2018 International Power Electronics Conference (IPEC-Niigata 2018 –ECCE Asia-)

Niigata, Japan
20-24 May 2018

Pages 1460-2152

IEEE Catalog Number: CFP1854I-POD
ISBN: 978-1-5386-4190-3

Copyright © 2018, IEEJ Industry Applications Society
All Rights Reserved

*** This is a print representation of what appears in the IEEE Digital Library. Some format issues inherent in the e-media version may also appear in this print version.

IEEE Catalog Number: CFP1854I-POD
ISBN (Print-On-Demand):   978-1-5386-4190-3
ISBN (Online):   978-4-88686-405-5

Additional Copies of This Publication Are Available From:

Curran Associates, Inc
57 Morehouse Lane
Red Hook, NY 12571 USA
Phone: (845) 758-0400
Fax: (845) 758-2633
E-mail: curran@proceedings.com
Web: www.proceedings.com

TABLE OF CONTENTS

THREE-PHASE INDUCTIVE POWER TRANSFER SYSTEM WITH 12 COILS FOR RADIATION NOISE REDUCTION 69
Keisuke Kusaka ; Jun-Ichi Itoh

SECONDARY-SIDE-ONLY CONTROL FOR SMOOTH VOLTAGE STABILIZATION IN WIRELESS POWER TRANSFER SYSTEMS WITH CONSTANT POWER LOAD 77
Giorgio Lovison ; Takehiro Imura ; Hiroshi Fujimoto ; Yoichi Hori

CONSTANT CURRENT CHARGING AND THE MAXIMUM SYSTEM EFFICIENCY TRACKING FOR WIRELESS CHARGING SYSTEMS EMPLOYING DUAL-SIDE CONTROL 84
Zhenjie Li ; Xiaoliang Huang ; Kai Song ; Jinhai Jiang ; Chunbo Zhu ; Zhijiang Du

ELECTRIC FIELD COUPLING TYPE HIGH POWER WIRELESS POWER TRANSFER WITH LEAKAGE ELECTRIC FIELD STRUCURE 88
Mitsuru Masuda

TRANSFER POWER ANALYSIS OF CAPACITIVELY ISOLATED OUTLET AND PLUG (CAPISOP) USING SERIES RESONANCE 94
Hirohito Funato ; Koki Amano ; Takuya Hatsumi ; Junnosuke Haruna

WIDE VOLTAGE GAIN RANGE LLC DC/DC TOPOLOGIES: STATE-OF-THE-ART 100
Qi Cao ; Zhiqing Li ; Haoyu Wang

DUAL HALF-BRIDGE LLC RESONANT CONVERTER WITH HYBRID-SECONDARY-RECTIFIER (HSR) FOR WIDE-OUPUT-VOLTAGE APPLICATIONS 108
Jae-Il Baek ; Chong-Eun Kim ; Keon-Woo Kim ; Min-Su Lee ; Gun-Woo Moon

A STUDY ON THE ANALYSIS AND CONTROL OF NO-LOAD CHARACTERISTICS OF LLC RESONANT CONVERTER FOR PLASMA PROCESS 114
Min-Jun Kwon ; Woo-Cheol Lee

MECHANISM OF CURRENT IMBALANCE IN LLC RESONANT CONVERTER WITH CENTER TAPPED TRANSFORMER 118
Mitsuru Sato ; Shingo Nagaoka ; Takeshi Uematsu ; Toshiyuki Zaitsu

PERFORMANCE STUDY OF HIGH-POWER HALF-BRIDGE INTERLEAVED LLC CONVERTER 123
Hung-I Hsieh ; Hui-Lung Chiu ; Guan-Chyun Hsieh

MULTI-CHIP SIC MOSFET POWER MODULES FOR STANDARD MANUFACTURING, MOUNTING AND COOLING 130
Alberto Castellazzi ; Asad Fayyaz ; Emre Gurpinar ; Abdallah Hussein ; Jianfeng Li ; Bassem Mouawad

AN ALTERNATIVE METHOD TO ACCURATELY DETERMINE THE THERMAL RESISTANCE OF SIC MOSFET STRUCTURES WITH DISCRETE DIODES 137
Andras Vass-Varnai ; Young Joon Cho ; Gabor Farkas ; Marta Rencz

HEAT-RESISTANT PACKAGING TECHNOLOGY FOR WIDE BANDGAP POWER DEVICES AND THERMAL RELIABILITY TESTING 142
K. Suganuma ; H. Zhang ; S. Nagao ; C. Chen ; T. Sugahara ; A. Shimoyama ; A. Suetake

VERIFICATION OF IDENTIFICATION ACCURACY OF LOSS CALCULATED BY INVERSE THERMAL ANALYSIS 148
Yuki Ikari ; Kazushige Nakao

PACKAGING ARCHITECTURES FOR SILICON CARBIDE POWER ELECTRONIC MODULES 153
H. Alan Mantooth ; Simon S. Ang

DEVELOPMENT OF A HOMO-POLAR BEARINGLESS MOTOR WITH CONCENTRATED WINDING FOR HIGH SPEED APPLICATIONS 157
Dai Suzuki ; Takaaki Oiwa

HIGH-SPEED SLOTLESS PERMANENT MAGNET MACHINES: MODELLING AND DESIGN FRAMEWORKS 161
S. Jumayev ; K.O. Boynov ; E.A. Lomonova ; J. Pyrhonen

DEVELOPMENT AND PERFORMANCE OF HIGH-SPEED SPM SYNCHRONOUS MACHINE 169
Kota Kawanishi ; Keisuke Matsuo ; Takayuki Mizuno ; Koji Yamada ; Takashi Okitsu ; Kouki Matsuse

1.2KW 100,000RPM HIGH SPEED MOTOR FOR AIRCRAFT 177
Takehiro Jikumaru ; Gen Kuwata

COMPARATIVE EVALUATION OF Y-INVERTER AGAINST THREE-PHASE TWO-STAGE BUCK-BOOST DC-AC CONVERTER SYSTEMS 181
Michael Antivachis ; Dominik Bortis ; David Menzi ; Johann W. Kolar

DC-POWERED OFFICE BUILDINGS AND DATA CENTRES : THE FIRST 380 VDC MICRO GRID IN A COMMERCIAL BUILDING IN GERMANY 190
Tilo Pueschel

RECENT TREND IN POWER ELECTRONICS FOR ICT SYSTEMS 196
Hiroshi Nakao ; Yu Yonezawa ; Yoshiyasu Nakashima

GREEN BASE STATION USING ROBUST SOLAR SYSTEM AND HIGH PERFORMANCE LITHIUM ION BATTERY FOR NEXT GENERATION WIRELESS NETWORK (5G) AND AGAINST MEGA DISASTER 201
M. Nakamura ; K. Takeno

OPTIMIZATION OF MAINTENANCE BY FAILURE PREDICTION CONSIDERING INSTANTANEOUS AND CUMULATIVE EFFECTS OF EXTERNAL ENVIRONMENTS 207
Kaisei Kanetani ; Masahiro Yamazaki ; Tadatoshi Babasaki ; Hideaki Kim ; Tatsushi Matsubayashi

HYBRID CONVERTERS WITH REDUCED INDUCTOR LOSS FOR INTEGRATABLE POWER CONVERSION 213
Gab-Su Seo ; Hanh-Phuc Le

ENERGY SAVING SYSTEM TREND FOR HARBOR CRANE WITH LITHIUM ION BATTERY 219
Hidemasa Yoshihara

INVERTER DRIVE OF DYNAMOMETERS FORAUTOMOTIVE EVALUATION SYSTEM 227
Shizunori Hamada ; Toshimichi Takahashi ; Nobutaka Kezuka ; Masaju Kouketsu ; Shingo Ishigaki

EXPERIMENTAL INVESTIGATION OF PROTOTYPE ALL-SIC CONVERTER FOR ULTRA-HIGH-SPEED ELEVATOR 233
Kazuhisa Mori ; Kaoru Katoh ; Yohei Matsumoto ; Tatsushi Yabuuchi ; Naoto Ohnuma

HIGH-VOLTAGE, LARGE-CAPACITY CONVERTER TECHNOLOGIES AND THEIR APPLICATIONS 238
Daisuke Yoshizawa ; Paul Bixel ; Masahiko Tsukakoshi

HIGHER RADIAL SUSPENSION FORCE OF MAGNETIC BEARING ON CENTRIFUGAL COMPRESSOR FOR HVAC 244
Yuji Nakazawa ; Yusuke Irino ; Atsushi Sakawaki ; Kazunobu Ohyama

NOVEL SWITCHING CONTROL METHOD FOR FULL-BRIDGE DC-DC CONVERTERS FOR IMPROVING LIGHT-LOAD EFFICIENCY USING REVERSE RECOVERY CURRENT 250
Fumihiro Sato ; Takae Shimada ; Takayuki Ouchi

A 800V/14V SOFT-SWITCHED CONVERTER WITH LOW-VOLTAGE RATING OF SWITCH FOR XEV APPLICATIONS 256
Byeongwoo Kim ; Kangsan Kim ; Sewan Choi

HIGH SPEED CONTROL METHOD FOR SUPERPOSING HIGH-FREQUENCY-HIGH-SINUSOIDAL-CURRENT WITH DC CURRENT TO ANALYZE BATTERY AC IMPEDANCE 261
Jin Xu ; Toshihiko Kishimoto ; Noboru Shimosato

EV BMS WITH TIME-SHARED ISOLATED CONVERTERS FOR ACTIVE BALANCING AND AUXILIARY BUS REGULATION 267
Z. Gong ; B.A.C. Van De Ven ; Y. Lu ; Y. Luo ; K. Gupta ; C. Da Silva ; H.J. Bergveld ; O. Trescases

A DRIVING CIRCUIT WITH PARTIAL POWER REGULATION FOR RGB LED LAMPS 275
You-Chun Huang ; Yu-Jen Chen ; Yong-Jyun Li ; Chin-Sien Moo

FPGA-BASED DYNAMIC DUTY CYCLE AND FREQUENCY CONTROLLER FOR A CLASS-E2DC-DC CONVERTER 282
Sanghyeon Park ; Juan Rivas-Davila

DESIGN METHODOLOGY OF 3 KW INDUCTION HEATING SYSTEM FOR BOTH LOW RESISTANCE AND HIGH RESISTANCE CONTAINERS IN A SINGLE BURNER 289
Si-Hoon Jeong ; Hwa-Pyeong Park ; Jee-Hoon Jung

MULTI-RESONANT INVERTER REALIZING DOWNSIZING AND LOSS REDUCTION FOR ALL-METALLIC IH COOKTOP 296
Takayuki Hirokawa ; Makoto Imai ; Atsushi Fujita

TEMPERATURE ESTIMATION OF ALUMINUM ELECTROLYTIC CAPACITOR UNDER ACTUAL CIRCUIT OPERATION 302
Kazuki Urata ; Toshihisa Shimizu

DESIGN AND EVALUATION OF CURRENT DISTRIBUTION IN POWER MODULE 309
Takaaki Ibuchi ; Eisuke Masuda ; Tsuyoshi Funaki

DEVELOPMENT OF IMPEDANCE-SOURCE INVERTER USING SIC-MOSFET 313
Ryuji IIjima ; Thilak Senanayake ; Takanori Isobe ; Hiroshi Tadano

CONTROL METHODOLOGY FOR REALIZATION OF 100KW HEECS CHOPPER WITH 99.5% EFFICIENCY 318
Yukinori Tsuruta ; Atsuo Kawamura

IRON LOSS REDUCTION IN THE CORES OF INDUCTION HEATING COILS FOR SMALL-FOREIGN-METAL PARTICLE DETECTOR WITH A 400-KHZ SIC-MOSFETS HIGH-FREQUENCY INVERTER 324

Takuya Shijo ; Yuki Uchino ; Yujiro Noda ; Hiroaki Yamada ; Toshihiko Tanaka

FREQUENCY TRACKING BURST-MODE PDM-CONTROLLED CLASS-D ZERO VOLTAGE SOFT-SWITCHING RESONANT CONVERTER FOR INDUCTIVE POWER TRANSFER APPLICATIONS 329

Yoichiro Tabata ; Tomokazu Mishima ; Tatsuya Kido

REDUCED-ORDER DYNAMICAL MODELS OF TUNED WIRELESS POWER TRANSFER SYSTEMS 337

Hongchang Li ; Jingyang Fang ; Yi Tang

DYNAMIC MODELLING AND CLOSED LOOP CONTROL OF TRANSMITTER PARALLEL AND RECEIVER SERIES COMPENSATED IPT TOPOLOGY FOR EV APPLICATIONS 342

Suvendu Samanta ; Akshay Kumar Rathore

DEVELOPMENT OF INDUCTIVE POWER TRANSFER SYSTEM FOR EXCAVATOR UNDER LARGE LOAD FLUCTUATION : CONSIDERATION OF RELATIONSHIP BETWEEN LOAD VOLTAGE AND RESONANCE PARAMETER 348

Jun-Ichi Itoh ; Kent Inoue ; Keisuke Kusaka

WIRELESS POWER TRANSFER SYSTEM USING THREE-PHASE TO SINGLE-PHASE MATRIX CONVERTER 356

Yuji Hayashi ; Hiromasa Motoyama ; Takaharu Takeshita

DESIGN OF A REDUCED-ORDER OBSERVER FOR SENSORLESS CONTROL OF DUAL-ACTIVE-BRIDGE CONVERTER 363

Nguyen Duy Dinh ; Goro Fujita

IMPROVED LOAD TRANSIENT RESPONSE OF A DUAL-ACTIVE-BRIDGE CONVERTER 370

Sheng-Zhi Zhou ; Chuan Sun ; Song Hu ; Guo Chen ; Xiaodong Li

MODULATION AND ACTIVE MIDPOINT CONTROL OF A THREE-LEVEL THREE-PHASE DUAL-ACTIVE BRIDGE DC-DC CONVERTER UNDER NON-SYMMETRICAL LOAD 375

Philipp Joebges ; Anton Gorodnichev ; Rik W. De Doncker

A NOVEL SWITCHING ALGORITHM TO IMPROVE EFFICIENCY AT LIGHT LOAD CONDITIONS FOR THREE-PHASE DAB CONVERTER IN LVDC APPLICATION 383

Hyun-Jun Choi ; Si-Hoon Jung ; Jee-Hoon Jung

DESIGN OF A HIGH-FREQUENCY DUAL-ACTIVE BRIDGE CONVERTER WITH GAN DEVICES FOR AN OUTPUT POWER OF 3.7 KW 388

Philipp Schülting ; Christian Winter ; Rik W. De Doncker

EXPLORATION OF THE DESIGN AND PERFORMANCE SPACE OF A HIGH FREQUENCY 166 KW/10 KV SIC SOLID-STATE AIR-CORE TRANSFORMER 396

Piotr Czyz ; Thomas Guillod ; Florian Krismer ; Johann W. Kolar

NOVEL CALCULATION METHOD OF IRON LOSS OF GAPPED INDUCTORS USING LOSS MAP 404

Yoshihiro Miwa ; Toshihisa Shimizu

VERIFICATION OF THE REDUCTION OF THE COPPER LOSS BY THE THIN COIL STRUCTURE FOR INDUCTION COOKERS 410

Morimasa Hataya ; Koki Kamaeguchi ; Eiji Hiraki ; Kazuhiro Umetani ; Takayuki Hirokawa ; Makoto Imai ; Hideki Sadakata

CONDITION MONITORING OF ELECTROLYTIC CAPACITOR BASED ON ESR ESTIMATION AND THERMAL IMPEDANCE MODEL USING IMPROVED POWER LOSS COMPUTATION 416

Sundararajan Prasanth ; Mohamed Halick ; Mohamed Sathik ; Firman Sasongko ; Tan Chuan Seng ; Peng Yaxin ; Rejeki Simanjorang

TEST SETUP FOR CHARACTERISATION OF BIASED MAGNETIC HYSTERESIS LOOPS IN POWER ELECTRONIC APPLICATIONS 422

Min Luo ; Drazen Dujic ; Jost Allmeling

A FAST OPEN-CIRCUIT FAULT DIAGNOSIS SCHEME FOR MODULAR MULTILEVEL CONVERTERS WITH MODEL PREDICTIVE CONTROL 428

Dehong Zhou ; Shunfeng Yang ; Yi Tang

AN ONLINE OPEN-CIRCUIT FAULT DIAGNOSIS AND FAULT TOLERANT SCHEME FOR THREE-PHASE AC-DC CONVERTERS WITH MODEL PREDICTIVE CONTROL 434

Dehong Zhou ; Yi Tang

THE LIFETIME ASSESSMENT OF A MICRO-INVERTER FOR PV APPLICATIONS 439

Tohihiro Shimao ; Koji Kato ; Youichi Ito ; Akio Iwabuchi ; Yongheng Yang ; Frede Blaabjerg

ONLINE HEALTH MONITORING OF MULTIPLE MOSFETS IN A GRID-TIED PV INVERTER USING SPREAD SPECTRUM TIME DOMAIN REFLECTOMETRY (SSTDR) 446
Sourov Roy ; Faisal Khan

AN IMPROVED EQUIVALENT MODEL FOR A LONG PV STRING UNDER PARTIAL SHADING CONDITIONS 453
Xiaoyang Wang ; Huiqing Wen ; Xingshuo Li

OPTIMIZED FLUX-WEAKENING CONTROL OF INDUCTION MOTOR FOR TORQUE ENHANCEMENT IN VOLTAGE EXTENSION REGION 459
Zhen Dong ; Yong Yu ; Bo Wang ; Qinghua Dong ; Dianguo Xu

IMPROVED PERFORMANCE OF CFTC-BASED DIRECT TORQUE CONTROL OF INDUCTION MACHINES BY INCREASING TORQUE LOOP BANDWIDTH 466
Ibrahim Mohd Alsofyani ; June-Hee Lee ; Byung-Moon Han ; Kyo-Beum Lee

μ-ANALYSIS EVALUATION OF A NOVEL COMBINED CURRENT-AND-SPEED CONTROL FOR INDUCTION MOTORS VIA ILQ DESIGN METHOD 471
Shuto Omori ; Hiroshi Takami ; Masashi Nakamura

LOSS MINIMIZATION CONTROL OF SENSORLESS SCALAR-CONTROLLED INDUCTION MOTOR DRIVES CONSIDERING IRON LOSS 478
Nguyen Anh Tan ; Dong-Choon Lee

TUNING OF INDUCTION MOTOR DRIVE WITH TORQUE SENSOR 483
Hajime Kubo ; Yugo Tadano

QUASI-TWO-LEVEL CONVERTER FOR OVERVOLTAGE MITIGATION IN MEDIUM VOLTAGE DRIVES 488
F. Bertoldi ; M. Pathmanathan ; R. S. Kanchan ; K. Spiliotis ; J. Driesen

A MEDIUM-VOLTAGE THREE-PHASE AC-DC CONVERTER CONSISTING OF CASCADED THREE-LEVEL BOOST-TYPE RECTIFIERS AND AN OPEN-END WINDING TRANSFORMER 495
Ryoji Tsuruta ; Hiromitsu Suzuki ; Ritaka Nakamura

A FAULT TOLERANT CONTROL STRATEGY FOR THE DELTA-CONNECTED CASCADED CONVERTER 503
Ping-Heng Wu ; Po-Tai Cheng

COOLING PERFORMANCE IMPROVEMENT OF HEAT SINK BY OSCILLATING HEAT PIPE ADDITION AND DESIGN FOR ENVIRONMENT OF OSCILLATING HEAT PIPE REFRIGERANT 511
Kuan-Chung Tey ; Kenichiro Suzuki

COMPACT LARGE CAPACITY GAS TURBINE STATIC STARTER 517
Hironori Kawaguchi ; Shigeyuki Nakabayashi ; Akinobu Ando ; Hiroshi Ogino ; Yasuaki Matsumoto ; Ikuto Udagawa ; Takahiro Ohta

VOLTAGE REFERENCE MODIFICATION SCHEME FOR RESONANCE SUPPRESSION IN LCL-FILTERED INVERTERS WITH DISCONTINUOUS PWM METHOD 521
Hyeon-Sik Kim ; Seung-Ki Sul

PARAMETRIC ROBUSTNESS ANALYSIS FOR PARALLEL FEEDFORWARD COMPENSATION BASED ACTIVE DAMPING OF LCL GRID CONNECTED INVERTER 528
Muhammad Talib Faiz ; Muhammad Mansoor Khan ; Xu Jianming ; Muhammad Ali ; Houjun Tang

OPEN-LOOP-BASED ISLAND-MODE VOLTAGE CONTROL METHOD FOR SINGLE-PHASE GRID-TIED INVERTER WITH MINIMIZED LC FILTER 534
Satoshi Nagai ; Jun-Ichi Itoh

EXPERIMENTAL VALIDATION OF ADAPTIVE CURRENT INJECTING METHOD FOR GRID-SYNCHRONIZATION IMPROVEMENT OF GRID-TIED REGS DURING SHORT-CIRCUIT FAULT 542
Shaokang Ma ; Hua Geng ; Geng Yang ; Bo Liu

ADAPTIVE CONTROL OF GRID-VOLTAGE FEEDFORWARD FOR GRID-CONNECTED INVERTERS BASED ON REAL-TIME IDENTIFICATION OF GRID IMPEDANCE 547
Roni Luhtala ; Tuomas Messo ; Tomi Roinila

MODEL BASED TUNING OF PROPORTIONAL RESONANT CONTROLLERS FOR VOLTAGE SOURCE INVERTERS 555
Stefan Almér ; Thomas Besselmann ; Mario Schweizer

AN SOC-BASED PLATFORM FOR INTEGRATED MULTI-AXIS MOTION CONTROL AND MOTOR DRIVE 560
Yongping Sun ; Ming Yang ; Yangyang Chen ; Wangpin He ; Dianguo Xu

VARIABLE SWITCHING FREQUENCY STRATEGY FOR ENHANCED SETTLING PERFORMANCE OF POSITION CONTROL WITHIN INVERTER LOSS LIMIT 565
Choongin Lee ; Jung-Ik Ha

TWO-WHEEL CANE FOR WALKING ASSISTANCE..571
Phi Van Lam ; Yasutaka Fujimoto

FALL PREVENTION AND VIBRATION SUPPRESSION OF WHEELCHAIR USING RIDER MOTION STATE..575
Isseki Takahashi ; Toshiyuki Murakami

STABILIZATION METHOD FOR RESIDENTIAL DC SYSTEM BASED ON PASSIVITY CRITERION..583
Hiroaki Kakigano

A NOVEL CONTROL APPROACH TO MULTI-TERMINAL POWER FLOW CONTROLLER FOR NEXT-GENERATION DC POWER NETWORK..588
Kenji Natori ; Yuta Nakao ; Yukihiko Sato

DC MICROGRID FOR TELECOMMUNICATIONS SERVICE AND RELATED APPLICATION..593
Keiichi Hirose

MVDC DISTRIBUTION GRIDS FOR ELECTRIC VEHICLE FAST-CHARGING INFRASTRUCTURE..598
Marco Stieneker ; Benedict J. Mortimer ; Arne Hinz ; Adolf Müller-Hellmann ; Rik W. De Doncker

REVIEW OF RESONANT GATE DRIVER IN POWER CONVERSION..607
Bainan Sun ; Zhe Zhang ; Michael A.E. Andersen

A LOW PROFILE HIGH FREQUENCY LED DRIVING SYSTEM BASED ON AIRCORE PLANAR INDUCTOR..614
Yueshi Guan ; Xihong Hu ; Shu Zhang ; Yijie Wang ; Dianguo Xu ; Wei Wang

ANALYSIS AND COMPENSATION OF DEAD-TIME EFFECT IN SIC-DEVICE-BASED HIGH-SWITCHING-FREQUENCY INVERTERS..619
Qingzeng Yan ; Xibo Yuan ; Xiaojie Wu ; Yiwen Geng

CONTROL AND PERFORMANCE OF NEW ASYMMETRICAL OPERATION FOR SWITCHED-CAPACITOR-BASED RESONANT CONVERTERS..626
Hadi Setiadi ; Hideaki Fujita

HIGH-FREQUENCY RESONANT CONVERTER WITH SYNCHRONOUS RECTIFICATION FOR HIGH CONVERSION RATIO AND VARIABLE LOAD OPERATION..632
Lei Gu ; Kawin Surakitbovorn ; Juan Rivas-Davila

SMART PV INVERTERS FOR SMART GRID APPLICATIONS..639
Cheng-Jhen Yang ; Terng-Wei Tsai ; Yi-Chan Li ; Cheng-Yu Tang ; Yaow-Ming Chen ; Yung-Ruei Chang

HIGH-VOLTAGE BI-DIRECTIONAL HALF-BRIDGE THREE-LEVEL SERIES RESONANT CONVERTER WITH FREQUENCY MODULATION CONTROL..645
Lee Sih-Yi ; Jhang Jynu-Jhe ; Lin Jing-Yuan ; Hsieh Yao-Ching ; Chiu Haung-Jen

A CONTROL STRATEGY FOR FLYING-START OF SHAFT SENSORLESS PERMANENT MAGNET SYNCHRONOUS MACHINE DRIVE..651
Zih-Cing You ; Sheng-Ming Yang

CONTACTLESS EV POWER TRACK SYSTEM WITH SEGMENT-EXCITED INDUCTIVELY COUPLED STRUCTURE..657
Jia-You Lee ; Yu-Chi Wang ; Chih-Yi Liao

DRIVING TEST EVALUATION OF SENSORLESS VEHICLE DETECTION METHOD FOR IN-MOTION WIRELESS POWER TRANSFER..663
Katsuhiro Hata ; Kensuke Hanajiri ; Takehiro Imura ; Hiroshi Fujimoto ; Yoichi Hori ; Motoki Sato ; Daisuke Gunji

A SYSTEM DESIGN METHOD OF HIGH-FREQUENCY CLASS-D INVERTER FOR WIDEBAND CURRENT CONTROL..669
Hiroki Kurumatani ; Seiichiro Katsura

ANALYSIS OF INTERIOR PERMANENT MAGNET TWO DEGREES OF FREEDOM MOTOR BASED ON CROSS-COUPLED STRUCTURE..675
Yoshiyuki Hatta ; Tomoyuki Shimono

STUDY COMPARISON BETWEEN FIREFLY ALGORITHM AND PARTICLE SWARM OPTIMIZATION FOR SLAM PROBLEMS..681
Mounia Janah ; Yasutaka Fujimoto

BANDWIDTH LIMITATIONS IN FORCE CONTROL OF A SERIES ELASTIC ACTUATOR WITH BACKLASH AND QUANTIZATION..688
Hanul Jung ; Chan Lee ; Sehoon Oh

ROTOR SHAPE OPTIMIZATION OF INTERIOR PERMANENT MAGNET SYNCHRONOUS MOTORS WITH CONCENTRATED WINDINGS BY CONSIDERING END-LEAKAGE FLUX..693
Katsumi Yamazaki ; Hiroki Narushima

LOSS ANALYSIS OF PERMANENT-MAGNET SYNCHRONOUS MACHINES CONSIDERING IN-PLANE EDDY CURRENT IN ELECTRICAL STEEL SHEETS..699

Hideki Ohguchi ; Satoshi Imamori ; Katsumi Yamazaki ; Haiyan Yui ; Masao Shuto

STUDY ON INFLUENCE OF DIFFERENCE IN STRUCTURE OF CONCENTRATED WINDING IPMSMS OBTAINED BY AUTOMATIC DESIGN 704

A. Ura ; M. Sanada ; S. Morimoto ; Y. Inoue

CARRIER HARMONIC LOSS REDUCTION TECHNIQUE ON DUAL THREE-PHASE PERMANENT-MAGNET SYNCHRONOUS MOTORS WITH PHASE-SHIFT PWM..711

Yoshihiro Miyama ; Haruyuki Kometani ; Kan Akatsu

FLUX INTENSIFYING PM-MOTOR WITH VARIABLE LEAKAGE MAGNETIC FLUX TECHNIQUE..718

Masahiro Aoyama ; Toshihiko Noguchi

CONTINUOUS OPERATION CONTROL OF PMSM IN THE CASE OF DC POWER SUPPLY LOSS..726

Jongwon Heo ; Keiichiro Kondo

MODEL PREDICTIVE CONTROL FOR MULTIPHASE MOTOR DRIVES – A TECHNOLOGY STATUS REVIEW..732

A. Tenconi ; S. Rubino ; R. Bojoi

INFLUENCE OF FAST SWITCHING SEMICONDUCTORS ON THE WINDING INSULATION SYSTEM OF ELECTRICAL MACHINES..740

Kay Hameyer ; Andreas Ruf ; Florian Pauli

CENTRALIZED CONTROL OF MODULAR MULTI RECTIFIER FOR MOTOR DRIVE APPLICATIONS UNDER UNBALANCED GRID..746

Yipeng Song ; Pooya Davari ; Frede Blaabjerg

VECTOR CONTROL OF MAGNETICALLY MODULATED MOTOR FOR POWER SPLITTING OF HEV APPLICATION..753

Toshihiko Noguchi ; Sawanth Krishna Machavolu ; Masahiro Aoyama ; Yuto Motohashi

IMPEDANCE-BASED STABILITY EVALUATION OF VIRTUAL SYNCHRONOUS MACHINE IMPLEMENTATIONS IN CONVERTER CONTROLLERS..759

Eneko Unamuno ; Atle Rygg ; Mohammad Amin ; Marta Molinas ; Jon Andoni Barrena

STABLE POWER SUPPLY METHOD FOR HOUSEHOLD APPLIANCES VIA VIRTUAL SYNCHRONOUS GENERATOR IN SINGLE-PHASE THREE-WIRE MICROGRID..767

Yuko Hirase ; Hidehiko Nakagawa ; Eiji Yoshimura ; Shogo Katsura ; Kensho Abe ; Osamu Noro ; Kazushige Sugimoto ; Kenichi Sakimoto

A NOVEL OSCILLATION DAMPING METHOD OF VIRTUAL SYNCHRONOUS GENERATOR CONTROL WITHOUT PLL USING POLE PLACEMENT..775

Jia Liu ; Yushi Miura ; Toshifumi Ise

OPERATION OF A MODULAR MULTILEVEL CONVERTER CONTROLLED AS A VIRTUAL SYNCHRONOUS MACHINE..782

Salvatore D'arco ; Giuseppe Guidi ; Jon Are Suul

ASSESSMENT OF VIRTUAL SYNCHRONOUS MACHINE BASED CONTROL IN GRID-TIED POWER CONVERTERS..790

Chi Li ; Igor Cvetkovic ; Rolando Burgos ; Dushan Boroyevich

RESEARCH ON THE BLOCKCHAIN-BASED INTEGRATED DEMAND RESPONSE RESOURCES TRANSACTION SCHEME..795

Shengnan Zhao ; Yang Li ; Beibei Wang ; Huiling Su

INDIRECT CURRENT CONTROL FOR SEAMLESS TRANSFER OF UTILITY INTERACTIVE INVERTER..803

Kyungbae Lim ; Injong Song ; Jaeho Choi

STUDY OF AC POWER INTERCHANGE AND DC POWER INTERCHANGE FOR MICRO GRID SYSTEMS..809

Kazuto Yukita ; Daiki Owaki ; Shunsuke Horie ; Toshiro Matsumura ; Yasuyuki Goto

STABILITY ENHANCEMENT STRATEGY FOR ISLANDING MICROGRID WITH MULTI-TYPE INVERTERS BASED ON HYBRID IMPEDANCE MODELLING..815

Meiqin Mao ; Yong Ding ; Yatao Shen ; Liuchen Chang

DC POWERED DATA CENTER WITH 200 KW PV PANELS..822

Keiichi Hirose

INFLUENCES OF DETERIORATION IN CAPACITOR AND INDUCTOR ON CURRENT SENSORLESS STATIC MODEL DC-DC CONVERTER..826

Fujio Kurokawa ; Masashi Taguchi ; Jizhe Wang ; Hidenori Maruta ; Nobumasa Matsui

CAPACITIVE DIVIDER BASED PASSIVE START-UP METHODS FOR FLYING CAPACITOR STEP-DOWN DC-DC CONVERTER TOPOLOGIES 831
Michael Halamicek ; Tom Moiannou ; Nenad Vukadinovic ; Aleksandar Prodic

HIGH VOLTAGE GAIN INTERLEAVED ACTIVE-CLAMP FORWARD (IACF) CONVERTER HAVING REDUCED PRIMARY CONDUCTION LOSS 838
Yeonho Jeong ; Mu-Hyun Park ; Gun-Woo Kim ; Byoung-Hee Lee ; Gun-Woo Moon

CONTROL OF SWITCHING-CAPACITOR BASED BUCK-BOOST CONVERTER 845
M. Veerachary ; Vasudha Khubchandani

IMPROVEMENT OF UPLOAD TRANSIENT RESPONSES FOR ULTRA HIGH STEP-DOWN CONVERTER 851
Y.T. Yan ; K.I. Hwu

POWER ELECTRONICS AND CONTROL TECHNOLOGIES FOR HOUSEHOLD WASHER 856
Toru Niki

DEVELOPMENT OF ROOM AIR CONDITIONER WITH TWIN-PROPELLER FANS 860
Takamasa Uemura ; Tomoya Fukui ; Kenichi Sakoda

ELECTROLYTIC CAPACITOR-LESS SINGLE-PHASE TO THREE-PHASE INVERTER WITH HARMONICS SUPPRESSION CONTROL FOR AIR CONDITIONER 866
Nobuo Hayashi ; Takuro Ogawa ; Tomoisa Taniguchi ; Morimitsu Sekimoto

LATEST DEVELOPMENT OF SIC POWER MODULE-BASED SINGLE-STAGE AC-AC RESONANT CONVERTER FOR HIGH-FREQUENCY INDUCTION HEATING APPLICATIONS 872
Tomokazu Mishima

AN OPTIMIZED CONTROL STRATEGY TO IMPROVE THE CURRENT ZERO-CROSSING DISTORTION IN BIDIRECTIONAL AC/DC CONVERTER BASED ON V2G CONCEPT 878
Lei Jing ; Xiaoqing Wang ; Bodong Li ; Maohang Qiu ; Bo Liu ; Min Chen

PER-PHASE CONTROL STRATEGY OF THE THREE-PHASE FOUR-WIRE INVERTER 883
Yi-Chan Li ; Terng-Wei Tsai ; Cheng-Jhen Yang ; Yaow-Ming Chen ; Yung-Ruei Chang

OPPORTUNITIES FOR PERFORMANCE IMPROVEMENT OF SINGLE-PHASE POWER CONVERTERS THROUGH ENHANCED AUTOMATIC-POWER-DECOUPLING CONTROL 889
Huawei Yuan ; Sinan Li ; Wenlong Qi ; Siew-Chong Tan ; S. Y. Ron Hui

ZERO VOLTAGE SWITCHING SCHEME FOR FLYBACK CONVERTER TO ENSURE COMPATIBILITY WITH ACTIVE POWER DECOUPLING CAPABILITY 896
Hiroki Watanabe ; Jun-Ichi Itoh

MODEL PREDICTIVE FAULT TOLERANT CONTROL OF BIDIRECTIONAL AC/DC CONVERTER WITH VOLTAGE BALANCE OF SPLIT CAPACITOR 904
Nan Jin ; Chongyan Zhao ; Leilei Guo

PWM STRATEGY FOR PARALLEL OPERATION OF THREE PHASE CONVERTERS TIED TO GRID 911
Hyun-Sam Jung ; Seung-Ki Sul

PRACTICAL ISSUES AND IMPLEMENTATION CIRCUITS OF THE DIGITAL-ANALOG HYBRID FULL FEED-FORWARD METHOD WITH UNIPOLAR AND BIPOLAR MODULATIONS 917
Xin Zhang ; Henry S. H. Chung ; Zhixun Ma

AN AC-DC POWER CONVERTER FOR ELECTROLYTIC CAPACITOR-LESS LED DRIVER WITH HIGH LUMINOUS EFFICACY 922
Kwon-Sik Park ; Byuong-Jun Seo ; Kyoung-Suk Kang ; Eui-Cheol Nho

AN IMPROVED CASCADED DUAL-BUCK INVERTER 927
Usman Ali Khan ; Honnyong Cha ; Ashraf Ali Khan ; Heung-Geun Kim ; Wilson Eberle ; Liwei Wang

A SINGLE-SWITCH INTEGRATED-STAGE LED DRIVER BASED ON CUK AND CLASS-E CONVERTER 934
Shu Zhang ; Yijie Wang ; Xiaosheng Liu ; Yan Zhou ; Dianguo Xu

A FAULT-TOLERANT PARALLEL INVERTER APPLIED TO MICRO-GRID 939
Xiangyue Shi ; Jinjie Peng ; Zhifeng Qiu ; Wei Xiong

STABILITY ANALYSIS OF GRID-CONNECTED CONVERTERS WITH ADD-ON VOLTAGE SUPPORT FUNCTIONALITY USING REPETITIVE CONTROL 946
Y. Zhang ; M. G. L. Roes ; M. A. M. Hendrix ; J. L. Duarte

ADAPTIVE SERIES STABILIZER MODULE FOR THE GRID CONNECTED INVERTER UNDER VARIABLE GRID CONDITIONS 953
Xin Zhang

AN IMPROVED DROOP CONTROL BASED SMOOTH TRANSFER CONTROL STRATEGY 957
Xin Meng ; Jinjun Liu ; Zeng Liu ; Ronghui An

FREQUENCY RESPONSE ANALYSIS OF LOAD EFFECT ON DYNAMICS OF GRID-FORMING INVERTER 963

Matias Berg ; Tuomas Messo ; Teuvo Suntio

A NEW CONTROL METHOD FOR TRIPLE-ACTIVE BRIDGE CONVERTER WITH FEED FORWARD CONTROL 971

Takanobu Ohno ; Nobukazu Hoshi

ANALYSIS OF PFM OPERATION MODEL FOR CAPACITOR CHARGER RESONANT TOPOLOGY WITH ENERGY DOSAGE 977

Pengyu Jia ; Yiqin Yuan ; Shengwen Fan ; Zhenyu Shan

AN ACTIVE-CLAMPED CURRENT-FED HALF-BRIDGE DC-DC CONVERTER WITH THREE SWITCHES 982

Truong-Duy Duong ; Minh-Khai Nguyen ; Young-Cheol Lim ; Joon-Ho Choi

A HIGH GAIN QUASI SINGLE STAGE LLC RESONANT DC/DC CONVERTER WITH COUPLED INDUCTOR AND PARTIAL ACTIVE CLAMP 987

Chongcan Huo ; Xiaogao Xie ; Shuai Jiang ; Hanjing Dong

SUPPRESSION OF RIPPLE CURRENT IN HIGH STEP-UP DC-DC CONVERTER UTILIZING COCKCROFT-WALTON CIRCUIT WITH INDUCTOR 992

Takumi Yasuda ; Masataka Minami ; Shin-Ichi Motegi ; Masakazu Michihira

AN OPTIMAL DESIGN METHOD CONSIDERING TRANSFORMER PARASITIC CAPACITANCE OF LLC RESONANT CONVERTERS 998

Naizeng Wang ; Xu Yang ; Mofan Tian ; Haiyang Jia ; Guangzhao Xu ; Zhenwei Li

COMPARISON OF HARMONIC LINEARIZATION AND HARMONIC STATE SPACE METHODS FOR IMPEDANCE MODELING OF MODULAR MULTILEVEL CONVERTER 1004

Jing Lyu ; Xin Zhang ; Jingjing Huang ; Jianwen Zhang ; Xu Cai

AN IMPROVED PHASE-SHIFTED PWM FOR A FIVE-LEVEL HYBRID-CLAMPED CONVERTER 1010

Kui Wang ; Nianzhou Liu ; Zedong Zheng ; Yongdong Li

INTEGRATED CONTROL METHODS FOR ASYMMETRICAL CASCADED H-BRIDGE RECTIFIER 1015

Wenjing Dai ; Jie Chen ; Xin Chen ; Chunying Gong

TRANSIENT VOLTAGE STRESS MODELING FOR SUBMODULES OF MODULAR MULTILEVEL CONVERTERS UNDER GRID VOLTAGE SAGS 1021

Zhijian Yin ; Yongheng Yang ; Huai Wang

SVPWM STRATEGY BASED ON MULTILEVEL 3LNPC-CR 1027

Xiaoqiong He ; Pengcheng Han ; Xiaolan Lin ; Yi Wang ; Xu Peng

THE MULTIPLE DEGREE OF FREEDOM BASED NEUTRAL POINT POTENTIAL CONTROL OF THREE LEVEL NEUTRAL POINT CLAMPED CONVERTERS 1032

Bo Guan ; Shinji Doki

A MODIFIED PHASE-SHIFTED PWM TECHNIQUE FOR THE GRID-CONNECTED HYBRID CASCADED CONVERTER 1038

Yu-Chen Su ; Po-Tai Cheng

NOVEL T-TYPE DUAL-BUCK INVERTER WITH MINIMUM NUMBER OF INDUCTORS 1046

Tien-The Nguyen ; Honnyong Cha ; Bang Le-Huy Nguyen ; Heung-Geun Kim

CONTROL OF DIRECT AC/AC MODULAR MULTILEVEL CONVERTER IN RAILWAY POWER SUPPLY SYSTEM 1051

Shuguang Song ; Jinjun Liu ; Shaodi Ouyang ; Xingxing Chen ; Baojin Liu

WIRELESS POWER TRANSFER: CRITICAL REVIEW OF RELATED STANDARDS 1062

Mohamad Abou Houran ; Xu Yang ; Wenjie Chen ; Mehdi Samizadeh

COMPARATIVE STUDY OF SINGLE-PHASE FUNDAMENTAL COMPONENT FREQUENCY ESTIMATION SCHEMES UNDER TIME-VARYING HARMONIC DISTORTION OPERATION 1067

E. B. Kapisch ; J. L. Duarte ; C. A. Duque

A COMPREHENSIVE DEAD-TIME COMPENSATION METHOD FOR A THREE-PHASE DUAL-ACTIVE BRIDGE CONVERTER WITH HYBRID MODULATION SCHEMES 1073

Jingxin Hu ; Zhiqing Yang ; Rik W. De Doncker

EVALUATION OF A HIGH-FREQUENCY REACTOR WITH A NEW WIRE GUIDE FOR A TOROIDAL CORE 1080

Hideki Ayano ; Akira Fujimura ; Yoshihiro Matsui

CORE LOSS EVALUATION IN POWDER CORES: A COMPARATIVE COMPARISON BETWEEN ELECTRICAL AND CALORIMETRIC METHODS 1087

Yuki Ishikura ; Jun Imaoka ; Mostafa Noah ; Masayoshi Yamamoto

MODELING, MAGNETIC DESIGN, AND SIMULATION METHODS CONSIDERING DC SUPERIMPOSITION CHARACTERISTIC OF POWDER CORES USED IN POWER CONVERTERS 1095

Jun Imaoka ; Kenkichiro Okamoto ; Masahito Shoyama ; Yuki Ishikura ; Mostafa Noah ; Masayoshi Yamamoto

MODELLING AND DESIGN OF A MEDIUM FREQUENCY TRANSFORMER FOR HIGH POWER DC-DC CONVERTERS 1103

Miloš Stojadinovic ; Jürgen Biela

EVALUATION OF INDUCTOR LOSSES ON Z-SOURCE INVERTER CONSIDERING AC AND DC COMPONENTS 1111

Ryuji IIjima ; Naoki Kamoshida ; Rene Alexander Barrera Cardenas ; Takanori Isobe ; Hiroshi Tadano

AN INTEGRATING STRUCTURE OF OUTPUT FILTER FOR GRID CONNECTED INVERTER BASED ON FMLF TECHNIQUE 1118

Jie Ma ; Yenan Chen ; Pingping Chen ; Wenxing Zhong ; Dehong Xu

NEW SCREENING METHOD FOR IMPROVING TRANSIENT CURRENT SHARING OF PARALLELED SIC MOSFETS 1125

Junji Ke ; Zhibin Zhao ; Peng Sun ; Huazhen Huang ; James Abuogo ; Xiang Cui

PSPICE MODELING AND APPLICATION FOR SIC POWER MOSFET TO EVALUATE THE POWER LOSS IN FULL-BRIDGE CONVERTER 1131

Juan Wei ; Fei Lin ; Zhongping Yang ; Xianjin Huang ; Chanjuan Xiao ; Hao Zhang ; Wencai Liang

ALL-SIC MODULE PACKAGING TECHNOLOGY 1137

Kento Shirata ; Norihiro Nashida ; Hideyo Nakamura ; Yoshitaka Nishimura

A NEW SMALLEST 1200V INTELLIGENT POWER MODULE FOR THREE PHASE MOTOR DRIVES 1141

Minsub Lee ; Miran Baek ; Junbae Lee ; Daewoong Chung

DESIGN AND ENHANCEMENT OF ESD RELIABILITY IN CIRCULAR UHV 300-V NLDMOS POWER COMPONENTS 1145

Shen-Li Chen ; Yi-Hao Chao ; Chih-Ying Yen ; Jen-Hao Lo ; Chun-Ting Kuo ; Yu-Lin Lin ; Yi-Hao Chiu ; Pei-Lin Wu ; Yu-Lin Jhou

A TECHNOLOGY ANALYSIS OF VOLTAGE SHARING IN SERIES CONNECTED POWER DEVICES 1149

Z Davletzhanova ; O Alatise ; R Bonyadi ; J Ortiz-Gonzalez ; T Dai ; M Jennings ; L Ran ; P Mawby

FAILURE MECHANISM ANALYSIS AND PHYSICS-OF-FAILURE LIFETIME PREDICTION METHOD FOR PRESS-PACK THYRISTOR OF CONVERTER VALVE 1157

Ning Liang ; Zhigang Zhang ; Yating Gou ; Cuicui Liu ; Zebin Yang ; Jiangnan Chen ; Fang Zhuo ; Feng Wang

SURGE VOLTAGE ABSORPTION BY A SILICON CARBIDE AVALANCHE-DIODE WITH P-N STRUCTURE 1162

K. Koseki ; Y. Tanaka

CALCULATION OF THYRISTOR RELIABILITY PARAMETER OF UHVDC CONVERTER VALVE IN HEMP ENVIRONMENT 1167

Zhigang Zhang ; Yating Gou ; Cuicui Liu ; Zebin Yang ; Xiaotong Du ; Jiangnan Chen ; Fang Zhuo ; Feng Wang ; Yuanliang Lan ; Caiwang Sheng

GENERALIZED STACKELBERG GAME-THEORETIC APPROACH FOR JOINTED ENERGY AND RESERVE COORDINATION OF ELECTRIC VEHICLES 1172

Tianyang Zhao ; Xuewei Pan ; Lei Li ; Fei Zhao ; Can Wang

IMPEDANCE INFLUENCE ANALYSIS OF PHASE-LOCKED LOOPS ON THREE-PHASE GRID-CONNECTED INVERTERS 1177

Yuncheng Wang ; Xin Chen ; Yang Zhang ; Jie Chen ; Chunying Gong

PULSE-INJECTION-BASED SENSORLESS CONTROL METHOD WITH IMPROVED DYNAMIC CURRENT RESPONSE FOR PMSM 1183

Hechao Wang ; Kaiyuan Lu ; Dong Wang ; Frede Blaabjerg

INFLUENCE OF PARAMETER VARIATIONS ON OPERATING CHARACTERISTICS OF MTPF CONTROL FOR DTC-BASED PMSM DRIVE SYSTEM 1189

Keisuke Fujii ; Yukinori Inoue ; Shigeo Morimoto ; Masayuki Sanada

A QUIET POSITION SENSORLESS CONTROL FOR AN IPMSM BASED ON EXTENDED EMF AND VOLTAGE INJECTION SYNCHRONIZED WITH PWM CARRIER 1196

Yuki Ishii ; Hiroki Yamashita ; Hisao Kubota

STUDY OF TORQUE RIPPLE REDUCTION AND TORQUE BOOST BY MODIFIED TRAPEZOIDAL MODULATION 1202

Satoshi Joryo ; Kazuto Tatsumi ; Toshimitsu Morizane ; Katsunori Taniguchi ; Noriyuki Kimura ; Hideki Omori

FAULT DIAGNOSIS METHOD OF CURRENT SENSOR FOR PERMANENT MAGNET SYNCHRONOUS MOTOR DRIVES 1206

Guoqiang Zhang ; Guoxin Wang ; Gaolin Wang ; Junya Huo ; Lianghong Zhu ; Dianguo Xu

SENSORLESS SPEED CONTROL OF DIESEL-GENERATOR SYSTEMS BASED ON MULTIPLE SOGI-FLLS 1212

Ngoc Dat Dao ; Dong-Choon Lee ; Dae-Sik Lim

ROBUSTNESS OF SIMPLIFIED SPEED-SENSORLESS VECTOR CONTROL FOR INDUCTION MOTOR 1217

Naoki Akao ; Mineo Tsuji ; Shin-Ichi Hamasaki

MAXIMUM TORQUE CONTROL REFERENCE FRAME BASED ON A TORQUE MAP FOR IPMSMS WITH LARGE INDUCTANCE VARIATION 1223

Kazuki Ohta ; Takumi Ohnuma ; Shinji Doki

PMSM MODEL DISCRETIZATION IN CONSIDERATION OF PARK TRANSFORMATION FOR CURRENT CONTROL SYSTEM 1228

Masamichi Inoue ; Shinji Doki

PSEUDO-RANDOM HIGH-FREQUENCY SINUSOIDAL VOLTAGE INJECTION BASED SENSORLESS CONTROL FOR IPMSM DRIVES 1234

Guoqiang Zhang ; Huiying Wang ; Gaolin Wang ; Junya Huo ; Lianghong Zhu ; Dianguo Xu

AT-NPC 3-LEVEL INVERTER-FED INDUCTION MOTOR VECTOR CONTROL WITH NEUTRAL POINT VOLTAGE CONTROL 1240

K. Sudo ; M. Tsuji ; S. Hamasaki ; T. Fukuoka ; H. Ichinose

INVESTIGATION OF VARIOUS POSITION ESTIMATION ACCURACY ISSUES IN PULSE-INJECTION-BASED SENSORLESS DRIVES 1246

Hechao Wang ; Kaiyuan Lu ; Dong Wang ; Frede Blaabjerg

POSITION SENSORLESS CONTROL OF SWITCHED RELUCTANCE MOTOR USING ESTIMATED PWM PHASE VOLTAGE 1253

Y. Nakazawa ; K. Ohyama ; H. Fujii ; H. Uehara ; Y. Hyakutake

EXPERIMENTAL CONFIRMATION OF THRUST AND ATTRACTIVE FORCE CONTROL OF LINEAR INDUCTION MOTOR BY TWO DIFFERENT FREQUENCY COMPONENTS 1259

Kenta Sannomiya ; Toshimitsu Morizane ; Noriyuki Kimura ; Hideki Omori

GA BASED OPTIMIZED TRAJECTORIES OF ROTATING SPEED AND D-Q AXIS CURRENTS FOR AN IPMSM 1264

Shuta Kumagai ; Kaoru Inoue ; Toshiji Kato

2-DEGREE-OF-FREEDOM DEADBEAT CONTROL WITH DISTURBANCE COMPENSATION FOR PMSM DRIVE SYSTEM USING FPGA 1270

Arata Takahashi ; Shotaro Takakura ; Tomoki Yokoyama

EXTENDED EMF-BASED SIMPLE IPMSM SENSORLESS VECTOR CONTROL USING COMPENSATED CURRENT CONTROLLER 1276

Takatoshi Inoue ; Yasumasa Hamabe ; Mineo Tsuji ; Shin-Ichi Hamasaki

FULL-BAND OUTPUT IMPEDANCE MODEL OF VIRTUAL SYNCHRONOUS GENERATOR IN DQ FRAMEWORK 1282

Li Wenbing ; Wang Jianhua ; Song Jingyu ; Luo Fangfang ; Gao Shang ; Wu Zaijun

AN MTPA CONTROL METHOD OF A PMSM AND A SYNRM BASED ON A DTC IN THE STATOR FLUX LINKAGE SYNCHRONOUS FRAME 1289

Gimpei Itoh ; Yukinori Inoue ; Shigeo Morimoto ; Masayuki Sanada

EEMFS EXCITED BY SIGNAL INJECTION FOR POSITION SENSORLESS CONTROL OF PMSMS AND THEIR PERFORMANCE COMPARISON BY USING IMAGINARY ELECTROMOTIVE FORCE 1295

Takumi Nimura ; Shota Kondo ; Shinji Doki ; Mutuwo Tomita

HARMONIC CURRENT CANCELLATION METHOD FOR PMSM DRIVE SYSTEM USING RESONANT CONTROLLERS 1301

Dongsheng Li ; Yoshitaka Iwaji ; Yasuo Notohara ; Ken Kishita

ESTIMATION ERROR ANALYSIS OF STATOR FLUX OBSERVER FOR DTC-BASED PMSM DRIVES 1308

Atsushi Shinohara ; Kichiro Yamamoto

APPLICATION OF FICTITIOUS REFERENCE ITERATIVE TUNING TO CONTROLLER DESIGN FOR VARIOUS MACHINES 1315

Hidehiro Ikeda ; Kazuya Goto ; Feili Zhang ; Kazuya Kayashima ; Tsuyoshi Hanamoto

HIGH EFFICIENCY CONTROL FOR PERMANENT MAGNET MOTOR DRIVE SYSTEM WITH FUEL CELLS CONNECTED IN SERIES WITH ELECTRIC DOUBLE-LAYER CAPACITORS 1322

Kichiro Yamamoto ; Fumiya Ohdera ; Atsushi Shinohara

COMPARATIVE STUDY OF SPEED RIPPLE REDUCTION BY VARIOUS CONTROL METHODS IN PMSM DRIVE SYSTEMS WITH PULSATING LOAD 1329

Yuma Komaru ; Yukinori Inoue ; Shigeo Morimoto ; Masayuki Sanada

ESTIMATION OF THE PARAMETERS OF THE SERVO DRIVE SYSTEM USING PARTICLE SWARM OPTIMIZATION ALGORITHM 1336

Helin Zhu ; Jae Hyuk Choi ; Sang Uk Park ; Jusuk Lee ; Hyong Gun Lee ; Hyung Soo Mok

A PROGRAMMABLE BATTERY TEST SYSTEM WITH ENERGY RECYCLING FEATURE BASED ON SINUSOIDAL LOADING TECHNIQUE 1341

Chang-Hua Lin ; Guan-Jung Chen ; Hwa-Dong Liu ; Kun-Feng Chen

DEVELOPMENT OF LARGE-CAPACITY CONVERTER FOR BATTERY ENERGY STORAGE SYSTEMS 1346

Hiroyoshi Komatsu ; Tatsuji Katayama ; Noriko Kawakami

ANALYSIS AND COMPARISON OF DC/DC TOPOLOGIES IN PARTIAL POWER PROCESSING CONFIGURATION FOR ENERGY STORAGE SYSTEMS 1351

Maria C. Mira ; Zhe Zhang ; A. E. Michael Andersen

TWO-STAGE PROTECTION FOR MULTI-CHANNEL POWER ELECTRONIC CONVERTERS FED LARGE ASYNCHRONOUS HYDRO-GENERATING UNIT 1358

R. R. Semwal ; Anto Joseph

CURRENT SHARING CONTROL FOR SERIES-PARALLEL CHANGEOVER USING BATTERY AND ELECTRIC DOUBLE-LAYER CAPACITOR BANK 1364

Taisei Nishino ; Keisaku Isozaki ; Naoki Kogai ; Kyungmin Sung

CONTROL METHOD OF ENERGY STORAGE SYSTEM TO IMPROVE OUTPUT POWER OF PCS 1370

Mikiya Ishibashi ; Hitoshi Haga ; Kenji Arimatsu ; Koji Kato

A CONTROL STRATEGY OF MMC BATTERY ENERGY STORAGE SYSTEM BASED ON ARM CURRENT CONTROL 1376

Liu Danqing ; Wang Guangzhu ; Ou Zhujian ; Liu Jiaxing

EQUIVALENT RESISTANCE CONTROL FOR MAXIMUM POWER TRANSFER METHOD OF PIEZOELECTRIC ELEMENT IN VIBRATION POWER GENERATION 1381

Kenya Takamura ; Hiroaki Yamada ; Toshihiko Tanaka ; Tomoharu Yada ; Hajime Fujiwara

DC BUS VOLTAGE STABILIZATION FOR CASCADED POWER CONVERTER BY INTEGRATING AN EXTRA PORT INTO LOAD SIDE PSFB 1386

Jiang You ; Weiyan Fan ; Mengyan Liao

COMMON MODE CURRENT REDUCTION OF THREE-PHASE CASCADED MULTILEVEL TRANSFORMERLESS INVERTER FOR PV SYSTEM 1391

Wenjie Wang ; Ke Chen ; Lijun Hang ; Anping Tong ; Yiliang Gan

CURRENT SHARING/VOLTAGE SHARING CONTROL STRATEGY FOR CASCADED DC/DC CONVERTER IN PHOTOVOLTAIC DC COLLECTION SYSTEM 1397

Bo Chen ; Yi Wang ; Yanjun Tian ; Shilei Wei

PCC VOLTAGE COMPENSATION OF PV INVERTER WITH ACTIVE POWER DECOUPLING CIRCUIT 1403

Duck-Hwan Hwang ; Jung-Yong Lee ; Younghoon Cho

A NOVEL PARTIAL SHADING DETECTION ALGORITHM UTILIZING POWER LEVEL MONITORING OF PHOTOVOLTAIC PANELS 1409

Thusitha Randima Wellawatta ; Sung-Jin Choi

BOOST INTEGRATED THREE-PHASE SOLAR INVERTER USING CURRENT UNFOLDING AND ACTIVE DAMPING METHODS 1414

N. Ha Pham ; Tomoyuki Mannen ; Keiji Wada

LINEAR ACTIVE DISTURBANCE REJECTION CONTROL FOR ISOLATED THREE-PORT CONVERTER 1421

Jiang You ; Mengyan Liao ; Weiyan Fan

STABILITY CONSTRAINED GAIN OPTIMIZATION OF DROOP CONTROLLED CONVERTERS IN DC NANOGRIDS 1426

Soumya Bandyopadhyay ; Laura Ramirez-Elizondo ; Pavol Bauer

SIC BASED SSPC FOR HIGH VOLTAGE SPACE APPLICATIONS 1435

D. Marroquí ; A. Garrigós ; José M. Blanes ; R. Gutiérrez

AN IMPROVED VOLTAGE-TYPE GRID-CONNECTED CONTROL STRATEGY FOR COMPENSATING UNBALANCED VOLTAGE 1442

Liu Hongpeng ; Zhou Jiajie ; Wang Wei

DUAL TWO-STAGE ISOLATED BIDIRECTIONAL DC-DC CONVERTER FOR DC GRID STORAGE 1447

Gabriel Tibola ; Jorge L. Duarte

MODULAR MULTILEVEL CONVERTER WITH CAPACITOR VOLTAGE SELF-BALANCING USING REDUCED NUMBER OF VOLTAGE SENSORS 1455

Taiyuan Yin ; Yue Wang ; Xiaolei Wang ; Shiyuan Yin ; Shumin Sun ; Guanglei Li

PLUG AND OUTLET IN HOUSEHOLD DC LOW VOLTAGE MICRO-GRID POWER DISTRIBUTION 1460
Worapong Pairindra ; Surin Khomfoi

PERFORMANCE PROGRAMMING TECHNIQUE FOR MULTI-STAGE DC POWER DISTRIBUTION SYSTEMS 1465
Syam Kumar Pidaparthy ; Hansang Kim ; Yeonjung Kim ; Byungcho Choi

COORDINATION CONTROL FOR PARALLELED INVERTERS BASED ON VSG FOR PV/BATTERY MICROGRID 1472
Meiqin Mao ; Cheng Qian ; Liuchen Chang ; Yan Du

ADAPTIVE VOLTAGE CONTROL SCHEME FOR DAB BASED MODULAR CASCADED SST IN PV APPLICATION 1478
Tao Liu ; Yang Xuan ; Xu Yang ; Peng Xu ; Yang Li ; Lang Huang ; Xiang Hao

SIX-STEP MMC-BASED HIGH POWER DC-DC CONVERTER 1484
Stefan Milovanovic ; Dražen Dujic

COMBINED DC POWER FLOW CONTROLLER FOR DC GRID 1491
Yongning Chi ; Xizhou Du ; Siqi Liu ; Xu Cai

AN APPROACH FOR THE EMULATION OF DC GRID ADMITTANCES: IMPLEMENTATION ON A BUCK CONVERTER 1498
Enrique Rodriguez-Diaz ; Fracisco D. Freijedo ; Drazen Dujic ; Juan C. Vasquez ; Josep M. Guerrero

A COMPOUND CONTROLLER FOR POWER FLOW AND SHORT-CIRCUIT FAULT IN DC GRID 1504
Han Ye ; Wu Chen ; Pengpeng Pan ; Xiaokun He

DESIGN PROCEDURE AND CONTROL OF A HYBRID CIRCUIT BREAKER WITH ADAPTABLE PULSE CURRENT INJECTION 1509
Andreas Jehle ; Jürgen Biela

A PRAGMATIC SOH AND SOC CO-ESTIMATOR FOR LITHIUM-ION BATTERIES IN SMART GRID APPLICATIONS 1517
Kaiyuan Li ; King Jet Tseng ; Feng Wei ; Boon-Hee Soong

MODELING AND STABILITY ANALYSIS OF PARALLEL DROOP-CONTROLLED AND CURRENT-CONTROLLED INVERTERS 1524
Shike Wang ; Zeng Liu ; Jinjun Liu ; Ronghui An

DIRECT WIRELESS BATTERY CHARGING SYSTEM 1530
Woo-Seok Lee ; Jin-Hak Kim ; Shin-Young Cho ; Il-Oun Lee

AN IMPROVED PWM SCHEME TO ACHIEVE ZERO-VOLTAGE SWITCHING FOR ALL DEVICES IN THREE-PHASE ISOLATED MATRIX RECTIFIER 1537
Xuerui Lin ; Yunwei Ryan Li ; Jahangir Afsharian ; Dewei David Xu

FIXED-FREQUENCY HF GATE DRIVER BY A PUSH-PULL SELF-EXCITATION LC OSCILLATOR HAVING A CAPACITANCE TRANSISTOR 1543
Naoyuki Ishibashi ; Takuya Mizushima ; Masahiko Hirokawa ; Akihiko Katsuki

A FLEXIBLE REDUCED CAPACITOR VOLTAGES STRATEGY FOR VARIABLE-SPEED DRIVES WITH MODULAR MULTILEVEL CONVERTER 1549
Fangzhou Zhao ; Guochun Xiao ; Daoshu Yang ; Zhiqian Wu ; Xin Meng

A LEAKAGE FLUX CANCELLATION TECHNIQUE FOR SERIES-PARALLEL COMBINED RESONANT CIRCUITS WITH ASYMMETRIC ROTARY TRANSFORMERS USED FOR ULTRASONIC SPINDLE DRIVE 1554
Jun Imaoka ; Masahito Shoyama

A NOVEL STRUCTURAL HEALTH MONITORING SYSTEM WITH WIRELESS POWER AND BI-DIRECTIONAL DATA TRANSFER 1562
Yujin Jangs ; Keon-Woo Kim ; Moo-Hyun Park ; Nayoung Lee ; Gun-Woo Moon

CONTROL STRATEGY FOR STARTER GENERATOR IN UAV WITH MICRO JET ENGINE 1567
Jun-Ichi Itoh ; Kazuki Kawamura ; Hiroyuki Koshikizawa ; Kazuyuki Abe

STUDY ON THE INFLUENCE OF VOLTAGE VARIATIONS FOR NON-INTRUSIVE LOAD IDENTIFICATIONS 1575
Yu-Hsiu Lin ; Shun-Kang Hung ; Men-Shen Tsai

BASIC EXPERIMENT OF A MAGLEV SYSTEM FOR A FLEXIBLE STEEL PLATE WITH CURVATURE: FUNDAMENTAL CONSIDERATION ON LEVITATION STABILITY UNDER DISTURBANCE 1580
Makoto Tada ; Kazuki Ogawa ; Takayoshi Narita ; Hideaki Kato ; Hiroyuki Moriyama

PERFORMANCE OF HYBRID MAGNETIC LEVITATION CONTROL SYSTEM FOR THIN STEEL PLATE BY EMS AND PMS: EXPERIMENTAL EVALUATION OF APPLYING OPTIMAL GAP AND ARRANGEMENT OF PMS..1586

Yasuaki Ito ; Yoshiho Oda ; Kengo Okuno ; Toshiki Suzuki ; Masahiro Kida ; Takayoshi Narita ; Hideaki Kato ; Hiroyuki Moriyama

A PRACTICAL LITHIUM-ION BATTERY MODEL BASED ON THE BUTLER-VOLMER EQUATION 1592

Kaiyuan Li ; King Jet Tseng ; Feng Wei ; Boon-Hee Soong

BONDING TECHNOLOGY USING COLD-ROLLED AG SHEET IN DIE-ATTACHMENT APPLICATIONS 1598

Seungjun Noh ; Chanyang Choe ; Chuantong Chen ; Hao Zhang ; Katsuaki Suganuma

HIGH-FREQUENCY SELF-DRIVEN SYNCHRONOUS RECTIFIER CONTROLLER FOR WPT SYSTEMS 1602

Akihiro Konishi ; Kazuhiro Umetani ; Eiji Hiraki

AUTOMATIC RESONANCE FREQUENCY TUNING METHOD FOR REPEATER IN RESONANT INDUCTIVE COUPLING WIRELESS POWER TRANSFER SYSTEMS....................1610

Masataka Ishihara ; Kazuhiro Umetani ; Eiji Hiraki

INDUCTIVE POWER TRANSFER FOR T5 FLUORESCENT LAMP LIGHTING SYSTEM 1617

Chung-Chuan Hou ; Tang-Jung Chen ; Ching-Chen Chen ; Chen-Wei Chang ; Po-Wei Wang

AN IMPLEMENT 1.5 MHZ OF INDUCTION HEATING FOR ALUMINUM BASED ON VACUUM TUBE OSCILLATOR CIRCUIT..1622

A. Bilsalam ; P. Chanmontree ; S. Supanyapong ; V. Chunkag

SINGLE-INDUCTOR MULTIPLE-OUTPUTS DIMMABLE LED DRIVER WITH BUCK CONVERTER...1626

Ta-Wei Huang ; Wei-Jing Tseng ; Jun-Xian Huang

A SOFT-SWITCHED THREE-LEVEL T-TYPE INVERTER WITH AUXILIARY COMMUTATED POLES...1634

Apollo Charalambous ; Xibo Yuan

CARRIER-BASED REALIZATION OF ARBITRARY SPACE-VECTOR PWM METHODS FOR THREE-LEVEL INVERTERS 1642

Somboon Sangwongwanich ; Supakorn Paiboon

MULTI-LEVEL TOPOLOGY BASED LINEAR AMPLIFIER FAMILY FOR REALIZATION OF NOISE-LESS INVERTERS...1649

Hidemine Obara ; Tatsuki Ohno ; Atsuo Kawamura

A NEW ZERO-VOLTAGE SWITCHING THREE-LEVEL CONVERTER WITH REDUCED RECTIFIER VOLTAGE STRESS 1655

Keon-Woo Kim ; Cheon-Yong Lim ; Dong-Kwan Kim ; Yu-Jin Jang ; Gun-Woo Moon

MODEL PREDICTIVE CONTROL OF A THREE-LEVEL NPC RECTIFIER WITH A SLIDING MANIFOLD TERM..1661

Xiaonan Gao ; Wei Tian ; Xicai Liu ; Zhenbin Zhang ; Ralph Kennel

H∞ CONTROL-BASED VIBRATION SUPPRESSION IN ROBOT ARM WITH STRAIN WAVE GEARING 1666

Tran Vu Trung ; Makoto Iwasaki

FINE FORCE SENSORLESS FORCE CONTROL BASED ON FRICTION-FREE DISTURBANCE OBSERVER 1673

Ohishi Kiyoshi ; Naoki Kamiya ; Toshimasa Miyazaki ; Yuki Yokokura

KINEMATICS AND TRACKING CONTROL OF A FOUR AXIS ANTENNA FOR SATCOM ON THE MOVE 1680

Oguz Kaan Hancioglu ; Mustafa Celik ; Ugur Tumerdem

POSITION SENSORLESS POSITION CONTROL FOR DUAL SOLENOID ACTUATOR..............1687

Sakahisa Nagai ; Atsuo Kawamura

CAE TECHNOLOGY APPLICATION TREND FOR LARGE-CAPACITY POWER ELECTRONICS DEVELOPMENT 1692

Teruo Yoshino ; Kuniaki Nagasaka ; Shigeaki Nakabayashi ; Ikuto Udagawa ; Isamu Tominaga ; Junya Konno

XILINX SYSTEM GENERATOR BASED MODELLING OF FINITE STATE MPC.....................1698

Vijay Kumar Singh ; Ravi Nath Tripathi ; Tsuyoshi Hanamoto

POWER HARDWARE-IN-THE-LOOP SETUP FOR STABILITY STUDIES OF GRID-CONNECTED POWER CONVERTERS 1704

Tommi Reinikka ; Henrik Alenius ; Tomi Roinila ; Tuomas Messo

PASSIVITY-BASED LCL FILTER DESIGN OF GRID-CONNECTED VSCS WITH CONVERTER SIDE CURRENT FEEDBACK 1711

Shih-Feng Chou ; Xiongfei Wang ; Frede Blaabjerg

ADAPTIVE CONTROL OF DC POWER DISTRIBUTION SYSTEMS: APPLYING PSEUDO-RANDOM SEQUENCES AND FOURIER TECHNIQUES.. 1719

Tomi Roinila ; Hessamaldin Abdollahi ; Silvia Arrua ; Enrico Santi

AN IMPROVED FINITE-SET MODEL PREDICTIVE TORQUE CONTROL FOR INTERIOR PERMANENT MAGNET SYNCHRONOUS MOTOR DRIVES.. 1724

Xinan Zhang ; Gilbert Foo ; Tung Ngo

PREDICTIVE TORQUE CONTROL FOR FIVE PHASE INDUCTION MOTOR DRIVE WITH COMMON MODE VOLTAGE REDUCTION 1730

Apekshit Bhowate ; Mohan Aware ; Sohit Sharma ; Yogesh Tatte

INDIRECT MATRIX CONVERTER FOR PERMANENT-MAGNET-SYNCHRONOUS-MOTOR DRIVES BY IMPROVED TORQUE PREDICTIVE CONTROL.. 1736

Yun Jang ; Yeongsu Bak ; Kyo-Beum Lee

PREDICTIVE DC-LINK CURRENT CONTROL BASED ON IPMSM DISCRETE STATE EQUATION FOR INVERTER WITHOUT INDUCTOR OR ELECTROLYTIC CAPACITOR 1741

Yousuke Akama ; Kodai Abe ; Kiyoshi Ohishi ; Yuki Yokokura ; Koji Kobayashi ; Tatsuki Kashihara

NEW SEARCH ALGORITHM OF MODEL PREDICTIVE CONTROL TO REDUCING CALCULATION AMOUNT FOR IMPROVING STEADY CURRENT CONTROL PERFORMANCE.. 1747

Masahiro Shimaoka ; Shinji Doki

DISTRIBUTED POWER SHARING STRATEGY FOR ISLANDED MICROGRIDS WITHOUT FREQUENCY AND VOLTAGE DEVIATIONS.. 1752

Tuan V. Hoang ; Hong-Hee Lee

LIFETIME-ORIENTED DROOP CONTROL STRATEGY FOR AC ISLANDED MICROGRIDS.................... 1758

Yanbo Wang ; Dong Liu ; Fujin Deng ; Dao Zhou ; Zhe Chen

EXPERIMENT ON HIERARCHICAL CONTROL BASED POWER QUALITY ENHANCEMENT FOR STANDALONE MICROGRID.. 1764

Darith Leng ; Sompob Polmai ; Kittichot Soontorntaweesub

A DISTRIBUTED PREDICTIVE CONTROL STRATEGY BASED ON STATE ESTIMATOR FOR ISLANDED MICROGRID 1771

Mi Dong ; Li Li ; Xiaoyu Tian

MAXIMUM POWER POINT TRACKING METHOD FOR PV MODULE UNDER WIDE RANGE VARYING IRRADIANCE LEVELS.. 1777

Hwa-Dong Liu ; Chang-Hua Lin

DUAL MPPT CONTROL AND FIELD TESTING FOR SWITCHED CAPACITOR-BASED CELL-LEVEL POWER BALANCING UTILIZING DIFFUSION CAPACITANCE OF PHOTOVOLTAIC CELLS.. 1782

Masatoshi Uno ; Yota Saito ; Masaya Yamamoto ; Shinichi Urabe

SERIES RESONANT DC-DC CONVERTER WITH DUAL-MODE RECTIFIER FOR PV MICROINVERTERS.. 1788

Yanfeng Shen ; Huai Wang ; Zhan Shen ; Yongheng Yang ; Frede Blaabjerg

VOLTAGE-REFERENCE ACTIVE POWER DECOUPLING BASED ON BOOST CONVERTER FOR SINGLE-PHASE BRIDGE INVERTER.. 1793

Shuang Xu ; Meiqin Mao ; Riming Shao ; Liuchen Chang

A SINGLE-PHASE COMMON GROUND BOOST INVERTER FOR PHOTOVOLTAIC APPLICATIONS 1799

Tan-Tai Tran ; Minh-Khai Nguyen ; Young-Cheol Lim ; Joon-Ho Choi

STUDY FOR FURTHER INTRODUCTION OF THE ELECTRONIC FREQUENCY CONVERTERS TO THE TOKAIDO SHINKANSEN 1803

Toshimasa Shimizu ; Ken Kunomura ; Masahiko Kai ; Hiroki Miyajima ; Teruhisa Matsui

COUNTERMEASURE FOR PARTIAL TURN-OFF OF THYRISTOR CHANGEOVER SWITCH INTRODUCED TO TOHOKU SHINKANSEN SHIN-YONO SECTIONING POST 1810

Yuki Mizumoto ; Nobuhito Kurosawa

HARDWARE–IN–THE–LOOP REAL–TIME SIMULATION EXPERIMENT PLATFORM FOR TRACTION POWER SUPPLY SYSTEM BASED ON DSPACE-XSIM...................................... 1816

Runze Zhang ; Fei Lin ; Zhongping Yang ; Hu Cao ; Yuping Liu

EVALUATING THE NON-SINUSOIDAL AND NON-SYMMETRIC REGIMES FROM A RAILWAY SUPPLYING SUBSTATION 1822

Ileana-Diana Nicolae ; Petre-Marian Nicolae ; Radu-Florin Marinescu

A FUNDAMENTAL TRAIN RUNNING EXPERIMENT FOR A BASIC PERFORMANCE VERIFICATION OF A TRAIN POWER DEMAND CONTROL SYSTEM BY DECENTRALIZED CONTROL ALGORITHM 1828

Yusuke Oki ; Tomoyuki Ogawa ; Yoko Takeuchi ; Tatsuhito Saito ; Jun'ichiro Kawaguchi

VERIFICATION OF SIC BASED MODULAR MULTILEVEL CASCADE CONVERTER (MMCC) FOR HVDC TRANSMISSION SYSTEMS 1834
Y. Ishii ; T. Jimichi

CONTROL OF A 6.6-KV TRANSFORMERLESS STATCOM BASED ON THE MMCC-SDBC USING SIC MOSFETS 1840
Laxman Maharjan ; Toshihisa Tajyuta ; Hiroshi Shinohara ; Akio Suzuki ; Akio Toba

ISOLATED THREE–PHASE AC/DC CONVERTER USING A SOFT–SWITCHING TECHNIQUE FOR BATTERY CHARGER 1847
Yuto Matsui ; Kazuma Suzuki ; Takaharu Takeshita ; Wataru Kitagawa

IMPLEMENTATION OF A MINIATURIZED SIC INVERTER 1854
Hideaki Fujita ; Cristian Andres Garces Guajardo

DESIGN CONSIDERATION OF FLYING CAPACITOR MULTILEVEL INVERTERS USING SIC MOSFETS 1860
Yukihiko Sato ; Kenji Natori

A CONTROL METHOD OF OVERVOLTAGE SUPPRESSION ACROSS THE DC CAPACITOR IN A GRID-CONNECTION CONVERTER USING LEG SHORT-CIRCUIT OF POWER MOSFETS DURING THE INITIAL CHARGE 1866
Tomoyuki Mannen ; Keiji Wada

THE ESSENTIAL RELATIONSHIP BETWEEN DEADBEAT PREDICTIVE CONTROL AND CONTINUOUS-CONTROL-SET MODEL PREDICTIVE CONTROL FOR PWM CONVERTERS 1872
Bi Liu ; Tao Chen ; Wensheng Song

DEADBEAT CONTROL FOR MULTI-LEVEL INVERTER USING 1MHZ MULTISAMPLING METHOD FOR UTILITY INTERACTIVE SYSTEM 1877
Ryosuke Kikuchi ; Ryunosuke Araumi ; Tomoki Yokoyama

1MHZ MULTISAMPLING DEADBEAT CONTROL WITH DISTURBANCE COMPENSATION METHOD FOR THREE PHASE PWM INVERTER 1883
Hiroaki Ueta ; Tomoki Yokoyama

MODULAR MULTILEVEL CONVERTER REPLACED ONE MODULE WITH HIGH VOLTAGE IGBT 1890
Kazunobu Oi ; Kenta Takasho ; Yugo Tadano

INCREASED EFFICIENCY AND REDUCED REALIZATION EFFORT OF DSBC AND DSCC MODULAR MULTILEVEL CONVERTERS (MMCS) 1896
A. Hillers ; J. Biela

COMMON-MODE VOLTAGE INJECTION TECHNIQUES FOR QUASI TWO-LEVEL PWM-OPERATED MODULAR MULTILEVEL CONVERTERS 1904
Jakub Kucka ; Axel Mertens

CURRENT TRACKING AND CELL-VOLTAGE LIMITATIONS OF MODULAR MULTILEVEL CONVERTERS WITH DIRECT DIGITAL CONTROL 1912
T.-F. Wu ; T.-C. Chou ; K.-E. Lin ; T.-Y. Li

SWITCHING LOSS ANALYSIS OF SIC-MOSFET BASED ON STRAY INDUCTANCE SCALING 1919
Keiji Wada ; Masato Ando

MODELING AND OPTIMIZATION OF DISPLACEMENT WINDINGS FOR TRANSFORMERS IN DUAL ACTIVE BRIDGE CONVERTERS 1925
Zhan Shen ; Yanfeng Shen ; Zian Qin ; Huai Wang

OPTIMIZED SELECTION AND UTILIZATION OF DC-LINK CAPACITOR IN A SINGLE-PHASE PV GRID INVERTER SYSTEM 1931
Caspar Collins ; Li Ran

AN EVALUATION CIRCUIT FOR DC-LINK CAPACITORS USED IN A HIGH-POWER THREE-PHASE INVERTER WITH CONDITION MONITORING 1938
Kazunori Hasegawa ; Ichiro Omura ; Shin-Ichi Nishizawa

RECENT MARKET AND TECHNICAL TRENDS IN COPPER ROTORS FOR HIGH-EFFICIENCY INDUCTION MOTORS 1943
Daniel Liang ; Victor Zhou

OVERVIEW OF THE LATEST RESEARCH AND DEVELOPMENT FOR COPPER DIE-CAST SQUIRREL-CAGE ROTORS 1949
Shu Yamamoto

A NOVEL HEAT-RESISTANT INSULATION-PROCESSING AGENT APPLICABLE TO COPPER DIE-CAST SQUIRREL-CAGE ROTORS 1955
Junichi Uchida ; Yuki Sueuchi ; Naosumi Kamiyama

INSULATION-PROCESSING OF COPPER DIE-CAST SQUIRREL-CAGE ROTOR ON MOTOR EFFICIENCY IN HIGH-SPEED OPERATION OVER 10,000 R/MIN 1960
Hideaki Hirahara ; Akira Tanaka ; Shu Yamamoto

HIGH-PRECISION ROTOR POSITION ESTIMATION FOR HIGH-SPEED SPMSM DRIVE BASED ON STATE OBSERVER AND HARMONIC ELIMINATION 1966
Peng Yang ; Xi Xiao ; Meng Zhang ; Shkodyrev Vyacheslav

HARMONIC LOSS REDUCTION IN HIGH SPEED MOTOR DRIVE SYSTEMS BY FLYING CAPACITOR MULTILEVEL INVERTER 1972
Anudari Tumurbaatar ; Sae Mochidate ; Koji Yamaguchi ; Tomohiro Matsuda ; Yukihiko Sato

CURRENT SOURCE TYPE PMSG WIND TURBINE SYSTEM WITH THREE-PHASE THREE-SWITCH BUCK-TYPE RECTIFIER FOR MACHINE-SIDE CONVERTER 1977
Beomseok Chae ; Tahyun Kang ; Yongsug Suh

A STUDY OF 10MW LOAD COMMUTATED INVERTER FOR GAS-TURBINE START-UP 1985
An Hyunsung ; Cha Hanju

PROTOTYPING OF 500 KVA MEDIUM FREQUENCY TRANSFORMER FOR OFFSHORE DIRECT-CURRENT COLLECTION GRID 1991
Tomoyuki Hatakeyama ; Naoyuki Kurita ; Mamoru Kimura

PSCAD/EMTDC AND RTDS SIMULATION ANALYSIS OF MULTIVENDOR MULTI-TERMINAL HVDC SYSTEM CONNECTED TO OFFSHORE WINDFARMS 1997
Hiroshi Suwa ; Takuro Arai ; Takahiro Ishiguro ; Tohru Yoshihara ; Mamoru Kimura ; Tsuneshisa Wachi ; Takahiro Horikoshi ; Tatsuhito Nakajima

INTEROPERABILITY OF MODULAR MULTILEVEL CONVERTERS AND 2-LEVEL VOLTAGE SOURCE CONVERTERS IN A LABORATORY-SCALE MULTI-TERMINAL DC GRID 2003
Salvatore D'arco ; Atsede G. Endegnanew ; Giuseppe Guidi ; Jon Are Suul

PRINCIPLE EXPERIMENT OF CURRENT COMMUTATED HYBRID DCCB FOR HVDC TRANSMISSION SYSTEMS 2011
Ryuta Hasegawa ; Kazuhisa Kanaya ; Yushi Koyama ; Toshiaki Matsumoto ; Takahiro Ishiguro

A THREE-INPUT CENTRAL CAPACITOR DC/DC CONVERTER 2016
Jiaxin Liu ; Feng Gao

SERIES/PARALLEL SWITCHING CIRCUITS USING POWER MOSFETS FOR PHOTOVOLTAIC MODULES 2022
Masamichi Tanemo ; Koki Matsudate ; Shinichi Nomura

MODULARIZED EQUALIZATION ARCHITECTURE BASED ON SWITCHED CAPACITOR CONVERTER TO VIRTUALLY UNIFY MISMATCHED PHOTOVOLTAIC PANEL CHARACTERISTICS 2030
Masatoshi Uno ; Masaya Yamamoto

BUCK-BOOST TYPE MPPT CIRCUIT SUITABLE FOR PHOTOVOLTAIC GENERATION OF VEHICLE INSTALLATION 2036
Fumihisa Kano ; Yuji Kasai ; Hideki Kimura ; Kouhei Sagawa ; Junnosuke Haruna ; Hirohito Funato

VERIFICATION TEST OF ENERGY-EFFICIENT OPERATIONS AND SCHEDULING UTILIZING AUTOMATIC TRAIN OPERATION SYSTEM 2042
Shoichiro Watanabe ; Yasuhiro Sato ; Takafumi Koseki ; Eisuke Isobe ; Jun Kawashita

THE DIRECT BENEFIT OF SIC POWER SEMICONDUCTOR DEVICES FOR RAILWAY VEHICLE TRACTION INVERTERS 2047
Shingo Makishima ; Kazuki Fujimoto ; Keiichiro Kondo

THE LOSS CHARACTERISTICS OF PSFB ZVS DC-DC CONVERTER APPLIED TO THE AUXILIARY POWER SYSTEM 2051
Xianjin Huang ; Juan Zhao ; Fei Lin

SURVEY ON ELECTROMAGNETIC INTERFERENCE ANALYSIS FOR TRACTION CONVERTERS IN RAILWAY VEHICLES 2058
Zhichang Yang ; Hong Li ; Chao Feng ; Yanfeng Jiang ; Fei Lin ; Zhongping Yang

DEVELOPMENT OF TRACTION MOTOR FOR NEW ZERO - EMISSION VEHICLE 2066
Akinobu Iwai ; Satoshi Honjo ; Hirofumi Suzumori ; Toshio Okazawa

EMC DESIGN AND DEVELOPMENT METHODOLOGY FOR TRACTION POWER INVERTERS OF ELECTRIC VEHICLES 2073
Isao Hoda ; Jia Li ; Hiroki Funato

SIMULATION-DRIVEN DESIGN OPTIMIZATION OF A MULTILAYER EMC INPUT FILTER 2078
Fatou Diouf ; Nadim Sakr ; Anna Gheonjian

EV TRACTION INVERTER EMPLOYING DOUBLE-SIDED DIRECT-COOLING TECHNOLOGY WITH SIC POWER DEVICE 2082
Takashi Hirao ; Masami Onishi ; Yusuke Yasuda ; Akihiro Namba ; Kinya Nakatsu

AN OVERVIEW OF STABILITY IMPROVEMENT METHODS FOR WIDE-OPERATION-RANGE FLYBACK CONVERTER WITH VARIABLE FREQUENCY PEAK-CURRENT-MODE CONTROL ... 2086

Ching-Hsiang Cheng ; Ching-Jan Chen ; Shinn-Shyong Wang

DESIGN AND IMPLEMENTATION OF A HIGH POWER DENSITY ACTIVE-CLAMPED FLYBACK CONVERTER ... 2092

Yu-Chen Liu ; Bing-Siang Huang ; Cheng-Hung Lin ; Katherine A. Kim ; Huang-Jen Chiu

OPTIMIZED VARIABLE ON-TIME CONTROL FOR LED LIGHTING DRIVER ... 2097

Jizhe Wang ; Haruhi Eto ; Fujio Kurokawa

DESIGN OF MULTIMODE BATTERY CHARGER WITH DYNAMIC VOLTAGE TRACKING CONTROL ... 2102

Pang-Jung Liu ; Lin-Hao Chien ; Song-Kai Lee ; Ang-Tung Chen

DUAL-SLOT POWER-PICKUP STRUCTURE FOR CONTACTLESS STRIP INDUCTIVE POWER TRACK SYSTEM ... 2107

Jia-You Lee ; I-Lin Chen ; Chien-Tzu Ko

DISCONTINUOUS SVM TECHNIQUE FOR THREE-LEG VSI FED BALANCED/UNBALANCED TWO-PHASE LOADS ... 2113

Supanut Charoensuksirikul ; Yuttana Kumsuwan

REDUCTION OF POWER LOSSES BASED ON GENERALIZED TWO-LEVEL PWM ALGORITHM FOR A NINE-SWITCH VSI ... 2121

Neerakorn Jarutus ; Yuttana Kumsuwan

SIC-BASED THREE-PHASE QUASI-Z-SOURCE INVERTER VERSUS THE TWO-STAGE TOPOLOGY - A COMPARISON ... 2129

Kornel Wolski ; Mariusz Zdanowski ; Jacek Rabkowski

DC-SIDE CIRCUIT IMPLEMENTATION OF A THREE-PHASE INVERTER FOR BALANCING PHASE-LEG CAPACITOR CURRENTS ... 2137

Takashi Hirao ; Keiji Wada ; Toshihisa Shimizu

A THREE-PHASE HYBRID SWITCHED-BOOST INVERTER ... 2145

Minh-Khai Nguyen ; Tan-Tai Tran ; Hoan-Tien Luong ; Kyoung-Won Lee ; Youn-Ok Choi ; Geum-Bae Cho

THE EFFECT OF BUILT-IN CR SNUBBER CAPACITOR INTO THE POWER MODULE ... 2149

Ryotaro Hata ; Shigeki Nishiyama

EVALUATION OF NOVEL HYBRID PROTECTION BASED ON PYROSWITCH AND FUSE TECHNOLOGIES ... 2153

Tomokazu Sakuraba ; Rémy Ouaida ; Song Chen ; Thibaut Chailloux

OPTIMAL DESIGN OF A MAGNETICALLY COUPLED FILTER FOR HIGH EFFICIENCY, LOW COST AND LOW VOLUME DC-DC BATTERY STORAGE CONVERTER ... 2158

Timothé Delaforge ; Robert Pasterczyk ; Mickaël Robert ; Hervé Chazal ; Jean-Luc Schanen ; Sébastien Mariethoz

HIGH POWER/CURRENT INDUCTOR LOSS MEASUREMENT WITH SHUNT RESISTOR CURRENT-SENSING METHOD ... 2165

Pin Yu Huang ; Toshihisa Shimizu

SENSITIVITY ANALYSIS OF MEDIUM FREQUENCY TRANSFORMER DESIGN ... 2170

Marko Mogorovic ; Drazen Dujic

STANDARD MODELS FOR POWER ELECTRONIC SYSTEM SIMULATION ... 2176

Koichi Shigematsu ; Hiroki Ishikawa ; Taku Noda ; Kentarou Fukushima ; Yoichi Sekiba ; Yusuke Kouno ; Takashi Abe ; Takayuki Sekisue ; Shinji Katoh

MODELING AND MODEL PARAMETER EXTRACTION OF WIDE BANDGAP POWER SEMICONDUCTOR DEVICE, PACKAGE, AND CIRCUIT FOR SIMULATING FAST SWITCHING BEHAVIOR ... 2181

Tsuyoshi Funaki

STABILITY ANALYSIS METHODS OF A GRID-CONNECTED INVERTER IN TIME AND FREQUENCY DOMAINS ... 2186

Toshiji Kato ; Kaoru Inoue ; Taiki Sakiyama

FINITE ELEMENT METHODS FOR MULTI-OBJECTIVE OPTIMIZATION OF A HIGH STEP-UP INTERLEAVED BOOST CONVERTER ... 2193

Wilmar Martinez ; Camilo Cortes ; Ahmad Bilal ; Jorma Kyyra

HIGH FIDELITY REAL-TIME SIMULATION OF MULTI-LEVEL CONVERTERS ... 2199

Jost Allmeling ; Niklaus Felderer ; Min Luo

AN ENHANCED HIGH FREQUENCY PULSATING VOLTAGE INJECTION METHOD BASED ON IMMUNE ALGORITHM FOR SENSORLESS IPMSM DRIVES ... 2204

Yanping Zhang ; Zhonggang Yin ; Chao Du ; Youyun Wang ; Xiangdong Sun

POSITION ESTIMATION ACCURACY IMPROVEMENT FOR MAGNETIC SALIENCY BASED SENSORLESS CONTROL INCLUDING CROSS-COUPLING FACTOR 2210
Keita Shimamoto ; Shinya Morimoto

SENSORLESS DRIVE IN THE LOW SPEED REGION AND AUTO-TUNING METHOD FOR PERMANENT MAGNET SYNCHRONOUS MOTORS 2216
Naofumi Nomura ; Shinichi Higuchi

HIGH STABILITY V/F CONTROL OF PMSM USING STATE FEEDBACK CONTROL BASED ON N-T COORDINATE SYSTEM 2224
Yosuke Matsuki ; Shinji Doki

STABILIZATION METHOD USING EQUIVALENT RESISTANCE GAIN BASED ON V/F CONTROL FOR IPMSM WITH LONG ELECTRICAL TIME CONSTANT 2229
Jun-Ichi Itoh ; Takato Toi ; Koroku Nishizawa

SINGLE-PHASE SOLID-STATE TRANSFORMER USING MULTI-CELL WITH AUTOMATIC CAPACITOR VOLTAGE BALANCE CAPABILITY 2237
Jun-Ichi Itoh ; Kazuki Aoyagi ; Keisuke Kusaka ; Masakazu Adachi

A DEVELOPED DUAL MMC ISOLATED DC SOLID STATE TRANSFORMER AND ITS MODULATION STRATEGY 2245
Yan Li ; Chao Liu ; Xu Cai

DC FAULT RIDE-THROUGH OF A THREE-PHASE DUAL-ACTIVE BRIDGE CONVERTER FOR DC GRIDS 2250
Jingxin Hu ; Shenghui Cui ; Rik W. De Doncker

A COMPOUND 10KV DVR SYSTEM BASED ON SOLID STATE TRANSFORMER STRUCTURE 2262
Yaqian Zhang ; Jianzhong Zhang ; Xing Hu ; Zakiud Din

A DUAL-ENERGY-SOURCE UNINTERRUPTIBLE POWER SUPPLY (UPS) 2270
Hao Wang ; Dehong Xu ; Binci Xu ; Haijin Li ; Ye Zhu

INFLUENCE OF WIND POWER FORECASTS ON EQUITABLE DISTRIBUTION METHOD OF WIND POWER CURTAILMENT 2278
Daisuke IIoka ; Hiroumi Saitoh

COMPARISON OF OPTIMIZED DEMAND OF EGS FOR MINIMIZING FUEL CONSUMPTION AND EGS MODEL WITH POWER GRID FREQUENCY USING A HPSPITAL LOAD WITH PV 2283
Yuji Mizuno ; Teppei Baba ; Fujio Kurokawa ; Nobumasa Matsui

COORDINATED DFIG WIND TURBINES AND SOLAR PV GENERATORS FOR INTER-AREA OSCILLATION DAMPING 2287
Tossaporn Surinkaew ; Issarachai Ngamroo

ENERGY MANAGEMENT USING A QUICK CHARGER WITH STORAGE BATTERIES FOR ELECTRIC VEHICLES 2292
Taku Ishibashi ; Toyonari Shimakage ; Norikazu Takeuchi ; Takaaki Kikuchi ; Midori Nonogaki

A METHOD FOR JUNCTION TEMPERATURE ESTIMATION UTILIZING TURN-ON SATURATION CURRENT FOR SIC MOSFET 2296
Hui-Chen Yang ; Rejeki Simanjorang ; Kye Yak See

FIELD BUS FOR DATA EXCHANGE AND CONTROL OF MODULAR POWER ELECTRONIC SYSTEMS WITH HIGH SYNCHRONISATION ACCURACY 2301
Stefan Rietmann ; Simon Fuchs ; André Hillers ; Jürgen Biela

ANALYTICAL INVESTIGATION ON ASYMMETRIC LCC COMPENSATION CIRCUIT FOR TRADE-OFF BETWEEN HIGH EFFICIENCY AND POWER 2309
Kodai Takeda ; Takafumi Koseki

PROBABILISTIC PCA-SUPPORT VECTOR MACHINE BASED FAULT DIAGNOSIS OF SINGLE PHASE 5-LEVEL CASCADED H-BRIDGE MLI 2317
Nagendra Vara Prasad Kuraku ; Yigang He ; Murad Ali

A STUDY ON EDGE SUPPORTED ELECTROMAGNETIC LEVITATION SYSTEM: FUNDAMENTAL CONSIDERATION ON LEVITATION PERFORMANCE OF THIN STEEL PLATE 2324
Yoshiho Oda ; Yasuaki Ito ; Kengo Okuno ; Masahiro Kida ; Toshiki Suzuki ; Takayoshi Narita ; Hideaki Kato ; Hiroyuki Moriyama

APPLICATION OF FACTS DEVICES FOR A DYNAMIC POWER SYSTEM WITHIN THE USA 2329
Jan Paramalingam ; Fuminori Nakamura ; Akihiro Matsuda ; Daisuke Yamanaka ; Taichiro Tsuchiya

CAPACITOR VOLTAGE BALANCING IN SEMI-FULL-BRIDGE SUBMODULE WITH DIFFERENTIAL-MODE CHOKE : (INVITEDPAPER) 2335
Kalle Ilves ; Yuhei Okazaki ; Nan Chen ; Muhammad Nawaz ; Antonios Antonopoulos

RESEARCH ON KEY TECHNOLOGY AND EQUIPMENT FOR ZHANGBEI 500KV DC GRID 2343
Hui Pang ; Xiaoguang Wei

WHAT LED TO SUCCESS IN ACADEMIC RESEARCH ON THE FAMILY OF MODULAR MULTILEVEL CASCADE CONVERTERS? 2352
Hirofumi Akagi

OPERATING PRINCIPLE OF CURRENT RESONANT CONVERTER USING AIR CORE TRANSFORMER FOR ISOLATED POWER SUPPLY ON CHIP 2360
Seiya Abe ; Hikaru Kaishakuji ; Satoshi Matsumoto

ANALYSIS FOR HIGH-FREQUENCY LLC RESONANT CONVERTER WITH PLANAR TRANSFORMER AT LIGHT-LOAD CONDITION 2365
Keon-Woo Kim ; Jae-Il Baek ; Yeonho Jeong ; Ki-Mok Kim ; Gun-Woo Moon

A NOVEL FULL DIGITAL CONTROL H-BRIDGE DC-DC CONVERTER FOR POWER SUPPLY ON CHIP APPLICATIONS 2370
Shigeki Nakano ; Toshiomi Oka ; Seiya Abe ; Satoshi Matsumoto

A HIGH-EFFICIENCY POWER SUPPLY FROM MAGNETIC ENERGY HARVESTERS 2376
Cheon-Yong Lim ; Yeonho Jeong ; Keon-Woo Kim ; Feel-Soon Kang ; Gun-Woo Moon

OPPORTUNITIES FOR LEVERAGING LOW-VOLTAGE GAN DEVICES IN MODULAR MULTI-LEVEL CONVERTERS FOR ELECTRIC-VEHICLE CHARGING APPLICATIONS 2380
Mojtaba Ashourloo ; Mohammad Shawkat Zaman ; Miad Nasr ; Olivier Trescases

A NEW CONTROL STRATEGY FOR MODULAR MULTILEVEL CONVERTER OPERATING IN QUASI TWO-LEVEL PWM MODE 2386
Chao Wang ; Kui Wang ; Zedong Zheng ; Yongdong Li

A CURRENT-SOURCE TYPE MMC WITH DELTA-CONNECTED ARMS FOR SMES 2393
Yushi Miura ; Toshifumi Ise

NEW MODULE WITH ISOLATED HALF BRIDGE OR ISOLATED FULL BRIDGE FOR MODULAR MEDIUM VOLTAGE CONVERTER 2400
Yunpeng Si ; Yifu Liu ; Qin Lei

DEVELOPMENT OF A 700-V-CLASS REVERSE-BLOCKING IGBT FOR ADVANCED T-TYPE NEUTRAL POINT-CLAMPED POWER CONVERSION SYSTEM 2404
Hiroki Wakimoto ; Haruo Nakazawa ; David H. Lu ; Takashi Matsumoto ; Yoichi Nabetani

CERAMIC EMBEDDING AS PACKAGING SOLUTION FOR FUTURE POWER ELECTRONIC APPLICATIONS 2410
Hoang Linh Bach ; Tobias Maximilian Endres ; Daniel Dirksen ; Sigrid Zischler ; Christoph Friedrich Bayer ; Andreas Schletz ; Martin März

MICROELECTROMECHANICAL SYSTEM (MEMS) RESONATOR: A NEW ELEMENT IN POWER CONVERTER CIRCUITS FEATURING REDUCED EMI 2416
A N M Wasekul Azad ; Sourov Roy ; Abu Saleh Imtiaz ; Faisal Khan

A LUMPED THERMAL MODEL INCLUDING THERMAL COUPLING EFFECTS AND BOUNDARY CONDITIONS FOR CAPACITOR BANKS 2421
Qiusheng Wang

HYSTERESIS MODELING OF MAGNETIC DEVICES BASED ON RELUCTANCE NETWORK ANALYSIS 2426
Yoshiki Hane ; Kenji Nakamura

OPTIMAL SIZING AND PLACEMENT OF SOLAR POWERED CHARGING STATION UNDER EV LOADS PENETRATION USING ARTIFICIAL BEE COLONY TECHNIQUE 2430
Yuttana Kongjeen ; Kulsomsup Yenchamchalit ; Krischonme Bhumkittipich

A COMPARISON OF AVERAGE MODEL, SAMPLED-DATA MODEL AND MULTI-FREQUENCY MODEL BASED ON DC/DC CONVERTERS 2435
Xiangpeng Cheng ; Jinjun Liu ; Zeng Liu ; Yiming Tu ; Danhong Xue

SMALL-SIGNAL DISCRETE-TIME MODELING AND DIGITAL CONTROL OF THE BI-DIRECTIONAL DC/DC CONVERTERS 2441
Jia Yaoqin ; Xu Yingchun ; Hou Yijie

ENERGY MANAGEMENT OF HYDROGEN-STORAGE PHOTOVOLTAIC GENERATION SYSTEM WITH A FUNCTION OF SUPPRESSING SHORT-PERIOD COMPONENTS 2449
Yuuki Machida ; Akihisa Goto ; Akiko Takahashi ; Shigeyuki Funabiki

A DYNAMIC BATTERY CHARGING APPROACH FOR ENERGY TRADING IN THE SMART GRID 2456
Avinash Sharma ; Akshay Kumar Rathore ; Rajesh Kumar

A FORCED COMMUTATION METHOD OF THE SOLID-STATE TRANSFER SWITCH IN THE UNINTERRUPTED POWER SUPPLY APPLICATIONS 2462
Meng-Jiang Tsai ; Jiuyang Zhou ; Po-Tai Cheng

ONLINE INTERNAL IMPEDANCE MEASUREMENTS OF LI-ION BATTERY USING PRBS BROADBAND EXCITATION AND FOURIER TECHNIQUES: METHODS AND INJECTION DESIGN..2470
Jussi Sihvo ; Tuomas Messo ; Tomi Roinila ; Roni Luhtala

A DC CURRENT FLOW CONTROLLER FOR MESHED HVDC GRIDS 2476
Viktor Hofmann ; Mark-M. Bakran

AN ISOLATED SOFT-SWITCHING HYBRID-SOURCE DC-DC CONVERTER FOR DC OFFSHORE WIND FARMS...2484
Shenghui Cui ; Jingxin Hu ; Marco Stieneker ; Rik W. De Doncker

A TRANSFORMERLESS MULTI-CELL SOLID-STATE FAULT CURRENT LIMITER FOR MEDIUM VOLTAGE POWER SYSTEM..2490
Pantarote Techama ; Sompob Polmai ; Chanin Bunlaksananusorn

A NOVEL DC POWER FLOW CONTROLLER FOR HVDC GRIDS WITH DIFFERENT VOLTAGE LEVELS...2496
Ya'nan Wu ; Han Ye ; Wu Chen ; Xiaokun He

DESIGN AND CONTROL OF SINGLE-PHASE GRID-CONNECTED PHOTOVOLTAIC MICROINVERTER WITH REACTIVE POWER SUPPORT CAPABILITY 2500
Geon-Hong Min ; Kyung-Hwan Lee ; Jung-Ik Ha ; Myong Hwan Kim

OPTIMAL SIZE AND MULTI-OBJECTIVE CONTROL OF BATTERY ENERGY STORAGES IN DISTRIBUTION SYSTEM WITH HIGH PENETRATION OF DISTRIBUTED PV GENERATORS 2505
Meiqin Mao ; Lei Zhou ; Yangyang Wang ; Liuchen Chang

MISSION PROFILE-ORIENTED CONTROL FOR RELIABILITY AND LIFETIME OF PHOTOVOLTAIC INVERTERS 2512
Ariya Sangwongwanich ; Yongheng Yang ; Dezso Sera ; Frede Blaabjerg

DISCONTINUOUS CURRENT MODE CONTROL FOR MINIMIZATION OF THREE-PHASE GRID-TIED INVERTER IN PHOTOVOLTAIC SYSTEM 2519
Hoai Nam Le ; Jun-Ichi Itoh

A THEORETICAL ANALYSIS ON STATIC CHARACTERISTICS OF VOLTAGE BASED CONTROL METHOD AND CURRENT BASED CONTROL METHOD FOR THE WAYSIDE ENERGY STORAGE SYSTEM IN DC-ELECTRIFIED RAILWAY 2527
Hiroyasu Kobayashi ; Keiichiro Kondo ; Diego Iannuzzi

IMPROVEMENT OF A DC ELECTRICAL RAILWAY SIMULATOR USING ARTIFICIAL INTELLIGENCE 2534
Alvaro J. Lopez-Lopez ; Ramon R. Pecharroman ; Antonio Fernandez-Cardador ; Asuncion P. Cucala

FEEDING-LOSS REDUCTION BY HIGHER-VOLTAGE DC RAILWAY FEEDING SYSTEM WITH DC-TO-DC CONVERTER...2540
Hidenori Shigeeda ; Hiroaki Morimoto ; Kazuhiko Ito ; Toshiyuki Fujii ; Naoki Morishima

MODELING AND SIMULATION OF NOVEL RAILWAY POWER SUPPLY SYSTEM BASED ON POWER CONVERSION TECHNOLOGY 2547
Minwu Chen ; Ruofei Liu ; Shaofeng Xie ; Xiaofang Zhang ; Yimin Zhou

COMPARATIVE STUDY ON FRONT-END PARAMETER IDENTIFICATION METHODS FOR WIRELESS POWER TRANSFER WITHOUT WIRELESS COMMUNICATION SYSTEMS...........................2552
Sinan Li ; S. Y. Ron Hui

A NEW TYPE OF WIRELESS V2X SYSTEM WITH A DUAL-ACTIVE BIDIRECTIONAL SINGLE-ENDED CONVERTER AND OPTIMIZED SIC-MOSFET 2558
Hideki Omori ; Aoto Yamamoto ; Naoki Mukaiyama ; Masahito Tsuno ; Kenji Fukuda ; Hisato Michikoshi ; Noriyuki Kimura ; Toshimitsu Morizane

METAL OBJECT DETECTION SYSTEM WITH PARALLEL-MISTUNED RESONANT CIRCUITS AND NULLIFYING INDUCED VOLTAGE FOR WIRELESS EV CHARGERS.............................2564
Seog Y. Jeong ; Van X. Thai ; Jun H. Park ; Chun T. Rim

WIRELESS EV CHARGING SYSTEM WITHOUT AIR-GAP AND MISALIGNMENT 2569
Wenxing Zhong ; Dehong Xu

FIXED SLOPE CARRIER PWM FOR INDIRECT MATRIX CONVERTER 2576
Tzung-Lin Lee ; Chun-Yao Hung ; Yen-Wen Chen ; Wen-Mei Huang

CARRIER-BASED OVERMODULATION STRATEGY FOR MATRIX CONVERTERS.......................................2581
Paiboon Kiatsookkanatorn ; Somboon Sangwongwanich

THREE-PHASE TO HIGH-FREQUENCY SINGLE-PHASE MATRIX CONVERTER : A FREQUENCY CONTROL SUITABLE FOR SOFT SWITCHING..2589
Wataru Kodaka ; Satoshi Ogasawara ; Koji Orikawa ; Masatsugu Takemoto ; Takashi Hyodo ; Hiroyuki Tokusaki

TWO-STEP COMMUTATION FOR ISOLATED DC-AC CONVERTER WITH MATRIX CONVERTER...2596
Shunsuke Takuma ; Jun-Ichi Itoh

A DC-LINK CAPACITOR VOLTAGE OSCILLATION REDUCTION METHOD FOR A MODULAR MULTILEVEL CASCADE CONVERTER WITH SINGLE DELTA BRIDGE CELLS (MMCC-SDBC) 2604
Takaaki Tanaka ; Huai Wang ; Frede Blaabjerg

OPTIMIZED DECOUPLING CONTROL OF FLYING CAPACITOR IN ANPC FIVE-LEVEL INVERTER 2611
Fusheng Wang ; Deyou Zheng ; Jianing Wang ; Fei Li ; Fang Liu ; Shuying Yang ; Zhen Xie

CASCADED DUAL-BUCK AC-AC CONVERTER USING COUPLED INDUCTORS 2619
Sanghun Kim ; Duekjin Jang ; Heung-Geun Kim ; Honnyong Cha

INSTANTANEOUS POWER LOSS CALCULATION FOR MMC BASED ON VIRTUAL ARM MATHEMATICAL MODEL 2625
Yin Shiyuan ; Wang Yue ; Yin Taiyuan ; Nie Cheng ; Duan Guozhao ; Wang Zhang

COMPARISON OF CURRENT CONTROL STRATEGIES IN MODULAR MULTILEVEL CONVERTER 2630
Jianzhao Wei ; Anirudh Budnar Acharya ; Lars Norum ; Pavol Bauer

MODEL PREDICTIVE CONTROL OF A MODULAR MULTILEVEL CONVERTER WITH AN IMPROVED CAPACITOR BALANCING METHOD 2638
Shichong Zhang ; Baodong Bai ; Dezhi Chen

HIGH STEP-UP DC-DC CONVERTER BASED ON MULTI-CELL COUPLED INDUCTOR DIODE-CAPACITOR NETWORK 2646
Xinying Li ; Yan Zhang ; Jinjun Liu ; Pengxiang Zeng

NOVEL ACTIVE CLAMPING STEP-DOWN DC-DC CONVERTER WITH LOWER VOLTAGE STRESS 2653
Chi-Hsuan Hsu ; Jun-Min Jian ; Jiann-Fuh Chen ; Hsuan Liao

DESIGN AND EVALUATION OF A MAGNETICALLY-LOOSELY-COUPLED INDUCTOR FOR A FOUR-PHASE INTERLEAVED BOOST CHOPPER 2660
Hiroki Kowatari ; Toshinori Kitamura ; Nobukazu Hoshi

A SYNCHRONOUS-REFERENCE-FRAME I-V DROOP CONTROL METHOD FOR PARALLEL-CONNECTED INVERTERS 2668
Mingshen Li ; Yonghao Gui ; Zheming Jin ; Yajuan Guan ; Josep M. Guerrero

TRANSIENT STABILITY IMPACT OF THE PHASE-LOCKED LOOP ON GRID-CONNECTED VOLTAGE SOURCE CONVERTERS 2673
Heng Wu ; Xiongfei Wang

COMPREHENSIVE ANALYSIS OF VIRTUAL IMPEDANCE-BASED ACTIVE DAMPING FOR LCL RESONANCE IN GRID-CONNECTED INVERTERS 2681
Teng Liu ; Zeng Liu ; Jinjun Liu ; Yiming Tu ; Zipeng Liu

A COMPARATIVE STUDY OF THE TRADITIONAL FS-MPC AND THE PROPOSED CSF-PCC FOR THE THREE-PHASE GRID-CONNECTED INVERTERS 2688
Zhixun Ma ; Xin Zhang ; Jingjing Huang

CONSTANT SWITCHING-FREQUENCY PREDICTIVE- CURRENT-CONTROL METHOD WITH A DICHOTOMY SOLUTION FOR THE GRID-TIED INVERTERS 2692
Zhixun Ma ; Xin Zhang ; Jingjing Huang ; Zhao Bin ; Lyu Jing

OBSERVER-BASED ACTIVE DAMPING FOR GRID-CONNECTED CONVERTERS WITH LCL FILTER 2697
Y. Zhang ; M. G. L. Roes ; M. A. M. Hendrix ; J. L. Duarte

CONDUCTION LOSS ANALYSIS AND OPTIMIZATION DESIGN OF FULL BRIDGE LLC RESONANT CONVERTER 2703
Yugang Yang ; Lifei Zhang ; Tianshu Ma

FULL-BRIDGE T-TYPE ISOLATED DC/DC CONVERTER WITH WIDE INPUT VOLTAGE RANGE 2708
Dong Liu ; Yanbo Wang ; Fujin Deng ; Zhe Chen

RESEARCH ON HIGH EFFICIENCY LLC DC-DC CONVERTER BASED ON SIC MOSFET 2714
Pengcheng Han ; Xiaoqiong He ; Haijun Ren ; Zhiqing Zhao ; Xu Peng

AN IMPROVED DUAL PHASE SHIFT CONTROL STRATEGY FOR DUAL ACTIVE BRIDGE DC-DC CONVERTER WITH SOFT SWITCHING 2718
Miao Hong ; Gao Xuanjie ; Zeng Chengbi ; Duan Shujiang

DEVELOPMENT OF AN SIC HIGH-FREQUENCY PWM INVERTER USING A THICK MULTILAYER PCB TO MINIMIZE STRAY INDUCTANCE 2725
Kohsuke Ishikawa ; Satoshi Ogasawara ; Masatsugu Takemoto ; Koji Orikawa

FAST SWITCHING PLANAR POWER MODULE WITH SIC MOSFETS AND ULTRA-LOW PARASITIC INDUCTANCE 2732
Arash Edvin Risseh ; Hans-Peter Nee ; Konstantin Kostov

EXPERIMENTAL EVALUATION OF INVERTER SYSTEM CONSISTING OF 4-PARALLEL GAN DEVICES UNIT 2738

Yoshiya Ohnuma ; Satoshi Miyawaki ; Fumiya Hattori ; Masayoshi Yamamoto

IMPACT OF THE THERMAL-INTERFACE-MATERIAL THICKNESS ON IGBT MODULE RELIABILITY IN THE MODULAR MULTILEVEL CONVERTER 2743

Yi Zhang ; Huai Wang ; Zhongxu Wang ; Yongheng Yang ; Frede Blaabjerg

NANOSCALE INVESTIGATION OF THE POWER MOSFET BY THE AFM/KFM/SCFM 2750

Mizuki Nakajima ; Yuuki Uchida ; Nobuo Satoh ; Hidekazu Yamamoto

SIMULATION ANALYSIS OF OPTIMUM GATE DRIVING CONDITIONS OF IGBTS 2756

Satoshi Sugahara ; Masaki Kawakami ; Kousuke Kamakura

IMPROVEMENT OF THE I2T CAPABILITY FOR XEV ACTIVE SHORT CIRCUIT PROTECTION BY COMBINATION OF RC-IGBT AND LEADFRAME TECHNOLOGIES 2764

Keiichi Higuchi ; Hayato Nakano ; Akihiro Osawa ; Akio Kitamura ; Shunji Takenoiri ; Daisuke Inoue ; Souichi Yoshida ; Hiromichi Gohara

INVESTIGATION OF SWITCHING BEHAVIOR OF AN IGBT UNDER SOFT TURN-OFF IN APPLICATION FOR DUAL-ACTIVE BRIDGE CONVERTERS 2768

Eri Ogawa ; Yuichi Onozawa ; Rik W. De Doncker

600 V HIGH VOLTAGE GATE DRIVER IC (HVIC) WITH 1.0 MHZ HIGH FREQUENCY OPERATION FOR LLC CURRENT RESONANT POWER SUPPLY 2774

Masaharu Yamaji ; Masashi Akahane ; Takahide Tanaka ; Akihiro Jonishi ; Hidetomo Ohashi ; Masahiro Sasaki ; Hitoshi Sumida

AN INTEGRATED VOLTAGE AND CURRENT BALANCING STRATEGY OF SERIES-PARALLEL CONNECTED IGBTS 2780

Xiaotong Du ; Fang Zhuo ; Haotian Sun ; Hao Yi ; Yanlin Zhu

THERMAL DESIGN AND ANALYSIS OF A CABLE CHARGER USED FOR PORTABLE ELECTRONICS 2785

Mofan Tian ; Xu Yang ; Naizeng Wang ; Yang Chen ; Laili Wang

PARASITIC INDUCTANCE DESIGN CONSIDERATIONS TO SUPPRESS GATE VOLTAGE OSCILLATION OF FAST SWITCHING POWER SEMICONDUCTOR DEVICES 2789

Yusuke Sugihara ; Kimihiro Nanamori ; Masayoshi Yamamoto ; Yasuki Kanazawa

THE EXAMINATION OF INCREASING OPERATION SPEED OF CONSEQUENT POLE TYPE AXIAL GAP MOTOR FOR HIGHER OUTPUT POWER DENSITY 2796

Toru Ogawa ; Tomohira Takahashi ; Masatsugu Takemoto ; Satoshi Ogasawara ; Hideaki Arita ; Akihiro Daikoku

BASIC STUDY OF PMASYNRM WITH BONDED MAGNETS FOR TRACTION APPLICATIONS 2802

Marika Kobayashi ; Shigeo Morimoto ; Masayuki Sanada ; Yukinori Inoue

STUDY ON ROTOR STRUCTURE SUITABLE FOR IMPROVING POWER DENSITY AND EFFICIENCY IN IPMSMS FOR AUTOMOTIVE APPLICATIONS 2808

R. Imoto ; M. Sanada ; S. Morimoto ; Y. Inoue

EXAMINATION OF THE DEMAGNETIZATION SUPPRESSION EFFECT OF PLACING FLUX BARRIERS IN AN IPMSM USING RARE-EARTH BONDED MAGNETS 2814

Takashi Umeda ; Masayuki Sanada ; Shigeo Morimoto ; Yukinori Inoue

A NOVEL POLE-CHANGING METHOD WITH A MULTIPLE THREE-PHASE INVERTER 2820

Yuki Hidaka ; Taiga Komatsu ; Hideaki Arita

STARTING CHARACTERISTICS OF AN ULTRA-LIGHTWEIGHT MOTOR USING MAGNETIC RESONANCE COUPLING 2826

Kenta Takishima ; Kazuto Sakai

DESIGN AND BASIC CHARACTERISTICS ANALYSIS OF TOROIDAL WINDING AXIAL GAP INDUCTION MOTOR 2832

Ryosuke Sakai ; Yukihiro Yoshida ; Katsubumi Tajima

MAGNET ARRANGEMENT SUITABLE FOR LARGE AIR GAP LENGTH IN LINEAR PM VERNIER MOTOR 2836

Tatsuya Ninomiya ; Abdulaziz Gasim ; Shoji Shimomura

MICRO ELECTROMAGNETIC VIBRATION ENERGY HARVESTER WITH MECHANICAL SPRING AND IRON FRAME FOR LOW FREQUENCY OPERATION 2842

Yecheng Shen ; Kaiyuan Lu ; Yongming Xia

MEASUREMENT OF TWO-LEVEL INVERTER INDUCED CURRENT SLOPES AT HIGH SWITCHING FREQUENCIES FOR CONTROL AND IDENTIFICATION ALGORITHMS OF ELECTRICAL MACHINES 2848

Simon Decker ; Andreas Liske ; Daniel Schweiker ; Johannes Kolb ; Michael Braun

A NEW TOPOLOGY OF SWITCHED-CAPACITOR MULTILEVEL INVERTER FOR SINGLE-PHASE GRID-CONNECTED WITH ELIMINATING LEAKAGE CURRENT...2854

Mehdi Samizadeh ; Xu Yang ; Bagher Karami ; Wenjie Chen ; Mohamad Abou Houran ; Adib Abrishamifar ; Abdolreza Rahmati

AN INTERLEAVED BUCK-CASCADED BUCK-BOOST INVERTER FOR PV GRID-CONNECTION APPLICATIONS...2860

Chien-Hsuan Chang ; Chun-An Cheng ; Hung-Liang Cheng

A NOVEL PV ARRAY CONNECTION STRATEGY WITH PV-BUCK MODULE TO IMPROVE SYSTEM EFFICIENCY...2866

Chi Shao ; Wenjie Wang ; Lijun Hang ; Anping Tong ; Shitao Wang

A COMMON-MODE VOLTAGE REDUCTION FOR TWO-STAGE THREE-PHASE TRANSFORMERLESS PV INVERTERS...2871

Adisak Promyoo ; Surapong Suwankawin

A GRID-CONNECTED PV-ENERGY STORAGE SYSTEM WITH SYNCHRONOUS GENERATOR CHARACTERISTICS...2877

Huadian Xu ; Jianhui Su ; Ning Liu ; Yong Shi ; Yan Du

A TRANSFORMERLESS BIDIRECTIONAL DC-DC CONVERTER BASED ON POWER UNITS WITH UNIPOLAR AND BIPOLAR STRUCTURE FOR MVDC INTERCONNECTION...2882

Lejia Sun ; Fang Zhuo ; Feng Wang ; Hao Yi ; Baohui Ma

NEW MODULATION CONTROL OF CONVERTER SYSTEM APPLIED FOR OFFSHORE WIND FARMS...2887

Naoki Kawabata ; Noriyuki Kimura ; Toshimitsu Morizane ; Hideki Omori

SPHERE DECODING BASED LONG-HORIZON PREDICTIVE CONTROL OF THREE-LEVEL NPC BACK-TO-BACK PMSG WIND TURBINE SYSTEMS...2895

Ferdinand Grimm ; Zhenbin Zhang ; Ralph Kennel

BASED ON PCHD AND HPSO SLIDING MODE CONTROL OF D-PMSG WIND POWER SYSTEM...2901

Lijun Hou ; Xuemei Zheng ; Chao Wang ; Yangman Li ; Haoyu Li

ESTABLISHMENT AND DYNAMIC CONTROL OF WIND INDUCTION GENERATOR...2907

M. Z. Lu ; V. K. Ganisetti ; C. M. Liaw

MIDDLE FREQUENCY SOLID STATE TRANSFORMER FOR HVDC TRANSMISSION FROM OFFSHORE WINDFARM...2914

Noriyuki Kimura ; Toshimitsu Morizane ; Isao Iyoda ; Kazushige Nakao ; Tomoki Yokoyama

SIMULATION OF WIND POWER GENERATION SYSTEM USING SWITCHED RELUCTANCE GENERATOR AND CAPACITOR-LESS AC-AC CONVERTER...2921

Guyuan Ji ; Kazuhiro Ohyama

VARIABLE FREQUENCY CONTROL AND FILTER DESIGN FOR OPTIMUM ENERGY EXTRACTION FROM A SIC WIND INVERTER...2932

Abdallah Hussein ; Alberto Castellazzi

EXPERIMENTAL VERIFICATIONS OF UPFC USING DEADBEAT CONTROL WITH 3-PHASE UNBALANCED COMPENSATION...2938

Shin-Ichi Hamasaki ; Hiroto Fukuda ; Syohei Tokumaru ; Mineo Tsuji

A CONTROL METHOD FOR TWO TYPES OF THREE-PHASE TRANSFORMERLESS UNIFIED POWER QUALITY CONDITIONER...2944

Fujian Li ; Guochun Xiao ; Fangzhou Zhao ; Shuai Zhang ; Baojin Liu

DESIGN OF CUSTOMER-END CONVERTER SYSTEMS FOR LOW VOLTAGE DC DISTRIBUTION FROM A LIFE CYCLE COST PERSPECTIVE...2948

A. Mattsson ; P. Nuutinen ; T. Kaipia ; P. Peltoniemi ; J. Karppanen ; V. Tikka ; A. Lana ; P. Pinomaa ; P. Silventoinen ; J. Partanen

A CONTROL METHOD OF DC CAPACITOR VOLTAGE IN MMC FOR HVDC SYSTEM USING NEGATIVE SEQUENCE CURRENT...2956

Hanis Afiqah Binti Jaffar ; Ahmad Arif Bin Abd Rahman ; Hiroaki Kakigano

A COORDINATE AND DISTRIBUTED CONTROL SCHEME FOR MULTILEVEL AND MULTI-STAGE MEDIUM VOLTAGE SOLID STATE TRANSFORMER...2963

Jintong Nie ; Liqiang Yuan ; Qing Gu ; Jianning Sun ; Zhengming Zhao

AN IMPROVED HARMONIC POWER SHARING SCHEME OF PARALLELED INVERTER SYSTEM...2969

Liu Hongpeng ; Liu Xiaoxi ; Zhang Wei ; Wang Wei

THE GRID IMPEDANCE ADAPTATION DUAL MODE CONTROL STRATEGY IN WEAK GRID...2973

Ming Li ; Xing Zhang ; Ying Yang ; Pengpeng Cao

TRANSMISSION POWER ANALYSIS AND CONTROL OF THE DC TRANSFORMER IN HYBRID AC/DC MICROGRID 2980
Jingjin Huang ; Xin Zhang ; Tengfei Zhang

A NOVEL FLEXIBLE INTERCONNECTION SCHEME FOR MICROGRID TO OPTIMIZE THE CAPACITY OF ENERGY STORAGE SYSTEM (ESS) 2986
Zhou Jianqiao ; Zhang Jianwen ; Cai Xu ; Li Zhuyong ; Wang Jiacheng ; Zang Jiajie

VSC CONTROL AND PARAMETERS DESIGN BASED ON VIRTUAL SYNCHRONOUS GENERATOR 2992
Fang Liu ; Meng Wang ; Zhen Xie ; Fusheng Wang ; Jinxin Deng ; Xing Zhang

MULTI-TARGET VIRTUAL RESISTANCE CONTROL STRATEGY IN A 400 HZ LOW VOLTAGE MICROGRID 2997
Yuze Li ; Xuejun Pei ; Zhi Chen ; Hanyu Wang ; Yong Kang

AN ADAPTIVE POWER COMPENSATION STRATEGY FOR THE VOLTAGE STABILIZATION OF LCL-VSC BASED MICROGRIDS 3002
Sheng Xu ; Wu Cao ; Dongchen Fan ; Jianfeng Zhao ; Shunyu Wang

RESONANCE DETECTION STRATEGY FOR MULTIPLE GRID-CONNECTED INVERTERS-BASED SYSTEM USING CASCADED SECOND-ORDER GENERALIZED INTEGRATOR 3010
Wu Cao ; Dongchen Fan ; Kangli Liu ; Jianfeng Zhao ; Liheng Ruan ; Xiaojun Wu

HARMONIC STABILITY ASSESSMENT BASED ON GLOBAL ADMITTANCE FOR MULTI-PARALLELED GRID-CONNECTED VSIS USING MODIFIED NYQUIST CRITERION 3015
Wu Cao ; Dongchen Fan ; Kangli Liu ; Jianfeng Zhao ; Liheng Ruan ; Xiaojun Wu

THE AC TRACTION POWER SUPPLY SYSTEM FOR URBAN RAIL TRANSIT BASED ON NEGATIVE SEQUENCE CURRENT COMPENSATOR 3020
Tianshu Zhao ; Xu Peng

GRID CONNECTED POWER GENERATION CONTROL METHOD FOR Z-SOURCE INTEGRATED BIDIRECTIONAL CHARGING SYSTEM 3025
Xu Jia ; Guoming Chuai ; Haonan Niu ; Qianfan Zhang

AN ISOLATED PFC CONVERTER WITH HARMONIC MODULATION TECHNIQUE FOR EV CHARGERS 3030
Byung-Kwon Lee ; Jun-Young Lee ; Dong-Hun Kang

HIGHLY DYNAMIC SWITCHING FREQUENCY-BASED CALCULATION OF POWER QUANTITIES, FUNDAMENTAL WAVEFORMS, AND RMS VALUES OF INVERTER-FED ELECTRICAL MACHINES 3034
Alexander Stock ; Johannes Teigelkötter ; Johannes Büdel

DESIGN AND ANALYSIS OF HIGH VOLTAGE POWER SUPPLY FOR INDUSTRIAL ELECTROSTATIC PRECIPITATORS 3040
Shengwen Fan ; Yiqin Yuan ; Pengyu Jia ; Zhigang Chen ; Haisi Li

LOAD SHARING OPERATION IN N+1 UPS SYSTEM BY USING HARMONIC SHARING CONTROL METHOD 3046
Prashant Patel ; Sagar Naina ; Utsav Patel ; Premal Patwa

RESEARCH ON CAPACITY OPTIMIZATION OF PV-WIND-DIESEL-BATTERY HYBRID GENERATION SYSTEM 3052
Cailing Zhu ; Furong Liu ; Sheng Hu ; Shu Liu

A NUMERICAL ANALYSIS AND IMPROVEMENT OF OUTPUT CHARACTERISTICS IN DIFFERENT PASSIVE RECTIFIERS BASED ON VIBRATION GENERATORS 3058
Tomoki Sakabe ; Masataka Minami ; Shin-Ichi Motegi ; Masakazu Michihira

CIRCUIT MODELING APPROACH FOR ANALYZING TRIBOELECTRIC NANOGENERATORS FOR ENERGY HARVESTING 3063
Bo-Kyung Yoon ; Jeong Min Baik ; Katherine A. Kim

GENERAL POWER ELECTRIC CONVERTER MODEL 3069
Jingwen Xie

A MODULAR CONVERTER- AND SIGNAL-PROCESSING-PLATFORM FOR ACADEMIC RESEARCH IN THE FIELD OF POWER ELECTRONICS 3074
Rüdiger Schwendemann ; Simon Decker ; Marc Hiller ; Michael Braun

CONTROL IC FOR BOOST-FLYBACK CONVERTER FOR ENERGY HARVESTING APPLICATIONS 3081
Jhih-Sian Li ; Kai-Hui Chen ; Jui-Hung Lai ; Jun-Xian Huang

NEW CONCEPT OF THE DC-DC CONVERTER CIRCUIT APPLIED FOR THE SMALL CAPACITY UNINTERRUPTIBLE POWER SUPPLY 3086
Dang Minh Huynh ; Yoichi Ito ; Shinji Aso ; Koji Kato ; Kenji Teraoka

COMPARATIVE STUDY ON THE PERFORMANCE OF DUAL-PHASE TAPPED-INDUCTOR BOOST CONVERTER AND INTERLEAVED BOOST PARALLEL-INPUT SERIES-OUTPUT CONVERTER IN 40 TO 400V APPLICATIONS 3092
Niño Christopher Ramos ; Tsuyoshi Funaki

A NEW STANDBY STRUCTURE INTEGRATED WITH BOOST PFC CONVERTER FOR SERVER POWER SUPPLY 3100
Jae-Il Baek ; Jae-Kuk Kim ; Jae-Bum Lee ; Moo-Hyun Park ; Gun-Woo Moon

NONISOLATED TWO-CHANNEL LED DRIVER WITH SIMPLE SNUBBER 3107
Jong-Woo Kim ; Jung-Kyu Han ; Jih-Sheng Lai

DESIGN AND IMPLEMENTATION OF SINGLE-PHASE ASYMMETRIC MULTILEVEL STATCOM 3112
Hao Chen ; Yang Han ; Ping Yang ; Congling Wang ; Josep M. Guerrero

SUBMODULE VOLTAGE BALANCING AND LOSS EQUALISATION IN ALTERNATE ARM CONVERTERS BASED ON VIRTUAL VOLTAGES 3117
Georgios Konstantinou ; Harith R. Wickramasinghe ; Salvador Ceballos ; Josep Pou

BALANCED CONDUCTION LOSS DISTRIBUTION AMONG SMS IN MODULAR MULTILEVEL CONVERTERS 3123
Zhongxu Wang ; Huai Wang ; Yi Zhang ; Frede Blaabjerg

SIMPLIFICATION OF MODEL PREDICTIVE CONTROL FOR MODULAR MULTILEVEL CONVERTER THROUGH DIRECT VOLTAGE LEVEL SELECTION 3129
Xingxing Chen ; Jinjun Liu ; Shaodi Ouyang ; Shuguang Song ; Rui Luo

FAMILY OF INTEGRATED MULTI-INPUT MULTI-OUTPUT DC-DC POWER CONVERTERS 3134
Bang Le-Huy Nguyen ; Honnyong Cha ; Tien-The Nguyen ; Heung-Geun Kim

LOW-COMPLEXITY STATE-SPACE BASED SYSTEM IDENTIFICATION AND CONTROLLER AUTO-TUNING METHOD FOR MULTI-PHASE DC-DC CONVERTERS 3140
Marc Kanzian ; Harald Gietler ; Christoph Unterrieder ; Matteo Agostinelli ; Michael Lunglmayr ; Mario Huemer

A PHASE-SHIFT DOUBLE FULL-BRIDGE (PSDB) CONVERTER WITH THREE SHARED LEADING-LEGS 3145
Junjie Zhu ; Qinsong Qian ; Shengli Lu ; Weifeng Sun ; Le Zhang

DUAL ACTIVE BRIDGE SYNCHRONOUS RECTIFIED STEP-DOWN CONVERTER 3151
Chien-Chun Huang ; Chang-Lin Tsai ; Tsung-Lin Tsai ; Yao-Ching Hsieh ; Huang-Jen Chiu ; Jing-Yuan Lin

ACCURATE IMPEDANCE MODEL OF GRID-CONNECTED INVERTER FOR SMALL-SIGNAL STABILITY ASSESSMENT IN HIGH-IMPEDANCE GRIDS 3156
Tuomas Messo ; Roni Luhtala ; Aapo Aapro ; Tomi Roinila

MODELING OF UNBALANCED THREE-PHASE GRID-CONNECTED CONVERTERS WITH DECOUPLED TRANSFER FUNCTIONS 3164
Wei Liu ; Xiongfei Wang ; Frede Blaabjerg

PREDICTING VOLTAGE CHARACTERISTIC OF CHARGING MODEL FOR LI-ION BATTERY WITH ANN FOR REAL TIME DIAGNOSIS 3170
Minella Bezha ; Naoto Nagaoka

IMPEDANCE MODELING AND STABILITY ANALYSIS OF THE CASCADED THREE-PHASE SYMMETRIC SYSTEMS USING COMPLEX TRANSFER FUNCTIONS 3176
Teng Liu ; Zeng Liu ; Jinjun Liu ; Yiming Tu ; Zipeng Liu

ACOUSTIC NOISE REDUCTION OF 12/8 POLES SRM WITHOUT EFFICIENCY DROP USING SIMPLE CURRENT WAVEFORMS 3182
Kyohei Kiyota ; Kenji Amei ; Takahisa Ohji ; Jun Jisaki ; Masanobu Nakai

STUDY OF SWITCHED RELUCTANCE MOTOR DIRECTLY DRIVEN BY COMMERCIAL THREE-PHASE POWER SUPPLY 3186
Masaki Takahashi ; Kohei Aiso ; Kan Akatsu

DOUBLE STATOR AXIAL-FLUX SWITCHED RELUCTANCE MOTOR FOR ELECTRIC CITY COMMUTERS 3192
Hiroki Goto

TORQUE RIPPLE REDUCTION USING ASYMMETRIC FLUX BARRIERS IN SYNCHRONOUS RELUCTANCE MOTOR 3197
Yuuto Yamamoto ; Shigeo Morimoto ; Masayuki Sanada ; Yukinori Inoue

ON-BOARD SINGLE-PHASE ELECTRIC VEHICLE CHARGER WITH ACTIVE FRONT END 3203
Theodore Soong ; Peter W. Lehn

A BIDIRECTIONAL BUFFERED CHARGING UNIT FOR EV'S (BBCU) 3209
Gabriel Fernandez

RECONFIGURABLE CONVERTER WITH MULTIPLE-VOLTAGE MULTIPLE-POWER FOR E-MOBILITY CHARGING 3215
Mohamed S A Dahidah ; He Liu ; Vassilios G. Agelidis

DEVELOPMENT OF A SERIES HYBRID ELECTRIC VEHICLE LABORATORY TEST BENCH WITH HARDWARE-IN-THE-LOOP CAPABILITIES 3223
Poria Fajri ; Nima Lotfi ; Mehdi Ferdowsi

NEW THREE-PHASE STATIC TRANSFER SWITCH USING AC SSCB 3229
Seung-Min Song ; Jin-Young Kim ; In-Dong Kim

HARMONICS COMPENSATION IN HIGH FREQUENCY RANGE OF ACTIVE POWER FILTER WITH SIC-MOSFET INVERTER IN DIGITAL CONTROL SYSTEM 3237
Shin-Ichi Hamasaki ; Kengo Nakahara ; Mineo Tuji

CONTROL OF BUCK-BOOST DIRECT MATRIX CONVERTER WITH LOW VOLTAGE RIDE-THROUGH CAPABILITY 3243
Nico Remus ; Martin Leubner ; Wilfried Hofmann

AN IMPROVED PLL BASED SEAMLESS TRANSFER CONTROL STRATEGY 3251
Xin Meng ; Jinjun Liu ; Zeng Liu ; Ronghui An

EFFICIENT URBAN RAILWAY DESIGN INTEGRATING TRAIN SCHEDULING, ONBOARD ENERGY STORAGE, AND TRACTION POWER MANAGEMENT 3257
Warayut Kampeerawar ; Takafumi Koseki ; Fulin Zhou

OPTIMAL CONTROL METHOD OF AN ENERGY STORAGE SYSTEM FOR ENERGY SAVING 3265
Yoko Takeuchi ; Tomoyuki Ogawa ; Keisuke Sato ; Hiroaki Morimoto ; Tatsuhito Saito

START-UP AND TRANSIENT OPERATION OF A BIDIRECTIONAL CHOPPER WITH AN AUXILIARY CONVERTER 3273
Hamzeh J. Ahmad ; Haruna Ohnishi ; Makoto Hagiwara

EXPERIMENTAL RESULTS OF QUASI-OPTIMAL CHARGING CURRENT PATTERNS TO REDUCE THE INTERNAL HEAT GENERATION OF THE LITHIUM-ION BATTERY 3280
Yoshiaki Taguchi ; Gaku Yoshikawa

DEVELOPMENT OF TEST METHODS AND EVALUATION RESULTS FOR 500KV HVDC CONVERTER 3286
Keisuke Hattori ; Asuka Ohtake ; Takayoshi Kamejima ; Haruhisa Wada

DISSIPATION LOOP FOR SHOOT-THROUGH FAULTS IN HVDC CONVERTER CELLS 3292
Keijo Jacobs ; Staffan Norrga ; Hans-Peter Nee

A SUPPRESSION METHOD OF HARMONIC INSTABILITY IN LINE-COMMUTATED CONVERTERS APPLYING ACTIVE HARMONIC FILTERS 3299
Kenichiro Sano ; Toshiaki Kikuma ; Tatsuhito Nakajima ; Junya Kanno

EXPERIMENT OF SEMICONDUCTOR BREAKER USING SERIES-CONNECTED IEGTS FOR HYBRID DCCB 3304
Kazuyasu Takimoto ; Hiroshi Takenaka ; Toshiaki Matsumoto ; Takahiro Ishiguro

STUDY OF EMI CAUSED BY BUCK CONVERTER ON CONTROLLER AREA NETWORK 3309
Ryo Shirai ; Toshihisa Shimizu

A STUDY ON REDUCTION TECHNIQUES OF A WIDEBAND COMMON-MODE VOLTAGE PRODUCED BY A PWM INVERTER 3315
Shotaro Takahashi ; Satoshi Ogasawara ; Masatsugu Takemoto ; Koji Orikawa ; Michio Tamate

A MODIFIED DISCONTINUOUS PWM FOR COMMON-MODE VOLTAGE ELIMINATION IN 3-LEVEL 4-LEG PWM CONVERTER SYSTEM 3323
Seon-Ik Hwang ; Jun-Hyung Jung ; In-Ho Cho ; Jang-Mok Kim ; Yung-Deug Son

EMI ANALYSIS OF FULL-SIC INTEGRATED POWER MODULE 3329
Xiliang Chen ; Wenjie Chen ; Yu Ren ; Liang Qiao ; Yilin Sha ; Xu Yang

EXPERIMENTAL VERIFICATION OF COUPLING EFFECT AND POWER TRANSFER CAPABILITY OF DYNAMIC WIRELESS POWER TRANSFER 3332
Chan Anyapo ; Nithiphat Teerakawanich ; Chowarit Mitsantisuk ; Kiyoshi Ohishi

NEIGHBORING EFFECTS ON THE DEACTIVATED INVERTER IN A SEGMENTED DYNAMIC WIRELESS EV CHARGING SYSTEM 3338
Qingwei Zhu ; Yanjie Guo ; Lifang Wang ; Shufan Li ; Chenglin Liao

MULTIPLE EXCITING VOLTAGE CONTROL FOR MAXIMIZATION OF MULTI-HOP WIRELESS POWER TRANSFER EFFICIENCY 3344
Masato Sasaki ; Masayoshi Yamamoto

GENERAL ANALYTICAL MODEL FOR INDUCTIVE POWER TRANSFER SYSTEM WITH EMF CANCELING COILS 3349
Keita Furukawa ; Keisuke Kusaka ; Jun-Ichi Itoh

STABILITY INFLUENCE OF FILTER COMPONENTS PARASITIC RESISTANCE ON LCL-FILTERED GRID CONVERTERS 3357
Hiroaki Matsumori ; Toshihisa Shimizu ; Frede Blaabjerg ; Xiongfei Wang ; Dongsheng Yang

REAL-TIME ESTIMATION CONTROL OF INDUCTANCE PARAMETERS USING DUST CORE MATERIALS FOR PWM INVERTER..3363

Kazu Imai ; Takuma Yoshino ; Ohasi Shunsuke ; Tomoki Yokoyama

CONTROL DESIGN OF OUTPUT-STAGE FILTERLESS SINUSOIDAL-WAVE INVERTER..........................3369

Shinichi Hiroshige ; Kenji Yamanaka ; Masahide Hojo

SERIES REACTIVE POWER COMPENSATOR WITH REDUCED CAPACITANCE FOR HYBRID TRANSFORMER..3375

Yuki Takahashi ; Takanori Isobe ; Hiroshi Tadano

AN INSIGHT INTO THE VOLTAGE RISING BEHAVIOR DURING TURN-OFF PROCESS OF SERIES CONNECTED SIC MOSFETS ON CIRCUIT LEVEL..3383

Panrui Wang ; Feng Gao ; Yang Jing ; Yufeng Chen ; Lei Zhang

PARALLELING SIX 320A 1200V ALL-SIC HALF-BRIDGE MODULES FOR A LARGE CAPACITY POWER STACK..3390

David Hongfei Lu ; Hiromu Takubo ; Sho Takano ; Yuhei Suzuki

3.3KV ALL-SIC MODULE FOR ELECTRIC DISTRIBUTION EQUIPMENT..3396

Ryohei Takayanagi ; Katsumi Taniguchi ; Satoshi Kaneko ; Naoyuki Kanai ; Keishirou Kumada ; Motohito Hori ; Yoshinari Ikeda ; Kouji Maruyama ; Itsuo Kawamura

PRESENT STATUS OF SIC BASED POWER CONVERTERS AND GATE DRIVERS – A REVIEW..............3401

Abhijit Choudhury

METHOD OF APPLYING FORCE DISTRIBUTION FUNCTION FOR LINEAR SWITCHED RELUCTANCE MOTOR DRIVEN BY CURRENT SOURCE INVERTER..3406

Tadashi Hirayama ; Shuma Kawabata

A NOVEL DRIVE CIRCUIT FOR SWITCHED RELUCTANCE MOTORS WITH BIPOLAR CURRENT DRIVE..3412

Hiroki Ishikawa ; Yuma Uesugi ; Seiya Sakurai

TORQUE RIPPLE MINIMIZATION CONTROL OF SRM BASED ON NOVEL MOTOR MODEL CONSIDERING MUTUAL COUPLING EFFECT..3418

Sungyong Shin ; Naruse Hikaru ; Takashi Kosaka ; Nobuyuki Matsui

COMPARISON OF HIGH FREQUENCY VOLTAGE INJECTION METHODS FOR SHAFT SENSORLESS CONTROL OF WOUND-FIELD FLUX SWITCHING MACHINE..3426

Hong-Quan Nguyen ; Sheng-Ming Yang

DESIGN AND EXPERIMENTAL VERIFICATION OF A DAB MEDIUM FREQUENCY TRANSFORMER FOR A 6.6KV/200V SOLID STATE TRANSFORMER..3431

Rene Barrera-Cardenas ; Takanori Isobe ; Terazono Katsushi ; Tadano Hiroshi

RESEARCH ON THE UNBALANCED COMPENSATION RANGE OF DELTA-CONNECTED CASCADED H-BRIDGE MULTILEVEL SVG..3439

Rui Luo ; Yingjie He ; Yiming Tu ; Xingxing Chen ; Jinjun Liu

STATIC SYNCHRONOUS COMPENSATOR TO STABILIZE GRID VOLTAGE FOR WIND AND PHOTOVOLTAIC POWER PLANT..3450

Ryota Okuyama ; Naoki Morishima ; Yusuke Ashizaki ; Yohei Itaya

LARGE EQUALIZATION CURRENT CONTROL STRATEGY FOR SERIES CONNECTED BATTERY PACKS BASED ON BUCK-BOOST CONVERTER..3455

Xinbo Liu ; Zhuo Gao ; Xuehao Huang ; Yaohan Zou

A MULTI-PORT BIDIRECTIONAL POWER CONVERSION SYSTEM FOR REVERSIBLE SOLID OXIDE FUEL CELL APPLICATIONS..3460

Xiang Lin ; Kai Sun ; Jin Lin ; Zhe Zhang ; Wei Kong

SELF-PREHEATING METHOD FOR LI-ION BATTERY USING BATTERY IMPEDANCE ESTIMATOR..3466

Dong-Kwan Kim ; Young-Dal Lee ; Sang-Hyun Ha ; Yu-Jin Jang ; Gun-Woo Moon

ACTIVE ANTI-ISLANDING TECHNIQUE WITH REDUCED NON-DETECTION ZONE FOR CENTRALIZED INVERTERS..3471

Prashant Jain ; Vivek Agarwal ; Bishnu Prasad Muni ; Eswar Rao ; Deepak Gehlot ; S. Gautam Kumar

DEVELOPMENT OF SIC APPLIED TRACTION SYSTEM FOR SHINKANSEN HIGH-SPEED TRAIN..3478

Kenji Sato ; Hirokazu Kato ; Takafumi Fukushima

DEVELOPMENT OF A HIGH POWER DENSITY AUXILIARY CONVERTER BASED ON 1700V 225A SIC MOSFET FOR TRAMS..3484

Liu Hao ; Fei Lin ; Zhongping Yang ; Hu Cao ; Meng Xia

EXPERIMENTAL TESTS RESULTS OF DAMPING CONTROL WITH OVER VOLTAGE RESISTOR FOR REGENERATIVE BRAKE CONTROL OF RAILWAY VEHICLE..3490

Natsuki Kawagoe ; Febry Pandu Wijaya ; Hiroyasu Kobayashi ; Keiichiro Kondo ; Tetsuya Iwasaki ; Akihiko Tsumura ; Takumi Nagashima ; Yoshinori Yamashita ; Ryota Gondo

COILS LAYOUT OPTIMIZATION OF DYNAMIC WIRELESS POWER TRANSFER SYSTEM TO REALIZE OUTPUT VOLTAGE STABLE...3495
Yi Wang ; Fei Lin ; Zhongping Yang ; Panpan Cai ; Zhiyuan Liu

QUICK CHARGER FOR A BATTERY USING MODULAR MATRIX CONVERTER (MMXC).......................3501
Kazuma Suzuki ; Takaharu Takeshita

VARIABLE OUTPUT VOLTAGE CONTROL OF AN ISOLATED BI-DIRECTIONAL AC/DC CONVERTER WITH A SOFT-SWITCHING TECHNIQUE...3507
Takumi Hamaguchi ; Kazuma Suzuki ; Wataru Kitagawa ; Takaharu Takeshita

A NEW MODULATION METHOD APPLYING OPTIMAL DUTY CYCLE AND PHASE SHIFT FOR BIDIRECTIONAL ISOLATED THREE-PHASE AC/DC CONVERTER BASED ON MATRIX CONVERTER...3514
Koji Shigeuchi ; Jin Xu ; Noboru Shimosato ; Yukihiko Sato

DECOUPLING CONTROL METHOD FOR ELIMINATING DC BIAS FLUX OF HIGH FREQUENCY TRANSFORMER IN A BIDIRECTIONAL ISOLATED AC/DC CONVERTER......................3522
Kensuke Sakuma ; Koji Shigeuchi ; Jin Xu ; Noboru Shimosato ; Yukihiko Sato

INTERLEAVED VOLTAGE-DOUBLER BOOST CONVERTER FOR POWER FACTOR CORRECTION...3528
Bo-Jia Huang

ZVS INTERLEAVED TOTEM-POLE BRIDGELESS PFC CONVERTER WITH PHASE-SHIFTING CONTROL..3533
Moo-Hyun Park ; Jae-Il Baek ; Jung-Kyu Han ; Cheon-Yong Lim ; Gun-Woo Moon

A ZERO-VOLTAGE-SWITCHING TOTEM-POLE BRIDGELESS BOOST POWER FACTOR CORRECTION RECTIFIER HAVING MINIMIZED CONDUCTION LOSSES.................................3538
Young-Dal Lee ; Chong-Eun Kim ; Jae-Il Baek ; Dong-Kwan Kim ; Gun-Woo Moon

POWER-FACTOR-CORRECTION WITH POWER DECOUPLING FOR AC-TO-DC CONVERTER...3544
Wan-Jung Chen ; Tsung-Hsi Wu ; Yao-Ching Hsieh ; Chin-Sien Moo ; Po-Hsiang Wen

DESIGN AND ANALYSIS OF THE DISTRIBUTED CONTROLLER FOR THE MODULAR MULTILEVEL CASCADED CONVERTER...3549
Ping-Heng Wu ; Yu-Chen Su ; Po-Tai Cheng

ASYMMETRIC MIXED MODULAR MULTILEVEL CONVERTER TOPOLOGY IN HYBRID BIPOLAR HVDC TRANSMISSION SYSTEMS...3557
Joon-Hee Lee ; Jae-Jung Jung ; Seung-Ki Sul

HIGH POWER MEDIUM VOLTAGE 10 KV SIC MOSFET BASED BIDIRECTIONAL ISOLATED MODULAR DC–DC CONVERTER...3564
Sayan Acharya ; Ritwik Chattopadhyay ; Anup Anurag ; Satish Rengarajan ; Yos Prabowo ; Subhashish Bhattacharya

MULTI-LEVEL POWER CONVERTER USING SERIES-CONNECTED SOLID-STATE TRANSFORMERS...3572
Yuichi Mabuchi ; Yuki Kawaguchi ; Kimihisa Furukawa ; Mitsuhiro Kadota ; Mizuki Nakahara ; Akihiko Kanoda

CAPACITOR VOLTAGE CONTROL OF MMC-STATCOM DURING UNBALANCED AC SYSTEM FAULT...3578
Kaho Nada ; Takeshi Kikuchi ; Tsuguhiro Takuno ; Toshiyuki Fujii ; Ryosuke Uda ; Takashi Sugiyama

SIC BASED POWER SEMICONDUCTOR IN APPLICATIONS - ASPECTS AND PROSPECTS.....................3584
Peter Friedrichs

ELECTROMAGNETIC MODELING APPROACHES TOWARDS VIRTUAL PROTOTYPING OF WBG POWER ELECTRONICS...3588
Ivana Kovacevic-Badstübner ; Daniele Romano ; Giulio Antonini ; Jonas Ekman ; Ulrike Grossner

SILICON BASED DEVICES FOR DEMANDING HIGH POWER APPLICATIONS.................................3596
A. Kopta ; J. Vobecky ; M. Rahimo ; T. Wikström ; U. Vemulapati ; C. Papadopoulos ; C. Corvasce ; M. Andenna ; F. Dugal ; F. Fischer ; S. Hartmann

RECENT PROGRESS IN HIGH TO ULTRA-HIGH-VOLTAGE SIC POWER DEVICES: DEVELOPMENT AND APPLICATION...3603
Y. Yonezawa

DYNAMIC DRIFT EFFECTS IN GAN POWER TRANSISTORS: CORRELATION TO DEVICE TECHNOLOGY AND MISSION PROFILE...3607
Joachim Würfl ; Eldad Bahat-Treidel ; Oliver Hilt ; Maria Troppenz ; Mihaela Wolf ; Jan Böcker ; Carsten Kuring ; Sibylle Dieckerhoff

COMPENSATION METHOD OF RADIAL UNBALANCE FORCE AT FAILURE OF A MOTOR SECTION IN A D-Q AXIS CURRENT CONTROL BEARINGLESS MOTOR.......................................3613
Masahide Ooshima

A BEARINGLESS SYNCHRONOUS RELUCTANCE SLICE MOTOR WITH ROTOR FLUX BARRIERS 3619

Thomas Holenstein ; Thomas Nussbaumer ; Johann W. Kolar

PARAMETER IDENTIFICATIONS OF CURRENT-FORCE FACTOR AND TORQUE CONSTANT IN SINGLE-DRIVE BEARINGLESS MOTORS 3627

Hiroya Sugimoto ; Akira Chiba

DAMPENING OF AXIAL VIBRATIONS IN A BEARINGLESS FLUX-SWITCHING SLICE MOTOR BY FIELD CURRENT REGULATION 3632

Bianca Klammer ; Karlo Radman ; Wolfgang Gruber

ANALYSIS AND DESIGN OF A BEARINGLESS AXIAL-FORCE/TORQUE MOTOR WITH FLEX-PCB WINDINGS 3640

Nobuyuki Kurita ; Walter Bauer ; Gerald Jungmayr ; Wolfgang Gruber ; Wolfgang Amrhein

A PLOTTER-BASED AUTOMATIC MEASUREMENT AND STATISTICAL CHARACTERIZATION OF MULTIPLE DISCRETE POWER DEVICES 3644

Michihiro Shintani ; Benjamin Dauphin ; Kazuki Oishi ; Masayuki Hiromoto ; Takashi Sato

A NOVEL HIGH-SPEED SIC MOSFET DRIVER WITH A LOW SWITCH-VOLTAGE STRESS 3650

Xiuqin Wei ; Yuchong Sun ; Hiroo Sekiya

ENHANCEMENT OF DRIVING CAPABILITY OF GATE DRIVER USING GAN HEMTS FOR HIGH-SPEED HARD SWITCHING OF SIC POWER MOSFETS 3654

Takafumi Okuda ; Takashi Hikihara

DESIGN AND EXPERIMENTAL VERIFICATION OF ROBOT ARM OPERATION FOR POWER PACKET DISPATCHING SYSTEM 3658

Tomoki Yokoyama ; Ryunosuke Araumi ; Kazunori Asada ; Takashi Ando

A RESOURCE SHARING MODEL IN A POWER PACKET DISTRIBUTION NETWORK 3665

H. Ando ; R. Takahashi ; S. Azuma ; M. Hasegawa ; T. Yokoyama ; T. Hikihara

DECOUPLED DSOGI-PLL FOR IMPROVED THREE PHASE GRID SYNCHRONISATION 3670

A. A. Nazib ; D. G. Holmes ; B. P. Mcgrath

A DEVIATION ELIMINATION CONTROL BASED ON AUTONOMOUS CURRENT-SHARING CONTROLLER FOR THE PARALLEL-CONNECTED INVERTERS IN AC MICROGRIDS 3678

Yajuan Guan ; Wei Feng ; Baoze Wei ; Wenzhao Liu ; Mingshen Li ; C. Juan Vasquez ; M. Josep Guerrero

SISO TRANSFER FUNCTIONS FOR STABILITY ANALYSIS OF GRID-CONNECTED VOLTAGE-SOURCE CONVERTERS 3684

Hongyang Zhang ; Lennart Harnefors ; Xiongfei Wang ; Jean-Philippe Hasler ; Hans-Peter Nee

A COMMUNICATION-INDEPENDENT REACTIVE POWER SHARING SCHEME WITH ADAPTIVE VIRTUAL IMPEDANCE FOR PARALLEL CONNECTED INVERTERS 3692

Ronghui An ; Zeng Liu ; Jinjun Liu ; Shike Wang

DESIGN AND INTEGRATION OF THE BI-DIRECTIONAL ELECTRIC VEHICLE CHARGER INTO THE MICROGRID AS EMERGENCY POWER SUPPLY 3698

Yang Song ; Pengcheng Li ; Yuanliang Zhao ; Shuai Lu

STABILITY IMPACT OF PV INVERTER GENERATION ON MEDIUM VOLTAGE DISTRIBUTION SYSTEMS 3705

Ye Tang ; Rolando Burgos ; Chi Li ; Dushan Boroyevich

1MW POWER CONDITIONING SYSTEM WITH MULTIPLE DC INPUTS FOR PVS AND BATTERIES 3711

Yasuaki Furusho ; Yasuyuki Noto ; Kansuke Fujii

A ROBUST AND FLEXIBLE DC-LINKED 3-PHASE ENERGY MANAGEMENT SYSTEM WITH ADAPTIVE DROOP CONTROL STRATEGY 3717

Yue Ma ; Yuki Ishikura ; Hitoshi Tsuji ; Kazuaki Mino

MAXIMUM POWER POINT TRACKING CONTROL FOR SMALL HYDROELECTRIC GENERATION 3723

Kazuya Azegami ; Masashi Takiguchi ; Junya Yano ; Hirohiko Tsutsumi ; Toshitake Masuko

DESIGN AND EXPERIMENTAL VERIFICATION OF A THREE-PHASE DUAL-ACTIVE BRIDGE CONVERTER FOR OFFSHORE WIND TURBINES 3729

Takushi Jimichi ; Murat Kaymak ; Rik W. De Doncker

OPTIMIZED BIDIRECTIONAL PFC RECTIFIERS & INVERTERS - SI VS. SIC VS. GAN IN 2L AND 3L TOPOLOGIES - 3734

Jonas Wyss ; Jürgen Biela

A STANDARD BLOCK OF "SERIES CONNECTED SIC MOSFET" FOR MEDIUM/HIGH VOLTAGE CONVERTER 3742

Qin Lei ; Chunhui Liu ; Yunpeng Si ; Yifu Liu

DESIGN AND TESTING OF 1 KV H-BRIDGE POWER ELECTRONICS BUILDING BLOCK BASED ON 1.7 KV SIC MOSFET MODULE 3749
Jun Wang ; Rolando Burgos ; Dushan Boroyevich ; Zeng Liu

A FLYBACK CONVERTER WITH SIC POWER MOSFET OPERATING AT 10 MHZ: REDUCING LEAKAGE INDUCTANCE FOR IMPROVEMENT OF SWITCHING BEHAVIORS 3757
Kazuki Hashimoto ; Takafumi Okuda ; Takashi Hikihara

A STUDY ON LOAD FLUCTUATION OF ISOLATED DC-DC CONVERTER WITH CLASS PHI-2 INVERTER USING GAN-HFET 3762
Yuta Yanagisawa ; Yushi Miura ; Hiroyuki Handa ; Tetsuzo Ueda ; Toshifumi Ise

SINGLE-INDUCTOR MULTIPLE-OUTPUT CURRENT-SOURCE CONVERTER WITH IMPROVED CROSS REGULATION AND SIMPLE CONTROL STRATEGY 3768
Zheng Dong ; Xiaolu Lucia Li ; Chi K. Tse

LIMIT OPERATING FREQUENCY OF PEAK CURRENT-MODE CONTROL DC-DC CONVERTER CONSIDERING TURN-OFF DELAY TIME 3773
Ryo Ute ; Kazuya Fujiwara ; Jun Imaoka ; Masahito Shoyama

A NOVEL SINGLE SWITCH HIGH FREQUENCY DC/DC CONVERTER AND ITS MATHEMATIC MODEL 3780
Yueshi Guan ; Xihong Hu ; Shu Zhang ; Yijie Wang ; Dianguo Xu ; Wei Wang

ANALYSIS OF CLOSED LOOP OPERATION OF AN ISOLATED BIDIRECTIONAL DAB DC-DC CONVERTER WITH LC COUPLING 3785
Bruno Yukio Enomoto ; Kelly C. M. Carvalho ; Lourenço Matakas Junior ; Wilson Komatsu

ISOLATED AC/DC CONVERTER USING SIMPLE PWM STRATEGY 3791
Naoki Hirose ; Yuto Matsui ; Takaharu Takeshita

ANALYSIS OF ONE PHASE LOSS OPERATION OF THREE-PHASE ISOLATED BUCK MATRIX-TYPE RECTIFIER WITH EIGHT-SEGMENT PWM SCHEME 3797
Jahangir Afsharian ; Dewei David Xu ; Bin Wu ; Bing Gong ; Zhihua Yang ; Jun-Ichi Itoh

NOVEL ISOLATED BIDIRECTIONAL INTEGRATED DUAL THREE-PHASE ACTIVE BRIDGE (D3AB) PFC RECTIFIER 3805
F. Krismer ; E. Hatipoglu ; J. W. Kolar

LOAD VOLTAGE REGULATION METHOD FOR AN ISOLATED AC-DC CONVERTER WITH POWER DECOUPLING OPERATION 3813
Shohei Komeda ; Hideaki Fujita

OPTIMAL DESIGN OF A LOW COST 20KW 99.1% EFFICIENCY ACTIVE ZCS ISOLATED DC-DC CONVERTER 3820
Timothé Delaforge ; Sébastien Mariéthoz

SOFT-SWITCHING ANALYSIS AND PFM CONTROL METHOD OF BIDIRECTIONAL DC/DC CONVERTER TOPOLOGY 3825
Yijie Wang ; Haoyu Wang ; Hongyu Song ; Dianguo Xu

A FULLY SOFT-SWITCHED PWM DC-DC CONVERTER USING AN ACTIVE-SNUBBER-CELL 3833
Hai N. Tran ; Adhistira M. Naradhipa ; Sunju Kim ; Ali Tausif

FLYING CAPACITOR RESONANT POLE INVERTER WITH DIRECT INDUCTOR CURRENT FEEDBACK 3840
Sjef J. Settels ; Jorge L. Duarte ; Jeroen Van Duivenbode

DESIGN OF A GAN-BASED WIRELESS POWER TRANSFER SYSTEM AT 13.56 MHZ TO REPLACE CONVENTIONAL WIRED CONNECTION IN A VEHICLE 3848
Kawin Surakitbovorn ; Juan Rivas-Davila

EFFICIENCY MAXIMIZATION OF INDUCTIVE POWER TRANSFER SYSTEM BY IMPEDANCE AND SWITCHING FREQUENCY CONTROL IN SECONDARY-SIDE CONVERTER 3855
Ryosuke Ota ; Dannisworo S. Nugroho ; Nobukazu Hoshi

ANALYSIS OF OPTIMAL OPERATION FREQUENCY RANGE FOR BATTERY CHARGING IN WPT SYSTEM 3863
Yongbin Jiang ; Min Wu ; Junwen Liu ; Yue Wang ; Laili Wang ; Hailong Zhang

INITIAL CURRENT INJECTION METHOD OF A DIRECT THREE-PHASE TO SINGLE-PHASE AC/AC CONVERTER FOR INDUCTIVE CHARGER 3870
Ferdi Perdana Kusumah ; Jorma Kyyrä

MISSION PROFILE EMULATOR FOR PERMANENT MAGNET SYNCHRONOUS MACHINE BASED ON THREE-PHASE POWER ELECTRONIC CONVERTER 3877
Yubo Song ; Ran Cheng ; Ke Ma

A VARIABLE DC BUS VOLTAGE BASED POWER HARDWARE-IN-THE-LOOP EMULATION OF ELECTRIC MOTORS WITH WIDE VARIATION IN INTERFACE FILTER INDUCTANCE 3884
Tsai-Fu Wu ; Mitradatta Misra ; Ying-Yi Jhang ; Chang-Jun Yang ; Yin-Chi Xu

COPPER LOSS MINIMIZATION CONTROL AT ZERO OUTPUT VOLTAGE FOR ELECTROLYTIC CAPACITOR-LESS INVERTER 3890
Kodai Abe ; Haruya Kada ; Kiyoshi Ohishi ; Hitoshi Haga ; Yuki Yokokura

ARMATURE TEMPERATURE ESTIMATION INSENSITIVE TO ROTOR FLUX VARIATION FOR SPMSM 3896
Toshiki Sano ; Kiyoshi Ohishi ; Yuki Yokokura ; Hiroki Iwata ; Yuji Ide ; Daigo Kuraishi ; Akihiko Takahashi

VIRTUAL SYNCHRONOUS GENERATOR CONTROL WITH RELIABLE FAULT RIDE-THROUGH CAPABILITY BY ADOPTING MODEL PREDICTIVE CONTROL 3902
Jonggrist Jongudomkarn ; Jia Liu ; Toshifumi Ise

RESHAPING QUADRATURE-AXIS IMPEDANCE OF THREE-PHASE GRID-CONNECTED CONVERTERS FOR LOW-FREQUENCY STABILITY IMPROVEMENT 3910
Yi Tang ; Jingyang Fang ; Xiaoqiang Li ; Hongchang Li

COMPARISON BETWEEN TRADITIONAL DROOP AND A NEW AUTONOMOUS CONTROL SCHEME FOR PARALLEL INVERTERS 3916
Mohammad Bani Shamseh ; Teruo Yoshino ; Atsuo Kawamura

A NOVEL MICROGRID POWER SHARING SCHEME ENHANCED BY A NON-INTRUSIVE FEEDER IMPEDANCE ESTIMATION METHOD 3924
Baojin Liu ; Zeng Liu ; Jinjun Liu ; Ronghui An ; Shuguang Song

DEVELOPMENT OF A 3.2MW PHOTOVOLTAIC INVERTER FOR LARGE-SCALE PV POWER PLANTS 3929
Naoya Shibata ; Tsuguhiro Tanaka ; Masahiro Kinoshita

IMPEDANCE-BASED STABILITY ANALYSIS OF LARGE-SCALE PV STATION UNDER WEAK GRID CONDITION CONSIDERING SOLAR RADIATION FLUCTUATION 3934
Yiming Tu ; Jinjun Liu ; Teng Liu ; Xiangpeng Cheng

EXPERIMENTAL VERIFICATION OF GRID-CONNECTION OF A PV CONVERTER USING A SYMMETRICALLY CONNECTED BOOST CONVERTER FOR A HIGH-LEG DELTA TRANSFORMER 3940
Daiki Yamaguchi ; Hideaki Fujita

A NOVEL SINGLE- STAGE HIGH-FREQUENCY BOOST INVERTER CASCADED BY RECTIFIER-INVERTER SYSTEM FOR PV GRID-TIE APPLICATIONS 3945
Hamdy Radwan ; Mahmoud A. Sayed ; Takaharu Takeshita ; Adel A. Elbaset ; G. Shabib

NINE SWITCHES MATRIX CONVERTER USING BI-DIRECTIONAL GAN DEVICE 3952
Takashi Hirota ; Kentaro Inomata ; Daisuke Yoshimi ; Masato Higuchi

A MODEL PREDICTIVE DUAL CURRENT CONTROL METHOD FOR INDIRECT MATRIX CONVERTER FED INDUCTION MOTOR DRIVES 3958
Mei Yang ; Chen Lisha ; Liang Wang ; Yunwei Li

FAULT TOLERANT PREDICTIVE CONTROL OF THREE-LEVEL NEUTRAL-POINT-CLAMPED BACK-TO-BACK POWER CONVERTERS 3965
Zhenbin Zhang ; Xicai Liu ; Kejun Cai ; Feng Gao ; Ralph Kennel

TWO-STAGE OPTIMIZATION BASED PREDICTIVE TORQUE CONTROL WITH REDUCED COMPLEXITY FOR A THREE-LEVEL INVERTER DRIVEN INDUCTION MOTOR 3971
Ilham Osman ; Dan Xiao ; Faz Rahman

DESIGN CHALLENGES OF SIC DEVICES FOR LOW- AND MEDIUM-VOLTAGE DC-DC CONVERTERS 3979
Georges Engelmann ; Alexander Sewergin ; Markus Neubert ; Rik W. De Doncker

DESIGN AND TESTING OF 6 KV H-BRIDGE POWER ELECTRONICS BUILDING BLOCK BASED ON 10 KV SIC MOSFET MODULE 3985
Jun Wang ; Slavko Mocevic ; Jiewen Hu ; Yue Xu ; Christina Dimarino ; Igor Cvetkovic ; Rolando Burgos ; Dushan Boroyevich

HIGH POWER MEDIUM VOLTAGE CONVERTERS ENABLED BY HIGH VOLTAGE SIC POWER DEVICES 3993
Sanket Parashar ; Ashish Kumar ; Subhashish Bhattacharya

SOFT-SWITCHING – THE KEY TO HIGH POWER WBG CONVERTERS 4001
Deepak Divan ; Zheng An ; Prasad Kandula

SIC: TECHNOLOGY ENABLER FOR MV DC/DC GALVANICALLY INSULATED MODULAR CONVERTERS 4009
S. Alvarez ; M. Bellini ; U. Vemulapati ; F. Canales ; M. Rahimo

A BEARINGLESS SLICE MOTOR WITH A SOLID IRON ROTOR FOR DISPOSABLE CENTRIFUGAL BLOOD PUMP 4016
Tadahiko Shinshi ; Ryo Yamamoto ; Yoshiki Nagira ; Junichi Asama

REDUCED HARDWARE PARALLEL DRIVE FOR NO VOLTAGE BEARINGLESS MOTORS 4020
Eric L. Severson

DUAL FIELD-ORIENTED CONTROL OF BEARINGLESS MOTORS WITH COMBINED WINDING SYSTEM 4028
  Wolfgang Gruber ; Siegfried Silber

OPEN-CIRCUIT FAULT TOLERANT STUDY OF BEARINGLESS MULTI-SECTOR PERMANENT MAGNET MACHINES 4034
  G. Valente ; L. Papini ; A. Formentini ; C. Gerada ; P. Zanchetta

BALANCE CONTROL OF SPLIT CAPACITOR POTENTIAL FOR MAGNETICALLY LEVITATED MOTOR SYSTEM USING ZERO-PHASE CURRENT 4042
  Takaaki Oiwa

ASYMMETRICAL HALF-BRIDGE CONVERTER WITH ZERO DC-OFFSET CURRENT IN TRANSFORMER USING NEW RECTIFIER STRUCTURE 4049
  Jung-Kyu Han ; Jong-Woo Kim ; Seung-Hyun Choi ; Jih-Sheng Lai ; Gun-Woo Moon

CIRCULATING CURRENT-LESS PHASE-SHIFTED FULL-BRIDGE CONVERTER WITH NEW RECTIFIER STRUCTURE 4054
  Jung-Kyu Han ; Gun-Woo Moon

A BI-DIRECTIONAL CURRENT DETECTION USING CURRENT TRANSFORMERS FOR BI-DIRECTIONAL DC-DC CONVERTER 4059
  Seiji Iyasu ; Yuji Hahashi ; Yuuichi Handa ; Kimikazu Nakamura ; Keiji Wada

A 10 MHZ GANFET BASED ISOLATED HIGH STEP-DOWN DC-DC CONVERTER 4066
  Prasanth Thummala ; Dorai Babu Yelaverthi ; Regan Zane ; Ziwei Ouyang ; Michael A. E. Andersen

ANALYSIS AND DESIGN OF A PARALLEL RESONANT CONVERTER FOR CONSTANT CURRENT INPUT TO CONSTANT VOLTAGE OUTPUT DC-DC CONVERTER OVER WIDE LOAD RANGE 4074
  Tarak Saha ; Hongjie Wang ; Baljit Riar ; Regan Zane

NOVEL SINUSOIDAL INPUT CURRENT SINGLE-TO-THREE-PHASE Z-SOURCE BUCK+BOOST AC/AC CONVERTER 4080
  M. Haider ; D. Bortis ; J. W. Kolar ; Y. Ono

SIMPLE PWM STRATEGY OF A MATRIX CONVERTER FOR MINIMIZING OUTPUT VOLTAGE HARMONICS 4088
  Takuya Oshima ; Takaharu Takeshita

NOVEL THREE-LEVEL BACK-TO-BACK CONVERTERS: STRUCTURE, MODULATION METHOD, AND EXPERIMENT 4096
  S. Sangwongwanich ; K. Niyomsatian ; S. Samermurn ; S. Nuchnoi ; S. Suwankawin

MODEL PREDICTIVE CONTROL USING SUBDIVIDED VOLTAGE VECTORS FOR CURRENT RIPPLE REDUCTION IN AN INDIRECT MATRIX CONVERTER 4104
  Keon Young Kim ; Yeongsu Bak ; Jin-Hyuk Park ; Kyo-Beum Lee

DC-LINK RIPPLE CURRENT REDUCTION IN BACK-TO-BACK CONVERTERS WITH DPWM 4109
  Anatolii Tcai ; Kyo-Beum Lee

AN ANALYSIS OF CLASS DE VOLTAGE-SOURCE PARALLEL RESONANT INVERTER 4114
  Takeshi Kondo ; Tsuyoshi Inaba ; Yoshikazu Sakai ; Hirotaka Koizumi

AN IMPROVEMENT ON EXTENDED IMPEDANCE METHOD TOWARDS EFFICIENT STEADY-STATE ANALYSIS OF HIGH-FREQUENCY CLASS-E RESONANT INVERTERS 4122
  Junrui Liang

OUTPUT POWER CAPABILITY COMPARISONS OF CLASS-E POWER AMPLIFIERS WITH HARMONIC RESONANCE 4127
  Hiroo Sekiya ; Xiuqin Wei ; Yuchong Sun

A CLASS Φ2 RESONANT BUCK CONVERTER WITH RIPPLE INJECTION BURST CONTROL METHOD 4133
  Min Lin ; Masahiko Hirokawa

PRACTICAL DESIGN TECHNIQUE FOR HIGH POWER DENSITY LLC RESONANT CONVERTER 4139
  Shingo Nagaoka ; Hiroyuki Onishi ; Koji Takatori ; Toshiyuki Zaitsu ; Takeshi Uematsu

OPERATIONAL STUDY AND PROTECTION OF A SERIES RESONANT CONVERTER WITH DC CURRENT INPUT APPLIED IN DC CURRENT DISTRIBUTION SYSTEMS 4145
  Hongjie Wang ; Tarak Saha ; Baljit Riar ; Regan Zane

A STUDY ON IMPROVEMENT OF POWER UTILIZATION RATE OF ENERGY SYSTEMS WITH PVS AND BATTERIES 4151
  Hiroaki Endo ; Masakatsu Kurisaka ; Tsutomu Ueno ; Yusuke Yoshioka ; Kaoru Inoue ; Toshiji Kato

A NOVEL DC DISTRIBUTION NETWORK WITH MULTI-LEVEL BUS VOLTAGES AND ITS ENERGY MANAGEMENT SYSTEM DESIGN 4157
  Jingjin Huang ; Xin Zhang ; Zhixun Ma ; Jianfang Xiao

A NOVEL DC-SIDE-PORT IMPEDANCE MODELING OF MODULAR MULTILEVEL CONVERTERS BASED ON HARMONIC STATE SPACE METHOD 4162
Jing Lyu ; Xin Zhang ; Zhixun Ma ; Xu Cai

AN IMPROVED MASTER-SLAVE CONTROL FOR THREE-PORT CONVERTER BASED DISTRIBUTED DC GRID-CONNECTED PV SYSTEM 4168
Siyue Jiang ; Kai Sun ; Hongfei Wu ; Haixu Shi ; Xiaofeng Dong ; Syed Muhammad Raza Kazmi

SENSORLESS POSITION ESTIMATION, PARAMETER IDENTIFICATION AND CONTROL INTEGRATION FOR PERMANENT MAGNET SYNCHRONOUS MACHINES USING CURRENT DERIVATIVE MEASUREMENTS 4174
M.X. Bui

DYNAMIC PERFORMANCE IMPROVEMENT OF BIDIRECTIONAL SWITCHED-CAPACITOR DC/DC CONVERTER BY RIGHT-HALF-PLANE ZERO ELIMINATION 4181
Ding Kaicheng ; Zhang Yan ; Liu Jinjun ; Zeng Pengxiang ; Zhang Jinshui

A MATRIX BASED ISOLATED BIDIRECTIONAL AC-DC CONVERTER WITH LCL TYPE INPUT FILTER FOR ENERGY STORAGE APPLICATION 4186
Prathamesh Pravin Deshpande ; Amit Kumar Singh ; Sanjib Kumar Panda

ON A STUDY OF VOLTAGE DIVIDING CLASS Φ AMPLIFIER 4193
Katsutoshi Hirayama ; Tadashi Suetsugu ; Yudai Furukawa ; Fujio Kurokawa

A DPWM BASED CONTROL STRATEGY TO INTEGRATE PHOTOVOLTAIC SYSTEM AND BATTERY STORAGE USING GRID CONNECTED THREE-LEVEL T-TYPE INVERTER 4198
Mohammad M. Hashempour ; Yue-Ting Tsai ; T. L. Lee

IMPEDANCE MEASUREMENT OF MEGAWATT-LEVEL RENEWABLE ENERGY INVERTERS USING GRID-FORMING AND GRID-PARALLEL CONVERTERS 4205
Matias Berg ; Tuomas Messo ; Tomi Roinila ; Henrik Alenius

IMPROVED VIRTUAL INDUCTANCE BASED CONTROL STRATEGY OF DFIG UNDER WEAK GRID CONDITION 4213
Ran Fang ; Wenjia Chen ; Xueguang Zhang ; Dianguo Xu

CONTROL OF VSC-HVDC FOR WIND FARM INTEGRATION WITH REAL-TIME FREQUENCY MIRRORING AND SELF-SYNCHRONIZING CAPABILITY 4220
Renxin Yang ; Chen Zhang ; Xu Cai ; Gang Shi ; Jing Lyu

A STUDY ON STEADY-STATE CHARACTERISTICS OF SERIES-CONNECTED WIND FARM USING AN EXPERIMENTAL SET OF LABORATORY SIZE 4227
Fujio Tatsuta ; Shoji Nishikata

A NOVEL ISLANDING DETECTION METHOD WITH TWO-PHASE MAGNIFICATION INSPECTION 4233
Jian-Tang Liao ; Shun-Hao Yeh ; Hong-Tzer Yang

Author Index

Plug and Outlet in Household DC Low Voltage Micro-grid Power Distribution

Worapong Pairindra[1] and Surin Khomfoi[2]

1 Dept. of Electrical Engineering Technology, Faculty of Industrial Technology,
Valaya Alongkorn Rajabhat University, Thailand,13180
2 Dept. of Electrical Engineering, Faculty of Engineering,
King Mongkut's Institute of Technology Ladkrabang. Bangkok ,Thailand,10502
*E-mail: worapong73@hotmail.com[1], kkhsurin@kmitl.ac.th[2]

Abstract- **The main idea of using a dc to dc plug and outlet with high efficiency for household appliances is proposed in this paper. A five-level, 300-W, 200-kHz diode-clamped multilevel technique with selective harmonic elimination is utilized as a power source with input from a 300-V DC link bus and output controlled at 48V. A contactless transformer with an air core and a ferromagnetic material core was chosen in this study for testing and evaluation. Electric shock elimination, galvanic isolation, and arc-free operation can be achieved with this scheme. Additionally, the design's low total harmonic distortion can reduce the size of the inductive power transformer core and increase the power density of the system. Moreover, an LLC series resonance technique with fundamental harmonic approximation (FHA) was designed and added to the rear side of the power converter to achieve the zero voltage switching condition. Finally, the outcome is illustrated by both MATLAB/Simulink and experimental results.**

Keywords— Contactless power transfer, Multilevel converter, Selective Harmonic Elimination, LLC series resonant converter

I. INTRODUCTION

Currently, the power generated from renewable or sustainable energy sources has captured public attention because of the potential global reduction of fossil fuel use. The use of dc power distribution for household appliances is a promising future for next-generation management of micro-grid systems [1] including distributed generators (DGs) such as wind, solar, diesel engines, or backup batteries. There are many researches in the micro-grid applications such as [2]-[4]. Therefore, power conversion in dc power distribution for higher efficiency has become a major challenge for many researchers recently. However, delivering power from the connected dc link bus source directly to the load causes a penalty for the conventional plug and outlet. Electric shock, arc discharge, and galvanic isolation issues are the main obstacles for power distribution because of the non-zero crossing of the dc signal. The main idea is the replacement of plug and receptacle for household dc power distribution: the old-fashioned direct metal contact will be replaced with an inductive power transfer (IPT) system as seen in [5]–[11] and will send power through a high switching-frequency converter.

Power conversion density has the attention of many researchers because of the sizes and capabilities of power devices. Generally, the coaxial transformer with an air core has been reviewed in [5][6] with different sizes of coreless transformers. The size can be reduced by increasing the switching frequency according to the transformer design equation. The advantages of using an air core transformer compared with a ferromagnetic material core (FMC) can be categorized as follows:

1. The power loss in the core material due to fluctuation of the switching frequency can be eliminated.

2. The cost-effectiveness of the IPT is increased.

3. The ferromagnetic material core must be designed in specific dimensions to meet the appropriate typical system design.

However, air-core IPT still has a problem compared with the FMC IPT, in terms of power conversion efficiency and power density.

The proposed solution introduces IPT with an air-core (solenoid) transformer that consists of two solenoid coils; the outer shelter coil is set as the primary coil, and the inner coil is designed as the secondary coil, as shown in Fig. 1. Maxwell's equation is used to calculate the parameters of the two solenoids in the IPT design. The new approach of using the FMC was setup and evaluated; a separable PQ40/40 ferrite core with the calculated IPT in turns functioned as a power conversion device for higher power density and higher efficiency compared with the air-core IPT. The advantages of using the FMC transformer can be described as follows:

1. The size reduction due to the high flux density of the FMC depends on the material type of each manufacturer.

2. The power density meets the standard requirements compared with the conventional plug and outlet.

Generally, IPT technology is based on sending a highly magnetic coupling over the media, inducing the secondary coil with high-frequency switching from the source converter. Conventional converters, such as half-bridge and full-bridge converters, can send voltage across the high-speed transformer. Unfortunately, the square wave signal has a high total harmonic distortion (THD), which affects the IPT transformer design. Therefore, a five-level diode-clamped inverter with high switching frequency is introduced for the IPT system in order to

1460

reduce the THD and increase the power density of the plug and receptacle. The ideal concept design of the IPT can be illustrated in Fig. 1. and Fig 2. The receptacle and plug in Fig. 1. has no physical metal contact for power transferring and Fig. 2. shows the structure of the PQ 40/40 core which left hand side is the primary coil and right hand side represents a secondary coil, respectively. Under the power transferring operation, these 2 cores should be pressed to keep the distance at minimal range.

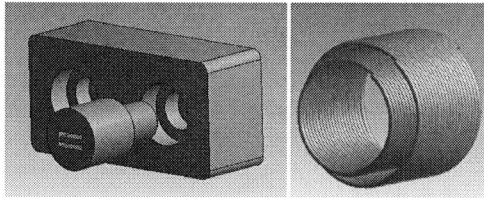

Fig. 1. Ideal air-core contactless transformer

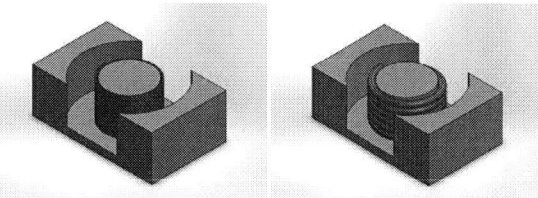

Fig. 2. Inductive power transfer with ferromagnetic material

II. MULTILEVEL INDUCTIVE POWER TRANSFER TECHNOLOGY

The multilevel converter as reviewed in [12]–[15] functions as a power source of the IPT system. A five-level DCML converter generates 200 kHz of switching pulses with different angles as shown in Fig. 4. Because the THD depends on the switching angles, selective harmonic elimination (SHE) is introduced to find the appropriate angles for the input of the IPT primary coil to meet the design with the sinusoid transformer calculation. Theoretically, a conventional converter, such as a half-bridge or full-bridge converter, has approximately 11% lower power density than the sinusoid waveform. Therefore, it is obvious that a low-THD converter can potentially transfer more power than a conventional converter with the same volume.

The connected LLC series resonant technique with FHA has been adapted to the system for reducing the switching losses due to the high number of operating switches. The combination of ZVS and the new MOSFET (R_{DS_ON}) technologies assure that the efficiency will still meet the requirement. Additionally, the output voltage will be controlled by the frequency variation algorithm as shown in the system gain versus nominal frequency curve.

The multilevel switching device voltage stress is reduced because of the characteristics of the staircase level of the converter. The applications of the multilevel can be from high power to low power with different switching frequencies. This study introduced the five-level DCML converter for the power source with 200-

kHz switching frequency; it is controlled by a microcontroller unit (MCU, Texas Instruments DSP TMS320F38335), which can generate eight non-inversing pulse-width modulation (PWM) and eight inverting PWM signals for the operating devices.

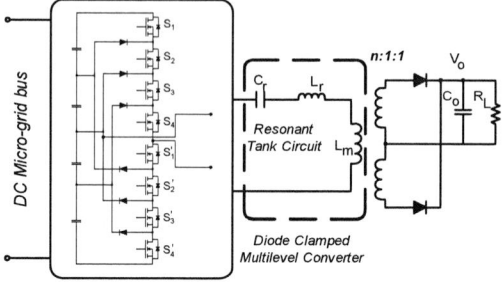

Fig. 3. Structure of the DCML with IPT

The DCML converter with SHE technique and its switching states are illustrated in Table 1.

TABLE I
DCML SWITCHING STATES

V_{out}	Switching state							
	S_1	S_2	S_3	S_4	S_1'	S_2'	S_3'	S_4'
150 V	1	1	1	1	0	0	0	0
75 V	0	1	1	1	1	0	0	0
0 V	0	0	1	1	1	1	0	0
−75 V	0	0	0	1	1	1	1	0
−150 V	0	0	0	0	1	1	1	1

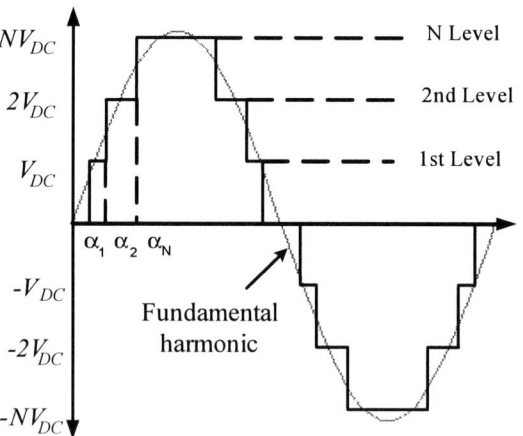

Fig. 4. N-level output voltage waveform

The operating staircase voltage is illustrated in Fig. 4. It can be written in terms of a Fourier series equation and can be described as following:

Fundamental component of the output voltage V_{f1} representing the fundamental harmonic:

$$V_{f1} = \frac{4V_{DC}}{\pi} \sum_{i=1}^{N} \cos(\alpha_i) \qquad (1)$$

The $(2l-1)$-th harmonic components of the output voltage can be found by

$$V_{h(2l-1)} = \frac{4V_{DC}}{(2l-1)\pi} \sum_{i=1}^{N} \cos((2l-1)\alpha_i) \qquad (2)$$

N level, l=1,2,3,4.....

The nonlinear polynomial equations can be solved by the Newton–Raphson method, and the third-harmonic elimination switching angles are calculated from the polynomial equations as shown in the following:

$$\cos(1\alpha_1) + \cos(1\alpha_2) = M_A \qquad (3)$$

$$\cos(3\alpha_1) + \cos(3\alpha_2) = 0 \qquad (4)$$

where

$$\cos(3\alpha) = 4\cos^3\alpha - 3\cos\alpha \qquad (5)$$

$$4\cos^3(\alpha_1) - 3\cos(\alpha_1) + 4\cos^3(\alpha_2) - 3\cos(\alpha_2) = 0 \qquad (6)$$

The DCML inverter with the SHE technique was simulated and validated with the calculated parameters by using MATLAB/Simulink. The Newton–Raphson method was utilized for solving the nonlinear mathematical equation angles. The switching angles were applied to the DCML, and the THD$_V$ output is illustrated in Fig. 5. The lowest THD$_V$ with appropriate switching angles is found at 12° and 48°, and the selected harmonic magnitudes were eliminated.

The five-level converter with a modulation index of 1.64 almost eliminates the third and fifth harmonic order from the output and transfers the power through the contactless transformer with a low THD of 17.54%. By using the resonant tank circuitry (RLC passive devices) without additional switching devices, the resonant current waveform approximates a sinusoidal shape and is shifted from the drain-source voltage of each switching device as shown in Fig. 6.

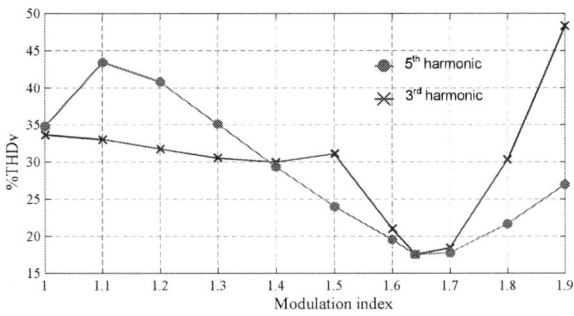

Fig. 5. Third and fifth harmonic elimination compared with THD

III. EXPERIMENTAL AND SIMULATION RESULTS

In this study, the implementation was based on power transfer by using an air-core or solenoid transformer and a ferrite core transformer. The IPT was implemented by using the DCML power converter with an IRF740 power switch and an FOD3184 high-frequency driver from Fairchild IC. The controlled PWM signals were generated by a Texas Instruments TMS320F23885 microcontroller. The DCML contained four individually non-inverse PWMs in the upper portion and four inverse PWMs for the lower-portion switching devices. The controlling block set controlled the switching algorithm via CCS 5.4.0 and enabled comprehension of the feedback analog-to-digital converter (ADC) signals in real time. The power converter generated the five-level voltage with proper angles and provided power to the contactless transformers. The prototype setup for IPT was calculated and substituted to the plant with the FHA technique; parameters are shown from Table 2 to Table 4.

TABLE II
SPECIFICATION OF THE PROTOTYPE CONVERTER

Prototype specification	Parameter
Input DC link voltage	300 V
Output voltage	48 V
Switching frequency	200 kHz

TABLE III
DMCL CONVERTER WITH LLC COMPONENTS

Component	Parameter
Resonant inductor	23.24 μH
Resonant capacitor	27.24 nF
Magnetized inductor	81.36 μH
Switching device	IRF740
High speed diode	MUR840
High speed driver	FOD3184

TABLE IV
AIR-CORE CONTACTLESS TRANSFORMER SPECIFICATION

Wire type	SWG 17	SWG 12
Turns	57	30
Layers per turn	1	
Diameter	≈ 65 mm	≈ 60 mm
Height	≈ 80 mm	

TABLE V
FERROMAGNETIC CORE CONTACTLESS TRANSFORMER
SPECIFICATION

Wire type	SWG 18	SWG 16
Turns	12	4
Layers per turn	1	
Ferrite core	PQ 40/40(Diameter 14.70mm)	
Height	≈30 mm	

Fig. 8. Diode clamped multi-level converter prototype

With the SHE technique applied to the switching algorithm, the ferromagnetic material core voltage harmonic spectrum is illustrated in Fig. 9. Obviously, the third and fifth harmonic orders with 200-kHz fundamental harmonic are almost eliminated from the total contents, and the THD is approximately 17%.

The primary and secondary coil size of the ferromagnetic material core are approximately 1/3 those of the air-core transformer. The secondary voltage still maintains the five-level waveform and can be rectified into a dc signal at the end of the stage.

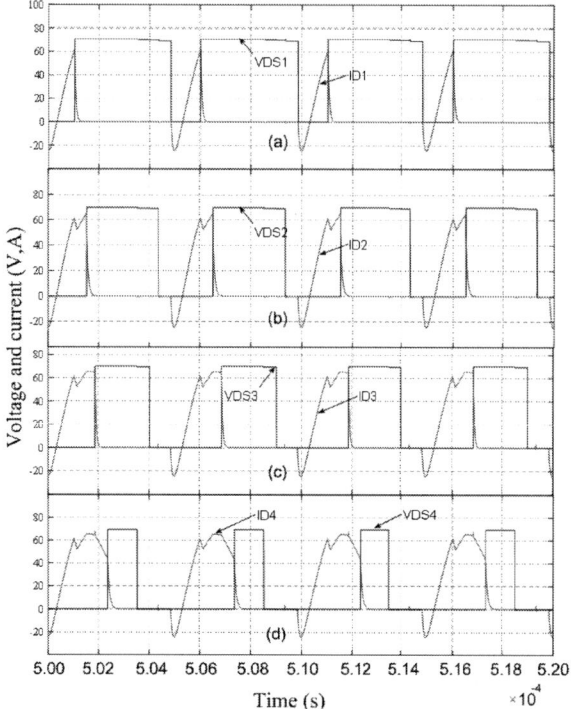

Fig. 6. Drain-source voltage and current of the DCML converter:(a) MOSFET$_1$, (b) MOSFET$_2$, (c) MOSFET$_3$, (d) MOSFET$_4$

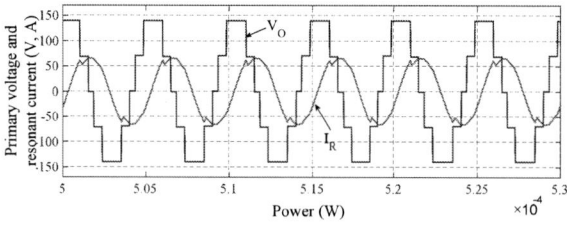

Fig. 7. Inverter output voltage and resonant current

A 12-kW Kikusui PCR4000LA digital dc power generator was utilized as a power unit for the dc link bus of the DCML converter in order to generate a five-level output voltage. The prototype setup for IPT was calculated and substituted to the plant with the FHA technique.

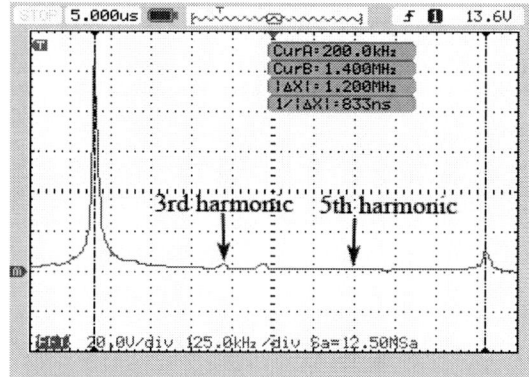

Fig. 9. DCML harmonic order content

(20.0V/DIV, 125 kHz/DIV, Sa 12.5 NSa)

The efficiencies of the ferromagnetic core and air-core are illustrated in Fig. 10. One of most important key factor for high effectiveness system is the coupling coefficient, k or the total line of flux of one coil cuts through the secondary coil. If k equal to 1 means all the line of the flux of one coil cuts all through the secondary coil, if $k > 0.5$ the two coils are called tightly coupled. Otherwise, if $k < 0.5$ the two coils are called loosely coupled. The distance of the primary coil and secondary coil could be improved by adjusting the gap with high precision fabrication to create a high efficiency power transfer via the contactless transformer.

1463

The 2018 International Power Electronics Conference

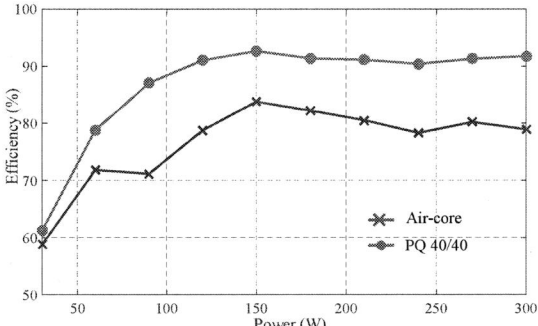

Fig. 10. Efficiency of the IPT with PQ40/40 core compared with air-core

The single-layer solenoid with an outer shelter coil and an inner coil transformer was measured with a Hewlett-Packard HP4194A impedance/gain phase analyzer. The coupling coefficient is equal to 0.91 with a tight coupling condition, assuming there is no distance between the two coils for air-core IPT. Fig. 11. shows the load fluctuation of DCML at rated power and 110%, apparently, the converter output still operates under ZVS condition.

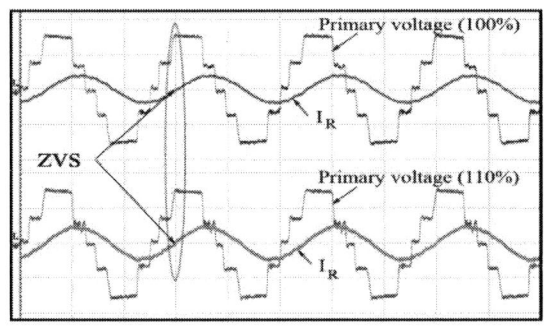

Fig. 11. DCML ZVS under load variation (I_r 10A/DIV, $V_{primary}$ 100V/DIV, T 2μS/DIV)

IV. CONCLUSION

A contactless power transfer prototype with different kinds of materials was implemented and evaluated in this study. Initially, the MATLAB/Simulink simulation was used for calculating and simulating the LLC series resonant current and the voltage; the hardware tests showed that the current was shifted from the drain-source voltage and the design performed soft switching (ZVS) for every switching device. The maximum efficiency for the air-core and the ferromagnetic material core were 83.8% and 92.6%, respectively. However, the cost effectiveness of the system was revealed by the reversal of the power density when comparing the two types of IPT. Obviously, the ferromagnetic material core power density is greater than air core by a factor of more than 3. Additionally, the IPT transformer size can be dramatically reduced with higher switching frequency that matches the designed resonant tank. Finally, this proposed method could lead us to a promising future for dc to dc conversion with renewable energy sources.

REFERENCES

[1] H. Kakigano, Y. Miura, T. Ise, T. Momose, H. Hayakawa, "Fundamental Characteristics of DC Microgrid for Residential Houses with Cogeneration System in Each House," *in Proceedings of Power and Energy Society General Meeting - Conversion and Delivery of Electrical Energy in the 21st Century, 2008, Pittsburgh, Pennsylvania, USA, July 2008.*

[2] Surin Khomfoi, Warachart Sae-Kok, and Issarachai Ngamroo, "An Open Circuit Fault Diagnostic Technique in IGBTs for AC to DC Converters Applied in Microgrid Applications," *Journal of Power Electronics,* vol. 11, no. 6, pp. 801-810, 2011.

[3] Surin Khomfoi, "A Cascaded Hybrid Multilevel Inverter Incorporating a Reconfiguration Technique for Low Voltage DC Distribution Applications," *Journal of Power Electronics,* vol. 16, no. 1, pp. 340-350, 2016.

[4] Worapong Pairindra, Surin Khomfoi, "An Optimization Design of the Diode Clamped Multi-Level Converter for Coaxial Inductive Power Transfer on the Low Voltage DC Micro-grid," *Journal of Electrical Engineering & Technology,* vol 13, no, 1 pp. 333-344, 2018.

[5] Yushi Miiura, Satoshi Ojika, Tomofumi Ise, "Voltage Control of Inductive Contactless Power Transfer System with Coaxial Coreless Transformer for DC Power Distribution," *in Proceedings of IEEE IPEC2014,* Hiroshima, Japan, page 1430 - 1437, May 2014.

[6] Satoshi Ojika, Yushi Miura and Toshifumi Ise, "Inductive Contactless Power Transfer System with Coaxial Coreless Transformer for DC Power Distribution," *in Proceedings of IEEE ECCE2013* Asia Downunder (ECCE Asia), Melbourne, Australia, page 1046 - 1051, June 2013.

[7] Ihssen Jabri, Fethi Ghodbane, Adel Bouallegue, Adel Khedher, "Modeling of the Coreless Transformer and Response to Misalignment and Gap Variation," *Proceedings of IEEE Electrical Sciences and Technologies in Maghreb (CISTEM), 2014,* Tunis, Tunisia, November 2014.

[8] Yusuke Hayashi, Hajime Toyoda, Toshifumi Ise, "Design Consideration for Contactless DC Connector in High Power Density Future 380 V DC Distribution System, " *Proceedings of IEEE Energy Conversion Congress and Exposition (ECCE), 2014 IEEE,* Pittsburgh, PA, USA, September 2014.

[9] Yusuke Hayashi, Hajime Toyoda, Toshifumi Ise and Akira Matsumoto, "Contactless DC Connector Based on GaN LLCConverter for Next-Generation Data Centers," *IEEE Trans. on Industry applications,* vol 51,No. 4, pp. 3244-3253, July/August 2015

[10] Sangwook Han, and David D. Wentzloff, "Wireless Power Transfer Using Resonant Inductive Coupling for 3D Integrated ICs," *in Proceedings of IEEE 3D Systems Integration Conference (3DIC) 2010,* Munich, Germany, Nov. 2010.

[11] Yusuke Hayashi, Hajime Toyoda, Toshifumi Iseand Akira Masumoto, "Contactless DC Connector based on ISOP–IPOS Topology for High Power Density 380 V DC Power Feeding System ," *in Proceedings of IEEE 2014 16th European Conference on Power Electronics and Applications,* Lappeenranta, Finland, Aug 2014.

[12] Said Barkati, Lotfi Baghlib,El Madjid Berkouk, Mohamed-Seghir Boucherit, " Harmonic elimination in diode-clamped multilevel inverter using evolutionary algorithms" *Electric Power Systems Research* 78, pp. 1736–1746, 2008.

[13] Sule Ozdemir, Engin Ozdemir, Leon M. Tolbert, and Surin Khomfoi, "Elimination of Harmonics in a Five-Level Diode-Clamped Multilevel Inverter Using Fundamental Modulation," *in Proceedings of IEEE 2007 7th International Conference on Power Electronics and Drive Systems,* Daegu, South Korea, October 2007.

[14] D. Graham Holmes, Thomas A. Lipo, *Pulse width modulation for power converter:* A John Wiley &Son, 2003, p. 440-449.

[15] S. Khomfoi and L. M. Tolbert, Multilevel Power Converters, Power Electronics Handbook, 2nd Edition, Elsevier, Chapter 17, pp. 451-482. 2007.

The 2018 International Power Electronics Conference

Performance Programming Technique for Multi-Stage Dc Power Distribution Systems

Syam Kumar Pidaparthy*, Hansang Kim, Yeonjung Kim, and Byungcho Choi

School of Electronics Engineering, Kyungpook National University, Daegu, South Korea

*E-mail: syamkumar.537@gmail.com

Abstract—**This paper presents a new design technique for multi-stage dc power distribution systems, which consistently offers the pre-programmed performance for dc distribution bus, irrespective of detrimental interactions among subsystem components. The paper demonstrates that the proposed design technique provides the aimed performance, even for cases in which the conventional design encounters instability due to adverse interactions from an ill-conditioned source subsystem. The performance of the proposed technique is theoretically validated and experimentally supported using a prototype two-stage dc power distribution system.**

Keywords—*Dc power distribution systems, distribution bus dynamics, impedance overlap, line filter design, performance programming, stability.*

I. INTRODUCTION

Dc distribution power systems employ several cascaded/paralleled stages of converters and line filters to achieve efficient and reliable power conversion [1]. Such multi-stage dc power systems are exposed to potential performance degradation or instability, due to adverse dynamic interactions among cascaded/paralleled subsystems [2]–[12]. Elaborate techniques have been proposed to resolve the aforementioned problem, such as active damping techniques [3], [4], controller tuning methods [5], [6], feedforward controls [7], [8], and virtual impedance implementations [9]–[12]. The prior techniques employed additional feedback controllers or dedicated control circuits, which complicates the controller design. The existing techniques mainly focused on stability, while paying limited attention to dynamic behaviors of dc distribution buses.

This paper proposes a new performance programming technique which allows dc power distribution systems to consistently secure both stability and pre-programmed performance for dc distribution bus, at the presence of considerable drifts in small-signal dynamics of subsystems. The proposed technique does not require any extra feedback controller or circuit component and only needs to redesign the pre-existing line filters. The new technique enables system designers/integrators to program the frequency- and time-domain performance of dc distribution bus. A performance-programmed system produces the aimed performance for most practical operational

conditions, even for cases where conventionally-designed systems result in instability.

II. TWO-STAGE DC POWER DISTRIBUTION SYSTEM

The experimental system shown in Fig. 1 is used in demonstrating both the problems of the conventional design and benefits of the proposed technique. The system consists of a practical voltage source − a voltage source with a finite output impedance, current-mode controlled boost front-end converter with an input filter, and load subsystem. The load subsystem is configured as a parallel connection of two current-mode controlled buck load converters. Each load converter has a separate line filter stage. The current source i_{bus} is employed in order to evaluate the transient performance of the dc distribution bus. A switch, **sw** in Fig. 1, is placed at the junction among the practical voltage source, ideal voltage source, and front-end converter, in order to contrast the system operation with an ideal voltage source to the operation with a practical voltage source. The system performance and stability are evaluated for the following two different cases:

1) **Case A**: the system is connected to the ideal voltage source with **sw** at **a** position.

2) **Case B**: the system is operated with the practical voltage source with **sw** at **b** position.

This paper uses the impedance based stability criterion [13], where the impedance ratio defined as a minor loop gain is utilized to evaluate the system stability. Two distinct minor loop gains are defined for the stability analysis. The first minor loop gain is defined as an impedance ratio at the input port of the front-end converter, $T_{mn1} = Z_s/Z_{iC}$, where Z_{iC} is the input impedance of the front-end converter, and Z_s is the source impedance seen by the front-end converter. The second minor loop gain is defined as the impedance ratio at the output port of the front-end converter, $T_{mn2} = Z_{oC}/Z_L$, where Z_{oC} is the output impedance of the front-end converter, and $Z_L = Z_{L1} \parallel Z_{L2}$ is the load impedance seen by the front-end converter.

A. Problems in Conventional Design Practice

This section addresses the problems of the dc power distribution system, designed based on the conventional approach [13], [14]. The system stability and performance of the conventionally-designed dc power distribution system are shown in Fig. 2. Two line filters in the load subsystem are designed to provide an impedance gap

---

This research was supported by Basic Science Research Program through the National Research Foundation of Korea (NRF) funded by the Ministry of Education (NRF- 2016R1D1A1B03931966).

1465

The 2018 International Power Electronics Conference

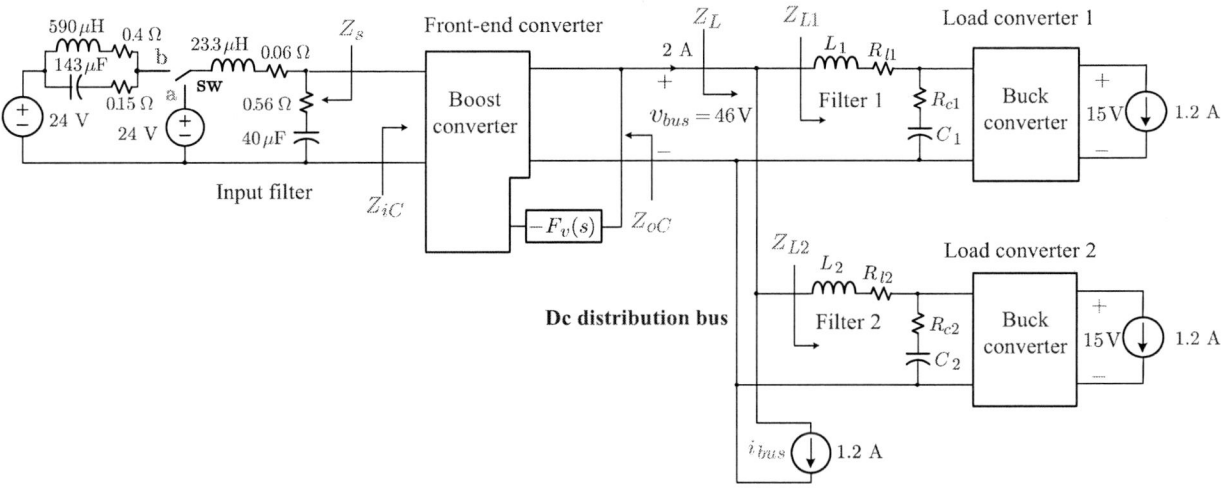

Fig. 1. Dc power distribution system.

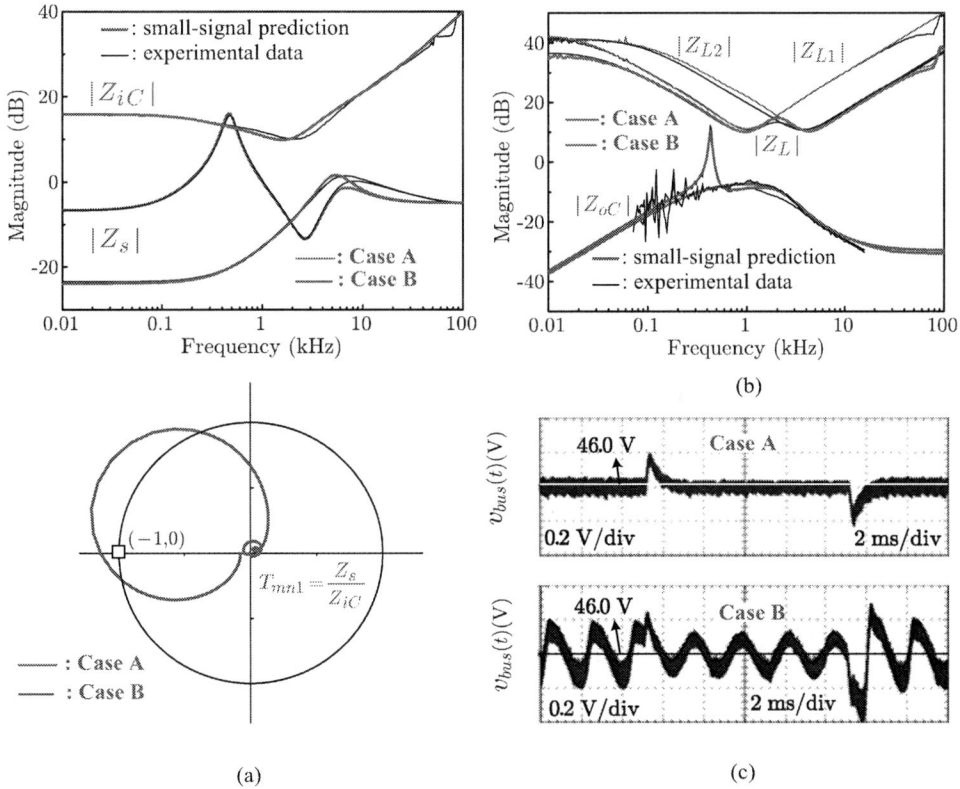

(a)

(b)

(c)

Fig. 2. Dc power distribution system with conventional design. (a) Impedances at input port of front-end converter and minor loop gain T_{mn1}. (b) Impedances at output port of front-end converter. (c) Transient response of bus voltage.

between $|Z_{oC}|$ and $|Z_L|$, as shown in Fig. 2(b). The circuit components of the line filters are

- Filter 1 for Load converter 1: $L_1 = 512\ \mu H$, $R_{l1} = 0.128\ \Omega$, $C_1 = 50\ \mu F$, and $R_{c1} = 3.5\ \Omega$.

- Filter 2 for Load converter 2: $L_2 = 138\ \mu H$, $R_{l2} = 0.093\ \Omega$, $C_2 = 13.9\ \mu F$, and $R_{c2} = 4\ \Omega$.

The stability and performance analyses of the conventionally-designed power conversion are summarized for the two different cases.

1) Case A with ideal voltage source: The first minor loop gain T_{mn1} in Fig. 2(a) indicates that the system is stable with sufficient margin. The system stability is well supported in Fig. 2(c), which shows the transient

1466

waveform of the bus voltage, v_{bus}, in response to 0.5 A step changes in i_{bus}.

2) Case B with practical voltage source: As shown in Fig. 2(a), the practical voltage source induces a peaking to $|Z_s|$, making $|Z_s|$ exceed $|Z_{iC}|$ and the polar plot of T_{mn1} encircle the $(-1, 0)$ point. This is an indication of the instability stimulated by the practical voltage source. This instability can be clearly seen in the form of large peaking in the output impedance of the converter, $|Z_{oC}|$, around the frequencies where $|Z_s|$ exceed $|Z_{iC}|$, as shown in Fig. 2(b). This system instability is also verified by the bus voltage waveform in Fig. 2(c).

III. IMPEDANCE OVERLAP AND SYSTEM PERFORMANCE

The conventional design provides an impedance gap between front-end converter and load subsystem, in order to suppress the interaction from the load subsystem. This design approach implicitly presumes that the interaction due to the load subsystem is detrimental to the system stability and performance. In contrary to general perception, a recent publication [15] proved that an *impedance overlap between front-end converter and load subsystem* can be exploited to enhance the system stability and performance. This paper further presents the application potentials of the impedance overlap to enhance the system performance. Moreover, the impedance overlap can be used an instrumental tool to pre-program the dynamics of dc distribution bus. This section investigates the impacts of the impedance overlap on the system dynamics.

A. Impact of Impedance Overlap on Converter Input Impedance

Figure 3 illustrates the impact of the impedance overlap on the input impedance of the front-end converter. The load impedance, $Z_L = Z_{L1} \parallel Z_{L2}$, is deliberately designed to create an impedance overlap at the output port of the front-end converter as shown in Fig. 3(a). The $|Z_L|$ falls below $|Z_{oC}|$ for the frequency range of $\omega_1 < \omega < \omega_2$. At this frequency range, the minor loop gain $|T_{mn2}|$ becomes positive and crosses the 0 dB line at ω_1 and ω_2. Figure 3(b) shows the polar plot of the minor loop gain, $T_{mn2} = Z_{oC}/Z_L$. The polar plot of T_{mn2} crosses the unit cirlce ω_1 and ω_2, where two different phase margins are marked on the unit circle: $\text{PM}_1 = 180° - \angle T_{mn2}(j\omega_1)$ at ω_1 and $\text{PM}_2 = \angle T_{mn2}(j\omega_2) - (-180°)$ at ω_2.

Figure 3(c) illustrates the input impedance of the converter with the impedance overlap, $|Z'_{iC}|$. The impedance overlap provides a boost for $|Z'_{iC}|$ from $|Z_{iC}|$ of the conventional design. Thus, the amount of the impedance overlap between $|Z_{oC}|$ and $|Z_L|$ is projected as a magnitude boost in $|Z'_{iC}|$ in the frequency range of the overlap, $\omega_1 < \omega < \omega_2$. As will be shown later, the boost in $|Z'_{iC}|$ provides protections from potential stability problems at the input port of the front-end converter.

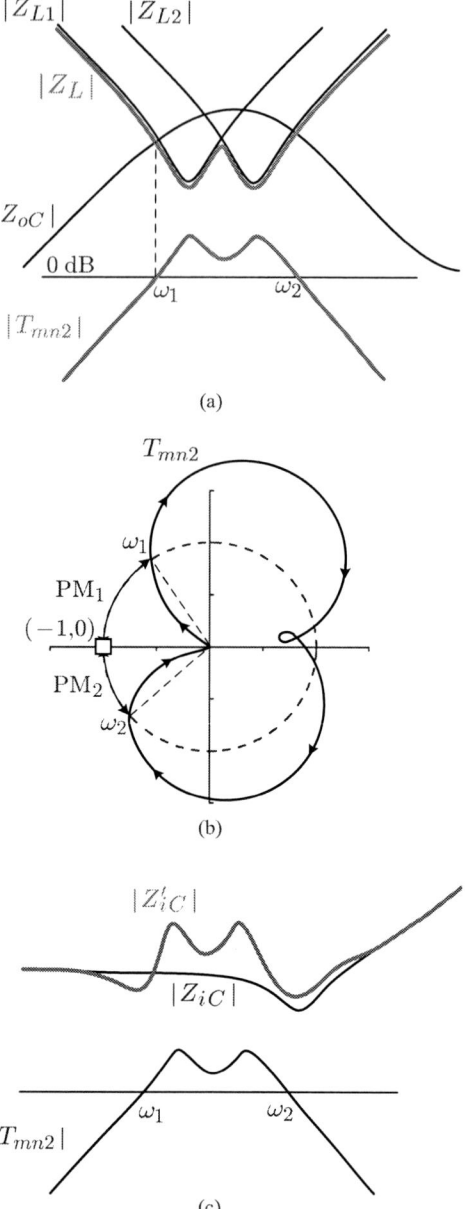

Fig. 3. Impact of impedance overlap on the converter input impedance. (a) Impedance overlap. (b) Polar plot of minor loop gain, T_{mn2}. (c) Minor loop gain and converter input impedance.

B. Bus Impedance and Distribution Bus Dynamics

Figure 4 depicts the bus impedance and transient response of dc bus voltage in the presence of the impedance overlap. The bus impedance, Z_{bus} in Fig. 4(a), is constructed using the relation $Z_{bus} = Z_{oC} \parallel Z_{L1} \parallel Z_{L2}$. The $|Z_{bus}|$ exhibits two isolated peaks at the crossover frequencies at ω_1 and ω_2. The two peaks transform into underdamped second-order terms in Z_{bus}, which will produce decaying oscillations in time-domain responses [16]. The low frequency oscillation at ω_1 usually dominates the dc bus voltage response, v_{bus}, due to the condition

The 2018 International Power Electronics Conference

TABLE I.   ASYMPTOTIC ANALYSIS OF BUS IMPEDANCE AND BUS VOLTAGE

Low-frequency approximation of bus impedance	Prediction of bus voltage
$$\bar{Z}_{bus}(s) = k_d s \dfrac{1}{1 + \dfrac{s}{Q_1 \omega_1} + \dfrac{s^2}{\omega_1^2}}$$	$$v_{bus} = \dfrac{\Delta i_{bus}\, k_d\, \omega_1}{\sqrt{1 - 1/(4Q_1^2)}}\, e^{-\frac{\omega_1}{2Q_1} t} \sin\left(\omega_1 \sqrt{1 - 1/(4Q_1^2)}\right) t$$
$$k_d = \dfrac{\lvert Z_{oC}\rvert_{@\omega_1}}{\omega_1}$$	$$\Delta \bar{v}_{bus\,max} = \Delta i_{bus}\, k_d\, \omega_1\, e^{-\left(\frac{1}{\sqrt{4Q_1^2 - 1}} \tan^{-1} \sqrt{4Q_1^2 - 1}\right)}$$
$$Q_1 = \dfrac{1}{\sqrt{(2 - 2\cos \mathrm{PM}_1)}}$$	$$t_s \approx \dfrac{6Q_1}{\omega_1}$$

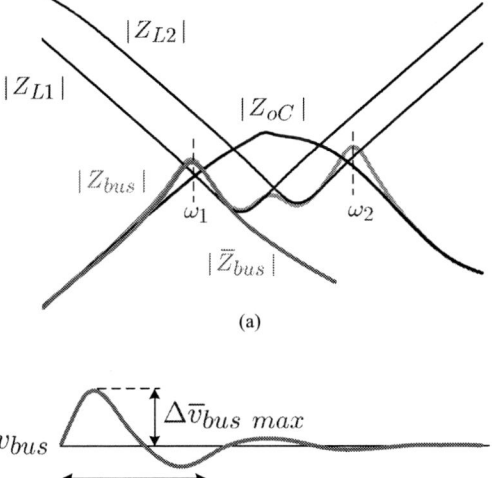

Fig. 4.   Bus impedance and step load response of bus voltage. (a) Formation of bus impedance. (b) Prediction of bus voltage.

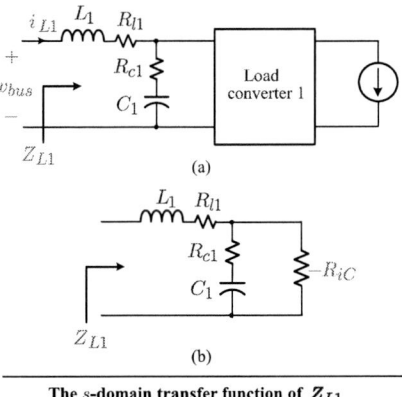

The s-domain transfer function of Z_{L1}

$$Z_{L1}(s) = -R_{iC} \dfrac{1 + \dfrac{s}{Q_o \omega_o} + \dfrac{s^2}{\omega_o^2}}{1 + \dfrac{s}{\omega_p}}$$

$$Q_o = \dfrac{1}{R_{l1} + R_{c1}} \sqrt{\dfrac{L_1}{C_1}} \quad \omega_p = -\dfrac{1}{C_1 R_{iC}} \quad \omega_o = \dfrac{1}{\sqrt{L_1 C_1}}$$

(c)

Fig. 5.   Evaluation of Z_{L1}. (a) Block diagram. (b) Circuit model. (c) s-domain transfer function of Z_{L1}.

$\omega_2 \gg \omega_1$. Therefore, the v_{bus} response, can be predicted from the low-frequency asymptotic approximation of the bus impedance. The v_{bus} expression in response to a step current change Δi_{bus} can be evaluated from

$$v_{bus} = \mathcal{L}^{-1}\left(\bar{Z}_{bus} \Delta i_{bus}/s\right) \qquad (1)$$

where $\bar{Z}_{bus}$ is low-frequency approximation of Z_{bus}. Table I gives the expressions for $\bar{Z}_{bus}$, and the dc bus voltage response of (1). Figure 4(b) illustrates the transient waveform v_{bus}. The expressions for the performance metrics of v_{bus}, such as maximum overshoot $\Delta \bar{v}_{bus\,max}$ and the 5%-boundary settling time t_s in Fig. 4(b), are given in Table I. These expressions indicate the performance can be determined from the phase margin PM_1 at ω_1.

IV.   PERFORMANCE PROGRAMMING AND LINE FILTER DESIGN

For the known Z_{oC}, Z_L can be purposely designed so that T_{mn2} has a specified phase margin PM_1 at ω_1. The PM_1 has the corresponding Q_1 value. The resulting Q_1 value, along with ω_1 and k_d, determines the maximum overshoot and setting time of the bus voltage.

Referring to Fig. 3(a), the aimed PM_1 is determined at the frequency where $\lvert Z_{oC}\rvert$ intersects with $\lvert Z_{L1}\rvert$. The Z_{L1} is the input impedance of the load subsystem branch, associated with the line filter having the larger capacitor. We presume that Filter 1, feeding Load converter 1 in Fig. 1(a), has the larger capacitor. This implies that the performance of the distribution bus can be programmed by properly selecting the circuit components of Filter 1, while maintaining $C_1 \gg C_2$.

1468

The 2018 International Power Electronics Conference

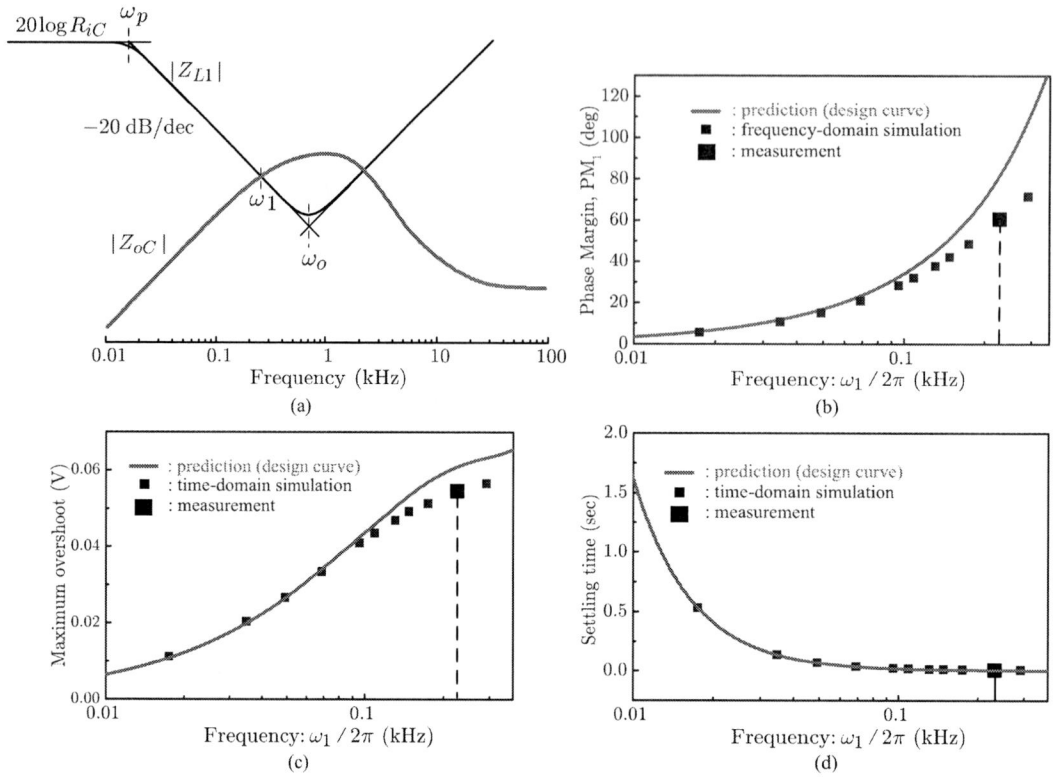

Fig. 6. Line filter design and system performance. (a) Line filter design. (b) ω_1 and phase margin. (c) ω_1 and maximum overshoot. (d) ω_1 and settling time.

Figure 5 describes the evaluation of Z_{L1}. Figure 5(a) is a block diagram of the load subsystem branch associated with Filter 1. Figure 5(b) is the circuit model for Z_{L1}, obtained by replacing the input impedance of Load converter 1 with its negative resistance approximation [15], [17], $-R_{iC} = -V_{bus}/I_{L1}$ with I_{L1} being the dc current flowing into Load converter 1. The s-domain expression of Z_{L1} is given in Fig. 5(c). Details about the load impedance evaluation and analysis can be found in [15], [17].

The design of Filter 1 and associated performance programming are illustrated in Fig. 6. The asymptotic plot of $|Z_{L1}|$ is drawn in Fig. 6(a), along with the Bode plot of $|Z_{oC}|$, generated with the ideal voltage source, i.e., **sw** at **a** position. The following relationship holds at ω_1

$$20\log R_{iC} \quad - \quad 20\log\left(\frac{\omega_1}{\omega_p}\right) = |Z_{oC}|_{@\omega_1} \quad (2)$$

From the s-domain expression of Z_{L1} shown in Fig. 5(c), the phase margin , $\mathrm{PM}_1 = 180° - \left(\angle Z_{oC@\omega_1} - \angle Z_{L1@\omega_1}\right)$, at ω_1 is derived as

$$\mathrm{PM}_1 = -\angle Z_{oC@\omega_1} \quad + \quad \tan^{-1}\left(\frac{\omega_1}{\omega_p}\right)$$
$$+ \quad \tan^{-1}\left(\frac{\omega_1/(Q_o\omega_o)}{1-\omega_1^2/\omega_o^2}\right) \quad (3)$$

From the expressions (2) and (3), a direct correspondence

between ω_1 and PM_1 is derived as

$$\mathrm{PM}_1 = -\angle Z_{oC@\omega_1} \quad + \quad \tan^{-1}\left(\frac{R_{ic}}{10^{|Z_{oC}|_{@\omega_1}/20}}\right)$$
$$+ \quad \tan^{-1}\left(\frac{\omega_1/(Q_o\omega_o)}{1-\omega_1^2/\omega_o^2}\right) \quad (4)$$

This relationship is converted into the $\omega_1 - \mathrm{PM}_1$ curve under the assumption that Q_o and ω_o are preselected, which constitutes the design curve for Filter 1.

Figure 6(b) is the $\omega_1 - \mathrm{PM}_1$ curve, constructed with $Q_o = 1.0$ and $\omega_o = 2\pi \cdot 375$ rad/s. The $\omega_1 - \mathrm{PM}_1$ curve specifies the attainable phase margin PM_1 and associated ω_1 value, for the given Z_{oC} characteristics and line filter structure. The $\omega_1 - \mathrm{PM}_1$ curve is plotted for the frequency range of $2\pi \cdot 10$ rad/s $< \omega_1 < 2\pi \cdot 375$ rad/sec. When ω_1 exceeds $2\pi \cdot 375$ rad/sec, the performance programming feature will be lost. For the experimental front-end converter, ω_1 is selected at $2\pi \cdot 229$ rad/s to obtain the phase margin of $\mathrm{PM}_1 = 75°$, as indicated with a vertical line in Fig. 6(b). The required pole frequency ω_p for this design is determined from (2)

$$\omega_p = 10^{\frac{|Z_{oC}|_{\omega_1}}{20}}\frac{\omega_1}{R_{ic}} = 10^{\frac{-11.78}{20}}\frac{2\pi \cdot 229}{115} = 2\pi \cdot 0.51 \text{ rad/s}$$

From the given ω_p, ω_o, and Q_o values, the filter components are determined as $L_1 = 66.7$ μH, $R_{l1} = 0.017$ Ω, $C_1 = 2.7$ mF, and $R_{c1} = 0.14$ Ω, based on the expressions shown in Fig. 5(c).

1469

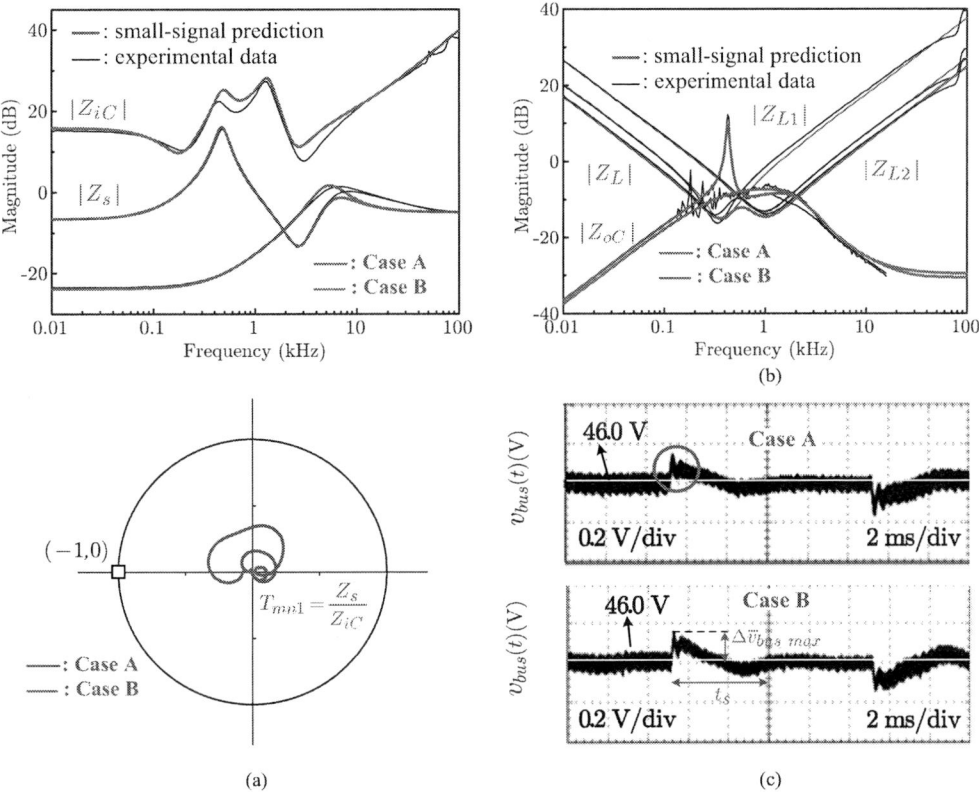

Fig. 7. System dynamics of performance programmed dc distribution system. (a) Impedances at input port of front-end converter and minor loop gain T_{mn1}. (b) Impedances at output port of front-end converter. (c) Transient response of bus voltage.

The accuracy of the $\omega_1 - \mathrm{PM}_1$ curve in Fig. 6(b), which served as the design curve for Filter 1, is assessed with computer simulations. The small squares represent the phase margins of the converter loop gain, obtained from frequency-domain simulations.

The phase margin curve in Fig. 6(b) provides good estimates for the maximum overshoot and settling time of the bus voltage, v_{bus}, as given in Table I. Figures 6(c) and 6(d) are the predictions and assessments of the maximum overshoot and 5%-boundary settling time for the case of a 0.5 A current change at the distribution bus. Time-domain simulations are displayed with small squares, in order to show the accuracy of the predictions.

V. DYNAMICS OF PERFORMANCE-PROGRAMMED TWO-STAGE DC DISTRIBUTION SYSTEM

Figure 7 shows the performance of the experimental dc power distributed system whose line filters are designed for $\mathrm{PM}_1 = 75°$ at $\omega_1 = 2\pi \cdot 229$ rad/s, $\Delta v_{bus\,max} = 60$ mV and $t_s = 3.2$ ms. Filter 1 was designed in Section IV for performance programming, but Filter 2 components are selected to meet the condition $C_1 \gg C_2$. The components for the line filters are
• Filter 1: $L_1 = 65.4~\mu$H, $R_{l1} = 0.02~\Omega$, $C_1 = 2.5$ mF, and $R_{c1} = 0.132~\Omega$.
• Filter 2: $L_2 = 36~\mu$H, $R_{l2} = 0.03~\Omega$, $C_2 = 0.75$ mF, and $R_{c2} = 0.18~\Omega$.

The line filters generate the aimed impedance overlap between $|Z_{oC}|$ and $|Z_L|$ as illustrated in Fig. 7(b). As demonstrated in Section III, the impedance overlap reshapes the input impedance of the front-end converter, as shown in Fig. 7(a). The reshaped input impedance provides a sufficient impedance gap at the input port of front-end converter. This impedance gap totally nullifies the detrimental interaction coming from the practical voltage source. As shown in Fig. 7(c), the bus voltage reveals the maximum overshoot of $\Delta v_{bus\,max} = 54$ mV which settles in $t_s = 4$ ms. The system exhibited the same programmed performance for both the ideal voltage source and practical voltage source. This shows the robustness of the proposed performance programming technique.

VI. CONCLUSIONS

This paper presented a new performance programming technique for multi-stage dc power distribution systems. The proposed technique constantly guarantees the stability and preprogrammed performance metrics for dc distribution bus. The implementation of the new technique only needs to redesign pre-existing line filters. The outcomes of this paper lay out general design methodology for multi-stage dc power distribution systems.

REFERENCES

[1] W. Tabisz, M. Jovanovic, and F. Lee, "Present and future of distributed power systems," in *Proc. IEEE Appl. Power Electron. Conf. Expo.*, Feb 1992, pp. 11–18.

[2] S. K. Pidaparthy and B. Choi, "Input impedances of pwm dc-dc converters: Unified analysis and application example," *Journal of Power Electronics (JPE)*, vol. 16, no. 6, November 2016.

[3] M. Wu and D. D. C. Lu, "A novel stabilization method of LC input filter with constant power loads without load performance compromise in dc microgrids," *IEEE Trans. Ind. Electron.*, vol. 62, no. 7, pp. 4552–4562, July 2015.

[4] M. Wu, D. D. C. Lu, and C. K. Tse. "Direct and optimal linear active methods for stabilization of LC input filters and DC/DC converters under voltage mode control," *IEEE Journal on Emerging and Selected Topics in Circuits and Systems*, vol. 5, no. 3, pp. 402–412, Sept 2015.

[5] G. Sulligoi, D. Bosich, G. Giadrossi, L. Zhu, M. Cupelli, and A. Monti, "Multiconverter medium voltage dc power systems on ships: Constant-power loads instability solution using linearization via state feedback control," *IEEE Trans. on Smart Grid*, vol. 5, no. 5, pp. 2543–2552, Sept 2014.

[6] Y. Gu, D. Zhang, X. Wu, and X. Zhang, "Research on stability improvement of the cascaded dc-dc converters based on ac signal sampling control method," *IEEE Trans. Power Electron.*, vol. 33, no. 5, pp. 4547–4559, May 2018.

[7] A. Riccobono, J. Siegers, and E. Santi, "Stabilizing positive feedforward control design for a dc power distribution system using a passivity-based stability criterion and system bus impedance identification," in *Proc. Applied Power Electronics Conference and Exposition (APEC), 2014 Twenty-Ninth Annual IEEE*, March 2014, pp. 1139–1146.

[8] J. Siegers, S. Arrua, and E. Santi, "Stabilizing controller design for multi-bus mvdc distribution systems using a passivity based stability criterion and positive feed-forward control," in *Proc. Energy Conversion Congress and Exposition (ECCE), 2015 IEEE*, Sept 2015, pp. 5180–5187.

[9] X. Zhang, X. Ruan, and Q. C. Zhong, "Improving the stability of cascaded dc/dc converter systems via shaping the input impedance of the load converter with a parallel or series virtual impedance," *IEEE Trans. Ind. Electron.*, vol. 62, no. 12, pp. 7499–7512, Dec 2015.

[10] X. Lu, K. Sun, J. M. Guerrero, J. C. Vasquez, L. Huang, and J. Wang, "Stability enhancement based on virtual impedance for dc microgrids with constant power loads," *IEEE Trans. Smart Grid*, vol. 6, no. 6, pp. 2770–2783, Nov 2015.

[11] X. Zhang, X. Ruan, and Q.-C. Zhong, "Improving the stability of cascaded dc/dc converter systems via shaping the input impedance of the load converter with a parallel or series virtual impedance," *IEEE Trans. Ind. Electron.*, vol. 62, no. 12, pp. 7499–7512, Dec 2015.

[12] X. Zhang, Q. C. Zhong, and W. L. Ming, "Stabilization of cascaded dc/dc converters via adaptive series-virtual-impedance control of the load converter," *IEEE Trans. Power Electron.*, vol. 31, no. 9, pp. 6057–6063, Sept 2016.

[13] R. D. Middlebrook, "Input filter considerations in design and application of switching regulators," in *Proc. IEEE Ind. Appl. Soc.*, October 1976, pp. 366 –382.

[14] B. Choi and B. Cho, "Intermediate line filter design to meet both impedance compatibility and EMI specifications," *IEEE Trans. Power Electron.*, vol. 10, no. 5, pp. 583–588, Sep 1995.

[15] S. K. Pidaparthy, B. Choi, H. Kim, and Y. Kim, "Stabilizing effects of load subsystem in multistage dc-to-dc power conversion systems," *IEEE J. Emerg. Sel. Topics Power Electron.*, vol. 5, no. 4, pp. 1589–1603, Dec 2017.

[16] B. Choi, *Pulsewidth Modulated DC-to-DC Power Conversion: Circuits, Dynamics, and Control Designs.* John Wiley & Sons, 2013.

[17] S. K. Pidaparthy and B. Choi, "Stability analysis of PWM converters connected to general load subsystems," in *Proc. IEEE 9th ICPE-ECCE Asia*, June 2015, pp. 1033–1040.

Coordination Control for Paralleled Inverters Based on VSG for PV/Battery Microgrid

Meiqin Mao，Cheng Qian，Liuchen Chang, Yan Du
Research Center for Photovoltaic System Engineering
Hefei University of Technology
Hefei 23009, China
mmqmail@163.com

Abstract— **With increasing penetration of the intermittent renewable energy resources (RESs), Virtual Synchronous Generator (VSG) technology is proposed as it can provide backup inertia to improve dynamic performance of frequency. Traditionally, power sharing among inverters is in proportion to their rated power, the output power of RESs can not been maximally used. In this paper, a parameter adaptive control strategy is proposed with the consideration of PV output power and load demand for a VSG-controlled islanded PV/battery microgrid. The proposed strategy can maximize the utilization of PV generation. Finally, the effectiveness of the proposed control strategy is verified through two different simulation conditions.**

Index Terms-- **Virtual synchronous generator (VSG), PV, parameter constraint, parameters adaptation.**

I. INTRODUCTION

Increased awareness of limited energy resources and deteriorative environment has driven a considerable effort toward the integration of renewable distributed generators (DGs) such as wind turbines and photovoltaic in the microgrid system [1], [2]. These DGs are generally connected to the power system through inverters, the inverter-based DGs can not provide enough inertia for the system [3]. Moreover, the output power of renewable energy sources (RESs) is intermittent and fluctuating due to the effects of the environment. These all pose a challenge to the safe and stable operation of the power system [4].

Virtual Synchronous Generator (VSG), which integrates the advantages of traditional synchronous generator and inverter, is proposed to address the above problems. VSG can enhance the stability of the frequency by simulating rotor motion equations [5]. As a result, VSG is helpful to improve the penetration of RESs and enhance power system stability [6].

At present, the researches of VSG cover the control implementation [7], parameter design or stability analysis [8]. These studies mainly focus on the control of inverter, so the DC side is usually replaced by an ideal voltage source without the consideration of the output characteristics of RESs [9]. In fact, RESs such as PV may not provide enough energy buffers for VSG. To ensure the stable operation of VSG, battery storage system is usually combined with PV at the DC side, where battery storage system is used to smooth the PV output power fluctuations and provide energy buffers for VSG. But in an islanded PV/battery microgrid, power sharing among the paralleled VSGs is usually in proportion to their rated power. The output power of some VSGs may be less than the maximum output power of the PV they connected, leading these PV generation can not be fully utilized, while the output power of some VSGs may be more than the maximum output power of the PV they connected, thus battery storage system is required to balance the power difference. Therefore, the inverter capability and PV output power should been taken into account when designing the controller parameters of VSG to optimize power sharing. The time-varying property and finiteness of PV are considered in [4], a control strategy for PV generation system is proposed to keep the DC voltage stable and reach the optimal power matching based on the load or scheduled power, but the topology doesn't include energy storage. In [10], the available PV power and battery conditions are both taken into account, realizing reasonable power sharing under different load demands by adjusting droop coefficient and reference power dynamically. But the control strategy is for droop control and the effect of line impedance is ignored, thus it is prone to power oscillation while load changing. Based on the aforementioned literatures, this paper proposes a parameter adaptive control strategy for VSG-controlled islanded hybrid PV/battery microgrid to maximize the utilization of PV and ensure reasonable power sharing between VSGs.

The rest of the paper is arranged as follows: the microgrid structure and the VSG control strategy are presented in Section II. In Section III, a constraint for parameters selection can be obtained through analyzing the transfer function models of paralleled VSGs. In Section IV, a control strategy is proposed and the criteria for selecting the power sharing coefficient under different operation modes are presented. In

Section V, the proposed method is validated by Matlab/Simulink through various operating conditions. Section VI is the conclusion.

II. MICROGRID STRUCTURE AND VSG CONTROL STRATEGY

A. Structure of the PV/Battery microgrid

The microgrid consisting of hybrid PV/battery units, VSG-controlled inverters and loads is depicted in Fig.1. The PV unit includes PV arrays and a dc–dc boost converter, the energy storage unit includes a battery storage and a bidirectional dc/dc converter. All the local loads in microgrid are centralized in a load Z_{Load}. The inverter output terminal connects to the local AC bus and the microgrid connects to the main grid via a solid state switch. The microgrid operates in grid-connected mode when the solid state switch at PCC is closed, otherwise it operates in islanded mode. The micro sources provide the required electrical energy for the local loads corporately.

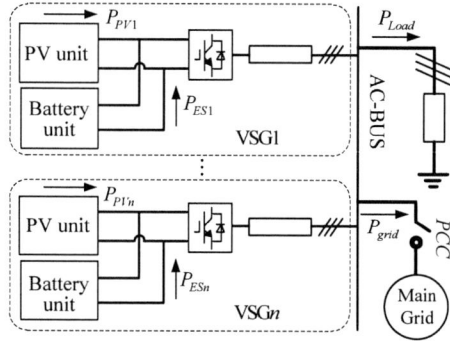

Fig. 1. Structure diagram of the studied PV/Battery microgrid

B. VSG Control Strategy

According to the simplified mechanical model of a synchronous generator, the swing equation is added in the power control loop to imitate inertia. Fig.2 shows the control block diagram of VSG-controlled inverter. The R/X ratio of line impedance in low voltage network is large, so introducing virtual impedance control by adding appropriate virtual inductance can realize the decoupling control of active power and reactive power, and improve power sharing accuracy between paralleled VSGs.

The swing equation of VSG is shown as:

$$P_m - P - D_P(\omega - \omega_0) = J\frac{d\omega}{dt} \quad (1)$$

Where, J is the moment of inertia, ω and ω_0 are rotor and reference angular frequency, P_m and P_e are the mechanical power of the rotor shaft and electromagnetic power of the stator respectively, D_P is the damping coefficient. Meanwhile, adding ω-P adaptive frequency control to emulate the droop characteristics of synchronous generator. P_{ref} is reference active power and m represents the droop coefficient.

$$P_m = P_{ref} + m(\omega_0 - \omega) \quad (2)$$

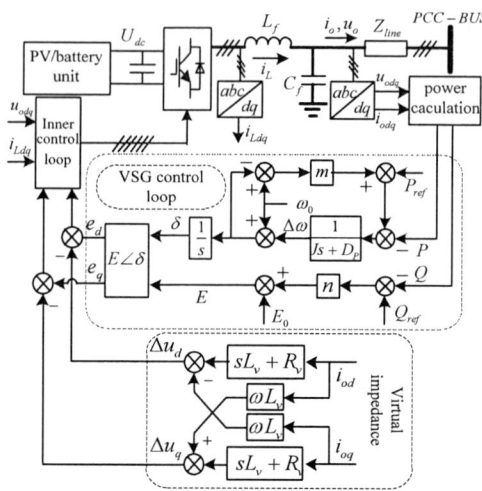

Fig. 2. Control block diagram of VSG-controlled inverter

III. DYNAMIC CHARACTERISTIC ANALYSIS OF PARALLELED VSGs

A. Transfer Function Models of Paralleled VSGs

One advantage of VSG control technology is that it can mimic the behavior of traditional synchronous generators. However, the introduction of swing equation increases the order of the system, which leads that VSG is prone to power oscillation during transient process [11]. In this section, transfer function models for multiple VSGs are established to analyze dynamic response characteristics of VSGs and the influence of parameters. P_i , P_{refi} is the output power and reference power of ith inverter respectively, J_i is the rotor inertia of ith inverter, D_i is the equivalent damping coefficient of ith inverter and $D_i=D_P+m$, the small signal model of power control loop can be presented as:

$$\Delta\omega_i = \frac{1}{J_i s + D_i}(\Delta P_{refi} - \Delta P_i) \quad (3)$$

The output impedance of VSG is considered to be inductive when virtual impedance control is applied, it can be derived from power transfer equation:

$$\Delta P_i = \frac{3}{2}\frac{E_i V}{X_i}\cos\delta_i\Delta\delta_i = k_i\Delta\delta_i \quad (4)$$

Where E_i represents output voltage amplitude of ith inverter, V represents voltage amplitude of PCC, if the virtual inductance is large enough, the total output reactance X_i of VSG can be replaced by virtual inductance $X_i=\omega L_{vi}$, the phase difference $\Delta\delta_i$ between VSG and PCC can be obtained as :

$$\Delta\delta_i = \frac{1}{s}(\Delta\omega_i - \Delta\omega) \quad (5)$$

For the system consisting of multiple VSGs, the balanced power flow among different VSGs and loads should satisfy (6) when the load demand changes:

$$\sum_{k=1}^{m} \Delta P_k = \Delta P_{Load} \tag{6}$$

According to (3) ~ (6), we can obtain the following transfer function models as shown in (7) ~ (9):

$$G_{pl} = \frac{\Delta P_i}{\Delta P_{Load}} = \frac{G_i\left(J_i s + D_i\right)}{\sum_{m=1}^{n} G_m\left(J_m s + D_m\right)} \tag{7}$$

$$G_{ri} = \frac{\Delta P_i}{\Delta P_{refi}} = \left(1 - \frac{G_i\left(J_i s + D_i\right)}{\sum_{j=1}^{n} G_j\left(J_j s + D_j\right)}\right) G_i \tag{8}$$

$$G_{rm} = \frac{\Delta P_i}{\Delta P_{refm}} = -\frac{G_i\left(J_i s + D_i\right)}{\sum_{j=1}^{n} G_j\left(J_j s + D_j\right)} G_m ; \ (m = 1, 2, \cdots n \ and \ m \neq i) \tag{9}$$

Where $\quad G_i = \dfrac{k_i}{J_i s^2 + D_i s + k_i} ; \ (i = 1, 2, \mathrm{L} \ n) \tag{10}$

G_{pl}, G_{ri}, G_{rm} represents the influence of load, reference power of ith inverter and reference power of other inverters on the output power of ith inverter respectively. From the transfer function models, the output power of paralleled VSGs in the islanded microgrid is not only related to the load demand and its own parameters, but also affected by the parameters of other VSGs, these effects are expressed through transfer functions.

B. Response Characteristic Analysis

Compared with traditional synchronous generator, the moment of inertia and damping coefficient of VSG can be chosen in real time to meet the demands of the microgrid [12]. Based on this advantage, the parameters can be changed adaptively to meet different demands.

By analyzing G_{pl}, an important conclusion can be found that if the parameters satisfy the following relations that is given by (11) (α_i is a introduced constant and $\sum_{i=1}^{n} \alpha_i = 1$), G_{pl} will become a proportional component as shown in (12), when load demand changes, the output power of VSG changes proportionally without dynamic oscillation process.

$$J_1 : J_2 : \cdots : J_n = D_1 : D_2 : \cdots : D_n = \frac{1}{X_1} : \frac{1}{X_2} : \cdots : \frac{1}{X_n} = \alpha_1 : \alpha_2 : \cdots : \alpha_n \tag{11}$$

Under this condition, G_{ri}, G_{rm} will become second order loops as shown in (13) and (14), when reference power changes, the output power of VSG will go into oscillation and the oscillation characteristic is related to G_i.

$$G_{pl} = \alpha_i \tag{12}$$

$$G_{ri} = (1 - \alpha_i) G_i \tag{13}$$

$$G_{ri} = -\alpha_i G_m ; (m \neq i) \tag{14}$$

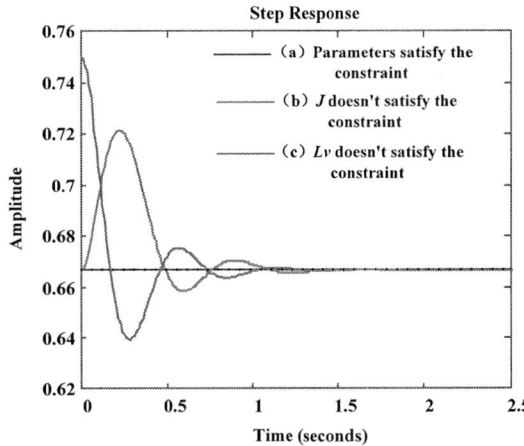

Fig. 3. Step response of G_{pl} for two VSGs system under different conditions of parameters

Fig.3 shows the step response curve of G_{pl} for two VSGs system under different conditions of parameters. When virtual inertia, equivalent damping coefficient and virtual inductance meet the constraint shown in (11), the two VSGs share load power proportionally without oscillation while load changing, as shown in Fig.3(a). When equivalent damping coefficient or virtual inductance doesn't meet the constraint, the output power of VSGs is gradually stabilizing after a period of oscillation as shown in Fig.3(b) and Fig.3(c). Differently, the initial value of G_{pl} is proportional to $1/X_i$, and the steady state value is proportional to D_i. So when the moment of inertia doesn't satisfy the constraint, the initial value of output power of VSG is equal to the steady state value, but when the virtual inductance doesn't satisfy the constraint, the initial value and the steady state value are not equal. Equation (11) can be regarded as a constraint for parameters selection.

For G_{ri} and G_{rm}, G_i represents the power response characteristic of a single VSG, the natural oscillation frequency ω_n and the damping ratio ξ_i is expressed as:

$$\begin{cases} \omega_n = \sqrt{k_i / J_i} \\ \xi_i = 0.5 D_i / \sqrt{J_i k_i} \end{cases} \tag{15}$$

Fig. 4. Step response of G_{ri} for two VSGs system under different damping

1474

When reference power changes, the dynamic process of output power of VSG is related to its own rotor inertia, equivalent damping coefficient, virtual inductance and so on. The steady state value of G_{ri} is less than unit, which means the output power can't increase to the given value. From the step response of G_{ri}, the output power oscillation decreases as the damping increases, so it can suppress power oscillation by choosing suitable J_i and D_i, Step response of G_{ri} for two VSGs system under different damping can be found in Fig.4.

IV. PROPOSED ADAPTIVE CONTROL STRATEGY FOR HYBRID PV/BATTERY UNITS-VSG

Through the analysis of transfer functions in the previous section, if the parameters meet the constraint shown in (11), G_i can be written as:

$$G_i = \alpha_i G_N, i = 1, 2, \cdots, n$$

Where $G_N = \dfrac{k_N}{J_N s^2 + D_N s + k_N}$, considering the paralleled VSGs as a whole, J_N, D_N is the total rotor inertia and total equivalent damping coefficient respectively, its damping ratio of the whole system is ξ. The reference values of rotor inertia J_N, equivalent damping coefficient D_N and virtual inductance L_N can be chosen based on the RoCoF, the steady state frequency deviation and damping ratio ξ. The coefficient α_i can be obtained as follows, then the parameters of ith inverter are $J_i = \alpha_i J_N$, $D_i = \alpha_i D_N$, $L_{vi} = L_N / \alpha_i$, and the output power of ith inverter P_i is equal to $\alpha_i P_L$, so α_i is called power sharing coefficient.

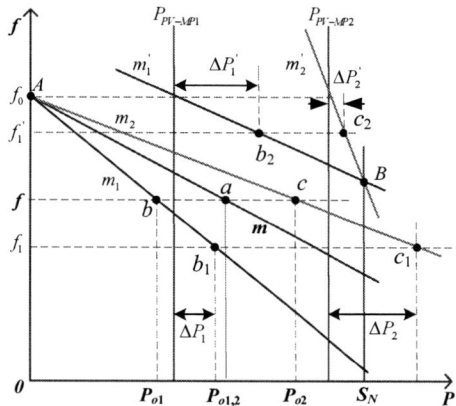

Fig. 5. *f-P* curves for VSG

Next, the analysis will be done for two paralleled VSGs shown in Fig.1, it is also applicable to multiple VSGs connected in parallel. Traditionally, power sharing is in proportion to the rated capacity of the inverter. Supposing the two VSGs have the same rated capacity, their *f-P* droop curves are shown as *Aa* in Fig.5, the VSGs share the load power equally.

The proposed control strategy takes the output power of photovoltaic into account. Assuming P_{PV-MP1}, P_{PV-MP2} is the PV maximum output power of the two VSGs respectively, the rated capacity of the inverters is S_N, the load power is P_{load}.

Depending on the total PV maximum output power and the total load power of the system, the microgrid can operate in two main modes:

Mode *I*: $\sum P_{PV-MPPi} > P_{Load}$, the total PV maximum output power is more than the total load power. If the equivalent damping coefficients of the two VSGs are equal, VSGs both work at point *a* and share the load evenly. Then the output power of VSG1 is more than P_{PV-MP1}, but the output power of VSG2 is less than P_{PV-MP2}. Under this situation, it can't make full use of PV output power and the power difference needs battery to provide, which may reduce the lifetime of the battery.

In this mode, if the power sharing coefficient α_i is chosen to be proportional to $P_{PV-MPPi}$ and set $P_{refi} = 0$, the load sharing will depend on the PV output capability, the two VSGs will work on line *Ab* and *Ac* respectively, VSG1 works at point *b* and VSG2 works at point *c*. The output power of VSGs is less than the maximum output power of PV they connected. The control method ensures the maximization utilization of PV power and extends the lifetime of battery storage.

Mode *II*: $\sum P_{PV-MPPi} < P_{Load}$, when the load power continues to increase until the total PV maximum output power is less than the total load power. If VSGs share the load equally, the situation is similar to Mode *I*. If the load sharing still follows line *Ab* and *Ac*, VSG1 will work at point b_1 and VSG2 will work at point c_1. Although the total load demand doesn't reach the output limit of the inverters, VSG2 may overload.

In this mode, the reference power of the inverters should be changed first, let $P_{refi} = P_{PV-MPi}$, it ensures that the inverters output the PV maximum output power first. Following α_i is chosen to be proportional to $S_N - P_{PV-MPi}$, then the *f-P* curves of VSGs are line $b_2 B$ and $c_2 B$, VSG1 works at point b_2, VSG2 works at point c_2. The remaining power of the two VSGs $\Delta P_1'$ and $\Delta P_2'$ will be allocated according to α_i, $\Delta P_1' \propto S_N - P_{PV-MP1}$, $\Delta P_2' \propto S_N - P_{PV-MP2}$. The total load demand doesn't reach the output limit of the inverters $P_{PV-MP1} + \Delta P_1' + P_{PV-MP2} + \Delta P_2' < 2S_N$, so $P_{PV-MP1} + \Delta P_1' < S_N$, $P_{PV-MP2} + \Delta P_2' < S_N$, VSGs will not overload. The control method ensures the maximization utilization of PV power, and avoids the overload operation of the inverters.

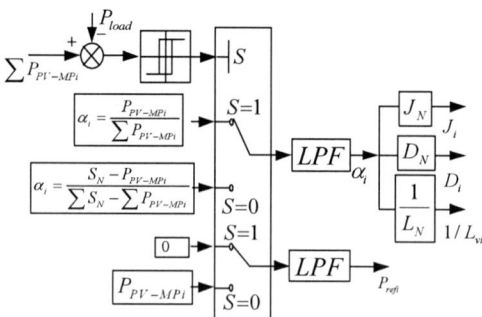

Fig. 6. Parameter adaptive control strategy for VSG

Through the analysis above, the proposed control strategy is shown in Fig.6. A hysteresis comparator is added during the switching processes, LPF is used to reduce the rate of change of α_i, P_{refi}, and improve system stability. If $\sum P_{PV-MPPi} > P_{Load}$,

S=1, it corresponds to Mode I, while if $\Sigma P_{PV\text{-}MPPi}<P_{Load}$, S=0, it corresponds to Mode II.

V. SIMULINK RESULTS

To verify the feasibility of the proposed control strategy, a microgrid model consisting of two paralleled VSGs is built in the Matlab/Simulink simulation environment. The simulated system structure is shown in Fig. 1, and the main simulation parameters are given in Table I.

TABLE I. MAIN SIMULATION PARAMETERS

Parameter	Values
Filter inductance L_s/mH	1
Filter capacitance C_s/μF	40
Line inductance L/mH	0.3
Line resistor R/Ω	0.01
Reference virtual inertia J_N	2
Reference damping coefficient D_N	63.7
Reference virtual inductance L_N/mH	2.5
Rated power of each VSG P_N/kW	15

Two different conditions are performed to verify the performance of the proposed control strategy. Fig.7 shows the active power of the load. Fig.8 shows the PV maximum output power.

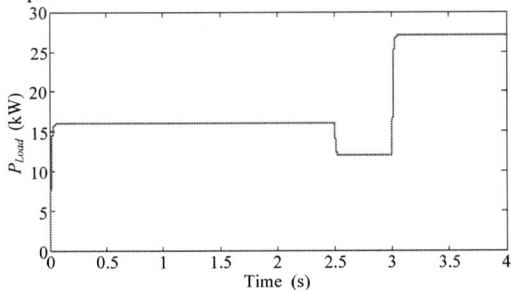

Fig. 7. Waveform for load demand

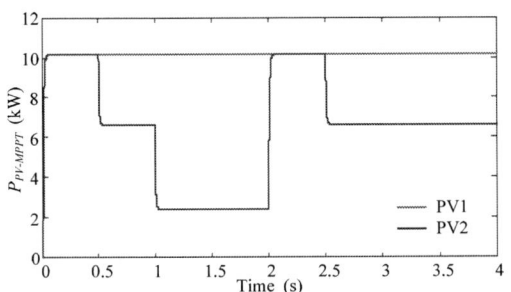

Fig. 8. Waveforms for PV maximum output power

A. PV Maximum Output Power Variations

The simulation results in response to PV maximum output power variations are present before 2 seconds. Under this scenario, the load power is kept at 16kW.

Initially, the PV maximum output power of the two VSGs is set at 10kW. The total maximum PV output power is more than the load and PV generation systems have the same output capacity, so the VSGs share the load power evenly, as shown in Fig.9 before 0.5s.

At t=0.5s, the PV maximum output power of one VSG is reduced to 7kW, while that of VSG2 is kept unchanged at 10kW. The total maximum PV output power is still more than the load, the system remains in Mode I. Depending on the proposed control strategy, α_i is proportional to $P_{PV\text{-}MPi}$ as shown in Stage I of Fig.10, then VSGs share the load power according to the output capacity of PV they connected, as shown in Stage I of Fig.9.

At t=1s, PV maximum output power of one VSG changes to 3kW, that of the other is remained at 10kW. The total PV maximum output power is less than the load, the system is in Mode II. Let $P_{refi}=P_{PV\text{-}MPi}$, as shown in Stage A of Fig.11, the inverters output the PV maximum output power first. α_i is adjusted to be proportional to S_N-$P_{PV\text{-}MPi}$, as shown in Stage II of Fig.10, the remaining power is distributed based on S_N-$P_{PV\text{-}MPi}$. The Stage II in Fig.9 shows the output power of VSGs.

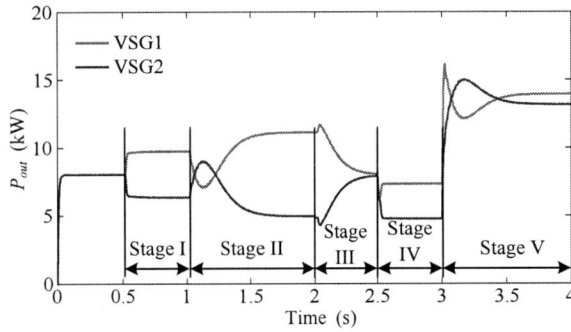

Fig. 9. Output powers of the two VSGs

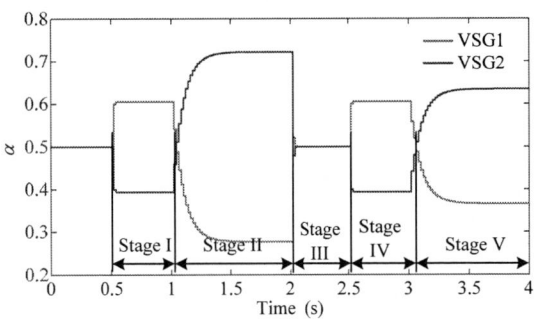

Fig. 10. Waveforms for power sharing coefficient α_i

The 2018 International Power Electronics Conference

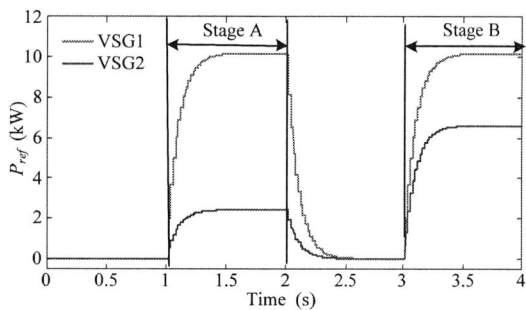

Fig. 11. Waveforms for reference powers P_{refi}

B. Load Demand Variations

The simulation results in response to load demand variations are present after 2.5 seconds. Under this scenario, the PV maximum output power of the two VSGs is 10kW and 7kW respectively.

At t=2.5s, the load is kept at 12 kW, the total PV maximum output power is more than the load. The system is in Mode *I*. the changed α_i is shown in Stage IV of Fig.10, which is proportional to $P_{PV\text{-}MPi}$, VSGs share the load power depending on the PV output capability. Stage IV of Fig.9 shows the output power of VSGs.

At t=3s, the load changes to 27kW, the total PV maximum output power is less than the load power. The system is in Mode *II*, if the load is allocated as the same as t=2.5s, the output power of VSG1 will be 15.8kW, which is more than the rated power of VSG. But according to the proposed control strategy, let $P_{refi}=P_{PV\text{-}MPi}$, as shown in Stage B of Fig.10, and α_i is adjusted to be proportional to S_N- $P_{PV\text{-}MPi}$ that is shown in Stage V of Fig.9. The method ensures PV works at MPP and avoids overloads of VSGs.

It is notable that three parameters need to be adjusted in the proposed strategy and they satisfy the constraint shown in (11). From Stage I and IV in Fig.9, we can see that the output power of two VSGs varies in proportional to the change of load without any oscillation. From Stage II and V of Fig.9, when changing the reference power, the existence of G_N leads to the output power oscillation during operation mode switching.

In conclusion, the proposed control strategy in this paper guarantees the optimal distribution of load in paralleled VSGs on the premise of giving priority to the use of PV.

VI. Conclusion

In this paper, a parameter adaptive control strategy is proposed for a VSG-controlled islanded hybrid PV/battery microgrid. The whole system can operate in two modes, the parameters in different modes is adjusted depends on different requirements, any transients can be eliminated when the load changes. So the method also enhances the stability of the system. The simulation results verify that the proposed method can maximize the use of PV and avoid the overload of VSG, which ensures the reasonable power sharing between VSGs. The battery conditions are not considered in this paper, while they also affect the operation of the microgrid. So in the future, battery conditions will be taken into account to cover all the possible operation modes of the microgrid.

Acknowledgment

This work was supported in part by the National Natural Science Foundation of China under Grant 51577047, in part by the International Science & Technology Cooperation Project of Anhui Province under Grant 1604b0602015.

References

[1] Mahmoud M S, Saif U R M, A. L.-Sunni F M, "Review of microgrid architectures – a system of systems perspective," *Renewable Power Generation IET* 2015, 9(8):1064-1078.

[2] Tuballa M L, Abundo M L, "A review of the development of Smart Grid technologies," *Renewable & Sustainable Energy Reviews*, 2016, 59:710-725.

[3] Liu J, Miura Y, Ise T, "Power Quality improvement of microgrids by virtual synchronous generator control," in *Electric Power Quality and Supply Reliability. IEEE*, 2016:119-124.

[4] Zheng Tianwen, Chen Laijun, Liu Wei, et al, "Multi-mode Operation Control for Photovoltaic Virtual Synchronous Generator Considering the Dynamic Characteristics of Primary Source," *Proceedings of the CSEE*, 2017,37(2):454-463(in Chinese).

[5] Bevrani H, Ise T, Miura Y, "Virtual synchronous generators: A survey and new perspectives," *International Journal of Electrical Power & Energy Systems*, 2014, 54(1):244-254.

[6] Serban I, Ion C P, "Microgrid control based on a grid-forming inverter operating as virtual synchronous generator with enhanced dynamic response capability," *International Journal of Electrical Power & Energy Systems*, 2017, 89:94-105.

[7] D'Arco S, Suul J A, Fosso O B, "A Virtual Synchronous Machine implementation for distributed control of power converters in SmartGrids," *Electric Power Systems Research*, 2015, 122(6):180-197..

[8] Wu H, Ruan X, Yang D, et al, "Small-Signal Modeling and Parameters Design for Virtual Synchronous Generators," *IEEE Transactions on Industrial Electronics*, 2016, 63(7):4292-4303.

[9] Lv Zhipeng, Sheng Wanxing, Zhong Qingchang et al, "Virtual Synchronous Generator and Its Applications in Micro-grid". *Proceedings of the CSEE*, 2014(16):2591-2603(in Chinese).

[10] Karimi Y, Oraee H, Golsorkhi M S, et al, "Decentralized Method for Load Sharing and Power Management in a PV/Battery Hybrid Source Islanded Microgrid," *IEEE Transactions on Power Electronics*, 2017, 32(5):3525-3535.

[11] Shintai T, Miura Y, Ise T, "Oscillation Damping of a Distributed Generator Using a Virtual Synchronous Generator," *IEEE Transactions on Power Delivery*, 2014,29(2):668-676.

[12] Alipoor J, Miura Y, Ise T, "Power System Stabilization Using Virtual Synchronous Generator With Alternating Moment of Inertia," *IEEE Journal of Emerging & Selected Topics in Power Electronics*, 2015, 3(2):451-458.

The 2018 International Power Electronics Conference

Adaptive voltage control scheme for DAB based modular cascaded SST in PV application

Tao Liu[1 2*], Yang Xuan[1], Xu Yang[1], Peng Xu[1], Yang Li[1], Lang Huang[1 2], Xiang Hao[2]

1 School of electrical engineering, Xian Jiaotong University, Xi'an, China

2 Renewable energy research institute, TBEA Xinjiang Sunoasis Co., LTD, Xi'an, China

*E-mail:75279071@qq.com

Abstract- Solid state transformer (SST) is able to interface PV panels and distributed grid, which could serve as a new structure for future PV plants. However, the wide operating voltage range of PV system reduces the efficiency of DAB based modular cascaded SST system. Thus, this paper analyzes the impact of DAB voltage deviation between input and output side under frequency component analysis. Besides adjusting primary and secondary side duty cycles, which is widely used in previous literature, this paper utilizes DAB output voltage as an extra control freedom in SST. Then, an adaptive voltage control (AVC) scheme is proposed, which changes DAB output voltage in accordance with input voltage. This paper presents the detailed implementation of AVC and thoroughly compares SST cell efficiency with previous modulation method. Finally, the proposed method is verified in 10kV/1MW SST system.

Keywords- SST; DAB; Adaptive voltage control; efficiency;

I. INTRODUCTION

Current central PV plants mostly feed the energy into distributed grid with the voltage no less than 10kV. The structure is composed of three-phase H-bridge inverter and line frequency transformer. However, in the application, this structure has several drawbacks. First, inverter is unable to work when input DC voltage is low, which reduces the equivalent working hours. Secondly, the bulky line frequency transformer is less efficient and intelligent [1]. According to the scenario pictured in the future renewable electric energy delivery and management system (FREEDM), the PV energy can be directly transferred into distributed grid with solid state transformer (SST) [2].

Among all the feasible topology for SST, cascaded modular multilevel converter based SST is most promising due to its advantage in scalability, modularity and redundancy [3]. In this topology, SST is composed of many cells, which are connected in parallel at DC side and in series at AC side. Inside each cell, dual active bridge (DAB) is used to achieve galvanic isolation due to its high efficiency performance and wide operation range capability [4]. The H-bridge inside SST cell is implemented to transfer DC energy into AC energy. The topology is shown in Fig. 1.

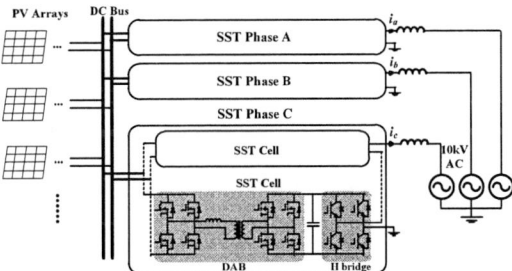

Fig. 1. SST topology in PV application

One of the most important concerns for PV converting system is efficiency. For SST in PV application, it has to deal with the problem of large input voltage range, which mostly affects the efficiency of DAB. Previously, numerous literatures have proposed methods to increase efficiency when DAB input voltage has big deviation with output voltage. Most of the studies derive DAB current in time domain and increases the efficiency with different methods. In [5], the modulation of DAB is categorized into triangular current mode and trapezoidal current mode. The choice of the modulation mode is based on different working condition. It is proven to obtain higher efficiency comparing with traditional phase shift modulation (PSM). To further reduce the conduction loss, closed form solution for modulation method is presented in [6], which achieves the minimum RMS current in the whole operation range. On the other hand, there are also many literatures analyzing the switching loss inside DAB and achieves ZVS of semiconductor device in wild operating range [7]. By considering both conduction and switching loss, the complete loss model of DAB and optimal efficiency trajectory is presented in [8]. Aforementioned methods analyze DAB in the time domain and achieve DAB model with high precision. However, the modeling process requires switching cycle to be divided into different periods. This requires huge on-line or off-line computation. Moreover, different voltage range and power rating exacerbate the calculation burden.

Alternatively, there are also methods, which build DAB model in the frequency domain and achieve high efficiency performance. Under this modeling method, DAB voltage and current is expressed in flourier series form and the physical model is more intuitive since the active power and reactive power can be expressed in different order. Under such mechanism, the reactive power suppression method is presented in [9]. To fully optimize fundamental component, [10] proposes a

1478

fundamental duty modulation (FDM), which is reported to achieve both small RMS current and wild ZVS range.

Till now, most of the control methods utilize the duty cycles of DAB H-bridges and phase shift angel as control freedom. There are other methods proposing control variables like variable frequency [11] and extra duty cycle [12]. However, the implementation is complex and could cause EMI problems.

Different from previous study in DAB modulation improvement, this paper focuses on the extra control freedom for DAB in SST application, which is the output voltage of DAB. This paper creates the DAB model under frequency domain and analyzes the impact of output voltage on DAB efficiency. Through adjusting the DAB output voltage based on input side voltage, DAB obtains higher efficiency than previous methods. This paper presents the detailed implementation method of adaptive voltage control (VAC) for SST system in PV application. Then, the complete loss model is implemented to verify the loss reduction under VAC. Finally, the method is implemented in 10kV/1MW SST system. The experiment results validate the effectiveness of proposed methods.

II. DAB Modeling

Previously, most of DAB modeling is achieved through the time domain. However, the modeling process is quite complex. Thus, this chapter utilizes the frequency component analysis (FCA) to achieve DAB modeling.

DAB converter topology and typical signals are shown in Fig. 2.

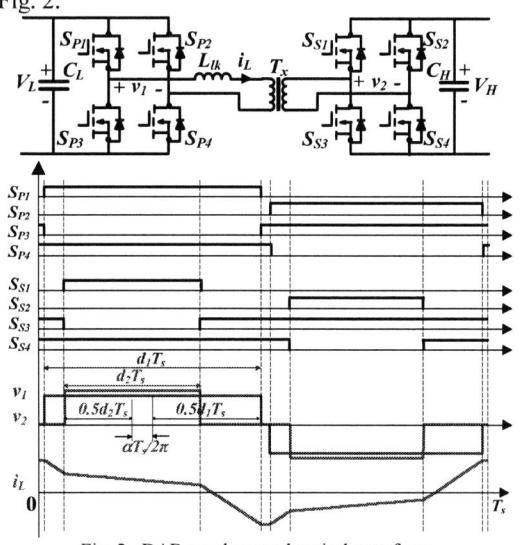

Fig. 2. DAB topology and typical waveform

According to the waveform, the output voltage in both H-bridges can be expressed in Flourier series form as below [9]:

$$v_1(t) = \sum_{n=1,3..} v_{1\_n}(t) = \sum_{n=1,3..} (-1)^{\frac{n-1}{2}} \frac{4V_L}{n\pi} \sin(n\pi d_p)$$

$$v_2(t) = \sum_{n=1,3..} v_{2\_n}(t) = \sum_{n=1,3..} (-1)^{\frac{n-1}{2}} \frac{4V_H}{n\pi} \sin(n\pi d_s) \angle n\alpha \tag{1}$$

Here, v_1 is the input side H-bridge voltage and v_2 is the output side H-bridge voltage. The voltage applies on the inductor and creates the DAB current, which can be expressed in (2)

$$i_L(t) - i_L(t_0) = \frac{1}{L} \int [v_1(t) - \frac{1}{K} v_2(t)] dt \tag{2}$$

Here, K is the turns ratio of transformer. Due to the symmetric nature in DAB, the current satisfies:

$$i_L(t) = -i_L(\frac{T_s}{2} + t) \tag{3}$$

Combining (1-3), DAB current can be derived accordingly in (4).

$$i_L(t) = \sum_{n=1,3..} i_{L\_n}(t) = \sum_{n=1,3..} I_{L\_n} \angle \theta_i^n \tag{4}$$

Where, the key parameters satisfy:

$$I_L^n = \sqrt{\left(\frac{4(V_H \sin(n\pi d_2)\cos(n\alpha) - KV_L \sin(n\pi d_1))}{Kn^2\pi\omega L}\right)^2 + \left(\frac{4V_H \sin(n\pi d_2)\sin(n\alpha)}{Kn^2\pi\omega L}\right)^2} \tag{5}$$

$$\theta_i^n = \arctan\left(\frac{V_H \sin(n\pi d_2)\cos(n\alpha) - KV_L \sin(n\pi d_1)}{V_H \sin(n\pi d_2)\sin(n\alpha)}\right)$$

Now that the current expression and input side voltage expression are derived, the value of active current and reactive current in each order can be obtained in (6).

$$i_{Ld\_n} = \frac{4V_H \sin(n\pi d_2)\sin(n\alpha)}{Kn^2\pi\omega L}$$

$$i_{Lq\_n} = \frac{4(V_H \sin(n\pi d_2)\cos(n\alpha) - KV_L \sin(n\pi d_1))}{Kn^2\pi\omega L} \tag{6}$$

Here, i_{Ld} presents the active current and i_{Lq} presents the reactive current. The total RMS current is derived in (7).

$$I_{L\_RMS} = \sqrt{\sum_{n=1,3..} \left((i_{Ld\_n})^2 + (i_{Lq\_n})^2\right)} \tag{7}$$

The power of DAB combines the active power in each order, which is presented in (8).

$$P = \sum_{n=1,3..} \frac{8V_L V_H}{Kn^3\pi^2\omega_s L} \sin(n\pi d_1)\sin(n\pi d_2)\sin(n\alpha) \tag{8}$$

III. DAB Analysis with Different Cases

Now that DAB modeling is achieved, we can analyze DAB under different working condition. The parameter of DAB converter is shown in Table I.

TABLE I
DAB KEY PARAMETERS

Symbol	Meaning	Value
P_o	DAB full power	31kW
V_L	Input voltage	400V~800V
V_H	Nominal output voltage	800V
f_s	Switching frequency	20kHz
L_r	DAB Inductor	40μH
K	Turns ratio	1:1

First, we consider the case when input voltage and output voltage is the same. The output power of DAB is 50% rated power. Here, PSM is used. The active current and reactive current is shown in different order in Fig. 3 case 1. It shows that when input and output voltage is the same, reactive current is extremely small, most of the current is active current. Thus, RMS current is small for this case. To exam the effect of smaller input voltage, the analytical result of V_L=700V with the same power is shown in Fig. 3 case 2. Due to smaller input voltage, the active current in each order is increased slightly. Nevertheless, the reactive current in fundament frequency increases significantly. The reason is because most of DAB operates with small phase shift angel [4]. Thus, the voltage difference mainly applies on the inductor and increases the reactive current. For higher order current, it reduces with higher order according to (6), which is ignored according to [9].

In order to suppress the RMS current, fundamental current optimization is proposed. It shows good performance in RMS current suppression and ZVS achievement [10]. Using this method, the result is shown in Fig. 3. The result shows that the fundamental reactive current is eliminated. However, this modulation scheme also causes problem especially at 3rd order. The reason is because 3rd order harmonic current helps change the current polarity in switching instance so that ZVS is more likely to be achieved [10].

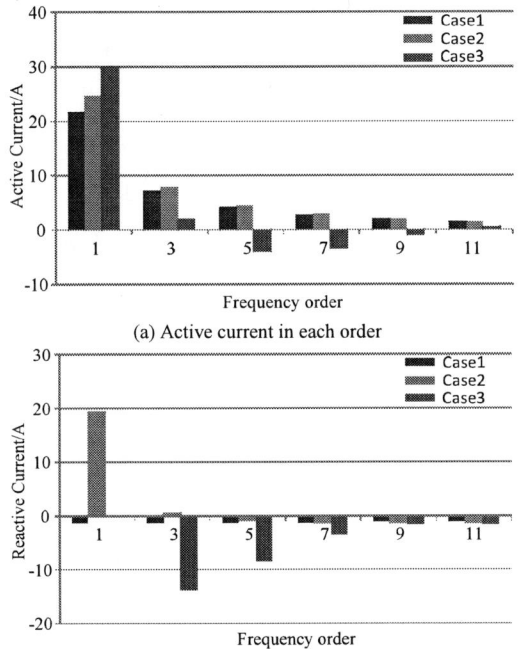

(a) Active current in each order

(b) Reactive current in each order

Fig. 3. Active current and reactive current in different cases (Case1: V_L=800V under PSM; Case 2 V_L=700V under PSM; Case 3 V_L=700V under FDM)

The analytical result shows that large reactive current also exists under FDM. There are other methods changing both d_1 and d_2. However, the reduced duty cycle means larger RMS current to produce required power. For these methods, they either have large loss (large RMS current or non-ZVS switching loss) or are more difficult to implement. So far, there is no modulation method, which is comparable with DAB operation with no voltage deviation.

IV. ADAPTIVE VOLTAGE CONTROL FOR DAB

When DAB operates in SST system, it not only realizes the galvanic isolation, but also serves as the reliable voltage source for DC-AC stage. Thus DAB

output voltage can be variable as long as it is constant in the short time scale. For MPPT control, it is implemented several hundred millisecond, which is much slower than H-bridge control, whose control is implemented every 50us. Thus, DAB output voltage can be adjusted according to the input side voltage.

Previous analysis illustrates that zero voltage deviation is preferable for DAB operation. Thus, this paper proposes to control output voltage according to input voltage. For this condition, the output voltage becomes smaller, which would require larger current for the output side. Thus, the RMS current comparison between AVC and FDM is required. Here, we list the case when V_L=700V and compare the RMS current between two methods for the whole operation range.

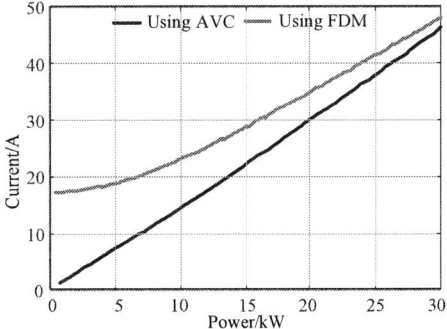

Fig. 4. RMS current comparison between AVC and FDM

The result shows that AVC achieves even smaller RMS current comparing with FDM. This is because FDM requires large reactive current at 3rd order to achieve ZVS. The RMS current reduction is even more significant at light load condition which is because 3rd order reactive power is more prominent at light load. Even though AVC is preferable, it has its own constraint. This is because DAB output voltage has its own inverting constraint, which means DAB output voltage must be larger than critical voltage V_{cr}. The critical voltage satisfies

$$NV_{cr} > \frac{\sqrt{2}}{\sqrt{3}} V_{grid} \qquad (9)$$

For SST in this paper, the cell number N is 13. So, DAB output voltage must be larger than 650V in the implementation. However, the PV output voltage ranges from 400V~800V. When it is smaller than 650V, FDM shall be implemented. This is due to its implementation simplicity and effectiveness [10]. However, we still need to determine the preferable DAB output voltage. Here, we list the case when V_L=500V and compare the RMS current with different V_H using FDM.

1480

The 2018 International Power Electronics Conference

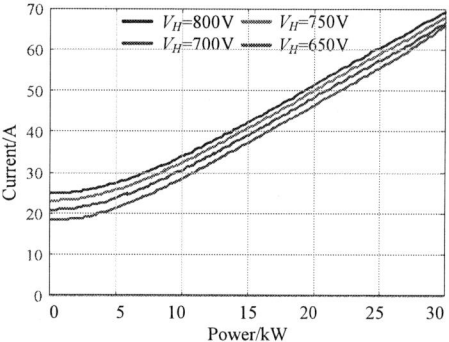

Fig. 5. . RMS current comparison with different V_H

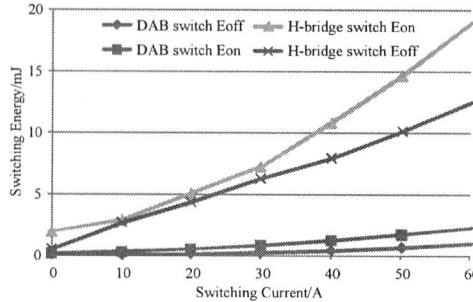

Fig. 7. Switching energy of DAB and H-bridge switches

It shows that RMS current is smaller with lower V_H. Thus, when V_L is smaller than 650V, V_H reference shall be close to V_{cr} in the application. Now that the AVC control scheme can be obtained with the control diagram depicted in Fig. 6. In the application, the minimum V_H is determined as 680V for enough margin.

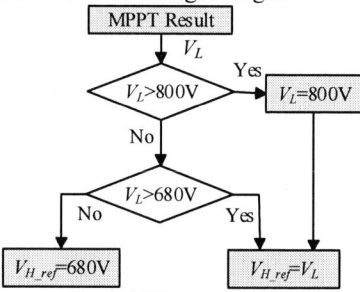

Fig. 6. AVC control diagram

V. LOSS ANALYSIS OF AVC

Previous chapter only compares the RMS current, which only presents the conduction loss of DAB. This chapter gives the detailed loss analysis of AVC which considers the loss of DC-DC stage as well as DC-AC stage.

The key parameters of SST cells are shown in Table II.

TABLE II
SST CELL KEY PARAMETERS

Symbol	Meaning	Value
R_{ds}	DAB switch resistance	40mΩ
R_{ac}	Transformer winding resistance	65mΩ
R_L	Inductor ESR	25mΩ
V_{ce}	DC-AC stage IGBT saturation voltage	1V(@Ice=5A) 1.3V(@Ice=25A)
f_{DAB}	DAB switching frequency	20kHz
f_H	H-bridge switching frequency	300Hz

Since the switching loss is one important concern for detailed loss analysis. The turn-on and turn-off loss of DAB switch and H-bridge switch needs to be determined first. The double pulse platform is built and the test result is shown in Fig.7.

With all these given parameters, the loss breakdown can be achieved for SST cells working under different cases. First, we compare AVC with FDM when V_L=700V. Here, we undertake the loss analysis under 20% and 70% rated power.

The detailed loss breakdown is shown in Fig. 8. The result proves the effectiveness of AVC. The efficiency improvement of AVC mostly lies in DAB conduction loss. It shall be noted that even though FDA achieves ZVS for the switch, the turn-off current is higher than that under AVC. Thus, it suffers higher turn-off loss.

Following similar method, the loss analysis is undertaken when V_L=500V. For PV application, when the PV panel output voltage is small, the output power is always limited. Thus, we only consider 20% output power. The loss breakdown is shown in Fig. 8. The result shows that with smaller voltage deviation in DAB, SST cell achieves better efficiency.

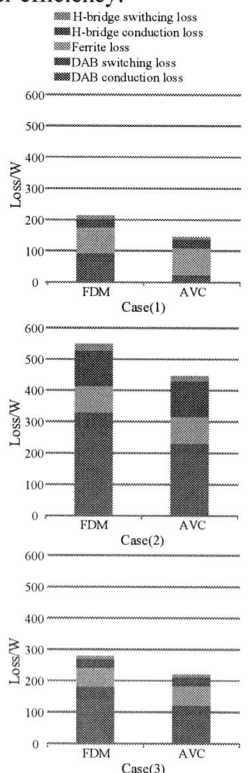

Fig. 8. Loss breakdown of SST cell
(Case(1):V_L=700V at 20% power; Case(2):V_L=700V at 70% power; Case(3):V_L=500V at 20% power)

1481

The 2018 International Power Electronics Conference

VI. EXPERIMENT RESULT

The experiment is conducted under 10kV/1MW SST prototype shown in Fig. 9.

Fig. 9. Picture of SST prototype

Under AVC, the systematic control structure is shown in Fig.10. Here, DAB controls its individual output voltage, which achieves good voltage balancing in each module. Cascaded H-bridge controls the output current and the current reference comes from input side voltage control loop. MPPT algorithm utilizes P&O method.

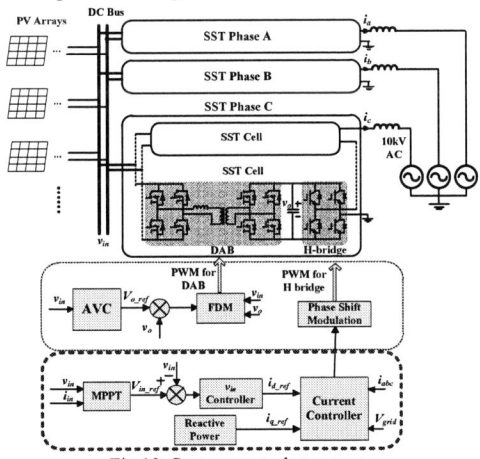

Fig.10. System control structure

First, the comparison between FDM and VAC is conducted. Here, SST operates with V_L=700V at 20% load. DAB waveforms under two methods are shown in Fig. 11. The comparison shows that RMS current as well as the switching current is significantly larger under FDM. Thus, the efficiency can be improved using VAC.

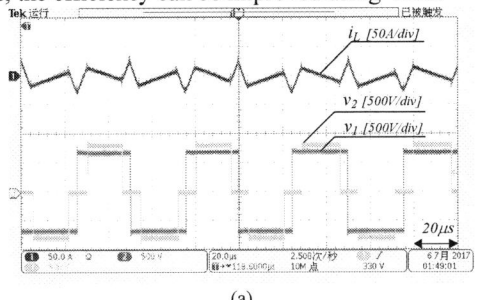

(a)

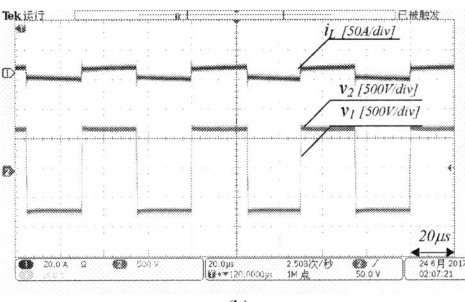

(b)

Fig. 11. Waveform comparison under FDM(a) and VAC(b)

Then AVC is implemented inside SST system. The efficiency is tested for different input voltage and operation power. The efficiency curve is shown in Fig. 12.

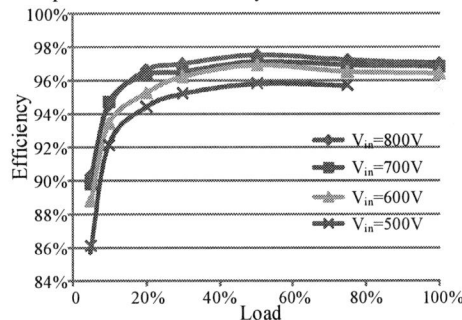

Fig. 12. Efficiency curve under AVC

It shows that SST achieves high efficiency at different input voltage. When V_L=700V, the efficiency is slightly smaller than the case when V_L=800V due to AVC control scheme. At even lower voltage, the output voltage is not able to follow input voltage. Thus, output voltage equals its limit value and FDM takes effect. The efficiency reduction is still small. Under proposed control scheme, SST shows high efficiency performance at whole operation range.

VII. CONCLUSION

SST is promising to subverting current power grid including distributed generation system. However, the output character of distributed generation is possible to reduce the SST system efficiency. This paper analyzes the wide voltage range impact on DAB based modular cascaded SST system. Then, an adaptive voltage control scheme for DAB output voltage is proposed. The detailed loss analysis and experiment result proves the effectiveness of proposed method. With AVC, SST system achieves high efficiency under different voltage range, which facilitates the reform of current PV plant structure.

REFERENCES

[1] B. Xiao, L. Huang, J. Mei, C. Riley, L. M. Tolbert and B. Ozpineci "Modular Cascaded H-Bridge Multilevel PV Inverter with Distributed MPPT for Grid-Connected Applications". *IEEE Trans. Ind. Appl.*, vol. 51, no. 2, pp. 1722-1731, Sep. 2015.

[2] X. She, A. Q. Huang, S. Lukic, and M. Baran, "On integration of solid state transformer with zonal DC

microgrid," *IEEE Trans. Smart Grid*, vol. 3, no. 2, pp. 975–985, Jun. 2012.

[3] B. Zhao, Q. Song, W. Liu and Y. Sun, "Overview of dual-active-bridge isolated bidirectional DC-DC converter for high-frequency-link power conversion system," *IEEE Trans. Power Electron.*, vol. 29, no. 8, pp. 4091–4106, Aug. 2014.

[4] J. M. Carrasco, L. G. Franquelo, J. T. Bialasiewicz, E. Galvan, R. C. P. Guisado, M. A. M. Prats, J. I. Leon and N. Moreno-Alfonso, "Power electronic systems for the grid integration of renewable energy sources: A survey," *IEEE Trans. Ind. Electron.*, vol. 53, no. 4, pp. 1002–1016, Jun. 2006.

[5] Y. Xie, J. Sun and J. S. Freudenberg, "Power Flow Characterization of a Bidirectional Galvanically Isolated High-Power DC/DC Converter Over a Wide Operating Range," *IEEE Trans. Power Electron.*, vol. 25, no. 1, pp. 54–66, Jan. 2010.

[6] F. Krismer and J. W. Kolar, "Closed form solution for minimum conduction loss modulation of DAB converter," *IEEE Trans. Power Electron.*, vol. 27, no. 1, pp. 174–188, Jan. 2012.

[7] J. Everts, F. Krismer, J. Van den Keybus, J. Driesen, and J. W. Kolar, "Optimal ZVS modulation of single-phase single-stage bidirectional DAB AC/DC converters," *IEEE Trans. Power Electron.*, vol. 29, no. 8, pp. 3954–3970, Aug. 2014.

[8] D. Costinett, D. Maksimovic and R. Zane "Design and Control for High Efficiency in High Step-Down Dual Active Bridge Converters Operating at High Switching Frequency," *IEEE Trans. Power Electron.*, vol. 28, no. 8, pp. 3931–3940, Aug. 2013.

[9] B. Zhao, Q. Song, W. Liu, G. Liu and Y. Zhao, "Universal high-frequency-link characterization and practical fundamental-optimal strategy for dual-active-bridge dc-dc converter under PWM plus phase-shift control," *IEEE Trans. Power Electron.*, vol. 30, no. 12, pp. 6488 - 6494. Dec. 2015.

[10] W. Choi, K. M. Rho, and B. H. Cho, "Fundamental Duty Modulation of Dual-Active-Bridge Converter for Wide-Range Operation," *IEEE Trans. Power Electron.*, vol. 31, no. 6, pp. 4048–4064, Jun. 2016.

[11] G. Oggier and M. Ordonez, "High efficiency DAB converter using switching sequences and burst-mode," *IEEE Trans. Power Electron.*, vol. 31, no. 3, pp. 2069–2082, Jun. 2015.

[12] T. Hirose, M. Takasaki and Y. Ishizuka, "A Power efficiency improvement technique for a Bidirectional Dual Active Bridge DC-DC Converter at Light Load," *IEEE Trans. Ind.Appl.*, vol. 50, no. 6, pp. 4047–4055, Dec.2014

Six-Step MMC-Based High Power DC-DC Converter

Stefan Milovanović and Dražen Dujić
Power Electronics Laboratory - PEL
École Polytechnique Fédérale de Lausanne - EPFL
Station 11, CH-1015 Lausanne
stefan.milovanovic@epfl.ch, drazen.dujic@epfl.ch

Abstract—Both DC collection and transmission grids have enjoyed a lot of popularity lately. In order to successfully take advantage of the benefits offered by DC systems, reliable connection between two networks of different voltage levels has to be obtained. This paper presents high step-down ratio, isolated DC-DC converter intended to connect high/medium voltage grid with a low voltage DC grid. Proposed converter combines two Modular Multilevel Converters in series at high/medium voltage side with conventional Six-Step Converter at low voltage side, interfaced by means of Three Winding Three-Phase transformer. Basic operating principles, along with design and sizing rules, are presented and supported by simulation results.

Index Terms—HVDC, MVDC, Modular Multilevel Converter (MMC), High Power DC-DC Conversion

I. Introduction

Steadily increasing electrical energy consumption demands have led to the need for the modern power systems to expand. However, limited ampacity of existing power lines is said to represent one of the most significant bottlenecks towards expanding power transfer capabilities of the existing systems [1]–[4]. High Voltage Direct Current (HVDC) systems have shown certain advantages over their High Voltage Alternating Current (HVAC) counterparts, some of which are reduced transmission losses, alleviated stability problems, no limitations in terms of distance at which the energy can be transferred through a line, etc. On the other hand, stringent ecological requirements have led to renewable energy resources proliferation. Consequently, HVDC transmission and Medium Voltage Direct Current (MVDC) distribution grids are expected to expand since they represent promising solutions for integration of renewable energy sources into the existing grid. Additionally, success of HVDC technology is expected to be replicated within Medium Voltage (MV) domain. Further, large off-shore wind power plants usually require MVDC connection with existing distribution grids due to higher transmission efficiency and lower overall system cost compared to AC systems [5], [6]. Additional advantage provided by the DC systems is the fact that bulky Low Frequency Transformer (LFT) can be replaced by smaller Medium Frequency Transformer (MFT), leading to high power densities and decreased system volume and weight.

Ideas of increasing power transfer capacity of existing AC power lines by converting them into DC, appeared during 1970s [1]. It was stated in [1] that conversion of an AC line into DC shows potential of power increase by the factor of almost 3.5. Having an AC line converted into DC might result in different line configurations [3], [4], however this paper focuses on converters operating within bipolar networks with neutral conductor.

Within the aforementioned systems, DC-DC power converter can be labeled as a key component providing the means for interfacing two DC networks of different voltage levels. However, certain requirements have to be met in order to provide reliable connection between two networks, some of which are high step-up/down voltage ratio, high efficiency, isolation due to safety reasons, redundancy, modularity, etc. Galvanic isolated power converters are quite often referred to as Solid State Transformers (SSTs). SST concept has been envisioned within AC grids a long time ago, however research interests including SST have only increased over time. So far, majority of SST configurations have involved multiple conversion stages connected by means of multiple MFTs, whereas this paper focuses on bulk power processing concept implying the existence of one single MFT in the isolation stage. Due to the fact that High Voltage (HV)/MV has to be handled at either side of the converter, Modular Multilevel Converter (MMC) is used.

MMC-based high power DC-DC converters have been subject to various research projects. In [6] MMC-based Single-Phase (1PH) Dual-Active Bridge (DAB) was proposed, with aim of obtaining electronic tap changer. However, problems with high voltage stress imposed on isolation stage, originating from instantaneous MMC cells insertion, had not been addressed. To overcome the aforementioned problems, Quasi Two-Level (Q2L) operation was proposed [7], [8]. Such an operation implies sequential insertion of MMC-alike cells in equidistant time steps within a fraction of fundamental converter operating period. On the other hand, topologies operating with sinusoidal currents within the isolation stage were also reported [9]. However, keeping HV/MV semiconductor devices switching frequency equal to MFT operating frequency tends to be quite challenging in these cases. Therefore, topologies operating similarly to conventional DAB were preferred by the authors. None of the aforementioned references has analyzed the employment of MMC-based DAB within bipolar DC grids with neutral conductor, where redundancy principle can be utilized providing suitable converter structure is chosen.

The 2018 International Power Electronics Conference

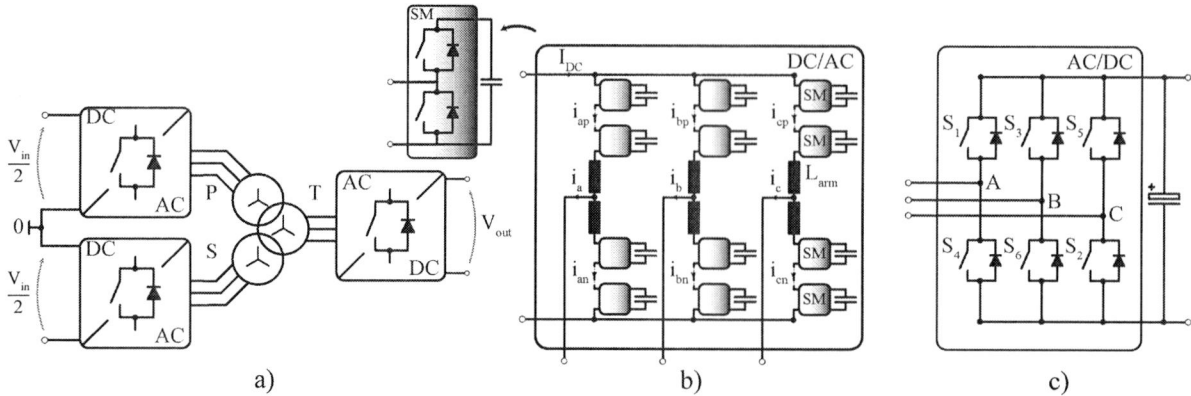

a)   b)   c)

Figure 1: a) Proposed topology b) MV MMC stage c) LV Six-Step Converter

II. CONVERTER STRUCTURE

Fig. 1 presents the proposed topology consisting of series connection of two MMCs at HV/MV side and Six-Step Converter (SSC) at Low Voltage (LV) side. With aim of interfacing HV/MV grid, both MMCs employ Half-Bridge (HB) cells whose number depends on available HV/MV along with voltage class of semiconductors to be used. HV/MV and LV stages are interconnected by means of three winding Three-Phase (3PH) transformer. According to Fig. 1a, HV/MV side windings are referred to as primary (P) and secondary (S) windings, whereas LV side winding is referred to as tertiary (T) winding. Analyzed system ratings are presented in Tab. I. In order to maintain switching frequency within reasonable limits, both MMCs as well as SSC, operate in six-step mode. Therefore, in terms of operation principles, the proposed topology corresponds to 3PH DAB presented in [10]. LV side semiconductor devices switch at MFT operating frequency, thus creating the possibility for employing low conduction losses switching devices such as Insulated Gate-Commutated Thyristor (IGCT). In addition, Zero-Voltage Switching (ZVS), a property inherent to DAB within a restricted operating range, leads to reduction of converter switching losses. Another advantage offered by the proposed topology is its redundancy in case of failure of either MMC. On these terms, faulty MMC gets isolated from the rest of the circuit, whereas converter can continue to operate with half the rated power, without the need for system reconfiguration. The same applies if any of the voltage poles were to be lost.

SSC operates with square-wave voltages, therefore creating four level voltage waveform at transformer T-winding. However, driving HV/MV converter stage with square-wave voltages would have detrimental effect on transformer's insu-

lation. Consequently, with aim of alleviating problems related to high dV/dt stress, MMC cells were inserted sequentially during the fundamental period fraction labeled with θ_d, as suggested by Fig. 2. Hence, voltage waveforms generated by MMC arms correspond to the ones introduced with Q2L converter [7], [8]. Another convenient feature offered by the sequential cells insertion is the ability to balance the MMC capacitor voltages since every cell receives different amount of charge depending on the time instant at which it receives the switching signal [11].

Primary and secondary MMCs are connected in series, nonetheless due to galvanic isolation, existence of transformer leakage inductance as well as MMCs' arm inductances, they are independent from each other. Hence, only one MMC can be analyzed, bearing in mind that obtained results apply for the other one providing converter operates with no faults.

Power flow analysis can be conducted using the circuit presented in Fig. 3a. In order to derive relevant equations, voltages V_{an}, V_{bn} and V_{cn} are the ones of interest. Voltages V_{si} generated at a MMC AC pole are defined by (1), where V_n and V_p denote MMC lower and upper arm voltages, respectively.

$$V_s = \frac{V_n - V_p}{2} \tag{1}$$

According to Fig. 1, maximum AC voltage generated by a single MMC branch equals $V_s^{max} = V_{in}/4$. Analyzing the circuit from Fig. 3a, along with Fig. 3b, voltage V_{an} can be calculated as (2), according to which its maximum equals

Table 1: Analyzed system ratings

Input Voltage	V_{in}	± 20kV
Output Voltage	V_{out}	1.5kV
Rated Power	P	10MW
Operating frequency	f	400Hz

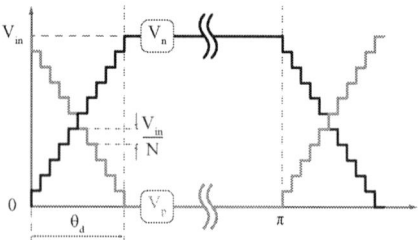

Figure 2: Sequential MMC cell insertion

1485

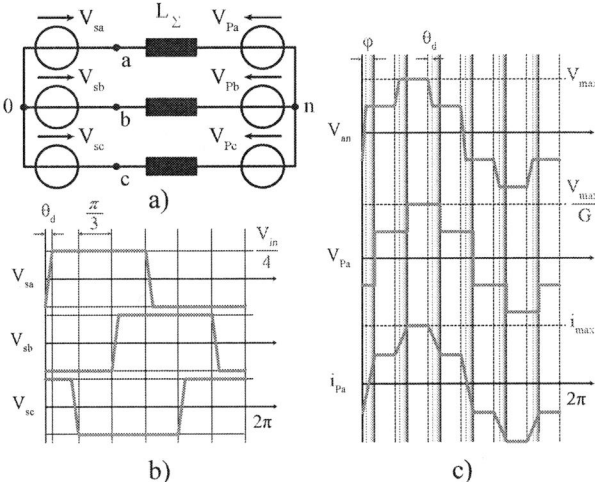

Figure 3: a) Equivalent circuit of one MMC connected to the MFT b) Voltages generated at MMC AC poles c) Idealized waveforms relevant for power flow analysis (G=1)

$V_{max} = V_{in}/3$. Even though MMC cells insertion occurs in discrete time instants, AC pole voltage waveform was linearized with aim of alleviating converter sizing procedure as well as relevant equations derivation. Normally, number of cells employed within a MMC-based system is quite high, therefore linearization should not significantly hinder precision of the obtained results. Further, transformer EMF seen from P/S side depends on its turns ratio m_t. Ratio $G = V_{in}/2m_t V_{out}$ can be referred to as converter voltage gain. It can be shown that this parameter plays a major role in defining converter current shape, as well as that it does not necessarily have to be equal to one. Fig. 4 presents the influence of converter voltage gain G on MMC arm currents. Phase A of both MMCs will be observed. At time instant $t = 0$ upper arm of the observed phase ramps its voltage down. In order to achieve ZVS, prior to switching an IGBT on, its free-wheeling diode needs to

conduct. Therefore, MMC upper arm current needs to remain negative in order to achieve ZVS for all the cells within an arm. It can be seen Fig. 4 that the higher the converter voltage gain, the more negative the MMC arm current at time instant $t = 0$.

Relevant per-phase voltage and current waveforms are presented in Fig. 3c. Similarly to conventional DAB, adjustment of the phase angle φ between voltage V_{an} and transformer EMF denoted by V_{Pa}, enables the control of power at which the energy is transferred through the converter. Observing given waveforms, power equation for either MMC can be derived (3), whereas full power of the system can be obtained by multiplying derived expression by two. In (3) MFT operating frequency was denoted by f, whereas L_Σ accounts for the sum of transformer leakage inductance, half the MMC arm inductance and, eventually added, external inductance.

$$V_{an} = \frac{2V_{sa} - V_{sb} - V_{sc}}{3} \tag{2}$$

$$P_{mmc} = \frac{V_{in}^2}{3\omega L_\Sigma G}\left\{\varphi\left(\frac{1}{2} - \frac{3\varphi}{8\pi}\right) - \frac{\theta_d^2 + (2\pi - 3\varphi)\theta_d}{8\pi}\right\} \tag{3}$$

It is noteworthy that MMCs' phase-shifts can be controlled separately due to the fact that they operate independently.

III. SYSTEM SIZING

In order to ensure proper converter operation, passive components such as MMC cell capacitances, arm inductances, output capacitor as well as transformer turns ratio have to be correctly determined. However, sizing procedure depends on converter nominal operating points such as phase angle φ, voltage gain G, arm voltage ramp-up time described by angle θ_d, etc.

Firstly, reasonable range of the angles φ and θ_d should be discussed. Converter operation was simulated employing 3.3kV rated IGBTs. Considering that MMC cell rated voltage should be around 55% of employed semiconductors voltage class, one might calculate that number of cells per MMC arm equals $N = 20kV/(0.55 \cdot 3.3kV) \approx 11$, according to Tab. I. Considering standard dead times of devices falling into this

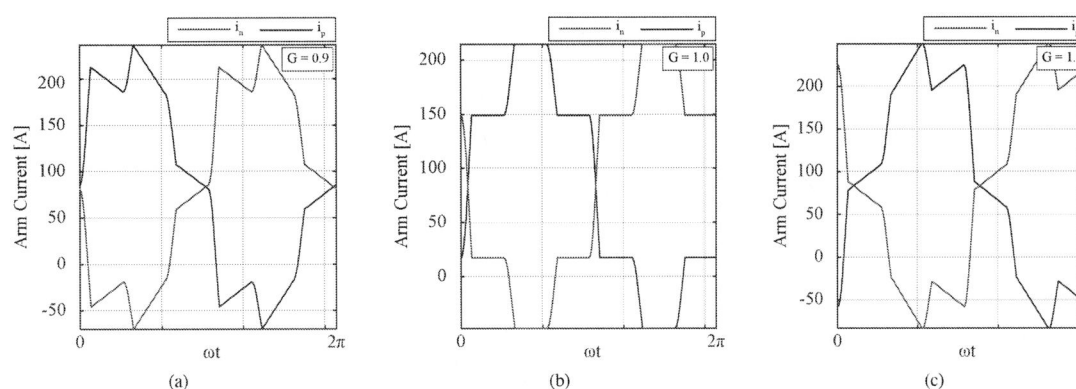

Figure 4: Influence of converter voltage gain G on MMC arm currents. It can be seen that the higher the gain G the lower the upper arm current at time instant t=0. Therefore, ZVS conditions might be provided for the HV/MV side devices.

voltage class being around $5\mu s$, it can be assumed that ramp-up time of an arm voltage equals $50\mu s$ at least. If converter operating frequency was set too high, possibility that arm voltage ramp-up time would be higher than nominal phase angle would exist. With aim of avoiding such a phenomenon so that converter operation resembles conventional DAB as much as possible, operating frequency was set as 400Hz. Therefore, angle θ_d equals approximately $7.2°$. This leaves the possibility to set nominal phase angle φ as $15°$, which is expected to provide solid output voltage regulation flexibility while keeping AC stage reactive power moderate. With aim of achieving ZVS at HV/MV side, converter voltage gain was set as G=1.1. Tab. II summarizes the design of passive components used for system simulations.

A. Arm inductor design

MMC arm inductors were sized so that power transfer can be controlled without the need for adding any external inductances in series with the MFT, while providing the possibility to limit converter common-mode currents ripple. It was assumed that MFT leakage inductance can be considered negligible compared to the MMC arm inductance. Once the nominal power, nominal phase angle, voltage gain, and arm voltage ramp-up time are known, arm inductance can be determined using (3).

B. Output capacitor design

Output capacitance was determined so to improve converter dynamic response during abrupt load changes, as well as to alleviate output voltage oscillations originating from the SSC output current nature. It was reported in [12], [13] that phase angle abrupt changes should be treated properly in order to avoid SSC current offset and oscillations. However, methods proposed in [12], [13] rely on the prevention of phase angle change during a fraction of fundamental converter operating period. Hence, if abrupt load change occurred during the period at which the phase angle change is forbidden, output voltage drop has to be prevented by determining output filter capacitance properly.

The most critical case can be perceived as the abrupt full load connection to the converter LV bus in the time instant at which output voltage sampling occurred slightly prior to the load connection. It will be assumed that phase-angle changes will be forbidden until the next sampling instant, leaving the system operating as if no load was connected (namely, no power is transferred through the converter). Therefore, until the next sampling instant, energy demanded by the load must be supplied by the output capacitor. Allowed voltage drop, system sampling time and reference output voltage were denoted as ΔV, T_{samp} and V_{ref}, respectively. Unless converter starts providing the load with current, system can be modeled as a simple RC circuit, at which capacitor initial voltage equals V_{ref}. Hence (4), can easily be derived.

$$C_{out} \geq -\frac{T_{samp}P_{nom}}{V_{ref}^2 ln\left(1-\frac{\Delta V}{V_{ref}}\right)} \qquad (4)$$

C. MMC cell capacitance design

MMC cells voltage balancing method presented in [11] was used. Consequently, every cell was inserted/bypassed with duty cycle matching 0.5, however instants at which switching signals were passed to the cells differed by $\beta_d = \theta_d/(N-1)$. In order to perform MMC capacitor sizing, converter current waveforms need to be known. Suitable current waveform integration leads to amount of charge which causes MMC cell voltage ripple. Cell capacitance can be chosen according to desired voltage ripple caused by arm current flowing through it providing the cell is inserted into the circuit.

D. Transformer turns ratio determination

As already mentioned, converter voltage gain G can be chosen different to unity in order to provide conditions for ZVS at HV/MV side. It can be shown that, on these terms, converter voltage gain should be greater or equal to some value G^* (however, this analysis falls out of this paper scope). Having this constraint as well as voltage gain definition in mind, transformer turns ratio can be determined as (5).

$$m_t = floor\left[\frac{V_{in}}{2V_{out}G^*}\right] \qquad (5)$$

Table II: Passive components design

MMC arm inductance	L_{arm}	7.2mH
Submodule capacitance	C_{sm}	1mF
Output capacitance	C_{out}	12mF
Number of MMC cells per arm	N	11
Transformer turns ratio	m_t	12

IV. OVERALL CONVERTER CONTROL

Observing the topology depicted in Fig. 1, it can be concluded that output voltage can be controlled by means of SSC output current, whose mean value can be derived from (3) if system losses were neglected. However, power equation is nonlinear, therefore linearization might be performed with aim of relieving system computational burden. Taking system ratings from Tab. I into account, power equation (3) can be linearized around nominal phase shift. However, it can be shown that linearization around $\pi/6$ does not significantly hinder power calculation precision, leading to (6).

$$\varphi = \frac{1}{A}\left(\frac{V_{out}\overline{i_R}^*}{2} + B\right) \qquad (6)$$

where

$$A = \frac{V_{in}^2}{48\omega LG\pi}\left(6\theta_d + 7\pi\right)$$

$$B = \frac{V_{in}^2}{24\omega LG\pi}\left(\theta_d^2 + 2\pi\theta_d\right)$$

As presented in Fig. 5, output voltage is sensed and fed through the Low-Pass Filter (LPF) in order to diminish the influence of higher order harmonics on the regulation loop. Error signal is forwarded to PI regulator, whereas directly measurable output current is used as a feed-forward in order to

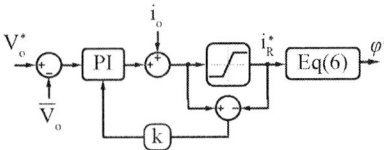

Figure 5: Output voltage control structure

keep regulator actions moderate and therefore improve system dynamic response. SSC mean current reference $\overline{i_R}^*$ is passed to the block which extracts reference phase shift according to (6). Thereafter, phase shift information is being used by modulator which generates switching signals for both MMCs, as well as SSC.

V. SIMULATION RESULTS

Previously discussed system was simulated in PLECS and steady state results are presented in Fig. 6. System was controlled so that output voltage remains equal to the reference value defined in Tab. I. To avoid abrupt output power changes, which might lead to either instabilities or require oversizing the output capacitor, full load (10MW) was applied sequentially in time steps of 0.75s. Firstly, half the rated load was connected to the LV bus at time instant denoted by t_1, whereas the other half was connected 0.75s after, at time instant denoted by t_2. It can be seen from Fig. 6 that control system manages to maintain the output voltage V_o at desired reference. At time instant denoted by t_3, ideal current source with current ramping-up from zero to twice the nominal output current over the period of 0.25s was connected to the LV bus in order to inspect

converter behavior in case the energy needs to be routed in the opposite direction. It can be seen from Fig. 6 that converter output current i_o changes according to the ramp-up time of the ideal current source connected to the LV bus.

Fig. 7 presents converter operating waveforms under full load during four fundamental periods. It can be seen that output voltage V_o consists of mean value being equal to the reference defined by Tab. I along with ripple occurring at six times converter operating frequency. As expected, SSC output current i_{SSC} proves to behave the same. Transformer P, S and T winding currents (i_{Pi}, i_{Si} and i_{Ti}, respectively) are of the same shape, which is logical considering that both MMCs were controlled so that powers delivered to the LV side match. Steady state MMCs' input currents i_{in1} and i_{in2} consist of mean value with small ripple originating from the ripple over MMC cell capacitors superimposed. Neutral conductor current i_n can be neglected, however during faults it is actually the neutral conductor which is going to overtake the current of the faulty voltage pole (MMC). However, one might also notice that, even thought they are symmetrical, transformer currents shape slightly differs compared to the ideal 3PH DAB due to the fact that voltage gain G was selected different than unity with aim of achieving ZVS at the HV/MV side.

Fig. 8 presents MMC arm currents, which consist of a DC component along with half the transformer current. Owing to the fact that transformer's currents sum up to zero, as well to the existence of a DC component in MMC arm currents, one might conclude that, in steady state, MMC input current should be flat ideally. Therefore, filtering capacitors in the HV/MV stage can be omitted.

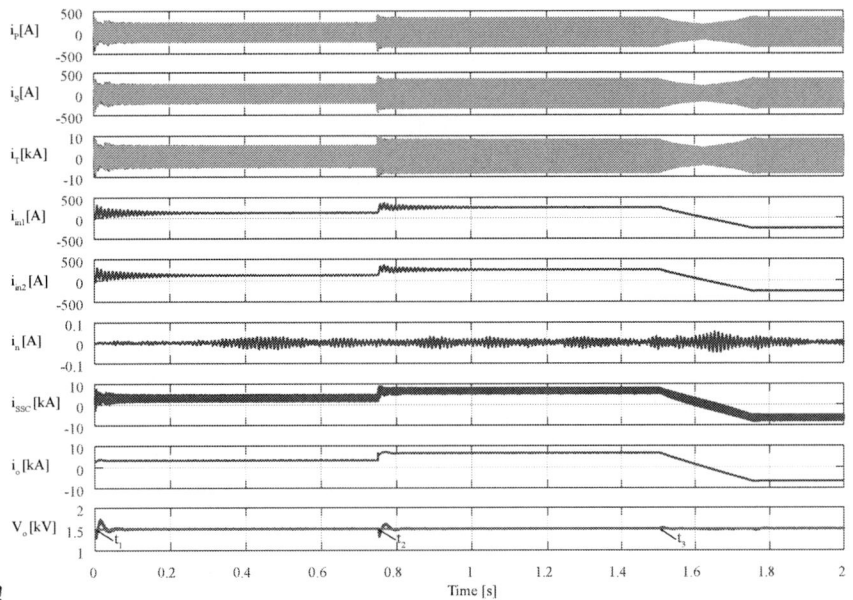

Figure 6: Converter operating waveforms after stepwise load changes ($t = t_1$) and ($t = t_2$) as well as power reversal ($t \geq t_3$)

The 2018 International Power Electronics Conference

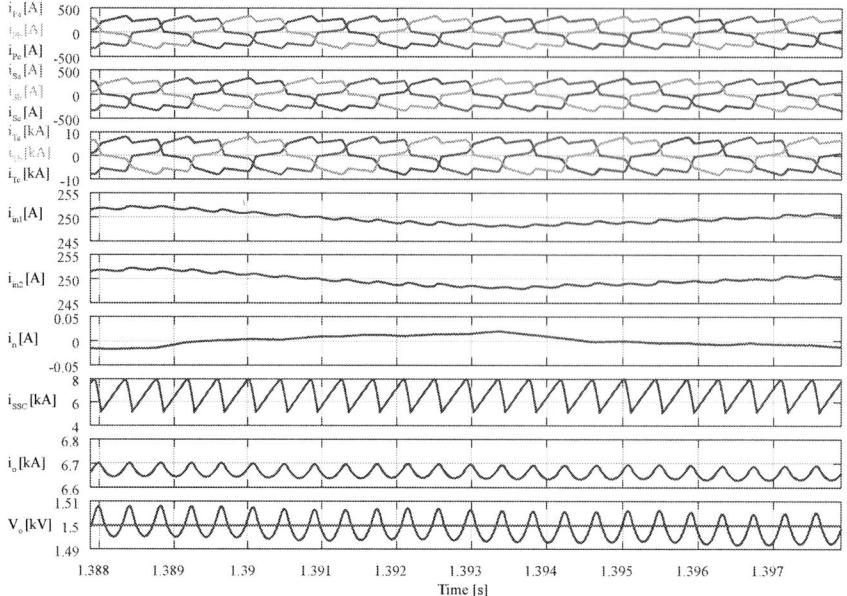

Figure 7: Converter operating waveforms under full load during four fundamental periods

Fig. 9 presents voltages across a MMC arm over several fundamental periods. It can be seen that voltage ripple equals around 3% of a cell nominal voltage. Moreover, zoomed part of the graph indicates that switching pulses were indeed delayed with respect to each other by $\beta_d = \theta_d/(N-1)$.

In order to verify that HV/MV side devices experience ZVS, Fig. 10 depicts currents of upper (i_H) and lower (i_L) switches within an upper arm of either MMC. If converter voltage gain G was correctly determined, all the switches within an arm should experience ZVS, meaning that during the arm voltage transition period, determined by the angle θ_d, observed arm current should remain negative. A cell being inserted into the circuit first was labeled as "Cell 1", whereas the one being inserted the last was labeled as "Cell 11" (please notice that gate signal was multiplied by 100 for the sake of presenting the analyzed situation on a single graph). When MMC upper arm cells receive inserting signal, current should divert from lower switches to the upper ones. In order to achieve ZVS, free-wheeling diode of the upper switch should start conducting during the dead-time interval, which means that upper switch current should be negative, which is easy to verify from Fig. 10. On the other hand, when an upper arm gate signal equals zero, lower arm cells get inserted into the circuit. Once again, observed arm current has to be negative in order to achieve ZVS. It can be seen from Fig. 10 that for both the first and the last cell inserted into the lower arm, its current is still negative, meaning that ZVS is obtained for both MMC arms.

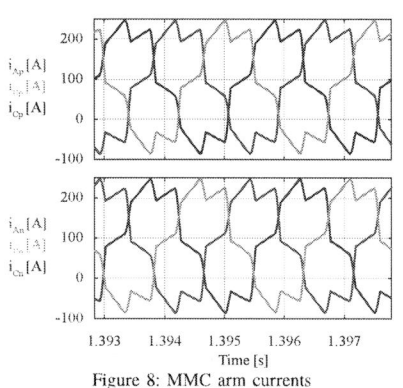

Figure 8: MMC arm currents

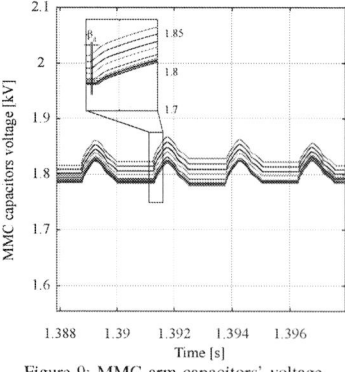

Figure 9: MMC arm capacitors' voltage

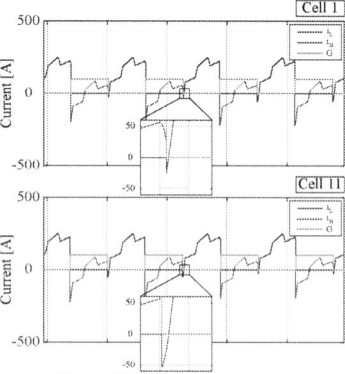

Figure 10: MMC arm current

VI. Conclusion

This paper proposed six-step MMC-based high power DC-DC converter intended for connection of HV/MV bipolar grid with LV DC grid. In order to provide the possibility to successfully utilize redundancy principle inherently offered within bipolar grids with neutral conductor, proposed converter employs Three Winding 3PH transformer. Both converter stages operate at MFT operating frequency, therefore reducing switching losses and providing the means for system operating frequency increase with respect to conventional AC systems. Operating principles resemble the ones introduced with the appearance of the DAB. Additionally, correct determination of transformer turns ratio can provide soft switching at both converter stages.

Acknowledgment

This work is part of the Swiss Competence Center for Energy Research (SCCER) initiative which is supported by the Swiss Commission for Technology and Innovation (CTI) with focus on Future Swiss Electrical Infrastructure (FURIES).

References

[1] A. Clerici, L. Paris, and P. Danfors, "Hvdc conversion of hvac lines to provide substantial power upgrading," *IEEE transactions on Power Delivery*, vol. 6, no. 1, pp. 324–333, 1991.

[2] M. Häusler, G. Schlayer, and G. Fitterer, "Converting ac power lines to dc for higher transmission ratings," *ABB REVIEW*, pp. 4–11, 1997.

[3] D. M. Larruskain, I. Zamora, O. Abarrategui, and Z. Aginako, "Conversion of ac distribution lines into dc lines to upgrade transmission capacity," *Electric Power Systems Research*, vol. 81, no. 7, pp. 1341–1348, 2011.

[4] D. M. Larruskain, I. Zamora, O. Abarrategui, and A. Iturregi, "Vsc-hvdc configurations for converting ac distribution lines into dc lines," *International Journal of Electrical Power & Energy Systems*, vol. 54, pp. 589–597, 2014.

[5] S. Lundberg, *Wind farm configuration and energy efficiency studies: series DC versus AC layouts*. Chalmers University of Technology, 2006.

[6] S. Kenzelmann, A. Rufer, D. Dujic, F. Canales, and Y. R. De Novaes, "Isolated dc/dc structure based on modular multilevel converter," *IEEE Transactions on Power Electronics*, vol. 30, no. 1, pp. 89–98, 2015.

[7] I. Gowaid, G. P. Adam, S. Ahmed, D. Holliday, and B. W. Williams, "Analysis and design of a modular multilevel converter with trapezoidal modulation for medium and high voltage dc-dc transformers," *IEEE Transactions on Power Electronics*, vol. 30, no. 10, pp. 5439–5457, 2015.

[8] I. Gowaid, G. Adam, A. M. Massoud, S. Ahmed, D. Holliday, and B. Williams, "Quasi two-level operation of modular multilevel converter for use in a high-power dc transformer with dc fault isolation capability," *IEEE Transactions on Power Electronics*, vol. 30, no. 1, pp. 108–123, 2015.

[9] S. Cui, N. Soltau, and R. W. De Doncker, "A high step-up ratio soft-switching dc-dc converter for interconnection of mvdc and hvdc grids," *IEEE Transactions on Power Electronics*, 2017.

[10] R. W. De Doncker, D. M. Divan, and M. H. Kheraluwala, "A three-phase soft-switched high-power-density dc/dc converter for high-power applications," *IEEE transactions on industry applications*, vol. 27, no. 1, pp. 63–73, 1991.

[11] S. Shao, M. Jiang, J. Zhang, and X. Wu, "A capacitor voltage balancing method for a modular multilevel dc transformer for dc distribution system," *IEEE Transactions on Power Electronics*, 2017.

[12] S. P. Engel, N. Soltau, and R. W. De Doncker, "Instantaneous current control for the three-phase dual-active bridge dc-dc converter," in *Energy Conversion Congress and Exposition (ECCE), 2012 IEEE*. IEEE, 2012, pp. 3964–3969.

[13] S. P. Engel, N. Soltau, H. Stagge, and R. W. De Doncker, "Improved instantaneous current control for the three-phase dual-active bridge dc-dc converter," in *ECCE Asia Downunder (ECCE Asia), 2013 IEEE*. IEEE, 2013, pp. 855–860.

The 2018 International Power Electronics Conference

Combined DC Power Flow Controller for DC Grid

Xu ZHONG[1], *Student Member, IEEE*, Miao ZHU[1*], *Senior Member, IEEE*, Yongning CHI[2], Xizhou DU[3], Siqi LIU[1] and Xu CAI[1]

1 School of Electronic Information and Electrical Engineering, Shanghai Jiao Tong University, Shanghai, China

2 China Electric Power Research Institute, Beijing, China

3 State Grid Shanghai Municipal Electric Power Company, Shanghai, China

*E-mail: miaozhu@sjtu.edu.cn

Abstract— The utilization of DC power flow controller can increase the control freedom in the complex DC gird, which will improve both the active power distribution and coordination control performance of DC grid. The general concept frame of Combined DC power flow controller has been proposed in this paper, as well as a detailed topology. The operation principles, circuit characteristics and control strategy of the proposed DC power flow controller are analyzed in detail. The simulation results proved that the two lines under various conditions, and good application prospect.

Keywords—DC grid, DC power flow, DC power flow controller

I. INTRODUCTION

With the rapid construction of large-scale PV power stations/wind farms, DC transmission technology of high-efficiency and large-scale renewable energy stations has drawn greater attention. HVDC transmission with the advantages of long transportation distance and flexible operation, suitable to connect remote renewable energy stations, has become a hotspot in the smart grid. Compared to two-terminal HVDC system, multi-terminal HVDC system is more economical and flexible to solve the problems of large-scale renewable energy integration, large-capacity long-distance power transmission and transmission corridor shortage and other issues. The converter station with the core technology of the voltage source converter (VSC), can effectively control the input power into the multi-terminal DC gird while the power flow of each DC line is passively determined by its own line resistance [1-3].

The factors that determine the current in DC power grid are only the line resistance and the line terminal voltage. Therefore, the power flow of the DC power grid can only be changed by changing the resistance of the line and the line terminal voltage. DC power flow controller (DCPFC), is divided into resistance-type DC power flow controller and voltage-type DC power flow controller. Resistance-type DC power flow controller topology and its control are relatively simple, equivalent to increase the equivalent resistance of transmission lines, but with the shortcomings of large loss and one-way current adjustment.

From the view of the topology, voltage type DC power flow controller are divided into DC transformers, series

adjustable voltage source (SAVS) and interline DC power flow controller (IDCPFC). The input and output sides of DC transformer [4-5] are connected to positive and negative lines of different voltage-level DC grid, equivalent an adjustable voltage source in series with the transmission line. But the device needs to withstand system-level voltage with complex design and high cost. SAVS [6-8] is proposed to adjust the current in series with positive polarity line or negative polarity line. Although the device requires lower voltage-level and power-level, it needs external power supply.

IDCPFC [2] is proposed, which uses capacitance as the energy transfer buffer to transfer partial power in one line to another line and achieves power flow control by the power exchange of two lines with no external power supply. IDCPFC is simplified for topology, circuit modeling and control strategy analysis in [9,12]. But the larger voltage ripples and current ripples are the problem because the working effect of IDCPFC is equivalent to two voltage sources, frequently strung and bypassed in two lines. A novel IDCPFC [10] with the inductance as the energy transfer buffer is proposed with the capacitance in series with lines, which greatly reduces voltage and current ripples. In [11], IDCPFC is improved to be applied to the situation of line power reversal to meet the needs of different power flow direction with the introduction of the coupling inductance.

However, IDCPFC can only actively control the current of one line with the passive current control of the other line. That is, IDCPFC is unable to control the current of two lines actively with the limited application. This paper proposes a novel DC power flow controller which can meet current control multi-targets. As this DC power flow controller can also function as a separate SAVS or IDCPFC, it is so called the Combined DC power flow controller (CDCPFC). This paper describes the operation principles and circuit characteristics of CDCPFC, proposes a double-target completed-decoupling control strategy with simulation verification.

II. DCPFC COMBINATION MECHANSIM

To adjust the power flow of two transmission lines, there should be three modes: meanwhile increase the currents of the two transmission lines; meanwhile decrease the currents of the two transmission lines; increase the current of one line and decrease the current of the other line. IDCPFC can only achieve the active control of one line's current while the other line's current is controlled passively. That is, IDCPFC can't achieve the active control of two lines, classifying as the degraded mode of controlling the current of two lines. Thus, there is the need for a novel DC power flow controller to ensure the diversity of DC system power flow control. To cope with all kinds of potential power flow scheduling situation, the aim of power flow controller is adjusting the DC power flow flexibly to achieve safe transmission and decrease the loss. To achieve fast, flexible, multi-target DC power flow control, combined construction of a new DC power flow controller is in need. In order to achieve fast, flexible and multi-target DC power flow control, we need to combine the construction of a new DC power flow controller.

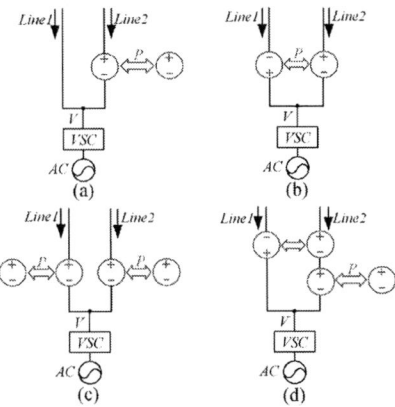

Fig.1 Combined mechanism of DCPFC. (a) SAVS. (b) IDCPFC. (c) Combination Scheme 1. (d) Combination Scheme 2.

Fig.1(a) and Fig.1(b) are schematic diagram of series adjustable voltage source and IDCPFC, which are all single-target DC power flow controllers. Two series adjustable voltage sources are used in combination, called combination scheme 1, as shown in Fig.1(c); one IDCPFC and one series adjustable voltage source are used in combination called combination scheme 2, as shown in Fig.1(d). Combining two single-target DC power flow controllers can achieve the power flow control of two DC lines. Thus, the two combination schemes can achieve three current adjustment modes of two lines. Compared with the combination scheme 1, the combination scheme 2 requires less external power supply with more simple circuit structure and lower equipment cost. But more series voltage sources need to be strung in lines. When the series adjustable voltage source is bypassed, the combination scheme 1 function as IDCPFC function, so the combination scheme 2 retains IDCPFC function with better economical and practical features.

Considering that DC power flow controller of combination scheme 2 needs to connect in series with more voltage sources, the combination scheme 2 is simplified to obtain the schematic diagram of the composite DC power flow controller in Fig.2(a). The device can be improved by external auxiliary circuit connected with DC bus, without an external voltage source, whose schematic diagram is shown in Fig.2(b) below. If the equipment only exchange energy with one line, it achieves series adjustable voltage source function, as shown in Fig.2(c); if it achieves energy exchange without the external voltage source, it can achieve IDCPFC function, as shown in Fig.2(d). The composite DC power flow controller inherits the function of series adjustable voltage source and IDCPFC, and can also realize three current adjustment modes of two lines, to be full-function, so called Combined DC Power Flow Controller.

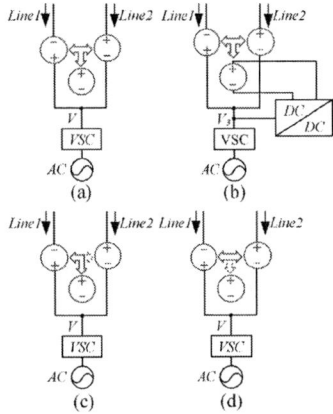

Fig.2 Principles and Function of CDCPFC. (a) CDCPFC with an external voltage source. (b) CDCPFC with external auxiliary circuit. (c) CDCPFC as SAVS. (d) CDCPFC as IDCPFC.

III. CDCPFC AND ITS OPERATION PRINCIPLES

Based on the working principles and topological structure analysis of CDCPFC above, this paper presents a CDCPFC circuit topology, shown in Fig.3.

A. Topology Description

DC power flow controller connects two transmission lines and an external DC voltage source to realize the power flow regulation of two transmission lines. CDCPFC consists of eight IGBTs (Q_1 to Q_8) and reverse parallel diodes (D_1 to D_8), four series diodes (D_{1b} to D_{4b}), one coupling inductance, two series capacitors (C_1 and C_2) of transmission lines and two bypass switches (S_1 and S_2), wherein the bidirectional switch consists of two IGBTs in series. The reference direction of each electrical quantity is shown in Fig.3

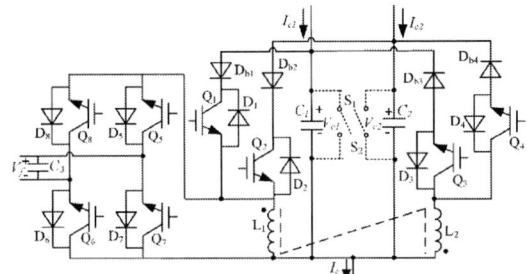

Fig. 3 Topology of CDCPFC

When the bypass switches S_1 and S_2 are closed, the capacitors C_1 and C_2 are shorted, CDCPFC is in the bypass maintenance status and does not involved in the system power flow control. When the bypass switches S_1 and S_2 are disconnected, there are six equivalent circuits 1-6 by

1492

controlling the ONs and OFFs of the eight IGBTs to control the power flow control of the two transmission lines. The operation principles of CDCPFC is analyzed by three different flow conditions as follows.

B. Operation Principles

There are only four kinds of power flow adjustment objectives: 1) increase I_{c1} and I_{c2}; 2) decrease I_{c1} and I_{c2}; 3) decrease I_{c1} and increase I_{c2}; 4) increase I_{c1} and decrease I_{c2}. The energy transfer path's principles of four objectives are consistent. If both Q_1/Q_2 are off, both V_{c1} and V_{c2} will continue to rise. Therefore, in order to keep the balance of V_{c1} and V_{c2}, it is necessary to transfer the capacitive energy to L_1 and L_2. Meanwhile, in order to keep the balance of I_{L1} and I_{L2}, it is necessary to transfer the energy of L_1 and L_2 to output capacitor C_3 of the external voltage source.

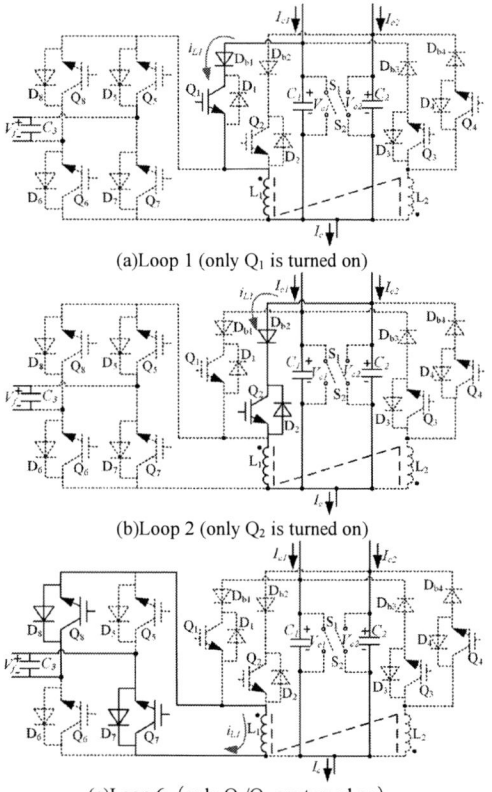

(a)Loop 1 (only Q_1 is turned on)

(b)Loop 2 (only Q_2 is turned on)

(c)Loop 6 (only Q_7/Q_8 are turned on)

Fig.4 Operation modes when I_{c1}/I_{c2} are positive

Case1: I_{c1} and I_{c2} are positive: In this case, I_{c1} and I_{c2} are decreased, as an example, that is, equivalent negative DC voltage sources are in series into line1 and line2 (voltage directions of C_1 and C_2 shown in Fig.5). The whole work process can be divided into three stages. Stage 1, first Q_1 is turned on, C_1-L_1-Q_1-D_{b1} form loop 1, as shown in Fig.4(a). C_1 transfers energy to L_1, and I_{L1} increases linearly. Stage 2, Q_1 is turned off / Q_2 is turned on, then C_2-L_1-Q_2-D_{b2} form loop 2, as shown in Fig.4(b). The energy is shifted from C_2 to L_1, and I_{L1} increases linearly. Stage 3, Q_2 is turned off / Q_7 and Q_8 are turned on, then C_3-L_1-Q_7-Q_8 form loop 6, as shown in Fig.4(c). L_1 charges C_3, and I_{L1} decreases linearly. When Q_7/Q_8 are turned off and Q_1 is turned on, another switch period begins. The power flow

controller transfers some energy of C_1 and C_2 to C_3 to decrease I_{c1} and I_{c2}, whose specific switching mode is shown in Fig.4. Q_1, Q_2 and Q_7/Q_8 are complementary to each other. Q_3, Q_4, Q_5 and Q_6 are always in the off status. Based on the analysis above, Q_1, Q_2 and Q_7/Q_8 need to be controlled.

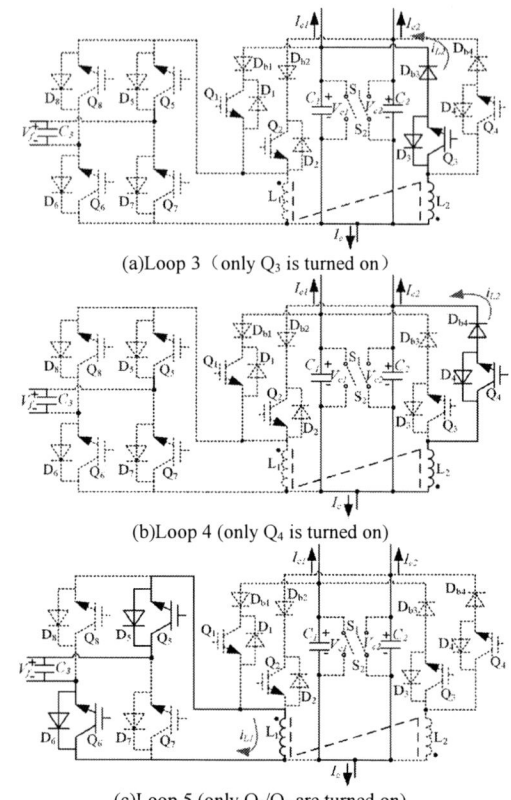

(a)Loop 3 (only Q_3 is turned on)

(b)Loop 4 (only Q_4 is turned on)

(c)Loop 5 (only Q_5/Q_6 are turned on)

Fig.5 Operation modes when I_{c1}/I_{c2} are negative

Case2: I_{c1} and I_{c2} are negative: here, I_{c1} and I_{c2} are to be increased, that is, equivalent negative DC voltage source string into line1 and line2 (voltage directions of C_1 and C_2 shown in Fig.5). Stage 1, first open Q_3, C_1-L_2-Q_3-D_{b3} form loop 3, as shown in Fig.5(a). L_2 transfers energy to C_1, and I_{L2} decreases linearly. Stage 2, Q_3 is turned off / Q_4 is turned on, then C_2-L_2-Q_4-D_{b4} form loop 4, as shown in Fig.5(b). The energy is shifted from L_2 to C_2, and I_{L2} decreases linearly. Stage 3, Q_4 is turned off / Q_5 and Q_6 are turned on, then C_3-L_1-Q_5-Q_6 form loop 5, as shown in Fig.5(c). The L_2's energy is transferred to L_1 because of the coupling inductance. C_3 charges L_1, and I_{L1} increases linearly. When Q_5/Q_6 are turned off and Q_3 is turned on, another switch period begins. The power flow controller transfers some energy of C_3 and C_1 to C_2 to increase I_{c1} and I_{c2}, whose specific switching mode is shown in Fig.5. Q_3, Q_4 and Q_5/Q_6 are complementary to each other. Q_1, Q_2 Q_7 and Q_8 are always in the off status. Based on the analysis above, Q_3, Q_4 and Q_5/Q_6 need to be controlled.

Case3: I_{c1} and I_{c2} are in different directions: there are two conditions: I_{c1} is positive and I_{c2} is negative or I_{c1} is negative and I_{c2} is positive. According to the symmetry, I_{c1} is decreased and I_{c2} is increased, as an example to illustrate

1493

the condition where I_{c1} is positive and I_{c2} is negative, that is, one equivalent negative and positive DC voltage sources are strung into line1 and line2 (voltage directions of C_1 and C_2 shown in Fig.6). Stage 1, first open Q_1, C_1-L_1-Q_1-D_{b1} form loop 1, as shown in Fig.6(a). C_1 transfers energy to L_1, and I_{L1} increases linearly. Stage 2, Q_1 is turned off/Q_4 is turned on, then C_2-L_2-Q_4-D_{b4} form loop 4, as shown in Fig.6(b). The energy of L_1 is transferred to L_2 because of the coupling inductance. The energy is shifted from L_2 to C_2, and I_{L2} decreases linearly. If the increasing energy of L_1 in loop 1 is less than the decreasing energy of L_2 in loop 4, the decreasing energy of the inductance needs to be obtained from C_3. Stage 3, Q_4 is turned off/Q_5 and Q_6 are turned on, then C_3-L_1-Q_5-Q_6 form loop 5 in Fig.5(c). The L_2's energy is transferred to L_1 for the coupling inductance. C_3 charges L_1, and I_{L1} increases linearly. Otherwise, if the increasing energy of L_1 in loop 4 is more than the decreasing energy of L_2 in loop 4. L_1 charges C_3, and Q_7/Q_8 need to turn on and I_{L1} decreases linearly in Fig.4(c). When Q_5/Q_6 are turned off and Q_1 is turned on, another switch period begins.

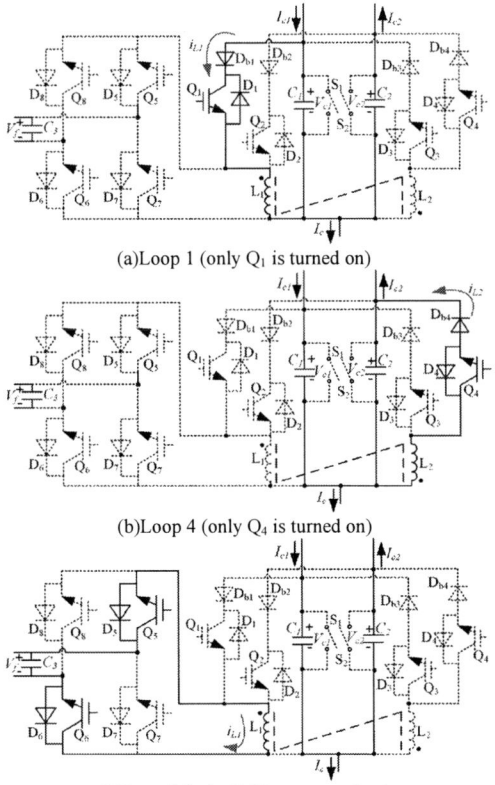

(a)Loop 1 (only Q_1 is turned on)

(b)Loop 4 (only Q_4 is turned on)

(c)Loop 5 (only Q_5/Q_6 are turned on)

Fig.6 Operation modes when I_{c1}/I_{c2} are in different direction

The power flow controller balances the difference between C_1 and C_2 with C_3 to decrease I_{c1} and increase I_{c2}, whose specific switching mode is shown in Fig.6. Q_1, Q_4 and Q_5/Q_6 are complementary to each other. Q_2, Q_3 Q_7 and Q_8 are always in the off status. Based on the analysis above, Q_1, Q_4 and Q_5/Q_6 need to be controlled.

According to the analysis above, combined with the current direction and the need for power flow control, the

CDCPFC has 17 kinds of conditions totally. Table 1 shows the switching statuses of each switch under different operation condition. The switching action rules can be summarized as below:

● While I_{c1} is positive, Q_1 works, Q_3 shuts; while I_{c1} is negative, Q_3 works, Q_1 lockouts;
● While I_{c2} is positive, Q_2 works, Q_4 shuts; while I_{c2} is negative, Q_4 works, Q_2 lockouts;
● While control objectives I_{c1}/I_{c2} are increased, Q_5/Q_6 work, Q_7/Q_8 shut; while control objectives I_{c1}/I_{c2} are decreased, Q_7/Q_8 work, Q_5/Q_6 shut;
● When I_{c1}/I_{c2} control objectives are different, the capacitance energy of one line is partially transferred to the inductor; the capacitor of the other line is required to obtain part energy from the inductor. While the inductor requires energy from the capacitor, Q_5/Q_6 work, Q_7/Q_8 shut. While the inductor needs to transfer energy to the capacitor C_3, Q_7/Q_8 works, Q_5/Q_6 shut.

Tab.1 Operation statuses of the CDCPFC

I_{c1}/I_{c2} direction	I_{c1}/I_{c2} control targets	V_x/V_y polarity	Controlled switching device	case
+/+	↑/↑	-/-	Q_1、 Q_2、 Q_5/Q_6	1
	↓/↓	+/+	Q_1、 Q_2、 Q_7/Q_8	2
	↑/↓	-/+	Q_1、 Q_2、 Q_5/Q_6 orQ_7/Q_8	3
	↓/↑	+/-	Q_1、 Q_2、 Q_5/Q_6 orQ_7/Q_8	4
-/-	↑/↑	+/+	Q_3、 Q_4、 Q_5/Q_6	5
	↓/↓	-/-	Q_3、 Q_4、 Q_7/Q_8	6
	↑/↓	+/-	Q_3、 Q_4、 Q_5/Q_6 orQ_7/Q_8	7
	↓/↑	-/+	Q_3、 Q_4、 Q_5/Q_6 orQ_7/Q_8	8
+/-	↑/↑	-/+	Q_1、 Q_4、 Q_5/Q_6	9
	↓/↓	+/-	Q_1、 Q_4、 Q_7、 Q_8	10
	↑/↓	-/-	Q_1、 Q_4、 Q_5/Q_6 orQ_7/Q_8	11
	↓/↑	+/+	Q_1、 Q_4、 Q_5/Q_6 orQ_7/Q_8	12
-/+	↑/↑	+/-	Q_2、 Q_3、 Q_5/Q_6	13
	↓/↓	-/+	Q_2、 Q_3、 Q_7/Q_8	14
	↑/↓	+/+	Q_2、 Q_3、 Q_5/Q_6 orQ_7/Q_8	15
	↓/↑	-/-	Q_2、 Q_3、 Q_5/Q_6 orQ_7/Q_8	16
——	——	——	S_1、 S_2	17

C. Control Strategy

The control strategy of CDCPFC is shown in Fig.7. The dispatch system sends control commands to CDCPFC according to the current system power flow, that is, the current reference $I_{c1\text{-ref}}$ and $I_{c2\text{-ref}}$ of the controlled line. Meanwhile, according to the corresponding specific conditions, CDCPFC will receive command signal of the condition sequence from the dispatch system, select the switching device which participates in the operation through the logic unit, and lock the switching device which does not participate in the operation. $I_{c1\text{-ref}}$ and $I_{c2\text{-ref}}$ are compared with the actual current I_{c1} and I_{c2}. The difference obtained by the PID regulator are compared with sawtooth carrier. Finally, the three groups of four drive signals are obtained through the logic operations.

In this control strategy, the power flow control target of the two lines are completely decoupled, without constrained relationship. The signal waveforms above are complementary in each switching cycle. In order to construct three sets of complementary drive signals, the concept of "dynamic upper limit value" is introduced in the control strategy, that is, the duty cycle of a group of input signals is limited with set value. In Fig.7, for the PID unit, the upper limit of output limiter unit is determined by D_1. Thus, three sets of complementary drive signals can be obtained in the control strategy.

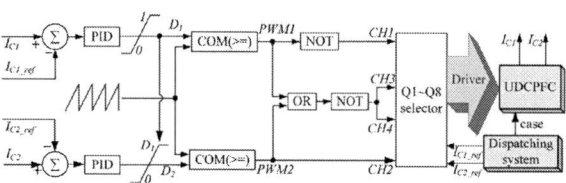

Fig.7 Block diagram of control strategy

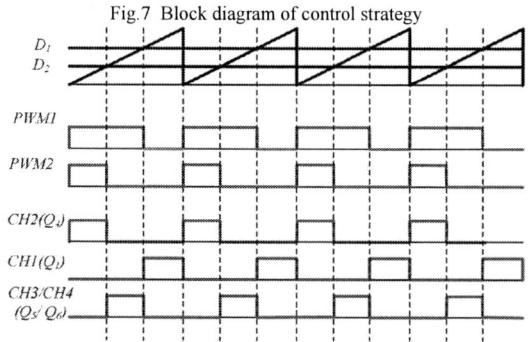

Fig.8 Waveforms of control strategy

D. Charateristic Analysis

In the case of the technical situation of Fig.5, for example, a detailed control strategy can be obtained on the basis of Tab.1 and Fig.7. Due to the symmetry of CDCPFC topology and 17 cases described above, case 12 is taken as an example to analyze the control signal of CDCPFC in Fig.8. Set the duty cycle of Q_1 as D_1, Q_4 as D_2 and Q_5/Q_6 as $1-D_1-D_2$. While the current of coupling inductance is continuous, the equation (1) can be obtained by the principles of voltage-second balance.

$$V_{c1}D_1T_s - V_{c2}D_2T_s - V_f(1 - D_1 - D_2)T_s = 0 \qquad (1)$$

Where, T_s is the switching period.

While the voltages of C_1 and C_2 are continuous, according to the principles of amperes-second balance, we can obtain:

$$I_{c1}(1 - D_1)T_s + (I_{c1} - I_L)D_1T_s = 0 \qquad (2)$$

$$(I_{c2} - I_L)D_2T_s + I_{c2}(1 - D_2)T_s = 0 \qquad (3)$$

The equations are simplified above, we can obtain:

$$I_{c1} = D_1 I_L \qquad (4)$$

$$I_{c2} = D_2 I_L \qquad (5)$$

Where, I_L is average inductor current.

Equation (4) and (5) stand for the average model of CDCPFC. The duty cycle D_1 and D_2 have a linear relationship with the line current, which indicates that D_1 and D_2 can control the current of two lines. According to (1)(4)(5), we can obtain:

$$V_{c1}I_1 - V_{c2}I_2 - V_f I_L(1 - D_1 - D_2) = 0 \qquad (6)$$

$$I_L = I_{c1} + I_{c2} + (V_{c1}I_{c1} - V_{c2}I_{c2})/V_f \qquad (7)$$

Equation (6) is the power conservation equation while CDCPFC efficiency is 100%. In particular, when $D_1+D_2=1$, the energy exchange exists only between C_1 and C_2, which means that CDCPFC functions as IDCPFC. When $D_1=0$ or $D_2=0$, the energy exchange exists only between C_1 and C_3 or C_2 and C_3, which means CDCPFC functions as a single series adjustable voltage source.

IV. SIMULATION VERIFICATION

In order to verify the feasibility and validity of the proposed CDCPFC topology, an equivalent three-terminal ring-type monopole earth loop DC transmission system model is constructed in MATLAB/SIMULINK environment, as shown in Fig.9. VSC1 and VSC2 work in constant power mode, respectively, deliver 160MW and 80MW power to the system; VSC3, as the system power output terminal, operates in constant voltage mode, $V_3 = 150kV$; the transmission line parameters are shown in Table 2; CDCPFC is installed at VSC3, the capacitors C_1 and C_2, respectively, are in series with the line2 and line3. The external power supply required by CDCPFC uses a DAB-based voltage source [8], which acquires the energy from the DC bus of VSC3. The simulation parameters are set to: $C_1=35mF$, $C_2=8.5mF$, $L_1=L_2=L=2mH$, and the switching frequency is 1kHz.

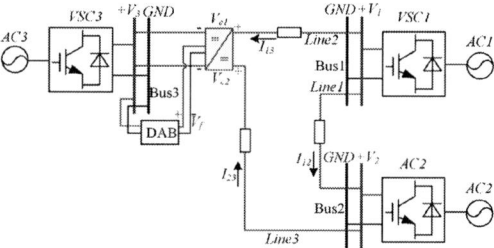

Fig. 9 Three-terminal Ring HVDC System Simulation Model

Tab.2 transmission line parameters

Parameter	Line 1	Line 2	Line 3
Distance/km	200	300	100
Resistor/Ω	2	3	1
Inductance/mH	80	120	40

A. Put into operation

t=0~2s, CDCPFC is bypassed, the initial steady-status currents I_{13} and I_{23} of lines 2 and 3 are 0.615kA and 0.968kA, respectively; t=2s, CDCPFC get instructions to control I_{13} and I_{23}, whose current reference values were 0.85kA and 0.7kA, then CDCPFC is put into operation. The simulation waveform of the whole input process is shown in Fig.10. Fig.10(a) shows the line current quickly responds when CDCPFC puts into operation (red curve corresponding to I_{13}; blue curve corresponding to I_{23}). Fig.10(b) shows the terminal voltage curve of VSCs. The line capacitance voltage curves are shown in Fig.10(c).

1495

Fig.10(d) shows the three groups of complementary switching signal waveform. The simulation results are consistent with the theoretical analysis which verifies the correctness of the operation principles.

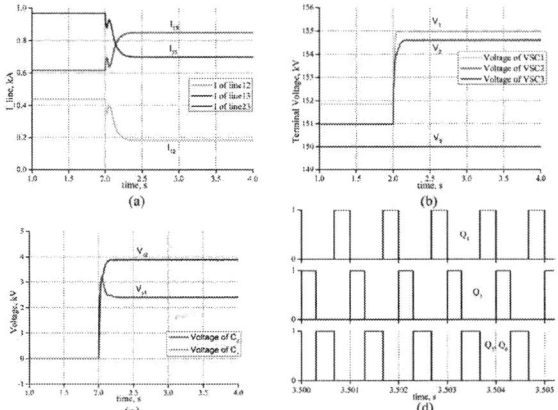

Fig.10 Simulation waveforms when putting into operation

(a)I_{12}, I_{13} ,I_{23} (b)V_1,V_2 (c)V_{c1} , V_{c2}(d)Q_1, Q_3 , Q_7/Q_8

B. Power flow reversal

This simulation is used to verify the power flow reversal process of line 3. Before the current reverses, CDCPFC is bypassed and I_{23} = 0.968kA. t=2s, CDCPFC is instructed to control I_{13} and I_{23} with current reference values of 1.56kA and -0.06kA, respectively. The simulation results of the whole process are shown in Fig.11. According to Fig.11(a), the direction of I_{23} in line 3 is changed from the positive direction to the negative direction and the direction of I_{13} in line 2 is not changed, thereby indicating the CDCPFC can cope with power flow reversal occasion. In Fig.11(b) and (c), the capacitor voltages V_{C1} and V_{C2} of the line 2 and line 3 are 4.81kV and 10.69kV, respectively, and the terminal voltages of VSC1 and VSC2 rise to 159.50kV and 160.63kV to meet the requirements of the power flow reversal.

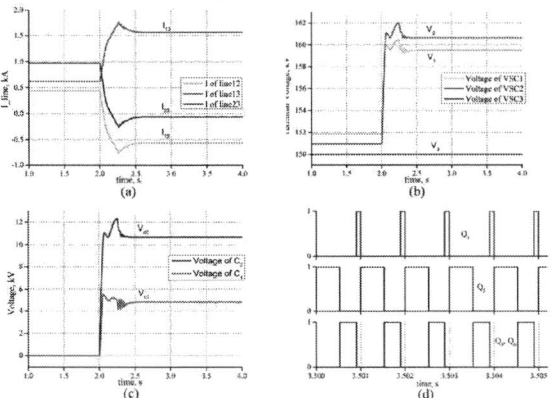

Fig. 11 Simulation waveforms when power is reversed

(a)I_{12}, I_{13} ,I_{23} (b)V_1,V_2 (c)V_{c1} , V_{c2}(d)Q_1, Q_3 , Q_7/Q_8

C. Power jump of VSC2

Set the CDCPFC to control I_{13} and I_{23} to reach 0.85kA and 0.7kA respectively (same as part A). At t = 3s, the output power of VSC2 changes stepwise from 80MW to 90MW, VSC1 output power remains unchanged, then the

control reference value of I_{13} and I_{23} remains unchanged. Fig.12(a) shows that the controlled line current had the rapid return to the desired current value by about the transient process of 0.3s. VSC terminal voltage and line string capacitance voltage shown in Fig.12(b) and (c) returned to steady status by the transient process of about 0.2s. The simulation shows that in the event of external disturbances (eg, terminal output power changes), CDCPFC can quickly respond to external power changes to make the system reach a new stable status quickly. Thus, the proposed CDCPFC control strategy is valid.

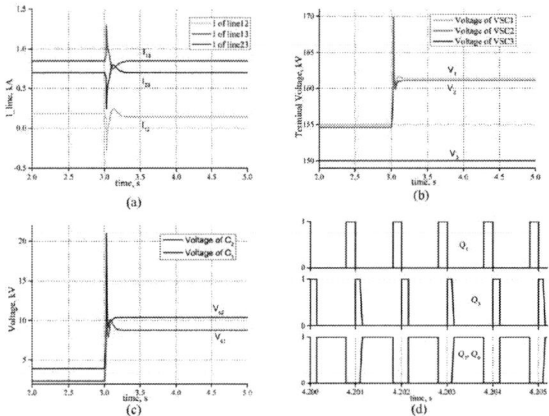

Fig. 12 Simulation waveforms when power jump of VSC2

(a)I_{12}, I_{13} ,I_{23} (b)V_1,V_2 (c)V_{c1} , V_{c2}(d)Q_1, Q_3 , Q_7/Q_8

V. CONCLUSION

This paper analyzes the shortcomings of IDCPFC in 2-line power flow control. Based on the structure of IDCPFC, this paper presents a schematic diagram of Combined DC power flow controller with the ability to realize the flexible control of two-line currents. Based on the schematic diagram, a topology circuit is proposed, and its operation principles, control strategy and working characteristic are analyzed. A simulation model of three - terminal HVDC transmission system is established in MATLAB / SIMULINK, which verifies the function of the power flow controller. The principle analysis and simulation results show that the DC power flow controller can effectively realize the multi-line power flow control function.

ACKNOWLEDGEMENT

The authors' work is financially supported by Research Program of State Grid Shanghai Municipal Electric Power Company (Unified Power Flow Controller Technology and Its Application in Shanghai Power System).

REFERENCES

[1] D. Jovcic, M. Hajian, H. Zhang, G. Asplund, "Power Flow Control in DC Transmission Grids Using Mechanical and Semiconductor Based DC/DC Devices," in 10th IET Int. Conf. AC and DC Power Transmission, Dec. 2012, pp. 1-6.

[2] C D Barker, R S Whitehouse, "A Current Flow Controller for Use in HVDC Grids," in 10th IET Int. Conf. AC and DC Power Transmission, Dec. 2012, pp. 1-5.

[3] H. Diab, S. Tennakoon, C. Gould, et al., "An investigation of power flow control methods in multi terminal high voltage DC grids," in 50th Int. Universities Power Engineering Conference, Sep. 2015, pp. 1-5.

[4] K. Natori, H. Obara, K. Yoshikawa, et al., "Flexible power flow control for next-generation multi-terminal DC power network," in 6th IEEE Int. Conf. Energy Conversion Congress and Exposition(ECCE), Nov. 2014, pp. 778-784.

[5] D. Jovcic., "Bidirectional, High-power DC Transformer," IEEE Trans. Power Del., vol. 24, no. 4, pp. 2276-2283, Oct. 2009.

[6] Q. Mu, J. Liang, Y. Li, X. Zhou, "Power Flow Control Devices in DC Grids," in Proc. of the 2012 IEEE Power and Energy Society Meeting, Jul. 2012, pp. 1-7.

[7] N. Deng, P. Wang, X. Zhang, et al., "A DC current flow controller for meshed modular multilevel converter multiterminal HVDC grids," IEEE CSEE Journal of Power and Energy System, vol. 1, no. 1, pp. 43-51, Feb. 2015.

[8] L. Yao, H. Cui, J. Zhuang, et al., "A DC Power Flow Controller and Its Control Strategy in the DC Grid," in 8th Power Electronics and Motion Control Conference, May. 2016, pp.1-6.

[9] J. Sau-Bassols, E. Prieto-Araujo, O. Gomis-Bellmunt, "Modelling and Control of an Interline Current Flow Controller for Meshed HVDC Grids," IEEE Trans. Power Del., vol. 32, no. 1, pp. 11-22, Feb. 2017.

[10] W. Chen, X. Zhu, L. Yao, et al., "An Interline DC Power-Flow Controller (IDCPFC) for Multiterminal HVDC System", IEEE Trans. Power Del., vol. 30, no. 4, pp. 2027-2036, Aug.2015.

[11] W. Chen, X. Zhu, L.Yao L Z, et al., "A Novel Interline DC Power Flow Controller (IDCPFC) for Meshed HVDC Grids," IEEE Trans. Power Del., vol. 31, no. 4, pp. 1719-1727, Aug. 2016.

[12] X. ZHONG, M. ZHU, R. HUANG, X. CAI, "Combination strategy of DC power flow controller for multi-terminal HVDC system," in 6th IET Int. Conf. Renewable Power Generation, Oct. 2017, pp. 1-7.

An Approach for the Emulation of DC Grid Admittances: Implementation on a Buck Converter

Enrique Rodriguez-Diaz, Fracisco D. Freijedo*, Drazen Dujic*, Juan C. Vasquez, and Josep M. Guerrero
Department of Energy Technology, Aalborg University. Emails:{erd, juq, joz}@et.aau.dk
*Ecole Polytechnique Federale de Lausanne - EPFL, Power Electronics Laboratory - PEL
Emails:{francisco.freijedo, drazen.dujic}@epfl.ch

Abstract—**This work proposes a method to actively emulate dc grids, including their equivalent admittance, by means of a modified current control structure of a dc/dc converter. The stability of power converter units (PCUs), interconnected with electrical grids is frequently assessed by the characteristics of the equivalent admittance/impedance of the grid, at the point of connection, and the PCU. Several approaches have been proposed to shape the PCU equivalent admittance, by means of the control structure, in order to fulfil the requirements about stability, robustness and dynamic response. On top of this, the integration of dc distribution systems is steadily growing, due to its simpler implementation and the potential cost and losses reduction. The presence of PCUs, in dc grids, is inherently higher, which complicates the overall stability assessment. In order to verify an impedance/admittance stability analysis in a laboratory scaled system, the characteristics of the impedance/admittance needs to be emulated. A suitable control structure, implemented on a buck converter, to actively emulated dc grid admittance has been proposed and tested. The analytical expressions of the emulator converter admittance, with the modified control structure, are presented. The admittances of a MVDC collection network for offshore wind farms, and a MVDC drive, working with a vector control scheme, have been calculated to be used for the emulation. The experimental validation has shown the overall feasibility of the proposed emulation approach.**

I. Introduction and Motivation

The integration of dc distribution systems has grown significantly in recent years due to the potential efficiency improvement, cost reduction and simplicity. The specific benefits of implementing a dc distribution system varies significantly with the application [e.g. ac grid availability, high presence of energy storage systems (ESS) or electronics loads, etc.], however, for isolated electrical power systems, the use of dc voltage is, in general, the most convenient option [1], and in the telecommunication and automotive industries, dc distribution system are a mature solution [2], [3].

Lately, this trend can be observed in higher power electric systems, as the ones used for maritime and aircraft applications [4]–[7], but also in grid connected systems with a high presence of dc loads, as datacenters [8].

Furthermore, a high penetration of renewable energy sources (RES) and ESS at consumption level, allows for a significant efficiency improvement, by avoiding unnecessary conversion stages, since distribution between mostly dc-based devices [e.g. photovoltaic (PV) panels, batteries, and electronic loads] is performed more efficiently if dc voltage is employed [1], [9]. For higher power distribution, through longer distances, the

absence of reactive power, makes dc voltage a more suitable candidate for distribution in low-voltage (LV) [10], medium-voltage (MV) [11], [12], and high-voltage (HV) [13].

Nevertheless, the use of dc, always relies on power converter units (PCUs) to adapt the voltage levels, making the presence of PCUs inherently higher in dc distribution systems, which can compromise the overall stability of the system [14]–[18].

Impedance-based methods are frequently used to assess the stability of the electrical system, and to design the control structures of the power converters [6], [7], [16]–[22]. The electrical power system can be systematically analysed by studying the interaction of the output impedance of the source converter (e.g. a MVDC collector converter of an offshore wind farm [12], or a front-end rectifier for supplying a datacenter [23], etc.), and the equivalent admittance of the rest of the system, aggregated in the total grid admittance.

The grid admittance depends on several factors (i.e. distribution architecture, individual admittance of the connected power converters, cables...) and replicating the specific grid conditions for testing the source converter can be challenging.

In this work, an approach to actively emulate a given dc grid admittance, is shown. The emulation is performed by means of a modified current control structure that sets a relation between the input voltage and current of the PCU. The concept of the active admittance emulator is shown in Fig. 1. In many applications, the output impedance of the source converter $Z_s(s)$ is shaped, so a set of design requirements are fulfil when interacting with the equivalent admittance of the rest of the dc grid, $Y_g(s)$, as seen from the point of connection of the source converter. Therefore, in order to perform the experimental validations, the approach shown in Fig. 1, can be followed, where the dc grid admittance is emulated by a second PCU connected at the terminal of the sources converter.

The same reasoning has been applied for emulating the behaviour of devices, where a dc/dc converter is used to replicate different components such us PV panels and batteries [24], [25], for a given emulated conditions (e.g. solar irradiation, temperature, state of charge, etc.).

II. Modelling and Control of Buck Converter for Admittance Emulation

The proposed approach, for emulation of an equivalent dc grid admittance, is based on a modified current control structure, that shapes the equivalent input admittance of the

The 2018 International Power Electronics Conference

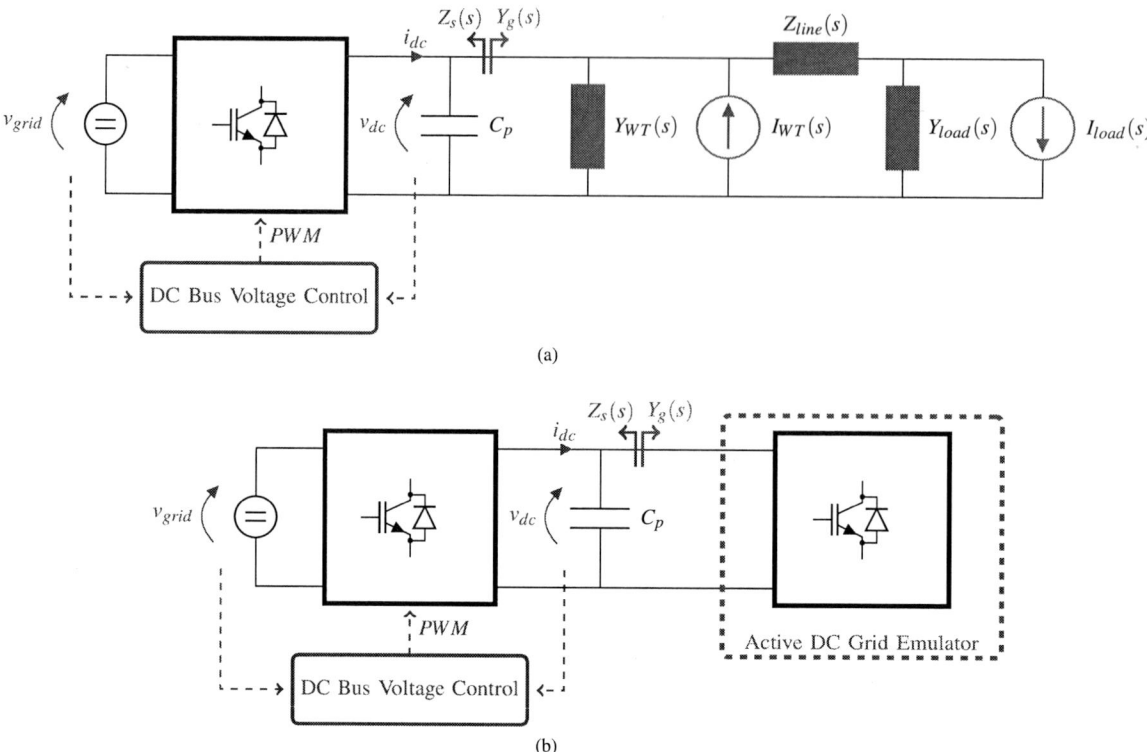

Fig. 1: Concept of dc grid admittance emulator.

emulator converter (Y_{em} in the rest of the paper), in order to match the objective dc grid admittance (Y_g in the rest of the paper).

In the scenario shown in Fig. 1, the dc grid admittance defines the grid current responses against voltage variations at the terminal of the source converter, therefore, it can be seen that the converter needs to regulate the input current as function of the input voltage, in order to emulate the dc grid behaviour.

In this work, a buck converter is modelled for the application. The current control is performed on the inductor of the converter, as shown in Fig. 2. The current references are obtained from the input voltage (v_s) measurements and the objective grid admittance (e.g $i_s^*(s) = Y_g(s)v_s(s)$), therefore it is expected that, within the current controller bandwidth, $Y_{em}(s) \approx Y_g(s)$. The use of the buck topology, allows for a lower converter output voltage (i.e. $V_o < V_s$), nevertheless, the inductor is placed at the output, and therefore, i_s is not directly controlled (i.e. the control is performed on i_o, and i_o^* needs to be estimated form i_s^*).

Alternatively, using a boost converter, would allow to directly control i_s, however, it requires a higher voltage on the output side, ideally twice as big, which can compromise the implementation of this topology due to hardware limitations, especially for MVDC or HVDC applications. Therefore, for this work, the buck topology has been analysed.

For the admittance calculation of the emulator, the current and voltage waveforms have been averaged during one switch-

ing period, therefore the equations that described the system, are as follows:

$$L\frac{di_o(t)}{dt} = [v_s(t) - v_o(t)]d(t) - v_o(t)[1 - d(t)] \quad (1)$$

$$i_s(t) = i_o(t)d(t) \quad (2)$$

where $v_s(t)$, $v_o(t)$, $i_s(t)$ and $i_o(t)$ are the voltages and currents on the high voltage and low voltage sides, respectively, L is the inductance value of the inductor in the buck converter, and $d(t)$ is the duty cycle.

It can be seen that equation (1) and (2) are non-linear, therefore linearazing the equations around a given operating point, and assuming that V_o is tightly regulated ($\hat{v}_o(t) \approx 0$), results in the following linear expressions in the s-domain:

$$Lsi_o(s) = d(s)V_s + v_s(s)D \quad (3)$$

$$i_s(s) = i_o(s)D + d(s)I_o \quad (4)$$

From the control structure shown in Fig. 2, the expression of $d(s)$ can be obtained.

$$d(s) = \frac{e^{-st_L}H_{zoh}(s)}{V_s}K(s)[\frac{v_s(s)Y_g(s)}{D} - i_o(s)] \quad (5)$$

where $Y_g(s)$ is the equivalent admittance of grid that is being emulated, $K(s)$ is the main current controller, D and V_s are the dc quantities of the duty cycle and the input voltage respectively, and e^{-st_L} and $H_{zoh}(s)$ are the system delays due

1499

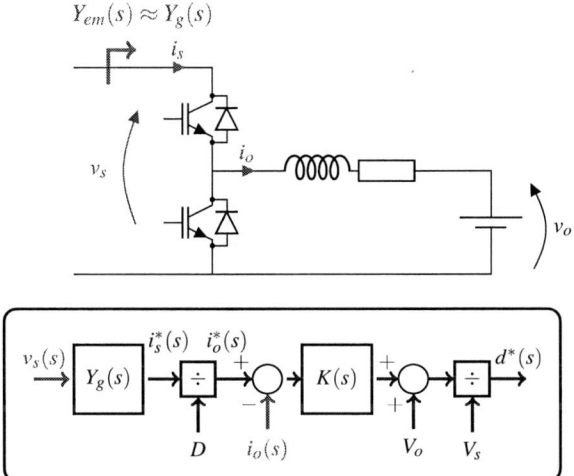

Fig. 2: Control structure of the active dc grid admittance emulator based on a buck converter.

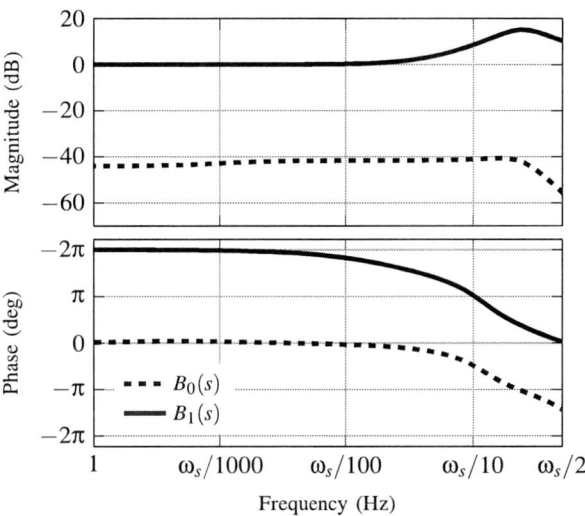

Fig. 3: Frequency responses of $B_0(s)$ and $B_1(s)$.

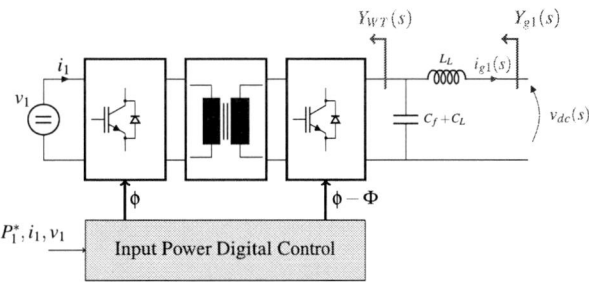

Fig. 4: Studied System of a DC Collection Network for Offshore Wind Farms.

to the sampling and PWM action.

The model of the system is shown in Fig. 2. From equations (1), (2), and (5), the emulator admittance can be derived, and the expression is the following:

$$Y_{em}(s) = \frac{i_s(s)}{v_s(s)} = B_0(s) + Y_g(s)B_1(s) \qquad (6)$$

where,

$$B_0(s) = \frac{Y_o/Y_s - Y_sC}{Ls + C} \qquad (7)$$

$$B_1(s) = \frac{C + Y_oC(Ls + C) - Y_oC^2}{Ls + C} \qquad (8)$$

where, $Y_o = I_o/V_o$, $Y_s = I_s/V_s$, and $C = K(s)e^{-st_L}H_{zoh}(s)$.

Equation (6) shows that $Y_{em}(s)$ is formed by two terms, a term $B_0(s)$ independent of the objective admittance, and a second term, $B_1(s)Y_g(s)$, dependent on the objective admittance. It is easily appreciated that an ideal admittance emulator (i.e. $Y_{em}(s) = Y_g(s)$) is achieved when $B_0(s) = 0$, and $B_1(s) = 1$ for the whole frequency range. This is not achievable in practise due to finite controller gains and time delays.

As it can be seen from (7), and (8), the tuning of the current regulator, $K(s)$, and the converter parameters (L, V_s, V_o, etc.) allows to shape the frequency response of $B_0(s)$, and $B_1(s)$. It has been found that, tuning the current controller to cancel the pole of the converter inductor ($K_i = R/L$) and a close loop bandwidth ($K_p = 2\pi Lbw$) slightly below $\omega_s/10$, gives the best match of both magnitude and phase of the objective admittance, $Y_g(s)$.

Fig. 3 show the frequency responses of $B_0(s)$, and $B_1(s)$, where the e^{-st_L} and $H_{zoh}(s)$ have been approximated by their first order expressions [20]. It can be seen that, first, $B_0(s)$ has a magnitude below -40 dB, and second, $B_1(s)$ has a magnitude of 0 dB up to $\omega_s/10$ approximately, while introducing some phase delay after $\omega_s/100$.

III. EQUIVALENT ADMITTANCES OF DC GRIDS

In order to assess the feasibility of the proposed emulation approach, two representatives cases of dc distribution systems have been selected and modelled, in order to use the calculated dc grid admittances [$Y_{g1}(s)$ and $Y_{g2}(s)$] in the emulator. The dc grid admittances, calculate in this Section, include the effect of constant power sources (CPSs), constant power loads (CPLs), and resonances introduced by the distribution line parameters. The presence of this elements y a dc distribution grid can compromise the stability of the system, therefore they are of most interest to test the emulation approach. Also, both dc grid admittances have been calculated in p.u. values.

A. MVDC collection Network for Offshore Wind Farms

The first case consist on a dc distribution line that connects the output of a WT, interfaced by a dc-dc converter, with a collector converter, that it is connected to the HVAC or HVDC grid. This configuration, shown in Fig. 4, has been widely proposed in the literature for WT farms collectors [11], [12], [26]–[29].

In this particular case, a WT interfaced by a dual-active-bridge (DAB) converter, working under power control oper-

The 2018 International Power Electronics Conference

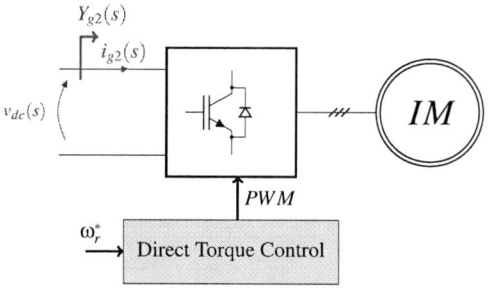

Fig. 5: Analysed system of a MV drive supplied from a DC source.

Table I: Physical parameters used for the study cases.

Parameter	Value
DAB	
Nominal Power (P_1)	1 MW
Nominal Input Voltage (V_1)	1000 V
Nominal Output Voltage (V_2)	10 kV
Controller Bandwidth (α)	$2\pi 10$ rad/s
Line	
Line inductance (L_L)	0.347 mH/km
Line resistance (R_L)	0.089 Ω/km
Line Capacitance (C_L)	0.307 uF/mk
Filter capacitance (C_f)	6.5 mF
Length	1 km
Drive	
Nominal Power	3 MW
Nominal DC bus Voltage	10 kV
Switching Frequency	250 Hz
Control Bandwidth	$2\pi 15$ rad/s

ation, has been modelled. The low voltage (LV) side of the DAB converter receives the output power of the rectifier (P_1), connected to the WT generator. Therefore the DAB acts a physical actuator of a higher level mechanical controller, that regulates the power that is extracted from the turbine.

In WT applications, the requirements, for the controller design, are the specified maximum allowable time constant, and no overshoot in the power tracking response. The parameters used for this study case are shown in table I. Linearazing the system at a given operating point, the output admittance of the WT is seen as a CPS, up to the bandwidth of the power controller. The analytical derivation of the WT equivalent admittance is as follows:

$$Y_{WT}(s) = \frac{I_2}{V_2} \frac{\alpha}{s+\alpha} \qquad (9)$$

where, I_2 and V_2 are the output current and voltage of the DAB converter, and α is the close loop bandwidth of the power controller.

From Fig. 4, the equivalent admittance of the grid (Y_{g1}), as

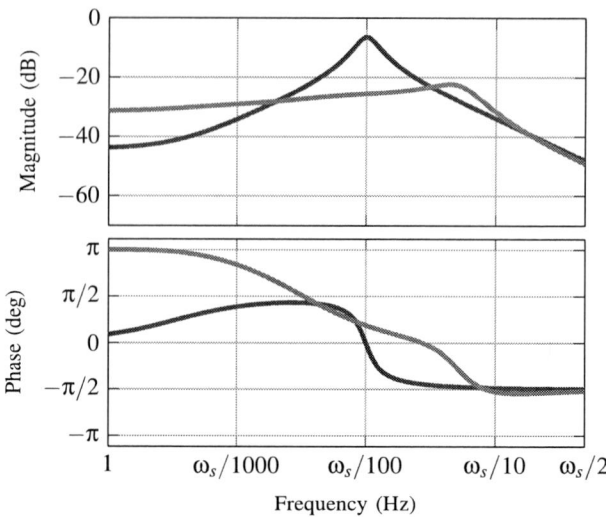

Fig. 6: Calculated equivalent dc grid admittances. DC collection network for offshore wind farms, $Y_{g1}(s)$ (blue), and for a drive working with DTC, $Y_{g2}(s)$ (red).

seen from the collector, can be calculated as follows:

$$Z_{g1}(s) = \frac{1}{Y_{WT}(s)+Y_C(s)} + Z_L(s) \qquad (10)$$

$$Y_{g1}(s) = \frac{i_{g1}(s)}{v_{dc}(s)} = \frac{1}{Z_{g1}(s)} \qquad (11)$$

where, $Y_C(s)$ is the admittance of the combination of the DAB filter capacitance (C_f) and parasitic line capacitance (C_L), and $Z_L(s)$ is the impedance of the parasitic line inductance.

The frequency response of $Y_{g1}(s)$ is shown in Fig. 6. As expected, at low frequencies, the admittance is dominated by the power control of the WT [$Y_{WT}(s)$], while the filter capacitance, together with the parasitic line parameters, create a resonance. The dc value of $Y_{g1}(s)$ is given the operating point of the WT (i.e. $P_2 = V_2 I_2$ in steady state operation), while for higher frequencies, beyond the controller bandwidth, the line parameters dominate the shape of $Y_{g1}(s)$.

B. MVDC Drive

The second case is based on a drive connected to a MVDC electrical distribution system for maritime applications. In this application, the stability of the electrical system can be compromise by high presence of high-power motor drives (propulsion motors can make up to 80% of the electrical load of the ship [6]), which generally show a CPL behaviour [18].

Vector controls of motors are used as the control of the propulsion drives [e.g. field-oriented control (FOC) and DTC]. The specific parameters used to derived the equivalent admittance of the propulsion drive, $Y_{g2}(s)$, are shown in Table I, however an analitycal expression for the equivalent input admittance has not been obtained. The system has been simulated using PLECS, and $Y_{g2}(s)$ has been measured.

1501

The 2018 International Power Electronics Conference

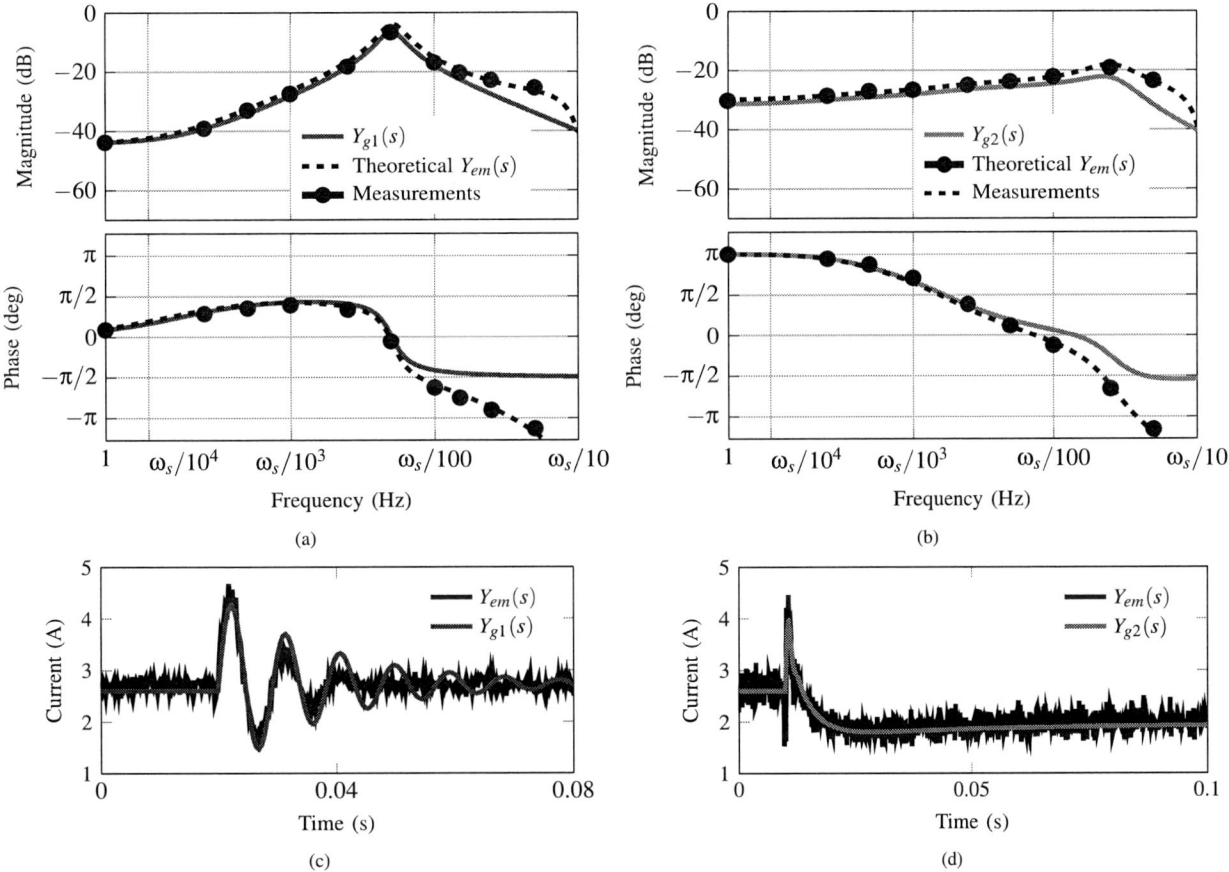

Fig. 7: Emulator admittance measurement tests. (a)-(b) Admittances measured in a specific set of frequencies. (c)-(d) Input voltage steps.

The frequency response of $Y_{g2}(s)$ is shown in Fig. 6. It can be seen that at low frequencies, inside the controller bandwidth, the admittance effectively shows a CPL behaviour.

IV. EXPERIMENTAL VERIFICATION

A prototype, with the configuration shown in Fig. 2, has been build for the experimental verification of the proposed active dc grid admittance emulator. The physical parameters of the experimental setup are shown in Table II.

The dc grid is created by a REGATRON TopCon TC.ACS, used as controllable dc source, that has also been employed to introduce a perturbation in the dc voltage, in order to measure the emulator admittance in a pre-defined set of frequencies. The dc grid admittances, shown in Fig. 6, have been implemented as the objective admittances [$Y_g(s)$] in the control structure, shown in Fig. 2. The results from the experimental verification are shown in Fig. 7, as it can be seen, the results are in a good agreement with the theoretical analysis and modelling.

As expected from Fig. 3, there is a magnitude amplification of the objective admittances for frequencies above $\omega_s/10$. For the emulation of most grid admittances this is not an issue,

since in practise, the magnitude of grid admittance naturally tends to decrease as the frequency increases. This magnitude amplification is created by the current reference estimation used in this work [$i_o^*(s) \approx i_s^*(s)/D$, see Fig. 2], since at high frequencies the term [$d(s)I_o/D$] start to be non-negligible.

The admittance of the emulator has been measured for an specific set of frequencies [see Figs 7(a) and 7(b)]. Since there is a maximum frequency at which the admittances can

Table II: Physical parameters used for the experimental validation.

Parameter	Value
Nominal Power (P_1)	1 kW
Nominal DC Grid Voltage (V_s)	400 V
Nominal Output Voltage (V_o)	200 V
Converter Inductor	$L = 14$ mH, $R = 1$ Ω
PI regulator	$132(1 + 71.5/s)$
Switching frequency	20 kHz
Theoretical closed loop bandwidth	$2\pi1500$ rad/s

1502

The 2018 International Power Electronics Conference

be measure with the available hardware, a second set of experiment have been performed. Figs 7(c) and 7(d) show the response of the emulator for a dc voltage step (v_s). It can be seen that the measured current responses well matches the theoretical ones.

V. CONCLUSION

In this work a novel method for the active emulation of dc grid admittances has been proposed. A modified current control structure, that effectively establishes a relation between the input voltage and current responses, defined by a given grid admittance, is shown. The buck-based emulator is able to effectively emulate the objective grid admittances, nevertheless it has been seen that the emulation performance is inherently limited by the switching frequency.

As the results have shown, the analytical expressions derived, for the emulator admittance's, accurately matches the measurements. Also, both admittances' phases and magnitudes are well emulated within the bandwidth of the current controller. The delays introduced by the inner current control loop limit the emulation capability to one tenth of the switching frequency approximately, where the phase and magnitude errors start to be significant.

REFERENCES

[1] E. Rodriguez-Diaz, J. C. Vasquez, and J. M. Guerrero, "Potential energy savings by using direct current for residential applications: A Danish household study case," in *2017 IEEE Second Int. Conf. DC Microgrids*. IEEE, jun 2017, pp. 547–552.

[2] A. T. Ghareeb, A. a. Mohamed, and O. a. Mohammed, "DC microgrids and distribution systems: An overview," *IEEE Power Energy Soc. Gen. Meet.*, vol. 119, pp. 407–417, 2013.

[3] F. Blaabjerg, A. Consoli, J. Ferreira, and J. VanWyk, "The Future of Electronic Power Processing and Conversion," *IEEE Trans. Power Electron.*, vol. 20, no. 3, pp. 715–720, may 2005.

[4] R. M. Cuzner and V. Singh, "Future Shipboard MVdc System Protection Requirements and Solid-State Protective Device Topological Tradeoffs," *IEEE J. Emerg. Sel. Top. Power Electron.*, vol. 5, no. 1, pp. 244–259, mar 2017.

[5] R. Soman, M. M. Steurer, T. A. Toshon, M. O. Faruque, and R. M. Cuzner, "Size and Weight Computation of MVDC Power Equipment in Architectures Developed Using the Smart Ship Systems Design Environment," *IEEE J. Emerg. Sel. Top. Power Electron.*, vol. 5, no. 1, pp. 40–50, mar 2017.

[6] U. Javaid, F. D. Freijedo, D. Dujic, and W. van der Merwe, "Dynamic Assessment of Source-Load Interactions in Marine MVDC Distribution," *IEEE Trans. Ind. Electron.*, vol. 64, no. 6, pp. 1–1, 2017.

[7] F. Gao, S. Bozhko, A. Costabeber, G. M. Asher, and P. W. Wheeler, "Control Design and Voltage Stability Analysis of a Droop-Controlled Electrical Power System for More Electric Aircraft," *IEEE Trans. Ind. Electron.*, vol. 0046, no. c, pp. 1–1, 2017.

[8] H. Yajima, K. Usui, T. Hayashi, Y. Ooi, Y. Yoshida, K. Hirose, and A. Tadahito, "Energy-saving effects of super computers by using onsite solar power and direct HVDC feeding systems," *INTELEC, Int. Telecommun. Energy Conf.*, vol. 2, pp. 31–34, 2016.

[9] B. Glasgo, I. L. Azevedo, and C. Hendrickson, "How much electricity can we save by using direct current circuits in homes? Understanding the potential for electricity savings and assessing feasibility of a transition towards DC powered buildings," *Appl. Energy*, vol. 180, pp. 66–75, oct 2016.

[10] E. Rodriguez-Diaz, F. Chen, J. C. Vasquez, J. M. Guerrero, R. Burgos, and D. Boroyevich, "Voltage-Level Selection of Future Two-Level LVdc Distribution Grids: A Compromise Between Grid Compatibiliy, Safety, and Efficiency," *IEEE Electrif. Mag.*, vol. 4, no. 2, pp. 20–28, jun 2016.

[11] M. Stieneker, Nurhan Rizqy Averous, N. Soltau, H. Stagge, and R. W. De Doncker, "Analysis of wind turbines connected to medium-voltage DC grids," in *2014 16th Eur. Conf. Power Electron. Appl.* IEEE, aug 2014, pp. 1–10.

[12] J. Robinson, D. Jovcic, and G. Joos, "Analysis and Design of an Offshore Wind Farm Using a MV DC Grid," *IEEE Trans. Power Deliv.*, vol. 25, no. 4, pp. 2164–2173, oct 2010.

[13] R. Adapa, "High-Wire Act: HVdc Technology: The State of the Art," *IEEE Power Energy Mag.*, vol. 10, no. 6, pp. 18–29, nov 2012.

[14] Z. Shuai, D. Liu, J. Shen, C. Tu, Y. Cheng, and A. Luo, "Series and Parallel Resonance Problem of Wideband Frequency Harmonic and Its Elimination Strategy," *IEEE Trans. Power Electron.*, vol. 29, no. 4, pp. 1941–1952, apr 2014.

[15] X. Wang, F. Blaabjerg, and W. Wu, "Modeling and analysis of harmonic stability in an AC power-electronics- based power system," *IEEE Trans. Power Electron.*, vol. 29, no. 12, pp. 6421–6432, 2014.

[16] M. Cespedes, L. Xing, and J. Sun, "Constant-Power Load System Stabilization by Passive Damping," *IEEE Trans. Power Electron.*, vol. 26, no. 7, pp. 1832–1836, jul 2011.

[17] L. Harnefors, R. Finger, X. Wang, H. Bai, and F. Blaabjerg, "VSC Input-Admittance Modeling and Analysis Above the Nyquist Frequency for Passivity-Based Stability Assessment," *IEEE Trans. Ind. Electron.*, vol. 64, no. 8, pp. 6362–6370, aug 2017.

[18] A. Emadi, A. Khaligh, C. H. Rivetta, and G. A. Williamson, "Constant power loads and negative impedance instability in automotive systems: Definition, modeling, stability, and control of power electronic converters and motor drives," *IEEE Trans. Veh. Technol.*, vol. 55, no. 4, pp. 1112–1125, 2006.

[19] L. Harnefors, M. Bongiorno, and S. Lundberg, "Input-Admittance Calculation and Shaping for Controlled Voltage-Source Converters," *IEEE Trans. Ind. Electron.*, vol. 54, no. 6, pp. 3323–3334, dec 2007.

[20] F. D. Freijedo, E. Rodriguez-Diaz, M. S. Golsorkhi, J. C. Vasquez, and J. M. Guerrero, "A Root-Locus Design Methodology Derived from the Impedance/Admittance Stability Formulation and Its Application for LCL Grid-Connected Converters in Wind Turbines," *IEEE Trans. Power Electron.*, vol. 8993, no. c, pp. 1–1, 2017.

[21] L. Harnefors, A. G. Yepes, A. Vidal, and J. Doval-Gandoy, "Passivity-Based Controller Design of Grid-Connected VSCs for Prevention of Electrical Resonance Instability," *IEEE Trans. Ind. Electron.*, vol. 62, no. 2, pp. 702–710, feb 2015.

[22] X. Zhang, X. Ruan, and C. K. Tse, "Impedance-Based Local Stability Criterion for DC Distributed Power Systems," *IEEE Trans. Circuits Syst. I Regul. Pap.*, vol. 62, no. 3, pp. 916–925, mar 2015.

[23] L. Schrittwieser, J. W. Kolar, and T. B. Soeiro, "99buck-type SiC MOSFET PFC rectifier minimizing life cycle cost in DC data centers," *INTELEC, Int. Telecommun. Energy Conf.*, vol. 2016-November, no. 1, pp. 47–58, 2016.

[24] J. P. Ram, H. Manghani, D. S. Pillai, T. S. Babu, M. Miyatake, and N. Rajasekar, "Analysis on solar PV emulators: A review," *Renew. Sustain. Energy Rev.*, vol. 81, no. July 2017, pp. 149–160, jan 2018.

[25] T. Baumhofer, W. Waag, and D. Sauer, "Specialized battery emulator for automotive electrical systems," in *2010 IEEE Veh. Power Propuls. Conf.* IEEE, sep 2010, pp. 1–4.

[26] S. Vogel, T. W. Rasmussen, and W. Z. El-Khatib, "Investigation of DC Collection Networks for Offshore Wind Farms," Technical University of Denmark, Tech. Rep. August, 2014.

[27] D. Jovcic, "Step-up DCDC converter for megawatt size applications," *IET Power Electron.*, vol. 2, no. 6, p. 675, 2009.

[28] Y.-H. Chen, C. Gabriel Dincan, R. J. Olsen, M.-C. Schimmelmann, P. Kjaer, and C. Leth Bak, "Study for Characterization of Electricla Propoerties of DC Collection System in Offshore Wind Farms," in *CIGRE 2016, B4-301 2016*, jul 2016, pp. 1–13.

[29] F. Deng and Z. Chen, "Operation and control of a DC-grid offshore wind farm under DC transmission system faults," *IEEE Trans. Power Deliv.*, vol. 28, no. 3, pp. 1356–1363, 2013.

A Compound Controller for Power Flow and Short-circuit Fault in DC Grid

Han Ye, Wu Chen*, Pengpeng Pan, Xiaokun He

Center for Advanced Power-Conversion Technology and Equipment, School of Electrical Engineering
Southeast University, Nanjing, China
*E-mail: chenwu@seu.edu.cn

Abstract-With the rapid development of HVDC transmission and the increasingly complex structure of DC networks, DC power flow control and short circuit fault problems become prominent. Based on the technology of interline direct current power flow controller and fault suppression, this paper proposes a new compound controller for power flow and short-circuit fault in DC grid. The device can not only realize the flexible adjustment of power between two lines, but also curb the short current and cut off the fault line in time. The operating principles and control strategy are introduced in a three-terminal meshed grid. The performance is validated by simulation in PLECS.

Keywords— HVDC transmission; power flow control; short-circuit fault; compound device

I. INTRODUCTION

Multi-terminal HVDC technology is an effective measure for renewable energy and one of the important directions of future power system [1]-[3]. With the continuous expansion of DC power grid, the further increase in the number of DC nodes and the increasingly complex grid structure, power flow control and DC fault will be more prominent. When power is not adjusted effectively, it may lead to the overload of some lines, and even breakdown of the entire DC grid [4]. In addition, due to the low resistance of DC grid, once short-circuit fault occurs, fault current rises rapidly in a short time. Meanwhile, the parallel capacitor on the DC side discharges rapidly. Large short-circuit current goes into the fault point, reaching several times or even hundred times of the rated current within a few milliseconds [5]-[6]. Besides, DC fault may not only damage converter station devices, but also spread to the entire system immediately, making the system out of work. Taking the technical level of fault location and DC circuit breaker into account, it is urgent to limit the short-circuit current effectively and remove the fault lines timely [7].

DC power flow is determined by DC line voltage and resistance, so it can be controlled by changing resistance and voltage. [8]-[9] proposed a variable resistor scheme. Variable resistance controller has the advantages of simple structure and flexible control, but it also has some problems such as large on-state loss, only one-way adjustment, limited range of adjustment. A DC

This work is supported by the State Key Laboratory of Advanced Power Transmission Technology Beijing China (Grant No. GEIRI-SKL-2017-010).

transformer in [10]-[11] can change the line voltage by controlling the duty cycle of the switch. It can regulate system power effectively and have the ability of fault isolation. In addition, the power flow controller has to withstand system-level power and voltage. Due to the low resistance of the DC grid, [12] proposed an adjustable voltage source structure based on the thyristors to achieve power flow regulation. The method has a large adjustment range, but it requires more devices and may inject harmonics into the AC system [13]-[14]. Interline power flow controller in [15]-[16] does not need external source. It can exchange power between the lines and can be equivalent to the voltage source stringed into the line. It has low cost and low loss, but it may introduce voltage ripple. As DC short-circuit fault current rises rapidly to a large peak value in a short time and has no natural zero-crossing, DC fault circuit breaker has become the main challenge of the development of DC grid. The existing DC circuit breakers are divided into three categories -- mechanical, solid-state and hybrid circuit breakers. Mechanical circuit breaker is generally the transformation of the traditional AC circuit breaker. It increases the oscillation circuit, making short-circuit current have zero-crossing. Solid-state breaker mainly uses the high power electronic device. Compared with the mechanical circuit breaker, it has the advantages of fast switching speed, short-circuit current inhibition, high reliability [17], but its on-state loss is large. Hybrid DC circuit breaker combines the advantages of mechanical circuit breaker and solid-state circuit breaker, making improvement on them. It has the advantages of low loss and fast breaking. ABB first proposed a hybrid DC circuit breaker topology and verified its feasibility [18].

At present, DC power flow controller and fault current limiting device can be used in grid, respectively. However, power flow control is relatively frequent, while the suppression and removal of DC fault only occurs in few moment, so it is not frequent. The existing devices have no capacity of both power flow control in steady state and transient fault current suppression. Based on the technology of power flow controller and DC circuit breaker, this paper proposes a novel compound device-power flow controlling DC circuit breaker (PFCCB), which combines power control with DC fault breaking. Compared with the previous independently operated equipment, this scheme can effectively reduce the number, volume, loss and cost of devices.

II. TOPOLOGY AND OPERATING PRINCIPLES

Based on the characteristics of interline power flow controller and hybrid DC circuit breaker, this paper presents a compound controller, as shown in Fig.1. It is composed of a pair of coupling inductors ($L_1 \sim L_2$), two capacitances ($C_1 \sim C_2$), four IGBTs ($Q_1 \sim Q_4$), four diodes ($D_{b1} \sim D_{b4}$), two fast mechanical switches ($UFD_1 \sim UFD_2$) and three fault-state switches ($MB_1 \sim MB_3$) and arresters ($SA_1 \sim SA_3$) parallel to them. The main functions of the compound device are power flow regulation of DC grid, short circuit current suppression and the removal of fault line.

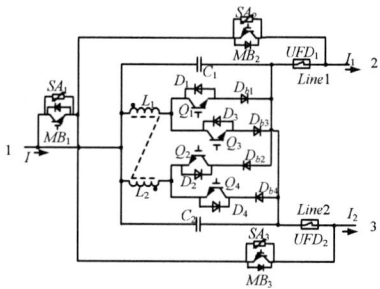

Fig. 1 Topology of the compound controllers for power flow and DC short-circuit fault

Taking three typical operating conditions as an example, the characteristics of the compound device for DC grid are analyzed (The direction in Fig. 1 is the positive reference direction of each electrical quantity).

Mode 1: I_1 and I_2 have the same direction

a) Power flow control: In the case that short circuit fault does not occur, the short circuit control part of system is locked, so $MB_1 \sim MB_3$ are turned off, while power flow control part is in work. Since both I_1 and I_2 are in positive direction, according to the symmetry of the system, take reducing I_1 and increasing I_2 to adjust the power flow as an example. It is equivalent to increase in the resistance of line1 and decrease in the resistance of line2. The voltage direction of the capacitor is consistent with the reference direction in Fig.1. Under normal regulation, C_1 is charged and C_2 discharges. In order to maintain the voltage balance in one cycle, the energy in C_1 needs to be transferred to C_2.

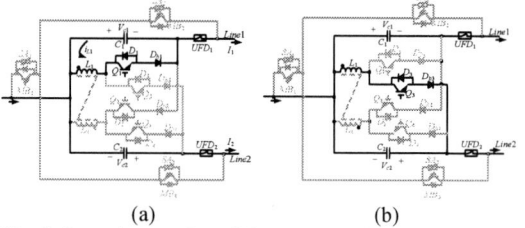

Fig. 2 Operation modes of the compound device with I1 and I2 the same direction. (a) Only Q_1 is on. (b) Only Q_3 is on

According to the energy transmission path, when Q_1 is on, Q_1, D_{b1}, C_1 and L_1 form a current loop, so energy in L_1 increases. After a period of time, turn off Q_1 and turn on Q_3, then L_1, Q_3, D_{b3} and C_2 form a current loop, so energy in L_1 is transferred to C_2. The specific switching mode is shown in Fig.2.

b) Short-circuit fault control: As direction of I_1 and I_2 is the same, according to the symmetry, take the short circuit at line1 as an example.

When line1 fails, turn on the switches Q_1 and Q_3 immediately and turn off the switches Q_2 and Q_4, then the branch with switches Q_1 and Q_3 are the commutation branch. Delay for a while, turn on the switches $MB_1 \sim MB_3$, then most current transfers to the branch with $MB_1 \sim MB_3$ switches. Delay for some time again and then turn off switch Q_1 on the fault branch, and turn off fast mechanical switch UFD_1 to cut off the fault line in zero current state. After mechanical switch is fully turned off, the switches $MB_1 \sim MB_3$ are turned off so that the residual fault current flows through the arrester SA_1 and the anti-parallel diode of the switch MB_2 and is consumed by them. The specific switching mode is shown in Fig.3. At this point, the fault branch has been removed from the system, and non-fault branches resume to operation.

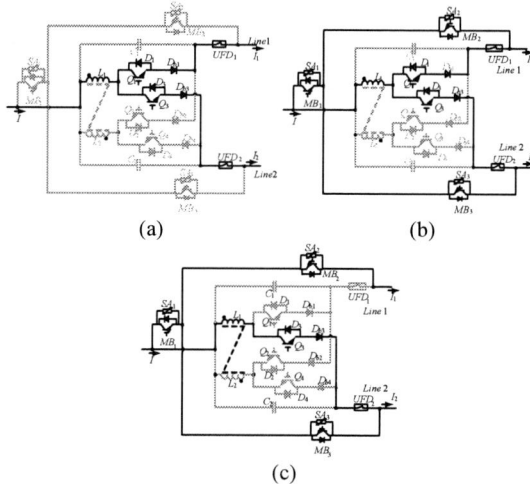

Fig. 3 Operation modes of the compound device with I_1 and I_2 the same direction. (a) Turn on Q_1 and Q_3. (b) Turn on $MB_1 \sim MB_3$. (c) Remove the fault line.

Mode 2: I_1 and I_2 have opposite directions

a) Power flow control. Due to the symmetry, take I_1 in positive direction, I_2 in negative direction, increasing I_1 and increasing I_2 to adjust the flow as an example. Then the resistance in line1 need to reduce and resistance in line2 need to increase. The direction of C_1 is contrary to Fig.1, C_2 is the same with Fig.1 and the energy of C_2 is transferred to C_1 in one cycle. According to the energy transmission path, turn on Q_4 and energy in L_2 will increase. Then turn off Q_4 and turn on Q_1. The energy will transfer from C_2 to C_1 by the coupling inductance. The specific switch mode shown in Fig.4.

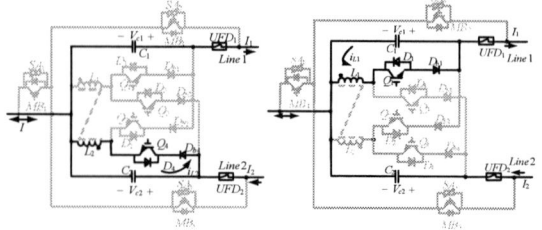

(a)   (b)

Fig. 4 Operation modes of the compound device with I_1 and I_2 opposite direction. (a) Only Q_4 is on. (b) Only Q_1 is on.

b) Short circuit control. Take I_1 in positive direction, I_2 in negative direction, and short circuit occurs at line1 as an example. When line1 fails, turn on Q_1 and Q_4 immediately and turn off Q_2 and Q_3. Then the branch with Q_1 and Q_3 is the fault commutation branch. Delay for a while, turn on the switches $MB_1 \sim MB_3$, then most of the fault current transfers to the branch with $MB_1 \sim MB_3$. After a delay, turn off Q_1 on the fault branch and then turn off the fast mechanical UFD_1 to cut off the fault line. Continue to delay for a period of time, and then turn off the switch $MB_1 \sim MB_3$, so that the residual fault current flows through the arrester SA_1 and the anti-parallel diode of the switch MB_2 and consumed by them. The specific switching mode is shown in Fig.5. By this time, the fault line has been removed and the system resumes to work.

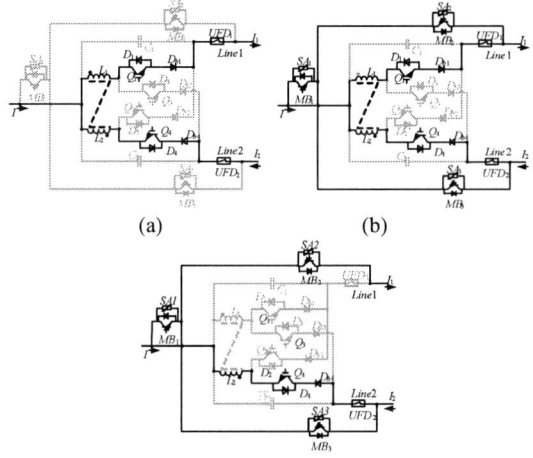

(a)   (b)

(c)

Fig. 5 Operation modes of the PFCCB with I_1 and I_2 opposite direction. (a) Turn on Q_1 and Q_3. (b) Turn on $MB_1 \sim MB_3$. (c) Remove the fault line

III. CONTROL STRATEGY

Based on above analysis, control strategy of the proposed PFCCB can be divided into two parts: power flow control in normal state and fault line isolation in short-circuit state.

TABLE I
OPERATING MODES OF THE PFCCB

I_1/I_2 direction	Requirements		Controlled switches				
	I_1	I_2	Q_1	Q_2	Q_3	Q_4	$MB_1 \sim MB_3$
P/ P	↑	↓	1	0	D	0	0
	↓	↑	D	0	1	0	0
N/ N	↑	↓	0	1	0	D	0
	↓	↑	0	D	0	1	0
P/ N	↑	↓	1	0	0	D	0
	↓	↑	D	0	0	1	0
N/ P	↑	↓	0	1	D	0	0
	↓	↑	0	D	1	0	0

In this table, P and N mean positive direction and

negative direction, respectively; D means duty cycle of the controlled switch.

Operating modes and their controlled switches are shown in Table I and operating timing in fault state is described in Fig. 6.

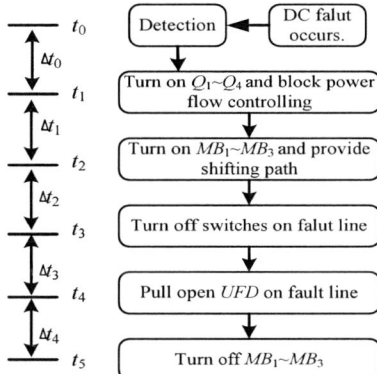

Fig.6 Operation timing of the PFCCB

IV. SIMULATION RESULTS

In this paper, the three-terminal system and its reference direction and parameters [19], [20] are shown in Fig.7. The line parameters are shown in Table II. The converter stations VSC1 and VSC2 operate in the constant power, P_1=300MW and P_2=120MW. The converter station VSC3 operates in constant voltage and the voltage U_3=200kV. The parameters of the transmission line are shown in Table II. The compound device is connected to the outlet side of the VSC3 converter station, which connects the line 2 and the line 3, that is, the capacitor C_1 is cascaded into the line 3, and the capacitor C_2 is connected to line 2. The PFCCB parameters are: C_1=1600μF，C_2=4000μF，L_1=L_2=80mH.

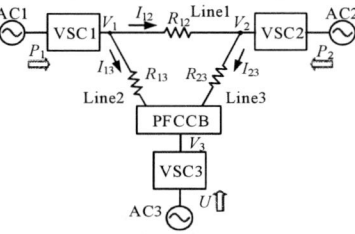

Fig.7 Three-terminal meshed DC grid

TABLE II
LINE PARAMETERS OF THE DC GRID

Transmission line parameters	Line1	Line2	Line3
Length/km	100	200	150
Resistance/Ω	1	2	1.5
Induction/mH	40	80	60

Next, this paper performs simulation verification mainly from the three aspects: power flow control, current limiting characteristics in case of fault, fault removal.

A. Normal power flow control

The compound device works in normal operation, controlling the line current I_{23}=0.25kA.

1506

According to the power flow equation of the DC system, and the power and voltage of three converter stations, the current and voltage can be calculated: $I_{12}=$ -0.33kA , $I_{13}=1.80$kA , $I_{23}=0.25$kA, , $V_1=204.76$kV , $V_2=205.26$kV , $V_{c1}=4.63$kV , $V_{c2}=0.64$ kV. Power flow control module of DC system begins to work at 4s, controlling the line current $I_{23}=0.25$kA, and the simulation results are shown in Fig.8. From Fig.8 (a) to (c), it can be seen that the simulation results are basically consistent with the theoretical analysis. The current I_{23} is controlled to the preset target 0.25kA in a short time, and the transient process is smoother. Converter stations are working in the normal voltage level.

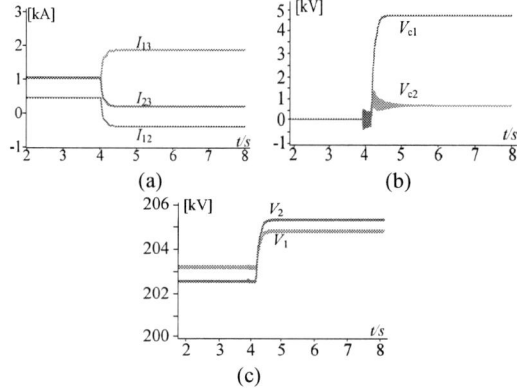

(a)   (b)

(c)

Fig.8 The waveforms in normal power flow control operation. (a) Line current waveforms. (b) Capacitor voltage waveforms. (c) Converter station voltage waveforms.

B. Fault line removal under short circuit fault

At the initial moment, the system is in normal working condition. Before 8s, the compound device works to control the line current $I_{23}=0.5$kA. The system operates stably. At 8s, a short circuit fault occurs on the line2, the short circuit controller puts into the system, and removes the fault line from the system. The simulation results are shown in Fig.9.

The low resistance of DC system makes the current I_{short} of short circuit point increase rapidly, and the capacitor voltage V_c on the fault line charge quickly in the event of a short circuit. It can be seen from Fig.9, the line current I_{13} increases rapidly. At t_1, the fault is detected and the current control part is blocked immediately so the fault current goes into the fault commutation branch and then commutation branch current I_{PFCCB} rises. At t_2, the device turn on the transmission branch switches MB_1 ~MB_3, and then the fault current goes into the MB branch, and the fault line capacitor voltage V_C began to decline to avoid overvoltage. At t_3, turn off the IGBTs on the fault branch, and cut off the residual current I_{PFCCB} of the fault line. At t_4, cut off the mechanical switch UFD in the zero current state, making the fault branch completely removed from the line. The compound device removes the faulty branch within a few milliseconds, and the non-faulty branches resume to normal operation after the removal of short circuit fault. From the V_1 waveform in the picture, it can be seen that, the system after the removal of a short circuit, the converter station quickly

control the voltage to 200kV and avoid the collapse of the DC system.

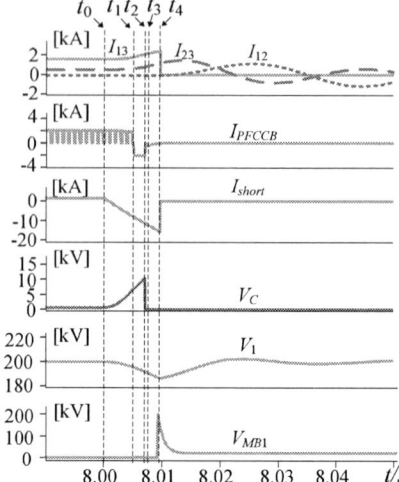

Fig.9 The waveforms in DC fault control operation.

C. Study on current limiting under short-circuit fault

When the short-circuit fault occurs and the system turns on the switch Q_1~Q_4 as the fault commutation branch. At this time the coupling inductance plays the role of limiting current. In the case of short-circuit fault, in order to study the effect of current limitation, comparative simulation about whether there is an inductance in the system is made, and the results are shown in Fig.10. It can be seen that the presence of the coupling inductance significantly slows the rising rate of the fault current. Unlike a separate inductor, the current through coupling inductance rises rapidly to a much lower current level than the final peak at first and then rises slowly, due to the effect of other windings' normal operating current. From the short period after the short circuit occurs in Fig.10 (b), the peak of the short-circuit current is significantly reduced by the presence of the coupling inductance within a few milliseconds of the fault removal process.

To study the current limiting characteristics and value selection of the coupling inductance in the compound device, simulation research for different coupling inductance values is carried out, with inductance 80mH, 400mH and 1600mH, respectively. The simulation results are shown in Fig 11. It can be seen from Fig.11 (a) that the rising speed of the short-circuit current slows down as the coupling inductance increases. The larger the inductance value is, the better the effect of suppressing the short-circuit current. However, the short-circuit current rising rate and peak value at different inductance values are close to each other within a few milliseconds, as shown in Fig. 11(b), when the compound device is in fault control mode. Besides, if coupling inductance value is too large, it will lead to worse dynamic performance of DC power flow control and the increase in the device volume. Therefore, it is advantageous for the improvement of the device performance to select a

The 2018 International Power Electronics Conference

relatively small inductance as far as possible in the case of satisfying the current limiting condition.

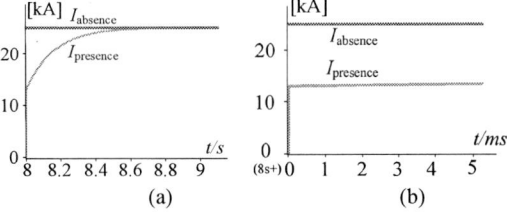

(a)   (b)

Fig. 10 The waveforms comparison between the presence and absence of inductor. (a) Within 1s after the short circuit occurs. (b) Within 5ms after the short circuit occurs.

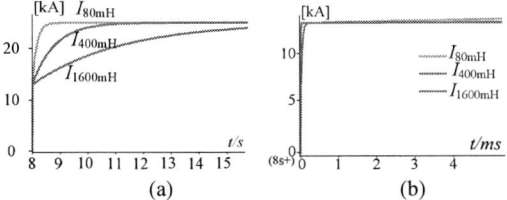

(a)   (b)

Fig.11 The waveforms comparison between different inductance. (a) Within 7s after the short circuit occurs. (b) Within 5ms after the short circuit occurs.

V. CONCLUSION

This paper presents a compound device with power flow control and short circuit control for DC grids. It can not only realize the effective regulation of the power flow in the DC power grid, but also can effectively suppress the rising rate and peak value of the short circuit current and remove short circuit fault immediately when a short circuit occurs in the DC power network. Compared with the traditional DC power flow controller and DC circuit breaker operating independently, the device effectively reduces the use of power electronic devices and the system costs.

REFERENCES

[1] Sun Wei, Yao Liangzhong, Li Yan. "Study on operation control strategies of DC grid with multi-voltage level considering large offshore wind farm grid integration". *Proceedings of the CSEE*, vol.35, no.4, pp.776–785,2015.

[2] Yao Liangzhong, Wu Jing, Wang Zhibing. "Pattern analysis of future HVDC grid development". *Proceedings of the CSEE*, vol.34, no.34, pp.6007–6020, 2014.

[3] Tang Guangfu, Luo Xiang, Wei Xiaoguang. "Multi-terminal HVDC and DC-grid technology". *Proceedings of the CSEE*, vol.33, no.10, pp.8–17, 2013.

[4] Jovcic D, Hajian M, Zhang H. "Power flow control in DC transmission grids using mechanical and semiconductor based DC/DC devices." //*Proceedings of the 10th IET International Conference on AC and DC Power Transmission (ACDC)*. Birmingham, UK: IEEE, pp. 1-6, 2012.

[5] Chang B, Cwikowski O, Barnes M. "Point-to-point two-level converter system faults analysis." //*7th IET International Conference on Power Electronics, Machines and Drives*. Manchester, UK: IET, pp.1-6, 2014

[6] Yang J, Fletcher J E, O'Reilly J. "Short-circuit and ground fault analyses and location in VSC-based dc network cables." *IEEE Transactions on Industrial Electronics*, vol.59, no.10, pp.3827-3837, 2012.

[7] Zhang Xuena, Zhao Chengyong, Pang Hui. "A control and protection scheme of multi-terminal DC transmission system based on MMC for DC line fault." *Automation of Electric Power Systems*, vol.37, no.15, pp.140-145, 2013.

[8] Q. Mu, J. Liang, Y. Li, X. Zhou. "Power flow control devices in DC grids." *IEEE Power and Energy Society General Meeting*, 2012.

[9] K. Rouzbehi, A. Miranian, A. Luna. "Towards fully controllable multi-terminal DC grids using flexible DC transmission systems." *IEEE Energy Conversion Congress and Exposition (ECCE)*, pp.5312–5316, 2014.

[10] K. Natori, H. Obara, K. Yoshikawa. "Flexible power flow control for next-generation multi-terminal DC power network." *IEEE Energy Conversion Congress and Exposition (ECCE)*, pp. 778–784, 2014.

[11] B. P. Kumar, J. Tomas, M. Subhasish. "DC-DC converter based load flow control in HVDC grids." PCT WO 2012/037957 A1.

[12] E. Veilleux, B. T. Ooi. "Multiterminal HVDC with thyristor power-flow controller." *IEEE Transactions on Power Delivery*, vol.27, no.3, pp.1205–1212, 2012.

[13] B. P. Kumar, M. Subhasish, J. Tomas. "Series-connected dc/dc converter for controlling the power flow in a HVDC power transmission system." PCT WO 2012/037964 A1, 2012.

[14] S. Balasubramaniam, J. Liang, C. Ugalde-Loo. "An IGBT based series power flow controller for multi-terminal HVDC transmission." *International Universities' Power Engineering Conference (UPEC)*, pp.1–6, 2014.

[15] C. D. Barker, R. S. Whitehouse. "A current flow controller for use in HVDC grids." *10th IET International Conference on AC and DC Power Transmission (ACDC)*, pp.1–5, 2012.

[16] R. S. Whitehouse, C. D. Barker. Current flow controller, PCT EP 2013/061313 A1.

[17] P Zhendong, R Zhigang, J Nan, Yang Chenguang. "Design and analysis for a novel DC solid state current limiting circuit breaker." *Proceedings of the CSEE*, vol.37, no.4, pp.1028-1037, 2017.

[18] Callavik M, Blomberg A, Häfner J. "The hybrid HVDC breaker." *ABB Grid Systems Technical Paper*, 2012.

[19] Chen W, Zhu X, Yao L. "An interline DC power-flow controller (IDCPFC) for multiterminal HVDC system." *IEEE Transactions on Power Delivery*, vol.30, no.4, pp.2027-2036, 2015.

[20] Chen W, Zhu X, Yao L. "A Novel Interline DC Power Flow Controller (IDCPFC) for Meshed HVDC Grids." *IEEE Transactions on Power Delivery*, vol.31, no.4, pp. 1719-1727, 2016.

1508

Design Procedure and Control of a Hybrid Circuit Breaker with Adaptable Pulse Current Injection

Andreas Jehle and Jürgen Biela

Laboratory for High Power Electronic Systems, ETH Zürich

Email:jehle@hpe.ee.ethz.ch, URL:http://www.hpe.ee.ethz.ch

Abstract—Hybrid circuit breaker are key components for high voltage DC transmission as they are the only possibility to avoid a complete shutdown of a DC grid in case of faults. In this paper, the focus is on the design of a new hybrid circuit breaker with improved current injection, which allows a fast and reliable arc extinction in the mechanical switch that conducts the line current under normal operation. After presenting the turn-off procedure, the design of the switching modules and the optimization of the pulse circuit is shown. Additionally, a reliable control is presented, which adapts the pulse current to the line current and ensures a successful current interruption.

I. INTRODUCTION

One of the major remaining problems of HVDC systems is to turn lines quickly off in case of a fault [1], where e.g. the semiconductors in VSC can be destroyed due to the fast increasing currents. Besides turning off the complete DC grid in case of a fault [2], DC circuit breaker can be used to disconnect only the faulty part of a (multi-terminal) grid. Especially hybrid circuit breakers (HCB) are an interesting option, since HCB combine a mechanical circuit breaker (MCB) with a power electronic circuit enabling a fast turn-off, while having low on state losses [3], [4]. For disconnecting a faulty line, the HCB must fulfill three basic requirements [1]:

1) Generate a Zero Current Condition (ZCC) in the MCB
2) Dissipate the stored energy in the (dis-)connected lines
3) Withstand the system voltage

For generating a ZCC in the MCB, several concepts have been proposed in the past. A concept is for example to commutate the current with semiconductor devices from the MCB to a parallel auxiliary current branch and open the MCB without current [5]–[7]. However, the semiconductor devices for commutating the current generate additional on-state losses since they are in series to the MCB. Alternatively, the MCB can be opened under current resulting in an arc, which is extinguished by a ZCC generated with an increasing resonant current [8] or an injected counter-current [9]–[11]. However, a successful arc extinction in the MCB without reignition of the arc depends on several parameters as e.g. the gap distance at the ZCC [12], the arc duration [13], the dv/dt across the MCB shortly after the arc extinction and the di/dt of the current in the MCB shortly before the arc extinction [14]–[16]. While the gap distance and the arc duration depend on the MCB design and the opening duration of the MCB, the di/dt and dv/dt significantly depend on the power electronic circuit of

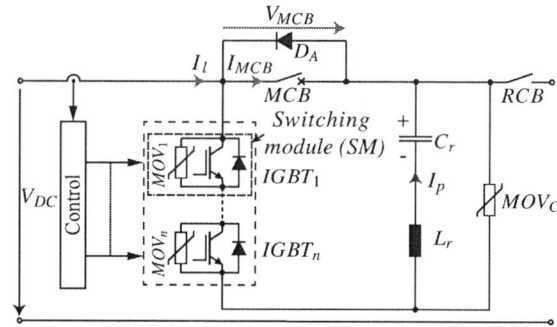

Fig. 1. Single pulse HCB with adaptable pulse current (HCB-APC): By changing the number of turned off IGBTs, the number of varistors in the pulse current path is changed in order to adapt the pulse current I_p to the line current I_l.

the HCB. Depending on the MCB, the HCB must generate a maximum di/dt between approximately $20A/\mu s$ and $200A/\mu s$ [13], [15] for a successful arc extinction.

However, most HCB concepts with current injection do generally not enable a significant control of the injected pulse current and a low di/dt and dv/dt for all possible line currents can only be achieved with large passive components. Therefore, a new concept for a HCB with adaptable pulse current (HCB-APC) has been proposed in [11], which is able to achieve the required di/dt and dv/dt with relatively small passive components and a low number of semiconductors.

While in [11] only the basic concept has been shown, this paper presents the design and control of the HCB-APC for $400kV$ and $12kA$. With the optimization of the HCB presented in this paper, the HCB-APC requires only passive components with a relatively small volume for a reliable turn-off due to the adaption of the pulse current. Compared to controls of other HCB concepts, the considered control not only opens the MCB and triggers a pulse current, but also adapts the pulse current to the fault current amplitude.

First, the topology and the operation principle of the HCB-APC are explained in section II. Thereafter, the influence of the grid on the HCB design is explained in section III-A. In section III-B, the focus is on designing the IGBTs and varistors, which are used to control the pulse current. The optimization of the pulse circuit with its trade-off between the passive components and the number of semiconductors is shown in section III-C. Finally, the control structure and the required measurements are explained in the sections III-D and III-E. The focus in this case is especially on the reliable operation.

The 2018 International Power Electronics Conference

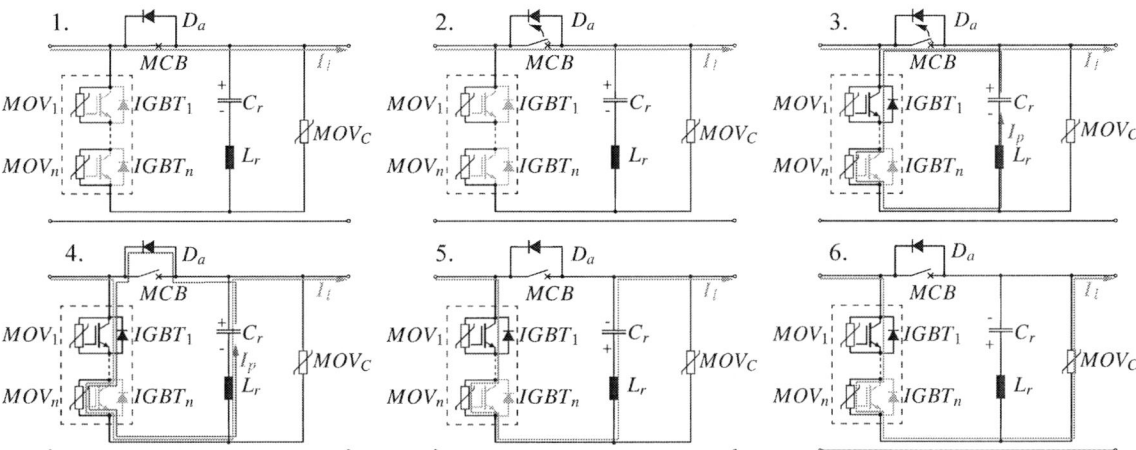

Fig. 2. Operation principle of the HCB-APC divided in 6 steps: 1) The fault is detected by the HCB. 2) The MCB is opened resulting in an arc. 3) A pulse current is injected in the MCB to generate the ZCC. 4) Diode D_A conducts and prohibits the ITIV. 5) The TIV is increased. 6) The remaining energy in the lines is dissipated.

II. OPERATION PRINCIPLE OF THE PROPOSED HCB-TOPOLOGY

In this section, the operation principle of the HCB-APC shown in Fig.1 is described. To generate the pulse current, a series connection of a precharged capacitor C_r, an inductor L_r and switching modules (SM) is used [11]. Each SM consists of an IGBT and a parallel varistor. The turn-off procedure of the HCB in case of a fault, can be divided into 6 steps (Fig. 2):

- Fault detection (Fig. 2-1): First the line current through the MCB increases until the fault is detected.

- MCB opening (Fig. 2-2): After the fault is detected, the MCB is opened resulting in an arc, while the line current increases further.

- Interruption (Fig. 2-3): Shortly before the MCB is completely open, a pulse current is triggered by turning on a number of IGBTs depending on the line current to generate a ZCC to extinguish the arc. The ideal timing is that the current in the MCB becomes zero when the MCB just has reached its maximum contact distance / is fully open.
  To generate a low di/dt before the arc extinction, the pulse current is adapted to the line current. This is achieved by adapting the number of turned off IGBTs depending on the fault current, respectively by adapting the number of MOVs in the $C_r L_r$ circuit. The general strategy is to increase the pulse current fast at the beginning by including only a small number of MOVs and then to achieve a pulse current with a low di/dt shortly before the peak pulse current by including additional MOVs.

- Conduction of diode D_A (Fig. 2-4): After the arc is extinguished, diode D_A starts to conduct and prohibits a negative initial transient interruption voltage (ITIV) across the MCB until the pulse current becomes lower than the line current.

- TIV increase (Fig. 2-5): After diode D_A blocks, the MCB blocks a current in the MCB branch and the line current continues to charge capacitor C_r while the MCB regains its full blocking capability. During this and the next step, the transient interruption voltage (TIV) across the MCB can be influenced by inserting or removing MOVs in the line current path by turning some IGBTs on or off. This allows to share the TIV between capacitor C_r and the series connection of IGBTs so that neither of them must be able to block the full TIV.

- Energy dissipation (Fig. 2-6): As soon as the voltage across the capacitor C_r is high enough to commutate the line current to MOV_C, the energy of the lines is dissipated in MOV_C and varistors $MOV_1 - MOV_n$ (depending on turned on IGBTs).

In Fig. 4, the waveforms for a turn-off at a line current of $10.7kA$ is exemplary shown. After an assumed detection time of $2ms$, the MCB is opened within 16ms. The pulse current is generated by turning first a high number of IGBTs on and after a short time several IGBTs are turned off. For the shown $10.7kA$, the optimization results in a turn-on of all but one IGBTs and a turn-off of 9 additional IGBTs after $20\mu s$. Due to the oscillations of the grid current, the ZCC occurs shortly after the MCB is completely open.

III. DESIGN OF THE HCB

In the following, the design of the HCB is shown for the four terminal grid with $V_{DC} = 400kV$ presented in [17] (Fig. 3) as a benchmark case. In the first part (section III-A, III-B and III-C), the hardware of the turn-off procedure is designed in three main steps (Fig. 5). First, the impact of the grid on the HCB design is evaluated. In the second step, the SMs are designed. In the third step, the pulse circuit ($C_r L_r$-circuit)

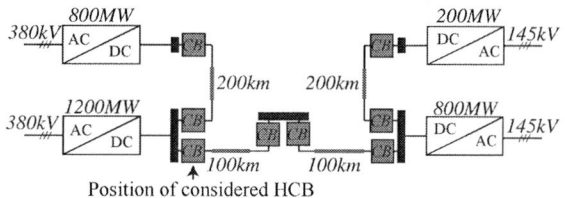

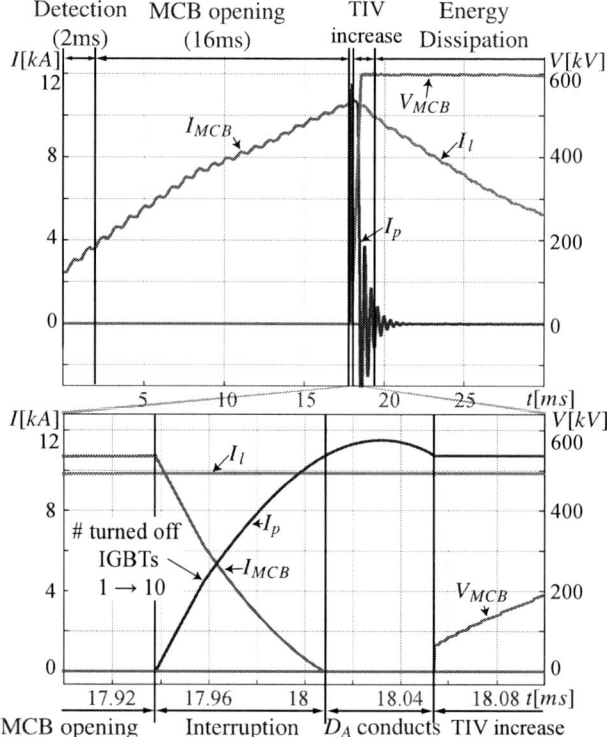

Fig. 3. Exemplary block diagram of a 4-terminal symmetric monopolar DC grid [17] including cables (green) and overhead lines (red)(OHL), which is used as case study for designing the HCB.

Fig. 4. The MCB I_{MCB}, the pulse I_p and the line current I_l as well as the MCB voltage V_{MCB} are shown during the turn-off process for a fault at $t = 0ms$ for: $V_{DC} = 400kV$, $I_{fault,max} = 12kA$, $t_{det} = 2ms$, $t_{open} = 16ms$. The upper figure shows the complete turn-off process, while the lower shows the pulse generation in detail. The pulse current is generated by turning first all but one IGBTs on and after $20\mu s$ 9 additional IGBTs off.

is optimized. The second part presents the control of the HCB. The general control strategy and the possible adaptations during the pulse current generation to increase the robustness are presented in section III-D and III-E. In the third part, the charging system of the HCB is presented (section III-F).

A. Impact of the grid

The specifications, the configuration (symmetric monopolar with high impedance grounding at each converter station) and the components (MMC, current limiting inductors, cables and overhead lines (OHL)) of the grid determine the basic design parameters of the HCB.

First of all, the DC operating voltage $V_{DC} = 400kV$ deter-

mines the maximum blocking voltage across the HCB and therefore also across the MCB $V_{HCB,max} = V_{MCB,max}$. Generally, the typical maximum blocking voltage is chosen to be 50% higher than the DC voltage [18]. This already limits the choice of the MCB to gas circuit breaker, since vacuum circuit breaker are limited to relatively low voltages [16]. However, the opening time of the MCB increases with the required contact distance, which results for gas circuit breaker in relatively long opening times even if multiple MCBs are series connected. During this opening time and the detection time, the current through the MCB increases. Therefore, current limiting reactors must be installed in series to the HCB to limit the maximum current [18]. For the exemplary design of the HCB, the SF6-MCB presented in [15] is assumed. To block the maximum voltage across the HCB $V_{HCB,max} = 600kV$, four MCBs are connected in series. Additionally, grading capacitors are connected in parallel to the MCB to provide an equal voltage distribution.

Since the opening time of 16ms is relatively long, current limiting reactors $L_{lim} = 300mH$ are used [19], which limit the maximum line current $I_{l,max}$ in case of a fault. While $I_{l,max}$ can be approximated for a constant DC voltage by $I_{l,max} = \frac{V_{DC}}{L_{lim}}(t_{open} + t_{det})$, $I_{l,max}$ is hard to predict in a grid with converter since multiple sources with varying voltage supply the increasing line current. Therefore, $I_{l,max}$ is predicted numerically with simulations ($I_{l,max} = 11.8kA$). With the simulations also the maximum change of the line current amplitude $\Delta I_{f,max}(\Delta t)$ in a short time interval Δt and the maximum energy to dissipate E_{dis} is determined. $\Delta I_{f,max}(\Delta t)$ is used to predict the possible line current trajectories for the control. E_{dis} defines the required energy dissipation capability of the HCB and depends on the position of the HCB in the grid and therefore varies for each HCB in the grid [18]. For the considered design, a $E_{dis} = 97MJ$ is determined (Fig. 3).

B. Switching module (SM) design

The design of the SMs and the modeling of the varistors are an important aspect of the design, since these have a major impact on the pulse current. Three SM parameters are important for the design: The maximum SM current $I_{mod,max}$, the maximum blocking voltage $V_{mod,max}$ and the maximum energy $E_{mod,max}$, which the varistors of a SM must dissipate. The maximum SM current and therefore the maximum IGBT current is defined by the maximum pulse current, which is equal to the maximum line current $I_{mod,max} = I_{p,max} = I_{f,max}$. Although a single varistor can conduct the current $I_{mod,max}$, $N_{MOV,par}$ varistors must be parallelized to dissipate $E_{mod,max} = N_{MOV,par}E_{MOV,max}$. Additionally, the current sharing between the varistors vary due to tolerances of the varistors so that only approximately 70% of the energy dissipation capability of the varistors can be used [20]. Therefore, the number of varistors must be increased by a factor 1.43. For $E_{dis} = 97MJ$ this results in 24 parallel connected varistors with each an energy dissipation capability of $9.7kJ/kV_{VC}$.

The maximum voltage across the SMs is equal to the maximum varistor voltage $V_{MOV,max} = V_{mod,max}$, which depends on

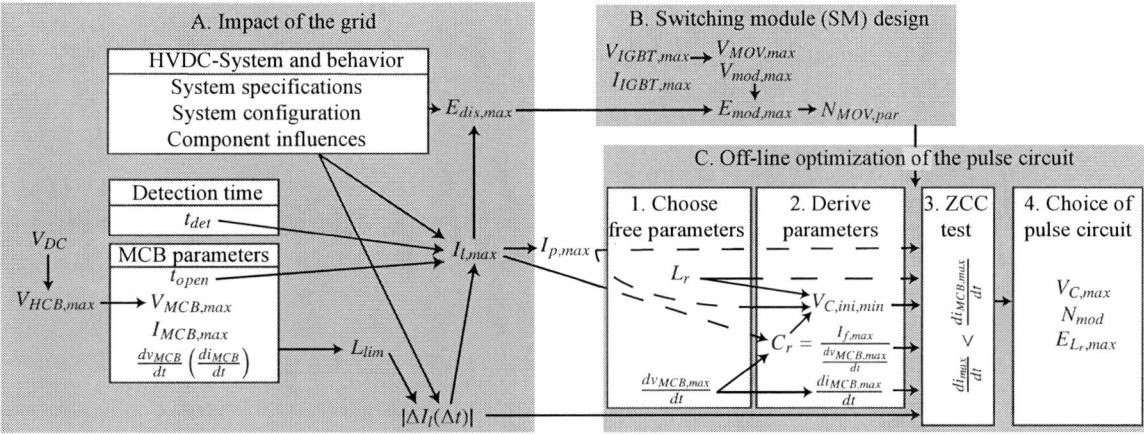

Fig. 5. Design and optimization procedure for the HCB: In a first step, the impact of the grid on the design of the HCB is determined. The MCB opening time and the detection time determine with the grid and the current limiting inductance L_{lim} the maximum line current in case of a fault $I_{l,max}$ and the energy $E_{dis,max}$, which must be dissipated in the HCB varistors. Additionally, the maximum line current change in a time interval $\Delta I_l(\Delta t)$ is determined. In a second step, the SM design is determined. While the maximum SM voltage and current is limited by the available IGBTs, the number of varistors depend on the energy to dissipate. In a third step, the pulse circuit is optimized. By varying inductance L_r and the allowed $\frac{dv}{dt}$ across the MCB after the arc extinction, a set of parameters for the pulse circuit can be found, which generate the required low di/dt.

the number of parallel connected varistors $N_{MOV,par}$. In order to protect the IGBTs from overvoltages, the maximum varistor voltage must be below the maximum IGBT voltage $V_{MOV,max} < V_{IGBT,max}$. This limits the maximum SM voltage $V_{mod,max}$ for $4.5kV$ IGBTs to values below $2.8kV$. Besides the impact of the tolerances on the number of varistors per SM, the tolerances of the varistors and IGBTs also influence the generated pulse current. These tolerances must be taken into account during the optimization, since they must be compensated by increased capacitance and inductance values. The conventional model for varistors using a piecewise linear resistor is sufficient for a first rough design [21], but neglects the hysteresis and the influence of the varistor inductance, which must be included in a detailed design [22]. The hysteresis differs by several hundreds of volts, which can be modeled with a voltage source depending on rise time and dissipated energy [23]. The inductance is mainly important for determining the maximum voltage across the IGBT during the current commutation from the IGBT to the varistors and is otherwise only added to the inductor inductance. Beside the mentioned effects, in the optimization additionally a total tolerance of 5% is utilized for the different tolerances of the varistor [20].

For the IGBTs, the turn-on and turn-off delay times must be compensated by adapting the timing of the switching signals. The delay times depend on the switched current and must therefore be estimated to some degree depending on the ideal pulse current (without tolerances). However, the variations in the delay times change the current amplitude, especially when the number of turned off IGBTs changes, and is therefore as additional jitter taken into account.

C. Off-line optimization of the pulse circuit ($C_r L_r$-circuit)

The general design of the pulse circuit is a compromise, which aims for:

- Minimal stored energy in inductor L_r
- Minimal stored energy in capacitor C_r
- Low number of SMs
- Low di/dt in the MCB (depends on the TIV after the arc extinction)

To generate a ZCC, the maximum pulse current $I_{p,max}$ must be higher than the maximum line current $I_{l,max} = 11.8kA$ (section III-A). Therefore, a maximum required pulse current $I_{p,max} = 12kA$ is chosen. The optimization of the pulse circuit is basically done in four steps (Fig.5 Part C):

1) Choose free parameters (Fig.5 Part C 1): Two parameters can be varied and lead to different pulse circuits. The first parameter is the maximum dv/dt across the MCB after the arc extinction. The second parameter is the inductance value of L_r.

2) Derive parameters (Fig.5 Part C 2): In the second step, additional values are derived. The minimum capacitance value $C_r = I_{l,max}/\frac{dv}{dt}_{max}$ depends on the maximum dv/dt across the MCB after the arc extinction. This value is generally used as capacitance value for the pulse circuit to minimize the stored capacitive energy. Higher capacitance values are only used if the initially stored energy is not high enough to generate the required pulse current amplitude. Additionally, the maximum dv/dt across the MCB after the arc extinction defines the maximum allowed di/dt in the MCB for a successful arc extinction. A high di/dt results in a low maximum dv/dt and requires a high capacitance value. Therefore, the highest dv/dt is chosen for which the required low di/dt can be generated, to achieve a low stored energy in the capacitance.

The inductance value in combination with the pulse current and the capacitance value defines the minimum initial capacitor voltage $V_{C,ini,min} = I_{p,max}\sqrt{\frac{L_r}{C_r}}$. However, the actually required minimum initial capacitor voltage, which results from the optimizations, is slightly higher due to the component tolerances.

3) ZCC test (Fig.5 Part C 3): In the third step, it is checked if the chosen free and derived parameters are suitable to generate a ZCC for all possible line current values.

a) Pulse current: First, the pulse currents for all switching times t_2 (in steps of $1\mu s$) and numbers of turned off IGBTs (N_1, N_2) are calculated. Due to the deviations of the initial capacitor voltage $V_{C,ini} \pm \Delta V_{C,ini}$, tolerances in the SMs and component tolerances (assumed $L_{tol} = 5\%$, $C_{tol} = 10\%$, the generated pulse currents can differ from the calculated pulse current waveforms. Also, the arc voltage of the gas circuit breaker, which is relatively high compared to the arc voltage of vacuum circuit breakers, influences the pulse current. Thus, a conservative value for the arc voltage between $0V$ and $6kV$ is assumed. Another possible source for deviations is the possibility that an IGBT fails.

In order to account for all these effects, maximum and minimum pulse current envelopes ($i_{p,min}$ and $i_{p,max}$) are calculated apart from the ideal pulse currents ($i_{p,ideal}$). Fig. 6a) shows one exemplary ideal pulse current $i_{p,ideal}$ and its pulse current envelopes ($i_{p,min}$ and $i_{p,max}$) for $t_2 = 20\mu s$, $N_1 = 1$ and $N_2 = 9$.

In addition to the currents, also the maximum di/dt of the pulse currents plus the maximum fault current di/dt are calculated, which is equal to the maximum MCB current slope. Fig. 6b) shows the maximum MCB current slope for the pulse current of Fig. 6a).

b) For checking if a ZCC is achieved, also the fault current is required. Therefore, for all fault current measurements I_l with current measurement resolution $\pm\Delta I_l/2$ the possible fault current ranges are evaluated. The procedure is exemplary shown in Fig. 6c) for a single fault current measured $70\mu s$ before the MCB is open and the pulse current shown in Fig. 6a). During the generation of the pulse current, the fault current is changing. Therefore, the fault current range is increasing for the time Δt after the current measurement.

c) Two important ZCC exist. The earliest possible time instance at which a ZCC can occur $t_{ZCC,min}$ is when the maximum pulse current envelope $i_{p,max}$ is equal to the minimum of the fault current range $I_{l,70} - \Delta I_l/2 - \Delta I_{f,max}(\Delta t)$. This defines the maximum time t_1 at which the pulse current can be triggered before the MCB is completely open. Additionally, the minimum pulse current $i_{p,min}$ at a time instance $t_{ZCC,max}$ must be equal to the maximum of the fault current range $I_{l,0} + \Delta I_l/2 + \Delta I_{f,max}(\Delta t)$ so that certainly a ZCC exists. In between of these two time instances ($t_{ZCC,min}$ and $t_{ZCC,max}$), the

maximum MCB current slope must be below the maximum allowed di/dt (Fig. 6b) for a successful arc extinction.

If several pulse currents exist, which fulfill this requirement, the one with the lowest di/dt is chosen.

The steps b) and c) are repeated for different measurement times (in $5\mu s$ steps before the MCB is open). If for all fault currents pulse currents exist, which fulfill these requirements, the chosen free parameters of step 1) can be used for the pulse circuit.

4) Choice of pulse circuit (Fig.5 Part C 4): In the fourth step, the optimum pulse circuit of the third step is chosen and the number of SMs (and IGBTs), respectively the maximum voltage of the capacitor during the line current interruption, are determined. For the benchmark case an optimum is found for an inductance value of $L_r = 400\mu H$, a capacitance value of $C_r = 10\mu H$ and a minimum capacitor voltage $V_{C,ini} = 80kV$, which allows to generate a maximum di/dt below $114\frac{A}{\mu s}$.

The number of SMs and the maximum voltage of the capacitor depend on each other, since after the arc extinction, the IGBTs block together with the MOV_C the TIV. Therefore, this is a trade-off between the stored energy in the capacitor and the number of SMs. The minimum for both is determined by the minimum initial capacitor voltage $V_{C,ini,min}$, which is required for the pulse current. For the benchmark case, the number of IGBTs is minimized. With a continuous blocking voltage of $1kV$ per SM ($2.5kV$ at peak current), 80 series connected SMs are required to block the minimum initial capacitor voltage. However, the $V_{C,ini,min} = 80kV$ represents the minimum voltage for a successful ZCC and for the passive charging circuit presented in section III-F a higher number of series connected SMs are required to decrease the charging time.

D. On-line control of the HCB

The structure of the HCB with the control and measurement signals is shown in Fig. 7 and the control procedure is shown in Fig. 8. The HCB is opened either if the measured line current is higher than a preset current limit ($I_l > I_{DC,max}$) or if an external detection circuit sends an opening signal. After the detection of the fault, the MCB is opened. While the MCB opens, the line current is measured.

Shortly before the MCB is completely open, a pulse current is generated to extinguish the arc as soon as the MCB is open after the duration t_{open}. The pulse current is generated with two switching steps as described in section II. The switching times (t_1/t_2) and the number of turned off IGBTs/active varistors (N_1/N_2) in the pulse circuit are taken from a look-up table. The data in the look-up table is generated off-line within the optimization (section III-C). The look-up table lists the switching times and number of IGBTs for different measured fault currents $I_{l,meas}$ and initial capacitor voltages $V_{c,meas}$. With these values a pulse current is generated that

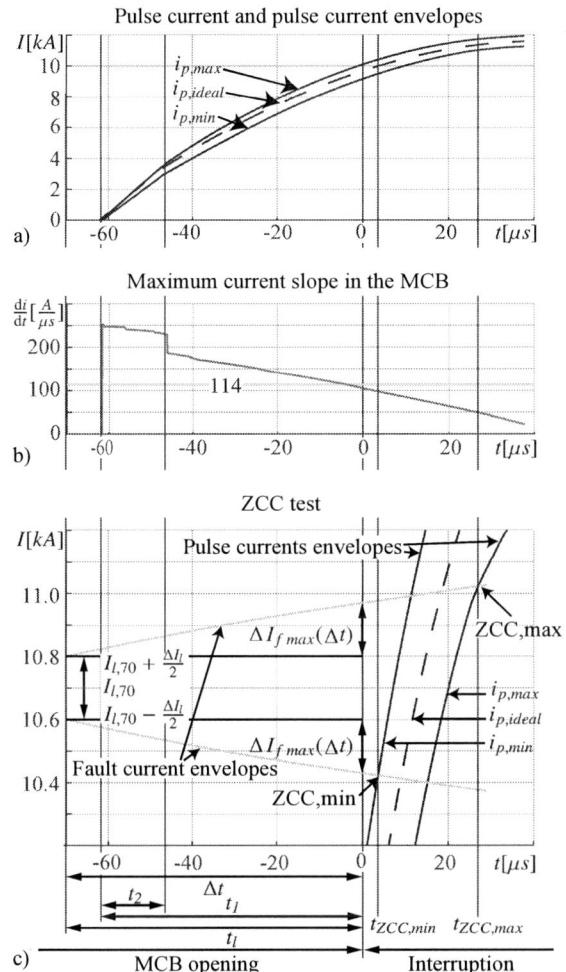

a) *Pulse current and pulse current envelopes*

b) *Maximum current slope in the MCB*

c) *ZCC test*

Fig. 6. Exemplary pulse current used for a line current $I_l = 10.7kV$ measured $70\mu s$ before the MCB is completely open: a) The ideal pulse current and its envelopes due to turning on all but one IGBT $62.3\mu s$ before the MCB is completely open ($t = 0$) and turning at $t = -42.3\mu s$ 9 additional IGBTs off $42.3\mu s$. b) di/dt of the MCB current. c) The ZCC of the current in the MCB is between two times instances ($t_{ZCC,min}$ and $t_{ZCC,max}$), which mark the time instances when the maximum pulse current envelope is equal to the minimum of the fault current range and when the minimum pulse current envelope is equal to the maximum of the fault current range. In between these time instances, the MCB current slope must be below the maximum allowed $di/dt = 114A/\mu s$.

extinguishes the arc in t_l with a low di/dt.

However, the tolerances of the components and measurement values result in some deviation from the ideal pulse current. Although some deviation can be tolerated and is included in the optimization, higher deviations would generate a too high di/dt or do not generate a ZCC. Therefore, the pulse current is continuously measured and compared to the ideal pulse current slope $I_{ideal}(N_1, N_2, t_1, t_2)$. As soon as the deviation is too high, an additional IGBT is turned on or off.

After the ZCC, diode D_A starts to conduct until the line current is equal to the current in the pulse circuit. The line current charges capacitor C_r and increases the voltage across

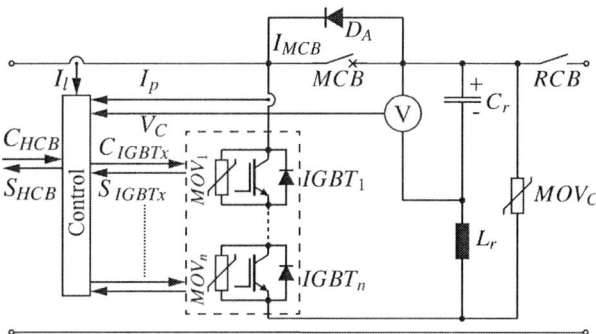

Fig. 7. Control structure of the HCB: With the line current measurement I_l the fault is detected and tracked to predict the required pulse current. For generating the pulse current, the IGBTs are controlled with the signals C_{IGBT}. The pulse current measurement I_p and the status signals of the IGBTs S_{IGBT} are used as feedback. The capacitor voltage measurement V_C is used for measuring the initial capacitor voltage and for controlling the MCB voltage after the arc extinction. Additional signals for status S_{HCB} and control C_{HCB} can be used for communication with the overall control of the grid.

the MCB. However, this voltage increase can be too slow for low current values. Therefore, additional IGBTs can be turned off and the voltage slope is increased, which results in a lower energy to dissipate. The voltage controller can either use the voltage measurement to determine the capacitor voltage or approximate the voltage based on the pulse current measurement. With the pulse current measurement also the voltage across the inductor and the IGBTs and therefore the voltage across the MCB can be approximated.

After the current decreased to zero, the RCD is opened, which completes the turn-off procedure.

E. Measurements

Three measurements are important for controlling the injected pulse current. Since the required isolation voltages are high, the communication to the main control is performed via optical fibers. As alternative, current measurements based on the Faraday effect [24] and voltage measurements based on the Pockels effect can be used.

1) Line current measurement: The most important measurement is the line current measurement. First, it can be used to detect the fault due to an overcurrent or a fast current increase. Second, it determines the line current amplitude during the fault for the control to adapt the pulse current. Since the line current changes only slowly due to the current limiting inductances ($30A$ in $10\mu s$) only a relatively small bandwidth is required. Therefore, the same current measurement as for the pulse current is used.

2) Pulse current measurement: Several non-idealities can influence the pulse current, especially if the possibility of IGBT failures is included. Therefore, the pulse current must be measured

The 2018 International Power Electronics Conference

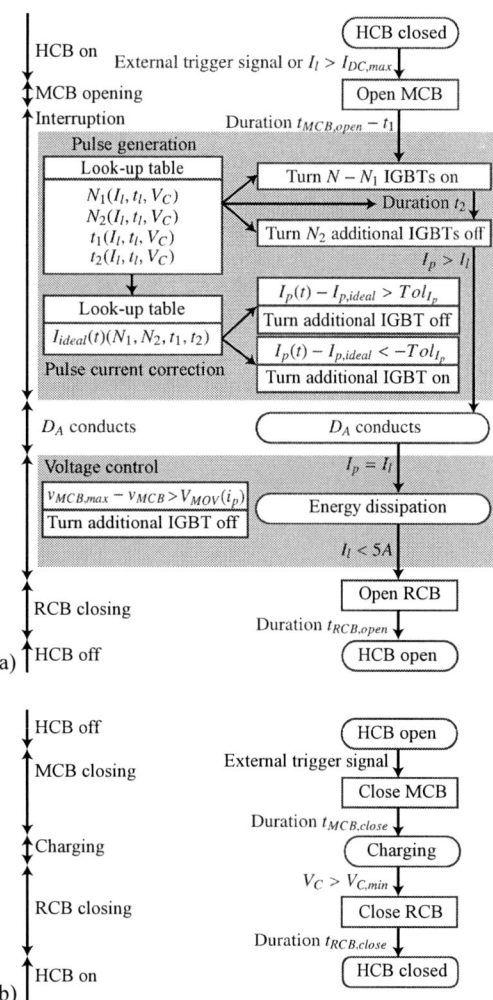

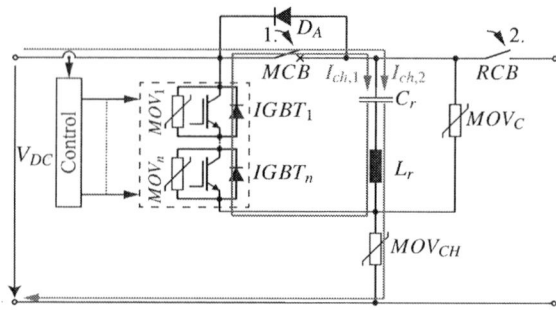

Fig. 9. Passive charging circuit for C_r: To allow an immediate turn-off after closing the HCB, the capacitor must be precharged before closing the RCB. This is done by closing first the MCB (1.), which generates a polarity reversal of the capacitor voltage due to the current $I_{ch,1}$. The capacitor is then completely charged by the current $I_{ch,2}$ due to the DC voltage. Finally, the RCB is closed (2.).

can generate a ZCC for all possible line currents. A deviation from this voltage results in a deviation of the pulse current. To avoid a too high deviation of the pulse current, two possibilities exist:

a) A precise charging system, which would require a current measurement and the possibility to control actively the capacitor voltage.

b) The voltage of the capacitor is measured and different switching times and numbers of turned off IGBTs are used for the different initial capacitor voltages, which is the chosen possibility for the considered HCB.

In both cases only the initial voltage must be known and the resolution can be relatively low ($\Delta V_{C,ini} = 100V$) as long as it is included as tolerance in the optimization. After the arc extinction, the measurement can be used together with the number of turned off IGBTs and the current in the pulse circuit to determine and control the TIV.

F. Passive Charging Circuit

The HCB uses the precharged capacitor C_r to generate the pulse current. Therefore, a charging system for the capacitor is required, which can be either a passive charging circuit via the line voltage or an additional isolated converter. To avoid an increased complexity of the system, a passive charging circuit is chosen for the considered HCB.

For precharging C_r, a varistor MOV_{ch} is used, which basically forms a voltage divider with the varistors of the SMs (Fig.9). As long as the capacitor voltage is low, the complete DC voltage is applied to the charging varistor MOV_{ch} and a current flows through capacitor C_r. As soon as the capacitor is charged, only a part of the DC voltage is applied so that the varistor does only conduct a negligible current. To block the DC voltage, the sum of the continuous blocking voltage of the charging varistor MOV_{ch} and the continuous blocking voltage of the series connected SMs must have a continuous blocking voltage of at least $400kV_{V_c}$.

The design of the charging circuit with numerical simulations shows that a continuous blocking voltage of MOV_{ch} of $300kV$

Fig. 8. Control procedure for the HCB: a) Turn off: After the detection of the fault, the MCB is opened. Depending on the line current measurement, the switching times and the number of turned off IGBTs are chosen to generate a pulse current, which generates a ZCC after the MCB is completely open. If the generated pulse current deviates too much from the ideal pulse current, additional IGBTs are turned on/off to compensate the deviation. After the zero current crossing, the voltage across the MCB is controlled until the line current decreased to a low current level, which the RCD can interrupt. b) Turn on: For charging the HCB, the MCB is first closed before the RCD is closed.

and potentially adapted. However, high voltage IGBTs have relatively long turn-on and turn-off delays. Therefore, a complete control of the pulse current is not an option and the pulse current is only corrected if the deviation from the ideal pulse current (without tolerances) is too high. The chosen measurement resolution is therefore relatively low with a resolution of $\pm 100A$, but requires a sample frequency of $200kHz$.

3) Capacitor voltage measurement: The capacitor of the pulse circuit must initially be charged to a minimum voltage $V_{C,ini,min}$ to generate a pulse current, which

TABLE I. RESULTING DESIGN OF A BENCHMARK CASE FOR THE GRID SHOWN IN FIG. 3.

V_{DC}	kV	400
$V_{MCB,max}$	kV	600
N_{GCB}		4
L_{lim}	mH	300
$I_{f,max}$	kA	11.8
di_{MCB}/dt_{max}	$A/\mu s$	114
dv_{MCB}/dt_{max}	$V/\mu s$	600
$L_r \pm 5\%$	μH	400
$V_{C,ini,min}$	kV	80
$C_r \pm 1\%$	μF	10
$V_{C,max}$	kV	300
$N_{SM} = N_{IGBTS}$		100
$N_{VAR,SM}$ $(1kV_{V_C}, 9.7kJ)$		24
N_{VAR,MOV_C} $(1kV_{V_C}, 9.7kJ)$		3600
$N_{VAR,MOV_{CH}}$ $(1kV_{V_C}, 9.7kJ)$		300

with 100 SMs results in a capacitor voltage of $85kV$. In addition to the continuous charging during the on-state, the capacitor must be charged before the HCB is turned on. This is performed by closing the MCB first, which allows to charge the capacitor before the RCB is closed in a second step and the connection to the line is established.

IV. CONCLUSION

In the first part of the paper, the design of the HCB procedure is presented. First, the important parameters of the grid and their impact on the HCB design are shown. Then, the switching module design and the optimization of the pulse circuit is presented, including the tolerances, which must be considered in a design process to ensure a reliable current interruption. The results of a exemplary design are shown and summarized in Tab.I. Finally, the control procedure of the HCB and determining the switching times and signals is shown, which allows to generate an optimal pulse current. Therefore, the paper gives a good overview how a HCB with adaptable pulse current injection can be constructed and controlled for a reliable interruption of DC currents.

ACKNOWLEDGMENTS

This project is carried out in the frame of the Swiss Centre for Competence in Energy Research on the Future Swiss Electrical Infrastructure (SCCER-FURIES) with the financial support of the Swiss Commission for Technology and Innovation (CTI - SCCER program).

REFERENCES

[1] C. Franck, "HVDC Circuit Breakers: A Review Identifying Future Research Needs," *IEEE Trans. on Power Delivery*, vol. 26, no. 2, pp. 998–1007, April 2011.

[2] D. Schmitt, Y. Wang, T. Weyh, and R. Marquardt, "DC-side fault current management in extended multiterminal-HVDC-grids," in *9th Int. Multi-Conf. on Systems, Signals and Devices (SSD)*, March 2012.

[3] J.-B. Curics, J. Descloux, S. Nguefeu, P. Rault, L. Violleau, F. Colas, X. Guillaud, W. Grieshaber, and B. Raison, "DEMO 3 testing results from DC network mock-up and DC breaker prototype status report for European Commission Deliverable: D11.3," *Twenties Deliverable*, Feb 2013.

[4] M. Bucher and C. Franck, "Fault current interruption in multiterminal HVDC networks," *IEEE Trans. on Power Delivery*, no. 99, 2015.

[5] J. Häfner and B. Jacobson, "Proactive hybrid HVDC breakers–A key innovation for reliable HVCD grids," in *Symp. on the electric power system of the future–Integrating supergrids and microgrids*, Sept 2011.

[6] R. Sander, M. Suriyah, and T. Leibfried, "A Novel Current-Injection Based Design for HVDC Circuit Breakers," in *Conf. for Power Electronics, Intelligent Motion, Renewable Energy and Energy Management (PCIM)*, May 2015.

[7] W. Grieshaber, J.-P. Dupraz, D.-L. Penache, and L. Violleau, "Development and test of a 120 kV direct current circuit breaker," in *45th CIGRE Session*, Aug. 2014, pp. 24–29.

[8] L. Ängquist and S. Norrga, "Arrangement, system, and method of interrupting current," WO Patent 2 016 003 357, Jan. 7, 2016.

[9] Y. Wang and R. Marquardt, "A fast switching, scalable DC-Breaker for meshed HVDC Super Grids," in *Int. Exhibition and Conf. for Power Electronics, Intelligent Motion, Renewable Energy and Energy Management (PCIM)*, May 2014.

[10] C. D. M. Oates, E. K. Chukaluri, and W. R. Crookes, "Circuit breaker apparatus," WO Patent 2 012 100 831, Aug. 2, 2012.

[11] A. Jehle and J. Biela, "Hybrid circuit breaker for HVDC grids with controllable pulse current shape," in *19th European Conf. on Power Electronics and Applications (EPE ECCE Europe)*, Sept 2017.

[12] Y. Shan, T. C. Lim, B. W. Williams, and S. J. Finney, "Successful fault current interruption on DC circuit breaker," *IET Power Electronics*, vol. 9, no. 2, pp. 207–218, 2016.

[13] J. Kaumanns, "Influence of the arcing time on the interruption behaviour and current zero conditions of vacuum circuit breakers," in *Proc. 8th Int. Symposium on Discharges and Electrical Insulation in Vacuum (ISDEIV)*, Aug 1998.

[14] R. P. P. Smeets and V. Kertesz, "Evaluation of high-voltage circuit breaker performance with a validated arc model," *IEE Proc. - Generation, Transmission and Distribution*, vol. 147, Mar 2000.

[15] K. Arimatsu, Y. Yoshioka, S. Tokuyama, Y. Kato, and K. Hirata, "Development and interrupting tests on 250kV 8kA HVDC circuit breaker," *IEEE Transactions on Power Apparatus and Systems*, vol. PAS-104, no. 9, pp. 2452–2459, Sept 1985.

[16] T. Schultz and C. M. Franck, "Interruption capability investigation of a model gas circuit-breaker for HVDC switching applications," in *21st Int. Conf. on Gas Discharges and their Applications*, Sept 2015.

[17] *Guide for the Development of Models for HVDC Converters in a HVDC Grid*. CIGRE, 2014.

[18] *Technical requirements and specifications of state-of-the-art HVDC switching equipment*. CIGRE, 2017.

[19] *High Voltage Direct Current Transmission Proven Technology for Power Exchange*. Siemens AG - Energy Sector.

[20] F. Hohmann and M. M. Bakran, "Impacts on the current distribution of metal oxide varistors for overvoltage protection in IGBT modules," in *18th European Conference on Power Electronics and Applications (EPE ECCE Europe)*, Sept 2017.

[21] G. V. N. Bezerra, L. A. M. M. Nobrega, J. F. S. Jnior, G. R. S. Lira, V. S. Brito, E. G. Costa, and M. J. A. Maia, "Evaluation of surge arrester models for overvoltage studies," in *Int. Conf. on High Voltage Engineering and Application (ICHVE)*, Sept 2014.

[22] F. Hohmann and M. M. Bakran, "IGBT switching behavior with parallel surge arrester for medium voltage applictation," in *International Exhibition and Conference for Power Electronics, Intelligent Motion, Renewable Energy and Energy Management*, May 2017.

[23] T. Hagiwara, T. Funabashi, H. Watanabe, N. Takeuchi, and T. Ueda, "A metal-oxide surge arrester model with active V-I characteristics," in *Electrical Engineering in Japan*, vol. 121, 1997.

[24] A. J. Rogers, "Optical technique for measurement of current at high voltage," *Proceedings of the Institution of Electrical Engineers*, vol. 120, no. 2, pp. 261–267, February 1973.

A Pragmatic SOH and SOC Co-Estimator for Lithium-ion Batteries in Smart Grid Applications

Kaiyuan Li[1], King Jet Tseng[2*], Feng Wei[2], Boon-Hee Soong[3]

1 Energy Research Institute @ NTU, Nanyang Technological University, Singapore
2 Electrical Power Engineering, Singapore Institute of Technology, Singapore
3 School of Electrical and Electronic Engineering, Nanyang Technological University, Singapore
*E-mail: KingJet.Tseng@SingaporeTech.edu.sg

Abstract- Reliable online co-estimation of state of health (SOH) and state of charge (SOC) of Li-ion batteries were of paramount importance for the realistic battery management system (BMS). This work aimed at bridging laboratory test and real-life battery operation data with a comprehensive analysis to provide a coherent and non-invasive approach based on probability density function (PDF), for building a pragmatic model to co-evaluate SOH and SOC of Li-ion batteries for smart grid applications. PDF results based on practical applications revealed that there was a prominent regularity of the voltage probabilities with regards to the SOH, which were exploited for setting up an online SOH evaluation scale gauge. Utilization of the load current of realistic smart grid further improve the practical generality of the proposed algorithm. The battery online SOC was determined afterwards based on the online OCV variation and the extracted SOH values. Commercial Li-ion batteries at arbitrary SOH and SOC level were tested to validate the effectiveness and robustness of the proposed algorithm, and the test results showed high accuracy and reliability of the proposed algorithm for co-evaluating SOH and SOC.

I. INTRODUCTION

Since this decade, there has been an ever-growing tendency towards smart grid (SG) to address the increasingly stringent concerns on greenhouse gas emission and fuel economy, where Li-ion batteries are among the predominant candidates for energy storage systems (ESSs) [1]. From the view of practical application, an advanced battery management system (BMS) enabling safe and optimal battery energy utilization becomes desirably sought. As core functions of BMS, reliable co-estimation of state of health (SOH) and state of charge (SOC), however, still remained challenging since they could be hardly measured by any sensors. Our research is based on a practical island SG where Li-ion batteries are employed as ESS, and we are facing a situation that the Li-ion battery data in SG was inaccessible until early this year, when the batteries had been working for circa 200 days. Accessible variables are the current, voltage, and ambient temperature of Li-ion batteries. As a result, in this work, we aim to devise a pragmatic approach to forecast the SOH and SOC of Li-ion batteries on the fly without interrupting the normal battery operation regimes in the smart grids applications, and the online SOH estimation is emphasized.

SOH was expressed by ratio of the battery capacity compared to that of fresh ones. To date, the most popular method for SOH identification was post-mortem analyses such as electrochemical impedance spectrum (EIS) [2] and cyclic voltammogram (CV) scan [3]. As to the online SOH estimation method, it could be further divided into two basic classes, namely, model-based [4-12] and data-driven approach [13-27]. Model-based approach included the capacity loss models built by Han et al. [4] and Ouyang et al. [5] predicting the degradation extent of Li-ion batteries, where SOH dependence on the stressing variables were reflected with chemical kinetics, and the cell capacity evolution at different conditions were predicted with acceptable accuracy; however, estimation accuracy was greatly dependent on the quantity and quality of the data provided. Meanwhile, the parameter/state observer approach offered a viable solution for SOH identification where SOH was cast as a variable to be determined. Key scenarios compromised observers based on recursive total-least-squares [6], sliding-mode [7], extended Kalman filter [8], particle filter [9], recursive least square (RLS) [10, 11], proportional integral [12], etc. Due to their desirable features of closed-loop and online estimation, remarkable adaptability and effectiveness in the co-evaluations of SOH together with SOC (sometimes with model parameters) were guaranteed, but the accuracy was highly sensitive to the robustness and credibility of the models prescribed.

Data-driven approaches were flexible and model-free. Take for example, researchers exploited neural networks [13, 14], support vector machines (SVMs) [15], fuzzy logic [16], and monotonic echo state network (MONESN) [17] to acquire the online SOH, where the battery operational variables such as current, voltage, and temperature acted as inputs into these black-box models. However, cycling and ageing history of battery were needed in these models, which turned out to be unrealistic for our SG situations. Other data-driven approaches were also utilized for non-invasive degradation diagnosis of Li-ion batteries, including differential thermal voltammetry (DTV) [18], differential voltage analysis (DVA) [19], incremental capacity analysis (ICA) [19-22], probability density function (PDF) [23-25]. These methods could be easily implemented; however, for the application of ICA and DVA, it was sometimes difficult to obtain the differential voltage from the sampled data in circumstances when current was low or voltage plateaus. The PDF-based approach was proposed as an extension of ICA/DVA, and satisfying SOH estimation results had been achieved [23-25]. Nevertheless, the rigorous prerequisites that the battery must be charged/discharged at constant current (CC) confined its application in practical applications. Compared to the battery cycling in laboratory [26, 27], the operation regimes in practical applications were inevitably volatile and hard to predict.

After the literature review, it could be summarized that all the cited SOH evaluation approaches were inappropriate for our case, and the common drawbacks were: (1) most studies used capacity degradation or resistance increase as SOH metric, where SOH was defined under relatively constrained situations, i.e., fully cycling at static current. Unfortunately, this was too strong an assumption for SG application. (2) almost all the online approaches for SOH estimation were based on the constant charge/discharge regimes of Li-ion batteries, which was difficult to realize in realistic SG. Further (3), in seeking to address our predicament, SOH had to be traced online from an arbitrary state rather than fresh, whereas all the cited works desired the cycle history right from fresh state.

In seeking to solve the problem we encountered and fully address the above-mentioned demerits, a pragmatic and non-invasive analyzing approach was proposed in this work to predict the online SOH and SOC of the Li-ion batteries used in real-life smart grid applications. The remainder of this paper is organized as follows: Section II introduces the proposed SOH and SOC co-evaluation approach based on PDF. The experimental tests and results, and model validation and discussion are presented in Section III. Finally, conclusions are summarized, and future work is addressed in Section IV.

II. APPROACH FOR SOH AND SOC CO-EVALUATION

SG load profile during a random 72 hours (from September 7th 00:00 a.m. till September 10th 00:00 a.m.) was shown in Fig. 1, where current for charge was indicated as negative and positive for discharge. In Fig. 1, there was a conspicuous daily charge period with dynamic rate from 9:00am to 12:30pm approximately, as indicated by C_n (n=1,2,3), and an everyday recurrent varying discharge period from circa 12:45pm till 15:45pm, marked by D_n (n=1,2,3). Actually, during the recorded battery operation of 8 months in the smart grid applications, C_n and D_n windows could be spotted as a daily routine. The sampled data of batteries during C_n and D_n were sorted out and exploited in this study for setting up the SOH scale gauge and predicting the online SOH of the on-site Li-ion batteries.

A. Probability Density Function Based Approach

The PDF method was introduced in [23-25] as a statistical analysis of the sampled battery data for SOH estimation. As derived in [23], PDF substantially had the equivalent basis and characteristic with ICA/DVA regardless of different techniques, thus, the theoretical utility of PDF could be guaranteed. The differentiation of the proposed method in this study from that presented in [23] was that, our proposed approach was more practical-application-oriented. Compared to the firm prerequisites that the battery must be charged from 0% SOC at C/3 CC, and discharged from 100% SOC at C/3 CC in [23], this work substantially extended the PDF applicability in pragmatic applications, where the battery initial SOC, charge/discharge

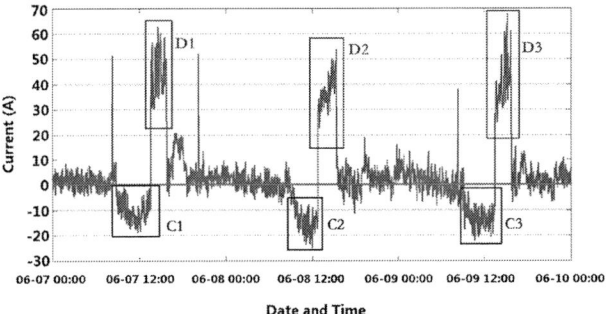

Fig. 1. Typical load current in smart grid applications.

current were not constrained. The only requirement was that the battery to be operated for at least 20 minutes for the PDF to be performed, which was quite pervasive and easy to meet in the practical smart grid applications, as implied in Fig. 1.

The procedure of the proposed approach could be summarized as follows: the BMS in the smart grid kept monitoring the battery parameters including current (I), voltage (V) and temperature (T) during normal duty regimes, and the battery online OCV was obtained simultaneously via the multi-timescale OCV observer which would be specified later. Once the continuous charging/discharging (C_n/D_n) regimes took place on the batteries on-site in the smart grid, the same load current and temperature profile was applied on several batteries of the same type off-site in the laboratory from the same initial OCV of the on-site batteries when C_n/D_n started. The reason that we regulated the OCVs of the off-site batteries same as that of the on-site batteries was that, the OCV was the only controllable variable of the off-site batteries that could be adjusted same as that of the on-site ones. In cases of continuous charge/discharge regimes, the terminal voltage characteristics of batteries with same initial OCV values but various SOH levels would deviate from each other gradually. Based on the experimental test results, there exhibited a similar tendency among the voltage responses of the batteries with different SOH levels, and after a period of at least 20 minutes the discrepancy become pronounced and discernible. More significantly, the deviated terminal voltage responses of these batteries could be utilized to predict the SOH value of a certain battery, as long as the SOH levels of other batteries were known, and a SOH gauge based on the terminal voltage responses was established. Accordingly, the off-site batteries were of different SOH level of N% ($0 < N \leq 100$ and signified that the capacities of the aged batteries equaled to N% of that of fresh ones). The unified initial OCV was obtained through discharging the fully charged batteries at rated current of 0.33 C for a certain period and relaxing for at least 2 hours. After the C_n/D_n regimes were completed both on-site and off-site, the terminal voltage characteristics of both the on-site and off-site batteries were obtained, where the off-site voltage responses were utilized as SOH scale gauge and the on-site voltage responses were substituted and fitted into the SOH gauge after PDF, thus, the online SOH could be calculated accordingly. After the SOHs of the on-site batteries were determined, the online SOC could be acquired by substituting the online SOH and OCV and the instantaneous operating temperature of the batteries into the SOC estimation model, which would be exemplified in the next section.

B. Proposed SOH and SOC Co-evaluation Model

Based on the above discussion, the SOH evaluation boiled down to extracting the online variation of OCV and the terminal voltage responses during C_n/D_n of Li-ion batteries. In this study, the 2nd order RC electrical circuit model (ECM) was applied, as shown in Fig. 2, since it insured sufficient precision while remained acceptable simplicity of the model configuration. The voltage source V_{OC} denoted the OCV, and the remainder of the model included an ohmic resistor (R_i) and two RC combination networks (R_s and C_s for the slow dynamic, and R_f and C_f for the fast dynamic), which characterized different battery transient dynamics. When battery was cycled, its electrical behaviors could be expressed by the following state-space equations:

$$C_s \frac{dV_s}{dt} + \frac{V_s}{R_s} = I_L \tag{1}$$

$$C_f \frac{dV_f}{dt} + \frac{V_f}{R_f} = I_L \tag{2}$$

$$V_t = V_{OC} - V_s - V_f - R_i \cdot I_L \tag{3}$$

where I_L is the load current, which is positive for discharge and negative for charge; V_t, V_s, and V_f are the battery terminal voltage, the voltage of the slow and fast dynamics, respectively. The following expressions could be derived from (1) and (2):

$$V_s(t) = e^{-\frac{nT}{R_s C_s}} V_s(t - nT) + (1 - e^{-\frac{T}{R_s C_s}}) R_s \sum_{j=1}^{n} e^{-\frac{(j-1)T}{R_s C_s}} I_L(t - jT) \tag{4}$$

$$V_f(t) = e^{-\frac{nT}{R_f C_f}} V_f(t - nT) + (1 - e^{-\frac{T}{R_f C_f}}) R_f \sum_{j=1}^{n} e^{-\frac{(j-1)T}{R_f C_f}} I_L(t - jT) \tag{5}$$

where T is the onboard sampling time for OCV estimation and n is the sample interval. Based on (3), (4) and (5), the state-space model of the ECM could be written as:

$$\begin{bmatrix} V_s(t) \\ V_f(t) \end{bmatrix} = \begin{bmatrix} \alpha_1 & 0 \\ 0 & \alpha_2 \end{bmatrix} \cdot \begin{bmatrix} V_s(t - nT) \\ V_f(t - nT) \end{bmatrix} + \begin{bmatrix} \beta_1 \\ \beta_2 \end{bmatrix} \cdot I_L(t - jT)$$

$$V_{OC}(t) - V_t(t) = \begin{bmatrix} 1 & 1 \end{bmatrix} \cdot \begin{bmatrix} V_s(t) \\ V_f(t) \end{bmatrix} + R_s \cdot I_L(t) \tag{6}$$

where $\alpha_1 = \exp(-nT/R_s C_s)$, $\alpha_2 = \exp(-nT/R_f C_f)$,

$\beta_1 = (1 - \exp(-\frac{nT}{R_s C_s})) R_s \sum_{j=1}^{n} e^{-\frac{(j-1)T}{R_s C_s}}$,

And $\beta_2 = (1 - \exp(-\frac{nT}{R_f C_f})) R_f \sum_{j=1}^{n} e^{-\frac{(j-1)T}{R_f C_f}}$.

The transfer function of (6) in z-domain could be calculated:

$$V_{OC}(z) - V_t(z) = \begin{bmatrix} 1 & 1 \end{bmatrix} \cdot \left(z \cdot I_{2 \times 2} - \begin{bmatrix} \alpha_1 & 0 \\ 0 & \alpha_2 \end{bmatrix} \right) + R_s \cdot I_L(z) \tag{7}$$

Through the inverse z-transform, the following can be derived:

$$V_{OC}(t) = (\alpha_1 + \alpha_2) V_{OC}(t - nT) - \alpha_1 \alpha_2 V_{OC}(t - (n + 1)T) + V_t(t) - (\alpha_1 + \alpha_2) \cdot V_t(t - nT) + \alpha_1 \alpha_2 V_t(t - (n + 1)T) + R_s \cdot I_L(t) + [\beta_1 + \beta_2 - (\alpha_1 + \alpha_2) R_s] \cdot I_L(t - nT) + (\alpha_1 \alpha_2 R_s - \alpha_1 \beta_2 - \alpha_2 \beta_1) \cdot I_L(t - (n + 1)T) \tag{8}$$

which can be written in a linear form alternatively as:

$$y(t) = \varphi^T(t) \theta(t) + e(t) \tag{9}$$

with parameters $\varphi^T(t) = (\alpha_1 + \alpha_2 - \alpha_1 \alpha_2 - 1)$,

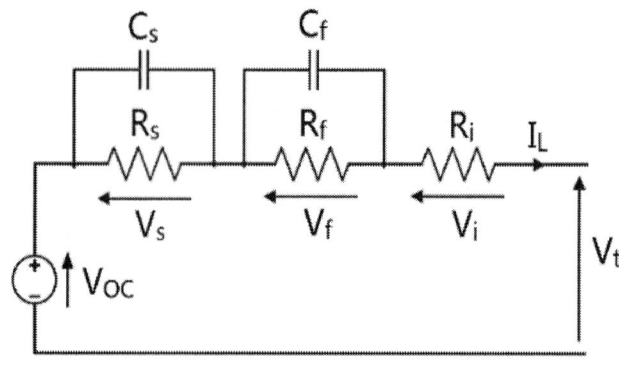

Fig. 2. Schematic of the 2nd order ECM.

$\theta(t) = V_{OC}(t)$,

$y(t) = V_t(t) - (\alpha_1 + \alpha_2) \cdot V_t(t - nT) + \alpha_1 \alpha_2 V_t(t - (n + 1)T) + R_s \cdot I_L(t) + [\beta_1 + \beta_2 - (\alpha_1 + \alpha_2) R_s] \cdot I_L(t - nT) + (\alpha_1 \alpha_2 R_s - \alpha_1 \beta_2 - \alpha_2 \beta_1) \cdot I_L(t - (n + 1)T)$,

$e(t) = \alpha_1 \alpha_2 [V_{OC}(t - nT) - V_{OC}(t - (n + 1)T)] + (\alpha_1 \alpha_2 - \alpha_1 - \alpha_2) [V_{OC}(t) - V_{OC}(t - nT)]$

V_{OC} is the parameter to be determined. $\varphi^T(t)$ can be obtained by the parameter values and the output $y(t)$ is combined both battery terminal voltage, load current and parameter values. The error term $e(t)$ is negligible if the sampling time Δt_1 is properly chosen. In this study, Δt_1 is chosen to be 1.0 s.

The stability and sensitivity studies revealed that the value of R_i has more effects on the OCV estimator; however, these study details are not included in this digest due to the size limit. To this end, the identification of R_i is separated out of the parameter estimation with an independent timescale. The following expression could be drawn according to (3):

$$\Delta V_t(t) = -R_i \cdot \Delta I_L(t) + \varepsilon(t) \tag{10}$$

and $\varepsilon(t) = \Delta V_{OC} - \Delta V_s - \Delta V_f$. $\Delta V_t(t)$ denoted the voltage difference between the sampling interval and so does other terms. When the sampling time Δt_2 is selected to be a sufficient short period, the erroneous term $\varepsilon(t)$ approaches zero and it is plausible to obtain the value of R_i through the voltage difference and the current difference. Whereas this approach is deficient in the cases when $\Delta I_L(t)$ is a quite small value, thusly, a constraint is applied that R_i is only updated when $|\Delta I_L(t)| \geq \xi$ and ξ is a triggering threshold determined from the battery specification.

Let Δt_3 be the sampling time for the transient parameters. Equation (4) and (5) can be deformed as:

$$V_t(t) + R_s I_L(t) = V_{OC}(t) - (\alpha_1 + \alpha_2) V_{OC}(t - \Delta t_3) + \alpha_1 \alpha_2 V_{OC}(t - 2\Delta t_3) + (\alpha_1 + \alpha_2) V_t(t - \Delta t_3) - \alpha_1 \alpha_2 V_t(t - 2\Delta t_3) + [(\alpha_1 + \alpha_2) R_s - \beta_1 - \beta_2] I_L(t - \Delta t_3) + [\alpha_1 \beta_2 + \alpha_2 \beta_1 - \alpha_1 \alpha_2 R_s] I_L(t - 2\Delta t_3) \tag{11}$$

By taking the difference transformation, (11) could be expressed:

$$\Delta[V_t(t) + R_s I_L(t)] = \Delta[V_{OC}(t) - (\alpha_1 + \alpha_2)V_{OC}(t - \Delta t_3) +$$

$$\alpha_1\alpha_2 V_{OC}(t - 2\Delta t_3)] + (\alpha_1 + \alpha_2)\Delta V_t(t - \Delta t_3) -$$

$$\alpha_1\alpha_2\Delta V_t(t - 2\Delta t_3) + [(\alpha_1 + \alpha_2)R_s - \beta_1 - \beta_2]\Delta I_L(t -$$

$$\Delta t_3) + [\alpha_1\beta_2 + \alpha_2\beta_1 - \alpha_1\alpha_2 R_s]\Delta I_L(t - 2\Delta t_3) \qquad (12)$$

Alternatively, the regression form of (11) could be built as:

$$p(t) = m^T(t)q(t) + \omega(t) \qquad (13)$$

where $p(t) = \Delta[V_t(t) + R_s I_L(t)]$,

$$m^T(t) = [\Delta V_t(t - \Delta t_3) \quad \Delta V_t(t - 2\Delta t_3) \quad \Delta I_L(t - \Delta t_3)$$

$$\Delta I_L(t - 2\Delta t_3)],$$

$$q(t) = [\alpha_1 + \alpha_2 \quad -\alpha_1\alpha_2 \quad (\alpha_1 + \alpha_2)R_s - \beta_1 - \beta_2$$

$$\alpha_1\beta_2 + \alpha_2\beta_1 - \alpha_1\alpha_2 R_s]^T,$$

and

$$\omega(t) = \Delta[V_{OC}(t) - (\alpha_1 + \alpha_2)V_{OC}(t - \Delta t_3) + \alpha_1\alpha_2 V_{OC}(t -$$

$$2\Delta t_3)] .$$

Similar to (9) and (10), (13) could be solved with Δt_3. Value of the transient parameters was identified after $q(t)$ was obtained:

$$R_s = \frac{\beta_1}{1 - \alpha_1}, \qquad C_s = -\frac{\Delta t_3(1 - \alpha_1)}{\beta_1 \ln\alpha_1},$$

$$R_f = \frac{\beta_2}{1 - \alpha_2}, \qquad C_f = -\frac{\Delta t_3(1 - \alpha_2)}{\beta_2 \ln\alpha_2}.$$

The stability and precision tests indicated that a large Δt_3 is more suitable for the transient parameter estimation, however, the stability and precision testes were not shown in this summary due to size limits. In this study, the three timescales are chosen as $\Delta t_1 = \Delta t_2 = 1$ s, and $\Delta t_3 = 5$ s. In order to solve the regression model in (9), (10) and (13), the U-D factorization based recursive least square (UDRLS) was applied, where the covariance matrix P was factorized as $P = U \cdot D \cdot U^T$. Detailed procedure of UDRLS was listed in Table I [11]. Compared to the conventional method such as EKF and RLS, UDRLS offered enhanced stability and less computational burden.

TABLE I
UDRLS ALGORITHM DETAILS

Initialization: $\theta_0 = E[\theta_0]$, $P_0 = UDU^T$, forgetting factor λ
Step 1: Define two functions f and v:
$f = [f_1, ..., f_n]^T = U^T(k-1)\varphi^T(k)$, $v = [v_1, ..., v_n]^T = D(k-1)f$
Step 2: for j=1,2,3, execute Step 2.1-2.2
  Step 2.1: Compute the following:
  $\alpha_j = \alpha_{j-1} + f_j v_j$, $D(k)_{jj} = (\alpha_{j-1}D(k-1)_{jj})/(\alpha_j\lambda)$, $b_j = v_j$, $c_j = -f_j/\alpha_{j-1}$
  Step 2.2: For i=1,2, ... j-1, Compute the following:
  $U(k)_{ij} = U(k-1)_{ij} + b_j c_j$, $b_i = b_i + U(k-1)_{ij}b_j$
  Update gain: $K(k) = [b_1, ..., b_n]^T/\alpha_n$
Step 3: Posteriori parameter update: $\theta(k) = \theta(k-1) + K(k)\beta(k)$

III. EXPERIMENTAL TESTS AND MODEL VALIDATION

To exemplify the proposed SOH and SOC co-estimator, a kind of commercial Li-ion battery with rated capacity of 20.0 Ah and rated voltage of 2.3 V was chosen for testing. A fresh cell and 3 aged cells each at SOH of 95%, 90% and 85% were selected for setting up the off-site SOH scale gauge in order to predict the SOH of 3 arbitrary cells with unknown SOH of the same type and brand, which were deemed to perform as the on-site batteries in smart grid. The various SOHs were acquired via the accelerated ageing test on fresh batteries. It had been verified in one of our recently published work that the accelerated ageing test was applicable for the type and brand of battery we used in the current work, but not for some others. The initial OCVs of the on-site batteries before C_n and D_n were founded to be overall below 2.2 V and above 2.4 V, respectively. In light of the facts that SOCs of the on-site batteries before C_n and D_n were highly dependent on the previous load conditions, in this study, we chose 2.2V (~30% SOC for fresh cell) and 2.1V (~10% SOC for fresh cell) as the initial OCV for batteries underwent C_n, and 2.4V (~80% SOC for fresh cell) and 2.5V (~90% SOC for fresh cell) as the initial OCV values for batteries underwent D_n. However, in real-life applications, the initial OCVs of the off-site batteries were tuned same as that of the on-site batteries as detected by the online OCV observer before C_n and D_n started. The reason we chose 2.2 V and 2.1 V for C_n and 2.4 V and 2.5 V for D_n as initial OCV values was to validate the effectiveness and applicability of the proposed method at various scenarios. Due to size limit, only the test results with ~2.2 V initial OCV for batteries under C_n was illustrated and analyzed as an instance. In supplement, the operation temperature of the on-site batteries in the smart grid was founded to be 26±1.5 °C, and this subtle temperature fluctuations would basically cause no characteristic changes of the battery, therefore in the off-site tests, the batteries were placed in the laboratory with room temperature of ~26°C.

The on-site Li-ion batteries with arbitrary ageing levels had been cycled under the I profile collected from the smart grid for more than 3 weeks to fully simulate the batteries performance in real-life smart grid applications, in light of which the online battery SOC and SOH were totally unpredictable. Fig. 3 showed the I profile during a random C_n period. One could observe that I was of negative values all the time, indicating that the battery was always being charged during C_n as aforementioned. The OCVs of on-site batteries were found to be 2.22 V according to the online OCV estimator. The four off-site batteries with various SOH levels (100%, 95%, 90% and 85%) but unified OCV of 2.22 V were loaded under the I profile in Fig. 3 and the corresponding terminal voltage responses were recorded and shown in Fig. 4. A consistent voltage tendency could be perceived from the 4 curves in Fig. 4 that, the voltage was lower as the capacity faded. To quantify the battery SOH with regards to the voltage responses, the PDF results of the 4 voltage curves were illustrated in Fig. 5. The horizontal axis of Fig. 5 was voltage and the vertical axis was the corresponding probability value of the battery terminal voltage obtained via the *ksdensity* function in Matlab. The integration areas of each the 4 PDF curves all equaled to 1 and the changing trends of the probability magnitude as the battery aged were indicated by arrows. Concretely, the beginning and ending parts of the PDF curves exhibited a distinct tendency of movements to the left as the battery SOH decreased and the capacity faded. It can be

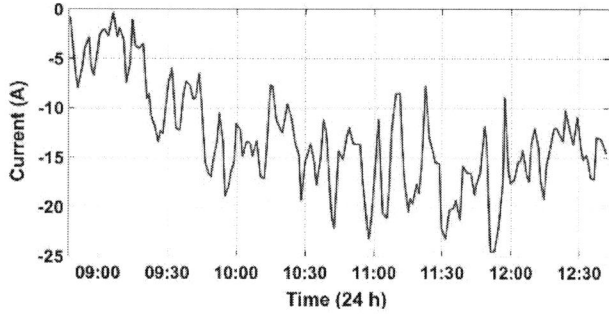

Fig. 3. Load current of Li-ion batteries during C_n.

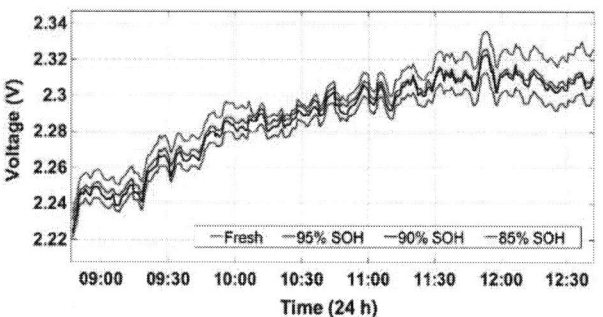

Fig. 4. Voltage responses of Li-ion batteries with various SOH in C_n.

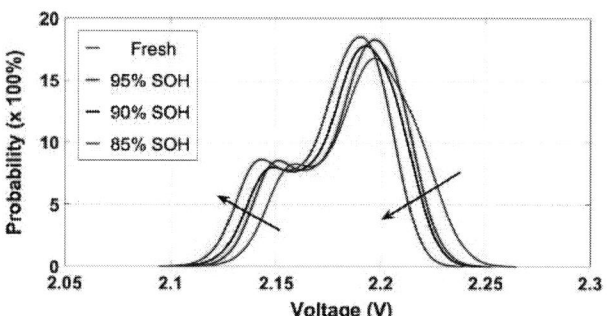

Fig. 5. PDF result of the 4 voltage responses.

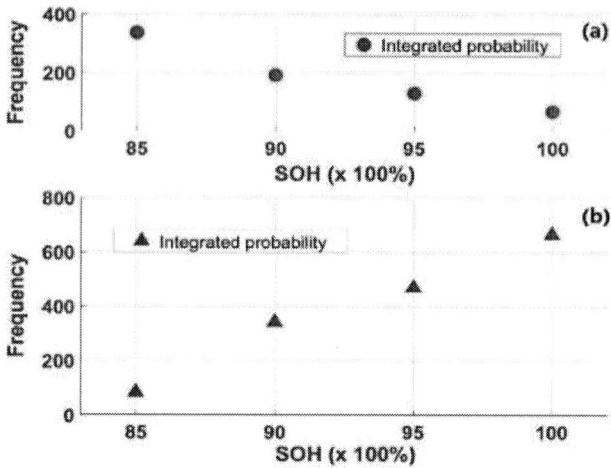

Fig. 6. Relation between the integrated probability and the battery SOH during beginning and ending voltage thresholds.

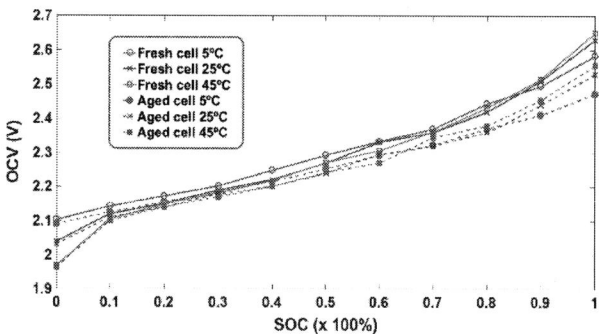

Fig. 7. OCV-SOC relation incorporating T and SOH effects.

concluded that the probability magnitude at the beginning and ending parts of the PDF curves displayed a consistent regularity as the battery aged. In the sequel, if the probability values within a certain voltage threshold of the beginning and ending parts were integrated separately, the integrated frequency should have the same trend as that of the probability magnitude. To be specific, the voltage threshold in this test was 2.11 V - 2.16 V for the beginning and 2.19 V - 2.24 V for the ending part. The voltage thresholds and the corresponding integrated probability of the 4 batteries with various SOH level were listed in Table II. Fig. 6 showed the trajectory of the integrated probability, which was indicated as frequency, with respect to the battery SOH. Based on the test results and the discussion above, the extracted PDF curves were competent to reveal the relationship between the statistics and the battery SOH. As a result, the online SOH

evaluation of the on-site batteries in the smart grid applications was able to be performed. When the integrated probability of the on-site batteries were available, the frequency results were fitted into the SOH scale gauge model illustrated in Fig. 6, where the red dots represented the frequency results during the beginning threshold and the blue triangles stood for the frequency values of the ending threshold .The SOH obtained from the beginning and ending voltage thresholds were further averaged with a forgetting factor λ, which was determined empirically based on experimental results. After the online SOH of on-site batteries was determined, the SOH, OCV and temperature values were substituted into the SOC estimation model, which was shown in Fig. 7 as an illustration, and the online SOC values could be evaluated accordingly. Detailed SOC estimation would be provided in the full paper.

To verify the proposed algorithm for SOH and SOC co-estimation, cycling data of the 3 on-site batteries were collected and the integrated probability during the two thresholds during C_n was obtained and transformed into the evaluated SOH values. The results were shown in Table III, and SOH_B and SOH_E denoted the estimated SOH values for the beginning and ending

The 2018 International Power Electronics Conference

TABLE II
STATISTICAL INTEGRATED PROBABILITY VALUES

SOH	2.11 V – 2.16 V	2.19 V – 2.24 V
100%	68.5	665.7
95%	129.6	470.1
90%	191.5	341.2
85%	335.1	82.7

TABLE III
SOH EVALUATION ERROR OF THE PROPOSED ALGORITHM

Battery	1	2	3
Actual SOH	93%	87%	84.1%
SOH_B	91.5%	87.6%	83.0%
Error%	-1.5%	0.6%	-1.1%
SOH_E	93.3%	86.2%	85.3%
Error%	0.3%	-0.8%	1.2%

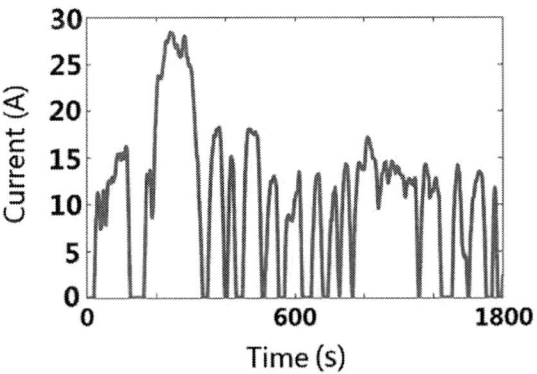

Fig. 8. Load current profile for validating the SOC evaluation results.

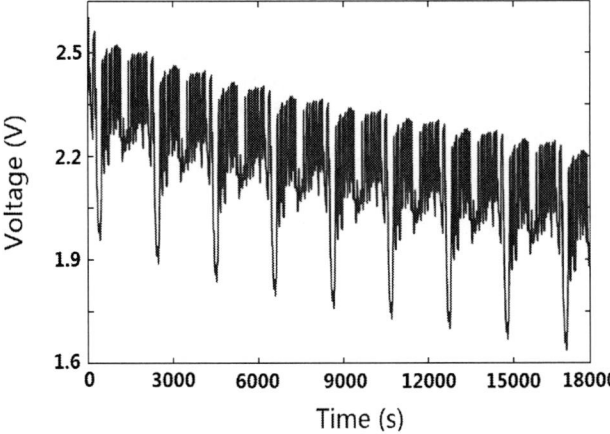

Fig. 9. Terminal voltage response of the Li-ion battery in the model validation test.

thresholds of C_n, respectively. Compared with the actual battery SOH values, the estimation results were confined within ±1.5% percentage error bound. The forgetting factor λ between SOH_B and SOH_E during C_n was hence determined accordingly.

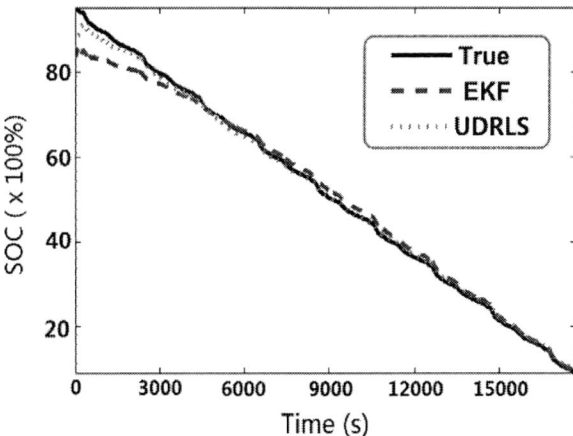

Fig. 10. Comparison of the actual SOC values with the estimated results, via UDRLS and EKF approach.

Furthermore, another forgetting factor λ between the SOH determined from the C_n and D_n was also utilized in order for achieving further improvements on the online SOH evaluation fidelity.

The battery online SOC was afterwards evaluated and advantages of UDRLS compared to the conventional Extend Kalman Filter (EKF) is also illustrated. A cell with 95% SOH was loaded respectively under the real-time current profile as shown in Fig. 8, which is extracted from a smart grid set up on a small island in Singapore. The terminal voltage response of the battery is shown in Fig. 9. The SOC estimation results based on the proposed algorithm is illustrated in Fig. 10 and the true SOC based on the CC approach and the estimation results of EKF is also shown in Fig. 10. It can be observed clearly from Fig. 10 that the proposed UDRLS algorithm outperforms EKF in the sense of comparable estimation accuracy and a faster convergence. What's more, the simulation time of the proposed algorithm is 0.175 second, while that of EKF is 1.92 second. Thus, the proposed algorithm has a lower computation cost compared to the pervasively exploited EKF.

Based on the above experimental test results and discussion, the proposed SOH and SOC co-estimator is able to achieve accurate SOH and SOC evaluation results with an accelerated processing time, compared to the conventional EKF algorithm.

IV. CONCLUSION

Compared to the traditional PDF method, this study extended its practicability to the pragmatic smart grid applications, for analyzing the sampled battery cycling data to realize the online SOH and SOC co-evaluation. To be specific, the stringent prerequisites for PDF in the literature were eliminated and the real-life battery load profiles were exploited in the laboratory directly without interrupting the battery normal operation regimes. In addition, the introduction of forgetting factor for both the beginning/ ending threshold, and charging/discharging regimes would help to further improve the online SOH

The 2018 International Power Electronics Conference

estimation accuracy. After the SOH was determined by the proposed algorithm, the online SOC of the batteries would be derived directly from the SOC model, incorporating the effects of battery SOH, OCV, and temperature into thorough consideration. Based on the testing results of commercial Li-ion batteries, the proposed SOH and SOC co-estimator was capable of providing a viable and pragmatic solution for the online battery status determination in realistic BMS applications. More details and figures would be provided in the full paper.

REFERENCES

[1] B. Dunn, H. Kamath, and J.M. Tarascon, "Electrical energy storage for the grid: A battery of choices," *Science*, vol. 334, pp. 928–935, 2011.

[2] F. Huet, "A review of impedance measurements for determination of the state-of-charge or state-of-health of secondary batteries," *J. Power Sources*, 70 (1998), pp. 59-69.

[3] Southampton Electrochemistry Group, "Instrumental Methods in Electrochemistry," Ellis Horwood, Chichester, London (1985) pp. 178–227.

[4] X. Han, M. Ouyang, L. Lu, J. Li, "A comparative study of commercial lithium ion battery cycle life in electric vehicle: Capacity loss estimation," *J. Power Sources*, 268 (2014), pp. 658–669.

[5] M. Ouyang, X. Feng, X. Han, L. Lu, Z. Li, X. He, "A dynamic capacity degradation model and its applications considering varying load for a large format Li-ion battery," *Appl. Energy*, 165 (2016), pp. 48-59.

[6] T. Kim, Y. Wang, Z. Sahinoglu, T. Wasa, S. Hara, W. Qiao, "A rayleigh quotient-based recursive total-least-squares online maximum capacity estimation for lithium-ion batteries," *IEEE Trans. Energy Convers.*, vol. 30, no. 3, pp. 842-850, 2015.

[7] M. Gholizadeh, F. R. Salmasi, "Estimation of state of charge, unknown nonlinearities, and state of health of a lithium-ion battery based on a comprehensive unobservable model," *IEEE Trans. Ind. Electron.*, vol. 61, no. 3, pp. 1335-1344, 2014.

[8] C. Zou, C. Manzie, D. Nesic, A.G. Kallapur, "Multi-time-scale observer design for state-of-charge and state-of-health of a lithium battery," *J. Power Sources*, 335 (2016), pp. 121–130.

[9] Z. Liu, G. Sun, S. Bu, J. Han, X. Tang, M. Pecht, "Particle learning framework for estimating the remaining useful life of lithium-ion batteries," *IEEE Trans. Instrum. Meas.*, vol. 66, no. 2, pp. 280-293, 2016.

[10] Z. Wei et al. "A multi-timescale estimator for battery state of charge and capacity dual estimation based on an online identified model," *Appl. Energy* (2017), http://dx.doi.org/10.1016/j.apenergy.2017.02.016.

[11] X. Tang, X. Mao, J. Lin, B. Koch, "Capacity estimation for Li-ion batteries," *in Proc. Am. Control Conf.*, Jun./Jul. 2011, pp. 947-952.

[12] L. Zheng, L. Zhang, J. Zhu, G. Wang, J. Jiang, "Co-estimation of state-of-charge, capacity and resistance for lithium-ion batteries based on a high-fidelity electrochemical model," *Appl. Energy*, 180 (2016), pp. 424-434.

[13] H. Chaoui, C.C. Ibe-Ekeocha, "State of charge and state of health estimation for lithium batteries using recurrent neural networks," *IEEE Trans. Veh. Technol.*, Article in Press, DOI 10.1109/TVT.2017.2715333.

[14] G. W. You, S. Park, D. Oh, "Real-time state-of-health estimation for electric vehicle batteries: A data-driven approach," *Appl. Energy*, 176 (2016), pp. 92-103.

[15] V. Klass, M. Behm, G. Lindbergh, "A support vector machine-based state-of-health estimation method for lithium-ion batteries under electric vehicle operation," *J. Power Sources*, 270 (2014), pp. 262-272.

[16] M. Landi, G. Gross, "Measurement techniques for online battery state of health estimation in vehicle-to-grid applications," *IEEE Trans. Instrum. Meas.*, vol. 63, no. 5, pp. 1224-1234, 2014.

[17] D. Liu, W. Xie, H. Liao, Y. Peng, "An integrated probabilistic approach to lithium-ion battery remaining useful life estimation," *IEEE Trans. Instrum. Meas.*, vol. 64, no. 3, pp. 660-670, 2015.

[18] B. Wu, V. Yufit, Y. Merla, R. F. Martinez-Botas, N. P. Brandon, G. J. Offer, "Differential thermal voltammetry for tracking of degradation in lithium-ion batteries," *J. Power Sources*, 273 (2015), pp. 495–501.

[19] C. Pastor-Fernández, K. Uddin, G. H. Chouchelamane, W. D. Widanage, J. Marco, "A comparison between electrochemical impedance spectroscopy and incremental capacity-differential voltage as li-ion diagnostic techniques to identify and quantify the effects of degradation modes within battery management systems," *J. Power Sources*, 360 (2017), pp. 301-318.

[20] M. Dubarry, V. Svoboda, R. Hwu, B. Y. Liaw, "Incremental capacity analysis and close-to-equilibrium OCV measurements to quantify capacity fade in commercial rechargeable lithium batteries," *Electrochem. Solid State Lett.*,9 (10) A454-A457 (2006).

[21] C. Weng, X. Feng, J. Sun, H. Peng, "State-of-health monitoring of lithium-ion battery modules and packs via incremental capacity peak tracking," *Appl. Energy*, 180 (2016), pp. 360-368.

[22] X. Li, J. Jiang, L. Y. Wang, D. Chen, Y. Zhang, C. Zhang, "A capacity model based on charging process for state of health estimation of lithium ion batteries," *Appl. Energy*, 177 (2016), pp. 537-543.

[23] X. Feng, J. Li, M. Ouyang, L. Lu, J. Li, X. He, "Using probability density function to evaluate the state of health of lithium-ion batteries," *J. Power Sources*, 232 (2013), pp. 209-218.

[24] X. Si, "An adaptive prognostic approach via nonlinear degradation modeling: application to battery data," *IEEE Trans. Ind. Electron.*, vol. 62, no. 8, pp. 5082-5096, 2015.

[25] X. Feng, J. Li, L. Lu, J. Hua, L. Xu, M. Ouyang, "Research on a battery test profile based on road test data from hybrid fuel cell buses," *J. Power Sources*, 209 (2012), pp. 30-39.

[26] *Battery Test Manual for Plug-in Hybrid Electric Vehicles*, Idaho National Laboratory (2008).

[27] PNGV Battery Test Manual (2001).

The 2018 International Power Electronics Conference

Modeling and Stability Analysis of Parallel Droop-Controlled and Current-Controlled Inverters

Shike Wang*, Zeng Liu, Jinjun Liu and Ronghui An

State Key Lab of Electrical Insulation and Power Equipment, Xi'an Jiaotong University, Xi'an, Shaanxi, China
*E-mail: wangshikepe@gmail.com

Abstract— As an integrated system consists of different sources and loads, stability issue is of great importance to microgrids. Existing works neglect the interaction stability issue between parallel droop controlled inverter and current-controlled inverter in islanded microgrids. This work is presented to fill this gap. Firstly, the small-signal terminal-characteristic model of inverters are conducted. Not only conventional output impedance and admittance but also the transfer functions representing the dynamic of fundamental angular frequency are considered in this model. Then the interaction stability of parallel inverters are studied based on generalized Nyquist criterion. Finally, the simulation results are presented to verify the analysis.

Keywords—droop control; phase-locked loop; terminal characteristics; stability analysis

I. INTRODUCTION

Since the Paris Agreement on climate change was signed by 175 countries on 22 April 2016 [1], as an effective solution of mitigating greenhouse gas emissions, renewable energy and distributed power generation have attracted increasing attention worldwide. Generally, renewable energy and other distributed energy resources (DERs) need power electronic interfaces to connect to the utility grid [2]. With increasing number of DERs penetrating the utility networks, microgrid was developed as an integrated system to manage control problems when different power electronic interfacing DERs are interconnected [3-4].

As a complex system consists of different types of sources and loads, research works on AC microgrids always attach great importance to stability issues. In state-space based approaches, system stability can be accessed by observing the eigenvalues of state matrix [5-7]. As the number of paralleled converters increasing, the order of system state matrix will be multiplied up, which challenges the feasibility of this method. The impedance-based modeling approach provides better practicability for three-phase AC systems. After obtaining the small-signal output impedance and admittance of source and load in the synchronous reference frame (SRF), generalized Nyquist criterion (GNC) can be applied to predict the stability of the parallel system [8-9]. It is widely used to study the stability of grid-connected three-phase converters [10-12], cascaded voltage source inverter and constant power loads [13-14], as well as

parallel inverters with active load sharing scheme [15]. Stability analysis becomes more complicated in terms of droop controlled inverters. Since the system fundamental angular frequency varies with inverter output active power based on the droop scheme [16], inverters also have coupling in frequency. In [17], Liu *et al.* proposed a novel small-signal terminal characteristic model for droop-controlled inverters, representing the dynamic interactions between inverter output current and system fundamental angular frequency. Furthermore, a stability criterion was proposed in [18].

Aforementioned works didn't cover a general case in islanded microgrids: the droop controlled inverter and current-controlled inverter (CCI) with phase-locked loop (PLL) parallel together. Different from connecting to the main grid, the variable fundamental angular frequency at point of common coupling (PCC) jeopardize operation of PLL and CCI. At the same time, the droop controlled voltage source is also influenced by CCI. Therefore, the interaction instability may occur between parallel inverters. This issue is crucial to operation of islanded microgrids, which is seldom studied in existing published works. To fill this gap, this paper first conducts the terminal-characteristic model of droop controlled inverter and PLL synchronized CCI. Besides conventional output impedance and admittance, the transfer functions representing the dynamic of fundamental angular frequency are covered in this model. Then, based on GNC, the system stability are studied. Finally, the simulation results are presented to verify the modeling and stability analysis.

II. TERMINAL-CHARACTERISTICS MODELING

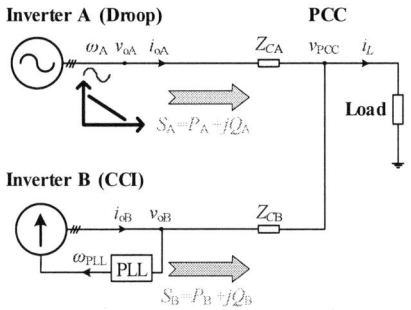

Fig. 1. Configuration of studied system.

This work was supported by the National Natural Science Foundation of China under Grant 51437007, and the Power Electronics Science and Education Development Program of Delta Environmental & Educational Foundation under Grant DREM2014002.

1524

The 2018 International Power Electronics Conference

In order to simplify the following analysis, as Fig.1 shows, a parallel system composed of one droop controlled inverter (A) and one PLL synchronized CCI (B) is studied. These two inverters are connected to the PCC through distribution cable Z_C. The system fundamental angular frequency and amplitude of PCC voltage is defined by Inverter A. Based on the droop scheme, ω_A and v_{oA} vary with its output active and reactive power. Inverter B synchronizes with the voltage at its output terminal through a PLL, delivering certain amount of power to load. The two inverters are not only coupled in voltage and current, but also in fundamental angular frequency. In following sections, the small-signal terminal characteristics of inverter A and B are first conducted separately then combined together to analysis the stability of the overall system.

A. Droop Controlled Inverter

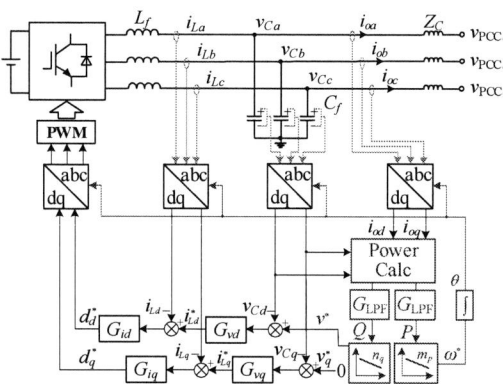

Fig. 2. Block diagram of a droop-controlled inverter.

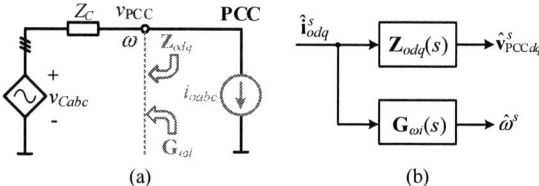

(a) (b)

Fig. 3. Terminal characteristic model of droop controlled inverter.

The power stage and control block diagram of a typical three-phase droop controlled inverter is depicted in Fig. 2. Based on the droop equations shown in (1):

$$\omega^* = \omega_0 - m_p(P - P_0)$$
$$v^* = V_0 - n_q(Q - Q_0) \tag{1}$$

where ω^* and v^* are calculated references for fundamental angular frequency and output voltage amplitude. P and Q are output active and reactive power. m_p and n_q are droop slopes. P_0 and Q_0 are initial output power setting points at rated frequency ω_0 and voltage amplitude V_0.

Since PCC voltage and system frequency is defined by this inverter, as Fig.3 (a) shows, it can be regarded as a controlled voltage source. The small-signal terminal-characteristic model is shown in Fig.3 (b). The output impedance $\mathbf{Z}_{odq}(s)$ describes the relationship between the small-signals of PCC voltage $\hat{\mathbf{v}}_{PCCdq}$ and output

current $\hat{\mathbf{i}}_{odq}$. The distribution cable impedance Z_C is included in $\mathbf{Z}_{odq}(s)$. The transfer function $\mathbf{G}_{\omega i}(s)$ is the novel terminal characteristic proposed by [17], describing the interaction from the excitation of $\hat{\mathbf{i}}_{odq}$ to its response of fundamental angular frequency $\hat{\omega}$ in small-signal sense. In order to represent the small-signal terminal characteristics of all paralleled inverters in one common SRF, the SRF of this droop-controlled voltage source is chosen to be system reference SRF, which is remarked by the superscript s. Since detailed modeling process of $\mathbf{Z}_{odq}(s)$ and $\mathbf{G}_{\omega i}(s)$ has already been discussed in [17], no more repetition will be presented in this paper.

B. PLL Synchronized CCI

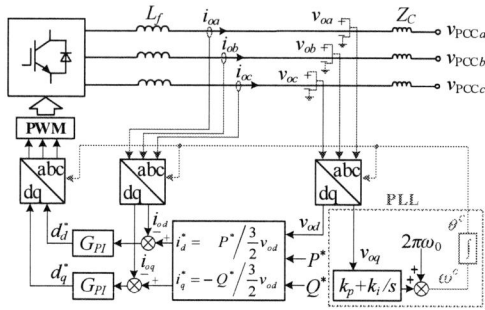

Fig. 4. Block diagram of a PLL synchronized CCI.

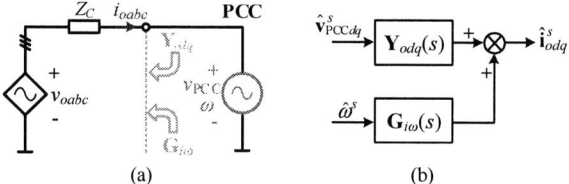

(a) (b)

Fig. 5. Terminal characteristic model of CCI.

As Fig. 4 shows, the CCI measures the voltage amplitude and frequency at its output terminal through a SRF-PLL. The angular position of this dq reference frame is controlled to drive v_{oq} to zero. Therefore, only v_{od} is needed to calculate the output current references according to P^* and Q^*. PI controllers G_{id} and G_{iq} ensure that the output current i_{od} and i_{oq} track the calculated references i_d^* and i_q^*. Since oscillation in PCC voltage and system frequency may influence the operation of CCI, in small-signal modeling, v_{PCC} and ω^s are considered as disturbances to the output current of CCI. In Fig.5, the output admittance $\mathbf{Y}_{odq}(s)$ represents the dynamic interaction between $\hat{\mathbf{i}}_{odq}$ and $\hat{\mathbf{v}}_{PCCdq}$. The transfer function $\mathbf{G}_{i\omega}(s)$ describes the relationship between $\hat{\mathbf{i}}_{odq}$ and $\hat{\omega}$. Due to the dynamic of PLL, there exists a phase angle difference between CCI's individual SRF and system SRF. All small-signal variables are converted to system SRF in Fig.5.

According to the power stage shown in Fig.5 (a), the equations of v_{odq}, v_{pccdq}, i_{odq} and ω can be derived as (2) in system SRF. After applying small-signal linearization to all circuit variables including system fundamental

1525

angular frequency, the small-signal model can be obtained in Fig. 6, whose corresponding transfer functions are in (3).

$$L_C \frac{di_{od}}{dt} + r_C i_{od} - \omega i_{oq} = v_{od} - v_{PCCd}$$

$$L_C \frac{di_{oq}}{dt} + r_C i_{oq} + -\omega i_{od} = v_{oq} - v_{PCCq} \tag{2}$$

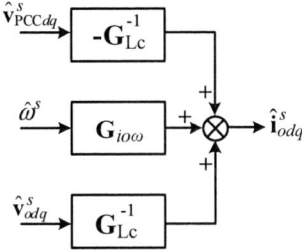

Fig. 6. Small-signal model of the power stage.

$$\mathbf{G}_{Lc} = \begin{bmatrix} sL_C + r_C & -\Omega_0 L_C \\ \Omega_0 L_C & sL_C + r_C \end{bmatrix}$$

$$\mathbf{G}_{io\omega} = \mathbf{G}_{Lc}^{-1} \cdot \begin{bmatrix} L_C I_{oq} \\ -L_C I_{od} \end{bmatrix} \tag{3}$$

As for the modeling of inner control loop, the variables should be represented in CCI's individual SRF. As Fig.4 shows, the fundamental angular frequency of this SRF ω^c is generated by PLL. The block diagram in Fig.7 shows the small-signal model between $\hat{\mathbf{v}}_{odq}^c$ and $\hat{\mathbf{i}}_{odq}^c$, whose transfer functions are in (4).

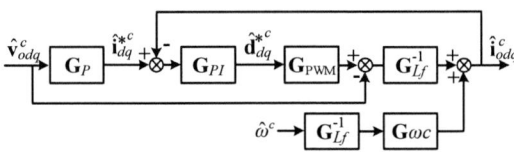

Fig. 7. Small-signal model of inner control loop.

$$\mathbf{G}_P = \frac{2}{3} \begin{bmatrix} -\dfrac{P^*}{V_{od}^2} & 0 \\ \dfrac{Q^*}{V_{od}^2} & 0 \end{bmatrix} \quad \mathbf{G}_{PI} = \begin{bmatrix} k_{pc} + k_{ic}/s & 0 \\ 0 & k_{pc} + k_{ic}/s \end{bmatrix}$$

$$\mathbf{G}_{PWM} = \begin{bmatrix} \dfrac{V_{dc}}{2} & 0 \\ 0 & \dfrac{V_{dc}}{2} \end{bmatrix} \quad \mathbf{G}_{Lf} = \begin{bmatrix} sL_f + r_f & -\Omega_0 L_f \\ \Omega_0 L_f & sL_f + r_f \end{bmatrix} \tag{4}$$

$$\mathbf{G}_{\omega c} = \begin{bmatrix} L_f I_{oq} \\ -L_f I_{od} \end{bmatrix}$$

Since terminal characteristics $\mathbf{Y}_{odq}(s)$ and $\mathbf{G}_{io\omega}(s)$ are all defined in system SRF, it is necessary to derive the quantitative relationship between these two SRFs for building the complete small-signal model of CCI [19]. As Fig.8 (a) shows, the $\hat{\mathbf{v}}_{odq}^s$ represented in system SRF are

first transformed to *abc* frame, then converted to CCI's SRF. After combing the inverse Park transformation and Park transformation together, as well as two integrators, Fig.8 (a) can be simplified as Fig.8 (b).

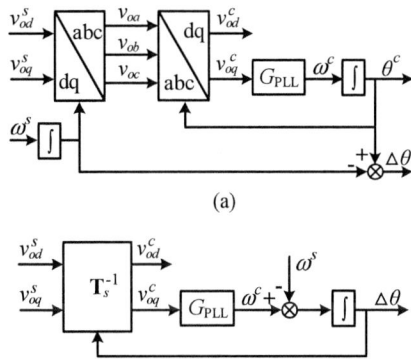

(a)

(b)

Fig. 8. Block diagram of PLL. (a) Orignial; (b) Simplified.

In Fig. 8, the phase angle difference of two SRFs is represented by $\Delta\theta$, whose expression is in (5). The transform matrix from system SRF to CCI's SRF is expressed in (6).

$$\Delta\theta = \int \left(G_{PLL} v_{oq}^c - \omega^s \right) dt \tag{5}$$

$$\begin{bmatrix} v_{od}^c \\ v_{oq}^c \end{bmatrix} = \mathbf{T}_s^{-1}(\Delta\theta) \begin{bmatrix} v_{od}^s \\ v_{oq}^s \end{bmatrix} = \begin{bmatrix} \cos(\Delta\theta) & \sin(\Delta\theta) \\ -\sin(\Delta\theta) & \cos(\Delta\theta) \end{bmatrix} \begin{bmatrix} v_{od}^s \\ v_{oq}^s \end{bmatrix} \tag{6}$$

After small-signal linearization, the small-signal model of $\Delta\theta$ is expressed in (7).

$$\Delta\hat{\theta} = G_{\theta v} \hat{\mathbf{v}}_{odq}^s + G_{\theta\omega} \hat{\omega}^s \tag{7}$$

where

$$G_{\theta v} = \begin{bmatrix} 0 & \dfrac{G_{PLL}}{s + G_{PLL} \cdot V_{od}} \end{bmatrix} \quad G_{\theta\omega} = -\dfrac{1}{s + G_{PLL} \cdot V_{od}}$$

$$G_{PLL} = k_p + k_i/s$$

Based on (6), the transformation equation from $\hat{\mathbf{v}}_{odq}^c$ to $\hat{\mathbf{v}}_{odq}^s$ is obtained as (8). Similarly, the transformation equation from $\hat{\mathbf{i}}_{odq}^s$ to $\hat{\mathbf{i}}_{odq}^c$ is obtained as (9). The $\Delta\theta_0$ represents the steady-state phase angle difference of two SRFs.

$$\begin{bmatrix} \hat{v}_{od}^s \\ \hat{v}_{oq}^s \end{bmatrix} = \mathbf{T}_s(\Delta\theta_0) \begin{bmatrix} \hat{v}_{od}^c \\ \hat{v}_{oq}^c \end{bmatrix} + \mathbf{T}_v \Delta\hat{\theta}$$

where

$$\mathbf{T}_s = \begin{bmatrix} \cos(\Delta\theta_0) & -\sin(\Delta\theta_0) \\ \sin(\Delta\theta_0) & \cos(\Delta\theta_0) \end{bmatrix} \tag{8}$$

$$\mathbf{T}_v = \begin{bmatrix} -V_{od}\sin(\Delta\theta_0) - V_{oq}\cos(\Delta\theta_0) \\ V_{od}\cos(\Delta\theta_0) - V_{oq}\sin(\Delta\theta_0) \end{bmatrix}$$

$$\begin{bmatrix} \hat{i}_{od}^c \\ \hat{i}_{oq}^c \end{bmatrix} = \mathbf{T}_s^{-1}(\Delta\theta_0)\begin{bmatrix} \hat{i}_{od}^s \\ \hat{i}_{oq}^s \end{bmatrix} + \mathbf{T}_c\Delta\hat{\theta}$$

where   (9)

$$\mathbf{T}_c = \begin{bmatrix} -I_{od}\sin(\Delta\theta_0) + I_{oq}\cos(\Delta\theta_0) \\ -I_{od}\cos(\Delta\theta_0) - I_{oq}\sin(\Delta\theta_0) \end{bmatrix}$$

Based on (3) to (9), the algebraic expression of CCI's terminal characteristics $\mathbf{Y}_{odq}(s)$ and $\mathbf{G}_{i\omega}(s)$ can be derived in (10) and (11).

$$\mathbf{Y}_{odq}(s) = \left[\mathbf{I} - \mathbf{G}_{Lc}^{-1}\left[\left[\mathbf{I} - (\mathbf{T}_s\mathbf{K}\mathbf{T}_c + \mathbf{T}_v)\mathbf{G}_{\theta v}\right]^{-1}\cdot \mathbf{T}_s\mathbf{K}\mathbf{T}_s^{-1}\right]\right]^{-1}\cdot\left(-\mathbf{G}_{Lc}^{-1}\right) \quad (10)$$

$$\mathbf{G}_{i\omega}(s) = \left[\mathbf{I} - \mathbf{G}_{Lc}^{-1}\left[\left[\mathbf{I} - (\mathbf{T}_s\mathbf{K}\mathbf{T}_c + \mathbf{T}_v)\mathbf{G}_{\theta v}\right]^{-1}\cdot \mathbf{T}_s\mathbf{K}\mathbf{T}_s^{-1}\right]\right]^{-1}$$
$$\cdot\left[\mathbf{G}_{i\omega\omega} + \mathbf{G}_{Lc}^{-1}\left[\mathbf{I} - (\mathbf{T}_s\mathbf{K}\mathbf{T}_c + \mathbf{T}_v)\mathbf{G}_{\theta v}\right]^{-1}\cdot(\mathbf{T}_s\mathbf{K}\mathbf{T}_c + \mathbf{T}_v)\mathbf{G}_{\theta\omega}\right] \quad (11)$$

where

$$\mathbf{K} = (\mathbf{G}_{PWM}\mathbf{G}_{PI}\mathbf{G}_P + \mathbf{G}_{oc}\begin{bmatrix} 0 & G_{PLL}\end{bmatrix} - \mathbf{I})^{-1}\cdot\left(\mathbf{G}_{Lf} + \mathbf{G}_{PWM}\mathbf{G}_{PI}\right)$$

III. STABILITY CRITERION FOR PARALLEL INVERTERS

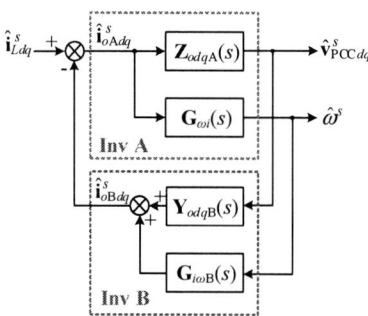

Fig. 9. Small-signal model of parallel inverters.

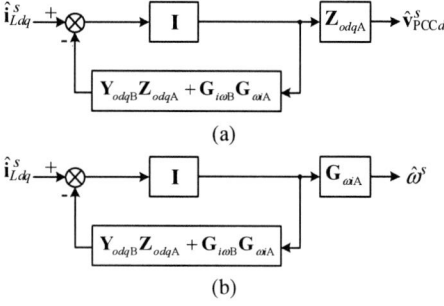

(a)

(b)

Fig. 10. Equivalent small-signal model of parallel inverters.

$$\mathbf{Z}_{sodq}(s) = \frac{\hat{\mathbf{v}}_{PCCdq}^s}{\hat{\mathbf{i}}_{Ldq}^s} = \mathbf{Z}_{odqA}(\mathbf{I} + \mathbf{Y}_{odqB}\mathbf{Z}_{odqA} + \mathbf{G}_{i\omega B}\mathbf{G}_{\omega iA})^{-1}$$

$$\mathbf{G}_{\omega is}(s) = \frac{\hat{\omega}^s}{\hat{\mathbf{i}}_{Ldq}^s} = \mathbf{G}_{\omega iA}(\mathbf{I} + \mathbf{Y}_{odqB}\mathbf{Z}_{odqA} + \mathbf{G}_{i\omega B}\mathbf{G}_{\omega iA})^{-1} \quad (12)$$

For parallel droop controlled inverters, covering the interactions of fundamental frequency, a stability

criterion is proposed in [18] to analyze system stability. This stability criterion is applied to parallel droop-controlled Inverter A and current-controlled Inverter B in Fig.1. As Fig.9 shows, the small-signal terminal characteristic models of inverter A and B are combined together. In islanded microgrids, the parallel system is required to provide stable voltage and frequency to load. Therefore, the small-signal model of parallel inverters in Fig.9 represents the relationship from the excitation in load current $\hat{\mathbf{i}}_{Ldq}^s$ to its response in PCC voltage $\hat{\mathbf{v}}_{PCCdq}^s$ and system fundamental angular frequency $\hat{\omega}^s$, which can be divided into two equivalent block diagrams in Fig.10. The transfer functions are presented in (12). As (12) shows, the overall system transfer functions $\mathbf{Z}_{sodq}(s)$ and $\mathbf{G}_{\omega i}(s)$ can be expressed as series of a feedback loop and terminal characteristic transfer functions of inverter A.

Normally, the design of inner control parameters of single inverter should ensure the stable operation in standalone mode. Thus, there is no right-half-plane (RHP) poles in $\mathbf{Z}_{odqA}(s)$, $\mathbf{G}_{\omega iA}(s)$ and $\mathbf{Y}_{odqB}(s)$ and $\mathbf{G}_{i\omega B}(s)$. The overall system stability is determined by the feedback loop. According to GNC, the interaction stability criterion of parallel inverters in Fig.9 can be expressed as:

$$Z_{RHP} = -N_L \quad (13)$$

where Z_{RHP} denotes the number of RHP poles in $\mathbf{Z}_{oSdq}$ and $\mathbf{G}_{\omega S}$. N_L is the net sum of anticlockwise encirclements of the $(-1+j0)$ point by characteristic loci of the return ratio matrix $\mathbf{L}(s)$ shown in (14).

$$\mathbf{L}(s) = \mathbf{I} + \mathbf{Y}_{odqB}\mathbf{Z}_{odqA} + \mathbf{G}_{i\omega B}\mathbf{G}_{\omega iA} \quad (14)$$

The parallel system is stable if and only if N_L is equal to zero, which means no RHP poles in system transfer functions $\mathbf{Z}_{sodq}(s)$ and $\mathbf{G}_{\omega i}(s)$. Therefore, the interaction stability can be accessed by terminal characteristics of single droop-controlled and current-controlled inverters.

IV. STABILITY ANALYSIS

In this section, the impact of key factors including bandwidth of PLL in CCI and voltage control bandwidth of droop controlled inverter on system interaction stability are studied based on terminal characteristics of individual inverters and the stability criterion proposed in section III.

The calculation of transfer functions are conducted in MATLAB. In accordance with Fig.1, a simulation prototype of one droop droop-controlled inverter and one current- controlled inverter is built in PSCAD/EMTDC. The circuit and control parameters in both modeling and simulation are listed in Table I.

For grid-tied CCI with current feedback control and PLL, Wen *et al.* proved that its q-q channel impedance behaves as a negative resistor. High bandwidth of PLL increases the frequency range of this negative resistor, which could destabilize the system [12]. In accordance with this conclusion in previous work, as Fig.11 shows, the same negative resistor behavior, i.e. negative admittance, could also be observed in inverter B output admittance Y_{oqq}. The bode plots of $\mathbf{G}_{i\omega B}(s)$ are also

1527

significantly changed with PLL bandwidth as Fig.12 shows.

According to Fig.11, the overall system is under worst stability condition when the bandwidth of CCI's PLL increases to case III: 50Hz. In time-domain simulation results shown in Fig.13, the parallel system lose stability after 2s when the parameters of PLL regulator change from case I: k_p=0.31, k_i=0.12 to case III: k_p=1.57, k_i=15.5.

Although increasing PLL bandwidth in CCI might incur instability, for parallel inverters, expanding the bandwidth of voltage control loop in droop-controlled inverter can effectively sustain stable operation. When CCI's PLL parameters are fixed as case III, the initial voltage control parameters in Table I (case A: k_{pv}=0.1, k_{iv}=100) of Inverter A are changed to case B: k_{pv}=0.5, k_{iv}=120 and case C: k_{pv}=0.7 k_{iv}=170. From characteristic loci of $\mathbf{L}$(s) in Fig.14, it can be seen that as the increasing of Inverter A's voltage control bandwidth, the system changes from unstable to stable state. This conclusion is verified by simulation results in Fig. 15. Under the case C design of Inverter A's voltage controller, despite larger noises on locked system frequency, the system maintains stable operation even after the CCI's PLL bandwidth changes from case I: 10Hz to case III: 50Hz.

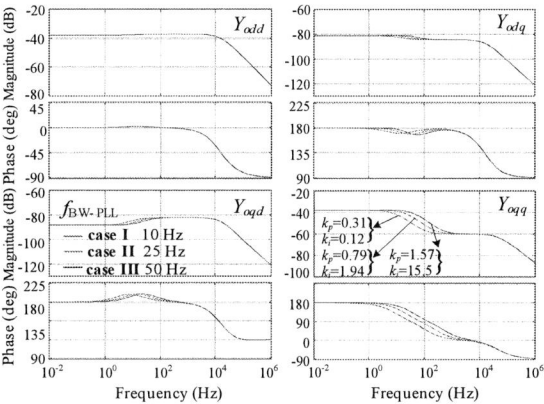

Fig. 11. Bode plot of $\mathbf{Y}_{odq0}(s)$ with different PLL bandwidth.

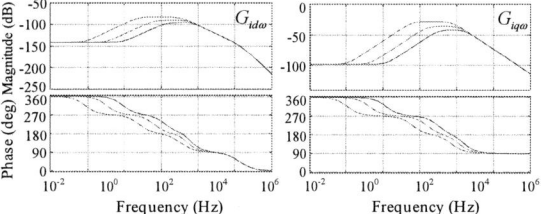

Fig. 12. Bode plot of $\mathbf{G}_{iB}(s)$ with different PLL bandwidth.

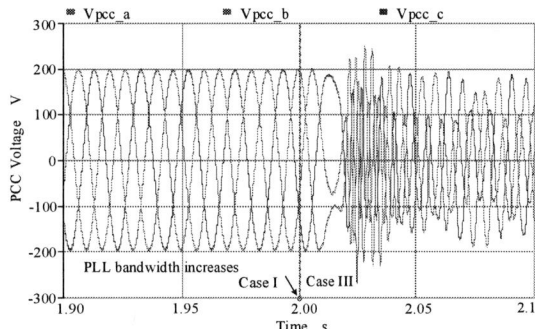

Fig. 13. PCC voltage waveform when PLL bandwidth changes from case I: 10Hz to case III: 50Hz.

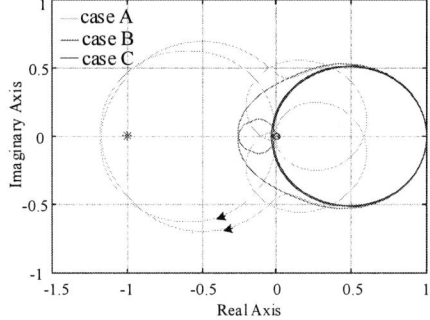

Fig. 14. Characteristic loci of $\mathbf{L}$(s) with different Inverter A's voltage control bandwidth.

TABLE I   CONTROL AND CIRCUIT PARAMETERS

Inverter A	
Parameters	Value
Inverter rated capacity S	6 kVA
Rated fundamental angular frequency ω_0	2π·50 rad/s
Rated voltage magnitude V_0	200 V
Active power bias P_0	1500 W
Reactive power bias Q_0	0 Var
P-ω^* droop slope m_p	1×10^{-4} rad/(s·W)
Q-V^* droop slope n_q	1×10^{-3} V/Var
Filter inductor L_{fA}	3 mH
Parasitic resistor of filter inductor r_{fA}	0.02Ω
Filter capacitor C_f	30 μF
Distribution cable inductance L_{CA}	2 mH
Distribution cable inductance r_{CA}	0.1 Ω
Voltgae proportional coefficient k_{pv}	0.1
Voltgae integral coefficient k_{iv}	100

Inverter B	
Parameters	Value
Inverter rated capacity S	6 kVA
Active power reference P^*	500 W
Reactive power reference Q^*	0 Var
PLL proportional coefficient k_p	0.31
PLL integral coefficient k_i	0.12
Filter inductor L_{fB}	3 mH
Parasitic resistor of filter inductor r_{fB}	0.02Ω
Distribution cable inductance L_{CB}	1 mH
Distribution cable inductance r_{CB}	0.05Ω
Current proportional coefficient k_{pc}	2
Current integral coefficient k_{ic}	100

The 2018 International Power Electronics Conference

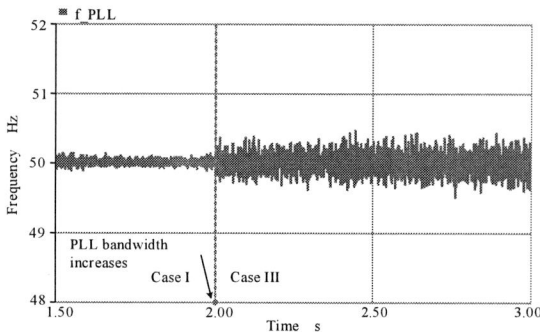

Fig. 15. CCI's locked frequency with case C design of voltage controller in droop-controlled inverter.

V. CONCLUSION

Considering the dynamic of system fundamental angular frequency, this paper conducts the small-signal terminal-characteristics model of droop-controlled inverter and current-controlled inverter. Not only conventional output impedance and admittance, but also the terminal-characteristics representing the dynamic interactions of transmitted current and system fundamental angular frequency are covered in this model. Based on GNC, the interaction stability of the parallel inverters are studied. The analysis results suggest that increasing PLL bandwidth of current-controlled inverter could destabilize the system while expanding the bandwidth of voltage control loop in droop controlled inverter could help sustain system stability.

REFERENCES

[1] "PARIS AGREEMENT Signature Ceremony" [Online]. United Nations Framework Convention on Climate Change (UNFCCC). Available:http://newsroom.unfccc.int/media/632121/list-of representatives-to-high-level-signature-ceremony.pdf

[2] B. Kroposki, C. Pink, R. DeBlasio, H. Thomas, *et al.*, "Benefits of power electronic interfaces for distributed energy systems," *IEEE Trans. on Energy Conversion*, vol. 25, no. 3, pp. 901-908, 2010.

[3] R. Lasseter et al. White Paper on Integration of Distributed Energy Resources—The CERTS MicroGrid Concept. [Online]. Available: http://certs.lbl.gov/pdf/LBNL_50 829.pdf.

[4] R. H. Lasseter, "Microgrids and distributed generation," *Journal of Energy Engineering*, vol. 133, no.3, pp. 144-149, 2007.

[5] E. Coelho, P. Cortizo, and P. Garcia, "Small-signal stability for parallelconnected inverters in stand-alone AC supply systems," *IEEE Trans. Industry Applications*, vol. 38, no.2, pp. 533-542, 2002.

[6] N. Pogaku, M. Prodanovic, and T. C. Green, "Modeling, Analysis and Testing of Autonomous Operation of an Inverter-Based Microgrid," *IEEE Trans. on Power Electronics*, vol. 22, no.2, pp. 613-625, 2007.

[7] K. Yu, Q. Ai, S. Wang, J. Ni, and T. Lv, "Analysis and Optimization of Droop Controller for Microgrid System Based on Small-Signal Dynamic Model," *IEEE Trans. on Smart Grid*, vol. 7, pp. 695-705, 2016.

[8] A. MacFarlane and I. Postlethwaite, "The generalized Nyquist stability criterion and multivariable root loci," *International Journal of Control*, vol. 25, no.1, pp. 81-127, 1977.

[9] M. Belkhayat, "Stability criteria for ac power systems with regulated loads," Ph.D. dissertation, Purdue University, West Lafayette, IN, USA, Dec. 1997

[10] J. Sun, "Impedance-Based Stability Criterion for Grid-Connected Inverters," *IEEE Trans. Power Electronics*, vol. 26, no.11, pp. 3075-3078, 2011.

[11] X. Wang, L. Harnefors, and F. Blaabjerg, "Unified Impedance Model of Grid-Connected Voltage-Source Converters," *IEEE Trans. Power Electronics*, vol. 33, no. 2, pp. 1775-1787, 2018.

[12] B. Wen, D. Boroyevich, R. Burgos, P. Mattavelli, and Z. Shen, "Analysis of D-Q Small-Signal Impedance of Grid-Tied Inverters," *IEEE Transactions on Power Electronics*, vol. 31, pp. 675-687, 2016.

[13] Z. Liu, J. Liu, W. Bao, and Y. Zhao, "Infinity-Norm of Impedance-Based Stability Criterion for Three-Phase AC Distributed Power Systems With Constant Power Loads," *IEEE Transactions on Power Electronics*, vol. 30, no. 6, pp. 3030-3043, 2015.

[14] B. Wen, D. Dong, D. Boroyevich, R. Burgos, P. Mattavelli, and Z. Shen, "Small-Signal Stability Analysis of Three-Phase AC Systems in the Presence of Constant Power Loads Based on Measured d-q Frame Impedances," *IEEE Trans. Power Electron.*, vol. 30, no.10, pp. 26-38, 2015.

[15] Z. Liu, J. Liu, X. Hou, Q. Dou, and T. Liu, "Output Impedance Modeling and Stability Prediction of Three-Phase Paralleled Inverters with Master-Slave Sharing Scheme Based on Terminal Characteristics of Individual Inverters," *IEEE Transactions on Power Electronics*, vol.31, pp. 5306-5320, 2016.

[16] M. C. Chandorkar, D. M. Divan, and R. Adapa, "Control of parallel connected inverters in standalone AC supply systems," *IEEE Trans. Ind. Appl.*, vol. 29, no. 1, pp. 136–143, Jan./Feb. 1993.

[17] Z. Liu, J. Liu, D. Boroyevich, R. Burgos, *et al.*, "Small-signal terminal-characteristics modeling of three-phase droop-controlled inverters," in *IEEE Energy Conversion Congress and Exposition (ECCE)*, 2016, pp. 1-7.

[18] Z. Liu, J. Liu, D. Boroyevich, and R. Burgos, "Stability criterion of droop-controlled parallel inverters based on terminal-characteristics of individual inverters," in *IEEE 8th International Power Electronics and Motion Control Conference (IPEMC-ECCE Asia)*, 2016, pp. 2958-2963.

[19] Z. Liu, J. Liu, and D. Boroyevich, "Small-Signal Terminal Characteristics Modeling of Three-Phase Boost Rectifier with Variable Fundamental Frequency", in *31st Annual IEEE Applied Power Electronics Conference and Exposition*, 2016, pp. 739-745.

Direct Wireless Battery Charging System

Woo-Seok Lee, Jin-Hak Kim, Shin-Young Cho, and Il-Oun Lee*
Department of Electrical Engineering, Myongji University, Yong-in, South Korea
*E-mail: leeiloun@mju.ac.kr

Abstract— **This paper presents a direct wireless battery charging system. The output current of the series-series compensated wireless power transfer (SS-WPT) system is used as a current source, and the output voltage of AC-DC converter controls the current source. Therefore, the proposed wireless battery charging system needs no battery charging circuit carrying out charging profiles, and can solve space constraints and thermal problem in many battery applications. In addition, the proposed wireless battery charging system can implement easily most other charging profiles. In this paper, the proposed wireless battery charging system is implemented and the feasibility is verified experimentally according to constant-current constant-voltage charging profile.**

Keywords— *Wireless barttery charging, wireless power transfer, and charging profile.*

I. INTRODUCTION

The widespread use of portable electronic devices and electric vehicles has begun recently, the many electronic devices and electronic vehicles are plugged into the wall outlet via the power cable for many hours a day, and the use of wired charging apparatuses becomes a part of daily life. In these life, a large number of wire-chargers are discarded as E (electronics)-waste by contact failure such as broken wires or short-circuit problem, etc. Due to the problem of the increasing E (electronic)-wastes, there has been increasing interest in research and development of wireless power transfer (WPT) technology that can be utilized to transfer power to batteries without requiring expensive failure-prone connector.

The WPT technology provides charging system with low maintain cost, high reliability and ability to operate even in the extreme environment. However, the wireless battery charging system requires more power stages than the wired battery charging system [1]-[4]. Concretely, the wireless battery charging system needs WPT system that consists of power transmitter and power receiver. An exclusive controller is also required to regulate the output of WPT system since the power transferred to the receiver of WPT system is not regulated whenever the load changes. The inverter or converter in Power Electronics is usually controlled by three methods: pulse width modulation (PWM), frequency modulation (FM), and amplitude modulation (AM). The AM requires an additional stage for the DC-DC converter in order to control the amplitude of the input voltage. On the other hand, both PWM and FM need no additional stage since

the inverter or converter uses power semiconductor switches for the power conversion. For this reason, when the PWM or FM is applied to the power transmitter of WPT system, the power transferred to the receiver can be easily regulated. Yet, high current stress and large power loss are generated since the voltage and current in the power transmitter are not in phase. Due to this problem, regulation circuits such as synchronous rectifiers [1]-[2] or impedance tuner [3] are necessary in the receiver of WPT system. Furthermore, battery-charging circuits such as low-dropout (LDO) regulator [1]-[2] or synchronous buck converter [3] are required for the battery charging. Fig. 1 shows a conventional wireless battery charging system. As abovementioned, the conventional system consists of the following five key power stages.

1. AC-DC converter
2. Power transmitter of WPT system
3. Power receiver of WPT system
4. Regulation circuit
5. Battery charging circuit

For the wireless battery charging system, the power receiver, regulation circuit, and battery charging circuit must be embedded inside portable electronic devices and electric vehicles, but there is not enough space for these power stages. In addition to this problem, the regulation and battery-charging circuits generate huge heat and raise the thermal stress of electronic devices while it is being charged.

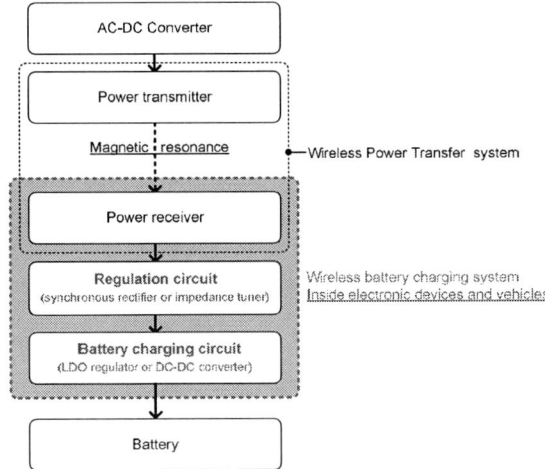

Fig. 1. Conventional wireless battery charging system.

The 2018 International Power Electronics Conference

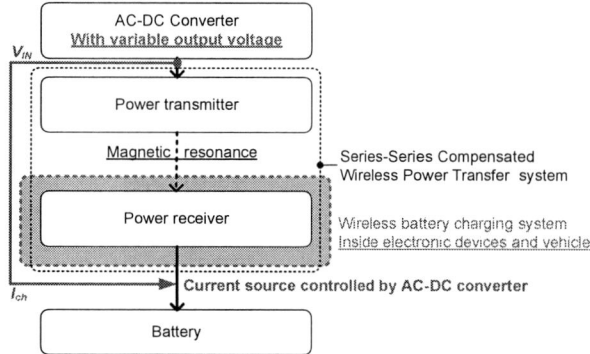

Fig. 2 Proposed direct wireless battery charging system.

In this paper, a direct wireless battery charging system is proposed. The regulation and battery-charging circuits in the conventional wireless battery charging system are removed, and the battery is charged directly from WPT system. Fig. 2 shows the proposed direct wireless battery charging system. As shown in Fig. 2, the proposed system is only made of AC-DC converter, power transmitter, and power receiver. The type of the applied WPT system is a series-series compensated wireless power transfer (SS-WPT) system, and it is connected directly to the battery. Generally, the output of SS-WPT system has the inherent characteristic as a current source. Hence, without the help of the exclusive regulation and battery-charging circuits, the battery can be charged directly from WPT system by adjusting the output voltage of the existed AC-DC converter in front of WPT system under constant-current constant-voltage (CC-CV) charging profile [5]-[11] or multi-step current charging profile [12]-[18], etc. Like this, the proposed system is simpler than the conventional system, and can be more valuable in many battery applications such as portable electronics devices and electronics vehicles, etc.

The paper is organized as follows. In Section II, the inherent current-source characteristic of SS-WPT system is described. And, the implementation of the CC-CV charging profile in the proposed system is explained In Section III. Experimental verification is presented in Section IV, and then, Section V draws the conclusion.

II. CURRENT SOURCE CHARACTERISTICS OF SERIES-SERIES COMPENSATED WIRELESS POWER TRANS

In the proposed wireless battery charging system, the SS-WPT system followed by AC-DC converter consists of half-bridge inverter, resonant tank, full-bridge rectifier, filter, and load as the battery, as shown in Fig. 3. The half-bridge inverter applies a square voltage into the resonant tank. The square voltage, V_{square} can be described as

$$V_{square}(t) = \begin{cases} V_{IN} & (0 < \omega t \le \pi) \\ 0 & (\pi < \omega t \le 2\pi) \end{cases} \tag{1}$$

The resonant tank consists of transmitter coil, receiver coil, transmitter capacitor, and receiver capacitor. The capacitors in the resonant tank resonate with coils and improve the conversion efficiency of system. Since the capacitors are in series with coils, the structure is called series-series (SS) compensation [19]-[20]. Since the series-series compensated resonant tank acts as a band pass filter, the effect of any harmonic components in the input square voltage V_{square} can be neglected, except for its fundamental component. Then, the transmitter and receiver voltages in the resonant tank are sinusoidal and expressed as V_T and V_R, respectively. The transmitter and receiver currents are also sinusoidal and expressed as I_T and I_R, respectively. The load represented as a load resistance R_L can be described with an equivalent load resistance R_{eq} [21].

$$R_{eq} = \frac{8}{\pi^2} R_L \tag{2}$$

Fig. 4(a) shows the equivalent circuit of the SS-WPT system in Fig. 3, which consists of a sinusoidal transmitter voltage V_T, capacitors, a coupled-inductor model, and an equivalent load resistance R_{eq}. The components in the receiver side are transformed into the transmitter side by the effective transformer turns-ratio, n. Here, the n can be expressed as

$$n = \sqrt{L_T / L_R}, \tag{3}$$

where L_T and L_R are the transmitter coil's self-inductance and receiver coil's self-inductance, respectively.

The modified equivalent circuit of SS-WPT system can be represented with coupling coefficient κ, L_T, and C_T as shown in Fig. 4(b). Fig. 5 shows the modified equivalent

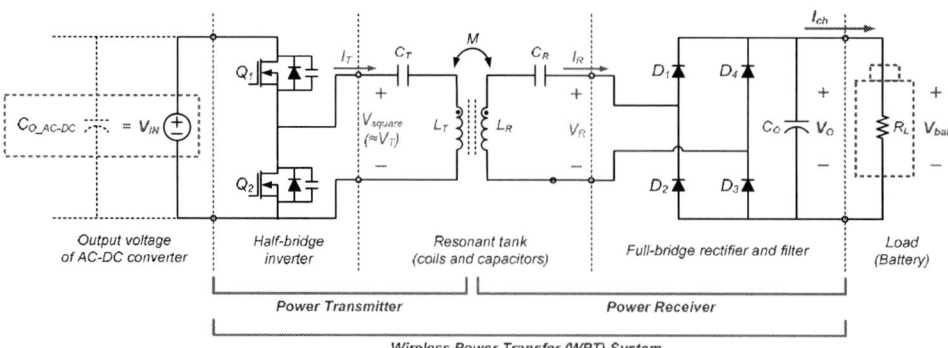

Fig. 3. Series-series compensated wireless power transfer circuit in the proposed system.

1531

circuit of SS-WPT system in frequency-domain. The series impedance Z_S and parallel impedance Z_P are expressed as:

$$Z_S = j\omega L_T\left(1-\frac{\omega^2}{\omega_r^2}\right) - j\kappa\omega L_T = -j\kappa\omega_r L_T = -j\kappa Z_O, \quad (4)$$

$$Z_P = j\kappa\omega L_T = j\kappa\omega_r L_T = j\kappa Z_O, \quad (5)$$

where ω_r is the resonant angular frequency and Z_O is the characteristic impedance. The Z_S and Z_P have the same magnitude, but different sign. To define the characteristics between input and output parameters, the resonant tank can be described as the transmission matrix in eq. (6).

$$\begin{bmatrix} V_T \\ I_T \end{bmatrix} = \begin{bmatrix} 0 & -j\kappa Z_O \\ -j\dfrac{1}{\kappa Z_O} & 0 \end{bmatrix} \begin{bmatrix} nV_R \\ I_R/n \end{bmatrix} \quad (6)$$

From eq. (6), the relationship between input and output parameters can be obtained as

$$V_T = -j\kappa Z_O I_R / n, \quad (7)$$

$$I_T = -j\kappa\frac{1}{Z_O}nV_R. \quad (8)$$

In eq. (7), it is known that the sinusoidal receiver current, I_R is proportional to the sinusoidal transmitter voltage V_T. This means that SS-WPT system driven by the fixed output voltage of AC-DC converter behaves as a constant current-source. In addition, if the magnitude of the output voltage of AC-DC converter is adjusted, the SS-WPT system is able to be a voltage controlled current source at the resonant frequency. From eq. (7), the battery charging current I_{ch} in the SS-WPT system in Fig. 3 can be described with the input voltage V_{IN} as follows.

$$I_{ch} = \frac{4}{\pi^2}\cdot\frac{nV_{IN}}{\kappa Z_O}. \quad (9)$$

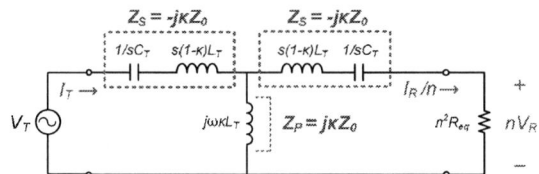

(a)

(b)

Fig. 4. Equivalent circuits of series-series compensated wireless power transfer system: (a) equivalent circuit and (b) modified equivalent circuit.

Fig. 5. Equivalent circuit of series-series compensated wireless power transfer system in frequency domain.

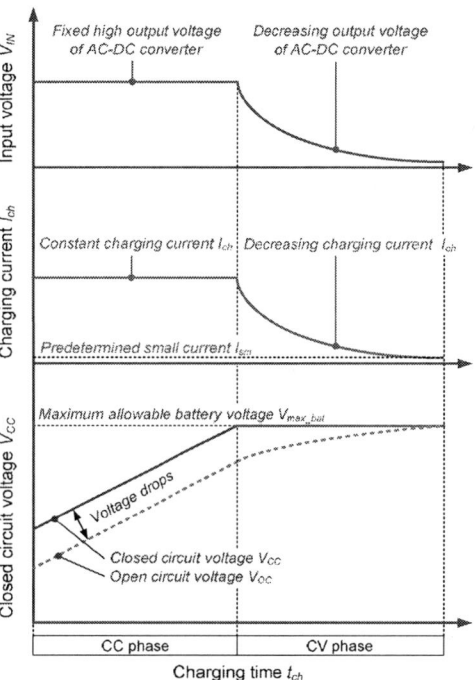

Fig. 6. Profile of constant-current constant-voltage charging in the proposed direct wireless battery charging system.

Then, the output voltage V_O can be expressed with the I_{ch} as in eq. (10).

$$V_O = I_{ch}R_L = \frac{4}{\pi^2}\cdot\frac{nV_{IN}}{\kappa Z_O}R_L. \quad (10)$$

Because the output voltage V_O of SS-WPT system is clamped at the battery voltage V_{bat}, the load resistance R_L can be determined from the V_{bat} and I_{ch}.

$$R_L = \frac{V_{bat}}{I_{ch}}. \quad (11)$$

III. IMPLEMENTATION OF CC-CV CHARGING

When the battery is not charged, it can be approximately modeled as only a capacitor with high capacitance C_{bat}. At this time, the battery voltage V_{bat} is called as open circuit voltage V_{OC}. On the other hand, when the battery is charged by the charging current I_{ch}, the closed loop is formed. Due to the IR voltage drop V_η across the overpotential resistance R_η and the voltage drop V_P by polarization, the V_{bat} deviates from the open circuit voltage V_{OC}. In this case, the V_{bat} is called as closed circuit voltage V_{CC} and is expressed as

$$V_{CC} = V_{OC} + V_\eta + V_P. \quad (12)$$

To charge the battery fully, the open circuit voltage V_{OC} should arrive at the maximum allowable battery voltage V_{max_bat}. However, the two voltage drops of V_η and V_P prevent the open circuit voltage V_{oc} from reaching the V_{max_bat}. To solve this, the charging current I_{ch} should be reduced to a predetermined small value I_{sm} at the end-of-charge because the V_η and V_P are proportional to the I_{ch}.

A. CC-CV charging profile

On the CC phase of the CC-CV charging, the relative high current is required for I_{ch} for the fast charge. Since the charging current I_{ch} of SS-WPT system is proportional to the input voltage V_{IN} as in eq. (9), this means that the output voltage of AC-DC converter in the proposed direct wireless charging system (or V_{IN}) must be set high. According to this principle, a fixed high V_{IN} is requited for a constant high I_{ch} in the proposed system and then, the I_{ch} will charge the battery directly until the closed circuit voltage V_{CC} reaches the maximum allowable battery voltage V_{max_bat}, as shown in Fig. 6.

On the CV phase, the I_{ch} begins to decreasing and the decreasing I_{ch} reduces the two voltage drops of V_η and V_P to allow the open circuit voltage V_{OC} to reach the maximum allowable battery voltage V_{max_bat}. At the end of CV phase, the I_{ch} reaches the predetermined small current I_{sm}. For this mechanism, the proposed system must reduce the output voltage of AC-DC converter on CV phase. As a result, the I_{ch} of SS-WPT system will decrease and the open circuit voltage V_{OC} will continues to increase. The two voltage drops V_η and V_P will decrease gradually on the CV phase. On the other hand, the closed circuit voltage V_{OC} will be hold at the maximum allowable battery voltage V_{max_bat} due to the constant sum of the increased open circuit voltage V_{OC} and the decreased two voltage drops V_η and V_P.

When CC-CV charging is implemented in the conventional system, any transient phase may occur during the switching time from CC control mode to CV control mode [7]. However, in the proposed wireless battery charging system, the current source characteristic of SS-WPT system is used on the CC phase and only the voltage feedback loop is required on the CV phase. This means that in the proposed system, the CC phase is always switched to the CV phase smoothly.

B. Other charging profile

In the proposed wireless charging system, the SS-WPT system is used as a current source. The current source can be controlled by adjusting the output voltage of AC-DC converter. That is, the proposed wireless battery charging system acts as a voltage controlled current source. Since the voltage controlled current source makes easily various magnitudes of current, it is noted that most other charging techniques such as multi-step current charging [12]-[18], constant power charging [22-24], boost charging [25], varying current decay [26-27] can be implemented easily by the use of the proposed system.

IV. EXPERIMENTAL VERIFICATION

A. Experimental conditions

A prototype is made to verify the proposed wireless battery charging system. Symmetric wire-wound spiral coils (inner diameter: 30mm and outer diameter: 70mm) with ferrite sheet (width: 90mm, length: 75mm, and height: 500μm) are fabricated as in Fig. 7. Symmetric wire-wound spiral coils are separated by the air gap of 15mm. The thickness of transmitter case is 1mm, and the thickness of receiver case is 1mm. The parameters for the experiment are listed in Table I. The lithium-ion batteries used for the experiment are Samsung prismatic 3100mAh batteries, which have the minimum allowable battery voltage V_{min_bat} of 3.4V and the maximum allowable battery voltage V_{max_bat} of 4.3V. The battery is charged according to the CC-CV charging profile. The input voltage V_{IN}, the charging current I_{ch}, and the closed circuit voltage V_{CC} are recorded by digital recorder MV1000 simultaneously.

Fig. 8 shows the measured I_{ch} by changing the input voltage V_{IN} of SS-WPT system. As shown in Fig. 8, the charging current I_{ch} of the proposed system increases linearly from 100mA to 1000mA when adjusting the V_{IN} from 3.61V to 27.46V. From this result, it is confirmed that the SS-WPT system behaves as a voltage controlled current source at the resonant frequency. The voltage controlled current source will charge the two batteries from the minimum allowable battery voltage V_{min_bat} of 3.4V to the maximum allowable battery voltage V_{max_bat} of 4.3V. And the I_{ch} at the end of CV phase is set to the predetermined small current I_{sm} of 100mA.

B. CC-CV charging profile

Fig. 9 shows the measured curves of CC-CV charging in the proposed system. The input voltage V_{IN} is set at 27.46V to obtain the charging current I_{ch} of 1000mA on the CC phase. Then, the I_{ch} of 1000mA starts to charge the battery. The closed circuit voltage V_{CC} increases sharply due to the two voltage drops of V_η and V_P at the start point of the CC phase, and then reaches the maximum allowable battery voltage V_{max_bat} of 4.3V from 3.9V, as shown in Fig. 9. Fig. 10 shows the measured key waveforms in the SS-WPT system while the closed circuit voltage V_{CC} moves from 3.9V to 4.3V. As shown in Fig. (10), the key waveforms at the V_{CC} of 3.9V is nearly similar to that at 4.3V. At the end of CC phase, the V_{CC} reaches the V_{max_bat} of 4.3V, but the open circuit voltage V_{OC} will be not 4.3V due to the two voltage drops V_η and V_P.

TABLE I. PARAMETERS FOR EXPERIMENT

Symbol	Descriptions	Values/Part name
L_T	Self-inductance of transmitter coil	50μH
L_R	Self-inductance of receiver coil	50μH
κ	Coupling coefficient	0.4
n	Turn ratio	1 (27:27)
C_T	Transmitter capacitance	78nF
C_R	Receiver capacitance	78nF
Q_1-Q_2	Switches	50CN10N
D_1-D_4	Diodes	PMEG4030ER
f_r	Resonant frequency	80kHz

1533

The 2018 International Power Electronics Conference

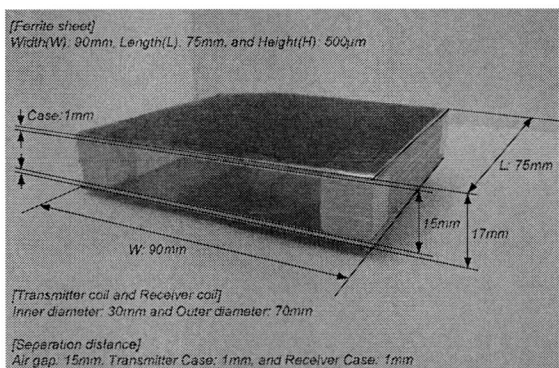

[Ferrite sheet]
Width(W): 90mm, Length(L): 75mm, and Height(H): 500μm

Case: 1mm

L: 75mm

15mm 17mm

W: 90mm

[Transmitter coil and Receiver coil]
Inner diameter: 30mm and Outer diameter: 70mm

[Separation distance]
Air gap: 15mm, Transmitter Case: 1mm, and Receiver Case: 1mm

Fig. 7. Symmetric wire-wound spiral coils with ferrite sheet.

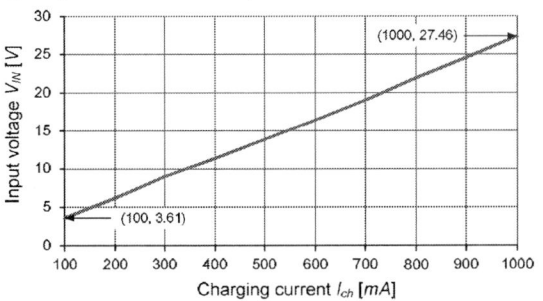

Fig. 8. Measured relationship between input voltage and charging current in series-series compensated wireless power transfer system.

On the CV phase after the CC phase, the V_{IN} is reduced in order to decrease the I_{ch} for charging the battery fully. Then, because the two voltage drops of V_η and V_P is reduced, the battery continues to be charged by the gradually decreasing. Fig. 11 shows the key waveforms in SS-WPT system on the CV phase. Since the WPT system operates at the resonant frequency, the input square voltage V_{square} and the transmitter resonant current

I_T are almost in phase. Also, the receiver resonant current I_R leads the I_T by about $\pi/2$. When the I_{ch} reaches the predetermined small current I_{sm} of 100mA, the CC-CV charging profile ends. The total charging time t_{ch} is 13540 seconds. Concretely, the charging times on the CC phase is 9238 seconds and the CV phase needs a time of 4302 seconds. With this experiment, it is also confirmed that the total charging time t_{ch} measured by the proposed direct wireless battery charging system is almost equal to that of the conventional wired battery charging system (13200 seconds) under the same condition.

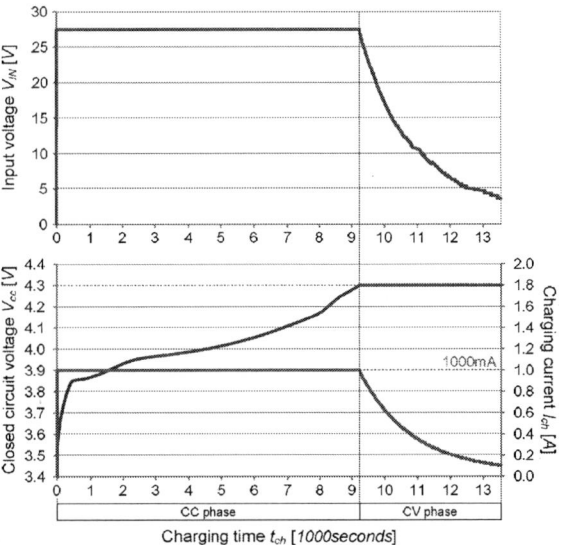

Fig. 9. Measured curves of CC-CV charging in the proposed wireless battery charging method.

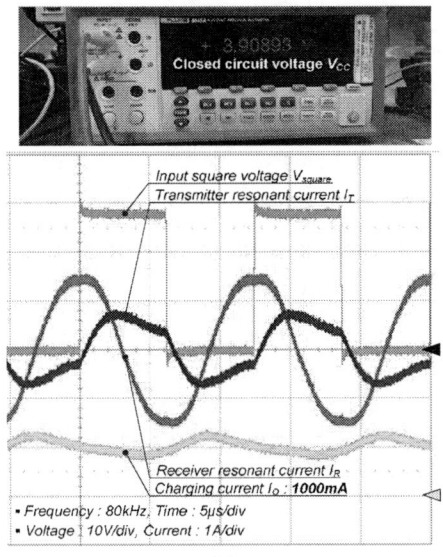

(a)

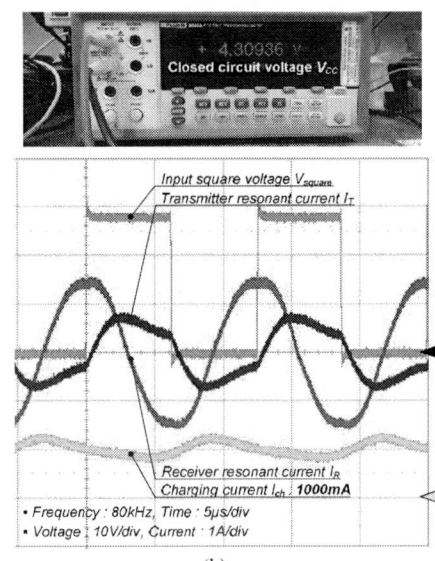

(b)

Fig. 10. Key waveforms of series-series compensated wireless power transfer system in constant-current phase: (a) closed circuit voltage of 3.9V and load resistance 3.9Ω at the middle of constant-current phase and (b) closed circuit voltage of 4.3V and load resistance 4.3Ω at the end of constant-current phase.

1534

The 2018 International Power Electronics Conference

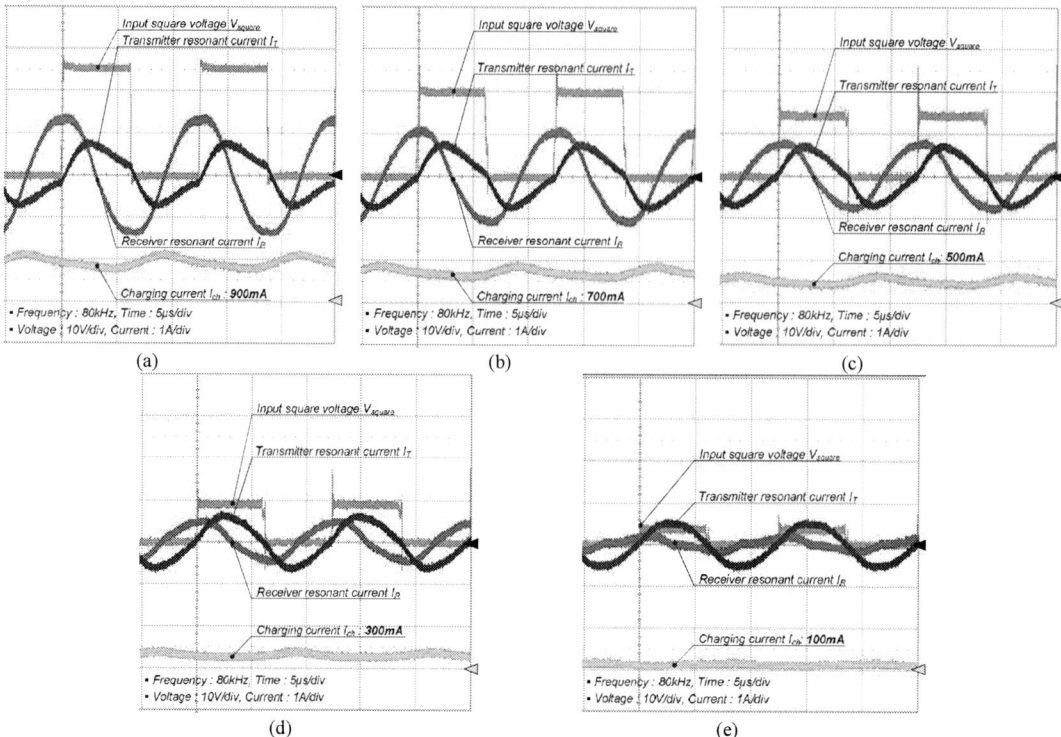

Fig. 11. Key waveforms of series-series compensated wireless power transfer system in constant-voltage phase: (a) charging current of 900mA, (b) charging current of 700mA, (c) charging current of 500mA, (d) charging current of 300mA, and (e) charging current of 100mA.

V. CONCLUSIONS

This paper proposes a direct wireless battery charging system. Because the proposed system takes advantage of the inherent current-source characteristic of SS-WPT system, the regulation and battery charging circuits in the conventional system are removed and hence it can be implemented easily and simply in many battery applications such as portable electronics devices and electric vehicles, etc. In addition, a voltage controlled current source implemented by adjusting the output voltage of AC-DC converter in the proposed system can charge directly the battery. This enables to implement easily various charging profiles such as CC-CV, MCC, etc. without an exclusive current feedback loop. In this paper, the abovementioned merits of the proposed system was analyzed in details and the feasibility was verified with the experiment based on a prototype of the proposed system, lithium-ion batteries of Samsung prismatic 3100mAh, and CC-CV charging profile. From the results, it can be insisted that the proposed system will be more competitive than the conventional system.

REFERENCES

[1] Why Not A Wire? The case for wireless power, Texas Instruments [Online] Available: http://www.wirelesspowerconsortium.com/technology/why-not-a-wire-the-case-for-wireless-power.html

[2] Texas Instruments (TI), "bq51050B and bq51051B Datasheet", http://www.ti.com/lit/ds/symlink/bq51050b.pdf, Jan 2014

[3] Linear Technology, "LTC4120 Datasheet", http://cds.linear.com/docs/en/

datasheet/4120fa.pdf, 2013

[4] Pengfei Li and Bashirullah, R., "A Wireless Power Interface for Rechargeable Battery Operated Medical Implants," *Circuits and Systems II: Express Briefs, IEEE Transactions on*, vol.54, no.10, pp.912,916, Oct. 2007

[5] S.S. Zhang, K. Xu, and T.R. Jow, "Study of the charging process of a LiCoO2-based Li-ion battery", *Journal of Power Sources*, vol. 160, no. 2, pp1349-1354, October 2006

[6] S. Dearborn, "Charging Li-ion batteries for maximum run times," *Power Electron. Technol. Mag.*, pp. 40–49, Apr. 2005.

[7] Chen, Min, and Gabriel A. Rincón-Mora, "Accurate, Compact, and Power-Efficient Li-Ion Battery Charger Circuit," *Circuits and Systems II: Express Briefs, IEEE Transactions on*, vol.53, no.11, pp.1180-1184, Nov. 2006

[8] Lin, Chia-Hsiang, Chun-Yu Hsieh, and Ke-Horng Chen., "A Li-Ion Battery Charger With Smooth Control Circuit and Built-In Resistance Compensator for Achieving Stable and Fast Charging," Circuits and Systems I: Regular Papers, IEEE Transactions on , vol.57, no.2, pp.506-517, Feb. 2010

[9] Do Valle, Bruno, Christian T. Wentz, and Rahul Sarpeshkar, "An Area and Power-Efficient Analog Li-Ion Battery Charger Circuit," Biomedical Circuits and Systems, IEEE Transactions on , vol.5, no.2, pp.131-137, April 2011

[10] Inoa, E.; Jin Wang, "PHEV Charging Strategies for Maximized Energy Saving," *Vehicular Technology, IEEE Transactions on*, vol.60, no.7, pp.2978,2986, Sept. 2011

[11] Bo-Yuan Chen; Yen-Shin Lai, "New Digital-Controlled Technique for Battery Charger With Constant Current and Voltage Control Without Current Feedback," *Industrial Electronics, IEEE Transactions on*, vol.59, no.3, pp.1545,1553, March 2012

[12] Liu, Yi-Hwa, Jen-Hao Teng, and Yu-Chung Lin. "Search for an optimal rapid charging pattern for lithium-ion batteries using ant colony system algorithm."*Industrial Electronics, IEEE Transactions on* 52, no. 5 pp.1328-1336, 2005

[13] Ikeya, Tomohiko, et al. "Collaborative investigation on charging electic-vehicle battery systems for night-time load levelling by Japanese electric power companies." *Journal of power sources* 69, no. 1 pp.103-111, 1997.

The 2018 International Power Electronics Conference

[14] Ikeya, Tomohiko, et al. "Multi-step constant-current charging method for electric vehicle, valve-regulated, lead/acid batteries during night time for load-levelling."*Journal of power sources* 75, no.1, pp.101-107, 1998

[15] Ikeya, Tomohiko, et al. "Charging operation with high energy efficiency for electric vehicle valve-regulated lead–acid battery system." *Journal of power Sources* 91, no.2, pp.130-136, 2000.

[16] Ikeya, Tomohiko, et al. "Multi-step constant-current charging method for an electric vehicle nickel/metal hydride battery with high-energy efficiency and long cycle life." *Journal of power sources* 105, no.1, pp.6-12, 2002

[17] Liu, Yi-Hwa, Ching-Hsing Hsieh, and Yi-Feng Luo, "Search for an Optimal Five-Step Charging Pattern for Li-Ion Batteries Using Consecutive Orthogonal Arrays," *Energy Conversion, IEEE Transactions on* , vol.26, no.2, pp.654-661, June 2011

[18] Liu, Yi-Hwa, and Yi-Feng Luo, "Search for an Optimal Rapid-Charging Pattern for Li-Ion Batteries Using the Taguchi Approach," *Industrial Electronics, IEEE Transactions on* , vol.57, no.12, pp.3963-3971, Dec. 2010

[19] Wang, Chwei-Sen, Grant A. Covic, and Oskar H. Stielau. "Power transfer capability and bifurcation phenomena of loosely coupled inductive power transfer systems." *Industrial Electronics, IEEE Transactions on,* vol. 51.no.1, pp.148-157, 2004

[20] Moradewicz, Artur J., and Marian P. Kazmierkowski. "Contactless energy transfer system with FPGA-controlled resonant converter." *Industrial Electronics, IEEE Transactions on,* vol. 57.no.9, pp.3181-3190, 2010

[21] L.Steigerwald, "A comparison of Half-Bridge Resonant Converters," *Power Electron. IEEE Transactions on,* vol.3, no.2, April 1988

[22] Poon, N. K., Bryan MH Pong, and Chi K. Tse. "A constant-power battery charger with inherent soft switching and power factor correction." *Power Electronics, IEEE Transactions on,* vol.18, no.6, pp.1262-1269, 2003

[23] Yan, Xinxiang, and Dean Patterson. "A high-efficiency on-board battery charger with unity input power factor." *International Journal of Renewable Energy Engineering* 2,.no.1 pp.141-147, 2000

[24] Kuperman, A., U. Levy, J. Goren, A. Zafransky, and A. Savernin. "Battery Charger for Electric Vehicle Traction Battery Switch Station." *IEEE Transactions on Industrial Electronics, vol.* 99, no. 1, 2012.

[25] Notten, P. H. L., J. H. G. Veld, and JRG van Beek. "Boost charging Li-ion batteries: A challenging new charging concept." *Journal of power sources,* vol. 145, no. 1, pp. 89-94, 2005.

[26] Chung, S. K., A. A. Andriiko, A. P. Mon'ko, and S. H. Lee. "On charge conditions for Li-ion and other secondary lithium batteries with solid intercalation electrodes." *Journal of power sources* 79, no. 2, pp.205-211, 1999

[27] Sikha, G., P. Ramadass, B. S. Haran, R. E. White, and Branko N. Popov. "Comparison of the capacity fade of Sony US 18650 cells charged with different protocols." *Journal of power sources* 122, no. 1,pp.67-76, 2003

The 2018 International Power Electronics Conference

An Improved PWM Scheme to Achieve Zero-Voltage Switching for All Devices in Three-Phase Isolated Matrix Rectifier

Xuerui Lin[1], Yunwei (Ryan) Li[1], Jahangir Afsharian[2] and Dewei (David) Xu[2]

1 Electrical and Computer Engineering, University of Alberta, Edmonton, Canada
2 Electrical and Computer Engineering, Ryerson University, Toronto, Canada
E-mail: {lin2, yunwei.li}@ualberta.ca, {jafshari, dxu}@ryerson.ca

Abstract-The zero-voltage switched (ZVS) technique can lower the converter switching loss and increase the conversion efficiency and power density. This paper proposed an improved PWM scheme with ZVS operation for three-phase isolated matrix rectifier. The presented PWM scheme removed two hard-switch actions at existing PWM scheme and permits all 12 MOSFET devices to be turned on under ZVS. The operation principles and steady-state analysis of the converter are addressed. The proposed ZVS operation of MOSFET switches is analyzed and operating states during switch transitions is stated. The switching loss comparison between existing and proposed PWMs is discussed. The ZVS operation of proposed PWM scheme is verified by both simulation and experimental results.

Keywords—Matrix rectifier, ZVS, SVM, MOSFET.

I. INTRODUCTION

Matrix rectifier is derived from traditional matrix converter, and it also inherits all the advantages of traditional matrix converters. Isolated high frequency three-phase matrix based rectifier is designed based on a three-phase to single-phase matrix converter, which is proposed in [1–4] as shown in Fig. 1, capable of implementing ZVS by utilizing resonance between MOSFET parasitic capacitances and transformer leakage inductance.

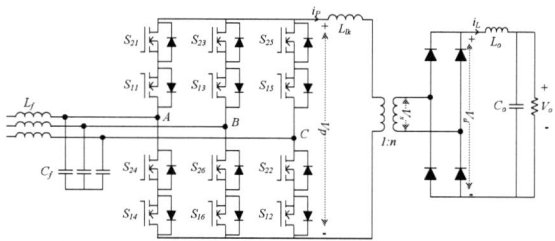

Fig. 1. Three-phase PWM rectifier.

For matrix converter, one of the most challenging problems is modulation methods, which can be considered as the basis of designing a matrix converter. Years of continuous effort on matrix converter have been dedicated to the development of different modulation and control methods. The basic idea of modulation strategy for matrix converter is to use the input voltage to

synthesize output voltage, while using the output current to synthesize the input current. According to different synthesis methods, there are several major modulation methods, including switching function modulation strategy [5], space vector modulation (SVM) strategy [6], carrier-based modulation strategy [7]. Many other modulation methods can be derived from these main schemes.

The modulation methods for traditional matrix converter also can be applied to isolated matrix rectifier. The current based SVM is especially suited for three-phase isolated matrix rectifier. The study in [2] proposed a PWM scheme for three-phase isolated matrix rectifier, and realized ZVS operation for eight out of ten switch actions in each switching cycle. Within any 60° interval, since the converter switching frequency is much higher than the line frequency, the line voltages can be treated as constants during each switching cycle. Therefore, the converter is analyzed as two full-bridge phase-shift (FB-PS) converter sub-topologies operating alternatively within a switching cycle. Phase *A* and *B* forms sub-topology *x*, and phase *A* and *C* forms sub-topology *y*. These two sub-topologies work as two FB-PS converters independently. The detailed ZVS operation for full-bridge converter has been described in reference [8–10], and SVM technique used to achieve unity power factor is also discussed in many references [11–13].

Depending on how the transformer primary side voltage is synthesized by the three-phase voltages, different switching pattern can be generated. In order to maintain volt-sec balance, sub-topology *x* and *y* are used to generated positive and negative voltage pulses interactively in high frequency. Each two pulses are separated by a zero-volt interval with same duration. For the same switching pattern, by using different pre-turn-on switches, ZVS results will also be different. In this paper, a switching method for eight-segment switching pattern by removing two hard-switches is proposed. With this proposed PWM method, the switching loss of converter is reduced, and the conversion efficiency and power density are improved. The rest of this paper is organized as follows. The operation principle of the PWM is demonstrated in Section II. Steady-state and ZVS

1537

operation of MOSFETs are described in Section III. Then, switching loss comparison between two PWMs is presented in Section IV. Simulation and experimental results are presented in Section V, followed by the conclusion in Section VI.

II. OPERATION PRINCIPLE

Within each sector, by rearrangement the operation sequence of sub-topology x and y, different PWM schemes can be generated. Due to transformer isolation, the transformer primary side voltage V_p must be alternating positive and negative in high frequency to maintain volt-sec balance. During Sector I, switches S_{11}, S_{21}, S_{16}, and S_{26} in sub-topology x are turned on to create a positive voltage pulse v_{AB}. To keep transformer flux balanced, switches S_{14}, S_{24}, S_{15}, and S_{25} in sub-topology y are turned on to create a negative voltage pulse v_{AC}. Then, switches S_{11}, S_{21}, S_{12}, and S_{22} in sub-topology y and switches S_{14}, S_{24}, S_{13}, and S_{23} in sub-topology x are turned on to generate another two voltage pulses.

Fig. 2 shows the circuit operation waveforms during -30°≤ θ ≤30° interval, with excessively decreased switching frequency f_{sw} so that the PWM details can be observed. Different shades are used to represent different sub-topologies, which are used to generate different parts of the waveforms.

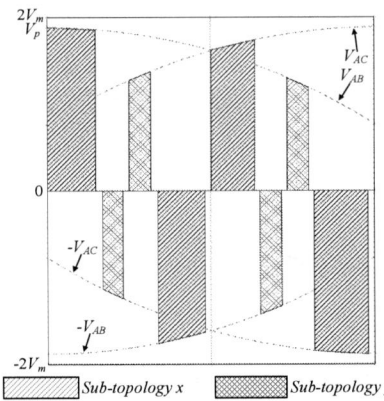

Fig. 2. PWM waveforms: primary side voltages.

III. STEADY-STATE AND ZVS OPERATION ANALYSIS

A. Steady-State Analysis

Paper [2] proposed the eight-segment PWM sequence $\overrightarrow{I_{x+}}, \overrightarrow{I_{0+}}, \overrightarrow{I_{x-}}, \overrightarrow{I_{0-}}, \overrightarrow{I_{y+}}, \overrightarrow{I_{0+}}, \overrightarrow{I_{y-}}, \overrightarrow{I_{0-}}$. This PWM scheme consists two hard-switched actions. In this paper, the vector sequence in each switching period is divided into the sequence of $\overrightarrow{I_{x+}}, \overrightarrow{I_{0+}}, \overrightarrow{I_{y-}}, \overrightarrow{I_{0-}}, \overrightarrow{I_{y+}}, \overrightarrow{I_{0+}}, \overrightarrow{I_{x-}}, \overrightarrow{I_{0-}}$, and the dwell time of each vector is $T_x/2, T_0/4, T_y/2, T_0/4, T_y/2, T_0/4, T_x/2, T_0/4$ respectively.

In this method, the analysis is based on assumption that transformer is ideal, and forward voltage drop across all diodes is zero. Fig. 3 shows the complete operation of the three-phase converter during one switching period T_S.

It shows the circuit waveforms of primary voltage V_P and current i_P, the rectified secondary voltage V_d and output inductor current i_L and the corresponding switch gate signals in sector I. Leakage inductance caused duty loss (shaded areas) can be observed when primary side current changes directions.

In this switching method, all switch states are decided by space vector and primary current direction. For example, in sector I, voltage potential v_A is always higher than v_B and v_C. Switches S_{14} and S_{21} can be kept on all the time since the body diodes are forward biased. If $i_P>0$, switches S_{13} and S_{15} can be kept on because the body diodes are forward biased. For the same reason, if $i_P<0$, switches S_{26} and S_{22} can be kept on, as shown in Table I. The states of synchronous rectification switches are only decided by voltage of one phase. This method not only avoid short circuit problem, also can easily achieve ZVS for all switches. The switching states and space vectors in sector I are summarized in Table I.

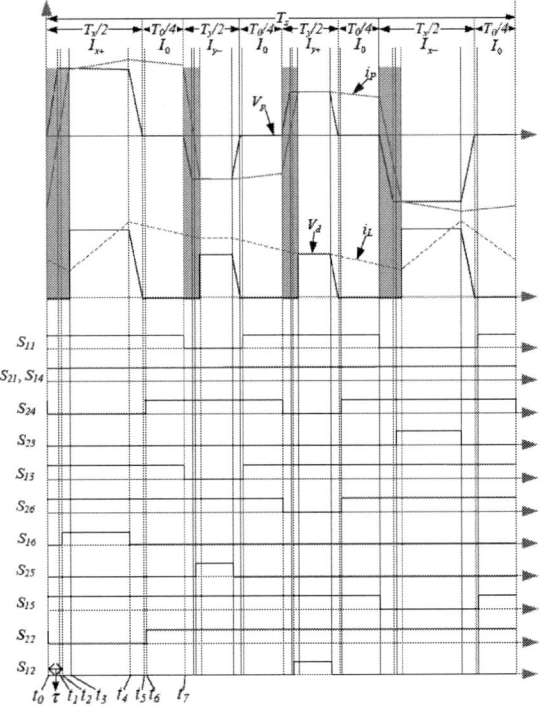

Fig. 3. Circuit waveforms: primary voltage and current, rectified secondary voltage and output inductor current and corresponding switch gate driver in sector I.

TABLE I
SWITCHING STATES AND SPACE VECTORS IN SECTOR I

Space vectors	S_{21}	S_{11}	S_{24}	S_{14}	S_{23}	S_{13}	S_{26}	S_{16}	S_{25}	S_{15}	S_{22}	S_{12}
$\overrightarrow{I_{1+}}$	1	1	0	1	0	1	1	1	0	1	0	0
$\overrightarrow{I_{0+}}$	1	1	1	1	0	1	1	0	0	1	1	0
$\overrightarrow{I_{2-}}$	1	0	1	1	0	0	1	0	1	1	1	0
$\overrightarrow{I_{0-}}$	1	1	1	1	0	1	1	0	0	1	1	0
$\overrightarrow{I_{2+}}$	1	1	0	1	0	1	0	0	0	1	1	1
$\overrightarrow{I_{0+}}$	1	1	1	1	0	1	1	0	0	1	1	0
$\overrightarrow{I_{1-}}$	1	0	1	1	1	1	1	0	0	0	1	0
$\overrightarrow{I_{0-}}$	1	1	1	1	0	1	1	0	0	1	1	0

1538

B. ZVS Operation Analysis

Since in the proposed PWM scheme, the vector transition only consists of transitions between active vectors and zero vectors, the ZVS operation analysis in this section only gives one example of vector transition from zero vector to active vector ($\overrightarrow{I_{0-}}$ to $\overrightarrow{I_{1+}}$) and one example from active vector to zero vector ($\overrightarrow{I_{1+}}$ to $\overrightarrow{I_{0+}}$) in sector I. The same analysis can be applied to the other vector transitions and the rest of other five sectors.

1. Vector transition from zero vector $\overrightarrow{I_{0-}}$ to active vector $\overrightarrow{I_{1+}}$ (t_0 to t_3).

At t_{0-}, transformer primary side current i_P is circulating through phase A. The primary side voltage v_P is clamped to zero. Current vector is $\overrightarrow{I_{0-}}$ in this mode, as shown in Fig. 4 (a).

At t_0, S_{24} and S_{22} are turned off. Energy stored in leakage inductance L_{lk} starts transferring to output capacitances of S_{24}, S_{16}, S_{22}, and S_{12} due to resonance between them as shown in Fig. 4 (b). The primary current i_P also starts resonating down to zero.

At t_1, when voltage across S_{16} reaches zero, D_{16} starts conducting as shown in Fig. 4(d), and the voltage of switch S_{16} is clamped to zero. At t_2, before i_P cross zero, S_{16} is turned on at zero voltage. In order to achieve ZVS of S_{16}, energy stored in L_{lk} should be enough to charge the total equivalent capacitance of C_{24}, C_{16}, C_{22} and C_{12}. The simplified circuit of resonant process is shown in Fig. 4(c). The transformer capacitance is also considered as part of total equivalent capacitance. The total equivalent capacitance can be calculated as (1).

$$C_{eq} = \frac{5}{2}C_o + C_{TR} \qquad (1)$$

where C_o is output capacitance of MOSFETs. C_{TR} is the transformer capacitance. As stated in [1], the optimum deadtime τ (t_0 to t_2) can be estimated as (2).

$$\tau = \frac{\pi}{2}\sqrt{L_{lk}C_{eq}} \qquad (2)$$

At t_3, primary side current i_P reaches secondary side current i_L and as a result line voltage v_{AB} appears across the transformer primary side and vector transition from $\overrightarrow{I_{0-}}$ to $\overrightarrow{I_{1+}}$ complete. At this moment, the voltage across S_{24}, S_{16}, S_{22}, and S_{12} are v_{AB}, 0, $\frac{1}{2}v_{AB}$, and $(v_{AC} - \frac{1}{2}v_{AB})$ respectively. Other vector transitions from zero vector to active vector is the same with the presented transition process.

2. Vector transition from active vector $\overrightarrow{I_{1+}}$ to zero vector $\overrightarrow{I_{0+}}$ (t_4 to t_6).

At t=$[t_3, t_4]$, the current vector is $\overrightarrow{I_{1+}}$ (Fig. 4 (f)). At t_4, S_{16} is turned off, and transformer primary side current i_P starts charging capacitances C_{24}, C_{16}, C_{22} and C_{12} as shown in Fig. 4 (g).

At t_5, voltage across C_{24} reduces to zero and D_{24} starts conducting. At the same time, the voltage across C_{22} reaches zero, and D_{24} starts conducting, as shown in Fig. 4 (h). S_{24} and S_{22} are turned on at zero voltage at t_6, as

shown in Fig. 4 (i). The primary voltage v_P is clamped to zero. Vector transition from active vector to zero vector operation is completed. At this moment, the voltage across S_{24}, S_{16}, S_{22}, and S_{12} are 0, v_{AB}, 0, and v_{AC} respectively. The total equivalent capacitance to be charged to achieve ZVS is same as that from zero vector to active vector, and the combined energy stored in both L_{lk} and L_o is used to charge/discharge equivalent capacitance. ZVS can be easily achieved.

During t=$[t_6, t_7]$, the primary voltage v_P is clamped to zero. Primary current freewheeling through S_{11}, D_{21}, S_{14} and D_{24}. The current vector is $\overrightarrow{I_{0+}}$.

The rest vector transitions between active vector and zero vector operation is similar with the analyzed part and will not be discussed in here.

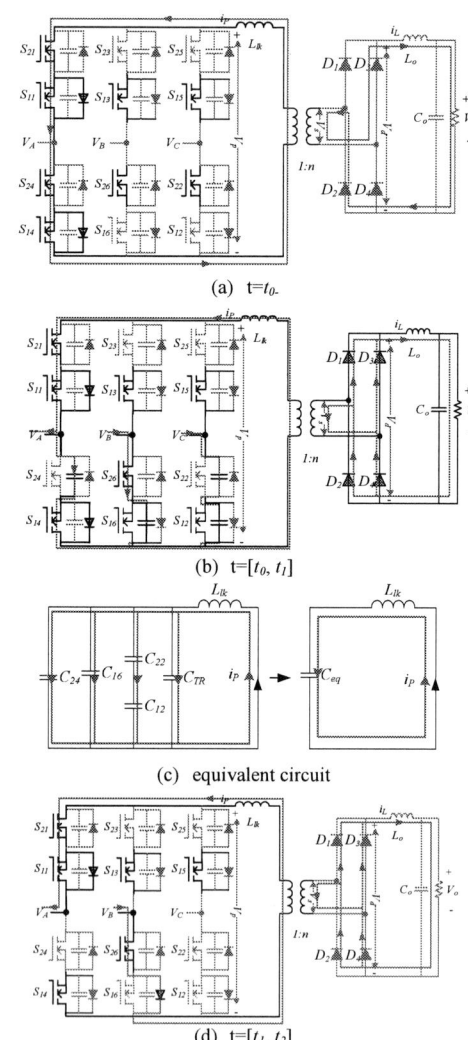

(a) t=t_0.

(b) t=$[t_0, t_1]$

(c) equivalent circuit

(d) t=$[t_1, t_2]$

The 2018 International Power Electronics Conference

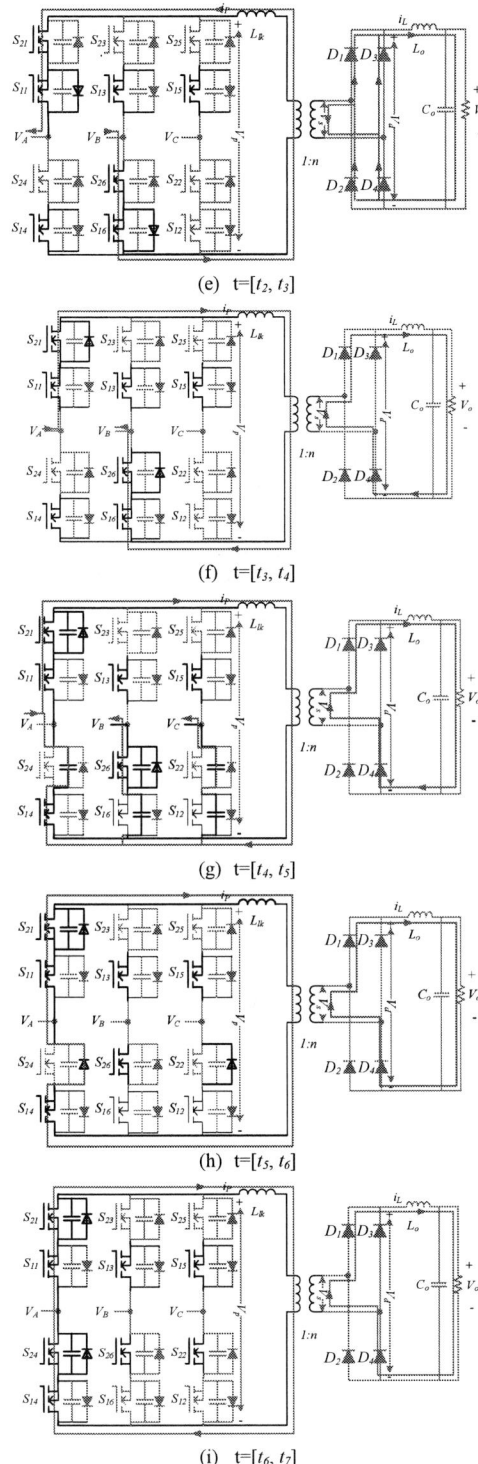

(e) t=[t_2, t_3]

(f) t=[t_3, t_4]

(g) t=[t_4, t_5]

(h) t=[t_5, t_6]

(i) t=[t_6, t_7]

Fig. 4. ZVS operation between zero vector and active vector.

IV. SWITCHING LOSS COMPARISON BETWEEN TWO PWMS

The analysis in this section is based on the switch actions in first half sector of sector I, and the results are

the same for the other sectors due to the symmetrical of the switching patterns.

In the existing PWM scheme, there are ten turn-on actions and ten turn-off actions in each switching cycle, among which, the switches S_{22} and S_{15} are turned on under hard-switched condition. Before the MOSFET S_{22} is turned on the drain-source voltage V_{ds} is $\frac{1}{2}v_{AB}$, and is varies from $\frac{3}{4}V_m$ to $\frac{\sqrt{3}}{2}V_m$ during -30° ≤ θ ≤ 30° interval. The output capacitance loss during switch turn-on process also contribute the switching loss. The output capacitance loss incurred by charging and discharging the MOSFETs' output capacitances C_{oss}. For hard-switched cases, all energy stored in C_{oss} is lost and discharged into MOSFET channel during the switching cycle. However, for ZVS cases, the energy stored in C_{oss} is transferred to other capacitors prior to turning on the MOSFET [14]. During hard-switched turn-on transitions, the reverse recovery process of MOSFET parasitic diode also causes turn-on loss. Two out of ten turn-off actions are turned off under zero current, thus have no impact on the switching loss. In summary, for the existing PWM, the two turn-on actions and eight turn-off actions contribute to the switching loss.

For the proposed PWM scheme, there are twelve turn-on actions and twelve turn-off actions in each switching cycle. All twelve turn-on actions are under ZVS condition. So the turn-on loss for the proposed PWM is zero. Among the twelve turn-off actions, four switches are turned off under zero current. So totally eight turn-off actions contribute to the switching loss in this PWM scheme. The switching loss comparison between two PWMs is summarized in Table II.

TABLE II
SWITCHING LOSSES COMPARISON BETWEEN TWO PWMS.

PWM scheme	Existing PWM	Proposed PWM
Number of turn-on actions with ZVS	8	12
Number of turn-on actions with non-ZVS	2	0
Number of no impact turn-off actions (ZCS)	2	4
Number of turn-off actions with full current	8	8

V. SIMULATION AND EXPERIMENTAL RESULTS

The simulation model and experiment prototype is setup at rated power of 3.3 kW and m_a =0.8. Detail parameters are listed in Table III.

TABLE III
SIMULATION AND EXPERIMENTAL PARAMETERS

L_f	90µH	L_o	450µH
C_f	10µF	C_o	220µF
f_{sw}	100kHz	f_{grid}	60Hz
V_o	330V	$V_{ll,rms}$	180V
L_{lk}	5.7µH	n	2

To observe the ZVS operation of MOSFET switches, one switch S_{11} on phase A is selected, thereby showing the ZVS of all active switches. Fig. 5 shows the

The 2018 International Power Electronics Conference

experimental waveforms of phase A current i_a, the transformer primary side voltage v_p and the drain-source voltage V_{ds} of MOSFET S_{11} in six sectors operation. For traditional eight-segment PWM scheme, S_{11} is turned on under non-ZVS condition in sector III, so the MOSFET turn-on actions with proposed sequence are analyzed at three locations of t_1, t_2, and t_3 as shown in Fig. 6 for comparison.

At t_1, S_{11} turn-on action represents the ZVS turn-on action for transitions within sectors other than sector III. At t_2 and t_3, S_{11} turn-on represents the ZVS turn-on actions during Sector III.

Fig. 6. (a), (c), and (e) present the simulation waveforms of drain-source voltage V_{ds} and gating signals of S_{11} at t_1, t_2, and t_3 respectively. Fig. 6. (b), (d), and (e) represent the corresponding experiment waveforms at t_1, t_2, and t_3.

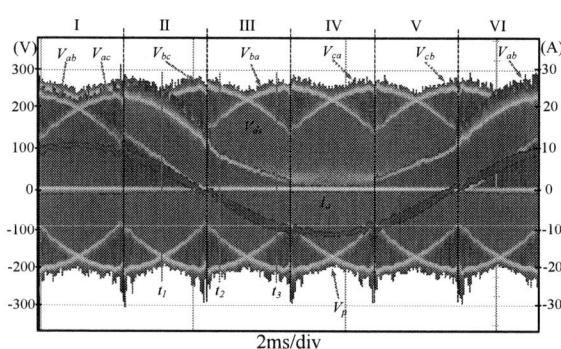

Fig. 5. Waveform of S_{11} in six sectors.

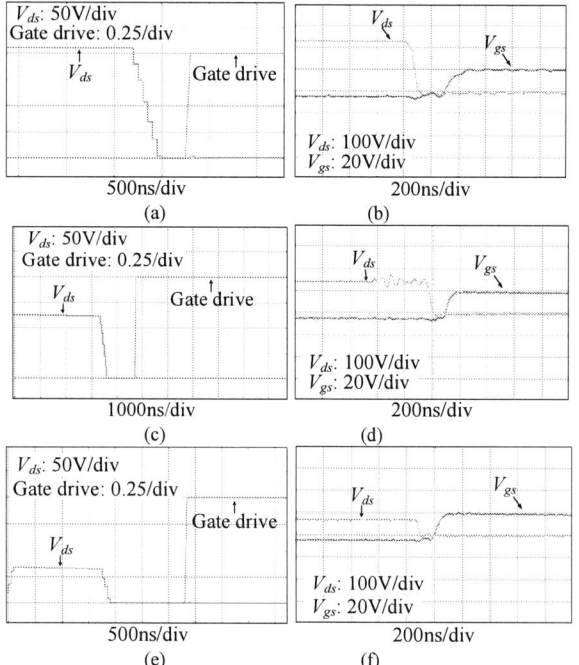

Fig. 6. Simulation and experimental results of S_{11} turn-on actions during six sectors.

In the simulation results, as shown in Fig. 6 (a), (c), and (e), drain-source voltage V_{ds} tends to zero before

gate-drive signal approaches high, which means that the body diode is on before the switch is turned on such that ZVS is realized. The results are verified by experiment results as shown in Fig. 6 (b), (d), and (f).

Fig. 7 shows the input phase current waveforms with proposed PWM scheme implementation. The total harmonic distortion (THD) of the input phase current is 2.4%.

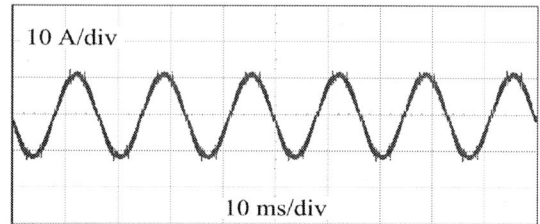

Fig. 7. Experimental waveform of input phase current.

VI. CONCLUSIONS

A PWM method for three-phase isolated matrix based rectifier is proposed in this paper. Due to ZVS operation for all twelve semiconductors, the matrix rectifier can operate at higher switching frequency. The analysis of ZVS operation during vector transitions is discussed. ZVS operation for all 12 semiconductors is obtained by using proposed switching method. The switching loss comparison between the existing PWM and proposed PWM is stated. The ZVS operation for all devices of proposed PWM scheme is verified by simulation and experiment results.

REFERENCES

[1] V. Vlatkovic, D. Borojevic, X. Zhuang, and F. C. Lee, "Analysis and design of a zero-voltage switched, three-phase PWM rectifier with power factor correction," in *Proc. 23rd Annual IEEE Power Electronics Specialists Conference (PESC)*, Article no. 4518235, Toledo, Spain, Jun.-Jul, 1992.

[2] V. Vlatkovic, D. Borojevic, and F. C. Lee, "A zero-voltage-switched, three-phase isolated PWM buck rectifier," *IEEE Trans. on Power Electron.*, vol. 10, pp. 148–157, Mar. 1995.

[3] J. W. Kolar, U. Drofenik, and F. C. Zach, "VIENNA Rectifier II— A novel single-stage high-frequency isolated three-phase PWM rectifier system," *IEEE Trans. on Ind. Electron.*, vol. 46, no. 4, pp. 674–691, Aug. 1999.

[4] D. S. Greff and I. Barbi, "A single-stage high-frequency isolated threephase ac/dc converter," in *Proc. 32nd IEEE Industrial Electronics Society Annual Conference*, pp. 2648–2653, Paris, France, Nov. 2006.

[5] L. Zhang, C. Watthanasaran, and W. Shepherd, "Control of AC-AC matrix converter for unbalanced and/or distorted supply voltage in *Proc. 2001 IEEE 32nd Annual Power Electronics Specialists Conference (PESC)*, Vancouver, Canada, Jun. 17–21, 2001, vol. 2, pp. 1108–1113.

[6] L. Huber and D. Borojevic, "Space vector modulator for forced commutated cycloconverters," in *Proc. Industry Applications Society (IAS) Annual Meeting*, San Diego, USA, Oct. 1–5, 1989, vol. 1, pp. 871–876.

[7] Y. D. Yoon, and S. K. Sul, "Carrier based modulation technique for matrix converter," *IEEE Trans. on Power Electron.*, vol. 21, no.6, pp. 1691–1703, Nov. 2006.

[8] J. A. Sabate, V. Vlatkovic, R.B.Ridley, F. C. Lee, and B. H. Cho, "Design considerations for high-voltage high-power full-bridge zero-voltage-switched PWM converter," in *Proc. Fifth Annual IEEE Applied Power Electronics Conference and Exposition (APEC)*, Article no. 3788172, Los Angeles, USA, Mar. 1990.

[9] L. Zhu, "A novel soft-commutating isolated boost full-bridge ZVS-PWM DC-DC converter for bidirectional high power applications," *IEEE Trans. Power Electron.*, vol. 21, no. 2, pp. 422-429, Mar. 2006.

[10] B. Gu, C. Y. Lin, B. Chen, J. Dominic, and J. S. Lai, "Zero-voltage-switching PWM resonant full-bridge converter with minimized circulating losses and minimal voltage stresses of bridge rectifiers for electric vehicle battery chargers," *IEEE Trans. on Power Electron.*, vol. 218, no. 10, pp. 4657-4667, Oct. 2013.

[11] T. Zhao, X. Guo, J. Su, and D. Xu, "Improved space vector modulation for matrix converter based isolated rectifiers," in *Proc. 40th IEEE Industrial Electronics Society Annual Conference*, Article no. 14950912, Dallas, USA, Oct.-Nov, 2014.

[12] J. Afsharian, D. Xu, B. Gong, and Z. Yang, "Space vector demonstration and analysis of zero voltage switching transitions in three-phase isolated PWM rectifier," in *Proc. 2015 IEEE Energy Conversion Congress and Exposition (ECCE)*, Article no. 15569397, Montreal, Canada, Sept, 2015.

[13] B. Wu, *High-power converters and ac drives*. Hoboken, NJ, USA: John Wiley & Sons, 2006.

[14] J. B. Fedison, M. Fornage, M. J. Harrison, and D. R. Zimmanck, "Coss related energy loss in power MOSFETs used in zero-voltage-switched applications," in *Proc. Twenty-Ninth Annual IEEE Applied Power Electronics Conference (APEC)*, Fort Worth, USA, Mar. 16–20, 2014, pp. 150–156.

The 2018 International Power Electronics Conference

Fixed-Frequency HF Gate Driver by a Push-Pull Self-Excitation LC Oscillator Having a Capacitance Transistor

Naoyuki Ishibashi [1], Takuya Mizushima [1], Masahiko Hirokawa [2] and Akihiko Katsuki [1*]

1 Next Generation Switching Power Circuit Course, Graduate School of Engineering, Nagasaki University,
1-14, Bunkyo-machi, Nagasaki-shi, Nagasaki, 852-8521 Japan
2 Advanced Products Development Center, Technology & Intellectual Property HQ, TDK Corporation,
2-15-7, Higashi-Ohwada, Ichikawa-shi, Chiba, 272-8558 Japan
*E-mail: katsuki@nagasaki-u.ac.jp

Abstract— **The size of many electronic products becomes smaller due to the improvement of integration technologies year by year. The miniaturization of convertors is realized by using high-frequency switching. Resonant converters are effective because of the low switching loss in semiconductor switches. At high frequency (HF), the phase-lag in feedback circuit increases and has to be considered. This lag leads to unstable operation. Due to this reason, we have examined a self-excitation LC oscillator as gate drive circuit. Resonant converters can be controlled by the frequency or the duty ratio. The driver to be discussed here can be used for both control methods. Usually, the oscillation frequency varies by the Miller effect in main switch of the outer converter. This paper proposes a fixed frequency driver using self-excitation LC oscillator. To compensate the oscillation frequency, a capacitance transistor is used.**

Keywords— Capacitance transistor, dual-gate MOSFET, frequency compensation, push-pull self-exciation LC HF gate driver.

I. INTRODUCTION

Nowadays, almost electronic circuits are small-sized because the technology of integrated circuits has been developing day after day. Since the volume of switching power supply depends on that of reactance components, high-frequency switching is effective for the miniaturization of power supply. However, the switching loss increases in proportion to the switching frequency. Therefore, we should use resonant converters because switching loss is reduced. In resonant converters, output voltage depends on the switching frequency. It is preferable that the switching frequency is kept constant. Furthermore, the lag in the feedback circuit has to be small for the system stability. Push-pull self-excitation LC oscillators [1-3] are suitable for the high frequency (HF, 3-30 MHz) band use.

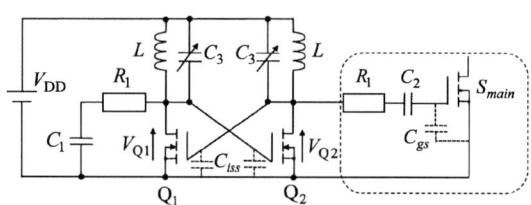

Fig. 1. Gate driver using a push-pull self-excitation LC oscillator [3].

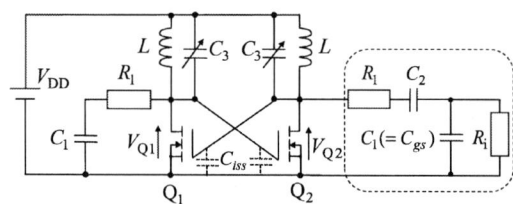

Fig. 2. Experimental circuit of a push-pull LC oscillator.

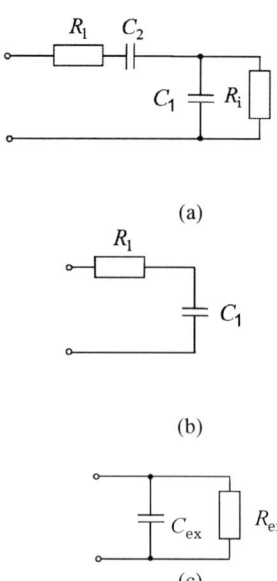

Fig. 3. The coupling circuit in Fig. 2 : (a) the circuit encircled by a dotted line, (b) the idealized equivalent circuit, (c) the series-to-parallel converted circuit.

1543

Self-excitation LC oscillators are classified to two types. One is single type and the other is push-pull type. The single-type oscillators are suitable for the very high frequency (VHF, 30-300 MHz) band use. However, the single-type driver stops when the oscillation conditions, i.e., the frequency condition and the amplitude condition are not satisfied. To the contrary, the push-pull typedriver stably operates even if parameters change due to external factor. However, the push-pull type driver needs the commutation time between two MOSFETs and then is not suited for VHF band use.

In this paper, we present an analysis on the push-pull LC self-excitation gate driver suitable for about 10-MHz oscillation. At first, the oscillation frequency is analyzed on the case of symmetrical operations. Next, asymmetrical operations in actual use are researched in detail. The time period in which one MOSFET operates and the time period in which the other MOSFET operates are calculated separately. Combining these time periods, we obtain the frequency of oscillation. Finally, utilizing a capacitance transistor, we propose a constant-frequency self-excitation gate driver by the use of frequency compensation.

II. Circuit and Fundamental Characteristics of LC Oscillator

A. Circuit

The object of this research is the gate driver shown in Fig. 1. Replacing the gate-source capacitance C_{gs} of the main switch S_{main} by the capacitance C_1, we obtain Fig. 2, where R_i is the input resistance of S_{main}. Here, we assume that C_1 equals C_{gs}. Fig. 3(a) is the partial circuit encircled by a dotted line in Fig. 2. When the capacitance C_2 and R_i is assumed to infinite, Fig. 3(a) is simplified to Fig. 3(b). In this case, a symmetrical operation can be performed. The experimental circuit is Fig. 2 with the coupling circuit of Fig. 3(b).

B. Oscillation frequency

According to [3], the oscillation frequency f_{osc} is expressed as follows:

$$f_{osc} = \frac{1}{2\pi\sqrt{L(C_1 + C_{iss} + C_3)}} \tag{1}$$

The MOSFETs used in this experiment are L8821, whose C_{iss} equals 33 pF from its datasheet. Since the inductance $L = 1$ µH and the capacitance $C_3 = 10$ pF, f_{osc} can be obtained as follows: $f_{osc} = 24.0$ MHz at $C_1 = 1$ pF, $f_{osc} = 21.9$ MHz at $C_1 = 10$ pF, $f_{osc} = 19.7$ MHz at $C_1 = 22$ pF, and $f_{osc} = 18.3$ MHz at $C_1 = 33$ pF.

Fig. 4 presents the measured characteristics on the relation between f_{osc} and the resistance R_1. It is seen that these values of oscillation frequency are different from the measured ones. Moreover, Eq. (1) does not contain R_1, though the measured oscillation frequency much depends on R_1. Thus, we convert the series circuit of Fig. 3 (b) into the parallel circuit of Fig.3 (c) and use C_{ex}

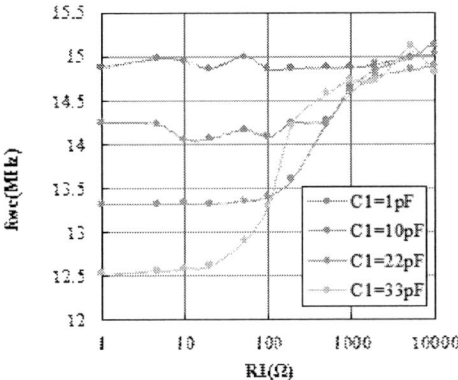

Fig. 4. Measured characteristics on the oscillation frequency f_{osc} against the resistance R_1.

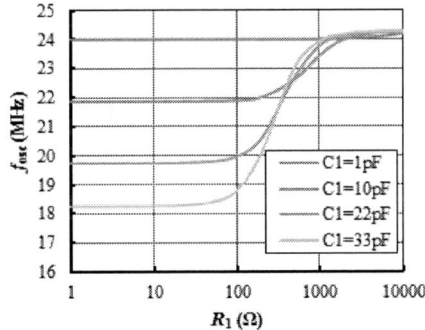

Fig. 5. Calculated characteristics on the oscillation frequency f_{osc} from Eq. (2) and Eq. (3) against the resistance R_1.

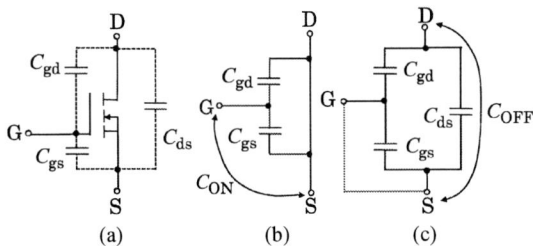

Fig. 6. Capacitances of MOSFETs Q_1 and Q_2: (a) Internal capacitances, (b) ON state, (c) OFF state.

instead of C_1. The oscillation frequency f_{osc} can be obtained from the following equations.

$$f_{osc} = \frac{1}{2\pi\sqrt{L(C_{ex} + C_{iss} + C_3)}} \tag{2}$$

where

$$C_{ex} = \frac{C_1}{1 + 4\pi^2 f_{osc}^2 C_1^2 R_1^2} \tag{3}$$

The result on calculation is depicted in Fig. 5. This graph is qualitatively similar to Fig. 4. However, the values of Fig. 5 are much higher than those of Fig. 4.

The 2018 International Power Electronics Conference

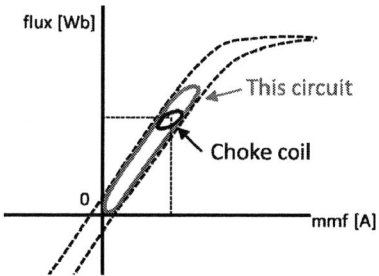

Fig. 7. Magnetizing curve of the inductor L.

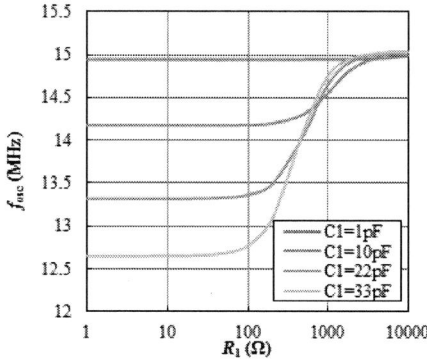

Fig. 8. Calculated characteristics on the oscillation frequency f_{osc} from Eq. (3) and Eq. (7) against the resistance R_1.

Next, we consider C_{iss}. With respect to the MOSFET at ON-state, C_{iss} means the capacitance between the gate and the drain. In a push-pull oscillator, an ON-state MOSFET and an OFF-state MOSFET are connected. Fig. 6(a) shows the capacitances between terminals. The ON-state MOSFET and the OFF-state MOSFET are expressed as Fig. 6(b) and Fig. 6(c), respectively. Therefore, C_{iss} should be corrected to C_{MOS} as follows:

$$C_{MOS} = C_{ON} + C_{OFF} \qquad (4)$$
$$= 42 \text{ pF} + 28 \text{ pF} = 70 \text{ pF}$$

where

$$C_{ON} = C_{gs} + C_{gd} \; (= C_{iss}) \qquad (5)$$
$$= 34 \text{ pF} + 8\text{pF} = 42 \text{ pF}$$

$$C_{OFF} = C_{ds} + C_{gd} \; (= C_{oss}) \qquad (6)$$
$$= 26 \text{ pF} + 2 \text{ pF} = 28 \text{ pF}$$

We need to correct not only the capacitance of MOSFETs but also the inductance of coils. In this experiment, choke coils for dc power supplies are used as the inductor L. Fig. 7 explains the magnetizing curve of these inductors. When these coils are set in this oscillator, an effective inductance should be corrected from L to L', where $L' > L$. We calculated that $L' = 1.4L$. As a result, the oscillation frequency f_{osc} can be obtained from Eq. (3) and the following equation.

$$f_{osc} = \frac{1}{2\pi\sqrt{L'\left(C_{ex} + C_{MOS} + C_3\right)}} \qquad (7)$$

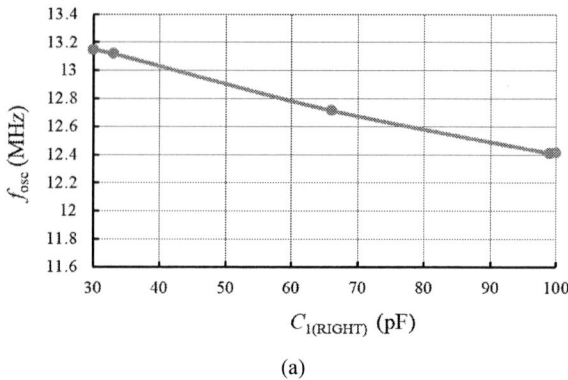

(a)

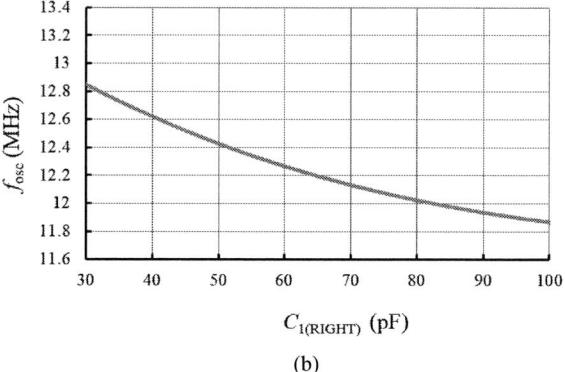

(b)

Fig. 9. Oscillation frequency f_{osc} vs. the right-hand side capacitance $C_{1(RIGHT)}$ characteristics, where the left-hand side capacitance $C_{1(LEFT)} = 33$ pF, $C_3 = 10$ pF, $R_1 = 100 \ \Omega$, $L' = 1.4$ μH : (a) measured result, (b) analyzed result.

Fig. 8 presents analyzed characteristics of oscillation frequency f_{osc}. Calculated values comport well with measured ones.

III. INFLUENCE OF THE MILLER CAPACITANCE IN MAIN SWITCH

Here, we consider the case that two capacitances C_1 in Fig. 2 have different values. Therefore, two MOSFETs Q_1 and Q_2 conduct during different time period, respectively. In this case, asymmetrical operation is observed. Fig. 9(a) shows the measured oscillation frequency f_{osc} vs. the capacitance $C_{1(RIGHT)}$ characteristics, where $C_{1(RIGHT)}$ is the capacitance of the right-hand side capacitor C_1. Here, the capacitance of the left-hand side capacitor C_1, $C_{1(LEFT)} = 33$ pF. Next, Fig 9(b) presents the results by analysis. This time, we analyzed each half-period from Eq. (3) and Eq. (7).

In the actual operation, the left-handed side capacitance is fixed. However, the right-handed side capacitance is varied according to the condition of the outer converter. The Miller effect in the main switch S_{main} is unavoidable.

1545

IV. REACTANCE TRANSISTOR

To compensate the above-mentioned unbalance of $C_{1(LEFT)}$ and $C_{1(RIGHT)}$, we use a reactance-transistor as a variable capacitor or variable inductor. The reactance transistor have two types which are capacitance transistor and inductance transistor.

The circuit of capacitance transistor is shown in Fig. 10. And Fig. 11 shows equivalent circuit of a capacitance transistor. This output impedance Z_0 is shown by the parallel circuit of drain resistance r_d and equivalent capacitance C_{eq}. This impedance Z_0 can show equivalent capacitance C_{eq} where $1/r_d \ll \omega (1+g_m R)C$. The equivalent capacitance C_{eq} is given by the following equation.

$$Z_0 \doteqdot \frac{1}{j\omega(1+g_m R)C} \triangleq \frac{1}{j\omega C_{eq}} \qquad (8)$$

where

$$C_{eq} = (1+g_m R)C \qquad (9)$$

$$R \ll \frac{1}{\omega C} \qquad (10)$$

It is difficult to design circuit from the condition (10) if capacitance transistor is used at high frequency like over 100 MHz.

The circuit of inductance transistor is shown in Fig. 12. And Fig. 13 shows equivalent circuit of an inductance transistor. This output impedance Z_0 is shown by the parallel circuit of drain resistance r_d and equivalent inductance L_{eq}. This impedance Z_0 can show equivalent inductance L_{eq} where $1/r_d + 1/R \ll g_m / \omega CR$. The equivalent inductance L_{eq} is given by the following equation.

$$Z_0 \doteqdot \frac{j\omega CR}{g_m} \triangleq j\omega L_{eq} \qquad (11)$$

where

$$L_{eq} = \frac{CR}{g_m} \qquad (12)$$

$$R \gg \frac{1}{\omega C} \qquad (13)$$

It is difficult to design circuit from the condition (13) at low frequency like under 1 MHz because capacitance C is requested to be lower than the gate-source capacitance of MOSFET. Therefore, we use the capacitance transistor as compensation circuit in this paper.

V. EXPERIMENT OF CAPACITANCE TRANSISTOR

The capacitance transistor used in our experiments is shown in Fig. 14. Here, a dual-gate MOSFET used in this experiment is 3SK291 whose the Miller capacitance is extremely small. Even at high frequency, the impedance looking into the drain-source terminals, Z_0 is almost pure capacitive. The mutual conductance g_m is adjustable by the gate-source bias voltage V_{G1S} or V_{G2S}. Fig. 15 indicates characteristics of the mutual conductance in [4]. Forward transfer admittance in this graph is almost equal to the mutual conductance g_m.

Here, we need the maximum capacitance of at least 66 pF. Therefore, we design the capacitance C is 10 pF and

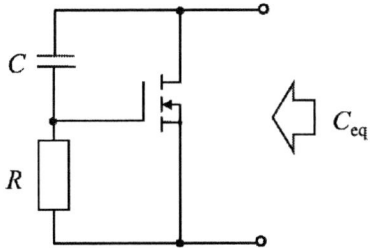

Fig. 10. Basic circuit of a capacitance transistor.

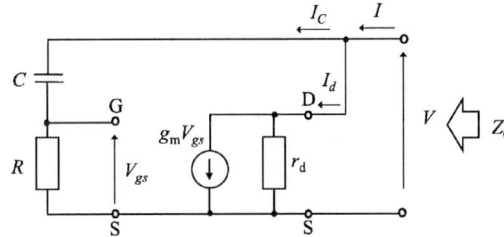

Fig. 11. Equivalent circuit of a capacitance transistor.

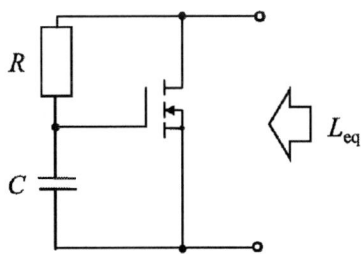

Fig. 12. Basic circuit of an inductance transistor.

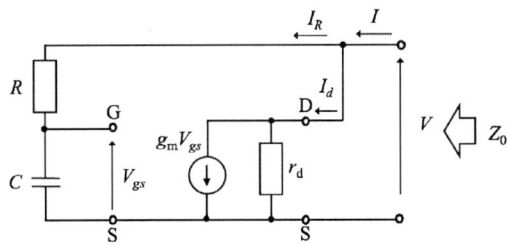

Fig. 13. Equivalent circuit of an inductance transistor.

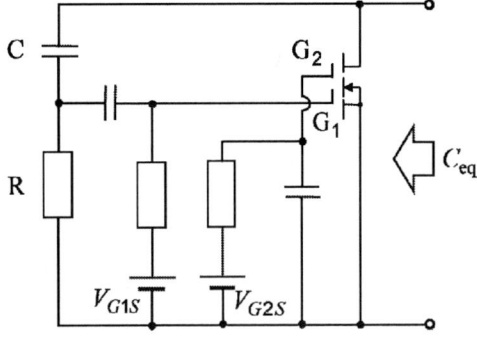

Fig. 14. Experimental circuit of a capacitance transistor.

The 2018 International Power Electronics Conference

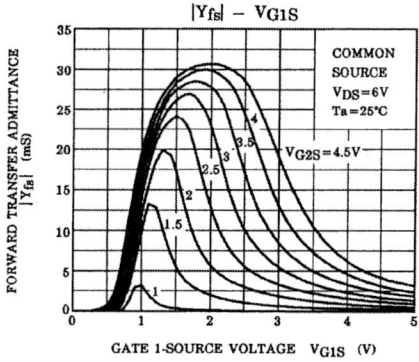

Fig. 15. Mutual conductance g_m of 3SK291 from its datasheet [4].

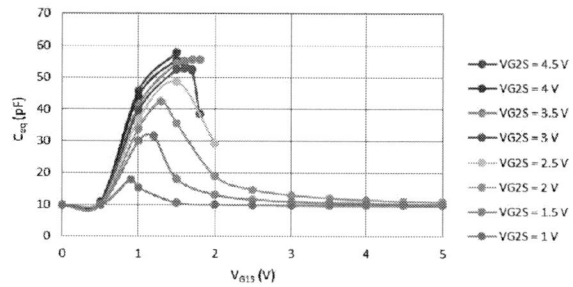

Fig. 16. Measured equivalent capacitance C_{eq} of capacitance transistor.

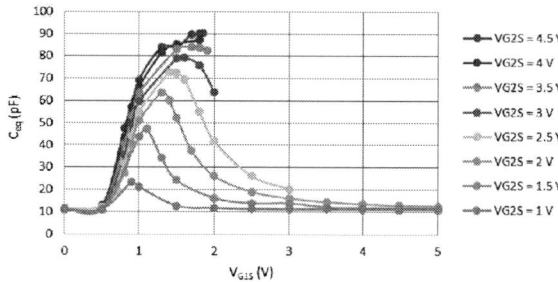

Fig. 17. Measured equivalent capacitance C_{eq} using two-parallel MOSFETs.

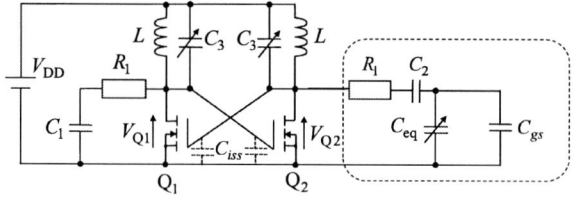

Fig. 18. Proposed push-pull LC oscillator with frequency compensation by capacitance transistor.

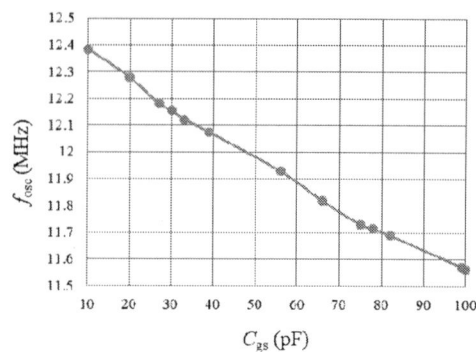

Fig. 19. Measured oscillation frequency f_{osc} without compensation.

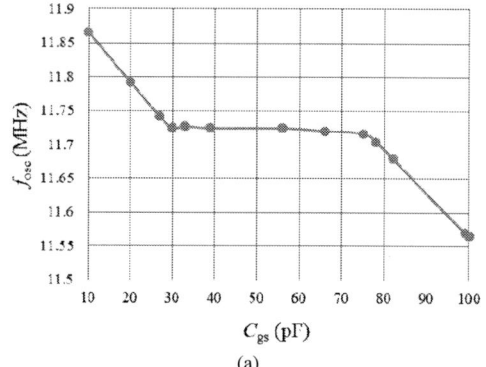

(a)

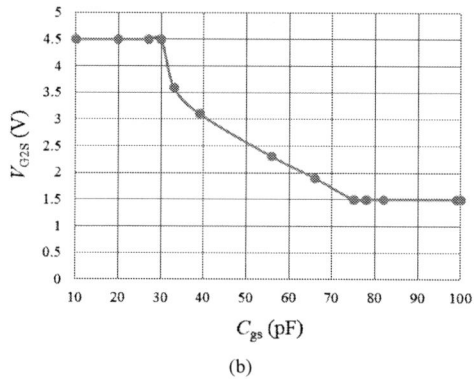

(b)

Fig. 20. Experiment results on frequency compensated driver using capacitance transistor: (a) Oscillation frequency f_{osc}, (b) Control voltage V_{G2S}, where $V_{G1S} = 1.5$ V.

the resistance R is 200 Ω at the experiment circuit. The equivalent capacitance C_{eq} becomes almost 70 pF by Eq. (9) when the mutual conductance g_m is maximum about 30 mS from Fig. 15. In this time, the relation (10) becomes $200 \ll 1360 \,[\Omega]$ at the frequency $f = 11.7$ MHz.

Therefore, these parameters satisfy the condition of Eq. (10).

Fig. 16 shows the measured characteristics of the capacitance C_{eq} at the condition of the drain-source dc voltage $V_{DS} = 6$ V. When the voltage V_{G1S} is changed from 0 to 6 V and the voltage V_{G2S} is changed from 1 V to 4.5 V, about 10 pF (the minimum capacitance) to about 58 pF (the maximum capacitance) can be obtained. We estimated the characteristics of the mutual conductance g_m from Eq. (9) and Fig. 16. When the voltage V_{G1S} is changed from 0 to 6 V and the voltage V_{G2s} is changed

1547

from 1 V to 4.5 V, about 0 (the minimum conductance) to about 24 mS (the maximum conductance) can be obtained.

We need peak capacitance of over 66 pF at compensation circuit. However the maximum equivalent capacitance was about 58 pF at this result. Therefore, we used two-parallel MOSFETs because the mutual conductance g_m is larger than that of one MOSFET. Fig. 17 shows the measured characteristics of the capacitance C_{eq} using two-parallel MOSFETs at the condition of V_{DS} = 6 V. When the voltage V_{G1S} is changed from 0 to 6 V and the voltage V_{G2S} is changed from 1 V to 4.5 V, about 11 pF (the minimum capacitance) to about 90 pF (the maximum capacitance) can be obtained. In this time, from Eq. (9) and Fig. 17, the measured value of the equivalent mutual conductance g_m becomes about 0 (the minimum conductance) to about 40 mS (the maximum conductance) when the voltage V_{G1S} is changed from 0 to 6 V and the voltage V_{G2S} is changed from 1 V to 4.5 V. In this case, we obtain about 1.5 times the equivalent capacitance and the mutual conductance. The equivalent capacitance peaks to about 72 pF at V_{G1S} = 1.5 V and V_{G2S} = 2.5 V. Therefore, it seems that this capacitance transistor is enough to operate as a variable capacitance device for this purpose.

VI. FREQUENCY-COMPENSATED GATE DRIVER

Fig. 18 shows an example of the frequency-compensated push-pull gate driver. Fig. 19 shows the characteristics of frequency without using the capacitance transistor as compensation circuit, where C_1 = 100 pF, C_2 = 1000 pF and R_1 = 100 Ω.

The oscillation frequency with compensation is presented at Fig. 20 (a). Fig. 20 (b) shows the gate-source voltage V_{G2S} as control voltage at Fig. 20 (a). In this time, the gate-source voltage V_{G2S} of capacitance transistor was changed from 1.5 V to 4.5 V at V_{G1S} = 1.5 V. The reference point is 11.8191 MHz at C_{gs} = 66 pF. Therefore, capacitance transistor is effective for compensation circuit.

VII. CONCLUSION

At first, we have reported the detailed analysis of oscillation frequency. Moreover, we have proposed a oscillation frequency becomes almost constant at about 11.72 MHz against from C_{gs} = 30 pF to C_{gs} = 76 pF. The error rate is +0.05 % and -0.03 % when reference point is 11.7202 MHz at C_{gs} = 66 pF.

In Fig. 19 with no compensation, the error rate is +2.9 % (at C_{gs} = 30 pF) and -0.7 % (at C_{gs} = 76 pF) when frequency-compensated gate driver by the use of a push-pull self-excitation LC oscillator. We have considered the capacitance transistor and the inductance transistor as compensation circuit for this driver.

In the proposed circuit with compensation by two-parallel-MOSFET capacitance transistor, the variation of the oscillation frequency reduced to almost one-forty-third in comparison with the original driver. The equivalent capacitance of a capacitance transistor effectively compensates the change in the gate-source capacitance by Miller effect in the main switch.

REFERENCES

[1] J.-H. C. Zhan, K. Maurice, J. Duster, and K. T. Kornegay, "Analysis and Design of Negative Impedance LC Oscillators Using Bipolar Transistors," *IEEE Trans. Circuits Syst.*, vol. 50, no. 11, pp. 1461-1464, November 2003.

[2] H. Moon, S. Kang, Y. T. Kim, and K. Lee, "A Fully Differential LC-VCO Using a New Varactor Control Structure," *IEEE Microw. Wireless Compon. Lett.*, vol. 14, no. 9, pp. 410-412, September 2004.

[3] P. Shamsi and B. Fahimi, "Design and Development of Very High Frequency Resonant DC-DC Boost Converters," *IEEE Trans. Power Electron.*, vol. 27, no. 8, pp. 3725-3733, August 2012.

[4] 3SK291, Datasheet from TOSHIBA, March. 2014.

A Flexible Reduced Capacitor Voltages Strategy for Variable-Speed Drives with Modular Multilevel Converter

Fangzhou Zhao*, Guochun Xiao, Daoshu Yang, Zhiqian Wu and Xin Meng
School of Electrical Engineering, Xi'an Jiaotong University, Xi'an, Shaanxi, China
*E-mail: zfzhaha@stu.xjtu.edu.cn

Abstract— Modular multilevel converter (MMC) has developed into an attractive topology for the variable-speed motor drives. However, the main challenge for the applications is the large low-frequency capacitor voltage fluctuation. This paper proposes a novel control strategy that is able to reduce the average capacitor voltages in a satisfying range when the motor operates below the base speed. Thus, the capacitor voltages can be well controlled under the nominal voltage for much better safety and reliability, and in the meantime the motor is able to operate with lower speed, larger load torque, and lower circulating current in the MMC for the wider range of operation conditions and greater control performances. The effectiveness of the proposed control strategy are demonstrated by the simulation results in PSCAD/EMTDC.

Keywords— *Modular Multilevel Converter (MMC), Variable-speed driver, Low-frequency fluctuation, Capacitor voltages balancing.*

I. INTRODUCTION

The Modular Multilevel Converter (MMC) has become one of the most popular multilevel topologies for the medium and high voltage applications in recent years [1]. Due to its unique advantages, such as modularity, low switching frequency and harmonic distortion, the MMC has been widely applied in the High-Voltage Direct Current transmission technology (HVDC) [2]. The MMC based medium-voltage application for variable-speed motor drives also shows meaningful prospect. However, the major technical challenge of the MMC based variable-speed drives that limits its development is the large fluctuation over the submodule (SM) capacitor voltages during the low-speed operation of the motor, also called low-frequency fluctuation [3].

Obviously, the low-frequency ripples over the capacitor voltages have a distinct negative impact on the system and may lead to the capacitor breakdown when the peak ripple voltage exceeds the rated value. What's more, the low-frequency component gradually becomes even larger in the lower speed operation.

Therefore, to solve this problem, the traditional high-frequency common-mode voltage and circulating current injection methods to suppress the low-frequency ripple

are proposed in [4][5], and comparisons of different injected waveforms are presented in [6]. In addition, the hybrid injection methods are also introduced in [7][8]. These injection methods can effectively suppress the low-frequency ripples over the capacitor voltages even if the output frequency is very low, which ensures the start-up of the motor. However, the injected common-mode voltage definitely accelerates the insulation aging of the winding in the motor and increase the risk of the winding deterioration. Besides, the peak value of the arm current is sharply risen thanks to the injected circulating current, and the tradeoff between the injected voltage and current exists in terms of the device rating and cost. In fact, the new improved MMC topologies to suppress the low-frequency fluctuation are presented in [9]-[11], but they inevitably cause more device costs and complicate the control structures. The optimal selection of the average capacitor voltages is proposed in [12]. It uses the energy margin to reduce the average capacitor voltages, and thus a greater low-frequency voltage ripple is allowed. However, the speed range using this method is limited (approximately 0.3-1.0 p.u.) and the margin to reduce the average voltage is small.

This paper proposes a novel control strategy that is able to reduce the average capacitor voltages in a large range. In this way, larger low-frequency fluctuation can be accept, since the average voltage is decreased and the peak voltage certainly will not exceed the rated value of the capacitors for superior safety and reliability. In the meantime, combining with the injection method, the motor is able to operate with wider speed range, larger load torque, and lower injected common-mode voltages. The proposed method changes the common rules of inserting and bypassing the SMs, which clearly differs from the optimal selection of the average voltage in [12]. Thus, the reduced voltage range of the proposed strategy is much larger than [12] and the speed is well controlled from zero to the nominal, not 0.3-1.0 p.u. in [12]. What's more, larger load torque can also be added applying the proposed method, comparing with the 40% rated torque in [6]. Besides, since the larger low-frequency ripple is allowed in the proposed strategy, lower injected high-

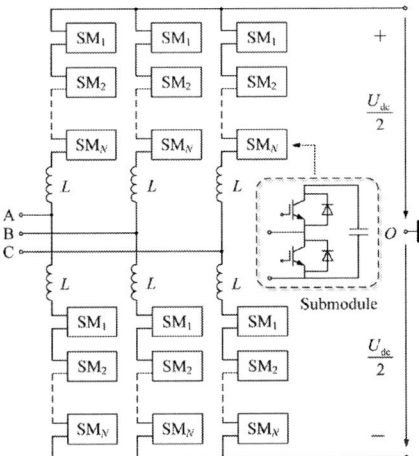

Fig. 1. The MMC basic structure.

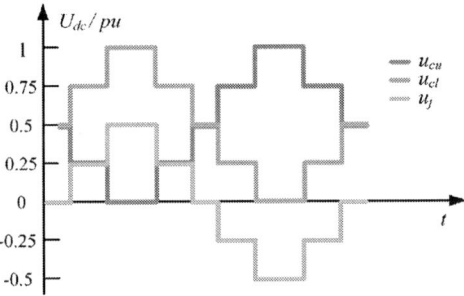

Fig. 2. Example of the voltages waveforms with 4 SMs.

frequency common-mode voltages are required, hence the risks of the insulation breakdown and large peak arm current are comparatively decreased, improving the efficiency and reliability of the system.

The outline of this paper is as follows. The basic operating principles of the MMC are presented in Section II. The proposed reduced capacitor voltages control method and its analysis are introduced in Section III. Simulation results in PSCAD/EMTDC are given in Section IV. Finally, conclusions are drawn in Section V.

II. BASIC OPERATING PRINCIPLES OF THE MMC

The MMC basic structure is shown in Fig. 1. Clearly, each arm of the MMC contains an inductor L and N series-connected SMs, and the half-bridge topology is widely applied in the SMs. Thus, one phase of the MMC has $2N$ SMs, and totally N SMs in the upper and lower arms are inserted to maintain the DC-link voltage. As a result, the nominal capacitor voltage U_{CN} is obtained as

$$U_{CN} = U_{dc} / N \qquad (1)$$

where V_{dc} is the DC-link voltage.

The output voltage u_j ($j=a,b,c$) of the MMC can be realized by controlling the upper and lower arm as [13]

$$u_{cu} = \left(U_{dc} - u_j\right)/2 \qquad (2a)$$

$$u_{cl} = \left(U_{dc} + u_j\right)/2 \qquad (2b)$$

where u_{cu} and u_{cl} are the control voltages for the upper and lower arms. Using the Phase-Shifted Carrier PWM (PSC-PWM) [14] or Nearest Level Control (NLC) [15] modulation, the output voltage u_j ($j=a,b,c$) of the MMC can be realized. To illustrate this relationship, an example of voltages with 4 SMs is shown in Fig. 2. To make it simple, the NLC is applied in Fig. 2. It can be seen that the output voltage has up to 5 levels, and totally 4 SMs are inserted in the upper and lower arms at any moment.

III. PROPOSED REDUCED CAPACITOR VOLTAGES METHOD

A. Basic Principles of the Proposed Strategy

As mentioned in the Section I, the MMC has totally N SMs in the upper and lower arms to maintain the DC-link voltage, which is also the principle to design the nominal average capacitor voltage as (1). Unlike this traditional principle, the proposed method applies more SMs to maintain the DC-link voltage, and the average capacitor voltages decrease naturally. To be detailed, suppose the nominal number of SMs to support the DC-link voltage N_s, and N_s is definitely larger than N to reduce the average voltage. Hence, the following relationship of the average capacitor voltage U_{CNs} can be obtained

$$U_{CNs} = U_{dc} / N_s < U_{CN} \qquad \left(N_s > N\right) \qquad (3)$$

Obviously, the average capacitor voltage is evidently reduced, and it gets gradually lower with larger N_s. However, normally the MMC only has N SMs in each arm, not counting the redundant SMs. When the proposed method is applied, N_s in upper and lower arm should support the DC-link voltage together, and thus there is no moment that all the SMs in an arm are bypassed since $N_s > N$. As a result, waveforms of the upper, lower arm and the output voltages are changed accordingly. To illustrate it, an example of voltages using the proposed strategy ($N_s = 5$) with 4 SMs is shown in Fig. 3. Clearly, since $N_s = 5$, the upper and lower arms have 5 inserted SMs in all at any moment, and the average capacitor voltage reduces to 0.2pu, comparing with the 0.25pu in the Fig. 2. However, the voltage levels of the waveforms decrease from 5 (Fig. 2) to 4, and the amplitude of the output voltage also gets smaller. In fact, the largest peak value of the output voltage using the proposed method can be calculated as

$$U_{am} = \frac{U_{CNs}\left(2N - N_s\right)}{2} = \frac{U_{dc}}{2}\frac{2N - N_s}{N_s} < \frac{U_{dc}}{2} \qquad (4)$$

Equation (4) indicates that the largest peak voltage of the MMC that is able to output gets lower. In other words, the modulation index decreases in the proposed strategy. Generally, the decline of the modulation index is not allowed in the normal MMC applications, such as the MMC-HVDC, and it means the operation range becomes smaller. However, in the MMC based variable-speed motor drives, the amplitude of the output voltage is approximately proportional to the motor speed, and it is correspondingly small when the speed is low, which

1550

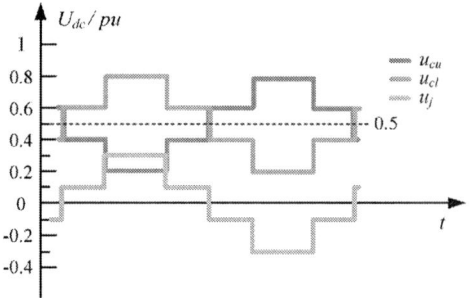

Fig. 3. Example of the voltages waveforms using the proposed method ($N_s = 5$) with 4 SMs.

exactly applies to the proposed method. Therefore, although the proposed method leads to a decrease of the modulation index, it only has a small impact on the system because the requirement of the modulation index is small anyway in a low-speed operation. What's more, the proposed method is able to change the N_s to the nominal value N and make the system back to normal smoothly for the high-speed operation or large load torque condition.

In conclusion, by applying the N_s ($N_s > N$), the average capacitor voltages can be reduced evidently. Since the output voltage in low-speed operation is small, the decrease of the modulation index of the proposed method is also acceptable.

B. Proposed Control Strategy

According to the analysis above, with the reduced average capacitor voltage, larger low-frequency ripple can be allowed, and the peak capacitor voltage is designed to be around the nominal value. Therefore, combining with the high-frequency common-mode voltage and circulating current injection method [6], the proposed control strategy can develop its advantages significantly. Because larger low-frequency ripple is accommodated, it's acceptable to inject much lower voltage components, and in the meantime lower speed with larger load torque can be realized. The complete block diagram of the proposed control strategy is shown in Fig. 4. Fig. 4(a) indicates that the N_s is larger than N in the low-speed operation, and at a certain moment it gradually decreases to N for the high-speed operation. Fig. 4(b) shows the averaging control that the capacitor voltage reference is controlled by N_s, and u_{Cave} is the average capacitor voltages of the phase. The traditional balancing and circulating current control in [6] is applied in Fig. 4(c). As for the injected high-frequency components i_{inj} and u_{com}, the 3-order harmonic [6] is used, but much lower injected common-mode voltages are required due to the reduced the average capacitor voltages. However, to apply the proposed method with N_s, the outputs of the controller in Fig. 4 should be transformed in terms of the N_s. Therefore, according to the control structure, the final control voltages for each SM in the upper and lower arms are calculated as

$$u_{cui}^* = u_{con} - \frac{u_j + u_{com}}{2N - N_s} + \frac{N_s}{2N^2}U_{dc} \qquad (5a)$$

$$u_{cli}^* = u_{con} + \frac{u_j + u_{com}}{2N - N_s} + \frac{N_s}{2N^2}U_{dc} \qquad (5b)$$

From (5), it can be seen that output voltage u_j and injected high-frequency voltage u_{com} should be equally separated into $2N-N_s$ SMs, because N_s SMs are used to maintain the DC-link voltage. Besides, the control voltages are lifted to the third terms in (5) for the PSC-PWM, and the control voltage with the carrier wave ($N=4$, $N_s=5$) is shown in Fig. 5. In fact, the lifted value in (5) is essential to the proposed method, because it exactly demonstrates that the N_s (not N) SMs are supporting the DC voltage in the low-speed operation. For the high-speed condition, with N_s gradually decrease to N, the lifted value in (5) equals to $U_{dc}/2N$, which is half of the carrier wave, and the system is back to normal. Clearly, the range of the output voltage is narrowed due to the rise of the DC term as shown in Fig. 4, and thus it is easier to the meet the peak value of the triangular carrier wave and get saturated. However, as claimed before, the required output voltage is comparatively small during low-speed operation and the beginning of the start-up, and consequently the decrease of the operation range of the proposed method is totally acceptable.

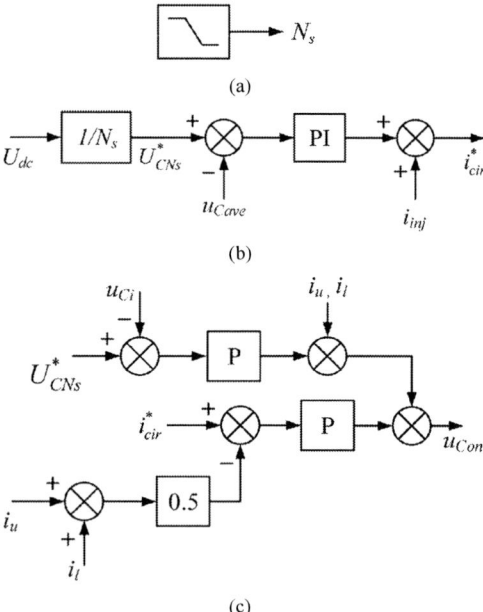

Fig. 4. Block diagrams of the proposed control strategy: (a) N_s control, (b) averaging control, (c) balancing and circulating current control.

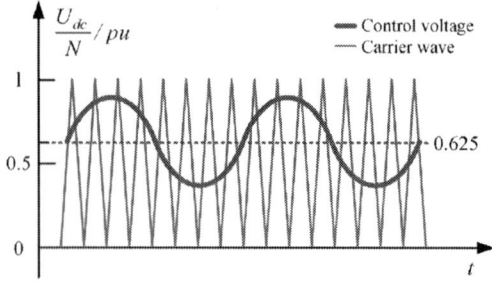

Fig. 5. Control voltage with the carrier wave ($N=4$, $N_s=5$).

IV. SIMULATION RESULTS

To verify the proposed method, the MMC based variable-speed motor drive simulation has been built in PSCAD/EMTDC. The steady-state and start-up control performances of the proposed method are verified properly with details to prove the validity and superiority. The main parameters of the MMC stage and the motor are respectively listed in Table. I and Table. II. As for the motor, the Field-Oriented Control (FOC) is applied to control the motor speed with required load torque for better dynamic performances.

The simulation results of the proposed control strategy at N_s=5, rotor speed ω_m=4.71rad/s and load torque T_L=10Nm are shown in Fig. 6. It can be seen that the average capacitor voltages in Fig. 6(a) are reduced to 120V (U_{dc}=600V, N_s=5) from the nominal value 150V using the proposed method. Therefore, the capacitor voltages can be well controlled far below the rated value for much better safety and reliability with over half of the rated load torque. The arm currents in Fig. 6(b) and line currents in Fig. 6(c) all indicate great performances of the proposed method. What's more, much lower injected common-mode voltages are required due to the decrease of average voltage. Fig. 7 shows the simulation results with only one quarter of the injected common-mode voltages. Obviously, the capacitor voltages ripples in Fig. 7(a) are evidently increased because far lower common-mode voltages are injected. However, although the ripples become much larger, the peak voltage is still far below the rated value 150V, and the arm currents in Fig. 7(b) and line currents in Fig. 7(c) show little differences from Fig. 6. As a result, in virtue of the reduction of the average capacitor voltage, lower injected common-mode voltages are required, and it certainly helps to decrease the disadvantages of the injection, such as the winding insulation deterioration. In that sense, the proposed strategy further enhances the reliability of the system. The proposed method can be well applied to the steady-state of the low speed situation with a desirable range of the load torque, and the reduced average capacitor voltage enables larger voltage fluctuation with comparatively lower injected components to lessen the disadvantageous influences on the motor.

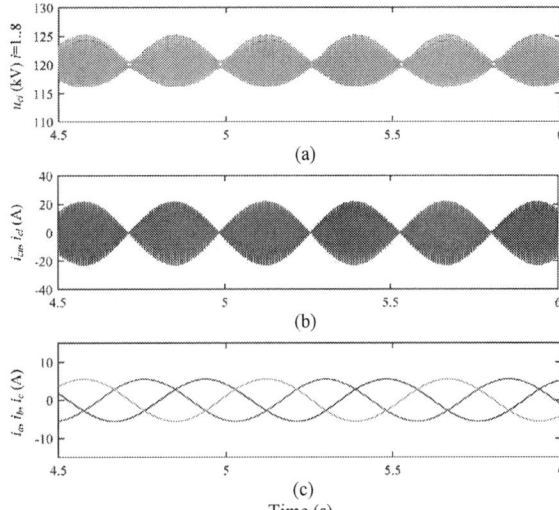

Fig. 6. Simulation waveforms of the proposed control strategy at N_s=5, ω_m=4.71rad/s and T_L=10Nm: (a) capacitor voltages of all SMs in phase A, (b) arm currents, (c) line currents.

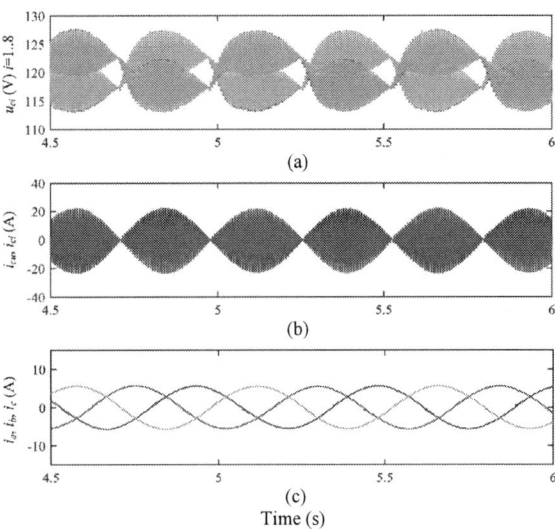

Fig. 7. Simulation waveforms of the proposed control strategy at N_s=5, ω_m=4.71rad/s and T_L=10Nm with only one quarter of the injected common-mode voltages: (a) capacitor voltages of all SMs in phase A, (b) arm currents, (c) line currents.

The start-up waveforms of the proposed method from starting to half of the rated angular speed are given in Fig. 8. The average capacitor voltage increases at 2s (N_s gradually decreases to 4) and gets back to the nominal value at 5s. Clearly, the capacitor voltages in Fig. 8(a) are controlled well without any overshoot, indicating the satisfying tracking performance of the proposed method. In the meantime, the angular speed in Fig. 8(b) demonstrate great start-up performances of the MMC based motor drives applying the proposed strategy. What's more, the arm currents and three phase line currents are given in Fig. 8(c) and Fig. 8(d) - the frequencies are gradually increasing with the motor speed, and the amplitudes are evidently controlled with satisfying results. As a result, the proposed strategy can be well utilized in the start-up process.

TABLE I
PARAMETERS OF MMC

Symbol	Meaning	Value
P	Rated power	4kW
U_s	Rated AC voltage (l-l)	380V
U_{dc}	Rated DC voltage	600V
N	Number of SMs per arm	4
C	Capacitance of SM	3.3mF
U_{CN}	Rated capacitor voltage	150V
L	Arm inductor	6mH
f_c	Switching frequency	2kHz

TABLE II
PARAMETERS OF MOTOR

Symbol	Meaning	Value
P	Rated power	2.2kW
U_s	Rated voltage (l-l)	380V
f	Rated frequency	50Hz
n	Rated speed	1430rpm
T	Rated torque	14.7Nm
p	Pole-pair number	2

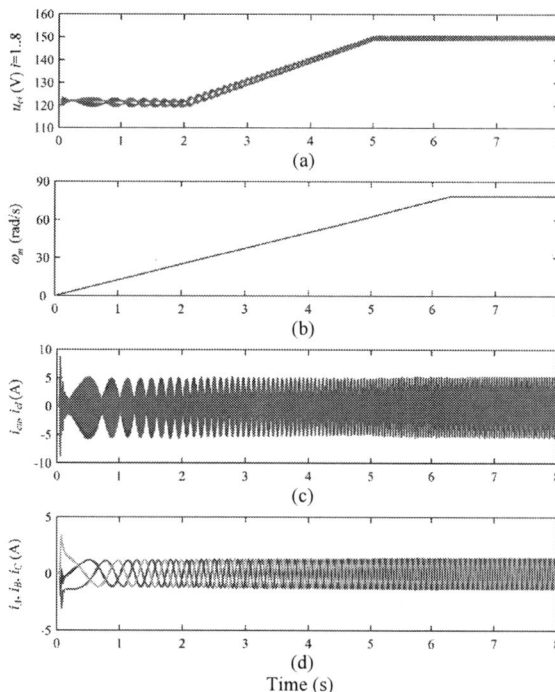

Time (s)

Fig. 8. Simulation waveforms of the proposed control strategy from starting to half of the rated angular speed: (a) capacitor voltages of all SMs in phase A, (b) angular speed, (c) arm currents, (d) line currents.

V. CONCLUSIONS

This paper proposes a novel control strategy that could effectively reduce the average capacitor voltages for much higher reliability and safety in terms of the large low-frequency voltage fluctuation. Since the average capacitor voltage can be reduced in a comparatively large range by applying the proposed method, greater low-frequency ripples are thusly accommodated, and in the meantime the peak voltage is ensured to be under the rated value for safety. Therefore, comparing with the traditional injection, a large proportion of the injected common-mode voltage is eliminated with only one quarter of the injected component is required in the proposed control method. Hence, the disadvantages caused by the common-mode voltage are significantly minimized, such as the winding insulation deterioration. What's more, the capacitor voltages is able to track the reference and get back to the nominal value smoothly in the start-up operation of the motor. The simulation results of the MMC based variable-speed motor drive in PSCAD verifies the proposed control method and demonstrate the effectiveness of the reduced average capacitor voltages with much lower injected common-mode voltage and the satisfying performances of the motor start-up operation. In conclusion, the proposed control strategy is a great choice for the MMC based variable-speed motor drive in terms of the low-speed operation.

REFERENCES

[1] S. Debnath, J. Qin, B. Bahrani, M. Saeedifard, and P. Barbosa, "Operation, control, applications of the modular multilevel converter: A review," *IEEE Trans. Power Electron.*, vol. 30, no. 1, pp. 37–53, Jan. 2015.

[2] F. Zhao, G. Xiao, M. Liu and D. Yang, "A Fast Sorting Strategy Based on Two-Way Merge Sort for Balancing Capacitor Voltages in Modular Multilevel Converter," *Journal of Power Electron.*, vol. 16, no. 4, pp. 1256-1267, Mar. 2017.

[3] M. Hagiwara, K. Nishimura, and H. Akagi, "A medium-voltage motor drive with a modular multilevel PWM inverter," *IEEE Trans. Power Electron.*, vol. 25, no. 7, pp. 1786–1799, Jul. 2010.

[4] K. Wang, Y. Li, Z. Zheng, and L. Xu, "Voltage balancing and fluctuation-suppression methods of floating capacitors in a new modular multilevel converter," *IEEE Trans. Ind. Electron.*, vol. 60, no. 5, pp. 1943–1954, May. 2013.

[5] A. Antonopoulos, L. Ångquist, S. Norrga, K. Ilves, L. Harnefors, and H.-P. Nee, "Modular multilevel converter AC motor drives with constant torque from zero to nominal speed," *IEEE Trans. Industry Applications*, vol. 50, no. 3, pp. 1982–1993, May/Jun. 2014.

[6] M. Hagiwara, I. Hasegawa, and H. Akagi, "Start-up and low-speed operation of an electric motor driven by a modular multilevel cascade inverter," *IEEE Trans. Industry Applications*, vol. 49, no. 4, pp. 1556–1565, Jul./Aug. 2013.

[7] S. Debnath and M. Saeedifard, "Optimal control of modular multilevel converters for low-speed operation of motor drives," *in Proc. 29th IEEE-APEC*, Fort Worth, TX, USA, Mar. 2014, pp. 247–254.

[8] L. He, K. Zhang, J. Xiong, S. Fan, and Y. Xue, "Low-frequency ripple suppression for medium-voltage dives using modular multilevel converter with full-bridge submodules," *IEEE J. Emerg. Sel. Topics Power Electron.*, vol. 4, no. 2, pp. 657–667, Jun. 2016.

[9] S. Du, B. Wu, N. Zargari, and Z. Cheng, "A Flying-Capacitor Modular Multilevel Converter for Medium-Voltage Motor Drive," *IEEE Trans. Power Electron.*, vol. 32, no. 3, pp. 2081-2089, Mar. 2016.

[10] S. Du, B. Wu, K. Tian, N. Zargari, and Z. Cheng, "An Active Cross-Connected Modular Multilevel Converter (AC-MMC) for a Medium-Voltage Motor Drive," *IEEE Trans. Ind. Electron.*, vol. 63, no. 8, pp. 4707-4717, Aug. 2016.

[11] B. Li, S. Zhou, D. Xu, S. Finney, and B. Williams, "A Hybrid Modular Multilevel Converter for Medium-Voltage Variable-Speed Motor Drives," *IEEE Trans. Power Electron.*, vol. 32, no. 6, pp. 4619-4630, Jun. 2017.

[12] A. Antonopoulos, L. Angquist, L. Harnefors, and H.-P. Nee, "Optimal selection of the average capacitor voltage for variable-speed drives with modular multilevel converters," *IEEE Trans. Power Electron.*, vol. 30, no. 1, pp. 227–234, Jan. 2015.

[13] D. Wu and L. Peng, "Analysis and Suppressing Method for the Output Voltage Harmonics of Modular Multilevel Converter," *IEEE Trans. Power Electron.*, vol. 31, no. 7, pp. 4755-4765, Jul. 2016.

[14] M. Hagiwara and H. Akagi, "Control and experiment of pulsewidth modulated modular multilevel converters," *IEEE Trans. Power Electron.*, Vol. 24, No. 7, pp. 1737–1746, Jul. 2009.

[15] Q. Tu and Z. Xu, "Impact of sampling frequency on harmonic distortion for modular multilevel converter," *IEEE Trans. Power Delivery*, vol. 26, no. 1, pp. 298–306, Jan. 2011.

The 2018 International Power Electronics Conference

A Leakage Flux Cancellation Technique for Series-Parallel Combined Resonant Circuits with Asymmetric Rotary Transformers Used for Ultrasonic Spindle Drive

Jun Imaoka[1], Masahito Shoyama[1]
1 Department of Electrical Engineering, Kyushu University, Fukuoka, Japan
Email: imaoka@ees.kyushu-u.ac.jp, shoyama@ees.kyushu-u.ac.jp

Abstract— This paper proposes a novel leakage flux cancellation technique suitable for series-parallel combined resonant circuits with asymmetric rotary transformer used for ultrasonic spindles drive. An asymmetric rotary transformer is utilized in the ultrasonic spindle in order to realize automated tool change. However, an asymmetric rotary transformer is likely to occur much leakage fluxes in secondly core due to asymmetric core structures. This large leakage inductance affects a drive circuit using a series-parallel combined resonant circuit, and the resonant characteristics in the drive circuit become very steep. In this state, the resonance circuit does not realize the desired function when the drive frequency fluctuates or the error of resonance parameters occurs. Therefore, this paper proposes a novel leakage flux cancellation technique used for ultrasonic spindle drive. The effectiveness of the proposed method is discussed from the theoretical analysis and experimental test.

Keywords— *Series-Parallel Combined Resonant Circuits, Asymmetric Rotary Transformer, Spindle, Leakage Flux Cancellation, Automatic Tool Changer*

I. INTRODUCTION

Precision machining equipment which is operated by automatic control programs has gained much attention in order to realize factory automation capable of reducing labor cost and eliminating human error. Among various pieces of machining equipment, spindles using an ultrasonic transducer are well-known as one of the outstanding processing equipment for difficult-to-machine materials or hard brittle materials. Ultrasonic spindles have outstanding machining performance by utilizing both of turning forces by a high-speed motor and high-speed vibration by the ultrasonic transducer [1-5]. Fig. 1 shows a configuration of the ultrasonic spindle using a symmetric rotary transformer. In general, the ultrasonic spindle is composed of a rotary transformer, a rotary axis, an ultrasonic transducer, and tools such as drills, end mills, and grinders. The ultrasonic transducer can vibrate based on the inverse piezoelectric effect and is effective to expand the lifetime of tools by improving machining performance in comparison with spindles without the ultrasonic transducer. In addition, in order to drive the ultrasonic transducer located on the rotary axis, the sinusoidal voltage with 40 kHz and sufficient

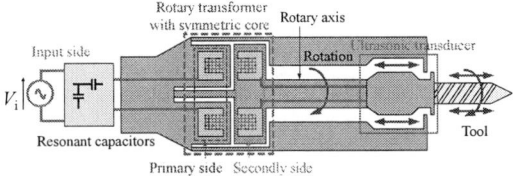

Fig. 1. The sectional view of the ultrasonic spindle using a symmetric rotary transformer.

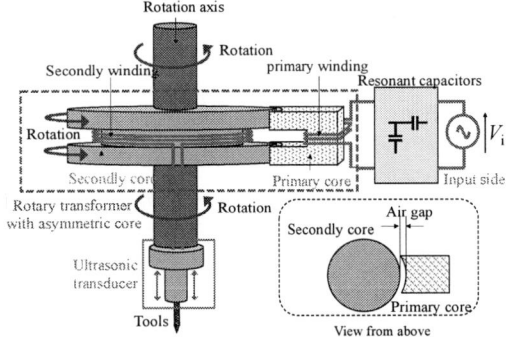

Fig. 2. The three-dimensional view of the ultrasonic spindle using an asymmetric rotary transformer.

amplitudes have to be supplied to the ultrasonic transducer from the input terminal of the fixed lateral shown in Fig. 1.

As conventional voltage transfer method to rotating object such as the ultrasonic spindle, the voltage transfer method using the combination of a slip ring and a brush is well known. However, the conventional voltage transfer method using a slip ring and a brush is needed regular maintenance work due to wear of the brush and deterioration of voltage transfer characteristic. Therefore, factories have to be stopped during changing a deteriorated brush and slip ring. Therefore, in order to improve upon this problem, the rotary transformer is equipped instead of a blush and a slip ring to realize maintenance free and reduction of running cost of the machining equipment.

However, two technical issues are encountered by replacing the conventional voltage transfer method to the voltage transfer method using the rotary transformer.

The 2018 International Power Electronics Conference

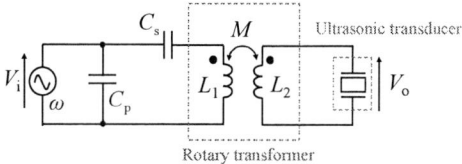

(a) Circuit configuration of Type A.

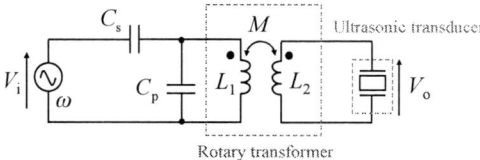

(b) Circuit configuration of Type B.

Fig. 3. Circuit configurations of series-parallel combined resonant circuits used for ultrasonic spindle drive.

First, a voltage drop caused by leakage inductances in the primary and secondary windings occurs, and sufficient voltage to drive the ultrasonic transducer cannot be transferred to the ultrasonic transducer. Second, the power factor of the input voltage source deteriorates due to the magnetizing inductance of the rotary transformer.

Therefore, to solve the above two technical issues, series-parallel combined resonant circuits was proposed in [1]-[2]. In these resonant circuits mentioned in [1]-[2], the series resonant capacitor is utilized to cancel the leakage inductances. In addition, the magnetizing inductance is canceled by the parallel resonant capacitor. Therefore, by using both of series and parallel resonances, the two technical issues can be solved. The reference [1] was proposed two resonant circuits using a symmetric rotary transformer by changing the location of the resonant capacitors. In addition, the circuit design methods for both of series-parallel combined resonant circuits were also proposed.

On the other hand, an ultrasonic spindle using an asymmetric rotary transformer was proposed in [5]. This system configuration of a series-parallel combined resonant circuit using an asymmetric transformer is shown in Fig. 2. The attractive feature of the ultrasonic spindle using asymmetric rotary transformers is that a function of automatic tool change can implement. The automatic tool change means automatically changing the type of tools located on a spindle according to a machining procedure. The automatic tool change can be realized by that the unit which is combined the secondary core, the ultrasonic transducer, and a tool is exchanged by mechanical exchange devices. Therefore, personnel for tool replacement in factories is also unnecessary and this contributes to a reduction of manufacturing cost. Thus, it enables an automatic, comfortable and complex processing of complicated workpieces. However, when the asymmetric rotary transformer is used in ultrasonic spindles as the voltage transfer method, the resonance characteristics become very steep because of low coupling coefficients of the transformer, and there is a high possibility that the resonance will come off due to

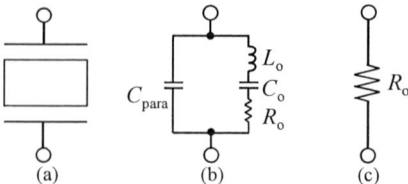

Fig. 4. Equivalent circuit models of the ultrasonic transducer. (a) is the symbol of the ultrasonic transducer, (b) is the equivalent circuit model, (c) is the equivalent circuit model at series resonant frequency decided by L_o and C_o.

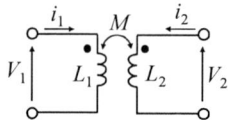

(a) Simple circuit model of the transformer using the mutual inductance.

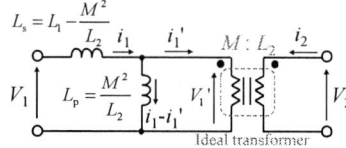

(b) Equivalent circuit model of the transformer used in Type A.

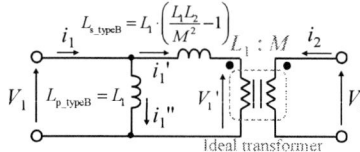

(c) Equivalent circuit model of the transformer used in Type B.

Fig. 5. Equivalent circuit models of the two winding transformer.

increasing the leakage inductance on the secondary side.

Therefore, this paper proposes a performance improvement method utilizing a novel leakage flux cancellation technique which uses copper tape or plate. Appling this technique, even if the drive frequency of the input voltage is slightly shifted or the resonant frequencies of both series and parallel resonant circuits differ due to tolerance, reduction of voltage drop and power factor improvement effect is maintained. The effectiveness of the proposed method is confirmed by theoretical analysis and experimental results.

II.   SERIES-PARALLEL COMBINED RESONANT CIRCUIT USED FOR ULTRASONIC SPINDLE DRIVE

A. Circuit Configuration and Equivalent Circuit Models

Fig. 3 shows the circuit configurations of the series-parallel combined resonant circuits. These circuit configurations are respectively called type A and B in [1]. where V_i and V_o are the input and output voltages with the angular frequency ω. Thereafter, f means the frequency of the input voltage source. L_1, L_2, and M mean the self-inductances on the primary and secondary windings and the mutual inductance, respectively. C_s and C_p are the series resonant and the parallel resonant capacitances, respectively. The difference between the Type A and B is the location of the resonant capacitors. In addition, these resonant capacitors cannot insert to the secondary side of the rotary transformer because the secondary side rotates by rotating force by a motor, and there is the possibility

1555

TABLE I
EQUIVALENT CIRCUITS OF SERIES-PARALLEL COMBINED RESONANT CIRCUIT AND THEIR DESIGN EQUATIONS

	Type A		Type B	
Configuration				
Combined inductance related to series resonance L_s	$L_{s\_typeA} = L_1 - \dfrac{M^2}{L_2}$	(6)	$L_{s\_typeB} = L_1 \cdot \left(\dfrac{L_1 L_2}{M^2} - 1 \right)$	(11)
Combined inductance related to parallel resonance L_p	$L_{p\_typeA} = \dfrac{M^2}{L_2}$	(7)	$L_{p\_typeB} = L_1$	(12)
Series resonant capacitance C_s	$C_{s\_typeA} = \dfrac{1}{\omega_0^2 \cdot L_{s\_typeA}}$	(8)	$C_{s\_typeB} = \dfrac{1}{\omega_0^2 \cdot L_{s\_typeB}}$	(13)
Parallel resonant capacitance C_p	$C_{p\_typeA} = \dfrac{1}{\omega_0^2 \cdot L_{p\_typeA}}$	(9)	$C_{p\_typeB} = \dfrac{1}{\omega_0^2 \cdot L_{p\_typeB}}$	(14)
Inductance ratio to transfer the same voltage from input to the output side	$L_2 = M$	(10)	$L_1 = M$	(15)

that it will come off due to centrifugal force.

In addition, the equivalent circuit model of the ultrasonic transducer is shown in Fig. 4. The equivalent circuit of the ultrasonic transducer consists of an intrinsic resistance R_o, an intrinsic capacitor C_o, an intrinsic inductance L_o and a parasitic capacitor C_{para}. In general, C_{para} is a very small value in comparison with other values such as L_o, C_o, R_o, and the equivalent circuit model at the series-resonant frequency decided by L_o and C_o can be shown by only R_o.

Then, the equivalent circuit of the two winding transformer is described. The simple circuit model of the two winding transformer using the mutual inductance is shown in Fig. 5 (a). However, the circuit model shown in Fig. 5 (a) cannot understand intuitively the inductances related to the series and parallel resonances. For intuitive understanding, converting the circuit model of the transformer is important. Fig. 5 (b) shows an equivalent circuit model of the transformer used in Type A which is separated into the inductances related to series and parallel resonance. where L_{s_typeA} and L_{p_typeA} mean the combined inductances related to series and parallel resonances, respectively. The turns ratio of the ideal transformer can show by $M : L_2$. Then, the reason why this equivalent circuit model can be obtained is described. The meaning of the transformer equivalent circuits is to fulfill the following equation:

$$\begin{cases} V_1 = j\omega L_1 \cdot i_1 + j\omega M \cdot i_2 \\ V_2 = j\omega L_2 \cdot i_2 + j\omega M \cdot i_1 \end{cases} \quad (1)$$

Then, the equivalency of the circuit model shown in Fig. 5 (b) is validated. From the primary side of the transformer, the following equation can be obtained:

$$\begin{aligned} V_1 &= j\omega \cdot \left(L_1 - \frac{M^2}{L_2} \right) \cdot i_1 + V_1' \\ &= j\omega \cdot \left(L_1 - \frac{M^2}{L_2} \right) \cdot i_1 + j\omega \left(\frac{M^2}{L_2} \right) \cdot \left(i_1 - i_1' \right) \\ &= j\omega \cdot \left(L_1 - \frac{M^2}{L_2} \right) \cdot i_1 + j\omega \left(\frac{M^2}{L_2} \right) \cdot \left(i_1 + i_2 \cdot \frac{L_2}{M} \right) \\ &= j\omega L_1 \cdot i_1 + j\omega M \cdot i_2 \end{aligned} \quad (2)$$

Similarly, from the secondary side of the transformer model shown in Fig. 5 (b), the following equation can be obtained:

$$\begin{aligned} V_2 &= V_1' \cdot \frac{L_2}{M} = j\omega \left(\frac{M^2}{L_2} \right) \cdot \left(i_1 - i_1' \right) \cdot \frac{L_2}{M} \\ &= j\omega \left(\frac{M^2}{L_2} \right) \cdot \left(i_1 + i_2 \cdot \frac{L_2}{M} \right) \cdot \frac{L_2}{M} \\ &= j\omega L_2 \cdot i_2 + j\omega M \cdot i_1 \end{aligned} \quad (3)$$

Therefore, the equivalency of the circuit model of the transformer for Type A shown in Fig. 5 (b) is proven theoretically because the voltage equations in the primary and secondary sides are the same as (1).

Then, the equivalency of the circuit model of the transformer for Type B shown in Fig. 5 (c) is described. Similarly, from the primary side of the transformer shown in Fig. 5 (c), the following equation can be obtained:

$$\begin{aligned} V_1 &= j\omega L_1 \cdot i_1^* = j\omega L_1 \cdot \left(i_1 - i_1' \right) \\ &= j\omega L_1 \cdot \left(i_1 + i_2 \cdot \frac{M}{L_1} \right) \\ &= j\omega L_1 \cdot i_1 + j\omega M \cdot i_2 \end{aligned} \quad (4)$$

From the secondary side, the following equation yield:

$$\begin{aligned} V_2 &= V_1' \cdot \frac{M}{L_1} = \left(j\omega L_1 \cdot i_1^* - j\omega L_1 \cdot \left(\frac{L_1 L_2}{M^2} - 1 \right) \cdot i_1' \right) \cdot \frac{M}{L_1} \\ &= j\omega M \cdot \left(i_1 - i_1' \right) - j\omega \cdot M \cdot \left(\frac{L_1 L_2}{M^2} - 1 \right) \cdot i_1' \\ &= j\omega M \cdot i_1 - j\omega \frac{L_1 L_2}{M} \cdot i_1' = j\omega M \cdot i_1 + j\omega \frac{L_1 L_2}{M} \cdot i_2 \frac{M}{L_1} \\ &= j\omega L_2 \cdot i_2 + j\omega M \cdot i_1 \end{aligned} \quad (5)$$

Therefore, the equivalency of the transformer circuit model for Type B is also proven.

B. Circuit Design for Series-Parallel Combined Resonant Circuits

Then, using both of equivalent circuit models of the ultrasonic transducer and the transformer, the equivalent circuit models of the series-parallel combined resonant circuits used in Type A and B are shown in Table I. As shown in Table I, the difference between the equivalent circuit of Type A and B is the position of the resonant capacitors. Type A is a circuit which series resonant circuit is buried between the parallel resonant circuit. Type B is a circuit which parallel resonant circuit is buried between the series resonant circuit. The series resonance between L_s and C_s contributes to reducing the voltage drop caused by the leakage inductances, and the sufficient voltage to drive the ultrasonic transducer can be transferred from the input voltage source. The parallel resonance between L_p and C_p contributes to improving the power factor at the input side because the impedance of the parallel resonance becomes the highest at the parallel resonant frequency. Therefore, the two problems mentioned in section I can be solved by using series-parallel combined resonant circuits.

The design equations for both of resonant circuits are also shown in Table I. It notes that the resonant frequencies of series and parallel resonant circuits and ultrasonic transducer has to match each other. In these equations shown in Table I, ω_0 means the resonance angular frequency, and its frequency is represented as f_0. In addition, to transfer the same voltage from the input to the output sides, the equations (10) and (15) in each circuit must be fulfilled. In addition, when (5)-(9) or (11)-(14) in each circuit are fulfilled, the series-parallel combined resonant circuits for the ultrasonic spindle can be operated in the same way as voltage transfer method using a blush and a slip ring.

The next section, a magnetic design method that can satisfy the constraints of $L_2=M$ or $L_1=M$ in both circuits is proposed by using a magnetic circuit model of transformers.

III. Transformer Design Method Using Equivalent Magnetic Circuit Model

In this section, the magnetic design methods for fulfilling the constraints of $L_2=M$ or $L_1=M$ are proposed by applying a magnetic circuit model. Fig. 6 shows the magnetic circuit model of the transformer with two windings. By using this equivalent magnetic circuit model, the self-inductance L_1, L_2, and the mutual inductance M can be modeled by

$$
\begin{cases}
L_1 = N_1^2 \cdot \dfrac{R_m + R_{lk1}}{R_m \cdot R_{lk1}} \\[2mm]
L_2 = N_2^2 \cdot \dfrac{R_m + R_{lk2}}{R_m \cdot R_{lk2}} \\[2mm]
M = \dfrac{N_1 \cdot N_2}{R_m} \\[2mm]
k = 1 \bigg/ \sqrt{\dfrac{(R_{lk1} + R_m) \cdot (R_{lk2} + R_m)}{R_{lk1} \cdot R_{lk2}}}
\end{cases}
\tag{16}
$$

where N_1 and N_2 are the number of turns on the primary and secondary windings, respectively. R_{lk1}, R_{lk2}, and R_m mean the magnetic reluctances on the magnetic paths where leakage flux flows and the magnetic path where magnetizing flux flows, respectively. In addition, these magnetic reluctance values are decided by magnetic core structure. Furthermore, the coupling coefficient is the

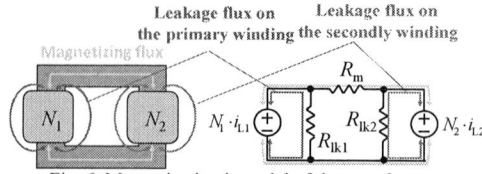

Fig. 6. Magnetic circuit model of the transformer.

function of magnetic reluctances and is not related to the number of turns. Therefore, using (16), the turn ratio which can fulfill the constraints of $L_2=M$ in case of the type A can be shown by

$$
\frac{N_{1\_typeA}}{N_{2\_typeA}} = \frac{R_m + R_{lk1}}{R_{lk1}} \quad (\text{Type A})
\tag{17}
$$

Therefore, when using (17) and the magnetic reluctance values R_{lk1}, R_{lk2}, and R_m decided by the magnetic core structure, the number of turns which can fulfill constraint of $L_2=M$ can design easily.

On the other hand, the turn ratio which can fulfill the constraints of $L_1=M$ in case of type B can be shown by

$$
\frac{N_{1\_typeB}}{N_{2\_typeB}} = \frac{R_m + R_{lk2}}{R_{lk2}} \quad (\text{Type B})
\tag{18}
$$

Using (17) or (18), the turn ratio of the winding in each resonant circuit can be decided easily.

IV. The Problem of the Asymmetric Magnetic Core Structure and the Performance Improvement Utilizing a Leakage Flux Cancellation Technique

A. Technical Issues of Series-Parallel Combined Resonant Circuit Using Asymmetric Rotary Transformer

In this section, the technical issues of using series-parallel combined resonant circuits in ultrasonic spindles are introduced. If these resonant components are designed to fulfill resonant conditions shown in Table I, the input voltage source looks connected directly to load R_o.

However, fulfilling simultaneously these resonant conditions is very difficult in practical use because there are three components having frequency characteristics. The resonant components are summarized as follows:
1. Series resonance in the resonant circuits
2. Parallel resonance in the resonant circuits
3. Series resonance in the ultrasonic transducer

Among these resonant components, matching the resonant frequency of the ultrasonic transducer with the frequency of input voltage source is a priority design requirement. This is because the ultrasonic transducer does not oscillate when the resonance frequency of the ultrasonic transducer and the frequency of the input voltage do not match.

From the resonant circuit perspective, when matching the frequencies of the input voltage source and the ultrasonic transducer under the condition that resonant frequency of the ultrasonic transducer and the resonant frequencies of the resonant circuits differ, the robustness of the resonant circuit against frequency fluctuation is required. In addition, the asymmetric rotary transformer shown in Fig. 2 tend to have lower coupling coefficient

The 2018 International Power Electronics Conference

TABLE II
VOLTAGE GAIN AND INPUT IMPEDANCE OF RESONANT CIRCUITS

	Type A		Type B	
Voltage gain T	$T_{typeA} = \dfrac{\lvert V_o \rvert}{\lvert V_{in} \rvert} = \dfrac{C_s \cdot M \cdot R_o \cdot \omega^2}{\sqrt{R_o^2 \cdot \left(C_s \cdot L_1 \cdot \omega^2 - 1\right)^2 + \left(L_2 \cdot \omega - C_s \left(L_1 \cdot L_2 - M^2\right) \omega^3\right)^2}}$	(19)	$T_{typeB} = \dfrac{\lvert V_o \rvert}{\lvert V_{in} \rvert} = \dfrac{C_s \cdot M \cdot R_o \cdot \omega^3}{\sqrt{R_o^2 \cdot \left(\left(C_s + C_p\right) \cdot L_1 \cdot \omega^2 - 1\right)^2 + \left(L_2 \cdot \omega - \left(C_s + C_p\right)\left(L_1 \cdot L_2 - M^2\right) \omega^3\right)^2}}$	(21)
Input impedance Z_{in}	$Z_{in\_typeA} = \lvert Z_{in\_typeA} \rvert e^{-j\theta_{typeA}}$ $\lvert Z_{in\_typeA} \rvert = \sqrt{\dfrac{a_A{}^2 + b_A{}^2}{c_A{}^2 + d_A{}^2}}, \quad \theta_{typeA} = \tan^{-1}\left(\dfrac{b_A \cdot c_A - a_A \cdot d_A}{a_A \cdot c_A + b_A \cdot d_A}\right)$ $\begin{cases} a_A = R_o\left(1 - C_s \cdot L_1 \cdot \omega^2\right) \\ b_A = L_2 \cdot \omega - C_s \cdot \left(L_1 \cdot L_2 - M^2\right) \cdot \omega^3 \\ c_A = \omega^2\left(C_s \cdot C_p \cdot \left(L_1 \cdot L_2 - M^2\right) \cdot \omega^2 - \left(C_s + C_p\right) \cdot L_2\right) \\ d_A = R_o \cdot \left(\left(C_s + C_p\right) \cdot \omega - C_s \cdot C_p \cdot L_1 \cdot \omega^3\right) \end{cases}$	(20)	$Z_{in\_typeB} = \lvert Z_{in\_typeB} \rvert e^{-j\theta_{typeB}}$ $\lvert Z_{in\_typeB} \rvert = \sqrt{\dfrac{a_B{}^2 + b_B{}^2}{c_B{}^2 + d_B{}^2}}, \quad \theta_{typeB} = \tan^{-1}\left(\dfrac{b_B \cdot c_B - a_B \cdot d_B}{a_B \cdot c_B + b_B \cdot d_B}\right)$ $\begin{cases} a_B = R_o - R_o \cdot L_1 \cdot \left(C_p + C_s\right) \cdot \omega^2 \\ b_B = L_2 \cdot \omega - \left(C_s + C_p\right) \cdot \left(L_1 \cdot L_2 - M^2\right) \cdot \omega^3 \\ c_B = -C_s \cdot L_2 \cdot \omega^2 + C_s \cdot C_p \cdot \left(L_1 \cdot L_2 - M^2\right) \cdot \omega^4 \\ d_B = C_s \cdot R_o \cdot \omega \cdot \left(1 - C_p \cdot L_1 \cdot \omega^2\right) \end{cases}$	(22)

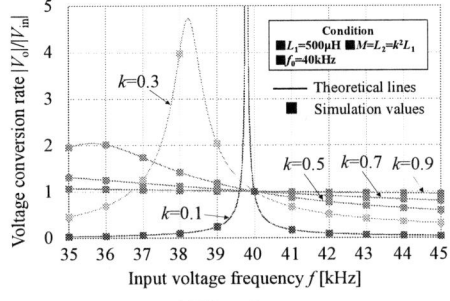

(a) Type A

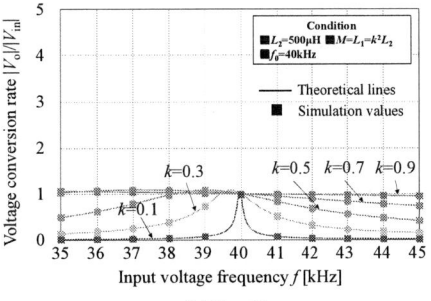

(b) Type B

Fig. 7. Voltage gain characteristics of the both resonant circuits.

than symmetric rotary transformer shown in Fig. 1 due to asymmetric core structures. In the asymmetric magnetic core structure, much leakage fluxes generate at the secondary side core which does not face the primary side core. This leakage inductance affects the series-parallel combined resonant circuit, and resonant characteristics become very steep.

Although the resonant circuit is inserted to suppress the voltage drop and improve the power factor, and these effects should be maintained in wide frequency band because the resonant frequency of the ultrasonic transducer has an individual difference or tolerance. Therefore, in next subsection, the resonance characteristics of the series-parallel combined resonance circuits investigates under the condition that input frequency and low coupling coefficient are respectively changed.

B. Voltage Gain and Input Impedance Characteristics between Resonant Frequency and in the Vicinity Thereof

First of all, voltage gain characteristics between the input and output voltages in both resonant circuit in type A and B are theoretically derived. From the equivalent circuit model for Type A shown in Table I, the ratio of the amplitude of V_o to that of V_i can be derived by (19). In addition, the power factor of the input voltage source is decided by the phase angle θ_{Zin_typeA} of the input impedance Z_{in_typeA} shown in Table I. As the analysis

results, the input impedance Z_{in_typeA} can be summarized in (20). When the resonant circuit of Type A and the transformer are designed on based on design method mentioned in Table I, the amplitude of the input impedance Z_{in_typeA} agrees with R_o value at the resonant frequency f_0. In addition, the phase of Z_{in_typeA} is 0 degree at the resonant frequency f_0. Likewise, the voltage gain of type A is 1 at resonant frequency f_0 because L_s is canceled by C_s and the turn ratio of the ideal transformer is 1:1.

Similarly, the voltage gain and the input impedance characteristics in case of Type B can be theoretically derived, the analysis results are shown by (21) and (22).

Then, using analysis results shown in Table II, the voltage gains and the input impedance characteristics in resonant circuits are investigated when the input frequency f and the coupling coefficient k of the rotary transformer are changed. The coupling coefficient is defined as follows:

$$k = \frac{M}{\sqrt{L_1 \cdot L_2}} \quad (23)$$

In addition, to confirm the validity of the analysis results, the analysis results are compared with simulation results. By using (19)-(23), the voltage gain characteristics of type A and B are respectively illustrated in Fig. 7 (a) and (b). The evaluation conditions such as circuit parameters are respectively shown in Fig. 7 (a) and (b). As seen in Fig. 7 (a) and (b), when the coupling coefficient k decreases and the input voltage frequency f deviates from

The 2018 International Power Electronics Conference

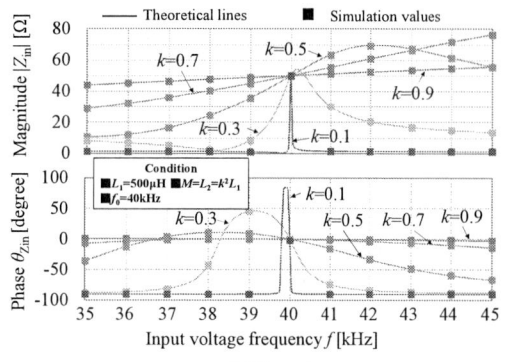
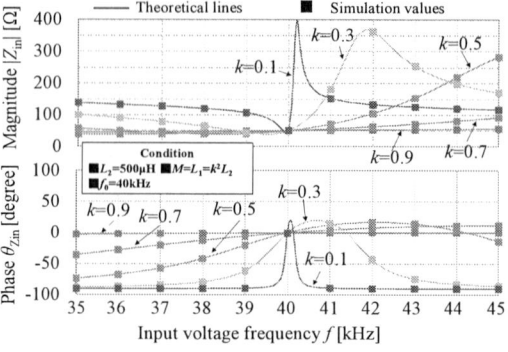

(a) Type A   (b) Type B

Fig. 8. Amplitude and phase of the input impedance of the both resonant circuits.

the resonance frequency f_0, the voltage gain at an optional frequency dramatically increase or decrease in comparison with the voltage gain at the resonant frequency f_0. As the difference between the voltage gain characteristics of Types A and B, that of type A is dramatically increased when f is in the range of below the resonant frequency f_0 and as coupling coefficient decrease. On the other hand, in the case of Type B, there is no such feature. On the basis of these analysis results, to improve the robustness of the resonant circuits against frequency fluctuation, the higher coupling coefficients are more effective since the voltage gain almost does not change. In addition, in Fig. 7 (a) and (b), dots and solid lines respectively mean the simulation results and the theoretical lines derived from (19) and (21). As shown in Fig. 7, the validity of these analyses are confirmed since these values agree with each other.

Then, the amplitudes and the phases of the input impedance of the resonant circuits are shown in Fig. 8 (a) and (b). As shown in Fig. 8 (a) and (b), the magnitude of the input impedance in the resonant circuits is greatly varied when f deviates from f_0 and as the coupling coefficient decreases. In addition, the phase θ_{Zin} of the input impedance is also dramatically changed as coupling coefficients decrease. Therefore, under the condition that coupling coefficient is lower, both of the magnitude and phase of input impedance sensitively change. Therefore, in order to improve the robustness of the frequency variation or that of mismatching the resonant parameters in the resonant circuits, the improving coupling coefficients of the rotary transformer is one of the effective ways.

C. A Leakage Flux Cancellation Technique to Improve Coupling Coefficient

A novel leakage cancellation technique to improve the coupling coefficient of asymmetric transformers is introduced in this subsection. Conceptual diagram of the proposed leakage flux cancellation technique is shown in Fig. 9. The proposed leakage flux cancellation technique is based on inserting copper tape or plate into the magnetic path where leakage flux flows. The proposed leakage flux cancellation technique function based on the following. First, original leakage flux generated from an asymmetric transformer cross the copper tape or plate

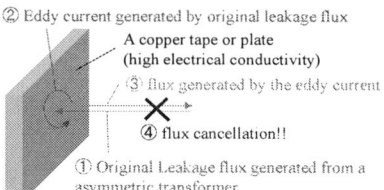

Fig. 9. Conceptual diagram of the proposed leakage flux cancellation technique.

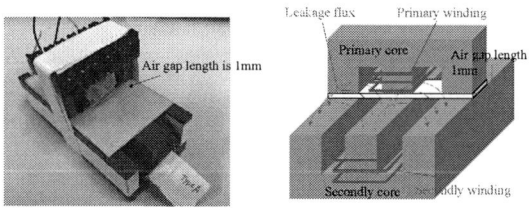

(a) Prototype.   (b) Image of the leakage flux path.

Fig. 10. Asymmetric core transformer for a basic experiment.

TABLE III
MEASURED VALUES OF MAGNETIC RELUCTANCE OF ASYMMETRIC TRANSFORMER WITHOUT PROPOSE TECHNIQUE

Magnetic reluctance on the leakage flux magnetic path on primary winding	R_{lk1}	17.04A/μWb
Magnetic reluctance on the leakage flux magnetic path on secondly winding	R_{lk2}	1.35A/μWb
Magnetic reluctance on the magnetizing flux magnetic path	R_m	4.84A/μWb
Coupling coefficient	k	0.413

having high electrical conductivity. Second, an eddy current is generated by original leakage flux on the basis of the Lenz's law. Third, flux is generated by the eddy current. Finally, the original leakage flux is canceled by the flux generated by the eddy current on the surface of the copper plate or tape. Therefore, the proposed leakage flux cancellation technique can improve coupling coefficients of an asymmetric rotary transformer by inserting properly the copper plate into the magnetic path where leakage flux flows.

D. Prototypes of Asymmetric Transformer with or without the Leakage Flux Cancellation Technique

In order to show the effectiveness of the proposed leakage flux cancellation technique, prototypes of the

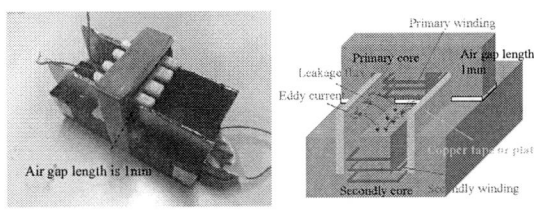

(a) Prototype.　　　(b) Image of the leakage flux path.

Fig. 11. Asymmetric core transformer with the proposed leakage flux cancellation technique.

TABLE IV
MEASURED VALUES OF MAGNETIC RELUCTANCE OF ASYMMETRIC
TRANSFORMER WITH PROPOSE TECHNIQUE

Magnetic reluctance on the leakage flux magnetic path on primary winding	R_{lk1}	32.67A/μWb
Magnetic reluctance on the leakage flux magnetic path on secondly winding	R_{lk2}	4.42A/μWb
Magnetic reluctance on the magnetizing flux magnetic path	R_m	4.66A/μWb
Coupling coefficient	k	0.652

asymmetric transformers are made by using general-purpose cores, and it is shown in Fig. 10. The matter that the secondary side core is larger than the primary core is the same essentially as the asymmetric rotary transformer shown in Fig. 2. The measured values of the magnetic reluctances are shown in Table III. As seen in this table, the magnetic reluctance of the leakage flux magnetic path on the secondary winding is very low, and leakage flux is easy to flow at the secondary core. In addition, the magnetic reluctance values shown in Table III are calculated by using (16) and the measured inductance value of the prototype.

On the other hand, the asymmetric transformer using the leakage flux cancellation technique to reduce leakage flux is also made, and it is shown in Fig. 11. The measured values of the magnetic reluctances are shown in Table IV. When comparing the measured values of the magnetic reluctances shown in Table III and IV, the magnetic reluctance values of the asymmetric transformer with proposed technique can improve in comparison with that of the asymmetric transformer without the proposed technique. Therefore, the coupling coefficient can improve from 0.413 to 0.652.

V. EXPERIMENTAL VALIDATIONS

A. Design of Resonant Circuit and Transformer

The series-parallel combined resonant circuit with the asymmetric transformer is designed in this subsection. In resonant circuit topologies, type A is selected. The design flowchart of the series-parallel combined resonant circuit of type A is shown in Fig. 12. As preconditions of this design, the self-inductance L_1 on the primary winding is designed as 500 μH. In addition, the resonant frequency f_o of the series-parallel resonant circuits of type A is designed as 40 kHz. The designed and measured values of each parameters are shown in Table V. As shown in Table V, the designed and measured values agree each other. Therefore, the validity of the transformer design method mentioned in section III is confirmed.

By using (19)~(20) and the designed values shown in

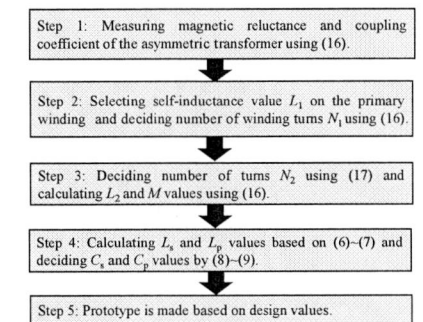

Fig. 12. Design flowchart of series-parallel combined resonant circuits (Type A).

TABLE V
DESIGNED AND MEASURED VALUES OF SERIES-PARALLEL
COMBINED RESONANT CIRCUIT

	Asymmetric transformer without proposed technique		Asymmetric transformer with proposed technique	
	Designed values	Measured values	Designed values	Measured values
N_1	43.4turns	44turns	45.2turns	45turns
N_2	9.5turns	9turns	22turns	22turns
L_1	500μH	514μH	500μH	506μH
L_2	85.3μH	84.1μH	213μH	212μH
M	85.3μH	82.0μH	213μH	215μH
L_s	414.7μH	434.1μH	286μH	287.4μH
L_p	85.3μH	79.9μH	213μH	218.7μH
C_s	38.1nF	36.5nF	55.2nF	55.1nF
C_p	185.4nF	195.7nF	74.3nF	72.4nF
R_o	50Ω			

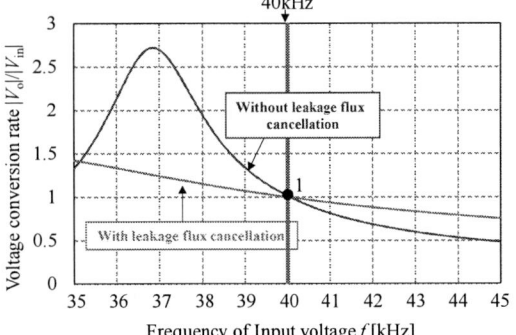

Fig. 13. Voltage conversion rate.

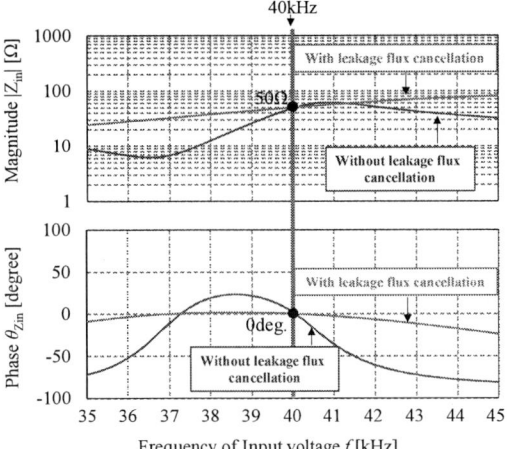

Fig. 14. Input impedance characteristics.

1560

The 2018 International Power Electronics Conference

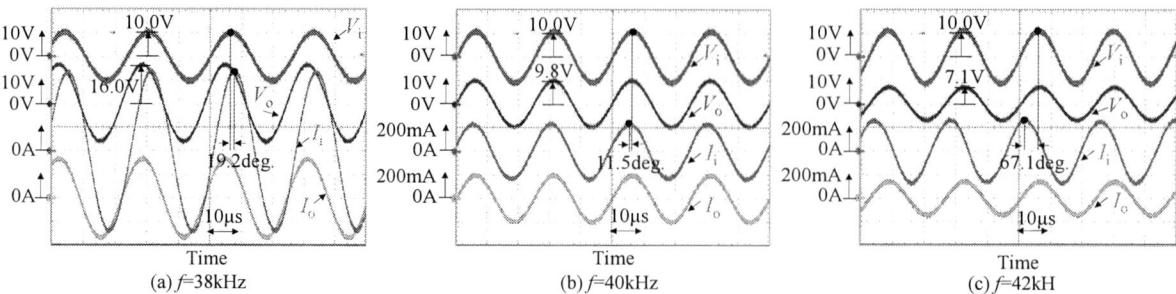

Fig. 15. Experimental waveforms of series-parallel combined resonant circuit with the asymmetric transformer without the proposed technique.

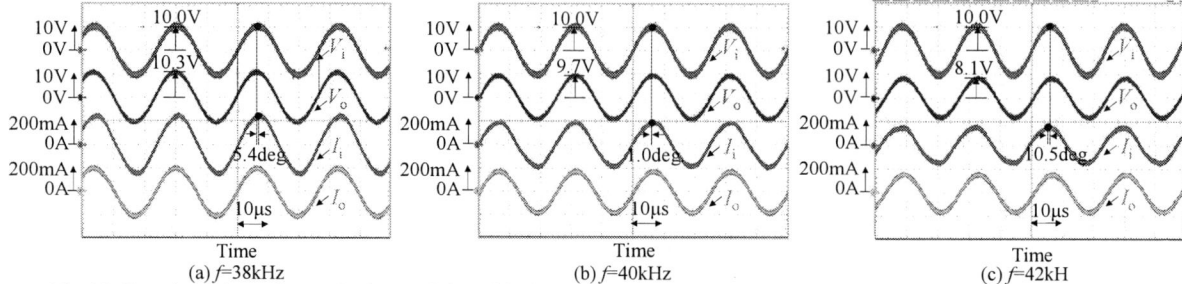

Fig. 16. Experimental waveforms of series-parallel combined resonant circuit with the asymmetric transformer with the proposed technique.

Table V, the voltage conversion rate, the amplitudes and phase of the input impedance when using the proposed technique and when not using it are shown in Figs. 13 and 14, respectively. As seen in Figs. 13 and 14, even if the frequency f of input voltage source changes for matching the resonant frequency of the ultrasonic transducer, the voltage conversion rate, the amplitude and phase of Z_{in} do not change dramatically in case of that with the cancellation technique. Therefore, the proposed method contributes to reducing the voltage drop caused by the leakage inductances and improving the power factor in input side even if the resonance parameter has the error or the frequency of input voltage changes in a wide frequency range.

B. Experimental Validations

Then, the experimental evaluations are carried out. Figs. 15 and 16 show the experimental results when the series-parallel combined resonant circuit using the asymmetric transformers with and without the proposed cancellation technique, respectively. As shown in Figs. 15 and 16, the amplitude of the output voltage V_o and the phase of the input current I_{in} when using the asymmetric transformer with the cancellation technique hardly change even if the frequency of the input voltage source varies between 38kHz and 42kHz. The resonant circuit without the proposed technique, the output voltage and phase of the input current I_{in} dramatically change with changing the frequency of the input voltage. Therefore, the proposed method is effective for improving the power factor and reducing the voltage drop caused by the leakage inductances of the transformer. In other words, the proposed method gives excellent performances in robustness to the error of the resonance parameters or that to a variation of the input voltage frequency.

VI. CONCLUSIONS

This paper proposed a novel leakage flux cancellation technique suitable for series-parallel combined resonant circuit with asymmetric rotary transformers used for ultrasonic spindle drive. The proposed method contributes to reducing the voltage drop caused by the leakage inductances and improving the power factor in input side even if the resonance parameter has the error or the frequency of input voltage changes in wide frequency ranges.

REFERENCES

[1] J. Imaoka, M. Nam, M. Shoyama, and H. Fujita, "Design of series-parallel combined resonant circuit with rotary transformer used for ultrasonic spindle drive," *in Proc. IEEE 3rd International Future Energy Electronics Conference and ECCE Asia (IFEEC 2017 - ECCE Asia),* Jun. 2017, pp. 2244-2249.

[2] Y. Inoue, Y. Gao, M. Shoyama, and H. Fujita, "Application of Wireless Electromagnetic Resonance Technology to Rotary Transformer Used in Ultrasonic Spindle," *in Proc. International Conference on Power Electronics (ICPE) - ECCE Asia,* Jun. 2015, pp. 1922-1928.

[3] D. Bortis, I. Kovacevic, L. Fassler, and J. W. Kolar, "Optimization of Rotary Transformer for High-Speed Applications," *in Proc. IEEE Workshop on Control and Modeling for Power Electronics (COMPEL),* Oct. 2013, pp. 1-6.

[4] D. Bortis, L. Fassler, A. Looser, and J. W. Kolar, "Analysis of Rotary Transformer Concepts for High-Speed Applications," *in Proc. IEEE Applied Power Electronics Conference and Exposition (APEC),* Mar. 2013, pp. 3262-3269.

[5] Y. Luan, B. Lin, X. Ma, and X. Zhu, "Innovative Contactless Energy Transfer Accessory for Rotary Ultrasonic Machining and Its Circuit Compensation Based on Coil Turns," *IEEE Trans. on Ind. Electron.,* Vol. 64, No. 10, pp. 7810-7818, Apr. 2017.

A Novel Structural Health Monitoring System with Wireless Power and Bi-directional Data Transfer

Yujin Jang[1*], Keon-Woo Kim[2], Moo-Hyun Park[2], Nayoung Lee[2] and Gun-Woo Moon[2]

1 Department of Future Vehicle, KAIST, Daejeon, Republic of Korea
2 School of Electrical Engineering, KAIST, Daejeon, Republic of Korea
E-mail: yujinjang@kaist.ac.kr

Abstract—This paper proposes a novel structural health monitoring (SHM) system for wireless sensor node (WSN) that enables power and physical data of WSN to be transmitted bi-directionally. The main point is system overview, the analysis, and implementation of proto-type SHM system. Conventional method of data transmission for SHM system is cable based sensor or WSN with battery. These result in high maintenance cost. To improve them, a novel SHM system with wireless power and data transfer (WPDT) is proposed. The proposed system has a WSN, primary control switch, load variation switch, and a modulation circuit for data transmission. It detects packetized digital data of the WSN by using load variation switch with WPDT method. Then, these demodulation circuit converts them to WSN data. Therefore, the proposed SHM system can reduce maintenance cost of the smart bridge by eliminating the communication module for data transmission. The WPDT method is implemented to assess validity. Wireless power is transferred to the WSN with a distance of 300 mm under concrete structure with steel rebar array. In addition, data transmission is properly conducted between primary resonator side and secondary resonator side with a 500bps data rate. Finally, a proto-type SHM system for the WSN has been built to experimentally evaluate the effectiveness. The results show that it operates the WSN by using wireless power and enables the WSN to transmit physical data of a smart bridge for damaged or undamaged conditions.

Keywords—wireless power and data transfer, load variation, concrete structure, digital packet transmission.

I. INTRODUCTION

Diagnosis of industrial application such as bridge is important issue. If diagnosis of the bridge is inaccurate, safety of the bridge is not guaranteed. Thus, structural health monitoring (SHM) of the bridge status is noticeable system for keeping in check of a damage. The SHM system for the bridge performs continuous monitoring of the concrete structure and offers accurate data. The SHM system do not substitute for visual estimation of human. Nevertheless, as for proper assignment of the SHM system, it is possible to monitor remotely condition of the bridge. This informs some change during inspection period to engineers. In addition, this enables them to plan and design maintenance activity. As a result, the SHM system leads safety and reduces maintenance cost of the bridge. To achieve the SHM

system, various sensors are mounted in the bridge, and measure various physical data of the bridges such as vibration, temperature, and the presence of concrete cracks. Then, status of the bridge can be detected from the data and transmitted to a control center that facilitates management of the bridge. Bridges that incorporate a maintenance system by using these sensors for condition measurement and data transmission are known as smart bridges [1]-[7].

In other words, the smart bridge needs to SHM system for a diagnosis of the physical state. A conventional SHM system for this comprises many WSNs that have cable based communication module and power supply. Hundreds of these WSNs are installed at the bottom or internal concrete structure of the bridge for measurement of the physical status. In addition, the cable based communication modules installed with the WSNs transmit the physical information to a data acquisition unit and store it to ground server system. Theses cables are connected to long distances in the smart bridge [8]-[9].

However, this SHM system has some problems. It is that wires for communication and power supply take up a lot of volume. In addition, these wires are exposed to external environment such as severe weather, which causes reliability problem. As a result, these conventional SHM systems result in high maintenance costs. To improve this, WSNs for SHM have been studied. These can improve the above-mentioned problems by transmitting data wirelessly. However, the power of these WSNs depends on the battery. This means that battery power cannot guarantee the life of the WSN. In addition, since the WSN is installed inside the bridge, it is very difficult to replace the battery. Moreover, a separate communication module is required to transmit the physical data of the bridges collected in the WSN. As a result, battery-powered WSNs still have high maintenance costs.

To solve this problem, a novel SHM system that wirelessly transfers power to a WSN and simultaneously transmits data is proposed. To do this, the proposed system supplies power to operate the WSN. Then, the signal for the operation of the WSN is transmitted from external side of the bridge to WSN. The physical data of the bridges detected at the WSN is then transmitted back to the external side of the bridge. In other words, the proposed system not only supplies power to the WSN wirelessly, but also can transmit data with bidirectional communication. The proposed system has the following advantages. Because it supplies power wirelessly to the WSN, there is no secondary battery. In addition, data can be transmitted using a wireless transfer system, so that additional communication modules for a data transmission can be eliminated.

II. SYSTEM OVERVIEW

Fig. 1 shows the concept diagram of the SHM system with the wireless power and data (WPDT) method There is an inspection vehicle on the outside of the bridge, and a WSN is placed inside the bridge. The operation method of the proposed system is as follows. The inspection vehicle arrives at the point where the WSN is installed to collect the physical data of the bridge. This inspection vehicle includes a resonator for wireless power transfer (WPT) and a data acquisition board. Then, the inspection vehicle supplies power to the WSN using a primary side resonator (PSR) and the secondary side resonator (SSR). Then, it transmits a request signal to the WSN by using the PSR and the SSR. At this time, the WSN collects the physical data inside the bridge. The collected data is converted into packetized digital data (PDD) by the WSN and then transmitted back to the inspection vehicle through the SSR and the PSR. When the data transfer is completed, the inspection vehicle moves to the another point where the other WSN is installed, and repeats the above-described process.

Fig. 2 shows an overall block diagram of the SHM system. The SHM system with WPDT technology encompasses three parts: a transmitter, a concrete structure, and a receiver. The transmitter is located in the primary side of the overall SHM system. It is composed of a power supply block and a data acquisition block by using MCU board. The power supply block has a series-series compensation topology for high efficiency of WPT and a PSR for the WPT. In addition, it includes a data acquisition block for processing PDD transmitted from the receiver. This block acquires the PDD generated by the transmitter resonator. It then substitutes the PDD for analog physical data about tendon force which is one of the physical data of the bridge. This is the primary transmitter side (PTS) of the SHM system. A concrete structure with a steel rebar array is situated between the primary transmitter and the secondary receiver. The steel rebar array is situated at the upper and lower sides of the concrete structure. The power supply generated by the PTS and the PDD generated by a secondary receiver side (SRS) have to penetrate this structure. The receiver is located in the secondary side of the overall SHM system. It is composed of a power management block, a data communication module, and a WSN. The power management block performs rectification of transferred power from the PTS with 12V/1A output. This output

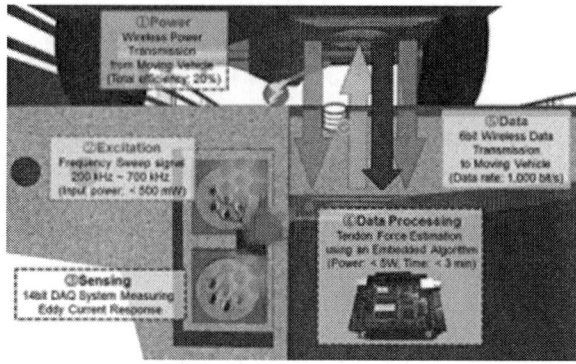

Fig. 1. Concept diagram of a wireless power and data transfer system using the inspection vehicle for SHM sensors

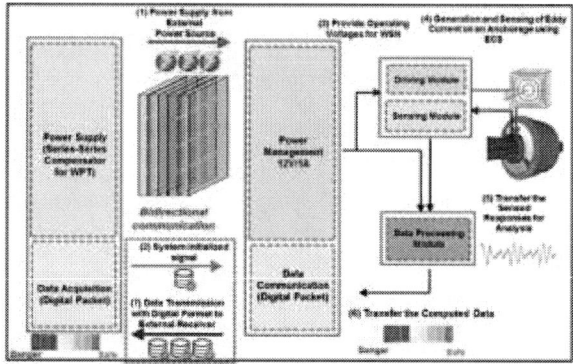

Fig. 2. Block diagram of the proposed scenario realization with the WPDT

power is stored in an internal battery and used for the WSN.

In addition, it includes a data communication circuit for transmitting acquired data from the WSN to the PTS. The SRS has a WSN that consists of driving, sensing, and processing modules. This is driven by wireless power transferred to the PTS. The data acquisition block then sends a request signal for WSN drive. The driving module generates and senses eddy current on an anchorage using the eddy current sensor (ECS) of the tendon force monitoring system. The sensing module then detects generated data to the driving module. At the same time, it transfers the sensed response for analysis to the data processing module by using a MCU board. It then analyzes the computed data to determine bridge damage. It outputs data in eight steps, from the most dangerous condition to the safest condition. The generated eight step data are converted into PDD by data communication module. The PDD is transmitted to the PTS through the concrete structure.

The proposed SHM system must satisfy several constraints. (1) The amount power and efficiency transferred wirelessly is 10 W and 30%, respectively. (2) The diameter of the PSR for the WPT is limited to 300mm since the space to be attached to the inspection vehicle is fixed. (3) The distance between the PSR and the SSR due to the bridge construction specifications is 300 mm. (4) The WPDT method for the proposed SHM should penetrate concrete structures with rebar arrays. (5)

1563

The 2018 International Power Electronics Conference

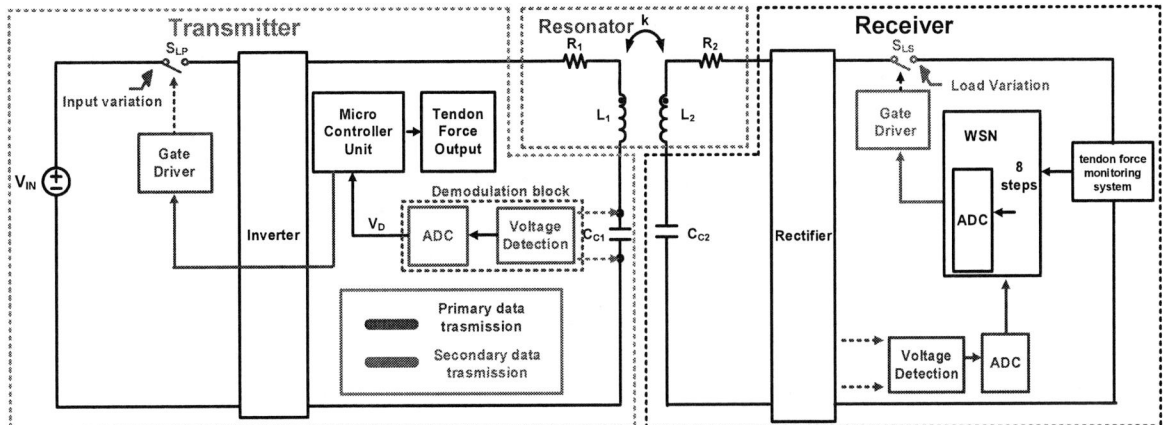

Fig. 3. Proposed wireless power and data transfer method with input variation and load variation swathes.

The WSN embedded in bridges collects physical data of bridges in 8 steps of the PDD. These five constraints should be considered in the proposed SHM system with the WPDT method for a smart bridge.

III. DESIGN CONSIDERATION

Fig.3 shows the proposed WPDT method for the SHM system. This is composed of a transmitter, a resonator, and a receiver. The transmitter has a power source, an inverter, a resonant capacitor for power transfer, an AC voltage-sensing block with data demodulation, and a MCU. And an input variation switch (S_{LP}) and a gate driver for primary side data transmission. The resonator is paired to the transmitter and the receiver. The receiver has a rectifier for converting AC to DC voltage, a load variation switch (S_{LS}) for data transmission, a gate driver, a ADC block, and a WSN. The basic idea for the WPDT is detection of input variation by S_{LP} and load variation by S_{LS}. The S_{LP} converts the voltage of the secondary rectifier. The voltage difference of the secondary rectifier is converted into digital information. Then, the WSN starts driving by this digital information. The WSN detects the physical data of the bridge with 8 step PDD. The digital data generated by WSN drive S_{LS}. The S_{LS} varies the circulation current of the PTS. The demodulation block then detects this information, and converts it to digital data. Finally, the digital data are translated by the MCU to physical data of the bridge. Fig. 4 shows a detailed block diagram for PDD detection. Fig. 4(a) shows a detection block diagram of the rectifier voltage that varies through S_{LP} of the PTS. The voltage of the rectifier block changes by S_{LP}. In other words, when S_{LP} is ON, the rectifier waveform is switched by a preset frequency, and when S_{LP} is OFF, it is zero. This waveform adjusts the output level of 0 to 5V by the next stage comparator. At this time, this is converted into a digital signal by the frequency-to-voltage converter of the final stage. In other words, when the switching waveform is applied, it outputs a pre-determined DC voltage and when 0V is applied, it outputs 0V. Fig. 4(b) shows a block diagram for detecting a variable voltage by the S_{LS}. The resonance voltage (V_{Cc1}) of the PTS by the S_{LS} is

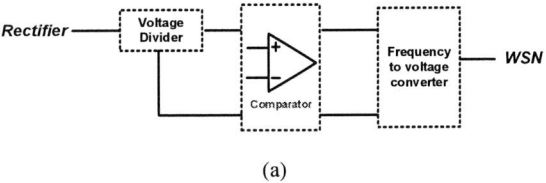

(a)

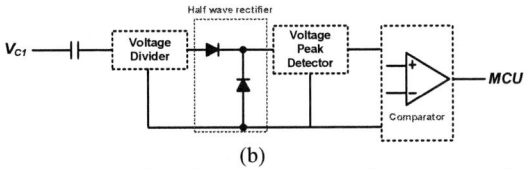

(b)

Fig. 4. (a) Schematic of the modulation circuit for data transfer for the PDD by S_{LP} (b) Schematic of the modulation circuit for data transfer for the PDD of S_{LS}

$$
\begin{aligned}
V_{Difference}(s) &= \left(\frac{1}{s^2 L_1 C_{C1} + s C_{C1} R_1 + 1} - \frac{1}{s^2 L_1 C_{C1} + s(C_{C1} R_1 + C_{C1}(\operatorname{Re} Z_{ref})) + 1} \right) V_{IN} \\
&= \left(\frac{s C_{C1}(\operatorname{Re} Z_{ref})}{\begin{array}{l} s^4 L_1^2 + s^3 (2L_1 C_{C1} R_1 + L_1 C_{C1}(\operatorname{Re} Z_{ref})) + \\ s^2 (C_{C1}^2 R_1^2 + 2L_1 + C_{C1}^2 (\operatorname{Re} Z_{ref}) R_1) + s(2C_{C1} R_1 + C_{C1}) + 1 \end{array}} \right)
\end{aligned}
\tag{1}
$$

changed. When S_{LS} is ON, the resonant voltage corresponding to the WPT is taken to V_{C1}. On the other hand, when the S_{LS} is OFF, the circulation current of the PTS abruptly increases and a very large resonance is detected. In other words, when the S_{LS} is ON, the secondary side impendence is projected on the PTS to form the resonance voltage of C_{C1}, but when the S_{LS} is OFF, the impedance of the SRS is not projected. As a results. a large voltage is generated in the C_{C1}. The difference between these resonance voltages can be expressed by the following equation [10].

When equation (1) is calculated by MATLAB, the difference in the V_{Cc1} is maximized by the L-C resonance frequency of the transmitter. In other words, when the WPDT method is operated at the resonance frequency of

1564

the transmitter and the receiver, the AC voltage variation of the resonant capacitor of the transmitter is easily detected. This resonance voltage is half-wave rectified through a voltage divider and a diode rectifier. This waveform is converted to DC by a voltage peak detector and converted to digital information by a final stage comparator. The reason why the primary side data transmission and the secondary side data transmission are different is that the detected voltage information is DC and AC.

IV. EXPERIMENTAL RESULTS

A prototype SHM system is designed to measure the tendon force of a bridge. As shown in Fig.4, the prototype SHM system with the proposed WPDT circuit supplies power to the WSN through a reinforced concrete structure. At the same time, the PTS sends WSN drive signal and the WSN measures physical data of a bridge and transmits the data to the PTS with transformed PDD through concrete structure. In addition, steel rebar arrays are located at a 150mm and 250mm distance from the bottom of the reinforced concrete structure respectively. The distance between steel rebar is 150mm and the overall height of the concrete structure is 300mm with the rebar array. Meanwhile, in terms of the data rate of this system, 500bit/s is set for transmitting physical data of a bridge. This is sufficient to transmit data for the safety diagnosis of a bridge.

Fig.5 shows waveforms of the prototype SHM system with the proposed WPDT method. Fig. 5(a) shows that DC voltage waveform is transformed by the frequency-to-voltage converter shown in Fig.4(a) from the input voltage of secondary rectifier. In (a) waveform, the input PDD of the PTS is yellow, input voltage of frequency-to-voltage converter is red, and the reproduced PDD of the SRS is green respectively. It has low voltage that is generated by the low signal of the input PDD (yellow line). In contrast, a high signal of the input PDD generates switching voltage of the secondary rectifier. The PDD (green line) are then reproduced at the SSR. Fig. 5(a) shows that the input PDD of the PTS is transmitted by the proposed input variation method to the PTS without any distortion.

Fig. 5(b) shows that DC voltage waveform is transformed by the demodulation circuit shown in Fig. 4(b) from the input voltage of half wave rectifier. In (b) waveform, the input PDD of the SSR is blue, input voltage of half wave rectifier is red, and the reproduced PDD is yellow, respectively. It has high voltage that is generated by the low signal of the input PDD (blue line). In contrast, a high signal of the input PDD means S_{LS} ON generates low voltage of the DC voltage. This is compared to the predetermined reference voltage. The PDD (yellow line) are then reproduced at the PSR. Fig. 5(b) shows that the input PDD is transmitted by the proposed load variation method to the PTS. As a result, the PDD that is input to the SRS is transmitted by resonant voltage variation with the proposed load variation method. In addition, the reproduced PDD is generated from the SSR without any distortion. This

Fig. 4. A proto-type SHM system to measure the physical data of a bridge

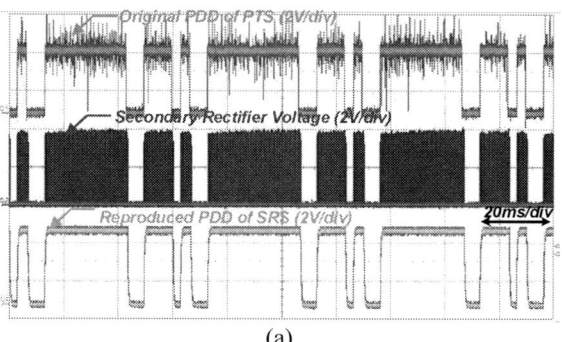

(a)

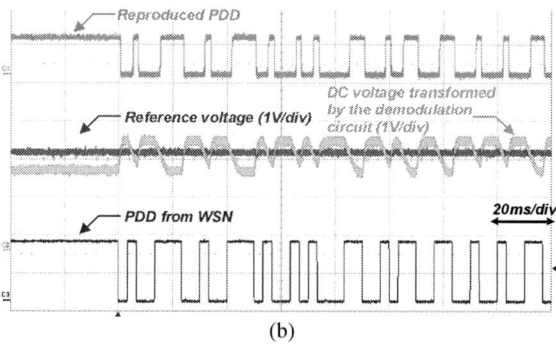

(b)

Fig. 5. Voltage waveforms under data transmission condition (a) Reproduced PDD by S_{LP} (b) Reproduced PDD by S_{LS}

means that data transmission in both directions is possible while the wireless power transfer is proceeding. Experiments confirmed that bidirectional data transmission is possible without any additional circuit.

V. CONCLUSION

In this paper, a novel SHM system with WPDT method is proposed. In addition, it is capable of bi-directional communication. As shown in the previous section, the proposed method drives the WSN with wireless power. Then, the WSN driving signal of the PTS is transmitted to the SRS, and at the same time, the physical data of the bridge generated in the WSN of the SRS is transmitted to the PTS. Moreover, it is confirmed that no signal distortion is occurred in this process. The proposed SHM system has simple data transmission principle and can eliminate additional communication module and battery. Therefore, the proposed SHM system is suitable for low cost bridge application.

ACKNOWLEDGMENT

This work was supported by the National Research Foundation of Korea (NRF) grant funded by the Korea government (MSIP) (No. 2016R1A2B2010328).

REFERENCES

[1] Chandrashekhara. K,Watkins. S.E., Nanni. A., and Prakash Kumar,. "Design and technologies for a smart composite bridge" in IEEE 2004 Intelligent Transportation Systems, 2004, pp. 954-959

[2] Sohn, H., Farrar, C. R., Hemez, F. M., & Czarnecki, J. J. (2002). A Review of Structural Health Review of Structural Health Monitoring Literature 1996-2001 (No. LA-UR-02-2095). Los Alamos National Laboratory.

[3] http://www.economist.com/node/17647603

[4] Setijadi, E.; Suwadi; Slamet, B.P.; Muntaqo, A.A.; In'am; Nur, A.E.; Suprobo, P.; Faimun; Febry, F.A. "Design of large scale structural health monitoring system for long-span bridges based on wireless sensor network" in IEEE 2013 Awareness Science and Technology and Ubi-Media Computing (iCAST-UMEDIA), 2013, pp. 169-174

[5] Li, A., Ding, Y., Wang, H., & Guo, T. "Analysis and assessment of bridge health monitoring mass data—progress in research/development of Structural Health Monitoring". Science China Technological Sciences, 2012, 55(8), 2212-2224.

[6] Ko, J. M., and Y. Q. Ni. "Technology developments in structural health monitoring of large-scale bridges." Engineering structures 27.12 (2005): 1715-1725.

[7] Curadelli, R. O., Riera, J. D., Ambrosini, D., & Amani, M. G. (2008). Damage detection by means of structural damping identification. Engineering Structures, 30(12), 3497-3504.

[8] Lynch, J. P., Law, K. H., Kiremidjian, A. S., Kenny, T., & Carryer, E.. "A wireless modular monitoring system for civil structures". In Proceedings of the International Modal Analysis Conference–IMAC , 2002, February

[9] Casas, Juan R. "A combined method for measuring cable forces: the cable-stayed Alamillo bridge, Spain." Structural Engineering International 4.4 (1994): 235-240.

[10] Y.J. Jang, J.K. Han, S.Y. Cho, G.W. Moon, J.M. Kim, H. Sohn, Wireless Power and Data Transfer System for Smart Bridge Sensors. Proceedings of Annual IEEE Applied Power Electronics Conference Exposition. 2016.

Control Strategy for Starter Generator in UAV with Micro Jet Engine

Jun-ichi Itoh[1*], Kazuki Kawamura[1*], Hiroyuki Koshikizawa[2] and Kazuyuki Abe[2]

1 Department of Electrical, Electronics and Information Engineering, Nagaoka University of Technology, Niigata, Japan
2 Development Department, YSEC Co., Ltd, Niigata, Japan
*E-mail: itoh@vos.nagaokaut.ac.jp, k_kawamura@stn.nagaokaut.ac.jp

Abstract— This paper proposes control strategy of a starter generator connected to a jet engine for an unmanned aerial vehicle system. Thrust is generated by both the jet engine and propellers which are powered by the jet engine through the starter generator. A flight range can be extended since energy density of the jet engine in the developed system is higher than battery energy density in the conventional system. Moreover, the starter generator directly connects to the jet engine and rotates at high speed for miniaturization. The proposed control strategy achieves the starting, the powering and the cooling operations with the starter generator. It is confirm through an experiment of a 3-kW prototype, that the prototype system achieves the maximum conversion efficiency of 92.7%. The minimum generator current THD is 16.5% at 70000 r/min. Further, the exhaust nozzle temperature is controlled within the maximum deviation of 2% regarding to the command value in study state.

Keywords— *Starter generator, Jet engine, Unmanned aerial vehicle(UAV), V/f control.*

I. INTRODUCTION

Recently, unmanned aerial vehicles (UAVs) have been actively studied for rescue activities in disaster [1–4]. In particular, the multicopter-type UAV has two advantages. First, it is easy to approach danger zones because of unmanned operation. Second, the multicopter-type UAV does not need a designated landing space. However, the multicopter-type is generally powered by batteries [2]. The flight range and carrying weight are limited because of the battery energy density [5]. Therefore, UAV with a jet engine has been developed [6]. In the developed UAV system, thrust is generated by both the jet engine and propellers which are powered by the jet engine through the starter generator. The flight range can be extended since the energy density of the jet engine is higher than the battery energy density. Furthermore, the developed UAV system is also be used as an emergency power supply owing to the starter generator.

An auxiliary power unit (APU) is generally used for starting and cooling the jet engine [7–8]. However, the use of APU leads to the increase in cost and size of the system. Furthermore, the rotation speed of the generator in APU is low because the generator is connected to the jet engine through reduction gears [9–10]. Therefore, the

generator tend to be large in a high power capacity system.

In this paper, the UAV system with a jet engine and the control strategy of the starter generator are proposed. In the developed UAV system, only the starter generator is used for starting and cooling, which eliminates the use of APU. Furthermore, the starter generator connects directly to the jet engine and rotates at high speed for miniaturization. The challenge of this paper is the achievement of the stable operation through the proposed control strategy even when the starter generator transits among operation modes, i.e., starting mode, powering mode, and cooling mode without APU and reduction gears. In particular, the synchronous frequency command limiter and the output power limiter are used in the proposed control method. In addition, modulation method is modified by the estimated intersection phase based on synchronous PWM. Through the experiments, it is confirmed that the prototype achieves the maximum conversion efficiency of 92.7%, the minimum generator current THD of 16.5% at 70000 r/min. Further, the exhaust nozzle temperature is controlled within maximum deviation of 2% compared to the command value in the steady state.

II. DEVELOPED UAV SYSTEM

Figure 1 shows the configuration of the developed an UAV system. The jet engine and the starter generator are directly connected without reduction gears. The jet engine powers six propellers through the starter generator. In the aerial applications, weight reduction of the starter generator is required from the viewpoint of flight range. Thus, the starter generator is rotated at high speed for miniaturization and weight reduction.

Figure 2 shows the mode transition diagram of the developed UAV system. A host controller selects the operation mode. The operation modes are described as follows;

A. All Off Mode

This mode is a stationary state. The power converter is not operated(gate off).

B. Standby Mode

The DC/DC converter boosts the DC-link voltage from the battery voltage to 300 V. However load is

C. Startup Mode

The starter generator is driven by the AC/DC converter in order to assist both the ignition of the jet engine and the acceleration up to 50000 r/min. The host controller controls the starter generator speed in this state. Further, the jet engine controls exhaust nozzle temperature.

D. Run Mode

The powering operation is performed in the range of the rotation speed from 50000 to 70000 r/min, and the battery is charged. The jet engine controls the speed, whereas the starter generator controls the output power.

E. Stop Mode

The engine output is halted, whereas the starter generator decelerates and cools the jet engine. When the exhaust nozzle temperature of the jet engine is cooled to 50°C or less, the operation of the power converter is stopped. Then, the operation mode is shifted to the all off mode.

III. MODULATION METHOD FOR EVEN-ORDER HARMONIC COMPONENTS SUPPRESSION

A. Continuous PWM

The starter generator is driven at the rotational speed of the jet engine, the carrier frequency, and the fundamental frequency are close to each other. This leads to the low-order harmonic components and the beat components on generator current. Thus, in the Run Mode, the nine-pulse synchronous PWM technique is used.

Figure 3 shows the voltage command with a modulation index of 0.8 and a triangular carrier with a frequency ratio of nine when the continuous PWM is applied. As shown Fig. 3(a), the voltage command v_u^* is compared with the triangular carrier to generate a PWM signal in general. As shown Fig. 3(b), the modulation index command V_m^* and the red carrier u_{mc} are compared to generate a PWM signal. The deformed carrier u_{mc} is calculated by

$$u_{mc} = \frac{u_m}{\sin\theta} \quad (\theta \neq 0°, 180°) \tag{1}.$$

where θ is the phase of inverter voltage command and u_m is the triangular carrier.

The proposed modulation method estimates the intersection phase of the modulation signal and the carrier using a look-up table of the deformed carriers in the software, then outputs a voltage command according to the estimated intersection phase. It can be implemented in a micro-controller because this proposed modulation method is implemented without changing the hardware.

Table I shows the estimated intersection phase patterns of the continuous PWM. As shown in Table I, the estimated intersection phases of sectors zero and nine are determined to be 0° and 180° because of the synchronous PWM, respectively. The relationship between the phase and the modulation index command of the deformed carrier u_{mc} of the sectors 1, 2, 3, and 4 is tabulated as

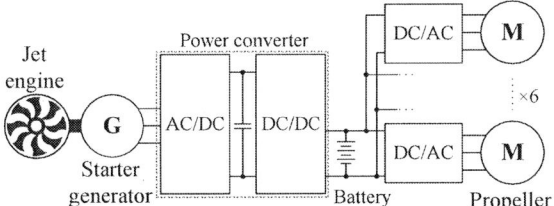

Fig. 1. Configuration of developed UAV system.

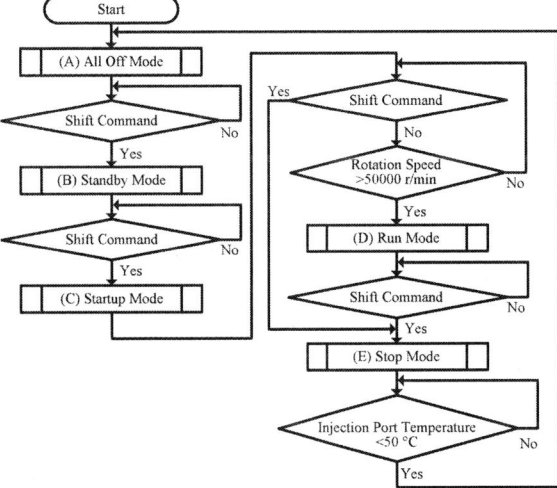

(a) Flowchart of developed UAV system operation.

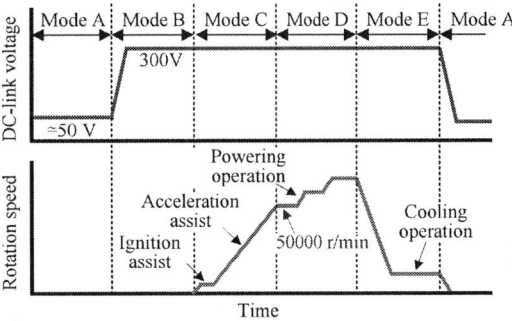

(b) Overview operation of jet generator.

Fig. 2. Mode transition diagram of developed UAV system.

look-up table 1, 2, 3, and 4. Note that α_1, α_2, α_3 and α_4 are the estimated intersection phases referred from these look-up tables. For the sector 5 and later, it is not necessary to prepare the tables in order to estimate the phase using the symmetry each 90°. The proposed modulation method acquires the estimated intersection phase in each sector using the phase referred to from look-up table 1 to 4 and the relationship in Table II, and outputs a voltage command according to the estimated intersection phase. The generated PWM signal is equivalent to that generated by analog control. Therefore, even-order harmonic components do not occur in PWM signal.

B. Discontinuous PWM

Figure 4 shows a modulation signal and a carrier with frequency ratio of 9 and a modulation index of 0.8 when

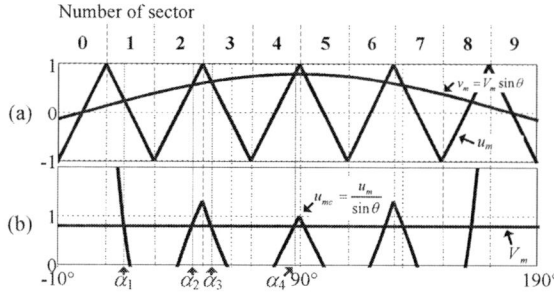

Fig. 3. Waveforms of voltage commands and carriers with continuous PWM.

TABLE I
Proposed estimated phase patterns of continuous PWM.

Sector	Look Up Table	Phase of Intersection Point	Sector	Look Up Table	Phase of Intersection Point
0		$0°$	9		$180°$
1	1	α_1	10	1	$180°+\alpha_1$
2	2	α_2	11	2	$180°+\alpha_2$
3	3	α_3	12	3	$180°+\alpha_3$
4	4	α_4	13	4	$180°+\alpha_4$
5	4	$180°-\alpha_4$	14	4	$360°-\alpha_4$
6	3	$180°-\alpha_3$	15	3	$360°-\alpha_3$
7	2	$180°-\alpha_2$	16	2	$360°-\alpha_2$
8	1	$180°-\alpha_1$	17	1	$360°-\alpha_1$

α_X : Phase by Look Up Table X

the discontinuous PWM is employed. The discontinuous PWM signal v_{xd}^* shown in Fig. 4(a) is calculated by adding the following offset to the three phase modulation signal v_x.

$$v_{xd}^* = v_x^* + v_{offset}^* , \quad (x = u, v, w)$$

$$v_{offset}^* = \begin{cases} 1-|v_{max}| & \text{if } |v_{max}| \ge |v_{min}|, \\ -1+|v_{min}| & \text{if } |v_{min}| < |v_{max}|, \end{cases} \quad (2).$$

$$\text{and} \begin{cases} v_{max} = \max[v_u^*, v_v^*, v_w^*] \\ v_{min} = \min[v_u^*, v_v^*, v_w^*] \end{cases}$$

As mentioned in Section A, a deformed carrier u_{md} in Fig. 4(b) is used. The intersection phases in sectors 0 and 4 are defined as $0°$ and $90°$ in advance, respectively. Therefore, the deformed carriers of sectors 1, 2 and 3 are only required to estimate the intersection phases. The deformed carrier u_{md} is calculated by

$$u_{md} = \frac{u_m+1}{\sin\theta - \sin(\theta - 120°)} \quad (10° \le \theta < 60°) \quad (3).$$

Note that in the section of more than $60°$ in sector 3, the intersection phase is set to $60°$ when the modulation index is 0.577 or less.

Table II shows the estimated intersection phase patterns of the discontinuous PWM. The relationship between the phase and the modulation index command of the deformed carrier u_{md} of the sector 1, 2, and 3 is defined as in look-up table 1, 2, and 3. By referring the phases α_1, α_2, α_3 using the modulation index from these look-up tables, the intersection phase of the carrier and the modulation signal in each sector is estimated by the relationship shown in Table II. By using this method, the symmetry of the PWM signal is secured even in the discontinuous modulation, and the even-order harmonic

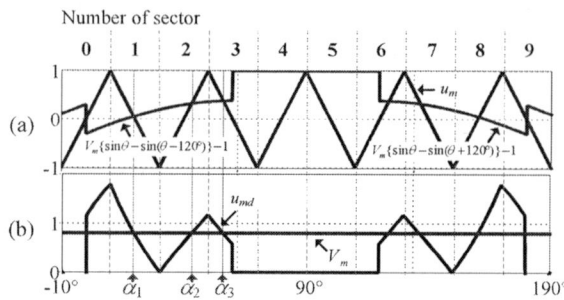

Fig. 4. Waveforms of voltage commands and carriers with discontinuous PWM.

TABLE II
Proposed estimated phase patterns of discontinuous PWM.

Sector	Look Up Table	Phase of Intersection Point	Sector	Look Up Table	Phase of Intersection Point
0		$0°$	9		$180°$
1	1	α_1	10	1	$180°+\alpha_1$
2	2	α_2	11	2	$180°+\alpha_2$
3	3	α_3	12	3	$180°+\alpha_3$
4		$90°$	13		$270°$
5		$90°$	14		$270°$
6	3	$180°-\alpha_3$	15	3	$360°-\alpha_3$
7	2	$180°-\alpha_2$	16	2	$360°-\alpha_2$
8	1	$180°-\alpha_1$	17	1	$360°-\alpha_1$

α_X : Phase by Look Up Table X

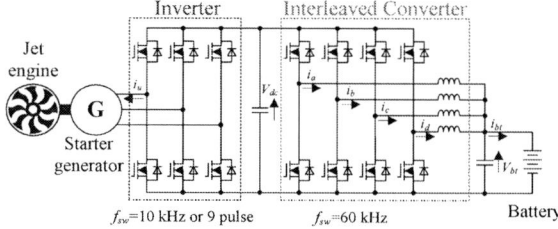

Fig. 5. Configuration of power converter.

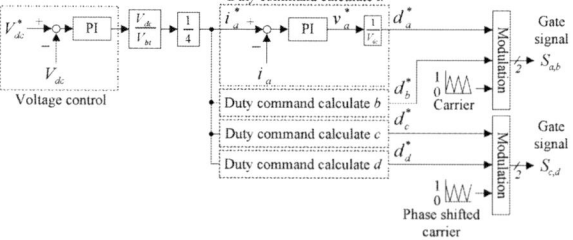

Fig. 6. Control block diagram of DC/DC interleaved converter.

components do not occur in the PWM signal.

IV. CONTROL STRATEGY FOR STARTER GENERATOR

Figure 5 shows the configuration of the power converter. This converter consists of a three-phase inverter and a four-leg interleave DC/DC converter. Since the battery voltage is approximately 50 V, the DC/DC converter is required to boost the voltage to 300 V in order to drive the inverter.

Figure 6 shows the control block diagram of DC/DC interleaved converter. The DC link voltage V_{dc} is regulated to the command value. The current imbalance among four legs is suppressed by the current control of

1569

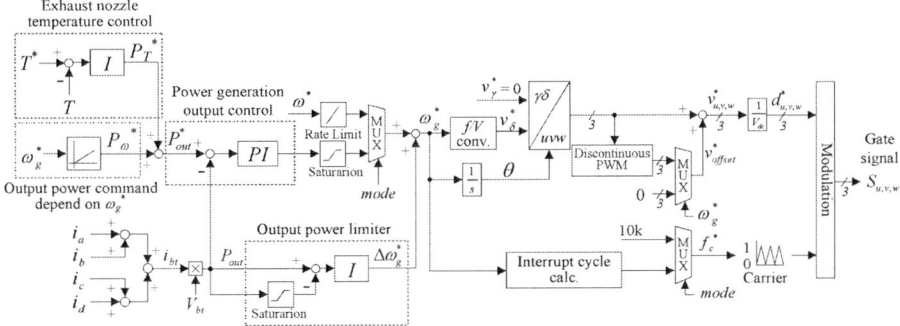

Fig. 7. Control block diagram of 3-phase inverter.

each leg. Moreover, the carrier of each two legs is phase shifted by half a period compared to the other two legs. As a result, the switching frequency is equivalently doubled and the current ripple is reduced to half [11].

Figure 7 shows the control block diagram of the three phase inverter. The power control is operated in the run mode, whereas the V/f control is employed in the other modes. The rotation speed of the jet engine is suddenly reduced because the output power of the jet engine is not sufficiently high when the inverter control switched from the V/f control to the output power control or exhaust nozzle temperature control for the run mode. In other words, the self-sustained operation of the jet engine is difficult at the low speed. In order to solve this problem, a synchronous frequency command limiter is applied. As a result, the speed is kept constant until the jet engine output becomes sufficiently high. During the startup mode, the generator torque suddenly changes. In order to prevent overcurrent in this operation, an output power limiter is introduced. Consequently, the synchronous frequency command is compensated in order to avoid the sudden change in the torque. Furthermore, around the rated speed, discontinuous PWM is employed to deal with overmodulation region.

V. STABILIZATION ANALYSIS OF POWER GENERATION OUTPUT CONTROL

Figure 8 shows the characteristics of jet engine thrust against the rotation speed. As shown Fig. 8, a thrust of 68.0 N is obtained at the rotation speed of 70000 r/min. Under the atmospheric pressure, the atmospheric temperature, and the air density are constant, the thrust of the jet engine depends only on the rotational speed regardless of the output power. Since the thrust of the jet engine is proportional to the cube of the rotational speed, the thrust F obtained by the measured value is approximated by the cube of the rotational speed as follows;

$$F = k_F \omega^3 \tag{4},$$

where ω is rotation speed of the jet engine and k_F is coefficient obtained from the measured value.

Figure 9 shows the block diagram of the output power control system with a jet engine. In this system, the generator synchronous angular frequency ω_g is produced

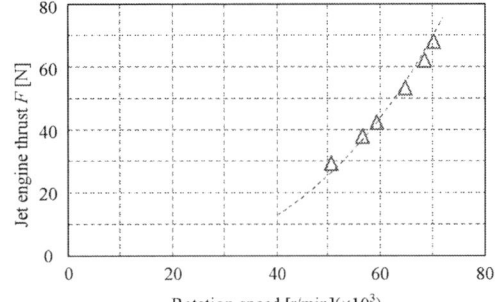

Fig. 8. Characteristics of jet engine thrust.

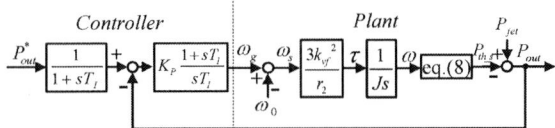

Fig. 9. Block diagram of output power control system with jet engine.

by the difference between the output power command P_{out}^* and the output power detection value P_{out}. Further, it is assumed that the response of the rotation speed control for the jet engine is sufficiently slower than that of the output power control. By ignoring the loss, the total power of the jet engine P_{jet} is calculated by

$$P_{jet} = P_{th} + P_{out} \tag{5},$$

where P_{th} is the thrust power of the jet engine. This thrust power is added to the shaft power P_{out} that drives the propeller. This shaft power is determined by the flight speed and the thrust of the aircraft. However, if the aircraft is stationary as in the test, the shaft power cannot be calculated from the flight speed. In this case, the stationary shaft power P_{th_s} is calculated by

$$P_{th\_s} = \frac{F}{11.2} \times 736 \tag{6}.$$

Substituting (4) into (6) and setting the coefficient as k_{th}, the stationary shaft power [12] is calculated by

$$P_{th\_s} = \frac{736k_F}{11.2} \omega^3 = k_{th} \omega^3 \tag{7}.$$

In order to analyze the stability of the control, the rotation angular velocity is linearized around the steady-state points.

$$P_{th\_s} = 3k_{th} \omega_0^2 \omega \tag{8}.$$

The 2018 International Power Electronics Conference

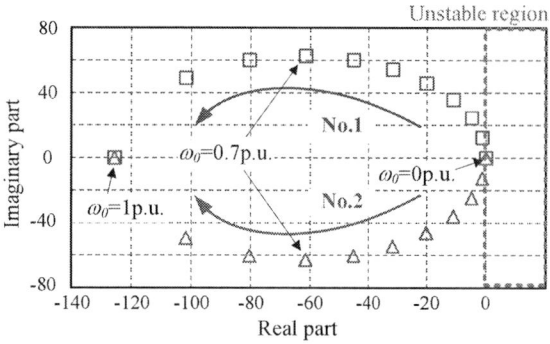

Fig. 10. Roots locus when the initial angular velocity ω_0 is increased.

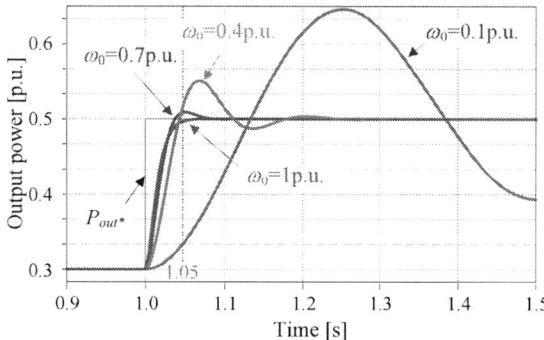

Fig. 11. Step response of output power when output power command is changed from 0.3 p.u. to 0.5 p.u.

Note that ω_o is the initial angular velocity ω_0 at the steady-state point. Consequently, the transfer function from input to output of this control system is expressed by

$$G(s) = \frac{-\dfrac{9K_p k_{vf}^2 k_{th} \omega_0^2}{K_i r_2 J}}{s^2 - \dfrac{9K_p k_{vf}^2 k_{th} \omega_0^2 s}{r_2 J} - \dfrac{9K_p k_{vf}^2 k_{th} \omega_0^2}{K_i r_2 J}} \quad (9),$$

where K_p is the proportional control gain, K_i is the integral control gain, k_{vf} is the voltage coefficient in the v/f control, J is the total inertia of the jet engine and generator, and r_2 is the secondary winding resistance of the generator. Furthermore, K_p and K_i are expressed as functions of the damping coefficient ζ and the response angular frequency ω_n.

$$K_p = -\frac{2\xi \omega_n r_2 J}{9 k_{vf}^2 k_{th} \omega_0'^2} \quad (10)$$

$$K_i = \frac{2\xi}{\omega_n} \quad (11)$$

Note that ω_0' is the initial angular velocity. This angular velocity should be set accordingly to the detection value of the angular velocity. However, the angular velocity detection is not employed in the test; therefore, this value is predetermined as following.

Figure 10 shows the roots locus when the initial

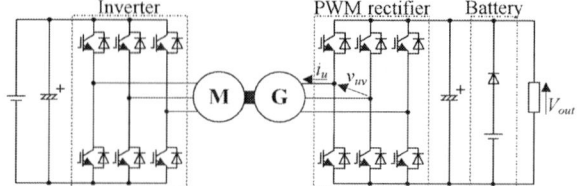

Fig. 12. Configuration of experimental system.

TABLE III
Specification of starter generator.

Parameter	Value
Poles	2
Rated rotary field speed	70000 r/min
Rated speed	68271 r/min
Rated voltage	200 V
Rated current	15.3 A
Rated power	4 kW
Rated torque	0.6 N·m
Weight	3.0 kg
Diagram	110 mm
Full length	192 mm

angular velocity ω_0 is increased. In this system, the powering operation is performed in the rotation speed range from 0.7 to 1.0 p.u. Therefore, the initial angular velocity setting value ω_0' is 0.7 p.u. The damping coefficient ζ is set to 0.7. The response angular frequency ω_n is set; thus, the overshoot time is 0.05 seconds, which is 1/10 of the jet engine control period of 0.5 seconds. As shown in Fig. 10, when the rotation speed is 0 p.u., the control system is at the stability limit because the poles locate on the imaginary axis. The control system becomes stable because the poles move to the negative half plane when the rotation speed is larger than 0 p.u.

Figure 11 shows the step response of output power when output power command is changed from 0.3 p.u. to 0.5 p.u. As shown in Fig. 11, at a rotation speed of 0.7 p.u. and 1.0 p.u., the response is equal to or larger than the design response time. The response time is delayed and a large overshoot occurs in output power at the rotation speed of 0.4 p.u. and 0.1 p.u., which is the low-speed range. However, such large overshoot does not occur since the power generation operation is performed only in the high-speed range in this system.

VI. EXPERIMENTAL RESULTS

A. Modulation method for even-order harmonic components suppression

Figure 12 shows the experimental system. Table III shows the specification of the starter generator. In this test, two motors shown in Table III are connected instead of the jet engine. In addition, a small capacity DC regulated power supply is connected to supply the excitation current at the time of starting since the starter generator is an induction generator.

Figure 13 shows a block diagram of the PWM

1571

converter. This control system is adjust the slip angle frequency and control the DC-link voltage. In the high speed range, switching from the asynchronous PWM to the synchronous PWM. Furthermore, around the rated speed, discontinuous PWM is employed to deal with the overmodulation region.

Figure 14 shows the operation waveforms of the continuous PWM at frequency ratio of nine and the rotation speed of 0.8 p.u. The modulation index is 0.871, and both the conventional method and the proposed method control the output voltage to be constant at 300 V.

Figure 15 shows the harmonic analysis results of the generator current of the continuous PWM. As shown Fig. 15(b), the proposed method suppresses low even-order harmonic components, such as second, eighth, and tenth

order, which are generated by the conventional method. Also, the eighth harmonic component was reduced by 99.2% compared to the conventional method. In addition, the generator current total harmonic distortion (THD) is reduced by 9.99% compared to the conventional method.

Figure 16 shows the operation waveforms of the discontinuous PWM at a frequency ratio of nine and a rotation speed of 1.0 p.u. The modulation index is 1.08, and both the conventional method and the proposed method control the output voltage to be constant at 300 V.

Figure 17 shows the harmonic analysis results of the generator current of the discontinuous PWM. As Fig. 17(b) shown, the proposed method suppresses low even-order harmonic components, such as second, eighth, and tenth order, which are generated by the conventional method. Also, the eighth harmonic component was reduced by 99.1% compared to the conventional method. In addition, the generator current THD is reduced by 7.14% compared to the conventional method. Therefore, this method is effective also in the discontinuous PWM.

B. Control strategy for starter generator

Figure 18 and Table IV shows the prototype of the jet

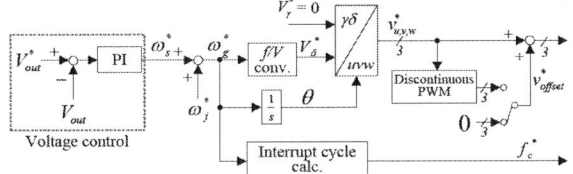

Fig. 13. Block diagram of PWM converter.

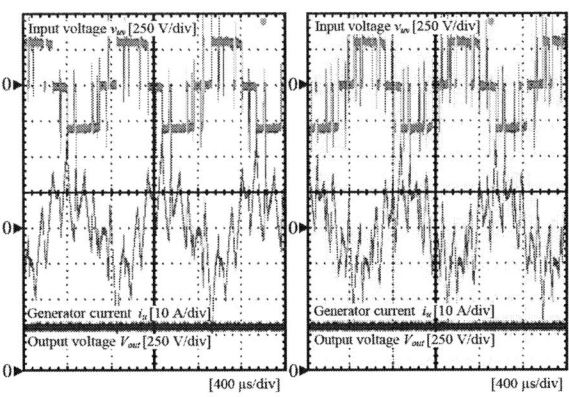

(a) Conventional method.   (b) Proposed method.
Fig. 14. Experimental results of discontinuous PWM.

(a) Conventional method.   (b) Proposed method.
Fig. 16. Experimental results of continuous PWM.

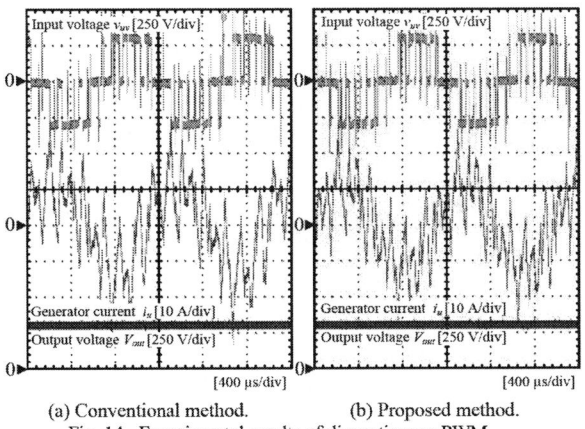

(a) Conventional method.

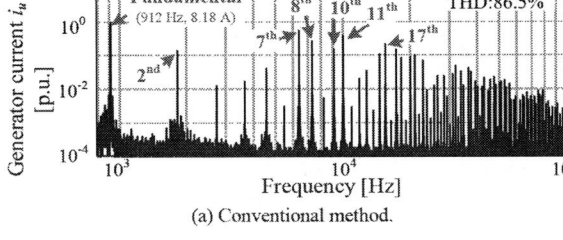

(b) Proposed method.
Fig. 15. Frequency analysis results of discontinuous PWM.

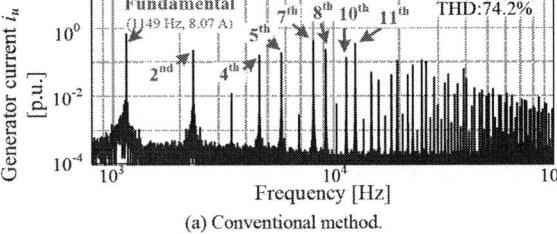

(a) Conventional method.

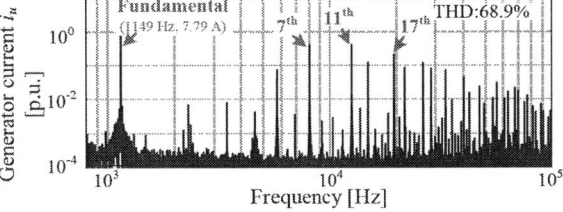

(b) Proposed method.
Fig. 17. Frequency analysis results of continuous PWM.

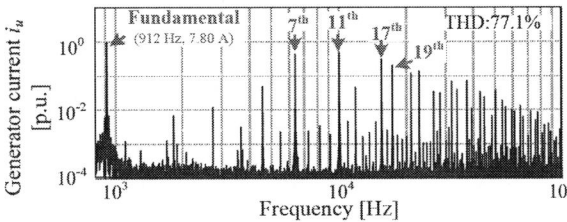

1572

generator and the specifications of the jet engine. As shown in Fig. 18, the starter generator is connected to the jet engine without a speed reduction gear.

Figure 19 shows the experimental waveforms of Run Mode at the rotation speeds of 70000 r/min, where the output power to the battery is 2.98 kW. As shown in Fig. 19, the DC link voltage is regulated to 300 V. Furthermore, the stable power generation operation is achieved since the battery current is constant at any rotation speeds.

Figure 20 shows the harmonic analysis results of the generator current in Fig. 19. As shown in Fig. 20, even-order low harmonic components are less than 0.2% which is sufficiently smaller than the fundamental component.

Figure 21 shows the characteristics of the generator current THD against output power. As shown in Fig. 21, the minimum current THD of 16.5% is achieved at the rotation speed of 70000 r/min and the output power of 2.98 kW. This is because the fundamental component of the generator current increases as the output power increases.

Figure 22 shows the efficiency characteristics of the power converter. As shown in Fig. 22, the maximum efficiency of 92.7% is achieved at the rotation speed of 70000 r/min and the output power of 2.98 kW.

Figure 23 shows the temperature characteristics of the exhaust nozzle against rotation speed and the output power when the ambient temperature is 26°C. As shown in Fig. 23, as the rotation speed of the jet engine increases, higher the output power is obtained at the same exhaust nozzle temperature. Further, this system has the highest efficiency when the exhaust nozzle temperature is around 800°C. Therefore, an output power command depends on angle frequency command ω_g^* is as shown in Fig. 23.

Figure 24 shows the experimental results the jet generator operation with the exhaust nozzle temperature control. The exhaust nozzle temperature command is 800°C. The jet engine is accelerated to 60000 r/min, 65000 r/min, and 70000 r/min in the run mode. Then the jet engine is decelerated to 60000 r/min. As shown in Fig 24, even when the rotation speed accelerates or decelerates, the exhaust nozzle temperature converges to the command value. The exhaust nozzle temperature can be controlled within the maximum deviation of 2% compared to the command value in the steady state. Furthermore, the exhaust nozzle temperature at 70000 r/min drops to 700°C because the inflow current of the battery is limited. In addition, the proposed control method achieves among the starting, the powering and the cooling operations. The transition without the deceleration is achieved by the synchronous frequency command limiter when the startup mode changes to the run mode. The output power gradually approaches zero after this transition, because the starter generator maintains the rotation speed until the output power of the jet engine becomes sufficiently high. The acceleration without overload is achieved in the startup mode by an

Fig. 18. Prototype of jet generator.

TABLE IV
Specification of jet engine.

Parameter	Value
Weight	2.9 kg
Diameter	131 mm
Full length	281 mm
Rated thrust	165 N
Rated speed	100000 r/min

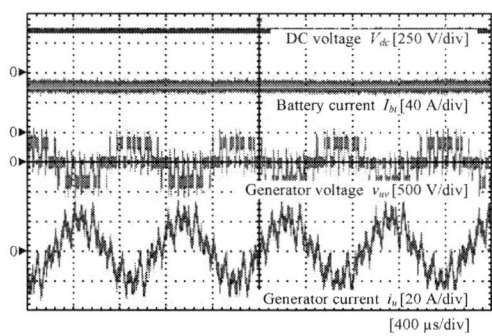

Fig. 19. Experimental waveforms of Run Mode.

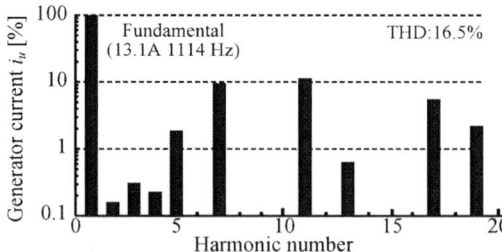

Fig. 20. Harmonic components on the generator current.

output power limiter, which limits a command up to 1 kW. Further, the starter generator decelerates by a free run when the operation mode transitions to the stop mode from the run mode. Then the inverter restarts at a rotation speed of 1000 r/min. The starter generator simultaneously performs the cooling operation.

VII. CONCLUSION

The control strategy for UAV with the jet engine were proposed in this paper. The stable transition without decelerating and overcurrent between the operation modes of the starter generator was achieved by the synchronous frequency command limit and the output

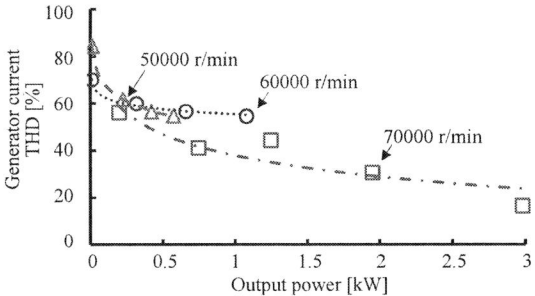

Fig. 21. Characteristics of generator current total harmonics distortion.

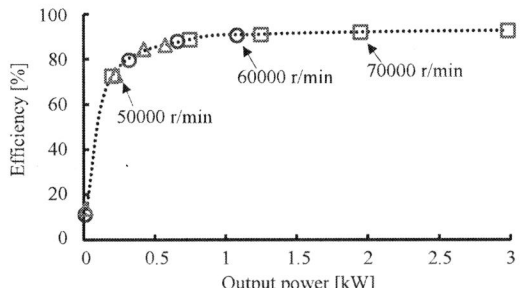

Fig. 22. Characteristics of efficiency of power converter.

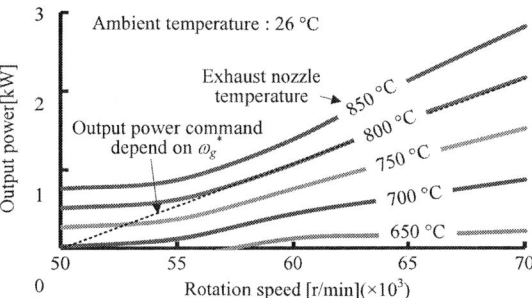

Fig. 23. Characteristics of exhaust nozzle temperature against rotation speed and output power, and output power command according to rotation speed.

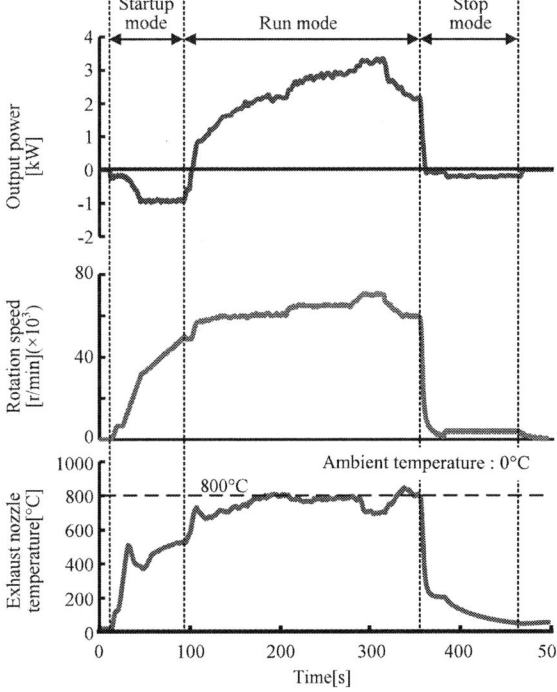

Fig. 24. Experimental results of operation of jet generator.

power limiter. In addition, the even-order low harmonic components are suppressed by the modulation method using the estimated intersection phase for the synchronous PWM. The 3-kW prototype system achieved the maximum conversion efficiency of 92.7%, the minimum generator current THD of 16.5% at 70000 r/min. Further, the exhaust nozzle temperature was controlled within the maximum deviation of 2% of the command value.

ACKNOWLEDGMENT

This paper is based on results obtained from a project subsidized by the New Energy and Industrial Technology Development Organization (NEDO) of Japan.

REFERENCES

[1] A. C. Satici, H. Poonawala, M. W. Sppong:"Robust Optical Control of Quadrotor UAVs", IEEE Access, vol. 1, pp. 79-93, 2013.

[2] N. Gageik, P. Benz, S. Montenegro:"Obstacle Detection and Collision Avoidance for a UAV With Complementary Low-Cost Sensors", IEEE Access, vol. 3, pp. 599-609, 2015

[3] J. Shiau, D. Ma, P. Yang, G. Wang, J. Gong:"Design of a Solar Power Management System for an Experimental UAV", IEEE Transactions on Aerospace and Electronic Systems, vol. 45, No. 4, pp. 1350-1360, 2009

[4] J. Shiau, D. Ma, P. Yang, G. Wang, J. Gong:"Predictor-Based Control of a Class of Time-Delay Systems and Its Application to Quadrotors", IEEE Transactions on Industrial Electronics, vol. 64, No. 4, pp. 459-469, 2016

[5] M. Whittingham "History, Evolution, and Future Status of Energy Storage", Proceedings of the IEEE, vol. 100, pp. 1518-1534, 2012

[6] "Small UAV turbojet engine developed in Japan". http://robohub.org/small-uav-tarbojet-engine-developed-in-japan/

[7] S. Chuangpishit, A. Tabesh, Z. Shahrbabak M. Saeedifard:"Topology Design for Collector Systems of Offshore Wind Farms With Pure DC Power Systems", IEEE Transactions on Industrial Electronics, vol. 61, No. 1, pp. 320-328, 2014

[8] Yosei Hirano, Takashi Yoshida, Kiyoshi Ohishi, Toshimasa Miyazaki, Yuki Yokokura, and Masataka Sato,"Vibration Suppression Control Method for Trochoidal Reduction Gears under Load Conditions", IEEJ J. Industry Applications, vol.5, no.3, pp.267-275, 2016

[9] Takashi Yoshioka, Thao Tran Phuong, Akinori Yabuki, Kiyoshi Ohishi, Toshimasa Miyazaki, and Yuki Yokokura," High-performance Load Torque Compensation of Industrial Robot using Kalman-filter-based Instantaneous State Observer", IEEJ J. Industry Applications, vol.5, no.3, pp.267-275, 2016.

[10] Takashi Yoshioka, Thao Tran Phuong, Akinori Yabuki, Kiyoshi Ohishi, Toshimasa Miyazaki, and Yuki Yokokura, "High-performance Load Torque Compensation of Industrial Robot using Kalman-filter-based Instantaneous State Observer", IEEJ J. Industry Applications, vol.5, no.3, pp.267-275, 2016.

[11] Toshiki Nakanishi, and Jun-ichi Itoh,"Control Strategy for Modular Multilevel Converter based on Single-phase Power Factor Correction Converter", IEEJ J. Industry Applications, vol.6, no.1, pp.46-57, 2017.

[12] E. Torenbeek: "Synthesis of Subsonic Airplane Design", Kluwer Academic Publishers, pp. 97-140, 2014

The 2018 International Power Electronics Conference

Study on the Influence of voltage variations for Non-Intrusive Load Identifications

Yu-Hsiu Lin[1], Shun-Kang Hung[2] and Men-Shen Tsai[3*]

[1] Department of Computer Science and Information Management, Providence University, Taichung City, Taiwan
[2] Avnet, Inc. www.avnet.com (Taiwan sales location), Taipei, Taiwan
[3] Graduate Institute of Automation Technology, National Taipei University of Technology, Taipei, Taiwan
*E-mail: mstsai@ntut.edu.tw

Abstract-In a smart grid, electricity energy demands requested from down-stream sectors continuously increase. One way to meet the energy demands is utilizing a Non-Intrusive Load Monitoring (NILM)-style system to monitor and manage residential and commercial electrical appliances in respond to demand response programs. NILM is as an electricity energy audit to energy-saving issues.

This paper aims at developing an NILM identification considering the influence of voltage variations. An NILM system normally consists of "Data Acquisition", "Event Detection and Feature Extraction", and "Load Classification." The goal of load classification in NILM is to identify the operation ON/OFF status of individual household appliances. The k-Nearest Neighbors Classifier is used as the load classifier of the NILM. In NILM, measured physical phenomena, voltage signals/power profiles, to load classification for long-term load monitoring vary. The variations affect the identification performance. In this paper, a cross-validation strategy is conducted and used to deal with sampled data with systematic errors from classified loads. Different types of loads are used to verify the influence of the voltage variations to load identifications. As the experimentation reported in this paper shows, a satisfactory recognition rate of 97.60% to the NILM load identification addressed in this paper is achieved. The proposed NILM system is able to classify loads with proper robustness.

I. INTRODUCTION

Nowadays, due to the global warming and climate change, it is very important to restrict the production of green-house gases through effective use of residential and commercial electrical appliances. One way to monitor and manage these electrical appliances used is utilizing the traditional intrusive-type load monitoring approaches that plug-load smart plugs need to be installed. It burdens customers with a high investment in construction of their load monitoring system. Therefore, researchers have developed Non-Intrusive Load Monitoring (NILM)-type load monitoring approaches [1-13] that only single set of non-plug-load voltage and current sensors is installed in the electrical panel of a load monitoring field. Power consumption on individual electrical appliances in the field can be identified through the analysis of NILM on

acquired composite electrical voltage and/or current signals.

In literature [11] and [12], to show the repeatability of the turn-on transient energy of the appliances where voltage variations from -10% to +10% in intervals of 1% were simulated and conducted for experimentation, the authors introduced the Coefficient of Variation of measuring the dispersion of a probability/frequency distribution. An adaptive load monitoring scheme actively analyzing real-time/on-line load data can also be used to identify appliances under voltage variations [13]. In this paper, we consider the influence of voltage variations on NILM identification, and a cross-validation strategy that produces combinatorial datasets to diminish systematic errors on sampled data is conducted and used to deal with sampled data/measured power profiles of appliances with voltage variations. This paper focuses on studying the influence of the voltage variations to the NILM load identification performance. A NILM consists of "Data Acquisition", "Event Detection and Feature Extraction", and "Load Classification." The load classification can be considered as a machine learning and pattern recognition task, the k-Nearest-Neighbor Classifier (k-NNC) is used to identify individual household appliances. In NILM, the acquired voltage signals are computed as power envelope with current signals. However, the voltage varies for long-term NILM execution. The variations affect classification performance of the NILM/k-NNC. Thus, a cross-validation strategy, a model-validation technique that shows how a statistical analysis generalizes to an independent dataset is conducted in this paper. It is also used to deal with data sampled with systematic errors, projected to representative features, and classified for load identification. Different types of electrical loads are used to verify the influence of the voltage variations to load identification of NILM.

The organization of this paper is as follows. Sec. I introduces NILM. The proposed NILM method including "Data Acquisition", "Event Detection and Feature Extraction", and "Load Classification" is discussed in Sec. II. The laboratory tests and experimental results are

The 2018 International Power Electronics Conference

shown in Sec. III. Discussion of influence of voltage variations to the identification performance is also discussed in Sec. III. Finally, Sec. IV concludes this study.

II. METHODOLOGY

Fig. 1 shows the workflow of a NILM system: "Data Acquisition", "Event Detection and Feature Extraction", and "Load Identification/Load Classification." Especially, during the training process of the NILM/load classification addressed in this study, a cross-validation procedure is conducted and used to overcome systematic bias from collected data samples. The cross-validation procedure dealing with sampled data with systematic errors from identified loads will be introduced in the end of this section.

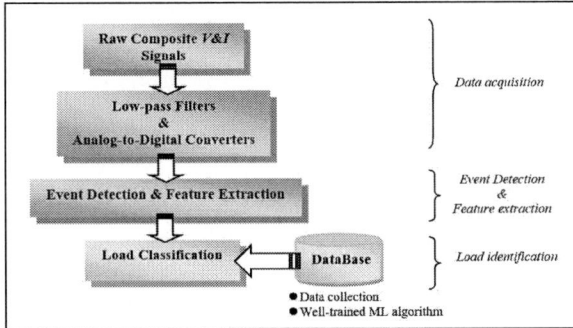

Fig. 1. Workflow of the a NILM system

Both composite current and voltage signals are simultaneously and continuously sampled through data acquisition system. Acquired data are conditioned, digitized, and then analyzed by the NILM system.

Event detection aims at detecting abrupt power changes that correspond to the state change of electrical appliance. The NILM system takes a *features* reading with a load event detected, analyzes the data through a cross-validation strategy (during the training process). The NILM system deduces the type of electrical appliance that is being energized or de-energized using the k-NNC [12]. The k-NNC is one of the most basic classifiers for data classification. Due to its simplicity, the k-NNC is often used as the baseline classification method in comparison with other sophisticated classification approaches in pattern recognition.

When a load is turned on, the transient of the power profile of the load continues for a certain period of time. The steady-state situation of the turned-on load occurs when the load settles down. Fig. 2 illustrates the event detection and feature extraction of the NILM in this paper. In order to extract features from a load (Fig. 2a) for different types of electrical appliances, a transient feature extraction scheme here works to identify transient response of monitored electrical appliances. As concluded in [2]-[4], [7], [9]-[11], transient current of different types of loads shows uniqueness and repeatability.

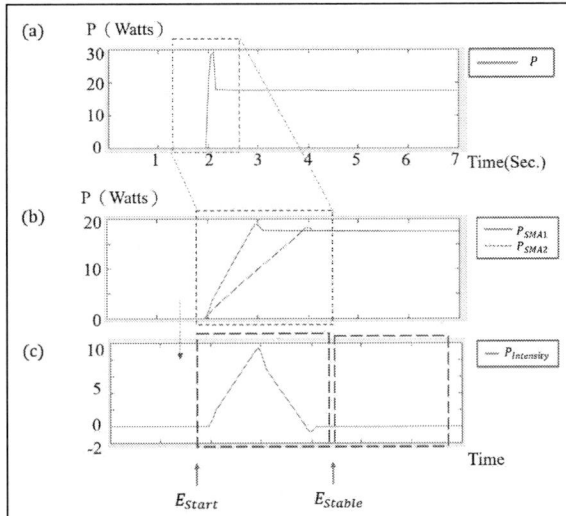

Fig. 2. Illustration of event detection and feature extraction of the NILM to a load operation scenario.

During the transient feature extraction process, a Simple Moving Average (SMA), the unweighted mean of previous n data points, is used to identify P_{SMA1} and P_{SMA2} (Fig. 2b). The power intensity, $P_{Intensity}$ (Fig. 2c), of P_{SMA1} against P_{SMA2} can be computed by (1). Finally, to retrieve the period of transient response (Fig. 2c) of the load (Fig. 2a), one can identify E_{start} and E_{stable}. For the event detection of the NILM in this paper to the load operation scenario (Fig. 2a), a pre-specified power threshold, $PWR_{Threshold}$, to detect a load event of a turned-on load can be computed by (2). The transient period can be measured through the detection of the transient response of the load (Fig. 2a).

$$P_{Intensity} = P_{SMA1} - P_{SMA2} \qquad (1)$$

In this paper, the data length, n, of P_{SMA1} in time duration is 2 (sec); it is equal to 4 for P_{SMA2}.

$$PWR_{Threshold} = \frac{P_{Intensity}}{\alpha} \qquad (2)$$

Where,

$\{PWR_{Threshold} \in N : PWR_{Threshold} > \beta\}$, α and β are constant.

The transient feature extraction scheme used in this paper and illustrated in Fig. 2 can be conducted in industry applications such as Prognostics and Health Management. To extract features from the load pattern(s) (Fig. 2a) for load classification, this paper uses (3)-(4) to transform the load pattern(s) into distinctive representatives as an input set of features of the classifier.

$$\Delta P = P_{SMA}(E_{Stable}) - P_{SMA}(E_{Start}) \qquad (3)$$

$$\Delta S = S_{SMA}(E_{Stable}) - S_{SMA}(E_{Start}) \qquad (4)$$

In (3)-(4), P stands for active power/real power (Watts); S represents apparent power (VA).

For load classification on distinctive representative features extracted from acquired data of the NILM, in this

1576

paper, the k-NNC, which is often used as a baseline method in comparison with other sophisticated approaches in machine learning and pattern recognition due to its simplicity [14], is used as the load classifier for NILM. In the machine learning and pattern recognition researches, the k-NNC is a non-parametric method used for classification. An example of the k-NNC is shown in Fig. 3. The test sample (the circle) can be classified either to the first class of square or to the second class of triangle. If the neighborhood size, k, is 3 (samples inside the solid-line circle), it is assigned to the second class because there are 2 triangles and only 1 square inside the inner circle. If k is 5 (samples inside the dashed-line circle), it is assigned to the first class since there are 3 squares, 2 triangles and 1 star inside the outer circle. In practice, one way to identify the best number of the neighborhood size, k, is to resort to an exhaustive search

In this paper, the k-NNC is used as the load classifier of the NILM for load identification to identify individual electrical appliances monitored. It learns from an on-site-data collected during the training process; once the training process has been completed, on-line load monitoring performs. Fig. 4 depicts the workflow of the training process and the tests of the model, k-NNC, by the NILM in this paper.

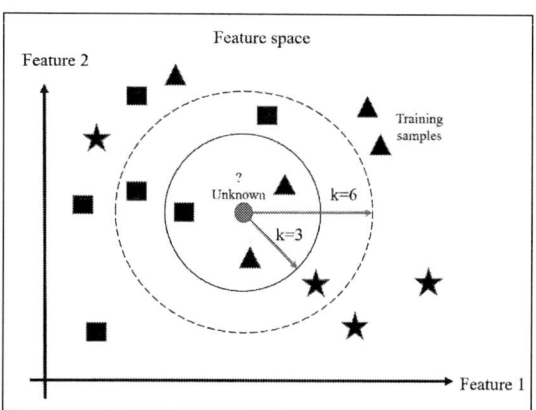

Fig. 3. Example of the k-NNC employed by the NILM.

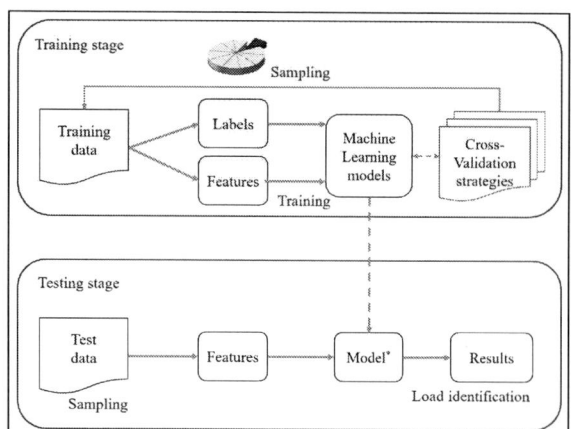

Fig. 4. Workflow of the training process and the tests of the model by the NILM in this paper. A cross-validation procedure is used to overcome systematic bias during the training process.

Sampling is biased if it systematically favors some observations over others. Sampling bias is sometimes called systematic bias. For NILM as predictive modeling, feature data being acquired through data acquisition and analyzed via event detection and feature extraction "evolve" over time with a systematic bias between separate training and validation datasets, as shown in Fig. 5. Therefore, in this paper, a cross-validation procedure in Fig. 4 is conducted and used to deal with data sampled with systematic errors.

III. EXPERIMENTATION

Fig. 6 shows the experiment set-up of the NILM proposed in this paper and demonstrated by different types of electrical loads. The electrical loads used to verify the proposed NILM include a table lamp, a fan, a hair dryer, an LED light, a spiral bulb, and an artificial electrical load simulated by an electronic load. Table I lists the electrical loads represented by their representative features, ΔP and ΔS, by (3)-(4) and used to verify the proposed NILM in this paper. During the data collection process, the data length, n, of P_{SMA1} in (1) in time duration is set to 2 (sec); it is set to 4 for P_{SMA2} in (1). Threshold β in (2) is set to 2 (Watts). In Table I, to each monitored electrical appliance with or without multi-state transitions, statistics is based on data samples with a total of 1,100 data points acquired, analyzed, and collected from each monitored electrical appliance. An HIOKI PW3390 high-precision and broad-range power analyzer is used to simultaneously and continuously acquire composite voltage and current signals at the power entrance. The TM6331HC digital timer switches offer timed On/Off control of the monitored electrical loads.

An active Current clamp that generates output voltage that is proportional to the current is used; its output voltage is wired to the power analyzer. The voltage of the AC power source is also wired to the power analyzer. The sampling rate of the power analyzer during the data collection process is 10 Hz (100ms/Sample). The power analyzer transmitting the digitized data to be analyzed to the NILM developed in this paper with a PC via a USB cable.

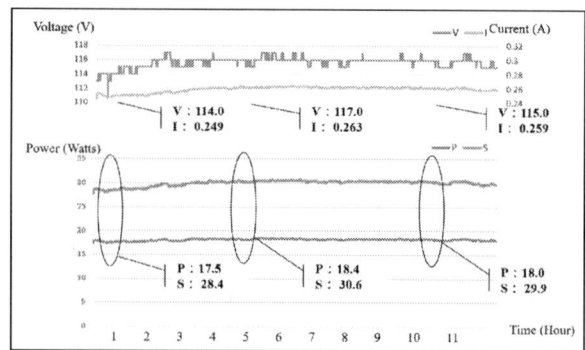

Fig. 5. Influence of voltage variations to NILM addressed in this paper.

The 2018 International Power Electronics Conference

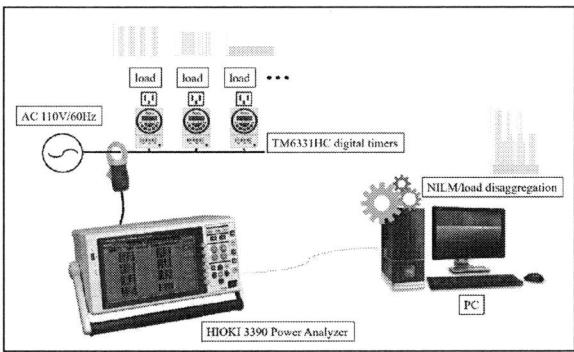

Fig. 6. Experimental set-up of the proposed NILM.

TABLE I
ELECTRICAL LOADS REPRESENTED BY THEIR FEATURES AND USED
TO VERIFY THE NILM IN THIS PAPER

Load	Active power (Watts)		Apparent power (VA)	
	Avg.	Std.	Avg.	Std.
Table lamp	0018.1	00.24	0030.0	00.65
Fan (Low)	0055.7	01.13	0055.7	01.13
Fan (High)	0075.3	01.65	0084.5	02.01
Hair dryer (Low)	0530.7	06.89	0569.4	08.61
Hair dryer (High)	1011.4	27.15	1011.4	27.16
LED light	0009.7	00.05	0010.0	00.06
Spiral bulb	0022.6	00.16	0032.6	00.45
Artificial load	0018.0	00.26	0033.0	00.48

The k-NNC used in this paper relies on an exhaustive search on k, in order to identify data samples with over-lapped features in the feature space. Sampling is biased if it systematically favors some observations over others. For the NILM as predictive modeling, feature data being acquired through data acquisition and analyzed via NILM evolve over time with a systematic bias between training and validation datasets. In this paper, a cross-validation strategy is used to deal with data sampled with systematic errors.

Table II shows the cross-validation procedure of the NILM in this paper, where totally 16 combinatorial datasets (there are 4 training sets and 4 test sets) are generated for training and tests of the k-NNC to the proposed NILM. In Table II, to each monitored electrical appliance, 1,100 data samples are acquired at different time. These samples are split into a training dataset and a test dataset for the NILM in this paper.

Fig. 7 shows the feature space of data acquired from and analyzed for the monitored table lamp in Table I. The k-NNC used in this experimentation resorts to an exhaustive search on k: the best value of k is 25. An overall recognition rate of 97.60% is achieved. To the model trained, a confidence level of 95.00% reflecting a significance level of 0.05 [15] and a systematic bias of ±0.75% interval are achieved. To get a more robust k-NNC of the NILM, one can use a k-Means clustering-combined distance-weighted k-NNC as the load classifier of the NILM.

TABLE II
THE CROSS-VALIDATION PROCEDURE, WHICH PRODUCES A TOTAL
OF 16 COMBINATORIAL DATASETS FOR THE K-NNC, ON DATA
SAMPLES OF EACH MONITORED ELECTRICAL LOAD

Time	Training samples	
	Training set/Test set	# of samples
the 1st hour	Training set 1	75
the 2nd hour	Training set 2	75
the 3rd hour	Training set 3	75
the 4th hour	Training set 4	75
	Test samples	
the 5th hour	Test set 1	200
the 7th hour	Test set 2	200
the 9th hour	Test set 3	200
the 11th hour	Test set 4	200
Total		1,100

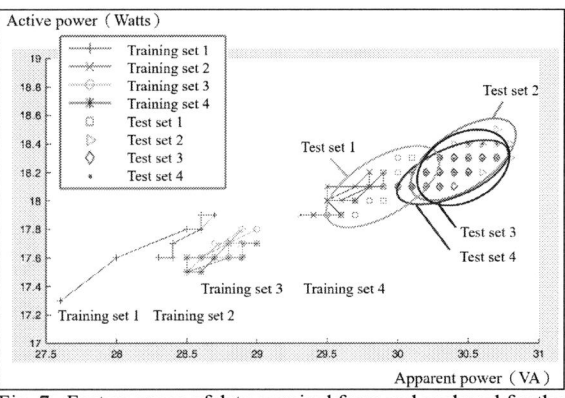

Fig. 7. Feature space of data acquired from and analyzed for the monitored table lamp.

IV. CONCLUSIONS

In this paper, an NILM system that applies a k-NNC as the load classifier to identify different types of electrical appliances is proposed. In NILM, measured physical phenomena, voltage signals/power profiles, to load classification for long-term load monitoring vary. The variations affect classification accuracy of the NILM. Thus, in this paper, a cross-validation strategy, which produces combinatorial datasets to diminish systematic errors on sampled data, is proposed and applied by the NILM. As the experimentation reported in this paper shows, the NILM proposed in this paper is feasible. Electricity energy consumption on each of monitored individual electrical appliances can be estimated by the proposed NILM system as an electricity energy audit, since the operation On/Off status of each monitored electrical appliance can be identified by the NILM.

In the future, to obtain a more robust k-NNC of the NILM, one can use a k-Means clustering-combined distance-weighted k-NNC as the load classifier of the NILM where the k-Means clustering selects representative data points for each load class and distances by the Euclidean distance metric are used as a weighting factor for the distance-weighted k-NNC.

The 2018 International Power Electronics Conference

Acknowledgment

This research was supported in part by the Ministry of Science and Technology, TAIWAN, under Contract numbers MOST 106-2221-E-027-086-MY2 and MOST 107-3113-E-006-007.

The authors would also like to thank the reviewers for their valuable suggestions on this paper.

References

[1] S. Drenker and A. Kader, "Nonintrusive Monitoring of Electric Loads," *IEEE Computer Applications in Power*, vol. 12, no. 4, pp. 47-51, 1999.

[2] C. Laughman, K. Lee, R. Cox, S. Shaw, S. B. Leeb, L. Norford, and P. Armstrong, "Power Signature Analysis," *IEEE Power & Energy Magazine*, vol. 99, no. 2, pp. 56-63, 2003.

[3] A. Shrestha, E. L. Foulks, R. W. Cox, "Dynamic load shedding for shipboard power systems using the non-intrusive load monitor," *IEEE Electric Ship Technologies Symposium*, pp. 412-419, Baltimore, MD, USA, 20-22 April 2009.

[4] H. H. Chang, C. L. Lin, H. T. Yang, "Load Recognition for Different Loads with the Same Real Power and Reactive Power in a Non-intrusive Load-morning System," *in Proc. of the 12th International Conference on Computer Supported Cooperative Work in Design (CSCWD 2008)*, pp. 1122-1127, Xi'an, China, 16-18 April 2008.

[5] A. Cole and A. Albicki, "Nonintrusive Identification of Electrical Loads in a Three-Phase Environment Based on Harmonic Content," *IEEE Conf. on Instrumentation and Measurement Technology*, pp. 24–29, Baltimore, MD, USA, 1-4 May 2000.

[6] Y. Nakano, H. Murata1, K. Yoshimoto1, S. Hidaka, M. Tadokoro, and K. Nagasaka, "Non-Intrusive Electric Appliances Load Monitoring System Using Harmonic Pattern Recognition - Performance Test Results at Real Households," *in Proc. of the 4th International Conference on Energy Efficiency in Domestic Appliances and Lighting*, pp. 477-488, London UK, 21-23 June 2006.

[7] S. B. Leeb, S. R. Shaw, and J. L. Kirtley Jr., "Transient Event Detection in Spectral Envelope Estimates For Nonintrusive Load Monitoring," *IEEE Trans. on Power Delivery*, vol. 10, no. 3, pp. 1200-1210, 1995.

[8] S. R. Shaw, S. B. Leeb, L. K. Norford and R. W. Cox, "Nonintrusive Load Monitoring and Diagnostics in Power Systems" *IEEE Trans. on Instrumentation and Measurement*, vol. 57, no. 7, pp. 1445-1454, 2008.

[9] H. T. Yang, H. H. Chang, C. L. Lin, "Design a Neural Network for Features Selection in Non-intrusive Monitoring of Industrial Electrical Loads," *in Proc. of the 11th International Conference on Computer Supported Cooperative Work in Design CSCWD 2007*, pp. 1022-1027, Melbourne, Vic., 26-28 April 2007.

[10] H. H. Chang, C. L. Lin, L .S. Weng, "Application of artificial intelligence and non-intrusive energy-managing system to economic dispatch strategy for cogeneration system and utility," *Computer Supported Cooperative Work in Design, 13th International Conference on (CSCWD 2009)*, 22-24 April 2009, pp.740-745.

[11] K. L. Chen, H. H. Chang, and N. M. Chen, "A new transient feature extraction method of power signatures for Nonintrusive Load Monitoring Systems," 2013 IEEE International Workshop on Applied Measurements for Power Systems (AMPS), pp. 79-84, Aachen, Germany, 25-27 September 2013.

[12] H. H. Chang, C. L. Lin, and J. K. Lee, "Load identification in nonintrusive load monitoring using steady-state and turn-on transient energy algorithms," *in Proc. of the 2010 14th International Conference on Computer Supported Cooperative Work in Design (CSCWD 2010)*, pp. 27-32, Shanghai, China, China, 14-16 April 2010.

[13] H. Shao, M. Marwah, and N. Ramakrishnan, "A Temporal Motif Mining Approach to Unsupervised Energy Disaggregation," *in Proc. of the Twenty-Seventh AAAI Conference on Artificial Intelligence (AAAI'13)*, pp. 1327-1333, Bellevue, Washington, 14-18 July, 2013.

[14] Jyh-Shing Roger Jang, "Data Clustering and Pattern Recognition," available at the links for on-line courses at the author's homepage at http://mirlab.org/jang.

[15] A. P. Field. Discovering Statistics Using IBM SPSS Statistics (4th Edition), SAGE Publications Ltd., 2013.

The 2018 International Power Electronics Conference

Basic Experiment of a Maglev System for a Flexible Steel Plate with Curvature: Fundamental Consideration on Levitation Stability under Disturbance

Makoto Tada[1], Kazuki Ogawa[2], Takayoshi Narita[2*], Hideaki Kato[2] and Hiroyuki Moriyama[2]
1 Course of Mechanical Engineering, Tokai University, Hiratsuka, Japan
2 Department of Prime Mover Engineering, Tokai University, Hiratsuka, Japan
*E-mail: narita@tsc.u-tokai.ac.jp

Abstract— These days, studies of electromagnetic levitation technology have been performed. When an ultrathin and flexible steel plate is to be levitated, levitation control becomes difficult because the ultrathin steel plate undergoes increased flexure. We have proposed the levitation of an ultrathin steel plate that is bent to an extent which does not induce plastic deformation. In this study to elucidate the levitation stability under disturbance, random disturbance was input into the levitation system and bending levitation experiments were carried out using the ultrathin steel plate with a thickness of 0.19 mm.

Keywords— *Electromagnetic levitation, Bending levitation control, Flexible steel plate, Vibration control, Random noise*

I. INTRODUCTION

Magnetic levitation technology capable of grasping and conveying objects in a noncontact manner attracts attention and vigorous studies utilizing the characteristics of magnetic force such as suction control method and induced repulsion method are being conducted [1-3]. Regarding the stability problem of the levitation body, a control method considering the mechanical characteristics of the structure is being studied [4-6]. It is expected that magnetic levitation technology can be applied to thin steel plate production processes such as cold rolling, which requires high surface quality. However, elastic vibration is induced because of the flexibility brought about by the plate thickness and area, so stability at the time of levitation is significantly impaired.

To solve this problem, our research group has constructed an electromagnetic levitation control system in which the relative distance between the electromagnet and a steel plate is constantly maintained, aiming to prevent the steel plate from falling from the conveyer or contacting the electromagnet during electromagnetic levitation conveyance. Furthermore, we have proposed a method of levitating a thin steel plate with a thickness of less than 0.3 mm by moderately bending it beforehand [7,

8]. The levitation stability was also improved by bending and the levitation stability could be maintained by the bending even if a disturbance was input to the control current [9]. However, considering a more practical situation, the electromagnet may be exposed to disturbances. In this study, we examined the bending levitation performance experimentally in which an external disturbance was applied to a steel plate by vibrating the frame where the electromagnet unit was installed.

II. SYSTEM FOR CONTROL EXPERIMENT

Figure 1 shows the outlines of the control system. Figure 2 shows a schematic illustration of the experimental apparatus. The object of electromagnetic levitation is a rectangular zinc- coated steel plate (SS400) with length $a = 800$ mm, width $b = 600$ mm and thickness $h = 0.19$ mm. To accomplish noncontact support of a rectangular ultrathin steel plate using five pairs of electromagnets (Nos. 1-5) as if the plate were hoisted by strings, the displacement of the steel plate was measured by five eddy-current gap sensors. Among the five pairs of electromagnets, the four pairs at the corners were inclined and a central electromagnet was

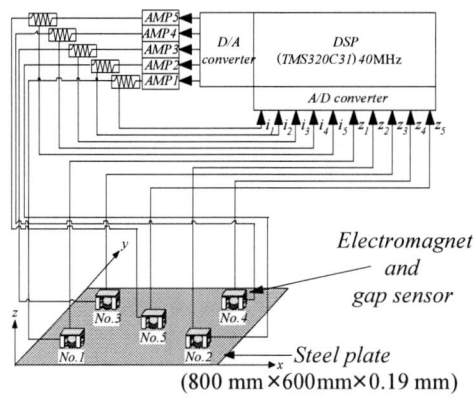

Fig. 1. Electromagnetic levitation control system.

moved in the vertical direction. In addition, the distance between the surfaces of electromagnets and the steel plate was controlled at 5 mm even when θ was changed. Figure 3 shows how the steel plate levitates with respect to each electromagnetic unit inclination angle θ. Thus, by moving the five electromagnets, bending magnetic levitation of the steel plate was realized.

In this system, independent control—in which information on the detected values of displacement, velocity, and coil current of the electromagnets under study at one position was fed back only to the same electromagnet— was performed, and the optimal control law was obtained on the basis of optimal control theory of the discrete time system [9]. As shown in Fig. 4, steel plates are divided into five virtual masses, and each is modeled as a lumped parameter system.

Furthermore, we model the steel plate as a one-dimensional beam seen from the x-axis direction and define the natural deflection angle θ_{na} which is the deflection angle at the support point when supporting the steel plate only with electromagnets No. 1 to No. 4.

$$\theta_{na} = \frac{\rho g l}{2Eh^2}\left(l^2 - 6d^2\right) \tag{1}$$

Here, h: thickness of the steel plate [m], ρ: density of the steel plate [kg/m3], g: gravitational acceleration [m/s2], l: electromagnet unit No. 1 and No. 3 viewed on the x- distance between No. 2 and No. 4 [m], d: distance from the electromagnet unit No. 1 or No. 2, No. 3, No. 4 seen on the x-axis [m], E: Young's modulus of the steel plate [N/m²]. According to eq. (1), when the electromagnet is installed as shown in Fig. 1, the natural deflection angle of the steel plate used in this paper is $h = 0.19$ mm and $\theta_{na} = 17.2$ °.

The vibrator shown in Fig. 5 was attached below the three frames on which the electromagnet unit was installed, so that the frame could be vibrated up and down. Table 1 shows the specifications of the vibrator. After adjusting the amplitude and phase of the frame to be constant by using a sine wave of constant frequency, it was possible to levitate the steel plate while vibrating the frame. In each frame, an eddy-current gap sensor was installed and the displacement of the frame during excitation was measured.

III. LEVITATION EXPERIMENT UNDER PULSE DISTURBANCE

In order to determine the frequency band of the waveform to be input in the frame vibration experiment, the steel plate was levitated with the frame vibrated using a pulse disturbance, and the resonance frequency of the steel plate was measured based on the amplitude spectrum acquired from the displacement gauge.

In addition, the amplitude of the frame excitation was set to the maximum amplitude at which the steel plate could levitate, but it could not levitate at the electromagnet angle $\theta = 0°$. Therefore, the experiment was carried out with an amplitude of about one third of the maximum amplitude only at $0°$.

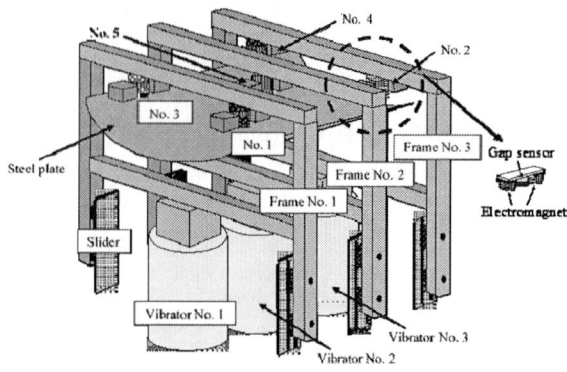

Fig. 2. Schematic illustration of the experimental apparatus.

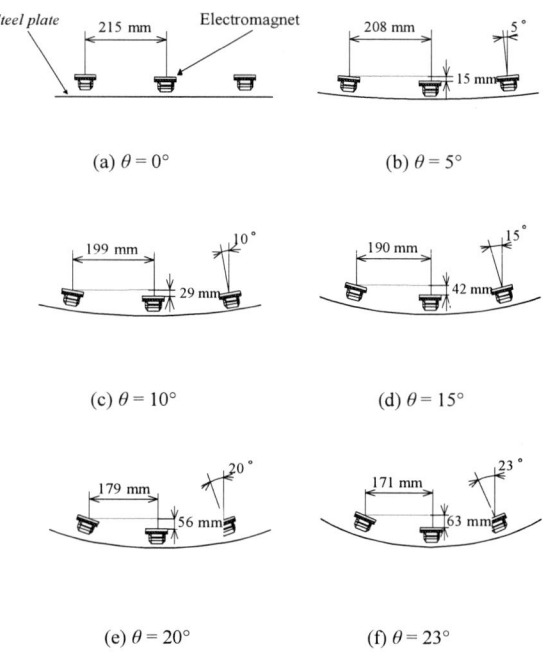

Fig. 3. Relationship between tilt angle of electromagnets θ and shape of steel plate.

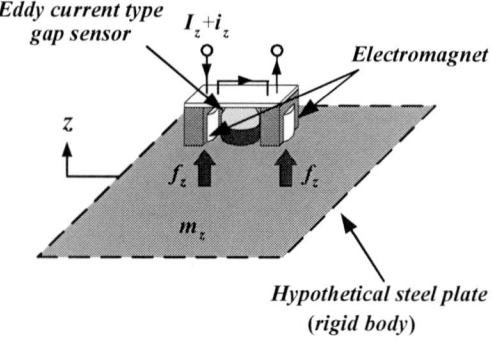

Fig. 4. Theoretical model of levitation control.

Figure 6 shows the time histories of displacement and amplitude spectrum of vibrating frames when the frame is excited using a pulse disturbance. Figure 7 shows the time histories of displacement and amplitude spectrum of the steel plate when the frame was excited using a pulse disturbance. From this spectrum waveform, the frequency with the largest spectrum value is defined as the resonance frequency. For example, in Fig. 7, the angle of the electromagnet is set to $\theta = 10°$. The maximum spectrum in the electromagnet 1ch at this time is 0.0534 mm, and the frequency at that time is 5.78 Hz, so this is taken as the resonance frequency. Representative values obtained by performing the same experiment ten times at each electromagnetic angle are shown in Table 2. From Table 2, it was confirmed that the resonance frequency exists in the band of 0 to 10 Hz.

IV. LEVITATION EXPERIMENT UNDER RANDOM DISTURBANCE

From the resonance frequency of the steel plate obtained in the previous chapter, random noise with a band set between 0 and 10 Hz was added to the frame, and the levitation performance was compared from the standard deviation of displacement and levitation probability for each electromagnetic angle. Experiments were conducted with three types of electromagnet angles of $0°$, $13°$, and $20°$. The standard deviation of displacement was measured 10 times for each electromagnetic angle, and the average value of these results was taken as the experimental value. At that time, in order to eliminate the influence of the transient state, measurement was carried out about 10 seconds after the start of levitation. In addition, the levitation probability

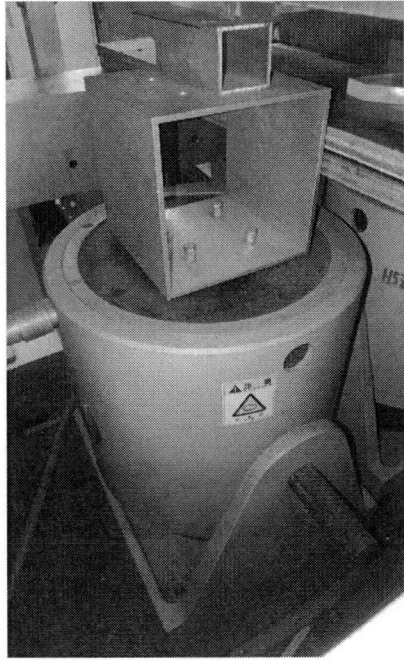

Fig. 5. Vibrator (EMIC 513-B).

TABLE I
SPECIFICATIONS OF VIBRATOR

Maximum excitation force	10 kgf$\pm$5 %
Maximum acceleration	28 G$\pm$5 %
Frequency range	3 Hz~13 kHz
Maximum input current	5.5 A$\pm$5 %
Maximum speed	177 cm/s
Maximum amplitude	10 mm

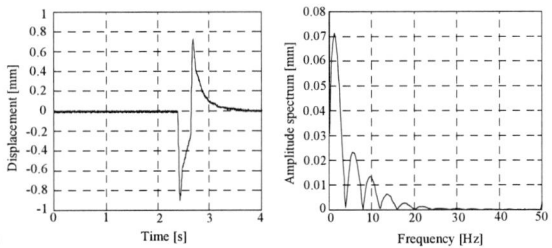

Fig. 6. Time histories of displacement and amplitude spectrums of vibrating frames by the pulse disturbance.

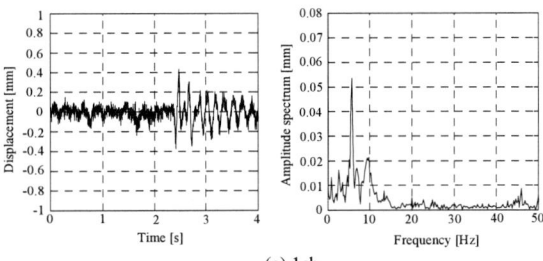

(a) 1ch

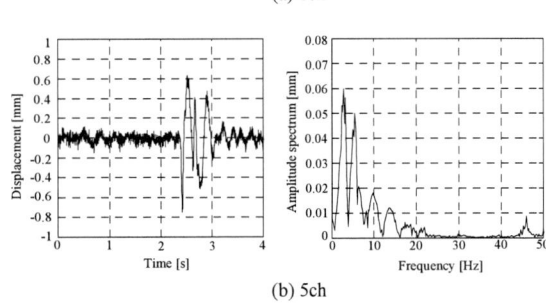

(b) 5ch

Fig. 7. Bending levitation result when vibrating frames by the pulse disturbance ($\theta = 10°$).

TABLE II
REPRESENTATIVE VALUE OF RESONANCE FREQUENCY

Tilt angle of electromagnets [deg]	Resonance frequency (1~4ch) [Hz]	Resonance frequency 5ch [Hz]
0	2.3	3.8
5	7.6	2.7
10	5.0	3.0
15	7.3	3.2
20	4.7	3.2

The 2018 International Power Electronics Conference

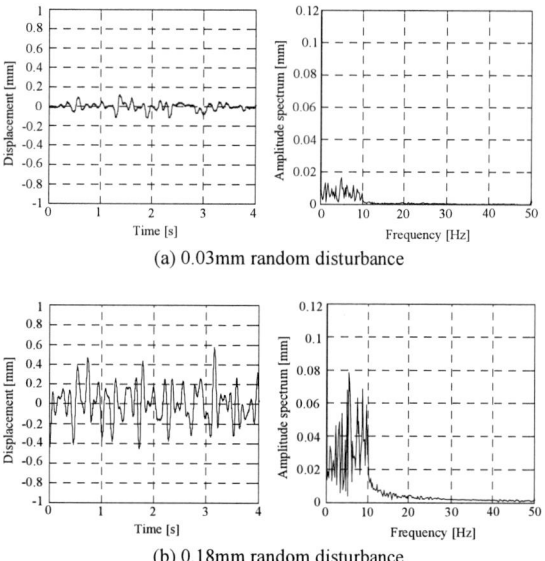

(a) 0.03mm random disturbance

(b) 0.18mm random disturbance

Fig. 8. Time histories of displacement and amplitude spectrums of vibrating frames by the random disturbance.

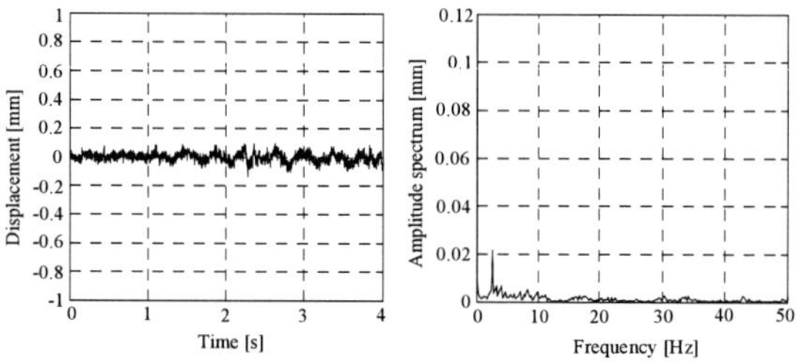

(a) $\theta = 0°$

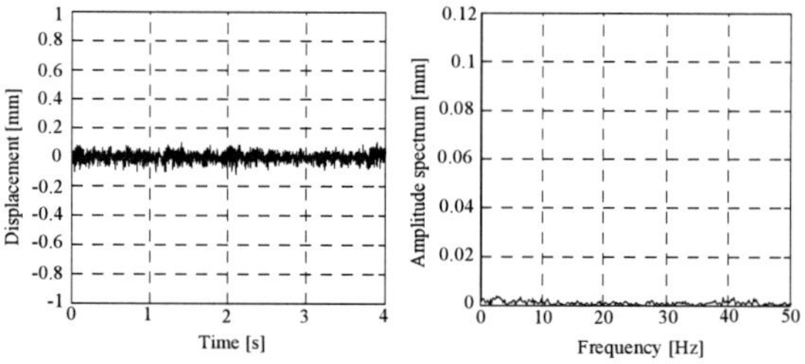

(b) $\theta \doteq 13°$

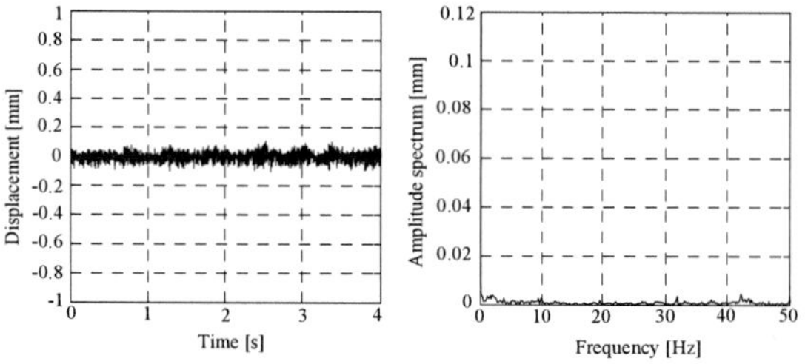

(c) $\theta = 20°$

Fig. 9. Bending levitation result when vibrating frames without disturbance.

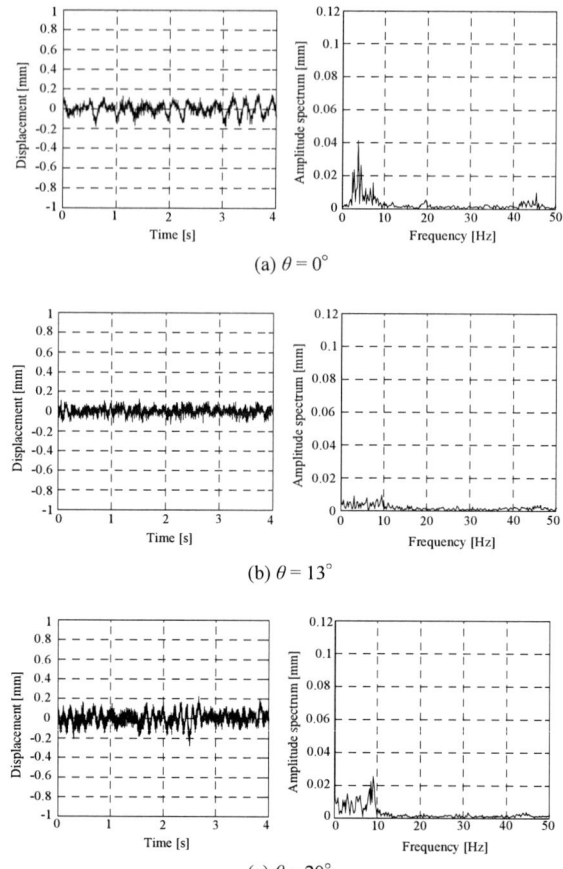

(a) $\theta = 0°$

(b) $\theta = 13°$

(c) $\theta = 20°$

Fig. 10. Bending levitation result when vibrating frames by the 0.03 mm random disturbance.

was regarded as a successful levitation when the levitation continued for 30 seconds after the input of the disturbance, and the number of times the levitation succeeded out of the 50 levitation experiments was calculated by a percentage. As for the amplitude of the frame vibration, as in the experiment using the pulse disturbance, the maximum amplitude at which the steel plate can levitate at the electromagnetic angle of 10° to 15° (standard deviation of displacement about 0.18 mm), and the amplitude of about one third of the maximum amplitude (standard deviation of displacement about 0.03 mm) were used.

Figure 8 shows the time histories of displacement and amplitude spectrums of vibrating frames by the random disturbance, Fig. 9 shows the time histories of displacement and amplitude spectrum of the steel plate without disturbance, and Fig. 10 and 11 shows the time histories of displacement and amplitude spectrum of the steel plate when the frame was excited using a random disturbance. From the spectra in Figs. 8, 9, 10 and 11, it is shown that the disturbance input in the 0 to 10 Hz band resonates with the steel plate. However, it can be seen by comparing Figs. 8, 9 and 10 that the spectrum when the electromagnetic angle of 13° is smaller than the spectrum when the other electromagnetic angles.

1583

Table 3 shows the result of standard deviation of displacement and Fig. 12 shows the result of levitation probability with respect to the difference of the electromagnet angle. From Table 3 and Fig. 12, it can be seen that when the frame is vibrated, the standard deviation of displacement increases for all the electromagnet angle. In the case of levitation without disturbance, electromagnet angles of 13° and 20°, which bent the steel plate, had almost the same standard deviation of displacement. However, when the disturbance was input, the standard deviation of displacement which the electromagnet angle of 13° has not deteriorated more than when electromagnet angles are 0° and 20°. For example, from Table 3, when the amplitude of the frame is large, the standard deviation of displacement when the electromagnet angle of 20° is deteriorated by about 60% compared with the case of the electromagnet angle of 13°, and it is impossible to levitate itself when the electromagnet angle of 0°.

The reason is the relationship between deflection and restoring force of a bending steel plate. By inclining the electromagnet and bending the steel plate, the deflection of each part of the steel plate is suppressed and the levitation stability is improved. However, if the degree of bending of the steel plate is too much, the restoring force is increased, the vibration readily occurs, and the levitation stability deteriorates.

As shown in Fig. 12, in the case of no disturbance, no large change in the levitation probability due to the electromagnet angle was observed, but when the disturbance was input, the levitation probability decreased greatly at 0° and 20° which are electromagnet angles at which the standard deviation of displacement is not suppressed. However, the decreasing rate of the levitation probability tended to be small at 13° which the standard deviation of displacement was suppressed.

From the above results, the bending steel plate by inclined at the optimum electromagnet angle levitates stable even if the frame on which the electromagnet unit was installed was vibrated.

V. CONCLUSIONS

In this study, we used a bending magnetic levitation system effective for non-contact transport of thin steel plate, inputs the disturbance assumed on the steel plate transport line into the experimental apparatus, and conducted experiments on the levitation performance at that time. As a result, by levitation at the optimum bending angle, vibration during levitation is suppressed even when external disturbance is input to the device, and it is confirmed that the good levitation state can be realized.

From now on, we would like to carry out a levitation experiment to verify the effectiveness of the bending magnetic levitation system even when using other plate thickness in the disturbance. We also plan to investigate the disturbance cancellation control that acquires the displacement of the frame with a sensor and calculates the absolute displacement of the steel plate and performs control.

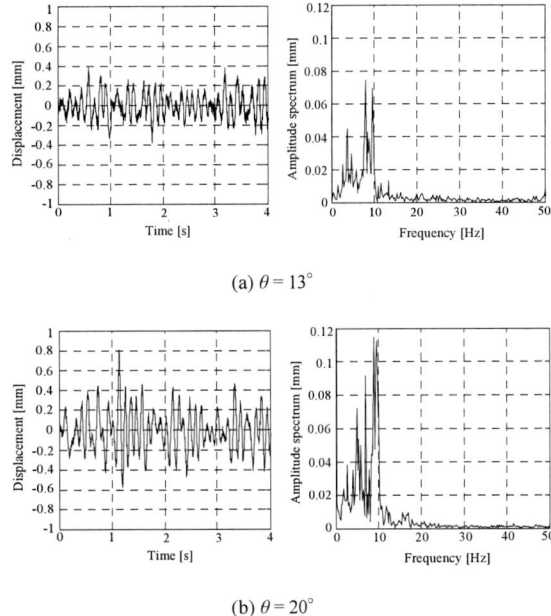

(a) $\theta = 13°$

(b) $\theta = 20°$

Fig. 11.   Bending levitation result when vibrating frames by the 0.18 mm random disturbance.

TABLE III
STANDARD DEVIATION OF DISPLACEMENT WHEN VIBRATING THE FRAMES BY THE RANDOM DISTURBANCE

Tilt angle of electromagnets [°]	Standard deviation of displacement [mm]		
	Without disturbance	0.03 mm random disturbance	0.18 mm random disturbance
0	0.0346	0.0596	-
13	0.0294	0.0363	0.1292
20	0.0299	0.0608	0.2164

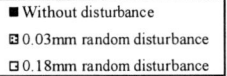

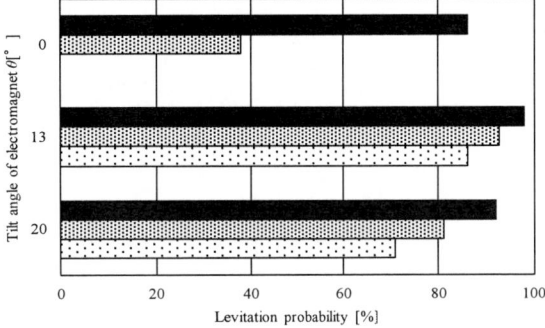

Fig. 12.   Levitation probability when vibrating the frames by the random disturbance.

REFERENCES

[1] Y. Chung, C. Lee, D. Kim, H. Kang, Y. Park and Y. Yoon, "Conceptual Design and Operating Characteristics of Multi-Resonance Antennas in the Wireless Power Charging System for Superconducting MAGLEV Train," *IEEE Trans. on Applied Superconductivity*, vol. 27, no. 4, pp. 3601805, 2017.

[2] A. Patel, V. Kalitka, S. C. Hopkins, A. Baskys, A. F. Albisetti, G. Giunchi, A. Molodyk and B. A. Glowacki, "Magnetic Levitation Between a Slab of Soldered HTS Tape and a Cylindrical Permanent Magnet," *IEEE Transactions on Applied Superconductivity*, vol. 26, no. 3, pp. 3601305, 2016.

[3] F. Sun and K. Oka, "Magnetic Suspension Using Variable Flux Path Control Mechanism with Permanent Magnet (Simultaneous Suspension Experiment of Two Iron Balls)," *Transactions of the Japan Society of Mechanical Engineers Series C*, vol. 78, no. 792, pp.2771-2780, 2012, (in Japanese).

[4] T. Ohji, Y. Katsuda, K. Amei and M. Sakui, "Structure of One-Axis Controlled Repulsive Type Magnetic Bearing System with Surface Permanent Magnets Installed and Its Levitation and Rotation Tests," *IEEE Transactions on Magnetics*, vol. 47, no. 12, pp. 4734-4739, 2011.

[5] Y. Takada, T. Kimura and T. Nakagawa, "Influence of Inductance Properties on a Magnetic Levitation for Thin-Steel Plates," *IEEE Transactions on Magnetics*, vol. 53, no. 11, pp. 2500705, 2017

[6] K. Ota, Y. Arita and T. Nakagawa, "Noncontact conveyance system in synchronization with a thin steel plate using feedforward correction," *43rd Annual Conference of the IEEE Industrial Electronics Society*, pp. 3063-3068, 2017.

[7] H. Yonezawa, H. Marumori, T. Narita, S. Hasegawa and Y. Oshinoya, "Bending Magnetic Levitation Control for Thin Steel Plate (Experimental Consideration Using Sliding Mode Control)," *2014 International Power Electronics Conference*, pp. 3055-3060, 2014.

[8] M. Tada, H. Yonezawa, H, Marumori, T. Narita and H. Kato, "Integrated Control of Bending Levitation for Flexible Steel Plate Using Sliding Mode Control," *Journal of JSAEM*, vol. 25, no. 2, pp.82-87, 2017, (in Japanese).

[9] H. Kato, H. Marumori, H. Yonezawa and T. Narita, "Vibration Suppression Effect in a Bending-Levitated Flexible Steel Plate by the Electromagnetic Force," *ASME Journal of Vibration and Acoustics*, vol. 138, no. 5, pp. VIB-15-1159, 2016.

Performance of Hybrid Magnetic Levitation Control System for Thin Steel Plate by EMs and PMs: Experimental Evaluation of Applying Optimal Gap and Arrangement of PMs

Yasuaki Ito[1*], Yoshiho Oda[1], Kengo Okuno[1], Toshiki Suzuki[1], Masahiro Kida[1],
Takayoshi Narita[2*], Hideaki Kato[2] and Hiroyuki Moriyama[2]
1 Course of Mech. Eng., Tokai University, 4-1-1 Kitakaname, Hiratsuka-shi, 259-1292, Japan
2 Dep. of Prime Mover Eng., Tokai University, 4-1-1 Kitakaname, Hiratsuka-shi, 259-1292, Japan
*narita@tsc.u-tokai.ac.jp

Abstract- **In the production line of thin steel plates which are widely used in various industrial products, steel plates are generally conveyed while in contact with rollers in the conveyance process. This leads to deterioration in the quality of the plate surface, such as flaws and peeling of the plated layer. As a solution, we have proposed a hybrid levitation control system for thin steel plate using the static magnetic force generated by permanent magnets, which have no operational costs. We investigated the levitation stability when changing the distance between the permanent magnet arrangement and the steel plate using a 0.24 mm thin steel plate. We investigated the levitation stability when 0.24 mm thin steel plate was used to change the distance, arrangement and number of permanent magnets and steel plate. We confirmed that they affect the levitation stability.**

Keywords— steel plate, magnetic levitation, genetic algorithm, permanent magnet.

I. INTRODUCTION

In the production line of thin steel plates which are widely used in various industrial products, steel plates are generally conveyed while in contact with rollers in the conveyance process. The contact leads to deterioration in the quality of the plate surface, such as flaws and peeling of the plated layer. As countermeasures to these problems, research on noncontact conveyance using magnetic levitation technology has been actively carried out recently [1,2]. The researchers in our laboratory have been involved in examining noncontact magnetic levitation control of the conveyance of a rectangular thin steel plate. Furthermore, we have proposed to use static attractive force of permanent magnet for levitation supplement. Though we considered optimal arrangement for stable levitation, it is hard to find optimal arrangement form a number of arrangement patterns of permanent magnets. Therefore, we applied genetic algorithm (GA) in search method for the optimal arrangement of permanent magnets considering the interactions that the magnets had with each other. We can improve the levitation stability using the obtained optimal arrangement of permanent magnets[3-5]. However, the levitation stability of the magnetically levitated steel plate using the optimal arrangement of the

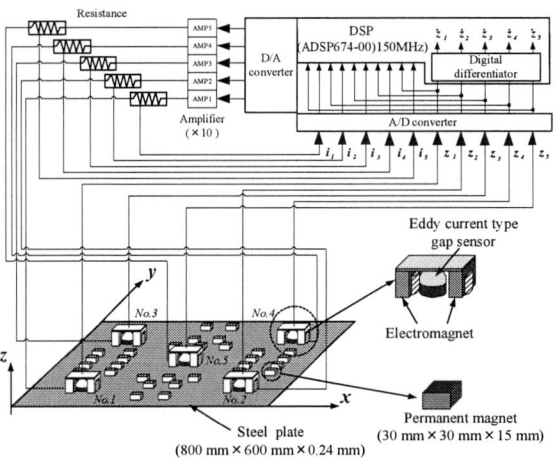

Fig. 1. Schematic illustration of the hybrid levitation system for thin steel plate using electromagnets and permanent magnets.

permanent magnet has not been sufficiently investigated. In this report, we report on the levitation stability of 0.24 mm thin steel plate when changing the arrangement of the permanent magnets and the distance between the steel plates.

II. NOMENCLATURE

The definitions of symbols used in this paper are as follows:

E : Young's modulus of the thin steel plate [N/m²]
f : vertical static magnetic force applied to the plate which is generated by the permanent magnets [N/m²]
f_z : *dynamic magnetic force [N]*
F_z : *magnetic force of the coupled magnets in the equilibrium state [N]*
g : acceleration due to gravity [m/s²]
h : plate thickness [m]
i_z : dynamic current of the coupled electromagnets [A]

The 2018 International Power Electronics Conference

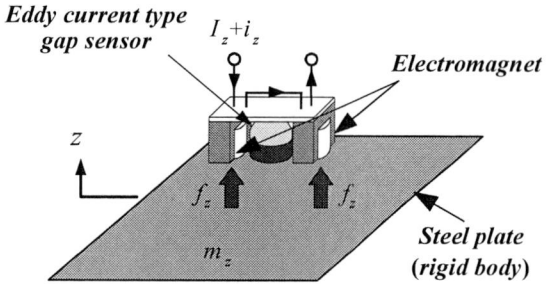

Fig. 2. Modeling of levitation control for one electromagnet unit by lumped parameter system.

I_z : current of the coupled electromagnets in the equilibrium state [A]

J_D : evaluation function of the maximum deflection [m]

J_z : evaluation function of the average absolute deflection [m]

L_{lea}: leakage inductance of the one magnet coil [H]

L_z : inductance of one electromagnet coil in the equilibrium state [H]

m_z : virtually divided steel plate [kg]

N : the total number of analysis points [-]

R_z : resistance of the coupled magnet coils [Ω]

T_s : sampling time [s]

v : Poisson ratio [-]

v_z : dynamic voltage of the coupled magnets [V]

x : coordinates in the width direction [m]

y : coordinates in the longitudinal direction [m]

z : vertical displacement from the equilibrium state [m]

z_i : displacement at each analysis point on the thin steel plate [m]

z_{max} : maximum deflection of the thin steel plate [m]

Z_0 : gap between the steel plate and electromagnet in the equilibrium state [m]

ρ : plate density [kg/m³]

III. HYBRID ELECTROMAGNETIC LEVITATION SYSTEM

A. Outline of System

A schematic illustration of the hybrid levitation system for thin steel plate using electromagnets and permanent magnets is shown in Fig. 1. The levitation object is a zinc-coated rectangle steel plate whose length is 800 mm, width is 600 mm and thickness is 0.24 mm. The hybrid levitation system consists of 5 electromagnet units which generates the attractive force for levitation control and several permanent magnets to generate the static attractive force for levitation supplement. The electromagnet unit has a pair of electromagnet and a eddy-current type noncontact displacement sensor is installed in the center of the pair of electromagnet. 4 electromagnet units are installed at near the corner of the steel plate (Nos. 1-4 in Fig. 1) and the other (No. 5 in Fig. 1) is installed at center of the steel plate. The pair of electromagnet is electrical connected in series.

The electromagnet is controlled to maintain a gap which is distance between the surface of the steel plate and eddy-current displacement sensor to be 5 mm. To levitate the steel plate, an output voltage is calculated considering the displacement of the steel plate, velocity and the current of the electromagnet by digital signal processor (DSP). The velocity is calculated from the displacement measured by the eddy-current type gap sensor using digital differentiation in DSP. The current is obtained from measured voltage of external resistance. Measured valued are input into DSP via an A/D converter. The calculated voltage by DSP is output via a D/A converter, and amplified by power supply amplifier to drive electromagnets.

Furthermore, permanent magnets are installed around electromagnet to support levitation control. In this paper, the material of permanent magnet is ferrite and size is 30 mm × 30 mm × 15 mm. The surface flux density of the permanent magnet is 0.12 T.

B. Modeling of Hybrid Electromagnetic Levitation System for Thin Steel Plate

A mathematical control model is formulated to construct the hybrid magnetic levitation control system. In this paper, each electromagnet unit is controlled by considering the displacement from gap sensor installed in itself, the velocity and the current of the electromagnet unit. Therefore, the control model for levitation is established in each unit. Figure 2 shows modeling of levitation control for one electromagnet unit by lumped parameter system. The steel plate is virtually divided into 5 hypothetical masses. In an equilibrium levitation state, the electromagnet need to generate a constant attractive force which is equal to weight of the virtually divided steel plate, and the change of the attractive force from the equilibrium state causes the motion of the virtually divided steel plate. The vertical motion of the steel plate is expressed as

$$m_z \ddot{z} = 2 f_z \qquad (1)$$

The material of the electromagnet is ferrite and the shape and size is E-type as shown in Fig. 3. The cross-sectional area of the electromagnet core where the magnetic flux of the central convex portion passes is 225 mm². Winding an enameled wire with a diameter of 0.5 mm is turned over the central protrusion 1005 times to make an electromagnet. The characteristics of the electromagnet are estimated on the basis of the following three assumptions. At first, the permeability of the core is infinite. Secondly, the eddy current inside the core is negligible. Finally, the inductance of the electromagnetic coil is expressed as the sum of the component inversely proportional to the gap between the steel plate and magnet and the component of leakage inductance. If deviation from the static equilibrium state is very small, based on the assumption, the characteristic equations of the electromagnet are linearized as

1587

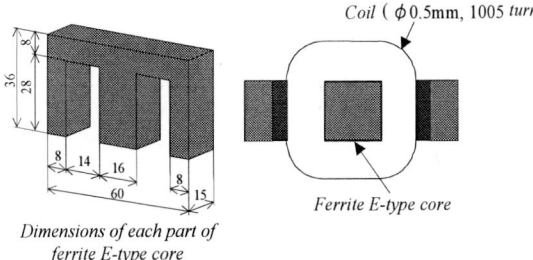

Fig. 3. Specifications of electromagnet.

$$f_z = \frac{2F_z}{Z_0} z + \frac{2F_z}{I_z} i_z \tag{2}$$

$$\frac{d}{dt} i_z = -\frac{L_{eff}}{L_z} \cdot \frac{I_z}{Z_0^2} \dot{z} - \frac{R_z}{2L_z} i_z + \frac{1}{2L_z} v_z \tag{3}$$

$$L_z = \frac{L_{eff}}{Z_0} + L_{lea} \tag{4}$$

Using the state vector, the eq. (1) - (4) are written as the following state equations:

$$\dot{z} = A_z z + B_z v_z \tag{5}$$

$$z = \begin{bmatrix} z & \dot{z} & i_z \end{bmatrix}^{\mathrm{T}}$$

$$A_z = \begin{bmatrix} 0 & 1 & 0 \\ \dfrac{2F_z}{m_z Z_0} & 0 & \dfrac{2F_z}{m_z I_z} \\ 0 & -\dfrac{L_{eff}}{L_z} \cdot \dfrac{I_z}{Z_0^2} & -\dfrac{R_z}{2L_z} \end{bmatrix}$$

$$B_z = \begin{bmatrix} 0 & 0 & \dfrac{1}{2L_z} \end{bmatrix}^{\mathrm{T}}$$

TABLE I

THE VALUE OF EACH PARAMETER

Parameters	Values
m	0.864 kg
E	206 GPa
v	0.3
Z_0	5×10^{-3} m
R_n	21.0 Ω
L_{eff}	2.55×10^{-4} Hm
L_{lea}	0.090 H
T_s	0.001 s

TABLE II

THE FEEDBACK GAIN OF LEVITATION CONTROL SYSTEM

Parameters	Values (Nos. 1-4 in Fig. 1)	Values (No. 5 in Fig. 1)
f_z	1.41×10^4	1.52×10^4
f_v	1.37×10^2	1.55×10^2
f_i	0.444×10^2	0.443×10^2

Furthermore, v_z is calculated by state feedback control of eq. (6).

$$v_z = -F_z x \tag{6}$$

$$F_z = \begin{bmatrix} f_z & f_v & f_i \end{bmatrix}$$

F_z is feedback gain of this control system and obtained by trial and error. Parameters used in control system are shown in Table 1 and 2.

To implement these electromagnetic levitation system in the DSP, it is necessary to construct the control model in discrete time system. Therefore, the state equations (eq. (5)) is transformed as

$$z_d(i+1) = \boldsymbol{\Phi} z_d(i) + \boldsymbol{\Gamma} v_d(i) \tag{7}$$

$$\boldsymbol{\Phi} = \exp(A_z T_s)$$

$$\boldsymbol{\Gamma} = \left\{ \int_0^{T_s} \left[\exp(A_z \tau) \right] d\tau \right\} B_z$$

IV. DETERMINATION OF OPTIMAL ARRANGEMENT

A. Search Method of Optimal Arrangement

In previous study, to placement of the permanent magnets suppresses the levitated steel plate and it is confirmed that arrangement of permanent magnets to decrease static deflection of the levitated steel plate improve levitation stability. To obtain the shape of the steel plate levitated by electromagnets and permanent magnets, the equation of the static deflection of the steel plate need to solve. The equations for the static deflection of the rectangular thin plate are expressed as

$$D \nabla^4 z = f - \rho h g \tag{8}$$

$$D = \frac{E h^3}{12(1 - v^2)}, \quad \nabla^4 = \frac{\partial^4}{\partial x^4} + 2 \frac{\partial^4}{\partial x^2 \partial y^2} + \frac{\partial^4}{\partial y^4}$$

To solve mathematically this equation is complex. Therefore, the deflection of the steel plate is numerical obtained using finite difference method. The calculated mesh size of the steel plate is 10 mm × 10 mm.

1588

The 2018 International Power Electronics Conference

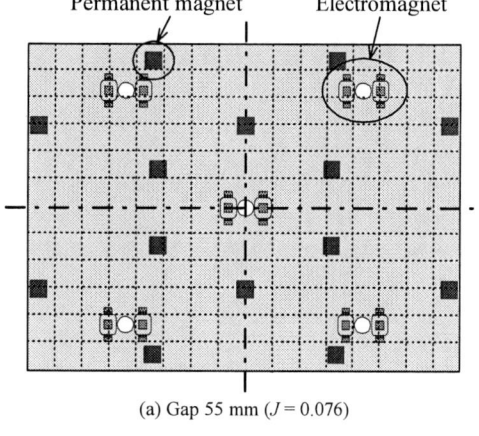

(a) Gap 55 mm ($J = 0.076$)

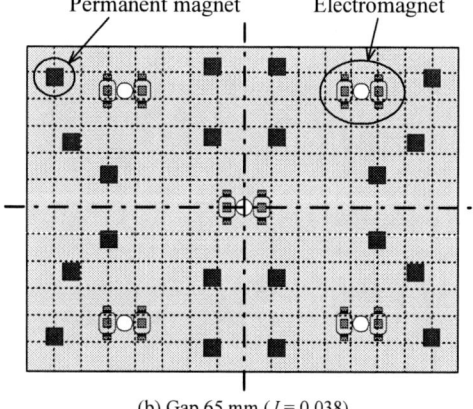

(b) Gap 65 mm ($J = 0.038$)

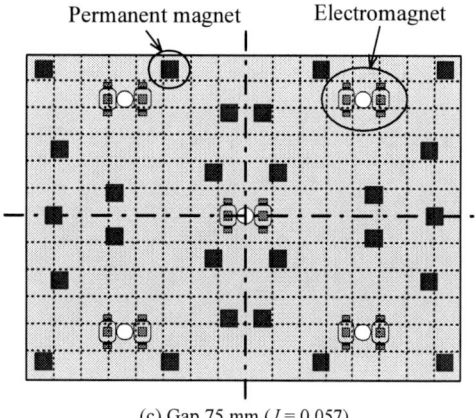

(c) Gap 75 mm ($J = 0.057$)

Fig. 4. Optimal arrangement of permanent magnets in each gap.
(plate thickness = 0.24 mm)

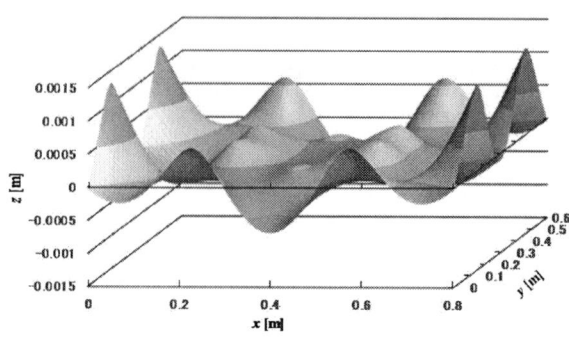

(a) Gap 55 mm ($J = 0.076$)

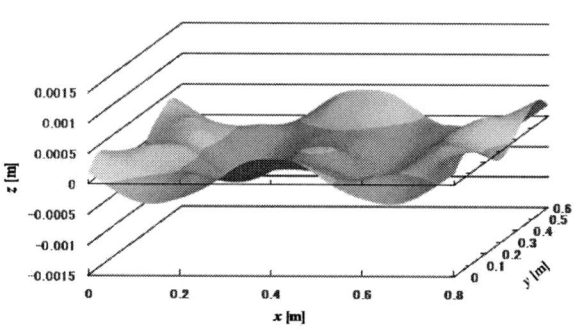

(b) Gap 65 mm ($J = 0.038$)

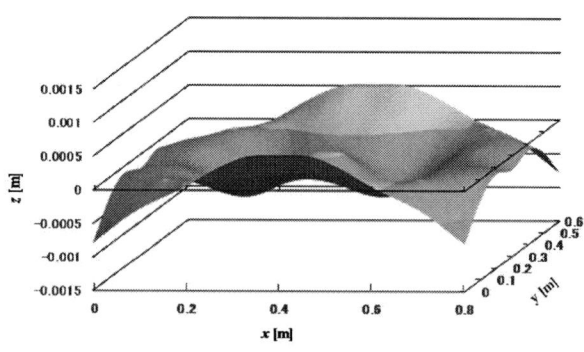

(c) Gap 75 mm ($J = 0.057$)

Fig. 5. The shape of steel plate as bird's eye view in the case of optimal arrangement in each gap.
(plate thickness = 0.24 mm)

In this study, the optimal number, arrangement and gap which is the distance from the steel plate to the surface of the permanent magnet are searched to reduce effectively deflection of the steel plate for optimization of the hybrid electromagnetic levitation system with considering the interaction. However, it is practically impossible to search for the optimal arrangement of the permanent magnets by experiment because combinations of search patterns are very numerous. Therefore, the optimal parameter is sought using a GA, as an optimization algorithm[4, 5]. In this search, the maximum number of permanent magnets was set to 15 in the area of 1/4 of the steel plate.

1589

B. Evaluation Function of GA

In order to obtain the optimum arrangement for suppressing the deflection of the steel plate using GA, a function for quantitatively evaluating how much the deflection was suppressed as compared with the shape of the steel plate which the permanent magnet is not installed due to the installation of the permanent magnet is defined.

The evaluation function J is defined as eq. (9),

$$J = \frac{J_Z}{J_{Z0}} \times w_Z + \frac{J_D}{J_{D0}} \times w_D \tag{9}$$

$$J_Z = \frac{\sum_{i=1}^{N} |z_i|}{N}, \ J_D = |z_{max}|, \ w_Z + w_D = 1$$

As reference evaluation values, J_{Z0} is the mean deflection and J_{D0} is the maximum deflection of the thin steel plate when no permanent magnets are used, and each term in eq. (9) is nondimensionalized. J_Z/J_{Z0} of eq. (9) in clause one is a ratio to which a permanent magnet sets up and the total deflection decreases, it is shown that the thin steel plate shape was improved by the value of J_Z/J_{Z0} small. J_D/J_{D0} of eq. (9) in clause two is a ratio to which a permanent magnet sets up and the regional deflection decreases, it is shown that the thin steel plate shape was improved by the value of J_D/J_{D0} small. w_Z and w_D are the weight coefficients for the mean deflection and maximum deflection of the thin steel plate, respectively. $w_Z = 0.5$ and $w_D = 0.5$, for which deflection is expected to be most greatly suppressed from the results of simulations, were adopted. The values of J are obtained at each gap, and we assume the gap of minimum value of J the optimal gap.

C. Result of Optimization Using GA

Figure 4 shows the result of optimal arrangement of permanent magnets. Figure 5 shows the plate shape of steel plate when optimal arrangement of permanent magnets is used. Figure 4, 5 (a) – (c) show results of gap 55 mm, gap 65 mm and gap 75 mm. Evaluation values J in gap 55 mm, 65 mm and 75 mm are 0.076, 0.038 and 0.057. These result shows the optimal arrangement of gap 65mm can suppress the deflection of the steel plate.

V. LEVITATION EXPERIMENT

To evaluate the optimal arrangement of gap 55 mm, gap 65 mm and gap 75 mm obtained by a GA, we carried out a levitation experiment of steel plate thickness 0.24 mm. In each of the above cases, after the thin steel plate is magnetically levitated, the y-axis direction elastic first order mode of the thin steel plate which is most susceptible to vibration is considered. A sinusoidal wave of 3.49 Hz, which is its natural frequency of 1st elastic mode in y axis, was input as a disturbance to the electromagnet unit No. 5 (see Fig. 1) in Fig. 1 to elastically vibrate the thin steel plate. The levitation experiment was conducted 50 times with gap 55 mm, gap 65 mm and gap 75 mm, respectively. Success if it keeps levitation for 30 seconds is successful,

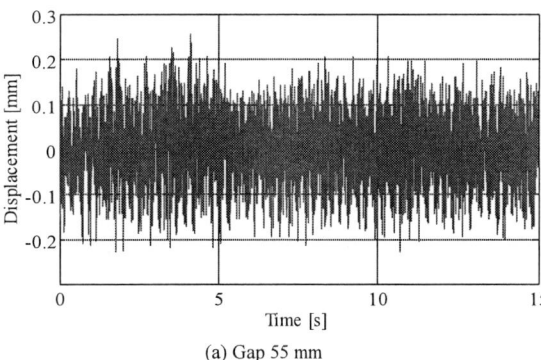

(a) Gap 55 mm

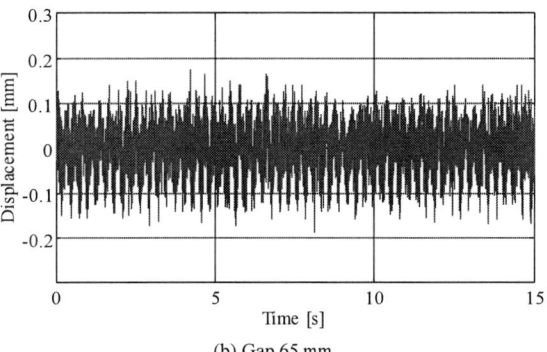

(b) Gap 65 mm

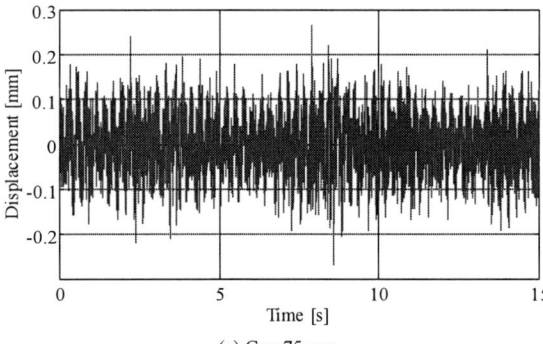

(c) Gap 75 mm

Fig. 6. Time histories of displacement of the steel plate in the case of optimal arrangement in each gap.

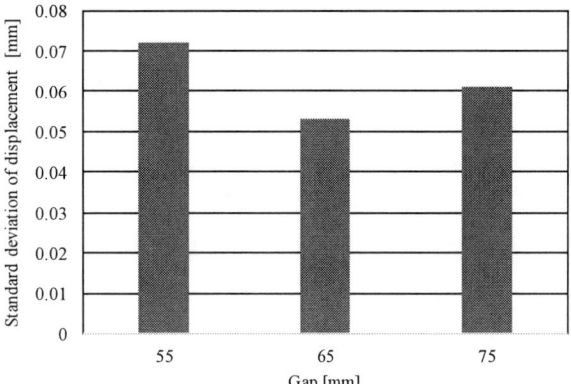

Fig. 7. Standard deviation of displacement in the case of optimal arrangement in each gap.

1590

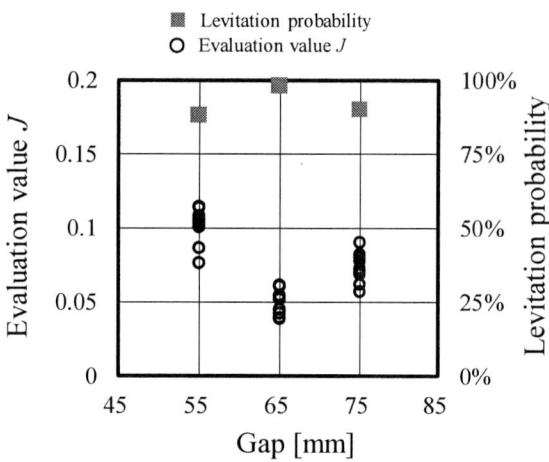

Fig. 8. Evaluation value and levitation probability for each gap.

it can't be lifted for 30 seconds, or it makes contact with the electromagnet and fails. Figure 6 and 7 shows the time history of displacement and the standard deviation of displacement in each condition. gap 65 has a displacement standard deviation of about 26% as compared with gap 55, Compared to gap 75, the displacement standard deviation was reduced by about 13%.

Figure 8 shows the relationship between J, levitation probability and the gap obtained by searching for the optimal positions. ■ plots in Fig. 8 mean the levitation probability. ● plots are the evaluation value J. It was confirmed that the standard deviation of displacement also decreases as the evaluation value J decreases. It is also confirmed that the levitation performance improves because the standard deviation of displacement decreases and the levitation probability increases.

VI. CONCLUSION

In this study, we investigated the levitation stability when changing the distance between the permanent magnet arrangement and the steel plate using a 0.24 mm thin steel plate. It was found that the levitation stability of the steel plate is affected by the optimal arrangement of each gap. In the future, we will also conduct experiments in other gaps and search for the optimal gap.

REFERENCES

[1] H. Marumori, H. Yonezawa, T. Narita, S. Hasegawa and Y .Oshinoya, Bending Levitation Control for Flexible Steel Plate —Fundamental Study on Elastic Vibration Control— , *Journal of the Japan Society of Applied Electromagnetics*, vol. 23, no.2, pp. 138-144, 2015.

[2] Y. Maruyama, T. Mizuno, M. Takasaki and Y. Ishino, Development of 3-DOF Active Control Type Magnetically Suspended Gyro —Mechanical System—, *The Japan Society of Mechanical Engineers Series C*, vol. 76, no. 766, pp. 1445-1451, 2010.

[3] Y. Oshinoya and K. Ishibashi, Development of Electromagnetic Levitation Control Device for a Rectangular Sheet Steel, *The Japan Society of Mechanical Engineers Series C*, vol. 67, no. 661, pp. 2855–2862, 2001.

[4] T. Narita, S. Hasegawa and Y. Oshinoya, Hybrid Electromagnetic Levitation System for Thin Steel Plates Using Permanent Magnets, *Journal of the Magnetics Society of Japan*, vol. 37, no. 2, pp. 29-34, 2013.

[5] H. Ishii, T. Narita and H. Kato, Hybrid Magnetic Levitation System for Thin Steel Plate by Electromagnets and Permanent Magnets —Basic Study on Optimal Placement Search Considering the Interaction of the Magnetic Field—, *Journal of the Japan Society of Applied Electromagnetics and Mechanics*, vol. 24, no. 3, pp. 149-154, 2016.

A Practical Lithium-ion Battery Model Based on the Butler-Volmer Equation

Kaiyuan Li[1], King Jet Tseng[2*], Feng Wei[2], Boon-Hee Soong[3]

1 Energy Research Institute @ NTU, Nanyang Technological University, Singapore
2 Electrical Power Engineering, Singapore Institute of Technology, Singapore
3 School of Electrical and Electronic Engineering, Nanyang Technological University, Singapore
*E-mail: KingJet.Tseng@SingaporeTech.edu.sg

Abstract-For Li-ion batteries based energy storage devices, the residual runtime is one of the most important indexes for the energy management and optimization in smart grid and electric vehicle applications. In this present study, the residual runtime of batteries is indicated by the residual available energy and the index of State of Energy (SOE) instead of the traditional State of Charge (SOC). To improve the accuracy of SOE estimation, a novel practical Li-ion battery model is built in this work against the dynamic load and the non-linear battery effects. To acquire advantages of both models, a Butler-Volmer equation-based electrochemical model is merged with an electrical model based on principles of electrochemistry from the discharge curve of the battery. Commercial Li-ion batteries are tested at single-cell level under dynamic loads for model validation. Simulation results verify the high precision and robustness of the proposed battery model.

I. INTRODUCTION

Energy storage technology is experiencing a booming development. Among all the available technologies, Li-ion battery has become the most promising solution, which is capable of meeting the application requirement imposed by smart grids and electric vehicles [1, 2]. This is due to their outstanding performance such as high volumetric and gravimetric energy density, prolonged cycle life, low self-discharge rate and enhanced safety [3].

Battery management system (BMS) has become a vital component in order to assure reliable, efficient, durable, and safe operations of Li-ion battery based energy storage systems (ESSs). Nowadays, modern BMSs rely on battery models to track key indexes of each cell and further the whole system. This requires the model to accurately capture the battery behavior under dynamic loads. As a critical parameter, the state of energy (SOE) has been replacing the traditional state of charge (SOC) to describe the residual available energy and runtime of batteries, on which the battery-based ESSs are dependent [4]. Thusly, proper management and operation of battery-based power systems/components require battery models able to predict the battery SOE under dynamic loads accurately.

In the literature, plenty of Li-ion battery models has been published. Generally, these models can be classified into three categories: electrochemical model, electrical model, and analytical model [5]. The electrochemical models are of the highest accuracy and are mainly derived from the electrochemical theory proposed by Doyle et al. [6]. However, thorough knowledge of the electrochemical processes in the battery is required to establish these models, causing them very complicated and computationally time-consuming. Electrical battery models use the equivalent circuit to simulate the electrical behaviors of the battery [5, 7]. Nevertheless, the existing electrical battery models seldom integrate the non-linear battery behavior such as the rate energy effects and recovery effects, thus, their accuracy is reduced especially under dynamic loads. Electrical models could capture the terminal voltage responses of the battery for co-design and co-operation with other devices in the system, but they are not feasible to be operated alone for BMS usage of real-time performance prediction under dynamic loads. Analytical models are generally the electrochemical models with reduced orders [8], where the intuitive kinetic battery model can capture the non-linear battery effects with simplified differential equations. However, these models cannot track the circuit characteristics of the battery, which is vital for co-simulation in the systems.

More comprehensively, several hybrid battery models have been introduced in the literature for higher estimation accuracy. Kim et al. [8] utilized a kinetic model to represent the non-linear battery effects and built a hybrid model composed of a kinetic model and an electrical model. However, this model did not consider the change of the model parameters due to load variations. Zhang et al. [9] presented a combination of the Rakhmatov diffusion model and an electrical model. However, this combined model was had to configure due to the complex structures. Juang et al. [10] proposed a Butler-Volmer (BV)-equation-based electrical battery model, and it is sufficient to capture the non-linear battery behaviors under dynamic loads. Unfortunately, applications of the proposed identification method were improper for Li-ion batteries because the voltage responses of the discharge current could hardly reach equilibrium, although this model exhibited promising results for lead-acid battery [11]. Liu et al. [11] employed BV equation on high-power Li-ion batteries and set up a battery model to capture the terminal voltage of batteries under varying loads. An empirical look-up table for SOE estimation was merged with an electrical battery model in [12]; however, popularization of this approach was deficient due to lack of theoretical derivations.

In conclusion, all the models in [5-12] have at least one of the following demerits: (1) complicated and intensive computational requirements; (2) not capable of involving the battery non-linear effects, and (3) inapplicable for integration with other electrical devices. Therefore, there is a desire to develop a comprehensive Li-ion battery model capable of predicting the battery SOE and runtime as well as capturing the battery voltage characteristics accurately under dynamic load conditions.

In this study, a hybrid Li-ion battery model based on the BV equation is presented to fulfill the demerits. The model combines an electrochemical model and an electrical model, and it is capable of predicting the SOE and runtime based on principles of electrochemistry from the discharge curve of the battery. In Section II, brief introduction of SOE and BV equation, and the proposed battery model is presented. The model extraction procedures and results are presented in Section III. The testing results of Li-ion batteries are compared with the simulation results of the proposed model in Section IV for model validation. Finally, Section V the conclusion and future work is addressed.

II. LI-ION BATTERY MODELING

A. State of Energy

Analogous to SOC, SOE described information of the battery residual energy. Only the SOE during discharging was investigated in this work, and it was calculated as [13-16]:

$$SOE(t) = SOE_0 - \frac{1}{E_{max}} \int_{t_0}^{t} P(\tau) d(\tau) \qquad (1)$$

where SOE(t) represented the SOE at time t similarly hereinafter, SOE_0 was the initial SOE; E_{max} denoted the total available energy inside the Li-ion battery, and $P(\tau)$ indicated the discharge power. When batteries operated in ESSs, one of the most straightforward parameters to monitor and measure was the operating load power. For instance, in application of smoothing the solar products, batteries were operated in the manner of providing/absorbing electrical power in accordance with the demand-supply profile. Actually, the operating power $p(t)$ could be considered as the discharge power of batteries. In the current study, the operating power rate was taken as the direct input index of control in the proposed battery model. Detailed study of the index of SOE and the influencing factors on SOE were referred to our previous work.

B. Butler-Volmer Equation

As presented in [5] and [7], the traditional electric battery models were insufficient to accurately capture the dynamic responses of the batteries due to lacking adequate representation of the electrochemical battery reactions. To address this problem, BV equation was introduced in this study to describe the battery electrochemical reactions and to characterize the over-potential phenomena.

Battery polarization was mainly composed of two components, which were the electrochemical polarization and the concentration polarization. The electrochemical polarization appeared once the electric field inside the battery was built, while the concentration polarization required a longer time to be built, resulting from the lithium ions transferring in the electrolyte when it was dominated by the molecule diffusion. In this work we took use of the conclusion drawn in [11], i.e., it was feasible to regard the short time constant polarization as electrochemical polarization since molecule diffusion were more tendency to build polarization with larger time constants, and the electrochemical polarization effects for power Li-ion batteries was much more significant than the diffusion effects. Therefore, in this work, we neglected the molecule diffusion effects inside the battery and focus on the electrochemical polarization during the model development. Additionally, according to the test results of [12], batteries used in this study had hysteresis effects less than 6mV thus the hysteresis effects were neglected without any noticeable influence on model accuracy. Consequently, the Thevenin battery model was chosen as the reference of the proposed battery model since it simplified the model configuration while remained sufficient precision.

With a thorough understanding of the ion transfer kinetics inside the battery, the ideal overpotential $\eta(t)$ under steady state could be expressed based on the BV equation as:

$$J(t) = J_0 \left[\exp\left(\frac{\alpha e}{K_B T}\eta(t)\right) - \exp\left(\frac{-(1-\alpha)e}{K_B T}\eta(t)\right) \right] \qquad (2)$$

where $J(t)$ and J_0 indicated the current density and the exchange current density. α denoted the transfer coefficient which was assumed to be a constant of 0.5 [11]. Parameters of e, K_B and T were the charge per electron, the Boltzmann constant and the temperature in Kelvin, respectively.

Taken into consideration that the current density was hard to be measured externally, the BV equation was rewritten as:

$$\frac{1}{J_0 \times S \times V(t)} P(t) = \left[\exp\left(\frac{\alpha e}{K_B T}\eta(t)\right) - \exp\left(\frac{-(1-\alpha)e}{K_B T}\eta(t)\right) \right] \qquad (3)$$

since

$$P(t) = J(t) \times S \times V(t) \qquad (4)$$

where S was the effective area and $v(t)$ was the battery terminal voltage. Assuming that the value of α hardly made a sudden change and basically equaled to 0.5, the overpotential $\eta(t)$ caused by the power variation could be solved as:

$$\eta(t) = \frac{2K_B T}{e} \ln \left(\frac{1}{2J_0 SV(t)} P(t) + \sqrt{\left(\frac{1}{2J_0 SV(t)} P(t)\right)^2 + 1} \right) \qquad (5)$$

Based on (5), we could acquire the specific change of the overpotential at a particular SOE, which was induced only by the dynamic operating power. When the discharge power increased, there was a sharp drop of the battery terminal voltage, which indicated that the over-potential deviated from the equilibrium state; however, when the battery was at equilibrium, the overpotential would become steady and approach to equilibrium after some rest time.

The 2018 International Power Electronics Conference

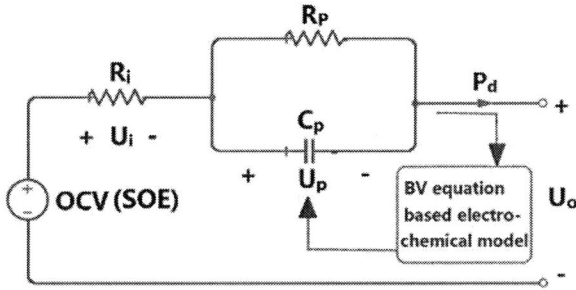

Fig. 1: Schematic of the proposed battery model.

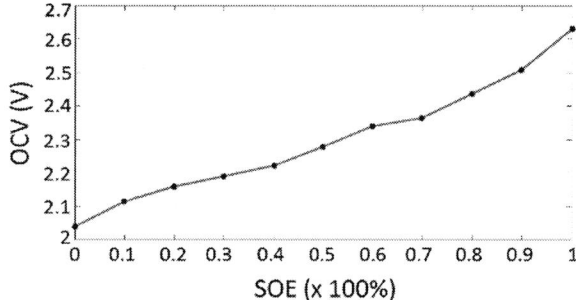

Fig. 2: Extracted OCV-SOE characteristics of Li-ion battery.

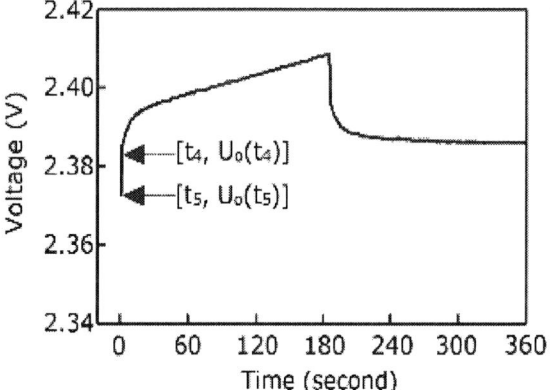

Fig. 3: Voltage response.

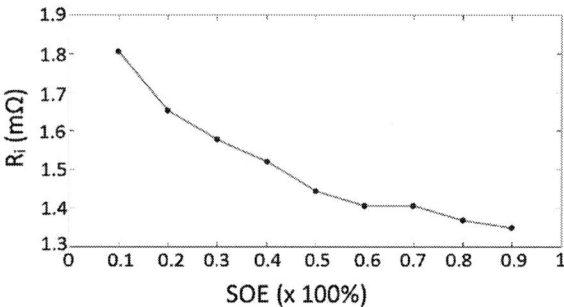

Fig. 4: R_i values at various SOE.

C. Proposed Li-ion Battery Model

A test bench had been built in the laboratory and commercial lithium titanate (LTO) battery (rated capacity of 20.0 Ah and rated voltage of 2.3 V) was utilized. The cut-off voltage threshold for the battery was 2.7 V for charging and 1.5 V for discharging. The proposed battery model was based on Thevenin model and embedded with a BV equation-based electrochemical model to capture the battery overpotential, as shown in Fig. 1. With the help of the 2 types of Li-ion battery models, the proposed model was capable of characterizing comprehensive battery behaviors and estimating the SOE under dynamic load conditions.

According to [17], assumed a battery discharge case with a power of P_d during $t_1 < t < t_2$, following relaxation period $t_2 < t < t_3$, then the model parameter was expressed as:

$$U_P = \begin{cases} \frac{R_P \, P_d(t)}{U_o} \cdot \left[1 - e^{-\frac{t-t_1}{\tau}}\right], & t_1 < t < t_2 \\ U_P(t_2)\left(e^{-\frac{t-t_2}{\tau}}\right), & t_2 < t < t_3 \end{cases} \quad (6)$$

$$U_O(t) = OCV(t) - U_P(t) - R_i \, P_d/U_o(t) \quad (7)$$

where $\tau = R_p \, C_p$. SOE reduction due to discharge power could be calculated from (1), and the initial SOE was obtained by the OCV-SOE characteristic curves. Based on (6), battery polarization could be separated into 2 components: polarization was established when there was a non-zero load, and it started to fall back to zero during battery relaxation. All the model parameters, OCV, R_i, R_p and C_p, were functions of SOE and the load power P_d, according to [17]. Parameter extraction method and results were detailed in Section III.

III. MODEL EXTRACTION

A. OCV-SOE Characteristics

OCV, as an indispensable component, represented the battery voltage when it had been rested for hours and at equilibrium. In this study, method in [19] was utilized to acquire the OCV-SOE characteristic, and that of 25°C was illustrated in Fig. 2.

B. Internal Resistance R_i

The internal resistance of the battery, R_i, representing the lumped resistance of the electrolyte, the electrodes, and other components inside battery, was responsible for the generation of internal energy loss and thermal heat of the battery. Fig. 3 illustrated a typical battery voltage response under pulse power test, where the terminal voltage rose immediately and continued to grow exponentially. The internal resistance of the battery was determined by (8). In Fig. 3, the power pulse test began at t_5, and t_4 was the time for the next sampling points, specifically, $t_4 = t_5 + \Delta t$, where Δt equaled to 1 second.

$$R_i = \frac{U_0(t_4) - U_0(t_5)}{I_{pulse}} = \frac{U_0(t_5 + \Delta t) - U_0(t_5)}{I_{pulse}} \quad (8)$$

The values of R_i were dependent on SOE and the load power rates; the detailed extraction results were referred to our

1594

The 2018 International Power Electronics Conference

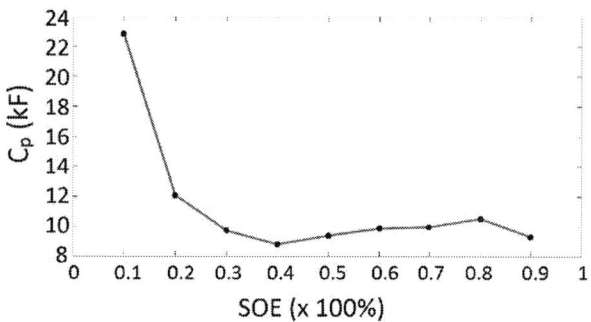

Fig. 5: R_p values at various SOE.

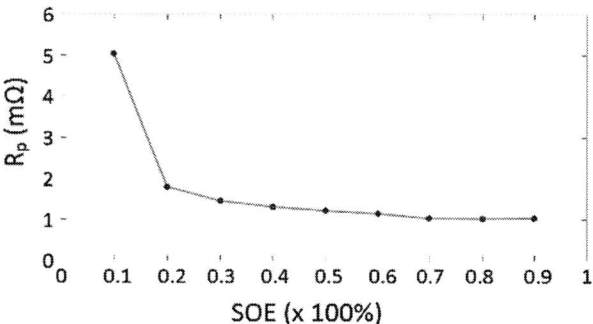

Fig. 6: C_p values at various SOE.

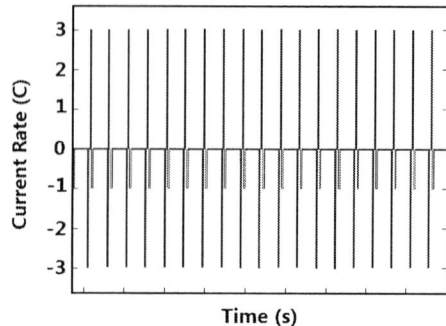

Fig. 7: Current profile for the model verification test.

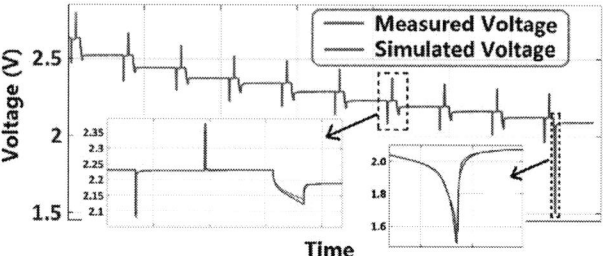

Fig. 8: Comparison of the measured and simulated voltage results.

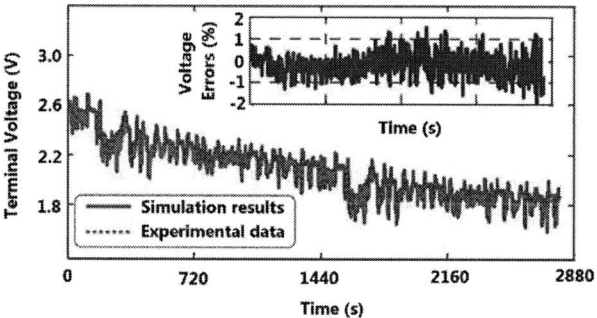

Fig. 9: Comparison of voltage results under FUDS.

previous study and were not provided here. R_i values used in the BV equation based models were illustrated in Fig. 4.

C. Polarization Components R_p and C_p

The battery polarization components was heavily dependent on the load power variation. Based on (5) and (6), battery polarization at a certain SOE could be described as:

$$U_p = \frac{R_P\,P_d(t)}{U_o} \cdot \left[1 - e^{-\frac{t-t_1}{\tau}}\right]$$

$$= \frac{2K_BT}{e}\ln\left(\frac{1}{2J_0SV(t)}P(t) + \sqrt{\left(\frac{1}{2J_0SV(t)}P(t)\right)^2 + 1}\right) \quad (9)$$

Nonetheless, the polarization resistance R_p during relaxation period was independent of the discharge power since no excitation was applied externally. Thus, R_p was calculated as:

$$R_P = \frac{\frac{2K_BT}{e}\ln\left(\frac{1}{2J_0SU_o(t)}P(t) + \sqrt{\left(\frac{1}{2J_0SU_o(t)}P(t)\right)^2 + 1}\right)}{\frac{P_d(t)}{U_o(t)}\left[1 - e^{-\frac{t-t_1}{\tau}}\right]} \quad (10)$$

R_p value during relaxation was constant. As to C_p, it was solely dominated by the time constant τ, and it was calculated from:

$$C_p = \frac{\tau}{R_p} \quad (11)$$

Based on (10) and (11), the values of polarization components were obtained, here the values of R_p and C_p used in the BV equation were shown in Fig. 5 and Fig. 6, separately.

Finally, all the parameters of the proposed model shown in Fig. 1 were obtained, and the values were input into Matlab.

When battery was used as power sources and discharged at dynamic load conditions, the terminal voltage response and the discharging load conditions of the battery was collectible; then the model parameters could be calculated separately from (8), (10) and (11). After the model parameters extracted, the online OCV values of the battery could be obtained, where the instantaneous SOE could be acquired according to the OCV-SOE characteristics illustrated in Fig. 2.

IV. MODEL VALIDATION

The proposed model was built in Matlab/Simulink and tests on commercial LTO battery cells were conducted for validating the efficacy of the proposed model. The measured results of an

1595

The 2018 International Power Electronics Conference

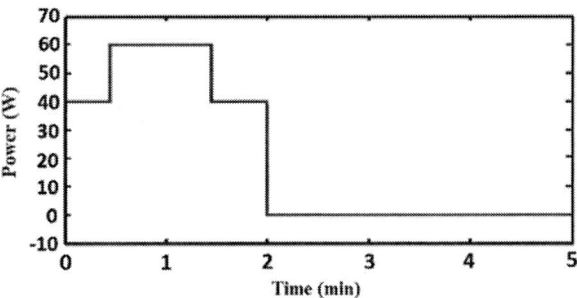

Fig. 10: Load profiles for the model validation tests.

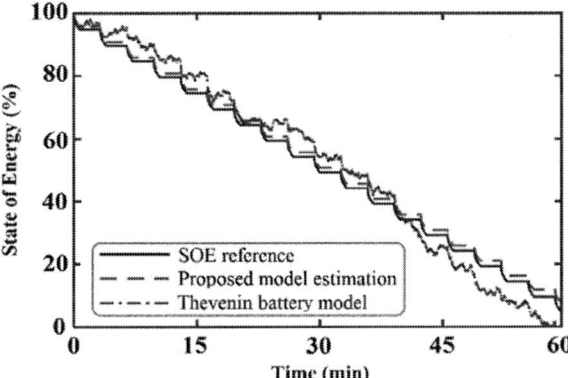

Fig. 11: Comparison of SOE estimation with the actual ones.

LTO battery cell were compared with the simulation results from the proposed model. Additionally, the SOE estimation results of the proposed model were compared with that of the Thevenin battery model so as to verify the superiority of the proposed model on SOE evaluation.

A. Verification of the terminal voltage estimation

To validate the prediction of the terminal voltage of the proposed model, a battery cell was loaded from 100% SOC at the discharge profile shown in Fig. 7 until the lower cut-off voltage was reached. The measured voltage results were compared with the simulated results of the proposed battery model. The results were illustrated in Fig. 8, where the red line was the simulated voltage and the green line represented the measured one. A close agreement between the two results could be observed, and the maximum discrepancy was at the end of discharge when the cell was nearly depleted (SOC 0-10%). The Maximum Absolute Percentage Error (MAPE) was 4.5%. In the SOC range from 95% to 10%, the Root Mean Square Percentage Error (RMSPE) was less than 2%.

To validate the prediction of terminal voltage under varying loads, a cell was loaded under the FUDS profile until 1.5V was reached. The measured terminal voltage was compared to the simulated ones, and the results were drawn in Fig. 9. The simulation results matched closely with the measured ones, with

RMSPE less than 1.5% and the MAPE was found when the battery was almost depleted.

In ESS applications, batteries would rarely be operated below 10% SOC, thus deviations in this region were of tiny importance. Some BMSs were programmed to discharge the battery no lower than 10% SOC to extend its lifespan. Even though the battery was discharged below 10%, the errors were acceptable small. Therefore, the proposed model was able to sufficiently model the battery voltage behaviors under various load conditions.

B. Verification of the battery SOE estimation

To check the effectiveness of the SOE prediction, a cell and the model were both discharged at the loads in Fig. 10 repetitively until the battery was completely exhausted. The power for discharge was indicated as positive values and negative for charge. The comparison results were shown in Fig. 11, where the SOE estimation results were compared to the actual ones. The SOE estimation results were acquired via loading the proposed model at the load profiles in Fig. 7 repetitively, and the actual SOE was acquired by power integration approach. The black lines represented the actual SOE, and the red lines depicted the SOE estimation results from the proposed model. To check the superiority of the proposed model on SOE estimation over the traditional Thenevin battery model, the estimated SOE results via Thevenin model were also indicated in Fig. 11, which was illustrated in blue. It could be observed from Fig. 11 that, when the discharge power underwent a dramatical variation, the SOE results from the Thevenin model deviated from the actual SOE values heavily, while the SOEs of the proposed model exhibited a much closer match to the actual ones. The RMSPE of the proposed model was 2.96%, whereas that of the Thevenin model was as high as 12.02%. After comparing the SOE results from two different models with the actual SOE, conclusion could be drawn that our model proposed in this study had a superior performance and was able to achieve a significant improvement compared to the Thevenin battery model when estimating the SOE of the battery, and our model was capable of predicting the SOE of Li-ion batteries with high fidelity and reliability under dynamic load conditions.

V. CONCLUSION

In this study, a novel practical battery model is presented, where a Butler-Volmer equation based electrochemical battery model is embedded with a Thevenin electrical model to realize high-fidelity terminal voltage and SOE prediction based on electrochemical principles from the voltage curves of the battery during discharge regimes. The proposed model is set-up in Matlab/Simulink, and the validation test results reveal that the proposed model is capable of achieving high-fidelity terminal voltage and SOE prediction of the battery under dynamic loads. Compared to the typical Thevenin battery model, the proposed model realizes improved accuracy on SOE estimation. In conclusion, this work offers a feasible solution for the estimation and prediction on terminal voltage, SOE and runtime of Li-ion batteries under dynamic load conditions for the realistic BMS applications. Study on the temperature effects and aging of Li-ion batteries are anticipated in the future work with experimental results.

REFERENCES

[1] B. Dunn, H. Kamath, and J.M. Tarascon, "Electrical energy storage for the grid: A battery of choices," *Science*, vol. 334, pp. 928–935, 2011.

[2] S. Vazquez, S.M. Lukic, E. Galvan, L.G. Franquelo, J.M. Carrasco, "Energy Storage Systems for Transport and Grid Applications," *IEEE Trans. on Ind. Electron.* vol. 57, no.12, pp. 3881-3895, 2010.

[3] DOE/EPRI 2013 Electricity Storage Handbook in Collaboration with NRECA, Tech. Rep. SAND 2013-5131, Sandia National Laboratories.

[4] X. Liu, J. Wu, C. Zhang, Z. Chen, "A method for state of energy estimation of lithium-ion batteries at dynamic currents and temperatures", *J. Power Sources* 270 (2014) 151-157.

[5] K. Li, K. J. Tseng, "An equivalent circuit model for state of energy estimation of lithium-ion battery", *IEEE APEC* Mar 2016, LA, US.

[6] M. Doyle, T.Fuller, J.Newman, "Modelling of galvanostatic charge and discharge of the lithium/polymer/insertion cell", *Journal of the Electrochemical society*, 140 (1993) 1526-1533.

[7] X. Hu, S. Li, H. Peng, "A comparative study of equivalent circuit models for Li-ion batteries", *J. Power Sources* 198 (2012) 359-367.

[8] T. Kim, W. Qiao, "A hybrid battery model capable of capturing dynamic circuit characteristics and nonlinear capacity effects", *IEEE Trans. Energy Convers.*, vol. 26, no. 4, pp. 1172-1180, Dec. 2011.

[9] J. Zhang, S. Ci, H. Sharif, and M. Alahmad, "An enhanced circuit-based model for single-cell battery," in *Proc. 25th Annu. IEEE APEC*, 2010, pp. 672–675.

[10] L. W. Juang, P. J. Kollmeyer, T. M. Jahns, and R. D. Lorenz, "Improved nonlinear model for electrode voltage-current relationship for more consistent online battery system identification," *IEEE Trans. Ind. Appl.*, vol. 49, no. 3, pp. 1480-1488, May/Jun. 2013.

[11] S. Liu, J. Jiang, W. Shi, Z. Ma, L.Y. Wang, and H. Guo "Butler–Volmer-equation-based electrical model for high-power lithium titanate batteries used in electric vehicles," *IEEE Trans. on Ind. Electron.* vol. 62, no.12, pp. 7557-7568, 2015.

[12] K. Li, K. J. Tseng, L. Moraleja, "Study of the influencing factors on the discharging performance of lithium-ion batteries and its index of state-of-energy," presented at *IECON2016*, Florence, Italy, 2016.

[13] C. Lin, H. Mu, R. Xiong, J. Cao, "Multi-model probabilities based state fusion estimation method of lithium-ion battery for electric vehicles: State-of-energy," *Appl. Energy* (2016).

[14] G. Dong, Z. Chen, J. Wei, C. Zhang, P. Wang, "An online model-based method for state of energy estimation of lithium-ion batteries using dual filters", J. Power Sources 301 (2016) 277-286.

[15] K. Mamadou, A. Delaille, E. Lemaire-Potteau, Prof. Y. Bultel, "New method for the energetic performances evaluation of electrochemical storage devices", ECS Trans. 2010 volume 25, issue 35, 105-112.

[16] Mamadou K, Lemaire E, Delaille A, Riu D, Hing SE, Bultel Y. "Definition of a state-of-energy indicator (SOE) for electrochemical storage devices: application for energetic availability forecasting." *J. Electrochem. Soc.* 2012 159(8): A1298-1307.

[17] K. Li, B.H. Soong, K. J. Tseng, "An Electrical Model Capable of Estimating the State of Energy for Lithium-ion Batteries used in Energy Storage Systems," presented at *IEEE SPEC 2016*, New Zealand, 2016.

[18] S. Abu-Sharkh, D. Doerffel, "Rapid test and non-linear model characterisation of solid-state lithium-ion batteries", *J. Power Sources* 130 (2004) 266-274.

Bonding technology using cold-rolled Ag sheet in die-attachment applications

Seungjun Noh[1,2*], Chanyang Choe[1,2], Chuantong Chen[2], Hao Zhang[2], and Katsuaki Suganuma[2]

[1]Department of Adaptive Machine Systems, Osaka University, Osaka, Japan
[2]The Institute of Scientific and Industrial Research, Osaka University, Osaka, Japan
*Email-Address: sjnoh@eco.sanken.osaka-u.ac.jp

Abstract—This work introduce a solid-state bonding technology, which uses cold-rolled Ag sheet as a bonding material for die-attachment applications. The process of cold rolling led to grain refinement of Ag sheet from the micron to sub-micron level, which provides a driving force for hillock formation on the Ag sheet. Cross-section observation revealed that cold rolled Ag sheets well bonded with the Si chip and Cu substrate by the grown hillocks. Since the Ag sheet can provide excellent electrical and thermal performances as die-attachment, this novel bonding technology can be applied to wide band-gap power devices that require high-performance interconnections.

Keywords— solid-state bonding technology, WBG semiconductor, cold-rolling Ag, grain refinement.

I. INTRODUCTION

To satisfy demands for high energy efficiency and reliability in extreme environments, wide band-gap semiconductor materials including silicon carbide (SiC) and gallium nitride (GaN) have been adopted in power electronic systems, due to its excellent electrical and physical properties [1]–[3]. Moreover, the development of SiC power devices allows for the use of power electronic systems at high temperatures (>250 °C). The commonly used Sn-based solder alloys for die-attachment are not reliable for high power density and high-temperature power electronic systems due to crack formation at the intermetallic compound [4]. In order to address the increasing challenge of reliability problems for high operating temperature of WBG power devices, sintered silver (Ag)-paste as the bonding layer has been gaining much attention due to its superior electrical performance, thermal conductivity, high melting temperature [5], [6].

However, since Ag sinter paste consists of Ag powder and organic solvent, which results in decreasing of shear strength due to residual organic solvent after the sintering process, and large-scale are also difficult [7]. In addition, the sintered Ag joining layer has a micro- or nano-porous structure, which may degrade the electrical properties of the bonding material and induce crack nucleation, thus affecting the reliability of die-attachment structure in service [8]. It has been reported that large voids are generated within the porous structure of the sintered Ag joining layer under thermal stress, which results in failure of power devices [9]–[11]. Furthermore, the thermomechanical stress caused by severe mismatch in the coefficients of thermal expansion (CTE) between the bonding chip and the substrate is more critical to porous structure leading to local thermomechanical stress [12].

On the other hand, previous research has revealed that interface between sputtering Ag layers can be bonded together in the Si/Ag/Ag/Si structure with a heating process [13]. In the heating process, Ag films can bond together by converting hillock formation into abnormal grains growth because of stress induced migration. Stress induced migration has been studies for many years and this bonding technology was called as stress migration bonding (SMB) [14]. In SMB, hillock formation is induced by stress migration, which serves as driving force for interface bonding; thus, it is hypothesized that two surfaces of Ag metallization layers can be bonded together by using bulk Ag sheet if hillocks can grow on the its surface by optimizing the growth conditions. Because Ag sheet has low-yield strength and ductility, it is a challenge to use this as a replacement to the porous sintered Ag and achieve the desired bonding for power device electronics.

In this work, a bonding technology based on cold-rolled Ag sheets is adopted to bond Si chips to Cu substrates by using the SMB theory. Cold rolling was performed on the bulk Ag to induce driving force for hillock growth that is rarely reported. This study also optimized the hillocks growth conditions on the surface of Ag sheet by changing the thickness reduction ratio to achieve interface bonding in a die attachment structure.

II. TECHNICAL WORK PREPARATION L

Four kinds of Ag sheets were used as starting materials. The initial thickness for cold-rolling process varied from 1mm to 0.1 mm to provide different microstructure and the residual stress of the cold-rolled Ag sheets, as shown Fig. 1(a). Before the cold rolling process, annealing was carried out at 400 °C for 1 h to obtain a uniform microstructure. The Ag sheets with a thickness of 0.3, 0.5 and 1 mm were cold-rolled up to 0.1 mm, respectively, which corresponds to reduction ratio of 66, 80 and 90%. The cold rolling was performed at room temperature without lubrication by a draft of 0.1 mm per pass. After cold rolling, Ag sheets were cut into 4.0 mm × 4.0 mm pieces.

The 2018 International Power Electronics Conference

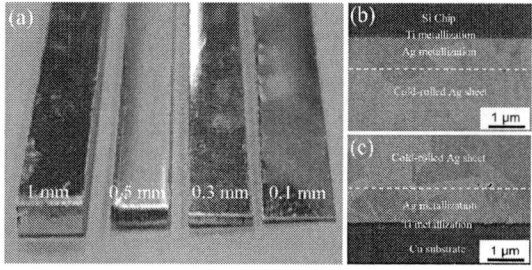
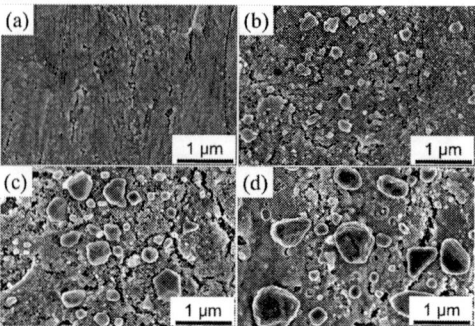

Fig. 1. (a) Samples of Ag sheets for cold rolling. Cross-section SEM image of bonding structures at (b) Si/ sheet, (c) sheet/ Cu.

The Si chips and Cu substrates were prepared with the dimensions 4.0 mm × 4.0 mm and 8 mm × 16 mm, respectively. After Ti metallization layer with thickness of 0.1 μm deposited on the surfaces of both the Si chips and Cu substrates, Ag metallization layer with thickness of 1.0 μm were deposited by radiofrequency (RF) sputtering. The Ti metallization layer can be as a barrier layer to prevent the formation of Cu oxide into the Ag sheet [15]. After cleaning all the specimens by ultrasonic cleaning in acetone for 5 min, the Ag sheets were sandwiched between Si chip and Cu substrate and heated on a hot plate under 1 MPa pressure at 300 °C in air for 1 h. The cross-section microstructure of the joints as well as the morphological changes in cold-rolled Ag sheets after heating at 300 °C were observed by field-emission scanning electron microscopy (FE-SEM, SU8020, HITACHI). By this process, Si chip and Cu substrate were successfully bonded with cold-rolled Ag sheet. Figs. 1(b) and (c) show example of SEM images after the bonding with 90% reduction ratio of cold-rolled Ag sheet, where the Ag sheet is well bonded with Ag metallization layer on both the Si chip and the Cu substrate, respectively. Die-shear test (XD-7500, DAGE) was performed on five bonded specimens at a shear rate of 50 μm/s.

III. RESULTS AND DISCUSSION

Fig. 2 shows the surface morphology of Ag sheets with various reduction ratios after heating at 300°C. Fig. 2(a) shows a surface morphology of initial Ag sheet after heating, featuring flat surface without any hillock formation. With increasing rolling reduction in thickness, tiny hillocks were observed on the surface of cold-rolled Ag sheet after heating, as shown Fig. 2(b) and (c). When the cold-rolling was conducted to a 90 % reduction in thickness, the largest size of hillock was observed among all specimens, as shown Fig. 2(d). The reason for formation of hillock on the cold-rolled Ag sheet after heating can be explained by migration of Ag atoms induced by residual stress release [16]. It was reported that the compressive residual stress was generated on the surface after cold rolling by large deformation [17], which was a driving force for the migration of Ag atoms, resulting in hillock formation after heating [18]. In addition, decreased grain size by cold rolling from micro to sub-micro provides a pathway for the migration Ag atoms along the grain boundary for hillock formation.

Fig .2. Surface morphology of Ag sheets with various reduction ratios after heating at 300 °C: (a) 0 %, (b) 66 %, (c) 80 %, and (d) 90 %.

The size of hillocks was measured with various reduction ratios in thickness after heating, as shown in Fig. 3. With increasing reduction ratio, the hillock size was increased. This result agrees well with the evolution of microstructure after heating (Fig. 2). In addition, the shear strength changes of bonded specimens with various reduction ratios are shown in Fig. 3. The error bar in shear strength is contributed from the non-uniform contacts due to the surface roughness between those samples. The die-attachment structure bonded with cold-rolled Ag sheet was achieved even using initial Ag sheet that did not form hillock during heating (Fig. 2(a)). It was reported that hillocks easily grown on the both surface of Si and Cu substrate coated by Ti/Ag metallization layers because a large compressive stress occurrence due to CTE mismatch between Si and its coated layers and also Cu substrate and coated layers. Therefore, it is supposed that the sputtered Ag metallization/Ag sheet interface could be bonded by hillock formation from the both surface of Si and Cu substrate. . With increasing rolling reduction in thickness, the shear strength of the bonded specimens increased to about 20 MPa by using the cold-rolled Ag sheet with 90% reduction ratio. The results indicate that contact interface between Ag sheet and substrates can be well bonded together when large quantity of hillock formed on the surface, leading to a solid-state bonding interface.

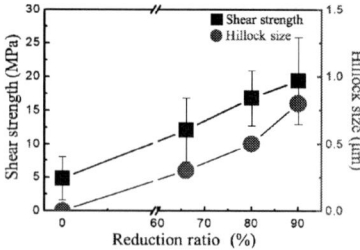

Fig. 3. Correlation between hillocks size and shear strength with various reduction ratios.

The evolution of microstructure during cold-rolling was observed, as shown in Fig. 4. Before cold rolling, the average grain size of Ag sheet was about 4.82 μm, as shown in Fig. 4(a). After cold rolling, the grain size decreased from the micro-scale to sub-micro-scale,

1599

resulting in increasing of grain boundary density (Fig. 4(b), (c)). When the cold-rolling was conducted to a 90 % reduction in thickness, average grain size decreased up to 0.32 μm. It has been reported that Ag grain boundary plays a key role in the formation of hillocks [18]. In the present work, by cold-rolling process, grain size of Ag sheet was decreased, leading to the increasing of pathway to migrate Ag atoms, which resulted in a larger hillock formation. Thus, cold-rolled Ag sheet with 90% showed the highest shear strength because of its better migration ability of Ag atoms among all specimens.

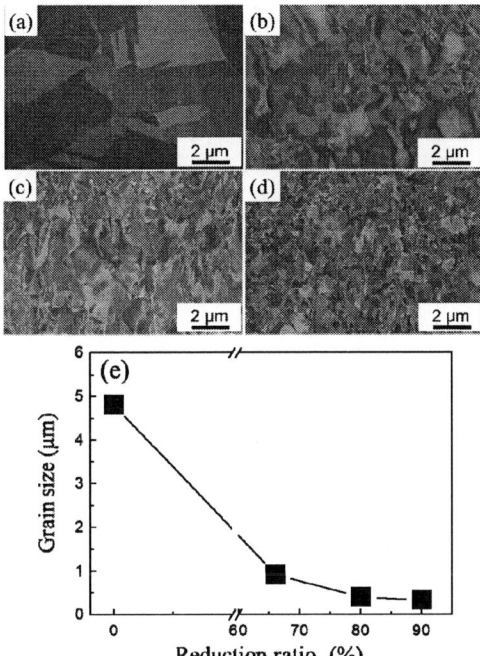

Fig .4. Microstructure evolution of Ag sheets with various reduction ratios: (a) 0 %, (b) 66 %, (c) 80 %, and (d) 90 %. Variations in grain size (e).

IV. CONCLUSION

In this study, a bonding technology based on cold-rolled Ag sheets is adopted to bond Si chips to Cu substrates by using the SMB theory, which uses hillock growth caused by stress migration. Various reduction ratios such as 0, 66, 80 and 90% have been evaluated. The highest die-shear strength was achieved by using the cold-rolled Ag sheet with 90% reduction ratio due to its better migration ability of Ag atoms among all specimens. Our results revealed the potential applications of highly thermal-stable SMB joints such as multi-chip integrations and next-generation power devices.

ACKNOWLEDGMENT

This work was supported by the JST Advanced Low Carbon Technology Research and Development Program (ALCA) project "Development of a high frequency GaN power module package technology" (Grant No. J165101047)

REFERENCES

[1] J. Biela, M. Schweizer, S. Waffler, and J. W. Kolar, "SiC versus Si - Evaluation of potentials for performance improvement of inverter and DCDC converter systems by SiC power semiconductors," *IEEE Transactions on Industrial Electronics*, vol. 58, no. 7, pp. 2872–2882, 2011.

[2] H. S. Chin, K. Y. Cheong, and A. B. Ismail, "A Review on Die Attach Materials for SiC-Based High-Temperature Power Devices," *Metallurgical and Materials Transactions B*, vol. 41, no. 4, pp. 824–832, 2010.

[3] Y. Gao, A. Q. Huang, S. Krishnaswami, J. Richmond, and A. K. Agarwal, "Comparison of static and switching characteristics of 1200 V 4H-SiC BJT and 1200 V Si-IGBT," *IEEE Transactions on Industry Applications*, vol. 44, no. 3, pp. 887–893, 2008.

[4] K. S. Kim, C. H. Yu, N. H. Kim, N. K. Kim, H. J. Chang, and E. G. Chang, "Isothermal aging characteristics of Sn-Pb micro solder bumps," *Microelectronics Reliability*, vol. 43, no. 5, pp. 757–763, 2003.

[5] K. Suganuma, S. Sakamoto, N. Kagami, D. Wakuda, K. S. Kim, and M. Nogi, "Low-temperature low-pressure die attach with hybrid silver particle paste," *Microelectronics Reliability*, vol. 52, no. 2, pp. 375–380, 2012.

[6] Y. H. Mei, J. Y. Lian, X. Chen, G. Chen, X. Li, and G. Q. Lu, "Thermo-mechanical reliability of double-sided IGBT assembly bonded by sintered nanosilver," *IEEE Transactions on Device and Materials Reliability*, vol. 14, no. 1, pp. 194–202, 2014.

[7] K. Qi, X. Chen, and G.-Q. Lu, "Effect of interconnection area on shear strength of sintered joint with nano-silver paste," *Soldering & Surface Mount Technology*, vol. 20, no. 1, pp. 8–12, 2008.

[8] V. Caccuri, X. Milhet, P. Gadaud, D. Bertheau, and M. Gerland, "Mechanical Properties of Sintered Ag as a New Material for Die Bonding: Influence of the Density," *Journal of Electronic Materials*, vol. 43, no. 12, pp. 4510–4514, 2014.

[9] P. O. Quintero and F. P. McCluskey, "Temperature cycling reliability of high-temperature lead-free die-attach technologies," *IEEE Transactions on Device and Materials Reliability*, vol. 11, no. 4, pp. 531–539, 2011.

[10] Y. Tan, X. Li, G. Chen, Y. Mei, and X. Chen, "Three-Dimensional Visualization of the Crack-Growth Behavior of Nano-Silver Joints During Shear Creep," *Journal of Electronic Materials*, vol. 44, no. 2, pp. 761–769, 2014.

[11] H. Zhang, C. Chen, S. Nagao, and K. Suganuma, "Thermal Fatigue Behavior of Silicon-Carbide-Doped Silver Microflake Sinter Joints for Die Attachment in Silicon/Silicon Carbide Power Devices," *Journal of Electronic Materials*, pp. 1–6, 2016.

[12] C. Chen *et al.*, "Self-healing of cracks in Ag joining layer for die-attachment in power devices," *Applied Physics Letters*, vol. 109, no. 9, 2016.

[13] T. Kunimune, M. Kuramoto, S. Ogawa, T. Sugahara, S. Nagao, and K. Suganuma, "Ultra thermal stability of LED die-attach achieved by pressureless Ag stress-migration bonding at low temperature," *Acta Materialia*, vol. 89, pp. 133–140, 2015.

[14] S. Lin *et al.*, "Nano-volcanic Eruption of Silver," *Scientific Reports*, vol. 6, p. 34769, 2016.

[15] H. Zhang, S. Nagao, K. Suganuma, H. J. Albrecht, and K. Wilke, "Thermostable Ag die-attach structure for high-temperature power devices," *Journal of Materials Science: Materials in Electronics*, vol. 27, no. 2, pp. 1337–1344, 2016.

[16] S. Noh, C. Choe, C. Chen, and K. Suganuma, "Heat-resistant die-attach with cold-rolled Ag sheet," *Applied Physics Express*, vol. 11, no. 1, 2018.

[17] E. Brinksmeier, J. T. Cammett, W. König, P. Leskovar, J. Peters, and H. K. Tönshoff, "Residual Stresses - Measurement and Causes in Machining Processes," *CIRP Annals - Manufacturing Technology*, vol. 31, no. 2, pp. 491–510, 1982.

[18] C. Oh, S. Nagao, T. Kunimune, and K. Suganuma, "Pressureless wafer bonding by turning hillocks into abnormal grain growths in Ag films," *Applied Physics Letters*, vol. 104, no. 16, 2014.

High-frequency Self-Driven Synchronous Rectifier Controller for WPT Systems

Akihiro Konishi[*], Kazuhiro Umetani and Eiji Hiraki

Graduate school of Natural Science and Technology, Okayama University, Okayama, Japan

*E-mail: p3448o58@s.okayama-u.ac.jp

Abstract - **Resonant inductive coupling wireless power transfer (RIC-WPT) is the promising wireless power supply method to mobile applications. It is well-known that the efficiency and the power transfer capability of RIC-WPT systems deeply depend on the quality factor of the receiver-side resonator, which tends to decline by the conduction loss at rectifying diodes. Therefore, the synchronous rectification is one of the effective solutions. However, the unique characteristics of RIC-WPT, that is the weak magnetic coupling between the transmitter-side and the receiver-side resonator, tend to make the introduction of the synchronous rectification difficult. In order to address this issue, this paper newly proposes a self-driven synchronous rectifier controller suitable for RIC-WPT. This synchronous rectifier controller consists of basic analog elements and gate drivers, and no additional power supply is necessary for driving itself. In addition, the controller doesn't require any information from the transmitter-side resonator. Experiments successfully verified the operating principle as well as the efficiency improvement and clarified the usefulness of the synchronous rectifier with proposed controller for practical application of RIC-WPT.**

Keywords— synchronous rectifier, wireless power transfer, magnetic resonant , propagation delay

I. Introduction

Wireless Power Transfer (WPT) systems have advantages such as no bulky mechanical connecter and diversity of charging location. So, a variety of mobile applications applied WPT have been proposed such as railway vehicles, electric vehicles' chargers, biomedical implant devices and mobile phones [1]–[8]. Especially, Resonant Inductive Coupling WPT (RIC-WPT) systems have great advantages such as high tolerance for a misalignment and longer transmission distance compared with an electromagnetic induction based WPT. Therefore, this technique is a promising wireless power supply method to the future mobile applications. For the practical application of WPT systems, it is necessary to increase an efficiency and a power transfer capability.

The efficiency of typical RIC-WPT systems [9][10] shown in Fig. 1 can be defined as;

$$\eta = \frac{(\omega M)^2 R_L}{(R_R+R_L)\{R_T(R_R+R_L)+(\omega M)^2\}},$$ (1)

where, R_T, R_R and R_L are an AC resistance of transmitter-side resonator, receiver-side resonator and the load, respectively. M is mutual-inductance between the transmitter-side and receiver-side resonator. ω is the operating frequency. The optimum load resistance which can receive a maximum power is derived from (1) as;

$$R_{Lopt} = R_L\sqrt{1+k^2 Q_T Q_R},$$ (2)

where, Q_T and Q_R are the quality factor of the transmitter-side and receiver-side resonator, respectively. k is coupling coefficient. The maximum efficiency in this optimum load condition is defined as

$$\eta_{max} = \frac{k^2 Q_T Q_R}{\left(1+\sqrt{1+k^2 Q_T Q_R}\right)^2}.$$ (3)

(2) and (3) indicates that optimization of the load impedance [11]–[16] and increasing Q_T and/or Q_R [17] are effective techniques to improve the efficiency and the power transfer capability. In particular, the latter approach contributes to both efficacies.

In many case, Q_R is smaller compared to Q_T, since the receiving coil tends to be much smaller than the transmitter-side coil caused by space limitation of mobile receiving targets. Furthermore, the voltage drop of rectifying diodes connected to the receiving resonant path in series apparently behaves similarly to decreasing Q_T. Because, decreasing Q_T is equivalent to the increase of the voltage drop in the AC resistance of the receiver-side resonator R_R. Thus, the suppression of the voltage drop in the rectifying diode will increase the efficiency and the power transfer capability.

The synchronous rectifier using MOSFETs [18]–[20] is a well-known solution to reduce the voltage drop of the rectifying diode, which has been widely used to the receiver-side of the isolated DC-DC converter. A typical synchronous rectifier is shown in Fig. 2. The rectifying MOSFETs are driven synchronously with the voltage supplied by the transmitter-side; i.e. control signals for the inverter. In order to apply this method to the RIC-WPT, a kind of communication systems is required between the transmitter-side and the receiver-side, which may not be effective for costs and space factors.

Furthermore, this method can be applied under the condition that the transmitter-side voltage and the receiver-side voltage are in phase. However, in the RIC-WPT systems, the input voltage to the transmitter-side resonator and the induced voltage of the receiver-side resonator are not in phase, because of a weak magnetic coupling between the transmitter-side and receiver-side resonators. Due to this phase difference, the control signals for rectifying MOSFETs become out of synchronization with the rectified current. Therefore, the synchronized rectification with the voltage supplied by the transmitter-side inverter does not lead to effective conversion.

One of the helpful knowledge for this issue may lie in self-driven synchronous rectifiers proposed in [21]–[28]. These self-driven synchronous rectifiers are driven based on the detecting technique of the polarity of the voltage or current of the rectifier input. In either case, the synchronous rectifier is controlled with the receiver-side information only. Therefore, no communication method is necessary. However, many of synchronous rectifier controllers (SRC) need auxiliary winding coils [24] or current transformers [25]–[28] to detect the polarity of the voltage or current. These additional components may not be effective for costs and space factors. Certainly, the self-driven rectifier introduced in [21] is free from

additional windings, though the introduced rectifier forced the Pch-MOSFET based rectification.

The purpose of this paper is to propose a simple SRC specialized for RIC-WPT systems. The proposed SRC has following advantages,

(1) Self-driven; without any external information and external power source.
(2) Low power consumption; without expensive processers such as DSP and FPGA, and without extra windings to obtain the current information.
(3) Possibility for high frequency operation; miniaturization of the propagation delay by analog controls.

The following discussion consists of four sections. Section □ presents the circuit configuration and the operation principle of the proposed SRC. Section □ presents the experiment to verify the effectiveness of the proposed SRC. Finally, section □ gives the conclusions.

II. PROPOSED SYNCHRONOUS RECTIFIER CONTROLLER

A. Circuit configuration

This subsection presents the circuit configuration of the proposed SRC employed to RIC-WPT with a self-driven synchronous rectifier. This SRC generates the synchronous gate signals based on the receiver-side resonant current. The direction of receiver-side resonant current is directly derived from the time-differential of the detected voltage applied to the receiver-side resonant capacitor. Therefore, extra windings or shunt resisters are not necessary which tend to decrease Q_R.

Fig. 3 shows the circuit diagram. The proposed SRC consists of the voltage generator, comparator, start-up controller, dead time controller, and isolated gate drivers. The regulator U_1 is employed for the power supply to logic and analog ICs by generating the V_{cc} from the output voltage of rectifier. Control ICs should be

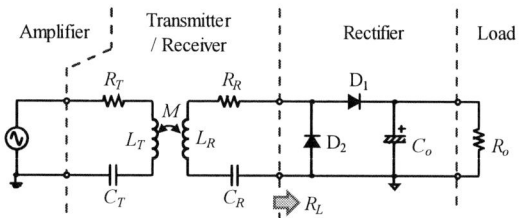

Fig. 1. Typical resonant inductive coupling wireless power transfer terminated by the rectifier.

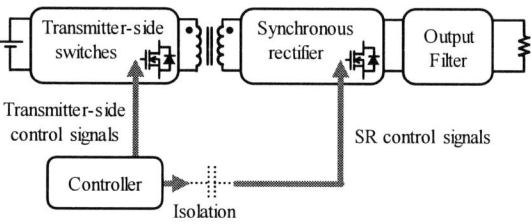

Fig. 2. Isolated converter block diagram

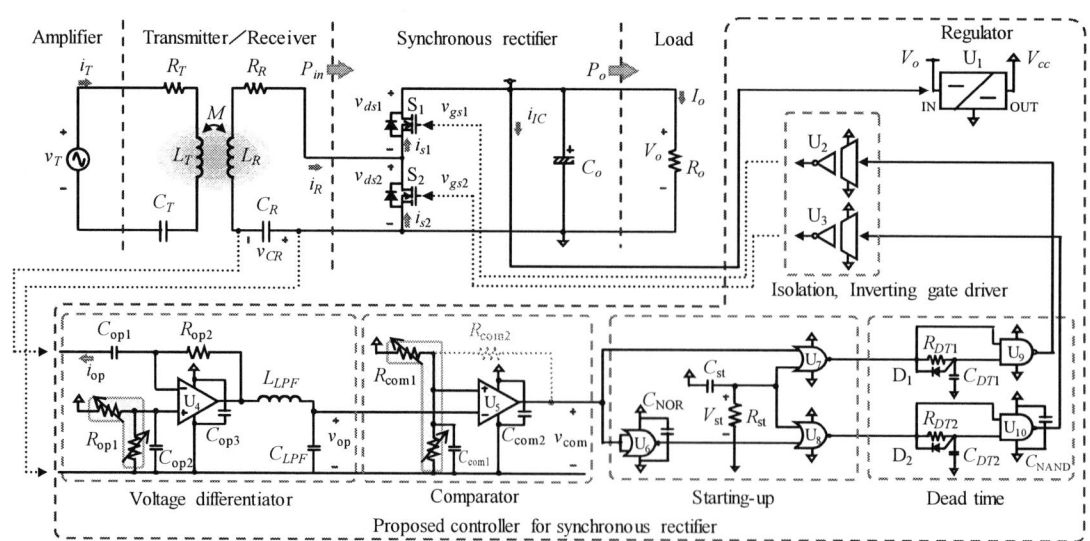

Fig. 3. Proposed controller for synchronous rectifier specialized for RIC-WPT systems.

1603

carefully selected by balancing performance and power consumption

B. Operation principle

This subsection presents how to generate gate signals and the operation principle in RIC-WPT systems. The proposed SRC generates gate signals as follows.

1) OP amplifier U_4, R_{op2} and C_{op1} are employed to find out the current i_{op} by differentiating the receiver-side resonant capacitor voltage v_{CR}. A differentiator outputs v_{op} when an intermediate potential bias is applied. v_{op} is defined as:

$$v_{op} = \left(C_{op1} R_{op2} / C_R\right) i_R + V_{cc} / 2 , \qquad (4)$$

where, C_{op1} and R_{op2} are the constants of the differentiator, i_R is the receiver-side resonant current, C_R is the receiver-side resonant capacitance, respectively.

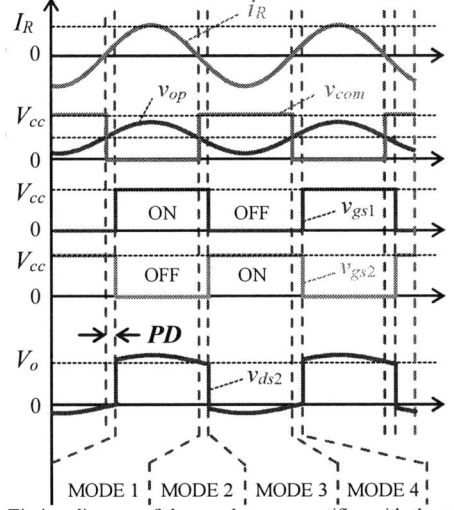

Fig. 4. Timing diagram of the synchronous rectifier with the proposed SRC.

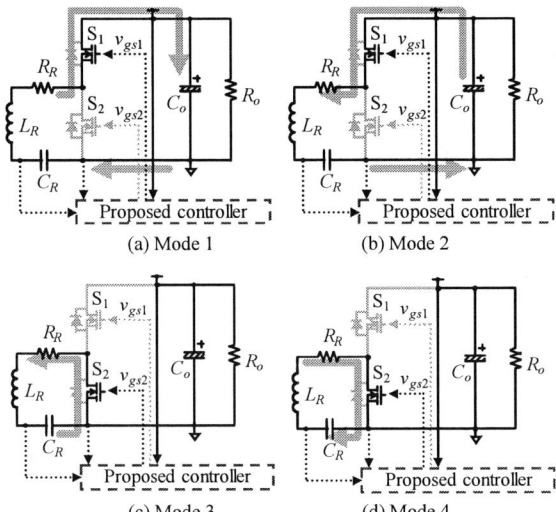

(a) Mode 1   (b) Mode 2

(c) Mode 3   (d) Mode 4

Fig. 5. Operation modes of the synchronous rectifier with the proposed SRC.

2) The comparator with the hysteresis derives a direction of a differentiated current i_{op}. The comparator outputs "Low" when i_R flows via R_R-S_1-C_o-C_R, and outputs "High" when i_R flows via C_R-S_2-R_R. The additional external hysteresis added by R_{com2} strongly suppresses an unintended operation.

3) Gate signals pass through starting-up circuits with the large time constant with milli-seconds to suppress an unintended operation in the transient state such as a start-up.

4) Gate signals including dead times are generated by NAND gates with the nano-second order time constant.

A time lag from the moment when the direction of the resonant current i_R changes from negative to positive to the moment when the drain-source voltage v_{ds2} rises is defined as propagation delay (*PD*). The *PD* is mainly caused by the response of the control ICs.

According to the direction of the receiver-side resonant current and the state of MOSFETs, the operation mode of the self-driven synchronous rectifier with the proposed SRC can be divided into four modes. Fig. 4 shows a timing diagram of the synchronous rectifier with the proposed SRC, and Fig. 5 shows the operation modes of the rectifier, where i_R is the receiver-side resonant current. v_{op} and v_{com} are the output voltage of the OP amplifier U_4 and the comparator U_5; v_{gs1} is the gate-source voltage of MOSFET S_1; v_{gs2} and v_{ds2} are the gate-source voltage and the drain-source voltage of MOSFET S_2; respectively. In this subsection, dead times are ignored.

MODE 1; S_1 turns on and S_2 turns off. i_R flows from the receiver-side resonator to the smoothing capacitor C_o. The conduction loss is suppressed, because the voltage drop of S_1 is far smaller than that of SBD.

MODE 2; Direction of i_R changes to negative. However, S_1 is still on-state due to the *PD*. Therefore, a part of the receiving power is returned to the transmitter-side via resonators.

MODE 3; S_2 turns on and S_1 turns off. i_R flows through S_2 to form the circulating current path. Therefore, it is impossible to receive a transmitting power in this mode. The conduction loss is suppressed as with mode 1.

MODE 4; Direction of i_R changes to positive. S_2 is still on-state due to the *PD*. Then, i_R flows through S_2 to form the circulating current path.

According to the mode transition, the received power is released during the *PD*. Furthermore, the switching loss of S_1 and S_2 increase as the *PD* increases. Therefore, the *PD* should be designed to be as short as possible.

III. EXPERIMENT AND EVALUATION

A prototype circuit was built to verify the self-driven synchronous rectifier with the proposed SRC. Table I presents the specification of the experimental set-up. Fig. 6 depicts the diagram of the experiment. A high-frequency AC amplifier (nF HSA4014) was employed for the AC voltage source. The operating frequency of the RIC-WPT systems will be higher to several MHz to reduce the size of resonators. However, 200 kHz was selected for the experiment to suppress influence of parasitic components. The resonant frequency of the transmitter-side and receiver-side resonators were also set at 200 kHz. The resonant capacitance was selected so as to satisfy the following relation,

$$f_r = 1/2\pi\sqrt{L_T C_T} = 1/2\pi\sqrt{L_R C_R}, \qquad (5)$$

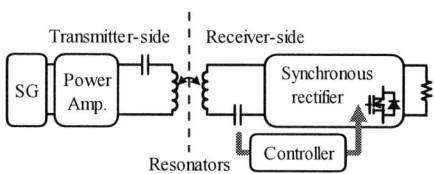

Fig. 6. Block diagram of the experiment.

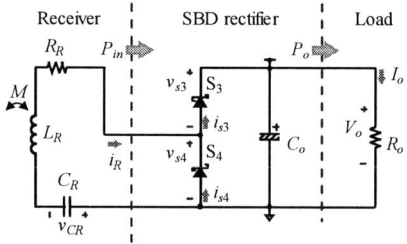

Fig. 7. Circuit diagram of the SBD rectifier in comparison.

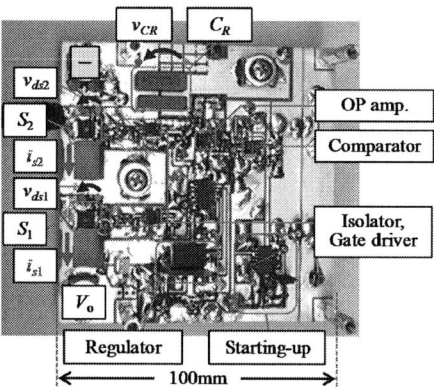

(a) Synchronous rectifier with proposed controller

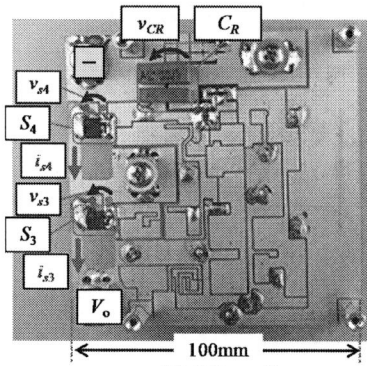

(b) SBD rectifier

Fig. 8. Photographs of PCBs' layout.

where, L_T and L_R are the self-inductance of the resonant coils, C_T and C_R are the resonant capacitance, respectively.

The spiral transmitter-side and receiver-side coils were made of a litz-wire to suppress a conduction loss by the skin effect. High voltage polypropylene film capacitors with a low equivalent series resistance (ESR) were selected for the resonant capacitors to reduce an ESR loss. The receiving power was adjusted by changing the AC power source voltage v_p and the DC load resistance R_o. To compare with the proposed approach, an SBD rectifier was prepared in Fig. 7. Fig. 8 shows the photographs of (a) synchronous rectifier and (b) SBD rectifier. These rectifiers were mounted on almost the same PCB layout each other.

A. Steady state operation

We observed operating waveforms of the rectifiers. Fig. 9 shows the observed waveforms of (a) S_1, (b) S_2, and (c) SRC at $P_o = 41$ W. Fig. 10 shows observed waveforms of S_4 of the SBD rectifier. As can be seen in Fig. 9 and 10, the drain-source voltage of MOSFETs S_1 and S_2 in the on-state was smaller than the cathode-anode

TABLE I
SPECIFICATION OF THE EXPERIMENTAL SET-UP

Component	Symbol	Value
Resistance of resonator	R_T	0.12 Ω
	R_R	0.12 Ω
Self-inductance of coils	L_T	36.4 μH
	L_R	36.3 μH
Resonant capacitance	C_T	17.41 nF
	C_R	17.47 nF
Mutual-inductance	M	3.635 μH
Coupling coefficient	k	0.1
Outer diameter of resonant coil	-	210 mm
Transmitting gap	-	100 mm
Number of turns	N_p, N_s	12
Resonant frequency	f_r	200 kHz
Operating frequency	f_o	200 kHz
Smoothing capacitance	C_o	330 μF / 50 V
		47 μF / 50 V
		1 μF / 250 V
Load resistance	R_o	7.71–100.6 Ω
@Differentiator	R_{op1}	100k Ω
	R_{op2}	2.4k–13k Ω
	C_{op1}	2 pF
@LPF (f_c = 1.61MHz)	L_{LPF}	4.7 μH × 4
	C_{LPF}	510 pF
@Comparator	R_{com1}	100k Ω
	R_{com2}	-
@Starting-up	R_{st}	100k Ω
	C_{st}	0.1 μF
Dead time of synchronous rectifier	-	50 ns
MOSFET (40V, 50A)	S_1, S_2	FDD8447LF085 ($r_{ds(on)}$ = 7.9mΩ@V_{GS} = 5V)
SBD (45V, 15A)	S_3, S_4	STPS15L45C-Y (V_F = 0.48V@ I_F = 5A)
Isolator	-	ADUM3100BRZ
Gate driver	-	MIC4417
	R_{g_int}	7.8 Ω @V_{cc} = 5V
OP amplifier	U_4	CLC1005ISO8MTR
Comparator	U_5	LT1719IS8
Regulator	U_1	TSR 0.5-2450 (V_{in} : ~32V)

1605

voltage of SBDs. Therefore, the synchronous rectifier with the proposed SRC suppressed the conduction loss as the theory. Also, the definite time lag between the zero-crossing point of the resonant current and the commutation timing of the MOSFETs were observed as shown in Fig. 9(a) and (b). Due to the PD, S_1 and S_2 were forced the hard-switching. As a result, the observed voltage surge in the synchronous rectifier was larger compared with the SBD rectifier.

B. Frequency Followability

Even if the operating frequency of the AC power source fluctuates, the frequency of gate signals must synchronize with the frequency of the receiver-side resonant current. Therefore, we observed the gate signals when the operating frequency was fluctuated from 180 kHz to 220 kHz. As shown in Fig. 11, the proposed SRC generated the gate signals synchronized with the resonant current. The observed PD was almost constant at 140ns and obviously independent from the operation frequency.

C. Start-up Response

To avoid the unintentional operation during the transient state such as a start-up, SRC should stop generating "High" gate signals until the output voltage V_o of the rectifier sufficiently drops below the UVLO threshold of the ICs in the SRC. A starting-up circuit with

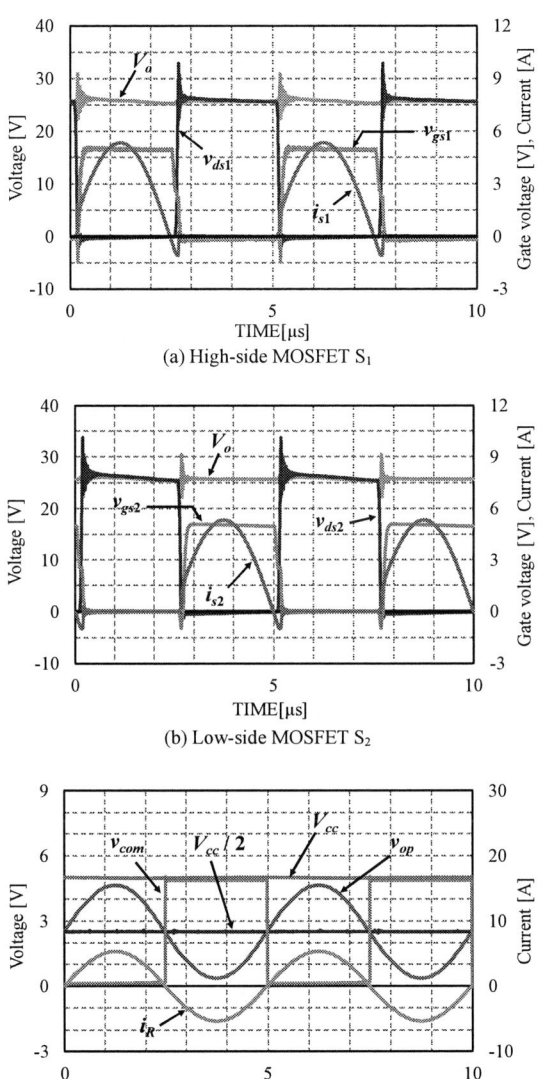

(a) High-side MOSFET S_1

(b) Low-side MOSFET S_2

(c) Synchronous rectifier controller

Fig. 9. Waveforms of the synchronous rectifier with the proposed SRC under $R_o = 15.61\Omega$, $P_o = 41$W, $I_R = 3.8$Arms.

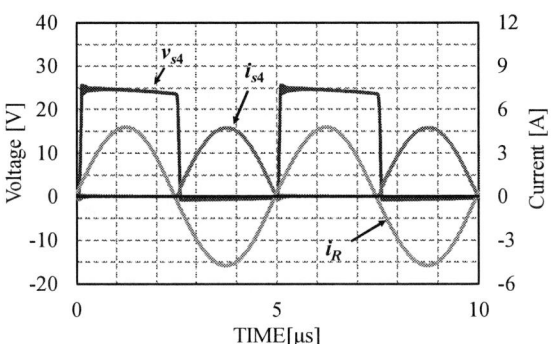

Fig. 10. Waveforms of the SBD rectifier under $R_o = 15.61\Omega$, $P_o = 35$W, $I_R = 3.4$Arms.

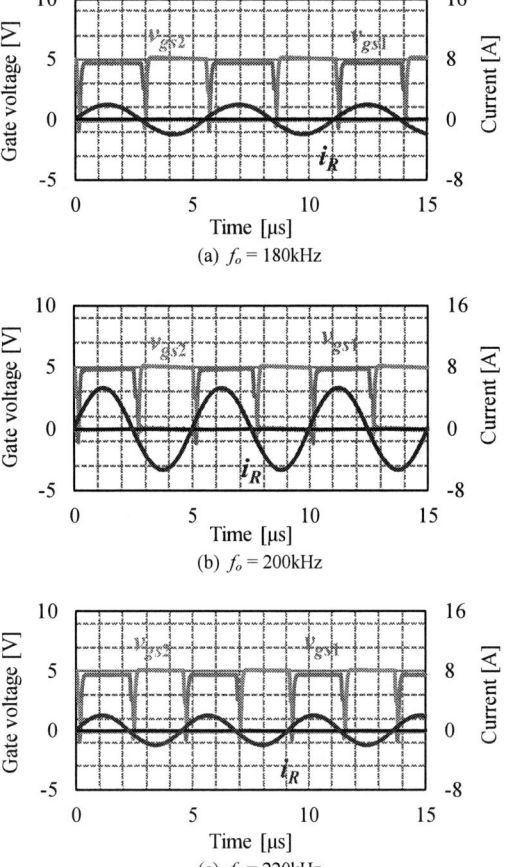

(a) $f_o = 180$kHz

(b) $f_o = 200$kHz

(c) $f_o = 220$kHz

Fig. 11. Operation waveforms of synchronous rectifier with different operating frequencies.

the large time constant is inserted between the comparator and gate drivers of the SRC, so that "High" gate signals are not transmitted to MOSFETs during the start-up. This idling time is roughly determined by the time constant $R_{st}C_{st}$. Fig. 12 shows observed waveforms of the start-up response. When the output voltage V_o exceeded around 8V, the SRC started up. The SRC waited for a while keeping the gate signals "Low". After that, the SRC started generating the gate signals at the timing that V_{st} dropped below the threshold of the NAND gates U_7 and U_8 in the start-up circuit.

D. Rectification Loss Analysis

Loss Analysis to verify a conduction loss suppression of the synchronous rectifier, we analyzed the power loss for the different output power in comparison with the synchronous rectifier with the SRC and the SBD rectifier. The output DC power P_o was measured by digital multi-meters (Keysight 34461A), and the input AC power P_{in} was measured by a 12-bit oscilloscope (Lecroy HDO4034A, PP026) and a current sensor (Tektronix TCP A300, 305A). We adjusted the output power by changing the output load R_o and the input AC voltage. In this experiment, we kept the output voltage V_o to 25.6 V which is approximately 80% of the maximum input voltage 32 V of the regulator U_1 in the SRC.

Fig. 13 (a) shows the measured the power loss at the different output power, Fig. 13 (b) shows the loss analysis results at the high and low output power. The loss reduction effect of the synchronous rectifier in all power regions. As shown in Fig.13 (b), the conduction

loss of the synchronous rectifier was drastically suppressed in comparison with the SBD rectifier. On the other hand, the huge switching loss was generated due to the hard-switching of the MOSFETs by the PD of the proposed SRC. To focus on the power consumption of the SRC, it was roughly constant regardless of the output power. Especially, in the low output power region, the power consumption of the SRC was dominant.

Fig. 14 shows the detailed loss distribution in the proposed SRC at $P_o = 11$W. The power consumption of

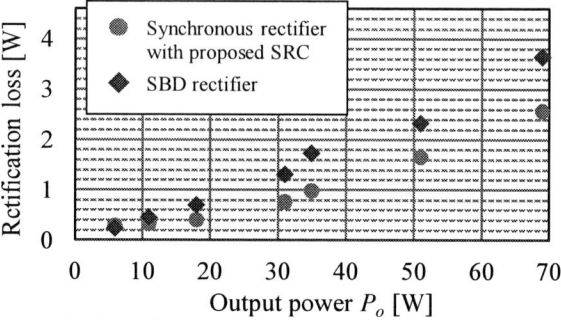

(a) Observed rectification loss in different output powers

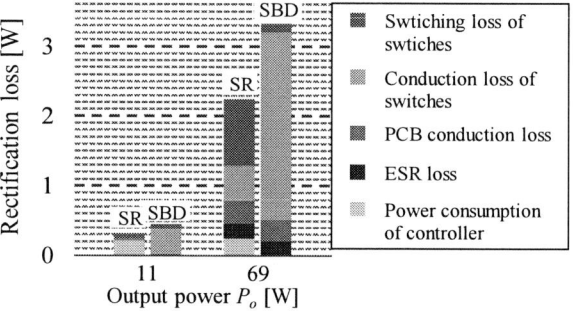

(b) Detail loss analysis in high/low power region
Fig. 13. Loss analysis of rectifiers.

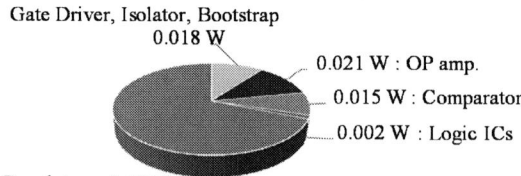

Fig. 14. Detail loss analysis of the proposed SRC under $P_o = 11$W.

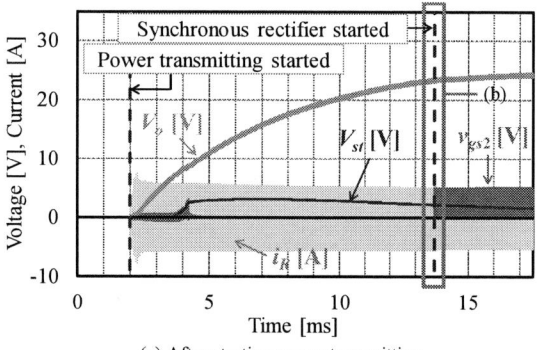

(a) After starting power transmitting

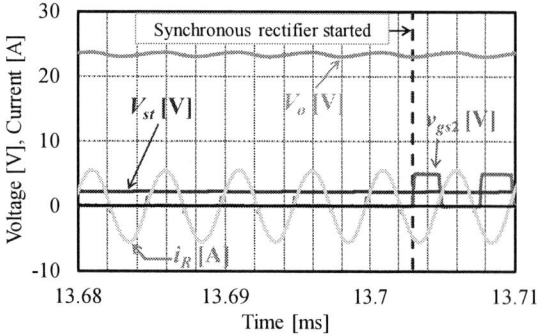

(b) At starting synchronous rectification
Fig. 12. Observed waveforms of the start-up response.

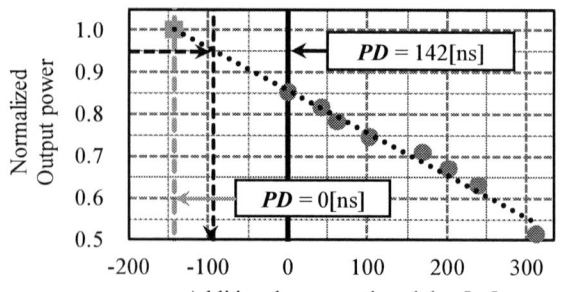

Fig. 15. Relationship between normalized output power P_o and propagation delay.

1607

each ICs was measured from the observed V_{cc} voltage and one's input current. A small shunt resister was inserted into V_{cc} terminal of ICs for detecting the input current. The major loss element of the SRC was a switching-regulator U_1. Therefore, the power loss will be further reduced by replacing the switching regulator to the charge pomp circuit. For the smaller power application such as the order of several watts small mobile apparatus, the OP amplifier and gate drivers with the low power consumption are indispensable to reduce power consumption in the SRC.

E. Influence of propagation delay

The PD generated by SRC involves not only a reverse power flow in mode 2, but also increase in the switching loss by the hard-switching. This means that the PD is the important parameter which involve to the output power of the synchronous rectifier. Therefore, we evaluated the influence of the PD on the output power of the synchronous rectifier. In order to add the intentional PD to the initial value 142 ns, the variable RC-delay circuit was inserted between the comparator and the starting-up circuit in the SRC. Fig. 15 shows measured relationship between an intentionally added PD and the normalized output power when the AC power source voltage v_p and the load resistance R_o are fixed to 9.0 Vrms and 15.61 Ω, respectively. As the PD increased from the initial value, the output power decreased linearly. Here, the ideal output power with the $PD = 0$ was estimated by the linear approximation.

Fig. 15 indicates that the PD should be designed to 50 ns in order to suppress the power attenuation within 5 % from the ideal output power. This result indicates that the MHz-class RIC-WPT with the proposed SRC require that the PD should be 10ns or less in order to suppress the power attenuation within 5 %.

IV. CONCLUSION

The RIC-WPT is the promising wireless power supply method to mobile applications. It is well-known that the efficiency and the power transfer capability of RIC-WPT systems deeply depend on the quality factor of the receiver-side resonator, which tends to decline by the conduction loss at rectifying diodes. Therefore, the synchronous rectification is one of the effective solutions. However, the unique characteristics of RIC-WPT, that is the weak magnetic coupling between the transmitter-side and the receiver-side resonator, tend to make the introduction of the synchronous rectification difficult.

In order to address this issue, this paper newly proposed a self-driven synchronous rectifier controller suitable for RIC-WPT. The proposed SRC consists of basic analog elements and gate drivers, and no additional power supply is necessary for driving itself. In addition, the SRC doesn't require any information from the transmitter-side resonator.

Experiments successfully verified the operating principle as well as the efficiency improvement and clarified the usefulness of the synchronous rectifier with the proposed SRC for practical applications of RIC-WPT.

REFERENCES

[1] R. Haldi, K. Schenk, "A 3.5 kW wireless charger for electric vehicles with ultra high efficiency", Proc. IEEE Energy Convers. Congr. Expo. (ECCE), pp. 668-674, 2014.

[2] F. Musavi, W. Eberle, "Overview of wireless power transfer technologies for electric vehicle battery charging", IET Power Electron., vol. 7, no. 1, pp. 60-66, Jan. 2014.

[3] C. M. Apostoaia, M. Cernat, "The inductive power transfer system for electric vehicles", OPTIM & 2017 ACEMP, May, 2017

[4] J. H. Kim et al., "Development of 1-MW inductive power transfer system for a high-speed train", IEEE Trans. Ind. Electron., vol. 62, no. 10, pp. 6242-6250, Oct. 2015.

[5] M. P. Kazmierkowski, R. M. Miskiewicz, A. J. Moradewicz, "Inductive coupled contactless energy transfer systems—A review", Proc. Sel. Problems Elect. Eng. Electron. (WZEE), pp. 1-6, 2015.

[6] J. Yungtaek, M. M. Jovanovic, "A contactless electrical energy transmission system for portable-telephone battery chargers", IEEE Trans. Ind. Electron., vol. 50, no. 3, pp. 520-527, Jun. 2003.

[7] L. Li, H. Liu, H. Zhang, W. Xue, "Efficient Wireless Power Transfer System Integrating With Metasurface for Biological Applications", IEEE Transactions on Industrial Electronics, vol. 65, Issue. 4, pp. 3230-3239, 2018.

[8] Y. Li, R. Mai, T. Lin, Y. Liu, Y.g Li, Z. He, J. Yu, "Design and implementation of a novel WPT system for railway applications", Emerging Technologies: Wireless Power Transfer, IEEE PELS Workshop, 2017.

[9] M. Kato, T. Imura, Y. Hori, "New characteristics analysis considering transmission distance and load variation in wireless power transfer via magnetic resonant coupling", INTELEC 2012 IEEE 34th International.

[10] S. Li, C. C. Mi, "Wireless Power Transfer for Electric Vehicle Applications", IEEE Journal of Emerging and Selected Topics in Power Electronics, vol. 3, no. 1, Mar, 2015

[11] T. Hiramatsu, X. Huang, M. Kato, T. Imura, Y. Hori, "Wireless Charging Power Control for HESS Through Receiver Side Voltage Control", The Applied Power Electronics Conference and Exposition, pp. 1614-1619, 2015.

[12] D. Ahn, S. Hong, "Wireless power transfer resonance coupling amplification by load-modulation switching controller", IEEE Trans. Ind. Electron., vol. 62, no. 2, pp. 898-909, 2015.

[13] H. Li, J. Li, K. Wang, W. Chen, X. Yang, "A maximum efficiency point tracking control scheme for wireless power transfer systems using magnetic resonant coupling", IEEE Trans. Power Electron., vol. 30, no. 7, pp. 3998-4008, 2015.

[14] Y. Li, R. Mai, et al., "Efficiency optimising strategy for dual-coupled transmitters based WPT systems", in Electronics Letters, Vol. 52, No. 22, pp. 1877-1879, 2016.

[15] R. Mai, Y. Liu, Y. Li, P. Yue, G. Cao, Z. He, "An Active Rectifier Based Maximum Efficiency Tracking Method Using an Additional Measurement Coil for Wireless Power Transfer", IEEE Transactions on Power Electronics, No. 99, pp. 1-1.

[16] W. K. Seol, S. K. Chung, "Current vector control of wireless power transfer system with 2D transmitting coils", Electronics Letters, vol. 54, Issue. 2, pp.91-93, 2018.

[17] H. Yu, G. Zhang, G. Liu, L. Jing, Q. Liu, "Asymmetry in Wireless Power Transfer Between a Superconducting Coil and a Copper Coil", IEEE Transactions on Applied Superconductivity, vol. 28, Issue. 3, 2018.

[18] P. Hurtuk, R. Radvan, M. Frivaldsky, "Full bridge converter with synchronous rectifiers for low output voltage application", Proc. Applied Electronics (AE), 2011.

[19] K. Murata, K. Harada, T. Harimoto, "Self turn-on loss of MOSFET as synchronous rectifier in dc/dc buck converter-In case of a low driving impedance", Proc. IEEE Power Electron. Spec. Conf., pp. 3348-3353, Jun. 2008.

[20] B.-H. Kwon, J.-H. Youm, J.-W. Lim, "A line-voltage-sensorless synchronous rectifier", IEEE Trans. Power Electron., vol. 14, no. 5, pp. 966-972, Sep. 1999.

[21] T. Koyama, T. Honjo, M. Ishihara, K. Umetani, E. Hiraki, "Simple self-driven synchronous rectifier for resonant inductive

coupling wireless power transfer", Telecommunications Energy Conference (INTELEC), Oct. 2017.

[22] W. A. Tabisz, F. C. Lee, D. Y. Chen, "A Mosfet Resonant Synchronous Rectifier for High-Frequency DC/DC Converters", IEEE Power Electronics Specialists Conference (PESC), 1990.

[23] E. Sakai, K. Harada, "Synchronous Rectifier for Low Voltage Switching Converter", Proc. IEEE INTELEC'95, pp. 471-475, 1995.

[24] P. Alou, J. A. Cobos, O. Garcia, R. Prieto, J. Uceda, "A new driving scheme for synchronous rectifiers: single winding self-driven synchronous rectification", IEEE Trans. Power Electron., vol. 16, no. 6, pp. 803-811, Nov. 2001.

[25] X. Xie, J. C. P. Liu, F. N. K. Poon, M. H. Pong, "A novel high frequency current-driven synchronous rectifier for low voltage high current applications", Proc. Applied Power Electronics Conference and Exposition (APEC), vol. 1, pp. 469-475, Mar. 2001.

[26] G.-Y. Jeong, "High efficiency asymmetrical half-bridge converter using a self-driven synchronous rectifier", IET Power Electron., vol. 1, no. 1, pp. 62-71, Mar. 2008.

[27] W. X. Zhong, S. Y. Hui, W. C. Ho, Xun Liu, "Using self-driven AC-DC synchronous rectifier as a direct replacement for tradiional power diode rectifier", IEEE Trans. Ind. Electron., vol. 59, no. 1, pp. 392-401, Jan. 2012.

[28] G. K. Y. Ho, R. Yu, B. M. H. Pong, "Current driven synchronous rectifier for LLC resonant converter with a novel integrated current transformer", in PEMD 2012, Mar. 2012.

The 2018 International Power Electronics Conference

Automatic Resonance Frequency Tuning Method for Repeater in Resonant Inductive Coupling Wireless Power Transfer Systems

Masataka Ishihara[1*], Kazuhiro Umetani[1] and Eiji Hiraki[1]
1 Graduate School of Natural Science and Technology, Okayama University, Okayama, Japan
*E-mail: p4wv0vf6@s.okayama-u.ac.jp

Abstract— Intermediate resonator (repeater) in resonant inductive coupling wireless power transfer (RIC-WPT) systems can increase the transmission distance between the transmitting and receiving coils. The induced current in the repeater is easily affected by the frequency splitting phenomenon because the quality factor (*Q*-factor) of the repeater is generally as high as several hundreds. If this phenomenon occurs, induction of large current is often difficult in the repeater because the one peak characteristic of the induced current is no longer expected and the single peak of resonance splits into multiple peaks which shift the resonant frequency corresponding to the circuit parameters. In this paper, we approach this difficulty by applying an auxiliary circuit to the RIC-WPT system with the repeater. As a result, under the fixed operating frequency, the induced current in the repeater can be kept large even if the frequency splitting phenomenon occurs. The effectiveness of proposed system was supported by the simulation and experiment.

Keywords— *wireless power transfer, repeater, automatic tuning, frequency splitting phenomenon*

I. INTRODUCTION

Resonant inductive coupling wireless power transfer (RIC-WPT) has recently been actively researched as a high efficiency wireless power transfer technique for various applications such as electric vehicles [1, 2], mobile devices, implanted biomedical devices [3], etc. The RIC-WPT systems utilize the magnetic induction between the magnetically coupled transmitting and receiving coils to transfer the power.

However, the transmission distance between the transmitting and receiving coils is relatively short because the magnitude of the magnetic coupling becomes significantly weak as the transmission distance increases [4, 5]. Therefore, in order to apply RIC-WPT technique to various practical applications, increasing the transmission distance is essential.

As a method of improving the transmission distance, adding the intermediate resonator (repeater) in addition to the transmitting and receiving coils has been widely studied [2, 4–6]. As shown in Fig. 1, the repeater is inserted at an arbitrary place between the transmitting resonator (transmitter) and the receiving resonator

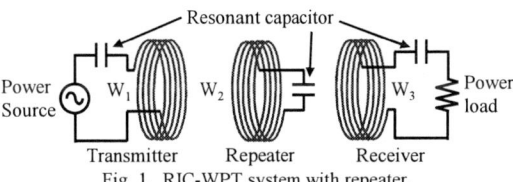

Fig. 1. RIC-WPT system with repeater.

(receiver), where W_1–W_3 represent the transmitting, repeating, and receiving coils, respectively.

As shown in Fig. 1, because the repeater has a passive configuration with only a coil and a capacitor, the quality factor (*Q*-factor) is as high as about several hundreds. Therefore, the repeater is easily excited by the little main flux generated by the transmitter. As a result, the transmission distance between the transmitter and the receiver can be increased because the repeater can generate the magnetic field by its resonance even in the location far from the transmitter.

However, the repeater tends to have low robustness against the misalignment due to its high *Q*-factor. The reason is that the high *Q*-factor easily leads to the frequency splitting phenomenon [7–12] even under weak magnetic coupling. When the frequency splitting phenomenon occurs, the RIC-WPT system has multiple resonance modes [10]. In other words, the one peak characteristics of the induced current is destroyed and the single peak of resonance splits into multiple peaks with different resonance frequencies. Especially in the RIC-WPT system configured by two coupled resonators, two resonance frequencies appear when the frequency splitting phenomenon occurs [7–12]. Among these two resonance frequencies, the lower frequency corresponds to the odd mode while the higher frequency corresponds to the even mode [8, 10].

The resonance frequencies of these multiple resonance modes are shifted according to the shift of the circuit parameters such as the change in the magnetic coupling factor [8, 11] due to the tolerance of the location of the resonators and the natural tolerance of the inductance or capacitance in the resonators. Therefore, optimum operation is commonly difficult, if the parameters of the

1610

resonator or the operating frequency cannot be adjusted to follow the resonance frequencies. However, if the frequency of the main magnetic flux generated by the transmitter deviates from the resonance frequencies corresponding to the resonance modes, the induced current in the repeater significantly decreases. As a result, the repeater does not operate effectively, causing difficulty in increasing the transmission distance.

There can be two possible approaches to solve this problem: One is dynamical adjustment of the operating frequency [11, 12]; and the other is dynamical adjustment of the resonance frequency of the resonator [9, 13–17], according to the condition of the magnetic coupling.

The former approach can be easily implemented because no additional circuit component is necessary [13]. However, in some case, this approach may not be effective, because the operating frequency may exceed frequency regulation for a RIC-WPT system allowed in a narrow bandwidth [9, 14]. In addition, if multiple resonators in the RIC-WPT system have different resonance frequency due to the manufacturing error, only adjusting the operating frequency is difficult for the practical applications [13]. Therefore, in these cases, the latter approach, i.e. adjusting the resonance frequency of the resonator, may be preferable.

Several methods have been previously proposed as a technique for actively adjusting the resonance frequency [13–17]. Many of these techniques, however, require an additional controller for the adjustment, which may lead to complexity in the circuit implementation. In order to overcome this drawback, [17] has proposed a simple circuit named as the Automatic Tuning Assist Circuit (ATAC), which can adjust the resonance frequency without any controller.

The ATAC is originally proposed to application for the transmitting resonator. According to [17], the ATAC can automatically tune the resonance frequency at the operating frequency. The ATAC can adjust the current phase of the resonator to achieve the unity power factor, which indicates the adjustment of the resonance frequency, only by driving the ATAC in synchronization with the operating frequency.

The previous study [17] has successfully elucidated effectiveness of the ATAC in the transmitting resonator under the one peak characteristic. However, effectiveness of the ATAC in the other resonators or under the frequency splitting phenomenon is still unknown. Particularly, as discussed above, adjustment of the resonance frequency at the repeater, which may easily be operated under the frequency splitting phenomenon, may be important for practical WPT systems. Therefore, the effectiveness of the ATAC should also be investigated in the application to the repeater.

The purpose of this paper is to elucidate the effectiveness of the ATAC for improving the robustness of the performance of the repeater. This paper focuses on the effect of the misalignment of the repeater and elucidates robust power transfer performance regardless to the resultant change in the coupling coefficient. Particularly, this paper verifies the effectiveness of the ATAC for the improvement of the robustness of the repeater when the frequency splitting phenomenon occurs.

The following discussion consists of three sections. Section II reviews the basic operation principle of the ATAC. Section III analyzes the RIC-WPT system with the repeater which applied the ATAC. Section IV presents simulation and experimental results to verify and discuss the effectiveness of the proposed system. Finally, section V presents the conclusions.

II. OPERATION PRINCIPLE OF ATAC

In this section, the basic operating principle of the ATAC is explained using a circuit shown in Fig. 2 which a single resonator and the ATAC are connected, where r_r is the series parasitic resistance of the resonator, L_r is the self-inductance of the coil, C_r is the resonating capacitance, C_A is the capacitance of the smoothing capacitor, and V_s is the voltage of source.

As shown in Fig. 2, the ATAC consists of the four switches Q_1–Q_4 and a capacitor C_A which has the capacitance sufficiently larger than the resonance capacitor C_r. In some cases, the ATAC having half-bridge configuration which consists of two switches may be used as reported in [17]. Two switches are turned on and off in pair, Q_1 and Q_4 are always opposite to Q_2 and Q_3. Q_1 and

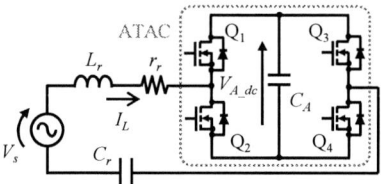

Fig. 2. Single resonator with ATAC.

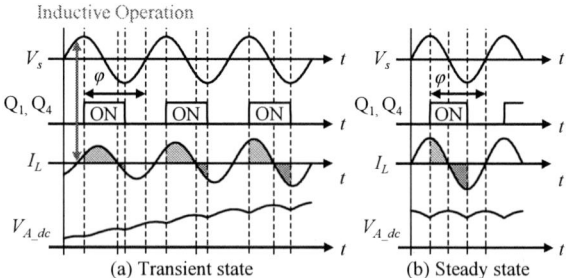

(a) Transient state   (b) Steady state

Fig. 3. Operating waveform of Fig. 2 in inductive operation.

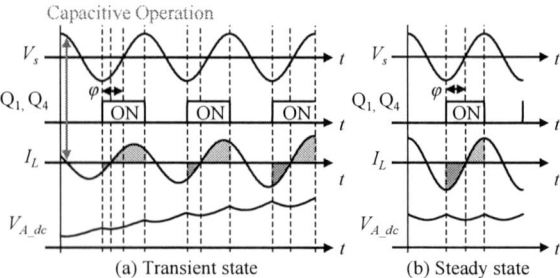

(a) Transient state   (b) Steady state

Fig. 4. Operating waveform of Fig. 2 in capacitive operation.

The 2018 International Power Electronics Conference

Q_4 are driven with an arbitrary phase difference φ with respect to the voltage source V_s. The switching frequency of Q_1–Q_4 is equal to the frequency of V_s.

Fig. 3 and Fig. 4 show waveforms in the transient and steady states when automatically adjusting the resonance frequency of the resonator which operates as the inductive or capacitive resonator. As shown in Fig. 3 (a) and Fig. 4 (a), the time integrations of the current charging and discharging the C_A are unbalanced in the transient state. As a result, the voltage of C_A increases. Because the capacitance of C_A is sufficiently larger than the C_r, C_A operates as a voltage source and supplies current to the resonator. A part of the current in the resonator generated by C_A operates to cancel the reactive current generated by V_s. In other words, the ATAC injects the reactive current in the resonator. Therefore, the ATAC equivalently adjusts the resonance frequency of the resonator. Then, when the time integration of the current charging the C_A matches the time integration of the current discharging the C_A by adjusting the resonance frequency, it becomes the steady state shown in Fig. 3 (b) and Fig. 4 (b).

As shown in Fig. 3 (b) and Fig. 4 (b), the phase difference between Q_1 (and Q_4) and I_L is always π or $3\pi/2$ in the steady state. Therefore, the phase of the resonant current can be arbitrarily adjusted by changing φ. Especially, in case of Fig. 3, when φ is set to $3\pi/2$, the reactive current generated by V_s can be canceled completely. Therefore, the resonator operates purely as resistive circuit. On the other hand, in case of Fig. 4, when φ is set to $\pi/2$, the reactance of the resonator can be equivalently canceled completely.

In case of the inductive operation (Fig. 3), the ATAC does not operate unless the phase difference φ is set to $\pi<\varphi<2\pi$ because it is necessary for C_A to generate a negative voltage in order for the ATAC to operate at $0<\varphi<\pi$. However, because Q_1–Q_4 have antiparallel diodes, negative voltage is not generated in C_A. For the same reason, in case of the capacitive operation (Fig. 5), the ATAC does not operate unless the phase difference φ is set to $0<\varphi<\pi$.

Moreover, in case of the inductive operation (Fig. 4), Q_1–Q_4 can achieve the ZVS turn-on. On the other hand, in the case of the capacitive operation (Fig. 5), Q_1–Q_4 operate hard switching. Therefore, compensating the inductive resonator is more efficient than compensating the capacitive resonator.

III. THEORETICAL ANALYSIS

In this section, the RIC-WPT with repeater applying the ATAC is analyzed. In order to compare the robustness of the repeater of the both case wherein the ATAC is applied to the RIC-WPT system or does not, we derive the induced current in the repeater and the current in the transmitter. First, we analyze the RIC-WPT system to which the ATAC is not applied. Then, we suggest the method which can improve the robustness of the repeater using the ATAC.

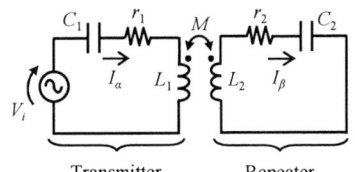

Fig. 5. Simplified equivalent circuit of Fig. 1.

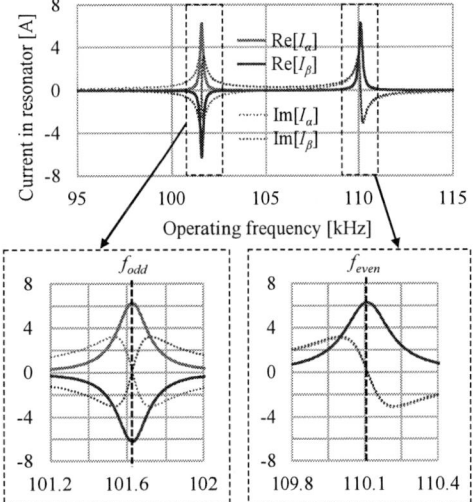

Fig. 6. Real and imaginary parts of current of fig. 5.

A. RIC-WPT without ATAC

Firstly, Fig. 5 illustrates the simplified equivalent circuit of Fig. 1, where L_1 and L_2 are the self-inductance of W_1 and W_2, C_1 and C_2 are the capacitance of the resonance capacitor, r_1 and r_2 are the series parasitic resistance of the transmitter and the repeater, and M is mutual inductance between W_1 and W_2.

In this study, we target the RIC-WPT system for small-sized low-power applications such as small mobile devices and implanted biomedical devices. In general, in case of these applications, the receiving coil has the relatively small magnetic coupling to the other coils [18]. Hence, for the simplification, it is assumed that the influence of the receiving resonator is ignored. The RIC-WPT systems which are easily affected by the receiving resonator will be addressed in other future paper.

In addition, for convenience, we assume that the transmitter and repeater have the same parameter, i.e. $L_1=L_2=L$, $C_1=C_2=C$, $r_1=r_2=r$.

According to the Kirchhoff's voltage law, the system of Fig. 5 can be described as

$$\begin{cases} V_i = rI_\alpha + j(\omega L - 1/\omega C)I_\alpha - j\omega M I_\beta, \\ -j\omega L I_\beta + j\omega M I_\alpha = rI_\beta - jI_\beta/\omega C. \end{cases} \quad (1)$$

Then, we define the impedance as

$$j(\omega L - 1/\omega C) = jX. \quad (2)$$

By substituting (2) into (1), the currents of the transmitter and the repeater can be derived accordingly as

$$I_\alpha = V_i(N_{1R} + jN_{1I})/D_p, \quad (3)$$

$$I_\beta = V_i(N_{2R} + jN_{2I})/D_p, \quad (4)$$

1612

where, N_{1R}, N_{1I}, N_{2R}, N_{2I}, and D_p are defined as

$$N_{1R}=r\{(r^2+X^2)+\omega^2M^2\},$$
$$N_{1I}=X(\omega^2M^2-r^2-X^2),$$
$$N_{2R}=2\omega MrX,\qquad\qquad\qquad (5)$$
$$N_{2I}=\omega M(\omega^2M^2+r^2-X^2),$$
$$D_p=2\omega^2M^2\{r^2-X^2+\omega^2M^2/2\}+(r^2+X^2)^2.$$

Next, we discuss the method which can improve the robustness of the induced current in repeater by applying the ATAC to Fig. 5. From (3), (4), and (5), the real and imaginary parts of the current in each resonator can be described as shown in Fig. 6, where f_{odd} and f_{even} are the odd and even mode resonance frequencies, respectively. Parameters V_i, L, M, C, and r are set at 2.25V, 140.70μH, 11.26μH, 16.14nF, and 0.18Ω, respectively.

As shown in Fig. 6, $\mathrm{Re}[I_\alpha]$ and $\mathrm{Re}[I_\beta]$ are in opposite phases at f_{odd}. Furthermore, the magnitude of $\mathrm{Im}[I_\alpha]$ and $\mathrm{Im}[I_\beta]$ are almost zero. On the other hand, at f_{even}, $\mathrm{Re}[I_\alpha]$ and $\mathrm{Re}[I_\beta]$ are in phases and $\mathrm{Im}[I_\alpha]$ and $\mathrm{Im}[I_\beta]$ are almost zero.

Therefore, if the ATAC is operated to cancel the $\mathrm{Im}[I_\alpha]$ or $\mathrm{Im}[I_\beta]$, it may be possible to drive equivalently the RIC-WPT as either of two resonance modes even if the operating frequency deviates from the frequency corresponding to the resonance mode. Therefore, we propose Fig. 7 which the ATAC is applied to the repeating side in order to cancel $\mathrm{Im}[I_\beta]$ and Fig. 8 which the ATAC is applied to the transmitting side in order to cancel $\mathrm{Im}[I_\alpha]$. Next, we analyze Fig. 7 and Fig. 8.

B. RIC-WPT with ATAC of Repeating Side (Fig. 7)

In order to derive I_{L1} and I_{L2}, we derive V_A in the steady state, where V_A is the effective value of the fundamental wave component of the rectangular wave voltage generated by the ATAC.

The phase diagram of Fig. 9 shows V_i and V_A of Fig.7. Because the phase of Q_1 and Q_4 is advanced by φ with respect to V_i, the phase difference between V_A and V_i is φ. To simplify the analysis, the phasor diagram with V_i as a reference shown in Fig. 9 (a) is rotated clockwise by $\varphi-\pi/2$. As a result, the phasor diagram with jV_A as a reference shown in Fig. 9 (b) is obtained.

I'_α, I'_β are defined as I_α and I_β after rotating the phasor diagram, respectively. From (3), I'_α and I'_β in the phasor diagram of Fig. 9 (b) can be expressed as

$$\mathrm{Re}[I'_\alpha]=V_i(N_{1R}\sin\varphi+N_{1I}\cos\varphi)/D_p,$$
$$\mathrm{Im}[I'_\alpha]=V_i(-N_{1R}\cos\varphi+N_{1I}\sin\varphi)/D_p,$$
$$\mathrm{Re}[I'_\beta]=V_i(N_{2R}\sin\varphi+N_{2I}\cos\varphi)/D_p,\qquad (6)$$
$$\mathrm{Im}[I'_\beta]=V_i(-N_{2R}\cos\varphi+N_{2I}\sin\varphi)/D_p.$$

In the steady state, the imaginary part of I_{L2} in Fig. 9 (b) is zero because the imaginary part of the currents generated by V_A and V_i cancel each other. Therefore, $\mathrm{Im}[I_{L2}]$ expressed as

$$\mathrm{Im}[I_{L2}]=\mathrm{Im}[I'_\beta]+\mathrm{Im}[-jV_A/Z_o]=0,\qquad (7)$$

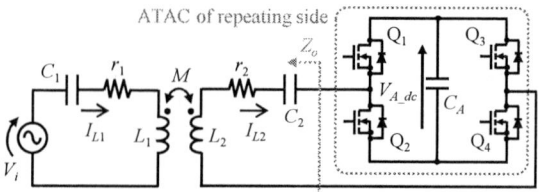

Fig. 7. RIC-WPT with ATAC of repeating side.

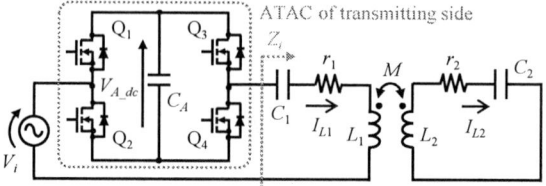

Fig. 8. RIC-WPT with ATAC of transmitting side.

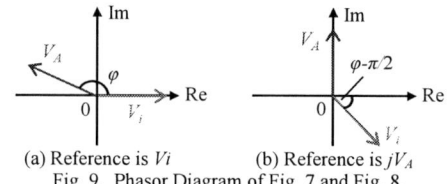

(a) Reference is V_i (b) Reference is jV_A

Fig. 9. Phasor Diagram of Fig. 7 and Fig. 8.

where, Z_o expressed as

$$Z_o=r\left(1+\frac{\omega^2M^2}{r^2+X^2}\right)+jX\left(1-\frac{\omega^2M^2}{r^2+X^2}\right).\qquad (8)$$

By substituting (6) and (8) into (7), V_A can be derived accordingly as

$$V_A=2\sqrt{2}V_{A\_dc}/\pi=V_i(-N_{2R}\cos\varphi+N_{2I}\sin\varphi)/N_{1R}.\qquad (9)$$

I_{L2} is represented by the superposition of the current by V_A and the current by V_i. Therefore, I_{L2} expressed as

$$\begin{aligned}I_{L2}&=\mathrm{Re}[I'_\beta]+\mathrm{Re}[-jV_A/Z_o]\\&=\{(N_{2R}\sin\varphi+N_{2I}\cos\varphi)V_i+N_{1I}V_A\}/D_p.\end{aligned}\qquad (10)$$

Then, we derive I_{L1}. Similar to equation (10), I_{L1} is also represented by the superposition of the currents generated by V_i and V_A. The real and imaginary parts of I_{L1} generated by V_i are expressed by (6). Moreover, the real and imaginary parts of I_{L1} generated by V_A can be derived using the Kirchhoff's voltage law similar to (3) and (4). Therefore, I_{L1} can be expressed as

$$I_{L1}=\sqrt{\left(\mathrm{Re}[I'_\alpha]+\frac{N_{2I}}{D_p}V_A\right)^2+\left(\mathrm{Im}[I'_\alpha]-\frac{N_{2R}}{D_p}V_A\right)^2}.\qquad (11)$$

C. RIC-WPT with ATAC of Transmitting Side (Fig. 8)

In case of Fig. 8, in order to derive I_{L1} and I_{L2}, we also derive V_A similar to the case of Fig. 7. Fig.9 (b) is used for the phasor diagram for analyzing Fig. 8.

In the steady state, the imaginary part of I_{L1} equal to zero because the phase difference between V_A and I_{L1} is always π or $3\pi/2$ as explained in previous section. Therefore, $\mathrm{Im}[I_{L1}]$ can be described as

$$\mathrm{Im}[I_{L1}]=\mathrm{Im}[I'_\alpha]+\mathrm{Im}[-jV_A/Z_i]=0 \ , \qquad (12)$$

where, Z_i expressed as

$$Z_i=Z_o=r\left(1+\frac{\omega^2M^2}{r^2+X^2}\right)+jX\left(1-\frac{\omega^2M^2}{r^2+X^2}\right). \qquad (13)$$

By substituting (6) and (13) into (12), V_A can be derived accordingly as

$$V_A=2\sqrt{2}V_{A\_dc}/\pi=V_i(-N_{1R}\cos\varphi+N_{1I}\sin\varphi)/N_{1R}. \ (14)$$

I_{L1} is represented by the superposition of the current by V_A and the current by V_i. Therefore, I_{L1} expressed as

$$\begin{aligned}I_{L1}&=\mathrm{Re}[I'_\alpha]+\mathrm{Re}[-jV_A/Z_i]\\&=\{(N_{1R}\sin\varphi+N_{1I}\cos\varphi)V_i+N_{1I}V_A\}/D_p\,.\end{aligned} \qquad (15)$$

Finally, similar to (11), I_{L2} is obtained as

$$I_{L2}=\sqrt{\left(\mathrm{Re}[I'_\beta]+\frac{N_{2I}}{D_p}V_A\right)^2+\left(\mathrm{Im}[I'_\beta]-\frac{N_{2R}}{D_p}V_A\right)^2}\,. \ (16)$$

IV. DISCUSSION AND EXPERIMENTAL RESULTS

Based on the analysis in the previous section, we evaluate the effectiveness of the proposed systems (Fig. 7 and Fig. 8) when the frequency splitting phenomenon occurs. In addition, we confirm the appropriateness of the analysis results of the previous section by experiment and simulation.

Table I shows the circuit parameters for the experiment and simulation. Fig. 10 shows the photographs of experimental setup of the RIC-WPT system with the ATAC of the repeating side. In the experiment, V_i of Fig. 5, Fig. 7, and Fig. 8 are generated by the half-bridge inverter. The gate signals for the ATAC of Fig. 7 and Fig. 8 are generated using the function generator. Moreover, as pointed out in the previous section, the phase difference φ is set to $\pi/2$ or $3\pi/2$ in order to cancel $\mathrm{Im}[I_\alpha]$ or $\mathrm{Im}[I_\beta]$.

Then, we define frequency regions as shown in Fig. 11, where f_0 is natural resonance frequency of the transmitter and the repeater. As pointed out in Section II, negative voltage cannot be generated in C_A of the ATAC because Q_1–Q_4 have antiparallel diode respectively. Therefore, φ is selected as $\pi/2$ or $3\pi/2$ so that V_{A_dc} does not become negative. Specifically, in the case of the RIC-WPT with the ATAC of the repeating side (Fig .7), φ is set to $\pi/2$ in

TABLE I
CIRCUIT PARAMETERS FOR EXPERIMENT AND SIMULATION

Input AC voltage	V_i	1.80V
Self-inductance of W_1	L_1	55.35μH
Parasitic resistance of W_1	r_1	0.16Ω
Capacitance of transmitter	C_1	12.15nF
Self-inductance of W_2	L_2	55.59μH
Parasitic resistance of W_2	r_2	0.16Ω
Capacitance of repeater	C_2	12.13nF
Mutual inductance	M	5.55μH
Phase difference	φ	$\pi/2$ or $3\pi/2$

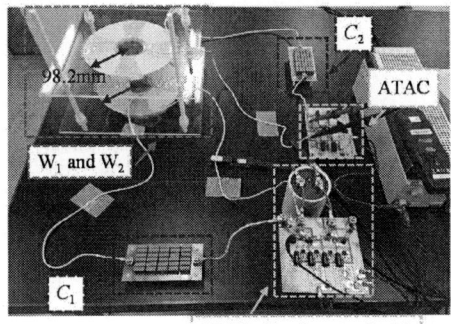

Fig. 10. Experimental setup.

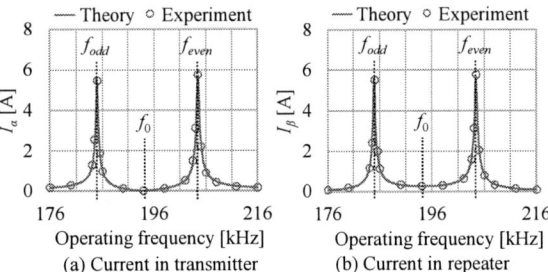

Fig. 11. Definition of frequency regions.

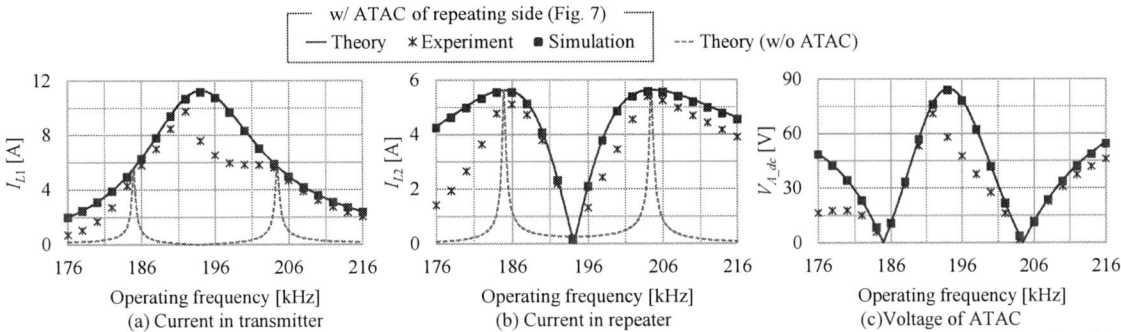

(a) Current in transmitter (b) Current in repeater

Fig. 12. Calculated and experimental results of frequency characteristics.

the frequency regions II and III, and φ is set to $3\pi/2$ in the frequency regions I and IV. On the other hand, in the case of the RIC-WPT with the ATAC of the transmitting side

— Theory　*Experiment　■ Simulation　--- Theory (w/o ATAC)

w/ ATAC of repeating side (Fig. 7)

(a) Current in transmitter　(b) Current in repeater　(c)Voltage of ATAC

Fig. 13. Calculated, simulated, and experimental results of frequency characteristics for the RIC-WPT which applied ATAC to repeating side.

1614

The 2018 International Power Electronics Conference

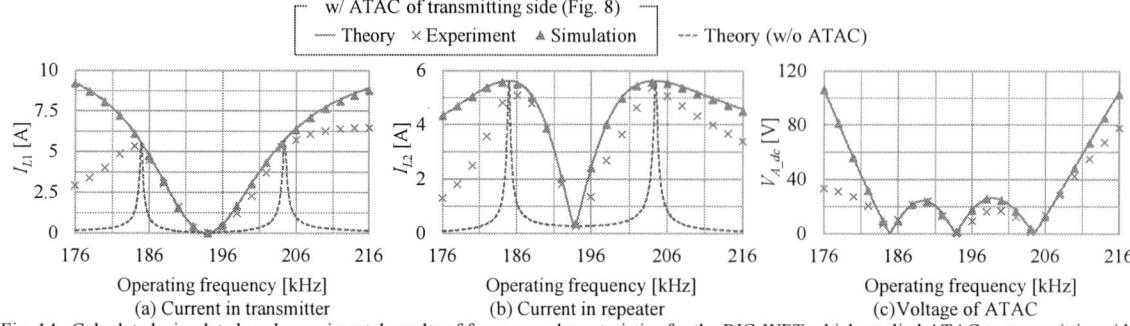

Fig. 14. Calculated, simulated, and experimental results of frequency characteristics for the RIC-WPT which applied ATAC to transmitting side.

(Fig. 8), φ is set to $\pi/2$ in the frequency regions I and III, and φ is set to $3\pi/2$ in the frequency regions II and IV.

In order to confirm the appropriateness of the analysis results of the previous section, we evaluate the frequency characteristic of Fig. 5, Fig. 7, and Fig. 8. Firstly, Fig. 12 shows the calculated and experimental results of the frequency response of the current in each resonator. As can be seen form Fig. 12, the experimental results correspond well with the calculated values in case of the RIC-WPT system without the ATAC.

Then, calculated, simulation and experimental results of the frequency characteristics of the RIC-WPT which the ATAC is applied to the repeating side (Fig. 7) and the RIC-WPT which the ATAC is applied to the transmitting side (Fig.8) are shown in Fig. 13 and Fig. 14, respectively. In cases of the RIC-WPT with the ATAC, the experimental results have some error with respect to the calculated value compared with the RIC-WPT without the ATAC. These errors occurred because the Q-factor of the resonator used in the experiment is as large as about several hundreds, so that the slight conduction loss and switching loss of Q_1–Q_4 in the ATAC are relatively easily influence to experimental results. In fact, the simulation results that the conduction and switching losses are not included are corresponded well with the calculated values. Therefore, the appropriateness of the theoretical analysis in the previous section can be confirmed.

As can be seen Fig. 13 and Fig. 14, the error is particularly large in the frequency regions I and III because the switches (Q_1–Q_4) of the ATAC cannot be realized ZVS (Zero Voltage Switching) turn-on in these frequency regions as shown in Table II, where HS represents Hard

TABLE II
SOFT SWITCHING ANALYSIS OF ATAC

Frequency region	I	II	III	IV
Q_1–Q_4 of ATAC (repeating side)	HS turn-on	ZVS turn-on	HS turn-on	ZVS turn-on
Q_1–Q_4 of ATAC (transmitting side)	HS turn-on	ZVS turn-on	HS turn-on	ZVS turn-on

Switching. Therefore, when the ATAC is applied to the RIC-WPT, it is desirable to apply it to the frequency range which ZVS turn-on can be realized.

From Fig. 14 (b), the RIC-WPT with the ATAC of the transmitting side can increase I_{L2} in all frequency regions compared to the RIC-WPT without the ATAC. Similarly, as can be seen Fig. 13 (b), the RIC-WPT with the ATAC of the repeating side can increase I_{L2} in a wide frequency range as compared with the RIC-WPT without the ATAC except frequency near f_0. Moreover, in the cases of RIC-WPT systems with the ATAC, the frequency characteristic of I_{L2} can be made gentle near f_{odd} and f_{even} compared to the RIC-WPT without the ATAC. Therefore, I_{L2} can be kept large even at frequencies a little deviated from the frequency at which I_{L2} reaches a peak.

As can be seen from Fig. 13 (b) and Fig. 14 (b), the frequency characteristics of I_{L2} are almost same between the two kinds of the RIC-WPT with the ATAC. On the other hand, the frequency characteristics of I_{L1} and V_{A_dc} are significantly different between the two kinds of the RIC-WPT with ATAC. In the frequency regions I and IV, the RIC-WPT with the ATAC of the repeating side (Fig .7) is superior to the RIC-WPT with the ATAC of the transmitting side (Fig. 8) because I_{L1} and V_{A_dc} can be made

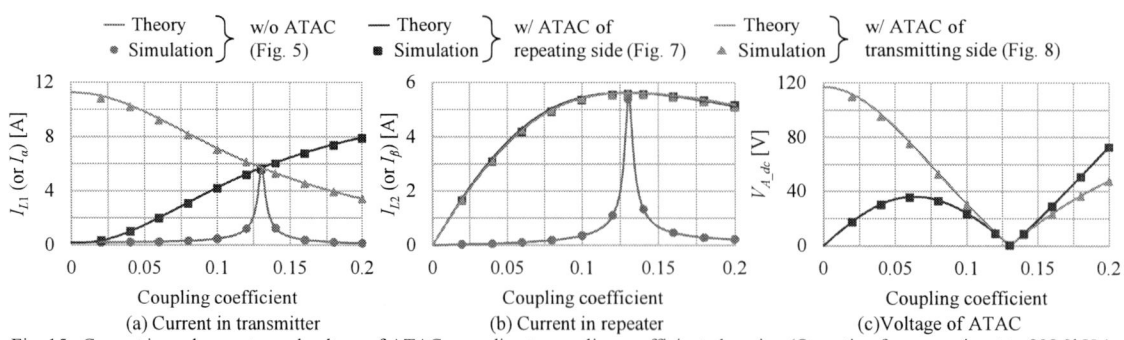

Fig. 15. Current in each repeater and voltage of ATAC according to coupling coefficient changing (Operating frequency is set to 208.0kHz).

1615

smaller. Meanwhile, in the frequency regions II and III, the RIC-WPT with the ATAC of the transmitting side (Fig. 8) is more advantageous than the RIC-WPT with the ATAC of the repeating side (Fig .7).

Next, under the operating frequency is arbitrarily fixed, we evaluate the characteristics when the coupling coefficient k is changed, where k is defined as $k=M/L$. The operating frequency is set to 208.0 kHz. Fig. 15 shows the evaluation results. From Fig. 15, when the ATAC is not applied, a large current cannot be induced to the repeater unless the coupling coefficient is set to a specific value. This means that the robustness of the repeater against the misalignment is low. On the other hands, when the ATAC is applied, a relatively large current can be induced to the repeater in a wide coupling coefficient range. Therefore, the robustness of the repeater can be improved by applying the ATAC.

V. CONCLUSIONS

The frequency splitting phenomenon easily occurs in the induced current of the repeater because the Q-factor of the repeater is high. Therefore, the repeater has low robustness against misalignment. In order to improve the robustness of the repeater, we applied the ATAC to the RIC-WPT with the repeater. As a result, it is clarified that the induced current in the repeater can be kept large in a relatively wide operating frequency range which is slightly different from the resonance frequency. Moreover, experiment and simulation successfully verified effectiveness of the proposed system as well as appropriateness of the theoretical analysis.

REFERENCES

[1] C.-S. Wang, G. A. Covic, and O. H. Stielau, "Power transfer capability and bifurcation phenomena of loosely coupled inductive power transfer systems," *IEEE Trans. Ind. Electron.*, vol. 51, issue. 1, pp. 148–157, Feb. 2004.

[2] D. H. Tran, V. B. Vu, and W. choi, "Design of a high-efficiency wireless power transfer system with intermediate coils for the on-board chargers of electric vehicles," *IEEE Trans. power. Electron.*, vol. 33, issue. 1, pp. 175–187, Jan. 2018.

[3] D. Ahn and S. Hong, "Wireless power transmission with self-regulated output voltage for biomedical implant," *IEEE Trans. Ind. Electron.* vol. 61, issue. 5, pp. 2225–2235, May. 2014.

[4] D. Ahn, S. Hong, "A study on magnetic field repeater in wireless power transfer," *IEEE Trans. on Industrial Electronics*, vol. 60, no. 1, pp. 360–371, 2013.

[5] K. Lee and S. H. Chae, "Power transfer efficiency analysis of intermediate-resonator for wireless power transfer," *IEEE Trans. power. Electron.*, vol. 33, issue. 3, pp. 2484–2493, Mar. 2018.

[6] P. K. S. Jayathurathnage, and D. M. Vilathgamuwa, "Optimization of a wireless power transfer system with a repeater against load variations," *IEEE Trans. Ind. Electron.*, vol. 64, issue. 10, pp. 7800–7809, Oct. 2017.

[7] W. Q. Niu, J. X. Chu, W. Gu, and A. D. Shen, "Exact analysis of frequency splitting phenomena of contactless power transfer systems," *IEEE Trans. on Circuit and Systems*, vol.60, no. 6, pp. 1670–1677, Jun. 2013.

[8] R. Huang, B. Zhang, D. Qiu, and Y. Zhang, "Frequency splitting phenomena of magnetic resonant coupling wireless power transfer," *IEEE Trans. on Magnetics*, vol. 50, no. 11, 8600204, Nov. 2014.

[9] D. -W. Seo, J. -H. Lee, and H. -S. Lee, "Optimal coupling to achieve maximum output power in a WPT system," *IEEE Trans. power. Electron.*, vol. 31, issue. 6, pp. 3994–3998, Jun. 2016.

[10] O. Karaca, F. Kappeler, D. Waldau, R. M. Kennel, and J. Rackles, "Eigenmode analysis of a multiresonant wireless energy transfer system," *IEEE Trans. Ind. Electron.*, vol. 61, issue. 8, pp. 4134–4141, Aug. 2014.

[11] D. -W. Seo and J. -H. Lee, "Frequency-tuning method using the reflection coefficient in a wireless power transfer system," *IEEE Microw. Compon. Lett.*, vol. 27, issue. 11, pp. 959–961, Nov. 2017.

[12] Y. Li, C. Zhang, Q. Yang, J. Li, Y. Zhang, X. Zhang, and M. Xue, "Improved ant colony algorithm for adaptive frequency-tracking control in WPT system," *IET Microw. Antennas Propag*, vol. 12, issue. 1, pp. 23–28, Jan. 2018.

[13] R. Mai, P. Yue, Y. Liu, Y. Zhang, and Z. He, "A dynamic tuning method utilizing inductor paralleled with load for inductive power transfer," *IEEE Trans. power. Electron.*, to be published.

[14] A. Trigui, S. Hached, F. Mounaim, A. C. Ammari, and M. Sawan, "Inductive power transfer system with self-calibrated primary resonant frequency," *IEEE Trans. on Power Electronics*, vol. 30, no. 11, pp. 6078–6087, Nov. 2015.

[15] N. Liu, and T. G. Habetler, "An active-rectifier-based maximum efficiency tracking method using an additional measurement coil for wireless power transfer," *IEEE Trans. Power Electron.*, vol. 33, issue. 1, pp. 716–728, Jan. 2018.

[16] T. Isobe, K. Kobayashi, K. Wakasugi, and R. Shimada, "Efficiency improvement of contactless energy transfer systems using series compensation device named MERS," in *Proc. 14th European Conf. Power Electron. Applicat. (EPE)*, 2011, Birmingham, pp. 1–10.

[17] Y. Endo and Y. Furukawa, "Proposal for a new resonance adjustment method in magnetically coupled resonance type wireless power transmission," in *Proc. 2012 IEEE MTT-S International Microwave Workshop Series on Innovative Wireless Power Transmission: Technologies, Systems, and Applications (IMWS-IWPT)*, 2012, Kyoto, pp. 263–266.

[18] S. Mao, J. Zhang, K. Song, G. Wei, and C. Zhu, "Wireless power transfer using a field-enhancing coil and a small-sized receiver with low coupling coefficient," *IET power electron.*, vol. 9, issue. 7, pp. 1546–1552, Jun. 2016.

The 2018 International Power Electronics Conference

Inductive Power Transfer for T5 Fluorescent Lamp Lighting System

Chung-Chuan Hou*, Tang-Jung Chen, Ching-Chen Chen, Chen-Wei Chang and Po-Wei Wang
Dept. of Electrical Engineering, Chung Hua University, Hsinchu, Taiwan, R.O.C.
*E-mail: bird@chu.edu.tw

Abstract—This study discusses the characteristics of the inductive power transfer (IPT) for T5 fluorescent lamps (FL) lighting system. The advantages of the IPT technology are electrical isolation and capability of operating in dirt environment. The transfer function and frequency response of the IPT for T5 FL lighting system are derived. The experimental results under varied air gap and load are discussed to evaluate the property of the IPT for T5 FL lighting system.

Keywords— Inductive Power Transfer, Fluorescent Lamp, Series-Parallel Resonant, Frequency Response.

I. INTRODUCTION

The T5 fluorescent lamp (FL) becomes more extensive because of low cost and high lumen/power. The analysis and design of the ballast for T5 FL lighting system are presented in the previous researches [1-6]. In the recent years, the technology of the inductive power transfer (IPT) develops rapidly [7-10]. The advantages of the IPT technology are electrical isolation and capability of operating in dirt environment. Therefore, this study discusses the characteristics of the IPT for T5 FL lighting system [11].

Figure 1 shows the inductive power transfer for T5 FL lighting system. A single phase full bridge inverter is designed to produce an AC voltage (v_{in}) as the input of the IPT system. The technology of the IPT system adopts the series-parallel resonant (SP) topology. The transformer consists of two layers. One layer is planar spiral coils; the outer diameter and inner diameter of the planar spiral coils are 43 mm and 20 mm. The turns of spiral coils are 80. The other layer is a square ferrite (53mm x 53mm x 5mm). The primary inductance is L_p (258 µH), secondary inductance is L_s (258 µH), and mutual inductance is M ($= L_m$), respectively. Equation (1) defines the coupling coefficient (k) as follow.

$$k = \frac{M}{\sqrt{L_p L_s}} \tag{1}$$

The capacitances C_P (0.22 µF) and C_S (0.033 µF) are placed at primary side and secondary side of the IPT system. The voltage and current of the T5 FL are v_L and i_L. The equivalent resistance of the T5 FL is R_L defined as in (2).

$$R_L = v_L / i_L \tag{2}$$

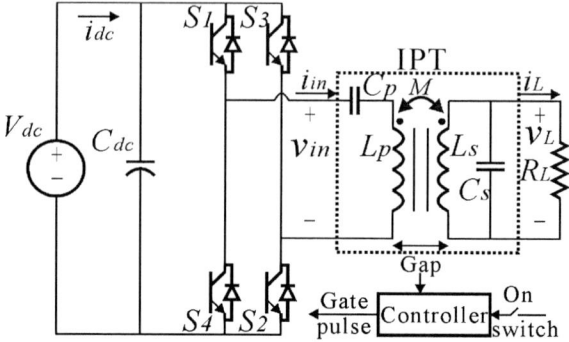

Fig. 1. Inductive power transfer for T5 FL lighting system.

The R_L is a nonlinear function. When lamp is inactive, the R_L is high impedance; when lamp is active, the R_L is low impedance [6]. The gap is defined as the distance between two planar spiral coils. The gap signal imports to the controller and decides the operating range (gap: 2mm - 8mm). When switch is turn on, the controller generates the gate signals for MOSFET (IPW65R041CFD) switches (S_1, S_2, S_3, and S_4).

II. TRANSFER FUNCTION OF IPT SYSTEM

Figure 2 shows the simplified circuit of the IPT for T5 FL lighting system. The oscillation frequency ω_o is defined as ($\omega_o = 1/\sqrt{L_s C_s}$; $f_o = \omega_o /2\pi = 55$kHz). The primary resistance and secondary resistance of planar spiral coils are r_P and r_S (0.6Ω). The equivalent resistance of transformer loss is R_m. The primary leakage inductance is L_{lp} and secondary leakage inductance is L_{ls} as given in (3).

$$L_{lp} = (1-k) \cdot L_p, \quad L_{ls} = (1-k) \cdot L_s \tag{3}$$

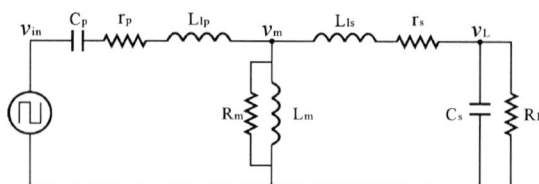

Fig. 2. Simplified circuit of the IPT for T5 FL lighting system.

As shown in figure 2, according the simplified circuit of the IPT for T5 FL lighting system, the transfer function is simplified as two zeros (s = 0, 0) and five poles as in (4).

$$\frac{v_L(s)}{v_{in}(s)} = \frac{s^2 \times C_P \times L_m \times R_L / [R_m \times (R_L + r_S)]}{(1 + \frac{s}{p_H}) \times (1 + \frac{2\zeta_1}{\omega_{n1}}s + \frac{s^2}{\omega_{n1}^2}) \times (1 + \frac{2\zeta_2}{\omega_{n2}}s + \frac{s^2}{\omega_{n2}^2})} \quad (4)$$

where high frequency pole (s = -p_H) is defined as in (5).

$$p_H = R_m \times (\frac{1}{L_{lp}} + \frac{1}{L_{ls}} + \frac{1}{L_m}) \quad (5)$$

The low frequency complex-conjugate poles (s= -p_{11}, -p_{12}) are defined as in (6).

$$1 + \frac{2\zeta_1}{\omega_{n1}}s + \frac{s^2}{\omega_{n1}^2} = 0 \Rightarrow s = -(\zeta_1 \pm j\sqrt{1 - \zeta_1^2})\omega_{n1} \quad (6)$$

The damping ratio ζ_1, natural undamped frequency ω_{n1}, and resonant frequency ω_{r1} are as in (7).

$$\zeta_1 = r_p C_p \omega_{n1} / 2$$
$$\omega_{n1}^2 = 1 / C_p (L_{lp} + L_m)$$
$$\omega_{r1} = \sqrt{1 - \zeta_1^2}\ \omega_{n1} \quad (7)$$

The middle frequency complex-conjugate poles (s = -p_{21}, -p_{22}) are defined as in (8).

$$1 + \frac{2\zeta_2}{\omega_{n2}}s + \frac{s^2}{\omega_{n2}^2} = 0 \Rightarrow s = -(\zeta_2 \pm j\sqrt{1 - \zeta_2^2})\omega_{n2} \quad (8)$$

The damping ratio ζ_2, natural undamped frequency ω_{n2}, and resonant frequency ω_{r2} are as in (9).

$$\zeta_2 = 1 / 2R_L C_S \omega_{n2}$$
$$\omega_{n2}^2 = (1 + r_S / R_L) / C_S (L_{ls} + \frac{L_{lp}L_m}{L_{lp} + L_m})$$
$$\omega_{r2} = \sqrt{1 - \zeta_2^2}\ \omega_{n2} \quad (9)$$

The frequency response of the SP resonant IPT system for T5 FL lighting system (gap: 2mm - 8mm) is as shown in figure 3. The parameters of the SP resonant IPT system are as follow. The DC side and AC side voltages of the single phase full bridge inverter are V_{dc} (50V) and v_{in}. The switching frequency (fsw) for SP resonant topology IPT system is from 22 kHz to 100 kHz and the oscillation frequency ω_o (fo) is designed as 55 kHz. Figure 3(a) (R_L: 1k Ω) shows the resonant frequency ω_{r1} (21 kHz) and ω_{r2} is moved to the oscillation frequency (fo) as air gap increased. Figure 3(b) (R_L: 400 Ω) shows the output voltages near resonant frequency ω_{r2} are decreased as load increased (R_L from 1kΩ to 400Ω). As shown in figure 3, the frequency response of the SP topology IPT system for T5 FL lighting system agrees with the transfer function given in (4). The inductances and poles of the IPT for T5 FL lighting system (gap: 2mm - 8mm) are as table I (R_L: 400 Ω).

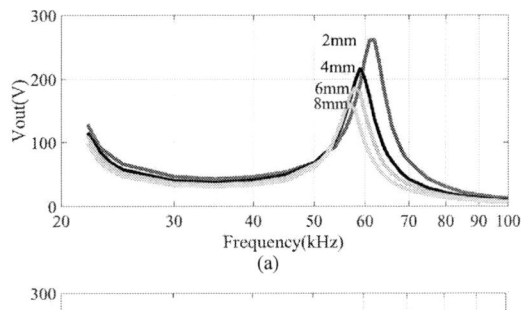

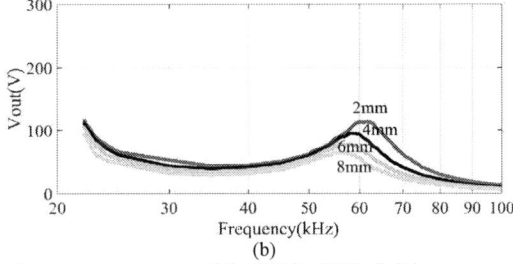

Fig. 3. Frequency response of the IPT for T5 FL lighting system under varied air gap (a) R_L (1k Ω) (b) R_L (400 Ω).

TABLE I
INDUCTANCES AND POLES OF THE IPT FL LIGHTING SYSTEM

Gap	2mm	4mm	6mm	8mm
L_{lp}=L_{ls}	149.64uH	159.96uH	170.28uH	180.6uH
L_m	108.36uH	98.04uH	87.72uH	77.4uH
k	0.42	0.38	0.34	0.30
ω_{r1}	21 kHz	21 kHz	21 kHz	21 kHz
ω_{r2}	60 kHz	59 kHz	58 kHz	57 kHz
P_H	2.3MHz	2.3 MHz	2.3MHz	2.3MHz

III. EXPERIMENTAL RESULTS

The experimental platform is design as follow:
- IPT: Lp (258 μH) and Ls (258 μH); transformer 80 turns, ferrite material 3C90; C_P (0.22 μF) and C_S (0.033 μF); gap 2-8mm.
- Load: R_L 400Ω - 1kΩ; T5/28W fluorescent lamp.
- Inverter: V_{dc} : 50V; MOSFET (IPW65R041CFD); fsw: 22-100 kHz.
- Controller: Atmel ATMEGA 328P.

Figure 4 shows the input power, output power and efficiency of the IPT for T5 FL lighting system under load R_L (1kΩ). The input power and output power are decreased as air gap increasing as shown in figure 4(a) and 4(b). The efficiency is higher than 90% operated at switching frequency range 51-60 kHz (gap: 2mm - 8mm) as shown in figure 4(c).

As shown in figure 3(a) and figure 4 (R_L : 1k Ω), the output voltage, input power, output power and efficiency operated at fsw 62 kHz and gap 2mm are 260.7 Vrms, 73.2W, 67.6W, and 93% ; operated at fsw 59 kHz and gap 4mm are 217.0 Vrms, 50.4W, 46.8W, and 93% ; operated at fsw 58 kHz and gap 6mm are 186.2 Vrms, 36.9W, 34.4W, and 93% ; operated at fsw 57 kHz and gap 8mm are 160.8 Vrms, 27.9W, 25.6W, and 91%, respectively.

The 2018 International Power Electronics Conference

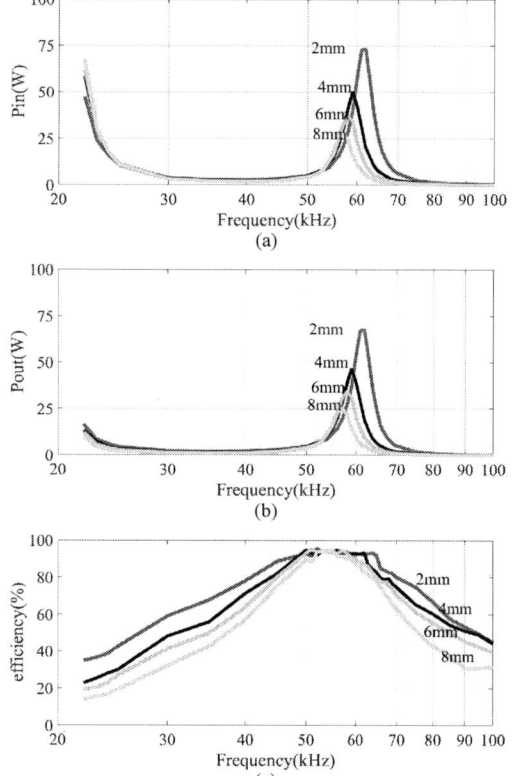

Fig. 4. Input-output powers and efficiency of the IPT for T5 FL lighting system under load R_L (1k Ω) (a) input power (b) output power (c) efficiency.

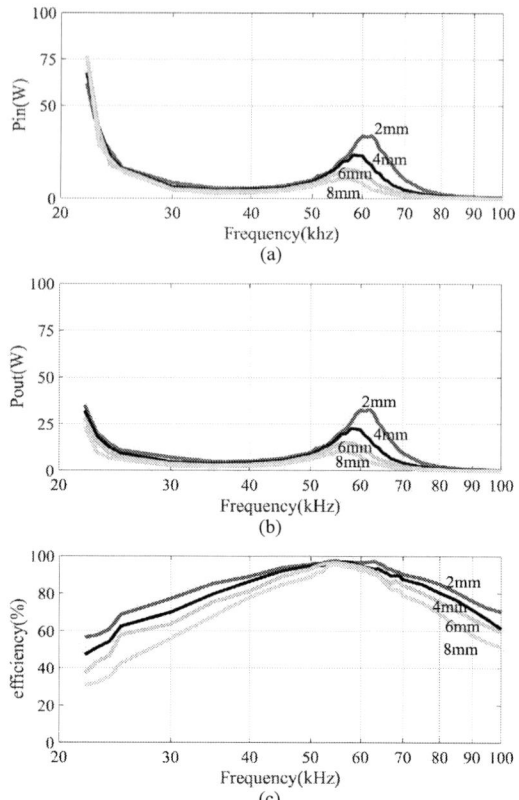

Fig. 5. Input-output powers and efficiency of the IPT for T5 FL lighting system under load R_L (400 Ω) (a) input power (b) output power (c) efficiency.

Figure 5 shows the input power, output power and efficiency of the IPT for T5 FL lighting system under load R_L (400 Ω). The input power and output power are decreased as air gap increasing as shown in figure 5(a) and 5(b). The efficiency is higher than 90% operated at switching frequency range 51-64 kHz (gap: 2mm - 8mm) as shown in figure 5(c).

As shown in figure 3(b) and figure 5 (R_L : 400 Ω), the output voltage, input power, output power and efficiency operated at fsw 60 kHz and gap 2mm are 113.2 Vrms, 33.6W, 32.4W, and 96% ; operated at fsw 59 kHz and gap 4mm are 94.5 Vrms, 23.4W, 22.4W, and 95.5% ; operated at fsw 58 kHz and gap 6mm are 77.0 Vrms, 15.5W, 14.9W, and 95.5% ; operated at fsw 57 kHz and gap 8mm are 64.4 Vrms, 11.0W, 10.4W, and 94%, respectively.

As shown in figure 4 and figure 5, the input and output powers of the SP resonant topology IPT for T5 FL lighting system agree with the frequency response as shown in figure 3.

Figure 6 demonstrates the voltage and current of T5 FL operated at transient state and steady state (fsw: 60 kHz; gap: 2mm). The transient state waveforms of the T5 FL during ignition as shown in figure 6(a). The controller generates the gate pulse at time (t_0) and the FL illuminates at time (t_1) (t_1-t_0: 100 ms). The steady state waveforms of the T5 FL are $v_{L,peak}$ (200V) and $i_{L,peak}$ (0.3A) as shown in figure 6(b).

Figure 7 demonstrates the voltage and current of T5 FL operated at transient state and steady state (fsw: 59 kHz; gap: 4mm). The transient state waveforms of the T5 FL during ignition as shown in figure 7(a). The controller generates the gate pulse at time (t_0) and the FL illuminates at time (t_1) (t_1-t_0: 220 ms). The steady state waveforms of the T5 FL are $v_{L,peak}$ (190V) and $i_{L,peak}$ (0.3A) as shown in figure 7(b).

Figure 8 demonstrates the voltage and current of T5 FL operated at transient state and steady state (fsw: 58 kHz; gap: 6mm). The transient state waveforms of the T5 FL during ignition as shown in figure 8(a). The controller generates the gate pulse at time (t_0) and the FL illuminates at time (t_1) (t_1-t_0: 315 ms). The steady state waveforms of the T5 FL are $v_{L,peak}$ (200V) and $i_{L,peak}$ (0.25A) as shown in figure 8(b).

Figure 9 demonstrates the voltage and current of T5 FL operated at transient state and steady state (fsw: 57 kHz; gap: 8mm). The transient state waveforms of the T5 FL during ignition as shown in figure 9(a). The controller generates the gate pulse at time (t_0) and the FL illuminates at time (t_1) (t_1-t_0: 280 ms). The steady state waveforms of the T5 FL are $v_{L,peak}$ (220V) and $i_{L,peak}$ (0.2A) as shown in figure 9(b).

As shown in figure 6, 7, 8, and 9, the illuminated time of T5 FL is increased as gap increased. The v_L and i_L of T5 FL are in phase operated at steady state. The IPT for T5 FL lighting system exhibits superior characteristics and agrees with the frequency response.

1619

The 2018 International Power Electronics Conference

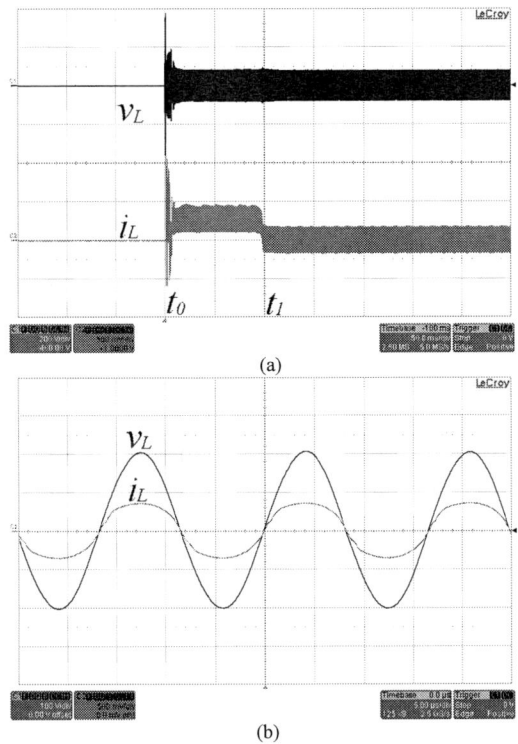

(a)

(b)

Fig. 6. Waveforms of fluorescent lamp operated at fsw 60 kHz (gap: 2mm) (a) transient state (v_L : 200V/div.; i_L : 0.5A/div.; t : 50ms/div.) (b) steady state (v_L : 100V/div.; i_L : 0.5A/div.; t : 5 μs/div.).

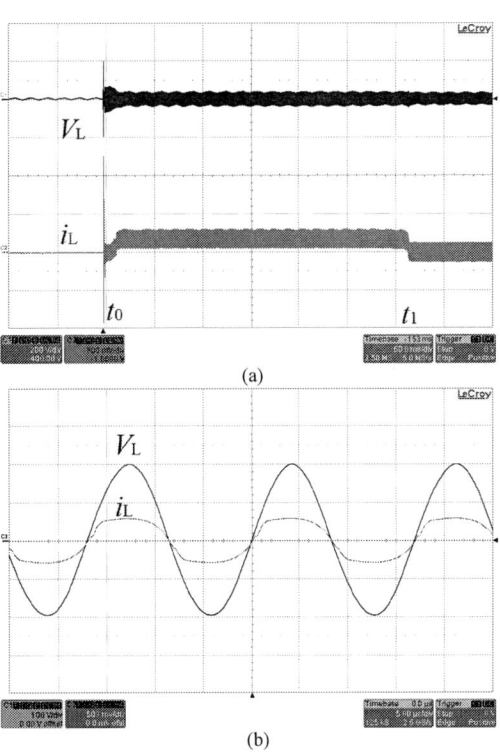

(a)

(b)

Fig. 8. Waveforms of fluorescent lamp operated at fsw 58 kHz (gap: 6mm) (a) transient state (v_L : 200V/div.; i_L : 0.5A/div.; t : 50ms/div.) (b) steady state (v_L : 100V/div.; i_L : 0.5A/div.; t : 5 μs/div.).

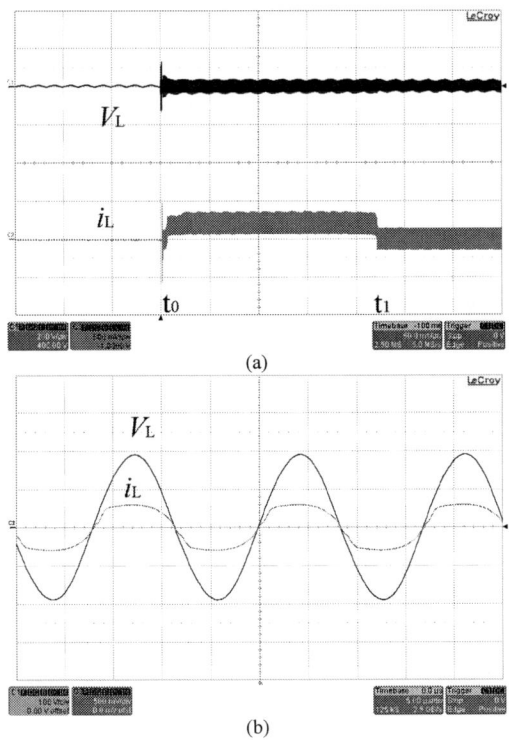

(a)

(b)

Fig. 7. Waveforms of fluorescent lamp operated at fsw 59 kHz (gap: 4mm) (a) transient state (v_L : 200V/div.; i_L : 0.5A/div.; t : 50ms/div.) (b) steady state (v_L : 100V/div.; i_L : 0.5A/div.; t : 5 μs/div.).

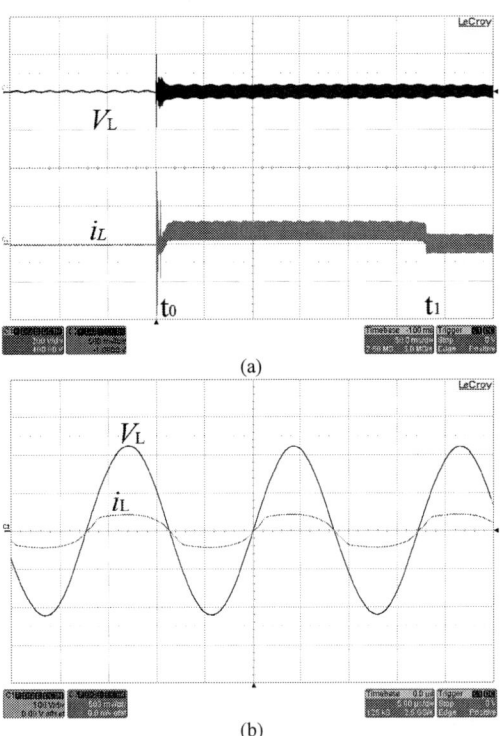

(a)

(b)

Fig. 9. Waveforms of fluorescent lamp operated at fsw 57 kHz (gap: 8mm) (a) transient state (v_L : 200V/div.; i_L : 0.5A/div.; t : 50ms/div.) (b) steady state (v_L : 100V/div.; i_L : 0.5A/div.; t : 5 μs/div.).

IV. CONCLUSIONS

The characteristics of inductive power transfer for T5 fluorescent lamp lighting system are presented. The technology of the inductive power transfer adopts the series-parallel resonant (SP) topology. The transfer function and frequency response of the SP topology inductive power transfer for T5 fluorescent lamp lighting system are derived and measured. The efficiency of the SP topology inductive power transfer is higher than 90% operated at switching frequency range 51-60 kHz and gap range 2-8mm. The illuminated time of T5 fluorescent lamp is 100-315 ms. The voltage and current waveforms of T5 fluorescent lamp exhibit superior performances adopted SP topology inductive power transfer technology.

REFERENCES

[1] R. Gules and W. M. dos Santos, "An auxiliary self oscillating preheating system for self oscillating fluorescent lamp electronic ballasts," *IEEE Trans. on Industrial Electronics*, vol. 59, no. 4, pp. 1859–1868, April 2012.

[2] W. M. Ng and S. Y. Hui, "Design of a single ultra-low-loss magnetic ballast for a wide range of T5 high-efficiency fluorescent lamps," *IEEE Trans. on Industrial Electronics*, vol.59, no.4, pp.1849-1858, April 2012.

[3] H. V. Marques and J. M. Alonso, "Constant-frequency magnetically controlled universal ballast with SoS compliance for TL5 fluorescent lamps," *IEEE Trans. on Power Electronics*, vol.27, no.4, pp.2163-2175, April 2012.

[4] H. C. Yen, "Analysis of balancing driver for multiple cold cathode fluorescent lamps," *IEEE Trans. on Industrial Electronics*, vol.57, no.4, pp.1354-1359, April 2010.

[5] Guan-Chyun Hsieh, "Group-asymmetrical PWM control for dimmable fluorescent lamp ballast without striation and thermostat effect," *IEEE Trans. on Power Electronics*, vol.24, no.5, pp.1293-1303, May 2009.

[6] Tsai-Fu Wu, Yung-Chun Wu and Zing-Ying Su, "Design considerations for single-stage electronic ballast with dimming feature," *IEEE Trans. on Industry Applications*, vol.37, no.5, pp.1537-1543, Sept./Oct. 2001.

[7] Su Y. Choi, Beom W. Gu, Seog Y. Jeong, and Chun T. Rim, "Advances in wireless power transfer systems for roadway-powered electric vehicles," *IEEE Journal of Emerging and Selected Topics in Power Electronics*, vol. 3, no. 1, pp. 18-36, March 2015.

[8] Liang Chen, Ganesh R. Nagendra, John T. Boys, and Grant A. Covic, "Double-coupled systems for IPT roadway applications," *IEEE Journal of Emerging and Selected Topics in Power Electronics*, vol. 3, no. 1, pp. 37-49, March 2015.

[9] John M. Miller, Omer C. Onar, and Madhu Chinthavali, "Primary-side power flow control of wireless power transfer for electric vehicle charging," *IEEE Journal of Emerging and Selected Topics in Power Electronics*, vol. 3, no. 1, pp. 147-162, March 2015.

[10] Siqi Li and Chunting Chris Mi, "Wireless power transfer for electric vehicle applications," *IEEE Journal of Emerging and Selected Topics in Power Electronics*, vol. 3, no. 1, pp. 4-17, March 2015.

[11] Xiaohui Qu, Wei Zhang, Siu-Chung Wong, and Chi K. Tse, "Design of a current-source-output inductive power transfer LED lighting system," *IEEE Journal of Emerging and Selected Topics in Power Electronics*, vol. 3, no. 1, pp. 306-314, March 2015.

The 2018 International Power Electronics Conference

An Implement 1.5 MHz of Induction Heating for Aluminum Based on Vacuum Tube Oscillator Circuit

A.Bilsalam[1*], P.Chanmontree [1], S.Supanyapong[1] and V.Chunkag[2]

[1]Department of Electrical Engineering Technology, College of Industrial Technology (CIT)
King Mongkut's University of Technology North Bangkok, Bangkok, 10800, Thailand
[2]Department Electrical and Computer Engineering, Faculty of Engineering
King Mongkut's University of Technology North Bangkok, Bangkok, 10800, Thailand
E-mail: anusak.b@cit.kmutnb.ac.th

Abstract— This paper presents an implement 1.5 MHz induction heating for aluminum based on vacuum tube oscillator circuit. It is operated by supplying input voltage 220 V/50 Hz to transformer with 220 V/2.4 kV then rectify form AC to DC voltage, which is then supplied to vacuum tube oscillator circuit. It produce an AC voltage, which has the waveform with high frequency close to the resonance frequency. This paper shows the heat up for aluminum with 0.95 kilowatt power and switching frequency at 1.5 MHz, consequently the circulating current at the induction coil and the work piece, causing heat at the work piece. Experimental results, induction hardening had a **high power factor equal to 0.993 in a heat up aluminum rod with 10-mm diameter from room temperature to 600 ºC within time about 60 second at the maximum full load power.**

Keywords— *Vacuum Tube; Oscillator circuit; Induction heating; Aluminium*

I. INTRODUCTION

Nowadays, the government of Thailand economy policy is encouraging such as in food agriculture, health wellness, smart devices robotics mechatronic, digital convergence and creative culture high value services to the development roadmap Thailand 4.0 by the office of industrial economics [1]. Smart devices robotics mechatronic group is consist development old and new technology to create value. An induction heating industry is important to Thailand economy, which widely used such as in car assembly, garbage and ore mine etc. From the industries mentioned above the induction heating participate in production process. The induction heating can heat up specific tools and metal products which added require heat for transform such as melting, hardening and forging. The advantages of this method of heating compared with old fuel method are heat can be given to a specific work piece, work piece and energy source are not touched because heat occur on skin work piece, no smoke from heating and can be start and stop easily, lower energy used at result, a capital cost will decrease also and fast response for heat up work piece. High frequency induction heating principle is that a high frequency AC voltage apply to a coil with work piece is placed in the coil, from a coil which receiving AC causing magnetic field coil will induce and heat would occur at cross-section area of work

piece due to skin effect [2] showed in Fig.1 following can be determine by $\sqrt{\rho/\mu\pi f}$ which ρ, μ is specific resistance and magnetic permeability respectively.

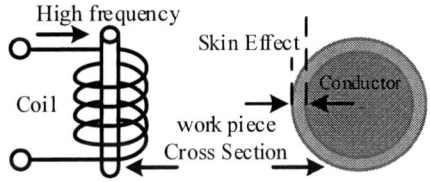

Fig. 1. Phenomenon skin effect of work piece is placed in the coil

From the above mentioned high frequency inverters are widely used. The popular topologies are class-E, half-bridge, full-bridge and multi-level inverter. Class-E inverter is an interesting topology because it comprise only single switch, easy gate driven, and all parameters can be calculated. However, the stress voltage across the power switch is a limitation due to this voltage is around three range higher than the input voltage [3-6]. While the medium to high power level used half-bridge, which consist of two switches and two capacitors input side. It can be achieve operation under the zero voltage and zero current switching condition as a result high efficiency of half-bridge compare with full-bridge inverter as a less than the number switches. The disadvantage of this power state is used very large capacitor input side operated regulation voltage. It not can be possible operation high frequency range. When the imperative high current and density can be used full-bridge inverter due to structure combination four switches and operation alternate are both. The relationship of output current and power density bounden dead time are importance as suitability and sufficient with next period [7-8]. In addition, can be connected multi-level inverter is a front-end of the power state in high power and frequency range is using more switch numbers and request a complicate gate driven control circuit. From over all power state with the induction heating mostly common connected matching impedance tank in the increase gain current output with heat up of work piece [9-10]. As then mentioned is importance issue semiconductor-discrete

This research support for finance by College of Industrial Technology (CIT), King Mongkut's University of Technology North Bangkok.

device, range operation switch frequency , power density and the fuel consumption of energy for heat up of work piece.

The proposed circuit is described in Section II. Section III presents experimental and discussion result and the overall conclusion is finally given in Section IV

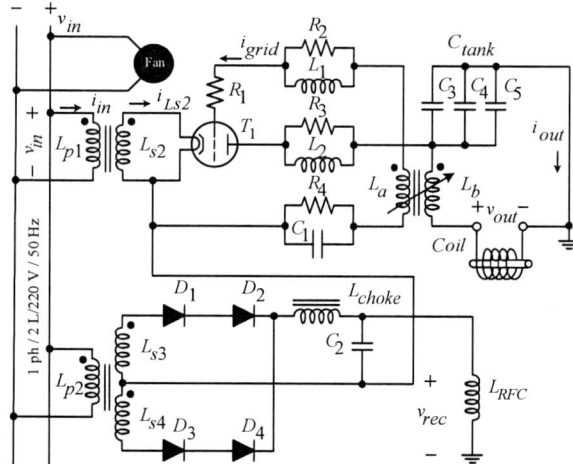

Fig. 2. Propose induction heating base on tube oscillate circuit

II. TECHNICAL WORK PREPARATION

A circuit of the proposed an implement 1.5 MHz of induction heating for aluminium base on vacuum tube oscillate circuit is shown Fig. 1. The circuit consists of center-tap full wave rectifier D_1- D_2- D_3- D_4, and LC filters L_{choke}- C_2. Axially adjust current grid circuit R_2- R_4- L_1- C_1 and main adjust current grid L_a- L_b. The Capacitors tank C_3 - C_4- C_5 are matching impedance power stage circuit.

A. Vacuum Tube Oscillater Circuit

There are survey type semiconductor switching devices such as Silicon Control Rectifier (SCR), Metal - Oxide-semiconductor Field - Effect Transistor (MOSFET), Insulated – Gate Bipolar Transistor (IGBT) etc. However, these devices cannot be used for high power density at the range of radio frequency except some auxiliary circuit complexity are added for example quasi-resonant and full resonant topology. These, vacuum tube is considered because it can be served the high frequency and high power application. For example in this paper used vacuum tube 7T84RB from TOSHIBA has some parameter as follow: is forced air cooling. The most important parameter are frequency 110 MHz, DC plate voltage 7 kV, plate dissipation 1.5 kW and output power Class C 3.8 kW at shown in fig.1. Vacuum tube consist of 3 connection: the cathode is used as a filament, anode is used plate terminal and finally terminal is grid it can be served increase control and acceleration of electrons be shaped into a wire mesh thousand around a vacuum tube.

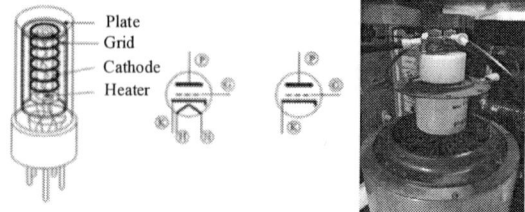

Fig. 3. Propose induction heating base on tube oscillator circuit

B. Resonant Oscillator and Grid Current Circuit

The main generate frequency circuit are used LC filter, in the first stage it is caused by discharge of a capacitor tank though anode and cathode vacuum tube. Inductor La coupling voltage from Inductor Lb pass to Inductor La that was connected to resistance R_4 and capacitor C_1 are function self-driven switching frequency, it can calculate following (1-2)

$$f_r = \frac{1}{2\pi\sqrt{(L_b + L_a)C_{tank}}} \tag{1}$$

$$k = \frac{M}{\sqrt{(L_a L_b)}}, \ a = \frac{L_b}{M} \tag{2}$$

When f_r = Switching frequency (Hz)

$L_b + L_a$ = Summation of inductance

C_{tank} = Capacitors tank resonant

a = Ration mutual inductance

k, M = Coupling coefficient and mutual inductance

C. Center-Tap Full Wave Rectifier Circuit

When the supplying AC input 220 V/50 Hz of line frequency transformer, it can be step-up voltage from 220 V to high voltage 2.4 kV. After that AC high voltage through center-tap full wave rectifier circuit with the covers into full wave. Power diode series with the increase the voltage rating from peak inverse voltage of center-tap transformer. The DC quality wave from depends on in inductor output, which saves block AC component in output size and in the voltage wave from depends on size capacitor. Moreover, choose to add RF choke will help reduce in-rush and high ripple current of capacitor rectifier in practice would bulky chokes have eliminated main frequency power supplies.

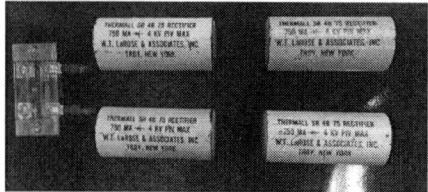

Fig. 4. High voltage power diode of center-tap full wave rectifier circuit

In this paper selected by using air core to create chokes and positive ground with decrease wear of a structural material.

III. CALCULATD HEAT EFFICIENCY

The work pieces is an aluminum weight 4 grams (0.004 kg) heat up from room temperature 33 °C to maximum temperature 600 °C use heating time 60 seconds. The specific heat capacity of aluminum is equal to 900 J/kg can be calculated from (3).

$$Q = mc\,\Delta T \qquad (3)$$

$$= \ 0.004\,kg \times \left(900\,{}^{J}\!/\!{}_{kg} \times 0.004\,kg\right) \times \left(600\,{}^{\circ}c - 33\,{}^{\circ}c\right)$$

When Q = Energy heat rate (kcal)

m = Mass (kg)

c = Specific heat capacity (J/kg)

ΔT = Rate of change temperature (s)

The energy heat, which is calculated from (3), is Q = 8.164 kcal or equal to 34181.03 J. The time is used to provide heat for 60 second. Therefore, output power can be obtained by

$$Power = \frac{Joules}{Second} = \frac{34181.03}{60} = 569.68\,\text{W} \qquad (4)$$

The measured input power equal to 950 W. Hence, total efficiency of a prototype is equal 60 %

IV. EXPERIMENTAL RESULTS

A prototype of the circuit was implement and built. The circuit parameters listed in Table 1. The main power stage operating in self-driven 1.5 MHz fix switching frequency. The heat up aluminum work pieces a size diameter 10 mm. The line voltage was 220 V_{rms}, the line frequency 50 Hz. The voltage and current input size were measured from digital oscilloscope. The measured about output power 570 W, when the heat up work pieces at a full power.

TABLE I.   EXPERIMENTAL RESULTS

Current Grid 40 mA							
Time (Sec)	20	30	40	50	60	70	80
Temp (°C)	307	402	479	545	588	596	603
Current Grid 50 mA							
Time (Sec)	20	30	40	50	60	70	80
Temp (°C)	320	399	481	543	598	604	613
Current Grid 60 mA							
Time (Sec)	20	30	40	50	60	70	80
Temp (°C)	295	378	455	514	562	570	578
Current Grid 70 mA							
Time (Sec)	20	30	40	50	60	70	80
Temp (°C)	290	374	453	511	561	577	582

Remake: All Temperature start up approximate equal to 32 °C room temperature set-up

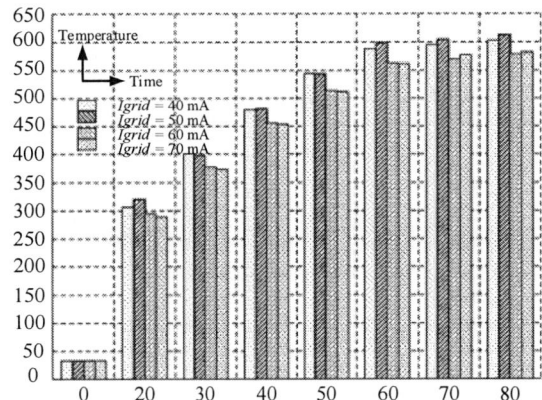

Fig. 5.   Experimental result relationship temperature and time

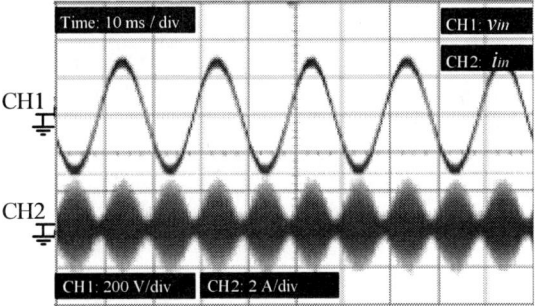

Fig. 6.   Measured voltage and current full load waveforms at the input side

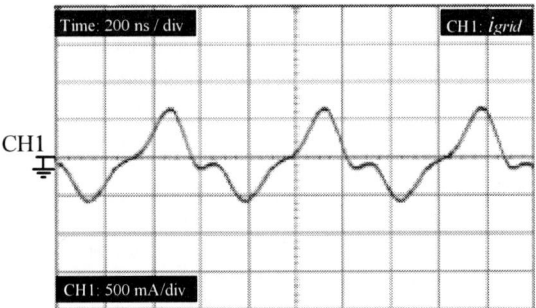

Fig. 7.   Measured current grid

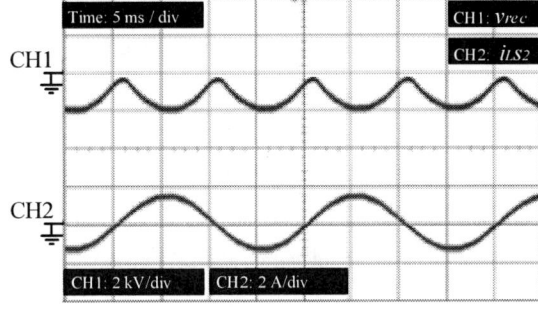

Fig. 8.   Measured voltage and current waveforms at the rectifier circuit

The 2018 International Power Electronics Conference

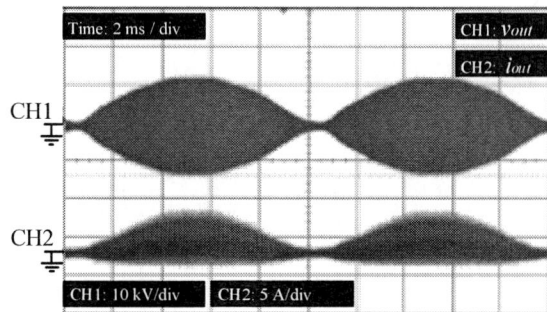

Fig. 9. Measured voltage and current full load waveforms at the output side

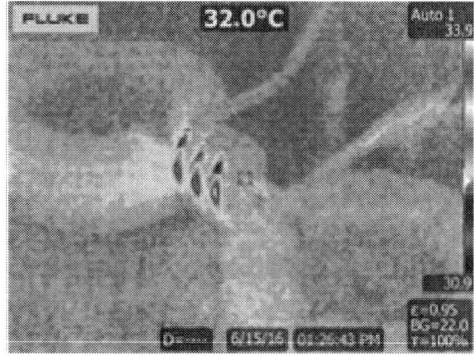

Fig. 10. Experimental result heat up work piece room temperature

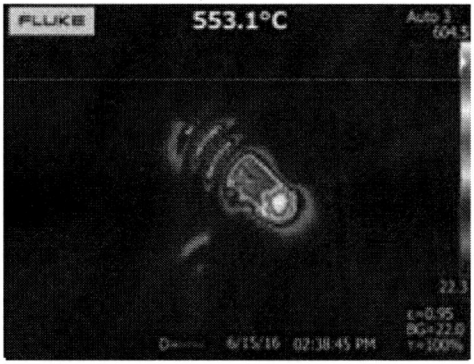

Fig. 11. Experimental result heat up work piece high temperature 600 ºC

The voltage and current at an input side power of 950 W as shown Fig. 5. Fig. 6 show the current control grid waveform similarity high frequency sinusoidal of vacuum tube that they operated full load about value 50 mA. The negative voltage waveform of a full-wave center tap rectifier about 2000 volts and current pass to tube is a sinusoidal wave form. When, the heat up work piece use only a little power about value 570 W and just take 60 minute shown in Fig.8. Fig.9 and Fig.10-11, show the experimental result from thermal infrared camera in the heat up work piece from room to high temperature are approximately value 32 ºC – 600 ºC, respectively

V. CONCLUSION

The paper presented an implement 1.5 MHz of induction heating for aluminium base on vacuum tube oscillate circuit. The main power switch using tube oscillate 7T84RB TOSHIBA it can be switching frequency 1.5 MHz are receive high frequency AC from full wave center tap rectifier. From experimental results in a heat up work pieces size diameter 10-mm from room temperature 33 to 600ºC within time about 60 second and 60% efficiency at maximum full power. The energy consumption which use to generate heat at sample work is 0.01583 kWh. This is considered to be a huge energy saving in the industry. In addition, it is the development of research induction group to reduce the import of machinery from abroad.

APPENDIX: SYSTEM PARAMERTERS

Diode rectifier D_1, D_2, D_3, D_4 SR 4B75, Inductor chock L_{choke} = 125 mH, Inductor Radio Frequency L_{RFC} = 12.1 µH, Capacitor filter C_2 = 1.2 pF/13 kV, Resistors R_1, R_2, R_3, R_4 = 150 Ω, 50 Ω, 50 Ω, 6 kΩ, Inductors L_1, L_2 = 6.9 µH, 6.9 µH, Capacitor C_1 = 1000 pF/13 kV, Capacitors Matching C_3, C_4, C_5 = 5600 pF/13 kV, Inductors Matching L_a, L_b = 3.5074 µH, 1.03 µH

REFERENCES

[1] Report, thailand Industrial Development Strategy 4.0 of Ministry of industry, October,2016

[2] E. 1. Davies, 1. and Simpson, P.,1979, Induction Heating Handbook. ,McGraw-Hili, UK

[3] E.J. Dede, J. Jordán, V. Esteve, A.E. Navarro and A. Ferreres, "On The Design of a High Power IGBT Sesies Resonant Inverter for Induction Forging Applications," *AFRICON, 1996., IEEE AFRICON 4th* ,pp. 206-208.

[4] C. Ekkaravarodome, P. Charoenwiangnuea, K. Jirasereeamornkul, "The Simple Temperature Control for Induction Cooker based on Class-E inverter," *In Proc. IEEE 10th Electrical Engineering/ Electronics, Computer Telecommunications and information Technology (ECTI 2013)*, pp. 1-6.

[5] H. Sugimura, S.P. Mun, S. K. Kwon, E. Hiraki and M. Nakaoka, "Active Voltage Clamped Edge-Resonant Soft Switching PWM High Frequency Cyclo converter Using Bidirectional Switches," *IEEE Power Electronics Specialists Conference*,2008, pp. 3917-3923.

[6] P. Charoenwiangnuea, C. Ekkaravarodome, I. Boonyaroonate, P. Thounthong and K.Jirasereeamornkul, "Design of Domestic Induction Cooker based on Optimal Operation Class - E inverter with parallel Load Network under large – Signal Excitation," *Journal of Power Electronic*, Vol. 17, No. 4, pp. 892-904

[7] V. Pichetjamroen, Y. Naras and Kouki Matsuse, "Analysis of Two Continuous Control Regions of Conventional Phase Shift and Transition Phase Shift for Induction heating Inverter under ZVS and NON-ZVS Operation," ," *IEEE Trans. Power electron*, Vol. 23, No. 6, 2008, pp. 2794-2805.

[8] A.Bilsalam and J.Haema, "a high frequency isolated asymmetrical Class-D resonant inverter for induction heating saw blades application," *In Proc. IEEE 10th Electrical Engineering/ Electronics, Computer Telecommunications and information Technology (ECTI 2015)*, pp. 1-4.

[9] F. Canales, P. Barbosa and F.C.Lee, "A Zero - Voltage and Zero – Current Switching Three-Level dc/dc Converter," *IEEE Trans. Power electron*, Vol. 17, No. 6, 2002, pp. 898-904.

[10] A. Jangwanitlert, J. Songboonkaew, W. Thammasiriroj and J.C. Balda, "Analysis of Three-Level ZVS PWM Inverter for Induction Heating Applications," *CES/IEEE Power Electronics and Motion Control Conference*, pp. 1-5, 2006

The 2018 International Power Electronics Conference

Single-Inductor Multiple-Outputs Dimmable LED Driver with Buck Converter

Ta-Wei Huang[1], Tsorng-Juu Liang[2], *Fellow, IEEE* and Wei-Jing Tseng[3], and Jun-Xian Huang[4]

Department of Electrical Engineering/Green Energy Electronics Research Center (GREERC)

/ Hierarchical Green-Energy Materials Research Center

National Cheng Kung University, Tainan, Taiwan

Email: tjliang@mail.ncku.edu.tw

Abstract— **Individual LED dimming control is required for RGB light mixing applications by controlling LED currents independently with multiple-output LED drivers. In this thesis, a single-inductor multiple-output dimmable LED driver with buck converter is designed. According to the overall output power and input voltage conditions, the buck converter with the primary-side control can be operated in QR mode, frequency limitation and frequency reduction mode. The varied frequency controlled signal method is proposed to control the dimming sequence and the conducting times of the RGB LED strings. Finally, the buck-type SIMO LED driver with the universal input voltages 90 - 264 Vrms to drive Red (25 V/0.2 A), Green (40 V/0.2 A) and Blue (40.5 V/0.2 A) LED lamps is implemented. The dimming range is from 5 % to 100 % individually. The highest error of the dimming is 10% and the highest efficiency is 94%.**

Keywords— *Dimming, LED driver, buck converter, single-inductor multiple-outputs.*

I. INTRODUCTION

Generally, the control schemes for the multi-string LED systems can be generally classified into two types, as shown in Fig. 1. and Fig. 2. Fig. 1.shows an AC-DC stage which produces a single common output bus V_o connected with all LED strings. This topology is not proper for LED applications because the peak value of the output current for each LED string is N times larger than the desired current for the N number of LED strings. Therefore, the LED strings may be burned out .

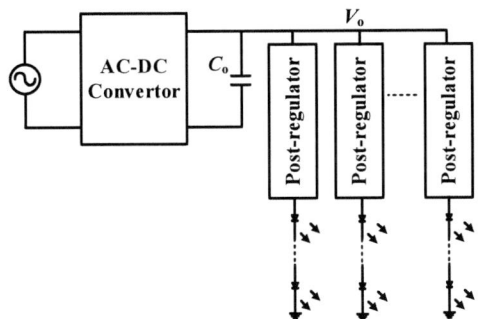

Fig. 1. Common output bus voltage.

Fig. 2. shows an AC-DC stage with separate output voltages for each LED string. The output voltage of the multi-string can be different due to the separated output voltage buses .

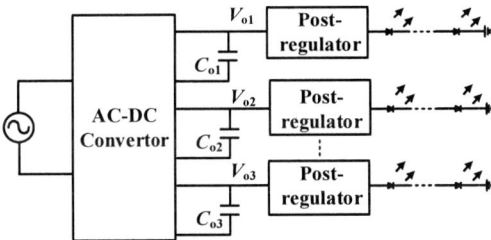

Fig. 2. Separate output voltage for each LED string.

A single-inductor multiple-output (SIMO) [1]-[4] dimmable LED driver with low-side switch buck converter is used in this thesis. The varied frequency controlled signal method is proposed to control the sequences and conducting times of the output-side switches. With the different dimming level, the LED driver can be operated in three modes, including quasi-resonant (QR) mode, frequency limitation with valley switching (FL) mode, and frequency reduction (FR) mode to improve efficiency and control accuracy. By using digital control, the buck-type LED driver can achieve 5% - 100% dimming range for each output in the universal input voltage range [5].

II. TECHNICAL WORK PREPARATION

A flyback LED driver with the primary-side regulation is shown in Fig. 3. The flyback LED driver with primary-side regulation (PSR) control method estimates the output current by the auxiliary winding and the current sensing resistor Rcs. Without using optocoupler and feedback circuit to feedback the output current of the LED modules, the cost and efficiency of the circuit can be better than the topology of the secondary-side regulation circuit. By reducing the components of the secondary side, not only can lower the costs of the components but also upgrade the efficiency [6]-[8].

1626

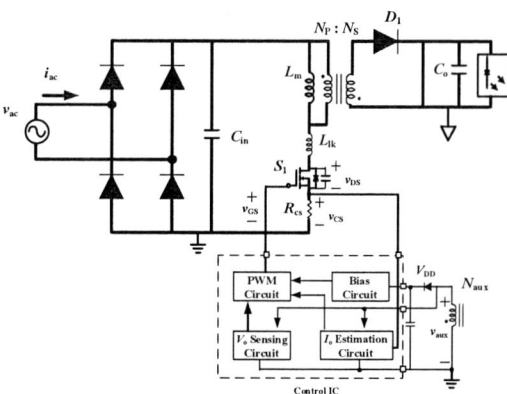

Fig. 3. Flyback LED driver with primary-side regulation.

A buck-type single-stage AC-DC single-inductor multiple-output (SIMO) LED driver is shown in Fig. 4. N number of switches $(S_{O\_1} - S_{O\_N})$ is in series with output diodes $(D_{O1} - D_{ON})$ to determine which channel will be conducted. Compared to the conventional topology, the single-inductor multiple-output topology has compact volume and higher efficiency [6].

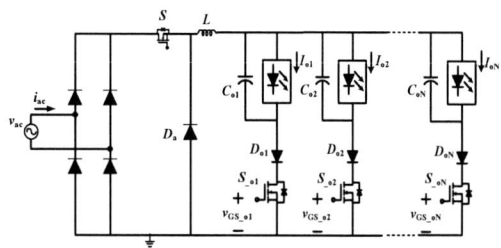

Fig. 4. Buck-type single-stage AC-DC SIMO LED driver.

Fig. 5 shows the system diagram of the buck-type LED driver with the output current estimation. As shown, a DSP is used to perform the output current estimation algorithm by sensing the peak current of the switch through the ADC module and detecting the diode conduction period. The valley detection mechanism is also contained in the control algorithm to turn on the switch at the valley voltage.

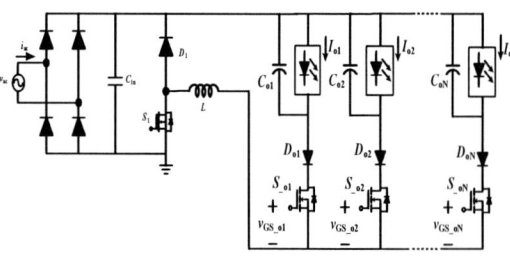

Fig. 5. Buck-type single-stage SIMO LED driver with low-side switch estimation

III. ANALYSIS OF PROPOSED CONTROL SCHEME FOR DIMMABLE SINGLE-INDUCTOR MULTIPLE-OUTPUT LED DRIVER

Fig. 6 shows the system diagram of the buck-type SIMO LED driver for RGB LED outputs. As shown in Fig. 6, the primary-side of the buck type LED driver is with a solid-ground switch. Based on the output power of the LEDs and input voltage, the buck converter can be operated in QR mode, FL mode, and FR mode [5]. A DSP is used to perform the output current estimation algorithm, and send the signal which consists of three different kinds of frequency (40 kHz, 150 kHz, and 400 kHz) sequentially to the low-pass filter, and the two band-pass filters to control RGB control switches. The switching frequency range of the main switch is 20 kHz - 100 kHz, and the switching frequency of the RGB control switches is 1 kHz, which means 3 kHz per switch.

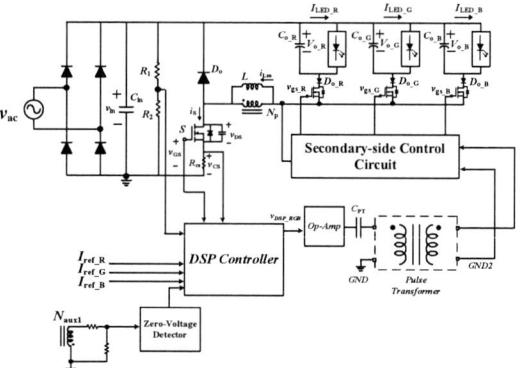

Fig. 6. Overall system topology of the buck-type SIMO LED driver.

RC Passive Low-Pass Filter

Fig. 7 shows an RC (Resistor-Capacitor) passive low-pass filter. The low-pass filter only allows low-frequency signals from 0 Hz to its cut-off frequency to pass [9]-[10]. The cut-off frequency and phase shift angle can be designed by using the following equation (1) and (2):

$$fc = \frac{1}{2\pi R_{f\_low} C_{f\_low}} = \frac{1}{2\pi \times 4k \times 1 \times 10^{-9}} = 40kHz \quad (1)$$

$$\text{Phase Shift } \varphi = -\arctan(2\pi R_{f\_low} C_{f\_low}) \quad (2)$$

After having the filtered signal that passes through C_{f_low}, D_{O_low40k} and C_{O_low40k} these two components can make sure the V_{out} will be a DC signal that sends to the comparator.

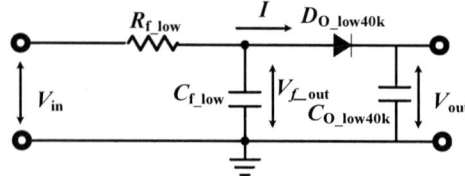

Fig. 7. RC passive low-pass filter

RLC Passive Band-Pass Filter

Fig. 8. shows an RLC (Resistor-Inductor-Capacitor) passive band-pass filter. Unlike a wide frequency range band-pass filter, a narrow frequency range band-pass filter is required in this thesis [11]. To avoid the cut-off frequencies of the three filters being too close to each other, the narrow range band-pass filters are required. The equations and the designed process of choosing the proper components are explained as follows .

By defining the resistance of the capacitor $C_{f\_band150k}$ as :

$$X_C = \frac{1}{2\pi f_c C_{f\_band150k}} \tag{3}$$

The reactance of the inductor $L_{f\_band150k}$ can be defined as :

$$X_L = 2\pi f_c L_{f\_band150k} \tag{4}$$

The net reactance for an AC circuit with an inductor and capacitor in series can be defined as (5):

$$X = X_C - X_L \tag{5}$$

As letting the net reactance to be zero, which means the $X_C = X_L$. By placing equations (3) and (4) into this equation, the following equations can be derived as (6):

$$f_c = \frac{1}{2\pi\sqrt{L_{f\_band150k}C_{f\_band150k}}} \tag{6}$$

The f_c is the resonant frequency, this happens when the reactance of the inductor and the capacitor are equal. Under the resonant frequency f_c, the signal can pass through the RLC band-pass filter.

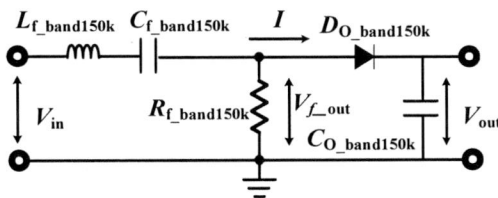

Fig. 8. RLC passive band-pass filter

Varied Frequency Controlled Signal Method

In this thesis, a method is proposed to control the gate signals of the RGB control switches (S_R, S_G and S_B). Fig. 9. shows the whole system of the control topology and the details are described as follows. First, the DSP sends a signal which is composed of three sequential frequencies to the op-amplifier to amplify the voltage. To ensure the output signal of the op-amplifier can pass through the pulse transformer, a capacitor C_{PT} is placed between the op-amplifier and pulse transformer to block the DC signal.

Second, by sending the AC output signals from the pulse transformer to the three filters (Low-pass filter with cut-off frequency 40 kHz, Band-pass filters with cut-off frequencies 150 kHz and 400 kHz), the three filters can

determine the correct timing for the signals to pass through the filters automatically. The diodes and capacitors (D_{O_low40k}, $D_{O\_band150k}$, $D_{O\_band400k}$, C_{O_low40k}, $C_{O\_band150k}$, and $C_{O\_band400k}$) are the components that can ensure the output signals from the filters to become small ripple DC signals that are sending to the comparators (Comp_R, Comp_G, Comp_B).

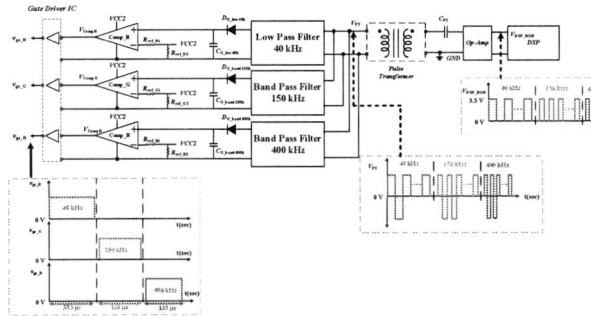

Fig. 9. Topology and signals for variable frequency controlled signal method.

Operational Modes of Primary-Side Switch

Fig. 10 shows the system operational modes of the buck-type SIMO LED driver. As the dimming percentage decreases from 100% to 5%, the mode of the circuit changes from Quasi-resonant (QR) mode, to frequency limitation with valley switching (FL) mode, then to the frequency reduction (FR) mode.

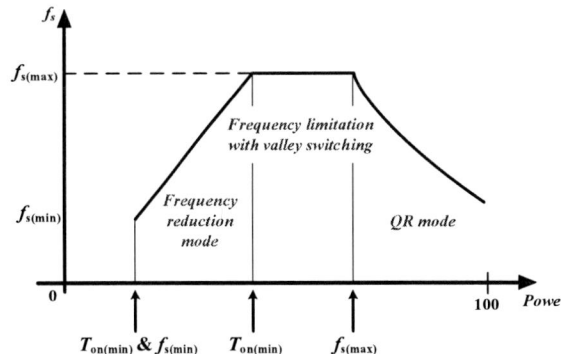

Fig. 10. Circuit operational modes.

As the percentage of dimming is 100%, the operation mode begins from QR mode. When the dimming percentage decreases, the turn-on time of the primary-side switch decreases. Thus, the switching frequency of the primary-side switch will rise until the frequency reaches the maximum limitation, and the circuit mode will change into the frequency limitation with valley switching mode. In frequency limitation with valley switching mode, the turn-on time of the primary-side switch still decreases when the dimming percentage is decreased. As the turn-on time of the primary-side switch reaches the minimum value, the LED driver will enter FR mode [5]

Output Current Estimation

To estimate the output current, the resonant period of the inductor current and dead time between each output are ignored to simplify the analysis. As the primary-side switch turns on, the inductance L can receive the energy from the power supply through the rectifier linearly. When the switch is turned off, the energy that stored in the inductor will then transform to the three output LED strings sequentially.

.

To estimate the LEDs output current, the resonant period of the inductor current is ignored to simplify the analysis. Unlike the single output circuit, the calculation of the three outputs circuit has to be divided by three.

$$I_{L,avg} = \frac{1}{3} \cdot (\frac{1}{2} \cdot I_{S,pk} \cdot \frac{T_{on} + T_{dsc}}{T_s}) \tag{7}$$

By using the amp-second balance law on the output capacitor C_o, the output LED current I_{LED} is same as the average of the inductor current $I_{L,avg}$, which can be written as (8).

$$I_{LED} = I_{L,avg} = \frac{1}{3} \cdot (\frac{1}{2} \cdot I_{S,pk} \cdot \frac{T_{on} + T_{dsc}}{T_s}) \tag{8}$$

To estimate the LED current under the PSR method, the sensing resistor R_{cs} and the auxiliary winding N_{aux1} is essential to the circuit. As shown in Fig. 11, the peak current $I_{S,pk}$ of the switch can be sensed by the sensing resistor R_{cs}. The diode conduction time T_{dsc} can be detected through the auxiliary winding. Thus, the LED current can be estimate as follows.

$$I_{LED,est} = \frac{1}{3} \cdot (\frac{1}{2} \cdot \frac{V_{cs,pk}}{R_{cs}} \cdot \frac{T_{on} + T_{dsc}}{T_s} \cdot \tag{9}$$

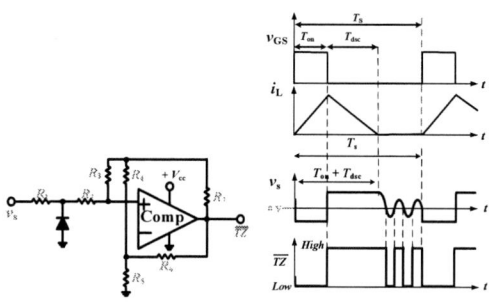

Fig. 11. The circuit and key waveforms of zero-voltage detector.

Quasi-Resonant (QR) Mode

The corresponding key waveforms of the buck-type SIMO LED driver operated in DCM is shown in Fig. 12. Fig. 12. shows one of the turn on period of the secondary-side switching period; thus, the red LED string is chose to be the example waveform. The PWM module of DSP not only can perform the digital to analog (DAC) function but also can control the duty cycle. One of the most important functions of PWM Counter is active counter-

compare registers CMPA and CMPB. In Fig. 12, CMPA is used to set the turn-off instant of the switch, CMPB is used to set the ADC sampling instant [5].

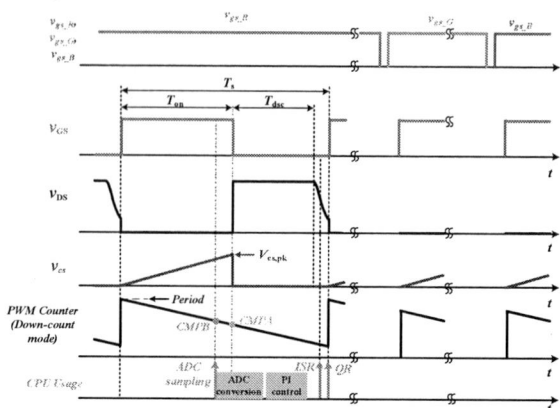

Fig. 12. Control timing diagram.

Fig. 13 shows the PI control function block. The function of PI control is important for the feedback and compensation for the circuit by using DSP to control. As shown in Fig. 13, I_{REF} is the reference signal, and I_{FB} is the feedback signal of the PI controller. Assuming that the steady-state error is zero, equation (10) can be obtained.

$$V_{REF} = V_{cs,pk} \cdot \frac{T_{on} + T_{dsc}}{T_S} \tag{10}$$

Hence, the reference current I_{REF} can be expressed in (11).

$$I_{REF} = \frac{1}{3} \cdot (\frac{1}{2} \cdot \frac{V_{REF}}{R_{cs}}) \tag{11}$$

According to (10), $V_{cs,pk}(T_{on}+T_{dsc})/T_s$ is the feedback term of the PI controller. By the feedback control, the controller output Io is the key to decide the on-time of the switch in the next switching cycle. To set the value of PI module as Fig. 13, suitable KI and KP is essential. With the suitable KI, the steady-state error will be zero and the system can reach the steady state in a short period of time.

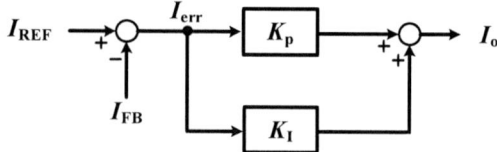

Fig. 13. PI control function block3.

Frequency limitation with valley-switching (FL) mode

In this section, the difference between QR mode and frequency limitation with valley switching mode will be discussed. Fig. 13(a) shows the LED driver operated in QR mode, Fig. 13 (b) shows the LED driver operated in frequency limitation with valley switching. As the discussion in the previous section, when the operating frequency of the primary-side switch is lower than the maximum frequency limitation, the switch can be turned on at the first valley voltage, which is known as the QR control, as shown in Fig. 13 (a). Different from the QR mode, since the switching frequency at the first valley voltage is higher than the maximum frequency limitation and the PI control is not done, the switch will not be turned on at this instant. From Fig. 13 (b), the turn-on instant of the primary-side switch is decided by examining the switching frequency in zero-voltage detection ISR. As the PI control finishes its calculation in a few resonant period, if the primary-side switching frequency is equal or below the maximum frequency limitation, the primary-side switch can be turned on with valley voltage .

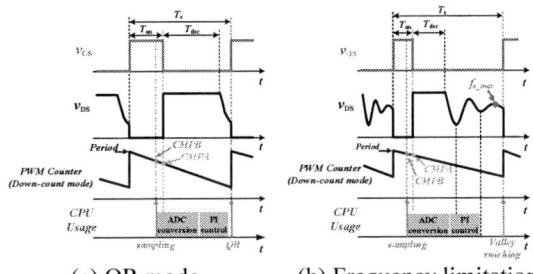

(a) QR mode   (b) Frequency limitation

Fig. 13 Valley switching in QR mode and frequency limitation mode.

Frequency reduction (FR) mode

If the turn on time reaches the minimum edge, the frequency limitation mode is going to change to the next mode, the frequency reduction (FR) mode. The frequency reduction mode can regulate the LED current and improve the efficiency at low dimming condition by reducing the switching loss. Under the control algorithm of frequency reduction mode, the primary-side switching frequency can keep reducing until the switching frequency reaches to a proper level during the low percentage of dimming. The performance period of frequency reduction (FR) mode is shown in Fig. 14, there is a boundary between the valley switching and non-valley switching in this mode. Based on the damping voltage of the primary-side switch, the FR mode can be divided into valley switching phase and non-valley switching phase.

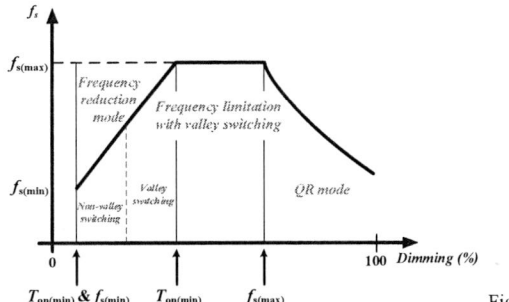

Fig. 14. Boundary of valley switching and non-valley switching in FR mod

Fig. 15 shows the valley switching phase in frequency reduction mode. Based on (11), as the feedback term has to follow the reference current I_{REF} to regulate the LEDs current to the desired value, the valley switching instant is determined by the PI control loop. As the I_{REF} is smaller than the I_{FB}, i.e. the steady-state error is not zero, then the controller will skip a few valley voltages to increase the switching period; thus, the feedback term will gradually follow the reference signal. On the opposite side, as I_{REF} is greater than I_{FB}, the controller has to make sure the switching period is shorter, by neglecting a short period of switching period, the resonant period of the primary-side switch can be shorter. From this, the I_{FB} will gradually follow the reference value.

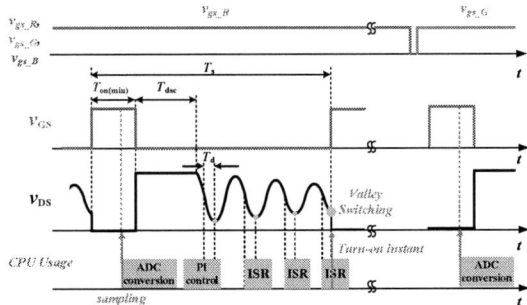

Fig. 15. Valley switching phase in frequency reduction mode.

IV. EXPERIMENTAL RESULTS

Table I is the specifications of the buck-type SIMO LED driver, which is designed for 60 Hz universal input voltage range, i.e. 90 − 264 V_{rms}. Table II shows the specifications of RGB LEDs and the LED strings.

TABLE I
SPECIFICATIONS OF THE BUCK-TYPE SIMO LED DRIVERT

Parameter		Value
Input line voltage	v_{ac}	90-264Vrms
Input line frequency	f_{line}	60 Hz
Switching frequency	f_s	20 kHz – 100 kH
Dimming frequency	f_{dim}	1 kHz
LED current		200mA rated
LED module		12 LEDs a string

TABLE II
Specifications of RGB LEDs

LED colors	Red	Green	Blue
LED in series	12	12	12
LED voltage, V_{LED}	25 V	40 V	41 V

TABLE III
KEY COMPONENTS

Components	Parameter	Value
Inductor	L	400 μH
Output capacitor	C_o	1 x 1000 μF / 50 V Per channel
Input capacitor	C_{in}	100 nF / 630 V
Primary-side Switch	S	UTC1N60/ 600 V, 1 A
Output switches	S_R, S_G, S_B	DMG1012UW/ 20 V, 1A
Output diode	D_o	MUR160/ 600 V, 1A per channel
DSP controller		TMS320F28335

Fig. 17 shows the waveform of the primary-side v_{GS} signal and the output side v_{gs_R}, v_{gs_G}, v_{gs_B} gate signals. The switching frequency of the output gate signals is 1 kHz, which means the turn-on time of each gate signals is 333 μs. Fig. 18 shows the waveform of the DSP output signal for the three filters, and output switches signals

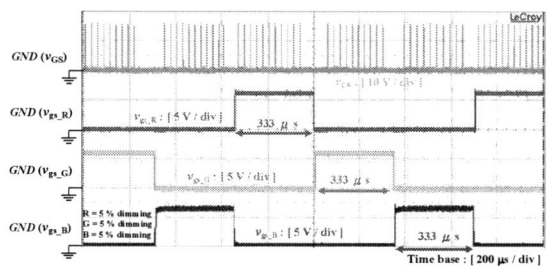

Fig. 17 Primary-side switch signal under 5% dimming conditions at input voltage 90 V_{rms} (b) v_{ac} = 220 V_{rms}

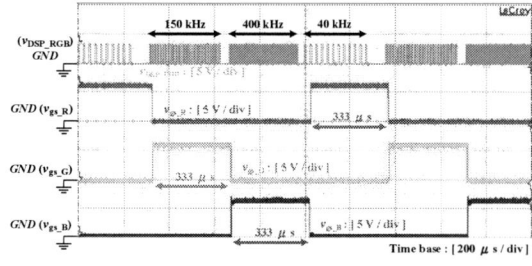

Fig. 18 DSP output signal and the gate driver signals for RGB controlling switches

Fig. 19, and Fig. 20 show the waveforms of the PSR buck-type SIMO LED driver operating in different modes with 110 V_{rms} and 220 V_{rms} input voltages with RGB LEDs. As shown in Fig. 4.19(a), Fig. 4.19(b), Fig. 4.20(a), and Fig. 4.20(b), the valley-voltage-switching is achieved to verify the theoretical analysis.

Fig.19. shows the measured waveforms of different dimming conditions for RGB LEDs at 110 Vrms. As shown in Fig. 19(a), the circuit is operated in FR mode. When the dimming percentage increases, the mode enters into FL mode. The valley switching is achieved in QR mode and FL mode. Fig.20. shows the measured waveforms of different dimming conditions for RGB LEDs at 220 Vrms. As shown in Fig.20(c), when the circuit is operated in the minimum frequency, the valley switching is not achieved.

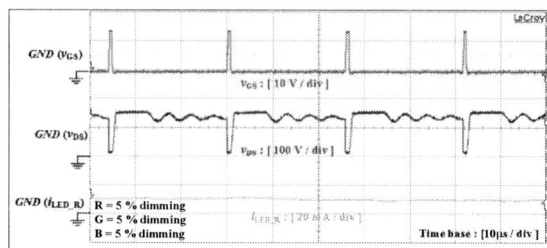

(a) i_{LED_R} waveform at R(5%), G(5%), B(5%) dimming levels (FR mode)

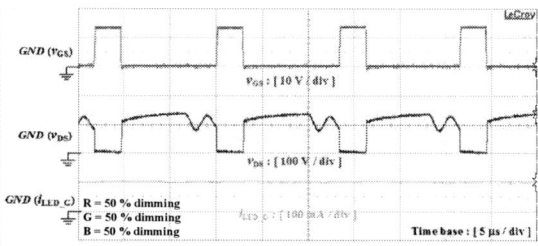

(b) i_{LED_G} waveform at R(50%), G(50%), B(50%) dimming levels (FL mode)

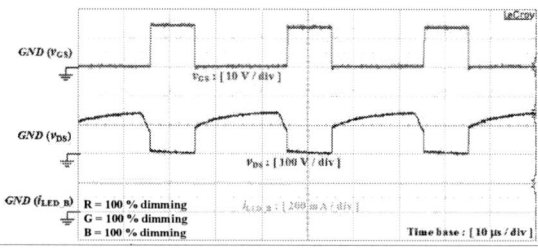

(c) i_{LED_B} waveform at R(100%), G(100%), B(100%) dimming levels (QR mode)

Fig. 19 v_{GS}, v_{DS}, and i_{LED} waveforms with various dimming conditions at 110 V_{rms}

Table IV lists the experimental results in the universal input voltage and full load dimming for multiple outputs in Red, Green, and Blue LEDs. The results show the output current almost remains at the rated current when changing the input voltage from 90 ~ 264 V_{rms}.

Table IV
Experimental results of RGB LEDs under 100 % dimming with universal input voltages

v_{ac} (Vrms)	90	110	130	200	220	240	264
V_{LED_R} (V)	25.2	25.4	25.1	25	24.8	24.6	24.8
I_{LED_R} (mA)	197.2	195	199	200	204.3	204	202
Error (%)	1.4	2.5	0.5	0	2.15	2	1
V_{LED_G} (V)	41	39.8	39.7	40.1	40	40.6	41
I_{LED_G} (mA)	195	201	201.7	199.3	199	196	195
Error (%)	2.5	0.55	0.85	0.35	0.5	2	2.5
V_{LED_B} (V)	43	40.9	40.76	41.3	41.3	41.4	42
I_{LED_B} (mA)	195	200	200.9	197.8	198.5	197.8	196
Error (%)	2.5	0	0.45	1.1	0.75	1.1	2

Fig. 21. shows the efficiency curves in the universal input voltage range. As shown, the efficiency decreases with the increasing input voltage since the circuit is operated at high switching frequency when the input is the high line voltage. As a result, the switching loss dominates at high line condition. The highest efficiency is 94%.

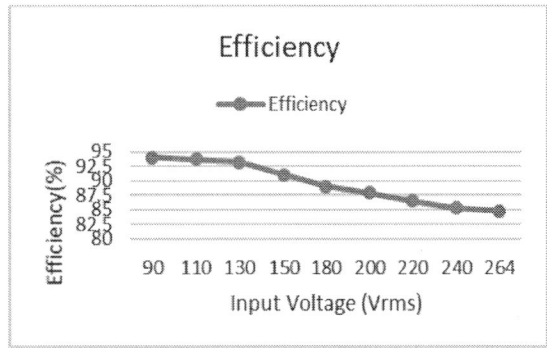

Fig. 21. Measured efficiency curves under RGB LEDs 100 % dimming conditions

V. CONCLUSIONS

The single-stage AC-DC PSR buck converter single-inductor multiple-output (SIMO) LED driver with the output current estimation and varied frequency controlled signal method for 5% - 100% dimming range is proposed in this thesis. The hardware prototype for the Red (25 V/0.2 A), Green (40 V/ 0.2 A) and Blue (40.5 V/ 0.2 A) LED modules have been designed and implemented. The hardware prototype can verify the feasibility of the proposed control strategies.

The topology of single-inductor multiple-output buck converter has the advantages of lower costs and compact volume. To control the outputs of the proposed circuit, the varied-frequency controlled signal method is proposed in this thesis. By sending sequential constant time of three kinds of frequencies from the DSP to the three different filters, which can determine the turn on time and the dimming sequence of the output switches. Thus, the dimming sequence of the LED strings can be controlled. Based on the dimming levels, the LED driver can be operated in three modes, including QR mode, frequency limitation with valley switching, frequency reduction mode. In QR mode, the switch turns on with the lowest voltage, which can reduce the switching loss. As the frequency reaches the maximum, the circuit operates under frequency limitation mode and the valley-voltage-switching mode can be achieved. When the turn-on time of the primary-side switch is decreased to the minimum, the circuit begins to operate under frequency reduction mode.

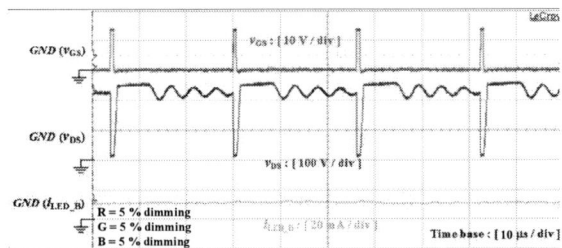

(a) i_{LED_R} waveform at R(5%), G(5%), B(5%) dimming levels (QR mode)

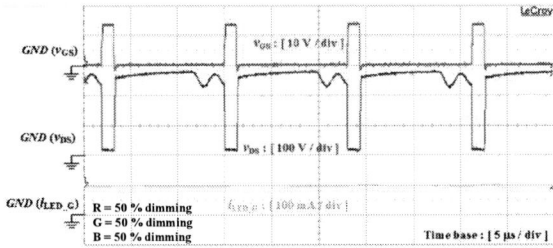

(b) i_{LED_G} waveform at R(50%), G(50%), B(50%) dimming levels (FL mode)

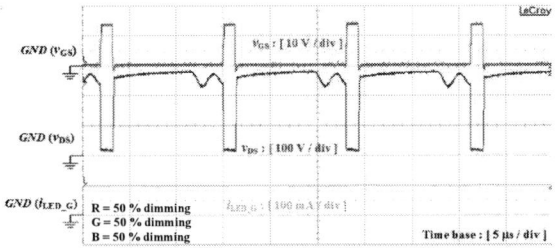

(c) i_{LED_B} waveform at R(100%), G(100%), B(100%) dimming levels (QR mode)

Fig. 20. v_{GS}, v_{DS}, and i_{LED} waveforms with various dimming conditions at 220 V_{rms}

REFERENCES

[1] S. Li, Y. Guo, S. C. Tan and S. Y. Hui, "An Off-line Single-Inductor Multiple-Output LED Driver With High Dimming Precision and Full Dimming Range," in IEEE Transactions on Power Electronics, vol. 32, no. 6, pp. 4716-4727, June 2017.

[2] Y. Guo, S. Li, A. T. L. Lee, S. C. Tan, C. K. Lee and S. Y. R. Hui, "Single-Stage AC/DC Single-Inductor Multiple-Output LED Drivers," in IEEE Transactions on Power Electronics, vol. 31, no. 8, pp. 5837-5850, Aug. 2016.

[3] M. H. Huang and K. H. Chen, "Single-Inductor Multi-Output (SIMO) DC-DC Converters With High Light-Load Efficiency and Minimized Cross-Regulation for Portable Devices," IEEE Journal of Solid-State Circuits, vol. 44, no. 4, pp. 1099-1111, Apr. 2009.

[4] M. G. V. Bautista, W. R. Liou, and M. L. Yeh, "Dimmable Multi-Channel RGB LED Driver," in Proc. IEEE ECCE Asia Downunder, Jun. 2013, pp. 1259-1262.

[5] P. Y. Lin, T. J. Liang, C. W. Chang, K. H. Chen and B. K. Huang, "Buck-type wide-range dimmable LED driver," *2017 IEEE Applied Power Electronics Conference and Exposition (APEC)*, Tampa, FL, pp. 1563-1569, 2017.

[6] J. S. Li and T. J. Liang, K. H. Chen, and Y. J. Lu, "Primary-Side Controller IC Design for Quasi-Resonant Flyback LED Driver," in Energy Conversion Congress and Exposition (ECCE), pp. 5308 - 5315, 2015.

[7] Y. C. Chuang, Y. L. Ke, H. S. Chuang, and C. C. Hu, "Single-Stage Power-Factor-Correction Circuit with Flyback Converter to Drive LEDs for Lighting Applications," in IEEE IAS, Oct. 2010, pp. 1-9.

[8] America (IFNA) Corp. "Design Guide for QR Flyback Converter" Jan, 2013

[9] Texas Instruments, "Input Filter Design for Switching Power Supplies" 2010

[10] Electronics Tutorials, "Passive Low Pass Filter". [Online]. Available: http://www.electronics-tutorials.ws/filter/filter_2.html

[11] Electronics Tutorials, "Passive Band Pass Filter". [Online]. Available:http://www.electronics-tutorials.ws/filter/filter_4.html

The 2018 International Power Electronics Conference

A Soft-switched Three-level T-type Inverter with Auxiliary Commutated Poles

Apollo Charalambous and Xibo Yuan*

Department of Electrical and Electronic Engineering, University of Bristol, Bristol, United Kingdom
*E-mail: xibo.yuan@bristol.ac.uk

Abstract-A 3-level T-type inverter that employs the auxiliary commutated pole inverter (ACPI) concept for fully soft-switched operation, is presented in this paper for the first time. This way, switching loss and dv/dt stress can be reduced, with little auxiliary loss and small resonant components. Importantly, the high-frequency content of the EMI source, i.e., the output voltage, can be attenuated in a controlled and predictable manner, with the help of the ACPI concept. This way, applications with stringent EMI standards such as the engine starter/generator drive for more electric aircraft (MEA), can doubly benefit from the superior AC waveform quality and the attenuated high-frequency EMI of the novel inverter, with increased power density. The 3-level ACPI resonant operation is described, and the analytical expressions that dictate it are derived through an equivalent LC circuit. SABER simulations are provided for proof of concept. It is observed that in the high-frequency region above 2 MHz, the 3-level ACPI enjoys a 16-dB reduction in harmonic content, compared to the hard-switched T-type converter.

Keywords— Soft-switching, T-type, Three-level, Auxiliary commutated pole inverter.

I. INTRODUCTION

The T-type inverter was first presented in [1], and has smaller component counts and a simpler structure than most multilevel (ML) topologies [2-4]. A 3-level (3L) T-type phase leg is shown in Fig. 1, where a midpoint branch consisting of the S_2/D_2 and S_3/D_3 device pairs, complements a 2-level (2L) phase leg consisting of the S_1/D_1 and S_4/D_4 pairs. This way, it can switch between 3 DC potentials, P, O, and N, for generating a stepped voltage pulse-train.

Unlike other ML converters, the T-type is unsuitable for medium/high-voltage applications because S_1/D_1 and S_4/D_4 have to be rated for the full DC-link voltage. However, it still enjoys the other benefits that come with the ML concept. The

stepped voltage pulse-trains lead to AC waveforms that have substantially lower total harmonic distortion (THD), compared to the 2L voltage-source converter (VSC) [5-8]. Moreover, since a driver for high switching frequencies is pushing the harmonics of the AC waveforms to higher frequencies for more effective filtering, the ML concept incentivises the reduction of the switching frequency, which improves efficiency [6,7].

Importantly, the stepped pulse-trains exhibit reduced harmonic content. For example, if a voltage-source inverter (VSI) is switched between 3 instead of 2 levels, the output voltage amplitude is halved, and its harmonic content drops by 6 dB, in the entire frequency spectrum. This way, the ability of this voltage for conducted EMI deteriorates. VSI-based motor drives can especially benefit from reduced common-mode (CM) EMI, as issues that plague the controlled motor like shaft voltages and bearing currents, can be mitigated. In addition, the reduced dv/dt rate can alleviate the device stress and the motor terminal overvoltages [9-12]. The T-type can bring all these benefits to low-voltage applications [4,6], such as the engine starter motor drive used in more electric aircrafts (MEA) [13-15].

Still, the T-type suffers from switching loss, which is why snubber circuits have been used for zero-voltage (ZVS) and zero-current switching (ZCS). In [16], a 30-kW T-type prototype employing a passive snubber, achieves 98.3% maximum efficiency, 0.3% more than with hard-switching. However, the resonant inductor, which is placed in series with the midpoint devices, has to carry the full load current at certain intervals, which can increase its losses and size, and the snubber capacitors do not charge or discharge fully at light load, compromising ZVS. Moreover, the midpoint devices must be rated for the full DC-link voltage. In [17], passive snubbers and auxiliary voltage sources are employed, but with imperfections like the snubber capacitors sometimes discharging into the incoming switches and provoking current spikes, and the main devices having to withstand the sum of the DC-link and the auxiliary source voltages.

Instead, a fully PWM-compatible ZVS topology with independent phase-leg control could be used that can be readily employed for VSC-based applications, such as the auxiliary commutated pole inverter (ACPI) [18,19]. The ACPI is fully soft-switched, with low auxiliary circuit loss, small component counts, and a fairly simple operational concept. For these reasons, it is also suitable for high-power applications, which is why, it has been extended to traditional ML topologies like the

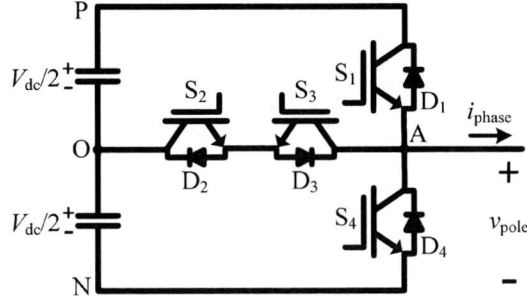

Fig. 1. Phase-leg schematic of the 3-level T-type inverter.

1634

cascaded H-bridge in a 1.1-GW multi-cell implementation for high voltage direct current transmission [20], and the neutral point clamped converter in a 10-MVA implementation in [21], and in a lower-rated 3-kW implementation in [22], where it is seen that the switching loss is halved with respect to hard switching. Further, the ACPI was combined with a 3L flying capacitor converter in [23].

To the knowledge of the authors however, the ACPI has not been combined with the T-type inverter. While both the ML and the soft-switching concepts are inherently more complicated in execution, and come with considerable component counts when compared to the 2L VSC, an opportunity is created here for conceiving a fairly simple, fully soft-switched circuit that can offer all the advantages of the ML concept to low-voltage applications, while at the same time, achieving switching loss reduction.

More than increased efficiency and smaller THD however, the main driver for this work is the high-frequency content attenuation of the output voltage in VSI-based motor drives. None of the aforementioned work examines ML soft switching from a harmonic content point of view. In [24], the ability of soft switching for tackling conducted EMI at source, in an optimized and predetermined manner, is examined by fully analysing and designing a 2L ACPI. The work presented here is a direct extension [24], applied to a 3L T-type inverter.

II. THE 3-LEVEL T-TYPE CONVERTER WITH ACPI BRANCHES

The 3L T-type has two switching circuits: the top one that switches between the P and O rails, and involves S_1/D_1 and S_3/D_3, and the bottom one between O and N that involves S_4/D_4 and S_2/D_2. This realization leads to the novel 3L T-type inverter with two ACPI branches, presented in Fig. 2(a). For comparison, the 2L ACPI, which employs only one auxiliary branch, is depicted in Fig. 2(b). Like the 2L version, the 3L ACPI is fully PWM-compatible, with independent phase control, and is also readily extendible to 3 phases.

The top branch facilitates the transitions between S_1 and S_3, and is formed by the L_{rP} inductor and the auxiliary devices S_{a1}/D_{a1} and S_{a3}/D_{a3}. The bottom branch facilitates the transitions between S_4 and S_2, and is formed by L_{rN}, S_{a2}/D_{a2} and S_{a4}/D_{a4}. The 3 snubber capacitors, C_{r1}, C_{r4}, and C_{r23} complete the setup. Similar to [23], both branches could be connected to only one, more highly-rated resonant inductor. Also, the four capacitors that split the DC link into equal parts of $V_{dc}/4$, might experience voltage imbalance, as discussed in [22], but this is not explored here. The resonant process transpires exactly like in the 2L ACPI [18], with all the main devices switched under ZVS, and all the auxiliary ones under ZCS. A description of the S_1 transitions for positive i_{phase} follows.

Fig. 3 depicts the switching currents i_{S1} and i_{S3}, the upper resonant inductor current i_{LrP}, the three snubber capacitor voltages v_{Cr23}, v_{Cr1} and v_{Cr4}, and the gate signals involved with the turn-on transition from D_3 to S_1 (O to P), along with the circuits formed during the old steady state, resonance, and the new steady state.

As D_3 is conducting with S_2 (S_2 is always on when switching between P and O), S_{a1} turns on under ZCS, and i_{LrP} starts ramping up for the accumulation of resonant energy. Meanwhile, i_{S3} decreases at the same rate, until it reverses direction. This way, S_3 starts conducting along with D_2, and i_{LrP} and i_{S3} keeps ramping towards their initial conditions of I_{trip} and I_{boost}, respectively. At the end of t_{ramp_on}, S_3 turns off, and L_{rP} starts resonating with the three capacitors.

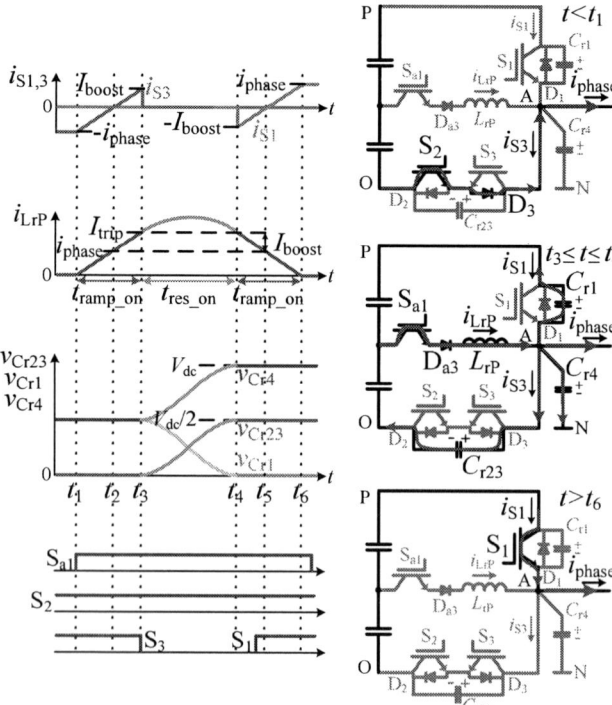

Fig. 3. Switching currents i_{S1}, i_{S3}, resonant waveforms i_{LrP}, v_{Cr23}, v_{Cr1}, v_{Cr4}, gate signals, and sub-circuits for the turn-on transition of S_1.

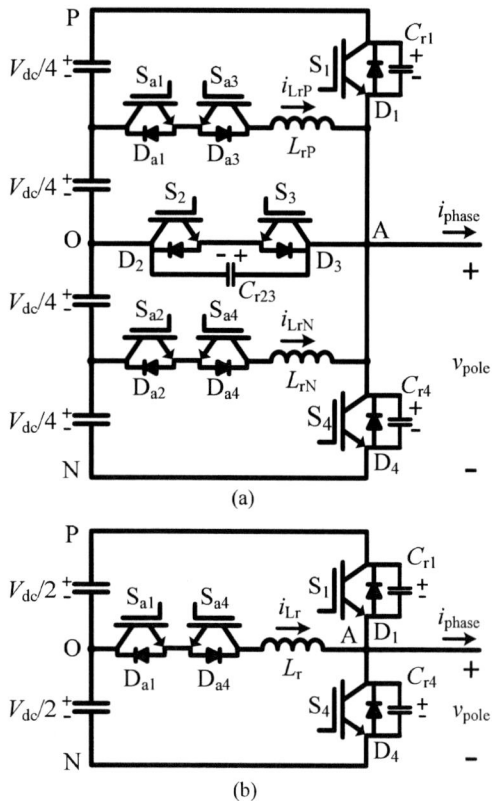

Fig. 2. Phase-leg schematic of (a) the 3L ACPI and (b) the 2L ACPI.

During t_{res_on}, the three voltages swing in a sinusoidal manner, with v_{Cr1} discharging from $V_{dc}/2$ to 0, at which point D_1 clamps and resonance ends. While i_{LrP} starts ramping down, a grace period is created for gating on the incoming S_1 under

The 2018 International Power Electronics Conference

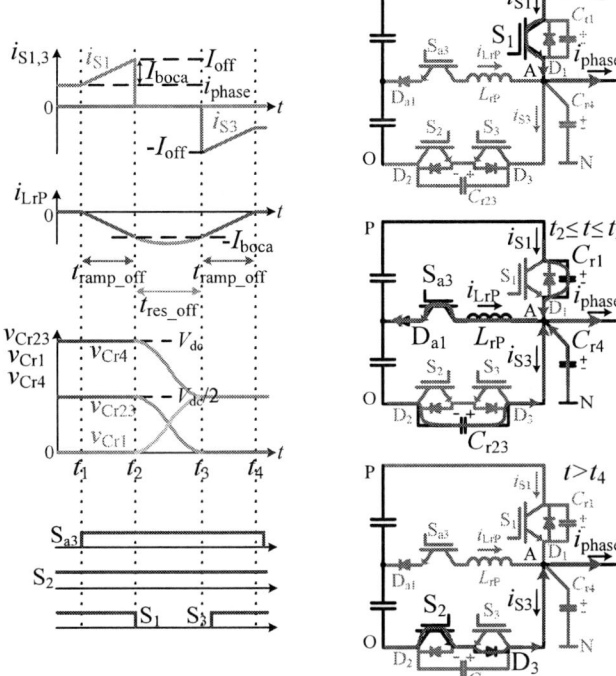

Fig. 4. Switching currents i_{S1}, i_{S3}, resonant waveforms i_{LrP}, v_{Cr23}, v_{Cr1}, v_{Cr4}, gate signals, and sub-circuits for the turn-off transition of S_1.

ZVS. As soon as i_{LrP} reaches i_{phase} again, S_1 starts conducting. ZVS turn-on and the shaping of the voltage edges are achieved. Finally, i_{LrP} reaches zero, S_{a1} turns off naturally, and the new steady state is reached.

Fig. 4 illustrates the turn-off transition from S_1 to D_3 along with the circuits formed during the old steady state, resonance, and the new steady state. With S_1 conducting, S_{a3} turns on under ZCS, and i_{LrP} starts ramping down. Meanwhile, S_1 keeps conducting until i_{S1} reaches I_{off} and i_{LrP} reaches $-I_{boca}$, at the end of t_{ramp_off}. Then, S_1 turns off under ZVS, and resonance elapses. During t_{res_off}, the three voltages swing back in a sinusoidal manner. When v_{Cr23} becomes 0, it is clamped by D_3. ZVS turn-off and the shaping of the voltage edges is once again achieved. Finally, i_{LrP} decreases to zero and S_{a3} turns off naturally.

III. EQUIVALENT SERIES LC CIRCUIT ANALYSIS

A. Extraction of LC Equivalents

In motor drive applications, the displacement power factor is inductive and usually close to unity. This way, the 3L T-type experiences 8 transitions during the fundamental cycle T_1: two turn-ons and two turn-offs under the two i_{phase} polarities in the top switching circuit, and the same for the bottom circuit. The resonant circuits of the soft-switched version will then have to assist all 8 transitions. As in the 2L ACPI [24], all the transitions can be analysed with the help of the series LC equivalent circuit, and can facilitate the design of the 3L ACPI for high-frequency content attenuation of its output voltage.

Fig. 5 shows the top resonant circuit for the transitions between P and O, where the black arrows mark the default resonant current polarities. For $i_{phase} > 0$, the green arrows denote the polarities for the turn-on transition, and the red ones denote the polarities for the turn-off transition. The combination of the i_{LrP} and i_{phase} polarities determines the kind of transition

to be realised. During turn-ons, i_{phase} impedes the resonant action, and during turn-offs, it facilitates it. From here, the top LC equivalent is extracted, with the equivalent capacitor being the sum of the 3 snubbers capacitors. The LC equivalent is driven by a voltage of $V_{dc}/4$, a quarter of the DC-link voltage, and i_{phase} acts as a parallel DC current source. The solutions to the step response of the top LC equivalent are $v_{AO} = v_{Cr23}$ and i_{LrP}, and, informed by the appropriate initial conditions seen in Fig. 5, are distinguished between the turn-on and the turn-off transition.

Similarly, Fig. 6 depicts the bottom resonant circuit for the transitions between O and N. For $i_{phase} < 0$, marked by the yellow arrow, the green arrows are for the turn-on transition of D_2 to S_4, and the red ones are for the turn-off transition of S_4 to D_2. Once again, the combination of the i_{LtN} and i_{phase} polarities determine the transition to be realised, with them concurring during turn-ons, and opposing each other during turn-offs. From here, the bottom LC equivalent is extracted, with A and L_{rN} moved to the bottom of the schematic. The default polarities though reversed with regards to the top LC equivalent, still reflect the positive currents and voltages. The solutions to the step response of the bottom LC equivalent are now v_{AO} and i_{LtN}, and are informed by the appropriate initial conditions for the description of the turn-on and the turn-off transition.

Figs. 5 and 6 describe four T-type transitions. The other four can be equally described by the appropriate LC equivalents, with the correct current and voltage polarities and initial conditions.

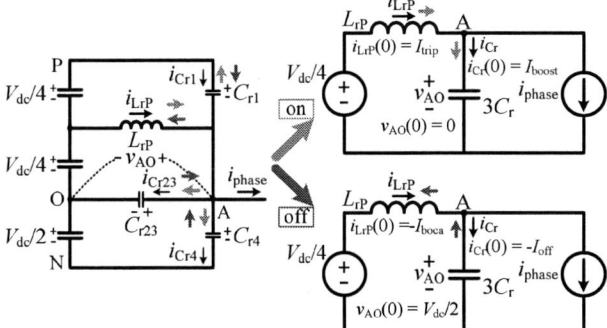

Fig. 5. Top resonant circuit and its LC equivalent when $i_{phase} > 0$.

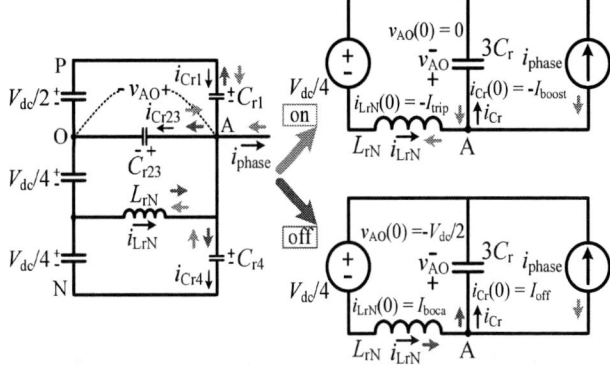

Fig. 6. Bottom resonant circuit and its LC equivalent when $i_{phase} < 0$.

B. Top LC Circuit Step Response for Positive Phase Current

The top circuit solutions i_{LrP} and v_{AO}, while $i_{phase} > 0$, are presented in this section. Once v_{AO} is found, v_{pole}, the voltage between the output node and the negative DC rail is simply

1636

$v_{\text{pole}} = v_{\text{Cr4}} = V_{\text{dc}}/2 + v_{\text{AO}}$. The turn-on solutions are described by (1) and (2), and the turn-off ones by (3) and (4):

$$i_{\text{LrP}}(t) = i_{\text{phase}} + I_{\text{boost}} \cos(\omega_0 t) + \frac{V_{\text{dc}}/4}{Z_0} \sin(\omega_0 t) \quad (1)$$

$$v_{\text{AO}}(t) = \frac{V_{\text{dc}}}{4}[1 - \cos(\omega_0 t)] + Z_0 I_{\text{boost}} \sin(\omega_0 t) \quad (2)$$

$$i_{\text{LrP}}(t) = i_{\text{phase}} - I_{\text{off}} \cos(\omega_0 t) - \frac{V_{\text{dc}}/4}{Z_0} \sin(\omega_0 t) \quad (3)$$

$$v_{\text{AO}}(t) = \frac{V_{\text{dc}}}{4}[1 + \cos(\omega_0 t)] - Z_0 I_{\text{off}} \sin(\omega_0 t) \quad (4)$$

Z_0 and ω_0 are the resonant impedance and the resonant frequency, respectively, and are described by (5) and (6):

$$Z_0 = \sqrt{L_{\text{rP}}/3C_r} = \sqrt{L_{\text{rN}}/3C_r} \quad (5)$$

$$\omega_0 = 1/\sqrt{3L_{\text{rP}}C_r} = 1/\sqrt{3L_{\text{rN}}C_r} \quad (6)$$

Solving (1) for $i_{\text{LrP}} = I_{\text{trip}}$ and (3) for $i_{\text{LrP}} = -I_{\text{boca}}$, yields the expression for $t_{\text{res\_on}}$ in (7), and for $t_{\text{res\_off}}$ in (8):

$$t_{\text{res\_on}} = \frac{2}{\omega_0} \tan^{-1}\left[\frac{V_{\text{dc}}/4}{Z_0(I_{\text{trip}} - i_{\text{phase}})}\right] = \frac{2}{\omega_0} \tan^{-1}\left(\frac{V_{\text{dc}}/4}{Z_0 I_{\text{boost}}}\right) \quad (7)$$

$$t_{\text{res\_off}} = \frac{2}{\omega_0} \tan^{-1}\left[\frac{V_{\text{dc}}/4}{Z_0(I_{\text{boca}} + i_{\text{phase}})}\right] = \frac{2}{\omega_0} \tan^{-1}\left(\frac{V_{\text{dc}}/4}{Z_0 I_{\text{off}}}\right) \quad (8)$$

Expressions (7) and (8) describe the resonant intervals during any given turn-on and turn-off transition, at any switching cycle during T_1, as long as the correct i_{phase} level and initial condition values are plugged in.

C. High-frequency Response of the Output Voltage

The resonant intervals are very important as they are the actual rise and fall times of the snubber capacitor voltages, and it is only during then that the resonant waveform expressions of (1) – (4) hold. Essentially, the ACPI is a quasi-resonant topology, resorting to resonance only for facilitating the transitions of the main devices. This way, ZVS is achieved and the switching loss is reduced, without excessive loss in the auxiliary components. Crucially though, as witnessed in Figs. 3 and 4, the capacitor voltages, among them v_{pole}, the voltage treated as the source of conducted EMI, have their edges shaped in a sinusoidal manner.

This is important for the high-frequency content of v_{pole}, as it can dictate the EMI performance of the inverter. The frequency spectrum of a pulse-train with linear edges has a roll-off of -40 dB/dec in the higher frequency range, and the only way to dictate where in the spectrum this roll-off begins is by altering the duration of those linear edges [25]. Sinusoidal and S-shaped edges on the other hand have smooth corners. This property introduces a -60 dB/dec roll-off [26,27].

IV. DESIGNING THE 3-LEVEL ACPI

A. Fixed-timing and Variable-timing Control

There is now an opportunity to influence both the duration and the shape of the v_{pole} edges. For specific circuit conditions V_{dc} and i_{phase}, the resonant components L_{rP}, L_{rN}, and C_r can be designed specifically for high-frequency content attenuation. As seen, the initial condition of the capacitor current i_{Cr}, defined here as the boost currents I_{boost} and I_{off}, appear in (1), (2), (3), (4), (7) and (8). By properly controlling these boost currents, v_{pole} can be profiled with slow and smooth edges for reduced high-frequency content, in a predetermined way.

As explained in [28], the two fundamental ACPI control schemes are fixed timing and variable timing, and they refer to the t_{ramp} intervals during which the inductor currents build up to their initial conditions, I_{trip} and I_{boca}, termed here as the trip currents. The t_{ramp} intervals dictate both the boost and the trip currents. With fixed timing, I_{trip} and I_{boca} are fixed throughout T_1. This forces the boost currents to vary, meaning that the t_{res} intervals of (7) and (8) will vary in duration as well. Hence, each v_{pole} pulse will be dissimilar to its adjacent ones in terms of edge duration and shape, and the frequency response will become unpredictable. Conversely, variable timing can manipulate the t_{ramp} intervals in a way that the boost currents stay fixed throughout T_1. The v_{pole} edges will then be shaped in a consistent manner from one switching cycle to the next, and the frequency response will become more predictable. Further, it can be demanded that $t_{\text{res\_on}} = t_{\text{res\_off}}$, for symmetrical v_{pole} pulses [24].

Fixed-timing control is easy to implement, and in its simplest form requires no current and voltage sensing whatsoever. However, excessively large trip currents, even when not warranted for the phase current level at which a transition takes place, lead to excessive auxiliary loss and current stress in both the main and the auxiliary devices. Variable timing can build the trip currents according to the phase current level and maintain the boost currents fixed, only by sensing the i_{phase} and V_{dc} levels. It also has to distinguish which auxiliary switch is to assist in a given transition. However, aside from the superior v_{pole} frequency response [24], the auxiliary loss and the current stress are reduced.

B. Design Procedure

For proof of concept, the intended 3L ACPI is designed for the circuit conditions of $V_{\text{dc}} = 500$ V, and a peak phase current of $I_{\text{phase\_pk}} = 18$ A. Sinusoidal PWM (SPWM) is employed with $m_A = 0.83$ modulation index, $f_s = 20$ kHz switching frequency and $f_1 = 400$ Hz fundamental frequency. For a direct comparison of the v_{pole} frequency response achieved by the 2L ACPI in [24], and the 3L ACPI designed here, the resonant intervals should last the same. They are chosen to be $t_{\text{res\_on}} = t_{\text{res\_off}} = 1.2$ µs, substantially slower than what a hard-switched inverter would achieve, which is desirable if the frequency response is to be enhanced. This way it is also demanded that the variable timing scheme produces symmetrical v_{pole} pulses. For this to happen the following boost current condition is demanded:

$$I_{\text{boost}} = I_{\text{off}} = I_{\text{phase\_pk}} = 18 \text{ A} \quad (9)$$

This informs the value of the resonant inductors:

$$L_{\text{rP}} = L_{\text{rN}} = \frac{V_{\text{dc}} \cdot t_{\text{ramp\_on\_max}}}{8 I_{\text{phase\_pk}}} \quad (10)$$

Where $t_{\text{ramp\_on\_max}}$ is the maximum ramp interval that appears during T_1, under this variable-timing scheme. It is emphasized that soft switching should not violate the timing constraints imposed by the modulation scheme, i.e., it should be well over before the next transition is commanded. Hence, the t_{ramp} and t_{res} intervals must comply with and not be too long. This way, for $t_{\text{ramp\_on\_max}} = 400$ ns, $L_{\text{rP}} = L_{\text{rN}} = 1.39$ µH. From (7), (6), and (5), the snubber capacitors are estimated as $C_r = 62$ nF.

This variable-timing scheme is expected to perform as well as the corresponding 2L design, with regards to high-frequency response, as seen in [24]. However, only simulations under fixed timing are available. A fixed-timing scheme though, derived from the variable-timing scheme described here, is useful for estimating the current stress of the devices, and sizing them properly. As such, by maintaining the values of the

The 2018 International Power Electronics Conference

resonant components, the longest resonant interval is demanded to be $t_{res\_on\_max} = 1.2$ μs, and since t_{res} varies all the time, all the other intervals will have a shorter duration. For a fixed $t_{ramp\_on} = t_{ramp\_off} = 400$ ns, the trip currents will be constant at $I_{trip} = I_{boca} = 36$ A. While this is an indicator of how current stress and power loss can become excessive, it renders fixed timing very simple in its implementation, as there is no need to distinguish between turn-ons and turn-offs.

V. SIMULATION RESULTS

This section presents the results of a 3L ACPI phase leg, simulated in SABER under the simplest fixed-timing scheme. Its time- and frequency-domain results are compared to those of a 2L ACPI simulated under the same fixed-timing scheme. The ACPI circuits are also compared to the results of two hard-switched inverters, a 3L T-type and a 2L VSI. Ideal switch models and piece-wise linear diode models are used in all the inverters. The DC-link voltage and the sinusoidal phase current are simulated as ideal sources, with the latter having a phase difference of -30° with regards to the sinusoidal reference.

A. Time-domain Results

Fig. 7 compares the 3L and the 2L ACPI, over two fundamental cycles. The i_{LrP} and i_{LrN} currents behave like the two halves of the 2L i_{Lr} current. As a result, i_{LrP} and i_{LrN} are 7.6 Arms each, almost $\sqrt{2}$ times smaller than the RMS value of i_{Lr}. This means a thinner winding for L_{rP} and L_{rN}. However, the maximum peaks remain the same. Thus, the cores of the two inductors will have the same peak magnetic flux density rating

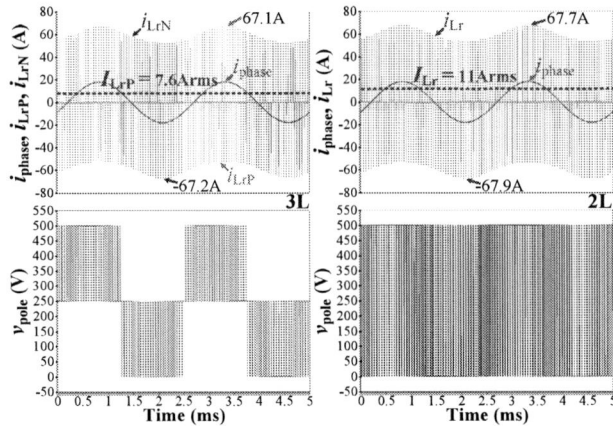

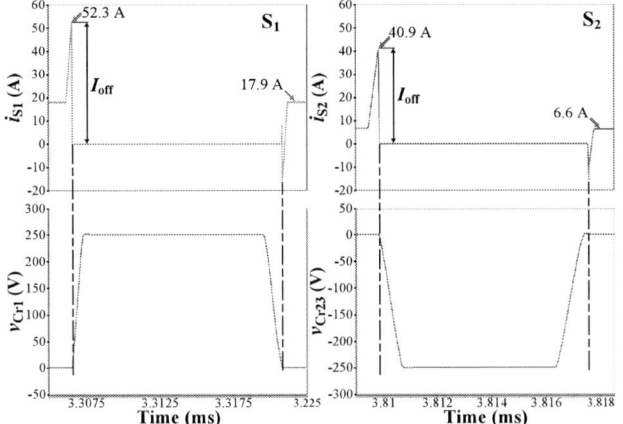

Fig. 7. 3L ACPI waveforms of i_{phase}, i_{LrP}, i_{LrN} and v_{pole}, and 2L ACPI waveforms of i_{phase}, i_{Lr} and v_{pole}, over two fundamental cycles.

Fig. 8. ZVS operation of S_1 at I_{phase_pk}, and of S_2 at low positive i_{phase}.

as L_r. If only one inductor is built, its winding should be rated for the same RMS current as in the 2L ACPI. Fig. 7 also shows the 3L and 2L SPWM-modulated v_{pole} pulse-trains. The constituent pulses of the 3L v_{pole} have an amplitude of $V_{dc}/2$.

Fig. 8 shows the transitions of the outer switch S_1 at I_{phase_pk}, and of the midpoint switch S_2 at a low positive load. They both transition under ZVS, with their currents falling and rising swiftly while the voltage across them is zero. Due to $\cos\varphi < 1$, S_2 transitions with D_4 at positive light load, for a limited time during T_1. The fixed-timing simulation indicates that the outer devices should be rated for a maximum I_{off} current of 52 A. The midpoint devices should have the same pulsed current rating as well, because always at the end of a given turn-off transition, an

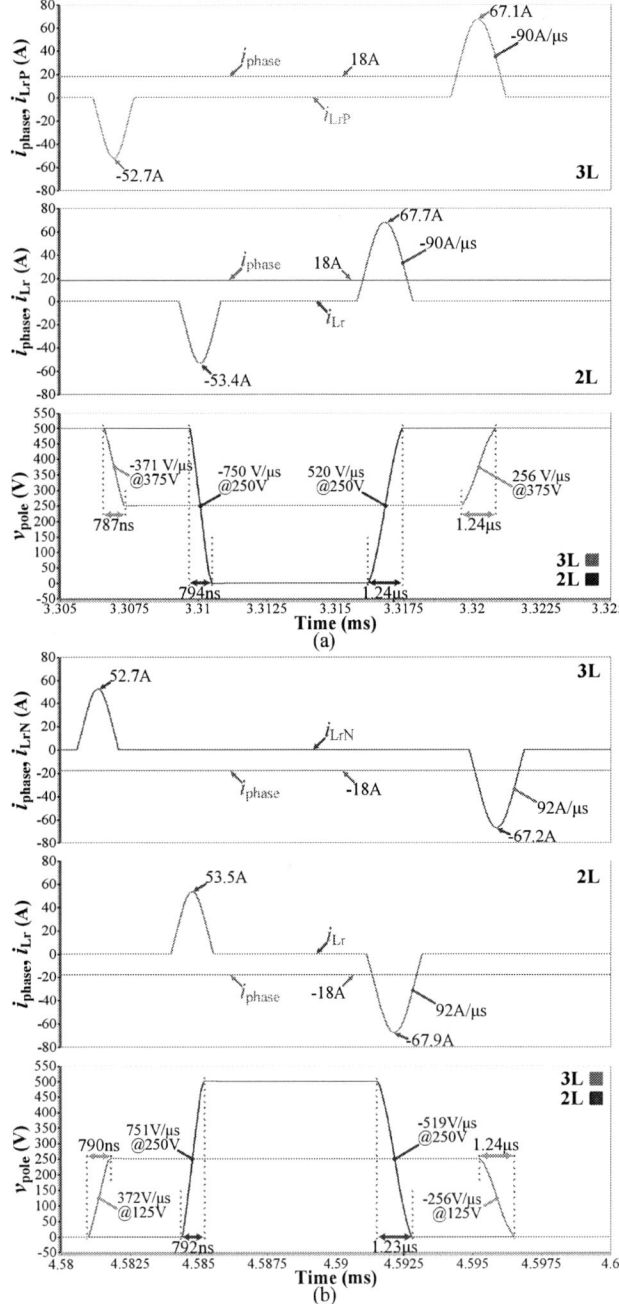

Fig. 9. Resonant transitions of the 3L and the 2L ACPI at (a) $+I_{phase\_pk}$ and at (b) $-I_{phase\_pk}$.

1638

The 2018 International Power Electronics Conference

incoming midpoint diode (D_2 or D_3) will conduct I_{off}, regardless of cosφ. At peak phase current, that will be 52 A.

Fig. 9(a) shows the top resonant circuit performing the transitions of S_1 (turn-off followed by a turn-on) at $+I_{phase\_pk}$. Fig. 9(b) depicts the bottom resonant circuit transitions at $-I_{phase\_pk}$. They transpire in the same manner and involve S_4 turning off and then on again.

These switching cycles are compared to their corresponding 2L ones. Since in both inverters the transitions occur at $|I_{phase\_pk}|$ and the fixed-timing schemes are designed in the same way, the inductor current peaks are very similar in the two inverters. The slope of the inductor current ramps is also unchanged, because, with respect to the 2L ACPI, the i_{LrP} and i_{LrN} currents are built by half the driving voltage, while the inductance is also halved. The 2L inductor in [24] is 2.71 µH.

The length of the v_{pole} edges is also maintained when changing from 2 over to 3 levels, but due to the fixed-timing scheme, the edge durations are dissimilar, and the pulses are asymmetrical. By halving the amplitude of the 3L v_{pole} pulses, while retaining the edge duration, the slope at the midpoint is halved with respect to the 2L pulse-train, and the main devices can benefit from greatly reduced dv/dt stress.

B. Frequency-domain Results

The spectra presented in this section were generated with the FFT function provided by SABER, where a rectangular window and 500,000 time-domain samples of the time-domain v_{pole} waveforms were used over 2 fundamental cycles.

Following the analysis of the time-domain pulses in Fig. 9, Fig. 10 compares the v_{pole} frequency response of the 3L and the 2L ACPI. While the fundamental harmonic of 400 Hz is identical in the two inverters, the 3L spectrum has a harmonic content that is between 5 and 7 dB lower than the 2L one. This trend however, breaks down at about 5 MHz onwards. Normally, halving the amplitude of the pulse-train should halve the harmonic content, and a uniform 6 dB reduction should be observed, in the entire frequency spectrum. For reference, the simulated spectra of the hard-switched 3L T-type and the 2L VSI are depicted in Fig. 11, along with the time-domain pulses at I_{phase_pk}. It is seen that the hard-switched spectra agree better with theory, as a more or less 6 dB harmonic content reduction is universally observed. This could be explained by the fact that SABER has difficulty in drawing the soft-switched spectra, which under fixed timing, are derived from pulse-trains with asymmetrical pulses and with edges that vary in duration from one switching cycle to the next.

Despite the erratic behaviour that results from fixed timing, Fig. 12 illustrates how soft-switching the 3L T-type can greatly reduce the high-frequency content of v_{pole}. While the low-

frequency spectrum remains unchanged, the -40 dB/dec roll-off appears earlier in the soft-switched spectrum, which leads to approximately a 16-dB harmonic content reduction in the high frequencies. This is expected since the longest soft-switched edge during the turn-on transition, is by 5 times longer than the hard-switched ones, as seen in the simulated time-domain pulses at I_{phase_pk}. However, no -60 dB/dec region is seen in any of the ACPI spectra. This might be because the simulated edges retain a somewhat linear shape with no smooth corners. This, coupled with the always-varying duration of the edges makes it

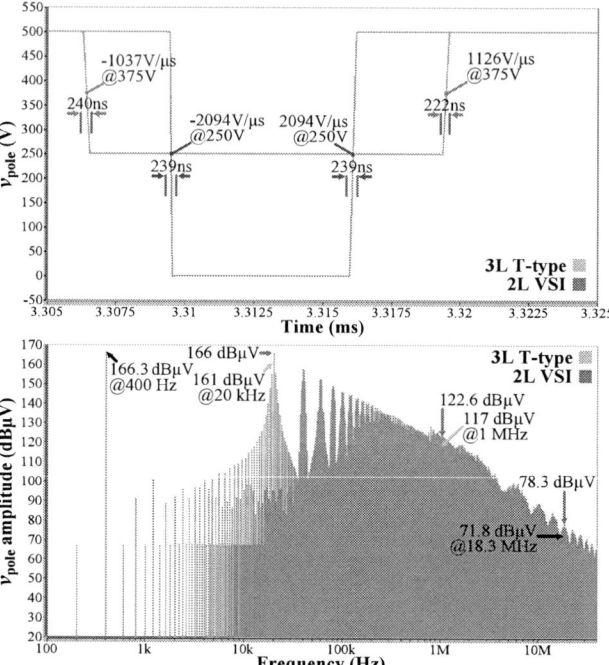

Fig. 11. Simulated v_{pole} pulses at I_{phase_pk} and spectra of the hard-switched 3L T-type and the hard-switched 2L VSI.

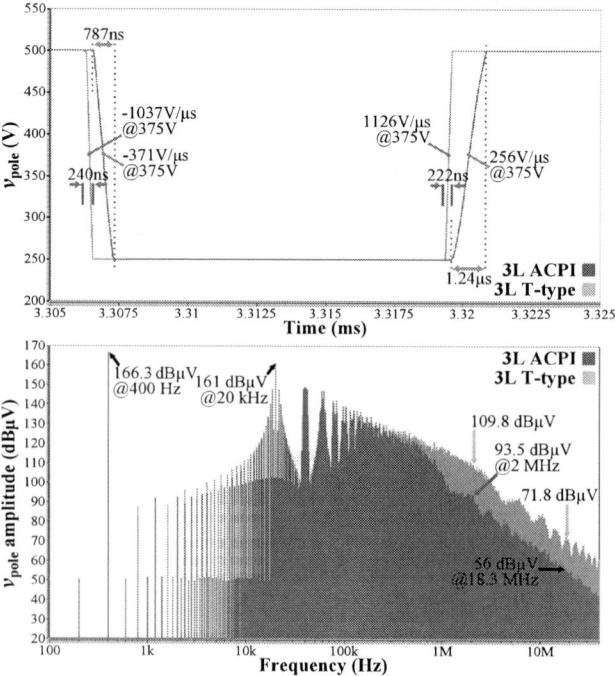

Fig. 12. Simulated v_{pole} pulses at I_{phase_pk} and spectra of the 3L T-type (hard-switching) and the 3L ACPI under the simplest fixed-timing control.

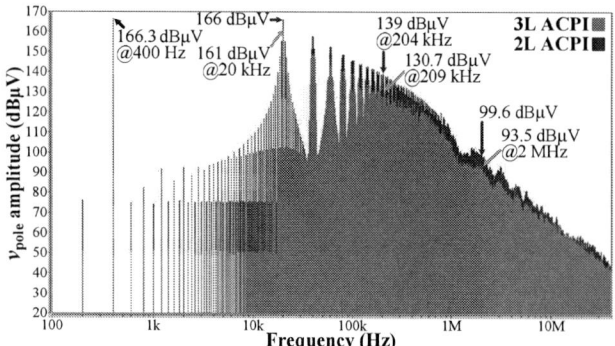

Fig. 10. Simulated v_{pole} spectra of the 3L and 2L ACPI under the simplest fixed-timing control.

1639

The 2018 International Power Electronics Conference

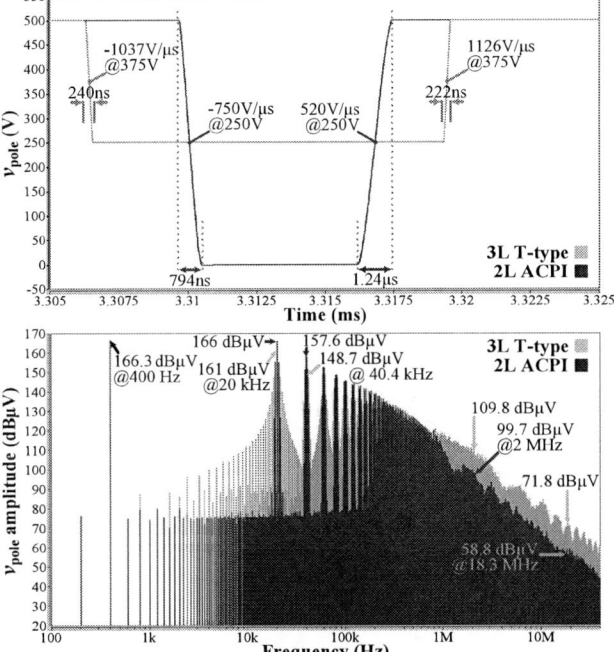

Fig. 13. Simulated v_{pole} pulses at I_{phase_pk} and spectra of the 3L T-type and the 2L ACPI under the simplest fixed-timing control.

hard to precisely predict the frequency response. The simulations indicate that the 3-level ACPI can attenuate the high-frequency harmonics at least thanks to the long duration of the edges, if not for their smoother shape.

Lastly, Fig. 13 compares the v_{pole} pulses of the 2L ACPI and the 3L hard-switched T-type. It is clear how the 2-level ACPI leads to slower transitions, with smaller dv/dt, though not as small as when the T-type is soft-switched. The frequency response of the 2L ACPI, also shown in Fig. 13, has a 5 to 9-dB higher harmonic content in the low-frequency region than the 3L T-type. This is to be expected, since the amplitude of the 3L pulse-train is halved, despite its faster transitions. The effect of the transitions starts to matter beyond 700 kHz. The -40 dB/dec roll-off starts early in the 2L ACPI spectrum, resulting in an approximately 13-dB reduction in harmonic content, with regards to the hard-switched T-type.

This is an interesting result. If Fig. 10 is thought to be completely correct, then not much is gained in high-frequency attenuation when the T-type is soft-switched. However, according to theory, the 3L ACPI should be able to deliver a further 6-dB reduction of harmonic content, for a total of 19 dB less content in the high frequencies, with respect to the hard-switched T-type. Then, the other benefits that the T-type has over the 2L VSI can be retained, such as potentially higher efficiencies and better-quality output waveforms for a harmonic content that can be attenuated across the frequency spectrum. Moreover, of all the inverters examined here, the 3L ACPI imposes the lowest dv/dt stress.

On the other hand, the component count of the 3L ACPI is large. It requires 4 main and 4 auxiliary device pairs for a total of 16, 3 snubber capacitors and 2 resonant inductors per phase. Note that, the four auxiliary device pairs will only have to block $V_{dc}/4$. Hence, for a 500-V DC link, faster and cheaper 250-V MOSFETs can be used, instead of IGBTs. In addition, their continuous current rating will be smaller than that of the 2L ACPI auxiliary devices, since the i_{LrP} and i_{LrN} waveforms have a smaller RMS value than the 2L i_{Lr} waveform. The pulsed

current rating however will have to be retained, and MOSFETs with quite high pulsed current ratings will be necessary.

As far as high-frequency attenuation is concerned, it is reasonable to suggest that the 2L ACPI can still perform quite well, without the added control complexity and device count of the 3L ACPI, and without further splitting the DC link and possibly exacerbating any voltage balance issues. The 2L ACPI only adds two more auxiliary devices per phase, albeit of twice the voltage rating, since they have to block a voltage of $V_{dc}/2$. However, only one more isolated gate driver supply is added per phase, making for 3 supplies in total. The 3L ACPI would require 5 isolated gate driver supplies per phase.

VI. CONCLUSIONS

A 3L T-type inverter with ACPI branches has been presented in this paper, with a focus on attenuating the high-frequency content of its output voltage, for enhanced conducted EMI performance. For this reason, the resonant operation of the 3L ACPI has been studied, with the help of the equivalent LC circuits, and the analytical expressions of the resonant magnitudes and intervals have been extracted. By controlling the resonant intervals, the edges of the output voltage can be actively shaped to be long and with smooth corners, for enhanced high-frequency response, and as such, a variable-timing scheme has been presented for a design primarily focused on high-frequency content attenuation.

Fixed-timing simulations have been employed for proving the operational concept, that is, the full ZVS of the main devices and the realization of the resonant transitions. It has been observed that the 3L ACPI output voltage exhibits a high-frequency content that is attenuated by 16 dB, compared to the hard-switched T-type. The limitations of the simulation however, and perhaps of the novel inverter as well, are highlighted by the fact that the 2L ACPI can also result in a substantial high-frequency content attenuation when compared to the hard-switched T-type, albeit exhibiting increased content in the PWM region.

Conclusively, an overall system evaluation that considers component counts and ratings, efficiency, superior-quality output waveforms, reduced dv/dt stress, and harmonic content reduction across the spectrum, all potential benefits offered by the 3L ACPI, should be undertaken with the help of variable-timing simulations and importantly, an experimental prototype. All these traits can benefit VSI-based applications, especially where stringent EMI standards apply, such as the engine starter motor drive on the MEA, because the combination of superior AC waveform quality and attenuated conducted EMI can help to increase the power density by reducing the required passive filters.

ACKNOWLEDGMENT

This project was supported by Safran Electrical & Power UK and the Royal Academy of Engineering under the Research Chairs and Senior Research Fellowships scheme.

REFERENCES

[1] J. Holtz, "Selbstgefuhrte weschselrichter mit treppenformiger ausgangsspannung fur grose leistung und hohe frequenz," *Siemens Forschungs-und Entwicklungsberichte,* vol.6, no.3, pp.164-171, 1977.

1640

[2] M. Schweizer and J. W. Kolar, "High eciency drive system with 3-level T-type inverter," in *Proc. EPE'11 Conf.*, pp.1-10, Aug 2011.

[3] R. G. d. Almeida Cacau, R. P. Torrico-Bascop, J. A. F. Neto, and G. V. Torrico-Bascop, "Five-level T-type inverter based on multistate switching cell," *IEEE Transactions on Industry Applications*, vol. 50, no. 6, pp. 3857-3866, Nov 2014.

[4] M. Schweizer and J. W. Kolar, "Design and implementation of a highly efficient three-level T-type converter for low-voltage applications," *IEEE Transactions on Power Electronics*, vol. 28, no. 2, pp. 899-907, Feb 2013.

[5] L. G. Franquelo, J. Rodriguez, J. I. Leon, S. Kouro, R. Portillo, and M. A. M.Prats, "The age of multilevel converters arrives," *IEEE Industrial Electronics Magazine*, vol. 2, no. 2, pp. 28-39, June 2008.

[6] I. Staudt, *3L NPC & TNPC Topology*, 5th ed., Semikron, October 2015, Application Note AN-11001.

[7] J. Rodriguez, J.-S. Lai, and F. Z. Peng, "Multilevel inverters: a survey of topologies, controls, and applications," *IEEE Transactions on Industrial Electronics*, vol. 49, no. 4, pp. 724-738, Aug 2002.

[8] P. C. Loh, D. G. Holmes, and T. A. Lipo, "Implementation and control of distributed PWM cascaded multilevel inverters with minimal harmonic distortion and common-mode voltage," *IEEE Transactions on Power Electronics*, vol. 20, no. 1, pp. 90-99, Jan 2005.

[9] C. Jaeger, I. Grinbaum and J. Smajic, "Numerical simulation and measurement of common-mode and circulating bearing currents," in *Proc. IEEE ICEM '16 Conf.*, pp. 486-491, 2016.

[10] J. Erdman, R. J. Kerkman, D. Schlegel and G. Skibinski, "Effect of PWM inverters on AC motor bearing currents and shaft voltages," in *Proc. IEEE APEC '95 Conf.*, vol.1, pp. 24-33, 1995.

[11] A. von Jouanne, P. Enjeti, and W. Gray, "The effect of long motor leads on PWM inverter fed AC motor drive systems," in *Proc. IEEE APEC '95 Conf.*, vol.2, pp. 592-597, 1995.

[12] S. Karugaba, A. Muetze, and O. Ojo, "On the common-mode voltage in multilevel multiphase single- and double-ended diode-clamped voltage-source inverter systems," *IEEE Transactions on Industry Applications*, vol. 48, no. 6, pp. 2079-2091, Nov 2012.

[13] B. Sarlioglu and C. T. Morris, "More Electric Aircraft: Review, Challenges, and Opportunities for Commercial Transport Aircraft," *IEEE Transactions on Transportation Electrification*, vol. 1, no. 1, pp. 54-64, June 2015.

[14] M.E. Elbuluk and M.D. Kankam, "Potential starter/generator technologies for future aerospace applications," *IEEE Aerospace and Electronic Systems Magazine*, vol. 12, no. 5, pp. 24-31, May 1997.

[15] M. Sinnett, "787 no-bleed systems: Saving fuel and enhancing operational efficiencies," *Boeing Aero Mag.*, vol. 4, pp. 6-11, 2007.

[16] M. W. Gekeler,"Soft switching three level inverter (S3L inverter)," in *Proc. EPE'15 Conf.*, pp. 1-10, Sept 2013.

[17] F. G. Stein and Y. R. de Novaes, "Analysis of a snubber for the T-type NPC converter," in *Proc. IEEE ISIE'15 Conf.*, pp. 239-244, June 2015.

[18] R. W. De Doncker, and J. P. Lyons, "The auxiliary resonant commutated pole converter," in *Proc. IEEE IAS Annual Meeting' 90*, vol. 2, pp. 1228-1235, 1990.

[19] W. McMurray, "Resonant snubbers with auxiliary switches," in *Proc. IEEE IECON' 89 Conf.*, vol.1, pp. 289-834, 1989.

[20] M. Heuvelmans, T. Moder, and S. Norrga, "Soft-switching cells for high-power converters," in *Proc. IEEE IECON'14 Conf.*, pp. 1806-1812, Oct 2014.

[21] J. G. Cho, J. W. Baek, D. W. Yoo, and C. Y. Won, "Three level auxiliary resonant commutated pole inverter for high power applications," in *Proc. IEEE PESC'96 Conf.*, vol. 2, pp. 1019-1026, Jun 1996.

[22] X. Yuan, G. Orglmeister, and I. Barbi, "ARCPI resonant snubber for the neutral-point-clamped inverter," *IEEE Transactions on Industry Applications*, vol. 36,no. 2, pp. 586-595, Mar 2000.

[23] A. Radan and M. Samaram, "The ARCP flying capacitor inverter," in *Proc. EPE'05 Conf.*, pp.1-8, Sept 2005.

[24] A. Charalambous, X. Yuan, and N. McNeill, "High-frequency EMI attenuation at source with the auxiliary commutated pole inverter," *IEEE Transactions on Power Electronics*, 2017 (Early access online).

[25] Michigan State University. Electromagnetic Compatibility (EMC) Course Notes. http://www.egr.msu.edu/em/research/goali/notes/ (2007).

[26] A. Karvonen, T. Thiringer, P. Futane, T. Tuveson, and H. Holst, "Reduction of EMI in switched mode converters by shaped pulse transitions," *Society of Automotive Engineers (SAE) World Congress*, 2007, Paper 2007-01-0361.

[27] N. Oswald, B. H. Stark, D. Holliday, C. Hargis, and B. Drury, "Analysis of shaped pulse transitions in power electronic switching waveforms for reduced EMI generation," *IEEE Transactions on Industry Applications*, vol. 47, no. 5, pp. 2154-2165, Sept.-Oct. 2011.

[28] K. Ma, D. Xu, T. Zhang, and S. Igarashi, "The evaluation of control strategies for auxiliary resonant commutated pole inverter," in Proc. *IEEE ECCE'09 Conf.*, pp. 810-816, 2009.

The 2018 International Power Electronics Conference

Carrier-Based Realization of Arbitrary Space-Vector PWM Methods for Three-Level Inverters

Somboon Sangwongwanich[*] and Supakorn Paiboon
Dept. of Electrical Eng., Faculty of Engineering, Chulalongkorn University, Bangkok, Thailand
*E-mail: somboona@chula.ac.th

Abstract— **Various space-vector modulation methods developed for three-level inverters require complicated calculation. Some researches attempt to solve this problem using a carrier-based modulation with the addition of appropriate zero voltages. However, to find the equivalent carrier-based realization is not always easy, especially for those space-vector modulations which include dipolar switching mode. This paper proposes a systematic procedure to derive an equivalent carrier-based modulation for an arbitrary space-vector modulation by decomposition of voltage commands and addition of generalized zero voltages. The proposed procedure is based on two key mathematical concepts 1) the barycentric coordinates for space-vector calculation and 2) the generalized dipolar modulation concept. Experimental results are given to prove the correctness of the proposed carrier-based algorithm.**

Keywords— *Three-level inverters, space-vector PWM, carrier-based realization, dipolar PWM.*

I. INTRODUCTION

The three-level inverter in Fig. 1 is now one of the basic converters used in several applications [1]. Various space-vector modulation (SVM) [2]-[16] and carrier-based modulation (CBM) [17]-[28] methods have been developed for three-level inverters. As compared to SVM, CBM is much easier to implement, especially when the dc bus voltages are unbalanced. Several attempts have been done to relate the SVM to the CBM [29]-[37]. However, two obstacles exist when one wants to find the equivalent CBM for a given SVM.

First, calculation of duty cycles or dwell times in the SVM is usually done on the 2-dimensional space vector coordinates instead of the original three-phase coordinates. Therefore, the duty cycles of the switching vectors are not expressed explicitly in terms of the phase voltages. This makes it difficult to analyze how the final output voltages generated by the SVM are related to the commanded phase voltages. Since CBM is based on the comparison of the three-phase voltage references and the carrier waveform, the equivalent CBM can be derived if the generated output voltages of the given SVM are analytically known as an explicit function of the three-phase voltage commands.

The above difficulty can be avoided by considering another way around, i.e. calculating instead the equivalent SVM of a given CBM. And it is shown that

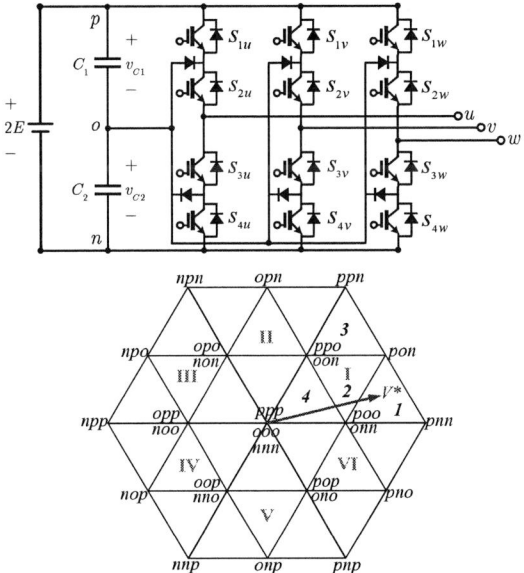

Fig. 1. Space vector diagram of a three-level inverter.

certain nearest-three-vector SVM methods are realizable by the unipolar CBM method by adding a zero voltage to the voltage commands [29]-[37]. However, from the space-vector viewpoint, the unipolar modulation covers only the SVM methods which use 3 or 4 switching vectors [2]-[8], while there are several SVM methods which use 5, 6 or 7 vectors [9]-[16]. Also, the selected vectors may not be the nearest vectors as well. Therefore, limitation to the unipolar modulation is the second obstacle in finding the CBM equivalent to the SVM.

The objective of this paper is to derive a CBM method which is equivalent to a given (arbitrary) SVM method. The derivation procedure presented in this paper solves the previous obstacles by introducing two key mathematical frameworks for modulation of the three-level inverter. First, barycentric-coordinate technique is proposed to compute the duty cycles of the SVM directly in terms of phase coordinates. Second, the generalized dipolar modulation concept is adopted to make it possible to analyze the characteristics of the SVM methods which use 3 to 7 switching vectors. Feasibility of all the analytical results will be verified by experiment on a neutral-point-clamped (NPC) three-level inverter.

1642

II. Space Vector Diagram of Three-Level Inverter and Barycentric Coordinates Concept

Fig. 2 is the space vector diagram of the switching space vectors generated by the three-level inverter. It should be noted that the coordinates of each vector in this diagram are not the 2-dimensional space vector coordinates as usually seen in the literatures. Here, the switching space vectors are expressed in terms of three-phase coordinates instead, whose values are equal to the three-phase voltages generated by the switching state minus the zero-voltage term. These switching vectors are coplanar, i.e. lie on the 2-dimensional space vector plane embedded in the 3-dimensional space of the three phase voltages.

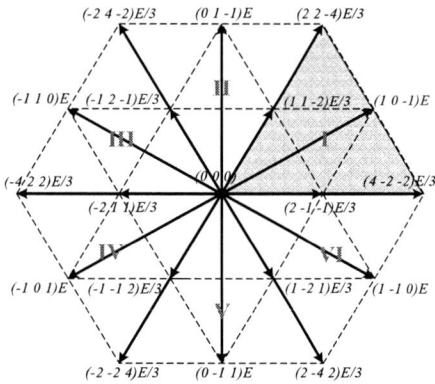

Fig. 2. Three-phase coordinates of hexagonal vector diagram of the three-level inverter.

In the nearest-three-vector (NTV) and the virtual-nearest-three-vector (NTV²) modulations, the 1st sector of Fig. 2 is divided into several sectants (triangles) as shown in Figs. 3(a) and 4(a).

By SVM, a given voltage command $\mathbf{V}^* = (v_u^*, v_v^*, v_w^*)$ belonging to a specific sectant is decomposed in terms of the three vertex vectors $\mathbf{V}_1, \mathbf{V}_2, \mathbf{V}_3$ as shown by (1) and (2) where $\lambda_1, \lambda_2, \lambda_3$ are the duty cycles of the three vertex vectors.

$$\mathbf{V}^* = \lambda_1\mathbf{V}_1 + \lambda_2\mathbf{V}_2 + \lambda_3\mathbf{V}_3 \tag{1}$$

$$\lambda_1 + \lambda_2 + \lambda_3 = 1 \tag{2}$$

Normally the SVM does this calculation in 2-dimensional space vector (α, β) coordinates based on the space vector transformation (3).

$$\mathbf{V}_{\alpha-\beta} = \begin{bmatrix} v_\alpha \\ v_\beta \end{bmatrix} = \sqrt{\frac{2}{3}} \begin{bmatrix} 1 & -1/2 & -1/2 \\ 0 & \sqrt{3}/2 & -\sqrt{3}/2 \end{bmatrix} \begin{bmatrix} v_u \\ v_v \\ v_w \end{bmatrix} = V\angle\theta \tag{3}$$

Therefore, the voltage command vector, the vertex vectors, and the duty cycles $\lambda_1, \lambda_2, \lambda_3$ are not explicitly expressed in terms of the original three-phase coordinates (v_u^*, v_v^*, v_w^*), but as functions of (v_α, v_β) or (V, θ) instead. This makes it difficult to relate the final output voltages generated by the SVM to the original commanded voltages.

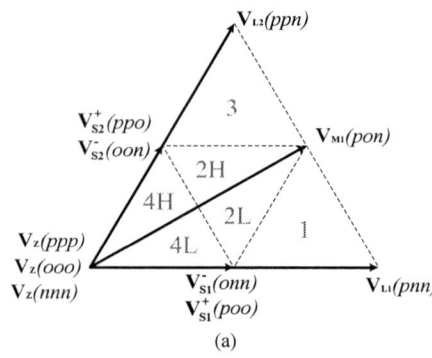

(a)

	1st Sectant		2nd Sectant
λ_{S1}	$2 + (v_w^* - v_u^*)/E$	λ_{S1}	$-1 + (v_w^* - v_v^*)/E$
λ_{L1}	$-1 + (v_u^* - v_v^*)/E$	λ_{M1}	$-1 + (v_u^* - v_w^*)/E$
λ_{M1}	$(v_v^* - v_w^*)/E$	λ_{S2}	$1 + (v_v^* - v_u^*)/E$
	3rd Sectant		4th Sectant
λ_{S2}	$2 + (v_w^* - v_u^*)/E$	λ_Z	$1 + (v_w^* - v_u^*)/E$
λ_{M1}	$(v_u^* - v_v^*)/E$	λ_{S1}	$(v_u^* - v_v^*)/E$
λ_{L2}	$-1 + (v_v^* - v_w^*)/E$	λ_{S2}	$(v_v^* - v_w^*)/E$

(b)

Fig. 3. Nearest-Three-Vector (NTV) modulation. (a) Sectants of 1st sector. (b) Corresponding duty cycles (Barycentric coordinates).

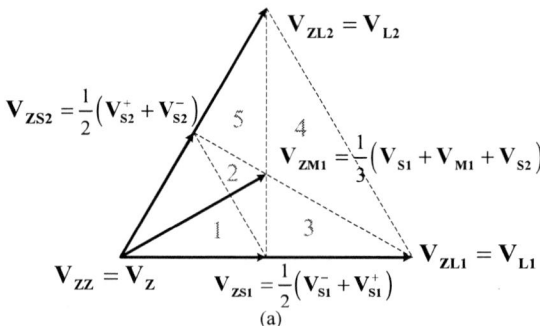

(a)

	2nd Sectant		3rd Sectant
λ_{ZS1}	$2 + 3v_w^*/E$	λ_{ZS1}	$2 + 3v_w^*/E$
λ_{ZM1}	$-3 + 3(v_u^* - v_w^*)/E$	λ_{ZL1}	$-1 + 3v_u^*/2E$
λ_{ZS2}	$2 - 3v_u^*/E$	λ_{ZM1}	$3(v_v^* - v_w^*)/2E$
	4th Sectant		5th Sectant
λ_{ZL1}	$-1 + 3v_u^*/2E$	λ_{ZM1}	$3(v_u^* - v_v^*)/2E$
λ_{ZL2}	$-1 - 3v_w^*/2E$	λ_{ZL2}	$-1 - 3v_w^*/2E$
λ_{ZM1}	$3 + 3(v_w^* - v_u^*)/2E$	λ_{ZS2}	$2 - 3v_u^*/E$

* duty cycles for 1st Sectant (NTV²) = duty cycles of 4th Sectant (NTV).

(b)

Fig. 4. Nearest-Three-Virtual-Vector (NTV²) modulation. (a) Sectants of 1st sector. (b) Corresponding duty cycles (Barycentric coordinates).

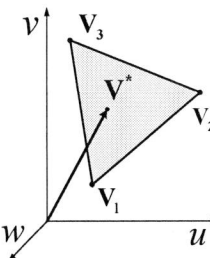

Fig. 5. Barycentric coordinates in 3-dimensional phase space.

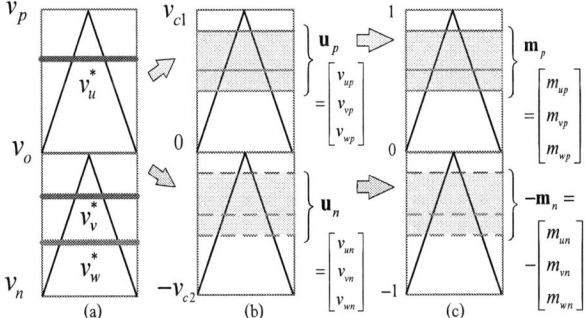

Fig. 6. Generalized dipolar PWM. (a) Commanded voltages. (b) Upper and lower-bus references. (c) Normalized references (duty cycles).

From the geometry viewpoint, vector decomposition in Figs. 3 and 4 is in fact the so-called barycentric coordinate calculation as shown in Fig. 5, which can be done on 3-dimensional phase space instead of 2-dimensional space vector plane. The corresponding coordinates (duty cycles) can be calculated directly from the phase coordinates by (4) [38], wherein "×" is the vector product and $[\,]_j$ is the j-axis component.

$$
\begin{aligned}
\lambda_1 &= \frac{[(\mathbf{V}^* - \mathbf{V}_3) \times (\mathbf{V}_2 - \mathbf{V}_3)]_j}{[(\mathbf{V}_1 - \mathbf{V}_3) \times (\mathbf{V}_2 - \mathbf{V}_3)]_j} \\[4pt]
\lambda_2 &= \frac{[(\mathbf{V}^* - \mathbf{V}_3) \times (\mathbf{V}_1 - \mathbf{V}_3)]_j}{[(\mathbf{V}_2 - \mathbf{V}_3) \times (\mathbf{V}_1 - \mathbf{V}_3)]_j} \\[4pt]
\lambda_3 &= 1 - \lambda_1 - \lambda_2 \quad (j = u, v, w)
\end{aligned}
\tag{4}
$$

Using (4), the duty cycles λ_k for each switching (vertex) vector $\mathbf{V}_k$ of the selected sectant can be found in terms of three-phase voltage commands (v_u^*, v_v^*, v_w^*) for the NTV and NTV² modulations as shown in Figs. 3(b) and 4(b), respectively, wherein $2E$ denotes the dc bus voltage. Another benefit of using barycentric coordinates is that duty cycle calculation (4) can be applied to any triangular shape (e.g. irregular triangular shape as in NTV² scheme). Therefore, duty cycles for other kinds of sectant division [9][11][16] can be found in the same manner.

III. DIPOLAR MODULATION THEORY FOR THREE-LEVEL INVERTER

Although most of the SVM methods are equivalent to the unipolar carrier-based PWM, there are still many other SVM methods [9]-[16] which result in both unipolar and dipolar mode in the modulation. Therefore, in this paper the dipolar modulation is adopted and is considered in a broad sense to include also unipolar or bipolar modulation as a special case. In the following, the mathematical framework for dipolar modulation will be briefly reviewed [22][27].

First, the output voltages v_u, v_v, v_w of the three-level inverter can be represented in the form of modulation or duty-cycle matrix $[m_{ij}]$ as shown in (5).

$$
\begin{bmatrix} v_u \\ v_v \\ v_w \end{bmatrix} = \begin{bmatrix} v_u^* + v_z \\ v_v^* + v_z \\ v_w^* + v_z \end{bmatrix} = \begin{bmatrix} m_{up} & m_{uo} & m_{un} \\ m_{vp} & m_{vo} & m_{vn} \\ m_{wp} & m_{wo} & m_{wn} \end{bmatrix} \begin{bmatrix} v_p \\ v_o \\ v_n \end{bmatrix}
\tag{5}
$$

$$
0 \le m_{ij} \le 1, \ \sum_j m_{ij} = 1, \ i = \{u, v, w\}, \ j = \{p, o, n\}
$$

where v_p, v_o, v_n are the dc-bus pole voltages, and v_z is the zero voltage added to the commanded voltages.

Under the generalized dipolar modulation concept, the output voltages referred to the neutral point 'o' can be written as a sum of the upper- and lower-bus voltage references $\mathbf{u}_p, \mathbf{u}_n$ as shown in (6)-(8) and Fig. 6.

$$
\overbrace{\begin{bmatrix} v_{uo} \\ v_{vo} \\ v_{wo} \end{bmatrix}}^{\mathbf{u}} \triangleq \begin{bmatrix} v_u - v_o \\ v_v - v_o \\ v_w - v_o \end{bmatrix} = \begin{bmatrix} v_u^* + v_z - v_o \\ v_v^* + v_z - v_o \\ v_w^* + v_z - v_o \end{bmatrix}
\tag{6}
$$

$$
= \overbrace{\begin{bmatrix} m_{up} \\ m_{vp} \\ m_{wp} \end{bmatrix}}^{\mathbf{u}_p} (v_p - v_o) + \overbrace{\begin{bmatrix} m_{un} \\ m_{vn} \\ m_{wn} \end{bmatrix}}^{\mathbf{u}_n} (v_n - v_o)
\tag{7}
$$

$$
\therefore \ \overbrace{\begin{bmatrix} v_{uo} \\ v_{vo} \\ v_{wo} \end{bmatrix}}^{\mathbf{u}} \triangleq \overbrace{\begin{bmatrix} v_{up} \\ v_{vp} \\ v_{wp} \end{bmatrix}}^{\mathbf{u}_p} + \overbrace{\begin{bmatrix} v_{un} \\ v_{vn} \\ v_{wn} \end{bmatrix}}^{\mathbf{u}_n}
\tag{8}
$$

where $v_{c1} = v_p - v_o$ and $v_{c2} = v_o - v_n$ are the upper- and lower-bus voltages, respectively.

In general, the commanded voltages $\mathbf{V}^* = (v_u^*, v_v^*, v_w^*)$ and the zero voltage v_z are also decomposed into the upper- and lower-bus components as shown in (9)-(11).

$$
\overbrace{\begin{bmatrix} v_u^* \\ v_v^* \\ v_w^* \end{bmatrix}}^{\mathbf{V}^*} \triangleq \overbrace{\begin{bmatrix} v_{up}^* \\ v_{vp}^* \\ v_{wp}^* \end{bmatrix}}^{\mathbf{V}_p^*} + \overbrace{\begin{bmatrix} v_{un}^* \\ v_{vn}^* \\ v_{wn}^* \end{bmatrix}}^{\mathbf{V}_n^*}
\tag{9}
$$

$$
v_{zo} = v_z - v_o \triangleq v_{zp} + v_{zn}
\tag{10}
$$

$$
v_{zp} = \tfrac{1}{3} \sum_{j=u,v,w} v_{jp}, \quad v_{zn} = \tfrac{1}{3} \sum_{j=u,v,w} v_{jn}.
\tag{11}
$$

Therefore, the upper- and lower-bus voltage references $\mathbf{u}_p, \mathbf{u}_n$ can finally be written as (12).

$$\overbrace{\begin{bmatrix} v_{up} \\ v_{vp} \\ v_{wp} \end{bmatrix}}^{\mathbf{u}_p} = \begin{bmatrix} v_{up}^* + v_{zp} \\ v_{vp}^* + v_{zp} \\ v_{wp}^* + v_{zp} \end{bmatrix}, \quad \overbrace{\begin{bmatrix} v_{un} \\ v_{vn} \\ v_{wn} \end{bmatrix}}^{\mathbf{u}_n} = \begin{bmatrix} v_{un}^* + v_{zn} \\ v_{vn}^* + v_{zn} \\ v_{wn}^* + v_{zn} \end{bmatrix}. \quad (12)$$

For unipolar modulation,

$$[\mathbf{u}_p]_j = \max(v_j^* + v_z, 0),$$

$$[\mathbf{u}_n]_j = \min(v_j^* + v_z, 0). \quad (13)$$

For dipolar modulation,

$$[\mathbf{u}_p]_j = v_{jp}^* + v_{zp},$$

$$[\mathbf{u}_n]_j = v_{jn}^* + v_{zn}. \quad (14)$$

In summary, as depicted in Fig. 6, to realize a carrier-based modulation of any SVM, the corresponding two voltage references $[\mathbf{u}_p]_j, [\mathbf{u}_n]_j$ or equivalently the upper/lower-bus commands $\mathbf{V}_p^*, \mathbf{V}_n^*$ and the zero voltages v_{zp}, v_{zn} must be found.

IV. Derivation Procedure of The Equivalent CBM Method for a Given Arbitrary SVM Method

The procedure to find the equivalent CBM of an SVM is described by reversing the process of Fig. 6 as follows.

Step 1) Determine the sectant of the commanded phase voltages $\mathbf{V}^*$ and find the duty cycles λ_i of the three vertex vectors of the selected sectant.

Step 2) Choose the redundant switching states and calculate the duty cycle (d_i) of each switching state from the duty cycles λ_i.

Step 3) From the switching sequence, draw the PWM waveforms for three-phase output voltages as shown in Fig. 7(a).

Step 4) From the PWM waveforms, calculate the average upper- and lower-bus voltage references $\mathbf{u}_p, \mathbf{u}_n$ as shown in Fig. 7(b).

Step 5) Using (9)-(12), find the zero voltages v_z, v_{zp}, v_{zn} and the two voltage references $\mathbf{V}_p^*, \mathbf{V}_n^*$.

Owing to the use of barycentric coordinate technique, all the calculations in Steps 1)-5) are done in terms of the commanded phase voltages $\mathbf{V}^* = (v_u^*, v_v^*, v_w^*)$. Tables 1-3 give the equivalent CBM methods for three examples of SVM methods. The results are shown for the 1st sector but can be repeated for other sectors as well. Similar correlation can be derived also for other SVM schemes [2]-[16].

V. Experimental Results and Discussion

Experiment is done on an NPC inverter to verify the analysis results. Parameters of the experimental setup are: dc bus 540 V, dc-bus capacitor C=560 uF, series RL load

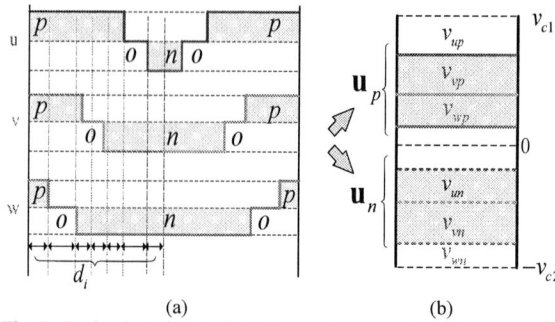

(a)   (b)

Fig. 7. Derivation of upper/lower-bus reference voltages. (a) PWM waveforms of the output voltages. (b) Averaged upper/lower-bus reference voltages.

(R = 52 Ω, L = 69 mH), switching frequency 4 kHz, output voltage 230 V 50 Hz. Figs. 8-10 compare the simulation results of SVM methods (shown in Tables 1-3) with the experimental results of the equivalent CBM methods. Here, $\Delta v_c = v_{c1} - v_{c2}$ denotes the dc-bus imbalance.

Table 1 NTV modulation in [6] and its zero voltage.

Sectant	zero voltage (v_z)	4 switching vectors
1	$\mathrm{mid}(v_u^*, v_v^*, v_w^*)/2$	oon,pnn,pon,poo
3	$\mathrm{mid}(v_u^*, v_v^*, v_w^*)/2$	oon,pon,ppn,ppo
2H	$(\min(v_u^*, v_v^*, v_w^*) + E)/2$	oon,pon,poo,ppo
2L	$(\max(v_u^*, v_v^*, v_w^*) - E)/2$	onn,oon,pon,poo
4H	$\max(v_u^*, v_v^*, v_w^*)/2$	oon,ooo,poo,ppo
4L	$\min(v_u^*, v_v^*, v_w^*)/2$	onn,oon,ooo,poo

Table 2 NTV modulation in [7] and its zero voltage.

Sectant	zero voltage (v_z)	3 switching vectors
1	$-E - \min(v_u^*, v_v^*, v_w^*)$	onn,pnn,pon
3	$-E - \min(v_u^*, v_v^*, v_w^*)$	oon,pon,ppn
2H	$-E - \min(v_u^*, v_v^*, v_w^*)$	onn,oon,pon
2L	$-E - \min(v_u^*, v_v^*, v_w^*)$	onn,oon,pon
4H	$-\max(v_u^*, v_v^*, v_w^*)$	ooo,oon,onn
4L	$-\max(v_u^*, v_v^*, v_w^*)$	ooo,oon,onn

Table 3 NTV² modulation in [10], its upper/lower-bus reference voltages, and upper/lower-bus zero voltages.

Sectant	Upper/lower-bus commands and zero voltages	5 switching vectors
1	$v_{jp}^* = v_j^*/2,$	ppo,ooo,poo,ooo,oon
2	$v_{jn}^* = -v_j^*/2$	ppo,poo,pon,oon,onn
3	$v_{zp} = -\min(v_u^*, v_v^*, v_w^*)/2$	ppo,poo,pon,pnn,onn
4	$v_{zn} = -\max(v_u^*, v_v^*, v_w^*)/2$	ppo,ppn,pon,pnn,onn
5	$v_z = \mathrm{mid}(v_u^*, v_v^*, v_w^*)/2$	ppo,ppn,pon,oon,onn

The 2018 International Power Electronics Conference

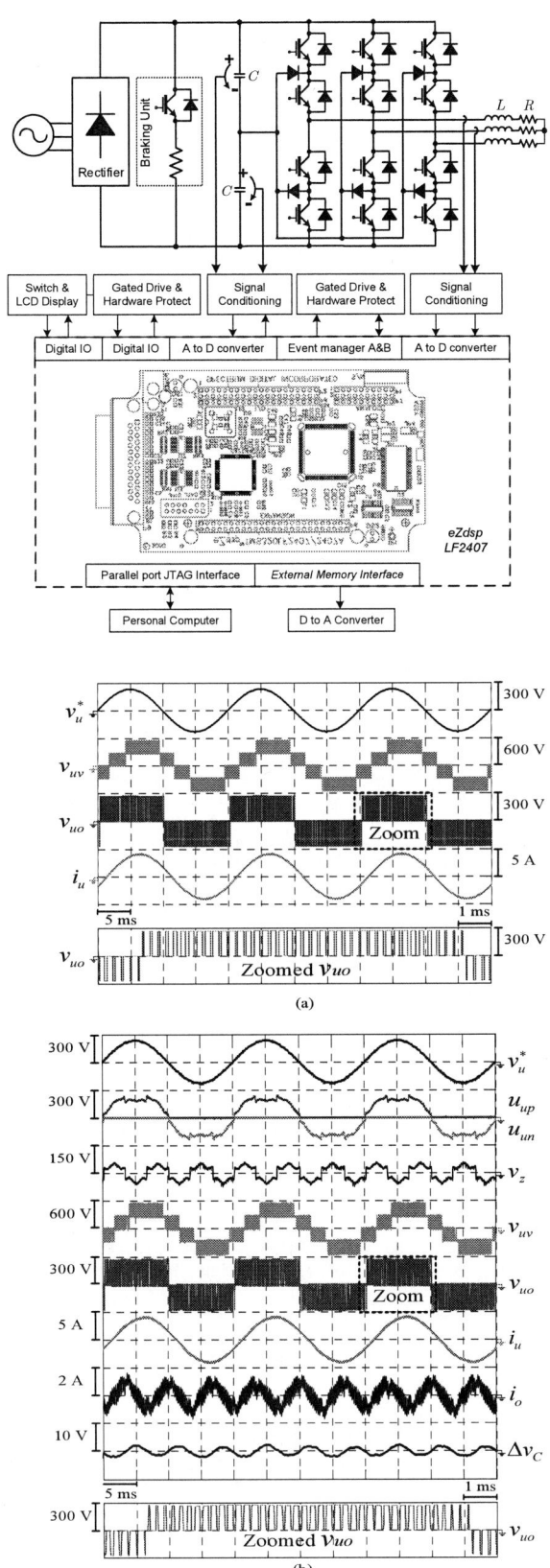

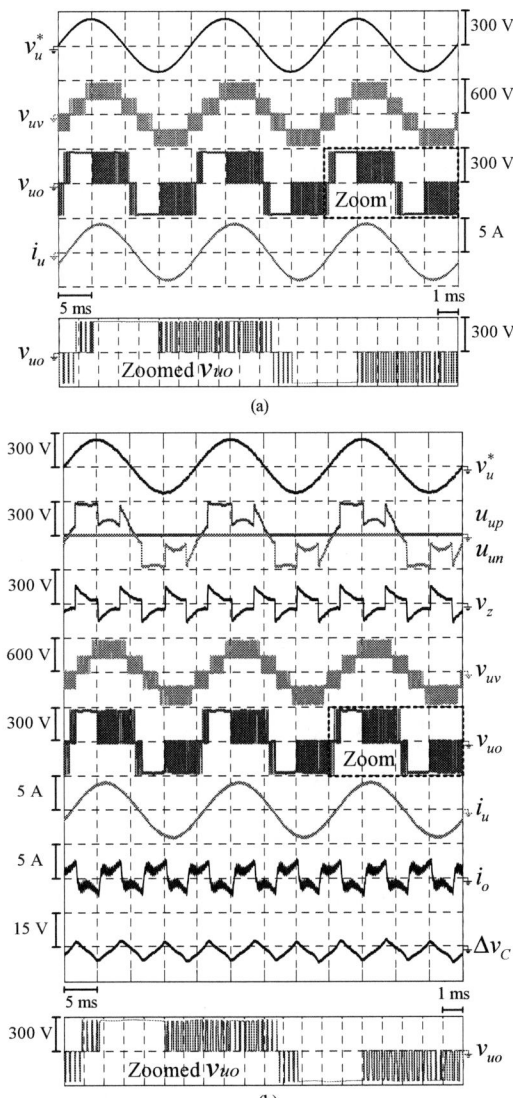

Fig. 10. (a) Simulation results with NTV SVM in [7]. (b) Experimental results with the equivalent CBM of (a).

Fig. 8. Experimental setup of the three-level inverter.

From the output voltage waveforms and also the upper/lower-bus reference voltages in Fig. 9, it is clear that the NTV modulation in [6] is a unipolar PWM. This conforms with the switching vectors shown in Table I. Simulation results of the SVM in Fig. 9(a) matches well with the experimental results of its realization by CBM, confirming that both are equivalent. Fluctuation in the neutral-point current and voltage is noted because of the inherent nature of the unipolar modulation.

As for the NTV SVM in [7], the output voltage waveforms show clearly the discontinuous PWM characteristic of this SVM. Waveforms of the upper/lower-bus reference voltages shown in its equivalent CBM reflects also the clamping of the output phase to the positive(p) or negative(n) dc bus. It is also seen that the zero voltages used in Figs. 9 and 10 are

Fig. 9. (a) Simulation results with NTV SVM in [6]. (b) Experimental results with the equivalent CBM of (a).

1646

The 2018 International Power Electronics Conference

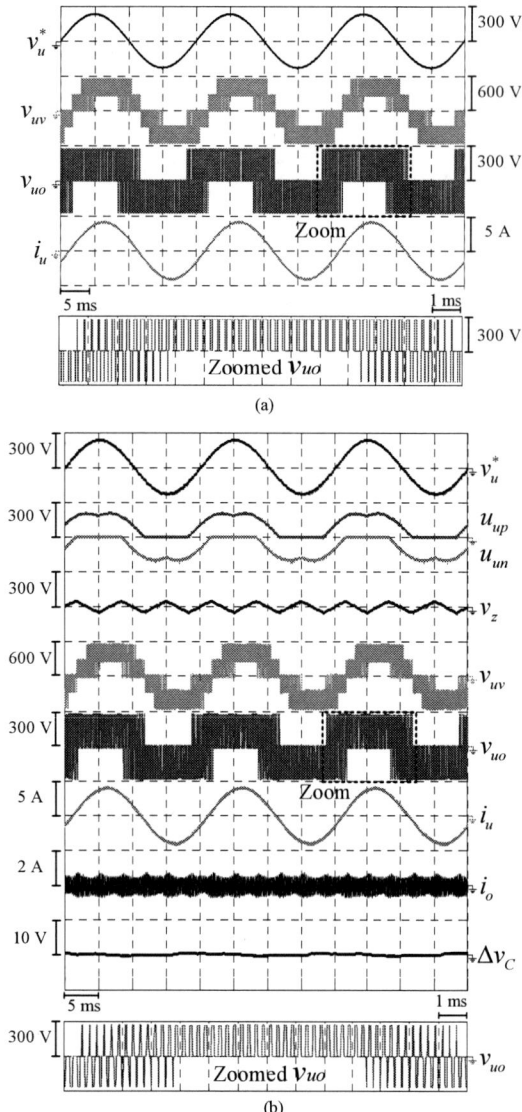

Fig. 11. (a) Simulation results with NTV² SVM in [10]. (b) Experimental results with the equivalent CBM of (a).

different. Again, perfect equivalent is obtained between the NTV SVM and the CBM.

Finally, Fig. 11 is the results for NTV² SVM in [10]. As can be seen from the simulation and experimental waveforms, this modulation scheme is a mix of unipolar and dipolar switching. Two output phases are in unipolar mode and the remaining phase is in dipolar mode. From Table III and the experimental waveform of the zero voltage v_z, this modulation scheme adds the same zero voltage as that of two-level inverter. The key features of the NTV² SVM are the zero averaged neutral-point current and the balanced dc bus, which is confirmed by the experimental waveforms of its equivalent CBM.

VI. CONCLUSIONS

This paper presents carrier-based realization of all space-vector modulation methods for three-level inverters using dipolar modulation theory, which greatly simplify the practical implementation. Algorithm to derive reference voltages for comparison with double carrier waves is given using only voltage commands and appropriate zero voltages. As examples, the proposed algorithm is applied to two space-vector modulation methods, i.e. the nearest-three-vector modulation (NTV) and the nearest-three-virtual-vector (NTV²) modulation. The simulation and experimental results show that the proposed unified algorithm can be realized very easily, and all the theoretical results are correct and feasible in practice.

REFERENCES

[1] J. Rodriguez, S. Bernet, P. K. Steimer, and I. E. Lizama, "A survey on neutral-point-clamped inverters," *IEEE Trans. on Industrial Electronics*, vol. 57, no. 7, pp. 2219–2230, 2010.

[2] J. H. Seo, C. H. Choi, and D. S. Hyun, "A New Simplified Space–Vector PWM Method for Three-Level Inverters," *IEEE Trans. on Power Electronics*, vol. 16, no. 4, pp. 545-550, 2001.

[3] M. Koyama, T. Fujii, R. Uchida, and T. Kawabata, "Space voltage vector-based new PWM method for large capacity three-level GTO inverter," in *Proc. of IECON 1992*, pp.271-276.

[4] B. Kaku, I. Miyashita, and S. Sone, "Switching loss minimized space vector PWM method for IGBT three-level inverter," *IEE Proc. on Electric Power Applications*, vol. 144, no. 3, pp. 182-190, 1997.

[5] N. Celanovic, and D. Borojevich, "A Comprehensive Study of Neutral-Point Voltage Balancing Problem in Three-Level Neutral-Point-Clamped Voltage Source PWM Inverters," *IEEE Trans. on Power Electronics*, vol. 15, no. 2, pp. 242-249, 2000.

[6] J. Pou, R. Pindado, D. Boroyevich, and P. Rodriguez, "Evaluation of the Low-Frequency Neutral-Point Voltage Oscillations in the Three-Level Inverter," *IEEE Trans. On Industrial Electronics*, vol. 52, no. 6, pp. 1582-1588, 2005.

[7] L. Helle, S. Munk-Nielsen, and P. Enjeti, "Generalized Discontinuous DC-Link Balancing Modulation Strategy for Three-Level Inverters," in *Proc. of the Power Conversion Conference (PCC)*, 2002, pp. 359-366.

[8] A. Choudhury, P. Pillay, and S. S. Williamson, "DC-Link Voltage Balancing for a Three-Level Electric Vehicle Traction Inverter Using an Innovative Switching Sequence Control Scheme," *IEEE Journal of Emerging and Selected Topics in Power Electronics*, vol. 2, no. 2, pp. 296-307, 2014.

[9] K. Ma, and F. Blaabjerg, "Unifying and Generating of Space Vector Modulation Sequences for Multilevel Converter," in *Proc. of Applied Power Electronics Conference and Exposition (APEC)*, 2014, pp. 196-201.

[10] S. Busquets-Monge, J. Bordonau, D. Boroyevich, and S. Somavilla, "The Nearest Three Virtual Space Vector PWM-A Modulation for the Comprehensive Neutral-Point Balancing in the Three-Level NPC Inverter," *IEEE Power Electronics Letters*, vol. 2, no. 1, pp. 11-15, 2004.

[11] C. Hu, X. Yu, D. G. Holmes, W. Shen, Q. Wang, F. Luo, and N. Liu, "An Improved Virtual Space Vector Modulation Scheme for Three-Level Active Neutral-Point-Clamped Inverter," *Trans. on Power Electronics*, vol. 32, no. 10, pp. 7419-7434, 2017.

[12] K. Tian, J. Wang, B. Wu, D. Xu, Z. Cheng, and N. R. Zargari, "A Virtual Space Vector Modulation Technique for the Reduction of Common-Mode Voltages in Both Magnitude and Third-Order Component," *Trans. on Power Electronics*, vol. 31, no. 1, pp. 839-848, 2016.

[13] C. J. Zhang, Y. Tang, D. Han, H. Zhang, X. Zhang, and K. Wan, "A Novel Virtual Space Vector Modulation Strategy for the Neutral-Point Potential Comprehensive Balance of Neutral-Point-Clamped Converters," *Journal of Power Electronics*, vol. 16, no. 3, pp. 946-959, 2016.

[14] S. B. Monge, S. Somavilla, J. Bordonau, and D. Boroyevich, "Capacitor Voltage Balance for the Neutral-Point Clamped Converter using the Virtual Space Vector Concept With Optimized Spectral Performance," *IEEE Trans. on Power Electronics*, vol. 22, no. 4, pp. 1128-1135, 2007.

[15] S. Busquets-Monge, J. D. Ortega, J. Bordonau, J. A. Beristáin, and J. Rocabert, "Closed-Loop Control of a Three-Phase Neutral-Point-Clamped Inverter Using an Optimized Virtual-Vector-Based Pulsewidth Modulation," *IEEE Trans. on Industrial Electronics*, vol. 55, no. 5, pp. 2061-2071, 2008.

[16] K. Ma, and F. Blaabjerg, "Modulation Methods for Neutral-Point-Clamped Wind Power Converter Achieving Loss and Thermal Redistribution Under Low-Voltage Ride-Through," *IEEE Trans. on Industrial Electronics*, vol. 61, no. 2, pp. 835-845 2014.

[17] A. Nabae I. Takahashi, and H. Akagi, "A New Neutral-Point-Clamped PWM Inverter," *IEEE Trans. on Industry Applications.*, vol. 17, no. 5, pp. 518–523, 1981.

[18]] S. Ogasawara, and H. Akagi, "Analysis of variation of neutral point potential in neutral-point-clamped voltage source PWM inverters," in *Conf. Rec. IEEE/IAS Annu. Meeting*, vol. 2, 1993, pp. 965-970.

[19] Q. Song et al., "A Neutral-Point Potential Balancing Algorithm for Three-Level NPC Inverters Using Analytically Injected Zero Sequence Voltage," in Proc. of APEC, 2003, vol. 1, pp. 1572-1580.

[20] D. Zhou, "A self-balancing space vector switching modulator for three-level motor drives," *IEEE Trans. on Power Electronics*, vol. 17, no. 6, pp. 1024-1031, 2002.

[21] T. Bruckner and D. G. Holmes, "Optimal pulse-width modulation for three-level inverters," *IEEE Trans. on Power Electronics*, vol. 20, pp. 82-89, 2005.

[22] B.Velaerts, P. Mathys, E. Tatakis, and G. Bingen, "A novel approach to the generation and optimization of three-level PWM wave forms," in *Proc. of PESC*, 1988, vol. 2, pp. 1255-1262.

[23] R. M. Tallam, R. Naik, and T. A. Nondahl, "A Carrier-Based PWM Scheme for Neutral Point Voltage Balancing in Three-Level Inverters," *IEEE Trans. on Industry Applications.*, vol. 41, no. 6, pp. 1734-1743, 2005.

[24] F. Bauer, and H. D. Heining, "Quick response space vector control for a high power three-level-inverter drive system," *Archiv fur Elektrotechnik*, Springer-Verlag, vol. 74, pp. 53-59, 1990.

[25] J. Pou, J. Zaragoza, P. Rodriguez, S. Ceballos, V. M. Sala, R. P. Burgos, "Fast-Processing Modulation Strategy for the Neutral-Point-Clamped Converter With Total Elimination of Low-Frequency Voltage Oscillations in the Neutral Point," *IEEE Trans. on Industrial Electronics*, vol. 54, no. 4, pp. 2288-2294, 2007.

[26] J. Zaragoza, J. Pou, S. Ceballos, V. M. Sala, E. Robles, and C. Jaen, "Voltage-Balance Compensator for a Carrier-Based Modulation in the Neutral-Point-Clamped Converter," *IEEE Trans. on Industrial Electronics*, vol. 56, no. 2, pp. 305-314, 2009.

[27] A. Saengseethong, and S. Sangwongwanich, "A new modulation strategy for capacitor voltage balancing in three-level NPC inverters based on matrix converter theory," in *Proc. of International Power Electronics Conference (IPEC) 2010*, pp. 2358-2365.

[28] S. Sangwongwanich, "Double-carrier-based modulation theory of three-level inverters and a new discontinuous PWM for neutral-point voltage balancing," in Proc. of *IECON 2012*, pp. 4961-4966.

[29] Vladimir Blasko, "Analysis of a Hybrid PWM Based on Modified Space-Vector and Triangle-Comparison Methods," *IEEE Trans. on Industry Applications*, vol. 33, no. 3, pp. 756-764, 1997.

[30] S. R. Bowes, and Y. S. Lai, "The Relationship Between Space-Vector Modulation and Regular-Sampled PWM," *IEEE Trans. on Industrial Electronics*, vol. 44, no. 5, pp. 670-679, 1997.

[31] F. Wang, "Sine-Triangle versus Space-Vector Modulation for Three-Level PWM Voltage-Source Inverters," *IEEE Trans. on Industry Applications*, vol. 38, no. 2, pp. 500-506, 2002.

[32] H. Wu, and X. He, "Inherent Correlation between Multilevel Carrier-based PWM and Space Vector PWM: Principle and Application," in *Proc. of the 4th IEEE International Conference on Power Electronics and Drive Systems*, pp. 276-281, 2001.

[33] W. Yao, H. Hu, and L. Zhengyu, "Comparisons of Space-Vector Modulation and Carrier-Based Modulation of Multilevel Inverter," *IEEE Trans. on Power Electronics*, vol. 23, no.1, pp. 45–51, 2008.

[34] C. Da-peng, S. Wen-xiang, X. Hui, C. Guo-cheng, and C. Chen, "Research on Zero-sequence Signal of Space Vector Modulation for Three-level Neutral-point-clamped Inverter Based on Vector Diagram Partition," in *Proc. of IEEE 6th International on Power Electronics and Motion Control Conference*, 2009. pp. 1435-1439.

[35] J. Pou, J. Zaragoza, S. Ceballos, M. Saeedifard, and D. Boroyevich, "A carrier-based PWM strategy with zero-sequence voltage injection for a three-level neutral-point-clamped converter. *IEEE Trans. on Power Electronics*, vol. 27, no.2 , pp. 642–651, 2012.

[36] K. Zhou, and D. Wan, "Relationship Between Space-Vector Modulation and Three-Phase Carrier-Based PWM: A Comprehensive Analysis," *IEEE Trans. on Industrial Electronics*, vol. 49, no. 1, pp. 186-196, 2002.

[37] B. P. McGrath, D. G. Holmes, and T. Lipo, "Optimized Space Vector Switching Sequences for Multilevel Inverters," *IEEE Trans. on Power Electronics*, vol. 18, no. 6, pp. 1293-1302, 2003.

[38] C. Koblbauer, "Barycentric Coordinates,", 2012 (Available at: http://koblbauermath.weebly.com).

The 2018 International Power Electronics Conference

Multi-Level Topology Based Linear Amplifier Family for Realization of Noise-Less Inverters

Hidemine Obara, Tatsuki Ohno, and Atsuo Kawamura

Department of Electrical and Computer Engineering, Yokohama National University, Yokohama, Japan
*E-mail: obara-hidemine-mh@ynu.ac.jp, oono-tatsuki-dy@ynu.ac.jp, kawamura@ynu.ac.jp

Abstract— Nowadays, all of power converters utilize switching operation to reduce losses of semiconductor power devices. However, the switching operation causes high harmonics, high electromagnetic interference (EMI), and enlarging passive components in a circuit. These are the typical issues in the power electronics, because dv/dt and di/dt in a power converter tend to high due to practical use of SiC and GaN devices with high-speed switching capability. In this paper, multi-level topology based linear amplifiers (MLLA) without the switching operation are proposed for realization of noise-less inverters. Although the conventional linear amplifier such as the class-B amplifier generate large loss due to a linear operation, MLLA can reduce the loss by multiple series connection of MOSFETs. Three basic topologies, diode-clamped linear amplifier (DCLA), flying-capacitor linear amplifier (FCLA), and modular-cascaded linear amplifier (MCLA) are introduced and summarized as the MLLA family. The operation principles of the three basic topologies are quite similar, and loss generated during a linear operation in the MOSFETs become smaller, as the number of series connected MOSFET increases. These features enable to realize a noise-less inverter with high efficiency even compared with the PWM inverters. Finally, it is verified that 4-series circuits of the DCLA, FCLA, and MCLA achieve higher efficiency of 82% despite of the linear amplifier without the harmonics and EMI.

Keywords— *high efficiency, linear amplifier, low electromagnetic interference, multi-level converter.*

I. INTRODUCTION

In general power converter, switching operation of semiconductor power devices is widely utilized to realize high efficiency. However, the switching operation causes harmonic distortion and electromagnetic interferences (EMI) [1][2]. These are typical issues on the power electronics. Meanwhile, some kinds of linear amplifiers such as class-B amplifier have been practically used to achieve low distortion waveforms. However, their efficiency is quite low than that of the PWM power converters using the switching operation.

As a solution of this problem, diode-clamped linear amplifier (DCLA) has been proposed to realize high-efficiency and low distortion, simultaneously [3]-[5]. The circuit topology of the DCLA is configured based on diode-clamped multi-level converter [6]. As other type topology, flying-capacitor linear amplifier (FCLA) has been reported based on the topology of flying-capacitor

multi-level converters [7][8].

This paper proposes and summarizes such multi-level topology based linear amplifier (MLLA) family including a novel proposed topology, modular-cascaded linear amplifier (MCLA) based on the modular multi-level cascaded converter (MMC) [9]. The three circuits, DCLA, FCLA, and MCLA are called basic topologies of MLLA. This means that other kind of such MLLA can be realized by using the same analogy based on the various multi-level converter topologies [10]. A distinguished feature of the MLLA is the reduction of the loss generated in the linear operation by using multiple series connected power devices. It enables to realize an inverter with high efficiency, less harmonics and EMI, and less output filter, simultaneously. In this paper, the circuit configuration, operation principle, and conversion efficiency on the basic topologies are summarized as MLLA family.

In the chapter II, the fundamentals of three basic topologies are described. Theoretical efficiency of MLLA is also derived. The chapter III shows the experimental waveforms of each prototype circuit. Finally, the chapter IV conclude the contents of this paper.

II. COMPARISON OF BASIC TOPOLOGIES OF MULTI-LEVEL TOPOLOGY BASED LINEAR AMPLIFIER FAMILY.

The operation principles of the three basic topologies for the reduction of loss generated by the linear operation of the power devices are quite similar. As mentioned above, so far, the DCLA and FCLA have been reported first in [4] on 2008 and [8] on 2017, respectively. This paper proposes the MCLA firstly, and summarize the three topologies as the MLLA. Moreover, any other topologies of the MLLA with the similar operation principle can be proposed, because many topologies of the multi-level converters have been reported so far [10]. The three basic topologies become the origin of all the MLLA topologies.

A. Diode-Clamped Linear Amplifier (DCLA)

Fig. 1 (a) shows the circuit configuration of 4-series connected DCLA per phase. In this regard, the number m of series connected MOSFETs per arm is defined in the MLLA like the number of the levels in the multi-level converter. The circuits shown in Fig. 1 are the 4-series connected circuits of each topology as an example. The DCLA is configured $2m$ MOSFETs and $2(m-1)$ diodes. Moreover, m DC voltage dividing capacitors including ($m-1$) floating capacitors are needed to keep multi-level

1649

The 2018 International Power Electronics Conference

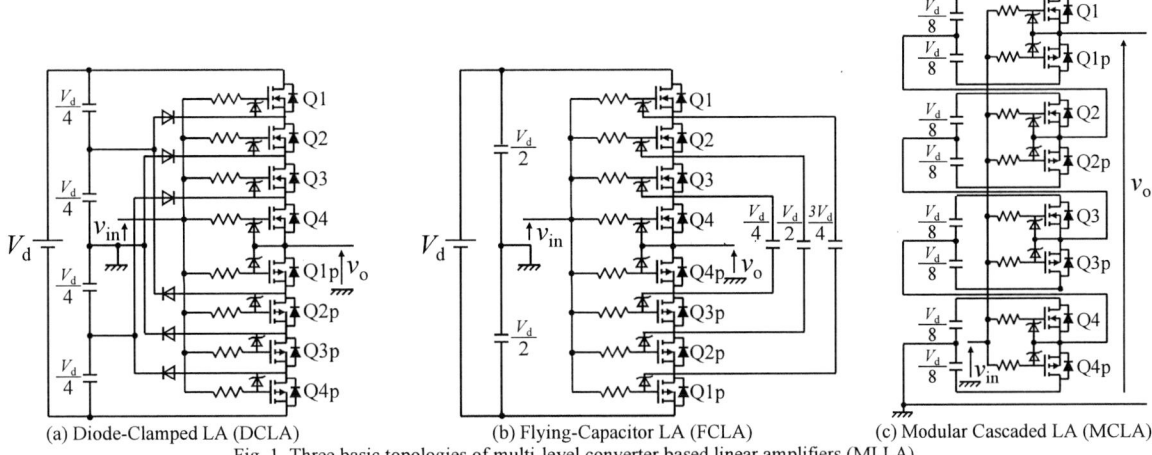

(a) Diode-Clamped LA (DCLA)   (b) Flying-Capacitor LA (FCLA)   (c) Modular Cascaded LA (MCLA)

Fig. 1. Three basic topologies of multi-level converter based linear amplifiers (MLLA).

TABLE I
COMPARISON OF THREE BASIC TOPOLOGIES OF MLLA.

Topology	Diode-Clamped Linear Amplifier (DCLA)	Flying-Capacitor Linear Amplifier (FCLA)	Modular-Cascaded Linear Amplifier (MCLA)
The number of MOSFETs	$2m$	$2m$	$2m$
The number of diodes	$2(m-1)$	0	0
The number of capacitors	m (shared with DC-link capacitors)	$(m-1)+2$	$2m$
Advantage	Small capacitor	Modularized topology	Modularized topology
Drawback	• Capacitor voltage unbalance • Diodes loss	• Large flying capacitor • Capacitor loss	Large number of isolated DC power supplies for each bridge
For capacitors voltage balance	Any additional circuits are required.	Voltage balancing control can be realized by individual linear gate circuit through load current.	Voltage balancing control for each cell can be realized by individual linear gate circuit through circulating current.

voltages in the circuit. Because, the MLLA reduces the loss by decreasing the applied voltage of each MOSFET during the linear operation as shown in Fig. 2. This operation needs the multiple DC voltages in the circuit. Moreover, these capacitor voltages have to be kept at the prescribed values during circuit operation. In the diode-clamped multi-level converter, any voltage balancing circuit is required to keep the voltage balance of the DC capacitors. In the same way, the DCLA also needs any voltage balancing circuit. It is a main drawback of the DCLA as indicated in Table I.

B. Flying-Capacitor Linear Amplifier (FCLA)

Fig. 1 (b) shows the circuit of the 4-series FCLA per phase. FCLA is configured by (m-1) flying-capacitors instead of the diodes and series capacitors in DCLA to realize the multiple voltage in the circuit. It has ladder and modularized circuit configuration without the cross-line in whole circuit. Generally, the linear amplifier does not need to pay attention the stray inductance, because there are no pulsed voltage and current. Although the stray inductance tends to be larger in such modularized configuration, it does not become a problem in the linear amplifier unlike the PWM power converters with the switching operation.

Although the capacitor voltages balance cannot be achieved in the described circuit configuration with the common gate circuits in Fig. 1 (b), it can be realized by preparing individual gate circuits for each MOSFET. This means that the voltage balancing control can be realized as the same with the flying capacitor multi-level converters

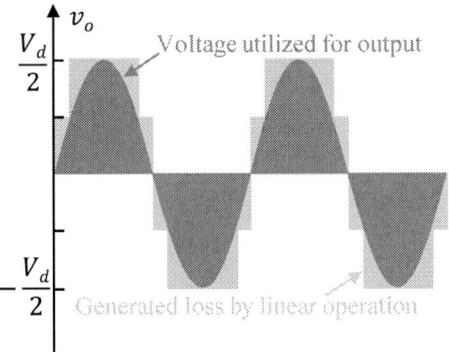

Fig. 2. Mechanism of loss reduction in 4-series MLLA.

in the future. However, the capacitor volume becomes larger than that of the PWM converters, because the charge and discharge frequency is the same to the output frequency due to the no switching operation.

C. Modular Cascaded Linear Amplifier (MCLA)

Fig. 1 (c) shows the 4-series MCLA per phase. This topology is based on the modular multi-level cascaded converter (MMC) and proposed in this paper. The multiple cascaded connection of the conventional class-B amplifiers is utilized. Thus, it is comparatively advantageous to design and implementation. However, the large number of the isolated DC power supplies are needed in the described circuit configuration. The voltage balancing control for each cell can be realized like the

1650

MMC utilized the circulating current by using the individual gate circuits for each MOSFET in the future.

D. Operating Waveforms of Three Topologies

Figs. 3 to 6 show the simulation waveforms of the three topologies when the total input DC voltage V_d is 200 V and pure resistive load is 47 Ω. As can be seen in these waveforms, all of three basic topologies of MLLA operate as a current amplifier. It means that the output voltage becomes the same with the input voltage, and the output current can be amplified. In Figs. 4 to 6, the voltage v_{QX} indicates the drain-source voltage of the switch QX of each 4-series topology shown in Fig. 1. It is seen that each

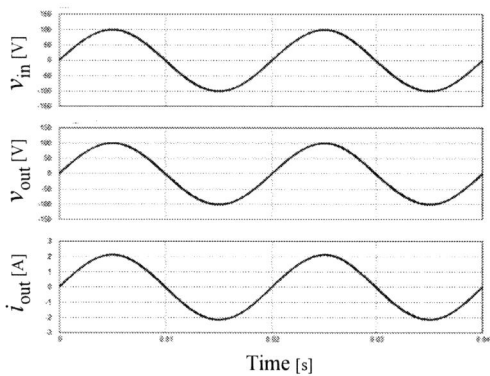

Fig. 3. Input voltage, output voltage and current in the three basic topologies.

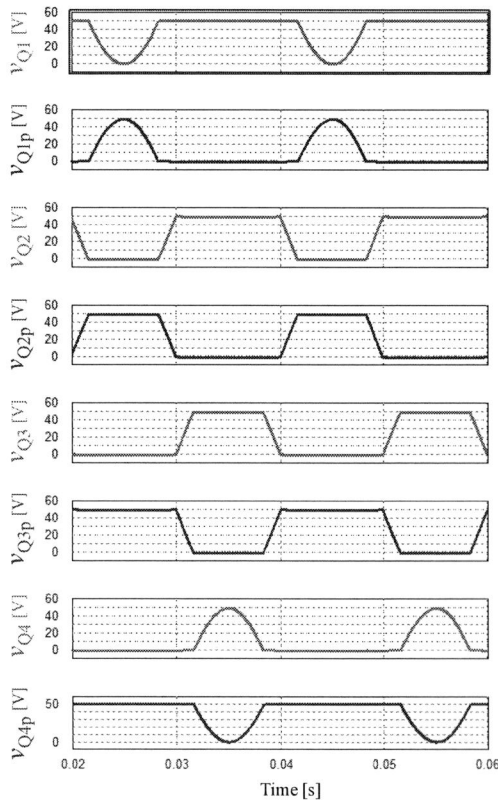

Fig. 4. Drain-source voltages of switches Q1-Q4p in DCLA.

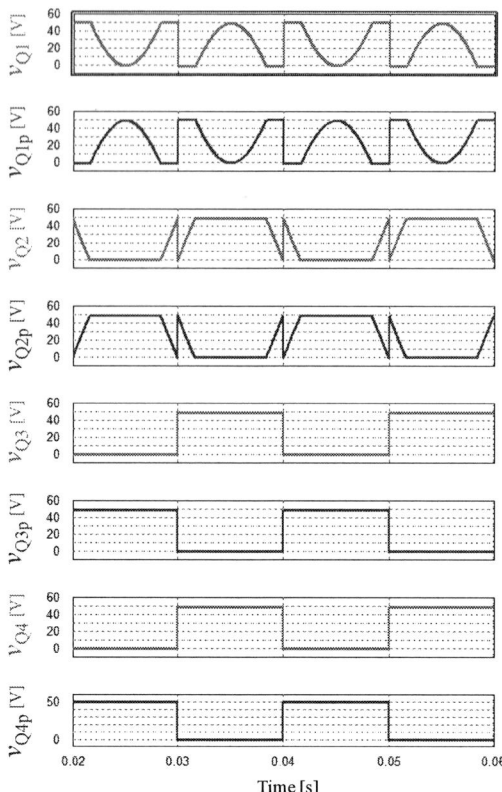

Fig. 5. Drain-source voltages of switches Q1-Q4p in FCLA.

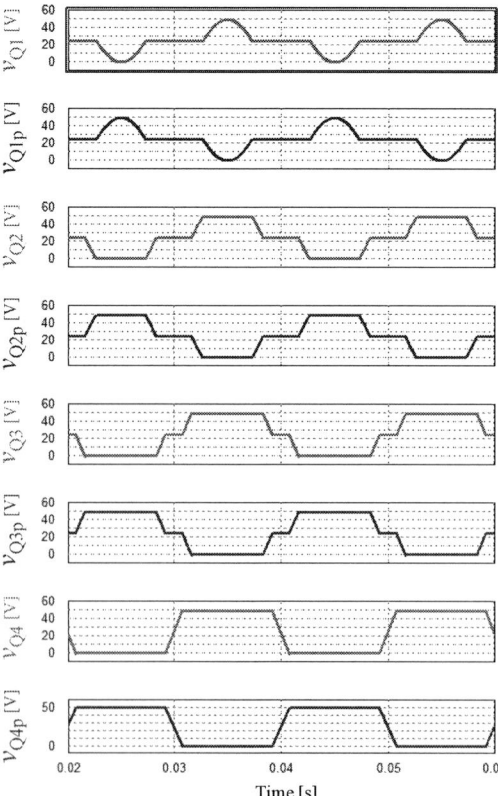

Fig. 6. Drain-source voltages of switches Q1-Q4p in MCLA.

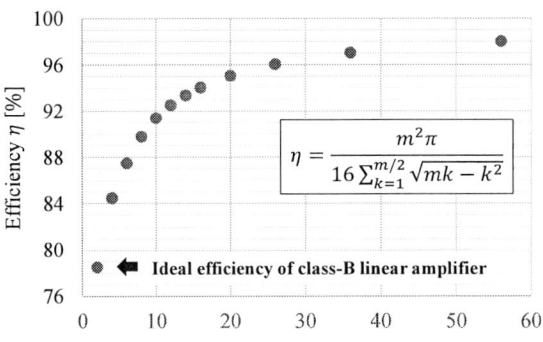

Fig. 7. Ideal efficiency of the three basic topologies of MLLA.

drain-source voltage becomes lower than 50 V which is one-fourth the total input DC voltage V_d. It means that the applied voltage of each MOSFET during the linear operation becomes lower compared with the class-B amplifier as the number m of series MOSFETs increases. Meanwhile, it is confirmed that these waveforms have a little different among the three topologies. The operating schemes depend on the connection of the multiple voltage sources including the capacitors, MOSFETs, and diodes. The mechanism for the loss reduction shown in Fig. 2 is basically the same in the three topologies. The all three topologies can decrease the applied voltage of the MOSFETs during the linear operation.

E. Theoretical Efficiency

When the output voltage v_o and current i_o are given by (1) and (2), respectively, the ideal efficiency of the three basic topologies can be calculated by the same equation, because the mechanism of the loss reduction of the MOSFETs is the same as mentioned above [4][8]. The ideal efficiency of the MLLA is derived as (3).

$$v_o = \frac{V_d}{2}\sin\theta \qquad (1)$$

$$i_o = I_{max}\sin\theta \qquad (2)$$

$$\eta = \frac{m^2\pi}{16\sum_{k=1}^{m/2}\sqrt{mk-k^2}} \qquad (3)$$

Fig. 7 shows the ideal efficiency of the MLLA by (3). It can be confirmed that the efficiency of MLLA can be improved by increasing the number of the series MOSFETs. However, as the number of series MOFETs increases, the progress of the efficiencies is gradually saturated. If the number increases to 36 MOSFETs, the efficiency reaches over 97% in principle. Thus, MLLA with achieves the same level on the efficiency for the general PWM inverters despite of the linear amplifier without the switching operation.

III. Experimental Investigation of FCLA

A. Prototype Circuits and Experimental Condition

Figs. 8 (a) to 8 (c) show the overview of developed prototype circuits of the 4-series DCLA, FCLA, and MCLA, respectively. In the prototypes, the circuit board

(a) 4-series DCLA

(b) 4-series FCLA

(c) 4-series MCLA

Fig. 8. Prototype circuit of three topologies of MLLA.

TABLE II
ELECTRICAL CHARACTERISTIC OF MOSFETS AND GATE CIRCUIT.

Parameters	Symbol	FQA28N15 (n-channel)	FQA36P15 (p-channel)
Drain to source maximum voltage	V_{DS}	150 V	-150 V
Rated current of MOSFETs	I_D	33 A	-36 A
On-state resistance	R_{on}	0.067 Ω	0.076 Ω
Gate to source threshold voltage	V_{th}	2.0 – 4.0 V	-2.0 – -4.0 V
Input capacitance	C_{in}	1250 pF	2550 pF
Gate resistance	R_{gate}	25 kΩ	11.8 kΩ
Zener voltage of zener diode	V_{zener}	15 V	15 V

The 2018 International Power Electronics Conference

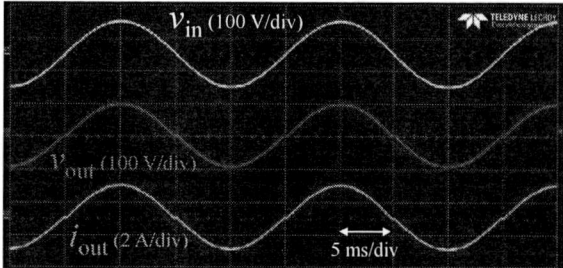

(a) Input voltage, output voltage and current.

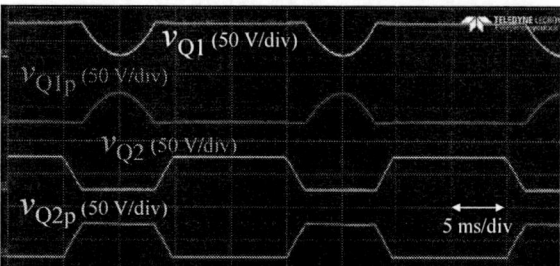

(b) Drain-source voltage of switches Q1-Q2p.

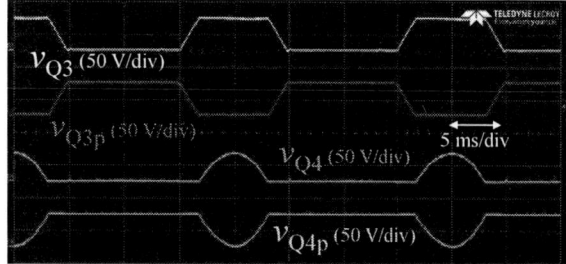

(c) Drain-source voltage on switches Q3-Q4p.

Fig. 9. Experimental waveforms of 4-series DCLA

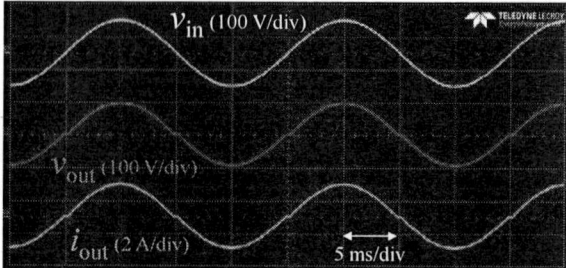

(a) Input voltage, output voltage and current.

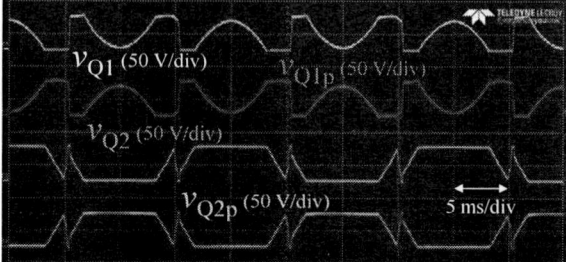

(b) Drain-source voltage of switches Q1-Q2p.

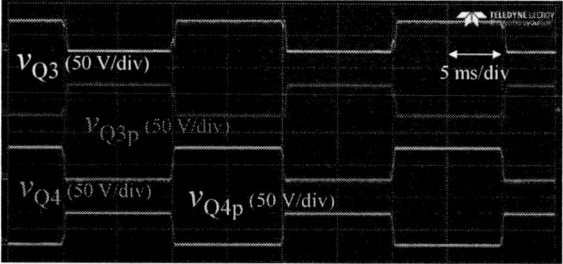

(c) Drain-source voltage of switches Q3-Q4p.

Fig. 10. Experimental waveforms of 4-series FCLA

of the DCLA and FCLA is the same. The FCLA can be realized by replacing the diodes in the DCLA to the flying capacitors as can be seen the circuit configurations shown in Fig.1 (a) and (b). In the MCLA, a unit module is implemented as the class-B amplifier, the number m of series MOSFETs can be flexibly changed by combining plural unit modules.

Table II lists the electrical characteristic of the MOSFETs and the specification of the components in the gate circuit. The experimental condition is the same with the simulation condition described in the chapter II and Figs. 3 to 6. The total input DC voltage V_d is 200 V, the input voltage is sinusoidal wave with the amplitude of 100 V and frequency of 50 Hz. The pure resistive load is 47 Ω. Isolated DC power supplies are used for the multiple voltages in each circuit.

B. Experimental Waveform

Figs. 9 to 11 show the experimental waveforms of the input voltage v_{in}, output voltage v_o, output current i_o, and drain-source voltage waveforms of the MOSFETs in the prototype DCLA, FCLA, and MCLA, respectively. It can be seen that the output voltage follows the input voltage

adequately. From this figure, it is confirmed that the MLLA operates as a current amplifier with the same input and output voltages. Although a little distortion is appeared in the output voltage when the polarity of the output current is changed. This phenomenon is the same with the well-known cross-over distortion caused by the threshold voltage of the MOSFETs in the conventional class-B amplifier. It can be compensated by adding a DC offset for the input voltage or applying a feedback control [8]. In the drain-source voltage waveforms, it is confirmed that each MOSFET separately operates as the active-state for the reduction of the loss generated in the linear operations. Moreover, it is verified that each drain-source voltage of the MOSFETs becomes always lower than 50 V which is one-fourth of the DC input voltage of 200 V. It is verified that the prototypes of the DCLA, FCLA, and MCLA operate according to the principle.

C. Measured Efficiency

The efficiency of the prototypes is measured by using the power analyzer HIOKI PW6001. The comparison between the ideal and measured efficiencies in each topology is shown in TABLE III. The 4-series DCLA,

1653

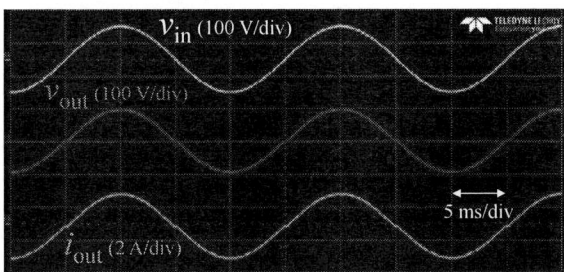

(a) Input voltage, output voltage and current.

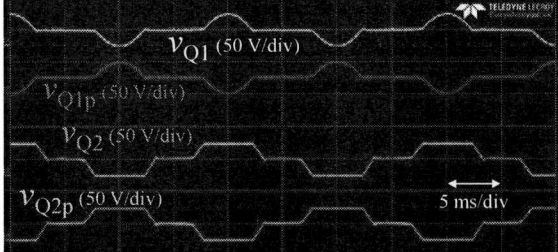

(b) Drain-source voltage of switches Q1-Q2p.

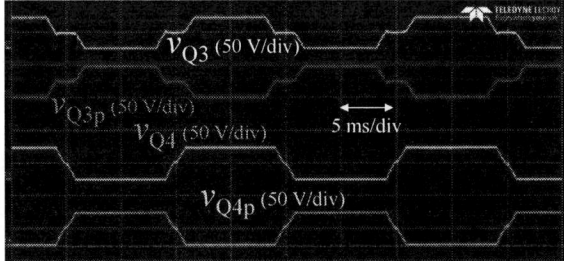

(c) Drain-source voltage of switches Q3-Q4p.

Fig. 11. Experimental waveforms of 4-series MCLA

TABLE III
COMPARISON OF IDEAL AND MEASURED EFFICIENCIES IN THREE-TOPOLOGIES OF 4-SERIES MLLA.

Topology	Conventional Class-B LA	4-series DCLA	4-series FCLA	4-series MCLA
Ideal efficiency	78.5%	84.5%		
Measured efficiency	-	82.0%	82.2%	81.3%

FCLA and MCLA achieve higher efficiency around 82%. It is approximate to the ideal efficiency of 84.5 % and improved over 3 points compared to the ideal efficiency of the conventional class-B linear amplifier. It is considered that the difference between measured and ideal efficiencies are mainly caused by the cross-over distortion. It is expected that the number of the series MOSFETs increases, the efficiency becomes higher according to the principle as shown in Fig. 7 [4][8].

IV. CONCLUSIONS

In this paper, the multi-level topology based linear amplifier family has been proposed. The three basic topologies, the diode-clamped linear amplifier (DCLA),

flying-capacitor linear amplifier (FCLA), and modular-cascaded linear amplifier (MCLA) have been summarized and compared from the viewpoint of the circuit configuration, operation principle, and efficiency. Finally, the prototype of the 4-series connected circuits of DCLA, FCLA, and MCLA have been realized. It has been verified that the 4-series amplifiers achieve higher efficiency of 82% improved 4 points for the ideal efficiency of the conventional class-B linear amplifier. It has been shown that each topology of MLLA family is an advantageous candidate for realization of a noise-less inverter with high efficiency and no EMI by increasing the number of the series connected MOSFETs.

ACKNOWLEDGMENT

This work was supported in part by JSPS KAKENHI Grant Number 16K18064.

REFERENCES

[1] Y. Murai, T. Kubota, and Y. Kawase, "Leakage current reduction for a high-frequency carrier inverter feeding an induction motor," *IEEE Trans. Ind. Appl.*, Vol. 28, No. 4, pp.858-863, 1992.

[2] G.L. Skibinski, R.J. Kerkman, and D. Schlegel, "EMI emissions of modern PWM ac drives," *IEEE Ind. Appl. Mag.*, Vol. 5, No. 6, pp.47-80, 1999.

[3] H. Fujita, N. Yamashita, "Performance of a Diode-Clamped Linear Amplifier," *IEEE Trans. Power Electron.*, Vol. 23, No. 2, pp.824-831, 2008.

[4] H. Fujita, "A Single-Phase Utility-Interface Circuit Without Any AC Inductor nor EMI Filter," *IEEE Trans. Ind. Appl.*, Vol. 45, No. 5, pp.1860-1867, 2008.

[5] Y. Yokokura, K. Ohishi, "Fine Sensorless Force Control using Diode-Clamped Linear Amplifiers," *IEEJ Journal IA*, Vol. 3, No. 3, 2014.

[6] J. Lai, F. Z. Peng, "Multilevel Converters - A New Breed of Power Converters", *IEEE Trans. Ind. Appl.*, Vol. 32, No. 3, pp.509-517, 1996.

[7] A. Meynard, H. Foch, "Multi-Level Conversion: High Voltage Choppers and Voltage-Source Inverters" *IEEE PESC'92*, pp.397-403, 1992.

[8] T. Ohno, M. Katayama, H. Obara, and A. Kawamura, "Flying Capacitor Linear Amplifier to Realize Both High-efficiency and Low Distortion for Power Conversion Applications Requiring High Quality Waveforms", *IEEE PEDS2017*, pp.907-912, 2017.

[9] H. Akagi: "Classification, Terminology, and Application of the Modular Multilevel Cascade Converter (MMCC)", *IEEE Trans. Power Electron.*, Vol.26, No.11, pp.3119-3130, 2011.

[10] S. Kouro, M. Malinowski, K. Gopakumar, J. Pou, L. G. Franquelo, B. Wu, J. Rodriguez, M. A. Pérez, and J. I. Leon: "Recent Advances and Industrial Applications of Multilevel Converters", *IEEE Trans. Ind. Electron.*, Vol.57, No.8, pp.2553-2580, 2010.

The 2018 International Power Electronics Conference

A New Zero-Voltage Switching Three-Level Converter with Reduced Rectifier Voltage Stress

Keon-Woo Kim[1]*, Cheon-Yong Lim[1], Dong-Kwan Kim[2], Yu-Jin Jang[2] and Gun-Woo Moon[1]
1 School of Electrical Engineering, KAIST, Daejeon, Republic of Korea
2 Department of Future Vehicle, KAIST, Daejeon, Republic of Korea
*E-mail: rainbowdot@kaist.ac.kr

Abstract- **Three-level converters are an attractive structure when the high input voltage is required. Although the voltage stress of switches in the three-level converter becomes half of the input voltage, zero-voltage switching cannot be achieved at light load conditions, which degrades the light load efficiency. This paper proposes a new three-level converter with achieving zero-voltage switching over entire load conditions. In addition, in the proposed converter, the voltage stress of rectifier diodes is low, leading to a reduction of conduction loss in the rectifier circuit. Therefore, by reducing the switching losses in primary switches, light load efficiency increases, and by decreasing the conduction loss in the rectifier circuit, heavy load efficiency is improved. The proposed converter is tested by the prototype with 540-600 VDC input and 750W (48 V/15.625 A) output.**

I. INTRODUCTION

Over the past several years, the high input voltage dc-dc converter has become a very important issue in medium to high power applications. When the input of a power factor correction (PFC) is three-phase ac voltage, rectified dc voltage becomes high as much as 700V. To utilize the switches which have the small voltage stress, many studies have been investigated [1]-[6]. Among them, three-level (TL) converter is one of the most attractive structures as shown in Fig. 1 [5]-[6]. However, in TL converters, switching losses because of its high input voltage degrade the efficiency under light load conditions. Also, conduction losses due to its circulating currents flowing through primary circuit and currents flowing through the rectifier circuit decrease the efficiency under heavy load conditions. Thus, the switching losses and conduction losses should be minimized for high efficiency.

To decrease the switching losses, many research have been applied zero-voltage-switching (ZVS) technique. [7]-[10]. The works in [7]-[8], ZVS operation for primary switches can be performed by using the leakage inductor energy of the transformer. The soft-switching of leading switches can be easily performed, but it of lagging switches is difficult to be achieved under light load conditions. Research in [9]-[10] extends the ZVS range of lagging switches. However, many components are required such as inductors or diodes, and they are still suffering from large conduction loss due to circulating current during the freewheeling period. In [11]-[13], all primary switches can be softly switched under entire load conditions. However, the converters have large circulating resonant current in the primary circuit.

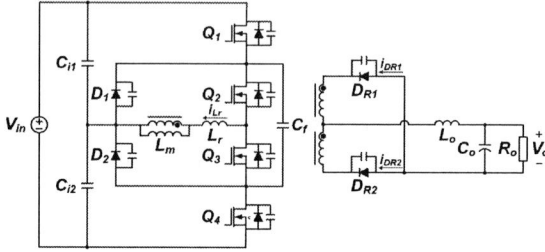

Fig. 1 Conventional three-level converter.

Although soft switching technique can increase the light load efficiency due to reduced switching losses, large circulating resonant current increases the conduction loss which decreases the heavy load efficiency.

The aim of reducing conduction losses of TL converters, many papers have been studied [14]-[18]. The works in [14]-[16], rectifier diodes with small voltage stress can be utilized, and circulating current in primary circuit is eliminated. However, many components are required resulting in a complex structure, and lagging switches cannot achieve the soft-switching which degrades the light load efficiency. Research in [17] eliminates circulating current in the primary circuit, and achieves ZVS condition of primary switches. However this converter requires two clamping diodes in primary side circuit, and it has large voltage stress on rectifier diodes. In [18], dual half-bridge cascaded TL converter is suggested. Clamping diode and flying capacitor are eliminated by combining two half-bridge structures. This converter can utilize the rectifier diodes with small voltage stress, and reduce the circulating current in primary circuit. However, ZVS of lagging switch cannot be achieved under light load conditions, and only half of the circulating current is eliminated.

This paper proposes a new TL converter by using a coupled inductor in the rectifier circuit. Circulating current in primary circuit is eliminated, and the rectifier diodes with small voltage stress can be used. Because circulating current in primary circuit is eliminated, and the rectifier diodes with small forward voltage drop can be utilized, the conduction losses are decreased. In addition, switching losses can be decreased by ZVS operation under overall load conditions. Therefore, because of the smaller conduction losses and switching losses, the proposed converter can have higher efficiency under overall load conditions.

1655

The 2018 International Power Electronics Conference

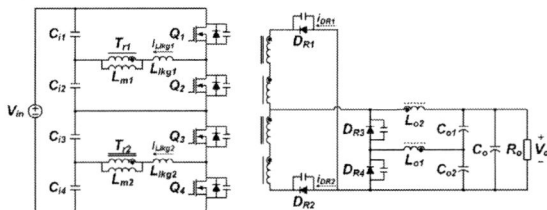

Fig. 2 Proposed three-level converter.

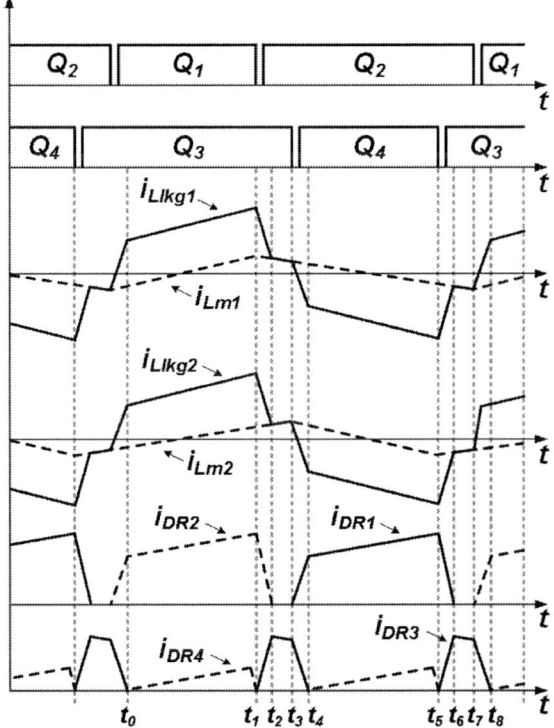

Fig. 3 Operating waveforms of the proposed three-level converter in steady state.

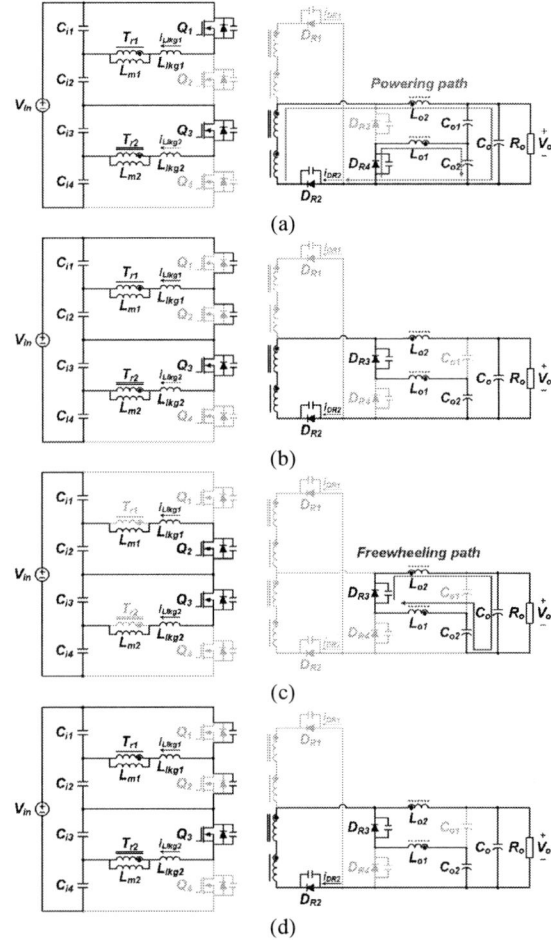

Fig. 4 Operating circuits of the proposed TL converter. (a) Mode 1. (b) Mode 2. (c) Mode 3. (d) Mode 4.

II. MODE ANALYSIS

The primary circuit of the proposed TL converter is same with the dual half-bridge cascaded TL converter, and the secondary side is composed of rectifier diodes and coupled inductor as shown in Fig. 2. The turn-on time of Q_1 is complementary to that of Q_2 with a dead time. The gate signals of Q_3 and Q_4 are also same with 180-degree phase difference. Operating waveforms at steady state is shown in Fig. 3. Due to the symmetric operation of two half cycles, only one of the half cycles is explained, and the equivalent circuits in each mode are shown in Fig. 4. For a simple analysis, the following assumptions have been made as follows.

1) Input capacitances C_{i1}, C_{i2}, C_{i3}, and C_{i4} are large enough. The voltage across input capacitors V_{Ci1}, V_{Ci2}, V_{Ci3}, and V_{Ci4} are calculated as follow:

$$V_{C1} = V_{C4} = \frac{(1-D)V_{in}}{2},\qquad(1)$$

$$V_{C2} = V_{C3} = \frac{DV_{in}}{2}.\qquad(2)$$

2) The output capacitance is large enough that output voltage, V_o, is constant.

3) Primary switches are all MOSFETs.

Mode 1 [t_0-t_1]: When Q_1 and Q_3 are turned on, mode 1 starts. During mode 1, power is transferred from primary circuit to secondary circuit. The voltage across the primary side of T_{r1} is $(1-D)V_{in}/2$, and it of T_{r2} is $DV_{in}/2$, where D is duty ratio and V_{in} is the input voltage. Output current flows through D_{R2} and D_{R4}.

Mode 2 [t_1-t_2]: When Q_1 is turned off, mode 2 starts. During this mode, ZVS turn-on of Q_2 is achieved. Until the leakage inductor current of T_{r1}, i_{lkg1}, becomes same with the magnetizing current of T_{r1}, i_{Lm1}, leakage inductor energy helps ZVS, and after that magnetizing inductor energy is involved in ZVS operation. Because D_{R2} becomes reverse-biased, the current flowing through D_{R2}, i_{DR2} is decreased, and the current flowing through D_{R3}, i_{DR3} is increased.

Mode3 [t_2-t_3]: When Q_2 is turned on, and i_{DR2} become zero, mode 3 starts. Because D_{R1} and D_{R2} are reverse-biased, the output current does not transfer to the primary

1656

circuit, which means no freewheeling current in the primary circuit. The output current flows through the coupled inductor and D_{R3}.

Mode4 [t_3-t_4]: When Q_3 is turned off, mode 4 starts. During this mode, ZVS turn-on of Q_4 is achieved. The voltage across Q_4 is discharged with the magnetizing inductor until the voltage across Q_4 become ($V_{in}/2$-$nV_o/2$). After that, ZVS is performed with leakage inductor energy.

III. CHARACTERISTIC OF THE PROPOSED CONVERTER

A. DC Voltage Gain

Ignoring the duty cycle loss, by using the voltage-second balance of the output inductor, the voltage gain can be calculated. The average voltage across the output inductor, $v_{Lo,avg}$, is calculated as follow:

$$v_{Lo,avg} = \left(\frac{V_{in}}{2n} - V_o\right)DT_s + \left(-\frac{n_c}{1+n_c}V_{o1}\right)\frac{(1-2D)T_s}{2} = 0, \quad (3)$$

where V_o is output voltage, n is turns ratio of the transformer, D is the duty ratio of Q_1 and Q_4, T_s is switching period and n_c is turns ratio of the coupled inductor.

Therefore, dc voltage gain of the proposed TL converter, M_{prop}, is obtained as follow:

$$M_{prop} = \frac{1+2Dn_c}{2n(1+n_c)}. \quad (4)$$

The voltage gain of the proposed TL converter becomes larger according to a decrease of n_c. In the same way, dc voltage gain of the converter in [18], M_{conv}, is calculated as follow:

$$M_{conv} = \frac{D(3-2D)}{2n}. \quad (5)$$

B. Voltage Stress of Rectifier Diodes.

In Fig. 5, the equivalent circuit at the secondary side of the proposed TL converter is represented. When switch Q_1 and Q_3 is turned on, the output current flows through D_{R2} and D_{R4} as shown in Fig. 5(a). Because the rectifier circuit is center tapped rectifier, the voltage across D_{R1}, v_{DR1} is twice as the voltage across D_{R3}, v_{DR3}. Assuming junction capacitance of the rectifiers is C_D and the secondary leakage inductance is $L_{lkg,c}$, $v_{DR1}(t)$ and $v_{DR3}(t)$ are calculated as follow:

$$v_{DR1}(t) = 2\left(V_o + n_c V_{o2}\cos(\omega_c t)\right), \quad (6)$$

$$v_{DR3}(t) = V_o + n_c V_{o2}\cos(\omega_c t), \quad (7)$$

where $\omega_c = (L_{lkg,c}C_D)^{1/2}$.

When Q_2 and Q_3 are turned on, the freewheeling current flows through D_{R3} as shown in Fig. 5(b). The voltage across D_{R4}, v_{DR4} is calculated as follow:

$$v_{DR4}(t) = V_{o2} + \frac{V_{o1}}{n_c + 1}\cos(\omega_c t). \quad (8)$$

Without considering the reverse recovery of the rectifier diodes, the maximum voltage of D_{R1} and D_{R2} is ($2V_{in}/n$-$2V_o$), the maximum voltage of D_{R3} is (V_{in}/n-V_o), and the maximum voltage of D_{R4} is ((n_c-

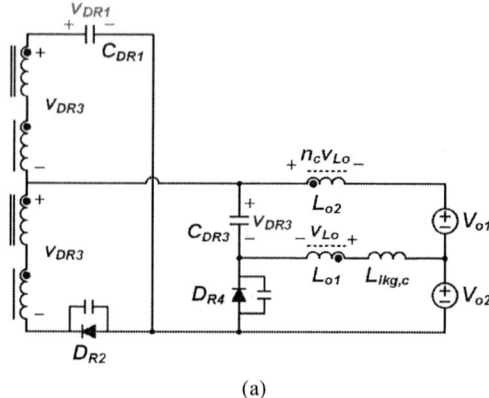

(a)

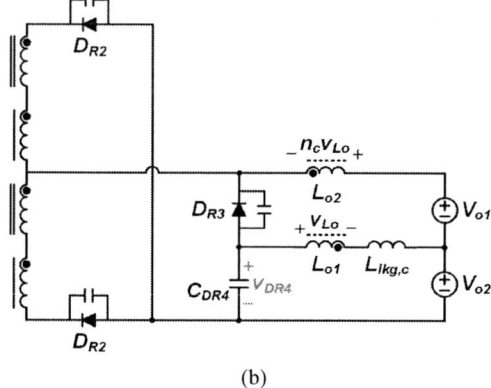

(b)

Fig. 5 Equivalent circuit at the secondary side of the proposed TL converter. (a) When Q_1 and Q_3 are turned on. (b) When Q_2 and Q_3 are turned on.

1)$V_{in}/(2nn_c(n_c+1))+V_o/n_c$). In the same way, the maximum voltage across D_{R1} and D_{R4} of the converter in [18] is ($2V_{in}/n$-$AV_{in}/8n$), and the maximum voltage across D_{R2} and D_{R3} of the converter in [18] is ($2V_{in}/n$-$AV_{in}/4n$) where $A=(3-(9-16nV_o/V_{in})^{1/2})$.

C. ZVS condition

In the proposed TL converter, leading switches can easily achieve the ZVS turn-on, because the required ZVS energy for output capacitor of leading switches can be provided from energy stored in the output inductor and the magnetizing inductor. Thus, leading switches can achieve ZVS turn-on under overall load conditions.

In the proposed TL converter, required ZVS energy for lagging switches can be provided from the energy stored in the magnetizing inductor and leakage inductor. Until the voltage across the lagging switch becomes ($V_{in}/2$-$nV_o/2$), magnetizing inductor current discharges the output capacitor of lagging switches. After that, the energy stored in the leakage inductor involves the soft-switching operation. Thus, the leakage inductor current at the starting point of ZVS operation should be large enough. At this time, the leakage inductor current decreases until the leakage current become same with the magnetizing current. In order to perform the ZVS operation of the lagging switch, the converter should satisfy following conditions:

1657

$$\frac{1}{2}\left(L_{lkg1}+L_{lkg2}\right)I_{Lm,peak}^{2}\geq C_{oss}\left(\frac{V_{in}}{2}\right)^{2}\frac{1}{2D+1}, \qquad (9)$$

where L_{lkg1} is the leakage inductor of T_{r1}, L_{lkg2} is the leakage inductor of T_{r2}, $I_{Lm,peak}$ is the peak current of the magnetizing inductor current, and C_{oss} is the output capacitance of the primary switches.

Using (9), to achieve soft switching of lagging switches under entire load conditions, the maximum value of the magnetizing inductance can be obtained as follows:

$$L_{m}\leq\sqrt{\frac{\left(L_{lkg1}+L_{lkg2}\right)\left(2D+1\right)}{2C_{oss}}\frac{D(1-D)}{2f_{s}}}, \qquad (10)$$

where L_{m} is the magnetizing inductance and f_{s} is the switching frequency.

IV. DESIGN GUIDELINE OF THE PROPOSED CONVERTER

The proposed TL converter is designed with 540-600 V input voltage, and 48 V/15.625 A output. The switching frequency is chosen as 100kHz.

A. Turns ratio

Using (4) and (5), the voltage gain of the converter in [18] and the proposed TL converter are expressed in Fig. 6. Because even at the minimum input voltage, the output voltage of the converter should be regulated, turns ratio of the transformer is designed considering the maximum voltage gain. Due to duty cycle loss, the converter cannot utilize the full duty, 0.5. Assuming that the converter can utilize the duty ratio as 0.4, the maximum voltage gain of the proposed TL converter is same with it of the converter in [18] when n_{c} is 1.5 as shown in Fig. 6. At this time, the minimum input voltage of proposed TL converter is 540 V, and turns ratio of the transformer can be obtained as follow:

$$n=\frac{1+2Dn_{c}}{2(1+n_{c})}\frac{V_{in}}{V_{o}}=\frac{1+2\cdot0.4\cdot1.5}{2(1+1.5)}\cdot\frac{540}{48}=4.95. \qquad (11)$$

Therefore, turns ratio of transformer should be designed smaller than 4.95.

B. Magnetizing inductance

Although small magnetizing inductance allows soft-switching operation, large magnetizing inductor current causes large conduction loss. Using (11), the magnetizing inductance can be calculated to perform ZVS of primary switches under overall load conditions. With the assumption that $L_{lkg1}=L_{lkg2}=8\mu H$, and $C_{oss}=170pF$, L_{m} should be designed smaller than $341\mu H$.

C. Rectifier diodes

Based on the analysis in section III-B, the voltage stress of the rectifier diodes according to the input voltage is shown in Fig. 7. In the proposed TL converter, the voltage stress of D_{R1} and D_{R2} is 154V, it of D_{R3} is 77V, and it of D_{R4} is 40V. Thus, 200 V voltage rating diode can be used for D_{R1} and D_{R2}, 100V voltage rating diode

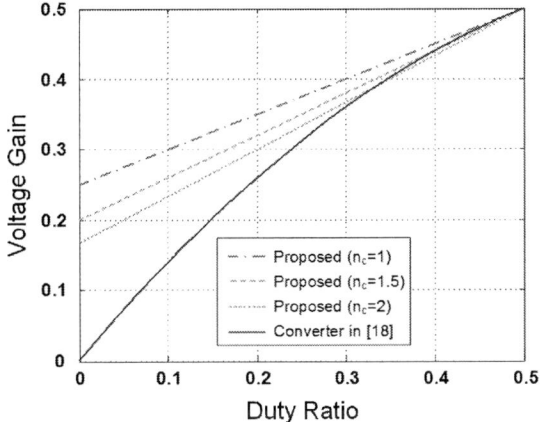

Fig. 6 Voltage gain versus duty ratio

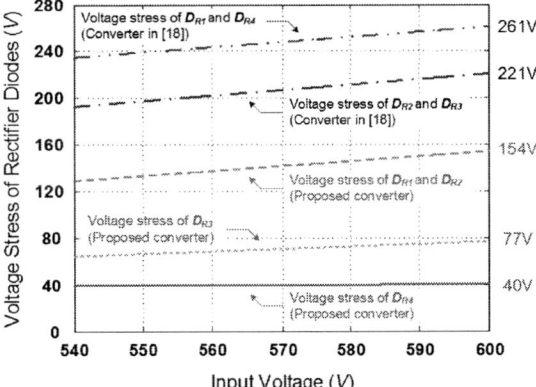

Fig. 7 Rectifier diodes voltage stress according to input voltage

can be used for D_{R3}, and 80V voltage rating diode can be used for D_{R4}. As shown in Fig. 7, the proposed TL converter can utilize the rectifier diodes with smaller voltage stress compare with the converter in [18].

V. EXPERIMENTAL RESULTS

In order to confirm the effectiveness of the proposed TL converter, a 750 W prototype converter is built, and detailed specification is listed in Table I.

Fig. 8 shows the measured experimental waveforms of the converter in [18] and proposed converter at 50% load conditions. The converter in [18] eliminates freewheeling current only one of them as shown in Fig. 8(a). Therefore, the converter in [18] still has conduction loss in primary circuit during the freewheeling period. However, the proposed converter has no circulating current during freewheeling periods as shown in Fig. 8(b). Thus, the proposed converter has smaller conduction loss in primary circuit. ZVS operation of Q_{4} which is lagging leg switch is shown in Fig. 9. It shows that Q_{4} is turned on after drain source voltage of Q_{4} becomes zero, and soft-switching is performed under overall load conditions.

1658

The 2018 International Power Electronics Conference

TABLE I
COMPONENTS LIST OF PROTOTYPE

Items	Converter in [18]	Proposed
Input voltage	540 - 600 V	
Output voltage	48V	
Magnetizing inductance (L_m)	720 μH	320 μH
Resonant inductance (L_r)	Leakage = 6 μH External = 6μH	Leakage = 6 μH
Primary switch (Q_1, Q_2, Q_3, Q_4)	SPA12N50C3 (500 V, 12 A)	
Rectifier diodes	D_{R1}, D_{R4} FFA60UP30DN (300V, V_F=1.5V)	D_{R1}, D_{R2} MBR20200CT (200V, V_F=0.8V)
	D_{R2}, D_{R3}, MBRB40250TG (250V, V_F=0.86V)	D_{R3} MBR20H100CTG (100V, V_F=0.77V) D_{R4} MBR20L80CTG (100V, V_F=0.45V
Output inductance	L_o 17 μH	L_o 14 μH

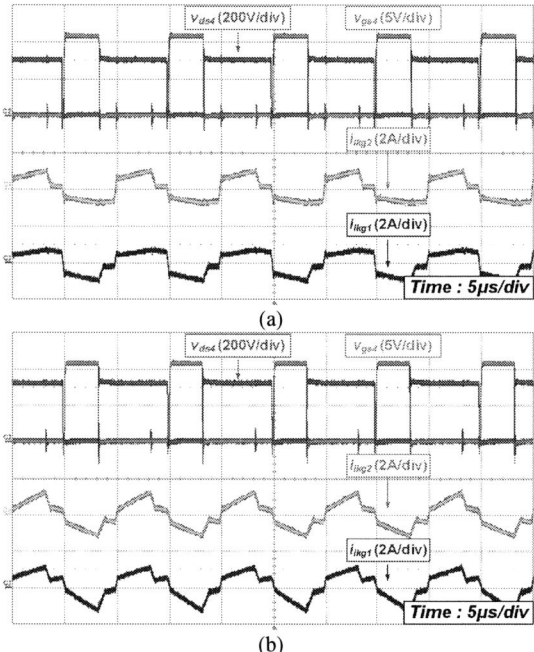

(a)

(b)

Fig. 8 Measured key waveform at 50% load condition when V_{in}=600V. (a) For the converter in [18]. (b) For the proposed converter.

The measured efficiency of the converter in [18] and proposed converter when the input voltage is 600 V is presented in Fig. 10. Under heavy load conditions, the proposed converter can obtain high efficiency due to small conduction loss in primary circuit and rectifier circuit. Under light load conditions, the proposed converter can obtain high efficiency because of ZVS operation of primary switches.

VI. CONCLUSION

This paper proposes a new TL converter which can achieve zero-voltage switching operation and have decreased voltage stress of rectifier diodes. In rectifier

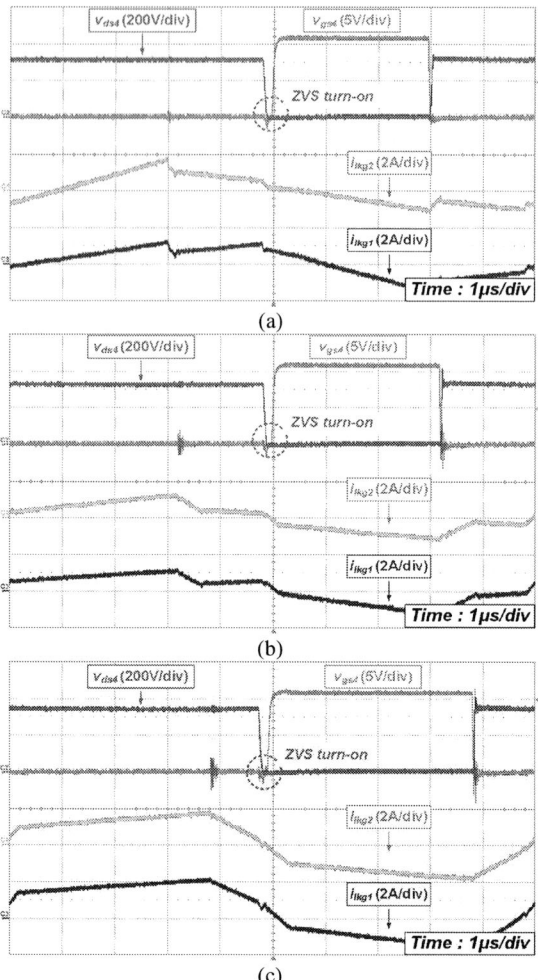

(a)

(b)

(c)

Fig. 9 ZVS waveform of the proposed converter when V_{in}=600V. (a) Under 10% load conditions. (b) Under 50% load conditions. (c) Under 100% load conditions

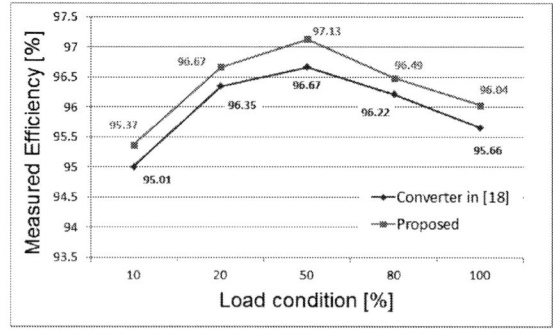

Fig. 10 Measured efficiency.

circuit, two diodes are added and output inductor is replaced with a coupled inductor. Because output current flows through the coupled inductor and additional diode during the freewheeling period, circulating current does not transfer to the primary circuit. Also, due to smaller voltage stress at rectifier diodes, the diode with smaller forward voltage drop, which can decrease the conduction loss, can be utilized. Furthermore, ZVS operation can be

1659

achieved under overall load conditions, and switching loss is decreased. Therefore, the proposed converter can obtain high efficiency because of small conduction loss and switching loss. As a result, the proposed converter is expected to be widely used in high input voltage applications.

ACKNOWLEDGMENT

This work was supported by the National Research Foundation of Korea (NRF) grant funded by the Korea government (MSIP)(No. 2016R1A2B2010328).

REFERENCES

[1] T. C. Lim, B. W. Williams, S. J. Finney, and P. R. Palmer, "Series connected IGBTs using active voltage control technique," *IEEE Trans. Power Electron.*, vol. 28, no. 8, pp. 4083–4103, Aug. 2013.

[2] S. Ji, T. Lu, Z. Zhao, H. Yu, and L. Yuan, "Series-connected HV-IGBTs using active voltage balancing control with status feedback circuit," *IEEE Trans. Power Electron.*, vol. 30, no. 8, pp. 4165–4174, Aug. 2015.

[3] D. Sha, K. Deng, and X. Liao, "Duty cycle exchanging control for input series-output-series connected two PS-FB DC-DC converters," *IEEE Trans. Power Electron.*, vol. 27, no. 3, pp. 1490–1501, Mar. 2012.

[4] W. Chen and G. Wang, "Decentralized voltage-sharing control strategy for fully modular input-series–output-series system with improved voltage regulation," *IEEE Trans. Ind. Electron.*, vol. 62, no. 5, pp. 2777–2787, May 2015.

[5] J. R. Pinheiro and I. Barbi, "The three-level ZVS-PWM DC-to-DC converter," *IEEE Trans. Power Electron.*, vol. 8, no. 4, pp. 486–492, Oct. 1993.

[6] S. Han, G. Moon, and M. Youn, "A voltage-balanced phase-shifted three level DC/DC converter operating from high-input voltage," *IEEE Power Electron. Lett.*, vol. 1, no. 3, pp. 74–77, Sep. 2003.

[7] X. B. Ruan, D. Y. Xu, L. Q. Zhou, B. L. Li, and Q. H. Chen, "Zero voltage-switching PWM three-level converter with two clamping diodes," *IEEE Trans. Ind. Electron.*, vol. 49, no. 4, pp. 790–799, Aug. 2002.

[8] K. Jin, X. Ruan, and F. Liu, "An improved ZVS PWM three-level converter," *IEEE Trans. Ind. Electron.*, vol. 54, no. 1, pp. 319–329, Feb. 2007.

[9] Y. Jang and M. M. Jovanovic, "A new three-level soft-switched converter," *IEEE Trans. Power Electron.*, vol. 20, no. 1, pp. 75–81, Jan. 2005.

[10] Y. Shi and X. Yang, "Wide load range ZVS three-level dc-dc converter: Four primary switches, capacitor clamped, two secondary switches, and smaller output filter volume," *IEEE Trans. Power Electron.*, vol. 31, no. 5, pp. 3431–3443, May 2016.

[11] F. Canales, P. Barbosa, C. Aguilar, and F. C. Lee, "A high-power-density DC/DC converter for high-power distributed power systems," *in Proc. IEEE PESC*, 2003, pp. 11–18.

[12] I.-O. Lee and G. W. Moon, "Analysis and design of a three-level LLC series resonant converter for high- and wide-input-voltage applications," *IEEE Trans. Power Electron.*, vol. 27, no. 6, pp. 2966–2979, Jun. 2012.

[13] I.-O. Lee, S.-Y. Cho, and G.-W. Moon, "Three-level resonant converter with double LLC resonant tanks for high-input-voltage applications," *IEEE Trans. Ind. Electron.*, vol. 59, no. 9, pp. 3450–3463, Sep. 2012.

[14] T. T. Song, N. Huang, and A. Ioinovici, "A zero-voltage and zero-current switching three-level DC–DC converter with reduced rectifier voltage stress and soft-switching-oriented optimized design," *IEEE Trans. Power Electron.*, vol. 21, no. 5, pp. 1204–1212, Sep. 2006.

[15] F. X. Liu, J. J. Yan, and X. B. Ruan, "Zero-voltage and zero-current switching PWM combined three-level DC/DC converter," *IEEE Trans. Ind. Electron.*, vol. 57, no. 5, pp. 1644–1654, May 2010.

[16] E. Chu, X. Hou, H. Zhang, M. Wu, and X. Liu, "Novel zero-voltage and zero-current switching (ZVZCS) PWM three-level DC/DC converter using output coupled inductor," *IEEE Trans. Power Electron.*, vol. 29, no. 3, pp. 1082–1093, Mar. 2014.

[17] Z. Guo, K. Sun, and D. Sha, "Improved ZVS three-level DC–DC converter with reduced circulating loss," *IEEE Trans. Power Electron.*, vol. 31, no. 9, pp. 6394–6404, Sep. 2016.

[18] Z. Guo, K. Sun, and L. Zhang, "Analysis and Evaluation of Dual Half-Bridge Cascaded Three-Level DC-DC Converter for Reducing Circulating Current Loss," *IEEE J. Emerg. Sel. Topics Power Electron.*, vol. 5, no. 1, pp. 351-362, Mar. 2017

The 2018 International Power Electronics Conference

Model Predictive Control of a Three-Level NPC Rectifier with a Sliding Manifold Term

Xiaonan Gao[1], Wei Tian[1,*], Xicai Liu[1], Zhenbin Zhang[2], Ralph Kennel[1]

1: Chair of Electrical Drive Systems and Power Electronics,Technical University of Munich, Munich, Germany

2: School of Electrical Engineering, Shandong University, Jinan, China

* corresponding author, E-mail: wei.tian@tum.de

Abstract—In this paper, a finite control set model predictive control (FCS-MPC) method for three-level neutral-point-clamped (3L-NPC) rectifiers without the outer control loop is presented. The proposed control method is very simple as only a sliding manifold term is added into the cost function to regulate the dc-bus voltage directly. Instead of calculating the suitable dynamic references, only the steady-state references are required in this method. Besides, we have compared the conventional proportional-integration controller based MPC method (PI-MPC), MPC with dynamic references design method (DR-MPC), and the proposed MPC with a sliding manifold term method (SM-MPC) with simulation. The simulation results show that the proposed SM-MPC outperforms the PI-MPC in the condition of the step change in load and has advantage over the DR-MPC when the voltage reference changes.

Index Terms—three-level neutral-point-clamped (3L-NPC) rectifier, direct dc-bus voltage control, model predictive control (MPC), sliding manifold.

I. INTRODUCTION

Recently, the three-level neutral-point clamped (NPC) active front-end (AFE) rectifier has been widely used in industrial applications because it is more suitable for the high power rating and has lower total harmonic distortion.

The main control objectives of AFEs are to obtain sinusoidal input currents, regulate the dc-bus voltage and eliminate the neutral point voltage. In order to achieve these control objectives, many control strategies have been presented in the last decades. The most popular strategies are classified into three categories: 1) voltage-oriented control (VOC), 2) direct power control (DPC), and 3) model predictive control (MPC) [1]–[3]. The VOC method uses the indirect scheme which controls dc-bus voltage through regulating d-axis current. In the internal loop (denoted as current loop), two PI controllers are used to track their current references and generate dq voltage commands. DPC is another high performance control strategy for the three-level NPC rectifier due to its fast dynamic response. Different from VOC, DPC uses a look-up table to obtain the switching actions. However, DPC is also an indirect scheme for the dc-bus voltage control. Model predictive control (MPC) has been a very promising control method for power electronics, motor drives, and renewable energy systems because it can easily handle additional nonlinear constraints and take the switching characteristics into consideration. Since the dynamic model of the NPC AFE rectifier include nonlinear nature, MPC is more suitable for such converter.

In [4], an external PI controller is designed to regulate the dc-bus voltage and generate the power reference for inner predictive control. By using cascaded control structure, this method realizes the decoupling of the d-axis current (active power) and dc-bus voltage. However, the tuning of the PI controller is not an easy task because of the nonlinear relationship between the d-axis current and dc-bus voltage. In [5], the authors proposed a way to directly design the suitable references of the d-axis current and dc-bus voltage. Therefore, the external control loop is not required in this case. The key idea of this method is to use a smoothed dc-bus voltage reference to control the d-axis current and dc-bus voltage in the same control loop. In [6], the MPC with dynamic references design (DR-MPC) is further extended into the back-to-back converter and a comparison of the conventional PI-MPC and DR-MPC is presented. However, because of the smoothed voltage reference, the dynamic response of the system slows down and the system may even become unstable when a fast dynamic response of dc-bus voltage is required. This is because the method does not adequately consider the strong coupling between the dc-bus voltage and the d-axis current. In the strong coupling system, it is not reasonable to just track one variable but not consider the dynamic response of the coupled variable. Therefore, it is better to design a trajectory which takes both the dc-bus voltage and the d-axis current into account. The similar idea has been used in the sliding mode control by enforcing the state variables in the sliding manifold to achieve the decoupling control. Inspired by this, we propose a novel MPC with the sliding manifold (SM-MPC) for NPC AFE rectifiers. In this method a sliding manifold term, consisting of the dc-bus voltage and the d-axis current, is added into the cost function, so that the two coupling state variables will approach or restrain on the sliding manifold and reach their references at last.

The remainder of this paper is organized as follows. In Section II, the dynamic model of the NPC AFE rectifier is presented. In Section III, PI-MPC, DR-MPC, and the proposed SM-MPC are presented. Simulation results are presented in Section IV. Finally, Section V concludes this paper.

II. SYSTEM DESCRIPTION AND MODELING

The main circuit of a three-level NPC rectifier is shown in Fig.1, where S_{in} ($i = a, b, c$, $n = 1, 2, 3, 4$) represents the semiconductor switch of the rectifier, e_i, i_i and u_i ($i = a, b, c$) are the three-phase grid voltages, currents and input voltage of the rectifier, L and R are inductance and resistance of the input filter, C_1 and C_2 are capacitors of the dc-bus, U_{dc1} and U_{dc2} are the voltages of C_1 and C_2 respectively, U_{dc} ($U_{dc} = U_{dc1} + U_{dc2}$) represents the dc-bus voltage, i_o is the current of neutral point. Note that, the model of the three-level NPC AFE rectifier are derived in rotating dq frame by

1661

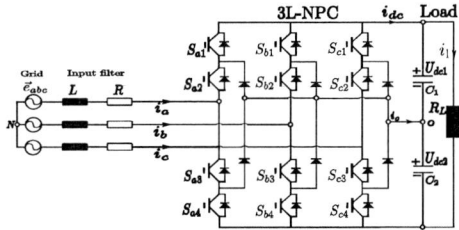

Fig. 1. Topology of three-level NPC rectifier

using Park-Transformation

$$\frac{2}{3} \begin{bmatrix} \cos(\theta) & \cos\left(\theta - \frac{2}{3}\pi\right) & \cos\left(\theta + \frac{2}{3}\pi\right) \\ \sin(\theta) & \sin\left(\theta - \frac{2}{3}\pi\right) & \sin\left(\theta + \frac{2}{3}\pi\right) \\ 0.5 & 0.5 & 0.5 \end{bmatrix}. \quad (1)$$

The dynamic of grid currents in the rotating dq frame can be expressed as

$$\begin{cases} \frac{di_{\mathrm{d}}}{dt} = \frac{1}{L}(e_{\mathrm{d}} - u_{\mathrm{d}} - Ri_{\mathrm{d}} + \omega Li_{\mathrm{q}}) \\ \frac{di_{\mathrm{q}}}{dt} = \frac{1}{L}(e_{\mathrm{q}} - u_{\mathrm{q}} - Ri_{\mathrm{q}} - \omega Li_{\mathrm{d}}) \end{cases}. \quad (2)$$

In this equation, u_{d} and u_{q} are related to the switching states, dc-bus voltages U_{dc1} and U_{dc2}. We define the switching state S_i ($i = a, b, c$) as

$$S_i = \begin{cases} 1 & \text{if } S_{i1} = 1 \text{ and } S_{i2} = 1 \\ 0 & \text{if } S_{i1} = 0 \text{ and } S_{i2} = 1 \\ -1 & \text{if } S_{i1} = 0 \text{ and } S_{i2} = 0 \end{cases} \quad (3)$$

for phase i. Therefore, there are 27 available switching states for the NPC rectifier. The rectifier output voltage can be obtained as

$$\begin{bmatrix} u_{\mathrm{a}} \\ u_{\mathrm{b}} \\ u_{\mathrm{c}} \end{bmatrix} = \frac{U_{\mathrm{dc1}} + U_{\mathrm{dc2}}}{6} \cdot \begin{bmatrix} 2 & -1 & -1 \\ -1 & 2 & -1 \\ -1 & -1 & 2 \end{bmatrix} \begin{bmatrix} S_{\mathrm{a}} \\ S_{\mathrm{b}} \\ S_{\mathrm{c}} \end{bmatrix}$$
$$+ \frac{U_{\mathrm{dc1}} - U_{\mathrm{dc2}}}{6} \cdot \begin{bmatrix} 2 & -1 & -1 \\ -1 & 2 & -1 \\ -1 & -1 & 2 \end{bmatrix} \begin{bmatrix} |S_{\mathrm{a}}| \\ |S_{\mathrm{b}}| \\ |S_{\mathrm{c}}| \end{bmatrix}. \quad (4)$$

Hence, the corresponding voltages u_{d} and u_{q} can be obtained by using the Park-Transmission given in Eq. (1).

The dynamic of the dc-bus voltage, which are related to the switching states and the current flow of the rectifier, can be modeled as follows

$$\frac{dU_{\mathrm{dc}}}{dt} = \frac{dU_{\mathrm{dc1}}}{dt} + \frac{dU_{\mathrm{dc2}}}{dt} = \frac{1}{C}(i_{\mathrm{dc}} - i_{\mathrm{l}}) + \frac{1}{C}(i_{\mathrm{dc}} - i_{\mathrm{l}} + i_{\mathrm{o}}) \quad (5)$$

where $i_{\mathrm{dc}} = (i_{\mathrm{a}}, i_{\mathrm{b}}, i_{\mathrm{c}}) \cdot (S_{\mathrm{a}}, S_{\mathrm{b}}, S_{\mathrm{c}})^{\mathrm{T}}$ and $i_{\mathrm{o}} = -(i_{\mathrm{a}}, i_{\mathrm{b}}, i_{\mathrm{c}}) \cdot (|S_{\mathrm{a}}|, |S_{\mathrm{b}}|, |S_{\mathrm{c}}|)^{\mathrm{T}}$ are related to the switching states and grid currents.

The difference between upper and lower capacitor voltage can be written as

$$\frac{dU_{\mathrm{o}}}{dt} = \frac{dU_{\mathrm{dc1}}}{dt} - \frac{dU_{\mathrm{dc2}}}{dt} = -\frac{1}{C}i_{\mathrm{o}}. \quad (6)$$

From Eq.(2) and Eq.(5), we can find that grid currents and the capacitor voltages (U_{dc1} and U_{dc2}) are coupled to each other. Moreover, the number of independent control inputs (the switching states) is less than the number of system independent states (d-axis current i_{d}, q-axis current i_{q} and dc-bus voltage). These features make such rectifier difficult

to control. This is the reason why the cascade structure is often used to control the rectifier.

For applying MPC, all dynamic equations will be derived in discrete-time format by using the forward Euler method in order to obtain the prediction model. Due to the length and focus of this paper, the discrete equations will not be given.

III. CONTROL METHODS

A. Classical PI-MPC

A PI controller and an inner predictive controller are used in this scheme. The PI controller is the outer control loop to regulate the dc-bus voltage. The predictive controller, as the inner controller, is used to control the grid currents and the error between the upper and lower capacitor voltage.

To design the PI controller for the dc-bus voltage, its dynamics under perfect tracking in the inner loop, i.e., $i_{\mathrm{d}} \to i_{\mathrm{d}}^*$, $i_{\mathrm{q}} \to 0$ and $U_{\mathrm{o}} \to 0$ are first derived. According to the power balance between the input and output, the dc-bus satisfies

$$\frac{d\left(U_{\mathrm{dc}}^2\right)}{dt} = \frac{2}{R_L C}\left(E_{\mathrm{m}} R_L i_{\mathrm{d}}^* - (i_{\mathrm{d}}^*)^2 R_L R\right) - \frac{2}{R_L C}U_{\mathrm{dc}}^2$$
$$= \frac{2}{R_L C}\phi\left(i_{\mathrm{d}}^*\right) - \frac{2}{R_L C}U_{\mathrm{dc}}^2 \quad (7)$$

where E_{m} is the amplitude of the grid voltage and ϕ is the nonlinear function of i_{d}^*. Note that, a nonlinear relationship between the input i_{d}^* and output U_{dc}^2 can be seen from this equation, which is the obstacle when we design the PI controller. In order to simplify the design of the controller, we assume that the nonlinear function is approximately equal to the linear one, i.e., $\phi\left(i_{\mathrm{d}}^*\right) \cong E_{\mathrm{m}} R_L i_{\mathrm{d}}^*$. This assumption is reasonable when E_{m} is much larger than R. Then the usual linear control methods can be used to design the PI control law.

For the inner predictive controller, three state variables should be controlled: i_{d}, i_{q} and U_{o}. Therefore, the cost function can be written as

$$J_{\mathrm{PI-MPC}} = \lambda_{\mathrm{Uo}}(U_{\mathrm{o}}(k+1))^2 + \lambda_{\mathrm{id}}(i_{\mathrm{d}}^* - i_{\mathrm{d}}(k+1))^2$$
$$+ \lambda_{\mathrm{iq}}(i_{\mathrm{q}}^* - i_{\mathrm{q}}(k+1))^2 \quad (8)$$

where λ_{id}, λ_{iq} and λ_{Uo} are weighting coefficients, $i_{\mathrm{d}}(k+1)$, $i_{\mathrm{q}}(k+1)$ and $U_{\mathrm{o}}(k+1)$ can be obtained by the aforementioned prediction model. i_{d}^* is the output of the PI controller and i_{q}^* is set to 0 to achieve the unit power control. Then the MPC algorithm will evaluate all the switching combinations and choose the optimal one which can minimize the cost function $J_{\mathrm{PI-MPC}}$.

B. MPC with dynamic references design (DR-MPC)

For this scheme, the dc-bus voltage, grid currents and the voltage difference of the dc-bus capacitors control are realized in one control loop. The cost function of this scheme should include an extra term for the dc-bus voltage control compared with $J_{\mathrm{PI-MPC}}$

$$J_{\mathrm{DR-MPC}} = \lambda_{\mathrm{id}}(i_{\mathrm{d}}^*(k+1) - i_{\mathrm{d}}(k+1))^2$$
$$+ \lambda_{\mathrm{iq}}(i_{\mathrm{q}}^* - i_{\mathrm{q}}(k+1))^2 + \lambda_{\mathrm{Uo}}(U_{\mathrm{o}}(k+1))^2 \quad (9)$$
$$+ \lambda_{\mathrm{Udc}}(U_{\mathrm{dc}}^*(k+1) - U_{\mathrm{dc}}(k+1))^2.$$

Note that, because the outer PI controller does not exist in this scheme, the suitable reference for the active current $i_\mathrm{d}^*(k+1)$ should be provided. The authors in [5] proposed a method to design the compatible references. Based on the fact that the capacitor current cannot be made arbitrarily large, hence the dc-bus voltage reference is given in a gradually approaching manner. Therefore, instead of U_dc^* in $J_\mathrm{PI-MPC}$, $U_\mathrm{dc}^*(k+1)$ is used in $J_\mathrm{DR-MPC}$

$$U_\mathrm{dc}^*(k+1) = U_\mathrm{dc}(k) + \frac{1}{N}\left(U_\mathrm{dc}^* - U_\mathrm{dc}(k)\right). \qquad (10)$$

It can be seen that this equation is a filter essentially. By regulating N within a certain band, the tracking response will be influenced. More specifically, if a faster tracking response is required, N should be given to a small value.

The next step is to find a compatible reference of the active current $i_\mathrm{d}^*(k+1)$ for the $U_\mathrm{dc}^*(k+1)$. This active current value can be solved from a binary equation which represents the power balance of the input and output of the rectifier. For the filtered reference $U_\mathrm{dc}^*(k+1)$, the output power can be written as

$$P_\mathrm{out}^* = P_\mathrm{Cdc1}^* + P_\mathrm{Cdc2}^* + P_\mathrm{RL}^* \qquad (11)$$

where P_Cdc1^*, P_Cdc2^* and P_RL^* represent the power of C_dc1, C_dc2 and R_L respectively. The precise expressions are given by

$$
\begin{aligned}
P_\mathrm{Cdc1}^* &= \frac{C}{T_\mathrm{s}}\left(\frac{U_\mathrm{dc}^*(k+1)}{2} - U_\mathrm{dc1}(k)\right)\left(\frac{U_\mathrm{dc}^*(k+1)}{2}\right)\\
P_\mathrm{Cdc2}^* &= \frac{C}{T_\mathrm{s}}\left(\frac{U_\mathrm{dc}^*(k+1)}{2} - U_\mathrm{dc2}(k)\right)\left(\frac{U_\mathrm{dc}^*(k+1)}{2}\right)\\
P_\mathrm{RL}^* &= \frac{(U_\mathrm{dc}(k)+U_\mathrm{dc}^*(k+1))^2}{4R_\mathrm{L}}.
\end{aligned}
\qquad (12)
$$

The input power of the rectifier needs to satisfy

$$P_\mathrm{in}^* = \frac{3}{2}E_\mathrm{m}i_\mathrm{d}^*(k+1) - \frac{3}{2}i_\mathrm{d}^*(k+1)^2 R = P_\mathrm{out}^*. \qquad (13)$$

Then the active current reference $i_\mathrm{d}^*(k+1)$ can be calculated by

$$i_\mathrm{d}^*(k+1) = \frac{E_\mathrm{m}}{2R}\left(1 - \sqrt{1 - \frac{8RP_\mathrm{out}^*}{3E_\mathrm{m}^2}}\right). \qquad (14)$$

C. Proposed MPC with a sliding manifold (SM-MPC)

As mentioned in the introduction, the MPC with dynamic references scheme does not adequately consider the strong coupling between the dc-bus voltage and the active current. It makes the system reach the true reference (dc-bus voltage reference U_dc^*) by slowly changing the dc-bus voltage reference ($U_\mathrm{dc}^*(k+1)$) which is given to the predictive controller. When a fast response of the dc-bus voltage is required, the oscillations of the dc-bus voltage and currents will be observed.

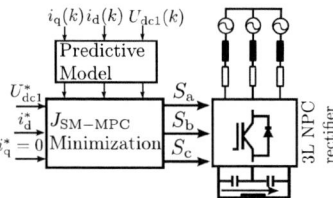

Fig. 2. Control structure of the SM-MPC scheme

TABLE I
SYSTEM CONFIGURATION.

Parameters	Values
Source voltage E_m[V]	120
Line inductor L[H]	20×10^{-3}
Inductor resistance R [Ω]	0.2
DC-bus voltage U_dc [V]	300
DC capacitance $C_1 = C_2$ [F]	1000×10^{-6}
Load resister $R_\mathrm{L}, R_\mathrm{L1}$[$\Omega$]	50/150
Sampling time T_s [μs]	20
PI-MPC parameters $K_\mathrm{p}/K_\mathrm{i}$, $\lambda_\mathrm{id} = \lambda_\mathrm{iq}$	0.00009/0.012, 1
DR-MPC parameters $N, \lambda_\mathrm{Udc}, \lambda_\mathrm{Uo}, \lambda_\mathrm{id} = \lambda_\mathrm{iq}$	500, 0.05, 0.01, 1
SM-MPC parameters $c, \lambda_\mathrm{iq}, \lambda_\mathrm{Udc1}, \lambda_\mathrm{Uo}, \lambda_\mathrm{S}$	6.4, 6, 0.01, 1

The proposed scheme (shown in Fig. 2) does not use the reference design manner because it is difficult to give suitable references of the two strong coupling state variables. Inspired by the sliding mode control, we expect the motion of the two coupling state variables to be tied (or gradually approach) to a pre-designed linear curve (sliding manifold). In order to achieve this goal, a sliding manifold term should be included in the cost function:

$$
\begin{aligned}
J_\mathrm{SM-MPC} &= \lambda_\mathrm{Udc1}(U_\mathrm{dc1}^* - U_\mathrm{dc1}(k+1))^2 + \lambda_\mathrm{S}S^2\\
&\quad + \lambda_\mathrm{iq}\left(i_\mathrm{q}^* - i_\mathrm{q}(k+1)\right)^2 + \lambda_\mathrm{Uo}(U_\mathrm{o}(k+1))^2
\end{aligned}
\qquad (15)
$$

where S is the sliding manifold term and it is defined as

$$S = c\left(i_\mathrm{d}^* - i_\mathrm{d}(k+1)\right) + \left(U_\mathrm{dc1}^* - U_\mathrm{dc1}(k+1)\right). \qquad (16)$$

Note that, $U_\mathrm{dc1}^*(\frac{1}{2}U_\mathrm{dc}^*)$ is the set value and i_d^*, corresponding to the U_dc^*, is the reference value when the system reaches the steady state.

$$i_\mathrm{d}^* = \frac{E_\mathrm{m}}{2R}\left(1 - \sqrt{1 - \frac{8R(U_\mathrm{dc}^*)^2}{3E_\mathrm{m}^2 R_\mathrm{L}}}\right) \qquad (17)$$

It can be seen from the cost function (15), the first and the fourth term are used to control the the dc-bus voltage and balance the neutral point potential. The third term can control the reactive current which we often set its reference to 0 in order to achieve the unit power control. The second term is the sliding manifold term which is used to make the motion of the coupling state variables (the active current and capacitors voltage) approach to or reach the sliding manifold. If this term is equal to 0, the active current and dc-bus voltage are decoupled as one variable can be expressed by the other one.

IV. SIMULATION RESULTS

The parameters of the simulated system are listed in Table I. The performance comparison of PI-MPC, DR-MPC and SM-MPC under the step change in load are displayed in Fig. 3 - Fig.5. It can be seen that both the DR-MPC and the SM-MPC exhibit the low dc-bus voltage drop and fast dynamic response. Compared with these two methods, the PI-MPC scheme shows the longer regulation time and larger dc-bus voltage drop. In such condition (step change in load), the outer PI controller prevents the fast dynamic response performance of the dc-bus voltage and cannot take the full advantage of MPC: fast dynamic response.

The comparison results on the step changes in the dc-bus voltage reference are also given in Fig.6 - Fig.8. All of these

The 2018 International Power Electronics Conference

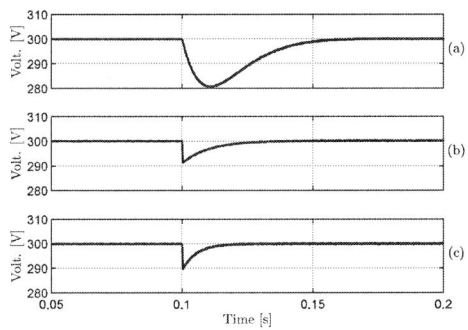

Fig. 3.   DC-bus voltage responses to step change in load. (a) PI-MPC. (b) DR-MPC. (c) SM-MPC.

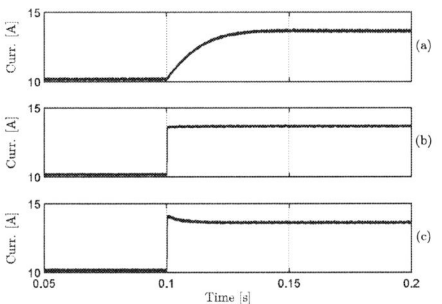

Fig. 4.   D-axis current responses to step change in load. (a) PI-MPC. (b) DR-MPC. (c) SM-MPC.

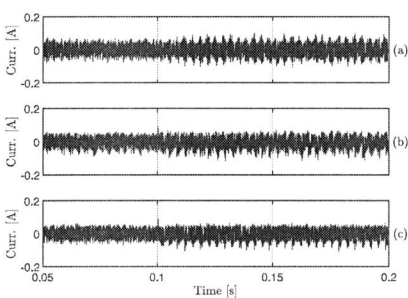

Fig. 5.   Q-axis current responses to step change in load. (a) PI-MPC. (b) DR-MPC. (c) SM-MPC.

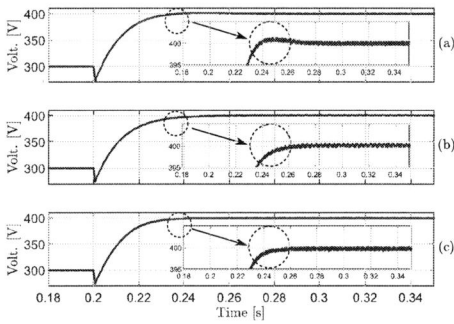

Fig. 6.   DC-bus voltage responses to step change in voltage reference. (a) PI-MPC. (b) DR-MPC. (c) SM-MPC.

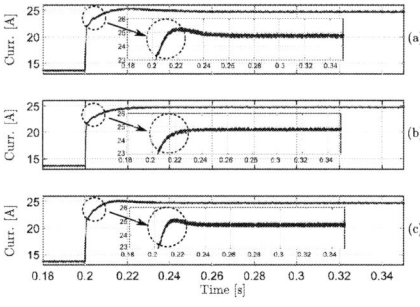

Fig. 7.   D-axis voltage responses to step change in voltage reference. (a) PI-MPC. (b) DR-MPC. (c) SM-MPC.

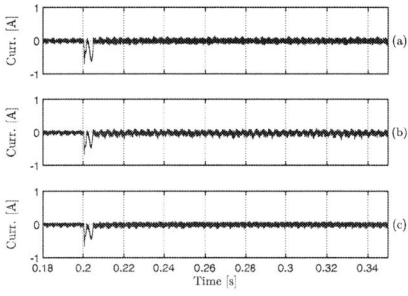

Fig. 8.   Q-axis voltage responses to step change in voltage reference. (a) PI-MPC. (b) DR-MPC. (c) SM-MPC.

three methods show fast tracking performance. However, a very small oscillation can be observed on the dc-bus voltage and currents when DR-MPC is applied. Although increasing the N can reduce (or even eliminate) this oscillation, the rising time will become longer. Actually, we can see in Fig. 6 that the rising time of dc-bus voltage for DR-MPC is the longest one but the voltage still has a small oscillation. This means that we cannot continue to reduce the value of N to achieve the fast dynamic response of the dc-bus voltage otherwise it will generate a larger oscillation and further reduce the power quality on the input side. As we discussed above, DR-MPC does not consider the strong decoupling relationship between the dc-bus voltage and the active current adequately.

The SM-MPC exhibits fast tracking and strong disturbance rejection capability in these two cases. However, by using this method a very small steady-state error (less than 0.5V) can be observed on the dc-bus voltage. This is because the actual efficiency of the rectifier is not considered when we calculate the steady-state reference of the active current.

V.   Conclusion

In this paper, the MPC with a sliding manifold term scheme (SM-MPC) has been proposed. Within this scheme, the dc-bus voltage is directly controlled without a outer PI controller. The control structure of SM-MPC is very simple as only a sliding manifold term is added into the cost function. Also, the calculation of this scheme is not complicated because only the steady-state references are needed in the cost function. Besides, a comparison of the classical PI-MPC, MPC with dynamic references design scheme (DR-

MPC) and SM-MPC is given through simulation results. The results show that the MPC schemes without outer

PI control loop have the better tracking and disturbance rejection performance when a step change in load occurs. However, when the dc-bus voltage reference changes, the dynamic response of DR-MPC is slow compared with the other two methods. The SM-MPC performs best but has a very small bias on the dc-bus voltage in steady state. The PI-MPC scheme and DR-MPC scheme can guarantee the zero steady-state error.

REFERENCES

[1] M. P. Kazmierkowski, M. Jasinski, and G. Wrona, "Dsp-based control of grid-connected power converters operating under grid distortions," *IEEE Transactions on Industrial Informatics*, vol. 7, no. 2, pp. 204–211, May 2011.

[2] A. Bouafia, J. P. Gaubert, and F. Krim, "Predictive direct power control of three-phase pulsewidth modulation (pwm) rectifier using space-vector modulation (svm)," *IEEE Transactions on Power Electronics*, vol. 25, no. 1, pp. 228–236, Jan 2010.

[3] Z. Song, W. Chen, and C. Xia, "Predictive direct power control for three-phase grid-connected converters without sector information and voltage vector selection," *IEEE Transactions on Power Electronics*, vol. 29, no. 10, pp. 5518–5531, Oct 2014.

[4] J. Rodriguez, J. Pontt, P. Correa, P. Lezana, and P. Cortes, "Predictive power control of an ac/dc/ac converter," in *Fourtieth IAS Annual Meeting. Conference Record of the 2005 Industry Applications Conference, 2005.*, vol. 2, Oct 2005, pp. 934–939 Vol. 2.

[5] D. E. Quevedo, R. P. Aguilera, M. A. Perez, P. Cortes, and R. Lizana, "Model predictive control of an afe rectifier with dynamic references," *IEEE Transactions on Power Electronics*, vol. 27, no. 7, pp. 3128–3136, July 2012.

[6] Z. Zhang, F. Wang, T. Sun, J. Rodrguez, and R. Kennel, "Fpga-based experimental investigation of a quasi-centralized model predictive control for back-to-back converters," *IEEE Transactions on Power Electronics*, vol. 31, no. 1, pp. 662–674, Jan 2016.

H_∞ Control-based Vibration Suppression in Robot Arm with Strain Wave Gearing

Tran Vu Trung, Makoto Iwasaki*

Department of Electrical and Mechanical Engineering, Nagoya Institute of Technology, Nagoya, Japan
*E-mail: iwasaki@nitech.ac.jp

Abstract—The paper proposes a robust controller design using H_∞ synthesis for a two-degree-of-freedom (2-Dof) full-closed control system, aiming to suppress the mechanical vibration due to elasticity in a robot arm. The proposed design has been verified by experiments using a prototype.

Keywords—flexible robots, H_∞ controller design, robust control, vibration suppression.

I. INTRODUCTION

Strain Wave Gearing (SWG) is widely implemented in industrial manipulators thanks to their compact mechanism, high gear ratios, and no backlash [1]. In the robot arms with SWG, mechanical resonant vibrations are generally excited due to the angular transmission errors and flexibility. Robust vibration suppression problem is required to deal with various sources of disturbances, such as load variation, interacting force between joints, and nonlinear properties in SWG. This paper concentrates on suppressing the vibration due to the low stiffness of the robot link and SWG, as well as improving the robustness against load inertia variation in general.

Robust control designs using H_∞ synthesis for robot arms have been investigated in many publications. A survey of linear H_∞ schemes for robot manipulators with rigid model was introduced in [2]. Recently, flexible joints with gearing are getting more interests from researches on H_∞ controller design. References [3] and [4] proposed H_∞ approaches for the flexible robots that were modeled as two-mass systems. In the fact, there exists not only a vibration mode due to the wave gear's flexibility, but the elasticity in mechanical structures also causes more vibration modes. The physical model of a robot arm is then a three-mass or four-mass system, instead of the traditional two-mass models [5]-[6].

In this paper, a 2-Dof control system with robust feedback controller designed using H_∞ synthesis, together with feedforward compensators constructed based on co-prime factorization method [7], for an actual horizontal one-axis robot is proposed to suppress the vibration, as well as to ensure the robust control performance against load parameter variations.

II. CONVENTIONAL CONTROL OF ROBOT ARM

A. Configuration and Modeling of Experimental Robot

The experimental robot arm is a horizontal one-axis robot. It is composed of an AC servo motor with a strain wave gearing. Figure 1 shows a schematic configuration of the experimental setup. Encoders are installed at both motor and load sides, which enable a double feedback control. The motor-side encoder is mounted on the motor shaft to feed back the motor angular position, while the load-side encoder measures the position of the gearing's output. The load weight at the tip of arm can be changed by adding or removing some stacked cylindrical loads of 2.5 kg, up to 15 kg (maximum of 6 cylindrical loads). Specifications of motor, wave gearing and supplementary equipment are listed in Table I.

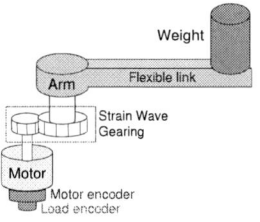

Fig. 1. Configuration of experimental robot arm.

Figure 2 shows experimental frequency responses from the input torque to the motor and arm positions with different weights of load: 15, 12.5, and 10 kg. They all exhibit two vibration modes: the primary one is considered to result from elastic deformation of the link, while the secondary one is caused by the elastic deformation of the gearing. Notice here that the phase characteristics of the motor and load sides are different at the two modes: in-phase at the first vibration frequency, while out-of-phase at the second one. Therefore, the robot arm can be modeled as a 3-mass system as defined in Figure 3(a), where τ is the input torque; N is the gear ratio; θ_m, θ_a, and θ_l are the angular positions of motor, gearing's output, and load tip, respectively; J_m, J_a, and J_l are the moments of inertia; D_m, D_a, and D_l are the viscous damping coefficients; K_1 and K_2 are the gearing stiffness and link stiffness, respectively.

Except for the parameters J_m, D_m, and K_1 that are

TABLE I.   SPECIFICATIONS OF EXPERIMENTAL SETUP

Rated Torque	1.27 Nm
Maximum instant torque	3.8 Nm
Rated speed	3000 rpm
Moment of inertia	1.04×10^{-4} kgm^2
Motor encoder resolution	131072 pulse/rev
Load encoder resolution	1048576 pulse/rev
Gear ratio	50

The 2018 International Power Electronics Conference

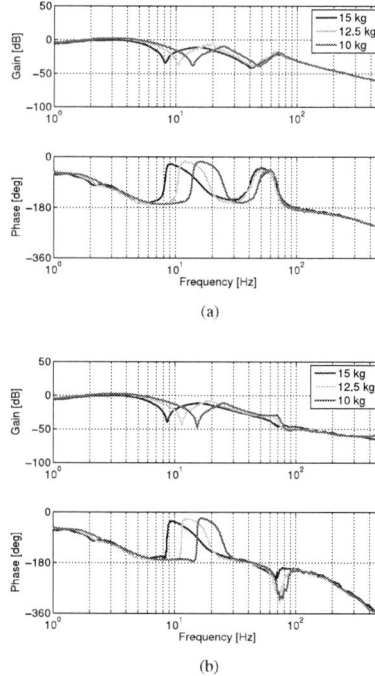

(a)

(b)

Fig. 2. Bode characteristics from torque to motor position θ_m (a), and to arm position $N\theta_a$ (b), with different weights of load.

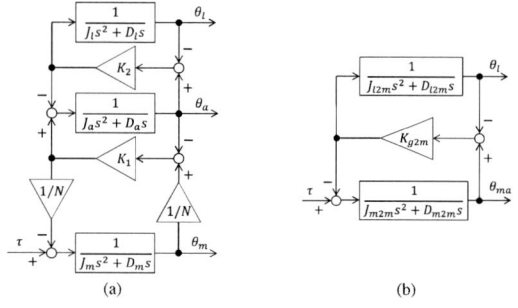

(a)   (b)

Fig. 3. Three-mass model (a), and two-mass model (b) of robot arm

given by mechanical design sheets, it is difficult to obtain exact values of the linear 3-mass model's parameters due to effects of nonlinear friction on frequency response analysis. The moments of inertia J_a and J_l can be calculated from the shape and weight of arm and load as well as the place where the elastic deformation occurs. The other parameters D_a, D_l, and K_2 are then determined by curve fitting. The identified parameters for the full-load (15 kg) case, which is considered as nominal model, are listed in Table II. For the two other cases of 12.5 and 10 kg load, variations of load parameters J_l and K_2 are shown in Table III.

B. 2-Dof full-closed control system

A conventional 1-degree-of-freedom (1-Dof) full-closed control system of the robot arm includes two cascaded feedback loops: a motor velocity control loop

TABLE II.   PARAMETERS OF NOMINAL THREE-MASS MODEL

N	50
J_m	1.04×10^{-4} kgm^2
D_m	1.4×10^{-4} Nms/rad
K_1	40000 Nm/rad
J_a	0.65 kgm^2
D_a	5 Nms/rad
K_2	6200 Nm/rad
J_l	2.1 kgm^2
D_l	3 Nms/rad

TABLE III.   LOAD PARAMETERS WITH DIFFERENT WEIGHTS

Load weight	J_l	ΔJ_l	K_2	ΔK_2
15 kg	2.1 kgm^2	0%	6200 Nm/rad	0%
12.5 kg	1.64 kgm^2	-22%	8250 Nm/rad	$+33\%$
10 kg	1.47 kgm^2	-30%	13650 Nm/rad	$+120\%$

inside and an arm position loop outside. In general, the position controller is a proportion (P) and the velocity controller is a proportion-integrator (PI) compensator:

$$C_p(s) = K_{pp}, \tag{1}$$

$$C_v(s) = K_{vp}\left(1 + \frac{1}{T_{vi}s}\right), \tag{2}$$

where the controller parameters for the experimental setup are listed in Table IV.

In order to improve motion performance, a 2-degree-of-freedom (2-Dof) control framework with feedback (FB) control system and feedforward (FF) compensators should be a promising approach [8]. Figure 4 shows a block diagram of the 2-Dof full-closed control system with coprime factorization-based FF compensators.

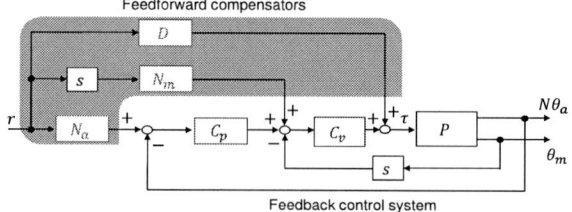

Fig. 4. Block diagram of full-closed 2-Dof control system.

Assume that the transfer functions from torque to motor position and arm position can be expressed by

$$P_m(s) = \frac{\theta_m}{\tau} = \frac{P_{Nm}(s)}{P_d(s)}, \tag{3}$$

$$P_a(s) = \frac{N\theta_a}{\tau} = \frac{P_{Na}(s)}{P_d(s)}. \tag{4}$$

By using the above functions, the FF compensators

TABLE IV.   PARAMETERS OF P-PI CONTROLLER

K_{pp}	K_{vp}	T_{vi}
25.1327 s^{-1}	0.0391 Nms/rad	0.03 s

are determined as follows:

$$N_m(s) = \frac{P_{Nm}(s)}{F_c(s)}, N_a(s) = \frac{P_{Na}(s)}{F_c(s)}, D(s) = \frac{P_d(s)}{F_c(s)},$$
(5)

where $F_c(s)$ is a low-pass filter that ensures the transfer functions of FF compensators to be proper.

The closed-loop transfer function from reference r to arm position $N\theta_a$ then becomes

$$T_{yr}(s) = N_a(s) = \frac{P_{Na}(s)}{F_c(s)}.$$
(6)

In practice, it is hard to implement 3-mass-model-based FF compensators because of their high order and excessive gains. Therefore, an approximate 2-mass model, as shown in Figure 3(b), is used for FF design. In this approximation, the joint is assumed rigid, hence we get

$$\theta_m = N\theta_a,$$
(7)

$$J_{m2m} = J_m + \frac{J_a}{N^2}, D_{m2m} = D_m + \frac{D_a}{N^2},$$
(8)

$$J_{l2m} = \frac{J_l}{N^2}, D_{l2m} = \frac{D_l}{N^2}, K_{g2m} = \frac{K_2}{N^2}.$$
(9)

By using the 2-mass model, the denumerator $P_d(s)$ of plant is a 4th-order polynomial, while the numerators $P_{Nm}(s)$ and $P_{Na}(s)$ are assigned the same 2nd-order polynomial. Thus, the low-pass filter $F_c(s)$ should be 4th-order one:

$$F_c(s) = \left(\frac{s + \omega_c}{\omega_c} \right)^4,$$
(10)

where the cut-off frequency ω_c is chosen such that the control bandwidth is as wide as possible. Parameters of the actual FF compensators are listed in Table V.

Figure 5 shows a comparison of frequency responses for experimental data, the 3-mass model, and the 2-mass model used for FF design. While the 3-mass system with friction model fits the actual plant well, the approximate 2-mass model represents only the primary vibration mode that is excited due to the flexibility of link.

C. Vibration suppression performance of conventional 2-Dof control system with load variation

Figures 6(a), (b), and (c) show comparative waveforms of arm position between 1-Dof control system and 2-Dof control system for different cases of load weight. As for the case of 15 kg, which the nominal model is

TABLE V.   PARAMETERS OF TWO-MASS MODEL-BASED FEEDFORWARD DESIGN

J_{m2m}	3.64×10^{-4} kgm^2
D_{m2m}	21×10^{-4} Nms/rad
K_{g2m}	2.48 Nm/rad
J_{l2m}	8.4×10^{-4} kgm^2
D_{l2m}	12×10^{-4} Nms/rad
ω_c	$2\pi \times 40$ rad/s

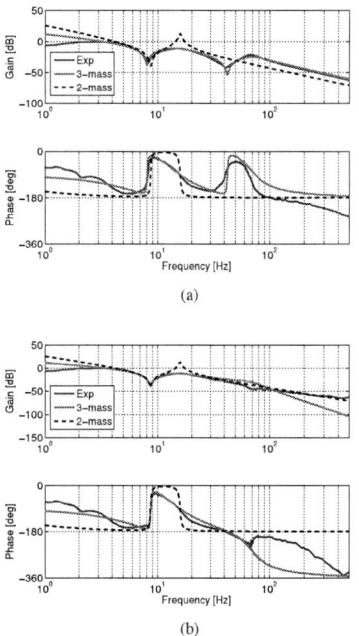

Fig. 5.   Comparative Bode characteristics from torque to motor position θ_m (a), and to arm position $N\theta_a$ (b), with different models for 15 kg load (nominal case).

constructed based on, 2-Dof control shows an effective vibration suppression with faster response than 1-Dof control. However, when the load weight decreases to 12.5 kg and 10 kg while FF compensators are fixed, the 2-Dof control performance deteriorates, even worse than 1-Dof control. The performance degradation due to load variation of 2-Dof control system is illustrated with an enlarged view, Figure 6(d), where the horizontal dashed lines denote the error band of ± 0.02 deg. The cases of 12.5 kg and 10 kg show a significantly higher overshoot and much slower settling time than the nominal case.

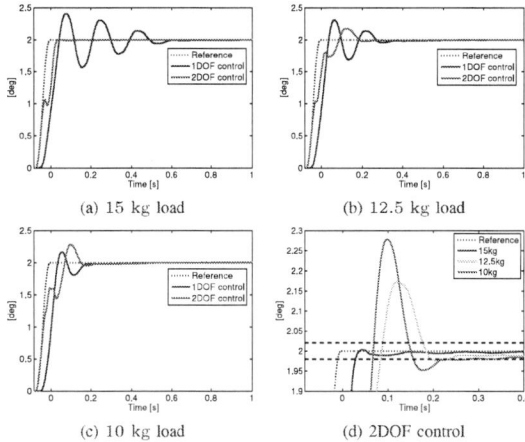

Fig. 6.   Experimental response waveforms of arm position (θ_a) with conventional P-PI control

A natural solution to this problem is to apply the

variable FF compensators that updates J_l and K_2's values when the load changes. Nevertheless, it is not always possible to obtain sufficiently precise values of the load-side parameters because of nonlinearities and complexity of mechanical structure. The identification error is inevitable, therefore, a robust FB controller design is expected to ease the deterioration due to the model error between FF compensators and actual plant for 2-Dof control system.

III. H_∞ CONTROL-BASED FEEDBACK CONTROL DESIGN

A. Problem of shaping four transfer functions for full-closed 2-Dof control system

A common H_∞ mixed-sensitivity synthesis aims to shape the frequency responses of sensitivity function $S(s)$ and complementary sensitivity function $T(s)$ [9]. In this research, two more transfer functions are considered to shape the gain characteristics.

Denote the differences between the ideal FF compensators that are calculated as Equation (5), and the actual FF compensators, by $\Delta N_a(s)$, $\Delta N_m(s)$, and $\Delta D(s)$. Then, the influence of these differences on control system are respectively characterized as the following disturbances, as shown in Figure 7:

$$d_1 = r\Delta D(s), d_2 = rs\Delta N_m(s), d_3 = r\Delta N_a(s). \quad (11)$$

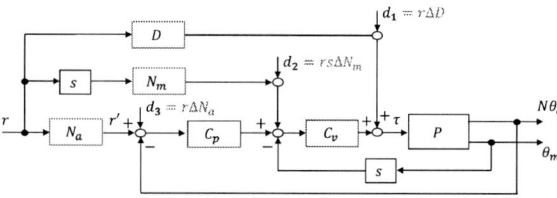

Fig. 7. Block diagram of 2-Dof control system with presence of model errors in FF compensators.

From the viewpoint of FB control system design shown in Figure 8, the gains of transfer functions from every disturbances to the output ($N\theta_a$) should be limited. Denote the transfer functions of the output for d_1, d_2, and d_3 by $G_{d1}(s)$, $G_{d2}(s)$, and $G_{d3}(s)$, respectively. Notice that since $G_{d3}(s)$ is identical to the complementary sensitivity function, only four transfer functions $S(s)$, $T(s)$, $G_{d1}(s)$, and $G_{d2}(s)$ should be considered in the gain-shaping problem for the H_∞ design.

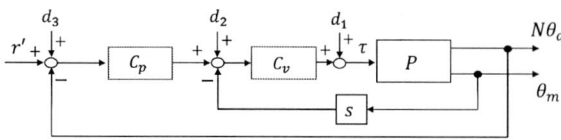

Fig. 8. Block diagram of FB control system with disturbances.

B. Structure of the proposed H_∞ synthesis

Figure 9 shows a block diagram of the H_∞ synthesis proposed for the problem of shaping four transfer functions $S(s)$, $T(s)$, $G_{d1}(s)$, and $G_{d2}(s)$. The position and velocity controllers are combined into a controller $\mathbf{K}(s)$ whose two inputs are the arm position error ($r' - N\theta_a$) and the motor velocity ($\omega_m = \dot\theta_m$), with one output is the torque τ_{ref}, as follows:

$$\tau_{ref} = \mathbf{K}(s) \begin{bmatrix} r' - N\theta_a \\ \omega_m \end{bmatrix} \quad (12)$$

$$= \begin{bmatrix} K_1(s) & -K_2(s) \end{bmatrix} \begin{bmatrix} r' - N\theta_a \\ \omega_m \end{bmatrix} \quad (13)$$

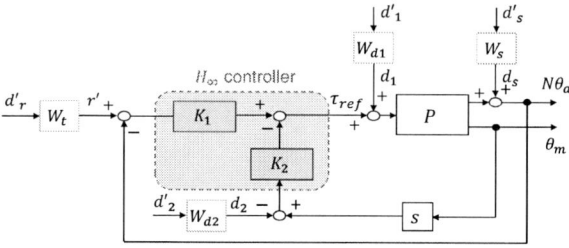

Fig. 9. Block diagram of H_∞ synthesis.

The formulations of $S(s)$, $T(s)$, $G_{d1}(s)$, and $G_{d2}(s)$ are then given by

$$S(s) = \frac{1 + K_2(s)sP_2(s)}{1 + K_2(s)sP_2(s) + K_1(s)P_1(s)}, \quad (14)$$

$$T(s) = \frac{K_1(s)P_1(s)}{1 + K_2(s)sP_2(s) + K_1(s)P_1(s)}, \quad (15)$$

$$G_{d1}(s) = \frac{P_1(s)}{1 + K_2(s)sP_2(s) + K_1(s)P_1(s)}, \quad (16)$$

$$G_{d2}(s) = \frac{P_1(s)K_2(s)}{1 + K_2(s)sP_2(s) + K_1(s)P_1(s)}. \quad (17)$$

Four weighting functions $W_s(s)$, $W_t(s)$, $W_{d1}(s)$ and $W_{d2}(s)$ are applied to output's disturbance d_s, reference r', FF-generated disturbances d_1 and d_2, respectively. Then, the closed-loop function T_{zw} from the new disturbances d'_s, d'_r, d'_1, and d'_2 of the generalized plant to the evaluation output $N\theta_a$ can be given by

$$T_{zw} = \begin{bmatrix} W_s S & W_t T & W_{d1}G_{d1} & W_{d2}G_{d2} \end{bmatrix}. \quad (18)$$

The task of design is to find the controller $\mathbf{K}(s)$ that makes the H_∞ norm of T_{zw} smaller than 1, then $S(s)$, $T(s)$, $G_{d1}(s)$, and $G_{d2}(s)$ satisfy the following gain-shaping inequalities:

$$|S(s)| < \left| \frac{1}{W_s(s)} \right|, \quad (19)$$

$$|T(s)| < \left| \frac{1}{W_t(s)} \right|, \quad (20)$$

$$|G_{d1}(s)| < \left| \frac{1}{W_{d1}(s)} \right|, \quad (21)$$

$$|G_{d2}(s)| < \left| \frac{1}{W_{d2}(s)} \right|. \quad (22)$$

Thus, by choosing the appropriate weighting functions, the desired shapes of $S(s)$, $T(s)$, $G_{d1}(s)$, and $G_{d2}(s)$ satisfying the requirements on robustness and tracking performance can be achieved.

C. Selection of weighting functions

Selecting weighting functions is the most difficult and important task in the H_∞ controller design. Since the gain-shaping problem is often multi-objective, it consumes time to obtain a set of weighting functions making a good balance between objectives. In this design, weighting functions for the sensitivity gain $|S(s)|$ and the complementary sensitivity gain $|T(s)|$ take priority because of their importance in stability and robustness of FB control system in general. While the weighting function for $|S(s)|$ specifies the performance of closed-loop system with the nominal plant, the weighting function for $|T(s)|$ guarantees the robust stability against model error.

In details, $|W_s(s)|$ should have a low upper-limit in order to keep $|S(s)|$ small, aiming to obtain a nominally stable closed-loop system. Additionally, $W_s(s)$ should have an integral part at low-frequency range to eliminate steady error. Nonetheless, H_∞ design techniques generally require that the weighting functions has no unstable poles, which results in adding a small positive number to the denumerator of $W_s(s)$, i.e. we only need $|S(0)|$ to be not greater than a very small value, instead of perfect tracking ($|S(0)| = 0$) [10]. Since these specifications, the weighting function for $|S(s)|$ is chosen as follows:

$$W_s(s) = 0.32 \frac{s + 2\pi \cdot 2}{s + 0.001}.$$ (23)

Regarding the complementary sensitivity function, $|W_t(s)|$ should be greater than the gain of multiplicative uncertainty for every frequency:

$$|\Delta W_t(s)| > |\Delta_m(s)| = \left| \frac{P(s) - P_n(s)}{P_n(s)} \right|,$$ (24)

where $P(s)$ is the actual plant and $P_n(s)$ is the nominal model, as shown in Figure 10.

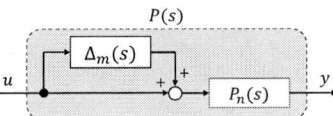

Fig. 10. Actual plant $P(s)$ consisting of nominal model $P_n(s)$ and multiplicative uncertainty $\Delta_m(s)$.

The nominal complementary sensitivity function then satisfies the following inequality:

$$|T(s)| < \left| \frac{1}{W_t(s)} \right| < \left| \frac{1}{\Delta_m(s)} \right|.$$ (25)

Thus, according to the small gain theorem, the closed-loop is robustly stable with all designated cases of multiplicative uncertainty. The following weighting function is selected in order to meet the condition (24):

$$W_t(s) = \left(90 \frac{s + 2\pi \cdot 20}{s + 2\pi \cdot 2000} \right)^4 \cdot \frac{s^2 + 240.5s + 2988}{s^2 + 1.093s + 2988}.$$ (26)

Its satisfaction with the requirement (24) is illustrated in Figure 11. The line representing for 15 kg-load case can not be seen in this figure because $\Delta_m(s) = 0$, i.e. $|\Delta_m(s)|$ in dB-unit is too small for nominal case. As for the two other cases, the gain of $\Delta_m(s)$ has a peak at the anti-resonant frequency of the nominal plant. These peaks are dealt with by a Notch filter that is just the second factor of $W_t(s)$ in expression (26).

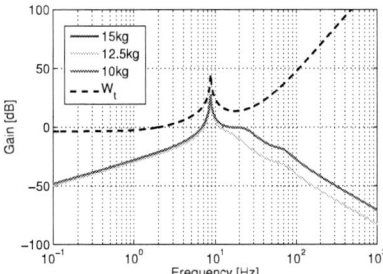

Fig. 11. Bode gain plots of weighting function $W_t(s)$ and multiplicative uncertainty $\Delta_m(s)$ in different cases of load weight.

Finally, the weighting functions $W_{d1}(s)$ and $W_{d2}(s)$ are selected such that the gains $|G_{d1}(s)|$ and $|G_{d2}(s)|$ do not exceed -14 dB, as follows:

$$W_{d1}(s) = 5 \frac{s + 2\pi \cdot 0.5}{s + 0.0001},$$ (27)

$$W_{d2}(s) = 5.$$ (28)

D. Calculation of H_∞ controller

An H_∞ controller including two stable 3rd-order transfer functions $K_1(s)$ and $K_2(s)$ are obtained by using MATLAB *Robust Control Toolbox*'s function **Hinfstruct**. In order to implement the H_∞ controller based on the structure shown in Figure 4, the equivalent position and velocity controllers are computed by

$$C_p(s) = \frac{K_1(s)}{K_2(s)},$$ (29)

$$C_v(s) = K_2(s).$$ (30)

Figure 12 shows Bode plots of $C_p(s)$ and $C_v(s)$ with P-PI control and H_∞ control for comparison.

IV. RESULTS

A. Evaluations of Robustness

The robustness of the proposed H_∞ control-based design is compared with that of the conventional P-PI control. Figure 13 shows the Bode gain plots of four transfer functions $S(s)$, $T(s)$, $G_{d1}(s)$, and $G_{d2}(s)$, with the selected corresponding weighting functions. It has proved the ability to suppress the peaks of sensitivity

The 2018 International Power Electronics Conference

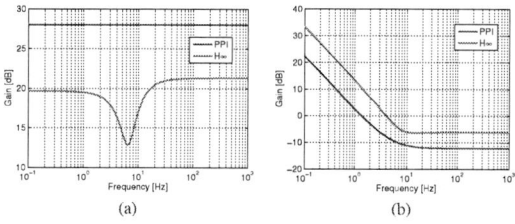

Fig. 12. Bode gain plots of position controller $C_p(s)$ (a), and velocity controller $C_v(s)$ (b).

gains $|S(s)|$ and $|T(s)|$ of the proposed H_∞ design. Although the sensitivity gain $|S(s)|$ for H_∞ control is higher than that for P-PI control in low-frequency range, it is fine because the tracking performance of 2-Dof control system is actually decided by FF compensators. As for complementary sensitivity gain $|T(s)|$, which is crucial to robustness, the proposed design shows a lower characteristic than the conventional one for every frequency. Moreover, the robust stability of closed-loop system is proven by the Nyquist plots of open-loop systems as shown in Figure 14. The proposed design (b) shows Nyquist loci further away from the point $(-1; 0)$ than those of the conventional design (a).

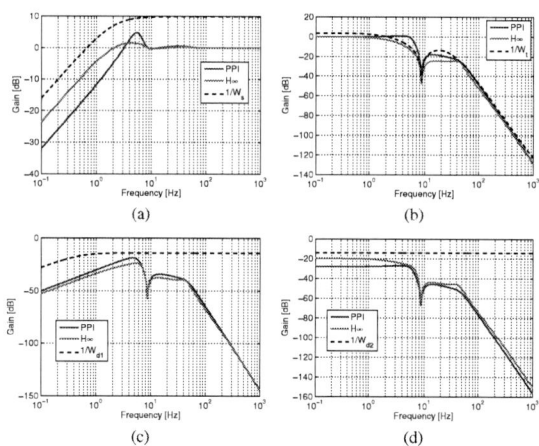

Fig. 13. Bode gain plots of four transfer functions: (a) Sensitivity function $S(s)$, (b) Complementary sensitivity function $T(s)$, (c) d_1-disturbance rejection function $G_{d1}(s)$, (d) d_2-disturbance rejection function $G_{d2}(s)$.

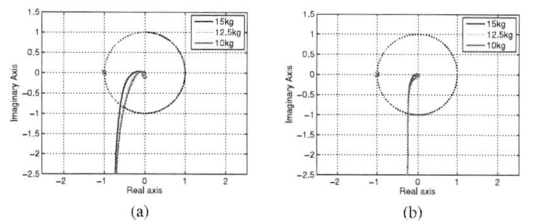

Fig. 14. Nyquist plots: (a) P-PI control, (b) H_∞ control.

To clarify the robust performance improvement, some numerical comparative evaluations are listed in Table VI. With H_∞ control, the gain and phase margins are

TABLE VI. COMPARISON OF ROBUSTNESS

	Conventional P-PI control	Proposed H_∞ control
Gain margin [dB]	11.26	23.59
Phase margin [deg]	52.75	75.60
$|S|_{\max}$ [dB]	4.74	1.58
$|T|_{\max}$ [dB]	1.05	0.001

improved noticeably, while the peak gains of sensitivity functions are also suppressed considerably.

B. Experimental results

The proposed controllers have been discretized by a linear transformation with sampling time of $T_s = 250\mu s$ for implementation. Figure 15 shows comparative experimental waveforms of arm position (θ_a) between the conventional design based on P-PI control and the proposed design based on H_∞ control, for different weights of load. FF compensators are kept unchanged for all experiments.

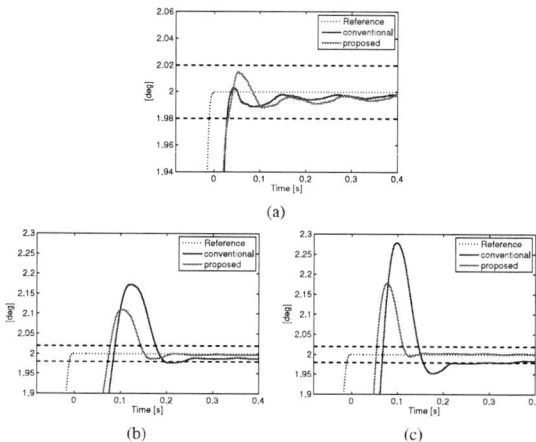

Fig. 15. Experimental response waveforms of arm position (θ_a) for 2-Dof control system: (a) 15 kg load, (b) 12.5 kg load, (c) 10 kg load.

In the nominal case of 15 kg load, the model error is insignificant; accordingly, the difference in response waveforms between the proposed method and the conventional one is not considerable. When the model error between FF compensators and the actual plant becomes significant, the H_∞ control-based system shows better performance with smaller overshoot and faster response than the conventional one. In details, in case of 12.5 kg load, overshoot drops from 8.6% to 5.5%, while settling time (for the error band of ± 0.02 deg) is reduced from 0.23 s to 0.14 s. As for 10 kg load, the overshoot declines from 13.9% to 8.9%, and the settling time decreases from 0.35 s to 0.11 s.

V. CONCLUSION AND FUTURE WORK

In this paper, a 2-Dof full-closed control system consisting of static coprime factorization-based FF compensators and H_∞ control-based robust FB controllers has been proposed to suppress the mechanical vibration due to low stiffness of robot link and joint for an experimental robot arm with strain wave gearing. The controller was

1671

designed by shaping the gains of four transfer functions. Two of them are basic components of traditional H_∞ mixed-sensitivity synthesis, while the two others are derived from the rejection of disturbances caused by FF compensators. Robustness improvement of the proposed method has been proven by frequency responses as well as the increase in gain and phase margins. In addition, effectiveness of the proposed design has been verified by experiments with different weights of load. In comparison with the conventional P-PI control system, overshoot and settling time are remarkably reduced for the cases in which the model error becomes appreciable.

In the fact, the paper neglected some nonlinear properties of the gearing, e.g. Angular Transmission Errors (ATEs). The vibration caused by ATEs can be suppressed using H_∞ control by shaping the gain of the transfer function from ATEs to the arm position besides sensitivity functions.

Furthermore, there are various methods to design FF compensators apart from the coprime factorization. The proposed H_∞ control-based FB compensator is expected to combine successfully with other FF designs, such as the traditional computed-torque FF [11], and the FF compensator for ATEs [12]. Evaluation on vibration suppression and robustness of such combinations is one of future directions of research.

REFERENCES

[1] M. Iwasaki, "Modeling and Compensation for Angular Transmission Errors for Precision Positioning Devices with Harmonic Drive Gearings," Proc. of 6th International Conference on Positioning Technology, pp. 13-20, 2014.

[2] H. G. Sage, M. F. Mathelin, and E. Ostertag, "Robust control of robot manipulators: A survey," International Journal of Control, vol. 72, no. 16, pp. 1498-1522, 1999.

[3] J. S. Yeon, J. Yim, and J. H. Park, "Robust control using recursive design method for flexible joint robot manipulators," Journal of Mechanical Science and Technology, vol. 25, no. 12, pp. 32053213, 2011.

[4] P. Axelsson et al., "H_∞ Synthesis Method for Control of Nonlinear Flexible Joint Models," Proc. of the 19th IFAC World Congress, pp. 210-216, 2014.

[5] M. Ostring, S. Gunnarsson, and M. Norrlof, "Closed-loop identification of an industrial robot containing flexibilities," Control Engineering Practice, vol. 11, no. 3, pp. 291-300, 2003.

[6] S. Moberg, J. Ohr, and S. Gunnarsson, "A Benchmark Problem for Robust Feedback Control of a Flexible Manipulator," IEEE Trans. on Control Systems Technology, vol. 17, no. 6, pp. 1398-1405, 2009.

[7] K. Itoh, M. Iwasaki, and N. Matsui, "Optimal design of robust vibration suppression controller using genetic algorithms," IEEE Trans. on Industrial Electronics, vol. 51, no. 5, pp. 947-953, 2004.

[8] M. Iwasaki, K. Seki, and Y. Maeda, "High-Precision Motion Control Techniques: A Promising Approach to Improving Motion Performance," IEEE Industrial Electronics Magazine, vol. 6, no. 1, pp. 32-40, 2012.

[9] S. Skogestad and I. Postlethwaite, "Multivariable Feedback Control: Analysis and Design, 2nd Edition," John Wiley & Sons, 2005.

[10] K. Zhou, "Essentials of Robust Control," Prentice Hall, 1999.

[11] Y. D. Song, A. T. Alouani, and J. N. Anderson, "Robust path tracking control of industrial robots: An H_∞ approach," Proc. of First IEEE Conference on Control Applications, pp. 25-30, 1992.

[12] M. Iwasaki et al., "Modeling and compensation for angular transmission error of harmonic drive gearings in high precision positioning," Proc. of IEEE/ASME International Conference on Advanced Intelligent Mechatronics, pp. 662-667, 2009.

The 2018 International Power Electronics Conference

Fine Force Sensorless Force Control Based on Friction-Free Disturbance Observer

Ohishi Kiyoshi, Naoki Kamiya, Toshimasa Miyazaki and Yuki Yokokura

Nagaoka University of Technology

1603-1 Kamitomioka, Nagaoka, Niigata 940-2188, JAPAN

Email:ohishi@vos.nagaokaut.ac.jp, s153123@stn.nagaokaut.ac.jp,

miyazaki@vos.nagaokaut.ac.jp, yokokura@vos.nagaokaut.ac.jp

Abstract—This paper proposes force sensorless force control by using a notch-type friction free disturbance observer. The force control is structured by using acceleration and speed controllers. The reaction force estimation system reduces the effect of the static friction using a dither signal. A conventional friction free disturbance observer becomes unstable when a low-frequency dither signal is applied. In order to solve this problem, a notch-type friction free disturbance observer is employed, which allows the use of low-frequency dither signals. This paper explains the difference between the conventional design method and the proposed design method. The validity of the proposed notch-type friction free disturbance observer is verified through the experimental results by using an actual industrial robot.

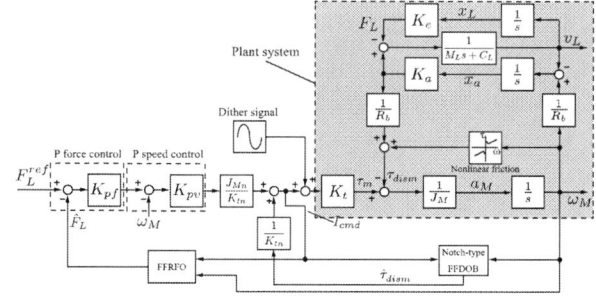

Fig. 1. Block diagram of Force sensorless force control.

I. INTRODUCTION

Recently, robots are required to work as substitutes for humans, and force control is an important element technology for robots to perform these tasks. In normal force control, a force sensor is attached to a robot and the sensor is subsequently controlled. However, the force sensor is extremely expensive and poses a problem wherein the mechanical stiffness decreases when it is attached to a robot. Therefore, force control without the use of a force sensor is desirable. It is important to estimate the force with high accuracy in force sensorless force control. The performance of the force estimation deteriorates due to the nonlinear friction of the controlled object. With respect to the ball screws and harmonic gears used in general industrial robots, a significant effect of static friction is observed in extremely low-speed operations, thus deteriorating the force estimation performance[1]. A method of using a dither signal to reduce the effect of the static friction has been proposed [2], [3], [4]. The dither signal is input to the current command value as a sine wave. The dither signal vibrates the motor side, and thereby reduces the effect of the static friction of the robot. A conventional force sensorless force control using a dither signal is used as a friction free disturbance observer (FFDOB) in a minor loop [3]. A FFDOB is a disturbance observer capable of outputting the estimated disturbance value from which the vibration corresponding to the frequency of the dither signal has been removed. During extremely low-speed operation such as a pushing motion of the robot, the steady-state error remains between the estimated and true value of the reaction force due to the effect of the static friction, leading to the deterioration in performance. The dither signal is important in reducing the effect of the static friction. In order to further reduce the effect

of static friction, it is necessary to increase the vibration of the robot. In general, mechanical vibrations are easily transmitted at lower frequencies, and a low-frequency dither signal is an optimal choice for this purpose. However, the FFDOB limits the frequency of the dither signal that is used and diverges at a low frequency. Therefore, the proposed method enables control even with a low- frequency of the dither signal.

This paper proposes a force sensorless force control by using a notch-type FFDOB. Furthermore, experiments with respect to pushing motion are performed by using an industrial robot, and the effectiveness of the notch-type FFDOB is validated.

II. FORCE SENSORLESS FORCE CONTROL

Fig.1 shows the block diagram of the force sensorless force control system. The force control is structured by acceleration and speed controllers [5]. The force sensorless force control is realized by using a friction free reaction force observer (FFRFO) [3], [6]. The FFRFO estimates the reaction force of the load side. The estimated value is feedbacked to the force control system as opposed to the force sensor. The system assumes an extremely low-speed pushing motion. It is considered as a single inertia model due to its extremely low speed. The speed controller gain is designed as 1/5th of the pole of the disturbance observer, by idealizing the acceleration control system. The gain of the force controller is determined through trial and error.

Specifically, $J_M, K_t, R_b, K_a, M_L, C_L,$ and K_e denote the motor inertia, torque constant, gear ratio, stiffness, load-side mass, viscosity friction, and environmental stiffness, respectively, and $I_{cmd}, \omega_M, x_a, x_L, v_L$, and F_L represent the current reference, motor-side velocity, deformation, load-side position, load-side velocity, and load force, respectively.

III. CONVENTIONAL METHOD (MULTIPLE ROOTS DESIGN)

A. Friction Free Disturbance Observer

The FFDOB is designed by using the model in Fig.2. The disturbance torque τ_{dism} is input to a second-order transfer function. The second-order transfer function expresses the vibration caused by the dither signal. This is because the disturbance torque includes the vibration of the dither signal.

The pole design of the observer is:

$$(s + \alpha)^n \qquad (1)$$

and designs by Pole Assignment Method(n: order of observer, α: pole design of observer). In this paper, this design is called the multiple roots design.

1) Zero-Order FFDOB: The model of disturbance torque τ_{dism} is defined as a step signal. The disturbance torque is expressed as follows:

$$\frac{d}{dt}\tau_{dism} = 0. \qquad (2)$$

The FFDOB is designed by using Eq.1. The pole design is expressed as follows:

$$(s + \alpha)^4 = s^4 + 4\alpha s^3 + 6\alpha^2 s^2 + 4\alpha^3 s + \alpha^4. \qquad (3)$$

The state equations of the zero-order FFDOB are expressed as follows:

$$\dot{\hat{\mathbf{x}}}_{zd}(t) = \hat{\mathbf{A}}_{zd}\hat{\mathbf{x}}_{zd}(t) + \hat{\mathbf{B}}_{zd}\mathbf{u}_{zd}(t) \qquad (4)$$

$$\hat{\mathbf{y}}_{zd}(t) = \hat{\mathbf{C}}_{zd}\hat{\mathbf{x}}_{zd}(t) \qquad (5)$$

$$\hat{\mathbf{A}}_{zd} = \begin{bmatrix} -g_{zd1} & -\frac{1}{J_{Mn}} & 0 & 0 \\ -g_{zd2} & 0 & 1 & 0 \\ -g_{zd3} & 0 & 0 & 1 \\ -g_{zd4} & 0 & -\omega_0^2 & -2\zeta_1\omega_0 \end{bmatrix},$$

$$\hat{\mathbf{B}}_{zd} = \begin{bmatrix} g_{zd1} & \frac{K_{tn}}{J_{Mn}} \\ g_{zd2} & 0 \\ g_{zd3} & 0 \\ g_{zd4} & 0 \end{bmatrix}, \hat{\mathbf{C}}_{zd} = \begin{bmatrix} 1 & 0 & 0 & 0 \\ 0 & 1 & 0 & 0 \\ 0 & 0 & 1 & 0 \\ 0 & 0 & 0 & 1 \end{bmatrix},$$

$$\hat{\mathbf{x}}_{zd}(t) = \begin{bmatrix} \hat{\omega}_M(t) & \hat{\tau}'_{dism}(t) & \dot{\hat{\tau}}'_{dism}(t) & \ddot{\hat{\tau}}'_{dism}(t) \end{bmatrix}^T,$$

$$\mathbf{u}_{zd}(t) = \begin{bmatrix} \omega_M(t) & I_{cmd}(t) \end{bmatrix}^T.$$

where ω_0 denotes the angular frequency of the dither signal and ζ denotes the attenuation coefficient. Observer gains $g_{zd1} \sim g_{zd4}$ are expressed as follows:

$$
\begin{aligned}
g_{zd1} &= -2\zeta_1\omega_0 + 4\alpha \\
g_{zd2} &= J_{Mn}(-6\alpha^2 + 8\zeta_1\omega_0\alpha - 4\zeta_1^2\omega_0^2 + \omega_0^2) \\
g_{zd3} &= J_{Mn}(-4\alpha^3 + 12\zeta_1\omega_0\alpha^2 + (-16\zeta_1^2\omega_0^2 + 4\omega_0^2)\alpha \\
&\quad + (8\zeta_1^3\omega_0^3 - 4\zeta_1\omega_0^3)) \\
g_{zd4} &= J_{Mn}(-\alpha^4 + 8\zeta_1\omega_0\alpha^3 + (-24\zeta_1^2\omega_0^2 + 6\omega_0^2)\alpha^2 \\
&\quad + (32\zeta_1^3\omega_0^3 - 16\zeta_1\omega_0^3)\alpha - 16\zeta_1^4\omega^4 + 12\zeta_1^2\omega_0^4 - \omega_0^4).
\end{aligned}
$$

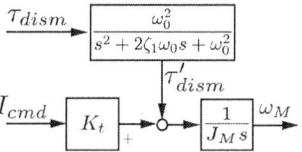

Fig. 2. Design model of FFDOB and Notch-type FFDOB.

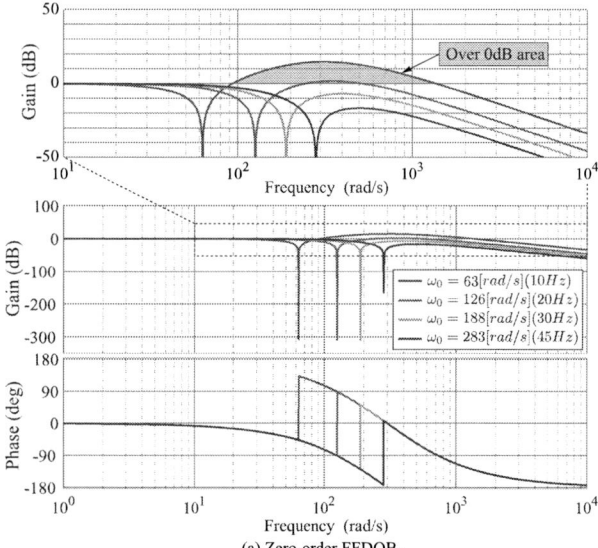

(a) Zero-order FFDOB.

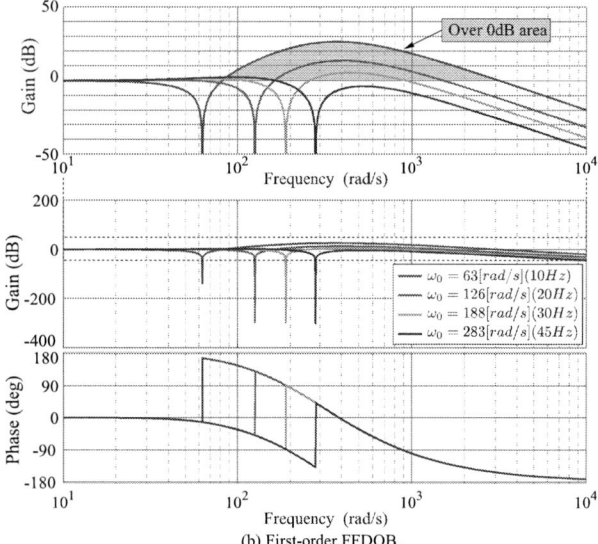

(b) First-order FFDOB.

Fig. 3. Estimation characteristics of FFDOB when changing dither frequency ω_0. $(\hat{\tau}_{dism}/\tau_{dism})$

The estimated disturbance torque $\hat{\tau}_{dism}$ is obtained by substituting the estimated values of the disturbance torque

$\ddot{\hat{\tau}}'_{dism}$, $\dot{\hat{\tau}}'_{dism}$ and $\hat{\tau}'_{dism}$ yielded by the observer into Eq.6,

$$\hat{\tau}_{dism} = \frac{\ddot{\hat{\tau}}'_{dism} + 2\zeta_1\omega_0\dot{\hat{\tau}}'_{dism} + \omega_0^2\hat{\tau}'_{dism}}{\omega_0^2} \quad (6)$$

and $\hat{\tau}_{dism}$ denotes the estimated disturbance torque of τ_{dism}.

The Bode diagram spanning from the disturbance torque τ_{dism} to the estimated disturbance $\hat{\tau}_{dism}$ is shown in Fig.3-(a). The parameters of the Bode diagram are listed in Table.I. The transfer function spanning from the disturbance torque τ_{dism} to estimated disturbance torque $\hat{\tau}_{dism}$ is expressed as follows:

$$\frac{\hat{\tau}_{dism}}{\tau_{dism}} = \frac{\alpha^4\left(2s\zeta_1\omega_0 + s^2 + \omega_0^2\right)}{(\alpha + s)^4\,\omega_0^2}. \quad (7)$$

From the Bode diagram, the frequency ω_0 of the dither signal is not transmitted to the estimated value. When the dither signal frequency of the FFDOB is lowered, the area beyond 0 dB increases, and thus divergence easily occurs. Therefore, high frequency is used for the dither signal of the FFDOB.

2) First-Order FFDOB: The model of disturbance torque τ_{dism} is defined as a ramp signal. The disturbance torque is expressed as follows:

$$\frac{d}{dt}\tau_{dism} = \dot{\tau}_{dism}, \quad (8)$$

$$\frac{d}{dt}\dot{\tau}_{dism} = 0. \quad (9)$$

The FFDOB is designed by using Eq.1. The pole design is expressed as follows:

$$(s+\alpha)^5 = s^5 + 5\alpha s^4 + 10\alpha^2 s^3 + 10\alpha^3 s^2 + 5\alpha^4 s + \alpha^5. \quad (10)$$

The state equations of the first-order FFDOB are expressed as follows:

$$\dot{\hat{\mathbf{x}}}_{fd}(t) = \hat{\mathbf{A}}_{fd}\hat{\mathbf{x}}_{fd}(t) + \hat{\mathbf{B}}_{fd}\mathbf{u}_{fd}(t) \quad (11)$$

$$\hat{\mathbf{y}}_{fd}(t) = \hat{\mathbf{C}}_{fd}\hat{\mathbf{x}}_{fd}(t) \quad (12)$$

$$\hat{\mathbf{A}}_{fd} = \begin{bmatrix} -g_{fd1} & -\frac{1}{J_{Mn}} & 0 & 0 & 0 \\ -g_{fd2} & 0 & 1 & 0 & 0 \\ -g_{fd3} & 0 & 0 & 1 & 0 \\ -g_{fd4} & 0 & -\omega_0^2 & -2\zeta_1\omega_0 & \omega_0^2 \\ -g_{fd5} & 0 & 0 & 0 & 0 \end{bmatrix},$$

$$\hat{\mathbf{B}}_{fd} = \begin{bmatrix} g_{fd1} & \frac{K_{tn}}{J_{Mn}} \\ g_{fd2} & 0 \\ g_{fd3} & 0 \\ g_{fd4} & 0 \\ g_{fd5} & 0 \end{bmatrix}, \hat{\mathbf{C}}_{fd} = \begin{bmatrix} 1 & 0 & 0 & 0 & 0 \\ 0 & 1 & 0 & 0 & 0 \\ 0 & 0 & 1 & 0 & 0 \\ 0 & 0 & 0 & 1 & 0 \\ 0 & 0 & 0 & 0 & 1 \end{bmatrix},$$

$$\hat{\mathbf{x}}_{fd}(t) = \begin{bmatrix} \hat{\omega}_M(t) & \hat{\tau}'_{dism}(t) & \dot{\hat{\tau}}'_{dism}(t) & \ddot{\hat{\tau}}'_{dism}(t) & \dot{\hat{\tau}}_{dism}(t) \end{bmatrix}^T,$$

$$\mathbf{u}_{fd}(t) = \begin{bmatrix} \omega_M(t) & I_{cmd}(t) \end{bmatrix}^T.$$

Observer gains $g_{fd1}\sim g_{fd5}$ are expressed as follows:

$$g_{fd1} = 5\alpha - 2\zeta_1\omega_0$$

$$g_{fd2} = J_{Mn}(-10\alpha^2 + 10\zeta_1\omega_0\alpha - 4\zeta_1^2\omega_0^2 + \omega_0^2)$$

$$g_{fd3} = J_{Mn}(-10\alpha^3 + 20\zeta_1\omega_0\alpha^2 + (-20\zeta_1^2\omega_0^2 + 5\omega_0^2)\alpha + 8\zeta_1^3\omega_0^3 - 4\zeta_1\omega_0^3)$$

$$g_{fd4} = J_{Mn}(-5\alpha^4 + 20\zeta_1\omega_0\alpha^3 + (-40\zeta_1^2\omega_0^2 + 10\omega_0^2)\alpha^2 + (40\zeta_1^3\omega_0^3 - 20\zeta_1\omega_0^3)\alpha - 16\zeta_1^4\omega_0^4 + 12\zeta_1^2\omega_0^4 - \omega_0^4)$$

$$g_{fd5} = -\omega_0^{-2}J_{Mn}\alpha^5.$$

The estimated disturbance torque $\hat{\tau}_{dism}$ is obtained by substituting the solution of the state equation of the first-order FFDOB in Eq.6.

The Bode diagram spanning from the disturbance torque τ_{dism} to the estimated disturbance $\hat{\tau}_{dism}$ is shown in Fig.3-(b). The transfer function spanning from the disturbance τ_{dism} to the estimated disturbance torque $\hat{\tau}_{dism}$ is expressed as follows:

$$\frac{\hat{\tau}_{dism}}{\tau_{dism}} = \frac{\alpha^4\left(2s\zeta_1\omega_0 + s^2 + \omega_0^2\right)(\alpha + 5s)}{\omega_0^2\,(s+\alpha)^5}. \quad (13)$$

When the frequency of the dither signal is lowered in a manner similar to the zero-order FFDOB, the area beyond 0 dB is high, and thus the first-order FFDOB also easily diverges.

Therefore, in the multiple roots design, it is impossible to lower the frequency of the dither signal.

IV. PROPOSED METHOD (NOTCH-TYPE DESIGN)

A. Notch-type Friction Free Disturbance Observer

The notch-type FFDOB is also designed by using the model in Fig.2. The pole design of the observer is:

$$(s^2 + 2\zeta_2\omega_0 s + \omega_0^2)(s+\alpha)^{n-2} \quad (14)$$

and designs by Pole Assignment Method(n: order of observer, α: pole design of observer, ζ_2: attenuation coefficient). In this paper, this design is called the notch-type design.

1) Notch-type Zero-Order FFDOB: The form of the state equation is similar to that of the zero-order FFDOB. Therefore, the state equation of the notch-type zero-order FFDOB is expressed in Eqs.4 and 5. However, observer gains $g_{zd1}\sim g_{zd4}$ are not similar. The notch-type FFDOB is designed by using Eq.14. The pole design is expressed as follows:

$$\begin{aligned}&(s^2 + 2\zeta_2\omega_0 s + \omega_0^2)(s+\alpha)^2 \\ &= s^4 + 2(\zeta_2\omega_0 + \alpha)s^3 + (4\alpha\zeta_2\omega_0 + \alpha^2 + \omega_0^2)s^2 \\ &\quad + 2\alpha\omega_0(\alpha\zeta_2 + \omega_0)s + \alpha^2\omega_0^2. \end{aligned} \quad (15)$$

The gains of the notch-type zero-order FFDOB are expressed as follows:

$$\begin{aligned} g_{zd1} &= 2\omega_0(\zeta_2 - \zeta_1) + 2\alpha \\ g_{zd2} &= -J_{Mn}\left\{4\zeta_1\omega_0^2(\zeta_1 - \zeta_2) + 4\alpha\omega_0(\zeta_2 - \zeta_1) + \alpha^2\right\} \\ g_{zd3} &= 2J_{Mn}\omega_0\left(-2\zeta_1\omega_0 + \alpha + \omega_0\right)\left(-2\zeta_1\omega_0 + \alpha - \omega_0\right) \\ &\quad (\zeta_1 - \zeta_2) \\ g_{zd4} &= -4\omega_0^2 J_{Mn}\left(-2\zeta_1^2\omega_0 + \alpha\zeta_1 + \omega_0\right) \\ &\quad (-2\zeta_1\omega_0 + \alpha)\left(\zeta_1 - \zeta_2\right). \end{aligned}$$

The estimated disturbance torque is obtained by substituting the solution of the state equation of the notch-type zero-order FFDOB in Eq.6. The transfer function spanning from the disturbance torque τ_{dism} to the estimated disturbance torque $\hat{\tau}_{dism}$ is expressed as follows:

$$\frac{\hat{\tau}_{dism}}{\tau_{dism}} = \frac{s^2 + 2\zeta_1\omega_0 s + \omega_0^2}{s^2 + 2\zeta_2\omega_0 s + \omega_0^2}\frac{\alpha^2}{(s+\alpha)^2}. \quad (16)$$

The nominal values are $K_{tn} = K_t$ and $J_{Mn} = J_M$. From Eq.16, the notch-type zeroth-order FFDOB is proved to include

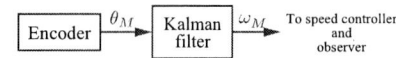

Fig. 5. Block diagram of Kalman filter for velocity estimation.

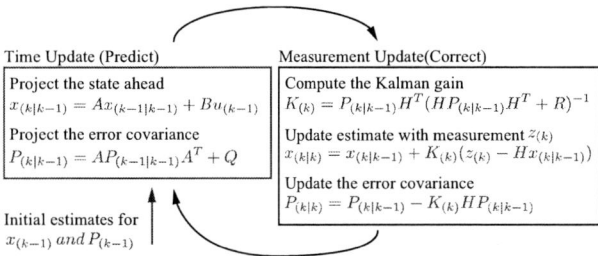

Fig. 6. Calculation step of Kalman filter[7].

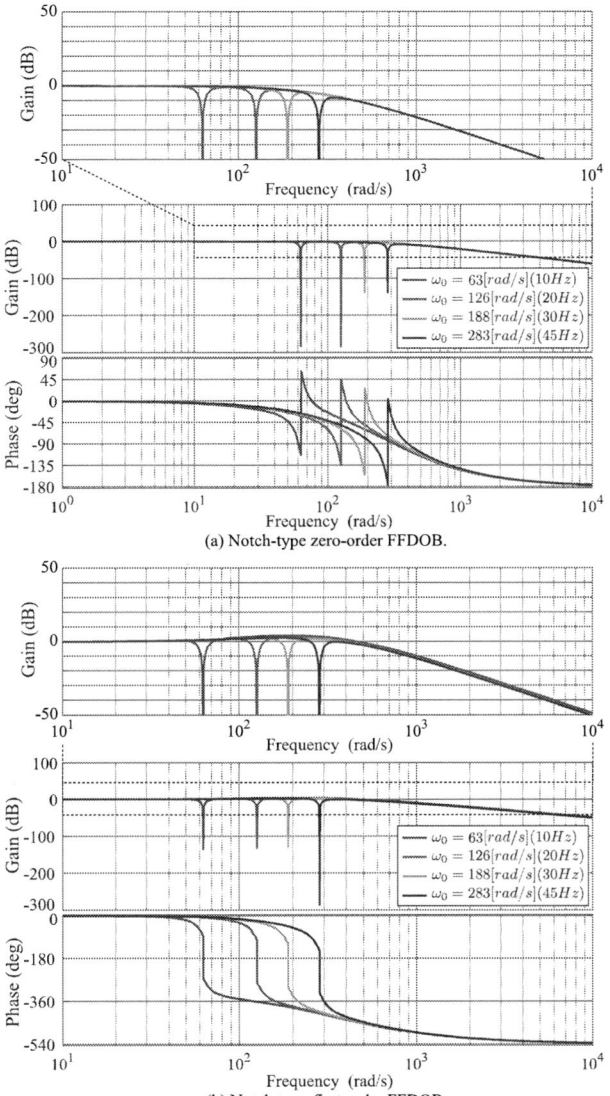

(a) Notch-type zero-order FFDOB.

(b) Notch-type first-order FFDOB.

Fig. 4. Estimation characteristics of notch-type FFDOB when changing dither frequency ω_0.$(\hat{\tau}_{dism}/\tau_{dism})$

the notch characteristics. The Bode diagram spanning from the disturbance torque τ_{dism} to the estimated disturbance torque $\hat{\tau}_{dism}$ is shown in Fig.4-(a). It is observed that a region beyond 0 dB does not appear even when the frequency of the dither signal decreases.

2) Notch-type First-Order FFDOB: The form of the state equation is similar to that of the first-order FFDOB. Therefore, the state equation of the notch-type first-order FFDOB is expressed in Eqs.11 and 12. However, observer gains $g_{fd1}{\sim}g_{fd5}$ are not identical. The notch-type FFDOB is designed by using

Eq.14. The pole design is expressed as follows:

$$
\begin{aligned}
&(s^2 + 2\zeta_2\omega_0 s + \omega_0^2)(s + \alpha)^3 \\
&= s^5 + (2\,\zeta_2\omega_0 + 3\,\alpha)\,s^4 \\
&\quad + (6\,\alpha\,\zeta_2\omega_0 + 3\,\alpha^2 + \omega_0{}^2)\,s^3 \\
&\quad + (6\,\alpha^2\zeta_2\omega_0 + \alpha^3 + 3\,\alpha\,\omega_0^2)\,s^2 \\
&\quad + (2\,\alpha^3\zeta_2\omega_0 + 3\,\alpha^2\omega_0^2)\,s + \alpha^3\omega_0{}^2
\end{aligned}
\tag{17}
$$

The gains of the notch-type first-order FFDOB are expressed as follows:

$$
\begin{aligned}
g_{fd1} &= 2\omega_0(\zeta_2 - \zeta_1) + 3\,\alpha \\
g_{fd2} &= -2J_{Mn}\omega_0(\zeta_2 - \zeta_1)(-2\zeta_1\omega_0 + 3\alpha) - 3J_{Mn}\,\alpha^2 \\
g_{fd3} &= -2J_{Mn}\omega_0(\zeta_2 - \zeta_1)\{\omega_0^2(4\zeta_1^2 - 1) \\
&\quad + 3\alpha(-2\zeta_1\omega_0 + \alpha)\} - J_{Mn}\alpha^3 \\
g_{fd4} &= -2J_{Mn}\omega_0(\zeta_2 - \zeta_1)\{4\zeta_1\omega_0^3(1 - 2\zeta_1^2) \\
&\quad + 3\alpha\omega_0^2(4\zeta_1 - 1) + \alpha^2(\alpha - 6\zeta_1\omega_0)\} \\
g_{fd5} &= -\alpha^3 J_{Mn}.
\end{aligned}
$$

The transfer function spanning from the disturbance torque τ_{dism} to the estimated disturbance torque $\hat{\tau}_{dism}$ is expressed as follows:

$$
\frac{\hat{\tau}_{dism}}{\tau_{dism}} = \frac{s^2 + 2\zeta_1\omega_0 s + \omega_0^2}{s^2 + 2\zeta_2\omega_0 s + \omega_0^2} \frac{\{(2\alpha\zeta_2 + 3\omega_0)s + \alpha\omega_0\}\,\alpha^2}{(s + \alpha)^3\omega_0}. \tag{18}
$$

The nominal values are $K_{tn} = K_t$ and $J_{Mn} = J_M$. From Eq.18, the notch-type first-order FFDOB is also proven to include the notch characteristics. The Bode diagram spanning from the disturbance torque τ_{dism} to the estimated disturbance torque $\hat{\tau}_{dism}$ is shown in Fig.4-(b). When the frequency of the dither signal is decreases, a region beyond 0 dB emerges although, its area is low.

The system does not diverge even when the frequency of the dither signal decreases. Therefore, in the notch-type design, it is possible to lower the frequency of the dither signal.

V. KALMAN FILTER FOR VELOCITY ESTIMATION

Velocity information of motor side is important for the observer and speed control. The velocity information is obtained by pseudo-differentiating position information of an encoder. However, the position information includes observation noise.

The 2018 International Power Electronics Conference

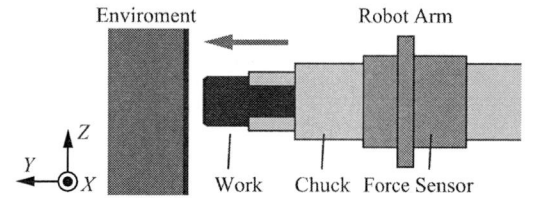

Fig. 7. Pushing motion.

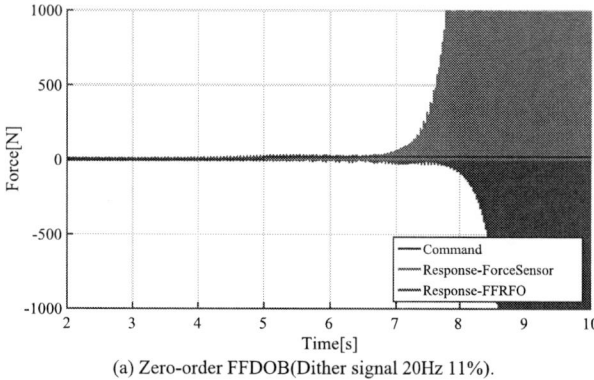

(a) Zero-order FFDOB(Dither signal 20Hz 11%).

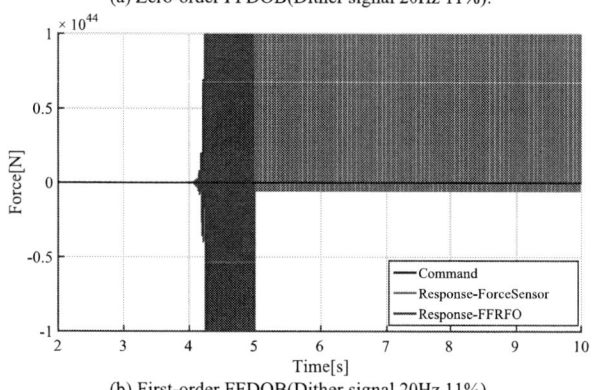

(b) First-order FFDOB(Dither signal 20Hz 11%).

Fig. 8. Numerical simulation results of force sensorless force control using FFDOB.

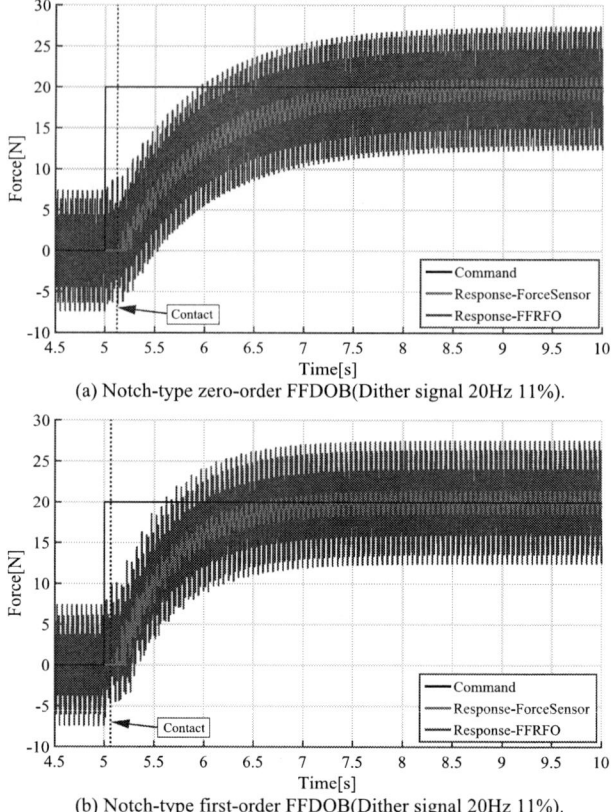

(a) Notch-type zero-order FFDOB(Dither signal 20Hz 11%).

(b) Notch-type first-order FFDOB(Dither signal 20Hz 11%).

Fig. 9. Numerical simulation results of force sensorless force control using notch-type FFDOB.

on data up to time k, $\mathbf{Q}$ denotes covariance matrix of process noise, $\mathbf{R}$ denotes the covariance matrix of measurement noise, and $\mathbf{Q}$ and $\mathbf{R}$ are expressed as follows:

$$\mathbf{Q} = E[ww^T] \tag{21}$$
$$\mathbf{R} = E[vv^T]. \tag{22}$$

where $\mathbf{R}$ is calculated from experimental results, and $\mathbf{Q}$ is obtained by an iterative calculation through off-line processing. The effect of the Kalman filter is confirmed by the experiment.

VI. SIMULATIONS AND EXPERIMENTS

A. Simulation and Experimental Conditions

This paper confirms the effectiveness of the proposed method by pushing a robot against an environment. A six-axis industrial robot is used for the experiments. The XYZ(1-3) axis of the robot use a linear motion mechanism with a ball screw. The WRC(4-6) axis of the robot is used as a rotation mechanism by using a harmonic gear. In the simulation and experiment, the robot arm is moved in the **Y** axis direction and pushes against the environment. Table.I lists the parameters of the control system. The frequency and amplitude of the dither signal are determined through trial and error.

Therefore, the velocity information includes observation noise of the encoder and calculation noise. The noise affects the estimation accuracy of the reaction force. This paper utilizes a Kalman filter for velocity estimation as a means to decrease the noise(Fig.5). The Kalman filter is expressed as follows:

$$x_{(k+1)} = \mathbf{A}x_{(k)} + \mathbf{B}u_{(k)} + w_{(k)} \tag{19}$$
$$z_{(k)} = \mathbf{H}x_{(k)} + v_{(k)} \tag{20}$$

$$\mathbf{A} = \begin{bmatrix} 1 & T_s \\ 0 & 1 \end{bmatrix}, \mathbf{B} = \mathbf{0}, \mathbf{H} = \begin{bmatrix} 1 & 0 \end{bmatrix}$$

where $x_{(k)}$ denotes the state variable($[\theta_M \ \omega_M]^T$), Ts denotes the sampling time, $z_{(k)}$ denotes the measured data corrupted by noise, $w_{(k)}$ and $v_{(k)}$ denote process noise and measurement noise, respectively, and are calculated by the flow in Fig.6[7]. Additionally, $(k|k-1)$ denotes a value that is calculated based on data up to time k - 1, $(k|k)$ denotes a value calculated based

1677

The 2018 International Power Electronics Conference

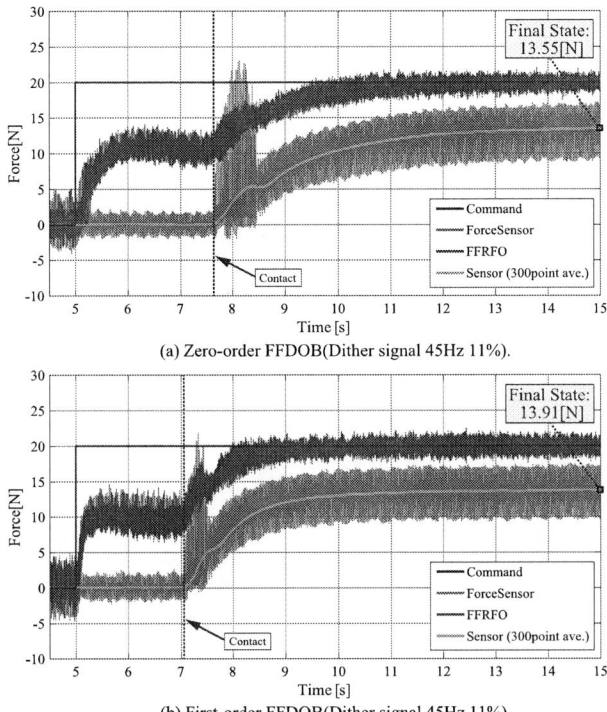

(a) Zero-order FFDOB(Dither signal 45Hz 11%).

(b) First-order FFDOB(Dither signal 45Hz 11%).

Fig. 10. Experimental results of force sensorless force control using FFDOB.

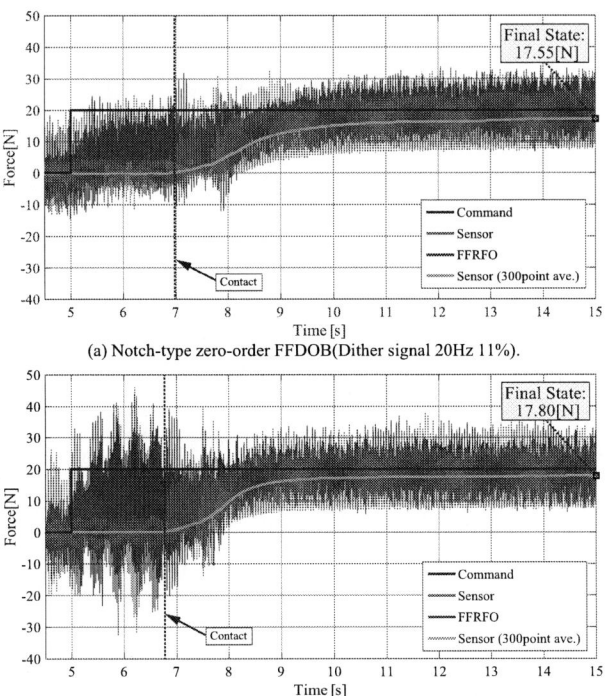

(a) Notch-type zero-order FFDOB(Dither signal 20Hz 11%).

(b) Notch-type first-order FFDOB(Dither signal 20Hz 11%).

Fig. 11. Experimental results of force sensorless force control using Notch-Type FFDOB.

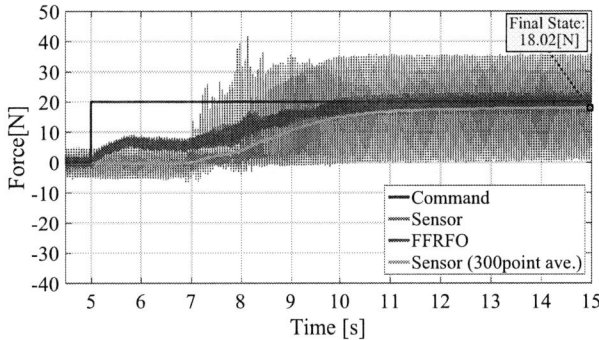

Fig. 12. Experimental results of force sensorless force control using Notch-Type Zero-order FFDOB and Kalman filter.(Dither signal 20Hz 11%)

TABLE I. EXPERIMENTAL PARAMETERS.

Proportional gain of speed ??controller	K_{pv}	60 rad/s
Proportional gain of force controller	K_{pf}	0.006 rad/s
Poles of FFRFO, FFDOB and Notch-type FFDOB	α	300 rad/s
Attenuation coefficient 1	ζ_1	0
Attenuation coefficient 2	ζ_2	0.1
Dither angular frequency	ω_0	$40\pi, 90\pi$ rad/s
Dither amplitude (Rated motor torque 0.64)	A	11%(0.0704[Nm])

Fig.7 shows the condition of the experiment. A force sensor is attached to the tip of the robot arm. However, the force sensor is used for evaluation and not for control. The environment and robot arm commence from a non-contact state.

B. Simulation Results

This section simulates both a conventional and the proposed method. The frequency of the dither signal is set as low to validate the difference between the proposed and conventional method. Figs.8-(a) and 8-(b) show the FFDOB simulation results. In the FFDOB results, the divergence occurs irrespective of whether it is zero-order or first-order. In the multiple roots design, it is impossible to lower the frequency of the dither signal.

Figs.9-(a) and 9-(b) show the notch-type FFDOB simulation results. The notch-type FFDOB results do not diverge. With respect to the results of rise time, the zero-order is 0.89 [s] and first-order is 0.56 [s]. Thus, the performance of the notch-type first-order FFDOB exceeds that of the notch-type zero-order FFDOB. Therefore, in the notch-type design, it is possible to lower the frequency of the dither signal.

C. Experimental Results

The experimental results of the force sensorless force control by using the FFDOB are shown in Figs.10-(a) and 10-(b). Conventional methods diverge with low frequency dither signals. Therefore, a conventional method experiments at a high frequency does not cause divergence. Figs. 10-(a) and 10-(b) show the experimental results of the zero-order and first-order FFDOB, respectively. With respect to the force command

1678

value of 20 [N], the steady-state error of the zero-order FFDOB is 6.45 [N] and that of the first-order FFDOB is 6.09 [N].

The experimental results of the force sensorless force control using the notch-type FFDOB are shown in Figs.11-(a) and 11-(b). Figs.11-(a) and 11-(b) show the experimental result of the notch-type zero-order FFDOB and notch-type first-order FFDOB, respectively. A steady-state error exists between the estimated and true value due to the effect of nonlinear friction. With respect to the force command value of 20 [N], the steady-state error of the notch-type zero-order FFDOB is 2.72 [N] and the notch-type first-order FFDOB is 2.20 [N]. With respect to the results of rise time, the notch-type zero-order FFDOB is 1.88 [s] and the notch-type first-order FFDOB is 1.56 [s]. Therefore, the notch-type first-order FFDOB exhibits a high performance in both in the steady-state error and rise time.

Thus, the notch-type FFDOB permits accurate controls even when the dither signal exhibits a low frequency. The reaction force estimation accuracy improved by 28%. By using the notch-type FFDOB, it is possible to lower the frequency of the dither signal. The proposed method is validated through the improvement in the stability.

The experimental results of the force sensorless force control by using a Kalman filter for velocity estimation are shown in Fig.12. The force sensorless force control by using notch-type first-order FFDOB and Kalman filter diverge. Therefore, it is impossible to use the Kalman filter in the notch-type first-order FFDOB. Nevertheless, it is possible to use the Kalman filter in the notch-type zero-order FFDOB. The vibration of the force estimation is improving by using Kalman filter. However, the force sensor value exhibits a high vibration. The high vibration is not optimal. The use of the Kalman filter allows improvements in the vibration and accuracy of the force estimation.

VII. CONCLUSION

This paper proposes a force sensorless force control system by using a notch-type FFDOB. It is possible to lower the frequency of the dither signal by using the proposed method. The effectiveness of the proposed method is confirmed in the experiment using an industrial robot. A force sensor less force control is achieved without divergence even when a low frequency is used for the dither signal. By using the notch-type FFDOB, it is possible to lower the frequency of the dither signal, which validate the improvement in the stability.

A future task will include maintaining of the force estimation accuracy and reduction in the vibration of the force sensor.

REFERENCES

[1] M.Iwasaki, Y.Maeda, M.Kawafuku, H.Hirai: "Improvement of Precise Positioning Performance by Modeling and Compensation for Nonlinear Friction.", IEEJ Transactions on Industry Applications, 126(6), 732-740, 2006

[2] S. Anayama, K. Ohishi, Y. Yokokura, T. Miyazaki and A. Tsukamoto: "Sensor-less Force Control Using Resonance Ratio Control and Friction Free Reaction Force Observer", IEEE SAMCON2015, TT5-3-3, 2015.

[3] H. Nakamura, K.Ohishi, Y. Yokokura, T. Miyazaki and A. Tsukamoto: "Sensorless Force Control System Using Friction-Free First Order Disturbance Observer", IEEE SAMCON2016, TT8-1-3, 2017.

[4] N. Kamiya, K. Ohishi, Y. Yokokura and T. Miyazaki: "Force Sensorless Force Control Using Notch-Type First-Order Friction Free Disturbance Observer", IEEJ Industry Applications Society Conference, 2-20, 136-138, 2017

[5] S. Komada, T. Murakami and K. Ohnishi: "Force feedback control of multi-degrees-of-freedom robot based on acceleration.", IEEJ Journal of Industry Applications, 109(5), 325-332, 1989

[6] R. Furusawa, T. Asai, K. Ohishi, K. Majima, K. Kageyama, M. Takatsu and S. Urushihara: "Development of Sensorless Force Control System Based on Nonlinear Friction Phenomenon for Electric Injection Molding Machine.", IEEJ Journal of Industry Applications, 131(3), 311-318, 2011

[7] T.Phuong, K.Ohishi, C.Mitsantisuk, Y.Yokokura, K.Ohnishi, R.Oboe, A.Sabanovic: "Disturbance Observer and Kalman Filter Based Motion Control Realization.", IEEJ Journal of Industry Applications, 7(1), 1-14, 2018

Kinematics and Tracking Control of a Four Axis Antenna for Satcom on the Move

Oguz Kaan Hancioglu[1,2*], Mustafa Celik[1,3], Ugur Tumerdem[4]

1 Control Systems, PROFEN Communication Technologies & Services, Inc., Istanbul, Turkey
2 Mechatronics Engineering, Istanbul Technical University, Istanbul, Turkey
3 Physics Engineering, Ankara University, Ankara, Turkey
4 Mechanical Engineering, Marmara University, Istanbul, Turkey
*E-mail: oguz.hancioglu@profen.com

Abstract—**In this paper, we propose a 4 axis antenna which is capable of tracking satellites, through the precise real-time control of azimuth, cross-level, elevation and polarization angles. Due to the well-known gimbal lock problem in 2-3 axis antenna systems, our design has a redundant degree of freedom, and this redundancy is utilised to solve the issue of inverse kinematics around singularities. In the paper, we derive the kinematics of the 4 DOF redundant orientation mechanism and propose a method to avoid the gimbal lock issue. Also, utilising satellite position data (from TLE Database) and GPS localisation, we derive the necessary kinematics equations to position the antenna towards any visible satellite from any position on earth. Furthermore, we propose a PI-based cascade controller for tracking the satellites. To verify our analytical results we present experimental results on a 4 axis antenna test platform that we have built and compare our results with the feedback from a satellite in the form of guidance RF signals. The proposed system can find and track satellites with a 1-degree error without satellite signal feedback.**

Keywords—**Kinematics Equation, Redundant Robot, Servo Control, SOTM Antenna**

I. INTRODUCTION

The Internet has a become a vital necessity for mankind, and the need to access internet on mobile platforms has resulted in the development of automatically controlled SOTM (Satellite Communication on the Move) Antennas. The purpose of these antenna systems is to point to a satellite among the visible constellation of satellites, lock on to the incoming satellite signals and sustain the communication link when there are disturbances acting on the mobile platforms (such as a ship moving in rough seas). For this purpose the most commonly used SOTM platform types are 2 axis and 3 axis platforms. 2 axis platforms of are often the choice when designing an mobile antenna platform as they can perform rotations in azimuth and elevation axes which provide a horizon to zenith hemispheric satellite coverage [1][2]. However there is a major caveat with 2 axis platforms as they can not operate when the elevation angle is about 90 degrees from the horizon due to the gimbal lock problem [3]. This is the problem known as a workspace singularity in the robotics literature [4]. As

a solution to the gimbal lock/singularity problem of the two axes platforms, a third axis is often added to the mobile robot antenna platforms as a redundant axis to avoid singularities [5]. However there is still a problem with these platforms if the antennas used in the SOTM system make use the linear polarization of the signal such as Ku-Band, Ka-Band, C-Band and X-Band antennas. In these cases the polarization, as well as the elevation and azimuth angles, have to be controlled for successful satellite communications [6]. This corresponds to roll-pitch-yaw orientation control in robotics. In this paper, to achieve gimbal-lock/singularity and linear polarization, we propose a 4 axis mobile antenna platform that can be utilized in satellite tracking with polarization. The proposed antenna has 4 active axes (degrees of freedom) which are the azimuth, cross-level, elevation and polarization axes. Cross-level or x-level axis, which is tilted at 30 degrees with the horizontal, is utilized as a redundant axis to solve the problem of gimbal lock. In order to receive communications from the satellite, the antenna has to have the correct orientation in 3 dimensions, which include roll, pitch and yaw angles. In this paper, we provide for the first time a kinematics solution for such an antenna system that can track a satellite signal with polarization near singular configurations.

The organization of the paper is as follows: in section II the kinematic description of the satellite coordinate frame relative to the antenna frame is determined by the use of satellite coordinates obtained from the TLE database and the antenna coordinates received from GPS. This description determines the desired orientation for the antenna. In section III, the kinematic equations, forward and inverse, are obtained for the redundant 4 DOF antenna platform. The gimbal lock-singularity problem is described and a solution to that problem using kinematics is proposed. Section IV describes the proposed control system and the architecture for a prototype antenna that we have built. Section V presents the experiment results that we have performed on the developed prototype test bed and satellite guidance signal intensities have been measured show that the proposed technique is successful in finding and tracking with 1 degree pointing errors. This can be further improved by utilizing satellite signal feedback in the loop.

This work is funded by PROFEN Communication Technologies & Services, Inc. and TUBITAK (The Scientific and Technological Research Council of Turkey) with a project code 1130278.

II. SATELLITE AND ANTENNA FRAME DESCRIPTION

In this section, we find the description of the satellite frame in relative to the antenna frame so that we can compute the target orientation for the antenna. When the antenna orientation is correctly computed it should be pointing to the selected satellite. The algorithm we present here can enable the antenna to point to any satellite in its visible space, given that the satellite coordinates with respect to the antenna are known. This algorithm uses coordinate frame descriptions as well as matrix transformations to find the desired orientation as a rotation matrix. The Earth-Centered, Earth-Fixed (ECEF) coordinate system is used as an inertial frame.

ECEF XY-Plane is the equatorial plane with X-Axis pointing in longitude 0 direction. The algorithm uses three stages to obtain the desired rotation matrix:

Step 1: The position vector and rotation matrix of the given satellite is calculated with respect to ECEF coordinate system. In this step, the positon vector is calculated using satellite radius, latitude and longitude with respect to ECEF.

$$\vec{r}_{Sat}^{ECEF} = \begin{bmatrix} \vec{r}_{Sat}\cos\left(Lat_{Sat}\right)\cos\left(Long_{Sat}\right) \\ \vec{r}_{Sat}\cos\left(Lat_{Sat}\right)\sin\left(Long_{Sat}\right) \\ \vec{r}_{Sat}\sin\left(Lat_{Sat}\right) \end{bmatrix} \quad (1)$$

The rotation of the satellite is calculated using satellite longitude, latitude and polarization from TLE database.

$$R_{Sat}^{ECEF} = R_z(Long_{Sat})R_y(Lat_{Sat})R_x(Pol_{Sat}), \quad (2)$$

Where $\vec{r}_{Sat}^{ECEF}$ represent the position of given satellite with respect to ECEF. $\vec{r}_{Sat}$ represent the radius of given satellite. $Lat_{Sat}, Long_{Sat}, Pol_{Sat}$ angles represent the latitude, longitude and polarization of the satellite respectively. R_{Sat}^{ECEF} represent the rotation of the given satelltie. R_z, R_y, R_x are respresent the rotation around z axis, rotation around y axis, rotation around x axis respectively.

Step 2: The position vector and rotation matrix of the antenna is calculated with respect to ECEF coordinate system assuming earth shape is spherical. In this step, the positon vector is calculated using antenna radius, latitude and longitude with respect to ECEF as is given in (3). The rotation of the antenna is calculated using antenna longitude, latitude and polarization as:

$$\vec{r}_{Ant}^{ECEF} = \begin{bmatrix} \vec{r}_{Ant}\cos\left(Lat_{Ant}\right)\cos\left(Long_{Ant}\right) \\ \vec{r}_{Ant}\cos\left(Lat_{Ant}\right)\sin\left(Long_{Ant}\right) \\ \vec{r}_{Ant}\sin\left(Lat_{Ant}\right) \end{bmatrix} \quad (3)$$

$$R_{Ant}^{ECEF} = \begin{bmatrix} \vec{e}_{xAnt}^{ECEF} & \vec{e}_{yAnt}^{ECEF} & \vec{e}_{zAnt}^{ECEF} \end{bmatrix} \quad (4)$$

Where $\vec{r}_{Ant}^{ECEF}$ represent the position of given antenna with respect to ECEF. $\vec{r}_{Ant}$ represent the radius of given antenna which in here radius of antenna is equal to radius of earth. $Lat_{Ant}, Long_{Ant}$ angles represent the latitude and longitude of the antenna respectively. R_{Ant}^{ECEF} represent the rotation of the antenna. $\vec{e}_{xAnt}^{ECEF}, \vec{e}_{yAnt}^{ECEF}, \vec{e}_{zAnt}^{ECEF}$ describe the antenna frame vectors w.r.t. ECEF x,y,z coordinates respectively.

Step 3: The position vector and rotation matrix between satellite and antenna is calculated using the satellite and antenna position vectors and orientation matrices obtained in step 1 and step 2:

$$\vec{r}_{Sat}^{Ant} = R_{Ant}^{ECEF'}\left[\vec{r}_{Sat}^{ECEF} - \vec{r}_{Ant}^{ECEF}\right] \quad (5)$$

$$R_{Sat}^{Ant} = \begin{bmatrix} \vec{e}_{xSat}^{Ant} & \vec{e}_{ySat}^{Ant} & \vec{e}_{zSat}^{Ant} \end{bmatrix} \quad (6)$$

Where $\vec{r}_{Sat}^{Ant}$ represent the position vector between the antenna and satellite. R_{Sat}^{Ant} represent the rotation matrix between the antenna and satellite. $\vec{e}_{xSat}^{Ant}, \vec{e}_{ySat}^{Ant}, \vec{e}_{zSat}^{Ant}$ represents the column vectors of the rotation matrix from the satellite to the antenna.

The rotation matrix found in step 3 describes the orientation of the satellite with respect to antenna [7]. The Pointing Orientation Algorithm we use in this paper as a result, is a function block which takes satellite radius, satellite longitude, satellite latitude, satellite polarization, earth radius, antenna longitude, antenna latitude and creates the pointing rotation matrix as seen in Fig. 5.

After the satellite orientation is calculated, desired orientation matrix for the antenna is calculated as:

$$R_{desired} = [R_{XYZ}]^T R_{Sat}^{Ant} \quad (7)$$

Here R_{XYZ} is the orientation matrix of the antenna base with respect to the world frame. The desired matrix will be the input of the inverse kinematics algorithm in the next section.

III. KINEMATIC MODELLING OF SOTM ANTENNA

The antenna has 4 rotational joints which are azimuth, cross level or x-level, elevation and polarization axis which are shown in Fig. 1. In order to obtain the kinematic model, Denavit-Hartenberg (DH) frame assignment convention was used.

The kinematic structure of antenna is redundant because it has 4 active joints while the rotational workspace of the antenna is 3 dimensional (roll, pitch, yaw). It is possible to obtain a desired roll, pitch, yaw orientation with azimuth, elevation and polarization axes of the antenna. However a singularity will occur when the azimuth and polarization axes are aligned. To accomodate for singularities in the inverse kinematics, cross-level is utilised as a redundant axis. During normal operation (when the antenna is not near a singularity) there is no need for the cross level axis. However, when the elevation approaches 90 degrees, the singularity occurs. The cross-level axis is used to avoid the singularity.

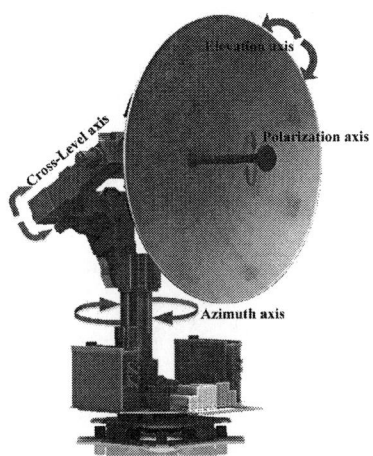

Fig. 1. The isometric view of the antenna CAD model and rotational axes

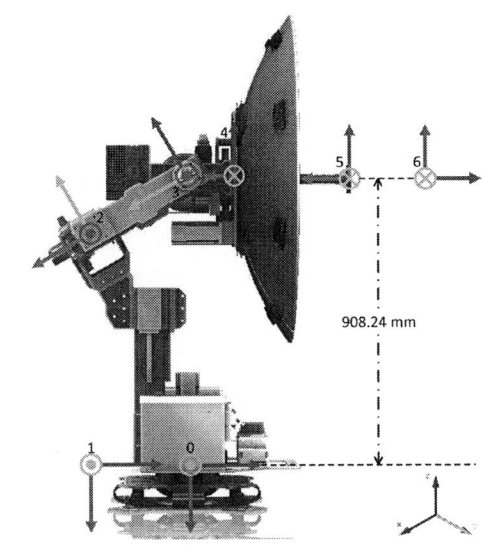

Fig. 2. The prototype antenna CAD model and axes' frame

A. Forward Kinematic Model of SOTM Antenna

The CAD drawing of the prototype antenna system can be seen in Fig. 2 and the coordinate frame assignment can be seen on Fig. 3. The $\{x_0, y_0, z_0\}$ frame is placed according to the North-East-Down (NED) coordinate frame convention and the z axis of this frame represents azimuth axis. β is an offset angle representing the mechanical tilt between cross-level and azimuth axes placed there in order to increase the stiffness of the system [3]. While this offset improves the mechanical properties, the kinematic modelling of the antenna becomes harder. In most robotic systems, the axes of the robots are usually selected with 0 or 90 degrees between sequential axes to simplify inverse kinematics solutions. Due to the use of this offset angle, a virtual $\{x_1, y_1, z_1\}$ frame is assigned to correct the relationship between tip and base of the antenna. $\{x_2, y_2, z_2\}$, $\{x_3, y_3, z_3\}$, $\{x_4, y_4, z_4\}$ and $\{x_5, y_5, z_5\}$ frames describe the cross level, elevation, polarization and end effector frames and the z axes of the cross level, elevation and polarization frames are the actuator rotation axes . The $\{x_6, y_6, z_6\}$ frame is a virtual frame added to the antenna system to accommodate for GPS errors. It is always a fixed value for each GPS sensor. Therefore, the orientation matrix which is created through collected data from GPS sensor can be used in the antenna system directly.

The DH representation of the antenna kinematics model with respect to the assigned coordinates is shown in Table I. The DH parameters are created omitting GPS frame.

Where, $\beta = 30$, $l_1 = 355.42mm$, $l_2 = 703mm$ and $l_3 = 410.4mm$ for our system and are obtained from the CAD drawings. The homogeneous transformation matrices between the frames can then be found as follows:

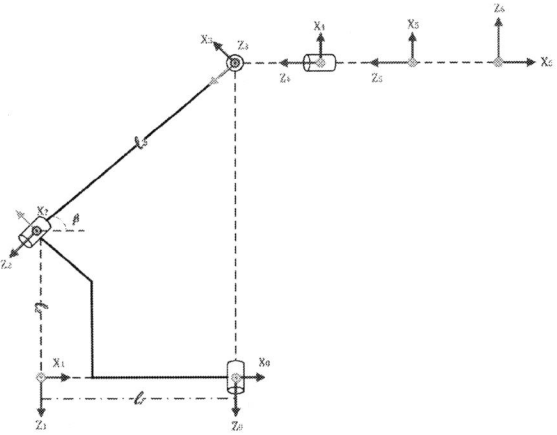

Fig. 3. The kinematic model of antenna and frame placement

$$A_1 = \begin{bmatrix} \cos(\theta_1) & -\sin(\theta_1) & 0 & l_1\cos(\theta_1) \\ \sin(\theta_1) & \cos(\theta_1) & 0 & l_1\sin(\theta_1) \\ 0 & 0 & 0 & 0 \\ 0 & 0 & 0 & 1 \end{bmatrix}$$

$$A_2 = \begin{bmatrix} 0 & -\cos(90-\beta) & -\sin(90-\beta) & 0 \\ 1 & 0 & 0 & 0 \\ 0 & \sin(90-\beta) & \cos(90-\beta) & -l_2 \\ 0 & 0 & 0 & 1 \end{bmatrix}$$

$$A_3 = \begin{bmatrix} \cos(\theta_3-90) & 0 & -\sin(\theta_3-90) & 0 \\ \sin(\theta_3-90) & 0 & \cos(\theta_3-90) & 0 \\ 0 & -1 & 0 & -l_3 \\ 0 & 0 & 0 & 1 \end{bmatrix}$$

TABLE I. THE DH TABLE OF THE ANTENNA

i	α_i	a_i	d_i	θ_i
1	0	l_1	0	$\theta_1{}^*$
2	$90 - \beta$	0	$-l_2$	90
3	-90	0	$-l_3$	$\theta_3{}^* - 90$
4	90	0	0	$\theta_4{}^* + \beta$
5	0	0	0	$\theta_5{}^*$

$$A_4 = \begin{bmatrix} \cos(\theta_4 + \beta) & 0 & \sin(\theta_4 + \beta) & 0 \\ \sin(\theta_4 + \beta) & 0 & -\cos(\theta_4 + \beta) & 0 \\ 0 & 1 & 0 & 0 \\ 0 & 0 & 0 & 1 \end{bmatrix}$$

$$A_5 = \begin{bmatrix} \cos(\theta_5) & -\sin(\theta_5) & 0 & 0 \\ \sin(\theta_5) & \cos(\theta_5) & 0 & 0 \\ 0 & 0 & 0 & 0 \\ 0 & 0 & 0 & 1 \end{bmatrix}$$

$$A_6 = \begin{bmatrix} 0 & 0 & -1 & 0 \\ 0 & 1 & 0 & 0 \\ 1 & 0 & 0 & 0 \\ 0 & 0 & 0 & 1 \end{bmatrix}$$

The forward kinematics model can be calculated using as:

$$T_6^0 = A_1 A_2 A_3 A_4 A_5 A_6 \tag{8}$$

B. Inverse Kinematic Model of SOTM Antenna

In the inverse kinematics, the necessary joint angles are computed that will result in a desired orientation for the antenna. Only the rotation parts of the transformation matrices are used as the tool position is not important in this case. The desired orientation matrix for the antenna was obtained in section II. Then we should have:

$$R_{desired}^0 = R_1^0 R_2^1 R_3^2 R_4^3 R_5^4 R_6^5 \tag{9}$$

During normal operation, the antenna should not be singular. The antenna has a singularity when the elevation axes approaches to 90 degrees. So during normal operation we assume that the antenna is non-singular, in non-singular cases we do not need the redundant cross-level axis. As a result, during normal operation we assume that the cross level axis is fixed: $\theta_3 = 0$. Therefore, R_3^2 matrix is constant. The required angles for the active joints, which are in this case elevation, azimuth and polarization, can be found from equation 9 as a function of the desired antenna orientation.

The inverse kinematic solution for the elevation angle (θ_4) is then:

$$\theta_4 = atan2(-0.866, 0.5) + atan2(\pm\sqrt{1 - r_{31}^2}, r_{31}) \tag{10}$$

The inverse kinematic solution for the azimuth angle (θ_1) is:

$$\cos(\theta_1) = \frac{r_{11}}{0.5\sin(\theta_4 + \beta) + 0.866\sin(\theta_4 + \beta)} = a \tag{11}$$

$$\sin(\theta_1) = \frac{r_{21}}{0.5\sin(\theta_4 + \beta) + 0.866\sin(\theta_4 + \beta)} = b \tag{12}$$

$$\theta_1 = atan2(b, a) \tag{13}$$

And the inverse kinematic solution for the polarization angle (θ_5) is found as:

$$\cos(\theta_5) = \frac{r_{33}}{0.5\sin(\theta_4 + \beta) + 0.866\sin(\theta_4 + \beta)} = a \tag{14}$$

$$\sin(\theta_5) = \frac{r_{32}}{0.5\sin(\theta_4 + \beta) + 0.866\sin(\theta_4 + \beta)} = b \tag{15}$$

$$\theta_5 = atan2(b, a) \tag{16}$$

There are two solution sets for the inverse kinematic solutions (a,b) due to the positive and negative solutions of θ_4. The correct solution set in our case is selected such that θ_4 must be between 0 and 90 degrees.

The singularity will occur when the azimuth and polarization axes are aligned. This depends on the antenna location, satellite location and base orientation of the antenna. In our design, when the antenna is in near singular configurations, while the cross-level axis is being used the azimuth axis is not used. Thus, cross-level, elevation and polarization axes are used to obtain the desired antenna orientation. In the singular region, the azimuth axis is fixed and the azimuth angle (θ_1) is known. Therefore, R_1^0 matrix is constant. Then, we have to solve the following equation:

$$R_2^{1^T} R_1^{0^T} R_{desired}^0 = R_3^2 R_4^3 R_5^4 R_6^5 \tag{17}$$

To calculate elevation, cross-level and polarization angles. Elevation angle (θ_4) is calculated as:

$$\cos\theta_4 + \beta = a \tag{18}$$

Where $a = 0.5r_{11} + 0.87r_{11}\cos(\theta_1) + 0.87r_{21}\sin(\theta_1)$.

$$\theta_4 = atan2(\pm\sqrt{1 - a^2}, a) - \beta \tag{19}$$

Cross-level angle (θ_3) can be determined from:

$$\cos(\theta_3 - 90) = \frac{r_{21}\cos(\theta_1) - r_{11}\sin(\theta_1)}{\sin(\theta_4 + \beta)} = a \tag{20}$$

$$\sin(\theta_3 - 90) = \frac{0.87r_{31} - 0.5r_{21}\sin(\theta_1) - 0.5r_{11}\cos(\theta_1)}{\sin(\theta_4 + \beta)} = b \tag{21}$$

$$\theta_3 = atan2(b, a) + 90 \tag{22}$$

And the polarization angle (θ_5) inverse kinematics solution is given as:

$$\cos(\theta_5) = \frac{0.5r_{33} + 0.87r_{13}\cos(\theta_1) + 0.87r_{23}\sin(\theta_1)}{\sin(\theta_4 + \beta)} = a \tag{23}$$

$$\sin(\theta_5) = \frac{0.5r_{32} + 0.87r_{12}\cos(\theta_1) + 0.87r_{22}\sin(\theta_1)}{\sin(\theta_4 + \beta)} = b \tag{24}$$

$$\theta_5 = atan2(b, a) \tag{25}$$

Again, there are two solution sets (a,b) for inverse kinematics in the singular region for positive and negative θ_5 values. In the mechanically correct solution set, θ_5 must be between 0 and 180 degrees.

IV. CONTROL SYSTEM ARCHITECTURE

We use the robust cascade PI control architecture often used in motion control for the orientation control of the antenna as shown in Fig. 4. The position controller has a cascade structure with a PI controller that provides a reference for a velocity controller which also uses a PI controller [8]. The controller creates a torque reference, and the torque control is achieved by the current control of the Servo Drive which is a Beckhoff system in our case.

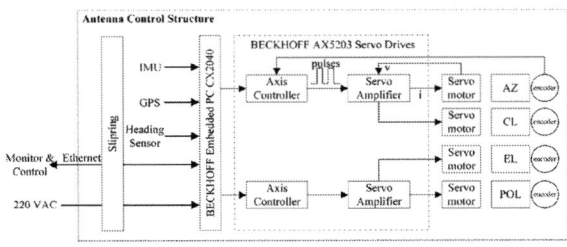

Fig. 4. The control structure diagram of the antenna

The position feedback is directly obtained from motor absolute encoders and the velocity value is estimated through the use of position and current measurements. Set point generator generates a LSPB (Linear Segments with Polynomial Blends) trajectory in order to obtain smooth movement.

Matlab and Twincat software are used for control the antenna. Twincat support Matlab/Simulink. In order to control antenna movement Simulink control algorithm is used via Twincat. This algorithm takes the IMU (Inertial Measurement Unit) and GPS sensor data, calculates the desired position and control the position with specified velocity and acceleration as it is shown in Fig. 5.

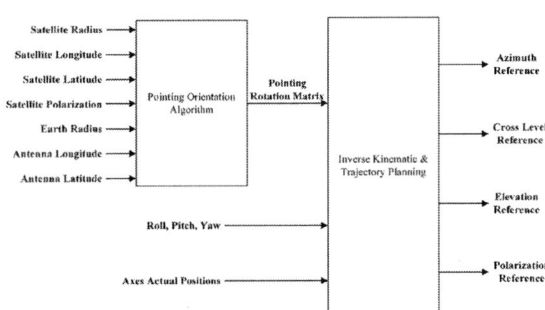

Fig. 5. The control algorithm diagram of the antenna

V. THE TESTBED AND EXPERIMENTS

The prototype antenna used as a test bed is shown in Fig. 6. RF signals from the satellite and obtained through the Ku-band antenna was measured with a Spectrum Analyzer (Rohde&Schwarz, FPC1000). The RF signal which comes from LNB (Low Noise Block) was connected to the Spectrum Analyzer using DC splitter.

Fig. 6. The test setup of the antenna

In the experiments, the initial position of the antenna has been calibrated with respect to the kinematic model of antenna. After initialisation, the desired rotation matrix was calculated using GPS values of TURKSAT 4A and Istanbul. The calculated orientation matrix was used for the inverse kinematics and the joint angles were obtained. R_{4A}^{IST} matrix was calculated with respect to TURKSAT 4A and Istanbul GPS values as:

$$R_{4A}^{IST} = \begin{bmatrix} -0.7148 & 0.1664 & 0.6792 \\ 0.252 & 0.9673 & 0.02814 \\ -0.6524 & 0.1913 & -0.7334 \end{bmatrix} \tag{26}$$

The inverse kinematic solution of the antenna was calculated with the obtained rotation matrix R_{4A}^{IST}:

$$q = \begin{bmatrix} 160.6 & 90 & 0 & 40.72 & 165.4 \end{bmatrix}^T \tag{27}$$

The 2018 International Power Electronics Conference

Validity of the antenna pointing algorithm was verified through the analysis of the satellite guidance signals obtained via the antenna with the Spectrum Analyzer. The maximum signal strength was measured -61.26 dBm at 11.7 GHz as in Fig. 7. In Fig. 8, Fig. 9, Fig. 10 and Fig. 11 the actual position and velocity azimuth, cross-level, elevation and polarization axis is shown respectively. It is clear that all axes follow their position references with zero steady state error. It is obtained that the peak of the polarization is caused by the high friction. In order to compensate the high friction the controller applies more torque and this causes the velocity peak. Fig. 12 shows the RF signal strength when the axes approach their set points.

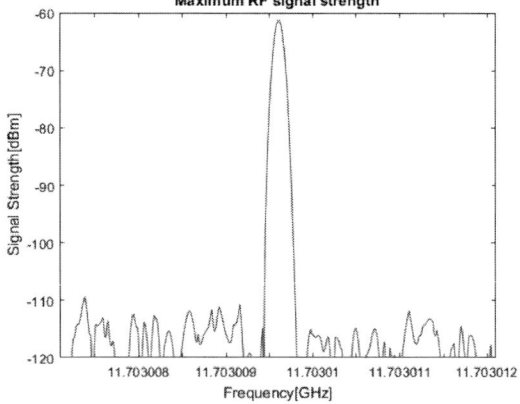

Fig. 7.   Maximum RF signal strength

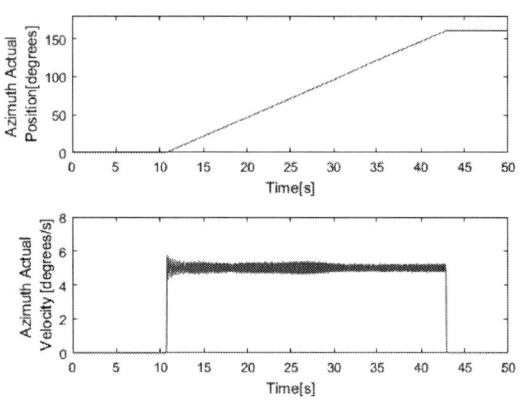

Fig. 8.   The actual position and velocity of the azimuth axis

In the experiments 4 consecutive tests were performed, and at each test antenna starts from the initial position and tries to find the satellite. At each test, the strength of the signal received by the antenna was recorded.

The test results are shown in Table II. In the tests, error values defined as the maximum signal strength minus measurement signal are computed. This difference is caused by a combination of azimuth and elevation axis

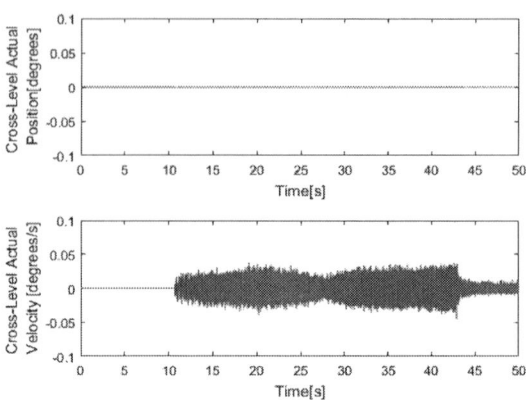

Fig. 9.   The actual position and velocity of the cross-level axis

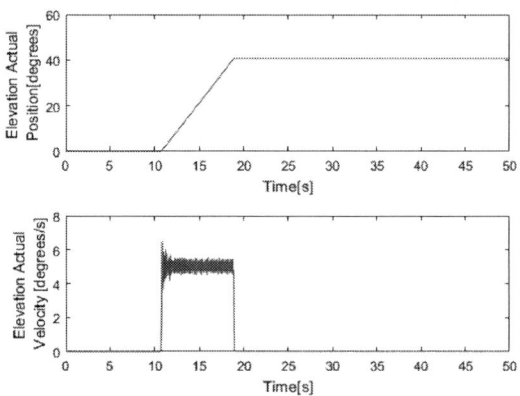

Fig. 10.   The actual position and velocity of the elevation axis

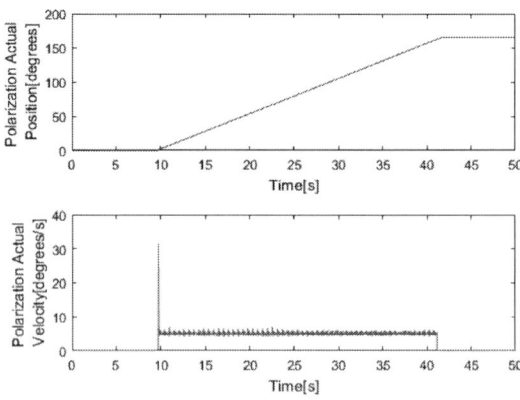

Fig. 11.   The actual position and velocity of the polarization axis

errors. The table shows that if the effect of the azimuth and elevation axes error are assumed equal, there is a 0.685 deflection angle in azimuth axis and there is a 0.668 deflection angle in elevation axis. The total error can be calculated using Euclidean geometry. The results of the each tests are given respectively on the table. We

1685

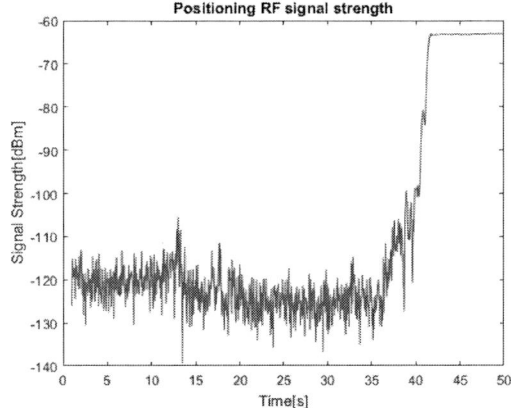

Fig. 12. Positioning RF signal strength from Spectrum Analyzer

TABLE II. THE TEST RESULT OF THE ANTENNA

Test Number	Error	Az deflection angle	El deflection angle	Total
1	$2.185dB$	0.685	0.668	0.956
2	$2.610dB$	0.763	0.722	1.050
3	$2.480dB$	0.744	0.708	1.027
4	$2.385dB$	0.724	0.695	1.003

have observed that these errors are mostly caused by gearbox backlash in the mechanism and other mechanical inaccuracies. This can be further improved in our next iterations.

VI. CONCLUSIONS

To conclude, while the test results show us the pointing of the antenna has an accuracy of 1 degree, which is adequate in locking on to the satellite and as a result tracking is successful. Furthermore, satellite feedback can be utilized in the control algorithm to correct for these errors. This leads us to believe that the proposed system can achieve more accurate tracking. Furthermore this algorithm can be used to achieve pointing of the antenna under dynamic disturbances (stabilization).

ACKNOWLEDGEMENT

The authors are greatful to R. Tekay and M. Tekin for useful comments. The authors would like to thank N. Durmaz for technical drawings.

REFERENCES

[1] Ekstrand B., "Equations of motion for a two-axes gimbal system", *IEEE Transactions on Aerospace and Electronic Systems*, vol. 37, no. 3, pp. 1083-91, 2001.

[2] Abdo MM, Vali AR, Toloei AR, Arvan MR, "Stabilization loop of a two axes gimbal system using self-tuning PID type fuzzy controller", *ISA transactions*, vol. 31, no. 53, pp. 591-302, 2014.

[3] Debruin J, "Control systems for mobile satcom antennas", *IEEE Control Systems*, vol. 28, no. 1, pp. 86-101, 2008.

[4] Craig JJ., "Introduction to robotics: mechanics and control", *Upper Saddle River, Pearson Prentice Hall*, 2005.

[5] Jiang J, Yao B, Guo J, Chen QW, "Modeling and nonlinear computed torque control of ship-mounted mobile satellite communication system", *International Journal of Automation and Computing*, vol. 9, no. 5, pp. 459-466, 2012.

[6] Suzuki Y, Yamashita F, Kobayashi K, Takeda Y, "A Novel Digitally Polarization Tracking Antenna for Ku-band Mobile Satellite Communication Systems", *In28th AIAA International Communications Satellite Systems Conference(ICSSC-2010)* , pp. 8721, 2010.

[7] Chair of Astronautics, Technical University of Munich, Eutelsat Communication, "Calculation of Azimuth, Elevation and Polarization for non-horizontal aligned Antennas", 2015, Retrieved from http://www.eutelsat.com/files/contributed/support/pdf/azimuth-elevation-polarization.pdf.

[8] BECKHOFF, "Functional description for servo drives of the series AX5000", 2017, Retrieved from https://download.beckhoff.com/ download/document/motion/ax5000_function_description_hw2_en.pdf

Position Sensorless Position Control for Dual Solenoid Actuator

Sakahisa Nagai[1], Atsuo Kawamura[1]

1 Department of Electrical and Computer Engineering, Yokohama National University,
79-5, Tokiwadai, Hodogaya, Yokohama, Japan
*E-mail: nagai-sakahisa-yf@ynu.jp, kawamura@ynu.ac.jp

Abstract—This study aims to realize a compact actuator system. Solenoid actuators have some advantages such as low cost and high output with small volume. Additionally, the structure is simple, therefore the size can be easily miniaturized. In general, the solenoid actuator can generate the driving force in only one direction, hence a spring is necessary to return the position at the initial position. To realize quick motion, dual solenoid actuators (DSA) which can generate the driving force in both directions are studied. In this study, sensoeless technique is used to realize further miniaturization of the system. This paper proposes position sensorless position control for DSA. In the proposed control, a disturbance observer is used to achieve robust position control. The control performance was confirmed by experiments. As a result, the position was completely followed to 0.1 Hz sinusoidal command. The frequency characteristics of the proposed control were measured. The proposed control can achieve that the gain is from 0 dB to 1 dB and the total harmonic distortion is less than 10 % when the frequency is up to 5 Hz.

Keywords—Dual solenoid actuator, position control, sensorless estimation

I. INTRODUCTION

Solenoid actuators are widely used in many applications such as switches and valves [1][2]. Their advantages are low cost, simple structure, and high output with small volume. In general, the solenoid actuator can generate the driving force in one direction by inputting current. A spring is required to return the mover at the initial position. Therefore, the force in the reverse direction depends on the mover position. In order to return the position quickly, the driving force should be generated in the reverse direction as well.

Dual solenoid actuators (DSA) which can generate the driving force in both directions are studied [3][4]. The structure of the DSA is shown in Fig. 1. The DSA consists of two solenoid actuators: an upper solenoid actuator and a lower solenoid actuator. In order to generate the reverse direction force, the lower solenoid actuator is attached to the upper solenoid actuator in the opposite direction. The movers are connected together. In [3], zero vibration on-off position control was proposed to improve the power efficiency and switching/chattering phenomenon. Fast and small overshoot position control was achieved. In [4], feedforward controller, which generates the flux trajectory, was proposed.

This study aims to realize a compact actuator system by applying a sensorless drive technique to the DSA.

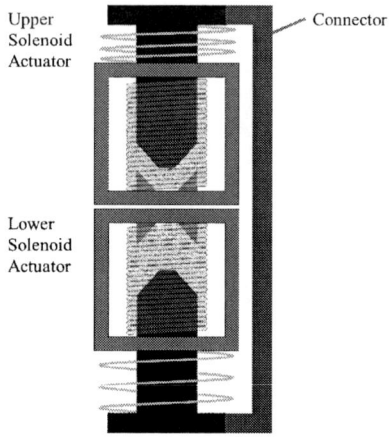

Fig. 1. Structure of DSA.

Sensorless drive technique is effective to make the system small and inexpensive [5]. When the mover position of the DSA changes, the inductance also changes. Hence the position can be estimated by using the inductance characteristics. In this paper, the mover of the DSA is light, therefore the sensing signal should have less harmonics, which causes small oscillation and unpleasant sound. We have proposed position sensorless estimation method for compact solenoid actuators [6], [7]. In their proposed method, the input signal consists of a DC driving signal and an AC sensing signal. Therefore, the input signal has less harmonics. In this paper, the sensorless position estimation method for the DSA is based on [6], [7]. The sensing signal is input to only the upper solenoid actuator. The inductance of the upper solenoid actuator is calculated from the measured voltage and current. From the calculated inductance, the position is estimated by using the inductance characteristics measured in advance.

This paper proposes position sensorless position control for DSA. The estimated position is used in the position feedback control. In the proposed control, a disturbance observer (DOB) is used to achieve robust control [8]. Experiments were conducted to confirm the control performance.

This paper is organized as follows: Section II explains the modelling of the DSA. Section III describes the sensorless position estimation for the DSA. Section IV describes the measurement results of the inductance characteristics. Section V explains the sensorless position control. The experimental results of the sensorless position

The 2018 International Power Electronics Conference

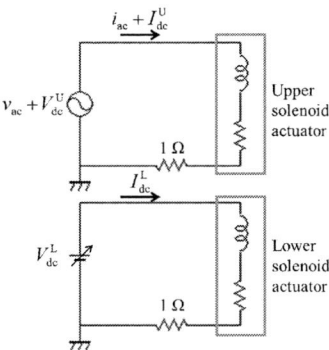

Fig. 2. Driving circuit of DSA.

control are shown in Section V. The conclusions are described in Section VII.

II. DUAL SOLENOID ACTUATOR

This section describes the modeling of the DSA. Fig. 1 shows the structure of the DSA. A DSA consists of two solenoid actuators. The lower solenoid actuator is attached to the upper solenoid actuator in the opposite direction. The movers are connected together. When current flows through the upper solenoid actuator, the magnetic force is generated in a downward direction. On the other hand, when current flows through the lower solenoid actuator, the magnetic force is generated in an upward direction. Therefore, a DSA can generate the driving force in both directions by changing the current path. A spring is mounted to each solenoid actuator, in order to set the mover position at the center of the working range when the current does not flow.

Fig. 2 shows the driving circuit of the DSA. The circuits of the lower solenoid actuator and the upper solenoid actuator are separated. A shunt resistor is used in each circuit to measure the current. The superscript "L" and "U" denote the lower solenoid actuator and the upper solenoid actuator. The subscript "ac" means the AC component, which is a high frequency sinusoidal wave and used to estimate the mover position. The subscript "dc" means the DC component, which is used to drive the solenoid actuators. v_{ac} and i_{ac} denote the AC voltage and current. V_{dc} and I_{dc} denote the DC voltage and current. The sensing AC signal is input to only the upper solenoid actuator. The signals of the upper and lower solenoid actuators are directly generated by using D/A converters.

Fig. 3 shows the force applied to the mover. The motion equation can be derived as

$$M\ddot{x} = F_{\mathrm{mag}} - F_{\mathrm{spr}} + F^{\mathrm{dis}} \tag{1}$$

where x, M, and K denote mover position, mover mass, and elastic coefficient of the spring. F_{mag}, F_{spr}, and F^{dis} denote magnetic force, elastic force, and disturbance force, respectively. F_{mag} is calculated as

$$F_{\mathrm{mag}} = \begin{cases} \frac{\alpha}{2}\frac{\partial L}{\partial x}I_{\mathrm{dc}}^2 & \text{when } I_{\mathrm{dc}} \geq 0 \\ -\frac{\alpha}{2}\frac{\partial L}{\partial x}I_{\mathrm{dc}}^2 & \text{when } I_{\mathrm{dc}} < 0 \end{cases} \tag{2}$$

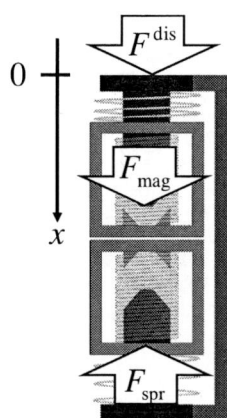

Fig. 3. Force applied to mover.

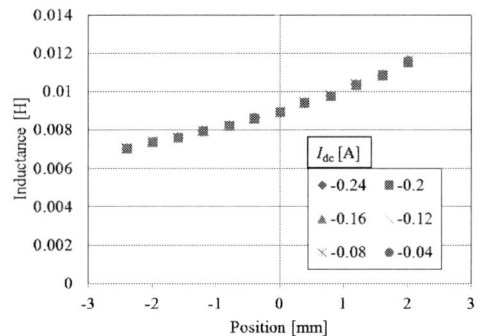

(a) $I_{\mathrm{dc}} < 0.0$.

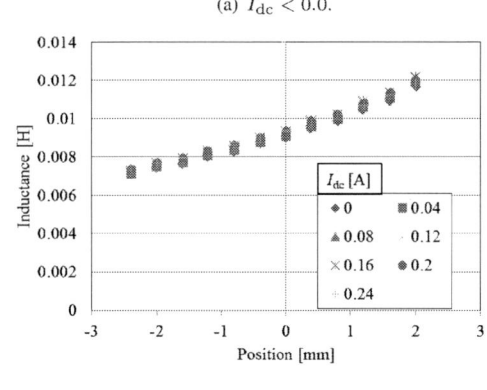

(b) $I_{\mathrm{dc}} \geq 0.0$.

Fig. 4. Measuremnt results of relation among inductance, position, and DC current.

where α and L denote adjustment coefficient and inductance. The AC current also generates the magnetic force. However, the AC current is much smaller than the DC current. Therefore, the magnetic force caused by the AC current can be ignored. I_{dc} is defined as

$$I_{\mathrm{dc}} = I_{\mathrm{dc}}^{\mathrm{U}} - I_{\mathrm{dc}}^{\mathrm{L}}. \tag{3}$$

III. SENSORLESS POSITION ESTIMATION

This section describes the sensorless position estimation. The estimation method is based on the method

1688

The 2018 International Power Electronics Conference

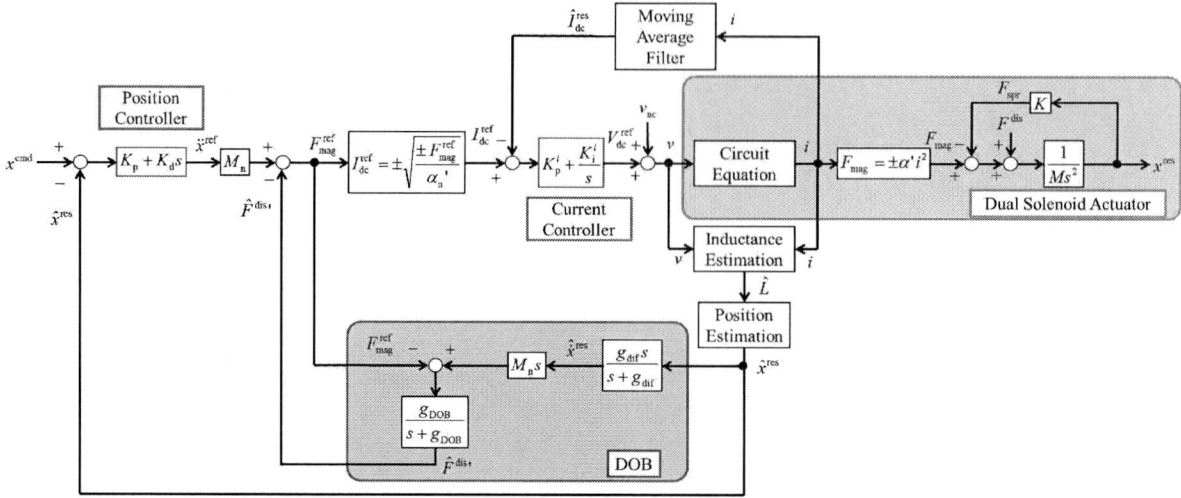

Fig. 5. Block diagram of sensorless position control.

that we have proposed in [6], [7]. The inductance of the solenoid actuator varies when the mover position is changed. Therefore, the position can be estimated from the inductance by using the relation between the position and inductance that is measured in advance. The circuit equation of the upper solenoid actuator can be calculated as

$$v_{ac} + V_{dc}^{U} = R\left(i_{ac} + I_{dc}^{U}\right) + \frac{d\Psi}{dt} \tag{4}$$

where R and Ψ denote internal resistance of the upper solenoid actuator and magnetic flux. Ψ is calculated by multiplying the inductance and the current. Solving (4) into the inductance yields

$$L = \frac{1}{2\pi f}\sqrt{\left(\frac{\overline{v}_{ac}}{\overline{i}_{ac}}\right)^2 - \left(\frac{V_{dc}^{U}}{I_{dc}^{U}}\right)^2} \tag{5}$$

where f, $\overline{v}_{ac}$, and $\overline{i}_{ac}$ denote frequency of the sensing signal, amplitude of the sinusoidal voltage and current, respectively. By using (5), the inductance of the upper solenoid actuator can be obtained from the measured voltage and current. When I_{dc}^{U} is very small, V_{dc}^{U}/I_{dc}^{U} becomes very large. In order to calculate the inductance accurately when I_{dc}^{U} is very small, V_{dc}^{U}/I_{dc}^{U} is set at the value of the internal resistance R.

IV. INDUCTANCE MEASUREMENT

This section describes the measurement results of the relation between the mover position and the inductance. The inductance was measured using (5) with change of the position and the driving current. Fig. 4 shows the measurement results of the relation among the inductance, position, and DC current. The inductance characteristics are assumed as

$$L = \frac{1}{a\left(I_{dc}\right)x + 1/b\left(I_{dc}\right)} \tag{6}$$

where $a(I_{dc})$ and $b(I_{dc})$ are coefficients which are functions of I_{dc}. Solving (6) into x yields

$$\hat{x} = \frac{1}{a\left(I_{dc}\right)}\left(\frac{1}{\hat{L}} - \frac{1}{b\left(I_{dc}\right)}\right). \tag{7}$$

By using (7), the position can be estimated from the inductance.

From (2), the value of $\frac{\partial L}{\partial x}$ is required to calculate the magnetic force. As a result of the inductance characteristics shown in Fig. 4, the slope $\frac{\partial L}{\partial x}$ almost does not change, therefore $\frac{\partial L}{\partial x}$ can be assumed as a constant. The magnetic force characteristics can be rewritten as

$$F_{mag} = \begin{cases} \alpha' I_{dc}^2 & \text{when } I_{dc} \geq 0 \\ -\alpha' I_{dc}^2 & \text{when } I_{dc} < 0 \end{cases} \tag{8}$$

where α' is a constant.

V. POSITION CONTROL

This section explains the sensorless position control. Fig. 5 shows the block diagram of the sensorless position control. A PD controller (whose gains are K_p and K_d) is used for the position control. A PI controller (whose gains are K_p^i and K_i^i) is used for the current control. The measured current includes the AC and DC components, therefore the driving current should be separated from the measured current. A moving average filter is utilized to calculate the driving current. The controller gains were experimentally determined.

The upper right block denotes the plant model of the DSA. The lower block denotes a Disturbance Observer (DOB) which is used to achieve robust position control [8]. The superscript "cmd", "ref", and "res" means command, reference, and response, respectively. The subscript "n" means nominal value. The symbol with "^" denote estimated value. s is Laplace operator. A pseudo differentiation is used to calculate the velocity from the estimated position. g_{dif} and g_{DOB} denotes the cutoff frequencies of the pseudo differentiation and the DOB.

1689

TABLE I. PARAMETERS.

Position controller gains	$K_\mathrm{p}, K_\mathrm{d}$	2500, 100
Current controller gains	$K_\mathrm{p}^i, K_\mathrm{i}^i$	9.3, 10000
Cutoff frequency of pseudo differentiation	g_dif	100 rad/s
Cutoff frequency of DOB	g_DOB	100 rad/s
Mover mass	M_n	10.41 g
Elastic coefficient	K	113 N/m
Sampling frequency		8 μs
Points of moving average		125
Frequency of sensing AC signal		1 kHz
Internal resistor of solenoid actuator	R	14.2 Ω

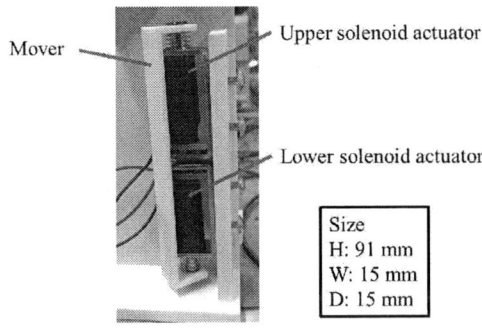

Fig. 6. Dual solenoid actuator.

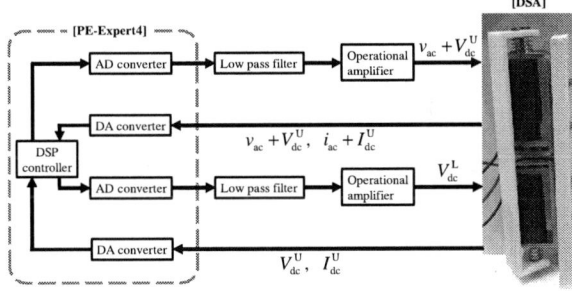

Fig. 7. Experimental system.

The magnetic force can be calculated by (8). Therefore, the reference driving current can be calculated as

$$I_\mathrm{dc}^\mathrm{ref} = \begin{cases} \sqrt{\dfrac{F_\mathrm{mag}^\mathrm{ref}}{\alpha'}} & \text{when } F_\mathrm{mag}^\mathrm{ref} \geq 0 \\ -\sqrt{\dfrac{-F_\mathrm{mag}^\mathrm{ref}}{\alpha'}} & \text{when } F_\mathrm{mag}^\mathrm{ref} < 0 \end{cases} \quad (9)$$

The motion equation can be expressed as (1). In this paper, the elastic force is also regarded as the disturbance. The DOB estimates the disturbance force as

$$\hat{F}^{\mathrm{dis}'} = M\hat{\ddot{x}}^\mathrm{res} - F_\mathrm{mag}^\mathrm{ref}. \quad (10)$$

The DOB in Fig. 5 is constructed using (10) [8].

VI. EXPERIMENTS

This section describes the experimental results of the sensorless position control. Experiments were conducted to confirm the control performance. Fig. 6 shows the DSA used in the experiments. Two compact solenoid actuators "DS-05A, J100-001" produced by CKD Corporation are combined in the DSA. Fig. 7 shows the experimental

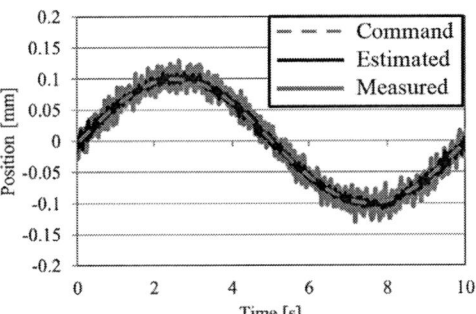

(a) 0.1 Hz sinusoidal command.

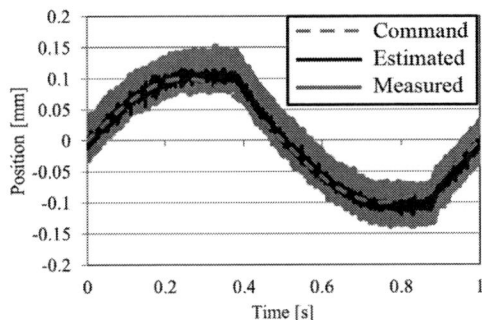

(b) 1.0 Hz sinusoidal command.

Fig. 8. Position response.

system. "PE-Expert4" made by Myway Plus Corporation was used as a digital controller. The output signal of the DA converter is input to the DSA through a low pass filter and an operational amplifier. The voltage and current of the upper and lower solenoid actuators are measured by the AD converter. Table I shows the parameters in the control.

Fig. 8 shows the position response when 0.1 Hz and 1 Hz sinusoidal commands were applied. In each figure, the thick line, black thin line, and red thin line denote position measured by a laser sensor, estimated position, and position command, respectively. As a result, the measured position and estimated position is almost same, therefore the position was accurately estimated without a position sensor. From Fig. 8(a), it was confirmed that the position completely followed the 0.1 Hz sinusoidal position command. From Fig. 8(b), it was confirmed that the position almost followed the position command. However, the response waveform was distorted.

The frequency characteristics of the sensorless position control were measured. Fig. 9 shows the experimental results. The vertical axis in Fig. 9 expresses the gain which is calculated as

$$G = 10 \log_{10} \frac{\overline{\hat{x}^\mathrm{res}}}{\overline{x^\mathrm{cmd}}}. \quad (11)$$

$\overline{\hat{x}^\mathrm{res}}$ and $\overline{x^\mathrm{cmd}}$ denote the amplitude of the estimated position response and position command, respectively. As a result, the position was controlled with the gain from

1690

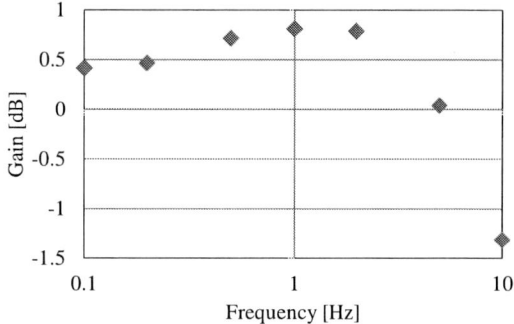

Fig. 9. Experimental results of frequency characteristics.

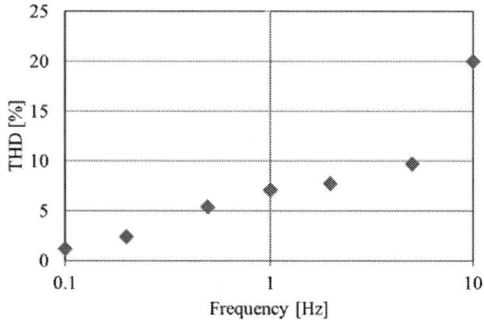

Fig. 10. Experimental results of total harmonic distortion.

0 dB to 1 dB when the command frequency is less than 5 Hz.

In order to confirm the waveform distortion of the response, total harmonic distortion (THD) was calculated. The THD was calculated as

$$\text{THD} = \frac{\sqrt{\overline{\hat{x}_2^{\text{res}}}^2 + \overline{\hat{x}_3^{\text{res}}}^2 + \overline{\hat{x}_4^{\text{res}}}^2 + \overline{\hat{x}_5^{\text{res}}}^2}}{\overline{\hat{x}_1^{\text{res}}}} \times 100 \ [\%]. \quad (12)$$

The subscript means the harmonic order. Therefore, $\overline{\hat{x}_1^{\text{res}}}$ denotes the fundamental component. In addition, the harmonic component whose order is more than 6 is assumed to be very small and ignored in this calculation. Fig. 10 shows the calculation results of the THD. As a result, the THD becomes larger when the frequency increases. However, it was confirmed that the THD is less than 10 % when the frequency is less than 5 Hz. The reason why the THD got worse when the frequency increases is that the DSA has non-linearity such as the magnetic force characteristics and friction force caused by the mechanical structure.

VII. Conclusions

This paper has proposed position sensorless position control for DSA. The position estimation was conducted by calculating the inductance from the measured voltage and current and using the inductance characteristics measured in advance. In the position control, the DOB was applied to achieve robust control. The control performance was confirmed by experiments. As a result, the position completely followed the 0.1 Hz sinusoidal position command. From the measurement results of the frequency characteristics and the THD, the proposed control achieved that the gain was from 0 dB to 1 dB and the total harmonic distortion was less than 10 % when the frequency was up to 5 Hz.

The proposed sensorless control is useful because the size and cost of the DSA system can be reduced by the sensorless drive technique.

References

[1] X. Zhao, L. Li, J. Song, C. Li, and X. Gao, "Linear control of switching valve in vehicle hydraulic control unit based on sensorless solenoid position estimation," *IEEE Trans. Ind. Electron.*, vol. 63, no. 7, pp. 4073–4085, Jul. 2016.

[2] S. Wu, X. Zhao, C. Li, Z. Jiao, and Y. Qu, "Multi-objective optimization of a hollow plunger type solenoid for high speed on/off valve," *IEEE Trans. Ind. Electron.*, to be published.

[3] L. Yu and T. N. Chang, "Zero vibration on-off position control of dual solenoid actuator," *IEEE Trans. Ind. Electron.*, vol. 57, no. 7, pp. 2519–2526, Jul. 2010.

[4] T. Braun and J. Reuter, "A distributed parameter approach for dual solenoid valve control with experimental validation," in *Proc. 18th International Conference on Mechods and Models in Automation and Robotics*, pp. 235–240, Aug. 2013.

[5] P. Mercorelli, "A motion-sensorless control for intake valves in combustion engines," *IEEE Trans. Ind. Electron.*, vol. 64, no. 4, pp. 3402–3412, Apr. 2017.

[6] S. Nagai, T. Nozaki, and A. Kawamura, "Environmental robust position control for compact solenoid actuators by sensorlesss simultaneous estimation of position and force," *IEEE Trans. Ind. Electron.*, vol. 63, no. 8, pp. 5078–5086, Aug. 2016.

[7] S. Nagai, T. Nozaki, and A. Kawamura, "Real time sensorless estimation of position and force for solenoid actuators," *IEEJ Journal of Industry Applications*, vol. 5, no. 2, pp. 32–38, Mar. 2016.

[8] S. Katsura, Y. Matsumoto, and K. Ohnishi, "Modeling of force sensing and validation of disturbance observer for force control," *IEEE Trans. Ind. Electron.*, vol. 54, no. 1, pp. 530–538, Feb. 2007.

CAE Technology Application Trend for Large-Capacity Power Electronics Development

Teruo Yoshino[1*], Kuniaki Nagasaka[1], Shigeaki Nakabayashi[1], Ikuto Udagawa[1], Isamu Tominaga[1] and Junya Konno[1]

1 Power Electronics Systems Division, Toshiba Mitsubishi-Electric Industrial Systems Corporation, Tokyo, Japan

*E-mail: YOSHINO.teruo@tmeic.co.jp

Abstract— Large-capacity power electronics in the range of MW or more has large physical dimensions, heavy weight and high power dissipation although the efficiency is high enough. These features sometimes put difficulties into the development only by the hardware prototypes. Then, CAE (Computer Aided Engineering) technology is now going to be applied to overcome some of difficult and complicated designs of parasitic inductance, thermal performance, mechanical strength and so on. The paper summarized trend of CAE technology application by introducing actual examples. The simulation results of CAE tools are evaluated by experiments with actual prototypes. In summary, the CAE technology of these days is found contributing effectively to the development process of large-capacity power electronics equipment.

Keywords— *CAE: Computer Aided Design, 3D: Three Dimentional Design, Product development process, Digital mockup, Front loading, Parastic inductance, Thermal design and cooliing, Mechanical strength*

I. INTRODUCTION

Different from the small power electronics such like those which can be put on desks, large-capacity power electronics for industries in the range of MW or more has features as follows;

· high current ratings of kA range,
· high power dissipation of 10kW range despite of high efficiency,
· large physical dimensions in 10m range and
· heavy weight in 1000kg range.

These features require advanced design technologies from following viewpoints;

· parasitic inductance,
· cooling and
· mechanical strength.

The parasitic inductance affects both the voltage overshoot at turn-off and current balance among semiconductor devices. However, the parasitic inductance is not a visible lumped component like a resistor or a capacitor but an invisible distributed parameter.

Cooling design is not so simple for large-capacity equipment since it includes complicated cooling fluid flow paths among components of various shapes.

The mechanical structure of the large-capacity power electronics should support heavy components. It is required to withstand dynamic forces in transportation and earthquakes especially in Japan. Then, advanced mechanical design on the dynamic forces is required in addition to the ordinary design on static forces.

This paper summarizes trends of technologies for large-capacity power electronics development mainly from viewpoints of parasitic inductance, cooling and mechanical strength mentioned above.

Section II overviews the development process for large-capacity power electronics and compares between the old process and the new process. In the new process, CAE and 3D (3-Dimensional) design technologies are actively used to improve the quality of development[1]. They also contribute to shorten the lead time. Section III describes technology trend on the parasitic inductance design showing a typical example of the current balance among semiconductor power devices[2]. Section IV describes technology trend on the cooling design with a typical example of the forced air cooling[3][4]. The self-convection cooling[5][6] is also introduced very briefly. Section V describes technology trend on the mechanical strength[7] with a typical example of the dynamic analysis considering the vibration.

For large-capacity power electronics with high voltage ratings of 10 to 100kV range, the other technologies for series connection and insulation are required. However, this paper focuses on the three factors mentioned above.

II. DEVELOPMENT PROCESS OF LARGE-CAPACITY POWER ELECTRONICS

The development process of the large-capacity power electronics itself is the same to that of the small power electronics. The process starts from the prototype design stage then continues to the manufacturing stage and performance evaluation stage. Some trial and error loops of these three stages are repeated until the expected performances are achieved.

A. Conventional development process

In case of the large-capacity power electronics, a trial and error loop may take much resources and long time for weeks or months. For example, small change in shape of a metal plate needs many works, the drawing revision, metal material purchase, manufacturing the new plate by a vendor, assemble again the equipment with the new plate and test again.

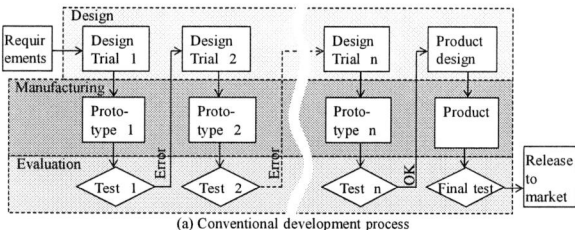

(a) Conventional development process

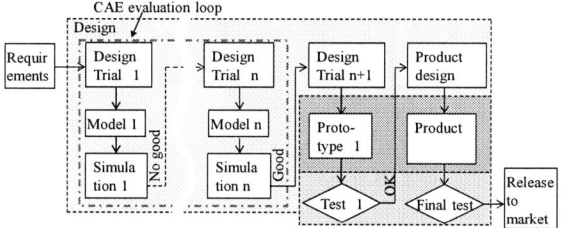

(b) Development process with CAE

Fig. 1. Comparison between old and new processes.

In old times, the design on the factors described in Section I could not be done precisely enough. The engineers roughly designed a prototype based on their experiences and with simple calculations. Then, many trial and error loops might have been necessary to complete the development when it includes unknown factors out of experiences. Such old development process took long time in manufacturing and evaluation processes as shown in Fig. 1 (a). It was difficult to reach the optimized design through the conventional process.

B. Advanced development process

However, these days, advanced CAE tools and 3D design tools are available and desk top PCs (Personal Computers) have sufficient computation performances for the tools to work in short time, minutes or hours. Such technology environment makes possible for the engineers to repeat trial and error loops just in their PCs as shown in Fig. 1(b).

The 3D design tools make possible to visualize the arrangement of components inside of the equipment before an actual prototype is made. The data in 3D design tool, dimensions of conductors for example, can be exported to the CAE tools and can make so-called "digital mock-up" efficiently. Then, with CAE tools, the parasitic inductances, for example, can be evaluated in the PCs. Namely, the trial and error loops can be done in PCs of engineers. This approach contributes to improve the quality of design before the actual prototype is made. Although the lead time of design stage may take longer, this approach, so-called "front loading", can reduce the number of trial and error loops of the actual hardware prototype resulting in short lead time for total development process.

III. PARASITIC INDUCTANCE AND CURRENT BALANCE

Some power electronics products are rated at high voltage in the range of kV. In such case, parasitic inductance design is important from viewpoint of turn-off voltage overshoot[8]. However, this section focuses on the parasitic inductance calculation from viewpoint of the current balance among semiconductor devices connected in parallel in high current power electronics products.

A. Inductance calculation methods and features

The parasitic inductance calculation is usually performed by electro-magnetic field analysis tools. The major methods of analysis are based on FEM (Finite Element Method) or BEM (Boundary Element Method). The other method, PEEC (Partial Element Equivalent Circuit) is also used and its references concerning power electronics are found[9][10][11].

The features of FEM, BEM and PEEC are listed and compared in this subsection.
a) FEM

In the FEM, the finite space including the conductor is divided into small cubes. In each small cube, the electro-magnetic equations are solved one by one. This type of analysis tool is now available in the market. Some of them can be obtained with reasonable cost. And some of them also offer the data import function from 3D tools. Then, the FEM tools are useful in practice. However, the FEM has restrictions when calculating the parasitic inductance considering its nature.

The relationship between the inductance L and electromagnetic energy W made by current I in a conductor is given by Equation (1) [12]. The equation requires magnetic fields including those up to infinite distance from the conductor in order to obtain accurately the total magnetic energy W.

$$W = \frac{1}{2}LI^2 \tag{1}$$

On the other hand, the FEM analysis tool, as its name shows, the calculation is done within a finite volume of space. Then, the space volume surrounding the conductor should be carefully selected when the FEM tool is applied from viewpoint of calculation accuracy.
b) BEM

In the BEM, the equations are made for the surface of the conductor. The surface of the conductor is divided into small areas. This approach is different from that of FEM where small cubes in the space are considered. The equations are solved in each area and provide solutions including those at the infinite distance automatically. Then, in the case of BEM, the space volume is not necessary to be considered.

However, the cost of analysis tool based on the BEM is still expensive and is not practical as development design tools.
c) PEEC

In the PEEC, the conductor is divided into cubes or strings. Inductance of each cube or string can be calculated by simple theoretical equations[13]. For calculating the conductor inductance, the simultaneous linear equations of cubes or strings are made and solved.

However, this type of analysis tools is not available in

the market. The engineers are required to make and to solve the complicated equations by themselves. Then, this type of analysis method is not practical for development design works.

B. Analysis on parastic induactance and current balance

In this subsection, an analysis and its evaluation result is introduced as an example for FEM calculation. The analysis was done on the complicated bus bar design for the circuit, a single phase full bridge converter consisting of 4-paralled IGBTs shown in Fig. 2.

In order to connect four IGBTs in parallel, bus bars for AC output to the load are designed. The shapes of the bus bars are shown in Fig. 3. Each bus bar has a single terminal to the load and four branches to IGBTs.

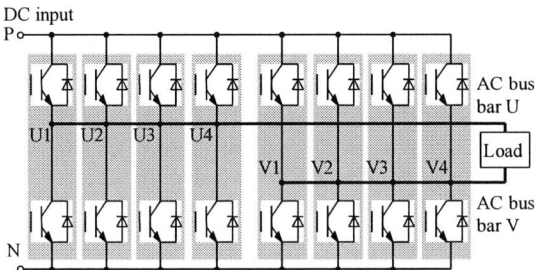

Fig. 2. Single phase full bridge with 4-paralled IGBTs

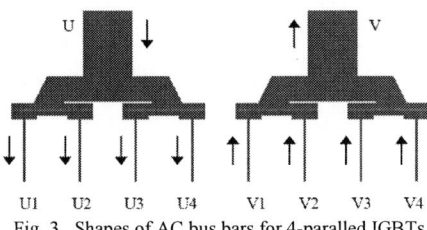

Fig. 3. Shapes of AC bus bars for 4-paralled IGBTs

The current sharing analysis is performed on this AC bus bar considering the parasitic inductances as shown in Fig. 4. In the analysis, the frequency of the current is taken as parameter. The frequency is scanned from DC to 5000Hz. As easily understood from symmetrical shape in Fig. 3, the current balance for DC is good among 4 IGBTs of U1 to U4 since it is determined only by the resistance. However, for high frequencies, the unbalance is observed because of parasitic inductance differences among four current paths.

C. Evaluation of current balance alalysis and disxussion

The current sharing analysis result in Fig. 4 is evaluated by the test with the actual bus bars as shown in Fig. 5. Two tests are performed with the DC current and the pulse current. The pulse current is used to measure the current balance for high frequency region. The analysis and the evaluation test results show similar tendency that the current to U4 takes the maximum value.

The results can be evaluated to be reasonable from the qualitative discussions below.

Considering the line-symmetrical shape of the AC busbars in Fig. 3, it is understood that the current flowing in U1 branch is almost equal to current in V4 branch. And the current flowing U4 branch is also is almost equal to current in V1 branch. The other observation indicates that the current path from U terminal, via U1 branch and V4 branch to V terminal makes a large loop while another current path U terminal, via U4 branch and V1 branch to V terminal makes a small loop. It is expected that the inductance of the large loop is greater than that of the small loop.

When the frequency of current is high, the impedance is dominated by the inductance. Then, it is expected that smaller current flows in the large loop and that large current in the small loop. It means that the current in U1 branch is smaller than in U4 branch.

The analysis like those shown here now can be done in the PCs of the engineers who themselves design the bus bars. Then, they can determine optimized shape of the bus bars and make drawings for actual manufacturing.

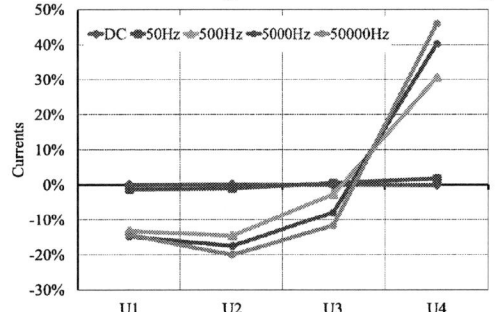

Fig. 4. Current balance calculation considering parasitic inductance

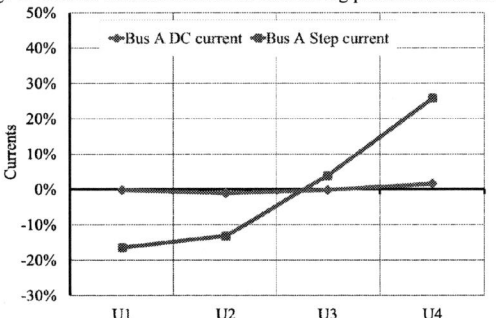

Fig. 5. Current balance measured with actual bus bars

IV. COOLING AND THERMAL DESIGN

The large-capacity power electronics of MW range dissipates power of 10 kW range when the efficiency is 99% which is sufficiently high for practical application. For large power dissipation of this range, usually, the forced air cooling is applied. Self-convection cooling is sometimes applied responding to requirements such like small audible noises and less maintenance works.

A. Heat sink thermal resistance analysis

As a typical example for such power electronics, a high power semiconductor device of press pack type is assembled with heat sinks as shown in Fig. 6. In old

1694

times, in order to design the heat sink, several prototypes are made and the thermal resistances are measured in experiments. In this example, new design process to the thermal design is introduced here.

In the new process, 8 heat sink models are made by 3D design tool with different fin thickness, pitch between fins and number of fins in the heat sink as shown in Fig. 6. The 8 heat sink models are listed in Table I. Each model is designed by the 3D design tool. The dimensions of the model are transferred to a CAE tool for thermal fluid analysis and the cooling performance of each model is evaluated by the simulation. Fig. 7 shows one of simulation results. The simulation results are also listed in Table I for each model.

From the simulations, an optimal design is selected. In this example, it is the model No. 6, which has the smallest number of fins among those with the thermal resistance of 0.016K/W. A pair of prototype is made and the thermal resistance is measured. The thermal resistance from the analysis is found to be in good agreement with the measured value in the experiment as shown in Table II. Through this verification, the accuracy of the simulation is proved to be sufficient for practical application.

B. Cooling design of semiconductor stack in enclosure

Then, the evaluated heat sinks are applied to the converter as shown in Fig. 8. The heat sinks are located within an enclosure with the power semiconductor devices and cooled by forced air flow. In the enclosure, resistors are also placed in the downstream of the heat sink. The thermal fluid analysis can be performed to this size of equipment these days.

The selected heat sink model in the first step of simulation is used to make a semiconductor stack model for the next step of simulation as shown in Fig. 9. The stack includes four semiconductor devices connected in series. In the simulations, the types of fan and the number of fan are surveyed evaluating the temperature of the semiconductor device and the air flow rate of the heat sinks. Examples of such simulation are shown in Fig. 9. From the simulations, as the cooling fans, those listed in Table III are confirmed to fulfil the requirements. The air flow rate unbalance among heat sinks results in the cooling performance unbalance among semiconductors. Then, the air flow rate variations among heat sinks are also calculated and found to be less than 10%.

The thermal resistances of the semiconductor stack are compared with the measured value of the actual prototype stack as listed in Table IV. The simulation results are confirmed in good agreement with the measured values.

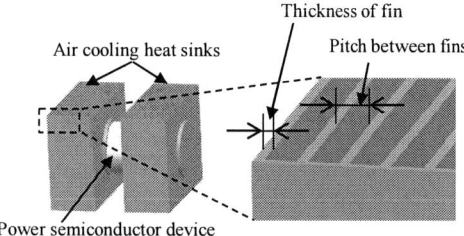

Fig. 6. An example of heat sink model.

TABLE I
HEAT SINK PARAMETERS AND SIMULATION RESULTS

Model No.	1	2	3	4
Thickness of fins	0.5mm	0.5mm	0.5mm	0.5mm
Pitch between fins	2.5mm	3mm	3.5mm	4mm
Number of fins	25	21	19	17
Thermal resistance	0.016K/W	0.017K/W	0.017K/W	0.017K/W
Model No.	5	6	7	8
Thickness of fins	1.0mm	1.0mm	1.0mm	1.0mm
Pitch between fins	2.5mm	3mm	3.5mm	4mm
Number of fins	21	19	17	15
Thermal resistance	0.016K/W	0.016K/W	0.017K/W	0.017K/W

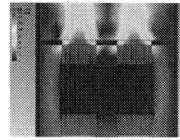 

(a) Air flow rate distribution   (b) Temperature distribution

Fig. 7. Examples of heat sink cooling performance simulation.

TABLE II
THERMAL RESISTANCES FROM ANALYSIS AND EXPERIMENT

	Simulation result	Measured value
Thermal resistance	0.016 K/W	0.016 K/W

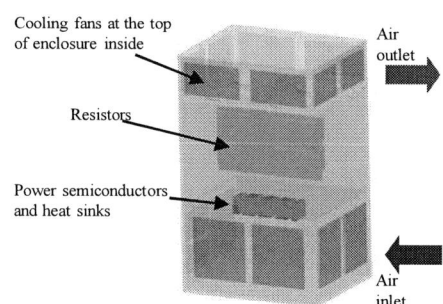

Fig. 8. An example of converter model for thermal fluid analysis.

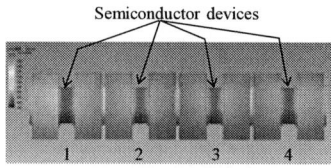

(a) Temperature analysis

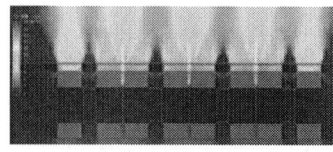

(b) Air flow rate analysis

Fig. 9. Semiconductor stack thermal fluid analysis.

TABLE III
SELECTED COOLING FANS THROUGH SIMULATIONS

Type of fan	Maximum air flow rate	Number of fans	Thermal resistance of stack
Axial flow fan	138m³/min	6	0.016K/W
Centrifugal fan	194m³/min	4	0.014K/W

1695

TABLE IV
COMPARISON OF THERMAL RESISTANCE BETWEEN SIMULATION AND
ACTUAL MEASUREMENT

Location of semiconductor device	Simulation	Measurement
1	0.015K/W	0.014K/W
2	0.014K/W	0.014K/W
3	0.014K/W	0.013K/W
4	0.014K/W	0.015K/W

C. Total cooling design of enclosure

After the cooling design of the semiconductor stack, the total cooling design of the enclosure is performed as the third step of simulation. The enclosure contains many components of various shapes in addition to the semiconductor devices. Some of them also dissipate loss and requires the cooling.

An example of the simulation result is shown in Fig. 10. Fig. 10(a) shows the flow rate of the cooing air and that the resistors are efficiently cooled by the air flow. Fig. 10(b) shows the temperature rise. From the analysis the hottest point is expected to be observed at the top of the resistors. The actual hardware prototype is made to evaluate the accuracy of analysis. The resistor temperature is measured and found to be in good agreement with the simulated value as shown in Table V.

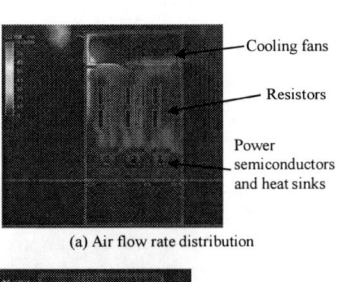

(a) Air flow rate distribution

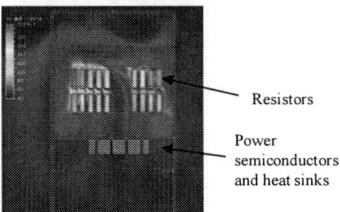

(b) Temperature distribution

Fig. 10. An example of thermal fluid analysis on all equipment.

TABLE V
TEMPERATURE RISE FROM ANALYSIS AND EXPERIMENT

	Simulation result	Measured value
Temperature rise	156.2K	146.5K

D. Self-convection cooling

In the forced air cooling system, the air flow is externally driven by the pressure made by the cooling fans. However, in case of the self-convection cooling, the pressure itself is also necessary to be calculated from the density difference of the air between cool place and warm place[5][6].

The temperature varies place by place in actual power electronics products and the air density is also varies. Then, it is considered that such simulation is more difficult than those of the forced cooling. However, these days, the analysis tools are available and can simulate the self-convection cooling systems.

V. MECHANICAL STRENGTH DESIGN

A. Purpose of the mechanical anaysis

The large-capacity power electronics uses large and heavy components. Fig. 11 is one of examples of the heavy component. The dimensions are of several tens of cm and the weight is several tens of kg. It is called a stack which includes high-power semiconductor devices, cooling heat sinks, insulators and mechanical supports. The stacks are placed in an enclosure of large-capacity power electronics.

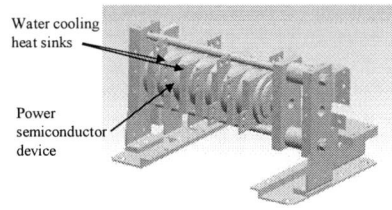

Fig. 11. Mechanical analysis model of a stack with high-power semiconductor devices.

When the enclosure is transported, the enclosure and the stacks are required to withstand the vibrations. In such case, usually, analysis on the eigenvalue, the intrinsic mechanical resonance frequency, is performed to find which frequency in the vibration, the equipment is sensitive to.

B. Mechanicalviberation anaysis example

An example of displacement of the stack obtained by the eigenvalue analysis is shown in Fig. 12. The displacement is magnified by 50 times of the original value for easier visual recognition of which part the displacement is significant. It is noted that this analysis is not static but dynamic where all the mechanical components in the model are oscillated by the resonance frequency.

Fig. 12 shows the displacement due to the first mode of oscillation of the stack. The first mode appears as right-left mode of oscillation seen from the front of the stack. The resonance frequency is analyzed to be 40Hz as listed in Table VI. From the analysis it is also found that the second mode appears as front-back mode of oscillation and has resonance frequency of 71Hz.

The mechanical stress for the first mode of oscillation is shown in Fig. 13. In Fig. 13, the parts colored red shows the places where the mechanical stress is analyzed to be large.

From these analyses, the engineer can see the weak points of the mechanical structure. Then, the structure can be reinforced with appropriate design change before making the hardware prototype.

C. Verification with actual mechanical structure

In order to verify the effectiveness of the analysis, prototypes are made for the designs before and after the reinforcement. The frequency response of the stacks is measured by a pick-up sensor placed in the stack. The results are shown in Fig. 14 and in Table VI.

Before reinforcement, from the measurement of the actual mechanical structure, it is found that the stack has a resonance frequency at around 30Hz. After the reinforcement, it is increased to around 60Hz. As shown in Table VI, the analyzed values and the actually measured values show good agreement.

In the transportation vibration, it is usually said that the lower frequency components are included much more than the higher frequency components. Then, the reinforcement is evaluated to effective to improve the mechanical strength of the stack in the practical vibration environment.

Displacement magnified by 50 times for easier visual recognition

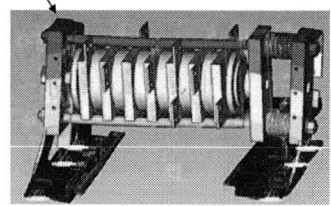

Fig. 12. Eigenvalue analysis of high-power semiconductor stack.

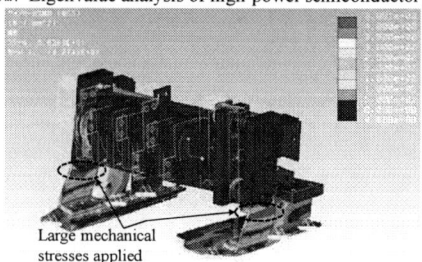

Large mechanical stresses applied

Fig. 13. Mechanical stress distribution in high-power semiconductor stack.

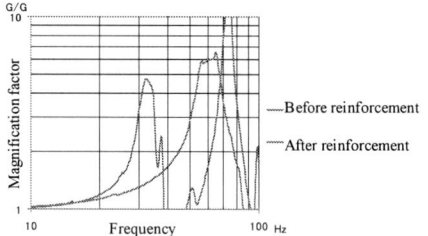

Fig. 14. Measured frequency responses of prototype stacks.

TABLE VI
TEMPERATURE RISE FROM ANALYSIS AND EXPERIMENT

	Before reinforcement	After reinforcement
Simulation results	40Hz	59Hz
Measured values	33Hz	58Hz

D. Impact response analysis

These days, the dynamic mechanical analysis of the impact response is going to be available to the power electronics engineers in addition to the eigenvalue analysis. In the eigenvalue analysis, the sinusoidal oscillation is assumed and the steady state response to the vibration can be analyzed. On the contrary, the impact response analysis is the transient analysis in the time region. With this analysis, the transient displacement and mechanical stress can be analyzed.

VI. CONCLUSIONS

The paper summarized the trend of CAE technology applied to develop the large-capacity power electronics from viewpoint of practical equipment design. The progress in CAE tools and PC calculation performances enable engineers to do a kind of "desk top trial and error loops". From comparisons between simulations and the experiments, it is proven that the CAE analysis tools now show sufficiently good accuracy for practical power electronics equipment designs. The CAE tools are considered to making a kind of revolution in the power electronics development process.

At the same time, however, it is noted that, in actual applications of CAE tools, careful considerations and evaluations based on facts are necessary when making analysis models and evaluating the simulation results.

REFERENCES

[1] S. Nakabayashi et.al, "CAE application to the power converters development," Conference of IAS of IEE of Japan, 1-64, (2008) (in Japanese)

[2] J. Konno et.al, "Study for relationship between the design of bus bars and the balance of currents in IGBTs connected in parallel," Conference of IAS of IEE of Japan, 1-81, (2018) (in Japanese)

[3] I. Udagawa et.al, "Optimal thermal design of the power converters by CAE," Conference of IAS of IEE of Japan, 1-139, (2018) (in Japanese)

[4] Y. Tanaka et.al, "Optimal Thermal Design of UPS by CAE," General Meeting of IEE of Japan, 4-126, (2017) (in Japanese)

[5] Y. Jin, K. Matsuoka, T. Takahashi and N. Takahashi, "100kW PV PCS with natural convection cooling for outdoor installation," 2014 International Power Electronics Conference (IPEC-Hiroshima 2014 - ECCE ASIA), Hiroshima, 2014, pp. 3207-3212.

[6] J. Li et.al, "High Efficiency PCS with Self Convection Cooling for Photovoltaic Systems Into the world," Conference of IAS of IEE of Japan, 1-61, (2013) (in Japanese)

[7] M. Mukunoki et.al, "The vibration evaluation by eigenvalue analysis," General Meeting of IEE of Japan, 4-073, (2014) (in Japanese)

[8] D. Yoshizawa, M. Mukunoki, K. Omote, M. Hayashi and T. Isida, "A large capacity 3-level IEGT inverter," 2014 International Power Electronics Conference (IPEC-Hiroshima 2014 - ECCE ASIA), Hiroshima, pp. 1950-1955 , (2014)

[9] K. Tsuboi et.al, "A Simplified Calculating Method of Bus Bar Inductances and Its Application for Stray Resonance Analysis in Inverter DC link," IEEJ Trans. on Industry Applications, vol. 117, No.11, pp.1364-1374, (1997) (in Japanese)

[10] T. Matsumoto et.al, "Technology of IEGT application," Toshiba Review, Vol. 55, No. 7, pp.11-14 (2000)

[11] M. Ando et.al, "Design of Wiring Structure by Considering Bus Bar Inductance.," IEEJ Trans. on Industry Applications, vol. 132, No.4, pp.510-517, (2012) (in Japanese)

[12] L. Popova et al., "Modelling of low inductive busbars for medium voltage three-level NPC inverter," 2012 IEEE Power Electronics and Machines in Wind Applications, Denver, CO, (2012), pp. 1-7

[13] H. B. Dwight, "Geometric Mean Distances for Rectangular Conductors," in Transactions of the American Institute of Electrical Engineers, vol. 65, no. 8, pp. 536-538, Aug. (1946)

Xilinx System Generator Based Modelling of Finite State MPC

Vijay Kumar Singh[1*], Ravi Nath Tripathi[2] and Tsuyoshi Hanamoto[1]

1 Graduate School of Life Science and Systems Engineering, Kyushu Institute of Technology
Kitakyushu, Fukuoka, Japan

2 Next Generation Power Electronics Research Center, Kyushu Institute of Technology
Kitakyushu, Fukuoka, Japan

*E-mail: vijay.ec1985@gmail.com

Abstract— Finite state-model predictive control (FS-MPC) has gained considerable interest for implementing the current control of power converters. The increase in complexity level of FS-MPC is a concern for design and prototyping of the controller. Xilinx system generator (XSG) provides model-based design platform to realize rapid design and prototyping of digital controller through analysis of intermediate outputs. The discrete nature of FS-MPC readily suits the XSG environment for controller implementation. This paper proposes the design and development of FS-MPC controlled three phase voltage source inverter (VSI) in integrated platform of XSG and MATLAB-Simulink. The system performance is analyzed through simulation results.

Keywords— *Finite state-model predictive control, voltage source inverter, xilinx system generator.*

I. INTRODUCTION

The control of voltage source inverters (VSI) has been extensively studied by the power electronics researchers for development and operation of inverter based system. Current control methods developed for VSI can be categorized as conventional and modern controllers, or linear and nonlinear controllers [1]. Linear controllers like sinusoidal pulse width modulation (SPWM) and space vector modulation (SVM) require carrier-based modulators. Nonlinear controllers like hysteresis current controllers (HCC) which are based on comparators and requires high switching frequency. In addition to these conventional current control techniques, advance controllers such as model predictive control (MPC) based on the system model has been recently gained attraction for VSIs among researchers [2]-[5]. The growing interest in MPC is due to its attractive characteristics of handling multi input multi output (MIMO) cases, system constraints and nonlinearity cases effectively, although it suffers from the problem of high computation requirement.

Finite state-model predictive control (FS-MPC) is a particular classification of MPC that exploits the finite number of switching states for the controlled operation of power converter. The control action is based upon an optimization problem that does not require any kind of external modulator [6]. This method reduces the computational loads as compared to generalized MPC method by considering discretized model of the power converter. Possible switching states are used for discretized model of a power converter for FS-MPC. The future behavior is being predicted from the discretized model of the power converter in one prediction horizon that reduces the online computations.

In general, predictive current controllers are implemented using the microprocessors/digital signal processors (DSPs) due to high computational requirements [7-8]. However, these micro-programmable solutions have some major disadvantages such as computation delay for selection of optimum switching state. And, selected switching state applied to the converter lags one sample period. The delay problem has been clearly recognized and some modifications has been given for compensation [7]-[9].

Considering the above mentioned delay problem, field programmable gate array (FPGA) can be used for implementation as an alternative because of parallel processing nature [10]. However, real-time controller implementation using an FPGA is a tedious process and requires the knowledge of hardware description language (HDL). Hence, prototype designing is complex and time consuming task even for the skilled researchers or engineers with increase in the level of controller complexity [11-12].

Xilinx system generator (XSG) platform provides a virtual FPGA environment for the designing of controller. In addition, the integrated platform of MATLAB-Simulink with XSG provides the functionality of automatic HDL code generation that can be further utilized for straightforward implementation of FPGA based controller [13-15]. Therefore, design and modelling of controller in XSG is an essential part at the initial stage for the development of real-time system using FPGA.

This paper proposes the design and modelling of FS-MPC for the current control of a three phase VSI system using integrated platform of XSG and MATLAB-Simulink. A discrete-time model of the system is used to predict the behavior of the states (variables) at each sampling interval that suits the XSG environment because of discrete nature of the controller.

The 2018 International Power Electronics Conference

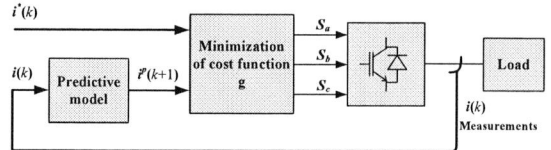

Fig. 1. Block diagram of current control using FS-MPC

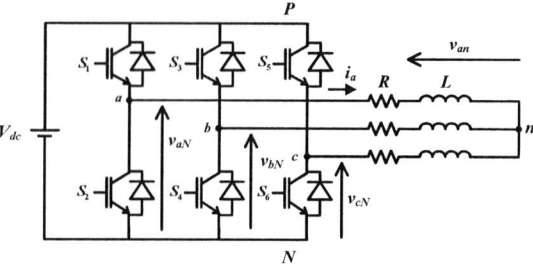

Fig. 2. Power circuit of three phase VSI with RL load

II. FS-MPC OPERATING PRINCIPLE

FS-MPC needs discrete-time model to formulate the control algorithm by using possible number of switching states of a power converter. Discrete-time model of the converter system is used to predict future variables corresponding to each switching state. For the optimum switching state selection corresponding to present sampling interval, a selection criteria (cost function) should be defined and calculated based on the future predicted variables to be controlled. The switching state corresponding to the minimum cost function will be selected in each sampling interval and applied to the converter in the next sampling time.

FS-MPC applied to the three phase VSI for the control of load currents is represented as block diagram in Fig. 1. The steps of current control using FS-MPC is as follows:

- The load side three phase current $i(k)$ at the kth sampling interval is measured and given to predictive model for the prediction of the future variables.
- The load current for the sampling interval $(k+1)$ is predicted corresponding to the possible switching states of the inverter.
- Further, the cost function g is calculated corresponding to possible switching state using the predicted load current $i^P(k+1)$ and reference current $i^*(k)$.
- The cost function g is optimized and optimum switching state is selected corresponding to the minimum cost function.
- The optimum switching state is further used for switching of the inverter in next sampling interval.

To minimize the load current error in the next sampling interval, cost function g is expressed in as follows:

$$g = \left| i_\alpha^* - i_\alpha^p (k+1) \right| + \left| i_\beta^* - i_\beta^p (k+1) \right| \qquad (1)$$

where i_α^* and i_β^* are reference current considering the real and imaginary parts, i_α^p and i_β^p are the predicted load current vector for sample $(k+1)$ considering the real and imaginary parts.

The $\alpha\beta$ stationary reference domain is preferred for the computation of cost function g considering computational complexity of FS-MPC algorithm. The conversion from three phase abc quantities to stationary $\alpha\beta$ domain is performed using Clark transformation relation given as:

TABLE I
SWITCHING SIGNALS OF VSI

Gating signals	S_a	S_b	S_c
1	S_1 ON, S_2 OFF	S_3 ON, S_4 OFF	S_5 ON, S_6 OFF
0	S_1 OFF, S_2 ON	S_3 OFF, S_4 ON	S_5 OFF, S_6 ON

$$\begin{bmatrix} i_\alpha \\ i_\beta \end{bmatrix} = \frac{2}{3} \begin{bmatrix} 1 & -\frac{1}{2} & -\frac{1}{2} \\ 0 & \frac{\sqrt{3}}{2} & -\frac{\sqrt{3}}{2} \end{bmatrix} \begin{bmatrix} i_a \\ i_b \\ i_c \end{bmatrix} \qquad (2)$$

III. THREE PHASE VSI SYSTEM

A. Inverter Model

The power circuit of three phase VSI with RL load is shown in Fig. 2. The switching states of VSI in Table I are determined based on the switching signals S_a, S_b and S_c corresponding to phase a, b, and c respectively. The switching states S can be expressed in vector form as

$$S = \frac{2}{3}\left(S_a + aS_b + a^2 S_c \right) \qquad (3)$$

where $a = e^{i(2\pi/3)}$

The output voltage vectors of VSI are represented as

$$v_i = \frac{2}{3}\left(v_{aN} + av_{bN} + a^2 v_{cN} \right) \qquad (4)$$

where v_{aN}, v_{bN}, and v_{cN} are the phase voltages of VSI. The load voltage vector v_i is defined corresponding to the switching state vector S as

$$v_i = SV_{dc} \qquad (5)$$

Considering all possible switching combinations of the switching signals, eight switching states and hence eight voltage vectors are obtained as shown in Table II where two voltage vectors v_0 and v_7 are same with value 0.

B. Load Model

For the balance three phase RL load, the current can be represented as

1699

TABLE II
VOLTAGE VECTORS AND SWITCHING STATES WITH INDEX NUMBER

Index number	v_i	$[S_a\ S_b\ S_c]$	$[v_{real}\ v_{imag}]$
0	v_0	[0 0 0]	[0, 0]
1	v_1	[1 0 0]	$[2V_{dc}/3, 0]$
2	v_2	[0 1 0]	$[-V_{dc}/3, \sqrt{3}\ V_{dc}/3]$
3	v_3	[1 1 0]	$[V_{dc}/3, \sqrt{3}V_{dc}/3]$
4	v_4	[0 0 1]	$[-V_{dc}/3, -\sqrt{3}V_{dc}/3]$
5	v_5	[1 0 1]	$[V_{dc}/3, -\sqrt{3}V_{dc}/3]$
6	v_6	[0 1 1]	$[-2V_{dc}/3, 0]$
7	v_7	[1 1 1]	[0, 0]

$$i = \frac{2}{3}\left(i_a + a i_b + a^2 i_c\right) \tag{6}$$

where i_a, i_b, and i_c are the three phase currents through RL load. By applying Kirchhoff's voltage law, the load current dynamics can be described by the vector equation

$$v_i = Ri + L\frac{di}{dt} \tag{7}$$

where R and L represents a load resistance and inductance respectively and v_i represents output voltage of the inverter.

C. Discrete-time Model

The discrete-time model of the load current in (7) represents predictive model for sampling time T_S. It will be used to predict the future value of load currents considering the voltage vectors and the measured load currents at the sampling instant k. The discrete-time model is defined using forward Euler approximation considering the load current derivative di/dt and represented as

$$\frac{di}{dt} \approx \frac{i(k+1) - i(k)}{T_S} \tag{8}$$

After substituting (8) into (7), an expression of predicted future load current at sampling interval $k+1$, can be obtained for each of the eight voltage vectors $v_i(k)$ generated by the inverter as

$$i^p(k+1) = \left(1 - \frac{RT_S}{L}\right)i(k) + \frac{T_S}{L}v_i(k) \tag{9}$$

where $i^p(k+1)$ denotes the predicted future load currents

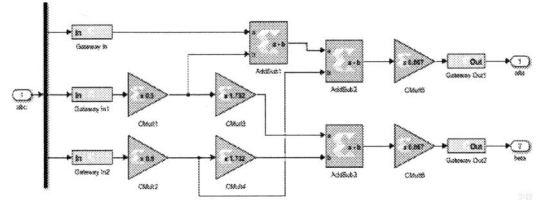

Fig. 3. abc to $\alpha\beta$ transformation in XSG

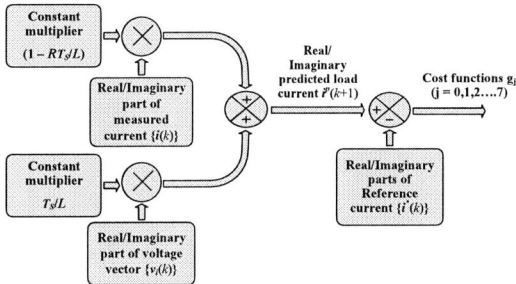

Fig. 4. Block diagram for cost function computations

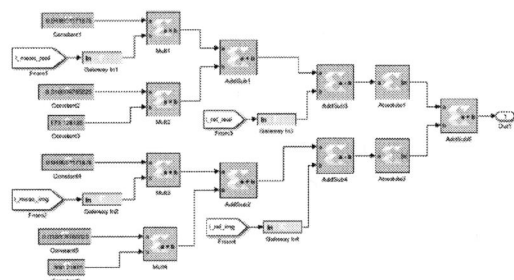

Fig. 5. Computation of cost function in XSG

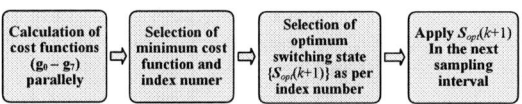

Fig. 6. Block diagram for selection of optimum switching state

at time $k+1$, $i(k)$ is the measured load currents at instant k and the inverter voltage $v_i(k)$ represents the output voltage of the inverter at instant k.

IV. IMPLEMENTATION OF FS-MPC IN XSG

The implementation of the FS-MPC is based upon the computation of cost functions and selection of the switching signals corresponding to optimum cost function. The implementation steps for modelling of controller is explained in the following sections:

A. Transformation of abc to αβ

The cost functions can be computed using the three phase abc quantities but to reduce the computational burden, the abc quantities are transformed into $\alpha\beta$ domain. Transformation of sampled three phase load currents $i_a(k)$, $i_b(k)$, $i_c(k)$ into real and imaginary components $i_\alpha(k)$, $i_\beta(k)$ is modelled using (2) in XSG environment as shown in Fig. 3. $i_\alpha(k)$ and $i_\beta(k)$ are used to compute the cost functions.

B. Computation of Cost Function

The computation of cost function is explained and represented in Fig. 4 through block diagram. Future

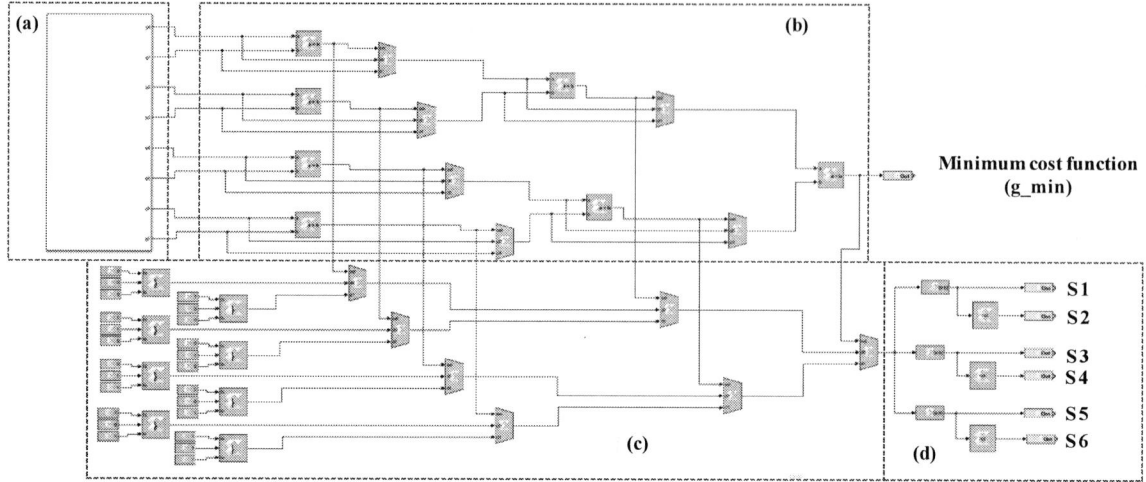

Fig. 7. Modelling of the FS-MPC in XSG with (a) computation of cost function, (b) selection of minimum of cost function, (c) selection of optimum switching state and (d) application of switching state to inverter

currents $i^p_\alpha(k+1)$ and $i^p_\beta(k+1)$ are predicted for computation of the cost function using $i_\alpha(k)$ and $i_\beta(k)$ and inverter model as represented in (9). The cost function need to be computed for each inverter voltage vectors (v_0 – v_7) defined for corresponding switching states (S_a, S_b, S_c) as given in Table II to select minimum cost function. The measured currents $i_\alpha(k)$, $i_\beta(k)$ and the reference currents $i^*_\alpha(k)$, $i^*_\beta(k)$ are fed to XSG model (using input block of Xilinx blockset) as shown in Fig. 5 for computation of cost function corresponding to voltage vector v_3.

C. Selection of Optimum Switching State

The block diagram for selection of optimum switching state corresponding to minimum cost function is shown in Fig. 6. The optimum switching state is selected corresponding to index number of the minimum cost function for each sampling interval. And, the index number is defined for each voltage vectors and switching states as mentioned in Table II. The XSG modelling for the selection of optimum switching state shown in Fig. 7 is developed based upon the block diagram in Fig. 6. The switching signals (S1-S6) are generated for selected optimum switching state.

V. SIMULATION RESULTS AND DISCUSSION

The power circuit of three phase VSI is designed in MATLAB-Simulink environment using floating-point representation, whereas the controller is implemented in XSG environment using fixed-point representation. The parameters considered for the simulation are as follows; dc link voltage (V_{dc}): 520 V, load resistor (R): 10 Ω, load inductor (L): 10 mH, sampling time (T_S): 50 μs.

A. System Performance

The performance of three phase VSI system is analyzed for three phase load current and inverter output

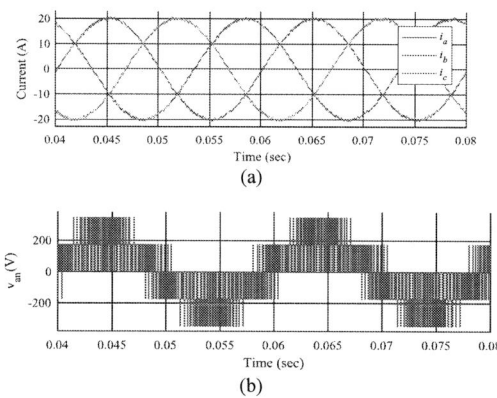

Fig. 8. Steady state (a) three phase load currents, (b) phase voltage in phase 'a', for sampling time T_S = 50 μs

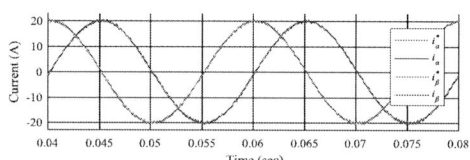

Fig. 9. Steady state load current i_α, i_β for T_S = 50 μs

voltage. The performance of the VSI system is demonstrated as shown in Fig. 8 for T_S = 50 μs. The system is also investigated considering tracking performance and change in T_S for FS-MPC. The tracking performance for i_α and i_β is demonstrated corresponding to i^*_α and i^*_β respectively for T_S = 50 μs in Fig. 9 and T_S = 25 μs in Fig. 10.

The system performance can be improved by selecting lower T_S leads to higher sampling frequency f_S however, higher f_S results in increase in maximum switching frequency f_{sw} defined as half of the f_S in case of FS-MPC. It is required to mention that even though the latest

1701

The 2018 International Power Electronics Conference

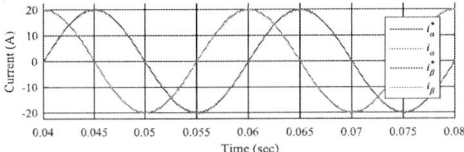

Fig. 10. Steady state load current i_α, i_β for $T_S = 25$ µs

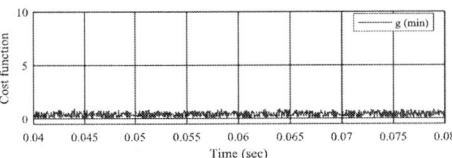

Fig. 11. Intermediate response: Cost function (minimum current error) for $T_S = 50$ µs

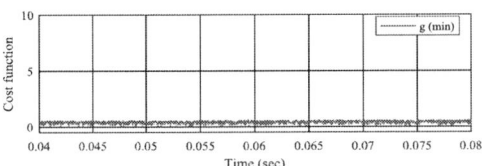

Fig. 12. Intermediate response: Cost function (minimum current error) for $T_S = 25$ µs

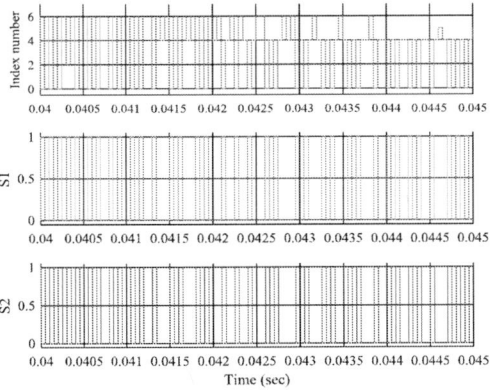

Fig. 13. Index number of selected optimum switching state with switching signals for leg 'a' with $T_S = 50$ µs

microprocessors having high fundamental clock frequency, are capable of handling high f_S but the maximum f_{sw} selection have been limited by maximum possible f_{sw} operation of a device.

B. Intermediate Response

The computational steps of FS-MPC are: computation of minimum cost function, selection of index number and selection of optimum switching states. The intermediate responses are demonstrated for computational steps mentioned above. The selection of minimum cost function results in minimum current error shown in Fig. 11 and 12 for T_S of 50 µs and 25 µs respectively. The FS-

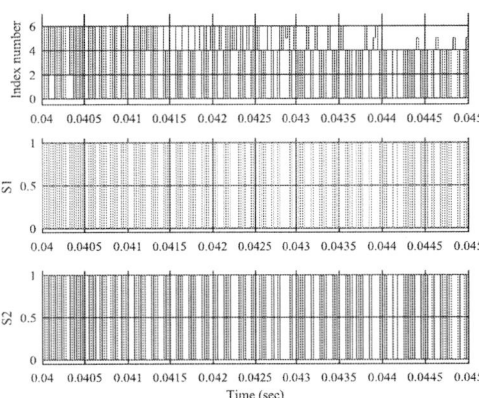

Fig. 14. Index number of selected optimum switching state with switching signals for leg 'a' with $T_S = 25$ µs

MPC performance is validated for current error corresponding to change in T_S. The selection of index number corresponding to minimum cost function consecutively the selection of optimum switching state corresponding to selection of index number are shown in Fig. 13 for $T_S = 50$ µs and Fig. 14 for $T_S = 25$ µs respectively. In addition, the generated switching signals of leg 'a' are demonstrated according to selected optimum switching state corresponding for each T_S.

The responses for intermediate computational steps authenticate the operation of FS-MPC through step-by-step analysis. The rapid development and prototyping of modern controllers and improvement in existing controller can be performed through investigation and verification of the intermediate outputs.

VI. CONCLUSIONS

FS-MPC is developed in XSG through modelling for the output current control of three phase VSI system modelled in MATLAB-Simulink environment. In this paper, modelling based approach is adopted to analyze and substantiate the computational steps of FS-MPC. The performance of FS-MPC is validated considering change in sampling time. The XSG platform can be used for digital controller design adopting model based approach. In future, integrated platform of MATLAB-Simulink-XSG can be used for Hardware-in-the-Loop (HIL) verification of FS-MPC and straightforward implementation of real-time experimental system through FPGA using automatic HDL code generation functionality.

REFERENCES

[1] N. Mohan, T. M. Undeland, and W. P. Robbins, Power Electronics, 2nd ed. New York: Wiley, 1995.

[2] J. Rodríguez and P. Cortes, Predictive Control of Power Converters and Electrical Drives, 1st ed. New York: Wiley-IEEE Press, 2012.

[3] J. Rodríguez, J. Pontt, C. Silva, P. Correa, P. Lezana, P. Cortés, and U. Ammann, "Predictive current control of a voltage source inverter," *IEEE Trans. Ind. Electron.*, vol. 54, no. 1, pp. 495–503, Feb. 2007.

1702

The 2018 International Power Electronics Conference

[4] S. Kouro, M. A. Perez, J. Rodriguez, A. M. Llor and H. A. Young, "Model Predictive Control: MPC's Role in the Evolution of Power Electronics," *IEEE Ind. Electron. Mag.*, vol. 8, no. 1, pp. 16–31, Mar. 2014.

[5] S. Vazquez, J. I. Leon, L. G. Franquelo, J. Rodriguez, H. A. Young, A. Marquez, and P. Zanchetta, "Model predictive control: A review of its applications in power electronics," *IEEE Ind. Electron. Mag.*, vol. 9, no. 4, pp. 8–21, Dec. 2015.

[6] J. Rodríguez, M. P. Kazmierkowski, J. R. Espinoza, P. Zanchetta, H. Abu-Rub, H. A. Young, and C. A. Rojas, "State of the art of finite control set model predictive control in power electronics," *IEEE Trans. Ind. Informat.*, vol. 9, no. 2, pp. 1003–1016, May 2013.

[7] H. Abu-Rub, J. Guzinski, Z. Krzeminski, and H. A. Toliyat, "Predictive current control of voltage-source inverters," *IEEE Trans. Ind. Electron.*, vol. 51, no. 6, pp. 585–593, Jun. 2004.

[8] P. Lezama, R. Aguilera, and D. E. Quevedo, "Model predictive control of an asymmetric flying capacitor converter," *IEEE Trans. Ind. Electron.*, vol. 56, no. 6, pp. 1839–1846, Jun. 2009.

[9] O. Kukrer, "Discrete-time current control of voltage-fed three-phase PWM inverters," *IEEE Trans. Power Electron.*, vol. 11, no. 3, pp. 260–269, Mar. 1996.

[10] M. W. Naouar, A. A. Naassani, E. Monmasson, and I. Slama-Belkhodja, "FPGA-based predictive current controller for synchronous machine speed drive," *IEEE Trans. Power Electron.*, vol. 23, no. 7, pp. 2115–2126, Jul. 2008.

[11] A. Myaing, M.O. Faruque, V. Dinavahi, and C. Dufour, "Comparison of insulated gate bipolar transistor models for FPGA-based real-time simulation of electric drives and application guideline", *IET Power Electron.*, vol. 5, no. 3, pp. 293–303, March 2012.

[12] S. Mekhilef, and A. Masaoud, "Xilinx FPGA based multilevel PWM single phase inverter," *IEEE Int. Conf. on Ind. Tech. (ICIT 2006)*, 15–17 December 2006, pp. 259–264.

[13] Vivado Design Suite User Guide: Model-Based DSP Design Using System Generator, June 7, 2017.

[14] Vivado Design Suite Tutorial: Model-Based DSP Design Using System Generator, June 7, 2017.

[15] Vivado Design Suite User Guide: Release Notes, Installation, and Licensing, April 20, 2017.

Power Hardware-in-the-Loop Setup for Stability Studies of Grid-Connected Power Converters

Tommi Reinikka[1], Henrik Alenius[2], Tomi Roinila[1], Tuomas Messo[2]

1 Laboratory of Automation and Hydraulics, Tampere University of Technology, Tampere, Finland
2 Laboratory of Electrical Energy Engineering, Tampere University of Technology, Tampere, Finland
*E-mail: tommi.reinikka@tut.fi

Abstract—Interaction of grid-connected inverter and power grid can be analyzed through their impedances. The impedances can be used to evaluate stability at the point of common coupling and to define how the system reacts to harmonic distortion. Because the grid conditions have capacitive characteristics and vary over time a voltage source connected through inductors is not sufficient. Power-hardware-in-the-loop method can be used to produce a controlled research environment. The power grid is simulated on a real-time digital system simulator (RTDS) and emulated using voltage amplifier. However, the method causes issues with accuracy and stability of the test bench. Instability can be hazardous as large overcurrent and voltages may appear. In this paper the reliability of the test bench is evaluated. Stability and accuracy of the test bench are tested by simulation and experiments. Suitability of the test bench for performing experiments is evaluated.

Keywords—*Power Hardware-in-the-loop, Power grid emulation, Ideal transformer method, Emulation interface*

I. INTRODUCTION

The power grid and power generation is going through a paradigm change in the near future as more distributed generation is connected to the grid. Virtually all of these power sources, such as photovoltaic power plants and wind farms, are connected to the power grid by power electronics. The increased use of the active devices causes significant effects in the grid dynamics [1]. Large-scale stability problems have occurred in real systems due to impedance-based interaction between multiple inverters and the grid [2], [3]. As the grid conditions vary significantly, different scenarios have to considered and tested to guarantee the stability of the inverters.

The effect of the grid impedance at the point of common coupling (PCC) has been under research. In many studies the impedance has been assumed to be ideally inductive. However, the actual grid impedance is time varying and has also capacitive characteristics [4]. To be able to study the effects of these characteristics, models that are more complex are required to represent the effects of the impedance at the PCC.

The dynamics of the power grid have become more complex as the active components in the power grid are becoming more common, and the direction of power flow may change due to distributed energy resource. Therefore, a method is needed to study the interaction of the power grid and the inverter in time-varying grid conditions. Power hardware-in-the-loop (PHIL) setup allows running tests with a real device connected to a realistic grid instead of pure software environment [5]. The method is commonly used in the auto mobile industry, and recently, in the studies of the dynamics of grid-connected power converters.

PHIL test-bench consists of the hardware under test, interface, and the simulation. The power grid interface is electrically emulated using either voltage amplifier or power converter. The bandwidth of the grid emulation on the hardware side depends on the interface and the use of the power converter limits the bandwidth to a few hundred Hertz [6], [7]. In this paper a linear voltage amplifier is used for the grid side emulation to allow studying the impedance-interactions over a wide range of frequencies.

Power hardware-in-the-loop testing is used when a realistic simulation model is hard or impossible to make and pure hardware test is not reasonable due to cost or complexity. Accuracy of PHIL testing has been under study in [8] as usually there is no possibility to compare the real system to the one made for PHIL-testing. A method to evaluate the accuracy has been studied in [9]. This allows simulations, PHIL-experiments and knowledge of the laboratory equipment to be used as the basis for analyzing the stability of the PHIL-setup and the accuracy of the emulated impedance.

The PHIL system requires thorough testing on its stability since the method is prone to instability. The instability is caused by error in the amplification and the A/D and D/A conversions [8]. The simulation is discrete time and as such causes delay into the control loop of the system. Furthermore the amplifier may cause switching noise and has finite response time. The effects caused by interface algorithms and the actual interface have been studied in [8], [10]. Fast nature of the phenomena present in the power electronic device limits choices for the interface algorithm.

In this paper, a PHIL setup with an RTDS-based grid impedance emulation is presented. Thus, the hardware-under-test can be connected to complex and versatile grid conditions instead of inductance and stiff voltage source. Moreover, such a setup would allow studying novel concepts such as virtual inertia in weak grids. A pure inductance will fail in this matter since it does not deteriorate the robustness of the grid frequency. The model used in this paper consists of an upstream medium-voltage grid and a group of 50 households.

The 2018 International Power Electronics Conference

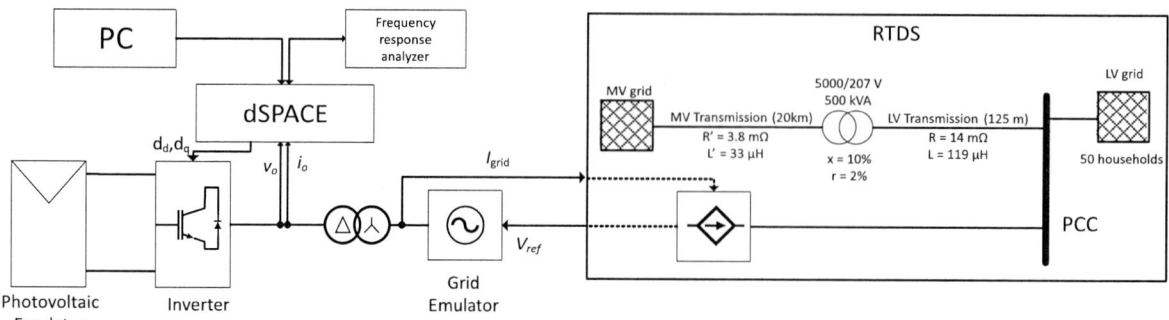

Fig. 1: Layout of the test bench

II. MODELING OF GRID IMPEDANCE

Impedance-based interaction between the grid-connected converter and the grid may cause increase in harmonic content or even instability [11]. Thus, the grid impedance is an important design parameter for grid-connected converters. However, grid impedance of a weak grid is often modeled as a plain series inductor or a series connection of resistance and inductor with inductance of a few mH. Thorough impedance analysis of a real grid are rare, but the acquired results show very different impedance behavior compared to impedance of an inductor [12], [4], [13]. Particularly, series and parallel resonances cause drastic changes in grid conditions, which may not be included in simple grid impedance models. In addition, grid impedance experiences often a strong time-variation during day, as the load profile of the grid changes.

The grid impedance is always related to a certain PCC, and is affected by the characteristics of upstream grid, nearby distributed generation and loads in the grid. Upstream grid is usually mostly inductive due to inductance from transmission lines and transformers. However, grid-connected filters and current control of converters or loads in distribution grid add capacitive element to the grid impedance. Parallel-connected capacitances are cumulative, and thus, the resulting equivalent capacitance may reach significant value. According to [14] a typical household has an equivalent capacitance of 0.6-6 μF, and consequently, the resulting total capacitance may shift the resonant frequency to as low as 250 Hz, which corresponds to the 5th harmonic. The households are depicted as a lumped parallel RLC load, where the value of the equivalent resistance is approximated from the total actual power consumption and capacitance based on 6 μF per household. The PCC is located within the low-voltage distribution grid.

A. HARDWARE-UNDER-TEST

Fig. 1 shows the layout of the test bench. It consists of a PV-emulator, inverter, a 3-phase linear voltage amplifier and a Real Time Digital Power System (RTDS) -simulator. The inverter is controlled by a dSPACE real-time simulator. The test bench is capable of emulating

a PV-generator connected to the power grid through an inverter up to 15 kW. The current PHIL test bench allows fast modifications to the inverter control algorithms and passive filters without having to build a new prototype.

Formerly, the voltage amplifier was used as a stiff voltage source, which had fixed voltage and frequency, and the grid interaction was modeled by connecting inductors between the grid emulator and the inverter. In this paper, the grid emulator is connected to the RTDS for realistic grid side emulation. Current measurements are used as inputs to the RTDS to simulate the response of the power grid. In addition, emulating the grid impedance allows fast changes to be made to the power grid layout, and consequently, the operating point of the grid can be changed while the system is online. For example, the system can emulate tripping a part of the grid in order to change the power flow in the grid. However, the coupling of the test bench to a real-time simulator causes delay to the grid emulation, which can be weaken the accuracy and stability of the test setup. The setup may become unstable even if the system emulated in the test bench would be stable in reality. The instability is caused by error in the voltage amplification.

Fig. 2 shows the control diagram of the RTDS-PHIL test bench. The stability of the system can be evaluated through the information of the laboratory equipment and the characteristics of the applied interface algorithm. For accuracy of the experiments, the best-suited interface algorithms for testing the power electronics are either ideal transformer method (ITM) for its simplicity or damping impedance method (DIM) for its stability [9]. DIM requires matching of the damping impedance in the simulation with the impedance of the hardware-under-test [15], which is complicated due to active nature of the inverter. In this paper ITM has been chosen as the interface algorithm because of its simplicity and sufficient stability. Using the DIM is a possible future improvement.

The delay of the simulation and the conversions between analog and digital signals causes error to the grid emulator output. The error is amplified in the simulation if the impedance ratios between the simulated and hardware parts of the system are not considered. When using voltage type ITM, the impedance of the simulation may never

1705

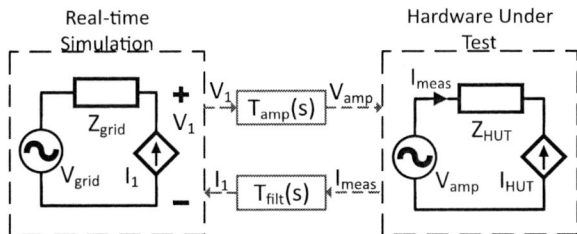

Fig. 2: Control diagram of the PHIL grid emulation with ITM interface algorithm

TABLE I: Voltage amplifier specifications

Voltage Amplifier	PAS 15000
Nominal Voltage (V_{ac})	135 V_{rms} / 270 V_{rms}
Frequency Range (f_{bw})	30 kHz (-3 dB)
Continuous Power	5000 VA per phase
Slew Rate	$< 52 V/\mu s$

be higher than the actual impedance of the hardware side [9]. Otherwise the simulation causes a positive feedback loop for the amplification error and the system becomes unstable.

III. EMULATION INTERFACE

The grid emulation is done by interfacing the power grid simulation to the hardware through a 3-phase linear voltage amplifier. The stability and accuracy of the test setup is affected by the voltage amplifier. The system is affected by the voltage amplifier because there is a small delay between the simulation calculating the correct reference values and the voltage amplifier being able to follow the set reference values. The use of a PHIL-test bench requires exact knowledge of the applied amplifier. In this paper the applied voltage amplifier is a three-phase mains simulation system DM 15000/PAS manufactured by Spitzenberger&Spies. The rated output power of the voltage amplifier is 15 kVA and it is capable of operating as a power sink up to 4.5 kVA. Table I shows the details of the amplifier.

For the stability and accuracy analysis of the system the transfer function of the voltage amplifier needs to be determined. A step test in combination with the curve fitting is used to approximate the amplifier with a second-order transfer function. The transfer function attained through curve fitting is

$$T_{amp} = e^{-3.7*10^{-6}s} \frac{1}{1 + 2.70*10^{-6}s + 3.57*10^{-12}s^2} \tag{1}$$

In Fig. 3 a step response using (1) is compared to the actual step response measured from the amplifier. The calculated step response follows the measured step response accurately and, as such, gives the functionality of the amplifier accurately for the purposes of this paper.

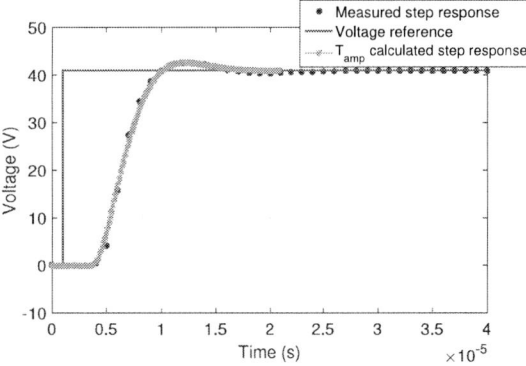

Fig. 3: Measured (blue) step response from the voltage amplifier and calculated (orange) step response from the transfer function T_{amp}

IV. SYSTEM STABILITY

The stability of the system can be evaluated by approximating the transfer functions of the interface algorithm and the real-time simulation. The loop gain caused by the grid emulation can be approximated by

$$G = e^{-sT_d} T_{amp} T_{filt} \tag{2}$$

$$Z_s = Z_g G \tag{3}$$

where e^{-sT_d} is the delay caused by the sampling frequency of the grid simulation, T_{amp} is the voltage amplifier transfer transfer function, T_{filt} is the current measurement low-pass filter, Z_g is the simulated impedance and Z_s is the output impedance of the grid emulation.

$$T_{filt} = \frac{1}{1 + \frac{s}{2*\pi*f_{co}}} \tag{4}$$

where f_{co} is the cutoff frequency of the filter. The approximation of the system transfer function was tested by comparing a simulated impedance of the grid emulation and a measured output impedance of the voltage amplifier. A conventional frequency-response analyzer was used to measure the impedance of the inverter. The voltage amplifier has a finite bandwidth and time delay. The amplifier transfer function is approximated as a second-order transfer function in (1). The measurements show that the system has an extra delay of 83 μs from an unaccounted source. The delay caused by the simulation is 2 times the sampling frequency, which is with 40 kHz sampling frequency 50 μs. The delay of the voltage amplifier is 3.6 μs. Total delay T_d with 40 kHz sampling frequency is 136.6 μs

Fig. 4 shows the verification of the transfer function by comparing the calculated impedance to the actual measured impedance. The test was done by injecting a perturbation signal with the inverter and measuring the simulated impedance at the output of the voltage amplifier. The transfer function for the emulated impedance was tested with two tests. In the first experiment, a 1.0 mH simulated inductance was applied with a sampling frequency of 40 kHz, and a low-pass filter with a cutoff frequency of 1.6 kHz. In the second experiment, a 0.2 mH simulated inductance was applied with a sampling frequency of 80 kHz, and a low-pass filter with a cutoff frequency of 10.6 kHz. The measured output impedance of the voltage amplifier corresponds to the value of the approximated impedance. Thus, the evaluation of the stability can be done based on the approximated impedance emulation.

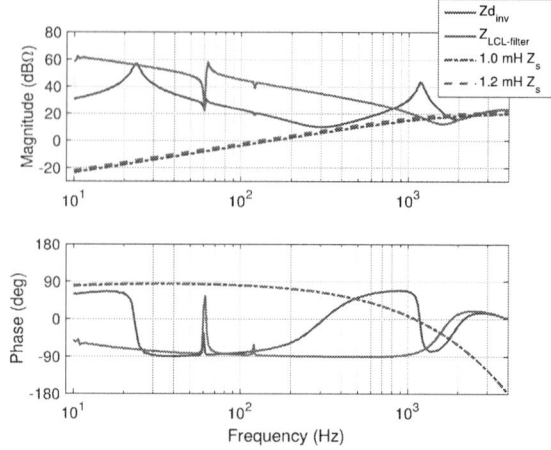

Fig. 5: Measured impedance output impedances of the inverter, the LCL-filter when the inverter is offline and the impedance simulated with the transfer function approximation G

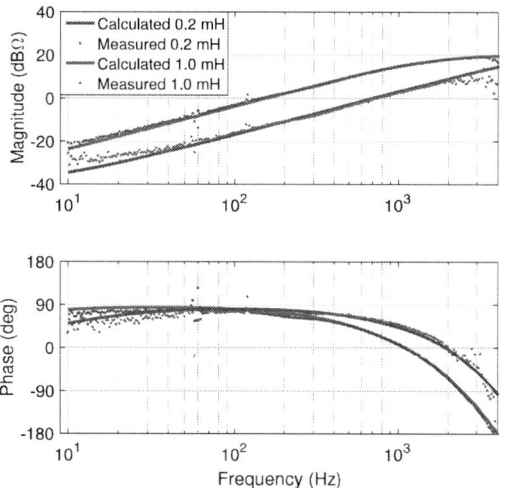

Fig. 4: Measured impedance from the output of the voltage amplifier compared to the approximated impedance emulation.

Fig. 5 shows the measured d-component output impedances of the inverter and the LCL-filter when the inverter is offline. The impedances are measured from the output of the voltage amplifier. These are compared to the approximated grid impedance emulation of 1.0 mH and 1.2 mH. Due to the delay present in the system the phase shift of the impedances is below -90 degrees above 2.5 kHz. The magnitudes of 1.2 mH emulated impedance and the hardware impedance overlap above 2.5 kHz, which is a concern for the stability.

Fig. 6 shows the ratio of grid and inverter impedance on a complex-plane. Stability can be evaluated by applying the Nyquist stability criterion. Thus, the system is unstable if the contour encircles the -1,0 point. The stability analysis is done when the inverter is offline as the LCL-filter becomes dominant in the inverter output impedance above its resonant frequency. The setup is stable when the emulated inductance is 1 mH and unstable when the inductance is increased to 1.2 mH.

Figs. 7-9 show the PHIL-grid emulation in stable and unstable operation. When the emulated inductance is increased from 1.0 mH to 1.2 mH the test bench is no longer able to attenuate the error caused by the emulation, and harmonic resonance occurs. When the instability

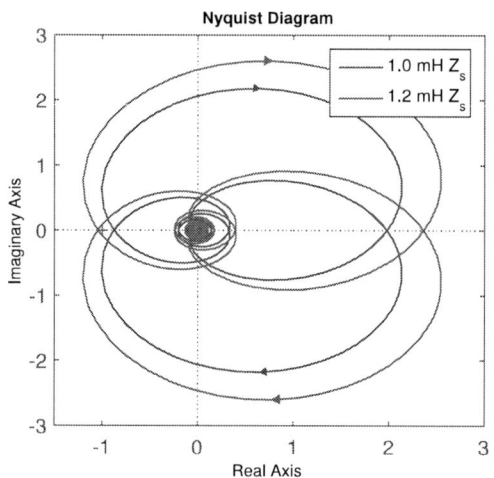

Fig. 6: Nyquist stability analysis with 1.0 mH and 1.2 mH inductances

occurs, the protection of the PHIL-setup disconnects the current measurements from the simulation. The current measurements in Fig. 9 shows that the oscillation is occuring around 2.7 kHz as predicted by the bode plot in Fig. 5. Attaining the stability for the system has three different options. The most straightforward option is to increase the hardware impedance i.e. in this setup the LCL-filter grid side inductor has lower inductance than commonly used in commercial devices. It could be switched to a larger inductance. The second option is to choose another interface algorithm such as the DIM. The third option is to tune the low-pass filter so that the magnitude of the simulated impedance is always lower than the limit for stability without affecting the phase in the tested frequency range. This can be achieved with a first-order filter or more sophisticated filters such as Butterworth or FIR-filter. The case in Figs. 8 and 9 is

1707

stable when the lowpass-filter cutoff frequency is lowered from 1.6 kHz to 1.45 kHz.

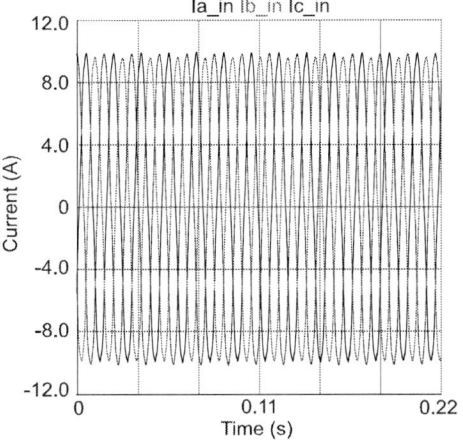

Fig. 7: Currents in each phase in stable operation. 1.0 mH emulated inductance

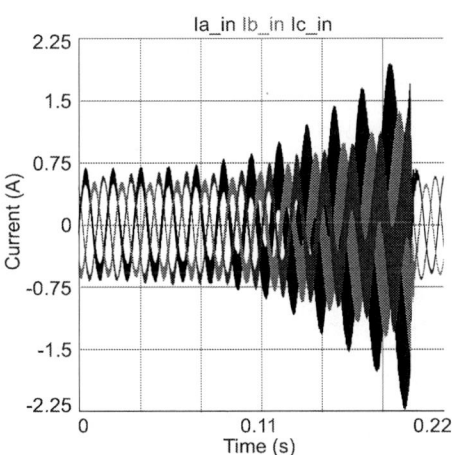

Fig. 8: Currents in each phase in unstable operation. 1.2 mH emulated inductance

V. ACCURACY OF THE PHIL SETUP

The accuracy of the system can be evaluated by comparing the transfer functions of the modeled system with and without the transfer functions caused by the PHIL-setup interface. The delay and slight distortion caused by the amplifier is compared to the assumed real system. Fig. 10 shows the block diagram of the transfer function when the effect of the voltage amplifier and the simulator are taken into account.

The accuracy is analyzed by comparing the voltage ratio between the grid connection and the inverter output voltage between an ideal system and the PHIL-test bench. The transfer functions can be formed by voltage division over simulated and hardware impedances. The ideal system can be shown as

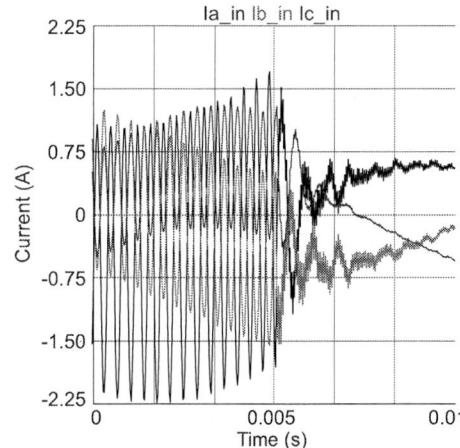

Fig. 9: Zoomed in view of the unstable case with 1.2 mH emulated inductance

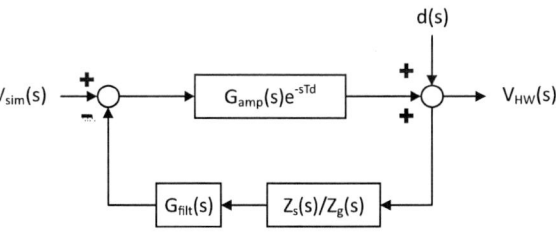

Fig. 10: Transfer function of the emulation interface

$$T_{\text{ideal}} = \frac{Z_{\text{HW}}(s)}{Z_{\text{sim}}(s) + Z_{\text{HW}}(s)} \qquad (5)$$

The transfer function of the test bench with the effect of delays and finite response time of the amplifier can be taken in to account as

$$T_{\text{PHIL}} = \frac{Z_{\text{HW}}(s)}{Z_{\text{sim}}(s)T_{\text{amp}}(s)e^{-sT_{\text{d}}} + Z_{\text{HW}}(s)} * T_{\text{amp}}(s)e^{-sT_{\text{d}}} \qquad (6)$$

From (5)-(6) we can calculate the approximated error for a certain frequency range can be calculated by [16]

$$a(w) = 20 * log \left| \frac{T_{\text{PHIL}}(s) - T_{\text{ideal}}(s)}{T_{\text{ideal}}(s)} \right| \qquad (7)$$

Fig. 11 shows the error caused by the interface algorithm with the filter cutoff-frequencies used in Fig. 4 for determining the accuracy of the formed transfer functions. It can be determined that the PHIL-testing requires high cutoff frequency of the current measurement low-pass filter and small delay in the system when the tests require accuracy at high frequency range.

The maximum relative error of the output voltage of the voltage amplifier is 0.06 at 60 Hz and 0.8 at 800

The 2018 International Power Electronics Conference

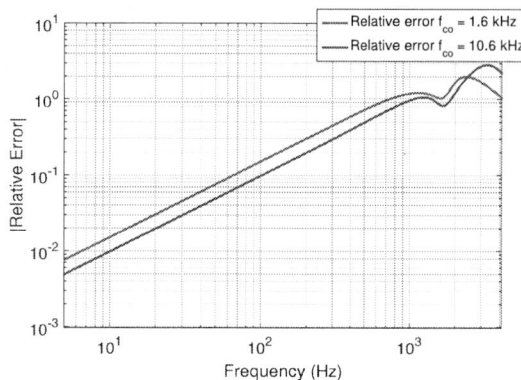

Fig. 11: Relative error in inverter output voltage with different current measurement filter-cutoff frequencies

TABLE II: Power-grid parameters

Emulated Grid Impedance	R_{load}	C_{eq}	L_{grid}
Value	15 Ω	300 μF	180 μH

Hz for cutoff frequency of 10.6 kHz, and 0.09 at 60 Hz and 1 at 800 Hz for cutoff frequency of 1.6 kHz. The emulation does not achieve exact replication of the impedance values of a real power grid. If the exact grid impedance is not required the test bench can reproduce the correct waveform for the impedance.

VI. TEST SCENARIO

The system is tested in a scenario where the PCC is located within a low-voltage distribution system consisting of group of households and capacitive loads. The system is shown in Fig. 1 and the corresponding grid parameters are shown in Table II

The grid is modeled with low power flow i.e. during night-time. Therefore, the equivalent resistance R_{load} is relatively high. The low-voltage system is supplied from the medium-voltage upstream grid with inductance L_{grid}. A resonance between capacitive low-voltage-branch with equivalent capacitance C_{eq} and inductive medium- voltage branch can be seen in the grid impedance at the PCC. Fig. 12 shows the measurement points in the setup. The impedance is measured at the output of the inverter and the output of the voltage amplifier.

Figs. 13 and 14 represents the impedance measurements done with the inverter connected to the grid. The measurements are performed at the output of the inverter. The measurement includes the effect of the grid side LCL-filter inductor and the isolation transformer. Above 500 Hz the delay in the grid emulation begins to affect the results. The low-pass filter is set to 10.6 kHz, which affects the phase of the measurement above 1 kHz. The series and parallel resonance in the grid model can be seen in the impedance measurements performed on the hardware side. The measurements are done in the dq-domain which causes the peak of the parallel resonance

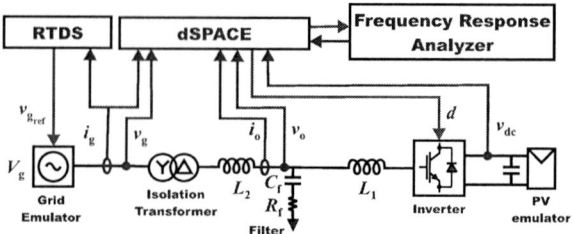

Fig. 12: Impedance is measured at the output of the inverter and at the output of the voltage amplifier

to be divided into two peaks at 60 Hz and -60 Hz of the actual resonant frequency. The impedance follows the calculated results. Difference is observed at the low frequency, which is caused by the lack of detail in simulation of the isolation transformer.

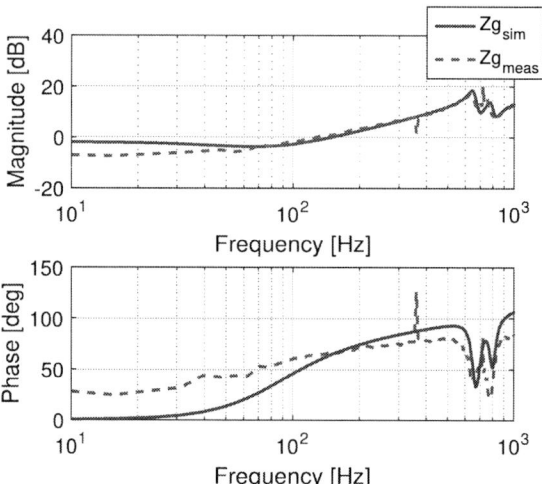

Fig. 13: Measured (orange) and calculated (blue) impedances at the output of the inverter

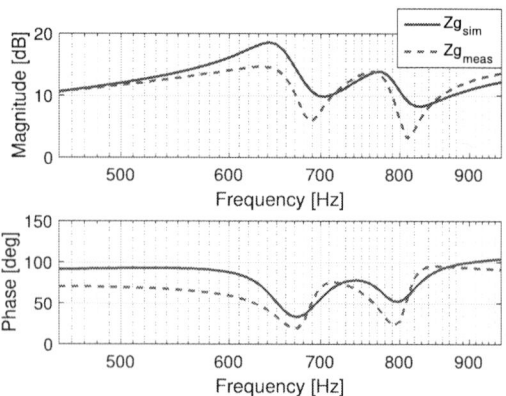

Fig. 14: Measured (orange) and calculated (blue) impedances around the frequency range of the parallel and series resonance at the output of the inverter

The 2018 International Power Electronics Conference

Fig. 15 shows the impedance measured at the output of the voltage amplifier. The system follows the waveform of the simulated impedance but has error in the magnitude as shown in Fig. 11. The error is relative to 10 percent error in the output voltage of the amplifier. The error can be reduced if the overall delay present in the system is decreased.

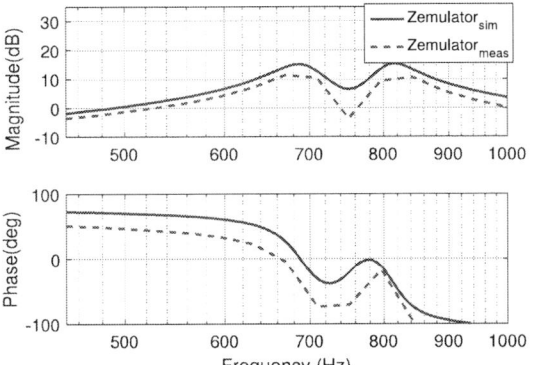

Fig. 15: Measured (orange) and calculated (blue) impedances of the parallel and series resonance at the output of the voltage amplifier

VII. CONCLUSIONS

PHIL grid emulation provides a powerful tool for analyzing the interactions between the grid and inverter. The proposed PHIL setup is shown to be capable of producing an accurate grid impedance since the impedance measured from the hardware side follows reasonably well the calculated frequency response. In this paper, a power grid with a resonant spike in its impedance was modeled. The hardware-side impedance measurement shows that the test setup is capable of producing resonance in the grid impedance in the high frequency range. However, the accuracy of the emulated impedance is decreasing when the frequency is higher. However, for the use of studying the inverter side load-interaction the method is accurate enough. The waveform of the impedance follows the predicted curve and with this setup the error was minimized to around 10 percent. The test bench can be used for studying the interaction of the inverter and the power grid when the exact impedance magnitude is not required.

REFERENCES

[1] C. Li, "Unstable Operation of Photovoltaic Inverter from Field Experiences," in IEEE Transactions on Power Delivery, vol. PP, no. 99, pp. 1-1, doi: 10.1109/TPWRD.2017.2656020, Jan. 2017.

[2] J. Liu and S. Q. Zhao, "Study on Sub-Synchronous Control Interaction (SSCI) of Wind-Power Generator", Advanced Materials Research, Vols. 953-954, pp. 518-521, June 2014.

[3] C. Buchhagen, C. Rauscher, A. Menze, and J. Jung, "BorWin1 - First Experiences with harmonic interactions in converter dominated grids", Die Energiewende - Blueprints for the new energy age, pp. 1-7, Nov. 2015.

[4] L. Jessen, S. Gunter, F. W. Fuchs, M. Gottschalk, and H. J. Hinrichs, "Measurement results and performance analysis of the grid impedance in different low voltage grids for a wide frequency band to support grid integration of renewables," 2015 IEEE Energy Conversion Congress and Exposition (ECCE), pp. 1960-1967, 2015.

[5] de Jong, E, de Graff, R, Vassen, P, Crolla, P, Roscoe, A, Lefuss, F, Lauss, G, Kotsampopoulos, P and Gafaro, European White Book on Real-Time Power Hardware in the Loop Testing: DERlab Report No. R- 005.0. 1 edn, Jan 2012.

[6] A. S. Mkinen, T. Messo and H. Tuusa, "Power hardware in-the-loop laboratory test environment for small scale wind turbine prototype," 2014 16th European Conference on Power Electronics and Applications, Lappeenranta, pp. 1-10, Aug. 2014.

[7] C. Yin et al., "Virtual impedance method of the Power hardware-in-the-loop simulation to improve its stability and accuracy," 2016 IEEE 8th International Power Electronics and Motion Control Conference (IPEMC-ECCE Asia), Hefei, pp. 2752-2758, May 2016.

[8] W. Ren, "Accuracy Evaluation of Power Hardware-in-the-Loop (PHIL) simulation", Ph.D. Dissertation, Florida State University, Tallahassee, USA, ISBN: 9780549467694, July 2007.

[9] D. Barakos, P. Kotsampopoulos, A. Vassilakis, V. Kleftakis and N. Hatziargyriou, "Methods for stability and accuracy evaluation of Power Hardware In the Loop simulations," MedPower 2014, Athens, pp. 1-5, Nov. 2014.

[10] F. Lehfuss, G. Lauss, P. Kotsampopoulos, N. Hatziargyriou, P. Crolla and A. Roscoe, "Comparison of multiple power amplification types for power Hardware-in-the-Loop applications," 2012 Complexity in Engineering (COMPENG). Proceedings, Aachen, pp. 1-6, July 2012.

[11] J. Sun, "Impedance-based stability criterion for grid-connected inverters", IEEE Transactions on Power Electronics, vol. 26, no. 11, pp. 3075-3078, April 2011.

[12] A. Knop and F. W. Fuchs, "High frequency grid impedance analysis by current injection," 35th Annual Conference of IEEE Industrial Electronics, pp. 536-541, Nov. 2009.

[13] J. Xie, Y. X. Feng and N. Krap, "Network Impedance Measurements for Three-phase High-voltage Power Systems," 2010 Asia-Pacific Power and Energy Engineering Conference, Chengdu, pp. 1-5, March 2010.

[14] J. H. R. Enslin and P. J. M. Heskes, "Harmonic interaction between a large number of distributed power inverters and the distribution network," IEEE Transactions on Power Electronics, vol. 19, no. 6, pp. 1586-1593, Nov. 2004.

[15] A. Riccobono, E. Liegmann, M. Pau, F. Ponci and A. Monti, "Online Parametric Identification of Power Impedances to Improve Stability and Accuracy of Power Hardware-in-the-Loop Simulations," IEEE Transactions on Instrumentation and Measurement, vol. 66, no. 9, pp. 2247-2257, Sept. 2017.

[16] A. Viehweider, G. Lauss and L. Felix, "Stabilization of Power Hardware-in-the-Loop simulations of electric energy systems", Simulation Modelling Practice and Theory, Volume 19, Issue 7, pp. 1699-1708, Aug. 2011,

Passivity-based LCL Filter Design of Grid-Connected VSCs with Converter Side Current Feedback

Shih-Feng Chou, Xiongfei Wang and Frede Blaabjerg
Department of Energy Technology, Aalborg University, Aalborg, Denmark
shc@et.aau.dk, xwa@et.aau.dk, fbl@et.aau.dk

Abstract— **This paper presents a design method for LCL filters used in grid-connected voltage source converters (VSCs). The converter-side current feedback and the filter-capacitor voltage feedforward control methods are used. Besides the harmonic current injection limit, the passivity of the input admittance of VSCs is also considered for a co-design of the controller gains and passive filter components. Thus, the stability of current control can be ensured against the grid impedance variation. The effectiveness of the method is confirmed by simulations and experimental tests.**

Keywords— *Input Admittance, LCL Filter, Passivity-based Design, Voltage Source Converters.*

I. INTRODUCTION

Voltage-source converters (VSCs) are commonly used with renewable energy generations, flexible ac and/or dc transmission systems [1]. The switching characteristics of VSCs generate unwanted voltage and current harmonics, which may pollute the power grid and give lower power quality. The LCL filter is the most popular and mature solution to attenuate switching ripples at the point of common coupling (PCC) [2]. Due to its higher gain roll-off, the LCL filter can provide a more efficient attenuation at the switching frequency with a smaller total inductance in comparison with the L filter. Following the same idea, the higher order LLCL filter is introduced [3], which enables to reduce the filter size further by using an LC trap tuned at the switching frequency. Nevertheless, the resonance of the LCL and LLCL filters may also affect the stability of current control, and there by deteriorating the injected current [4].

With the increase use of VSCs in the power system, the assumption of a stiff power grid no longer holds and the interaction between converters is becoming a critical issue [5]. The multi-paralleled VSCs are commonly used in large-scale renewable energy generation systems, e.g. photovoltaic power plants and wind farms. The research works on the resonances and stability of multi –paralleled VSCs have been continuously reported [6]-[8].

The passivity-based control, compared to other active damping approaches, has recently gained more attention, due to the unknown grid impedance variation [9]. The method enables the VSC to behave like a passive component and thus no negative damping will be contributed from the VSCs. Moreover, in the frequency domain, the passivity concept implies that the VSC input admittance has a non-negative real part [10]. This property provides a sufficient, yet not necessary, stability condition. Hence, for the VSC-based power systems, the system will be kept stable if the input admittances of all the VSCs are designed passive [11].

A large number of research works have thus been reported on the frequency-domain passivity-based control [12]-[15]. In [14] and [15], the grid side current is used as feedback control input. However, the safe operating area (SOA) of the power semiconductor devices is the characteristic that ensured that the device can be operated without self-damage. It means the information of the converter side current is usually required for protection. The additional current sensor will increase the cost of converter, which is drawback. The research work [12] is using the converter side current feedback control. The active damping gain is designed according to the current control bandwidth and the system time-delay, which is varied with the current controller. The research paper [13] separates the input admittance in two parts, one is related to the passive components and the other one is time delay and controller dependent. Since the passivity of the passive components is no need to doubt, the passivity can be designed according to the time delay. However, the real part of two parallel complex number form impedance is not only related to the real part of each complex number but also to the imaginary part of the complex number.

This paper presents a co-design method of controller gains and the passive components. The converter-side current feedback control and the filter-capacitor voltage feedforward control are used. When the passivity is guaranteed as discussed [12], the harmonic current emission is considered and limited by the filter capacitor and grid side inductor design. The effectiveness of the method is validated by simulations and experimental tests.

II. PASSIVITY-BASED LCL FILTER DESIGN

The generic equivalent circuit of the LCL filter is shown in Fig. 1(a). For a three-phase system, the current control and the filter capacitor voltage feedforward control are implemented in the stationary frame. The current of the converter side inductor and the voltage of the filter capacitor are measured.

The 2018 International Power Electronics Conference

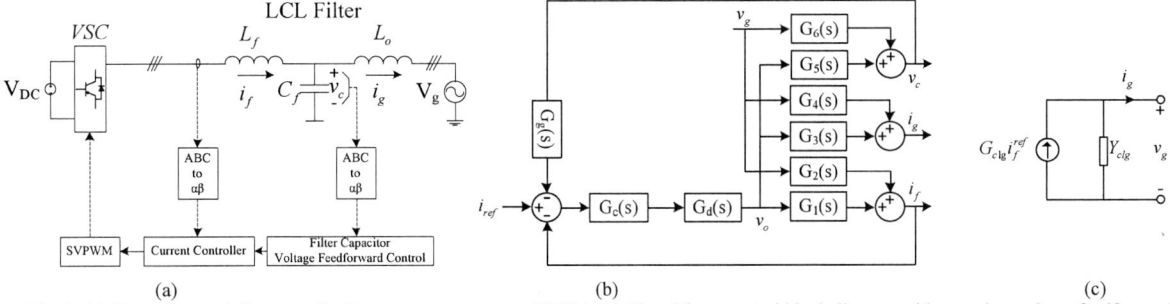

Fig. 1. (a) Overall control diagram of voltage source converters (VSCs) (b) Closed-loop control block diagram with capacitor voltage feedforward control (c) Equivalent admittance Model.

The dc-link voltage is assumed to have a constant value. As the controller is implemented in the stationary frame, the phase-locked loop (PLL) is not used in the system and there by is the low-frequency oscillation and the unexpected harmonics avoided [17].

A. Design for Passivity

Since the LCL filter is used here, there are three state variables with two inputs. V_o is the converter output voltage and V_g is the grid voltage. The open-loop transfer functions can be derived using superposition theorem. In the control block diagram Fig. 1(b), the plant and the controller transfer functions are presented in (1). K_p equals to $2\pi f_c L_t$, f_c is the crossover frequency of the open-loop current loop gain and L_t is the total inductance, which is the sum of L_f and L_o. T_d is the system time delay that is given by (2). Where T_s, f_s are the sampling period and sampling frequency, respectively [16].

$$G_1 = \frac{i_f(s)}{v_o(s)} = \frac{C_f L_o s^2 + 1}{C_f L_f L_o s^3 + (L_o + L_f)s}$$

$$G_2 = \frac{i_f(s)}{v_g(s)} = \frac{-1}{C_f L_f L_o s^3 + (L_o + L_f)s}$$

$$G_3 = \frac{i_g(s)}{v_o(s)} = \frac{1}{C_f L_f L_o s^3 + (L_o + L_f)s}$$

$$G_4 = \frac{i_g(s)}{v_g(s)} = -\frac{C_f L_f s^2 + 1}{C_f L_f L_o s^3 + (L_o + L_f)s}$$

$$G_5 = \frac{v_c(s)}{v_o(s)} = \frac{L_o}{C_f L_f L_o s^2 + (L_o + L_f)}$$

$$G_6 = \frac{v_c(s)}{v_g(s)} = \frac{L_f}{C_f L_f L_o s^2 + (L_o + L_f)}$$

$$G_c(s) = K_p$$

$$G_d(s) = e^{-T_d s}$$

$$G_g(s) = K_f s$$

$$T_d = 1.5T_s, \quad T_s = \frac{1}{f_s} \tag{2}$$

(1)

The filter-capacitor voltage feedforward control gain K_f will be studied and designed in the following section. The control block diagram of the converter is shown in Fig. 1(b). The grid side current can be presented as (3). G_{clg} is the closed-loop current loop gain and Y_{clg} is the input admittance of the converter, respectively. The equivalent admittance model is drawn in Fig. 1(c).

$$i_g = G_{clg} i_f^{ref} - Y_{clg} v_g \tag{3}$$

$$G_{clg} = \frac{G_c G_d G_5}{1 + G_c G_d G_g G_3 + G_c G_d G_1} \tag{4}$$

$$Y_{clg} = \frac{-G_6 - G_c G_d G_g G_3 G_6 - G_c G_d G_1 G_6 + G_c G_d G_g G_4 G_5 + G_c G_d G_2 G_5}{1 + G_c G_d G_g G_3 + G_c G_d G_1} \tag{5}$$

Taking the real part of (5) that yields (6) that is presented in next page. The passivity is related to the numerator of (6), which depends on system delay, converter side inductors and the feedforward gain of filter capacitor voltage feedforward path because the denominator is always positive.

B. Design Procedure of Passive Components

The flow chart of the design procedure is shown in Fig. (2).

1. Converter Side Inductor:

The instantaneous output current ripple in a three-phase space vector pulse width modulation (SVPWM) converters has been identified and has been analyzed in [18]. The maximum peak-to-peak current ripple amplitude is almost a linear function of modulation index and inverse proportional to the converter side inductors.

For a given maximum of peak-to-peak current ripple Δi_{pp}^{max}, the inductor is designed as (7). m is the modulation index, V_{dc} is the dc bus voltage and T_{sw} is the switching period.

$$L_f = \frac{V_{dc} T_{sw}}{2\sqrt{3} \Delta i_{pp}^{max}} m \tag{7}$$

2. Filter Capacitor Voltage Feedforward Control Gain:

The passivity depends on system delay, converter side inductors and the feedforward gain of the integration of the filter capacitor voltage that is depicted in (6).

1712

$$\mathrm{Re}\left(Y_{\mathrm{clg}}\right)=\frac{K_p\left(1+K_fL_f\omega^2\right)\cos\left(T_d\omega\right)}{\left[\left(L_o+L_f\right)-C_fL_fL_o\omega^2\right]^2\omega^2-2K_p\omega\left[\left(L_o+L_f\right)-C_fL_fL_o\omega^2\right]\left[1-\left(K_fL_o+C_fL_o\right)\omega^2\right]\sin\left(T_d\omega\right)+K_p^2\left[1-\left(K_fL_o+C_fL_o\right)\omega^2\right]^2} \tag{6}$$

$$\left|G_{\mathrm{clg}}\right|=\frac{\left|K_p\right|}{\sqrt{\left[\left(L_o+L_f\right)-C_fL_fL_o\omega^2\right]^2\omega^2-2K_p\omega\left[\left(L_o+L_f\right)-C_fL_fL_o\omega^2\right]\left[1-\left(K_fL_o+C_fL_o\right)\omega^2\right]\sin\left(T_d\omega\right)+K_p^2\left[1-\left(K_fL_o+C_fL_o\right)\omega^2\right]^2}} \tag{10}$$

$$\left|Y_{\mathrm{clg}}\right|=\frac{\sqrt{K_p^2\left(K_f+C_f\right)^2\omega^2-2K_p\omega\left(K_f+C_f\right)\left(1-C_fL_f\omega^2\right)\sin\left(T_d\omega\right)+\left(1-C_fL_f\omega^2\right)^2}}{\sqrt{\left[\left(L_o+L_f\right)-C_fL_fL_o\omega^2\right]^2\omega^2-2K_p\omega\left[\left(L_o+L_f\right)-C_fL_fL_o\omega^2\right]\left[1-\left(K_fL_o+C_fL_o\right)\omega^2\right]\sin\left(T_d\omega\right)+K_p^2\left[1-\left(K_fL_o+C_fL_o\right)\omega^2\right]^2}} \tag{11}$$

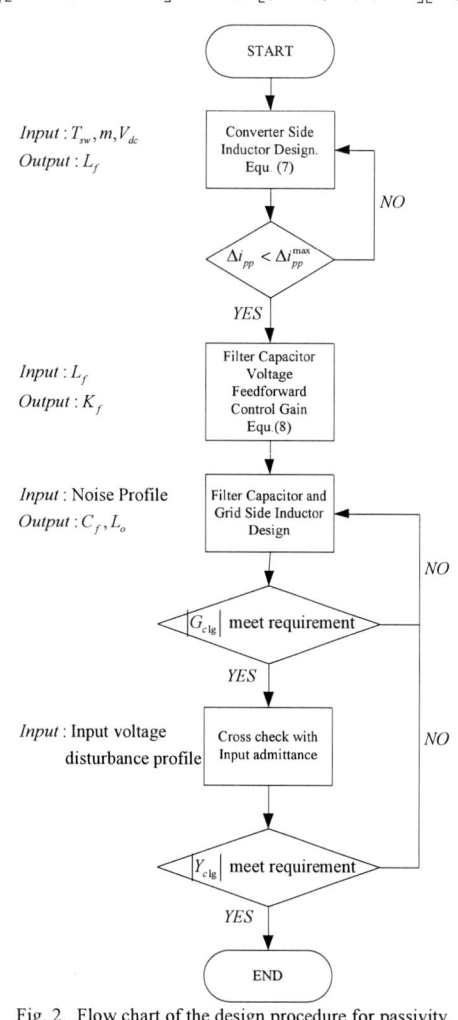

Fig. 2. Flow chart of the design procedure for passivity.

For guaranteeing the passivity of the converter, the numerator of (6) must stay positive in the frequency range under consideration. Due to the system delay described in (2) and the inductance of the converter side inductor is chosen according to (7). The feedforward control gain can be designed as (8) then the input admittance of the converter will remain passive up to (9).

$$K_f=-\frac{4T_d^2}{L_f\pi^2} \tag{8}$$

$$\omega=\pi f_s \tag{9}$$

3. Filter Capacitor and Grid Side Inductor Design:

The converter output current is related to two transfer functions with two inputs according to equation (3). One is the closed-loop current loop gain G_{clg} with current command i_f^{ref} and the other one is the input admittance Y_{clg} with input voltage v_g. The magnitude of the closed-loop current loop gain and the input admittance are presented as (10) and (11), respectively.

Firstly, the closed-loop current loop gain is considered as noise amplifier if the input voltage is ideal and sinusoidal. For simplifying equation (10), the resonance frequency of the LCL filter and the resonance frequency of the filter capacitors with grid side inductors are assumed to be high enough. This assumption means the capacitance of the filter capacitors cannot be too large. It is a common design guideline because the reactive power absorbed by the converters depends on the filter capacitors [20].

The peak value of the closed-loop current loop gain will appear close to but not equal to the LCL resonant frequency. By designing the filter capacitor and the closed-loop current loop gain at LCL resonant frequency, the peak value of the closed-loop current loop gain can be restricted effectively.

Equation (12) is the gain at the LCL resonant frequency. Define $C_f=-aK_f$ when K_f is designed as (8). Equation (13) shows that the closed-loop current loop gain at the LCL resonant frequency relates to the inductors used in the converter and the ratio between the capacitance of the filter capacitor C_f and K_f.

TABLE I
SYSTEM PARAMETERS

Symbol	Meaning	Value
V_{DC}	DC bus voltage	700 V
V_g	Grid voltage	400V_{rms}, 50 Hz
S_{base}	Rated power	7.5 kVA
f_s	Switching frequency	10 kHz
K_p	Proportional gain	9, f_c =540 Hz
K_f	Feedforward gain	5.77E-06
L_f	Converter side inductor	1.58 mH
L_o	Grid side inductor	1.06 mH
L_t	Total inductance	2.64 mH(0.053 pu)
C_f	Filter capacitor	4.50 uF,5.77 uF and 19.8 uF

$$\left|G_{c\lg}\right|_{\omega=\sqrt{\frac{L_o+L_f}{C_f L_f L_o}}}=\cfrac{1}{1-\left(K_f L_o+C_f L_o\right)\cfrac{L_o+L_f}{C_f L_f L_o}} \qquad (12)$$

$$\left|G_{c\lg}\right|_{\omega=\sqrt{\frac{L_o+L_f}{C_f L_f L_o}}}=\cfrac{1}{1+\cfrac{(1-a)}{a}\cfrac{L_o+L_f}{L_f}} \qquad (13)$$

Fig. 3 shows the closed-loop current loop gain with three different filter capacitors, which a equals to 0.78, 1 and 3.43, respectively. The other system parameters are shown in Table I.

To explain the meaning of the closed-loop current loop gain further, the dashed line in Fig. 3 is drawn according to the current distortion limits given in the IEEE 519-2014 [19], which assume there is 10 mA white noise in the current loop. This dashed line varies with different noise profile. The noise profile can be a uniform probability distribution or not. The system needs to be identified to determine the source of the noise, e.g. noise from the switching power supply, coupling effect of the PCB layout.

For a 19.8 uF filter capacitor, the requirement is hardly satisfied with 10mA white noise. If the noise is under controlled, 19.8 uF filter capacitor can still meet the requirement. The switching of the VSC is the main source of the noise.

The current harmonic at the switching frequency can be limited and be attenuated by using larger filter capacitor, as the example $a=3.43$ shown in Fig. (2). The increase of the grid side inductor can also improve the attenuation at the switching frequency. On the other hand, the harmonic current at the switching frequency will have less attenuation with a small filter capacitor like the example $a=0.78$ shown in Fig. (2), which is not acceptable in some applications and will increase the cost of the EMI filter [2].

Secondly, the grid voltage will never be ideal in practical power system. The grid voltage may contain some harmonics.

According to IEEE 519-2014, the individual voltage harmonics distortion limits are specified that the individual voltage harmonics need to be below 5% of the nominal voltage.

The magnitude of the input admittance is presented in (11). The phase of the input admittances is between negative 90 degrees to positive 90 degrees that means the passivity is guaranteed by designing the filter capacitor voltage feedforward control gain. With lower magnitude of the input admittance, the converter has a better immunity to the grid voltage harmonics.

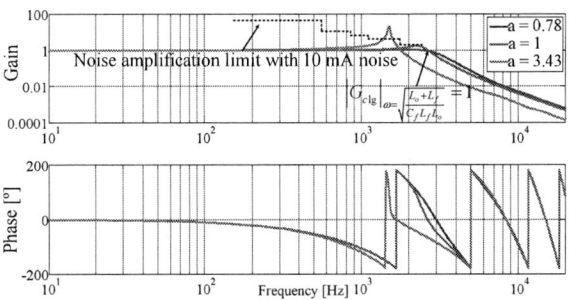

Fig. 3. Magnitude and phase of the closed-loop current control loop gain with different filter capacitors.

Fig. 4 presents the input admittance with three different filter capacitors. The dashed line is drawn according to 5% individual input voltage harmonics, which is the strictest condition. If the current harmonics need to fulfill the requirement under this distorted voltage then the inductance of the inductors must increase, which will increase the cost of the converter and the volume of the converter. The solid line is drawn according to 1% individual input voltage harmonics and the converter can meet most part of the requirement under this condition but if the input voltage harmonics occur between 23th and 50th of the base frequency then the current harmonics will exceed the limits.

According to (10) and (11), all the passive components and the parameters of the controller affect the closed-loop current loop gain and the input admittance. The design procedure 1 and 2 guarantee the converter passivity and make sure the converter side current ripple is limited that is the safety issue of the switching components.

The converter side inductor is designed in procedure 1 and the filter capacitor voltage feedforward control gain only depends on the converter side inductor. The design of the filter capacitor and the grid side inductor is independent of procedure 1 and 2. Procedure 3 decides the filter capacitor and the grid side inductor. The magnitude of the closed-loop current loop gain and the magnitude of the input admittance are studied.

There are many options to determine the filter capacitor and the grid side inductor according to equation (13). The options are reduced when a cross check with the magnitude of the input admittance is done, which is presented in (11). The iterations are required to meet the current harmonic emission limit.

In this paper, the IEEE 519-2014 is used to determine the performance of the converter. From the passivity point of view, the model depicted in Fig. 1(c) is the most important model. Even though there are different types of filters and different controller designs, the design procedure can be applied to different topologies that can help to estimate the performance of the converter.

1714

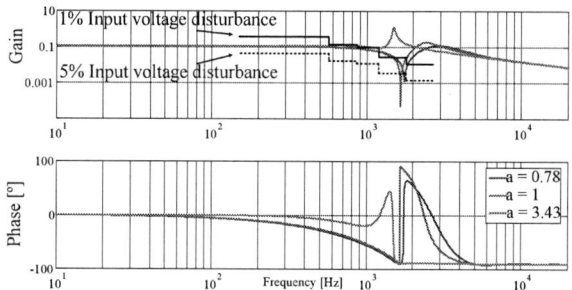

Fig. 4. Magnitude and phase of the input admittance with different filter capacitors.

III. SIMULATION RESULTS

The system parameters are listed in Table I. The capacitance of the filter capacitors are chosen to be 4.5 uF and 19.8 uF, respectively. The inductance of the inductors are 1.58 mH and 1.06 mH, respectively. The values of the passive components are measured by an LCR meter and used in the following simulation and experimental sections. According to the design procedure, the value of the passive components is optimized but the choice of the passive components is limited, e.g. the capacitance of the filter capacitor has certain values and as well as the tolerance of the passive components. The simulation results are shown from Fig. 5 to Fig. 7. The time-domain current waveforms are not much different. The Fast Fourier Transform (FFT) of the grid side current can be used to verify the theoretical results. The solid line in the FFT waveforms of the grid side current is drawn according to the current distortion limits in IEEE 514-2014.

In Fig. 5, a 4.5 uF filter capacitor is used. The simulation result shows the current harmonics fulfill the requirement that is recommended by IEEE 519-2014. As mentioned in the previous section, the peak value of the current harmonics at the switching frequency is about 20 mA, which may cause EMI problems.

On the other hand, when using 19.8 uF filter capacitor as shown in Fig. 6 the attenuation of the switching frequency harmonic current is improved. The drawback of using larger filter capacitor is that the current harmonics are close to 1.6 kHz and exceed the IEEE 519-2014 recommended current harmonics limit, which is unwanted.

Since the voltage source used in the simulation is ideal, the FFT spectrum of the grid side current shows that the noise in the simulation circuit is less than 10 mA but large than 1 mA. The noise profile has very different characteristics in the practical system compare with the simulation conditions.

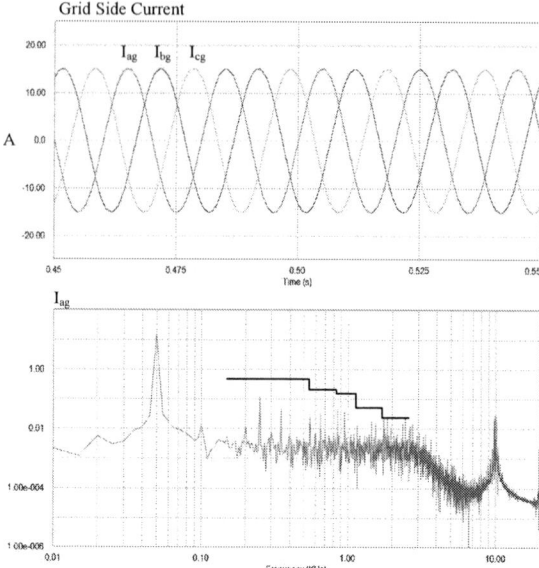

Fig. 5. Simulation result (upper) and the FFT waveform (lower) of the grid side current with filter capacitor 4.5 uF.

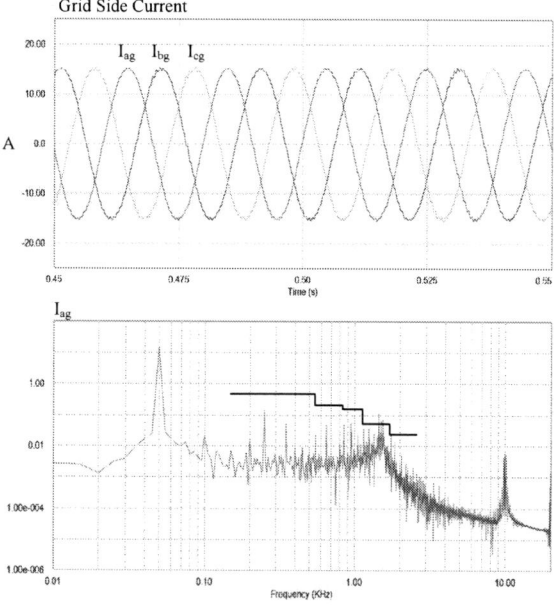

Fig. 6. Simulation result (upper) and the FFT spectrum (lower) of the grid side current with filter capacitor 19.8 uF.

In Fig. 7, the 4.5 uF filter capacitor is used. The result shows the simulation result with 7.8 V, 1250 Hz line voltage harmonic injection. The induced current harmonics is about 250 mA, which is above the limits depicted in IEEE 519-2014. The injection voltage is the 25th order voltage harmonics and the value is 1.4% of the nominal voltage. The simulation results validate the theoretical results.

1715

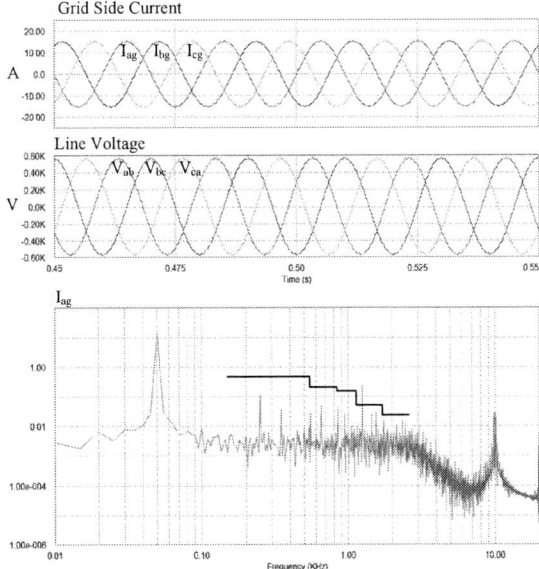

Fig. 7. Simulation result (upper) and the FFT spectrum (lower) of grid side current with 4.5 V, 1250 Hz voltage injection.

IV. EXPERIMENTAL RESULTS

The system parameters are listed in Table I. The value of the passive components are measured by LCR meter. A regenerative Grid Simulator, Model 61800 Series, which is designed by Chroma generate the grid voltage. As shown in Fig. 8(b), the grid voltage is not as good as shown in the simulations. The controller of the converter is implemented by using a dSPACE system.

A 4.5 uF filter capacitor is used in Fig. 8. The harmonics current fulfil the requirement that is recommended in IEEE 519-2014. There are two groups of current harmonics that are noticeable in the frequency spectrum. One is the switching frequency that is depicted at the right circle in fig. 8(a). These current harmonics even influence the grid simulator, which induce voltage harmonics at these frequencies. The other group of current harmonics is closed to 3.3 kHz that is influenced by the voltage harmonic generate by grid simulator. The voltage harmonic is circled out in Fig. 8(b). The magnitude of this voltage harmonic is about 0.4 V and the induced current harmonics are about 30 mA.

In Fig. 9, 19.8 uF filter capacitor is used. The frequency spectrum of the line voltage shows in Fig. 9(b) that the grid simulator is just slightly influenced by the converter because the attenuation of the current harmonics closed to switching frequency is improved. On the other hand, the current harmonics closed to 30th order of the base frequency in Fig. 9(b) are increased. The reason for this result is that the closed-loop current loop gain has a larger magnitude compare with the result in Fig. 8, which is discussed in section II.

As Fig. 9(a) shows, the current harmonics are closed to the boundary but do not exceed it.

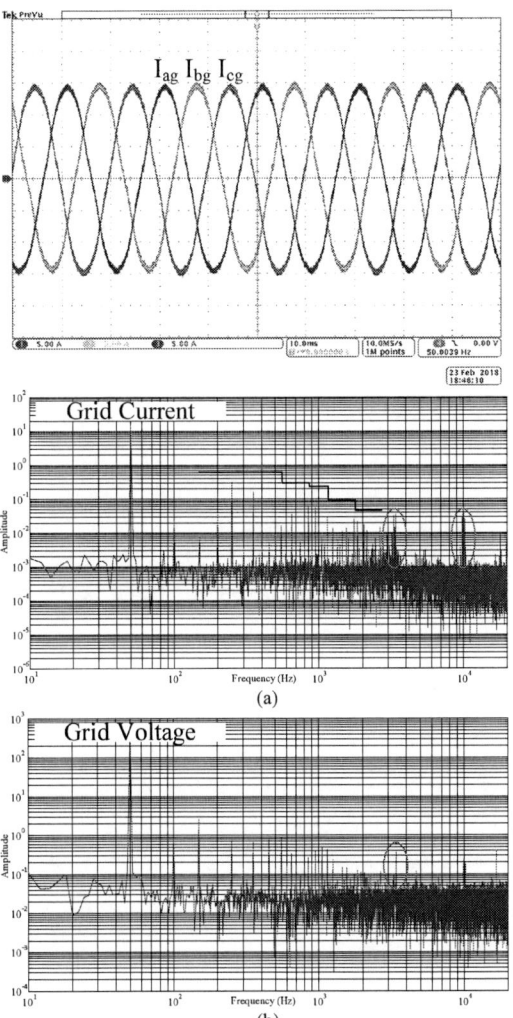

Fig. 8. (a) Experimental results and the FFT spectrum of the grid side current with filter capacitor 4.5 uF. (b) The FFT spectrum of the line voltage.

There are many facts that can lead to the result. First, the non-ideal characteristics of the inductor and filter capacitor might reduce the harmonic currents, e.g. skin effect of the inductors, the equivalent series resistance (ESR) of the filter capacitor. Secondly, the noise profile in the system has a lower value compared with the simulation results.

In Fig. 10 the 4.5uF filter capacitor is used. The result shows the experiment result with 7.8 V, 1250 Hz line voltage harmonic injection. The induced current harmonics is about 650 mA, which is above the limit depicted by IEEE 519-2014. The injection voltage is the 25th order voltage harmonics and the value is 1.4% of the nominal voltage. The dashed line shown in Fig. 9(b) depicts 1% voltage harmonic limit. The result shows that even though the value of the voltage harmonics is almost nothing but the induced harmonic current can still exceed the current harmonics limit.

The 2018 International Power Electronics Conference

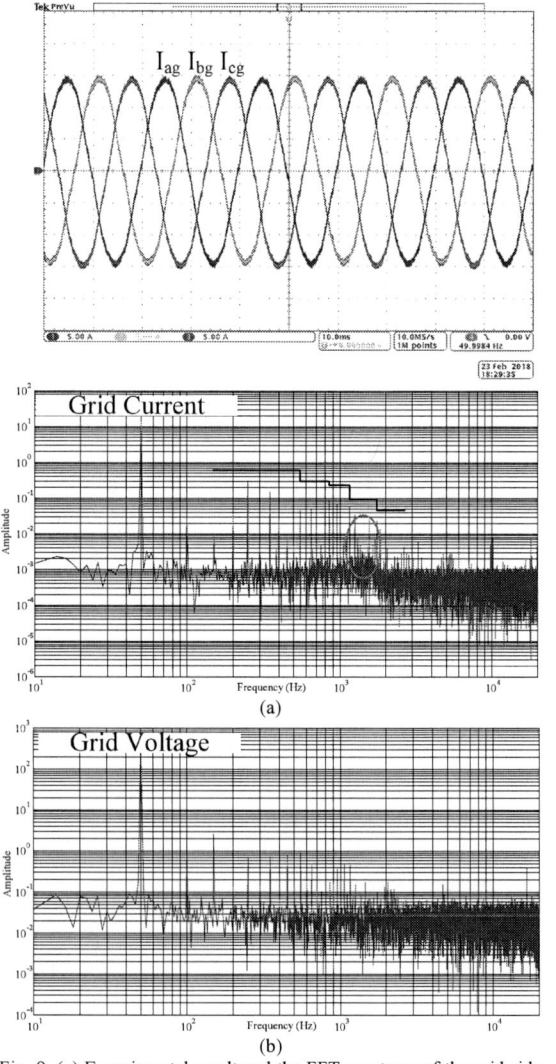

Fig. 9. (a) Experimental result and the FFT spectrum of the grid side current with filter capacitor 19.8 uF. (b) The FFT spectrum of the line voltage.

V. CONCLUSIONS

Designing the filter capacitor voltage feedforward control gain guarantees the passivity of the VSCs with the converter side current feedback. This paper presents a co-design method of the controller gains and the passive components.

The filter capacitor and the grid side inductor are designed according to the magnitude of the closed-loop current loop gain and the input admittance. The harmonic current emission is considered and limited. The simulation results and the experimental results validate the theoretical deduction.

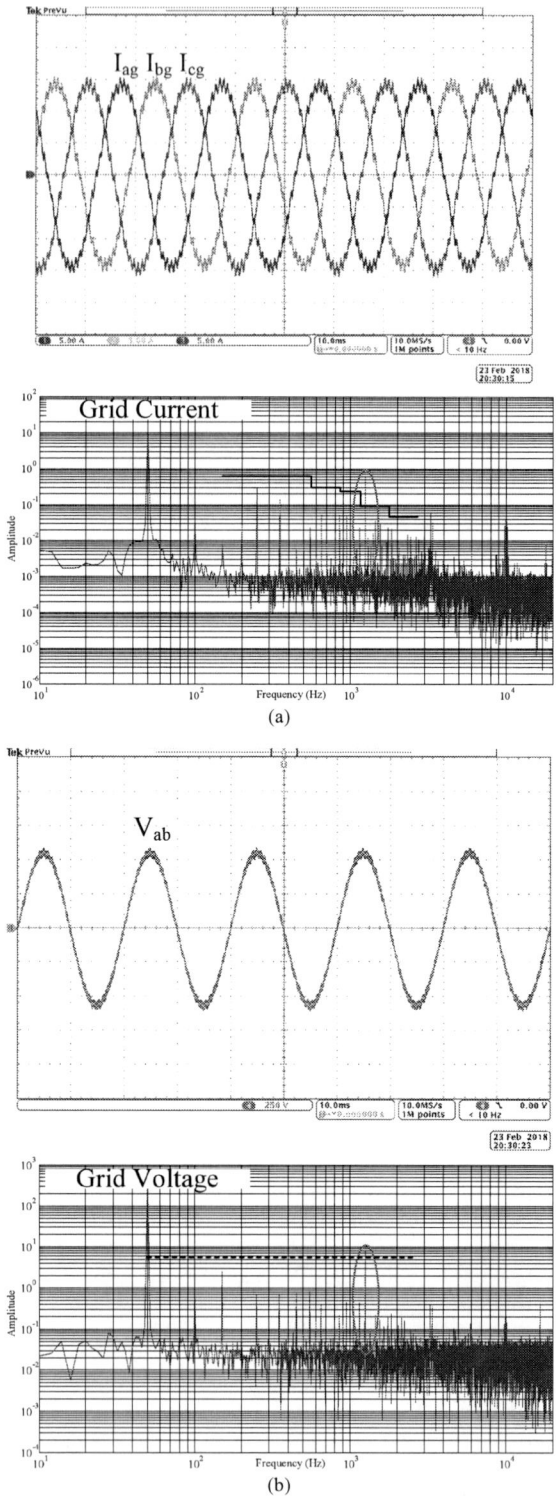

Fig. 10. (a) Experimental result and the FFT spectrum of the grid side current with 4.5 V, 1250 Hz voltage injection. (b) The waveform and the FFT spectrum of the line voltage.

REFERENCES

[1] F. Blaabjerg, Z. Chen, and S. Kjaer, "Power electronics as efficient interface in dispersed power generation systems," *IEEE Trans. Power Electron.*, vol. 19, no. 5, pp. 1184–1194, Sep. 2004.

[2] M. Liserre, F. Blaabjerg, and S. Hansen, "Design and control of an LCL-filter-based three-phase active rectifier," *IEEE Trans. Ind. Appl.*, vol. 41, no. 5, pp. 1281–1291, Sep-Oct. 2005.

[3] W. Wu, Y. He, and F. Blaabjerg, "An LLCL power filter for single-phase grid-tied inverter," *IEEE Trans. Power Electron.*, vol. 27, no. 2, pp. 782–789, 2012.

[4] M. Huang, X. Wang, P. C. Loh and F. Blaabjerg, "LLCL-filtered grid converter with improved stability and robustness." *IEEE Trans. on Power Electron.*, vol. 31, no. 5, pp. 3958–3967, May 2016 .

[5] X. Wang, F. Blaabjerg, M. Liserre, Z. Chen, J. He, and Y. Li, "An active damper for stabilizing power-electronics-based AC systems," *IEEE Trans. Power Electron.*, vol. 29, no. 7, pp. 3318–3329, Jul. 2014.

[6] X. Wang, F. Blaabjerg, Z. Chen, and W. Wu, "Resonance analysis in parallel voltage-controlled distributed generation inverters," in *Proc. 28th Annu. IEEE Appl. Power Electron. Conf. Expo.*, 2013, pp. 2977–2983.

[7] X. Wang, F. Blaabjerg, and P. C. Loh, "An impedance-based stability analysis method for paralleled voltage source converters," in *Proc. IEEE Int. Power Electron. Conf. (IPEC-Hiroshima 2014 - ECCE ASIA)*, 2014, pp. 1529–1535.

[8] X. Wang, F. Blaabjerg, and P. C. Loh. "Passivity-Based Stability Analysis and Damping Injection for Multiparalleled VSCs with LCL Filters." *IEEE Trans. on Power Electron.*, vol. 32, no. 11, pp.8922–8935, Nov. 2017.

[9] A. Riccobono and E. Santi, "A novel passivity-based stability criterion (PBSC) for switching converter DC distribution systems," in *Proc. IEEE Appl. Power Electron. Conf.*, 2012, pp. 2560–2567.

[10] L. Harnefors, L. Zhang, and M. Bongiorno, "Frequency-domain passivity-based current controller design," *IET Power Electron.*, vol. 1, no. 4, pp. 455–465, Dec. 2008.

[11] L. Harnefors, X. Wang, A. G. Yepes, and F. Blaabjerg, "Passivity-based stability assessment of grid-connected VSCs—An overview," *IEEE J. Emerg. Sel. Topics Power Electron.*, vol. 4, no. 1, pp. 116–125, Mar. 2016.

[12] L. Harnefors, A. G. Yepes, A. Vidal, and J. Doval-Gandoy, "Passivity-based controller design of grid-connected VSCs for prevention of electrical resonance instability," *IEEE Trans. Ind. Electron.*, vol. 62, no. 2, pp. 702–710, Feb. 2015.

[13] X. Wang, F. Blaabjerg, and P. C. Loh, "Proportional derivative based stabilizing control of paralleled grid converters with cables in renewable power plants," in *Proc. IEEE Energy Convers. Congr. Expo. (ECCE)*, 2014, pp. 4917–4924.

[14] X. Wang, R. Beres, F. Blaabjerg and P. C. Loh, "Passivity-based design of passive damping for LCL-filtered voltage source converters." in *Proc. IEEE Energy Convers. Congr. Expo. (ECCE)*, 2015, pp. 3718–3725.

[15] X. Wang, F. Blaabjerg, and P. C. Loh, "High-performance feedback-type active damping of LCL-filtered voltage source converters," in *Proc. IEEE Energy Convers. Congr. Expo.*, 2015, pp. 2629–2636.

[16] D. M. Van de Sype, K. D. Gusseme, F. D. Belie, A. P. Van den Bossche, and J. A. Melkebeek, "Small-signal z-domain analysis of digitally controlled converters," *IEEE Trans. on Power Electronics*, vol. 21, no. 2, pp. 470–478, Mar. 2006.

[17] X. Wang, L. Harnefors, and F. Blaabjerg. "Unified Impedance Model of Grid-Connected Voltage-Source Converters." *IEEE Transactions on Power Electronics*, vol. 33, no. 2, pp. 1775–1787, Feb. 2018.

[18] G. Grandi, and J. Loncarski, "Evaluation of current ripple amplitude in three-phase PWM voltage source inverters," In *Compatibility and Power Electronics (CPE), 2013 8th International Conference on*, 2013, pp. 156–161.

[19] "IEEE Recommended Practice and Requirements for Harmonic Control in Electric Power Systems," in *IEEE Std 519-2014 (Revision of IEEE Std 519-1992)*, pp. 1–29, June 2014.

[20] M. Liserre, F. Blaabjerg, A. Dell Aquila, "Step-by-step design procedure for a grid-connected three-phase PWM voltage source converter," *International Journal of Electronics*, vol. 91, no. 8, pp. 445–460, Aug. 2004.

The 2018 International Power Electronics Conference

Adaptive Control of DC Power Distribution Systems: Applying Pseudo-Random Sequences and Fourier Techniques

Tomi Roinila[1*], Hessamaldin Abdollahi[2], Silvia Arrua[2], Enrico Santi[2]

1 Laboratory of Automation and Hydraulics, Tampere University of Technology, Tampere, Finland

2 Department of Electrical Engineering, University of South Carolina, Columbia, United States

*E-mail: tomi.roinila@tut.fi

Abstract—**Interconnected AC and DC power-distribution systems are prone to interactions that can reduce system stability. Since the system parameters usually vary over time and with operating conditions, adaptive controllers are required to guarantee desired system operation. Recent studies have presented adaptive methods based on online measurements of converter small-signal transfer functions. This paper proposes a computationally efficient strategy for adaptive control that enables each converter in a system to adjust its own local control. In the paper, a pseudo-random-binary-sequence (PRBS) injection is applied, and the converter control-to-output frequency response is measured online. The controller loop gain is quickly computed based on the measured control-to-output response. A loop-shaping technique is then utilized for adaptive control design to obtain desired stability margins. Advantages of the presented method compared to previously presented techniques are that it does not require a parametric system model for control design, and that the frequency-response measurements do not require opening the feedback loop. The online adaptive control method is fully implementable in the converter digital controller. Experimental results are presented and used to demonstrate the effectiveness of the proposed method.**

Keywords—*Power converters, frequency response, adaptive control, measurements, DC system stabilization*

I. INTRODUCTION

Dc power distribution systems consisting of multiple feedback-controlled switching-power converters have become extensively used in various industrial applications, such as in telecommunication systems, aircrafts, and electric cars [1], [2]. Advantages of such systems include power interface flexibility, reduced weight and size, highly efficient energy conversion, and easy incorporation of dc-type renewable resources [3]. One of the problems of dc distribution systems is the possible stability degradation that occurs when multiple converters are connected to a common bus [4].

Recent studies have presented wideband identification techniques for measuring converter small-signal transfer functions required for stability analysis and adaptive-control design [5]–[10]. In these methods, a broadband signal such as pseudo-random binary sequence (PRBS) is injected, for example, on top of the controller reference voltage. Then, the output voltage and/or current is measured, and the desired small-signal transfer function

is computed through Fourier techniques. The authors in [5] applied this PRBS technique and measured the converter control-to-output transfer function. A parametric model was then computed and used for control design. The authors in [6] applied an adaptive control strategy based on converter open-loop frequency-response measurements. Converter impedance-measurement techniques were shown in [8], [10], and [11], and a direct loop-gain-measurement technique in [9], but the methods were not applied for adaptive control.

This work extends previous studies and proposes a general, easy-to-implement adaptive-control approach, that can be used along with conventional control schemes. In the paper, the converter small-signal control-to-output transfer function is measured online using the PRBS injection and Fourier techniques. The measured response is then used for control design, and for computationally obtaining the converter loop gain. The proposed method is computationally efficient, does not require any curve-fitting methods, such as least-square, to obtain parametric models, and can be fully embedded in the converter control platform. The wideband measurement technique provides a fast transfer-function measurement, no external signal generator or data-acquisition device are required, and all the transfer functions can be measured without opening the feedback loop. The methods are versatile, and can be used in a number of applications including system monitoring, stability analysis and adaptive control design.

The remainder of the paper is organized as follows. Section II reviews the theory and methods applied in the paper; the theory of nonparametric system identification and maximum-length binary sequence is briefly reviewed, and an adaptive control method based on loop-shaping technique is introduced. Section III presents experimental results based on a custom-designed switching converter utilizing an adaptive digital controller. Section IV draws conclusions.

II. THEORY AND METHODS

A. System Identification

A power-converter system can be considered as a linear time-invariant system for small disturbances. According to basic control theory, this type of system can be fully characterized by its impulse response, which can

1719

be transformed into frequency domain and represented by a frequency-response function.

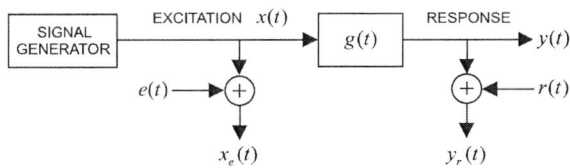

Fig. 1: Typical measurement set up.

Fig.1 shows a typical setup where the device under test, presented by an impulse-response function $g(t)$, is to be identified. The system is perturbed by the excitation $x(t)$, which yields the corresponding output response $y(t)$. The measured signals are corrupted by noise, as presented by $e(t)$ and $r(t)$. The measured excitation and output response can now be denoted by $x_e(t)$ and $y_r(t)$. The noise signals are assumed to resemble white noise and are uncorrelated with $x(t)$ and $y(t)$. All of the signals are assumed to be zero mean sequences.

B. Maximum-length binary sequence (MLBS)

Pseudo-random binary sequence (PRBS) is a periodic broadband signal based on a sequence of length N. The most commonly used signals are based on maximum-length binary sequences (MLBS). Such sequences exist for $N = 2^n - 1$, where n is an integer. They are popular because they can be easily generated using feedback shift register circuits. [12]

Fig. 2 shows a 12-bit-length MLBS in the time domain and in the frequency domain. The sequence is generated at 20 kHz and has signal levels ± 1V. The energy spectrum has an envelope and drops to zero at the generation frequency and its harmonics.

The MLBS x has the lowest possible peak factor $|x|_\text{peak}/x_\text{rms} = 1$ regardless of its length, which means that the sequence is well suited for sensitive systems that require small-amplitude perturbation. Due to the deterministic nature of the sequence, the signal can be repeated and injected precisely and the signal-to-noise ratio (SNR) can be increased by synchronous averaging of the response periods.

C. Frequency-Response Computation

There are several techniques for computing the frequency-response function from input-output data generated by a broadband injection [13]. The simplest method is to use straight Fourier transform of the measured voltage and current responses although this method does not cancel out the effect of external noise. In this paper, a logarithmic averaging procedure is proposed to obtain the frequency response $G(j\omega)$ as

$$G(j\omega) = \left(\prod_{k=1}^{P} \frac{Y_{rk}(j\omega)}{X_{ek}(j\omega)} \right)^{1/P} \qquad (1)$$

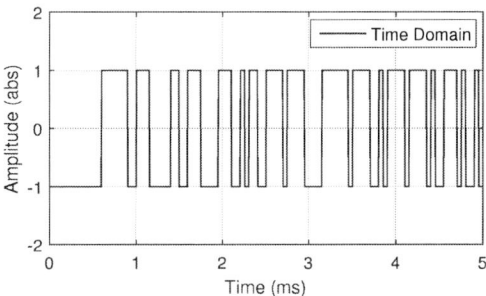

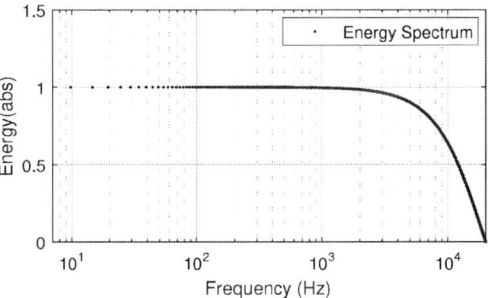

Fig. 2: MLBS in the time and frequency domain.

where P denotes the number of injected excitation periods, and Y_r and X_e are the measured output and input sequences, respectively. In this method, the measurements from both input and output sides are segmented and Fourier transformed according to the length of excitation, after which (1) is applied. The method tends to cancel out the effect of noise (that is uncorrelated with the measured input and output), both from input and output sides [13].

D. System Model and Adaptive Control

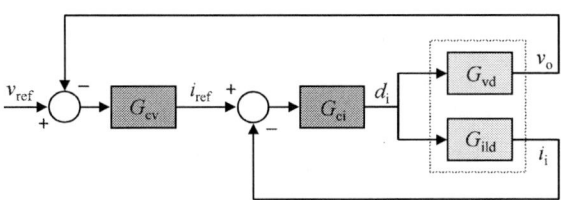

Fig. 3: Block diagram of the converter under study.

In this work, the system under study is a DC/DC buck converter shown by a block diagram in Fig. 3. The converter operates under nested control: an inner current loop regulates the inductor current i_i with respect to the reference i_ref that is produced by an outer voltage-loop. The open-loop converter model is represented by the transfer functions i) from control to output voltage G_vd, and ii) from control to inductor current G_ild. The transfer functions G_ci and G_cv represent the PI controllers used for current and voltage loops. The detailed small-signal closed-loop model of the converter can be found in [14]. Having the current-loop closed, the control-to-output transfer function with respect to the outer voltage loop is the transfer function G_{co} from the reference inductor current i_{ref} to the output voltage v_o, given by

$$G_{co}(s) = G_{vd}(s)\frac{G_{ci}(s)}{1 + G_{ci}(s)G_{ild}(s)} \qquad (2)$$

The voltage loop gain can be given as

$$T_v(s) = G_{cv}(s)G_{co}(s) \qquad (3)$$

In this work, the controller adaptation is implemented on the voltage controller G_{cv}. The design for the controller is based on the following conditions:

1) $\parallel T_v(j2\pi f_c) \parallel = 1$

2) $\sphericalangle T_v(j2\pi f_c) = PM - \pi$

where f_c and PM denote the desired cross-over frequency and phase margin, respectively. The conditions simply state that the loop gain has a gain of 0 dB at desired cross-over frequency f_c (condition 1), and the phase value has the desired margin (condition 2).

The adaptive control structure is based on the identification of the control-to-output frequency response $G_{co}(s)$ of the converter. In the structure, only the control-to-output frequency response G_{co} is measured, and the loop gain is calculated as in (3). The overall adaptive routine is summarized in Fig. 4.

The inputs to this routine are the MLBS injection, the desired bandwidth and phase margin. The routine starts by perturbing the control input which is the reference current and measuring the output voltage. This procedure yields the control-to-output transfer function $G_{co}(s)$. The loop-gain is then computed using (3) with the most recently updated controller transfer function from the previous cycle of adaptation. Then, the gain cross-over frequency and phase margin are computed. If the specified conditions are not met, $G_{co}(s)$ will be utilized for controller redesign. Finally, the controller transfer function gets updated with the calculated gains of the PI controller.

The proposed routine guarantees that the controlled converter has a bandwidth and phase margin that are approximately equal to the specified desired values. Therefore, the dynamic behavior of the system can be kept fixed and very close to the desired performance. It is emphasized, that in the proposed routine no model-fitting technique is used. The non-parametric results of the identification procedure are directly utilized for stability verification and controller design. This is in fact very advantageous from a practical standpoint, as it guarantees fast controller adaptation, while only limited computational resources are required. Equally important, the proposed structure makes it possible to perform the monitoring and stability verification by computationally obtaining the loop gain while measuring only the control-to-output frequency response. Therefore, no direct loop gain measurement or loop opening are required, and thus the converter can continue its normal operation. It should also be emphasized that the proposed algorithm

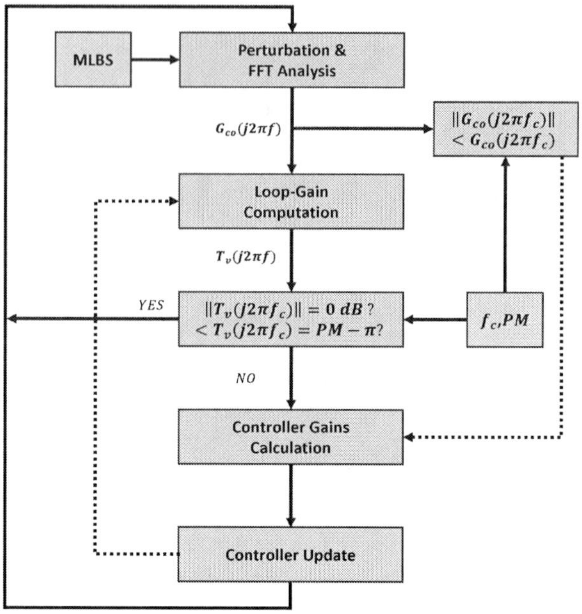

Fig. 4: Adaptive control routine.

is a general method that can be used to enhance conventional control schemes, so that they can perform on-the-fly adjustments with respect to variations occurring in the operating conditions of the converter. For the most common cases of PI and PID controllers, this algorithm can be directly applied, as is done in this work for the case of a PI controller.

III. EXPERIMENTAL RESULTS

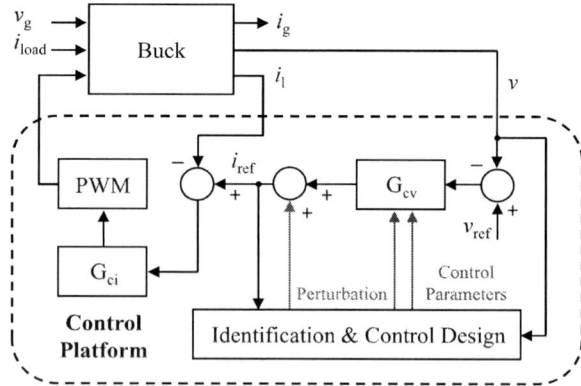

Fig. 5: Block diagram of adaptively controlled Buck converter.

The DC-DC switching converter with an adaptive digital controller depicted in Fig. 5 was constructed in the laboratory using a custom designed IGBT-based switching converter. The converter (Buck) operates under feedback control with an inner current loop and an outer voltage loop based on PI control strategy. The parameters of the outer loop voltage controller are regularly updated using the proposed method. The voltage controller is designed such that the outer loop gain has a crossover frequency of 150 Hz and a phase margin of 65 degrees.

To verify the operation of the adaptive controller an interconnected system shown in Fig. 6 was built in which the buck converter acts as source converter and supplies a load voltage source inverter (VSI). The source converter is supplying the load VSI with 500 W at 100 V. The load VSI is supplying a resistive load with a peak voltage of 45 V. The switching frequency for both converters is 10 kHz.

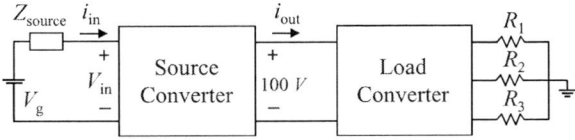

Fig. 6: Interconnected system under test.

A 4095-bit-length MLBS was designed using a shift register with XOR feedback [15]. The PRBS was generated at 20 kHz, and thus, the frequency resolution was 20 kHz/4095 $\approx$ 4.88 Hz. The designed MLBS was injected into the nominal current reference and the converter output voltage was measured. The data was collected after which Fourier methods were applied and the control-to-output response $G_{co}(s)$ was computed. The controller transfer function $G_{cv}(s)$ is known at the beginning of each adaptation cycle, and the online computation of the loop gain is done as shown in (3). The MLBS was applied for 5 periods, and, hence, the measurement time was approximately 1.02 s. The injection amplitude was adjusted such that converter output voltage deviated from its nominal value no more than 5 percent.

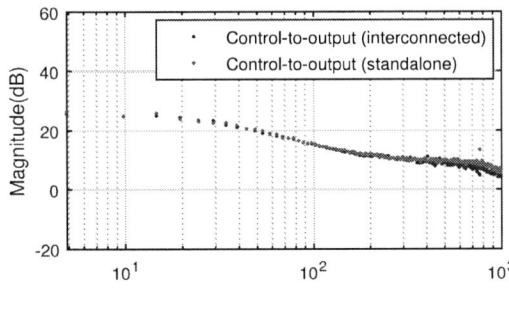

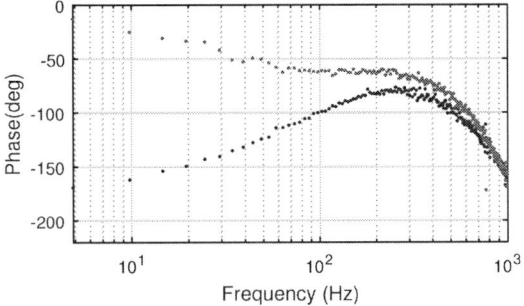

Fig. 7: Measured control-to-output transfer function.

Fig. 7 shows the measured control-to-output responses (with and without the load inverter). The figure clearly

shows how the constant power load nature of the tightly regulated load converter causes a significant phase deterioration. Fig. 8 shows the computed loop gain (with the load inverter). As a reference, the loop gain is also experimentally measured by directly injecting the MLBS into the loop. As the figure shows, the computationally obtained loop gain accurately follows the reference in a wide frequency band.

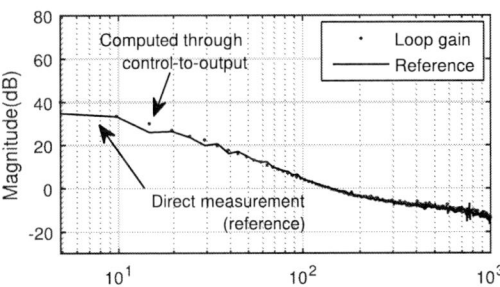

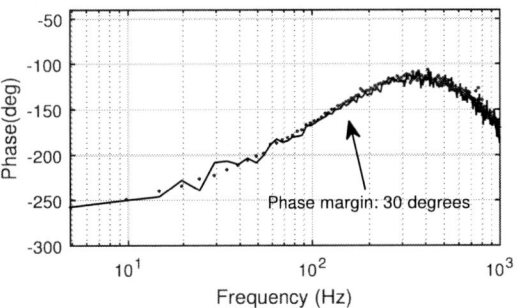

Fig. 8: Measured loop gain.

The controller design for the buck converter is originally done for a resistive load (standalone case). However, in this test case the source buck converter is supplying a tightly regulated converter, which exhibits constant-power-load behavior. This can be verified in the control-to-output frequency response (see Fig. 7) where the phase is different for the two cases: the phase approaches zero at low frequencies for the standalone resistive case and -180 degrees for the interconnected case. This demonstrates the importance of control adaptation in an interconnected system in order for the source subsystem to preserve sufficient stability margins. In fact, the actual phase margin is deteriorated to as low as 30 degrees instead of the desired 65 degrees as shown in Fig. 8. At higher power levels, the situation would become worse and the deviation in effective stability margins with respect to the design values would further increase. The adversely affected stability margin of the source can easily drive the overall system into poorly performing or unstable modes.

Fig. 9 shows the loop gain after the control adaptation procedure; the phase margin is returned to the desired value of 65 degrees with the desired 150 Hz crossover frequency. The time-domain step response in the bus voltage, shown in Fig. 10, shows improvement: with adaptive

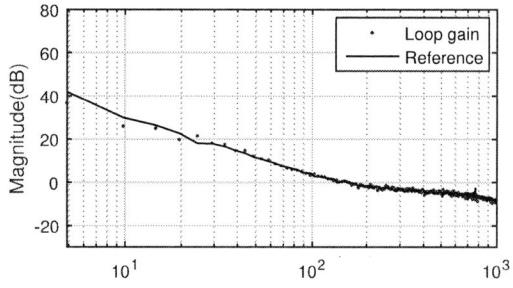

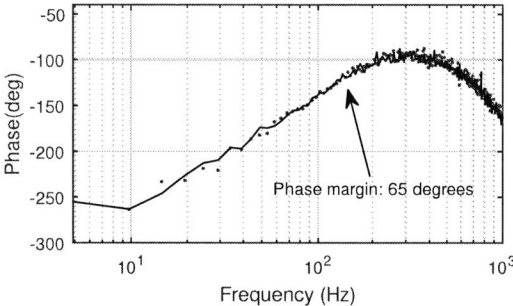

Fig. 9: Computed and measured loop-gain after controller update.

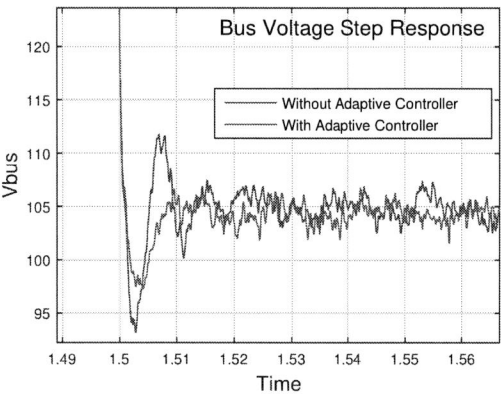

Fig. 10: Step in the bus voltage for both cases.

control the bus voltage oscillations are well damped. The results confirm the effectiveness of the proposed method in improving the stability and performance status of the interconnected system.

IV. CONCLUSIONS

An adaptive-control algorithm based on integrated wideband system-identification implemented in the converter's controller is proposed. Unlike most previously presented methods, the technique does not require parametric small-signal models or opening the feedback loop. In the proposed method, the load-affected control-to-output transfer function is measured. The measurement result and the most recently updated controller transfer function are then used for the computation of the loop-

gain, evaluation of stability margins, and controller redesign. The proposed method is well suited for online stability enhancement of DC power distribution systems in which the stability margins are affected by the interactions of converter feedback loops. Unlike most previously presented methods, the technique does not require any parametric fitting, and therefore, the method is computationally highly efficient. The proposed procedure for the computation of loop-gain eliminates the need for loop opening, and therefore, the control adaptation does not disturb the normal system operation. The effectiveness of the proposed method was experimentally validated in an interconnected system.

REFERENCES

[1] M. Baran and N. Mahajan, "DC distribution for industrial systems: Opportunities and challenges," *IEEE Trans. on Industry Applications*, vol. 39, no. 6, pp. 1596–1601, 2003.

[2] Q. Zhou and D. Yuan, "Prediction on future dc power systems," in *Proc. IEEE International Power Electronics and Motion Control Conference*, 2009, p. 11921195.

[3] A. Riccobono and E. Santi, "Comprehensive review of stability criteria for DC power distribution systems," *IEEE Trans. on Industry Applications*, vol. 50, no. 5, pp. 3525–3535, 2014.

[4] A. Emadi, A. Khaligh, C. Rivetta, and G. Williamson, "Constant power loads and negative impedance instability in automotive systems: definition, modeling, stability, and control of power electronic converters and motor drives," *IEEE Trans. on Vehicular Technology*, vol. 55, no. 4, pp. 1112–1125, 2006.

[5] B. Miao, R. Zane, and D. Maksimovic, "Automated digital controller design for switching converters," in *Proc. IEEE Power Electronics Specialists Conference*, 2005, pp. 2729–2735.

[6] M. Shirazi, R. Zane, and D. Maksimovic, "An autotuning digital controller for DC-DC power converters based on online frequency-response measurement," *IEEE Trans. on Power Electronics*, vol. 24, no. 11, pp. 2578–2588, 2009.

[7] A. Barkley and E. Santi, "Improved online identification of a DC/DC converter and its control loop gain using cross-correlation methods," *IEEE Trans. on Power Electronics*, vol. 24, pp. 2021–2031, 2009.

[8] A. Barkley, R. Dougal, and E. Santi, "Adaptive control of power converters using digital network analyzer techniques," in *Proc. Applied Power Electronics Conference and Exposition*, 2011, pp. 1824–1832.

[9] T. Roinila, M. Vilkko, and T. Suntio, "Fast loop gain measurement of switched-mode converter using binary signal with specified Fourier amplitude spectrum," *IEEE Trans. on Power Electronics*, vol. 17, no. 6, pp. 2746–2755, 2009.

[10] ——, "Frequency-response measurement of switched-mode power supplies in the presence of nonlinear distortions," *IEEE Trans. on Power Electronics*, vol. 25, pp. 2179–2187, 2010.

[11] T. Roinila, H. Abdollahi, S. Arrua, and E. Santi, "Online measurement of bus impedance of interconnected power electronic systems: Applying orthogonal sequences," in *Proc. IEEE Energy Conversion Congress and Exposition*, 2017, pp. 5783–5788.

[12] K. Godfrey, *Perturbation Signals for System Identification*. Prentice Hall, UK, 1993.

[13] R. Pintelon and J. Schoukens, *System Identification - A Frequency Domain Approach*. The Institute of Electrical and Electronics Engineers, Inc. New York, 2001.

[14] J. Siegers, S. Arrua, and E. Santi, "Stabilizing controller design for multi-bus mvdc distribution systems using a passivity based stability criterion and positive feed-forward control," *IEEE Journal of Emerging and Selected Topics in Power Electronics*, vol. 5, pp. 14–27, 2017.

[15] L. Ljung, *System Identification-Theory for the User*. Prentice Hall PTR, USA, 1999.

An Improved Finite-set Model Predictive Torque Control for Interior Permanent Magnet Synchronous Motor Drives

Xinan Zhang[1]*, Gilbert Foo[2] and Tung Ngo[2]

1 School of Electrical and Electronic Engineering, Nanyang Technological University, Singapore, Singapore
2 School of Engineering, Computer and Mathematical Sciences, Auckland University of Technology, Auckland, New Zealand
*E-mail: zhangxn@ntu.edu.sg

Abstract— This paper presents a deadbeat based finite-set model predictive torque control (FS-MPTC) for interior permanent magnet synchronous motor drives. The proposed method contributes to significantly reduce the comparatively high torque ripples experienced in conventional FS-MPTC by using two voltage vectors (VVs) in one sampling cycle. The duty ratio of active VV is determined based on deadbeat torque regulation principle, which equivalently increases the weighting of torque control in FS-MPTC. This may lead to flux de-regulation under certain circumstances. To overcome this obstacle, a simple but effective trade-off scheme is proposed based on flux errors. The validity of the proposed method is verified through simulations.

Keywords— *interior permanent magnet synchronous motor, predictive torque control, flux de-regulation, duty ratio.*

I. INTRODUCTION

Interior permanent magnet synchronous motor (IPMSM) has attracted extensive research attentions owing to its high torque and power density in a wide speed range [1]. It is one of the most promising candidates for high speed electrical drives. Two of the most popular control strategies for high performance IPMSM are known as field-oriented control (FOC) [2] and direct torque control (DTC) [3-4]. FOC is famous for its smooth torque and flux response under different operating conditions and constant switching frequency. Nonetheless, the usage of coordinate transformation, PI current controllers and space vector modulation (SVM) leads to complex control structure. Moreover, FOC produces sluggish dynamic responses. In comparison, DTC features at simple control structure, good parameter robustness, less rotor position dependency and fast dynamics. The main drawbacks of DTC are high torque ripples and variable switching frequency [5]. In the past decade, a substantial amount of research efforts have been made to improve the torque response of DTC while maintaining its desirable features.

Among these works, finite-set model predictive torque control (FS-MPTC) is highly recommended for IPMSM drives due to its capability of selecting the optimal voltage vector (VV) dynamically and overcoming one cycle delay in digital control through prediction [6-7]. Compared with the conventional DTC, FS-MPTC maintains the features of fast dynamic response and reduced rotor position dependency while achieving smaller torque ripples with reduced stator current harmonics. This is realized through model based prediction and real-time cost function minimization. Generally, the cost function of FS-MPTC consists of weighted torque and flux errors, from where the optimal VV that produces the minimum torque and flux errors is selected and applied in the next sampling cycle.

Despite of its superior performance, the conventional FS-MPTC still suffers from comparatively high torque ripples because of the usage of only one VV in an entire sampling cycle. This effect is especially significant at low sampling frequencies. To further attenuate the torque ripples in FS-MPTC, new research proposals are published recently [8-11]. In [8], a new MPTC approach is proposed to improve both transient and steady-state performances of PMSM drives. Although good torque and flux regulations are obtained by this approach, its nature of offline torque ripple minimization makes it lose the capability of online optimization. Thus, both the one step ahead prediction for counteracting one cycle digital delay and the dynamic balancing of torque and flux errors become infeasible since they require the information from last control cycle. Some other methods, such as [9] and [10], provide more control flexibility in FS-MPTC by enlarging the finite-set using virtual VVs. With virtually more available VVs and SVM, these proposals attained excellent torque regulation with much smaller ripples for PMSM drives. Nevertheless, the computational complexity is increased to a large extent compared to conventional MPTC. Furthermore, precise rotor angular position is used in the torque and flux control by both methods, indicating that their rotor position dependency is high. In 2017, a discrete duty cycle control method [11] is proposed for MPTC of induction motors. It improves the steady-state torque response by firstly predicting the effect of zero VV on torque control and subsequently compensating for the possible deficiency of zero VV via a set of discretized new VVs. Except for the increased computational burden, the duty ratio of active VV in this approach is not determined based on torque or flux error. This implies a degradation on its accuracy

The 2018 International Power Electronics Conference

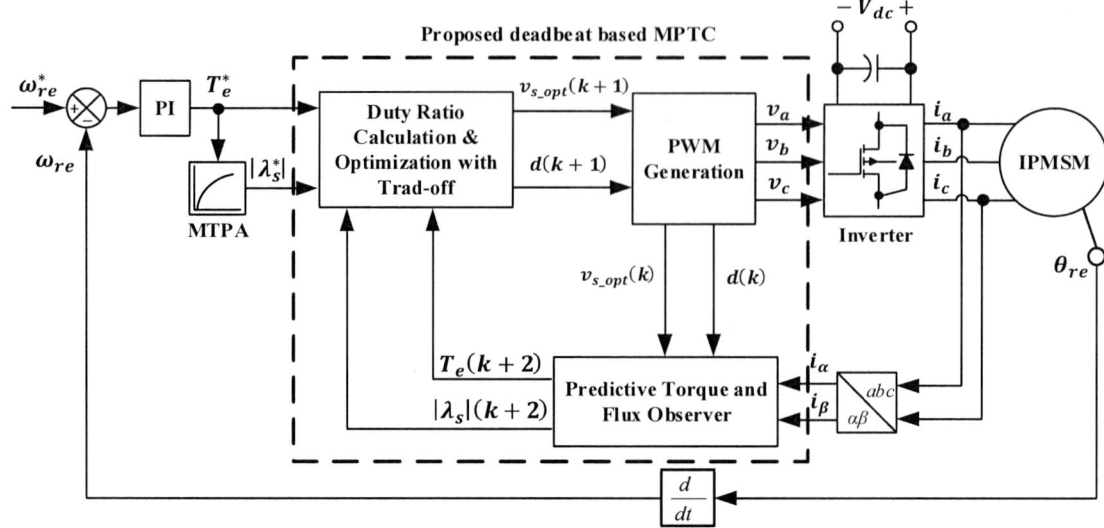

Fig. 1. Block diagram of the proposed model predictive torque control with improved stability.

To effectively reduce torque ripples experienced in conventional FS-MPTC while retaining its features of fast response and less rotor position dependency, this paper proposes a deadbeat based method with two VVs (including one active and one zero VV) applied in each sampling cycle. In addition, the possible flux de-regulation problem caused by the increased emphasis on torque control is discussed. A simple but effective trade-off between the conventional FS-MPTC and the proposed method is provided to ensure that flux regulation is not

deteriorated by torque ripple minimization. The block diagram of the proposed method is illustrated in Fig. 1.

The rest of this paper is organized as follows: Firstly, the mathematical model of IPMSM and the conventional MPTC are discussed in Section II. Then, the proposed deadbeat based duty ratio calculation is presented in Section III. After that, the problem of flux de-regulation is also discussed and a trade-off improvement is given to solve this problem in the same Section. Simulation results are demonstrated in Section IV to verify the validity of the proposed method. Finally, a conclusion is drawn in Section V.

To estimate $T_e(k+1)$ and $\lambda_s(k+1)$, where $(k+1)$ represents the next sampling cycle, the value of $\lambda_s(k)$ must be obtained first. In DTC, the stator flux can be estimated by (3),

$$\lambda_s(k) = \lambda_s(k-1) + T_s\big(v_s(k-1) - R_s i_s(k-1)\big) \quad (3)$$

where T_s is the sampling period. Based on the measured stator currents $i_s(k)$ at the current sampling cycle and $\lambda_s(k)$, $\lambda_s(k+1)$ is derived as

$$\lambda_s(k+1) = \lambda_s(k) + T_s\big(v_s(k) - R_s i_s(k)\big) \quad (4)$$

By referring to (2), it is seen that $i_s(k+1)$ is also need for the prediction of $T_e(k+1)$. Theoretically, stator current is predicted based on (5),

$$\frac{di_s}{dt} = -\frac{R_s}{L_q}i_s + \frac{\omega_{re}}{L_q}\big(\lambda_s - L_q i_s\big) + \frac{1}{L_q}v_s \quad (5)$$

where L_q denotes the stator inductance in q-axis and ω_{re} is rotor electrical speed. By using the Euler forward method, $i_s(k+1)$ is derived as:

$$i_s(k+1) = \left(1 - T_s\frac{R_s}{L_q}\right)i_s(k) + T_s\frac{\omega_{re}}{L_q}\big(\lambda_s(k) - $$

$$L_q i_s(k)\big) + \frac{T_s}{L_q}v_s(k) \quad (6)$$

II. IPMSM MODEL AND CONVENTIONAL FS-MPTC

A. Mathematical model of the IPMSM

The mathematical model of IPMSM in the stationary $(\alpha - \beta)$ reference frame is given in (1) and (2).

$$v_s = R_s i_s + \frac{d\lambda_s}{dt} \quad (1)$$

$$T_e = \frac{3}{2}P(\lambda_s \times i_s) \quad (2)$$

where

R_s		stator resistance
$v_s = [v_{s\alpha} \quad v_{s\beta}]^T$		stator voltage vector
$i_s = [i_{s\alpha} \quad i_{s\beta}]^T$		stator current vector
$\lambda_s = [\lambda_{s\alpha} \quad \lambda_{s\beta}]^T$		stator flux vector with
T_e		electromagnetic torque
P		number of pole pairs.

B. Conventional FS-MPTC

The conventional FS-MPTC is implemented in two steps: 1) torque and flux prediction based on predicted current, 2) iterative cost function minimization by comparing the numerical value of cost function with all the possible voltage vectors. This can be further illustrated as:

1) T_e and λ_s prediction

1725

Subsequently, $T_e(k+1)$ is predicted as

$$T_e(k+1) = \frac{3}{2}P(\boldsymbol{\lambda}_s(k+1) \times \boldsymbol{i}_s(k+1)) \quad (7)$$

To compensate for one cycle digital delay, the torque and flux have to be further predicted for one more step ahead by shifting forward one sampling cycle:

$$T_e(k+2) = \frac{3}{2}P(\boldsymbol{\lambda}_s(k+2) \times \boldsymbol{i}_s(k+2)) \quad (8)$$

where the predictions of $\boldsymbol{\lambda}_s^{k+2}$ and $\boldsymbol{i}_s^{k+2}$ are obtained as:

$$\boldsymbol{\lambda}_s(k+2) = \boldsymbol{\lambda}_s(k+1) + T_s(\boldsymbol{v}_s(k+1) - R_s\boldsymbol{i}_s(k+1)) \quad (9)$$

$$\boldsymbol{i}_s(k+2) = \left(1 - T_s\frac{R_s}{L_q}\right)\boldsymbol{i}_s(k+1) +$$
$$T_s\frac{\omega_{re}}{L_q}\left(\boldsymbol{\lambda}_s(k+1) - L_q\boldsymbol{i}_s(k+1)\right) + \frac{T_s}{L_q}\boldsymbol{v}_s(k+1) \quad (10)$$

2) Iterative cost function minimization

To obtain optimal torque and flux control, the value of cost function (11) has to be evaluated with all the possible VVs $\boldsymbol{v}_s(k+1)$. In (11), notation k_f represents the weighting factor of stator flux error, For three-phase, two-level inverter, the number of possible VVs is 7. It is noted that parameter k_f needs to be tuned to achieve the best dynamic response. A good starting point to select k_f is (12), where T_{rated} and λ_{rated} are the rated torque and stator flux, respectively.

$$J = \left|T_e^{ref} - T_e^{k+2}\right| + k_f\left||\lambda_s^{ref}| - |\lambda_s^{k+2}|\right| \quad (11)$$

$$k_f = \frac{T_{rated}}{\lambda_{rated}} \quad (12)$$

III. Proposed Deadbeat based Duty Ratio Calculation and Improvement on Flux Regulation

A. Deadbeat based Duty Ratio Calculation

In labeling figure axes please use words rather than As mentioned above, the application of only one VV in an entire sampling cycle inevitably generates high torque ripples. To further diminish the torque ripples, two voltage vectors, including one active VV and one zero VV, are used in the proposed method. This means VV $\boldsymbol{v}_s(k+1)$ becomes the combination of two VVs $\boldsymbol{v}_{s1}(k+1)$ and $\boldsymbol{v}_{s2}(k+1)$. Define the time duration of active VV $\boldsymbol{v}_{s1}(k+1)$ as t_1. Then, the time duration of zero VV $\boldsymbol{v}_{s2}(k+1) = \boldsymbol{0}$ is $(T_s - t_1)$. In this way, $T_e(k+2)$ can be re-derived from (8), (9), and (10) as

$$T_e(k+2) = \frac{3}{2}P\left[(\boldsymbol{K}_1 + t_1\boldsymbol{V}) \times \left(\boldsymbol{K}_2 + t_1\frac{1}{L_q}\boldsymbol{V}\right)\right] \quad (13)$$

where $\boldsymbol{K}_1 = \boldsymbol{\lambda}_s(k+1) - T_sR_s\boldsymbol{i}_s(k+1)$ $\quad (14)$

$$\boldsymbol{K}_2 = \left(1 - T_s\frac{R_s}{L_q}\right)\boldsymbol{i}_s(k+1) + T_s\frac{\omega_{re}}{L_q}\left(\boldsymbol{\lambda}_s(k+1) - L_q\boldsymbol{i}_s(k+1)\right) \quad (15)$$

$$\boldsymbol{V} = \boldsymbol{v}_{s1}(k+1) \quad (16)$$

To minimize the torque ripples, deadbeat principle is applied for torque regulation. It imposes the relation of $T_e(k+2) = T_e^*$, where T_e^* denotes the reference torque. Theoretically, this means the controller expects the produced torque at $(k+2)$-*th* sampling cycle can be equivalent to the reference torque. Such a relation leads to the determination of time duration t_1:

$$t_1 = \frac{T_e^*\left(\frac{2}{3P}\right) - (\boldsymbol{K}_1 \times \boldsymbol{K}_2)}{\left(\boldsymbol{K}_1\frac{1}{L_q} - \boldsymbol{K}_2\right) \times \boldsymbol{V}} \quad (17)$$

Using (17), the potentially optimal time duration for each active VV can be calculated. The resultant time duration will be substituted into cost function (11) for iterative evaluation to find out the overall optimal active VV.

B. Potential Flux De-regulation and Improvement

It is noteworthy that the deadbeat based torque control produces the potentially optimal time duration by considering only the minimization of torque error. It ignores the flux regulation. Even though flux error is still taken into account in cost function (11), the selected optimal VV, which originally satisfies both torque and flux control as in the conventional MPTC, is already adjusted by deadbeat based torque control. By giving preference to torque control, flux de-regulation might be caused since the adjusted optimal VV can be insufficient to satisfy both torque and flux control under some circumstances. In other words, the application of deadbeat principle equivalently increases the weighting of torque control in the optimization. Moreover, the increased weighting is varying dynamically with respect to the changing of operating conditions, such as load and speed. Consequently, it is difficult to ensure proper flux regulation under all operating conditions.

To solve this problem, we propose a trade-off scheme based on flux error. The amplitude of flux error is compared with a pre-defined threshold that guarantees acceptable flux regulation. If the amplitude of flux error is within such threshold, the control priority is given to torque ripple minimization based on deadbeat. Otherwise, the full optimal VV in the conventional MPTC is used to drive the flux error back into the area below such threshold. It is simple since only flux error is used in the scheme. Meanwhile, it is effective because balanced torque and flux control is proven to be achievable in conventional MPTC [7]. The threshold of flux error amplitude can be selected in a similar manner as DTC. In this paper, this threshold is chosen to be 3% of the rated flux.

C. Pseudo Algorithm of the Proposed Method

Based on the previous discussions in Parts A and B, the proposed control algorithm can be summarized as follows:

Step 1) Measure the stator current $\boldsymbol{i}_s(k)$ and rotor speed $\omega_{re}(k)$, estimate the stator flux $\boldsymbol{\lambda}_s(k)$ using (3), and set $\boldsymbol{v}_s(k)$ as the optimal voltage vector applied in the previous sampling cycle.

Step 2) Predict the stator flux $\boldsymbol{\lambda}_s(k+1)$ and stator current $\boldsymbol{i}_s(k+1)$ using (4) and (6), respectively.

Step 3) Compare flux error amplitude with the threshold:
1) if smaller than threshold, for each VV $\boldsymbol{v}_s^i(k+$

1), with $i = 0,1,\cdots,6$, calculate the optimal time duration using (17);

2) if larger than threshold, for each voltage vector $v_s^i(k+1)$, with $i = 0,1,\cdots,6$, set $t_1 = T_s$.

Step 4) Predict the torque $T_e(k+2)$ and stator flux $\lambda_s(k+2)$ using (8) to (9). Then, evaluate the value of cost function J_i using (11) for each VV $v_s^i(k+1)$.

Find the minimum cost function among J_i. Then, determine the corresponding optimal VV $v_s(k+1)$ with its time duration. Return to Step 1.

IV. SIMULATION RESULTS

To demonstrate the validity of the proposed method, simulation results are presented in this Section. The parameters of IPMSM used in the simulations are given in Table I. The sampling frequency of both conventional MPTC and the proposed MPTC is set at 5kHz.

TABLE I. PARAMETERS OF THE TEST IPMSM

Rated torque	4 Nm
Number of pole pairs	2
Stator resistance	5.5 Ω
Permanent magnet flux linkage	0.297 Wb
d-axis inductance	0.0868 H
q-axis inductance	0.1697 H
DC link voltage	200 V

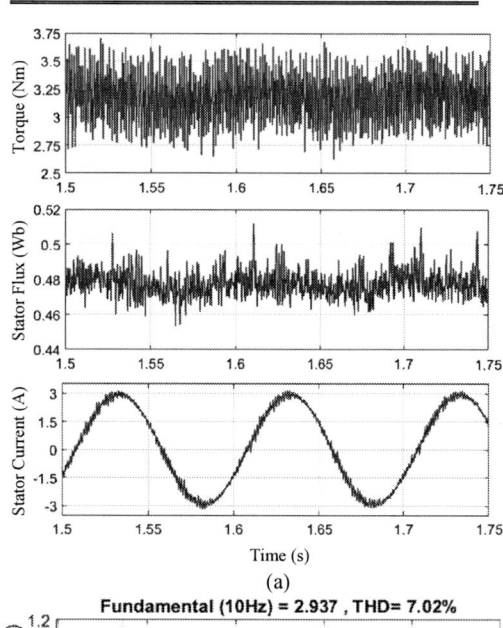

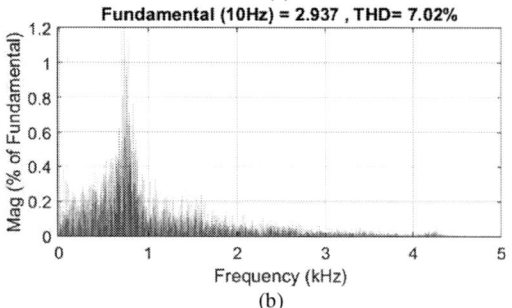

(a)

Fundamental (10Hz) = 2.937 , THD= 7.02%

(b)

Fig. 2. Steady-state response of Conventional MPTC at 300rpm with 3.2Nm load.

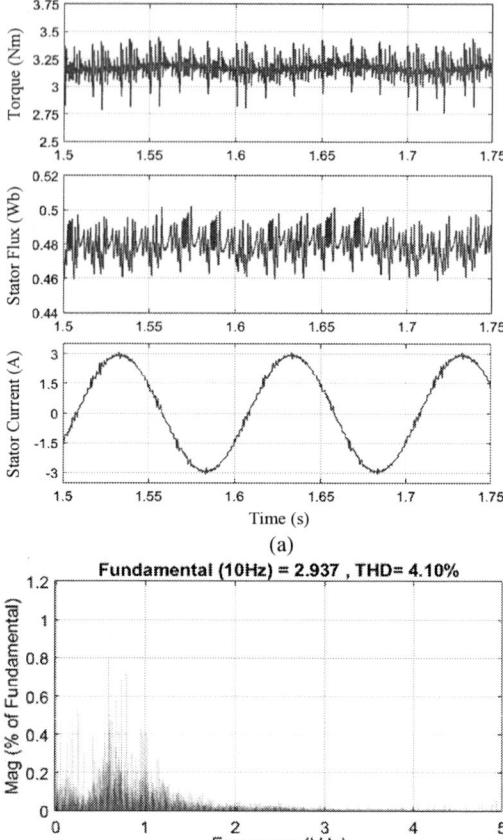

Fig. 3. Steady-state response of the proposed improved MPTC at 300rpm with 3.2Nm load.

By comparing Fig. 2(a) with Fig. 3(a), it is clearly seen that the proposed method contributes to significantly reduce torque ripples and stator current harmonics while obtaining a smoother flux response compared to the conventional MPTC. The first subplots of Fig. 2(a) and Fig. 3(a) indicate an approximately 50% torque ripple reduction by using the proposed method. Meanwhile, the flux ripples are reduced by nearly 27% as shown in the second subplots of Fig. 2(a) and Fig. 3(a). To investigate the total harmonic distortion (THD) of stator currents produced by different methods, the frequency spectrums of stator currents are plotted in Fig. 2(b) and Fig. 3(b). Based on this, the THD of stator current is estimated as 7.02% and 4.1% for conventional MPTC and the proposed method, respectively. Hence, the THD is attenuated by roughly 41% using the proposed method. From the above discussions, it can be concluded that the proposed method significantly improves the steady-state performance of IPMSM drives.

Furthermore, the transient speed and torque responses of the proposed method are also excellent and comparable to those of the conventional MPTC. This is illustrated in Figs. 4 and 5. By comparing Fig. 4(a) and Fig. 4(b), it is seen that the proposed method obtains a fast step speed response of approximately 0.1s for a step speed change from 300 rpm to 600 rpm. This is nearly equivalent to the step speed response obtained by the conventional MPTC. Nonetheless, the proposed method is superior in torque ripple reduction as shown by the comparison of subplots 2 of Fig. 4(a) and Fig. 4(b). In addition, performance of the proposed method is also examined in the scenario of dynamic torque reversal when the motor is held standstill. The results are given in Fig. 5. By comparing subplots 2 of Fig. 5(a) and Fig. 5(b), it is visible that the dynamic torque response is nearly the same under two different control approaches. However, the proposed method generates smaller torque ripples when steady-state is reached.

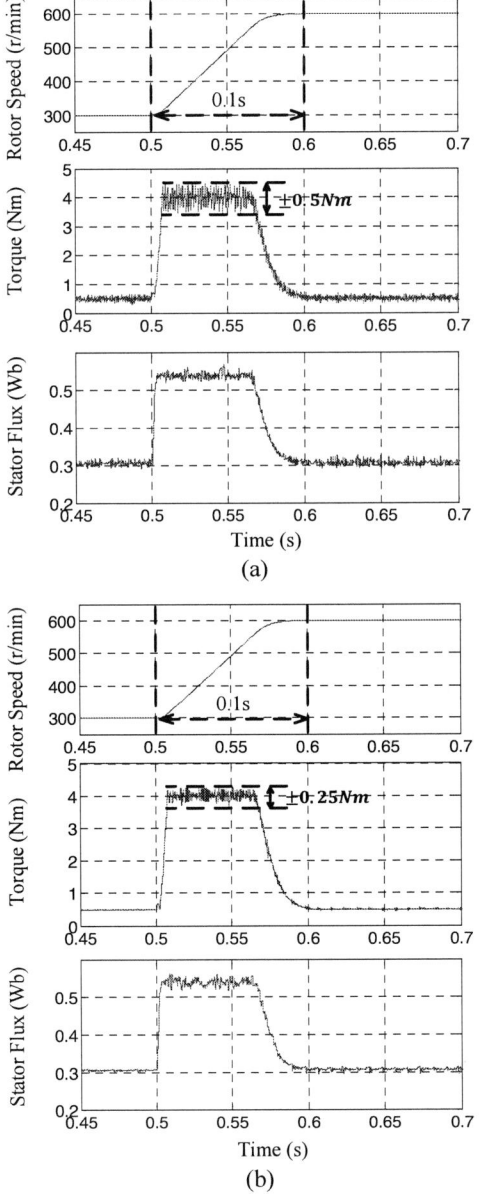

Fig. 4. Performance of step speed test: (a) speed, torque and flux responses of IPMSM controlled by classical MPTC during step speed test, (b) speed, torque and flux responses of IPMSM controlled by the proposed method during step speed test.

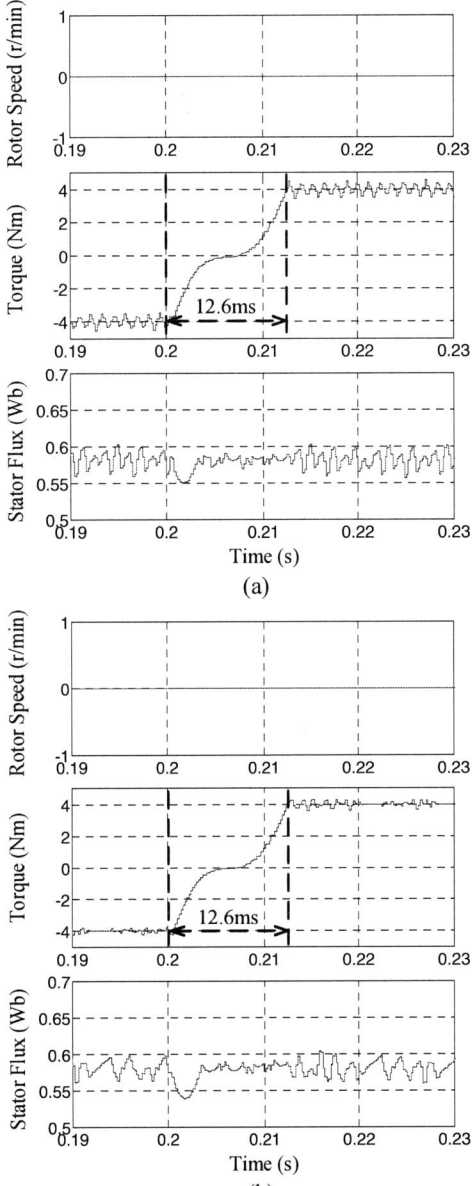

Fig. 5. Performance of torque reversal test: (a) speed, torque and flux responses of IPMSM controlled by classical MPTC during torque reversal, (b) speed, torque and flux responses of IPMSM controlled by the proposed method during torque reversal.

The problem of flux de-regulation without the proposed trade-off scheme is illustrated in Fig. 6. It is observed from subplot 3 of Fig. 6 that the stator flux goes out of control frequently under step load and step speed changes. This is mainly caused by the neglecting of flux regulation in the deadbeat based duty ratio determination. By using the proposed trade-off scheme, a well-regulated stator flux is obtained together with excellent torque and speed responses as shown in Fig. 7. Therefore, it can be concluded that the proposed method is capable of reducing torque ripples without sacrificing the flux regulation performance.

The 2018 International Power Electronics Conference

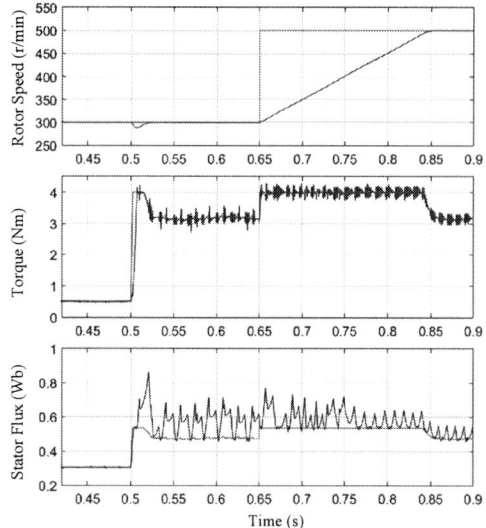

Fig. 6. Illustration of flux de-regulation problem.

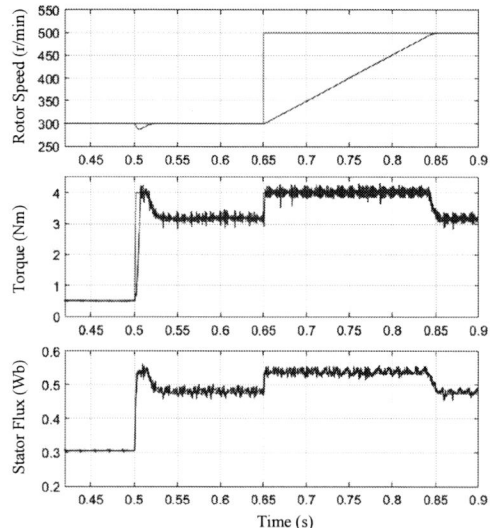

Fig. 7. Overcoming of flux de-regulation problem by using the proposed trade-off scheme.

V. CONCLUSIONS

This paper proposes an improved model predictive torque control (MPTC) approach which contributes to significantly reduce the torque ripples and stator current harmonics for IPMSM drives. The proposed method shows much better drives performance compared to the conventional MPTC while maintaining the advantages of fast response and less rotor position dependency.

ACKNOWLEDGMENT

This work is supported by the Nanyang Technological University Start-up-grant, under M4082085.040 NTU.

REFERENCES

[1] M. A. Rahman, "History of interior permanent magnet motors," *IEEE Ind. Applicat. Magazine,* vol. 19, pp. 10-15, 2013.

[2] F. Niu, B. Wang, A. S. Babel, K. Li, and E. G. Strangas, "Comparative Evaluation of Direct Torque Control Strategies for Permanent Magnet Synchronous Machines," *IEEE Trans. Power Electron.,* vol. 31, pp. 1408-1424, 2016.

[3] M. F. Rahman, L. Zhong, and L. Khiang Wee, "A direct torque-controlled interior permanent magnet synchronous motor drive incorporating field weakening," *IEEE Trans. Ind. Applicat.,* vol. 34, pp. 1246-1253, 1998.

[4] L. Zhong, M. F. Rahman, W. Y. Hu, and K. W. Lim, "Analysis of direct torque control in permanent magnet synchronous motor drives," *IEEE Trans. Power Electron.,* vol. 12, pp. 528-536, 1997.

[5] N. R. N. Idris, A. H. M. Yatim, "Direct torque control of induction machines with constant switching frequency and reduced torque ripple," *IEEE Trans. Ind. Electron.,* vol. 51, no. 4, pp. 758-767.

[6] W. Xie, X. Wang, F. Wang, W. Xu, R. M. Kennel, D. Gerling, R. D. Lorenz, "Finite-control-set model predictive torque control with a deadbeat solution for PMSM drives," *IEEE Trans. Ind. Electron.,* vol. 62, no. 9, pp. 5402-5410.

[7] Y. Zhang, J. Zhu, W. Xu, "Analysis of one step delay in direct torque control of permanent magnet synchronous motor and its remedies," 2010 *International Conference on Electrical Machines and Systems,* pp. 792-797, 2010.

[8] M.H. Vafaie, et al. "A new predictive direct torque control method for improving both steady-state and transient-state operations of the PMSM," *IEEE Trans. Power Electron.,* vol. 31, no. 5, pp. 3738-3753, 2016.

[9] Z. Zhou, C. Xia, Y. Yan, Z. Wang, T. Shi, "Torque Ripple Minimization of Predictive Torque Control for PMSM with Extended Control Set," *IEEE Trans. Ind. Electron., early access,* 2017.

[10] Y. Wang, X. Wang, W. Xie, F. Wang, M. Dou, R.M. Kennel, D. Gerling, "Deadbeat Model-Predictive Torque Control With Discrete Space-Vector Modulation for PMSM Drives," *IEEE Trans. Ind. Electron.,* vol. 64, no. 5, pp. 3537-3547, 2017.

[11] B. Asaei, S.O. Ahmadi, "Discrete Duty-Cycle-Control Method for Direct Torque Control of Induction Motor Drives with Model Predictive Solution," *IEEE Trans. Power Electron., early access,* 2017.

Predictive Torque Control for Five Phase Induction Motor Drive with Common Mode Voltage Reduction

Apekshit Bhowate[1*], Mohan Aware[1], Sohit Sharma[1] and Yogesh Tatte[2]

1 Electrical Engineering Department, Visvesvaraya National Institute of Technology, Nagpur, India
2 Electrical Engineering Department, SBJITMR, Nagpur, India
*E-mail: apekshit.bhowate@students.vnit.ac.in

Abstract- **Multiphase induction motors are used in application requiring high reliability, fault tolerance and reduced per phase power. Common mode voltage (CMV) in induction motor (IM) drives causes winding insulation failure, bearing leakage current and acts as source for electromagnetic interference. This paper presents finite state predictive torque control (PTC) for two level five phase induction motor in order to reduce the CMV. A new cost function is formulated to reduce CMV and total harmonic distortion (THD) and increase efficiency retaining the dynamic performance and variable speed operation of the drive. The performance of drive is analyzed with different voltage vectors applied to aforementioned constraints (CMV, current THD and efficiency) in cost function. Experimental results and analysis confirm the validity of the applied algorithm.**

Keywords—Predictive torque control, five phase induction motor, cost function, common mode voltage, total harmonics distortion.

I. INTRODUCTION

The hardware implementation of multiphase drive system is possible due to the rapid invention in the semiconductor technology. The direct torque control (DTC) technique which has shown its capability compared to scalar and vector control schemes is implemented in [1]. The different PTC methods are proposed in [2]-[4] for the performance improvement in the drive. In [5], the PTC technique is compared with the DTC technique and has overcome it. In [5] it is suggested that the PTC method can be used for controlling the five-phase induction motor in a better way. In the inverter controlled motor drive the CMV has always appeared. In [6] and [7] the different space vector pulse width modulation schemes are proposed for controlling the CMV. As far as the CMV in DTC methods are concerned, the research papers [8] and [9] are proposed for controlling the CMV.

In this paper, the PTC [10] method with new cost function is proposed for controlling the five-phase induction motor. Along with torque and flux, three more constraints of *xy*- current, CMV and switching frequency are added in the proposed cost function. In multiphase motors presence of *xy*-current causes distortion in stator current and increases the losses of motor. The efficiency of motor depends on the average switching of the drive.

Less the number of switching, better is the efficiency of the machine. Three cases with different choices of voltage vectors are considered and analyzed by using proposed cost function.

The paper is structured as: Section 2 describes the modeling of five phase induction machine and its vector mapping in *dq* and *xy*-subspace. CMV of the five phase induction motor drive with different voltage vectors is shown in Section 3. Section 4 presents the finite state predictive torque control technique with three cases applied to proposed cost function. Section 5 presents and discuss the experimental results. While the last section summarizes the conclusion.

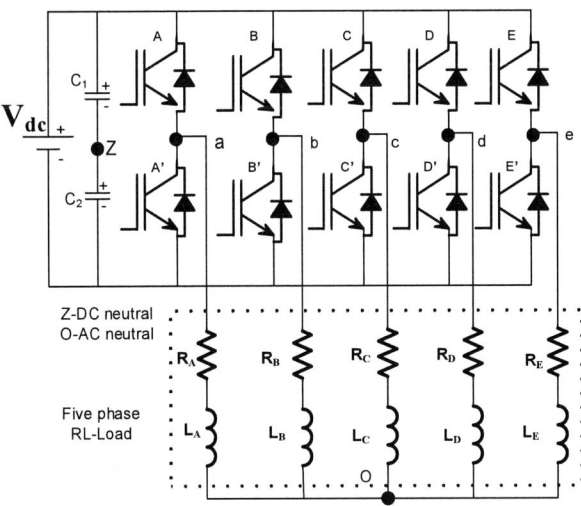

Fig. 1. Schematic diagram of five phase induction motor.

II. FIVE PHASE INDUCTION MOTOR MODELING AND VECTOR MAPPING

The schematic diagram of five phase induction motor drive is shown in Fig. 1. There are 2^5=32 voltage vectors available in five phase drive. The induction motor used for the analysis is having distributed winding with 72^0 spatially displaced voltage vectors between the stator windings. The voltage space vector in the two orthogonal subspaces can be defined as.

$$V_{dq} = V_d + jV_q$$

$$= \frac{2}{5}\left(V_a + V_b e^{j\phi} + V_c e^{j2\phi} + V_d e^{j3\phi} + V_e e^{j4\phi}\right) \quad (1)$$

$$V_{xy} = V_x + jV_y$$

$$= \frac{2}{5}\left(V_a + V_b e^{j2\phi} + V_c e^{j4\phi} + V_d e^{j8\phi} + V_e e^{j12\phi}\right) \quad (2)$$

The five-phase symmetrical physical quantities in natural coordinate system can be projected to two orthogonal subspace *dq* and *xy* as shown in Fig. 2.

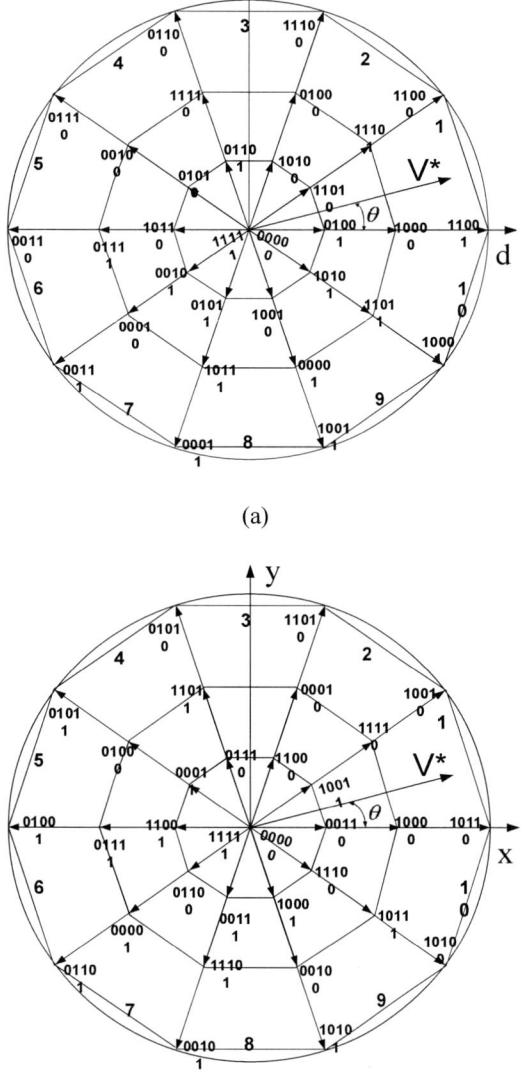

(a)

(b)

Fig. 2 Voltage vectors in (a) *d-q subspace* and (b) *x-y subspace* plane.

The *d-q* stator flux (λ^s_{dq}), rotor flux (λ^r_{dq}) and *x-y* stator flux (λ^s_{xy}), rotor flux (λ^r_{xy}) of the motor considering

stationary reference frame can be written as

$$\lambda^s_{dq} = \int (v^s_{dq} - i^s_{dq} r_s) dt \quad (3)$$

$$\lambda^r_{dq} = \int (v^r_{dq} - i^r_{dq} r_r) dt \quad (4)$$

$$\lambda^s_{xy} = \int (v^s_{xy} - i^s_{xy} r_s) dt \quad (5)$$

$$\lambda^r_{xy} = \int (v^r_{xy} - i^r_{xy} r_r) dt \quad (6)$$

$$\lambda^s_{dq} = (L_{ls} + L_m) i^s_{dq} + L_m i^r_{dq} \quad (7)$$

$$\lambda^r_{dq} = (L_{lr} + L_m) i^r_{dq} + L_m i^s_{dq} \quad (8)$$

$$\lambda^s_{xy} = L_{ls} i^s_{xy} \quad (9)$$

$$\lambda^r_{xy} = L_{lr} i^r_{xy} \quad (10)$$

where, v^s_{dq}, v^r_{dq}, v^s_{xy}, v^r_{xy} and i^s_{dq}, i^r_{dq}, i^s_{xy}, i^r_{xy} represent stator *d-q x-y*, rotor *d-q x-y* voltage and current complex vectors. The r_s, r_r, L_{ls}, L_{lr} and L_m are the stator resistance, rotor resistance, stator leakage inductance, rotor leakage inductance and mutual inductance. The electromagnetic torque can be expressed as:

$$T_e = \frac{5}{2} P \frac{L_m}{\sigma L_s L_r} (|\lambda^s_{dq}| * |i^s_{dq}|) sin(\theta) \quad (11)$$

where, leakage coefficient $(\sigma) = 1 - \dfrac{L_m^2}{L_s L_r}$, P is the number of pole pairs, '*' denotes the complex conjugate, L_s is the stator inductance and L_r is the rotor inductance.

III. COMMON MODE VOLTAGE IN FIVE PHASE DRIVE

The common-mode voltage (u_{cm}) in the star connected five-phase induction motor is given as

$$u_{cm} = \frac{u_{az} + u_{bz} + u_{cz} + u_{dz} + u_{ez}}{5} \quad (12)$$

where, u_{az}, u_{bz}, u_{cz}, u_{dz}, and u_{ez} are the inverter output phase voltages with respect to midpoint of the dc-link voltage. From (12), it can be seen that the CMV impressed can go upto $u_{cm} = V_{dc}/2$. CMV seen in the drive is of three different level depending on the voltage vector applied. From (12) following values of CMV can be extrapolated: $\pm V_{dc}/10$, $\pm 3V_{dc}/10$ and $\pm 5V_{dc}/10$. Table I describe the CMV related to different type of voltage vectors. The CMV of switching vectors can be clustered as following:

1) Large voltage vector: When two phases are connected to positive of dc bus and three phases are connected to negative of dc bus or vice versa.

2) Medium voltage vector: When one phases is connected to positive of dc bus and four phases are connected to negative of dc bus or vice versa.

3) Small voltage vector: When two phases are connected to positive of dc bus and three phases are connected to negative of dc bus or vice versa.

4) Zero voltage vector: When all phases are connected to positive of dc bus or all five phases are connected to negative of dc bus.

As in three phase drive, to reduce CMV use of zero voltage vector is avoided. Similarly here the use of both zero and

The 2018 International Power Electronics Conference

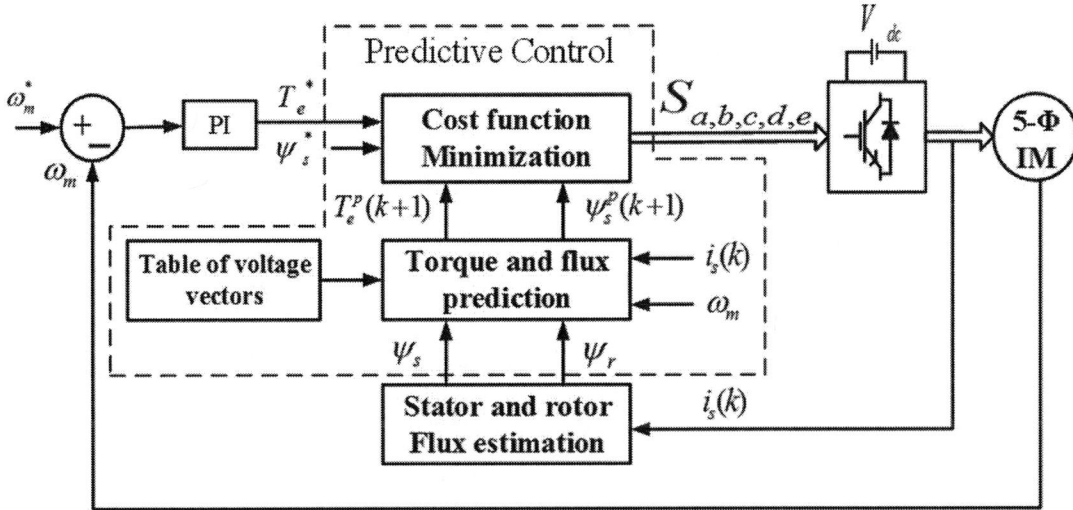

Fig. 3   Block diagram of proposed PTC scheme

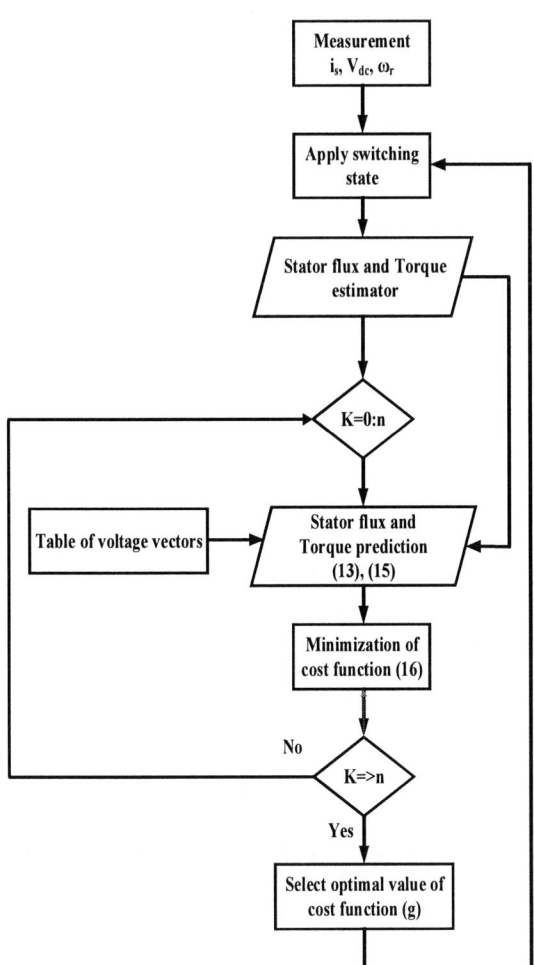

Fig. 4 Flowchart of PTC algorithm

medium voltage vector can be avoided as both produces more CMV as shown in Table I

Table I: CMV in Five phase induction motor drive with different voltage vectors.

Voltage vectors	Type	CMV Value
V_L	large	$\pm V_{dc}/10$
V_M	medium	$\pm 3V_{dc}/10$
V_S	small	$\pm V_{dc}/10$
V_Z	zero	$\pm V_{dc}/2$

IV.   FINITE STATE PREDICTIVE TORQUE CONTROL

For implementing PTC the values of the future are need to be predicted so as to select the optimized voltage vector according to the constraints defined in the cost function. To do this, the motor drive is modelled and the using Euler's forward transformation, it is discretized to formulate the control algorithm. The assumptions taken in to consideration are: winding are perfectly symmetrical, MMF is sinusoidally spatially distributed, negligible core loss and magnetic saturation and air gap is uniform throughout. To formulate the estimators and prediction models, discretizing (3)-(11) using sampling time of T_s.

The proposed scheme is shown in Fig. 3. The stator and rotor flux are estimated from the measured values of stator phase current, dc bus voltage and the speed using sensors. Using the prediction model the stator flux and motor electromagnetic torque is estimated for the instant (k+1). The predictive controller is formulated by discretizing the motor model equations given in Section II. The predictive model at every instant calculates the future value of torque, flux, CMV and THD. The estimated value and the predicted value is compare inside the cost function. This process is repeated for all the voltage vectors. The voltage vector which gives the minimum value of the cost function

1732

is chosen and fired to the inverter. The proposed algorithm is shown in the flowchart Fig. 4.The reference torque is calculated from the speed error by a PI feedback loop. The reference flux is given with nominal value.

$$\psi_s^p(k+1) = \psi_s(k) + T_s\, v(k) - R_s\, T_s\, i_s(k) \qquad (13)$$

$$i_s^p(k+1) = \left(1 + \frac{T_s}{t_\sigma}\right) i_s(k) + \left(\frac{T_s}{t_\sigma + T_s}\right)$$
$$\times \left\{\frac{1}{R_\sigma}\left[\left(\frac{k_r}{t_r} - j\, k_r\omega\right)\psi_r(k) + v_s(k)\right]\right\} \qquad (14)$$

$$T^p(k+1) = \frac{5}{2}p\,\mathrm{Im}\left\{\overline{\psi_s^p}(k+1) i_s^p(k+1)\right\} \qquad (15)$$

A cost function is formulated which finds the error between the present value and predicted value given by (16).

$$g_h = \left|T_e^* - T_e^p(k+1)\right|_h + \lambda_\psi \left|\psi_s^* - \psi_s^p(k+1)\right|_h$$
$$+ \lambda_{xy}\left|0 - i_{xy}^p(k+1)\right|_h$$
$$+ \left|\lambda_{CMV} V_{CMV}\right|_h + \left|\lambda_{SW} C_{SW}\right|_h \qquad (16)$$

where $\lambda_\psi = T_n/\psi_{s_n}$ - flux weighing factor:

$\lambda_{xy} - xy$ current weighing factor

λ_{cmv} - CMV weighing factor

λ_{sw} - Switching weighing factor

The constraints are given different weighing factor which penalizes cost function according to the predicted value. The cost function is evaluated for all 32 voltage vectors. The voltage vector with minimum value of cost function is selected and applied to the inverter. Along with torque and flux error three more constraints i.e xy-current, CMV and number of switching are included in the proposed cost function. In multiphase motor drives xy-current is source of harmonics. So by proper combination of voltage vectors it can be reduced. Different voltage vectors produces different CMV. CMV is computed from the voltage vector fired as given in Table I. The efficiency of the drive depends on the average switching frequency.

This is calculated from variable C_{sw}. This variable calculates the number of switching transitions from present state to other states. Accordingly, the switching state with minimum switching transitions is selected. Thus by varying the value of weighing factors desired performance of drive can be obtained. In this paper the performance of drive is analyzed by combination of different sets of voltage vectors as categorized in cases below in Table II.

Table II: Cases under consideration of five phase induction motor drive with different voltage vectors

	Type of voltage vectors	CMV Value
Case 1	$V_L,\ V_M,\ V_S,\ V_Z$	$\pm V_{dc}/2$
Case 2	$V_L,\ V_M$	$\pm 3V_{dc}/10$
Case 3	$V_S,\ V_M$	$\pm 3V_{dc}/10$

Case 1: In this case due to application of zero voltage vector the CMV is maximum ($\pm 0.5\ V_{dc}$). But as zero voltage vectors are used so the THD of the stator current is minimum.

Case 2: In this case as only large and medium voltage vectors are used the CMV is restricted to ($\pm 0.3\ V_{dc}$). The reflection of combination of large and medium voltage vector in xy - subspace is less. So the THD of the stator current is less.

Case 3: In this case only small and medium voltage vectors are used which makes the CMV upto ($\pm 0.1\ V_{dc}$). CMV is minimum in this case. However the reflection of combination of small and medium voltage vector in xy - subspace is large. So the THD of the stator current is more as compared to case 2.

Accordingly, different voltage vectors defined for above cases are used for evaluating minimum value of cost function so as to achieve the desired performance. The cost function and corresponding weighing factor used for all the three cases is same for the fair comparison of all the three cases.

V. Experimental Results

The proposed PTC technique is validated through the experimental results. For experimentation the five-phase induction motor of rating 1 kW has taken. The DC bus is set at 240V. Semikron IGBT modules is to build the five phase two level inverter. The current is sensed by LEM LA 25P. The dc bus voltage is sensed by the LEM LV − 25. The experimental results of CMV, current waveform and current THD are presented for all the three cases as shown in Fig. 5 − 7. The experimental results presented for the CMV in all the PTC schemes have CMV equals to $V_{dc}/2$, $3V_{dc}/10$ and $3V_{dc}/10$ as mentioned in the previous section. The experimental results of current waveform are also presented in all the PTC schemes along with their current THD. It is noted that the current THD waveform presented in all the PTC schemes has the special trend that the PTC having lower CMV has higher current THD and vice-versa.

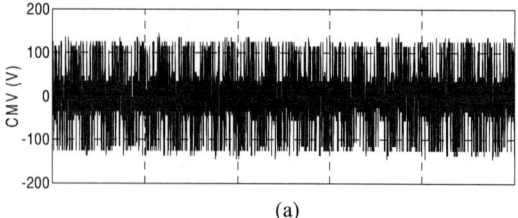

(a)

The 2018 International Power Electronics Conference

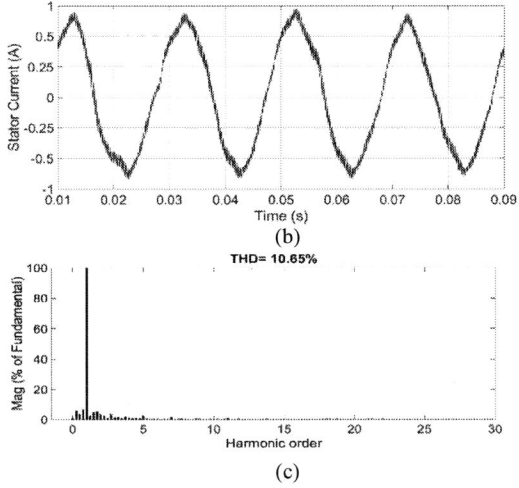

(b)

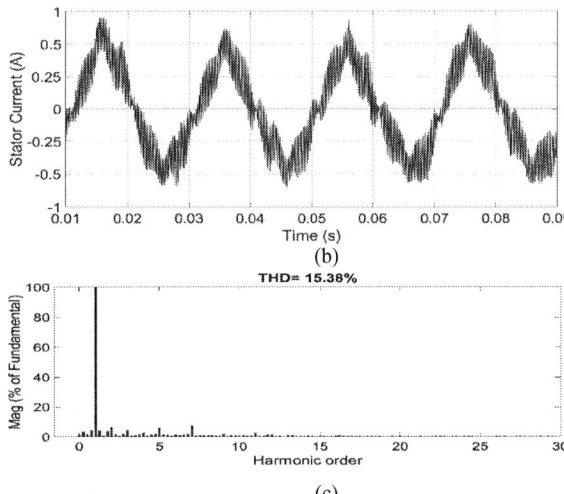

(b)

(c)

Fig. 5. Case 1: With V_L-V_M-V_S-V_Z vectors (a) CMV (20 ms/div), (b) Stator current of phase a and (c) THD spectrum of stator current

(c)

Fig. 7. Case 3: With V_M-V_S vectors (a) CMV (20 ms/div), (b) Stator current of phase a and (c) THD spectrum of stator current

VI. CONCLUSION

In this paper the PTC technique for the five-phase induction motor with new cost function is proposed. With the proposed cost function three different cases are analyzed with three different combination of voltage vectors. The different combinations are compared in context of CMV and current THD. It is concluded that the new cost function has worked effectively for the CMV reduction. It is also marked that by employing different combinations of voltage vectors the generated CMV has different values. The case with lower CMV has higher current THD. This is due to the xy – subspace harmonics present in each cases.

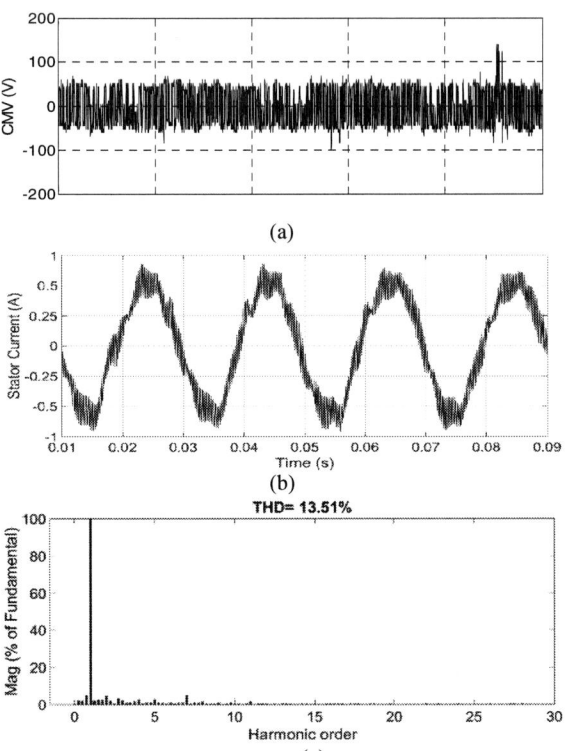

(a)

(b)

(c)

Fig. 6. Case 2: With V_L-V_M vectors (a) CMV (20 ms/div), (b) Stator current of phase a and (c) THD spectrum of stator current

REFERENCES

[1] Y. Tatte; M. Aware, "Torque Ripple and Harmonic Current Reduction in Three-Level Inverter Fed Direct Torque Controlled Five-Phase Induction Motor", *IEEE Transactions on Industrial Electronics*, vol. 64, no. 7, pp. 5265-5275, July 2017 .

[2] H. Miranda, P. Cortés, J. Yus, and J. Rodríguez, "Predictive torque control of induction machine based on state-space model," *IEEE Trans. Ind. Electron.*, vol. 56, no. 6, pp. 1916–1924, Jun. 2009.

[3] M. R. Arahal, F. Barrero, S. Toral, M. J. Durán, and R. Gregor, "Multiphase current control using finite-state model-predictive control," *Control Eng. Pract.*, vol. 17, no. 5, pp. 579–587, May 2009.

[4] J. Riveros, B. Bogado, J. Prieto, F. Barrero, S. Toral, and M. Jones, "Predictive torque control for five-phase induction motor drives," in *Proc. 36th Annu. Conf. IECON*, 2010, pp. 2467–2472.

[5] J. Riveros, F. Barrero, E. Levi, M. Duran, S. Toral, and M. Jones, "Variable-speed five-phase induction motor drive based on predictive torque control," *IEEE Trans. Ind. Electron.*, vol. 60, no. 8, pp. 2957–2968, Aug 2013.

[6] M. Duran, J. Prieto, and F. Barrero, "Space vector pwm with reduced common-mode voltage for five-phase induction motor drives operating in overmodulation zone," *IEEE trans. Power Electron.*, vol. 28, no. 8, pp. 4030–4040, Aug 2013.

[7] M. J. Durán, J. Prieto, F. Barrero, J. A. Riveros and H. Guzman, "Space-Vector PWM With Reduced Common-Mode Voltage for Five-Phase Induction Motor Drives," in *IEEE Transactions on Industrial Electronics*, vol. 60, no. 10, pp. 4159-4168, Oct. 2013.

[8] Y. N. Tatte and M. V. Aware, "Direct Torque Control of Induction Motor with Common-Mode Voltage Elimination," *Journal of Electric Power Components and Systems*, vol. 44, no. 20, pp. 2310-2324, November 2016.

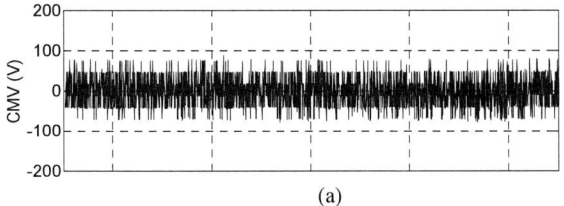

(a)

[9] Y. Tatte; M. Aware, "Direct Torque Control of Five-Phase Induction Motor with Common-Mode Voltage and Current Harmonics Reduction," *IEEE Transactions on Power Electronics*, vol. 32, no. 11, pp. 8644-8654, November, 2017.

[10] A. Bhowate and M. Aware, "CMV suppression using a new predictive direct torque control for induction motor," *2016 7th India International Conference on Power Electronics (IICPE)*, Patiala, India, 2016, pp. 1-6.

Indirect Matrix Converter for Permanent-Magnet-Synchronous-Motor Drives by Improved Torque Predictive Control

Yun Jang, Yeongsu Bak, and Kyo-Beum Lee*

Department of Electrical and Computer Engineering, Ajou University, Suwon, Korea

*E-mail: kyl@ajou.ac.kr

Abstract— This paper presents indirect-matrix-converter (IMC) for permanentmagnet synchronous-motor (PMSM) drives using improved torque predictive control (TPC). The IMC does not require capacitors in the DC-link, which makes it possible to reduce the volume of system unlike typical system of the AC-DC-AC power conversion. The IMC has advantages which are sinusoidal waveform in input-output current and high-power density. Therefore, researches in the IMC for PMSM drives using various torque control methods have been actively studied. In this paper, the IMC for the PMSM drives using the improved TPC is proposed. The ripple component of the torque of the PMSM is declined by the proposed method. also, the dynamic torque response is improved compared with the conventional torque control methods. The performance of the proposed method is verified by the results of the PSIM simulation tool.

Keywords— *Indirect Matrix Converter (IMC), Permanent-Magnet-Synchronous-Motor (PMSM), Torque Contorl, Torque Predictive Control (TPC).*

I.   INTRODUCTION

Recently, many researches have been studied on AC-AC converters because interest in renewable energy and electric vehicles have increased. The fundamental AC-AC converter is composed an inverter stage, DC-link, and a rectifier stage. The DC-link is composed of electrolytic capacitors with large volume. The electrolytic capacitors have disadvantages of increasing the volume and short lifetime of the entire system. In order to overcome these drawbacks, an indirect matrix converter (IMC) can be used for the AC-AC converters [1]–[3].

The IMC is not required the DC-link electrolytic capacitors [4], [5]. Therefore, it is able to overcome the disadvantages caused by the electrolytic capacitors with large volume of the general AC-AC converter. It has sinusoidal waveform in the input-output current and high-power density [6]. Therefore, the IMC can be used in renewable energy systems, uninterrupted power supply (UPS), and motor driving systems. Recently, researches in the IMC for a permanent-magnet-synchronous-motor (PMSM) drives using various torque control methods such as a direct-torque-control (DTC) or a torque predictive-control (TPC) have been studied [7], [8].

In the torque control method using the DTC, a hysteresis controller and a look-up table are used to control the torque of the PMSM. It has advantages of being easy to implement and having a fast dynamic torque response [9]. However, it has disadvantage of increasing ripple components in the torque and current waveform of the PMSM because a fixed voltage vector selected by the look-up table is applied during a control period [10]–[13]. Contrary to the torque control method using the DTC, in torque control method using the TPC, the voltage vectors are calculated for torque control of the PMSM using the relation with torque, flux, and voltage equation instead of the look-up table [14], [15]. It has fast dynamic torque response and simple controller configuration. However, similar to the torque control method using the DTC, it has increasing ripple components in the torque and current waveform of the PMSM [16], [17].

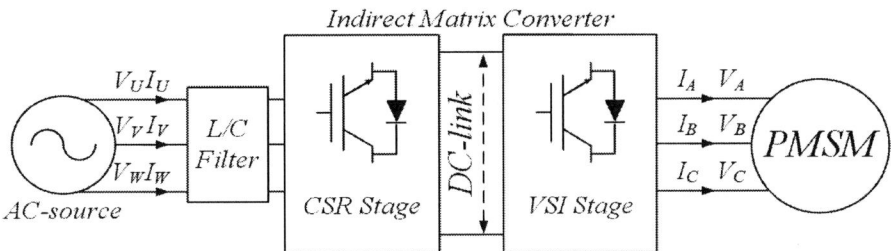

Fig. 1. Circuit configuration of IMC for PMSM drives.

The 2018 International Power Electronics Conference

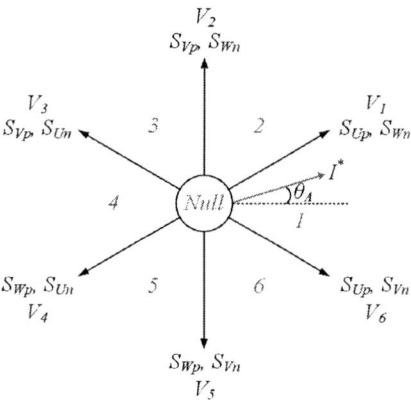

Fig. 2. Space vector diagram of CSR stage.

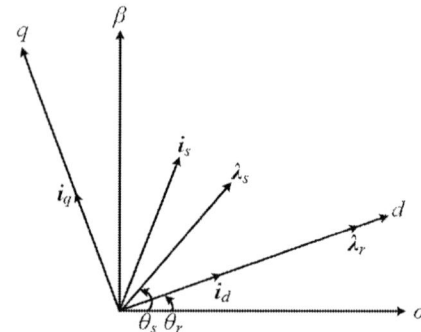

Fig. 3. Space vector diagram of the PMSM with different coordinate axes.

Therefore, in this paper, the IMC for the PMSM drives using the improved TPC is proposed. The ripple component of the torque of the PMSM is declined by the proposed method. The performance of the proposed method is verified by the results of the PSIM simulation tool.

II. TOPOLOGY AND MODULATION METHOD OF IMC

A. Overall topology of the IMC

Fig. 1 shows a configuration of the IMC for the PMSM drives. It is composed of three stages which are an input stage connected to the AC source, the power conversion system using the IMC, and an output stage connected to the PMSM. In addition, the L/C filter is used to increase the power quality of AC source input voltage. The IMC consists of the two stages and one hypothetical DC-link. The two stages are mainly composed of voltage source inverter (VSI) and current source rectifier (CSR). The VSI and CSR are connected through the DC-link. because DC-link is not required DC-link capacitors. Finally, the PMSM can be controlled through the VSI and CSR.

B. CSR Modulation method of the IMC

The highest voltage of the AC source can be transferred to the DC-link through the modulation in the CSR stage. Through the modulation of the CSR stage, the input stage current is maintained perfectly sinusoidal and unity power factor. Fig. 2 shows the diagram of the space vector in the CSR stage. The space vector of the CSR stage is composed of the six active states and three null states. When the pair of the switches are operated ON states at the same time, the null states have occurred and the DC-link voltage is shorted to zero. On the other hand, the power produced from the AC source is transmitted to the output stage in the six active states.

In Fig. 2, the reference current phasor (I^*), which is reproduced by the nearest vectors, is located in sector 1. It is reproduced by the V_1 and V_6 using the duty ratios (d_X and d_Y). The duty ratios for modulation of the CSR are obtained as follows. The reference phase currents (I^*_U, I^*_V, I^*_W) are expressed as in (1).

$$I^*_U = I_m \cos(\omega_i t) = I_m \cos\theta_U,$$
$$I^*_V = I_m \cos\left(\omega_i t - \frac{2}{3}\pi\right) = I_m \cos\theta_V, \qquad (1)$$
$$I^*_W = I_m \cos\left(\omega_i t - \frac{4}{3}\pi\right) = I_m \cos\theta_W,$$

where I_m is the phase current amplitude, ω_i is the angular frequency of the AC source, and θ_U, θ_V, and θ_W are individual phase angles. Besides, the d_X and d_Y are expressed as in (2).

$$\cos\theta_U + \cos\theta_V + \cos\theta_W = 0,$$
$$d_X = -\frac{\cos\theta_V}{\cos\theta_U}, \quad d_Y = -\frac{\cos\theta_W}{\cos\theta_U}. \qquad (2)$$

Furthermore, the DC-link voltage average (V_{DC-av}) is able to calculate as in (3) through the line voltage, d_X, and d_Y.

$$V_{DC-av} = d_X V_{UV} - d_Y V_{WU}$$
$$= \frac{3V_m}{2\cos\theta_A} \cdot \cos\phi_i, \quad \left(-\frac{\pi}{6} \le \theta_A \le \frac{\pi}{6}\right), \qquad (3)$$

where V_m is phase voltage amplitude and ϕ_i is the power-factor angle.

The modulation method is equal to the common inverter of the VSI stage. The space vector of the VSI stage is categorized by null states (V_0, V_7) and active states ($V_1 \sim V_6$). The carrier based PWM method is used for the modulation of the VSI stage. Additionally, in the VSI stage, the compensation of V_{dc-av} is required to produce sinusoidal input-output currents.

III. TORQUE PREDICTIVE CONTROL METHOD

A. Conventional TPC

In the conventional TPC, in order to calculate the voltage vector for torque control of the PMSM, it is necessary to analyze the interaction formula of the torque, flux, and voltage. Fig. 3 shows the space vector diagram with different coordinate axes. The reference frames of

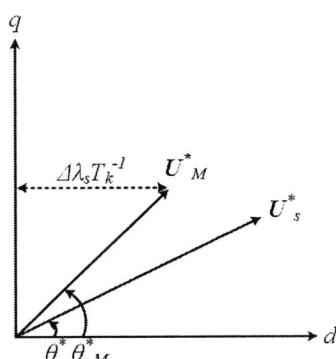

Fig. 4. Improved TPC strategy.

the α-β and the d-q are stationary reference frame and rotating frame synchronized to the rotor of the PMSM, respectively. The θ_s and θ_r indicate the location in the flux vector of stator and flux vector of rotor. The voltage vector in stationary reference frame can be expressed as in (4).

$$U_s = R_s i_s + \frac{d}{dt}\lambda_s, \tag{4}$$

where U_s ($=[U_{\alpha s}, U_{\beta s}]^{\mathrm{T}}$) is voltage vector in the stator, i_s ($=[i_\alpha, i_\beta]^{\mathrm{T}}$) is current vector in the stator, λ_s ($=[\lambda_\alpha, \lambda_\beta]^{\mathrm{T}}$) is flux vector in the stator, and R_s is resistance in the stator. The λ_s can be expressed as in (5).

$$\lambda_s = L_s i_s + \lambda_r, \tag{5}$$

where L_s is self-inductance in the stator and λ_r is flux vector in the rotor. The electromagnetic-torque can be expressed as in (6).

$$T_e = \frac{3}{2} p_n \left(\lambda_s \times i_s\right), \tag{6}$$

where p_n is the number of pole pairs. The changing rate T_e as in (7) is expressed by substituting differential form of (4), (5), and (6).

$$\Delta T_e = \frac{3}{2}\frac{p_n}{L_s}\lambda_r \times U_s T_k$$
$$+ \left(-\frac{R_s}{L_s}T_e - \frac{3}{2}\frac{p_n}{L_s}\lambda_s \times \frac{d}{dt}\lambda_r\right)T_s, \tag{7}$$

where T_s is control period and T_k is voltage vector sustainment time. From (7), the torque of the PMSM can be controlled using the us. Additionally, the angle of between the λ_s and the us can be calculated as in (8).

$$\theta^* = \sin^{-1}\left(\frac{\Delta T_e + \left(\dfrac{R_s}{L_s}T_e + \dfrac{3}{2}\dfrac{p_n}{L_s}\lambda_s \times \dfrac{d}{dt}\lambda_r\right)T_s}{\dfrac{3}{2}\dfrac{p_n}{L_s}|\lambda_r\|T_k U_s|}\right). \tag{8}$$

B. Improved TPC

In the conventional TPC, the voltage vector magnitude for torque control of the PMSM is fixed to M $(0 < M \leq 1)$ in order to simplify the control structure. However, it causes that the ripple components of the torque and current waveform of the PMSM are increased by the fixed M. Additionally, it is difficult that choose the proper voltage vector by the torque control of the PMSM in the low-speed section because the voltage vector magnitude fixed to the M influences on voltage vector sustainment time.

Therefore, in this paper, the improved TPC is used to overcome these disadvantages of the conventional TPC. In the improved TPC, the modified voltage-vector with the appropriate voltage vector magnitude is used for torque control of the PMSM. It can be obtained by using the variation of stator-flux and output-torque of the PMSM. The appropriate magnitude of the voltage vector for the stator flux control is calculated by using the stator flux error as in Fig. 4. As a results, through the improved TPC, the ripple components of the torque and current waveform of the PMSM can be effectively decreased.

From (5), if the voltage drop caused by the stator resistance does not be considered, the magnitude of the α-axis stator voltage ($U_{\alpha s}$) for control the λ_s as the flux in the stator can be expressed as in (9).

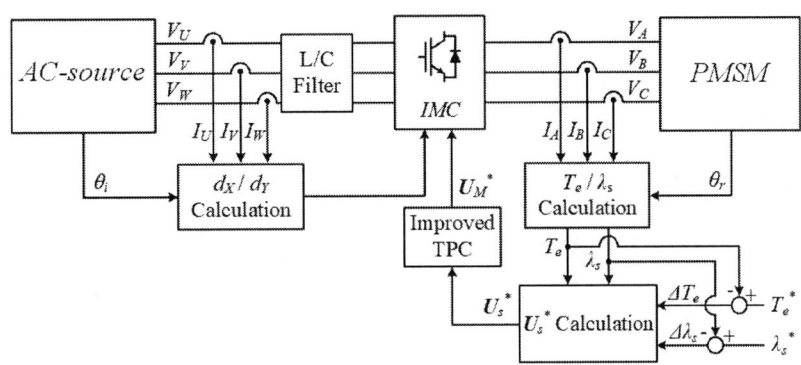

Fig. 5. Control block diagram of the IMC for PMSM drives using the proposed method.

The 2018 International Power Electronics Conference

TABLE I
SIMULATION PARAMETERS OF PERMANENT-MAGNET
SYNCHRONOUS MOTOR

Parameters	Values
Rated power	11 [kW]
Rated current	19.9 [A]
Rated speed	1750 [rpm]
Rated torque	60 [Nm]
Stator resistance	0.349 [Ω]
Stator inductance	15.6 [mH]
Permanent-magnet flux	0.554 [Wb]
Number of poles	6

$$\Delta\lambda_s = \left(U_s - R_s i_s\right)T_s, \quad U_{\alpha s} = \Delta\lambda_s T_s^{-1}. \tag{9}$$

In addition, in the improved TPC, the appropriate the voltage vector magnitude (M) should be calculated for decreasing the ripple components of the torque and current waveform of the PMSM. It is calculated by using a ratio of the maximum reference of the voltage-vector (U_s^*) and magnitude of the modified reference voltage vector (U_M^*) as in (10). Additionally, it can be expressed by using the $U_{\alpha s}$ as in (9) and trigonometric function.

$$M = \frac{\left|U_M^*\right|}{U_s^*} = \frac{\sqrt{U_{\alpha s}^2 + U_{\beta s}^2}}{U_s^*}$$
$$= \frac{\sqrt{\left(\Delta\lambda_s T_s^{-1}\right)^2 + \left(U_s^* \sin\theta^*\right)^2}}{\left|U_s^*\right|}. \tag{10}$$

The phase angle of the U_M^* as the modified reference voltage vector can calculate as in (11) from (9) and (10).

$$\theta_M^* = \sin^{-1}\left(\frac{\left|U_s^*\right|}{\left|U_M^*\right|}\sin\theta^*\right). \tag{11}$$

As a results, through the M and θ_M^*, the U_M^* for torque control of the PMSM in the improved TPC is selected. The ripple components of the torque and current waveform of the PMSM can be reduced by the improved TPC using the U_M^*.

C. Control Method of the IMC for PMSM drives

Fig. 5 shows the control block diagram in the IMC for the PMSM drives using the proposed method. The stator flux (λ_s) and output torque (T_e) of the PMSM are calculated using the output current (I_A, I_B, I_C) and the phase angle (θ_r) of the PMSM rotor. The reference voltage vector (U_s^*) is calculated by the errors of the stator flux ($\Delta\lambda_s$) and the output torque (ΔT_e) of the PMSM. Additionally, the modified reference voltage-vector can be calculated by the appropriate voltage-vector magnitude (U_M^*). Finally, the modified reference voltage for the torque control of the PMSM is produced by the modified reference voltage vector, flux error, and phase angle of the PMSM.

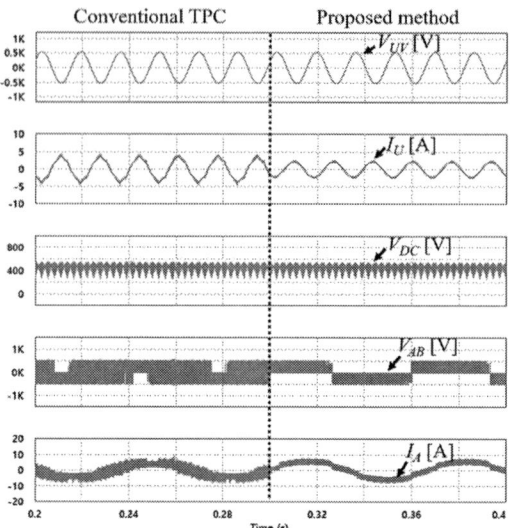

Fig. 6. Simulation results of input-output line to line voltages (V_{UV} and V_{AB}), input-output phase currents (I_U and I_A), and DC-link voltage (V_{DC}) of the IMC.

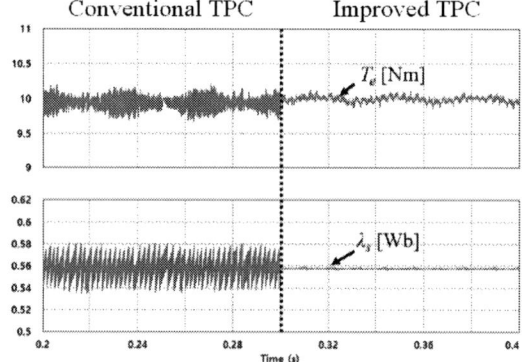

Fig. 7. Simulation results of output torque and stator-flux with load at 300 rpm depending on the control method.

IV. SIMULATION

The IMC for the PMSM drives using the proposed method was simulated by the PSIM in order to verify the effectiveness of the improved TPC method comparing with the conventional TPC. The parameters of the PMSM in the simulation are presented in Table I. The line-to-line voltages of the AC source are 330 V$_{rms}$ and frequency is 60 Hz. Additionally, the sampling time of the control scheme was 100 μs and the switching frequency was 10 kHz.

Fig. 6 shows the simulation results of the input-output line to line voltages (V_{UV} and V_{AB}), input-output phase currents (I_U and I_A), and the DC-link voltage (V_{DC}) of the IMC depending on the control method. In Fig 6, the output torque of the PMSM is controlled 10 Nm with load at 300 rpm. The control method is changed to the proposed method from the conventional TPC at 0.3 s. The I_U and I_A as the input-output currents of the IMC are obtained to sinusoidal waveform by the characteristic of the IMC. In the conventional TPC, the I_U and I_A have the

1739

ripple components. Contrary to the conventional TPC, in the proposed method using the improved TPC, the ripple components of the I_U and I_A are effectively reduced.

Fig. 7 shows the simulation results of the output torque (T_e) and stator flux (λ_s) with load at 300 rpm depending on the control method. The T_e and λ_s of the PMSM are controlled to 10 Nm and 0.56 Wb, respectively. The proposed method using the improved TPC is applied at 0.3 s. In Fig. 7, the ripple components of the T_e and λ_s are decreased by the proposed method from 0.66 Nm to 0.23 Nm and from 0.03 Wb to almost 0 Wb, respectively.

In other words, in the IMC for the PMSM drives using the improved TPC, the dynamic torque response is improved and the ripple components are decreased.

V. CONCLUSION

This paper presents the IMC for PMSM drives using the improved TPC. The IMC for control of the PMSM guarantees the sinusoidal waveform in the input-output current and high-power density. Fundamentally, the various torque control methods such as DTC and TPC are used to control the PMSM using the IMC. However, these control methods cause the ripple components of the output-torque and phase-current of the PMSM. Therefore, in this paper, the IMC for the PMSM drives using the improved TPC is proposed. It guarantees the decrease of the ripple components in the torque and current waveform. Additionally, the dynamic torque response in the proposed method is improved. The performance of the proposed improved TPC method is verified by the PSIM simulation results.

REFERENCES

[1] Y. Bak, E. Lee, and K.-B. Lee, "An indirect matrix converter for dual output AC-drive system with reduced number of power transistors," *in Proc. CENCON Conf.*, 2014, pp. 360–364.

[2] Y. Bak, J.-S. Lee, and K.-B. Lee, "Balanced current control strategy for current source rectifier stage of indirect matrix converter under unbalanced grid voltage conditions," *Energies*, vol. 10, no. 1, Jan. 2017.

[3] M. Rivera, J. Rodriguez, B. Wu, J. R. Espinoza, and C. A. Rojas "Current control for an indirect matrix converter with filter resonance mitigation," *IEEE Trans. Ind. Electron.*, vol. 59, no. 1, pp. 71–79, Jan. 2012.

[4] M. Jussila and H. Tuusa "Comparison of simple control strategies of space-vector modulated indirect matrix converter under distorted supply voltage," *IEEE Trans. Power Electron.*, vol. 22, no. 1, pp. 139–148, Jan. 2007.

[5] M. Rivera, J. Rodriguez, J. R. Espinoza, and H. Abu-rub "Instantaneous reactive power minimization and current control for an indirect matrix converter under a distorted ac supply," *IEEE Trans. Ind. Info.*, vol. 8, no. 3, pp. 482–490, Aug. 2012.

[6] Y. Bak, E. Lee, and K.-B. Lee, "Indirect matrix converter for hybrid electric vehicle application with three-phase and single-phase outputs," *Energies*, vol. 8, no. 5, pp. 3849–3866, Apr. 2015.

[7] C. Xia, J. Zhao, Y. Yan, and T. Shi, "A novel direct torque control of matrix converter-fed PMSM drives using duty cycle control for torque ripple reduction," *IEEE Trans. Ind. Electron.*, vol. 61, no. 6, pp. 2700–2713, June 2014.

[8] P. Correa, J. Rodriguez, M. Rivera, J. R. Espinoza, and J. W. Kolar, "Predictive control of an indirect matrix converter," *IEEE Trans. Ind. Electron.*, vol. 57, no. 6, pp. 1847–1853, June 2009.

[9] D. Casadei, F. Profumo, G. Serra, and A. Tani, "FOC and DTC: Two viable schemes for induction motors torque control," *IEEE Trans. Power Electron.*, vol. 17, no. 5, pp. 779–787, Sep. 2002.

[10] K.-B. Lee and F. Blaabjerg, "Simple power control for sensorless induction motor drives fed by a matrix converter," *IEEE Trans. Energy Convers.*, vol. 23, no. 3, pp. 781–788, Sep. 2008.

[11] K.-B. Lee, F. Blaabjerg, and T.-W. Yoon, "Speed-sensorless DTC-SVM for matrix converter drives with simple nonlinearity compensation," *IEEE Trans. Ind. Appl.*, vol. 43, no. 6, pp. 1639–1649, Nov. 2007.

[12] K.-B. Lee, J.-H. Song, I. Choy, and J.-Y Yoo, "Improvement of low-speed operation performances of DTC for 3-level inverter-fed induction motors," *IEEE Trans. Ind. Electron.*, vol. 48, no. 5, pp. 1006–1014, Oct. 2001.

[13] K.-B. Lee, J.-H. Song, I. Choy, and J.-Y. Yoo, "Torque ripple reduction in DTC of induction motor driven by 3-level inverter with low switching frequency," *IEEE Trans. Power Electron.*, vol. 17, no. 2, pp. 255–264, Mar. 2002.

[14] H. Zhu, X. Xiao, and Y. Li, "Torque ripple reduction of the torque predictive control scheme for permanent-magnet synchronous motors," *IEEE Trans. Power Electron.*, vol. 59, no. 2, pp. 871–877, Feb. 2012.

[15] M. Siami, D. A. Khaburi, and J. Rodriguez, "Torque ripple reduction of predictive torque control for PMSM drives with parameter mismatch," *IEEE Trans. Power Electron.*, vol. 32, no. 9, pp. 7160–7168, Sep. 2017.

[16] M. Pacas and J. Weber, "Predictive direct torque control for the PM synchronous machine," *IEEE Trans. Ind. Electron.*, vol. 52, no. 5, pp. 1350–1356, Oct. 2005.

[17] M. Lopez, J. Rodriguez, C. Silva, and M. Rivera, "Predictive torque control of a multidrive system fed by a dual indirect matrix converter," *IEEE Trans. Ind. Electron.*, vol. 62, no. 5, pp. 2731–2741, May 2015.

Predictive DC-link Current Control Based on IPMSM Discrete State Equation for Inverter without Inductor or Electrolytic Capacitor

Yousuke Akama[1],Kodai Abe[2]. Kiyoshi Ohishi[3], Yuki Yokokura[4], Koji Kobayashi, Tatsuki Kashihara
Nagaoka University of Technology, JAPAN
Email: [1]s143097@stn.nagaokaut.ac.jp, [2]abe_nagaoka@stn.nagaokaut.ac.jp,
[3]ohishi@vos.nagaokaut.ac.jp, [4]yokokura@vos.nagaokaut.ac.jp

Abstract—**In this paper, predictive control of DC-link current is proposed based on the interior permanent-magnet synchronous motor (IPMSM) discrete equation for electrolytic capacitorless inverters. In electrolytic capacitorless inverters, the electrolytic capacitor and power factor correction circuit are not used. Therefore, control of the DC-link current of the inverter to improve the input power factor of direct DC-link current control has been proposed as a method of controlling the DC-link current of the inverter. However, control errors occur in the DC-link current due to sampling error in the motor current. Due to this control error, a resonant current is generated in the inductance and the DC-link capacitor. The predictive control of the DC-link current proposed in this paper reduces this control error. The effectiveness of the proposed method is confirmed by experiments.**

Keywords—*Electrolytic capacitorless inverter, IPMSM, Predictive Control*

I. INTRODUCTION

In recent years, a variable-speed drive system for motors has been widely used to realize energy saving[1], [2], [3]. In general, single-phase to three-phase inverters are used for variable speed operation of motors. To use a single-phase to three-phase inverter, a power factor correction (PFC) circuit and a large electrolytic capacitor are required[4]. However, large electrolytic capacitors increase the size and shorten the life of the device. In order to solve these problems, a motor drive system using a film capacitor instead of an electrolytic capacitor has been proposed [5]. In this system, because there is no energy buffer in the DC-link part, power ripple is absorbed by the kinetic energy accumulated in the moment of inertia of the motor. In addition, we have proposed an direct DC-link current control (DDCCC) that improves the power factor by directly controlling the DC-link current toward electrolytic capacitorless inverter [6]. However, because the direct DC-link current control controls the DC-link current in a feedforward manner, a control error occurs in the DC-link current due to the motor current that changes during the control period. As this control error depends on the output voltage vector, if the voltage vector is steeply manipulated, the control error has a high-frequency component. As a result, a resonance current flows. Therefore, we propose a predictive control based on the IPMSM discrete equation that can reduce

the control error of the DC-link current. In addition, this study confirms the effectiveness of the proposed method by experiment.

II. DIRECT DC-LINK CURRENT CONTROL

In the electrolytic capacitorless inverter system, the DC-link electrolytic capacitor is replaced with a small-capacity film capacitor, and the PFC circuit is omitted. Therefore, the current flowing into the capacitor decreases, and by controlling the DC-link current, the input power factor improves in the electrolytic capacitorless inverter. However, distortion occurs in the grid current due to the delay in the motor current control system and the induced harmonic voltage of the IPMSM. Therefore, we proposed an direct DC-link current control (DDCCC) that improves the power factor by directly controlling the DC-link current of the electrolytic capacitorless inverter. DDCCC is a method of directly controlling the DC-link current by correcting the output voltage of the inverter in a feedforward manner.

$$I_{dc} = \frac{1}{2}i_{\alpha}v_{\alpha n}^{*} + \frac{1}{2}i_{\beta}v_{\beta n}^{*} \tag{1}$$

where $v_{\alpha n}^{*}$ and $v_{\beta n}^{*}$ are the normalized α and β axis voltage references and i_{α} and i_{β} are the α and β axis current, respectively. The process of normalization involves dividing the voltage command value by the DC-link voltage V_{dc} and then multiplying by 2. The normalized voltage is a dimensionless quantity, like the duty ratio. By substituting the average DC-link current reference I_{dc}^{*} into the left-hand side of Eq.(1), the following equations indicating that the average DC-link current that is equal to its reference I_{dc}^{*} are obtained:

$$v_{\beta n}^{*} = -\frac{i_{\alpha}}{i_{\beta}}v_{\alpha n}^{*} + \frac{2I_{dc}^{*}}{i_{\beta}} \tag{2}$$

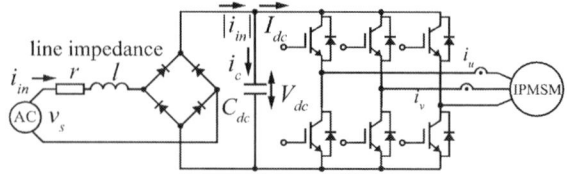

Fig. 1. Electrolytic capacitorless inverter

Equation (2) indicates that the DC-link current, which is equal to its reference I_{dc}^*, is obtained by outputting the voltage vector on the line. The DDCCC method controls the DC-link current by correcting the output voltage command $v_o^{'}$ of the current controller on the DC-link current line, as shown in Fig.2. Fig.2 (a) shows the state of DDCCC in which voltage saturation does not occur, and the voltage vector is corrected to a point on the DC-link current line with the same phase as $v_o^{'}$. Fig.2 (b) shows the state of the DDCCC at voltage saturation, and the voltage vector is corrected to the intersection of the voltage-limiting circle and the DC-link current line.

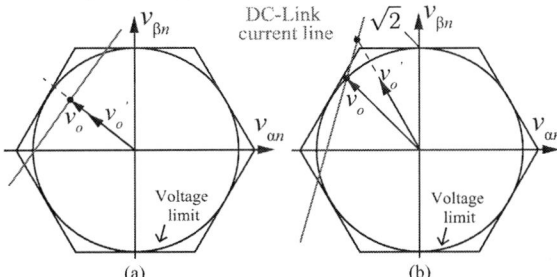

(a)   (b)

Fig. 2.   direct DC-link current control

III.   CONTROL ERROR OF DC-LINK CURRENT BY IVC

In DDCCC, the voltage vector is manipulated in a feedforward manner so as to obtain the desired DC-link current based on the sampling value of the continuously changing phase current. However, as shown in Fig.3 , a control error occurs in the control of the DC-link current because it is delayed by one control period from the time when the voltage vector is determined by DDCCC until it is actually reflected at the inverter. Even during the period when the inverter outputs the desired voltage, because the motor current changes during the control period depending on the output voltage, this also increases the control error in the DC-link current. Because of these factors, DDCCC using the DC-link current line always produces a control error in the DC-link current, and the error amount varies depending on the voltage output from the inverter and the speed and current of the motor. Therefore, in DDCCC, which may cause the output voltage to abruptly change, there is a problem in that the control error of the DC-link current is generated in a pulse shape due to the change in the output voltage.

In DDCCC, the cause of the control error of the DC-link current is that the one control period delay of the voltage output is not taken into consideration in Eq.(1), and the motor current is controlled depending on the output voltage. Based on the voltages $\overline{v_{qn}^*}[k+1]$ and $\overline{v_{dn}^*}[k+1]$ outputted from the inverter at time [k+1] and [k], the relationship between the current actually flowing through the DC-link and the motor current is described in Eq.(3).

$$I_{dc\_ave}[k+1] = \frac{i_d[k+1]+i_d[k+2]}{4}\overline{v_{dn}^*}[k+1]$$
$$+ \frac{i_q[k+1]+i_q[k+2]}{4}\overline{v_{qn}^*}[k+1] \quad (3)$$

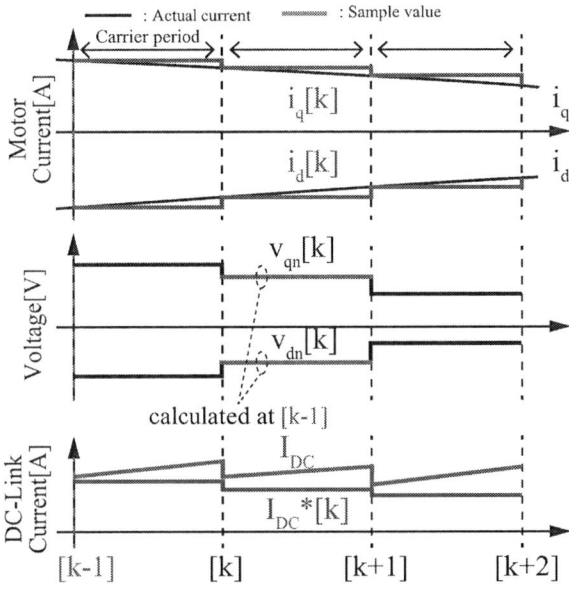

Fig. 3.   Sampling time series

Eq.(3) shows that the current flowing through the DC-link during one control cycle period is based on the average value of the motor current and the output voltage. Because $i_d[k+2]$ and $i_q[k+2]$ in Eq.(3) change with $\overline{v_{dn}^*}[k+1]$ and $\overline{v_{qn}^*}[k+1]$, the average value of the DC-link current varies depending on the output voltage. Moreover, the control error of the DC-link current by DDCCC can be obtained based on Eq.(3). In DDCCC, the voltage is output on the DC-link current line under all circumstances, so that the only degree of freedom of control is the phase of the output voltage θ. In addition, the phase of the voltage vector is defined as follows.

$$\frac{v_{qn}^{'}[k+1]}{v_{dn}^{'}[k+1]} = \tan\theta \quad (4)$$

By these definitions, when a point on the DC-link current line is expressed using the phase of the voltage vector, the following is obtained.

$$v_{dn}[k+1] = \frac{2I_{dc}^*[k]}{i_q[k]\tan\theta + i_d[k]} \quad (5)$$

$$v_{qn}[k+1] = \frac{2I_{dc}^*[k]\tan\theta}{i_q[k]\tan\theta + i_d[k]} \quad (6)$$

By substituting Eq.(7) and Eq.(8) into Eq.(3), it is possible to obtain the response of the DC-link current by DC-link current control using DDCCC.

$$\begin{aligned} i_d[k+2] &= A_{11}i_d[k+1] + A_{12}i_q[k+1] \\ &+ b_{11}\frac{V_{dc}[k]}{2}\overline{v_{dn}^*}[k+1] \\ &+ b_{12}\frac{V_{dc}[k]}{2}\overline{v_{qn}^*}[k+1] - b_{12}\omega_{re}[k]\phi_a \quad (7) \end{aligned}$$

$$
\begin{aligned}
i_q[k+2] \;=\; & A_{21}i_d[k+1] + A_{22}i_q[k+1] \\
& + \; b_{21}\frac{V_{dc}[k]}{2}\overline{v_{dn}^{*}}[k+1] \\
& + \; b_{22}\frac{V_{dc}[k]}{2}\overline{v_{qn}^{*}}[k+1] - b_{22}\omega_{re}[k]\phi_a \quad (8)
\end{aligned}
$$

The coefficients A and b used for Eq.(7) and Eq.(8) will be described later. When the response value of the DC-link current represented by Eq.(3) is defined as Eq.(9), the error rate of the control error of the DC-link current can be defined by Eq.(10).

$$
\overline{I_{dc}}[k+1] = \frac{I_{dc}[k+1] + I_{dc}[k+2]}{2} \tag{9}
$$

$$
I_{dcError} = \left| \frac{I_{dc}^{*}[k] - \overline{I_{dc}}[k+1]}{I_{dc}^{*}[k]} \right| \tag{10}
$$

The error rate distribution of the DC-link current in the voltage-limitation hexagon under the conditions shown in Table1 and Table2 is shown Fig.4; the control error rate of the DC-link current at the point on the DC-link current line at the phase θ^{*} is also shown.

TABLE I.   Calculation parameters

parameter	value
Stator resistance R_a	1.034 Ω
d-axis inductance L_d	14.15 mH
q-axis inductance L_q	23.40 mH
Linkage flux ϕ_a	0.0869 V/(rad/s)
Number of pole pairs P	3 pole
Input voltage	200 V_{rms}

TABLE II.   Motor condition

parameter	value
d-axis current $i_d[k+1]$	-3.76 [A]
q-axis current $i_q[k+1]$	2.48 [A]
DC-Link current ref I_{dc}^{*}	1.38[A]
Electrical angle θ_e	2.98 [rad]
Electrical angular velocity ω_e	472.1[rad/s]

As can be seen from Fig.4, the control error of the DC-link current differs depending on the location on the DC-link current line. In addition, when the harmonic of the magnetic flux density distribution of the rotor is large, the harmonic of the motor current increases, such that the motor current changing during one control period increases and the control error increases. Such a control error arises as long as a method of correcting the voltage vector based on the sampled current is used.

IV. Predictive Control of DC-Link Current Based on IPMSM Discrete Equation

DDCCC is based on Eq.(2). However, there is an error in Eq.(2), and the relationship between the DC-link current actually flowing and the motor current is described by Eq.(3). Therefore, in order to control the DC-link current more strictly, it can be seen that it is better to lead the DC-link current lines from Eq.(3). Here, we define [k] as the current time point and define the voltage command calculated at [k] as $\overline{v_{dn}^{*}}[k+1]$ and $\overline{v_{qn}^{*}}[k+1]$. The future values of the motor currents included in Eq.(3) $i_d[k+1]$

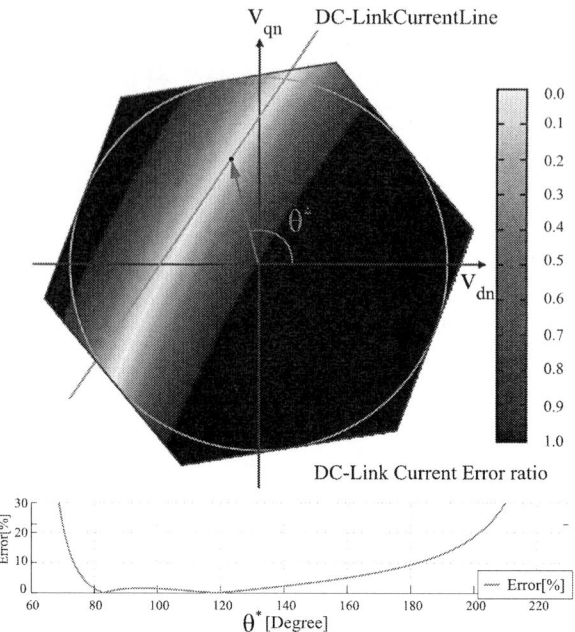

Fig. 4.   DC-link current-control error rate distribution

and $i_q[k+1]$ are calculated from the discrete IPMSM equations by Eq.(11) and Eq.(12).

$$
\begin{aligned}
i_d[k+1] \;=\; & A_{11}i_d[k] + A_{12}i_q[k] + b_{11}\frac{V_{dc}[k]}{2}\overline{v_{dn}^{*}}[k] \\
& + \; b_{12}\frac{V_{dc}[k]}{2}\overline{v_{qn}^{*}}[k] - b_{12}\omega_{re}[k]\phi_a \quad (11)
\end{aligned}
$$

$$
\begin{aligned}
i_q[k+1] \;=\; & A_{21}i_d[k] + A_{22}i_q[k] + b_{21}\frac{V_{dc}[k]}{2}\overline{v_{dn}^{*}}[k] \\
& + \; b_{22}\frac{V_{dc}[k]}{2}\overline{v_{qn}^{*}}[k] - b_{22}\omega_{re}[k]\phi_a \quad (12)
\end{aligned}
$$

The elements included in Eq.(11) and Eq.(12) all have known values at the time point [k], so they can be immediately calculated as numerical values. By substituting Eq.(11), Eq.(12), Eq.(7), and Eq.(8) into Eq.(3) and solving for $\overline{v_{qn}^{*}}[k+1]$, we obtain the equation of the DC-link current line considering the motor current that varies with one sample delay and the output voltage. The solved expression is expressed in Eq.(13).

$$
\begin{aligned}
\overline{v_{qn}^{*}}[k+1] \;=\; & \frac{n_1 \pm \sqrt{n_2}}{2a_3} \tag{13} \\
n_1 \;=\; & -a_1 - a_5\overline{v_{dn}^{*}}[k+1] \\
n_2 \;=\; & \left(a_1 + a_5\overline{v_{dn}^{*}}[k+1]\right)^2 \\
& - 4a_3(a_2\overline{v_{dn}^{*}}[k+1] + a_4\overline{v_{qn}^{*}}[k+1]^2 - 4I_{dc}^{*}[k])
\end{aligned}
$$

1743

$$a_1 = i_q[k+1] + A_{21}i_d[k+1] + A_{22}i_q[k+1]$$
$$+ \; b_{22}\left(-\omega_{re}\phi_{fa}\right) \tag{14}$$

$$a_2 = i_d[k+1] + A_{11}i_d[k+1] + A_{12}i_q[k+1]$$
$$+ \; b_{22}\left(-\omega_{re}\phi_{fa}\right) \tag{15}$$

$$a_3 = b_{22}\frac{V_{dc}[k+1]}{2} \tag{16}$$

$$a_4 = b_{11}\frac{V_{dc}[k+1]}{2} \tag{17}$$

$$a_5 = (b_{21}+b_{12})\frac{V_{dc}[k+1]}{2} \tag{18}$$

Equation (13) is an equation of an ellipse passing through the vicinity of the conventional DC-link current line. This equation is called the predictive DC-link current curve.

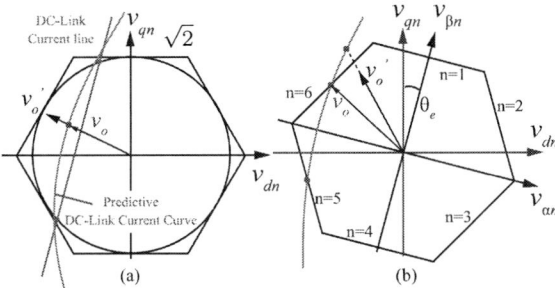

Fig. 5.   DC-link current line and predictive DC-link current curve

The proposed method predictively controls the DC-link current using Eq.(13) . The predictive DC-link current curve is used in the same manner as the conventional DDCCC, as shown in (a) and (b) of Fig.5.

A. IPMSM discrete equation

The predictive control of the DC-link current is based on the IPMSM discrete equation. The discrete equation of IPMSM used here is obtained by discretizing the state equation of IPMSM by the zero-order hold of the sampling time T_s. The state equation of IPMSM is as follows:

$$\frac{d}{dt}\begin{bmatrix} i_d \\ i_q \end{bmatrix} = \begin{bmatrix} -\frac{R_a}{L_d} & \omega_{re}\frac{L_q}{L_d} \\ -\omega_{re}\frac{L_d}{L_q} & -\frac{R_a}{L_q} \end{bmatrix}\begin{bmatrix} i_d \\ i_q \end{bmatrix} + \begin{bmatrix} \frac{1}{L_d} & 0 \\ 0 & \frac{1}{L_q} \end{bmatrix}\begin{bmatrix} v_d \\ v_q \end{bmatrix}$$
$$+ \begin{bmatrix} \frac{1}{L_d} & 0 \\ 0 & \frac{1}{L_q} \end{bmatrix}\begin{bmatrix} 0 \\ -\omega_{re}\phi_a \end{bmatrix}$$

$$\frac{d}{dt}\boldsymbol{x}(t) = \boldsymbol{A}_c x(t) + \boldsymbol{b}_c u(t) + \boldsymbol{b}_c d \tag{19}$$

The equation obtained by discretizing Eq.(19) by the zero-order hold of the sampling time T_s is given below:

$$\boldsymbol{x}[i+1] = \boldsymbol{A}x[i] + \boldsymbol{b}u[i] + \boldsymbol{b}d \tag{20}$$

The coefficient matrix A in Eq.(13) is as follows:

$$\boldsymbol{A} = \begin{bmatrix} A_{11} & A_{12} \\ A_{21} & A_{22} \end{bmatrix} = e^{\boldsymbol{A}_c T_s} = \text{Ł}^{-1}\left[(sI - A_c)^{-1}\right] \tag{21}$$

Similarly, matrix b can be obtained as follows:

$$\boldsymbol{b} = \begin{bmatrix} b_{11} & b_{12} \\ b_{21} & b_{22} \end{bmatrix} = \int_0^{T_s} e^{\boldsymbol{A}_c \tau} d\tau \cdot \boldsymbol{b}_c \tag{22}$$

V.   EXPERIMENTAL RESULT

Table1 and Table3 show the motor parameters and inverter circuit constants used for the experiment. The experimental results at a control frequency of 16 kHz are shown in Fig. 6 − 9. In the experimental result with a torque load of 0.6 Nm at an operation speed of 2000 rpm, the DC-link current line becomes unstable due to the low motor current, and the voltage vector is steeply manipulated. For this reason, because the control error of the DC-link current has a high-frequency component, a resonance circuit parasitic on the input portion is excited and a harmonic current is generated within the input current. In the results using the proposed method, as the control error of the DC-link current is reduced, the harmonic current of the input current is also reduced. Similarly, the experimental result at a torque load of 1 Nm at an operation speed of 3000 rpm reduces the harmonic of the input current by the proposed method. The FFT analysis results of the input current waveform are shown in Fig.14 and Fig.15. In the FFT analysis of the proposed method, the harmonic content is lower than the harmonic regulation value. Furthermore, the experimental results when the control is changed to 10 kHz under similar conditions are shown in Fig. 1017. In the conventional method, the control error of the DC-link current increases as the time of one control cycle increases, and the harmonic amount increases compared with that of the case of 16 kHz. In the proposed method, even when the control period is 10 kHz, the harmonic content is lower than the harmonic regulation value.

TABLE III.   EXPERIMENTAL CONDITIONS

parameter	value
DC-link capacitance C_{dc}	11 μF
Line inductance l	0.4 mH
Line resistance r	0.5 Ω
Poles of current control system w_c	4000 rad/s

TABLE IV.   HARMONIC REDUCTION BY THE PROPOSED METHOD

Operating speed	Decrease amount of 2-40 th order harmonic	Decrease amount of 41-60 th order harmonic
2000rpm(16kHz)	2.62%	44.51%
3000rpm(16kHz)	19.03%	32.68%
2000rpm(10kHz)	2.62%	44.51%
3000rpm(10kHz)	19.03%	32.68%

The 2018 International Power Electronics Conference

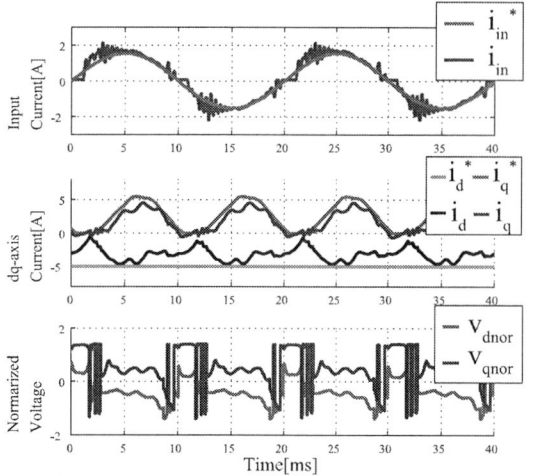

Fig. 6. Experimental result 2000rpm, 0.6Nm 16kHz(Conventional method)

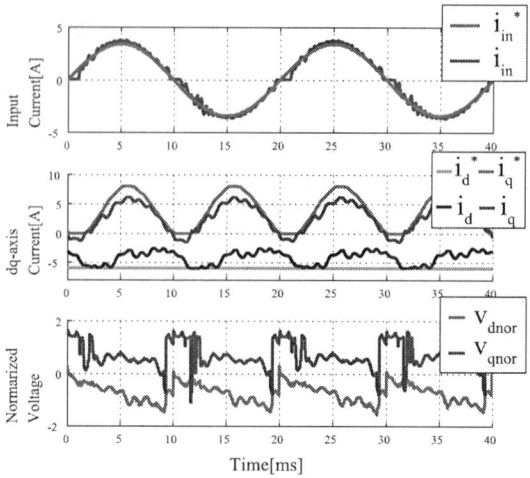

Fig. 9. Experimental result 3000rpm 1Nm 16kHz (Proposed method)

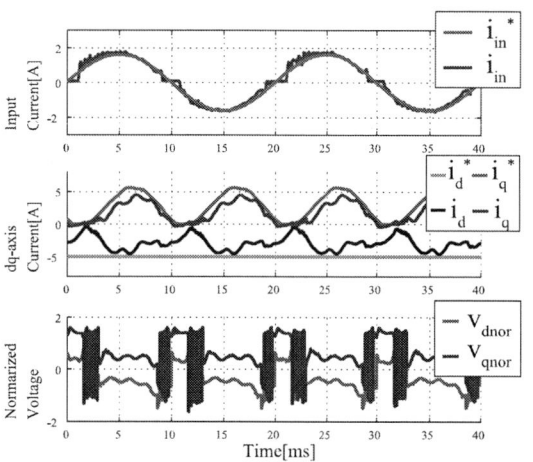

Fig. 7. Experimental result 2000rpm 0.6Nm 16kHz(Proposed method)

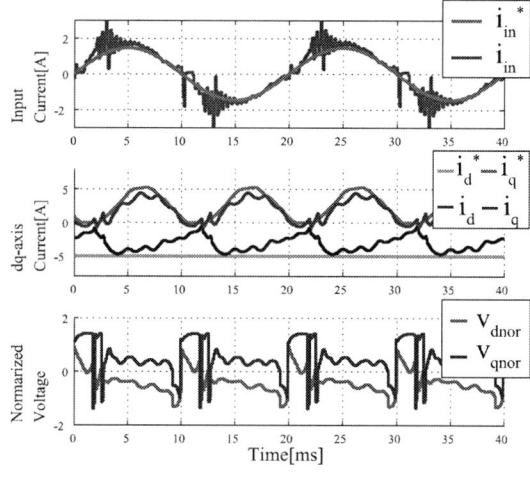

Fig. 10. Experimental result 2000rpm 0.6Nm 10kHz(Conventional method)

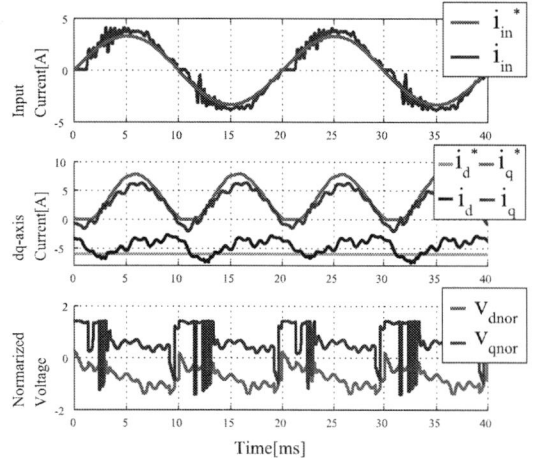

Fig. 8. Experimental result 3000rpm 1Nm 16kHz (Conventional method)

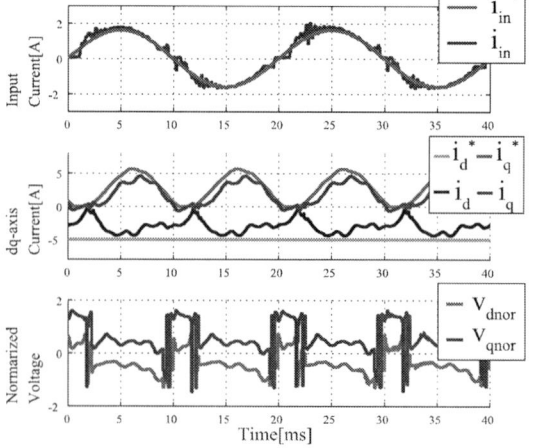

Fig. 11. Experimental result 2000rpm 0.6Nm10kHz(Proposed method)

1745

The 2018 International Power Electronics Conference

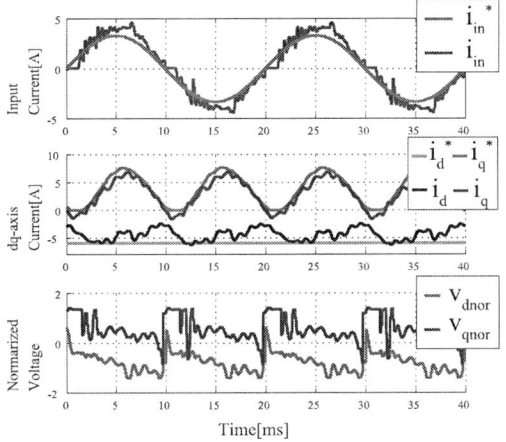

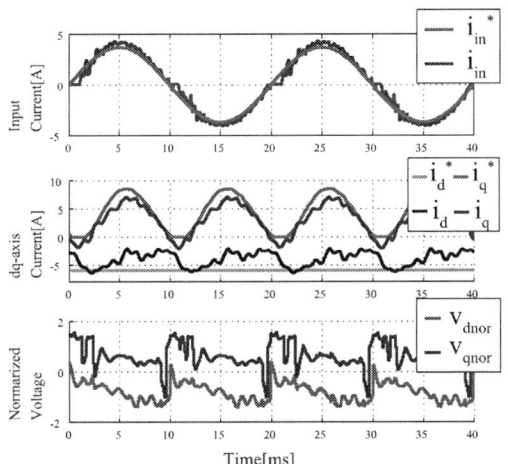

Fig. 12. Experimental result 3000rpm 1Nm 10kHz(Conventional method)

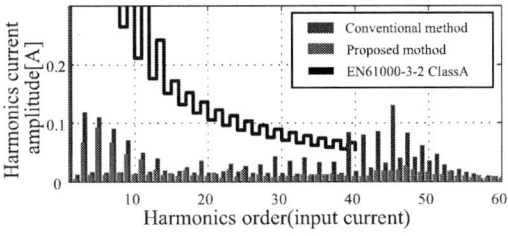

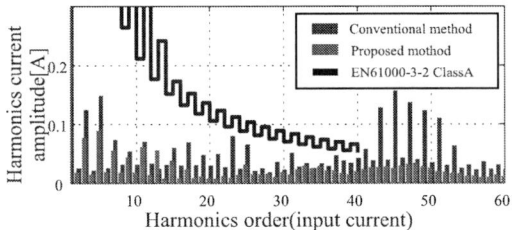

Fig. 13. Experimental result 3000rpm 1Nm 10kHz(Proposed method)

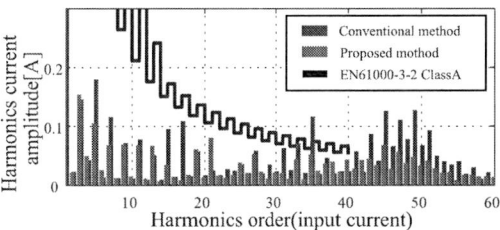

Fig. 14. FFT analysis of input current at 2000rpm 0.6Nm16kHz

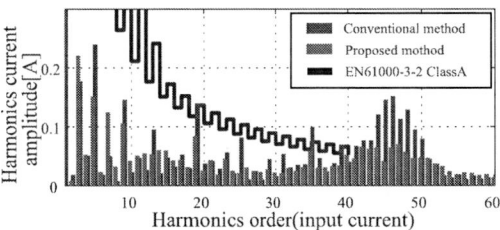

Fig. 16. FFT analysis of input current at 2000rpm 0.6Nm10kHz

Fig. 15. FFT analysis of input current at 3000rpm 1Nm16kHz

Fig. 17. FFT analysis of input current at 3000rpm1Nm10kHz

VI. CONCLUSION

A predictive control method for DC-link current was proposed based on the IPMSM discrete equation. In the conventional DDCCC, a control error occurs in the DC-link current, generating an input current harmonic by the high frequency component of the control error due to steep voltage operation. In the proposed method, the harmonic of the input current is reduced because the control error of the DC-link current is reduced. The effectiveness of the proposed method is confirmed by experiments.

REFERENCES

[1] S. Shao, E. Abdi, and R. McMahon, "Low-Cost Variable Speed Drive Based on a Brushless Doubly-Fed Motor and a Fractional Unidirectional Converter", *IEEE Trans. Ind. Electron.*, vol. 59, no. 1, pp. 317-325, 2012.

[2] J. Kolb, F. Kammerer, M. Gommeringer, and M. Braun, "Cascaded Control System of the Modular Multilevel Converter for Feeding Variable-Speed Drives", *IEEE Trans. Power Electron.*, vol. 30, no. 1, pp. 349–357, Jan., 2015.

[3] H. K. Samitha Ransara and U. K. Madawala, "A Torque Ripple Compensation Technique for a Low-Cost Brushless DC Motor Drive", *IEEE Trans. Ind. Electron.*, vol. 62, no. 10, pp. 6171–6182, Oct., 2015.

[4] M. K. H. Cheung, M. H. L. Chow, and C. K. Tse, "Design and Performance Considerations of PFC Switching Regulators Based on Noncascading Structures", *IEEE Trans. Ind. Electron.*, vol. 57, no. 11, pp. 3730-3745, 2010.

[5] K. Inazuma, K. Ohishi, and H. Haga, "High-Power-Factor Control for Inverter Output Power of IPM Motor Driven by Inverter System without Electrolytic Capacitor", *IEEE Int. Symp. Ind. Electron. (ISIE)*, pp. 619-624, 2011.

[6] K. Abe, K. Ohishi, H. Haga, Y.Yokokura "Instantaneous voltage vector control and d-axis current reference calculation to improve source current waveform for an electrolytic capacitor-less single-phase to three-phase inverter", *19th International Conference on Electrical Machines and Systems (ICEMS)*, pp1-6, 2016.

[7] Yeongrack.Son, Jung-Ik.Ha"Direct Power Control of a Three-Phase Inverter for Grid Input Current Shaping of a Single-Phase Diode Rectifier With a Small DC-Link Capacitor", *IEEE Transactions on Power Electronics*, Volume: 30, Issue: 7, July 2015

New Search Algorithm of Model Predictive Control to Reducing Calculation Amount for Improving Steady Current Control Performance

Masahiro Shimaoka[1*], Shinji Doki[1]

1 Department of Information and Communication Engineering
Graduate School of Engineering, Nagoya University, Aichi, Japan
*E-mail: m-shimaoka@nagoya-u.jp

Abstract—We have proposed Model predictive Control which can manage the voltage phase resolution. This method decides the optimum permutation of voltage vector by predicting current behaviors for all permutations and evaluating behaviors. Our method can realize same steady current control performance of FOC, however, there is the issue which calculation amount become very huge. In this paper, we propose new search algorithm of our MPC which can reduce calculation amount. In Propose algorithm, we take a notice that control set of our MPC can be classified by type and number of the voltage vector. Calculation amount is reduced by selecting search space depending on operation points. We show that new algorithm can be greatly reduced calculation amount while keeping steady current control performance.

Keywords—Model Predictive Control, Motor Drive System, Reducing Calculation Amount, Search Algorithm

I. INTRODUCTION

Model Predictive Control (MPC) has been used in various fields against the background of the advancement of a processor's performance[1]. MPC is one of optimum control that predicts future object's behaviors using object's mathematical motel for possible inputs and selects a optimum input by evaluating behaviors. In addition, MPC can consider the system constraints in mathematical model, and it is easy to tune and intuitive.

Motor drive system has been taken notice as an application of MPC due to the further improvement of a processor's performance[2] - [8]. In motor drive systems, controller considers motor and inverter as mathematical models, it determines input voltage by predicting and evaluating behaviors. In some approach of MPC, the controller treats the input voltage as disintegration quantity such as VSI's output voltage vector. It predicts the behavior of current, torque or flux in each inverter ' s output and select the optimum input voltage(Finite Control Set MPC : FCS-MPC) [3] - [8]. Many of these approach is used for reducing current ripple, torque ripple or switching frequency in the same operation range as the conventional control method, such as Field Oriented Control (FOC) using PWM (Pulse Width Modulation). Meanwhile, the current vector control based on FCS-MPC can also use overmodulation and square wave drive because the controller select the optimum output voltage vector directly. This method realize faster current

response and wider drive range of motor drive system[9]. Thus, the current vector control based on FCS-MPC can realize higher performance of motor drive systems compared with ordinary methods.

However, conventional Model Predictive current vector control based on FCS-MPC(CMP-CVC) has the issue that is inferior in steady current control performance compared with conventional control methods. This is caused by the voltage phase resolution. The voltage phase resolution of CMP-CVC is determined by control period in a characteristic. Therefore the voltage phase resolution is nearly 1000 times large compared with FOC using PWM. It needs shorting of control period for improving the voltage phase resolution of CMP-CVC, however, control period can't be shortened enough by the restriction of the power device. So, it needs to design the voltage phase resolution and switching frequency independently. To improve the voltage phase resolution, several approaches that define new voltage as control set such as Circular Dichotomy-based Voltage or Virtual Voltage Vector have been proposed[7], [8]. Control sets of these methods are not voltage vector of the inverter but voltages which the amplitude and the phase are changed, the amount of control set increase greatly. Because the control period of the current controller of motor drive systems is tens microseconds, there is the issue that implementation is prevented by calculation amount. To solve this problem,. some approaches which compatible the improvement of the voltage phase resolution and the reduction of calculation amount have been proposed by limiting the control set using the knowledge of power electronics. However, we think there are possibility that performance of CMP-CVC may be impaired by these limitation based on the knowledge of power electronics.

To improve the voltage phase resolution, we have proposed new model predictive control, Model Predictive Modulated Current Vector Control (MPM-CVC). Our method defines the prediction period dividing the control period anew. The controller makes voltage patterns which are permutations of voltage vector, predicts a current behavior in every prediction period and determines optimum voltage pattern by searching. By imposing the limitation that number of switching during control period is maximum one time, propose method is able to improve the voltage phase resolution keeping the average

The 2018 International Power Electronics Conference

switching frequency constant by this approach[10], [11]. Furthermore, our method can also keep the advantages of FCS-MPC that are fast response and wide operation range because the controller determines the voltage sequence directly[12].

However, MPM-CVC has ssame issue that calculation amount becomes more enormous likewise other approach which define a new control set. Therefore, we have verified by setting rough voltage phase resolution or only confined operation point.

To overcome this problem, we propose new search algorithm of MPM-CVC to reducing calculation amount. Proposed search algorithm focuses on the driving state of PMSM and that the voltage patterns of MPM-CVC are able to be classified. New algorithm can reduce the calcuration amount by limiting the search space. In this paper, we describe new search algorithm and show the effect of proposed algorithm by the simulation.

II. SYSTEM MODEL

A. PMSM Model

The mathematical model of PMSM is give as follows;

$$
\boldsymbol{v}_{dq} = \begin{bmatrix} R + pL_d & -\omega_{re}L_q \\ -\omega_{re}L_d & R + pL_q \end{bmatrix} \boldsymbol{i}_{dq} + \boldsymbol{e}_{dq} \quad (1)
$$

The symbols of PMSM model are listed in Table.I.

TABLE I: Symbols of PMSM Model

$\boldsymbol{v}_{dq} = [v_d, v_q]^T$	stater voltage vector in dq-axis
$\boldsymbol{i}_{dq} = [i_d, i_q]^T$	stater current vector in dq-axis
$\boldsymbol{e}_{dq} = [0, \omega_{re}Ke]^T$	electric force vector in dq-axis
R	stater resistance
L_d, L_q	stater inductance of dq-axis
Ke	electric force constant
ω_{re}	electric angular velocity of rotor

We can obtain the current state equation as equation.(2) by transforming from equation.(1).

$$
\frac{d}{dt}\boldsymbol{i}_{dq} = \boldsymbol{A}\boldsymbol{i}_{dq} + \boldsymbol{B}(\boldsymbol{v}_{dq} - \boldsymbol{e}_{dq}) \quad (2)
$$

$$
\boldsymbol{A} = \begin{bmatrix} -\frac{R}{L_d} & \frac{\omega_{re}L_q}{L_d} \\ -\frac{\omega_{re}L_d}{L_q} & \frac{R}{L_q} \end{bmatrix}, \boldsymbol{B} = \begin{bmatrix} \frac{1}{L_d} & 0 \\ 0 & \frac{1}{L_q} \end{bmatrix}
$$

The discrete-time current state equation converted from equation.(2) is follows;

$$
\boldsymbol{i}_{dq}(n+1) = \boldsymbol{A_d}\boldsymbol{i}_{dq}(n) + \boldsymbol{B_d}(\boldsymbol{v}_{dq}(n) - \boldsymbol{e}_{dq}(n)) \quad (3)
$$
$$
\boldsymbol{A_d} = exp(\boldsymbol{A}t_s), \ \boldsymbol{B_d} = [\boldsymbol{A_d} - \boldsymbol{I}]\boldsymbol{A}^{-1}\boldsymbol{B}
$$

where, t_s is discrete-time. We can calculate a current behavior after t_s by equation.(3)

B. 2-level 3-phase VSI

2-level three-phase voltage source inverter like Fig.1 is widely used in motor drive systems. One phase has two device switches as shown in Fig.1, two switch states exist at each phase. Therefore, this type inverter has eight different voltage vectors in all like Table.II and Fig.2(V_0 to V_7),which are determined by the combination of the switch state at each phase. Looking as the moment, the input of motor can be regarded as sequence of voltage vector.

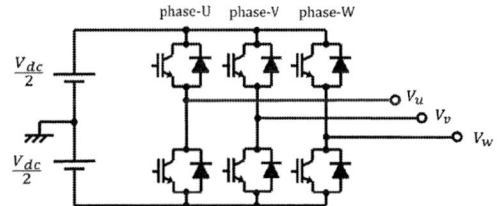

Fig. 1: 2-level 3-phase VSI

TABLE II: Voltage Vector of Inverter

	U	V	W
V_0	$-V_{dc}/2$	$-V_{dc}/2$	$-V_{dc}/2$
V_1	$V_{dc}/2$	$-V_{dc}/2$	$-V_{dc}/2$
V_2	$V_{dc}/2$	$V_{dc}/2$	$-V_{dc}/2$
V_3	$-V_{dc}/2$	$V_{dc}/2$	$-V_{dc}/2$
V_4	$-V_{dc}/2$	$V_{dc}/2$	$V_{dc}/2$
V_5	$-V_{dc}/2$	$-V_{dc}/2$	$V_{dc}/2$
V_6	$V_{dc}/2$	$-V_{dc}/2$	$V_{dc}/2$
V_7	$V_{dc}/2$	$V_{dc}/2$	$V_{dc}/2$

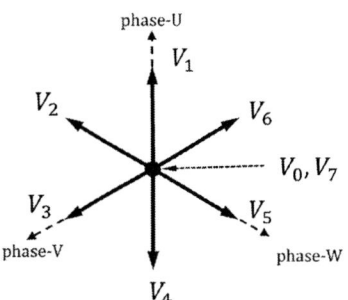

Fig. 2: Voltage Vector of the VSI

III. CURRENT VECTOR CONTROL BASED ON FCS-MPC

The current control systems of PMSM based on FCS-MPC is shown in Fig.3. The controller determines input voltage directly without using a modulator.

A. Control Flow of CMP-CVC

In CMP-CVC, the controller predicts future current behaviors during control period for each voltage vector

1748

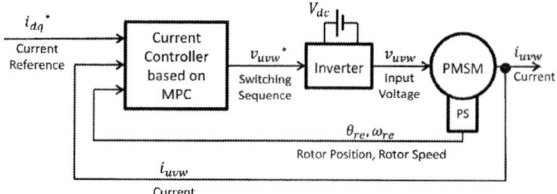

Fig. 3: Current Control System based on FCS-MPC

by using equation.(3). And then, it selects the optimum voltage vector which cost function(equation.(4)) is smallest.

$$ J = W_{i_d} * |i_d^* - \hat{i}d| + W_{i_q} * |i_q^* - \hat{i}_q| \qquad (4) $$

where, i_d^* and i_q^* are dq-axis current reference, $\hat{i}_d$ and $\hat{i}_q$ are predicted current behavior, W_{i_d} and W_{i_q} are weighting factor. Incidentally, the prediction horizon is the same as the control period for simplicity.

Fig.?? shows the control flow of CMP-CVC. Basic method controls PMSM by repeating the prediction of future current behaviors and the evaluation by cost function. The controller of CMP-CVC decides the input voltage vector for a motor during the control period directly without using a modulator, so it can use overmodulation and square-wave drive[9].

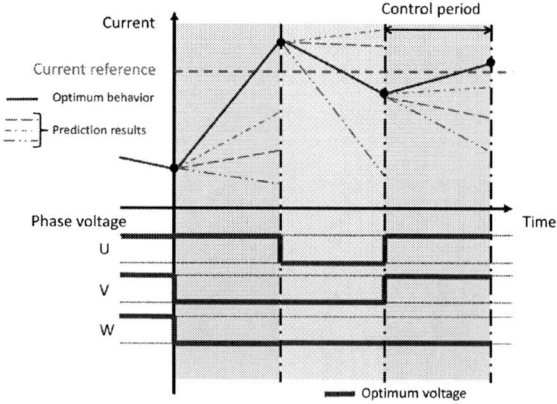

Fig. 4: The Control Flow of CMP-CVC

B. Provlem of CMP-CVC

As is well known, the voltage phase resolution affects the steady current control performance. The voltage phase resolution of CMP-CVC is the control period in a principle. This is 1000times larger than that of FOC using PWM. For the reason, CMP-CVC is inferior in steady current control performance compared with ordinary methods using a modulator. For improving the performance, the subdivision of the voltage phase resolution is needed. In CMP-CVC, this is realized by shortening the

control period. However, the abbreviation of the control period leads to rising the average switching frequency. The restrictions of the equipment such as a performance of a controller or a switching component prevent to shorten the control period. Therefore, a manage method of the voltage phase resolution without shortening the control period is required in current vector control based on FCS-MPC.

C. Control Flow of MPM-CVC

For this problem, we have proposed new current vector control method for motor drive systems based on FCS MPC. We call propose method, Model Predictive Modulated Current Vector Control(MPM-CVC). Our method introduces anew the prediction period in which the control period is devided by the integer Nc. As control set, the controller makes voltage patterns which are permutations of voltage vector like equation.(5) under the limitation that the number of switching in each phase is maximum one time during the control period.

$$ \hat{V}(m) = [V(1), V(2), \ldots, V(k), \ldots, V(N_c)] \qquad (5) $$
$$ V(k) \in V_0 \sim V_7 (k : 1 \sim N_c) $$

where m is pattern number, N_c is Control period / Prediction period.

Propose controller predicts future current behaviors at each prediction period for voltage patterns (equation.(5)). After that, Controller decides the optimum voltage pattern which cost function value(equation.(6)) is minimum.

$$ J = W_{i_d} * \sum_{k=1}^{N_c} |i_d{}^* - \hat{i}_d(k)| + W_{i_q} * \sum_{k=1}^{N_c} |i_q{}^* - \hat{i}_q(k)| $$
$$ (6) $$

where $\hat{i}_d(k)$ and $\hat{i}_q(k)$ are predicted dq-axis current behavior in k step ($k : 1 \sim N_c$).

Fig.5 shows the control flow of MPM-CVC. The average switching frequency is kept by the limitation of switching number during control period, so our propose method can improve the voltage phase resolution from the control period to the prediction period independently of the control period. Furthermore, our method don't use any modulator and consider the permutation of voltage vector, the controller decides the voltage sequence directly, MPM-CVC can also realize fast response and wide drive range [12].

IV. REDUCING CALCULATION AMOUNT OF MPM-CVC

MPM-CVC can design the voltage phase resolution and the average swithing frequency independently by introducing the prediction period. However, MPM-CVC has the issue that the calculation amount is very huge because the controller predicts behaviors for all possible

The 2018 International Power Electronics Conference

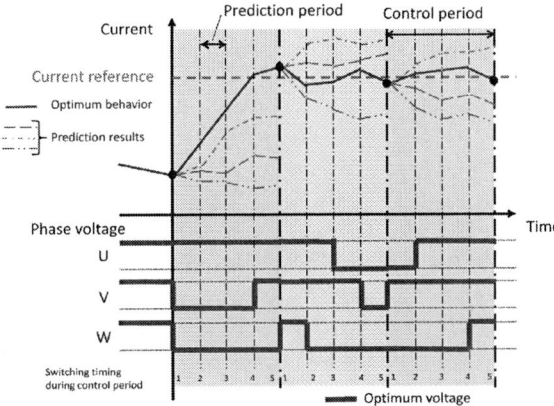

$N_c = 4 = $ Control period / Prediction period

Fig. 5: The Control Flow of MPM-CVC

voltage patterns. Quantity of voltage patterns represents equation.(7), those is more than thousands for realizing the same steady current control performance as the conventional method. Reduction of calculation amount is needed for improving the steady current control performance.

$$(N_c + 1)^3 \qquad (7)$$

We propose new search algorithm to reduce calculation amount of MPM-CVC. As can be easily imagined, the limitation of voltage patterns to be predicted is effective for reducing the calculation amount. Prediction time can be reduced by reducing the number of voltage patterns, that is, restricting the search space like Fig.6. In Fig.6, Orange area shows the search space. At this time, it is important not to miss the optimum voltage pattern in space limit. We pay attention to steady drive of PMSM for not missing the optimum voltage pattern.

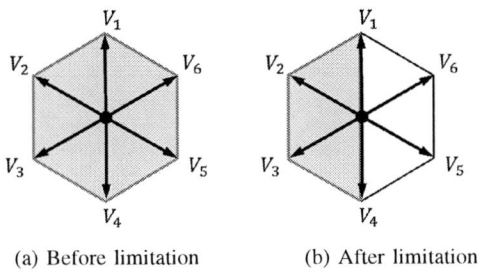

(a) Before limitation   (b) After limitation

Fig. 6: Example of Search Space

The voltage that is necessary for driving of PMSM in steady state can be expressed as the equation.(8). In this paper, we define this voltage as virtual voltage reference.

$$v_{dq}^* = \begin{bmatrix} R & -\omega_{re}L_q \\ -\omega_{re}L_d & R \end{bmatrix} i_{dq}^* + e_{dq} \qquad (8)$$

It can be considered that the optimum voltage pattern exists around virtual voltage reference in steady drive. Therefore, it can be considered that computational complexity can be greatly reduced by setting only the vicinity of the virtual voltage reference as the search space.

At this point, we focus on the characteristic of MPM-CVC. MPM-CVC create voltage patterns as control set in advance, voltage patterns can be classified by type and number of voltage vector as shown in Fig.7. In Fig.7, Nc is set 4 and a trapezoid represents one class. Examples of voltage patterns in red area is presented in equation.(9). Unlike SVM, voltage patterns of MPM-CVC is permutation of voltage vectors, not combinations. MPM-CVC cannot prevent an increase of calculation amount because it consider permutation of voltage vector, but it can realize fast response and wide drive range.

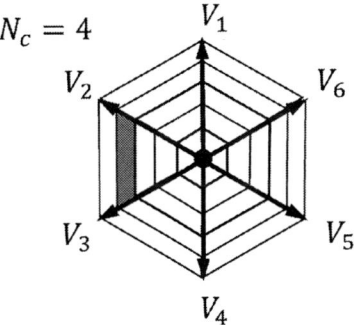

Fig. 7: Classification of Voltage Patterns

$$[V_0, V_2, V_2, V_2], [V_2, V_2, V_2, V_7], [V_0, V_2, V_2, V_3],$$
$$[V_2, V_2, V_3, V_7], [V_0, V_2, V_3, V_3], [V_2, V_3, V_3, V_7], \quad (9)$$
$$[V_0, V_3, V_3, V_3], [V_3, V_3, V_3, V_7]$$

The controller of new search algorithm selects some classes as search space from the amplitude and phase of virtual voltage reference like Fig.8. Propose search algorithm drastically reduce calculation time by limiting search space.

V. SIMULATION

We confirmed the effect of new search algorithm by the simulation. The simulation condition is listed in Table.III. We compared the execute time of controller and steady current control performance.

Table.IV shows execute time of the whole of simulation and the controller. The each execute time is measured by Simulink Profile Report. Proposed search algorithm could greatly reduce execute time of the controller by limiting search space.

Fig.9 and Fig.10 shows simulation results. The proposed algorithm can realized the same steady current control performance compared with conventional algorithm as shown in Fig.9 and Fig.10. We confirmed the

1750

The 2018 International Power Electronics Conference

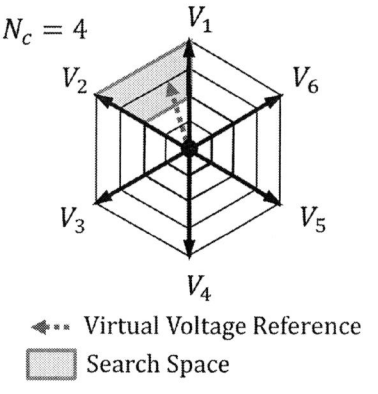

$N_c = 4$

◄·· Virtual Voltage Reference

▨ Search Space

Fig. 8: Selection of Search Space

effectiveness of proposed search algorithm from this simulation.

TABLE III: Simulation Condition

Control Period	40 [μs]
Division Number(N_c)	40
Simulation time	100 [ms]
Speed	1000 [rpm]
Torque Reference	1 [Nm]

TABLE IV: Calculation Time of Each method

Execute Time	Simulation	Controller	Rate
Conventional algorithm	641.27[s]	168.88[s]	100[%]
Proposed algorithm	471.13[s]	7.48[s]	4.4[%]

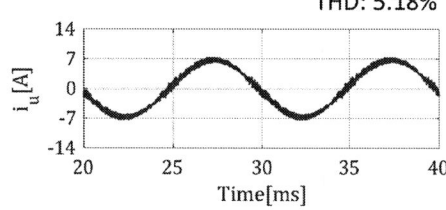

THD: 5.18%

Fig. 9: Simulation Result of Conventional Algorithm

VI. CONCLUSIONS

This paper describes new search algorithm of MPM-CVC for reducing calculation amount and we show the effect of new algorithm by the simulation. MPM-CVC cam manage the voltage phase resolution, however, our method has the issue which calculation amount becomes huge when the voltage phase resolution is set sufficiently fine. We noticed that the control set of MPM-CVC, voltage patterns, can classify by the voltage vector. Propose

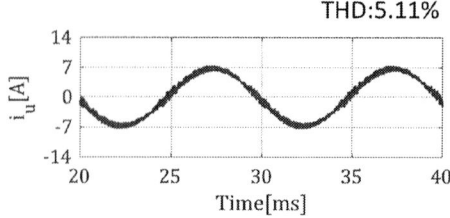

THD:5.11%

Fig. 10: Simulation Result of Proposed Algorithm

search algorithm limits search space by selecting some class according to the drive situation. We showed that the propose algorithm can reduce calculation time greatly and achieve the steady current control performance as conventional algorithm by the simulation.

REFERENCES

[1] Jan M. Maciejowski, "Predictive Control with Constraints",Pearson Education, 2002.

[2] G. Cimini, D. Bernardini, A. Bemporad S. Levijok, "Online Model Predictive Torque Control for Permanent Magnet Synchronous Motors", IEEE International Conference on Industrial Technology, 2015.

[3] M. Preindl, S. Bolognani, "Model predictive direct torque control with finite control set for PMSM drive systems, part 2: Field weakening operation", IEEE Transactions on Industrial Informatics, vol. 9, no. 2, pp. 648657, May 2013.

[4] M. Preindl, S. Bolognani, "Model predictive direct torque control with finite control set for PMSM drive systems, part 1: Maximum torque per ampere operation", IEEE Transactions on Industrial Informatics, vol. 9, no. 4, pp. 19121921, Nov. 2013.

[5] T. Zanma, M. H. M. Kawasaki, K. Liu, and A. Imura, "Model Predictive Direct Torque Control for PMSM with Discrete Voltage Vectors", IEEJ Journal of Industry Applications, vol. 3, no. 2, 2014. pp.121-130.

[6] Y. Cho, K. B. Lee, J. H. Song, and Y. I. Lee, "Torque-Ripple Minimization and Fast Dynamic Scheme for Torque Predictive Control of Permanent-Magnet Synchronous Motors", IEEE TRANSACTIONS ON POWER ELECTRONICS, vol. 30, no. 4, 2015, pp.2182-2190.

[7] X. Mei, F. Wang, R. Kennel, "A Circular Dichotomy-based Method for Model Predictive Control with Fixed Switching Frequency for Electric Drives", The 42nd Annual Conference of the IEEE Industrial Electronics Society, 2016.

[8] Y. Wang, X. Wang, W. Xie, F. wang, M. Dou, R. Kennel, R. Lorenz, D. Gerling, "Deadbeat Model-Predictive Torque Control With Discrete Space-Vector Modulation for PMSM Drives", IEEE Transactions on Industrial Electronics, vol64 No.5, May 2017.

[9] M. Kadota, S.Lerdudomsak, S. Doki, and S. Okuma: A Novel Current Control System of IPMSM Operating at High Speed Based on Model Predictive Control, The 4th Power Conversion Conference, April, pp.1315-1319, 2007.

[10] T. Maeda, S. Doki, "Improvement of Torque Control System of PMSM based on Model Predictive Control", The 37th Annual Conference of the IEEE Industrial Electronics Society, 2011

[11] H. Ohata, T. Sakamoto, and S. Doki, "Expansion of Operating Range and Improvement of Torque Response of PMSM Drive by Using Model Predictive Control", The 2014 International Power Electronics Conference, 2014.

[12] M. Shimaoka, S. Doki, "Current control system of PMSM based on Model Predictive Control for seamless drive between PWM mode and Square-wave mode of inverter", The 42th Annual Conference of the IEEE Industrial Electronics Society, 2016.

1751

The 2018 International Power Electronics Conference

Distributed Power Sharing Strategy for Islanded Microgrids without Frequency and Voltage Deviations

Tuan V. Hoang and Hong-Hee Lee
School of Electrical Engineering, University of Ulsan, Ulsan, Korea
E-mail: hoangtuan0404@gmail.com, hhlee@mail.ulsan.ac.kr

Abstract— This paper introduces a distributed power sharing strategy for the droop-controlled microgrids in which the power sharing errors and frequency/voltage deviations are completely eliminated based on the distributed cooperative control. The proposed strategy requires only a spare communication network in which the DG unit uses local and its neighbors' information to execute the control algorithm. In the proposed strategy, a virtual-impedance-regulator based on the consensus algorithm is developed to eliminate the power sharing errors in the islanded microgrid in spite of the load condition changes, and a modified droop controller based on the neighbors' power information is also introduced to cope with the frequency and voltage deviation issue. The proposed method does not use any information such as the feeder impedances, or the detailed microgrid configuration. The feasibility and effectiveness of the proposed method are also verified through digital simulation on an islanded microgrid system.

Keywords— *Droop control, distributed generation (DG), microgrid, power sharing.*

I. INTRODUCTION

Recently, distributed generation (DG) has been widely installed to the power distribution systems to supply the growing load demand for the utility with reduced effects on the environment and distribution infrastructures. And, the microgrid concept has been introduced to provide superior cooperation among multiple DG units. The microgrid can flexibly operate in grid-connected or islanded mode, and provide a more trustworthy power to the customers [1].

In the islanded operation, the load power demand must be accurately shared by multiple DG units to avoid the overload condition of a certain DG unit. Conventionally, the active power-frequency (P-ω) and the reactive power-voltage magnitude (Q-V) droop control methods have been adopted to realize the power sharing with a decentralized method [2]. Even though the system with the droop method is simple since only local measurement and no communication links are required, the reactive power sharing is highly vulnerable to the mismatch in feeder impedances [3].

To cope with the inaccurate power sharing issues, the virtual impedance-based control method has been considered as one of the most effective solutions [4]–[6]. However, these methods usually require the information about the DG feeder impedances which is hard to be available. Furthermore, due to the "plug-and-play" character of loads and DG units, the microgrid configuration has changed frequently and randomly. Consequently, the real-time feeder impedance is key information for the virtual impedance-based methods, which is very difficult to detect in practice. Furthermore, the frequency and voltage deviations are inevitable in the droop-controlled microgrids. In order to remove these problems, a centralized secondary control has been utilized [7]. Nevertheless, to carry out the centralized-based methods, some additional sensors are needed to measure the voltage at the point of common coupling (PCC), and also, several PI-type controllers are required to compensate voltage and frequency deviations. Furthermore, since the centralized secondary methods include a central controller and communication links between the central controller and DG units, they remain defenseless against a failure of the central controller or the single-point communication link. Recently, the distributed secondary controller has been proposed as a preferable solution [8]–[10] because only a sparse communication network are required without central controller and. And also, the additional sensors used to measure the PCC voltage are unnecessary as frequency and the voltage magnitude from the droop controller are used. However, even though the distributed control methods can solve the voltage and frequency deviation issue in the microgrid system, it increases computation burden on the local controllers due to the integral operation at each DG unit.

In this paper, we propose a distributed power-sharing strategy for the islanded microgrids which provides both accurate power sharing and no frequency/voltage deviations in the microgrid system. The contributions of this paper are summarized follows:

1) A virtual impedance based on the consensus algorithm is developed to deal with the power sharing problem in islanded microgrid. The proposed methodology can be directly implemented without any information about feeder impedances, load

1752

powers, and microgrid structure which is favorable in practice.

2) The traditional droop controllers are modified by using the neighbors' power information to eliminate the voltage/frequency deviations once the accurate power sharing is achived. Therfore, the controller can be implemented without the PCC voltage measurement as well as the integral operation such as PI-type controller either in the central controller or in the local controllers.

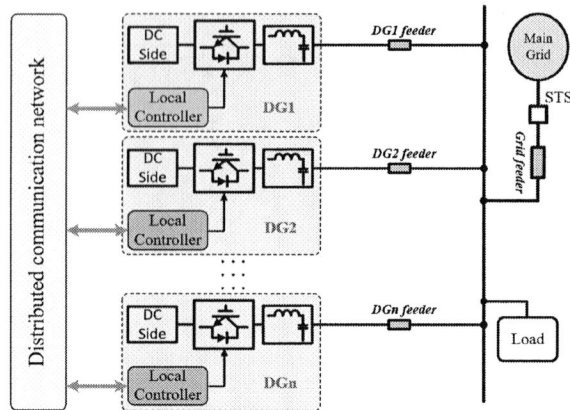

Fig. 1. Configuration of the islanded microgrid.

II. MICROGRID OPERATION

Fig. 1 illustrates an islanded microgrid configuration. In Fig. 1, the microgrid consists of n DG units and a number of loads. Each DG unit includes a DC source, inverters, LC filter, and local controller. The DG unit can communicate with others through the distributed communication network. According to the operation requirements, the microgrid can be in islanded mode or grid-connected mode by means of controlling the static transfer switch (STS) at PCC.

During the islanded operation, DG units are conventionally operated via the P-ω and Q-V droop controller in which the angular frequency and voltage magnitude set point of the DG output are given as

$$\omega_i = \omega_0 - m_i P_i , \qquad (1)$$

$$E_i = E_0 - n_i Q_i , \qquad (2)$$

where ω_0 is the DG nominal angular frequency; E_0 is the DG nominal voltage magnitude; P_i and Q_i are respectively the active and reactive powers of ith DG unit after low-pass filter (LPF); m_i and n_i are the slopes of P-ω and Q-V droop controllers of the ith DG unit, respectively. From (1) and (2), the voltage reference $V_{droop,i}$ for inner voltage controller is derived as

$$V_{droop,i} = E_i \sin\left(\int \omega_i dt\right) . \qquad (3)$$

However, the application of droop controller in (1) and (2) exhibits some limitations:

- The deviations of frequency and voltage magnitude from the nominal values,
- Poor reactive power sharing among DG units because of the feeder impedance mismatch and local loads.

III. PROPOSED CONTROL SCHEME

A. Communication Network Model

To perform the distributed control methods, the communication network to exchange information among DG units is required. Based on graph theorem [11], we describe the distributed communication network as an undirected and connected graph $G = (\zeta, \Gamma)$ as shown in Fig. 2, in which $\zeta = \{1,2,...,n\}$ is the non-empty finite set of n DG nodes, and $\Gamma \subseteq \zeta \times \zeta$ is the set of undirected

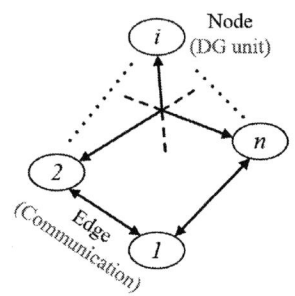

Fig. 2. Graphical representation of communication network.

edges. We call two nodes i and k as neighbors if there exists a communication link between them, i.e. $(i, k) \in \Gamma$. Then, the neighbor set of node i is established by $N_i = \{k \in \zeta \mid (i,k) \in \Gamma\}$, and $|N_i|$ is denoted as its cardinality. The adjacency matrix of the graph G is an $n \times n$ matrix A defined as $A = (a_{ik})_{N \times N}$ with $a_{ik} = 1$ if $(i, k) \in \Gamma$, otherwise $a_{ik} = 0$. Assuming that the graph contains no self-edge, i.e., $a_{ii} = 0, \forall i \in \zeta$, the degree matrix is denoted by $D = diag(d_i, d_2,...,d_N)$ where d_i is the in-degree of node i defined by $d_i = \sum_{j \in N_i} a_{ij}$. From these definitions, we obtain the Laplacian matrix of the graph by $\mathcal{L} = D - A$.

Finally, for the undirected and connected graph, the following important results for Laplacian matrix $\mathcal{L}$ are used in this paper: $\mathcal{L}$ is a symmetric and positive semi-definite matrix with zero as a simple eigenvalue, and $\underline{1}_n$ is right eigenvector corresponding to the zero-eigenvalue, i.e., $\mathcal{L}\underline{1}_n = 0$.

B. Proposed Control Scheme

In this paper, we develop a distributed control scheme thanks to the aid of communication network as shown in Fig. 3. The proposed strategy includes three main control parts: virtual impedance control to regulate power sharing, modified droop equations to remove voltage and frequency deviations, and double-loop voltage controller

1753

The 2018 International Power Electronics Conference

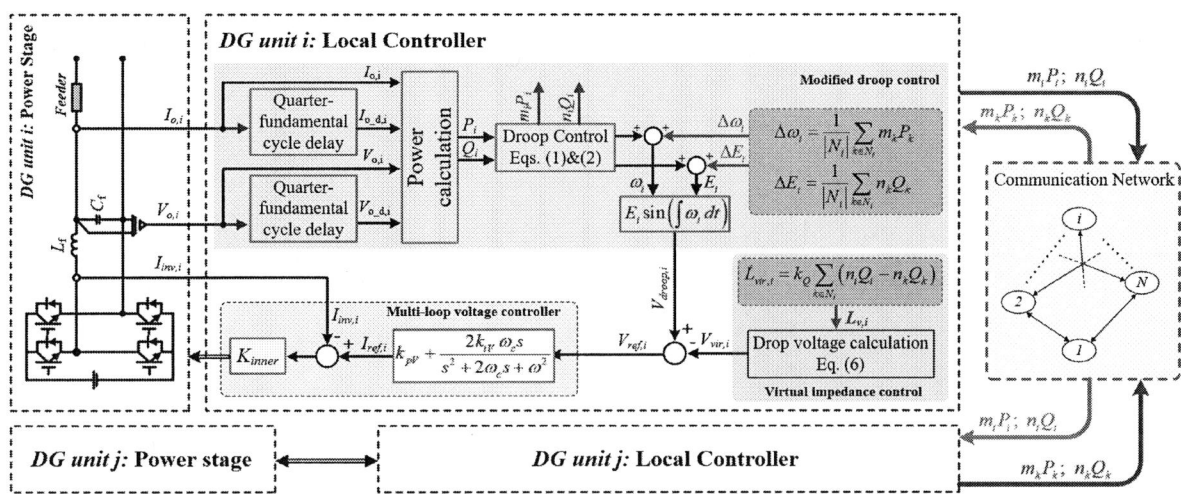

Fig. 3. Proposed distributed power sharing control scheme.

to output designed DG voltage.

1) Virtual Impedance Control: In order to achieve accurate reactive power sharing ($n_iQ_i = n_kQ_k$) by using the consensus control [12], the n_iQ_i should be chosen as DG state. However, in the consensus control, selected DG state is required to be directly controllable while n_iQ_i cannot. To cope with this problem, we use the controllable quantity $\dot{L}_{vir,i}$ instead of $n_i\dot{Q}_i$, and the consensus protocol for i^{th} DG unit is designed as follows:

$$\dot{L}_{vir,i} = k_Q \sum_{k \in N_i} \left(n_iQ_i - n_kQ_k \right), \qquad (4)$$

where k_Q is the positive control gain.

From (4), the update rule of virtual impedance at i-th DG unit is then expressed as

$$L_{vir,i} = k_Q \int \sum_{k \in N_i} \left(n_iQ_i - n_kQ_k \right) dt. \qquad (5)$$

And, the voltage drop on the virtual impedance, $V_{vir,i}$, is calculated as

$$V_{vir,i} = -\left(\omega_i L_{vir,i} \right) I_{o\_d,i}, \qquad (6)$$

where $I_{o\_d,i}$ is the conjugated signal of the DG line current $I_{o,i}$ obtained by delaying $I_{o,i}$ for a quarter fundamental cycle. Then, the reference voltage $V_{ref,i}$ for the inner voltage controller is modified as

$$V_{ref,i} = V_{droop,i} - V_{vir,i} = V_{droop,i} + \omega_i L_{vir,i} I_{o\_d,i}. \qquad (7)$$

Theorem 1: The reactive power sharing errors in the system are eliminated with virtual impedance control in (4).

Proof: Firstly, some global variables for entire n-DG system are defined as

$$\dot{L}_{vir} = \left[\dot{L}_{vir,1} \; \dot{L}_{vir,2} \cdots \dot{L}_{vir,n} \right]^T,$$

$$nQ = \left[n_1Q_1 \; n_2Q_2 \cdots n_nQ_n \right]^T.$$

Then, the whole n-DG system is expressed as (8)

$$\dot{L}_{vir} = k_Q \mathcal{L} nQ. \qquad (8)$$

From an undirected and connected graph, $\mathcal{L}\mathbf{1}_n = 0$. Hence, by setting $\dot{L}_{vir} = 0$ in (8) at the steady-state, (9) is obtained:

$$0 = k_Q \mathcal{L} nQ \Leftrightarrow nQ = \alpha \mathbf{1}_n \Leftrightarrow n_iQ_i = n_kQ_k, \forall i,k \in \zeta, \quad (9)$$

where $\alpha \in \mathbb{R} \setminus \{0\}$.

□

2) Modified Droop Equations: To deal with the frequency and voltage magnitude deviation issue, the droop controllers in (1) and (2) are modified as

$$\omega_i = \omega_0 - m_iP_i + \delta\omega_i, \qquad (10)$$

$$E_i = E_0 - n_iQ_i + \delta E_i, \qquad (11)$$

where $\delta\omega_i$ and δE_i are secondary compensation signals which are obtained as follows:

$$\delta\omega_i = \frac{1}{|N_i|} \sum_{k \in N_i} m_kP_k; \quad \delta E_i = \frac{1}{|N_i|} \sum_{k \in N_i} n_kQ_k. \quad (12)$$

Theorem 2: The frequency deviation issue is solved with the control law in (10) without any effect to the accurate active power sharing.

Proof: We define the matrix $N \in \mathbb{R}^{n \times n}$ as $N := diag\left(|N_i|^{-1} \right)$; and two column vectors as $e^\omega := col\left(e_i^\omega \right)$ with $e_i^\omega = \omega_i - \omega_0$ and $mP := col\left(m_iP_i \right)$. Then, the equation (10) is rewritten for n-DG system as

$$e^\omega = -N\mathcal{L}mP. \qquad (13)$$

Because all DG units operates at the same frequency at

1754

The 2018 International Power Electronics Conference

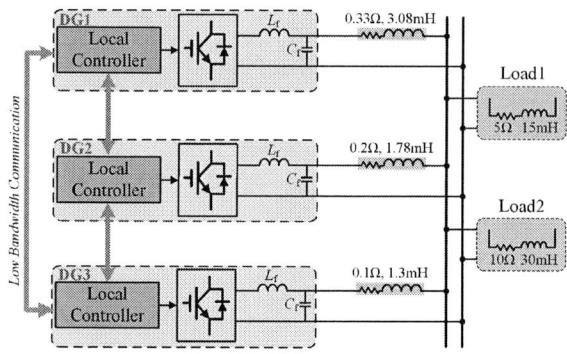

Fig. 4. Islanded microgrid used in simulation.

TABLE I
PARAMETERS IN SIMULATION

System parameter	Values		
LC filter, L_f/C_f	1.5 mH/ 15 µF		
DC link voltage, V_{dc}	250 V		
Nominal operating voltage, E_0/f_0	120 V, 60 Hz		
Switching frequency, f_{sw}	10 kHz		

Droop control	Case I	Case II		
	DG1=DG2=DG3	DG1	DG2	DG3
m (rad/W·s)	0.0015	0.0045	0.003	0.0015
n (V/Var)	0.0055	0.0165	0.011	0.0055

Double-Loop Voltage Control	Values
k_{pV}	0.1
k_{iV}	30
K_{inner}	20
ω_c	12.5 rad/s

Virtual Impedance Control	Values
k_Q	0.01

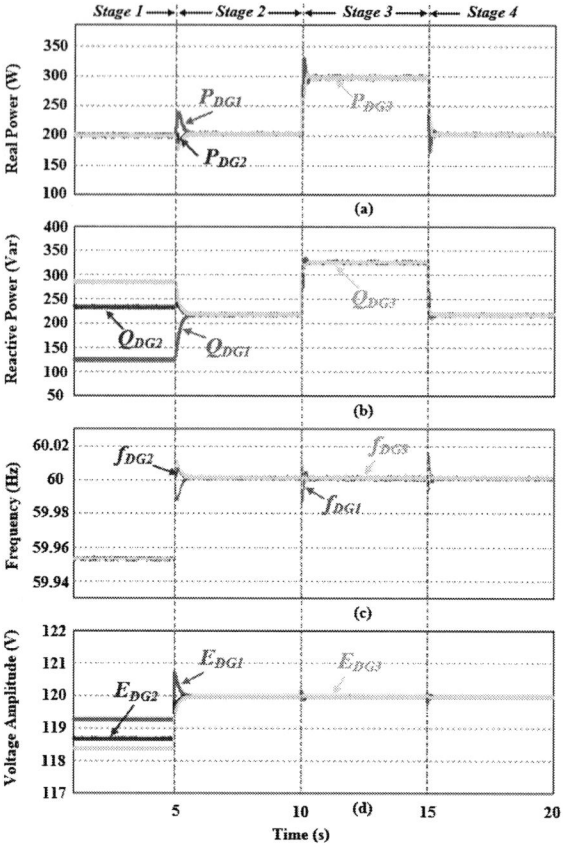

Fig. 5. Performance with the proposed controller: (a) active power; (b) reactive power; (c) frequency; (d) voltage amplitude.

the steady state, for $\beta \in \mathbb{R}$ the following relationship is achieved:

$$e^\omega = \beta \underline{1}_n. \tag{14}$$

By substituting (14) into (13), and multiplying both sides with $\underline{1}_n^T N^{-1}$, the following new equation is obtained:

$$\underline{1}_n^T N^{-1} \beta \underline{1}_n = \underline{1}_n^T N^{-1} N \mathcal{L} m P \Leftrightarrow \beta \underline{1}_n^T N^{-1} \underline{1}_n = \underline{1}_n^T \mathcal{L} m P \tag{15}$$

From the balanced Laplacian matrix, $\underline{1}_n^T \mathcal{L} = 0$, the equation (15) becomes

$$\beta \underline{1}_n^T N^{-1} \underline{1}_n = 0. \tag{16}$$

Since $\underline{1}_n^T N^{-1} \underline{1}_n = |N_1| + |N_2| + ... + |N_n| > 0$, the following result is obtained from (16):

$$\beta = 0 \Rightarrow e^\omega = 0 \Leftrightarrow \omega_i = \omega_0, \forall i \in \zeta. \tag{17}$$

Moreover, when $e^\omega = 0$ in (13), then,

$$0 = N \mathcal{L} m P \Leftrightarrow m P = \gamma \underline{1}_n \Leftrightarrow m_i P_i = m_k P_k, \forall i,k \in \zeta, \tag{18}$$

where $\gamma \in \mathbb{R} \setminus \{0\}$.

From (17) and (18), it is clear that both the frequency restoration and perfect active power sharing are realized with the control law in (10).

Theorem 3: The voltage deviation is eliminated with the control law in (11) when the accurate reactive power sharing is obtained by the virtual impedance in (4).

Proof: As proven previously, the reactive power errors are removed by the virtual impedance in (4). Therefore,

$$n_i Q_i = n_k Q_k, \forall i,k \in \zeta. \tag{19}$$

Thus, from (12) and (19), it can be derived as following:

$$\Delta E_i = n_i Q_i, \forall i \in \zeta. \tag{20}$$

By substituting (20) into (11), the following equality is driven:

$$E_i = E_0, \forall i \in \zeta. \tag{21}$$

□

3) Double-Loop Voltage Controller: From (10) and (11), the voltage reference in (7) becomes

$$V_{ref,i} = E_i \sin\left(\int \omega_i dt\right) + \omega_i L_{vir,i} I_{o\_d,i}. \tag{22}$$

1755

The 2018 International Power Electronics Conference

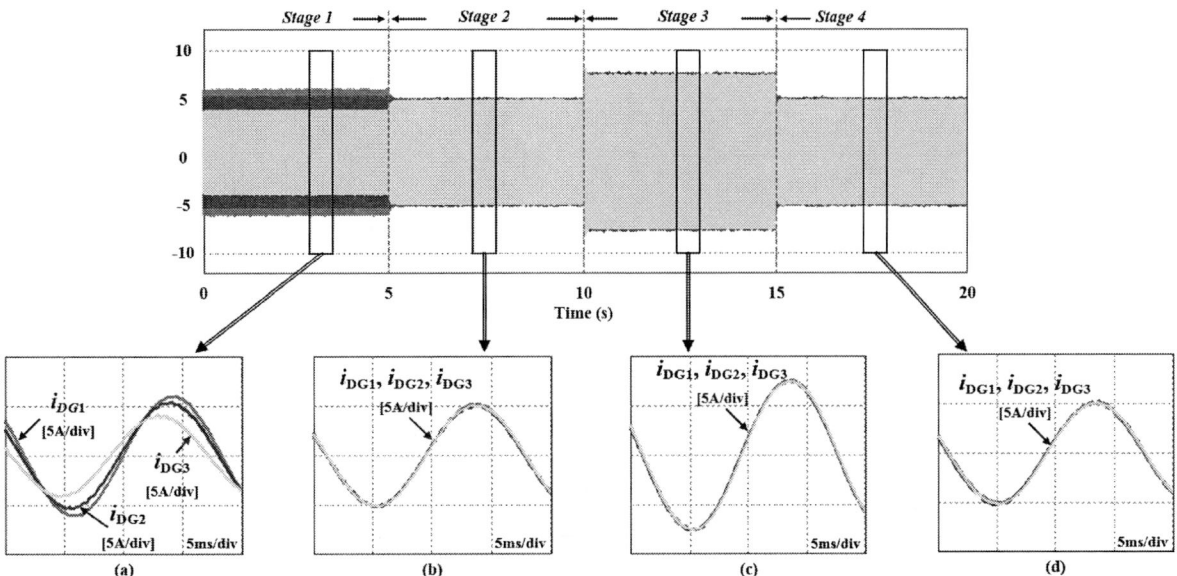

Fig. 6. Zoom-in DG current waveforms: (a) conventional droop method with load 1 at *t = 0s*, (b) proposed method with load 1 at *t = 5s*, (c) proposed method with load1 and 2 at *t = 10s*, (d) proposed method with load 1 at *t = 15s*.

To generate DG output voltage as reference in (22), the double-loop voltage controller is well adopted. In the double-loop voltage controller, the outer loop is a non-ideal proportional-resonant (PR) controller and the inner loop is a simple proportional controller as shown in (23) and (24), respectively:

$$G_{outer}(s) = k_{pv} + \frac{2k_{iv}\omega_c s}{s^2 + 2\omega_c s + \omega_{DG}^2}, \qquad (23)$$

$$G_{inner}(s) = K_{inner}. \qquad (24)$$

IV. SIMULATION VALIDATION

The proposed power sharing strategy is applied to the islanded microgrid in Fig. 4, which consists of three DG units and several loads at PCC, and it is simulated thanks to the aid of the PSIM environment. The system parameters are listed in Table I. A 1ms delay is included in the communication links of DG units.

For performance evaluation, we investigate two cases of the DGs in Fig. 4: the DG units have the same power ratings (Case I) and different power rating (Case II).

A. Case I: Same Power Ratings

The system performance with the proposed control method is shown in Fig. 5; the microgrid is originally operated with the conventional droop method with *Load1*. At *t = 5s*, the proposed method is applied. In order to verify the effectiveness of the proposed method according to the scenarios of step load change, *Load2* is attached to and detached from the system at *t = 10s* and *15s*, respectively.

From the power flows of DG units in Figs. 5(a) and 5(b), even though the active power demand is accurately

shared among DG units, the accurate reactive power sharing is not achieved with the conventional method in the stage 1 due to the mismatched feeder impedances. Once the proposed method is activated at *t = 5s*, the total reactive power demand is shared accurately among DG units after small transient duration. The accurate reactive power sharing is remained continuously regardless of the load power demand change in stage 3. Furthermore, the proposed control scheme always maintains the perfect active power sharing at the steady-state.

Figs. 5(c) and 5(d) show the frequency and voltage amplitude set points produced by the modified droop control. As can be seen, with the conventional droop control in stage 1, the frequency and voltage amplitude are deviated from the nominal values. However, those deviations are completely eliminated with the proposed method after *t = 5s* regardless of the load conditions.

Fig. 6 shows the DG output current waveforms corresponding to the stages in Fig. 5. As can be seen, both phase and magnitude of DG currents are different from each other in stage 1 with the conventional droop method (See Fig. 6(a)). However, with the proposed method from *t = 5s*, the DG currents are almost the same irrespective load condition changes (See Figs. 6 (b)-(d)).

B. Case II: Different Power Ratings

In this case, the ratio of power rating for DG units in Fig. 4 are designed as 1:1.5:3. The control performance with the proposed control method is shown in Fig. 7, wherein the waveforms correspond to those in Fig. 5. As clearly shown, irrespective of different power ratings, the proposed method provides proportional power sharing, and a good performance in frequency and voltage regulation similar to those with the same DG power ratings.

1756

The 2018 International Power Electronics Conference

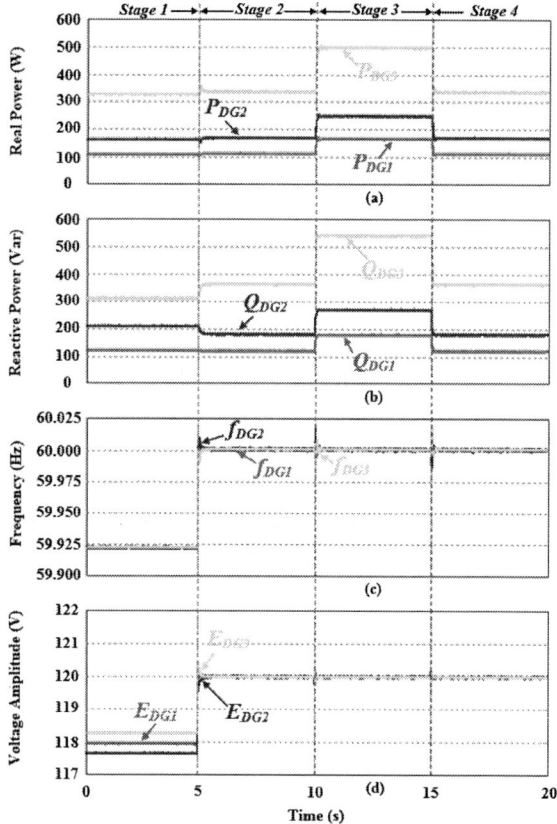

Fig. 7. Performance with the proposed controller (Case II): (a) active power; (b) reactive power; (c) frequency; (d) voltage amplitude.

V. CONCLUSIONS

In this paper, an adaptive virtual impedance control approach has been proposed by means of the consensus algorithm to achieve accurate power sharing for the islanded microgrids. The virtual impedance control can be implemented directly without any information about physical feeder impedance, detailed microgrid configuration, and load powers. Furthermore, when the accurate power sharing is realized, the traditional droop controller is modified to eliminate the frequency/voltage deviations without any additional sensor. The proposed approach uses only a spare communication network which is less vulnerable to the communication link failure. Therefore, the system becomes more reliable. Simulation results are also provided to validate the feasibility and effectiveness of the proposed control method.

ACKNOWLEDGMENT

This work was partly supported by the National Research Foundation of Korea Grant funded by the Korean Government (NRF-2015R1D1A1A09058166) and the Korea Institute of Energy Technology Evaluation and Planning (KETEP) and the Ministry of Trade, Industry & Energy (MOTIE) (No. 20174030201490).

REFERENCES

[1] A. Mehrizi-Sani and R. Iravani, "Potential-function based control of a microgrid in islanded and grid-connected modes," *IEEE Trans. Power Syst.*, vol. 25, no. 4, pp. 1883 - 1891, Nov. 2010.

[2] K. De Brabandere, B. Bolsens, J. Van den Keybus, A. Woyte, J. Driesen, and R. Belmans, "A voltage and frequency droop control method for parallel inverters," *IEEE Trans. Power Electron.*, vol. 22, no. 4, pp. 1390 - 1420, Apr. 2007.

[3] J. He and Y. W. Li, "Analysis, design and implementation of virtual impedance for power electronics interfaced distributed generation," *IEEE Trans. Ind. Appl.*, vol. 47, no. 6, pp. 2525 - 2538, Nov./Dec. 2011.

[4] J. He, Y. W. Li, J. M. Guerrero, F. Blaabjerg, and J. C. Vasquez, "An islanding microgrid power sharing approach using enhanced virtual impedance control scheme," *IEEE Trans. Power Electron.*, vol. 21, no. 1, pp. 193 - 201, Jan. 2013.

[5] Y. Zhu, F. Zhou, F. Wang, B. Liu, and Y. Zhao, "A wireless load sharing strategy for islanded microgrid based on feeder current sensing," *IEEE Trans. Power Electron.*, vol. 30, no. 12, pp. 6706 - 6719, Dec. 2015.

[6] Y. Zhu, F. Zhou, F. Wang, B. Liu, R. Gou, and Y. Zhao, "A virutal impedance optimization method for reactive power sharing in networked microgrid," *IEEE Trans. Power Electron.*, vol. 31, no. 9, pp. 2890 - 2904, Apr. 2016.

[7] J. M. Guerrero, J. Vasquez, L. G. de Vicuna, et al., "Hierarchical control of droop-controlled AC and DC microgrids – a general approach toward standardization," *IEEE Trans. Ind. Electron.*, vol. 58, no. 1, pp. 158 - 172, Jan. 2011.

[8] Q. Shafiee, J. M. Guerrero, and J. C. Vasquez, "Distributed secondary control for islanded microgrids – a novel approach," *IEEE Trans. Power Electron.*, vol. 29, no. 2, pp. 1018 - 1031, Feb. 2014.

[9] J. W. Simpson-Porco, Q. Shafiee, F. Dorfler, et al., "Secondary frequency and voltage control of islanded microgrids via distributed averaging," *IEEE Trans. Ind. Electron.*, vol. 62, no. 11, pp. 7025 - 7038, Nov. 2015.

[10] J. Schiffer, T. Seel, J. Raisch, and T. Sezi, "Voltage stability and reactive power sharing in inverter-based microgrids with consensus-based distributed voltage control," *IEEE Trans. Control Syst. Technol.*, vol. 24, no. 1, pp. 96 - 108, Jun. 2016.

[11] C. Godsil and G. F. Royle, *Algebraic Grapth Theory*. Berlin, Germany: Spinger-Verlag, 2001.

[12] Q. Shafiee, V. Nasirian, J. C. Vaquez, J. M. Guerrero, and A. Davoudi, "A multi-functional fully distributed control framework for ac microgrids," *IEEE Trans. Smart Grid*, to be published.

Lifetime-Oriented Droop Control Strategy for AC Islanded Microgrids

Yanbo Wang[1*], Dong Liu[1], Fujin Deng[2], Dao Zhou[1], Zhe Chen[1]

1. Department of Energy Technology, Aalborg University, Aalborg, Denmark
2. School of Electrical Engineering, Southeast University, Nanjing, China
*E-mail: ywa@et.aau.dk

Abstract— This paper presents a lifetime-oriented droop control approach for AC islanded microgrids, which is able to perform equal thermal stress distribution among paralleled inverters. An electro-thermal model and temperature estimation model is first established to observe the thermal stress of inverters, and then the temperature-power droop characteristic are built according to the electro-thermal model. Finally, the lifetime-oriented droop controller is proposed. Simulation results show that the proposed droop control strategy is able to perform automatically thermal stress sharing by regulating output powers of paralleled inverters, which thus mitigates the effect of unequal thermal-mechanical stresses on long term lifetime of paralleled inverters. The proposed droop control approach preserves the advantages of conventional droop control methods, and also implements the active thermal control for islanded microgrids. It thus increases the average lifetime and enhances long-time reliability of microgrids.

Keywords— *Droop control, lifetime, paralleled inverter, thermal sharing, AC microgrid.*

I. INTRODUCTION

With the increasing penetration of renewable energy sources, distributed power systems such as microgrids and active distribution networks [1]-[2] are becoming promising solutions to integrate various distributed energy sources and local loads [1], which are able to improve the reliability and flexibility of electricity utilization. Microgrids can be operated either in grid-connected mode or in islanded mode according to the power system demands.

For islanded microgrids, droop control schemes [1]-[3] have been intensively developed to perform power sharing among paralleled inverters. The attractive advantage of droop control is to automatically assign output power among distributed generators (DG) without using critical communication links, which thus improves reliability and flexibility of distributed power generation. So far, existing droop control strategies mainly focus on proportional power sharing according to maximum rating of DG inverters. However, the accuracy of power sharing tends is deteriorated due to unbalanced line impedance. Hence, the improved droop control schemes with adaptive virtual impedance [4], state estimation [2] and

disturbance injection [3] have been proposed to improve the accuracy of power sharing. In addition, several novel droop control strategies with consideration of optimized cost [5] and intermittent effect of renewable energy [6] have been presented.

However, the microgrids tend to be employed to supply reliable electricity for critical loads in safe-critical systems such as ship microgrid, aircraft power system and remote microgrid [7]-[8]. Hence, the reliability and safety are essential concerns in microgrids dominated by critical loads. And the efforts toward reliability enhancement of microgrids should be further explored. However, this research aspect has not ever been considered in droop-controlled microgrids.

It is well-known that operating temperatures of power modules have significant impacts on the long-term reliability of power electronic system, where over-temperature and temperature fluctuations caused by the losses of power semiconductors are the main factors causing power modules failures [9]. Fig. 1 shows the typical junction temperature profile of power module. Failure of power modules may occur if power modules are operated continuously at a junction temperature (T_j) higher than the maximum temperature (T_{j_max}) [9].

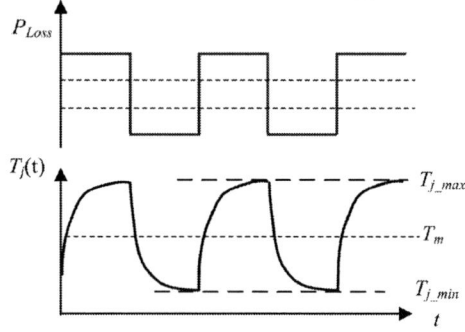

Fig. 1. Junction temperature profile of semiconductor device [9].

In practical microgrids, thermal operating points of inverters may be different due to the differences of power modules and thermal impedance, etc. [10]-[12]. The unequal thermal stress distribution among paralleled inverters may happen due to the difference of inverter parameters, which thus mitigates long term lifetime of microgrids. Thermal management is becoming an

1758

important aspect of power electronic systems with the increasing demands of reliability and high power density [13]. Active thermal control strategies by reducing or redistributing load power have been introduced to improve system efficiency and increase the average lifetime cycle of inverters. To enhance long-term reliability of power converters, an active thermal control-based current sharing scheme for paralleled inverters is proposed in [10], where the effect of parameters variation on temperature is investigated and load current is redistributed among paralleled inverters according to their temperature difference. A dynamic electro-thermal model is presented in [12], which is able to estimate the transient junction temperature of the semiconductor devices, and power sharing among paralleled converters is facilitated according to the estimated junction temperature. The proposed method is able to increase the overall system efficiency and reliability. However, the communication link is employed to define reference voltage for the control method. Hence, the existing droop control strategies fail to perform active thermal control in microgrids, which weakens system long-term reliability.

To enhance system reliability, this paper presents a lifetime-oriented droop control scheme. An electro-thermal model is first built, and temperature estimation model is presented to observe thermal distribution of paralleled inverters. Then, the dynamic droop control strategy is proposed to perform active thermal balance. The proposed method is able to redistribute thermal dissipation and mitigate the thermal-mechanical stresses, which also preserves the inherent advantage of conventional droop control. The main contributions of this paper are: (1) A lifetime-oriented droop control scheme with active thermal balance is proposed. (2) To enhance the sharing accuracy of thermal stress, virtual impedance-based temperature-power droop control strategy is developed.

II. CONVENTIONAL DROOP CONTROL SCHEME

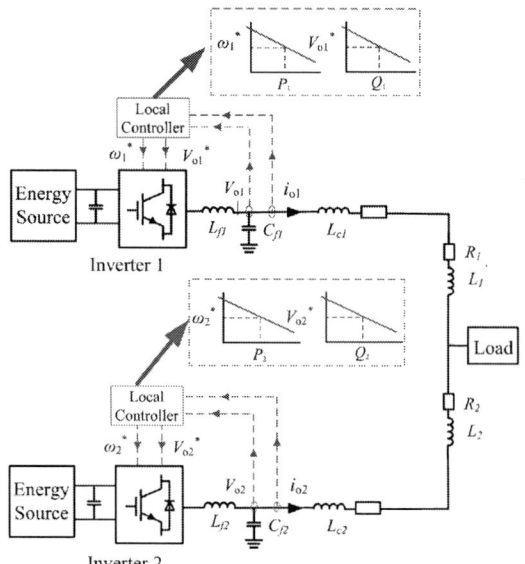

Fig. 2 The circuit configuration of droop-controlled microgrids

Droop control schemes have been widely proposed to perform proportionally power sharing in islanded microgrids. Fig. 2 shows the circuit configuration of droop-controlled microgrids, where active power-frequency (P-ω) and reactive power-voltage (Q-V) droop control strategy is employed to perform power sharing. The principle of droop control strategy is illustrated in Fig. 3 [2].

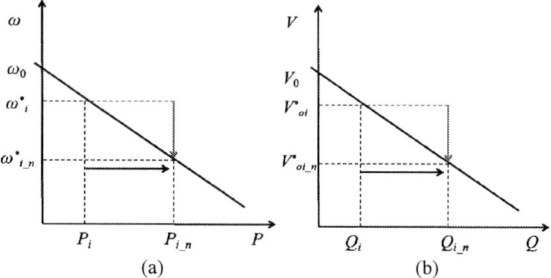

(a)   (b)

Fig. 3 The diagram of droop control strategy

The mathematical relationship of droop controller may be represented as (1) and (2) [2]

$$\omega_i^* = \omega_0 - m_i P_i \qquad (1)$$

$$V_{oi}^* = V_0 - n_i Q_i \qquad (2)$$

The initial operating point of power controller are (P_i, ω_i^*) and (Q_i, V_{oi}^*). In the presence of load disturbances, the terminal frequency and voltage of DG inverter will be changed to track output power. Then, power controller is operated in new operating point (P_{i_n}, $\omega_{i\_n}^*$) and (Q_{i_n}, $V_{oi\_n}^*$). ω_i^* and V_{oi}^* are output angle frequency and voltage of i-th inverter. P_i and Q_i are output active power and reactive power of i-th inverter. m_i and n_i are droop coefficients. The droop coefficients tend to be tuned according to respective maximum ratings of DG units as (3) and (4).

$$m_1 P_{\max 1} = m_2 P_{\max 2} = ... = m_i P_{\max i} = \omega_{\min} - \omega_{\max} \qquad (3)$$

$$n_1 Q_{\max 1} = n_2 Q_{\max 2} = ... = n_i Q_{\max i} = V_{\min} - V_{\max} \qquad (4)$$

where $P_{\max i}$ and $Q_{\max i}$ are maximum active power and reactive power capacity of i-th inverter.

However, the difference of power modules in inverters may cause different thermal operating points, so that the unequal thermal stress may occur due to the difference of these parameters, which weakens long-term reliability of microgrids. Thus, this paper presents a lifetime-oriented droop control method.

III. ELECTRO-THERMAL MODEL AND TEMPERATURE ESTIMATION MODEL

In this section, electro-thermal model is first built to estimate the temperature-current characteristic. Then, the temperature-current characteristic is incorporated into droop controller. Afterwards, the power sharing can be performed by the temperature-prioritized droop control strategy. The aim of electro-thermal modeling is to compute the average junction temperatures of power modules, and derive temperature-power relationship, which will be incorporated into droop controller.

A. Electro-thermal modeling

Power loss in voltage-source converter is composed of conduction loss and switching loss of power modules [13]. The average conduction losses of IGBT (P_{con_IGBT}) and diode (P_{con_Diode}) during the fundamental period may be represented as (5) and (6) [14].

$$P_{con\_IGBT} = \frac{1}{T_0}\int_0^{T_0/2}(V_{ce0}+rI_m\sin(\omega t))*I_m\sin(\omega t)*(\frac{1}{2}(1+m\sin(\omega t+\varphi)))dt$$
$$=\frac{1}{2}(V_{ce0}\frac{I_m}{\pi}+r\frac{I_m^2}{4})+m\cos\varphi(V_{ce0}\frac{I_m}{8}+\frac{1}{3\pi}rI_m^2)$$

$$(5)$$

$$P_{con\_Diode} = \frac{1}{T_0}\int_0^{T_0/2}(V_{ce0}+rI_m\sin(\omega t))*I_m\sin(\omega t)*(\frac{1}{2}(1+m\sin(\omega t+\varphi)))dt$$
$$=\frac{1}{2}(V_{T0}\frac{I_m}{\pi}+r\frac{I_m^2}{4})-m\cos\varphi\cdot(V_{T0}\frac{I_m}{8}+\frac{1}{3\pi}rI_m^2)$$

$$(6)$$

where T_0 is fundamental period, V_{ce0} is the constant voltage drop and r is the current magnitude dependent voltage in output characteristic of on-state voltage and collector current from datasheet [14]. I_m is amplitude of load current. m is modulation factor as the relation of AC voltage amplitude to DC-link voltage. φ is the phase angle between voltage and current. ω is fundamental angle frequency. The average switching losses of IGBT [14] over a fundamental period can be given as (7).

$$P_{sw\_IGBT} = \frac{1}{T_0}\sum_1^n E_{sw}(i) \qquad (7)$$

E_{sw} is the sum of turn-on and turn-off energy dissipation, which is given as (8).

$$E_{sw}(i) = (E_{on}(I_{nom},V_{nom})+E_{off}(I_{nom},V_{nom}))\frac{i}{I_{nom}}\cdot\frac{V_{dc}}{V_{nom}} \quad (8)$$

The turn-off recovery losses of diode may be calculated [14] as (9).

$$P_{sw\_diode} = \frac{1}{\pi}f_{sw}*E_{rec}(I_{nom})*(0.45\frac{i}{I_{nom}}+0.55)*\frac{V_{dc}}{V_{nom}} \quad (9)$$

E_{on} and E_{off} are measured turn-on and turn-off energy dissipation of IGBT at each switching pulse from datasheet. i is phase leg current. I_{nom} and V_{nom} are nominal current and voltage of power module. V_{dc} is DC-link voltage in practical application, which may be different from the nominal DC voltage. E_{off_diode} is turn-off recovery losses of diode.

Then, the total power loss can be represented as (10)-(11) by combining (5)-(9).

$$P_{loss\_IGBT} = P_{con\_IGBT}+P_{sw\_IGBT} \qquad (10)$$

$$P_{loss\_Diode} = P_{con\_diode}+P_{sw\_diode} \qquad (11)$$

B. Temperature estimation model

To reveal influence of load current on temperature, the temperature estimation model is established in this section. Junction temperature of power modules in DG inverters can be estimated by RC equivalent thermal models [15]-[16], which provide an effective solution to estimate junction temperature without direct measurements. Reduced order models of the thermal conduction path can be adopted to get an estimate of junction temperature. Real-time junction temperature estimation requires calculation of the instantaneous loss

in each power module of the voltage source inverter. This approach neglects the detailed behavior of the semiconductor devices within one switching cycle but gives sufficient details to include power cycling at fundamental frequency [17]. The temperature gradients and temperature swings caused by the losses of the power semiconductor affect mechanical strain within the models. Thermal resistance is a critical parameter to determine the thermal behavior of power electronic devices. In practical application of inverters, the thermal distribution may be different due to different thermal resistance. Fig.4 shows the thermal resistance chain of power modules applied in this work [15], where each current source represents the thermal power losses of power module. R_{thJC_IGBT} and R_{thJC_D} are thermal resistance of IGBT and diode from junction to case. R_{thCH_IGBT} and R_{thCH_D} are thermal resistance of IGBT and diode from case to heat sink. R_{thHA} is thermal resistance from heat sink to ambient temperature.

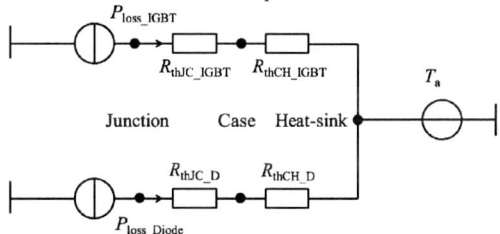

Fig. 4. Equivalent thermal model of semiconductor power module [15].

Then, the junction temperature of IGBT and diode can be derived from the equivalent thermal model [15] as shown in Fig. 4 as (12) and (13).

$$T_{jun\_T} = P_{loss\_IGBT}\cdot(R_{thJC\_IGBT}+R_{thCH\_IGBT})+T_a \qquad (12)$$

$$T_{jun\_D} = P_{loss\_Diode}\cdot(R_{thJC\_D}+R_{thCH\_D})+T_a \qquad (13)$$

where T_{jun_T} and T_{jun_D} are junction temperature of IGBT and diode, T_a is heat-sink temperature.

IV. THE PROPOSED LIFETIME-ORIENTED DROOP CONTROL STRATEGY

In this section, the lifetime-oriented droop control strategy is proposed according to the established electro-thermal model in Section III. With the assumption that three-phase load is balanced, each IGBT and diode in three legs has same thermal distribution. The temperature-current relationship of inverters can be derived from (12) and (13). In this work, a microgrid with two inverters is employed to exemplify the proposed droop control strategy, where different IGBT power modules are applied in different inverters. The parameters of power modules applied to calculate electro-thermal characteristics are given in Table I, which can be seen in datasheets [17]-[18].

As shown in Fig. 5, the temperature-current relationship of different inverters can be obtained by combining (12)-(13) and parameters in Table I. It can be observed that inverters have different temperature-current characteristics due to differences of semiconductor power modules. Furthermore, the mathematical representation of temperature-current relationship is obtained from Fig.5 by the least square fitting, which is given as (14)-(15).

TABLE I
PARAMETERS APPLIED IN ELECTRO-THERMAL MODEL

Inverter1		Inverter2	
Power module	FP10R06K L4	Power module	FS6R06VE3_B2
R_{thJC_IGBT}	2.2 K/w	R_{thJC_IGBT}	3.3 K/w
R_{thCH_IGBT}	0.29 K/w	R_{thCH_IGBT}	1.3 K/w
R_{thJC_D}	3.5 K/w	R_{thJC_D}	4.5 K/w
R_{thCH_D}	1.1 K/w	R_{thCH_D}	2.1 K/w
Current (Based value)	12A	Temperature (Based value)	75°
Frequency (Based value)	50Hz	Voltage (Based value)	110V

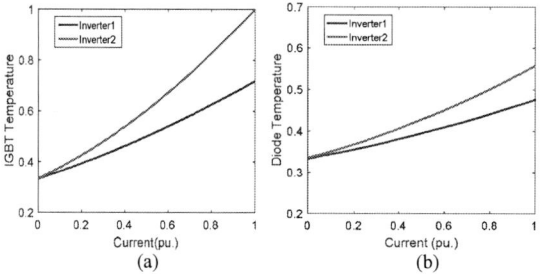

Fig. 5. Temperature-current relationship of inverter 1 and inverter 2. (a) Output current-dependent IGBT junction temperature. (b) Output current-dependent diode junction temperature.

$$T_{IGBT}(I_c) = a_1 I_c^2 + b_1 I_c + c_1 \tag{14}$$

$$T_{Diode}(I_c) = a_2 I_c^2 + b_2 I_c + c_2 \tag{15}$$

where $T_{IGBT}(I_c)$ and $T_{Diode}(I_c)$ are load current-dependent junction temperature of IGBT and diode. $a_1, b_1, c_1, a_2, b_2, c_2$ are coefficients of fitted equations, which can be obtained according to Fig.5.

The temperature-current characteristic (14)-(15) is incorporated into droop curves to perform power sharing control. Then, new active power-frequency and reactive power-voltage droop relationships with consideration of active thermal balance are presented as (16)-(20).

$$f_i^* = f_{max} - \Delta f_i \tag{16}$$

$$\Delta f_i = \frac{f_{max} - f_{min}}{P_{max}(T_{IGBT\_max})} \times T_{IGBT}\left(\frac{P_i}{V_{nom}}\right) \tag{17}$$

$$V_{oi}^* = V_{max} - \Delta V_{oi} \tag{18}$$

$$\Delta V_{oi} = \frac{V_{max} - V_{min}}{Q_{max}(T_{IGBT\_max})} \times T_{IGBT}\left(\frac{Q_i}{V_{nom}}\right) \tag{19}$$

$$\text{or} \quad \Delta V_{oi} = \frac{V_{max} - V_{min}}{Q_{max}(T_{Diode\_max})} \times T_{Diode}\left(\frac{Q_i}{V_{nom}}\right) \tag{20}$$

It can be seen that the new droop controller (16) and (18) are nonlinear equations associated with the temperature characteristics. The proposed P-f droop curves and Q-V droop curves are shown in Fig. 6. Compared with the conventional linear droop control strategies, the proposed droop control scheme is able to perform power sharing according to different temperature characteristics. The inverter with lower temperature will generate the more powers. Then, the thermal stress is redistributed, which thus achieves equal thermal distributions among paralleled inverters. The

temperature-dependent droop characteristic can be defined according to thermal models of different power modules.

Fig. 7 shows the diagram of the proposed droop control scheme. Power loss and thermal characteristic are calculated according to offline electro-thermal model established in section III. Then, the temperature-dependent droop controller is formulated by incorporating thermal characteristics.

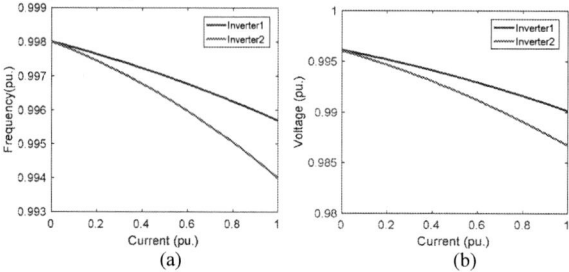

Fig. 6. The proposed lifetime-oriented droop control. (a) The temperature-dependent P-f droop curve. (b) The temperature-dependent Q-V droop curve.

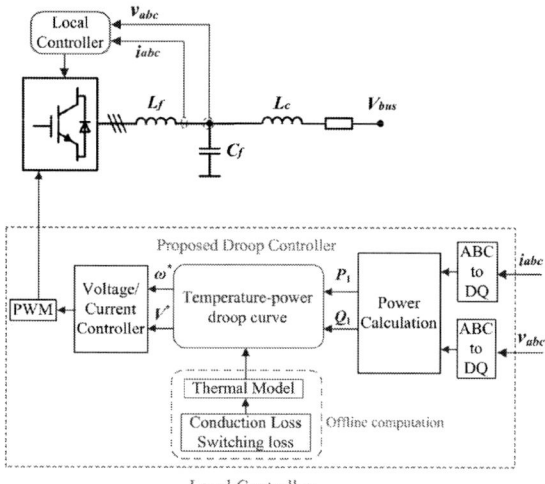

Fig. 7. The implementation of the proposed droop controller.

For reactive power-voltage droop controller, either IGBT or diode junction temperature may be selected as control objective. It is well-known that the accuracy of reactive power sharing is commonly affected by unbalanced feeder impedances [3], [18]. To eliminate the reactive sharing errors, inverter equivalent fundamental impedance can be reshaped by virtual impedance methods. For the lifetime-oriented droop controller, the accuracy of thermal sharing may be affected if only reactive power loads are fed. To guarantee the accuracy of thermal sharing, the following constraint (21) on inverter equivalent fundamental impedances is performed, where virtual impedance is configured into droop controller to reshape the terminal output performance as (22).

$$X_{f1} \cdot Q_{max1}(T_{jun1}) = X_{f2} \cdot Q_{max2}(T_{jun2}) \tag{21}$$

$$\frac{Q_{max1}(T_{jun1})}{Q_{max2}(T_{jun2})} = \frac{X_{f2}}{X_{f1}} = \frac{X_{o2} + X_{v2}}{X_{o1} + X_{v1}} \tag{22}$$

$Q_{maxi}(T_{juni})$ is junction temperature-dependent reactive power capacity of i-th inverter. X_{fi} is the desired inverter equivalent fundamental impedance. X_{oi} is the closed-loop output impedance of i-th inverter at fundamental frequency. X_{vi} is the virtual fundamental impedance of i-th inverter. Then, the reference voltage incorporating virtual impedance can be derived as (23) and (24). The control diagram is shown in Fig.8.

$$V_{refi\_d} = V_{odi}^* - (R_{vi} \cdot I_{odi} - \omega_0 L_{vi} I_{oqi}) \tag{23}$$

$$V_{refi\_q} = V_{oqi}^* - (R_{vi} \cdot I_{oqi} + \omega_0 L_{vi} I_{odi}) \tag{24}$$

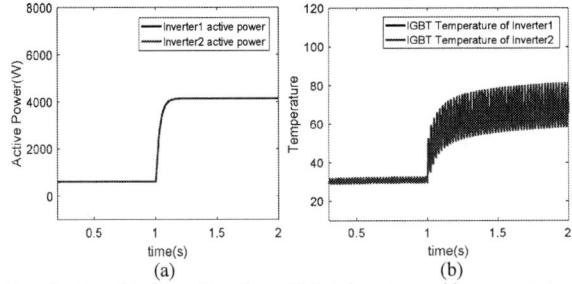

Fig.8. the control diagram of proposed droop controller with virtual impedance.

V. SIMULATION VERIFICATION

In order to validate the effectiveness of the proposed droop control strategy, the simulations in MALAB with PLECS blockset are performed in a scaled-down islanded microgrid with two inverters. The circuit diagram of exemplified microgrid is shown in Fig.2. The system parameters applied in simulation verification are given in Table II.

TABLE II
SIMULATION PARAMETERS

Parameters	Value	Parameters	Value
Inverter Rate	10kw	R_1/L_1	0.1Ω/2mH
f_s	10kHz	R_2/L_2	0.1Ω/2mH
L_f	1.5mH	Base Voltage (Phase RMS)	110V
C_f	25μF		
L_c	2mH	f_{max}/f_{min}	50.1Hz/49Hz

Case I Conventional droop control

Fig. 9. Simulated results of paralleled inverters with same power modules. (a) Active power sharing of inverters. (b) IGBT average junction temperature of inverters.

Fig. 9 shows simulation results of paralleled inverters with same power modules under conventional droop control method, where the Fig. 9(a) shows output active power of inverters and Fig. 9(b) shows IGBT average

junction temperature of inverters. It can be seen that the proportional power sharing and equal thermal dissipation are performed.

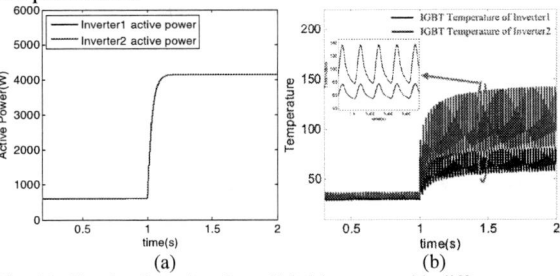

Fig. 10. Simulated results of paralleled inverters with different power modules. (a) Active power sharing of inverters. (b) IGBT Average junction temperature of inverters.

Fig. 10 shows the simulation results of paralleled inverters with different power modules under conventional droop control method. Although the proportional power sharing can be achieved, the thermal stress is unequally distributed once different thermal impedance characteristics of inverters happens, which causes unbalanced thermal dissipation as shown in Fig.10(b).

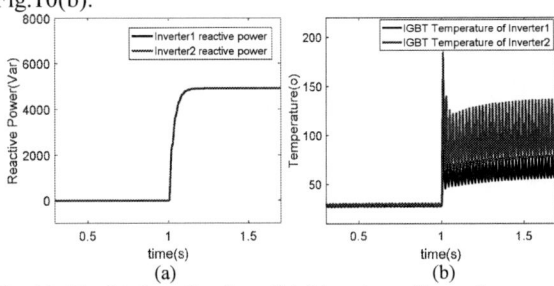

Fig. 11. Simulated results of paralleled inverters with reactive power loads. (a) Reactive power sharing of inverters. (b) IGBT Average junction temperature of inverters.

Fig. 11 shows the simulation results of paralleled inverters with reactive power loads. Although the reactive power can be equally shared, the thermal stress is unequally distributed due to difference of thermal impedance characteristics as shown in Fig. 11(b). Hence, the conventional droop control method fails to deal with unequal thermal stress in microgrid.

Case II Proposed lifetime-oriented droop controller

Fig. 12 shows simulation results of paralleled inverters with same power modules under proposed lifetime-oriented droop controller. Output active power and IGBT average junction temperature of inverters are shown in Fig.12(a) and Fig.12(b), respectively. It can be seen that the proportional power sharing and equal thermal dissipation are performed. Compared with conventional droop control approach, the proposed lifetime-oriented droop control strategy is still able to perform same results if power modules of inverters are same. It thus preserves advantage of conventional droop control strategies.

Fig. 13 shows the simulation results of paralleled inverters with different power modules under the proposed lifetime-oriented droop controller. It can be seen from Fig. 13(b) that the proposed droop control scheme is able to share automatically thermal dissipation by regulating active powers of inverters, so that the thermal dissipation can be equally distributed between inverters.

Fig.14 shows the simulation results of paralleled inverters with reactive power loads, where virtual impedance loop is incorporated into the proposed droop controller. It can be seen from Fig. 14(b) that the droop controller is able to equally redistribute thermal stress between paralleled inverters by regulating reactive powers of inverters.

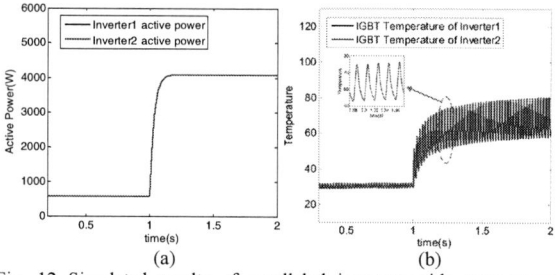

(a)   (b)

Fig. 12 Simulated results of paralleled inverters with same power modules under the proposed lifetime-oriented droop controller. (a) Active power sharing of inverters. (b) IGBT average junction temperature of inverters.

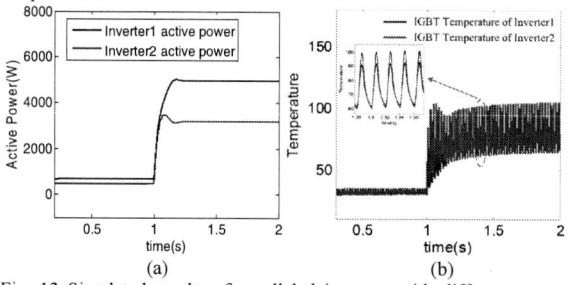

(a)   (b)

Fig. 13 Simulated results of paralleled inverters with different power modules under the proposed lifetime-oriented droop controller. (a) Active power sharing of inverters. (b) IGBT average junction temperature of inverters.

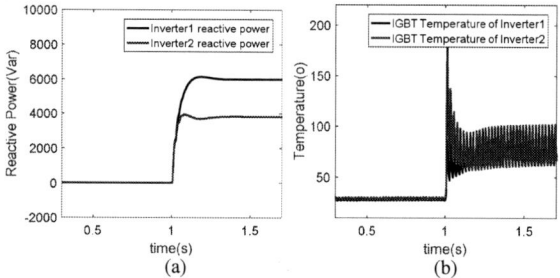

(a)   (b)

Fig. 14 Simulated results of paralleled inverters with different power modules. (a) Reactive power sharing of inverters. (b) IGBT average junction temperature of inverters.

VI. Conclusion

A lifetime-oriented droop control approach is developed in this paper. Electro-thermal model is first established to estimate temperature of power modules of inverters. Then, the temperature-current characteristic is incorporated into droop controller, and the lifetime-oriented droop control strategy is proposed. Virtual impedance is incorporated into temperature-power droop controller to enhance sharing accuracy of thermal stress. Simulation verifications show the proposed lifetime-oriented droop control approach is able to perform equal thermal stress distribution by regulating output powers of inverters, which also preserves inherent advantage of conventional droop control strategy. The proposed method thus enhances the long term lifetime and improves reliability of islanded microgrids.

References

[1] R. Lasseter, "Smart distribution: Coupled microgrids, " *Proc. IEEE*, vol. 99, no. 6, pp. 1074-1082, Jun. 2011.

[2] Y. Wang, Z. Chen, X. Wang, Y. Tian, Y. Tan, C. Yang, "An estimator-based distributed voltage predictive control strategy for AC islanded microgrids." *IEEE Trans. Power Electron.*, vol. 30, no. 7, pp. 3934-3951, July. 2015.

[3] J. He, Y. W. Li, and F. Blaabjerg, "An enhanced islanding microgrid reactive power, imbalance power, and harmonic power sharing scheme." *IEEE Trans. Power Electron.*, vol. 30, no. 6, pp. 3389-3400, Jun. 2015.

[4] J. Kim, J. M. Guerrero, P. Rodriguez, R. Teodorescu, and K. Nam, "Mode adaptive droop control with virtual output impedances for an inverter-based flexible AC microgrid," *IEEE Trans. Power Electron.*, vol. 26, no. 3, pp. 689-701, Mar. 2011.

[5] I. U. Nutkani, P. C. Loh, P. Wang, and F. Blaabjerg, "Cost-priorized droop schemes for autonomous AC microgrids." *IEEE Trans. Power Electron.*, vol. 30, no. 2, pp. 1109-1119, Feb. 2015.

[6] Y. Wang, Z. Chen, and F. Deng, "Dynamic droop scheme considering effect of intermittent renewable energy source". *in Proc. 2016 IEEE 7th International Symposium on Power Electronics for Distributed Generation Systems*, 2016, pp. 1-6.

[7] M. R. Hossain, and H. L. G III, "Real-time distributed coordination of power electronic converters in a DC shipboard distribution system." *IEEE Trans. on Energy Conversion*, vol. 32, no. 2, pp. 770-778, Jun. 2017.

[8] F. Gao, B. Bozhko, and S. Yeoh, G. Asher, and P. Wheeler. "Stability of multi-source droop-controlled electrical power system for more-electric aircraft." *in Proc. 2014 IEEE International Conference on Intelligent Energy and Power Systems*, 2014, pp. 122-126.

[9] D. A. Murdock, J. E. R. Torres, J. J. Connors, and R. D. Lorenz, "Active thermal control of power electronic modules." *IEEE Trans. on Ind. Applications*, vol. 42, no. 2, pp. 552-558, March. 2006.

[10] C. J. J. Joseph, M. R. Zolghadri, A. Homaifa, F. C. Lee, and R. D. Lorenz, "A novel thermal based current sharing control of parallel converters". *in Proc. 2004 26th Annual International Telecommunications Energy Conference*, 2004, pp. 647-653.

[11] K. Ma, M. Liserre, F. Blaabjerg and T. Kerekes, "Thermal loading and lifetime estimation for power device considering mission profiles in wind power converter." *IEEE Trans. Power Electron.*, vol. 30, no. 2, pp. 590-602, Feb. 2015.

[12] X. Yu, H. Wang, A. M. Khambadkone. "Control of parallel connected power converters for low voltage microgrid—Part I: A hybrid control architecture." *IEEE Transactions on Power electronics*. vol. 25, no. 12 (2010): 2971-2980.

[13] H. Wang, A. M. Khambadkone, and X. Yu. "Control of parallel connected power converters for low voltage microgrid—Part II: Dynamic electro-thermal modeling." *IEEE Transactions on Power electronics*. vol. 25, no. 12 (2010): 2971-2980.

[14] Infineon Technical Documentation. "Dimensioning program IPOSIM for loss and thermal calculation of infineon IGBT modules."

[15] S. Rohner, S. Bernet, M. Hiller, and R. Sommer, "Modulation, losses, and semiconductor requirements of modular multilevel converters." *IEEE Trans. on Ind. Electronics*. vol. 57, no. 8. 2010. pp: 2633-2642.

[16] N. Baker, M. Liserre, L. Dupont, and Y. Avenas, "Improved reliability of power modules: A review of online junction temperature measurement methods." *IEEE Industrial Electronics Magazine*, vol. 8, no. 3, pp. 17-27, Sep. 2014.

[17] "IGBT modules technical information", FS6R06VE3_B2, *Infineon datasheet*, 2007.

[18] "IGBT modules technical information", FP10R06KL4, *Infineon datasheet*, 2007.

[19] J. He, Y. W. Li, J. M. Guerrero, F. Blaabjerg, and J. C. Vasquez, "An islanding microgrid power sharing approach using enhanced virtual impedance control scheme." *IEEE Trans. Power Electron.*, vol. 28, no. 11, pp. 5272-5282, Nov. 2013.

Experiment on Hierarchical Control Based Power Quality Enhancement for Standalone Microgrid

Darith Leng[1], Sompob Polmai[2], Kittichot Soontorntaweesub[3]

Faculty of Engineering, King Mongkut's Institute of Technology Ladkrabang, Bangkok, Thailand

[1]E-mail: vothny.darith@yahoo.com, [2]E-mail: kpsompob@kmitl.ac.th, [3]E-mail: brightly.ks@gmail.com

Abstract—A microgrid is a new paradigm of power grid which has triggered advancement toward a green, intelligent and power efficiency. Due to the intermittent nature of microgrid's resources and the unpredictable power demand, the microgrid's voltage and frequency may experience deviation. Therefore, the decentralized control which could provide a faster response is a must. In this paper, the decentralized hierarchical control is implemented to handle the microgrid issues and enhance the power quality. The first level deals with power sharing between parallel generator. The frequency restoration and energy storage control are implemented in the second level to handle the frequency deviation which results from the high integration of PV. The third level is based on the multiagent system to ensure supply-demand balancing, autonomous decision-making and real-time energy monitoring. A laboratory scale microgrid is build up to verify the proposed control. The experimental results prove the effectiveness of the proposed control throughout various operating condition.

Keywords—Battery energy storage system , hierarchical control , internet of thing, multiagent system

I. INTRODUCTION

The electricity consumption in the world is constantly increasing, and our lives become more and more dependent on electricity. Due to environmental concern and the limit of fossil fuels, the energy production becomes more crucial and lead to deploying of renewable energy. The 21st century trend drives to the major change of electrical system from the ways it has ever been to be smarter and autonomous. It is a witness of a gradual transition with more sustainable and more decentralized towards a smarter power system which is a so-called smartgrid. The smartgrid represents a vision for the future power distribution systems, which integrates advanced sensing technologies, control methodologies and communication technologies into the current electricity grid. A microgrid is a new form of power system with the innovative control and management architecture at the distribution level. It is a promising technology for smartgrid with high integration of renewable energy, distributed generator, and energy storage system towards a more environmentally friendly and sustainable energy paradigm [1].

The strong dependency to meteorological of microgrid's sources and variation of power demand, it leads to voltage and frequency deviation and eventually resulting in an unstable system. To overcome these issues, vari-

ous control method has been proposed recently. In [2], the virtual generator that mimics the power-frequency relationship of a synchronous machine is proposed to solve the frequency oscillation during sudden load change which caused by lacking physical moment of inertia. The author [3] proposed a two-layers techno-economic energy management to optimize the operation, emission cost and to improve frequency deviation. In [4] Markov decision process based method is proposed to obtain the optimal load shedding strategy in an islanded microgrid to realize the supply-demand balancing.

The centralized control in current power system faces many challenges such as the increasing of the control variable, the complexity of the system, the alter behavior of loads as well as the uncertainty of environment. Hence, it is no longer suitable for the system control, especially for the future smart grid. The new trend of control and operation management pushes the conventional centralized control which is based on Supervisory Control and Data Acquisition (SCADA) to adopt the decentralized approach. Thus far, several standards (IEC 61499 and IEC 61131-3) are introduced to migrate from centralized to decentralized. The remarkable of these standards are platform independent, generic and special designed for distributed control applications such as agent-based technology [5]. In addition, with the new standard of substation automation system (IEC 61850), it provides the agent the ways to intercommunicate with the IED such as protection, control, and measurement unit. That is really convenient for the agent to be applicable in industries application.

As MAS is one of the most popular in system automation and smart control in terms of decentralized scheme, a well-designed MAS is flexible, reliable, and less expensive to implement, and it has a better chance of surviving single-point failures. Recently, MAS-based approaches have been applied to various power system applications, such as restoration plan [6], economics approach for residential electricity consumption [7], voltage regulation reactive power control [8], and demand response management [9].

Although many research works have been conducted to address the microgrid issues, most of them are realized in simulation only [10], [11]. In this paper, three levels hierarchical control is implemented and verified with a laboratory microgrid system. The first level is responded for power-sharing between the parallel generator, the second level is managed the system's voltage and frequency variation and the third level is used to realize autonomous decision-making, web-remote, and monitoring.

The rest of the article is organized as follows. The architecture of microgrid is described in Section II. In Section III, the proposed hierarchical control is presented. The experimental results and discussion are in Section IV. Finally, concluding remarks are given.

II. MICROGRID DESCRIPTION

The verification of proposed control only with software simulation, it can not guarantee the effectiveness while applying to the real system. Therefore, a real microgrid is required. Fig. 1 is a laboratory scale microgrid testbed which uses in this experiment to assess the proposed control.

Fig. 1: Laboratory scale Microgrid

The system's components are listed below:

- Two inverter-based generators which are direct control using DSP TMS320F28335 from Texas Instruments.

- PV simulator: The PV simulator is modeled based on a distributed generator ($DG3$). A GUI is implemented using Microsoft $C\#.net$ to interface between rooftop PV profile and controller DSP for controlling the $DG3$.

- Battery energy storage system: It comprises of eight batteries connected in series and it is cooperative control between DSP TMS320F28335 and agent.

- Loads: As this research focuses on a standalone microgrid, a better demand respond is important. Therefore, loads in this research are classified in two group such as critical load and non-critical load. A non-critical load is a group of loads which has low priority and it could shed or shift in case of the supply shortage.

The major parameters of the microgrid are provided in Table I.

TABLE I: Microgrid's parameters

Microsource 1 and 2 (Gen 1, Gen 2)	
Model	Kikusui PCR2000M
Power Rating	$220V, 50Hz, 2kVA$
P-Angle Droop Coefficient (m)	$2\pi(0.000185)$
Q-V Droop Coefficient (n)	$1/0.002592725$
Digital Controller (DSPF28335)	Sampling Rate: $12.78kHz$
Xd	$4mH$
Load	
Critical Load	$1500W, 1275VA$
Non-critical load	$1300W$
PV Simulator	
Model	Kikusui PCR2000M
Simulate	Based PV profile (Max: $1247W$)
Energy storage system	
SOLAR 12-65	8 * (12V, 65AH)
Multiagent system	
SPADE	Python (FIPA compliant)
Execute	Raspberry Pi 3 (Ubuntu Mate 16.04)

III. HIERARCHICAL CONTROL

Stemming from the conventional hierarchical control of large power systems [12] and IEC 62264, a microgrid's hierarchical control structure has been defined which consists of three levels as depicted in Fig. 2.

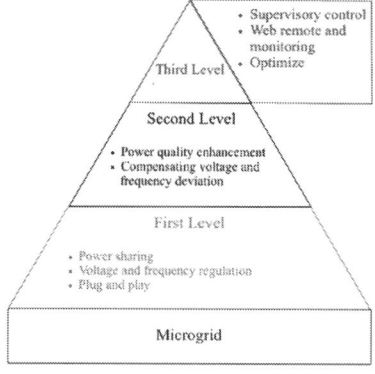

Fig. 2: Concept of proposed hierarchical control

In contrast to centralized controls, the emerging smart grid concept compels microgrid to adopt distributed cooperative methods as a result of the highly dynamic behavior of microgrids. Therefore, three-level decentralized hierarchical control is proposed. The detail of each control level is described below:

A. First Level

The primary level is also knowing as the local control which has the ability to perform the fastest respond with plug-and-play. Droop control is a well-known decentralized technique that mostly uses in the first level to realize the power-sharing between parallel generator or to avoid circulating current. According to the author [13] controlling converter interface generator via angle-droop can provide better stability margin compare to frequency-droop. As a result, the modified droop control is presented and expressed in Eq. 1.

$$\theta_E = \theta_{Eref} - K_f * P_{fb}$$
$$E = E_{ref} - K_v * Q_{fb} \qquad (1)$$

Where θ_E and E are the operating frequency and voltage, respectively. θ_{Eref} is the power angle for desire frequency and E_{ref} is the desired voltage.

Fig. 3 is the block diagram which is used to determine the control signals θ_E and E.

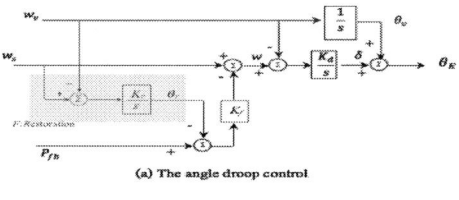

(a) The angle droop control

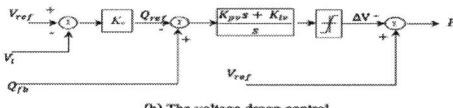

(b) The voltage droop control

Fig. 3: Frequency and voltage droop characteristic

B. Second Level

The main purpose of the secondary control is to handle the voltage and frequency deviation in the system. One major drawback of droop control [13] is that the frequency may operate at a new value which could be out of acceptable range. Therefore, the frequency restoration is implemented to keep the operating frequency close to the desired frequency ($\theta_E \simeq \theta_{Eref}$). The frequency restoration is described in Eq. 2 and expressed in color area of Fig. 3 to restore the frequency back to normal value.

$$\theta_r = K_r \int (\omega_s - \omega_v) \qquad (2)$$

Where K_r is a controller gain which is proportional to generator's angle droop coefficient. The restoration frequency (θ_r) is added with Eq. 1 to correction the local droop controller. Therefore, the Eq. 1 can be rewritten as below:

$$\theta_E = \theta_{Eref} - K_f * P_{fb} + \theta_r \qquad (3)$$

Furthermore, the intermittent nature of microgrid's source (PV) and the alter of loads could result in frequency deviation. Therefore, the fluctuations from PV power is needed to remove out. Fig. 4 shows the block diagram of PV power fluctuations mitigates by using the battery energy storage system (BESS). Two different control methods are used in BESS, so-called P_{fast}

and P_{slow}. P_{fast} contains the high-frequency component which is deployed to damp the power oscillation from PV and P_{slow} responds to BESS management which gets the control command from tertiary level. In P_{slow}, the constant K determines the charge-and-discharge rate and $Switch$ manages the time frame of BESS's charge-and-discharge with anticipating of the economic factor.

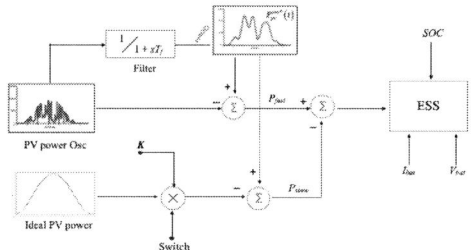

Fig. 4: PV power fluctuation mitigation

C. Third Level

The functionality of the third level is to provide the microgrid a smart operation, autonomous decision-making and real-time system monitoring. In this level; instead of using conventional SCADA, a multiagent system (MAS) is used. MAS is hardware or software based computer system which has design objective to handle the complex task in decentralized technique. Unlike SCADA, the agent can execute in low-level hardware (Embedded system) like Raspberry Pi3 that could reduce the cost of implementation, and it is usable in field control. Fig. 6 (Color area) is the group of agents which assigned to each component of Microgrid. Those agents execute in Raspberry Pi3 and their responsibility is described in Table II. Each agent has the different responses, and the $Algorithm 1$ is a sample of LOAD agent's task. The MAS is implemented using an open-source agent building toolkit, called SPADE [14].

TABLE II: AGENTs and Task

AGENT	TASKs
DG agent	It contains the information of DGs such as DG's capacity, actual power, status (On/Off) .., etc.
	It connects or isolates the reserve generator
	It executes the command from **web user interface** (wui) for manual control.
LOAD agent	It keeps tracking the energy consumption.
	It performs Demand Respond Management such as load-shedding and demand shift.
	It receives the command from **wui** for control the demand in manual.
PV agent	It holds the information of PV such as actual power and prediction power.
ESS agent	It communicates and tracks the storage level (SOC).
	It makes the decision when the ESS need to charge or discharge.
CONTROL agent	It works as a backup agent in case of any agent is down.
	It gathers the data from various agents and feed to the IoT platform for storage and monitoring.

The 2018 International Power Electronics Conference

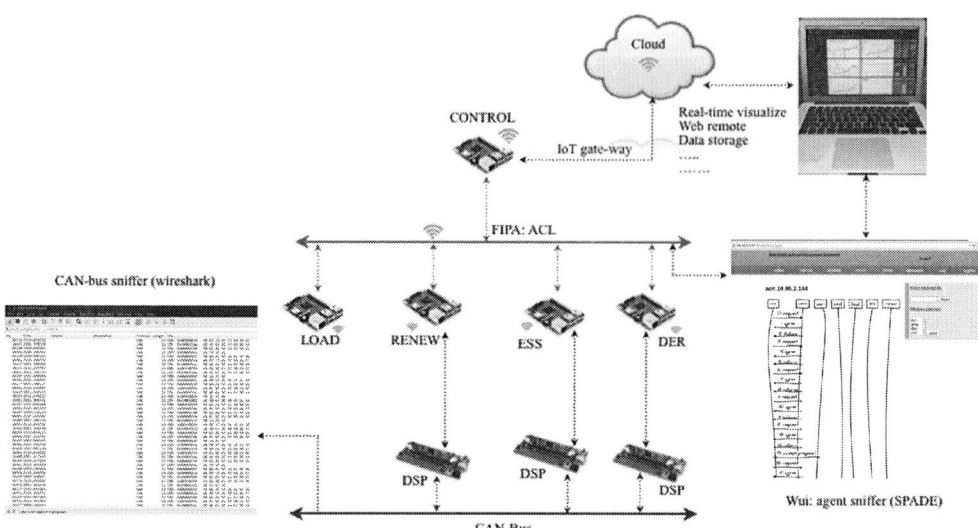

Fig. 5: Communication protocol

LOAD agent

Algorithm 1: Load-Shedding
1: Acquire Generator's data:
2:      $DG : P_{actual}, Q_{actual}, Capacity, Status$
3: Acquire Load's data:
4:      $Critical\ load : P_{crit}, Q_{crit}, Status$
5:      $Non - critical\ load : P_{nc}, Q_{nc}, Status$
6:      $P_{dm} = P_{crit} + P_{nc}$
7: $if\ \ P_{dm} > P_{dg}\ \ \ \ then$
8:      $Checking\ actual\ over\ power\ demand$
9:      $Checking\ non - critical\ load's\ status$
10:      $Minimize\ load\ to\ shed\ based\ on\ over\ demand$
11: $else$: no load-shedding
12: end

1) Communication Protocol: The agents are work in cooperate to form a multiagent system. The information and control commands from each agent are distributed in the system. Two standard communication mediums are used in this system to realize local-and-inter information exchange as demonstrated in Fig 5. The CAN-bus protocol is used for data exchange between the agent and local controller DSP and the TCP/IP protocol is used for data exchange between agent-agent based on FIPA-ACL standard, agent-to-web and agent-to-cloud. In Fig. 5 (Left-side) is the CAN-bus sniffer which used to monitor the data flow in CAN-bus and Fig. 5 (Right-side) is the agent-sniffer which is used to keep track of MAS.

IV. RESULTS AND DISCUSSION

The microgrid testbed is implemented in PEARL laboratory to evaluate the responses of the proposed control. Fig. 6 is a single-line diagram of laboratory scale microgrid assigning with the decentralize hierarchical control. The experiment was conducted for 500s and data recording is performed through power meter *Yokogawa*

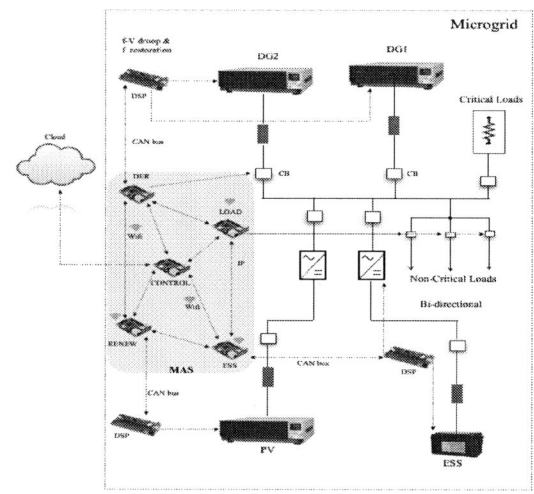

Fig. 6: Single-line diagram

WT 1600. The various scenarios are carried out for testing each control level and the results are discussed below.

A. Primary Level

As mentioned before, the functionality of the first level is to realize power-sharing of the parallel generator. In Fig. 7, the element 1 ($E1$) represents the power consumption which yellow-graph and red-graph are the total active power and reactive power, respectively. The element 2 ($E2$) consists of two graphs that the green-graph is the active power of $Gen\#1$ and the orange-graph is the active power of $Gen\#2$. In element 3 ($E3$), the ping-graph and blue-graph are the reactive power of $Gen\#1$ and $Gen\#2$, respectively. During $55s - 500s$, the

1767

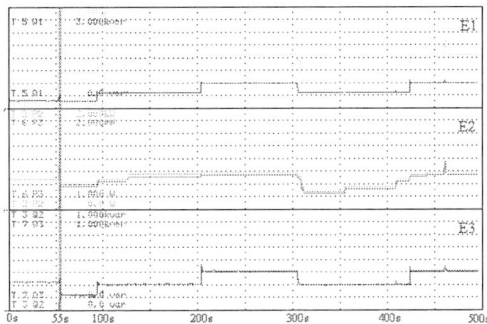

Fig. 7: Droop control: (E1) Active and reactive power of demands; (E2) Active power of Gen#1(Green) and Gen#2(Orange); (E3) Reactive power of Gen#1(Ping) and Gen#2(Blue)

active power output of Gen#1 (E2: Green) and Gen#2 (E2: Orange) and reactive power output of Gen#1 (E3: Ping) and Gen#2 (E3: Blue) are nearly equal. This result proves the effectiveness of modified droop control.

B. Secondary Level

The responsible of the second level is to correct or handle the impact of the first level or stochastic nature of renewable sources in the system. In this level, two different controls are implemented to ensure that system's voltage and frequency are operated in the standard range.

1) Frequency restoration: The frequency restoration is used to address the drawback of droop control, and it will take action while Gen#1 and Gen#2 are operating together. From $170s - to - 340s$ and $395s - to - 500s$, as shown in Fig. 8, the frequency ($E3 : Ping$) is restored to a normal value while the Gen#2 was taking into account.

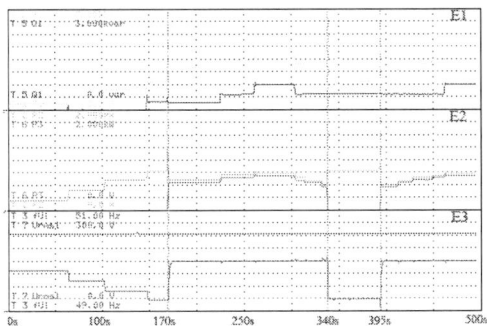

Fig. 8: Frequency restoration: (E1) Active and reactive power of demands; (E2) Active power of Gen#1(Green) and Gen#2(Orange); (E3) Ping (Frequency) and Cyan (Voltage)

2) Frequency Deviation Mitigation: The battery energy storage system in this paper is mainly used to smooth the power fluctuations of PV. Fig 9, the green-graph is the system's frequency without the BESS and, the red one is the system's frequency with presenting of the BESS. It

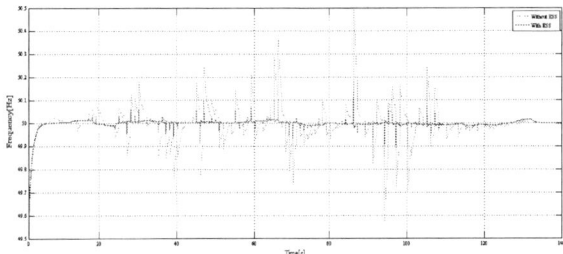

Fig. 9: Frequency deviation suppression with P_{fast}: Green without BESS; Red with BESS

is verified the influent of BESS to suppress the frequency deviation.

C. Tertiary Level

The third level is the core of hierarchical control to realize the smart microgrid concept. The state-of-art of control like the multiagent system and internet of thing (IoT) are utilized. Unlike JADE, SPADE provides the better way to develop and control the agents. Fig. 10 is a web user interface (wui) engaged with SPADE container. From this wui, the owner or operator can monitor and control the agent from anywhere that can access to the internet.

The multiagent system is work incorporate and share the data resources with each other in order to complete the goal. Below are some case studies which conduce to demonstrate the agent's performance :

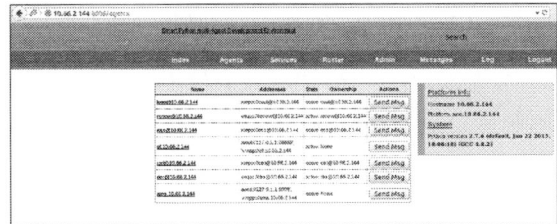

Fig. 10: SPADE: Web User Interface

1) Generator Control: The entire power demand is fully supported by the local generator when the microgrid is operated in standalone. The unpredictable of the power output of renewable sources may result in power supply-demand unbalancing. Therefore, the reserve generator is needed. The reserve generator in this system is referred to the fossil-based generator such as diesel generator which has high production cost, and it is allowed to operate in critical condition only to maintain the stability of the system.

In this case, the generator#2 works as a reserve generator. Fig. 11 demonstrated the effectiveness of MAS to control the performance of $Gen\#2$. The element 3 ($E3$) represents to $Gen\#2$. In case of power supply shortage, the DER agent will put the $Gen\#2$ into operation as it can clearly be seen during $25s - 170s, 230s -$

1768

The 2018 International Power Electronics Conference

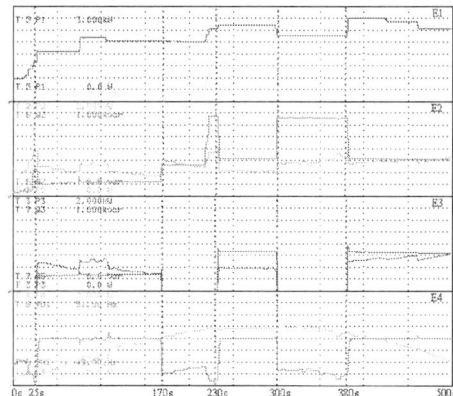

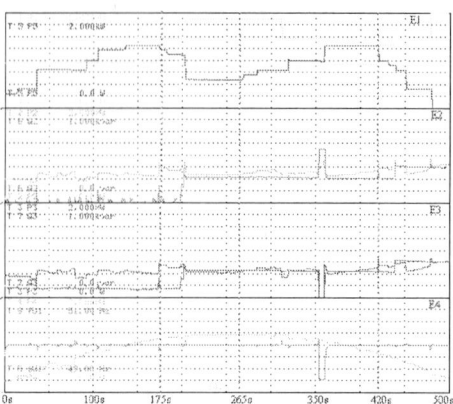

Fig. 11: Control reserve generator: (E1) Yellow (Critical load) and Red (Non-critical load); (E2) Power of Gen#1; (E3) Power of Gen#2; (E4) Cyan (PV) and Gray (f-Hz)

Fig. 12: Load-shedding: (E1) Yellow (Critical load) and Red (Non-critical load); (E2) Power of Gen#1; (E3) Power of Gen#2; (E4) Cyan (PV) and Gray (f-Hz)

$300s$ and $380s - 500s$. Otherwise, $Gen\#2$ will be disabled $(0 - 25s, 170s - 230s$ and $300s - 380s)$ as it is obvious in Fig. 11.

2) Load-Shedding: The load shedding is defined as a coordinated set of controls which results in decreasing of the electric consumption in a microgrid. It is conducted in case of insufficient supply of electricity from the local resources and the reserve generator. An optimal load shedding scheme finds the best stable equilibrium operating point for the system with the minimum amount of load to be shed. It could be realized by giving more priority to the critical load and managed the non-critical loads to minimize the total energy consumption. In Fig. 12, the element1 ($E1$) represents to the critical load's power ($Yellow$) and non-critical load's power (Red). From $175s - to - 265s$ and $420s - to - 500s$, the power supply is not fulfilled to the consumption, to prevent the system from collapse, the LOAD agent performs load shedding by cutting some non-critical load from the system. Consequently, the power of non-critical load is reduced during $175s - 265s$ and $420s - 500s$ ($E1 : red - graph$).

3) Web-Based Remote and Monitoring: Real-time monitoring plays an important rule in energy management system. It allows the utility as well as customer to make a better respond to any problem right away.

In this experiment, the CONTROL agent holds the whole system's data which receives from the various agents. The IoT gateway is implemented in this agent to profit the IoT technology. With the integration of IoT, the system's data are transmitted to the IoT platform called $THINGSBOARD$ [15] for real-time visualize, data storage & playback, alert and web remote. Fig 13 is some parts of the laboratory scale microgrid web-monitoring. It shows the necessary information of the $Gen\#1$, $Gen\#2$, loads and system's voltage and frequency, and the operator and the user can access it from anywhere using PC or $Mobile - phone$.

V. CONCLUSION

The microgrid is a promising technology for a future smart grid. The high integration of renewable sources, it may impact the system stability. In this research, the three-level hierarchical control is proposed to handle the microgrid issues. The First level used modified droop control for power-sharing between the parallel generator and eliminate circulating current. The Second level, the frequency restoration, and battery energy storage are implemented to handle the frequency deviation which results from the first level and the system. The Third level works like supervisory control which provides the system smart ability and autonomous operation. The real microgrid system is built up, and various scenarios are carried out. The experimental results prove the effectiveness of the proposed hierarchical control to ensure the system stability, improve the power quality, and provide the autonomous decision-making ability.

REFERENCES

[1] D. Chunxia, Y. Dong, L. Xinbin, and X. Yusheng, "Mas-based management and control strategies for integrated hybrid energy system," *IEEE Transactions on Industrial Informatics*, vol. 4, pp. 1332–1349, Aug 2016.

[2] T. Ise, T. Shintai, and Y. Miera, "Reactive power control for load shedding with vitual synchronous generator control," *Internation Conference on Electric Utility Deregulation and Restructuring and Power Technologies (DRPT)*, July 2012.

[3] A. khaledian, A. Ahmadian, and M. Aliakbar-Golkar, "Optimal droop gains assignment for real-time energy management in an islanding microgrid: a two-layer techno-economic approach," *IET*, vol. 11, pp. 2292–2304, July 2017.

[4] H. Gao, Y. Chen, Y. Xu, and C.-C. Liu, "Dynamic load shedding for an islanded microgrid with limited generation resources," *IET Generation, Transmission & Distribution*, vol. 10, pp. 2953–2961, August 2016.

[5] G. Frey and T. Hussain, "Modeling techniques for distributed control systems based on the iec 61499 standard - current approaches and open problems," *Proceedings of the 8th International Workshop on Discrete Event Systems (WODES 2006), Ann Arbor, Michigan, USA*, pp. 176–181, July 2006.

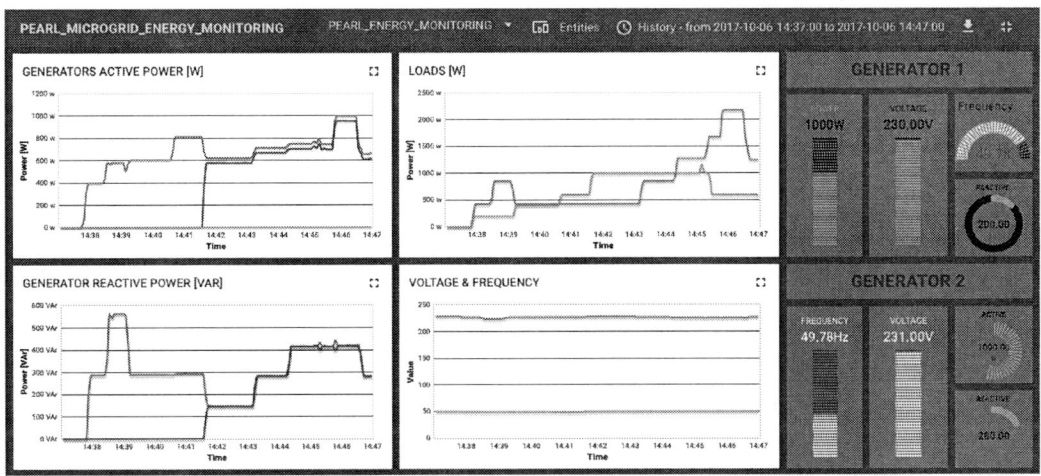

Fig. 13: Real-time web remote and monitoring

[6] E. Rokrok, M. Shafie-khah, P. Siano, and P. S. Catalão, "A decentralized multi-agent-based approach for low voltage microgrid restoration," *Energies*, vol. 10, p. 1491, Sep 2017.

[7] L. Siebert, A. Sbicca, A. Aoki, and G. Lambert-Torres, "A behavioral economics approach to residential electricity consumption," *energies, MDPI*, vol. 10, no. 6, p. 768, June 2017.

[8] A. Khatibzadeh, M. Besmi, A. Mahabadi, and M. Haghifam, "Multi-agent-based controller for voltage enhancement in ac/dc hybrid microgrid using energy storages," *energies, MDPI*, vol. 10, no. 2, p. 169, Feb 2017.

[9] V. Bui, A. Hussain, and H. Kim, "A multiagent-based hierarchical energy management strategy for multi-microgrids considering adjustable power and demand response," *IEEE Transactions on Smart Grid*, no. 99, June 2016.

[10] D. Li, B. Zhao, Z. Wu *et al.*, "An improved droop control strategy for low-voltage microgrids based on distributed secondary power optimization control," *energies*, vol. 10, no. 9, p. 1347, Sep 2017.

[11] L. Raju, R. S. Milton, and S. Mahadevan, "Multiagent systems based modeling and implementation of dynamic energy management of smart microgrid using macsimjx," *The Scientific World Journal-Hindawi*, vol. 2016, March 2016.

[12] J. M. Guerrero, J. C. Vasquez, J. Matas, L. G. D. Vicuna, and M. Castilla, "Hierarchical control of droop-controlled ac and dc microgrids — a general approach toward standardization,," *IEEE Trans. Ind. Electron*, vol. 58, no. 1, pp. 158–172, 2011.

[13] Q. Shafiee, J. Guerrero, and J. Vasquez, "Distributed secondary control for islanded microgrids—a novel approach," *IEEE Transactions on Power Electronics*, vol. 29, no. 2, Feb 2014.

[14] J. Palanca, G. Aranda, M. Escriva *et al.*, "Spade," *https://pypi.python.org/pypi/SPADE*.

[15] thingsboard.github.io, "Thingsboard - open-source iot platform," *Available at: https://thingsboard.io*.

The 2018 International Power Electronics Conference

A Distributed Predictive Control Strategy Based on State Estimator for islanded microgrid

Mi Dong, Li Li and Xiaoyu Tian

School of Information Science and Engineering

Central South University

Changsha, China

e-mail: mi.dong@csu.edu.cn; lili112209@163.com; txy15388967580@163.com.

Abstract—The traditional droop control is usually used to share real and reactive power without communication for fully decentralization power system. Because of that the traditional droop control has its inherent limitations, this paper presents a modified droop controller based on Kalman filter to reduce load effect, and then utilizes a distributed consensus algorithm to transmit error voltage to other distributed generation (DG) with sparse communication network, and last make use of distributed model predictive control (MPC) to achieve accurately load power sharing between multi-parallel inverters. Simulation results is verified the reliability of the proposed algorithm in realizing load sharing and the voltage-frequency restoration.

Keywords—droop control; Kalman filter; load effect; distributed consensus algorithm; distributed model predictive control.

I. INTRODUCTION

Microgrid, usually consisted of solar, tidal power and other distributed generations (DGs), plays an important role in the development of smart grid technology [1, 2]. A microgrid acts as a single controllable entity with respect to the maingrid, it can connect and disconnect from the maingrid to enable it to operate in both grid-connected or island-mode according to the instructions of the superior controller [3, 4]. Usually, a droop control is universally used to achieve power allocation automatically between DGs without communication in islanded mode [2, 5-7]. However, this control strategy has some inherent limitations, such as, it is easy to be affected by line impedance mismatch and variation in local load, thus leading voltage-frequency drop and poor reactive power sharing performance.

To address these problem, some significant researches have already been in progress for improving the performance of droop control, such as centralized voltage control [7, 10-12], the decentralized voltage control [8-9, 13-14], and distributed voltage control [16-20]. The centralized control mode needs complex and robust communication network [4, 17, 20] to collect global information and process huge amounts of data. Despite the communication technology has greatly improved, the communication networks still face great challenges in practical applications, especially for cost and stability [18, 20-21]. While the decentralized control mode is a wireless manner, it is enabled the "plug-and-play" function for DGs, which is help for to enhances the flexibility and reliability of system. But

This work was supported by the National Natural Science Foundation of China (No. 51677194)

this method cannot solve the load sharing problem appropriately, which prompts more and more scholars to add the communication to solve this problem. Currently, distributed control mode has become a potential control mode due to that DG unit in the distributed communication network only needs to transmit data messages to its adjacent DG unit. Driven by above motivation, this paper introduces a fully decentralized estimation method to estimate the rms-voltage in common bus based on Kalman filter. And then, a modified droop voltage-predictive control is used to address voltage-frequency drop and load sharing problem. The contributions are listed as follows:

(1) The proposed modified droop control based on Kalman filter can ensure the voltage in common bus maintaining at the rated value, which compensates the limitation of traditional droop control, and effectively restrain influence of noise, disturbance and other factors on the measurement results.

(2) A voltage-frequency recovery based on distributed consensus algorithm can achieve global consistency only with spare communication network.

(3) A distributed voltage predict control based on state estimation is investigated in detail, which can bring about accurate load sharing, and be against parameters perturbations.

The rest of the paper is arranged as below. In Section II, a mathematic model for system is introduced, and the limitations of traditional droop control is revealed. The novel control strategy, which is consisted of a modified droop voltage-predictive control based on Kalman filter and a voltage-frequency recovery algorithm, is described in Section III. Lastly, some simulation examples based on Matlab/Simulink via an islanded microgrid with multi-parallel DGs are given in Section IV, followed by conclusions made in Section V.

II. CONVENTIONAL DROOP CONTROL FOR MICROGRID

For an AC microgrid in the islanded operation mode, multi-DGs with inverters are connected to the common bus in the manner of parallel connection. Figure 1 shows a simplified circuit of a parallel system with two DGs connected to a common load across line impedances. L_{fi} and C_{fi} in Fig.1 are the filter inductance and filter inductance of i-th inverter, R_{li} and L_{li} are the line resistance and line inductance of i-th inverter. R_{load} and L_{load} are the resistance and inductance of load.

1771

The 2018 International Power Electronics Conference

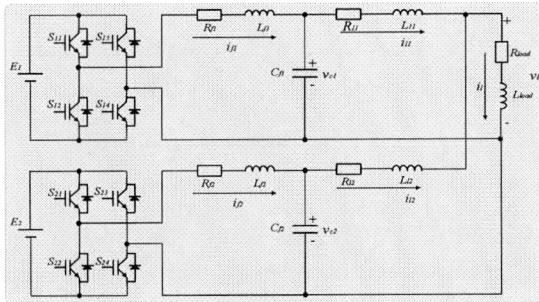

Figure 1. Circuit of two parallel single-phase voltage inverters.

In the islanded mode, the droop control [8-9] are employed to adjust output active and reactive power according to their frequency and voltage amplitude. When the line impedance is inductive, and the expressions of conventional droop control are as follows:

$$\begin{cases} f_i = f^* - m_i P_i \\ E_i = E^* - n_i Q_i \end{cases} \qquad (1)$$

where f^* and f_i are respectively the rated and reference angular frequency for i-th DG. E^* and E_i are the rated and reference value of output voltage. P_i and Q_i are the measured active and reactive power, which are calculated by following formulas:

$$\begin{cases} P_i \approx \dfrac{v_{ci} v_b}{R_{li,i} + jX_{li,i}} \sin \delta_n \\ Q_i \approx \dfrac{v_{ci} \left(v_{ci} - v_b \right)}{R_{li,i} + jX_{li,i}} \end{cases} \qquad (2)$$

where $X_{li,i} = \omega L_{li}$ is the line impedance at fundamental frequency of i-th DG, v_b is the load voltage in common bus:

$$v_b = v_{c1} - z_{l1} i_{l2} = v_{c2} - z_{l2} i_{l2} \qquad (3)$$

where $z_{li} = R_{li} + j\omega L_{li}$. To achieve load demand sharing, the droop coefficients should satisfy:

$$\begin{cases} m_1 P_{rated,1} = \cdots = m_i P_{rated,j} = \cdots = m_n P_{rated,n} \\ n_1 Q_{rated,1} = \cdots = n_i Q_{rated,j} = \cdots = n_n Q_{rated,n} \end{cases} \qquad (4)$$

where $P_{rated,i}$ and $Q_{rated,i}$ are the rated active and reactive power for i-th DG, respectively. Besides, the load sharing is also affected by line impedances of DGs [11]. Therefore, the load sharing should also satisfy:

$$\begin{cases} R_{li,1} P_{rated,1} = \cdots = R_{li,i} P_{rated,j} = \cdots = R_{li,n} P_{rated,n} \\ X_{li,1} Q_{rated,1} = \cdots = X_{li,i} Q_{rated,j} = \cdots = X_{li,n} Q_{rated,n} \end{cases} \qquad (5)$$

Only when the above two conditions are satisfied, can the accurate sharing of load demand be come true at steady-state. However, it is not easy to make the line impedance completely equal or in a fixed proportion. Furthermore, because of inherent characteristics of droop control, there is always a deviation between the output voltage and the rated voltage. The load voltage in common bus v_b is always affected by load Therefore, it is necessary to take measures to improve the traditional droop control.

III. PROPOSED CONTROL STRATEGY BASED ON STATE ESTIMATOR

In this section, a modified droop control algorithm as illustrated in Fig. 2 is introduced to solve the above problems, which consists of a KF-based voltage local estimator and a frequency-voltage recovery algorithm. The proposed KF-based voltage local estimator can estimate the rms-voltage of common bus through local measurements of output voltage and line current, which avoids delay effects due to remote communication. Furthermore, the distributed voltage-frequency recovery algorithm can maintain the output voltage and frequency keeping at the nominal value, even if load change happened. Next, the detail algorithm is depicted as below.

A. The modified voltage droop algorithm

To compensate the drop voltage caused by line impedances, many control algorithms have been presented to address this issue, among which the virtual impedance design has been widely used in [12-13,16]. However, the virtual impedance is not easily got because the line impedance is not fixed, and easy to be affected by the external environment, such as temperature.

Motivated by the control algorithm in [7], a robust and nature control algorithm is proposed. The voltage difference between nominal and common bus as a feedback, and the expression is written as:

$$n_i Q_i = K_v \left(E^* - E_b \right) \qquad (6)$$

where K_v is feedback gain, and E_b is rms-voltage in common bus. In the steady state, the right side of (3) always keeps the same for multi-parallel inverters. As long as the K_v is chosen properly, the system can easily meet the following condition:

$$n_1 Q_1 = n_2 Q_2 = \ldots = n_i Q_i = \text{constant} \qquad (7)$$

This control algorithm effectively controls the voltage of common bus near the constant voltage as desired no matter how the load changes. The modified droop control algorithm is illustrated in Figure 2.

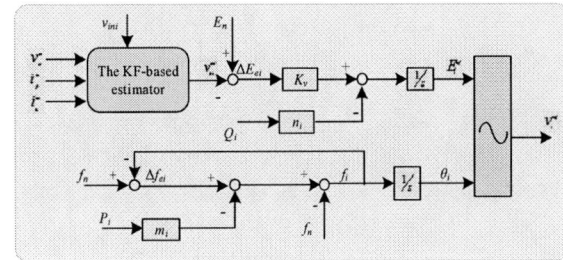

Figure 2. The modified droop control algorithm.

However, due to the introduction of E_b, the system should increase measuring instruments, which will increase the calculation error and even other instability factors. Besides, voltage drop ($E^* - E_b$) is always the same, therefore, there is no necessary to compute it for each inverter controller. In view of the above reasons, a Kalman filter is employed to estimate rms-voltage of common bus in some controllers, and the estimated value is transmitted to other controllers in the network through the cooperative control algorithm, which will be described in detail below.

B. Rms-Voltage Estimator Based on Kalman Filter

Kalman filters are one of the most popular dynamic state estimation techniques. In this paper, the state equation for i-th inverter can be written as (9)

$$C_{fi}\frac{dv_{ci}}{dt} = i_{fi} - i_{li}$$
$$L_{fi}\frac{di_{fi}}{dt} = v_{ci} - R_{fi}i_{fi} + v_{ini} \qquad (8)$$
$$L_{li}\frac{di_{li}}{dt} = v_{ci} - v_{bi} - R_{li}i_{li}$$

where C_{fi}, L_{fi} are filter capacitor and filter inductance, R_{li}, L_{li} are line impedance and line inductance, and R_{fi} is resistance of filter. State variable is defined as $x=[v_{ci}\ v_{bi}\ i_{fi}\ i_{li}]^{\mathrm{T}}$, where v_{ci}, v_{bi}, i_{fi}, i_{li} are voltage of filter capacitor and common bus, current of filter inductance and line inductance, respectively. v_{ini} is input voltage. The complete model including the Gaussian noise is expressed in the form of state space by

$$\begin{bmatrix} \frac{dv_{ci}}{dt} \\ \frac{dv_{bi}}{dt} \\ \frac{di_{fi}}{dt} \\ \frac{di_{li}}{dt} \end{bmatrix} = \begin{bmatrix} 0 & 0 & \frac{1}{C_{fi}} & -\frac{1}{C_{fi}} \\ 0 & 0 & 0 & 0 \\ -\frac{1}{L_{fi}} & 0 & -\frac{R_{fi}}{L_{fi}} & 0 \\ \frac{1}{L_{li}} & -\frac{1}{L_{li}} & 0 & -\frac{R_{li}}{L_{li}} \end{bmatrix} \begin{bmatrix} v_{ci} \\ v_{bi} \\ i_{fi} \\ i_{li} \end{bmatrix} + \begin{bmatrix} 0 \\ 0 \\ \frac{1}{L_{fi}} \\ 0 \end{bmatrix}[v_{ini}] + \begin{bmatrix} w_{vci} \\ w_{vbi} \\ w_{ifi} \\ w_{ili} \end{bmatrix} \qquad (9)$$

where w_{vci}, w_{vbi}, w_{ifi}, w_{ili} are the process noise of v_{ci}, v_{bi}, i_{fi}, i_{li} for i-th DG, respectively.

For estimation of parameters, the state variable v_{ci}, i_{fi}, i_{li} are measured. The output equation is defined by

$$\begin{bmatrix} v_{ci} \\ i_{fi} \\ i_{li} \end{bmatrix} = \begin{bmatrix} 1 & 0 & 0 & 0 \\ 0 & 0 & 1 & 0 \\ 0 & 0 & 0 & 1 \end{bmatrix} \begin{bmatrix} v_{ci} \\ v_{bi} \\ i_{fi} \\ i_{li} \end{bmatrix} + \begin{bmatrix} v_{vci} \\ v_{ifi} \\ v_{ili} \end{bmatrix} \qquad (10)$$

where v_{vci}, v_{ifi}, v_{ili} are the measurement noise. And the covariances matrix $\mathbf{Q}$ and $\mathbf{R}$ of process and measurement noise are chosen as

$$Q = diag[1e-4,1e-5,1e-3,1e-4] \qquad (11)$$
$$R = diag[0.1,0.1,0.1] \qquad (12)$$

The proposed voltage estimator based on Kalman filter has a good ability to respond to the parameters or noise perturbations.

C. Distributed voltage-frequency recovery algorithm

For the whole system, a cooperative and multi-agent control method is developed to reduce the amount of calculation for system. Firstly, some DGs, which are chosen as leader DGs, calculate the voltage-frequency deviation between rating and common bus:

$$\begin{cases} \Delta f_e = f^* - f_{pcc} \\ \Delta v_e = E^* - E_b^{est} \end{cases} \qquad (13)$$

where E_b^{est} is an estimated value for rms-voltage of common bus. A continuous-time consensus algorithm is used to deliver the voltage-frequency deviation to other follower inverter controllers. The specific expressions are as follows:

$$\begin{cases} k_f\Delta\dot{f}_i = -\sum_{j\in N, j\neq i} a_{ij}(\Delta f_i - \Delta f_j) + b_{fi}\Delta f_e \\ k_v\Delta\dot{v}_i = -\sum_{j\in N, j\neq i} a_{ij}(\Delta v_i - \Delta v_j) + b_{vi}\Delta v_e \end{cases} \qquad (14)$$

where k_f and k_v are the positive gains for the i-th DG. If the i-th DG is a leader, then $b_i=1$; otherwise, $b_i=0$. The parameter a_{ij} is an entry of the adjacency matrix.

D. Model Predictive Control based on State Estimation

With the emergence of a large network of industrial production, the distributed predictive control for real-time control algorithm appears to replace the traditional centralized predictive control [29-30]. The distributed MPC method establishes local mathematical model and objective function for each subsystem, which is helpful to reduce computational complexity. The local mathematical model of subsystem for inverters is:

$$\begin{bmatrix} \dot{v}_{ci} \\ \dot{i}_{fi} \end{bmatrix}_{x_i} = \begin{bmatrix} 0 & \frac{1}{C_{fi}} \\ -\frac{1}{L_{fi}} & -\frac{R_{fi}}{L_{fi}} \end{bmatrix}_{A_i} \begin{bmatrix} v_{ci} \\ i_{fi} \end{bmatrix}_{x_i} + \begin{bmatrix} 0 \\ \frac{1}{L_{fi}} \end{bmatrix}_{B_i}[v_{ini}]_{u_i} + \begin{bmatrix} -\frac{1}{C_{fi}} \\ 0 \end{bmatrix}_{B_{di}}[i_{li}]_{u_{di}} \qquad (15)$$

$$[v_{ci}]_{y_i} = [1\ 0]_{C_i}\begin{bmatrix} v_{ci} \\ i_{fi} \end{bmatrix}_{x_i}$$

where is i_{li} seen a measurable disturbance input, $\mathbf{B}_{di}$ is its measurable disturbance input matrix, $\mathbf{B}_i$ is an input-to-state matrix and $\mathbf{A}_i$ is a state matrix. The objective for control of parallel inverters is to ensure the voltage and frequency in common bus remain at the set point and enhance power sharing ability within limits for the input disturbances. Fig. 3 shows a distributed model predictive control scheme, plant represents the subsystem model of parallel inverters.

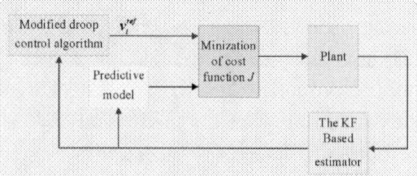

Figure 3. Distributed model predictive control scheme for i-th inverter subsystem.

The cost function J that reflects the control objects, which can be written as:

$$J = w_i^1 \left(v_{ci}^{ref} - v_c \right)^2 + w_i^2 \left(\Delta u_i \right) \qquad (16)$$

where w_i^1 and w_i^2 are the penalty factors to be adjusted, and their range is $0 \leq w_i^1, w_i^2 \leq 1$. The proper values for w_i^1 and w_i^2 can be chosen by simulation. The control object is to minimize the error between the set-point and the predicted output by finding the optimized control input vector Δu. The whole block diagram of proposed control strategy for multi-parallel inverters system is shown as Fig.4.

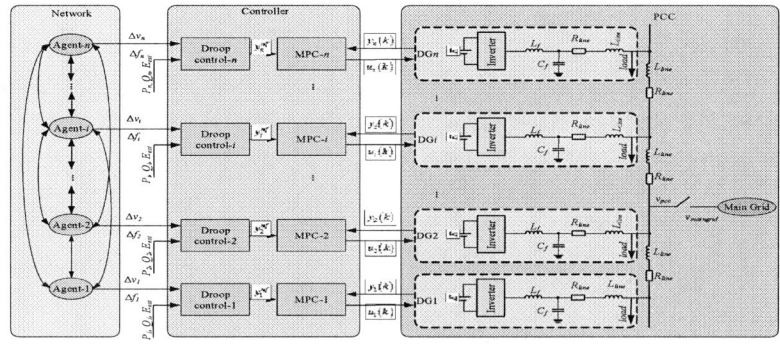

Figure 4. Block diagram of proposed control strategy for multi-parallel inverters system.

IV. SIMULATION RESULTS

To verify its effectiveness, the proposed control method is simulated in MATLAB/Simulink for an AC microgrid with two DGs connected across feeder lines as given in Fig. 2. The single-phase is designed with 48V-50Hz rating voltage. The line impedances are set as R_1=0.3Ω, L_1=3mH, R_2=0.6Ω, L_2=6mH, R_3=0.4Ω, L_3=4mH. The communication between DGs is undirected and ring connected network, and the system parameters are considered as given in Table I.

TABLE I.   SPECIFICATIONS OF THE MICROGRID SYSTEM

Name	Parameters	DG1-2
P-f droop coefficient	m	5×10^{-4}
Q-V droop coefficient	n	1×10^{-4}
Integrator gain Kv	k_v	10
Control horizon	Nc	4
Prediction horizon	Np	10

A. Case 1: Result under Proposed Control Algorithm with Distributed MPC and Traditional PI Control

In this case, a distributed MPC is used to come out the power sharing and keep voltage of common bus in rating value compared with the PI control with conventional droop control. It is can be seen form Fig. 5 that, no matter what kind of controller, the active power is shared precisely because of the inductive line impedance. While the reactive power cannot be shared under the traditional droop control when t<5s. Only when t>=5s the virtual impedance is added, the reactive power can realize divided accurately. However, the virtual impedance is not got easily in practical engineering. Therefore, a more adaptive virtual impedance method is needed for further investigations. Besides, the voltage in common bus and frequency drop is not solved by conventional droop control as shown in the left column of Fig. 5(c-d).

Compared to conventional droop control, the proposed distributed MPC method control each subsystem, which greatly reduce the amount of calculation. The simulation results are given in right of Fig. 5, and the results show the high accuracy of power sharing and slight fluctuation. In addition, the rms voltage and frequency in common bus are kept in the nominal value, 48V and 50Hz under the proposed algorithm with distributed MPC.

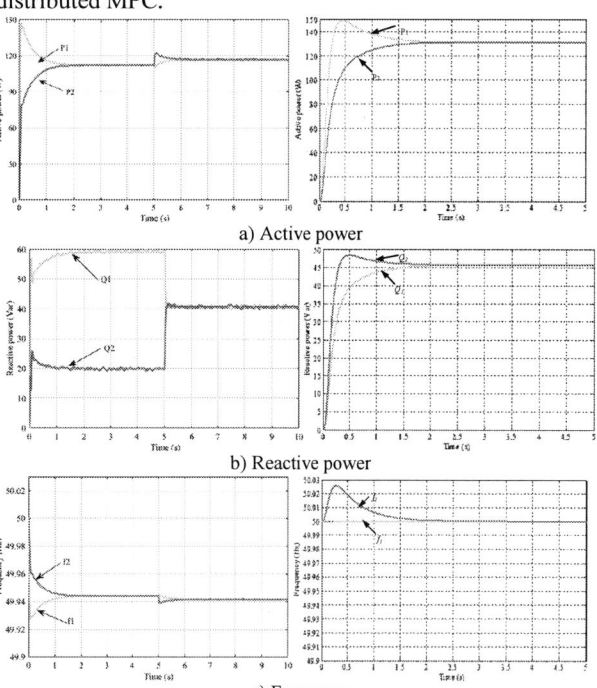

a) Active power

b) Reactive power

c) Frequency

The 2018 International Power Electronics Conference

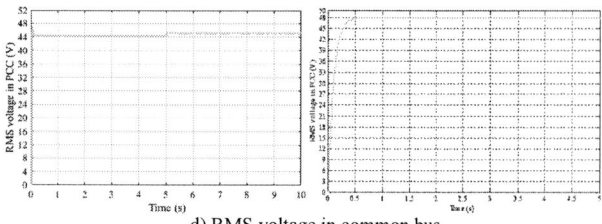

d) RMS voltage in common bus

Figure 5. Simulation results under the conventional droop control (left column) and proposed control with distributed MPC (right column).

B. Case 2: Power Sharing Performance to Load Change

Load is constantly changing in practice, thus the proposed control algorithm is needed to be robust to the change of the load. The simulation results for loads change are shown in Fig. 6(a-d). In the beginning, the total active and reactive power are 540W and 240Var. When t=5s, the total load is reduced to 240W and 90Var. The two DGs can quickly respond to load changes and ensure that the voltage in common bus is maintained at the rated value under the proposed distributed MPC.

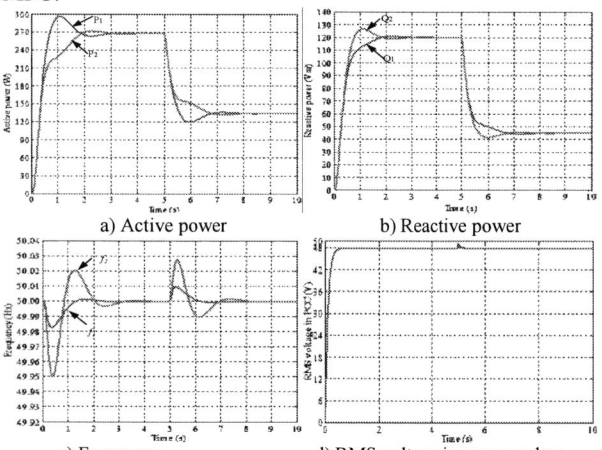

a) Active power　　　　b) Reactive power

c) Frequency　　　　d) RMS voltage in common bus

Figure 6. Simulation results of sharing performance to load change under the proposed control algorithm

C. Case 3: Proposed Control Algorithm with Plug-and-Play Function

The plug-and-play function is a necessary function of microgrid, which is simulated in this case. As shown in Fig.7, three DGs supply power for load at first, they achieve load sharing under the proposed distributed MPC. While when t=5s, DG3 is disconnected from the microgrid, and the rest two DGs update their states respectively through state estimators and redistributed active and reactive power to supply load. Because of the sudden switch off DG3, voltage in common bus fluctuate transitorily and then recover to the rated value.

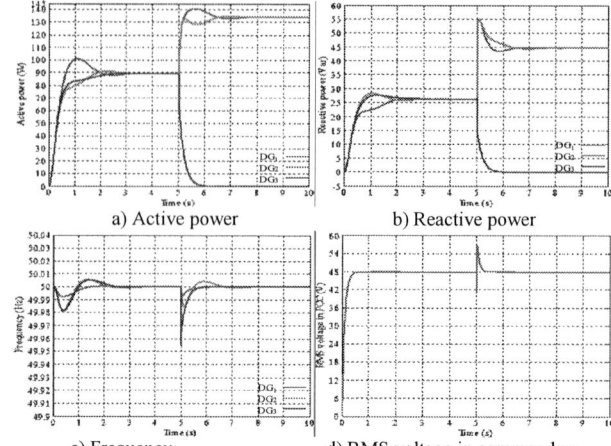

a) Active power　　　　b) Reactive power

c) Frequency　　　　d) RMS voltage in common bus

Figure 7. Simulation results of plug-and-play function under the proposed control algorithm.

V.　CONCLUSION

In this paper, the inaccurate load sharing problem for multi-parallel DGs respect to the different line impedances, and the voltage-frequency drop problem has been researched for islanded microgrids. A distributed modified droop control algorithm based on state estimator is introduced to solve these problems. The proposed control method uses Kalman filter to estimate the RMS voltage in common bus, and then transmit the voltage error between common bus and nominal value to follower DGs through communication network based on consensus distributed algorithm. Finally, a distributed MPC is put into effect, which is implemented on-line with low cost and maintained a good control performance, helps to make the system more reliable and flexible to plug and play function. Lastly, the proposed control method with sparse communication network is tested by simulation, which show the effectiveness of the control strategy.

REFERENCES

[1] R. H. Lasseter, "Smart distribution: Coupled microgrids," in *Proceedings of the IEEE*, Jun. 2011, vol. 99, No. 6, pp. 1074-1082.

[2] N. Pogaku, M. Prodanovic, T. C. Green, "Modeling, analysis and testing of automous operation of an inverter-based on microgrid," *IEEE Trans. Power Electron.*, vol. 22, No. 2, pp. 613-625, Mar. 2007.□

[3] Y. Abdel-rady, I. Mohaned, E. F. EI-Saadany, "Adaptive decentralized droop controller to preserve power shanring stability of paralleled inverters in distributed generation microgrids," *IEEE Trans. Power Electron.*, vol. 23, No. 6, pp. 2806-2816, Nov. 2008.

[4] C. L. Chen, Y. Wang, J. S. Lai, Y. -S. Lee, and D. Martin, "Design of parallel inverters for smooth mode transfer microgrid applications," *IEEE Trans. Power Electron.*, vol. 25, no. 1, pp. 6-15, Jan. 2010.

[5] R. Kamel, A. Chaouachi, and K. Nagasaka, "Three control strategies to improve the microgrid transient dynamic response during islated mode: A comparative study", *IEEE Trans. Ind. Electron.*, vol. 60, no. 4, pp. 1314-1342, Apr. 2013.

[6] J. M. Guerrero, J. C. Vasquez, J. Matas, L. G. d. Vicuna and M. Castilla, "Hierarchical control of droop-controlled AC and DC microgrids- A general approach toward standardization," *IEEE Trans. Ind. Electron.*, vol. 58, no. 1, pp. 158–172, Jan. 2011.

[7] Q. C. Zhong, "Robust droop controller for accurate proportional load sharing among inverters operated in parallel," *IEEE Trans. Ind. Electron.*, vol. 60, no. 4, pp. 1281–1290, Apr. 2013.

[8] J. M. Guerrero, J. C. Vasquez, and J. Matas, "Control strategy for flexible microgrid based on parallel lineinteractive UPS systems," *IEEE Trans. Ind. Electron.*, vol. 56, no. 3, pp. 726-736, Mar. 2009.

[9] J. M. Guerrero, J. Matas, L. Garcia de Vicuna, M. Castilla, and J. Miret, "Wireless-control strategy for parallel operation of distributed-generation inverters," *IEEE Trans. Ind, Electron.*, vol. 53, no. 5, pp. 1461-1470, Oct. 2006.

[10] J. Rocabert, A. Luna, F. Blaabjerg, and P. Rodriguez, "Control of power converters in AC microgrids," *IEEE Trans. Power Electron.*, vol. 27, no. 11, pp. 4734–4749, Nov. 2012.

[11] J. He and Y. W. Li, "An enhanced microgrid load demand sharing strategy," *IEEE Power Electron.*, vol. 27, no. 9, pp. 3984–3995, Sep. 2012.

[12] H. Mahmood, D. Michaelson and J. Jiang, "Accurate reactive power sharing in an islanded microgrid using adaptive virtual impedances," *IEEE Power Electron.*, vol. 30, no. 3, pp. 1605–1617, Mar. 2015.

[13] A. H. Etemadi, E. J. Davision, and R. Iravani, "A decentralized robust control strategy formulti-Dermicrogrids-Part I: Fundamental concepts," *IEEE Trans. Power Del.*, vol. 27, no. 4, pp. 1843–1853, Oct. 2012.

[14] A. L. Dimeas and N. D. Hatziargyriou, "Operation of a multiagent system for microgrid control," *IEEE Trans. Power Syst.*, vol. 20, no. 3, pp. 1447-1455, Aug. 2005.

[15] J. W. Simpson-Porco, Q. Shafiee, F. Dorfler, J. C. Vasquez, J. M. Guerrero and F. Bullo, "Secondary frequency and voltage control of islanded microgrids via distributed averaging," *IEEE Trans. Ind. Electron.*, vol. 62, no. 11, pp. 7025-7037, Nov. 2015.

[16] H. Zhang, S. Kim, Q. Sun and J. Zhou, "Distributed Adaptive virtual impedance control for accurate reactive power sharing based on consensus control in microgrids," *IEEE Trans. Smart Grid.*, vol. 8, no. 4, pp. 1749 – 1761, Jan. 2016.

[17] Q. Shafiee, J. M. Guerrero and J. C. Vasquez, "Distributed secondary control for islanded microgrids-A novel approach," *IEEE Power Electron.*, vol. 29, no. 2, pp. 1018–1031, Feb. 2014.

[18] A. Bidram, A. Davoudi, F. L. Liews and J. M. Guerrero, "Distributed cooperative secondary control of microgrid using feedback linearization," *IEEE Trans. Power Syst.*, vol. 28, no. 3, pp. 3462–3470, Aug. 2013.

[19] P. H. Divshali, A. Alimardani, S. H. Hosseinian and M. Abedi, "Decentralized cooperative control strategy of microsources for stabilizing autonomous VSC-based microgrids," *IEEE Trans. Power Syst.*, vol. 27, no. 4, pp. 1949-1959, Nov. 2012.

[20] A. Mehrizi-Sani and R. Iravani, "Potential-function based control of a microgrid in islanded and grid-connected models," *IEEE Trans. Power Syst*, 2010, vol. 25, pp. 1883–1891.

[21] Xin H. Qu Z, Seuss J, Maknouninejad A, "A self-organizing strategy for power flow control of photovoltaic generators in a distribution network," *IEEE Trans. Power Syst*, 2011, vol. 26, pp. 1462-1473.

[22] X.Wang, F. Blaabjerg, Z. Chen, and J. M. Guerrero, "A centralized control architecture for harmonic voltage suppression in islanded microgrids," in *Proc. IEEE Ind. Electron. Soc. Conf.*, 2011, pp. 3070–3075.

[23] Y. B. Wang, Z. Chen, X. F. Wang, Y. J. Tian, Y. D. Tan, C. Yang, "An Estimator-Based Distributed Voltage-Predictive Control Strategy for AC Islanded Microgrids," *IEEE Power Electron.*, vol. 30, No. 7, pp. 3934–3951, Jul. 2015.

[24] H. Moussa, J. P. Martin, S. Pierfederic, N. Moubayed, "Power sharing enhancement for islanded microgrid based on sate estimation of PCC rms-voltage," *IEEE, Ind, Electron Society*, 42nd annual conference, pp. 3878-3883, IECON, Oct. 2016.

[25] G. Farrai-Trecate, L. Galbusera, M. P. E. Marcinadi, and R. Scattolini, "Model predictive control schemes for consensus in multi-agent systems with single- and double-integrator dynamics," IEEE Trans. Autom. Control, vol. 54, no. 11, pp. 2560–2572, Nov. 2009.

[26] T. J. Besselmann, S. Almer, and H. J. Ferreau, "Model predictive control of load-demand inverter-fed synchronous machines," IEEE Trans. Power Electron., vol. 31, no. 10, pp. 7384–7393, Oct. 2016.

Maximum Power Point Tracking Method for PV Module Under Wide Range Varying Irradiance Levels

Hwa-Dong Liu[1], Chang-Hua Lin[1*]

1 Department of Electrical Engineering, National Taiwan University of Science and Technology, Taipei, Taiwan
*E-mail: link@mail.ntust.edu.tw

Abstract-This study proposes a novel algorithm for maximum power point tracking (MPPT) in solar-power-generating systems, and compares it with the conventional methods of hill climbing (HC). These conventional algorithms are prone to divergence under low irradiance levels (<150 W/m²), resulting in MPPT difficulties and a limited effective MPPT range; the HC algorithm in particular often entraps the actuating point near an inescapable local minimum and causes divergence. The proposed algorithm uses the angle between the sun and the horizon to develop a novel MPPT technique that extends the MPPT range to 100 W/m², effectively mitigating the divergence problems of the HC algorithms when the irradiance level is low (<150W/m²). The performance of the proposed algorithm was compared with that of the HC algorithm at varying irradiance levels, and the experimental results confirm the superiority of the proposed algorithm.

I. INTRODUCTION

Solar power output is poor on cloudy days (defined as having an irradiance level below 150W/m²) [1], and solar power output is dependent on climatic factors (e.g., irradiance level and temperature). Therefore, a MPPT controller could considerably enhance solar power efficiency.

Numerous MPPT algorithms are available for solar energy [1]–[5] and have been extensively investigated. The HC and perturbation and observation (P&O) algorithms are the most widely used because they are simple and cost-effective [2]. However, they have several major drawbacks; they converge slowly near the maximum power point (MPP) [3], and when the irradiance is steady, the tracked power point oscillates around the MPP and causes the system to generate less than optimal power [3], [4]. Moreover, on cloudy days, both algorithms have difficulties performing MPPT [3]. In addition, when the irradiance varies rapidly, both algorithms are prone to divergence [4], [5].

The present study proposes a novel MPPT algorithm for solar power systems that uses the angle between the sun and the horizon to improve the tracking. This algorithm can quickly and precisely detect the MPP and enhance the efficiency of a PV module on cloudy days. The results of an experimental comparison with a conventional HC algorithm under irradiance levels of 800 W/m², 500 W/m², 250 W/m², and 100 W/m², confirmed that the proposed algorithm performed more favorably than the conventional HC algorithm. The proposed algorithm extended the lower limit of the MPPT range to

100 W/m², remedied the HC algorithm's defect of entrapping the actuating point near an inescapable local minimum, and exhibited high reliability, rendering it suitable for applications in various irradiance conditions.

II. PV MODULE PROPERTIES AND HC ALGORITHM

A PV module consists of numerous solar power cells. Fig. 1 is a diagram of a circuit equivalent to a single solar power cell [3].

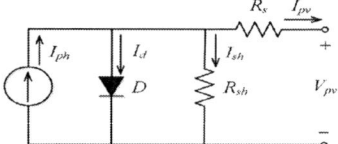

Fig. 1. Circuit equivalent to a single solar power cell.

$$I_{pv} = n_p \cdot I_{ph} - n_p \cdot I_d \left\{ \exp\left[\frac{q}{A \cdot k \cdot T \cdot n_s} \left(V_{pv} + I_{pv} \cdot R_s \right) \right] - 1 \right\} - n_p \frac{\left(V_{pv} + I_{pv} \cdot R_s \right)}{n_s \cdot R_{sh}} \quad (1)$$

The characteristics of the voltage V_{pv} and current I_{pv} in a PV module are expressed in the following equation [3], which is derived from this circuit and is suitable for various types of PV modules. Here, V_{pv} is the output voltage and I_{pv} is the current in the PV module; R_s and R_{sh} are the series and shunt resistances, respectively, in the PV module; q is the elementary charge (1.6×10^{-19} C); I_{ph} is the light-generated current; I_d is the reverse saturation current; A is the dimensionless junction material factor; k is the Boltzmann constant; T is the temperature (in Kelvin); and n_p and n_s are the numbers of solar power cells in parallel and in series, respectively.

The HC algorithm, which is commonly used for its simplicity and cost-effectiveness [3], involves disrupting the duty cycle of the power converter to affect the output power of the PV module (P_{pv}) for MPPT. However, the disruptive characteristic of the algorithm causes the actuating point to oscillate around the MPP, resulting in suboptimal power output. Moreover, the algorithm only involves comparing the power points with adjacent power points. When an actuating point is trapped near a local minimum without escape, a divergence [6] may occur.

III. PROPOSED ALGORITHM

Fig. 2 depicts the position of the PV module and the sun–horizon angle in this study. The experimental module was positioned at 25° N and 121° E, facing south, at a set-up angle (θ_1) of 23.5°. The sun–horizon angle was expressed as θ_2 (which is not the elevation angle of the sun). Because the sun's apparent path varies at each position on Earth, a generic path in which the sun rises in the east and sets in the west was adopted in this study [7]. Fig. 3. illustrates the locus of the sun at the location in summer of the experiment according to data provided by the Central Weather Bureau [8]. The sun generally rises at approximately 5:00 am; early morning is between 5:00 am and 8:00 am; morning runs from 8:00 am to 11:00 am; the noontime hours are 11:00 am to 1:00 pm; the afternoon is 1:00 pm to 6:00 pm; the evening runs from 6:00 pm to 7:00 pm; and the sun sets at approximately 7:00 pm. At noon, under ideal conditions (the absence of both shadows and rain), the PV module generated its maximum power. In the other periods (i.e., early morning, morning, afternoon, and evening), the PV module could not generate maximum power. Because the changing sun–horizon angle (θ_2) over the course of the day affected the likelihood for the PV module to generate its maximum power, a new MPPT technology was developed that accounted for θ_2.

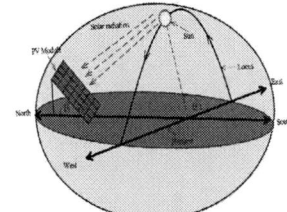

Fig. 2. Positioning of the PV module and the sun–horizon angle.

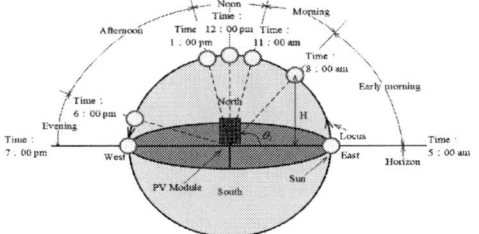

Fig. 3. Diagram of solar trajectory, time, and the sun–horizon angle θ_2.

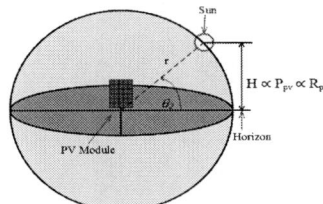

Fig. 4. Relationships between the sun–horizon angle (θ_2), the power of the PV module (P_{pv}), and the equivalent conductance of the PV module (R_{pv1}^{-1}).

Fig. 4, which is based on Fig. 3, illustrates the relationships between θ_2, the power of the PV module (P_{pv}), and the equivalent conductance of the PV module (R_{pv1}^{-1}), where r represents the distance between the sun and the PV module. According to Fig. 3, under ideal conditions the PV module generates its maximum power at noon, when the sun–horizon height (H) was maximal. Therefore, H is positively associated with P_{pv}. According to (2), P_{pv} is also positively related to R_{pv1}^{-1}. Fig. 4 is the reference to Solar trajectory data [7] when the sun is in motion the sun-horizon height (H) and θ_2 is a positive

relationship. Because θ_2 affects H, P_{pv}, and R_{pv1}^{-1}, θ_2 was regarded as a crucial reference factor for MPPT in the proposed algorithm. The value of θ_2 in radians (rad)=S/r; the value of θ_2=(rad·180°)/π.

$$P_{pv} = \frac{V_{pv}^2}{R_{pv}} = V_{pv}^2 \cdot R_{pv}^{-1} \qquad (2)$$

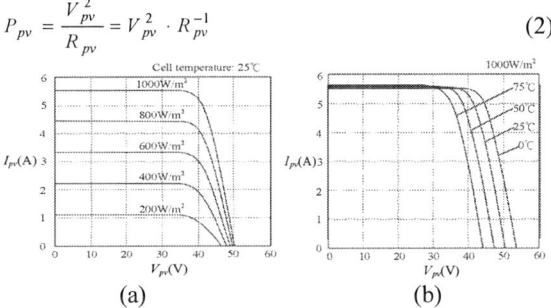

Fig. 5. I_{pv}–V_{pv} characteristic curves for a single PV module (Sanyo, model number HIP-205NKHB5). (a) Temperature 25 °C; irradiance levels 200 W/m², 400 W/m², 600 W/m², 800 W/m², and 1,000 W/m². (b) Irradiance level 1000 W/m²; temperatures 0 °C, 25 °C, 50 °C, and 75 °C.

According to Fig. 3 and Fig. 4, the PV module generates its maximum power at noon under ideal conditions because of the changes in the irradiance level and temperature. Conversely, in the early morning and evening, the PV module cannot generate its maximum power. From early morning to evening, changes in the irradiance level and temperature are substantially correlated to the time and θ_2. Fig. 5 illustrates the I_{pv}–V_{pv} characteristic curves of the PV module [9], which reveal that I_{pv} and V_{pv} change according to the changes in the irradiance level and temperature. R_{pv} also changed according to the irradiance level and temperature. Therefore, R_{pv} was considered a crucial factor connected to θ_2.

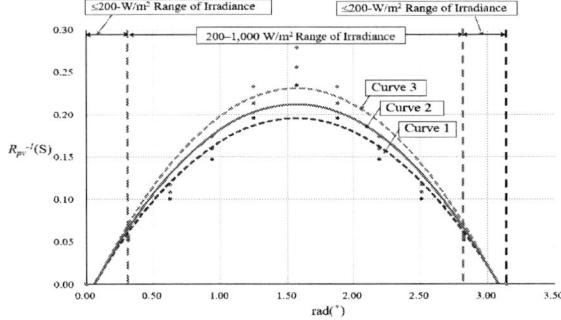

Fig. 6. Relationship between rad and R_{pv}^{-1}.

Curve fitting was conducted (in Microsoft Excel) on the relationship between rad and R_{pv}^{-1}, as indicated by Figs. 3 and 5. Accordingly, and three curves were drawn (Fig. 6) to display the relationship between rad and R_{pv}^{-1}: Curve 1 for a temperature of 0 °C and irradiance level of 0–1000W/m²; Curve 2 for an average temperature between 0°C and 75°C and an irradiance level of 0–1000 W/m²; and Curve 3 for a temperature of 75 °C and irradiance level of 0–1000W/m². The three curves could be approximated by the following quadratic equation to simplify the mathematical model:

$$R_{pv}^{-1} = a(rad)^2 + b(rad) + c \qquad (3)$$

Equation (3) can be used to obtain rad.

1778

$$rad = \frac{-b + \sqrt{b^2 - 4\left[a \cdot \left(c - R_{pv}^{-1}\right)\right]}}{2a} \qquad (4)$$

In Fig. 6, Curve 1 was drawn with (3) using a=−0.0852, b=0.2674, and c=−0.0146. Curve 2 was drawn with a=−0.0929, b=0.2918, and c=−0.0169, and Curve 3 with a=−0.1016, b=0.319, and c=−0.0194. A comparison of all three curves revealed that the mean deviations of Curves 1 and 3 from Curve 2 were both <5%; therefore, Curve 2 was chosen as the model for the relationship between rad and R_{pv}^{-1}. Moreover, because low irradiance (150W/m²) was already within the scope of Curve 2, this enabled a substantial expansion in the irradiance level within which the proposed algorithm could achieve MPPT.

The MPPT mathematical model was derived according the output voltage of the boost converter in a steady state V_o can be expressed as follows:

$$V_o = \frac{V_{pv}}{1 - D} \qquad (5)$$

Both the load resistance, which is defined as $R_o = V_o/I_o$, and the PV module's equivalent resistance, which can be defined as $R_{pv} = V_{pv}/I_{pv}$, are critical factors. In ideal condition, $P_{pv} = P_o$. Hence, the relationship between V_{pv}, V_o, R_{pv}, and R_o can be expressed as

$$\frac{V_{pv}^2}{R_{pv}} = \frac{V_o^2}{R_o} \qquad (6)$$

Further, from (5) and (6), the relationship between V_{pv}, R_{pv}. R_o, and the duty cycle D can be expressed as

$$\frac{V_{pv}^2}{R_{pv}} = \frac{\left(\dfrac{V_{pv}}{1 - D}\right)^2}{R_o} \qquad (7)$$

Moreover, according to (7), the load R_o can be expressed as

$$R_o = \frac{R_{pv}}{(1 - D)^2} \qquad (8)$$

The duty cycle D can be obtained through (8):

$$D = 1 - \sqrt{\frac{1}{R_{pv}^{-1} \cdot R_o}} \qquad (9)$$

Applying (3) and (4) in (9) yields the following equation on the relationship of the duty cycle (D), R_o, and rad:

$$D = 1 - \sqrt{\frac{1}{\left[a\left(rad\right)^2 + b\left(rad\right) + c\right] \cdot R_o}} \qquad (10)$$

The most important parameters a, b, c in (10) are easy to calculate. Generally, the installation of PV system seldom changed and have long lifespan up to 20 years. Hence, these said parameters are nearly constant values after PV system is installed. In other words, the parameters in the proposed algorithm almost do not need any modification unless the specification of the PV system or the topology of the power converter have been changed. Moreover, the proposed MPPT algorithm can be executed only when the output has load ($I_o \neq 0$). When the irradiance and PV cell temperature are known, the proposed algorithm can obtain the only R_{pv} at cast point by the MCU, and then Curve 2 in Fig. 7 can correspond to rad, the MPPT duty cycle D can be obtained through (10). Fig. 7 is the flowchart of the proposed algorithm,

where V_{pv} is the present voltage, I_{pv} the present current, and P_{pv} the present power of the PV module; V_o is the present output voltage and I_o the present output current of the boost converter; R_{pv} is the PV module's equivalent resistance and R_{pv}^{-1} its equivalent conductance; R_o is the load; rad is the radian equivalent of θ_2; a, b, and c are the parameters of the quadratic equation expressing the relationship between rad and R_{pv}^{-1}; and D is the PWM duty cycle during MPPT.

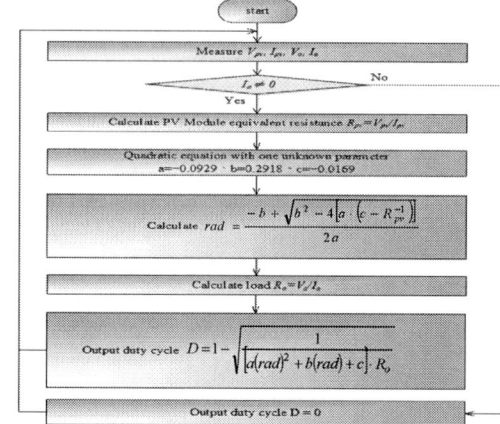

Fig. 7. Flowchart of the proposed algorithm.

Fig. 8 illustrates a boost converter typically employed for generating solar power [3], the system framework of which consists of two optical couplers, two current transducers, an inductor, a power MOSFET, diodes, a capacitor, a gate driver, and an MCU. Optical coupler 1 detects the output voltage of the PV module V_{pv}, and optical coupler 2 detects the output voltage of the boost converter V_o. Current transducer 1 detects the output current of the PV module I_{pv}, and current transducer 2 detects the output current of the boost converter I_o. The detected voltage and current signals are transmitted to the MCU. Employing the proposed algorithm, the MCU outputs a set of PWM signals at a frequency of 30 kHz and facilitates solar power MPPT by controlling power MOSFET S_1 with the gate driver.

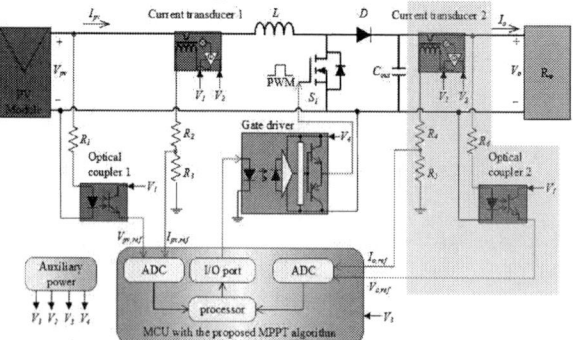

Fig. 8. Diagram of the boost converter with the proposed algorithm embedded.

IV. EXPERIMENT RESULTS

The performance of the system in Fig. 8, which incorporates the proposed algorithm, was compared with that of a system employing a conventional HC algorithm. Two PV modules (Sanyo, model number HIP-

205NKHB5) [9] were used in parallel during the experiment, hence, V_{MPP}=41V, I_{MPP}=10A, and P_{MPP}=410W. The booster converter had an inductance L=1mH and an output capacitance C_{out}= 220μF. The MCU used was the model 18F452 (Microchip Technology). In addition, to verify the suitability of the proposed algorithm under various irradiance conditions, experiments were conducted under irradiance levels of 800W/m², 500W/m², 250W/m², and 100W/m². The results reveal that the proposed algorithm yielded more favorable results than the conventional HC algorithm did (Figs. 10–11). At a constant temperature, increasing the irradiance level increased the P_{pv} of the PV module. A low irradiance level (<150 W/m²) led to a low P_{pv} in the PV module. When R_o was overloaded, P_{pv} could also be overloaded, thus nullifying MPPT. Therefore, different values of R_o were selected for different irradiance levels to prevent overloading.

Fig. 9 compares the result of the proposed algorithm and that of the HC algorithm at 10:20 am, when the irradiance level and temperature were respectively 800 W/m² and 25℃. As shown in Fig. 9(a), MPPT was activated at t=t_0, and the proposed algorithm operated consistently at the MPP. During this period, the equivalent resistance of the PV module (R_{pv}) was 4.76 Ω. According to (4), rad was 1.415° (θ_2=81°) and R_o=10 Ω, the optimal duty cycle (D) was 0.31 according to (10). Fig. 9(b) illustrates the test results of the HC algorithm. MPPT was activated when t=t_0 and diverged for 0.4s when t=t_1; when t=t_2, the algorithm operated at the MPP. This happened because the HC algorithm only compared adjacent power points, trapping the actuating point near a local minimum. The experimental results (Table I) verified that the proposed algorithm yielded a higher convergence time than did the HC algorithm.

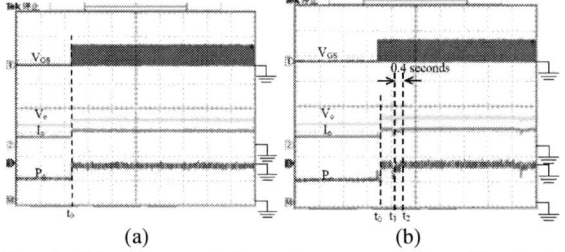

(a)   (b)

Fig. 9. Waveforms of the gate–source voltage V_{GS} in the MOSFET, output voltage V_o, output current I_o, and output power P_o for a PV module under irradiance 800 W/m² for (a) the proposed algorithm and (b) a HC algorithm (Ver: 20 V/div for V_{GS}; 50 V/div for V_o; 5 A/div for I_o; 200 W/div for P_o; and Hor: 2 s/div).

Fig. 10 illustrates a comparison between the results of the proposed and HC algorithms at 6:10 pm, when the irradiance level and temperature were respectively 100 W/m² and 25 °C. At t=t_0, MPPT was activated, and the proposed algorithm operated consistently at the MPP [Fig. 10.(a)]. During this period, R_{pv}=39.8 Ω. According to (4), rad=2.99° (θ_2=171°) and load R_o=150 Ω; according to (10), D=0.49. Fig. 10(b) illustrates the test results of the HC algorithm. At t=t_0, MPPT was activated but diverged for 80s, preventing the algorithm from operating at the MPP; at t=t_1, MPPT continued to

diverge. Because of the difficulty of completing the MPPT procedure under low-irradiance conditions (100W/m²) and the actuating point becoming trapped at a local minimum. The experimental results (Table I) verify that the proposed algorithm yielded a higher MPPT efficiency, both the output power P_o and the convergence time than did the HC algorithm.

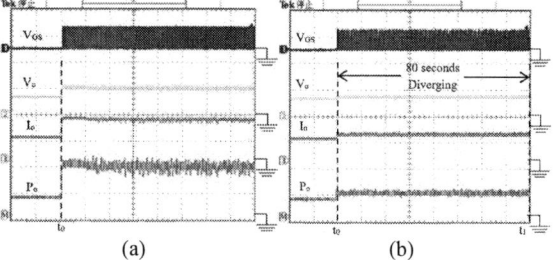

(a)   (b)

Fig. 10. V_{GS}, V_o, I_o, and P_o waveforms for a PV module under an irradiance of 100 W/m² for (a) the proposed algorithm and (b) an HC algorithm (Ver: 20 V/div for V_{GS}; 50 V/div for V_o; 250 mA/div for I_o; 50 W/div for P_o; and Hor: 10 s/div).

TABLE I
COMPARISON OF MPPT EFFICIENCIES, OUTPUT POWERS AND CONVERGENCE TIMES

Irradiance level (W/m²)	MPPT efficiencies		MPPT output powers		MPPT convergence times	
	Proposed	HC	Proposed	HC	Proposed	HC
800	99%	99%	280W	280W	0s	0.4s
500	96%	92%	90W	50W	0s	0s
250	93%	Divergent	60W	Divergent	0.8s	33.6s
100	92%	Divergent	20W	Divergent	0s	80s

V. CONCLUSIONS

As a basis for the proposed algorithm, rad revealed the MPPT as related to the changes in the irradiance level and temperature at different time points. R_{pv} was used to directly reflect the changes in the temperature and irradiance level, and the relationship between *rad* and R_{pv}^{-1} was illustrated through curve fitting. The proposed algorithm can adapt to a change in R_o to calculate the optimal D for MPPT. This is unlike the conventional HC algorithm, in which the actuating point may easily be entrapped near an inescapable local minimum and result in divergence. Moreover, the proposed algorithm prevented power loss arising from the oscillation of the actuating point near the MPP in the HC and P&O algorithms, and precisely facilitated MPPT by calculating the optimal trajectory of the duty cycle using the load R_o and rad. The proposed algorithm used the model corresponding to *rad* and R_{pv}^{-1} at 0 °C to 75 °C and irradiance level 0–1000W/m². Experimental results show that the MPPT efficiency was 99% at 800 W/m², and at least 92% at low irradiance (100 W/m²), superior to that of the HC algorithm. Therefore, the proposed algorithm enabled steady operation at the MPP, expanded the MPPT range to 100W/m², quickly and precisely facilitated MPPT for a module under 100W/m², and substantially improved the performance of the PV module in poor weather conditions, effectively remedying the persistent problem of conventional MPPT failure when irradiance level was below 150W/m^2. Therefore, the proposed algorithm was confirmed as highly reliable for PV modules under various irradiance levels.

ACKNOWLEDGMENT

This work was sponsored by the National Science Council, Taiwan, R.O.C., Project number: MOST 106-2221-E-011 -094 –MY3.

REFERENCES

[1] Mutoh, N.; Ohno, M.; Inoue, T., "A method for MPPT control while searching for parameters corresponding to weather conditions for PV generation systems," *IEEE Trans. Ind. Electron.*, vol. 53, no. 2, pp. 1055–1065, 2006.

[2] Kjær, S. B., "Evaluation of the "Hill Climbing" and the "Incremental Conductance" maximum power point trackers for photovoltaic power systems," *IEEE Trans. Energy Convers.*, vol. 27, no. 4, pp. 922–929, 2012.

[3] Alajmi, B.N.; Ahmed, K.H.; Finney, S.J.; Williams, B.W., "Fuzzy-logic-control approach of a modified hill-climbing method for maximum power point in microgrid standalone photovoltaic system," *IEEE Trans. Power Electron.*, vol. 26, no. 4, pp. 1022 –1030, 2011.

[4] Esram, T.; Chapman, P.L., "Comparison of photovoltaic array maximum power point tracking techniques," *IEEE Trans. Energy Convers.*, vol. 22, no. 2, pp. 439–449, 2007.

[5] Hong, Y.; Pham, S.; Yoo, T.; Chae, K.; Baek, K.; Kim, Y., "Efficient maximum power point tracking for a distributed PV system under rapidly changing environmental conditions," *IEEE Trans. Power Electron.*, vol. 30, no. 8, pp. 4209–4218, 2015.

[6] Pragallapati, N.; Sen, T.; Agarwal, V., "Adaptive velocity PSO for global maximum power control of a PV array under nonuniform irradiation conditions," *IEEE J. Photovoltaics*, vol. 7, no. 2, pp. 624–639, 2017.

[7] Solar trajectory data, (2017). [Online]. Available: http://www.itacanet.org/the-sun-as-a-source-of-energy/part-3-calculating-solar-angles

[8] Central Weather Bureau data, (2017). [Online]. Available: http://www.cwb.gov.tw/V7/astronomy/sunrise.htm

[9] PV module data, (2017). [Online]. Available: https://zh.scribd.com/document/37009862/Sanyo-Hip-205-200nkhb5

Dual MPPT Control and Field Testing for Switched Capacitor-Based Cell-Level Power Balancing Utilizing Diffusion Capacitance of Photovoltaic Cells

Masatoshi Uno[1*], Yota Saito[1], Masaya Yamamoto[1], and Shinichi Urabe[2]

1 Ibaraki University, Hitachi, Japan
2 Toyota Motor Corporation, Shizuoka, Japan
*E-mail: masatoshi.uno.ee@vc.ibaraki.ac.jp

Abstract—Capacitorless switched capacitor converter (SCC)-based voltage equalizers utilizing diffusion capacitance of photovoltaic (PV) cells have been proposed to increase the energy yield from partially-shaded PV modules. Although this equalizer offers simple and miniaturized circuit thanks to the capacitorless topology, the performance optimization is an elusive task as the optimal duty cycle to achieve the lowest power conversion loss is dependent on shading conditions and module configurations. This paper proposes a dual MPPT algorithm that controls duty cycles of not only the capacitorless equalizer but also a front-end boost converter. With the dual MPPT control, in addition to the operation at an MPP, the proposed equalizer can operate at an optimal duty cycle, at which the total Joule loss in the equalizer is minimized. A prototype of the proposed equalizer for six-cell module was built. The experimental results demonstrated the efficacy of the proposed equalizer as well as the dual MPPT.

Keywords—*Diffusion capacitance, photovoltaic cell, partial shading, switched capacitor converter, voltage equalizer.*

I. INTRODUCTION

Partial shading on a photovoltaic (PV) modules/strings comprising multiple cells connected in series is known to cause serious issues, such as significant decrease in energy yield and occurrence of multiple maximum power points (MPPs). For instance, 10% equivalent area of partial shading on a PV module reportedly results in 30% reduction in power generation [1]. Similar phenomena occur in curved PV panels, such as solar roofs for plug-in hybrid electric vehicles (PHEVs), due to uneven irradiance on the panel at all times, as depicted in Fig. 1. The impact of the reduced energy yield will be of serious concern in mobile and vehicular applications because of limited space for PV panels.

To address the partial shading issues mentioned above, various kinds of differential power processing (DPP) converters or voltage equalizers (hereafter, simply call voltage equalizers) have been developed. Bidirectional PWM converters [2]–[7] are the most straightforward topology, through which adjacent substrings exchange power depending on shading conditions. Single-input multi-output converters, such as multi-winding flyback converter [8], multi-stacked buck-boost converters [9], [10], and resonant voltage multiplier [11], [12] offer simpler circuit with reduced switch count. Isolated bidirectional flyback converters allow flexible power transfer even between remote substrings [13]–[15].

Among most popular and viable voltage equalizers is a switched capacitor converters (SCCs) [16]–[18]. In addition to their simple circuit, SCCs are advantageous over any other equalizers in terms of circuit volume because of their magnetic-less topology. In general, energy densities of capacitors are within a range of more than three orders of magnitude over those of inductors [19].

To achieve even simpler and more compact circuit, capacitorless SCC-based cell-level voltage equalizer utilizing diffusion capacitance of PV cells has been proposed, as shown in Fig. 2 [20]. A diffusion capacitance C_d, a parasitic element originating from a P-N junction of PV cells (see the inset of Fig. 2), is used as a capacitor of SCCs, achieving even simpler and more compact circuit design. Although cell-level equalizers are costly due to

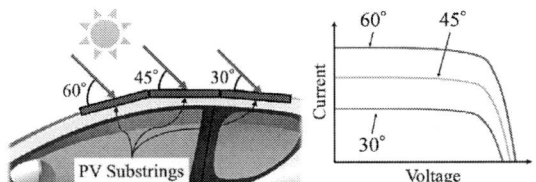

Fig. 1. Image of uneven irradiance and mismatched substring characteristics in solar roof for PHEVs.

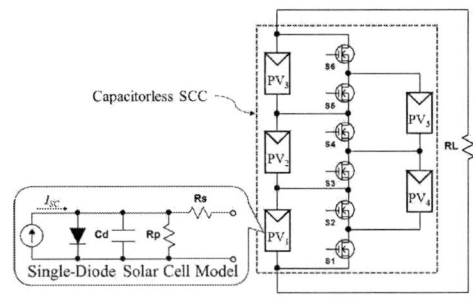

Fig. 2. Conventional capacitorless SCC equalizer utilizing diffusion capacitance of PV cells.

numerous switches necessary, an energy yield reportedly improves compared with substring- and panel-level equalizations [1].

The conventional capacitorless cell-level SCC equalizer, however, has to operate at a high switching frequency even under unshaded conditions, naturally increasing losses associated with the equalizer at high-frequency operations. In addition to the efficiency penalty under unshaded conditions, this capacitorless equalizer is only applicable to modules comprising odd number cells, likely limiting its applications because most PV modules consist of even number cells (e.g., 24, 36, and 72 cells).

This paper proposes a novel SCC-based equalizer utilizing diffusion capacitance of PV cells. Although an extra capacitor and inductor are necessary, losses associated with the equalizer can be reduced especially under unshaded conditions. Section II describes the proposed SCC-based equalizer using diffusion capacitances. Joule losses of the conventional and proposed equalizers are quantitatively compared in Section III. Section IV introduces a novel MPPT algorithm for the proposed equalizer (dual MPPT), followed by experimental verification in Section V.

II. PROPOSED SCC-BASED EQUALIZER UTILIZING DIFFUSION CAPACITANCE OF PV CELLS

A. Circuit Description and its Advantage

The proposed capacitorless SCC equalizer can be derived from the combination of a PWM buck converter and capacitorless SCC, as depicted in Fig. 3. Since an inductor L and capacitor C are additionally necessary, the proposed SCC-based equalizer is not completely capacitorless and is slightly larger than the conventional capacitorless equalizer. But the impact of the increased volume due to these added elements is infinitesimal as PV panels are far larger than these elements.

Similar to the conventional capacitorless equalizer, the odd- and even-numbered switches are alternately driven so that the equalizer opeates under partial shading conditions. In the ideal case, all capacitor voltages are automatically unified at any duty cycle. Under unshaded conditions, on the other hand, the equalizer operates with 0% duty cycle (i.e., $d = 0\%$ where d is the duty cycle of even-numbered switches), realizing the 3-series–2-parallel configuration. No pulsating currents flow with 0% duty cycle, and hence,

losses associated with high-frequency operation can be significantly reduced, as will be detailed in Section III.

Another advantage is that the proposed equalizer can be applied to PV panels consisting of even number cells. Although the topology shown in Fig. 3 is for six cells, it can be scaled to the arbitrary even number of cells by simply adding cells and switches. Since most PV panels comprise even number cells (e.g., 24, 60, 72, etc.), the proposed equalizer would be more suitable than conventional one.

B. Voltage Conversion Ratio

The added capacitor C and PV_6 correspond to an output and input of a PWM buck converter in the proposed equalizer, as can be seen from Fig. 3. Therefore, assuming all the cell voltages are ideally equal to V_{PV}, the voltage of C, V_C, is given by

$$V_C = dV_{PV}. \tag{1}$$

Hence, the module voltage, V_{mod}, is

$$V_{mod} = (3 + d)V_{PV}. \tag{2}$$

This suggests that module characteristics seen from the output are dependent on d. In other words, I–V and P–V characteristics of the module seen from the output change with d.

III. QUANTITATIVE JOULE LOSS COMPARISON

A. Charge Vector Analysis

The operation modes of the proposed SCC-based equalizer are shown in Fig. 4. By defining charge flows in each mode as designated in Fig. 4, the Kirchhoff's current law (KCL) at Nodes A–D in Mode A yields the following set of equations;

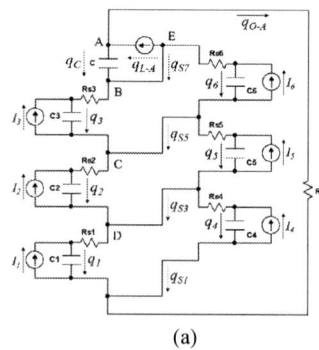

(a)

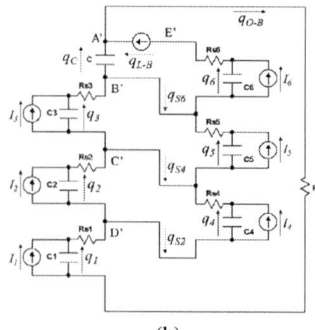

(b)

Fig. 4. Operation modes: (a) Mode 1, (b) Mode 2.

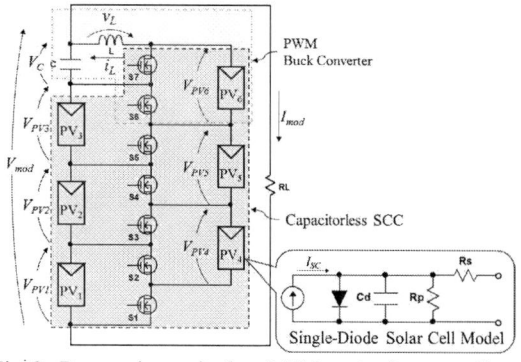

Fig. 3. Proposed capacitorless SCC-based voltage equalizer utilizing diffusion capacitance of PV cells.

1783

$$\begin{cases} 0 = -q_C + q_{L.A} - q_{O.A} \\ 0 = -q_3 - q_6 + q_C - q_{L.A} + d'(I_3 - I_6) \\ 0 = -q_2 + q_3 - q_5 + q_6 + d'(I_2 - I_3 + I_5 - I_6) \\ 0 = -q_1 + q_2 - q_4 + q_5 + d'(I_1 - I_2 + I_4 - I_5) \end{cases} \quad (3)$$

where $d' = (1 - d)$. Similarly, from the KCL at Nodes A'–E' in Mode B,

$$\begin{cases} 0 = q_C + q_{L.B} - q_{O.B} \\ 0 = q_3 + q_5 - q_6 - q_C + d(I_3 + I_5 - I_6) \\ 0 = q_2 - q_3 + q_4 - q_5 + d(I_2 - I_3 + I_4 - I_5), \\ 0 = q_1 - q_2 - q_4 + d(I_1 - I_2 - I_4) \\ 0 = q_6 - q_{L.B} \end{cases} \quad (4)$$

The charge delivered through L is proportional to the mode length, and therefore,

$$0 = dq_{L.A} - d'q_{L.B}. \quad (5)$$

For the sake of convenience, the voltage of R_L is assumed constant, so is the sum voltage of C_1–C_3 and C. Neglecting voltage drops across R_{s1}–R_{s3} gives

$$0 = q_{C1} + q_{C2} + q_{C3} + q_C. \quad (6)$$

Equations (3)–(6) can be arranged in the matrix form as

$$(7)$$

The last six rows refer to currents of cells, I_1–I_6. From the obtained matrix, the unique amount of charge delivered in each operation mode can be determined.

B. Joule Loss of Series-Resistance of Cells

A current flowing through the series resistance, R_{si} ($i = 1\ldots6$), is pulsating due to the switching operation of the equalizer, increasing Joule losses to some extent. The charge flowing through R_{si} in Modes A and B, $q_{Rsi.A}$ and $q_{Rsi.B}$, are expressed as

$$\begin{cases} q_{Rsi.A} = -q_i + d'I_i \\ q_{Rsi.B} = q_i + dI_i \end{cases}. \quad (8)$$

Currents of R_{si} in Modes A and B are

$$\begin{cases} I_{Rsi.A} = \dfrac{q_{Rsi.A}}{d'} \\ I_{Rsi.B} = \dfrac{q_{Rsi.B}}{d} \end{cases}. \quad (9)$$

The total Joule loss of R_{si}, $P_{Rs.tot}$, is

$$P_{Rs.tot} = \sum_{i=1}^{6} (d'I_{Rsi.A}^2 + dI_{Rsi.B}^2)R_{Si}. \quad (10)$$

Cells generate Joule losses in the form of $I_i^2 R_{si}$ when without the equalizer. Joule losses due to pulsating currents, $P_{Rs.tot.p}$, can be expressed as

$$P_{Rs.tot.p} = P_{Rs.tot} - \sum_{i=1}^{6} I_i^2 R_{Si}, \quad (11)$$

TABLE I
PARAMETERS USED FOR QUANTITATIVE COMPARISON

Switching Frequency, f_s	40 kHz
Series Resistance, R_{si}	5.0 mΩ
On-Resistance, R_{on}	13.4 mΩ
ESR of Inductor, r_L	3.6 mΩ
ESR of Capacitor, r_C	1.0 mΩ

where the second term on the right-hand side is the Joule loss when without the equalizer.

C. Joule Loss of Switches

Unique amount of charge flowing through switches can be similarly derived from Figs. 4(a) and (b), as

$$\begin{cases} q_{S1} = q_4 - d'I_4 \\ q_{S2} = q_4 + dI_4 \\ q_{S3} = -q_4 + q_5 + d'(I_4 - I_5) \\ q_{S4} = q_2 - q_3 + d(I_2 - I_3) \\ q_{S5} = -q_5 + q_6 + d'(I_5 - I_6) \\ q_{S6} = q_3 - q_C + dI_3 \end{cases}. \quad (12)$$

Switch current, i_{Si} ($i = 1\ldots7$), is

$$i_{Si} = \begin{cases} \dfrac{q_{Si}}{d'} & (i = 1,3,5,7) \\ \dfrac{q_{Si}}{d} & (i = 2,4,6) \end{cases}. \quad (13)$$

The total Joule loss of switches, $P_{S.tot}$, is given by

$$P_{S.tot} = R_{on}\left(\sum_{i=1,3,5,7} d'i_{Si}^2 + \sum_{i=2,4,6} di_{Si}^2\right), \quad (14)$$

where R_{on} is the on-resistance of switches.

D. Loss Comparison between Conventional and Proposed Equalizers

Based on the derived Joule loss models in the previous subsections, the total Joule losses of the proposed equalizer are compared with those of the conventional capacitorless equalizer. Parameters used for the comparison are listed in Table I. In this comparison, all cells generate 5.0 A (i.e., $[I_1, I_2, I_3, I_4, I_5, I_6] = [5, 5, 5, 5, 5, 5]$), and total losses are normalized by the number of cells because the proposed and conventional equalizers contain different numbers of cells;

$$Normalized\ Loss = \frac{Total\ Loss}{Number\ of\ Cells}. \quad (15)$$

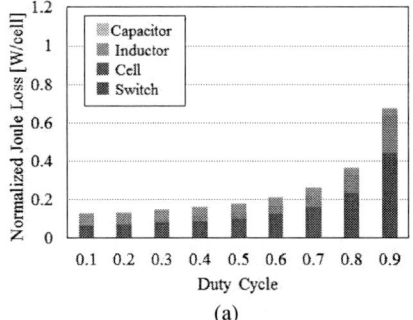

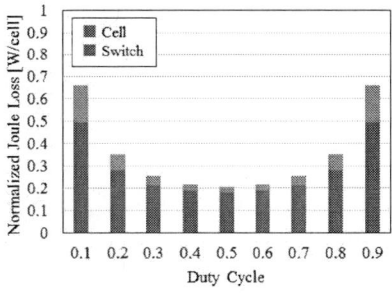

Fig. 5. Normalized total Joule losses as a function of duty cycle under unshaded condition: (a) Proposed SCC-based equalizer, (b) conventional capacitorless SCC equalizer.

1784

The 2018 International Power Electronics Conference

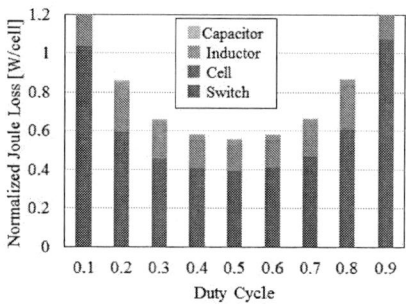

Fig. 6. Normalized total Joule losses of the proposed equalizer as a function of duty cycle under shaded condition of $[I_1, I_2, I_3, I_4, I_5, I_6] = [1, 5, 5, 1, 5, 5]$.

Theoretical normalized Joule losses under the unshaded condition are compared in Fig. 5. The normalized Joule losses in the conventional capacitorless SCC equalizer became lowest at $d = 0.5$, as shown in Fig. 5(b), because of its symmetry circuit (see Fig. 2). The proposed equalizer, on the other hand, showed the different tendency that the losses decreased at small d, as shown in Fig. 5(a). As d neared 0, the Joule loss of cells disappeared because the operation of $d = 0$ was equivalent to the 3-series–2-parallel configuration, as discussed in Section II—note that Joule losses of R_{si} associated only with ac current components due to switching were considered in order to fairly compare the losses due to equalizers [see (11)]. Losses of the conventional equalizer were quite larger because the conventional capacitorless equalizer must operate even under unshaded conditions (as mentioned in Section I), unnecessarily generating large Joule losses. These results revealed the superior characteristics of the proposed equalizer under unshaded conditions. And even under shaded conditions, the proposed equalizer exhibited better characteristics, although results are not shown for the sake of limited pages.

Figure 6 shows the normalized Joule losses of the proposed equalizer under a severely-shaded condition of $[I_1, I_2, I_3, I_4, I_5, I_6] = [1, 5, 5, 1, 5, 5]$. Losses under this shaded condition were larger than those under the unshaded condition because of the relatively large power transfer among cells for voltage equalization. More importantly, the duty cycle achieving the lowest loss under the shaded condition of Fig. 6 was different from that under the unshaded condition of Fig. 5(a). These results suggest that optimal duty cycle for the lowest Joule loss varies depending on shading conditions. In the next section, a novel dual MPPT control algorithm is proposed to realize the lowest Joule loss for the proposed SCC-based voltage equalizer.

IV. DUAL MPPT CONTROL

Module characteristic with the proposed SCC-based voltage equalizer varies depending on duty cycle d as (2) indicates; the module voltage tends to be high as d increases. In addition, an optimal duty cycle is dependent on shading conditions, as discussed in the previous section. It is noted that module's operating point cannot be determined with d but an external circuit that is

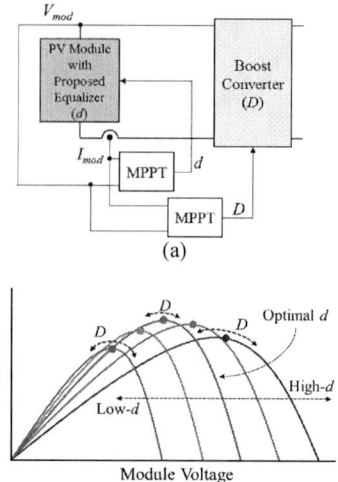

Fig. 7. (a) Dual MPPT control block diagram, (b) notional module characteristics with dual MPPT control.

traditionally a boost converter for ordinary PV panels. In other words, a PV module with the proposed SCC-based voltage equalizer operates in conjunction with an external boost converter.

A dual MPPT technique manipulating not only duty cycle d of the voltage equalizer but also duty cycle D of an external boost converter is proposed so that the module can operate at its MPP with minimizing Joule loss. The control block diagram for the proposed dual MPPT control is illustrated in Fig. 7(a). Notional module characteristics with the proposed dual MPPT are depicted in Fig. 7(b). An optimal module characteristic is sought by the voltage equalizer adjusting d while a maximum power point is tracked by the boost converter manipulating its duty cycle D. To decouple two MPPT control loops, a sampling interval of MPPT for d should be approximately 10 times longer than that for D.

V. EXPERIMENTAL RESULTS

A. Prototype

A prototype of the proposed equalizer using a 10 µH inductor and 300 µF capacitor for six cells was built. PV cells (125 ×125 mm) with a diffusion capacitance of 6–13 mF, depending on irradiance, were used. The equalizer was operated at 40 kHz.

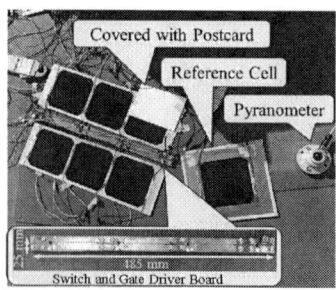

Fig. 8. Field testing for proposed SCC-based equalizer emulating a partial shading condition.

1785

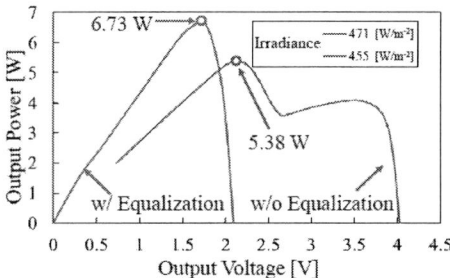

Fig. 9. Measured module characteristics with/without proposed SCC-based equalizer with $d = 0.1$.

B. Module Characteristics with/without Equalizer in Field Test

The field test at a fixed $d = 0.1$ was performed emulating a partial shading condition, as shown in Fig. 8. Half of PV_1 was covered with a postcard so that its short-circuit current was reduced to half those of other cells. Under this partial shading condition, the module characteristic of the proposed equalizer was manually swept using an electronic load. As a reference, a characteristic of a string comprising six cells connected in series without equalization (i.e., with bypass diodes) was also measured under the same shading condition.

The measured module characteristics are shown and compared in Fig. 9. The characteristic without equalization exhibited two power point maxima (one global and one local MPPs), and its maximum power was merely 5.38 W at 2.1 V. Meanwhile, the local MPP disappeared with the voltage equalization, and the maximum power increased to as high as 6.73 W, demonstrating the efficacy of the proposed SCC-based voltage equalizer.

C. Dual MPPT Control

In order to simulate the partial shading condition of $[I_1, I_2, I_3, I_4, I_5, I_6] = [1, 5, 5, 1, 5, 5]$ used for the quantitative analysis in Section III-D, solar cell emulation circuits shown in Fig. 10(a) were used instead of real PV cells. The cell emulators were based on an equivalent circuit model of PV cells, and their measured characteristics are shown in Fig. 10(b). The prototype of the equalizer and boost converter were controlled using TMS320F28335 (Texas Instruments) control card. An experimental setup for Dual MPPT control is shown in Fig. 11.

Firstly, module characteristics of the proposed equalizer under the severely-shaded condition of $[I_1, I_2, I_3, I_4, I_5, I_6] = [1, 5, 5, 1, 5, 5]$ were manually swept at various duty cycle to grasp the d-dependent module characteristics, as shown in Fig. 12. Measured characteristics significantly varied depending on d, as (2) indicated—the open-circuit voltage of the module increased with d. At extreme duty cycles, such as $d = 0.1$ and 0.9, measured characteristics showed a local MPP, indicating the cells were not satisfactorily equalized. This is attributable to insufficient equalization performance at extreme duty cycles, with which an equivalent resistance of SCCs reportedly soars [18]. The largest MPP was observed at $d = 0.5$, and these results agreed with the theoretical Joule loss tendency shown in Fig. 6.

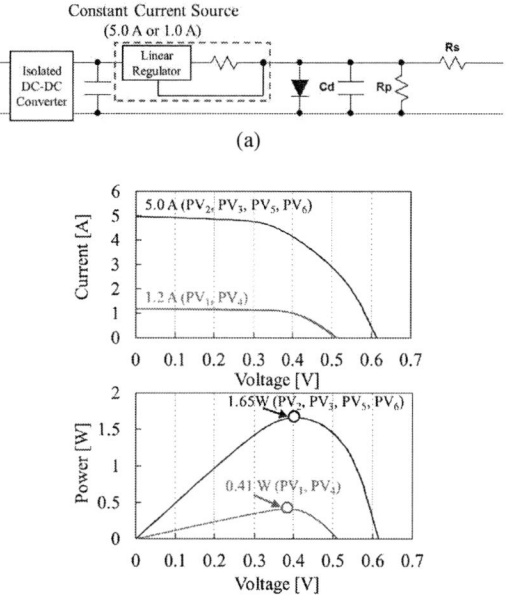

Fig. 10. (a) Solar cell emulator circuit and (b) its measured characteristics.

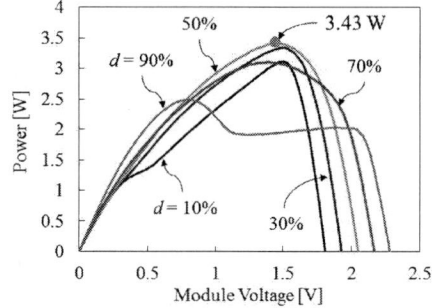

Fig. 11. Experimental setup for dual MPPT control.

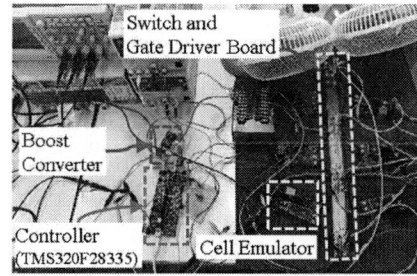

Fig. 12. Measured module characteristics under severely-shaded condition of $[I_1, I_2, I_3, I_4, I_5, I_6] = [1, 5, 5, 1, 5, 5]$.

Under the shaded condition of $[I_1, I_2, I_3, I_4, I_5, I_6] = [1, 5, 5, 1, 5, 5]$, the SCC-based equalizer was operated with the proposed dual MPPT control algorithm. Sampling intervals for MPPTs of the equalizer and boost converter were set to be 0.1 and 1.0 s, respectively.

Measured module voltage and power are shown in Fig. 13. Approximately 100 s after the beginning of the test, the

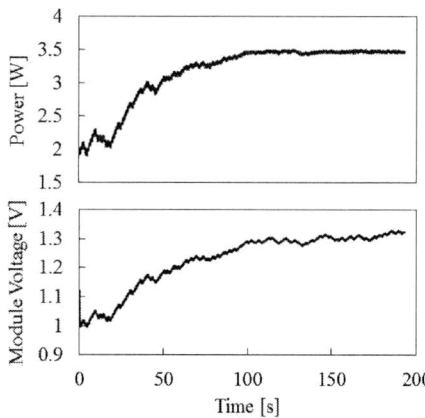

Fig. 13. Experimental results of module voltage and power during dual MPPT under severely-shaded condition.

operation came to the steady-state condition, and duty cycle fluctuated around 0.5. Extracted power under the steady-state was about 3.5 W and agreed very well with the results shown in Fig. 12, verifying the proposed dual MPPT technique.

VI. Conclusions

The SCC-based voltage equalizer utilizing diffusion capacitances of PV cells has been proposed to prevent partial shading issues of PV modules. The quantitative analysis revealed that the proposed voltage equalizer achieved lower Joule loss than did the conventional capacitorless SCC equalizer. Since an optimal duty cycle of the proposed equalizer varies depending on shading conditions, the dual MPPT technique has also been proposed to operate the module at its MPP with the optimal duty cycle minimizing the Joule loss in the equalizer.

The field test using the six-cell prototype has been performed. The local MPP successfully vanished, and the extractable maximum power significantly increased by the proposed voltage equalizer. With the proposed dual MPPT control, the module with the equalizer could operate at its MPP with the optimal duty cycle minimizing the Joule loss in the equalizer.

References

[1] S. M. MacAlpine, R. W. Erickson, and M. J. Brandemuehl, "Characterization of power optimizer potential to increase energy capture in photovoltaic systems operating under nonuniform conditions," *IEEE Trans. Power Electron.*, vol. 28, no. 6, pp. 2936–2945, Jun. 2013.

[2] H. J. Bergveld, D. Büthker, C. Castello, T. Doorn, A. D. Jong, R. V. Otten, and K. D. Waal, "Module-level dc/dc conversion for photovoltaic systems: the delta-conversion concept," *IEEE Trans. Power Electron.*, vol. 28, no. 4, pp. 2005–2013, Apr. 2013.

[3] P. S. Shenoy, K. A. Kim, B. B. Johnson, and P. T. Krein, "Differential power processing for increased energy production and reliability of photovoltaic systems," *IEEE Trans. Ind. Power Electron.*, vol. 28, no. 6, pp. 2968–2979, Jun. 2013.

[4] S. Qin, S.T. Cady, A.D.D. García, and R.C.N.P. Podgurski, "A distributed approach to maximum power point tracking for photovoltaic submodule differential power processing," *IEEE Trans. Power Electron.*, vol. 30, no. 4, pp. 2024–2040, Apr. 2015.

[5] S. Qin, C. B. Barth, Student, and R. C. N. P. Podgurski, "Enhancing microinverter energy capture with submodule differential power processing," *IEEE Trans. Power Electron.*, vol. 31, no. 5, pp. 3575–3585, May 2016.

[6] T. Shimizu, O, Hashimoto, and G. Kimura, "A novel high-performance utility-interactive photovoltaic inverter system," *IEEE Trans. Power Electron.*, vol. 18, no. 2, pp. 704–711, Mar. 2003.

[7] T. Shimizu, M. Hirakata, T. Kamezawa, and H. Watanabe, "Generation control circuit for photovoltaic modules," *IEEE Trans. Power Electron.*, vol. 16, no. 3, pp. 293–300, May 2001.

[8] J. Du, R. Xu, X. Chen, Y. Li, and J. Wu, "A novel solar panel optimizer with self-compensation for partial shadow condition," in Proc. *IEEE Applied Power Electron. Conf. Expo.*, APEC, pp. 92–96, 2013.

[9] M. Uno and A. Kukita, "Single-switch voltage equalizer using multi-stacked buck-boost converters for partially-shaded photovoltaic modules," *IEEE Trans. Power Electron.*, vol. 30, no. 6, pp. 3091–3105, Jun. 2015.

[10] M. Uno and A. Kukita, "Current sensorless equalization strategy for a single-switch voltage equalizer using multistacked buck–boost converters for photovoltaic modules under partial shading," *IEEE Trans. Ind. Appl.*, vol. 53, no. 1, pp. 420–429, Jan./Feb. 2017.

[11] M. Uno and A. Kukita, "Two-switch voltage equalizer using an LLC resonant inverter and voltage multiplier for partially-shaded series-connected photovoltaic modules," *IEEE Trans. Ind. Appl.*, vol. 51, no. 2, pp. 1587–1601, Mar./Apr. 2015.

[12] M. Uno and A. Kukita, "Single-switch single-magnetic PWM converter integrating voltage equalizer for partially-shaded photovoltaic modules in standalone applications," *IEEE Trans. Power Electron.*, vol. 33, no. 2, pp. 1259–1270, Feb. 2018.

[13] C. Olalla, D. Clement, M. Rodríguez, and D. Makisimović, "Architectures and control of submodule integrated dc-dc converters for photovoltaic applications," *IEEE Trans. Power Electron.*, vol. 28, no. 6, pp. 2980–2997, Jun. 2013.

[14] C. Olalla, C. Deline, D. Clement, Y. Levron, M. Rodríguez, and D. Makisimović, "Performance of power limited differential power processing architectures in mismatched PV systems," *IEEE Trans. Power Electron.*, vol. 30, no. 2, pp. 618–631, Feb. 2015.

[15] G. Chua, H. Wena, L. Jiangb, Y. Hub, and X. Lia, "Bidirectional flyback based isolated-port submodule differential power processing optimizer for photovoltaic applications," *Solar Energy*, vol. 158, pp. 929–940, Oct. 2017.

[16] J. T. Stauth, M. D. Seeman, and K. Kesarwani, "Resonant switched-capacitor converters for sub-module distributed photovoltaic power management," *IEEE Trans. Power Electron.*, vol. 28, no. 3, pp. 1189–1198, Mar. 2013.

[17] Z. Qiu and K. Sun, "A photovoltaic generation system based on wide voltage-gain dc-dc converter and differential power processors for dc microgrids," *Chinese J. Electrical Eng.*, vol. 3, no. 1, pp. 84–95, Jun. 2017.

[18] M. Uno and A. Kukita, "PWM converter integrating switched capacitor converter and series-resonant voltage multiplier as equalizers for photovoltaic modules and series-connected energy storage cells for exploration rovers," *IEEE Trans. Power Electron.*, vol. 32, no. 11, pp. 8500–8513, Nov. 2017.

[19] S. R. Sanders, E. Alon, H. P. Le, M. D. Seeman, M. Jhon, and V. W. Ng, "The road to fully integrated dc–dc conversion via the switched-capacitor approach," *IEEE Trans. Power Electron.*, vol. 28, no. 9, pp. 4146–4155, Sep. 2013.

[20] A. H. Chang, A. T. Avestruz, and S. B. Leeb, "Capacitor-less photovoltaic cell-level power balancing using diffusion charge redistribution," *IEEE Trans. Power Electron.*, vol. 30, no. 2, pp. 537–546, Feb. 2015.

Series Resonant DC-DC Converter With Dual-Mode Rectifier for PV Microinverters

Yanfeng Shen, Huai Wang, Zhan Shen, Yongheng Yang, and Frede Blaabjerg
Department of Energy Technology,
Aalborg University, Denmark
Email: yaf@et.aau.dk, hwa@et.aau.dk, zhs@et.aau.dk, yoy@et.aau.dk, fbl@et.aau.dk

Abstract— A new series resonant dc-dc converter is proposed for PV microinverters. Compared to the conventional series resonant converter (SRC), a dual-mode rectifier (DMR) is configured on the secondary side, which allows the proposed converter to achieve a wider gain range with a fixed-frequency phase-shift control. Moreover, the active switches and diodes are turned on with zero-voltage switching (ZVS) and off with zero-current switching (ZCS), respectively, thereby minimizing the switching losses. The operation principle is firstly detailed and then the converter characteristics are analyzed and compared. Finally, a 400-kHz 250-W converter prototype is built and tested to verify the advantages of the proposed converter.

I. INTRODUCTION

PV inverters can be classified into the central inverters, string inverters, and microinverters. Compared to the central and string inverters, microinverters have features in low power applications such as module-level (maximum power point tracking) MPPT, low PV-system installation effort, easy monitoring and failure detection of the PV-system [1], [2]. Nevertheless, there are still some challenges for microinverters: 1) The power conversion efficiency of microinverters is relatively low compared with transformerless string inverters (efficiency > 98%); thus, the efficiency performance needs to be improved [3]; 2) Microinverers are typically mounted on the backside of PV modules, and thus they will inevitably suffer from high temperature, causing an accelerated operational degradation [4]; 3) Due to the higher number of converters, the microinverter-based PV power systems have a big overhead of control- and auxiliary electronics [3].

Nevertheless, a multitude of PV microinverter topologies with MPPT have been proposed in the literature. In general, three types of structures can be identified: the high-frequency-link microinverter [5], the pseudo microinverter [6], and the DC-link microinverter [7]-[8]. The DC-link microinverters have the advantages of simple structure and easy performance optimization for each stage; therefore they have attracted much popularity.

The flyback-based two-stage microinverter is one of the most commonly adopted topologies because of its merits in cost and wide voltage regulation range [3], [7]-[8]. However, Leuenberger *et al* stated in [3] that the transformer of the dc-dc stage (i.e., the flyback converter) limits the dc-dc stage efficiency to 96.5% even with low-loss nanocrystalline core materials; thus the overall microinverter efficiency is limited to 94.5% in [3]. Therefore, it is necessary to improve the power

conversion efficiency at the dc-dc stage. The LLC resonant converter is a promising topology in terms of high efficiency [9],[10]. The problem with this topology is the narrow voltage gain range; thus hybrid control schemes are required, which increases the implementation complexity of MPPT. For instance, a full-bridge LLC resonant converter is designed for PV applications in [11]; however, a burst mode control should be used in addition to the variable frequency control. In [12], a hybrid control combining the pulse-frequency modulation (PFM) and phase-shift pulse-width modulation (PS-PWM) is employed on a full-bridge LLC resonant converter; the efficiency is improved, but the control complexity is significantly increased as well. In addition, various modified topologies have been proposed to extend the gain range of the LLC resonant converter [13]-[15]. In [13], a bidirectional switch is added on the secondary resonant tank, whereas the full-bridge diode rectifier is replaced with a semi-active rectifier in [14]-[15]. In this case, by controlling the secondary-side active switches, the boost operation is enabled. However, the primary-side switches have to suffer from high turn-off currents and thus high off-switching losses when the converters operate in the buck mode.

This paper proposes a new series resonant converter, as shown in Fig. 1, where a dual-mode rectifier is configured on the secondary side: either a full-bridge rectifier or a half-bridge rectifier (voltage doubler) can be formed. Thus, the proposed converter can handle a wide voltage gain range with the fixed-frequency phase-shift control. All active switches and diodes can achieve zero-voltage switching (ZVS) and zero-current switching (ZCS), respectively, leading to low switching losses. Moreover, the primary and secondary root-mean-square (RMS) currents do not vary significantly with respect to the voltage gain. Therefore, the proposed converter can achieve high efficiency over a wide gain range.

II. TOPOLOGY DERIVATION AND OPERATING PRINCIPLES

As shown in Fig. 1, the proposed Dual-Mode Rectifier based Series Resonant Converter (DMR-SRC) is realized by adding a pair of anti-series transistors (S_5-S_6) between the midpoints of the diode leg (D_3-D_4) and the output capacitor leg (C_{o1} and C_{o2}). Thus, a dual-mode rectifier can be formed by controlling the anti-series transistors S_5-S_6:

1) Half-Bridge Rectifier (HBR) mode: when the anti-series transistors S_5 and S_6 are switched on, a half-bridge rectifier

The 2018 International Power Electronics Conference

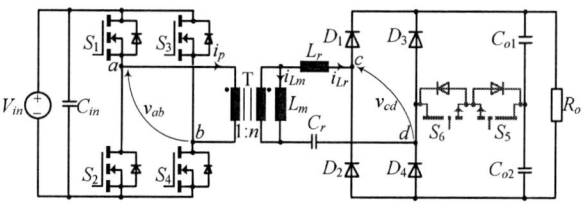

Fig. 1. Schematic of the proposed dual-mode rectifier based series resonant dc-dc converter.

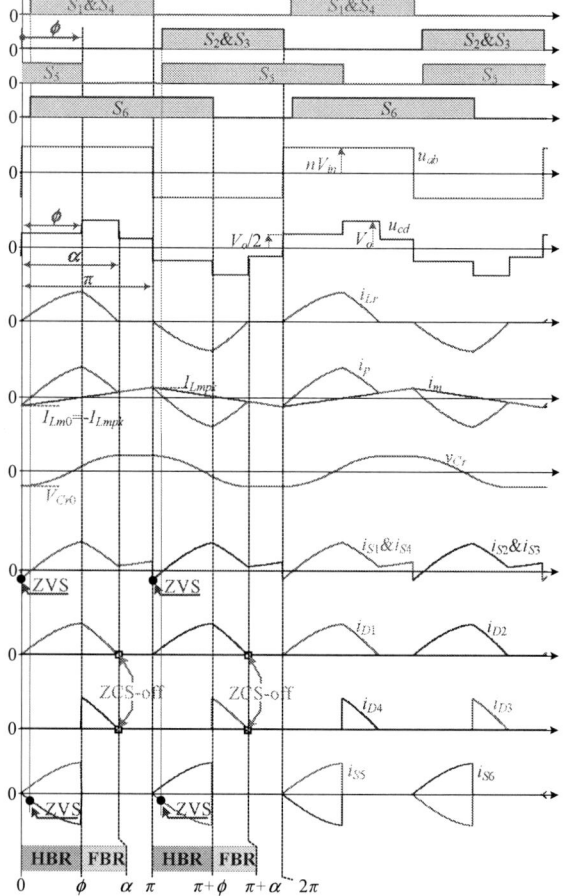

Fig. 2. Operating waveforms of the proposed converter.

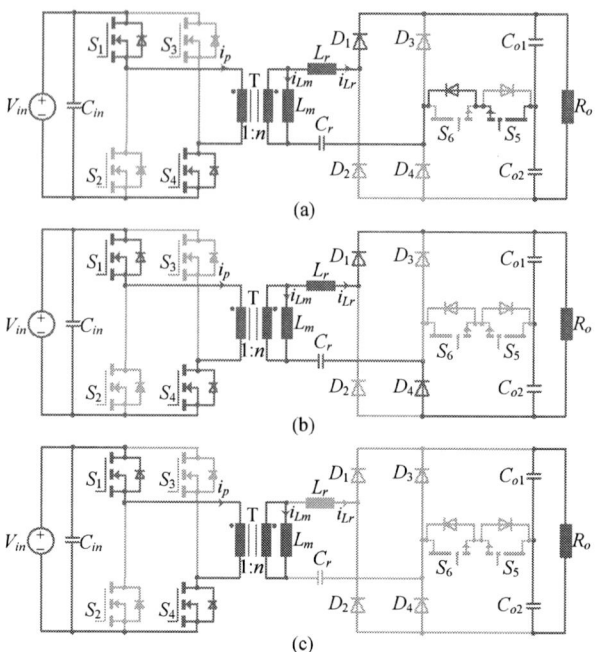

Fig. 3. Equivalent circuit of each operation stage. (a) Stage I: $[0, \phi]$; (b) Stage II: $[\phi, \alpha]$; (b) Stage III: $[\alpha, \pi]$;

The voltage gain of the converter is $G = V_o/(nV_{in})$, and the inductors ratio of L_m to L_r is defined as $m = L_m / L_r$. To simplify the analysis, the voltages and currents are referred to the secondary side. The quality factor is denoted as $Q = Z_r/R_o = P/(V_o^2/Z_r)$, where the characteristic impedance $Z_r = \sqrt{L_r / C_r}$. From (1), the initial capacitor voltage V_{Cr0} can be obtained: $V_{Cr0} = -\pi GQV_o/2$. The initial magnetizing current is denoted as I_{Lm0}, and the peak magnetizing current is represented as I_{Lmpk}. Due to the odd symmetry of the magnetizing current i_{Lm}, we have $I_{Lm0} = -I_{Lmpk} = -\pi V_o/(2mZ_rG)$.

The operating waveforms of the proposed DMR-SRC are shown in Fig. 2. The deadtime is neglected, and thus six stages are included in each switching cycle. The operating stages 1-3 over the first half switching cycle are illustrated in Fig. 3.

III. CONVERTER CHARACTERISTICS

A. Voltage Gain

The voltage gain can be derived as

$$G = \frac{1}{2\pi Q(\cos\phi + 3)} \times$$
$$\left(\sqrt{8\pi Q(2\pi Q - \cos^2\phi - \cos\phi + 2) + (1 - \cos\phi)^2} \right.$$
$$\left. + \frac{4\pi Q(4\pi Q + \sin^2\phi) - \cos^3\phi + 3\cos\phi - 2}{4\pi Q - \cos^2\phi - \cos\phi + 2} \right) \quad (2)$$

Then the voltage gain curves can be plotted, as shown in Fig. 4. It can be seen that the range of the normalized voltage gain is always from 1 to 2 regardless of the quality factor (i.e., the load). It should be noted that the inductors ratio m has no impact on

(voltage doubler) consisting of D_1, D_2, S_5, S_6, C_{o1} and C_{o2} presents on the secondary side;

2) Full-Bridge Rectifier (FBR) mode: when S_5 and S_6 are disabled, there is a full-bridge rectifier consisting of D_1-D_4 on the secondary side.

A fixed-frequency phase-shift modulation is applied to the DMR-SRC, as presented in Fig. 2. The voltage across the midpoints of the two primary-side switch legs, i.e., v_{ab}, is an ac square wave with an amplitude of V_{in}. The resonant inductor L_r and capacitor C_r resonate at frequency $f_r = 1 / \left(2\pi\sqrt{L_rC_r}\right)$. The charge variation of the resonant capacitor over half a switching cycle is

$$q_{hs} = -2V_{Cr0}C_r = \int_0^{T_s/2} i_{Lr}(t)\mathrm{d}t = \frac{T_sP}{2nV_{in}} \quad (1)$$

The 2018 International Power Electronics Conference

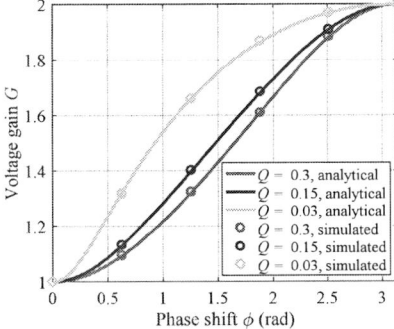

Fig. 4. Analytical and simulated curves of the normalized voltage gain for different quality factors.

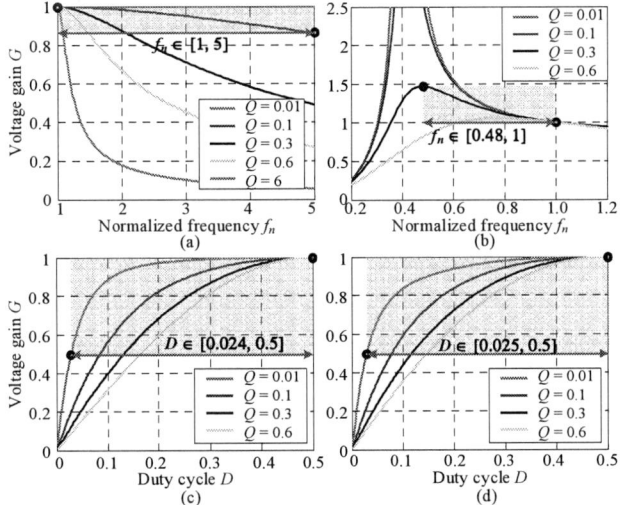

Fig. 5. Voltage gain characteristics of the full-bridge SRC and the full-bridge LLC resonant dc-dc converter. (a) SRC with variable frequency control (VFC); (b) LLC resonant converter with VFC; (c) SRC with duty cycle control; (d) LLC resonant converter with duty cycle control.

the voltage gain. Therefore, the magnetizing inductance can be designed large under the condition of satisfying ZVS conditions of the primary-side switches. The simulated results are also depicted in Fig. 4, which validates the analytical gain model given in (2).

The voltage gain characteristics of the full-bridge SRC and the full-bridge LLC resonant converter are depicted in Fig. 5. For the variable-frequency controlled SRC, the light-load gain range is very narrow even within a wide normalized frequency range $f_n \in [1, 5]$. For the variable-frequency controlled LLC resonant converter, its heavy-load gain range is narrow. In order to have a high full-load gain peak, the characteristic impedance has to be decreased, resulting in a wide frequency range but high conduction losses. With the fixed-frequency duty cycle control, the gain ranges of the SRC and the LLC resonant converter are extended. However, one problem is that the variation of the duty cycle D is wide. When the duty cycle D is small, the conduction loss rises and the soft-switching condition is lost because the peak magnetizing current is dramatically reduced in this case. In addition, when using the duty cycle

control, the primary-side switches of SRC and LLC converter have to turn off with a large current, and thus the off-switching

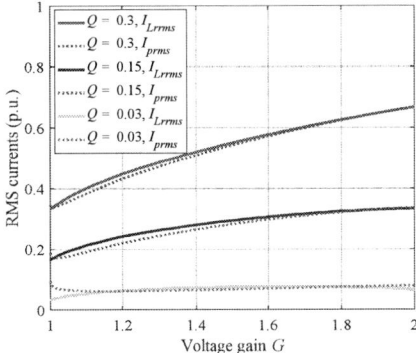

Fig. 6. Primary and secondary transformer RMS currents at $m = 10$.

loss is large. By contrast, the peak magnetizing current of the proposed resonant converter does not vary significantly with respect to the gain G. Thus, ZVS-on of the primary-side switches can be easily achieved while keeping the magnetizing inductance large. Moreover, the primary-side switches in the proposed converter are turned off at the peak magnetizing current, which is small in value; thus, the off-switching loss is small as well.

B. RMS Currents

The secondary and primary transformer RMS currents, I_{Lrrms} and I_{prms}, are shown in Fig. 6. It can be seen that at heavy loads, the RMS resonant current I_{Lrrms} is increasing with respect to the voltage gain. The reason is that when the gain increases, the input voltage decreases and thus the RMS current increases if the power is fixed. However, at light loads, the RMS current firstly increases and then decreases. This is because the angle α is small at light loads when the voltage gain is in the middle area. A smaller α means a larger RMS current if the power is fixed. For the primary transformer RMS current I_{prms}, it is overall rising as the voltage gain G increases. However, I_{prms} becomes flat with respect to G at light loads. Overall, the RMS current difference between the primary and secondary sides is small due to the large magnetizing inductance in the proposed converter.

IV. EXPERIMENTAL RESULTS

A 250-W converter prototype is built and its parameters are as follows: the input voltage $V_{in} = 20\text{-}40$ V, output voltage $V_o = 400$ V, nominal power $P_N = 250$ W, transformer turns ration 1 : $n = 1 : 10$, switching frequency $f_s = 400$ kHz, $L_r = 72$ µH, $C_r = 2.2$ nF, $S_1\text{-}S_4$: EPC2031, $S_5\text{-}S_6$: GS66504B, $D_1\text{-}D_4$: VS-8EWL06FN-M3.

All the primary-side eGaN FETs, i.e., $S_1\text{-}S_4$, can achieve ZVS-on and are turned off at the peak magnetizing current I_{Lm0} that is small in value. Thus, the switching loss of the primary-side switches is small and can be neglected. Only conduction loss is taken into account for the primary-side switches.

For the secondary switches $S_5\text{-}S_6$, they can achieve ZVS-on. However, they are turned off with a relatively large current. Therefore, there may be certain turn-off losses if the turn-off

1790

The 2018 International Power Electronics Conference

speed is not fast enough. In this research, two GaN HEMTs, GS66504B, from GaN Systems™ are used as the secondary-side

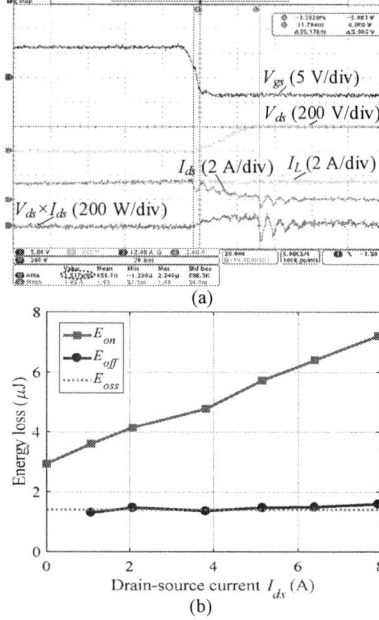

(a)

(b)

Fig. 7. Double-pulse test results. (a) Turn-off waveforms. (b) Turn-on and turn-off energy losses with respect to the drain-source current.

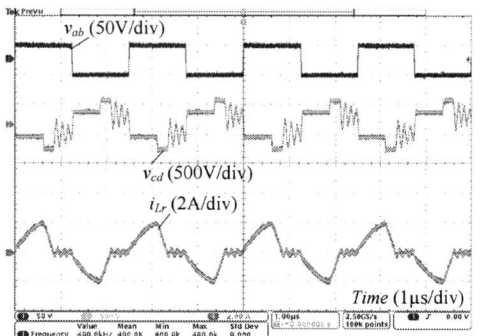

Fig. 8. Experimental waveforms at V_{in} = 26 V, V_o = 400 V and P_o = 250 W.

switches. To explore the turn-off power loss of GS66504B, a double-pulse test setup was built. The switching-off waveforms at V_{ds} = 200 V and I_{ds} = 1.43 A are shown in Fig. 7(a). When the gate voltage V_{gs} falls to below the threshold voltage, the channel will be cut off, but the drain-source voltage V_{ds} has not significantly risen. Therefore, the drain-source current is almost flowing into the output capacitor of the GaN HEMT, causing V_{ds} to rise from 0 to 200 V. As a result, the calculated turn-off energy loss is equal to the energy stored in the output capacitor C_{oss}, as shown in Fig. 7(b). It should be noted that the calculated turn-off energy loss E_{off} (= E_{oss}) is not dissipated during the turn-off period. If the switch is subsequently turned on under hard switching, the energy stored in the output capacitor C_{oss}, E_{oss}, is dissipated on the channel. However, the switch in this converter can achieve ZVS-on, which means that the energy stored in the output capacitor is transferred instead of being dissipated.

Therefore, for the proposed converter, the secondary-side switches implemented with GaN HEMTs can achieve almost lossless turn-off as well as ZVS-on.

The full-load waveforms at V_{in} = 26 V and V_o = 400 V are shown in Fig. 8. As can be seen, the steady-state waveforms match well with the analysis except for the high-frequency oscillations of resonant current i_{Lr} and voltage u_{cd} when each resonant interval ends. The oscillations are caused by the resonance between the resonant inductor, parasitic inductance of printed circuit board (PCB) traces and the output capacitance of rectifier diodes.

V. CONCLUSION

In this paper, a new fixed-frequency phase-shift controlled series resonant converter was proposed. The operating principles and characteristics were analyzed. A comparison between the proposed converter and the conventional SRC and LLC resonant converter was conducted: the proposed converter is advantageous in terms of maintaining high efficiency over a wide gain range. A 400-kHz 250-W converter prototype was tested, and high full- and half-load efficiencies were achieved over an input voltage range from 20 V to 40 V. Therefore, the proposed resonant converter is a good candidate for PV microinverter applications.

REFERENCES

[1] H. Oldenkamp and I. de Jong, "The return of the ac-module inverter," *in 24th European Photovoltaic Solar Energy Conference and Exhibition/4th World Conference on Photovoltaic Energy Conversion*, Hamburg, Germany, 21-24. Sept., 2009.

[2] S. B. Kjaer, J. K. Pedersen, and F. Blaabjerg, "A review of single phase grid-connected inverters for photovoltaic modules," *IEEE Trans. Ind. Appl.*, vol. 41, no. 5, pp. 1292–1306, Oct. 2005.

[3] D. Leuenberger, and J. Biela, "PV-Module Integrated AC Inverters (AC Modules) with Subpanel MPP-Tracking," *IEEE Trans. Power. Electron.*, vol. 32, no. 8, pp. 6150–6118, Aug. 2017.

[4] B. Gu, "Power converter and control design for high-efficiency electrolyte-free microinverters," Ph.D dissertation, Blacksburg, VA, Nov. 2013.

[5] A. Trubitsyn, B. J. Pierquet, A. K. Hayman, G. E. Gamache, C. R. Sullivan, and D. J. Perreault, "High-efficiency inverter for photovoltaic applications," in *Proc. IEEE Energy Conversion Congress and Exposition (ECCE)*, 2010, pp. 2803–2810.

[6] A. C. Kyritsis, E. C. Tatakis, and N. P. Papanikolaou, "Optimum design of the current-source flyback inverter for decentralized grid-connected photovoltaic systems," *IEEE Trans. Energy Convers.*, vol. 23, no. 1, pp. 281–293, 2008.

[7] J.-M. Kwon, B.-H. Kwon, and K.-H. Nam, "High-efficiency module integrated photovoltaic power conditioning system," *IET Power Electron.*, vol. 2, no. 4, pp. 410–420, 2009.

[8] S.H. Lee, W.J. Cha, J.M. Kwon, and B.H. Kwon, B.H., "Control Strategy of Flyback Microinverter With Hybrid Mode for PV AC Modules," *IEEE Trans. Ind. Electron.*, vol. 63, no. 2, pp.995-1002, Feb. 2016.

[9] H. Wu, Y. Li, and Y. Xing, "LLC resonant converter with semiactive variable-structure rectifier (SA-VSR) for wide output voltage range application," *IEEE Trans. Power Electron.*, vol. 31, no. 5, pp. 3389–3394, May 2016.

[10] Q. Zhang, C. Hu, L. Chen, A. Amirahmadi, N. Kutkut, Z. J. Shen, and I. Batarseh, "A center point iteration MPPT method with application on the frequency-modulated LLC micro-inverter", *IEEE Trans. Power Electron.*, vol.29, no.3, pp.1261 -1274, Mar. 2014.

[11] S. Abdel-Rahman, "Resonant LLC Converter: Operation and Design," available online [2016], http://www.infineon.com/dgdl/Application_Note_Resonant+LLC+Converter+Operation+and+Design_Infineon.pdf?fileId=db3a30433a047ba0013a4a60e3be64a1.

[12] X. Mao, Q. Huang, Q. Ke, Y. Xiao, Z. Zhang, and A. E. Andersen, "Grid-connected Photovoltaic Micro-inverter with New Hybrid Control LLC Resonant Converter," *in Proc. 42nd Annual Conference of the IEEE Industrial Electronics Society*, pp. 2319-2324, 2016.

[13] T. LaBella, W. Yu, J.-S. Lai, M. Senesky, and D. Anderson, "A bidirectional-switch-based wide-input range high-efficiency isolated resonant converter for photovoltaic applications," *IEEE Trans. Power Electron.*, vol. 29, no. 7, pp. 3473–3484, Jul. 2014.

[14] H. Wu, T. Mu, X. Gao, and Y. Xing, "A secondary-side phase-shiftcontrolled LLC resonant converter with reduced conduction loss at normal operation for hold-up time compensation application," *IEEE Trans. Power Electron.*, vol. 30, no. 10, pp. 5352–5357, Oct. 2015.

[15] X. Zhao, L. Zhang, R. Born, and J.-S. Lai, "A High-Efficiency Hybrid Resonant Converter With Wide-Input Regulation for Photovoltaic Applications," *IEEE Trans. Ind. Electron.*, vol. 64, no. 5, pp. 3684–3695, May 2017.

Voltage-Reference Active Power Decoupling Based on Boost Converter for Single-Phase Bridge Inverter

Shuang Xu[1], Meiqin Mao[2], Riming Shao[1], Liuchen Chang[1]

1. Emera and NB Power Research Centre for Smart Grid Technologies, University of New Brunswick, Fredericton, Canada
2. Research Center for Photovoltaic System Engineering, Hefei University of Technology, Hefei, China

Abstract—With the development of renewable energy systems such as photovoltaic (PV) systems, single-phase inverters are popular in low power distributed generations. But the single-phase voltage-source bridge inverters has low DC voltage utilization and power mismatch between the input and output sides. To address these issues, additional circuits and modulation techniques have been applied to the bridge inverters, but those methods generally add extra power electronics or complicate the control strategy. In this paper, a voltage-reference active power decoupling method without additional power electronics is applied to the typical boost converter and the voltage-source bridge inverter. Simulation and experimental results verified the feasibility of the proposed power decoupling method on the two-stage single-phase bridge inverter with boost converter.

Keywords— boost converter; bridge inverter; power decoupling; second-order ripple.

I. INTRODUCTION

Single-phase voltage source inverters (VSIs) are widely used for interconnecting distributed generation units that are rated up to 10 kW. In single-phase inverter systems, particularly photovoltaic (PV) single-phase inverters, constant input power to the inverter is desired at the DC side; whereas a pulsating instantaneous power is required by a single-phase AC load as produced by the sinusoidal voltage and current. In this case, the pulsating power can induce a ripple component on the DC voltage or current at double-line frequency, which makes the PV conversion efficiency low when the ripple components are high [1].

Passive power decoupling method generally places an electrolytic capacitor at DC side of the inverter, which not only shortens the PV system lifetime, but also enlarges the prototype volume.

Some current-reference active power decoupling methods are recently proposed in literature [2-5], which are commonly used in flyback-type inverters or current source inverters. The basic idea of the current-reference method is to charge the decoupling capacitor when the DC power is higher than the demanded instantaneous power at the AC side, and discharge the decoupling capacitor when the DC power is less than the the demanded instantaneous power at the AC side.

Since flyback inverter is not suitable for power higher than 1kW due to the difficulty of the flyback transformer design, the bridge inverters have been widely used in the distribution systems such as PV systems. The bridge inverter alone has low DC voltage utilization so that a boost converter is usually connected in front of the bridge circuit, as shown in Fig. 1. The

power decoupling techniques for bridge inverters have also been investigated [6-12], which utilize the voltage-reference active power decoupling method, diverting the pulsating power into decoupling capacitors by controlling the capacitor voltages.

PWM rectifier with a bidirectional buck converter controls the decoupling capacitor voltage as a rectified sine wave or a DC-biased sine wave in [6]. The rectified sine wave requires a complicated controller due to the sharp turns at the bottom. The single-phase inverter in [7] parallels a bidirectional boost converter to control the decoupling capacitor voltage as a DC-biased sine wave fluctuating at double line frequency. The DC-biased sine wave has a DC offset, which means the decoupling capacitor is never totally discharged so that the decoupling capacitor is under- utilized. Moreover, the bridge inverter still needs an extra DC-DC conversion stage such as a boost chopper to achieve a better DC voltage utilization.

To integrate the boost converter with power decoupling circuit, a decoupling capacitor is added to the boost converter in front of the bridge inverter in [2], [3]. The boost converter and bridge stage are connected after an isolated DC/DC converter with a very small capacitor, whose output is more a current source than a voltage source. The inductor has a high inductance to mitigate the current ripple because the inductor current is used as a control variable in the boost converter stage [3].

In this paper, the single-phase bridge inverter is connected with a voltage source (batter or PV panel) and the boost inductor requirement is flexible with the applied voltage-reference active power decoupling techniques instead of the previously used current-reference active power decoupling technique. Compared with conventional boost inverter in Fig. 1, the new single-phase bridge inverter has active power decoupling capability with only one extra decoupling capacitor. The inductance is also smaller because the inductor current is not a control variable for the operation of the inverter with the voltage-reference active power decoupling technique. The second-order ripple power is diverted into the film capacitor C_D, and the DC offset in C_D is used to help

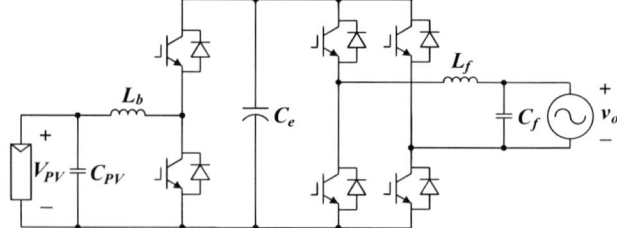

Fig. 1. Conventional single-phase bridge inverter with typical boost converter.

The research has been supported by National Natural Science Foundation of China under Grant 51677050.

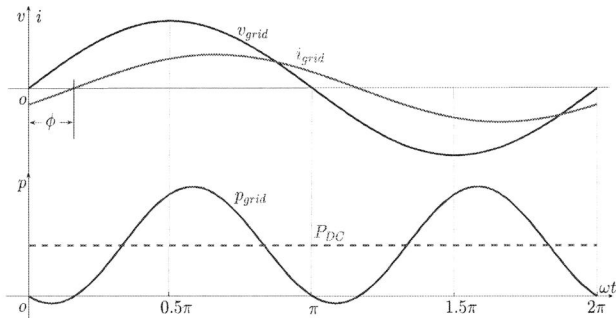

Fig. 2. Instantaneous voltage, current, and power at AC (grid) side.

support the DC-link voltage, as opposed to being under-utilized in previous DC voltage-reference power decoupling techniques.

II. PASSIVE POWER DECOUPLING

The conventional bridge inverter with a typical boost converter is shown in Fig. 1. As stated previously, the conventional bridge inverter needs a large electrolytic capacitor to buffer the second-order power mismatch between the DC and the AC sides.

In single-phase inverter systems, particularly PV systems, constant input power is desired at DC side whereas pulsating instantaneous power is required by the grid side, as shown in Fig. 2. The grid voltage and current can be expressed as:

$$v_{\text{grid}}(t) = \sqrt{2}V_{\text{grid}} \cdot \sin(\omega t) \tag{1}$$

$$i_{\text{grid}}(t) = \sqrt{2}I_{\text{grid}} \cdot \sin(\omega t + \varphi) \tag{2}$$

where V_{grid} is the rms value of the grid voltage, I_{grid} is the rms values of the grid current, ω is the angular frequency, and φ is the phase difference. The instantaneous AC power is presented as:

$$
\begin{aligned}
p_{\text{grid}}(t) &= v_{\text{grid}}(t) \cdot i_{\text{grid}}(t) \\
&= \underbrace{V_{\text{grid}}I_{\text{grid}} \cdot \cos\varphi}_{P_{DC}} \underbrace{-V_{\text{grid}}I_{\text{grid}} \cdot \cos(2\omega t + \varphi)}_{p_r} \tag{3}
\end{aligned}
$$

The first term in (3) represents the DC component P_{DC}, and the second term represents the second-order ripple component p_r.

Most of the second-order ripple power component p_r should be absorbed by the DC-link capacitor, as shown in the schematic diagram of conventional single-phase bridge inverter with boost converter in Fig. 3, but the DC-link capacitor has the same percentage of voltage oscillation as that of PV side. In order to minimize the power pulsation of the input PV panel, a large electrolytic capacitor C_e is required at PV side or DC link to minimize the voltage oscillation and absorb the ripple power.

Assume that the input DC voltage oscillation should be less than ±2% and the second-order ripple power provided by the PV panel is less than 5%, then the voltage oscillation at DC link should also be less than ±2% and the remaining 95% of the ripple power will be absorbed by the DC-link capacitor C_e. Thus the DC-link capacitance meeting the design requirement can be calculated by:

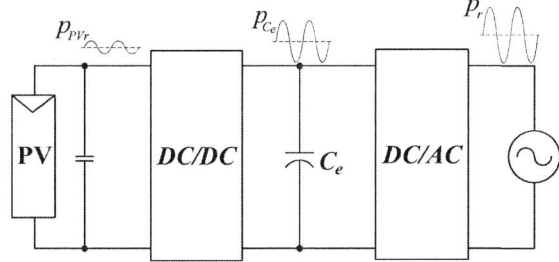

Fig. 3. Schematic diagram of conventional single-phase bridge inverter with boost converter showing ripple power at the DC side, DC-link, and AC side.

$$C_e = \frac{0.95P_r}{V_{\text{link}} \cdot 2\omega \cdot 0.02V_{\text{link}}} \tag{4}$$

where P_r is the magnitude of the second-order component of the ripple power, and V_{link} is the average DC-link voltage. With 400W rated power and 100V input DC voltage, the capacitance of C_e needed for power decoupling is 2.52mF. When the DC-link voltage is boosted up to 200V, the minimum capacitance of C_e can be reduced to 0.63mF, but still not enough to avoid the use of a large electrolytic capacitor. Therefore, the active power decoupling method that adds only a decoupling capacitor at boost converter stage is proposed to significantly reduce the capacitance requirement in single-phase inverter systems.

III. VOLTAGE-REFERENCE ACTIVE POWER DECOUPLING

To avoid using the bulky and short lifespan electrolytic capacitor, a voltage-reference active power decoupling technique is applied to the boost converter with adding a decoupling capacitor as shown in Fig. 4. The power decoupling analysis of the system is investigated. For the boost converter, the relationship between the input DC voltage V_{DC} and DC-link voltage v_{link} is:

$$v_{\text{link}} = \frac{1}{1 - d_c} \cdot V_{DC} \tag{5}$$

where d_c is the duty cycle of switch S_{c2}. Then the decoupling capacitor voltage v_{C_D} is:

$$v_{C_D} = \frac{d_c}{1 - d_c} \cdot V_{DC} \tag{6}$$

Suppose the single-phase VSI is operated with unity power factor, then the voltage and current of the AC load are as follows:

$$v_o = V_o \cdot \sin(\omega t) \tag{7}$$

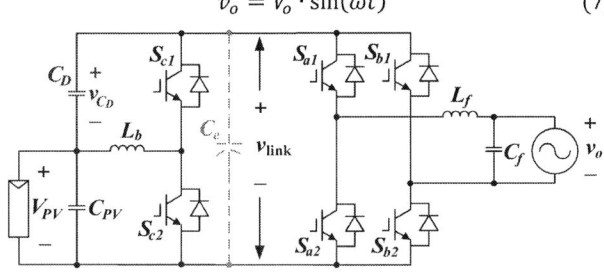

Fig. 4. Single-phase bridge inverter with voltage-reference active power decoupling based on boost converter.

$$i_o = I_o \cdot \sin(\omega t) \qquad (8)$$

and the current through filtering capacitor C_f can be calculated as:

$$i_{C_f} = C_f \cdot \frac{dv_o}{dt} = I_{C_f} \cdot \cos(\omega t) \qquad (9)$$

where $I_{C_f} = \omega C_f V_o$.

The instantaneous power absorbed by the AC load and through filtering capacitor are, respectively,

$$p_o = v_o i_o = \frac{V_o I_o}{2} - \frac{V_o I_o}{2} \cdot \cos(2\omega t) \qquad (10)$$

$$p_{C_f} = v_o i_{C_f} = \frac{V_o I_{C_f}}{2} \cdot \sin(2\omega t) \qquad (11)$$

from which we can see that the second-order pulsating power comes from both the AC load and the filtering capacitor. To balance the second-order pulsating power, the front end will be controlled to divert the pulsating power into the film capacitor C_D. The decoupling capacitor voltage v_{C_D} is controlled such that the instantaneous power p_{C_D} absorbed by C_D becomes:

$$p_{C_D} = P_o \cdot \cos(2\omega t) - P_{C_f} \cdot \sin(2\omega t) \qquad (12)$$

where $P_o = \frac{V_o I_o}{2}$ and $P_{C_f} = \frac{V_o I_{C_f}}{2}$.

The voltage across the decoupling capacitor is a DC-biased sine wave. In waveform control [8],[9], or Solution II in [10], the voltage-reference active power decoupling techniques directly add a second-order component to the DC offset. But these power decoupling techniques simultaneously introduced a fourth-order ripple power due to the multiplication of the second-order components in the capacitor voltage and current. To theoretically eliminate the fourth-order ripple power, the decoupling capacitor voltage is assumed as a DC offset V_d and an AC component $v_{C_{AC}}$ for further calculation. Suppose the voltage across the decoupling capacitor is:

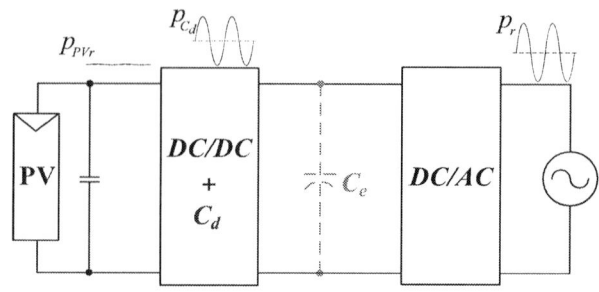

Fig. 5. Schematic diagram of conventional single-phase bridge inverter with boost converter showing ripple power at the DC side, DC-link, and AC side.

$$v_{C_D} = V_d + v_{C_{AC}} \qquad (13)$$

Then the current flowing through the decoupling capacitor is solved as:

$$i_{C_D} = C_D \dot{v}_{C_D} = C_D \dot{v}_{C_{AC}} \qquad (14)$$

where the above dot indicates the derivative of the variable.

The instantaneous power p_{C_D} absorbed by C_D is calculated as:

$$p_{C_D} = v_{C_D} i_{C_D} = C_D V_d \dot{v}_{C_{AC}} + C_D v_{C_{AC}} \dot{v}_{C_{AC}} \qquad (15)$$

Equaling the right hand sides of the equations (12) with (15), it can be obtained that:

$$C_D V_d \dot{v}_{C_{AC}} + C_D v_{C_{AC}} \dot{v}_{C_{AC}} = P_o \cdot \cos(2\omega t) - P_{C_f} \cdot \sin(2\omega t) \qquad (16)$$

Take the integral for both sides, a quadratic equation with respect to $v_{C_{AC}}$ is obtained:

$$C_D v_{C_{AC}}^2 + 2 C_D V_d v_{C_{AC}} = \frac{V_o I_o}{2\omega} \cdot \sin(2\omega t) + \frac{V_o^2 C_f}{2} \cdot \cos(2\omega t) \qquad (17)$$

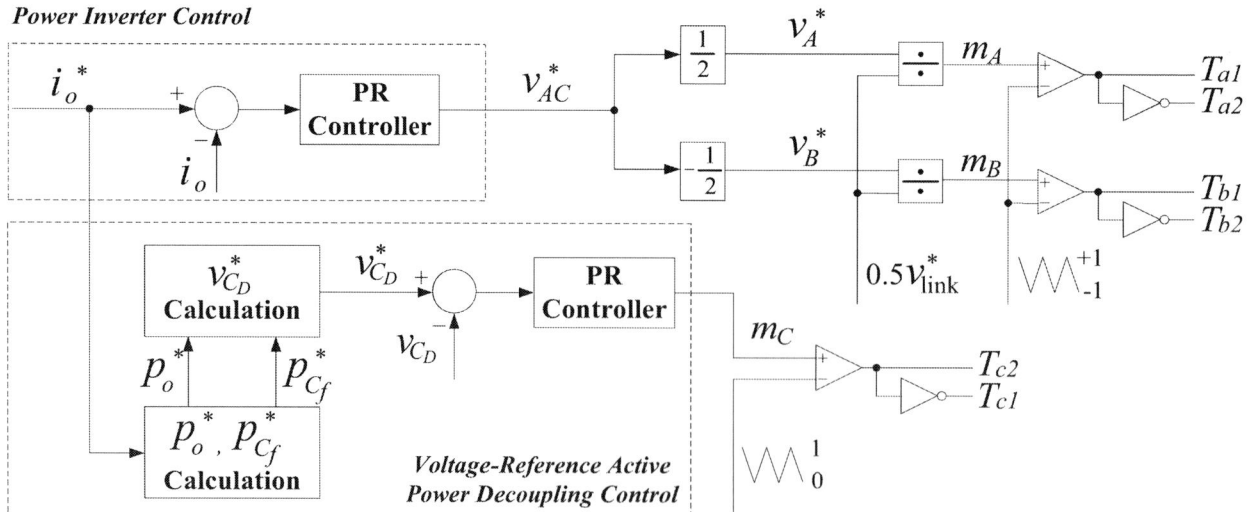

Fig. 6. Control diagram of the single-phase bridge inverter with voltage-reference active power decoupling based on boost converter.

The 2018 International Power Electronics Conference

From the quadratic function, the AC component $v_{C_{AC}}$ of the decoupling capacitor voltage is solved as:

$$v_{C_{AC}} = -V_d + \sqrt{V_d^2 + \frac{V_o I_o}{2\omega C_D}\sin(2\omega t) + \frac{C_f}{2C_D}V_o^2\cos(2\omega t)}$$

(18)

Then the decoupling capacitor voltage reference can be expressed as:

$$v_{C_D} = \sqrt{V_d^2 + \frac{V_o I_o}{2\omega C_D}\sin(2\omega t) + \frac{C_f}{2C_D}V_o^2\cos(2\omega t)} \quad (19)$$

And the duty cycle reference d_c for the boost converter can be determined by:

$$d_c = \frac{v_{C_d}}{v_{C_d} + V_{DC}} \quad (20)$$

With the voltage-reference active power decoupling technique, the schematic diagram of the single-phase bridge inverter with active power decoupling based on boost converter is shown in Fig. 5. Most of the second-order ripple power generated by the AC side has been absorbed by the film capacitor C_D. Theoretically, the second-order ripple power at DC side can be totally eliminated, as opposed to be partially eliminated by the passive power decoupling as described in Fig. 3.

The whole control algorithm is shown in Fig. 6, including the power inverter control and the voltage-reference active power decoupling control. In the power inverter control, a proportional-resonant (PR) controller is employed instead of a traditional proportional-integral (PI) controller to track the reference AC current. The power decoupling control algorithm stated in this section has been utilized to achieve the desired

capacitor voltage, which is also added with DC input voltage to reach higher DC-link voltage to ensure proper operation in the bridge inverter stage control.

IV. SIMULATION AND EXPERIMENTAL RESULTS

To confirm the validity of the voltage-reference active power decoupling technique on the single-phase voltage-source bridge inverter with boost converter, a 400W prototype is built and tested in PSIM 11.0.3 and experimental setup. The key parameters are listed in Table I.

TABLE I
PARAMETERS OF SINGLE-PHASE BRIDGE INVERTER

DC voltage V_{DC}	100V
AC voltage V_o	156V
AC current I_o	5.1A
Rated power P_o	400W
Filter Inductance L_f	0.8mH
Boost Inductance L_b	300μH
Filter Capacitance C_f	10μF
Power decoupling capacitance C_D	160μF
AC frequency f_{AC}	60Hz
Switching frequency f_s	12kHz

The simulation results are shown in Fig. 7, which contain the input DC voltage V_{DC}, decoupling capacitor voltage v_{C_D}, DC-link voltage v_{link}, output AC voltage v_o and DC current I_{DC}. The input DC voltage is much less than the maximum output AC voltage. Without the voltage-reference active power

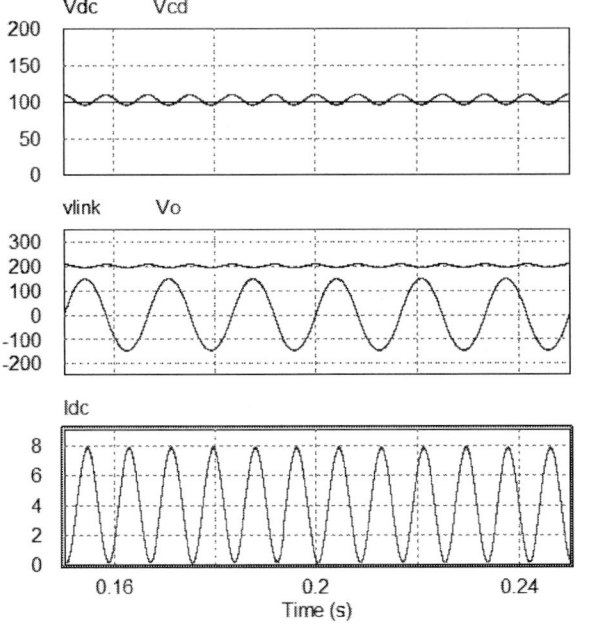 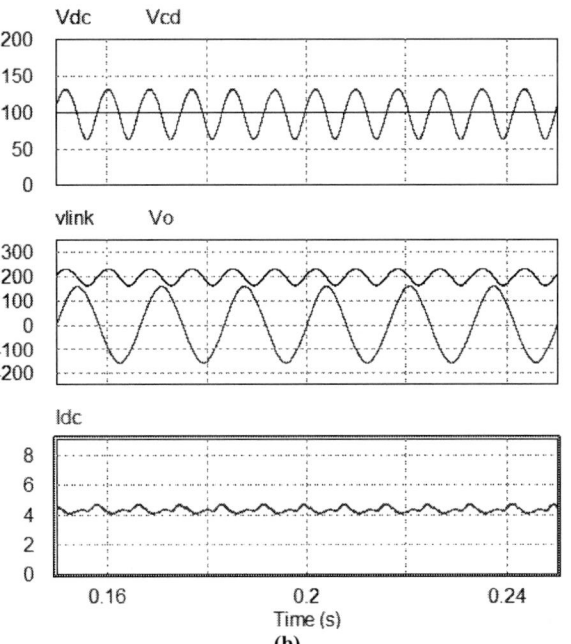

Fig. 7. Simulation results of the single-phase inverter with boost converter: (a) without power decoupling control. (b) with power decoupling control.

1796

The 2018 International Power Electronics Conference

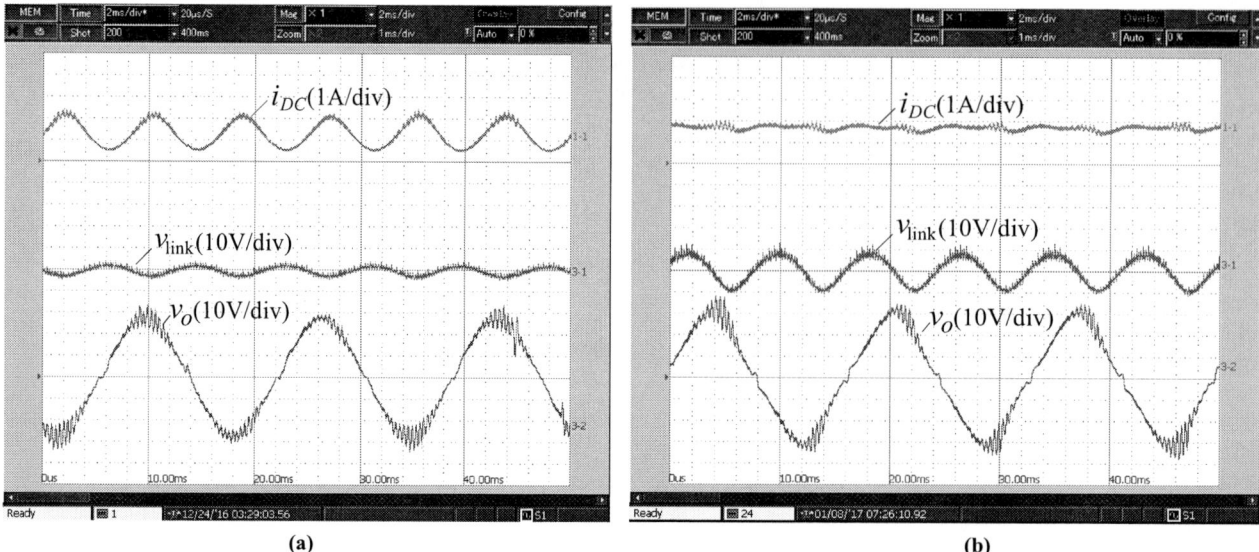

Fig. 8. Experimental results of the single-phase inverter with boost converter: (a) without power decoupling; (b) with power decoupling.

decoupling control, the boost stage before the bridge inverter successfully boost the voltage from 100V to around 200V, and the voltage across the decoupling capacitor has small ripple However, the input DC current contains a prominent second-order component, which can decrease the maximum power point tracking of the PV array and conversion efficiency of the inverter system.

With the proposed voltage-reference active power decoupling control method, the voltage across the decoupling capacitor has higher ripple because it absorbs the second-order ripple power. The DC offset in decoupling capacitor voltage helps support the DC link voltage. As it is illustrated in the DC current waveform in Fig. 7(b), the second-order ripple in DC current has been eliminated, achieve successful power decoupling.

During practical test, the DSP TMS320F28335 is employed to achieve the control functions in the simulation. The preliminary results are shown in Fig. 8, containing waveforms of input DC current, DC-link voltage, output AC voltage and output AC current. The DC-link voltage is always higher than the maximum output AC voltage, ensuring the bridge inverter never reach over-modulation region. In Fig. 8(a), the input DC current contains a prominent second-order ripple, indicating that the pulsating power has not been decoupled. In comparison, the second-order ripple has been eliminated in the input DC current in Fig. 8(b), indicating the successful power decoupling.

V. CONCLUSION

This paper presented a voltage-reference active power decoupling technique based on the boost converter before the single-phase bridge inverter. The power decoupling control technique reuses the DC offset in the decoupling capacitor voltage, and theoretically enable remove the second-order ripple power without introducing other-order harmonics such as fourth-order harmonic. Simulation and experimental results

verified the successful power decoupling control with substantial reduction of second-order ripple current at the DC side. With the voltage-reference active power decoupling technique, the topology also has small boost inductance compared with that of the current-reference active power decoupling, which can increase the power density of the topology.

REFERENCES

[1] C. R. Sullivan, J. J. Awerbuch and A. M. Latham, "Decrease in photovoltaic power output from ripple: Simple general calculation and the effect of partial shading," *IEEE Trans. Power Electron.*, vol. 28, no. 2, pp. 740-747, Feb. 2013.

[2] J. i. Itoh, H. Watanabe, K. Koiwa and Y. Ohnuma, "Experimental verification of single-phase inverter with power decoupling function using boost-up chopper," in *Proc. 15th Euro. Conf. Power Electron. Appl. (EPE)*, Sept. 2013, pp. 1-10.

[3] H. Watanabe, K. Koiwa, J. i. Itoh, Y. Ohnuma and S. Miyawaki, "Miniaturization of the boost-up type active buffer circuit in a single-phase inverter," in *Proc. Int. Power Electron. Conf. (IPEC- ECCE ASIA)*, May 2014, pp. 84-91.

[4] H. Hu, S. Harb, X. Fang, D. Zhang, Q. Zhang, Z. Shen and I. Batarseh, "A three-port flyback for PV microinverter applications with power pulsation decoupling capability," *IEEE Trans. Power Electron.*, vol. 27, no. 9, pp. 3953-3964, Sept. 2012.

[5] H. Hu, S. Harb, N. H. Kutkut, Z. J. Shen and I. Batarseh, "A single-stage microinverter without using eletrolytic capacitors," *IEEE Trans. Power Electron.*, vol. 28, no. 6, pp. 2677-2687, June, 2013.

[6] R. Wang, F. Wang, D. Boroyevich, R. Burgos, R. Lai, P. Ning and K. Rajashekara, "A high power density single-phase PWM rectifier with active ripple energy storage," *IEEE Trans. Power Electron.*, vol. 26, no. 5, pp. 1430-1443, May 2011.

[7] G. C. Christidis, A. C. Kyritsis, N. P. Papanikolaou and E. C. Tatakis, "Investigation of Parallel Active Filters' Limitations for Power Decoupling on Single-Stage/Single-Phase Microinverters," *IEEE J. Emerg. Sel. Topics Power Electron.*, vol. 4, no. 3, pp. 1096-1106, Sept. 2016.

[8] G. Zhu, S. Tan, Y. Chen and C. K. Tse, "Mitigation of low-frequency current ripple in fuel-cell inverter systems through waveform control," *IEEE Trans. Power Electron.*, vol. 28, no. 2, pp. 779-792, Feb. 2013.

[9] W. Yao, X. Wang, P. C. Loh, X. Zhang and F. Blaabjerg, "Improved Power Decoupling Scheme for a Single-Phase Grid-Connected Differential Inverter With Realistic Mismatch in Storage Capacitances," *IEEE Trans. on Power Electron.*, vol. 32, no. 1, pp. 186-199, Jan. 2017.

[10] H. Zhang, X. Li, B. Ge and R. S. Balog, "Capacitance, dc Voltage Utilizaton, and Current Stress: Comparison of Double-Line Frequency Ripple Power Decoupling for Single-Phase Systems," *IEEE Ind. Electron. Magazine*, vol. 11, no. 3, pp. 37-49, Sept. 2017.

[11] Y. Xia, J. Roy and R. Ayyanar, "A Capacitance-Minimized, Doubly Grounded Transformer less Photovoltaic Inverter With Inherent Active-Power Decoupling," in *IEEE Transactions on Power Electronics*, vol. 32, no. 7, pp. 5188-5201, July 2017.

[12] Y. Xia, Z. Yu and R. Ayyanar, "Power command compensation structure to improve the dynamic performance for single phase transformer-less photovoltaic inverters with dynamic power decoupling," in *Proc. IEEE Energy Convers. Congress and Expo. (ECCE)*, Oct. 2017, pp. 455-460.

The 2018 International Power Electronics Conference

A Single-Phase Common Ground Boost Inverter for Photovoltaic Applications

Tan-Tai Tran[1], Minh-Khai Nguyen[2*], Young-Cheol Lim[1], and Joon-Ho Choi[1]
1 Department of Electrical Engineering, Chonnam National University, Gwangju, Korea
2 Department of Electrical Engineering, Chosun University, Gwangju, Korea
* E-mail: khaibk@ieee.org.

Abstract- The paper introduces a new single-phase boost inverter for photovoltaic applications. The introduced inverter merges a boost converter and a single-phase inverter in a single-phase single-stage inverter. The proposed inverter has a simpler topology and uses the minimum number of power electronic components. A high ac output voltage of the introduced topology is generated from a low dc voltage source by implementing one cycle control. Furthermore, the introduced topology shares a common ground between input voltage source and load. The operating principles, simulation results and circuit analysis of the single-phase boost inverter with common ground are illustrated. Also, a simple pulse width modulation technique is shown. The experimental results are also shown to validate the precise performance of the introduced inverter.

Keywords— Single-phase inverter, common ground, boost inverter, PV applications.

I. INTRODUCTION

The question of how to deal with the phenomenon of energy crisis still remains a source of constant debate in our contemporary society. Fossil fuels are non-renewable, that is, they become too environmentally damaging or too expensive. Meanwhile, renewable energy sources such as fuel cell, Photovoltaic, and wind energy will never run out. So, the use of renewable energy sources is productive approaches to deal with the energy requirement of our social development. Nevertheless, most renewable energy sources produce an output dc voltage. As a result, an inverter is required to invert this dc electric energy into the ac electric energy. The inverters can operate in either grid-connected or stand-alone.

The voltage-source inverters (VSIs) are popularly used in applications such as uninterruptible power supplies (UPS), hybrid electric vehicles, and distributed power systems [1]-[3]. Nevertheless, the traditional voltage source inverters are buck dc-ac power conversion, where the total DC source voltages are always higher than the ac output voltage peak. Therefore, the most traditional popular dc-ac power inverters for renewable energy source integration are based on the two-stage topology, which consists an inverter and a front-end dc-dc converter. An additional dc-dc boost converter [4]-[10] is required to achieve the desired value of output voltage. As a result, the additional power converter performs two-stage power conversion with low efficiency and high cost. The existing inverter topologies which can be used as boost inverter have some the following demerits: (a)

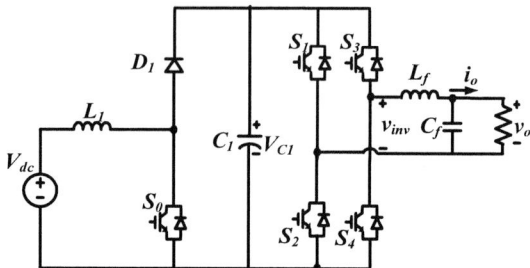

Fig. 1. Conventional two-stage inverter topology.

complex control technique [11]-[13], (b) high semiconductor device count [14]-[15], and (c) high number of capacitors and inductors [15]-[16]. In addition, a transformer is used to boost voltage and supply isolation in Photovoltaic applications. As a result, the cost, volume, and weight of the inverter were risen. So, the inverter with Common ground can constructively limit the leakage current of the Photovoltaic system.

To alleviate solve the aforementioned demerits of the conventional inverters, this paper discusses a new topology of a single-stage single-phase boost inverter with common ground. The introduced inverter is composed of one diode, one inductor one dc-link capacitor, and four controllable switches. The proposed modulation scheme is simple. According to aforementioned characteristics, the introduced inverter is applicable to photovoltaic or fuel cells applications. The circuit analysis, operating principles and simulation results of the single-phase boost inverter with common ground are presented. Also, a simple control strategy for the introduced topology is illustrated. Experimental results are shown to confirm the precise performance of the introduced inverter. In section II, introduced topology, control method and principles of inverter operation are presented. Section III shows Simulation results. Afterward, experimental results are indicated in section IV. Finally, the conclusion is shown section V.

II. SINGLE-PHASE BOOST INVERTER WITH COMMON GROUND

Fig. 1 illustrates a conventional two-stage inverter topology for the system when the input dc voltage is less than the output voltage peak. The converter combine with a dc-dc boost converter is composed of capacitors C_1, MOSFET S_0 and boost inductor L_1. The current flows back the input source can be blocked by Diode D_1. The

1799

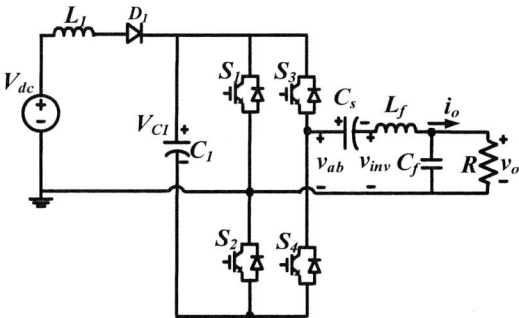

Fig. 2. The single-phase boost inverter with common ground.

inverter is a H–bridge circuit using four MOSFET (S_1–S_4). The merit of the conventional two-stage inverter topology is that the H–bridge circuit and boost converter are worked separately and the controller design is pretty easy. Nevertheless, a two-stage topology leads to high cost and low efficiency.

The configuration of the single-phase boost inverter with common ground is illustrated in Fig. 2. As shown in Fig. 2, one of the H-bridge circuit legs is participated in boost converter. Therefore, the proposed system consists of one DC sources, four active switches, one diode, one dc-link capacitor, one inductor, two capacitor filter and an inductor filter connected to the load.

In boost state, S_1 and diode D_1 are switched on, while S_2 are switched off as shown in Fig. 3(a) and Fig. 3(b). The inductor stores energy while the capacitor is discharged. In this state, the time interval is $D.T$, where T and D is a switching period and the duty cycle of each cycle, respectively. We obtain:

$$L_1 \frac{di_{L_1}}{dt} = V_{dc} . \qquad (1)$$

In boost state, the inverted voltage, v_{ab}, equals zero as shown in Fig 3(a) and a negative capacitor C_1 voltage as shown in Fig 3(b).

In non-boost state, S_2 and diode D_1 are switched on, while S_1 is switched off as shown in Fig. 3(c) and Fig. 3(d), the inductor discharged while the capacitor is charged. We obtain:

$$L_1 \frac{di_{L_1}}{dt} = V_{C1} - V_{dc} . \qquad (2)$$

Applying the voltage balance law to L_1 in steady state, (1) and (2) yield:

$$V_{C1} = \frac{V_{dc}}{1-D} . \qquad (3)$$

The DC-link voltage is expressed as:

$$V_{PN} = V_{C1} = \frac{V_{dc}}{1-D} . \qquad (4)$$

The boost factor of the introduced topology is defined as follows:

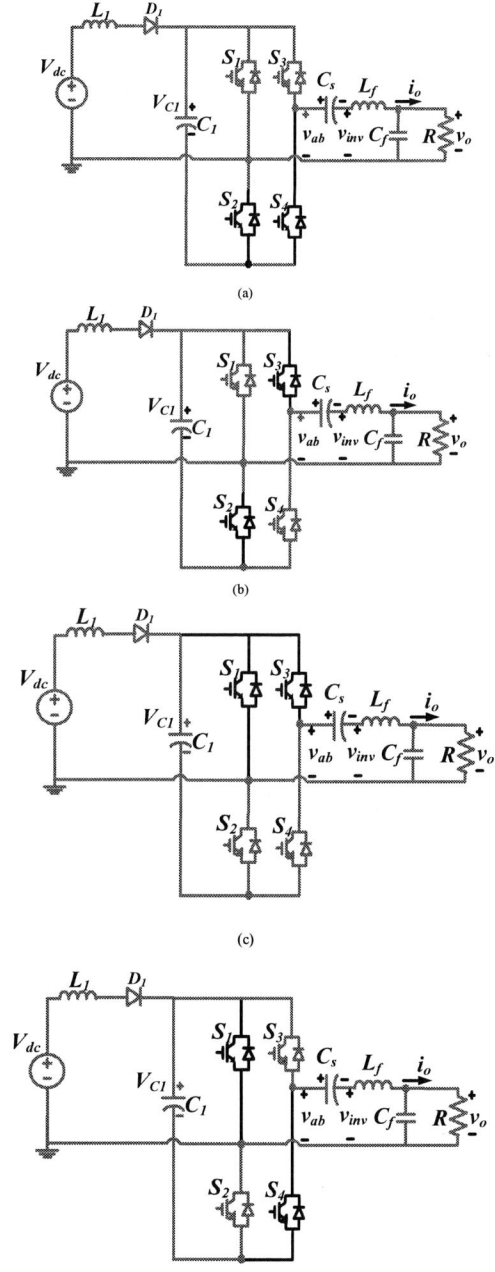

Fig. 3. Operating states of introduced inverter; (a)-(b) boost state, and (c)-(d) non-boost state.

$$B = \frac{V_{PN}}{V_{dc}} = \frac{1}{1-D} . \qquad (5)$$

Fig. 4 shows a sinusoidal pulse-width modulation (PWM) strategy for the proposed topology. With the modulation in Fig. 4, a control waveform, $V_{control}$ is used to produce control signals for S_3 and S_4 switches by comparing to a high-frequency carrier waveform, V_{tri}. A fixed voltage $-V_{SH}$ is compared to another high-frequency carrier waveform (dashed line) with double frequency

The 2018 International Power Electronics Conference

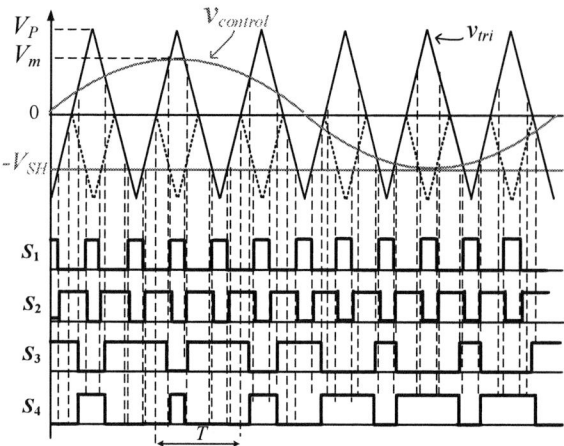

Fig. 4. PWM strategy for the proposed inverter.

TABLE I: PARAMETERS OF THE INTRODUCED INVERTER

Parameter	Value
Output voltage	110 Vrms / 50Hz
Inductor (L_1) and Capacitor (C_1)	1 mH and 1000 μF
Filter Capacitor (C_s)	1000μF
Filter inductor (L_f)	3 mH
Filter Capacitor (C_f)	10μF
Resistive load (R)	60 Ω
Switching frequency	10KHz

and half of the amplitude of that of high-frequency carrier waveform V_{tri} to provide control signals for S_1 and S_2 switches. The output voltage of inverter is a three-level as $-V_{PN}$, 0 and V_{PN}.

The output voltage peak is achieved by:

$$V_o = M.\frac{V_{PN}}{2}, \tag{6}$$

where M is the modulation index. From (5) and (6), the ac output voltage peak is rewritten as:

$$V_o = M.\mathrm{B}.\frac{V_{dc}}{2}. \tag{7}$$

III. SIMULATION RESULTS

In order to confirm the operating principle of the introduced topology, PSIM simulation is used. Table I indicates the simulation parameters for the single-phase boost inverter with common ground. The input dc voltage is set to 96 V in the proposed inverter in order to test properties of introduced topology.

Fig. 5 and Fig.6 illustrate the simulation waveforms of the introduced topology at V_{dc} = 96 V. From Fig. 5, we can see that the ac output voltage of the introduced inverter is 110 Vrms with three-level. The DC-link voltage is 315 V as shown in Fig. 6(a). The average

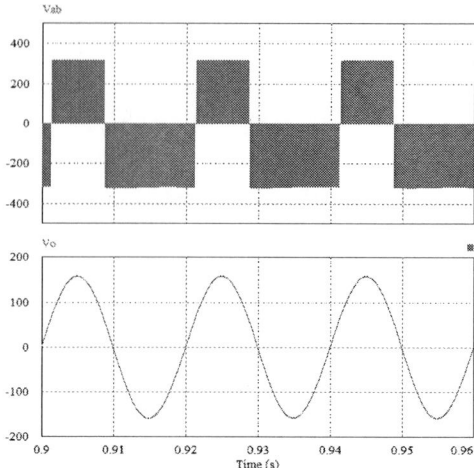

Fig. 5. Simulation results for the introduced inverter. (a) 3-level output voltage; (b) output voltage with using Filter.

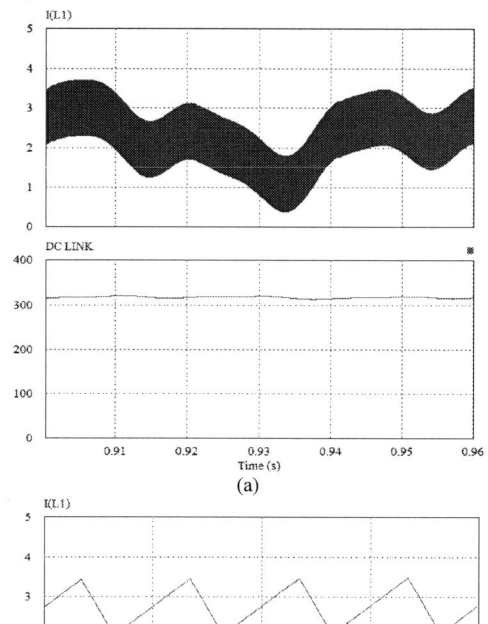

(a)

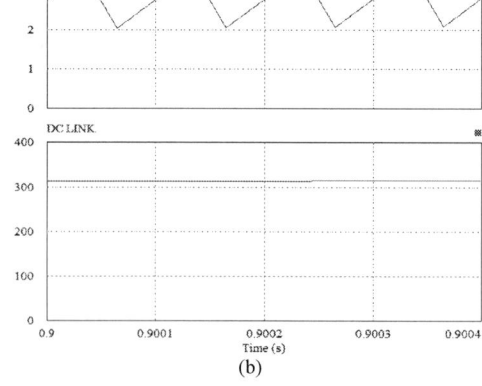

(b)

Fig. 6. Simulation results for the introduced inverter. From top to bottom: (a)-(b) input current and dc link voltage.

inductor current is 2.28A. As shown in Fig 6(b), the high-frequency peak-to-peak inductor ripple current is 1.37 A.

1801

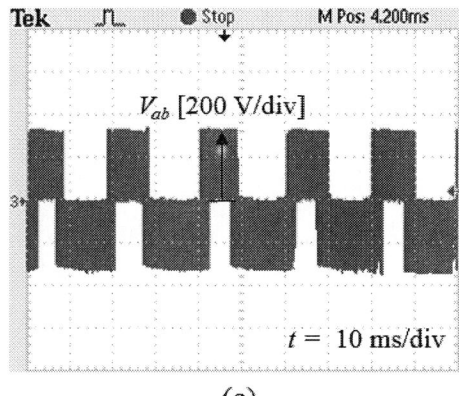

(a)

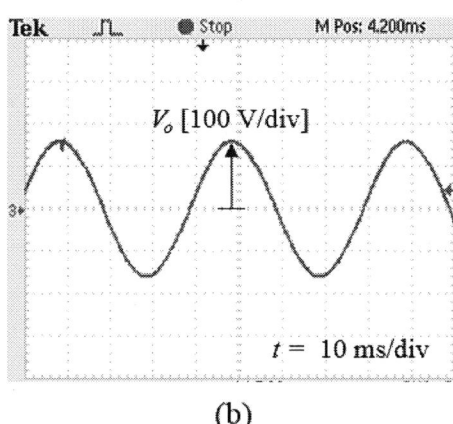

(b)

Fig. 7. Experimental waveforms for the introduced inverter when V_{dc} = 96 V. (a) Three-level output voltage; (b) output voltage with using filter.

IV. EXPERIMENTAL RESULTS

Based on a TMS320F28335 DSP, a 200 W laboratory prototype is built to confirm the performance of the introduced inverter. The parameters of experimental prototype are the same as the simulation. The input dc voltage and output voltage are 96 V and 110 Vrms at 50 Hz, respectively.

Fig. 7 shows the experimental waveform of the introduced inverter at V_{dc} = 96 V. The measured output voltage of the introduced inverter is 110 Vrms with three-level as illustrated in Fig. 7.

V. CONCLUSION

This paper shows circuit analysis and operating principles of the single phase single-stage boost inverter with common ground. In the introduced inverter, one of the H-bridge circuit legs participates in a boost converter. Furthermore, the ground of dc input voltage source is directly connected to ground of ac load. The introduced inverter requires only one inductor, one dc-link capacitor, one diode, and four switches. In addition, a simply pulse width modulation strategy is presented to control the introduced inverter. Since ac output voltage of the introduced inverter is higher than dc input voltage, the introduced topology suits PV and fuel-cell applications.

Simulation and experimental results are given to validated the precise performance of the introduced inverter.

REFERENCES

[1] H. Demomi, L.J. Bome, and C. Nayar, "Design and implementation of a low cost sine wave inverter," *IEEE – ISIE International Symposium on Industrial Electronics,* pp. 280-285, 2003.

[2] T. Kerekes, R. Teodorescu, M. Liserre, C. Klumpner, and M. Sumner, "Evaluation of three-phase transformerless photovoltaic inverter topologies," *IEEE Trans. Power Electron.,* vol. 24, no. 9, pp. 2202–2211, 2009.

[3] J.M. Carrasco, L.G. Franquelo, J.T. Bialasiewicz, E. Galvan, R.C. PortilloGuisado, M.A.M. Prats, J.I. Leon, and N.M. Alfonso, "Power-electronic systems for the grid integration of renewable energy sources: a survey," *IEEE Trans. Ind. Electron.,* vol. 53, no. 4, pp. 1002–1016, Jun. 2006.

[4] S.B. Kjaer, J.K. Pedersen and F. Blaabjerg, "A Review of single-phase grid-connected inverters for photovoltaic modules", *IEEE Trans. Ind. Appl.,* vol. 41, vo. 5, pp. 1292–1306, 2005.

[5] F. Schimpf and L. E. Norum, "Grid connected converters for photovoltaic, state of the art, ideas for improvement of transformerless inverters," *Nordic Workshop on Power and Industrial Electronics,* 2008.

[6] H. Matsuo and F. Kurokawa, "New solar cell power supply system using a boost type bidirectional dc–dc converter", *IEEE Trans. Ind. Electron,* vol. 31, no. 1, pp. 1118–1126., 1984.

[7] W. Hu, H. Wu, Y. Xing and K. Sun, "A full-bridge three-port converter for renewable energy application." *Proc. IEEE Applied Power Electron. Conf.,* 2014, pp. 57–62.

[8] Y.M. Chen, Y.C. Liu and F.Y. Wu, "Multi-input dc/dc converter based on the multiwinding transformer for renewable energy applications," *IEEE Trans. Ind. Appl.,* vol. 38, no. 4, pp. 1096–1104, 2002.

[9] C. Zhao, S.D. Round and J.W. Kolar, "An isolated three-port bidirectional dc–dc converter with decoupled power flow management," *IEEE Trans. Power Electron.,* vol. 23, no.5, pp. 2443–2453, 2008.

[10] B. Singh, D.T. Shahani and A.K. Verma, "Neural network controlled grid interfaced solar photovoltaic power generation," *IET Power Electron.,* vol. 7, no. 3, pp. 614–626, 2014.

[11] M.K. Nguyen and T.T. Tran, "A single-phase single-stage switched-boost inverter with four switches," *IEEE Trans. Power Electron.,* vol. PP, no. 99, 2017.

[12] H. Ribeiro, B. Borges and A. Pinto, "Single-stage DC–AC converter for photovoltaic systems," *Proc. IEEE Energy Conversion Congress and Exposition,* 2010, pp. 604–610

[13] F.Z. Peng, "Z-source inverter," *IEEE Trans. Ind. Appl.,* vol. 39, no. 2, pp. 504–510, 2003.

[14] M.K. Nguyen, T.V. Le, S.J. Park, Y.C. Lim and J.Y. Yoo, "Class of high boost inverters based on switched-inductor structure," *IET Power Electron.,* vol. 8, no. 5, pp. 750–759, 2015.

[15] H. Oleksandr, R.C. Carlos, R.C. Enrique, V. Dmitri and S.Serhii, "Single phase three-level neutral-point-clamped quasi-Z-source inverter," *IET Power Electron,* vol. 8, no. 1, pp. 1–10., 2015.

[16] R.O. Caceres and I. Barbi, "A boost dc–ac converter: analysis, design, and experimentation," *IEEE Trans. Power Electron.,* vol. 14, no. 1, pp. 134–141, 1999.

The 2018 International Power Electronics Conference

Study for further introduction of the Electronic Frequency converters to the Tokaido Shinkansen

Toshimasa Shimizu[1][*], Ken Kunomura[1], Masahiko Kai[2], Hiroki Miyajima[3] and Teruhisa Matsui[4]

1 General Technology Division Technology Research and Development Department, Central Japan Railway Company, Aichi, Japan
2 Construction Department, Central Japan Railway Company, Kanagawa, Japan
3 Railway system Division, Toshiba Infrastructure systems & Solutions Corporation, Kanagawa, Japan
4 Power System Solution and Distribution System R&D Department, Toshiba Corporation, Tokyo, Japan
*E-mail: toshimasa.shimizu@jr-central.co.jp

Abstract— In the Tokaido Shinkansen, frequency conversion systems are used for unification of the feeding electric power frequency. For the base power, rotary frequency changers are used for these tolerances against inrush current or temporary heavy load current. Then, the frequency conversion equipment are mainly composed by the rotary frequency changers. In this report, studies are done for further introduction of the electronic frequency converters, replacing rotary frequency changers. And, progresses for replacements are realized by investigating controls, blocking and re-drive system, with reducing special rating.

Keywords— *Shinkansen; Frequency converter; Inrush current*

I. INTRODUCTION

Although the Tokaido Shinkansen railway line extends over both 50 Hz and 60 Hz electric power system regions in Japan, the rolling stocks are driven by 60 Hz single phase electric power supply. Therefore, in the 50 Hz region, the electric power frequency has to be converted from 50 Hz to 60 Hz. For this purpose, the Tokaido Shinkansen has equipped frequency conversion systems, using the Rotary Frequency Changers (the followings are referred to as "RFC"), from 1964: the Tokaido Shinkansen operation was started. And from 2003, the Electronic Frequency Converters (the followings are referred to as "EFC") are introduced with technical development. Now, the 6-units of the RFCs and 3-units of the EFCs are operating (every set consisted for rating 60 MVA), and realize high density operation with stable electric power supply[1]-[7]. Especially, the EFCs introduction brought us many merits: low-loss, low-cost, flexibility in operation, layout easiness, and so on.

But, the three EFCs are used limited purpose. The two EFCs, called the Tsunashima 4th EFC and the Numazu EFC, were introduced for the enforcement of the RFCs' power system. The last EFC, called the Oi EFC, was introduced to feed a large railyard independently.

For this situation that the existing three EFCs are used in limited purpose, the tolerances against inrush current or temporary heavy load current are the key

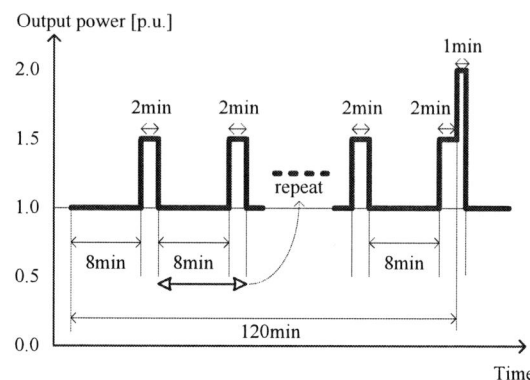

Fig. 1. Special rating of the RFC. The RFC have special rating for load power (1p.u. = 60 MVA) and negative sequence power (1p.u. = 30 Mvar) showed above.

characteristics. As shown in the Fig. 1, the RFC has the tolerances against inrush current or temporary heavy load current, because of the special rating. The RFC can output 1.5 p.u. for 2 minutes with 8 minutes interval, and 2.0 p.u for 1 minutes with 2 hours interval. On the other side, the EFC has no overcurrent tolerances, and output upper 1 p.u. in normal operation. Then, by the comparison of the introduction costs, the EFC haven't been able to take place of the RFC, without the typical situation in feeding railyard (whose load current is milder compared to feeding main line).

But, the further EFCs introductions are continuously wanted, because of the EFC's merits, especially in the maintenance. The stop term for the EFC maintenance only needs one sixth of the stop term for the RFC maintenance. In comparison for the maintenance cost, the EFC has advantage against the RFC for reducing 70%. These are from the difference between the RFC and the EFC, for having moving elements or not.

In Table 1, the comparison between the RFC and the EFC is shown.

Therefore, replacement from the RFC to the EFC are tried, with having the same rated capacity. And, in this report, we made progress, as replacing the one-third of the RFCs to EFCs. Details are described below.

1803

TABLE I
COMPARISON BETWEEN THE RFC AND THE EFC

RFC	Merit	1.	Having tolerance against inrush current or temporary heavy load current which over the normal ratings
	Demerit	1.	Higher maintenance cost and longer term for maintenance
		2.	Need huge water and oil for cooling
		3.	Larger loss in operation
		4.	Difficulty in connecting other power source
		5.	Restriction in layout: Need for large squared space
EFC	Merit	1.	Lower maintenance cost (cutting 70%) and shorter term for maintenance (about 1/6)
		2.	Lower loss in operation
		3.	Flexibility in connecting other power sources
		4.	Flexibility in layout
	Demerit	1.	Little tolerance against inrush current and temporary heavy current which over the normal ratings

II. THE POWER SYSTEM OF THE TOKAIDO SHINKANSEN FREQUENCY CONVERSION AREA

In Fig. 2, the power system of the Tokaido Shinkansen frequency conversion substations area is shown. As described before, the RFCs are the base power sources with the tolerances against inrush current or temporary heavy load current. Three EFCs are installed for the reinforcement of the RFCs' power system. With this policy, there are nine FCs, classified into six sets of the RFCs' power sources and three sets of the EFCs'. Each RFC and EFC unit is rated 60 MVA, so the full rates of the RFCs and EFCs is the 540 MVA.

FC power supply systems are divided into three groups, with one specialized power system of the Oi EFC, for the

Oi railyard power supply. The other two FC power supply systems are for the main line, basically consisted of three RFCs. We call these two FC power supply systems as the Tsunashima and the Nishisagami FC power systems. And, these two FC power systems are reinforced by EFCs, for realize high density operation of trains. For the simplification, these two FC power systems are called as the FC power systems in following sentences.

These setups of the FC power systems in the Tokaido Shinkansen are supposed to operate with one unit maintenance. So, in the each FC power system, operation with two RFCs and one EFC reinforcement are normal situation, and the two RFCs support against inrush current or temporary heavy load current.

III. THE LOADING POWER SITUATION OF THE FC POWER SYSTEMS

In Fig. 3, Fig. 4, Fig. 5, and Fig. 6, estimated positive-phase-sequence current and negative-phase-sequence current are shown, calculated by the density of the train operation. In these figures, 1p.u. means the normal ratings of the each RFC's positive-phase-sequence current and negative-phase-sequence current. The EFC's normal ratings are basically the same as the RFC's ratings in load current, twice above in negative sequence current for output flexibility.

As you can see, because of the temporary heavy load, estimated positive-phase-sequence current and negative-phase-sequence current show the difficulty to operate in normal ratings range. In operation with one set maintenance (two RFCs and one EFC operation), the normal ratings range are refined +3p.u. and -3p.u. line. Although, the power conditioners for negative sequence

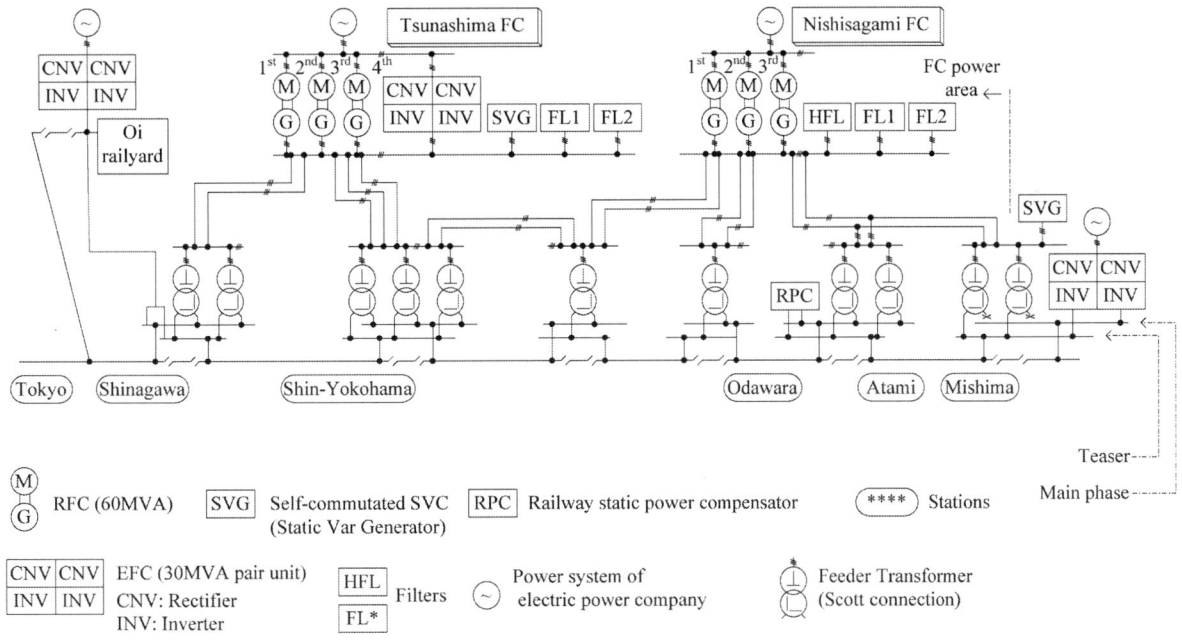

Fig. 2. the power system of the Tokaido Shinkansen frequency conversion substations area.

The 2018 International Power Electronics Conference

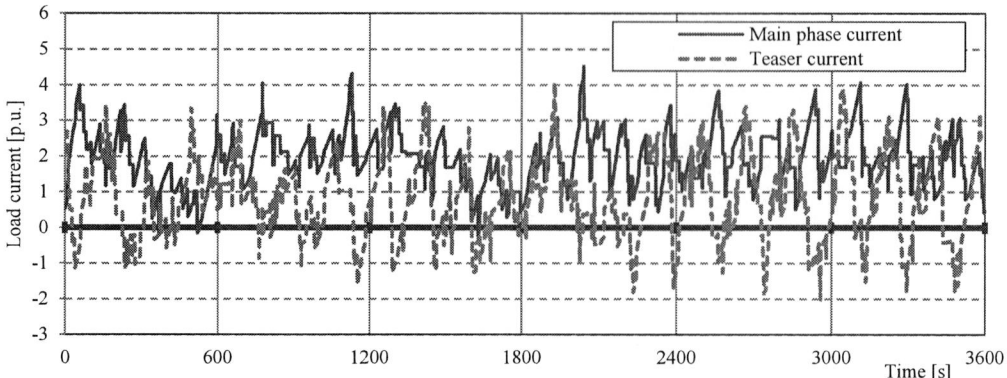

Fig. 3. Estimated positive-phase-sequence current of the Tsunashima FC power system.

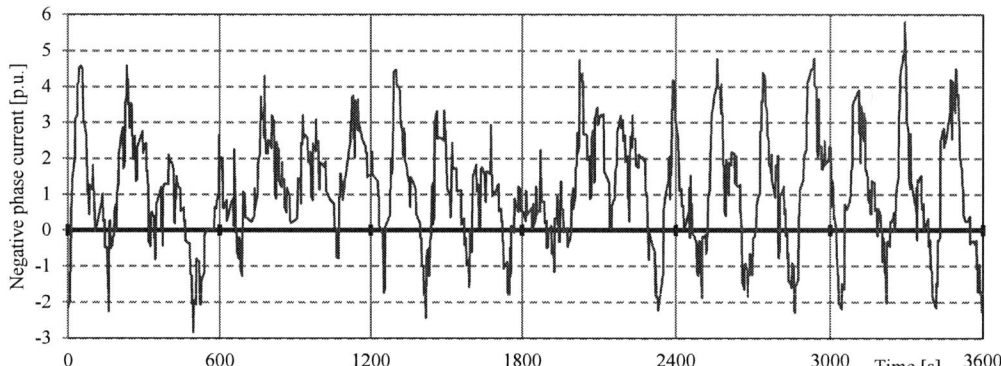

Fig. 4. Estimated negative-phase-sequence current of the Tsunashima FC power system.

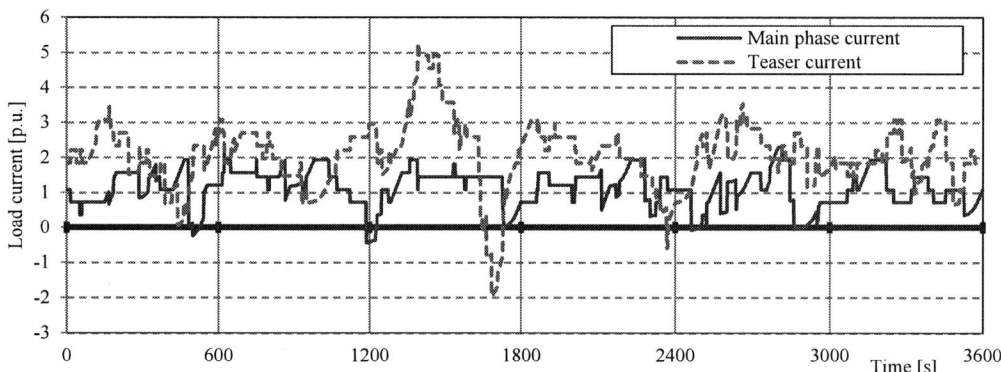

Fig. 5. Estimated positive-phase-sequence current of the Nishisagami FC power system.

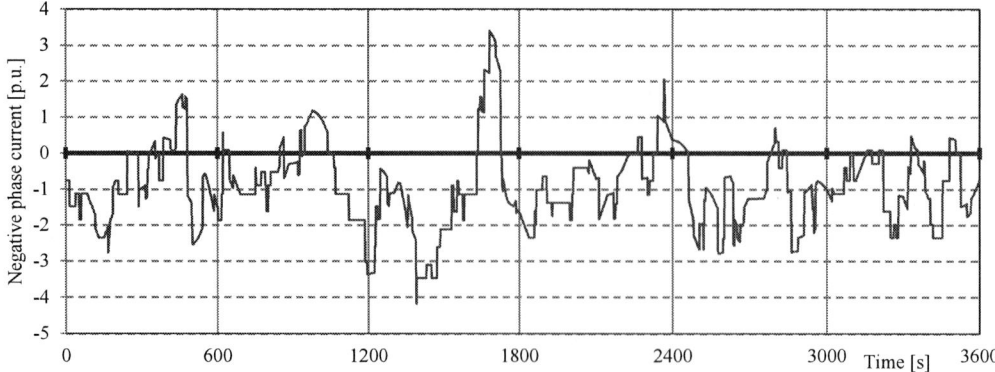

Fig. 6. Estimated negative-phase-sequence current of the Nishisagami FC power system.

1805

power are installed, the capacities of the power conditioners are under 1.7 p.u.. So, the special ratings of the RFCs are necessary in the FC power systems operation. In the Tsunashima FC power system, the power conditioner is consisted by the one self-commutatted SVC (the followings are referred to as "SVG"). And, in the Nishisagami FC power system, the power conditioners are consisted by one railway static power conditioner (the followings are referred to as "RPC") and one SVG in feeding substations.

Then, the special ratings of positive-phase-sequence and negative-phase-sequence current of EFCs must be considered, when we try to replace RFCs to EFCs.

IV. THE INRUSH CURRENT SITUATION OF THE FC POWER SYSTEMS

In another viewpoint, static converters are weak against inrush current. Inrush current mainly occurs by transformer magnetizing inrush. In the Tokaido Shinkansen, static converters including the EFCs are set to avoid inrush current by finding inrush current and protect by gate block operation.

But, the FC power systems have to continue electric power supply against inrush current. And, the stability of the FC power system relies on the RFCs' tolerances against inrush current.

Then, for the replacement of the EFCs from RFCs, the way for continuous electric power supply against transformer magnetizing inrush must be realized.

V. OVERCOMING PLAN

As referred in Chapter III and IV, there are two points to overcome in the replacement from the RFCs to the new EFCs. The first point is the special ratings for temporary heavy positive-phase-sequence current or negative-phase-sequence current, and the second point is the continuous electric power supply against transformers magnetizing inrush currents. These overcoming points are very difficult, and we decided the limitation of the replacement: replace one RFC to one EFC in each FC power system. This means the new FC power system is consisted of two RFC and one new EFC combination, enforced by existing EFCs. The image of the replacement is shown in Fig. 7. And, improvements below are based on this situation. Especially, improvements are done under hard situation, and the combination of one RFC and one new EFC (in one RFC maintenance, and enforced by the existing EFCs) is set for the basic investigating situation.

For the first point to overcome, easy strategy may result to set the new EFC to higher special rating which can output as Fig. 1. But, doubling rating in maximum output power directly causes the converters size up. For the lower cost and downsizing, this situation is not wanted. Then, the best combination of the RFCs and the new EFC output is studied, for the reduction of the new

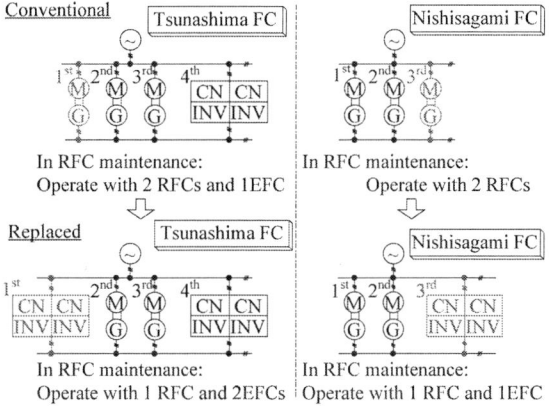

Fig. 7.   Image of the replacement from one RFC to EFC.

EFC's ratings involving special rating, described in Chapter VI.

For the second point, overcoming plans are divided to 2 measures. The first measure is the reduction of the inrush current. The second measure is the protection and re-drive system in hard inrush current situation. The details are described in Chapter VII.

VI. STUDY FOR THE BEST COMBINATION OF THE RFCs AND THE NEW EFC OUTPUTS FOR REDUCTION OF THE NEW EFC'S RATINGS

In the study for the best combination of the RFCs and the new EFC outputs for reduction of the new EFC's special rating, the output apportionment is focused point. And, the control of the equalized output of positive-phase-sequence and negative-phase-sequence power is concluded for suitable. And, this study is based on the operations in special rating range of the new EFC and the RFCs, because of the heavy load situation. The details are below.

Fig. 8 shows the comparison of the ways of the new EFC's output, with the new EFC's special rating shown in Fig. 9. This EFC's special rating is based on the basic rating of 60 MVA, the same ratings of the other RFCs. The ways for the combination of the RFCs and the new EFC were shown in Fig. 10. The comparison in Fig. 8 is done under these situations: In the Nishisagami FC power system, the new EFC operates with one RFC, without the RPC in Atami substation, and with the SVG in the Numazu substation and half unit of the Numazu EFC.

When the new EFC operates with equalized output apportionment of positive-phase sequence power and negative-phase-sequence power, the new EFC basically outputs equally as the RFC's output. On the other hand, when the new EFC operates with priority in negative sequence power output, the new EFC mainly outputs negative-phase-sequence power.

And, in this study, comparison mainly focused on RFC's output current. This is because the new EFC's output should be strictly limited and controlled by special

1806

The 2018 International Power Electronics Conference

rating for the protection of the converters, although the existing RFCs are not. But, the RFCs also should be operate within the special ratings shown in Fig. 1, without instantly heavy load as inrush current.

Investigated result is this: As in Fig. 8, when the new EFC operate with priority in negative-phase-sequence power, the RFC can't operate within the special rating. On the other hand, when the new EFC operate with equalized output apportionment, the RFC can operate

within the special rating. This means that the way to operate with priority in negative-phase-sequence power needs more power, as having the more high ratings of the new EFCs.

These results means that, for reduction of the new EFC's ratings involving special rating, efficient usage of the RFCs ratings in both positive-phase-sequence and negative-phase-sequence power are important. In priority in negative-phase-sequence power output of the EFC, the

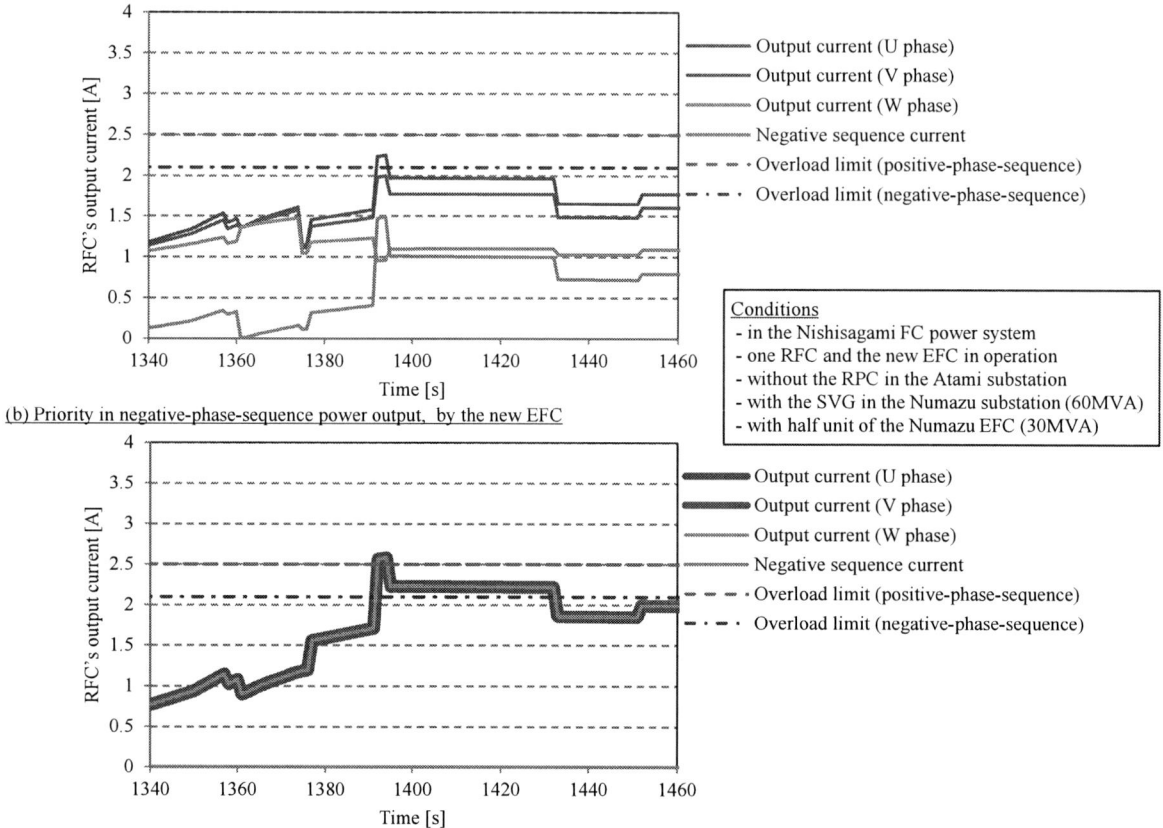

(a) Equalized output apportionment of positive-phase sequence power and negative-phase-sequence power, by the new EFC

Conditions
- in the Nishisagami FC power system
- one RFC and the new EFC in operation
- without the RPC in the Atami substation
- with the SVG in the Numazu substation (60MVA)
- with half unit of the Numazu EFC (30MVA)

(b) Priority in negative-phase-sequence power output, by the new EFC

Fig. 8. Comparison of the output policy with the new EFC's special rating shown in Fig. 9.

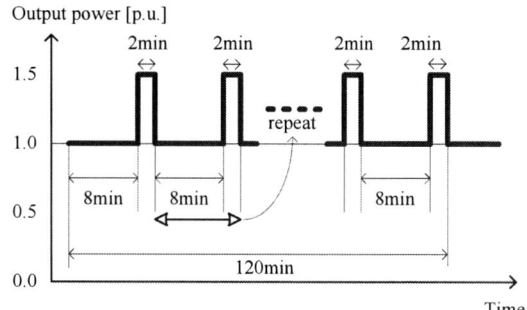

Fig. 9. Special rating of the new EFC. The new EFC have special rating for load power (1p.u. = 60 MVA) and negative sequence power (1p.u. = 30 Mvar) showed above.

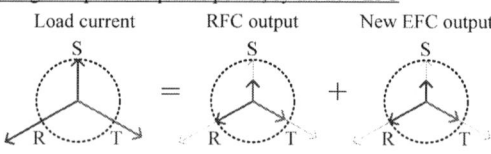

Fig. 10. Ways for the combination of the RFCs and the new EFC output.

1807

RFC's rating of the negative-phase-sequence is used little. On the other hand, in equalized output apportionment in positive-phase-sequence and negative-phase-sequence power, the RFC's output of both positive-phase-sequence and negative-phase-sequence power are used near the ratings.

From this, the new EFC's equalized output apportionment of positive-phase-sequence power and negative-phase-sequence power enables the efficient utilization of the RFCs ratings involving special rating. And, with this way, the new EFC's special rating can be reduced as 90 MVA, shown in Fig. 9.

VII. IMPROVEMENTS FOR THE CONTINUOUS ELECTRIC POWER SUPPLY AGAINST TRANSFORMERS MAGNETIZING INRUSHES

In the study for the continuous electric power supply against transformer magnetizing inrushes, overcoming plans are divided into 2 measures.

For the first measure of the overcoming plans, phase controlled closing systems are introduced for all of the feeder transformers in the FC power systems. With these apparatuses, magnetizing inrush current by the feeder transformers are critically reduced, and the new EFC can keep continuous operation under the reduced magnetizing inrush current.

Although this improvement clear the largest magnetizing inrush currents caused by feeder transformers in the FC power systems, there remains other magnetizing inrush factors by auto transformers and train transformers. For these magnetizing inrush factors, the second overcoming measure is introduced: the blocking and re-drive system.

In new EFC's protection system, the protection policy is based on the RFC operation continuing. This new EFC's protection system are basically operate in main phase and teaser converted from 3 phase. This means to protect in actual load changing, realizing effective and immediate operation. With this protection system, the new EFC will operate in safety range. And when the large magnetizing inrush current occurs, the gate signals of the new EFC's converter will be blocked and the RFC keep the electrical power supply.

The re-drive system of the new EFC is the needed function for continuous operation of whole FC power system. When the blocking of the gate signals of the new EFC operates by a magnetizing inrush current, the RFC should keep operation alone. This situation is the only way now, but this situation can endure for little time. Then, having ten seconds from the blocking of the gate signals of the new EFC's converter, the gate signals are de-blocked and the new EFC re-drives.

In fig. 11, the developed FC power system image for the continuous electric power supply with the new EFC is shown. And, with these improvements, replacement plan of the one RFC to the new EFC in each FC power system is realized.

VIII. CONCLUSIONS

In this report, studies for the further introduction of the EFCs replacing RFCs in the Tokaido Shinkansen are done. The replacement of the one RFC to the new EFC in each FC power system, the Tsunashima FC power system and the Nishisagami FC power system, is realized with reducing the new EFC's special rating which results practically low-cost for introduction.

In the study for the best combination of the RFCs and the new EFC outputs for reduction of the new EFC's special rating, equalized output apportionment of positive-phase-sequence power and negative-phase-sequence power and priority in negative-phase-sequence power are compared. And, the new EFC's equalized output apportionment enables the efficient utilization of the RFCs ratings involving special rating.

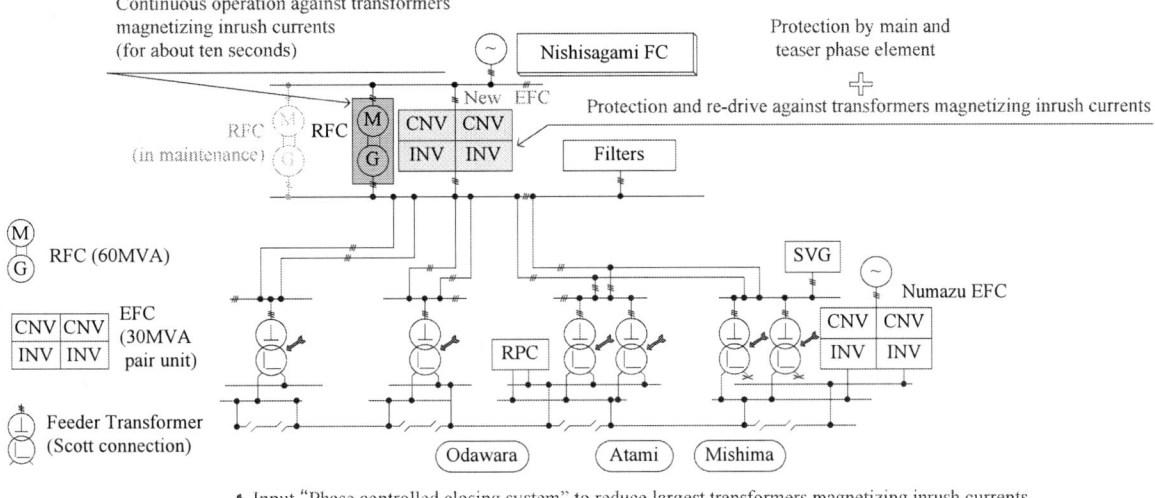

Fig. 11.　Image of the FC power system for the continuous electric power supply with the new EFC.

In the improvements for the continuous electric power supply against transformer magnetizing inrush currents, two overcoming measures are developed. The first is the introduction of the phase controlled closing systems for the largest magnetizing inrush currents by feeder transformers. And, the second is the new EFC's protection system of the blocking and re-drive system.

With these improvements, further EFC usage is stepped ahead, reducing electrical loss, maintenance cost and term. 2 sets of the new EFCs reported here are under constructing, and they must contribute safety and stability operation of the Tokaido Shinkansen.

REFERENCES

[1] K. Ito, et al., "Electronic Frequency Converter", Proc. of the 2004 JIAS Conf., III, pp.347-352(2004)

[2] K. Kunomura, et al., "Electronic Frequency Converter", The 2005 International Power Electronics Conference (IPEC-Niigata 2005), pp. 2187-2191 (2005)

[3] K. Kunomura, et al.: "Electronic Frequency Converter Controls A.C. Voltage Using Fixed Power Factor Method", IEE Japan Trans. IA, Vol. 129, No. 7, 2009, pp. 768-774 (2009)

[4] K. Kunomura, et al., "Electronic Frequency Converter Feeding Single Phase Circuit and Controlling Feeder Voltage with Fixed Power Factor Method for Shinkansen," IEEE Trans. on Power Electronics, vol.27, No.9, pp.3888-3896(2012)

[5] K. Kunomura, et al., "Electronic Frequency Converter that Controls A.C. Voltage using Fixed Power Factor Method", Proc. of Technical Meeting on Power Engineering, IEE Japan, PE-08-171, pp. 37-42 (2008)

[6] K. Kunomura, et al., "Stability analysis of A.C. Voltage Controls for Shinkansen Feeding System with Multiple Power Converters", IEE Japan Trans. IA, Vol. 130, No.4, pp.536-543(2010)

[7] T. Shimizu, et al, "Application of Electronic Frequency Converter to the Shinkansen Railyard Power Supply", IEEJ Journal IA, Vol.4, No.4 pp.315-322(2015)

The 2018 International Power Electronics Conference

Countermeasure for Partial Turn-off of Thyristor Changeover Switch Introduced to Tohoku Shinkansen Shin-Yono Sectioning Post

Yuki Mizumoto[1*] and Nobuhito Kurosawa[1]
1 Tokyo Electrical Construction & System Integration Office, East Japan Railway Company, Tokyo, Japan
*E-mail: y-mizumoto@jreast.co.jp

Abstract— For the first time as East Japan Railway Company, we adopted the thyristor changeover switch in the changeover section of Shin-Yono sectioning post. However, Shin-Yono SP has a characteristic that the stray capacitance of the neutral section is large. Therefore, if a "partial turn-off" phenomenon occurs in the thyristor constituting the changeover switch, in the worst case, it may lead to failure of the equipment. In this paper, we analyzed the conditions under which partial turn-off occurs, focusing on the voltage phase displacement. We also examined the settling and sequence of asynchronous detection relays introduced as a countermeasure.

Keywords— *Shinkansen, neutral section, thyristor change-over switch, partial turn-off*

I. INTRODUCTION

In the Shinkansen (high-speed railway) power supply system, three-phase AC power is transformed at traction substations into single-phase AC power with two phases differing by 90 degrees, and power is supplied to the trains by direction. In substations and sectioning posts, different power sources are butted. Therefore, in order to ensure high-speed operation with keeping the train powering, "changeover section" is installed which switches the power supply to the train instantaneously (within 300ms) with changeover switch. Since the changeover switch performs opening/closing every time it passes through the train, it requires high reliability with frequent operation. Conventionally, a vacuum changeover switch (hereafter, VS) that mechanically opens/closes an electric circuit in a vacuum valve has been used for a changeover section. Currently, a thyristor changeover switch (hereafter, THS) using a thyristor element has been developed, and introduced to the substations/sectioning posts of the Tokaido Shinkansen since 2005 [1], [2]. Figure 1 shows the operation of THS in the changeover section. The THS electrically opens/closes the circuit by 12 series thyristors. Therefore, it has the advantage of being maintenance free and usable for a long time since THS does not cause mechanical abrasion.

We introduced THS for the first time in East Japan Railway Company (hereafter, JR East) in Shin-Yono sectioning post (hereafter, Shin-Yono SP) where we

renewed substation equipment, and it has been used since August 2017. In introducing THS to Shin-Yono SP, there was concern that the "partial turn-off" phenomenon of the thyristor leads to damage of equipment. This study aims to analyze the condition of occurrence of partial turn-off for THS introduced to Shin-Yono SP and to examine the interlocking relay and sequence as the measures against equipment failure.

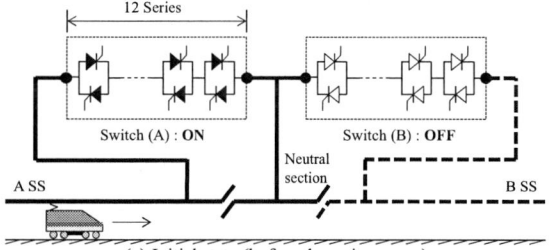
(a) Initial state (before the train passes)

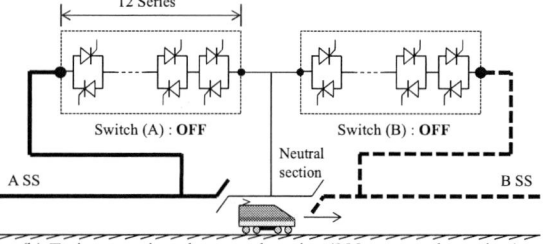
(b) Train enters into the neutral section (300ms no-voltage time)

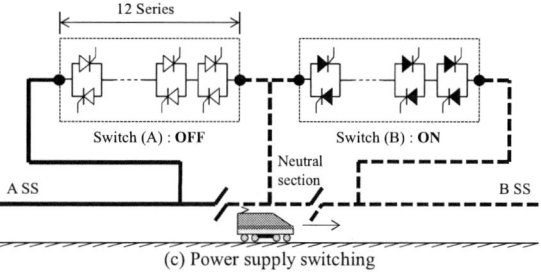
(c) Power supply switching

Note: Conventional type consists of two pairs of VS (A, B).
Fig. 1. Operation of thyristor changeover switch in the changeover section.

1810

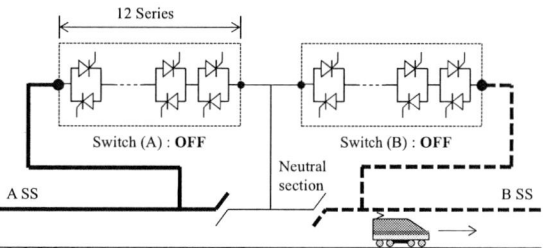

(d) Train leaves from the neutral section (The switch returns to the initial state)

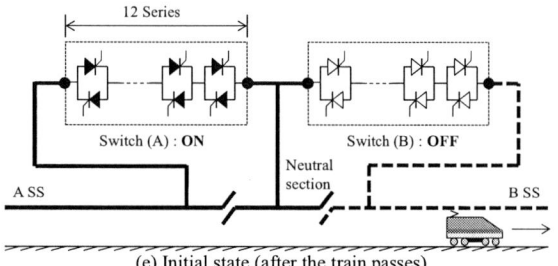

(e) Initial state (after the train passes)

Note: Conventional type consists of two pairs of VS (A, B).
Fig. 1. Operation of thyristor changeover switch in the changeover section. (Continuation)

II. INTRODUCTION OF THS TO SHIN-YONO SP

Figure 2 shows the Shinkansen route map of JR East, and Figure 3 shows the feeding system diagram around Shin-Yono SP. Shin-Yono SP is the sectioning post that all the Shinkansen lines (Tohoku/Joetsu/Hokuriku /Yamagata/Akita) of the JR East area pass, and is an overcrowded section where about 15 trains (one way) pass each hour in the peak time. Therefore, we spend much effort on VS maintenance. For example, we inspect the mechanism parts and breaking parts of the VS in about 1.6 years (when the VS operation is 50,000 times). Next, we replace the vacuum valve of the VS in about 3.2 years (when the VS operation is 100,000 times). Finally, we replace the VS of Shin-Yono SP in about 6.3 years (when the VS operation is 200,000 times). On the other hand, the THS that is electrically opened/closed by thyristors has an expected life of 30 years irrespective of the number of operations, and it is maintenance free except for some replacement parts. Therefore, THS was adopted for the first time in the equipment replacement of Shin-Yono SP.

III. STRAY CAPACITANCE OF THE NEUTRAL SECTION

THS is designed on the premise that the "stray capacitance" of the neutral section of the installation site is small enough. This is because that when the THS is used in a section where the stray capacitance of the neutral section is large, partial turn-off phenomenon may occur due to the dispersion of individual thyristor element characteristics.

Measurement results of the stray capacitance of the feeding circuit at the test line before the opening of the Tohoku Shinkansen were reported in reference [3].

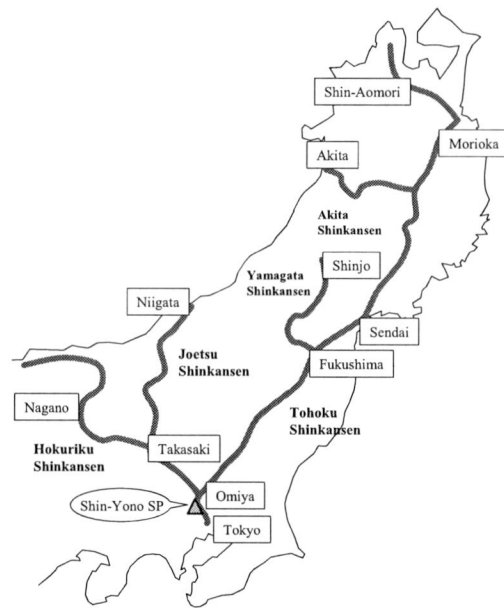

Fig. 2. Shinkansen route map of East Japan Railway Company

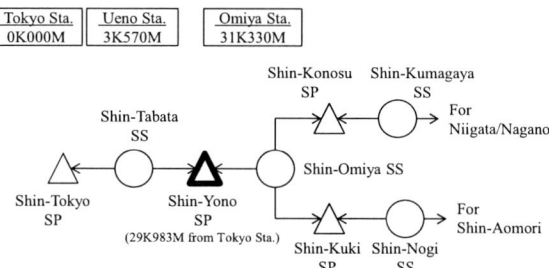

Fig. 3. Feeding system diagram. (From Shin-Tokyo SP to Shin-Nogi SS / Shin-Kumagaya SS)

However, Shin-Yono SP which renewed substation equipment was supposed to have a large stray capacitance of the neutral section for reason as follows;

(1) The locations of the neutral section and the SP are about 2km apart.
(2) It is composed of GIS and cable (20m or more) in order to reduce the live part of SP.

Therefore, we applied voltage to the neutral section with a generator and measured the leakage current to estimate the stray capacitance. As a result, the stray capacitance of the neutral section of Shin-Yono SP was estimated to be 0.052μF, and it turned out that it exceeded the assumed value of THS equipment design. Therefore, based on the measurement result, it was necessary to verify whether partial turn-off occurs.

IV. ABOUT PARTIAL TURN-OFF PHEOMENON

The THS is composed of 12 series thyristors connected in inverse-parallel. The thyristor has a characteristic that it does not maintain the ON state after the gate signal is turned off unless the gate signal is turned on and current is supplied that is stronger than the "latching current". This latching current inevitably varies due to individual differences for each thyristor.

Here, in order to briefly explain the partial turn-off phenomenon, we assume a case where two thyristors are connected in series (Th1, Th2). The latching current of Th1 is defined as I_{Th1} and the latching current of Th2 is defined as I_{Th2} ($I_{Th1} < I_{Th2}$). Whether partial turn-off occurs or not depends on the current value I and the timing of turning off the gate signal. Figure 4 shows the thyristor current at gate signal off.

(1) As shown in Figure 4(a), when the gate signal turns off at the timing of I_{Th1}, $I_{Th2} > I$, both Th1 and Th2 are turned OFF immediately since the current value is less than the latching current.

(2) As shown in Figure 4(b), when the gate signal turns off at the timing of $I_{Th1} < I < I_{Th2}$, Th2 turns OFF immediately because the current value is less than the latching current I_{Th2}. On the other hand, Th1 continues the ON state because the current value is not less than the latching current I_{Th1}. As described above, a phenomenon in which a part of the thyristors connected in series is turned OFF and the remaining thyristors are kept ON is called partial turn-off.

(3) As shown in Figure 4(c), when the gate signal turns off at the timing of I_{Th1}, $I_{Th2} < I$, both Th1 and Th2 continue the ON state because the current value is not less than the latching current, and no partial turn-off occurs .

(4) As shown in Figure 4(d), when the gate signal turns off after $I_{Th1} < I$ peak value $< I_{Th2}$ and the peak value has passed, Th2 turns OFF immediately because the current value is less than the latching current I_{Th2}. On the other hand, since the current value is greater than the latching current I_{Th1}, Th1 continues to be ON, and partial turn-off occurs.

(5) As shown in Figure 4(e), in the case of the current I peak value $< I_{Th1}$, I_{Th2}, partial turn-off does not occur regardless of the timing of the gate signal off.

Normally, the switches of THS share the feeding voltage by 12 thyristors. However, if a partial turn-off occurs, total voltage is applied to the thyristor turned OFF earlier. As a result, it exceeds the specified withstand voltage of the thyristor, which may lead to breakage of the equipment in the worst case.

V. ABOUT PARTIAL TURN-OFF PHEOMENON

A. Calculation of Partial Turn-off

When there is a train load, the train load current occupies most of the thyristor current, which is very large compared to the latching current. If the gate signal is turned off near the zero point of the train load current, partial turn-off may occur. Therefore, the off phase of gate signal is set so as to avoid the zero point of the train load current, and partial turn-off is prevented.

When there is no train load (Figure 1 (d) switch (B) ON → OFF) and the stray capacitance of the neutral section is sufficiently small, partial turn-off does not occur at any timing off since the peak value of the

thyristor current is less than the latching current. However, since the stray capacitance of the neutral section of the Shin-Yono SP is large, even if there is no train load, there is a possibility that the passing current peak value may exceed the latching current.

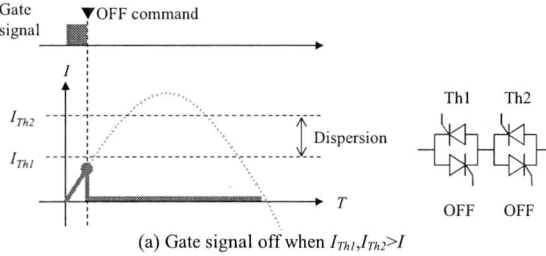

(a) Gate signal off when I_{Th1},I_{Th2}>I

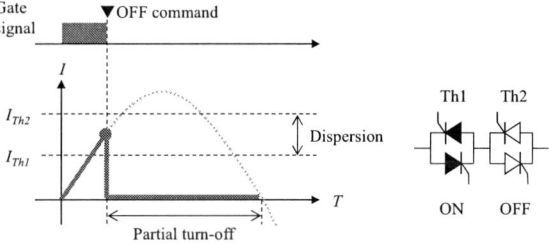

(b) Gate signal off when I_{Th1}<I<I_{Th2} (Partial turn-off)

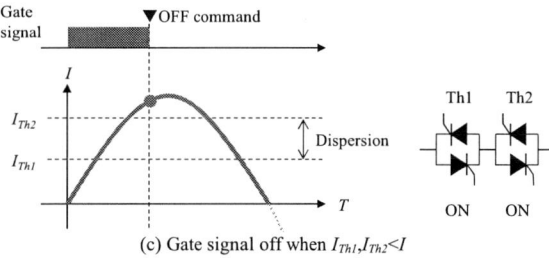

(c) Gate signal off when I_{Th1},I_{Th2}<I

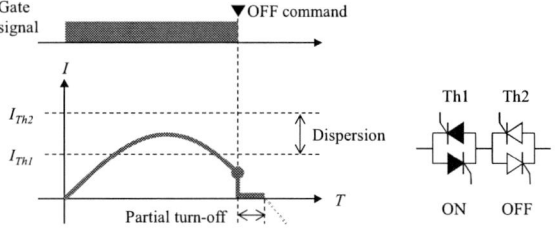

(d) Gate signal off after passing the peak value (Partial turn-off)

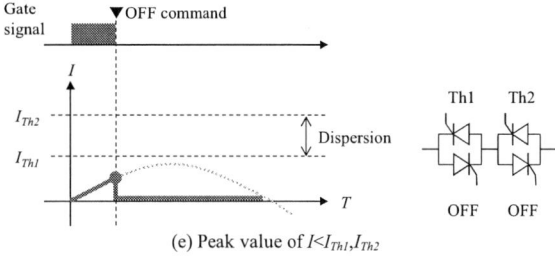

(e) Peak value of I<I_{Th1},I_{Th2}

Fig. 4. Thyristor current at gate signal off.

The 2018 International Power Electronics Conference

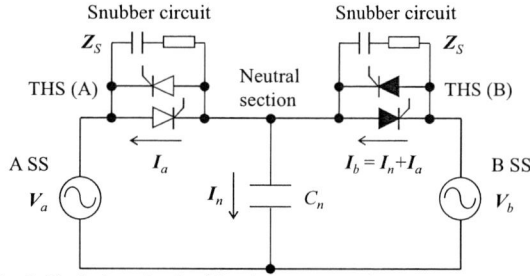

Fig. 5. Equivalent circuit of the changeover section.

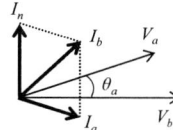

Fig. 6. Vector diagram of V and I.

Therefore, we analyzed the partial turn-off occurrence condition by calculating the thyristor current using the equivalent circuit of the section. Figure 5 shows the equivalent circuit when the train leaves the section and switch (B) turns from ON to OFF. Here, V_a and V_b are feeding voltages from substations on both sides of SP, C_n is stray capacitance of the neutral section, and Z_s is snubber circuit impedance. In Figure 5, the current I_b flowing through the thyristor of switch (B) is expressed by the formula (1), as the sum of the leakage current I_n flowing through the stray capacitance of the neutral section and the current I_a flowing through the snubber circuit of switch (A).

$$I_b = I_n + I_a = j\omega C_n V_b + \frac{V_b - V_a e^{j\theta_a}}{Z_s} \quad (1)$$

Figure 6 shows the vector diagram of V and I. Here, θ_a is the phase displacement of V_a with reference to V_b (lead is positive). In the case where the calculation result of the instantaneous current I_b at gate signal off corresponds to Figure 4 (b) or (d), it is determined that the partial turn-off occurs in the THS.

In formula (1), stray capacitance of the neutral section C_n and snubber circuit impedance Z_s are fixed constants. Therefore, we analyze the partial turn-off occurrence condition with the feeding voltage and the voltage phase difference as parameters. Table I shows the main calculation conditions.

B. Calculation Results

Table II shows the calculation results of the phase displacement θ_a where partial turn-off occurs for the combination of V_a, V_b and f. Calculation results are summarized below.

TABLE I
MAIN CALCULATION CONDITIONS

Symbol	Meaning	Value
V_a, V_b	Feeding voltage	22.5-30.0 kVrms
θ_a	Phase displacement of V_b to V_a	0 ± 120 deg
f	Frequency	50 Hz ± 1%
C_n	Stray capacitance of neutral section	0.052μF (measured value)
I_L	Latching current (Max/Min)	Specification of THS
Z_S	Snubber circuit impedance	Specification of THS
θ_{OFF}	Gate signal off phase	Specification of THS

TABLE II
CALCULATION RESULTS OF THE PHASE DISPLACEMENT WHERE PARTIAL TURN-OFF OCCURS

No.	V_a [kV]	V_b [kV]	f [Hz]	Phase displacement θ_a where partial turn-off occurs [deg] (Lead+/Lag-)		
				-7.5 deg error in gate signal off phase	No error in gate signal off phase	+7.5 deg error in gate signal off phase
1	22.5	22.5	49.5	None	None	None
2	22.5	22.5	50	None	None	None
3	22.5	22.5	50.5	None	None	None
4	22.5	25.5	49.5	None	None	None
5	22.5	25.5	50	None	None	None
6	22.5	25.5	50.5	None	None	None
7	22.5	30.0	49.5	None	None	None
8	22.5	30.0	50	None	None	None
9	22.5	30.0	50.5	None	None	-120～-105
10	25.5	22.5	49.5	111～116	None	None
11	25.5	22.5	50	109～112	None	None
12	25.5	22.5	50.5	107	None	None
13	25.5	25.5	49.5	94～108	None	None
14	25.5	25.5	50	93～103	None	None
15	25.5	25.5	50.5	91～97	None	None
16	25.5	30.0	49.5	73～92	None	-120～-108
17	25.5	30.0	50	71～83	None	-120～-99
18	25.5	30.0	50.5	None	None	-120～-93
19	30.0	22.5	49.5	103～120	None	None
20	30.0	22.5	50	101～120	None	None
21	30.0	22.5	50.5	100～117	None	None
22	30.0	25.5	49.5	89～118	89	None
23	30.0	25.5	50	88～114	None	-120～-109
24	30.0	25.5	50.5	87～110	None	-120～-102
25	30.0	30.0	49.5	72～108	None	-120～-90
26	30.0	30.0	50	71～103	-120～-112	-120～-86
27	30.0	30.0	50.5	69～97	-120～-99	-120～-82

1813

(1) From the relationship between I_b and I_L, partial turn-off occurs when -120deg $\leqq \theta_a \leqq$ -99deg when the gate signal turns off. Considering the error (±7.5deg) of the gate signal turns off timing, partial turn-off occurs when -120deg $\leqq \theta_a \leqq$ -82deg and +69deg$\leqq \theta_a \leqq$+120deg.

(2) In the range of -67deg $< \theta_a <$ +68deg, partial turn-off does not occur regardless of the gate signal off phase since the peak value of I_b is smaller than the I_L minimum value. However, if the gate signal off phase is not controlled due to a control power supply failure, partial turn-off may occur within the range of -120deg$\leqq \theta_a \leqq$ -67deg and +68deg$\leqq \theta_a \leqq$ +120deg since the peak value of the thyristor current exceeds the minimum latching current value.

As an example of the calculation result, Figure 7 shows a combination of V_b and θ_a where partial turn-off occurs when V_a=30kV (fixed), V_b=22.5-30kV and θ_a=0±120deg. Figure 7 shows that partial turn-off occurs when the feeding voltage and phase displacement is located in the gray area. From the calculation results, partial turn-off occurrence depends largely on the phase displacement of feeding voltage. Therefore, in order to judge occurrence of partial turn-off, it is necessary to pay attention to the phase displacement of the feeding voltage.

C. Voltage Phase Measurement in Shin-Yono SP

In the previous section, it was shown that the partial turn-off depends on the phase displacement of feeding voltage. Therefore, we measured the phase to grasp the phase displacement of the actual Shin-Yono SP. Table III shows the measurement condition, and Figure 8 shows the result of voltage and phase displacement in Shin-Yono SP.

In the case of normal feeding, the power sources of the same phase of the traction transformer (scott connected transformer or roof-delta connected transformer) are butted at SP so that no phase displacement occurs. In the measurement result, the phase displacement was about 10 degrees at the maximum due to the influence of the train load and the upper power supply system.

On the other hand, if Shin-Omiya SS drops out in on accident, it is assumed that the power supply is secured by extended feeding from Shin-Nogi SS or Shin-Kumagaya SS. In this case, since the power sources of different phases are butted at the Shin-Yono SP, the phase displacement is around 90 degrees.

In consideration of the calculation result, the partial turn-off does not occur during normal feeding operation because the phase displacement is small, so THS can be used without problems. However, when an accident causes extended feeding from Shin-Nogi SS or Shin-Kumagaya SS, phase displacement of around 90 degrees occurs at Shin-Yono SP, and there is a possibility that partial turn-off occurs in THS.

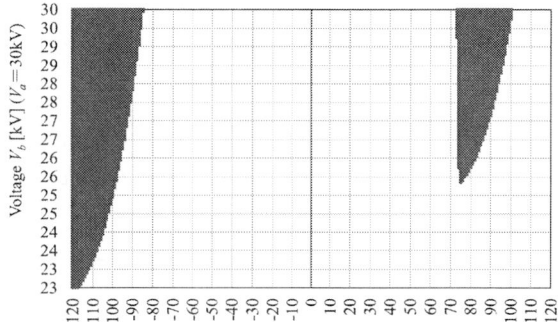

Fig. 7. The area of partial turn-off occurrence. (V_a=30kV, V_b=22.5-30kV, θ_a=0±120deg)

TABLE III
PHASE DISPLACEMENT MEASUREMENT CONDITION IN SHIN-YONO SP

No.	System	For Tokyo	For Omiya	Frequency
1	Normal	Shin-Tabata SS (M)	Shin-Omiya SS (A)	Everyday
2	Extend I	Shin-Tabata SS (M)	Shin-Nogi SS (T)	Accident (Very rare)
3	Extend II	Shin-Tabata SS (M)	Shin-Kumagaya SS (T)	Accident (Very rare)

*1 "Normal" is the 30-second interval measurement data from December 15, 2016 to January 10, 2017.
*2 "Extend" is data measured for 30 minutes during the night of March 10, 2017. Therefore, the phase difference may fluctuate more during the daytime with a train load.
*3 Phase difference is based on the Omiya side. (Lead+/Lag-)
*4 M: Main phase of scott connected transformer
   T: Teaser of scott connected transformer
   A: A phase of roof-delta connected transformer

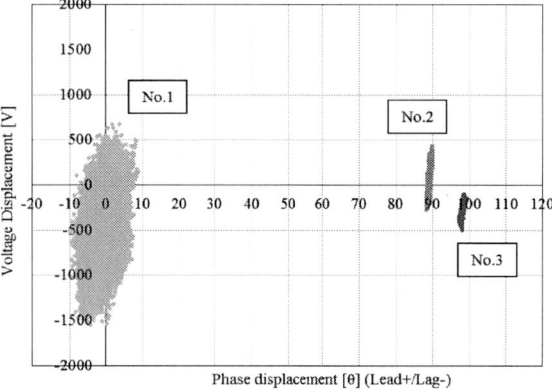

Fig. 8. Configuration of active/standby changeover switch in the changeover section. (Reference : Shin-Omiya SS side)

VI. MEASURES TO AVOID PARTIAL TURN-OFF

Neutral section is standby duplex system, and when failure occurs in the THS, it switches to the standby unit VS to ensure redundancy. In addition, when both the active unit THS and the standby unit VS become unusable, extended feeding system is constructed by automatic sequence, and power supply is secured. Figure 9 shows the configuration of the active/standby unit in the changeover section.

As mentioned above, in order to prevent THS failure due to partial turn-off, it is necessary to put the THS into disuse mode at the time of extended feeding from Shin-Nogi SS or Shin-Kumagaya SS generating a phase displacement of 90 degrees, and to use the standby VS.

Therefore, we installed a protection relay (asynchronous detection relay, 25F) on the control panel for monitoring the voltage phase as shown in Figure 9. And we set up a sequence to disable THS when the phase displacement of butted power supply detected by 25F relay exceeds a certain level.

Figure 10 shows the 25F relay characteristics and settling. First, the requirement of the phase displacement detection element 25F-θ is as follows;

(1) To reliably detect and operate the phase displacement range where partial turn-off occurs.

(2) Do not unnecessarily detect during normal operation. The settling value of 25F-θ is 60 degrees based on the calculation and measurement results. Next, the settling value of 25F-UV is 60V (corresponding to 18kV in 30kV system) in consideration of operation only when there is a certain voltage. Finally, the settling value of 25F-T is 3.0sec in order to prevent instantaneous unnecessary detection. This is because partial turn-off occurs in the case of extended feeding from Shin-Nogi SS or Shin-Kumagaya SS, and it is assumed that the phase displacement will not instantaneously increase to the partial turn-off area during normal feeding operation.

The main sequences related to 25F relay are as follows. It is possible to prevent the use of the THS under the condition that the partial turn-off occurs, and to prevent the failure of the thyristor.

(1) Power supply by normal feeding;

Since 25F relay does not operate, THS (active unit) can be closed. If the switching failure of the THS occurs, it automatically switches to the VS (standby unit).

(2) Power supply by extended feeding from Shin-Nogi SS or Shin-Kumagaya SS;

Since 25F relay is in operation, THS cannot be closed and VS is closed. In this state, if switching failure of the VS occurs, both changeover switches are disabled.

(3) Power supply by extended feeding from Shin-Nogi SS or Shin-Kumagaya SS, but THS has already closed;

Since 25F relay is in operation, THS is regarded as failure and it switches automatically to VS.

(4) Power supply by extended feeding from Shin-Nogi SS or Shin-Kumagaya SS, but VS has already closed;

25F relay is in operation, but it continues to operate with VS as it is.

(5) When the phase displacement exceeds the set value in normal feeding (this will not occur normally);

As in (3), it automatically switches to the VS.

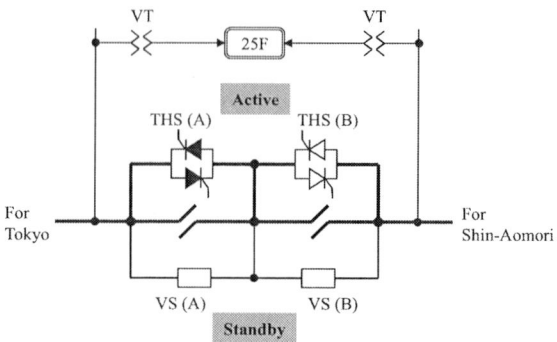

Fig. 9. Configuration of active/standby changeover switch in the changeover section.

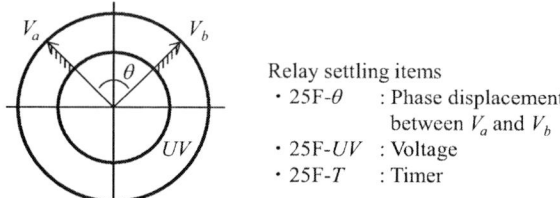

Relay settling items
- 25F-θ : Phase displacement between V_a and V_b
- 25F-UV : Voltage
- 25F-T : Timer

Fig. 10. 25F relay characteristics and settling.

VII. Conclusions

In this paper, in order to introduce THS into Shin-Yono SP with large stray capacitance of the neutral section, we analyzed the condition under which the partial turn-off phenomenon occurs. And we showed that the partial turn-off occurrence condition can be distinguished by the feeding voltage phase displacement.

In addition, we measured the phase displacement of Shin-Yono SP by field measurements. As a result, although partial turn-off cannot occur in normal feeding operation, it was found that there is a possibility that the phase difference is large and the partial turn-off may occur in extremely rare operation such as in an accident. Therefore, we installed an asynchronous detection relay to monitor the phase displacement on the control panel, and examined relay setting and sequence. THS started to be used at Shin-Yono SP in August 2017, and it has contributed to the stable transportation of JR East's Shinkansen.

References

[1] Ken Kunomura et al., "Development of Static Changeover Switch for Shinkansen" 2006 National convention record, IEEJ, 5-184, 2006.

[2] Naoki Suzuki et al., "Introduction of the Thyristor Changeover Switch to Tokaido Shinkansen Whole Line" 2015 IEEJ Industry Appl. Soc. Conference, 5-5, 2015.

[3] Mochinaga et al., "Tohoku Shinkansen Oyama General Test Line Electrical Circuit Constant Measurement Test [Published in Japanese.]" Railway Technical Research Institute, 78-181, 1978.

Hardware-in-the-loop Real-time Simulation Experiment Platform for Traction Power Supply System Based on dSPACE-Xsim

Runze Zhang[1*], Fei Lin[1], Zhongping Yang[1], Hu Cao[2] and Yuping Liu[2]
1 School of Electrical Engineering, Beijing Jiaotong University, Beijing, China
2 CCRC Qingdao Sifang Rolling Stock Reseaarch Institute Co. Ltd., Qingdao, China
*E-mail: 16126069@bjtu.edu.cn

Abstract- **Pulse Width Modulation technology is widely used in High-Speed Railway. The harmonic distribution caused by this technology is extensive, the characteristic harmonics are quite obvious, and it influences the stability of grid-train coupling relationship. In the study of the relationship between traction grid and train, matlab off-line simulation technology is mostly adopted. But this method has the disadvantages of slow simulation speed, it can not reflect the control algorithm and operation state of the train. This paper adopts the hardware-in-the-loop real-time simulation technology, which contacts the dSPACE simulator with actual controller to simulate the train and verify the control algorithm, and the Xsim simulator is used to build the traction grid model at the same time. Finally, two devices are combined to simulate the train operation state on traction power supply system in real time, so as to prepare for the development of the train algorithm.**

Keywords—traction power supply system（TPSS）; real-time simulation; dSPACE;Xsim

I. INTRODUCTION

Sinusoidal pulse width modulation (PWM) technology has been used in electrified railway system for more than 20 years, this technology produce high frequency harmonics in the train traction power supply system and it will be injected into the traction grid through the pantograph [1],[2]. The traction power supply system (TPSS) can be equivalent to a distributed parameter system, when the transformer leakage inductance is matched with the traction grid impedance and the frequency of harmonic current which injects into the traction grid is equal to the resonance frequency of the traction grid, it will lead to the phenomenon of harmonic amplification, which will cause the distortion of the grid voltage. For example, there has been a resonance accident when China CRH-380 was running in the Beijing-Shangai pilot test, which resulted in train arrester burned and tranction grid lossd voltage with a short time.

In the offline simulation modeling and analysis of traction grid，references [4]-[7] proposed several models, such as Norton, crossed frequency-coupled matrix approaches to describe the external features of nonlinear loads for harmonic studies. Different load types will affect the impedance analysis of TPSS. There are many analysis methods for TPSS, such as simulation analysis [8], frequency scan [9], Modal assessment [10],

etc. Reference[11] establishes the distribution parameter model of the traction grid and the equivalent Norton model of the train, and it uses Modal analysis theory to evaluate the resonance of TPSS. The above methods have achieved some progresses in off-line simulation, but it is limited by the simulation speed and can not simulate the operation state of the train on the line. There are disadvantages in the verification of the train algorithm. In this paper, an equivalent five conductor model of traction grid is established by using multi conductor chain grid model on Matlab, and it is transplanted to the hardware in the loop simulation machine--Xsim. After that, it will be combined with the existing dSPACE high speed train traction drive system simulation equipment to finish the grid-train coupling experiment.

In the second part of this paper, it introduces the mathematical modeling method of traction grid on the basis of the multi conductor chain grid model theory and analyzes the impedance characteristics of the system. The third part will introduce the joint simulation experiment platform based on dSPACE-Xsim and the logic mode which the train has real-time operation on the system. The fourth part will carry out experimental verification of the experimental platform. Finally, it will provide a summary of the whole paper.

II. MATHEMATICAL MODELING OF TRACTION POWER SUPPLY SYSTEM

A. Traction Grid Mathema tical Modeling[13]

In high speed railway system, the traction grid consists of several parallel transmission lines, such as contact wire, positive feeder, rail, bearing cable and so on. The AT(auto transformer) double track traction power supply system is shown in Fig. 1.

For facilitating the analysis, many scholars proposed multi conductor chain grid model to separate the traction grid for different lengths. In order to investigate the harmonic amplification and resonance conditions, it is necessary to use a sufficiently accurate distribution parameter model, which is related to the distributed capacitance of the line. According to the traction grid structure, the system can be divided into different sections. In this paper, the length of line is selected into 1km to divide section and the equivalent chain circuit is shown in Fig. 2. In Fig.2, Z and Y are the admittance and impedance matrixes in multi conductor chain grid model, their parameters are determined by the material and sectional area of the conductors. The longitudinal current I is a current source in the system, such as a high speed train.

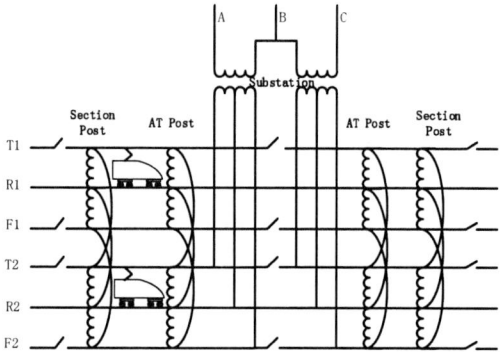

Fig. 1. AT traction power supply system diagram

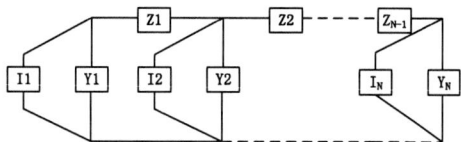

Fig. 2. Equivalent chain circuit

According to the node position of chain grid model and electrical network theory, the node admittance equation can be obtained as follows:

$$Y = \begin{bmatrix} Y_1 + Z_1^{-1} & -Z_1^{-1} \\ -Z_1^{-1} & Z_1^{-1} + Y_2 + Z_2^{-1} \\ & & \ddots & & \ddots \\ & & & Z_{N-2}^{-1} + Y_{N-1} + Z_{N-1}^{-1} & -Z_{N-1}^{-1} \\ & & & -Z_{N-1}^{-1} & Z_{N-1}^{-1} + Y_N \end{bmatrix} \quad (1)$$

The above calculation parameters can be applied to the simulation model.

Assuming that the number of parallel conductors in the traction grid is m, and the spatial distribution of the traction grid is known, the impedance and admittance matrix of the unit length can be calculated according to the complex depth theory and Carson theory. The matrices are m-order, the unit length π type equivalent circuit is shown in Fig. 3.

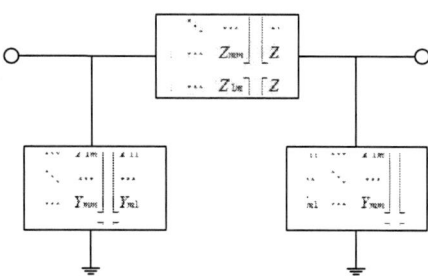

Fig.3. π type equivalent circuit with m conductor

B. Impedance Characteristic Analysis of Traction Grid

In high speed railway system, many traction grids have adopted parallel 2×27.5kV 50Hz autotransformer (AT)-fed system. According to the parameters of a power supply interval, the impedance and admittance matrix of the traction power supply system can be calculated, and the 12-conductor model of the system is equivalent to 5-conductor simplified model [12] [13]. In this paper, the power supply interval is 48km, and the traction substation is located at the left of the line. The autotransformer is set at every 12km. The whole line consists of 48 1km 5-conductor π type equivalent circuits. Combining with the above calculation results, the equivalent model of traction grid can be constructed in Matlab-simulink, which can effectively and accurately reflect the impedance characteristic. In the absence of special statement, this paper will use the above line conditions.

According to the traction grid model, the impedance and phase angle analysis of the entire traction grid can be obtained from every kilometer point by using the frequency sacn method which is showed in Fig. 4 and Fig. 5.

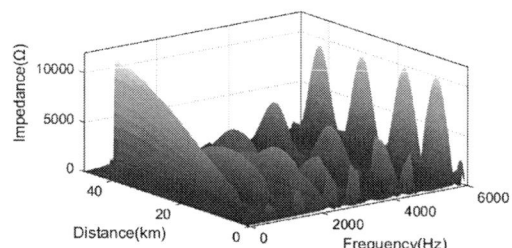

Fig. 4. Impedance analysis of traction grid

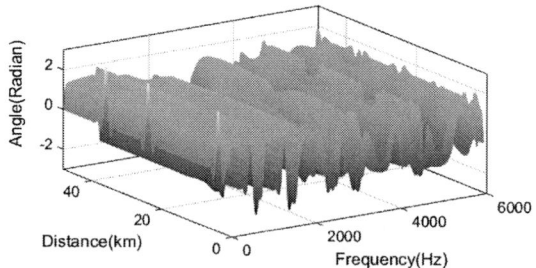

Fig. 5. Phase angle analysis of traction grid

It can be seen from the above two figures that the whole system is a multi resonance peak system from any point of traction grid (there are six obvious resonant peaks in the system). From the frequency axis side of Fig. 4, the frequency-impedance relation diagram is shown in Fig. 6.

1817

The 2018 International Power Electronics Conference

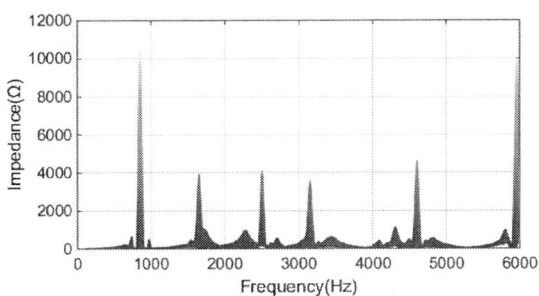

Fig. 6. Frequency-impedance relation diagram

From the above figure, the resonant frequency is independent of the line position and the maximum impedance value of each resonance point is different. When the phase angle changes from the positive value to the negative value, it indicates that the parallel resonance is occurred in this frequency. For the same resonant frequency, the impedance value varies with the distance from the substation, and the times which reaches the maximum impedance value of system are different.

III. JOINT SIMULATION EXPERIMENTAL PLATFORM BASED ON DSPACE-XSIM

A. Introduction of joint simulation experiment platform

In the TPSS, the train receives electricity from the traction grid through the pantograph, and the train traction drive system injects the current into the traction grid through it which is showm in Fig.7. In the actual train, the transformer usually contains multiple secondary windings, and only one group is drawn in the diagram.

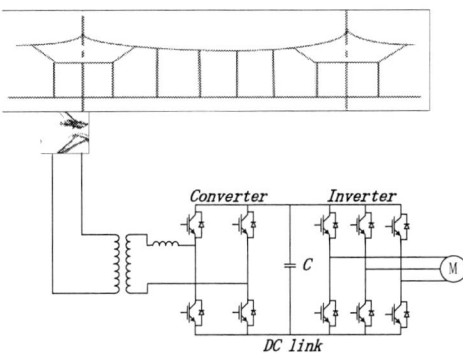

Fig. 7. Diagram of traction drive system

In the simulation modeling of traction grid, an equivalent 5 conductor model is built for every 1km. Because of the demand of the line length, the CPU sub-core parallel computing method is needed in the simulator to complete the simulation modeling of the traction grid. The performance of Xsim real-time simulator is better than dSPACE in parallel calculation, therefore, the traction grid model is set up in the Xsim simulator, and the train traction drive system model is built in dSPACE. The analog of the train running on the line is completed through the signal transmission between two devices.

According to the signal transmission relationship between the train and the traction grid in Fig. 7, when the joint simulation experiment is carried out, the Xsim simulator needs to transmit the traction grid voltage signals to the dSPACE through the AO(Analog Output) channel, and dSPACE receives voltage signals through the AI(Analog Input) channel, then, the voltage signal will input to the controlled voltage source to restore traction grid voltage. The transmission of the current signals is similar to the voltage. The above procedure is shown in Fig. 8. The interface signal delay of the device is in microsecond level, so it has minimal influence on the joint simulation.

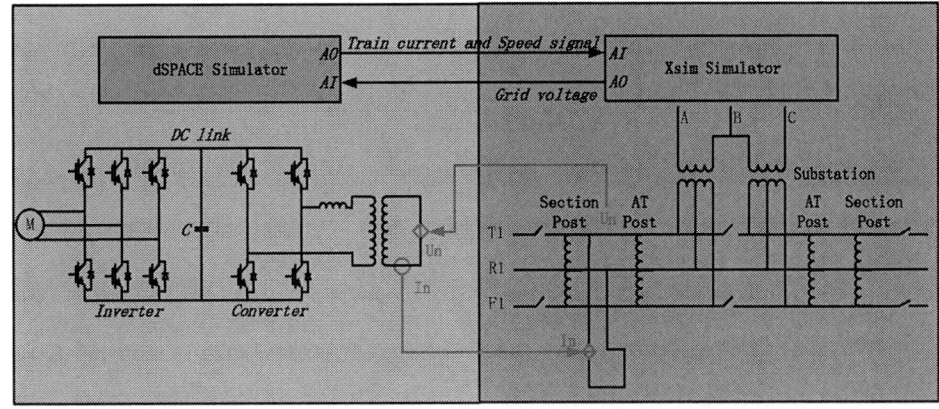

Fig. 8. Signal flow diagram of grid-train coupling experiment platform

B. Train real-time operation logic method

In off-line simulation, it is difficult to simulate the real-time operation of the train on the actual line due to the limitation of the simulation speed. With the help of Xsim parallel calculation and the real-time operation of dSPACE train model, the above real-time experiments can be accomplished through the mutual transmission of signals in Fig. 8.

1818

In Xsim simulation model, controlled current source is used to simulate the train operation. Therefore, when simulating the different positions of the train on the line, the controlled current source can be placed between any two unit length 1km π equivalent circuit, and it can use the position signal to determine which controlled current source receives current signals. Xsim simulator can obtain the speed signal transmitted by dSPACE through the IO port at first, and the position signal can be calculated by integrating the speed signal with time.

Due to the limitation of simulation resources, the traction grid can only be separated by 1km. Suppose that the interval number of π type circuit in Xsim model is m, the currents which is sent to the controlled current source are as follows:

1) Extend the obtained current signal into current matrix $I=[\ I_{n1}\ I_{n2}\cdots I_{nm}]$;

2) Generating m-order unit diagonal matrix;

3) Select 'position_signal' row matrix according to the position signal;

4) Use the obtained matrix in 3) '. * ' current matrix I, get the current signals matrix $I' = [0\quad 0\cdots I_{position\_signal}\cdots 0]$;

The elements in the above matrix I' are sequentially transferred to the controlled current sources in the model. Due to the rest of the elements are zero addition to the $I_{position\_signal}$, it is equivalent to a break circuit when the current is zero. It means that the controlled current sources are not connected to the circuit except the 'position_signal' current source.

The voltage signals are handled in a manner similar to the current signals. Based on the above signals transfer logic method, the flow chart is shown in Fig. 9. The logic algorithm can also be used for the subsequent multi train real-time online operation.

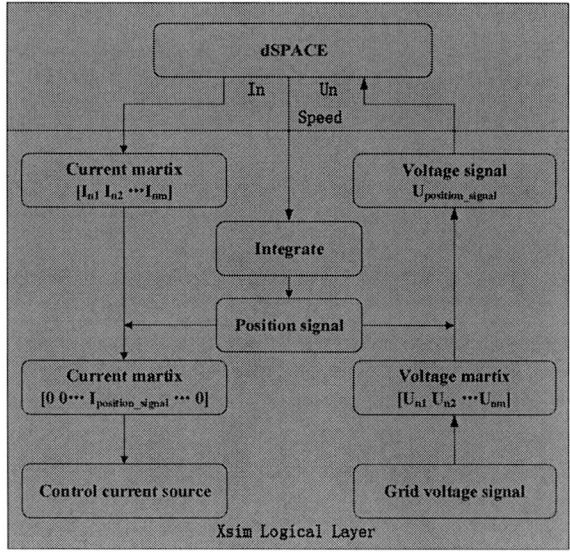

Fig. 9. Logic method flow diagram of dSPACE-Xsim

Combined with Fig.8, in Xsim, the traction grid simulation model is established by using the equivalent multi conductor model. The traction grid per unit length is equivalent to a π type equivalent circuit. The length and parameters of the model can be changed according to the specific line requirements, and then the actual operating conditions of the train can be simulated. The train model in the dSAPCE is built according to the topology of the actual train circuit, and then compiled in the FPGA to ensure the simulation speed of the model. The dSPACE connects to the TCU through hardware in the loop mode, ensuring that the operation environment of the algorithm is the same as the actual train operation enviroment. Through the above model, the consistency of the train simulation condition and the actual operating condition can be ensured.

IV. EXPERIMENT VERIFICATION

A. Introduction of experiment platform

What is shown in Fig.10 is the dSPACE-Xsim joint simulation cabinet, the simulator and controller are placed by layer structure, which is convenient for the rear alignment and data test.

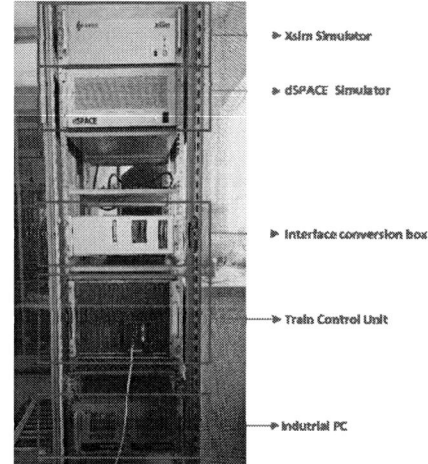

Fig. 10. dSPACE-Xsim Simulator experiment platform

Xsim and dSPACE work as traction grid and traction drive system, respectively. It uses the Train Control Unit(TCU) to verify the control strategy of converter and inverter. The interface conversion box is responsible for the signal conversion between the controller and the simulator.

According to the combined logic algorithm in the third part above, the real-time simulation and the offline simulation time are shown in table 1 under the simulation step 50us of grid-train coupling model.

TABLE I
TIME COMPARISON OF DIFFERENT SIMULATION MODELS

Simulation time(s)	Time for off-line simulation(s)	Time for real-time simulation(s)
1	20.2	1
5	102.7	5
10	208.3	10
20	423.1	20

From the above results, it can be seen that, in case of larger calculation of the simulation model, the offline simulation takes long time and cannot meet the demand of the real-time operation of the train. While in real-time simulation, when the amount of computation is not overrun, the running time of the model is strictly consistent with the wall clock, and the simulation calculation speed is significantly improved compared with the off-line simulation, which can meet the needs of real-time operation of the grid-train coupling simulation.

B. Analysis of experimental results

When the train is operating normally on the line and the position changes, the traction grid voltage and the train current waveform are shown as follows.

Fig. 11 is grid voltage and current waveform. It can be seen that voltage and current waveform is stable.

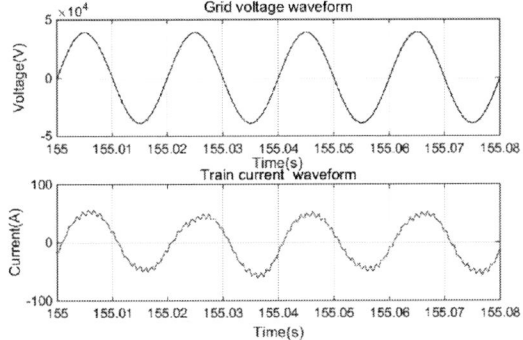

Fig. 11. Grid voltage and train current waveform

In Fig. 12, the FFT analysis of train current shows that the harmonic current is concentrated near 1000Hz and confirms to the harmonic characteristics of the four quadrant converter. Fig.13 is the FFT analysis of grid voltage. It is widely distributed, in which the content of 1000Hz is rich because the harmonic current which injected traction grid is concentrated near the 1000Hz. The harmonic content of grid voltage near 4000Hz is rich because the impedance of this position bearby 4000Hz is large.

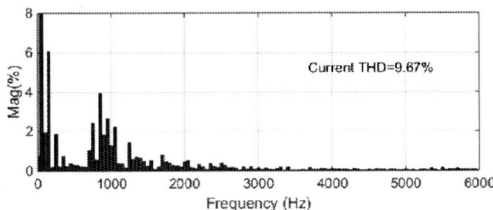

Fig. 12. Fourier analysis of transformer primary current

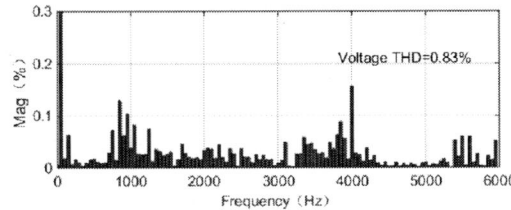

Fig. 13. Fourier analysis of Grid voltage

Fig. 14 is the partial enlargement of the voltage and current waveform when the train position changes. From this figure, it can be seen that the waveform of the position changing process is continuous, and there is no instability of the grid voltage due to the changing position.

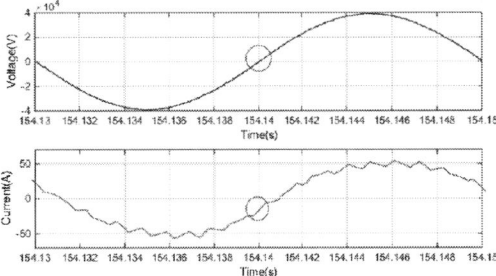

Fig. 15. Partial enlargement of the voltage and current waveform

V. CONCLUSION

In this paper, dSPACE and Xsim real-time simulator are used to simulate traction grid and traction drive system of the train, and a joint experiment platform for high-speed railway traction power supply system is set up. The external algorithm and network controller which connect to the platform can verify the control strategy of the actual train. At the same time, this paper introduces the logic control mode of the train real-time on-line operation. Finally, the feasibility of the experimental platform is verified by experiments, and it can be used for the train algorithm development subsequent.

REFERENCES

[1] Z. He, H. Hu, Y. Zhang and S. Gao, "Harmonic Resonance Assessment to Traction Power-Supply System Considering Train Model in China High-Speed Railway," in IEEE Transactions on Power Delivery, vol. 29, no. 4, pp. 1735-1743, Aug. 2014.

[2] Song K, Konstantinou G, Wu M, et al. Windowed SHE-PWM of Interleaved Four-Quadrant Converters for Resonance Suppression in Traction Power Supply Systems[J]. IEEE Transactions on Power Electronics, 2017, PP(99):1-1.

[3] H. Cui, W. Song, H. Fang, X. Ge and X. Feng, "Resonant harmonic elimination pulse width modulation-based high-frequency resonance suppression of high-speed railways," in IET Power Electronics, vol. 8, no. 5, pp. 735-742, 5 2015.

[4] "Modeling and simulation of the propagation of harmonics in electric power networks. I. Concepts, models, and simulation techniques," in IEEE Transactions on Power Delivery, vol. 11, no. 1, pp. 452-465, Jan 1996.

[5] E. Thunberg and L. Soder, "A Norton approach to distribution network modeling for harmonic studies," in IEEE Transactions on Power Delivery, vol. 14, no. 1, pp. 272-277, Jan 1999

[6] M. Fauri, "Harmonic modelling of non-linear load by means of crossed frequency admittance matrix," in IEEE Transactions on Power Systems, vol. 12, no. 4, pp. 1632-1638, Nov 1997.

[7] P. W. Lehn and K. L. Lian, "Frequency Coupling Matrix of a Voltage-Source Converter Derived From Piecewise Linear Differential Equations," in IEEE Transactions on Power Delivery, vol. 22, no. 3, pp. 1603-1612, July 2007.

[8] Hanmin Lee, Changmu Lee, G. Jang and Sae-hyuk Kwon, "Harmonic analysis of the korean high-speed railway using the eight-port representation model," in IEEE Transactions on Power Delivery, vol. 21, no. 2, pp. 979-986, April 2006.

[9] Shen Dong, Wang Zhonghong, J. Y. Chen and Y. H. Song, "Harmonic resonance phenomena in STATCOM and relationship to parameters selection of passive components," 2001 IEEE Power Engineering Society Winter Meeting. Conference Proceedings (Cat. No.01CH37194), Columbus, OH, 2001, pp. 1097 vol.3-.

[10] Y. Cui and W. Xu, "Harmonic Resonance Mode Analysis Using Real Symmetrical Nodal Matrices," in IEEE Transactions on Power Delivery, vol. 22, no. 3, pp. 1989-1990, July 2007.

The 2018 International Power Electronics Conference

[11] Z. He, H. Hu, Y. Zhang and S. Gao, "Harmonic Resonance Assessment to Traction Power-Supply System Considering Train Model in China High-Speed Railway," in IEEE Transactions on Power Delivery, vol. 29, no. 4, pp. 1735-1743, Aug. 2014.

[12] X. Chu, F. Lin and Z. Yang, "The analysis of time-varying resonances in the power supply line of high speed trains," 2014 International Power Electronics Conference (IPEC-Hiroshima 2014 - ECCE ASIA), Hiroshima, 2014, pp. 1322-1327.

[13] Mingli Wu. "Uniform Chain Circuit Model for Traction Networks of Electric Railways, "in Proceedings of the CSEE, vol. 30, no. 28,pp.52-58,October 2010(in Chinese)

Evaluating the Non-sinusoidal and Non-symmetric Regimes from a Railway Supplying Substation

Ileana-Diana Nicolae[1], Petre-Marian Nicolae[2] and Radu-Florin Marinescu[2]

1 Dept. of Computer Science and Information Technology, University of Craiova, Craiova, ROMANIA

2 Dept. of Electrical Engineering, Energetic and Aeronautics

*E-mail: nicolae_ileana@software.ucv.ro, pnicolae@elth.ucv.ro, rmarinescu@elth.ucv.ro

Abstract— Single phase currents and voltages were acquired from a substation used to supply power lines for railway transport. Different operational contexts were addressed (variable types of locomotives and trains, some of them being simultaneously nearby the substation). Firstly an evaluation of the background noise was performed by using either the mean signal or a thrashing tree. Fast Fourier and Wavelet Packet Transforms were used to evaluate the harmonic spectra and the major types of Root Mean Square quantities. The computed THD revealed significant distortions of the analyzed waveforms. The results yielded by different methods were compared. A good convergence of methods was reached in most of the cases and explanations for the differences recorded for the rest are proposed. Instantaneous values of the fundamental harmonic and global distortions were evaluated with the Stationary Wavelet Transform, allowing the evaluation of maximum/RMS values for currents/voltages to be injected when using active filtering.

Keywords— Harmonic analysis, Stationary Wavelet analysis, Wavelet Packet Transform analysis..

I. INTRODUCTION

Extending the railway transportation systems which use intensively the power converters and electric drives with various types of motors electrically driven bring a lot of benefits related to the transportation safeness, environment pollution diminishing and reliability of systems used for trains driving. Nevertheless there are some disadvantages associated to the electric driving systems [1]. Of them, those related to the harmonic pollution, associated to the non-symmetries due to the supplying in single-phase mode are extremely harmful. The harmonic currents induce unpleasant effects over the supplying cables, motors with electric driving and others. The non-sinusoidal and non-symmetric voltages induced in the supplying voltage affect other consumers connected to the same network as well. A correct analysis of the associated non-sinusoidal regimes (NR) associated to the non-symmetrical regimes induced by the electric network in the distribution network from which the railway network is supplied is a precondition for a correct compensation of NR.

II. OPERATIONAL CONTEXT

The non-sinusoidal voltages and currents which occur in the railway supplying network are caused by the driving with static converters of certain motors. Their supplying is made by means of a single phase transformer with special design which supplies a converter with a d.c. intermediate circuit. The three-phase inverter supplies the asynchronous motors.

The data addressed by this study were acquired from a distribution substation which supplies a Romanian electric railway network. In Romania, the electric power generated by power stations is carried to electric railway substations by 110kV three-phase transmission lines. The electrical supply system of a railway line provides electric power of the desired characteristics (AC, single phase, 25 kV) to the trains from the high-voltage network by single phase 110kV/25kV transformers. Each transformer is connected to two phases belonging to the three-phase system [2].

Dedicated programs, realized by the authors, were used to process the acquired data. The 1-st one relies on Fast Fourier Transform, the 2-nd one implements the Wavelet Packet Transform (WPT) [3]-[7] and the 3-rd one implements the Stationary Wavelet Transform [8]. Data were acquired with a portable equipment of type Chauvin Arnoux, used at the catenary's supplying.

Data were acquired from 2 feeders (F1 and F2) characterized by different operating regimes. F1 supplies both the substation with the auxiliary services and a railway station, whilst F2 supplies only the aerial electric line. Therefore recordings were acquired at different moments across a day and for different locomotives and trains, considering also cases when two trains were nearby the substation in the same time. Seven datasets were considered representative for this study. Fig.1 represents the 1-st period from all the analyzed datasets.

III. EVALUATING THE MEAN SIGNAL AND THE BACKGROUND NOISE

A first step prior to start the analysis with FFT, WPT and SWT was to evaluate the level of background noise, such as to make sure that the analysis is not affected by it. The method relying on average signal [9] could be used for currents in 5 out of the 7 cases. The exact number of acquired samples per period (NS), evaluated by using 2 crossings through 0 on ascending slopes (C1 and C2 - where C1 and C2 are separated by 3 periods), sometimes

The 2018 International Power Electronics Conference

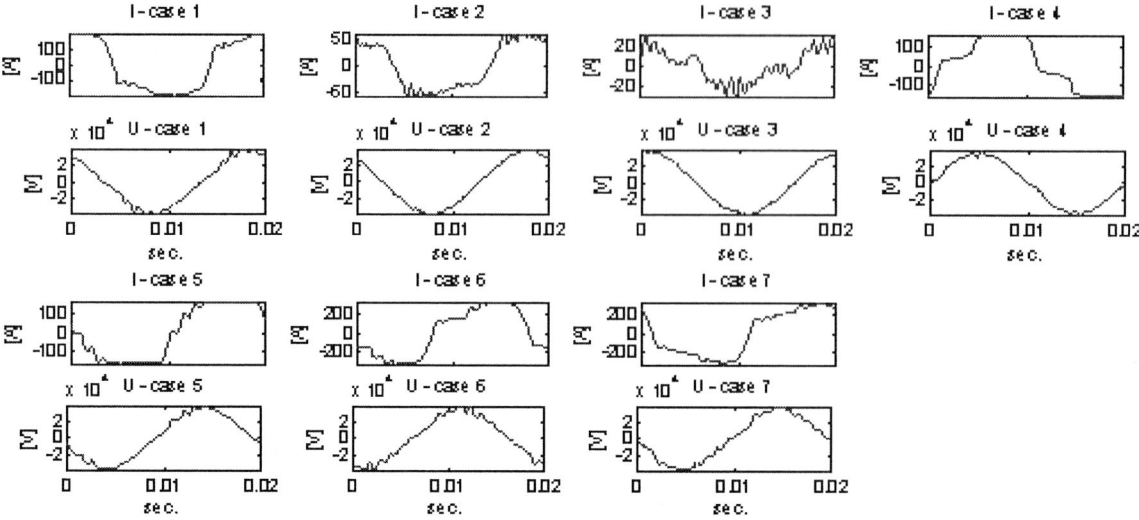

Fig. 1. 1-st period from the analyzed datasets.

was proved to be non-integer. Therefore, when evaluating the mean signal (MS) across a 30 periods length sequence from a dataset, MS tends to overlap best with the 15-th (central) period.

Fig. 2 depicts the overlapped 30 periods considered when evaluating the background noise for the 1-st dataset.

The estimated background noise (computed as difference between the central period and MS), is depicted in Fig. 3 for the 1-st dataset.

The cases where the method relying on average signal could not be used (namely the 3-rd and 4-th datasets) did not exhibit the required stationarity for 30 consecutive periods (Figs. 4 and 5).

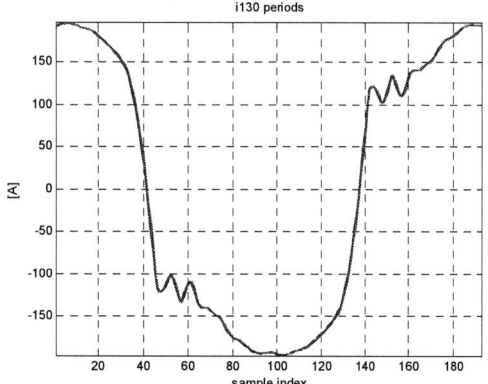

Fig. 2. 30 overlapped periods if the current considered when evaluating the background noise for the 1-st dataset.

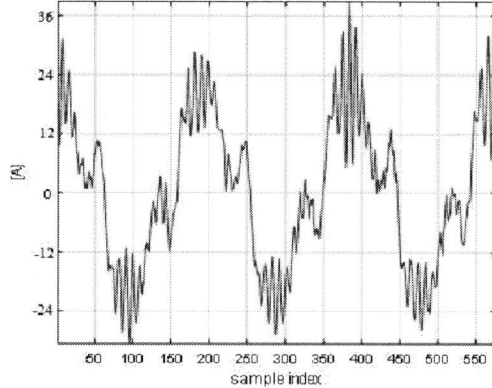

Fig. 4. The first 3 periods from current of the 3-rd dataset.

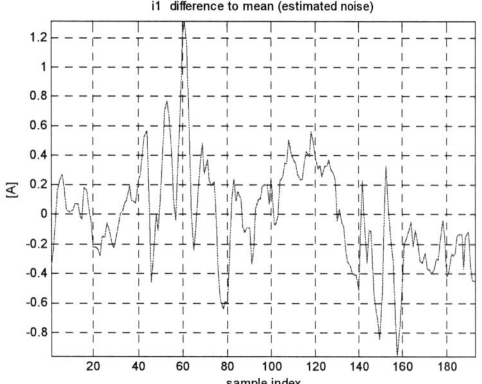

Fig. 3. Estimated noise for the current, 1-st dataset.

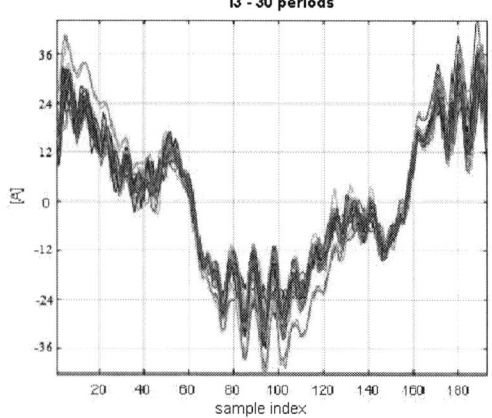

Fig. 5. 30 overlapped periods from the current of the 3-rd dataset.

1823

For the currents associated to the 3-rd and 4-th datasets, a different „per-period" technique of denoising, relying on a wavelet thrashing tree, was used. The noise was assumed to have a Gaussian nature (as in most of the industrial applications) in both denoising methods.

Table I gathers the main parameters used to characterize the noise presenting interest for this application: the „Signal to noise" ratio (SNR) and the mean value (MV) of the noise. MV was computed with:

$$MV = \left(\sum_{i=1}^{per} ES_i / per \right) / RMS \cdot 100 \qquad (1)$$

where: ES represents the estimated noise, per stands for the number of samples in a period and RMS represents the Root Mean Square (RMS) of the dataset associated to a specific noise.

TABLE I
PARAMETERS OF THE ESTIMATED NOISE

Id set	SNRI (p.u.)	SNRU (p.u.)	MVI (%)	MVU (%)
1	$1.92 \cdot 10^5$	$7.08 \cdot 10^4$	3.92	9.41
2	$485.13 \cdot 10^4$	$8.01 \cdot 10^3$	5.97	3.45
3	$148.33 \cdot 10^4$	$1.54 \cdot 10^4$	2.95	-3.12
4	$1.9 \cdot 10^4$	$8.57 \cdot 10^4$	-0.63	0.54
5	$1.19 \cdot 10^4$	$5.57 \cdot 10^4$	2.67	-6.19
6	$9.5 \cdot 10^3$	$8.8 \cdot 10^3$	-0.75	0.58
7	$4.51 \cdot 10^4$	$4.58 \cdot 10^4$	4.18	0.7

SNR was found to be at least $9.5 \cdot 10^3$, which is highly acceptable and therefore no additional measures had to be taken relative to the noise. Moreover, the absolute percent relative mean values of all estimated noises were at most 9.41%, proving that the assumption relative to the noise nature (Gaussian white) was true.

IV. ANALYSIS OF ACQUIRED DATA

A. Analysis methods

Three methods, implemented in MATLAB, were used to process the acquired data: FFT, WPT [7] and SWT [8]. WPT was implemented by using a binary tree relying on the „dwt" function from MATLAB. It has 6 levels, 4 components in each of the final nodes and an underlying wavelet mother of type Daubechies, length 40 [8]. This configuration was selected considering the good properties related to the harmonics selectivity relying on the final nodes energies. SWT has the same configuration, but it uses an unbalanced tree, decomposing only the vectors of approximations [8].

B. Evaluating the harmonic spectra

Figs. 6 and 7 depict the harmonic spectra evaluated with FFT for currents, respectively for voltages, grouped across the harmonic orders considering all 7 analyzed datasets.

Better emphasized by Fig. 8, the weight for the 2-nd harmonic exceeds the value of 1% for the 3-rd dataset. In this particular case a richer harmonic content was noticed, harmonic orders up to 49 exceeding the threshold of 1%. This dataset corresponds to a higher load of the substation (it was the case when a cargo train

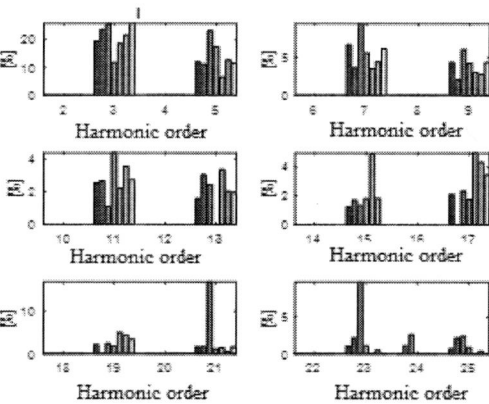

Fig. 6. Harmonic weights of currents.

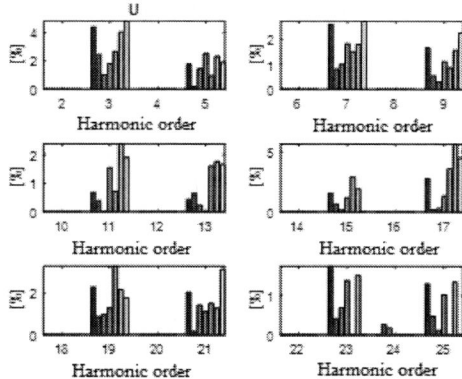

Fig. 7. Harmonic weights of voltages.

entered in the substation vicinity simultaneously with a passenger train).

Similar spectra were yielded by WPT considering the method presented in [3] for all datasets. An exception is related to the 3-rd dataset, where the conditions imposed by the WPT based analysis for harmonic order separation are not met.

In agree with the general picture of practical power applications, in all the analyzed cases the most significant harmonic orders were those with uneven ranks. Excepting the 3-rd dataset, both currents and voltages exhibited harmonic orders higher than 1% up to the 25-th harmonic order.

Variable weights were obtained for the same harmonic order for different datasets. Most obvious are the variations corresponding to the harmonic orders 3, 5 and 7.

C. Evaluating the Root Mean Square Values

Tables II-IV gather the root means square (RMS) indices evaluated with both FFT and WPT, according to the theories from [10] and [11]. Table V gathers the values of the Total Harmonic Distortions (THD).

The results yielded by the different analysis methods are similar, more significant relative percent differences being recorded with respect to the RMS of the distorting regime. A first explanation for this relies on the inherent errors introduced by numerical calculations.

The 2018 International Power Electronics Conference

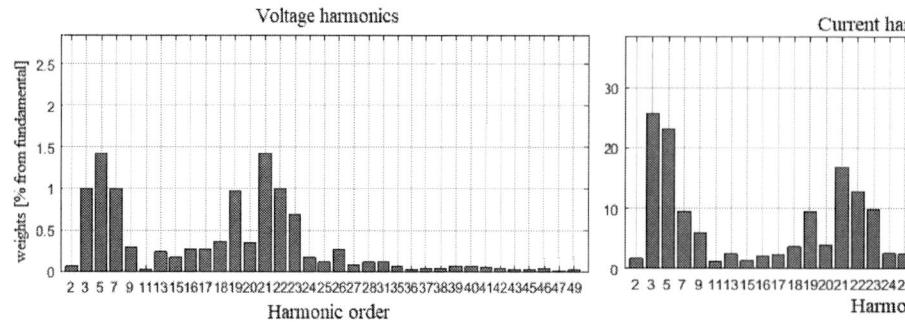

Fig. 8. Harmonic spectra for the 3-rd dataset. Left-voltages, right-currents.

TABLE II
RMS FOR FUNDAMENTAL

Dataset ID	For current [A]		For voltage [V]	
	FFT	WPT	FFT	WPT
1	152.08	153.12	25427.38	25467.5
2	41.29	41.08	26041.88	26020.23
3	13.81	16.29	26354.55	26774.48
4	133.14	133.76	25732.93	25237.9
5	132.6	131.04	25803.98	25348.93
6	224.4	220.6	25304.8	25710.65
7	224.4	220.6	25304.8	25710.65

TABLE IV
TOTAL RMS

Dataset ID	For current [A]		For voltage [V]	
	FFT	WPT	FFT	WPT
1	156.6	157.7	25500	25561.38
2	42.78	42.73	26053.8	26058.9
3	15.17	17.94	26366.75	26787.03
4	136.56	137.15	25767.73	25336.68
5	135.83	133.92	25867.23	25459.35
6	227.56	229.34	25767.73	25807.8
7	156.6	157.7	25500	25561.38

TABLE III
RMS FOR DISTORTIONS

Dataset ID	For current [A]		For voltage [V]	
	FFT	WPT	FFT	WPT
1	37.36	37.74	1926.83	2188.6
2	11.26	11.76	787.93	1419.23
3	6.30	7.55	801.9	820.04
4	30.41	30.36	1338.6	2234.93
5	29.43	27.62	1807.88	2368.48
6	59.09	62.72	2339.95	2237.3
7	37.36	37.74	1926.83	2188.6

TABLE V
TOTAL HARMONIC DISTORTIONS

Dataset ID	THDI [%]	THDU [%]
1	24.57	7.58
2	27.26	3.03
3	45.67	3.04
4	22.84	5.20
5	22.19	7.01
6	26.33	9.25
7	29.72	9.02

Fig. 9 depicts a 3D representation of the percent relative errors introduced by the numeric processing for the mentioned sampling rate in synthetic data. They are more significant for smaller weights and lowest harmonic orders [12], [13].

D. Evaluating the Instantaneous Values

The original analysis with SWT allowed an estimation of the instantaneous values for the fundamental harmonic [8]. The instantaneous values of distortions could be estimated as a difference between the acquired signals and the estimated values of the fundamental harmonic.

In some cases the half-periods were symmetric to a certain extent (Fig. 10), revealing proper applicability conditions for the evaluation with FFT. In other cases (Fig. 11) significant differences were noticed between the half-periods.

The maximum absolute values (acquired signals versus distortions) estimated with SWT across the 1-st period from each dataset are depicted in Fig. 12.

Their numerical values, along with the representations

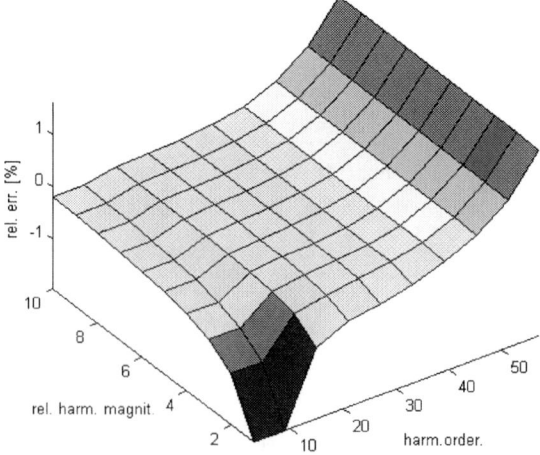

Fig. 9. Minimum percent relative errors recorded by using FFT when evaluating the RMS for distortions.

of instantaneous values, bring essential information to an overall picture of the operating regimes characteristics.

1825

The 2018 International Power Electronics Conference

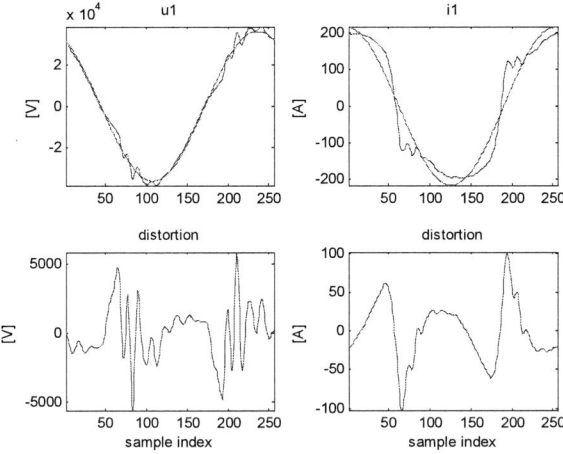

Fig. 10. Top – 1-st period from the 1-st data set, with acquired signal overlapped with the fundamental component extracted with SWT. Down –estimated distortions.

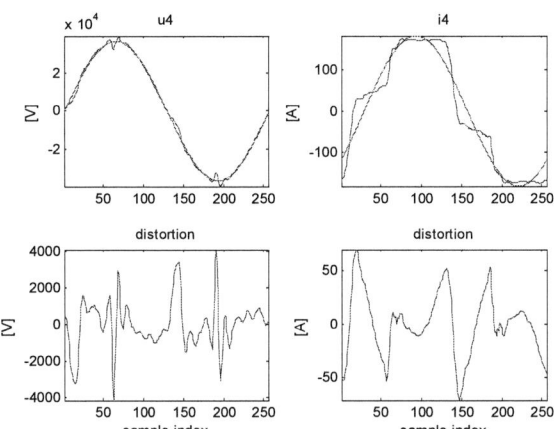

Fig. 11. Top – 1-st period from the 4th data set, w-ith acquired signal overlapped with the fundamental component extracted with SWT. Down –estimated distortions.

V. PUTTING ALL TOGETHER

FFT and WPT revealed a very "rich" harmonic spectrum, with highly variable weights of harmonics. For datasets corresponding to "lighter" loads, 13 harmonic orders proved to be more significant. For harder loads, more than 20 significant harmonic orders (with the upper limit of 49) were revealed (even harmonic orders becoming significant as well!).

The THD for currents exceeded the imposed limits for all datasets. At voltages the same limits were not obeyed for 4 out of 7 selected datasets.

A passive filtering cannot be considered as a solution to diminish the harmonic content considering the operating regimes characteristics.

An active filtering seems to be the only reliable solution to compensate the highly non-sinusoidal regime [14]. The maximum/RMS values for currents and voltages required for the designing (selection) of an active filter can be estimated by using SWT in a preliminary stage. Its usability in a real-time environment

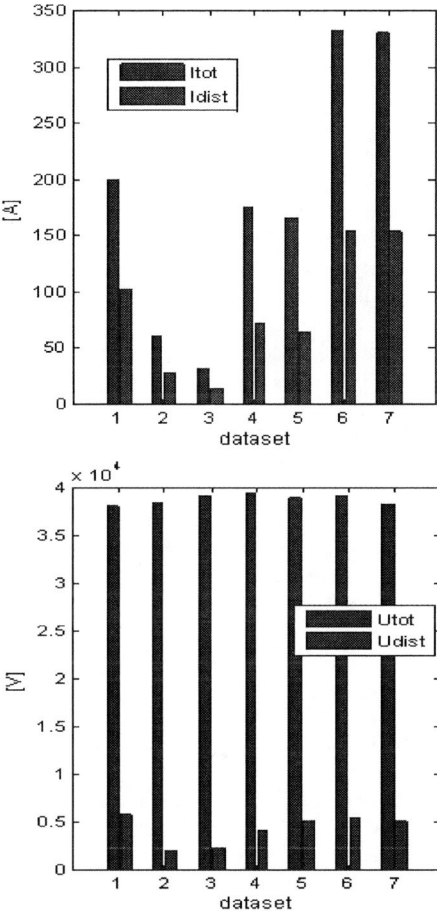

Fig. 12. Maximum absolute values (acquired signals versus distortions). Top – currents; bottom- voltages.

is still questionable because it requires long runtimes. Another problem is related to the accuracy of evaluating the components nearby the edges of the currently analyzed period.

On the other hand, the use of FFT for the extraction of the fundamental components relies on the symmetry between the half-periods which is not always fulfilled.

VI. CONCLUSIONS

The joint (FFT and WPT) analysis revealed that the THD values for currents varied within the range [21.26...41.54]% whilst those for voltages varied within the range [3.02...9.2]%. It means that the imposed limits for currents were exceeded for all datasets. The THD-related limits at voltages were exceeded in 4 out of 7 cases.

Very close values (FFT vs WPT) were obtained relative to the RMS for the fundamental harmonic. The absolute percent relative differences (PRD) FFT vs WPT were under 0.21% for currents and 1.9% for voltages.

With few exceptions, good agreement was obtained between the values of RMS associated to distortions. Variations within the ranges [-19.74 ...6.11]% for currents and respectively [-66.96 ... 4.38]% for voltages were obtained.

1826

A similar discussion can be made for the total RMS. Reference values provided by Riemann sums [15] were used. FFT provided worse evaluations (maximum absolute PRD of 4.41 % for currents and respectively 1.7% for voltages) whilst the PRD provided by WPT were only 1.08% for currents 0.01% for voltages.

In our opinion, besides the inherent errors introduced by numerical calculus, other possible causes for the recorded differences can be:
a) Unlike WPT and SWT, FFT considers the distortions generated only by harmonics, assuming symmetrical halves of one period;
b) The harmonic content of voltages is less significant, the low harmonic orders being preponderant. Studies made on synthetic waveforms [13] revealed that PRD-s depend on phase differences too. This allows for cumulated errors when they have opposite signs for the same phase shift in a FFT respectively WPT (SWT) approach.

Still, as long as the absolute values of the RMS for voltage distortions are small, the PRD are practically translated into acceptable real values for both transforms.

WPT does not provide the phase shifts associated to harmonics and therefore the harmonic powers cannot be estimated! Moreover, it involves long runtimes. This aspect can be improved by a multithreading approach or by truncating the wavelet coefficients. 25% runtime savings can be achieved with the price of an accuracy penalty equal to 0.5% [16].

The above mentioned techniques for runtime diminishing are valid for the SWT analysis too. Therefore it makes sense using Stationary Wavelet Transform. At this analysis the decimation phenomena is counteracted with the price of longer runtimes.

SWT offers a very useful and more accurate estimation of instantaneous values associated to the fundamental component and distortions. Instantaneous values of signals across frequency ranges can be obtained too by this transform, but their evaluation should require additional runtimes and will not help the real-time active filtering operation when only the fundamental (or distortions) components are needed in a first place.

Examples for the SWT based evaluation of the maximum/RMS absolute values, for acquired signals and distortions respectively, were provided. Non-symmetries between half-periods were better revealed by SWT. At this moment the authors make efforts to improve the accuracy of the SWT based analysis when evaluating more accurately the components nearby the edges of the currently analyzed period.

FFT and WPT revealed a very "rich" harmonic spectrum, with highly variable weights of harmonics. For datasets corresponding to "lighter" loads, 13 harmonic orders proved to be more significant. For harder loads, over 20 significant harmonic orders (with the upper limit of 49) were revealed (even harmonic orders becoming significant as well!) Therefore an active filtering seems to be the only reliable solution to compensate the highly non-sinusoidal regimes. The maximum/RMS values for currents and voltages required during the designing

(selection) of an active filter can be better estimated by using SWT in a preliminary stage.

ACKNOWLEDGMENT

This paper got support from the grant no. 59/05.09.2016 under the frame of POC-A1-A1.2.3-G-2015 and by a grant of the Romanian National Authority for Scientific Research and Innovation, CNCS/CCCDI – UEFISCDI, project number PN-III-P2-2.1-BG-2016-0240, within PNCDI III.

REFERENCES

[1] Z. Lijun, C. Ying, H. Zhijie, Q. Qipei, H. Jian and L. Yiuan, "Calculation Method for Harmonic Impedance of Traction Substation Grounding Grid", *Proceed. of POWERCON* Wollongong, NSW, Australia 2016.

[2] M. Popescu, A. Bitoleanu and M. Dobriceanu, "Harmonic Current Reduction in Railway Systems",*WSEAS Trans. on Systems*, vol.7, pp. 689-698, 2008.

[3] Nicolae, I.D. and Nicolae, P.M., " Practical Aspects Related to Paired Nodes and Paired Harmonics in WPT Analysis", *Proceedings of IECON 2016*, Florence, pp. 1-6, 2016

[4] J. Barros and R. Diego, "Analysis of harmonics in power systems using wavelet packet transform". *IEEE Trans. Instrumentation and Measurement*, vol. 57, pp. 63-69, Jan. 2008.

[5] E.Hamid and Z. Yokoyama Kawasaki , „Rms and Power Measurements: A Wavelet Packet Transform Approach". *Trans. Institute of Electrical Engineers of Japan*, vol. 122-B , no. 5, pp.599-606, May, 2002.

[6] J. Barros, R. Diego and M. Apraiz, "Applications of wavelet transform for analysis of harmonic distortion in power systems: A review. ", *IEEE Trans. on Instr. and Measurement*, 61 (10), pp. 2604 – 2611, Sept. 2012.

[7] I.D. Nicolae, P.M. Nicolae, D.C. Maria and L. Scărlătescu, Evaluating RMS of Linearly Variable Magnitude Waveforms by Using FFT and WPT. Theory and Practice., *Annals of the University of Craiova, Electrical Engineering series*, No. 40, 2016, pp. 33-38.

[8] I.D. Nicolae, R.F. Marinescu, D.C. Marinescu and P.M. Nicolae, Analyzing Signals from a Primary Winding of a Locomotive Transformer with Three Types of Wavelet-Based Transforms, *Annals of the University of Craiova, Electrical Engineering series*, pp. 98-106, No. 41, 2017.

[9] Pace University Webspace, "Signals and noise". available at goo.gl/QCAyk3, retrieved 2018.

[10] W.G. Morsi and M.E. El-Hawary, "Reformulating Power Compo-nents Definitions Contained in the IEEE Standard 1459-2000 Using Discrete Wavelet Transform", *IEEE Trans. on Power Delivery*, vol. 22, no. 3, pp.1910-1916, July 2007.

[11] A. Tugulea, "Criteria for the Definitions of the Electric Power Quality and its Measurement Systems," *ETEP*, vol. 6, no.5, pp.357-363, 1996.

[12] I.D. Nicolae, P.M. Nicolae and M.S. Nicolae, "Tunning the Parameters for the FFT Analysis of Waveforms Acquired from a Power Plant", *Proceed. of MPS* 2015 Cluj-Napoca, pp.1-6, 2015.

[13] I.D. Nicolae, P.M. Nicolae, I.D.Smărăndescu and M.Ş. Nicolae, "Wavelet Packet Transform, a Reliable and Fast Method to Obtain the Fundamental Components Required for Active Filtering in Power Plants", *Proceed. of PEMC* 2016 ,Varna, Bulgaria, 2016.

[14] H. Yap, A. Mohd , R. Mohd, K. H. Mohd and F. M. Nashiren, „Control Algorithms of Shunt Active Power Filter for Harmonics Mitigation: A Review", *Energies* vol 10(12), 2017, available at http://www.mdpi.com/1996-1073/10/12/2038/htm, retrieved 2018.

[15] K. Cartwright, „Determining the effective or RMS voltage of various waveforms without calculus", Technology Interface/Fall 2007, pag. 1-20, available at http://tiij.org/issues/issues/fall2007/30_Cartwright/Cartwright-Waveforms.pdf , 2007, retrieved 2018.

[16] I.D. Nicolae, P.M. Nicolae and I.D. Smărăndescu, "Diminishing the Computational Burden when Analyzing Electrical Signals with Long Wavelet Filters", *Annals of the University of Craiova, Electrical Engineering series*, pp. 34-39, No. 41, 2017..

A Fundamental Train Running Experiment for a Basic Performance Verification of a Train Power Demand Control System by Decentralized Control Algorithm

Yusuke Oki[1], Tomoyuki Ogawa[2], Yoko Takeuchi[2], Tatsuhito Saito[2], and Jun'ichiro Kawaguchi[3]

1 Department of Aeronautics and Astronautics, The University of Tokyo, Bunkyo, Japan
2 Railway Technical Research Institute, Kunitachi, Japan
3 Japan Aerospace Exploration Agency, Sagamihara, Japan
*E-mail: oki.yusuke@ac.jaxa.jp

Abstract— It is greatly important to restrict power consumption of running trains for reducing power consumption cost. This study aims to control the train's "power demand", average power consumption in 30 minutes, utilizing decentralized control scheme. This control system is a server-less system and is advantageous from the view point of plug & play and capital investment. In this work, it was verified that this control scheme can achieve the power demand control by the fundamental train running experiment.

Keywords— Power saving, Decentralized control, Broadcast, Internal model control, Transient response

I. INTRODUCTION

In the DC electric railway system it is significantly important to restrict the power consumption of trains. It cannot only reduce the contract power cost but also contribute to environment friendly preservation by energy saving. Since especially the contract power cost with the power company is determined by average power consumption in 30 minutes, called "power demand", there are great needs to reduce the power demand. Although it is required to restrict the power demand of the substation in order to reduce the power demand, it is also required to run all trains not to delay as much as possible even under power demand restriction. If the power demand consumption is kept to the restricted value with minimizing the delay time of trains, the robust railway system is provided to the area where the power provision is unstable such as emerging countries.

The power required by each train is different because the delayed trains must consume more power than other trains in order to recover their delays. Therefore, this theme is considered as the problem that the restricted power is distributed preferentially to trains which require a lot of power. In order to solve this problem we proposed "decentralized control scheme" which was developed from the power control method which was used in spacecraft for deep space exploration [1-3]. Decentralized control is a server-less system and does not require to gather the power consumption information of all components. Since the conservative way, "server-client", requires to allocate addresses to all components

to gather information of them, it is difficult that a new component joins the control system and a component gets out from the system (Not plug & play). The server-client is not fit to railway system because trains often enter and exit the area fed by the substation and other companies' trains enter the area. On the other hand, in the decentralized control scheme only the total power consumption information is broadcasted to all components and each component calculates its own power consumption based on the broadcasted power individually according with feedback control law. Since the calculations are conducted in parallel, the system does not require a server and the investment cost is dramatically lower than server-client. Moreover, it is free to enter and exit the control system because it does not require to allocate their addresses (Plug & play). Therefore, it is fit to railway system. Figure 1 and 2 show the control schemes of server-client and decentralized control.

In several previous works it has been studied that controlling the trains speed or running time decreases the energy consumption [4-6]. However, these studies have not aimed to reduce the power to the restricted value imposed on the substation because they do not use the feedback control unlike this work. Although there are some studies utilizing the server-client system to control the power consumption, the server-client control has several demerits described above [7].

We have developed train power demand control algorithm utilizing this decentralized control scheme using the numerical simulation [2, 3]. It was verified that keeping the restricted power demand imposed on the substation and averaging all trains' delay time by minimizing the delay time of which the most delayed train has are achieved at the same time in numerical simulation. However, the experiment by actual train running has not been conducted yet.

Therefore, this study aims to apply the decentralized control to actual trains and verify convergence of the power demand to the restricted value imposed on the substation by the experiment. Since when the ideally decentralized control is applied, the transient response of power demand has no over shoot and is converged to the restricted value in 30 minutes [3], this experiment examined the convergence in the transient response. As mentioned above, although this control scheme achieves compatible controlling the power demand and averaging trains' delay time, only the convergence of power demand is evaluated in this fundamental experiment.

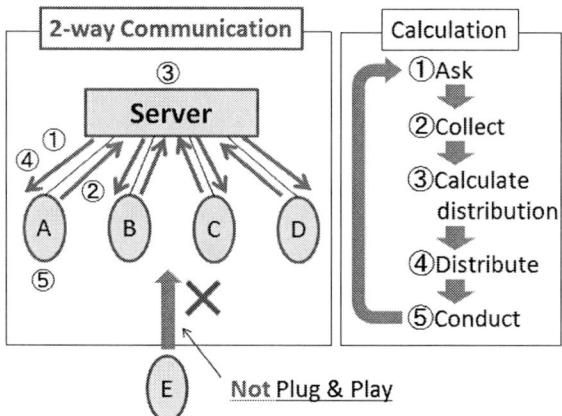

Fig. 1. Server-client control.

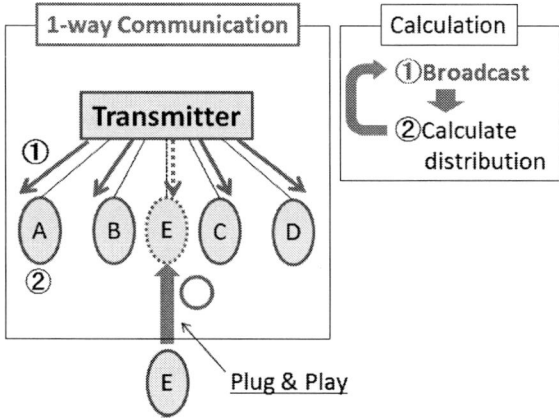

Fig. 2. Decentralized control.

II. DECENTRALIZED CONTROL

In this paper the "notch-off" velocity is set to be a control input. Notch-off velocity indicates the speed of changing the driving operation from powering to coasting. Changing the notch-off velocity adjusts the average power consumption of each train between stations and controls the substation's power demand. At the timing of each train arrival at the station, the train calculates its own notch-off velocity referring to the current power demand broadcasted from the substation. Figure 3 shows the concept of these calculation schemes. The feedback control law of each train is expressed by Eq. (1).

$$V_i(k+1) = \frac{1}{n+1}\left(\sum_{j=0}^{n} V_i(k-j)\right) - c \times \frac{S_t}{Q_{ii}} \times (P_d - P_d^*) \quad (1)$$

where V_i indicates a notch-off velocity of the train set i, k indicates the number starting one at the start of control, n

is a number of stations which a train set runs through in power demand averaging time, c is a conversion factor, S_t indicates the system sensitivity, Q_{ii} is a priority of train set i, P_d is the current power demand broadcasted from the substation, and P_d^* is a restricted value of power demand imposed on the substations. The characteristic of this control law is that the next V_i is determined the average V_i in power demand averaging time instead of the one step before V_i. This is the result of utilizing the internal model control. Although the control system is unstable because the power demand is moving average and is the phase delay element, the internal model control is able to stabilize this system. System sensitivity S_t is a common gain among all trains and Q_{ii} is a different value depending on the delay time. This study does not include weighing depending on the delay time and set same values of all trains as expressed by Eq. (2).

$$Q_{ii} = N \text{ (The number of all trains)}, \ S_t = 1 \quad (2)$$

The conversion factor c converts power demand to notch-off velocity. In this study c is determined by numerical simulation before the experiment.

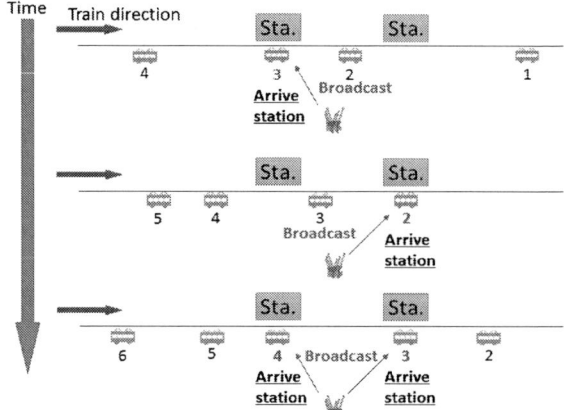

Fig. 3. Broadcast from substations to each train every departure from stations.

III. EXPERIMENT CONFIGURATION

A. Experiment Condition

This section describes the experiment configuration. Fundamental experiment condition is shown in Table. I. In the experiment two trains run between Nagatsuda station and Chuorinkan station on Tokyu Den'entoshi line. The overall view of this experiment is shown in Fig. 4. The instructor operates the software which calculates the notch-off velocity with Eq. (1) and delay time of the train at each station. After that, he tells the notch-off velocity to the driver. In order to set more detailed experiment conditions, we estimated this train running experiment using "Train Operation Power Simulator" developed for estimating train operation power, and multipurpose energy simulator "Hybrid-Speedy" developed for calculating speed profiles and vehicle energy [8-10].

Although it was verified that the decentralized control is valid even if there is re-acceleration, in this experiment it is considered that the operation flow is simply accelerate, coasting, and brake. The trains are not controlled in the following part of sections because these are the terminal stations.

(1) From Tsukimino station to Chuorinkan station
(2) From Tsukushino station to Nagatsuda station
(3) From Nagatsuda station to Tsukushino station

The experiment started at 1:45, but the control start time was set to be 2:25, 40 minutes after the start, in order to observe the power demand under no control after the 30-minutes power demand information was accumulated. In addition, since when the restricted value of the power demand is too low, the delays of the trains become too large and it is not practical, the restricted value of power demand is set to be 1000 kW which corresponds to about 70 % of the power demand under no control situation, 1471 kW. This value is the result of the preliminary experiment in the condition that the all trains are not controlled and run according to the diagram.

Since it is enough that the broadcast interval of power demand is sufficiently shorter than 30 minutes of demand integration time, it is set to be one minute in this experiment (Table I). It is also an advantage of this control law that the low speed communication of one minute intervals is effective.

B. Diagram

The diagram used in this experiment is shown in Fig.5. The precise conditions are showed in the following way.
(1) One roundtrip takes 30 minutes.
(2) Planned stopping time was set 20 seconds.

(3) Trains are able to recover their own delay time 5 seconds at most at each station.
(4) Trains are able to recover their own delay time 20 seconds at most at the terminal stations.

The reason of condition (1), if one roundtrip time is not completely 30 minutes, the profile of the power demand oscillates because of the small number of trains in this experiment and it becomes difficult to observe the profile. There is the sufficient number of trains in the actual railway system, this condition is not required. The reason of condition (2), the stopping time was set equivalent to the actual condition. In order to satisfy condition (1) and (2), planned running times are shortage in the range of 1 to 5 seconds between each stations, so that we set the condition (3). The condition (4) makes to keep the delay time of the train sets because the finish time of this experiment is determined.

C. Conversion Coefficient

Conversion coefficient is required for converge the power demand broadcasted from the substation to the velocity of each train. In order to determine the conversion coefficient in Eq. (1), the simulation was conducted. The simulation assumes that designating each train's running time form one station to the next station generates the train's speed profile using the speed profile generator for energy estimation [10]. Therefore, it is required to replace the notch-off velocity to the average velocity from one station to the next in Eq. (1). The relation of each train's running time and the average velocity is expressed by Eq. (3).

$$T_i = \frac{D}{V_i} \qquad (3)$$

where T_i indicates each train's running time, V_i indicates

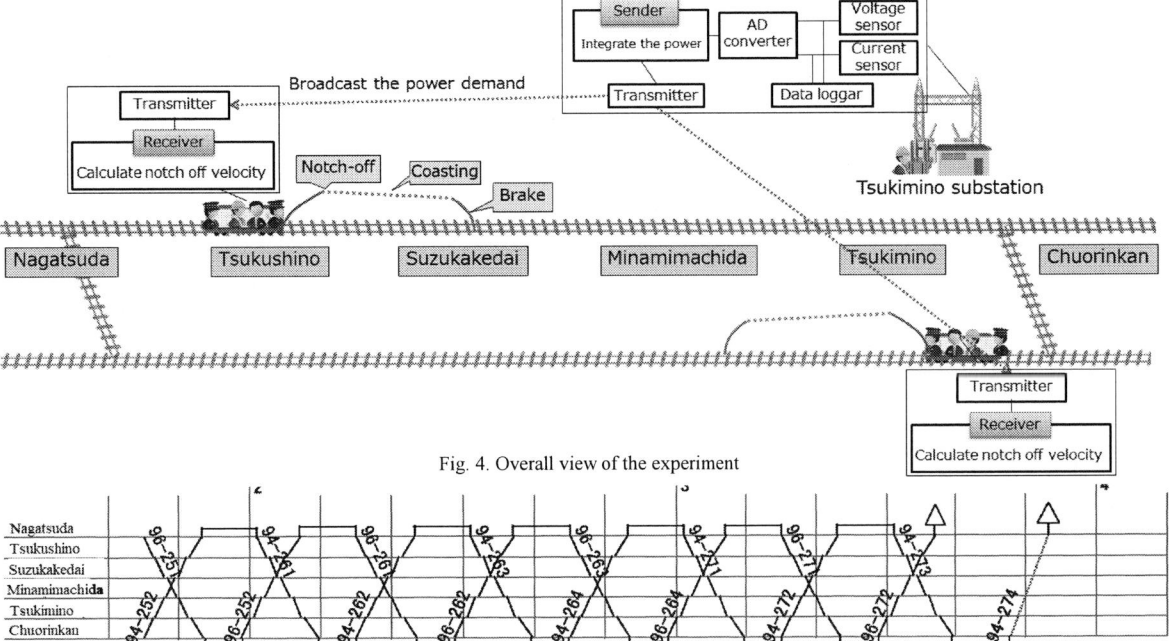

Fig. 4. Overall view of the experiment

Fig. 5. Diagram

the average velocity, D indicates the distance between stations. Therefore, it is necessary to obtain conversion coefficients for both notch-off velocity and average velocity. In order to obtain the two conversion coefficients several cases of simulation that trains run on the scheduled timetable and run 5, 10, 15, 20 seconds slower than the scheduled timetable were conducted. Figure 6 shows the relationship between the power consumption of the substation per one train and the velocity in the above cases. Since the slope in Fig. 6 corresponds to the conversion coefficient c, the two conversion coefficient c is obtained as shown in Table II. For the experiment, we use the conversion factor 0.067 for the notch-off velocity and compare this result with the result with the conversion factor 0.023 in the simulation.

TABLE I
FUNDAMENTAL CONDITION

Experiment section	Nagatsuda – Chuorinkan
Substation	Tsukimino substation
Train type	Tokyu 5000, 2 train set
Time	1:45 – 3:50
Power demand integration time	30 minutes
Broadcast span	1 minute
Power demand restricted value	1000 kW

TABLE II
CONVERSION FACTOR

Velocity	Value
Notch-off velocity	0.067
Average velocity	0.023

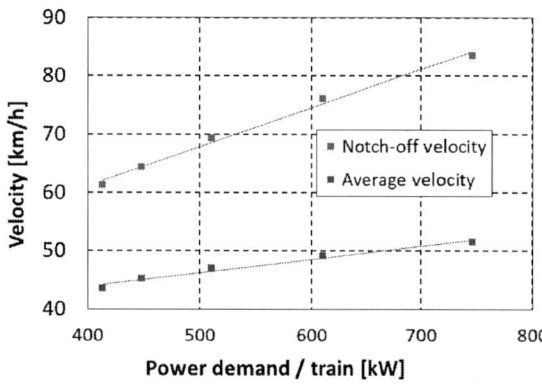

Fig. 6. Notch-off velocity and average velocity against power demand per one train

IV. EXPERIMENT RESULT

A. Power Demand

We describe the power demand of the substation in the result of running experiment. Figure 7 shows the power demand profile of the experiment data measured at substation, the power demand of the simulation, and the restricted value of the power demand. First, it is found that the experimental power demand in Fig. 7 is converged to the restricted value, 1000 kW at around 2:55, 30 minutes after the start of control at 2:25. Furthermore, there is almost no undershoot, and the

transient response of power demand when the internal model control expressed by Eq. (1) ideally works. It is considered that the power demand is converged stably and quickly. Since the average value of the power demand after convergence is 1020 kW, the error against the restricted value is only 2%.

It is found that the power demand profiles of the simulation and of the experiment are similar (Fig. 7). Note that the setting of the conversion coefficient in the simulation is in accordance with Table II. At first, Table III shows the power demand when trains are not controlled before 2:25. Comparing the experiment and simulation, we found that the simulation calculates the power demand with an accuracy of 1.7% against the experimental value. Table III also shows the root mean square error (RMSE) between the power demand and the restricted value during control. The definition formula of RMSE is expressed by Eq. (4) where i is the data number of the sampled power demand, i_0 is the data number at the control start time, and m is the number of data while controlling.

$$\text{RMSE} = \sqrt{\frac{1}{m} \sum_{i=i_0}^{m+i_0} \left(P_{d,i} - P_{d,i}^*\right)^2} \qquad (4)$$

Since the RMSE's error between the simulation and experiment is 1.4%, which is close to the accuracy of the power demand at when trains are not controlled before 2:25, it is found that the validity of the convergence and stability of the power demand of the experiment results was accurately reproduced by the numerical simulation. Therefore, it was verified that the decentralized control scheme enables to satisfy the power demand restricted value of the substation.

B. Delay

Finally, we describe the delays of two trains. Figure 8 shows the delay profiles of two trains. It is found that there is 40 seconds difference between train 1 and train 2 at the end of the experiment. It is possible that this differences caused by the difference of the timing when the instructors push the button of the software which calculates the notch-off velocity because the stopping time of trains at stations are also calculated when the same button is pushed. However, the observation of the tendency of each train set delay is natural because this experiment does not consider the delay time in the feedback control. Although the priority does not include the delay time in Eq. (2) in this fundamental experiment, if the delay time is included in the setting of the priority, the delay times of all trains is equalized. Both controlling the power demand of the substation and leveling the delay times of all trains are achieved at the same time [3]. In the next train running experiment, the equalizing of the delay times of all trains is going to be verified.

C. Speed Profile

Figure 9 and 10 show the actual speed profiles of two trains, the actual notch-off velocity, and the designated notch-off velocity which is the result of the calculation solving Eq. (1). At first, it is found that the actual notch-off velocities decrease after starting control 2:25 to satisfy the power demand restricted value. Furthermore, it is found that all actual notch-off velocities are near to the designated notch-off velocity and the differences between the actual notch-off velocity and the designated one are only 2.2 km/h on average. Therefore, it was verified that the drivers are able to operates the notch-off velocity told by the instructors.

TABLE III
COMPARISON BETWEEN EXPERIMENT AND SIMULATION

	Power demand under no control	RMSE against restricted value
Experiment	1493 kW	155.3 kW
Simulation	1467 kW	157.5 kW
Error between experiment and simulation	1.7 %	1.4 %

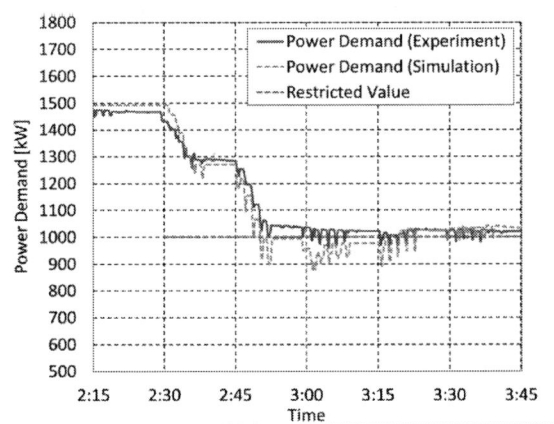

Fig. 7. Comparison of experimental data and simulation data about the power demand profile

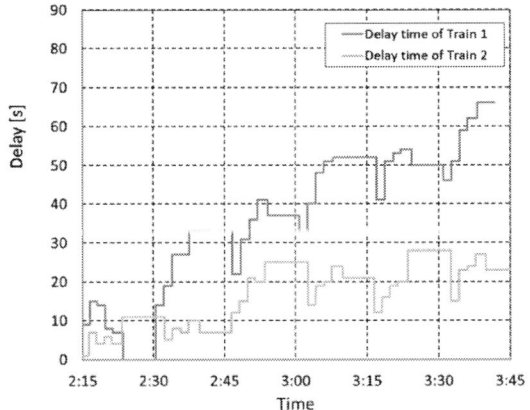

Fig. 8. Delay of two trains

V. CONCLUSIONS

This study conducted the experiment verifying that the power demand consumption of the substation can be controlled to the restricted value utilizing decentralized control scheme. As a fundamental train running experiment, the convergence of the power demand of the substation was evaluated without consideration of delay. The power demand profile of experiment data converges to the restricted value in 30 minutes non-oscillatory and stably. This behavior is the same transient response when this control ideally operates. Moreover, the validity of the convergence and stability of the power demand of the experiment results was reproduced by the simulation result because the error of the power demand after convergence between simulation and experiment was only 1.4%. Finally, it was verified that the drivers are able to operate the notch-off velocity told by the instructor with only 2 km/h error. In the future, we aim to simultaneously achieve both the convergence of power demand and the leveling of the delay time by the next experiment.

A part of development of the Train Operation Power Simulator was funded by the Railway Technology Development Grant from the Ministry of Land, Infrastructure, Transport and Tourism.

ACKNOWLEDGMENT

I would like to thank Tokyu Corporation and Tokyu Techno System Corporation for carrying out the train running experiment.

REFERENCES

[1] S. Hagino, "Development and Result of Hayabusa Spacecraft", *NEC Technical Report*, vol 64, no. 1, pp.130-138, 2011 (in Japanese)

[2] J. Kawaguchi, "Power Control system and Method, and Information Communication Ability Control System and Method", WO/2015/115385, 2015.

[3] Y. Oki, T. Ogawa, J. Kawaguchi, "Train Power Demand Control Using Decentralized and Parallel Control Scheme", *IEEJ Journal of Industry Applications*, vol.6, no. 6, 2017

[4] N. Kimura and M. Miyatake, "Strategy of Speed Restriction Allowing Extended Running Times to Minimize Energy Consumption and Passenger Disutility", *WIT Transactions on The Built Environment*, vol. 135, pp.733-743, 2014.

[5] M. Miyatake, R. Kuwahara, and S. Nakasa, "A simple adjustment of runtimes between stations for saving traction energy by means of mathematical programming", *Computers in Railways XIII: Computer System Design and Operation in the Railway and Other Transit Systems*, Vol. 127, pp. 451-459, 2013

[6] S. Watanabe and T. Koseki: "Train group control for energy-saving DC-electric railway operation", Power Electronics Conference, *IPEC-Hiroshima* 2014-ECCE-ASIA, 2014 International. IEEE, 2014

[7] A. U. Raghunathan, T. Wada, K. Ueda, and S. Takahashi: "Minimizing Energy Consumption in Railways by Voltage Control on Substations", *WIT Transactions on The Built Environment*, Vol. 135, pp. 697-708, 2014

[8] Y. Takeuchi, T. Ogawa, H. Morimoto, Y. Imamura, S. Minobe and S. Sugimoto, "Development of a Train Operation Power Simulator Using the Interaction between the Power Supply Network, Rolling Stock Characteristics and Driving Patterns, as Conditions", *Quarterly Report of RTRI*, Vol.58, No.2, 2017.

[9] T. Ogawa, K. Sato, Y. Imamura, S. Minobe, J. Kawamura, M. Kageyama, N. Shimada, "Speed Profile Generator for Energy Estimation of a Train Running Simulator", *IEEJ Transactions on Industry Applications*, vol.135, no.5, pp.563-570, 2015 (in Japanese).

[10] Tomoyuki Ogawa, Minoru Kondo, Kazumasa Kumazawa, Yoichi Imamura, Shingo Minobe, Junya Kawamura, Naoto Shimada, Tadashi Soeda, and Yoshikazu Sugiyama: "Development of a Multipurpose Energy Simulator for a Running Train", Joint Technical Meeting on "Linear Drives" and "Transportation and Electric Railway", *I.E.E. Japan*, LD-14-067/ TER-14-030 2014. (in Japanese)

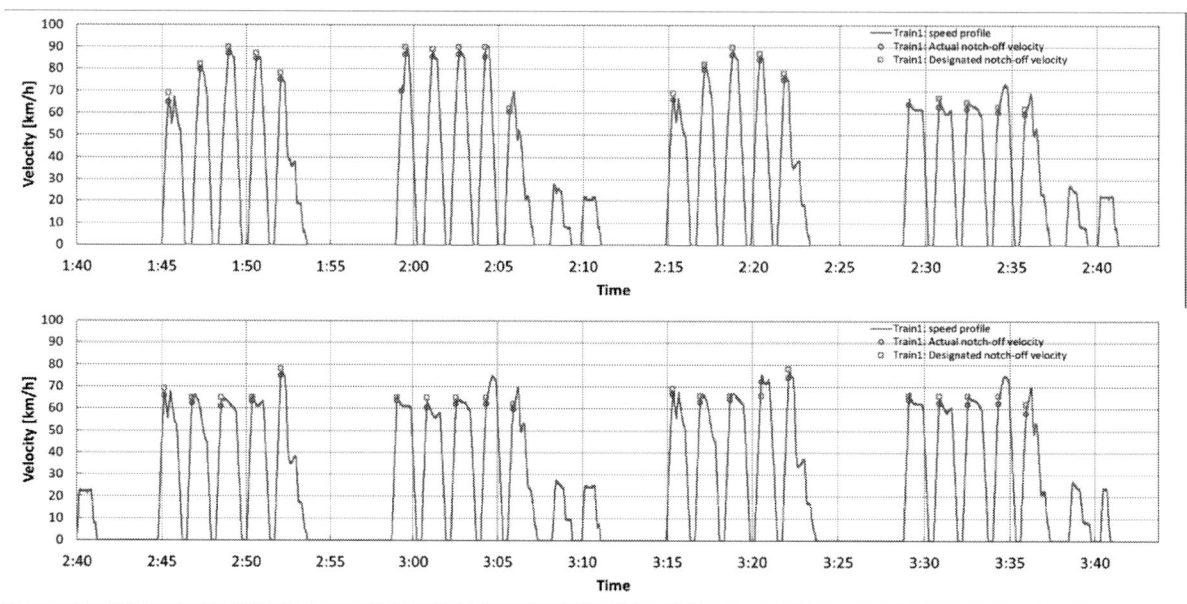

Fig. 9. Speed profile of train 1

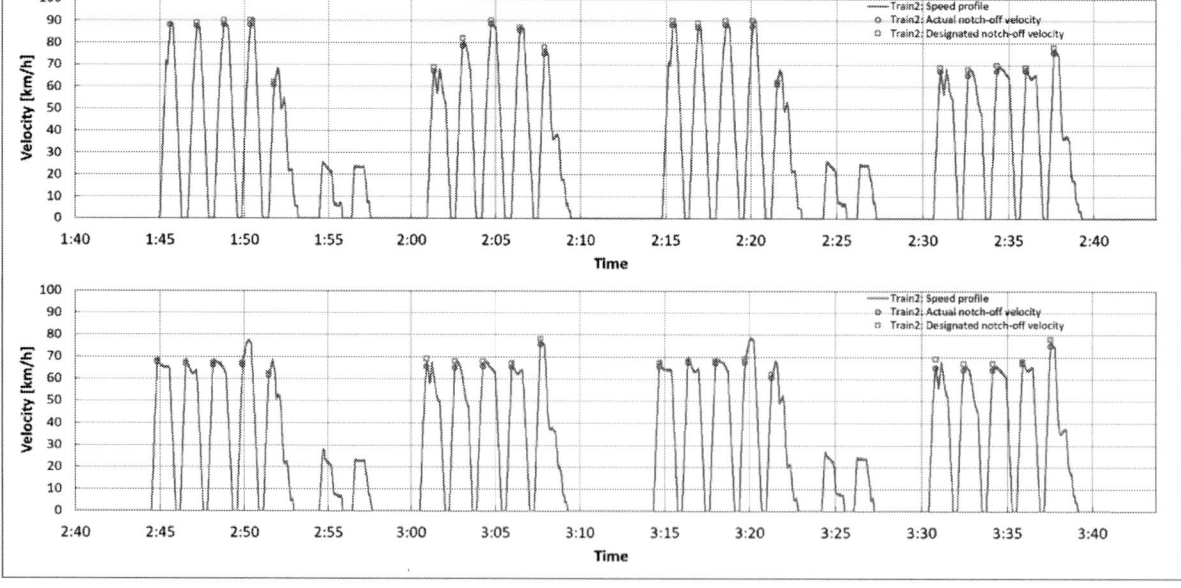

Fig. 10. Speed profile of train 2

The 2018 International Power Electronics Conference

Verification of SiC based Modular Multilevel Cascade Converter (MMCC) for HVDC Transmission Systems

Y. Ishii[1], and T. Jimichi[1]

1 Advanced Technology R&D Center, Mitsubishi Electric Corporation, Amagasaki, Japan

*E-mail: Ishii.Yuki@cb.MitsubishiElectric.co.jp

Abstract— This paper presents a verification result of SiC based modular multilevel cascade converter (MMCC) for HVDC transmission systems. The paper designs and evaluations SiC based MMCC for an assumed 576MW/±250kV HVDC system. The SiC devices' specification is 3300V 1500A SiC-MOSFET/SiC-schottky barrier diode. Because of using SiC devices, capacitance of the chopper cells decrease by 17%. In addition, loss calculation using device characteristics shows that more 50% semiconductor loss reduction is achieved. And also, a 300kVA SiC based prototype submodule is assembled. Compared to that using conventional Si IGBT modules, the volume of submodule decrease by 21% and the weight decrease by 14%.

Keywords— HVDC, MMCC, SiC

I. INTRODUCTION

In recent years, the introduction of offshore wind farm mainly in Europe has been increasing [1]. High-voltage direct current transmission (HVDC) capable of reducing power transmission loss and cost for transmission to land is adopted when offshore wind farm has a long shore distance. Until 2000, as a converter applied to HVDC, line-commutated HVDC with thyristor as the switching device was mainstream [2]. In the 2000s, the introduction of self-commutated HVDC using IGBT increased. As a self-commutated HVDC, 2 or 3 level converter system in which switching devices are connected in series in the 2000s appeared on the market [3]. After the 2010s, the modular multilevel cascade converter (MMCC) has appeared on the market [4].

MMCC has a configuration in which a plurality of unit converters called cell or submodule, are connected in series and has the advantages as follows:

1) Output is multi-level waveform, harmonic filter is unnecessary,

2) High voltage output can be easily configured by increasing the number of cells,

3) Standard specification transformer can be applied.

Because of such advantages, various studies on MMCC are being conducted [5]-[7].

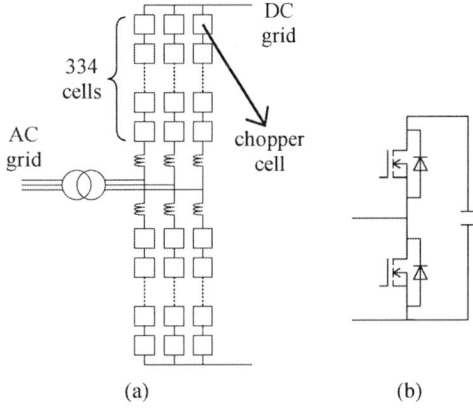

Fig. 1. Circuit configuration of MMCC: (a) whole circuit of MMCC, and (b) chopper cell.

TABLE I
SPECIFICATION OF ASSUMED HVDC TRANSMISSION SYSTEM

Capacity	576 MW
DC voltage	±250 kV
AC voltage	380kV, 50Hz

On the other hand, next generation switching devices have been actively developed. In particular, application of silicon carbide (SiC) is the mainstream for switching devices for high voltage / high power applications [8]. The application of SiC has already advanced to products such as electric railway and contributes to further reduction of power loss [9].

In this paper, we examine the effect of applying SiC-MOSFET / SiC-SBD to the switching devices of MMCC. The rating of the SiC devices used for this study is 3300 V / 1500 A. Examination shows that by increasing the switching frequency by applying SiC devices, the cell capacitor capacity is reduced by 17%. In addition, we evaluate the semiconductor loss using device characteristics and show that the semiconductor loss is reduced to 50% or less compared with the case using the conventional Si-IGBTs. Furthermore, we fabricated a cell using SiC devices, which shows that the cell volume was reduced by 21% and the weight was reduced by 14% as compared with the conventional case using Si devices.

II. CIRCUIT CONFIGURATION

Fig. 1(a) shows the circuit configuration of MMCC for HVDC transmission systems. Each leg consists of an inductor and a series connected chopper cells drown in Fig. (b).

Table I shows the specifications of the HVDC transmission system assumed in this verification. This specification refers to the value of the actual HVDC link, HelWin1, which connects offshore wind farm to the power grid of the German mainland. In addition, SiC modules' ratings are 3300V 1500A SiC-MOSFET/SiC-schottky barrier diode. Therefore, chopper cell capacitor voltage is set to 1500V in consideration of derating. Since the series connection number of chopper cells necessary for each leg N_{cell_leg} is obtained by dividing the rated DC voltage of HVDC by DC voltage each chopper cell outputs as follows:

$$N_{cell\_leg} = \frac{250kV}{1.5kV \times 0.5} = 333.3 \cdots \approx 334 \qquad (1)$$

In the above calculation, the chopper cell output voltage is calculated as half of the cell capacitor voltage. In an actual HVDC system, the number of chopper cells is decided in consideration of the HVDC voltage range and redundancy. Because of MMCC consisting of 6 legs, the number of total chopper cells N_{cell_total} is given by

$$N_{cell\_total} = N_{cell\_leg} \times 6 = 334 \times 6 = 2004 \qquad (2)$$

The power capacity per a chopper cell is calculated by dividing the power capacity of MMCC by N_{cell_total} as follows:

$$P_{cell} = 576MW / 0.95 / 2004 \approx 303kW \qquad (3)$$

Here, the power factor is assumed to be 0.95. As calculated above, the rated power capacity of chopper cell become about 300kW. In the following chapters, cell capacitor and semiconductor loss of this cell is analyzed.

III. DESIGN OF CHOPPER CELL'S CAPACITOR

This chapter describes the design of chopper cell capacitor. In MMCC, cell capacitor occupies a large volume and weights in chopper cell. Therefore, the reduction of cell capacitance leads significant advantages.

The cell capacitance is determined by the allowable value of cell capacitor voltage ripple ΔV_{cell}. The power flowing into chopper cell p_{cell} is calculated by

$$p_{cell} = v_{cell} \times i_{cell} \qquad (4)$$

where v_{cell} is chopper cell output voltage, i_{cell} is chopper cell input current. The energy fluctuation of chopper cell

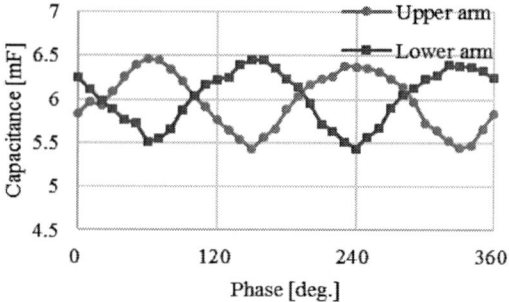

(a)

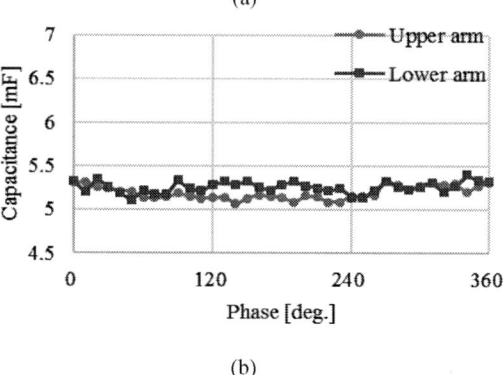

(b)

Fig. 2. Cell capacitance vs. carrier phase: (a) fc = 175Hz, and (b) fc = 350Hz.

ΔW_{cell} is given by

$$\Delta W_{cell} = \max\left(\int p_{cell} dt\right) - \min\left(\int p_{cell} dt\right) \qquad (5)$$

Whereas, ΔW_{cell} is obtained from Δv_{cell} as follows:

$$\Delta W_{cell} = \frac{1}{2} C_{cell} \left(V_{cell} + \Delta V_{cell}\right)^2$$

$$- \frac{1}{2} C_{cell} \left(V_{cell} - \Delta V_{cell}\right)^2$$

$$= 2 C_{cell} V_{cell} \Delta V_{cell} \qquad (6)$$

Here, V_{cell} is average cell capacitor voltage. Equation (5) and (6) gives the following equation about C_{cell}.

$$C_{cell} = \frac{\max\left(\int p_{cell} dt\right) - \min\left(\int p_{cell} dt\right)}{2 V_{cell} \Delta V_{cell}} \qquad (7)$$

By performing numerical calculation using the equation (7), it is possible to calculate the required cell capacitor capacity. Because the necessary capacitor capacitance fluctuates depending on the PWM carrier phase of the cell, it is necessary to obtain the maximum value of each required capacitor capacity when the PWM carrier phase

The 2018 International Power Electronics Conference

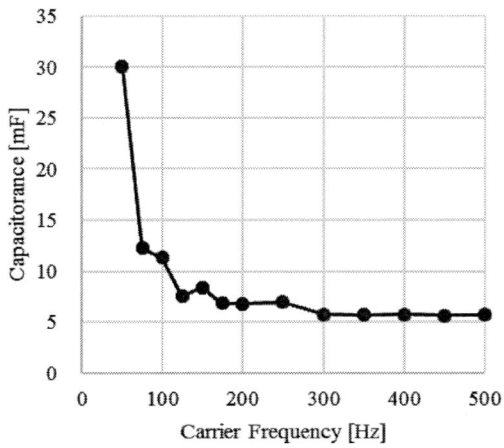

Fig. 3. Calculation result of the relationship between C_{cell} and f_c.

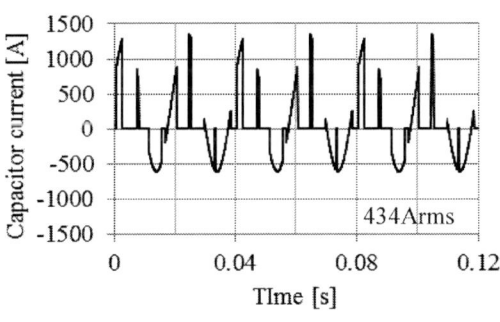

(a)

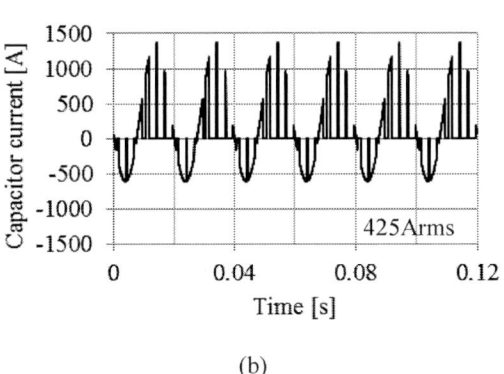

(b)

Fig. 4(a). Waveform of capacitor current when rms value is maximum: (a) f_c = 175Hz, and (b) f_c = 350Hz.

is shifted. In addition, the operating condition where the required capacitor capacity become the largest is as follows:

1) Rectifier end operation,
2) Maximum reactive power output.

Fig. 5. Appearance of cell capacitors.

Fig. 2(a) shows the calculation result of relationship between C_{cell} and PWM carrier phase at f_c = 175Hz. Fig. 2(b) shows also the result at f_c = 350Hz. In this analysis, V_{cell} is set to 1500V, and ΔV_{cell} is set to ±19%. From these results, C_{cell} for each fc is determined as follows:

$$f_c = 175 \text{Hz}: \; 6.5mF / 0.95 = 6.9mF \tag{8}$$
$$f_c = 350 \text{Hz}: \; 5.4mF / 0.95 = 5.7mF \tag{9}$$

Here, the tolerance of C_{cell} is assumed +10%/-5%. As above, increasing f_c by using SiC device, C_{cell} is reduced by about 17%.

Fig. 3 shows the result with other f_c calculating as above equations. Because reduction of C_{cell} is saturated into a certain value which is obtained by calculation in continuous waveform, the optimum value of f_c is around 350 Hz. Fig. 4(a) shows waveform of capacitor current with fc = 175Hz when rms value is maximum. Since the switching frequency is set to a non-integral multiple of 50 Hz, the waveform includes a waveform including the 0.5th order (25Hz) component. The root mean square value is 434 Arms. Fig. 4(b) shows waveform of capacitor current with fc = 350Hz. In this case, since fc is integral multiple of 50Hz, the waveform contains only the integer order components. The root mean square value is 425 Arms.

Fig. 5 shows the prototype of cell capacitor based on the above calculations, when the capacitance is 6.9mF, the external dimension is H470×W190×D450, and the weight is 46.5g. On the other hand, in the case of 5.7mF, the dimension is H400×W190×D450, and the weight is 39.5kg. The volume decreases by 15% ant the weight deceases by 15% compared with the Si device.

IV.   SEMICONDUCTOR LOSS EVALUATION

In this chapter, numerical analysis of steady state semiconductor loss in assumed HVDC system is performed. The characteristics of SiC devices and Si devices used for numerical analysis are as follows:

SiC: Developed device (3300 V / 1500 A),
Si: Mitsubishi CM1500HC-66 R (3300 V / 1500 A).

1836

The 2018 International Power Electronics Conference

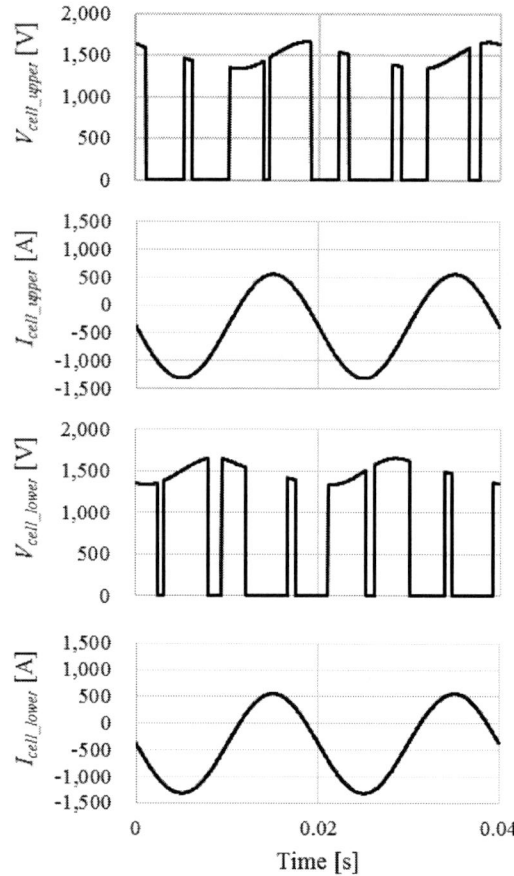

Fig. 6. Example waveform of V_{cell} and I_{cell}.

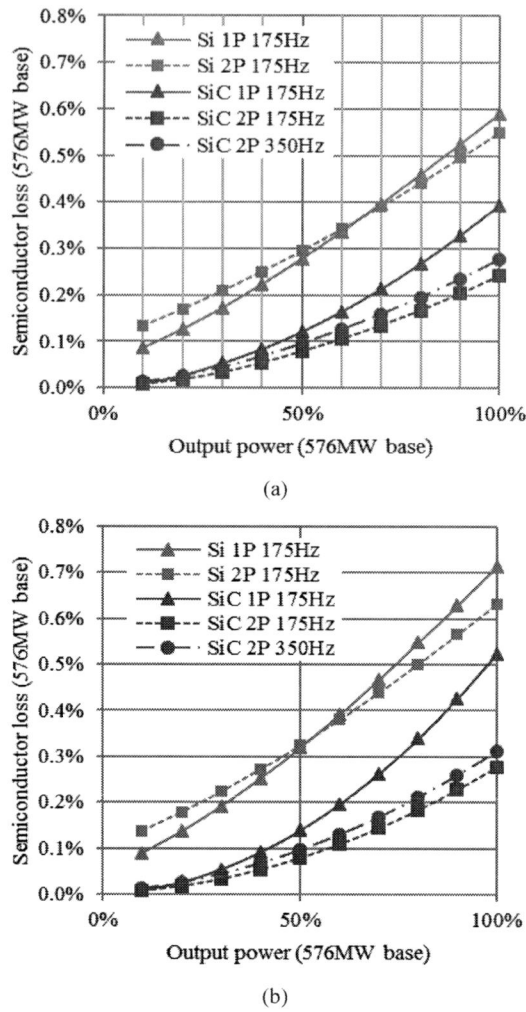

(a)

(b)

Fig. 7. Numerical analysis result of Semiconductor loss: (a) Rectifier side operation, and (b) Inverter side operation.

Loss calculation is carried out using data on V-I characteristics and switching loss characteristics when the junction temperature is 125℃. In the case of the SiC device, V-I characteristics is considering the synchronous rectification of MOSFET. The voltage and current of chopper cell is calculated from the operation principle of MMCC. Regarding the PWM control, a phase shift triangular comparison method that shifts the carrier phase for each cell is applied. Fig. 6 shows example waveform of V_{cell} and I_{cell}. Because these waveforms contain DC component, there is a difference in the amount of loss between REC side and INV side.

Fig. 7(a) shows the calculation result of semiconductor loss at rectifier side operation. Fig. 7(b) also shows the calculation result at inverter side operation. When using one Si-device per each arm, the semiconductor loss based on the rated power of MMCC is 0.59% at REC side and 0.71% at INV side. On the other hand, when the SiC-devices is used, the loss is 0.39% at REC side and 0.52% at INV side. When using two Si-devices in pararell per each arm, the semiconductor loss is 0.55% at REC side and 0.63% at INV side. Moreover, in case of SiC-devices, the loss is 0.24% at REC side and 0.28% at INV side. As described above, the semiconductor loss is reduced by 50% or more. Furthermore, the loss reduction effect is

further increased in the power range below the rated operating point. The reason is that the ON-voltage characteristic of IGBT has a built-in voltage, whereas the ON-voltage characteristic of MOSFET has no built-in voltage. Since SiC-devices' switching loss is rather small, the loss when the switching frequency is increased to 350Hz is also calculated. In this case, the loss is 0.28% at REC side and 0.31% at INV side.

Fig. 8(a) shows the itemized semiconductor losses of REC side at the rated operating point. Fig. 8(b) also shows that of INV side. Here, P_on_Tr is conduction loss of transistor device, P_sw_on is turn-on loss of transistor device, P_sw_off is turn-off loss of transistor device, P_on_Di is conduction loss of diode, and P_sw_rec is recovery loss of diode. In the HVDC system, since the chopper cell current include the direct current component, the proportion of each item of semiconductor loss varies greatly between REC side and INV side. At REC side,

1837

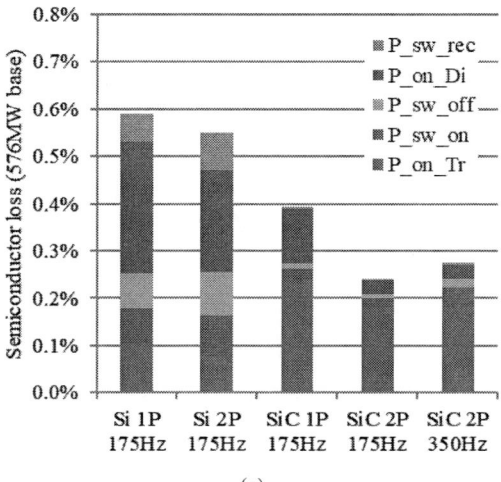

(a)

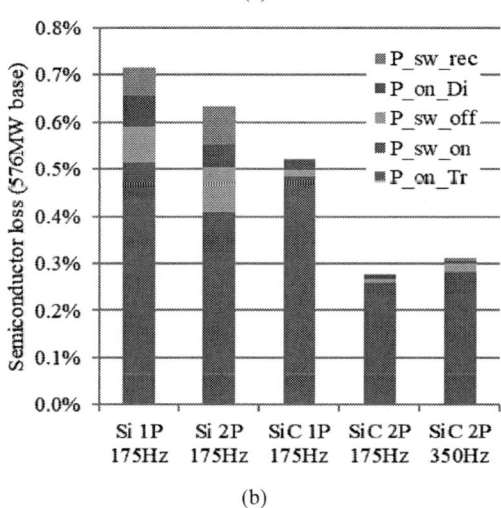

(b)

Fig. 8. Numerical analysis result of Semiconductor loss: (a) Rectifier side operation, and (b) Inverter side operation.

the conduction loss of diodes occupies a large proportion. On the other hand, the conduction loss of transistor devices becomes dominant at INV side. Since the conduction loss occupies a large proportion in both cases, reduction of conduction loss by application of SiC has the large effect as described above.

V. PROTOTYPE OF 300KVA SIC BASED CELL

Fig. 9 shows the appearance of the 300kVA cell prototype. The switching frequency of Si based cell is designed to be 175Hz and that of SiC based cell is to be 350Hz. This prototype is designed to evaluate full-bridge cell in addition to chopper cell. Aluminum water-cooling heatsink are adopted for cooling the semiconductor devices. Laminated copper busbars are applied to the main circuit conductor. Note that cell bypass means is not installed to this prototype cell. The external dimensions of Si based cell is H580×W420×D1070.

Fig. 9. Prototype of SiC based chopper cell.

TABLE II
COMPONENT WEIGHT OF CHOPPER CELL

Parts	Weight [kg] Si	SiC	Diff. [kg]	Ratio
Cell capacitor	46.5	39.5	7	84.9%
Bus bar	23.2	19.3	3.9	83.2%
Heatsink	10.0	7.2	2.8	72.0%
Module	9.6	9.2	0.4	95.8%
Others	50.9	45.1	5.8	88.6%
Total	120.3	140.2	19.9	85.8%

On the other hand, the dimensions of SiC based cell is H530×W420×D930 which means the volume of SiC based cell is reduced by 21% than Si based cell. The volume reduction by applying SiC devices mainly obtained by the followings.

1) Downsizing of water-cooling heatsink by reducing loss,
2) Downsizing of cell capacitor by reducing the capacitance.

Table 2 shows the weight composition of the cell prototype component. The cell capacitor has most large weight reduction. Besides, the reduction of the busbar and heatsink are large. The weight reduction of the entire cell is 14% by applying SiC devices.

VI. CONCLUSION

This paper examined the effect of applying SiC device to MMCC. The 576MW HVDC transmission system is assumed in this analysis. Moreover, 3300C 1500A SiC-MOSFET/SiC- schottky barrier diode is applied to the chopper cells. The analysis shows that the capacitance of cell capacitor of the SiC based cell is reduced by 17% because of increasing the switching frequency. Numerical calculation of semiconductor loss when SiC device is applied confirmed that the loss is reduced by 50% or more as compared with the conventional Si device. Furthermore, by fabricating SiC based cell prototype, it showed that the volume is reduced by 21% and weight by 14%.

ACKNOWLEDGMENT

This work was supported by Council for Science, Technology and Innovation (CSTI), Cross-ministerial Strategic Innovation Promotion Program (SIP), "Next-generation power electronics"(funding agency: NEDO).

REFERENCES

[1] GWEC, "Global Wind Report 2015," Available at http://www.gwec.net/wp-content/uploads/vip/GWEC-Wind-2015-Report_April-2016_22_04.pdf.

[2] W. long, and S. Nilsson, "HVDC transmission: Yesterday and today," *IEEE Power Mag.,* vol. 2, no. 5, pp. 22-31, 2007.

[3] N. Flourentzou, V. G. Agelidis, and G. D. Demetriades, "VSC-based HVDC power transmission systems: An overview," *IEEE Trans. Power Electron.,* vol. 24, no. 3, pp. 592-602, 2009.

[4] H. J. Knaak, "Modular multilevel converters and HVDC/FACTS: A success story," in *Proc. Conf. EPE, Birmingham, U.K.,* 2011, pp. 1–6.

[5] H. Akagi, "Classification, terminology, and application of the modular multilevel cascade converter (MMCC)," *IEEE Trans. Power Electron.,* vol. 26, no. 11, pp. 3119–3130, Nov. 2011.

[6] A. Lesnicar and R .Marquardt, "An innovative modular multilevel converter topology suitable for a wide power range," in *IEEE Conf. Rec. EPE 2003.*

[7] M. Hagiwara and H. Akagi, "Control and experiment of pulsewidthmodulated modular multilevel converters," *IEEE Trans. Power Electron.,* vol. 24, no. 7, pp. 1737–1746, Jul. 2009.

[8] M. Ostling, R. Ghandi and C. Zetterling, "SiC power devices - Present status, applications and future perspective," in *Proc. 23rd Int. Symp. Power Semiconductor Devices & IC's, San Diego, CA,* 2011, pp. 10-15.

[9] Mitsubishi Electric, "Mitsubishi Electric's Railcar Traction Inverter with All-SiC Power Modules Achieves 40% Power Savings," Available at http://www.mitsubishielectric.com/news/2015/pdf/0622-a.pdf.

Control of a 6.6-kV Transformerless STATCOM Based on the MMCC-SDBC using SiC MOSFETs

Laxman Maharjan[*], Toshihisa Tajyuta, Hiroshi Shinohara, Akio Suzuki and Akio Toba

Fuji Electric Co., Ltd., Tokyo, Japan
[*]Email: m-laxman@fujielectric.com

Abstract-This paper describes a transformerless STATCOM (static synchronous compensator) intended for 6.6-kV power distribution system in Japan. The STATCOM is characterized by the use of MMCC-SDBC (modular multilevel cascade converter – single-delta bridge-cells) topology and SiC MOSFETs. The paper discusses a control method developed for the 6.6-kV system for reactive-power and negative-sequence compensation with focus on dc-capacitor voltage control. The dc-capacitor voltage control is comprised of (a) inter-phase balancing control and (b) intra-phase balancing control. The latter uses a new technique that is based on the control of dead times of bride cells. A three-phase 200-V 6-kVA experimental system is designed, constructed and tested to verify the effectiveness of the control method.

Keywords— DC-capacitor voltage control, MMCC (modular multilevel cascade converters), reactive power, STATCOM (static synchronous compensator)

I. INTRODUCTION

In recent years, MMCC (modular multilevel cascade converters) have gained significant interest for grid applications [1], [3]-[12], [14]-[23]. Modular topology of these converters allows direct connection to medium- and high-voltage grids without using step-up line-frequency transformers. There exist different circuit configurations of MMCC [15]. However, for STATCOM applications, the most investigated configurations are:

(a) MMCC-SSBC (single-star bridge-cells), and
(b) MMCC-SDBC (single-delta bridge-cells).

The basic circuit unit of both configurations of MMCC is a single-phase H-bridge converter equipped with a dc capacitor, and is generally referred to as a bridge cell. For the same voltage rating, the MMCC-SSBC requires fewer bridge cells than the MMCC-SDBC and, as such, is the first choice as long as the STATCOM is required to control only positive-sequence reactive power [1], [9], [12], [23]. STATCOMs intended for voltage regulation in utility application fall into this category. However, when the STATCOM is required to control both positive- and negative-sequence reactive powers, the MMCC-SDBC is preferred over the MMCC-SSBC [5], [16], [19]. STATCOMs intended for flicker compensation in industry application, e.g. electric arc furnace, fall into this category [2], [13].

Fuji Electric is currently developing a 6.6-kV 200-kVA transformerless STATCOM based on the MMCC-SDBC as a demonstration project in Japan. The aim of the paper is discuss the control method developed for the 6.6-kV system for reactive-power and negative-sequence compensation with focus on dc-capacitor voltage control. The dc-capacitor voltage control presented in this paper is comprised of (a) inter-phase balancing control and (b) intra-phase balancing control. The former uses a technique based on the circulation of line-frequency zero-sequence current in delta loop. It has been widely reported in literatures [5], [7], [15]-[22]. The latter, on the other hand, uses a new technique that is based on the control of dead times of bridge cells. This paper provides a detailed explanation of the technique. Experimental results based on a three-phase 200-V 6-kVA downscaled system are presented to validate the efficacy of the control method.

II. THE 6.6-KV TRANSFORMERLESS STATCOM

Fig. 1 shows circuit configuration of the 6.6-kV 200-kVA transformerless STATCOM based on the MMCC-SDBC.

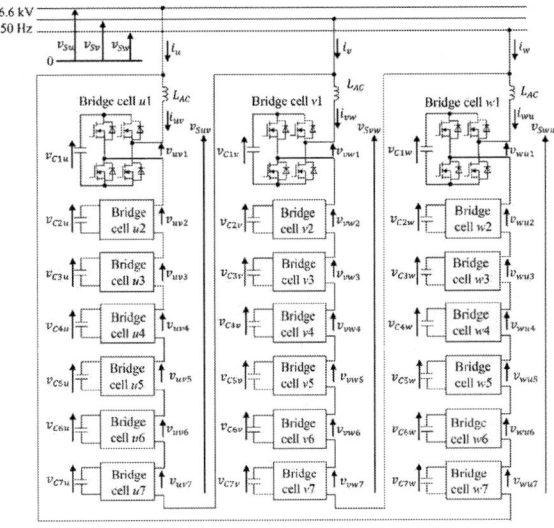

Fig. 1. Circuit configuration of the 6.6-kV 200-kVA STATCOM based on the MMCC-SDBC and using 3.3-kV SiC MOSFETs.

The STACOM is currently under development as a demonstration project in Japan. The aim of the project is to demonstrate the feasibility of a high-efficiency medium-voltage transformerless grid inverter by combining the MMCC-SDBC topology with SiC power switching devices.

Each phase of the STATCOM consists of a series connection of an ac inductor L_{AC} and seven single-phase H-bridge cells. The three phases are connected in delta. This configuration allows a zero-sequence current to be circulated in the delta loop, thus enabling the control of negative-sequence reactive power.

TABLE I
CIRCUIT PARAMETERS OF THE 6.6-KV 200-KVA SYSTEM

	Nominal line-to-line voltage	V_S	6.6 kV
	Capacity rating	P	200 kVA
	Line frequency	f	50 Hz
	Carrier frequency	f_C	5 kHz
	Equivalent carrier frequency		70 kHz
	Number of bridge cells per phase	N	7
Bridge cell	Nominal dc voltage	V_C	1700 V
	Nominal output voltage		942 V
	Nominal output current		10 A
	Capacity rating		9.5 kVA

Table I shows the circuit parameters of the 6.6-kV system. The nominal dc voltage is 1.7 kV. SiC MOSFETs rated at 3.3 kV are used as power switching devices. The use of SiC MOSFETs enables to reach high grid voltage with reduced number of bridge cells in series and achieve high efficiency. The 21 bridge cells will be controlled by a phase-shifted unipolar sinusoidal PWM (pulsewidth modulation) with a carrier frequency of 5 kHz. The equivalent carrier frequency will be 70 kHz and the resulting line-to-line ac voltages will be 15-level waveforms.

Fig. 2 shows a general view of one of the 21 power converter stacks of the 6.6-kV system. The stack consists of a single-phase H-bridge converter and four gate drive circuits. The stack also contains a self-fed dc power supply circuit that supplies a non-isolated power from the main circuit to the gate drive circuits. The H-bridge employs 3.3-kV SiC MOSFET modules from Fuji Electric.

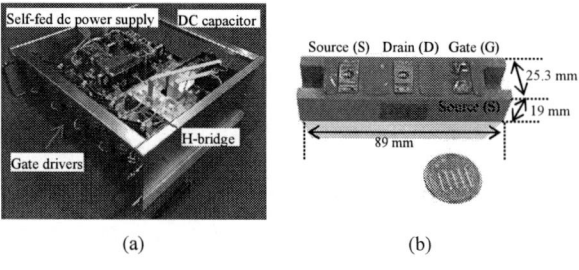

(a)   (b)

Fig. 2. General view of one of the 21 power converter stacks of the 6.6-kV system. (a) Power converter stack, and (b) Sample of the 3.3-kV SiC MOSFET module used in the H-bridge of the stack.

III. CONTROL METHOD

Fig. 3 shows block diagram of the control system. The control system can be broadly divided into:

(a) AC current control
(b) Feed-forward control of circulating current, and
(c) DC-capacitor voltage control.

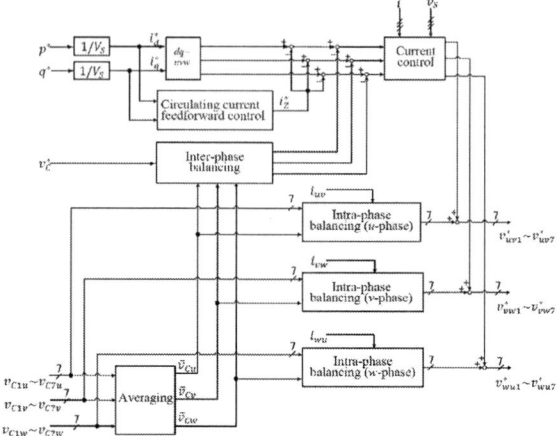

Fig. 3. Block diagram of the control system.

In contrast to a conventional STATCOM based on a two-level voltage-source inverter, a STATCOM based on the MMCC-SDBC has no common dc capacitor, but multiple floating dc capacitors as shown in Fig. 1. The dc-capacitor voltage control is, therefore, indispensable. The dc-capacitor voltage control in Fig. 3 is comprised of:

(a) Inter-phase balancing control, and
(b) Intra-phase balancing control.

The inter-phase balancing control regulates the mean dc voltages in the u-, v- and w-phases ($\bar{v}_{Cu}$, $\bar{v}_{Cv}$ and $\bar{v}_{Cw}$) at the dc voltage command (v_c^*). Here $\bar{v}_{Cu}$, $\bar{v}_{Cv}$ and $\bar{v}_{Cw}$ are given by

$$\bar{v}_{Cu} = \frac{1}{7}\sum_{j=1}^{7} v_{Cju} \tag{1}$$

$$\bar{v}_{Cv} = \frac{1}{7}\sum_{j=1}^{7} v_{Cjv} \tag{2}$$

$$\bar{v}_{Cw} = \frac{1}{7}\sum_{j=1}^{7} v_{Cjw} \tag{3}$$

The control is achieved by circulating a line-frequency zero-sequence current (i_Z) in the delta loop. The circulating current is given by

$$i_Z = \frac{1}{3}(i_{uv} + i_{vw} + i_{wu}) \tag{4}$$

The feed-forward control of circulating current in Fig. 3 can also be considered as a part of the inter-phase balancing control.

The intra-phase balancing control, on the other hand, regulates the dc voltages of the bridge cells in a phase (e.g. $v_{C1u} \sim v_{C7u}$) at the mean dc voltage in that phase ($\bar{v}_{Cu}$). This paper presents a new technique of intra-phase balancing that is based on the control of dead times of bridge cells. It is explained in detail in the following section.

IV. INTER-PHASE BALANCING CONTROL

A. Basic Principle

The basic principle of the presented intra-phase balancing is based on the fact that increasing the dead time of a bridge cell results in the rise in dc-capacitor voltage of the bridge cell.

Fig. 4 shows four main modes of operation of a single-phase bridge cell: (a) Mode 1, (b) Mode 2, (c) Mode 3 and (d) Mode 4. For a current direction from leg YV to leg UX,

- Mode 1 charges the dc capacitor,
- Mode 2 discharges the dc capacitor, and
- Modes 3 and 4 do not charge or discharge the dc capacitor.

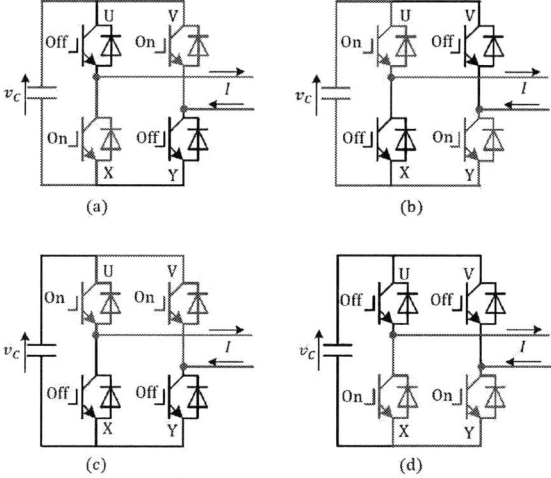

Fig. 4. Main modes of operation of a single-phase bridge cell. (a) Mode 1, (b) Mode 2, (c) Mode 3, and (d) Mode 4.

Fig. 5 shows four additional modes that occur in dead times between two main modes of Fig. 4. Modes A and B occur during the dead time of leg UX, while modes C and D occur during the dead time of leg VY. For a current direction from leg VY to leg UX,

- Modes A and D charge the dc capacitor, whereas

- Modes B and C do not charge or discharge the dc capacitor.

When the dead time of a bridge cell is increased, the durations of main modes 1 through 4 are shortened, while the duration of dead-time modes A through D are elongated.

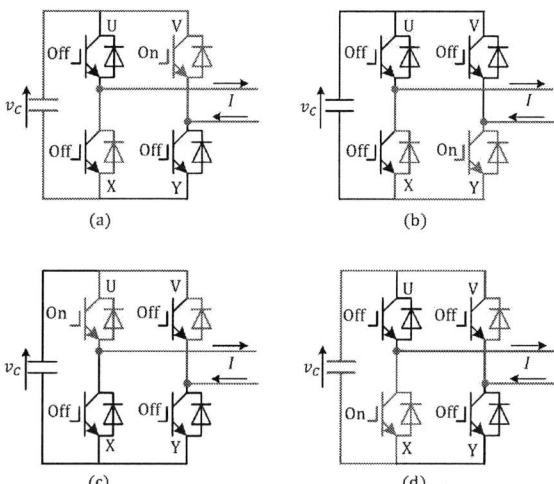

Fig. 5. Modes of operation each of which occur during the dead time between two main modes of Fig. 4. (a) Mode A, (b) Mode B, (c) Mode C, and (d) Mode D.

TABLE II
NET EFFECT OF AN INCREASED DEAD-TIME ON DC VOLTAGE AS COMPARED TO THE ONE WITHOUT DEAD-TIME CONTROL

Main mode[1]	Subsequent dead-time mode[1]	Net effect on dc-capacitor voltage[2]
Mode 1 (Charge)	Mode A (Charge)	No change
	Mode D (Charge)	No change
Mode 2 (Discharge)	Mode B (No change)	Rise
	Mode C (No change)	Rise
Mode 3 (No change)	Mode A (Charge)	Rise
	Mode C (No change)	No change
Mode 4 (No change)	Mode B (No change)	No change
	Mode D (Charge)	Rise

Table II summarizes net effects on dc-capacitor voltage of each shortened main mode and its subsequent elongated dead-time mode. It is seen that the net effect, as compared to the one without dead-time control, is either rise or no change in dc-capacitor voltage. However, when all the main modes and their subsequent dead-time modes are considered, the overall net effect will be rise in

---

[1] Indications in ( ) assume current direction from leg VY to leg UX.
[2] Net effect is as compared to the one without dead-time control.

dc-capacitor voltage. That means, the dc-capacitor voltage of a bridge cell rises with increase in dead time of the bridge cell.

For opposite current direction, the indications in parenthesis (charge, discharge or no-change) in Table II in each main mode and its subsequent dead-time mode may vary, but the overall net effect of all main modes and their subsequent dead-time modes will always be rise in dc-capacitor voltage for an increase in dead time. Note that the rise in dc-capacitor voltage here is as compared to the one without the control of dead time.

B. Block diagram

Fig. 6 shows the block diagram of inter-phase balancing control, taking u-phase bridge cell 1 as an example. For a particular bridge cell, the PI (proportional-integral) control calculates the dead-time offset $t_{d(offset)}$ based on the difference between the dc voltage in that cell (v_{C1u}) and the mean dc voltage of the phase which the cell belongs to ($\bar{v}_{Cu}$). The dead-time offset is then added to the minimum dead-time $t_{d(min)}$ to generate the dead time t_d to be used in that bridge cell. Note that minimum value of $t_{d(offset)}$ is limited to zero using a lower limiter, because t_d is not supposed to be less than $t_{d(min)}$.

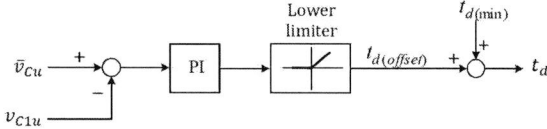

Fig. 6. Block diagram of the intra-phase balancing control, taking u-phase bridge cell 1 as an example.

If the dc-capacitor voltage of a bridge cell in a phase is lower than the mean dc voltage of the phase, the dead time of the bridge cell is increased as compared to that of other bridge cells in the phase. As explained earlier, this increases the dc-capacitor voltage of the bridge cell, resulting in dc-voltage balancing between bridge cells in the phase.

V. THE 200-V 6-KVA DOWNSCALED SYSTEM

Fig. 7 shows circuit configuration of the three-phase 200-V 6-kVA downscaled system.

Table I summarizes the circuit parameters. The number of bridge cells in a phase is four. The capacity rating of each bridge cell is 500 VA and the nominal dc voltage V_C is 100 V. The 12 bridge cells are controlled by phase-shifted unipolar sinusoidal PWM with a carrier frequency of $f_C = 5$ kHz. The resulting converter voltages (v_{uv}, v_{vw}

and v_{wu}) will thus be nine-level waveforms with an equivalent carrier frequency of 40 kHz.

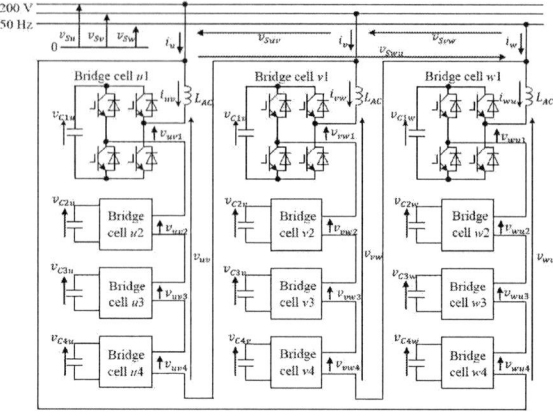

Fig. 7 Circuit configuration of the three-phase 200-V 6-kVA downscaled STATCOM based on the MMCC-SDBC.

TABLE III
CIRCUIT PARAMETERS OF THE 200-V 6-KVA DOWNSCALED SYSTEM

Nominal line-to-line voltage		V_S	200 V
Capacity rating		P	6 kVA
Line frequency		f	50 Hz
Carrier frequency		f_C	5 kHz
Equivalent carrier frequency			40 kHz
Number of bridge cells per phase		N	4
Bridge cell	Nominal dc voltage	V_C	100 V
	Nominal output voltage		50 V
	Nominal output current		10 A
	Capacity rating		500 VA

Fig. 8 shows a general view of the experimental system. The front view shows the control hardware while the rear view shows a total of six stacks of power converters. Each stack of power converter consists of two series connected single-phase H-bridge cells together with their gate-drive circuits and dc-voltage detection boards.

 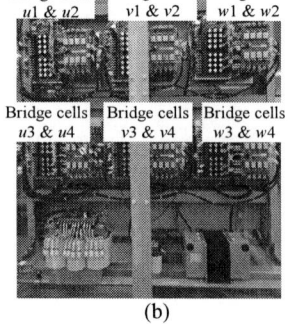

Fig. 8. General view of the three-phase 200-V 6-kVA downscaled system. (a) Front view, and (b) Rear view.

1843

VI. Experimental Results

Figs. 8 through 13 show experimental results obtained from the 200-V 6-kVA downscaled system. The voltage and current waveforms correspond to those in Fig. 7.

A. Steady-state performance

Fig. 9 shows experimental waveforms when the rated positive-sequence reactive power of 6 kVA (leading) was controlled. The input commands, not shown in the figure, were given in terms of the instantaneous active- and reactive-power commands, p^* and q^*.

The converter voltages v_{uv}, v_{vw} and v_{wu} were multilevel (nine-level) PWM waveforms with a voltage step of 100 V. The low-voltage step in the converter voltages made a significant contribution to reducing THD (total harmonic distortion) values of the output currents (e.g. THD value of i_u was as low as 3.1%).

The 12 dc-capacitor voltages, including the three v_{C1u}, v_{C1v} and v_{C1w} shown in the figure, were well balanced at 100 V, due to the dc-voltage control. It is clear from Fig. 9 that the dc-capacitor voltages contain both dc and ac (100-Hz) components. The dc-voltage control, however, regulates the dc component only.

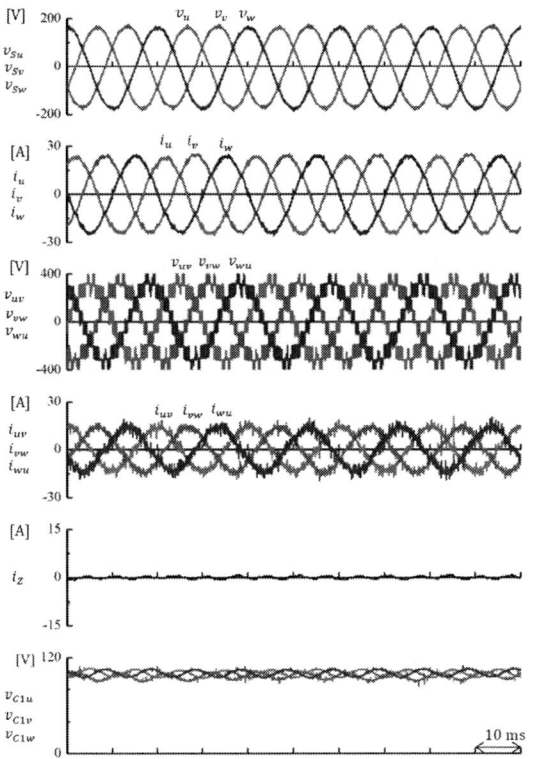

Fig. 9. Experimental waveforms when the rated positive-sequence reactive power of 6 kVA (leading) was controlled.

Fig. 10 shows experimental waveforms when a negative-sequence reactive power of 3 kVA was controlled. The output currents i_u, i_v and i_w were in uwv order. Here, the u-phase acted as an open circuit because i_{uv} was zero. The v-phase acted as a capacitor because i_{vw} led v_{vw} by $90°$, while the w-phase acted as an inductor because i_{wu} lagged v_{wu} by $90°$. The amplitudes of v_{uv}, v_{vw} and v_{wu} were different due to different operating modes.

The 12 dc-capacitor voltages, including the three v_{C1u}, v_{C1v} and v_{C1w} shown in the figure, were well balanced at 100 V. Also shown in the figure is the zero-sequence current i_Z that was circulated within the delta loop of the converter. It helped maintain the balancing of dc-capacitor voltages between the phases even during the control of negative-sequence reactive power.

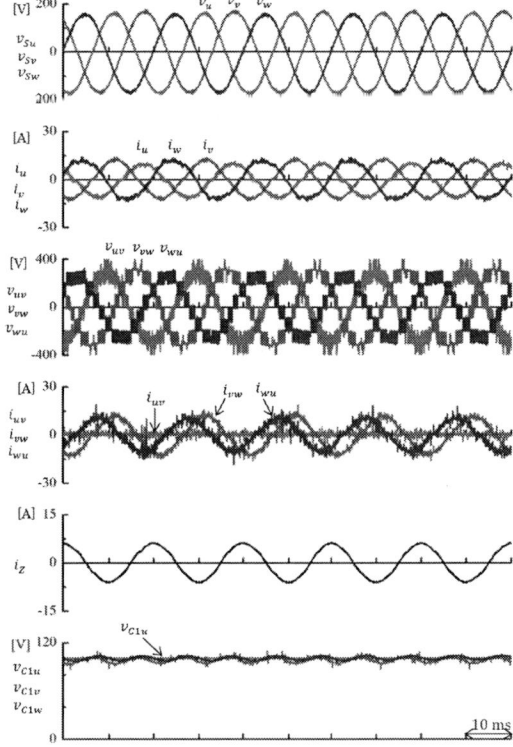

Fig. 10. Experimental waveforms when a negative-sequence reactive power of 3 kVA was controlled.

Fig. 11 shows experimental waveforms when a positive-sequence reactive power of 3 kVA and a negative-sequence reactive power of 3 kVA were controlled simultaneously.

The 2018 International Power Electronics Conference

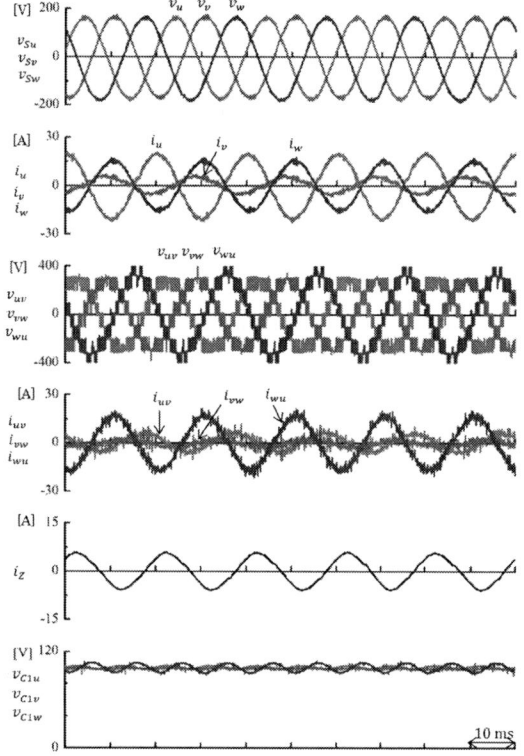

Fig. 11. Experimental waveforms when a positive-sequence reactive power of 3 kVA and a negative-sequence reactive power of 3 kVA were controlled simultaneously.

B. Transient-state performance

Fig. 12 shows experimental waveforms with a step change in positive-sequence reactive-power reference from 0 to 6 kVA (leading). The input commands were given in terms of the instantaneous active- and reactive-power commands, p^* and q^*. The waveforms of 12 dc-capacitor voltages ($v_{C1u} \sim v_{C4u}$, $v_{C1v} \sim v_{C4v}$ and $v_{C1w} \sim v_{C4w}$) show that a maximal dc voltage error of 10% (10 V) occurred during the transient state. As the dc capacitors were designed to have a dc voltage ripple of ±5% at rated power in steady state, the maximal ripple of 10% in such stringent transient state demonstrates an excellent performance.

Fig. 13 shows experimental waveforms with a step change in negative-sequence reactive-power reference from 0 to 3 kVA. As in Fig. 12, the waveforms of 12 dc-capacitor voltages ($v_{C1u} \sim v_{C4u}$, $v_{C1v} \sim v_{C4v}$ and $v_{C1w} \sim v_{C4w}$) show that a maximal dc voltage error of 10% (10 V) occurred during the transient state. Note that the u-phase dc voltages ($v_{C1u} \sim v_{C4u}$) contained no 100-Hz ripple because the u-phase inverter current i_{uv} was zero.

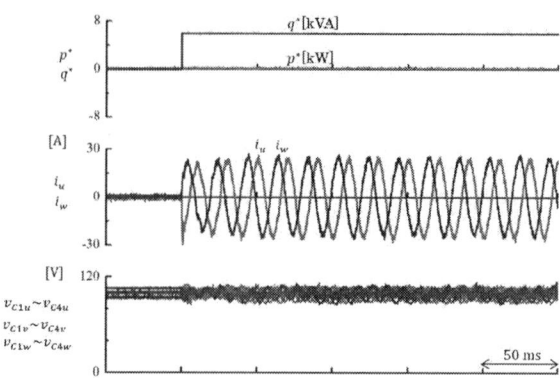

Fig. 12. Experimental waveforms with a step change in positive-sequence reactive power from 0 to 6 kVA (leading).

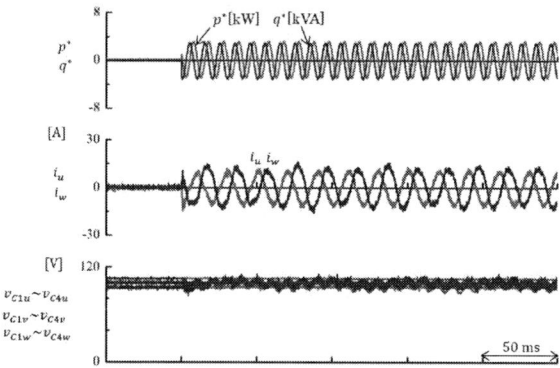

Fig. 13. Experimental waveforms with a step change in negative-sequence reactive power of 0 to 3 kVA.

C. Performance of intra-phase balancing control

Fig. 14 shows experimental waveforms when the intra-phase balancing was intentionally disabled during the control of positive-sequence reactive power of 3 kVA (leading). The dc capacitor voltages in each phase started to deviate from the mean dc voltage in the phase. Soon the dc over-voltage protection was activated. Note that the mean dc-capacitor voltages in the three phases remained balanced because the inter-phase balancing control was not disabled in this experiment. The waveforms of Fig. 14 conclude that the presented intra-phase balancing, which is based on the control of dead times of bridge cells, is effective in balancing the dc-capacitor voltages in a phase.

VII. CONCLUSION

The paper has discussed a 6.6-kV transformerless STATCOM based on the MMCC-SDBC and its control method with focus on dc-capacitor voltage balancing. The paper has provided a detailed description of a new intra-phase balancing technique which is based on the control of dead times of bridge cells. Experimental results based on a three-phase 200-V 6-kVA downscaled system

1845

has shown that the control method can achieve a rapid and independent control of positive- and negative-sequence reactive power, and regulate the dc-capacitor voltages within ±10% of the nominal dc voltage. The control method will be easily extended to the 6.6-kV 200-kVA demonstration system that is currently under development in Japan.

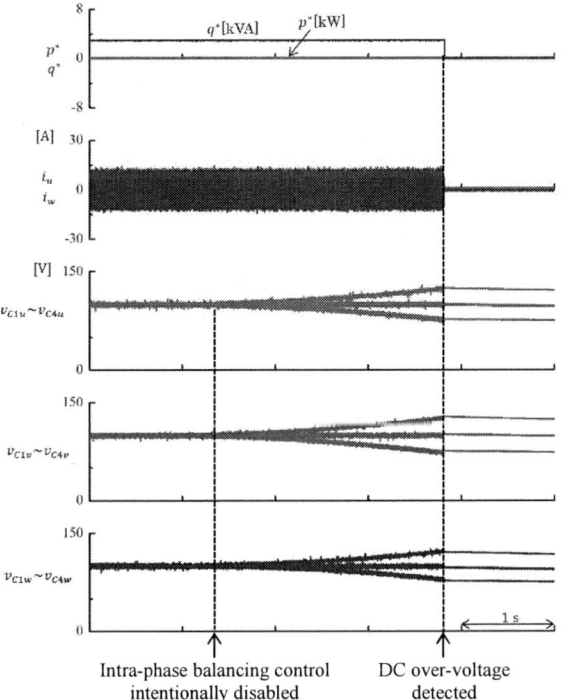

Fig. 14. Experimental waveforms when the intra-phase balancing was intentionally disabled during the control of a positive-sequence reactive power of 3 kVA (leading).

ACKNOWLEDGMENT

This work was supported by Council for Science, Technology and Innovation (CSTI), Cross-ministerial Strategic Innovation Promotion Program (SIP), "Next-generation power electronics" (funding agency: NEDO).

REFERENCES

[1] F. Z. Peng, J. S. Lai, J. W. Mckeever and J. VanCoevering, "A multilevel voltage-source inverter with separate dc sources for static VAr generation," *IEEE Trans. Ind. Appl.*, vol. 32, no. 5, pp. 1130-1138, Sep./Oct. 1996.

[2] C. Schauder, "STATCOM for compensation of large electric arc furnace installations," in *Proc. IEEE PES Summer Meeting*, Jul. 1999, pp. 1109-1112.

[3] Y. Liang and C. O. Nwankpa, "A new type of STATCOM based on cascading voltage-source inverters with phase-shifted unipolar SPWM," *IEEE Trans. Ind. Appl.*, vol. 35, no. 5, pp. 1118-1123, Sep./Oct. 1999.

[4] A. Lesnicar and R. Marquardt, "A new modular voltage source inverter topology," in *Proc. EPE*, Sep. 2003.

[5] F. Z. Peng and J. Wang, "A universal STATCOM with delta-connected cascade multilevel inverter," in *Proc. IEEE PESC*, Jun. 2004, pp. 3529-3533.

[6] K. Fujii, U. Schwarzer and R. W. De Doncker, "Comparison of hard-switched multi-level inverter topologies for STATCOM by loss-implemented simulation and cost estimation," in *Proc. IEEE PESC*, 2005, pp. 340-346.

[7] R. E. Betz, T. Summers, and T. Furney, "Symmetry compensation using a H-bridge multilevel STATCOM with zero sequence injection," in *Proc. IEEE IAS Annu. Meeting*, Oct. 2006, pp. 1724-1731.

[8] C. Han, Z. Yang, B. Chen, A. Q. Huang, B. Zhang, M. R. Ingram and A. Edris, "Evaluation of cascade-multilevel converter-based STATCOM for arc furnace flicker mitigation," *IEEE Trans. Ind. Appl.*, vol. 43, no. 2, pp. 748-755, Mar./Apr. 2007.

[9] H. Akagi, S. Inoue and T Yoshii, "Control and performance of a transformerless cascade PWM STATCOM with star configuration," *IEEE Trans. Ind. Appl.*, vol.43, no.4, pp.1041-1049, Jul./Aug. 2007.

[10] L. Maharjan, S. Inoue, and H. Akagi, "A transformerless energy storage system based on a cascade multilevel PWM converter with star configuration," *IEEE Trans. Ind. Appl.*, vol. 44, no. 5, pp. 1621-1630, Sep./Oct. 2008.

[11] L. Maharjan, S. Inoue, H. Akagi and J. Asakura, "State-of-charge (SOC)-balancing control of a battery energy storage system based on a cascade PWM converter," *IEEE Trans. Power Electron.*, vol.24, no.6, pp.1628-1636, Jun. 2009.

[12] B. Gultekin *et al.*, "Design and implementation of a 154 kV, ±50 MVAr transmission STATCOM based on 21-level cascaded multilevel converter," in *Proc. IEEE ECCE*, Sep. 2010, pp. 3936-3948.

[13] K. Usuki, F. Aoyama and M. Hanamatsu, "Development of SVC control for suppressing voltage fluctuations," in *Proc. IEEE ICPE*, May/Jun. 2011, pp. 2073-2080.

[14] H. J. Knaak, "Modular multilevel converters and HVDC/FACTS: a success story," in *Proc. EPE*, Aug./Sep. 2011.

[15] H. Akagi, "Classification, terminology, and application of the modular multilevel cascade converter (MMCC)," *IEEE Trans. Power Electron.*, vol. 26, no. 11, pp. 3119-3130, Nov. 2011.

[16] M. Hagiwara, R. Maeda and H. Akagi, "Negative-sequence reactive-power control by a PWM STATCOM based on a modular multilevel cascade converter (MMCC-SDBC)," *IEEE Trans. Ind. Appl.*, vol. 48, no. 2, pp. 720-729, Mar./Apr. 2012.

[17] P. Sochor, and H. Akagi, "Theoretical comparison in energy-balancing capability between star- and delta-configured modular multilevel cascade inverters for utility-scale photovoltaic systems," *IEEE Trans. Power Electron.*, vol. 31, no. 3, pp. 1980-1992, Mar. 2016.

[18] P. H. Wu, H. C. Chen, and P. T. Cheng, "Delta-connected cascaded H-bridge converter application in unbalanced load compensation," *IEEE Trans. Ind. Appl.*, vol. 53, no. 2, pp. 1254-1262, 2016.

[19] E. Behrouzian, and M. Bongiorno, "Investigation of negative-sequence injection capability of cascaded H-bridge converters in star and delta configuration," *IEEE Trans. Power Electron.*, vol.32, no.2, pp.1675-1683, Feb. 2017.

[20] H. Akagi, "Multilevel converters: Fundamental circuits and systems," *Proc. IEEE,* early access article.

[21] L. Maharjan, T. Tajyuta, A. Suzuki, A. Toba, Y. Matsumoto, and H. Akagi, "Control of a transformerless STATCOM based on the MMCC-SDBC (modular multilevel cascade converter – single-delta bridge-cells)," in *Proc. EPE*, Sep. 2017.

[22] J. J. Jung, J. H. Lee, S. K. Sul, G. T. Son, and Y. H. Chung, "DC capacitor voltage balancing control for delta-connected cascaded H-bridge STATCOM considering unbalanced grid and load conditions," *IEEE Trans. Power Electron.*, early access article.

[23] Y. Koyama, Y. Nakazawa, H. Mochikawa, A. Kuzumaki, K. Sano, and N. Okada, "A transformerless 6.6-kV STATCOM based on hybrid cascade multilevel converter using SiC devices," *IEEE Trans. Power Electron.*, early access article.

The 2018 International Power Electronics Conference

Isolated Three–Phase AC/DC Converter Using a Soft–Switching Technique for Battery Charger

Yuto Matsui, Kazuma Suzuki, Takaharu Takeshita, and Wataru Kitagawa

Nagoya Institute of Technology, Gokiso, Showa, Nagoya, Japan

E-mail: cjh14113@nitech.jp, 28513004@stn.nitech.ac.jp, take@nitech.ac.jp, kitagawa.wataru@nitech.ac.jp

Abstract—**This paper presents a high-frequency isolated three-phase AC/DC converter using a soft-switching technique, which consists of primary matrix converter, high-frequency link transformer, and secondary diode rectifier circuit. The authors propose the control method of the high-frequency isolated three-phase AC/DC converter to satisfy the soft switching conditions. The effectiveness of the proposed PWM strategy and soft switching in the high-frequency isolated three-phase AC/DC converter have been verified by experiments.**

Keywords—*high–frequency isolated three-phase AC/DC converter, soft–switching technique, matrix converter*

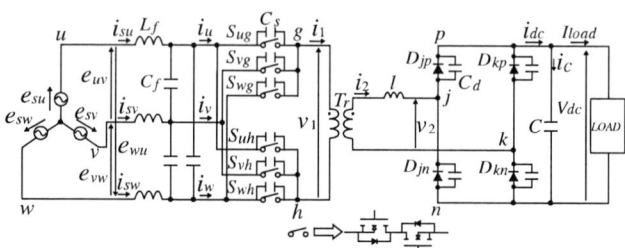

Fig. 1.   Isolated three-phase AC/DC converter

I. INTRODUCTION

In recent years, the conservation of resources and energy is required because the global environmental problems are serious, which are the exhaustion of resources and the global warming. The electric vehicle is one of solutions of these problems. Carbon dioxide emissions which are a cause of the global warming can be reduced because the electric vehicles don't emit carbon dioxide while driving. The electric vehicles can also be used as a household battery and an emergency power source. So, the development and spread of the electric vehicles are promoted. Moreover, more and more people will begin to use the electric vehicles because the travel distance is extended by being improved the performances of the batteries. The electric vehicles are charged with DC, but the electrical systems are operated on AC. An AC/DC converter to convert AC to DC is required. Also, it is necessary to prevent electric shocks during charging and prevent breaking down the equipments by accidents. The isolated AC/DC converter which has electric isolation by using transformer is required for protection and safety. The isolated AC/DC converter using a transformer of commercial frequency is composed of AC/DC rectifier circuit and DC/DC chopper circuit. The circuit system is large and heavy because it is used the transformer of commercial frequency and the large electrolytic capacitor. Therefore, the isolated AC/DC converters using a high-frequency transformer which is small and light are proposed.

One of the circuits has configuration using the inverter [1]-[4]. In this circuit, the AC/DC converter converts three phase AC voltage into DC voltage, the inverter converts DC voltage into high–frequency AC voltage, and the AC/DC converter converts high–frequency AC voltage into DC voltage. The size and the power loss of the power converters are increased because this circuit has three power converters. Another circuit has configuration using three–phase to single–phase matrix

converter [5]-[8]. In this circuit, the three–phase to single–phase matrix converter converts three phase AC voltage into high–frequency AC voltage, and the AC/DC converter converts high–frequency AC voltage into DC voltage. The size and the power loss of the power converters are reduced because this circuit has two power converters. The sizes of the transformer and the LC filters of the input and output sides can be reduced by using higher frequency. However, the higher switching frequency causes the larger switching loss because of using hard switching in these converters. The soft switching technique in the converters is required to reduce the switching loss [8]-[13].

This paper presents a high-frequency isolated three-phase AC/DC converter using a soft-switching technique, which controls the input current and the primary voltage, simultaneously. In order to reduce the converter loss, the authors propose the simple converter configuration composed primary three–phase to single–phase matrix converter, high–frequency transformer and secondary diode rectifier circuit. Also, the control method to satisfy the soft–switching conditions which are expressed by using the primary voltage and the primary current is proposed. In the proposed control method, the primary current reference is approximated by the sawtooth AC waveform to suppress the distortion of the source current. The effectiveness of the proposed circuit and control method which satisfies the soft-switching conditions have been verified by experiments.

II. ISOLATED THREE–PHASE AC/DC CONVERTER

A. Main circuit

Fig.1 shows an isolated three-phase AC/DC converter. The primary matrix converter and the secondary diode rectifier circuit are magnetically coupled through the high–frequency transformer with the turn ratio of $N_1/N_2 = 1$. The primary three–phase to single–phase matrix converter is composed

1847

The 2018 International Power Electronics Conference

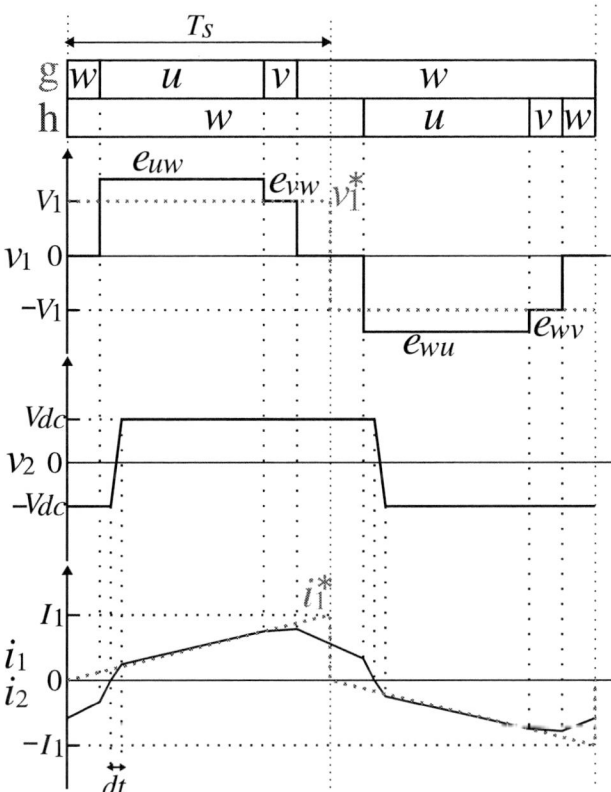

Fig. 2. Transformer voltage and current

TABLE I. SOFT–SWITCHING CONDITIONS BETWEEN PRIMARY VOLTAGE AND CURRENT

Switch	Primary voltage v_1	Primary current i_1
$S_{wg} \to S_{ug}$	$0 \to e_{uw}$	$i_1 < 0$
$S_{ug} \to S_{vg}$	$e_{uw} \to e_{vw}$	$i_1 > 0$
$S_{vg} \to S_{wg}$	$e_{vw} \to 0$	$i_1 > 0$
$S_{wh} \to S_{uh}$	$0 \to e_{wu}$	$i_1 > 0$
$S_{uh} \to S_{vh}$	$e_{wu} \to e_{wv}$	$i_1 < 0$
$S_{vh} \to S_{wh}$	$e_{wu} \to 0$	$i_1 < 0$

voltage reference v_1^* is given in the square–wave AC voltage with the amplitude V_1 in one period $2T_s$, and the primary current reference i_1^* is given in the sawtooth AC waveform with the amplitude I_1 in one period $2T_s$. The primary voltage waveform v_1 is determined by the switching patterns of the phase g and the phase h and the instantaneous value of the source line voltage. When the primary voltage reference v_1^* is positive, the switching patterns of each phase are shown in Fig.2. The switches in the phase g are turned on in the order of the minimum voltage phase w, the maximum voltage phase u, the intermediate voltage phase v, and the minimum voltage phase w. In the phase h, the switch of the minimum voltage phase w is always turned on. When the primary voltage reference v_1^* is negative, the primary voltage v_1 is realized as shown in Fig.2 by replacing the switching patterns of the phase g and the phase h in each other under the positive primary voltage reference ($v_1^* > 0$). Then, the average value of the primary voltage $\overline{v_1}$ during half–period of high–frequency T_s is equal to the primary voltage reference v_1^*. The secondary voltage v_2 is obtained the trapezoidal AC voltage with the amplitude of DC voltage V_{dc}. The secondary voltage v_2 is stepping up and down when the transformer current is equal to zero. The parallel capacitor of the diodes C_d are charged and discharged. The minute time dt in Fig.2 is the time when the secondary voltage v_2 is changing the sign, and is changed by the value of the parallel capacitors of the diodes C_d.

C. Soft switching

Table I shows the soft–switching conditions between the primary voltage v_1 and the primary current i_1. The soft–switching condition has two patterns. One pattern, when the primary voltage v_1 changes from lower voltage to higher voltage, the primary current i_1 should be negative. As am example, at the switching from S_{wg} to S_{ug} in Fig.2, the primary voltage v_1 changes from zero to higher voltage e_{uw}, then the primary current i_1 should be negative. On the contrary, when the primary voltage v_1 changes from higher voltage to lower voltage, the primary current i_1 should be positive. For example, at the switching from S_{ug} to S_{vg} in Fig.2, the primary voltage v_1 changes from higher voltage e_{uw} to lower voltage e_{vw}, then the primary current i_1 should be positive. All of the switching in Fig.2 satisfy the soft–switching conditions in Table I.

Fig.3 shows the soft–switching commutation of the primary converter from the switch S_{wg} to S_{ug} in Fig.2. In all modes in Fig.3, the primary current i_1 is negative, and the

of the six bi–directional switches $S_{ug} - S_{wh}$, and converts the three–phase AC voltage into single–phase high–frequency AC voltage. The bi–directional switch consists of two SiC–MOSFETs connected to each other in series and opposite direction. The secondary diode rectifier circuit is composed of the four diodes $D_{jp} - D_{kn}$, and converts the single–phase high–frequency AC voltage into DC voltage. The filter reactor L_f and the filter capacitor C_f are connected to the input side of the primary converter in order to suppress the outflow of harmonic current to the power supply. Also, the reactor l is connected to the high–frequency transformer for suppressing change of the transformer current, and the capacitors C_d are connected to the diodes in parallel for suppressing the distortion of the source current. The capacitors C_s are connected to the bi–directional switches of the primary converter in parallel for the soft–switching.

B. Operation of High–Frequency Transformer

Fig.2 shows the voltage and current waveforms of the high–frequency transformer in one period $2T_s$ in the condition of the source voltage $e_{su} > e_{sv} > 0 > e_{sw}$ under the source angle $\pi/6 \le \theta \le \pi/3$. The excitation current is ignored as sufficiently small compared with the primary and secondary currents i_1 and i_2. In Fig.2, the primary current i_1 and the secondary current i_2 are the same waveforms. The primary

1848

The 2018 International Power Electronics Conference

phase h is always connected to the switch S_{wh}. In the mode1, the primary current i_1 flows through the switch S_{wg}. Then, the parallel capacitors of the maximum voltage phase u and the intermediate voltage phase v are positively charged on the side of the power source because the output phase g and the minimum voltage phase w are the same potential. The parallel capacitor of the minimum voltage phase w is not charged. When the switch S_{wg} is turned off, the mode changes into the mode2. In the mode2, the primary current i_1 flows through three parallel capacitors in the phase g. Then, the switch S_{wg} can be realized zero voltage switching (ZVS) because the parallel capacitor of the minimum voltage phase w is not charged. The gate signal of the leftward switch of the switch S_{ug} has to be turned on by being completed the discharge of the parallel capacitor of the maximum voltage phase u. When the discharge of the parallel capacitor of the maximum voltage phase u is completed, the mode changes into the mode3. In the mode3, the primary current i_1 flows through the switch S_{ug}. Then, the switch S_{ug} can be realized zero voltage switching (ZVS) because the parallel capacitor of the maximum voltage phase u is not charged. The parallel capacitors of the intermediate voltage phase v and the minimum voltage phase w are positively charged on the side of the transformer because the output phase g and the maximum voltage phase u are the same potential. Also, the gate signal of the rightward switch of the switch S_{ug} has to be turned on by changing the primary current i_1 from negative to positive.

Fig.4 shows the soft–switching commutation of the primary converter from the switch S_{ug} to S_{vg} in Fig.2. In all mode in Fig.4, the primary current i_1 is positive, and the phase h is always connected to the switch S_{wh}. In the mode1, the primary current i_1 flows through the switch S_{ug}. Then, the parallel capacitors of the intermediate voltage phase v and the minimum voltage phase w are positively charged on the side of the transformer because the output phase g and the maximum voltage phase u are the same potential. The parallel capacitor of the maximum voltage phase u is not charged. When the switch S_{ug} is turned off, the mode changes into the mode2. In the mode2, the primary current i_1 flows through three parallel capacitors in the phase g. Then, the switch S_{ug} can be realized zero voltage switching (ZVS) because the parallel capacitor of the maximum voltage phase u is not charged. The gate signal of the rightward switch of the switch S_{vg} has to be turned on by being completed the discharge of the parallel capacitor of the intermediate voltage phase v. When the discharge of the parallel capacitor of the intermediate voltage phase v is completed, the mode changes into the mode3. In the mode3, the primary current i_1 flows through the switch S_{vg}. Then, the switch S_{vg} can be realized zero voltage switching (ZVS) because the parallel capacitor of the intermediate voltage phase v is not charged. The parallel capacitor of the maximum voltage phase u is positively charged on the side of the power source and the parallel capacitor of the minimum voltage phase w is positively charged on the side of the transformer because the output phase g and the intermediate voltage phase v are the same potential.

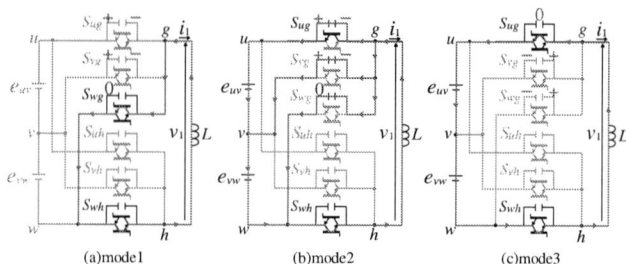

Fig. 3.   Soft–switching commutation from S_{wg} to S_{ug}

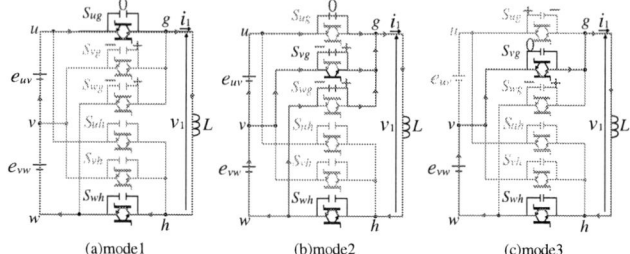

Fig. 4.   Soft–switching commutation from S_{ug} to S_{vg}

III. Control Method

A. Input Current References

The capacitor voltages can be approximated by the source voltages e_{su}, e_{sv}, and e_{sw} because the voltage drops across the input reactors L_f in Fig.1 are ignored as enough small compared with the source voltages. The source voltages e_{su}, e_{sv}, and e_{sw} are given by using the source line voltage effective value E and the phase angle θ as follows;

$$\begin{bmatrix} e_{su} \\ e_{sv} \\ e_{sw} \end{bmatrix} = \sqrt{\frac{2}{3}}E \begin{bmatrix} \cos\theta \\ \cos(\theta - 2\pi/3) \\ \cos(\theta - 4\pi/3) \end{bmatrix} \quad (1)$$

Also, the input current references i_u^*, i_v^*, and i_w^* are given by using the effective value of the input current references I and the power factor angle reference φ^* as follows;

$$\begin{bmatrix} i_u^* \\ i_v^* \\ i_w^* \end{bmatrix} = \sqrt{2}I \begin{bmatrix} \cos(\theta + \varphi^*) \\ \cos(\theta + \varphi^* - 2\pi/3) \\ \cos(\theta + \varphi^* - 4\pi/3) \end{bmatrix} \quad (2)$$

The input instantaneous power p_{in} is expressed by using (1) and (2) as follows;

$$p_{in} = e_{su}i_u^* + e_{sv}i_v^* + e_{sw}i_w^* = \sqrt{3}EI\cos\varphi^* \quad (3)$$

The effective value of the input current references I is determined by the balance between the input power and the output power. Assuming that the power loss in the primary side and secondary side power converters and the high–frequency transformer are nothing, the input instantaneous power p_{in} is equal to the output instantaneous power p_{out}. Therefore, the effective value of the input current references I is obtained by

1849

using (3) as follows;

$$I = \frac{p_{out}}{\sqrt{3}E\cos\varphi^*} \tag{4}$$

B. Duty Cycles

Only one of the three switches on the each output phase g and h has to be always turned on for preventing the current discontinuous and the short circuit between the switches. From this condition, the duty cycles $d_{ug} - d_{wh}$ of the six bi–directional switches $S_{ug} - S_{wh}$ during half–period T_s of the high–frequency transformer are given as follows;

$$d_{ug} + d_{vg} + d_{wg} = 1 \tag{5}$$
$$d_{uh} + d_{vh} + d_{wh} = 1 \tag{6}$$

For example, the duty cycles in the condition of the source voltage $e_{su} > e_{sv} > 0 > e_{sw}$ under the source angle $\pi/6 \le \theta \le \pi/3$ in Fig.2 are derived. The primary current reference i_1^* is given as follows;

$$i_1^* = \frac{I_1}{T_s}t \tag{7}$$

The duty cycles are determined uniquely by using the equation that two of the six duty cycles are equal to zero. When the primary voltage reference v_1^* is positive, the switch S_{wh} of the minimum voltage phase w is always turned on in the phase h. The duty cycles d_{uh}, d_{vh}, and d_{wh} are given as follows;

$$d_{uh} = 0, \ d_{vh} = 0, \ d_{wh} = 1 \tag{8}$$

The average values of the input currents $\overline{i_u}$, $\overline{i_v}$ during half–period T_s of high–frequency transformer are given as follows;

$$\overline{i_u} = \frac{1}{T_s}\int_{0.5d_{wg}T_s}^{(0.5d_{wg}+d_{ug})T_s} i_1^* dt$$
$$= \frac{I_1}{2}d_{ug}(d_{ug} + d_{wg}) \tag{9}$$

$$\overline{i_v} = \frac{1}{T_s}\int_{(0.5d_{wg}+d_{ug})T_s}^{(1-0.5d_{wg})T_s} i_1^* dt$$
$$= \frac{I_1}{2}(1 - d_{wg} - d_{ug}d_{wg} - d_{ug}{}^2) \tag{10}$$

The following equation regarding the input current reference and the average values of the input currents is obtained in order to control the source current as the sinusoidal current.

$$i_u^* : i_v^* = \overline{i_u} : \overline{i_v} \tag{11}$$

Also, the following equation regarding the duty cycles is obtained because the primary voltage reference v_1^* is equal to the average value of the primary voltage during half–period T_s of high–frequency transformer.

$$v_1^* = e_{uw}d_{ug} + e_{vw}d_{vg} \tag{12}$$

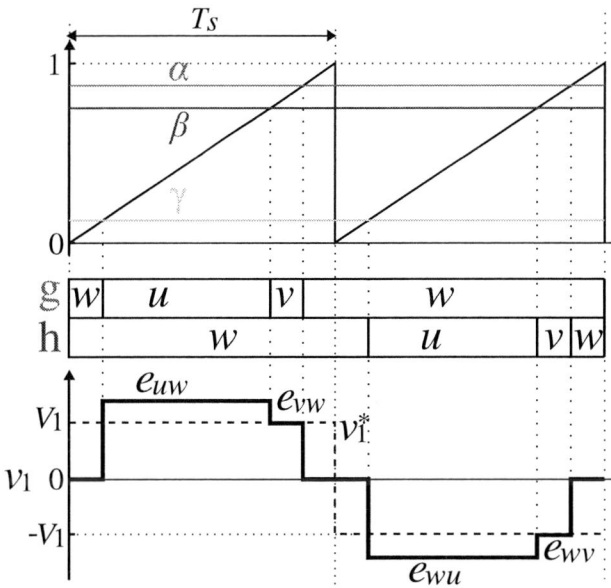

Fig. 5. Sawtooth wave comparison and switching pattern

Therefore, the duty cycles d_{ug}, d_{vg}, and d_{wg} are obtained by using (5), (6), (8) (12) as follows;

$$\left. \begin{aligned} d_{ug} &= \frac{-A + \sqrt{B^2 - 4AC}}{2A} \\ d_{vg} &= \frac{v_1^* - e_{uw}d_{ug}}{e_{vw}} \\ d_{wg} &= 1 - \frac{-A + \sqrt{B^2 - 4AC}}{2A} - \frac{v_1^* - e_{uw}d_{ug}}{e_{vw}} \end{aligned} \right\} \tag{13}$$

The coefficients A, B, and C in (13) are expressed as follows;

$$\left. \begin{aligned} A &= (i_u^* + i_v^*)e_{uw} \\ B &= (e_{uw} - v_1^*)i_u^* + (e_{vw} - v_1^*)i_v^* \\ C &= -i_u^* v_1^* \end{aligned} \right\} \tag{14}$$

The duty cycles d_{ug}-d_{wh} when the primary voltage reference v_1^* is negative are obtained by replacing the duty cycles of the phase g in (13) and the duty cycles of the phase h in (8). Also, the input currents i_u, i_v, i_w and the primary voltage v_1 can be obtained coincident with each reference by the PWM control to use these duty cycles.

C. PWM Strategy

The sawtooth wave comparison and the switching pattern in the condition of the source voltage $e_{su} > e_{sv} > 0 > e_{sw}$ under the source angle $\pi/6 \le \theta \le \pi/3$ in Fig.2 are explained. Fig.5 shows the sawtooth wave comparison and the switching pattern. The gate signals of each switch S_{ug}-S_{wh} are generated by comparing the comparison lines α, β, and γ with the sawtooth wave which changes from zero to one during half–period T_s of high–frequency transformer. The input currents i_u, i_v, i_w and the primary voltage v_1 are controlled coincident with thier references by using the gate

1850

The 2018 International Power Electronics Conference

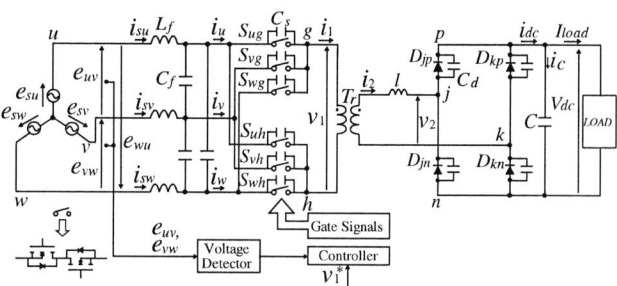

Fig. 6. Experimental system configuration

TABLE II. SPECIFICATIONS FOR EXPERIMENTS

Source voltage E, ω	200 V, $2\pi \times 60$ rad/s
DC voltage V_{dc}	180 V
Output power P_{out}	1200 W
Load R	27.0 Ω
Reactor l	300 μH
Capacitor of switch C_s	1 nF
Capacitor of diode C_d	20 nF
Input filter L_f, C_f	1.0 mH, 7.1 μF
Capacitor C	30 μF
Frequency of transformer	10 kHz
Ratio of transformer $N_1 : N_2$	1 : 1

signals which are generated by the sawtooth wave comparison. Also, three comparison lines α, β, and γ are expressed by using the duty cycles as follows;

$$\left.\begin{aligned} \alpha &= 1 - 0.5d_{wg} \\ \beta &= d_{ug} + 0.5d_{wg} \\ \gamma &= 0.5d_{wg} \end{aligned}\right\} \quad (15)$$

When the primary voltage reference v_1^* is positive, the gate signals of the switches S_{ug}-S_{wg} connected to the phase g are generated by comparing the comparison lines α, β, and γ with the sawtooth wave. The gate signal of the switch S_{wg} is turned on when the comparison line γ is larger than the sawtooth wave or the comparison line α is smaller than the sawtooth wave. The gate signal of the switch S_{ug} is turned on when the comparison line β is larger than the sawtooth wave and the comparison line γ is smaller than the sawtooth wave. The gate signal of the switch S_{vg} is turned on when the comparison line α is larger than the sawtooth wave and the comparison line β is smaller than the sawtooth wave. In the phase h, the gate signal of the switch S_{wh} is always turned on, and the gate signals of the switches S_{uh} and S_{vh} are always turned off. On the other hand, when the primary voltage reference v_1^* is negative, the switching pattern can be obtained by replacing the switching pattern of the phase g and the phase h during half–period T_s when the primary voltage reference is positive.

IV. EXPERIMENTAL RESULTS

A. Experimental Conditions

Fig.6 shows the experimental system configuration of the isolated thee-phase AC/DC converter. The bi–directional switches for the primary matrix converter are used the SiC–MOSFETs (CREE, C2M0040120D) and the diodes for the secondary diode rectifier circuit are used the SiC-SBDs (ROHM, SCT2120AF). The controller is a DSP. The duty cycles of the switches are calculated by using the detected values of the source line voltages e_{uv} and e_{vw} which are obtained by the detectors. The gate signals of the switches based on the calculated duty cycles are generated by using the FPGA. Table II shows the experimental conditions. The effective value of the source line voltage E is 200 V and the frequency is 60 Hz. The frequency of high-frequency transformer is 10 kHz. The parallel capacitor C_s to realize soft switching is 1 nF.

B. Experimental Waveforms

Fig.7 shows the experimental waveforms. The experimental waveforms are the source voltage e_{su}, the source current i_{su}, the input current i_u, the transformer voltages v_1, v_2, the transformer currents i_1, i_2, the DC voltage V_{dc}, and the load current I_{load}. In Fig.7, the input current i_u is includes the harmonic current, but the source current is controlled to the sinusoidal current by connecting the input LC filter to the input side of the primary converter. Also, the DC voltage is controlled to 180 V. Fig.8 shows the magnification waveforms of the transformer voltage and current in Fig.7. From Fig.8, the transformer voltage and current are controlled to the theoretical waveforms in Fig.2. The excitation current is ignored as sufficiently small because the primary current i_1 and the secondary current i_2 are the same waveforms. At the commutation from S_{wg} to S_{ug} on (a) in Fig.8, the soft–switching condition is satisfied because the primary current i_1 is negative. At the commutation from S_{ug} to S_{vg} and the commutation from S_{vg} to S_{wg} on (b) and (c) in Fig.8, the soft–switching condition is satisfied because the primary current i_1 is positive. The all of the commutations in Fig.8 satisfy the soft–switching conditions in Table I.

C. Switching Loss and Efficiency

Fig.9 shows the waveforms of the voltage and current of switch during the commutation from the switch S_{wg} to the switch S_{ug} in both circuits with and without the parallel capacitors of the switches C_s. The switching loss is reduced by connecting the parallel capacitors C_s to the switches of the primary converter because the overlap of the voltage and the current of the switches is reduced.

The efficiency is measured by the power analyzer (Yokogawa Electric, WT3000). Fig.10 shows the efficiency measurement result in both circuits with and without the parallel capacitors of the switches C_s under the constant load current I_{load} of 6.67 A. The soft switching is performed in the all of the condition in Fig.10. From Fig.10, when the output power P_{out} is 1200W, the efficiency of the circuit with the parallel capacitors C_s is 95.0%, and the efficiency of the circuit without the parallel capacitors C_s is 94.9%.

V. CONCLUSIONS

This paper presents the high–frequency isolated three-phase AC/DC converter using soft switching technique. The soft

The 2018 International Power Electronics Conference

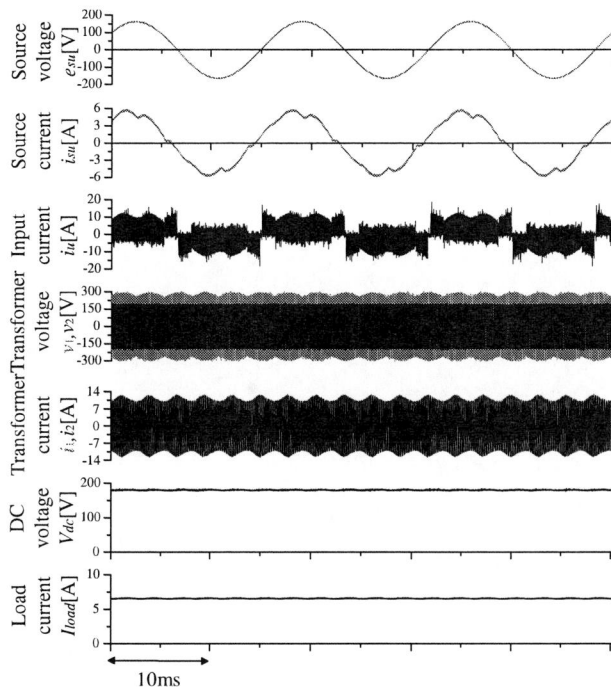

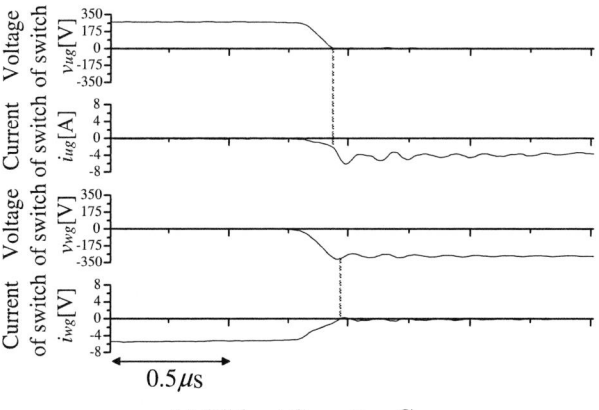

(a) Without Capacitor C_s

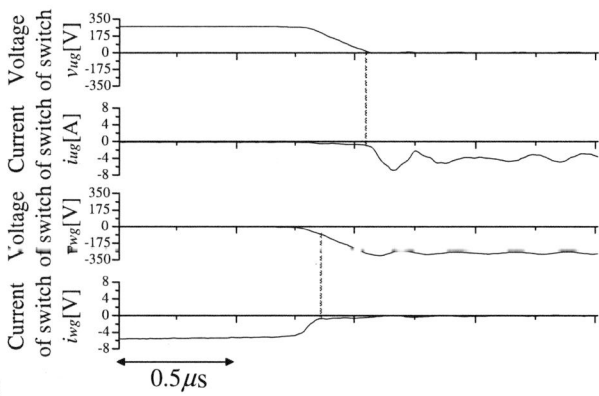

Fig. 7. Experimental waveforms

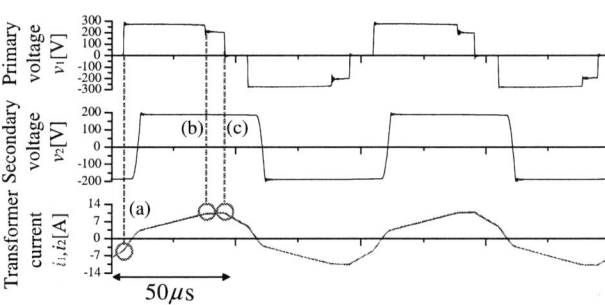

(b) With Capacitor C_s

Fig. 9. Waveforms of switch

Fig. 8. magnification waveforms

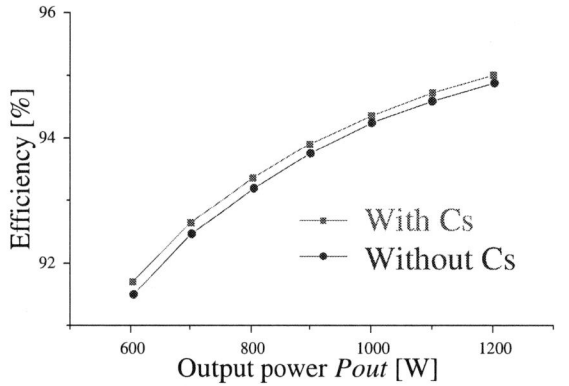

Fig. 10. Efficiency

switching in the primary matrix converter can be achieved in all of the commutaions, and the switching loss is reduced by connecting the parallel capacitors C_s to the switches of the primary matrix converter because the overlap of the voltage and the current of the switches is reduced. The effectiveness of the proposed control method and soft–switching in the high–frequency isolated three-phase AC/DC converter have been verified by experiments.

This work was supported by Council for Science, Technology and Innovation (CSTI), Cross-ministerial Strategic Innovation Promotion Program (SIP),"Next-generation power electronics" (funding agency: NEDO).

Electronics Conference and Exposition (APEC), pp.761-767 (2007)

[2] M. J. Erfani, T. Thiringer and S. Haghbin, "Performance and losses analysis of charging and discharging mode of a bidirectional DC/DC fullbridge converter using PWM switching pattern", *2011 IEEE Vehicle Power and Propulsion Conference*, pp.1-6 (2011)

[3] J. Everts, J. Van den Keybus, F. Krismer, J. Driesen and J. W. Kolar,

REFERENCES

[1] S. Inoue and H. Akagi, "A Bi-Directional DC/DC Converter for an Energy Storage System", *Twenty-Second Annual IEEE Applied Power*

"Integrated capacitor for common-mode EMI mitigation applicable to high frequency planar transformers used in electric vehicles DC/DC converters", *2012 Twenty-Seventh Annual IEEE Applied Power Electronics Conference and Exposition (APEC)*, pp.1048-1055(2012)

[4] D. Hamza and M. Pahlevaninezhad, "Switching control strategy for full ZVS soft-switching operation of a Dual Active Bridge AC/DC converter", *2014 IEEE Energy Conversion Congress and Exposition (ECCE)*, pp.4809-4814(2014)

[5] S. Manias and P. D. Ziogas, "A Novel Sinewave in AC to DC Converter with High-Frequency Transformer Isolation", *IEEE Transactions on Industrial Electronics*, Vol.IE-32, No.4, pp.430-438(1985)

[6] K. Inagaki, T. Furuhashi, A. Ishiguro, M. Ishida and S. Okuma, ""A New PWM Control Method for ac to dc Converters with High–Frequency Transformer Isolation", *IEEE Transactions on Industrial Electronics*, Vol.29, No.3, pp.486-492(1993)

[7] Amit K Singh, Pritam Das, S K Panda, "A novel matrix based isolated three phase AC-DC converter with reduced switching losses", *2015 IEEE Applied Power Electronics Conference and Exposition (APEC)*, pp.1875-1880(2015)

[8] Chushan Li, Yu Zhang, David Xu: "Soft-switching three-phase matrix based isolated AC-DC converter for DC distribution system" *2015 IEEE Energy Conversion Congress and Exposition (ECCE)*, pp.6755-6761(2015)

[9] Y. Zhaoyang, Z. Kun, J. Minli and W. Weiyang: "SPWM Modulation Strategy Study of AC/DC Matrix Converter with High Frequency Isolation" *International Conference on Electrical Machines and Systems (ICEMS)*, pp.3789-3792(2008)

[10] A. Ambikapathy, Gajendra Singh, Ashish Shrivastava: "Efficient soft-switching dc-dc converter for MPPT of a grid connected PV system" *2016 International Conference on Computing, Communication and Automation (ICCCA)*, pp.934-938(2016)

[11] Chushan Li, Yu Zhang, David Xu: "Soft-switching single stage isolated AC-DC converter for single-phase high power PFC applications" *2015 9th International Conference on Power Electronics and ECCE Asia (ICPE-ECCE Asia)*, pp.1103-1108(2015)

[12] Khaled A. Mahafzah, Klaus Krischan, Annette Muetze: "Efficiency enhancement of a three phase Soft Switching Inverter under light load conditions" *IECON 2016 - 42nd Annual Conference of the IEEE Industrial Electronics Society*, pp.3378-3383(2016)

[13] Jordi Everts, Jeroen Van den Keybus, Florian Krismer, Johan Driesen, Johann W. Kolar: "Switching control strategy for full ZVS soft-switching operation of a Dual Active Bridge AC/DC converter" *2012 Twenty-Seventh Annual IEEE Applied Power Electronics Conference and Exposition (APEC)*, pp.1048-1055(2012)

Implementation of a Miniaturized SiC Inverter

Hideaki Fujita and Cristian Andres Garces Guajardo
Tokyo Institute of Technology
hf@ieee.org

Abstract—**This paper presents analysis, design, and implementation of a miniaturized inverters using SiC-MOSFETs. Theoretical analysis and experimental verification in this paper reveals the relationship between the time-domain switching properties and the frequency-domain impedance characteristics. Based on the analysis, damping resistors are inserted into the dc side of the inverter. The damping resistors can effectively reduce the impedance at resonant frequencies in frequency domain, and they also enable to suppress the surge voltage and oscillation across the switching devices in time domain. Moreover, it has theoretically and experimentally been confirmed that the damping resistors also have the capability of reducing the switching power losses in the circuit.**

I. INTRODUCTION

Currently, SiC devices are available in the market, and it is possible to reduce the power losses of inverters by replacing an Si-IGBT module with an SiC-MOSFET module. The SiC-MOSFETs have a low on-state resistance with a relatively high voltage rating of more than 1200 V, which can be used instead of 1200-V class IGBTs. Moreover, since the SiC-MOSFETs have the fast switching capability, it is expected to increases the switching frequency to a higher frequency than IGBTs. In this case, the components used in the circuit has to suffer a very high dv/dt and di/dt and to suppress the induced electromagnetic interferences (EMI).

Various analysis and design methods are proposed to reduce the EMI issue induced by a PWM inverters, such as application of a common-mode chokes [4], active common mode cancelers [5], and multilevel topologies [6]. They mainly focuses on the common-mode leakage current because it produces a relatively strong conduction EMI. The frequency band of the conductive EMI is essentially in a range from 150 kHz to 30 MHz in a standard or regulation. The SiC-MOFET has a very fast turn-on and turn-off times of a few tens nano seconds or less at a relatively high dc-link voltage. This quick switching operation would cause a serious EMI issues at a higher frequency band.

For these reasons, it becomes attractive to improve the layout of the switching devices and passive components on printed circuit boards (PCBs)[7]-[9]. In these PCB design, it is discussed various problems to make the circuit small, eg., crosstalk between the power lines, reduction of the radiated EMI, temperature and thermal design, location of the cooling devices, and so on. In the miniaturized design, it is very important to pay attention to the size of components. Especially, the dc capacitors occupies a relatively large footprint in the power circuit PCB. Recent polymer film capacitors have a large current ripple ratings and are suitable as a dc

capacitor. However, they have a relatively large equivalent series inductance (ESL), and thus, it is required to add a capacitor having a better high-frequency response. Ceramic capacitors seems to be suitable for this purpose although its capacitance is relatively small. Therefore, the combination of film and ceramic capacitors are often used in the dc side of the power circuit PCB.

This paper presents analysis, design, and implementation of a miniaturized inverters using SiC-MOSFETs. In this paper, theoretical analysis and experimental verification has been conducted to reveal the relationship between the time-domain switching properties and the frequency-domain impedance characteristics. Based on the analysis, an inverter was designed and implemented to drive a 400-V, 15-kW induction motor using 1200-V SiC-MOSFETs. The newly-designed inverter circuit is equipped with damping resistors on its dc side to suppress the resonance between the dc capacitors and stray inductance. The added damping resistors make it possible to reduce the impedance at resonant frequencies in frequency domain successfully, and thus, they also make it possible to suppress the surge voltage, overshoot, and oscillation across the drain-to-source terminals of the switching devices in time domain. Moreover, it has theoretically and experimentally been confirmed that the damping resistors also have the capability of reducing the switching power losses in the circuit.

II. CIRCUIT CONFIGURATION

Fig. 1 shows circuit configuration of a traditional three-phase voltage-source inverter, which consists of six switching devices and a dc capacitor. In general, a three-phase inductive load should be connected to the three-phase ac output of the inverter. A practical inverter is usually would be connected to a three-phase rectifier with a relatively-large electrolytic capacitor as the power source. In this paper, the dc input is assumed to be regulated at a constant voltage E instead of a rectifier, as shown in Fig. 1 to make the following analysis simple.

The experimental setup also employs the same circuit configuration essentially. The main circuit is constructed on a printed circuit board (PCB) as well as the gate drive circuit. SiC-MOSFET (C2M0025120D: 1200 V, 90 A, Wolfspeed) is used as the switching devices. Two SiC-MOSFETs are connected in parallel and used as an arm of the inverter, and thus, the inverter uses twelve MOSFETs totally. Six separated heat sink are attached to the MOSFETs. No insulator is inserted between the heat sink and the MOSFET because each heat sink is isolated from the other arm. This makes it possible to reduce the thermal resistance.

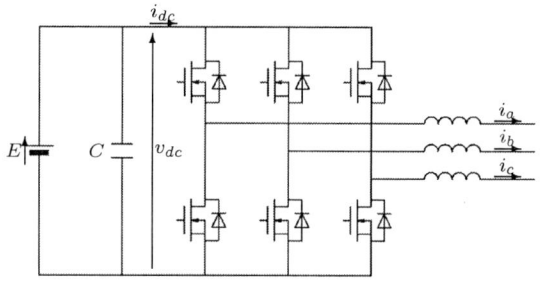

Fig. 1. Circuit configuration of the developed inverter.

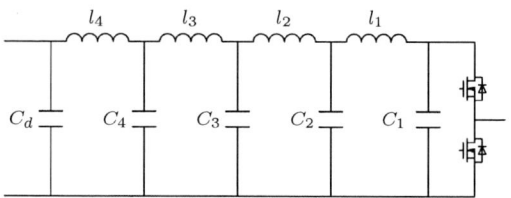

Fig. 2. Equivalent dc-link circuit of the inverter.

Fig. 2 shows the configuration of the dc-link capacitors, and the circuit parameter of the dc-link components are shown in TableI. The inverter is equipped with ceramic capacitors of $C_1 = 20$ nF, $C_2 = 33$ nF, and $C_3 = 300$ nF, and a film capacitor of $C_4 = 1$ μF. In general, it is better to attach the dc capacitor to the inverter bridge as close as possible for reduction of stray inductance in the connection. However, the size of the film capacitor C_4 is relatively larger than C_1 through C_3. Thus, the ceramic capacitors are required to be put closer to the MOSFETs than the film capacitor C_4. In the following experiment, an electrolytic capacitor of $C_4 = 3300$ μF was inserted between the film capacitor and a dc power supply to eliminate the internal impedance of the power supply. Note that l_4 in Table I represents the equivalent series inductance (ESL) in the electrolytic capacitor C_d. The stray inductance between C_4 and C_d are disregarded because the ESL in C_d is much larger than the actual stray inductance between C_4 and C_d.

TABLE I. PARAMETERS OF THE DC LINK COMPONENTS

C_1	ceramic capacitor	$10 \times 2 = 20$ nF
	ESL	1.1 nH
C_2	ceramic capacitor	33 nF
	ESL	1.6 nH
C_3	ceramic capacitor	$100 \times 3 = 300$ nF
	ESL	1.2 nH
C_4	film capacitor	1 μF
	ESL	73 nH
C_d	electrolytic capacitor	3000 μF
l_1	stray inductance	1.3 nH
l_2	stray inductance	1.9 nH
l_3	stray inductance	10 nH
l_4	ESL in C_d	100 nH

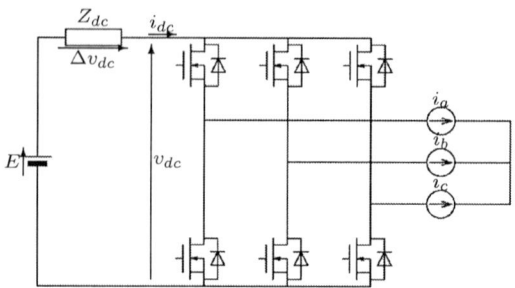

Fig. 3. Equivalent Circuit of the inverter.

III. TRANSIENT ANALYSIS OF THE SWITCHING TRANSITION

Fig. 3 is the equivalent circuit of a three-phase voltage-source inverter. At first, the the-phase load current can be considered as a constant three-phase current source I_a, I_b, and I_c because the load should be inductive and its current change can be negligible in a short switching duration. As the Kirchhoff's current laws, the following relation should be considered in the load current:

$$I_a + I_b + I_c = 0 \tag{1}$$

The dc side circuit of the inverter in Fig. 1 is replaced with an equivalent series impedance Z_{dc} in Fig. 3. The equivalent series impedance Z_{dc} is the dc-side impedance seen from the terminals of the switching devices. Note that the equivalent series impedance Z_{dc} should be considered to include the effect of all components, such as the capacitance, the series equivalent resistance (ESR), and the series equivalent inductance (ESL) of the dc capacitors, and stray inductance on the PCB pattern. Assuming that no leak current flows through the dc capacitor C in Fig. 1, the terminal voltage v_{dc} is equal to the dc supply voltage E when all the MOSFETs are turned off. Thus, the dc supply voltage should be E also in Fig. 3 as the Thévenin's theorem. Then, the voltage drop across the equivalent series impedance, Δv_{dc} is given by

$$\Delta v_{dc} = v_{dc} - E, \tag{2}$$

which represents the voltage ripples or the ac component included in the terminal voltage v_{dc} in time domain.

A. Time Domain Analysis

Fig. 4 shows two conduction states of the three-phase inverter assuming the switching states in the a-phase leg. In Fig. 4(a), the a-phase upper MOSFET is conducting, and the lower MOSFETs are in on-state in the b- and c-phase legs. In this state, the dc-side current is equal to the a-phase current $i_{dc} = i_a$. On the other hand, the lower MOSFETs are turned on in all the three legs in Fig. 4(b). Then, the load current circulates the lower arms, an then no current flows through the dc side as $i_{dc} = 0$. Therefore, the initial and final value of the dc current can be considered as $i_{dc}(0) = I_a$ and $i_{dc}(\infty) = 0$ in a turn-on transition, respectively. And they need to be modified as $i_{dc}(0) = 0$ and $i_{dc}(\infty) = I_a$ for a turn-off transition.

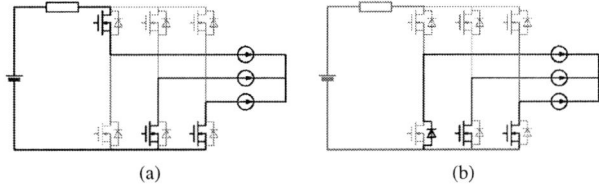

Fig. 4. Switching mode and current loop.

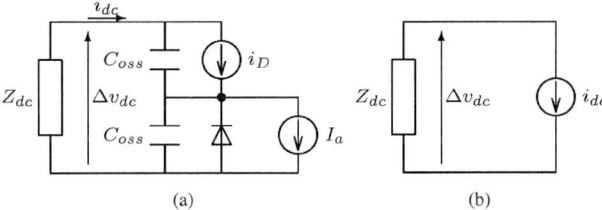

Fig. 5. Simplified circuit for the dynamic analysis.

Fig. 5 is a simplified equivalent circuit paying attention to the switching transition in Fig. 4. The current sources I_a and i_D represent the a-phase load current and the drain current through the upper MOSFET. If the equivalent series impedance Z_{dc} were expressed by a simple inductor L_{dc}, the applied voltage Δv_{dc} can also be expressed in time domain by

$$\Delta v_{dc}(t) = L_{dc} \frac{di_{dc}(t)}{dt}. \tag{3}$$

This implies that the equivalent series impedance strongly affects the voltage ripple in the dc side of the inverter. And thus, it is very important to reduce stray inductance existing in the dc side of the inverter. However, the total impedance Z_{dc} has a complicated network consisting of capacitors and stray inductance among them in a real circuit, for example as shown in Fig. 2.

B. Frequency Domain Analysis

Fig. 5 (a) considers the effect of the output stray capacitance C_{oss} in both upper and lower MOSFETs. The upper MOSFET is represented by the parallel connection of a current source i_D and C_{oss}, and the body diode and C_{oss} are only considered in the lower device as shown in Fig. 5 (a). Moreover, the dc power supply E in Fig. 4 is removed from Fig. 5 (a), and thus, this equivalent circuit is only effective for ac analysis to derive the surge and/or oscillating component Δ. It can also be expressed in frequency domain by using the Fourier transform as,

$$\Delta V_{dc}(\omega) = Z_{dc}(\omega)I_{dc}(\omega). \tag{4}$$

where $Z_{dc}(\omega)$, $\Delta V_{dc}(\omega)$, and $I_{dc}(\omega)$ are frequency responses of the impedance, surge component, and drain current, respectively. In this case, the frequency response of the impedance, $Z_{dc}(\omega)$ is require to evaluate the surge voltage. Although the equivalent series impedance Z_{dc} has complicated characteristics due to the dc capacitors and stray inductance, it can fortunately be measured by using an impedance analyzer

and/or a network analyzer. In addition, it is also required to estimate the spectrum of the drain current $I_{dc}(\omega)$.

Fig. 6(a) shows an equivalent circuit including two dc capacitors and stray inductance between the capacitors. Here, the capacitor C_d is assumed to be large enough to neglect its impedance at the resonant frequency of the circuit. The equivalent series resistance r is considered in the capacitor C. It is usually small, and the the parallel resonant angular frequency ω_r is simply calculated by

$$\omega_r = \frac{1}{\sqrt{lC}}. \tag{5}$$

The impedance seen from the current source is calculated as follows:

$$
\begin{aligned}
Z_{dc}(\omega) &= j\omega l \parallel \left(r + \frac{1}{j\omega C} \right) \\
&= R_{dc}(\omega) + jX_{dc}(\omega),
\end{aligned} \tag{6}
$$

where $R(\omega)$ is the transformed equivalent series resistance and $X(\omega)$ is the series reactance. The equivalent series resistance is given by

$$
\begin{aligned}
R_{dc}(\omega) &= \frac{\omega^4 l^2 C^2 r}{(1 - \omega^2 lC)^2 + \omega^2 C^2 r^2} \\
&= \frac{\left(\frac{\omega}{\omega_r} \right)^4 r}{\left[1 - \left(\frac{\omega}{\omega_r} \right)^2 \right]^2 + \left(\frac{\omega}{\omega_r} \right)^2 \left(\frac{r}{Z_0} \right)^2},
\end{aligned} \tag{7}
$$

and the series reactance is

$$
\begin{aligned}
X_{dc}(\omega) &= \omega l \frac{1 + \omega^2 C(Cr^2 - l)}{(1 - \omega^2 lC)^2 + \omega^2 C^2 r^2} \\
&= \omega l \frac{1 - \left(\frac{\omega}{\omega_r} \right)^2 + \left(\frac{\omega}{\omega_r} \right)^2 \left(\frac{r}{Z_0} \right)^2}{\left[1 - \left(\frac{\omega}{\omega_r} \right)^2 \right]^2 + \left(\frac{\omega}{\omega_r} \right)^2 \left(\frac{r}{Z_0} \right)^2}.
\end{aligned} \tag{8}
$$

As shown in (7), the transformed equivalent series resistance and reactance $R_{dc}(\omega)$ and $X_{dc}(\omega)$ have a largest value around the parallel resonant angular frequency, $\omega \approx \omega_r$. Thus, the most significant surge voltage or oscillation would usually be induced around the parallel resonant angular frequency ω_r. The transformed equivalent series resistance and reactance is simply obtained at ω_r as

$$R_{dc}(\omega_r) = \frac{l}{Cr} = \frac{Z_0^2}{r}, \tag{9}$$

$$X_{dc}(\omega_r) = \sqrt{\frac{l}{C}} = Z_0, \tag{10}$$

where Z_0 is the characteristic impedance between the stray inductance l and capacitor C, given by

$$Z_0 = \sqrt{\frac{l}{C}}. \tag{11}$$

Equation (9) implies that it is possible to reduce the transformed equivalent series resistance $R_{dc}(\omega_r)$ by suppressing

The 2018 International Power Electronics Conference

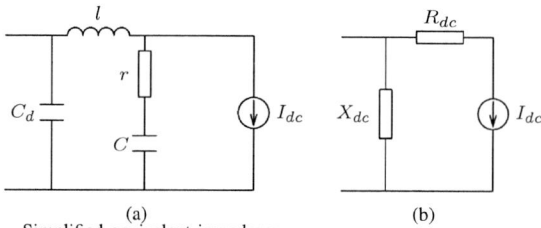

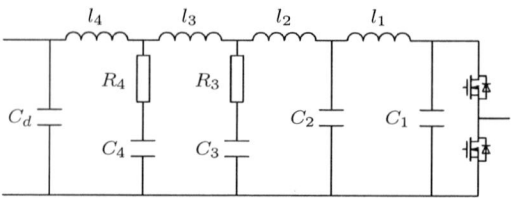

Fig. 6. Simplified equivalent impedance.

Fig. 7. Equivalent dc-link circuit of the inverter with damping resistors.

TABLE II. DAMPING RESISTORS

R_3	chip resistors	$1 \parallel 3 = 0.3\ \Omega$
R_4	carbon resistor	$0.6\ \Omega$

the stray inductance l. This well agrees with the result of the transient analysis. On the other hand, the transformed equivalent series resistance $R_{dc}(\omega_r)$ is inversely proportional to the real equivalent series resistance (ESR) r. Thus, a non-negligible surge voltage would be induced when the capacitor C has a very small ESR r. The estimated voltage ripple or oscillation across the dc link can be calculated as

$$
\begin{aligned}
\Delta \dot{V}_{dc}(\omega_r) &= [R_{dc}(\omega) + jX_{dc}(\omega)] \\
&= \left(\frac{Z_0^2}{r} + jZ_0 \right) I_{dc}(\omega_r).
\end{aligned} \tag{12}
$$

The total power loss in the transformed equivalent series resistance $R_{dc}(\omega)$ can be obtained by

$$
P = \int_0^\infty P(\omega)d\omega = \int_0^\infty R(\omega)I_{dc}^2(\omega)d\omega. \tag{13}
$$

In fact, this calculated power loss in the transformed equivalent series resistance $R_{dc}(\omega)$ occurs in the ESR of the capacitor C because no other lossy element is considered in the equivalent circuit in Fig. 6(a).

As mentioned above, the most significant oscillation appears around ω_r. Therefore, the dominant power lose is also assumed to be caused by the resonant frequency component in the equivalent current source, $I_{dc}(\omega_r)$. The power loss can be estimated by

$$
P(\omega_r) = R(\omega_r)I_{dc}^2(\omega_r) = \frac{Z_0^2}{r}I_{dc}^2(\omega_r). \tag{14}
$$

As shown in (14), the expected power loss is essentially in inverse proportion to the capacitor ESR r. Thus, relatively large power loss would occur and cause a non-negligible heat in the capacitor if the circuit uses dc capacitors with a low equivalent series resistance r.

IV. EFFECT OF DAMPING RESISTORS

Fig. 7 shows the circuit configuration of the dc link where damping resistors are applied to the capacitors. Two damping resistors R_3 and R_4 are connected in parallel with ceramic capacitor C_3 and the film capacitor C_4. The aim of the damping resistors are mainly to suppress the peak of the impedance Z_{dc}. Thus, these resistance values R_3 and R_4 are designed based on the characteristic impedance $Z_3 = \sqrt{l_3/C_3}$ and $Z_4 = \sqrt{l_4/C_4}$, respectively.

It is assumed to connect a damping resistor $R = Z_0$ to the circuit Fig. 5 (a). The induced voltage ripple or oscillation

across the dc link can be reduced as

$$
\Delta \dot{V}_{dc}(\omega_r) = Z_0(1 + j)I_{dc}(\omega_r). \tag{15}
$$

Assuming that the damping resistor is much higher than the ESR in the capacitor, the power loss in the damping resistance $R_{dc}(\omega)$ is also estimated as

$$
P_{dc}(\omega_r) = Z_0 I_{dc}^2(\omega_r). \tag{16}
$$

The total power loss in the damping capacitor and the ESR seems to be reduced because the ESR r is usually lower than the characteristic impedance Z_0.

V. EXPERIMENTAL RESULTS

Fig. 8 shows the measured impedance of the dc link seen from one of the three inverter legs. Four difference conditions are examined in Fig. 8. In case of no damping resistor, the impedance plot has three large peaks in the measured impedance at frequencies of 300 kHz, 1.2 MHz, and 20 MHz. The peak value of impedance reaches 6 Ω at 1.2 MHz which is caused by the parallel resonance between the ceramic capacitor $C_3 = 300$ nF and the film capacitor $C_4 = 1$ μF. And the resonance at 20 MHz is induced between ceramic capacitors of $C_2 = 33$ nF and $C_3 = 300$ nF. The damping resistor R_3 is required to eliminate the The resonance at 1.2 MHz is damped by the resistor R_3 inserted in series with either film or ceramic capacitor. The better impedance plot is obtained when both damping resistors are inserted to both film or ceramic capacitors.

Fig. 9 shows experimental waveforms of the experimental setup. The input dc voltage was 600 V, the switching frequency was set to 20 kHz. The practical dead time was 300 ns, no dead time compensation was implemented. A inductive load was connected to the inverter. The output line-to-line was modulated properly, and the line current has a sinusoidal waveform. Moreover, almost no overshoot was observed in the voltage waveform, irrespective of phase angle of the line current.

Figs. 10-13 show experimental turn-on and off waveforms measure with/without the damping resisters. In this measurement, the source current i_S was measured instead of the drain current because of the pin arrangement of the MOSFETs. Although i_S included the gate current i_G, which would be

1857

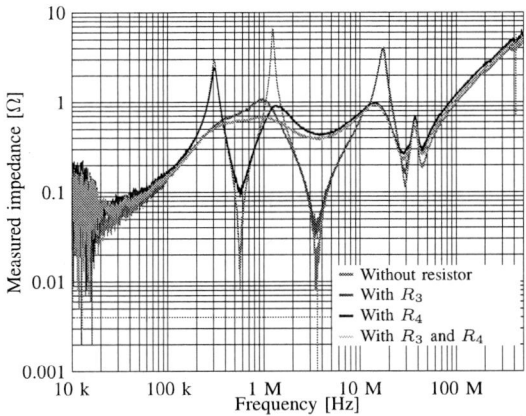

Fig. 8. Measured impedance of the dc link seen from one of the inverter legs.

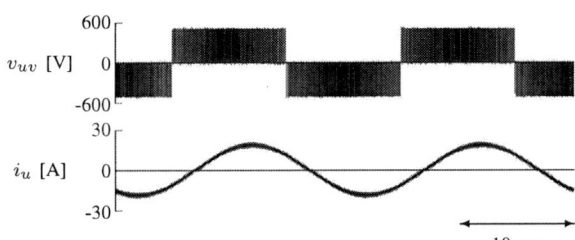

Fig. 9. Experimental Waveforms of

negligible. Figs. 10 and Figs. 11 included a continuous oscillation in Δv_{dc}, while it was well damped in Figs. 12 and Figs. 13.

Fig. 14 shows the measured power loss with/without the damping resisters. In this measurement, one of the three-phase leg was only used as a buck converter to improved the measurement accuracy. The input dc voltage was $E = 600$ V, and the switching frequency was set to 20 kHz. As shown

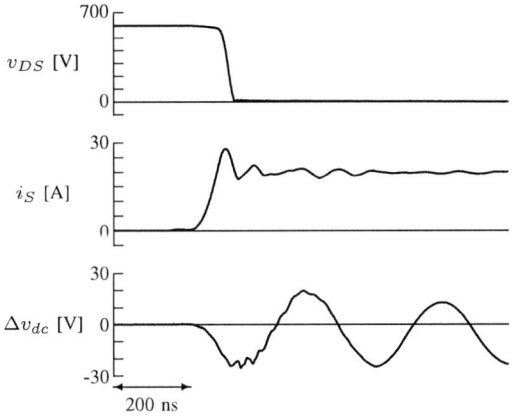

Fig. 10. Turn-on waveforms without damping resistor.

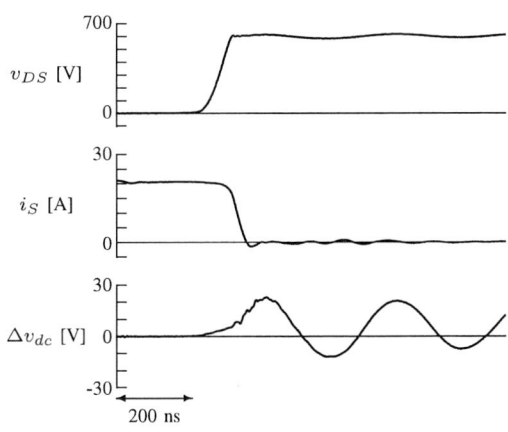

Fig. 11. Turn-off waveforms without damping resistor.

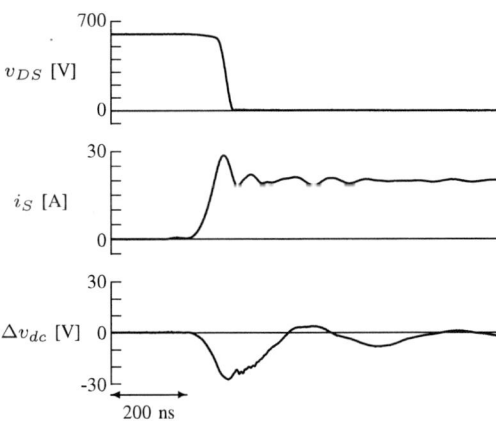

Fig. 12. Turn-on waveforms with damping resistors.

in Fig. 14, the damping resister reduced the power loss by 2 W. Although the difference is not so large, it is consumed in the small capacitors in case of the circuit without damping resistors. For this reason, the capacitor may causes undesired heat and make its lifetime short if no damping resistor is connected.

VI. CONCLUSION

This paper have discussed the analysis, design, and implementation of a miniaturized inverters using SiC-MOSFETs. An inverter was designed and implemented to drive a 400-V, 15-kW induction motor using 1200-V SiC-MOSFETs based on the analysis. The newly-designed inverter circuit is equipped with damping resistors on its dc link capacitors to suppress the resonance between the dc capacitors and stray inductance. The effect of the damping resistor has been evaluated in theoretical analysis and experimental verification. As a result, it has been clarified that the added damping resistor makes it possible to reduce the impedance at the resonant frequency, and to suppress the surge voltage, overshoot, and oscillation across the drain-to-source terminals. Moreover, it has also been

The 2018 International Power Electronics Conference

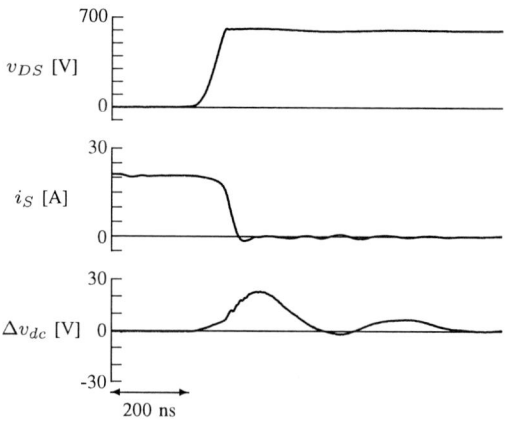

Fig. 13. Turn-off waveforms with damping resistors.

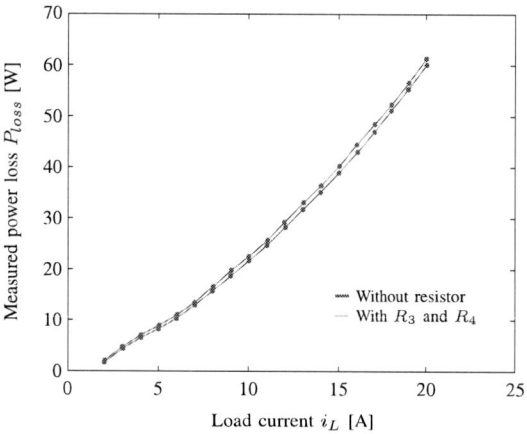

Fig. 14. Measured power loss.

confirmed that the damping resistor can reduce the power losses.

REFERENCES

[1] M. Bhatnagar, B. J. Baliga, "Comparison of 6H-SiC, 3C-SiC, and Si for power devices," IEEE Transactions on Electron Devices, vol. 40, no. 3, pp. 645-655, 1993.

[2] Woongje Sung, Kijeong Han, B. Jayant Baliga, "A comparative study of channel designs for SiC MOSFETs: Accumulation mode channel vs. inversion mode channel," International Symposium on Power Semiconductor Devices and IC's (ISPSD), 2017

[3] Woongje Sung, B. J. Baliga, "On Developing One-Chip Integration of 1.2 kV SiC MOSFET and JBS Diode (JBSFET)," IEEE Transactions on Industrial Electronics, vol. 64, no. 10, pp. 8206-8212, 2017.

[4] H. Akagi, H. Hasegawa, T. Doumoto, "Design and performance of a passive EMI filter for use with a voltage-source PWM inverter having sinusoidal output voltage and zero common-mode voltage," IEEE Transactions on Power Electronics vol. 19, no. 4, pp. 1069-1076, 2004.

[5] S. Ogasawara, H. Ayano, H. Akagi, "An active circuit for cancellation of common-mode voltage generated by a PWM inverter," IEEE Transactions on Power Electronics, vol. 13, no. 5, pp. 835-841, 1998.

[6] Haoran Zhang, A. Von Jouanne, Shaoan Dai, A. K. Wallace, Fei Wang, "Multilevel inverter modulation schemes to eliminate common-mode voltages," IEEE Transactions on Industry Applications vol. 36, no. 6, pp. 1645-1653, 2000.

[7] Handy Fortin Blanchette, Kamal Al-Haddad, "Solving EMI-Related Problems for Reliable High-Power Converters Design With Precomputed Electromagnetic Models," IEEE Transactions on Power Electronics, vol. 25, no. 1, pp. 219-227, 2010.

[8] Bingyao Sun, Rolando Burgos, Dushan Boroyevich, Remi Perrin, Cyril Buttay, Bruno Allard, Nicolas Quentin, Marwan Ali, "Two comparison-alternative high temperature PCB-embedded transformer designs for a 2 W gate driver power supply," IEEE Energy Conversion Congress and Exposition (ECCE), pp. 1-7, 2016.

[9] Qingzeng Yan, Xibo Yuan, Xiaojie Wu, "A 100kHz 95.91% efficiency SiC-device-based split output converter with EMI reduction," IEEE International Power Electronics and Motion Control Conference (IPEMC-ECCE Asia), pp. 13-20, 2016.

Design Consideration of Flying Capacitor Multilevel Inverters Using SiC MOSFETs

Yukihiko Sato and Kenji Natori
Graduate School of Science and Engineering
Chiba University, Chiba, Japan
*E-mail: ysato@ieee.org

Abstract— SiC-MOSFETs are expected as promising power switching devices from the view point of high-speed switching and lower on-state resistance. In the multilevel inverters, the number of switching devices in the main current path increases. Thus, lower on-state resistance of the SiC-MOSFETs is very useful to reduce conduction loss of the inverter. In addition, to reduce required capacitance of the flying capacitors, higher switching frequency operation is indispensable. From this point of view, the higher switching frequency in high power applications realized by the SiC-MOSFETs is very attractive. In this way, the SiC-MOSFETs can dramatically enhance the advantages of the flying capacitor multilevel inverters. In this paper, as a basis of design consideration for flying capacitor multilevel inverters using SiC-MOSFETs, various experimental investigations employing four different prototypes are carried out. The effect of the number of the output levels and thermal management are demonstrated. In addition, key issues in extension to higher voltage applications are investigated.

Keywords— *Multilevel Inverters, Flying Capacitor, SiC MOSFETs*

I. INTRODUCTION

Widely used 2-level inverters need output filters to improve pulsed output waveform especially in grid-connected applications. In addition, AC side inductors are required to mitigate sudden change in AC side current due to the switching operation. However their volume and weight are critical problems to realize high power-density converters. On the other hand, the multilevel inverter can realize multilevel waveform with lower harmonics and lower voltage step change [1]. Thus, the size of the output filters and AC side inductors can be reduced dramatically. As the multilevel inverters, various topologies have been proposed [2-4]. In this research, flying capacitor multilevel inverter is dealt with from the viewpoint of high power-density.

In general, the multilevel inverters have internal sources of multilevel voltage such as capacitors or floating voltage sources. The size of the multilevel voltage sources affects the practicability of the multilevel inverters significantly. In the flying capacitor inverters,

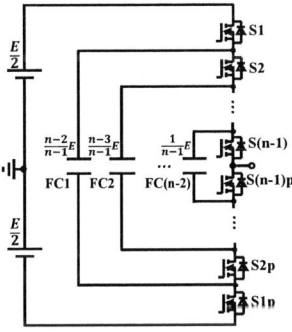

Fig. 1. Circuit configuration per phase of generalized n-level flying capacitor multilevel inverters.

flying capacitors used as the multilevel voltage sources are charged and discharged equally in each PWM carrier period. Thus, the flying capacitor inverter can inherently maintain the average capacitor voltages at a predetermined value without feedback control of the flying capacitor voltage [5]. Consequently, the capacitance of the flying capacitors can be reduced when the PWM switching frequency is high [6].

In this research, usefulness of the SiC-MOSFETs in the flying capacitor multilevel inverters is investigated. The SiC-MOSFETs have lower on-state resistance and higher speed switching capability in high power operating range. The lower on-state resistance can reduce conduction loss which tends to large due to a large number of switching devices in the main current path of the multilevel inverters. The higher speed switching capability enables higher switching frequency operation resulting in smaller required capacitance of the flying capacitors. These properties are very attractive to realize flying capacitor multilevel inverters with higher power density.

In this paper, the effectiveness of the SiC-MOSFETs in the flying capacitor multilevel inverters is demonstrated by experimental investigations employing four different prototypes of the flying capacitor multilevel inverters. First of all, the effect of the number

The 2018 International Power Electronics Conference

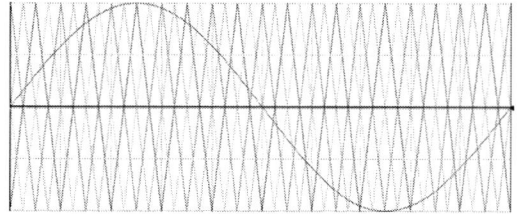

Fig. 2. Carrier phase shifted modulation.

of output levels and thermal management are investigated experimentally with two different prototypes as a basis of design considerations. Furthermore, extension of applicable power range of the flying capacitor inverter to higher voltage applications is investigated with another two different prototypes.

II. FEATURES OF FLYING CAPACITOR MULTILEVEL INVERTERS

A. Operation Principle

As the multilevel inverters, various topologies have been proposed such as a diode clamp type, a cascade connected type, and a flying capacitor type [3-5]. In this research, the flying capacitor topology is treated because it is most suitable for downsizing of the capacitors and their associated balancing circuits.

Fig.1 shows circuit configuration in a leg of the generalized n-level flying capacitor inverters. There are (n-1) power switches in each of the upper and lower half of the leg. The (n-2) flying capacitors hold voltages that are different by E/(n-1) each other. The output voltage levels except both the maximum and minimum voltages (E/2 and -E/2) are generated by combinations of the capacitor voltages and source voltage. In addition, it should be noted that an intermediate output voltages can be generated by several different switching states of the 2(n-1) power devices.

B. Characteristic and Advantages

In the flying capacitor topology, the voltage ripple of the capacitors can be reduced by introducing high switching frequency operation because each capacitor has charge and discharge modes in a PWM carrier period. In other words, it is possible to reduce the required capacitance of the flying capacitors for a given permissible ripple voltage with higher switching frequency.

When phase-shifted-carrier PWM is applied, the durations of different switching states are equal in a PWM carrier period and the charging and discharging of each flying capacitor are balanced in each carrier period. As the result, the average voltage of the flying capacitors can be inherently stable without feedback control [6].

In carrier phase shifted modulation shown in Fig.2 the phase of (n-1) carrier waves are shifted each other by 2π /(n-1) [rad]. These carrier waves are compared with the voltage reference signal. As a result, all the switching

Table.1. Specification of prototypes.

	Prototype-I	Prototype-II
Number of output levels	3, 5, 7 (Selectable)	7 (Fixed)
Maximum DC Input Voltage	300V	300V
MOSFET	SCT3030AL (Rohm)	
On resistance of MOSFET	30mΩ	
Rated voltage of MOSFET	650V	
Dead time	200ns	200ns

Fig.3 Prototype flying capacitor inverter with selectable number of output levels. (Prototype-I)

devices in the circuit perform switching equally in a whole output fundamental period. When the carrier period is short enough with respect to the fundamental period, it can be assumed that the output current does not change in one carrier period. Thus, the current flows through each switching device equally and the losses in switching devices become uniform. In other multilevel topologies, the current may concentrate on specific switching devices depending on the operating conditions. Consequently, the overall cooling efficiency decreases due to heat concentration. In contrast, the losses of switching devices in the flying capacitor topology is uniform. Thus, volume reduction of required cooling devices can be realized.

From the above discussion, it is concluded that the flying capacitor topology has advantages in the realization of high power density compared to other multilevel circuit topologies. In this context, the flying capacitor topology has been put into commercial products for medium voltage motor drive [7].

III. EXPERIMENTAL INVESTIGATIONS FOR DESIGN CONSIDERATION

In this section, the influence of the number of output levels on performance of the inverter such as quality of output waveforms and efficiency is investigated experimentally employing a prototype with selectable number of output levels. (Prototype-I) In addition, effect

1861

The 2018 International Power Electronics Conference

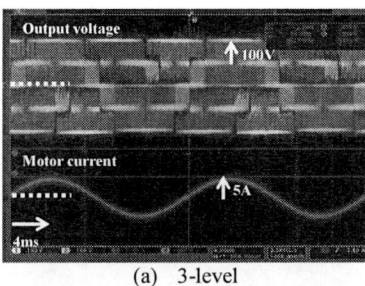

(a) 3-level

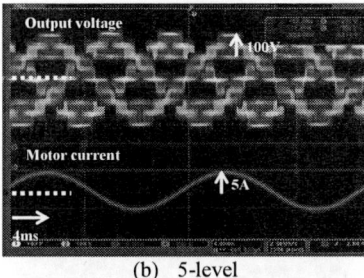

(b) 5-level

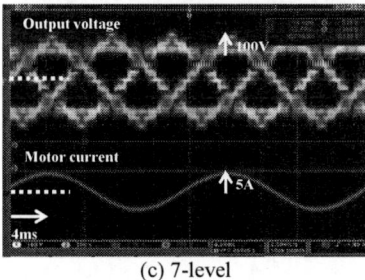

(c) 7-level

Fig.4. Experimental waveforms of line-to-line voltage and motor current for different number of output levels (Carrier frequency: 20kHz).

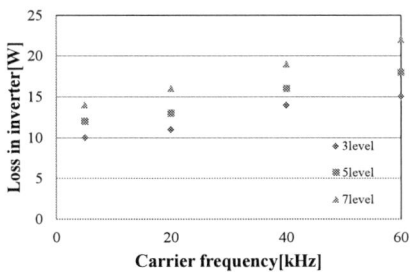

Fig.5. Total loss in inverter for three different output levels.

of parasitic parameters, heat distribution and thermal management are investigated employing another prototype with fixed number of the output levels. (Prototype-II) Specifications of these prototypes are summarized in Table 1.

A. Influence of Number of Output Levels

Fig.3 shows a prototype (Prototype-I) of the flying capacitor multilevel inverter specially constructed so that

Fig.6. Prototype seven level flying capacitor multilevel inverter. (Prototype-II)

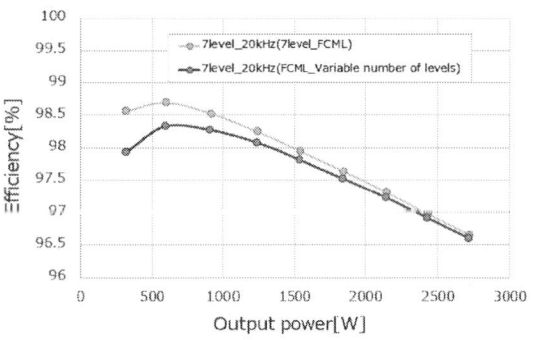

Fig.7. Comparison of efficiency. (Prototype-I and Prototype-II)

the number of the output levels can be selectable by changing connection of multiple circuit units. Each circuit unit is a three level multilevel inverter. A five and seven level inverters can be realized by combination of two and three circuit units, respectively. The circuit units is assembled on an aluminum-core printed circuit board. The heat generated in the circuit spreads through the aluminum core.

Fig.4(a)-(c) show experimental waveform of the inverter output voltage and current for three different numbers of levels. Fig.5 shows measured relationship between the power losses and the carrier frequency. In these experiments, the DC input voltage is set to 200V. As a load of the inverter, an induction motor is connected. The output power and frequency of the inverter are 450W and 50Hz, respectively. The carrier frequency is varied from 5kHz to 60kHz. In each of Fig.4(a)-(c), the upper and lower traces show the line-to-line voltage waveform and one of the line current waveforms, respectively. From Fig.4(a)-(c), we can see that the voltage waveform becomes closer to sinusoidal waveform as the number of the output levels increases. Although, any harmonic filter is not connected to the output of the inverter, the load current becomes purely sinusoidal regardless of the number of output levels due to the effect of multilevel output voltage.

From Fig.5, we can see that the power loss in the

1862

inverters increases as the number of output levels increases. In the prototype multilevel inverter, the number of series connected power devices in the current path increases as the number of the output level increases. In addition, the number of output level is changed by multiple combination of the three level multilevel inverter circuit units. Thus, the conduction loss becomes dominant when the number of output level increases. However, the voltage applied to the power devices is reduced as the number of output level increases. Thus, the required voltage rating of the power devices becomes low when the number of output level is high. In general, on-state resistance of power devices with lower voltage rating is low. Therefore, with proper selection of rated voltage of the power devices, the increase in the total on-state resistance is not a serious problem even when the number of output level is high.

B. Efficiency and Thermal Management

Fig.6 shows a prototype of seven level flying capacitor multilevel inverter for investigation of efficiency and thermal management under continuous operation. (Prototype-II) Fig.7 shows a comparison of efficiencies of Prototype-I and Prototype-II under the same operating condition. In this experiment, the DC input voltage and output frequency are set to 300V and 50Hz, respectively. The efficiency of Prototype-II is higher than that of Prototype-I especially in the low power range. In these prototypes, the same SiC-MOSFETs are used. However, the length of current path of Prototype-I is longer due to the interconnection between the circuit units. Thus, the increases in parasitic inductance and resistance result in additional losses due to the switching transient oscillation and conduction loss.

Fig.8(a) and (b) show temperature distribution of a circuit unit of Prototype-II at the rated load operation for two different switching frequencies. In these results, the maximum temperature is lower than 80 degree Celsius. Although the switching frequency of (b) is twice that of (a), the increase in temperature rise is not significant. From this result, we can see that the switching loss is not dominant due to the lower applied voltage of SiC-MOSFETs. In contrast to this, conduction loss due to the on-state resistance of the SiC-MOSFETs is dominant. In this prototype, the rated voltage 650V is excessive for actual applied voltage of 50V. It should be noted that the temperature rise can be permissible only with aluminum plates attached to the SiC-MOSFETs. Because of structure without heat sinks, the power density of 2.35W/cc can be attained in the Prototype-II.

IV. EXPERIMENTAL INVESTIGATIONS FOR HIGHER VOLTAGE APPLICATIONS

The advantage of SiC-MOSFET can be fully utilized in higher voltage applications due to the trend in specifications of commercially available products. In this section, some of key issues for extension of operating voltage is investigated experimentally employing down-scale prototype of higher voltage 7-level flying capacitor multilevel inverter (Prototype-III). In addition, operating performance of the flying capacitor inverters for grid-

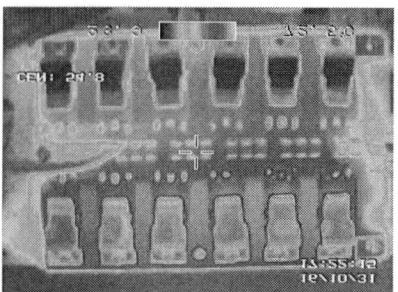

(a) Switching frequency: 20kHz

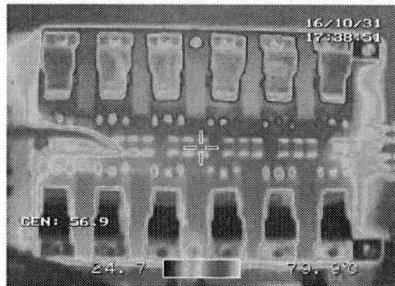

(b) Switching frequency: 40kHz

Fig.8. Temperature distribution at rated load operation. (Prototype-II)

Table.2. Specification of down-scale prototypes for higher voltage applications.

	Prototype-III	Prototype-IV
Number of output levels	7 (Fixed)	17 (Fixed)
Maximum DC Input Voltage	600V	300V
MOSFET	SCT3030AL (Rohm)	
On resistance of MOSFET	30mΩ	
Rated voltage of MOSFET	650V	
Dead time	200ns	200ns

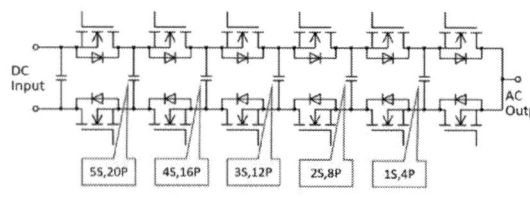

Fig.9. Circuit configuration of down-scale prototype for higher voltage applications. (Prototype-III)

Fig.10. Prototype flying capacitor multilevel inverter for higher voltage applications. (Porotype-III)

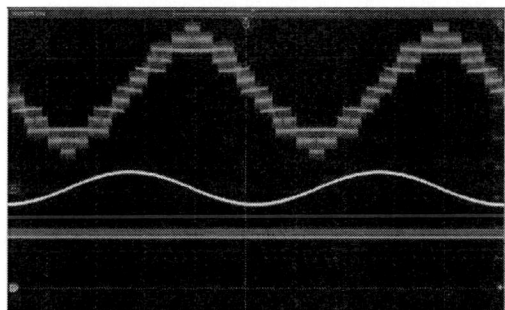

Fig.11. An example of experimental waveforms. (Prototype-III) (Top: Line-to-line voltage 100V/div, Middle: Line current 50A/div, Bottom: Flying capacitor voltages 20V/div)

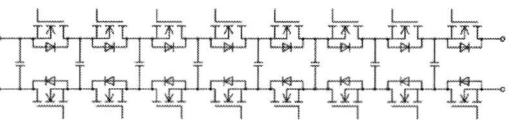

(a) 9-level inverter unit.

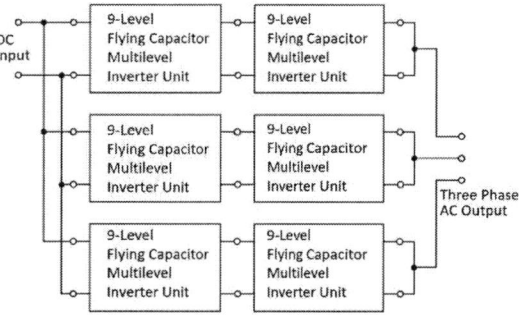

(b) Total circuit connection.

Fig.12. Circuit configuration of down-scale prototype for grid-connected applications. (Prototype-IV)

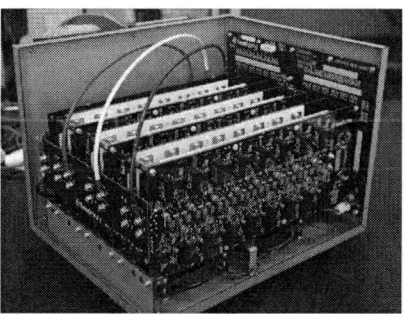

Fig.13. Prototype flying capacitor multilevel inverter for grid-connected application. (Porotype-IV)

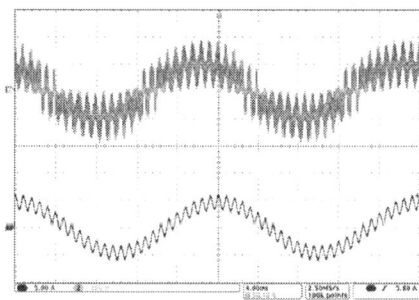

Fig.14. An example of experimental waveforms. (Porotype-IV) (Upper: Line-to-line voltage 100V/div, Lower: Line current 5A/div)

connected applications that are typical expected applications for higher voltage applications is investigated employing a prototype of 17-level inverter (Porotype-IV). The specifications of two prototypes for higher voltage applications are listed in Table 2.

A. Extension to Higher Voltage Applications

Fig.9 shows circuit diagram per phase of Prototype-III for higher voltage operation. Fig.10 shows photograph of Prototype-III. In the flying capacitor topology, the voltages applied to the individual flying capacitors are different according to their position in the circuit. The voltage applied to the flying capacitor nearest to the DC input terminal is the maximum. In Fig.9, all the flying capacitors are implemented by combination of same capacitors. According to the applied voltage, the numbers of series and parallel connection are selected as shown in Fig.9. In this case, the voltage balance in the series connected capacitors is one of the concerns. Fig.11 shows an example of experimental waveform of Prototype-III. In this figure, the upper, middle and bottom trances are the output line-to-line voltage, output line current, and voltages of series connected five flying capacitors nearest to the DC input terminal. From this result, we can see that the proper balance in the voltages of the flying capacitors is obtained.

B. Grid-Connected Applications.

Fig.12 shows circuit configuration of down-scale prototype 17-level flying capacitor multilevel inverter for grid-connected applications. The number of levels is determined assuming DC side voltage of 10kV for applications in 6.6kV distribution lines and voltage rating of SiC-MOSFET of 1.2kV. This prototype is constructed with six 9-level flying capacitor multilevel inverter units with open connection shown in Fig.12(a). The total connection diagram is shown in Fig.12(b). Introducing the unit configuration, size of the prototype can be reduced as shown in Fig.13. As promising applications in higher voltage range, grid-connected converters such as reactive power compensator and active filter for harmonic compensation without isolation transformers are expected. Fig.14 shows an example of experimental results. The upper and lower traces show the output line-to-line voltage and line current, respectively. In this experiment, a passive load that consists of an inductor (1mH) and a resister (12.3 Ω) per phase is connected. The modulation indexes are 0.6 and 0.2 for the fundamental and 25th harmonic components, respectively. The PWM carrier frequency is set to 10kHz. From the results in Fig.14, 25th harmonic component can be generated precisely without any unnecessary frequency components. Thus, we can confirm the current control performance required for active filters for harmonic compensation.

V. CONCLUSIONS

This paper has dealt with applications of the SiC-MOSFETs to flying capacitor multilevel inverters. The effectiveness of the SiC-MOSFETs has been demonstrated experimentally with two different prototypes for design consideration. As the result, power density of 2.35W/cc can be attained by the prototype 7-level flying capacitor multilevel inverter.

In addition, technical issues in the extension to higher voltage application such as voltage balance in series connected flying capacitors and controllability of harmonic compensation current have been demonstrated experimentally with another different prototypes.

ACKNOWLEDGMENT

The authors wish to thank Mr. Atsushi Toriumi and Mr. Tatsuya Yokoyama, Master Course Students at Graduate School of Science and Engineering, Chiba University, for their helpful assistance in the experimental work.

This work is supported by Cross-ministerial Strategic Innovation Promotion Program of Cabinet Office, Government of Japan.

REFERENCES

[1] Takumi Ogura, Jun-ichi Itoh: "Evaluation of Total Loss of Both an Inverter and Motor Depending on Modulation Strategy" SPC-09-184 (2009), in Japanese

[2] Jih-Sheng Lai, and Fang Zheng Peng: "Multilevel Converters-A New Breed of Power Converters", IEEE Trans. Ind. Appl., Vol. 32, No. 3, pp. 509-517 (1996)

[3] Samir Kouro, Mariusz Malinowski, K. Gopakumar, Josep Pou, Leopoldo G. Franquelo, Bin Wu, Jose Rodriguez, Marcelo A. Pérez, and Jose I. Leon: "Recent Advances and Industrial Applications of Multilevel Converters", IEEE Trans. Ind. Electron., Vol. 57, No. 8, pp.2553-2580 (2010)

[4] J. Rodriguez, Jih-Sheng Lai, F. Z. Peng: "Multilevel Inverters: A Survey of Topologies, Controls, and Applications", IEEE Trans. Ind. Appl., Vol. 49, No, 4, pp. 724-748 (2002)

[5] Haitham Abu-Rub, Joachim Holtz, Jose Rodriguez, and Ge Baoming: "Medium-Voltage Multilevel Converters-State of the Art, Challenges, and Requirements in Industrial Applications", IEEE Trans. Ind. Electron., Vol. 57, No. 8 pp.2581-2596 (2010)

[6] Hidemine Obara, Masamu Kamaga, Takumi Ito, Yukihiko Sato: "An Investigation of Capacitors for Flying Capacitor Converters" Vol.131 No.12 pp.1393-1400 (2011), in Japanese

[7] Alstom:Catalog "Symphony Air Cooled Medium Voltage AC Drives"

The 2018 International Power Electronics Conference

A Control Method of Overvoltage Suppression across the DC Capacitor in a Grid-Connection Converter Using Leg Short-Circuit of Power MOSFETs during the Initial Charge

Tomoyuki MANNEN[1]*, Keiji Wada[1]
Tokyo Metropolitan University, Tokyo, Japan
*E-mail: mannen@ieee.org

Abstract—This paper proposes a control method of initial charging of the dc-capacitor in voltage source power converters. The proposed method is suitable for the power converter equipped with a small dc capacitor and is capable of removing the initial charge circuit. The proposed method makes a leg short-circuit by using power devices in the power converter, which discharges the dc capacitor and suppresses its overvoltage due to the initial charge. The short-circuit operation is a short time period and is used one time in each leg for distributing the loss to each power devices. The experimental verification using a 200-V, 5-kVA three-phase converter shows that the proposed method can reduce the peak voltage of the dc capacitor from 520 V to 420 V. The experimental results exhibit a good capability of the proposed method of suppressing the overvoltage due to the initial charge without fatal damage to the power devices. As a result, the proposed method makes it possible to remove the initial charge circuit resulting in a reduction in the size and cost of power converters.

Keywords—*DC Capacitor, Grid-Connection Converter Initial charge, Short-circuit*

I. INTRODUCTION

Improvement of power semiconductor devices and application of new circuit topology and/or control method suitable for specific application make it possible to reduce passive components in power converters [1]. Recent power converters tend to employ a small film capacitor as the dc capacitor instead of a large electrolytic capacitor and try to reduce its volume and cost [1]–[3]. In addition, power converters equipped with wide-band-gap (WBG) semiconductor devices, such as SiC and GaN, are being put into practical use. The WBG devices can reduce a loss in the power devices resulting in size reduction of the heat sink [4]. According to the volume reduction of the main circuit in the power converter, the volume ratio of the collateral circuit occupied in the power converters increases.

When the power converter connects to the utility, an initial charge of the dc capacitor causes inrush current. For preventing the inrush current, the power converter employs an initial charge circuit, which usually consists of a resistor and a mechanical conductor. On the other hand, the initial charge circuit is an obstacle to reduce the size and cost of power converters because the design of

the initial charge circuit is based on the rated power. For example, when the dc capacitor is reduced, it is possible to reduce the charging resistor. However, it is impossible to reduce the mechanical conductor because the capacitor current in steady states operation is independent to the capacitance value. The initial charge circuit becomes large volume ratio occupied in the power converter when the dc capacitor reduced. Therefore, the initial charge circuit is a bottleneck to the power converter downsizing even though the initial charge circuit is essentially unnecessary to operate the power converter in steady states.

A small resonant capacitor installed in the dc side can reduce the inrush current [5]. This method needs additional passive components for the resonant circuit. On the other hand, thyristors are often used to prevent inrush current [6]–[9]. Reference [6] proposed a power converter equipped with thyristors instead of anti-parallel diodes have a capability for controlling the inrush current. However, the loss in the switching devices increases because one semiconductor switch consists of two power devices connected in series. For example, since it is impossible for MOSFETs to avoid anti-parallel diode, the thyristors are connected with MOSFETs in series.

If the power converter has no initial charge circuit, the inrush current accompanied with the initial charge is regulated by LC resonance between ac inductors and the dc capacitor in the power converter. When the small dc capacitor applied, since inrush current also becomes small, it is possible to charge the dc capacitor without power devices destruction by overcurrent. On the other hand, the dc capacitor voltage is increased to double of the source voltage. Since the rated voltage of the dc capacitor is usually designed as 150–160% of the source voltage, overvoltage occurs in the dc capacitor.

A discharge resistor and switch, which are connected with the dc capacitor in parallel have a capability of reducing the dc capacitor voltage and suppressing its peak. This method also needs additional components in the power converter. Here, many papers reported the results of short-circuit test [10]–[13], which indicate that power devices work as a resistor while both a large voltage and current are applied to the devices. In this case, quite a large amount of loss occurs in power devices.

1866

The 2018 International Power Electronics Conference

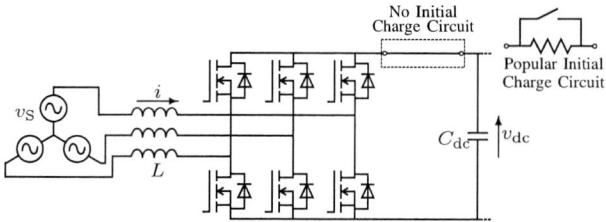

Fig. 1. Circuit diagram of a voltage source converter.

Fig. 2. Picture of the experimental setup.

This loss heats the power devices and may destructively damage them. Reference [11] indicates that a short time period of short-circuit can apply to power devices without damage.

This paper proposes a control method of suppressing overvoltage across the dc capacitor for the power converter equipped with a small dc capacitor. The proposed method makes a leg short-circuit for a short time period to discharge the dc capacitor by using the power devices and suppress its peak voltage. The short-circuit operation is used one time in each leg for distributing the loss to each power devices. The experimental verification confirms that the proposed method can suppress the overvoltage across the dc capacitor. The experimental results exhibit a good capability of the proposed method to suppress the peak dc-capacitor voltage due to the initial charge without fatal damage to the power devices.

II. CIRCUIT CONFIGURATION

Fig. 1 shows a circuit configuration of the power converter and its circuit parameters are listed in Table I. This configuration assumes a rectifier part of an electrolytic-capacitorless power converter for a three-phase motor drive system rated at 5 kVA. Fig. 2 is a picture of the experimental setup. The converter circuit is a three-phase voltage source converter consists of SiC-MOSFETs (Cree C2M0040120D) and SiC-SBDs (Cree C4D10120D). The power converter has ac inductors L as a switching ripple filter in its ac side. Only a small film capacitor C_{dc} is installed in its dc side as a dc capacitor. There is no initial charge circuit such as a resistor and switch. The circuit parameters are designed enough to operate the power converter at a switching frequency of 20 kHz [1], [2].

When a voltage source converter without any initial charge circuit connects to the voltage source at $t = 0$, the

TABLE I.   CIRCUIT PARAMETERS OF THE CONVERTER.

Source Voltage	V_S	200 V	
Source Frequency	f_S	50 Hz	
Rated Power	S	5 kVA	
AC Inductor	L	0.43 mH	(1.7%)
DC Capacitor	C_{dc}	50 μF	(0.5 ms)

dc capacitor starts to be charged through the ac inductors. While the dc capacitor voltage is lower than the highest line-to-line voltage of the voltage source, the anti-parallel diodes connect the power source, the ac inductors, and the dc capacitor in series regardless of gate signals of MOSFETs. Therefore, the highest line-to-line voltage of the voltage source, v_h, the source current i, and the dc capacitor voltage v_{dc} are given by following equations:

$$2L\frac{d}{dt}i(t)=v_h(t) - v_{dc}(t), \qquad (1)$$

$$C_{dc}\frac{d}{dt}v_{dc}(t)=i(t). \qquad (2)$$

Therefore, the dc capacitor voltage is derived from

$$v_h(t)=2LC_{dc}\frac{d^2}{dt^2}v_{dc}(t) + v_{dc}(t). \qquad (3)$$

Here, it is assumed that the charging time of the dc capacitor is sufficiently short and the source voltage has a constant value V_h during the initial charge. The source current i and the dc capacitor voltage v_{dc} are derived from (3) and the initial conditions, $v_{dc}(0) = 0$ and $i(0) = 0$, as follows:

$$v_{dc}(t)=V_h \left(1 - \cos \frac{t}{\sqrt{2LC_{dc}}} \right), \qquad (4)$$

$$i(t)=V_h \sqrt{\frac{C_{dc}}{2L}} \sin \frac{t}{\sqrt{2LC_{dc}}}. \qquad (5)$$

Therefore, the source current is limited to

$$i_{max}=V_h \sqrt{\frac{C_{dc}}{2L}} \qquad \left(t = \frac{\pi}{2}\sqrt{2LC_{dc}} \right). \qquad (6)$$

In this case, the dc capacitor voltage exceeds the source voltage because the ac inductor has a resonant energy and keeps the source current. Finally, when the source current reaches to zero, the dc capacitor also reaches to the maximum voltage given by

$$v_{max}=2V_h \qquad \left(t = \pi\sqrt{2LC_{dc}} \right), \qquad (7)$$

which is the double of the source voltage. As a result, overvoltage across the dc capacitor may occur when the converter has no initial charge circuit.

III. OVERVOLTAGE SUPPRESSION METHOD USING A LEG SHORT CIRCUIT OF MOSFETS

When the dc capacitor voltage is less than the source voltage, the source current increases resulting in increasing of the energy stored in the ac inductor during the initial dc capacitor charge. After the dc capacitor voltage reaches the source voltage, the source current starts to decrease and the stored energy is released from the ac inductor. Both the energy supplied from the source and the energy stored in the ac inductor flow into the dc capacitor resulting in the overvoltage across the dc capacitor. Therefore, in order to suppress the overvoltage,

1867

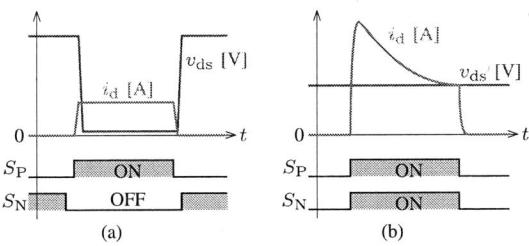

Fig. 3. Waveforms of the drain-source voltage v_{ds} and the source current i_d in a MOSFET under (a) a switching operation and under (b) a short-circuit operation

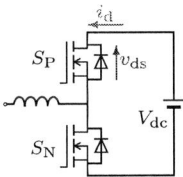

Fig. 4. A circuit diagram of a half-bridge leg.

the converter reduces the energy flow into the dc capacitor and releases the stored energy from the dc capacitor by consuming surplus energy in the converter, when the dc capacitor voltage exceeds the source voltage. However, the ac inductor has no capability to consume energy. Resistors can consume the energy but additional discharging resistor is required. On the other hand, the MOSFETs in the power converter can work as resistor [12], [13].

Fig. 3 shows drain-source voltage and drain current waveforms of a MOSFET in a half-bridge circuit shown in Fig. 4. Fig. 3(a) is a general switching operation. When a large voltage applies to the MOSFET, the current is almost zero. On the other hand, the large current flows though the MOSFET, the voltage across the MOSFET is quite low. Fig. 3(b) shows waveforms under a short circuit operation. A large current flows through the MOSFET but the voltage applied to the MOSFET is also large. In this case, the MOSFET acts as a resistor and the device current is limited by its resistance value. The current waveform has a peak value just after the MOSFET turns on and then the device current reduces. The resistance value of the MOSFET is almost proportional to the junction temperature [14]. Therefore, the device current yields a power loss and causes a junction temperature increasing and the resistance value rising resulting in the device current reduction. However, a high junction temperature may cause fatal damage to the MOSFET. The short circuit duration should be managed accurately to prevent from the destruction of the MOSFET.

A. Control method

This paper proposes an overvoltage suppression method during the initial charge of the dc capacitor. The proposed method makes a leg short circuit by using MOS-FETs in the power converter. In this case, the MOSFETs work as resistors for discharging the dc capacitor. The dc capacitor voltage is controllable because the MOSFETs are controlled by the power converter.

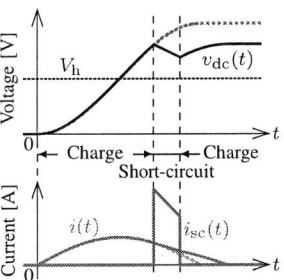

Fig. 5. Voltage and current waveforms in the proposed initial charge operation for the dc capacitor.

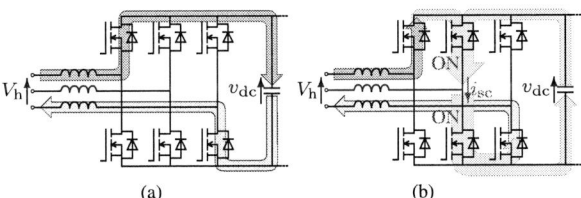

Fig. 6. A current path of the three-phase power converter under (a) a switching operation and (b) a short-circuit operation

Figs. 5–6 shows an operation principle of the overvoltage suppression method proposed in this paper. When the dc capacitor voltage is lower than the source voltage, the gates of all MOSFETs are always off. In this case, the dc capacitor charges through some anti-parallel diodes and the source current increases.

On the other hand, there are two operating modes when the dc capacitor voltage is higher than the source voltage. One is a charging mode shown in Fig. 6(a): all MOSFETs are off. In the charging mode, the dc capacitor voltage increases because the source current still remains and charges the dc capacitor through some anti-parallel diodes. The other is a leg short-circuit mode illustrated in Fig. 6(b): both upper and lower MOSFETs in one of three legs are turned on simultaneously. In the leg short-circuit mode, the dc capacitor voltage decreases steeply as if a small resistor is connected to the dc capacitor in parallel.

Only the leg short-circuit mode has a capability of reducing the dc capacitor voltage. However the source current may not decrease in the leg short-circuit mode though the source current must be reduced to zero to complete the initial charging. On the contrary, a negative voltage always applies to the ac inductor and the source current decreases in the charging mode. Therefore, both the charging mode and the leg short-circuit mode are necessary for the overvoltage suppression. The proposed method employs these two methods alternately to suppress the overvoltage across the dc capacitor.

B. Circuit analysis

Fig. 7 shows equivalent circuit of the initial charge for dc capacitor in Fig. 6. Fig. 7(a) is under charging mode. The charge current flows through two ac-inductors and the dc capacitor. The source current i and the dc

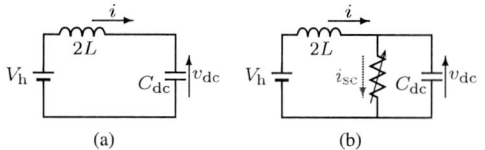

Fig. 7. An equivalent circuit diagram of the power converter under (a) charging mode and (b) short-circuit mode.

capacitor voltage v_{dc} are given by (1) and (2). Here, it is assumed that the charging time of the dc capacitor is sufficiently short and the source voltage has a constant value V_h during the initial charge. The source current i and the dc capacitor voltage v_{dc} are derived from (3) as follows:

$$v_{dc}(t) = k_1 \sin \omega t + k_2 \cos \omega t + V_h, \qquad (8)$$

$$i(t) = \frac{1}{Z}\left(k_1 \cos \omega t - k_2 \sin \omega t\right), \qquad (9)$$

where k_1 and k_2 are a constant, and ω and Z are the characteristic angular frequency and the characteristic impedance, respectively, given by

$$\omega = \frac{1}{\sqrt{2LC_{dc}}}, \qquad (10)$$

$$Z = \sqrt{\frac{2L}{C_{dc}}}. \qquad (11)$$

After the short-circuit duration, the circuit switches into the charge mode immediately. Here, it is assumed the source current and the dc capacitor voltage reach to I_{ne} and V_{ne} at the end of the n-th short-circuit duration $t = T_{ne}$, respectively. After the n-th short-circuit duration, the dc capacitor voltage v_{dcn} and the source current i_n are given by

$$v_{dcn}(t) = A_{ne}\cos\left(\omega t - \phi_{ne}\right) + V_h, \qquad (12)$$

$$i_n(t) = -A_{ne}\frac{1}{Z}\sin\left(\omega t - \phi_{ne}\right), \qquad (13)$$

where the amplitude A_{ne} and the phase angle ϕ_{ne} are calculated as follows:

$$A_{ne} = \sqrt{\left(V_{ne} - V_h\right)^2 + Z^2 I_{ne}^2}, \qquad (14)$$

$$\phi_{ne} = \omega\frac{T_{ne}}{2} + \\ \text{atan}\frac{\left(V_{ne} - V_h\right)\sin\frac{\omega T_{ne}}{2} + Z I_{ne}\cos\frac{\omega T_{ne}}{2}}{\left(V_{ne} - V_h\right)\cos\frac{\omega T_{ne}}{2} - Z I_{ne}\sin\frac{\omega T_{ne}}{2}}. \qquad (15)$$

In addition, (12)–(15) are applicable before the first short circuit duration. In this case, I_{0e} and V_{0e} are considered as the initial source current and the dc capacitor voltage at $t = 0$, resulting in $I_{0e} = 0$, $V_{0e} = 0$ and $T_{0e} = 0$.

Fig. 8 shows a calculated waveforms of the amplitude, the dc capacitor voltage, and the source current. The dc capacitor voltage and source current at T_{ne}, V_{ne} and I_{ne}, decide the maximum dc capacitor voltage in the charging mode. Only the timing changing into short circuit mode can control the peak voltage. The dc capacitor voltage peak is changed to $A_{ne} + V_h$ after the n-th short circuit operation.

Fig. 7(b) is the equivalent circuit diagram under short-circuit mode. Fig. 7(b) has an additional variable resistor

which assumes the power devices under short circuit operation. Operating conditions of the power devices decides the resistance value of the variable resistor and junction temperature strongly affects the resistance value [14], [15]. The resistance value usually increases under a short-circuit operation because the short-circuit power loss in the power devices heats its junction. To simplify the following analysis, the resistor voltage, or the dc capacitor voltage, is directly discussed instead of considering the variable resistor.

Here, it is assumed the dc capacitor voltage decreases linearly during the short-circuit mode. The dc capacitor voltage during the short-circuit operation is given by

$$v_{dcn}(t) = -K_V(t - T_{ns}) + V_{ns}, \qquad (16)$$

where K_V is voltage rate of change during short-circuit operation and T_{ns} and V_{ns} are the time and the dc capacitor voltage at the start of the n-th short-circuit duration, respectively. The source current is also derived from (16) as follows:

$$i_n(t) = \frac{K_V}{4L}(t - T_{ns})^2 + \frac{(V_h - V_{ns})}{2L}(t - T_{ns}) + I_{ns}. \qquad (17)$$

Therefore (16) and (17) are also expressed by using (12) and (13). The amplitude A_n and the phase angle ϕ_n are given by

$$A_n = \sqrt{\begin{aligned}&\left\{Z I_{ns} + \omega(t - T_{ns})\frac{(t - T_{ns})K_V + 2(V_h - V_{ns})}{2}\right\}^2 \\ &+ \left\{(t - T_{ns})K_V + (V_h - V_{ns})\right\}^2\end{aligned}}, \qquad (18)$$

$$\phi_n = \frac{\omega T_{ne}}{2} + \text{atan}\frac{(V_{ns} - V_h)\sin\frac{\omega t}{2} + Z I_{ns}\cos\frac{\omega t}{2}}{(V_{ns} - V_h)\cos\frac{\omega t}{2} - Z I_{ns}\sin\frac{\omega t}{2}}. \qquad (19)$$

The amplitude and phase angle change during the short circuit operation, only the short circuit operation can suppress the maximum dc capacitor voltage during the charging mode.

C. Timing and duration of the leg short-circuit mode

The voltage source provides energy for the power converter during the initial charge. The total amount of the provided energy W_s is given by

$$W_s = \int V_h i(t)\mathrm{d}t. \qquad (20)$$

On the other hand, after finishing the initial charge, the dc capacitor voltage reaches to V_{dcf}. At the same time, the energy stored in the dc capacitor W_{cf} is given by

$$W_{cf} = \frac{1}{2}C_{dc}V_{dcf}^2. \qquad (21)$$

The difference between the provided energy W_s and the stored energy W_{cf} is consumed in the short-circuit mode. In the leg short-circuit mode, the sum of the source current and the capacitor current flows through the MOSFETs which are turned on resulting in power loss. Since the power loss heats and damages the MOSFETs, the leg short-circuit mode should be used in a short time.

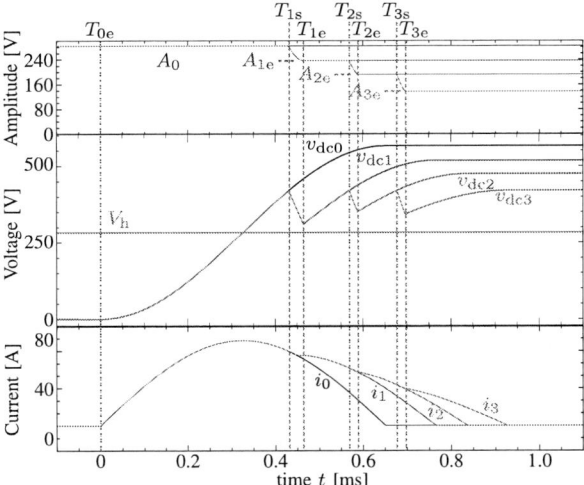

Fig. 8. The calculated the amplitude A, dc capacitor voltage v_{dc}, source current i.

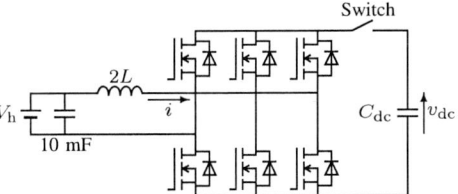

Fig. 9. A circuit diagram used in the experimental verification.

Therefore, the total amount of short-circuit capability of six MOSFETs must be larger than $W_s - W_{cf}$. The maximum short-circuit duration is provided in the data sheet of the MOSFET or measured in some paper [10].

To reduce the required short-circuit capability, the converter switches from the charging mode to the leg short-circuit mode when the dc capacitor voltage reaches the set maximum voltage V_{max}^*. The final dc capacitor voltage V_{dcf} is also set to the maximum voltage V_{max}^*. In addition, the higher dc capacitor voltage applying, the source current decreases faster. The initial charge duration and the total amount of the provided energy W_s reduce. For this reason, it is better to make the short-circuit three times with small duration rather than make the short-circuit once with large duration.

After the dc capacitor voltage reaches the set maximum voltage V_{max}^* in the charging mode, $v_{dcn}(t) = V_{max}^*$, the next short-circuit mode starts, The time when the n-th short-circuit mode starts, T_{ns}, is derived from (16) as follows:

$$T_{ns} = \frac{1}{\omega}\left(\phi_{(n-1)e} - \mathrm{acos}\frac{V_{max}-V_h}{A_{(n-1)e}}\right). \quad (22)$$

The initial value of the amplitude A is V_h. After the final short-circuit mode, the amplitude A becomes $V_{max}^* - V_h$ because the final dc capacitor voltage is set to be V_{max}^*. Therefore, the n-th short circuit mode decreases A to A_n and, finally, $A_3 - V_{max}^*$ V_h. The time when he n-th short-circuit mode ends, T_{ne}, can be derived from (18).

IV. EXPERIMENTAL VERIFICATION

FIg. 9 shows experimental circuit configuration for the proposed method. In the following experiments, a dc voltage power supply is used instead of a three-phase ac power source so as to set the same condition in these measurements. A large capacitor of 10 mF helps the voltage supply to keep the constant voltage at V_h. An air-core inductor $2L$ is connected to the ac side instead of the three-phase ac inductor L. The switch in the dc side connects the dc capacitor to the power source, as if it works like a breaker in the ac side. The voltage source was set to 283 V, the maximum line-to-line voltage in ac 200 V. The switch turns on at $t = 0$ which imitates what the power converter connected to the ac source.

Figs. 10–11 show experimental results of the initial charge of the dc capacitor. Fig. 10 is measured without the overvoltage suppression method. In this case, all gates of the MOSFETs were OFF and the dc capacitor is charged through only the anti-parallel diodes. The dc capacitor voltage v_{dc} and the source current i are increased from $t = 0$. The dc capacitor voltage was reached to the supply voltage at $t = 0.344$ ms and the peak of the source current was 64 A at the same time. After that, the source current decreases. When the source current decreased to zero, the dc capacitor voltage increased to 520 V.

Fig. 11 is measured with the proposed overvoltage suppression method. The waveforms of the dc capacitor voltage v_{dc} and the source current i are similar to those in Fig. 10 during $v_{dc} < V_h$. The proposed method keeps all gates of the MOSFETs off when the dc capacitor voltage is less than the source voltage. On the other hand, some gates of the MOSFETs are ON after the dc capacitor voltage reached the source voltage. When the dc capacitor voltage is reached to a set point of maximum voltage, 420 V, the proposed method turns on two MOSFETs belonging to one of three leg. In Fig. 11, the short-circuit occurs in the v-phase at $t = 0.463$ ms, the u-phase at $t = 0.555$ ms, and the w-phase at $t = 0.621$ ms. The short-circuit current i_{sc} has peaks at the short-circuit starting, and then the short-circuit current decreases because the junction temperature of the MOSFETs increases resulting in increasing its resistance value. Finally, the dc capacitor voltage reached to 420 V after the source current became 0. So, the proposed method reduced the peak of the dc capacitor voltage from 520 V to 420 V accompanying to the initial charge for the dc capacitor.

A large amount of power loss occurs in each MOSFET in the short-circuit operation. The power loss in two v-phase MOSFETs was 1.49 J. In the v-phase, the power loss in the two MOSFETs are balanced because the device current is the same. Thus, the loss in each MOSFET was 0.75 J in the v-phase. On the other hand, the loss in u- and w-phase MOSFETs are unbalanced because the device current has difference due to the source current. The upper MOSFET in the u-phase and the lower MOSFET in the w-phase have larger loss than the others. The power loss in the u- and w-phase MOSFETs were 0.90 and

The 2018 International Power Electronics Conference

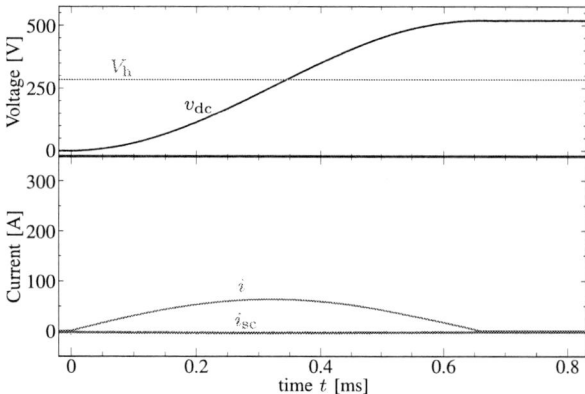

Fig. 10. The experimental waveforms during the initial charge for the dc capacitor without the initial charge circuit.

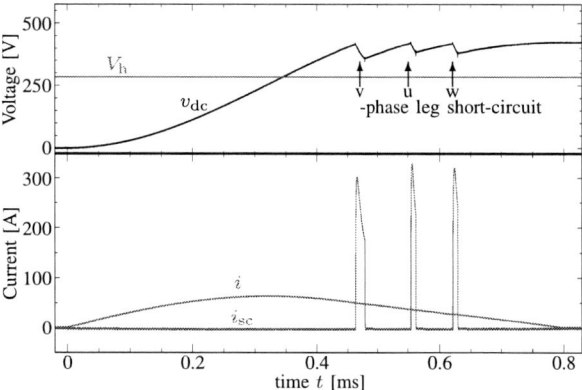

Fig. 11. The experimental waveforms during the initial charge for the dc capacitor with the proposed method.

0.91 J, respectively. Since the short circuit capability of the MOSFET is 1.4 J [10], the proposed method can apply to the power converter without destruction on MOSFETs.

V. CONCLUSION

This paper proposed a control method of suppressing overvoltage across the dc capacitor during its initial charge. The proposed method makes a leg short-circuit in the power converter by using its power devices and discharge the dc capacitor. Short time periods of the short-circuit operation can suppress the peak voltage across the dc capacitor.

The validity of the proposed method is confirmed in the experimental verification. The proposed method can suppress the peak of the dc capacitor voltage from 520 V to 420 V without fatal damage to the MOSFETs. Therefore, the proposed method can remove the initial charge circuit in the power converter and makes it possible to reduce the size and cost of the power converter.

REFERENCES

[1] I. Takahashi, H. Haga, "Inverter control method of IPM motor to improve power factor of diode rectifier," *IEEE Proceedings of the Power Conversion Conference-Osaka 2002*, pp. 142–147, 2002.

[2] H. Yoo, S. Sul, "A Novel Approach to Reduce Line Harmonic Current for a Three-phase Diode Rectifier-fed Electrolytic Capacitor-less Inverter," *IEEE APEC 2009*, pp. 1897–1903, 2009.

[3] T. Shimizu, K. Wada, N. Nakamura, "Flyback-Type Single-Phase Utility Interactive Inverter With Power Pulsation Decoupling on the DC Input for an AC Photovoltaic Module System," *IEEE Trans. Power Electron.*, vol. 21, no. 15 pp. 1264–1272, 2006.

[4] J.W. Kolar, D. Neumayr, D. Bortis, "Google Little Box Reloaded: How to Achieve 200W/in^3 & Beyond? Concepts - Evaluation - Barriers - Future," *IEEE APEC 2017*, 2017.

[5] M. Mohammadi, M. Ordonez, "Inrush Current Limit or Extreme Start-Up Response for LLC Converters Using Average Geometric Control," *IEEE Trans. Power Electron.*, vol. 3, no. 1, pp. 777–792, 2017.

[6] L. Gonthier, B. Renard, "AC/DC reversible mixed inverter with built-in inrush-current limitation and cut-off stand-by losses," *IEEE EPE'15 ECCE-Europe*, pp. 1–10, 2015.

[7] L. Wei, Z. Jankovic, Y. P. Patel, J. Hu, "Single phase precharge control method for active front end rectifier," *IEEE ECCE 2016*, pp. 1–6, 2016.

[8] M. Alam, W. Eberle, N. Dohmeier, "An Inrush Limited, Surge Tolerant Hybrid Resonant Bridgeless PWM AC-DC PFC Converter," *IEEE Energy Conversion Congress and Exposition (ECCE) 2014*, pp. 5647–5651, 2014.

[9] M. Kumar, L. Huber, M. M. Jovanovic, "Startup procedure for three-phase three-wire isolated ac-dc converter implemented with three single-phase converter modules," *IEEE APEC 2016*, pp. 46–53, 2016.

[10] A. E. Awwad, S. Dieckerhoff, "Short-circuit evaluation and overcurrent protection for SiC power MOSFETs," *IEEE EPE'15 ECCE-Europe*, pp. 1–9, 2015.

[11] C. Ionita, M. Nawaz, K. Ilves, F. Iannuzzo, "Short-circuit Ruggedness Assessment of a 1.2 kV/180 A SiC MOSFET Power Module," *IEEE ECCE 2017*, pp. 1982–1987, 2017.

[12] T. Oeder, A. Castellazzi, M. Pfost, "Electrical and thermal failure modes of 600V p-gate GaN HEMTs," Microelectronics Reliability, Vol. 76–77, pp. 321–326, 2017.

[13] L. Ceccarelli, P.D. Reigosa, F. Iannuzzo, F. Blaabjerg, "A survey of SiC power MOSFETs short-circuit robustness and failure mode analysis," Microelectronics Reliability, Vol. 76–77, pp. 272–276, 2017.

[14] Y. Du, R. Rodrigues, T. Jiang, "A Measurement Method to Extract the Transient Junction Temperature Profile of Power Semiconductors at Surge Conditions," *IEEE ECCE 2017*, pp. 2512–2519, 2017.

[15] F. Stella, G. Pellegrino, E. Armando, D. Dapra, "On-line temperature estimation of SiC power MOSFET modules through on-state resistance mapping," *IEEE Energy Conversion Congress and Exposition (ECCE) 2017*, pp. 5907–5914, 2017.

The essential relationship between deadbeat predictive control and continuous-control-set model predictive control for PWM converters

Bi Liu[1], Tao Chen[2], and Wensheng Song[1*]

1. Department of Electrical Engineering, Southwest Jiaotong University, Chengdu, 610031, China
2. CRRC Zhuzhou Institute Co., Ltd, Zhuzhou 412001, China

*E-mail: songwsh@swjtu.edu.cn

Abstract-This paper presents the equivalent principle of deadbeat predictive direct power control (DP-DPC) scheme and conventional continuous-control-set model predictive direct power control (MP-DPC) scheme of single-phase pulse width modulation (PWM) ac-dc converters, respectively, and reveals their inherent relationship from the view of the expressions of modulated voltage vector in *dq* rotating frame. The expression of the modulation index can be derived from the expression of modulated voltage vector in conventional MP-DPC scheme, which is in accordance with the result in the MP-DPC scheme with modulation function optimization. On this basis, the relationship of these CCS MP-DPC schemes with the same cost function is discussed in this paper. A comprehensive experimental comparison has been conducted, experimental results confirm the theoretical study and the effectiveness of the three DPC schemes.

Keywords-direct power control, model predictive control, continuous-control-set, single-phase PWM converter.

I. INTRODUCTION

Due to the excellent characteristics, such as bidirectional energy flow, high power factor, low current harmonics and good dc-link voltage regulation, single-phase pulse width modulation (PWM) ac-dc converters have been widely used in railway locomotive traction [1]. Therefore various control schemes have been reported in recent decades. Direct power/current control is an alternative control solution for single-phase PWM converters, which can realize good steady-state and rapid dynamic performance. For the similar characteristics between direct current control and direct power control (DPC), DPC is taken as a case study in this paper.

For the classical DPC scheme choosing the relevant converter switching state through a look-up table for the hysteresis comparators, it has many excellent properties: simple system configuration, high executing efficiency and fast dynamic response [2]. However, a high sampling frequency is necessary in this DPC scheme, and the switching frequency is variable [3]. To overcome those drawbacks, variable DPC schemes have been proposed to stabilize switching frequency, such as PI-based DPC [4], deadbeat predictive DPC (DP-DPC) [5][6], and continuous-control-set model predictive DPC (MP-DPC) schemes with PWM stage or vector sequences with optimized duty cycle control [7][8]. Recently, DP-DPC and MP-DPC schemes

become two well-known approaches in power converters application. In order to track the desired reference active and reactive powers, the DP-DPC with PWM scheme uses the system predictive model to calculate the desired modulation voltage vector during each sampling interval and adopts PWM stage to generate the gating signals [6]. And the conventional MP-DPC with PWM scheme defines a cost function as a criterion, selects the optimal modulation voltage vector while the minimization of the cost function is achieved [7]. Due to the accurate steady characteristic and fast dynamic response, much effort has been done on the advanced MP-DPC scheme. [8] presents a MP-DPC scheme with modulation function optimization, which applies the cost function to predict the optimal modulation index in two-axis stationary reference frame. The MP-DPC with duty cycle optimization is proposed in [9], which calculates the optimal duty cycle of the nonzero vector and a zero vector within a control interval.

However, there are few literatures for discussing the inherent relationship between DP-DPC and continuous-control-set MP-DPC with PWM stage. Thus, in this paper, the principles of the former two DPC schemes in single-phase PWM converters and the inherent relation between them are presented. The rest of this paper is organized as follows. Section II describes a power mathematical model of single-phase two-level PWM converters. The principle of DP-DPC and the conventional MP-DPC scheme is introduced in Section III, then the inherent relationship among the DP-DPC and continuous-control-set MP-DPC schemes, and the MP-DPC scheme with modulation function optimization scheme is revealed in Section IV. An experimental test of the conventional MP-DPC scheme is conducted in Section V, followed by a conclusion in Section VI.

II. SYSTEM DESCRIPTION AND MATHEMATICAL MODEL

Fig. 1 shows the topology of a single-phase two-level PWM converter, where u_s and i_s are the main voltage and the line current, respectively; u_{ab}, i_L and u_{dc} represent the input voltage of H-bridge, the dc-link current and dc-link voltage of the converter, respectively; L_s and R_s are symbols for the equivalent resistance and inductance of ac-side inductors; C_d

and R_L are the dc-link capacitor and equivalent resistive load; S_1, S_2, S_3 and S_4 represent IGBT modules with anti-parallel diodes on the converter's leg-a and leg-b, respectively.

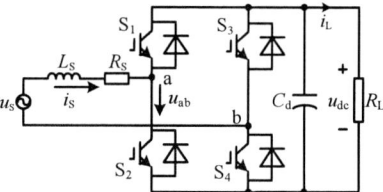

Fig. 1 The topology of a single phase two-level converter

As shown in Fig. 1, the mathematical model of the adopted converter can be expressed as

$$L_s \frac{di_s}{dt} = u_s - R_s i_s - u_{ab} \tag{1}$$

On the basis of three-phase instantaneous power theory, the instantaneous active power and reactive power of the adopted converter can be shown as [10]

$$\begin{bmatrix} P \\ Q \end{bmatrix} = \frac{1}{2} \begin{bmatrix} u_{s\alpha} & u_{s\beta} \\ u_{s\beta} & -u_{s\alpha} \end{bmatrix} \begin{bmatrix} i_{s\alpha} \\ i_{s\beta} \end{bmatrix} = \frac{1}{2} \begin{bmatrix} u_{sd} & u_{sq} \\ u_{sq} & -u_{sd} \end{bmatrix} \begin{bmatrix} i_{sd} \\ i_{sq} \end{bmatrix} \tag{2}$$

where P and Q are the instantaneous active and reactive powers, respectively. $u_{s\alpha}$, $u_{s\beta}$ represent the α-axis and β-axis components of the main voltage vector $\boldsymbol{u_s}$; $i_{s\alpha}$, $i_{s\beta}$ represent the α-axis and β-axis components of the line current vector $\boldsymbol{i_s}$ in $\alpha\beta$ stationary coordinate frame. u_{sd}, u_{sq} are the d-axis and q-axis components of $\boldsymbol{u_s}$; i_{sd}, i_{sq} are the d-axis and q-axis components of $\boldsymbol{i_s}$ in dq rotating frame.

According to (2), the fictitious β-axis components of $\boldsymbol{u_s}$ and $\boldsymbol{i_s}$ need to be generated to estimate the instantaneous active and reactive powers in single-phase converter system. Then the second-order generalized integral (SOGI) method is adopted to estimate the fictitious β-axis components of $\boldsymbol{u_s}$ and $\boldsymbol{i_s}$ in this paper, which can present advantages such as noise and harmonic filtering, simple structure, the adaptive characteristic and fast dynamic response [11].The transfer function expression of SOGI in s-domain can be written as

$$\begin{cases} x_\alpha (s) = \dfrac{k\omega s}{s^2 + k\omega s + \omega^2} x(s) \\ x_\beta (s) = \dfrac{k\omega^2}{s^2 + k\omega s + \omega^2} x(s) \end{cases} \tag{3}$$

where ω represents angular frequency of the main voltage u_s, k is the damping factor, and the outstanding performance of SOGI depends on the selection of damping factor k. [11] gives an optimal value of k to gain the fastest response while $k=1.57$.

Applying (3) to the main voltage u_s and the line current i_s, respectively, the block diagram of calculating $u_{s\alpha}$, $u_{s\beta}$, $i_{s\alpha}$, $i_{s\beta}$, P and Q is shown in Fig. 2.

The $\alpha\beta/dq$ transformation in Fig. 2, can be expressed as

$$\begin{bmatrix} d \\ q \end{bmatrix} = \begin{bmatrix} \cos \omega t & \sin \omega t \\ -\sin \omega t & \cos \omega t \end{bmatrix} \begin{bmatrix} \alpha \\ \beta \end{bmatrix} \tag{4}$$

From (1), the mathematical model in d-q frame of the adopted converters can be written as

$$\begin{cases} L_s \dfrac{di_{sd}}{dt} = u_{sd} - R_s i_{sd} - u_{abd} + \omega L_s i_{sq} \\ L_s \dfrac{di_{sq}}{dt} = u_{sq} - R_s i_{sq} - u_{abq} - \omega L_s i_{sd} \end{cases} \tag{5}$$

where u_{abd}, u_{abq} represent the d-axis and q-axis components of the modulated voltage vector $\boldsymbol{u_{ab}}$.

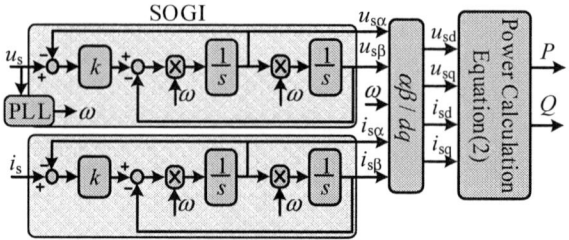

Fig. 2 Block diagram of instantaneous powers estimation of the adopted converter

Assuming that there is no fluctuation in the main voltage u_s, then the d-axis and q-axis voltage components u_{sd}, u_{sq} are constant values in normal condition. Generally, the main voltage u_s is orientated to d axis, in the d-q reference frame, where $u_{sd} = U_{sm}$, $u_{sq} = 0$. And U_{sm} is the peak values of the fundamental components in the main voltage u_s. Then substituting (2) into (5), the power mathematical model of single-phase PWM converter is shown as

$$\begin{cases} L_s \dfrac{dP}{dt} = \dfrac{1}{2} U_{sm}^2 - \dfrac{1}{2} u_{abd} U_{sm} - R_s P - \omega L_s Q \\ L_s \dfrac{dQ}{dt} = \dfrac{1}{2} u_{abq} U_{sm} - R_s Q + \omega L_s P \end{cases} \tag{6}$$

III. CONVENTIONAL DP-DPC AND MP-DPC OF THE ADOPTED CONVERTER

A. Principle of DP-DPC scheme

The equivalent resistance R_s is very small and can be ignored. Then, applying the first-order discrete approximation to the power mathematical model expressed in (6), the $(k+1)$th power sampling values are expressed as

$$\begin{cases} P(k+1) = P(k) + \dfrac{T_s}{L_s} \left(\dfrac{1}{2} U_{sm}^2 (k) - \dfrac{1}{2} u_{abd} (k) U_{sm} (k) - \omega L_s Q(k) \right) \\ Q(k+1) = Q(k) + \dfrac{T_s}{L_s} \left(\dfrac{1}{2} u_{abq} (k) U_{sm} (k) + \omega L_s P(k) \right) \end{cases} \tag{7}$$

where T_s is the sampling interval.

In order to eliminate the power errors at the $(k+1)$th sampling interval successfully, in DP-DPC $P(k+1)$ and $Q(k+1)$ can be expressed as

$$\begin{cases} P(k+1) = P^* \\ Q(k+1) = Q^* \end{cases} \tag{8}$$

By replacing (8) into (7), the reference values of the modulation voltages that satisfy the control law in (8) are obtained as

$$\begin{cases} u_{\text{abd}}(k) = U_{\text{sm}}(k) - \dfrac{2L_s}{U_{\text{sm}}(k)T_s}\left[P^* - P(k)\right] - \dfrac{2\omega L_s Q(k)}{U_{\text{sm}}(k)} \\[3mm] u_{\text{abq}}(k) = \dfrac{2L_s}{U_{\text{sm}}(k)T_s}\left[Q^* - Q(k)\right] - \dfrac{2\omega L_s P(k)}{U_{\text{sm}}(k)} \end{cases} \quad (9)$$

Then, transforming the reference voltages calculated from (9) to the $\alpha\text{-}\beta$ reference frame, which can be expressed as

$$\begin{bmatrix} u_{\text{ab}\alpha}(k) \\ u_{\text{ab}\beta}(k) \end{bmatrix} = \begin{bmatrix} \cos\omega t & -\sin\omega t \\ \sin\omega t & \cos\omega t \end{bmatrix}\begin{bmatrix} u_{\text{abd}}(k) \\ u_{\text{abq}}(k) \end{bmatrix} \quad (10)$$

where $u_{\text{ab}\alpha}$, $u_{\text{ab}\beta}$ represent the α-axis and β-axis components of the modulated voltage vector $\boldsymbol{u_{\text{ab}}}$ in two-phase stationary coordinate system. The fictitious signal $u_{\text{ab}\beta}$ is discarded and $u_{\text{ab}\alpha}$ will be accepted as the reference value for the PWM generator. The modulated voltage $u_{\text{ab}\alpha}$ is limited by dc-link voltage of the adopted converter, and the amplitude of $|u_{\text{ab}\alpha}|$ is lower than u_{dc} in normal condition. If the $|u_{\text{ab}\alpha}(k)|$ is larger than u_{dc}, the optimized $u_{\text{abd}}(k)$ and $u_{\text{abq}}(k)$ are out of the linear operation range. In this condition, the modulated voltage $u_{\text{ab}\alpha}$ remains u_{dc} or $-u_{\text{dc}}$, which can be expresses as

$$u_{\text{ab}\alpha} = \text{sgn}\left(u_{\text{ab}\alpha}(k)\right)u_{\text{dc}} \quad (11)$$

where $\text{sgn}(u_{\text{ab}\alpha}(k))$ represents the polarity of $u_{\text{ab}\alpha}(k)$. Fig. 3 shows the control block of the DP-DPC scheme.

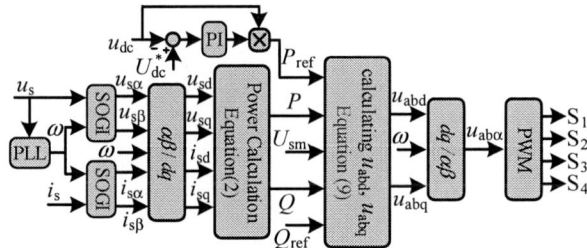

Fig. 3 Block diagram of the DP-DPC scheme

B. Principle of MP-DPC scheme

The control goal of MP-DPC is that the active and reactive powers are forced to track their references, respectively. Therefore, the cost function is usually defined as

$$J(k) = \left[P^* - P(k+1)\right]^2 + \lambda\left[Q^* - Q(k+1)\right]^2 \quad (12)$$

where λ represents the weighting factor of cost function. Substituting $P(k+1)$ and $Q(k+1)$ of (7) into (12), the cost function can be rewritten as

$$\begin{aligned} J(k) = & \left(\dfrac{U_{\text{sm}}(k)T_s}{2L_s}\right)^2\left[u_{\text{abd}}(k) - \begin{pmatrix} U_{\text{sm}}(k) - \dfrac{2\omega L_s Q(k)}{U_{\text{sm}}(k)} \\ -\dfrac{2L_s}{U_{\text{sm}}(k)T_s}\left(P^* - P(k)\right) \end{pmatrix}\right]^2 \\ & + \lambda\left(\dfrac{U_{\text{sm}}(k)T_s}{2L_s}\right)^2\left[u_{\text{abq}}(k) - \begin{pmatrix} -\dfrac{2\omega L_s P(k)}{U_{\text{sm}}(k)} \\ +\dfrac{2L_s}{U_{\text{sm}}(k)T_s}\left(Q^* - Q(k)\right) \end{pmatrix}\right]^2 \end{aligned} \quad (13)$$

The optimization control object is to minimize the cost function $J(k)$ in (13) by calculating appropriate control variables $u_{\text{abd}}(k)$ and $u_{\text{abq}}(k)$, which yields

$$\begin{cases} \dfrac{\partial J(k)}{\partial u_{\text{abd}}(k)} = 0 \\[3mm] \dfrac{\partial J(k)}{\partial u_{\text{abq}}(k)} = 0 \end{cases} \quad (14)$$

Solving the simultaneous (13) and (14), the optimized voltage variables $u_{\text{abd}}(k)$ and $u_{\text{abq}}(k)$ can be derived as

$$\begin{cases} u_{\text{abd}}(k) = U_{\text{sm}}(k) - \dfrac{2L_s}{U_{\text{sm}}(k)T_s}\left[P^* - P(k)\right] - \dfrac{2\omega L_s Q(k)}{U_{\text{sm}}(k)} \\[3mm] u_{\text{abq}}(k) = \dfrac{2L_s}{U_{\text{sm}}(k)T_s}\left[Q^* - Q(k)\right] - \dfrac{2\omega L_s P(k)}{U_{\text{sm}}(k)} \end{cases} \quad (15)$$

where the weighting factor λ is eliminated during the procedure of solving (14). Also the expression and limitation of $u_{\text{ab}\alpha}$ in MP-DPC can be deduced, which is the same as (10) and (11) in DP-DPC.

IV. RELATIONSHIP BETWEEN DP-DPC AND MP-DPC

A. Inherent relationship of continuous-control-set MP-DPC schemes

Due to those excellent characteristics, several continuous-control-set MP-DPC schemes, from different ways of implementation, have been proposed to improve the control performance of the adopted converter, such as modulation function optimization or duty cycle optimization. Although these schemes are realized in different ways, which adopt the same cost function and mathematical model, the ultimate essence is the same. Taking the MP-DPC with modulation function optimization as an example, the detail analysis is shown as follow.

According to (10) and (15), the expression of $u_{\text{ab}\alpha}$ can be written as

$$\begin{aligned} u_{\text{ab}\alpha}(k) = & \, U_{\text{sm}}(k)\cos\omega t - \dfrac{2L_s}{U_{\text{sm}}(k)T_s}\left[P^* - P(k)\right]\cos\omega t \\ & - \dfrac{2L_s}{U_{\text{sm}}(k)T_s}\left[Q^* - Q(k)\right]\sin\omega t \\ & + \dfrac{2\omega L_s}{U_{\text{sm}}(k)}\left[P(k)\sin\omega t - Q(k)\cos\omega t\right] \end{aligned} \quad (16)$$

On the basis of (3), the fundamental component expressions of the main voltage u_s in $\alpha\text{-}\beta$ reference frame can be defined as

$$\begin{cases} u_{s\alpha} = U_{\text{sm}}\cos\omega t \\ u_{s\beta} = U_{\text{sm}}\sin\omega t \end{cases} \quad (17)$$

Then substituting (17) into (16), the expression of modulation index m can be expressed as

1874

$$m(k) = \frac{u_{ab\alpha}(k)}{u_{dc}(k)} = \left\{ 2\omega L_s T_s \left(P(k) u_{s\beta}(k) - Q(k) u_{s\alpha}(k) \right) \right.$$
$$-2L_s u_{s\alpha}(k) \left(P^* - P(k) \right) - 2L_s u_{s\beta}(k) \left(Q^* - Q(k) \right) \quad (18)$$
$$\left. + u_{s\alpha}(k) U_{sm}^2(k) T_s \right\} \bigg/ u_{dc}(k) U_{sm}^2(k) T_s$$

From (18), the expression of m is the same with that of the MP-DPC with modulation function optimization scheme in [8]. Thus, it can be concluded that the inherent character of MP-DPC with modulation function optimization is accordant with that of the conventional MP-DPC. Similarly, the same conclusion also can be obtained between MP-DPC with duty cycle optimization and the conventional MP-DPC. So the internal essence of these continuous- control-set MP-DPC schemes with same cost function and mathematical model are in accordance.

B. Relationship between DP-DPC and continuous-control-set MP-DPC schemes

From (9) and (15), expressions of the control variables $u_{abd}(k)$ and $u_{abq}(k)$ in DP-DPC and MP-DPC schemes are the same. In DP-DPC, in order to eliminate the power errors, $P(k+1)$ and $Q(k+1)$ are defined as their reference values P^* and Q^*. With this definition, (12) will reach the minimized value as zero, which is the main control object of MP-DPC. It is more intuitive that $J(k)$ equals zero while substituting $u_{abd}(k)$ and $u_{abq}(k)$ of (9) into (13).

And in MP-DPC, the purpose of calculating the partial derivatives of $u_{abd}(k)$ and $u_{abq}(k)$ to $J(k)$ in (14) is to select the appropriate $u_{abd}(k)$ and $u_{abq}(k)$ which will predict $P(k+1)$ and $Q(k+1)$ to track their reference values without errors, which is the key definition in DP-DPC. (14) always has a global solution, or a set of $u_{abd}(k)$ and $u_{abq}(k)$ always exists that adjusts $J(k)$ to reach the extreme point. Although the powers $P(k+1)$ and $Q(k+1)$ predicted by the optimal $u_{abd}(k)$ and $u_{abq}(k)$ may not reach their references during one control interval, the prediction trend of powers is correct, and these predicted powers will reach their own reference values in several control intervals.

According to the above analysis, the essence of DP-DPC and continuous- control-set MP-DPC scheme is similar while the kind of implementation is different.

V. CONTROL PERFORMANCE OBSERVED FROM CIRCUIT EXPERIMENT

A scale-down single-phase PWM converter experimental platform has been implemented to verify the effectiveness of DP-DPC and continuous-control-set MP-DPC schemes. Fig. 4 shows a photo of the experimental hardware prototype platform, which is consist of the main power circuit, signal processing circuit, TMS320F2812 controller and gate signal driving circuit and so on.

Fig. 4 The photo of the experimental hardware platform

TABLE I
THE PARAMETERS OF EXPERIMENTAL SYSTEM

Parameters	Value
The main voltage(rms)U_s/V	60
The dc-link reference voltage U_{dc}^*/V	120
The dc-link rated load R_L/Ω	30
The ac-side inductor L_s/mH	4.76
The dc-link capacitor C_d/mF	1.65

Due to the same character of these DPC schemes, it just needs to verify the effectiveness of one scheme in those DPC schemes. Table I shows the experimental system parameters, and switching frequency f_s is set as 5kHz.

Fig. 5 a shows steady-state experimental waveforms of the main voltage u_s, line current i_s and dc-link voltage u_{dc} of MP-DPC scheme. Fig. 5 b shows the FFT analysis results of i_s in MP-DPC scheme in experimental test. Fig. 5 c shows experimental waveforms for dynamic response test when the reference active power steps up from 75% to 100% rating value. From Fig. 5, it can be noticed that MP-DPC scheme can achieve perfect steady-state and fast dynamic performance.

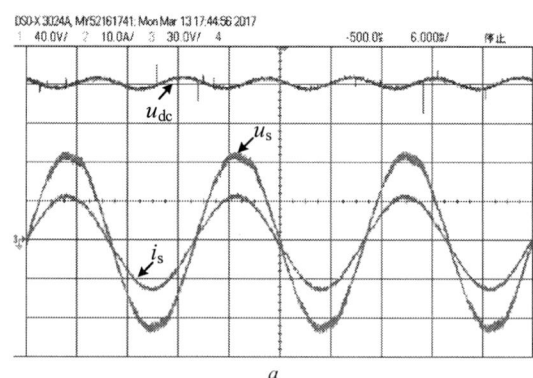

a

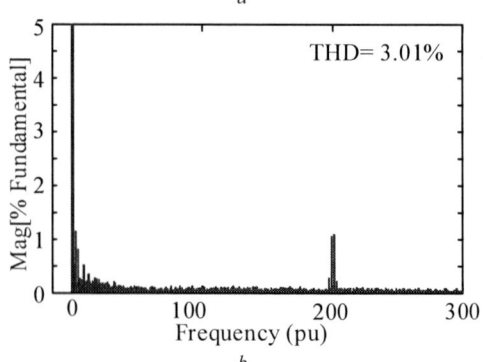

b

The 2018 International Power Electronics Conference

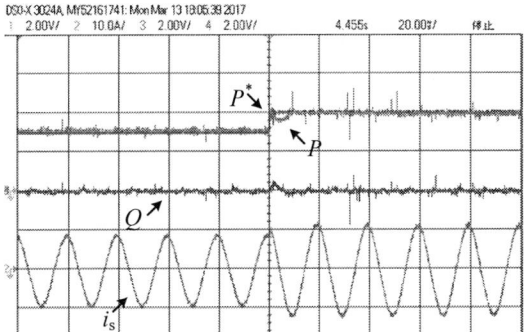

Fig. 5 Experimental results of MP-DPC scheme
a   Waveforms of the main voltage, line current and dc-link voltage in steady-state (u_s: 40V/div, i_s: 10A/div, u_{dc}:30V/div, Time: 6ms/div)
b   FFT analysis result of the line current
c Waveforms of the instantaneous powers while the active reference power steps up from 75% to 100% rating value (P, Q and P^*: 240W/div, i_s: 10A/div, Time: 20ms/div)

VI. CONCLUSION

This paper presents two kinds of DPC scheme for single-phase PWM converters, which are DP-DPC and conventional continuous-control-set MP-DPC schemes, and verifies their inherent similar relationship. Also, the inner characters of different continuous-control-set MP-DPC schemes with the same cost function and mathematical model are in accordance, such as MP-DPC with modulation function optimization or duty cycle optimization.

Although the design concepts of these schemes are absolutely different, both have the same expressions of reference voltages $u_{abd}(k)$ and $u_{abq}(k)$. The experimental results confirm the good performance of the MP-DPC scheme. Furthermore, this conclusion also applies equally to the direct current control (DCC), that is, deadbeat predictive DCC and continuous-control-set model predictive DCC schemes with PWM stage have the same inherent characteristic.

ACKNOWLEDGMENT

This work was supported in part by the National key R & D program of China under grants 2017YFB1200901.

REFERENCES

[1]   W. Song, S. Wang, C. Xiong, et al, " Single phase three-level SVPWM algorithm for grid-side railway traction converter and its relationship of carrier-based PWM," *IET Electrical Systems in Transportation*, vol.4, no. 3, pp. 78–87, 2014.

[2]   J. Norniella, J. Cano, G. Orcajo, et al, " Improving the dynamics of virtual-flux-based control of three-phase active rectifiers", *IEEE Trans. on Industrial Electronics*, vol. 61, no.1, pp. 177-187, Jan. 2014.

[3]   Chen, S. and Joos, G., "Direct power control of active filters with averaged switching frequency regulation," *IEEE Trans. on Power Electronics*, vol. 23, no. 10, pp. 2729-2737, 2008.

[4]   M. Malinowski, M. Jasinski, and M. Kazmierkowski, "Simple direct power control of three-phase PWM rectifier using space-vector modulation (DPC-SVM)," *IEEE Trans. on Industrial Electronics*, vol. 51, no.2, pp. 447-454, 2004.

[5]   W. Song, J. Ma, L. Zhou, et al, "Deadbeat predictive power control of single-phase three-level neutral-point-clamped converters using space-vector modulation for electric railway traction," *IEEE Trans. on Power Electronics*, vol. 31, no. 1, pp. 721–732, 2016.

[6]   M. Monfared, M. Sanatkar, and S. Golestan, "Direct active and reactive power control of single-phase grid-tie converters," *IET Power Electronics*, vol.5, no. 8, pp. 1544-1550, 2012.

[7]   H. Eskandari-Torbati and D. Khaburi, "Direct power control of three phase PWM rectifier using model predictive control and SVM switching," *Drive Syst. and Technol. Conf. on Power Electronics, Iran*, pp. 193-198, April 2013.

[8]   W. Song, Z. Deng, S. Wang, et al, "A simple model predictive power control strategy for single-phase PWM converters with modulation function optimization," *IEEE Trans. on Power Electronics*, vol. 31, no. 7, pp. 5279–5289, 2016.

[9]   Y. Zhang, W. Xie, Z. Li, et al, "Model predictive direct power control of a PWM rectifier with duty cycle optimization," *IEEE Trans. On Power Electronics*, vol. 28, no. 11, pp. 5343-5351, 2013.

[10]   J. Ma, W. Song, S. Jiao, et al, "Power calculation for direct power control of single-phase three-level rectifiers without phase-locked loop," *IEEE Trans. on Industrial Electronics*, vol.36, no.5, pp. 2871-2882, 2016.

[11]   A. Kulkarni and V. John, "A novel design method for SOGI-PLL for minimum settling time and low unit vector distortion," *Proc. IEEE Ind. Electron. Soc. Conf., Vienna*, pp. 274-279, Nov. 2013.

The 2018 International Power Electronics Conference

Deadbeat control for Multi-level Inverter using 1MHz Multisampling Method for Utility Interactive System

Ryosuke Kikuchi[1*], Ryunosuke Araumi[1], Tomoki Yokoyama[1]

1 Tokyo Denki University, 5, Asahicho, Senju, Adachiku, Tokyo, Japan
*E-mail: yoko@fr.dendai.ac.jp

Abstract—Recently, multi-level inverters were widely applied to grid connect system to downsize the filter component and reduce the harmonic distortions of the output voltage and the current. In these systems, the conventional PI control is widely used for the inverter control, and the dynamic performance should be considered in the case of FRT conditions. In this paper, a new deadbeat control method for multi-level inverters is proposed based on PWM hold technique. Three level inverters are investigated to achieve the fast transient response in the FRT conditions. 1MHz multisampling method was also applied to improve the dynamic response. The advantages of the proposed method was verified through simulations.

Keywords—*Multi-level inverter, Deadbeat control, Utility Interactive Inverter, Multisampling, PWM hold*

I. INTRODUCTION

Large distributed power systems such as mega solar powe plant and wind farm have much attention due to the environmental problems. In the case of many distributed power plants are connected to the power grid, the stability of the power grid should be considered[1]-[3]. For the utility interactive inverter system, the control robustness is required to suppress the uncertainty of the power grid. Fault ride through requirement should be also satisfied for the grid failure.

Recentry, multi-level inverters such as Neutral-Point Clamped (NPC) inverters, Flying Capacitor (FC) inverters or Modular Multilevel Converter (MMC) were widely applied for the large scale power systems because of low-power-loss and high-voltage-rating switching devices have been developed[4][5]. In multi-level inverters, the harmonics of the output voltage can be supressed by increasing the number of the voltage levels[6]-[8]. In the utility interactive inverter systems, the conventional PI control is widely used for the inverter control. In the utility interactive inverter system, the precise control performance is required to achieve both of the steady state characteristics and the transient characteristics[9]. In the conventional control method, it is hard to satisfy both of the characteristics because of the limitation of low carrier frequency in the large scale power converter. In the previous studies, the deadbeat control for two-level inverters was proposed and good transient response was achieved while the FRT conditions[1]-[3]. This study focused to construct the precise control strategy of multi-level inverteres by adopting the new deadbeat control.

In order to derive the discrete time model for two level PWM inverters, a PWM hold technique was used in [13]. Adopting a PWM hold technique for multi-level inverters, a new discrete model is proposed. To improve the control accuracy, 1MHz multisampling method was also adopted. The advantage of the proposed method was verified through simulations.

II. CONTROL METHOD

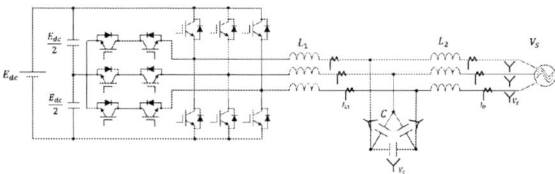

Fig. 1.   Main circuit of three-phase utility interactive inverter system

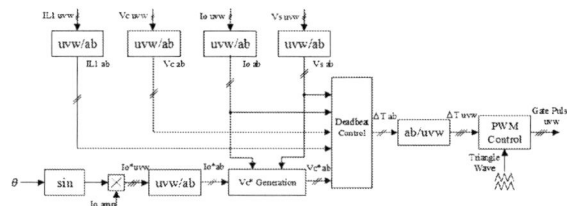

Fig. 2.   Control block diagram of the proposed method

Fig.1 shows the main circuit of three phase utility interactive system with the T-NPC inverter. The DC voltage is supplied from the distributed power supply. E_{DC} was connected to the inverter bridge, then transformed into the square wave voltages with the amplitude of 0[V], $\pm E_{dc}/2$[V] or $\pm E_{dc}$[V]. Then connected to the power grid via LC filter and the inductor L_2. The output current I_O is the control target of the utility interactive inverter system. Since the output current I_O is determined by the difference voltage between the grid voltage V_S and the capacitor voltage V_C, the output current I_O can be controlled by the proper operation of the capacitor voltage V_C. In addition, the capacitor voltage V_C can be controlled by the inverter voltage V_{inv} which is determined by the control input ΔT. Fig.2 shows the control block diagram of the proposed methase detected by the PLL(Phase Locked Loop) control. Adopting three-phase

1877

to two-phase transformation for eahod. The current reference is generated based on the utility pch state variables, three phase system can be converted to two independent single phase system. Then the control inputs ΔT can be derived in each single phase systems. Adopting two-phase to three-phase transformation for each control inputs ΔT_α and ΔT_β, three phase control inputs ΔT_U, ΔT_V, ΔT_W can be derived.

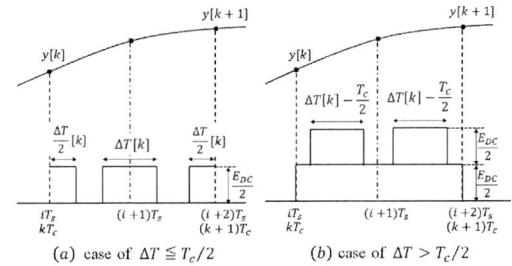

(a) case of $\Delta T \leqq T_c/2$ (b) case of $\Delta T > T_c/2$

Fig. 5.   Output line voltage of three level NPC inverter

III.   MODELING

A. *Modeling of multilevel PWM pulse*

Three phase system can be converted to two independent single phase system by three-phase to two-phase transformation. In this section, the modeling method for single phase multi-level system was described[14].

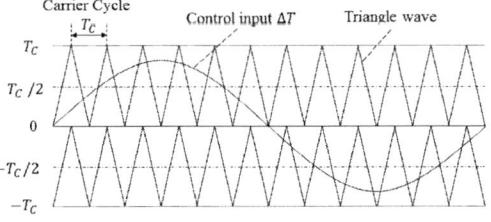

Fig. 3.   Level shifted PWM

Fig.3 shows the level shifted PWM method which is for the NPC inverter. In the level shifted multi carrier modulation, two triangular carrier waveforms are used as the PWM control method for three-level inverters. PWM control is carried out by comparing two triangular carrier waveforms and the control input ΔT[15]. Fig.4 shows the contol input and inverter line voltage in the three level inverter. The output level of the line voltage varied due to the value of the contol input.

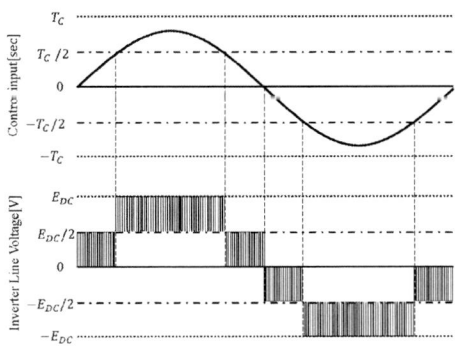

Fig. 4.   Contol input and Inverter line voltage

In the conventional PWM hold method, the centered pulse model was applied by equalizing the sampling period T_S and the carrier period T_C [1]. Fig.5(b) shows the inverter output voltage in the case that the control input exceeds half of the carrier period T_C. In this case, the centered pulse model can be adopted within one sampling period by doubling the sampling frequency f_S. Fig.5(a) shows the inverter output voltage in the case that the control input is within the half of the carrier period T_C. In this case, the pulse is divided into two pulses at both ends of the sampling period.

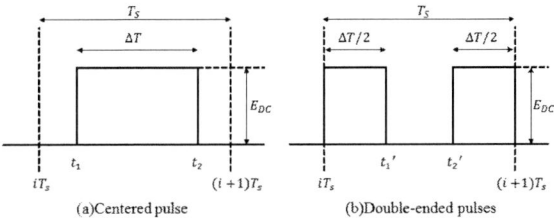

(a)Centered pulse   (b)Double-ended pulses

Fig. 6.   The centered pulse and the double-ended pulse within the sampling period

Fig.6 shows the centered pulse and the double-ended pulse in two level inverter. The discretization procedure of these pulses is described. The state equation for the discrete time system in two level invereters can be obtained as follows. (1) shows the continuous time system and (2) shows the discrete time system. B_{v1} and G_v are the matrix of the inverter output.

$$\dot{x_v}(t) = A_v x_v(t) + B_{v1} u_v(t) + B_{v2} V_s(t). \quad (1)$$

$$x_v[k+1] = F_v x_v[k] + G_v u_v[k] + H_v V_s[k]. \quad (2)$$

In Fig.6(a), t_1 and t_2 are obtained as follows,

$$t_1 = \frac{T_s - \Delta T}{2}, t_2 = \frac{T_s + \Delta T}{2}. \quad (3)$$

Integrating the output pulse over time from t_1 to t_2, the equation for the discretization with the centered pulse model is obtained as follows.

$$G_v = \int_{(T_s - \Delta T)/2}^{(T_s + \Delta T)/2} e^{A_v(T_s - \tau)} B_{v1} E_{dc} d\tau, \quad (4)$$

$$= e^{A_v Ts/2} B_{v1} E_{dc} \Delta T$$

In Fig.6(b), t_1' and t_2' are obtained as follows.

$$t_1' = iT_s + \frac{\Delta T}{2}, t_2' = (i+1)T_s - \frac{\Delta T}{2}. \quad (5)$$

The equation for the discretization with the double-ended pulses model is obtained by subtracting the integral result from t_1' to t_2' from the integral result from 0 to Ts.

$$G_v = \int_0^{T_s} e^{A_v(T_s - \tau)} B_{v1} E_{dc} d\tau \quad (6)$$

$$- \int_{t_1'}^{t_2'} e^{A_v(T_s - \tau)} B_{v1} E_{dc} d\tau,$$

$$= e^{A_v Ts/2} B_{v1} E_{dc} (T_s - (t_2' - t_1')),$$

$$= e^{A_v Ts/2} B_{v1} E_{dc} \Delta T.$$

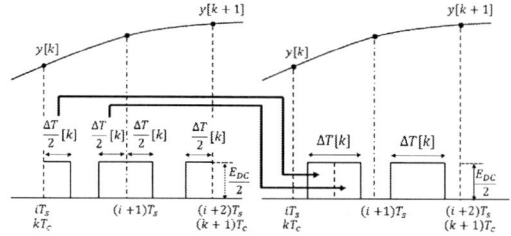

Fig. 7. Transformation to the centered pulse

To equivalent the centered pulse model and the double-ended pulse model, the double-ended pulse model can be converted to the centered pulse model by summing each pulse within one sampling period as shown in Fig.7.

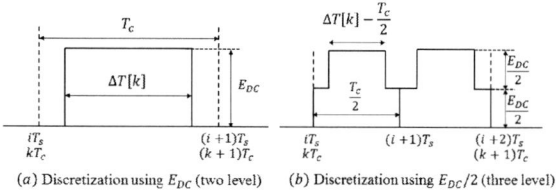

(a) Discretization using E_{DC} (two level)   (b) Discretization using $E_{DC}/2$ (three level)

Fig. 8. Discretization model of three level PWM pulse

Fig.8(a) shows the discretization of input matrix B_{v1} adopting the value of E_{dc} in two level inverters. The area of the output pulse $S_{E_{dc}}$ can be obtained as follows.

$$S_{E_{dc}} = \Delta T(k) E_{dc}. \quad (7)$$

As shown in Fig.8(a), the area inside the output pulse of two level inverters is $\Delta T(k) \cdot E_{dc}$ within one carrier period. In three level inverters, the area inside the output pulse should be $\Delta T(k) \cdot E_{dc}/2$ within one sampling period because the sampling frequency f_S of three level

inverters is twice of the the sampling frequency f_S for the two level inverters. Fig.8(b) shows the discretization of the input matrix B_{v1} using the half value of DC power supply $E_{dc}/2$. This area $S_{E_{dc}/2}$ can be obtained as follows.

$$S_{E_{dc}/2} = \left(\Delta T - \frac{T_c}{2} \right) \frac{E_{DC}}{2} + \left(\frac{T_c}{2} \right) \frac{E_{DC}}{2} = \frac{\Delta T(k) E_{DC}}{2}. \quad (8)$$

As shown in Fig.8(b), $S_{E_{dc}/2}$ is half of $S_{E_{dc}}$ within one sampling period. The area inside the output pulse of three level inverters equals to two level inverters within one carrier period. Therefore, the input matrix B_{v1} can be discretized using the half value of DC power supply $E_{dc}/2$ in three level invereters.

B. Reference generation

The capcitor voltage reference V_c is determined to match the output current I_o with the output current reference value I_O. It is assumed that the grid voltage V_s is the infinite utility line. Therefore, the output current I_o depends on the capacitor voltage V_c. The inductor voltage V_{L_2} is obtained as follows.

$$V_{L_2}(t) = V_c(t) - V_s(t). \quad (9)$$

The output current I_o is obtained as follows.

$$I_o(t) = \frac{1}{L_2} \int V_{L2} dt. \quad (10)$$

The first derivative of the output current I_o becomes;

$$\dot{I}_o(t) = \frac{V_c(t) - V_s(t)}{L_2}. \quad (11)$$

Since the grid voltage V_S is the ideal sinusoidal wave, the first and second derivative can be obtained as follows.

$$V_s(t) = sin\omega t. \quad (12)$$

$$\dot{V}_s(t) = \omega cos\omega t. \quad (13)$$

$$\ddot{V}_s(t) = -\omega^2 sin\omega t = -\omega^2 V_s(t). \quad (14)$$

The state variables are $I_{o\alpha,\beta}$, $V_{s\alpha,\beta}$ and $\dot{V}_{s\alpha,\beta}$. The state equation for the continuous time system becomes as follows[1]-[3];

$$\dot{x}_i = A_i x_i + B_i u_i, \quad (15)$$

$$x_i = \begin{bmatrix} I_{o\alpha,\beta} \\ V_{s\alpha,\beta} \\ \dot{V}_{s\alpha,\beta} \end{bmatrix}, A_i = \begin{bmatrix} 0 & -\frac{1}{L_2} & 0 \\ 0 & 0 & 1 \\ 0 & -\omega^2 & 0 \end{bmatrix},$$

$$B_i = \begin{bmatrix} \frac{1}{L_2} \\ 0 \\ 0 \end{bmatrix}, u_i = \begin{bmatrix} V_{c\alpha,\beta}(t) \end{bmatrix}.$$

The state equation based on the zero order hold for the discrete time system becomes as follows[1]-[3];

$$x_i[k+1] = F_i x_i[k] + G_i u_i[k], \qquad (16)$$

$$F_i = e^{A_i T_s} = \begin{bmatrix} f_{i11} & f_{i12} & f_{i13} \\ f_{i21} & f_{i22} & f_{i23} \\ f_{i31} & f_{i32} & f_{i33} \end{bmatrix},$$

$$, \quad G_i = \int_0^{T_s} e^{A_i \tau} B_i d\tau = \begin{bmatrix} g_{i11} \\ 0 \\ 0 \end{bmatrix}.$$

T_C is the carrier period. In addition, V_c^* is derived to solve the first line of the equation (17).

$$V_{c\alpha,\beta}^* = \frac{I_{o\alpha,\beta}^*[k] - f_{i11} I_{o\alpha,\beta}[k] - f_{i12} V_{s\alpha,\beta}[k] - f_{i12} \dot{V}_{s\alpha,\beta}[k]}{g_{i11}} \quad (17)$$

The capacitor voltage reference V_c is given by the equation (17), the output current I_o matches to the output current reference I_O at the next sampling instant.

C. Deadbeat Control

The derivation of the PWM pulse width ΔT in order to apply the single-rate deadbeat control method (SRDB) to the inverter output current control. The inductor voltage V_{L_1} is obtained as follows.

$$V_{L_1}(t) = L_1 \dot{I}_{L_1}(t). \qquad (18)$$

The first derivative of output current I_{L_1} becomes;

$$\dot{I}_{L_1}(t) = \frac{V_I(t) - V_c(t)}{L_1}. \qquad (19)$$

The capacitor current I_c can be obtained as follows.

$$I_c(t) = I_{L_1}(t) - I_o(t). \qquad (20)$$

The capacitor voltage V_c can be obtained as follows.

$$V_c(t) = \frac{1}{C} \int I_c dt. \qquad (21)$$

The first derivative of the capacitor voltage $\dot{V}_c$ becomes;

$$\dot{V}_c(t) = \frac{I_{L_1}(t) - I_o(t)}{C}. \qquad (22)$$

The state variables are $I_{L_1\alpha,\beta}$, $V_{c\alpha,\beta}$ and $I_{o\alpha,\beta}$. The state equation for the continuous time system becomes;

$$\dot{x}_v = A_v x_v + B_{v1} u_v + B_{v2} V_s \qquad (23)$$

$$x_v = \begin{bmatrix} I_{L_1\alpha,\beta} \\ V_{c\alpha,\beta} \\ I_{o\alpha,\beta} \end{bmatrix}, A_v = \begin{bmatrix} 0 & -\frac{1}{L_1} & 0 \\ \frac{1}{C} & 0 & -\frac{1}{C} \\ 0 & -\frac{1}{L_2} & 0 \end{bmatrix}$$

$$B_{v1} = \begin{bmatrix} \frac{1}{L_1} \\ 0 \\ 0 \end{bmatrix}, B_{v2} = \begin{bmatrix} 0 \\ 0 \\ -\frac{1}{L_1} \end{bmatrix}$$

The state equation for the discrete time system becomes;

$$x_v[k+1] = F_v x_v[k] + G_v u_v[k] + H_v V_s[k] \qquad (24)$$

$$F_v = e^{A_i T_s} = \begin{bmatrix} f_{v11} & f_{v12} & f_{v13} \\ f_{v21} & f_{v22} & f_{v23} \\ f_{v31} & f_{v32} & f_{v33} \end{bmatrix},$$

$$G_v = \int_{(T_s-\Delta T)/2}^{(T_s+\Delta T)/2} e^{A_v(T_c-\tau)} B_{v1} \frac{E_{DC}}{2} d\tau = \begin{bmatrix} g_{v11} \\ g_{v21} \\ g_{v31} \end{bmatrix},$$

$$H_v = \int_0^{T_s} e^{A_v(t-\tau)} B_{v2} d\tau = \begin{bmatrix} h_{v11} \\ h_{v21} \\ h_{v31} \end{bmatrix}.$$

In addition, ΔT is derived to solve the first line of the equation (25).

$$\Delta T_{\alpha,\beta}[k] = \frac{V_{c\alpha,\beta}^*[k] - f_{v21} I_{L1\alpha,\beta}[k] - f_{v22} V_{c\alpha,\beta}[k] - f_{v23} I_{o\alpha,\beta}[k] - h_{v21} V_{s\alpha,\beta}[k]}{g_{v21}}$$
$$(25)$$

The two phase control inputs ΔT_α and ΔT_β are given by the equation (25). Then the three phase control inputs ΔT_U, ΔT_V and ΔT_W can be derived by adopting three-phase to two-phase transformation for the two phase control inputs.

D. Multisampling control method

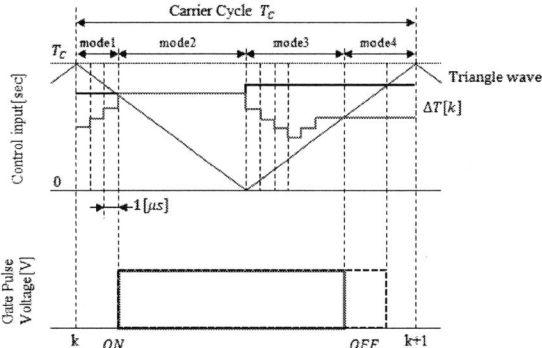

Fig. 9. Multisamplimg control method

Fig.9 shows the multisampling control method for the level shifted PWM. In the multisampling control method, one cycle of carrier is divided into four modes to calculate the gate pulse width[16]. In the mode1, the trun-on timing of the pulse is calculated every sampling period (1.0μsec). In the mode2, the value of ΔT is held to prevent the pulse from turning off. In the mode3, the trun-off timing of the pulse is calculated every sampling period (1.0μsec). In the mode4, the value of ΔT is held to prevent the pulse from turning on. 300 times sampling is carried out within one carrier interval in the proposed method. The PWM pulse is dynamically modified to improve the control accuracy.

1880

IV. SIMULATIONS

Simulations were carried out for the steady state condition and the power system disturbance condition for the conventional the PI control method and the proposed method. Grid voltage disturbance was settled to 20% of the steady state conditions. Three cases of grid conditions were evaluated. Grid voltage unbalance was settled to V_{sU}:80% and V_{sV}:20%, V_{sW}:50% of the steady state conditions. Short circuit of the power grid was settled to V_{sV}:0% of the steady state conditions. Simulation result for each condition was evaluated using THD(Total Harmonic distortion) of the output current I_O. Simulation result in the case of the power system disturbance condition was evaluated by observing the output waveform. Parameters for the simulations are indicated in TABLE I.

TABLE I. PARAMETER OF SIMULATION

Condition	Prameter	PI control	SRDB with multisampling
Sampling frequency	f_s	3.6[kHz]	1.0[MHz]
Carrier frequency	f_c	3.6[kHz]	3.6[kHz]
Nominal capacity	P		1.0[MVA]
Grid Voltage(RMS)	V_s		400[V]
Operating frequency	f		60[Hz]
DC power supply	E_{DC}		750[V]
Filter inductor	L_1		6.0%IZ(25.5[μ H])
Filter capacitor	C		10.0%IZ(555[μ F])
Interconnection inductor	L_2		2.04%IZ(8.5[μ H])
Proportational gain	K_p	0.08	-
Integration gain	K_i	0.001	-
Integration time	T_i	0.01	-

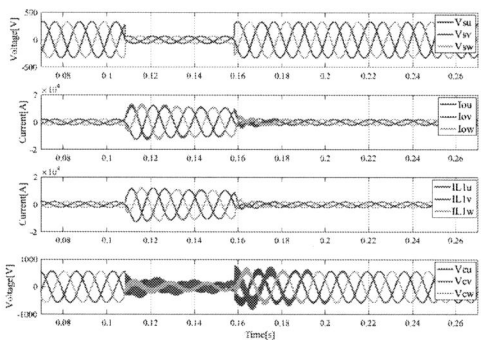

Fig. 10. Output waveforms for disturbance grid voltage conditions (PI control)

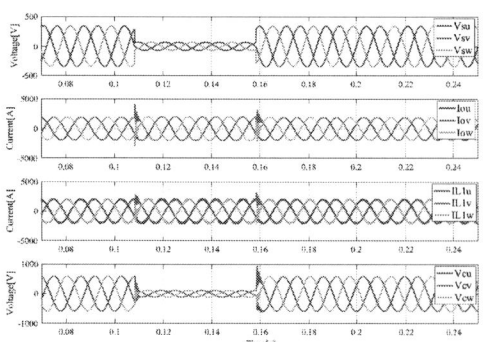

Fig. 11. Output waveforms for disturbance grid voltage conditions (SRDB)

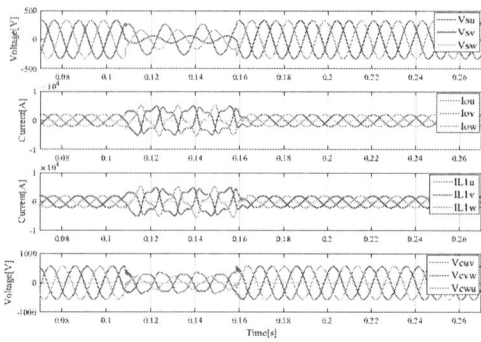

Fig. 12. Output waveforms for unbalanced grid voltage conditions (PI control)

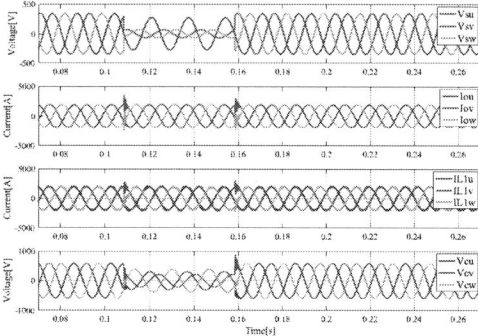

Fig. 13. Output waveforms for unbalanced grid voltage conditions (SRDB)

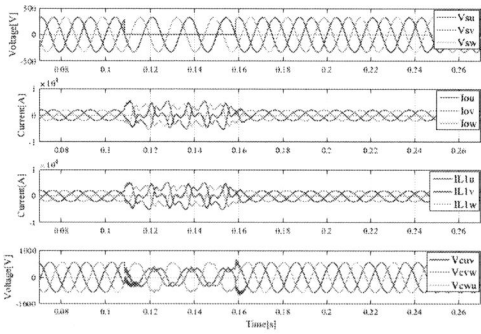

Fig. 14. Output waveforms for a short circuit conditions (PI contorol)

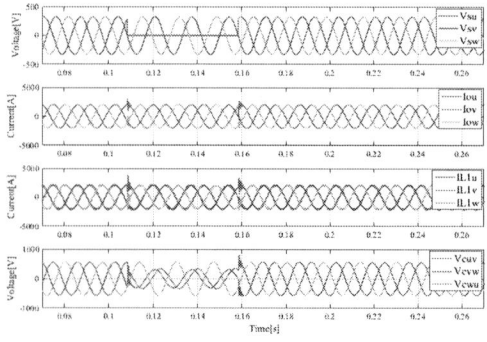

Fig. 15. Output waveforms for a short circuit conditions (SRDB)

Fig.10 and Fig.11 show the output waveforms in the case of the grid voltage disturbance conditions using PI control and SRDB. Fig.12 and Fig.13 shows in the case of grid voltage unbalance conditions using PI control and SRDB. Fig.14 and Fig.15 shows in the case of a short circuit conditions using PI control and SRDB. It was confirmed that the output current I_O can be controlled for the duration of grid voltage disturbance, grid voltage unbalance and a short circuit in the case of SRDB.

TABLE II shows the THD trend in the steady state condition.

TABLE II.   THD FOR STEADY STATE CONDITION

	amplitude	THD		
PI control	99.5[%](2031[A])	U:1.57[%]	V:1.46[%]	W:2.26[%]
SRDB	99.2[%](2025[A])	U:1.64[%]	V:1.64[%]	W:1.62[%]

TABLE III shows the THD trend in the case of the grid voltage failure conditions.

TABLE III.   THD FOR GRID VOLTAGE FAILURE CONDITIONS

conditions	control method	THD		
Disturbance	PI control	U:3.23[%]	V:3.52[%]	W:3.38[%]
Disturbance	SRDB	U:1.35[%]	V:1.36[%]	W:1.32[%]
Unbalance	PI control	U:26.3[%]	V:30.9[%]	W:31.6[%]
Unbalance	SRDB	U:1.55[%]	V:1.48[%]	W:1.46[%]
Short circuit	PI control	U:78.0[%]	V:48.6[%]	W:43.5[%]
Short circuit	SRDB	U:1.43[%]	V:1.49[%]	W:1.48[%]

It was confirmed that SRDB indicates superior control performance both of the steady state characteristics and the transient characteristics.

V.   CONCLUSION

Deadbeat control method based on PWM hold for three level invereters was proposed and verified through simulations. Deadbeat control method with multisampling was applied to improve the control performance in the case of grid voltage disturbance conditions. The control accuracy in the case of grid voltage disturbance can be improved using SRDB in the three level inverters. As a result, it was confirmed that the proposed method indicates good control performance both of the steady state characteristics and the transient responses.

REFERENCES

[1] Y.Hanashima, T.Yokoyama, "Fault Ride Through Capability of 100kHz Single Phase Utility Interactive Inverter with FPGA based Hardware Controller", EPE-PEMC, 2012

[2] M.Yoshida, T.Yokoyama, "Variable Carrier Frequency Deadbeat Control with Hysteresis Band using SoC-FPGA for Utility Interactive Inverter", EPE-PEMC, 2013

[3] S.Ohashi, T.Yokoyama, "Verification of estimation using hysteresis control for the variable carrier deadbeat control of single phase utility interactive inverter", IEEJ Journal of Industry Applications pp229-232, 2014

[4] Asmarashid Bin Ponniran, Koji Orikawa, and Jun-ichi Itoh,"Minimization of Passive Components in Multi-level Flying Capacitor DC-DC Converter",IEEJ Journal of Industry Applications, vol.5, no.1, pp.10-11, 2016.

[5] Ritwik Chattopadhyay, Subhashish Bhattacharya, Nicole C. Foureaux, Igor A. Pires, Helder de Paula, Lenin Moraes, Porfirio C. Cortizio, Sidelmo M. Silva, Braz Cardoso Filho, Jose A. de S. Brito," Low-Voltage PV Power Integration into Medium Voltage Grid Using High-Voltage SiC Devices",IEEJ Journal of Industry Applications, vol.4, no.6, pp.767-775, 2015.

[6] Toshiki Nakanishi, and Jun-ichi Itoh,"Control Strategy for Modular Multilevel Converter based on Single-phase Power Factor Correction Converter",IEEJ J. Industry Applications, vol.6, no.1, pp.46-57, 2017.

[7] Hideaki Fujita,"Emerging Technologies for Multilevel Converters in Japan",IEEJ Journal of Industry Applications, vol.1, no.2, pp.95-101, 2012.

[8] Asmarashid Bin Ponniran, Koji Orikawa, Jun-ichi Itoh," Minimization of Passive Components in Multi-level Flying Capacitor DC-DC Converter",IEEJ Journal of Industry Applications, vol.5, no.1, pp.10-11, 2016.

[9] Hitoshi Haga, Kenta Sayama, Kiyoshi Ohishi, and Takayuki Shimizu," Fine Voltage Control Based on Frequency Separation Two-Degrees-of-Freedom Control for Single-Phase Inverter",IEEJ J. Industry Applications, vol.5, no.6, pp.413-421, 2016.

[10] K.P. Gokhale, A. Kawamura, R.G. Hoft, "Dead Beat Microprocessor Control of PWM Inverter for Sinusoidal Output Waveform Synthesis", IEEE Trans. on Ind. Appl. Vol.IA-23, No.3, Sep/Oct. pp.901-910(1987).

[11] A.Kawamura, T.Haneyoshi, "Deadbeat Controlled PWM Inverter with Parameter Estimation Using Only Voltage Sensor" IEEE-trans. on PE, Vol.9, No.5, 1988

[12] T.Haneyoshi, A.Kawamura, "Waveform Compensation of PWM Inverter with Cyclic Fluctuating Loads" IEEEtrans. on INDUSTRY APPLICATIONS, Vol.24, No.4, 1988

[13] K.Sakata, H.Fujimoto:"Perfect Tracking Control of Servo Motor Based on Precise Model Considering Current Loop and PWM Hold",IEEJ Journal of Industry Applications Vol127, No6, 2007

[14] Shuhei Shimmyo, Katsutoku Takeuchi, Norio Takahashi, Makoto Matsushita, Kouhei Ohnishi," Multi-level Motor Drives for Torque Ripple Suppression Taking Control Sensitivity into Account",IEEJ Journal of Industry Applications, vol.5, no.2, pp.69-77, 2016.

[15] Natchpong Hatti, Surasak Nuilers, Jirayut Phontip, "Two-Triangle Carrier SPWM Modulation For a Five-Level Diode-Clamped PWM Inverter", IEEJ Journal of Industry Applications Vol4, No3, 2015

[16] M.Ito, R.Fujiwara, R.Araumi, T.Yoshino, "Robust digital control of single phase PWM inverter using 3MHz multi sampling method with FPGA based hardware controller", IPEMC, 2016.

1MHz multisampling deadbeat control with disturbance compensation method for three phase PWM inverter

Hiroaki Ueta*, Tomoki Yokoyama
Tokyo Denki University
5, Asahicho, Senju, Adachiku, Tokyo, Japan
E-mail: yoko@fr.dendai.ac.jp

Abstract—The demand of the UPS system is highly recommended for the power system of the important infrastructure. The precise output voltage control for the load variations are required. In the case of large scale UPSs, the power rating of the inverter becomes several hundred kVA or more, the switching frequency of the inverter is limited under several kHz range. The carrier interval becomes rather large, so the robustness for the load condition is deteriorated. The deadbeat control with disturbance compensation with 1MHz multisampling method for three phase PWM inverter system is proposed. Using FPGA based hardware controller, 1MHz sampling is realized while the carrier interval of the PWM pulse, the robustness for the load variation can be improved compared with the conventional deadbeat control method. Through simulations and experiments, the output characteristics were verified.

Keywords—FPGA, three phase PWM inverter, deadbeat control, UPS

I. INTRODUCTION

As the information technology growth in worldwide, the influence of the fluctuation of the grid power becomes huge[1]. Constant power supply system is required in the case of the grid power failure, especially for the large scale infrastructure like a data center, medical facilities, transportation system, industrial factories, and so on. In these systems, an uninterruptible power supply (UPS) is indispensable[2]. Recently, construction of hyper scale data center like the cloud system becomes widespread, the demands for the large scale and high reliable UPSs are increased[3]. In the case of large scale UPSs, the power rating of the inverter becomes several hundred kVA or more, the switching frequency of the inverter is limited under several kHz range to pretend the increasing of the switching loss[4][19]. In the lower carrier frequency range, percent impedance of the LC filter component should be designed rather large, which results in the cost up of the filter component. Also the output characteristics for the sudden load change or the nonlinear load condition have to satisfy the criteria, the precise control accuracy and the robustness are required for the UPSs.

The deadbeat control is the one approach to satisfy the precise control accuracy to the reference voltage or the current[6][7][8]. In these studies, multi-sampling method was combined with the deadbeat control using MHz range sampling approach using FPGA, the superior characteristics was obtained[9][10]. In this case, the response for the disturbance during the carrier interval was

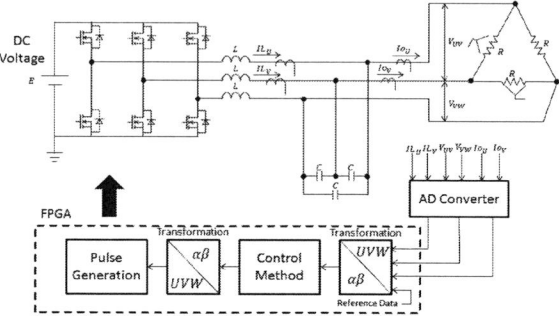

Fig. 1. Three-phase PWM inverter system

improved, the robustness for the parameter variations was also improved[4][11][12]. In this paper, the multisampling method combined with the deadbeat control for the three phase PWM inverter system is proposed. In the multi-sampling method, during the one carrier interval, plural data sampling is carried out and the recalculation of the control law is carried out and the PWM pulse width is updated in every sampling timing. As the result, the on timing and the off timing of the PWM pulse can be responsible to the disturbance of the load condition during the carrier interval. The multisampling method is extended to the three phase system, the design strategy is described for 20kHz carrier frequency and 5kHz carrier frequency. Through the simulations and the experiments, the superior characteristics of the proposed method is verified.

II. CONTROL METHOD

A. Modeling

The proposed three phase PWM inverter system is indicated in Fig.1. The three phase inverter bridge is connected to the LC filter component and the rated resistive load. Assuming that the three phase balanced system is considered, the three phase system can be converted to the independent two single phase systems. The control method for the single phase system can be applied to each independent single phase system[13]. Three phase to two phase conversion can be expressed as follows.

$$
\begin{bmatrix} x_\alpha \\ x_\beta \end{bmatrix} = \frac{2}{3} \begin{bmatrix} 1 & \frac{-1}{2} & \frac{-1}{2} \\ 0 & \frac{\sqrt{3}}{2} & \frac{-\sqrt{3}}{2} \end{bmatrix} \begin{bmatrix} x_a \\ x_b \\ x_c \end{bmatrix}. \tag{1}
$$

Three phase line to line voltage to two phase voltage conversion can be expressed as follows.

$$\begin{bmatrix} x_\alpha \\ x_\beta \end{bmatrix} = \frac{2}{3} \begin{bmatrix} \frac{1}{2} & 0 & \frac{-1}{2} \\ 0 & \frac{\sqrt{3}}{2} & 0 \end{bmatrix} \begin{bmatrix} x_{ab} \\ x_{bc} \\ x_{ca} \end{bmatrix}. \tag{2}$$

Using (1) and (2), three phase system can be converted to two independent systems, the state variables are settled as the output voltage V_α, V_β, the inductor current IL_α, IL_β and the disturbance current $Idis_\alpha$, $Idis_\beta$. The continuous time model of the proposed system becomes as follows.

$$\dot{x}(t) = Ax(t) + Bu(t). \tag{3}$$

$$A = \begin{bmatrix} \frac{-1}{RC} & \frac{1}{3C} & \frac{-1}{3C} \\ \frac{-1}{L} & 0 & 0 \\ 0 & 0 & 0 \end{bmatrix} B = \begin{bmatrix} 0 \\ \frac{1}{L} \\ 0 \end{bmatrix} x = \begin{bmatrix} v_\alpha \\ IL_\alpha \\ Idis_\alpha \end{bmatrix}, \begin{bmatrix} v_\beta \\ IL_\beta \\ Idis_\beta \end{bmatrix}$$

$, u = u_\alpha, u_\beta.$

$Idis$ in (3) can be derived from the sampled data of Io. The current for the rated resistive load is considered as the nominal current, the disturbance current can be described as follows.

$$\begin{aligned} Idis_U &= Io_U - \frac{V_{UV} - V_{VW}}{R}. \\ Idis_V &= Io_V - \frac{V_{VW} - V_{UV}}{R}. \\ Idis_W &= -Idis_U - Idis_V. \end{aligned} \tag{4}$$

As the disturbance current is included in the model, the robustness for the parameter variations or the load disturbance can be compensated. In the next section, the control method is constructed for the single phase system described in (3). The derived two independent PWM pulse width can be converted using two phase to three phase conversion, three phase PWM pulse width can be derived.

B. Deadbeat control

The sampling interval is settled as T, for the single phase PWM inverter system in (3), the discrete time model can be derived as follows[13][14].

$$x[k+1] = \Phi x[k] + \Gamma u[k]. \tag{5}$$

$$\Phi = e^{AT} = \begin{bmatrix} Z_{11} & Z_{12} & Z_{13} \\ Z_{21} & Z_{22} & Z_{23} \\ Z_{31} & Z_{32} & Z_{33} \end{bmatrix}, \Gamma = e^{\frac{AT}{2}} BE = \begin{bmatrix} G_1 \\ G_2 \\ G_3 \end{bmatrix},$$

$$x[k] = \begin{bmatrix} V_\alpha[k] \\ IL_\alpha[k] \\ Idis_\alpha[k] \end{bmatrix}, \begin{bmatrix} V_\beta[k] \\ IL_\beta[k] \\ Idis_\beta[k] \end{bmatrix}, u = u_\alpha[k], u_\beta[k].$$

Solve the first line of (5), the following equation can be derived.

$$u_{\alpha,\beta}[k] = \frac{V_{\alpha,\beta}[k+1] - Z_{11}V_{\alpha,\beta}[k] - Z_{12}IL_{\alpha,\beta}[k] - Z_{13}Idis_{\alpha,\beta}[k]}{G1}. \tag{6}$$

For the three phase PWM inverter, the derived PWM pulse width of the deadbeat control law can be converted using two phase to three phase conversion, three phase PWM pulse width can be derived as follows.

$$\begin{bmatrix} x_a \\ x_b \\ x_c \end{bmatrix} = \begin{bmatrix} 1 & 0 \\ \frac{-1}{2} & \frac{\sqrt{3}}{2} \\ \frac{-1}{2} & \frac{-\sqrt{3}}{2} \end{bmatrix} \begin{bmatrix} x_\alpha \\ x_\beta \end{bmatrix}. \tag{7}$$

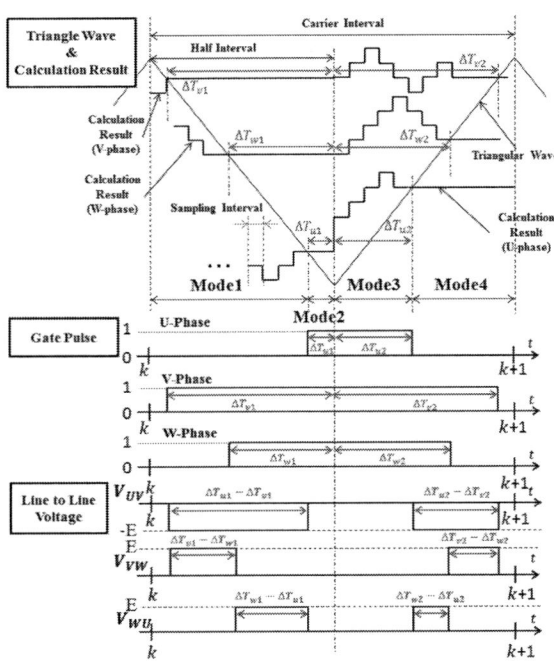

Fig. 2. PWM modulation with multisampling method

C. Multisampling method

The multisampling method can be realized using the very fast A/D converters and the FPGA based hardware controller[15][16]. MHz order sampling is carried out while the carrier interval, the PWM pulse width is re-calculated in every sampling timing and the switching timing is updated for the proper timing to respond to the disturbance of the load. Fig.2 shows the derivation procedure of the PWM pulse while one carrier interval. One carrier interval is divided into four mode. In mode 1, in every sampling timing, the control calculation is carried out and the pulse widths are compared with the triangular waveforms. When the pulse width becomes larger than the triangular waveforms, the gates are turned on, the mode is transient to the mode 2, and keep the on state to the center of the carrier period. Then the mode is transient to the mode 3, the control recalculation is carried out in every sampling timing. When the pulse width becomes smaller than the triangular waveforms, the gates are turned off. Then the mode is transient to mode 4, and the wait for the next carrier interval. Applying the multisampling method, the turn on timing and the turn off timing of the gate pulse can be determined independently, the PWM pulse widths are modified due to MHz order sampling data. If the load is the rated resistive load, the PWM pulse widths become same as the nominal deadbeat control output. And if the load is unnominal condition, the PWM pulse widths are modified to respond to the load conditions.

As the result, the 1MHz multisampling singlerate deadbeat control with disturbance compensation (1MHzMSS-RDBDC) can be realized.

The 2018 International Power Electronics Conference

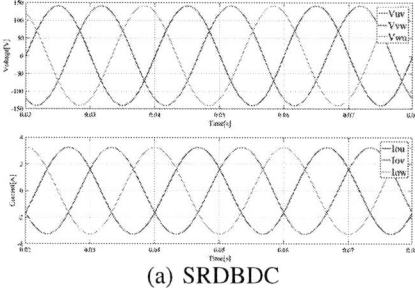
(a) SRDBDC

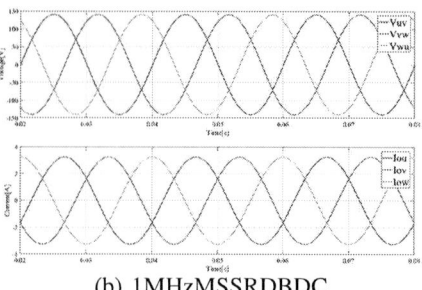
(b) 1MHzMSSRDBDC

Fig. 3. Simulation result(SRDBDC, 1MHzMSSRDBDC, 20kHz): rated load

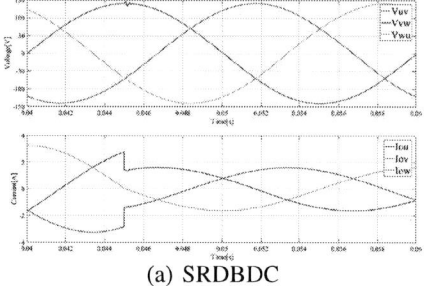
(a) SRDBDC

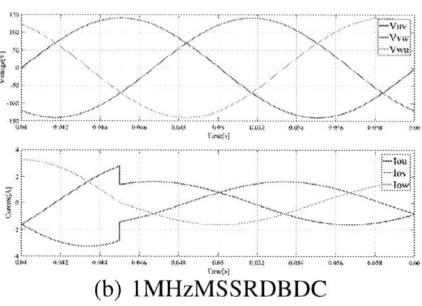
(b) 1MHzMSSRDBDC

Fig. 4. Simulation result(SRDBDC, 1MHzMSSRDBDC, 20kHz): step response

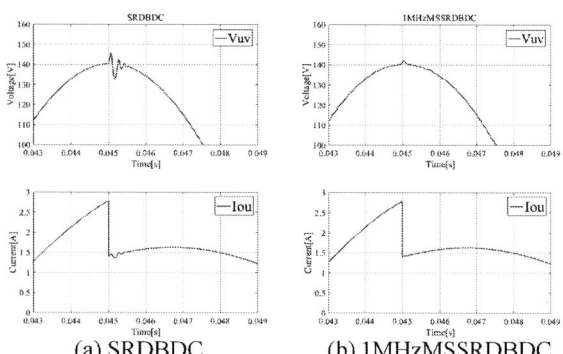
(a) SRDBDC   (b) 1MHzMSSRDBDC

Fig. 5. Enlargement waveforms of step response: 20kHz

III. SIMULATION

A. 20kHz carrier frequency

Simulations of the 1MHzMSSRDBDC is carried out for three phase PWM inverter system, the output characteristics were verified for the steady state error of the output voltage amplitude and the THD of the output voltage as the criteria. For the comparison, the single-rate deadbeat control with disturbance compensation method

TABLE I.   SIMULATION CONDITIONS

Output voltage frequency [Hz]	50
Output voltage amplitude [V_{rms}]	100
Input dc voltage [V]	200
Carrier frequency [kHz]	20
Reactor(L)[mH]	0.636(0.8%)
Capacitor(C)[μF]	4.00(9.4%)
Load Conditions	
Rated resistive load[Ω]	75.0
Load Change	
Load change conditions[Ω]	150.0

Fig. 6. Simulation result: 20kHz

TABLE II.   SIMULATION RESULT (STEADY STATE: 20KHz)

Control method	THD			Steady state error		
	U-Phase[%]	V-Phase[%]	W-Phase[%]	U-Phase[%]	V-Phase[%]	W-Phase[%]
SRDBDC	0.19	0.19	0.19	-0.66	-0.66	-0.66
1MHzMSSRDBDC	0.24	0.24	0.25	-0.70	-0.70	-0.71

TABLE III.   SIMULATION RESULT (STEP RESPONSE: 20KHz)

Control method	Maximum error			Settling time		
	U-Phase[V]	V-Phase[V]	W-Phase[V]	U-Phase[ms]	V-Phase[ms]	W-Phase[ms]
SRDBDC	15.49	7.56	8.18	0.32	0.31	0.23
1MHzMSSRDBDC	4.12	0.52	3.43	0.16	0.14	0.15

without multisampling method (SRDBDC) is also verified. The parameters used for the simulations are summarized in Table.I. The output characteristics of the rated resistive load is shown in Fig.3. The output characteristics of the step load change is shown in Fig.4. Enlargement waveforms of the U-phase voltage and the output current I_{ou} for the step load change is shown in Fig.5. The simulation result for the steady state conditions are summarized in Table.II. The load is the rated resistive load, so the output characteristics are almost same. The simulation result for the step load change conditions are summarized in Table.III. The maximum voltage error for the reference during the load change duration is summarized in the left side of the table, and the settling time for the voltage fluctuation is summarized in the right side of the

1885

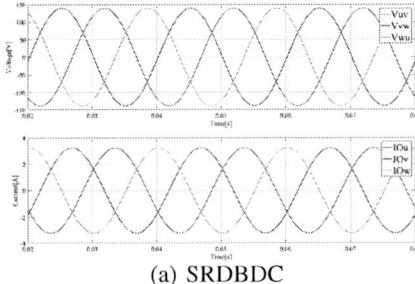

(a) SRDBDC

(b) 1MHzMSSRDBDC

Fig. 7. Simulation result(SRDBDC, 1MHzMSSRDBDC, 5kHz): rated load

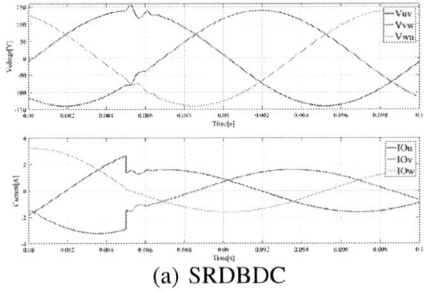

(a) SRDBDC

(b) 1MHzMSSRDBDC

Fig. 8. Simulation result(SRDBDC, 1MHzMSSRDBDC, 5kHz): step response

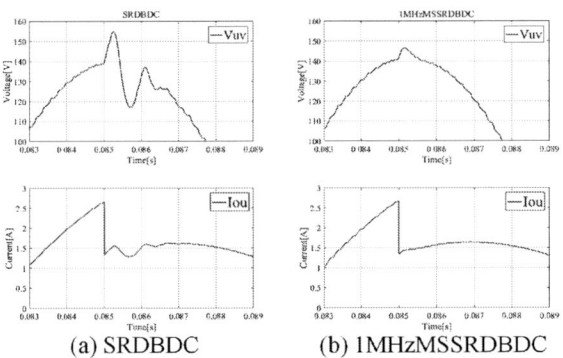

(a) SRDBDC   (b) 1MHzMSSRDBDC

Fig. 9. Enlargement waveforms of step response: 5kHz

table. 1MHzMSSRDBDC shows good transient response, the multisampling method is very effective for the load disturbance.

B. 5kHz carrier frequency

The simulations for a low carrier frequency was carried out. Parameters for the simulation of 5kHz carrier frequency are indicated in Table.IV. The output characteristics of the rated resistive load is shown in Fig.7. The output characteristics of the step load change is

TABLE IV.   SIMULATION CONDITIONS

Output voltage frequency [Hz]	50
Output voltage amplitude [V_{rms}]	100
Input dc voltage [V]	200
Carrier frequency [kHz]	5
Reactor(L)[mH]	3.180(4.00%)
Capacitor(C)[μF]	12.00(28.3%)
Load Conditions Rated resistive load[Ω]	75.0
Load Change Load change conditions[Ω]	150.0

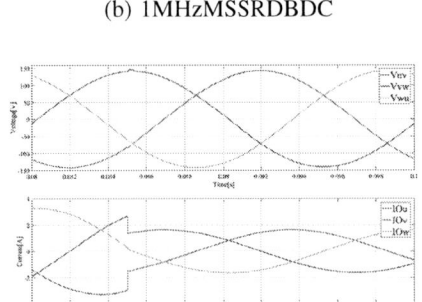

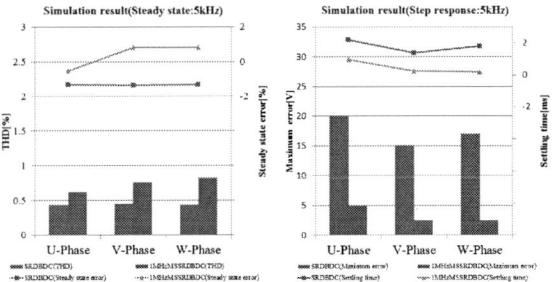

Fig. 10. Simulation result: 5kHz

TABLE V.   SIMULATION RESULT (STEADY STATE: 5kHz)

Control method	THD			Steady state error		
	U-Phase[%]	V-Phase[%]	W-Phase[%]	U-Phase[%]	V-Phase[%]	W-Phase[%]
SRDBDC	0.44	0.45	0.44	-1.34	-1.35	-1.34
1MHzMSSRDBDC	0.62	0.76	0.83	-0.53	0.80	0.82

TABLE VI.   SIMULATION RESULT (STEP RESPONSE: 5kHz)

Control method	Maximum error			Settling time		
	U-Phase[V]	V-Phase[V]	W-Phase[V]	U-Phase[ms]	V-Phase[ms]	W-Phase[ms]
SRDBDC	20.1	15.1	17.0	2.20	1.39	1.80
1MHzMSSRDBDC	4.99	2.56	2.55	0.96	0.24	0.20

shown in Fig.8. Enlargement waveforms of the U-phase voltage and the output current Iou for the step load change is shown in Fig.9. The simulation result for the steady state conditions are summarized in Table.II. The load is the rated resistive load, so the output characteristics are almost same. The simulation result for the step load change conditions are summarized in Table.VI. 1MHzMSSRDBDC shows good transient response in low carrier frequency, the multisampling method is very effective for the load disturbance.

The 2018 International Power Electronics Conference

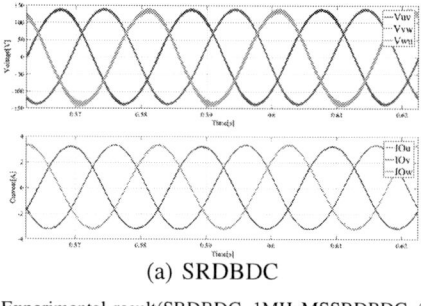
(a) SRDBDC

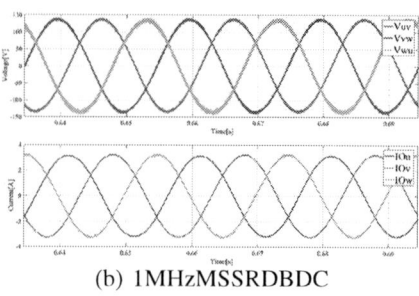
(b) 1MHzMSSRDBDC

Fig. 11.   Experimental result(SRDBDC, 1MHzMSSRDBDC, 20kHz): rated load

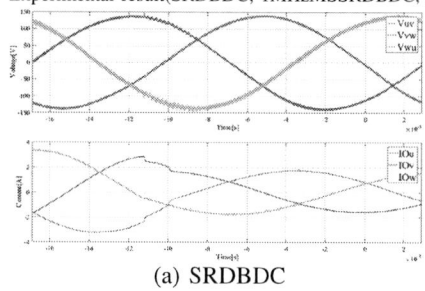
(a) SRDBDC

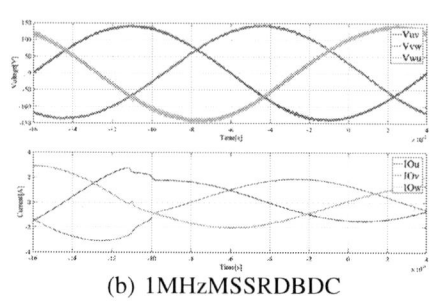
(b) 1MHzMSSRDBDC

Fig. 12.   Experimental result(SRDBDC, 1MHzMSSRDBDC, 20kHz): step response

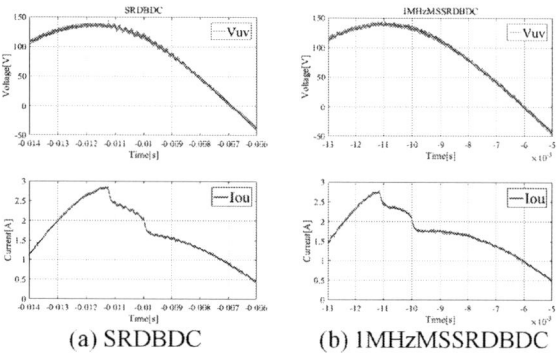
(a) SRDBDC   (b) 1MHzMSSRDBDC

Fig. 13.   Enlargement waveforms of step responce: 20kHz

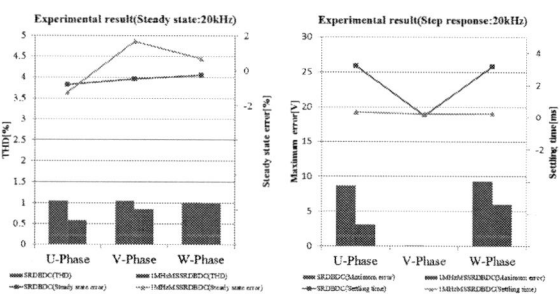

Fig. 15.   Experimental result: 20kHz

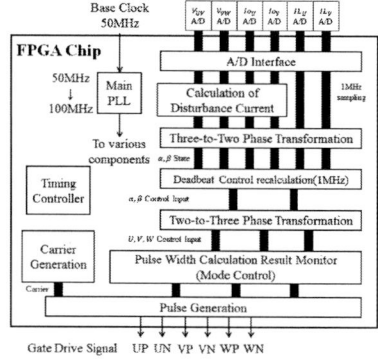

Fig. 14.   The Module Block Diagram of FPGA

IV.   EXPERIMENT

The experiments were carried out for the three phase inverter system as indicated in Fig.1, the FPGA based hardware controller was applied. The module block diagram of FPGA controller is indicated in Fig.14. In the FPGA controller, 50MHz oscillator was implemented, each control module was operated with 100MHz clock

TABLE VII.   EXPERIMENTAL RESULT(STEADY STATE:20KHZ)

Control method	THD			Steady state error		
	U-Phase[%]	V-Phase[%]	W-Phase[%]	U-Phase[%]	V-Phase[%]	W-Phase[%]
SRDBDC	1.05	1.05	1.02	-0.81	-0.48	-0.25
1MHzMSSRDBDC	0.58	0.86	0.99	-1.25	1.67	0.68

TABLE VIII.   EXPERIMENTAL RESULT(STEP RESPONSE:20KHZ)

Control method	Maximum error			Settling time		
	U-Phase[V]	V-Phase[V]	W-Phase[V]	U-Phase[ms]	V-Phase[ms]	W-Phase[ms]
SRDBDC	8.70	0.09	9.30	3.22	0.15	3.19
1MHzMSSRDBDC	3.09	0.03	5.99	0.36	0.23	0.27

generated by the main PLL circuit. The output voltage, the output current and the inductor current are converted in every 1MHz sampling timing using A/D converters. The disturbance currents were calculated and the three phase to two phase conversion was applied to the state variables, the independent two phase system can be derived. The deadbeat control law is applied to each single phase system, the derived pulse widths were converted to three phase pulse widths using two phase to three phase conversion. The derived pulse width were applied to the mode transition of the multisampling method, 1MHz multisampling control can be realized.

1887

The 2018 International Power Electronics Conference

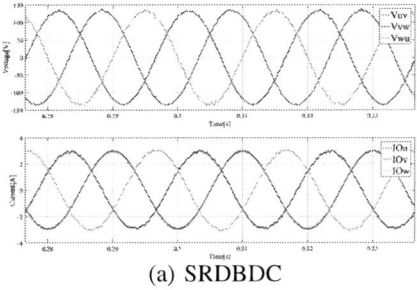 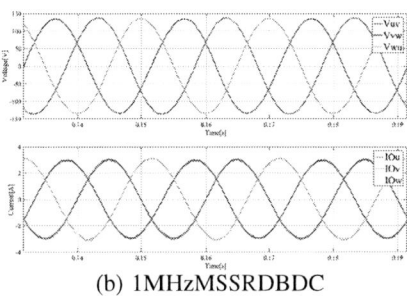

(a) SRDBDC   (b) 1MHzMSSRDBDC

Fig. 16. Experimental result(SRDBDC, 1MHzMSSRDBDC, 5kHz): rated load

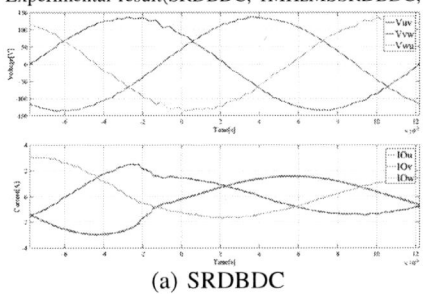 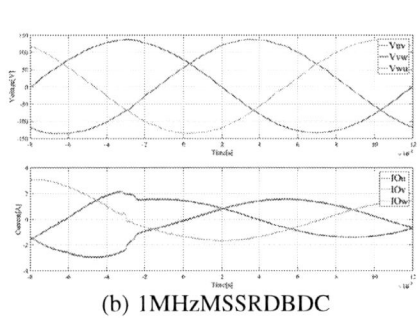

(a) SRDBDC   (b) 1MHzMSSRDBDC

Fig. 17. Experimental result(SRDBDC, 1MHzMSSRDBDC, 5kHz): step response

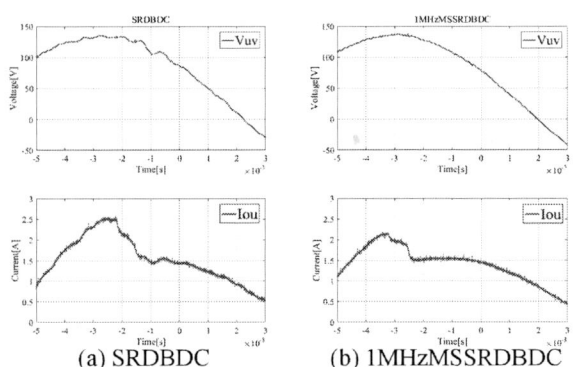

(a) SRDBDC   (b) 1MHzMSSRDBDC

Fig. 18. Enlargement waveforms of step responce: 5kHz

A. 20kHz carrier frequency

In the experiments, the output characteristics were verified for the steady state error of the output voltage amplitude and the THD of the output voltage as the criteria. The experimental result for the steady state conditions are summarized in Talbe.VII. The load is the rated resistive load, so the output characteristics are almost same, but the THD is improved in the case of 1MHzMSSRDBDC. The output characteristics of the rated resistive load is shown in Fig.11. The output characteristics of the step load change is shown in Fig.12. Enlargement waveforms of the U-phase voltage and the output current Iou for the step load change is shown in Fig.13. The experimental result for the step load change conditions are summarized in Table.VIII. The maximum voltage error for the reference during the load change duration is summarized in the left side of the table, and the settling time for the voltage fluctuation is summarized in the right side of the table. 1MHzMSSRDBDC shows good transient response, the multisampling method is very effective for the load disturbance.

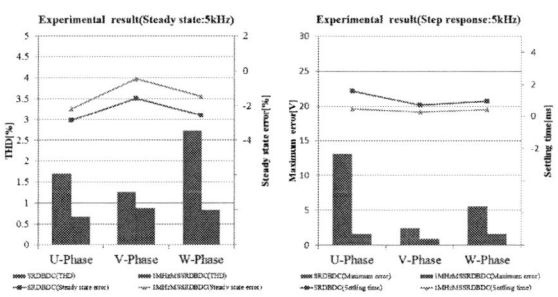

Fig. 19. Experimental result:5kHz

TABLE IX. EXPERIMENTAL RESULT (STEADY STATE:5KHZ)

Control method	THD			Steady state error		
	U-Phase[%]	V-Phase[%]	W-Phase[%]	U-Phase[%]	V-Phase[%]	W-Phase[%]
SRDBDC	1.70	1.27	2.73	-2.87	-1.58	-2.55
1MHzMSSRDBDC	0.67	0.88	0.83	-2.18	-0.46	-1.46

TABLE X. EXPERIMENTAL RESULT (STEP RESPONSE:5KHZ)

Control method	Maximum error			Settling time		
	U-Phase[V]	V-Phase[V]	W-Phase[V]	U-Phase[ms]	V-Phase[ms]	W-Phase[ms]
SRDBDC	13.1	2.46	5.57	1.61	0.75	0.99
1MHzMSSRDBDC	1.62	0.93	1.63	0.52	0.31	0.44

B. 5kHz carrier frequency

The experimental result in the 5kHz carrier frequency for the steady state conditions are summarized in Talbe.IX. From Table.IX, the THD and the steady state error is improved in the case of 1MHzMSSRDBDC. The output characteristics of the rated resistive load is shown in Fig.16. The output characteristics of the step load change is shown in Fig.17. Enlargement waveforms of the U-phase voltage and the output current Iou for the step load change is shown in Fig.18. The experimental result for the step load change conditions are summarized in Table.X. The maximum voltage error for the reference during the load change duration is summarized in the

1888

left side of the table, and the settling time for the voltage fluctuation is summarized in the right side of the table. 1MHzMSSRDBDC shows superior characteristics compared wth SRDBDC, 1MHz multi sampling method is very effective in the case of 5kHz carrier furequency.

V. CONCLUSION

For the three phase PWM inverter system, 1MHz multisampling single-rate deadbeat control with disturbance compensation method was proposed using FPGA based hardware controller. Through simulations and experiments, the output characteristics were evaluated. The robustness for the load condition was improved, the superior characteristics of the multisampling method is verified.

REFERENCES

[1] Shoishi Abe, Keisuke Ohnishi, "Development of High Reliability Three-Phase Four-Wire Modular Uninterruptible Power System", *JIASC*, (2015).

[2] Yoshihiro Hatakeyama, "The Latest Technology of Uninterruptive Power Supply", *IEIEJ*, Vol.26, No.3, pp185-189, (2006).

[3] Toshiki Nakamori, Xiaochen Zhang, Shoishi Abe, "Parallel System Uninterruptible Power Systems Development of Large Capacity 12MW Class", *JIASC*, (2017).

[4] Suguru Tahara, Tomoki Yokoyama, "1MHz Variable Sampling Deadbeat Control of Single Phase PWM Inverter in Low Carrier", *JIASC*, No.1-76,I435-440,(2007).

[5] Kent Taguri, Keiji Wada, "Experimental Verification of 3kVA Three-Phase PWM Inverter with 200kHz Switching", *JIASC*, (2014).

[6] K.Ishihara and A.Kawamura, "DSP Based Real Time Output Waveform Synthesis for Three Phase PWM Inverter", *T.IEE japan*, Vol.110-D, No.6, pp627-636, (1990-6).

[7] K.P.Gokhale, A.Kawamura, and R.G.Hoft, "Dead beat microprocessor control of PWM inverter for sinusoidal out-put waveform synthesis", *IEEE Trans. Industry Applications*, Vol23, No.5, pp901-910, (1987).

[8] Hironori Abe, Hiroshi Fujimoto, "Perfect Tracking Control of single Phase Inverter with Inter Sampling for Arbitrary Waveform",*JIASC*, Vol.128, No.2, pp110-116, (2008).

[9] T.Ishioka, and T.Yokoyama, "Verification of communication system for power electronics controller using FPGA", *JIASC*, (2009).

[10] M.Ito, R.Fujiwara, R.Araumi, T.Yoshino, "Robust digital control of single phase PWM inverter using 3MHz multi sampling method with FPGA based hardware controller", *IPEMC*, (2016).

[11] Mitsutoshi Ito, Reo Fujiwara, Tomoki Yokoyama, "3MHz multi sampling deadbeat control of single phase PWM inverter", *IFEEC*, (2015).

[12] Hiroki Uchida, Tomoki Yokoyama, "1MHz Variable Sampling Deadbeat Control of Single Phase PWM Inverter", *JIASC*, (2012).

[13] Atsuo Kawamura, "Modern Power electronics", SUURIKOUGAKU-SHA, (2005).

[14] K.Sakata, and H.Fujimoto, "Perfect Tracking Control of Servo Motor Based on PWM Hold Model", IIC-06-62, pp99-104, (2006).

[15] Toshihiro Ide, Tomoki Yokoyama, "Experiments of Deadbeat Contol for Three Phase PWM Inverter using FPGA based Hardware Controller", *JIASC*, (2004).

[16] T.Saigusa, T.Yokoyama, "Digital Control Method for 100kHz Single Phase Utility Interactive Inverter with FPGA based Hardware Controller", *EPE-PEMC*, (2010).

[17] Shin-ichi Hamasaki, Shinya Miyazaki, Tsuyoshi Takaki, Mineo Tsuji, "Verification of Performance of UPFC applying Deadbeat Control", *JIASC*, (2014).

[18] TingAn Lee, Masahiro Kinoshita, Nobuyuki Nagai, "High efficiency Large Capacity Uninterruptible Power Supply for 3-Phase 4-Wire Power System", *JIASC*, (2013).

[19] TingAn Lee, Masahiro Kinoshita, Nobuyuki Nagai, "Fine Voltage Control Based on Frequency Separation Two-Degrees-of-Freedom Control for Single-Phase Inverter", *IEEJ Journal of Industry Applications*, Vol.5, No.6, pp413-421, (2016).

Modular Multilevel Converter Replaced One Module with High Voltage IGBT

Kazunobu Oi[1*], Kenta Takasho[1] and Yugo Tadano[1]

1 Power Electronics Research Department, Meidensha Corporation, Numazu, Japan

*E-mail: ooi-ka@mb.meidensha.co.jp

Abstract — **Modular Multilevel Converters (MMCs) have advantages in application of high or medium voltage without transformers. They also can generate high quality voltage with fewer harmonics. Their unique merits make it possible as a leading technology in high/medium voltage applications. However, there still existed a concern that high conduction loss might limit its applications. In this paper, a MMC topology using high-voltage IGBT (HV-IGBT) as one cell module arm is proposed, by which conduction loss can be greatly reduced due to the output current flowing through the HV-IGBT in 1/3 of the cycle. Meanwhile, the proposed control stratagem can control the voltage of cell capacitor by circulating current. The effectiveness of the proposed circuit and control scheme is verified by simulation and experiment.**

Keywords — Capacitor Voltage Control, Circulating Current, Conduction Loss, Modular Multilevel Converter

I. INTRODUCTION

A Modular Multilevel Converter (MMC) [1]-[2] has a topology that its arms are configured with multiple chopper-cells connected in cascade. Therefore, it can easily produce higher voltage by just increasing the number of cells. Harmonics can also be reduced because its output voltage is actually composed by a series of tiny multi-level voltages. Through a stratagem of shifting switching timing of respective cells, a higher overall equivalent switching frequency of the unit, but lower individual switching frequency of each cell can be achieved. Thus, the total switching loss is reduced [3]. Today, MMCs are very popular in many applications [4].

One common topology of MMCs is shown in Fig. 1, in which two MMCs are connected in star as one phase. It can convert DC and AC in both directions [5]. However, in a higher voltage application, the number of switching elements or chopper cells should be increased. As a consequence, a higher conduction loss is caused. To solve the problem, a straight forward solution is to increase the capacitor voltage of each cell, so that the number of cells can be reduced. While it has a drawback that switching loss is significantly increased. Thus, it is necessary to modify the configuration of MMC to achieve low power losses in both conduction and switching [6].

In this paper, a solution of one MMC in star connection replaced by high-voltage IGBT (HV-IGBT) is proposed. Its topology and operating principle are presented. The circulating current required to control voltage of chopper cell is calculated. The operation performance and conduct loss reduction of the proposal are then verified by simulation. Through experiment, the feasibility of the proposed topology is investigated. The system is very stable during dynamic change between different operation modes.

II. TOPOLOGY AND OPERATING PRINCIPLE OF THE PROPOSED CIRCUIT

One phase of proposed circuit is presented in Fig. 2. In the upper arm, a HV-IGBT with a snubber capacitor C_P replaces a normal MMC, while the lower arm maintains the same configuration, e.g. multiple chopper cells connected in cascade. Each chopper cell consists of a DC capacitor and two low voltage IGBTs, as shown in Fig. 1. A buffer reactor L_B is connected in series in lower arm. The number of chopper cells in the lower arm and HV-IGBTs in the upper arm can be increased or decreased according to the system voltage and application requirements. As an example, four chopper cells and one HV-IGBT are used to configure a 5-level converter, as illustrated in Fig. 2.

The corresponding waveform of output voltage v'_U is presented in Fig. 3. This circuit requires 120-degree discontinuous PWM (DPWM) [7]. Fig. 4 illustrates current flowing during the interval of (a), (b), (c), and (c'). As shown in Fig. 3 and 4, during the interval (a), the HV-IGBT turns off. The output current comes from the lower arm. Once the HV-IGBT turns on during interval (b), the output current changes its path from the lower arm to the upper arm. Since there is only one HV-IGBT in the upper arm, the voltage drop of the upper arm is lower, thus conduction loss is lower. Also during (b), the circulating current is applied from the DC bus to the upper and lower arms to regulate the capacitor voltage of the lower arm cells.

It is seen that two intervals (c) and (c') inserted between intervals (a) and (b) to help HV-IGBT to achieve soft switching and avoid short circuit between HV-IGBT and snubber capacitor C_P. It can be explained as follows:

Firstly, in interval (c), HV-IGBT keeps off, there is current from lower arm to discharge C_P to zero to avoid short circuit between HV-IGBT and C_P. And then the freewheel diode of HV-IGBT is forced to conduct. Under this condition, the HV-IGBT turns on with zero voltage

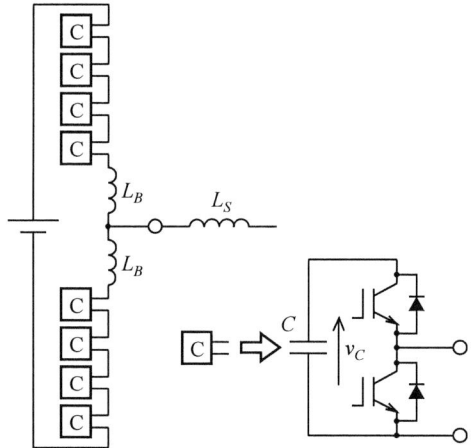

Fig. 1. Double star modular multilevel converter (One phase)

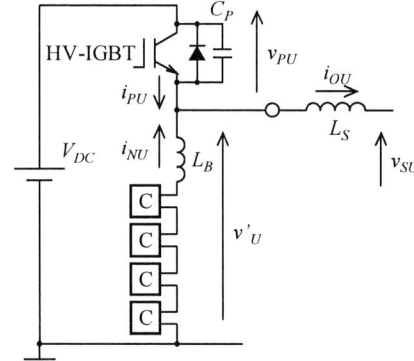

Fig. 2. Proposed modular multilevel converter (One phase)

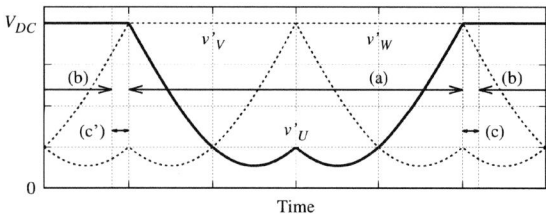

Fig. 3. Waveform of v'

and zero current.

Secondly, during interval (c'), the HV-IGBT turns off with zero voltage due to the effect of C_P. Therefore, zero voltage or zero current switching is achieved for both HV-IGBT turning-on and -off. It should be mentioned that to make C_P smaller and to help the zero voltage switching in interval (c'), the supporting current from lower arm is still necessary.

III. CIRCULATING CURRENT FOR CELL CAPACITOR VOLTAGE CONTROL

The circulating current required for capacitor voltage control of each cell is calculated as follows:

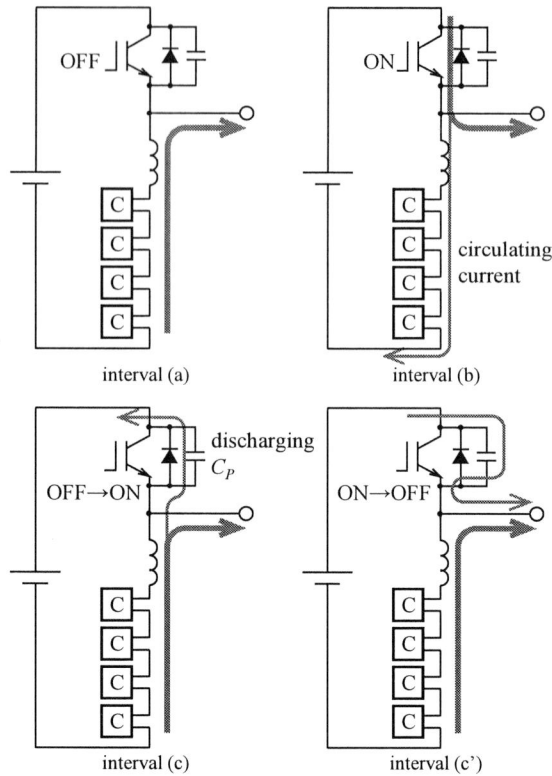

Fig. 4. Current flow in the proposed converter

Assume AC output voltage of phase U expressed as $v_{SU} = \sqrt{2/3}V \cos \omega t$, and its output current as $i_{OU} = \sqrt{2}I \cos(\omega t - \phi)$. While the voltage after DPWM, v'_U can be expressed as Eq. (1).

$$v'_U = \begin{cases} V_{DC} & \left(-\dfrac{\pi}{3} < \omega t < \dfrac{\pi}{3}\right) \\ \sqrt{2}V \cos\left(\omega t - \dfrac{\pi}{6}\right) + V_{DC} & \left(-\pi < \omega t < -\dfrac{\pi}{3}\right) \\ \sqrt{2}V \cos\left(\omega t + \dfrac{\pi}{6}\right) + V_{DC} & \left(\dfrac{\pi}{3} < \omega t < \pi\right) \end{cases} \quad (1)$$

where, $0 < V < V_{DC}/\sqrt{2}$.

During interval (a), the energy E_a, produced by the lower arm cells entirely, can be obtained by Eq. (2).

$$\begin{aligned} E_a &= \int_{-\pi/\omega}^{-\pi/3\omega} v'_U i_{OU}\, dt + \int_{\pi/3\omega}^{\pi/\omega} v'_U i_{OU}\, dt \\ &= \frac{I \cos\phi}{\sqrt{3}}\left(2\pi V - 3\sqrt{2}V_{DC}\right) \end{aligned} \quad (2)$$

If assuming the circulating current I_C as constant during interval (b), energy E_b charged by I_C can be obtained by Eq. (3).

$$E_b = \int_{-\pi/3\omega}^{\pi/3\omega} v'_U I_C\, dt = \frac{2\pi I_C V_{DC}}{3\omega} \quad (3)$$

As $E_a + E_b = 0$, the circulating current is expressed as Eq. (4).

$$I_C = \sqrt{3}I \cos\phi \left(\frac{3}{\pi\sqrt{2}} - \frac{V}{V_{DC}}\right) \quad (4)$$

Similarly, the circulating current I_{CD} of the double-star connection MMC can be expressed as Eq. (5) [8], and the circulating current I_{CN} of the NPC-MMC can be expressed as Eq. (6) [6].

$$I_{CD} = \frac{VI \cos \phi}{\sqrt{3} V_{DC}} \qquad (5)$$

$$I_{CN} = \frac{2I \cos \phi}{\sqrt{3}} \left(\frac{V}{V_{DC}} - \frac{\sqrt{3}}{\pi \sqrt{2}} \right) \qquad (6)$$

Compared with the conventional MMC, the proposed circuit seems to require the larger capacity for cell capacitor because the circulating current cannot flow during interval (a). However, during interval (b), as power is directly supplied from DC side to AC side, power loss can be reduced. Particularly, it can be seen from Eq. (4) that the required circulating current I_C is zero when $V = 3V_{DC}/\pi\sqrt{2} \cong 0.675V_{DC}$. It is implied that the proposed circuit can achieve high efficiency with smaller capacity of the cell capacitors in an application where the voltage is relatively high and the power output continues, such as a grid interconnection.

IV. CONTROL SCHEME OF THE PROPOSED CIRCUIT

In this circuit, the output current is regulated during interval (a), the balancing control of each cell capacitor is conducted during interval (b), and the discharge of the snubber capacitor C_P is controlled during intervals (c) and (c'). Thus the proposed control scheme varies in different intervals. While the control scheme in [2] and [8] is borrowed to control capacitor voltage of respective cells.

A. Output Current Regulation

Fig. 5 shows the block diagram of regulating the output current. It is a conventional current regulation scheme, where DPWM is used with dq conversion of 3 phase currents i_{OU}, i_{OV}, and i_{OW}. Two PI controllers are applied in dq axis to control output current closed to desired value. By inverse dq conversion and DPWM, the voltage commands v_U*, v_V*, and v_W* are obtained. For the HV-IGBT gate control, 'on' command is given only when corresponding v_U*, v_V*, or v_W* equals to 1. If their value is less than 1, the command to corresponding HV-IGBT should be 'off'. While the measured upper arm voltage v_{PU}, v_{PV}, and v_{PW} are also checked to make sure the command is given only when v_{PU}, v_{PV}, $v_{PW} <0$. Thus, the interval (c) and (c') are automatically generated to achieve soft switching of HV-IGBT.

B. Lower Arm Cell Capacitor Voltage Control

Fig. 6 shows the block diagram of capacitor voltage control for the lower arm cell in phase U. The average voltage of cell capacitor v_{Cavg} is measured and held at the beginning of interval (b). The deviation of measured voltage and voltage command V_C* is calculated. Here, V_C* is not equal to the value V_{DC} divided by number of cells 'n' but exceeded by α. Thus, the additional current for discharge of the snubber capacitor C_P during intervals (c) and (c') is guaranteed. Using Eq. (7), the required discharge current is obtained as:

$$i_{NU}^* = \frac{C}{\Delta t_{(b)}} \left(v_{CavgU} - V_C^* \right) = G_C \left(v_{CavgU} - V_C^* \right) \quad (7)$$

Where $\Delta t_{(b)}$ is the time duration of interval (b).

The control for compensating overs and shorts of charging due to error of capacitor's capacity is made in parallel. Deviation from V_C* after eliminating the ripple of the line frequency is applied to a P controller. Current command $i_{NU}*$ is calculated as a sum of the output of P controller and Eq. (7). Thus, the average of capacitor voltage is controlled by the current output from the lower arm cell module during interval (b).

During intervals (c) and (c'), the snubber capacitor in the upper arm needs to be discharged. In order to do so, besides of normal output current i_{OU}, the extra current should be provided by adding β in current command.

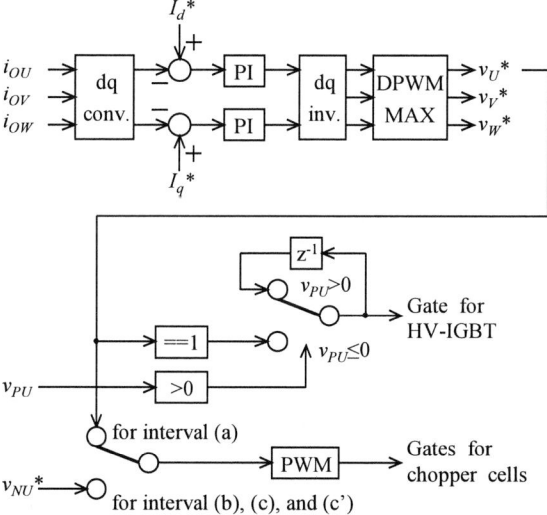

Fig. 5. Control block diagram for output current regulation

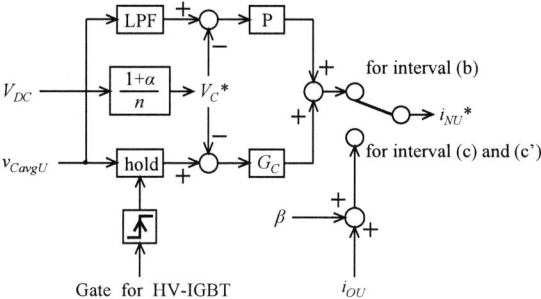

Fig. 6. Control diagram for cell capacitor voltage averaging control

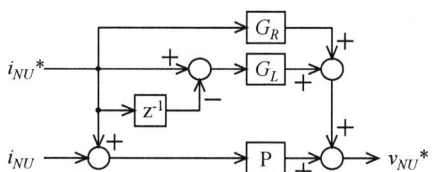

Fig. 7. Control diagram for current regulation of cell module in lower arm

C. Current Regulation Control for Lower Arm Cell Module

For the proposed circuit, discharge of the snubber capacitor C_P and voltage control of the cell capacity are required during intervals (b), (c), and (c'). When the line frequency is 50 Hz, it is only 6.67 ms, which requires high-speed current regulation. The voltage required to output $i_{NU}{}^*$ is obtained by Eq. (8), in which the inductance of L_B multiplies with the time differential of the current $i_{NU}{}^*$ as a feed-forward compensation.

$$v_{NU}^* = Ri_{NU}^* + \frac{L}{\Delta t_C}\Delta i_{NU}^* = G_R i_{NU}^* + G_L \Delta i_{NU}^* \qquad (8)$$

Where Δt_C is a half period of the equivalent carrier frequency. G_R is used to compensate for voltage drop due to the parasitic resistance.

Fig. 7 shows the block diagram for the current regulation of the lower arm cell module. Feedback control using P controller is added with feed-forward control as a voltage command $V_{NU}{}^*$.

V. VERIFICATION BY SIMULATION

Simulation using MATLAB was performed for verification, which has circuit parameters listed in Table I.

A. Operation Waveform

Fig. 8 shows the operating waveform in steady state condition under rated active power output, while Fig. 9 shows the operating waveform under rated active power input. It can be seen that a resonant current caused by the snubber capacitor C_P and buffer reactor L_B, as well as the switching ripple are superimposed on the upper arm current i_{PU} and the lower arm cell module current i_{NU}. However such ripple current is not superimposed on the output current i_O since L_S is larger than L_B. Thus, the waveform of i_O with small distortion was obtained. THD was 3.3% under rated active power output and 3.4% under input. i_{PU} becomes a negative value just before the upper arm switches and the discharge of the snubber capacitor occurs. Abrupt change of the current i_{PU} occurs twice during the interval of 6.67 ms with respect to i_{NU} regulation. For the voltage of the cell capacitor v_{CU}, although the ripple in one fundamental period is large, the average is controlled at 946.3 V after turning off the upper arm. The difference between the maximum and minimum values is less than 21 V.

B. Loss Comparison

In order to verify the conduction loss reduction of the proposed circuit, the loss in the IGBTs was compared with that of the traditional double-star MMC using PLECS. The data of CM400DY-34A (Mitsubishi) was referenced as the low voltage IGBTs for the lower arm cells and the data of MBN800H45E2 (Hitachi) was referenced as the HV-IGBTs for the upper arms. Table II shows the total number of IGBTs and cells. Two HV-IGBTs are connected in series in the proposed circuit. To make the rated voltage same, double-star MMCs with 16 cells per phase are used. To make the number of low voltage IGBTs same as the traditional double-star MMCs, two IGBTs are connected in parallel in the proposed circuit.

Results of power loss of proposed and traditional topology are listed in Table III. Conduction loss by the low voltage IGBT is significantly reduced in both active power input and output conditions. Switching loss of the HV-IGBT is almost zero which implies that soft switching is achieved. The difference of losses in total is about 3.5 kW so that effectiveness of power loss reduction in the proposed circuit is significant.

TABLE I
CIRCUIT PARAMETERS FOR SIMULATION

Power rating	2 MW
Line-to-line voltage	4.4 kV
Line frequency	50 Hz
AC current rating	262.4 A
Line inductance (L_S)	2.90 mH (9.4 %)
Buffer inductance (L_B)	0.31 mH (1 %)
DC link voltage (V_{DC})	7333.6 V
Voltage of chopper cells	946.3 V
Capacitance of chopper cells (C)	15 mF
Number of chopper cells per 1 phase	8
Carrier frequency	268.75 Hz
Equivalent switching frequency	2150 Hz
Snubber Capacitor (C_P)	1.2 μF
Unit capacitance time constant	80.6 ms

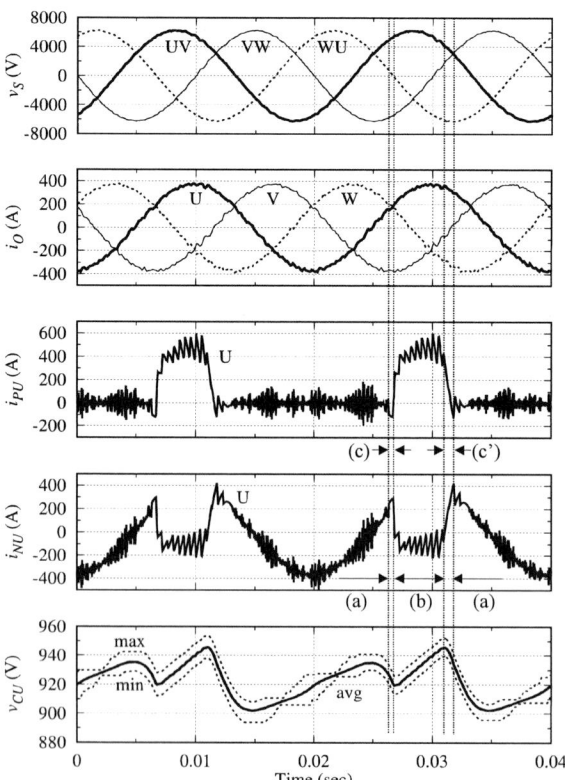

Fig. 8. Simulation results under active power output

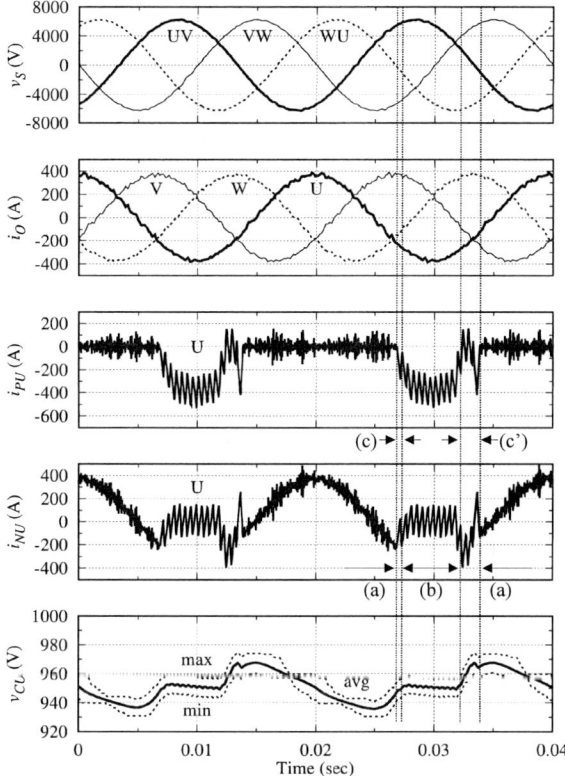

Fig.9. Simulation results under active power input

TABLE II
TOTAL NUMBER OF IGBTS FOR LOSS CALCULATION

total number	Proposed topology	Double star MMC
LV-IGBT	96	96
HV-IGBT	6	0
Chopper cell	24	48

TABLE III
CIRCUIT PARAMETERS FOR SIMULATION

Loss (W)		Proposed topology		Double star MMC	
		Output	Input	Output	Input
LV-IGBT	Cond.	5121.8	1279.1	9593.6	1584.8
	SW	938.8	967.6	1268.2	1268.6
LV-Diode	Cond.	1220.5	3771.8	1247.2	7622.9
	SW	353.7	294.8	516.3	509.0
HV-IGHT	Cond.	1555.0	21.6	-	-
	SW	0.0	3.0	-	-
HV-Diode	Cond.	23.0	1132.7	-	-
	SW	0.0	0.0	-	-
Total		9212.8	7470.6	12625.3	10985.3
		0.46 %	0.37 %	0.63 %	0.55 %

VI.   VERIFICATION BY EXPERIMENT

To confirm the feasibility of proposed topology, the experimental verification using 415 V and 15 kVA equipment was also conducted. Circuit parameters of the experiment are listed in Table IV. IGBT CM100DY-12A (Mitsubishi) was used for for both upper and lower arms, e.g. both HV-IGBT and the cell modules. In this experiment, only a capacitor was connected to the DC link, the experiment was carried out under no load condition. It meant the operation performance for reactive power was checked. As presented in Fig. 10, the equipment generates inductive reactive power of 1 p.u. before the time of −0.02 sec, while dynamically change to 1 p.u. of capacitive reactive power after 0.02 sec. A sinusoidal waveform with small distortion can be seen in output current i_O, in which 4.1% of THD is measured for inductive reactive power output and 3.7% for capacitive reactive power output.

The upper arm current i_{PU} is negative for the interval (c) and (c'), which indicates the discharging process of the parallel-connected snubber capacitor C_P during HV-IGBT switching. The cell module current i_{NU} is nearly zero in interval (b). This is because the output power is reactive. Thus, substituting $\phi=\pm\pi/2$ in Eq. (4) results in the required circulating current being zero. In fact, a slight charging current flows to supplement the lost portion. However, in interval (a), a current equivalent to i_O is required for i_{NU}. From Fig. 10, it is also seen that i_{NU} has fast changes in interval (c) and (c'), which means the high-speed current control is established.

For cell capacitor voltage v_{CU}, the command value α for maximum charging was set to 3% when turning off the upper arm IGBT. Experimental results also show that v_{CU} was properly controlled. The variation between maximum and minimum values of cell capacitor voltage was 1.9 V, which was very small.

In this experiment, the dynamic characteristic of proposed topology during changes of output power was also checked. From the time of −0.02 sec to 0.02 sec, the output power reference changed from rated inductive reactive power to rated capacitive one. The output followed the change and the operation of system was stable. During this period, the distortion of i_O was very small; v_{CU} followed the change of v_{DC} and the variation of v_{CU} kept small.

TABLE IV
CIRCUIT PARAMETERS FOR EXPERIMENT

Power rating	15 kW
Line-to-line voltage	415 V
Line frequency	50 Hz
AC current rating	20.87 A
Line inductance (L_S)	3.3 mH (9.0 %)
Buffer inductance (L_B)	0.38 mH (1.0 %)
DC link voltage (v_{DC})	733 V
DC link capacitor	4,700 μF
Voltage of chopper cells	1.03×v_{DC}/4
Capacitance of chopper cells (C)	5,600 μF
Number of chopper cells per 1 phase	4
Carrier frequency	768.75 Hz
Equivalent switching frequency	3075 Hz
Snubber Capacitor (C_P)	0.33 μF
Unit capacitance time constant	79.8 ms

The 2018 International Power Electronics Conference

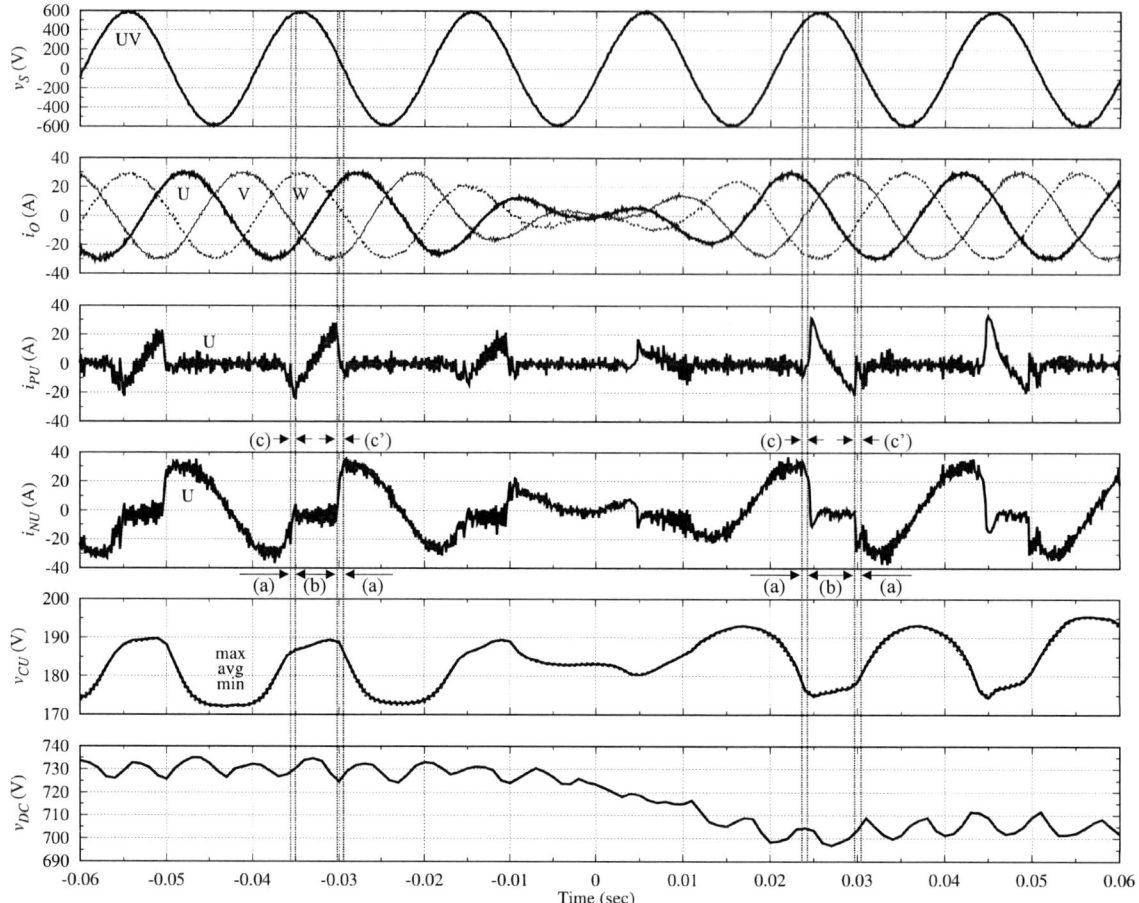

Fig.10. Experimental results under repower input

VII. CONCLUSIONS

In this paper, a circuit configuration where the upper arm of the double star MMC is replaced by HV-IGBT is proposed. The control scheme and conduction loss reduction are explained. Its operation performance was verified by simulation and experiment. As a result, it was confirmed that the capacitor voltage could be controlled correctly, and a slight distortion was obtained in output current. The operation was stable, even under the dynamic changes of output power. In addition, it had been proved that the proposed topology was superior in terms of efficiency to other circuit configurations, particularly under high voltage conditions. As estimated, 0.17% of IGBT loss reduction can be achieved compared to a conventional double-star MMC. In the future, more experiments will be conducted to have a complete investigation.

REFERENCES

[1] A. Lesnicar and R. Marquardt, "An innovative modular multilevel converter topology suitable for a wide power range," 2003 IEEE Bologna Power Tech Conference Proceedings, 2003, pp. 6 pp. Vol.3-.

[2] Makoto Hagiwara and Hirofumi Akagi, "PWM Control and Experiment of Modular Multilevel Converters," IEEJ Trans. IA, Vol.128, No.7, pp.957-965 (2008-7)

[3] Jiantao Liu; Jianguo Yao; Shengchun Yang; Ke Wang, "Loss analysis of two kinds of flexible HVDC converters," Power Electronics and Motion Control Conference (IPEMC), 2012 7th International , vol.3, no., pp.1669,1674, 2-5 June 2012

[4] Knaak, H.-J., "Modular multilevel converters and HVDC/FACTS: A success story," Power Electronics and Applications (EPE 2011), Proceedings of the 2011-14th European Conference on , vol., no., pp.1,6, Aug. 30 2011-Sept. 1 2011

[5] H. Akagi and M. Hagiwara: "Clasification and Terminology of modular multilevel cascade converters (MMCC)," IEEJ Annual Meeting Rec., No.4-043, pp.71-72 (2010-3)

[6] Ryuta Hasegawa and Daichi Suzuki, "Neutral Point Clamped Modular Multilevel Converter," Proceedings of the 2014 Japan Industry Applications Society Conference, 1-29, 2014, pp. I-155 - I-160

[7] K. Taniguchi, Y. Ogino, and H. Irie, "PWM technique for power MOS-FET inverter," IEEE Trans. Power Electron., July 1988, pp. 328-334.

[8] Hideaki Fujita, Makoto Hagiwara, Hirofumi Akagi, "Power Flow Analysis and DC-Capacitor Voltage Regulation for the MMCC-DSCC," IEEJ Trans. IA, Vol.132, No.6, pp.659-665 (2016-6)

The 2018 International Power Electronics Conference

Increased Efficiency and Reduced Realization Effort of DSBC and DSCC Modular Multilevel Converters (MMCs)

A. Hillers and J. Biela
Laboratory for High Power Electronic Systems
ETH Zurich, Physikstrasse 3, CH-8092 Zurich, Switzerland
Email: hillers@hpe.ee.ethz.ch

Abstract—The double-star chopper-cell (DSCC) and the double-star bridge-cell (DSBC) modular multilevel converters (MMCs) may be designed with a reduced number of modules when higher grid currents and an increased volume of the passive components are acceptable. The full-bridge modules in the DSBC even allow for designs with increased power conversion efficiency at the expense of additional modules. This paper presents an integrated optimal design methodology that exploits these additional degrees of freedom while at the same time maximizing power density and power conversion efficiency. To demonstrate the effectiveness of the proposed approach, a short-circuit-proof 250 kW, 35 kVDC bidirectional active rectifier prototype system is presented that has been designed with emphasis on reduced realization effort.

I. INTRODUCTION

Modular multilevel converters (MMCs) present an active area of research for a variety of applications such as HVDC [1], [2], grid-connected battery energy storage systems [3]–[5], variable frequency drives [6], [7], STATCOMs [8] and even self-contained dc ship grids [9]. The typical design process of an MMC consists of the following five basic steps:

1) Selection of the power semiconductors and therewith the critical module voltage V_{crit}
2) Selection of the switching frequency f_{sw}
3) Tradeoff between number of modules N and module capacitance C_{mod}.
4) Selection of the arm inductance L_{arm} based on the output current quality requirements and the maximum permitted fault current.
5) Optimal design of the inductors

Different optimization procedures have been presented for example in [10] and [11], which (partly) incorporate the listed steps into a comprehensive methodology.

In addition to steps 1) – 5) listed above, this paper proposes the consideration of the ac-dc voltage conversion ratio as a free parameter in the optimal design process: When designing the double-star bridge-cell (DSBC) or the double-star chopper-cell (DSCC) MMC to operate in boost mode, the number of modules can be reduced at the cost of an increased volume of the passive components and increased power losses. With the full-bridge modules in the DSBC, an operation in buck mode can even increase the power conversion efficiency at the expense of increasing the realization effort.

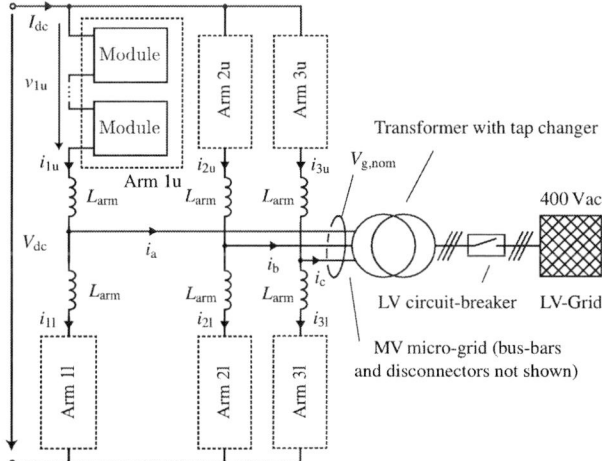

Fig. 1: Simplified circuit diagram of the MMC-based medium voltage rectifier prototype. The system connects directly to the 9 kV medium-voltage bus bar in the *Laboratory for High Power Electronic Systems* (HPE) at the ETH Zurich.

The paper is structured as follows: **Section II** explains the principles of operating double-star MMCs in boost mode and buck mode and presents a set of showcase specifications that is used to exemplify the calculations in this paper. **Section III** discusses the limitations that a conventional design entails. **Section IV** proposes an integrated optimal design methodology to overcome these limitations, allowing to reduce the realization effort and power losses. **Section V** presents a hardware prototype system that has been designed with emphasis on reduced realization effort.

II. OPERATING PRINCIPLES

In the following, the operating principles of the double-star chopper-cell (DSCC) and the double-star bridge-cell (DSBC) modular multilevel converters (MMCs) are introduced. At first, the electrical specifications of a 250 kW showcase system are discussed, which present the basis for the calculations performed in this paper. Afterwards, the steady-state operation in *buck mode* resp. *boost mode* is briefly recapitulated.

To prevent the excessive use of index variables, all considerations in this paper are exemplarily made for the first leg (e.g. i_{a}), the upper arm in the first leg (e.g. v_{1u}) or the

1896

TABLE I: Specifications of the DSBC hardware prototype system developed at the *Laboratory for High Power Electronic Systems* at ETH Zurich. These specifications present the basis for all numerical calculations performed in this paper.

Parameter		Value
Nominal dc voltage	$V_{dc,nom}$	35 kV
Maximum module voltage	V_{crit}	2.2 kV
Critical module current	I_{crit}	40 A
Nominal output power	P_{nom}	250 kW
Module switching frequency	$f_{sw,mod}$	1 kHz
Maximum dc current	$I_{dc,max}$	30 A
Pre-control dead-time	$T_{d,tot}$	50 μs

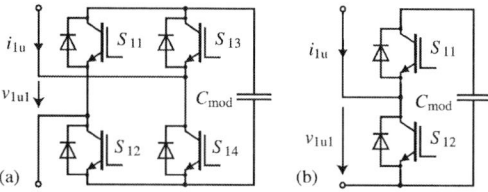

Fig. 2: Full-bridge module (a) and half-bridge module for the DSBC resp. DSCC, exemplarily shown for the first module in the first arm.

first module within the upper arm of the first leg (e.g. v_{1u1}) wherever possible. The explicit time-dependence of variables is omitted for the sake of readability (e.g. $v_{1u} \equiv v_{1u}(t)$).

A. Specifications of a 250 kW Showcase System

Fig. 1 shows the basic circuit diagram of a bidirectional active rectifier based on the modular multilevel converter (MMC). The system is designed to be part of the infrastructure at the *Laboratory for High Power Electronic Systems* at ETH Zurich and connects directly to the lab's own medium-voltage (MV) micro-grid. The voltage at the transformer clamps is adjustable between

$$V_{g,range} = 5\,\text{kV} \dots 25\,\text{kV}, \qquad (1)$$

which allows the MMC to be operated with different ac voltages. When the rectifier is not in use, the transformer may provide ac power to other equipment in the lab. For the sake of simplicity, the corresponding disconnectors and bus-bars are not shown.

The total dead-time of the voltage pre-control loop of the MMC is assumed to be smaller than $T_{d,tot} = 50\,\mu\text{s}$, including the delay of the grid-voltage measurement, the communication delay between the central controller and the modules and the time it takes to switch IGBTs. In case of a short-circuit, the system is specified to tolerate a maximum module current of $I_{crit} = 40\,\text{A}$ to prevent the arm inductors from saturating. The full electrical specifications are summarized in Tab. I.

In order to tolerate dc short-circuits, the system is based on the DSBC which uses full-bridge modules (see Fig. 2 (a)). For the sake of completeness, this paper also discusses the DSCC, which uses half-bridge modules (see Fig. 2 (b)) and is thus not short-circuit proof.

B. Steady-State Model

In the following, the steady-state operation in *boost mode* and *buck mode* are briefly recapitulated. The steady-state model presents the basis for the optimal design methodology proposed in section IV. The goal of the steady-state model is to calculate the arm voltages v_{1u}, v_{1l}, v_{2u}, v_{2l}, v_{3u}, v_{3l} and the respective arm currents i_{1u}, i_{1l}, i_{2u}, i_{2l}, i_{3u}, i_{3l} in steady-state, given the dc-link voltage V_{dc}, the dc-link current I_{dc}, the complex grid voltage $\underline{V}_g$ and the complex grid power $\underline{S}$. To simplify the calculations, it is assumed w.l.o.g. that $\underline{V}_g$ is real-valued ($\underline{V}_g = V_g$).

1) Continuous Assumption: Under typical operating conditions, the voltages of all module capacitors in each arm are kept approximately equal. This is commonly achieved with the sorting algorithm discussed e.g. in [2]. The short-term average of the output voltage of each arm can thus in general be controlled to an arbitrary value between the positive and negative limit imposed by the the *internal arm voltage*:

$$|v_{1u}| \leq v_{1u,int} = \sum_{n=1}^{N} v_{C1un}. \qquad (2)$$

The arm voltage v_{1u} is typically regarded as continuous [12], which is adopted for the following considerations.

2) Operating in Buck Mode and Boost Mode: In order to simplify the analysis of double-star connected MMCs, the arm-voltages are typically split into the parts $v_{1,grid}$, $v_{2,grid}$ and $v_{3,grid}$ that only have an influence on the grid currents, and the parts $v_{1,circ}$, $v_{2,circ}$ and $v_{3,circ}$, that only influence the circulating currents [13]:

$$v_{1u} = \frac{V_{dc}}{2} - v_{1,grid} + \frac{v_{1,circ}}{2}, \qquad (3)$$

$$v_{1l} = \frac{V_{dc}}{2} + v_{1,grid} + \frac{v_{1,circ}}{2}. \qquad (4)$$

The parts $v_{1,circ}$, $v_{2,circ}$ and $v_{3,circ}$ can be neglected for normal operation, because they are small compared to $v_{1,grid}$, $v_{2,grid}$ and $v_{3,grid}$. The operating factor

$$o = \frac{V_{dc,nom}}{2\hat{v}_{1,grid}} \overset{(13)}{\approx} \frac{V_{dc,nom}}{2\sqrt{\frac{2}{3}}V_{g,nom}} \qquad (5)$$

is introduced to describe whether the system is designed for operation in boost mode or buck mode. When the nominal peak-to-peak value of $v_{1,grid}$ is lower than the nominal dc voltage $V_{dc,nom}$, the system is designed for *boost mode*, corresponding to an operating factor of

$$o_{boost} > 1 \Rightarrow V_{dc,nom} > 2\sqrt{\frac{2}{3}}V_{g,nom}. \qquad (6)$$

According to (3), the arm voltages are not required to become negative in this case. The system is designed for *buck mode* when the operating factor is

$$o_{buck} < 1 \Rightarrow V_{dc,nom} < 2\sqrt{\frac{2}{3}}V_{g,nom}, \qquad (7)$$

which according to (3) can only be achieved with the full-bridge modules in the DSBC.

3) Ideal Arm Currents and Voltages: The ideal arm voltages during steady-state operation can be calculated using common phasor arithmetics as shown e.g. in [10]:

$$v_{1,\text{grid}} = \hat{V}_\alpha \cos(2\pi f t + \varphi_\alpha), \tag{8}$$

where

$$\underline{I}_a = \frac{S}{\sqrt{3}V_g}, \tag{9}$$

$$\hat{V}_\alpha = \sqrt{2}\left|\underline{I}_a\left(\frac{j\omega L_a}{2} + j\omega L_g\right) + V_g\right|, \tag{10}$$

$$\varphi_\alpha = \angle\left(\underline{I}_a\left(\frac{j\omega L_a}{2} + j\omega L_g\right) + V_g\right). \tag{11}$$

Notice how the calculations are independent of the operating factor. The angular frequency of the grid is $\omega = 2\pi f$. The grid inductance L_g is neglected in the following. Because $\frac{j\omega L_a}{2}$ is typically small compared to V_g, $v_{1,\text{grid}}$ can be approximated by

$$v_{1,\text{grid}} \approx \sqrt{\frac{2}{3}}V_g\cos(\omega t), \tag{12}$$

resulting in an arm voltage of

$$v_{1u} \approx \frac{V_{dc}}{2} - \sqrt{\frac{2}{3}}V_g\cos(\omega t). \tag{13}$$

The arm current can be expressed in the time domain as well:

$$i_{1u} = \frac{1}{2}\sqrt{2}I_a\cos(\omega t + \varphi_i) + \frac{I_{dc}}{3}. \tag{14}$$

In the above, $I_a = |\underline{i}_a|$ and $\varphi_i = \angle\underline{i}_a$ denote the RMS-value resp. the phase-angle of the output current. It is assumed that the currents are controlled ideally. It is further assumed that the converter is controlled in such a way, that the dc-link current

$$I_{dc} = \frac{|\underline{S}|}{V_{dc}} \tag{15}$$

is divided up equally among all three legs, which results in an average power transfer of zero to the module capacitors in steady-state (assuming the power-losses in the system are zero). Both assumptions are common for the design of MMCs operating in (quasi-) steady-state and lead to a negligible impact on the accuracy of the design.

4) Internal Arm Voltages: During normal operation, the instantaneous arm power

$$p_{1u} = \left(\frac{V_{dc}}{2} - \hat{V}_\alpha\cos(\omega t + \varphi_\alpha)\right)\left(\frac{I_a}{\sqrt{2}}\cos(\omega t + \varphi_i) + \frac{I_{dc}}{3}\right) \tag{16}$$

and therewith the instantaneous arm energy

$$w_{1u,\text{int}} = \int p_{1u}\,dt \tag{17}$$

fluctuate, leading to a fluctuation of the internal arm voltages. This fluctuation is buffered by the module capacitors; their dimensioning is one of the fundamental design steps as explained in section IV.

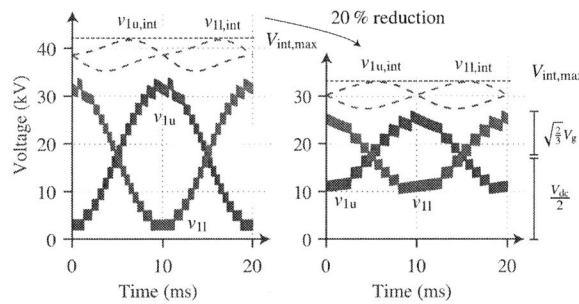

Fig. 3: By lowering the ac voltage, the maximum required internal arm voltage $V_{\text{int,max}}$ – and therewith the number of modules – can be reduced.

III. LIMITATIONS OF CONVENTIONAL DESIGNS

In the following, it is explained how the typical choice of the operating factor limits the maximum achievable power conversion efficiency and the realization effort.

A. Typical Choice of the Operating Factor

According to (13) and (6), the maximum arm voltage observed during normal operation is

$$v_{1u,\text{max}} \approx \frac{V_{dc,\text{nom}}}{2} + \sqrt{\frac{2}{3}}V_{g,\text{nom}} = \frac{V_{dc,\text{nom}}}{2}\cdot\left(1 + \frac{1}{o}\right) \overset{o=1}{=} V_{dc,\text{nom}}. \tag{18}$$

In conventional designs the operating factor is typically chosen to be

$$o_{\text{typ}} \approx 1 \Rightarrow V_{dc,\text{nom}} \approx 2\sqrt{\frac{2}{3}}V_{g,\text{nom}}, \tag{19}$$

which justifies the last equals sign[1].

The choice of $o_{\text{typ}} \approx 1$ is optimal in the sense that it maximizes the root-means-square (RMS) of both voltage components $\frac{V_{dc}}{2}$ and $\sqrt{\frac{2}{3}}V_g$ in (18) for a limited $v_{1u,\text{max}}$ (as defined by the maximum module voltage and the number of modules in (28)). In the case of the DSCC, where the arm-voltage has to be positive, this choice furthermore leads to the highest number of possible output voltage steps on the ac side (again assuming a limited $v_{1u,\text{max}}$) as can be inferred from Fig. 3 (a). However, this choice neither leads to the minimum number of modules nor the maximum power-conversion efficiency, as explained in the following.

B. Non-Optimal Realization Effort

The minimum number of modules per arm is determined by the maximum permissible module voltage V_{crit} and the maximum required peak value of the arm voltage $v_{1u,\text{max}}$ at any point in time [10]:

$$n_{\text{min}} = \frac{v_{1u,\text{max}}}{V_{\text{crit}}} \approx \frac{V_{dc,\text{nom}}\cdot\left(1 + \frac{1}{o}\right)\cdot 1.15}{2V_{\text{crit}}}, \tag{20}$$

The peak value of the arm voltage is calculated according to (18), assuming the system can be operated in (quasi-) steady-state at a constant grid-voltage. A dynamic control reserve of 15 % has been included.

---

[1]For actual designs of the DSCC, a margin for dynamic control of the dc and ac voltages has to be included to prevent the arm voltages from becoming negative (see section IV).

According to (20), the minimum required number of modules in each arm is reduced for designs with $o > 1$ (operation in boost mode). Fig. 3 shows the corresponding internal arm voltage and arm voltage waveforms during normal operation to illustrate this. Fig. 3 (a) shows the situation for the design with $o \approx 1.2$ discussed in section V-D. Owing to the dynamic control reserves, the effective operating factor is very close to one in this case. Fig. 3 (b) shows the situation for the design with $o = 2.38$ discussed in section V-E.

C. Non-Optimal Power Conversion Efficiency

While designing for boost mode decreases the minimum number of modules, designing for operation in buck mode may reduce the overall power losses. According to (5), (9) and (15), a decrease of the operating factor o leads to a decrease of the nominal ac current

$$I_{g,\text{nom}} = \frac{P_{\text{nom}}}{\sqrt{3}V_{g,\text{nom}}} = \frac{2\sqrt{2}}{3} \cdot o \cdot I_{\text{dc,nom}}, \qquad (21)$$

which in turn leads to a decrease of the nominal rms arm current

$$I_{1u,\text{rms,nom}} = \sqrt{\left(\frac{I_{g,\text{nom}}}{2}\right)^2 + \left(\frac{I_{\text{dc,nom}}}{3}\right)^2} = \frac{I_{\text{dc,nom}}}{3}\sqrt{2o^2+1}. \quad (22)$$

However, according to (20), the decrease of the operating factor at the same time leads to an increase of the minimum required number of series-connected modules, which has a negative impact on the power losses (more IGBTs in series).

The effect of increasing the power conversion efficiency by increasing the ac voltage while at the same time increasing the number of modules can be analyzed by calculating the derivative of the relative increase of the arm currents

$$\left.\frac{d\iota}{do}\right|_{o=1} = \frac{2}{3}, \quad \iota = \frac{I_{1u,\text{rms,nom}}}{I_{1u,\text{rms,typ}}} = \frac{\sqrt{2o^2+1}}{\sqrt{3}} \qquad (23)$$

as well as calculating the derivative of the relative increase of the minimum required number of modules

$$\left.\frac{d\nu}{do}\right|_{o=1} = -\frac{1}{2}, \quad \nu = \frac{n_{\min}}{n_{\min,\text{typ}}} = \frac{1}{2}\frac{1+o}{o}, \qquad (24)$$

with respect to o, where $I_{1u,\text{rms,typ}}$ and $n_{\min,\text{typ}}$ refer to the result of equations (22) resp. (20) when evaluated at $o = 1$.

The above calculations give an intuitive explanation for the power loss reduction: When assuming the power losses scale in a first approximation linearly with the current (the overall conduction losses $P_{\text{cond}} = NV_{\text{CE}} \cdot I_{\text{CE,avg}} + N\Delta R_{\text{CE,on}} \cdot I_{\text{CE,rms}}^2$ of the IGBTs within an arm even have a quadratic dependency of $I_{\text{CE,rms}}$), it becomes evident that the increase of the number of modules is overcompensated by the decrease of the arm currents. This effect is verified by the calculations performed in section V.

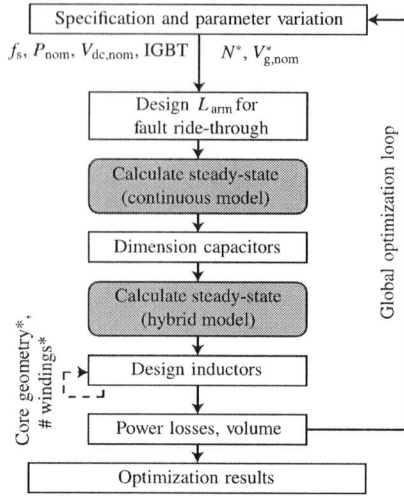

Fig. 4: Simplified flow-chart of the proposed global optimization procedure. The parameters marked with an asterisk (*) are the free parameters of the optimization.

IV. OPTIMAL DESIGN PROCEDURE

In the following, an integrated optimal design procedure is proposed that includes the operating factor into the design process. For the sake of simplicity, it is assumed that the nominal dc-link voltage $V_{\text{dc,nom}}$ is constant and that the ac voltage is chosen freely. However, the design could be performed for a fixed ac voltage and a freely chosen dc-link voltage or both in an analogous manner.

A. Global Optimization Loop

Fig. 4 shows the basic flow-chart of the proposed optimal design procedure. The optimal design procedure for the DSBC and DSCC is based on the optimal design procedure previously presented in [10] for the DSCC. The outermost *global* optimization loop iterates through all possible numbers of modules N and different ac voltages $V_{g,\text{nom}}$ for which the system can be designed. In every iteration, a full design of the modular multilevel converter is performed. In order to keep the analyses in this paper clear and concise, the following simplifications are made:

- The grid voltage is assumed to be constant at $V_{g,\text{nom}}$ at all times.
- The designed systems are assumed to operate in (quasi-) steady-state at all times.
- The switching frequency is fixed at $f_s = 1000\,\text{Hz}$ per module
- All considerations are made exemplarily for SEMIKRON SKM75GB17E4 1.7 kV IGBTs as used in the the prototype presented in section VI[2]

As presented in [5], the optimal design procedure can be used to compare systems operating at different switching frequencies and / or using different power semiconductors. The

---

[2]The modules of the prototype are realized with three-level full-bridges, but the neutral-point-clamped state is typically not used. For the simplified loss calculations, every switch is thus regarded as two series-connected IGBTs.

inclusion of these free parameters would have gone beyond the scope of this analysis. The possibility to increase the overall power conversion efficiency and decrease the realization effort is independent from the initial choice of the switching frequency and the choice of the power semiconductors, assuming the decrease of the arm currents leads to a likewise decrease of the power losses in the switches.

B. Power Losses

The switching losses and conduction losses in the IGBTs are calculated based on the respective loss curves given in [14]. Since the switching losses are only given for a certain collector-emitter voltage $V_{CE,ref}$, the loss curves are scaled with the instantaneous module voltage at the time of switching

$$E_{sw,IGBT}(V_{CE}, i_C) = E_{sw,IGBT}(i_C) \cdot \left(\frac{V_{CE}}{V_{CE,ref}}\right)^{1.3...1.4} \quad (25)$$

as suggested in [15]. All calculations have been performed for a junction temperature of $T_j = 125\,°C$ to capture the worst operating case and prevent thermal runaway.

The power losses in the inductors and the capacitors are small [10] compared to the power losses in the IGBTs and have been neglected in the following for the sake of simplicity.

C. Number of Modules

The minimum number of modules is calculated according to (20)

$$N_{min} = \lceil n_{min} \rceil \approx \left\lceil \frac{v_{1u,max}}{V_{crit}} \right\rceil = \left\lceil \frac{V_{dc,nom} \cdot \left(1 + \frac{1}{o}\right) \cdot 1.15}{2V_{crit}} \right\rceil, \quad (26)$$

including a margin of 15 % for dynamic control. In order to arrive at a physically feasible design, the actual number of modules N has to be higher than n_{min}. The overdimensioning factor

$$\lambda = \frac{N}{n_{min}} \quad (27)$$

makes designs with different specifications comparable.

D. Module Capacitors

In order to ensure a safe operation, the internal arm voltages must always be kept within their designated limits at all times. The maximum internal arm voltage is determined by the number of modules and the maximum module voltage:

$$V_{int,max} = NV_{crit}. \quad (28)$$

Perfect balancing of the module voltages within an arm is assumed. For an operation in (quasi-) steady state, it is sufficient to limit the minimum internal arm voltage to

$$V_{int,min} = v_{1u,max} = \frac{V_{dc,nom}}{2} \cdot \left(1 + \frac{1}{o}\right) \cdot 1.15 \quad (29)$$

which according to (3) ensures that the arms are at all times able to supply the nominal output voltage. A margin of 15 % has been included for dynamic control.

The limits of the internal arm voltages determine the size of the module capacitors [10]:

$$C_{mod} \geq \frac{2N\Delta w_{1u,int}}{v_{1u,max}^2 - V_{int,min}^2} = \frac{2}{NV_{crit}^2} \frac{\Delta w_{1u,int}}{1 - \frac{1}{\lambda^2}} \quad (30)$$

where Δw_{1u} denotes the energy that has to be buffered by all module capacitors in one arm during normal operation:

$$\Delta w_{1u,int} = w_{1u,int,max} - w_{1u,int,min} \quad (31)$$

The value of $\Delta w_{1u,int}$ can be determined analytically or numerically e.g. by evaluating (17) over the whole operating area.

E. System Volume

The design of the capacitors is optimal when the module capacitance (30) is minimal. The total design energy

$$W_{C,tot} = 6N \cdot \frac{1}{2} C_{mod} V_{crit}^2 = 6\frac{\Delta w_{1u,int}}{1 - \frac{1}{\lambda^2}} \quad (32)$$

of all $6N$ module capacitors together is used to calculate the overall volume of the module capacitors. *Electronicon* film capacitors have been taken as a reference, because they offer high energy densities ($\rho_{C,el} \approx 150\,J\,dm^{-3}$), excellent current handling capabilities and a long lifetime [16].

F. Arm Inductance

The arm inductance is determined by the maximum fault current during a dc or ac short-circuit as explained in [17]. In order to limit the overcurrent in case of an ac short-circuit, the arm inductance is required to be higher than

$$L_{arm} > L_{ac,min} = \frac{\sqrt{\frac{2}{3}} T_d V_{g,nom}}{\Delta \hat{i}_{1u}^{sc,ac}} = 19.7\,mH. \quad (33)$$

The maximum allowed current rise

$$\Delta \hat{i}_{1u}^{sc,ac} = I_{crit} - \hat{i}_{1u}^{sc,ac} = 18.7\,A \quad (34)$$

is the difference between the peak arm current

$$\hat{i}_{1u}^{sc,ac} = \frac{\sqrt{2}I_{g,nom}}{2} + \frac{I_{dc,max}}{3} = 11.3\,A + 10\,A = 21.3\,A. \quad (35)$$

at full output power and maximum dc-link current (reached at a reduced dc-link voltage, which is possible with the DSBC) and the maximum permissible module current I_{crit}. The numerical results are exemplarily given for the prototype system presented in section VI (see specifications in Tab. I).

In order to limit the dc-link current in case of a dc-link short-circuit, the arm inductance is required to be higher than [17]

$$L_{arm} > L_{dc,min} = \frac{\frac{2}{3} T_d V_{dc,nom}}{3\Delta \hat{i}_{1u}^{sc,dc}} = 14.8\,mH. \quad (36)$$

The maximum current rise

$$\Delta \hat{i}_{1u}^{sc,dc} = I_{arm,crit} - \hat{i}_{1u}^{sc,dc} = 26.3\,A \quad (37)$$

is the difference between the peak arm current when operating at full output-power and maximum dc-link voltage

$$\hat{i}_{1u}^{sc,dc} = \frac{\sqrt{2}I_{g,nom}}{2} + \frac{I_{dc,nom}}{3} 11.3\,A + 2.4\,A = 13.7\,A. \quad (38)$$

1900

The calculations have been performed for the nominal dc-link current of

$$I_{dc,nom} = \frac{P_{nom}}{V_{dc,nom}} = 7.1\,\text{A} \qquad (39)$$

and the nominal (maximum) dc-link voltage $V_{dc,nom}$, which presents the worst case situation. The numeric results are again exemplarily given for the prototype system presented in section VI.

G. Arm Inductors

The design of the inductors themselves presents a multi-objective optimization problem of its own whose consideration would have gone beyond the scope of this analysis. This optimization is represented by the *design inductors* step in the flow-chart shown in Fig. 4. In the prototype system presented in section VI, the arm inductors are split up among the individual modules which simplifies their isolation and cooling. An optimal design of the module inductors can been performed as proposed in [10].

V. Design Results

In order to demonstrate the effectiveness of the proposed design methodology, a 250 kW active rectifier system based on the DSBC has been designed according to the specifications listed in Tab. I. The results are presented in the following. At first, the influence that the choice of operating factor has on the design is discussed. Afterwards, the overall tradeoff between designing the system for highest power conversion efficiency and minimum system volume is discussed. Last but not least, three possible solutions are presented to illustrate how the MMC may by designed for highest power conversion efficiency, lowest realization effort or a compromise of both.

A. AC-Voltage Tradeoff

Fig. 5 illustrates the effect that a reduction (or increase) of the ac voltage has on the volume of the module capacitors and the power losses in the semiconductors. The results are shown for an ac voltage reduction

$$r_{ac} = 1 - \frac{2\sqrt{\frac{2}{3}}V_{g,nom}}{V_{dc,nom}} = 1 - \frac{1}{o} \qquad (40)$$

ranging from $r_{ac} = -20\% \dots 60\%$. This is more convenient than plotting the results as a function of the operating factor o directly. All designs are targeted for $\lambda \approx 115\%$[3].

It becomes evident, that a reduction of the grid voltage leads to a reduction of the number of modules but comes at the cost increased power losses. At the same time, the volume of the module capacitors is increased. According to (16) and (17), these have to buffer larger power fluctuations because of the increased grid currents. The reduction of the overall number of modules reduces the realization effort because it reduces the number of required communication channels, IGBTs, gate drivers, per-module control platforms, sensors, etc.

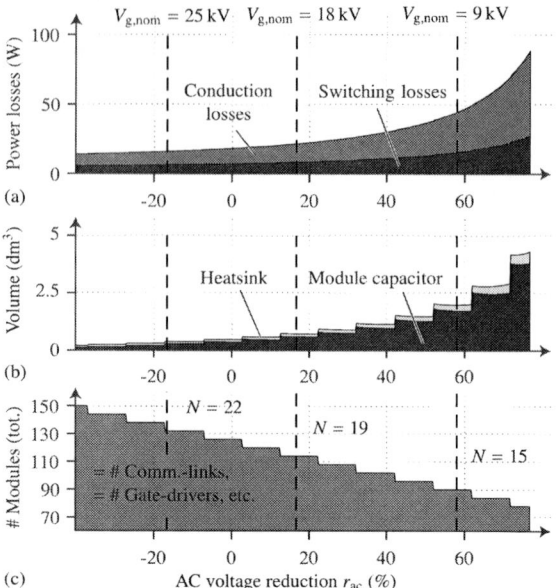

Fig. 5: An increase of the size of the module capacitors and an increase of the power losses is traded against a reduction of the overall number of modules and therewith a reduction of the number of gate drivers, switches, communication links, per-module control units, etc. The power losses and volumes are shown per module for designs with $\lambda \approx 115\%$.

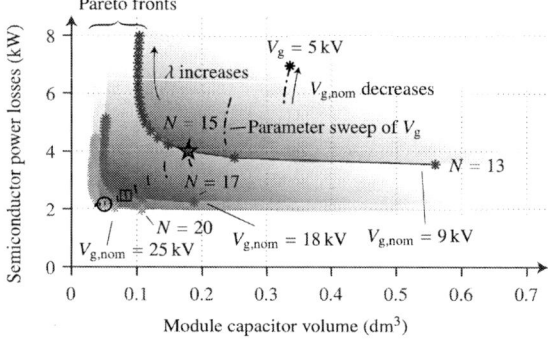

Fig. 6: The choice of the grid voltage affects the overall performance: All points of the same color denote pareto-optimal designs for different numbers of modules (and therewith different choices of λ) with the same nominal ac voltage $V_{g,nom}$. The black dashed-and-dotted curve denotes the results of the parameter sweep shown in Fig. 5

B. Pareto-Optimal Designs

Fig. 6 shows the results of the optimal design procedure for systems with $V_g = 25\,\text{kV}$, $V_g = 18\,\text{kV}$ and $V_g = 9\,\text{kV}$ when sweeping through different factors of $\lambda = 100\% \dots 250\%$. Each point marks a possible design. The lines connecting these points are only a visual aid.

The black dashed-and-dotted curve maps the results shown in Fig. 5 to the two-dimensional performance plane. It becomes evident, that the choice of the grid voltage affects the overall performance of the DSBC or DSCC without breaking

---

[3]The module number is an integer. Deviations from the ideal $\lambda = 115\%$ are caused by rounding.

with the optimal design principles. In particular, for every choice of $V_{g,nom}$, a variation of λ still allows to trade volume against power losses of the overall systems, as previously discussed in [10]. The yellow curve denotes systems operating in buck mode, which is only possible with the DSBC. The box, the circle and the star mark three possible solutions which are in the following discussed in detail.

C. Design for Increased Efficiency (Buck Mode)

When designing the DSBC for operation in buck mode, the power conversion efficiency can be improved. A corresponding design has been performed for a grid voltage of

$$V_{g,nom,buck} = 25\,kV. \tag{41}$$

The key design parameters for a design with $\lambda \approx 115\%$ are exemplarily given in Tab. II (a). The power-conversion efficiency of this system is calculated to be

$$\eta_{buck} = 99.2\% \tag{42}$$

at nominal load (based on the switching and conduction losses). The minimum number of modules per arm is

$$N_{min,buck} = 20. \tag{43}$$

The switching- and the conduction-losses are indicated by the leftmost dashed line shown in Fig. 5. The overall performance is denoted by the black circle in Fig. 6. The design adheres to the specifications presented in Tab. I.

Owing to the increased ac voltage, the required arm inductance (see Tab. II (a)) is comparatively large. For designs where the realization effort of the inductors is of particular concern, coupled inductors (or additional grid inductors) may be used [17]. However, the consideration of these options would have gone beyond the scope of this analysis.

D. Typical Design (Standard)

Typical system designs show a good compromise between power-conversion efficiency and realization effort. When designing the system with

$$V_{g,nom,typ} = 18\,kV \tag{44}$$

corresponding to an operating factor of $o \approx 1.2$. Because the operating factor does not include the margins for dynamic control, the effective operating factor of the presented typical solution is in fact very close to one, as can be inferred from Fig. 3. An overall power conversion efficiency of up to

$$\eta_{typ} = 99.0\% \tag{45}$$

can be achieved at nominal load (again calculated based on the switching- and the conduction-losses). The corresponding power losses are highlighted by the middle dashed line shown in Fig. 5. Again, the design has been performed for a target of $\lambda \approx 115\%$. The key design parameters are given in Tab. II (b) and the design adheres to the specifications presented in Tab. I. The minimum number of modules per arm is

$$N_{min,typ} = 17. \tag{46}$$

The black square shown in Fig. 6 represents the corresponding system design.

TABLE II: Key design parameters of a typical active rectifier system based on the modular multilevel converter (a) without considering the ac voltage as a free parameter (b) when optimizing for minimum realization effort.

	$V_{g,nom}$	C_{mod}	$L_{dc,min}$	$L_{ac,min}$	N_{min}	N	o
(a)	25 kV	18 μF	11.6 mH	39.4 mH	20	22	0.86
(b)	18 kV	36 μF	12.2 mH	30.0 mH	16	19	1.2
(c)	9 kV	104 μF	14.8 mH	19.7 mH	12	15	2.38

E. Design for Reduced Realization Effort (Boost Mode)

When allowing the DSBC or DSCC to operate in boost mode, the realization effort can be reduced. The black star shown in Fig. 6 represents a design with a grid voltage of

$$V_{g,nom,boost} = 9\,kV. \tag{47}$$

The minimum number of modules per arm is

$$N_{min,boost} = 13 \tag{48}$$

and the system features a maximum power-conversion efficiency of

$$\eta_{boost} = 98.4\% \tag{49}$$

(again calculated based on the switching- and the conduction-losses). While a design with $V_{g,nom} = 5\,kV$ would have been possible, the choice of $V_{g,nom} = 9\,kV$ presented the most attractive compromise between power losses, volume of the module capacitors and realization effort and thus has been chosen as the foundation of the hardware prototype discussed in the following section.

VI. PROTOTYPE SYSTEM

The proposed design method has been used to develop a full-scale DSBC hardware prototype that adheres to the specifications listed in Tab. I. Fig. 7 shows the assembly of an arm with three modules. The system is controlled via the synchronous converter control (SyCCo) bus presented in [18] and [19], which allows for simple and robust communication between the central controller and the modules while keeping the overall communication and control delay below the specified $T_{d,tot} = 50\,\mu s$. Auxiliary power is supplied to each module with the high voltage isolated multiport LLC resonant converter presented in [20].

Because the rectifier system is intended for supplying dc voltage to prototype power electronic converter systems at the *Laboratory for High Power Electronic Systems* at ETH Zurich, it is of utmost importance that the system itself is short circuit-proof, hence the DSBC topology. The protection and control system has been realized with a combination of central and local protection measures as proposed in [21].

Equation (20) suggests that the increasing the maximum module voltage, reduces the minimum number of modules in the system. It is hence desirable to use power-semiconductors with high blocking voltage capabilities. Unfortunately, low-current IGBT modules are only available off-the-shelf up to 1700 V. Therefore, NPC full-bridges are used in the modules to approximately double their maximum voltage. While this

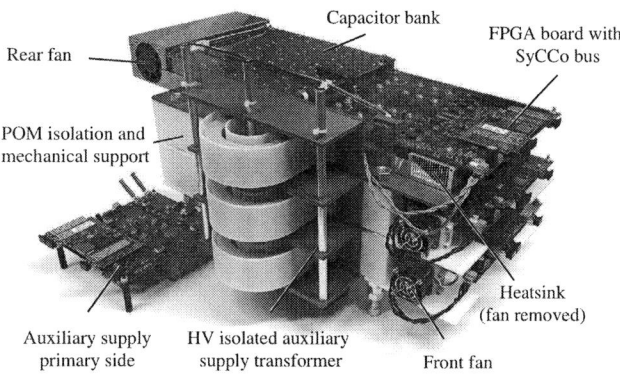

Fig. 7: Downscaled Hardware prototype with three modules arranged in one arm. The external wiring as well as some components have been removed for the sake of showing a more clear arrangement.

neither reduces the overall number of installed power semiconductors nor the overall number of installed gate drivers, it effectively halves the number of isolated auxiliary supplies, communication links, per-module control platforms as well as local measurement, protection and control circuitry, significantly reducing the overhead in build volume and the total bill-of-materials.

VII. Conclusion

When the double-star bridge-cell (DSBC) or double-star chopper-cell (DSCC) modular multilevel converters (MMC) are designed for operation in boost mode, their minimum required number of modules can be reduced. While this comes at the cost of increased power losses and an increase of the volume of the passive components, the benefits associated with the reduced assembly effort, the reduced number of power semiconductors and gate drivers and the reduced number of auxiliary components such as communication links, per-module control platforms and sensors may outweigh these drawbacks in many applications. In contrast to that, designing the DSBC for operation in buck mode allows to increase its power conversion efficiency at the expense of additional modules. With the design methodology presented in this paper, this additional degree of freedom may be exploited without breaking with the optimal design principles of double-star MMCs. To demonstrate the effectiveness of the proposed approach, a prototype system has been developed with emphasis on reduced realization effort. The system can provide an output voltage of 35 kV dc at a continuous rated power of 250 kW and uses favorable off-the-shelf low-current IGBT modules.

Acknowledgment

The authors would like to thank *ABB Switzerland Ltd.* and the *Bundesamt für Energie* (BFE) for their financial support of this very interesting project. This project is carried out within the frame of the *Swiss Centre for Competence in Energy Research on the Future Swiss Electrical Infrastructure* (SCCER-FURIES) with the financial support of the *Swiss Innovation Agency* (Innosuisse – SCCER program).

References

[1] A. Lesnicar and R. Marquardt, "An innovative modular multilevel converter topology suitable for a wide power range," in *IEEE Power Tech Conference, Bologna*, June 2003.

[2] D. Siemaszko, A. Antonopoulos, K. Ilves, M. Vasiladiotis, L. Aandngquist, and H.-P. Nee, "Evaluation of control and modulation methods for modular multilevel converters," in *International Power Electronics Conference (IPEC)*, June 2010.

[3] I. Trintis, S. Munk-Nielsen, and R. Teodorescu, "A new modular multilevel converter with integrated energy storage," in *Conference on IEEE Industrial Electronics Society (IECON)*, Nov. 2011, pp. 1075–1080.

[4] M. Vasiladiotis and A. Rufer, "Analysis and control of modular multilevel converters with integrated battery energy storage," *IEEE Transactions on Power Electronics*, vol. 30, no. 1, pp. 163–175, Jan 2015.

[5] A. Hillers and J. Biela, "Systematic comparison of modular multilevel converter topologies for battery energy storage systems based on split batteries," in *European Conference on Power Electronics and Applications (EPE)*, Sept. 2015.

[6] A. Korn, M. Winkelnkemper, and P. Steimer, "Low output frequency operation of the modular multi-level converter," in *IEEE Energy Conversion Congress and Exposition (ECCE)*, Sept 2010, pp. 3993–3997.

[7] W. Kawamura, K.-L. Chen, M. Hagiwara, and H. Akagi, "A low-speed, high-torque motor drive using a modular multilevel cascade converter based on triple-star bridge cells (mmcc-tsbc)," *IEEE Transactions on Industry Applications*, vol. 51, no. 5, pp. 3965–3974, Sept 2015.

[8] H. Mohammadi and M. Bina, "A transformerless medium-voltage statcom topology based on extended modular multilevel converters," *IEEE Transactions on Power Electronics*, vol. 26, no. 5, pp. 1534–1545, May 2011.

[9] V. Staudt, M. Jager, A. Rothstein, A. Steimel, D. Meyer, R. Bartelt, and C. Heising, "Short-circuit protection in DC ship grids based on MMC with full-bridge modules," in *Electrical Systems for Aircraft, Railway, Ship Propulsion and Road Vehicles (ESARS)*, Mar. 2015.

[10] A. Hillers and J. Biela, "Optimal design of the modular multilevel converter for an energy storage system based on split batteries," in *European Conference on Power Electronics and Applications (EPE)*, Sept. 2013.

[11] L. Baruschka and A. Mertens, "Comparison of cascaded h-bridge and modular multilevel converters for BESS application," in *Energy Conversion Congress and Exposition (ECCE)*, Sept. 2011.

[12] A. Antonopoulos, L. Angquist, and H.-P. Nee, "On dynamics and voltage control of the modular multilevel converter," in *European Conference on Power Electronics and Applications (EPE)*, Sept. 2009.

[13] S. Norrga, L. Angquist, K. Ilves, L. Harnefors, and H.-P. Nee, "Decoupled steady-state model of the modular multilevel converter with half-bridge cells," in *IET International Conference on Power Electronics, Machines and Drives (PEMD)*, Mar. 2012.

[14] SKM75GB17E4 datasheet. Online. SEMIKRON. [Online]. Available: www.semikron.com

[15] A. Wintrich, U. Nicolai, W. Tursky, and T. Reimann, *Applikationshandbuch Leistungshalbleiter*, E. Schoene and G. Stark, Eds. SEMIKRON International GmbH, 2010.

[16] E56/E59/SR17 Capacitors with large capacitances. Electronicon. [Online]. Available: www.electronicon.com

[17] N. Cherix and A. Rufer, "Dimensionnement formel des inductances de branches des convertisseurs modulaires multiniveaux (MMC)," in *Conference Electronique de Puissance du Futur*, July 2012.

[18] C. Carstensen and J. Biela, "A converter control field bus protocol for power electronic systems with a synchronization accuracy of +/- 5 ns," in *European Conference on Power Electronics and Applications (EPE)*, Sept. 2015.

[19] S. Rietmann, S. Fuchs, A. Hillers, and J. Biela, "Field bus for data exchange and control of modular power electronic systems with high synchronisation accuracy," in *International Power Electronics Conference (IPEC)*, May 2018.

[20] D. Peftitsis, M. Antivachis, and J. Biela, "Auxiliary power supply for medium-voltage modular multilevel converters," in *17th European Conference on Power Electronics and Applications (EPE), Geneva, Switzerland*, Sept. 2015.

[21] A. Hillers and J. Biela, "Central control and distributed protection of the DSBC and DSCC modular multilevel converters," in *IEEE Energy Conversion Congress and Exposition (ECCE-USA)*, Oct. 2016.

The 2018 International Power Electronics Conference

Common-Mode Voltage Injection Techniques for Quasi Two-Level PWM-Operated Modular Multilevel Converters

Jakub Kucka, Axel Mertens

Institute for Drive Systems and Power Electronics, Leibniz Universität Hannover, Hanover, Germany
E-mail: jakub.kucka@ial.uni-hannover.de

Abstract—This paper studies the application of the common-mode voltage injection techniques known from two-level inverters (carrier-based space-vector modulation and flat-top modulation in particular) to quasi two-level PWM-operated modular multilevel converters. Similar to two-level inverters, the available operating area is extended to higher modulation indices. The properties of the respective techniques are compared to each other and to the modular multilevel converter in normal operation mode. The results are based on simulations and the model is partly experimentally validated on a down-scaled converter prototype.

Keywords—modular multilevel converter, quasi two-level, PWM, capacitance reduction.

I. Introduction

The modular multilevel converter (MMC, Fig. 1, [1]) is a well-known topology used in HVDC and medium-voltage applications. Recently, many efforts have been made to apply this topology to electric drives, as the low output frequencies would lead to an excessively high module capacitance in normal operation mode.

The majority of the proposed solutions can be divided into two groups. In the first group, the converter topology is modified, adding new current paths (e.g. [2], [3]) or converter stages (e.g. [4]–[6]). In the second group, an additional high-frequency voltage component is injected to the common-mode voltage (e.g. [7]–[10]). While the first group increases the converter losses and the installed volume of semiconductor devices, the second group leads to converter oversizing, when the rated torque is demanded over the whole operating range [10]. Moreover, the amount of installed capacitance in the second group is still exceedingly high, when the rated output frequency is relatively low.

A more recent proposal addressing this problem is the quasi two-level PWM operation, first published in [11]. This operation mode mimics the standard two-level voltage source inverter (VSI). In order to massively reduce the amount of installed capacitance in modules, the multilevel property of the MMC has to be sacrificed. Although this might seem as a large drawback at first sight, the inductance of a low-speed machine should be sufficient for the current filtering despite the quasi two-level voltage shape. Moreover, as a trapezoidal (staircase) waveform is used instead of the rectangle form typical for a two-level VSI, the output voltage dv/dt is limited. This mitigates the critical problem of reflection due to long machine cables and thus significantly reduces the requirements on the machine isolation systems. Therefore, the quasi two-level operation of an MMC is expected to be highly beneficial for cost-sensitive low-speed medium-voltage drives, as various nominal voltages are achievable using standardized low-cost modules with a very low amount of installed capacitance.

This paper investigates the application of the common-mode voltage injection techniques, known from two-level VSIs, to quasi two-level PWM-operated MMCs. First, the principle of the quasi two-level operation is explained in Section II, and the trade-offs for maximum duty cycle, which have to be met during the design process, are derived in Section III. The CMV injection techniques and their application to quasi two-level PWM-operated MMCs are presented in Section IV. These are validated on a down-scaled prototype (Section V), and the simulations for different modulation indices and power factors are carried out. Finally, the results and their comparison are discussed in Section VI.

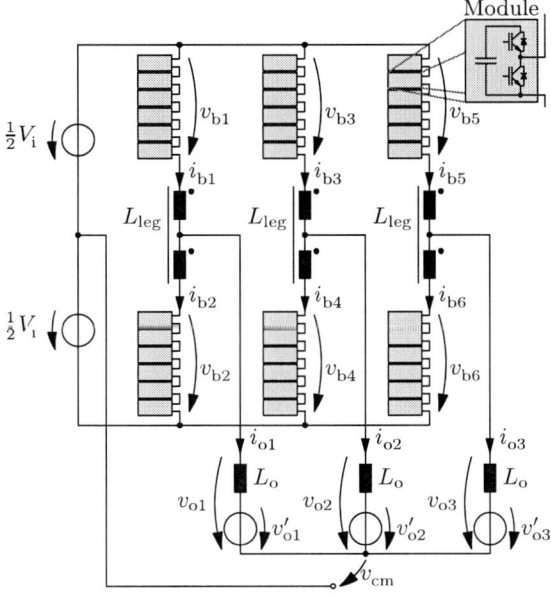

Fig. 1. Studied Modular Multilevel Converter.

1904

II. QUASI TWO-LEVEL PWM OPERATION

The principle of the quasi two-level PWM operation can be explained using Fig. 2. Because all three phase legs are identical and are controlled independently, it is sufficient to concentrate on the first phase leg consisting of branches 1 and 2, and (coupled) branch inductors.

As for two-level VSI, the output voltage is generated by a pulse-width modulation (PWM), by comparing the carrier c_1 to the duty cycle δ_1 (as shown in the first graph of Fig. 2). When the duty cycle signal is higher than the carrier, the voltage of branch 2 v_{b2} approximates the input direct voltage V_i with high-frequency modulation (HF modulation), and all the modules of branch 1 are short-circuited ($v_{b1} = 0$). When the duty cycle signal is lower, branch 1 approximates the input voltage and branch 2 is short-circuited.

Between these two described states, transition states are applied, where either all modules of both branches are switched on or all modules of both branches are short-circuited. The purpose of the transition states is to change the leg current

$$i_{\text{leg1}} = \frac{1}{2} \cdot (i_{b1} + i_{b2}) \tag{1}$$

as fast as possible.

In order to limit the dv/dt of the phase voltages and common-mode voltage v_{cm} (CMV), a minimum delay is asserted between every two switching instants of the modules within one branch. This leads to a staircase shape of the branch voltages and the phase voltage v_{o1} in Fig. 2.

The branch voltages determine the waveform of the output (phase) voltage v_{o1}, which determines the output current i_{o1} (with the load inductance L_o and the load voltage v'_{o1}).

The main idea for reducing the energy variation in the modules' capacitors is to split the output current i_{o1} between the branches 1 and 2, so that the branch with currently short-circuited modules handles most of the current and the current of the branch generating a high voltage is very low. This current division is adjusted by the leg current (Fig. 2).

However, because of the staircase form of the branch voltages and the limited di/dt of the leg current, the branch powers (buffered in the module capacitors) cannot always be instantaneously zero. Consequently, power peaks appear in the branch powers (last two graphs of Fig. 2). These power peaks are then compensated by a small branch current in the opposite direction while the branch voltage is high. This current will be further referred to as "compensation current".

III. DESIGN TRADE-OFFS FOR MAXIMUM DUTY CYCLE

The closer the duty cycle is to 1 or -1, the shorter is the time period available for the power-peak compensation in one of the branches, leading to higher branch currents. This means that the maximum possible duty cycle is

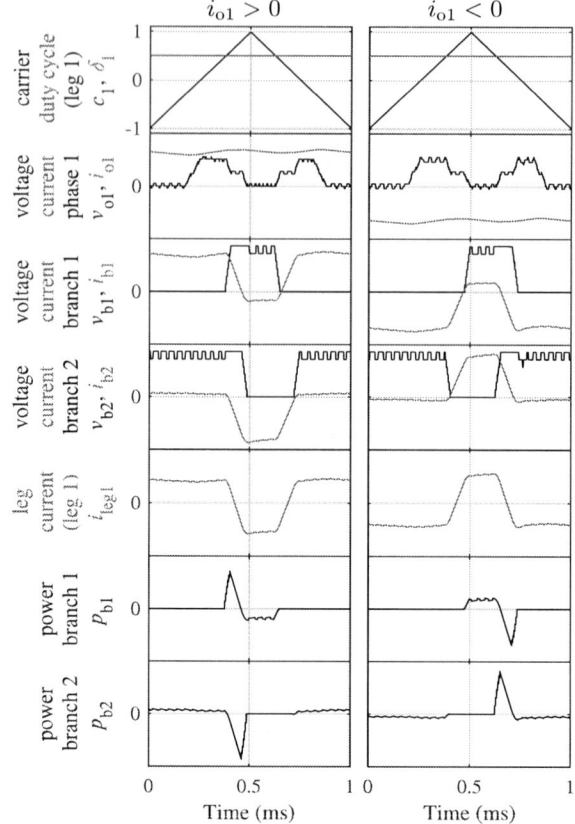

Fig. 2. Quasi two-level PWM operation example waveforms for positive and negative output current direction shown for one PWM period (1 ms).

a design parameter influencing the leg inductance, the module capacitance, and the HF-modulation frequency. Below, the derivation of these relationships (and trade-offs) will be presented.

The energy disturbance Δe_b caused by the mentioned power peak can be calculated as a triangle area under the branch power waveform according to (2). The length of the triangle is the time period T_T needed for the transition state to finish. The height of the triangle is the branch-power peak value $\hat{p}_b$.

$$\Delta e_b = \frac{1}{2} \cdot \hat{p}_b \cdot T_T = \frac{V_b \cdot \hat{i}_b}{2} \cdot T_T \tag{2}$$

The branch-power peak value can be calculated as a product of the peak branch current $\hat{i}_b$ and the maximum branch voltage V_b, when the staircase waveform of the branch voltage is simplified to a rectangle. The transition state duration

$$T_T = \frac{L_{\text{leg}} \cdot \hat{i}_b}{2 \cdot V_b - V_i} \tag{3}$$

is dependent on the leg inductance of the coupled inductor L_{leg}, peak branch current $\hat{i}_b$, and the difference between the input voltage V_i and the sum of both branch voltages $2V_b$, Fig. 1.

As the maximum branch current, which can occur, is approximately the output current amplitude $\hat{i}_b \approx \hat{i}_o$

and the maximum branch voltage can be roughly approximated as input voltage $V_b \approx V_i$, a worst-case estimation of the maximum branch energy deviation

$$\Delta e_{b,\max} \approx \frac{1}{2} \cdot L_{\text{leg}} \cdot \hat{i}_o^2 \quad (4)$$

can be derived using (2) and (3).

This energy deviation has to be compensated with a compensation current, while the corresponding branch voltage is high. Its value is the highest ($\hat{i}_c$), when the time period for the compensation is the shortest ($T_{c,\min}$):

$$\Delta e_{b,\max} = V_i \cdot \hat{i}_c \cdot T_{c,\min} \quad . \quad (5)$$

The time available for the compensation

$$T_{c,\min} = \frac{1 - \delta_{\max}}{2} \cdot \frac{1}{f_{\text{PWM}}} \quad (6)$$

is a function of the maximum duty cycle $\delta_{\max}$ and the PWM frequency f_{PWM}, when the duration of the transition state is neglected.

Substituting (4) and (6) into (5), the maximum compensation current

$$\hat{i}_c = \frac{L_{\text{leg}} \cdot i_o^2 \cdot f_{\text{PWM}}}{V_i} \cdot \frac{1}{1 - \delta_{\max}} \quad (7)$$

can be expressed as a function of the leg inductance and the maximum duty cycle. This equation shows that for a given input voltage V_i, PWM frequency f_{PWM}, and output current amplitude $\hat{i}_o$, the higher the maximum duty cycle $\delta_{\max}$ is required, the higher the maximum compensation current $\hat{i}_c$ or the lower the inductance of the leg inductor L_{leg} has to be. However, both scenarios are undesired, as a higher compensation current leads to higher semiconductor losses and a lower inductance requires a higher HF modulation frequency in order to keep the leg-current ripple limited. Moreover, when the maximum duty cycle approaches one, the HF modulation frequency or the compensation current would approach infinity. Therefore, the duty cycle must necessarily be limited to a value below one.

Considering the derived trade-offs, the presented converter has been designed for a maximum duty-cycle value limited to 0.9. This limits the maximum compensation current to 20 % of the nominal output-current amplitude.

Note that the duty cycles being exactly 1 or -1 are also achievable, as these do not cause any power peaks, and therefore, no compensation is necessary.

Equation (7) also implies that it is allowed for the duty cycle to exceed its designed maximum value, if the output current is reduced at these operating points accordingly. This option is not further studied in this paper.

IV. CMV INJECTION TECHNIQUES

The basis for the injection techniques is to modify the setpoint duty-cycle vector $\boldsymbol{\delta}' = [\delta_1', \delta_2', \delta_3']$, in order to improve some properties of the converter [12]. The resulting duty-cycle vector $\boldsymbol{\delta}$ used for the PWM is obtained by

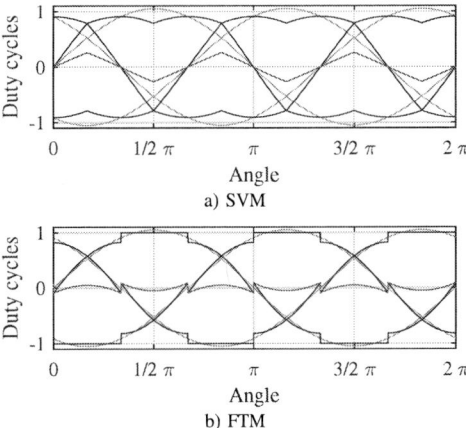

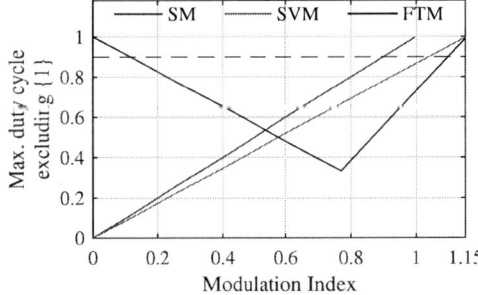

Fig. 3. Example duty-cycle waveforms, $M = 1.05$. Grey lines represent the setpoint sine duty cycles, red line the injected duty-cycle component, and the blue lines the resulting duty cycles used for the output PWM.

Fig. 4. Analytical comparison of the considered injection techniques.

adding the common-mode component δ_{cm} to the setpoint duty-cycle values for each phase δ_1', δ_2' and δ_3':

$$\boldsymbol{\delta} = [\delta_1, \delta_2, \delta_3] = [\delta_1' + \delta_{\text{cm}}, \delta_2' + \delta_{\text{cm}}, \delta_3' + \delta_{\text{cm}}] \quad . \quad (8)$$

One obvious option is not to inject any common-mode component:

$$\delta_{\text{cm}}^{\text{SM}} = 0 \quad . \quad (9)$$

This technique is also referred to as "Sine Modulation" (SM).

The second option is to apply the "Carrier-Based Space-Vector Modulation" (SVM) with

$$\delta_{\text{cm}}^{\text{SVM}} = -\frac{\max(\boldsymbol{\delta}') + \min(\boldsymbol{\delta}')}{2} \quad , \quad (10)$$

extending the maximum modulation index M without overmodulation. This technique is similar to the "Third-Harmonic Injection". However, it has a significant advantage, as the output angle does not have to be known to calculate the injected component. Therefore only SVM is investigated. Waveforms demonstrating the technique are plotted in Fig. 3a.

The third investigated technique is the "Flat-Top Modulation" (FTM), also called "Discontinuous Modulation", shown in Fig. 3b. This technique does not only extend the maximum modulation index, but it also reduces the

The 2018 International Power Electronics Conference

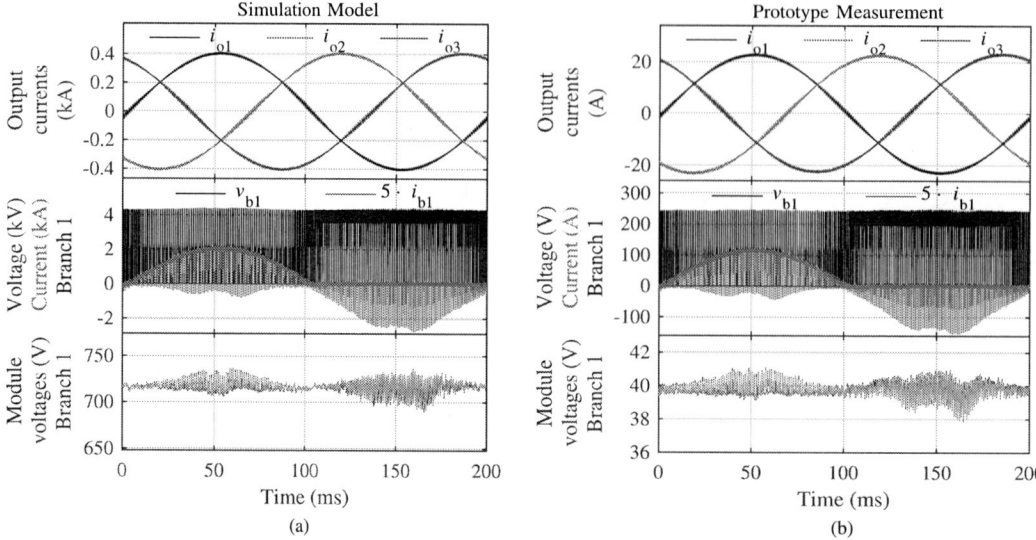

Fig. 5. Experimental validation of the quasi two-level PWM-operated modular multilevel converter with SVM, $M = 1.05$: a) simulated waveforms b) measured waveforms on down-scaled prototype. With down-scaled prototype, only three module capacitor voltages (out of six) were measured.

switching losses, as the converter does not have to switch, when the duty cycle is 1 or -1. The injected common-mode component can be calculated as:

$$\delta_{\mathrm{cm}}^{\mathrm{FTM}} = \mathrm{sign}\left(\max(\boldsymbol{\delta'}) + \min(\boldsymbol{\delta'})\right) \cdot \ldots \\ \left(1 - \max(\mathrm{abs}(\boldsymbol{\delta'}))\right). \quad (11)$$

In literature, modified variants of this technique can be found (e.g. [13]), which shift the position of the "flat top" according to the output current to further reduce the switching losses, when the output current is not aligned with the output voltage. However, the optimum shift has to be searched for each modulation index and power factor and its value changes in dependency on the maximum duty cycle achievable with the converter. Thus, only the described (simple) method will be further investigated in this paper.

As explained in section III, the maximum duty cycle of the quasi two-level PWM-operated converter is limited (e.g. to 0.9 with presented converter) except for the duty cycle being exactly 1 or -1. In Fig. 4, the maximum required duty cycle (excluding 1) is plotted in dependence on the modulation index. Using this figure and assuming a maximum duty cycle of 0.9, we can state that the maximum achievable modulation index with SM of 0.9 can be extended to ≈ 1.05 with SVM and even to ≈ 1.1 with FTM.

To improve the properties of the FTM with a quasi two-level PWM-operated MMC, it is recommended to inverse the carrier function for a particular converter phase x, if its required duty cycle δ_x is lower than zero.

This can be explained using Fig. 2. Observing branch voltage 1, it can be recognized that its values change from zero to high voltage and then back to zero during each PWM period. In contrary, the branch voltage 2 is firstly high, then zero, and afterwards high again. If the next required duty cycle is 1, the branch voltage 1 remains zero and the branch voltage 2 remains high, leading to a fluent

transition without any additional switching. However, if the next required duty-cycle value is -1, both branch voltages must be changed, leading to additional switching and significant branch energy disturbances. This can be prohibited simply, when the described PWM pattern is inverted for negative duty cycles, leading to a fluent transition to and from the state of duty cycle being -1. The simplest way to do so is by inverting the corresponding carrier.

V. Simulation Model and Its Experimental Validation

The simulation model used for the investigations was implemented in Matlab/Simulink, using Plecs toolbox for the electrical part of the model. The converter input source was assumed to be a direct-voltage source. The converter output was connected to an alternating-voltage source, controlling the first harmonic of the output current with a very slow proportional-resonant controller. The voltage THD of this voltage source was under 0.5 % in all operating points and therefore, the influence on the output current harmonics is negligible. The simulation parameters are summarized in Table I. The converter's energy storage time constant H (the total energy stored in the modules' capacitors divided by the maximum converter power) was calculated, assuming the maximum modulation index is

TABLE I. CONVERTER PARAMETERS

Parameter		Simulation	Prototype
Input voltage	V_{i}	4 kV	220 V
Output current amplitude	$\hat{\imath}_{\mathrm{o}}$	400 A	22 A
Output power factor	$\cos\varphi_{\mathrm{o}}$	1	0.996
Output frequency	f_{o}	5 Hz	5 Hz
PWM frequency	f_{PWM}	1 kHz	1 kHz
HF-modulation frequency	f_{HF}	25 kHz	25 kHz
Modules per branch	n_{mod}	6	6
Module capacitance	C_{mod}	0.2 mF	0.2 mF
Module setpoint voltage	V_C^*	720 V	40 V
Leg inductance (coupled)	L_{leg}	0.2 mH	0.2 mH
Load inductance	L_{o}	15 mH	15 mH
Energy storage constant	H	1.4 ms	1.4 ms

1907

The 2018 International Power Electronics Conference

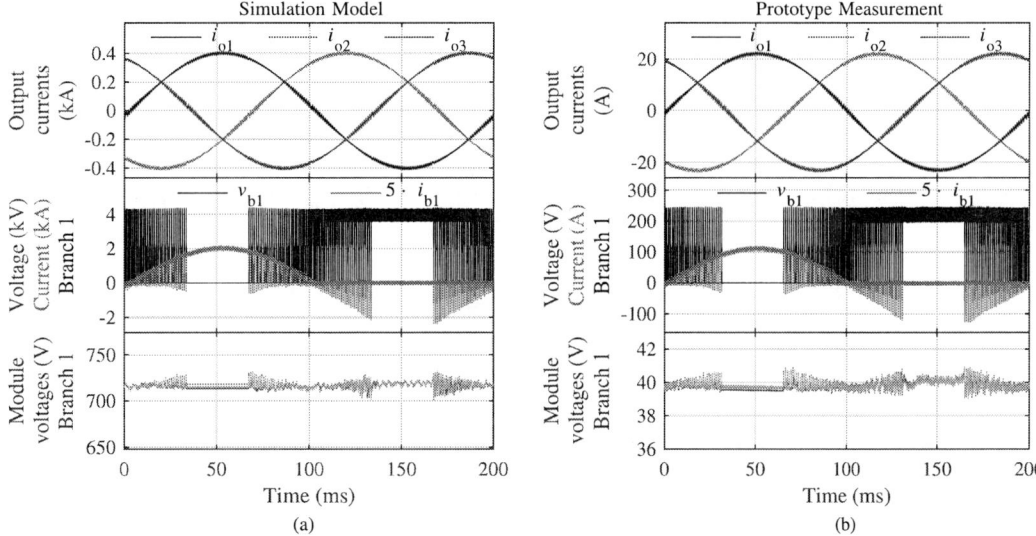

Fig. 7. Experimental validation of the quasi two-level PWM-operated modular multilevel converter with FTM, $M = 1.1$: a) simulated waveforms b) measured waveforms on down-scaled prototype. With down-scaled prototype, only three module capacitor voltages (out of six) were measured.

1.1. The loss calculation is based on the IGBT/Diode module `Infineon FZ400R12KE4` [14].

In order to validate the simulation model, a down-scaled experimental prototype was developed, Fig. 6. In reference to the simulation model, the prototype is scaled down by a factor of 18 for currents and voltages, Table I. In Fig. 5, the simulation results and prototype measurements are plotted, showing the operation with SVM. In Fig. 7, the operation with FTM is shown. The measured waveforms are zoomed in in Fig. 8. For a better correspondence, both the simulation and the measurements employ a resistor instead of the voltage source as a load. To reduce the signal noise in measurements due to EMI, the measured waveforms are filtered with an 1 MHz low-pass filter. In general, a very good correspondence between the waveforms can be stated and thus the simulation models are valid.

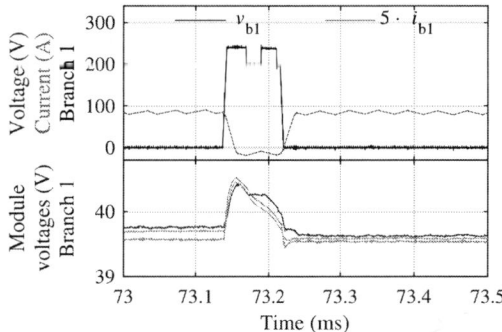

Fig. 8. Measured detail of the quasi two-level PWM operation (Fig. 7b zoomed in; the branch current and the branch voltage are not filtered).

VI. COMPARISON OF RESULTS

In Fig. 9a, the steady-state simulation results are plotted as a function of the modulation index for different injection techniques. The output current amplitude is kept constant at 400 A (see Table I) with unity power factor. In the first graph, the maximum duty cycles are plotted, confirming the findings from Fig. 4. It can also be seen that the maximum duty cycle was limited to 0.9. While the properties of the SVM and SM are almost identical (except for the extended modulation index with SVM), the FTM has a significant influence on all parameters. The first visible change is the significantly higher THD representing the output current ripple. Secondly, the semiconductor losses are lower, since less switching instants are necessary. Thirdly, the energy variation in the module capacitors and (consequently) the RMS value of module capacitor currents are lower.

However, the unity power factor, selected for the investigations in Fig. 9a, is the best case for the FTM. In Fig. 10a, the dependencies on the output-current phase angle φ_o for a constant modulation index $M = 0.8$ are plotted. As can be observed, the converter losses and the capacitor RMS currents are close to those of SM or SVM

Fig. 6. Photo of the down-scaled converter prototype.

1908

The 2018 International Power Electronics Conference

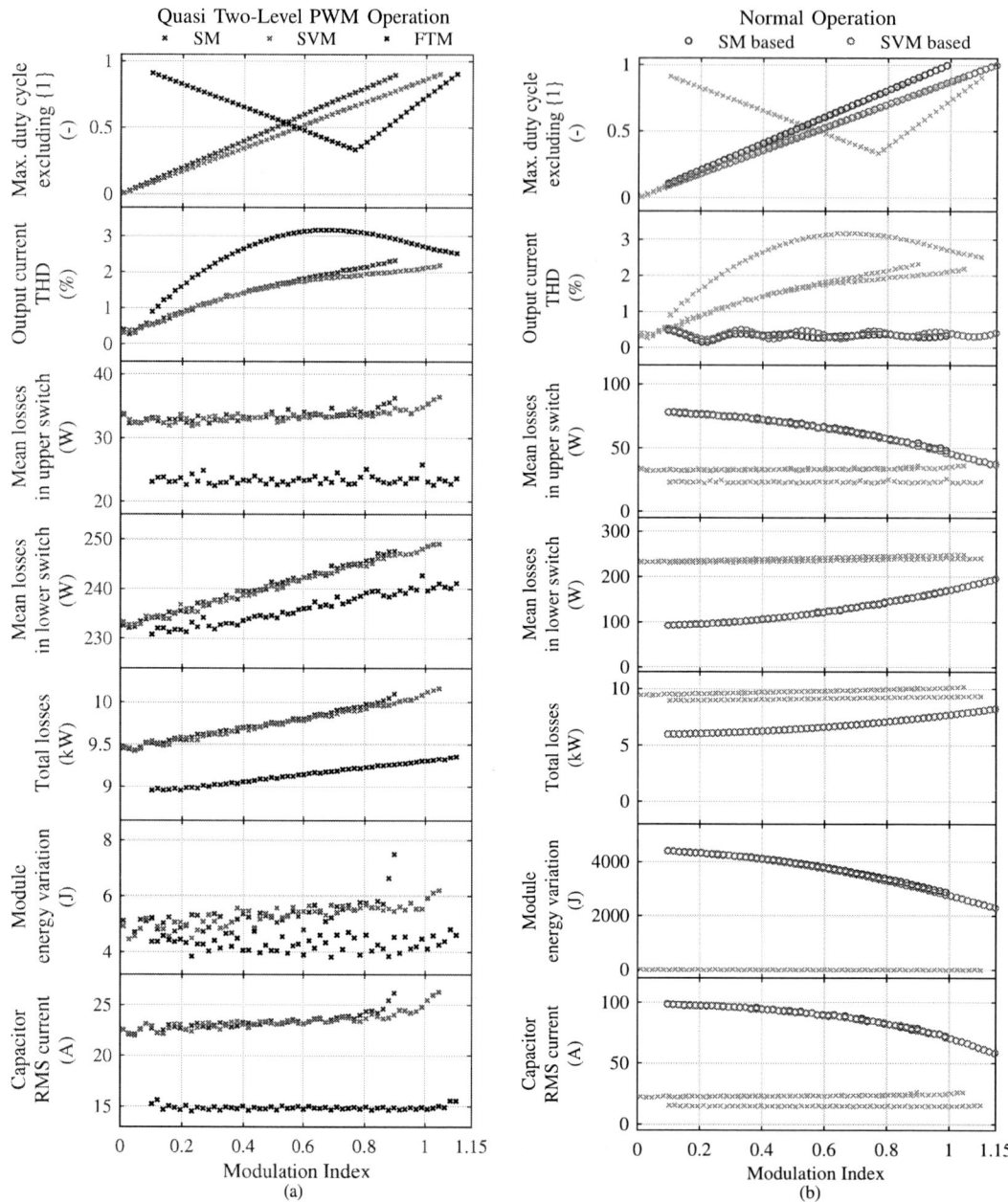

Fig. 9. Comparison of the injection techniques in dependence on modulation index with a constant output current and unity power factor: a) quasi two-level operation, b) normal operation mode – the grey "x" markers represent the quasi two-level operation data (for a better correspondence). The values of the mean losses, the energy variation and the capacitor current are given for the worst-case module during one simulated period (200 ms).

modulated converter, when reactive power is required at the converter output. Moreover, when only reactive power is required (φ_o is $\frac{1}{2}\pi$ or $\frac{3}{2}\pi$), the module energy variation with FTM becomes worse than the one of SM or SVM. This is caused by carrier signals being inverted while the branch current is the highest.

Furthermore, Fig. 10a shows that the quasi two-level PWM operation leads to lower losses in the module's lower IGBT/diode switch, when the active power is transmitted from the load. The cause is the shorter time in which the IGBTs conduct the branch current and the

longer time in which the diodes conduct it.

In general, considering both Fig. 9a and Fig. 10a, it can be stated that the upper IGBT/diode switch in the module has significantly lower losses than the lower switch and therefore could be designed smaller, thus reducing the semiconductor costs and switching losses.

The investigations from Fig. 9a and Fig. 10a have also been accomplished in normal converter operation mode. The results are shown in Fig. 9b and Fig. 10b, respectively. For a better correspondence to quasi two-

The 2018 International Power Electronics Conference

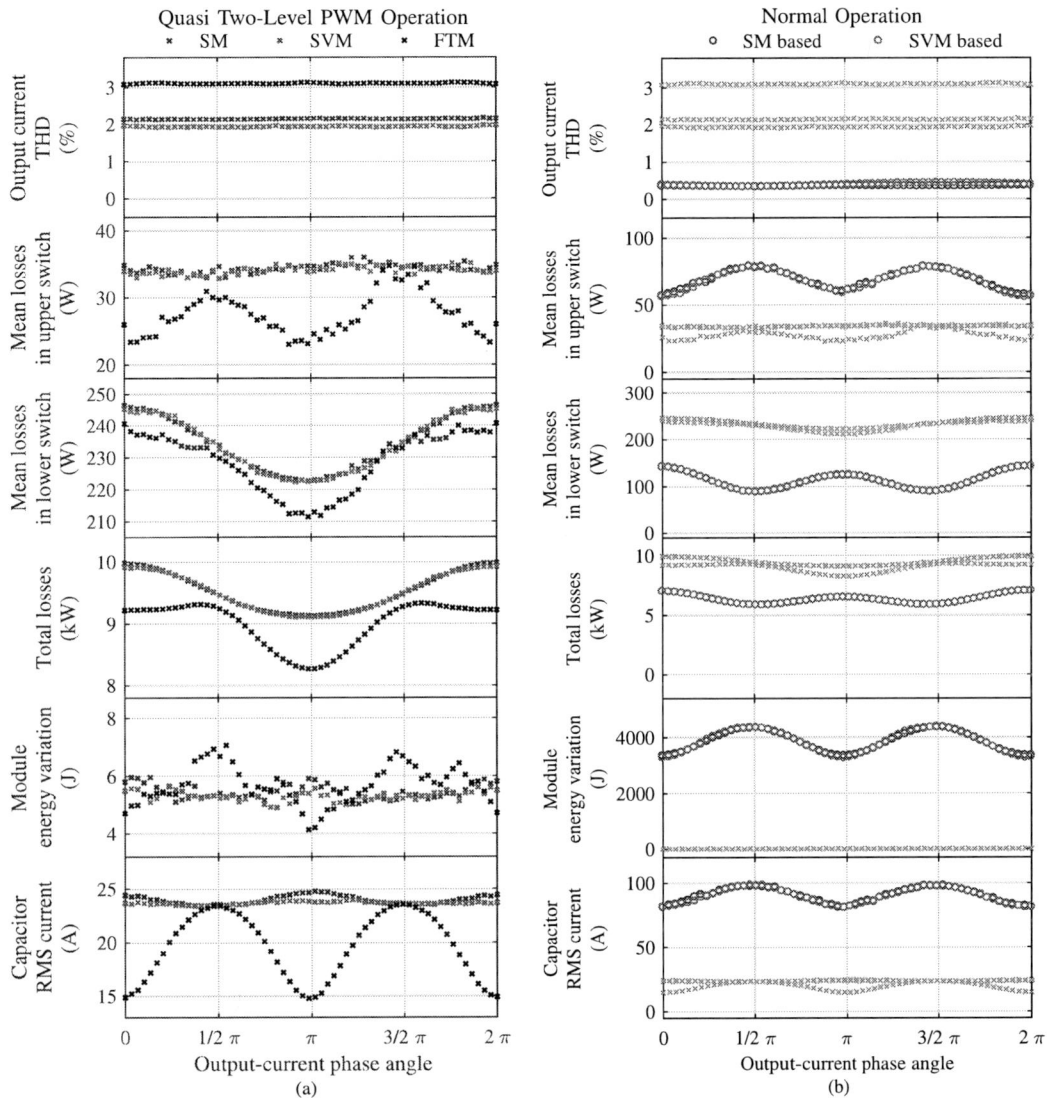

Fig. 10. Comparison of the injection techniques in dependence on the output-current phase angle φ_o with a constant output-current amplitude and modulation index $M = 0.8$: a) quasi two-level operation, b) normal operation mode – the grey "x" markers represent the quasi-two level operation data (for a better correspondence). The values of the mean losses, the energy variation and the capacitor current are given for the worst-case module during one simulated period (200 ms).

level PWM operation, the data from Fig. 9a and Fig. 10a are plotted as grey "x" markers. As significantly higher energy variation per module occurs in the normal operation mode, the module capacitance was increased to 200 mF (the energy storage constant was increased to 1.35 s, assuming the maximum modulation index is 1.15). The modulation (and control) frequency was chosen at 1 kHz, being identical with the PWM frequency of the quasi two-level PWM-operated MMC. The leg inductance was increased to 12 mH, accordingly.

While slightly higher modulation indices are reachable in normal operation mode, and the current distortion is significantly lower (as expected), the total energy variation, representing the amount of installed capacitance in the modules, is up to 850 times higher. Additionally, the module capacitors are loaded with significantly higher

RMS currents. The converter losses are generally lower in normal operation mode, as this leads to a lower number of switching instants. However, this advantage is expected to be diminished in drive applications during operation at machine speeds below nominal, when high-frequency CMV injection and additional circulating currents have to be applied.

VII. CONCLUSIONS

In this paper, three different injection techniques applied to the quasi two-level PWM-operated MMC were studied and compared. It can be concluded that all three techniques are feasible, which has also been validated on an experimental converter prototype.

As could be derived in this paper, the achievable modulation indices with the quasi two-level PWM operation

1910

are limited to some value below one, if no CMV injection is applied (the case of SM). The SVM technique extends the maximum achievable modulation index by a factor of ≈ 1.15 in comparison to SM, while the other important properties remain unchanged. The FTM technique can further extend the modulation index and additionally reduce the converter losses; however, it leads to higher output-current distortion. Therefore, it is recommended to use the SVM technique by default. The FTM technique can be optionally used for higher modulation indices, if the additional current distortion is acceptable.

The comparison to the normal operation of MMC confirmed the expected advantages of the quasi two-level PWM operation (i.e. significantly lower module capacitance requirements) and its disadvantages (i.e. higher current distortion and losses).

Additionally, the paper has shown that the upper IGBT/diode switch in modules has significantly lower losses than the lower switch, when the quasi two-level PWM operation is applied. In future work, the option to decrease the size of this switch will be investigated. Furthermore, the control implementation used for the converter prototype will be presented and the application of quasi two-level PWM-operated MMC to medium-voltage low-speed drives will be studied more extensively, drawing a comparison to other converter topologies and operation modes.

Acknowledgement

This work was supported by Deutsche Forschungsgemeinschaft (DFG, German Research Foundation) – project ME 1002/8-1.

References

[1] A. Lesnicar and R. Marquardt, "An innovative modular multilevel converter topology suitable for a wide power range," in *2003 IEEE Bologna Power Tech Conf. Proc.*, pp. 1–6, 2003.

[2] S. Du *et al.*, "An Active Cross-Connected Modular Multilevel Converter (AC-MMC) for a Medium-Voltage Motor Drive," in *IEEE Trans. Ind. Electron.*, vol. 63, no. 8, pp. 4707–4717, Aug. 2016.

[3] S. Du *et al.*, "A Flying-Capacitor Modular Multilevel Converter for Medium-Voltage Motor Drive," in *IEEE Trans. Power Electron.*, vol. 32, no. 3, pp. 2081–2089, March 2017.

[4] J. Kucka and L. Baruschka, "A hybrid modular multilevel DC/AC converter," in *17th Eur. Conf. on Power Electron. and Appl.*, pp. 1–10, Sep. 2015.

[5] Y. Okazaki *et al.*, "Multiple medium-voltage motor drives using modular multilevel cascade converters with medium-frequency transformers," in *2015 IEEE 2nd Int. Future Energy Electron. Conf.*, Taipei, pp. 1–6, 2015.

[6] B. Li *et al.*, "A Hybrid Modular Multilevel Converter for Medium-Voltage Variable-Speed Motor Drives," in *IEEE Trans. Power Electron.*, vol. 32, no. 6, pp. 4619–4630, June 2017.

[7] A. J. Korn *et al.*, "Low output frequency operation of the Modular Multi-Level Converter," in *2010 IEEE Energy Conversion Cong. and Expo.*, pp. 3993–3997, Sep. 2010.

[8] J. Kolb *et al.*, "Cascaded control system of the modular multilevel converter for feeding variable-speed drives," in *IEEE Trans. Power Electron.*, vol. 30, no. 1, pp. 349-357, Jan. 2015.

[9] Y. Okazaki *et al.*, "A speed-sensorless start-up method of an induction motor driven by a modular multilevel cascade inverter (MMCI-DSCC)," in *IEEE Trans. Ind. Appl.*, vol. 50, no. 4, pp. 2671–2680, Jul. 2014.

[10] A. Antonopoulos *et al.*, "Modular Multilevel Converter AC Motor Drives With Constant Torque From Zero to Nominal Speed," in *IEEE Trans. Ind. Appl.*, vol. 50, no. 3, pp. 1982–1993, May-June 2014.

[11] A. Mertens and J. Kucka, "Quasi two-level PWM operation of an MMC phase leg with reduced module capacitance," *IEEE Trans. Power Electron.*, vol. 31, no. 10, pp. 6765–6769, Oct. 2016.

[12] J. Holtz, "Pulsewidth modulation for electronic power conversion," in *Proc. of the IEEE*, vol. 82, no. 8, pp. 1194–1214, Aug. 1994.

[13] A. M. Hava *et al.*, "A high-performance generalized discontinuous PWM algorithm," in *IEEE Trans. Ind. Appl.*, vol. 34, no. 5, pp. 1059–1071, Sep./Oct. 1998.

[14] Infineon, "Technical Information: IGBT Module FZ400R12KE4," datasheet, revision 2.2, Nov. 2013.

Current Tracking and Cell-Voltage Limitations of Modular Multilevel Converters with Direct Digital Control

T.-F. Wu, T.-C. Chou, K.-E. Lin, and T.-Y. Li

Elegant Power Electronics Applied Research Laboratory (EPEARL)
Department of Electrical Engineering
National Tsing Hua University
Hsinchu, Taiwan, ROC
Email: tfwu@ee.nthu.edu.tw

Abstract—This paper presents current tracking and cell-voltage limitations of modular multilevel converters (MMC) with direct digital control to achieve real-power injection, rectification and low voltage ride through (LVRT) functions. MMC plays a pivotal role in either exchanging power between dc side and ac side or reducing power cable size for high voltage transmission systems. The direct digital control can track sinusoidal reference currents and regulate dc-bus and cell voltages regardless of step load change or ac voltage sagging. Moreover, cell-voltage expression has been derived to determine dynamics of cell-voltage regulation, as well as amplitude of cell-voltage ripple. In this paper, MMC control laws are reviewed first, and then limitations of cell-voltage range are discussed. Experimental and simulated results obtained from a 50 kW MMC have verified the desired functions.

Index Terms— *Direct Digital Control, Modular Multilevel Converter, Real-Power Injection Mode, Cell-Voltage Limitations, and Cell-Voltage Regulation.*

I.   INTRODUCTION

Nowadays, the role of power converters has become more and more important. According to demand, it needs to transform electricity to different forms with various types of converters, such as dc/dc [1], ac/dc, dc/ac [2] and ac/ac converters. Applications of dc/ac converters, for example, are suitable for uninterruptible power supply [3], active power filter [4], motor drive [5] and so on.

For high voltage applications, conventional converters [3]-[5] are not satisfied because of high voltage stress on both switches and capacitors. Accordingly, modular multilevel converters (MMCs) become more feasible in high voltage applications. The author of [6] has classified MMCs into four topologies.

In shaping a sinusoidal grid current, cell-voltage regulation or arm balancing between upper- and lower-arm cells are the important factors. Several methods to achieve either cell-voltage regulation or arm balancing control have been presented [7]-[9]. In [7], the authors proposed PI controllers combined with maximum power point tracker, redundant module controller, voltage sta-

bility controller and grid-connected controller to operate grid-connected PV systems. When the MMC operates under partial shading condition, the redundancy modules are used to compensate fluctuation voltage of photovoltaic panels. This control scheme is feasible for partial shading PV modules. The authors of [8] proposed a decoupled control between ac and dc sides to manipulate the circulating current reference to control the absorbed arm power, which can effectively balance arm voltages under pulsed dc load. Reference [9] proposed a voltage control combining averaging, individual-balancing and arm-balancing controls, and current tracking control was also conducted which was intentionally designed for grid connection and medium-voltage power systems.

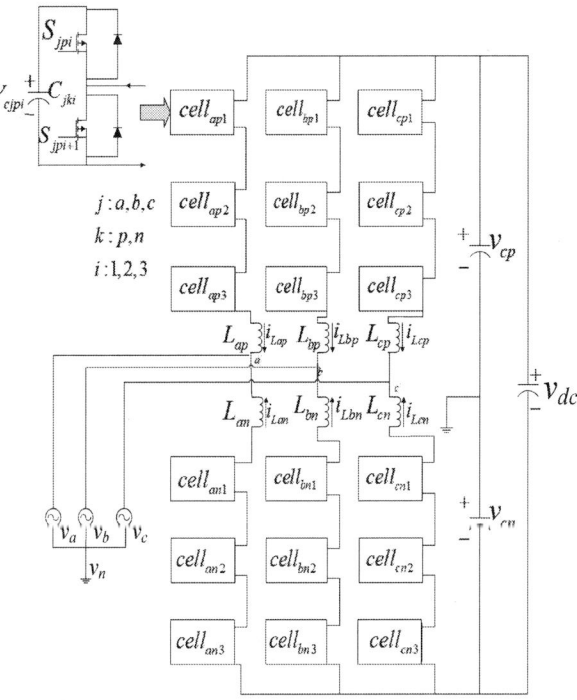

Fig. 1. Circuit configuration of an MMC.

All of the aforementioned control schemes [7]-[9] are typically based on *abc-to-dq* frame transformation and adopting PI controllers, which have limitations of taking into account filter-inductance variation and grid-voltage harmonics. Moreover, cell voltages cannot be expressed in terms of system parameters explicitly; thus, cell-voltage limitations and cell-ripple voltage reduction have not been discussed yet.

In this study, a direct digital control based on D-Σ process [10] is proposed for cell-voltage, dc-bus voltage regulation and balancing, and arm current tracking of an MMC. With the direct digital control approach, the duty-ratio control laws can be derived directly and the inverter can accommodate wide filter-inductance variation to achieve fast current tracking and voltage regulation. Furthermore, cell-voltage expression can be derived based on the proposed direct digital control to determine either dynamics of cell-voltage regulation or amplitude of cell-voltage ripple. Experimental and simulated results from a 50 kW MMC as shown in Fig. 1 are presented to confirm the discussed control scheme.

II. CONTROL LAWS FOR MMC

An MMC is shown in Fig. 1, where j represents phases, k is upper and lower arms, and i denotes the numbering of cell modules. In order to derive control laws directly, it is necessary to simplify Fig. 1 to single-phase model MMC. More details were presented in [10]. First of all, the control law for inductor current tracking can be derived based on voltage-averaging approach and Kirchhoff's voltage law:

$$D_k = \frac{1}{V_{cell}}(v_{ck} \mp v_{ac} \mp \frac{L_k \cdot \Delta i_{Lk}}{T_s}), \qquad (1)$$

$-$: upper arm $+$: lower arm

where V_{cell} is the total voltage of upper or lower arm cells and Δi_{Lk} is the inductor current variation over one switching cycle. For regulation of cell-capacitor voltage, the charge-balance principle is used. The following relationship can be obtained for cell voltage regulation:

$$\Delta d_{ki} = \pm \frac{C_{ki}}{T_s} \cdot \frac{(v_{cellref} - v_{cki})}{i_{Lk}}. \qquad (2)$$

$+$: upper arm $-$: lower arm

A control block diagram for current tracking is shown in Fig. 2. As can be seen from (1), control law duty ratio (D_k) is related to feedback signals, including inductor current, grid voltage, cell voltage, dc-bus voltage and inductance values. Thus, it can accommodate all of the discussed parameter variations. Additionally, it is worth noting that cell voltage regulation control signal is included in inductor current tracking control, resulting in large voltage variations in both switching cycle and different power levels. Thus, it will be a challenge in designing either an appropriate compensator or a limiter for Δd_{ki} to achieve fast dynamic response. Note that to keep

tight arm current tracking, Δd_{ki} should be properly limited, usually less than 10%.

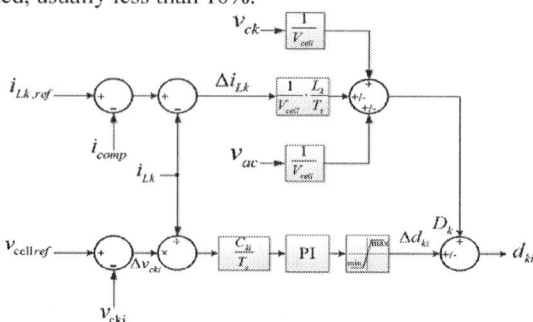

Fig. 2. Control block diagram for current tracking.

III. LIMITATIONS OF CELL-VOLTAGE RANGE

For sake of analyzing cell-voltage regulation dynamics, cell-voltage ripple and limitations of cell-voltage range, it needs to obtain cell-voltage expression explicitly, as shown below

$$
\begin{aligned}
v_{cpi}(t) = & \left(\frac{v_{cp}I_0}{2V_{cell}\omega C_{pi}} - \frac{v_{ac}I_{dc}}{V_{cell}\omega C_{pi}} - \frac{LI_0^2}{16V_{cell}\omega C_{pi}}\right) \\
& + \left(\frac{v_{cp}I_{dc}}{V_{cell}C_{pi}} - \frac{v_{ac}I_0}{4V_{cell}C_{pi}}\right)t \\
& + \left(\frac{v_{ac}I_{dc}}{V_{cell}\omega C_{pi}} - \frac{v_{cp}I_0}{2V_{cell}\omega C_{pi}}\right)\cos(\omega t) \\
& + \left(\frac{-LI_{dc}I_0}{2V_{cell}C_{pi}}\right)\sin(\omega t) \\
& + \left(\frac{LI_0^2}{16V_{cell}C_{pi}}\right)\cos(2\omega t) \\
& + \left(\frac{v_{ac}I_0}{8V_{cell}\omega C_{pi}}\right)\sin(2\omega t).
\end{aligned}
\qquad (3)
$$

In (3), the coefficient of the term "t" can be tuned by varying I_o and I_{dc} to decide dynamics of cell-voltage regulation. The coefficients of the terms "$cos(\omega t), sin(\omega t), cos(2\omega t), cos(2\omega t)$" represent the ripple of cell voltage. Cell-voltage ripple can be reduced effectively by injecting zero-mean current based on (3). In general, each cell voltage should be regulated to the value V_{dc}/N, in which V_{dc} represents dc-bus voltage, and N is the number of cell modules in the upper or lower arm. However, there is no need to regulate cell voltage to V_{dc}/N. It is possible to regulate cell voltage to be less than v_{dc}/N, or even can be different values. The effective range of cell voltage can be derived based on (1) and (2), and the duty-ratio range between 0 and 1. Thus, the following inequality can be derived

$$0 < \frac{1}{V_{cell}}(v_{ck} \mp v_{ac} \mp \frac{L_k \cdot \Delta i_{Lk}}{T_s}) \pm \frac{C_{ki}}{T_s} \cdot \frac{(v_{cellref} - v_{cki})}{i_{Lk}} < 1. \qquad (4)$$

The term "Δd_{ki}" in Fig. 2 for cell-voltage regulation can be neglected since the system is operated in the steady state. The upper arm is used for range calculation

1913

in this study. After rearranging (4), an inequality can be obtained below

$$V_{cell} > v_{cp} - v_{ac,\max} \sin(\omega t) - L \cdot \omega \cdot \frac{I_o}{2} \cdot \cos(\omega t), \qquad (5)$$

where I_o is the amplitude of grid current. Equation (5) shows the lower bound of cell voltage. The upper bound of cell voltage can be determined based on the amplitude of inductor current ripple, so that the cell voltage range can be rewritten as

$$V_{cell} > v_{cp} - v_{ac,\max} \sin(\omega t) - L \cdot \omega \cdot \frac{I_o}{2} \cdot \cos(\omega t), \qquad (6)$$

$V_{cell} <$ voltage ripple due to I_{MAX}.

In simulation waveforms, a single-phase MMC under 10 kW is used to verify the analysis of (6). The testing conditions are that N equals to 3 and dc-bus voltage is 2000 V. Simulation cases are shown in Table I.

Table I. Simulation cases

Case I	$V_{cell} = 2000$ V	$V_{cki} = 667$ V
Case II	$V_{cell} = 1500$ V	$V_{cki} = 500$ V
Case III	$V_{cell} = 1200$ V	$V_{cki} = 400$ V

Fig. 3(a) shows simulation waveform of cell voltage under **Case I**. The cell voltage can be regulated to 667 V. Fig. 3(b) shows simulation waveform of duty ratio. As can be seen, the duty ratio range is between 0.26 and 0.73.

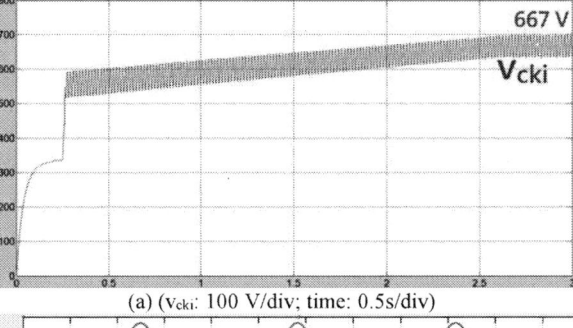

(a) (v_{cki}: 100 V/div; time: 0.5s/div)

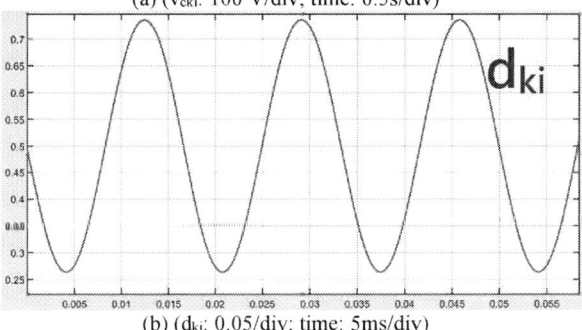

(b) (d_{ki}: 0.05/div; time: 5ms/div)

Fig. 3. Simulated waveforms: (a) cell voltage, and (b) duty-ratio range under **Case I**.

Fig. 4(a) shows simulation waveform of cell voltage under **Case II**. The cell voltage can be regulated to 500 V.

Fig. 4(b) shows simulation waveform of duty ratio. As can be seen, the duty ratio range is between 0.35 and 0.98.

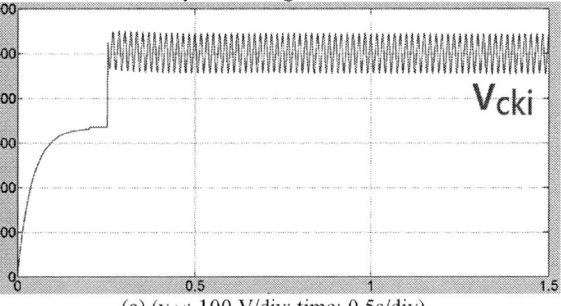

(a) (v_{cki}: 100 V/div; time: 0.5s/div)

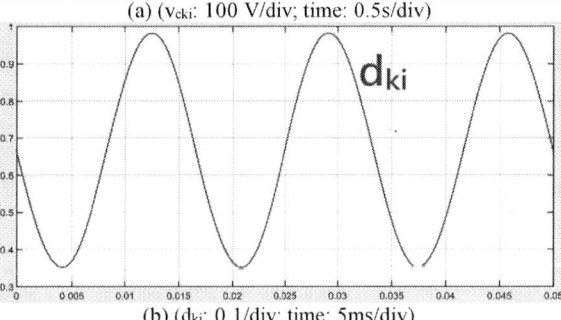

(b) (d_{ki}: 0.1/div; time: 5ms/div)

Fig. 4. Simulated waveforms: (a) cell voltage, and (b) duty-ratio range under **Case II**.

Fig. 5(a) shows simulation waveform of cell voltage under **Case III**. The cell voltage cannot be regulated to 400 V, because duty ratio has exceeded 1, as shown in Fig. 5(b).

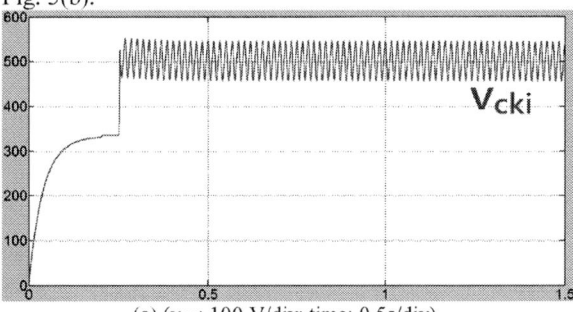

(a) (v_{cki}: 100 V/div; time: 0.5s/div)

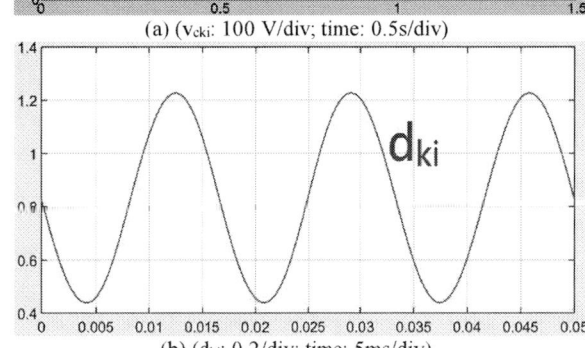

(b) (d_{ki}: 0.2/div; time: 5ms/div)

Fig. 5. Simulated waveforms: (a) cell voltage, and (b) duty-ratio range under **Case III**.

Among the three cases, the duty ratios of case I and case II are less than 1, so that cell voltage can be regulated to 667V and 500 V, respectively. On the contrary, the duty ratio of case III exceeds 1. Thus, cell voltage cannot be regulated to 400 V. With (6), it can select an appropriate cell voltage in accordance with system requirement. Although a lower cell voltage may cause higher inductor current ripple, it can be effective to reduce cell-voltage stress on capacitor. In addition, the voltage stress and switching loss of semiconductor components can be significantly reduced.

IV. EXPERIMENTAL AND SIMULATION RESULTS

Two 50 kW MMCs for recycling power test are shown in Fig. 6 and have been implemented in our lab, in which the Master module regulates dc-bus voltage and the Slave functions as a real power injector to ac grid. The three-phase line voltage is 580 V_{rms}. The MMC dc-bus voltage is ±1,000 V.

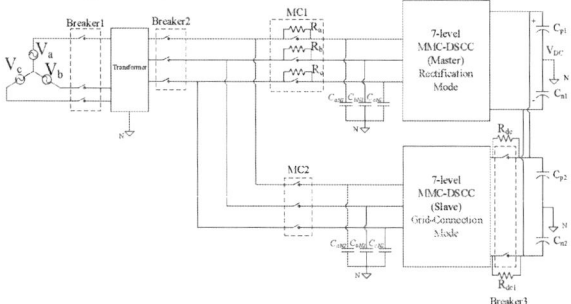

Fig. 6. Two MMCs for recycling power test.

Fig. 7(a) shows measured waveforms of grid currents operated in rectification mode under 50 kW. Fig. 7(b) shows zoom-in waveforms of Fig. 7(a). As can be seen, the grid current can be shaped sinusoidally and THD is 2.5 % under grid voltage distortion (6 %). The most important contribution is that the inductance value varying with arm current over 50 %, as shown in Fig. 8. It has been taken into account for generating lower current ripple with a smaller magnetic core.

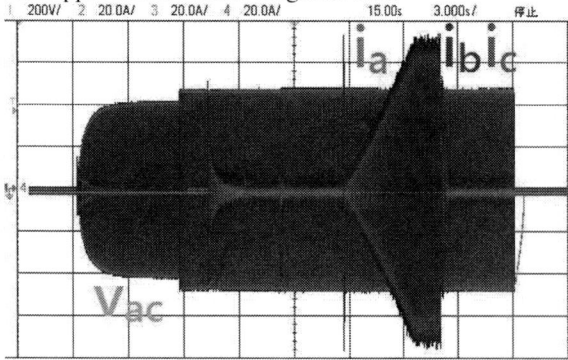

(a) (i_a, i_b, i_c: 20 A/div and v_{ac}: 200 V/div; time: 3s/div)

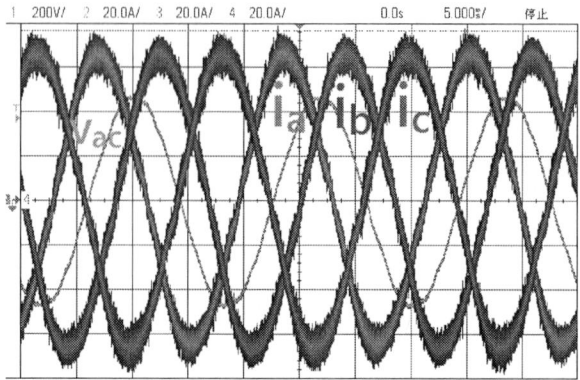

(b) (i_a, i_b, i_c: 20 A/div and v_{ac}: 200 V/div; time: 5ms/div)

Fig. 7. Measured waveforms: (a) grid current, and (b) zoom-in of grid current under 50 kW.

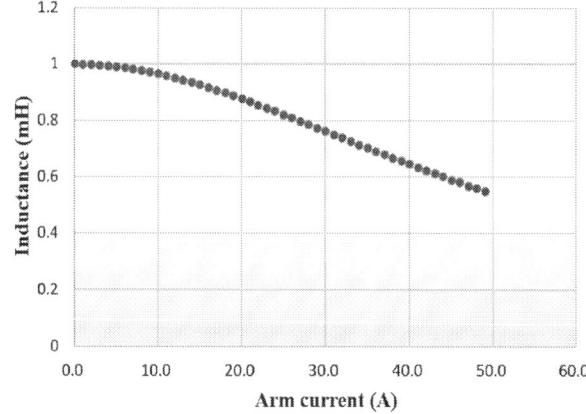

Fig. 8. Measured inductance of an inductor constructed with a Mega flux core (1 mH) corresponding to arm inductor currents 0–49 A.

Fig. 9(a) shows measured waveforms of grid current operated in real-power injection mode under 50 kW. Fig. 9(b) shows zoom-in waveforms of Fig. 9(a). As can be seen, the grid current can be shaped sinusoidally regardless of cell-voltage ripples. Moreover, grid-current (THD = 2.2 %) is low enough when taking into account inductance variation. Experimental results have verified current tracking with the proposed direct digital control.

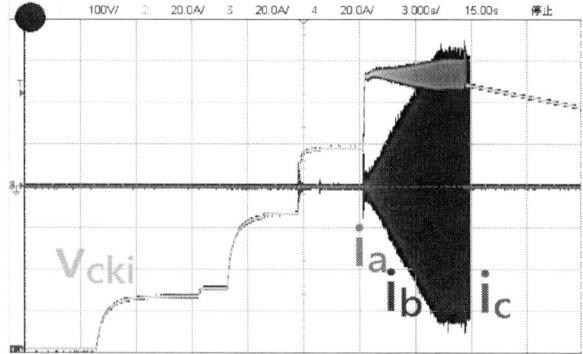

(a) (i_a, i_b, i_c: 20 A/div and v_{cki}: 100 V/div; time: 3s/div)

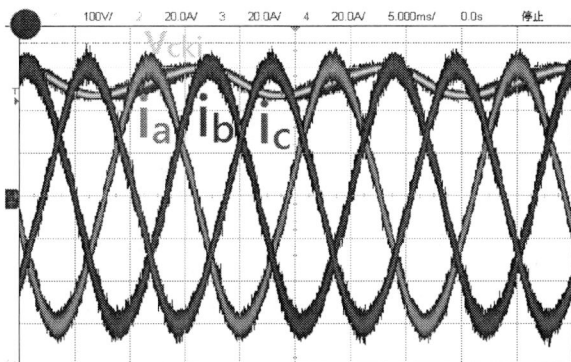

(b) (i_a, i_b, i_c: 20 A/div and v_{cki}: 100 V/div; time: 5ms/div)
Fig. 9. Measured waveforms: (a) grid current, and (b) zoom-in of grid current under 50 kW.

Fig. 10(a) shows measured waveforms under low-voltage ride through (LVRT) where ac voltage sags with 50 %. Fig. 10(b) shows zoom-in waveforms of Fig. 10(a). The grid current can be kept constant with the direct digital control when ac voltage drops 50%. With the coefficient term of "t" in (3), cell voltage can be fast regulated to reference voltage by tuning I_o and I_{dc}.

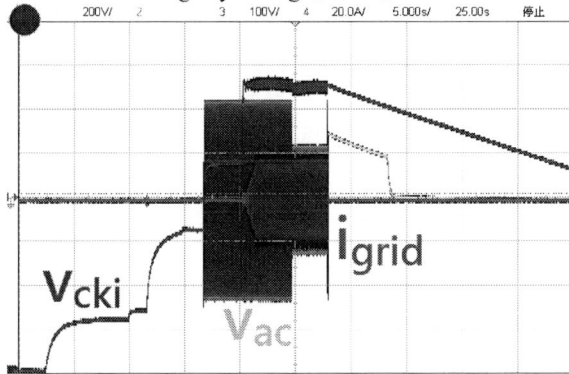

(a) (i_{grid}: 20 A/div, v_{ac}: 200 V/div and v_{cki}: 100 V/div; time: 5s/div)

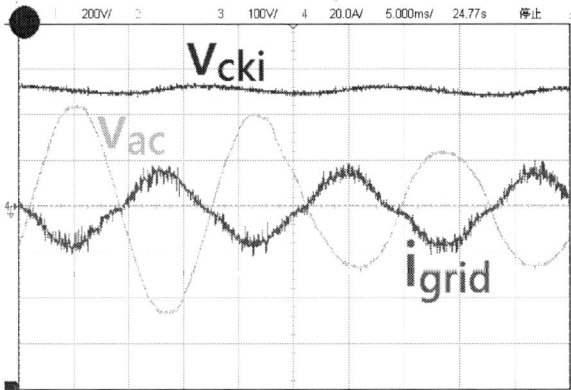

(b) (i_{grid}: 20 A/div, v_{ac}: 200 V/div and v_{cki}: 100 V/div; time: 5ms/div)
Fig. 10. Measured waveforms: (a) LVRT testing, and (b) zoom-in of LVRT testing.

Fig. 11(a) shows simulation waveforms of cell, grid voltages and grid current operated in rectification mode, in which (6) and Fig. 8 have been taken into account. Cell voltage is regulated to 500 V, resulting in higher lower-arm current ripple in Fig. 11(b). Lower-arm current ripple is higher than that of the upper-arm about 1.6 A. Grid current ripple will remain almost identical by considering the inductance variation shown in Fig. 8. With conventional *abc-to-dq* frame transformation method, it is hard to achieve the same current ripple. Most important of all, the grid current can be shaped sinusoidally under high THD (22.36 %) of grid voltage, as shown in the upper plot of Fig. 11(c). The THD of injected current is 3.28 %, as shown in the lower plot of Fig. 11(c).

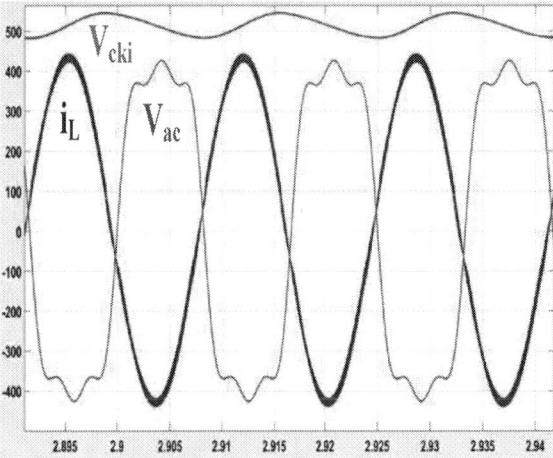

(a) (i_L: 100 A/10*div, V_{cki}: 100V/div and V_{ac}: 100 V/div; time: 5ms/div)

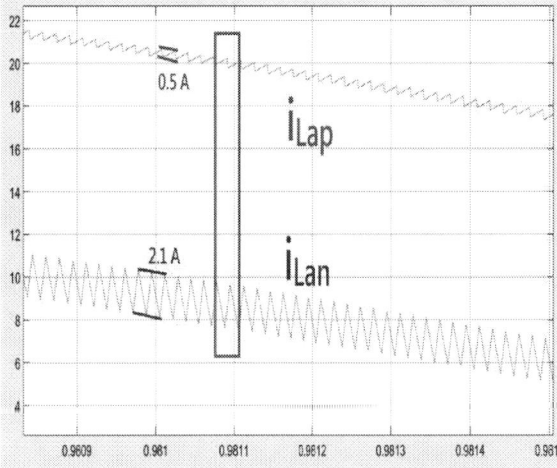

(b) (i_{Lap}: 2 A/div and i_{Lan}: 2 A/div; time: 0.1ms/div)

The 2018 International Power Electronics Conference

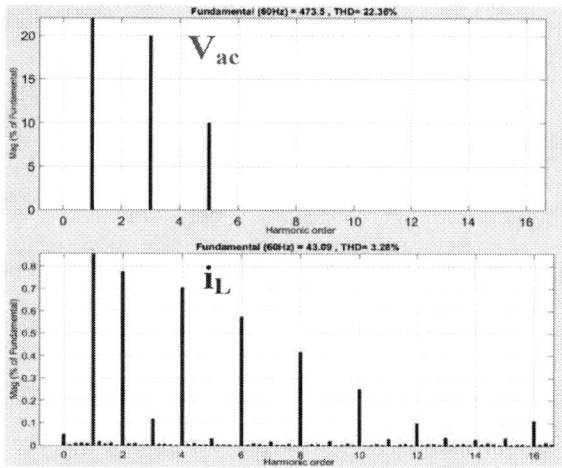

(c) (V$_{ac}$: 5 %/div and iL: 0.1 %/div; Harmonic order: 2/div)
Fig. 11. Simulation waveforms: (a) cell, grid voltages and grid current, (b) upper- and lower-arm inductor current, and (c) harmonic order of grid voltage and injected current.

In order to verify limitations of cell voltage, a single-phase MMC which is consisted of 7 cells (upper arm: 3 cells, lower arm: 4 cells) is shown in Fig. 12. Specifications of Fig. 12 are shown in Table II. On the whole, each cell voltage should be regulated to the value of V_{dc}/N. However, it is not necessary to regulate cell voltage to V_{dc}/N. On the contrary, it can regulate cell voltage to be less or higher than V_{dc}/N, or even can be different values.

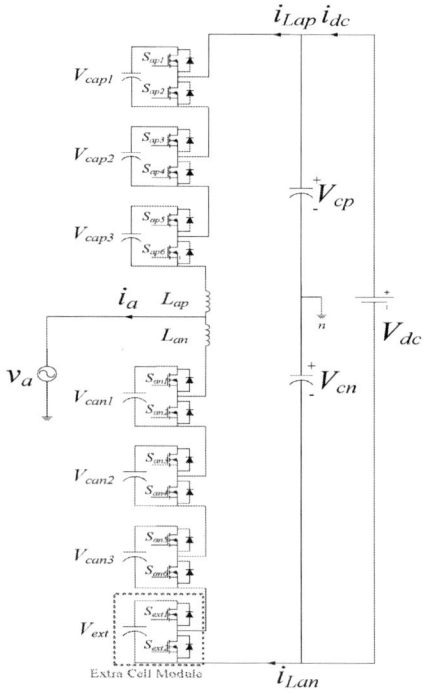

Fig. 12. Circuit configuration of a single-phase MMC.

Table II. Specifications of the single-phase MMC

V_{dc}	760 V
V_a	220 V$_{rms}$
V_{cjki} & V_{ext}	Depending on voltage command
Power rating	3 kW

Fig. 13(a) shows measured waveforms of cell, grid voltages and grid current operated in real-power injection mode, in which the upper- and lower-arm cells are regulated at different voltages. The grid-current ripples in positive-half and negative-half cycles are identical because the rule (V_{dc}/N) is used. Fig. 13(b) shows measured waveforms of cell, grid voltages and grid current, in which the upper- and lower-arm cells are regulated at the same voltage. It verifies that cell voltage can be higher than V_{dc}/N but grid current ripple in positive-half cycle will be larger than that in negative-half cycle.

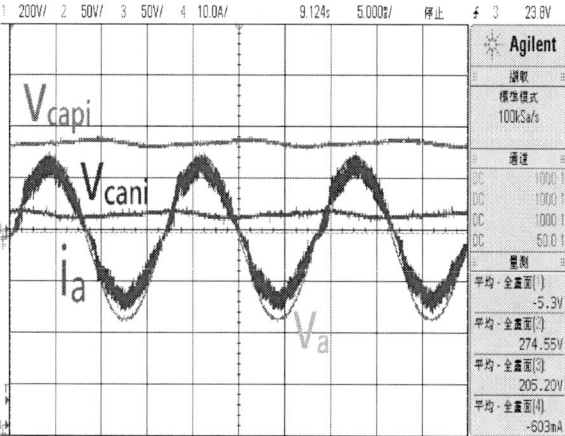

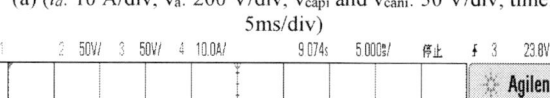

(a) (i_a: 10 A/div, v$_a$: 200 V/div, v$_{capi}$ and v$_{cani}$: 50 V/div; time: 5ms/div)

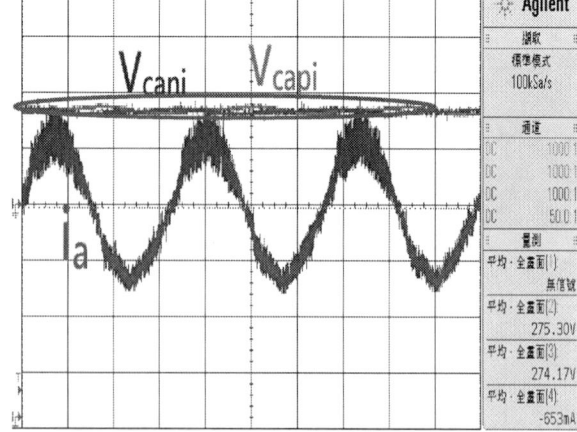

(b) (i_a: 10 A/div, v$_{capi}$ and v$_{cani}$: 50 V/div; time: 5ms/div)
Fig. 13. Measured waveforms: (a) cell, grid voltages and grid current (v$_{capi}$= 275 V, v$_{cani}$= 200 V), and (b) cell voltage and grid current (v$_{capi}$= 275 V, v$_{cani}$= 275 V).

Fig. 14 shows a photo of the single-phase MMC hardware implemented in our lab where the red box designates the extra cell.

Fig. 15 shows a single-cell module prototype hardware which consists of local controller and cell board. The distributed control method is adopted to achieve high reliability and scalability.

Fig. 14. Photo of the single-phase MMC prototype hardware.

Fig. 15. Photo of the single-cell module prototype hardware.

V. CONCLUSIONS

With the proposed direct digital control approach, the duty-ratio control law based on system parameters and feedback signals (inductor current, grid voltage, cell voltage, dc-bus voltage and inductance value) can be derived directly. Moreover, cell-voltage expression can be also derived explicitly to analyze cell-voltage regulation, cell-voltage ripple reduction, and the range of cell voltages. It is worthy of mentioning that cell voltage can be less or higher than V_{dc}/N, or even can be in different values. It can satisfy requirement of many applications and reduce the cost of cell-module capacitor.

REFERENCES

[1] H. Liu, L. Wang, Y. L. Ji, and F. Li, "A novel reversal coupled inductor high-conversion-ratio bidirectional dc-dc converter," *IEEE Transactions on Power Electronics*, vol. PP, no. 99, pp. 1–1, 2017.

[2] M. Wang, Q. Huang, S. Guo, X. Yu, W. Yu, and A. Huang, "Soft-switched modulation techniques for an isolated bi-directional dc-ac converter," *IEEE Transactions on Power Electronics*, vol. PP, no. 99, pp. 1–1, 2017.

[3] M. Aamir and S. Mekhilef, "An online transformerless uninterruptible power supply (UPSTabl) system with a smaller battery bank for low-power applications," *IEEE Transactions on Power Electronics*, vol. 32, no. 1, pp. 233–247, Jan 2017.

[4] C. S. Lam, L. Wang, S. I. Ho, and M. C. Wong, "Adaptive thyristor-controlled LC-hybrid active power filter for reactive power and current harmonics compensation with switching loss reduction," *IEEE Transactions on Power Electronics*, vol. 32, no. 10, pp. 7577–7590, Oct 2017.

[5] J. Li, M. Sumner, J. A. Padilla, and H. Zhang, "Fault signal propagation through the PMSM motor drive systems," *IEEE Transactions on Industry Applications*, vol. 53, no. 3, pp. 2915–2924, May 2017.

[6] H. Akagi, "Classification, terminology, and application of the modular multilevel cascade converter (mmcc)," *IEEE Transactions on Power Electronics*, vol. 26, no. 11, pp. 3119–3130, Nov 2011.

[7] F. Rong, X. Gong and S. Huang, "A Novel Grid-Connected PV System Based on MMC to Get the Maximum Power Under Partial Shading Conditions," in *IEEE Transactions on Power Electronics*, vol. 32, no. 6, pp. 4320-4333, June 2017.

[8] M. Jankovic, A. Costabeber, A. Watson and J. C. Clare, "Arm Balancing Control and Experimental Validation of a Grid Connected MMC with Pulsed DC Load," in IEEE Transactions on Industrial Electronics, vol. PP, no. 99, Jun. 2017, pp. 1-1.

[9] M. Hagiwara, R. Maeda, and H. Akagi, "Control and analysis of the modular multilevel cascade converter based on double-star chopper-cells (MMCC-DSCC)", IEEE Trans. Power Electronics, Vol. 26, no. 6, Jun. 2011, pp 1649-1658.

[10] T.-F. Wu, T.-C. Chou, Z.-C. Guo, C.-S. Wu, and C.-W. Huang, "D-Σ digital control based modular multilevel converter," in *2016 IEEE 8th International Power Electronics and Motion Control Conference (IPEMC-ECCE Asia)*, May 2016, pp. 3501–3507.

Switching Loss Analysis of SiC-MOSFET based on Stray Inductance Scaling

Keiji Wada and Masato Ando

Department of Electrical Electronics Engineering, Tokyo Metropolitan University, Hachioji, Japan
*E-mail: kj-wada@tmu.ac.jp

Abstract—Recently, high-speed switching circuits using SiC power modules have been developed for using the next-generation power converter circuits. The stray inductances inside the converter circuit caused by among DC capacitors, bus bars, and power module package are one of the most critical parameters, which influence not only the over-voltage but also the switching loss. This paper proposes a quantitative design procedure for the switching losses depending the stray inductances. This paper focuses on the effects of stray inductance on both the turn-on and turn-off losses, and the quantitative analyses are explained using a stray inductance scaling method. To verify the design procedure, experimental results rated at 500 V and 160 A are demonstrated using an all-SiC power module.

Keywords—Circuit design, Stray inductance, SiC-MOSFET, Switching loss

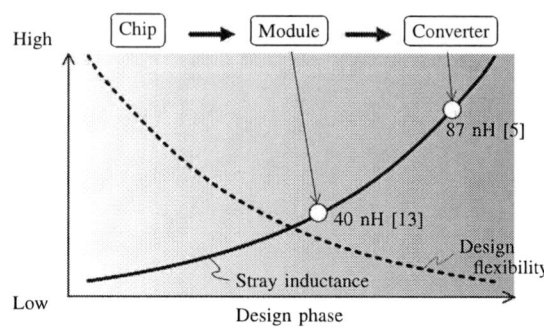

Fig. 1. Relationship between design flexibility and stray inductance.

I. INTRODUCTION

Wide bandgap power devices (SiC and GaN) have been developed to achieve higher-efficiency power converter circuits for traction, automotive, and industrial applications [1]-[5]. It is important to analyze and design the DC-side stray inductances of the power converter circuits used in these power devices, because these stray inductances affect the over-voltages under turn-off operations and the switching losses.

Many studies have demonstrated the relationships between the stray inductances and switching characteristics through experiments and circuit simulations [6]-[11]. In general, the stray inductance depends on the wiring structure between a DC capacitor and the power devices, and their internal structures. A low stray inductance structure has been proposed for reducing the over-voltage during turn-off operations [7]. Several papers have discussed the effects of stray inductances on both the turn-on and turn-off losses [8][9]. These papers show the experimental results of switching losses as variable stray inductances; however, quantitative analyses and design procedures of stray inductances have never been considered in detail. The influence of the stray inductance is especially critical for the switching losses in SiC-MOSFETs, body diodes of SiC-MOSFETs, and SiC-SBDs, because tail currents or recovery currents do not appear in these devices [12], as in Si-IGBTs or Si-PNDs. In addition, the experimental conditions of the stray inductance that exist while making the datasheets of the power devices may differ from that of the actual power converter geometry.

Fig. 1 shows the relationship between the design flexibility and the stray inductance. Power converter circuits consist of power modules, bus bars, gate drivers, and so on; therefore, the stray inductances of power converter become larger than that of a chip or a power module. For example, under experimental conditions, the stray inductance of a power module is set to 40 nH [13]; in contrast, that of a power converter which installs the power module is designed to be 87 nH [5]. Since the switching loss characteristics in the datasheet differ from those of a power converter, it is necessary to confirm the design procedure considering the influence of the stray inductance.

This paper proposes a quantitative design procedure for the switching losses depending the stray inductance, which can be explained using the scaling method for stray inductances. Scaling in this paper implies that the impedance of the stray inductance is represented as a percentage of the rated impedance in power converter circuits. The objective of this scaling method is to clarify the relationship between voltage and current ratings and the acceptable stray inductance. In order to clarify the influence of the stray inductance on the switching losses, the turn-on and turn-off switching waveforms obtained by considering stray inductances are shown. An analytical equation for the switching losses caused by the stray inductance is proposed to separate the switching waveforms. To verify the design procedure, analytical and experimental results are demonstrated using an all-SiC power module rated at 500 V, 160 A. In addition, the analytical result of the switching loss characteristics depending on the stray inductance is shown.

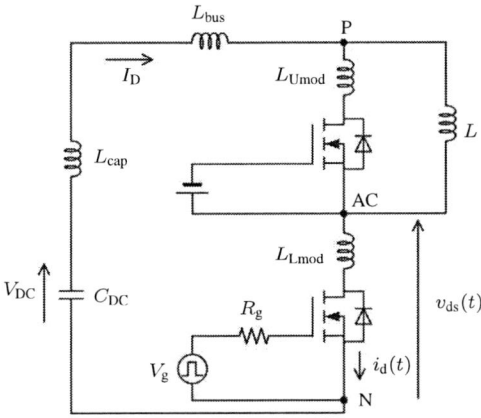

Fig. 2. Circuit configuration of the buck chopper circuit.

II. ANALYSIS OF INFLUENCE ON SWITCHING LOSSES

A. Definition of the scaling method for stray inductance

Fig. 2 shows the circuit configuration for the buck chopper circuit used in the analysis and experiment. In addition, Table I shows the definitions of the circuit parameters of Fig. 2. Electromagnetic force v_L (EMF) is generated during the turn-on and turn-off of the MOSFET, due to the stray inductance. Therefore, the drain-to-source voltage $v_{ds(t)}$ can be calculated by adding the DC voltage V_{DC} and the EMF v_L of the total loop stray inductance L_{loop}, as follows:

$$v_{ds}(t) = V_{DC} - v_L = V_{DC} - L_{loop} \times \frac{di_d(t)}{dt} \quad (1)$$

$$L_{loop} = L_{cap} + L_{bus} + L_{Umod} + L_{Lmod} \quad (2)$$

Then, dt and $di_d(t)$ can be replaced by I_D/δ_t, where δ_t is the switching period and I_D is the output current.

The rated impedance Z_{INV} of this circuit and the impedance Z_L of the stray inductance can be defined in volts per ampere ([V]/[A]) and henries per second ([H]/[s]), respectively.

$$Z_{INV} = \frac{V_{DC}}{I_D} \ [\Omega] \quad (3)$$

$$Z_L = \frac{L_{loop}}{\delta_t} \ [\Omega] \quad (4)$$

Therefore, the impedance ratio α is defined by the following equation [14]:

$$\alpha = \frac{Z_L}{Z_{INV}} = \frac{L_{loop}/\delta_t}{V_{DC}/I_D} = \frac{L_{loop} \times I_D/\delta_t}{V_{DC}} = \frac{v_L}{V_{DC}} \quad (5)$$

The impedance ratio α indicates the proportion of the DC voltage V_{DC} to the EMF v_L, as shown in (5). In addition, the impedance ratios of the turn-on and turn-off sides are defined as α_{on} and α_{off}, respectively.

B. Influence of stray inductance on turn-on loss

Fig. 3 shows a typical turn-on waveform considering the influence of stray inductances. The stray inductances L_{loop} of the broken line and solid line are set to L_1 and

TABLE I. DEFINITION OF CIRCUIT PARAMETERS

Symbol	Feature
L_{cap}	ESL of DC capacitor
L_{bus}	Stray inductance of the bus bar
L_{Umod} and L_{Lmod}	Upper- and lower-side ESLs of power device
R_g	External gate resistance
L	Output inductor
I_D	Output current of the circuit
V_{DC}	DC voltage of the circuit

L_2, respectively. The stray inductance L_2 is larger than L_1 in this paper.

$$L_2 > L_1 \quad (6)$$

Therefore, the stray inductances of L_1 and L_2 indicate the values from the datasheet and the actual power converter, respectively.

When the MOSFET is turned on, the drain current $i_d(t)$ rises from zero to the output current I_D. Then, the drain-to-source voltage $v_{ds}(t)$ falls from the DC voltage V_{DC}, because of the occurrence of EMF v_L. The turn-on losses of L_1 and L_2 during the period t_{on1} are defined as $E_{on1(L1)}$ and $E_{on1(L2)}$, respectively. The turn-on loss $E_{on1(L1)}$ of the stray inductance of L_1 during t_{on1} in Fig. 3 can be calculated using the following equation.

$$
\begin{aligned}
E_{on1(L1)} &= \int_0^{t_{on1}} v_{ds}(t) \times i_d(t) \ dt \\
&= (V_{DC} - v_{L1}) \times \int_0^{t_{on1}} i_d(t) \ dt \\
&= V_{DC} (1 - \alpha_{on1}) \times \int_0^{t_{on1}} i_d(t) \ dt \quad (7)
\end{aligned}
$$

The stray inductance of L_2 generates a larger EMF v_L compared to L_1; hence, the turn-on loss of L_2 decreases. The turn-on loss $E_{on1(L2)}$ of the stray inductance L_2 can be calculated to replace the on-side impedance ratio α_{on1} with α_{on2} as shown the following equation.

$$E_{on1(L2)} = V_{DC} (1 - \alpha_{on2}) \times \int_0^{t_{on1}} i_d(t) \ dt \quad (8)$$

Therefore, the ratio ΔE_{on1} between the turn-on losses $E_{on1(L1)}$ and $E_{on1(L2)}$, which occur during the period t_{on1} in Fig. 3, can be calculated using the on-side impedance ratios α_{on1} and α_{on2} as follows:

$$\Delta E_{on1} = \frac{E_{on1(L2)}}{E_{on1(L1)}} - 1 = \frac{\alpha_{on1} - \alpha_{on2}}{1 - \alpha_{on1}} < 0. \quad (9)$$

Here, α_{on1} and α_{on2} indicate the impedance ratios of L_1 and L_2, respectively.

After the drain current $i_d(t)$ attains the output current I_D, the drain-to-source voltage $v_{ds}(t)$ falls to the on-voltage of the MOSFET. The turn-on loss $E_{on2(L1)}$ of the stray inductance of L_1 during t_{on2} in Fig. 3 can be

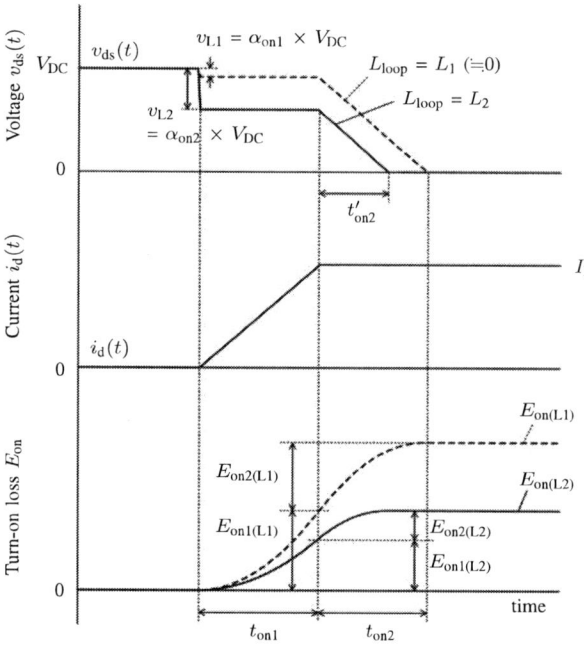

Fig. 3. Analytical turn-on waveforms of the MOSFET including the electromagnetic forces due to stray inductances.

calculated using the following equation.

$$
\begin{aligned}
E_{on2(L1)} &= \int_0^{t_{on2}} v_{ds}(t) \times i_d(t) \, dt \\
&= I_D \times \int_0^{t_{on2}} \frac{(V_{DC} - v_{L1})(t_{on2} - t)}{t_{on2}} \, dt \\
&= \frac{1}{2} V_{DC} (1 - \alpha_{on1}) \times I_D \times t_{on2} \quad (10)
\end{aligned}
$$

The switching speed of the drain-to-source voltage $v_{ds}(t)$ depends not on the stray inductance but on the gate-to-drain capacitance of the MOSFET [15]. In this analysis, the switching speed of the drain-to-source voltage $v_{ds}(t)$ is approximated to a linear function. In addition, the slope of the linear function for the stray inductance of L_1 and L_2 are set to the same value. Therefore, the stray inductance causes a ratio ΔE_{on2} in the turn-on loss, which can be calculated using the following equation.

$$
\begin{aligned}
\Delta E_{on2} &= \frac{E_{on2(L2)}}{E_{on2(L1)}} - 1 = \frac{(1 - \alpha_{on2}) \times t'_{on2}}{(1 - \alpha_{on1}) \times t_{on2}} - 1 \\
&= \left(\frac{1 - \alpha_{on2}}{1 - \alpha_{on1}} \right)^2 - 1 < 0 \quad (11)
\end{aligned}
$$

C. Influence of stray inductance on turn-off loss

Fig. 4 shows a typical turn-off waveform considering the influence of stray inductances. The stray inductances L_{loop} of the broken line and solid line are set to L_1 and L_2, respectively.

When the MOSFET is turned off, the drain-to-source voltage $v_{ds}(t)$ rises from zero to the DC voltage V_{DC}.

During this period t_{off1}, the drain current $i_d(t)$ does not change. The turn-off losses of L_1 and L_2 during the period t_{off1} are defined as $E_{off1(L1)}$ and $E_{off1(L2)}$, respectively. The turn-off loss $E_{off1(L1)}$ of the stray inductance of L_1 during t_{off1} in Fig. 4 can be calculated using the following equation.

$$
\begin{aligned}
E_{off1(L1)} &= \int_0^{t_{off1}} v_{ds}(t) \times i_d(t) \, dt \\
&= I_D \times \int_0^{t_{off1}} v_{ds}(t) \, dt \quad (12)
\end{aligned}
$$

The switching speed of the drain-to-source voltage $v_{ds}(t)$ during t_{off1} are set to the same value for the stray inductance of L_1 and L_2. Then, the ratio ΔE_{off1} between the turn-off losses $E_{off1(L1)}$ and $E_{off1(L2)}$, which occur during the period t_{off1} in Fig. 4, can be calculated using the on-side impedance ratios α_{off1} and α_{off2} as follows:

$$
\Delta E_{off1} = \frac{E_{off1(L2)}}{E_{off1(L1)}} - 1 = 0 \quad (13)
$$

After that, the drain current $i_d(t)$ falls from the output current I_D to zero. Then, the drain-to-source voltage $v_{ds}(t)$ exceeds the DC voltage V_{DC}, because of the occurrence of EMF v_L from (1). In this analysis, the slope of the drain current $i_d(t)$ is approximated as a linear function. The turn-off loss $E_{off2(L1)}$ of the stray inductance of L_1 during t_{off2} in Fig. 4 can be calculated using the following equation.

$$
\begin{aligned}
E_{off2(L1)} &= \int_0^{t_{off2}} v_{ds}(t) \times i_d(t) \, dt \\
&= (V_{DC} + v_{L1}) \times \int_0^{t_{off2}} i_d(t) \, dt \\
&= V_{DC} (1 + \alpha_{off1}) \times \int_0^{t_{off2}} i_d(t) \, dt \quad (14)
\end{aligned}
$$

The stray inductance of L_2 generates a larger EMF v_L compared to L_1; hence, the turn-off loss of L_2 increases. Therefore, the ratio ΔE_{off2} between the turn-off losses $E_{off1(L1)}$ and $E_{off1(L2)}$, which occur during the period t_{off2} in Fig. 4, can be calculated using the off-side impedance ratios α_{off1} and α_{off2} as follows:

$$
\Delta E_{off2} = \frac{E_{off2(L2)}}{E_{off2(L1)}} - 1 = \frac{\alpha_{off2} - \alpha_{off1}}{1 + \alpha_{off1}} > 0 \quad (15)
$$

III. EXPERIMENTAL VERIFICATION OF THE PROPOSED PROCEDURE

A. Experimental setup

The experiments are conducted using an SiC-MOSFET power module (CAS100H12AM1, Cree). Table II shows the parameters of the stray inductances and impedance ratios. The DC capacitor (PEH200YX4470M, KEMET) and power module are connected using a laminated bus bar. In addition, an external wire is used to insert the Rogowski coil for measuring the switching waveform. The voltage probe for measuring the drain-to-source voltage $v_{ds}(t)$ is connected between the terminal AC and N as shown in Fig. 2. Although the EMF of the

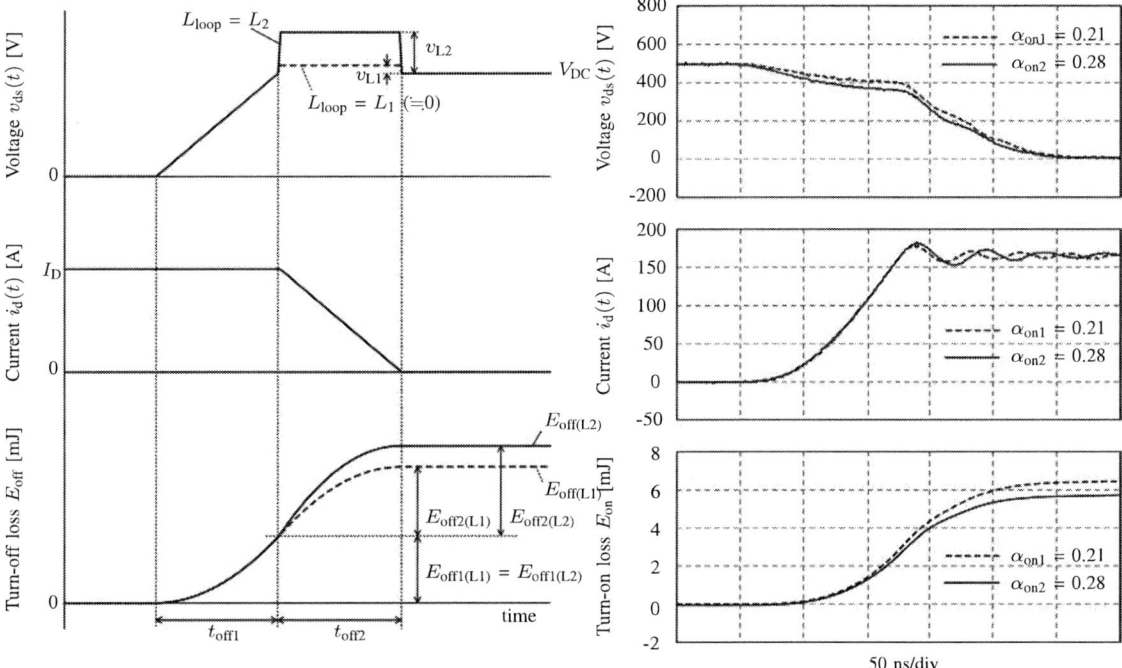

Fig. 4. Analytical turn-off waveforms of the MOSFET including the electromagnetic forces due to stray inductances.

Fig. 5. Experimental turn-on waveforms of different stray inductances when $I_D = 160$ A.

TABLE II. DETAILS OF THE STRAY INDUCTANCES AND IMPEDANCE RATIOS

Symbol	Type I	Type II
L_{cap}	5.7 nH	17 nH
L_{Umod}	10 nH	10 nH
L_{bus}	18.4 nH	18.4 nH
L_{loop}	34.1 nH	45.4 nH
α_{on}	0.21	0.32
α_{off}	0.32	0.43

TABLE III. ANALYTICAL AND EXPERIMENTAL RESULTS OF THE TURN-ON LOSS RATIO

	ΔE_{on1}	ΔE_{on2}
Experiment	−8.3%	−18.0%
Analysis	−9.0%	−17.2%

lower-side stray inductance L_{Lmod} exists inside the power module, the EMF of L_{Lmod} can not be measured at the voltage probe point.

The DC voltage V_{DC} is set to 500 V, and the external gate resistance is 5 Ω. The purpose of this experiment is to clarify the influence of the stray inductance on the turn-on and turn-off losses. Therefore, two types of stray inductance are applied in this experiment: one is 34.1 nH and the other is 45.4 nH. The on-side impedance ratios α_{on} are calculated to be 0.21 and 0.28, respectively, when the turn-on $di_d(t)/dt$ is set to 3.1 kA/μs. In addition, the off-side impedance ratios α_{off} are calculated to be 0.32 and 0.43, respectively, when the turn-off $di_d(t)/dt$ is set to 4.7 kA/μs.

B. Experimental result for turn-on loss

Fig. 5 shows the turn-on waveforms of the SiC-MOSFET, when the output current I_D is set to 160 A. The dashed and solid lines indicate the on-side impedance ratios α_{on} of 0.21 and 0.28, respectively. The switching speed $di_d(t)/dt$ of the drain current reaches 3.1 kA/μs for both the on-side impedance ratios.

Table III shows a comparison of the turn-on loss ratio ΔE_{on} from the experimental and analytical results. The turn-on loss is separated into E_{on1} and E_{on2}, as shown in Fig. 3. It is clear that E_{on1} and E_{on2} decrease with increase in the stray inductance. For example, E_{on1} is decreased from 2.95 mJ to 2.70 mJ, which is equal to a reduction of 8.3%. Moreover, the ratio ΔE_{on1} for the turn-on loss can be calculated to be −9.0%, from (9) using these on-side impedance ratios. Therefore, it is confirmed that the proposed method corresponds well with the experimental results. In addition, the ratio ΔE_{on2} from the analytical and experimental results corresponds well with each other.

C. Experimental result for turn-off loss

Fig. 6 shows the turn-off waveform of the SiC-MOSFET, when the output current I_D is set to 160 A. The dashed and solid lines indicate the off-side impedance ratios α_{off} of 0.32 and 0.43, respectively. The switching speed of the drain current $di_d(t)/dt$ reaches 4.7 kA/μs for both the off-side impedance ratios.

Table. IV shows a comparison of the turn-off loss ratio ΔE_{off} from the experimental and analytical results. The turn-off loss is separated into E_{off1} and E_{off2}, as shown in Fig. 4. The switching speed of the drain-to-

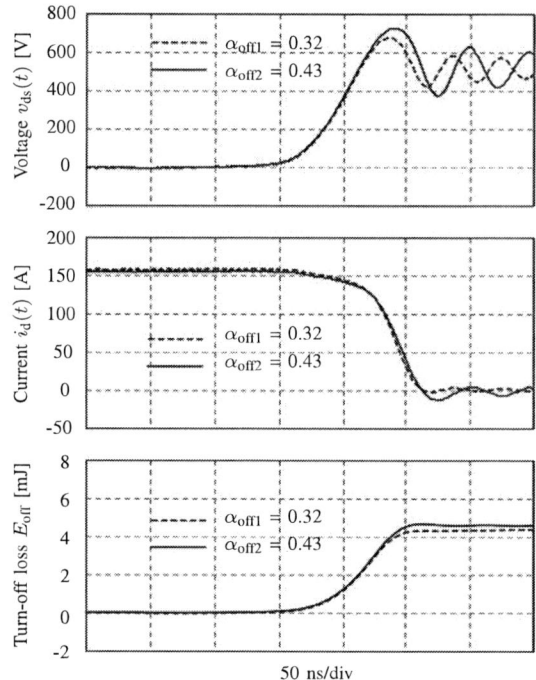

Fig. 6. Turn-off waveform for different stray inductances when $I_D =$ 160 A.

TABLE IV. ANALYTICAL AND EXPERIMENTAL RESULTS OF THE TURN-OFF LOSS RATIO

	ΔE_{off1}	ΔE_{off2}
Experiment	−0.4%	+9.8%
Analysis	0%	+8.1%

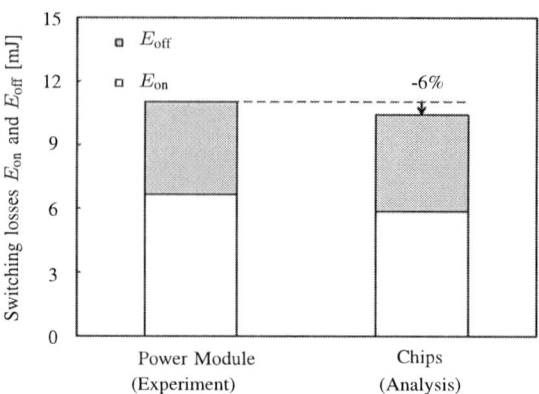

Fig. 7. Comparison of switching losses ratio for the power module and the chips.

source voltage $v_{\text{ds}}(t)$ during t_{off1} have the same value of 14.6 kV/μs for both the off-side impedance ratios of 0.32 and 0.43. Therefore, the turn-off loss of the off-side impedance ratios of 0.32 and 0.43 are 1.87 mJ and 1.86 mJ, respectively. This result corresponds to the analytical result as shown in (13).

In addition, it is clear that E_{off2} increases with increasing the stray inductance, because of the occurrence of higher EMF v_{L}. The E_{off2} increased from 2.52 mJ to 2.76 mJ, and the increasing ratio is 9.8%. Moreover, the ratio ΔE_{off2} in the turn-off loss can be calculated to be +8.1%, from (15) using these off-side impedance ratios. Therefore, it is confirmed that the proposed method corresponds well with the experimental results.

IV. INFLUENCE OF STRAY INDUCTANCE ON SWITCHING LOSS

A. Switching loss evaluation of SiC-MOSFET chips inside a power module

The switching losses of the SiC-MOSFET chips differ from the measured values of the power module, because an EMF v_{L} is generated by the stray inductance existing inside the power module. Then, the EMF v_{L} of the lower-side stray inductance L_{Lmod} inside the power module,

which is equal to 10 nH, is measured around the SiC-MOSFET chips.

Fig. 7 shows the switching loss comparisons between the power module and the chips. In this analysis, the DC voltage V_{DC} and output current I_{D} are set to 500 V and 160 A, respectively. In addition, the on-side impedance ratio α_{on} of the power module is set to 0.21, which indicates a total loop stray inductance of 34.1 nH. The on-side impedance ratio α_{on} of the chips is 0.28 considering the stray inductance L_{Lmod} inside the power module. Then, the turn-on loss E_{on} of the chip decreases by 12.0%, compared to that of the power module. On the other hand, the turn-off loss E_{off} of the chips increase by 4.1% compared to that of the power module, because the EMF v_{L} becomes larger. As a result, the switching loss of the chips can be reduced by 6%, compared to the value measured from the power module.

B. Switching loss minimization

This section discusses an optimal design from the viewpoint of the switching loss of the SiC-MOSFET. In this case, the gate resistances of the turn-on and turn-off sides are set to the same value. The external gate resistance R_{g}, DC voltage V_{DC}, and output current I_{D} are set to 10 Ω, 500 V, and 160 A, respectively. In the case of increasing stray inductance, the decrement of the turn-on loss exceeds the increment of the turn-off loss. Therefore, the switching loss can be reduced to design the stray inductance.

Fig. 8 shows the relationship between the stray inductance and the switching loss. The switching loss is defined as the sum of the turn-on and turn-off losses.

$$E_{\text{sw}} = E_{\text{on}} + E_{\text{off}} \quad (16)$$

This relationship is calculated by the experimental turn-on and turn-off losses of L_{loop} = 34.1 nH and (9), (11), (13), and (15). In the case of minimization of the stray inductance, which implies considering only the stray inductance of the power modules (20 nH), the switching loss increases by 8.6% compared to the stray inductance L_{loop} of 34.1 nH. In contrast, the greater the

1923

The 2018 International Power Electronics Conference

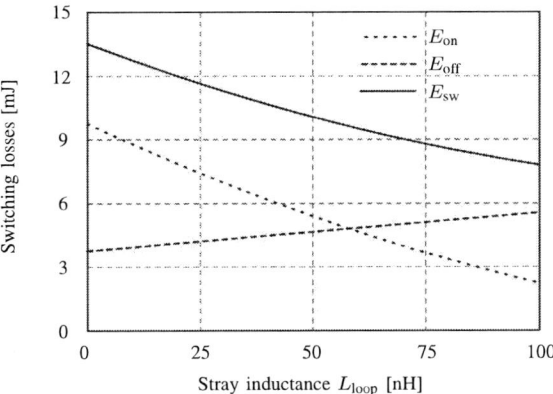

Fig. 8. Relationship between stray inductance and switching loss

stray inductance, the smaller the switching loss can be realized. For example, the tolerable stray inductance is defined to achieve an over-voltage below 70% of the voltage rating of the power modules. The tolerable stray inductance can be calculated using (1).

$$L_{\text{loop}} = \frac{1200 \text{ V} \times 70\% - 500 \text{ V}}{4.7 \text{ kA}/\mu s} = 74 \text{ nH} \qquad (17)$$

In the case of the stray inductance L_{loop} of 74.0 nH, the switching loss can be reduced by 20% compared to that of 34.1 nH. As a result, the switching loss can be calculated and minimized.

V. CONCLUSION

A scaling method for designing the stray inductance, considering its influence on the switching loss, was presented in this paper. The quantitative analyses and experimental results of the turn-on and turn-off losses were presented. The experiments were conducted using a SiC-MOSFET power module, for verifying the proposed method. In addition, the switching loss characteristics of the SiC-MOSFET chips and a design example for the minimization of switching losses using the scaling method were shown. Since the switching loss characteristics of the chips differ from those of the power module, the design accuracy of the cooling system can be improved by using the proposed procedure. The proposed method can be applied in future power converter circuits, which utilize SiC power devices.

REFERENCES

[1] D. Han, J. Noppakunkajorn, and B. Sarlioglu: "Comprehensive Efficiency, Weight, and Volume Comparison of SiC- and Si-Based Bidirectional DC-DC Converters for Hybrid Electric Vehicles" *IEEE Trans. on Vehicular Technology*, Vol. 63, No. 7, pp. 3001-3010 (2014)

[2] S. Shun and T. Shimizu: "A Study on Efficiency Improvement of High-frequency Current Output Inverter Based on Immittance Conversion Elemen", *IEEJ Journal of Ind. Appl.*, Vol. 4, No. 3, pp. 220-226 (2015)

[3] K. Aoyama, N. Motoi, Y. Tsuruta, and A. Kawamura: "High Efficiency Energy Conversion System for Decreaces in Electric Vehicle Battery Terminal Voltage", *IEEJ Journal of Ind. Appl.*, Vol. 5, No. 1, pp. 12-19 (2015)

[4] A. L-Masich, H. V-Blavi, J. M. B-Moncusf, and L. M-Salamero: "A High-Voltage SiC-Based Boost PFC for LED Applications", *IEEE Trans. on Power Electron.*, Vol. 31, No. 2, pp. 1633-1642 (2016)

[5] J. Fabre, P. Ladoux, and M. Piton: "Characterization and Implementation of Dual-SiC MOSFET Modules for Future Use in Traction Converters", *IEEE Trans. on Power Electron.*, Vol. 30, No. 8, pp. 4079-4090 (2015)

[6] K. Wang, X. Yang, H. Li, H. Ma, X. Zeng, and W. Chen, "An Analytical Switching Process Model of Low-Voltage eGaN HEMTs for Loss Calculation", *IEEE Trans. on Power Electron.*, Vol. 31, No. 1, pp. 635-647 (2016)

[7] M. C. Caponet, F. Profumo, R. W. De Doncker, and A. Tenconi: "Low Stray Inductance Bus Bar Design and Construction for Good EMC Performance in Power Electronic Circuits", *IEEE Trans. on Power Electron.*, Vol. 17, No. 2, pp. 225-231 (2002)

[8] J. Wang, H. S-h. Chung, and R. T-h. Li: "Characterization and Experimental Assessment of Effects of Parasitic Elements on MOSFET Switching Performance", *IEEE Trans. on Power Electron.*, Vol. 28, No. 1, pp. 573-590 (2013)

[9] Z. Chen, D. Boroyevich, and R. Burgos: "Experimental Parametric Study of the Parasitic Inductance Influence on MOSFET Switching Behavior", in *Proc. International Power Electronics Conf. (IPEC)*, pp. 164-169 (2010)

[10] J. Noppakunkajorn, D. Han, and B. Sarlioglu: "Analysis of High-Speed PCB With SiC Devices by Investigating Turn-OFF Overvoltage and Interconnection Inductance Influence", *IEEE Trans. on Power Electron.*, Vol. 1, No. 2, pp. 118-125 (2015)

[11] Y. Xiao, H. Shah, T. P. Chow, and R. J. Gutmann: "Analytical Modeling and Experimental Evaluation of Interconnect Parasitic Inductance on MOSFET Switching Characteristic", in *Proc. IEEE Applied Power Electronics Conf. and Exposition (APEC)*, Vol. 1, pp. 516-521 (2004)

[12] S. Jandi, O. Alatise, R. Bonyadi, P. Alexakis, C. A. Fisher, J. A. O. Gonzalez, L. Ran, and P. Mawby: "An Analysis of the Switching Performance and Robustness of Power MOSFETs Body Diodes: A Technology Evaluation", *IEEE Trans. on Power Electron.*, vol. 30, no. 5, pp. 2383-2394 (2015)

[13] Rohm, Appl. Note, "SiC Power Devices and Modules" (2014)

[14] M. Ando and K. Wada: "Design of Acceptable Stray Inductance based on Scaling Method for Power Electronics Circuits", *IEEE J. Emerg. Sel. Topics Power Electron.*, 2016.

[15] N. Mohan, T. M. Undeland, and W. P. Robbins: POWER ELECTRONICS Converters, Applications and Design SECOND EDITION, *JOHN WILEY & SONS, INC*, pp. 583-587 (1995)

Modeling and Optimization of Displacement Windings for Transformers in Dual Active Bridge Converters

Zhan Shen[*], Yanfeng Shen[*], Zian Qin[†], and Huai Wang[*]
zhs@et.aau.dk, yaf@et.aau.dk, Z.Qin-2@tudelft.nl, hwa@et.aau.dk
[*]Department of Energy Technology, Aalborg University, Aalborg, Denmark
[†]Department of Electrical Sustainable Energy, Delft University of Technology, Delft, Netherlands

Abstract— The transformer in the dual active bridge converter (DAB) is the key element which provides galvanic insulation and voltage conversion. The parasitic parameters, including winding capacitance, ac resistance, and leakage inductance, are the primary considerations in its winding design. Without proper consideration of those parameters could result in issues on current ringing, high power loss, and overheating. In this paper, a comprehensive study is devoted to those parameters. A winding design method is presented by taking all those parameters into consideration. Special attention is paid to the impact of displacement winding, which is quite often in the manufacture and especially in prototype design phase. Both the normal and displacement winding will be studied and compared, with analytical, simulation, and experimental methods. Through comparison, additional coefficients are introduced to the simple analytical equations so that they could also be applied for displacement windings. Several considerations are given to control those parameters within a reasonable range in the design and manufacture phase. Finally, the analysis and design method are verified by finite element method and the experimental results on a 120 kHz prototype, and can be extended to other high-frequency magnetic designs.

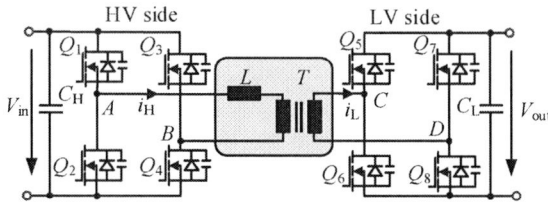

Fig. 1. A dual active bridge converter (DAB).

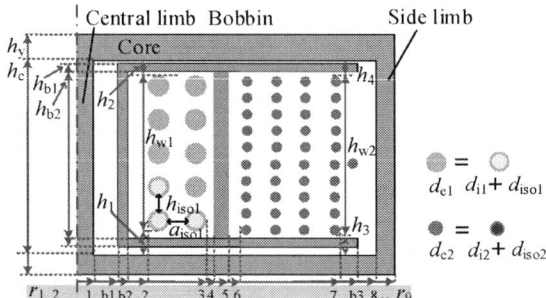

Fig. 2. Transformer cross-section and parameter definitions.

I. INTRODUCTION

The parasitic parameters of the transformer winding, e.g. capacitances, ac resistance, and leakage inductance, have significant impact on the performance of the dual active bridge converter (DAB) [1–3]. A high winding capacitance could cause severe current ringing and capacitive switching losses [4–6]. The efficiency of the transformer mainly depends on the core and resistive losses, the latter is directly related the efforts to minimizing the ac resistance [7, 8]. A certain leakage inductance is essential to the required power transfer and current shaping [1, 9]. Thus, a well predicted and optimized winding structure is important for the transformer design.

In this paper, a winding design method for the transformer is proposed considering capacitances, ac resistance and leakage inductance. The impact of the difference between the normal and displacement winding is investigated through the comparison of analytical, finite element method and experimental results. The comparison results can help engineers to use simple analytical model for displacement winding with modification coefficients, and to choose winding configurations from perspective of parasitic parameters control. Then followed by a pre-design of leakage inductance, the winding capacitance and ac resistance are optimized in a two-dimensional plane through Pareto optimization. Finally, the transformer for a 120 kHz DAB converter is designed to verify the proposed method.

II. IMPACT OF DISPLACE ORDER WINDING

The position of the winding in magnetics is difficult to fix in manufacture. Initially, the winding is designed as Fig 3. (a, b). With the tolerance errors, the winding is less likely to keep organized, and turns to the structure in Fig 3. (c, d). The displacement becomes serious with the increase of the

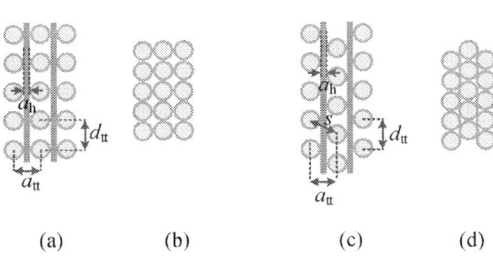

(a)　　　(b)　　　(c)　　　(d)

Fig. 3. Different winding placement positions (a) orthogonal winding, (b) compact orthogonal winding, (c) orthocyclic winding, (d) compact orthocyclic winding. The former two are defined as normal windings, and the latter two are defined as displacement windings.

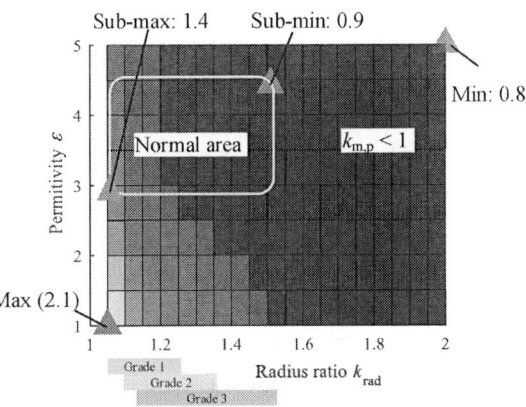

Fig. 4. The relation between radius ratio, permittivity, and displacement capacitance ratio

TABLE I
RADIUS AND DISPLACEMENT CAPACITANCE RATIO OF DIFFERENT WIRE GRADES

Grade	Radius ratio k_{rad}	Displacement capacitance ratio $k_{\text{m,p}}$
1	$1.05 \sim 1.24$	$1.48 \sim 0.99$
2	$1.09 \sim 1.37$	$1.28 \sim 0.94$
3	$1.13 \sim 1.54$	$1.19 \sim 0.89$

number of turns and layers, especially in high voltage ratio transformer. Moreover, sometimes in the compact winding design, the compact orthogonal (b) and compact orthocyclic windings (d), are applied on purpose to reduce the cost of layer insulation to achieve high window utilization factor and high power density. In the aspect of winding width (in a_{tt} direction), the width increment of the normal winding is d_e, while the orghocyclic winding is $\frac{\sqrt{3}}{2}d_e$, which is 13.4% less and thereby leads to a smaller volume of the winding. Besides, the impact of the displacement to various parameters is detailed discussed below.

A. Impact on winding capacitance

The cylindrical and parallel plate capacitance models $C_{\text{l,p}}$ are two most basic models for analytical calculation, the former is given as:

$$C_{\text{l,c}} = \varepsilon_0 \varepsilon_i \frac{2\pi h}{\ln(1 + d/r)}, \quad (1)$$

and the later is given as:

$$C_{\text{l,p}} = \varepsilon_0 \varepsilon_i \frac{2\pi h (r + d/2)}{d}, \quad (2)$$

where ε_0 and ε_i are vacuum permittivity and relative permittivity of insulation material, separately. They could also be used to model the capacitance between the winding layers by applying $d = d_{\text{eff}}$ as the effective insulation distance for orthogonal winding [10] (c.f. Fig 3 (a)):

$$d_{\text{eff}} = d_e + h - 2.3 \cdot r_i + 0.26 \cdot d_{\text{tt}}. \quad (3)$$

The displacement could introduce a large distribution of capacitance values [11]. The electric field of orthocyclic windings is different from the orthogonal winding, and the layer capacitance for those is given in [12]:

$$C_{\text{l,m}} = t\varepsilon_0 \frac{4\varepsilon_r \arctan\left[\frac{(\sqrt{3}-1)(2\varepsilon_r + k_{\text{ln}})}{(\sqrt{3}+1)\sqrt{k_{\text{ln}}(2\varepsilon_r + k_{\text{ln}})}}\right]}{\sqrt{2\varepsilon_r k_{\text{ln}} + k_{\text{ln}}^2}} \quad (4)$$

where $k_{\text{ln}} = \ln k_{\text{rad}}$, $k_{\text{rad}} = d_e/d_i$ is the radius ratio, t is the number of turns per layer.

The displacement capacitance ratio is defined for the comparison of the orthogonal and orthocyclic winding:

$$k_{\text{m,p}} = \frac{C_{\text{l,m}}}{C_{\text{l,p}}} \quad (5)$$

The relationship between $k_{\text{m,p}}$ and k_{rad} is illustrated in Fig. 4. The maximum value occurs at the low permittivity and radius ratio, while the minimum value vise verse. The upper right corner covered by purple color is those situations with $k_{\text{m,p}}$ smaller than 1. To have a further impression of the comparison, the enameled copper wire from Elektrisola is chosen to calculate the radius ratio range [13]. The diameter is from 0.01 mm to 0.5 mm according to IEC 60317. Normally, ε of insulation material is between 3 and 4.5. Combing those two boundaries, we could obtain a normal area in Fig. 4. The color bar under the axis illustrates the detailed radius ratio distribution of different grades. The detailed result of different displacement capacitance ratio $k_{\text{m,p}}$ is given in Table I. For large grade and thicker insulation, the displacement helps decrease the parasitic capacitance. For lower grade and thin insulation which is more common, the increase of capacitance due to displacement is very serious. Thus, an empirical coefficient could be introduced according to Table I when using simple parallel plate capacitance to predict the displacement orthocyclic winding. In order to reduce the capacitance due to the displacement, one solution is to add insulation layer between layers (Fig. 3 (a, c)), another is to use high-grade wires with thicker insulation. Although all of the three grades of wires have the similar lowest displacement capacitance ratio $k_{\text{m,p}}$ close to 0.9, using the high-grade insulation wire could ensure that the capacitance increase ratio due to the displacement could only reach 1.19, which is far below the Grade 1 wire. The above two solutions all focus on increasing

TABLE II
PARAMETERS OF INDUCTORS BUILT FOR COMPARISON

Parameters	Value	Units
Core type	ETD 59/31/22	
Core material	N97	
Bobbin type	B66398	
Pri. Number of layers p	3	
Sec. Number of layers p	3	
Turns per layer t	34	
Conductor diameter d	1	mm
Insulation coating r_{iso}	0.05	mm
Insulation distance a_{iso}	1.2	mm

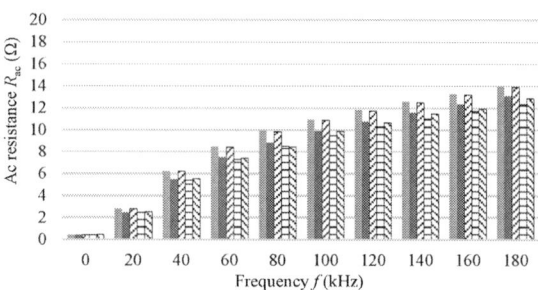

Fig. 6. Comparison of ac resistance R_{ac} in normal and displacement winding.

Fig. 5. The Prototype Td with displacement winding as Fig. 3 (d).

the thickness of the insulation system, and will have additional cost in large amount production.

B. Impact on ac resistance

The classic ac resistance model is derived with the assumption of ordered conductor placement and one-dimensional field distribution[16, 17], Dowell's equation for the winding ac resistance R_{ac} is:

$$R_{\mathrm{ac}} = R_{\mathrm{dc}}F_{\mathrm{r}}, \qquad (6)$$

with

$$F_{\mathrm{r}} = \triangle[\nu_3 + \frac{2}{3}(m^2 - 1)\nu_2] \qquad (7)$$

and

$$\nu_2 = \frac{\sinh\triangle - \sin\triangle}{\cosh\triangle + \cos\triangle}, \quad \nu_3 = \frac{\sinh(2\triangle) + \sin(2\triangle)}{\cosh(2\triangle) - \cos(2\triangle)}, \qquad (8)$$

where R_{dc} and R_{dc} are the winding ac and dc resistance, F_{r} is the winding resistance factor, ν_3, ν_2 are high-frequency coefficients related to skin and proximity effect, and m is the number of layers, the penetration ratio $\triangle$ is calculated by:

$$\triangle = \frac{d_{\mathrm{i}}}{\delta}, \qquad (9)$$

where d_{i} is the conductor diameter and δ is skin depth. With the equation above, there is no typical information of winding position included to predict the ac resistance. The difference

of mean length per turn (MLT) of normal and displacement winding can cause a small difference in their ac resistance. Normally, it takes more wire in the process of wiring for normal winding, so its MLT and ac resistance is slightly larger.

In order to illustrate the impact of displacement winding on the ac resistance and leakage inductance, two transformers Prototypes named Td and Tn are designed, analytically calculated and simulated. The former is with the displacement winding and the later is with the normal winding. They have three layers for primary and secondary, separately. Only Prototype Td is built due to that wiring the normal compact orthogonal winding by hand is very difficult. The specification of the core and bobbin are listed in Table II. The number of turns per layer t is 34 to reduce the impact of fringe effect [15]. Their comparison of ac resistance R_{ac} is given in Fig. 6. The results of the normal winding are given named Anal., FEM, and displacement winding are given named Anal. Dis., FEM Dis. and Meas. Dis.. All of the analytical, finite elemental model and experimental results shows that the displacement has no too much impact on the ac resistance in each frequency. The magnetic field keep straight in most places in both two cases due to their compact structure, thus no too much difference is introduced. The small difference in two cases is due to their small difference in MLT and hand-wiring process.

C. Impact on ac leakage inductance

The ac leakage inductance is to illustrate the energy storage ability of the winding structure [9, 16, 18, 19]:

$$L_\sigma = \mu_0 m_1^2 \frac{l_{\mathrm{w}}}{h_c}(\underbrace{d_{w1}\frac{m_1}{3}F_{L1} + d_{w2}\frac{m_2}{3}F_{L2}}_{\text{frequency dependent}}$$
$$\underbrace{+ d_g + d_{i1}\frac{m_1 - 1}{2m_1} + d_{i2}\frac{m_2 - 1}{2m_2}}_{\text{frequency independent}}). \qquad (10)$$

where m_1, m_2 is the layer number of primary and secondary winding; d_{w1}, d_{w2} is the thickness of two winding; d_g is the interwinding gap thickness; d_{i1}, d_{i2} is the inter-layer gap

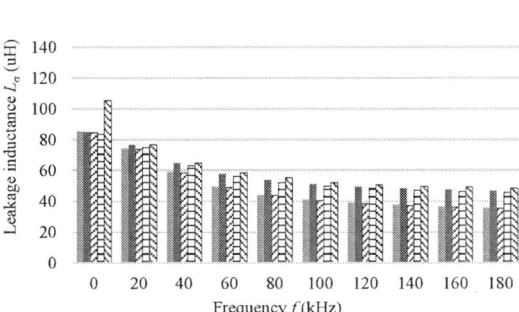

Fig. 7. Comparison of ac leakage inductance L_σ in normal and displacement winding.

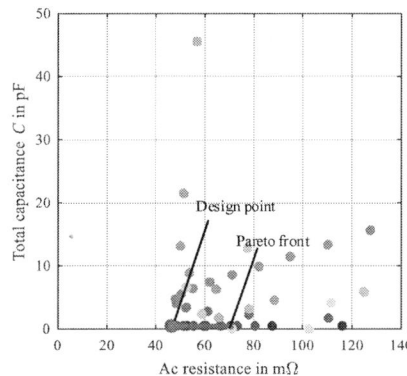

Fig. 8. High voltage winding design Pareto front and design point

Fig. 9. Inductor and transformer in a dual active bridge prototype.

thickness, F_L can be solved by:

$$F_{L1} = \frac{1}{2m_1^2\triangle}[(4m_1^2-1)\varphi_1 - 2(m_1^2-1)\varphi_2]. \quad (11)$$

with

$$\varphi_1 = \frac{\sinh(2\triangle) - \sin(2\triangle)}{\cosh(2\triangle) - \cos(2\triangle)} \quad \varphi_2 = \frac{\sinh\triangle - \sin\triangle}{\cosh\triangle - \cos\triangle} \quad (12)$$

Unlike the ac resistance value which is only dependent on the magnetic field inside the conductor, the inductance value is also dependent on the air or insulation between winding layers (frequency independent part).

With the two prototypes in the last section, a comparison of ac leakage inductance L_{ac} is given in Fig. 7. Analytically, the difference between the normal and displacement winding lies in the difference of the mean length per turn (MLT), the width of each winding, which are both very small. So the difference of the analytical results in both comparisons of the two cases is neglectable. Due to the accurate control of the leakage width between primary and secondary winding, the finite element simulation and experimental results also fit the analytical model very well.

III. PARASITIC PARAMETERS WINDING DESIGN

From the comparison of the parasitic parameters in the last section, the ac resistance and leakage inductance is position-independent in the reasonable range. The winding capacitance, on the other hand, is very sensitive to the position and the displacement. For the sake of convenience in practice, the simple analytical equations (1) and (2) are corrected with the empirical coefficients in Table I to predict the winding capacitance in displacement windings. For each wire grade, the larger displacement capacitance ratio is for very compact wiring and the small ratio is for the loose design. In this way, the use of complex equations can be omitted and the accuracy of the analytical model can be guaranteed.

In [20, 21], analytical methods are proposed to optimize the ac resistance with each number of layers p. It is later developed in [22] to optimize in the $(p, \triangle)$ domain. It is also possible to implement the capacitance in the $(p, \triangle)$ domain with some pre-optimization procedure. A series inductor will

be introduced for the leakage inductance, the initial design of the transformer requires only a minimum controlled leakage inductance 300 nH. The low voltage side capacitance and core capacitance is ignored due to the transfer of turn ratio, and only the high voltage side capacitance is considered. In the preliminary design, the leakage inductance is assumed using the dc leakage inductance formula, and a minimum insulation distance a_{psiso} then can be gained, and it is set as one of the boundary conditions for the design. Further, the winding capacitance is mainly decided by the layer insulation, thus maximum layer insulation a_{iso1} and a_{iso2} are chosen for the each design. Due to the power balance, the window width for the primary and secondary side winding is set equal. The thin insulation Litz wire conductor is chosen as the conductor. Under the assumptions above, the parameters now become a two-dimensional optimization problem in $(p, \triangle)$ domain, and a global optimization could be used with the method in [22] for both ac resistance and winding capacitance.

For verification, a transformer for a 120 kHz dual active bridge prototype is designed, and the design input and output parameters are given in Table III. The relationship between C_w and R_{ac} of the high voltage winding is plotted in Fig. 8 and the Pareto front is with red line. All the points on the Pareto front is with the $p-1$ situation, and is with no much difference in terms of capacitance. So the left side point of the Pareto front is chosen as the design point. The transformer prototype built with the optimal design is illustrated in Fig. 9. Due to the increase of chosen wire diameter, the real prototype has two layers of winding for the high voltage side. Alternatively, a comparison transformer was built with the same specification of high voltage side and the core, but with 0.2 mm foil as the

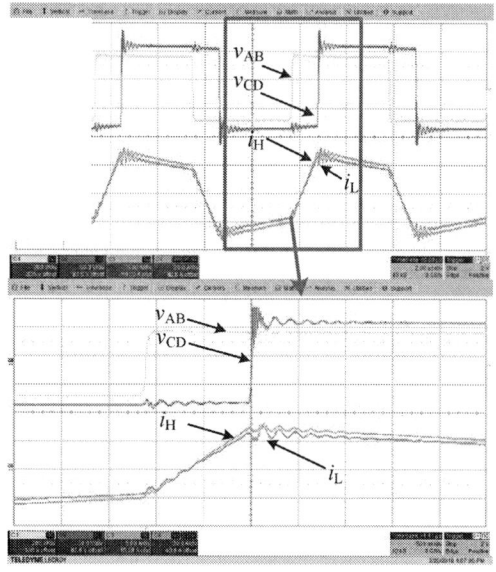

(a) Optimum designed transformer waveform.

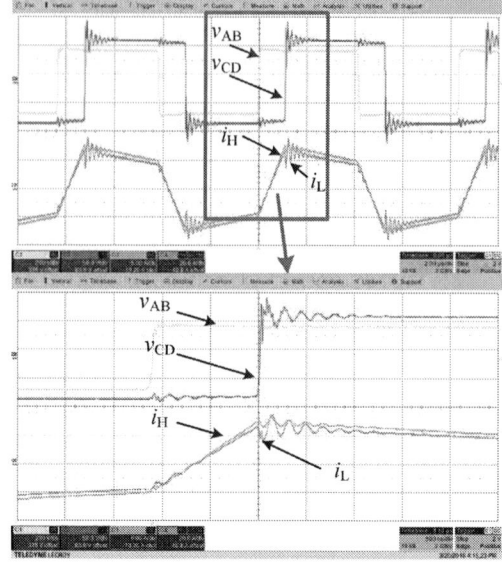

(b) Comparison transformer waveform.

Fig. 10. Optimum design and comparison transformer operation waveform.

TABLE III
TRANSFORMER WINDING DESIGN INPUTS AND OUTPUTS

Input Parameters	Value Pri.	Sec.	Units
Core type	ETD 59/31/22		
Core material	N97		
Bobbin type	B66398		
Number of turns p	21	6	
Number of strands k_{str}	50		
Output Parameters			
Number of layers p	1	1	
Turns in one layer t	21	6	
d_i output	0.15	0.23	mm
d_i chosen	0.20	0.20	mm
R_{ac} @ 100 kHz	118.52		mΩ
Primary Capacitance C	482		nF

low voltage winding. The comparison result is given in Fig. 10. v_{AB}, v_{CD}, i_H and i_L are HV and LV side voltage and current, separately. Both situations are with the same output voltage of 72V and output power of 1.1 kW, while the optimum one has the total system efficiency slightly improved (0.1%) to the comparison. The instrument used for the system efficiency measurement is N4L precision power analyzer. The current ringing of the optimum designed transformer is reduced, which is helpful for efficiency and EMI problems. Further reduction of the transformer capacitance requires a systematic

modeling and optimization of the capacitance network of the transformer [6].

IV. CONCLUSIONS

An improved transformer winding design method is proposed and verified in terms of ac resistance, leakage inductance, and parasitic capacitance. The impact of the winding displacement on those parasitic parameters are concretely discussed. With the comparison and understanding of the field distribution, the simple analytical equations originally for the normal winding can be applied for the displacement windings with the additional empirical coefficients. With a local optimization of capacitance, the proposed design method compresses a three-dimensional optimization to a two-dimensional problem, and achieves Pareto optimization for the ac resistance and winding capacitance. Finally, a transformer winding design case for a DAB converter is presented which leads to reduced current ringing and increased efficiency.

REFERENCES

[1] R. De Doncker, D. Divan, and M. Kheraluwala, "A three-phase soft-switched high-power-density DC/dc converter for high-power applications," vol. 27, no. 1, pp. 63–73, Jan. 1991.

[2] G. Ortiz, C. Gammeter, J. Kolar, *et al.*, "Mixed mosfet-igbt bridge for high-efficient medium-frequency dual-active-bridge converter in solid state transformers," in *Proc. IEEE Wksp. Control Model. Power Electron (COMPEL)*, 2013, pp. 1–8.

[3] S. Engel, M. Stieneker, N. Soltau, *et al.*, "Comparison of the modular multilevel DC converter and the dual-active bridge converter for power conversion in HVDC and MVdc grids," vol. 30, no. 1, pp. 124–137, Jan. 2015.

[4] J. Biela and J. W. Kolar, "Using transformer parasitics for resonant converters-a review of the calculation of the stray capacitance of transformers," vol. 44, no. 1, pp. 223–233, Jan. 2008.

[5] P. Thummala, H. Schneider, Z. Zhang, *et al.*, "Investigation of transformer winding architectures for high-voltage (2.5 kV) capacitor charging and discharging applications," vol. 31, no. 8, pp. 5786–5796, Aug. 2016.

[6] Z. Qin, Z. Shen, and F. Blaabjerg, "Modelling and analysis of the transformer current resonance in dual active bridge converters," in *Proc. IEEE Energy Conversion Congress and Exposition (ECCE)*, Sep. 2017, pp. 4520–4524.

[7] J.-P. Vandelac and P. Ziogas, "A novel approach for minimizing high-frequency transformer copper losses," vol. 3, no. 3, pp. 266–277, Jul. 1988.

[8] C. Sullivan, "Optimal choice for number of strands in a litz-wire transformer winding," vol. 14, no. 2, pp. 283–291, Mar. 1999.

[9] M. A. Bahmani and T. Thiringer, "Accurate evaluation of leakage inductance in high-frequency transformers using an improved frequency-dependent expression," vol. 30, no. 10, pp. 5738–5745, Oct. 2015.

[10] W. Schroder, "Berechnung der eigenschwingungen der doppellagigen langen spule," *Arch. Elektrotechnik*, vol. Band XI, Heft 6, pp. 203–229, 1922.

[11] M. Aghaei and S. Kaboli, "On the effect of disorder on stray capacitance of transformer winding in high-voltage power supplies," vol. 64, no. 5, pp. 3608–3618, May 2017.

[12] A. Massarini and M. K. Kazimierczuk, "Self-capacitance of inductors," vol. 12, no. 4, pp. 671–676, Jul. 1997.

[13] Elektrisola, *Technical data for enamelled copper wire by size acc. to IEC 60317*, http://www.elektrisola.com /fileadmin/webdata/english/Downloads/ELEKTRISOLA _EnCuWire_IEC_Datasheet_eng.pdff, Sep. 2017.

[14] Z. Shen, H. Wang, Y. Shen, *et al.*, "Winding design of series ac inductor for dual active bridge converters," in *Proc. IEEE Appl. Power Electron. Conf. Exposit.*, IEEE, 2018.

[15] F. Robert, P. Mathys, B. Velaerts, *et al.*, "Two-dimensional analysis of the edge effect field and losses in high-frequency transformer foils," vol. 41, no. 8, pp. 2377–2383, 2005.

[16] P. Dowell, "Effects of eddy currents in transformer windings," *Proc. Inst. Electr. Eng.*, vol. 113, no. 8, pp. 1387–1394, Aug. 1966.

[17] J. Ferreira, "Improved analytical modeling of conductive losses in magnetic components," vol. 9, no. 1, pp. 127–131, 1994.

[18] I. Villar, "Multiphysical characterization of medium-frequency power electronic transformers," PhD thesis, École Polytechnique Fédérale de Lausanne, Lausanne, Switzerland, 2010.

[19] M. Mogorovic and D. Dujic, "Medium frequency transformer leakage inductance modeling and experimental verification," in *Proc. IEEE Energy Conversion Congress and Exposition (ECCE)*, Sep. 2017, pp. 419–424.

[20] W. Hurley, E. Gath, and J. Breslin, "Optimizing the AC resistance of multilayer transformer windings with arbitrary current waveforms," vol. 15, no. 2, pp. 369–376, Mar. 2000.

[21] R. Wojda and M. Kazimierczuk, "Analytical optimization of solid–round-wire windings," vol. 60, no. 3, pp. 1033–1041, 2013.

[22] Z. Shen, Z. Li, L. Jin, *et al.*, "An ac resistance optimization method applicable for inductor and transformer windings with full layers and partial layers," in *Proc. IEEE Appl. Power Electron. Conf. Exposit.*, IEEE, 2017, pp. 2542–2548.

The 2018 International Power Electronics Conference

Optimized Selection and Utilization of DC-Link Capacitor in a Single-Phase PV Grid Inverter System

Caspar Collins and Li Ran*
School of Engineering, University of Warwick, Coventry, UK, CV4 7AL
*E-mail: l.ran@warwick.ac.uk

Abstract - Single-phase PV grid inverters usually require large reservoir capacitors on the DC-Link to absorb 2nd order harmonics. This paper shows in a design how to improve the utilization of metalized polypropylene film capacitor (MPFC) specifications in multiple dimensions simultaneously: breakdown voltage rating, capacitance, ripple current capability and warranted reliability performance. As a result, the system size and cost can be reduced, while the overall energy capturing efficiency is guaranteed. Under low-capacitance conditions the DC-Link voltage varies within a fundamental cycle; a boost converter is placed between the PV panel and the DC-Link to provide MPPT (maximum power point tracking). The modulation index of the output inverter can be dynamically varied for waveform compensation. Comparison is made with electrolytic capacitors and it is found that when using an optimized capacitor selection strategy, it is possible to mitigate the increased size and cost implications of film capacitors. Simulation demonstrates the design of a 2.5kW system, while experiment is conducted to prove the design concept.

Keyword — PV inverter, metalized polypropylene film capacitor, cost, size reduction, reliability.

I. INTRODUCTION

With the rise in popularity of domestic-scale PV systems comes increased research interest in small scale single-phase PV inverters. One of the major challenges of inverter design is reliability; a typical inverter has a lifetime of only 10-15 years, which is much shorter than the typical lifetime of PV modules which is approximately 25-30 years. A major cause of this short lifetime is the high failure rate of electrolytic capacitors [1] [2], and so it is desirable to replace these with high reliability metalized polypropylene film capacitors (MPFCs). However, MPFCs are considerably more expensive and are much larger than electrolytic capacitors per unit of capacitance. Through optimized capacitor selection and the utilization of the high current ripple capability of MPFCs it is possible to mitigate the negative cost and size implications by minimizing the required capacitance.

Minimizing the DC-Link capacitance results in a large double line frequency voltage ripple. If not compensated for in the control systems this has negative implications both on the maximum power point tracking (MPPT) and on the harmonic content of the output waveform. A dual-stage topology with boost capability is used; the gain of the boost converter can be adjusted such that the voltage across the PV array remains constant. The PV voltage can then be held at the maximum power point (MPP) despite the DC-

Link voltage ripple; thus, the power output of the PV array is not compromised. Methods of MPPT in cases of large DC-Link ripple voltage have been proposed in literature. These methods aim to use traditional MPPT techniques to vary the DC component of the duty ratio while live measurements of the DC-Link voltage are used to determine the AC component. This means that the MPP tracker does not need to produce the full time-varying duty ratio and so improves the efficiency. The magnitude of the H-bridge control signal can also be varied to successfully remove second order harmonics from the output signal by taking live measurement of the DC-Link voltage. Thus, it is possible to greatly reduce the DC-Link capacitance whilst maintaining output signal quality.

The DC-Link voltage is investigated and an optimized MPFC capacitor selection method is proposed. This is then applied to a 2.5 *kW* inverter design and the volume and cost implications are analyzed. Performance is then shown in simulation. An inverter operating under minimized capacitance is then shown experimentally.

II. DC-LINK VOLTAGE

A voltage ripple arises at the DC-Link as a result of the mismatch between the DC power supplied and the AC power output. Figure 1 shows the power flow in the studied topology.

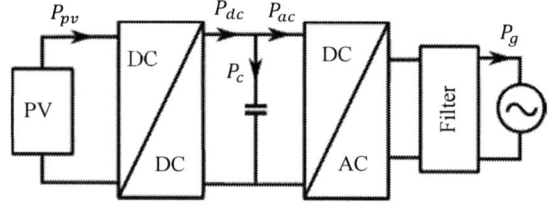

Figure 1. Power flow in a double-stage PV inverter with passive power decoupling.

For a grid current of $I_g = \hat{I}_g \sin(\omega t)$ and grid voltage $V_g = \hat{V}_g \sin(\omega t)$ the power drawn by the inverter is

$$P_{ac} = \hat{I}_g \hat{V}_g \sin^2(\omega t) \quad (1)$$
$$= P_{pv}(1 - \cos(2\omega t)), \quad (2)$$

assuming a system with ideal components and 100% efficiency in steady state. The power flow into the DC-Link capacitor can then be given by

$$P_c = P_{dc} - P_{ac} = P_{pv} \cos(2\omega t), \quad (3)$$

1931

From basic capacitor theory, it can be said that the energy stored in the DC-Link capacitor is given by

$$E = \frac{1}{2}CV_{DC}^2 = \int P_{pv}\cos(2\omega t)\, dt \qquad (4)$$

$$= \frac{P_{pv}}{2\omega}\sin(2\omega t) + const, \qquad (5)$$

where V_{DC} is the DC-Link voltage and C is the DC-Link capacitance. This gives

$$V_{DC} = \sqrt{\overline{V_{DC}}^2 + \frac{P_{pv}}{\omega C}\sin(2\omega t))}, \qquad (6)$$

Series expansion of this gives equation 8[1], where

$$k = \frac{P_{pv}}{2C\omega\overline{V_{DC}}^2}. \qquad (7)$$

For small values of k, k^2 and higher order terms can be ignored giving $\sqrt{\overline{V_{DC}}^2} \cong \overline{V}_{DC}$ and

$$V_{DC} \cong \overline{V}_{DC} + \Delta V_{DC,rip}\sin(2\omega t), \qquad (9)$$

where

$$\Delta V_{DC,rip} = k\overline{V}_{DC} \qquad (10)$$

$$\cong \frac{P_{pv}}{2C\omega\overline{V}_{DC}}. \qquad (11)$$

This estimation of voltage ripple is in agreement with that in literature [3]. However, for larger values of k it may be important to consider the higher order harmonic terms, particularly the 4th harmonic term. In this project, a k value of around 20% is used and so gives a 4th harmonic of around 1% of the mean DC-Link voltage and so does not need to be considered.

III. OPTIMIZED CAPACITOR SELECTION

Optimized capacitor selection is necessary to offset the lower capacitance-volume ratio and higher cost of MPFCs relative to electrolytic capacitors. The high current ripple capability of MPFCs allows for much greater capacitance minimization compared to electrolytic capacitors and so it is possible that given an optimized capacitor selection process the cost and size implications of MPFCs may be comparable to that of electrolytic capacitors. When selecting capacitors there are two main properties that need to be considered: breakdown voltage and capacitance. These two properties should be minimized as they both will have an impact on cost and size of the capacitors. The minimization of these properties is restrained by several limits; in order to utilize the capacitor effectively it is important to operate as close to these limits as possible.

A. DC-Link Ripple Voltage Limit

The DC-Link ripple voltage magnitude is limited by

$$\Delta V_{DC,rip} \lesssim k_{max}\overline{V}_{DC}. \qquad (12)$$

The limit is required to prevent 2nd order harmonics from propagating to the output waveform. k_{max} will depend on the control strategy used and typically varies from 0.1 to 0.25. In order to minimize capacitance, it is necessary to operate as close to this limit as possible. Rearranging

equation 7 gives

$$\overline{V}_{DC} \geq \sqrt{\frac{P}{2\omega C k_{max}}}, \qquad (13)$$

and this can be used to set the reference mean DC-Link voltage. For low capacitance values the required DC-Link voltage will be higher. Thus, there is a trade-off between capacitance minimization and the breakdown voltage requirements of the devices.

B. Breakdown Voltage Limit

The minimization of the breakdown voltage is restricted by two conditions. The first condition is that the breakdown voltage of the capacitor, V_{BD}, must be larger than the maximum instantaneous voltage across the DC-Link:

$$V_{BD} \geq (1 + k_{max})\sqrt{\frac{P}{2\omega C k_{max}}}. \qquad (14)$$

The second condition is given in the data sheet of the capacitor and states that

$$\Delta V_{DC,rip} \leq k_{BD}V_{BD}. \qquad (15)$$

k_{BD} typically ranges from $0.1 - 0.2$ for MPFCs depending on manufacturer, for example see [4] and [5]; however, some manufacturers do not provide a limit at all. To meet this limit either $\Delta V_{DC,rip}$ can be reduced by using a smaller value for k or the breakdown voltage can be increased. However, capacitor cost and size is generally more sensitive to capacitance than breakdown voltage so it is advisable to increase the breakdown voltage and operate at $\Delta V_{DC,rip} = k_{max}\overline{V}_{DC}$, this gives

$$V_{BD} \geq \frac{k_{max}}{k_{BD}}\sqrt{\frac{P}{2\omega C k_{max}}} \qquad (16).$$

The breakdown voltage must be chosen such that both conditions are met.

C. Minimum Voltage Limit

The third limit that must be considered is that the minimum DC-Link voltage must be greater than the peak grid voltage,

$$\overline{V}_{DC} - \Delta V_{DC,rip} \geq \hat{V}_g, \qquad (17)$$

this gives

$$\overline{V}_{DC} \geq \frac{\hat{V}_g}{1 - k_{max}}. \qquad (18)$$

This is the minimum DC-Link voltage that can be used for a given value of k.

D. Current Ripple Limit

The final limit to be considered is the limit that the current ripple rating must exceed the current ripple through the capacitor. The current ripple rating is approximately proportional to the capacitance; ~1 $A/\mu F$ for MPFCs. However, the full utilization of this limit cannot be achieved because the capacitance required to operate close

---

[1]

$$V_{DC} \cong \sqrt{\overline{V_{DC}}^2}\left[\left(1 - \frac{k^2}{4}\right) + \left(1 + \frac{3k^2}{8}\right)k\sin(2\omega t) + \frac{k^2}{4}\cos(4\omega t) - \frac{k^3}{8}\sin(6\omega t)\,...\right] \qquad (8)$$

to the current ripple rating is very small, and so corresponds to a very large breakdown voltage limit, which is not realistically achievable. Therefore, when selecting MPFCs the current ripple rating need not be considered as it does not limit the design.

IV. DC-LINK VOLTAGE RIPPLE COMPENSATION

Operating under capacitance minimized conditions results in a large voltage ripple at the DC-Link. This voltage ripple can reduce MPPT efficiency and distort the output waveform if not effectively compensated for.

A. MPPT

Modifications to traditional MPPT techniques that have been specifically designed for cases of high DC-Link voltage ripple are proposed in [6] [7] [8] [9]. [7] and [9] propose methods of using MPPT to control the DC component of the duty ratio and then use live measurement of the DC-Link voltage to calculate the desired duty ratio. This removes the requirement that the MPP tracker produces an output varying with double line frequency, therefore greatly reducing the dynamic performance requirements of the algorithm. Both methods show an improved efficiency in simulation.

Here an alternative method of implementation of the method proposed in [7] is shown. The desired duty ratio for a boost converter operating with continuous current is

$$D = 1 - \frac{V_{pv}}{V_{DC}}, \tag{19}$$

and the desired output from the MPPT as shown in [7] is

$$D_{mppt} = 1 - \frac{V_{pv}}{\bar{V}_{DC}}. \tag{20}$$

This gives the equation for the modifier which will be multiplied to the output of the MPPT to be

$$\frac{D}{D_{mppt}} = \frac{\bar{V}_{DC}}{V_{DC}} \left(\frac{V_{DC} - V_{pv}}{\bar{V}_{DC} - V_{pv}} \right). \tag{21}$$

This can easily be implemented in MATLAB using measurements of V_{DC} and V_{pv}, and by assuming that $\bar{V}_{DC} = \bar{V}_{DC,ref}$.

B. H-Bridge Control

Inverter control under capacitance minimized conditions for dual-stage topologies has been shown in the literature several times; [10], [11], [12], [13], and [14]. This is not the focus of this paper so is not covered in detail here. Most control systems involve an inner current control loop and an outer voltage control loop. The voltage control loop controls the DC-Link voltage at the desired value and produces the reference for the current control loop. This means that the response of the voltage loop must be relatively slow and so can result in large transients. The implementation of this method has been shown in [14] with a k value of 0.25. A similar method is employed in [10] but uses a ripple voltage estimator to determine the mean DC-Link voltage and operates with a k value of 0.18. [11] uses PR controllers and an indirect voltage control method such that the measurement of the DC-Link voltage is not required, a k value of 0.15 is used. [13] proposes a predictive current control method and is demonstrated with

a k value of 0.17. In order to effectively minimise DC-Link capacitance it is desirable to develop control systems capable of operating under as large of a voltage ripple as possible while maintaining good voltage control under transient conditions.

V. 2.5 kW DC TO 230 V, 50 Hz INVERTER DESIGN

The proposed capacitor selection method will now be applied to a 2.5 kW inverter design. The performance of the design is then shown in simulation. Several capacitors along the breakdown-voltage/capacitance trade-off are chosen and compared to determine the optimum design based on volume and cost. Volume is determined from data sheet values and cost is determined from values quoted by Mouser and Farnell for one component. Voltage spikes across the switching devices can lead to large instantaneous voltages across the devices, up to double the DC-Link voltage [15]. The cost of these devices must be considered during capacitor selection as the voltage ratings required of the SiC MOSFETs and the SiC schottky diode will vary depending on the chosen capacitor.

The cost of components varies greatly depending on market trends and only makes up a small portion of the total cost of an inverter. Therefore, there is a limit to the usefulness of a cost comparison, however it is important to ensure component costs of the various design are reasonably similar in order for a fair comparison to be made.

A. Capacitor Selection

Figure 2 shows the DC-Link capacitor limits for a 2.5 kW inverter in the voltage-capacitance space with $k_{max} = 0.2$ and $k_{BD} = 0.1$.

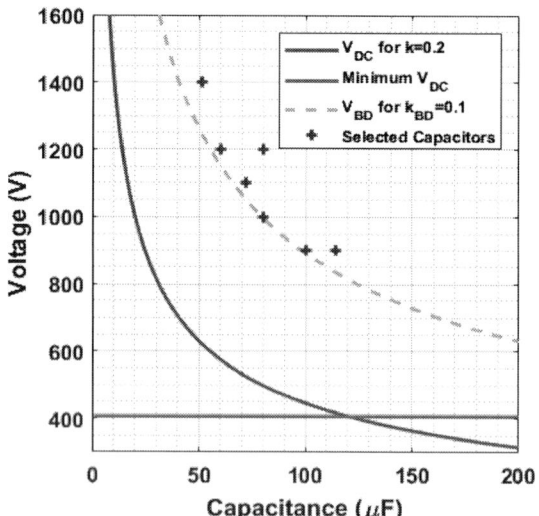

Figure 2. DC-Link Capacitor selection constraints for a 2.5 kW PV inverter with $k_{max} = 0.2$ and $k_{BD} = 0.1$.

Capacitors were chosen as close to the limit as market availability would allow. The capacitors selected for comparison are marked on Figure 2. In cases where several capacitors of the same rating were available the capacitor was chosen based on a balance of volume and cost minimization. Designs with just one capacitor were favored as these keep component count to a minimum and generally have a better capacitance-volume ratio.

Figure 3 shows a comparison of the volume of the chosen capacitors; the $114\mu F/900V$ and the $72\mu F/1100V$ capacitors are the most compact both with volumes of $187\ cm^3$.

Figure 4 shows a comparison of the cost of the selected designs. The three lowest capacitance designs are considerably more expensive than the other designs. This is because the breakdown voltage requirements of the switching devices exceed $1200\ V$ and therefore they require much more expensive components; a $1700\ V$ SiC MOSFET with a sufficient current rating costs £66.71 whereas a $1200\ V$ MOSFET costs just £9.74. High voltage Si IGBTs may be applicable for these designs; however, this will result in a reduced efficiency due to the higher switching losses in Si devices. It can be seen from Figure 4 that the component cost for designs under $1200\ V$ are reasonably similar, so a fair comparison can be made between the designs. The $114\mu F/900V$ capacitor is chosen as the final design as it is both the cheapest and the most compact of the compared designs.

B. Comparison with Electrolytic Capacitor Design

An equivalent circuit which utilizes electrolytic capacitors was designed. This design consists of two $1000\mu F/420V$ electrolytic capacitors in parallel, giving a total capacitance of 2000 μF. The large capacitance is required in order to provide a sufficiently large ripple current rating preventing further capacitance minimization, the rating of the capacitors chosen was 5.6 A.

The total cost of components for this design is £77, which is slightly cheaper than the chosen MPFC design which cost £84. However, the operational lifetime of the electrolytic capacitor is only 12,000 hrs and so will most likely require replacing at least once within the lifetime of the PV module. The MPFC capacitor has an operational lifetime of over 100,000 hrs. The cost of replacement will outweigh the price difference between the different types of capacitors. The total volume of the required electrolytic capacitors is 192 cm^3 which is actually marginally larger than the MPFC capacitor which has a volume of 187 cm^3. Therefore, under these conditions a properly optimized MPFC can both be more compact and of reasonable cost compared to electrolytic capacitors.

VI. SIMULSTION RESULTS

The design was then validated in simulation using Simulink. The inverter is simulated both with and without compensation for the DC-Link voltage ripple so that a comparison can be made and the effectiveness of the compensation techniques can be assessed. In the uncompensated case the duty ratio is constant and the 2nd harmonic is not removed from the output waveform; in the compensated case, the techniques outlined in Section IV are employed.

Figure 5 shows the simulated voltage waveforms for the uncompensated case. The harmonic distortion to the output voltage waveform is evident and is inadequate for grid connection. Furthermore, there is a large voltage ripple across the PV array which results in poor MPPT performance with an average MPPT efficiency of just 92.5%, this is shown in Figure 6.

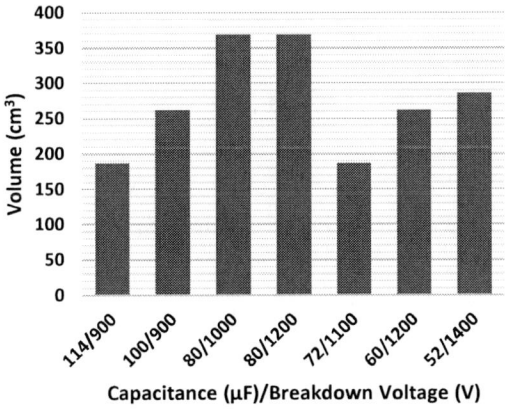

Figure 3. Comparison of DC-Link capacitor volume for the selected designs.

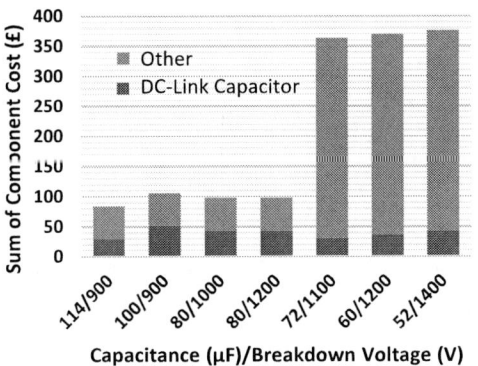

Figure 4. Comparison of DC-Link capacitor cost for the selected designs.

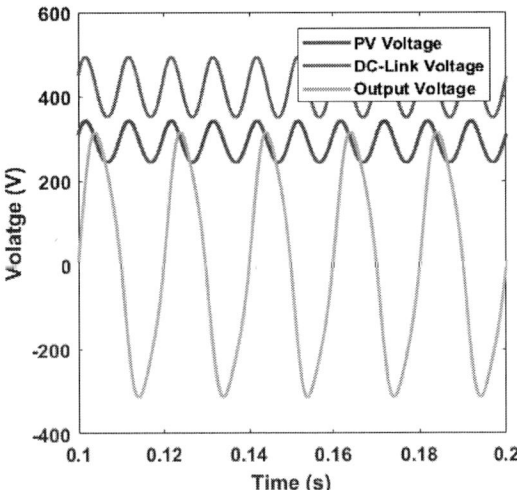

Figure 5. Simulated voltage waveforms for the uncompensated case.

The simulated voltage waveforms for the compensated case are shown in Figure 7. The distortion to the output waveform has successfully been removed and the voltage ripple across the PV array has been considerably reduced. Figure 8 shows the MPPT performance. It is clear that the output of MPPT only tracks the DC component whilst the duty ratio is able to vary such that the PV voltage remains constant despite large variations in the DC-Link voltage. The achieved average MPPT efficiency is 99.7% which is considerably higher than the case of no compensation. This highlights the importance of the boost converter stage in maintaining a high inverter efficiency and outweighs the potential losses associated with the boost converter stage.

The capacitor voltage ripple magnitude is $82\ V$ and average DC-Link voltage is $418\ V$, giving $k = 0.197$ and $k_{BD} = 0.092$. The maximum DC-Link voltage is $498\ V$ which is below the $900\ V$ rating of the capacitor and $1200\ V$ rating of the diode and MOSFET even when considering the potential for voltage spikes at double the DC-Link voltage. The minimum DC-Link voltage is $333\ V$ which is above the peak grid voltage of $325\ V$. The maximum RMS ripple current is $10\ A$ which is lower than the capacitor rating of $31\ A$.

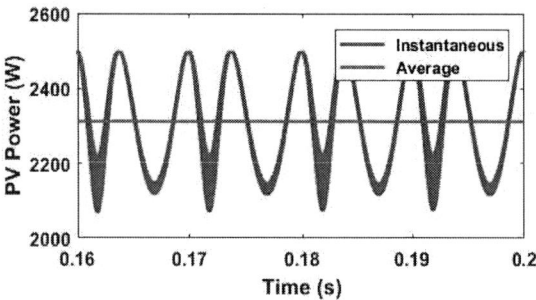

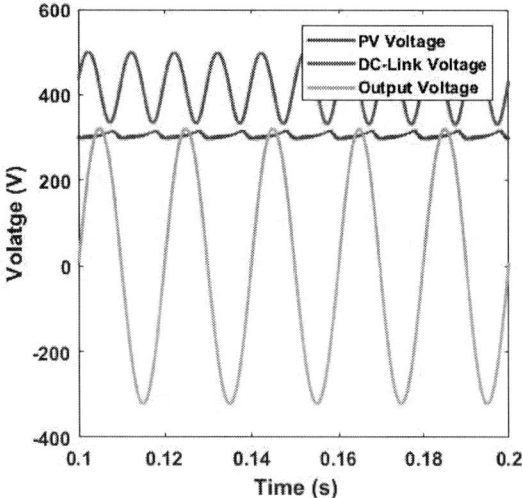

Figure 6. MPPT performance for a constant duty ratio (uncompensated case).

Figure 7. The simulated voltage waveforms for the compensated case.

Therefore, all the design limitations are met by this design. The design makes full use of the breakdown

voltage and voltage ripple ratings of the capacitor; however, only utilizes 32% of the ripple current rating. To achieve greater utilization of the current ripple rating further capacitance minimization is required, this can be achieved by using a larger value for k or by operating at a higher DC-Link voltage.

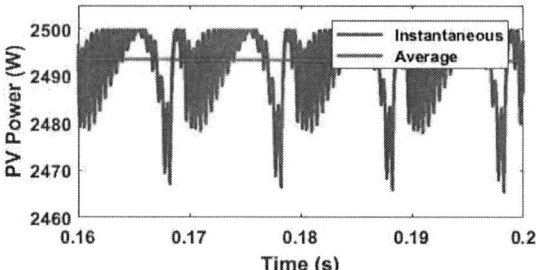

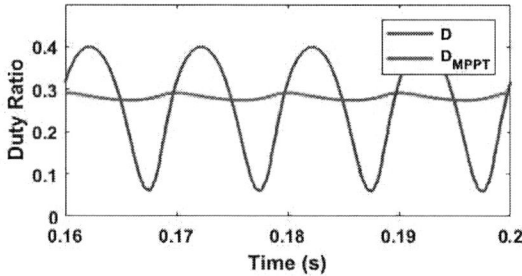

Figure 8. MPPT performance for a variable step P&O MPPT method with duty ratio modifier applied (compensated case).

VII. EXPERIMENTAL RESULTS

The phenomenon was then shown experimentally. An inverter was designed and tested under large ripple voltage conditions and the modulated control techniques were implemented. The designed inverter is shown in Figure 9. The EZPE1B406MTA $40\ \mu F$ film capacitor was used for the DC-Link. The PCB is supplied by a $4\ kW$ SL Series Magna Power DC Power supply and a load resistance of $100\ \Omega$ was used. The control strategy for the inverter was designed in Simulink but compiled and run in dSpace. The power supply was operated in fixed output voltage mode, this means that the MPPT could not be tested. Instead the duty ratio modifier was applied to a fixed duty ratio to prevent 2nd order harmonic from disrupting the input power supply. A constant voltage reference was used for the H-bridge controller and modulation was applied to remove the 2nd order harmonic component from the output.

The inverter was tested with and without compensation for the voltage ripple so that a comparison can be made and the effectiveness of the control techniques could be assessed. A mean DC Link voltage of $100\ V$ was used with a peak-to-peak ripple of $20\ V$ corresponding to a k value of 0.1.

The experimental results for the case of no compensation are shown in Figure 10. It is clear that there is a large second harmonic component in the PV voltage which would result in a reduced MPPT efficiency. Figure 11 shows the experimental results for the case when the

The 2018 International Power Electronics Conference

compensation techniques are applied. The PV voltage waveform is almost completely flat and so the

compensation technique shown in Section IV is successful in reducing the voltage ripple across the PV array.

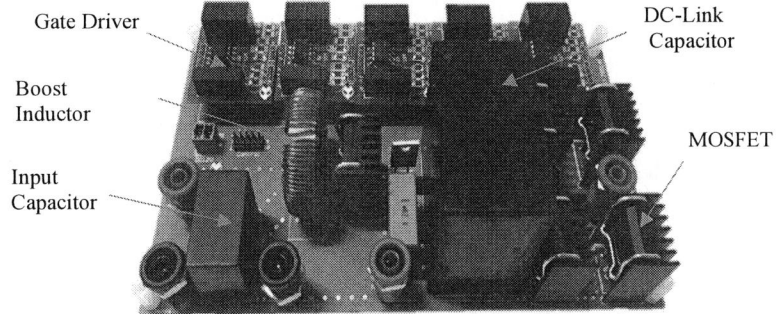

Figure 9. Photograph of the designed inverter used for experimentation.

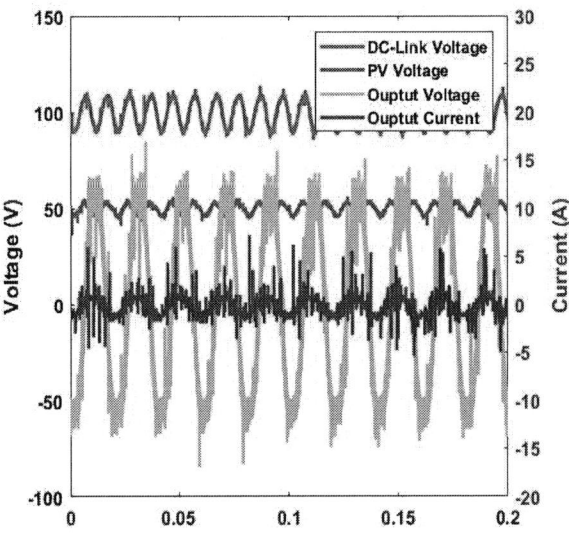

Figure 10. Experimental results for the case of no compensation.

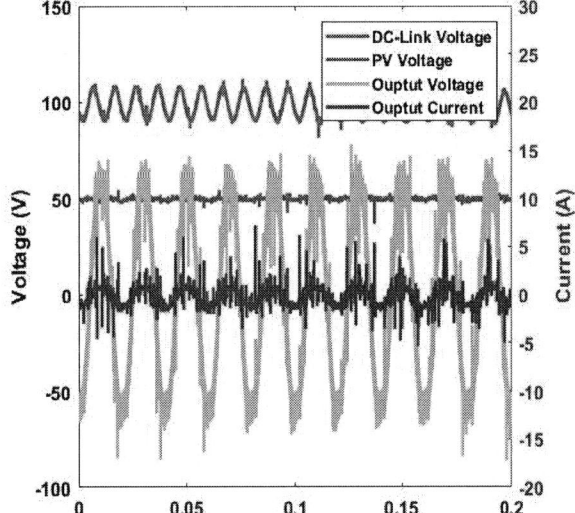

Figure 11. Experimental results for the case when compensation techniques are applied.

VIII. CONCLUSION

The utilization of MPFCs in domestic scale PV inverters has been investigated. This paper proposes an optimized capacitor selection technique to mitigate the cost and size implications of high reliability MPFCs. The proposed technique involves utilizing the high current ripple capability of MPFCs to minimize capacitance much further than is possible with electrolytic capacitors. It has been shown that the optimized capacitor selection method has the potential to fully mitigate the increased size and cost implications of MPFCs, therefore, allowing for the replacement of electrolytic capacitors in domestic-scale PV inverters. This will improve the reliability of such inverters and thus reduce downtime and maintenance costs; thereby reducing the contribution of the inverter to the LCOE for solar power. The operation of an inverter under reduced capacitance conditions has been shown and the control techniques have been verified experimentally.

IX. ACKNOWLEDGEMENT

The authors acknowledge the financial support of the EPSRC UK in project EP/K036327/1: Reconfigurable Distribution Networks.

X. REFERENCES

[1] J. Flicker, R. Kaplar, M. Marinella and J. Granata, "Lifetime Testing of Metallized Thin Film Capacitors for Inverter Applications," in *IEEE 39th Photovoltaic Specialists Conference (PVSC)*, Tampa, FL, 2013.

[2] "$1/W Photovoltaic Systems," U.S Department of Energy,Washington, 2010.

[3] H. Hu, S. Harb, N. Kutkut, I. Bataresh and Z. J. Shen, "A Review of Power Decoupling Techniques for Microinverters With Three Different Decoupling Capacitor Locations in PV Systems," *IEEE Transaction on Power Electronics*, vol. 28, no. 6, June 2013.

[4] Vishay Roederstein, "Metallized Polypropylene Film Capacitor DC-Link Capacitor," MKP1848C DC-Link Datasheet, March 2016.

[5] TDK, "Metallized Polypropylene Film Capacitors (MKP)," B32674 Datasheet, Feb. 2017.

[6] Y. Shi, R. Li, Y. Xue and H. Li, "High-Frequency-Link-Based Grid-Tied PV System With Small DC-Link Capacitor and Low-Frequency Ripple-Free Maximum Power Point Tracking," *IEEE Transaction of Power Electronics,* vol. 31, no. 1, Jan 2016.

[7] R. Kotti and W. Shireen, "An Efficient Robust MPPT Control for Grid-Connected Photovoltaic Systems with Reduced DC Link Capacitance," in *40th Annual Conference of the IEEE Industrial Electronics Society,* Dallas, TX, 2014.

[8] S. K. Yarlagadda and W. Shireen, "A Maximum Power Point Tracking Technique for Single-Phase PV Systems with Reduced DC-Link Capacitor," in *Applied Power Electronics Conference and Exposition (APEC),* Fort Worth, TX, 2014.

[9] J. Harsha Vardhana and B. Krishna Naick, "Optimization of PV MPPT performance and DC link voltage ripples compensation using a duty cycle modifier," in *Recent Advances and Innovations in Engineering (ICRAIE),* Jaipur, 2014.

[10] N. A. Ninad and L. Lopes, "Operation of Single-phase Grid-Connected Inverters with Large DC Bus Voltage Ripple," in *IEEE Canada Electrical Power Conference,* Montreal, 2007.

[11] F. Gao, D. Li, P. C. Loh, Y. Tang and P. Wang, "Indirect dc-link voltage control of two-stage single-phase PV inverter," in *Energy Conversion Congress and Exposition,* San Jose, 2009.

[12] Z. Luo, L. Lopes and H. Sun, "A Multi-Featured Single-Phase Utility Interface with Reduced DC Link Capacitor for Distributed Power Sources," in *IEEE Industrial Electronics Society,* Busan, 2004.

[13] J. Schonberger, "A single phase multi-string PV inverter with minimal bus capacitance," in *Power Electronics and Applications,* Barcelona, 2009.

[14] T. Brekken, N. Bhiwapurkar, M. Rathi, N. Mohan, C. Henze, and L. R. Moumneh, "Utility-Connected Power Converter for Maximizing Power Transfer From a Photovoltaic Source While Drawing Ripple-Free Current," in *IEEE Annual Power Electronics Specialists Conference,* Cairns, 2002.

[15] ABB, "Voltage Ratings of High Power Semiconductors," Application Note 5SYA 2051, Lenzburg, August, 2013.

An Evaluation Circuit for DC-Link Capacitors Used in a High-Power Three-Phase Inverter with Condition Monitoring

Kazunori Hasegawa, Ichiro Omura, and Shin-ichi Nishizawa

Abstract- This paper presents a condition monitoring method of dc-link capacitors used in a high-power three-phase PWM inverter intended for load life testing with an evaluation circuit consisting a low-voltage inverter. The evaluation circuit provides the equivalent ripple current waveform and dc-bias voltage to those of the high-power inverter. Hence, the monitoring method allows the load life testing with the equivalent condition to the high-power inverter at low cost. Experimental results verify that the monitoring method extracts both the ESR and capacitance changes of a capacitor under test.

I. INTRODUCTION

DC-link capacitors including electrolytic capacitors, film capacitors, and ceramic capacitors in power electronic converters are a major constraint on the improvement of power density as well as of reliability [1-3]. They usually have a shorter lifetime than the other components in power electronic converters. Reference [4] provides a survey of reliability in power electronic converters, which describes that capacitors were chosen by 18% of fragile components responsible for converter failure.

Condition monitoring plays an important role in estimating health condition of capacitors [5-9]. The so-called "ripple current tester" is developed, which supplies a sinusoidal ripple current and a dc-bias voltage into a capacitor under test [10]. The ripple current tester is designed for load life testing, application of which includes capacitor quality evaluation and quality assurance [10].

In general, however, characteristics of the capacitors are evaluated by a single sinusoidal current with frequencies of 120 Hz, 1 kHz, and so on [7, 8, 11]. Actual ripple current generated by the converter contains multiple frequency components [12], so that characteristics of the capacitor cannot be exactly evaluated by a single sinusoidal current. The authors of this paper have revealed that the power loss with the square-wave current injection cannot be estimated only by the root-mean-square (RMS) value of the capacitor current due to the nonlinear characteristic of the power loss [13]. The authors have proposed an evaluation circuit for testing dc-link capacitors used in a high-power three-phase inverter [14, 15]. The circuit presents an actual ripple current waveform that is equivalent to that of an existing full-scale inverter, although its power rating is smaller than that of the full-scale inverter by a factor of 10.

This paper presents a condition monitoring method of dc-link capacitors used in a high-power three-phase pulse-width modulated (PWM) inverter intended for load life testing with an evaluation circuit consisting of a low-voltage inverter. The monitoring method allows the load life testing with the equivalent condition to the high-power inverter at low cost.

II. EVALUATION CIRCUIT FOR DC-LINK CAPACITORS

The most effective method to test dc-link capacitors is measuring their characteristics with an existing converter in operation. This method, however, brings a high cost.

Fig. 1 shows an evaluation circuit that employs a small-power-rating low-voltage three-phase inverter [14, 15]. The low-voltage inverter is just used for producing the ripple current, while the high-voltage dc supply provides a dc-bias voltage into a capacitor under test. The bypassing capacitor C_{bypass} is used for circulating the ripple current. The choke inductor L_{LV} and resistor R_{HV} block the ripple current, through which only dc current flows. Hence, their impedances should be much larger than those of capacitors C_{UT} and C_{bypass} [14]. Note that the inductor and the two capacitors do not result in oscillation in practice [15].

The current rating of the low-voltage inverter is full-scale, while voltage rating is downscale. Hence, the circuit operates as a full-scale voltage-rating and full-scale current-rating inverter from the viewpoint of the dc bias voltage and ripple current although the power rating of the low-voltage inverter is greatly smaller than that of the full-scale inverter. This concept is similar to the circuits proposed in [7] and [11] in terms of the combination of a ripple current generator and a dc voltage source, while it produces the same current waveform as that generated by the inverter.

III. CAPACITOR CONDITION MONITORING METHOD

A. How to Extract ESR and Capacitance

Since the ripple current consists of multiple frequency components, it is given by

$$i_{CUT} = \sum_{k=1}^{n} I_{CUT}(\omega_k)\sin(\omega_k t + \varphi_k)$$

(1)

where $I_{CUT}(\omega_k)$ and ϕ_k are the amplitude and initial phase angle for each frequency, respectively. Fig. 2 shows an equivalent circuit of the capacitor under test for this analysis, where R_S is the ESR of the capacitor. The voltage across the capacitor under test is given by

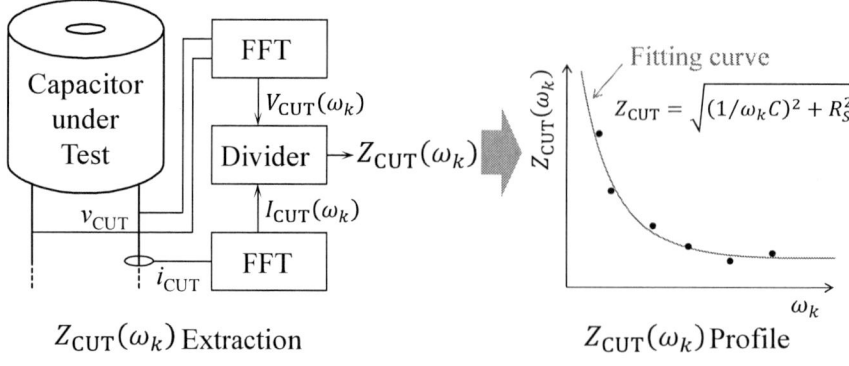

Fig.3 How to extract ESR and capacitance from the actual ripple current and voltage

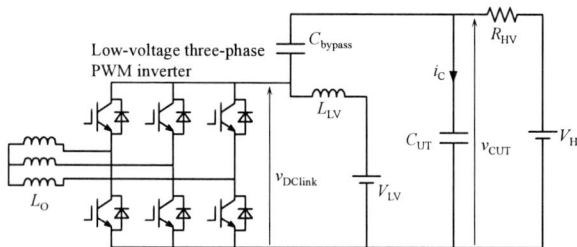

Fig. 1 Evaluation circuit consisting of a low-voltage three-phase inverter

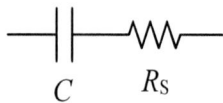

Fig . 2 Equivalent circuit of the capacitor for analysis

$$v_{CUT} = \frac{1}{C}\int i_{CUT}dt + R_S i_{CUT}$$

$$= \sum_{k=1}^{n} I_{CUT}(\omega_k)\left\{-\frac{1}{\omega_k C}\cos(\omega_k t + \varphi_k) + R_S \sin(\omega_k t + \varphi_k)\right\}$$

(2)

From Eq(2), the amplitude of v_{CUT} at each frequency is given by

$$V_{CUT}(\omega_k) = I_{CUT}(\omega_k)\sqrt{\left(\frac{1}{\omega_k C}\right)^2 + R_S^2}$$

(3)

Dividing (3) by $I_{CUT}(\omega_k)$ gives the impedance of the capacitor at each frequency as follows:

$$Z_{CUT}(\omega_k) = \frac{V_{CUT}(\omega_k)}{I_{CUT}(\omega_k)} = \sqrt{\left(\frac{1}{\omega_k C}\right)^2 + R_S^2}$$

(4)

Equations (4) suggest that the capacitance and ESR are independently calculated if impedances at different frequencies are obtained. Multiple frequency components of the ripple current help calculate the capacitance and ESR.

Fig. 3 shows a block diagram of ESR and capacitance extraction, which introduces the fast Fourier transform (FFT) that obtains $V_{CUT}(\omega_k)$ and $I_{CUT}(\omega_k)$. Dividing $V_{CUT}(\omega_k)$ by $I_{CUT}(\omega_k)$ yields the impedance at each frequency, $Z_{CUT}(\omega_k)$. One can plot an impedance profile of $Z_{CUT}(\omega_k)$ and draw a fitting curve by the least squares method, which indicates both the ESR and capacitance.

B. Capacitor Voltage Monitoring from DC-Link Voltage

A major concern to carry out the monitoring method is to measure the ripple component of the capacitor voltage because the voltage contains a much larger amount of dc-bias voltage than the ripple component.

Fig. 4 shows the relation between the dc-link voltage of the low-voltage inverter, v_{dclink}, and the voltages across the bypassing capacitor, v_{bypass}, and that across the capacitor under test, v_{CUT}. The dc-link voltage v_{dclink} is given by

$$v_{dclink} = v_{bypass} + v_{CUT} \tag{5}$$

The amplitude of the ripple component of the dc-link voltage, Δv_{dclink}, is also given by

$$\Delta v_{dclink} = \Delta v_{bypass} + \Delta v_{CUT} \tag{6}$$

where Δv_{bypass} and Δv_{CUT} are amplitudes of the ripple component of v_{bypass} and v_{CUT}, respectively. Hence, one can obtain the ripple component of the capacitor voltage by means of measuring the dc-link voltage. From equation (2), the ripple ratio of v_{dclink}, r_{dclink} is given by

$$r_{dclink} = \frac{\Delta v_{dclink}}{V_{dclink}} = \frac{\Delta v_{bypass} + \Delta v_{CUT}}{V_{dclink}}$$
$$= \frac{\Delta v_{bypass} + \Delta v_{CUT}}{V_{HV}} \times \frac{V_{HV}}{V_{dclink}}$$
$$= (r_{bypass} + r_{CUT}) \times n$$

(7)

where r_{bypass} and r_{CUT} are ripple ratios of v_{bypass} and v_{CUT} with respect to V_{HV}, respectively, and n corresponds to the ratio between voltage ratings of the full-scale inverter and the small inverter. Equation (7) suggests that the ripple ratio of the dc-link voltage is multiplied by n, so that the dc-link voltage has a large amount of ripple ratio even though the two capacitors contain a small amount of ripple component with respect to their dc-bias voltages. This improves voltage resolution of the ripple amplitude of the capacitor under test. For example, for $V_{HV} = 1000$ V and $V_{LV} = 100$ V, a 100-V class voltage sensor is available

The 2018 International Power Electronics Conference

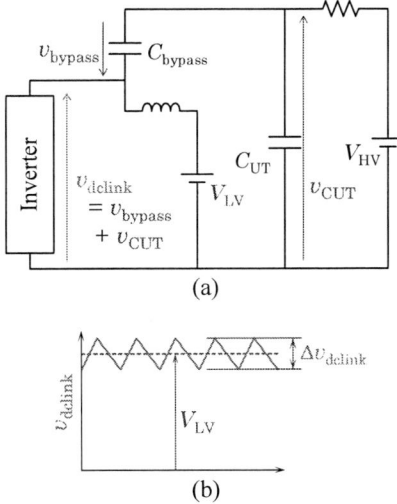

(a)

(b)

Fig. 4 Relation between dc-link voltage and capacitor voltages. (a) Positive direction of the voltages. (b) Ripple component appearing on the dc link.

Table I Ratings and circuit parameters of the evaluation circuit used in experiment.

System Power rating	P	69 kVA
Power rating of the inverter	P_{inv}	2.8 kVA
AC current rating	I_O	54 A
AC voltage rating	V_O	29 V
Low-voltage dc source	V_{LV}	48 V
High-voltage dc source	V_{HV}	1200 V
Load inductor	L_O	1 mH
Carrier frequency	f_{SW}	3 kHz
Output frequency	f_O	50 Hz
High-voltage choke resistor	R_{HV}	1 kΩ
Low-voltage choke inductor	L_{LV}	4 mH
Capacitors under test	C_{UT}	320 or 360 μF
Unit capacitance constant of the capacitors under test [16]	H_{CUT}	3.3 ms (320 μF)
		3.8 ms (360 μF)
Bypassing capacitor	C_{bypass}	320 μF

although the voltage across the capacitor under test is 1000 V.

IV. EXPERIMENT

Table I summarizes ratings and circuit parameters of the experimental circuit. It should be pointed out that the circuit acts as a 69-kVA full-scale inverter although the inverter power rating is 2.8 kVA that is 1/25 of that of the full-inverter. Sinusoidal PWM is applied to the low-voltage inverter with a unity modulation index. The capacitor under test C_{UT} employs metalized polypropylene capacitors. The inverter drives only reactive power in this experiment. The ripple current waveform somewhat changes according to the power factor of the output power but always contains multiple frequency components. Thus, the proposed method is effective even though the power factor changes.

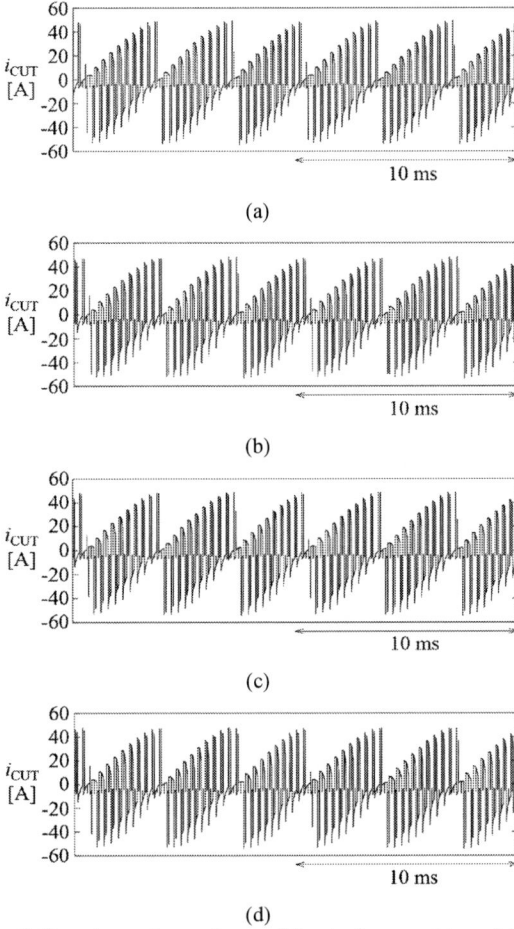

(a)

(b)

(c)

(d)

Fig. 5 Experimental waveforms of the ripple current i_{CUT}. (a) Condition A. (b) Condition B. (c) Condition C. (d) Condition D.

This paper introduces different capacitor conditions to present capacitance and ESR changes as follows:
1) Condition A: C_{UT} = 360 μF.
2) Condition B: C_{UT} = 360 μF with an intentionally-connected series resistance of 0.1 Ω, which shows the effect of an increase in ESR.
3) Condition C: C_{UT} = 320 μF.
4) Condition D: C_{UT} = 320 μF with a series resistance of 0.1 Ω.

Fig. 5 shows experimental waveforms of the ripple current, i_{CUT}. All the waveforms were almost the same because the inverter acts as a current source for the capacitor [14]. Fig. 6 shows experimental waveforms of the dc-link voltage, where the ripple amplitude of the voltage increased when the capacitance was reduced as well as the ESR was increased. Fig. 7 shows experimental waveforms of the voltage across the capacitor under test. All the waveforms stayed at 1200 V that were equal to V_{HV}.

Fig. 8 shows impedance profiles of the capacitor against each frequency component along with fitting curves. The impedance profiles are obtained from FFT results of the capacitor current shown in Fig. 5 and those of dc-link voltage shown in Fig. 6. The fitting curves are modeled by

1940

The 2018 International Power Electronics Conference

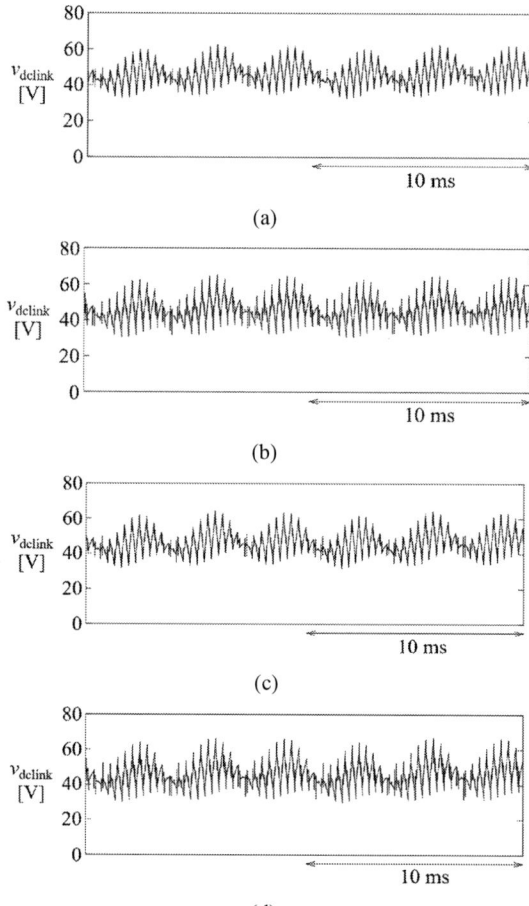

(a)

(b)

(c)

(d)

Fig. 6   Experimental waveforms of the DC-link voltage of the low-voltage inverter at V_{LV} = 48 V. (a) Condition A. (b) Condition B. (c) Condition C. (d) Condition D.

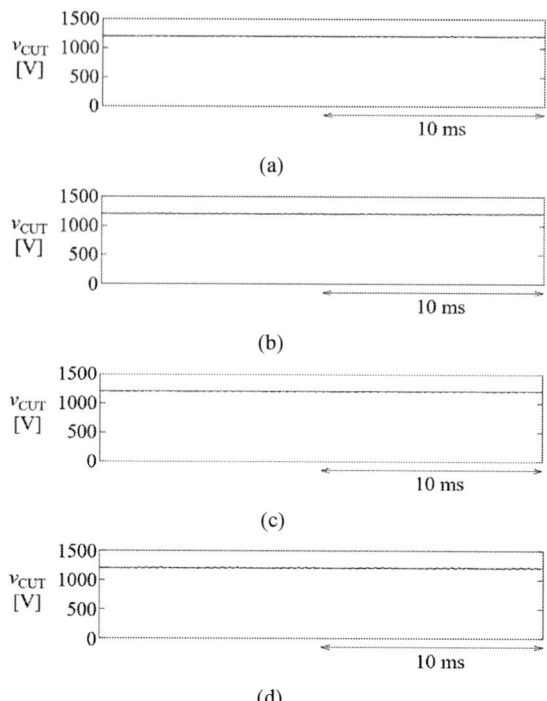

(a)

(b)

(c)

(d)

Fig. 7   Experimental waveforms of the voltage across the capacitor under test, v_{CUT} at V_{HV} = 1200 V. (a) Condition A. (b) Condition B. (c) Condition C. (d) Condition D.

analysis has confirmed that the monitoring method independently extracts the capacitance and ESR of the dc-link capacitor with the actual current generated by the low-voltage inverter. Experimental results verify the viability and effectiveness of the proposed method.

equation (4). Note that the impedance profile shows the sum of impedances of C_{bypass} and C_{UT}. Hence, C_{UT} is calculated from the sum of impedance under a condition of C_{bypass} = 320 µF. In Fig. 8(a), the capacitance was almost same as the series-connected capacitance of C_{bypass} and C_{UT}, and the ESR was 0 Ω, whereas the ESR in Fig. 8(b) increased to 0.1 Ω that was the same as the intentionally-connected resistance. Note that the series resistance is larger than the reactance of C_{UT} in a range more than 3 kHz, which makes an error in the capacitance C. For this reason, C_{UT} changed into 333 µF.

In Fig. 8(c), the impedance in a low-frequency region were slightly larger than that of Fig. 8(a) due to a smaller capacitance. In Fig. 8(d), the ESR increased to 0.1 Ω because of the intentionally-connected series resistance.

These results confirmed that the monitoring method well observed both the ESR and capacitance changes.

V. CONCLUSION

This paper has presented a condition monitoring method of dc-link capacitors used in a high-power three-phase PWM inverter with an evaluation circuit. Theoretical

REFERENCES

[1] J. W. Kolar, U. Drofenik, J. Biela, M. Heldwein, H. Ertl, T. Friedli, and S. Round, "PWM converter power density barriers," *IEE Japan Trans. Ind. Appl.*, vol. 128, no. 4, pp.468-480, 2008.

[2] H. Wang and F. Blaabjerg, "Reliability of capacitors for dc-link applications in power electronic converters—an overview," *IEEE Trans. Ind. Appl.* vol. 50, no. 5, pp. 3569-3578, 2014.

[3] K. Abe, H. Haga, K. Ohishi, and Y. Yokokura, "Current Ripple Suppression Control Based on Prediction of Resonance Cancellation Voltage for Electrolytic-Capacitor-Less Inverter," *IEEJ J. Ind. App.*, vol.6, no.1, pp.1-11, 2017.

[4] S. Yang, A. Bryant, P. Mawby, D. Xiang, L. Ran, and P. Tavner, "An Industry-Based Survey of Reliability in Power Electronic Converters," *IEEE Trans. Ind. Appl.*, vol. 47, no. 3, pp. 1441-1451, May/Jun. 2011.

[5] K. Harada, A. Katsuki, and M. Fujiwara, "Use of ESR for deterioration diagnosis of electrolytic capacitor," *IEEE Trans. Power Electron.*, vol. 8, no. 4, pp. 355-361, Oct. 1993.

[6] P. Venet, F. Perisse, M. H. El-Husseini, and G. Rojat, "Realization of a smart electrolytic capacitor circuit," *IEEE Ind. Appl. Mag.*, vol. 8, no. 1, pp. 16–20, Jan./Feb. 2002.

[7] A. M. R. Amaral, and A. J. M. Cardoso, "Estimating aluminum electrolytic capacitors condition using a low frequency transformer together with a dc power supply," *in Proc. of IEEE ISIE*, pp. 815-820, 2010.

[8] K. Abdennadher, P. Venet, G. Rojat, J. M. Retif, and C. Rosset, "A Real-Time Predictive-Maintenance System of Aluminum Electrolytic Capacitors Used in Uninterrupted Power Supplies," *IEEE Trans. Ind. Appl.* vol. 46, no. 4, pp. 1644-1652, Jul./Aug., 2010.

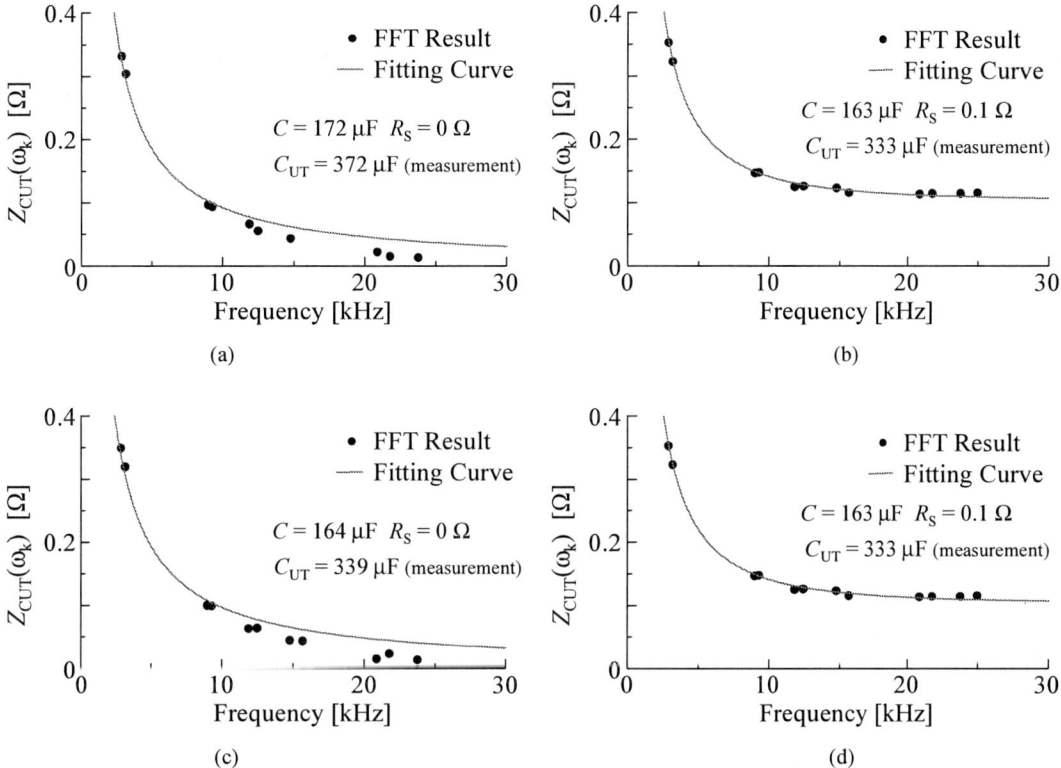

Fig. 8 Impedance profiles obtained from FFT results along with fitting curves indicating the ESR and capacitance. (a) Condition A. (b) Condition B. (c) Condition C. (d) Condition D.

[9] H. Soliman, H. Wang, and F. Blaabjerg, "A Review of the Condition Monitoring of Capacitors in Power Electronic Converters," *IEEE Trans. Ind. Appl.*, vol. 52, no. 6, pp. 4976-4989, Jul. 2016.

[10] *RIPPLE CURRENT TESTER MODEL 11800/11801/11810*, Chroma ATE Inc. 2014. [Online]. available: http://www.chromaate.com/File/DownLoad/42014

[11] M. Makdessi, A. Sari, P. Venet, P. Bevilacqua, and C. Joubert, "Accelerated Ageing of Metallized Film Capacitors Under High Ripple Currents Combined with a DC Voltage," *IEEE Trans. Power Electron.*, vol. 30, no. 5, pp. 2435-2444, May 2015.

[12] B. P. McGrath and D. G. Holmes, "A general analytical method for calculating inverter DC-link current harmonics," *IEEE Trans. Ind. Appl. vol. 45, no. 5*, Sep./Oct. 2009

[13] K. Hasegawa, K. Kozuma, K. Tsuzaki, I. Omura, and S. Nishizawa, "Temperature rise measurement for power-loss comparison of an aluminium electrolytic capacitor between sinusoidal and square-wave current injections," *Microelectron. Rel.*, vol. 64, pp. 98-100, 2016.

[14] K. Hasegawa, I. Omura, and S. Nishizawa, "Design and Analysis of a New Evaluation Circuit for Capacitors Used in a High-Power Three-Phase Inverter," *IEEE Trans. Ind. Electron.*, vol. 63, no. 5, pp. 2679-2687, May 2016.

[15] K. Hasegawa, I. Omura, and S. Nishizawa, "A New Evaluation Circuit with a Low-Voltage Inverter Intended for Capacitors Used in a High-Power Three-Phase Inverter," *IEEE APEC.*, pp. 3032-3037, Mar. 2016.

[16] H. Fujita, S. Tominaga, and H. Akagi, "Analysis and design of a dc voltage-controlled static var compensator using quad-series voltage-source inverters," *IEEE Trans. Ind. Appl.*, vol. 32, no. 4, pp. 970--977, 1996.

Recent market and technical trends in Copper Rotors for High-Efficiency Induction Motors

Daniel Liang[1*], Victor Zhou[1]

1 Motor & Motor System, International Copper Association, Beijing, China

*E-mail: Daniel.liang@copperalliance.asia

Abstract: **This paper will introduce the recent progress in Copper Rotor Motor in China. The development of IE4 and IE4+ copper rotor motors are introduced, and the detailed performance analysis is given. A compacted motor design which use copper rotor for pumps and escalators is introduced. The technical and economic analysis is also given. A pilot project which used rotor replacement technology to improve the installed motor efficiency will be presented from both technical and economical point of view. The total energy saving potential and the possibility to develop a new efficiency standard for the motor rebuilt will be discussed.—**

Keywords— High Efficiency Motor, Copper Rotor, Motor Rebuild

I. INTRODUCTION

Squirrel cage induction motors are the mostly used motors at present. Squirrel cage induction motors took a share of over 95% of the total middle and small motors output in 2016 in China[1]. Currently, squirrel cage induction motors are widely applied in industries, agriculture, commerce, public facilities , transportation and other applications. Motors are inseparable core equipment to transfer electrical energy into mechanical energy.

Currently, the rotors of the squirrel cage induction motors are produced by aluminum die casting technique with the advantages of low cost of rotors. However, for the small motors, the use of aluminum bars for rotors makes it difficult to ensure the motor to achieve the super efficiency requirements.

It is well known that the copper's electrical conductivity is around 40% higher than that of aluminum. If copper is used to replace the current widely used aluminum in rotors, the total loss of motors will be reduced, hence increasing the overall motor efficiency. Copper rotors can be used to produce high efficiency and even super premium efficiency motors without enlarge the motor frame size. Redesigning of motor's rotors and stators is required in line with the property of cast copper rotors which can maximize the motor efficiency to reach higher or even super premium energy efficiency.

The data show that by using copper rotors can reduce the motor energy consumption by 15%-25% and increase the efficiency of 2%-5%[2] .

Because copper's electrical resistivity is low, electric current in the rotors is bigger than that of aluminum rotors under the same induction voltage, that means fewer

materials can be used to produce motors with the same power and efficiency. It is because of the material used has been reduced, motors with copper rotors will be lower material cost, lighter weight and smaller volume than that of aluminum rotor motor[3]. At the same time, the excellent anti-corrosion features of copper can make copper rotor motors have better environmental tolerance and reliability. All these advantages provided a broader and more flexible design space for motor designers and producers, making them to pursue higher efficiency, lower cost, smaller volume and lower weight or find a balance among all these factors.

The above mentioned excellent features make the copper motors excellent choice for broad applications, including industrial motors, pump motors, compressor motors, electric vehicles, swimming pool circulation pump motors, crane motors, aviation motors, high speed motors, spindle motors, military used special motors, etc.

II. IE4 & IE4+ COPPER ROTOR MOTORS

The total motor production volume for small & medium motor (<=375KW) is 180,330 MW in 2016 in China. Table I gives the production volume from 2013 to 2016 for different efficiency motors[1].

TABLE I
PRODUCTION VOLUME FOR DIFFERENT EFFICIENCY MOTORS IN CHINA

Year	IE1	IE2	IE3	IE4
2013	62.13%	29.68%	8.00%	0.19%
2014	52.21%	38.37%	9.09%	0.33%
2015	52.12%	40.35%	7.30%	0.24%
2016	32.56%	56.79%	10.38%	0.27%

At present, IE2 and IE1 efficiency motor is still the major production in China. The high efficiency motor (IE3 & IE4) production volume is only 11% of the total motor market in 2016. The Chinese government has put a lot of efforts to improve the motor efficiency. The market share of high efficiency motor will be continuously increased in the coming years. The copper rotor technology is one of the easiest and economical solutions to meet the growing demand for high efficiency motors in the world.

In order to improve motor efficiency, motor manufacturers usually use more materials such as magnet wire and electrical steel. But using more material means

1943

increase the motor cost and meanwhile will increase the motor physical size[3]. There are standards to illustrate the motor physical size, there are limitations on motor center height and diameter. The materials that are needed to achieve the higher efficiency will force the motors size beyond the limits of the standards, especially in smaller motors. That enlarged motors can't be used for the situation where the space is limited. For small induction motors, in order to improve motor efficiency and reach the IE4 and even IE5 standards new methods and materials must be introduced.

There are several solutions to improve the motor efficiency to meet the requirements. The permanent magnet motors, the reluctance motors and the copper rotor motors. Each solution has the advantages and disadvantages. But considering the current market share of induction motors, above 95% is induction motors in Chinese small & medium motor market in 2016[1], the copper rotor solution is the most economical and feasible solution to achieve the high efficiency requirements.

Copper rotor is the most economical and easier way to improve the induction motor efficiency. Copper rotor motors make it easier to control the whole production process and lower material cost when compared to using high quality electrical steel. Copper rotor motors have a great cost advantage when compared to super-conduct material such as rare earth permanent magnet. Copper rotor technology makes it possible for motor manufacturers especially small & medium motor manufacturers to produce high or super efficiency motor without change their production line too much.

Die casting of pure copper is very challenge due to the high melt point of copper (1083°C). That was the obstacle for motor manufacturers to adopt copper rotor for their high efficiency motors. The melting point of pure copper (1083°C) is much higher than that of aluminum (660°C). That makes casting copper is much difficult compared with casting aluminum.

In the past decades, the international copper association (ICA) has done a lot of work in order to make casting copper rotor available by cooperated with several company worldwide. Current, copper rotor casting technology has been developed successfully and the commercialized production have been carried out in China. Last year, more than 200,000 copper rotors have been produced in China, which servers 5% of total high efficiency motor market in China.

Nanyang motor Co. Ltd. has used copper rotor technology to develop IE4 induction motor successfully (YZTE4 Series). The frame size is from 80 to 250, the output power is from 0.75KW to 55KW.

The VZTE4 series super premium efficiency three phase induction motors (Frame HB0-250) are developed by copper rotor technology in line with IEC60034-30-1 Standards. which ratings, mounting dimensions and electrical performances meet the requirements of latest IEC Standards. These motors possess many remarkable features such as IE4 super premium efficiency, low noise, little vibration, large margin of safety in temperature rise,

popular shape, safe and reliable operation. The detailed information of the YZTE4 series motor is as below table:

TABLE II
PARAMETERS FOR YZTE4 COPPER ROTOR MOTORS

Items	Value
Rated power	.75kW to 55kW
Frame Size	H80 – 250
Speed	3000,1500 or 1000rpm @ 50Hz
Insulation	Class F insulation with B rise @1.0 SF
Service factor	1.0
Phase	3
Voltage	380V ,415V,380/660V,440V @ 50Hz
Efficiency	IE4(IEC60034-30-1, Super-Premium efficiency)
Protection	IP55 Protection

In the design stage Nanyang motor optimized the whole design of copper rotor by considering the copper characteristics. The electromagnetic density of copper rotor motor was designed larger than that of aluminum by considering the higher efficiency and thermal conductivity of copper rotor motors. It reduced the motor size, total material usage and cost.

The results proved that copper rotor is one of the best ways to reach the IE4 efficiency. It makes it easy to improve the motor efficiency to reach higher level. For motor manufacturers, the copper rotors allow them to produce high efficiency motors without changing their current design too much with the cost advantage.

Besides Nanyang motor, more than 10 Chinese motor manufacturers have already developed their IE4 induction motors by using copper rotor technology.

Chinese government has already approved a standard draft for the IE4 copper rotor motors. The draft was developed by Shanghai Electrical Apparatus Research Institute, together with more than 20 top motor manufacturers as well as institutes and associations. The new IE4 motor standard named YZTE4(IP55) by using copper rotor, the standard covered the motor range from frame size 80-180, rated power from 0.55 to 22 kW. Table III to Table V gives the major technical parameters of this standard.

TABLE III
THE EFFICIENCY OF YETE4 STANDARD

Type by Poles	Average Efficiency (%)	
	IE4 Standard	Designed
2-6 P	88.81%	89.66%
2-8 P	89.73%	90.56%

TABLE IV
THE POWER FACTOR OF YETE4 STANDARD

Type by Poles	Average Power Factor	
	IE4 Standard	Designed
2-6 P	0.784	0.817
2-8 P	0.807	0.839

1944

TABLE V
THE STARTING TORQUE OF YETE4 STANDARD

Type	Average Starting Torque	
	IE4 Standard	Designed
1st	8.568	7.36
Tst	1.932	2.254
Tm	2.393	3.018

Compared with IE4 aluminum rotor motor design, the copper rotor design has smaller frame size, that means use less material and less weight.

Besides the IE4 motors, the IE4+ motors using copper rotor technology was also developed successfully. Nanyang Motor Co. ltd. by cooperated with Yunnan Copper Die Casting Co. Ltd and International Copper Association to develop IE4+ efficiency motor series based on copper rotor technology.

Copper rotor technology was selected in order to meet the IE4+ efficiency. As introduced, Nanyang motor has a lot of experience to develop high efficiency motor based on copper rotor technology. Nanyang has already developed IE4 series induction motor based on copper rotor technology.

Considering the copper rotor characteristic, the design for IE4+ series motor enlarged the magnetic density as well as electrical density of the motors. Meanwhile, the thermal duty was increased when compared with aluminum design.

The IE5 series motor has lower stray load loss. The stray load loss was only 0.6% of the input power by average based on the real testing results.

TABLE VI
TEST RESULTS OF IE4+ COPPER ROTOR MOTORS

Type	Tested Efficiency	Weighted Efficiency
YZTE4-112M-4P 4kW 415V 50Hz	91.37%	91.16%
YZTE4-160M-4P 11kW 415V 50Hz	93.99%	93.3%
NSPE184T-4P 5HP 460V 60Hz	91.9%	91.22%
NSPE254T-4P 15HP 460V 60Hz	93.62%	93.95%

This is a good example to prove the induction motor efficiency can be improved to IE4+ or IE5 level by adopting copper rotor technology.

III. COMPACT COPPER ROTOR MOTORS

Besides to improve motor efficiency, copper rotor could be used to reduce the motor physical size by 1 to 2 frame grade with the same power and efficiency, in another words, it could improve the motor power density[4][5].

This paper introduced 2 prototypes development for these copper rotor compact motors. These compact motors were used for gear box, pump, escalator, etc.

Compared with traditional aluminum rotor motors, these motors were 15-20% lighter and smaller. Table VII shows the comparison between traditional aluminum rotor motor and copper rotor motor for the 75KW-2 compressor driven motors.

TABLE VII
COMPARISON BETWEEN AL AND CU ROTOR MOTORS

	Efficiency（%）	Frame Size	Steel in Stator (kg)
Copper Rotor	94.91%	225	300
Aluminum Rotor	93.8%	280	375

From the table we can see that motor frame size could be decreased by 2 grades by using copper rotor. Meanwhile, the efficiency could be improved from original 93.8% to 94.91%, which means the efficiency grade could be improved from IE2 to IE3.

By cooperated with Nanyang motor, a 7.5kW IE3 compacted copper rotor has been developed and the performance has been tested. Table VIII gives the comparison between the copper rotor motor design and aluminum rotor design.

TABLE VIII
COMPARISON BETWEEN CU AND AL ROTOR 7.5KW MOTORS

	Cu rotors	Al rotors	Different
Rated Power (kw)	7.5	7.5	
Frame Size	132	132	
Efficiency Grade	IE3	IE3	
Starting Torque (Tst/TN)	2.51	2/43	
Starting Current (Ist/IN)	5.97	6.04	
Length (mm)	578	503	
Width (mm)	302	220	
Height(mm)	450	375	
Volume (dm³)	78.55	41.5	47%
Weight (kg)	98	65	34%

By using coper rotor, the motor' volume can be reduced by 47% and the weight can be reduced by 34% with the same efficiency.

Since less material used, the motor cost could be reduced also. Table IX gives the cost analysis by using copper rotor compared with aluminum rotor by RMB.

TABLE IX
COST COMPARISON BETWEEN CU AND AL ROTOR 7.5KW MOTORS

	Al rotors	Cu rotors	Different
Magnetic wire	369	365	4
Insulation material	37	35	2
Silicon steel	393	286	7
Rotor bar	36	168	-132
Shaft	28	13	15
House	256	167	89
Bearing	33	21	12
End Shield	11	8	3
Fan	50	33	17
Total	1213	1096	117

By using copper rotor, together with the reduce of the physical volume, the motor cost can be reduced by 10%.

For escalator driven motors, 2 types copper rotor motors were developed successfully. The torque is very important for escalator usage. Usually the locked rotor torque is 2 times than the rated torque for general purpose motors. But for escalator, the requirement is 2.2-2.5 times than rated torque. Based on copper rotor technology, we developed 2 types of motors, both achieved the IE3 efficiency, locked rotor torque is 2.9 times than rated torque. Meanwhile, the copper rotor helped to reduce the physical volume compared with aluminum design. For these 2 motors, we keep the same frame size but reduce the motor length to fit the space requirements of the escalator.

Table X and XI gives the technical parameters for these 2 types of motors.

TABLE X

PARAMETERS FOR 7.5KW ESCALATOR COPPER ROTOR MOTORS

	Tested	Standard
Frame	160M	
Rated Power (Kw)	7.5	
Poles	6	
Efficiency (%)	88.83	87.46
Power Factor	0.79	0.77-0.79
Locked Rotor Torque	2.85	>=2.0
Locked Rotor Current (A)	6.24	7.0-8.4
Max Torque	2.85	>=2.1
Min Torque	2.18	>=1.3
Vibration (m/s)	0.61	<2.2
Noise(dB)	54.1	73
Temperature Rise (K)	65.1	80

TABLE XI

PARAMETERS FOR 7.5KW ESCALATOR COPPER ROTOR MOTORS

	Tested	Standard
Frame	160M	
Rated Power (Kw)	11	
Poles	6	
Efficiency (%)	89.86	88.45
Power Factor	0.78	0.78-0.8
Locked Rotor Torque	2.95	>=2.0
Locked Rotor Current (A)	6.37	7.27-8.64
Max Torque	2.95	>=2.1
Min Torque	2	>=1.2
Vibration (m/s)	0.61	<2.2
Noise(dB)	54.4	73
Temperature Rise (K)	78.3	80

Besides the advantage of physical size and efficiency. The copper rotor motor has better performance on vibration as well as noise.

IV. COPPER ROTOR FOR ROTOR REPLACEMENT

Copper rotor technology can be used to improve the installed motor efficiency besides to improve the new built motor efficiency. By using the optimized-designed copper rotor to replace the aluminum rotor from old motors, the motor efficiency could be improved from 1 to 2 grades. That method just changed the motor rotor without changing the stator, the rotor replace word could be done during the motor repairmen stage. Compared with other motor refurbished methods, rotor replacement technology is the most simple and feasible and

economical way to improve the installed motor efficiency.

As reported, the prototype test has been done to verify the method. By cooperated with Yunnan copper die casting Co. Ltd and Nanyang Explosion Protection Group, 66 types of prototypes have been developed and tested the real performance, the old motor is Y series motor which is IE1 efficiency. The power range is from 0.55 to 22KW and include 2,4,6 pole motors. The test results are very satisfied. All these 66 types of motors have been improved the efficiency by 1 grade which is from original IE1 to IE2. Among these, 53 types of motor's efficiency have been improved by 2 grades, which is from original IE1 to IE3.

The prototypes development and the performance testing are very successfully. But there is still a lot of work to be done to make this method real practical in the industry. The challenge comes from the difference of the installed motors. Different from the new built motors, the installed motors are very different. The motors installed in a factory could be purchase from different manufacturers. Even purchased from same manufacturers, motors could be different after several years running. The rotor replacement sound simple and the prototype test proved it technical available. But a new and suitable process is needed to overcome the difference from the installed motors. Guiding by the process, the workers could be easier to identify which motor should be rebuild and how to rebuild the motor. In order to develop the suitable process, the field testing have conducted.

The field test has been conducted to test the real performance. We have rebuilt a IE1 motor located on a copper fabrication company. copper fabrication. The original motor is Y-series motor which is IE1 efficiency, after the re-built, the efficiency grade was improved to IE3. Table XII gives the test results.

TABLE XII

COMPARISON BETWEEN 22KW ORIGINAL AND REBUILD MOTOR

Item	Before Rebuild	After Rebuild
Active Power (kW)	21.33	19.47
Apparent Power (kW)	24.57	23.89
Voltage(V)	385.8	385.8
Current(A)	37.1	33.12
Power Factor	0.87	0.88

The energy saving is (21.33-19.47) / 21.33 = 8.72%, the payback period is less than 0.6 year.

We also conducted a field test in a paper making company. Totally 8 motors were selected and rebuilt by using rotor replacement technology.

Table XIII shows the test results. From the table we can find all motors' efficiency have been increased after conducted the rotor replacement, hence saving electrical energy during running.

The total annual energy saving for these 8 motors is 46,744.65kW after rebuilt. That means the annual energy saving is RMB 37,395 (The average electricity price is 0.8 RMB per kWh), and the total cost to rebuild these motors is less than RMB 35,000, the payback period is less than 1 year, which is very attractive for the motor end users to save the energy consumption.

TABLE XII
TEST RESULTS ON THE ENERGY SAVING FOR 8 MOTORS.

Equipment No.	Annual using hours	Average input power (kw)	Energy saving (%)	Annual Energy saving (kwh)
412	8,000	8.638	14.05	9,709.11
5311-2	8,000	9.985	10.27	8,203.68
P502	8,000	14.037	8.23	9,241.96
P506-B	8,000	6.952	3.84	2,135.65
5111	8,000	3.991	8.27	2,640.45
5110	8,000	3.690	6.69	1,974.89
402	8,000	20.995	6.82	11,454.87
409-1	8,000	8.782	1.97	1,384.04

After the field test, 2 pilot projects have been established in China supported by the local government, one located in a paper industry and another located in the mining company. Totally more than 1000 motors were rebuilt by using copper rotor technology to improve the efficiency, the average energy saving is around 5% and the payback period is less than 1 year.

V. CONCLUSION

Improving energy efficiency of motor is very important for sustainable development and saving energy cost for end users. Copper has excellent electrical conductivity. The usage of copper in motors is very important to improve the energy efficiency of the motor. As mentioned, copper rotor motor has many advantages, such as high efficiency, lower running temperature, light weight, smaller physical volume and less material usage.

Considering the larger market share of induction motors, copper rotor technology is the most easier and economical way to improve the motor efficiency. The paper introduced the IE4 an IE4+ motor prototype and standard development in China, which proved that copper rotor is workable to improve the motor efficiency.

For IE3 motors, copper rotor technology can improve the motor power density which means motor could be smaller and lighter while keep the same output power.

Besides the new build motors, copper rotor technology could be used to improve the installed motor efficiency, considering the huge amount of installed motors. The energy saving is remarkable by adopt this method.

REFERENCES

[1] All China Market Research Co. Ltd., "The report of Chinese motor market research 2016", pp 11.

[2] Zhang Limin, "The progress for high efficiency motors",Shanghai Large&Medium Motor, P38, 2012.

[3] Yu Jianbin, "The development of the super high efficiency motor based on copper rotor technology", Motor & motor control system, pp15-18, 2014.

[4] Deng Yaoqiang, "The development of high efficiency and power density induction motor", Electric Machine & Contral Application, pp53-56, 2016

[5] Huo Huafeng, "The development of high speed, high efficiency, high power density motors", Modern Business Trade Industry, pp186-187, 2017.

[6] Huang Jian, "The R&D for copper rotor high efficiency motors", Electrical Machinery Technology, pp1-5, 2012.

Overview of the Latest Research and Development for Copper Die-Cast Squirrel-Cage Rotors

Shu Yamamoto
Polytechnic University, Tokyo, Japan
E-mail: yamamoto@uitec.ac.jp

Abstract— This paper presents the overview of the latest research and development for copper die-cast squirrel-cage rotors. First, some examples comparing copper die-casting with aluminum die-casting are shown. Effects of copper die-cast on efficiency, temperature, size and weight are demonstrated. Second, the latest copper die-cast machines capable of mass-producing rotors with various dimensions are shown. Key assembling techniques, problems and countermeasures are discussed. Third, with reviewing some papers mainly related to manufacturing method, resent technical trends in copper die-cast and small-size fabricated copper bar rotors are summarized.

Keywords— aluminum, copper, die-cast squirrel-cage rotor, induction motor.

I. INTRODUCTION

Many aluminum die-cast rotors are used for squirrel cage induction motors. These efficiency and power factor are not superior to those of permanent magnet synchronous motors. However, induction motors have many merits that are not found in permanent magnet synchronous motors; open-loop drives and line starting are available, single inverter-fed parallel running and assemble/disassemble are easy. Thus, to enhance worth of induction motor, the efficiency must be improved more.

Since electrical conductivity of copper is 1.5 times or more than that of aluminum as shown in Table I, it is well-known that employing copper die-cast rotors instead of aluminum die casting leads to improving the efficiency of induction motors. In copper die-cast process, manufacturing cost is still high due to technical difficulties caused by high melting temperature of copper. However, recent copper die-cast technology has changed for the better so as to be applicable to not only prototype products but also mass-producing.

To address this situation, this paper presents the overview of the latest research and development for copper die-cast rotors. Many papers studying copper die-cast rotors have been presented, such as summarized in [1]. Among them, this paper focusses on manufacturing techniques and experimental studies and discusses usefulness, problems and countermeasures of copper die-cast rotors for ac motors. First, some examples

comparing copper die-casting with aluminum die-casting are shown. Effects of copper die-cast on efficiency, temperature, size and weight are demonstrated. Second, the latest copper die-cast machines capable of mass-producing rotors with various dimensions are shown. Key assembling techniques, problems and countermeasures are discussed. Third, with reviewing some papers mainly related to manufacturing method, resent technical trends in copper die-cast and small-size fabricated copper bar rotors are summarized.

TABLE I
DIFFERENCE OF MATERIAL PROPERTY BETWEEN ALUMINUM AND COPPER

Item	Aluminum	Copper
Electrical conductivity (%IACS)	65	102
Thermal conductivity (W/(mK))	236	398
Melting point (degrees Celsius)	660	1084

II. DIFFERENCE AMONG CAGE-ROTOR CONSTRUCTION

In [2], differences among four types of rotor construction, aluminum die-cast, copper die-cast, fabricated aluminum bar, and fabricated copper bar, are discussed in detail. The strong (and weak) points of copper bar can be summarized as follows.

➢ Efficiency and starting torque
Copper die-cast and fabricated copper bar rotors are the best to maximize the driving efficiency due to their higher electrical conductivity. But, the starting torque of the copper rotors is lower than that of the aluminum rotors. A trade-off problem between efficiency and starting torque exists.

➢ Cost
Aluminum die-cast is the best choice, especially for small motors. Copper die-cast and fabricated aluminum bar are the second best. Fabricated copper bar requires the highest cost due to the

- labor-cost associated with fabrication and material costs of copper.

> Rotor Design Flexibility
Copper rotor design is more flexible to changes in rotor bar resistivity because of an economic and practical point. Copper bars and alloys are available in many different levels of resistance, and no change in tooling is required.

> Rotational Stresses
Copper bars are approximately three times heavier than aluminum, so there must be sufficient strength in the laminations above the bars.

> Rotor Bar Heat Capacity
Copper bars are superior to aluminum bars. Due to difference of the material density and specific heat, aluminum bars will get much hotter, expand further, and generate much higher stresses while accelerating the same rotating inertia

> High-inertia load and multiple restart
Copper rotors can withstand high inertia loads better than aluminum rotors. Because high-inertia starts result in rapid bar heating.

From the consideration mentioned above, it is found that copper rotor bars have a lot of merits. However, it is also noticed that copper die-cast requires a large capital investment. It may be difficult to justify the expense of the copper bar rotor just for improving efficiency, especially on smaller machines. So, it is considered that, to widely spread the use of copper die cast, it is necessary to develop new technology which improves both manufacturing process and driving efficiency.

III. PERFORMANCE COMPARISON BETWEEN COPPER AND ALUMINUM DIE-CASTINGS

A. Temperature

Table II compares the temperature rise of the copper die-cast rotors [3]. One can see that the temperatures of the copper die-cast rotors are lower than those of the aluminum rotors.

B. Efficiency

Table III summarizes the change in the measured efficiency of prototype squirrel cage induction motor when aluminum die-cast is replaced to copper die-cast without changing the other specification and dimension [3]. Especially, it is confirmed that increment rate in efficiency of the small-capacity motor is higher than that of the large-capacity motor. A similar experimental result has been reported in [4].

C. Size and weight, and their impact

In [5], an aluminum die-cast induction motor and a copper die-cast induction motor are designed under the same baseline requirements, peak power rating of 50kW, maximum speed of 6,000 r/min, thermally-limited

maximum steady-state power capability of 22kW, motor diameter of 238mm, motor length of 173mm and efficiency at the base speed of 90%, for electric vehicle traction motor. When comparing between two designed motors, the copper die-cast motor became 23% lighter and 30% smaller than the aluminum die-cast motor. In addition, although the total losses of the copper die-cast motor was the same as that of aluminum die-cast one, the secondary cage loss of the aluminum die-cast motor was 37% of the total loss. Meanwhile that of the copper one was 17%. From the results, it is claimed in [5] that since this difference leads to a much higher heat transfer across the stator wall, the copper die-cast motor had more than a 30% increase in overall heat transfer capability than the equivalent aluminum die-cast motor.

TABLE II
EXAMPLE OF TEMPERATURE COMPARISON WHEN ALUMINUM DIE-CAST IS REPLACED TO COPPER DIE-CAST

Specification	Temperature (degrees Celsius)	
	Aluminum	Copper
11 kW	64.9	59.5
18.5 kW	59.4	47.2

TABLE III
EXAMPLE OF EFFICIENCY IMPROVEMENT WHEN ALUMINUM DIE-CAST IS REPLACED TO COPPER DIE-CAST

Specification	Efficiency (%)	
	Aluminum	Copper
3.7kW, 400V, 4-poles, 50Hz	85.9	87.3
3.7kW, 400V, 4-poles, 60Hz	86.0	87.6
0.75kW, 200V, 4-poles, 50Hz	80.6	82.8
0.75kW, 220V, 4-poles, 60Hz	82.5	84.5

IV. MANUFACTURE OF COPPER DIE-CAST ROTOR

A. Problem and countermeasure

In [6], problems and their countermeasures in achieving a cost-effective copper die-cast operation are quickly summarized as follow.

The first item is to prevent molten copper from forming CUPROUS OXIDE. The copper die-casting operation is in many ways similar to that of aluminum. However, unlike aluminum, once molten, copper reacts with oxygen in the air to form cuprous oxide. This leads to reducing electrical conductivity of the secondary cage, consequently, motor efficiency. Its countermeasure is to keep the oxygen content within a range of 300 to 600 ppm. To achieve this, carrying out gas-blanket protection while the copper melts is effective.

The second item is not to make the CASTING POROUS. Since the molten copper must be flow into the die cavity quickly, the gas entrained in the molten copper and the shrinkage during solidification lead to the casting porous. To minimize them, it is recommended that to search for the optimum operation condition by predicting where porosity defects arise on the basis of software-based simulation of the heat and fluid flows during casting.

The third is SHORT MOLD LIFE. It is explained that use of conventional materials and die designs would have resulted in a copper rotor die having a useful life of only 600 cycles compared to 50,000 cycles for a comparable aluminum die. This is because copper's high melting point also required copper rotors be cast at 1,200 degrees Celsius or higher. Preheating the die by 500 to 600 degrees Celsius, near the temperature of molten copper, contributes to boosting the life, which helps alleviate the thermal shock and subsequent cracks that normally reduce die life.

In [7], instead of H-13 tool steel which is often utilized for copper die-casting, mold materials suitable for high melting temperature in copper die-casting are explored experimentally. Seven prototypes of TZM, Anviloy 1200, INCONEL alloy 601, 617, 625, 718 and 754. Pros and cons of them are discussed with actual shot testing.

In [8], it is pointed out that copper's high melting temperature results in failure of conventional die steels, such as H-13 tool steel, by thermal fatigue of the surface (heat checking) in less than 100 shots. Moreover, high-temperature nickel-base alloy dies (e.g., INCONEL alloy 617) is recommended to increase die life [9]. It is also described that HAYNES alloy 230 has similar properties and is conventionally weld repairable.

B. Recent manufacturing method

In [10], a recent copper die-cast manufacturing process system applicable to not only prototype products but also mass-producing is reported. Fig. 1 shows main equipment: 690 ton die-casting machine and high frequency melting furnace (75 kW, 3,000 Hz). Fig. 2 indicates a process of copper die-casting with photo examples. The procedure consists of melting of pure copper, setting of rotor core, pouring and injecting of molten copper, ejecting of product and quality verification in turn. It is also reported that critical points of this process are management of material and temperature, technics to prevent the molten copper from oxidizing and mixing with dross, selection of optimum release agent and searching of the optimum die-casting condition based on simulation analysis of fluidity and solidifying. The quality verification is roughly divided into outside appearance inspection and interior failure diagnosis. In the former, lack of fluidity and crack of end ring are checked. In the latter, condition of bubbles can be observed whether they are separated into small pieces.

C. Items to increase the efficiency of copper die-cast rotor induction motor

In [11], it is investigated that the reason why the efficiency of a copper die-cast rotor induction motor for electrical vehicle traction drive, shown in Fig. 3, becomes lower than that of a fabricated copper bar induction motor in high speed range.

Here, five possibilities are explored by experiments on prototype products. In conclusion, the following two items are extracted as reasonable causes.

➢ Increase of short between cage bar and rotor core. It is considered that this is due to influence of inter bar current and excessive contact between cage bar and rotor core; it is found from outside appearance inspection of a die-cast copper bar pulled out from the rotor core, as shown in Fig. 4.

(a)

(b)

Fig. 1. Main equipment for copper die-casting, (a) 690 ton die-casting machine, (b) high frequency melting furnace (75 kW, 3,000 Hz) [10].

1st : melting of pure copper   2nd : setting of rotor core

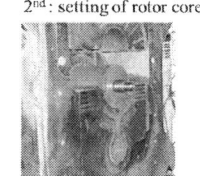

3rd : pouring and injecting   4th : ejecting of product

 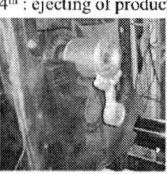

Fig. 2. Process of copper die-casting [10].

➢ Low rotor cage ac resistance, leading to increasing eddy current loss by spatial harmonics.

Meanwhile, it is concluded that the following items are not related with reduction of the efficiency.

➢ Impurities in molten copper.
➢ Deterioration of lamination steel sheet.
➢ Deterioration of the rotor slot insulation.

Thus, the following countermeasure is recommended to increase the efficiency in high speed range.

➢ Rotor slot insulation should be done.
➢ Water quenching of the die-cast rotor, immediately after pulling out from mold, should be done.

Fig. 3. Prototype of copper die-cast rotor induction motor for electrical vehicle traction drive [10].

Fig. 4. Die-cast copper bar pulled out from the rotor core. On the surface of the bar, one can see thin iron peeled off from the inner surface of the rotor slot [11].

As a result, it is described that the efficiency of the copper die-cast rotor induction motor employing the countermeasures mentioned above reaches to the same level as the fabricated copper bar induction motor at 10,000 r/min. In [11], however, specifications and ratings of the test motor, measurement method, type of high-frequency converter, etc. are not shown.

D. High-speed induction motor

In [12], mechanical design of a 200 kW, 30,000 r/min copper die-cast rotor induction motor for electrical vehicle traction drive is proposed. Mechanical issues, bearing, mechanical firmness, balancing and shaft-hub joint, are analyzed and optimized by means of FEA.

V. ALUMINUM DIE-CASTING WITH FABRICATED COPPER BAR

A. No tooling cost process

In [13], some methods to improve the efficiency of cage induction motor by no tooling cost process (that is a cost-effective process that minimizes economical influence in manufacture) are studied. As a concrete example, a cage rotor in which aluminum die-casting is carried out under the fabricated copper bars are inserted to the slots is investigated.

Four layouts of the copper bar in a double-cage slot are verified with evaluating secondary circuit parameters. Consequently, a layout in which three fabricated copper bars are inserted to a slot is employed for achieving both low loss steady-state operation and low starting current. Specifically, one is a bar having a slender rectangular cross section whose upper and lower bases are close to the top of the outer bar slot and the bottom of the inner bar slot. The other bars having a slender rectangular cross section (but are shorter than the former one) are both inserted to the upper side of the inner bar slot.

In FEA-based simulation, the efficiency was improved when the rotor designed optimally was employed. In experiments on a prototype motor, however, improvement of the efficiency was hardly seen. This reason is pointed out from observation of rotor cross section as follows.

➢ Bubbles in aluminum die-cast.
➢ Oxides produced by high temperature die-casting process aggravate the conductivity between copper bar and aluminum.
➢ Affinity between copper bar and aluminum is insufficient.

Thus, it is concluded that development of a new cost-effective countermeasure for these problems is one of important future works.

B. Downsizing of induction motor

In [1], an example of downsizing of induction motor without changing the driving performance by employing aluminum die-casting with fabricated rotor bar is shown. Fig. 5 shows the base induction motor with pure aluminum die-cast rotor. The nameplate rating is 2.2kW, 400V, 50Hz, 4P, 4.6A, 1435r/min, power factor is 0.81 and efficiency is 84.3%. Fig. 6 shows the tested induction motor with aluminum die-cast having fabricated rotor bars. Note that the rated current and rotational speed are 4.65A and 1463r/min respectively. The other nameplate like a general pure aluminum die-cast rotor. However, cross section of the rotor shown in Fig. 6(b) indicates that the outer side of the rotor slot is filled with high-resistive data are the same as the base induction motor. From Figs. 6(a) one can see that the rotor of the tested motor looks aluminum. Meanwhile in the inner side of the slot, low-resistive fabricated copper bar is inserted. It is considered that this design is employed for achieving higher starting

1952

torque and lower starting current such like double cage induction motors.

Table IV shows the dimension of the motors. From this table, it can be found that the weight, core stack length and volume are downsized successfully though the outer diameters of the rotor and stator increase a little.

Fig. 5. The tested motor with pure aluminum die-casting in [1].

(a)

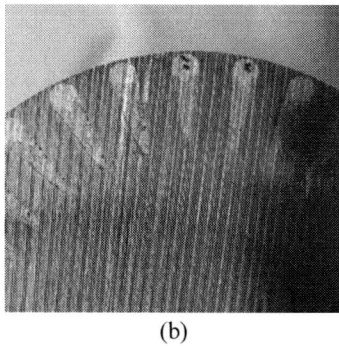

(b)

Fig. 6. The tested motor with aluminum die-casting having fabricated copper bars in [1], (a) rotor appearance, (b) cross section of the rotor core.

TABLE IV
COMPARISON BETWEEN PURE ALUMINUM DIE-CASTING AND ALUMINUM DIE-CASTING WITH FABRICATED COPPER BAR

Item	Pure aluminum die-casting	Aluminum die-casting with fabricated copper bar
Weight (kg)	28	21
Rotor outer diameter 2D (mm)	88	99.5
Stator outer diameter 2De (mm)	146	153
Rotor stack length (mm)	150	96
Volume $\pi D^2 L$ ($\times 10^{-3} m^3$)	9.21	7.46
Volume $\pi De^2 L$ ($\times 10^{-3} m^3$)	25.1	17.6
Rotor skew	use	no use

VI. PERMANENT MAGENT SYNCHRONOUS MOTOR WITH LINE-START WINDING USING FABRICATED COPPER BAR

In [14], improvement of the driving efficiency of an inverter-fed open-loop-controlled permanent magnet synchronous motor having line-start winding without deteriorating the starting performance is investigated. The key point is to use a fabricated copper bar whose diameter is smaller than a conventional line-start winding using fabricated aluminum bar. When this strategy is employed, it is possible to set permanent magnets close to periphery of rotor core. Since this leads to enlarge permanent magnet surface area, the driving efficiency can be improved. Experimental results on three prototype motors (model A: conventional aluminum bar, model B: copper bar having small diameter, and model C: aluminum bar having small diameter) demonstrate that the driving efficiency of the models B and C are higher than model A and the starting performance of the model B is better than model C.

VII. MAINTENANCE TECHNOLOGY

A. Fault detection of copper die-cast rotor

In [15] and [16], to detect various rotor faults, a rotor quality test system for die-cast copper rotors has been proposed with experimental verification. Since a copper die-casting operation is more difficult than an aluminum die-casting, the manufacturing problems involved in the die-cast copper rotors are higher. So, an exclusive system that is used in the testing of die-cast rotor in its manufacturing process is required. The rotor quality test for diagnosing faults is carried out according to the following three steps.

1) Weight test
2) End ring flaw detection test
3) Rotor quality test

The weight test is firstly performed to detect blow holes in rotor casting. Larger unacceptable voids in rotor bars and end rings after casting can be identified by weighting the rotor stack before and after casting. However, the weight test is not capable of giving indications to the position and magnitude of holes and lower porosity levels.

The end ring flaw detection test is secondary adapted to verify the porosity levels in the end rings by an ultrasonic tester. In this test, higher number of peaks detected by the ultrasonic tester indicates higher porosity levels in the end rings.

The rotor quality test is thirdly carried out to detect porosity, inter laminar shorting, bad skewing, lower conductivity, eccentric rotor cage, lamination property. This test is performed under the cast rotor (without stator) rotates in constant speed by a drive motor and a constant dc magnetic field is generated near the cast rotor bar. As the bar moves through the magnetic field, it cuts the field and induces a voltage in the bar. In this situation, a pickup coil is moved to be close to the cast rotor bar. Then, it acts as a secondary winding to detect the changes in the magnetic field due to this rotor bar current. By analyzing the signal of the pickup coil, discrimination of various rotor faults can be realized.

B. Defect prediction

In [17], an example to predict the porosity in edged-gated copper rotor die-casting using a casting simulation is shown with real die-cast verification. A casting simulation software ADSTEFAN is employed for the simulation on a 44-slot copper die-casting double cage rotor for a 15 HP induction motor. It is demonstrated that the porosities observed when the shot profile for pre-fill is changed can be properly appeared in both gate and ejector end rings.

VIII. CONCLUSIONS

This paper discussed usefulness, problems and countermeasures of copper die-cast rotors with referring overview of the latest research and development for copper die-cast squirrel-cage induction motor rotors.

The copper die-casting is surely a strong item to increase the efficiency with both reducing the rotor temperature and downsizing the volume of induction motors. Recent research and development would contribute to achieving high-quality copper die-casting in mass production.

Meanwhile, high production costs including of short mold life, low starting torque and high starting current must be improved to widely spread the use of copper die-cast induction motors.

ACKNOWLEDGMENT

The author thanks Mr. Y. Yuki (Japan Copper Development Association) and members of Investigating R&D Committee on latest trends of applications of AC motor drives in the users sights for their technical supports.

REFERENCES

[1] Investigating R&D Committee on latest trends of applications of AC motor drives in the users sights, "Trends of AC motor drive application from the users sights," IEE Japan, pp. 14-19, 2016 (in Japanese).

[2] M. Hodowanec, W. R. Finley, "Copper versus Aluminum Induction. Motor Rotors: Which construction is best?" IEEE IA Magazine, Vol.8, No.4, pp. 14-25, 2002.

[3] T. Akashi, Y. Yuki, "Current status and issues of Copper Die-Cast Motor Rotor," Proc. of JFS Meeting, Vol. 160th, p. 17, 2012 (in Japanese).

[4] A Modern Casting Staff Report, "Diecast Copper Rotors Improve Motor Efficiency," Modern Casting/April, pp. 33-35, 2006.

[5] James L. Kirtley Jr., Richard F Schiferl., Dale T. Peters, Edwin F. Brush Jr., "The Case for Induction Motors with Die-cast Copper Rotors for High Efficiency Traction Motors" SAE Technical Papers, SAE-2009-01-0956, 2009.

[6] T. Baldwin, "Industry could take a shine to COPPER-ROTOR MOTORS," Machine Design, Vol.84, No.13, pp. 70, 72, 81-82, 2012.

[7] John G. Cowie, Edwin F. Brush Jr., Dale T. Peters, Stephen P. Midson, "Materials & Modifications to Die Cast the Copper Conductors of the Induction Motor Rotor," Die Casting Engineer, Vol.45, No.5, pp. 38-42,44-47, 2001.

[8] J. Malinowski, J. Mccormick, K. Dunn, "Advances in Construction Techniques of AC Induction Motors: Preparation for Super-Premium Efficiency Levels," IEEE Transactions on IA, Vol. 40, No. 6, pp. 1665-1670, 2004.

[9] D. T. Peters, J. G. Cowie, E. F. Brush, and S. P. Midson, "Use of high temperature die material and hot dies for high pressure die casting pure copper and copper alloys," in Proc. 2002 Die Casting Congr., North American Die Casting Assoc., Rosemont, IL, Sept. 30–Oct. 2, 2002.

[10] T. Akashi, "Manufacturing Technology of copper die-cast rotor," JCDA Seminar document of die-cast copper rotor induction motors, 2014 (in Japanese).

[11] M. Burwell, "Efficiency improvement of high-speed copper rotor induction motors," JCDA Seminar document of die-cast copper rotor induction motors, 2014 (in Japanese).

[12] Epskamp Torsten, Butz Benjamin, Doppelbauer Martin, "Design and analysis of a high-speed induction machine as electric vehicle traction drive," IEEE Conference Proceedings, Vol. 2016, No. EPE'16 ECCE Europe, 2016.

[13] Aldo Boglietti, Andrea Cavagnino, Luca Ferraris, Mario Lazzari, and Giorgio Luparia, "No Tooling cost Process for Induction Motors Energy Efficiency Improvements," IEEE Transactions on IA, Vol. 41, No. 3, pp. 808-81, 2005.

[14] A. Takahashi, S. Kikuchi, "Line-Starting Permanent Magnet Synchronous Motor for IE4 Efficiency Classification," The Papers of Technical Meeting on Rotating Machinery, IEE Japan, RM-10-146, pp. 73-78, 2005 (in Japanese).

[15] Soby T. Varghese, Bhim Singh and K. R. Rajagopal, "Fault investigations on die-cast copper rotors," in Proc. IEEE Int. Conf. PEDES, Trivandrum, Dec. 2016.

[16] Soby T. Varghese, K.R. Rajagopal, "Design and development of rotor quality test system for die-cast copper rotors," IEEE Conference Proceedings, Vol. 2016, No. ICPEICES, 2016.

[17] H. Badarinarayan, I. Takahashi, J. Cowie, S. Midson, "Defect Prediction in Copper Motor Rotor Die Casting using Numerical Simulation," SAE Technical Papers, SAE-2005-01-1562, 2005

A Novel Heat-Resistant Insulation-Processing Agent Applicable to Copper Die-Cast Squirrel-Cage Rotors

Junichi UCHIDA [1*], Yuki SUEUCHI [1] and Naosumi KAMIYAMA [1]

1 Central Research Lab, NIHON PARKERIZING Co., Ltd, Hiratsuka, Kanagawa, Japan

*E-mail: juchida@parker.co.jp

Abstract— **Energy saving is one of the most important tasks in the world and a new type of induction motor has been developed to improve the efficiency for that purpose. The insulation coating layer between the copper and the rotor core is quite effective in improving the motor efficiency. However, the coating layer is easily removed when molten copper is flushed into the holes of the core. We evaluated several coating materials made of silicon compounds and pigments and succeeded in developing a coating giving good heat resistance and adhesion for high-efficiency copper die-cast cage induction motors.**

Keywords— **motor, insulation, coating, copper,**

I. INTRODUCTION

Nowadays, energy saving is one of the most important tasks and regulations have been tightened in the world[1-3]. A new type of induction motor has been developed to improve the efficiency for that purpose[4]. The most common form of induction motors utilizes a squirrel-cage rotor, which comprises a series of longitudinal conductor bars (usually made of aluminum or copper) inserted into slots located towards the periphery of a stack of steel laminations[5]. The conductor bars are connected at both ends by shorting end rings, producing what is known as a squirrel cage structure.

Induction motors are typically less efficient than permanent magnet motors[6]. To improve induction motors efficiency, it is necessary to re-design motors to reduce iron loss, copper loss and mechanical loss. However, the current design has already been optimized and further improvement would be limited with increasing the size or changing materials[7]. How can we improve the induction motor's efficiency?

The first step has been to improve the efficiency by making the squirrel cage from a highly conductive material such as copper. The squirrel cage for induction motors are typically produced by die casting aluminum, as shown in Figure 1. The rotors constructed from copper generally produce motors of higher overall efficiency, as the electrical conductivity of pure copper is more than 60% greater than that of aluminum[5].

A further development is an insulation coating between the copper and the rotor core which is made of steel lamination. Figure 2 shows an insulation coating between the copper and the rotor core. Theoretically, using insulation is quite effective to improve the motor efficiency. Specifically, it is important to focus on the leakage current between the squirrel cage of the lamination stack and copper bars[8].

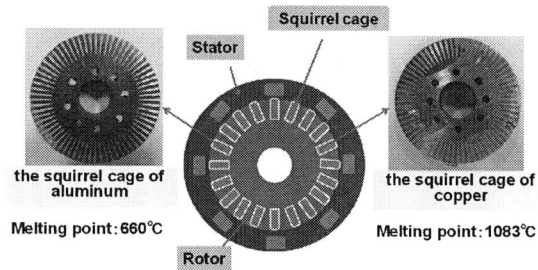

Fig. 1. The die cast rotors made of aluminum or copper

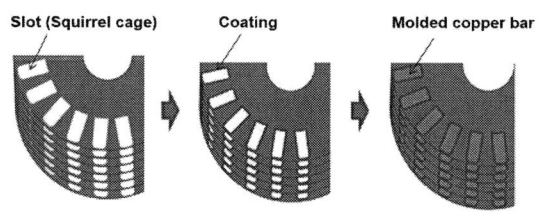

Fig. 2. The insulation coating between the copper and the rotor core

However even if the rotor core is coated with an insulating coating, the coating layer is easily removed when molten copper is flushed into the holes of the core in the production process. Even if the coating remains partly after the copper is casted, it is significantly damaged by extra-high temperatures[9],[10].

This study is an attempt to develop a novel heat-resistant insulation-processing agent applicable to copper die-cast squirrel-cage rotors. Therefore, the method we used for selecting an effective coating was dipping coated test panels into a pot filled with molten copper. We checked the appearance and electro-resistance after dipping. We evaluated several coating materials made of silicon compounds and pigments and developed a coating giving good heat resistance and adhesion for high-efficiency copper die-cast cage induction motors.

II. EXPERIMENTS

A. Chemical

The chemicals for insulation processing are as follows. The novel heat-resistant insulation chemicals ("HR-Series") consist of silicon compounds as main components. The comparisons used in tests are samples without coating.

Figure 3 shows the film model of the HR series. HR1 is an inorganic compound that prevents oxidation in a high temperature environment. HR8 is an inorganic/organic hybrid coating with inorganic insulating pigments that is resistant to continuous exposure to high temperatures. It provides high insulating properties due to the thickness of the coating film.

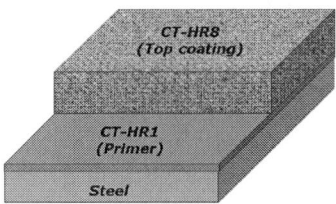

Fig. 3. The novel heat-resistant insulation film structure

B. Material and Processing

The test materials used for the studies are small rectangular steel samples (called here SPCC-SD).

Table 1 shows the coating process of the test panel. First, the bare samples were degreased by dipping in an isopropyl alcohol solution at room temperature. This was followed by drying using hot air blower. Finally, we coated the test panels with chemicals using bar-coat or roll-coat.

C. Evaluations

Effect to prevent oxidation of HR1

The steel samples coated by HR1 was exposed to the heat resistance test (300℃, 60 minutes). We observed the appearance of the test pieces. We analyzed the test pieces as follows.

We measured the electrical resistance of the coating layer on the test piece using surface insulation resistance testing apparatus. Depth profiles of the coating layer were determined using X-ray Photoelectron Spectroscopy (XPS).

Characteristics of HR8

The steel samples coated with both by HR1 and HR8 were exposed to the heat resistance test (600℃, 60 minutes). We observed the appearance of the test piece. We analyzed the test pieces as follows.

We measured the insulation breakdown voltage of the coating layer on the test piece using surface insulation resistance testing apparatus.

Test1

Simple testing was performed where small rectangular steel samples coated with both HR1 and HR8 were dipped into baths of molten copper. We analyzed the coating layer using a Scanning Electron Microscope (SEM-EDX). To further quantify this testing, the electrical resistance measurements were performed to determine the degree to which the coatings were still intact after immersion. Probes were used to perform the measurements at the following locations (see Figure 4):

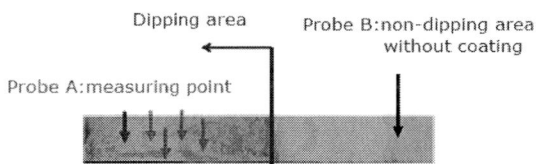

Probe A– Portion of sample immersed into molten metal
Probe B– Non-dipping area without coating

Fig. 4. Locations uses to test electrical resistivity

Test2

The details of the die casting trials are as follows. The rotors were cast by applying a coating of both HR1 and HR8 to the lamination stack. This coating was applied by dipping the assembled lamination stack into a container of the coating material. The coating was approximately 0.01mm to 0.02mm thick. This coating is used to provide a thermal barrier between copper and steel to prevent premature solidification during casting.

We analyzed the coating layer using a Scanning Electron Microscope (SEM-EDX).

TABLE I
Coating process of the test panel

Process	Chemical	Condition
Cleaning	IPA	Room temperature, 60s, Dip
Drying	–	Hot air blower
Coating	CT–HR1	Room temperature, Dip, Coating weight:0.1–0.5g/m^2
Drying		Peak metal temperature:70℃
Coating	CT–HR8	Room temperature, Dip, Coating weight:20–30 μ m
Pre–Drying		Peak metal temperature:100℃
Baking		180℃, 45min

The electrical resistance measurements were performed to determine the degree to which the coatings were still intact after immersion. Probes were used to perform the measurements at the following locations (see Figure 5):

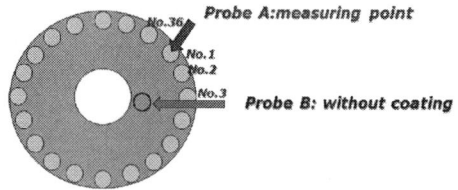

Fig. 5. Locations uses to test electrical resistivity

III. RESULTS AND DISCUSSION

A. Effect to prevent oxidation of HR1

The proprietary HR-1 has the main function of preventing oxidation. Figure 6 shows the appearance and the electrical resistance of the test piece after the heat resistance test. In the case of test pieces without coating, SPCC-SD turned blue after heating. On the other hand, the samples coated with HR-1 exhibited no change before and after the test.

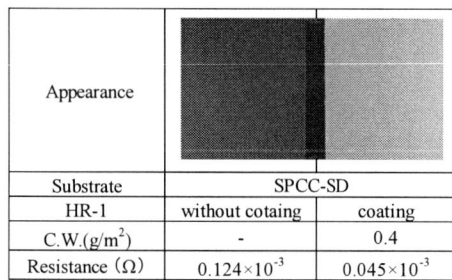

Appearance		
Substrate	SPCC-SD	
HR-1	without cotaing	coating
C.W.(g/m²)	-	0.4
Resistance (Ω)	0.124×10^{-3}	0.045×10^{-3}

Fig. 6. Result of heat resistance test on SPCC-SD

Figure 7 shows surface analytical results by XPS after heat resistance test on SPCC-SD without coating. By monitoring photoelectron peak position iron, it shows an iron peak shift. This means that an iron oxide formed on the surface. This is further supported by a slight rise in the electrical resistance of SPCC-SD without coating.

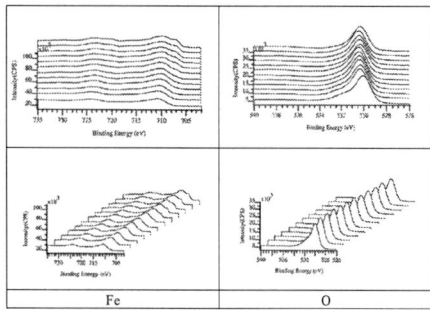

Fig. 7. Characteristics of SPCC-SD after heat resistance test by ESCA

Figure 8 shows surface analytical results by XPS after heat resistance test on SPCC-SD coated with HR-1. By monitoring photoelectron peak position iron, we can see that HR-1 prevents oxidation.

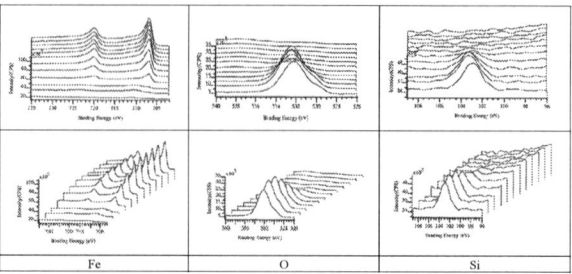

Fig.8. Characteristics of SPCC-SD with HR-1 coating after heat resistance test by ESCA

B. Characteristics of HR8

Figure 9 shows Insulation breakdown voltage of coating layer on the test piece after the heat resistance test.

As shown in Figure 9, the insulation breakdown voltage improves as the film thickness increases. The insulation breakdown voltage is not affected by the heat resistance test.

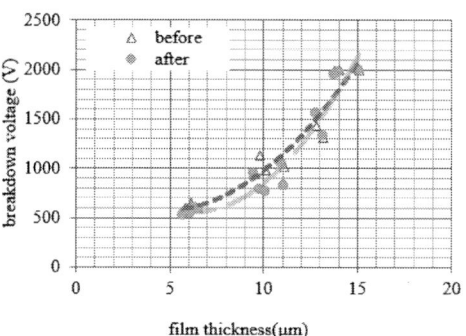

Fig. 9. Relationship between film thickness and breakdown voltage resistance after heat resistance test

C. Test 1

Figure 10 shows the surface appearance on coated SPCC-SD after it was dipped into baths of molten copper. The novel heat-resistant insulation coating provides excellent electrical resistance. The coating thickness of HR-8 does not influence electrical resistance between 0.01mm to 0.02mm.

It is possible to analyze the heat resistance by studying the electron intensity of the chemical elements in the cross section of coating using SEM-EDX, as shown in Figure 11. From the analysis results, Pigment A made the layer on the top of the coating and B made the layer between the coating and the substrate. It is expected that these three layers give excellent heat resistance.

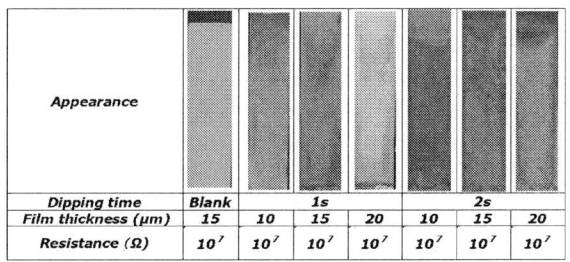

Appearance							

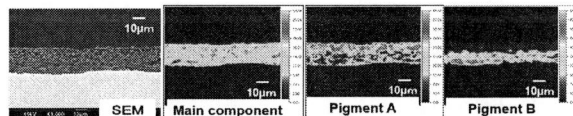

Dipping time	Blank	1s			2s		
Film thickness (µm)	15	10	15	20	10	15	20
Resistance (Ω)	10^7	10^7	10^7	10^7	10^7	10^7	10^7

Fig. 10. The surface appearance on coated SPCC-SD after it was the dipped into baths of molten copper

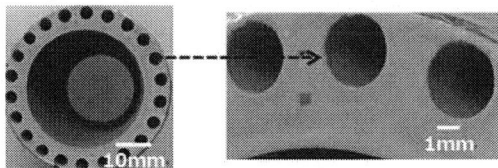

Fig. 11. The analytical results of the cross section of coating by SEM-EDX

D. Test 2

The rotors were cast applying the novel heat-resistant insulation coating to the whole stack. This coating was applied by dipping the assembled lamination stack into a container of the coating material, and a photograph of the typical coating thickness on the inside of the slots is shown in Figure 12.

Fig. 12. Coating appearance on lamination stack

Our major concern is whether the novel heat-resistant insulation coating can withstand high pressure when molten copper is flushed into the holes of the core in the production process.

To examine how well this coating was standing up to the injection of molten copper, several rotors were sectioned to directly examine the interface between the cast copper and the lamination stack.

The photograph of the core sectioned is shown in Figure 13. The surface appearance of copper and steel were turned white by the novel heat-resistant insulation coating, which indicates that some of the coating material did survive the copper die casting process.

Figure 14 shows surface analytical results by SEM-EDX. This means that the coating component, silicon and pigments, was also present on the surfaces of the conductor bar and the laminations.

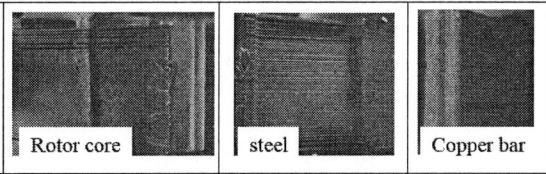

Fig. 13. The appearance rotors of sectioned after they were dipped into baths of molten copper

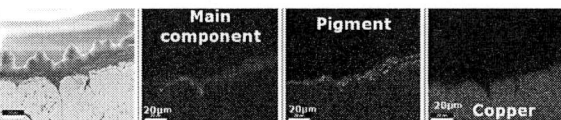

Fig. 14. The analytical results of the cross section of coating by SEM-EDX

Figure 15 shows the electrical resistance measurements. The novel heat-resistant insulation coating provides better electrical resistance at all the copper bars. We can see that the silicon compound has the good interface bonding to steel. Moreover, that good adhesion is achieved regardless of the high melting point of copper. This coating provides a thermal barrier between copper and steel to prevent solidification during casting.

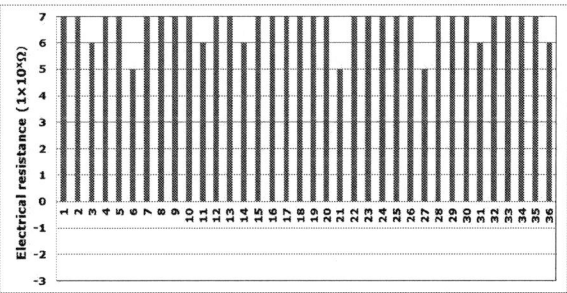

Fig. 15. The results of electrical resistance measurements

E. Mechanism of efficiency improvement

From the results of our investigation, it is important to mention that with respect to the sample without coating, focusing on the leakage current between the squirrel cages of the lamination stack and copper bars. We think that the motor efficiency of the sample without coating are adversely affected by the stray load loss as shown in Figure 16. The stray load loss is the leakage current between the squirrel cages of the lamination stack and copper bars. Especially, in case of the rotation speed higher, the motor efficiency is reduced. Because the contact pressure between copper bar and outside of slot increases and the stray load loss also increases as shown in Figure 17. It is difficult to achieve all-round performances.

By contrast, the effectiveness of a two-layer coating system prevents a contact between the squirrel cage of the lamination stack and copper bars. The coating component, silicon and pigments, has the interface bonding in between them even though the molten copper is flushed into the holes of the core. As the new coating is excellent barrier film of a polymer layer generated by the condensation of the coating component. It provides excellent performances to avoid stray load loss.

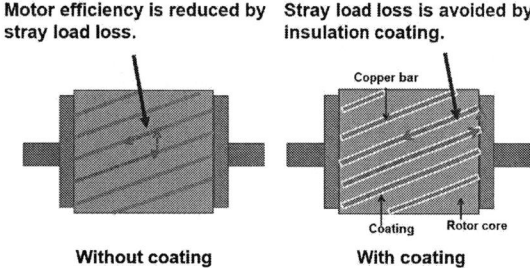

Fig. 16. Mechanism of efficiency improvement

Contact pressure between copper bar and outside of slot increases and stray load loss also increases.

Fig. 17. Estimated mechanism that motor efficiency is reduced under higher rotation

IV. SUMMARY

The novel heat-resistant insulation coating applies a 2-layer coating with both HR1 and HR8. HR1 can prevents oxidation in a high temperature environment. HR8 is resistant to continuous exposure to high temperatures. It provides high insulating properties due to the thickness of the coating film.

To summarize the results, the following conclusions can be drawn. It is important to mention the results of our investigation in comparison with to the core without coating. In the case of without coating, the copper does indeed make intimate contact with each individual lamination edge during the die casting. We think that the motor efficiency is adversely affected by the leak current by this contact point.

By contrast, the novel coating has the high properties of heat resistance and insulation regardless of the high temperature. The novel heat-resistant insulation coating consists of silicon compounds as main components. The silicon compound, which improves the interface bonding, has heat resistant.

We achieved the development of a coating giving good heat resistance and adhesion for high-efficiency copper die-cast cage induction motors. The novel heat-resistant insulation coating can be applied to a wide range of processing motor. This data shows that the novel heat-resistant insulation coating is successful in improving the efficiency of the die cast copper rotors, by providing a barrier between the copper and the steel laminations The novel heat-resistant insulation coating provides excellent electrical insulating performance for this purpose and also could be widely used on parts other than copper rotors.

REFERENCES

[1] Energy General Engineering Laboratory, "Energy commentary collection," 2007.
[2] Honda, "Matsushita's energy-saving motor development story," pp.15, 2007.
[3] Fleming Abrahamsen et al, "On the Energy Optimized Control of Standard and High Efficiency Induction Motors in CT and HVAC Applications", IEEE Trans, on Ind. Appl4, Vol.34, No4. , JUL/AUG, pp.822, 1998.
[4] Japan Standards Association, "High efficiency low pressure three phase cage induction motor JIS C4212 2000", 2000.
[5] M. Burwell, P. Caros, W. Rippel, "Improving the Efficiency of High Speed Induction Motors Using Die Cast Copper Rotors," Version1.5, 11 July, 2013.
[6] Ide, "Trend of latest technology in rotary machine field",126, 11, pp.718-737, 2006.
[7] K. Tsuboi, I. Hirozu, M.Hasegawa, "Efficiency Technology on Motors and Drive Systems", J.IEIE Jpa, Vol.29, No.3, pp.205-208, 2009.
[8] I. Oura, J. Uchida, Y. Kinoshita, I. Koiwa, "Nihon Parkerizing Coating Technology," Vol.28, 2016.
[9] ASM Handbook, "Aluminum and Aluminum Alloys", Ed: J.R. Davis, pp.25, 1993.
[10] ASM Handbook, "Copper and Copper Alloys", Ed: J.R. Davis, pp. 4, 2008.

Experimental Verification on the Effects of Insulation-Processing of Copper Die-Cast Squirrel-Cage Rotor on Motor Efficiency in high-speed operation over 10,000 r/min

Hideaki Hirahara, Akira Tanaka, and Shu Yamamoto
Polytechnic University, Tokyo, Japan
E-mail: hirahara@uitec.ac.jp

Abstract— **This paper verifies the effect of a new heat-resistant insulation-processing between rotor core and copper die-cast bar on motor efficiency in high-speed operations. Two prototypes of copper die-cast squirrel-cage rotors are assembled: an insulated-slot rotor and a non-insulated-slot rotor. These rotors are installed to the same stator whose ratings are 0.75 kW, 200 V, 2 poles and 12,000 r/min. Full-load driving tests over 10,000 r/min are implemented. The results suggest that to insulate between the copper die-cast rotor bar and rotor core by the proposed heat-resistant agent contributes to improving the motor efficiency in high-speed and heavy-load operations. In addition the impact of the slot insulation on loss segregation of the copper die-cast induction motor is clarified.**

Keywords— *copper die-cast, efficiency, high-speed induction motor, slot insulation.*

I. Introduction

To insulate rotor bars from rotor core is a well-known strategy to reduce stray-load loss mainly caused by inter bar current [1]-[3]. Such rotor slot insulation is often employed in production process of aluminum die-casting rotors. In case of copper die-casting, however, the rotor slot insulation is rarely used. This is because there is no heat-resistant coating material satisfying both low cost and sufficient insulation performance against high copper melting temperature that is about 1.5 times of aluminum.

To address this situation, a new coating material capable of overcoming the temperature (that is more than 1300 degrees Celsius) of die-casting molten copper has been developed [4]. Since this coating material is made with a ceramic agent mainly, it is relatively easy to coat the inner surface of the slot with low-cost tooling. So, it is considered that to employ this coating material is suitable for copper die-cast squirrel cage rotor. In [5], an on-load test on a copper die-cast cage induction motor with/without slot insulation is carried out. On the basis of this result, Ref. [5] claims that the efficiency of copper die-cast cage induction motor can be improved especially in high-speed range by coating the rotor slot by the heat-resistant agent. However, critical items, specifications

and ratings of the test motor, measurement method, type of high-frequency converter (sinusoidal, rectangular or pulse-width-modulation (PWM) waves?), etc., are not shown at all. So, it is difficult to sufficiently understand the effectiveness of the slot insulation on loss reduction and the behavior of loss segregation.

Thus, the aim of our paper is to quantitatively verify the effect of slot insulation using the high heat-resistant coating agent on copper die-cast squirrel cage induction motor in high-speed operation over 10,000 r/min.

Two prototypes of copper die-cast squirrel-cage rotors are assembled: an insulated-slot rotor and a non-insulated-slot rotor. These rotors are installed to the same stator whose ratings are 0.75 kW, 200 V, 2 poles and 12,000 r/min. Full-load driving tests over 10,000 r/min are implemented. The results suggest that to insulate between the copper die-cast rotor bar and rotor core by the proposed heat-resistant agent contributes to improving the motor efficiency in high-speed and heavy-load operations. In addition the impact of the slot insulation on loss segregation of the copper die-cast induction motor is clarified.

II. Properties of the Prototype Motor

The authors use two copper die-cast rotors: prototype-A without slot insulation and prototype-B with slot insulation. The dimension and external appearance of the prototype-B are the same as those of prototype-A except for rotor slot insulation. These rotors are installed to a common stator shown in Fig. 1. With respect to the motor design, the dimension and winding specification of a 0.75kW high speed induction motor with aluminum die-cast squirrel cage rotor are fully applied to the tested motor. The ratings and specifications of the tested motor are listed in Table I.

Fig. 2 show a typical cross section of rotor slots. The surface of the inside of the slots is coated by the high-temperature heat-resistant agent. Fig. 3 shows an example photo of an insulated slot with coating.

Fig. 1. Photo of the test motor.

TABLE I
SPECIFICATION OF THE TEST MACHINES: PROTOTYPE A WITHOUT SLOT INSULATION AND PROTOTYPE B WITH SLOT INSULATION.

Pole	2
Voltage	200 (V)
Output	0.75 (kW)
Class	F (155 °C)
Current	3.1 (A)
Frequency	200 (Hz)
Speed (max.)	12,000 (r/min)

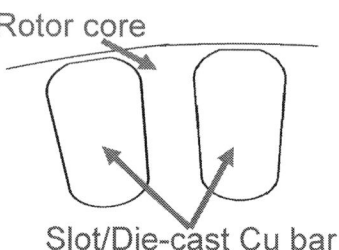

Fig. 2. Typical cross section of rotor slot.

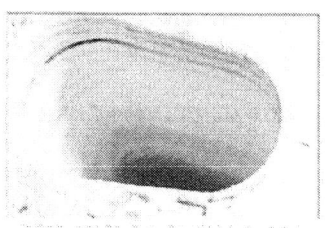

Fig. 3. Example of a rotor slot coated by high-heat-resistant material [5].

III. EXPERIMENTAL VERIFICATION

A. Experimetnal setup

Fig. 4 shows experimental setup. A loaded machine (IM), an inverter-fed four-guardant-speed-controlled induction generator load (G-IG) through a torque detector and a mechanical reduction gear, was coupled with the tested motor.

The test procedure of prototype-A is the same as that of prototype-B. First, after one hour warming operation, we carried out no-load tests, 12,000, 10,000 and 8,000 r/min, in turn. Next, we performed on-load tests in three frequencies whose synchronous speeds were 12,000, 10,000 and 8,000 r/min. During on-load test, the terminal voltage of the tested induction motor was kept in 200V/ 200Hz, 167V/167Hz and 133V/133Hz, respectively. In several load level from 25% to 150%, voltage, current, input, torque and rotor speed were measured by digital power meter and torque detector. Here, the level of the load was tuned by controlling the speed of the loaded induction generator so that the measured torque became just 0.90, 0.75, 0.60, 0.45, 0.30 and 0.15 Nm respectively. Namely, the tested induction motor was operated by open-loop V/F drive with no feedback control loop. Using a hot-line coil resistance meter [8], the stator resistance is measured in real time.

B. Efficiency

Fig. 5 shows efficiency versus torque characteristics in 8,000, 10,000 and 12,000 r/min operations. The results of Fig. 6 demonstrate that slot insulation of copper die-cast rotor contributes to improving the motor efficiency. At 0.6 Nm (100% load), the efficiencies of Fig. 5 (a), Fig. 5 (b) and Fig. 5 (c) rise by 2.2 %, 1.6 % and 1.9 % respectively by slot insulation. This result also suggests that the employed high-heat-resistant agent works well.

In Fig. 5 it can be also seen that, especially in heavy load range, the effectiveness of the slot insulation appears more clearly.

C. Loss

Fig. 6 shows stray load loss versus torque. One can see from Fig. 6 that the stray load loss of Prototype A is much larger than that of Prototype B, especially in heavy load operation range. The difference of stray load loss at 0.6 (Nm) were 10.4 (W) in 8,000 (r/min), 13.8 (W) in 10,000 (r/min), and 10.7 (W) in 12,000 (r/min), respectively.

Fig. 7 shows no-load rotational loss versus square of the terminal voltage. The no-load rotational loss is defined as $P_0 - 3I_0^2 r_1$. Here, P_0 is the input power at no-load test, I_0 is the stator current at no-load test, and r_1 is the stator resistance. At the rated voltages (133 (V) in 8,000 (r/min), 167 (V) in 10,000 (r/min), and 200 (V) in 12,000 (r/min)), the increment of core loss between two prototype motors were 6.95 (W) in 8,000 (r/min), 12.8 (W) in 10,000 (r/min), and 24.6 (W) in 12,000 (r/min), respectively.

Figs 8, 9 and 10 show stator current versus torque, stator copper loss versus torque, and rotor cage loss versus torque. In these figures, the difference between two prototype motors with respect to stator current, stator copper loss and rotor cage loss at 0.6 (Nm) were trivial.

The 2018 International Power Electronics Conference

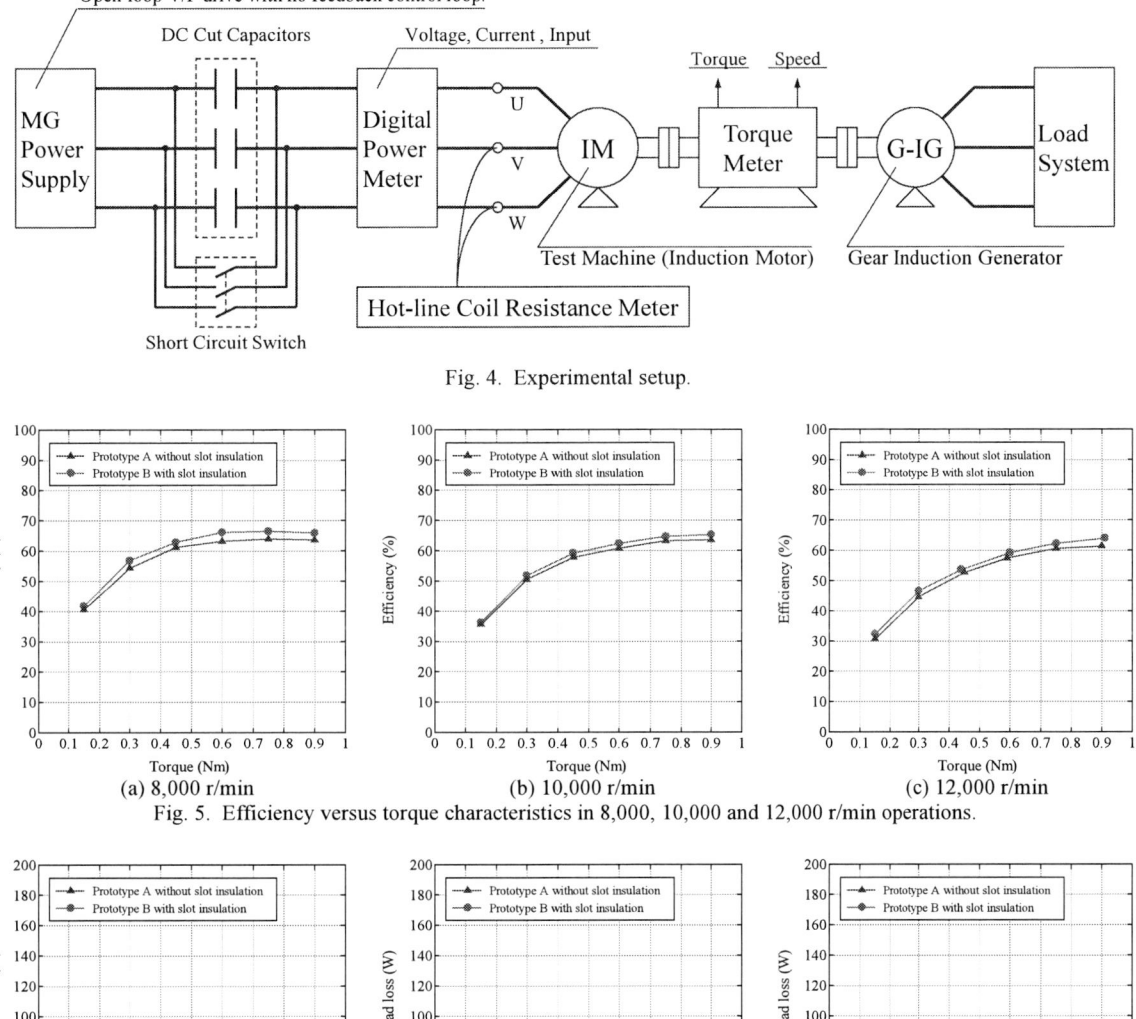

Fig. 4. Experimental setup.

Fig. 5. Efficiency versus torque characteristics in 8,000, 10,000 and 12,000 r/min operations.

Fig. 6. Stray load loss versus torque characteristics in 8,000, 10,000 and 12,000 r/min operations.

From these results, it can be seen that dominant factors that improves the efficiency of the prototype B with slot insulation are reductions of stray load loss and core loss.

The test is implemented by a pure sinusoidal power source. So, the stray load loss increasing with rise of load level in high speed over 10,000 r/min mainly takes account of inter-bar current loss, stator core loss caused by rotor slot harmonics, and rotor core loss caused by stator slot harmonics. It is assumed that the influence of phase band harmonics is much smaller than that of slot

harmonics [9]-[10]. Thus, it is considered that rotor slot insulation contributes to reducing the inter-bar currents since rotor slot insulation has little influence on increment of core loss by slot harmonics.

D. Rotor resistance

Figs. 11 and 12 show slip versus torque and power factor versus torque. From these figures it is confirmed that the performances of two motors are almost the same and no proportional shifting appears. This suggests that rotor insulation of the prototype B does not affect rotor

1962

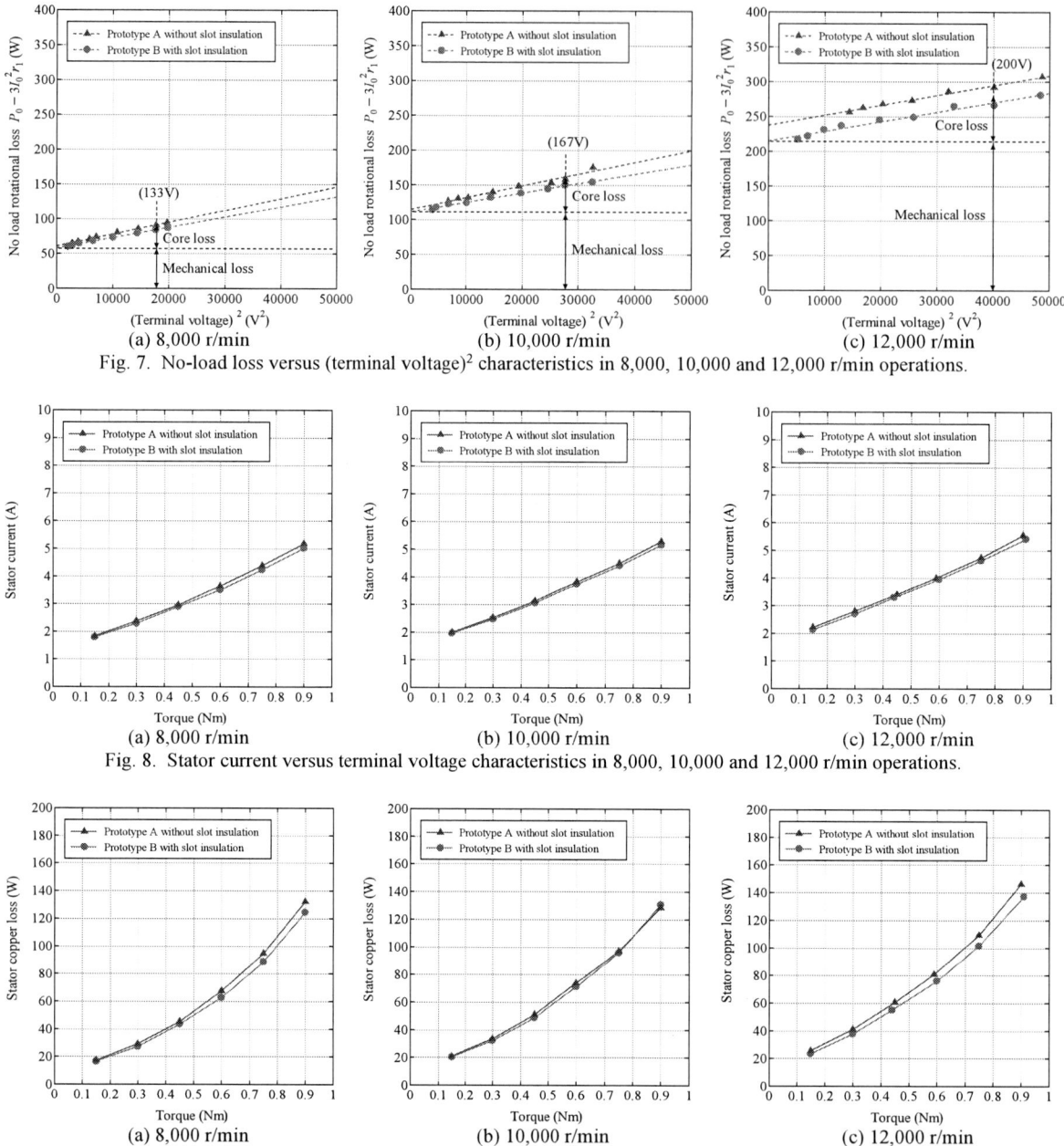

Fig. 7. No-load loss versus (terminal voltage)^2 characteristics in 8,000, 10,000 and 12,000 r/min operations.

Fig. 8. Stator current versus terminal voltage characteristics in 8,000, 10,000 and 12,000 r/min operations.

Fig. 9. Stator copper loss versus torque characteristics in 8,000, 10,000 and 12,000 r/min operations.

resistance.

In addition it is also found that trade-off between power factor and efficiency [11] is hardly seen.

IV. CONCLUSIONS

Effect of a new heat-resistant insulation-processing between rotor core and copper die-cast bar on motor efficiency in high-speed operations over 10,000 r/min was verified with experimental results on prototype induction motors having the same stator (0.75kW, 200V, 2P, 200Hz). One was a slot-insulated rotor, the other was a non-slot-insulated rotor. On the basis of no-load and full-load tests, the following results were confirmed and the validity of the new heat-resistant insulation-processing was demonstrated.

(1) The efficiency was improved by rotor slot insulation. When the rotor slot insulation was employed, increments of the efficiency were 2.2 % in 8,000 r/min, 1.6 % in 10,000 r/min and 1.9 % in 12,000 r/min.

(2) It was considered that one of dominant factors improving the efficiency was reduction of stray load loss by inter-bar currents.

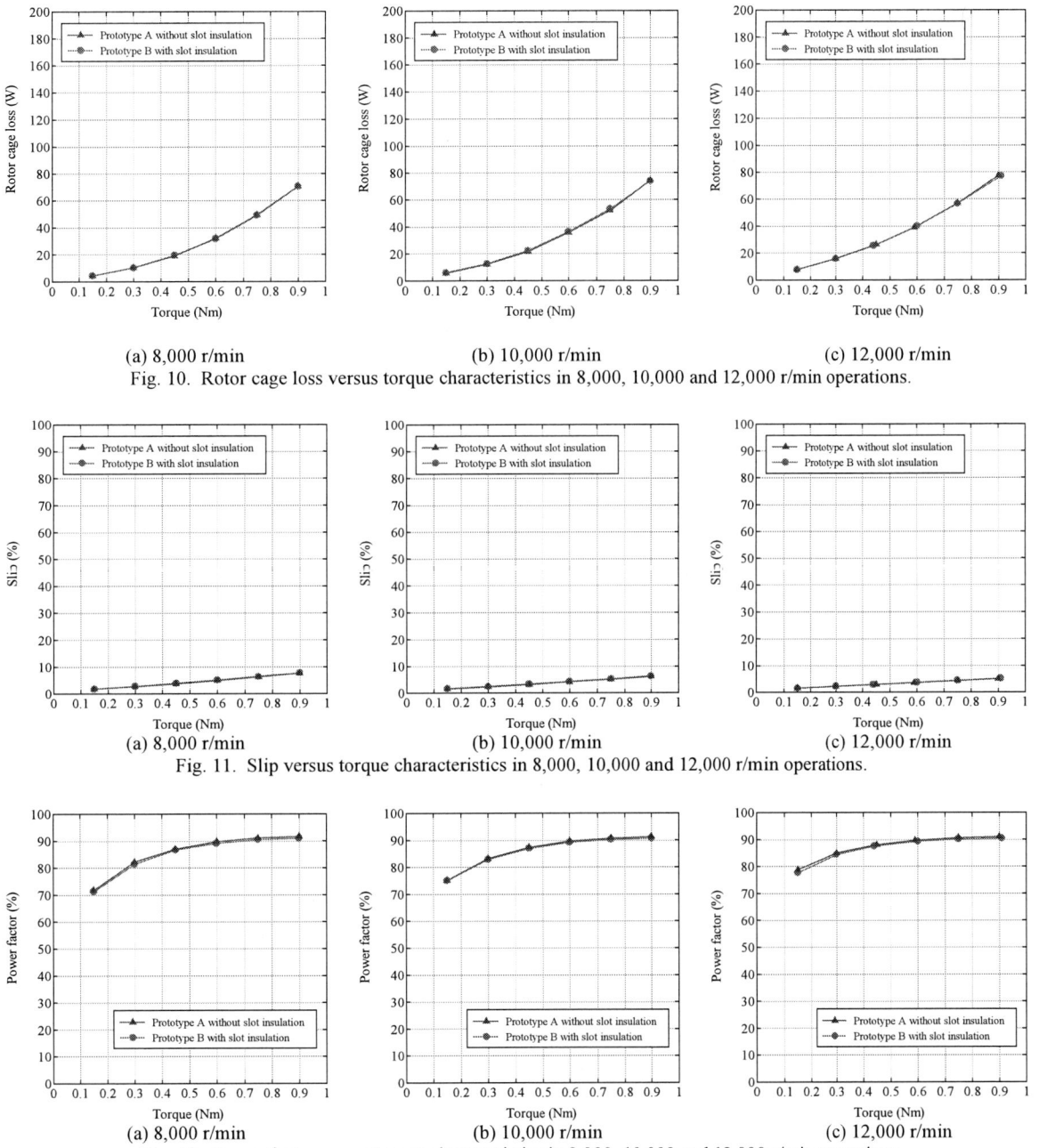

(a) 8,000 r/min   (b) 10,000 r/min   (c) 12,000 r/min

Fig. 10. Rotor cage loss versus torque characteristics in 8,000, 10,000 and 12,000 r/min operations.

(a) 8,000 r/min   (b) 10,000 r/min   (c) 12,000 r/min

Fig. 11. Slip versus torque characteristics in 8,000, 10,000 and 12,000 r/min operations.

(a) 8,000 r/min   (b) 10,000 r/min   (c) 12,000 r/min

Fig. 12. Power factor versus torque characteristics in 8,000, 10,000 and 12,000 r/min operations.

(3) Influence of the rotor slot insulation did not affect rotor resistance.

Experimental verification of pulse-width-modulation inverter-fed prototype induction motors is an important future work.

Acknowledgment

The author thanks International Copper Association for their technical supports.

References

[1] H. Nishizawa, "Study on Reliable Reduction of Stray Load Losses in Three-Phase Induction Motor for Mass Production," *IEEE Transactions on EC*, Vol. 2, No. 3, pp. 489-495, 1987.

[2] C. N. Glew, "Stray Load Losses in Induction Motors: A Challenge to Academia," Proc. of EMD 1997, Paper No. 444, 1997.

[3] S. Yasui, K. Ohkubo, and T. Obaru, "Study of Stray-Load Losses in Induction Motors," The Papers of Technical Meeting on Rotating Machinery, IEE Japan, RM-01-18, pp. 57-61, 2001 (in Japanese).

[4] I. Oura, J. Uchida, Y. Kinoshita, I. Koiwa, "Development of Heat Resistance Insulation Coating for Metal Materials of Electronics," Technical Report of Nihon Parkerizing, Vol. 28, No. 1, pp. 44-48, 2016 (in Japanese).

[5] M. Burwell, "Efficiency improvement of high-speed copper rotor induction motors," JCDA Seminar document of die-cast copper rotor induction motors, 2014 (in Japanese).

[6] The Institute of Electrical and Electronics Engineers, Inc., "IEEE Standard Test Procedure for Polyphase Induction Motors and Generators," IEEE Std. 112-1996, pp. 16-19, 1997.

[7] The Japanese Electrotechnical Committee, "Induction Machines," JEC-2137-2000, pp. 73-74, 2001 (in Japanese).

[8] "Inverter Hot Line Coil Resistance Meter DAC-HRI-3," Soken Electric Co. Ltd, Catalog, 2001 (in Japanese).

[9] K. Yamazaki, A. Suzuki, M. Ohto, and T. Takakura, "Harmonic loss and torque analysis of high speed induciton motors", IEEE Transactions on Industry Applications, vol. 48, no. 3, pp.933-941, (2012).

[10] K. Yamazaki, A. Suzuki, M. Ohto, and T. Takakura, "Circuit parameters determination involving stray load loss and harmonic torques for high-speed induction motors fed by inverters", *IEEE Transactions on EC*, vol. 28, no. 1, pp.154-163, 2013.

[11] A. H. Bonnettl, "An Overview of How AC Induction Motor Performance Has Been Affected by the October 24, 1977 Implementation of the Energy Policy Act of 1992," IEEE Trans. Ind. Appl., IA-36, No. 1, pp. 242-256, 2000.

High-precision Rotor Position Estimation for High-speed SPMSM Drive Based on State Observer and Harmonic Elimination

Peng Yang[1], Xi Xiao[1], Meng Zhang[2] and Shkodyrev Vyacheslav[3]

1 Department of Electrical Engineering, Tsinghua University, Beijing, China
2 Beijing Institute of Control Engineering, Beijing, China
3 Saint-Petersburg Polytechnic University, Russia
E-mail: xiao_xi@tsinghua.edu.cn

Abstract—A sensorless field-oriented control scheme (FOC) based on state observer is proposed. The estimated position and speed fluctuations caused by the high-order harmonics are analyzed. A harmonics filter is used to eliminate the harmonics. A high-speed surface-mounted permanent magnet synchronous motor (SMPSM) drive testbench is built to demonstrate the proposed control scheme. It is shown in the experimental results that the proposed method can significantly reduce the fluctuations of estimated position and speed.

Keywords— *harmonic elimination, high-speed surface permanent magnet synchronous motor (SPMSM), sensorless field-oriented control, state observer*

I. INTRODUCTION

Recently, the high-speed surface-mounted permanent magnet synchronous motor (SMPSM) is used in many industrial applications such as compressors and blowers. The operating speed of high-speed SPMSM is usually above 10000 rpm. Due to the high speed condition, an encoder cannot be installed on the rotor shaft [1]. Therefore, the sensorless field-oriented control (FOC) scheme is widely used in high-speed SPMSM drive. The scheme of typical sensorless FOC is shown in Fig. 1.

There are two kinds of sensorless estimation methods in the mainstream: the high frequency signal injection estimation methods (HFI) [2] [3] and the back-EMF based observers. As the HFI methods are only suitable for low speed condition, the back-EMF based observer is usually adopted in high-speed operation condition [4] [5]. The back-EMF based observers include state observer [6], the MRAS observer [7], sliding mode observer [8] and the Kalman filter observer, etc.

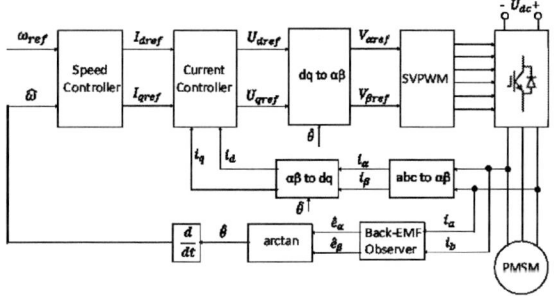

Fig.1. Scheme of typical sensorless FOC

Fig.1 shows the typical sensorless FOC for high-speed SPMSM. The sensorless observer estimates both position and speed of rotor shaft which are very important in FOC. Therefore, the accuracy of estimation will affect the performance of the control system. In high-speed sensorless PMSM drive, the estimation errors will affect the control stability and increase the losses. However, the estimated back-EMF signals inevitably contain many high-order harmonics, which are induced by various non-ideal factors [10]. The estimated rotor position error and speed fluctuation will occur because of the harmonics in back-EMF signals.

A state observer combined with harmonics filters is proposed to estimate the position and speed. The principle of the state observer which is used in this paper is presented and the back-EMF harmonics are analyzed in Section II. In Section III, the proposed harmonic filter is introduced to eliminate the harmonics. After that, Section IV proposes the sensorless FOC for high speed SPMSM. The experimental results are given in Section V.

II. PRINCIPLE OF SENSORLESS STATE OBSERVER

A. Model of PMSM

The stator voltage equations of the PMSM in αβ coordinates can be written as:

$$\begin{bmatrix} u_\alpha \\ u_\beta \end{bmatrix} = R_s \begin{bmatrix} i_\alpha \\ i_\beta \end{bmatrix} + \frac{d}{dt} \begin{bmatrix} L_\alpha & L_{\alpha\beta} \\ L_{\alpha\beta} & L_\beta \end{bmatrix} \begin{bmatrix} i_\alpha \\ i_\beta \end{bmatrix} + \omega\psi_f \begin{bmatrix} -\sin\theta \\ \cos\theta \end{bmatrix} \quad (1)$$

where

u_α , u_β	voltages in αβ axes;
i_α, i_β	currents in αβ axes;
R_s	stator resistance;
L_α, L_β, $L_{\alpha\beta}$	inductances;
Ψ_f	permanent magnetic flux linkage;
ω	electrical speed;
θ	electrical position.

For surface-mounted PMSM:

$$L_\alpha = L_\beta = L_s \quad (2)$$

$$L_{\alpha\beta} = 0 \quad (3)$$

where L_s is the stator inductance.

Therefore, (1) can be rewritten as:

$$L_s \frac{di_s}{dt} = u_s - Ri_s - e_s \quad (4)$$

where

$$i_s = \begin{bmatrix} i_\alpha & i_\beta \end{bmatrix}^T \quad (5)$$

$$u_s = \begin{bmatrix} u_\alpha & u_\beta \end{bmatrix}^T \quad (6)$$

$$e_s = \begin{bmatrix} e_\alpha \\ e_\beta \end{bmatrix} = \begin{bmatrix} -\omega\psi_f \sin\theta \\ \omega\psi_f \cos\theta \end{bmatrix} \quad (7)$$

The mechanical equation of SPMSM can be written as

$$J\frac{d\omega}{dt} = T_e - F\omega - T_L \quad (8)$$

$$T_e = 1.5 p\psi_f i_q \quad (9)$$

where

J	rotor moment of inertia;
p	pole pairs;
F	friction coefficient;
T_L	load torque;
T_e	electromagnetic torque;
i_q	q-axis current.

B. Rotor Position Estimation Using State Observer

The stator observer is designed as follows to obtain the back-EMF signals [6]:

$$\begin{cases} \dfrac{d\hat{i}_s}{dt} = \dfrac{R_s}{L_s}\hat{i}_s + \dfrac{1}{L}(u_s - \hat{e}_s) \\ \hat{e}_s = k(\hat{i}_s - i_s) = k\Delta i \end{cases} \quad (10)$$

where $\hat{i}_s$ and $\hat{e}_s$ represent the estimated current and back-EMF; k represents the observer gain; Δi_s represents the estimation error of current. By subtracting (4) from (10), the first-order differential equation is obtained.

$$\frac{d\Delta i_s}{dt} = \frac{e_s}{L} - \frac{R+k}{L}\Delta i_s \quad (11)$$

where

$$e_s = \begin{bmatrix} e_\alpha \\ e_\beta \end{bmatrix} = \begin{bmatrix} -\omega\psi_f \sin\omega t \\ \omega\psi_f \cos\omega t \end{bmatrix} \quad (12)$$

By solving the first-order differential equation in (11), Δi_s can be obtained as

$$\Delta i_s = \begin{bmatrix} \Delta i_\alpha \\ \Delta i_\beta \end{bmatrix} = \begin{bmatrix} -\omega\psi_f \sin(\omega t - \Delta\theta) \\ \omega\psi_f \cos(\omega t - \Delta\theta) \end{bmatrix} \quad (13)$$

where $\Delta\theta$ is the phase lag between estimated back-EMF and actual back-EMF:

$$\Delta\theta = \arctan(\frac{\omega L}{k+R}) \quad (14)$$

$\hat{e}_s$ can be obtained accordingly as

$$\hat{e}_s = \begin{bmatrix} \hat{e}_\alpha \\ \hat{e}_\beta \end{bmatrix} = k\Delta i_s = \begin{bmatrix} -k\omega\psi_f \sin(\omega t - \Delta\theta) \\ k\omega\psi_f \cos(\omega t - \Delta\theta) \end{bmatrix} \quad (15)$$

As can be seen in (15), $\hat{e}_s$ is synchronized with actual back-EMF with a phase lag. Therefore, the estimated rotor position can be calculated accurately:

$$\hat{\theta} = -\arctan(\frac{\hat{e}_\alpha}{\hat{e}_\beta}) + \Delta\theta = -\arctan(\frac{\hat{e}_\alpha}{\hat{e}_\beta}) + \arctan(\frac{\omega L}{k+R}) \quad (16)$$

The block diagram of the proposed state observer is shown in Fig.2.

C. Analysis of back-EMF harmonics

As is mentioned in [11], the 5th and 7th harmonics in the estimated back-EMF signals are mainly induced by the inverter nonlinearity and the gaps between permanent magnet pieces placed in the rotor.

Fig.3 and Fig.4 present the experimental results of the above sensorless state observer. The sinusoidal back-EMF signalsare distorted because of harmonics in Fig.3(a). In Fig.3(b), the Lissajous curve of the back-EMF signals is not a standard circle, which also shows the distortion caused by harmonics. Fig.4(a) shows that the calculated rotor positon fluctuates because of the large harmonic components. Fig.4(b) presents the estimated speed which is calculated by the estimated position. The estimated speed also fluctuates obviously.

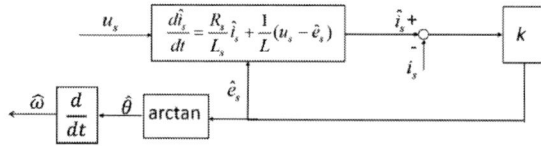

Fig.2. The proposed sensorless state observer

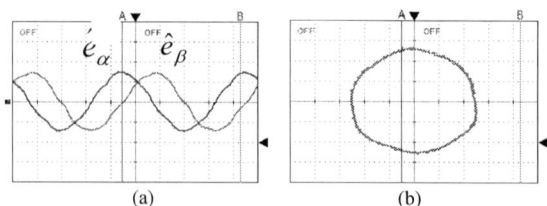

(a)   (b)

Fig.3. Analysis of back-EMF harmonics. (a) The estimated back-EMF by state observer. (b) Back-EMF Lissajous curve.

The 2018 International Power Electronics Conference

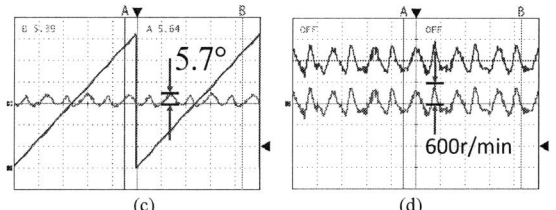

(c)   (d)

Fig.4. Estimated position and speed. (a) Estimated rotor position fluctuations. (b) Estimated speed ripples.

In the high-speed PMSM drive, the estimation errors of rotor position and speed caused by harmonics will affect the control stability and increase the losses. Therefore, the harmonics in the estimated back-EMF need to be reduced.

III. PROPOSED HARMONIC ELIMINATION METHOD

According to the analysis in Part II, the harmonics in the back-EMF estimation cause fluctuations of position and speed estimation. In this part, a harmonics filter is applied to reduce the harmonics.

A. Orthogonal PLL

PLL is widely used in sensorless control applications to acquire both rotor position and speed. Fig.5 shows the principle of orthogonal PLL which consists of the phase detector, the loop filter and the voltage controlled oscillator.

However, [12] points out that the performance of PLL will also be deteriorated by harmonics. So the harmonics filter is then introduced to filter the harmonics.

B. Harmonics Filter

A Synchronous Frequency-extract Filter (SFF) is proposed in [11]. The harmonics in the raw signal are filtered out by extracting the fundamental waves. The principle of SFF is presented in Fig.6.

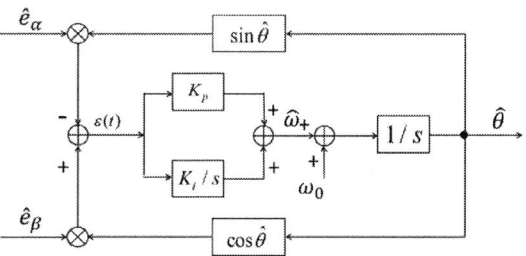

Fig.5. The block diagram of orthogonal PLL

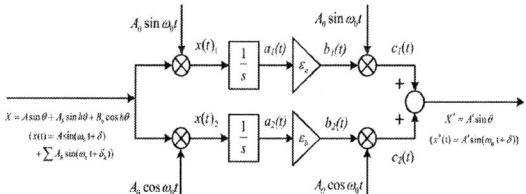

Fig.6. The block diagram of SFF

C. Proposed Harmonic Elimination Method

The proposed rotor position estimation method combined with PLL and SFFs is proposed to eliminate the harmonics in the back-EMF signals estimated by the aforementioned state observer. The proposed method is shown in Fig.7. The back-EMF signals are first input to a PLL to get the frequency and phase of fundamental wave. Then the output of the PLL is used in SFFs to eliminate the harmonics in the raw signals. Then both the position and speed are calculated using the filtered signals.

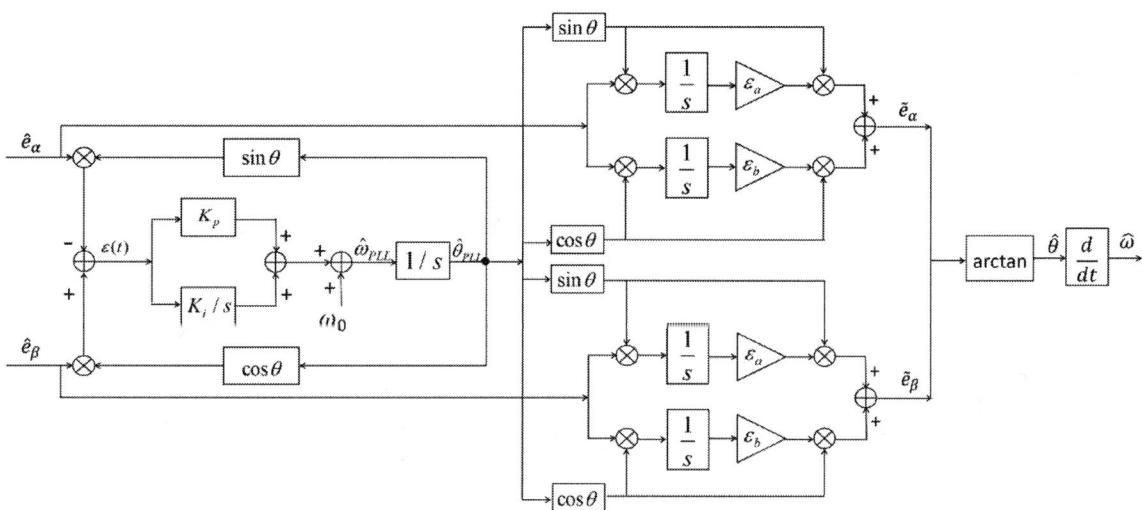

Fig.7. The proposed harmonic elimination method

1968

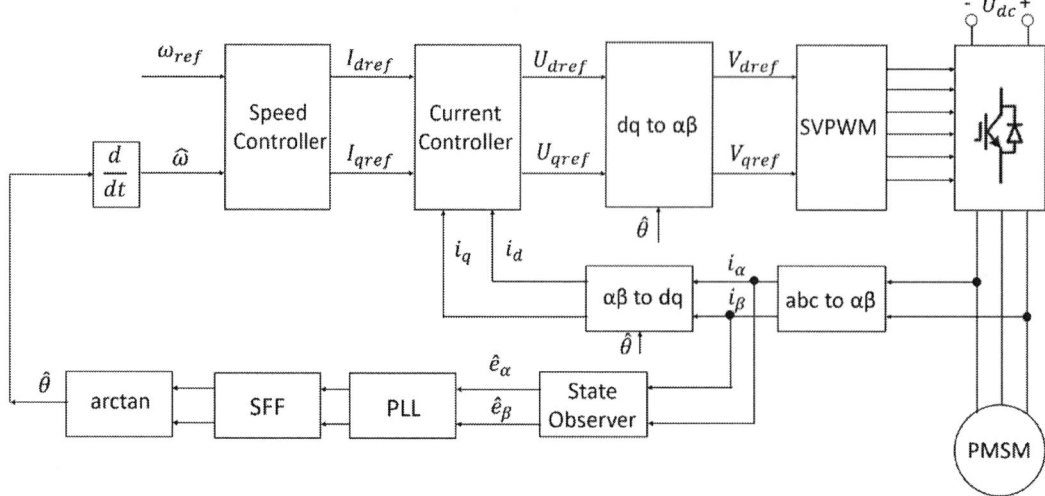

Fig.8. The proposed sensorless FOC

IV. Sensorless Field-Oriented Control Method for High Speed SPMSM

The sensorless FOC based on the proposed rotor position estimation method is shown in Fig.8. The high-speed SPMSM drive prototype is built, which consists of the frequency inverter and the high-speed SMPSM. The frequency inverter (MWINV-9R144) is a product of Myway Plus Corporation, which can output 15Arms AC current and 380 Vrms AC Voltage. The switching frequency is set to 10 kHz. A DSP (TMS320F28335) is used as the digital controller. A ball-bearing high-speed SPMSM which can achieve the speed of 15000 rpm is used. The photograph of the motor drive prototype is shown in Fig.9. The parameters of the high-speed SPMSM are shown in Table I.

The start-up from zero speed is especially considered because the back-EMF signals are hardly detected in low-speed region. The start-up consists of two stages: positioning stage and I/f control stage. In the positioning stage, the rotor position is set to 0° by generating a constant current in α axis. Then in the I/f control stage,

the amplitude of the current is set to be constant, and the frequency increases to accelerate the rotor. When the rotor speed accelerates to 800rpm, the control method will switch to sensorless field-oriented control. The back-EMF is observed by the aforementioned state observer. The proposed position estimation method is applied to obtain the estimated position $\hat{\theta}$ and the estimated speed $\hat{\omega}$. Both $\hat{\theta}$ and $\hat{\omega}$ are used in the field-oriented control scheme.

Table I. Experimental Parameters

Variable	Value
Pole pairs	2
Rated power	30kW
DC voltage	500V
Rated AC current	60Arms
Stator resistance	0.122Ω
Stator inductance	0.675mH
Flux linkage	0.0812 V.s
Rated speed	15000r/min

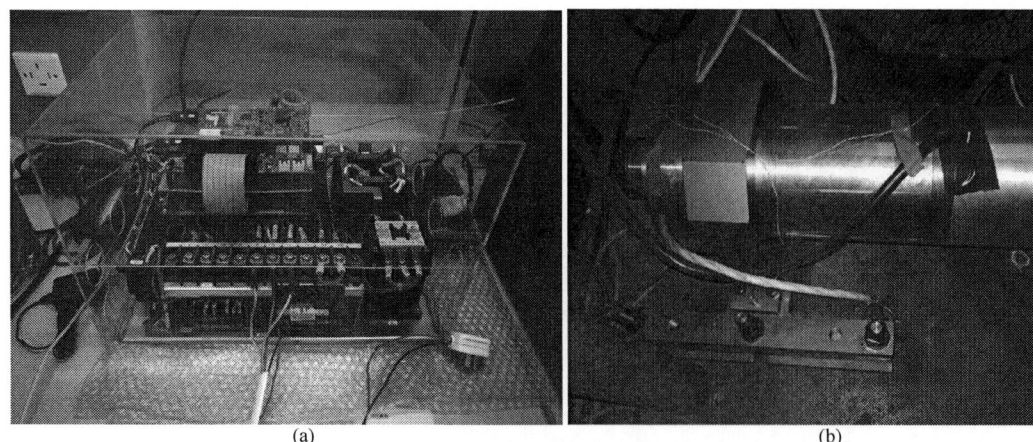

(a)   (b)

Fig.9. The photograph of high-speed SMPSM drive prototype. (a) Control system (MWINV-9R144). (b) High-speed SMPSM.

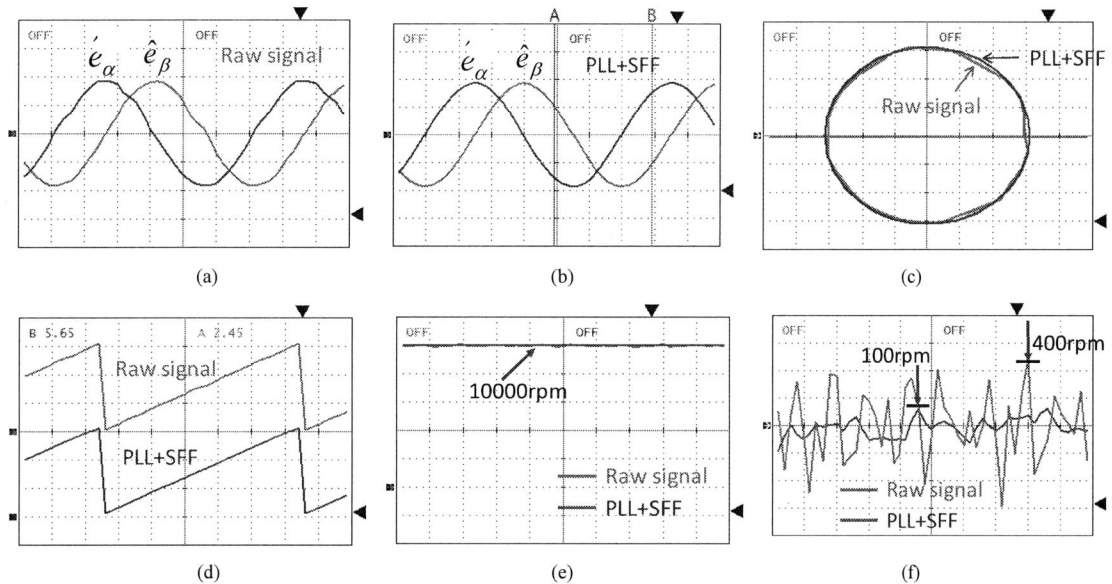

Fig.10. The experimental results at 10000rpm without load. (a) Raw signal of estimated back-EMF. (b) Back-EMF processed by PLL and SFF. (c) Back-EMF Lissajous curve. (d) Estimated rotor position. (e) Estimated rotor speed. (f) Estimated speed ripples.

V. EXPERIMENTAL RESULTS

The effectiveness of the proposed sensorless FOC scheme is validated in experiment without load. The estimation results are analyzed. The rotor speed is set to 10000 rpm. Fig.10 shows the experimental results at 10000 rpm without load.

Comparing Fig.10(a) and Fig.10(b), the back-EMF signals processed by PLL and SFF are more sinusoidal than the raw signals. And the Lissajous curve of the former is more close to a circle in Fig.10(c), which proves that the proposed harmonic elimination method is effective. In Fig.10(d) and Fig.10(f), the fluctuations of estimated rotor position and speed are reduced because of the harmonics are eliminated. In Fig.10(f), the speed ripples are significantly reduced.

VI. CONCLUSION

A sensorless field-oriented control scheme for high-speed SPMSM based on state observer and the proposed harmonic elimination method is proposed. The fluctuations of both rotor position and speed estimation are reduced due to the harmonic elimination method. The future work will focus on the efficiency of the proposed sensorless control scheme.

ACKNOWLEDGMENT

This work was supported by the National Natural Science Foundation of China under Grant 51577095.

REFERENCES

[1] Y. Jung-Sik, L. Wook-Jin, S. Seung-Ki, Y. Hyun-Sup, and K. Jin-Tae, "Sensorless vector control of super high speed turbo compressor," in Twentieth Annual IEEE Applied Power Electronics Conference and Exposition, 2005. APEC 2005, 2005, vol. 2, pp. 950-953 Vol. 2.

[?] J. M. Liu and Z. Q. Zhu, "Sensorless control strategy by square-waveform high-frequency pulsating signal injection into stationary reference frame," IEEE J. Emerg. Sel. Topics Power Electron, vol. 2, no. 2, pp. 171–180, Jun. 2014.

[3] T. Tuovinen and M. Hinkkanen, "Adaptive full-order observer with high frequency signal injection for synchronous reluctance motor drives," IEEE J. Emerg. Sel. Topics PowerElectron, vol. 2, no. 2, pp. 181–189, Jun. 2014.

[4] P. Yongsoon and S. Seung-Ki, "Sensorless control method for PMSM based on frequency-adaptive disturbance observer," IEEE J. Emerg. Sel. Topics Power Electron, vol. 2, no. 2, pp. 143–151, Jun. 2014.

[5] S. Shinnaka, "New "D-state-observer"-based vector control for sensorless drive of permanent-magnet synchronous motors," IEEE Trans. Ind. Appl., vol. 41, no. 3, pp. 825–833, Jun. 2005.

[6] L. Li, G. Tan, J. Liu and B. Kou, "Sensorless Control of PMSM Based on State Observer in the Steady and Transient State Operations without Phase Lag," in 2013 International Conference on Information Science and Technology, 2013.

[7] S. Kai, L. Kui, and H. Lipei, "Control strategy of PMSM drive in high speed operation for air-condition compressor," in 2008 34th Annual Conference of IEEE Industrial Electronics, 2008, pp. 1137-1142.

[8] W. Gaolin, Z. Hanlin, Z. Guoqiang, G. Xianguo, and X. Dianguo, "Adaptive compensation method of position estimation harmonic error for EMFbased observer in sensorless IPMSM drives," IEEE Trans. Power Electron., vol. 29, no. 6, pp. 3055–3064, Jun. 2014.

[9] Z. Zheng, Y. Li and M. Fadel, "Sensorless control of PMSM based on extended kalman filter," in 2007 European Conference on Power Electrionics and Applications, 2007.

[10] R. Wegener, F. Senicar, C. Junge, and S. Soter, "Low cost position sensor for permanent magnet linear drive," in Proc. IEEE Power Electron. Drives Syst., Nov. 2007, pp. 1367–1371.

[11] X. Song, J. Fang, B. Han, and S. Zheng, "Adaptive Compensation Method for High-Speed Surface PMSM Sensorless Drives of EMF-Based Position Estimation Error," IEEE Transactions on Power Electronics, vol. 31, no. 2, pp. 1438-1449, 2016.

[12] W. Liang, J. Qirong, H. Lucheng, Z. Chunpeng, and W. Yingdong, "A novel phase-locked loop based on frequency detector and initial phase angle detector," IEEE Trans. Power Electron., vol. 28, no. 10, pp. 4538– 4549, Oct. 2013.

Harmonic Loss Reduction in High Speed Motor Drive Systems by Flying Capacitor Multilevel Inverter

Anudari Tumurbaatar[1*], Sae Mochidate[1], Koji Yamaguchi[2], Tomohiro Matsuda[2] and Yukihiko Sato[1]

1 Graduate School of Science and Engineering, Chiba University, Japan
2 Products Development Center, IHI Corporation, Japan
*E-mail: afwa0111@chiba-u.jp

Abstract— In AC motor drive, the harmonic components in the output voltage of inverters generate additional iron losses in magnetic cores and additional copper losses in windings of the motor. Particularly in the case of high speed motors, the ripple of the current flowing through the motor increases, because the inductance is designed to be small. Therefore, the additional losses caused by the ripple of the current are often serious problems in the high speed motors.

In this research, as a solution to these problems, multilevel inverters are applied to high speed motor drive systems. Multilevel converters can generate multilevel output voltage using internal voltage sources. Thus, they can reduce harmonics and EMI.

In this paper, the loss reduction effect in the high speed motors by the multilevel inverters is confirmed by simulation and experiment.

Keywords—Multilevel inverter, High speed motor

I. Background

Voltage source inverters, which are widely used now, generate a great deal of harmonic contents in the output voltage because it is a pulse train consisting of two voltage levels. In the case of motor drive systems, which are typical examples of applications of the inverters, it is often reported that the iron loss and the copper loss may increase significantly due to the harmonic contents in the inverter outputs. These losses cause additional temperature rise in the motors and reduce motor efficiency considerably [1].

Particularly, in permanent magnet high speed motors, the inductance restricts upper limit of operating speed. Therefore, the inductance of the motor is designed to be small compared to ordinal motors. So, the ripple of the current flowing through the motor increases and the additional losses in the motor occurs significantly. In addition, the higher fundamental output frequency required for high speed drive makes the improvement of the output waveform difficult. Furthermore, poor heat dissipation capability of high speed motors due to the reduced volume makes limitation of power losses severe.

As a solution to these problems, harmonic filters are often connected to output terminal of the inverter to obtain high quality sinusoidal current waveform. But, in this case, the volume and weight of components of the harmonic filter such as inductors and capacitors are

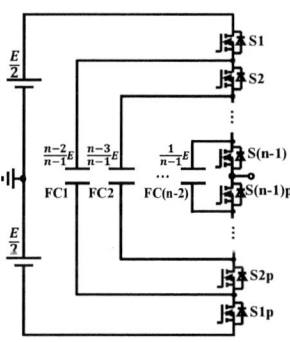

Fig. 1. Circuit configuration per phase of generalized n-level flying capacitor multilevel converters.

practical problem.

In this study, as a better solution to these harmonic problems, application of multilevel inverters to high speed motor drive systems is investigated. The multilevel inverters can generate multilevel voltage waveforms in the AC side with smaller instantaneous deviation from its reference. As a result, even without output filters, an almost sinusoidal output current can be obtained [2]. In this way, the redution of the harmonic loss in the high speed motor by multilevel inverter is expected.

In this paper, to clarify the advantage of multilevel inverters on harmonic loss reduction in high speed motor, as a first step, the harmonic copper loss is calculated for different number of output levels by simulation. Finally, the effectiveness in the reduction of harmonic loss is confirmed by experimental results employing a test machine.

II. Features Of Flying Capacitor Multilevel Inverters

A. Operation Principle

As the multilevel inverters, various topologies have been proposed such as a diode clamp type, a cascade connected type, and a flying capacitor type [3-5]. In this research, the flying capacitor topology is treated because it is the most suitable for downsizing of the capacitors and their associated balancing circuits.

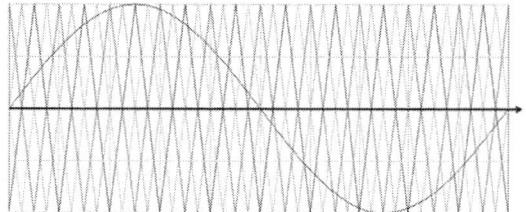

Fig. 2. Carrier phase shifted modulation.

Fig.1 shows circuit configuration in a leg of the generalized n-level flying capacitor converters. There are $(n-1)$ power switches in each of the upper and lower half of the leg. The $(n-2)$ flying capacitors hold voltages that are different by $E/(n-1)$ each other. The output voltage levels except both the maximum and minimum voltages $(E/2\ and\ -E/2)$ are generated by combinations of the capacitor voltages and source voltage. In addition, it should be noted that an intermediate output voltage can be generated by several different switching states of the $2(n-1)$ power devices.

B. Characteristic and Advantages

In the flying capacitor topology, the voltage ripple of the capacitors can be reduced by introducing high switching frequency operation because each capacitor has charge and discharge modes in a PWM carrier period. In other words, it is possible to reduce the required capacitance of the flying capacitors for a given permissible ripple voltage with higher switching frequency operation.

When phase-shifted-carrier PWM is applied, the durations of different switching states are equal in a PWM carrier period and the charging and discharging of each flying capacitor are balanced in each carrier period. As the result, the average voltage of the flying capacitors can be inherently stable without feedback control [6].

In carrier phase shifted modulation shown in Fig.2, the phases of $(n-1)$ carrier waves are shifted each other by $2\pi/(n-1)$ [rad]. These carrier waves are compared with the voltage reference signal. As the result, all the switching devices in the circut performs switching equally in a whole output fundamental period. When the carrier period is sufficiently short with respect to the fundamental period, it can be assumed that the output current does not change in one carrier period. Thus, the current flows through each switching device equally and the losses in switching devices become uniform. In other multilevel topologies, the current may concentrate on some specific switching devices depending on the operating conditions. Consequently, the overall cooling efficiency decreases due to heat concentration. In contrast, the losses of switching devices in the flying capacitor topology is uniform. Thus, volume reduction of required cooling devices can be realized.

From the above discussion, it is concluded that the flying capacitor topology has advantages in the realization of high power density compared to other multilevel circuit topologies.

In this context, the flying capacitor topology has been put into commercial products for midium voltage motor

drive [7].

III. SIMULATIONS ON HARMONIC COPPER LOSS REDUCTION IN HIGH SPEED MOTOR

In motor drive systems based on the multilevel inverter, proper selection of number of output levels is a key issue to reduce the power loss of motor drive system. As the number of output levels increase, the output waveform becomes sinusoidal and harmonics can be reduced.

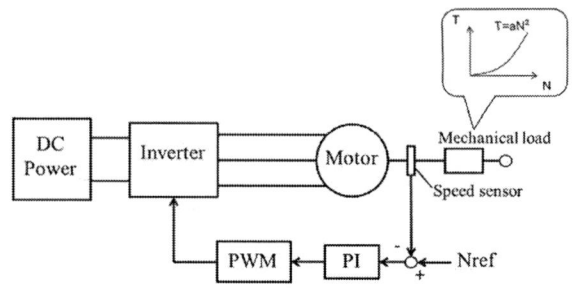

Fig. 3. Schematic block diagram in simulation.

TABLE I
SIMULATION CONDITIONS AND MOTOR PARAMETERS

Input DC voltage	270V
Motor output power	1.2kW
Rated speed	100krpm
Number of poles	2
Motor inductance (L_d, L_q)	20µH
Inverter switching frequency	20kHz

Consequently, the iron loss and copper loss in the motor due to the harmonics of inverter output will decrease. On the other hand, the losses in inverter tends to increase when the number of the output levels increases. Therefore, to clarify the effect of the number of output levels on total harmonic loss reduction in high speed motor drive systems, the investigation by simulation is carried out. Fig.3 shows a schematic block diagram of high speed motor drive system. In this simulation, the harmonic copper loss with different numbers of levels of inverter from 2 level to 9 level is calculated. The motor parameters and simulation conditions are shown in Table I. It should be noted that the motor inductance is very small compared to ordinal motors. As a load of the high speed motor, a mechanical load whose torque is proportional to speed squared is connected and vector control is applied.

Fig.4(a)-(c) show simulation waveforms of the inverter output line-to-line voltage and phase current for three different numbers of output levels. In Fig.4(a), the phase current contains very large harmonics due to the low motor inductance and high output fundamental frequency (1.6kHz) for high speed operation. From Fig.4(b) and (c), it is confirmed that the output voltage waveform becomes close to a sinusoidal wave as the number of output levels increases. In addition, although any harmonic filters are not connected between the high speed motor and the output terminal of the inverter, the load current becomes almost sinusoidal wave that has little distortion in the case of 9 level even when the motor inductance is so small and the fundamental output frequency is so high.

Fig.5 shows the result of FFT analysis of output line-to-line voltage and phase current waveforms. We can see that as the number of output levels increases, the amplitude of harmonic effective value can be decreased significantly.

Fig.6 shows the relative total harmonic distortion (THD) of the output voltage and current for different numbers of output levels with respect to that in the conventional 2 level inverter. From this figure, we can see that as the number of output levels increase, the THD can be decreased in both the voltage and current waveforms.

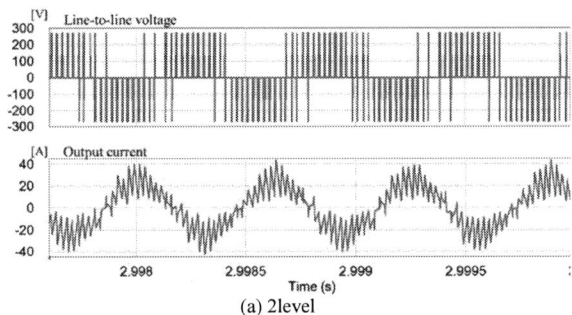

(a) 2level

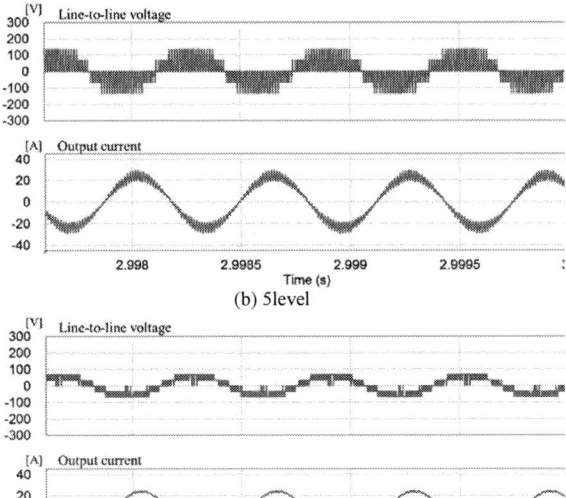

(b) 5level

(c) 9level

Fig. 4. Output line-to-line voltage and current waveform of inverter at each level by simulation.

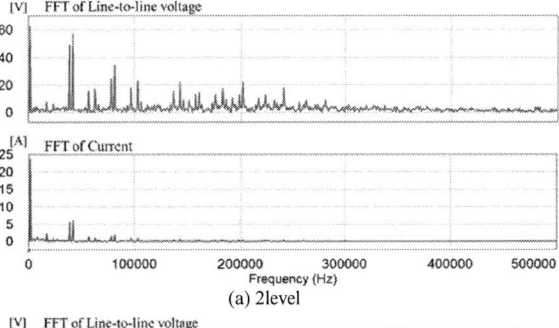

(a) 2level

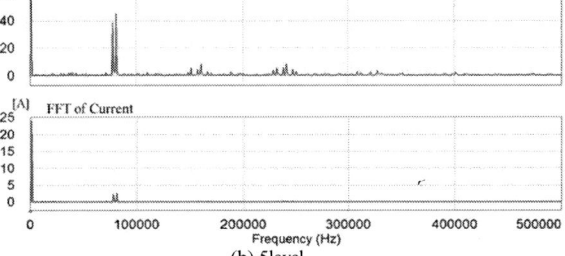

(b) 5level

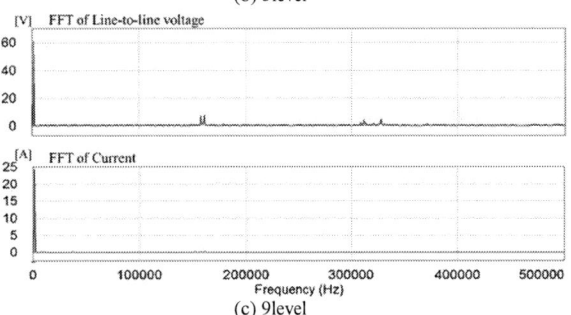

(c) 9level

Fig. 5. Fast Fourier transform of output waveforms of inverter at each level by simulation.

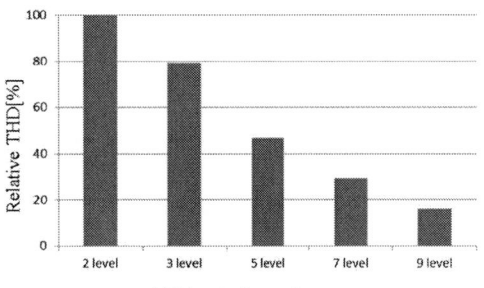

(a) Line-to-line voltage

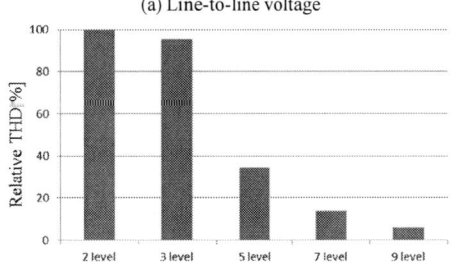

(b) Phase current

Fig. 6. Relative harmonic distortion of output waveforms for different numbers of output levels by simulation.

Under this simulation conditions, the THD have not changed constantly with increase in the number of output levels due to the low modulation factor. Fig.7 shows the harmonic copper loss in a high speed motor calculated by the result of FFT analysis. Table II shows the power losses in the high speed motor for five different numbers of output levels. The copper loss has been calculated from the following equation (1). Here, $R_{dc}, R_{ac}(f_n)$ are the DC and AC resistance in the high speed motor, $K_c(f_n)$ is the resistance coefficient that depends on frequency, and $I(f_n)$ is the effective value of current of frequency f_n obtained by the FFT analysis. In this simulation, the difference between the total power loss and the copper loss caused by the fundamental wave is defined as "harmonic copper loss".

$$W_{Cu} = \sum_{i=1} 3 * R_{ac}(f_n) * I(f_n)^2$$
$$= \sum_{i=1} 3 * k_c(f_n) * R_{dc} * I(f_n)^2 \qquad (1)$$

Fig. 7. Harmonic copper loss
for different numbers of output levels by simulation.

TABLE II
HARMONIC COPPER LOSS IN HIGH SPEED MOTOR BY SIMULATION

Number of output levels	Harmonic copper loss [W]
2	115.7
3	105.4 (-8.9% from 2 level)
5	30.7 (-73.5% ″)
7	8.7 (-92.5% ″)
9	1.3 (-98.9% ″)

From Fig.7 and Table II, as the number of output levels increase, the harmonic copper losses can be remarkably decreased in all cases. The reduction of harmonic copper loss from 3 level to 5 level is especially significant. However, the reduction over 5 level is small. Comparing 9 level to conventional 2 level, the harmonic loss can be reduced by 98.9% and we can confirmed that almost all the copper loss due to the harmonics components can be reduced. From these results, we can confirm that the multilevel inverter is very effective on the reduction of the harmonic loss in high speed motors.

IV. EXPERIMENTAL RESULT

A. Experimental System

In this experiment, the effect on reduction in the motor loss by the multilevel inverter is investigated. Fig.8 shows schematic diagram of the experimental system. The prototype of the flying capacitor multilevel inverter shown in Fig.9(a) is specially constructed so that the number of the output levels can be selectable by changing connection of multiple circuit units. Fig.9(b) shows the high speed motor coupled with a generator employed in the experiment. In this experiment, the number of levels of inverters was changed to 3, 5, and 7, and in each case the loss of the motor was measured. The experimental conditions are summarized in Table III.

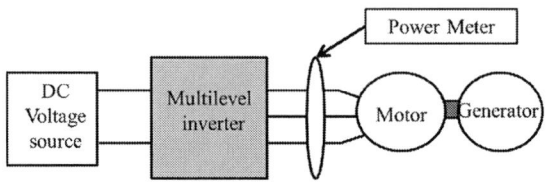

Fig. 8. Circuit connection for experiments.

(a) Prototype flying capacitor converter with
selectable number of output levels.

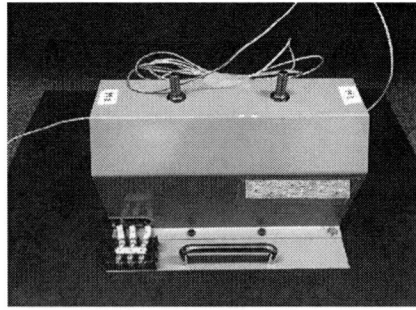

(b) High speed motor and Generator
Fig. 9. Experimental system.

TABLE III
EXPERIMENTAL CONDITIONS AND MOTOR PARAMETERS

Input DC voltage	40V
Rated speed	5.4krpm
Number of poles	10
Motor inductance (L_d, L_q)	20μH
Inverter switching frequency	20kHz

B. Experimental Results

Fig.10(a)-(c) shows the observed waveforms for three different numbers of output levels. Each figure shows the motor input line-to-line voltage and input current. It also shows the waveform of the induced voltage of the generator mechanically connected to the motor. From the observed waveform, it is confirmed that the input voltage waveform of the motor approaches sinusoidal and the distortion of the input current becomes smaller as the number of levels increases.

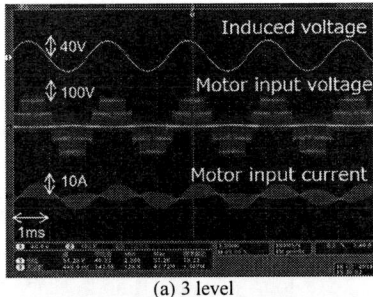

(a) 3 level

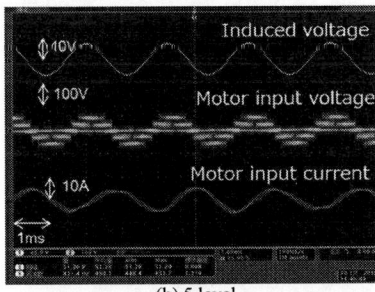

(b) 5 level

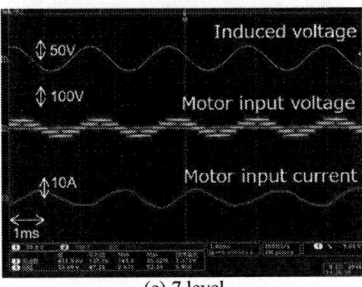

(c) 7 level

Fig. 10. Experimental waveforms.

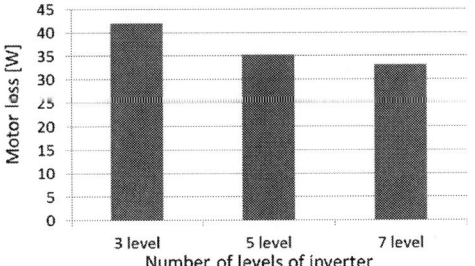

Fig.11. Total loss of motor by experiment.

Fig.11 shows motor loss for each level. In this experiment, the motor was driven at a rotational speed of 5400[rpm] with no load. Therefore, the power supplied to the motor by the inverter corresponds to total loss including copper loss, iron loss, mechanical loss, etc. When the 3-levels inverter is used, the motor loss is 42[W], whereas it is 35[W] at 5-levels and 33[W] at 7 levels. By changing the number of levels of the inverter from 3 to 7, motor loss was reduced by about 21.4%. From this result, it is predicted that a larger effect can be obtained by increasing the inverter level number from 2 to 7. And when the motor is driven in a higher speed range, this effect is expected to be significant.

V. CONCLUSION

This paper presents an investigation of loss reduction in high speed motor drive systems by introduction of multilevel inverters. First, the level of the inverter was changed from 2 to 9, and the harmonic loss of the motor was compared by simulation. In the simulation result, due to the significant reduction of harmonics included in the output current of inverter, the harmonic copper loss in high speed motor remarkably decreases. Next, the reduction in the harmonic losses due to the multilevel inverter has been demonstrated experimentally.

In this paper, the experimental investigation has been conducted under the limited conditions. The experimental investigations on operating characteristics under wide variety of load conditions, effect of the inverter output levels including conventional 2 level, reduction or elimination of a harmonic LC filter are required.

REFERENCES

[1] Takumi Ogura, Jun-ichi Itoh: "Evaluation of Total Loss of Both an Inverter and Motor Depending on Modulation Strategy" SPC-09-184 (2009), in Japanese

[2] Jih-Sheng Lai, and Fang Zheng Peng: "Multilevel Converters-A New Breed of Power Converters", *IEEE Trans. Ind. Appl.*, Vol. 32, No. 3, pp. 509-517 (1996)

[3] Samir Kouro, Mariusz Malinowski, K. Gopakumar, Josep Pou, Leopoldo G. Franquelo, Bin Wu, Jose Rodriguez, Marcelo A. Pérez, and Jose I. Leon: "Recent Advances and Industrial Applications of Multilevel Converters", *IEEE Trans. Ind. Electron.*, Vol. 57, No. 8, pp.2553-2580 (2010)

[4] J. Rodriguez, Jih-Sheng Lai, F. Z. Peng: "Multilevel Inverters: A Survey of Topologies, Controls, and Applications", *IEEE Trans. Ind. Appl.*, Vol. 49, No, 4, pp. 724-748 (2002)

[5] Haitham Abu-Rub, Joachim Holtz, Jose Rodriguez, and Ge Baoming: "Medium-Voltage Multilevel Converters-State of the Art, Challenges, and Requirements in Industrial Applications", *IEEE Trans. Ind. Electron.*, Vol. 57, No. 8 pp.2581-2596 (2010)

[6] Hidemine Obara, Masamu Kamaga, Takumi Ito, Yukihiko Sato: "An Investigation of Capacitors for Flying Capacitor Converters" Vol.131 No.12 pp.1393-1400 (2011), in Japanese

[7] Alstom:Catalog "Symphony Air Cooled Medium Voltage AC Drives"

The 2018 International Power Electronics Conference

Current Source Type PMSG Wind Turbine System with Three-phase Three-switch Buck-type Rectifier for Machine-side Converter

Beomseok Chae[1], Tahyun Kang[2], and Yongsug Suh[1]

1 Dept. of Electrical Engineering, Smart-Grid Research Center, Chonbuk National University, Korea

2 Research and Development Center, Milimsyscon Co, Seongnam, Korea

Abstract— This paper proposes a three-phase three-switch buck-type rectifier converter as a MSC of wind turbine system. Owing to a novel switching modulation scheme eliminating the unwanted diode rectifier mode switching state, the proposed system exhibits the satisfying ac voltage and current waveform quality and torque ripple up to the level of typical current source rectifier even under wide power factor range. The proposed system has been verified through the HILS test on the PMSG wind turbine model of 5MW/4160V. The proposed converter has been shown to provide the stator current THD of 3.9% and torque ripple of 1% under rated power condition. In addition to the inherent advantage of reduced switch count of three-phase three-switch buck-type converter, the proposed switching modulation technique can make this converter a quite viable solution for a MSC placed inside of nacelle, which is under severe volume, weight, and mechanical vibration design limitation.

Keywords— three-phase three-switch buck-type rectifier, current source rectifier, machine side converter

I. INTRODUCTION

Currently, wind energy is a significant part of global warming and greenhouse gas reduction. It is a very important field in research and development of sustainable energy. In terms of installation cost per energy generation, the larger system can achieve higher benefits. For this reason, the large-scale wind turbines are installed, i.e. medium-voltage and mega-watt systems. These large wind turbines have been installed in many places, and their effectiveness has been verified [1]-[3]. MV converter becomes more preferable as compared to the parallel connection of low voltage (LV) converters due to less number of component, high efficiency, and simple hardware design of power converters [4].

Various topologies and control methods have been developed for wind turbine system. Current source type converter has been regarded as one of many interesting circuit topologies in motor drives of MV class due to its inherent current limitation in fault condition and low dv/dt characteristic of ac line voltage in a long range cable connection [5]. Some work has been done for current

source rectifier based wind turbine system [6], [7]. Among many current source type converters, the three-phase three-switch buck-type converter is one candidate topology for the Machine Side Converter (MSC) of wind turbine systems. This three-phase three-switch buck-type converter consists of four diodes and only one active switch per phase, that is, a total of twelve diodes and three active switches in the MSC of three-phase resulting in a reduced component count as compared to the conventional current source converter. This reduced switch count property makes three-phase three-switch buck-type converter quite attractive solution for the MSC which is usually mounted within nacelle of wind turbine being subject to relatively critical design limitation of volume, weight, and mechanical vibration [8].

In spite of the advantage of less number of switches, three-phase three-switch buck-type converter has the inherent fatal problem of limited power factor operating range due to the structure of switch cell. Therefore, when the ac machine side voltage and current require the power factor angle beyond the possible operating range of converter, the switching modulation becomes disrupted leading to severe harmonic distortion of ac voltage and current. This property of limited power factor operating range prohibits the three-phase three-switch buck-type converter from being successfully applied to the MSC of wind turbine systems. This is because the Permanent Magnet Synchronous Generator (PMSG) or Doubly-Fed Induction Generator (DFIG) employed in typical wind turbine systems demand some reactive power to sustain magnetizing current of machines. There has been some previous work trying to increase the power factor operating range by various switching modulation techniques for three-phase three-switch buck-type converter [9]. However, none of them have paid enough attention to the practical switching modulation scheme satisfying the unique operating requirement of wind turbine systems of MW range.

The 2018 International Power Electronics Conference

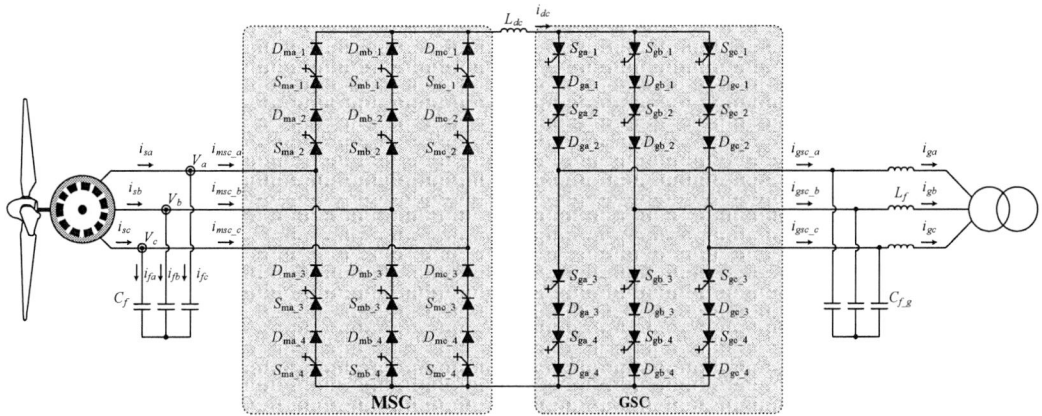

Fig. 1. Back-to-back type current source rectifier based converters for 5MW PMSG MV wind turbines.

This paper investigates the application of three-phase three-switch rectifier as a MSC in wind turbine systems. By employing a novel switching modulation scheme, the proposed system can be nicely fit into the wind turbine system requiring wide power factor operating range in the ac machine side. This novel switching modulation scheme avoids unwanted so called diode rectifier switching state by providing an accurate switching path of ac input current with respect to the reference command of ac input current under wide power factor operating range. The electromagnetic torque and driven current performance of the proposed system is compared with the current source type rectifier in place of the MSC of wind turbine system. The wind turbine of 5MW/4.16 kV Permanent Magnet Synchronous Generator (PMSG) is chosen as a common platform for the comparison work. In other words, back-to-back type current source rectifier based converter, which is regarded as the most popular current source type topology choice in this power range of 5MW, and the proposed back-to-back three-phase three-switch rectifier topology are analyzed and compared. The performance of two different types of converter system is studied with respect to the PMSG performance factor of stator currents and torque ripples.

The main objective of this paper is to show that the proposed three-phase three-switch buck-type rectifier can nicely work as the MSC of wind turbine system in a similar performance level of conventional current source

rectifier. This paper is structured in five main sections as the following. Section II provides the power semiconductor devices & system specification under comparison of 5MW PMSG Wind Turbine System (WTS). Section III explains a novel switching modulation technique of the three-phase three-switch buck-type rectifier. Section IV presents HILS test results of the proposed system in 5MW PMSG WTSs. Finally, the conclusion is given in Section V.

II. CURRENT SOURCE RECTIFIER AND THREE-PHASE THREE-SWITCH BUCK-TYPE RECTIFIER FOR WIND TURBINE SYSTEMS

A. Current Source Rectifier

Figure 1 shows the schematic of two-level current source rectifier with series connection of two IGCTs, i.e. ns=2. Series connection of two devices are required to meet the ac line input of 4.16kV. Each leg of the current source converters consists of four switches (Sm_x, Sg_x), and four reverse blocking diodes (Dm_x, Dg_x) [5]. The DC-Link has current Idc should be continuous. So zero switching state of the current source rectifier is equivalent to shorting one of three phase legs in the converter. In general, this placement of zero state vector complicates the control of current source rectifier compared to voltage source rectifier [10].

B. Three-phase Three-switch Buck-type Rectifier

In this paper, the new concept of wind turbine converter is proposed. Fig. 2 shows the schematic of three-phase three-switch buck-type rectifier with series connection of two IGCTs, i.e. ns=2. Series connection of two devices are required to meet the ac line input of 4.16kV. Each leg of the three-phase three-switch buck-type rectifier in the Machine Side Converter (MSC) consists of four reverse blocking diodes (Dm_x) and two IGCT switch (Sm_x) per leg. As a result, the proposed MSC as compared to that of conventional MSC in Fig. 1 has a reduced device count of active switch by half. The Grid Side Converter (GSC) is based on conventional current source inverter. Each leg of the GSC consists of

TABLE I
DEVICE PARAMETERS OF POWER SEMICONDUCTOR [5]

Device	Press-pack IGCT	Press-pack Diode
Manufacture	ABB	ABB
Code	5SHY 42L6500	5SDF 10H6004
Blocking Voltage	6.5 kV	6.0 kV
$I_{TGQM}/I_{F(AV)M}$	3.8 kA	1.1 kA

1978

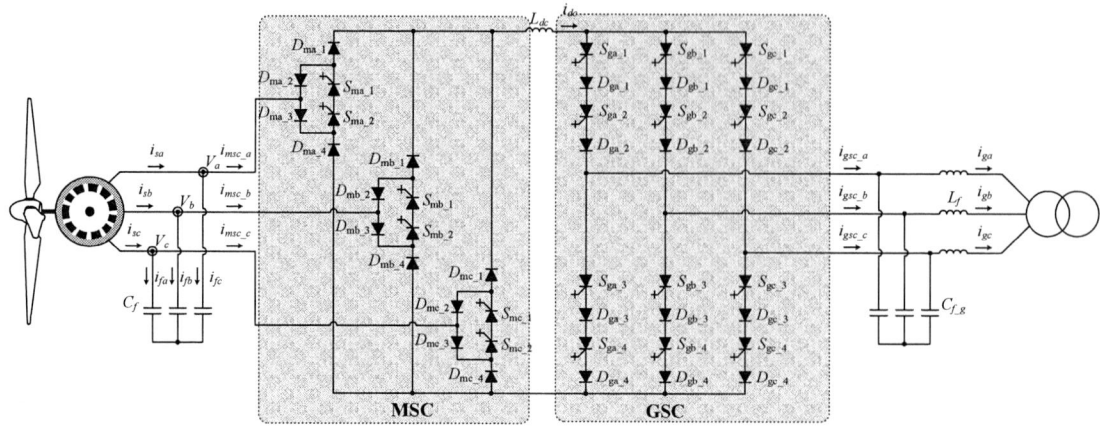

Fig. 2. Newly proposed structure of back-to-back type three-phase three-switch buck-type converters for 5MW PMSG MV wind turbines.

four switches (Sg_x), and four reverse blocking diodes (Dg_x). The conventional current source inverter is adopted in GSC in order to meet the grid code of wide power factor operation [11].

TABLE II
PARAMETERS OF 5MW PMSG WIND TURBINE SYSTEM

Parameter	Symbol	Value	Per unit
Rated power	P_m	5 MW	1.0
Rated wind speed	V_w	11.8 m/s	-
Rated stator voltage	V_m	4.16 kV	1.0
Rated q-axis current	I_{qs}	1,007 A	1.0
Rated stator frequency	f_{pm}	29.1 Hz	1.0
Rated rotor speed	ω_r	14.8 rpm	-
Rated rotor flux linkage	λ	18.0926 Wb	-
Rated mechanical torque	T_m	3.226 MN·m	-
Pole pairs	Z_p	118	-
Inertia	J	2.5e5 kg·m²	-
Stator resistance	R_s	80 mΩ	-
d-axis magnetizing inductance	L_{md}	5.5963 mH	-
q-axis magnetizing inductance	L_{mq}	5.5963 mH	-
PWM carrier frequency	f_{SW}	2,000 Hz	-
Grid frequency	f_{grid}	60 Hz	1.0
Grid side input voltage	V_{LL}	4.16 kV	1.0
Grid side input current	I_{AC_input}	708 A	1.0
Grid side filter inductance	L_f	0.98 mH	0.11
AC filter capacitance	C_{f_g}	0.26 mF	0.34
DC-link current	I_{DC}	997 A	-
DC-link inductance	L_{DC}	8.3 mH	-

C. Power Semiconductor Devices

In this paper, current source rectifier and three-phase three-switch buck-type rectifier employ the same switching devices of press-pack IGCT (ABB 5SHY 42L6500) and press-pack FRD (ABB 5SDF 10H6004) devices for the sake of consistent and fair comparison of two topologies. The main characteristics of employed power semiconductor devices are summarized in Table 1 [5].

D. System Specification

The system specifications of target current source rectifier and three-phase three-switch buck-type rectifier based wind turbine systems as shown in Fig. 1 and 2 are summarized in Table II, respectively. This study is conducted through simulations and HILS systems under 5MW output power and 4.16kV ac line voltage. Both power conversion systems employ the same filter networks in ac grid side and machine side for a fair comparison of current THD, power factor range, and electromagnetic torque as shown in Table II.

III. MODULATION STRATEGY OF THREE-PHASE THREE-SWITCH BUCK-TYPE RECTIFIER

A. General Modultion Strategy under Unity Power Factor Condition at Converter AC Input Side

The single leg of three-phase three-switch buck-type rectifier is driven by the single switch. This switch drives both positive and negative direction currents at ac side. For this reason, a rectified reference current of positive amplitude is compared with a unipolar triangular carrier at switching frequency in PWM block in contrary to the case of typical voltage source converter and current source converter. If the a-phase switch (Sa) is turned on, the a-phase converter input current (Ia) has a positive value under the condition of Va>Vc>Vb. In the same manner, Ib has a negative value when b-phase switch (Sb) is turned on under the same condition of Va>Vc>Vb. The c-phase is dependent on Sa and Sb. The *c*-phase converter input current (I*c*) has a negative value when the S*a* and S*c* are turned on at the same time. For the same reasoning, if S*b* and S*c* are turned on, then the I*c* becomes positive. When all switches are turned on; diode rectifier mode, then I*c* is zero because the *c*-phase voltage (V*c*) takes the median value between V*a* and V*b*.

In balanced 3-phase 3-wire system, the sum of 3-phase current is always zero. i.e. I*a*+I*b*+I*c*=0. As stated above, I*c* is dependent on S*a* and S*b*. But I*a* and I*b* are independent from the other phase switching condition when voltages are V*a*>V*c*>V*b* and current is in-phase with voltage at the converter ac input (I*a*>I*c*>I*b*, Sector 1). As a result, I*a* and I*b* are independently modulated according to the given reference signals. In the meanwhile, I*c* is naturally determined from the relationship of I*a*+I*b*+I*c*=0.

In general, switch of particular phase leg turns on when the rectified reference current of the corresponding phase becomes larger than the triangular carrier waveform. However, the switch of the phase leg whose current amplitude takes a median value among three-phase currents is kept turned on, e.g. S*c* in Sector 1, irrespective of other two switches. Once the switch turns on, the sign of converter input current flowing through that particular switch is determined by the sign of line voltage at the input capacitor.

B. Problem of Diode Rectifier Mode under Non-unity Power Factor Condition at Converter AC Input Side

The input current depends on the relative amplitude of the three-phase capacitor voltage under the diode rectifier mode with a switching state of (111). The current path changes according to the sector of the input capacitor voltage. In Sector 1-1, under the diode rectifier mode, *a*-phase current path is connected to the upper side of dc-link and *b*-phase current path is connected to the lower side of dc-link, but *c*-phase current path is not connected

to the dc-side. In Table III, these values of phase currents are shown under the column of in-phase condition. When the voltage condition changes from the in-phase to out-of-phase condition, the upper side and lower side of dc-link are connected to the different ac phases. For example, in Sector 1-1 under the leading angle condition between 0° and 30°, the values of phase currents during diode rectifier mode having the switching state of (111) are different from those under the in-phase condition. These different phase current values result in the failure of pulse width modulation where the localized average value of modulated current waveform is supposed to follow the reference signal of phase current. In the similar manner, the values of phase currents in Sector 1-2 become distorted under the lagging angle condition between 0° and 30° as shown in Table III. Therefore, changing the voltage condition can have a decisive influence on the current path to ac side. When there is a phase difference between the input current of the converter and the capacitor voltage, the current distortion is occurred at the converter input current [9].

C. Proposed Modultaion Scheme under Non-unity Power Factor Condition at Converter AC Input Side

In this study, the enhanced modulation scheme and its implemented modulation block for the back-to-back three-phase three-switch buck-type converters of Fig. 2 is proposed. This newly proposed modulation scheme prevents the failure of pulse width modulation of input phase current due to diode rectifier mode having the switching state of (111) for the MSC in PMSG wind turbines. This is made possible by changing the diode rectifier mode that switching state of (111) by its equivalent active switching state that generates the correct values of phase currents depending on the operating sector. Selection of switching combination from the improved modulation strategy is presented in

TABLE III
PHASE CURRENT UNDER VARIABLE POWER FACTOR CONDITIONS
IN SECTOR 1-1 AND SECTOR 1-2 WITH CONVENTIONAL METHOD

	Sector 1-1 I*a*>I*c*>I*b* (|I*b*|>|I*a*|>|I*c*|)								
	Vc>Va>Vb 0°< leading angle < 30°			Va>Vc>Vb In-phase			Va>Vc>Vb 0°< lagging angle < 30°		
	Ia	Ib	Ic	Ia	Ib	Ic	Ia	Ib	Ic
(110)	Idc	-Idc	0	Idc	-Idc	0	Idc	-Idc	0
(011)	0	-Idc	Idc	0	-Idc	Idc	0	-Idc	Idc
(111)	0	-Idc	Idc	Idc	-Idc	0	Idc	Idc	0

	Sector 1-2 I*a*>I*c*>I*b* (|I*a*|>|I*b*|>|I*c*|)								
	Vc>Va>Vb 0°< leading angle < 30°			Va>Vc>Vb In-phase			Va>Vc>Vb 0°< lagging angle < 30°		
	Ia	Ib	Ic	Ia	Ib	Ic	Ia	Ib	Ic
(110)	Idc	-Idc	0	Idc	-Idc	0	Idc	-Idc	0
(101)	Idc	0	-Idc	Idc	0	-Idc	Idc	0	-Idc
(111)	Idc	-Idc	0	Idc	-Idc	0	Idc	0	-Idc

TABLE IV
PHASE CURRENT UNDER VARIABLE POWER FACTOR CONDITIONS
IN SECTOR 1-1 AND SECTOR 1-2 WITH PROPOSED METHOD

	Sector 1-1 I*a*>I*c*>I*b* (|I*b*|>|I*a*|>|I*c*|)								
	Vc>Va>Vb 0°< leading angle < 30°			Va>Vc>Vb In-phase			Va>Vc>Vb 0°< lagging angle < 30°		
	Ia	Ib	Ic	Ia	Ib	Ic	Ia	Ib	Ic
(110)	Idc	-Idc	0	Idc	-Idc	0	Idc	-Idc	0
(011)	0	-Idc	Idc	0	-Idc	Idc	0	-Idc	Idc
(110)	Idc	-Idc	0	Idc	-Idc	0	Idc	-Idc	0

	Sector 1-2 I*a*>I*c*>I*b* (|I*a*|>|I*b*|>|I*c*|)								
	Vc>Va>Vb 0°< leading angle < 30°			Va>Vc>Vb In-phase			Va>Vc>Vb 0°< lagging angle < 30°		
	Ia	Ib	Ic	Ia	Ib	Ic	Ia	Ib	Ic
(110)	Idc	-Idc	0	Idc	-Idc	0	Idc	-Idc	0
(101)	Idc	0	-Idc	Idc	0	-Idc	Idc	0	-Idc
(110)	Idc	-Idc	0	Idc	-Idc	0	Idc	-Idc	0

The 2018 International Power Electronics Conference

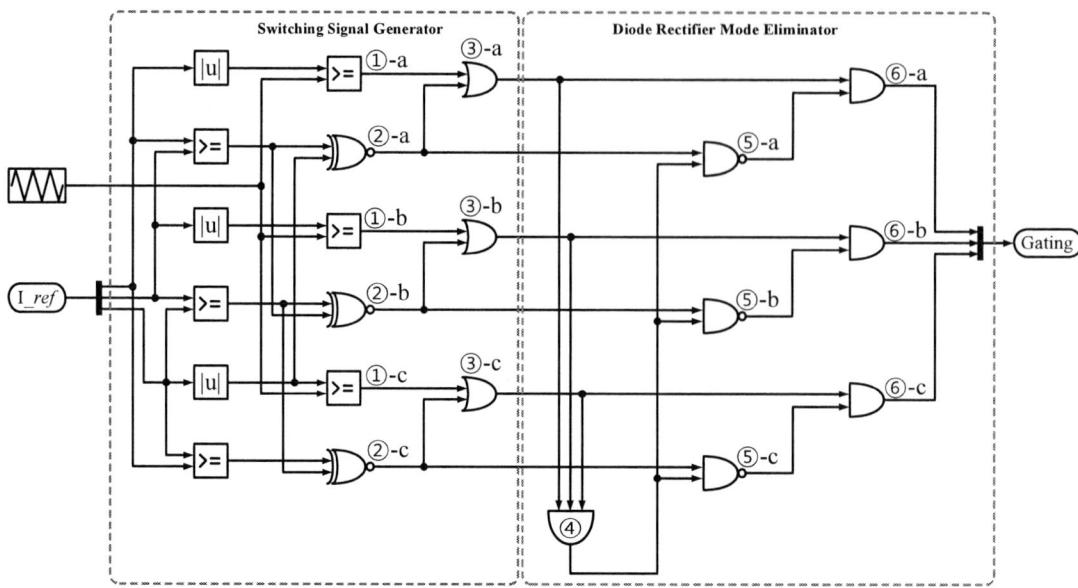

Fig. 3. Proposed modulation block of three-phase three-switch buck-type rectifier.

Table IV. For example, in Sector 1-1 the diode rectifier mode is replaced by the switching state of (110) so that the values of phase currents under the leading angle condition become equal to those of in-phase condition. In the same manner, in Sector 1-2 the diode rectifier mode is also replaced by the switching state of (110). This replacement corrects the values of phase currents under the lagging angle condition to match those of in-phase condition. As a result, this replacement of diode rectifier mode by the switching state of (110) in Sector 1 makes it possible that the current path is less affected by the variable power factor range of the converter [9].

In this paper, the advanced modulation scheme without diode rectifier mode is newly implemented in Carrier Based PWM (CBPWM). The practical implementation of this CBPWM using digital logic gates is presented in Fig. 3. In Fig. 3, *Switching Signal Generator Block* is the typical sine-triangle modulator (①) in which a rectified sine modulating waveform of positive value is compared with a unipolar triangular carrier waveform. In *Switching Signal Generator Block*, there are three NXOR gates; ②-a, ②-b, and ②-c. Each NXOR gate generates high value only when its corresponding phase current becomes a median phase, i.e. the amplitude of corresponding phase current is middle among three phase currents. For example, the output of ②-a becomes high when reference signal of *a*-phase converter input current becomes a median phase among three phase currents.

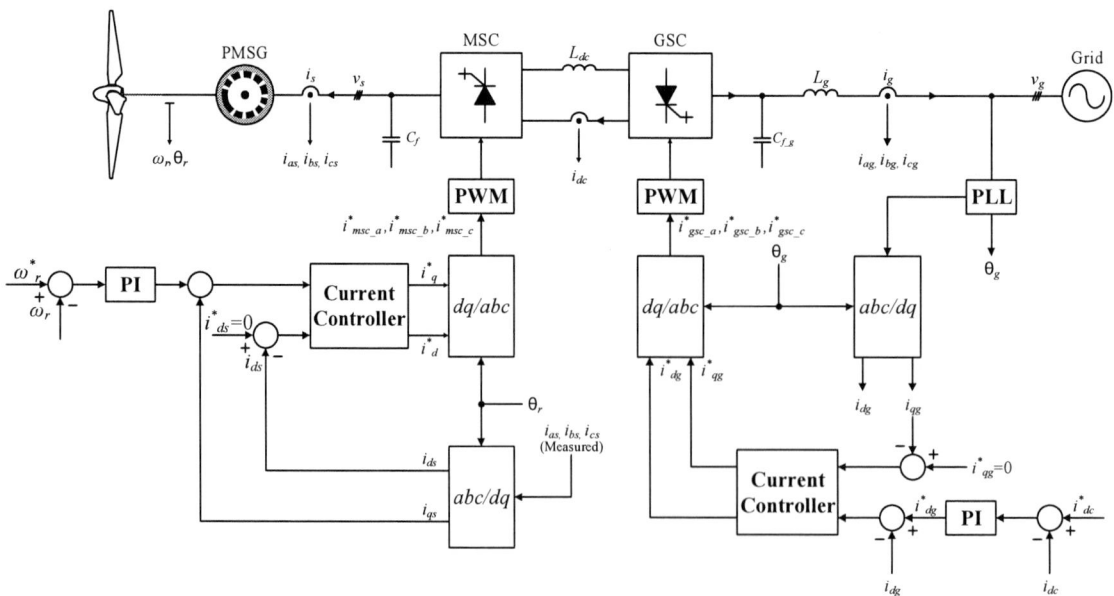

Fig. 4. Overall control block diagram of two different topologies of Fig. 1 and 2.

1981

The 2018 International Power Electronics Conference

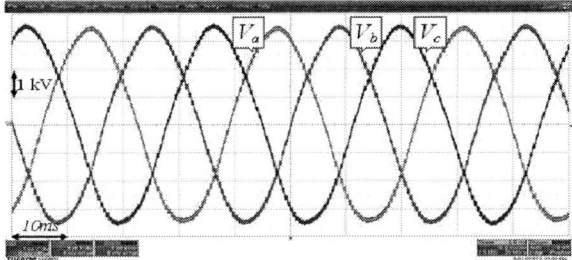

Fig. 5. Experimental waveforms of capacitor voltage (V*a*, V*b*, V*c*) in current source rectifier.

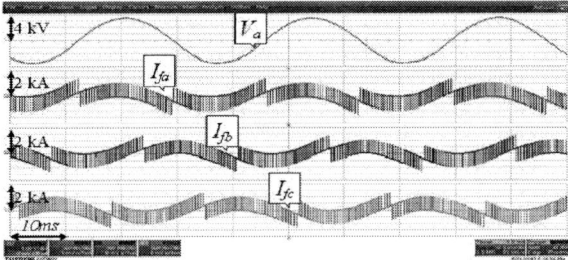

Fig. 6. Experimental waveforms of *a*-phase capacitor voltage and ripple current (V*a*, I*fa*, I*fb*, I*fc*) in current source rectifier.

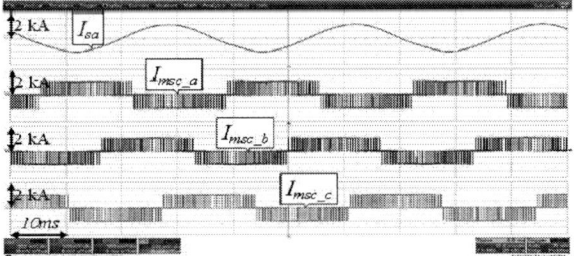

Fig. 7. Experimental waveforms of *a*-phase stator current and converter input current (I*sa*, I*msc_a*, I*msc_b*, I*msc_c*) in current source rectifier.

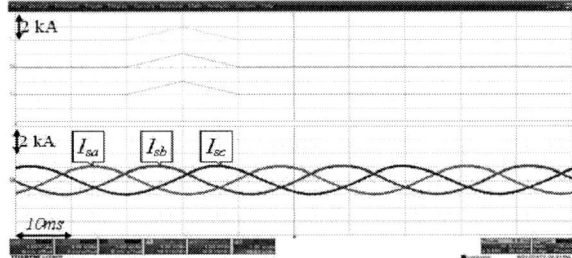

Fig. 8. Experimental waveforms of stator current (I*sa*, I*sb*, I*sc*) and frequency spectrum in current source rectifier.

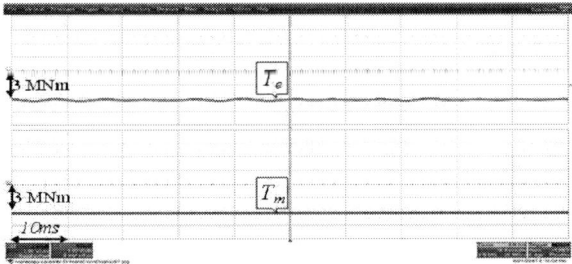

Fig. 9. Experimental waveforms of electromagnetic and mechanical torque (T*e*, T*m*) in current source rectifier.

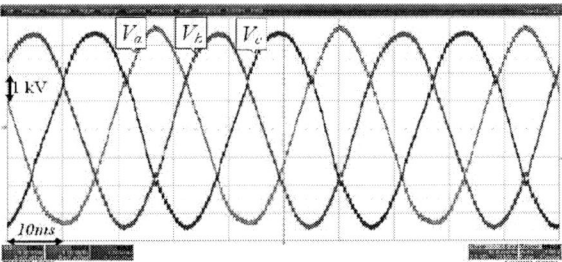

Fig. 10. Experimental waveforms of capacitor voltage (V*a*, V*b*, V*c*) in three-phase three-switch buck-type rectifier.

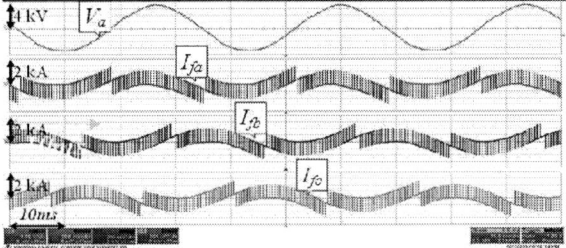

Fig. 11. Experimental waveforms of *a*-phase capacitor voltage and ripple current (V*a*, I*fa*, I*fb*, I*fc*) in three-phase three-switch buck-type rectifier.

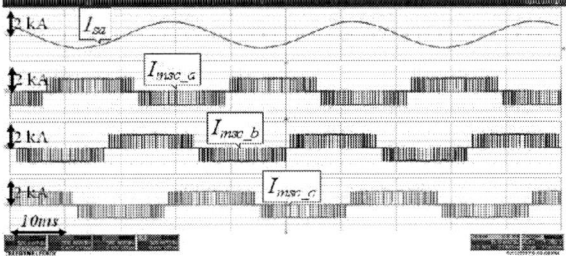

Fig. 12. Experimental waveforms of *a*-phase stator current and converter input current (I*sa*, I*msc_a*, I*msc_b*, I*msc_c*) in three-phase three-switch buck-type rectifier.

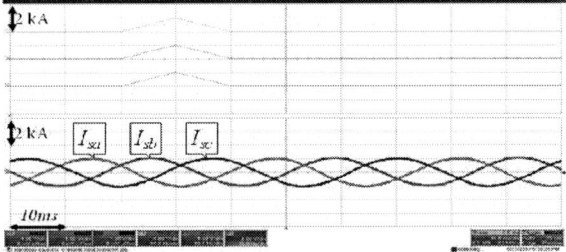

Fig. 13. Experimental waveforms of stator current (I*sa*, I*sb*, I*sc*) and frequency spectrum in three-phase three-switch buck-type rectifier.

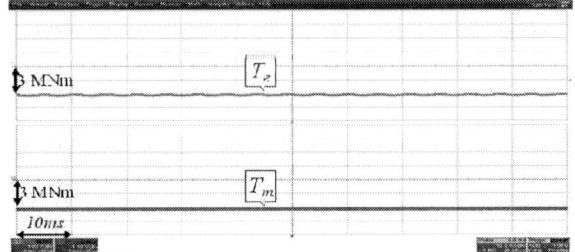

Fig. 14. Experimental waveforms of electromagnetic and mechanical torque (T*e*, T*m*) in three-phase three-switch buck-type rectifier.

The output signals of ①-a and ②-a are applied to OR gate of ③-a. This is to ensure that, in addition to the action of sine-triangle, the switch of *a*-phase is turned on whenever the *a*-phase becomes a median phase in order to provide the return path through converter switch of *a*-phase. In general, the OR gate of ③-a is indispensable for switching modulation of current source type converters. Usually, voltage source type converters do not require this OR gate in PWM logic since the each phase is modulated independent from other remaining phases.

In Fig. 3, *Diode Rectifier Mode Eliminator Block* implements the proposed modulation scheme in this paper. The output of OR gate (③-a) is then masked by AND gate of ⑥-a. This masking condition makes this modulation block different from state-of-the-arts solutions. The proposed masking condition is provided from NAND gate of (⑤-a). This masking condition is triggered when diode rectifier mode and median phase condition are met at the same time. In other words, under the condition of diode rectifier mode and a median phase, the turn-on signal for the switch of corresponding phase is forced to off, i.e. masking out. The diode rectifier mode of switching state (111) is detected when the output of AND gate (④) is high. Finally, the output of AND gate of ⑥-a is applied to *a*-phase switch of converter. This final output represents the switching table as given in Table IV. The actions of gates belonging to *b*-phase and *c*-phase are similar to those of *a*-phase.

IV. EXPERIMENT RESULTS OF THREE-PHASE THREE-SWITCH BUCK-TYPE RECTIFIER

The current source rectifier based WTS of Fig. 1 and three-phase three-switch buck-type rectifier based WTS of Fig. 2 have been verified through the real-time Hardware In the Loop (HIL) system of RT-Box. Two different topologies are compared with respect to two major performance factors; Total Harmonic Distortion (THD) of PMSG stator current and torque ripple under the rated power condition.

The real-time simulator is made based on the operating condition specified in Table II. Two different topologies considered in this paper adopt the same overall control scheme for MSC and GSC. Figure 4 describes the overall control block diagram of 5MW PMSG wind turbine system [12]. The employed real-time HIL system setup is illustrated in Fig. 15. Figure 5 through 9 present the experimental waveforms of current source rectifier based WTS, particularly for MSC. In this paper, the both converter operates under rectifier mode, i.e. power flows from the PMSG to the DC-link. The capacitor voltages are described in Fig. 5. In Fig. 6, along with the capacitor voltage of *a*-phase, the ripple currents of capacitor filter are also shown. The typical chopped converter input currents of MSC are described in Fig. 7 along with the filtered sinusoidal waveform of stator current of *a*-phase. The stator currents of all three phases

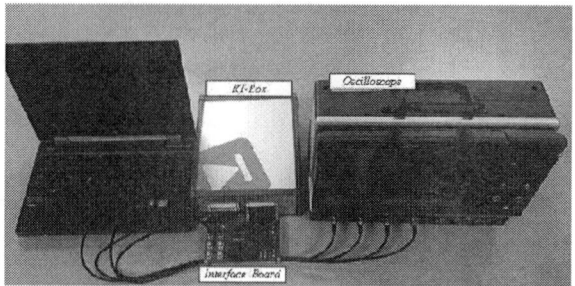

Fig. 15. System configuration of RT-Box.

TABLE V
PERFORMANCE FACTORS OF CURRENT SOURCE RECTIFIER AND THREE-PHASE THREE-SWITCH BUCK-TYPE RECTIFIER BASED WIND TURBINE SYSTEM UNDER THE RATED POWER CONDITION FROM HILS VERIFICATION

Factors	THD of stator current	Electromagnetic torque ripple (peak to peak)
Current Source Rectifier	3.0 %	23 kN·m
Three-phase Three-switch Buck-type Rectifier	3.9 %	31 kN·m

and its frequency spectrum are given in Fig. 8. As expected, the chopped converter input currents are effectively filtered out by the input capacitor filter network between the machine and current source rectifier of MSC. Finally, the electromagnetic torque generated by PMSG and mechanical torque applied through the prime mover, i.e. the blades of wind turbine, are illustrated in Fig. 9. The waveforms presented in Fig. 5 – 9 are regarded as benchmarking targets for the newly proposed three-phase three-switch buck-type rectifier based WTS.

Figure 10 through 14 present the experimental waveforms of three-phase three-switch buck-type rectifier based WTS, particularly for MSC. These waveforms are measured under the same operating condition of Table II. The capacitor voltages are described in Fig. 10. In Fig. 11, along with the capacitor voltage of *a*-phase, the ripple currents of capacitor filter are also shown. The typical chopped converter input currents of MSC are described in Fig. 12 along with the filtered sinusoidal waveform of stator current of *a*-phase. The stator currents of all three phases and its frequency spectrum are given in Fig. 13. Finally, the electromagnetic torque generated by PMSG and mechanical torque applied through the prime mover, i.e. the blades of wind turbine, are illustrated in Fig. 14. The power factor angle at the converter input node is measured to be 10°, leading. In Table V, two major performance factors; Total Harmonic Distortion (THD) of PMSG stator current and torque ripple under the rated power condition for two different topologies are obtained and presented.

It is noted that even under non-unity power factor condition, three-phase three-switch buck-type based WTS achieves the similar level of performance to those of current source rectifier based WTS. It is readily understood that the proposed modulation scheme as explained in Table IV is also successfully validated through experiment. As a result, three-phase three-switch buck-type rectifier can be effectively applied in PMSG wind turbines without suffering from the limited operating range of power factor in MSC. The successful utilization of three-phase three-switch buck-type rectifier as a MSC of PMSG wind turbines would give a rise to many potential engineering approaches to improve the performance of wind turbine and lower the system cost at the same time.

V. CONCLUSIONS

This paper proposes a three-phase three-switch buck-type converter as a MSC of wind turbine system. Owing to a novel switching modulation scheme eliminating the unwanted diode rectifier mode switching state, the proposed system exhibits the satisfying ac voltage and current waveform quality, i.e. THD, up to the level of typical current source rectifier even under wide power factor operating range. As shown in the paper, this stator current waveform of high quality leads to the torque ripple characteristics in the similar level of typical current source rectifier. These stator current waveform and torque ripple of high quality can be regarded as significant advantages for three-phase three-switch buck-type converter since this topology already has the inherent benefit of reduced switch count. Therefore, the proposed three-phase three-switch buck-type converter can be considered as a potential candidate for a MSC in wind turbine system overcoming severe design challenge due to relatively tough volume, weight, and mechanical vibration limit with respect to the converter placed inside of nacelle.

REFERENCES

[1]   H. J. Lee and S. K. Sul, "Wind power collection and transmission with series connected current source converters," in *Proc. 14th EPE*, pp. 1-10, Aug. 30-Sep. 1, 2011.

[2]   J. Sayago, T. Bruckner, and S. Bernet, "How to select the system voltage of MV drives: A comparison of semiconductor expenses," *IEEE Trans. Ind. Electron.*, Vol. 55, No. 9, pp. 3381-3390, Sep. 2008.

[3]   K. Lee, K. Jung, Y. Suh, C. Kim, H. Yoo, and S. Park, "Comparison of high power semiconductor devices losses in 5MW PMSG MV wind turbines," *IEEE Applied Power Electronics Conference and Exposition (APEC)*, pp. 2511-2518, Mar. 2014.

[4]   O. S. Senturk, L. Helle, S. Munk-Nielsen, and P. Rodriguez, "Power capability investigation based on electro-thermal models of press-pack IGBT three-level NPC and ANPC VSCs for multimegawatt wind turbines," *IEEE Trans. Power Electron.*, Vol. 27, No. 7, pp. 3196-3206, Jul. 2012.

[5]   T. Kang, T. Kang, B. Chae, K. LEE, and Y. Suh, "Comparison of efficiency for voltage source and current source based converters in 5MW PMSG wind turbine systems," *The Transactions of the Korean Institute of Power Electronics*, Vol. 20, No. 5, pp. 410-420, Oct. 2015.

[6]   J. Dai, D.D. Xu, and B. Wu, "A novel control scheme for current-source-converter-based PMSG wind energy conversion systems," *IEEE Trans. on Power Electronics*, Vol. 24, No. 4, pp. 963-972, Apr. 2009.

[7]   P. Tenca, A. A. Rockhill, T. A. Lipo, and P. Tricoli, "Current source topology for wind turbines with decreased mains current harmonics, further reducible via functional minimization," *IEEE Trans. Power Electron.*, Vol. 23, No. 3, pp. 1143-1155, May 2008.

[8]   T. Bruckner, S. Bernet, and P. K. Steimer, "Feedforward loss control of three-level active NPC converter," *IEEE Trans. Ind. Appl.*, vol. 43, no. 6, pp. 1588-1596, Dec. 2007.

[9]   B. Chae, T. Kang, T. Kang, and Y. Suh, "Reduced Current Distortion of Three-Phase Three-Switch Buck-Type Rectifier using Carrier Based PWM in EV Traction Battery Charging Systems," *The Transactions of the Korean Institute of Power Electronics*, Vol. 20, No. 4, pp. 375-387, Aug. 2015.

[10]   Y. Suh, J. K. Steinke, and P. K. Steimer, "Efficiency comparison of voltage-source and current-source drive systems for medium-voltage applications," *IEEE Trans. Ind. Electron.*, Vol. 54, No. 5, pp. 2521-2531, Oct. 2007.

[11]   P. E. Sutherland, "Ensuring stable operation with grid codes," *IAS Magazine*, pp. 60–67, Jan./Feb. 2016.

[12]   B. Wu, Y. Lang, N. Zargari, and S. Kouro, "Power conversion and control of wind energy systems," *NJ, IEEE Press*, 2011, ISBN 978-0-470-59365-3.

A Study of 10MW Load Commutated Inverter for Gas-Turbine Start-up

Hyunsung An and Hanju Cha

Electrical Engineering, Chungnam National University, Daejeon, Korea

Abstract— **This paper proposes a stable sensorless start-up method of the actual 10MW LCI system and analyzed the characteristic of the existing system for the simulation modeling. The 10MW LCI system is modeled to verify the performance of the proposed sensorless methods, controllers, operating modes and so on. The simulation model is designed by applying the operating conditions of the 10MW actual system and parameters of the 200MVA synchronous machine. The proposed sensorless methods for initial start-up of the gas-turbine is composed of the initial rotor position detector and methods for the low speed and high speed range. The operation mode of the LCI system is divided into the forced commutation mode and natural commutation mode according to the rotor speed, and the sensorless methods are applied differently by the operation mode. The sensorless methods are referred to as closed loop current control and flux estimator. The initial rotor position detector is applied to detect rotor position at the standstill of the machine using an induced back-emf by the injection of a field current, and the estimated error is less than 1%. After estimation of the initial rotor position, the LCI system is operated by the closed loop current control at the forced commutation mode, and it operates at less than 10% of the rated speed. The LCI system is operated from 10% to 90% of the rated speed through the flux estimator for the natural commutation mode, and the rotor speed/position is estimated by the phase locked loop. The performance of proposed sensorless start-up and the operation modes of LCI system is verified by the proposed simulation modeling.**

Keywords— *Gas-turbine start-up, Load commutated inverter, Static frequency converter, sensorless method.*

I. INTRODUCTION

An initial start-up system must be installed in the gas-turbine power station because a gas for turning the turbine starts internal combustion at approximately 30% of the rated speed [1]. Generally, a gas-turbine is difficult to start-up itself in the standstill state. In the past, the initial start-up system used dc motor or induction machines in spite of the maintenance and shaft vibration problems. A static frequency converter (SFC) using a thyristor switch has been studied [2]-[5], and it is used in a variety of the industrial applications like gas-turbine, pumped storage systems and start-up system for engines of variable capacity [6]-[10].

The SFC system is applied to a current source inverter (CSI) because of the large-scale system, and SFC system in the literature is a load commutated inverter (LCI) [11]-[14]. Typically, the gas-turbine over 200MVA requires the start-up system of about 10MW. The switching elements in the LCI system are the thyristor, and it has the uncontrollable turn-off characteristics. To turn off the thyristor, the current of the switch must be reduced to zero or a reverse voltage must be applied to the switch. Despite the turn-off characteristics, thyristor switches are highly reliable in the system of the power station with high current and high voltage. The thyristor is difficult to turn-off since the voltage of the machine is low during the initial operation of the LCI system. Therefore, a different operating method must be applied at the LCI system because of the insufficient induced voltage of the machine in the initial start-up. So, LCI systems apply the forced commutation mode (pulse mode) [15][16].

The rotor position and speed of the synchronous machine for the stable control must be obtained through the position sensors or the sensorless algorithms. However, the position sensors are limited in the power station because of the problems such as the maintenance and environment. The sensorless methods for LCI system have been studied, but the sensorless algorithms are limited due to the high power system and turn-off characteristic. Sensorless algorithms based on back-emf model and flux estimator are stable at high speed range, but it is very unstable at the standstill and extremely low speed range. Therefore, the motor drive system at the low speed range used sensorless methods like high frequency (HF) injection method or open-loop methods. As mentioned earlier, it is difficultly to be used to HF injection method in the [17] due to turn-off of the thyristor. So, LCI system based on thyristor needs methods for a stable sensorless control at the overall speed range. In addition, the large system such as LCI system for gas-turbine needs simulation modeling for verification of start-up algorithms. The analysis of existing systems and parameters of the gas-turbine is needed for modeling of LCI system.

In this paper, sensorless algorithm for stable start-up of the LCI system at the overall speed range is proposed through initial rotor position detector, closed loop current control and flux estimator. The gas-turbine and existing LCI system is analyzed by re-engineering, and the performance of sensorless algorithm, controllers and operation modes of the LCI system is evaluated by the 10MW simulation modeling.

II. START-UP SYSTEM FOR GAS TURBINE

Gas-turbine power station is composed of gas-turbine, LCI system, synchronous machine, excitation system and various control system, and it is shown as Fig. 1. The LCI system for initial start-up of the gas-turbine power station is important. The large synchronous machine operates as a motor or generator according to the operation mode of the gas-turbine and it operates as a generator when the rated speed is reached. A governor for generator, LCI system and excitation system are operated by a main control system according to command of control room in the power station. The LCI system is performed as a start-up system for the gas-turbine until 90% for the rated frequency (60Hz).

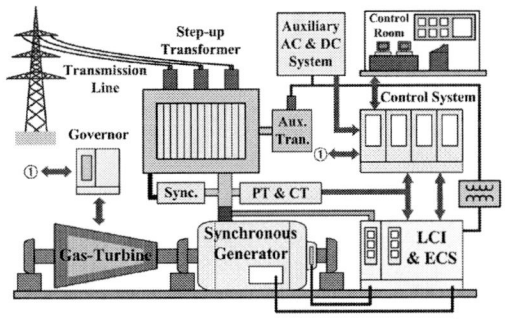

Fig. 1. Configuration of a gas-turbine power station with start-up based on the LCI

A. Load Commutated Inverter

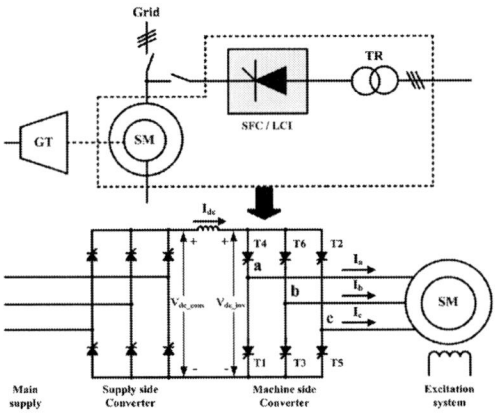

Fig. 2. Structure of a load commutated inverter

LCI system consists of a supply side converter (rectifier), machine side (inverter), synchronous machine, DC reactor and excitation system. The LCI system employs a DC reactor between the supply side converter and the machine side converter, and the torque of the synchronous machine is determined by firing angle of inverter and magnitude of the dc current. The supply side converter is a three phase rectifier circuit based on thyristor, where constant AC voltage of the grid is converted into variable DC voltage. Three phase AC power of the machine with a variable frequency and variable voltage is supplied by DC power generated through supply side converter.

The operation mode depends on the speed and voltage of the machine, and the LCI system has two modes. First, forced commutation mode (or pulse mode) exists in order to the low speed range (5%~10%). In case of the low speed,

commutation of the thyristor is difficult because the level of the machine voltage is low. Therefore, current of the machine must be reduced to zero by changing the firing angle of the supply side converter to the inversion mode. Second, LCI system is operated by the natural commutation mode after the appropriate voltage is secured. To turn off the thyristor reliably, the LCI system must be operated with a leading power factor to avoid commutation failure and it is very important.

B. Sensorless Methods according to the operation modes

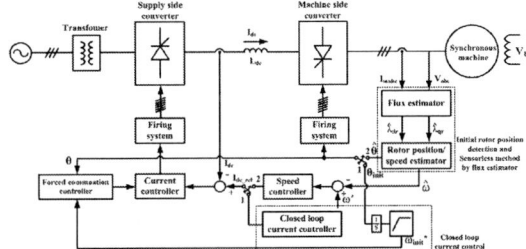

Fig. 3. Sensorless control block diagram of LCI system

The control method for stable start-up state is applied by speed control and current control based PI controller. In case of the less than 10%, current controller operates in conjunction with the forced commutation controller for turn-off of the thyristor switch such as Fig.3. DC voltage for start-up of the machine is generated by output of current controller. Flux estimator, closed loop current controller and rotor position/speed estimator are used for sensorless start-up, and rotor position/speed estimator based on the phase locked loop (PLL) estimate the rotor position/speed through inputting the estimated flux.

Sensorless methods are divided into closed loop current control and flux estimator depending on the rotor speed range or operation mode, and it is calculated by terminal voltage, current and so on. Modes of the sensorless are determined by switch 1 and 2 according to the speed. When switch 1 is closed, sensorless algorithm uses the closed loop current control and the initial rotor position detector, and flux estimator operate as information of the controller when switch 2 is closed after closed loop current control. Sequence of the sensorless start-up is shown in Fig. 4, and a purge, fire and speed-up mode are operation modes of the gas-turbine.

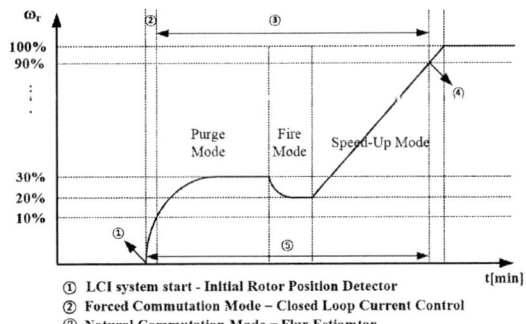

Fig. 4. Sequence of the sensorless methods

The sequence of sensorless methods consists of the initial rotor position detector, closed loop current control and flux estimator, sequentially. The initial rotor position detector is composed of a pure integrator and the PLL, and the rotor position is estimated by the induced back-emf through field current injection. The closed loop current control is used for stable start-up state at low speed range (forced commutation mode) within 10% after initial rotor position estimation. If the synchronous machine reaches 10% of the rated speed, start-up system is operated to the flux estimator because a back-emf of synchronous machine can be obtained sufficiently at the 10%, generally. The flux estimator is designed based on the phase difference of 90 degrees between voltage and flux. Namely, the flux estimator includes an integrator, basically. The detail contents of flux estimator are described in the [18].

III. ANALYSIS AND VERIFICATION

A. Load Commutated Inverter

The LCI system modeling is needed to verify the proposed algorithms, and analysis of the conventional system and the gas-turbine is essential for the LCI system modeling. In this paper, characteristics of the voltage, current and firing angle according to the speed and dc current are additionally analyzed.

1) Machine voltage and current of the LCI system

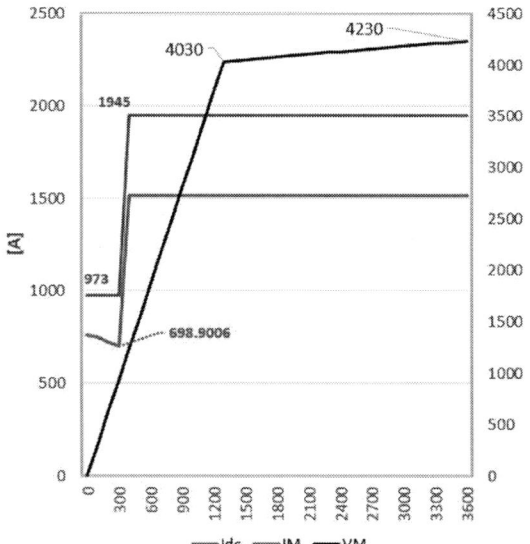

Fig. 5. Characteristic waveform of voltage and current of the synchronous machine

Rated voltage of the LCI system is 4230V, and a weak field area is 1350rpm. Rated DC current is 1945A, and DC current before the forced commutation mode (300rpm) is half-current of the rated current. Machine input current is reduced by '0'A range of the forced commutation mode as shown in Fig. 5. When the weak field control is start, field current by excitation system is decrease, and voltage of the machine is held constant even if the speed changes.

2) Firing angle of machine converter

The thyristor converter is operated by the leading factor to the prevention of the commutation failure, and it is related to the overlap angle. The overlap angle depends on the magnitude of the rotor speed and current of the machine as shown in fig. 6 (a) and (b), it shows the relationship between voltage and current of the machine. The phase difference (δ) between the electromotive force (e) and the terminal voltage (E_t) increases as the increment of the speed (ω) and current (I_t). On the other hand, the phase difference of between machine voltage and current is reduced. Therefore, when the LCI system is operating based on the rotor position, the increasing the speed of the synchronous machine can be cause commutation failure. A firing angle of the machine side converter must be reduced according to the increasing rotor speed and current.

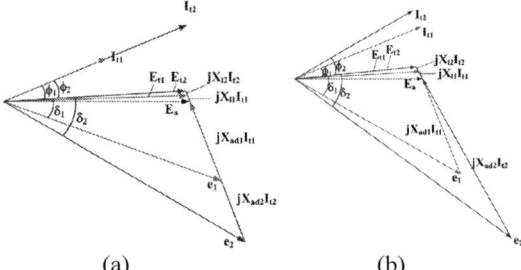

(a)   (b)

Fig. 6. Phase diagram of the LCI system (a) current (b) speed

The firing angle (α_{inv}) and overlap angle (μ) according to the variation of the speed and current of the machine is calculated by (1) and (2). Fig.7 and 8 show characteristic of the firing angle according to the driving conditions, and it show overlap angle of the actual system and calculated value. V_{LL}, L_s and I_{dc} are line voltage of the machine, synchronous inductance and DC current. The overlap angle is calculated by the actual parameters and it is compared by real value of the start-up system in the power station. Characteristics such as Fig. 7 and Fig. 8 should be reflected in the operation of the LCI system.

$$\mu = cos^{-1}\left(cos\alpha_{inv} - \frac{\sqrt{2}\Delta\omega L_s I_{dc}}{V_{LL}}\right) - \alpha_{inv} \quad (1)$$

$$\mu = cos^{-1}\left(cos\alpha_{inv} - \frac{\sqrt{2}\Delta I_{dc} L_s \omega}{V_{LL}}\right) - \alpha_{inv} \quad (2)$$

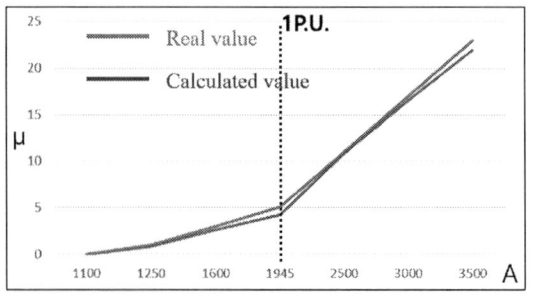

Fig. 7. Characteristic of firing angle according to machine current

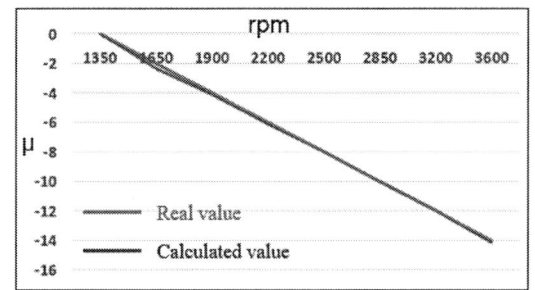

Fig. 8. Characteristic of firing angle according to machine speed

B. Modeling and Verification of the 10MW LCI system

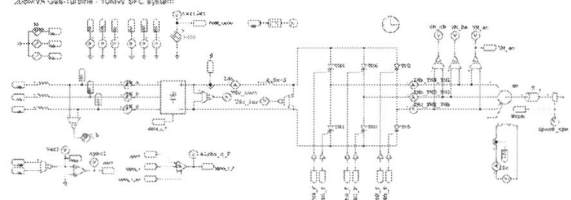

Fig. 9. Modeling of 10MW LCI system

A modeling of the 10MW LCI system consist of supply side converter, machine side converter, synchronous machine, load system, controller and so on as shown in Fig. 9. The simulation model is implemented by Psim in order to check precisely the performance such as the current controller, operation characteristic and so on. The simulation specification is the same as the Table I and Table II.

TABLE I
PARAMETERS OF MACHINE CONVERTER AND REACTOR

Parameters	Value	Unit
Rated power	10,000,000	W
Line to line voltage	4230	V
Rated frequency	60	Hz
Rated current (fund.)	1516	A
DC reactor	4.5	mH
Rated dc current	1945	A
Supply grid	4970	V

TABLE II
PARAMETERS OF SYNCHRONOUS MACHINE

Parameters	Value	Unit
Rated power	200,000,000	W
Line to line voltage	18,000	V
Rated frequency	60	Hz
Rated current	6450	A
Rated field current	1450	A
No load field current	552	A
Inertia	5360	Kgm^2
Pole	2	P

Fig. 10 shows the overall operation by the position sensor and the speed reference is set to 1350rpm. The DC current reference in the initial state is half-current of rated

current considering operation characteristic and it is changed to the rated current at the natural commutation mode. The operation modes (forced commutation mode and natural commutation mode) and constant speed control mode are sequentially operated in accordance with the speed range, and the forced commutation mode is performed up to 300rpm (8.33% of the rated machine speed). Before speed-up mode and internal combustion of gas-turbine, the purge mode for cleaning of the gas-turbine is performed by speed control, and it operate at the 1000rpm, approximately.

Fig. 11 (a) shows the operation of the forced commutation mode within the 100rpm and the firing angle of machine side converter was determined by considering start-up torque. The input current is decreased from the level of the DC current to 0A every 60 degree by changing of firing angle of the supply side converter as shown in Fig.11 (a). The firing angle of the supply side converter is converted to 150 degrees from the output value of the current controller according to command of forced commutation controller. Fig. 11 (b) shows the operation of the natural commutation mode, and the firing angle of machine side converter is determined by speed and current of the synchronous machine. The DC current is rated current (1945A), and the machine converter operates as a leading factor to prevention commutation failure. The basic performance by the position sensor of LCI system has been evaluated through the 10MW simulation results.

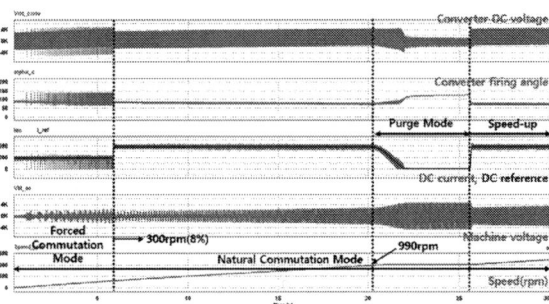

Fig. 10. Overall start-up of the LCI system by position sensor

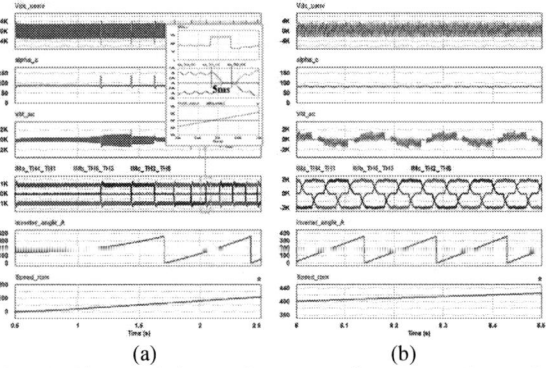

(a)　　　　　　　　(b)

Fig. 11. Characteristic waveforms according to operation modes (a) Forced commutation mode (b) Natural commutation mode

Fig. 12 shows overall start-up operation by the sensorless algorithm, and the performance of the sensorless algorithm, operation mode, speed/current

1988

controller (including forced commutation controller). The result waveforms show the DC voltage/current, output of controllers, estimated speed/angle and so on. The reference of the rotor speed and DC current are set the same as the operation conditions of the simulation by the position sensor. The sensorless start-up method is applied by the sequence depending on the speed range and the thyristor pair for the initial start-up of LCI system is determined by the initial rotor position detection. The TH 6 and TH 1 detected as thyristor pair for the initial operation in this paper.

Fig. 13 shows stable start-up state at the low speed range within 3% and the thyristor switches are turn off by the forced command of the forced commutation control. The LCI system in the low speed range performs stable control without the commutation failure and the firing angle of the supply side converter is instantaneously changed to 150 degrees by the forced command as shown in Fig. 13. Fig. 14 shows transition range through of the mode change from the forced commutation mode to the natural commutation mode. The LCI system has a shutdown time of 20ms depending on the operation condition of the actual system in the power station. The mode changing is performed at 300rpm (8.33%), and the estimated speed is also stabilized in the transition range. Fig. 15 shows natural commutation mode through the flux estimator at the 400rpm and the thyristors are converted naturally by terminal voltage in relation to the firing angle of the machine side converter. The LCI system operated stably with rated current without commutation failure.

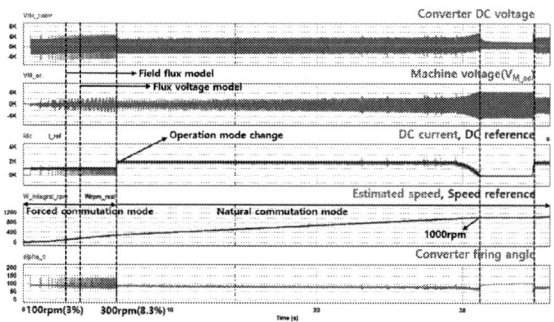

Fig. 12. Overall start-up of the LCI system by sensorless method

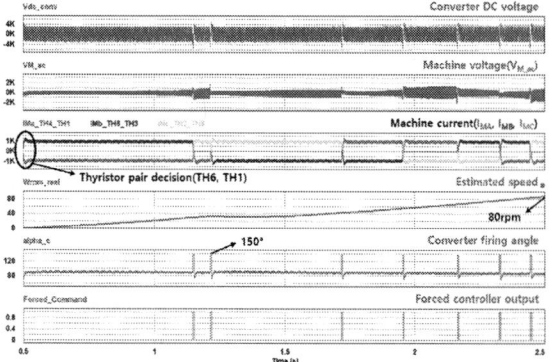

Fig. 13. Stable start-up at the low speed within 3% (Forced commutation mode)

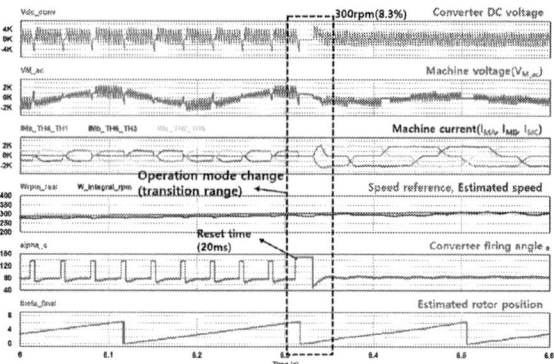

Fig. 14. Transition range by change of the operation mode

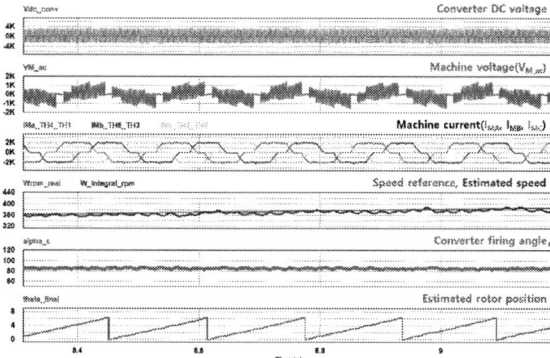

Fig. 15. Start-up operation at the natural commutation mode through flux estimator.

IV. CONCLUSIONS

This paper has been proposed a stable sensorless start-up method for the gas-turbine, and it consists of the initial rotor position detector, closed loop current control and flux estimator depending on the speed range of the machine. The initial rotor position detector has been composed simply by the integrator of the flux voltage model during initial build-up time of the field current, and the initial rotor position has been estimated around 1% quickly and accurately within 150ms. The sensorless methods have been applied differently by the operation mode, and the sensorless methods are referred to as the closed loop current control and the flux estimator. The LCI system has been operated by the closed loop current control at the forced commutation mode, and it operates at less than 10%. The LCI system is operated from 10% to 90% of the rated speed through the flux estimator for the natural commutation mode, and the rotor speed/position is estimated by the phase locked loop. The performance of the proposed sensorless start-up algorithm for the gas-turbine system has been evaluated by 10MW simulation modeling at the variable speed range, and the start-up system of actual capacity has been analyzed.

REFERENCES

[1] Seon-Hwan Hwang, Jang-Mok Kim, Ho-Seon Ryu and Gi-Gab Yoon, "Control Algorithm of Large Synchronous Machines for

Starting Gas Turbosets," *Journal of Power Electronics*, vol.9, no.2, pp.146-155, Mar. 2009.

[2] Ryota Okuyama, Yasuaki Matsumoto, Hiroshi Ogino, Shigeyuki Nakabayashi, Akinobu Ando and Yasuhiko Hosokawa, "Compact Static Starting Device for Gas Turbine," *in Proceedings of IEEE ECCE Asia 2013*, June, 2013

[3] S.D. Sudhoff, E.L. Zivi and T.D. Collins, "Start-up Performance of Load-Commutated Inverter Fed Synchronous Machine Drives," *IEEE Trans. Energy Conversion*, vol.10, no.2, pp.268-274, Jun. 1995.

[4] Hisanori Taguchi, Shinzo Tamai, Yasuhiko Hosokawa and Akinobu Ando, "APS Control Method for Gas Turbine Start-up by SFC," *in Proceedings of IEEE IPEC 2010 International Conference*, Jun, 2010.

[5] Jin Guangzhe, Zhou Changcheng, Gao Qiang and Xu Dianguo, "Research on System of Self-controlled Soft Start with Variable Frequency for Synchronous Motor," *in Proceedings of IEEE IPEMC 2012 7th international conference*, Jun. 2012.

[6] Wang Keshun, Zhang Lichun, Yang Bo, Li Guanjun, Tao Yibin, Fu Jianzhong, Li Jianfeng and Ji Liantao, "Developing and Simulation Research of the Control Model and Control Strategy of Static Frequency Converter," *in Prprounoceedings of IEEE ISDEA 2012 Second International Conference*, Jan, 2012.

[7] Jung-Chen Chiang, Chi-Jui Wu and Shih-shong Yen, "Mitigation of Harmonic Disturbance at Pumped Storage Power Station with Static Frequency Convertet," *IEEE Trans. Energy Conversion*, Vol.12, no.3, pp.232-240, Sept. 1997.

[8] Chihiro Hasegawa and Shoji Nishikata, "A Simple Starting Method for Self-Controlled Synchronous Motors in Electric Propulsion Systems for Ships," *in Proceedings of IEEE Power Electronics and Applications, 2007 European Conference*, Sept. 2007.

[9] Arun Kumar Datta, M.A. Ansari, N. R. Mondal, B. V. Raghavaiah, Manisha Dubey and Shailendra Jain, "Simulation of Static Frequency Converter for Synchronous Machine Operation and Investigation of Shaft Voltage," *International Journal of Electrical, Computer, Energetic, Electronic and Communication Engineering*, vol.8, no.3, 2014.

[10] Greg Magsaysay, Thomas Schuette and Russ J. Fostiak, "Use of a Static Frequency Converter for Rapid Load Response in Pumped-Storage Plants," *IEEE Trans. Energy Conversion*, vol.10, no.4, pp.1771-1782, Dec. 1995.

[11] kamalesh Hatua and V.T. Ranganathan, "A Novel VSI- and CSI-Fed Active-Reactive Induction Motor Drive with Sinusoidal Voltage and Currents," *IEEE Trans. Power Electronics*, vol.26, no.12, pp.3936-3947, Dec. 2011.

[12] Eduardo P. Wiechmann and Rolando Burgos, "On the Efficiency of Voltage Source and Current Source Inverters for High-Power Drives," *IEEE Trans. Industrial Electronics*, vol.55, no.4, pp.1771-1782, Dec. 2011.

[13] Amit Kumar Jain and V.T. Ranganathan, "Improved Control of Load Commutated Inverter Fed Salient Pole Wound Field Synchronous Motor Using Field Oriented Technique," *in Proceedings of IEEE ECCE 2012 Conference*, Sept. 2012.

[14] Bhim Singh, Sanjeev Singh and S.P. Hemanth Chender, "Harmonics Mitigation in LCI-Fed Synchronous Motor Drives," *IEEE Trans. Energy Conversion*, vol.25, no.2, pp.369-380, June. 2010.

[15] David Finney, "Variable frequency AC motor drive systems," Peter Peregrinus Ltd, 1988, p.202-274.

[16] Alberto Tessarolo and Roberto Menis, "On the Modeling of Commutation Transients in Split-Phase Synchronous Motors Supplied by Multiple Load-Commutated Inverters," *IEEE Trans. Industrial Electronics*, vol.57, no.1, pp.35-43, Jan. 2010.

[17] Ji-Hoon Jang, Seung-Ki Sul, Jung-Ik Ha, K Ide and M. Sawamura, "Sensorless Drive of Surface-Mounted Permanent Motor by High Frequency Signal Jnjection based on Magnetic Saliency", *IEEE Trans. Industry Applications*, vol. 39, no. 4, pp. 1031-1039, July/August. 2003.

[18] Hyunsung An and Hanju Cha, "A Robust Sensorless Start-up Method using Four Step Sequence for LCI system," *in Proceedings of IEEE ECCE 2016 Conference*, Sept. 2016.

Prototyping of 500 kVA Medium Frequency Transformer for Offshore Direct-Current Collection Grid

Tomoyuki Hatakeyama, Naoyuki Kurita, and Mamoru Kimura

Research & Development Group
Hitachi, Ltd.
2-1, Omika-cho 7-chome, Hitachi-shi, Ibaraki-ken, Japan
tomoyuki.hatakeyama.ff@hitachi.com

Abstract-Offshore wind farms are integrated to a power system via a long distance transmission system and an offshore platform which collects power from wind turbines. To reduce the size of the offshore platform, a medium voltage direct current (MVDC) collection grid based on modular DC-DC converters is a feasible solution. This paper deals with a medium frequency transformer (MFT), one of the most important components to realize compact and efficient modular DC-DC converters. To reduce large winding loss due to proximity effect, an interleaved winding arrangement is applied to the MFT with an amorphous core. MFTs with a rated power of 500 kVA with interleaved and non-interleaved winding arrangements are prototyped and experimentally evaluated. This study demonstrates that the interleaved winding arrangement provides higher efficiency because of the depressed proximity effect, at a capacity of 500 kVA. Our work indicates that the MFT with interleaved winding arrangement contributes to the reliability and miniaturization of the offshore platform.

I. INTRODUCTION

Offshore wind farms (OWFs) have already been operated commercially in Europe, and the development of OWFs are discussed and planned in many other region [1]. The location of the OWF needs to be paid attention in terms of wind velocity, landscape, and agreement of stakeholders, etc. One solution is to develop OWFs far from coasts. To reduce the transmission loss, a high voltage direct current (HVDC) technology is applied to the transmission system.

Conventional power system for OWFs is constructed with the HVDC transmission grid and a collection grid based on medium voltage alternative current (MVAC) as shown in Fig. 1(a) [2]–[5]. In this system the offshore platform collects the power of MVAC from the wind turbines and transfers the power of HVDC to the onshore power grid. The offshore platform consists of a utility frequency transformer and an AC to DC converter. The utility frequency transformer provides high voltage alternative current (HVAC), and the following AC to DC converter transforms HVAC to HVDC. In such system the offshore platform is costly due to its size.

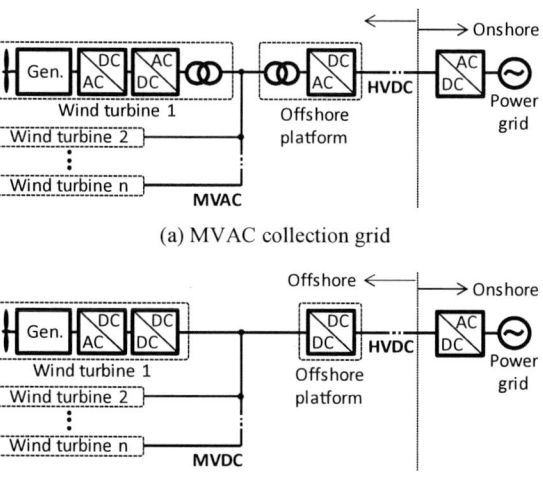

(a) MVAC collection grid

(b) MVDC collection grid

Fig. 1. Diagram of power systems for large scale offshore wind farms.

To reduce the size of offshore platform, a medium voltage direct current (MVDC) collection grid has been proposed as shown in Fig. 1(b) [6]–[10]. Each wind turbine has a DC-DC converter outputting the power of MVDC. The platform also contains of a DC-DC converter which boosts MVDC to HVDC. The utility frequency transformer is not needed in this system, therefore, reducing the size of the offshore platform is expected.

The DC-DC converter in the wind turbine is required to transform low voltage direct current (LVDC) to MVDC with sufficient insulation performance. To realize high voltage transformation ratio as well as insulation performance, a modular input-parallel output-series DC-DC converter has been proposed as shown in Fig. 2 [9][10]. This modular DC-DC converter is structured with several modules of DC-DC converter whose input terminals and output terminals are connected in parallel and in series, respectively. The module consists of an inverter, a medium frequency transformer (MFT), and a rectifier. The MFT is

The 2018 International Power Electronics Conference

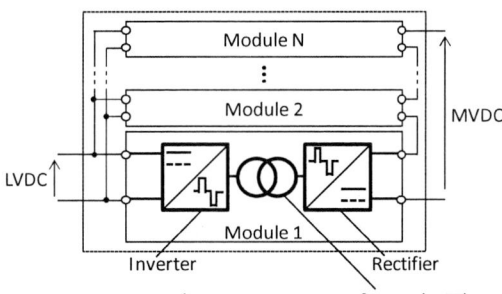

Fig. 2. Structure of a modular input-parallel output-series DC-DC converter.

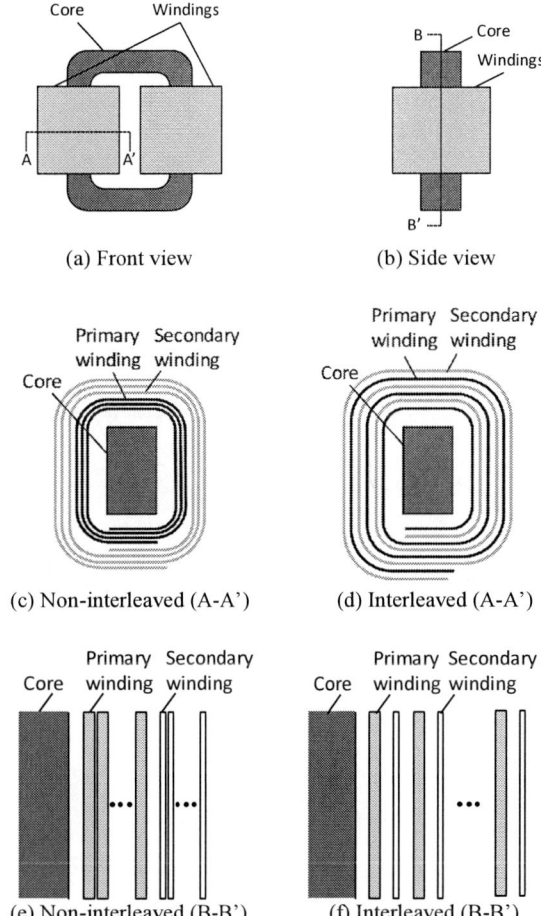

(a) Front view   (b) Side view

(c) Non-interleaved (A-A')   (d) Interleaved (A-A')

(e) Non-interleaved (B-B')   (f) Interleaved (B-B')

Fig. 3. Winding arrangement of the transformer. Cross sections A-A' and B-B' are indicated in (a) and (b), respectively.

one of the most important component among these, as it has large room to improve the efficiency and reduce the size. In general, the higher frequency provides the smaller size of MFTs, however, the larger undesired losses as well [11][12].

Literatures have presented that amorphous materials are suitable to be used as the core of the MFT because of their low core loss performance [13]–[16]. The other literatures have reported that the interleaved arrangement of the primary and secondary windings depresses the proximity effect, resulting in reducing copper loss [17][18]. However, large scale demonstration combining amorphous core and interleaved winding arrangement has not been shown in the literatures.

In this study, two amorphous MFTs are prototyped with different winding arrangements: non-interleaved and interleaved, to evaluate the effects of the winding arrangement. The rated power of the MFTs is set to be 500 kVA aiming at a practical wind turbine output, i.e., a 5 MW modular DC-DC converter is designed to be constructed with ten 500 kW modules for each wind turbine.

The rest of this paper is organized as follows: Section II provides information about specification of the prototyped MFTs. The loss evaluation method is given in Section III. The results are shown in Section IV. Section V and VI give the discussion and conclusion, respectively.

II. SPECIFICATION OF THE PROTOTYPED MEDIUM FREQUENCY TRANSFORMERS

The non-interleaved and interleaved arrangements are shown in Fig. 3. The primary winding and secondary winding are placed on both legs of the core as shown in Fig. 3 (a) and (b). Fig. 3 (c) and (d) show the cross sectional views of the non-interleaved and interleaved winding arrangements, respectively. This cross section corresponds to A-A' indicated in Fig. 3 (a). Following Fig. 3 (e) and (f) show the other cross sectional (B-B') views of the non-interleaved and interleaved winding arrangements, respectively, where B-B' line is indicated in Fig. 3 (b).

In the non-interleaved case, the primary winding is first wound around the leg of the core, then, the secondary winding is wound around the primary winding. Only the

outer-most turn of the primary winding and the inner-most turn of the secondary winding face each other. On the other hand, in the interleaved case, the primary winding and secondary winding face each other along all the turns. As the interleaved winding arrangement is expected to depress the proximity effect in all the turns, the copper loss is expected to reduce consequently.

Besides proximity effect, skin effect also increases copper loss. A litz wire is often used to depress skin effect [19]. However, the litz wire is not commercially available for a large power scale, such as several hundreds kVA. Instead, a copper sheet is chosen for this application because of ease to arrange and scalability.

Table I lists major specifications of the prototyped MFTs. The core is made of the amorphous ribbons with 25 µm thickness whose yoke is lap-jointed. The number of turn of the primary winding and secondary winding are set to be 30 and 60, respectively.

The appearances of the prototyped MFTs are shown in Fig. 4. The windings are molded whose upper and lower

1992

TABLE I
SPECIFICATIONS OF THE PROTOTYPED MFTs

Rated power (kVA)	500
Designed frequency (kHz)	3
Primary voltage (kV)	2.4
Secondary voltage (kV)	4.8
Primary current (A)	208
Secondary current (A)	104
Primary number of turns	30
Secondary number of turns	60
Core material	2605HB1M amorphous [20]

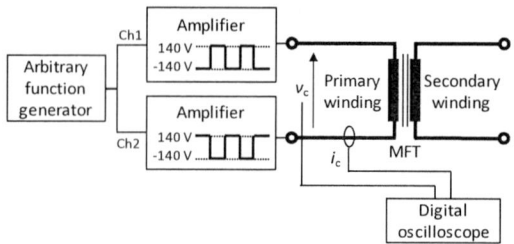

Fig. 5. Schematic diagram of the core loss evaluation system.

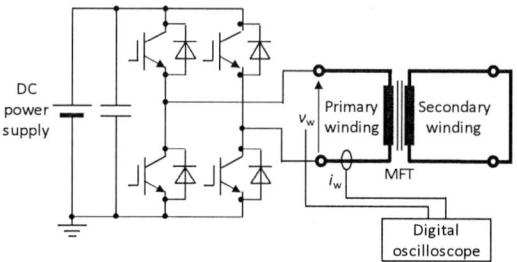

Fig. 6. Schematic diagram of the winding loss evaluation system.

(a) Non-interleaved

(b) Interleaved

Fig. 4. Appearance of the prototyped MFTs.

ends are sealed with a resin to realize sufficient insulation performance in air. A nonmagnetic stainless steel is used to fix the core and windings in order to suppress the stray loss due to the leakage magnetic field. A bus-bar connects the primary windings around both legs each other. The same applies to the secondary winding. A dedicated terminal is placed on the second turn of the winding around each leg to measure the core loss, as described in the following section.

Comparing Fig. 4 (a) and (b), the interleaved winding arrangement increases the size of MFT. This is because the insulation distance between the primary and secondary windings are longer than the insulation distance between each layers of the primary and secondary windings. Thus, the core in the case of interleaved arrangement is also larger than that of the MFT with non-interleaved winding.

III. LOSS EVALUATION METHOD

The core loss and winding loss of the assembled MFTs are experimentally evaluated. The measurement systems for the core loss and winding loss are configured differently as shown in Fig. 5 and 6, respectively, as the high voltage and large current outputs are needed, respectively.

A. Core loss evaluation method

The evaluation system of the core loss is shown in Fig. 5. The system is constructed with an arbitrary waveform generator and two bipolar-amplifiers. The maximum amplitude of the applied voltage is 280 V. Rectangular voltages with different frequencies are applied to the tap port placed on the second turn of the primary winding, where all the terminals of the secondary winding are opened.

The core loss (W_i) is calculated with the following equation using the measured v_c and i_c:

$$W_i = \frac{1}{M_i T} \int_0^T v_c(t) \cdot i_c(t) dt \tag{1}$$

where M_i and T are the mass of the core and a period, respectively.

W_i depends on the maximum flux density (B_m) and frequency (f). It can be approximated with the following equation [21]:

$$W_i = A_h B_m^2 f + A_e B_m^2 f^2 + A_a B_m^{1.5} f^{1.5} \tag{2}$$

where, A_h, A_e, A_a are the coefficients for hysteresis loss, classical eddy current loss, and anomalous eddy current loss, respectively.

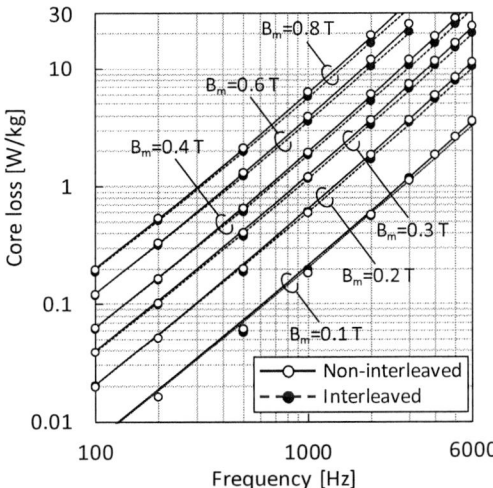

Fig. 7. Frequency performance of the core loss in the range of $B_m = 0.1$ T to 0.8 T. Symbols are the measured values. The solid and dashed lines are the approximation lines based on the eq. (2) for the non-interleaved and interleaved cases, respectively.

TABLE II
COEFFICIENTS OF CORE LOSS APPROXIMATION

	Non-interleaved	Interleaved
A_h	8.00×10^{-4}	8.00×10^{-4}
A_e	2.50×10^{-6}	1.65×10^{-6}
A_a	1.80×10^{-4}	1.80×10^{-4}

B. Winding loss evaluation method

The evaluation system of the winding loss is shown in Fig. 6. The winding loss of the assembled MFTs is measured by injecting the current to the primary winding up to 210 A$_{rms}$ using an IGBT full-bridged inverter system. The inverter applies the rectangular voltages with different frequencies to the primary winding where the secondary winding is shorted. In this test, the winding loss contains of primary winding loss and secondary winding loss. The winding loss (W_C) is calculated with the following equation.

$$W_C = \frac{1}{T}\int_0^T v_w(t)\cdot i_w(t)dt \qquad (3)$$

IV. LOSS CHARACTERISTICS OF THE ASSEMBLED MFTs

Firstly the core loss characteristics are shown. The frequency performance of the core loss in B_m range of 0.1 T to 0.8 T is shown in Fig. 7. Symbols are the measured values. The solid and dashed lines correspond to the approximation based on the eq. (2) for the MFTs with non-interleaved and interleaved winding arrangements, respectively. As shown in Fig. 7, the larger B_m and higher f increase the core loss. Comparing the non-interleaved and

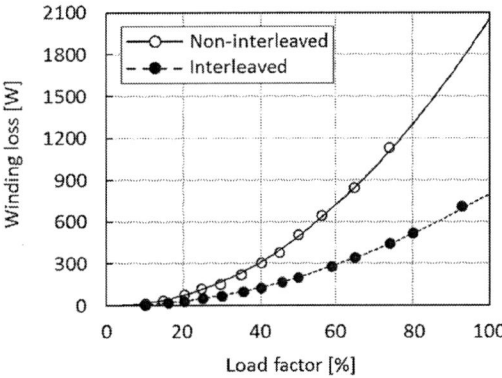

Fig. 8. Load factor performance of the winding loss at $f = 3$ kHz. The winding loss contains of both primary and secondary sides. The symbols and curves correspond to the measured values and cumulative approximation of them, respectively.

interleaved cases, the core loss of the former case is larger than that of the latter case. Table II lists the values of the coefficients for hysteresis loss, classical eddy current loss, and anomalous eddy current loss. Both MFTs have the same A_h and A_a. However, non-interleaved arrangement has approximately 1.5 times of A_e than that of interleaved arrangement. This seems to be because the core of the non-interleaved case is smaller than that of the interleaved case. The smaller core has the larger lap-jointed part, which causes in-plane eddy current loss [22].

Then, the winding loss characteristics are shown. Here the load factor is defined as the injected current divided by the rated current. The load factor performance of the winding loss is shown in Fig. 8. The symbols and curves correspond to the measured values and cumulative approximation of them, respectively. It is recognized that the winding loss in the case of interleaved arrangement is less than that in the case of non-interleaved arrangement for any load factor.

Likewise, the load factor performances of the winding loss and the approximated curves of them in the frequency range from 200 Hz to 10 kHz were also obtained. Using these approximated curves, the frequency performances of the winding losses at the rated load factor are derived. Fig. 9 shows the frequency performance of the winding loss. In the case of non-interleaved arrangement, the winding loss increases intensely as the frequency is increased. This is considered to be the effect of the proximity effect. On the other hand, in the case of interleaved arrangement, although the increase of the winding loss is recognized, the degree of increase is suppressed as compared with non-interleaved arrangement case. At the designed frequency of 3 kHz, the interleaved arrangement reduces the winding loss by 61% compared with non-interleaved arrangement case.

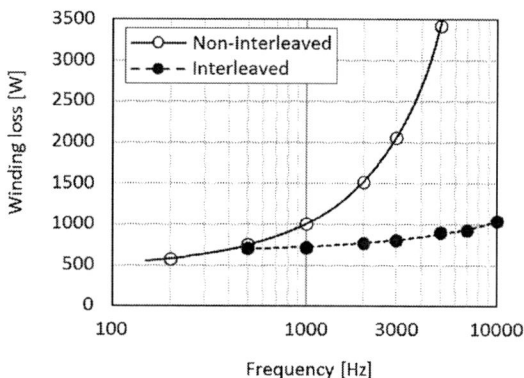

Fig. 9. Winding loss with frequency dependency at the rated load factor derived from the approximated load factor performance. The winding loss contains of both primary and secondary sides.

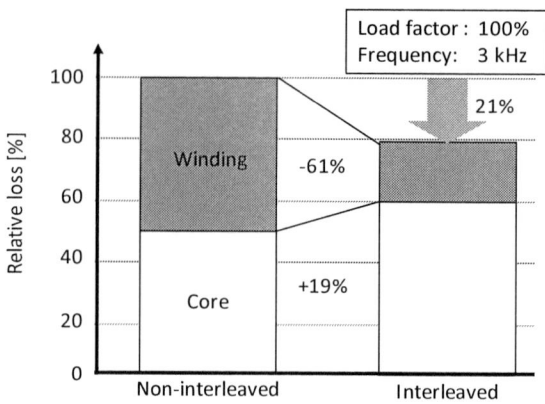

Fig. 10. Loss breakdown including core loss and winding loss.

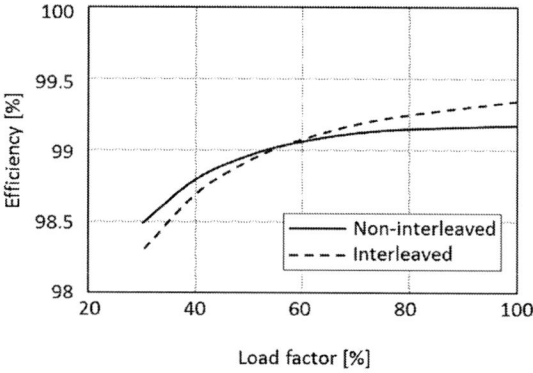

Fig. 11. Load factor performance of the efficiency derived from the core loss and winding loss.

V. DISCUSSION

The loss breakdown when the load factor is 100% is shown in Fig. 10. The core loss for the interleaved case is 19% larger than that for the non-interleaved case. This is because the core size for the interleaved case is larger than that for the non-interleaved case. However, the winding loss of the interleaved case is reduced by 61%, compared with the case of non-interleaved case. As a result, the sum of core loss and winding loss for the interleaved case is reduced by 21% compared with the non-interleaved case.

Fig. 11 shows the calculated load factor performance of the efficiency derived from the sum of core loss and winding loss for the interleaved and non-interleaved cases. When the load factor is more than approximately 60%, the interleaved case achieves higher efficiency because of the depressed proximity effect. The high efficiency at the heavy load simplifies a cooling system. Therefore, the MFT with interleaved winding arrangement is expected to be more reliable than the MFT with non-interleaved winding arrangement. Also, it is possible to make the MFT with interleaved winding arrangement smaller than the MFT with non-interleaved winding arrangement if the cooling system can be removed.

The larger capacity of the MFTs causes the larger winding loss due to the stronger proximity effect. Considering the efficiency of the MFT with capacity of more than 500 kVA, for example 1 MVA, based on the results of this study, the interleaved winding arrangement is expected to be even superior to the non-interleaved winding arrangement.

VI. CONCLUSION

The 500 kVA amorphous MFTs with interleaved and non-interleaved winding arrangements are prototyped and experimentally evaluated. The interleaved winding arrangement is confirmed to be more efficient than the non-interleaved winding arrangement by 21% at a rated power. Considering the load factor performance, the MFT with interleaved winding arrangement is more efficient when the load factor is more than 60%. High efficiency at the heavy load may dispense with the cooling system for the MFT with interleaved winding arrangement. Our work indicates that the MFT with interleaved winding arrangement contributes to the reliable and compact offshore platform.

ACKNOWLEDGMENT

This paper is based on the results obtained from a project commissioned by the New Energy and Industrial Technology Development Organization (NEDO).

REFERENCES

[1] Global wind energy council (GWEC), "Global Wind 2016 Report," 2017.

[2] N. M. Kirby, Lie Xu, M. Luckett and W. Siepmann, "HVDC transmission for large offshore wind farms," in *Power Engineering Journal*, vol. 16, no. 3, pp. 135-141, June 2002.

[3] A. B. Morton, S. Cowdroy, J. R. A. Hill, M. Halliday and G. D. Nicholson, "AC or DC? economics of grid connection design for

offshore wind farms," *The 8th IEE International Conference on AC and DC Power Transmission*, London, UK, 2006, pp. 236-240.

[4] P. Bresesti, W. L. Kling, R. L. Hendriks and R. Vailati, "HVDC Connection of Offshore Wind Farms to the Transmission System," in *IEEE Transactions on Energy Conversion*, vol. 22, no. 1, pp. 37-43, 2007.

[5] G. Ramtharan, A. Arulampalam, J. B. Ekanayake, F. M. Hughes and N. Jenkins, "Fault ride through of fully rated converter wind turbines with AC and DC transmission," in *IET Renewable Power Generation*, vol. 3, no. 4, pp. 426-438, December 2009.

[6] C. Meyer, M. Hoing, A. Peterson and R. W. De Doncker, "Control and Design of DC Grids for Offshore Wind Farms," in *IEEE Transactions on Industry Applications*, vol. 43, no. 6, pp. 1475-1482, 2007.

[7] W. Chen, A. Q. Huang, C. Li, G. Wang and W. Gu, "Analysis and Comparison of Medium Voltage High Power DC/DC Converters for Offshore Wind Energy Systems," in *IEEE Transactions on Power Electronics*, vol. 28, no. 4, pp. 2014-2023, April 2013.

[8] Y. Lee, G. Vakil, R. Feldman, A. J. Watson and P. W. Wheeler, "A high-power DC-DC converter based dual active bridge for MVDC grids on offshore wind farms," *2016 18th European Conference on Power Electronics and Applications (EPE'16 ECCE Europe)*, Karlsruhe, 2016, pp. 1-10.

[9] M. A. Bahmani, T. Thiringer, A. Rabiei and T. Abdulahovic, "Comparative Study of a Multi-MW High-Power Density DC Transformer With an Optimized High-Frequency Magnetics in All-DC Offshore Wind Farm," in *IEEE Transactions on Power Delivery*, vol. 31, no. 2, pp. 857-866, April 2016.

[10] H. Krishnamoorthy, M. Daniel, J. Ramos-Ruiz, P. Enjeti, L. Liu and E. Aeloiza, "Isolated AC–DC Converter Using Medium Frequency Transformer for Off-Shore Wind Turbine DC Collection Grid," in *IEEE Transactions on Industrial Electronics*, vol. 64, no. 11, pp. 8939-8947, Nov. 2017.

[11] W. J. Gu and R. Liu, "A study of volume and weight vs. frequency for high-frequency transformers," *Power Electronics Specialists Conference, 1993. PESC '93 Record., 24th Annual IEEE*, Seattle, WA, 1993, pp. 1123-1129.

[12] T. Hatakeyama and K. Onda, "Novel evaluation method for the volume of transformers with various magnetic cores," *2011 IEEE Ninth International Conference on Power Electronics and Drive Systems*, Singapore, 2011, pp. 480-485.

[13] L. Heinemann "An actively cooled high power, high frequency transformer with high insulation capability," in *Proc. IEEE Applied Power Electronics Conference and Exposition (APEC)*, pp. 352 – 357, 2002.

[14] M. Pavlovsky, S. de Haan, and J. A. Ferreira, "Reaching high power density in multikilowatt DC-DC converters with galvanic isolation," *IEEE Transaction on Power Electronics.*, vol. 24, no. 3, pp. 603–612, 2009.

[15] Z. M. Shafik, K. H. Ahmed, S. J. Finney, and B. W. Williams, "Nanocrystalline cored transformer design and implementation for a high current low voltage DC/DC converter," in *Proc. Power Electronics, Machines and Drives Conference (PEMD)*, 2010.

[16] T. Hatakeyama and K. Onda, "Core Loss Estimation of Various Materials Magnetized With the Symmetrical/Asymmetrical Rectangular Voltage," *IEEE Transaction on Power Electronics.*, vol. 29, no. 12, pp. 6628-6635, Dec. 2014.

[17] E. L. Barrios, A. Urtasun, A. Ursúa, L. Marroyo, and P. Sanchis, "High-Frequency Power Transformers With Foil Windings: Maximum Interleaving and Optimal Design," *IEEE Transaction on Power Electronics*, vol.30, pp. 5712–5723, Oct. 2015.

[18] H. Tanaka, K. Nakamura, and O. Ichinokura, "Winding Arrangement of High-Frequency Amorphous Transformers for MW-class DC-DC Converters," *Journal of Magnetic Society of Japan*, vol.40, pp.35-38, 2016.

[19] A. Roßkopf, E. Bär, and C. Joffe, "Influence of Inner Skin- and Proximity Effects on Conduction in Litz Wires," *IEEE Transaction on Power Electronics*, vol. 29, pp. 5454-5461, 2014.

[20] Amorphous Alloys for Transformer Cores, [Online] Available: http://www.metglas.com/assets/pdf/2605sa1.pdf

[21] G. Bertotti, "General Properties of Power Losses in Soft Ferromagnetic Materials," *IEEE Transaction on Magnetics*, vol. 24, pp. 621-630, 1988.

[22] R. U. Lenke, S. Rohde, F. Mura, and R.W. De Doncker, "Characterization of Amorphous Iron Distribution Transformer Core for Use in High-Power Medium-Frequency Applications," in *Proc. of IEEE Energy Conversion Congress and Exposition*, San Jose, CA, 2009, pp. 1060–1066.

PSCAD/EMTDC and RTDS Simulation Analysis of Multivendor Multi-terminal HVDC System Connected to Offshore Windfarms

Hiroshi Suwa[1], Takuro Arai[2], Takahiro Ishiguro[3], Tohru Yoshihara[4], Mamoru Kimura[4]
Tsuneshisa Wachi[5], Takahiro Horikoshi[5] and Tatsuhito Nakajima[6*]

1 Tokyo Electric Power Company Holdings, Incorporated, Tokyo, Japan
2 Toshiba Corporation, Tokyo, Japan
3 Toshiba Energy Systems and Solutions Corporation, Kawasaki, Japan
4 Hitachi, Ltd., Hitachi, Japan
5 JP Business Service Corporation, Tokyo, Japan
6 Tokyo City University, Tokyo, Japan
*E-mail: tnaka@tcu.ac.jp

Abstract- **A R&D project on a multi-terminal HVDC system is ongoing in Japan. This paper describes simulation analysis of a multivendor multi-terminal HVDC linked to offshore windfarms, by "PSCAD/ EMTDC" software and "RTDS" simulator for validating interoperability between multivendor converters. Satisfactory results were obtained.**

I. INTRODUCTION

A R&D project on a multivendor multi-terminal HVDC (high voltage dc transmission) system connected to off-shore windfarms is ongoing in Japan, jointly by Tokyo Electric Power Company Holdings, Inc. (Tepco), Toshiba Corp. and Hitachi, Ltd., as a contract research of NEDO (New Energy and Industrial Technology Development Organization). The HVDC systems connected to offshore windfarms, which are operated in commercial service in Europe, are two-terminal point-to-point HVDC types, and those ac/dc converters are provided by a single vendor.

In the R&D project in Japan, on the other hand, it is assumed that a multi-terminal HVDC system is applied for power transfer from the windfarms. It is expected that, by the multi-terminal HVDC, multiple windfarms at the locations with high wind potentials, and multiple onshore stations coupled to ac grids can be connected, and power transfer with high reliability can be realized consequently.

In the R&D project, it is also presumed that the HVDC station equipment is supplied by multiple vendors. This is because the capital cost of HVDC equipment will be reduced by competitive bidding between the vendors, and because large-scale windfarms will be developed one by one, and the HVDC system will be expanded in step by step. The multivendor multi-terminal HVDC system connected to the offshore windfarms is thus the project target.

The above-mentioned three companies engaged in the project have considered coordinated control schemes for the multivendor multi-terminal HVDC system, and challenged simulation analysis studies for validating inter-operability between multivendor converters, by making simulation models created by the vendors combined.

This paper describes simulation analysis studies of the multivendor multi-terminal HVDC system connected to the offshore windfarms, by using the "PSCAD/EMTDC" simulation software and "RTDS" digital real-time simulator, for validating interoperability between multivendor HVDC converters. A system configuration assumed for simulation analysis, and "PSCAD/EMTDC" and "RTDS" simulation results are reported in the followings.

II. SYSTEM CONFIGURATION OF MUTI-TERMINAL HVDC

A. Simulation Analysis Steps

Fig. 1 shows the way to proceed with the simulation analysis studies. The system configuration including the multi-terminal HVDC system, offshore windfarms and onshore ac grids were discussed and determined by the three companies involved in the project. Tentative requirement specifications for the multivendor multi-terminal HVDC were summarized and agreed by the companies. Then HVDC converter simulation models have been created, and basic simulation tests of single-vendor HVDC have been carried out by the vendors, Toshiba and Hitachi respectively. Preliminary simulation studies using a standard HVDC converter simulation model have been conducted by Tepco, which makes replacement to the vendor's models easier.

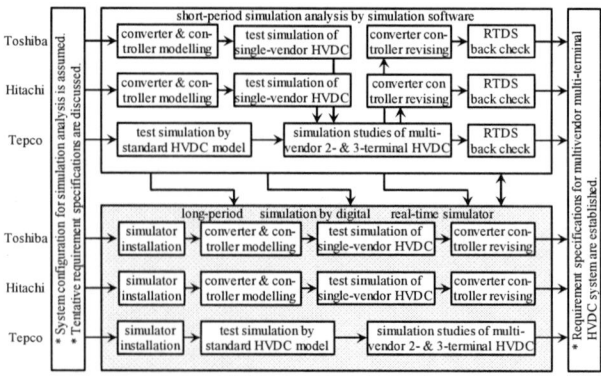

Fig. 1. Short-period and long-period simulation analysis.

B. System Configuration Assumed in NEDO Project

Fig. 2 shows a tentative system configuration for simulation analysis purposes. The assumed HVDC system is a five-terminal HVDC type, composed by two offshore terminals and three onshore terminals. The windfarms are modelled by simple current sources, by which windfarm mechanical and generator dynamics are neglected. AC grids consist of multiple generators so that effects on ac grid frequency by wind power variation and HVDC converter tripping operation can be considered. DC cables are used for bipole dc lines with a rated voltage ±500kV. Although the system configuration shown in Fig. 2 taken into account, the simulation studies have been started using a simplified two-terminal monopole HVDC system.

The tentative requirement specifications document describes steady-state control schemes, operating condition categories and sequences, faults assumed in the HVDC system and equipment responses. Control blocks, gains and time constants are not standardized in the document. These know-hows are preserved by each vendor, and not disclosed. The interfaces between a master controller and converter terminals are standardized on the other hand.

C. Simulation Analysis Tool

For the simulation analysis for validation, appropriate simulation tools need to be selected with the following requirements taken into consideration; (1) easiness of model date exchange with other companies, (2) preserve vendor's technical know-hows from other companies, (3) affinity with power electronic equipment for power grid applications, and (4) applicability to transient analysis of large scale systems. As the tools that meet these requirements, the power system transient analysis software "PSCAD/EMTDC" for short-period analysis, and the "RTDS" Technologies Inc. real-time simulator "RTDS" for long-period analysis were both selected. Fig. 3 shows the "RTDS" real-time simulator set installed in Tepco R&D center. The simulator hardware has 5 racks in the full-size cubicles, composed by print circuit board PB5 systems, and has 50 GTFPGA units in the middle-size cubicles specially designed for MMC models. This "RTDS" system enables simulation studies of a 5-terminal bipole HVDC system using MMC-type converters. The GTFPGA unit can simulate individual firing pulses for power devices in MMC in detail. Using

Fig. 3. Real-time digital simulator "RTDS".

Using this unit, a MMC model with 1024 submodules per arm can be created, and failures inside the converter can be simulated as well. The "RTDS" simulator is operated by an exclusive software "RSCAD", which is a multi-functional GUI software and easy to use.

III. MULTIVENDOR HVDC SYSTEM MODELS AND EXAMPLE SIMULATION ANALYSIS RESULTS

A. Short-period Simulation using "PSCAD/EMTDC"

The requirements specifications for the control and protection schemes of the multivendor multi-terminal HVDC system have been investigated. Short-period simulation models have been created, using "PSCAD/EMTDC" Ver 4.6, preceding "RTDS" simulation model creating. By "PSCAD/EMTDC", model creating and validation of interoperability of the multivendor HVDC are made easier.

A simulation model of multivendor two-terminal mono pole HVDC system, that is the minimum unit of HVDC system, was created at the beginning. The diagram of the two-terminal HVDC system is as shown in Fig. 4, and its "PSCAD/EMTDC" model is as shown in Fig. 5. The onshore and offshore converters are modelled as MMC. The simulation model interfaces of HVDC control signals and circuit terminals are standardized tentatively, based on discussions by the involved companies. Those model components that include vendor's know-hows are encapsulated by the vendor concerned, using the black boxing function of "PSCAD/EMTDC". Simulation studies of the multivendor two-terminal HVDC system were carried out, after each vendor conducted simulation tests for a single-vendor two-terminal HVDC system. The tests covered scenarios such as start/stop, power flow control, responses to AC grid fault. Fig. 5 shows an example model, in which the offshore converter model is the vendor A's model created and encapsulated by Toshiba, and the onshore one is the vendor B's model done by Hitachi. For validation of interoperability of the multivendor HVDC system, simulation studies with the opposite vendor combination (onshore by vendor B, offshore by vendor A) was also carried out.

In 2015, simulation studies to validate the scenarios using the multivendor two-terminal HVDC system model were carried out intensively. In 2016, the vendor-supplying HVDC converter models were upgraded by the vendors so that the HVDC system behaviors under onshore and offshore AC fault conditions can be calculated properly. One of the modified points was, they added a braking chopper to the onshore converter model

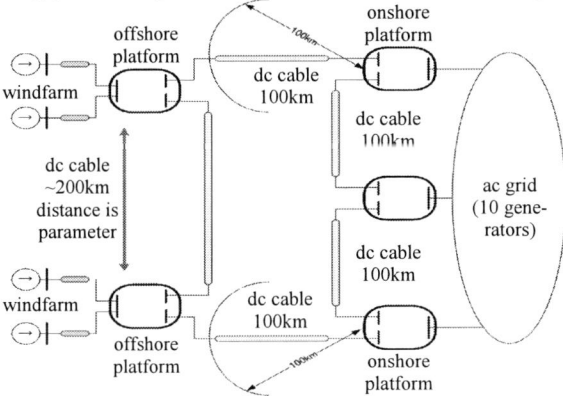

Fig. 2. Tentative system configuration for simulation purposes.

1998

to consume of surplus power in case of onshore grid AC fault and to avoid DC overvoltage. The other modified point was a fault-ride-through (FRT) function which was added to the current sources expressing wind generators, for preventing offshore AC overvoltages. The simulation studies for feasibility evaluation using the upgraded two-terminal HVDC model were successfully completed.

This paper shows two examples of simulation tests. One is of single-vendor's simulation tests under an offshore AC line fault [1], and another is the multivendor's simulation under an onshore grid fault [2].

At first, the single-vendor converters' results are presented. A converter switching control scheme for the offshore converter was examined to avoid converter tripping caused by overcurrents during the offshore AC line fault, and simulation studies were conducted with the HVDC model shown in Fig. 6. Simulation results under a three-phase-to-ground (3LG) fault condition are as shown in Fig. 7. Although the offshore ac bus voltages are almost completely lost as shown in the upper figure, the converter ac currents are sufficiently controlled to be small as shown in the lower figure. These results indicate that, by the applied control scheme, the offshore converter can continue operation during the offshore fault, consequently.

Next, the multivendor converters' simulation results, with an onshore grid fault scenario, are presented. During the onshore grid fault, the capacitor voltage of the onshore MMC may increase and stop the operation by the overvoltage due to the power injection from the offshore WF. In order to avoid the overvoltage of onshore MMC capacitor, the installation of the braking chopper at onshore side and its power consumption during the onshore grid fault are examined. Simulation studies under a 3LG fault were conducted with the HVDC model shown in Fig. 8. In this simulation, the offshore MMC is by vendor A), and the onshore one is by vendor B). In this simulation, the braking chopper is modeled as the series circuit of IGBT (insulated gate bipolar transistor) and the resistor. Simulation results are as shown in Fig. 9, which includes the waveforms of (a)VG, (b)PG and PDC, (c)VDC, (d)VCAVE and (e)Braking chopper operation, from top to bottom. In Fig. 9(e), "ON" means that the IGBT of the braking chopper is turned on, which is the power consumption mode of the braking chopper. "OFF" means that the IGBT of the braking chopper is turned off, which is the disconnection mode of the braking chopper from the HVDC system. VG drops when three-phase-to-ground fault occurred and VCAVE increases due to the power injection from offshore WF, although onshore

MMC cannot transmit this power to the onshore grid due to the VG drop. However, VDC and PDC are almost kept constantly and the VCAVE overvoltage can be avoided by the power consumption of the braking chopper. Finally, the onshore MMC operation can be recovered to the operation point before the 3LG fault without the stop of the MMC operation.

A multivendor three-terminal monopole HVDC model is being created currently. A master controller is provided for the model, with coordination between the three converter terminals, constraints by the onshore ac grids and those by the offshore windfarm taken into consideration.

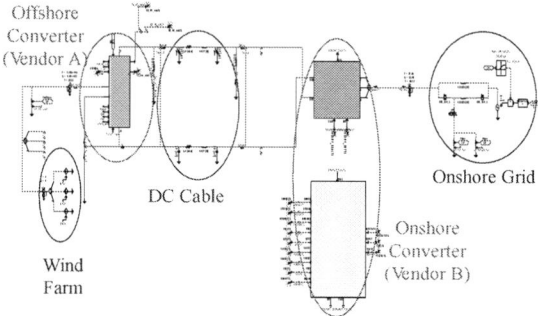

Fig. 5. Corresponding "PSCAD/EMTDC" model of Fig. 4.

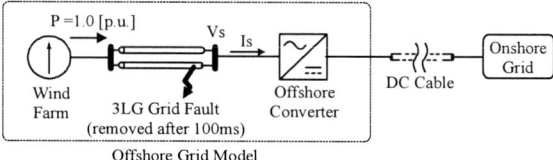

Fig. 6. HVDC simulation model for offshore ac line fault case.

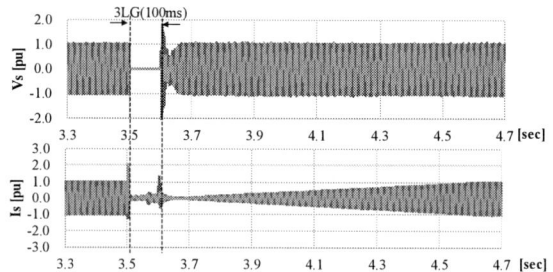

Fig. 7. Simulation results of offshore ac line fault case.
upper : offshore three-phase ac bus voltages [p.u.]
lower : offshore converter three-phase ac currents [p.u.]

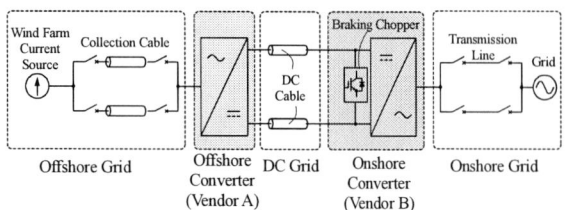

Fig. 4. Two-terminal monopole multivendor HVDC system.

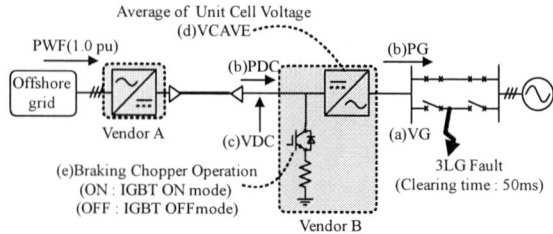

Fig. 8. HVDC simulation model for onshore AC line fault case.

1999

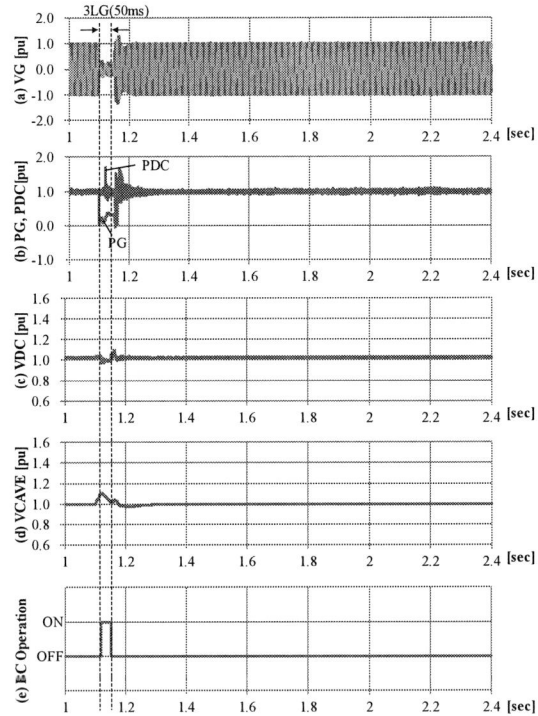

Fig.9 Simulation results of onshore AC line fault case.

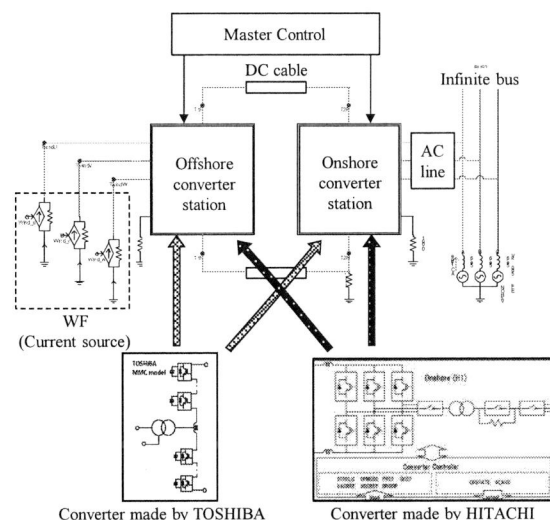

Fig. 10 Real-time simulator model using "RTDS".

Table 1. Converter control mode.

Converter mode	AC side		DC side
	Active power	Reactive power	
On-shore Converter Station	Capacitor voltage control	AQR or AC-AVR	P_{dc}-V_{dc} characteristic (DC-AVR)
On-shore Converter Station	Capacitor voltage control	AQR or AC-AVR	P_{dc}-V_{dc} characteristic (DC-APR)
Off-shore Converter Station	CVCF		Capacitor voltage control with DC-APR

B. Long-period Simulation using "RTDS"

In this section, the multivendor HVDC system is confirmed using "RTDS". The purposes of this section are not only showing converter operation designed by each vendor, but also verifying coordination between multiple converters under conditions such as initial charging of dc capacitors, starting operation, and rated power operation. Fig. 10 shows the overview of "RTDS" model for the multivendor two-terminal HVDC system. This system is so-called point-to-point HVDC. It consists of an offshore converter station and an onshore one. Generated power by the windfarm, which is expressed as current sources in the simulation studies, is transmitted to the onshore ac grid through the dc cables. In order to prevent leakage of the vendor's technical know-hows, the black boxing function of "RTDS" is used for creating the multivendor two-terminal HVDC model.

Table 1 shows the converter control mode. Onshore converter can control the active power and reactive power freely on the ac side provided that it does not deteriorate the power quality of AC system. Therefore the active power can be determined by the converter capacitor voltage control, and AQR (Auto reactive power regulator) or AC-AVR (Auto voltage regulator) can be applied for onshore converter using reactive power control. Furthermore, one or more onshore converters can operate as DC-AVR to maintain the voltage on the dc grid.

On the other hand, offshore converter is not able to control its active power because it is connected to the windfarm and operates CVCF (Constant voltage constant frequency). The active power of offshore grid is depend-

ing on the wind condition. Generated power by the windfarm flowing to the converter, it must be output to dc side using the converter capacitor voltage control. It means that offshore converter has the capacitor voltage control with DC-APR (Auto power regulator).

The converter control characteristic on the dc side is shown as Fig. 11. The P_{dc}-V_{dc} characteristic in Fig. 11(a) is referred to as a droop control. It is typically applied for the DC-AVR converter. As long as both V_{dc} and P_{dc} are between the upper limit and the lower limit, V_{dc} is proportional to P_{dc} with the slope $TxKdroop$. Actually, P_{dc} is often determined by other converter such as DC-APR converter. Moreover if $TxKdroop$ is set to be zero, V_{dc} can also be constant regardless of the P_{dc}. Using a droop control, it is also possible to contribute to the dc voltage control with two or more converter stations. In this case, if each converter has different K_{droop}, contribution to dc voltage control can be changed. Fig. 11(b) shows the characteristic of DC-APR. In the DC-APR mode, unless the output DC voltage exceeds the upper and lower limit, V_{dc} freely varies so that P_{dc} is constant.

The converter model specifications have been standardized tentatively based on the discussions by the companies. The converter models have common interfaces and are applicable to both offshore and onshore converters. Input / Output signals of the converter model are shown in Table 2. The master controller can control

The 2018 International Power Electronics Conference

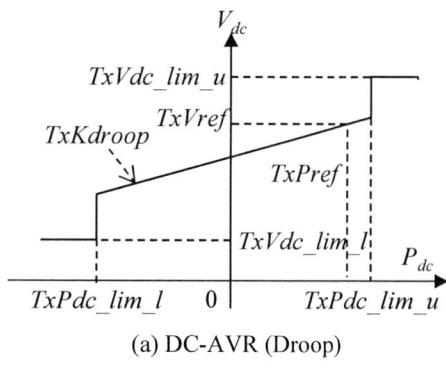

(a) DC-AVR (Droop)

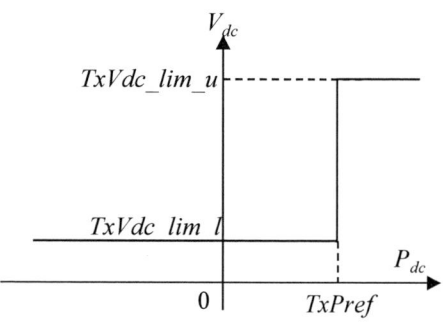

(b) DC-APR

Fig. 11. DC control characteristic.

Table 2 Input / Output signals of the converter model.

Category	Function	name	In / out
Converter control	Operation start	TxSTART	In
	Active power reference	TxPref	In
	Reactive power reference	TxQref	In
	DC voltage reference	TxVdcref	In
	AC voltage reference	TxVacref	In
	AC frequency reference	TxFref	In
	Droop	TxKdroop	In
	DC voltage upper limit	TxVdc_lim_u	
	DC voltage lower limit	TxVdc_lim_l	
	Converter power upper limit	TxPdc_lim_u	
	Converter power lower limit	TxPdc_lim_l	
	Control switching	TxACtap	In
	Output power	Tx_P / Tx_Q	Out
	Converter status	TxOV/TxOC	Out
Switch control	ACCB	TxACCB	In
	Bypass switch	TxBypass	In

each converter station by only using these signals. Each signal has an original name because signal name must not be duplicated in one model. A prefix "*T*" means converter made by Toshiba. The converter number is in "*x*".

Fig. 12 and Fig. 13 show an example of "RTDS" real-time simulation. These figures depict voltages, currents and power waveforms of the multivendor HVDC system from converter start-up to rated power operation. In Fig. 12, the onshore converter model is created by the vendor A (Toshiba), and the offshore model is by the vendor B (Hitachi). In Fig. 13, the offshore model is created by Hitachi, and the onshore model is by Toshiba, reversely.

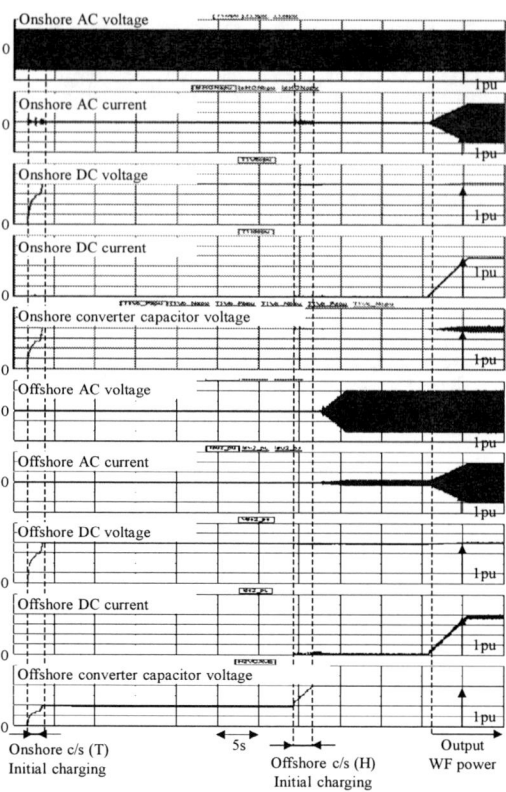

Fig. 12. "RTDS" simulation results of case from start-up to rated operation. (onshore : vendor A (T), offshore : vendor B (H))

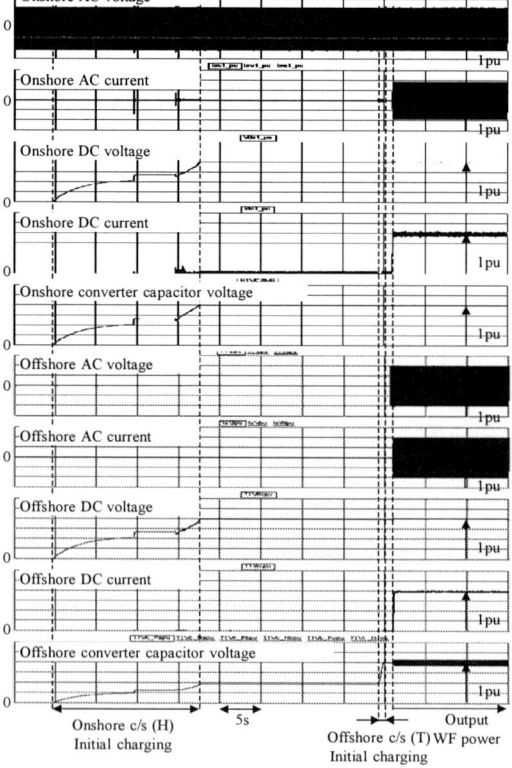

Fig. 13. "RTDS" simulation results of case from startup to rated operation. (onshore : vendor B (H), offshore : vendor A (T))

2001

First the onshore converter starts to charge its dc capacitors and the dc cables, and the offshore MMC converter dc capacitors are charged from the dc cable side, using the dc cable voltage. Then the offshore converter applies an ac voltage to the offshore ac lines, and finally, the windfarm modeled by the current sources starts to increase its output power to the rated power. The "RTDS" simulation results validate that, although the converter models provided by the two respective vendors have differences in the time constant of initial charging, startup timing and other control characteristics, the multivendor HVDC two-terminal system shows good interoperability, thanks to coordination between the two converters.

IV. CONCLUSIONS

This paper described simulation analysis studies of the multivendor multi-terminal HVDC system connected to the offshore windfarms, by using the "PSCAD/EMTDC" simulation software and "RTDS" digital real-time simulator, for validating interoperability between the multivendor HVDC converter terminals. A system configuration assumed for the simulation analysis, and "PSCAD/EMTDC" and "RTDS" simulation results were reported in the paper.

The tentative requirements specifications for the multivendor multi-terminal HVDC system were sufficiently discussed by the companies involved in the NEDO project. The interface signals between the master controller and converters were standardized based on the discussions. By using the black-boxing function of "PSCAD/EMTDC" and "RTDS", the simulation analysis studies of the multivendor two-terminal monopole HVDC system were successfully completed with preserving vendor's technical know-hows. Although waveforms of voltages, currents and power of various points in the HVDC circuits were differently observed, depending on the fact which vendor's converter is on the offshore side or onshore side, the operating behavior of multivendor HVDC system was not seriously affected by these differences. Only a limited number of items were standardized in the requirements specifications document, however, satisfactory results were obtained. These results validated interoperability between the multivendor HVDC converters.

Currently, simulation analysis studies of a multivendor three-terminal monopole HVDC system are being carried out. Requirements specifications for the master controller in steady-state and under transient conditions are being discussed by the project companies, and corresponding simulation studies will be started in the near future.

ACKNOWLEDGMENT

This paper is based on the results obtained from a project commissioned by the New Energy and Industrial Technology Development Organization (NEDO).

REFERENCES

[1] T. Yoshihara and M. Kimura, "Feasibility Study of Offshore Wind Farm Connected High Voltage Direct Current System under Grid Fault," *IEEJ Annual Conference 2017, 2-206, 2017 (Written in Japanese).*

[2] J. Maneiro, et. al., "Energy diverting converter topologies for HVDC transmission systems," *Power Electronics and Applications (EPE) 2013 15th European Conference on.,* 2013.

Interoperability of Modular Multilevel Converters and 2-level Voltage Source Converters in a Laboratory-Scale Multi-Terminal DC Grid

Salvatore D'Arco, Atsede G. Endegnanew, Giuseppe Guidi, and Jon Are Suul
SINTEF Energy Research, Trondheim, Norway
E-mail: salvatore.darco@sintef.no, Atsede.G.Endegnanew@sintef.no, Giuseppe.Guidi@sintef.no, Jon.A.Suul@sintef.no

Abstract-The prospect of future multivendor HVDC systems has led to concerns about interoperability between converter terminals based on different technologies. This paper investigates the interoperability of converter terminals in a small-scale experimental setup, consisting of two 2-level (2L) Voltage Source Converters (VSCs) and two Modular Multilevel Converters (MMCs) with different number of half-bridge sub-modules in each arm. The laboratory setup is utilized to demonstrate interoperability of a 2-level VSCs and an MMCs when operated in parallel on the dc-side as well as in parallel at a common connection point in an ac grid. Experimental results are presented for the investigated 4-terminal configuration with one 2L VSC controlling the dc voltage and all other converters controlling their active power flow. The presented results give an additional empirical confirmation that topological differences between properly designed converter terminals should have limited influence on system level operation and system interoperability of MT HVDC systems.

Keywords— Interoperability, HVDC Transmission, Modular Multilevel Converter, Power-Hardware-in-the-Loop

I. INTRODUCTION

HVDC transmission based on Voltage Source Converter (VSC) technology is presently an established solution for interconnecting weak ac networks and for grid connection of far-offshore wind farms [1], [2]. With the introduction of Modular Multilevel Converters (MMCs) [3], VSC-based HVDC transmission is also becoming a viable option for bulk power transfer [4], [5]. Due to the flexibility and controllability of VSC technology, is also suitable for operation of Multi-Terminal (MT) HVDC grids [6]-[8]. Two examples of MT HVDC grids based on VSC terminals are already in operation in China [9], [10]. However, these two installations have been designed, constructed and commissioned as MT HVDC systems by an integrated process with a predefined topology and functionality.

Future multi-terminal HVDC grids in Europe are expected to be a result of gradual expansion from point-to-point transmission schemes. Thus, European Transmission System Operators (TSOs) require that future MT HVDC systems should be based on a technical foundation allowing for multi-vendor solutions in a similar way as in traditional ac systems. For reliable operation of such multi-vendor systems it is critical to ensure interoperability between installations from different manufacturers [6], [7], [11]. This implies that HVDC converter stations commissioned separately at different times during the development of a MT HVDC system should be able to operate together without causing problems due to differences in topologies or control system implementations.

In principle, interoperability should be ensured in any power system where two or more converters can interact. Indeed, the term "interoperability" includes many different aspects, from selection of voltage levels and component interfaces to control system compatibility. The potential causes for operational incompatibility in a multi-vendor system are also numerous, and can have different implications in practical and theoretical terms. However, in the context of HVDC transmission, the concerns regarding converter interoperability seems to have increased substantially with the introduction of the MMC, which has expanded the number of possible configurations and topologies that can potentially coexist in the same system. An example of simulation-based analysis for confirming successful interoperability of HVDC converter terminals with different topologies was provided in [11].

Considering that interoperability between component ratings and interfaces can be easily ensured in MT HVDC systems, it is useful for practical convenience to differentiate between interoperability issues associated with the converter topology and interoperability aspects related to the control strategies. Indeed, the theoretical framework and the tools suitable for analysis of these two critical aspects of interoperability will differ significantly. For example, investigation of potential interoperability issues related to operation of different converter topologies in the same system would require very detailed models, which in practice restricts the analysis to time-domain simulations or experimental testing, as in [11], [12], [13]. By contrast, it can be expected that interoperability issues related to the control strategies of

---

This work was supported by the EU FP7 project, BestPaths – Beyond State-of-the Art Technologies for Repowering AC Corridors and Multi-Terminal HVDC Systems, under Grant Agreement No 612748.

the HVDC converter terminals, and especially interactions associated with the outer control loops and their tuning can be analyzed by average models, as discussed in [14]-[19].

Considering mainly the concerns of interoperability between different converter topologies in a MT HVDC system, this paper presents experimental results from a few selected cases that demonstrate interoperability of different VSC topologies. Thus, the paper attempts to advance from the experimental investigations presented in [12] and [13]. While successful operation of a point-to-point configuration combining a 2L VSC and an MMC terminal was demonstrated in [12], a multiterminal dc grid configuration based on 2L VSCs but with capability for Power-Hardware-in-the-Loop (P-HiL) simulation of the ac-side systems was presented in [13]. In this paper a four-terminal dc grid with two 2L VSCs and two MMCs with different number of half-bridge sub-modules, created within the Demo 1 of the BestPaths project is introduced [20]. According to the experience matured with the demonstrator and the test results presented in this paper, the VSC multiterminal configurations did not exhibit any interoperability issues due to the differences in the topologies. This gives an additional empirical evidence supporting that topological differences between properly designed converter terminals will have limited influence on system level operation and system interoperability of MT HVDC systems.

II. SYSTEM TOPOLOGIES AND CONTROL OF VSC HVDC CONVERTER TERMINALS

Although continuous research activities are being devoted to the improvement of design and control strategies for VSC-based HVDC converters, the basic control principles for HVDC terminals connected to large-scale ac systems are well established. For the investigations in this paper it is assumed that the converter stations will be operated by conventional vector-oriented control strategies that are synchronized to the ac grid voltage by a traditional Synchronous Reference Frame (SRF) Phase Locked Loop (PLL) [21]. A brief introduction to the assumed implementations for 2L VSCs and MMCs in the system configuration investigated in this paper is presented in the following.

A. Topology and control system for 2L VSCs

Fig. 1 shows an overview of the basic topology and control structure assumed for 2L VSCs. In addition to the 2-level topology, the system typically includes an LCL-filter (or another high order filter topology) on the ac-side and a relatively large capacitor on the dc-side [21]. The control system is usually based on an ac-side inner loop current controller in a synchronously rotating *dq* reference frame synchronized to the ac grid by a PLL [21]. The PLL detects the grid frequency, which is used in the decoupled PI current controllers, and the phase angle of the voltage measurements, which is needed for transformations between the stationary and synchronously rotating reference frames of the control system [21], [22].

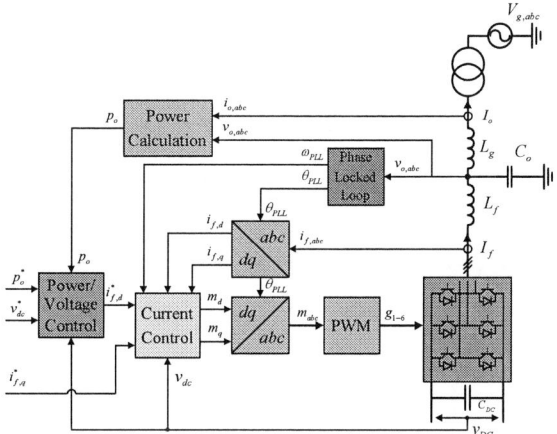

Fig. 1. Overview of assumed control system for 2L VSCs

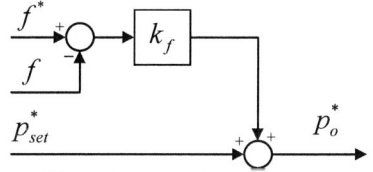

Fig. 2. Frequency droop controller

The control structure in Fig. 1 shows a general scheme where an outer loop controller regulates either the ac power or dc-side voltage by generating the active, d-axis, current reference to the inner loop current controller. In both cases, the outer loop is assumed to be based on a simple PI controller. It should be noted that an active power droop can easily be included in the dc voltage controller, or a dc voltage droop could be included in the power controller [16], [23]. For simplicity, operation with a single converter terminal controlling the dc voltage will be assumed in the following, while all other terminals will be operated with ac-side power control. However, it will be considered that an HVDC converter station can participate in the primary frequency control of the ac-grid [25], [24]. Thus, a frequency droop acting on the power reference for the control system can be introduced as shown in Fig. 2.

In this paper, the reactive q-axis current on the ac-side of the converter terminal is controlled to zero, implying unity power factor operation at the point of synchronization to the grid. However, a reactive power controller or an ac-voltage controller could easily be included in the control structure [21]. A reactive power controller could also be operated with an ac voltage droop, or an ac voltage controller could be operated with a reactive power droop. Such control system configurations would be most important in relatively weak ac systems and will not be explicitly studied in this paper.

B. Topology and control system for MMCs

An overview of the system topology and assumed control structure for MMCs is shown in Fig. 3. As can be seen from the figure, the MMC is assumed to be connected directly to the ac system without any capacitive filter. Furthermore, the MMC topology does

The 2018 International Power Electronics Conference

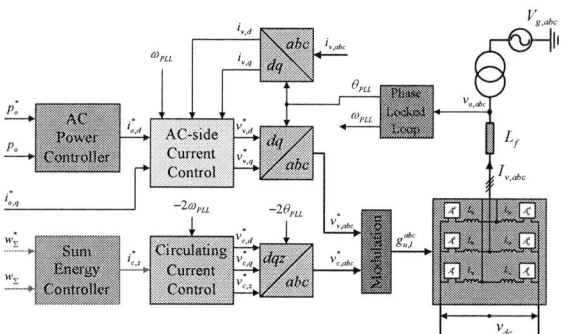

Fig. 3. Overview of assumed control system for MMCs

not depend on any capacitor at its dc-side terminals [3]. As also shown in Fig. 3, the assumed MMC control system has two parallel paths, for regulating the ac-side currents and the internal circulating currents, respectively. The ac-side control scheme is generally identical to the control scheme for the 2-L VSC from Fig. 1. Only an ac-side power controller will be considered in this paper as shown in Fig. 3 and it will be assumed that the reactive q-axis current component in the ac side will be controlled to zero. Nevertheless, the MMC could be also operated with the same combinations of control objectives and droop functions as outlined for the 2-L VSC.

From in Fig. 3, it can be seen that the ac-side current controllers generate the ac-side voltage references v^*_v, while the circulating current controller provides the internal voltage references v^*_c for each phase of the converter. These two voltage components are combined by a modulation strategy to generate a modulation signal for each arm of the MMC. The modulation strategy for an MMC can be implemented in several different ways, depending on whether the modulation signal is compensated for the continuous voltage oscillations in the total sum arm voltages or not [19]. In this paper, it is assumed that the modulation indices for upper (u) and lower (l) arm of the MMC are calculated by a division with the dc voltage as:

$$n_{u,k} \approx \frac{-v^*_{v,k} + v^*_{c,k} + v_{dc}}{v_{dc}}, \quad n_{l,k} \approx \frac{v^*_{v,k} + v^*_{c,k} + v_{dc}}{v_{dc}} \quad (1)$$

$$for \quad k \in \{a,b,c\}$$

The modulation signals resulting from (1) are used as input to a capacitor voltage balancing algorithm which will generate the gate signals of each individual sub-module of the MMC [3].

The circulating currents of the MMC are assumed to be controlled by a Circulating Current Suppression Controller (CCSC) according to [26]. Thus, the second harmonic component of the internal circulating currents of the MMC are controlled to zero by a set of decoupled PI current controllers in the negative sequence double frequency SRF, as shown in the upper part of Fig. 4.

If only the double frequency dq components of the circulating currents are controlled to zero, the dc components of the circulating currents and the total capacitor voltage in each arm of the MMC will be left uncontrolled. Thus, they will stabilize in steady-state operation according to the power flow, the dc-side

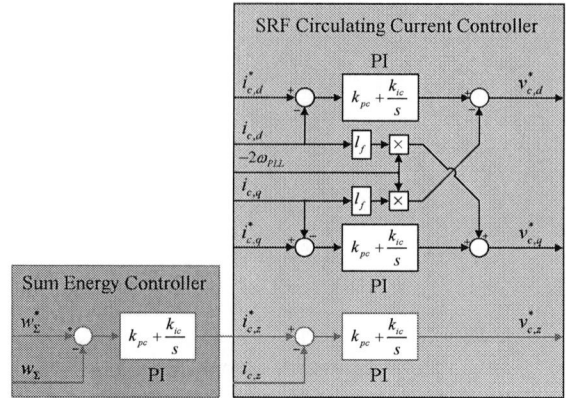

Fig. 4. Circulating current control of MMC

voltage and the equivalent resistance of each phase of the MMC. However, the dc-component of the circulating currents can be utilized to regulate the total energy stored in the internal capacitors of the MMC. Thus, the total arm voltage can be decoupled from the dc voltage of the converter.

A simple way to obtain control of the total energy stored in the MMC is to use a sum energy controller that provides a reference value for controlling the zero-sequence component of the circulating currents [27]. The corresponding controller implementation is shown in grey in the lower part of Fig. 4, where a PI controller is used to regulate the total energy stored in the MMC by providing a zero-sequence current reference. The feedback signal for the total energy can be calculated directly from the total capacitor voltages, as assumed in this paper, or it can be estimated from the currents and modulation signals of the MMC. If the sum energy controller is utilized, the zero-sequence component of the circulating currents must be regulated to its reference value. As shown in Fig. 4, this can be easily achieved with a PI current controller providing the zero sequence component of the internal voltage reference $v^*_{c,z}$.

III. MULTI-TERMINAL DC SYSTEM CONFIGURATION AND EXPERIMENTAL PROTOTYPES

The experimental setup utilized in this paper is based on results from a demonstration activity within the "BEST PATHS" project [20] supported by the European Union FP7 program. The studied configuration is installed in the National Smart Grid Laboratory of Norway [28]

A. System configuration of experimental setup

The investigated configuration represents a four-terminal dc grid interconnecting three asynchronous ac systems as represented in Fig. 5. Moreover, the setup includes a real-time simulator and a 200 kW high bandwidth grid emulator (EGSTON-COMPISO). This configuration allows for imposing controlled steady-state or transient conditions and offers the possibility for Power-Hardware-in-the-Loop (P-HiL) testing for investigating interoperability issues of MT HVDC systems. The two MMC-based terminals have different topologies and their control systems are based on

2005

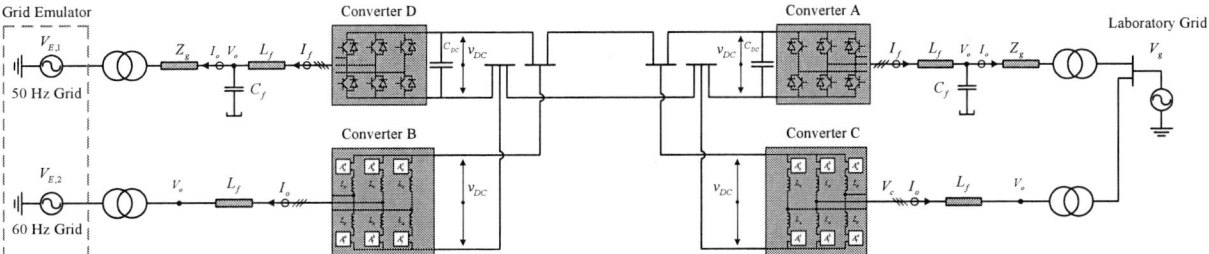

Fig. 5. Overview of MT DC grid configuration used for laboratory experiments

TABLE I AC GRID SPECIFICATION OF LABORATORY SETUP

AC grid	Rated voltage $V_{S,LL,RMS}$	Frequency	Converter Terminals
Laboratory grid, V_g	400 V	50 Hz (Nordic freq.)	A, C
Grid Emulator, V_{E1}	400 V	50 Hz (Regulated)	D
Grid Emulator, V_{E2}	400 V	60 Hz (Regulated)	B

different strategies for regulating the total energy stored within the internal capacitors.

The studied configuration is intended to trigger possible interoperability issues between the different converter topologies, on the dc-side and on the ac side. It consists of two 2L VSCs, denoted as Converter A and Converter D, and two MMCs denoted as Converter B and Converter C. All converters are connected in parallel on the dc side, forming a four-terminal dc grid. However, on the ac side, only Converter A (2L VSC) and C (MMC), with their associated transformers and filters, are connected in parallel to the same point in the power supply system available in the laboratory. Thus, these two converters operate at the frequency of the Nordic power system. The other two converters, Converter B (MMC) and D (2L VSC), are connected to independent ac grids established by the grid emulator to have a regulated grid frequency of 60 Hz and 50 Hz, respectively. Thus, three asynchronous ac systems are included in the experimental setup, as summarized in Table I.

For operating the system, Converter A acts as a slack-bus on the dc-side, since it is regulating its dc voltage, while the remaining three converters are power-controlled with references for the active (d-axis) current component generated by the outer loop active power controller. Furthermore, Converter B has a frequency droop controller that changes the power reference as a function of the deviation in the frequency of the associated ac grid according to Fig. 2. In all converters, the reactive (q-axis) current is controlled directly with a zero reference.

An OPAL-RT real-time simulator is utilized to implement the higher-level control of the converters and to coordinate their operation. The time step for both MMC and 2L converter controllers in the OPAL-RT was 100 µs. In addition, the real-time simulator controls the operation of the grid emulator, which was used to emulate two independent asynchronous ac grids with 50 Hz and 60 Hz frequency according to Fig. 5 and Table I.

B. Converter Prototypes

The converter units are custom-designed prototypes intended for laboratory-scale testing, and are rated for 60

Fig. 6. Overview of MMCs and view of a converter cell

TABLE II PARAMETERS OF THE VSCS AND ASSOCIATED GRID FILTER

Parameter	Value	Parameter	Value
Rated ac voltage $V_{S,LL,RMS}$	400 V	Filter inductance l_f	0.05 pu
Rated dc voltage V_{dc}	700 V	Filter resistance r_{Lf}	0.01 pu
Rated power S_b	60 kVA	Filter capacitance c_f	0.05 pu
Power controller gain: k_{pp}, k_{ip}	1, 10	Grid side inductance l_g	0.02 pu
DC voltage controller gain: k_{pvdc}, k_{ivdc}	10, 100	Grid side resistance r_g	0.01 pu
Current controller gain: k_{pc}, k_{ic}	0.5, 65	DC capacitance, c_{dc}	4 pu

kVA at 400 V ac (line-to-line RMS) and 700 V dc. The two 2L VSCs are identical in construction and are based on Semikron integrated IGBT modules. The converter terminals include an LCL filter on the ac side and a dc bus capacitor with parameters specified in the Table II. The control of the converters is implemented entirely in the OPAL-RT platform according to the scheme presented in section II. Indeed, the OPAL-RT platform contains a custom programmed FPGA, which is dedicated to sampling and conditioning of the measurements and to the generation of the gate signals. Conventional carrier-based PWM with third harmonic injection and 10 kHz switching frequency is applied.

The two MMC units are based on MOSFET half bridge cells and are connected without any external output filter or dc bus capacitor. The converters have been intentionally designed with different number of cells (i.e. Converter B and C have 18 and 6 cells per arm, respectively) to further expose potential interoperability issues associated with differences in the topology. This reflects in slight differences in the ratings of the power components mounted in each cell, due to the differences in the rated cell capacitor voltages A picture of main experimental setup showing the cabinets of the MMC prototypes and the power hardware of the MMC cells is shown in Fig. 6. The main design parameters for the two MMCs are summarized in Table III. Further information about the MMC prototypes and how they are designed to

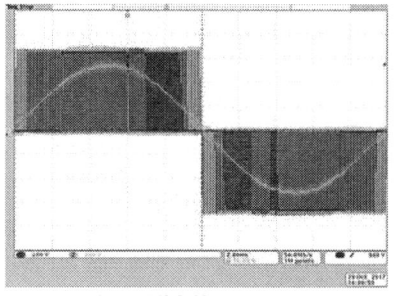 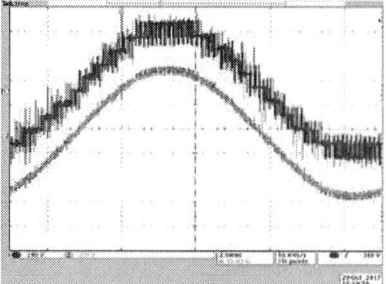 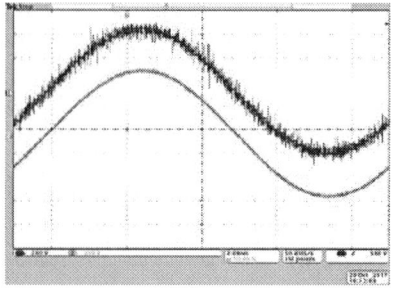

(a) 2L VSC (Converter D)   (b) 6 level MMC (Converter C)   (c) 18 level MMC (Converter B)

Fig. 7. Voltage wave forms before and after ac filter for all the different converter types

TABLE III

MAIN PARAMETERS OF THE REDUCED-SCALE MMC PROTOTYPE

Converter parameters	6 HB model	18 HB model
Name	*Converter C*	*Converter B*
Rated ac voltage $V_{S,LL,RMS}$	400V	400V
Rated dc voltage V_{dc}	700V	700V
Rated power S_b	60 kVA	60 kVA
Cells per arm	6 HB	18 HB
Nominal cell voltage	133V	44V
Arm inductance	1.4 mH	1,4 mH
Cell capacitance	5.9 mF	21.3 mF
Power controller gain: k_{pp}, k_{ip}	5e-4, 0.02	5e-4, 0.02
Current controller gain: k_{pc}, k_{ic}	0.5, 100	0.5, 100
Energy controller gain: k_{pc}, k_{ic}	-	1, 10
Circulating current controller gain: k_{pc}, k_{ic}	2, 100	2, 100

represent scaled models of a relevant HVDC terminal is available in [29].

The control hardware of the MMC prototypes is designed with a three-level hierarchical structure. At the lowest level, the cells are grouped into sets of 6, with each set connected to a local controller generating the gate signals of the 12 individual devices. These local controllers are daisy-chained with a fiber-optic link and connected to an upper layer controller responsible for controlling the two arms of a single phase. This controller implements the capacitor voltage balancing algorithms for selecting the cells to be switched within each arm. Finally, the three leg controllers are again daisy-chained and connected with an optical link to the OPAL-RT unit that implements the current controls, the grid synchronization and any other outer loop controllers.

IV. LABORATORY DEMONSTRATION OF AC- AND DC-SIDE INTEROPERABILITY OF 2L VSCS AND MMCS

This section presents experimental results obtained from the presented multiterminal configuration. As an example of converter interoperability, the configuration described in the previous sections was operated to transfer power between the three ac grids with the control modes and operating conditions already described in section III. It should be noted that for the following experiments, Converter B has an energy controller to regulate the total energy stored in the MMC, while Converter C has only a controller for the double frequency *dq* components of the circulating currents.

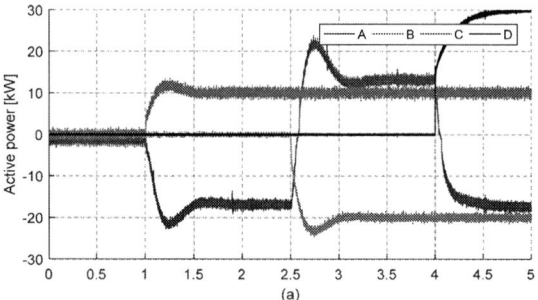

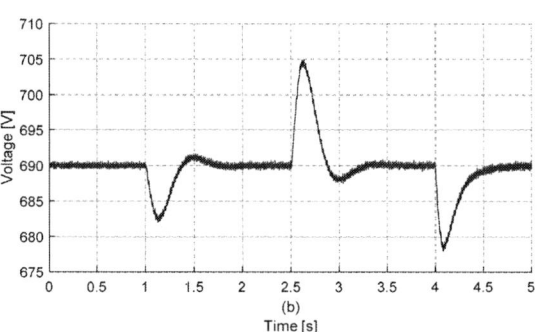

Fig. 8. (a) Measured active power on the ac side for all converters, and (b) Measured dc voltage

A. Steady-state operation

Fig. 7 shows oscilloscope screen-shots of the line-to-line voltage before the inductive output filter and after the filters, i.e. at the converter output terminals. The voltage in Fig. 7 (a) clearly highlights the characteristics of a 2L converter with the line-to-line voltage switching between the dc bus voltage and zero. In contrast, the voltages in Fig. 7 (b) and (c) present the behavior of a multilevel converter with clearly visible smaller steps corresponding to the cell voltage. It should be noted that the MMCs are not operated with nearest level modulation as commonly used for MMCs with hundreds of cells per arm: instead one cell in each arm is operated in PWM to reduce the THD of the voltage.

The expected steady state operation was observed, with power flow inside the dc grid and a net exchange of power between the different ac grids. For the specific choice of the references for the active power controlled converters (i.e. references equal to zero), the converter acting as a slack bus is operating with only a small active current to compensate for the system losses.

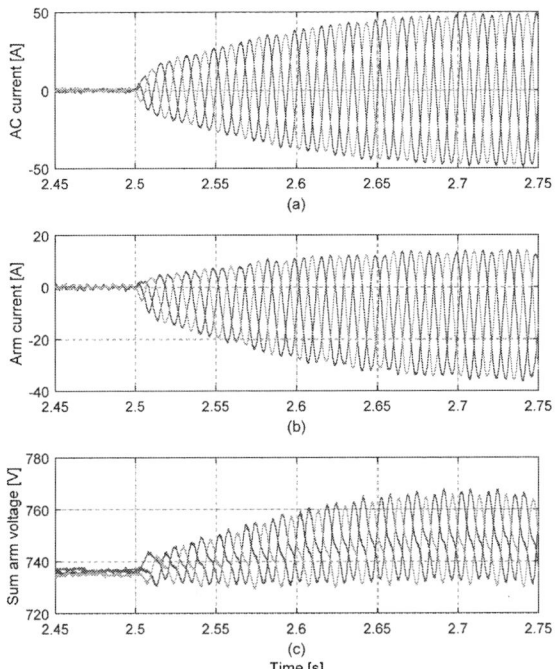

Fig. 9. MMC Converter B (a) AC currents, (b) upper arm currents in each leg, and (c) sum of upper arm voltages in each leg

B. Transient response

For further assessing potential interoperability issues, a test was performed to highlight the transient performances of the system in response to a step change in the active power references, as shown in Fig. 8 (a). Initially, all active power controlled converters were operated with zero reference, and the converter acting as a slack bus is operating with low active power to compensate for the system losses. The active power reference in MMC Converter C was changed from 0 to 10 kW at t = 1s. To compensate the power imbalance in the dc grid, the dc-voltage controlled Converter A changes its power flow. The power references in Converter B and Converter D were stepped from 0 to -20 kW and 30 kW at t = 2.5s and t = 4s, respectively. Note that power flowing out of a converter and into an ac grid is defined as positive. The measured powers reach steady state within about 0.5 s after the step of the power setpoint. The dc voltage disturbance caused by change in power flow is effectively rejected by the dc voltage controlling Converter A, as shown in in Fig. 8 (b), achieving smooth and stable operation. It can be noted that the transient response resulting from a change in the power reference for Converter D is different from the response when the power reference is changed for Converter C or D. This is mainly because the power controlled 2L VSC has different parameters and a different tuning of the active power controller than the MMCs.

Fig. 9 shows currents and voltages for Converter B. The transients in ac currents when the power reference of the converter is changed are shown in Fig. 9 (a). The arm currents and sum voltages for the upper arms of each leg/phase are shown in Fig. 9 (b) and (c), respectively.

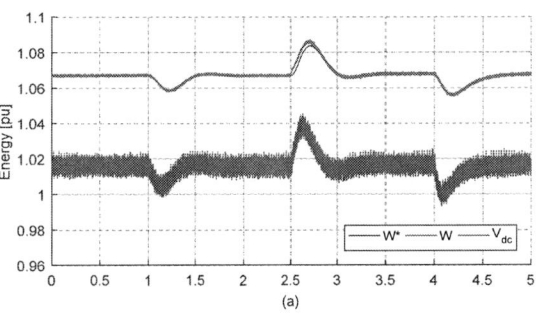

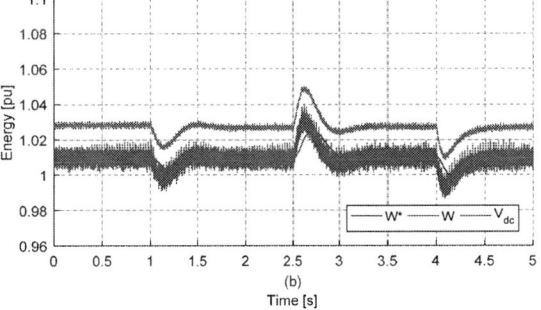

Fig. 10. Reference energy, measured energy, and dc voltage for MMC (a) Converter B, and (b) Converter C

The average value of the arm currents goes below zero as the power reference is negative. The sum of the arm voltages increases as the change of power flow causes the dc voltage to increase during the transient period (in Fig. 8 (b)). The currents and voltages in all phases remain balanced throughout the transient.

In Fig. 10, the total sum energy for Converter B and C MMCs are presented, scaled to per unit quantities by using the dc voltage of each converter as the base value. As mentioned, Converter B has an energy controller that generates a reference for zero sequence current as explained in [27], while Converter C only has circulating current suppression controller according to [26]. In this case, the energy reference for Converter B is set to be proportional to the square of the measured dc voltage. Thus, the sum energy reference follows the dc voltage dynamics, and the total energy stored in the MMC is accurately following this reference. This ensures a higher voltage margin for the control of the ac-side currents. If the energy reference was kept constant, this would also effectively decouple the total sum energy stored in the MMC from the dc-voltage in the multiterminal dc grid. The energy sum in Converter C is, however, following the dc voltage dynamics with only a small margin resulting from the peak value of the internal voltage oscillations of the cell capacitor voltages. The difference in the control strategy for the two MMCs do not have any noticeable influence on the other converters operating in the DC-grid.

C. Operation with grid frequency droop-control

As different power systems are being connected to each other through HVDC links, the frequency reserve needed to stabilize a synchronous system does not need

The 2018 International Power Electronics Conference

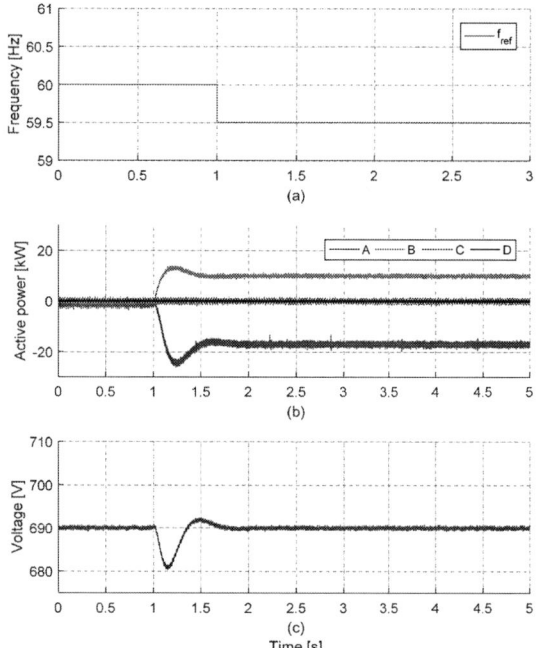

Fig. 11. (a) reference frequency, (b) active power at all converter terminals, and (c) dc grid voltage

to be only provided locally but can also partially supplied by the HVDC connection. Thus, an HVDC system can contribute to the primary frequency regulation by modifying the power flow at the converter terminal based on change in locally measured ac grid frequency. The active power controller of Converter B was modified as shown in Fig. 2 so that the converter participates in frequency control of the connected ac grid.

A frequency disturbance in the emulated ac grid V_{E2} was created by reducing the frequency reference of the grid emulator from 60 Hz to 59.5 Hz at t = 1s, as indicated in Fig. 11(a). This event caused the power flowing through Converter B to change from 0 to 15kW, resulting in increased power flow from the dc grid into the ac grid as shown in Fig. 11(b). To compensate for the change in power flow in the dc grid and the related change in dc voltage, Converter A changes the power flow and regulates dc grid voltage as shown in Fig. 11(b) and (c), respectively.

D. General observations

Considering the presented results, the experimental tests did not reveal any interoperability issues due to the differences in the converters topologies. Moreover, no undesired control interactions between the converters were observed for the simplified control schemes assumed for the different converter terminals. It should be noted that these results were obtained with four converters connected in parallel on the dc-side. This could be considered the potential worst case for direct interaction between the converter topologies since there is no significant impedance between the dc terminals of the different units. However, it should be kept in mind that long HVDC cables or lines can have internal resonance frequencies. Thus, any interoperability issues or stability

problems caused by the physical systems in such configurations would more likely result from interactions between a converter and the remaining electrical system than from any interaction caused by differences in topologies.

V. CONCLUSION

In this paper, the interoperability between converters in a multiterminal configuration has been considered from an empirical perspective based on the experience matured with a reduced-scale four-terminal dc grid with two 2L converters and two MMCs. First, the steady state operation of the dc grid with one 2L converter operating as a slack-bus in dc-voltage control mode has been presented. Then the transient behavior of the system in response to a change in active power flow or to a frequency step in the ac system was investigated. In general, no interoperability issues that could be associated to the differences in the converter topologies have been observed during the tests.

The configuration investigated in the paper included two identical 2L VSCs and two MMCs with different number of levels and inner controls. However, the experimental results do not indicate any noticeable interoperability aspects associated with the topology. The most noticeable differences in the response appeared to be related to the tuning of the voltage and current controller and to the presence of droop functionalities. Indeed, the topological characteristics are essential for the design of the hardware, the development of the internal controllers and all the aspects that are associated to the converter dynamics. However, in a system perspective the differences in the topology are less relevant except for elements like the presence of an output filter or of a bus capacitor. This would seem to justify the studies approaching the interoperability of outer loop control strategies by applying average models and techniques based on linearization for modal analysis.

ACKNOWLEDGEMENT

The authors would like to thank Kjell Ljøkelsøy, SINTEF Energy Research, for his efforts in designing and constructing the converter prototypes utilized in this paper. The authors also want to thank Salvador Ceballos, Tecnalia Research and Innovation, and Gilbert Bergna-Diaz, the Norwegian University of Science and Technology (NTNU), for their contributions to the implementation of the control systems for the MMC prototypes.

REFERENCES

[1] N. Flourentzou, V. G. Agelidis, G. D. Demetriades, "VSC-Based HVDC Power Transmission Systems: An Overview," in *IEEE Transactions on Power Electronics*, vol. 24, no. 3, March 2009, pp.592-602

[2] J. Glasdam, J. Hjerrild, Ł. H. Kocewiak, "Review on Multi-Level Voltage Source Converter Based HVDC Technologies for Grid Connection of Large Offshore Wind Farms," in *Proc of the 2012 IEEE POWERCON*, Auckland, New Zealand, 30 Oct. – 2 Nov. 2012, 6 pp.

[3] A. Lesnicar, R. Marquardt, "An innovative modular multilevel converter topology suitable for a wide power

range, in *Proceedings of the 2003 IEEE Bologna PowerTech Conference*, Bologna, Italy, 23-26 June 2003, vol.3, pp. 272-277

[4] J. Dorn, H. Gamback, J. Strauss, T. Westerweller, J. Alligan, "Trans Bay Cable – A Breakthrough of VSC Multilevel Converters in HVDC Transmission," in *Proceedings of the Cigré 2012 San Francisco Colloquium*, San Francisco, California, USA, 7-9 March 2012, 7 pp.

[5] P. L. Francos, S. S. Verdugo, H. F. Álvarez, S. Guyomarch, J. Loncle, "INELFE – Europe's first integrated onshore HVDC interconnection," *Proc. of the 2012 IEEE Power and Energy Society General Meeting*, PES GM 2012, San Diego, California, USA, 22-26 July 2012, 8 pp.

[6] D. Van Hertem and M. Ghandhari, "Multi-terminal VSC HVDC for the european supergrid: Obstacles," in *Renewable and Sustainable Energy Review*, vol. 14, no. 9, pp. 3156–3163, December 2010

[7] S. Cole, T. K. Vrana, O. B. Fosso, J.-B. Curtis, A.-M. Denis, C.-C. Liu, "A European Supergrid: Present State and Future Challenges," in *Proceedings of the 17th Power System Computation Conference*, PSCC 2011, Stockholm, Sweden, 22-26 August 2011, 7 pp.

[8] J. Beerten, O. Gomis-Bellmunt, X. Guillaud, J. Rimez, A. van der Meer, D. Van Hertem, "Modeling and control of HVDC grids: a key challenge for the future power system," in *Proceedings of the 18th Power Systems Computation Conference*, PSCC 2014, Wrocław, Poland, 18-22 August 2014, 21 pp.

[9] X. Li, Z. Yuan, J. Fu, Y. Wang, T. Liu, Z. Zhu, "Nanao Multi-terminal VSC-HVDC Project for Integrating Large-scale Wind Generation," in *Proceedings of the 2014 IEEE PES General Meeting, Conference & Exposition*, PES GM 2014, Washington D.C. USA, 27-31 July 2014, 5 pp.

[10] C. Li, X. Hu, J. Guo, J. Liang, "The DC Grid Reliability and Cost Evaluation with Zhoushan Five-Terminal HVDC Case Study," in *Proceedings of the 2015 50th Universities Power Engineering Conference*, UPEC 2015, Stoke-on-Trent, UK, 1-4 September 2015, 6 pp.

[11] G. P. Adam, S. J. Finney, B. Williams, "Interoperability of voltage source converters in dc grids," in *IET Generation, Transmission & Distribution*, Vol. 7, No. 11, November 2013, pp. 1310-1317

[12] N. Stanković, M. J. Carrizosa, A. Arzande, P. Egrot, J.-C. Vannier, "An HVDC experimental platform with MMC and two-level VSC in back-to-back configuration," in *Proceedings of the 2016 IEEE 25th International Symposium on Industrial Electronics*, ISIE 2016, 8-10 June 2016, Santa Clara, California, USA, pp. 436-441

[13] S.-A. Amamra, F. Colas, X. Guillaud, P. Rault, S. Nguefeu, "Laboratory Demonstration of a Multiterminal VSC-HVDC Power Grid," in *IEEE Transactions on Power Delivery*, Vol. 32, No. 5, October 2017, pp. 2339-2349

[14] G. O. Kalcon, G. P. Adam, O. Anaya-Lara, S. Lo, and K. Uhlen, "Small-signal stability analysis of multi-terminal VSC-based DC transmission systems," in *IEEE Transactions on Power Systems*, Vol. 27, No. 4, November 2012, pp. 1818 1830

[15] N. Chaudhuri, R. Majumder, B. Chaudhuri, and J. Pan, "Stability analysis of VSC MTDC grids connected to multimachine AC systems," *IEEE Transactions on Power Delivery*, Vol. 26, No. 4, October 2011, pp. 2774–2784

[16] J. Beerten, S. D'Arco, J. A. Suul, "Identification and Small-Signal Analysis of Interaction Modes in VSC MTDC Systems," in *IEEE Transactions on Power Delivery*, Vol. 31, No. 2, April 2016, pp. 888-897

[17] M. K. Zadeh, M. Amin, J. A. Suul, M. Molinas, O. B. Fosso, "Small-Signal Stability Study of the Cigré DC Grid Test System with Analysis of Participation Factors and Parameter Sensitivity of Oscillatory Modes," in *Proceedings of the 18th Power Systems Computation Conference*, PSCC 2014, Wrocław, Poland, 18-22 August 2014, 8 pp.

[18] G. Bergna Diaz, J. A. Suul, S. D'Arco, "Small-Signal State-Space Modeling of Modular Multilevel Converters for System Stability Analysis," in *Proceedings of the 2015 IEEE Energy Conversion Congress and Exposition*, ECCE 2015, Montreal, Quebec, Canada, 20-24 September 2015, pp. 5822-5829

[19] G. Bergna J. A. Suul, S. D'Arco, "State-Space Modeling of Modular Multilevel Converters for Constant Variables in Steady-State," in *Proceedings of the 17th IEEE Workshop on Control and Modeling for Power Electronics*, COMPEL 2016, Trondheim, Norway, 27-30 June 2016, 9 pp.

[20] EU FP7 project, "Beyond State-of-the Art Technologies for Repowering AC Corridors and Multi-Terminal HVDC Systems," BestPaths, http://www.bestpaths-project.eu/

[21] A. Yazdani, R. Iravani, "*Voltage-Sourced Converters in Power Systems – Modeling, Control, and Applications*," Hoboken, New Jersey, USA, IEEE Press / Wiley 2010

[22] S. D'Arco, J. Are Suul, M. Molinas, "Implementation and Analysis of a Control Scheme for Damping of Oscillations in VSC-based HVDC Grids," in *Proceedings of the 16th International Power Electronics and Motion Control Conference and Exposition*, PEMC 2014, Antalya, Turkey, 21-24 September 2014, pp. 586-593

[23] F. Thams, J. A. Suul, S. D'Arco, M. Molinas, F. W. Fuchs, "Stability of DC Voltage Droop Controllers in VSC HVDC Systems," in *Proceedings of the IEEE PowerTech Eindhoven 2015 – Towards Future Power Systems and Emerging Technologies*, Eindhoven, The Netherlands, 29 June – 2 July 2015, 7 pp.

[24] T. M. Haileselassie, K. Uhlen, "Primary frequency control of remote grids connected by multi-terminal HVDC," in *Proceedings of the IEEE PES General Meeting*, PES 2010, Minneapolis, Minnesota, USA, 25-29 July 2010, 6 pp.

[25] N. R. Chaudhuri, R. Majumder, B. Chaudhuri, "System Frequency Support Through Multi-Terminal DC (MTDC) Grids," in *IEEE Transactions on Power Systems*, Vol. 28, No. 1, February 2013, pp. 347-356

[26] Q. Tu, Z. Xu, L. Xu, "Reduced Switching-Frequency Modulation and Circulating Current Suppression for Modular Multilevel Converters," in *IEEE Transactions on Power Delivery*, Vol.26, No.3, pp.2009-2017, July 2011

[27] J. Freytes, G. Bergna, J. A. Suul, S. D'Arco, F. Gruson, F. Colas, H. Saad, X. Guillaud, "Improving Small-Signal Stability of an MMC with CCSC by Control of the Internally Stored Energy," in *IEEE Transactions on Power Delivery*, Vol. 33, No. 1, February 2018, pp. 429-439

[28] The National Smart Grid Laboratory of Norway, operated by NTNU and SINTEF in Trondheim, Norway, https://www.ntnu.edu/smartgrid

[29] K. Ljøkelsøy, G. Guidi, "Development of a scale model of a Modular Multilevel Converters," in *Proceedings of the 14th Deep Sea Offshore Wind R&D Conference – EERA DeepWind'2017, 18-20 January 2017, Trondheim Norway, Energy Procedia*, Vol. 137, October 2017, pp. 505-513

Principle Experiment of Current Commutated Hybrid DCCB for HVDC Transmission Systems

Ryuta Hasegawa[1*], Kazuhisa Kanaya[2], Yushi Koyama[2], Toshiaki Matsumoto[2] and Takahiro Ishiguro[3]

1 Toshiba Infrastructure Systems & Solutions Corporation, Kawasaki-shi, Kanagawa, Japan
2 Power and Industrial Systems Research and Development Center, Toshiba Corporation, Fuchu-shi, Tokyo, Japan
3 Transmission & Distribution Systems Div. Toshiba Energy Systems & Solutions Corporation, Kawasaki-shi, Kanagawa, Japan
*E-mail: ryuta.hasegawa@toshiba.co.jp

Abstract— We propose a current commutated hybrid Direct Current Circuit Breaker (DCCB) for HVDC transmission systems. The transmission losses of the developed DCCB are quite low because the normal current path includes only mechanical switches. Fast mechanical switches and a commutation circuit realize a fast current interruption. We also developed a test method for DCCBs, which allows for the use of existing facility such as AC generators. We tested the developed hybrid DCCB by the developed test method, therefore the test results prove the operation principle of the hybrid DCCB and the test method.

Keywords— Current commutation, HVDC, Hybrid DCCB, Semiconcuctor

I. INTRODUCTION

In recent years, optimizations of wide area systems which transmit the electric power of offshore wind power generation to each demand place of mainland have been considered. Since High Voltage Direct Current (HVDC) makes construction cost and losses lower in long transmission system, multi terminal HVDC systems with offshore wind power have been argued worldwide. Multi terminal HVDCs need self-excited converters which control power flow and Direct Current Circuit Breakers (DCCB) which interrupt fault currents within several milli-seconds for preventing voltage drops [1].

Hybrid DCCBs which include mechanical and semiconductor switches have been developed in several years. Low conduction loss characteristics of mechanical switches and high speed breaking performances of semiconductor switches are suitable for DCCBs. Because those hybrid DCCBs have a normal current path which includes auxiliary semiconductor breakers for current commutation, they carry transmission losses owing to voltage drops of the semiconductor breakers, and need cooling devices [2]-[6].

Meanwhile, DCCBs require equipment in order to test the performances of current interruption and voltage endurance. Although there is a test method which supplies the DCCBs with resonant currents and voltages using capacitors and inductors [4], DCCBs for several hundred megawatt HVDC enlarge the test equipment.

In this paper, we propose a hybrid DCCB with a normal current path including only mechanical switches, which allows for the fast interruption and minimized transmission losses. Moreover, we have developed a test method of DCCBs employing an AC generator which is used for testing AC circuit breakers (ACCB).

Firstly, we explain the operation principle of the proposed hybrid DCCB and the design method of a commutation circuit. Secondly, we introduce the developed test method using an AC generator. Finally, we report that the operation principle of the hybrid DCCB was proven by the test method.

II. OPERATION PRINCIPLE

A. Operation Principle

Fig.1 shows the proposed hybrid DCCB. The proposed hybrid DCCB includes a mechanical disconnector, a mechanical breaker, a commutation circuit, a semiconductor breaker, an arrester, a blocking diode, and a commutation inductor. Since the current path during normal transmission consists of a mechanical disconnector and breaker, transmission losses are quite low. The commutation circuit commutates a fault current from the mechanical breaker to the semiconductor breaker. However, the DCCB circuit of Fig.1 can interrupt only unidirectional faults, replacing the blocking diode to a semiconductor breaker enables bidirectional interruptions.

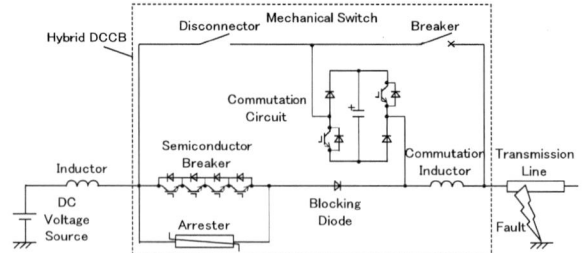

Fig. 1. Current commutated hybrid DCCB configuration.

Fig. 2 shows the operation principle of the commutation. (a): A system fault occurs, and a fault current increases through the inductor, the disconnector and the breaker. (b): When the fault is detected, the disconnector and breaker start to open. The commutation circuit discharges a capacitor to extinguish the arc of the breaker. A time when the commutation circuit starts to discharge is calculated so that a current zero crossing time corresponds to the opening time of the breaker. The blocking diode prevents the discharge current passing through the disconnector and the anti-parallel diodes of the semiconductor breaker. (c): The arc of the breaker is extinguished. (d): The commutation circuit stops to discharge the capacitor, and the current of the disconnector is commutated into the semiconductor breaker. (e): After waiting insulation recovery time of the disconnector, the semiconductor breaker turns off. The fault current is commutated into the arrestor, and the energy of the inductor is absorbed until the current interruption is completed.

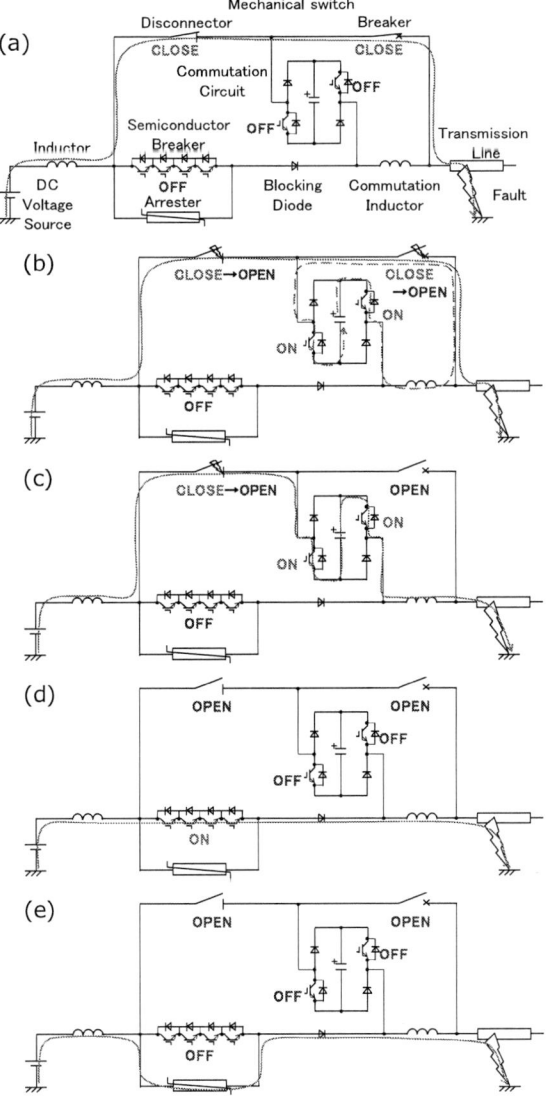

Fig. 2. Operation principle of hybrid DCCB

B. Design of Commutation Circuit

We designed a hybrid DCCB for a principle experiment. Table I shows specifications of the experiment. Because the withstand voltage of the disconnector is equal to the transient voltage after interruption, it needs longer opening time than that of the breker. We set the opening time of the disconnector to 3 ms and that of the breaker to 2 ms applying electromagnetic actuators to both. The fault detection time is defined by subtracting the disconnector opening time from the breaking time.

TABLE I
SPECIFICATIONS OF HYBRID DCCB FOR PRINCIPLE EXPERIMENT

Item	Value
Voltage rating	6.7 kV
Transient voltage after interruption	10 kV
Breaking current	9.0 kA
Breaking time	5 ms
Fault detection time	2 ms
Breaker opening time	2 ms
Disconnector opening time	3 ms

Since current zero crossing needs to be occurred between the opening time breaker and the disconnector, the time of current zero crossing is 4-5 ms after a fault. Therefore, a commutated current is approximately equal to the breaking current I_{brk}, and that leads to eq. (1).

$$I_{brk} \leq \frac{V_{com}}{\sqrt{\frac{L_{com}}{C_{com}}}} \sin \frac{T_{com}}{\sqrt{L_{com}C_{com}}} \qquad (1)$$

Where V_{com} is an initial voltage of the commutation circuit, L_{com} is an inductance of it, C_{com} is a capacitance of it, and T_{com} is a time after igniting it.

In order to analyze the drop voltage $\Delta \cdot V_{com}$ of the commutation circuit after a commutation, we calculated the discharge charge Q_{com} of the commutation circuit using eq. (2).

$$Q_{com} = \int_0^{T_{com}} \frac{V_{com}}{\sqrt{\frac{L_{com}}{C_{com}}}} \sin \frac{t}{\sqrt{L_{com}C_{com}}} dt =$$
$$C_{com}V_{com}\left(1 - \cos \frac{T_{com}}{\sqrt{L_{com}C_{com}}}\right) \qquad (2)$$

Substituting $Q_{com} \leq C_{com}\Delta V_{com}$ into eq.(2) yields eq. (3).

$$\Delta \cdot V_{com} \geq V_{com}\left(1 - \cos \frac{T_{com}}{\sqrt{L_{com}C_{com}}}\right) \qquad (3)$$

Arranging eq. (3), we obtain eq. (4) for solving the capacitor of the commutation circuit C_{com}.

$$C_{com} \geq \frac{1}{L_{com}}\left\{\frac{T_{com}}{\cos^{-1}(1-\Delta)}\right\}^2 \qquad (4)$$

Substituting eq. (4) into eq. (1) gives eq. (5) for solving the initial voltage of the commutation circuit V_{com}.

$$V_{com} \geq \frac{L_{com}I_{brk}\cos^{-1}(1-\Delta)}{T_{com}\sin\{\cos^{-1}(1-\Delta)\}} \qquad (5)$$

We can design the commutation circuit with these equations and analysis. The commutation time T_{com} should be set between the breaker opening time and the disconnector opening time. A parasitic loop inductance between the breaker and the commutation circuit can be utilized as the commutation inductance L_{com}, which is a minimized value. Low commutation inductance reduces the cost of the commutation circuit because the initial commutation voltage V_{com} is proportional to L_{com} as shown in eq. (5). We employed the parasitic loop inductance as the commutation inductance.

C. Mechanism of Fast Mechanical Switches

In order to commutate the current of mechanical switches into the semiconductor breaker within several milli-seconds, the disconnector and the breaker require a fast driving and reliable braking. We applied an electromagnetic actuator shown in Fig.3 to meet the performances [7]. Firstly, the coil current of an exciting circuit induces the current of a ring. Secondly, these currents generate electromagnetic force between the coil and ring, which drives a movable contact fast. Finally, a collision between a spring receiving part and an oil damper reduces a rebounding.

We operated an electromagnetic actuator for the principle experiment to open. Fig.4 shows contact strokes of it on experiment and 1D simulation. The results demonstrated the performances of the fast driving and the reliable braking, since the distance between contacts was more than 20 mm, which is enough for keeping 10-kV insulations, within 3 ms.

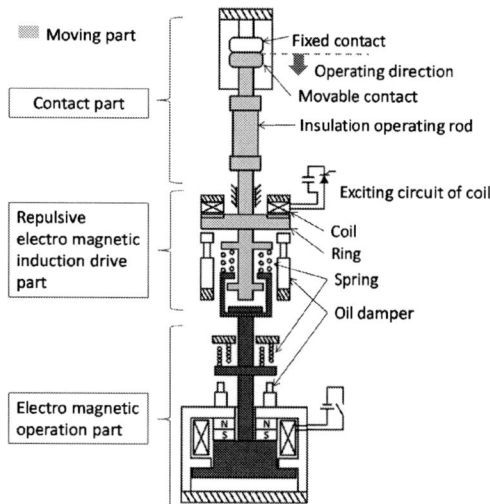

Fig. 3. Mechanical switch driven by electromagnetic actuator.

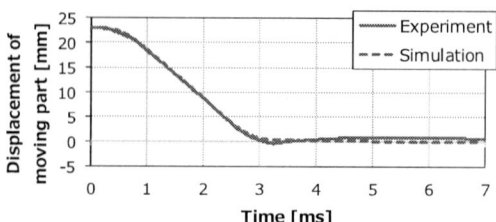

Fig. 4. Results of contact stroke in opening operation.

D. Semiconductor breaker

We applied 4-series press-pack Injection Enhanced Gate Transistors (IEGT; rating: 4.5 kV – 2.1 kA) to the semiconductor breaker for the principle experiment. The breaking current of 9 kA is more than 4 times of the rated current of 2.1 kA. Therefore, snubber circuits are added to the IEGTs as shown in Fig. 5. The addition of snubber circuits is more economical than an increase of IEGTs in parallel.

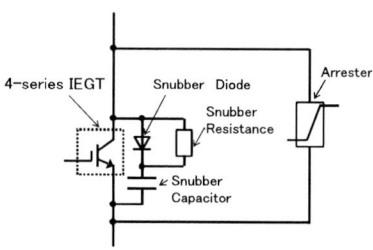

Fig. 5. Semiconductor breaker circuit.

III. TEST METHOD OF HYBRID DCCB

We have developed a test method of DCCBs for HVDC. Fig. 6 shows a test circuit. The circuit is configured with an AC generator for current source and DC capacitors for voltage source. The test circuit has a configuration allowing the use of existing equipment for testing ACCBs. The configuration reduces power and volume of test equipment because current source and voltage source are divided. The AC generator supplies a fault current and a transient voltage immediately after the current interruption, and the DC capacitor supplies a system voltage. SW2 is an auxiliary breaker to disconnect the AC generator after the interruption, and SW3 is an auxiliary switch to connect the DC capacitor during supplying the fault current.

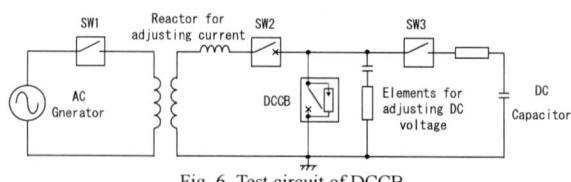

Fig. 6. Test circuit of DCCB.

Fig. 7 shows a test sequence of DCCB. (a): SW2 and the DCCB are closed for starting the test. (b): SW1 is closed to supply a fault current from the AC generator and the current starts to increase. (c): After detecting the

2013

fault, the DCCB starts to open. (d): SW3 is closed to connect the DC capacitor with the DCCB. (e): Although SW2 is opened, a generated arc continues to conduct it. (f): After DCCB interrupted the fault current, the current is commutated to the arrester connected to the DCCB in parallel. (g): The arrester absorbs the energy of the test circuit, and the arc of SW2 is extinguished. The DC source capacitor continues to apply a system voltage to the DCCB.

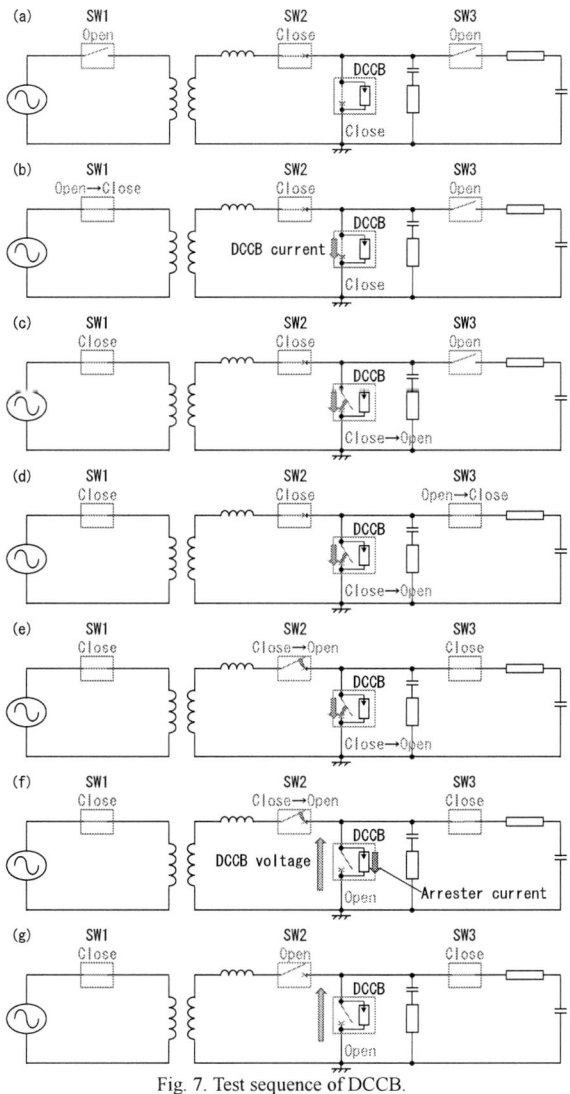

Fig. 7. Test sequence of DCCB.

IV. PRINCIPLE EXPERIMENT

We built a 10-kV experimental model of the proposed hybrid DCCB as shown in Fig.8, which was designed as shown in Table II. There is a feature that proposed hybrid DCCB can be realized with low commutation voltage.

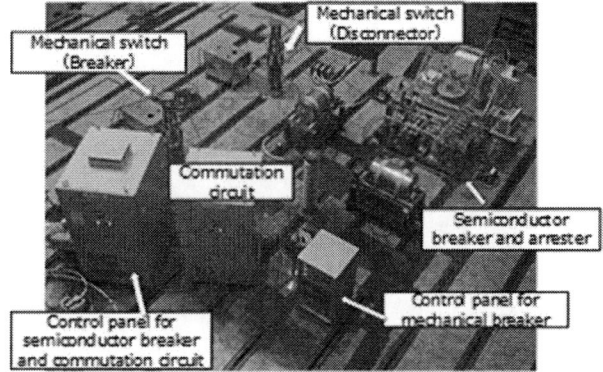

Fig. 8. 10-kV experimental model.

TABLE II
DESIGN PARAMETERS OF 10-KV EXPERIMENTAL MODEL

Item	Symbol	Value
Commutation voltage	V_{com}	400 V
Drop ratio of commutation circuit voltage after commutation	Δ	0.7
Commutation time	T_{com}	0.25ms

Fig. 9 shows the experimental waveforms of the experimental model using the test method. The full strokes of the mechanical switches are normalized to 1.

We define (a) in Fig. 8 as a fault time when the SW1 of Fig.4 is closed and AC generator starts to supply a fault current to the DCCB. A control device received a detected fault signal at (b). The control device operated the mechanical disconnector, breaker, commutation circuit, and semiconductor breaker in Fig.1 using the sequence introduced in Fig.2. The semiconductor breaker interrupted the fault current at (c) which was near the peak value of the sinewave. After the current interruption, the arrester absorbed inductance energy during the period between (c) and (d). SW2 disconnected the AC generator, and SW3 provided a DC voltage connecting the DC capacitor.

The experimental result demonstrated the performance of 9-kA current interruption and 10-kV voltage endurance.

Fig. 10 is an enlarged the waveforms of Fig. 9 for analyzing the commutation circuit. The commutation time met the specifications on Table II because the breaker current was crossed at 0.17 ms after starting commutation which was shorter than 0.25 ms. The drop ratio of the commutation circuit voltage satisfied the specifications too, because the value 53% of the experiment was lower than the 70% on Table II. Although the start and stop of commutation time was delayed by interface and gate drive circuits, the delays didn't affect to the performance of the current interruption.

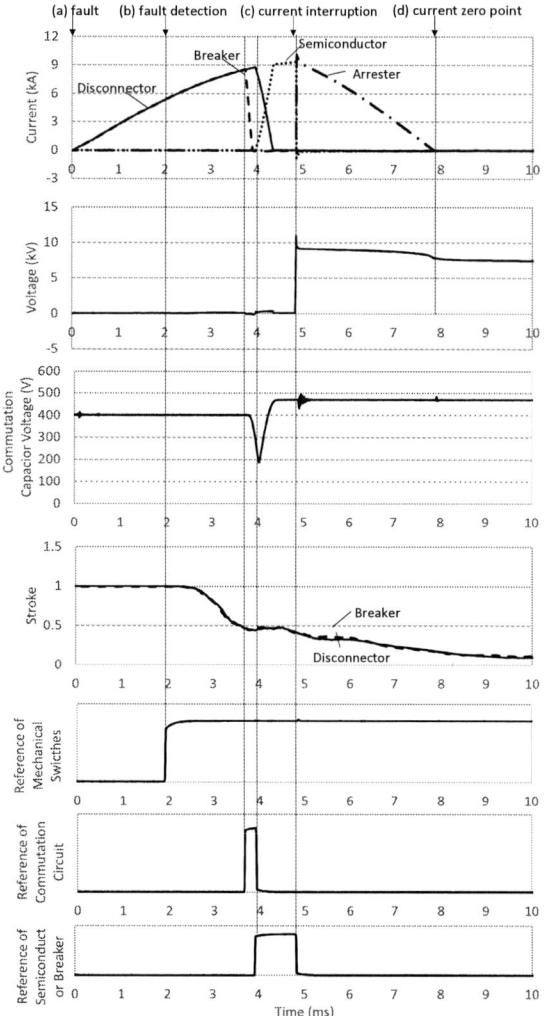

Fig. 9. Test waveforms of the 10-kV experiment model (0-10 ms).

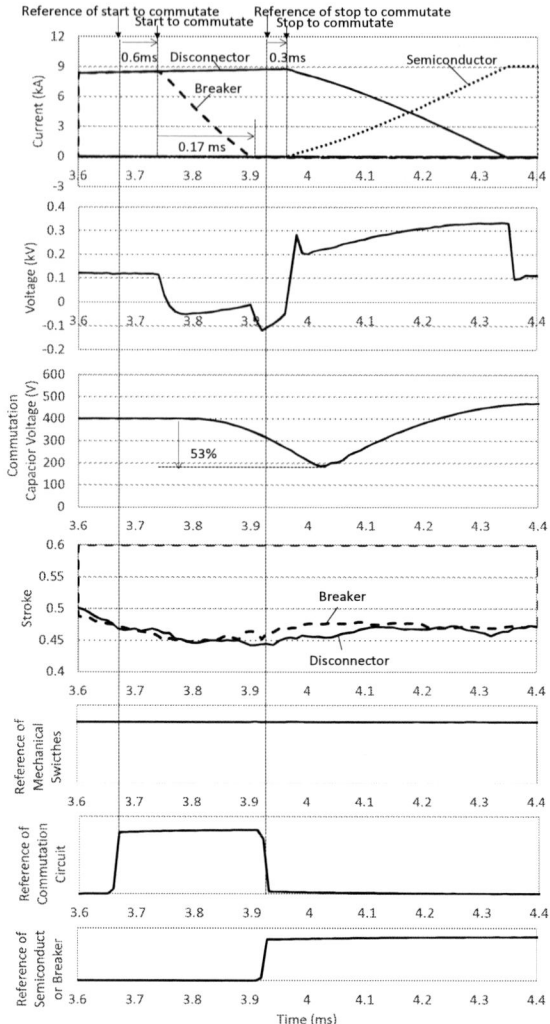

Fig. 10. Test waveforms of the 10-kV experiment model (3.6-4.4 ms).

V. CONCLUSIONS

This paper presented the current commutated hybrid DCCB for HVDC transmission systems. The proposed hybrid DCCB provides the low transmission losses, because the normal current path does not include semiconductor breakers but only mechanical breakers. Moreover, we proposed the design method of a commutation circuit and the test method of DCCBs using existing facility for ACCB such as AC generators.

The experimental results verified; the current commutated hybrid DCCB can interrupt a DC fault current by the proposed operation; the design method of a commutation circuit, the voltage of which is much lower than rated voltage of the DCCB, satisfies specifications set for the experiment; existing facility for ACCBs is able to apply to tests of DCCBs.

ACKNOWLEDGMENT

This paper is based on the results obtained from a project commissioned by the New Energy and Industrial Technology Development Organization (NEDO).

REFERENCES

[1] C. Franck, "HVDC circuit breakers: a review identifying future research needs," IEEE Trans. on Power Delivery, vol. 26, no. 2, pp. 998–1007, Apr. 2011.

[2] J. Hafner and B. Jacobson, "A key innovation for reliable HVDC grids", CIGRE symposium, Bologna, September 2011.

[3] A. Hassanpoor, J. Hafner, B. Jacobson, "Technical assessment of load commutation switch in hybrid HVDC breaker", pp. 3667–3673, IPEC, 2014.

[4] W. Grieshaber, J.-P. Dupraz, D.-L. Penache and L. Violleau, "Development and test of a 120kV direct current circuit breaker", B4-301, CIGRE, 2014.

[5] W. Zhou, X. Wei, S. Zhang, G. Tang, Z. He, J. Zheng, Y. Dan, C. Gao, "Development and test of a 200 kV full-bridge based hybrid HVDC breaker", EPE, 2015.

[6] G. F. Tang, X. G. Wei, W. D. Zhou, S. Zhang, C. Gao, Z. Y. He, J. C. Zheng, "Research and Development of a Full-bridge Cascaded Hybrid HVDC Breaker for VSC-HVDC Applications", A3-117, CIGRE, 2016.

[7] A. Daibo et al., "High-speed current interruption performance of hybrid DCCB for HVDC transmission system," *2017 4th International Conference on Electric Power Equipment - Switching Technology (ICEPE-ST)*, Xi'an, 2017, pp. 329-332.

The 2018 International Power Electronics Conference

A Three-input Central Capacitor DC/DC Converter

Jiaxin Liu[1][*] and Feng Gao[1]

1 School of Electrical Engineering, Shandong University, Jinan, China

*E-mail: sdu_liujiaxin@163.com

Abstract-This paper proposes a three-input central capacitor (TICC) dc/dc converter, which can control three input PV sources simultaneously with low semiconductor voltage/current stress and high efficiency. Meanwhile, it can help balance the dc-link voltage by properly regulating the transferred energy among three sources when connecting a rear-end three-level inverter. In addition, this paper elaborates the detailed operational principles of TICC converter. And Matlab simulations verified the performance of the proposed converter.

Keywords—DC/DC converter, photovoltaic converter, three level, transformer-less

I. INTRODUCTION

Transformer-less grid-connected PV systems present the merits of high efficiency and low cost to perform renewable energy production and power conversion [1]. Nowadays, the power ratings of PV plants are increasing remarkably and the insulation level of PV panel will increase to 1500V to improve the efficiency and reduce the cost. As the power rating increasing, it is often required to associate the converters in series or in parallel. For high current applications, interleaved boost converters shown in Fig. 1(a) are preferable, since the currents through the switches are just fraction of the input currents [2]. Besides, the size of the energy storage inductors and electromagnetic interference (EMI) filters can be reduced [3]. However, such circuit configuration doesn't reduce the voltage stress of the switches and diodes, which limits their high voltage applications. In [4], a dual-input central capacitor dc-dc converter shown in Fig. 1(b), with the capability of boosting the dual PV stings is presented, featuring the low semiconductor voltage/current stress and dissipation along with a wide operation range.

As the voltage rating increasing, multilevel converters should be employed to get a better power quality and reduce the dv/dt on switches. The three-level boost converter is presented to reduce the voltage stress of semiconductor devices to half of the total output voltage. However, its static gain is the same as that of the conventional boost converter [3], [6]. This paper proposes a three-input central capacitor (TICC) dc-dc converter as shown in Fig. 2 (a), which can track the tri-MPPs simultaneously and balance the dc-link voltage. Also the proposed converter features the low semiconductor voltage and current stress and high efficiency. This paper analyzes the operational principles of the proposed converter and makes a comparison with its counterparts. Matlab simulations verified the performance of the proposed dc-dc converter.

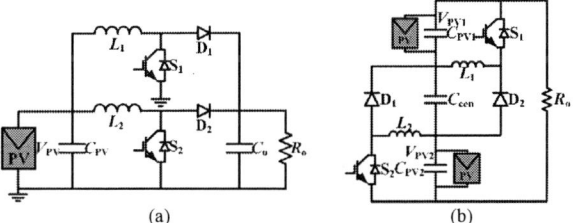

Fig. 1. (a) Interleaved boost converter, (b) Dual-input central capacitor converter.

II. OPERATIONAL PRINCIPLES OF TICC CONVERTER

The proposed three-input central capacitor (TICC) converter is drawn in Fig. 2 (a), where three independent PV sources and two central capacitors C_2 and C_5 are connected in series to power the inversion stage. The output dc voltage V_{bus} of the converter is equal to the sum of the voltage of six capacitors V_{Cx} (X=1~6), which can be written as

$$\begin{cases} V_{bus} = V_{up} + V_{down} \\ V_{up} = V_{C1} + V_{C2} + V_{C3} \\ V_{down} = V_{C4} + V_{C5} + V_{C6} \end{cases} \quad (1)$$

where V_{up} and V_{down} are half of the dc bus voltage.

The converter consists of four buck-boost converters, shown in Fig. 2 (b). The input side of each buck-boost converter is

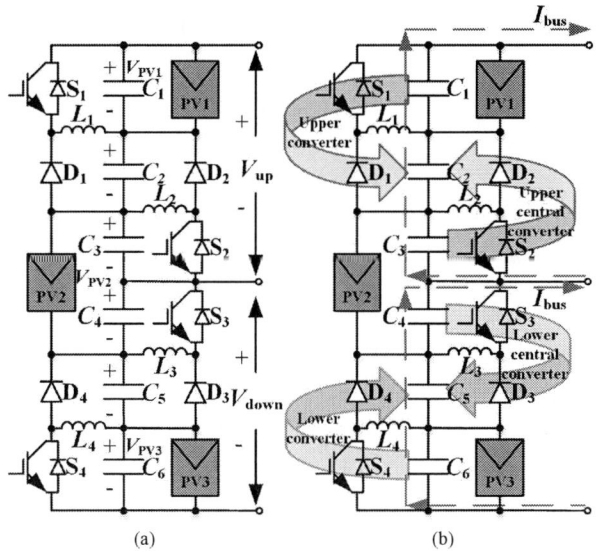

Fig. 2. (a) Three-input central capacitor converter, (b) power flow diagram of the proposed converter.

2016

connected to the PV source via the input capacitor and its output side is connected to another buck-boost converter via the output central capacitor.

As shown in Fig. 2 (b), the upper converter and lower converter are used to track the MPPs of the PV sources PV1 and PV3 and send their energy to the central capacitors C_2 and C_5 respectively. And the upper central converter and lower central converter need to track the MPPs of the PV source PV2 together and compensate the power difference between PV1 and PV3 to make V_{up} equal to V_{down}, which can be written as

$$V_{up} = V_{down} \qquad (2)$$

There are three independent PV sources in the proposed converter. The PV sources PV1 and PV3 usually have the same specification and their output power P_{pvx} ($x=1, 3$) can be expressed as

$$0 < P_{pvx} \leq P_{max} \quad (x = 1, 3) \qquad (3)$$

where P_{max} is the maximum output power of PV sources. The difference between their output power ΔP_{13} can be calculated as

$$0 < \Delta P_{13} = \left| P_{pv1} - P_{pv3} \right| \leq P_{max} \qquad (4)$$

In order to compensate the power difference between PV1 and PV3, the output power of PV2 at its MPP should be higher than the power difference, which can be expressed as

$$P_{pv2} \geq \Delta P_{13} \qquad (5)$$

The capacitors C_3 and C_4 have realized the decoupling of the upper central converter and lower central converter. So when the voltages on the capacitor C_3 and C_4 are different, they will absorb different power P_{C3} and P_{C4} from the PV source PV2, which can be calculated as

$$\begin{cases} \left| P_{C3} - P_{C4} \right| = \Delta P_{13} \\ P_{C3} + P_{C4} = P_{pv2} \end{cases} \qquad (6)$$

Then the upper central converter and lower central converter will send different energy to the capacitor C_2 and C_5 respectively to compensate the power difference between PV1 and PV3. In some extreme conditions, the output power of PV2 will be lower than the power difference between PV1 and PV3 due to partial shading. Then urgent control measures should be taken. The MPPT control of the PV source with higher output

TABLE I
OPERATIONAL STATE OF THE PROPOSED CONVERTER

Swiching mode				Operational state
S_1	S_2	S_3	S_4	
1	1	1	1	state 1
0	1	1	1	state 2
1	0	1	1	
1	1	0	1	
1	1	1	0	
1	1	0	0	state 3
0	0	1	1	
0	1	1	0	state 4
1	0	0	1	
1	0	1	0	
0	1	0	1	
0	0	0	1	state 5
0	0	1	0	
0	1	0	0	
1	0	0	0	
0	0	0	0	state 6

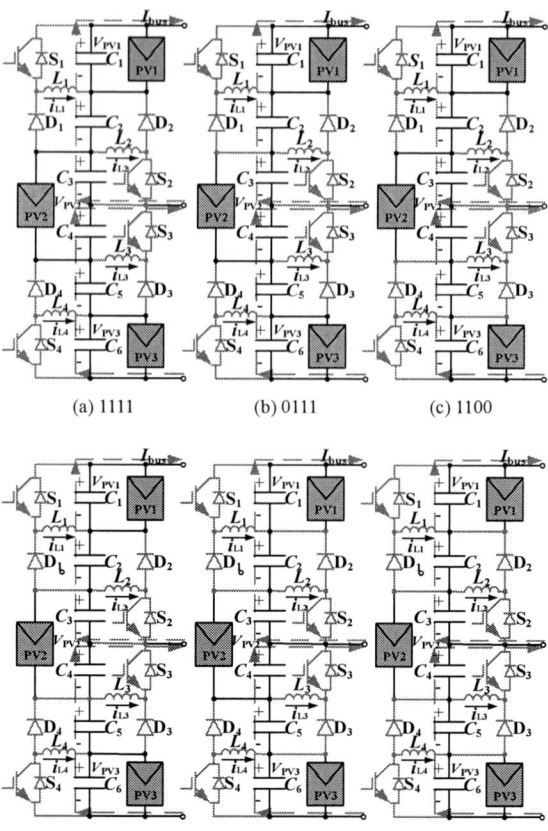

(a) 1111   (b) 0111   (c) 1100

(d) 0110   (e) 0001   (f) 0000

Fig. 3. Operational circuit diagrams of the proposed three-input central capacitor converter when the switch mode are (a) 1111, (b) 0111, (c) 1100, (d) 0110, (e) 0001 and (f) 0000.

power will be replaced by a non-maximum power point tracking (Non-MPPT) algorithm which reduces its output power to ensure the equation (2) to be satisfied [12], [13]. In this paper, three same PV sources have been used as the input sources.

In the proposed converter, four independent switches are employed, so there are sixteen switching modes in total as listed in Table I. It has been divided into six states according to the charging and discharging state of the inductors and capacitors.

State 1: Four input capacitors which are parallel with the PV sources charge the corresponding inductors through the switches respectively. And the load current is supplied by the series network which is composed of six capacitors, as shown in Fig. 3(a).

State 2: Three capacitors charge the corresponding inductors respectively. And one inductor discharges the energy that it has stored to the central capacitor through the diode. Fig. 3(b) shows one switching mode of the state 2.

State 3: Two capacitors charge the corresponding inductors respectively. And two inductors discharge their energy to the same central capacitor. Fig. 3(c) shows one switching mode of the state 3.

State 4: Two capacitors charge the corresponding inductors respectively. And two inductors discharge their energy to the different central capacitors respectively. Fig. 3(d) shows one switching mode of the state 4.

State 5: One capacitor charges the inductor. And three inductors discharge their energy to the central capacitors. Fig.

The 2018 International Power Electronics Conference

3(e) shows one switching mode of the state 5.

State 6: Four inductors discharge their energy to the central capacitors as shown in Fig. 3(f).

When the charging and discharging energy of the central capacitors C_2 and C_5 balance dynamically, the voltage of the central capacitors can be well regulated. When the converter operates at CCM condition, the relationship between the inductor current and bus current can be calculated as

$$\begin{cases} (1-d_1)i_{L1} + (1-d_2)i_{L2} = I_{bus} \\ (1-d_3)i_{L3} + (1-d_4)i_{L4} = I_{bus} \end{cases} \tag{7}$$

where d_x (x=1~4) is the duty ratio of the switch S_x (x=1~4) and the voltage gain under CCM can be obtained from the inductor volt-second balance, which can be calculated as

$$\begin{cases} V_{C2}/V_{C1} = d_1/(1-d_1) \\ V_{C2}/V_{C3} = d_2/(1-d_2) \\ V_{C5}/V_{C4} = d_3/(1-d_3) \\ V_{C5}/V_{C6} = d_4/(1-d_4) \end{cases} \tag{8}$$

III. PERFORMANCE COMPARISON

The proposed converter has a lower voltage and current stress of switches and diodes, which will bring high efficiency to the PV system. To analyze the properties of the proposed converter, its performance has been compared with its counterparts.

A. Analysis and Comparison of Device Quantity

The device quantity of the proposed converter and its counterparts has been compared from the aspects of total and average device quantity, which has been listed in TABLE II. The average device quantity is the ratio of total device quantity

TABLE II
COMPARISON OF DEVICE QUANTITY

	Boost	Interleaved Boost	3-level Boost	Dual-input Central Capacitor	Proposed
Quantity of PV sources	1	1	1	2	3
Total device quantity of transistors/diodes	1/1	2/2	2/2	2/2	4/4
Average device quantity of transistors/diodes	1/1	2/2	2/2	1/1	$\frac{4}{3}/\frac{4}{3}$
Total device quantity of inductors/capacitors	1/2	2/2	1/3	2/3	4/6
Average device quantity of inductors/capacitors	1/2	2/2	1/3	$1/\frac{3}{2}$	$\frac{4}{3}/2$

divided by quantity of PV sources.

B. Analysis and Comparison of Semiconductor Device Stress

When the proposed converter operated at CCM Mode and the output power of PV1 is equal to that of PV3, the relationship between the input PV voltage V_{pvx} (x=1, 3) and the output voltage V_{bus} can be written as

TABLE III
COMPARISON OF SEMICONDUCTOR DEVICE STRESS

	Boost	Interleaved Boost	3-level Boost	Dual-input Central Capacitor Converter	Proposed S1(4)/D1(4)	Proposed S2(3)/D2(3)
Peak voltage stress of transistors/diodes	V_{bus}	V_{bus}	$V_{bus}/2$	$V_{bus}-V_{PV}$	$(V_{bus}-V_{PV})/2$	$V_{bus}/2-V_{PV}$
Peak current stress of transistors and diodes (inductor current)	$\dfrac{V_{bus}\cdot I_{bus}}{V_{PV}}$	$\dfrac{V_{bus}\cdot I_{bus}}{2V_{PV}}$	$\dfrac{V_{bus}\cdot I_{bus}}{V_{PV}}$	$\dfrac{(V_{bus}-V_{PV})\cdot I_{bus}}{2V_{PV}}$	$\dfrac{(V_{bus}-V_{PV})I_{bus}}{3V_{PV}}$	$\dfrac{(V_{bus}-2V_{PV})I_{bus}}{3V_{PV}}$
Average current stress of transistors	$\dfrac{(V_{bus}-V_{PV})I_{bus}}{V_{PV}}$	$\dfrac{(V_{bus}-V_{PV})I_{bus}}{2V_{PV}}$	$\dfrac{(V_{bus}-V_{PV})I_{bus}}{V_{PV}}$	$\dfrac{(V_{bus}-2V_{PV})\cdot I_{bus}}{2V_{PV}}$	$\dfrac{(V_{bus}-3V_{PV})\cdot I_{bus}}{3V_{PV}}$	
Average current stress of diodes	I_{bus}	$I_{bus}/2$	I_{bus}	$I_{bus}/2$	$2I_{bus}/3$	$I_{bus}/3$
Total semiconductor rating	$\dfrac{2V_{bus}^2 I_{bus}}{V_{pv}}$	$\dfrac{2V_{bus}^2 I_{bus}}{V_{pv}}$	$\dfrac{2V_{bus}^2 I_{bus}}{V_{pv}}$	$\dfrac{2(V_{bus}-V_{pv})^2 I_{bus}}{V_{pv}}$	$\dfrac{2\left[(V_{bus}-V_{pv})^2+(V_{bus}-2V_{pv})^2\right]I_{bus}}{3V_{pv}}$	

$$\begin{cases} V_{PV1}\cdot d_1 = \dfrac{1}{2}(V_{bus}-3V_{PV1})\cdot(1-d_1) \\ \dfrac{V_{PV1}}{2}\cdot d_2 = \dfrac{1}{2}(V_{bus}-3V_{PV1})\cdot(1-d_2) \end{cases} \tag{9}$$

where d_1 and d_2 represents the duty ratio of the switch S_1 and S_2. The average currents through the switch S_1 and S_2 can be obtained by the method of power balance, which is calculated as

$$\begin{cases} I_{avg\_S1} = \dfrac{2}{3}\times\dfrac{\frac{1}{2}(V_{bus}-3V_{PV})\cdot I_{bus}}{V_{PV}} = \dfrac{(V_{bus}-3V_{PV})\cdot I_{bus}}{3V_{PV}} \\ I_{avg\_S2} = \dfrac{1}{3}\times\dfrac{\frac{1}{2}(V_{bus}-3V_{PV})\cdot I_{bus}}{\dfrac{V_{PV}}{2}} = \dfrac{(V_{bus}-3V_{PV})\cdot I_{bus}}{3V_{PV}} \end{cases} \tag{10}$$

Because the switches only conduct during the duty ratio of a period, the peak current stress of them can be calculated as

$$\begin{cases} I_{peak\_S1} = \dfrac{I_{avg,IGBT\_S1}}{d_1} = \dfrac{(V_{bus}-V_{PV})I_{bus}}{3V_{PVx}} \\ I_{peak\_S2} = \dfrac{I_{avg,IGBT\_S2}}{d_2} = \dfrac{(V_{bus}-2V_{PV})I_{bus}}{3V_{PVx}} \end{cases} \tag{11}$$

Here, the current ripple of the inductor has been neglected for simplicity. Similarly, the average current of the diodes D1 and D2 can be expressed as

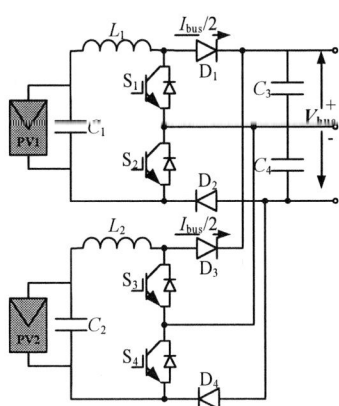

Fig. 4. Dual-input three level boost converter.

2018

$$\begin{cases} I_{avg\_D1} = \dfrac{2}{3} I_{bus} \\ I_{avg\_D2} = \dfrac{1}{3} I_{bus} \end{cases} \tag{12}$$

As a result, the semiconductor rating can be derived by multiplying the peak voltage stress by the peak current stress of the switches and diodes. The comparison of semiconductor device stress is listed in Table III. It can be seen that the voltage and current stress of the transistors and diodes of the proposed converter is lower than its counterparts. Although more devices have been used, the total semiconductor device rating of the converter is reduced.

C. Analysis and Comparison of Semiconductor Losses

To analyze the semiconductor losses of the proposed converter (TICC), it can be compared with the dual-input three level boost (DITHB) converter as shown in Fig. 4, considering the same quantities of transistors and diodes used in these two topologies.

The calculation process of this method is composed of two steps. First, the loss-calculation model for a single device (IGBT and diode) is established by using the method of curve fitting. Then the conduction and switching losses at each period are obtained from the loss-calculation model [10].

The semiconductor losses are comprised of transistor's switching losses and conduction losses, diode's reverse-recovery losses and conduction losses, which can be calculated as

$$\begin{cases} P_{switch\_IGBT} = \left(E_{on}\left(i_{peak}\right) + E_{off}\left(i_{peak}\right)\right) \cdot \dfrac{V_{sw}}{V_{rated}} \cdot f_{sw} \\ P_{con\_IGBT} = I_{avg\_IGBT} \cdot V_{CE}\left(i_{peak}\right) \\ P_{rec\_diode} = E_{rec}\left(i_{peak}\right) \cdot \dfrac{V_{sw}}{V_{rated}} \cdot f_{sw} \\ P_{con\_diode} = I_{avg\_diode} \cdot V_{F}\left(i_{peak}\right) \end{cases} \tag{13}$$

where,

P_{switch_IGBT}, P_{con_IGBT}: switching and conduction losses of IGBT,

$E_{on}\left(i_{peak}\right)$, $E_{off}\left(i_{peak}\right)$: turn-on and turn off losses of IGBT, which are the function of the collector current i_{peak} that passes through the IGBT,

V_{sw}: switching voltage of the switching process,

V_{rated}: switching voltage of the rated switching process,

f_{sw}: switching frequency,

I_{avg_IGBT}, I_{avg_diode}: average current that passes through the IGBT and diode,

$V_{CE}\left(i_{peak}\right)$: collector-emitter voltage of IGBT, which is the function of the collector current i_{peak} that passes through the IGBT,

P_{rec_diode}, P_{con_diode}: switching and conduction losses of diode,

$E_{rec}\left(i_{peak}\right)$: reverse recovery losses of diode, which are the function of the forward current i_{peak} that passes through the diode,

$V_{F}\left(i_{peak}\right)$: forward voltage of diode, which is the function of the forward current i_{peak} that passes through the diode.

The calculation of the semiconductor loss is based on the parameters listed in Table IV. Fig. 5 shows the proportion of the semiconductor loss in output power of the two converters. With the same semiconductors they have used, it can be seen that the semiconductor loss of proposed converter is much lower than

Table IV
PARAMETER FOR THE CALCULATION OF SEMICONDUCTOR LOSS

Parameter	value
Dc-bus voltage V_{bus}	800V
Dc-bus current I_{bus}	6A
Switching frequency f_{sw}	20kHz
IGBT module	FS3L30R07W2H3F_B11

Fig. 5. Proportion of semiconductor loss in output power.

the dual-input three level boost converter. In addition, the proposed converter with the capability of tracking the tri-MPPs simultaneously has a higher MPPT efficiency.

IV. CONTROL STRUCTURE OF THE PROPOSED CONVERTER

The flowchart of the switching of the control mode is shown in Fig. 6. The converter operates at the MPPT mode normally, tracking the MPPs of three PV strings simultaneously. Fig. 7 shows the block diagram of two double loop control strategies

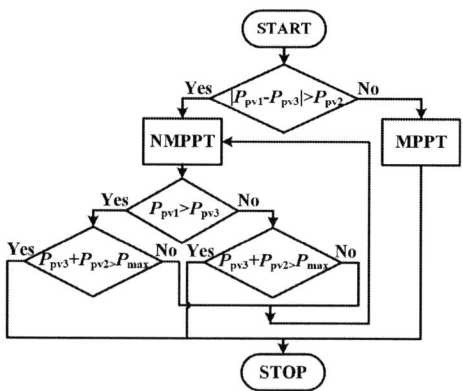

Fig. 6. The flowchart of the switching of the control mode.

[8]. For the upper and lower converter, the outer loop regulates the PV panel voltage and it provides the reference i_{Lx_ref} (x=1, 4) to the inductor current. And the inner loop controls the inductor current i_{Lx} (x=1, 4) by providing the duty cycle to the switch S_x (x=1, 4). For the upper central converter and lower central converter, the outer loop provides the current reference i_{Lx_ref} from the PV source voltage v_{PV2} and current i_{PV2}. The inner loop controls the average current of i_{L2} and i_{L3} and balances the dc-link voltage. The balancing duty ratio Δd is generated by the dc-link balancing control block as (14) [9]:

$$\Delta d = k_{p,balance}\left(V_{up} - V_{down}\right) \tag{14}$$

where $k_{p,balance}$ is the proportional gain of the dc-link voltage balancing controller. So the switches S_2 and S_3 are driven by the duty ratios d_2 and d_3.

However, in some extreme conditions where the output

The 2018 International Power Electronics Conference

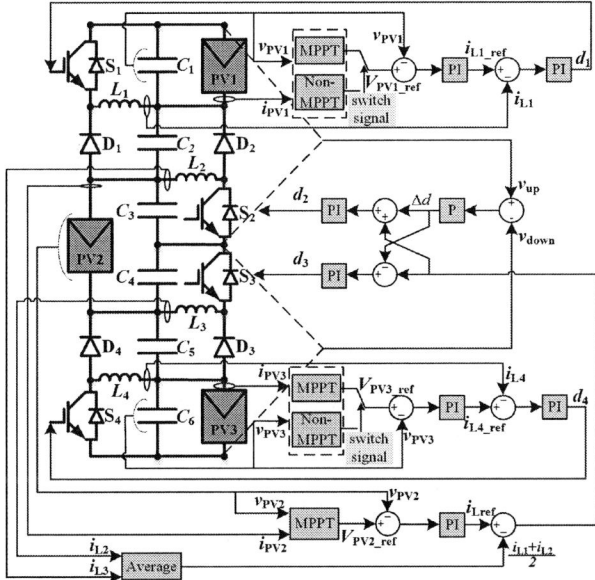

Fig. 7. Control diagram of three-input central capacitor converter.

power of PV2 at its MPP is lower than the power difference between the output power of PV1 and PV3, a Non-MPPT algorithm is introduced to balance the output voltage to ensure the equation (2) to be satisfied. For example, there is a sudden decrease of Ppv3, causing that equation (5) is not satisfied. The controller of PV3 will continue to track the new maximum power point of PV3, but the PV1 will be controlled to track a given power point to reduce its output power until the Ppv3 returns to a nominal power level. The determining condition whether the controller of PV3 can return to its MPPT mode is if the sum of output power of PV1 and PV3 is higher than the maximum output power of the PV2, which is written as

$$P_{pv1} + P_{pv3} \geq P_{max} \tag{15}$$

V. SIMULATION RESULTS

The Matlab simulation model has been built to verify the performance of the proposed three-input central capacitor converter. The corresponding circuit and control parameters are listed in Table V. To simulate the irradiation variation on the grid-tied PV system, the irradiation level of PV1 varies from 1000W/m² to 800W/m² at the time of 0.05s and recovers to 1000 W/m² at the time of 0.25s, while the irradiation level of PV2 and PV3 keep unchanged at 1000 W/m². Fig. 8(a) shows the simulated waveforms. It can be seen that half of dc-link voltages are regulated steadily to 400V and the grid currents are not deteriorated, because the upper central converter will send more energy to the capacitor C_2 to handle the power difference between PV1 and PV3. Thus, the proposed converter is able to handle the power difference during the unequal irradiance operation and guarantee the normal operation of the rear-end inversion stage.

Another case scenario is defined for verification of the proposed method, in which the irradiation level of PV1 varies from 1200W/m² to 400W/m² at the time of 0.05s and recovers to 1200 W/m² at the time of 0.25s, while the irradiation level of PV2 and PV3 keep unchanged at 800 W/m² and 1200 W/m². At the time of 0.05s, the output power of PV3 is higher than the sum of output power of PV1 and PV2, which enables the Non-MPPT control. So the output voltage of PV3 is improved to

TABLE V
PARAMETERS OF THR SIMULATION MODEL

Parameter	value
PV MPP voltage (1000W/m²,20□)	188V
PV MPP current (1000W/m²,20□)	6.57A
Rated dc bus voltage	800V
Grid voltage (line to neutral)	300V
Switching frequency	20kHz
Dc inductor / grid side inductor	3mH/5mH
PV capacitor / central capacitor	100uF/150uF

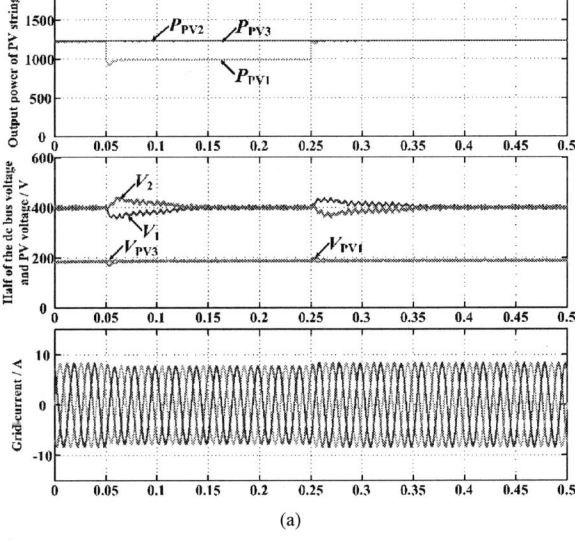

(a)

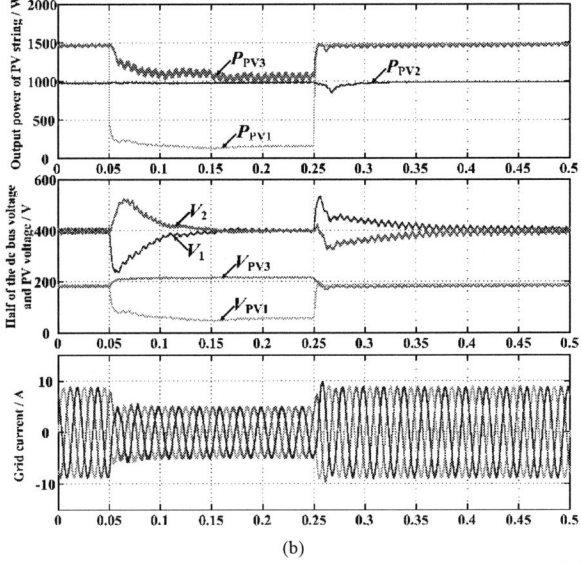

(b)

Fig. 8. (a) Simulated waveforms when PV1 irradiance declines to 800W/m² from 1000 W/m² at 0.05s and recovers to 1000W/m² at 0.25s, (b) Simulated waveforms when PV1 irradiance declines to 400W/m² from 1200 W/m² at 0.05s and recovers to 1200W/m² at 0.25s

reduce its output power. So the output voltage of the converter can keep balanced and the rear-end inversion stage can guarantee the normal operation.

VI. CONCLUSION

This paper proposes a three-input central capacitor converter for distributed PV system. Then, the detailed operational principles and the power difference handling capability are presented. Further, the performance of the proposed converter is analyzed through a comparative study. It can be seen that the semiconductor loss of the proposed converter has been reduced by reducing the semiconductor voltage and current stress. In addition, the corresponding control strategy with the capability of tracking the tri-MPPs simultaneously and balancing the dc-link voltage is presented. This three-input three level central capacitor converter along with the proposed strategy was verified through the simulation results.

REFERENCES

[1] J. F. Ardashir, M. Sabahi, S. H. Hosseini, F. Blaabjerg, E. Babaei and G. B. Gharehpetian, "A Single-Phase Transformerless Inverter With Charge Pump Circuit Concept for Grid-Tied PV Applications," in *IEEE Trans. Ind. Electron.*, vol. 64, no. 7, pp. 5403-5415, July 2017.

[2] R. Garth, W. J. Muldoon, G. C. Benson and E. N. Costague, "Multi-phase, 2-kilowatt, high-voltage, regulated power supply," *1971 IEEE Power Electronics Specialists Conference*, Pasadena, CA, 1971, pp. 110-116.

[3] F. L. Tofoli, D. De Castro Pereira, W. J. De Paula, and D. De Sousa Oliveira Junior, "Survey on non-isolated high-voltage step-up dc-dc topologies based on the boost converter," *IET Power Electron.*, vol. 8, no. 10, pp. 2044–2057, Oct. 2015.

[4] M. Chen, F. Gao, R. Li and X. Li, "A Dual-Input Central Capacitor DC/DC Converter for Distributed Photovoltaic Architectures," *IEEE Trans. Ind. Appl.*, vol. 53, no. 1, pp. 305-318, Jan.-Feb. 2017.

[5] R. Abdullah, N. A. Rahim, S. R. Sheikh Raihan and A. Z. Ahmad, "Five-Level Diode-Clamped Inverter With Three-Level Boost Converter," in *IEEE Trans. Ind. Electron.*, vol. 61, no. 10, pp. 5155-5163, Oct. 2014.

[6] J. M. Kwon, B. H. Kwon and K. H. Nam, "Three-Phase Photovoltaic System With Three-Level Boosting MPPT Control," in *IEEE Trans. Power Electron.*, vol. 23, no. 5, pp. 2319-2327, Sept. 2008.

[7] Z. Chen, L. Yuan, Z. Zhao and X. Sun, "Power losses in two- and three-level three phase photovoltaic inverters equipped with IGBTs," *2012 15th International Conference on Electrical Machines and Systems (ICEMS)*, Sapporo, 2012, pp. 1-6.

[8] F. D. Tan and R. D. Middlebrook, "A unified model for current programmed converters," *IEEE Trans. Power Electron.*, vol. 10, no. 4, pp. 397–408, Jul. 1995.

[9] J. S. Kim, J. M. Kwon and B. H. Kwon, "High-Efficiency Two-Stage Three-Level Grid-Connected Photovoltaic Inverter," in *IEEE Trans. Ind. Electron.*, vol. PP, no. 99, pp. 1-1.

[10] M. S. Agamy et al., "An efficient partial power processing DC–DC converter for distributed PV architectures," *IEEE Trans. Power Electron.*, vol. 29, no. 2, pp. 674–686, Feb. 2014.

[11] F. Blaabjerg, R. Teodorescu, M. Liserre and A. V. Timbus, "Overview of Control and Grid Synchronization for Distributed Power Generation Systems," in *IEEE Trans. Ind. Electron.*, vol. 53, no. 5, pp. 1398-1409, Oct. 2006.

[12] E. Afshari *et al.*, "Control Strategy for Three-Phase Grid-Connected PV Inverters Enabling Current Limitation Under Unbalanced Faults," *IEEE Trans. Ind. Electron.*, vol. 64, no. 11, pp. 8908-8918, Nov. 2017.

[13] M. Vekić, V. Porobić, S. Grabić, E. Adžić and C. Zogogianni, "Explicit active power reference tracking algorithm for photovoltaic converter," in *2017 International Symposium on Power Electronics (Ee)*, Novi Sad, Serbia, 2017, pp. 1-6.

Series/Parallel Switching Circuits Using Power MOSFETs for Photovoltaic Modules

Masamichi Tanemo[1], Koki Matsudate[1] and Shinichi Nomura[1*]
1 School of Science and Technology, Meiji University, Kawasaki, Japan
*E-mail: nomuras@meiji.ac.jp

Abstract— **This work discusses the feasibility of the series/parallel switching circuits using power MOSFETs in order to minimize the output power reduction of the photovoltaic power system due to the partial shading by connecting the shaded panels to the adjacent panel in parallel. From the experimental results: 1) When one of the photovoltaic panels is shaded at least, 50% of the reduction ratio of the solar radiation intensity becomes the lower limit of the threshold values for the series/parallel switching control. 2) The effect of the series/parallel switching control can be more expected when more than two of the photovoltaic panels are shaded. However, the sum of the solar radiation intensity applied to the panels in the parallel circuit should be less than 150% of the maximum intensity of the photovoltaic power system. 3) The shaded panel which should be connected to the adjacent panel in parallel can be detected by the bypass current measurements. The authors developed a switching circuit module prototype and carried out the operation tests using three photovoltaic panels. From the results, the switching circuit module successfully detects the partial shading conditions and switches from the series operation mode to the parallel operation mode.**

Keywords— *Partial shading detection, photovoltaic power system, power MOSFET, series/parallel switching control.*

I. INTRODUCTION

Photovoltaic power systems are expected to be one of the feasible options for renewable energy systems. However, the output power of the photovoltaic system is significantly reduced even when the photovoltaic panels are partially shaded. In order to extract the output power from the photovoltaic system and to prevent from the hot spot in the shaded panel due to the partial shading, the following feasible solutions have been mainly discussed:

1) Use of active bypass circuits instead of traditional antiparallel bypass diodes [1], or
2) Power assist by modular DC-DC converters incorporated into photovoltaic modules [2], [3], or
3) Circuit configuration change of photovoltaic panel strings [4].

In this work, the authors focus on the third solution and discuss the feasibility of series/parallel switching circuits using power MOSFETs. In general, although the shaded panel is disconnected by the bypass diode, this system tries to extract the output power from the shaded panel by

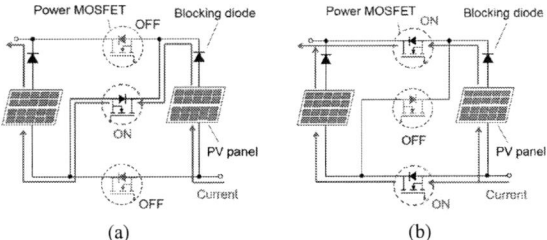

Fig. 1. Basic configuration of the series/parallel switching circuit for photovoltaic modules. In the normal operation, both panels are connected in series (a). If one of the panels is partially shaded, both panels are connected in parallel (b).

connecting the shaded panel to the adjacent panel in parallel. In the previous work, the authors investigated the effect of the series/parallel switching control by using solar cells [5]. In this work, the authors developed a prototype of a switching circuit module in order to investigate the control methods for the series/parallel switching circuit. This paper describes the threshold value for the series/parallel switching control and the detection method of the partial shading based on the experimental results and the circuit topology.

II. SERIES/PARALLEL SWITCHING CIRCUIT TOPOLOGY USING POWER MOSFETS

A. Basic Configuration

Fig. 1 shows the basic configuration of the series/parallel switching circuit using power MOSFETs. The switching circuit is connected between the photovoltaic panels. If both panels are uniformly illuminated by the solar radiation, both panels are connected in series as shown in Fig. 1(a). In the series operation mode, the center MOSFET is turned on and the other MOSFETs are turned off. Since the electrical potential of the left side panel is higher than that of the right side panel, the upper branch current can be blocked by the body diode of the MOSFET. The body diode of the MOSFET in the upper branch and that in the lower branch can be also used as a traditional antiparallel bypass diode for the left side panel and that for the right side panel, respectively.

On the other hand, if one of the panels is partially shaded, both panels are connected in parallel in order to extract the output power from the shaded panel. In the parallel operation mode, the center MOSFET is turned off and the other MOSFETs are turned on as shown in Fig. 1(b). Although the power MOSFETs in the upper and lower branches of the switching circuit can be replaced by diodes in the parallel operation mode, the authors try to use power MOSFETs in order to reduce the on-state power losses in the switching circuit. The external blocking diodes are composed of Schottky barrier diodes.

B. Series/Parellel Switching Circuit Module Prototype

In order to investigate the control methods of the series/parallel switching circuit for the photovoltaic system, the author develop a prototype of a switching circuit module for five photovoltaic panels. Fig. 2(a) shows a photograph of the switching circuit module prototype. As shown in Fig. 2(b), the number of the photovoltaic panels is selected by the mechanical selector switches.

The electric power load for the photovoltaic panels is controlled by the PWM control of the DC/DC boost converter. As shown in Fig. 2(c), the applied PWM signal with a duty ratio to the power MOSFETs of the boost chopper circuit is calculated from the microcomputer. By controlling the duty ratio, the output voltage of the solar cell module is changed from 0 V to the open circuit voltage and the output current is varied from 0 A to the short circuit current.

The gate signals applied to the power MOSFET of the series/parallel switching circuit is directly provided from the microcomputer. The gate drive circuit for each power MOSFET is insulated by using an isolated DC/DC converter and a photocoupler.

III. FEASIBLITY OF THE SERIES/PARALLEL SWITCHING CONTROL FOR PHOTOVOLTAIC MODULES

A. Experimental Conditions

The authors investigate the maximum power point dependence on the solar radiation disturbance between photovoltaic panels and evaluate the threshold value for the series/parallel switching control. A photograph and the specifications of the sample photovoltaic panel are shown in Fig. 3 and Table I, respectively. In this experiment, four photovoltaic panels are connected through the series/parallel switching circuits and the mechanical switches are used for the series/parallel switching instead of the power MOSFETs.

Fig. 4 shows the relationship between the solar radiation intensity and the output properties of the sample photovoltaic panel. From the results, the maximum power point and the short circuit current are proportional to the solar radiation intensity. Ignoring the internal resistance of the photovoltaic panel, the output current at the maximum power point is almost same as the short circuit current. Therefore, the threshold value of the series/parallel switching control can be evaluated by the solar radiation disturbance between the photovoltaic panels.

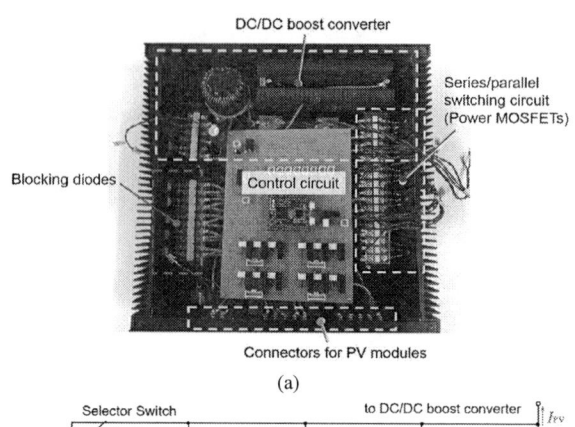

(a)

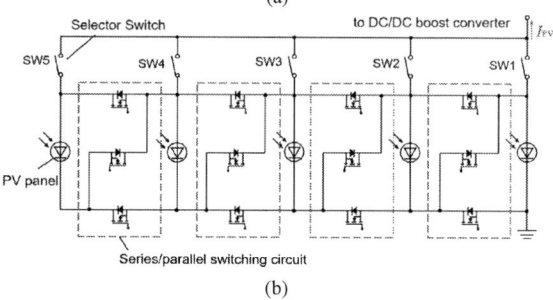

(b)

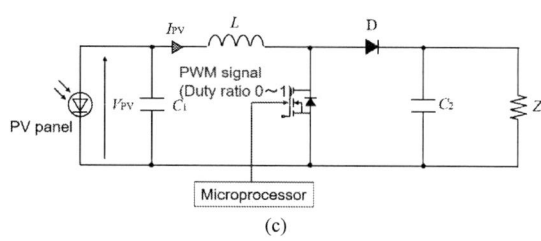

(c)

Fig. 2. A photograph of the series/parallel switching circuit module prototype for five photovoltaic panels (a). The number of the operating photovoltaic panels is selected by the selector switches (b). For instance, when the SW3 is turned on, three panels are connected to the DC/DC boost converter (c) through the series/parallel switching circuit.

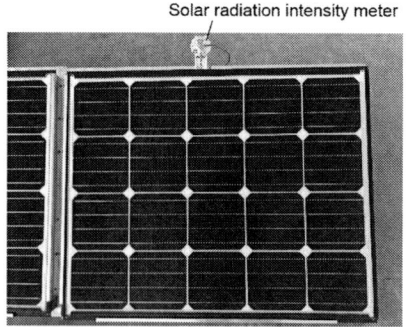

Fig. 3. A photograph of the sample photovoltaic panel.

TABLE I.
SPECIFICATIONS OF THE SAMPLE PHOTOVOLTAIC PANEL.

Maximum output	89 W
Open circuit voltage	12.7 V
Short circuit current	9.21 A
Maximum operating output voltage	10.3 V
Maximum operating output current	8.68 A

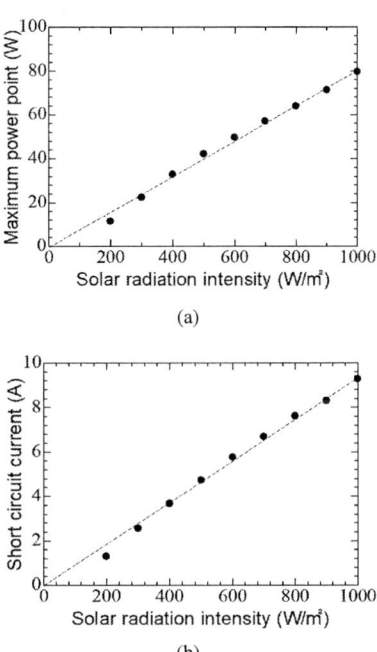

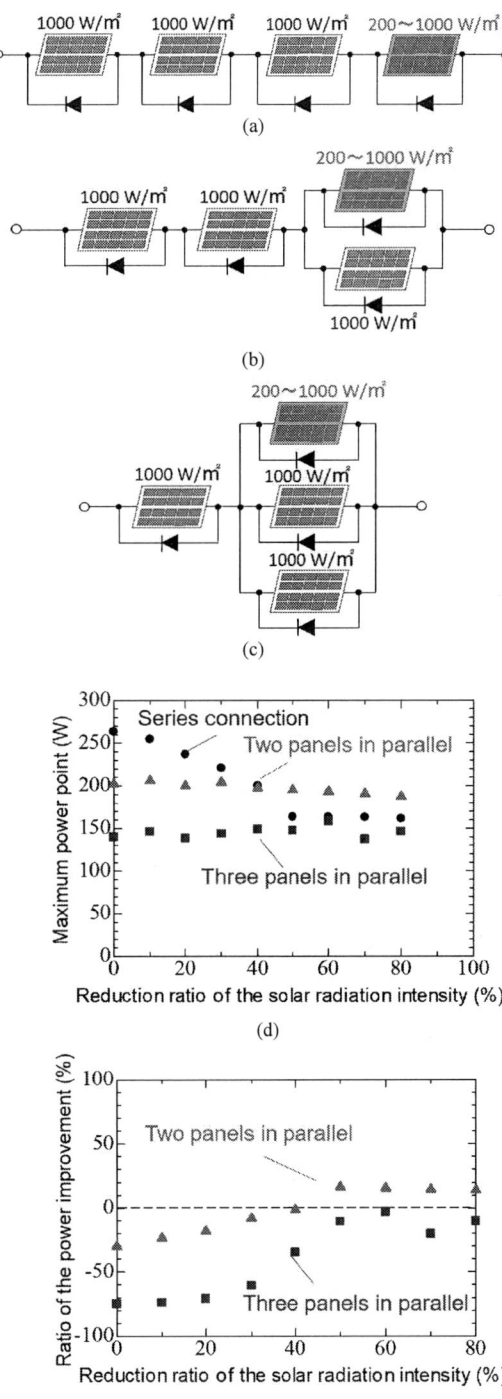

Fig. 4. The maximum power point (a) and the short circuit current (b) of the sample photovoltaic panel as a function of the solar radiation intensity. The surface temperature of the photovoltaic panel is 40 degrees centigrade.

The reduction ratio of the solar radiation intensity a_{ratio} is defined as

$$a_{\text{ratio}} = \frac{X_{\max} - X_{\min}}{X_{\max}} \times 100\%, \qquad (1)$$

where $X_{\max}$ is the maximum solar radiation intensity applied to the panel and $X_{\min}$ is the minimum solar radiation intensity of the shaded panels. If a_{ratio} is 0%, all photovoltaic panels are uniformly illuminated. If a_{ratio} is 100%, the photovoltaic panel is completely shaded (0 W/m²).

B. Case Studies of One Shaded Photovoltaic Panel

Fig. 5 compares the maximum power points between the series connection and the parallel connections of the photovoltaic panels when one of the panels is only shaded and the other panels are uniformly illuminated by the solar radiation. In the parallel connections, the shaded panel is connected to the adjacent panel in parallel as shown in Fig. 5(b) and Fig. 5(c). The maximum solar radiation $X_{\max}$ is 1000 W/m². The solar radiation of the shaded panel $X_{\min}$ varies from 200 W/m² to 1000 W/m².

From the results in Fig. 5(d), when the reduction ratio of the solar radiation intensity a_{ratio} is over 50%, the maximum power point in the series connection becomes a constant value. This property shows that the shaded panel is disconnected by the bypass diode. Then the maximum power point shifts to the lower voltage side.

Fig. 5. The maximum power point dependence on the solar radiation disturbance when one of the photovoltaic panel is shaded. The experimental conditions are all panels connected in series (a), two panels connected in parallel (b) and three panels connected in parallel (c). The maximum power points are compared between the series connection and the parallel connections of the photovoltaic panels (d). The ratio of the power improvement to the series connection case is compared between the case of two panels connected in parallel and the case of three panels connected in parallel under the same condition of the solar radiation disturbance (e).

On the other hand, the maximum power point in the parallel connections almost keeps the same maximum power which is obtained from three photovoltaic panels illuminated uniformly even when the solar radiation of the shaded panel is reduced. However, from the results in Fig. 5(e), when the reduction ratio of the solar radiation intensity a_{ratio} is over 50%, the maximum power in the case of two panels connected in parallel is 10% to 20% higher than that in the series connection case by the effect of the internal resistance reduction.

C. Case Studies of Two Shaded Photovoltaic Panels

Fig. 6 compares the maximum power points between the series connection and the parallel connections of the photovoltaic panels when two of the panels are shaded. In the parallel connections, the shaded panels are connected as shown in Fig. 6(b) and Fig. 6(c). Two of the photovoltaic panels are constantly illuminated by the solar radiation of 1000 W/m² whose value corresponds to the maximum intensity X_{max}. The other panels are shaded with a solar radiation rage from 200 W/m² to 1000 W/m² whose values correspond to X_{min}. All of the combinations of the solar radiation disturbance were evaluated.

From the results in Fig. 6(d) and Fig. 6(e), under the same condition of the solar radiation disturbance, the parallel connections can keep the higher output power from the photovoltaic power system compared with the series connection if the reduction ratio of the solar radiation intensity a_{ratio} is over 50%. Comparing the results in Fig. 6(e) to those in Fig. 5(e), the power improvement effect by the series/parallel switching control can be more expected when more than two of the photovoltaic panels are simultaneously shaded. Therefore, the 50% reduction ratio of the solar radiation intensity becomes one of the threshold values for the series/parallel switching control. However, especially in the case of three panels connected in parallel, there are some cases which can not improve the maximum power point. The validity of the threshold value is discussed in Section IV.

D. Case Studies of Three Shaded Photovoltaic Panels

Fig. 7 compares the maximum power points between the series connection and the parallel connections of the photovoltaic panels when three of the panels are shaded. In the parallel connections, the shaded panels are connected as shown in Fig. 7(b) and Fig. 7(c). One of the photovoltaic panels is constantly illuminated by the solar radiation of 1000 W/m² whose value corresponds to the maximum intensity X_{max}. The other panels are shaded with a solar radiation range from 200 W/m² to 1000 W/m² whose values correspond to X_{min}. All of the combinations of the solar radiation disturbance were evaluated.

From the results in Fig. 7(d) and Fig. 7(e), under the same condition of the solar radiation disturbance, the case of two panels connected in parallel can extract the output power from the shaded panels when the reduction ratio of

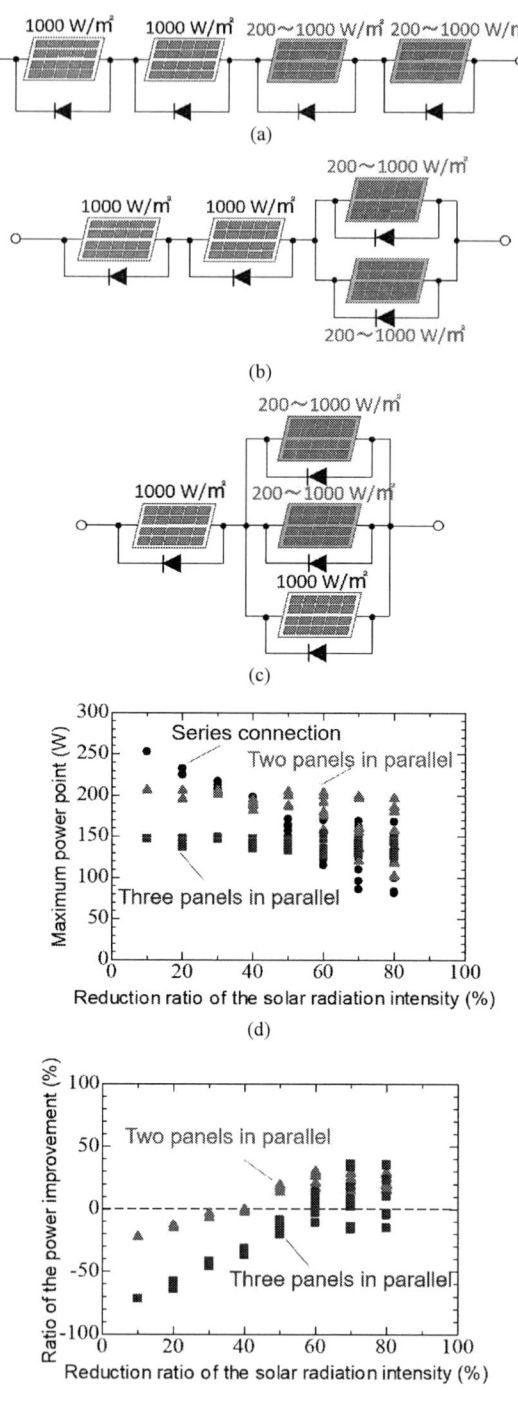

Fig. 6. The maximum power point dependence on the solar radiation disturbance when two of the photovoltaic panels are shaded. The experimental conditions are all panels connected in series (a), two panels connected in parallel (b) and three panels connected in parallel (c). The maximum power points are compared between the series connection and the parallel connections of the photovoltaic panels (d). The ratio of the power improvement to the series connection case is compared between the case of two panels connected in parallel and the case of three panels connected in parallel under the same condition of the solar radiation disturbance (e).

the solar radiation intensity a_{ratio} is over 50%. Compared to the series connection case, the maximum ratio of the power improvement becomes 30%. In the case of three panels connected in parallel, if the reduction ratio of the solar radiation intensity a_{ratio} is over 60%, there are some cases which can be expected to improve the maximum power point properties by the effect of the series/parallel switching control.

IV. Evaluation of the Threshold Values for the Series/Parallel Switching Control

A. Lower Limit of the Thereshold Value

From the results in the previous section, 50% of the reduction ratio of the solar radiation intensity becomes one of the threshold values for the series/parallel switching control. In order to evaluate the validity of this threshold value, the authors carried out the case studies using the switching circuit module prototype shown in Fig. 2.

Fig. 8 shows an example of the output properties for three photovoltaic panels with 50% of the reduction ratio of the solar radiation intensity. From Fig. 8(a), in the series connection, if the solar radiation intensity of one shaded panel is lower than 50% of that of the other panels, the maximum power point shifts to the lower voltage side. This property means that the shaded panel is disconnected by the bypass diode. When the shaded panel is connected to the adjacent panel in parallel, the maximum power point is increased by the effect of the reduction of the internal resistance. Therefore, 50% of the reduction ratio of the solar radiation intensity becomes the lower limit of the threshold value for the series/parallel switching control.

B. Limitation of Parallel Conditions

From the results in Section III, even when the reduction ratio a_{ratio} is over 50%, there are some cases which can not obtain the power improvement effect by the parallel connection. In this section, the authors discuss the limitation of the parallel conditions.

Fig. 9 shows the output properties with a solar radiation disturbance of 1000 W/m^2 – 500 W/m^2 – 300 W/m^2. Then the reduction ratio of the solar radiation intensity a_{ratio} becomes 70%. From the results in Fig. 9(a), since the maximum power point in the series connection case shifts to the lower voltage side, the shaded panel with 300 W/m^2 is disconnected by the bypass diode during the maximum power point tracking (MPPT) operation.

On the other hand, compared to the series connection, if the shaded panels are connected in parallel, the output current is increased as shown in Fig. 9(b). This result shows that the photovoltaic power system can extract the output power from the 300-W/m^2 shaded panel by the effect of the parallel connection. Since the maximum power point and the output current are proportional to the solar radiation intensity, the shaded panels in the parallel

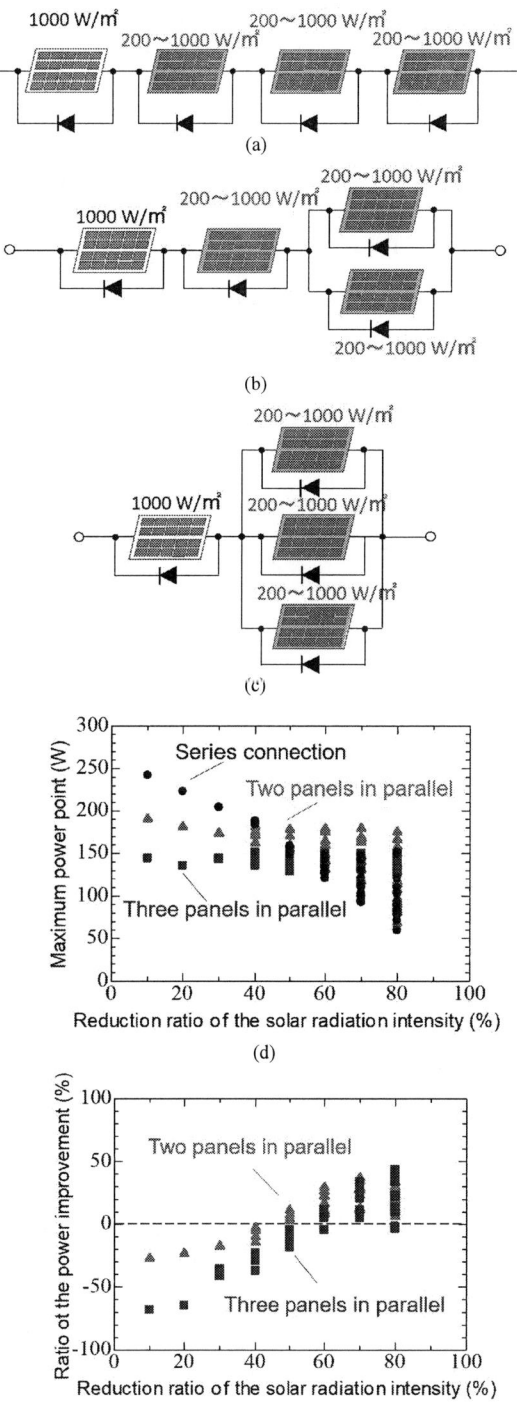

Fig. 7. The maximum power point dependence on the solar radiation disturbance when three of the photovoltaic panels are shaded. The experimental conditions are all panels connected in series (a), two panels connected in parallel (b) and three panels connected in parallel (c). The maximum power points are compared between the series connection and the parallel connections of the photovoltaic panels (d). The ratio of the power improvement to the series connection case is compared between the case of two panels connected in parallel and the case of three panels connected in parallel under the same condition of the solar radiation disturbance (e).

circuit generate the almost same maximum power from the panel with 800 W/m² of the solar radiation intensity. Therefore the photovoltaic power system can be equivalently evaluated as a series connected panel system with a solar radiation disturbance of 1000 W/m² − 800 W/m² as shown in Fig. 9(c). In this case, since the reduction ratio of the solar radiation intensity a_{ratio} can be reduced to 20%, the series/parallel switching control leads to the power improvement against the partial shading problem.

Fig. 10 shows an example of the series/parallel switching condition. In this example, the reduction ratio of the solar radiation intensity a_{ratio} is 80%. When the 200-W/m² shaded panel is connected to the 300-W/m² panel in parallel, the solar radiation disturbance is equivalently reduced to 1000 W/m² − 1000 W/m² − 500 W/m² which is the same condition discussed in Fig. 8. Therefore, in order to obtain the effect of the series/parallel switching control, the sum of the solar radiation intensity applied to the photovoltaic panels in the parallel circuit should be less than 150% of the maximum intensity of the photovoltaic power system as shown in Fig. 10.

V. CONTROL METHODS OF THE SERIES/PARALELL SWITCHING CIRCUIT MODULE

A. Detection Method of Partial Shading Conditions

As shown in Section II-A, if the partial shading occurs during the series operation mode, the shaded panel is disconnected by the body diode of the power MOSFET in the upper branch or in the lower branch of the series/parallel switching circuit. From this feature, the shaded panel which should be connected to the adjacent panel in parallel can be detected by the bypass current.

Using the series/parallel switching circuit module prototype, the short circuit test for three photovoltaic panels connected in series was carried out by setting the duty ratio of the DC/DC boost converter to be 1. The solar radiation intensity applied to one shaded panel varies from 100 W/m² to 1000 W/m² and the other panels are illuminated by 1000 W/m².

Fig. 11 shows the bypass current conditions in the cases of both ends of the series/parallel switching circuit module. When the photovoltaic panel in the left side end is shaded, the bypass current flows in the upper branch of the switching circuit as shown in Fig. 11(a). On the other hand, if the photovoltaic panel in the right side end is shaded, the bypass current flows in the lower branch of the switching circuit as shown in Fig. 11(b).

Fig. 11(c) shows the test result under the condition shown in Fig. 11(b). From the result, the bypass current I_{bypass} is proportional to the reduction ratio of the solar radiation intensity a_{ratio}. Since the output current through the photovoltaic panel is almost proportional to the solar radiation intensity, the relationship between the current through the shaded panel I_{shade} and the total current in the photovoltaic power system I_{PV} is given by

$$I_{shade} = \frac{X_{min}}{X_{max}} I_{PV}, \qquad (2)$$

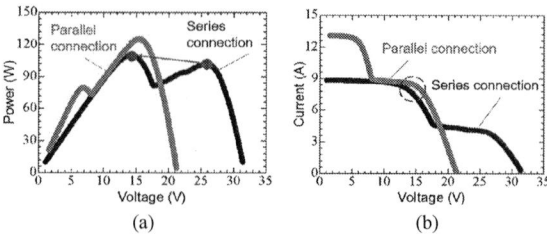

Fig. 8. Comparison of the *P-V* curves (a) and the *I-V* curves (b) between the series connection and the parallel connection with 50% of the reduction ratio of the solar radiation intensity using three photovoltaic panels. The solar radiation disturbance is 1000 W/m² − 1000 W/m² − 500 W/m². In the parallel connection, the shaded panel is connected to the adjacent panel in parallel.

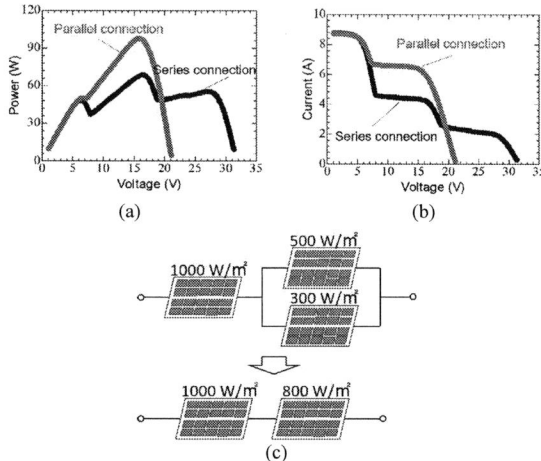

Fig. 9. Comparison of the *P-V* curves (a) and the *I-V* curves (b) between the series connection and the parallel connection using three photovoltaic panels. The solar radiation disturbance is 1000 W/m² − 500 W/m² − 300 W/m². After the shaded panels are connected in parallel, the photovoltaic power system can be equivalently evaluated as a series connected panel system with a solar radiation disturbance of 1000 W/m² − 800 W/m² (c).

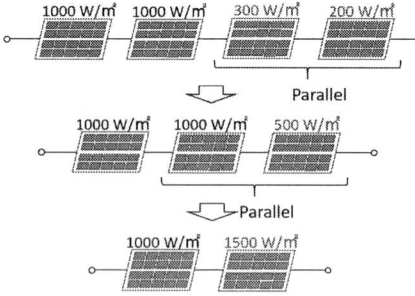

Fig. 10. Example of the series/parallel switching control and its equivalent condition.

where X_{min} is the solar radiation intensity applied to the shaded panel and X_{max} is the maximum solar radiation intensity. From Fig. 11(a) and Fig. 11(b), Kirchhoff's first law at the node *A* becomes

$$I_{PV} - I_{shade} - I_{bypass} = 0. \qquad (3)$$

2027

Therefore, using the reduction ratio of the solar radiation intensity a_{ratio} defined as (1), the bypass current is almost given by

$$I_{\text{bypass}} = \frac{a_{\text{ratio}}}{100} I_{\text{PV}}. \tag{4}$$

This equation means that the partial shading condition can be detected by measuring the bypass current. From the results in Section IV, when the bypass current is over 50% of the total current, the operation mode should be changed from the series operation mode to the parallel operation mode in order to extract the output power from the photovoltaic power system.

On the other hand, when the photovoltaic panel between the series/parallel switching circuits is shaded, the bypass currents flow in both upper and lower branches of the switching circuits as shown in Fig. 12(a). From the short circuit test with a range from 0% to 90% of the reduction ratio of the solar radiation intensity, the sum of the bypass currents $I_{\text{bypass1}} + I_{\text{bypass2}}$ is also proportional to the reduction ratio of the solar radiation intensity a_{ratio} as shown in Fig. 12(b). From Fig. 12(a), Kirchhoff's first law at the node A becomes

$$I_{\text{PV}} - I_{\text{shade}} - I_{\text{bypass1}} - I_{\text{bypass2}} = 0. \tag{5}$$

From (1), (2) and (5), the sum of the bypass currents is almost evaluated by

$$I_{\text{bypass1}} + I_{\text{bypass2}} = \frac{a_{\text{ratio}}}{100} I_{\text{PV}}. \tag{6}$$

This equation means that the partial shading conditions are detected by measuring the bypass current conditions. Therefore, when the sum of the bypass currents is over 50% of the total current, the operation mode should be changed from the series operation mode to the parallel operation mode in order to extract the output power from the photovoltaic power system.

B. Demonstration of the Series/Parallel Switching

The authors carried out the operation test of the partial shading detection and the series/parallel switching operation using the switching circuit module prototype. Fig. 13(a) shows the experimental circuit. The right side end of the photovoltaic panel (PV3) is reduced the applied solar radiation intensity. This condition corresponds to Fig. 11(b). The reduction ratio of the solar radiation intensity of one of three photovoltaic panels is 55%. The bypass currents in the upper and lower branches are measured by the current sensors (CS1 – CS4).

Fig. 13(b) shows the control procedure of the switching circuit module prototype. The control procedure for the MPPT operation is composed of three steps. The first step is the short circuit test to detect the partial shading conditions by measuring the bypass currents. In the short circuit test, the duty ratio of the PWM signals for the DC/DC boost converter is set to be 1. The DC/DC boost

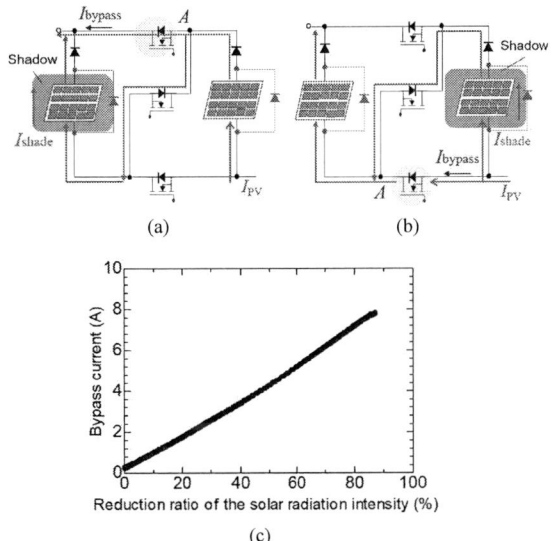

(a)   (b)

(c)

Fig. 11. Bypass current conditions in the left side end case of the switching circuit module (a) and the right side end case of that (b). The bypass current through the upper branch or the lower branch of the switching circuit module is proportional to the reduction ratio of the solar radiation intensity a_{ratio} (c)

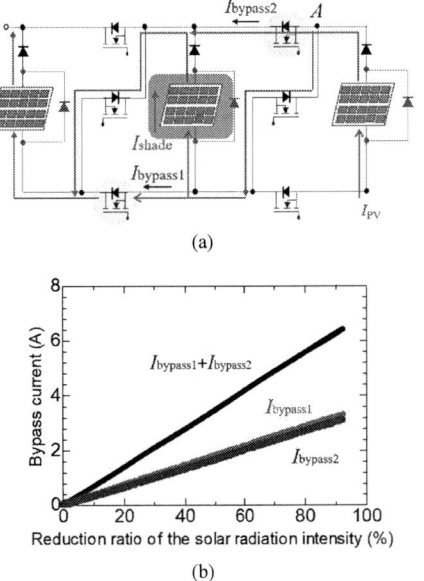

(a)

(b)

Fig. 12. Bypass current conditions when the photovoltaic panel between the series/parallel switching circuits is shaded (a). The total bypass current is proportional to the reduction ratio of the solar radiation intensity a_{ratio} (c)

converter circuit controls the output voltage and the output current. From (4), if the bypass current I_{bypass} is over 50% of the output current I_{PV}, the shaded panel (PV3) is connected to the adjacent panel (PV2) in parallel. The second step is the scanning of the maximum power point by varying the duty ratio of the PWM signals from 0 to 1. The third step is the MPPT control by the hill climbing method. In this demonstration, the authors confirm the fist and the second steps of the control procedure.

2028

Fig. 13(c) shows the test results of the partial shading detection and the series/parallel switching operation. From 0 to 1 seconds, the circuit condition is the open circuit for setting the offsets of the current sensors. In this period, the output voltage is 35 V. This voltage shows that three photovoltaic panels are connected in series. From 1 to 2 seconds, the short circuit test is carried out. In this period, the bypass current is over 50% of the output current. Then the switching circuit is changed from the series operation mode to the parallel operation mode at 2 seconds. From 2 to 4.5 seconds, the scanning of the maximum power point is carried out. After the scanning of the maximum power point, the output voltage becomes 23 V. This voltage shows that the shaded panel is connected to the adjacent panel in parallel.

VI. CONCLUSIONS

In order to minimize the output power reduction of the photovoltaic system due to the partial shading, the feasibility of the series/parallel switching circuits has been discussed. This system tries to extract the output power from the shaded panel by connecting the shaded panel to the adjacent panel in parallel. From the experimental results,

1) When one of the photovoltaic panels is shaded at least, 50% of the reduction ratio of the solar radiation intensity becomes the lower limit of the threshold values for the series/parallel switching control, and

2) The effect of the series/parallel switching control can be more expected when more than two of the photovoltaic panels are shaded. However, the sum of the solar radiation intensity applied to the photovoltaic panels in the parallel circuit should be less than 150% of the maximum intensity of the photovoltaic power system, and

3) The shaded panel which should be connected to the adjacent panel in parallel can be detected by the bypass current measurements.

The authors developed a prototype of a switching circuit module using power MOSFETs and carried out the operation tests using three photovoltaic panels. From the test results, the switching circuit module prototype successfully detects the partial shading conditions by measuring the bypass current and switches from the series operation mode to the parallel operation mode.

For the further steps of this work, the power improvement effect by the series/parallel switching control during the MPPT operation should be compared to that in the conventional photovoltaic power system. Addition to this, the number of photovoltaic panels connected in parallel should be optimized from the viewpoint of the required input DC voltage of the power conditioning system for the grid connection.

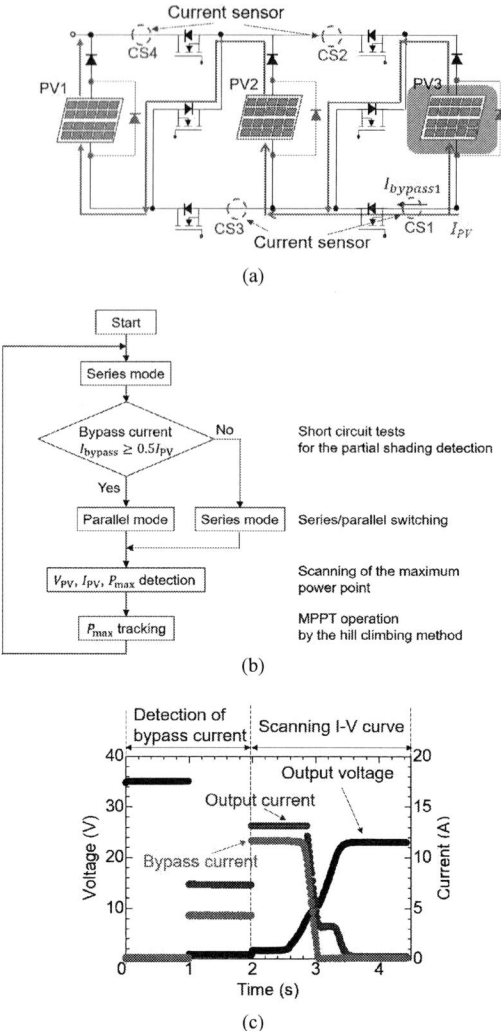

Fig. 13. Experimental circuit (a), the control procedure (b) and the test results (c) of the series/parallel switching circuit module prototype. V_{PV} and I_{PV} are the terminal voltage and the total current of the photovoltaic module, respectively. P_{max} is the maximum power point.

REFERENCES

[1] V. d'Alessandro, P. Guerriero, S. Daliento, "A simple bipolar transistor-based bypass approach for photovoltaic modules," *IEEE Journal of Photovoltaic*, vol. 4, no. 1, pp. 405-413, 2014.

[2] G. R. Walker, P. C. Serina, "Cascaded DC-DC converter connection of photovoltaic modules," *IEEE Trans. on Power Electronics*, vol. 19, no. 4, pp. 1130-1139, 2004.

[3] E. Roman, R. Alonso, P. Ibanez, S. Elorduizapatarietxe, D. Goitia, "Intelligent PV module for grid-connected PV systems," *IEEE Trans. on Industrial Electronics*, vol. 53, no. 4, pp. 1066-1073, 2006.

[4] T. Igarashi, T. Kumano, H. Hayashiya, T. Takino, "Efficiency improvement of rooftop photovoltaic system at railway station," *J. International Council on Electrical Engineering*, vol. 7, no. 1, pp. 41-50, 2017.

[5] K. Matsumoto, S. Nomura, "Performance improvement of photovoltaic power generation systems using on-off control methods," *in Proc. 2014 International Power Electronics Conference (IPEC-Hiroshima 2014 – ECCE ASIA)*, pp. 3218-3224, 2014.

Modularized Equalization Architecture Based on Switched Capacitor Converter to Virtually Unify Mismatched Photovoltaic Panel Characteristics

Masatoshi Uno and Masaya Yamamoto
Ibaraki University, Hitachi, Japan
*E-mail: masatoshi.uno.ee@vc.ibaraki.ac.jp

Abstract—Photovoltaic (PV) strings consisting of multiple panels connected in series are well known to suffer from partial shading or characteristic mismatch issues, such as significant reduction in power yield and occurrence of multiple power point maxima. Although various kinds of voltage equalizers have been developed to prevent the negative impacts of partial shading, conventional equalizers face variety of challenges, such as impaired extendibility and increased voltage stress of circuit elements. This paper proposes a novel switched capacitor converter (SCC)-based modular equalization architecture where PV panels are equalized in two layers—module and panel layers. A prototype for eight panels, which were grouped into two modules, was built, and equalization tests emulating a partial shading condition were performed. With the modular equalization system, the extractable maximum power was significantly increased, demonstrating the efficacy of the proposed modular equalization architecture.

Keywords—*Modularization, photovoltaic panel, partial shading, switched capacitor converter, voltage equalizer.*

I. Introduction

Applications of photovoltaic (PV) panels are rapidly expanding from residential rooftops to solar power plants. Electrical characteristic mismatch of substrings in a PV panel due to partial shading is well known to trigger serious issues. Total energy yield from a PV panel, whose substring characteristics are mismatched, is significantly reduced as a panel current detours through a bypass diode for the weakest substring [1]. In addition, the characteristic mismatch generates multiple power point maxima in its P–V characteristic curve, confusing ordinally maximum power point tracking (MPPT) algorithms.

To enhance energy yield from PV panels consisting of series-connected substrings, differential power processing (DPP) converters or voltage equalizers (hereafter, simply call equalizers) have been vigorously developed [2]–[20]. A fraction of unshaded substrings' power is transferred to shaded ones through voltage equalizers so that all substrings operate at the same voltage or even at each MPP, virtually unifying all substring characteristics.

In general, to obtain high voltage, multiple PV panels are connected in series to form a string. Characteristics of series-connected panels are often mismatched due to not only partial shading but also uneven ageing, generating the same issues as partial shading. The negative influence of the characteristic mismatch in PV strings can be precluded by DPP converters or voltage equalizers.

Most conventional voltage equalizers have been developed aiming for substring-level equalization—all substring characteristics in a panel are unified by voltage equalizers. For panel-level equalization, conventional voltage equalizers face variety of challenges, such as increased collective power conversion loss, impaired extendibility (or modularity), and increased voltage stress of circuit elements, as will be discussed in Section II.

This paper proposes a novel modular equalization architecture based on switched capacitor converters (SCCs). The proposed equalization architecture realizes good extendibility, mitigated voltage stress of circuit elements, and miniaturized circuit design at a time. Section II briefly reviews conventional equalization architectures. The concept of the proposed modular equalization architecture and its practical circuit implementation will be described in Section III, followed by simulation analysis for a dc equivalent circuit. Equalization tests emulating a partial shading condition was performed using a prototype for eight panels, and its results will be shown in Section V.

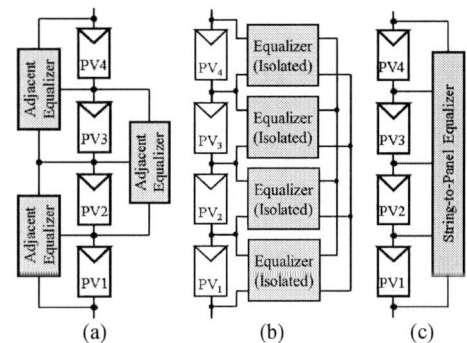

Fig. 1. Conventional equalization architectures: (a) Adjacent panel-to-panel equalization, (b) isolated direct panel-to-panel equalization, (c) string-to-panel equalization.

II. Voltage Equalization Architecture

Conventional DPP converters and voltage equalizers are roughly categorized into three groups: adjacent panel-to-panel equalizers, isolated direct panel-to-panel equalizers, string-to-panel equalizers.

The adjacent panel-to-panel equalization architecture shown in Fig. 1(a) is the most straightforward system. Equalizers transfer power only between adjacent two panels so that all panel characteristics are virtually unified. Bidirectional PWM converters [2]–[6], multi-stage choppers [7], [8], and ladder-type switched capacitor converters [see Fig. 2(a)] [9]–[12] are categorized into this architecture. The number of panels in this system can be arbitrary extended by simply adding panels as well as equalizers, hence offering good extendibility. A major drawback is the collective power conversion loss due to multiple power conversions, which tends to soar with a panel count in a string—for example, power from PV_4 must traverse three equalizers and two panels before reaching PV_1 in Fig. 1(a), collectively increasing the power conversion loss.

The isolated direct panel-to-panel equalization system [see Fig 1(b)] is based on the use of multiple isolated bidirectional converters [13]–[15]. Although this architecture allows flexible power transfer among panels without suffering from the cumulative power conversion loss, the need for numerous isolated converters is a major drawback as each equalizer contains a bulky and relatively expensive transformer.

A string-to-panel equalizer used in Fig. 1(c) is essentially a single-input multi-output converter, such as a multi-winding flyback converter [16], multi-stacked buck-boost converters [17], [18], and resonant voltage multipliers [19], [20]. The number of equalizers can be reduced to one, hence simplifying the system and reducing the cost. However, switches with high voltage rating are necessary for the string-to-panel equalization system because switches in these equalizers must be rated for full string voltage. In addition, since the input voltage of these equalizer is equal to a string voltage or sum voltage of series-connected panels, the equalizers need to be redesigned when the number of panels in series changes. In other words, poor extendibility is a drawback of this equalization system.

III. Proposed Modular Equalization Architecture Based on Switched Capacitor Converters

A. Switched Capacitor Converter

The proposed modular equalization architecture is based on modularized switched capacitor converters (SCCs). Although a variety of SCCs have been developed [21], [22], two SCC topologies shown in Fig. 2 can be used as voltage equalizers for series-connected PV panels. High- and low-side switches in both topologies operate with 50% duty cycle in a complementary mode.

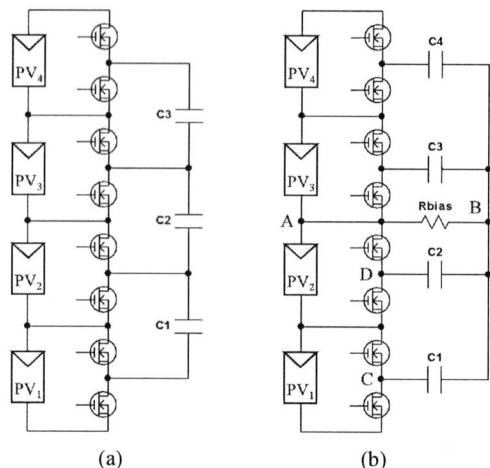

Fig. 2. Switched capacitor converters for (a) adjacent panel-to-panel equalization and (b) nonisolated direct panel-to-panel equalization.

The ladder-type SCC shown in Fig. 2(a) is one of the most popular adjacent equalizer. Voltage ratings of switches and capacitors are equal to a panel voltage V_{PV}. The number of panels connected in series can be arbitrary extended by simply stacking switches and capacitors, offering good extendibility. However, power transfer is limited between adjacent panels, and hence, power conversion loss would become collectively large due to multiple power conversions. Power from PV_1, for instance, has to be transferred via C_1–C_3 before reaching PV_4. The collective power conversion loss would be even significant in high-voltage string comprising numerous panels connected in series.

Meanwhile, the SCC shown in Fig. 2(b) is equivalent to a nonisolated direct panel-to-panel equalizer. Since capacitors are connected in series in this topology, the bias resistor R_{bias} is used to stabilize capacitor voltages. All capacitors are connected to the common node B, and therefore, power is transferred between any two panels through two capacitors. For example, power from PV_1 can reach PV_4 through only C_1 and C_4, reducing the power conversion stages compared to the SCC shown in Fig. 2(a).

Although this direct power transfer realizes efficient equalization, voltage stresses of capacitors tend to increase with the number of panels. Voltage stresses of C_1 and C_2 are briefly determined as an example. The average potential at node B is equal to that of node A thanks to R_{bias}. Average potentials at nodes C and D are $V_{PV}/2$ and $3V_{PV}/2$ (V_{PV} being an equalized panel voltage), respectively, when switches operate with 50% duty cycle. Hence, average voltages of C_1 and C_2 are determined to be $3V_{PV}/2$ and $V_{PV}/2$, respectively.

This tendency suggests that voltage stresses of outer capacitors (i.e., C_1 and C_4), which are placed far from the middle point B, are high, and vice versa for inner capacitors (i.e., C_2 and C_3). In addition, since capacitor

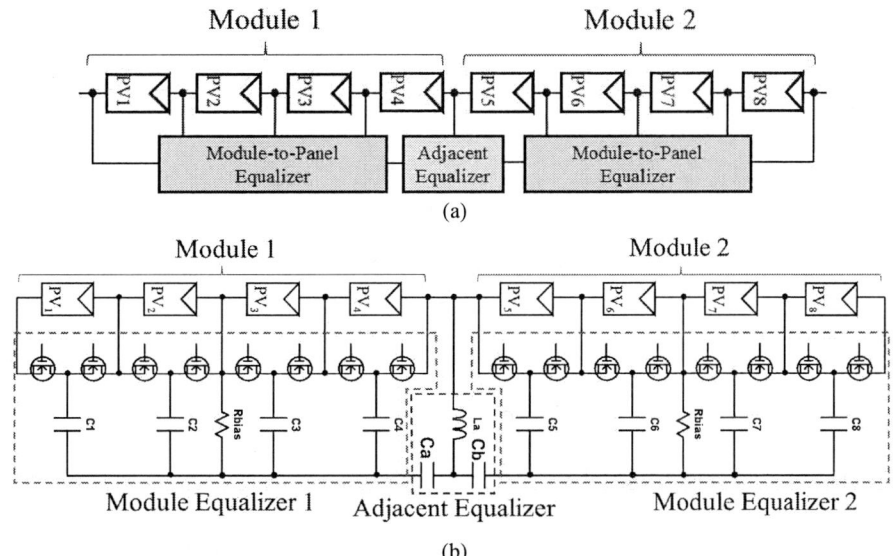

(a)

(b)

Fig. 3. Proposed modular equalization architecture: (a) Notional system, (b) practical implementation for two modules each comprising four panels connected in series.

voltages soar as the number of panels grows, reselection for capacitors is unavoidable for strings comprising a different number of panels. Thus, the number of panels cannot be readily changed with this SCC, impairing the extendibility of the system.

B. Proposed Modular Equalization Architecture

The notional system of the proposed modular equalization architecture is illustrated in Fig. 3. In this example, four panels are grouped as a module having a module-to-panel equalizer. Subsequently, modules are connected in series to form a string. Modules with a module equalizer are connected through an adjacent equalizer.

Characteristic mismatch in each module (i.e., mismatch in panel characteristics) is efficiently equalized by each module-to-panel equalizer, whereas the adjacent equalizer transfers power between two adjacent modules so that characteristic mismatch between modules is eliminated. In other words, PV panels in the modular architecture are equalized in two layers—module and panel layers.

The practical implementation of the proposed modular equalization architecture is shown in Fig. 3(b). Each module-to-panel equalizer employs the SCC shown in Fig. 2(b) that operates with a fixed 50% duty cycle. The adjacent equalizer, on the other hand, is a switchless topology consisting of two capacitors (C_a and C_b) and one inductor L_a. C_a and C_b are connected in series in order to

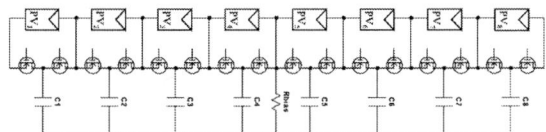

Fig. 4. SCC-based nonisolated direct panel-to-panel equalizer for eight panels.

halve their voltage stress, and their voltages are completely equalized by L_a at 50% duty cycle operation. The halved voltage stress of C_a and C_b realizes all-ceramic-capacitor circuit—if not, the voltage stress of capacitors might be as high as string voltage, and bulky film capacitors with high voltage rating would be necessary. Although one inductor is necessary in an adjacent equalizer, a small inductor with low current rating suffices because of a small bias current under steady-state conditions.

C. Voltage Stress of Capacitors

Voltage stresses of capacitors in the conventional SCC-based nonisolated direct panel-to-panel equalizer (Fig. 4) and proposed modular equalization system [Fig. 3(b)] are compared in Table I. All panel voltages are assumed to be equalized as 40 V in this comparison. Voltage stresses of outer capacitors of C_1 and C_8 in the conventional equalizer are the highest, while those of inner capacitors of C_4 and C_5 are low. This tendency is also true in each module in the proposed modular equalization system; outer capacitors in each module (C_1 and C_4 in Module 1, and C_5 and C_8 in Module 2) are exposed to relatively high voltage stress. However, their

TABLE I
VOLTAGE STRESS OF CAPACITORS

Capacitor	Conventional	Proposed
C_1	140 V	60 V
C_2	100 V	20 V
C_3	60 V	20 V
C_4	20 V	60 V
C_5	20 V	60 V
C_6	60 V	20 V
C_7	100 V	20 V
C_8	140 V	60 V
C_a, C_b	—	80 V

2032

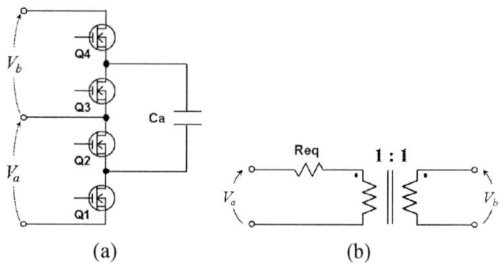

(a)   (b)

Fig. 5. (a) Basic SCC, (b) dc equivalent circuit.

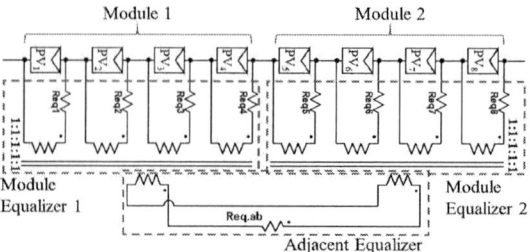

Fig. 6. DC equivalent circuit of proposed SCC-based modular equalization system for eight panels.

voltage stress is rather lower than that in the conventional equalizer thanks to C_a and C_b in the switchless adjacent equalizer.

Outer capacitors in the conventional SCC equalizer are exposed to high voltage stress of 140 V, and hence, bulky film capacitors are likely necessary. In the proposed modular equalization system, on the other hand, all capacitor voltage stresses are lower than 80 V. The reduced voltage stresses allow all-ceramic-capacitor circuit, achieving miniaturized circuit design.

IV. DC EQUIVALENT CIRCUIT AND ITS SIMULATION RESULTS

A. DC Equivalent Circuit

Simulation-based analysis for PV strings employing an MPPT algorithm takes a long stretch of time because of the huge difference between switching period and MPPT sampling interval—switching periods are around 10 μs (equivalent to 100-kHz switching frequency), whereas sampling intervals can be longer than hundreds of milliseconds.

In general, charge and discharge of a capacitor in SCCs can be equivalently expressed as an equivalent resistance that is inversely proportional to a capacitance and frequency. A basic SCC shown in Fig. 5(a) can be transformed into a dc equivalent circuit shown in Fig. 5(b) [12]. The value of R_{eq} [23] is given by

$$R_{eq} = \frac{1}{Cf_s} \frac{\exp\left(\dfrac{T}{\tau}\right) - 1}{\left\{\exp\left(\dfrac{dT}{\tau}\right) - 1\right\}\left\{\exp\left(\dfrac{(1-d)T}{\tau}\right) - 1\right\}}, \quad (1)$$

where C is the capacitance, f_S ($= 1/T$) is the switching frequency, d is the duty cycle, and τ is the time constant of capacitors. A dc equivalent circuit would significantly reduce the simulation burden and time for PV systems operating with MPPT.

The dc equivalent circuit of the proposed modular SCC equalization system is shown in Fig. 6. PV panels in each module are virtually connected in parallel through respective R_{eq} and an ideal multi-winding transformer. Meanwhile, the adjacent equalizer can also be expressed as $R_{eq.ab}$ that represents C_a and C_b. Overall, all panels are

virtually connected in parallel through R_{eq}, and therefore, their voltages are automatically nearly unified as long as voltage drops across R_{eq} are satisfactory small.

B. Simulation Results

A simulation-based equalization test was performed using the dc equivalent circuit shown in Fig. 7. The value of R_{eq} was determined to be 0.25 Ω, according to (1) and the conditions used for the experiments (see next section). Individual panel characteristics used for the simulation are shown in Figs. 7(a) and (b). In addition to the imbalanced panel characteristics in each module, module characteristics were also mismatched in order to verify the equalization performance of the proposed module equalization architecture—one and two panels in Modules A and B were mismatched, respectively.

Measured string characteristics with/without equalization are shown and compared in Fig. 7(c). Without the voltage equalizer, two power point maxima, including one global and one local MPP, were observed, and its extractable maximum power was merely 1020 W at 300 V. With the equalizer, on the other hand, the local MPP vanished, and the maximum power increased to as high as 1220 W, corresponding to 19.6% improvement.

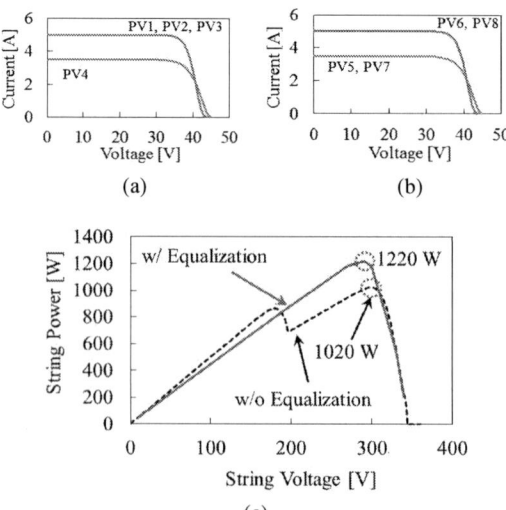

Fig. 7. DC equivalent circuit-based simulation results: (a) individual panel characteristics in Module 1, (b) individual panel characteristics in Module 2, (c) string characteristics with/without equalization.

2033

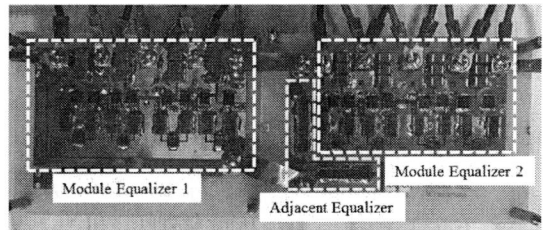

Fig. 8. Prototype of proposed modular equalization system for eight PV panels.

TABLE II
COMPONENT LIST

Component	Value
C_{out1}–C_{out8}	Ceramic Capacitor, 22 µF × 3, 50 V
C_1–C_8	Ceramic Capacitor, 10 µF × 5, 100 V
C_a, C_b	Ceramic Capacitor, 10 µF × 5, 100 V
L	1 mH
MOSFET	BSC320N20NS3G, R_{on} = 36 mΩ

V. EXPERIMENTAL RESULTS

A. Prototype

A prototype of the proposed modular SCC-based voltage equalizer for two modules, each comprising four panels, was built, as shown in Fig. 8. Components used for the prototype are listed in Table II, in which C_{out1}–C_{out8} are smoothing capacitors that are connected in parallel with panels (not illustrated in figures). Module equalizers were separately built and subsequently connected in series using copper plates. The switchless adjacent equalizer was mounted on one of the module equalizers (Module Equalizer 2). The prototype was operated at 100 kHz with 50% fixed duty cycle.

B. Equalization Test

The characteristic mismatch condition due to partial shading was emulated using solar array simulators (Keysight Technologies, E4361A). Individual panel characteristics in each module are shown in Figs. 9(a) and (b)—the characteristic of PV_8 was emulated using a constant-current–constant-voltage source with series and parallel resistors due to a lack of solar array simulators.

Measured string characteristics with/without the prototype are compared in Fig. 10. Similar to the simulation results shown in Fig. 7(c), the string characteristic without the equalization exhibited local MPPs, and its maximum power at the global MPP was 1050 W. Meanwhile, the voltage equalizer successfully eliminated the local MPPs, and the extractable maximum power increased to as high as 1281 W, corresponding to 22% improvement in power yield.

Measured capacitor voltages when the string operated at its MPP are shown in Table III. The observed tendency is very similar to that shown in Table I. Voltages of C_a and C_b were nearly identical, demonstrating the operation of the switchless adjacent equalizer.

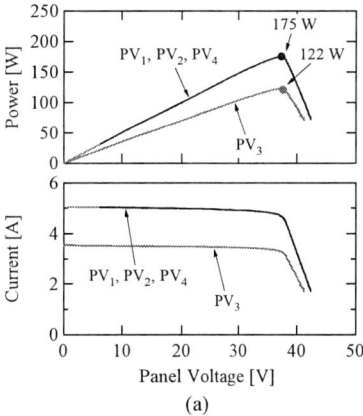

(a)

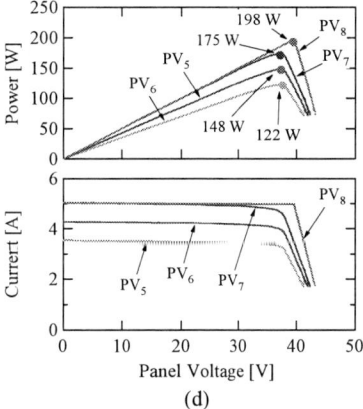

(d)

Fig. 9. Individual panel characteristics in (a) Module 1 and (b) Module 2.

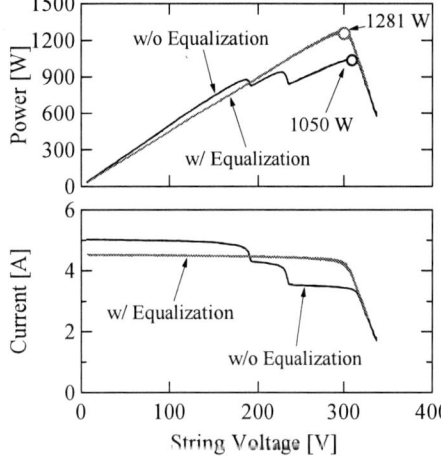

Fig. 10. Measured string characteristics with/without equalization.

VI. CONCLUSIONS

The SCC-based modular equalization architecture for PV strings has been proposed in this paper. SCC-based nonisolated direct panel-to-panel equalizers are connected through switchless adjacent equalizers,

TABLE III
MEASURED CAPACITOR VOLTAGES OF PROPOSED EQUALIZER
WHEN STRING OPERATED AT MPP

Capacitor	Voltage
C_1	57.1 V
C_2	19.2 V
C_3	18.3 V
C_4	55.9 V
C_5	56.2 V
C_6	18.8 V
C_7	18.0 V
C_8	54.5 V
C_a	75.1 V
C_b	74.3 V

achieving good modularity without suffering from high voltage stress of capacitors.

The prototype for the PV string consisting of two modules, each comprising four panels, was built, and the equalization test was performed emulating partial shading conditions. With the support of the proposed modular equalizer, local MPPs in the measured string characteristics successfully disappeared. In addition, the extractable maximum power dramatically increased, demonstrating the efficacy of the proposed modular voltage equalization system.

REFERENCES

[1] S. M. MacAlpine, R. W. Erickson, and M. J. Brandemuehl, "Characterization of power optimizer potential to increase energy capture in photovoltaic systems operating under nonuniform conditions," *IEEE Trans. Power Electron.*, vol. 28, no. 6, pp. 2936–2945, Jun. 2013.

[2] H. J. Bergveld, D. Büthker, C. Castello, T. Doorn, A. D. Jong, R. V. Otten, and K. D. Waal, "Module-level dc/dc conversion for photovoltaic systems: the delta-conversion concept," *IEEE Trans. Power Electron.*, vol. 28, no. 4, pp. 2005–2013, Apr. 2013.

[3] S. Qin, C.B. Barth, and R.C.N.P. Podgurski, "Enhancing microinverter energy capture with submodule differential power processing," *IEEE Trans. Power Electron.*, vol. 31, no. 5, pp. 3575–3585, May 2016.

[4] P. S. Shenoy, K. A. Kim, B. B. Johnson, and P. T. Krein, "Differential power processing for increased energy production and reliability of photovoltaic systems," *IEEE Trans. Ind. Power Electron.*, vol. 28, no. 6, pp. 2968–2979, Jun. 2013.

[5] S. Qin, S.T. Cady, A.D.D. García, and R.C.N.P. Podgurski, "A distributed approach to maximum power point tracking for photovoltaic submodule differential power processing," *IEEE Trans. Power Electron.*, vol. 30, no. 4, pp. 2024–2040, Apr. 2015.

[6] S. Qin, C. B. Barth, Student, and R. C. N. P. Podgurski, "Enhancing microinverter energy capture with submodule differential power processing," *IEEE Trans. Power Electron.*, vol. 31, no. 5, pp. 3575–3585, May 2016.

[7] T. Shimizu, O, Hashimoto, and G. Kimura, "A novel high-performance utility-interactive photovoltaic inverter system," *IEEE Trans. Power Electron.*, vol. 18, no. 2, pp. 704–711, Mar. 2003.

[8] T. Shimizu, M. Hirakata, T. Kamezawa, and H. Watanabe, "Generation control circuit for photovoltaic modules," *IEEE Trans. Power Electron.*, vol. 16, no. 3, pp. 293–300, May 2001.

[9] J. T. Stauth, M. D. Seeman, and K. Kesarwani, "Resonant switched-capacitor converters for sub-module distributed photovoltaic power management," *IEEE Trans. Power Electron.*, vol. 28, no. 3, pp. 1189–1198, Mar. 2013.

[10] A. H. Chang, A. T. Avestruz, and S. B. Leeb, "Capacitor-Less Photovoltaic Cell-Level Power Balancing using Diffusion Charge

Redistribution," *IEEE Trans. Power Electron.*, vol. 30, no. 2, pp. 537–546, Feb. 2015.

[11] Z. Qiu and K. Sun, "A Photovoltaic Generation System Based on Wide Voltage-Gain DC-DC Converter and Differential Power Processors for DC Microgrids," *Chinese J. Electrical Eng.*, vol. 3, no. 1, pp. 84–95, Jun. 2017.

[12] M. Uno and A. Kukita, "PWM Converter Integrating Switched Capacitor Converter and Series-Resonant Voltage Multiplier as Equalizers for Photovoltaic Modules and Series-Connected Energy Storage Cells for Exploration Rovers," *IEEE Trans. Power Electron.*, vol. 32, no. 11, pp. 8500‒8513, Nov. 2017.

[13] C. Olalla, D. Clement, M. Rodríguez, and D. Makisimović, "Architectures and control of submodule integrated dc-dc converters for photovoltaic applications," *IEEE Trans. Power Electron.*, vol. 28, no. 6, pp. 2980–2997, Jun. 2013.

[14] C. Olalla, C. Deline, D. Clement, Y. Levron, M. Rodríguez, and D. Makisimović, "Performance of power limited differential power processing architectures in mismatched PV systems," *IEEE Trans. Power Electron.*, vol. 30, no. 2, pp. 618–631, Feb. 2015.

[15] G. Chua, H. Wena, L. Jiangb, Y. Hub, and X. Lia, "Bidirectional flyback based isolated-port submodule differential power processing optimizer for photovoltaic applications," *Solar Energy*, vol. 158, pp. 929–940, Oct. 2017.

[16] J. Du, R. Xu, X. Chen, Y. Li, and J. Wu, "A novel solar panel optimizer with self-compensation for partial shadow condition," in Proc. *IEEE Applied Power Electron. Conf. Expo., APEC*, pp. 92–96, 2013.

[17] M. Uno and A. Kukita, "Single-switch voltage equalizer using multi-stacked buck-boost converters for partially-shaded photovoltaic modules," *IEEE Trans. Power Electron.*, vol. 30, no. 6, pp. 3091–3105, Jun. 2015.

[18] M. Uno and A. Kukita, "Current sensorless equalization strategy for a single-switch voltage equalizer using multistacked buck–boost converters for photovoltaic modules under partial shading," *IEEE Trans. Ind. Appl.*, vol. 53, no. 1, pp. 420–429, Jan./Feb. 2017.

[19] M. Uno and A. Kukita, "Two-switch voltage equalizer using an LLC resonant inverter and voltage multiplier for partially-shaded series-connected photovoltaic modules," *IEEE Trans. Ind. Appl.*, vol. 51, no. 2, pp. 1587–1601, Mar./Apr. 2015.

[20] M. Uno and A. Kukita, "Single-switch single-magnetic PWM converter integrating voltage equalizer for partially-shaded photovoltaic modules in standalone applications," *IEEE Trans. Power Electron.*, vol. 33, no. 2, pp. 1259–1270, Feb. 2018.

[21] Y. Ye, K.W.E. Cheng, Y.C. Fong, X. Xue, and J. Lin, "Topology, modeling, and design of switched-capacitor-based cell balancing systems and their balancing exploration," *IEEE Trans. Power Electron.*, vol. 32 no. 6, pp. 4444–4454, Jun. 2017.

[22] Y. Shang, B. Xia, F. Lu, C. Zhang, N. Cui, and C.C. Mi, "A switched-coupling-capacitor equalizer for series-connected battery strings," *IEEE Trans. Power Electron.*, vol. 32, no. 10, pp. 7694–7706, Oct. 2017.

[23] J. W. Kimball, P. T. Krein, and K. R. Cahill, "Modeling of capacitor impedance in switching converters," *IEEE Power Electron. Lett.*, vol. 3, pp. 136–140, Dec. 2005.

Buck-Boost Type MPPT Circuit Suitable for Photovoltaic Generation of Vehicle Installation

Fumihisa Kano [1,2*], Yuji Kasai [3], Hideki Kimura [4], Kouhei Sagawa [4], Junnosuke Haruna [1] and Hirohito Funato [1].

1 Dept. of Electr. & Electron. Eng., Utsunomiya Univ., Utsunomiya, Japan
2 Dept. of Innovative Electr. & Electron. Eng., Nat. Inst. of Tech., Oyama College, Oyama, Japan
3 Artificial Intelligence Research Center, Nat. Inst. of Advanced Industrial Science and Tech., Tsukuba, Japan
4 Dept. of Electr. & Electron. Eng., Tokai Univ., Hiratsuka, Kanagawa, Japan
*E-mail: kano@oyama-ct.ac.jp

Abstract— This paper proposes a buck-boost type MPPT circuit suitable for photovoltaic generation of solar car. By using an analog circuit for MPPT control, high conversion efficiency and weight reduction could be achieved by an accurate maximum power point tracking of photovoltaic generation. In order to reduce power conversion loss on wide voltage variation of the battery, buck-boost type dc-dc converter is used as a main power converter of the MPPT circuit. The proposed MPPT circuit which led to 12 sets of series-parallel was used for the solar car named "2017 Tokai Challenger" for the race and confirmed advantageous photovoltaic power generation.

Keywords— solar cell, photovoltaic generation, MPPT, solar car

I. INTRODUCTION

When installing on a moving vehicle such as the solar car, the solar power generation is affected by the change in direction and the shadow of obstacle. Compared with fixed type for industrial and residential use, the intensity of sunlight changes rapidly on the mobile object, and time change in power generation becomes faster[1,2]. When a Maximum Power Point Tracking (MPPT) controller circuit for photovoltaic generation fixed to the ground is applied to the photovoltaic generation of such a vehicle, the maximum power point tracking of the photovoltaic generation which suddenly fluctuates is difficult. Especially when solar cells are attached to a streamlined vehicle body like a solar car for race, since the normal plane orientation of the solar cells are different from each other, it is difficult to match the same power generation characteristics. Therefore, it is possible to obtain more power generation by connecting MPPT circuit to solar cells with the same plane orientation in order to minimize the influence of solar cells with different power generation characteristics[3,4]. However, use of several MPPT circuits to divide solar cells increases the power loss and weight of the circuit and affects the running performance of the vehicle. As a demand for the MPPT control circuit for installing the moving body, power saving, light weight, and compactness of the circuit are required together with accurate follow-up performance of the optimum

operating point[5]. In general, realizing an MPPT circuit by an analog control method which is not affected by quantization error can be realized by using many multipliers and bipolar elements, but power consumption of the control circuit becomes a problem[6,7,8,9].

In this paper, we have developed a simple MPPT circuit suitable for photovoltaic generation of solar cars combining power saving and weight saving with precise solar cell's optimal operating point tracking. The MPPT controller of the proposed circuit is composed of one analog switch and two operational amplifiers. A buck-boost converter is used as a main power converter in order to reduce power loss in high power region with wide variation of battery voltage. The proposed MPPT circuit was used for solar cars "2017 Tokai Challenger" on BWSC2017 race in order to confirm the effectiveness of solar power generation.

II. OPTIMUM OPERATING POINT DETECTION BY ANALOG METHOD

Fig. 1 shows the proposed buck-boost type MPPT circuit with analog control method. The circuit consists of a current detection circuit, a discrimination circuit, an integration circuit, a switching pulse generation circuit,

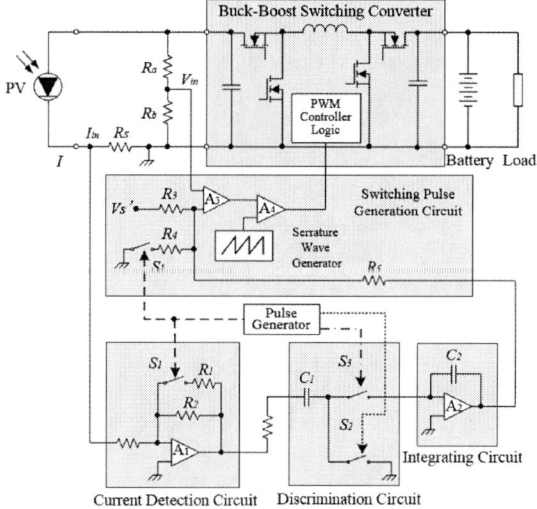

Fig. 1 Schematic Diagram of buck-boost type MPPT Circuit.

a pulse generation circuit of a modulation signal, and a buck-boost type switching converter.

Fig. 2 and Fig. 3 show representative characteristic curves of power-voltage (PV) and current-voltage (IV) characteristics of a solar cell. MPPT control operates by modulating the solar cell voltage with a modulation factor m of a switching converter. In two points of power comparison methods, the power Pb at voltage Vb is compared with the power Pa at voltage Va. Each power, Pa is power in voltage Va and current Ia, and the Pb is power in voltage Vb and current Ib. This Vb is voltage reduced for $k = 1 - m$ times by voltage Va. The target operating point voltage is increased when the power is $Pb > Pa$, and the voltage is reduced when the power is $Pa > Pb$. This power comparison is repeated and reaches the maximum power point of the PV curve top when power becomes $Pb = Pa$. The powers of Pa and Pb are calculated as equations (1) and (2) respectively.

$$Pa = Ia \times Va \quad\cdots\cdots\cdots\cdots\cdots\cdots\cdots\cdots\cdots\cdots (1)$$
$$Pb = Ib \times Vb = Ib \times k \times Va \quad\cdots\cdots\cdots\cdots\cdots (2)$$

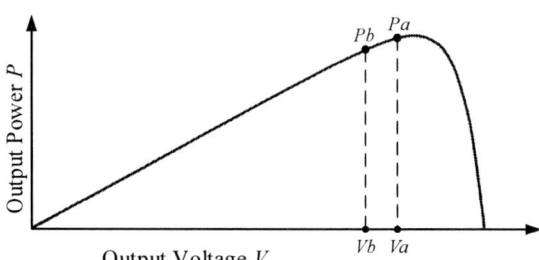

Fig. 2 Output Power vs Output Voltage of PV Generation.

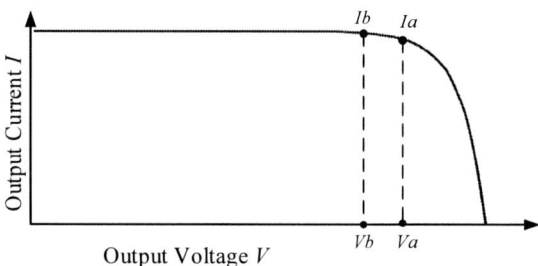

Fig. 3 Output Current vs Output Voltage of PV Generation.

From these equations, it can be seen that the current of optimum operating point is replaced Ia by with $Ib \times k$.

The analog switch in the circuit repeats ON / OFF at the timing signals of the $S1$, $S2$ and $S3$ by the pulse signal of 20 ms period shown in Fig. 4(a). Fig. 4(b) and Fig. 4(c) shows the modulation output voltage of the solar cell and the time change of the current, respectively. In the current detection circuit, the amplification factor of A1 changes according to the $S1$ signal. When the analog switch is OFF, the output voltage V_{OFF} of the operational amplifier A1 is a voltage proportional to the current flowing through Rs shown in the equation (3), assuming that the amplification factor is Aa and the current flowing through the current detection resistor Rs is Ia.

$$V_{OFF} = Aa \times Ia \quad\cdots\cdots\cdots\cdots\cdots\cdots\cdots\cdots\cdots (3)$$

The amplification factor of the operational amplifier A1 changes at the same magnification as the output voltage modulation factor k by switching between $R1$ and $R2$. When the signal of $S1$ is ON, the amplification factor is multiplied by $k \times Aa$ ($k = 1 - m$). Fig. 4 (d) shows the temporal change of the voltage modulation rate k synchronized with the $S1$ signal. If the current flowing in Rs is Ib, the output voltage V_{ON} of the operational amplifier A1 is obtained by the following equation.

$$V_{ON} = k \times Aa \times Ib \quad\cdots\cdots\cdots\cdots\cdots\cdots\cdots\cdots (4)$$

In the discrimination circuit, the output voltage of the current detection circuit is applied to the capacitor $C1$ in synchronization with the $S3$ pulse signal by the pulse signal. When $S1$ is ON, that is, the output voltage of A1 is V_{ON} and the analog switch of the discrimination circuit is ON by $S3$, $C1$ is charged up to V_{ON}.

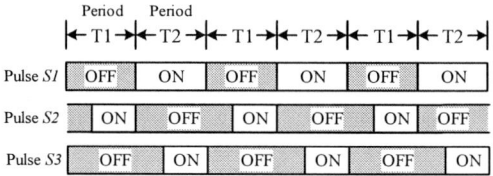

a) Time Sequence Chart for Pulse $S1$, $S2$ and $S3$.

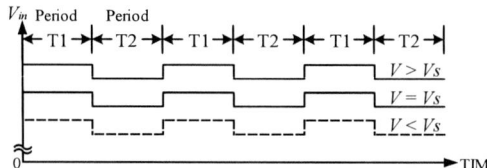

b) The Waveforms of The Photovoltaic Voltages.

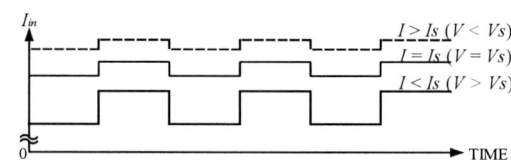

c) The Waveforms of The Photovoltaic Current.

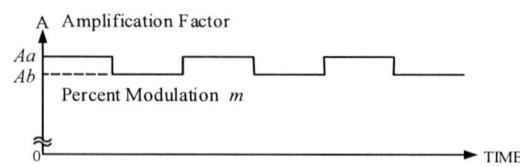

d) The Change of The Amplification Factor of the Current Detection Circuit.

Fig. 4 Time Sequence Chart for MPPT Circuit.

Next, when $S1$ is OFF, that is, $S2$ becomes ON when the output voltage is V_{OFF}, feedback is applied to the input of A2 of the integrating circuit through the capacitor $C2$, and the input voltage of the integrating circuit A2 becomes equal to V_{OFF}. When $S2$ of the analog switch is ON, the voltage of $C1$ is reset. By the ON / OFF operation of the discrimination circuit, a change ΔQ in

2037

the charge accumulated in the capacitor *C1* shown in equation (5) occurs. *ΔQ* becomes equal to the change of the charge in *C2*, and as a result, a voltage change *Vc* occurs at *C2* as shown in equation (6).

$$\Delta Q = V_{\text{OFF}} \times C1 - V_{\text{ON}} \times C1 \quad \cdots\cdots\cdots\cdots\cdots\cdots (5)$$

$$\begin{aligned} Vc &= \Delta Q / C2 = C1 / C2 \times (V_{\text{OFF}} - V_{\text{ON}}) \\ &= C1 / C2 \times (Aa \times Ib - k \times Aa \times Ia) \cdots\cdots\cdots (6) \end{aligned}$$

Vc becomes positive when *Pa* > *Pb* (*Ia* > *k* × *Ib*), negative when *Pa* <*Pb* (*Ia* < *k* × *Ib*), 0 when *Pa* = *Pb* (*Ia* = *k* × *Ib*) respectively. The voltage *Vc* is accumulated in the integrating circuit of the operational amplifier A2, and a voltage proportional to the current change flowing in *Rs* is output as shown in equation (7) according to the switch turn ON / OFF change by the pulse *S1*.

$$Vint = \int Vc \, dt \quad \cdots\cdots\cdots\cdots\cdots\cdots\cdots\cdots (7)$$

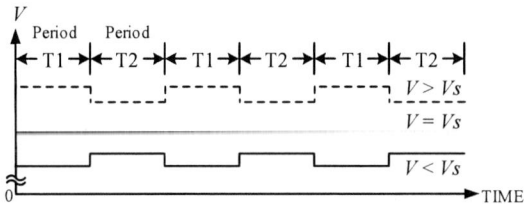

a) The Output Waveforms of The Current Detection Circuit.

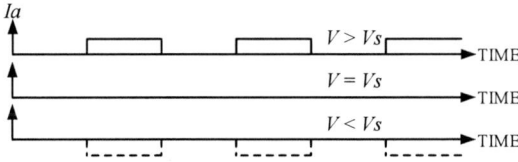

b) The Output Waveforms of The Discriminator Circuit.

Fig. 5 Time Sequence of Current Detection Circuit and Discriminator Circuit.

Fig. 5 (a) and 5 (b) show the output of the current detection circuit and output of the discrimination circuit respectively. The target input voltage *Vs* of the operational amplifier A3 in the switching pulse generating circuit is a voltage obtained by adding the reference voltage *Vs'* and the output voltage *Vint* of the integrating circuit. Voltages *Vs(ON)* and *Vs(OFF)* modulated at the modulation factor *m* shown in equations (8) and (9) are output by ON/OFF of the analog switch connected to *R3*, *R5* and *R4* by the *S1* signal.

$$Vs_{(\text{ON})} = (Vint + Vs') \times k \quad\cdots\cdots\cdots\cdots\cdots (8)$$
$$Vs_{(\text{OFF})} = Vint + Vs' \quad\cdots\cdots\cdots\cdots\cdots\cdots (9)$$

As a result, the control voltage signal for tracking the optimum operating point of the solar cell can be achieved by a simple circuit using two operational amplifiers and an analog switch by switching the amplification factor by the pulse signal. Since the elements necessary for circuit operation can be configured using CMOS elements, it is possible to reduce the power required for the control section operation.

III. IMPLEMENTATION OF MPPT CIRCUIT

When solar radiation strength in the daytime is strong, the optimum operating voltage of the solar cell is low, and it fluctuates highly when solar radiation strength in morning or in the evening. The battery voltage is high in the fully charged state and gradually decreases with the discharge. Fig. 6 shows the outline of voltage change over time for photovoltaic generation and battery. In the case of the boost type MPPT circuit, the boos ratio becomes large in a time zone where the generated power is large, that is, the loss of the circuit becomes large. Therefore, the buck-boost type dc-dc converter matching voltage is required to reduce a power conversion loss in daytime power generation.

Fig. 7 shows the MPPT circuit created as a non-insulated switching dc-dc converter with maximum input of 20V / 8A which performs buck-boost type operation. The weight of the MPPT circuit of analog control type produced is 95g, and the size of the circuit board is 100 × 100mm.

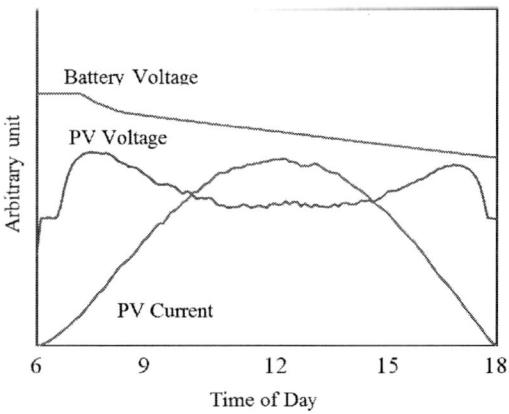

Fig. 6 Outline of PV and Battery Voltage Change.

Fig. 7 Analog Control System MPPT Circuit.

In the performance to follow the optimum operating point of the circuit, there is a correlation between the output voltage modulation factor m and the switching pulse frequency of the analog switch. As the output voltage modulation factor m is reduced and the switching pulse frequency is increased, the ideal follow-up operation becomes possible. On the other hand, due to the influence of the capacitive component of the solar cell and switching converter circuit, it is difficult to follow up exactly. Therefore, this time, the period of switching pulse signal was selected 20ms and the output voltage modulation rate was set to $m = 2\%$ ($k = 98\%$). With the setting voltage range of 9 to 20V, the minimum modulation voltage difference was 0.26V. Fig. 8 and Fig. 9 show the timing signal waveform of the analog switch, the voltage modulation waveform and the monitor current signal respectively.

The switching frequency of the converter was set to 120 kHz by calculation from the balance between the loss due to the FET and the saturation magnetic flux of the inductor.

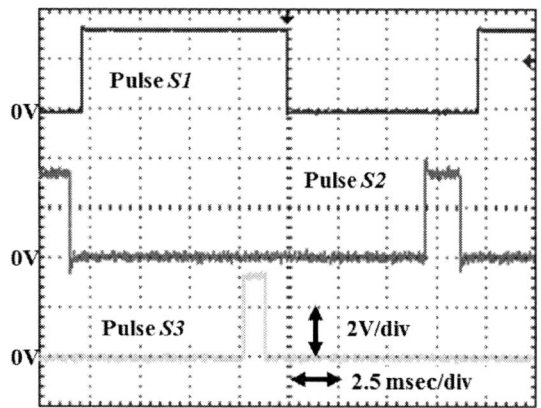

Fig. 8   Timing of Analog Switch Signal Waveforms.

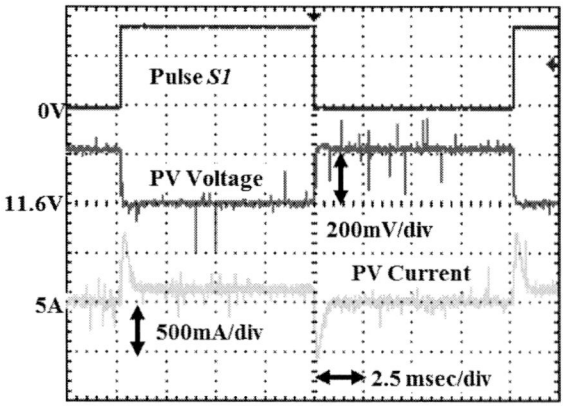

Fig. 9   MPPT circuit Voltage Modulation Waveform and Current Signal.

Fig. 10 shows the power conversion efficiency of the MPPT circuit which measured by varying the input power with the simulated solar cell as input. Power meter

(HIOKI PW3337) with accuracy of ± 0.1% was used for efficiency measurement.

In the case where the input power is low, the control drive power of the MPPT circuit is supplied from the input power, resulting in a decrease in efficiency. However, in the usual input power range of 50W or more, efficiency of 98.5% or more was confirmed. The power consumption of the controller including the FET driving power was measured as 156mW.

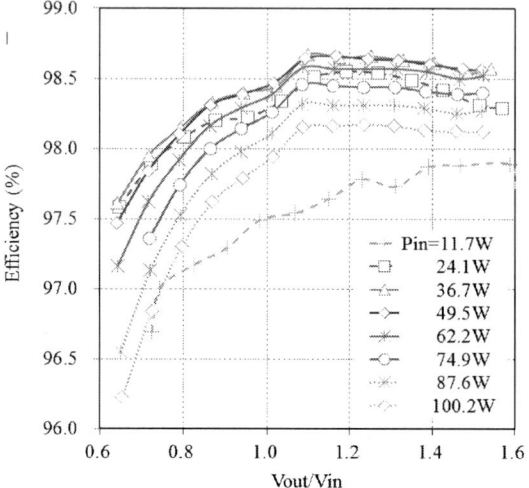

Fig. 10   Conversion Efficiency of MPPT Circuit.

Fig. 11 shows the result that measured tracking error properties of the MPPT control unit using a simulated solar module of the output of 60W. In the figure, maximum power point tracking by the MPPT control shows the result of a measurement that acted to P-V characteristic curve of the maximum power neighborhood that changed the target voltage of the MPPT circuit in manual operation.

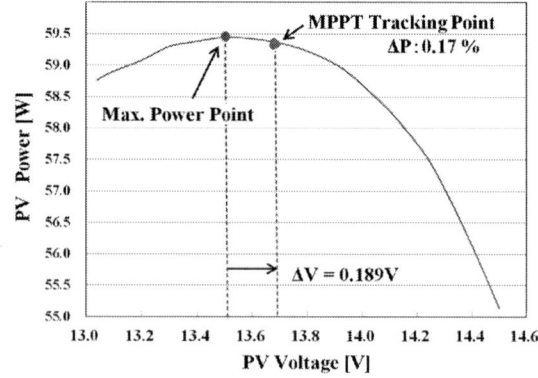

Fig. 11   Tracking Error Properties of MPPT Circuit.

As a result of having performed maximum power point tracking control using MPPT circuit, the voltage at actual operating point was 0.189V higher than the maximum power point voltage of P-V properties as show in a figure.

2039

However, a decrease in power caused by voltage tracking error is 0.17%, and the influence on power by the aim tracking error does not occur.

IV. APPLICATION TO THE SOLAR CAR FOR THE RACE

The solar car of Tokai University used the developed control circuit as MPPT controller, in the race of 2017 BWSC in Australia. Fig. 12 shows the 2017 Tokai Challenger.

Fig. 12 Solar car "2017 Tokai Challenger".

In Tokai Challenger equipped with Panasonic's photovoltaic module HIT (conversion efficiency 24.1%, total output 962W), the solar modules were divided and controlled by using 12 MPPT circuits. For conformity to the Li-ion battery voltage of DC 90V, MPPTs of similar generation properties were connected in 2 parallel 6 series.

The input photovoltaic voltage, current and the output voltage of each MPPT were recorded in sync with the control signal every 10 sec. from a controller by a CAN device on each MPPT circuit. Fig. 13 shows one-day change in the second day of the race. The photovoltaic generation power and the battery voltage were measured by using the data logger. The generation power at the race time was fluctuated by sun direction, cloud situation and shadow in running of the solar car. The battery voltage was changed by charge and discharge condition.

For 12 sets of MPPT circuits , PV power of each solar module is shown in Fig. 14. As for the time for battery charging until 8:00 or 17:00, the car body was turned to the right angle for a direction of the sun. The difference of the power output from each solar cell module controlled by MPPT circuit of this period was around 6W. However, in the running of the race, a difference occurred in the power output of each solar cell module by the motion of the direction to the sun and the influence by the partial shadow of the canopy. Especially, influence was large at the time of state of the angle of the low direction for the sun, and a difference of the generated output of around 30W occurred. Partial shadow generating of the solar cell module has a large influence to power output and shows efficiency of the distributed generation using MPPT circuits.

Fig. 15 shows the voltage change of the solar modules connected with each MPPT circuits in the second day of the race. The solar cell voltage is different since MPPT_01 and MPPT_02 have more number of the cells of the solar modules than other modules.

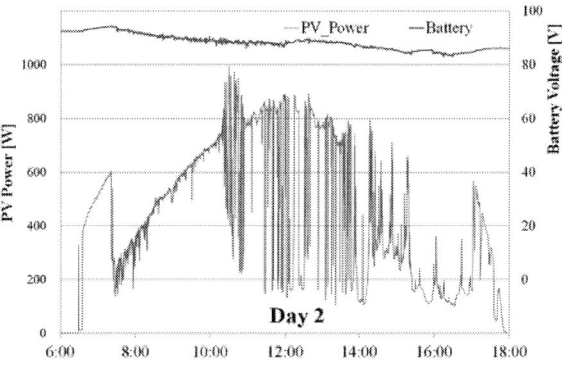

Fig. 13 PV Power and Battery Voltage (Day2).

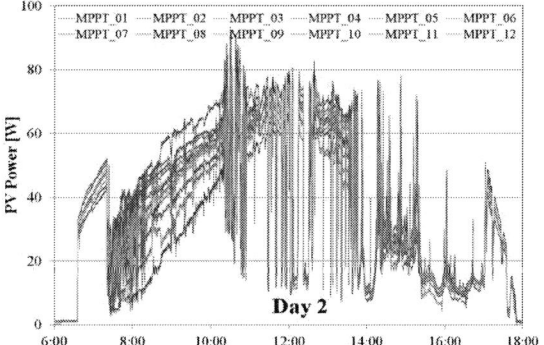

Fig. 14 PV Power of Each Module.

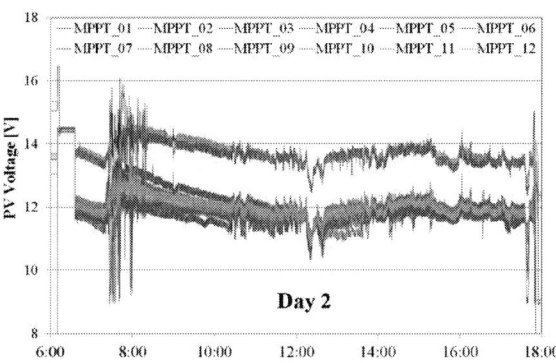

Fig. 15 Voltage of PV which Connected with Each MPPT.

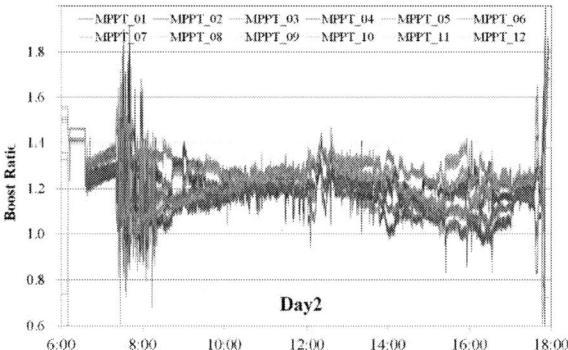

Fig. 16 Boost Ratio of Each MPPT Circuits.

2040

In this result, the difference of the solar cell module voltage occurs under the influence of a solar motion and the partial shadow of the canopy, and, as for the time zone of the sun angle of the high direction, there are few solar cell module voltage differences. The distributed control by MPPT circuit was able to suppress the influence of solar cell modules with different characteristics of solar car power generation.

Fig. 16 shows the boost ratio of the buck-boost operation of the power conversion circuit of each MPPT. The boost ratio was almost in the range from 1 time to 1.4 times. As the capacity of the Li-ion battery decreased, the voltage ratio in the power conversion circuit of MPPT gradually decreased. Furthermore, when there is a large difference in photovoltaic power generation of each solar cell module, the buck operation of several power conversion circuits of MPPT can be recognized. With respect to the power conversion efficiency of these MPPT circuits for solar power generation, a high power conversion efficiency of more than 98.5% for wide operating condition is confirmed by the measurement result of Fig. 10.

V. SUMMARY AND CONCLUSION

In this paper, we achieved MPPT circuit equipped to a vehicle which needed high efficiency, lightness and downsizing realized by adopting the analog circuit system using two op-amp and analog switches. The consumption power of the control circuit including the drive of the FET was 156mW by adopting a CMOS device to the main elements of the circuit.

12 sets of buck-boost type MPPT circuits were applied to the solar car "2017 Tokai challenger" at 2017 BWSC in Australia. Even if each solar module has different conditions of the power generation, output power of each modules was tracked to maximum power point using the proposed MPPT controller so that effective power generation was achieved. In addition, the MPPT circuit was able to act at power conversion efficiency of more than 98.5% in conditions of larger power more than about 50W.

The suitable MPPT circuit was able to accomplish enough power supply from a solar cell of the special quality of power generation which varied in a battery voltage by the combination with the high efficiency buck-boost type dc-dc converter to a vehicle.

REFERENCES

[1] Y. Haseo and T. Fujisawa, "Evaluation on Tracking Capability of MPPT for Running Solarcar," *International Conference on Control, Automation and Systems, Seoul*, Oct. 14-17, 2008, pp. 2933-2936.

[2] I. Nakir, A. Durusu, E. Ugur and M. Tanrıöven. "Performance Assessment of MPPT Algorithms for Vehicle Integrated Solar Systems," *Energy Conference and Exhibition (ENERGYCON), 2012 IEEE International*, 9-12 Sept. 2012, pp. 1034-1038.

[3] N. Femia, G. Lisi, G. Petrone, G. Spagnuolo, and M. Vitelli "Distributed Maximum Power Point Tracking of Photovoltaic Arrays: Novel Approach and System Analysis" *IEEE TRANSACTIONS ON INDUSTRIAL ELECTRONICS*, VOL. 55, NO. 7, JULY 2008, pp. 2610-2621

[4] T. Matsuyama, M. Yoda, et.all, "Experimental Introduction of Distributed Maximum Power Point Tracking Systems for PV System and Effect Applied to Solar Car,"*Journal of Japan Solar Energy Society 40(4)*, 2014, pp. 51-60.

[5] Hovens. M. G. P, and Prevoo, Y. K. L. M., "Maximum Power Point Tracking Topology, sensor and switch design," *Delft University of Technology*, 26 June 2012. https://repository.tudelft.nl/islandora/object/uuid:4f8b71a1-7766-462f-b6ec-6b1605bb5218?collection=education

[6] W. Stephen Woodward, "Maximum-Power-Point-Tracking Solar Battery Charger." *ELECTRIC DESIGN*, pp.114-118 September 14, 1998.

[7] Z. Liang, R. Guo and A. Huang, "A new cost-effective analog maximum power point tracker for PV systems," *Energy Conversion Congress and Exposition (ECCE), IEEE*, pp.624-631, 2010.

[8] Yuncong Jiang,Jaber A. Abu Qahouq, Ahmed Hassan, "Load current based analog MPPT controller for PV solar systems" *Applied Power Electronics Conference and Exposition (APEC), 2012 Twenty-Seventh Annual IEEE*, pp.911-914, 2012.

[9] F. Liu, Y. Han, Y. Gao and Jun Sun, "An Analog MPPT Controller IC together with its Application Circuit." *IEICE Electronics Express*, Vol. 11, N0. 20, pp.1-6, 2014.

Verification Test of Energy-Efficient Operations and Scheduling Utilizing Automatic Train Operation System

Shoichiro Watanabe[1*], Yasuhiro Sato[1], Takafumi Koseki[2], Eisuke Isobe[3] and Jun Kawashita[4]

1 National Traffic Safety and Environment Laboratory, Tokyo, Japan
2 The University of Tokyo, Tokyo, Japan
3 Japan Subway Association, Tokyo, Japan
4 Osaka Municipal Transportation Bureau, Osaka, Japan
*E-mail: sho-wata@ntsel.go.jp

Abstract— **It is well known that coasting and efficient use of notch command including a control of regenerating braking are useful for energy-saving operations. In addition, optimized scheduling can also contribute to a decrease in energy consumption. In this paper, verification tests of these energy-efficient operations and scheduling, which utilize Automatic Train Operation (ATO) system, are analyzed and the importance of time managements are explained.**

Keywords— *Energy-Saving, Hidden time, Optimization, Scheduling*

I. Introduction

In recent years, interest in global warming and environmental problems has been increased, and energy-saving technologies are expected to resolve them. In the field of DC-electric railway, energy-saving technologies are often discussed and categorized in hardware or software approaches. The main purposes of hardware strategies are to decrease catenary resistance loss, to decrease conversion loss and to use regenerative energy effectively. For example, energy storage system is installed in a substation and that capacity is optimized [1].

On the other hand, approaches for software strategies are different though the purposes are the same as hardware strategies. These software strategies have advantages that they do not cost much compared with hardware development and are able to obtain high energy efficiency which is reported in recent studies [2].

In this paper, energy-efficient operations and scheduling are focused on and methods to decrease energy consumption are explained by showing recent vehicle test results.

The authors are grateful to all staffs at the Ministry of Land, Infrastructure, Transport and Tourism, and Ministry of Economy, Trade and Industry, and related subway operators for their support to the Eco-Rail line project, and to Japan Society for the Promotion of Science.

II. Energy-Efficient Operations

A. The Importance of Time Managements

Energy-efficient operations are often discussed in previous studies. For example, optimized speed profiles [3-4] and effective use of regenerative braking with on-board assistance system [5] are studied. In these previous studies there were very severe time constraints because trains should reach the next station on time with an accuracy of seconds in order to follow timetable and provide comfortable traveling services to passengers. Previous study explained that running time was increased when energy-efficient operations such as effective use of regenerative braking control were carried out [6]. For this reason, these two strategies are important for energy-efficient operations as follows:

(a) Manage to allow a time margin by trying to cut down the running time, as much as possible, between the departing station and the arrival station under the time constraints.

(b) Use the time margin which is obtained by means of the strategy (a) and implement energy-efficient operations such as coasting or regenerative braking operations.

These strategies are simply explained in Fig.1. The strategy (a) is explained in Section III and (b) is explained in Section IV in detail.

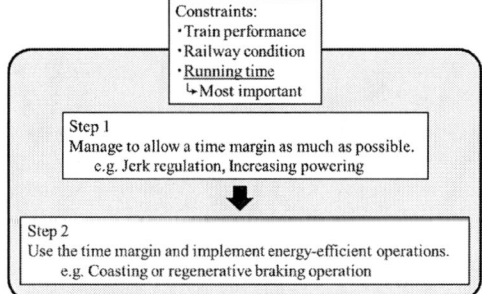

Fig.1. The main strategies of energy-efficient operations under running time constraints.

B. The Purpose of Effective Use of ATO

These operations are sometimes difficult for human drivers because notch off speed and braking command are complicated. Recently, Automatic Train Operation (ATO) is installed in rolling stocks and this system can control rolling stocks delicately. Researchers are interested in this ATO performance and currently are studying energy-efficient speed profiles considering real time analysis [7], reinforcement learning [8], CBTC (Communications-Based Train Control) [9] *etc.*

The authors focused on advanced control performance and applied the proposed methods to the ATO system, followed by experimental verification by on-track tests. Experimental results are shown and energy efficiency is also discussed.

III. THE TIME MARGIN, "HIDDEN TIME"

A. The Concept of Operation Time and Categories

Previous studies show that energy-efficient operations require additional time to implement their strategies [10]. For example, coasting speed profile is better than cruising speed profile from the energy-saving point of view. However, the average coasting speed is lower than the average cruising speed and additional running time is required when coasting is chosen instead of cruising. Figure 2 shows the categories of running time for train operations. A railway scheduling is designed based on the worst case scenario such as the lowest performance rolling stocks. In this situation, minimum running time to reach the next station is determined as marked (A) in Fig.2. Sometimes it is inconvenient for railway operators to use this time (A), and, therefore, recovery time (B) is added to round the number. The time (C) is generally used as a regular running time which is the sum of (A) and (B). Sometimes schedule is disrupted by troubles, and, therefore, margin time as marked (D) is considered. Railway services are acceptable when trains reach the next station in time (E) which is the sum of (C) and (D). However, rolling stocks can reach the next station usually in a shorter time compared with (A). The authors focused on time (F) and time (G) and time (G) was named "hidden time" in previous studies.

B. The Strategies for Hidden Time Management for Energy-Efficient Operations

Energy-efficient operations may be achieved if this hidden time is managed or increased because this hidden time allows energy-efficient strategies to be implemented. In order to increase the hidden time, authors proposed two methods below and carried out on-track tests.

(i) Increased acceleration in powering
(ii) Jerk regulation

The increased acceleration in powering (i) is one of the methods to manage the hidden time. Figure 3 shows comparison of speed profiles between conventional and proposed accelerations. This method mainly depends on inverter control but it is important for ATO to use this hidden time effectively because ATO system can control

notch off point and braking point with high accuracy compared with manual operations.

The jerk regulation (ii) is implemented in ATO control in order to provide comfortable quality for passengers in relation to powering. Figure 4 shows conventional and proposed methods of jerk regulation. ATO has many notch commands compared with human manual operations. For this reason, ATO can control acceleration delicately. However, conventional ATO uses 6-step notch command in powering as shown in Fig.4. Our proposed method is to use smoother notch steps in small increments; thus, conventional ride comfort is kept as well as the hidden time is obtained.

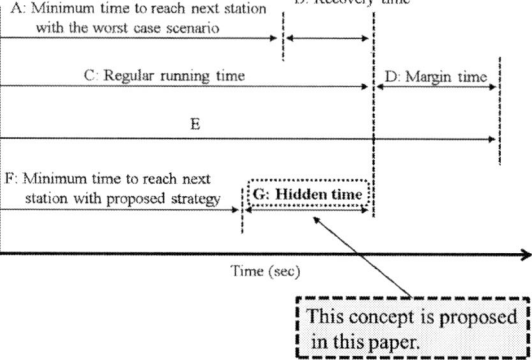

Fig.2. Categories of running time for train operations.

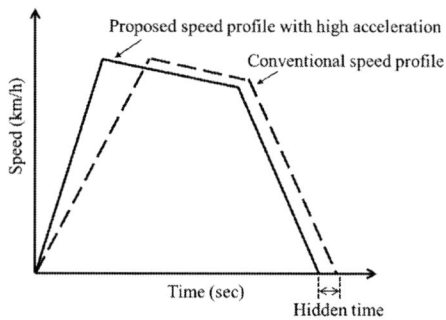

Fig.3. Management of the hidden time in high acceleration in powering.

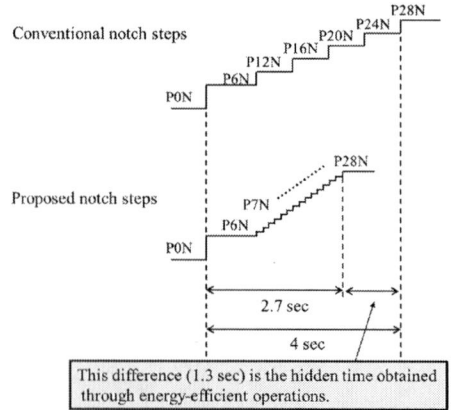

Fig.4. Management of the hidden time for jerk regulation.

IV. IMPREMENTATION OF ENERGY-EFFICIENT OPERATIONS

When the hidden time is obtained by means of the strategy which is explained in Section III, ATO can achieve energy-efficient driving. In this paper, three methods are explained [11].

A. Effective Use of Powering Notch Considering Motor Efficiency

The first method is the effective use of powering notch. Motor efficiency is not high when a command of low notch is executed. Recently, linear motor railway system is constructed as an urban transportation system. This rolling stock is activated by linear induction motor (LIM). The efficiency of this LIM is not high compared with rotary motor. For this reason, effects on energy consumption are also significant. Accordingly, in order to use powering notch effectively, use of low notch command is restricted for ATO.

B. Coasting Operations

The second method is the use of coasting operations, which can decrease energy consumption compared with cruising operations in urban railways. A starting point of cruising is determined based on mathematical analysis to be inputted into ATO control system.

C. Power-Limiting Braking for Increased Regeneration

The third method is power-limiting braking. Figure 5 shows that power of regenerative braking is decreased when speed is increased. Regenerative energy is increased when ATO controls braking operations in consideration of this regeneration performance. That is, as train braking power is limited in high speed area, regenerative braking can be used effectively. Mechanical braking is activated and kinetic energy is not reused as regenerative energy if braking power of train set exceeds the regenerative braking performance. In this case, heating loss is increased and, as a result, energy consumption is increased.

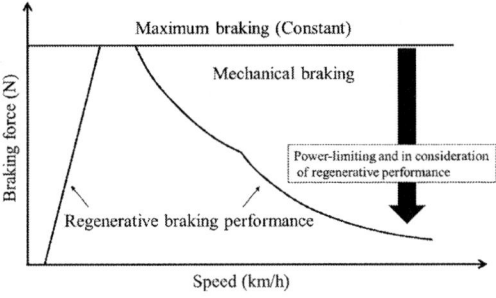

Fig.5. Performance of regenerative braking.

V. EVALUATION OF ENERGY CONSUMPTION

Energy consumption of rolling stocks has close relationships with running time. It is required to adjust running time so that running time is consistent all the time when energy-saving effect is evaluated. Though ATO can control rolling stocks with accuracy, sometimes it is difficult for ATO to ensure the same running time for conventional and proposed speed profiles. If that is the case, correction of energy consumption is required. However, this correction is difficult because;

(i) The relationships between energy consumption and running time are different in each section. Each section has different railway conditions such as curves and gradients.

(ii) The energy required to adjust running time is different for each of regular running times set for different sections between stations.

For these reasons, accurate numerical calculation and analysis of train dynamics are required. Figure 6 shows examples of the relationships between energy consumption and running time. In this calculation, railway conditions and constraints are satisfied and appropriate parameters of rolling stocks are chosen. When regular running time is 95 sec and actual running time is 100 sec, energy required for adjustments is 0.15 kWh. However, if regular running time is 68 sec and actual running time is 73 sec, required energy is increased up to 1.00kWh. ATO speed profile is designed in consideration of energy-saving effects while the regular running time is strictly followed, but sometimes differences are seen. Therefore, relationships between energy consumption and running time are analyzed to correct experimental results actually obtained so that we may be able to evaluate under the same running time.

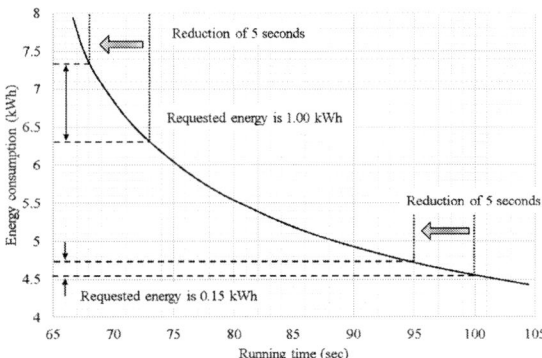

Fig.6. Relationships between energy consumption and running time.

VI. RESULTS OF ENERGY-EFFICIENT OPERATIONS

The results of energy-efficient operations are shown in the sections below. The methods which are explained in section III are implemented in ATO. The methods of increased acceleration and jerk regulations are analyzed in following chapters in A and B.

A. Strategy of Increased Acceleration in Powering

The experiments of this strategy were carried out in Nagahori Tsurumi-ryokuchi Line in Osaka in cooperation with Osaka Municipal Transportation Bureau. Figures 7 shows the conventional and proposed speed profiles. These two speed profiles have the same running time and, in the proposed speed profile, long coasting may be carried out by means of high acceleration. As a result, 19 % energy-saving effects are obtained as well as the regular running time is strictly followed.

The 2018 International Power Electronics Conference

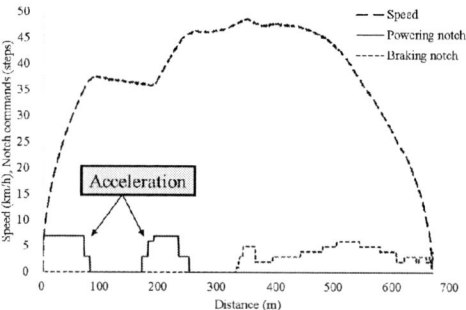

(a) Conventional speed profile.

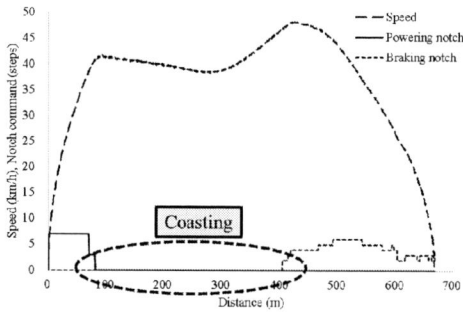

(b) Proposed speed profile with high acceleration.
Fig.7. Speed profiles in consideration of high acceleration tested in Osaka.

B. The Strategy of Jerk Regulation

The experiments of this strategy were carried out in cooperation with another railway company in Japan. Figures 8 also shows conventional and proposed speed profiles [11]. Figure 8 (b) shows jerk regulation to try to decrease the maximum speed, while (c) shows jerk regulation plus coasting operations. Energy-saving effects by decreasing the maximum speed are 6.1 % compared with (a) and (b). In addition, energy-saving effects by implementing coasting operation can also decrease energy consumption by 12.9 % compared with (a) and (c).

These experiments as shown in chapters A and B are carried out in urban railway and the distance between stations are not long. For this reason, hidden time is more effective for energy-saving operations.

VII. OPTIMIZED SCHEDULING

Previous studies have shown that energy consumption can be decreased by optimizing running time for all sections between stations [12]. Because factors, including curves, gradients and distances, are different for all sections, relationships between energy consumption and running time are also different according to such sections. Optimal running time can be calculated by choosing the same target point as shown in Fig.9. We try to optimize running time under restrictions where the total running time between the starting and terminal stations must be consistent. In addition, we pursue this optimization in a

unit of second, rather than minute. In these processes, ATO is considerably useful to attain accuracy.

VIII. RESULTS OBTAINED BY OPTIMIZED SCHEDULING AND ENERGY-SAVING OPERATIONS

The experiments of this strategy were carried out in cooperation with the railway company as mentioned in Chapter VI-B. Figure 10 shows energy consumption of conventional and proposed methods [10]. This proposed method includes energy-efficient operations, which are explained in Section IV, with jerk regulation, coasting operation and optimized scheduling. Train runs faster than in conventional speed profiles, but energy consumption is

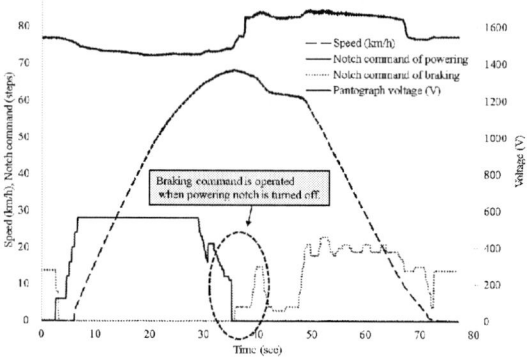

(a) Conventional speed profile.

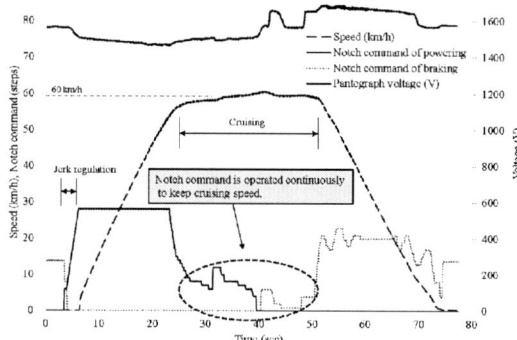

(b) Proposed speed profile with jerk regulation.

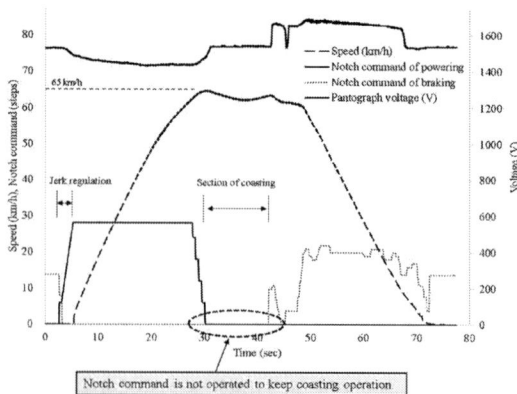

(c) Proposed speed profile with jerk regulation and coasting operations.
Fig.8. Speed profiles in consideration of jerk regulation.

2045

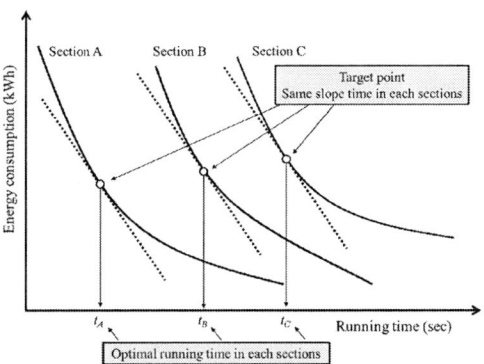

Fig.9. Target point and relationships between energy consumption and running time for each section.

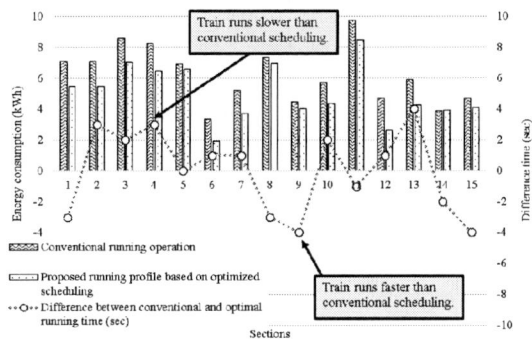

Fig.10. Energy consumption of conventional and proposed methods based on energy-efficient operations and optimized scheduling.

the same or decreased because energy-efficient operations contribute to energy saving. For this reason, energy-saving effects of optimal scheduling is yet to be separated from those of other methods, however, experiments and correction of running time is succeeded. Currently the authors are in the process of analysis to extract energy-saving effects of optimal scheduling. The results will be presented when available.

IX. CONCLUSIONS

In this paper, the importance of the hidden time in order to save energy consumption under the time constraints is explained. As a result of improvements in the conventional speed profiles, the hidden time is obtained, which contributes to the successful use of the techniques for energy-efficient operations.

The management methods utilizing high acceleration and jerk regulation are explained as the key factors. When the hidden time is obtained by these factors, the strategy of energy-efficient operations works including the effective use of notch commands, coasting operation and power-limiting braking.

The results show that these proposed methods can decrease energy consumption and that ATO system is useful for accurate operations. Thanks to this system, optimized scheduling methods may be used and energy-saving effects are gained.

ACKNOWLEDGMENT

The authors would like to express our cordial appreciation to all staffs at the Ministry of Land, Infrastructure, Transport and Tourism and Ministry of Economy, Trade and Industry of Japan and related subway operators for their support to the Eco-Rail line project. The authors also appreciate Japan Society for the Promotion of Science for their support by Grant-in-Aid of Research Fellow Number JP16J07015.

REFERENCES

[1] H. Lee, J. Song, H. Lee, C. Lee, and G. Jang : "Capacity Optimization of the Supercapacitor Energy Storages on DC Railway System Using a Railway Powerflow Algorithm", *International Journal of Innovative Computing, Information & Control(IJICIC)*, Vol.7, No.5B, pp.2739-2753 (2011)

[2] Shoichiro Watanabe, Masafumi Miyatake, Ryo Takagi, Tomoyuki Ogawa: "Analysis of Change in Energy Consumption by Enhancing the Driving Operation in DC-Electric Rolling Stocks", *IEEJ Industry Applications Society Conference*, 5-S1-5 (2016) (in Japanese)

[3] V. Doan, S. Watanabe, and T. Koseki : "The design of an optimal running curve for train operation based on a novel parameterization method aiming to minimize the total energy consumption", *Computers in Railways XIV: Railway Engineering Design and Optimization*, Vol.135, pp.175(2014)

[4] H. Ko, T. Koseki, and M. Miyatake: "Numerical study on dynamic programming applied to optimization of running profile of a train", *IEEJ Transactions on Industry Applications*, Vol.125, pp.1084-1092 (2005)

[5] S. Watanabe, Z. Yang, T. Koseki, K. Kondo, T. Mizuma, and Y. Hamazaki, "Experimental verification of power-limiting brake: Energy-saving train operation assistance," *IEEJ Transactions on Industry Applications*, vol. 134, no. 8, pp. 767–775, 2014.(in Japanese)

[6] Shoichiro Watanabe, Zhe Yang, Takafumi Koseki: "Analysis of the Relationship between Energy Consumption and Train Running Time Depending on Different Energy-Saving Train Operation Method", *IEEJ Joint Technical Meeting on Transportation and Electric Railway and Physical Sensors*,TER-13-005, PHS-13-009 (2013) (in Japanese)

[7] Q. Gu, T. Tang, F. Cao, and Y. Song : "Energy-Efficient Train Operation in Urban Rail Transit Using Real-Time Traffic Information", *IEEE Transactions on Intelligent Transportation Systems*, Vol.15, No.3, pp.1216-1233 (2014)

[8] J. Yin, D. Chen, and L. Li : "Intelligent Train Operation Algorithms for Subway by Expert System and Reinforcement Learning", *IEEE Transactions on Intelligent Transportation Systems*, Vol.15, No.6, pp.2561-2571 (2014)

[9] A. Fernández-Rodríguez, A. Fernández-Cardador, A. P. Cucala, M. Domínguez, and T. Gonsalves : "Design of Robust and Energy-Efficient ATO Speed Profiles of Metropolitan Lines Considering Train Load Variations and Delays", *IEEE Transactions on Intelligent Transportation Systems*, Vol.16, No.4, pp.2061-2071 (2015)

[10] Shoichiro Watanabe, Yasuhiro Sato, Takafumi Koseki, Takeshi Mizuma, Ryuji Tanaka, Yoshihiro Miyaji, Eisuke Isobe: "Energy-Saving Train Scheduling Based on Optimal Running Time Distribution and Verification Test Utilizing Automatic Train Operation System", *IEEJ Joint Technical Meeting on Vehicle Technology and Transportation and Electric Railway*,VT-17-020,TER-17-055 (2017) (in Japanese)

[11] Shoichiro Watanabe, Takafumi Koseki, Ryuji Tanaka, Eisuke Isobe: "Vehicle Test for Verifying Energy-Saving Train Control Based on Automatic Train Operation System", *IEEJ Joint Technical Meeting on Intelligent Transport Systems and Transportation and Electric Railway*,TER-16-070, ITS-16-049, pp. 35-40 (2016) (in Japanese)

[12] M. Miyatake : "A Simple Mathematical Model for Energy-saving Train Scheduling", *IEEJ Transactions on Industry Applications*, Vol.131, pp.860-861 (2011)

The Direct Benefit of SiC Power Semiconductor Devices for Railway Vehicle Traction Inverters

Shingo Makishima[1], Kazuki Fujimoto[1]*and Keiichiro Kondo[2]
1 Toyo Electric Mfg. Co., Ltd, Yokohama, Japan
2 , Faculty of Science and Engineering, Waseda University , Tokyo, Japan
*E-mail: kkondo@aoni.waseda.jp

Abstract- The only and the direct benefit by the loss reduction on SiC-SBD devices are studied in this paper. The loss of the Si-IGBT and SiC -SBD hybrid module type traction inverter for DC.600V tram car is calculated in this paper. The benefits of mass reduction and less dimensions of cooling system of the inverter are studied, compared with the case of Si-IGBT and Si-diode, and with the case of GTO and Si-diode. The thermal characteristics of the Si-IGBT and SiC -SBD hybrid module inverter are examined by the actual train operation.

I. INTRODUCTION

Around 30 years ago, 4.5kV Gate Turn Off thyristor (GTO) was developed and applied to voltage source type inverters (VSIs) to drive induction motors(IMs) for railway vehicle traction[1]. The Variable Voltage Variable Frequency (VVVF) inverter enabled DC 1.5kV powered electrical motive units (EMUs) to be driven by the induction motors. The benefits of IMs drive system for the railway vehicle traction, compared with DC motor traction systems, are,

-Reduction of the maintenance work with the brush and commutator on DC traction motor.
- Energy saving effect and less maintenance work with the mechanical brake parts by the wider speed range regenerative brake

The Self commutated function of GTO eliminates the bulky commutated circuit which is installed on the thyristor drive VVVF inverter, such as McMaray inverter in the late 1980's. 4.5kV GTOs initiate the "Era of inverter driven induction motor" in the railway vehicle traction technology field. However, as shown in Fig.1.[2], GTO still needs anord inductor to avoid the rush current at the turn on and needs snubber circuit to absorb the magnetic energy on the anord inductor and parasitic inductance at the turn off. The snubber circuit consists of bulky capacitor and resistance to absorb and consume the transient energy associated with turn on and turn off.

Therefore, in mid 1990's, Insulated Gate Bipolar Transistors (IGBTs) were developed and applied to ease the technical issue on the switching with GTOs. Switching duration of IGBT is around several times shorter than GTO and its switching loss is much lower than the one with GTO. In addition, gate is driven by the voltage signal with low loss of gate driver, compared with the current driven GTO[3]. At first of application, 1.7kV-400A IGBTs is utilized in the neutral point

cramped (NPC) 3 level inverter for DC 1.5kV EMUs[4]. This type of inverter was capable to drive an up to 200kW IM. Then 3.3kV-1.2kA IGBT reduce the size and dimension of inverter drastically because the snubber circuit itself is eliminated and the size of cooling system can be designed much smaller compared with the on for GTO inverter, because the heat sink do not need to deal with the loss from the snubber circuit.

This series of the innovation by the high voltage power semiconductor devices such as 4.5kW GTO and 3.3 kV IGBT acquire a "self-commuted function" on the high voltage power semi-conductor switching devices.

One of the latest technical topics on power electronics technology is the development of the wide band gap power semiconductor switching devices such as SiC devices. 3.3kV SiC-Metal Oxide Semiconductor Field Effect Transistor (MOSFET) is developed to focus on the

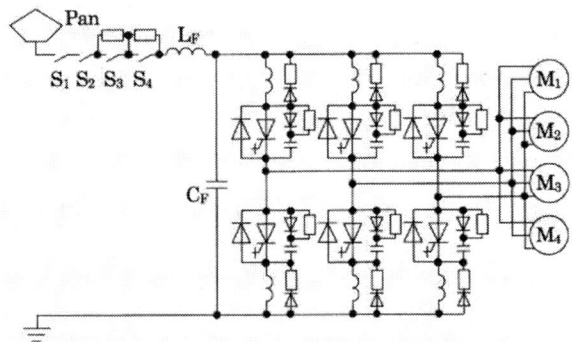

Fig.1 A example of circuit diagram of a traction inverter with 4.5kV GTO[2]

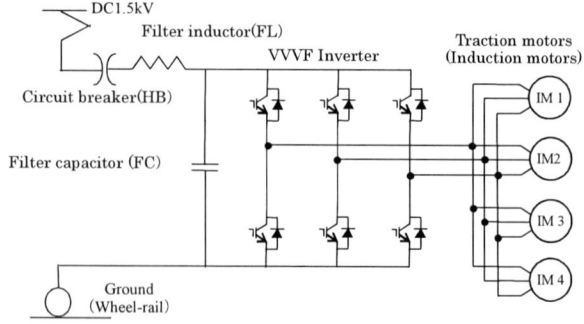

Fig.2 A circuit diagram of a typical traction system with a 3.3kV IGBT inverter.

railway vehicle traction applications. The SiC-MOSFET is associated with SiC shoot Key Barrier Diode (SBD) in the application. The SiC power semiconductor devices are featured in the high power range applications are as followings.
- Elimination of recovery current loss by SBD even with higher anti-voltage capability.
- Less conduction loss in proportion the drain current even on the unipolar devices such as MOSFET.
- Achieving both the lower switching loss even by the unipolar devices and higher anti-voltage capability.

The only and the direct benefit on the SiC devices for railway vehicle traction application is mainly to reduce the loss of the power converter. However, this direct befit has not been clearly discussed before in the railway vehicle application. Therefore, the benefit of loss reduction on SiC-SBD devices is studied in this paper. The loss of the Si-IGBT and SiC -SBD hybrid module type traction inverter for DC.600V tram car is calculated in this paper. The benefits of mass reduction and less dimensions of cooling system of the inverter are studied, compared with the case of Si-IGBT and Si-diode, and with the case of GTO and Si-diode. The thermal characteristics of the Si-IGBT and SiC -SBD hybrid module inverter are examined by the actual train operation. The only and the direct benefit on the SiC devices are clearly revealed through the discussion in this paper.

II. AN EXAMPLE OF MINIATURIZING PROPULSION INVERTER USING SiC DEVICES

This section presents an example of miniaturization of a propulsion inverter for an existing articulated tramcar using SiC devices. Propulsion inverter systems with GTO (1980~1990s) in the early days of their applications have reached the end of durable period, although the vehicle carbody still can be used. In this case, one of the options for the renovating the railway vehicles is to replace the conventional GTO inverter to a brand-new inverters on the existing carbody.

A case of an articulated tramcar is dealt with in this paper as one of the examples. Table 1 shows the specifications of the articulated tramcar. The articulated tramcar has 3 carbody sections and 4 bogies, as shown in Fig. 3. Two traction motors are mounted on the each bogie installed on the both ends of the vehicle. The first set of the articulated tramcar was constructed in 1990 and the inverter is designed for DC 600V input with GTO.

A. Constrains and development policy of the renovated inverter

The constraints on the design for the renovated inverter includes, as followings:
- continue to use the original traction motor; motor characteristics are unable to modify for the new inverter
- unable to increase the output power of auxiliary power supply unit; no additional power to cool the inverter

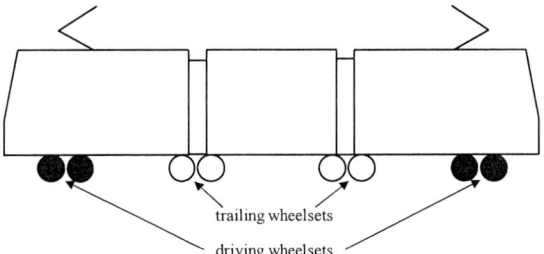

Fig. 3. Articulation formation of the tramcar

TABLE I
VEHICLE SPECIFICATION OF THE ARTICULATED TRAMCAR

Train length	27960 mm (3 carbody sections and 4 bogies)
Width	2496 mm
height	3820 mm
Track gauge	1435 mm
Electric system	600V DC overhead
Auxiliary power supply output	24V DC, 200V 60Hz 3phase AC
Maximum speed	60 km/h (operation) 80 km/h (design)
Acceleration	3.0 km/h/s (0.833 m/s^2)
Deceleration	5.0 km/h/s (1.39 m/s^2) (above 45km/h) 4.4 km/h/s (1.22 m/s^2) (below 45km/h)
Traction motors	Three-phase squirrel-cage induction motor 85kW, 440V, 144A, 1760 min^{-1}

- unable to change the brake system and characteristics; increasing regenerative power is impossible
- underfloor space is limited: the dimensions of the brand-new traction inverter must be the same or less than the conventional one.

Therefore, the new inverter needs to be installed with a natural-cooling system. The dimensions of cooler depend on the loss of semiconductor switching devices. This fact means that less loss of semiconductor switching devices much contributes to smaller sized inverter. Thus, SiC devices achieve lower losses than conventional Si devices and this contributes to reduce the dimensions of heat sink.

B. Losses and temperature rises of semiconductor switching devices

In order to estimate the mass and dimensions of the cooling system, losses associated with SiC-Hybrid and Si-IGBT switching devices are compared in the both performance inverter. The maximum voltage of both devices are 1700V and the rating current are 1200A. The only difference is the type of freewheeling diode, SiC-SBD or Si-Pin diode. Losses of the both type of propulsion inverters, with SiC-Hybrid and Si-IGBT switching devices, are estimated as shown in Table II, in the running condition with the heaviest load conditions.

TABLE II
LOSSES OF THE INVERTERS IN THE WORST RUNNING CONDITION

device	loss
SiC-Hybrid (CMH1200DC-34S)	900W
SI-IGBT (CM1200DC-34S)	1300W

TABLE III
TEMPERATURE RISES OF THE INVERTERS

device	cooler temperature	junction temperature
SiC-Hybrid module (CMH1200DC-34S)	95°C	125°C
SI-IGBT module (CM1200DC-34S)	125°C	140°C

TABLE IV
MASS AND SIZE COMPARISON OF INVERTERS

	SiC-Hybrid	Si-IGBT	GTO
mass	244 kg (65%)	255 kg (67%)	380 kg (100%)
occupied volume	0.5 m³ (42%)	0.7 m³ (60%)	1.2 m³ (100%)
power density	340 kW/m³ (240 %)	236 kW/m³ (167%)	141 kW/m³ (100%)

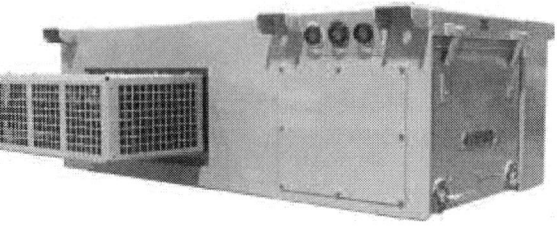

Fig. 4. The new propulsion inverter fitted with SiC-Hybrid propulsion inverter applied by SiC-Hybrid modules.

The loss reduction effect by changing the type of devices of Si to the one of SiC is generally considered as around 20% [5] [6]. These results in table II coincide with the general case..

Since the underfloor space for the traction inverter is limited and the thermal resistance of the heat sink must be less than 900W/55K. In this case, the maximum rise of temperature is shown in Table III, when the ambient temperature is 45°C.

There is not so much margin between the temperature of the heat sink and the limit temperature at the junction of the devices in case of applying Si-IGBT. Thus the SiC-Hybrid module should be applied to the brand new inverter.

C. Evaluation of the effect of SiC hybrid modules in the mass and Dimensions

In order to evaluate the effect of the application of SiC hybrid module in the brand-new inverter, Table IV shows the estimated design results of the dimensions and the mass and their reduction ratio of the the dimensions and the mass nominalized by the original GTO inverter. The mass and dimensions of the both brand-new inverters with Si-IGBT module and SiC hybrid modules are estimated under the above-mentioned assumptions.

Table IV presents that both the mass and the occupied volume are reduced by applying the less loss switching devices such as Si-IGBT module and SiC hybrid module. The output power density of the inverter with SiC-Hybrid module is 2.4 times higher than the conventional GTO inverter. The effect of miniaturisation by applying SiC-Hybrid and Si-IGBT inverters is not only due to semiconductor switching device difference, but also contributed by other factors such as smaller size gate controller. However, on the other hand, the effect of the 30% less volume of the inverter with SiC-Hybrid module, compared with Si-IGBT module inverter, mainly depends on the elimination of the recovery loss by the SiC-SBD in the SiC hybrid module.

Fig. 4 shows the outlook of the developed new

III. THERMAL TEST RESULTS

Figure 5 shows the thermal test ~~run~~ results of the experimental running by the articulated tramcar driven by the brand-new inverter with SiC-Hybrid modules. The test running is carried out with no load condition without any passengers under the condition of 800Hz switching frequency and synchronous switching mode with a 9 plus from the 30Hz of fundamental frequency. On the other hand, the temperature rise calculation in Table III is under the full load condition und the same switching frequency. In addition, the temperature rise is not saturated in any running section. Therefore, the maximum temperature rise in the test running is around 25 centidegree, though the rise of temperature in Table III is 50 centidegeree. According to the thermal resistance value of 99W/55K, the estimated loss heat generation in the test running is around 450W. The calculated temperature rise of the heat sink under the no load condition is around 440W. These results verify the design condition in Table III and the benefits of the application of SiC hybrid module is appropriately examined through this paper.

IV. CONCLUSIONS

In this paper, the only and the direct benefit on the SiC hybrid modules with Si-IGBT and SiC-SBD in the railway vehicle traction application is clearly discussed by compared with the loss of the SI-IGBT modules. The loss of the Si-IGBT module and SiC -SBD hybrid module type traction inverter for DC.600V tram car is calculated in this paper. The thermal characteristics of the Si-IGBT module and SiC-SBD hybrid module inverter are examined by the actual train operation. The only and the direct benefits on the SiC devices, which is downsizing of the heat sink of the traction inverter, are clearly revealed through the discussion in this paper.

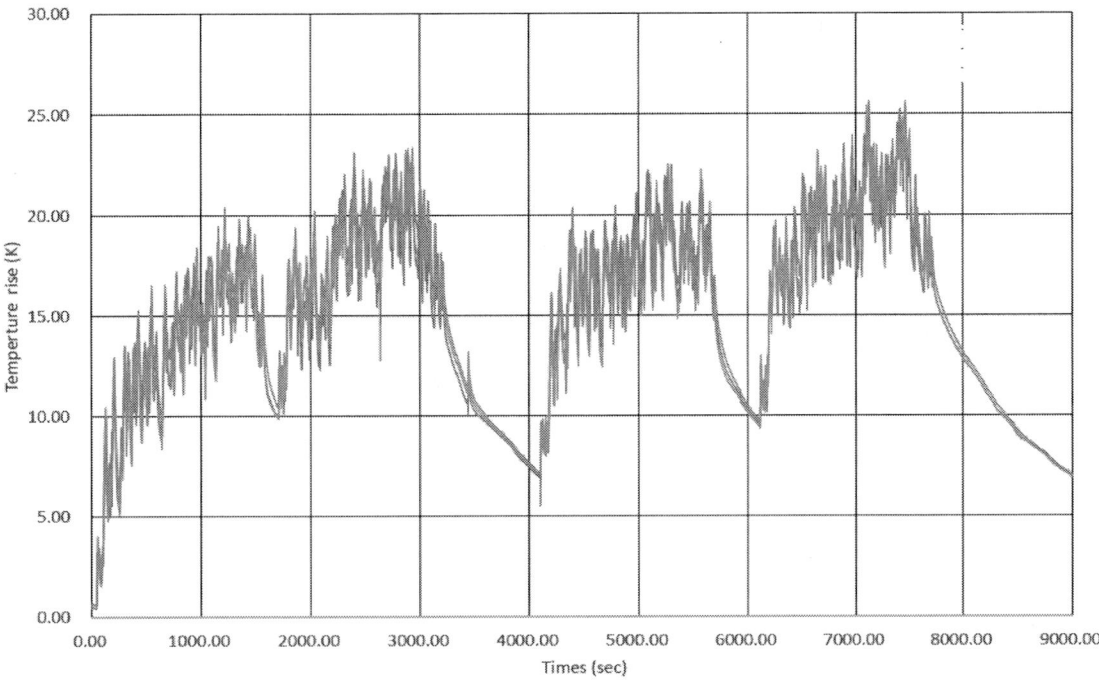

(--Temp. rise of heat sink for phase U, -- Temp. rise of heat sink for phase V, -- Temp. rise of heat sink for phase W)

Fig. 5 The measured temperature of the heat sink in no-load condition test running by the articulated tram can.

REFERENCES

[1] B. J. Wojtas,"The age of the squirrel cage (motors)", IEE Review, Vol.36, Issue 11, pp.435-438, 1990

[2] E Takahara, "A study of electrical railway vehicle traction control as control technology and power electronics," The Journal of The Institute of Electrical Engineers of Japan Vol. 122, No. 9 P 601-605, 2002.9

[3] Akiteru Ueda; Masahiko Ibamoto; Hiroshi Narita; Takamasa Hori; Takashi Tsuboi; Yukio Yamada "GTO Inverter for AC Traction Drives" , IEEE Transactions on Industry Applications、Volume: IA-19, Issue: 3, pp. 343-348,

[4] T. Ohmae;   K. Nakamura," Hitachi's role in the area of power electronics for transportation", IEEE,   International Conference on Industrial Electronics, Control, and Instrumentation, 1993. Proceedings of the IECON '93, pp.714-718, 1993

[5] Investigating R&D Committee on Technologies for high performance traction equipments, "Technologies for High Performance Traction Equipments," IEEJ Technical Report. Vol. 1227, 2011 (in Japanese)

[6] Katsumi Ishikawa;   Kazutoshi Ogawa;   Hidekatsu Onose; Norifumi Kameshiro; Masahiro Nagasu "Traction inverter that applies hybrid module using 3-kV SiC-SBD," *The 2010 International Power Electronics Conference - ECCE ASIA -, pp. 3266 – 3270,* 2010

The Loss Characteristics of PSFB ZVS DC-DC Converter Applied to the Auxiliary Power System

Xianjin Huang[1*], Juan Zhao[1] and Fei Lin[1]

1 School of Electric Engineering, Beijing Jiaotong University, Beijing, China

*E-mail: xjhuang@bjtu.edu.cn

Abstract-In this paper, the operation control and loss characteristics of a PSFB (Phase-Shifted Full-Bridge) ZVS DC/DC converter applied to auxiliary power system has been investigated. The working characteristics, device losses and system efficiency of the PSFB (Phase-Shifted Full-Bridge) circuit topology for 1500Vdc power supply using 3300V Si IGBT are studied. One type of ISOP (Input-Series Output-Parallel) double bridge topology using 1700V SiC MOSFET is proposed. The improved input voltage sharing control method is used to meet the requirements of the supply voltage and system capacity and it also improves the system's rapidity and stability. The 1700V Si IGBT and the 1700V SiC MOSFET used in ISOP PSFB converter are compared on the control, losses and efficiency. Simulation and experiment results show that the application of SiC devices exerts a positive effect on the volume reducing, weight decreasing and efficiency increasing.

Keywords— ISOP PSFB converter, loss characteristics, SiC MOSFET

I. INTRODUCTION

There are strict standards on the weighty and volume of auxiliary power system applied to the tram, LRT and metro train. Higher capacity density, smaller volume and slighter weight of the converter are beneficial to the vehicle layout design and transport capability. The power level of the latest domestic metro vehicle auxiliary inverter power system is generally 150kW. Most of the Shanghai, Guangzhou and Shenzhen subways are using the system composed of 1500Vdc bus power supply, 110Vdc battery charger and 380Vac three-phase inverter. The DC/DC converter can convert the 1500Vdc input into the 700Vdc output by bridge circuit, step-down transformer, diode rectifier and the LC filter. The half-bridge circuit can supply the 110Vdc with the 700Vdc input. The three-phase inverter generally uses bridge circuit topology, which converts the 700Vdc input into the 380Vac output. The advantages of this program are low cost and high reliability. The power density, volume and weight can be improved only by the output filter and the isolation. And the additional high-frequency DC/DC link is often used, which transports the output side industrial frequency transformer function to the former high-frequency transformer. The loss characteristics of the converter after increasing an additional high-frequency DC/DC link has been studied [1]. The different topologies of the converter fore-stage circuit has been studied [2-5]. The increasing topologies of the fore-stage circuit can improve the

converter weighty, volume and power density, but it also increases the cost of the device and the risk of failure. The DC/DC converter is generally composed of the bridge circuit, transformer isolation and output rectifier circuit. So there are 2 methods to improve the performance of the DC/DC converter, one is to improve the circuit topology, the other is to enhance the operating frequency, including the directly increase of the frequency and the use of new devices. An improved parallel circuit applied to the auxiliary power system has been investigated [6]. A train auxiliary power supply system based on PQ droop control algorithm has been designed [7]. There was a modular redundant design on the DC/DC converter to reduce the single module power [8]. The DC/DC converter with SiC MOSFET has been analyzed for the system loss and efficiency [9-10].

Soft-switching is the most common way to increase the switching frequency of the Si-based devices, thereby reducing device losses and ensuring system efficiency. Many literatures have carried out related research based on a certain circuit topology, including soft-switching circuit stability, soft-switching implementation and the efficiency of the circuit [11-15]. Wide bandgap device brings the characteristics of high-frequency and low-loss to enhance the circuit performance. Many circuit topologies with wide bandgap devices can be implemented under hard switches, reducing the design requirements for circuit parameters and simplifying control algorithms. The application of wide bandgap devices in aeronautical static converters using SiC MOSFETs has been studied [16]. Due to the commercialization process in the industry, the present common module of the SiC device is only at 1700V level, which also limits the applications of the SiC device in the subway auxiliary power supply system.

The ISOP (Input-Series Output-Parallel) PSFB (Phase-Shifted Full-Bridge) ZVS DC/DC converter circuit topology using 1700V SiC MOSFET suitable for 1500Vdc power supply is studied in this paper. The converter uses two power modules Input-Series Output-Parallel which is designed to meet the system voltage and power requirements. Each single power module uses full bridge structure which can achieve high-frequency isolation and phase-shifted control. The control method of the ISOP double bridge system is studied, and the input voltage and output current sharing are guaranteed by the improved input voltage sharing control algorithm. With two-level circuit structure using the traditional 3300V Si device and ISOP double bridge circuit structure using the ordinary 1700V Si device, the control characteristics, operating principles, loss and efficiency characteristics of the three structures are compared. Finally, the relationship between the loss and the operating frequency of the ISOP PSFB converter with SiC MOSFET is studied, and the design parameters of the converter in auxiliary power supply system are determined. The feasibility of the scheme is verified by simulation and experiment results.

II. ISOP PSFB ZVS DC/DC CONVERTER

A. A DC/DC converter applied to auxiliary power supply systems

The subway vehicle auxiliary power system often concludes the three parts. The first is a high power high voltage DC/DC converter which is supplied by the 1500V DC bus. In general, the output is 600Vdc or 700Vdc and the power is greater than 50kW. The second is a converter that charges the 110V battery charger. The power is generally 15 kW. The last is a 24Vdc or 48Vdc power supply module which supplies the control unit power. The power is generally not more than 500W. The high-voltage high-power DC/DC converter uses two-level structure in common. For 1500Vdc power supply, 3300V IGBT can meet the requirements. The conditional DC/DC converter structure is shown in Fig. 1.

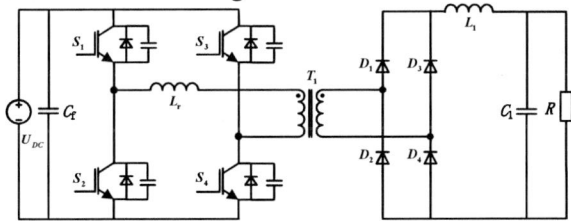

Fig. 1. Si IGBT PSFB converter

As the Fig. 1 shows, there are four IGBTs of the full bridge whose voltage level is 3300Vdc which can meet the requirements of the power supply level.

For 1700V SiC switching devices, the traditional two levels can not meet the voltage pressure requirements. In this situation, there will be the power modules in series or parallel structure to meet the voltage or current stress requirements. The input-series structure can reduce the voltage stress of the device, and the output-parallel structure can reduce the current stress of the output rectifier. With the input-series structure, 1700V SiC MOSFET can use in the full bridge structure. The ISOP PSFB converter using SiC MOSFET is shown in Fig. 2.

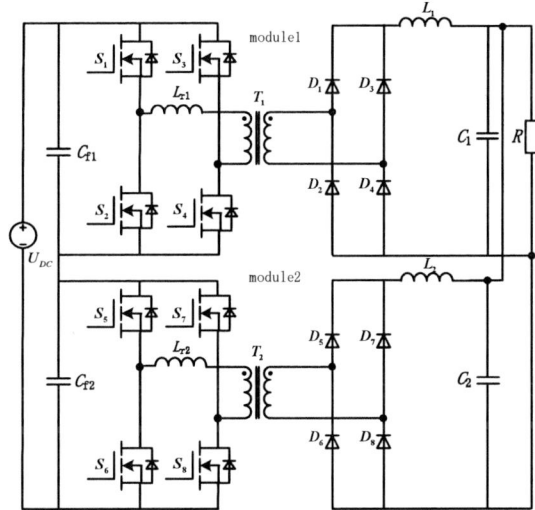

Fig.2. SiC MOSFET ISOP PSFB converter

Each power module is in full bridge structure and phase shift control in Fig. 2. The two full bridge modules share the input voltage so that the 1700V switching devices can meet the 1500V power supply requirements. And the improved input voltage sharing control is used in the ISOP structure to ensure the input voltage sharing and the output current sharing. The third converter is ISOP PSFB double circuit structure with the ordinary 1700V Si IGBT as shown in Fig. 2 which is only changed the switching device into the Si IGBT.

For a single full-bridge DC/DC converter, phase shift control is used to achieve ZVS. In phase shift control, the phase of S1 is ahead of S4 and the phase of S2 is ahead of S3, so it is often said that S1 and S2 make up the leading leg, and S3 and S4 make up the lagging leg. When there is not a shift-phase between the leading leg and the lagging leg, the primary side of the transformer obtains the largest volt-second product [17], allowing the converter to obtain the maximum output voltage. When the shift-phase is equal to or greater than $180°$, the primary side voltage of the transformer is 0 and there will be no output. In the dead zone, the resonant inductance L_r, the filter conductor L_f and switching device junction capacitance resonance to open the switches of the leading leg before the V_{ds} (V_{ce}) voltage drop to 0 to achieve ZVS. Besides, the resonant inductance L_r and its switching device junction capacitance resonance to open the lagging leg switches before the V_{ds} (V_{ce}) drop to 0 to achieve ZVS, too. The PSFB converter operating mode of an half period is summarized as Table 1.

TABLE I
PSFB CONVERTER OPERATING MODE OF AN HALF PERIOD

i_{Lr}	Pulse distribution	Conduction device
>0	S4(S1 OFF)	S1,S4,D1,D4
>0	S4	S4,D1,D4
>0	S4,S2	S4,SD2,D1,D4
<0	S2	SD2,SD3,D1,D2,D3,D4
<0	S2,S3	SD2,SD3,D1,D2,D3,D4
<0	S2,S3	S2,S3,D1,D2,D3,D4
<0	S2,S3	S2,S3,D2,D3

2052

Table 1 shows an half period of a PSFB converter operation mode, and the last half period is the similar. Sn represents the switching device of the full bridge in Fig. 1. SDn represents the corresponding switching device body diode. Dn is for the diode of the rectifier. i_L is the current flowing through the resonant inductance L_r.

Fig. 3 shows the converter waveforms during the different operation modes, the mode is correspond to the seven modes in table 1 from the time t0 to t7. The pluses of the switches in full bridge are in shifted-phase control.

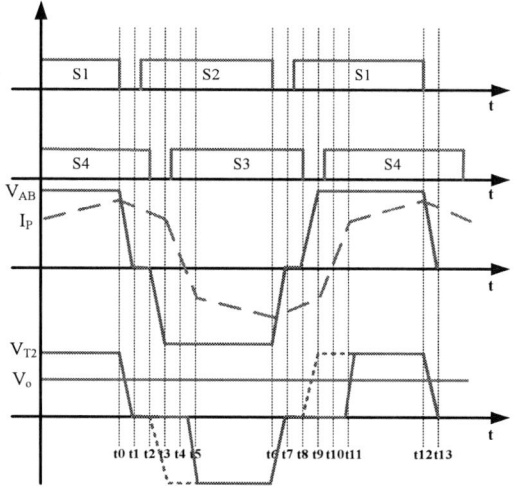

Fig. 3. ISOP PSFB converter operation waveforms

The pulses of switches, the shift-phase between the switches of the leading leg and the lagging leg, the current of the L_r, the voltage V_{AB} between the two midpoint of H bridge, the second side voltage of the transformer V_{T2}, V_o are shown in Fig. 3. As it also shows how the voltage V_{AB} and the current of the L_r change in the dead time. The dotted curve of the second side voltage of the transformer V_{T2} shows the duty cycle lost.

B. ISOP PSFB converter control

Due to the commercialization process in the industry, the present common module with the SiC device is only at 1700V level. In order to use the 1700V SiC device well, two power modules in series structure in the input which can ensure the voltage pressure is widely used. And the parallel structure in the output makes that one power module only needs meeting the half of the output current requirement. But there is a problem caused by the ISOP structure. The input voltage and the output current may not be shared equally by the two modules.

In order to achieve the goal of input voltage sharing and output current sharing. There is a normal control method as shown in Fig. 4.

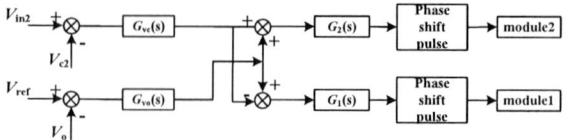

Fig.4. Commonly used input voltage sharing control

The amplitude of V_{in2} is half the input voltage U_{dc} in Fig. 2. V_{c2} is for the voltage of the capacitor C_{f2}. V_{ref} is the output voltage reference value. $G(s)$ is for the PI regulator of the various control loops. There are two control loops coupling in the common control method. The input voltage equalization loop and output current loop are coupled to ensure voltage and current sharing. If the control system can control the input voltage sharing and the output current can be shared, so the system can work stably [18] . As there is no current loop in the control system, the response of the system is slow.

Based on the commonly used input voltage sharing control method, in order to improve the rapidity and stability of the system, an improved input voltage sharing control method is proposed in Fig. 5.

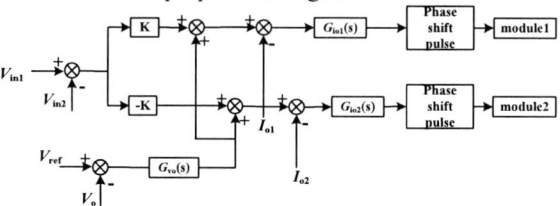

Fig. 5. An improved input voltage sharing control

As shown in Fig. 5, there is a common output voltage loop, individual average current mode inner current loop, whose references are adjusted by individual input voltage loops. And the difference between the input voltages of the two modules after multiplying a gain K, respectively plus or minus the value after the outside voltage loop regulator and then conducts as the output of the input voltage loops. Output control pulse is regulated by inner current loop regulator output to control the converter output.

The two voltage outer loops respectively ensure that the input voltage is equally divided and the output voltage is stable, and the inner current loop ensures the equalization of the output current. In the case of a suitable parameter setting, due to the inclusion of the inner current loop, the system's rapidity is improved and the stability of the system is also improved.

III. SIMULATION AND TESTS

There are 3 kinds of circuits to obtain the DC/DC converter objectives in auxiliary power system that is mentioned in section 2.1. These three kinds of structure simulation models are built by PSIM. The conventional converter is 10kHz IGBT PSFB. The control pulses and the voltages and currents of the switches are shown in Fig.6.

2053

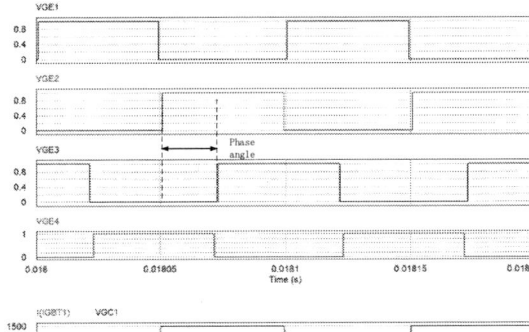

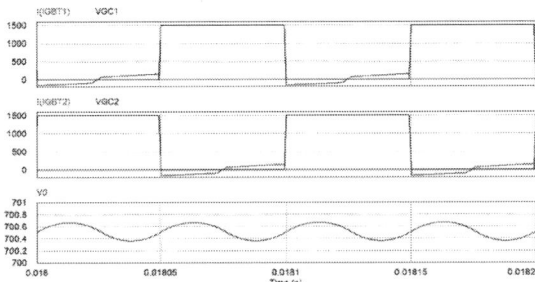

Fig. 6. 10kHz IGBT PSFB Simulation waveforms

The first converter uses FF400R33KF2C IGBT of Infineon. And most of the losses are conduction loss P_{con}, switching loss P_s, rectifier loss P_{rec}, circulation loss, P_{cir} transformer loss P_t and LC filter loss P_{Lf}. The formulas of the losses Calculation are as follows.

$$P_{con}=f_s\int_0^{T/2}U_{ce}(t)I_e(t)dt \tag{1}$$

In general, the conduction loss is calculated as formula (1) if there is the working waveforms. But in order to quickly and easily estimate the conduction loss, we will use data from the datasheets to estimate the value. So the conduction loss is

$$P_{con}(MOS)=I_{rms}^2R_{ds(on)}D_s \tag{2}$$

$$P_{con}(IGBT)=I_{rms}V_{ds(sat)}D_s \tag{3}$$

In the formula, $P_{con}(MOS)$ is the conduction loss for SiC MOSFET, $P_{con}(IGBT)$ is the conduction loss for IGBT. I_{rms} is the RMS value of the current flowing through the device. $R_{ds(on)}$ is on-state resistance of the device. D_s is the duty cycle.

We can determine that the switching energies at different power conditions are linearly related to the operating time voltage and current when we deal with high power applications. Based on the datasheet from manufacturer, we can know the single switching energies E_{on} and E_{off} of the device under specific test voltage and current. Then we can get the switching losses through single switching energy and switching frequency.

$$P_s=P_{turn-on}+P_{turn-off}=f_s(E_{on}+E_{off})\frac{V_d}{V_{dn}}\frac{I_d}{I_{dn}} \tag{4}$$

In the formula, E_{on} and E_{off} are the loss of turn on and turn off in the datasheet at the standard test condition, V_{dn} and I_{dn} are the blocking voltage and drain current in the datasheet at the standard test condition, V_d and I_d are the blocking voltage and drain current at the actual condition; and f_s is the switching frequency.

The rectifier loss concludes turn on loss and turn off loss

(reverse recovery loss) of rectifier diodes D_1-D_4.

$$P_{Don}=D_DI_pV_p \tag{5}$$

$$P_{Drr}=f_sE_{Drr}\frac{V_c}{V_{cn}}\frac{I_c}{I_{cn}} \tag{6}$$

$$P_D=P_{Don}+P_{Drr} \tag{7}$$

P_{Don} is the rectifier diode turn on loss, P_{Drr} is the rectifier diode reverse recovery loss, D_D is the average duty cycle, V_p is the turn on diode voltage, I_p is the the RMS value of the current flowing through the diode, R_D is equivalent resistance.

The circulating loss of the full bridge is about the time t_0-t_3, the switching device, the anti-parallel diode and the filter inductance conduction.

$$P_{cir}=f_s(\int_{t_1}^{t_3}U_F(t)I_{Lr}(t)dt+\int_{t_1}^{t_3}U_T(t)I_{Lr}(t)dt) \tag{8}$$

$U_F(t)$ is the turn on voltage of the switching device, $U_T(t)$ is the turn on voltage of the anti-parallel diode. We can calculate the loss according to the simulation waveforms data.

The loss of magnetic components is mainly composed of the core loss and winding losses.

$$P_{magnetic\ components}=P_{Fe}+P_{winding} \tag{9}$$

The Core loss P_{Fe} can generally be calculated by the following formula.

$$P_{Fe}=kf_s^\alpha B^\beta V_e \tag{10}$$

P_{Fe} is the core loss, B is the peak value of magnetic induction. k, α, β are constants, which depend on the grade of the material and they can be generally found from the manufacturers core manuals, V_e is for the volume of the core. The winding loss $P_{winding}$ can be expressed as (neglecting AC resistance losses):

$$P_{winding}=I_{rms}^2R_{coil} \tag{11}$$

R_{coil} is the DC resistance of winding.

The transformer magnetic intensity peak value can be expressed as

$$B_T=\frac{U_o}{4f_sA_eN_s} \tag{12}$$

Where A_e is the effective magnetic field of the transformer. And the RMS value of the current flowing through the transformer is as follows.

$$I_{T\_rms}=\sqrt{f_s\int_0^{DT_s}(\frac{I_o}{k})^2dt} \tag{13}$$

The loss value of the transformer P_t can be calculated by substituting the formula (12)-(13) into the formulas (9)-(11) respectively.

The peak value of the AC magnetic intensity of the filter inductor Lf can be expressed as

$$B_{m\_Lf}=\frac{\mu_r\mu_0N_{Lf}\Delta I_o}{2l_{Lf}} \tag{14}$$

Where ΔI_o is the output current pulsation, l_{Lf} is for the length of the effective magnetic circuit.

The current effective value of the filter inductor L_f is

$$I_{Lf\_rms}=I_o \tag{15}$$

The loss of the filter inductor can be obtained by substituting the formula (14)-(15) into the formulas (9)-(11) respectively.

According to the formulas, calculating the 10kHz IGBT PSFB converter with FF400R33KF2C, the losses percentage of the rated power 80kW is obtained. The

2054

proportion percentages of rated power P_{on}, P_t, P_{Lf}, P_{loop}, P_{rec}, P_s are 0.38%, 0.61%, 0.22%, 0.46%, 0.48%, 1.588% respectively.

The PSIM model simulation waveforms of ISOP PSFB converters with SiC MOSFET and Si IGBT are given as Fig. 7.

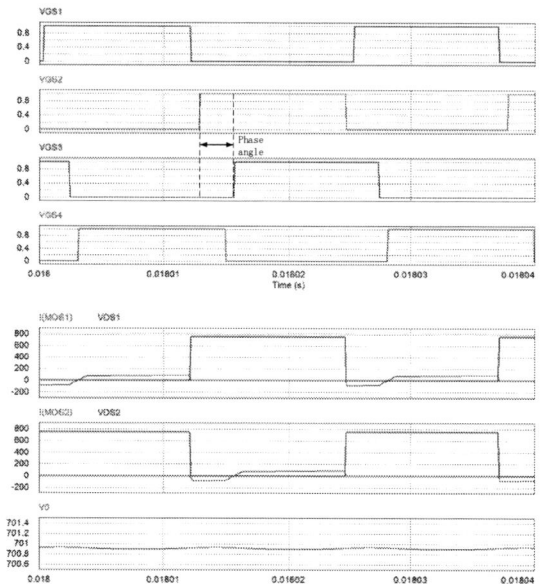

(1) 40kHz SiC MOSFET ISOP PSFB

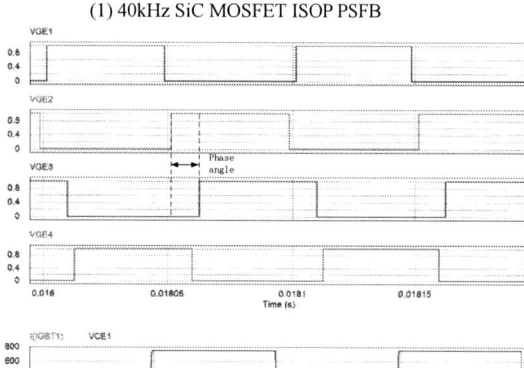

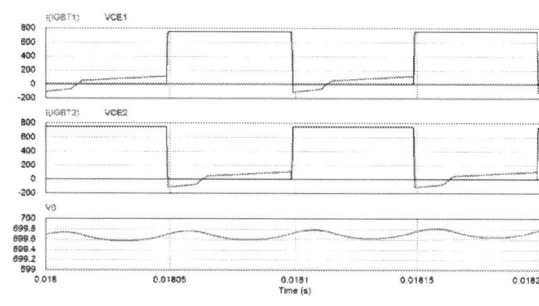

(2) 10kHz Si IGBT ISOP PSFB
Fig.7. Simulation waveforms

As the Fig.7 shows, all of the converters realize ZVS and the output voltage is obtained and kept stable.

A. Loss Characteristics Analysis of SiC ISOP Converter

Every converter is designed in the best conditions in which switching frequency is suitable, ZVS is realized and the output meets the requirements. The rated power of each converter is 80kW. The Si device operating frequency is 10kHz. The SiC device operating frequency is 40kHz. The main parameters and the switching device type of the converters are shown in the table 2.

TABLE II
THE MAIN PARAMETERS OF THE CONVERTERS

	switch type	P (kW)	fs (kHz)	Tn	Lr (uH)
Si PSFB	FF300R17KE4	80	10	1/1	31
SiC ISOP	CAS300M17BM2	80	40	1/1.4	4.65
Si ISOP	FF400R33KF2C	80	10	1/1.4	25.4

According to table 2, the loss and efficiency of each converter are calculated and is shown in Fig. 8.

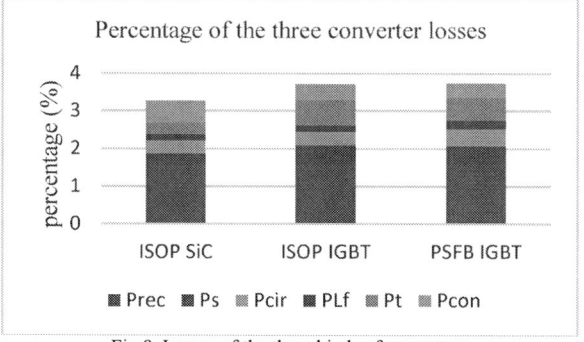

Fig.8. Losses of the three kinds of converters

In Fig.8, for 40kHz SiC MOSFET ISOP PSFB converter, The conduction loss P_{con} is the most, but all of the transformer loss P_t , the LC filter loss P_{Lf} , the circulation loss P_{cir} and the switching loss P_s are the least . For 10kHz Si IGBT ISOP PSFB converter, the transformer loss P_t and the switching loss P_s are the most, and the total loss is the most. For 10kHz Si IGBT PSFB converter, except the circulation loss P_{cir} is the most , the other losses are almost in the middle level. In all, the loss of the SiC MOSFET ISOP PSFB converter is the least, and the loss of the conventional Si PSFB converter is the most.

B. Loss Characteristics Analysis of SiC ISOP Converter

In order to analyze the SiC switching device advantages, the relationship between loss and frequency is studied and the system losses under full load conditions are also given.

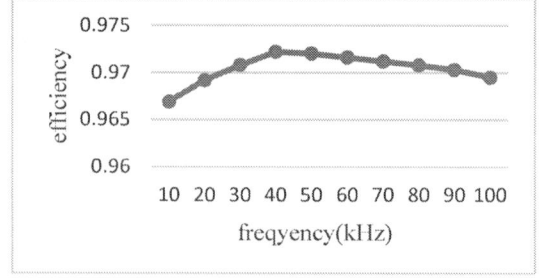

Fig.9. Efficiency of SiC ISOP PSFB with frequency

The system efficiency is almost reduced as the switching frequency increases as shown in Fig. 9. As the frequencies is up to 40kHz, the efficiency is almost at its highest. So if using the SiC ISOP PSFB converter, 40kHz frequency may be a good choice.

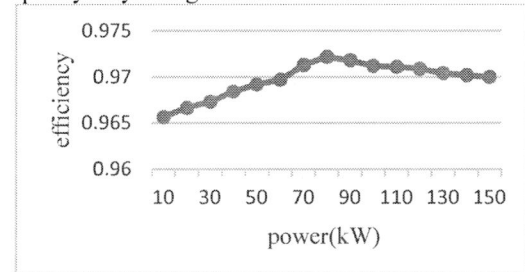

Fig.10. Effciency of SiC ISOP PSFB with power

The Fig. 10 above shows the loss of the converter at full load. The curve shows the efficiency is the lowest at 10kW. The loss increases as the power increases during 10-80kW. The efficiency decreases as the power increases after 80kW, but the downward trend is relatively stable and there is not a very large change. A high efficiency will be obtained by choosing 80kW as the rated power.

C. Single Power Module Test at Rated Power

A single SiC PSFB module to verify the correctness of the theory is built. The experiment module is shown as Fig. 11.

The topology is the same as in Fig. 1, or a module of the structure in Fig. 2, but the switching device is the SiC MOSFET. The rated frequency of the power supply is set as 40kW.

Fig. 11. A single phase shifted full bridge module

There is a transformer, 750V input module, control board, drive circuit and output module in Fig.11. At rated power (40kW) conditions, the voltages and currents of the leading leg switches and the lagging leg switches are tested as shown in Fig. 12.

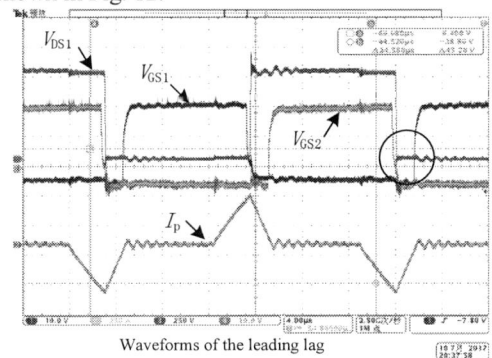

Waveforms of the leading lag

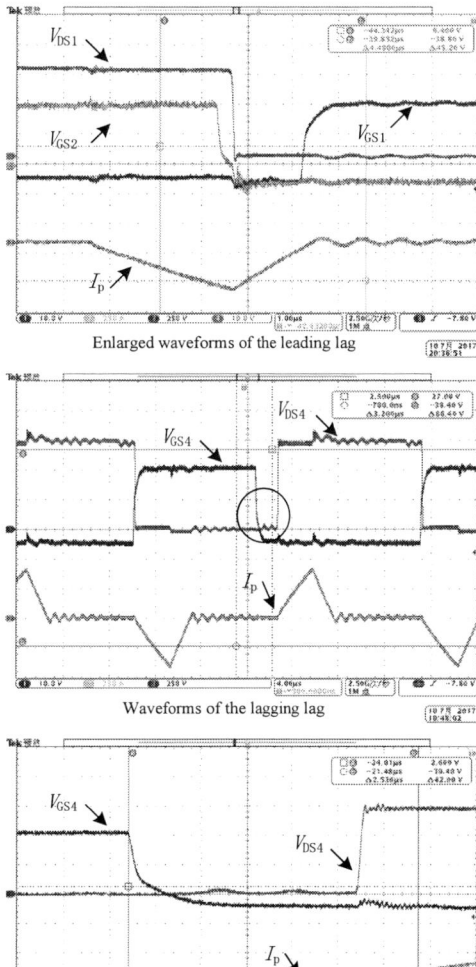

Enlarged waveforms of the leading lag

Waveforms of the lagging lag

Enlarged waveforms of the lagging lag

Fig. 12. Voltage and current waveforms of the switches

In the process of turning on the switches of the leading leg, the voltage and current do not overlap and ZVS is obtained. Due to the restrictions of the hardware parameters, the switches of the lagging leg do not achieve ZVS but achieve ZCS turn on and off to reduce loss.

IV. CONCLUSIONS

For the DC/DC converter in auxiliary power supply system of 1500Vdc input, the topology design of 1700V SiC MOSFET ISOP PSFB circuit is proposed. The output of the DC/DC converter is stable by an input voltage sharing control. Compared with 3300V Si IGBT PSFB converter and 1700V Si IGBT ISOP PSFB converter, 1700 V SiC MOSFET ISOP PSFB converter meets the design requirements and the good operating conditions, the lowest loss, the highest efficiency are obtained.

2056

ACKNOWLEDGMENT

First of all, I'd like to sincerely thank my mentor, Xianjin Huang and my teacher Fei Lin, for giving me guidance in theory and experiment. I am grateful to the School of Electrical Engineering of Beijing Jiaotong University for providing the experimental equipment and conditions. Finally, I also thank my laboratory partners for giving me help.

REFERENCES

[1] W. Zhou, S. Yang, X. Wu and K. Sheng, "Optimal design of SiC MOSFETs for 20kW DCDC converter," *2017 29th International Symposium on Power Semiconductor Devices and IC's (ISPSD)*, Sapporo, 2017, pp. 443-446.

[2] Deshang Sha, Bin Li, Wenqi Yuan, Zhiqiang Guo.,"An input-series-output-parallel PSFB ZVS DC-DC converter based on load adaptation with auxiliary LC network and wide voltage range." *Transactions of China Electrotechnical Society* 2016,36(13):3558-3564+3374. [2017-09-06]. DOI: 10.13334/j.0258-8013. pcsee. 150160

[3] S. R. Jang, S. H. Ahn, H. J. Ryoo and G. H. Rim, "Novel high voltage capacitor charger for pulsed power modulator," *2010 IEEE International Power Modulator and High Voltage Conference*, Atlanta, GA, 2010, pp. 317-321.

[4] I. O. Lee and G. W. Moon, "Phase-Shift PWM Converter With a Wide ZVS Range and Reduced Circulating Current," *IEEE Transactions on Power Electronics*, vol. 28, no. 2, pp. 908-919, Feb. 2013.

[5] Zhao, H. Li, X. Wu and J. Zhang, "An Improved Phase-Shift Full-Bridge Converter with Wide-Range ZVS and Reduced Filter Requirement," *IEEE Transactions on Industrial Electronics*, vol. PP, no. 99, pp. 1-1.

[6] Kun Shen, Jing Zhang, Xiaoyang Yao, Jian Wang. "An improved Parallel Inverter for Train Auxiliary Power Supply System". *Transactions of China Electrotechnical Society*, 2013,28 (05): 250-258. [2017-10-12]., 2013, 28(05) :250-258. [2017-10-12].

[7] Kun Shen, Jing Zhang ,Jian Wang. "Research on Auxiliary Power Supply System Based on PQ Droop Control Inverter Parallel Technology." *Transactions of China Electrotechnical Society*, 2011,26(07):223-229. [2017-10-12].

[8] Jin Wang, Xi Chen, Lan Xiao. "Study on a Novel Input Parallel Output Series DC/DC Converter". *Journal of Nanjing Normal University (Engineering & Technology Edition)*,2012,12(01):1-4. [2017-10-12].

[9] A. E. Awwad, N. Badawi and S. Dieckerhoff, "Efficiency analysis of a high frequency PS-ZVS isolated unidirectional full-bridge DC-DC converter based on SiC MOSFETs," *2016 18th European Conference on Power Electronics and Applications (EPE'16 ECCE Europe)*, Karlsruhe, 2016, pp. 1-10.

[10] S. Guo, P. Liu, R. Yu, L. Zhang and A. Q. Huang, "Analysis and loss comparison of megahertz high voltage isolated DC/DC converters utilizing integrated SiC MOSFET module," *2016 IEEE 4th Workshop on Wide Bandgap Power Devices and Applications (WiPDA)*, Fayetteville, AR, 2016, pp. 291-296.

[11] Zhao Lin Zhang, Ran Yan and Yong Luo, "A full-bridge ZVS converter applied to high power millimeter-wave gyrotron travelling wave tubes," *2016 IEEE International Conference on Microwave and Millimeter Wave Technology (ICMMT)*, Beijing, 2016, pp. 167-169.

[12] P. O'Neill, Jun Zhang and W. G. Hurley, "A Phase-shift Full-Bridge ZVS DC/DC Converter for Wireless Charging of Electric Vehicles," *2015 50th International Universities Power Engineering Conference (UPEC)*, Stoke on Trent, 2015, pp. 1-5.

[13] G. N. B. Yadav and N. L. Narasamma, "An Active Soft Switches Phase-Shift Full-Bridge DC–DC Converter: Analysis, Modeling, Design, and Implementation," in *IEEE Transactions on Power Electronics*, vol. 29, no. 9, pp. 4538-4550, Sept. 2014.

[14] I. O. Lee and G. W. Moon, "Phase-Shift PWM Converter With a Wide ZVS Range and Reduced Circulating Current," in *IEEE Transactions on Power Electronics*, vol. 28, no. 2, pp. 908-919, Feb. 2013.

[15] Zhang Yingwen, Guo Xinhua, Guo Baojia and Bian Yuanjun, "The simulation and design of PSFB ZVS DC-DC converter based on Saber," *2014 IEEE Conference and Expo Transportation Electrification Asia-Pacific (ITEC Asia-Pacific)*, Beijing, 2014, pp. 1-4.

[16] Xiaowei Ge, Zhao Zhang, Hong Cao, Shaojun Xie. "Applications of SiC MOSFET in Aeronautical Static Converter." *Chinese Journal of Power Supply*, 2016, 14 (04): 66-72. [2017-10-12]. DOI: 10.13234/j.issn.2095-2805. 2016 .4.66.

[17] Chunya Zhang, Lin He, Zhiguo Zhang. "Summary of development of switching power supply technology. " *Microelectronics*, 2016, (02):255-260+272.

[18] Lulu Cheng, Xinbo Ruan, Tao Zhang. Study on Control Strategy of DC-DC Converters Connected Input-Series Output-Parallel [J] .*Proceedings of the CSEE*, 2006 (22): 67-73.

The 2018 International Power Electronics Conference

Survey on electromagnetic interference analysis for traction converters in railway vehicles

Zhichang Yang, Hong Li[*], Chao Feng, Yanfeng Jiang, Fei Lin and Zhongping Yang
School of Electrical Engineering, Beijing Jiaotong University, Beijing, China
*E-mail: hli@bjtu.edu.cn

Abstract-In this paper, electromagnetic interference (EMI) analysis for traction converters in railway vehicles is surveyed. The mechanism of EMI generation is introduced firstly. Then the analysis and modeling methods for EMI are presented. After that, the suppression methods of EMI in traction converters are summarized. Finally, the development trend of EMI of traction converters is discussed.

Keywords— EMI, traction converters, railway vehicles.

I. INTRODUCTION

The rapid development of electrified railways has greatly improved the speed and convenience of transportation. However, the electromagnetic environment of electrified railway system, especially for high-speed railway, has increasingly become complex with the increasing application of information technology and advanced power electronics devices. Therefore, electromagnetic compatibility (EMC) is a critical issue, and it must be considered seriously in the design stage of railway system to ensure the functionality and safety of railway vehicles operation [1].

Through various generation mechanism, electrified railways can generate electromagnetic interference (EMI) noise over a wide frequency band in its power supply system, signaling and train control system, railway vehicles and other electrical/electronic equipment. Specifically, the main sources of EMI in railway vehicles include the power electronics harmonics brought by traction transformers, traction converters, motors and DC-DC converters; pantograph arcing; communication and antenna systems or other auxiliaries systems [2-4]. In particular, traction converters, which is one of the aforementioned sources EMI, play a very important role in railway vehicles as the high power energy flow control device and motor drive system [5]. However, the EMI caused by the operation of traction converters under high voltage level and current, cannot be ignored. The strong EMI noise will interfere communication system and control system and affect the functionality and safety of railway vehicles, which must be limited to meet EMC standards [6].

The EMI issue of traction converters in railway vehicles has been highly valued and widely studied in the past decades. The research topics about EMI mainly focus on the analysis of EMI generation mechanism, modeling and prediction of EMI and EMI suppression methods [7]. The EMI introduced by traction converters include that the high frequency harmonic noise caused by

pulse width modulation (PWM) control, the large di/dt and dv/dt caused by switching process of power switches, the leakage current caused by the common mode voltage and so on [8]. The EMI modeling is to model the power switches, establish the high frequency model of passive components and extract parasitic parameters in traction converter system, and constitute the final analysis model of EMI to analyze and predict the EMI level and provide guidance for the suppression of EMI in traction converters [7]. The EMI suppression methods are classified into two main types, by which EMI is suppressed in EMI sources and EMI coupling paths respectively. These two methods can be further subdivided into different categories according to the specific implement methods [7].

In this paper, EMI analysis for traction converters in railway vehicles is surveyed. The paper starts with an introduction to the mechanism of EMI generation. Then the modeling and predicting methods for EMI are presented. After that, the suppression methods of EMI in traction converters is summarized. Finally, the summary and conclusions are given.

II. MECHANISM OF EMI GENERATION

A. Structure of Traction converters

Fig. 1 shows the typical structure of traction drive system in high speed trains, which is made up of a traction transformer, traction converter including four-quadrant converters, a dc link, a three-phase inverter and motors [9]. Usually, the traction transformer has multiple secondary windings, four-quadrant converters and traction transformer are arranged as dual or multiple structure is adopted to reduce the current harmonics on grid side of traction transformer. Due to the same topology in the multiple structure, analysis can be focus on only single converter of them.

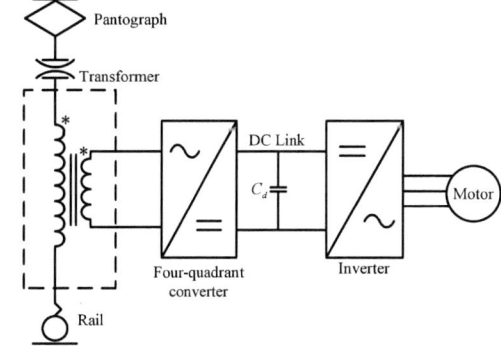

Fig. 1. The typical structure of traction drive system.

This work was supported by the Fundamental Research Funds for the Central Universities under Grant 2017JBM054 and the General Program of National Natural Science Foundation of China under Grant 51577010.

The 2018 International Power Electronics Conference

The typical structure of traction converter as shown in Fig.2, which contains a four quadrant converter with two levels, a dc link and a three phase inverter. The single phase ac voltage from pantograph is stepped down through traction transformer firstly, then is converted from ac to dc by the four quadrant converter, and the three phase ac power with adjustable voltage and frequency will be generated by three phase inverter to drive the motors. The control methods of four quadrant converters and inverter are usually sinusoidal pulse width modulation (SPWM).

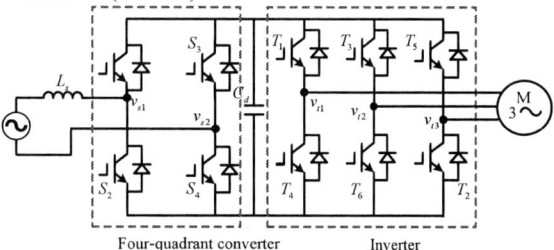

Fig. 2. Typical structure of traction converter.

B. EMI generation mechanism of traction converters

The EMI of traction converters can be classified into conducted EMI and radiated EMI, and the conducted EMI is the main interference noises, which is divided into the common-mode (CM) EMI and the differential mode (DM) EMI based on the EMI current flow paths. Thus in this paper, the conducted EMI is mainly introduced.

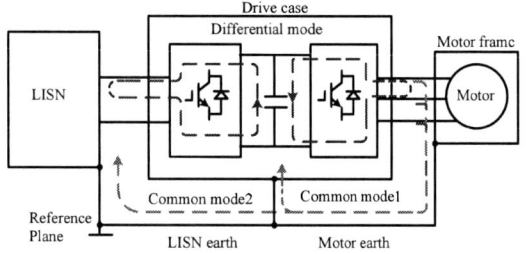

Fig. 3. Schematic of EMI noise propagation in a traction converter.

In general, the EMI introduced by traction converters include that the high frequency harmonic noise caused by SPWM control, the large di/dt and dv/dt caused by the switching process of power switches, the leakage current caused by the common mode voltage and so on [8]. The four quadrant converters and inverter controlled by SPWM contain rich high frequency harmonics while generating useful fundamental frequency signal, which will generate EMI noise in the form of conduction and radiation. The operation of high power switches in traction converters will also cause serious EMI problems because of the large di/dt and dv/dt are very large. The large di/dt and dv/dt will cause oscillation of the parasitic parameters of the circuits and generate high level of conducted EMI noise and radiated EMI noise. The pulsating CM voltage can charge and discharge to the parasitic capacitance and form the CM leakage current with high frequency harmonics, which will affect the electrical and electronic equipment with the same ground. The EMI noise propagation paths in a traction converter is shown in Fig. 3.

III. MODELING AND PREDICTING METHODS FOR EMI

The normal EMI analysis methods are based on testing in experiment, which can estimate EMI level of power converters. However, the experimental test cannot obtain the adequate information of EMI source, and cannot provide accurate guidance for EMI suppression. Therefore, it's meaningful to analyze the mechanism of EMI generation in traction converter and model the EMI equivalent circuit for predicting the EMI.

To analyze the EMI in traction converter, it is necessary to establish the high frequency model of power switches such as insulated gate bipolar translator (IGBT), passive elements, PCB wiring and other elements in traction converters [7].

A. Modeling of power switches

In [10], the model of IGBT and power diodes of traction converter is established accurately in PSpice to analyze the EMI characteristics of the traction converters. The corresponding IGBT model is given in Fig. 4. The IGBT model composed of a power MOSFET and a PNP transistor in Darlington structure. The parasitic parameter such as parasitic capacitor and stray inductance of IGBT, which could affect the turn on and turn off performance of IGBT, are also considered in this IGBT model, Based on the IGBT model in PSpice, the EMI characteristics and high frequency harmonics of railway traction system can be obtained in simulation.

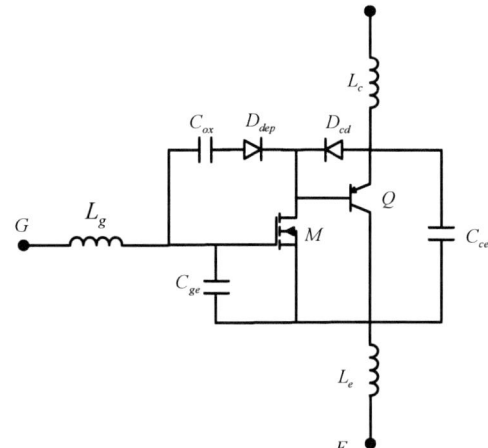

Fig. 4. Appropriate IGBT model in traction converter.

In [11-12], the IGBT module of power converters for traction application is considered as an EMI generation source. In [11], it indicates that the radiated EMI noise is dependent on the dV_{ce}/dt of IGBT module, followed by the driving parameters, and the value of dV_{ce}/dt is related to the value of di/dt directly. And, the EMI peaks appearing in high frequencies depend on the internal stray parameters, namely the parasitic inductances and capacitances of IGBT module, not the external parameters of circuit. The value of V_{ce} can only alter the internal capacitance and the resonance, which will further lead to the change of the resonant noise frequency. The current flowing through power switches can be related to the basic antenna, which leads to EMI noise emission.

2059

A high frequency model with lumped-charge approach for PIN diode is presented in [13]. The PIN diode model contains forward recovery and complete reverse recovery with sweeping-out and recombination effects. And the accurate parameter extraction method with curve-fitting approach is provided for the PIN diode model.

B. Modeling of passive components

In consideration frequency band range of EMI, passive components in traction converters including wire, cable, transformer, inductor, capacitor, resistor and so on can no longer be considered as ideal devices, since the parasitic parameters have great influence on their frequency characteristics of passive components. Thus, it is necessary to establish the high frequency model of passive components which contains their parasitic parameter effect in EMI model establishment of traction converters.

The modeling method of wire, cable, capacitor, resistor in high frequency band have been presented and the mathematical models of these components have been developed that consider the nonideal behavior of components are developed in [14]. The high frequency modeling method and high frequency characteristic analysis of magnetic components in power converters mainly adopt theoretical calculation method based on field analysis using finite element model [15] and the parameter extraction method based on impedance measurement using impedance analyzer [16].

C. Extracting of parasitic parameters

The conducted EMI will also transmit through parasitic parameters of traction converter system. Thus the parasitic parameters extraction is necessary for EMI modeling. Several methods for extraction of parasitic parameters have been developed but essentially they can be divided in two types: mathematical modeling techniques based on finite element model and impedance measurement with impedance analyzer. The parasitic capacitances between IGBT heat sink and ground was extracted by using measurement method in [17]. A lumped circuit model of wire coupling has been built in [18] based on the admittance blocks extracted from a mixed-potential integral equation formulation. Reference [19] present the extraction method of PCB parasitic parameters in power converters by using electromagnetic analysis software based on finite element method.

D. Modeling of EMI equivalent circuit

In [20], each pivotal subsystem of traction converter is modeled to simulate the system EMI accurately. In [21], the conducted EMI equivalent circuit of traction drive system is formed to analyze the influence rules of parasitic parameters on EMI spectrum distribution. The conducted EMI paths of traction drive system is shown in Fig. 5, and the equivalent circuit of conducted EMI is shown in Fig. 6. It considers the high frequency parameters of the traction transformer and the traction converter. According to the amplitude-frequency curve of

equivalent circuit, the influence rule of parasitic parameters on the EMI spectrum distribution can be analyzed. The above method can be applied to predict the conducted EMI noise and provide information for the EMC design.

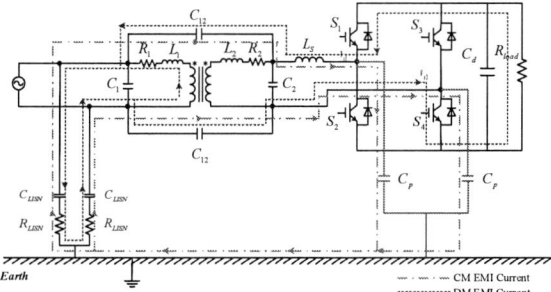

Fig. 5. The conducted EMI paths of traction drive system.

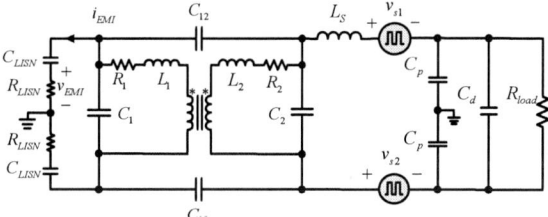

Fig. 6. Equivalent circuit of conducted EMI.

IV. SUPPRESSION METHODS FOR EMI IN TRACTION CONVERTERS

The three elements of EMI generation are interference source, interference propagation path and interfered equipment. Thus EMI suppression methods could be classified as two main types: one is to restrain the noise generation of interference source, and another is to modify interference propagation path by using filtering, shielding and other means.

According to above analysis, the classification of conducted EMI suppression techniques for power converters is shown in Fig. 7 [22]. The current EMI suppression methods along the propagation path is designing filters for power converters which can improve impedance characteristics of interference propagation paths and reduce EMI noise. The filters can be divided into external EMI filters and internal filters. Correspondingly, the external EMI filters can be separated into passive filters and active filters, and internal filters can be realized by designing of PCB layout and topology of converters reasonably. The EMI suppression at interference source, it can be carried out by circuit, layout and components design, better modulation methods and switching process improvement.

Although the power level, structure and other electrical characteristics of traction converters are different from that of other power converters, but the mechanism of EMI generation and main propagation paths are similar, which can provide a guidance for EMI suppression of traction converters in railway vehicles.

The 2018 International Power Electronics Conference

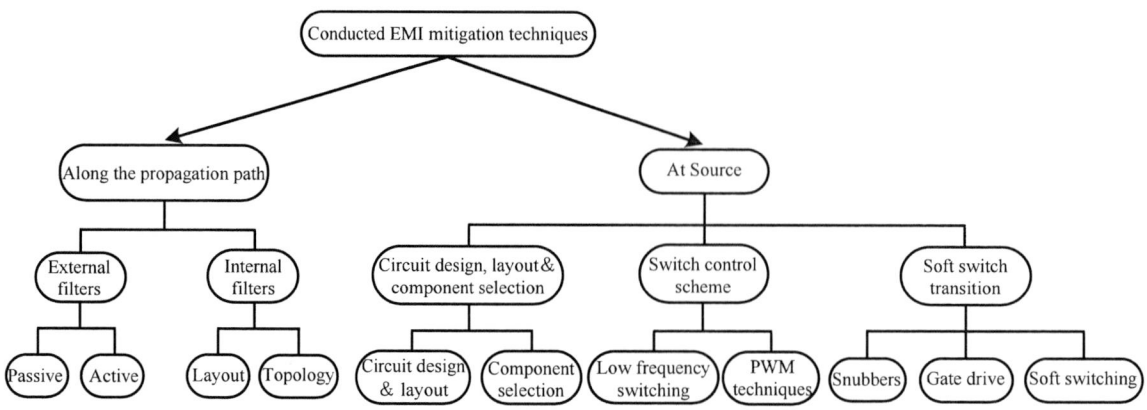

Fig. 7. Classification of EMI suppression techniques [22].

A. EMI suppression with filters

1) Passive filter

In [23], a passive method is proposed to suppress conducted EMI of AC drive system effectively according to the analysis and measurement results of CM currents. The star connected Y capacitances has been added in the input side of traction converter, as shown in Fig. 8. The value of Y capacitances has been obtained by the measurements in the time domain, while there is only a relative small CM voltage on the Y capacitances. Using this passive method, the CM current can be suppressed easily using a smaller CM choke, which will reduce the volume of filter. In [24], a detailed process for the design of filter in the input side of 3-phase traction inverter has been given with considering the system stability to alleviate dead-time power gap. An output passive filter of inverter has been proposed to reduce DM dv/dt and suppress CM voltage in [25]. The filter consists of a LC low-pass filter and a CM transformer to simplify the structure and design of the filter as shown in Fig. 9.

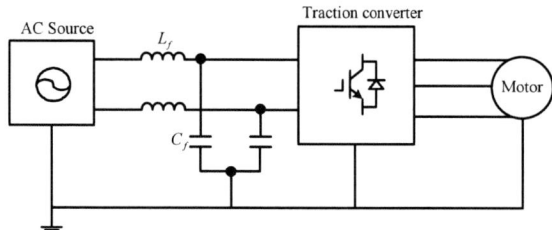

Fig. 8. A traction converter with the input passive filter.

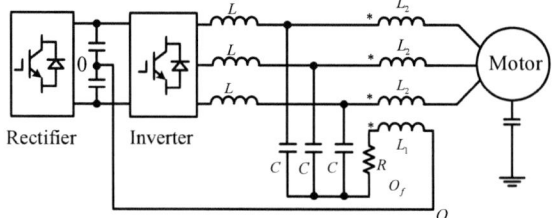

Fig. 9. Inverter with output passive filter.

2) Active filter

Active filter can suppress EMI using active electronic circuits and has the advantage of smaller size compared with the passive filter. An active filtering system to reduce the harmonics of railway power supply system has been presented in [26]. The active filtering system includes a transformer which is feed by a power amplifier and reinjects a voltage signal into dc line to suppress harmonics, the voltage signal is proportional to the line current. A CM active filter using in an induction motor drive system to suppress EMI has been described in [27]. The schematic of the CM active filter within the motor drive system is shown in Fig. 10, the CM active filter includes a detection circuit consists of three wye-connected capacitors and series connected capacitor, a transistor amplifier realize by a push–pull emitter follower and two complementary transistors in Darlington configuration, and a CM transformer. The operation principle of the CM active filter is that the CM voltage from detection system is transferred to the CM transformer by the transistor amplifier, and the compensation voltage reinjected to the transmission line between the inverter and motor through CM transformer to suppress CM voltage. In [28], a CM active filter applied in induction motor drives system is proposed. The active filter is designed based on CM voltage detection in the input side of inverter and a feedback scheme which is realized by using linear amplifiers. The CM active filter can increase the drive reliability and suppress EMI noise caused by high frequency CM current.

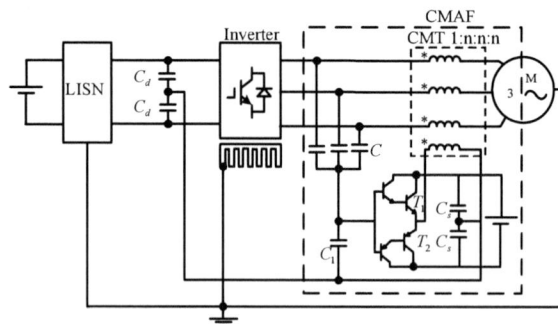

Fig. 10. Schematic of active filter in motor drive system.

B. EMI suppression with switching process optimization

High di/dt and dv/dt, voltage overshoots and oscillation occurred in switching process of s power switches will lead to serious EMI in traction converter system. Switching process optimization is the direct

2061

method to suppress EMI caused by the above cases. The switching process optimization methods involves Gate-drive improvements, soft-switching technology, snubbers.

1) Gate-drive improvements: The voltage and current waveforms of power switches in switching process are closely related to the characteristics of the gate-drive of power switches. The EMI suppression methods relative to gate-drive improvement contain slowing down switching transitions, reducing or eliminating overshoot and oscillation, which can be realized by increasing the drive resistance and adjusting gate voltage and current in switching process using active gate-drive [29-32].

In [29], an active gate drive circuit consisting of a gate drive power amplifier and active voltage clamping circuit is presented for IGBT based on two stage active gate control, which, as shown in Fig. 11. By using two capacitors and a transient voltage suppressor, different feedback can be gained at different turn-off instants for the gate drive circuit. Also, the voltage overshoots, oscillation and control the dv/dt of switching process is suppressed. In [30], an active gate drive with controlling gate current in the active region during turn off of IGBT is introduced to suppress overshoots. An active gate drive is presented for the improvement of switching performance of high power IGBT in [31]. The active gate drive has an adjustable complementary current source to give extra gate drive current which is helpful to increase the flexibility and controllability of active gate drive and reduce the overshoot in switching transients. Less EMI generation can also be achieved by optimizing the gate voltage when the collector voltage is rising or falling [32]. However, suppressing EMI with gate-drive improvement methods may increase the switching losses of power switches, and it is necessary to consider EMI suppression and the efficiency of system comprehensively when the gate-drive circuits are design.

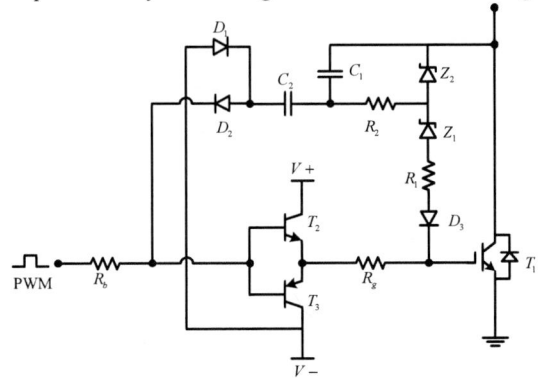

Fig.11. Schematic of an active gate drive circuit.

2) Soft-Switching Techniques: The resonant process of soft switching limits the rate of change of the voltage and current, and eliminates the spike of switching waveform, which can reduce EMI effectively in theory.

In [33], a comparison of the effect on is given when the soft switched and hard switched is are respectively employed into the PWM controlled inverter to suppress EMI suppression. As expected, the inverter with soft-switching works well to reduce EMI. However, in some cases, the effect of soft switching on EMI suppression is not obvious than hard-switch in PWM inverter because the additional circuit devices are applied to realize the soft switch. The functions and limitations of the soft-switching technology to improve the EMI of the AC-DC converter are analyzed in [34]. It is pointed out that the layout and routing of the circuit have an important influence on the suppression of EMI.

3) Snubbers: Snubber circuits are placed in parallel with the power switches to absorb the surge voltage thereby suppressing the EMI noise caused by voltage overshoot and oscillation in high frequency band.

A passive snubber circuit with resistor (R) and capacitor (C) filter is designed in [35] for a half-bridge module of power switches, which is used to eliminate voltage overshoot and oscillation, and suppress EMI simultaneously. In [36], a RC snubber circuit is proposed for a silicon carbide JFET inverter of motor drives as shown in Fig. 12, which is used to damp parasitic oscillations of power switches and achieve a better performance of EMC.

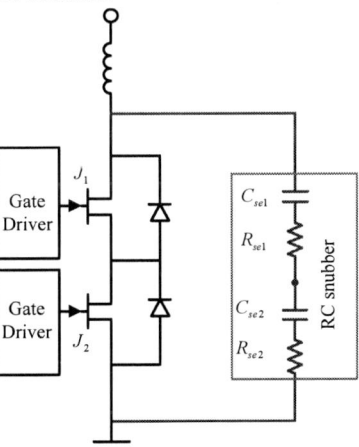

Fig.12. Schematic of snubber circuit.

C. EMI suppression with switching frequency modulation techniques

Traditionally, traction converters (including rectifiers and inverters) operate under PWM mode at a fixed switching frequency, in which the significant magnitudes of EMI occur at the switching frequency and its multiples frequencies. In order to suppress EMI from the source, many switching PWM strategies operating at variable frequency have been proposed, which are also known as the "spread-spectrum techniques". The principle is to extend the energy concentrated around the switching frequency harmonics before to a wider frequency range to reduce EMI peaks. At present, the main spread-spectrum methods contain periodic modulated PWM, random PWM and chaotic PWM.

In [37] and [38], different periodic modulated PWM schemes have been proposed to reduce EMI noise and achieve a well suppression effect. However, the disadvantage of periodic modulated PWM is its spectrum is discrete and limit the spectrum spread effect. To solve this problem, modulating carrier frequency twice approach is applied for the single-phase inverter in [39]. Compared with conventional periodic modulated PWM, the maximum suppression amplitude of EMI peaks can reach 5 dB or more with the scheme of modulating carrier frequency twice.

The previous researches on random PWM has been summarized and proposed an improved random PWM algorithm used for ac-drive in [40]. The comparison of experiment results between random PWM and fixed frequency PWM shows that the EMI under random PWM is 10 dB~20dB lower than that of fixed frequency PWM.

In practice, it is difficult to generate an ideal random signal and it is more difficult to generate a random signal with a specific distribution. Therefore, a pseudorandom sequence with finite length is actually used instead of an ideal random signal [41]. , by which the actual effect of random sequence will be greatly reduced. Due to the inherent randomness of chaotic systems, chaotic signals are used as the pseudorandom signals for PWM carrier modulation in many recent studies. The chaotic pulse position modulated PWM has been proposed and the chaotic amplitude frequency modulated PWM have been proposed in [42] and [43], which can significantly reduce the EMI peak. But for chaotic pulse position modulated PWM, some EMI peaks still exist, and for chaotic amplitude frequency modulated PWM, there are still some low-order harmonics. To overcome the drawbacks of the above two chaotic PWM schemes, a hybrid chaotic PWM control is proposed and adopted to an AC motor drive [44], which can beneficial to suppress the EMI peaks and reduce low frequency noises generation simultaneously. In [45] and [46], novel chaotic PWM modulations with a design of chaotic carrier have been proposed which are implemented in four-quadrant AC-DC converters, the realizing schematic is shown in Fig. 13, and the relationship between the changing rule of switching frequency and the EMI spectrum distribution is given to reveal the principle of EMI suppression by chaotic PWM based on deducing and analyzing the harmonic components of the four-quadrant AC-DC converter.

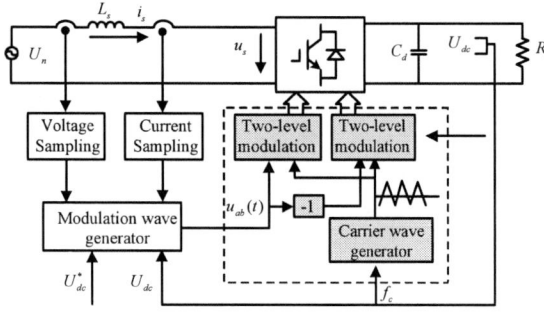

Fig.13. Four-quadrant AC-DC converter with Chaotic PWM.

Recently, another variable switching frequency PWM scheme based on current ripple prediction has been proposed [47], which not only can reduce the EMI noise, but also control peak or RMS value of the current ripple within a certain value.

In addition to adopting spread-spectrum techniques to suppress EMI, some researchers are seeking for new PWM strategies to eliminate common-mode interference. In [48], three PWM strategies have been employed to suppress common-mode interference: (1) Active vector PWM, (2) Three-dimensional PWM, (3) Active vector with memorizing PWM. The experiment result shows that the proposed methods can reduce EMI peaks about 20 dB at switching frequency, and suppress EMI in frequency range from few hundreds Hz to 600 kHz effectively.

The other methods of EMI suppression in traction converters include PCB layout designing, component selection and converter topology designing reasonably, well ground connection, shielding and so on. Actually, EMI suppression in traction converter is a complex issue. One suppression method can only solve the limit EMI problems, it is necessary to use multiple aforesaid methods comprehensively to meet EMC standards.

V. CONCLUSIONS

A comprehensive summary of the current literatures about EMI researches on traction converters in railway vehicles has been given in this paper. According to the results of survey, the current EMI researches of traction converters mainly focus on the EMI generation mechanism, EMI modeling and predicting methods, EMI suppression technologies.

Based on the topologies and operation principles of traction converters, the researches on EMI generation mechanism involve the analysis of EMI noise sources and EMI propagation paths. In the researches on EMI modeling and predicting methods, the IGBT and diode model, the high frequency model of passive components and parasitic parameters extraction have been studied deeply. What's more, the wide bandgap semiconductor devices like silicon carbide (SiC) device may be employed into traction converter in the future. Hence, it is necessary to establish the model of wide bandgap semiconductor to analyze the EMI further.

The EMI suppression technologies can be implemented at the EMI sources or along the coupling paths. Generally, it is preferable to suppress EMI at the EMI sources, compared with along the coupling paths. Since the switching frequency modulation techniques can suppress EMI effectively in low and medium frequency, and the gate-drive improvements can reduce the EMI in high frequency band, a reasonable combination of the two methods can suppress EMI in the whole frequency band, which has a significant advantage in operability for its not changing the power components or topology and little additional volume.

It should be noticed that the researches on the EMI of traction converters in railway vehicles are still inadequate. Particularly, with the increasing speed of railway vehicles, the development of EMC tractor is facing more and more serious challenges. Therefore, it's urgent for scholars and engineers to study the EMI comprehensively and deeply.

2063

REFERENCES

[1] A. Ogunsola, A. Mariscotti, "Electromagnetic compatibility in railways: analysis and management," *Springer Berlin*, 2013.

[2] C. A. Marshman, "EMC in railways," *IET 13th Professional Development Course on Electric Traction Systems*, London, UK, 2014, pp. 1-8.

[3] C. D. Xu, K. W. E. Cheng, Y. Zou, X. L. Wang, S. R. Raman, and X. D. Xue "Electromagnetic scattering of high power traction transformer in high speed railway based on FEM," *International Symposium on Electrical Engineering*, Hong Kong, China, 2016, pp. 1-5.

[4] S. Midya, D. Bormann, T. Schutte and R. Thottappillil, "Pantograph arcing in electrified railways—mechanism and influence of various parameters—Part II: with AC traction power supply," *IEEE Transactions on Power Delivery*, vol. 24, no. 4, pp. 1940-1950, 2009.

[5] W. Song, S. Jiao, Y. W. Li, J. Wang and J. Huang, "High-frequency harmonic resonance suppression in high-speed railway through single-phase traction converter with LCL filter," *IEEE Transactions on Transportation Electrification*, vol. 2, no. 3, pp. 347-356, 2016.

[6] J. Yu, Y. Lu and B. Chen, "Research on the EMI optimization of traction converter," *Railway Locomotive & Car*, vol. 36, no. 4, pp. 92-96, 2016.

[7] Z. Qian and H. Chen, "State of art of electromagnetic compatibility research on power electronic equipment," *Transactions of China Electrotechnical Society*, vol. 22, no. 7, pp. 1-11, 2007.

[8] X. L. Zhang, "Research on electromagnetic compatibility of traction converter System," *M.S. dissertation*, Beijing Jiaotong University, 2011.

[9] Z. He, H. Hu, Y. Zhang and S. Gao, "Harmonic resonance assessment to traction power-supply system considering train model in China high-speed railway," *IEEE Transactions on Power Delivery*, vol. 29, no. 4, pp. 1735-1743, 2014.

[10] K. Jia and R. Thottappillil, "EMC Assessment of the Railway Traction System by Using PSpice," *Asia-Pacific International Symposium on Electromagnetic Compatibility*, Beijing, China, 2010, pp. 598-601.

[11] G. Busatto, C. Abbate, F. Iannuzzo, L. Fratelli, B. Cascone and G. Giannini, "EMI characterisation of high power IGBT modules for Traction Application," *IEEE 36th Power Electronics Specialists Conference*, Recife, Brazil, 2005, pp. 2180-2186.

[12] G. Busatto, C. Abbate, L. Fratelli, F. Iannuzzo, G. Giannini and B. Cascone, "EMI analysis in high power converters for traction application," *European Conference on Power Electronics and Applications*, Dresden, Germany,2005, pp. 1-9.

[13] Y. Yuan and Z. Qian, "An improved lumped-charge model and parameter extraction approach of PIN diodes," *IEEE 33rd Annual Power Electronics Specialists Conference*, Cairns, Australia, 2002, pp. 1301-1304.

[14] C. R. Paul "Introduction to electromagnetic compatibility," *John Wiley & Sons, Inc.*, 2006.

[15] R. Asensi, J. A. Cobos, O. Garcia, R. Prieto and J. Uceda, "A full procedure to model high frequency transformer windings," *IEEE 1994 25th Annual IEEE Power Electronics Specialists Conference*, Taipei, Taiwan, 1994, pp. 856-863.

[16] J. Pleite, R. Prieto, R. Asensi, J.A. Cobos and E. Olias, "Obtaining a frequency-dependent and distributed-effects model of magnetic components from actual measurements," *IEEE Transactions on Magnetics*, vol. 35, no. 6, pp. 4490-4502, 1999.

[17] L. Arnedo and K. Venkatesan "Pspice simulation for conducted EMI and overvoltage investigations in a PWM induction motor drive system," *IEEE Workshop on Computers in Power Electronics*, Mayaguez, USA, 2002, pp. 132-137.

[18] D. Zhang, Y. Wen, Y. Wang, D. Liu, X. He and J. Fan, "Coupling analysis for wires in a cable tray using circuit extraction based on mixed-potential integral equation formulation," *IEEE Transactions on Electromagnetic Compatibility*, vol. 59, no. 3, pp. 862-872, 2017.

[19] J. Lai, X. Huang, E. Pepa, S. Chen and T.W. Nehl, "Inverter EMI modeling and simulation methodologies," *IEEE Transactions on Industrial Electronics*, vol. 53, no. 3, pp. 736-744, 2006.

[20] K. Jia and D. Ribbenfjard, "Prediction of the conducted emission generated by the electrified railway traction system," *IEEE International Symposium on Electromagnetic Compatibility*, Dresden, Germany, 2015, pp. 134-137.

[21] Z. Yang, H. Li, Z. Yang, Y. Xu and P. Lin, "Conducted electromagnetic interference analysis method for traction drive system in high speed trains," *IEEE Conference and Exp Transportation Electrification Asia-Pacific*, Harbin, China, 2017, pp. 1-5.

[22] K. Mainali and R. Oruganti, "Conducted EMI mitigation techniques for switch-mode power converters: A survey," *IEEE Transactions on Power Electronics*, vol. 25, no. 9, pp. 2344-2356, 2010.

[23] A. Kempski, R. Strzelecki, R. Smolenski and G. Benysek, "Suppression of conducted EMI in four-quadrant AC drive system," *IEEE 34th Annual Power Electronics Specialist Conference*, Acapulco, Mexico, 2003, pp. 1121-1126.

[24] M. Youssef, J. Abu-Qahouq and M. Orabi, "The electromagnetic compatibility design considerations of the input filter of a 3-phase inverter in a railway traction system," *IEEE Energy Conversion Congress and Exposition*, Atlanta, USA, 2010, pp. 4210-4216.

[25] Y. Jiang, D. Xu, Y. Liu and H. Zhao, "Research on a novel inverter output passive filter," *Proceedings of CESS*, vol. 24, no. 11, pp. 134-138, 2004.

[26] C. Broche, J. Lobry, P. Colignon and A. Labart, "Harmonic reduction in DC link current of a PWM induction motor drive by active filtering," *IEEE Transactions on Power Electronics*, vol. 7, no. 4, pp. 633-643, 1992.

[27] M. C. D. Piazza, A. Ragusa and G. Vitale, "Effects of common-mode active filtering in induction motor drives for electric vehicles," *IEEE Transactions on Vehicular Technology*, vol. 59, no. 6, pp. 2664-2673, 2010.

[28] M. C. D. Piazza, A. Ragusa and G. Vitale, "An optimized feedback common mode active filter for vehicular induction motor drives," *IEEE Transactions on Power Electronics*, vol. 26, no. 11, pp. 3153-3162, 2011.

[29] Q. Zheng, M. Du and K. Wei, "Research on electromagnetic interference suppression method of motor drive system," *Electric Drive*, vol. 47, no. 7, pp. 3-6, 2017.

[30] J. Kim, D. Park, J. Kim and B. Kwon, "An active gate drive circuit for high power inverter system to reduce turn-off spike voltage of IGBT," *7th International Conference on Power Electronics*, Daegu, Korea, 2007, pp. 127-131.

[31] F. Zhang, X. Yang, Y. Ren, L. Feng, W. Chen and Y. Pei "Advanced active gate drive for switching performance improvement and overvoltage protection of high-Power IGBTs," *IEEE Transactions on Power Electronics*, vol. 33, no. 5, pp. 3802-3815, 2018.

[32] X. Yang, X. Zhang and P. Palmer, "IGBT converters conducted EMI analysis by controlled multiple-slope switching waveform approximation," *IEEE International Symposium on Industrial Electronics*, Taipei, Taiwan, 2013, pp. 1-6.

[33] Y. Yang, H. Gao, X. Wang and J. Tang, "Study on soft switching technology to reduce electromagnetic interference of PWM inverter," *International Conference on Future Electrical Power and Energy Systems*, Qingdao, China, 2012, pp. 384-390.

[34] Y. Lu, M. Qian and T. Green, "Soft switched AC/DC power converter and their EMI property improvement," *Proceedings of the Csee*, vol. 20, no. 7, pp. 14-18, 2000.

[35] H. Kim, S. Ahn, I. Oh and J. Yook, "Reduction of radiated electromagnetic interference in 4-phases pulse width modulation converter system," *Electrical Design of Advanced Packaging and Systems Symposium*, Seoul, South Korea, 2008, pp. 39-42.

[36] X. Gong and J. A. Ferreira, "Investigation of Conducted EMI in SiC JFET Inverters Using Separated Heat Sinks," *IEEE Transactions on Industrial Electronics*, vol. 61, no. 1, pp. 115-125, 2014.

[37] A. Ruiz-Gonzalez, M. J. Meco-Gutierrez, F. Perez-Hidalgo, F. Vargas-Merino and J. R. Heredia-Larrubia, "Reducing acoustic noise radiated by inverter-fed induction motors controlled by a new PWM strategy," *IEEE Transactions on Industrial Electronics*, vol. 57, no. 1, pp. 228-236, 2010.

[38] Y. Xu, Q. Yuan, J. Zou and Y. Li, "Analysis of triangular periodic carrier frequency modulation on reducing electromagnetic noise of permanent magnet synchronous motor," *IEEE Transactions on Magnetics*, vol. 48, no. 11, pp. 4424-4427, 2012.

[39] J. Huang and R. Xiong, "Study on modulating carrier frequency twice in SPWM single-phase inverter," *IEEE Transactions on Power Electronics*, vol. 29, no. 7, pp. 3384-3392, 2014.

[40] K. Borisov, T. E. Calvert, J. A. Kleppe, E. Martin and A. M. Trzynadlowski, "Experimental investigation of a naval propulsion drive model with the PWM-based attenuation of the acoustic and electromagnetic noise," *IEEE Transactions on Industrial Electronics*, vol. 53, no. 2, pp. 450-457, 2006.

[41] B. Scheers and V. L. Nir, "Pseudo-random binary sequence selection for delay and add direct sequence spread spectrum modulation scheme," *IEEE Communications Letters*, vol. 14, no. 11, pp. 1002-1004, 2010.

[42] Z. Wang, K. T. Chau and C. Liu, "Improvement of electromagnetic compatibility of motor drives using chaotic PWM," *IEEE Transactions on Magnetics*, vol. 43, no. 6, pp. 2612-2614, 2007.

[43] Z. Wang and K. T. Chau, "Design and analysis of a chaotic PWM inverter for electric vehicles," *IEEE Industry Applications Annual Meeting*, 2007, pp. 1954-1961.

[44] Z. Zhang, K. T. Chau, Z. Wang and W. Li, "Improvement of electromagnetic compatibility of motor drives using hybrid chaotic pulse width modulation," *IEEE Transactions on Magnetics*, vol. 47, no. 10, pp. 4018-4021, 2011.

[45] H. Li, Z. Li, F. Lin, and B. Zhang, "Suppressing harmonics in four-quadrant AC-DC converters with chaotic SPWM control," *International Journal of Circuit Theory and Applications*, vol.42, no.4, 2014, pp. 331-342.

[46] H. Li, F. Lin, Z. Li, X. You and T. Q. Zheng and B. Zhang, "The application of chaotic PWM control for EMI suppression," *The International Journal for Computation and Mathematics in Electrical and Electronic Engineering*, vol. 32, no. 3, 2013, pp. 750-762.

[47] D. Jiang and F. Wang, "Variable switching frequency PWM for three-phase converters based on current ripple prediction," *IEEE Transactions on Power Electronics*, vol. 28, no. 11, pp. 4951-4961, 2013.

[48] C. Jettanasen, "Reduction of common-mode voltage generated by voltage-source inverter using proper PWM strategy," Asia-Pacific Symposium on Electromagnetic Compatibility, Singapore, Singapore, 2012, pp. 297-300.

Development of Traction Motor for New Zero - emission Vehicle

Akinobu Iwai[1][*], Satoshi Honjo[1], Hirofumi Suzumori[1], Toshio Okazawa[1]
1 Automobile R&D Center, Honda R&D Co., Ltd., Haga-machi, Haga-gun, Tochigi, Japan
*E-mail:Akinobu_Iwai@n.t.rd.honda.co.jp

Abstract- **This paper discusses structure that enhances both motive power performance and quietness of a newly developed FCV (Fuel Cell Vehicle) traction motor.**

To enhance motive power performance, the research focused on the stator lamination technique. An adhesive for fastening electric steel sheets inside dies is minimized. Reducing thickness of the adhesive layers raised maximum torque by 17% compared to the previous motor.

To enhance the quietness of the vehicle interior, it is important to lower the magnetic excitation force. To do this, a new magnetic circuit was developed and magnet position was optimized.

The newly developed traction motor to go on FCV provides quietness, high power and smooth driving feeling. Compared to the previous model, the new motor raises maximum output power from 100 kW to 130 kW, and maximum torque from 256 Nm to 300 Nm. It also lowers motor noise heard in the interior of the vehicle by 5-10 dB(A).

1. INTRODUCTION

In recent years, as automobile production volume in the world increases including the progression of motorization in emerging nations, societal demand for environmentally-friendly performance of automobiles is becoming severer every year. Under these circumstances, the automobile industries are introducing various next-generation clean cars while diversifying their energy sources, such as FCV, electric vehicles, and plug-in hybrid vehicles. Aiming to achieve the joy of free mobility and an enriched and sustainable society, Honda has promoted development toward the diffusion of FCV [1][2][3][4].

In March 2016, with a view toward the full-scale diffusion of fuel cell vehicles, the 2016 Model fuel cell vehicle was developed featuring a new revolutionary appeal that combines high "universal value" of utilization as a car with the "cutting-edge appeal" that is unique to fuel cell vehicles.

By shrinking the powertrain unit that includes the drive motor and fuel cell, etc. and which is equivalent to a conventional engine and transmission, and mounting them under the front hood, we realized the world's first five-seat sedan package of a fuel cell vehicle as a universal value.

As a cutting-edge appeal, we further enhanced the quiet, powerful and smooth drive feeling that is unique to fuel cell vehicles by improving the power performance of the drive motor.

In this paper, we report on a structure that both enhance high density power technology and quietness.

2. SYSTEM OUTLINE

Fig.1 shows the fuel cell powertrain unit of the new-type FCV. The fuel cell powertrain is primarily consist of a drive motor, a fuel cell stack, a fuel cell boost converter (FCVCU), a power control unit (PCU), and an electric turbo air compressor. Compactly consolidating the fuel cell powertrain to the same size as the V6 engine enables mounting these components under the front hood and achieving a comfortable interior space.

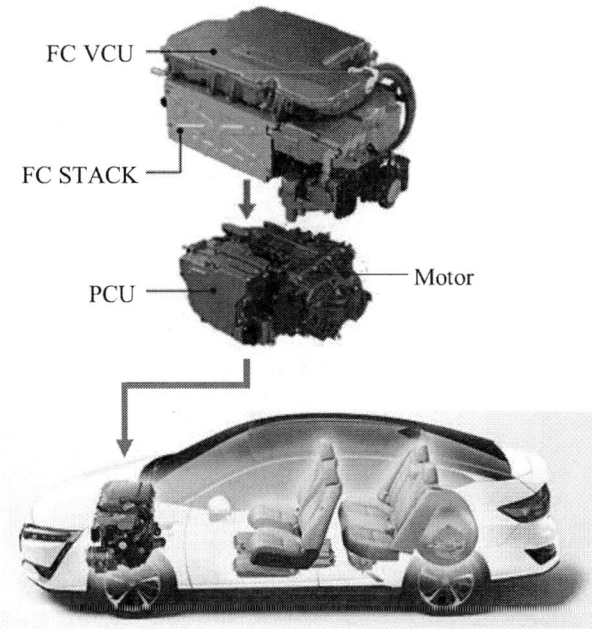

Fig.1 Fuel cell power train unit of new FCV

To consolidate the powertrain unit, the coaxial structure of motor is carried over from the previous model. Height is suppressed by structuring a hollow rotor shaft of the motor that allows the drive shaft to penetrate therein. (Fig.2) [5]

Furthermore, tilting the counter shaft arrangement of the drive motor forward by 90° from the previous model, and positioning the PCU in front of the motor secures the space for mounting the FCVCU and fuel cell stack above the motor and the PCU. (Fig. 3)

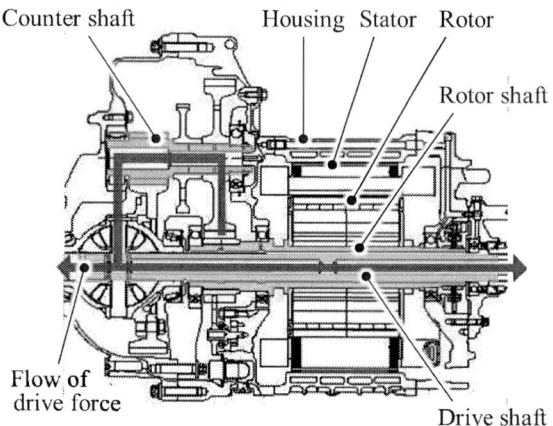

Fig.2 Section of drive unit

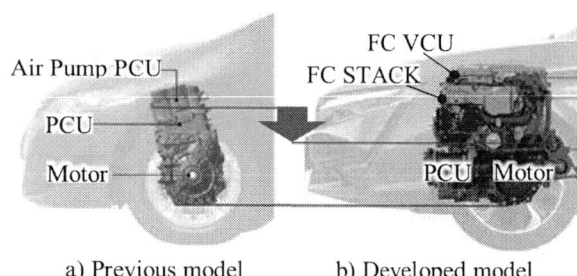

a) Previous model   b) Developed model

Fig.3 Layout of power train

3. AIM OF DRIVE MOTOR DEVELOPMENT

In the drive system of the new-type FCV, the acceleration performance and maximum speed of the vehicle depend on the torque, power characteristics, and the maximum rotation speed. Therefore, performance capable of high torque, high power, and high speed is required for the motor. In the new-type FCV, with the intention of achieving a seamless and endless acceleration feeling from the take-off unique to the motor drive, as well as powerful driving suitable for an upscale sedan, we are setting the power performance of the drive motor as shown in Table 1.

Moreover, in order to achieve the quietness unique to electric cars as targeted by the new-type FCV, suppression of own vibration and radiating noise as well as suppression of vibration transmission to vehicle parts are required of the drive motor.

Based on the above, we aimed to achieve the following two items in the new type motor development for fuel cell vehicles.

(1)   Enhancement of power performance
(2)   Enhancement of quietness

Table1 Specifications of motor

	Previous motor	Developed motor
Max. power	100kW	130kW
Max. torque	256Nm	300Nm
Max. rotation speed	12,500min^{-1}	13,000min^{-1}

4. APPLIED TECHNOLOGIES

4-1. Enhancement of power performance

4.1.1. Rotor magnetic circuit

To enhance the power performance of the drive motor, the interior permanent magnet rotor and distributed stator windings are adopted. A cross-section view of the motor is shown in Fig. 4.

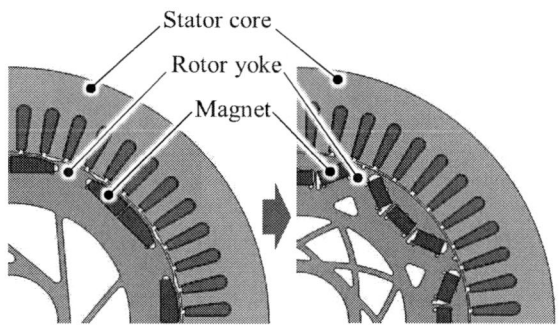

a) Previous model   b) Developed model

Fig.4 Section of motor

Securing the strength of the rotor yoke is an issue in enhancing the maximum generated torque, power, and rotation speed. Stress generated in rotor yoke depends on centrifugal force, press-fitting of the shaft and expansion difference due to temperature change of the magnet or the rotor yoke and resin for fixing the magnets. The measure for reducing these stresses of the rotor yoke are shown as follows.

At high rotation speed, a large centrifugal force is exerted on the magnet, and high stress is generated in the rotor yoke. If a simple magnet V-shaped arrangement is adopted in order to achieve the power performance, a maximum of 2.6 times the stress is generated in the yoke compared to the previous model.

2067

Therefore, three main changes of arrangement of rotor magnet were made from the previous model. The first changing point is to increase number of central ribs per each rotor yoke pole from 1 in the previous model to 2 in the developed model. The second changing point is to increase the thickness between the magnets and press-fit buffer section on the inside diameter. The third changing point is to configure the press-fit buffer section in a cross shape. (Fig.5)

These three changes made the rotor stress of the developed model equal to that of the previous model. (Fig.6)

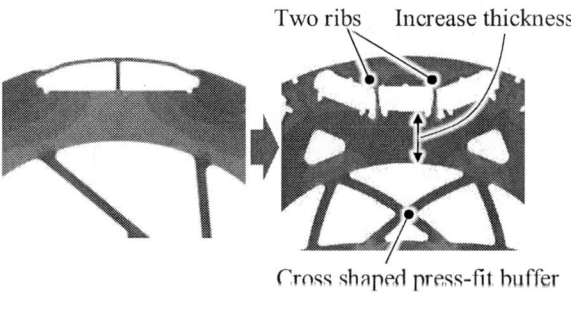

Two ribs   Increase thickness

a) Previous model   b) Developed model

Cross shaped press-fit buffer

Fig.5 Method for reducing rotor stress

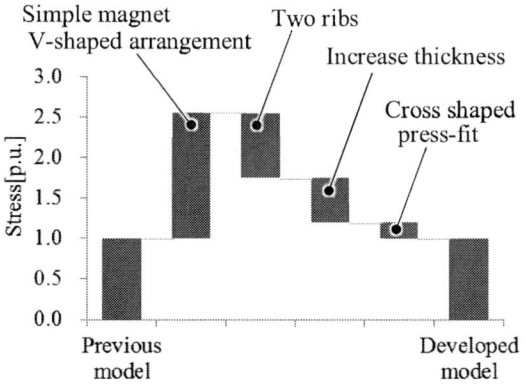

Fig.6 Rotor stress reduction effect

Also, when the temperature changes in the motor, the rotor yoke is deformed and stressed by a difference in expansion between the materials of the rotor yoke, the magnet and the magnet-fixing resin. Protrusions set in the slots suppress deformation and stress of the yoke by temperature change. (Fig.7)

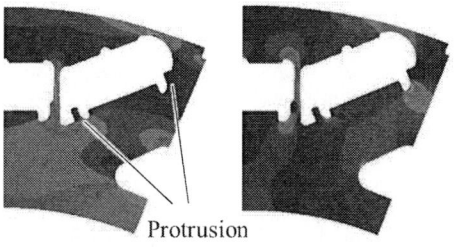

Protrusion

a) Centrifugal force   b) Thermal force

Fig.7 stress distribution of rotor

Furthermore, to reduce the stress resulting from deformation of the outer circumference of the rotor yoke when the shaft is press-fitted, we changed the shape of the spokes of the rotor yoke inner circumference. We were able to secure both rigidity to satisfy the necessary slip torque of the press-fitted section and reduce deformation of the peripheral section of the rotor yoke. (Fig.8)

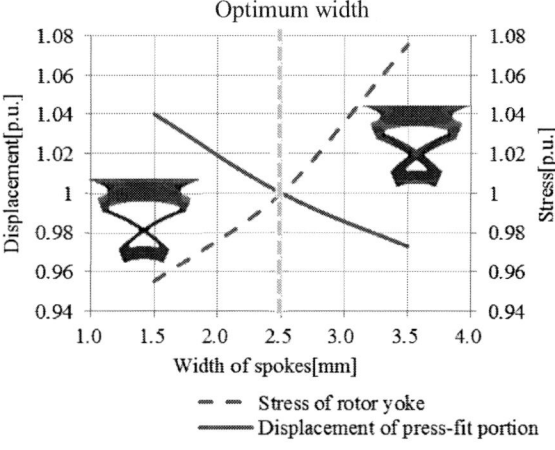

Fig.8 Relationship of Width of spokes with Displacement and Stress

4.1.2. Stator core lamination method

We focused on the lamination method of the stator core forming the magnetic circuit. Stator cores are made by laminating thin electromagnetic steel plates with film coating, and caulking is generally used as a method of bundling of the electromagnetic steel plates. However, caulking forms electrical short-circuit sections between the electromagnetic steel plates. When the motor is driven, short-circuits current flows due to the magnetic flux interlinking the closed circuit, resulting in loss. Therefore, instead of caulking, in the previous motor an adhesive layer is formed on the surface of the electromagnetic steel sheets, to bundle them together. Fig.9 shows manufacturing process of stator core.

In the previous motor, the electromagnetic steel sheets which surfaces are uniformly coated with the adhesive layer in the manufacturing process are used. The adhesive layer reduces the ratio (space factor) occupied by the electromagnetic steel sheet contributing to torque with respect to the entire lamination.

So, we adopted the new stator core production method shown in Fig.10 in the new-type motor to improve the space factor. It adds an adhesive application process in press equipment that punches the electromagnetic steel plate, and applies sufficient amount of adhesive to satisfy the required bundling ability for the stator plate. (Fig.11)

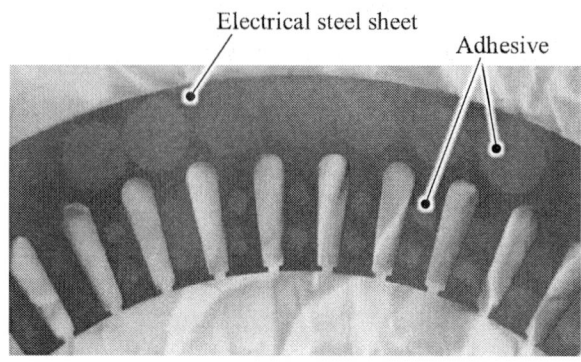

Fig.11 Adhesive application of stator plates

As a result, we reduced the thickness of the adhesive layer and improved the space factor by 4% from the conventional stator core. (Fig.12)

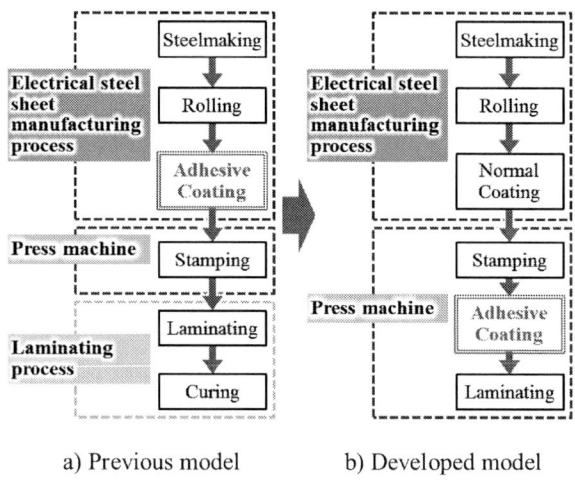

a) Previous model   b) Developed model

Fig.9 Comparison of stator core manufacturing process

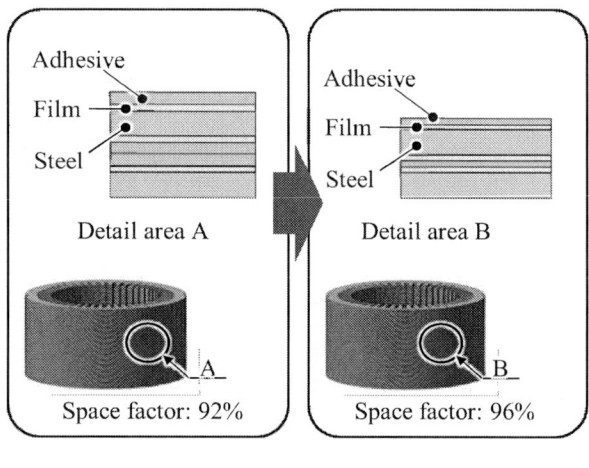

a) Previous model   b) Developed model

Fig. 12 Stator core lamination

In addition to the enhancement of space factor, by increasing the core laminate thickness and the input current, the maximum torque was increased by 17% as compared with the conventional model. (Fig.13)

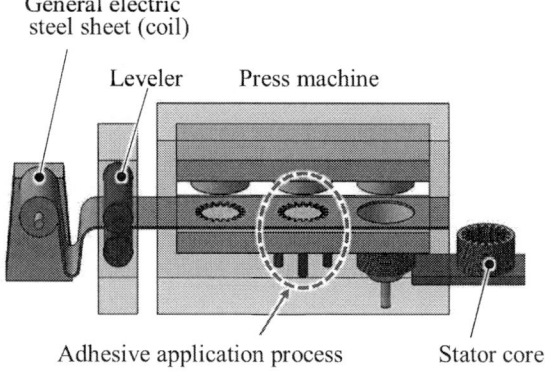

Fig.10 Product process of developed stator core

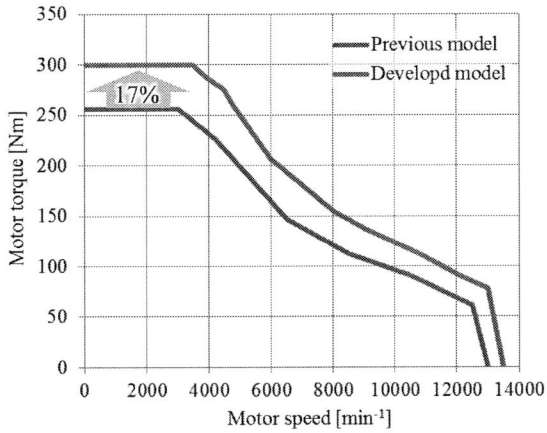

Fig.13 Motor torque performance

4-2. Enhancement of quietness

Fig.14 shows a typical example of cabin noise while driving for the previous motor. The horizontal axis shows motor speed, and the vertical axis shows the frequency. The noise levels at 6th order and 12th order harmonics frequency of motor electric angle are relatively high and pose a problem. The vibration generated by the structure of the drive motor is the main cause of the acoustic noise.

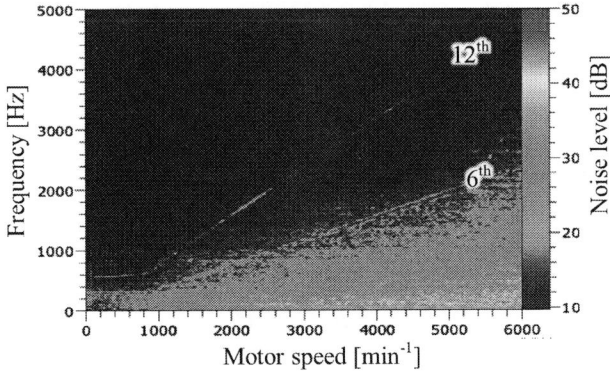

Fig.14 3D Map of cabin noise

There are two paths of noise propagation from the drive motor to the cabin: structure-borne and airborne. Structure-borne propagation is transmitted to the cabin via the bracket (point B), the mount (point C) and body (point D) from the fixed point (point A) for fastening the motor that is the source of vibration. Also, for airborne propagation, the radiating noise generated by the motor is attenuated by other parts and sound insulation materials, and propagates into the cabin (Fig.15). To achieve quietness of the motor, we focus on the 6th order noise

and 12th order noise, and implemented countermeasures to decrease the structure-borne propagation noise and airborne propagation noise.

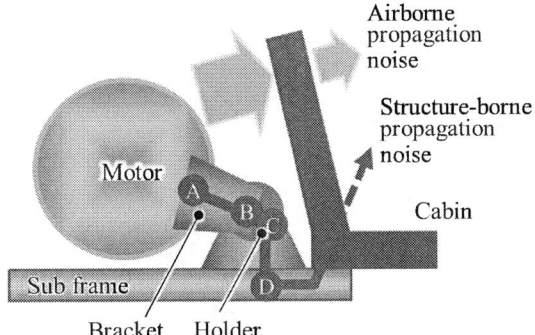

Fig.15 Paths of noise propagation

4.2.1 Reduction of torque ripple by magnet arrangement

To suppress the torque ripple that is a factor of motor vibration generation, we focus on the magnet arrangement of the rotor electromagnetic section. Fig.16 shows the relationship between the torque ripple and the layout angle b (electrical pole arc angle of magnet layout). As layout angle b becomes smaller, the torque ripple tends to decrease from 50deg to 44deg. Therefore, we selected layout angle b so as to minimize the torque ripple. [6]

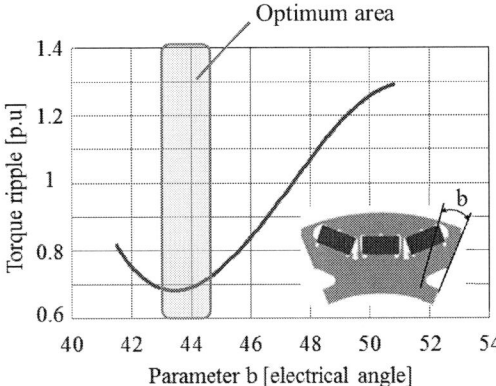

Fig.16 Relationship of magnet position and torque ripple

4.2.2 Vibration reduction by current control

To reduce vibration, we adopted harmonic current control. It is difficult to control high frequency harmonic currents. Therefore, we focused on 6th harmonic acoustic noise. To reduce 6th order harmonic acoustic noise, we

2070

intend to reduce 6th order harmonic torque ripple that cause 6th order harmonic acoustic noise. Additional 6th order currents command is superimposed to nominal current commands determined from required torque. Compensated harmonic currents are derived from required torque and motor speed.

Fig.17 shows the block diagram of harmonic currents control. Motor currents are controlled according to the superimposed d/q axis current commands. The d/q currents are converted to 3 phase motor currents as following formula. As a result, 7th order harmonic currents are superimposed to nominal 3 phase currents.[7]

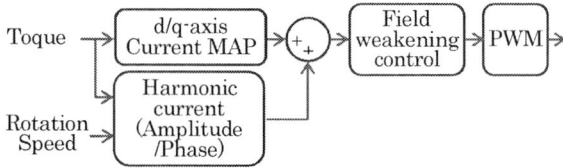

Fig.17 Block diagram of harmonic superposition

$$
\begin{cases}
i_d^{\,*} = i_d + A\cos(6n\theta + \delta) \\
i_q^{\,*} = i_q + A\sin(6n\theta + \delta)
\end{cases}
$$

$$
\sqrt{\frac{3}{2}}\,i_u = i_d^{\,*}\cos\theta - i_q^{\,*}\sin\theta
$$

$$
= i_q\sin\theta + i_q\cos\theta + A\sin\big[(6n+1)\theta + \delta\big]
$$

Fig.18 and Fig.19 show the current waveform and torque ripple waveform with and without harmonic current superposition. It is confirmed that superimposing the harmonic reduces the torque ripple compared to the torque waveform without superposition.

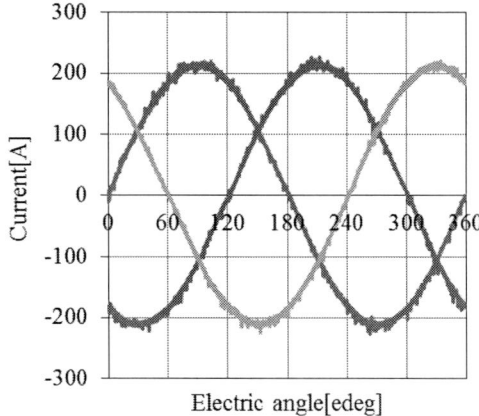

a) Without harmonic superposition

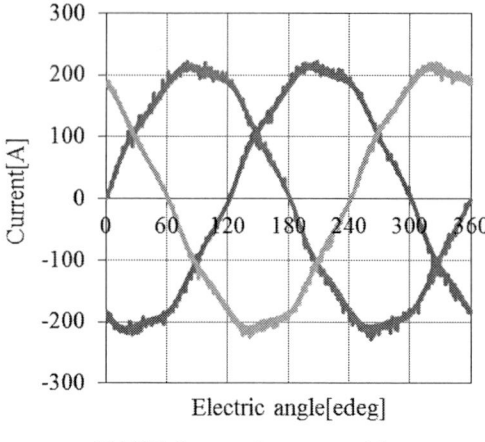

b) With harmonic superposition

Fig.18 Comparison of current wave form

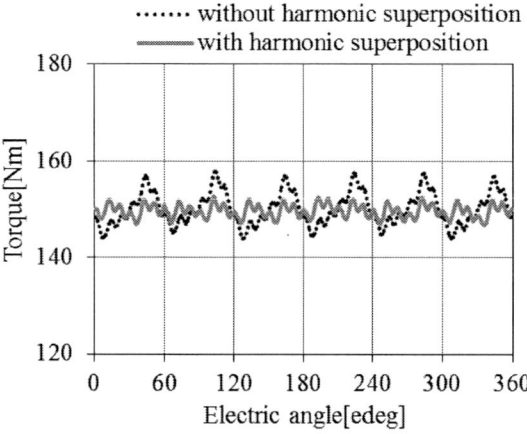

Fig.19 Comparison of torque ripple

4.2.3 Vibration reduction by multi-step skew

In the high speed area, the cabin noise caused by the motor is mainly 12th order noise by airborne propagation. Reduction of the excitation force in the radial direction of the motor is effective for reducing this noise. To reduce excitation force, we adopted a stepped skew that shifts the rotor yoke in the circumferential direction.

The conventional motor adopted 2-step skew, the new-type motor adopts 4-step skew to reduce the excitation force in the radial direction in the high frequency area. The divided four rotor yokes are press-fitted into the shaft with their phases shifted. (Fig.20)

2071

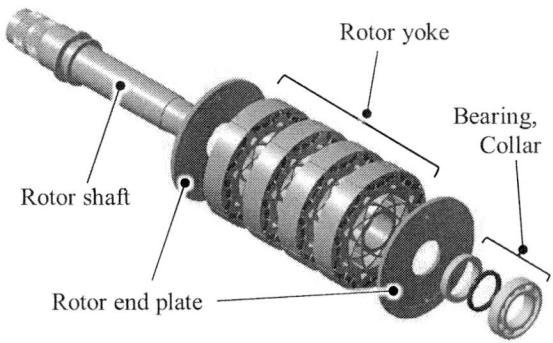

Fig.20 Rotor configuration

We enhanced quietness by applying these noise reduction technologies. Motor noise was reduced in the cabin by 10dB (A) in the low speed area (motor electrical angle 6th order noise) with motor speed at 3000rpm or less, and by 5dB (A) in the high speed area (motor electrical angle 12th order noise) with motor speed at 3000 rpm or more. (Fig.21)

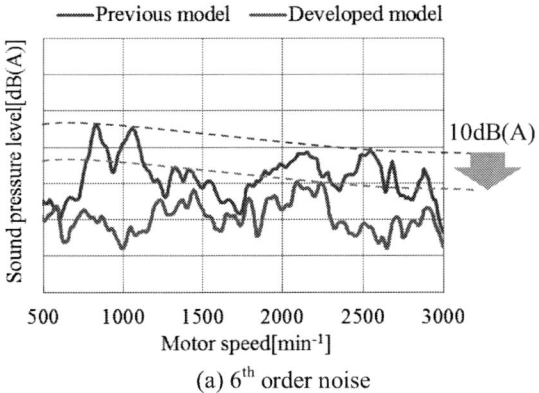

(a) 6th order noise

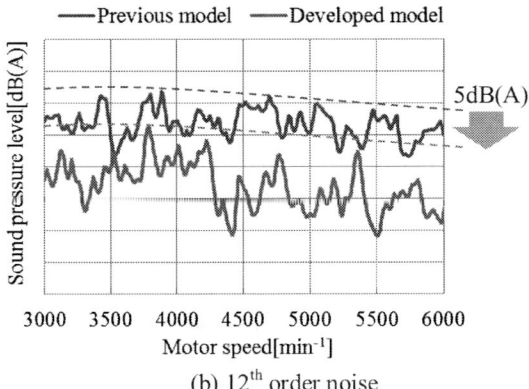

(b) 12th order noise

Fig.21 Motor sound pressure level inside cabin

5. CONCLUSION

We have newly developed a motor to be mounted on a new-type FCV. Appling new rotor magnetic circuit, stator core production method and current control enable a quiet, powerful and smooth drive feeling. Compared to the previous motor, the maximum power was increased from 100 kW to 130 kW and the maximum torque was increased from 256 Nm to 300 Nm, while maintaining the same volume. In addition, the motor noise inside the cabin was reduced by 5 to 10 dB(A) compared to that of the previous motor. As a result, the developed motor contributed to achieving of the seamless and endless acceleration feeling and the quietness unique to an electric vehicle.

REFERENCES

[1] Ogawa, T., Kimura, K., Uchibori, K., Fujimoto, S., Shimizu, K., *"Development of New Power Train for Honda FCX Fuel Cell Vehicle"* Honda R&D Technical Review, Vol.15, No.1, p.7-12(2003)

[2] Matsunaga, M., Fukushima, T., Ojima, K., Kimura, K., Ogawa, T., *"Fuel Cell Powertrain for FCX Clarity"* Honda R&D Technical Review, Vol.21, No.1, p.7-15(2009)

[3] Matsunaga, M., Fukushima, T., Ojima, K., *"Powertrain System of Honda FCX Clarity Fuel Cell Vehicle"* World Electric Vehicle Journal Vol.3

[4] Kimura, K., Kawasaki, T., Omura, T., Atsumi, T. and Shimizu, K., *"Development of New Fuel Cell Vehicle CLARITY FUEL CELL,"* Honda R&D Technical Review, Vol.28, No.1, p.1-7(2016)

[5] Iwai, A., Fukushima, T., Yamaguchi, N. and Shinoki, H., *"Development of Motor for 2009Model Fuel Cell Vehicle "* SAE Technical Paper, 20094591 Vol.40, No.5, p.1273-1278(2009)

[6] Ogami, M., Inoue, M., Takamatsu, H., Kamada, T. and Akai, K., *"Development of Motor for New Hybrid system"* SAE Technical Paper,No.45-16, p.1100-1105(2016)

[7] Yazaki, M., *"Vibration Reduction in Motors for the SPORT HYBRID SH-AWD"* SAE International Journal of Alternative Powertrains Vol. 4, No. 1 (May 2015), pp. 153-161

The 2018 International Power Electronics Conference

EMC Design and Development Methodology for Traction Power Inverters of Electric Vehicles

Isao Hoda[1]*, Jia Li[2], and Hiroki Funato[1]

1 Research & Development Group, Hitachi, Ltd., Yokohama, Japan
2 Automotive Products Research Laboratory, Hitachi America, Ltd., Farmington Hills, MI, USA
*E-mail: isao.hoda.tb@hitachi.com

Abstract — **A car manufacturer generally requires compliance with component-level electromagnetic compatibility (EMC) requirements. However, some countermeasures may be necessary due to differences in electromagnetic circumstances even if the requirements are complied with. The objective of this study is to establish a methodology for EMC design by predicting the vehicle-level radiated emissions in the early stages of component development. The technical challenge in performing vehicle-level EMC analysis is handling a wide dynamic range, from the inverter output to the emissions. In this study, the vehicle-level EMC analysis has been divided into three steps that calculate the output current, the common mode current, and the antenna voltage separately so that the each dynamic range can be reduced. Model validation achieved by comparing the measurement results using actual electric vehicles is also studied. With this analysis, designing the EMC of both vehicles and components can be realized.**

Keywords — *Common mode current, Electromagnetic compatibility, Inverter, Vehicle-level EMC analysis*

I. INTRODUCTION

The quantity of electronic components in automobiles has rapidly increased. Electromagnetic compatibility (EMC) has become a significant focus in the automotive industry as a way to confidently guarantee functionalities of these electronic components and keep down the negative effects on external systems. As the range of hybrid electric vehicles (HEV) and electric vehicles (EV) expands, EMC has become increasingly important because it uses high-voltage components with faster switching speeds.

Automotive component suppliers usually design and test their products at the component-level. However, the challenge with EMC is that components that have passed all their EMC tests on a component-level can still fail EMC testing at the vehicle-level after being loaded onto a vehicle. The biggest issue is that the test on the vehicle is usually performed at a very late stage of product development. Resolving an issue at a late stage of development is difficult, time consuming, and expensive. In order to avoid such issues, an automotive component supplier should consider the vehicle-level effects during the components' design phase.

Computer models provide good opportunities to simulate EMC testing at the vehicle-level. Chen [1] built a system-level EMC model to study the noise distribution mechanisms for the components embedded inside an engine block. Hubing [2] describes a general method of characterizing the component-level properties for system-level electromagnetic simulations. Zeng [3] created the vehicle-level analysis methodology to predict the noise propagation between the components and the vehicle antenna up to 2 MHz. However, there has not been a highly precise prediction of the resonance point of radiated noise around 6 MHz.

This work describes a vehicle-level EMC simulation methodology for estimating radiated emissions from a vehicle. The approach is carried out by running a combination of an equivalent circuit and a 3D full-wave electromagnetic field simulation for a simplified vehicle structure that includes inverter system components. Comparisons between simulations and measurement results between 500 kHz and 20 MHz are presented and discussed.

II. VEHICLE-LEVEL EMC MODELING

In this section, we discuss the difficulties in designing EMC for the automotive industry and the challenges in analyzing vehicle-level EMC. Then, we present a methodology for overcoming these challenges and describe analysis models for simulating vehicle-level emissions. Finally, we compare predicted results vs actual measurement results and show the accuracy of the simulations.

A. EMC Design in the Automotive Industry

In order to prevent potential EMC issues in the automotive industry, OEMs request that each electronic component meets their required specifications of component-level EMC before product shipment. Figure 1 shows an example of a component-level EMC test setup of an automotive inverter. The test determines the level of radiated emissions by placing inverters, cables, and other components on a metal plate that simulates the vehicle chassis and measures the voltage induced in an antenna positioned nearby.

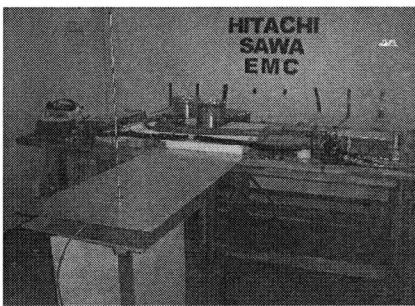

Fig. 1. Component-level EMC test setup

Test conditions such as the position of cables and the measurement antenna are stipulated in the International Special Committee on Radio Interference 25 (CISPR 25) international standards. In the case of radiated emissions testing, for example, this specifies measuring the induced voltage in an antenna placed 1 meter from the cable.

In contrast, vehicle testing is performed using a production-model vehicle. Vehicle EMC testing is broadly divided into radiated emissions testing that measures the induced voltage in an antenna positioned outside the vehicle and interference testing on vehicle-mounted antennas such as those used by car radios.

As described above, the measurement conditions of a component-level EMC test cannot imitate the real conditions of a vehicle-level EMC test. The risk of EMC issues increases at the vehicle prototype stage even though each component has passed its own requirements. If a vehicle-level EMC issue happens, the supplier may be required to revise the design to solve the issues. This results in considerable losses to both the automotive electronics manufacturer and the vehicle manufacturer.

Electromagnetic noise analysis at the beginning of the design stage (as shown in Figure 2) is effective in preventing such EMC issues. During the early design stage of both the components and the vehicle, a vehicle-level simulation is performed to ensure a good performance in minimizing the need to rework EMC after vehicle prototyping.

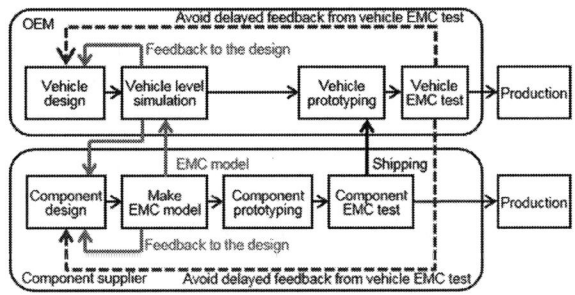

Fig. 2. Proposal of product development process utilizing vehicle-level EMC simulation

B. Mechanism of electromagnetic noise generation by an inverter and challenges of vehicle-level EMC analysis

The drivetrain inverter in a hybrid electric vehicle uses insulated-gate bipolar transistors (IGBTs) or similar semiconductors as switches that convert several hundred volts of direct current into the alternating current that drives the electric motor. Although the operation of semiconductors at a switching frequency is ordered by kHz, the high frequency components generated during voltage switching extend up to the MHz range. Furthermore, whereas an inverter has an output in the kilowatt range, it is standard practice to use shielded cables to connect the inverter, the motor, and high-voltage batteries because keeping the noise level considerably low (in the microwatt range or less) prevents interference from the leakage of electromagnetic noise.

The biggest challenge in vehicle-level simulation is implementing a very high dynamic power range from kilowatts of inverter output to microwatts of electromagnetic radiation with high accuracy. The vehicle-level simulation tool is continuously being refined to increase the accuracy of the predictions based on the previous generation of EMC data.

C. Methodology of vehicle-level analysis

The concept of an electric vehicle with a main inverter system is being studied to develop a system-level simulation methodology for predicting the radiated emissions in a vehicle. Major components of the inverter system include a power inverter, two motors, a high-voltage battery, and a cable connecting each component. In this system, the battery supplies direct current (DC) power to the inverter; that converts a DC to an alternating current (AC) using pulse width modulation (PWM) switching to rotate the motor. Due to the fast switching and high power, the inverter becomes the main source of noise in an electric vehicle., The components of the inverter system, the vehicle's body, and the antenna are modeled separately either by a Simulation Program with Integrated Circuit Emphasis (SPICE) simulator or by a 3D modeler to simulate this noise propagation.

The whole vehicle is modeled as a combination of these component models, as shown in Figure 3. In order to handle the high dynamic range while maintaining accuracy, electromagnetic radiation generating from the common mode current necessitates that the simulation be divided into three different analyses. First, time-domain circuit analysis determines the conductor current (I_N) of inverter output. Second, 3D electromagnetic analysis determines the conversion factor (K_1) from the conductor current (I_N) to the common mode current (I_C). Finally, 3D analysis determines the conversion factor (K_2) from the common mode current (I_C) to the antenna voltage (V_{ant}). From the above three analyses, the antenna voltage (V_{ant}) is calculated by:

$$V_{ant} = I_C \times K_2 = I_N \times K_1 \times K_2 \qquad (1)$$

The vehicle-level simulation is established with this analysis and can be applied to developing the inverter and shortening the development time by preventing vehicle-level countermeasures.

The 2018 International Power Electronics Conference

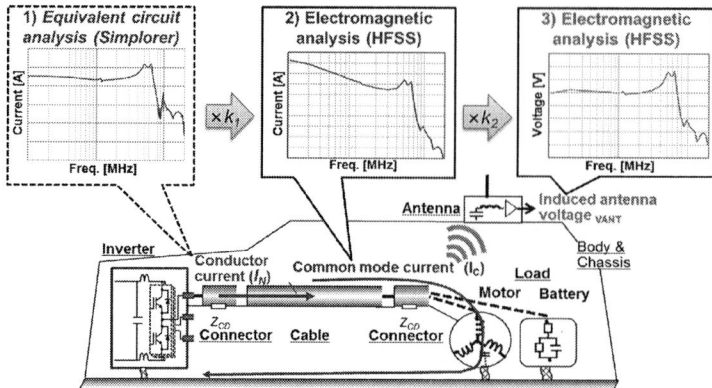

Fig. 3. Illustration of a system-level simulation methodology for calculating the radiated emissions of a vehicle

This combined analysis method has several merits, including saving computer resources (CPU and memory) and reducing time taken for analysis by calculating the common mode current analysis and the vehicle propagation analysis separately. Moreover, when this method is applied to the design of a product, the vehicle propagation analysis is only carried out if the antenna location has changed. There is no need to reanalyze the common mode current from the conductor current if the inverter system is not changed.

D. Analysis Model

In this study, ANSYS HFSS was used as the 3D modeler and ANSYS Simplorer was used for modeling the equivalent circuit. The common mode current across the cable harness and the induced voltage across the antenna were simulated and compared to the measurement results.

In the first stage of analysis, time-domain equivalent circuit analysis determined the conductor current (Icon) of inverter output. In order to expand the frequency range of the inverter model to 20 MHz, some insulated gate bipolar transistor (IGBT) and freewheeling diode (FWD) device parameters were modified. First, the parameters were adjusted with Simplorer's advanced dynamic IGBT model to fit static and dynamic characteristics to the specification sheet and to measured waveforms. Here, the static characteristic has an output characteristic (Ic-Vce curve) and a transmission characteristic (Ic-Vge curve). The dynamic characteristic has a capacitance curve of input (Cies), output (Coes), and reverse-bias. Then, the parameters were adjusted with Simplorer's dynamic diode

model to fit static and dynamic characteristics to the specification sheet and to measured waveforms. Here, the static characteristic has a forward characteristic (If-Vf curve), and the dynamic characteristic has a reverse-bias capacitance curve (Cj-Vf).

The equivalent circuit model of an inverter consists of a nonlinear active device (i.e., IGBT) and a passive device such as an AC bus-bar, DC bus-bar, smoothing capacitor, and filter. The parasitic inductance and the stray capacitance of the passive device are analyzed with the 3-D and 2-D parasitic extraction tool, the ANSYS Q3D extractor. Moreover, the gate driver circuit of IGBT is made to reflect the real product. The equivalent circuits of the HVAC cable, the HVDC cable, the high-voltage battery, and the motor are made from the impedance measurement of each component and fitting method.

Figure 4 is the schematic overview of the electromagnetic analysis model for calculating the conversion from the conductor current to the common mode current. The common mode current I_{CMN-AC} flowing on the HVAC cable is generated by the conductor current I_{CON-AC} through the thin shield thickness and the contact resistance between the shield and the connector between the inverter and motor. The common mode current I_{CMN-DC} flowing on the HVDC cable is also generated by the conductor current I_{CON-DC} through the thin shield thickness and the contact resistance between the shield and the connector between the inverter and motor. In this analysis model of HFSS, each conductor current is generated by the excitation source between the shield and the conductor in the inverter case.

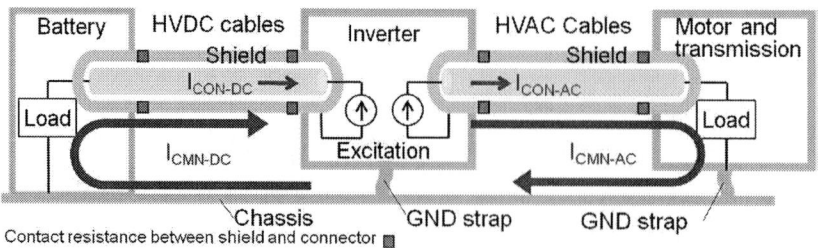

Fig. 4. The schematic overview of the electromagnetic analysis model for calculating the conversion from the conductor current to the common mode current

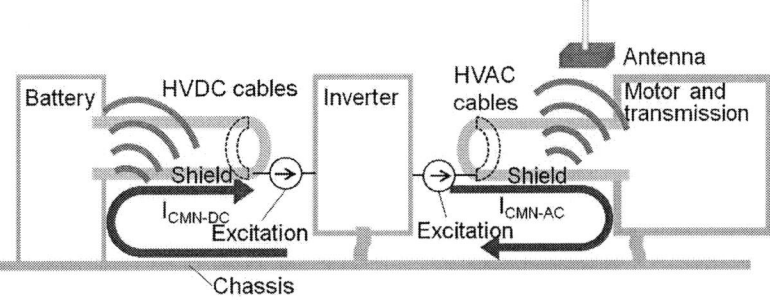

Fig. 5. The schematic overview of the electromagnetic analysis model for calculating the conversion from the common mode current to the antenna voltage

Figure 5 is the schematic overview of the electromagnetic analysis model for calculating the conversion from the common mode current to the antenna voltage. The electromagnetic field is generated from the loop of common mode current I_{CMN-AC} and I_{CMN-DC}, then it induces change in electrical voltage on the rod of the antenna. In this analysis model, the common mode currents I_{CMN-AC} and I_{CMN-DC} are generated by the excitation source between the shield and the inverter case.

Figure 6 shows the geometry of the electromagnetic analysis model for calculating the common mode current and the antenna voltage. All geometries of the full vehicle body are included in this analysis model. Two antenna locations are analyzed. The first location is on the roof, in the same location as the actual AM radio antenna. The second location is on the front bumper.

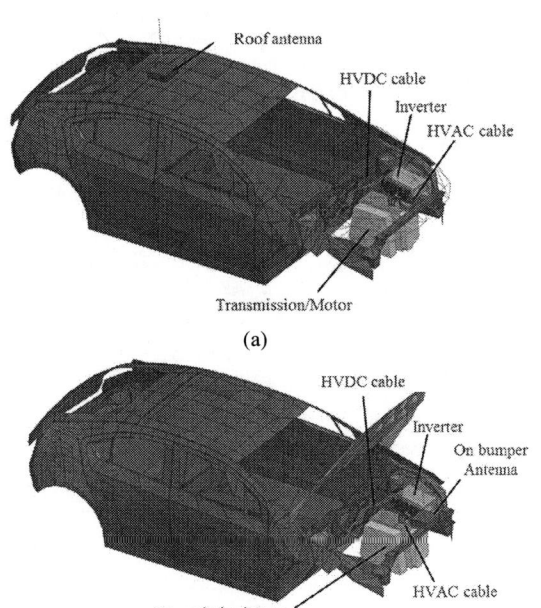

(a)

(b)

Fig. 6. Geometry of the electromagnetic analysis model.
(a) Roof antenna location and (b) front bumper antenna location.

E. Measurement method

The measurement results and the analysis results were compared in order to verify simulation accuracy. Figure 7 shows a simplified electric vehicle that was specifically prepared for conducting the electromagnetic noise measurements. In this vehicle, electrical components were removed, except for the drivetrain inverter (the subject of measurement and the most non-electrical component) in order to prevent the noise from the other components from influencing the measurement results.

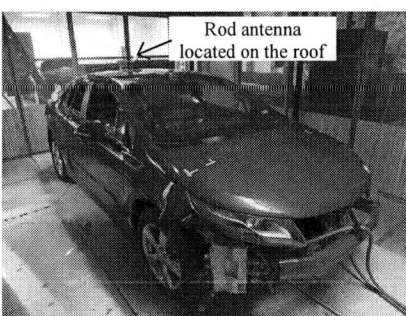

Fig. 7. Measurement setup of a simplified electric vehicle

The measurement items were common mode currents across the high-voltage AC and high-voltage DC cables and an induced voltage on the rod antenna. The antenna that measured radiated emissions was positioned at the same two locations, as mentioned in the Analysis Model section. All measurements were performed in a shielded room in order to prevent most disturbances from external noise.

F. Predicted versus actual measurement results and considerations

Figure 8 shows the analysis result of the common mode current across HVAC cables for comparing the measurement results. Here, we focused on the results of the HVAC cable that mainly contributed to the radiated emissions. The analyzed common mode current had two peaks to approximately 5 MHz and approximately 6.5 MHz within the measurement, and the frequency characteristics conformed to the measurement curve. The differences between the analysis and the measurements were within 6 dB at lower than two resonance points: 6.4 dB at the 5 MHz resonance point and 2.6 dB at the 6.5 MHz resonance point.

The 2018 International Power Electronics Conference

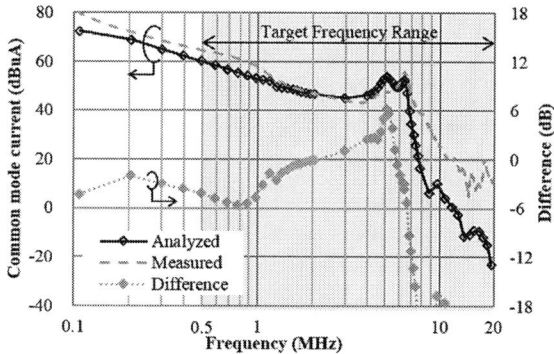

Fig. 8. Comparison of simulated and measured common mode current of the HVAC cable

Figure 9 and Figure 10 are the analysis results of the antenna voltage that was induced by the radiated electromagnetic field from the common mode current loop in order make comparisons with measurements. The antenna location shown in Figure 9 is on the roof. The antenna location shown in Figure 10 is on the front bumper.

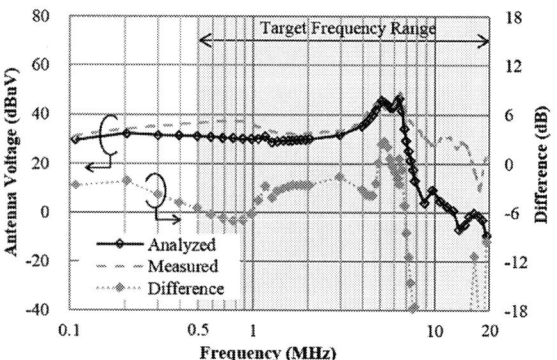

Fig. 9. Comparison of simulated and measured induced voltage at the rod antenna located on the roof

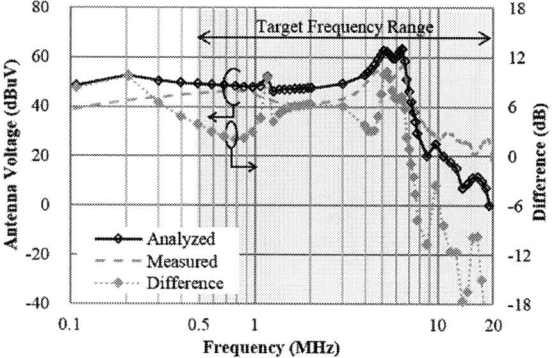

Fig. 10. Comparison of simulated and measured induced voltage at the rod antenna located on the front bumper

The analyzed antenna voltage had two peaks to approximately 5 MHz and approximately 6.5 MHz within the measurement, and the frequency characteristic conformed to the measurement curve. Regarding the roof antenna, the difference between the analysis and the measurement was within 7.1 dB, at lower than 2 MHz, 2.3

dB at the 5 MHz resonance point and 2.6 dB at the 6.5 MHz resonance point. Regarding the front bumper antenna, the difference between the analysis and the measurement was within 6.3 dB, at lower than 2 MHz, 7.1 dB at the 5 MHz resonance point and 7.0 dB at the 6.5 MHz resonance point.

III. CONCLUSION

We compared predicted and actual results for the frequency characteristics of induced voltages measured at the two antenna positions. When the antenna was located on the roof, there were errors up to 7.1 dB below 2 MHz and 2.6 dB at the resonance point, made by the common mode current loop. When the antenna was located on the front bumper, there were up to 9.7 dB of error below 2 MHz and errors of 9.9 dB at the resonance point, made by the common mode current loop. These results show that the actual measurement results and the simulated results have good agreement at a frequency range below the resonance point. A possible cause of errors in the high-frequency range above 10 MHz is simplification of the vehicle.

A full vehicle-level EMC analysis method to predict radio antenna noise from the inverter system of an electric vehicle has been developed. It can handle a wide dynamic range with practical accuracy by being divided into three steps. The measurements are carried out to verify simulation accuracy. The comparison results show that the maximum difference between the measurements and simulation was within 12 dB at up to a 20-MHz frequency peak. This result means that this methodology is applied to the actual design in a practical manner. This analysis makes it possible to achieve both vehicle and component design that takes EMC behavior into consideration in the early stages of product development.

ACKNOWLEDGMENT

This research was conducted in collaboration with General Motors Company. The authors wish to express their gratitude to everyone involved, especially to Mr. William Ivan and Mr. Andrew Baker of General Motors Company for making significant contributions.

REFERENCES

[1] C. Chen, "Predicting vehicle-level EMC performance utilizing on-bench component characterization results," Proc. of 1999 IEEE International Symposium on Electromagnetic Compatibility, Seattle, WA, USA, vol. 2, pp. 765–769, 1999.

[2] T. Hubing, "Component-Level Characterization for Vehicle-Level Electromagnetic Simulations," Proc. of the 2010 SAE Congress no. 2010-01-023, Detroit, 2010.

[3] H. Zeng, "Vehicle Level EMC Modeling for Electric Vehicle Application," Proc. of SAE World Congress 2015, no. 2015-01-0194, Detroit, 2015.

2077

Simulation-driven Design Optimization of a Multilayer EMC Input Filter

Fatou DIOUF[1*], Nadim SAKR[1] and Anna Gheonjian[2]
1 DEAMEE, Renault SA, Guyancourt, FRANCE
2 EMCoS Research Laboratory), EMCoS, Tbilisi, Georgia
*E-mail: fatou.diouf@renault.com

Power Electronics design in the automotive industry faces many challenges, it has to perform in a high currents environment, have a compact design, an easy fabrication and assembly process and finally be compliant with radio frequency conducted emission regulations. However, taking into account the challenges cited above, leads to a complex structure. The more complex the structure, the bigger the ElectroMagnetic Interferences (EMI), which can impair significantly the performances of the equipment.

In this paper we will present an optimization of a filter's design, thus a reduction of the EMIs and an improvement of its performances through simulation with CAD software. In order to analyze the performances of the filter, a 3D model of the filter has been developed and validated by comparing simulation and measurements results. Once the model is validated it is used to study the EMI issues, the impact of parasitic elements and possible improvements.

I. INTRODUCTION

Nowadays constraints for power electronics design in the automotive industry are increasing considerably and the structure of the power electronic systems are getting more and more complex.

The combination of higher constraints and more complex structures, raises the EMI level inside the equipment. Previous experiences have shown that EMIs due to parasitic elements are the main sources of degradation of equipment performances. It is the case in most of the filter designs in power electronics. In this paper, the case of an input filter presenting more than 50dB gap between the measurements and expected performances in a specific frequency band is studied.

In order to analyze the performances of the filter, a 3D model of the filter has been developed using EMCoS Studio [1]. The model is validated by comparing simulation and measurements results. Once the model is validated it is used to study the EMI issues, the impact of parasitic elements and improvement possibilities.

In section 2, the structure of the studied filter will be described, followed by a presentation of the first prototype model.

Section 3 presents the investigation and identification of the main EMI and performance issues.

In section 4, a comparison between simulation and measurement results of the improved filter will be presented.

In conclusion, the efficiency of CAD software as an investigation tool will be discussed as well as the gains it provides in the development phase of power electronic systems.

II. FILTER AND MODEL PRESENTATION

The filter is developed for a 3 phase charger. It has two stages with an LC/ LC structure and 4 layers. Fig. 1 and Fig. 2, present the filter, its first layout (V1) and circuit diagram.

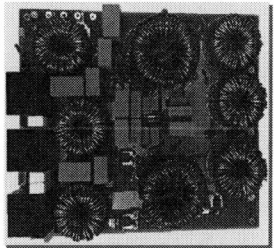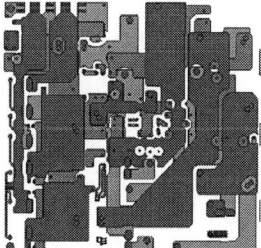

Fig. 1. 3 phase HV Filter and layout V1

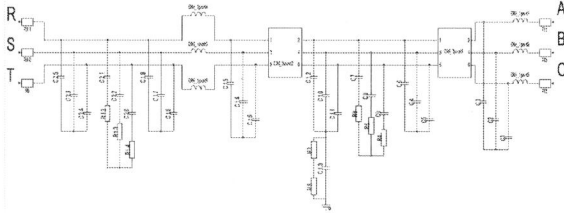

Fig. 2. 3 phase HV Filter circuit diagram

The first step is to model accurately the filter. Using [1], we are able to model the PCB of the filter using PEEC method and thus take into account the parasitic effects in the layout. The comparison between simulation and measurement in Fig. 3, offers a very good correlation. The frequency shift and the 6dB gap are due to the measurements conditions. They were performed on a 25Ω network instead of 50Ω in simulation, which explains the 6dB. The resonance frequency shift is due to the additional wires used on the measurement setup.

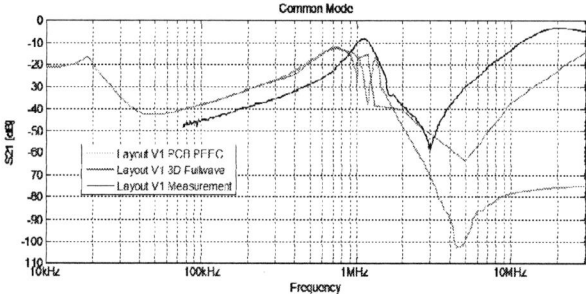

Fig. 3. Comparison between measurement and simulation results

Fig. 3 confirms the validity of the filter's model. The main issue, as stated in the introduction is the filter's performances which are below expectations and predictions of the SPICE model (see fig.4.).

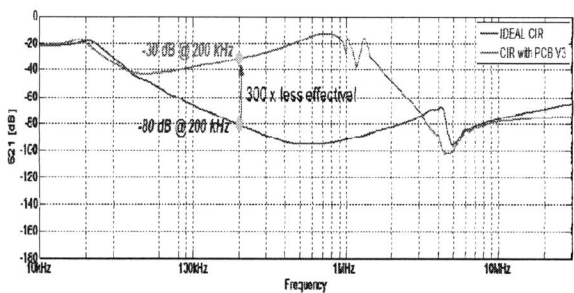

Fig. 4. Comparison between SPICE model (expectations *IDEAL CIR*) and PEEC model results (validated with measurement, *CIR with PCB V3)* of 3 phase HV Filter with layout V1

The low performances of the filter are due to the parasitic parameters of the PCB. With the PEEC model, we can identify the parasitic parameters and study their impact on the filter performances.

III. FILTER ANALYSIS

The PEEC model analysis is used to obtain the netlist of the PCB's equivalent circuit. The netlist includes parameters such as equivalent inductance of the traces and parasitic elements representing the coupling between the different parts of the PCB. Fig. 5 shows the final PEEC model and an overview of the PCB equivalent circuit netlist.

For example some high parasitic capacitances have been identified as shown in Fig. 6 and Fig. 7 by analyzing the different variables of the netlist. Two of those high parasitic capacitances are added to the ideal SPICE circuit in Fig. 8, and provides the same results as PCB measurements as presented in Fig. 9.

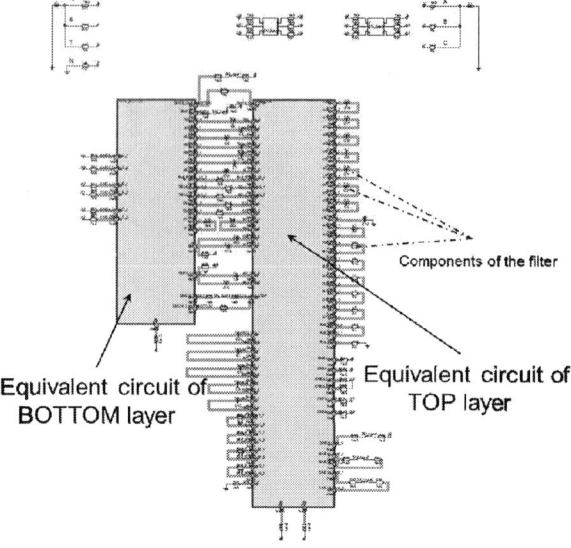

Fig. 5. PEEC model of 3 phase HV Filter and layout V1

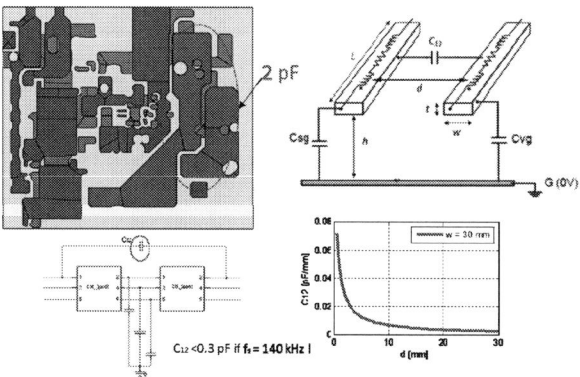

Fig. 6. Parasitic capacitance between two traces of the PCB on the same layer

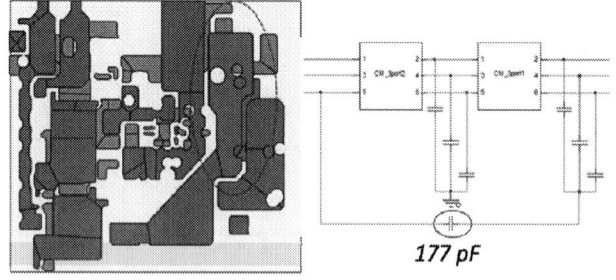

Fig. 7. Parasitic Capacitance between two traces of the PCB on two different layers

The 2018 International Power Electronics Conference

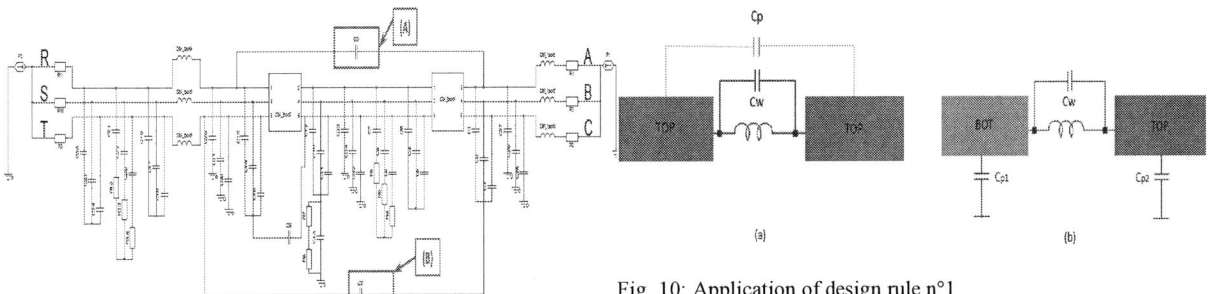

Fig. 8. Main parasitic capacitive couplings added to the 3 phase HV Filter circuit diagram

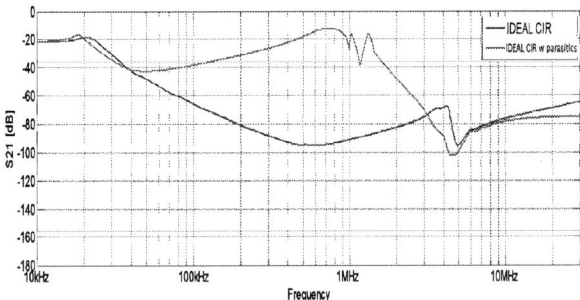

Fig. 9: Comparison between Ideal circuit and Ideal circuit with added parasitic capacitive couplings

The described investigative method allows the identification of the main coupling issues of the filter. After analysis of the parasitic elements with the PEEC method, the sources have been identified:

- Parasitic capacitive coupling between the two common mode chokes as indicated on Fig. 6. As shown in Fig. 8, the parasitic capacitance C_{35} (A) shortens the two common mode filter stages.
- Parasitic capacitive coupling between traces on two different layers, represented by capacitor C37 (B) in Fig. 8.

IV. FILTER'S IMPROVEMENTS WITH PEEC MODEL

After the identification of the main sources of parasitic coupling, the following design rules have been applied to the PCB:

- Connecting inputs and outputs of components respectively to the top and bottom layer, parasitic capacitive coupling between input and output (Cp in Fig. 10 (a)) is converted into common mode capacitances by having the ground plane between the layers (Cp1 and Cp2 in Fig. 10 (b)) which improves the common mode filtering.

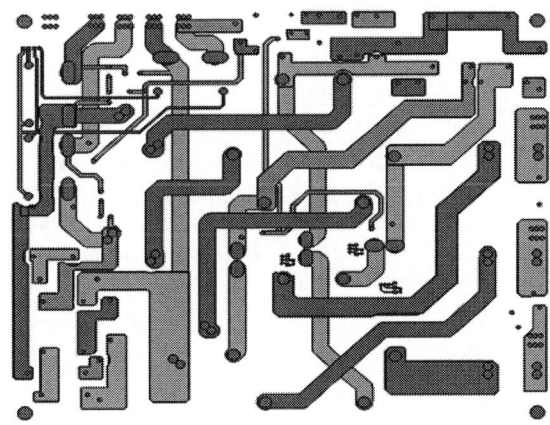

Fig. 10: Application of design rule n°1

- In the case of two components connected in series, if the input of one component and the output of the other are placed on the same layer, a minimum distance should be respected to reduce as much as possible the capacitive coupling between the input and output. (Ideally input and output on separate layers)

The changes made to the PCB layout V1 provided LayoutV2 (see Fig. 11.), and brought significant gain, as it can be seen on Fig. 12.

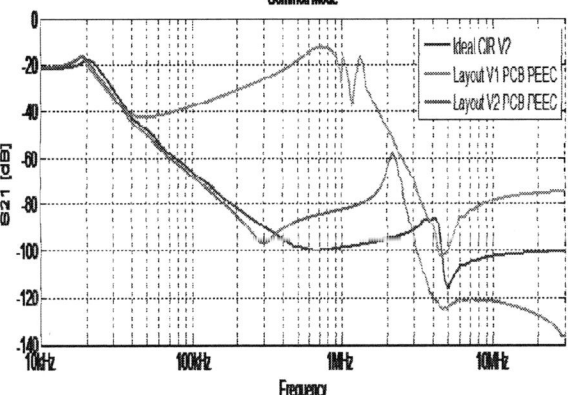

Fig. 11. Layout V2 model

Fig. 12. Comparison between the two layouts (V1 and V2) and ideal circuit

2080

V. FILTER'S IMPROVEMENTS WITH MoM

Following the layout improvements, using the PEEC method, the next phase: is the study of the interactions between PCB components and common mode current distribution.

This step requires the 3D model and the simulation with the method of moments (MoM) [2]. The simulation is done by replacing the radiating elements (chokes) of the filter by lumped equivalent elements. This configuration allows to isolate the impact of the interactions inside PCB.

Output

Frequency: 280 kHz

S21[dB] = -57 instead of -85 dB

Fig. 13. Current distribution of 3 phase HV Filter model (layout V2) at 280 kHz

The analysis of the current distribution, shows a rise of the current at the output of the filter. The analysis of the current flow on the ground plane, allows to understand the phenomenon which is described in Fig. 14.

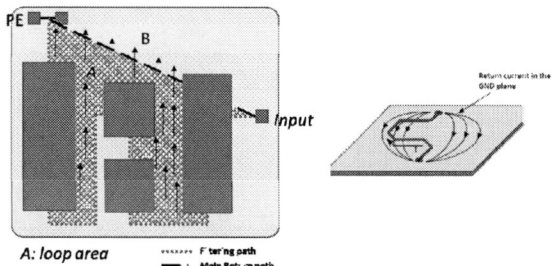

Fig. 14. Coupling of current between output and ground plane of the filter

The loop created between the traces of the filter and the ground plane, will induce a current on the output traces of the filter. The additional current at the output point reduces the filter's efficiency.

To solve the issue, the ground plane has been modified to reduce the size of the loop as in Fig. 15. The Filter with the modified ground presents better performances with less coupling as shown in Fig. 16. The presented findings with simulation are validated by measurements results in Fig. 17.

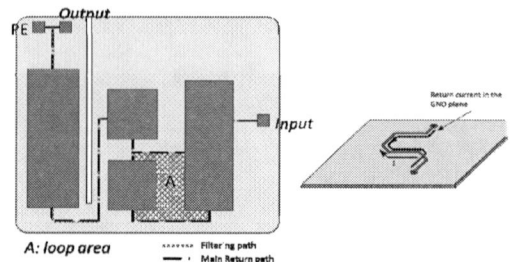

Fig. 15. Coupling of current between output and modified ground plane of the filter

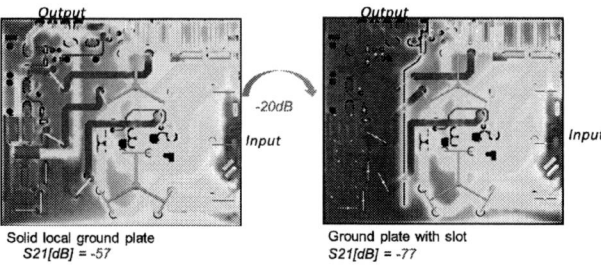

Solid local ground plate
S21[dB] = -57

Ground plate with slot
S21[dB] = -77

Fig. 16 Current distribution of 3 phase HV Filter model with modified ground plane at 280 kHz

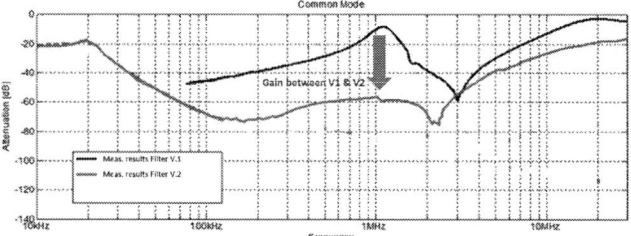

Fig. 17. Comparison of measurement results between the two version of the filter (layout V1 and layout V2)

VI. CONCLUSION

Using the CAD tool EMCoS Studio, we have been able to model and study the filter's performances. The investigation done with the PEEC models allowed the identification of parasitic couplings in the layout and its optimization. The following steps with the development of the 3D model, identifies the interferences due to magnetic and electric fields developed between the different layers of the filter.

The presented optimization method, allows significant gains in design & development time. With the simulation driven aspect different configurations can be tested simultaneously without having multiplying a number of prototypes.

REFERENCES

[1] EMCoS Ltd. EMCoS Studio 2017 - www.emcos.com
[2] K. O. Bowman and L. R. Shenton, "*Estimator: Method of Moments*", pp 2092-2098, Encyclopedia of statistical sciences, Wiley (1998).

EV Traction Inverter Employing Double-Sided Direct-Cooling Technology with SiC Power Device

Takashi Hirao[1]*, Masami Onishi[1], Yusuke Yasuda[1] Akihiro Namba[1], Kinya Nakatsu[1]

Research & Development Group, Hitachi, Ltd., Hitachi, Japan
*E-mail: takashi.hirao.nd@hitachi.com

Abstract—This paper demonstrates a high-power density automotive inverter with a SiC device. To enhance the inverter's performance, circuit implementation and double-sided direct-cooling technologies were developed. Use of an isometric circuit pattern, which connects the SiC MOSFET chips in parallel, suppresses the current imbalance between the parallel chips. The developed inverter reduces power loss by 60% compared with conventional silicon IGBT inverters.

Keywords—Electric vehicle, inverter, power module, SiC

I. INTRODUCTION

One of the key issues to building a sustainable society is the reduction of CO_2 emissions. Many industries, such as electrical power, home appliances, and transportation, are aimed at saving energy and reducing CO_2 emissions. In the transportation sector, electric vehicles (EVs) are an effective solution. EVs include power electronic components, such as inverters, motors, and batteries [1]. Downsizing the inverters has been strongly demanded, but the output power for EV traction is large. Therefore, both power density and power efficiency are key performance factors. Fig. 1 shows the power density trend of an automotive traction inverter [2]–[4], and we can see that the power density improved year by year. Furthermore, the use of SiC power devices are focused on enabling high efficiency and high-power density [5], [6].

This paper demonstrates a high-power density automotive inverter with a SiC device. To enhance the inverter's performance, circuit implementation and double-sided direct cooling technologies were developed. Use of isometric circuit pattern, which connects the SiC MOSFET chips in parallel, balances the current in each chip. The evaluation results indicate that the developed inverter is highly efficient.

II. KEY INVERTER TECHNOLOGIES

Fig. 2 shows the circuit configuration of the SiC inverter. As with the conventional silicon IGBT, the inverter is powered by a battery and supplies three-phase AC power to the motor. Several issues should be investigated to enhance the performance of the EV traction inverter with SiC devices. A large current output is required for EV traction, although the chip size of a SiC device is small. Furthermore, circuit implementation should be improved for high-speed switching in SiC devices. To

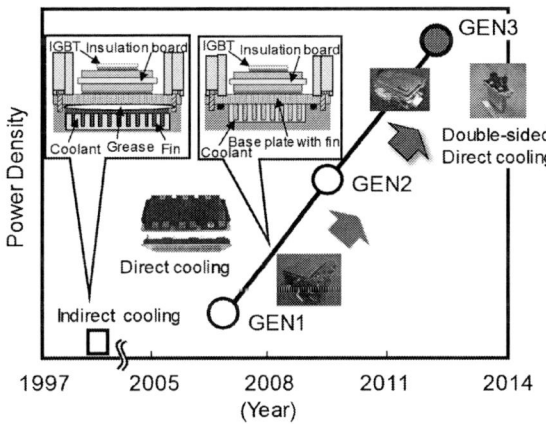

Fig. 1. Power density trend of automotive traction inverter.

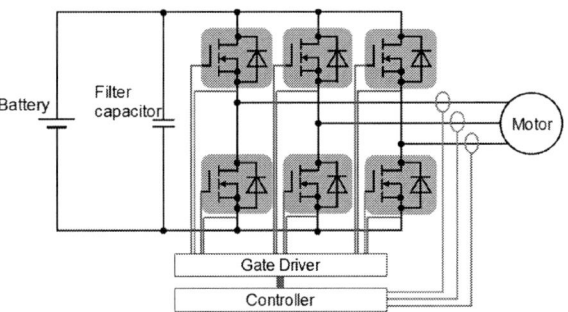

Fig. 2. Circuit configuration of SiC inverter.

address these issues, the following technologies were developed:

1) Double-sided direct-cooling technology
2) Improved circuit implementation
3) Balanced current sharing

A. Double-sided Direct-cooling Technology

To output the large current necessary for EV traction, even with a small chip area (such as SiC MOSFET), a reduction in thermal resistance is required. Figs. 3(a)–(c) show a cross-sectional view of the cooling structures. The conventional structure has thermal grease between the

The 2018 International Power Electronics Conference

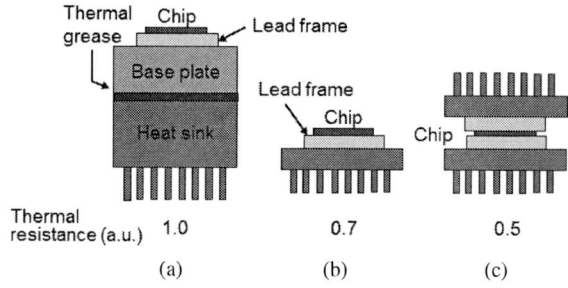

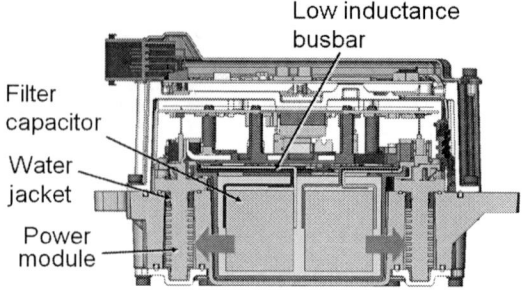

Fig. 5. Internal structure of developed inverter.

Fig. 3. Cross-sectional view of cooling structures of power modules. (a) Single-sided indirect cooling, (b) single-sided direct cooling, and (c) double-sided direct cooling.

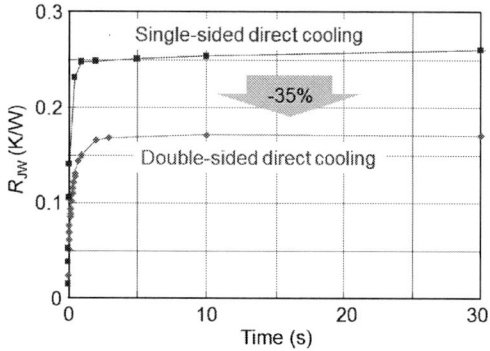

Fig. 4. Experimental results of thermal resistance using single-sided and double-sided direct-cooling structures.

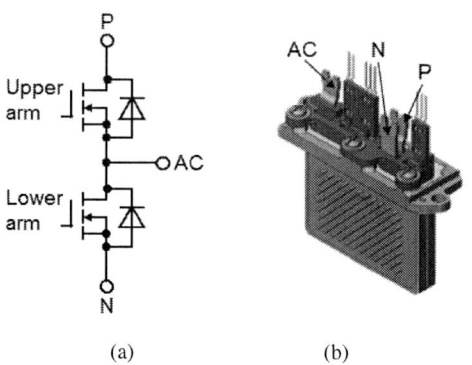

Fig. 6. Terminal structure of developed power module: (a) circuit configuration, and (b) terminal arrangement.

base plate and the heat sink, as shown in Fig 3(a), which limits the heat radiation performance. The direct-cooling structure eliminates the thermal grease because of a cooling fin on the bottom of the power module, as shown in Fig. 3(b). To further reduce the thermal resistance, a double-sided direct-cooling structure was developed, as shown in Fig. 3(c). In this structure, heat radiation passes through both the top and bottom of the chip. Fig. 4 shows the experimental results regarding thermal resistance with single-sided and double-sided direct-cooling structures. As shown in Fig. 4, the thermal resistance R_{JW} between the chip and the coolant was reduced by 35%. This fact is because of the large output current, even with the small chips of SiC MOSFETs.

B. Improved Circuit Implementation

A distinguishing characteristic of SiC MOSFETs is their capability of high-speed switching, which leads to a low switching loss. However, high-speed switching causes a large voltage surge depending on the stray inductance. Therefore, a circuit implementation with low stray inductance was developed. Fig. 5 shows the internal structure of the developed inverter. The stray inductance consists of a power module's internal inductance, a busbar inductance, and a filter capacitor's internal inductance. To reduce the busbar inductance, the terminal layout was designed to laminate the positive and the negative sides of the busbar. As a filter capacitor, a film capacitor was

used for high reliability. The cell and terminal layout was designed to reduce the capacitor's internal inductance. The capacitor cells are cooled by the water jacket, which is set around the capacitor.

The power module design also contributes to reducing the stray inductance. Figs. 6(a) and (b) show the circuit configuration in the power module and its terminal arrangement, respectively. As shown in Fig. 6(a), the power module consisted of upper and lower arm devices, that is, a 2 in 1 power module. Since the source electrode of the upper arm device and drain electrode of the lower arm device are connected inside the power module, there is little stray inductance. Furthermore, as shown in Fig. 6(b), the P and N terminals are located at adjacent ends, and the stray inductance on the terminals is reduced by maximizing the mutual inductance of the parallel terminals.

C. Balanced Chip Current Sharing

In accordance with the constraints of the wafer size and defects, the chips of the SiC MOSFETs are smaller than those of silicon IGBTs. However, the required output current for EV traction is large, therefore, multi-chip parallel connection is required. For a high-switching speed, the current imbalance between parallel chips may occur depending on the variation of the wire length.

Fig. 7(a) shows a cross-sectional view of the conventional circuit pattern for parallel chip connection. Each

2083

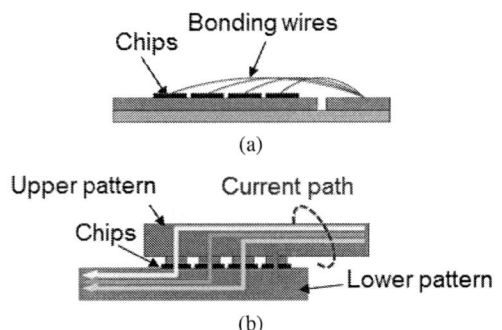

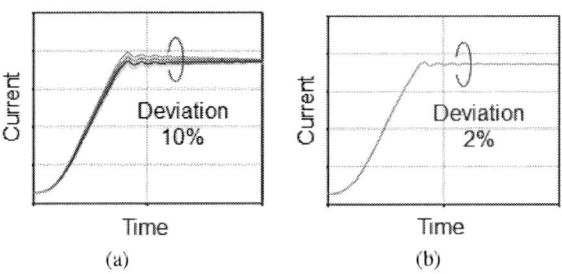

Fig. 7. Cross-sectional view of circuit pattern for parallel chip connection. (a) Conventional, and (b) developed structures.

Fig. 8. Simulated waveforms when using (a) conventional, and (b) developed structures.

Fig. 9. Prototype SiC power module.

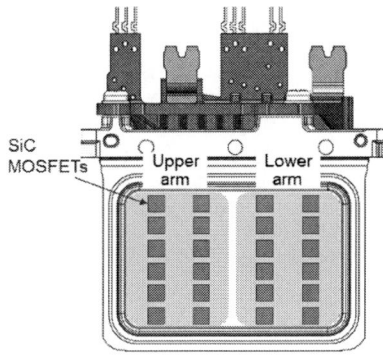

Fig. 10. Internal structure of prototype SiC module.

chip is connected by a bonding wire, and the length of the bonding wire is different for each chip. Fig. 7(b) shows a cross-sectional view of the developed circuit pattern for parallel chip connection. The main feature of the developed structure is that the chips are connected by a circuit pattern on both sides without needing bonding wire. Therefore, the current path on each chip is isometric.

Figs. 8(a) and (b) show the simulation waveforms of chip current sharing when using the conventional and developed structures. As shown in Fig. 6(a), when using the conventional structure, the deviation between each chip was 10%. As shown in Fig. 6(b), when using the developed structure, the deviation between each chip was suppressed to 2%.

III. PROTOTYPE SiC INVERTER

For the above investigation, a prototype SiC inverter was fabricated. Fig. 9 shows a prototype SiC power module. In this power module, double-sided direct-cooling technology is used for low thermal resistance. Fig. 10 shows the internal structure of the power module. The power module consists of upper and lower arm devices, that is, a 2 in 1 module. One arm has 12 chips in parallel, and a total of 24 chips are in the 2 in 1 module. These chips are connected as shown in Fig. 11. The SiC MOS-FET chips are connected by circuit patterns on both sides of them, which balances the chip current. Figs. 12(a), (b) and (c) show experimental switching waveforms at a junction temperature of 150°C, where V_{DS} represents the drain-source voltage, V_{GS} represents the gate-source

voltage, and I_D represents the drain current. As shown in the figure, high-speed switching was obtained at a large current of 400 A.

Fig. 13 shows a prototype SiC inverter. The internal structure was illustrated in Fig. 5, which has the feature of low stray inductance for high-speed switching. Fig. 14 shows the evaluation results of the inverter loss for silicon IGBTs and SiC MOSFETs. The developed inverter can reduce power loss by 60% compared with conventional silicon IGBT inverters.

IV. CONCLUSION

This paper demonstrated a high-power density and high-efficiency automotive inverter with a SiC device. To enhance the inverter's performance, circuit implementation and double-sided direct-cooling technologies were developed. Use of an isometric circuit pattern, which connects the SiC MOSFET chips in parallel, suppressed the current imbalance between the parallel chips. The developed inverter reduces the power loss by 60% compared with conventional silicon IGBT inverters.

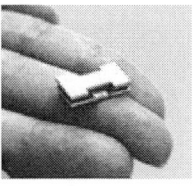

Fig. 11. Wiring structure for parallel connection of SiC MOSFETs.

The 2018 International Power Electronics Conference

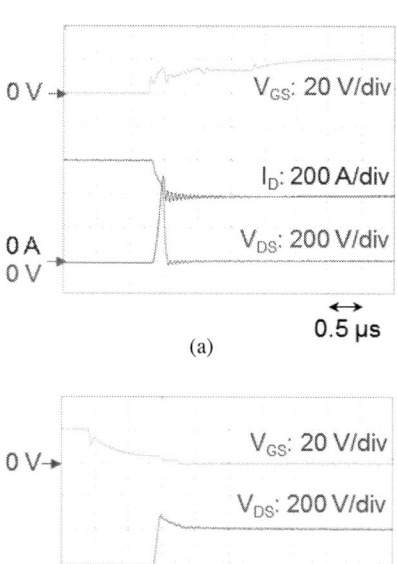

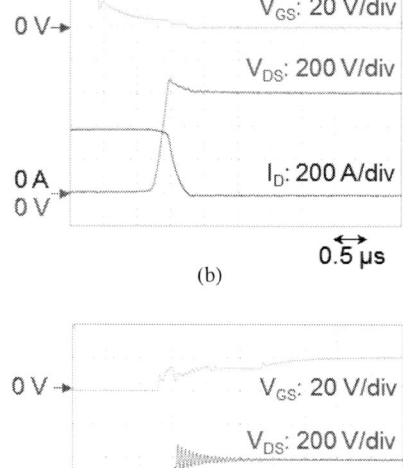

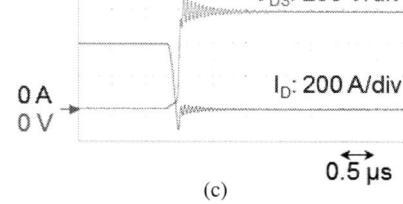

Fig. 12. Experimental switching waveforms of (a) turn on, (b) turn off, and (c) recovery.

Fig. 13. Prototype SiC inverter.

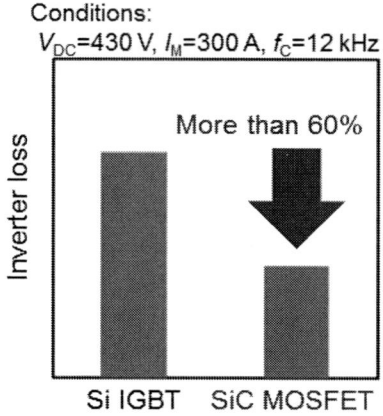

Fig. 14. Evaluation results of inverter loss for silicon IGBTs and SiC MOSFETs.

REFERENCES

[1] K. Ide, K. Nakatsu, H. Hombo, and K. Maki, "Power electronics component technologies for green mobility," *Hitachi Review*, vol. 60, pp. 310-315, 2011.

[2] K. Nakatsu, H. Suzuki, A. Nishihara, and K. Sasaki, "Next-generation inverter technology for environmentally conscious vehicles," *Hitachi Review*, vol. 61, pp. 254-258, 2012.

[3] K. Nakatsu, and R. Saito, "The next-generation high power density inverter technology for vehicle," in *Proc. IEEJ IPEC*, 2014, pp. 1925-1928.

[4] R. Saito, A. Matsushita, T. Tokuyama, K. Nakatsu, and T. Kimura, "Enhanced flexibility of inverter design by applying 2 in 1 double side power module," in *Proc. Automotive Power Electronics (APE)*, April 2015.

[5] K. Hamada, M. Nagao, M. Ajioka, and F. Kawai, "SiC-emerging power device technology for next-generation electrically powered environmentally friendly vehicles," *IEEE Trans. Electron. Devices*, vol. 62 no. 2, pp. 278-285, Feb. 2015.

[6] S. E. Schulz, "Exploring the High-Power Inverter: Reviewing critical design elements for electric vehicle applications," *IEEE Electrif. Mag.*, vol. 5 no. 1, pp. 28-35, Mar. 2017.

The 2018 International Power Electronics Conference

An Overview of Stability Improvement Methods for Wide-Operation-Range Flyback Converter with Variable Frequency Peak-Current-Mode Control

Ching-Hsiang Cheng[1], Ching-Jan Chen[1], *Member, IEEE*, Shinn-Shyong Wang[2]

[1]Department of Electrical Engineering, National Taiwan University, Taipei, Taiwan
[2]Richtek Technology Corporation, Chupei, Taiwan

Abstract—Recently, flyback converter with Universal Serial Bus Power Delivery (USB-PD) specification is widely used in low-power adaptor application. However, the wide output-voltage-range operation of USB-PD makes the stability issue of flyback converter more severe. This paper provides a brief overview of two stability improvement methods for variable-frequency peak-current mode (VFPCM) controlled flyback converter while possessing the variable frequency characteristic for high efficiency and completely interprets those detail techniques. Simulation validates the performances of the two methods. Furthermore, a control circuit and a flyback converter are built for experimental validation of NVFPCM control with constant crossover frequency. Finally, some comments are also summarized.

Key words—flyback converter, Universal Serial Bus Power Delivery (USB-PD), variable-frequency peak-current mode (VFPCM) control, stability improvement method.

I. INTRODUCTION

Flyback converter is widely adopted as cost-effective isolated power converter for low-power adaptor applications [1-2]. Recently, Universal Serial Bus Power Delivery (USB-PD) specification is proposed [3-4] for adaptor application. The adaptor with USB-PD specification has the advantage to supply various devices. For example, it provides 5V DC output voltage for cellphone load, and 19V DC output voltage for laptop computer load. However, the wide output-voltage-range operation of USB-PD application makes the stability issue of flyback converter more severe as can be seen later.

Variable-frequency peak-current-mode (VFPCM) control is a common control method for flyback converter to meet high efficiency specification of adaptor over the entire load range [5-8]. Fig. 1 shows a practical frequency characteristic of variable-frequency control. The switching frequency, f_s, decreases when the compensated voltage V_C decreases. Since V_C is equal to the sensed peak-current of switch and is proportional to output current, the switching frequency is reduced when the output current is reduced. Fig. 2 shows the simplified circuit diagram of VFPCM controlled flyback converter. A full bridge diode rectifier rectifies AC power to charge input capacitor of flyback converter and then the flyback converter generates a regulated DC output voltage V_O for load resistor R_L. L_m is the magnetizing inductor of transformer. I_D is the current of diode. The output capacitor is represented as C_o in series with equivalent series resistor (ESR) R_{Co}. The output voltage is sensed through an isolated type-II compensator consisting of an opto-coupler and a TL431 to generate compensated voltage V_C. In the controller, the compensated voltage V_C, the current signal

V_{CS} sensed by R_{CS}, and the slope compensation S_e determine the on-time. Compared with traditional constant-frequency peak-current-mode control, the compensated voltage V_C of VFPCM determines not only on-time but also switching frequency.

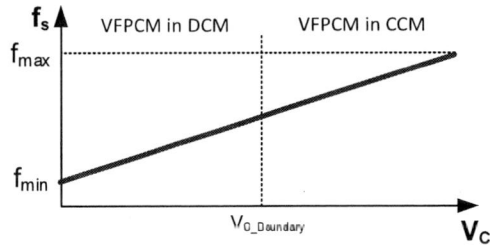

Fig. 1 Practical frequency characteristics of frequency reduction control

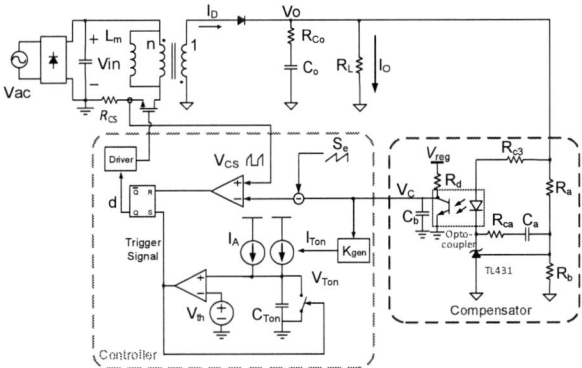

Fig. 2 Simplified circuit diagram of VFPCM control

Due to the wide operation range of USB-PD application, the stability issue of flyback converter with VFPCM control becomes severe. The converter has to be stable after compensation over the wide range of operation conditions including universal line voltage change, output voltage change, and switching frequency change due to VFPCM control. Fig. 3 shows the loop gains T(s) of VFPCM controlled flyback converter at various input voltages, output voltages, and load currents in 40W USB-PD application. The compensator is designed based on the small-signal model of VFPCM [9], and the converter is operated in continuous conduction mode (CCM). As shown in Fig. 3, the phase margins exist large difference at various operation conditions. In one extreme case, the phase margin is only 10° and leads to a near unstable condition for flyback converter, which is not acceptable for converter design. As shown in Fig. 3, the unstable issue is caused by cross-over frequency variation of loop gains and excessive phase drop since right-half-plane

2086

zero and high-frequency poles reduce phase margin for the loop gain with largest cross-over frequency.

In this paper, two stability improvement methods are reviewed and each of techniques is introduced, too. The paper is organized as follows. The reason for cross-over frequency variation of loop gain is analyzed in Section II. Section III and Second IV describes the first and second stability improvement methods, respectively. Section V demonstrates the experimental verifications. Finally, conclusions are made in Section VI.

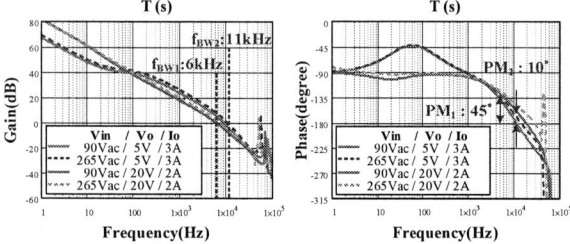

Fig. 3 Loop gains of VFPCM in various operation range of USB-PD application

II. EXPLANATION OF LOOP GAIN CROSS-OVER FREQUENCY VARIATION IN VFPCM CONTROL

Fig 4 shows a typical Bode plot of compensator design for flyback converter with VFPCM control. $G_{vc,spf}(s)$ is the control to output transfer function of flyback converter, and its cross-over frequency is defined as f_C. The cross-over frequency of loop gain is defined as f_{BW}. Since $G_{vc,spf}(s)$ is a transfer function with three poles and two zeros as (1) shown [9], the type II compensator is proper to compensate it [10-12]. In compensator design, the low-frequency zero of compensator is set to cancel the low-frequency pole of $G_{vc,spf}(s)$. The high-frequency pole of compensator is placed beyond f_{BW} to suppress high frequency noise. Besides, f_{BW} is lower than f_{ESR}, f_1, and f_{RHP}. Therefore, the loop gain is presented as -20dB/dec through cross-over frequency.

$$G_{VC,spf}(s) = \frac{v_o(s)}{v_c(s)} \cong G_{VC\_DC} \frac{\left(1+\dfrac{s}{\omega_{ESR}}\right)\left(1+\dfrac{s}{\omega_{RHP}}\right)}{\left(1+\dfrac{s}{\omega_{P1}}\right)\left(1+\dfrac{s}{Q_1\omega_1}+\dfrac{s^2}{\omega_1^2}\right)} \quad (1)$$

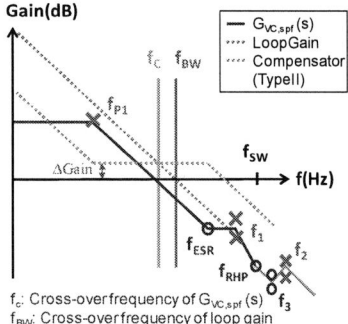

f_C: Cross-over frequency of $G_{VC,spf}(s)$
f_{BW}: Cross-over frequency of loop gain

Fig. 4 Bode plot of compesator design for flyback converter with VFPCM control

It is derived that the cross-over frequency of loop gain, f_{BW}, is only affected by f_C, when the frequency response of compensator is fixed. Based on the design of typical compensation in previous paragraph, the loop gain is

compensated as -20dB/dec line through cross-over frequency. Therefore, the relation of f_{BW} and f_C can be described as (2) from Fig. 4, where ΔGain is the middle-frequency gain of compensator. From (2), f_{BW} can be further presented as (3). Since ΔGain is constant for a fixed compensator, f_{BW} is only affected by f_C.

$$\frac{\Delta \text{Gain}}{\log f_{BW} - \log f_C} = -20 \left(dB / dec \right) \quad (2)$$

$$f_{BW} = 10^{\log f_C - \frac{\Delta \text{Gain}}{20}} \quad (3)$$

Based on the above results, f_{BW} can be observed by the equation of f_C. Based on the definition of Decibel unit, the unit gain of $G_{vc,spf}(s)$ happens at cross-over frequency ω_C as (4) shown. (4) can be approximated as (5) based on the compensated result in Fig. 4 that only ω_{P1} is below ω_C. (5) can be expressed as (6). From (6), f_C can be approximated as (7) since G_{VC_DC} is much greater than 1. From (7), it is clear that the f_C is determined by the multiplication of DC gain and low-frequency pole of $G_{vc,spf}(s)$.

$$\left| G_{VC,spf}(j\omega_C) \right| = 1 = G_{VC\_DC} \sqrt{ \frac{\left[1^2+\left(\dfrac{\omega_C}{\omega_{ESR}}\right)^2\right]\left[1^2+\left(\dfrac{\omega_C}{\omega_{RHP}}\right)^2\right]}{\left[1^2+\left(\dfrac{\omega_C}{\omega_{P1}}\right)^2\right]\left[\left(1-\dfrac{\omega_C^2}{\omega_1^2}\right)^2+\left(\dfrac{\omega_C}{Q_1\omega_1}\right)^2\right]} } \quad (4)$$

$$\left| G_{VC,spf}(j\omega_C) \right| = 1 \cong G_{VC\_DC} \sqrt{\left[1^2+\left(\frac{\omega_C}{\omega_{P1}}\right)^2\right]^{-1}} \quad (5)$$

$$\omega_C \cong \omega_{P1} \sqrt{G_{VC\_DC}^2 - 1^2} \quad (6)$$

$$f_C \cong \frac{\omega_C}{2\pi} = \frac{1}{2\pi} G_{VC\_DC} \omega_{P1} \quad (7)$$

The f_C variation of VFPCM control is investigated below. By substituting G_{VC_DC} and ω_{p1} of VFPCM [9] into (7), the f_C can be presented as (8). Then, (8) is further simplified as (9) since R_{Co} is usually much smaller than load resistor R_L and $k_2 \cdot R_{Co}$ is much smaller than 1 in practical application. From (9), it is shown that the f_C varies with different input and output voltages. Furthermore, it also changes with output current since switching period depends on output current.

$$f_C \cong \frac{G_{VC\_DC}\omega_{P1}}{2\pi}$$
$$= \frac{1}{2\pi}\left(\frac{DK_{gen}T_{SW}^2(2S_e+S_n)}{2V_{th}C_{Ton}}+1\right)\frac{1}{R_{CS}}\frac{DV_{in}}{V_O}\frac{R_L}{C_o(R_{Co}+R_L)}\frac{1}{\left(1-k_2\dfrac{R_L R_{Co}}{R_{Co}+R_L}\right)} \quad (8)$$

$$f_C \cong \frac{1}{2\pi}\left(\frac{DK_{gen}T_S^2(2S_e+S_n)}{2V_{th}C_{Ton}}+1\right)\frac{DV_{in}}{V_O}\frac{1}{R_{CS}}\frac{1}{C_o} \quad (9)$$

where

$$S_n = R_{CS} \cdot \frac{V_{in}}{L_m}, \; S_f = R_{CS} \cdot \frac{nV_O}{L_m}$$

The 2018 International Power Electronics Conference

As equation (9) presented, the f_{BW} and phase margin are variable since the fc varies with working conditions such as input voltage, output voltage and switching period. Therefore, if the variation of fc can be attenuated, the variations of f_{BW} and phase margin are also reduced and system becomes more stable. In order to decreases the variation of fc, there are two methods proposed in literature [13-14]. The two methods are introduced as following.

III. A PHASE MARGIN IMPROVEMENT METHOD FOR VFPCM CONTROLLED FLYBACK CONVERTER

The Fig.5 shows a simplified circuit diagram of VFPCM controlled flyback converter with phase margin improvement method (PMIM) [13]. Compared with original control structure, there is a first order low-pass filter inserted into control loop of switching frequency. The first order low-pass filter provides gain attenuation for high-frequency and keeps original gain for low-frequency.

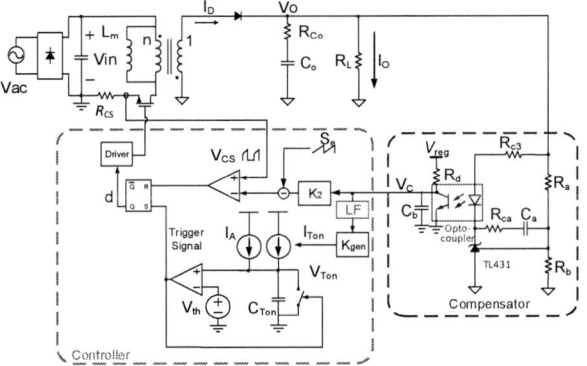

Fig. 5 Simplified circuit diagram of VFPCM control with PMIM

Based on the analysis results in Section II and literature [9], the fc of VFPCM is mainly affected by DC gain of $G_{vc,spf}(s)$ and DC gain of $G_{vc,spf}(s)$ is dominated by control to switching period function. Therefore, through a first order low-pass filter inserts into control loop of switching frequency, the high-frequency effect of control to switching period transfer function can be eliminated and variation of fc will be shrunk. Besides, the frequency reduction performance of steady-state operation still can be maintained same since low-frequency gain keeps same. As shown in equation (10), the first order low-pass filter provides a pole ω_{P_LF} to reduce the high-frequency gain for control to switching period transfer function.

$$\frac{T_S(s)}{v_C(s)} = -\frac{K_{gen}T_{SW}^2}{V_{th}C_{Ton}}\frac{1}{\left(1+\dfrac{s}{\omega_{P\_LF}}\right)} \quad (10)$$

A. Circuit implementation of PMIM

Fig 6 shows the circuit diagram of PMIM. By using current mirror, unity-gain buffer, and resistor R_1, R_2, R_A, the Vc can be transformed into currents to charge capacitors to determine switching frequency as shown in (11). Moreover, through the design of resistors R_1, R_2, R_A and capacitor C_{LF}, the gain of K_{gen} and first order low-pass filter can be easily realized as (12) shown.

$$I_A(s) = V_C(s)\cdot\frac{R_2}{R_A(R_1+R_2)}\cdot\frac{1}{1+sC_F(R_1//R_2)} \quad (11)$$

$$K_{gen} = \frac{R_2}{R_A(R_1+R_2)}, \quad \omega_{P\_LF} = \frac{1}{C_{LF}(R_1//R_2)} \quad (12)$$

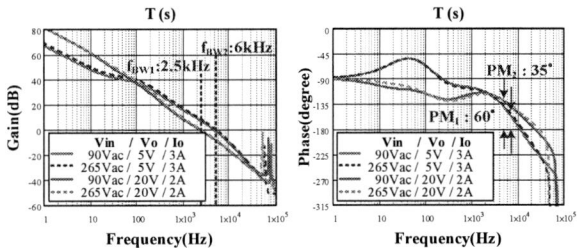

Fig. 6 Simplified circuit diagram of PMIM

B. Performance verification of PMIM

Fig. 7 shows the simulation results of loop gains of VFPCM controlled flyback converter with PMIM in various conditions. After adding the first order low-pass filter, the variations of f_{BW} is decreased and phase margin of VFPCM controlled flyback converter is increased, so system stability becomes better.

Fig. 7 Loop gains of VFPCM controlled flyback converter with PMIM in various operation range of USB-PD application

C. Comments

The PMIM method gives a simple, passive and low-cost solution to solve the stability issue of VFPCM controlled converter in USB-PD application. According to previous analysis, the f_{BW} and phase margin are variable since the fc variation. the fc is mainly affected by DC gain of $G_{vc,spf}(s)$ and DC gain of $G_{vc,spf}(s)$ is dominated by control to switching period function. In order to decrease the variations of f_{BW} and phase margin, a capacitor C_{LF} is inserted to original frequency control loop to eliminate the high-frequency gain of control to switching period transfer function. Thus, the phase margin has obvious improvement. However, the variation of phase margin caused by varied f_{BW} is still severe. In order to decrease the variation of phase margin, the literature [14] proposes a novel control for flyback converter which required frequency reduction control. The details of this control will be illustrated as below section.

IV. AN ADAPTIVE CONTROL TO ACHIEVE CONSTANT CROSS-OVER FREQUENCY FOR FLYBACK CONVERTER

The literature [14] proposes a novel variable-frequency peak-current mode (NVFPCM) control which can easily implement adaptive constant cross-over frequency control.

2088

The 2018 International Power Electronics Conference

Fig. 8 shows the simplified circuit diagram of NVFPCM controlled flyback converter. Compared with original VFPCM, the switching frequency is not varied by bias current I_A but by the threshold voltage V_{th}. When output current I_O decreases, V_C and I_{gen} also decrease. This results the same function of frequency reduction as VFPCM to increase light load efficiency. Moreover, an extra gain K_a is inserted into pulse-width modulation loop to add a degree of freedom for frequency reduction function design.

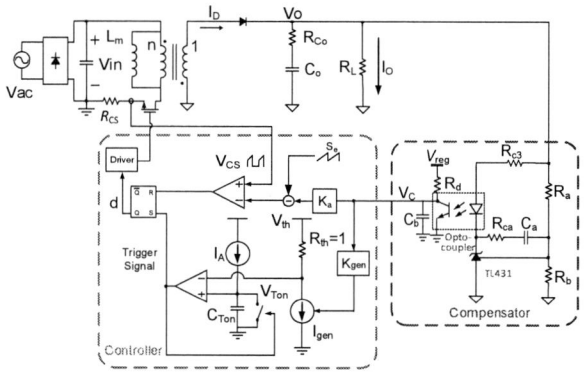

Fig. 8 Simplified circuit diagram of NVFPCM control

In order to investigate the f_C variation of NVFPCM control, the control to output transfer function $G_{vc,spf}(s)$ of NVFPCM controlled flyback converter needs to be derived out first. By same modeling strategy [9], the $G_{vc,spf}(s)$ is expressed as equation (11). The more detail mathematical equation is shown in literature [14].

$$G_{VC,spf}(s) = \frac{v_o(s)}{v_c(s)} \cong G_{VC\_DC} \frac{\left(1 + \dfrac{s}{\omega_{ESR}}\right)\left(1 + \dfrac{s}{\omega_{RHP}}\right)\left(1 + \dfrac{s}{Q_3\omega_3} + \dfrac{s^2}{\omega_3^2}\right)}{\left(1 + \dfrac{s}{\omega_{P1}}\right)\left(1 + \dfrac{s}{Q_1\omega_1} + \dfrac{s^2}{\omega_1^2}\right)\left(1 + \dfrac{s}{Q_2\omega_2} + \dfrac{s^2}{\omega_2^2}\right)} \quad (11)$$

where

$$G_{VC\_DC} = \left(\frac{K_{gen}C_{Ton}D(2S_e + S_n)}{2I_A} + K_a\right)\frac{1}{R_{CS}}\frac{DV_{in}}{V_O}\frac{R_L}{(1 - k_2 R_L)}$$

$$\omega_{P1} = \frac{1}{C_o(R_{Co} + R_L)}\frac{(1 - k_2 R_L)}{\left(1 - k_2\dfrac{R_L R_{Co}}{R_{Co} + R_L}\right)}$$

$$k_2 = -\frac{n^2(1-D)^3 T_S}{2L_m} - \frac{n(1-D)}{(V_{in} + nV_O)}\left(\frac{V_O}{(1-D)R_L} + \frac{n}{R_{CS}}S_e(1-D)T_S\right)$$

According to the discussion in Section II, the f_C of NVFPCM can be expressed as (12). From (12), it is found that if twice times of S_e is equal to S_f as (13), the f_C of NVFPCM can be simplified again as (14). Then, if the circuit parameters are designed to achieve equation (15), the term in (14) which times duty cycle is equal to zero. Thus, the f_C of NVFPCM can be kept constant as (16) and does not changed with input voltage, output voltage, switching frequency, and output current. That is, constant cross-over frequency over wide-operation-range can be achieved.

$$f_C \cong \frac{G_{VC\_DC}\omega_{P1}}{2\pi} \cong \frac{1}{2\pi}\left(\frac{K_{gen}C_{Ton}D(2S_e + S_n)}{2I_A} + K_a\right)\frac{DV_{in}}{V_O}\frac{1}{R_{CS}}\frac{1}{C_o} \quad (12)$$

$$2S_e = S_f \quad (13)$$

$$\begin{aligned}f_C &\cong \frac{G_{VC\_DC}\omega_{P1}}{2\pi} \\ &\cong \frac{1}{2\pi}\left(\frac{2I_A K_a n + nD(K_{gen}C_{Ton}S_n - 2I_A K_a)}{2I_A}\right)\frac{1}{R_{CS}}\frac{1}{C_o}\end{aligned} \quad (14)$$

$$K_{gen}C_{Ton}S_n - 2I_A K_a = 0 \quad (15)$$

$$f_C \cong \frac{G_{VC\_DC}\omega_{P1}}{2\pi} = \frac{1}{2\pi}\frac{nK_a}{R_{CS}}\frac{1}{C_o} \quad (16)$$

A. Circuit implementation of the proposed adaptive NVFPCM control

As mentioned above, constant f_c can be achieved if the control achieves equations (13) and (15). Equations (13) and (15) can be further derived as (17) and (18). In practical application, an auxiliary winding is employed to supply energy for control IC and this supply voltage, $V_{aux.}$, is usually proportional to output voltage by turns ratio, $n_{aux.}$, as equation (19) shown. Therefore, equation (17) can be expressed as (20).

$$S_e = \frac{S_f}{2} = \frac{nR_{CS}}{2L_m}V_O \quad (17)$$

$$I_A = \frac{K_{gen}C_{Ton}}{2K_a}\frac{R_{CS}}{L_m}V_{in} \quad (18)$$

$$V_O = \frac{V_{aux.}}{n_{aux.}} \quad (19)$$

$$S_e = \frac{S_f}{2} = \frac{nR_{CS}}{2L_m n_{aux.}}V_{aux.} \quad (20)$$

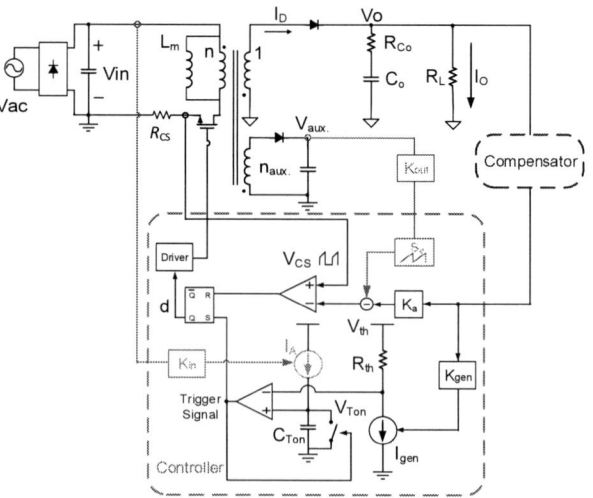

Fig. 9 Simplified circuit diagram of proposed adaptive NVFPCM control

Based on equations (19) and (20), the adaptive NVFPCM control is proposed as shown in Fig 9. K_{in} block achieves equation (21) to generate proper I_A, and K_{out} block achieves equation (22) to generates proper S_e. As seen in Fig.

2089

9, the output voltage information can be obtained through the auxiliary winding voltage without any extra isolated circuit. Besides, input voltage information can be obtained from input capacitor directly.

$$K_{in} = \frac{K_{gen}C_{Ton}}{2K_a}\frac{R_{CS}}{L_m} \quad (21)$$

$$K_{out} = \frac{nR_{CS}}{2L_m n_{aux.}} \quad (22)$$

B. Performance verification of cross-over frequency control by simulation

Fig. 10 shows simulated loop gains of adaptive control in various conditions The SIMPLIS simulation results present that the cross-over frequencies of loop gains are fixed at 5.5kHz and the variation of phase margin is reduced to 12°. In other words, the adaptive control achieves constant cross-over frequency which improves system stability effectively for USB-PD application. Compared to the minimum phase margin of 10° and phase margin variation of 35° in Fig. 3, the adaptive control improves stability over conventional VFPCM control for wide operating range.

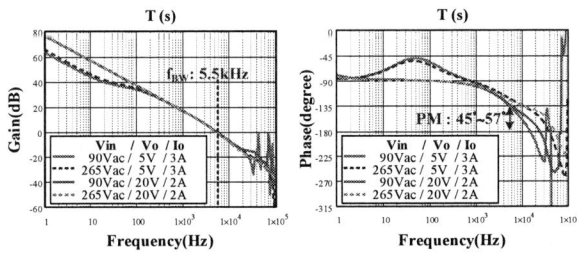

Fig. 10 Loop gain simulation of adaptive NVFPCM control in various conditions

C. Comments

The literature [13-14] proposes a novel variable-frequency peak-current mode (NVFPCM) control which possesses the variable frequency characteristic for high efficiency. Moreover, through small-signal mode analysis, an adaptive NVFPCM is proposed to achieve constant cross-over frequency control. This adaptive control utilizes I_A and S_e changed adaptively with input and output voltage to fix f_{BW} for various working conditions, thus the variation of phase margin is reduced obviously. Although the control circuit becomes more complex and the cost increases relatively while comparing with PMIM., this control achieves great performance for improving stability.

V. EXPERIMENTAL VERIFICATION

A prototype of flyback converter with adaptive NVFPCM control is built, as shown in Fig. 11, for validation of the control performance. The control circuit is built with an analog peak-current-mode controller, UC3843. The UC3843 provides a frequency setting pin for the user to adjust switching frequency. Thus, the variable-frequency control can be implemented by this pin as shown in Fig. 12. The K_{in} and K_{out} blocks of the proposed control are achieved by an ADL5315 from Analog Devices as the voltage controlled current source to implement bias current I_A and

slope compensation S_e, by an ADA4891 as the amplifiers to implement operational amplifier, and by an LM393 as the comparator to generate the trigger signal to determine switching frequency. The test parameters and working conditions for the experiments are the same as simulation parameters listed in Tables II.

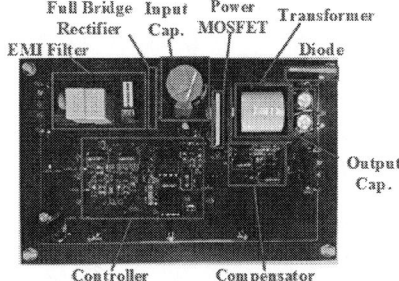

Fig. 11. A prototype flyback converter

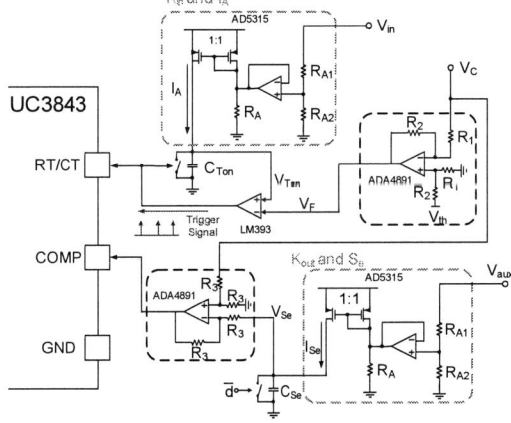

Fig. 12. Variable-frequency control implementation for UC3843
Circuit parameters of simulation

TABLE II
Circuit parameters of simulation

Item	NVFPCM in CCM			
Input voltage	127V		375V	
Output voltage	5V	20V	5V	20V
Output current	3A	2A	3A	2A
Switching frequency f_s	44kHz	80kHz	95kHz	114kHz
Magnetic inductor L_m	1.2mH			
Transformer turn ratio n	6.3			
Output capacitor C_o	1360µF			
ESR of capacitor R_{Co}	7mΩ			
Slope compensation S_e	6.7V/mS	26.8V/mS	6.7V/mS	26.8V/mS
Bias current I_A	40.5 uA	40.5 uA	119.5 uA	119.5 uA
Reference design	K_{gen}= 0,5A/V, K_n=0.33V/V Rcs = 0.51Ω, C_{Ton} = 1nF, V_{th} = 2.3V			
Compensator design	R_a= 68kΩ, R_{ca}=1kΩ R_{c3} = 600Ω, R_d=20kΩ C_a = 100nF, C_b = 1.8nF			
Low-side divided resistor R_b	68kΩ	9.7kΩ	68kΩ	9.7kΩ
Opto-coupler current gain	50%			

Figs. 13 verifies the model of $G_{vc,spf}(s)$ of adaptive NVFPCM control in various conditions. It indicates that the results of the proposed small-signal model are in excellent agreement with the measurements. Moreover, as SIMPLIS simulation results presented, the crossover frequency of $G_{vc}(s)$ in various conditions is the same. Namely, the adaptive NVFPCM control achieves constant crossover frequency. Fig. 14 shows the loop gains of the adaptive NVFPCM control. As shown in Fig. 14, both constant crossover frequency and proper phase margins are achieved at various working conditions, so the effectiveness of the proposed control method and system stability are verified.

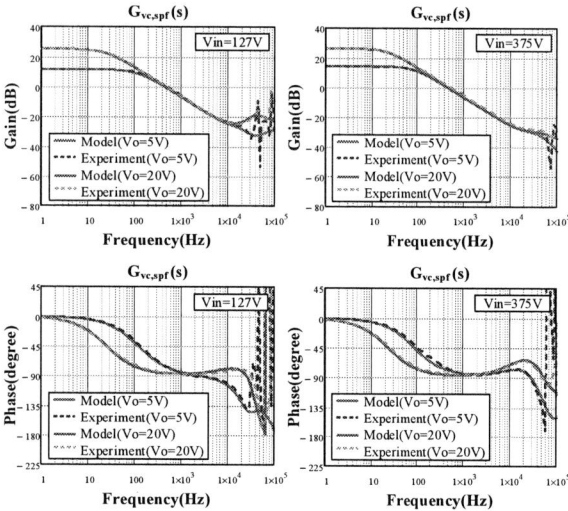

Fig. 13 Experimental verification of NVFPCM control model

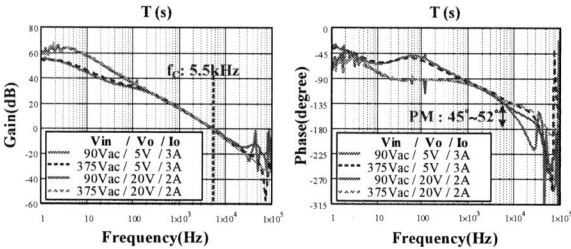

Fig. 14 Loop gain verification by experiment of the proposed NVFPCM control with constant crossover frequency at wide-range-operation

VI. CONCLUSIONS

Recently, VFPCM controlled flyback converter is widely used in wide-voltage-range operation such as USB-PD application. However, this application makes the stability issue of flyback converter more severe. In order to solve the issue, two stability improvement methods are suggested in literatures. The first method provides a simple, passive and low-cost solution, which utilizes the characteristic of a first order low-pass filter to eliminate the high-frequency gain of control to switching period transfer function. By using this method, the variations of f_{BW} is reduced and the phase margin of VFPCM controlled flyback converter is increased. Namely, the system stability becomes better. The second method proposes a constant cross-over frequency control which shows more great performance than PMIM in stability improvement. According to the small-signal model of NVFPCM, this adaptive control adjusts I_A and S_e adaptively to achieve constant f_{BW} for various working conditions. Thus the variation of phase margin is reduced effectively.

This paper presents a brief overview of the two stability improvement methods and completely interprets those detail techniques. Simulation validates the performances of the two methods and experiment presents that the adaptive NVFPCM control achieves constant crossover frequency. Finally, some comments are also summarized.

REFERENCES

[1] U. S. Padiyar, V. Kamath, "Design and implementation of a universal input flyback converter," in *Proc. IEEE ICEEOT*, 2016, pp. 3428-3433.

[2] Y. Li and J. Zheng, "A low-cost adaptive multi-mode digital control solution maximizing AC/DC power supply efficiency," in *Proc. IEEE Appl. Power Electron. Conf.*, 2010, pp. 349–354.

[3] Texas Instruments. USB Type-C and USB PD Source Controller , TPS25740 datasheet. Texas Instruments document. (2017). [Online]. Available: http://www.ti.com/

[4] Power Integrations. *20 W USB PD Power Supply Using Cypress CCG2 CYPD2134 and InnoSwitch™ –CP INN2215K, Design Example Report. Power Integrations document.* (2016). [Online]. Available: https://www.power.com/

[5] NXP semiconductor. *Green chip SMPS control IC, TEA1733A datasheet. NXP semiconductor document.* (2013). [Online]. Available: http://www.npx.com/

[6] ON semiconductor. *Fixed frequency current mode controller for flyback converters, NCP1237 datasheet. ON semiconductor document.* (2011). [Online]. Available: http://www.onsemi.com/

[7] Fairchild semiconductor. *Highly integrated green-mode PWM controller, FAN6862 datasheet. Fairchild semiconductor document.* (2012). [Online]. Available: http://www.fairchildsemi.com/

[8] Texas Instruments. *Constant-voltage constant-current flyback controller using opto-coupled feedback, UCC28740 datasheet. Texas Instruments document.* (2014). [Online]. Available: http://www.ti.com/

[9] C. H. Cheng, C. J. Chen and S. S. Wang, "Small-signal model of flyback converter in continuous-conduction mode with peak-current control at variable switching frequency," *IEEE Trans. Power Electron.*, vol. 33, no. 5, pp. 4145-4156, Sep. 2018.

[10] C. J. Chen, C. H. Cheng, B. S. Wu and S. S. Wang, "Unified small-signal model and compensator design of flyback converter with peak-current control at variable frequency for USB power delivery," *IEEE Trans. Power Electron.*, 201X.

[11] Fairchild semiconductor. *Application Note AN-4150: Green current mode pwm controller fan7602 Fairchild semiconductor document.* (2006). [Online]. Available: http://www.fairchildsemi.com/

[12] E. Wang, "Feedback control design of off-line flyback converter," App. note AN017, Jun. 2014. [Online]. Available: http://www.richtek.com/

[13] C. H. Cheng, C. J. Chen and S. S. Wang, "A phase margin improvement method for variable frequency peak-current-mode controlled flyback converter," *IEEE Trans. Power Electron.*, submitted.

[14] C. H. Cheng, C. J. Chen and S. S. Wang, "An adaptive variable frequency peak-current-mode control for wide-operation-range flyback converter," *IEEE Trans. Power Electron.*, submitted.

The 2018 International Power Electronics Conference

Design and Implementation of a High Power Density Active-Clamped Flyback Converter

Yu-Chen Liu[1], Bing-Siang Huang[2], Cheng-Hung Lin[2], Katherine A. Kim[3], Huang-Jen Chiu[2]

[1]Department of Electrical Engineering National Ilan University, Yilan, Taiwan, R.O.C

[2]Department of Electrical and Computer Engineering, National Taiwan University of Science and Technology, Taipei, Taiwan R.O.C

[3]School of Electrical and Computer Engineering Ulsan National Institute of Science and Technology Ulsan, South Korea

d10502205@mail.ntust.edu.tw

Abstract—**This paper presents a high power density power converter using an active-clamp flyback converter. The voltage spike in the conventional flyback converter can be eliminated to reduce electromagnetic interference and voltage stresses on power switches by using an active-clamped circuit. The relationship between zero-voltage-switching condition of high frequency power switches and magnetizing inductance and leakage inductance is analyzed and discussed. GaN devices are used both as primary and secondary switches to reduce the converter volume and switching losses, to achieve high power density. The planar transformer design reduces the conduction losses and current stresses of the secondary switches attributed to the winding configuration. With theoretical analysis and circuit simulation, a 65-W (19 V/ 3.3 A) active-clamped flyback converter with universal input voltage capability is developed and tested. The peak efficiency is up to 93% at low line voltage and the power density is 1.879 W/ cm³.**

Keywords—Gallium nitride device, active-clamped flyback converter, high frequency, high power density

I. INTRODUCTION

With environmental awareness, high efficiency is the main consideration for designing power supplies. One of the biggest markets of power supplies is the ac-dc adapter application, including smart phone, tablet, laptop, and game console, etc. Besides of the efficiency, circuit miniaturization is another market trend for adapter applications. Improving power density becomes the main design consideration for both efficiency and miniaturization. Magnetic component and capacitor volumes can be reduced by increasing switching frequency to further improve power density. Under high frequency conditions, large stray capacitance of silicon-based devices will lead to slow switching speed and high switching losses. By comparison, gallium nitride (GaN) devices [1-3] have low stray capacitance which can shorten switching time to produce lower switching losses and higher power density.

The flyback converter is usually used in lower power application because of its simple construction, isolation characteristic and low cost. An RCD snubber is usually applied in the circuit to eliminate the voltage spikes in the flyback converter and reduce the EMI noise. In the MHz frequency range, the loss that the snubber dissipates should not be omitted because the loss is proportional to the switching frequency. Several works [4, 5] achieved soft switching by using quasi-resonant switching. The active clamp technique [6-13] can not only achieve soft switching but also absorbs the leakage

inductance energy to reduce voltage stresses of primary switches and EMI noise [14, 15].

The leakage inductance is the main component that relates to zero-voltage-switching (ZVS) [16-18] in active-clamp flyback converters. In the MHz frequency range, the magnetic inductance also affects the ZVS condition. Both inductances need to be designed properly. To effectively control leakage inductance and minimize the circuit volume, a planar transformer [19] is implemented in the circuit.

In adapter applications, secondary winding loss and conduction loss is considerable compared to the primary side. Moreover, the winding loss will be higher because of the skin effect and proximity effect at MHz frequency. Two transformers can be applied and the secondary winding would be in parallel to widen the winding and reduce the secondary winding loss. Based on the parallel winding configuration, the matrix transformer discussed in [20-22] would further reduce the total core volume and core loss.

II. DESIGN OF ACTIVE-CLAMPED FLYBACK CONVERTER

A. Magnetic inductance

Fig. 1 shows the active-clamed flyback converter with two transformers. The basic operation principle of the active-clamped circuit is well known and will not be detailed in this paper. The two transformers reduce the secondary winding loss and the magnetic inductances L_{m1} and L_{m2} are in series.

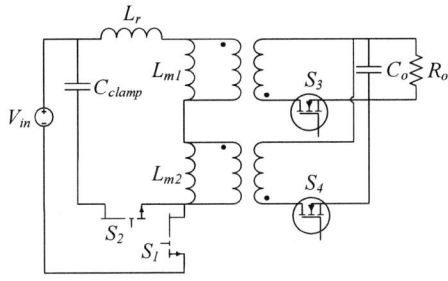

Fig. 1 Active clamp flyback converter with two transformers

Generally, leakage inductance is designed for ZVS of the main switch S_1 in active clamp circuit and affected by the parasitic capacitor of the switch. Whereas magnetic inductance

The 2018 International Power Electronics Conference

is designed based on the designated current ripple, which changes with switching frequency. Hence, leakage inductance will be larger than magnetic inductance at MHz frequency. Thus, ZVS relates to both inductances at MHz frequency.

Usually, the ZVS condition depends on energy stored in the inductance while the switch is turning on. At MHz frequency, the magnetic inductance is also designed to achieve soft switching or zero voltage switching of the main switch. Fig. 2 shows the waveforms of the current through the leakage inductance at the normal frequency (kHz) and MHz frequency. At lower frequency, leakage inductance current flows through the auxiliary switch S_2 from the positive to negative direction, and the magnetic inductance current is designed to flow in the positive direction. Only the negative current will assist the main switch to achieve soft switching, which means the magnetic inductance is not involved in ZVS at lower switching frequency. On the other hand, current of both inductances will flow from positive to negative direction at higher frequency. Thus, ZVS condition is related to both of the inductances. The more negative the current, the easier it is to achieve ZVS. Finally, the magnetic inductance is decided to be 19 μH to achieve ZVS at MHz frequency.

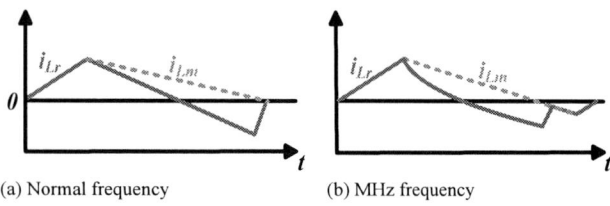

(a) Normal frequency   (b) MHz frequency

Fig. 2 Leakage inductance current

B. Turns Ratio

Design of the turns ratio affects the duty cycle of the primary and secondary side. The duty cycle will also affect the root mean square (rms) values of each current loop. The influence of rms values on efficiency is considerable. Fig. 3 shows the relationship of losses and turns ratio. The primary losses include conduction loss and primary winding loss. The secondary losses include the same loss types as the primary side. As shown, primary losses become smaller while the turns ratio increases because the primary side duty cycle is increasing, which results in decreasing of the rms value. Conversely, secondary losses increase with the turns ratio because the duty cycle of the secondary side is decreasing, which results in increasing of the rms value. The total losses can be combined with primary losses and secondary losses. The minimum of total losses is at a turns ratio of 5:1, which is the best choice for circuit efficiency. With turns ratio 5:1, the voltage stress is about 480 V for the primary switches and approximately 200 V for the secondary switches. Both conditions are acceptable and reasonable for GaN devices in adapter application.

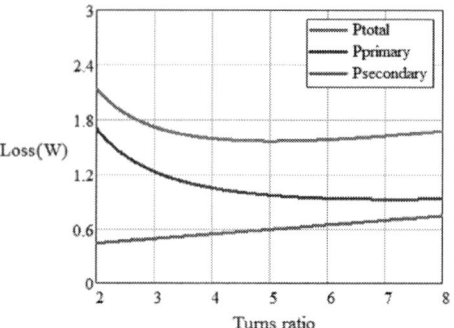

Fig. 3 Impact of turns ratio on converter loss

C. Core Cross-Sectional Area: Ae

The relationship between losses and core cross section area is shown in Fig. 4. Losses related to cross section area include primary winding loss and core loss. With a fixed turns ratio and winding area of transformer, increasing cross section area will reduce the core loss. Even though the core volume increases with larger cross section area. The core loss density which is the major factor of total core loss will effectively decrease with increasing cross section area. On the other hand, the primary winding loss is proportional to the cross section area since the primary winding length and winding resistance are also proportional to the cross section area. Considering the two losses in Fig. 4, core loss decreasing more rapidly than winding loss. Thus, cross section area should be chosen to be as large as possible. With emphasis on power density, the cross section area is designed to be 63 mm^2 for the transformer design.

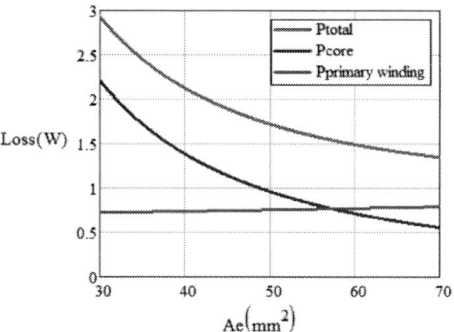

Fig. 4 Impact of core cross section area on converter loss

D. Clamping capacitor

Selection of the clamping capacitor is affected by the operating condition of the resonant tank, which includes the resonant frequency and current waveform of the resonant tank. Thus, current stress of the secondary synchronous rectifier is also determined by the selection of the clamping capacitor. The simulated current of the secondary side is shown in Fig. 5, showing three curves with different clamping capacitors. The smaller capacitor results in a higher resonant frequency and higher current stress on the switching device. There is current stress difference of about 4 A between 80 nF and 300 nF

2093

capacitance. Moreover, the higher current peak value means a higher current rms value, which is a major part of measuring the secondary conduction loss and secondary winding loss.

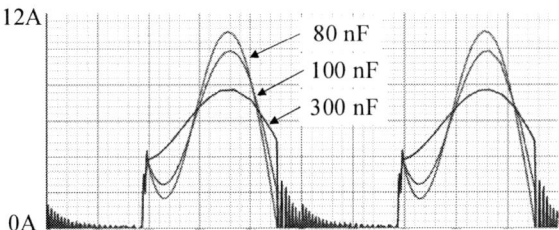

Fig. 5 Synchronous rectifier current with different clamping capacitors

Changing the clamping capacitor will directly impact the current of the resonant tank, which is constructed by leakage inductance and the clamping capacitor. The simulation current waveform is shown in Fig. 6. A larger clamping capacitor will also reduce the primary conduction loss and primary winding loss because of the lower current rms value. Further, ZVS of the main switch is also related to the negative current value of the resonant tank. With a larger clamping capacitor, the negative current value is higher when the current is cut off by the auxiliary switch. It is easier for the main switch to achieve ZVS condition with higher clamping capacitance. In this design, a 300 nF of clamping capacitor is used for relatively low conduction and switching loss of the main switch to gain higher efficiency.

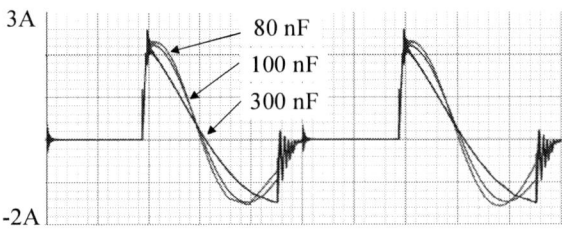

Fig. 6 Resonant tank current waveform with different clamping capacitor

III. WINDING CONFIGURATION

A. Secondary Winding

In adaptor applications, the secondary side suffers from high current stress and high conduction loss. Secondary conduction loss is even higher because of the skin effect and proximity effect at MHz frequency. Two transformers can be applied to reduce secondary current stress and winding loss. The equivalent resistance diagram of the two transformers is shown in Fig. 7. Compared to the conventional one, two transformers bring one more parallel winding to ease current stress and reduce winding loss. Especially for planar transformer applications, winding loss will generate heat and cause high temperatures in the circuit board.

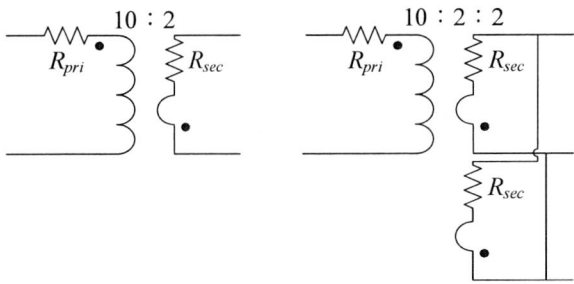

(a) one transformer   (b) two transformers

Fig. 7 Equivalent resistance diagram

However, two transformers will induce more core loss because of the doubled transformer volume. By applying a matrix transformer configuration, the number of cores can be reduced. There are four different matrix transformer configurations, as shown in Fig. 8. The upper left figure shows two separate core with a primary winding at each side. By changing the position, two transformers could be combined into one, which is shown in the upper right figure. The lower right figure shows the same core structure with different flux directions, where the center of the core would flow through the flux with both of the directions. Hence, the flux would be cancelled at the center of the core and the core loss at the center would also be cancelled. Furthermore, the center of the core could be removed since there is no flux flow through it. The core number can be reduced from two to one and still maintain the benefit of parallel secondary winding.

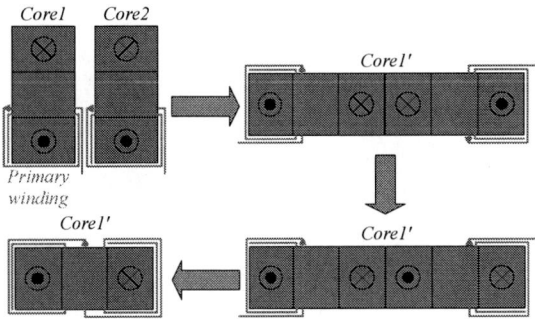

Fig. 8 Matrix transformer configuration

With the same number of primary turns, the layer of primary winding could be reduced by applying matrix transformer configuration. Fig. 9 shows the planar transformer where all the windings are wrapped around a single core. The red windings represent the primary side, where the width is 1.59 mm with ten turns. The blue windings represent the secondary side, where the width is 5.2 mm with two turns. There as six layers in total.

The 2018 International Power Electronics Conference

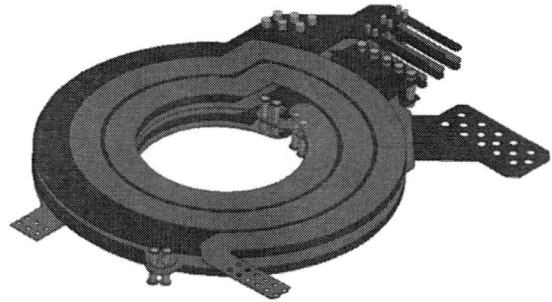

(a) Winding structure diagram

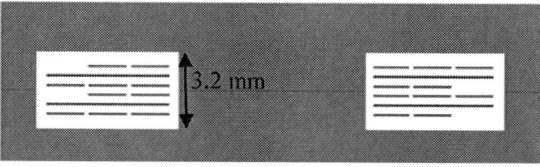

(b) Section view

Fig. 9 Winding without the matrix transformer configuration

With the same winding area of the core, the number of total layers can be reduced to four in the matrix transformer configuration. In planar transformer applications, the winding loss will cause heat generation in the circuit board. Attributed to the skin and proximity effect, winding loss is larger at MHz frequency. Thus, the reduction of layers will not only reduce the cost of the circuit board, but also reduce the heat generation.

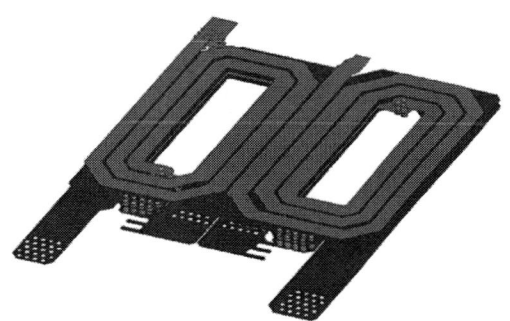

(a) Winding structure diagram

(b) Section view

Fig. 10 Winding with the matrix transformer configuration

IV. EXPERIMENTAL VERIFICATION

According to this design, a 65-W (19 V/ 3.4 A) prototype circuit is implemented, as shown in Fig. 11. The circuit is tested under open loop control testing and experimental testing is conducted to verifying the circuit operation at 1 MHz frequency.

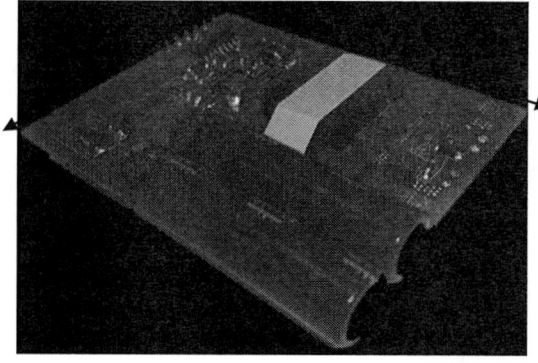

Fig. 11 1 MHz 65 W prototype circuit

GaN devices are used for both the primary and secondary side. The 650-V E-mode GaN from GaN Systems is used for primary side. The 200-V E-mode GaN from EPC is used for the secondary side to reduce conduction loss. The PCB board is 4-layer which is two layers fewer than the traditional winding construction. The power density is 1.88 W/cm^3.

Fig. 12 shows the waveforms at low line and high line input conditions. The voltage spike is eliminated by the active-clamped circuit. The ZVS condition is achieved at both input conditions. It reduces the switching loss and slows down the dv/dt at the turn on transient.

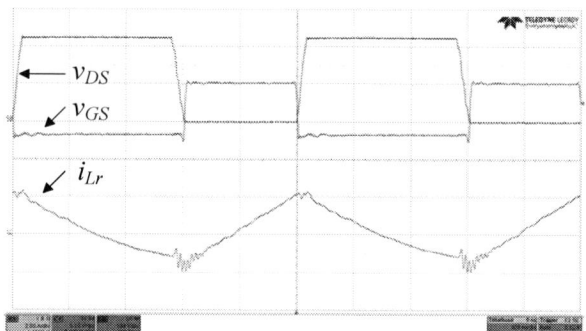

(a) Waveforms at 127 V$_{DC}$

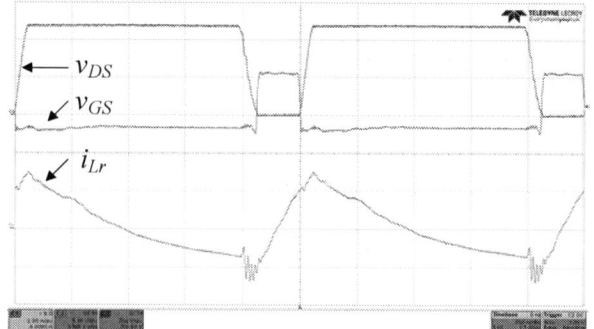

(b) Waveforms at 380 V$_{DC}$

Fig. 12 Circuit waveforms

2095

Fig. 13 shows the measured efficiency of the circuit. The synchronous rectifier is used in the circuit. And the peak of efficiency is over 93 % at low line input condition.

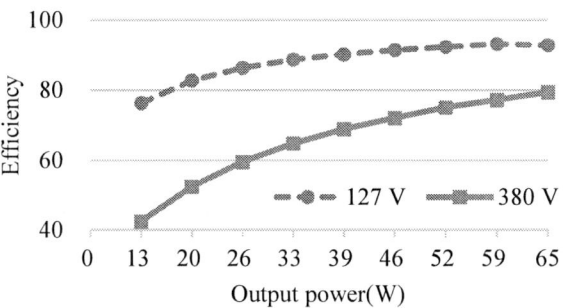

Fig. 13 Measured circuit efficiency

V. CONCLUSION

This paper presented the design method of active-clamped flyback converter at MHz frequency. The design is intended for higher efficiency and higher power density. The ZVS condition of the main switch is related to the magnetic inductance at higher frequencies. Two transformers in a matrix transformer configuration are applied to reduce the secondary winding loss caused by skin effect and proximity effect.

REFERENCES

[1] GaN Systems. (2016). How to drive GaN Enhancement mode HEMT [Online]. Available: http://www.mouser.com/pdfDocs/343654_GaNSystems__GN001_How_To_drive_GaN_EHEMT_Rev_20160426.pdf

[2] EPC. (2016). eGaN FET Drivers and Layout Considerations [Online]. Available: http://epc-co.com/epc/Portals/0/epc/documents/papers/eGaN%20FET%20Drivers%20and%20Layout%20Considerations.pdf

[3] E. A. Jones, F. Wang, D. Costinett, Z. Zhang, B. Guo, B. Liu, et al., "Characterization of an enhancement-mode 650-V GaN HFET," in 2015 IEEE Energy Conversion Congress and Exposition (ECCE), 2015, pp. 400-407.

[4] M. A. Bakar, K. Bertilsson, and R. Ambatipudi, "High frequency (MHz) soft switched flyback dc-dc converter using GaN switches and six-layered PCB transformer," in 8th IET International Conference on Power Electronics, Machines and Drives (PEMD 2016), 2016, pp. 1-6.

[5] Z. Zhang, K. D. T. Ngo, and J. L. Nilles, "A 30-W flyback converter operating at 5 MHz," in 2014 IEEE Applied Power Electronics Conference and Exposition - APEC 2014, 2014, pp. 1415-1421.

[6] X. Huang, J. Feng, W. Du, F. C. Lee, and Q. Li, "Design consideration of MHz active clamp flyback converter with GaN devices for low power adapter application," in 2016 IEEE Applied Power Electronics Conference and Exposition (APEC), 2016, pp. 2334-2341.

[7] R. Perrin, N. Quentin, B. Allard, C. Martin, and M. Ali, "High-Temperature GaN Active-Clamp Flyback Converter With Resonant Operation Mode," IEEE Journal of Emerging and Selected Topics in Power Electronics, vol. 4, pp. 1077-1085, 2016.

[8] J. h. Jung and S. Ahmed, "Flyback converter with novel active clamp control and secondary side post regulator for low standby power consumption under high-efficiency operation," IET Power Electronics, vol. 4, pp. 1058-1067, 2011.

[9] J. Zhang, X. Huang, X. Wu, and Z. Qian, "A High Efficiency Flyback Converter With New Active Clamp Technique," IEEE Transactions on Power Electronics, vol. 25, pp. 1775-1785, 2010.

[10] N. Quentin, R. Perrin, C. Martin, C. Joubert, B. Lacombe, and C. Buttay, "GaN Active-Clamp Flyback Converter with Resonant Operation Over a Wide Input Voltage Range," in PCIM Europe 2016; International Exhibition and Conference for Power Electronics, Intelligent Motion, Renewable Energy and Energy Management, 2016, pp. 1-8.

[11] Y. T. Yau, W. Z. Jiang, and K. I. Hwu, "Light-load efficiency improvement for flyback converter based on hybrid clamp circuit," in 2016 IEEE International Conference on Industrial Technology (ICIT), 2016, pp. 329-333.

[12] J. H. Kim, M. H. Ryu, B. D. Min, and E. H. Song, "A Method to Reduce Power Consumption of Active-Clamped Flyback Converter at No-Load Condition," in IECON 2006 - 32nd Annual Conference on IEEE Industrial Electronics, 2006, pp. 2811-2814.

[13] D. Murthy-Bellur and M. K. Kazimierczuk, "Active-clamp ZVS two-switch flyback converter," in 2011 IEEE International Symposium of Circuits and Systems (ISCAS), 2011, pp. 241-244.

[14] Y. Yang, D. Huang, F. C. Lee, and Q. Li, "Analysis and reduction of common mode EMI noise for resonant converters," in 2014 IEEE Applied Power Electronics Conference and Exposition - APEC 2014, 2014, pp. 566-571.

[15] X. Huang, J. Feng, F. C. Lee, Q. Li, and Y. Yang, "Conducted EMI analysis and filter design for MHz active clamp flyback front-end converter," in 2016 IEEE Applied Power Electronics Conference and Exposition (APEC), 2016, pp. 1534-1540.

[16] R. Watson, F. C. Lee, and G. C. Hua, "Utilization of an active-clamp circuit to achieve soft switching in flyback converters," IEEE Transactions on Power Electronics, vol. 11, pp. 162-169, 1996.

[17] S. Larousse, H. Razik, R. Cellier, N. Abouchi, and P. Volay, "Active Dead-time optimization for wide range Flyback active-clamp converter," in PCIM Europe 2016; International Exhibition and Conference for Power Electronics, Intelligent Motion, Renewable Energy and Energy Management, 2016, pp. 1-6.

[18] T. LaBella, B. York, C. Hutchens, and J. S. Lai, "Dead time optimization through loss analysis of an active-clamp flyback converter utilizing GaN devices," in 2012 IEEE Energy Conversion Congress and Exposition (ECCE), 2012, pp. 3882-3889.

[19] D. Fu, F. C. Lee, and S. Wang, "Investigation on transformer design of high frequency high efficiency dc-dc converters," in 2010 Twenty-Fifth Annual IEEE Applied Power Electronics Conference and Exposition (APEC), 2010, pp. 940-947.

[20] D. Reusch and F. C. Lee, "High frequency bus converter with integrated matrix transformers for CPU and telecommunications applications," in 2010 IEEE Energy Conversion Congress and Exposition, 2010, pp. 2446-2450.

[21] D. Reusch and F. C. Lee, "High frequency bus converter with low loss integrated matrix transformer," in 2012 Twenty-Seventh Annual IEEE Applied Power Electronics Conference and Exposition (APEC), 2012, pp. 1392-1397.

[22] D. Huang, S. Ji, and F. C. Lee, "LLC Resonant Converter With Matrix Transformer," IEEE Transactions on Power Electronics, vol. 29, pp. 4339-4347, 2014.

The 2018 International Power Electronics Conference

Optimized Variable On-Time Control for LED Lighting Driver

Jizhe Wang[1*], Haruhi Eto[1] and Fujio Kurokawa[2*]

1 Graduate School of Engineering, Nagasaki University, Nagasaki, Japan

2 Institute for Innovative Science and Technology, Nagasaki Institute of Applied Science, Nagasaki, Japan

1*E-mail: bb35211083@gmail.com

2*E-mail: kurokawa_fujio@nias.ac.jp

Abstract— This paper focus on how to implement variable on-time control to predictive boundary conduction mode boost PFC converter for driving LED lighting applications. Normally, the constant on-time control is implemented in predictive PFC converter, which leads to high THD of input line current. Varies of variable on-time control methods have been proposed to improve the THD of input line current. However, most of the proposed variable on-time calculation methods are too complex to be implemented in predictive control. In order to implement the variable on-time control to predictive control, a simple but very accurate variable on-time calculation method is applied. The optimized variable on-time control takes very short calculation time in DSP controller, hence it can be implemented in predictive control. The effectiveness of the proposed method which can reduce THD of line current is demonstrated on a 100 V input and 400 V/30 W output PFC prototype.

Keywords— *Predictive control; power factor correction; zero-crossing point; boundary conduction mode; power factor; total harmonic distortion.*

I. INTRODUCTION

Nowadays, about 20 % ~ 22 % of electrical energy is used for lighting. For instance, incandescent lighting represents about 2000 TWh/year. Since LED lights feature much higher energy efficiency, they are widely used instead of incandescent lights. The boost-type boundary conduction mode (BCM) power factor correction (PFC) converter, which is shown in Fig. 1, is commonly used as LED lighting drive controllers [1], [2]. PFC converters are required to achieve high power conversion efficiency and low total harmonic distortion (THD) of input line current [1]-[16].

In order to improve the efficiency, we proposed a predictive control method for boost-type BCM PFC converter for driving LED lighting applications [3]. In this method, the inductor current is predicted by analytical equations instead of being detected by a sensing-resistor. Therefore, the efficiency reduction caused by sensing-resistor can be eliminated.

Nonetheless, there is a problem exists in our previous method. In the previous method, the analysis of inductor current did not consider the effect of the parasitic capacitance C_j in diode and C_{oss} in MOSFET, as

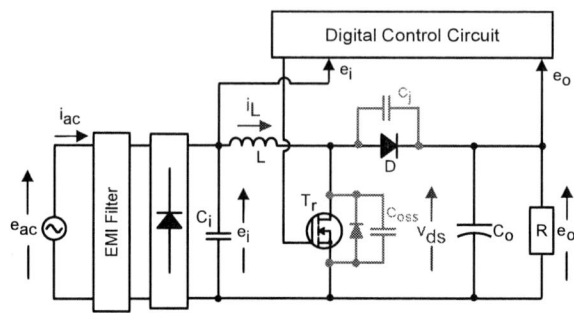

Fig. 1. Predictive digital controlled BCM PFC converter.

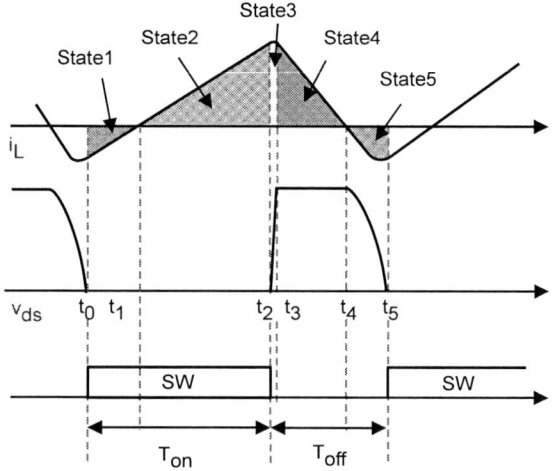

Fig. 2. Key waveforms of BCM PFC converter.

highlighted in Fig. 1. Actually, these parasitic capacitances resonant with the inductor after the demagnetization of inductor, thus the negative inductor current occurred [4], [9]. Figure 2 shows the key waveforms of BCM PFC converter when we consider the effect of parasitic capacitances. As a shown in Fig. 2, the PFC converter operates under 5 states, the inductor current is negative in state 1 and state 5.

In our previous method, since the negative inductor

2097

current is ignored, the switching on-time is calculated as a constant value and the constant on-time (COT) control is used [3]. However, the negative inductor current makes the average of inductor unable to be a sinusoidal curve and it is not able to follow the shape of input line voltage. As a result, the THD of line current is high in the COT control.

In order to decrease the distortion of line current and suppress THD, variable on-time (VOT) control methods have been studied by many researchers [9], [10]. These VOT control methods make the inductor current to build up more to cancel out the effect of negative inductor current. However, the derived equations of variable on-time in [9] and [10] include many times calculation of root and division. The root and division take a long calculation time in DSP controller. In predictive control, the total calculation time of one loop predictive control should be shorter than the switching on-time in the same switching cycle. Therefore, the derived equations of variable on-time in [9] and [10] are not suitable for predictive control.

[11] proposed a VOT control method for Totem-Pole BCM PFC converter. In this method, by approximating the very short resonance period of inductor current to a linearly change, the average inductor current is approximated to the average value of peak point and valley point of inductor current. Based on the approximation, a very simple but with high accuracy equation of variable on-time is derived. This derived equation is very simple, so it is suitable for the predictive control.

This approximation method for deriving the equation of variable on-time is applied in this paper. The equation of variable on-time for boost-type BCM PFC converter is derived first. The derived equation of variable on-time can then be implemented in the predictive control directly with a new predictive diagram. In this way, the VOT control can be combined with the predictive control. The high efficiency can be achieved and the THD of line current can be decreased.

II. OPERATION PRINCIPLE

2.1. Analysis of BCM Boost PFC Converter

Figure 2 shows the waveforms of i_L when we consider the impacts of parasitic resistance. In the proposed method, the equations of i_L in state 2 and state 4, the equation of v_{ds} in state 5 are most important in determining the switching on-time and switching off-time. They should be derived for predicting the zero-crossing point of inductor current.

In order to derive these equations accurately, the impacts of the internal resistance and the parasitic capacitance of each component are considered. The detailed modeling of the elements in the boost-type PFC converter is shown in Fig. 3. Based on the modeling, the equivalent circuits of BCM PFC converter in operating state 2, state 4 and state 5 can be shown as Fig. 4.

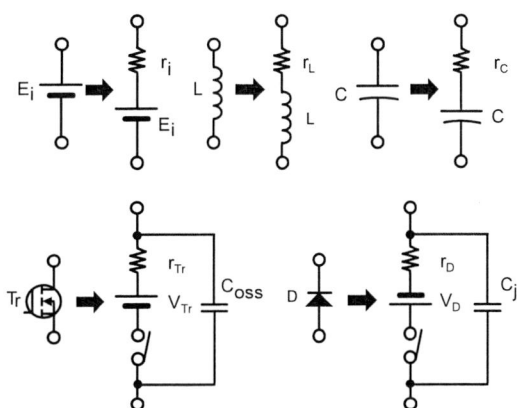

Fig. 3. Equivalent circuit models of boost converter elements.

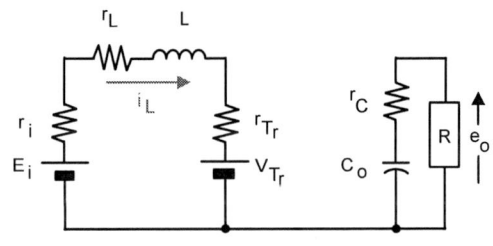

(a) Operating state 2

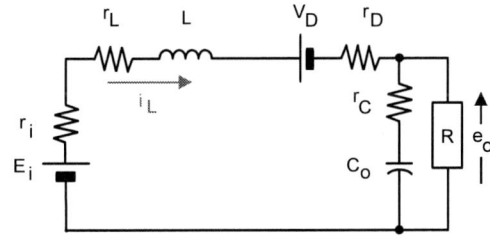

(b) Operating state 4

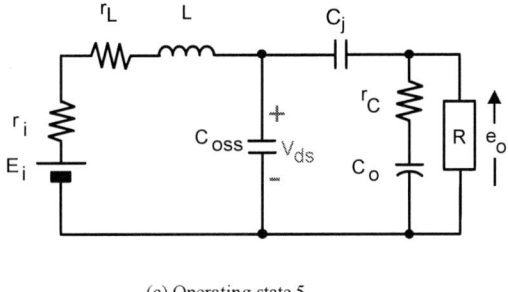

(c) Operating state 5

Fig. 4. Equivalent circuits of BCM PFC converter.

The equation of inductor current in state 2 $i_{L2}(t)$ is analyzed as follow:

$$i_{L2}(t) = K_1 + K_2(t_1) exp\{\alpha_2(t-t_1)\} \qquad (1)$$

where the constants K_1, $K_2(t_1)$ and α_2 in (1) are expressed by the equations as follows:

$$r_1 = r_i + r_L + r_{Tr} \tag{2}$$

$$E_1 = E_i - V_{Tr} \tag{3}$$

$$\alpha_2 = -\frac{r_1}{L} \tag{4}$$

$$K_1 = \frac{E_1}{r_1} \tag{5}$$

$$K_2(t_1) = i_{L2}(t_1) - \frac{E_1}{r_1}. \tag{6}$$

The equation of inductor current in state 4 $i_{L4}(t)$ is derived as follow:

$$i_{L4}(t) = G_1 + [G_2(t_3)\cos\{\omega_4(t-t_3)\} \\ + G_3(t_3)\sin\{\omega_4(t-t_3)\}]\exp\{\alpha_4(t-t_3)\} \tag{7}$$

where the constants G_1, $G_2(t_3)$, $G_3(t_3)$, ω_4 and α_4 in (7) are expressed by the equations as follows:

$$r_2 = r_i + r_L + r_D \tag{8}$$

$$E_2 = E_i - V_D \tag{9}$$

$$b_1 = \frac{1}{C(R+r_C)} + \frac{r_2 R + r_2 r_C + R r_C L}{L(R+r_C)} \tag{10}$$

$$b_2 = \frac{R+r_2}{LC(R+r_C)} \tag{11}$$

$$b_3 = \frac{E_2}{LC(R+r_C)} \tag{12}$$

$$b_4(t_3) = \frac{1}{L}\{E_2 + e_o(t_3)\} + \frac{L+CRr_C}{LC(R+r_C)}i_{L4}(t_3) \tag{13}$$

$$\alpha_4 = -\frac{b_1}{2} \tag{14}$$

$$\omega_4 = \frac{\sqrt{-(b_1^2 - 4b_2)}}{2} \tag{15}$$

$$G_1 = \frac{b_3}{b_2} \tag{16}$$

$$G_2(t_3) = i_{L4}(t_3) - \frac{b_3}{b_2} \tag{17}$$

$$G_3(t_3) = \frac{b_4(t_3)}{\omega_4} + \frac{\alpha_4 i_{L4}(t_3)}{\omega_4} - \frac{b_1 b_3 + b_3 \alpha_4}{b_2 \omega_4}. \tag{18}$$

The equation of voltage across the drain and source of MOSFET in state 5 $v_{ds5}(t)$ is derived as follow:

$$v_{ds5}(t) = E_i + [M_2(t_4)\cos\{\omega_5(t-t_4)\} \\ + M_3(t_4)\sin\{\omega_5(t-t_4)\}]\exp\{\alpha_5(t-t_4)\} \tag{19}$$

where the constants $M_2(t_4)$, $M_3(t_4)$, ω_5 and α_5 in (19) are expressed by the equations as follows:

$$\omega_5 = \sqrt{\frac{1}{L(C_j + C_{oss})} - (\frac{r_L + r_i}{2L})^2} \tag{20}$$

$$\alpha_5 = -\frac{r_L + r_i}{2L} \tag{21}$$

$$M_2(t_4) = E_o - E_i \tag{22}$$

$$M_3(t_4) = \frac{(E_o - E_i)(r_L + r_i)}{2L\omega_5}. \tag{23}$$

2.2. Variable On-time Control

Figure 5 shows the waveform of real inductor current and the waveform of approximated inductor current. As shown in Fig. 5, during the state 5, the inductor resonates with parasitic capacitances and parasitic resistances and the i_L is negative during this period. The inductor current during this period is non-linear. Since the period of this state is very short. Therefore, it can be approximated to linearly change. In this case, the average inductor current can be considered as the average value of peak point i_{L_peak} and valley point i_{L_valley}.

The peak point and valley point of inductor current can be expressed as follow:

$$i_{L\_peak} = \frac{e_i}{L}T_{on}(t) \tag{24}$$

$$i_{L\_valley} = \frac{e_o - e_i}{\sqrt{\frac{L}{C_j + C_{oss}}}}. \tag{25}$$

The average of inductor current i_{L_avg} can be obtained as follow:

$$i_{L\_avg} = \frac{i_{L\_peak} + i_{L\_valley}}{2}. \tag{26}$$

The sinusoidal reference current i_{ac_ref} is derived as follow:

$$i_{ac\_ref} = \frac{\sqrt{2}P_o}{\eta \cdot PF \cdot E_{in,rms}^2}\sin(2\pi f_s t). \tag{27}$$

where $E_{in,rms}$ indicates the RMS value of input line voltage.

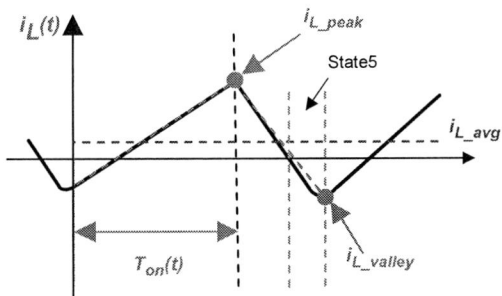

Fig. 5. Waveform of approximated inductor current.

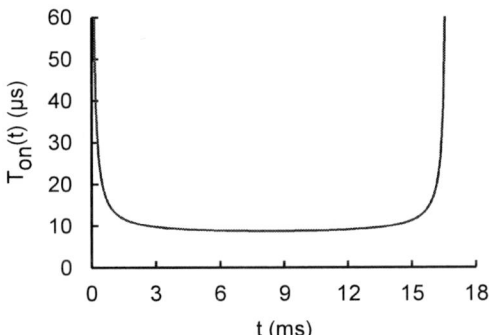

Fig.6. Calculated variable on-time $T_{on}(t)$ curve.

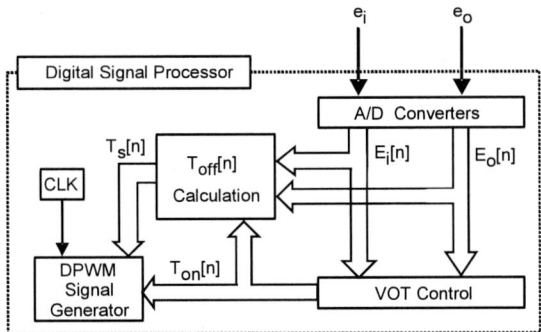

Fig. 7. Diagram of proposed VOT control combined with predictive control.

The equation of variable on-time $T_{on}(t)$ can be derived by making the i_{ac_ref} equal to i_{L_avg}. The $T_{on}(t)$ is derived as follow:

$$T_{on}(t) = \frac{2LP_o}{\eta \cdot PF \cdot E_{in,rms}^2} + \frac{e_o - e_i}{e_i}\sqrt{L(C_j + C_{oss})} . \tag{28}$$

Focusing on (28), we can see that the derived $T_{on}(t)$ only includes a division calculation of $(e_o-e_i)/e_i$. Compared with the equations of $T_{on}(t)$ in [9] and [10], it

is much more simplifier. Therefore, the derived $T_{on}(t)$ can be implemented to the DSP directly.

Figure 6 shows the equation curve of the $T_{on}(t)$, which is derived as (28). In the prototype of PFC converter, the alternating input voltage is 100 Vrms / 60 Hz, the output voltage is 400 V and the output power P_O is 30 W. The inductance L is 1.2 mH, the $C_j + C_{oss}$ is 100 pF, the PF is considered as 1, the efficiency is considered as 90 %. It is found that, the switching on-time is increased rapidly near the AC line zero-crossing point. Therefore, the inductor current can be built up more to cancel the effect of negative inductor current.

2.3. Diagram of Proposed VOT Control Combined with Predictive Control

Figure 7 shows the diagram of proposed VOT control combined with the predictive control for BCM PFC converter. First, the rectified input voltage e_i and the output voltage e_o are converted to the digital values $E_i[n]$ and $E_o[n]$. The switching on-time $T_{on}[n]$ is determined by the derived equation of variable on-time, which is (28). Then, the peak value of inductor current is obtained. The peak value is subsequently substituted to (7), the switching off-time $T_{off}[n]$ is calculated based on the (7) and (19). In this way, the optimal switching on-time and switching off-time for predicting the zero-crossing point of can be obtained. The combination of VOT control and predictive control can be achieved.

IV. EXPERIMENTAL RESULTS

To verify the performance of the proposed method, a 100 V input and 400 V/30 W output single-phase boost-type BCM PFC converter was implemented and tested. The switching frequency is 80 kHz to 120 kHz. The digital controller selects TI DSP TMS320F28335.

Figure 8 shows the measured experimental waveforms of input line voltage e_{ac} and line current i_{ac} in the COT control and VOT control, respectively. In the COT control, the switching on-time is constantly 9.5 μs while the whole phase of line voltage. As shown in Fig. 8(a), the period which line current keeps zero value in COT control is 0.97 ms. The THD of line current is 11.4 % and PF is 0.993 in the COT control.

By applying the VOT control, the switching on-time is increased and the inductor current is built up more near the AC line zero-crossing point. The switching on-time near the zero-crossing of line voltage is increased to 30.3 μs. It is much higher than that near the peak of line voltage, which is 9.4 μs. Therefore, the period which line current keeps zero value in VOT control is shorten to 0.50 ms, as shown in Fig. 8 (b). The THD of line current is 5.9 % and PF is 0.997 in the VOT control. The improvement in THD of line current is 48.2 %.

Figure 9 shows the measured harmonic currents together with the IEC 61000-3-2 Class C standard. It is

2100

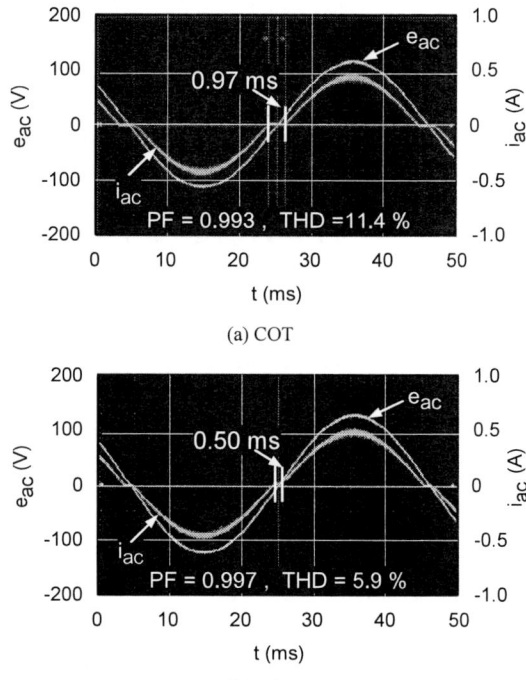

(a) COT

(b) VOT

Fig. 8. Experimental waveforms of line voltage and line current.

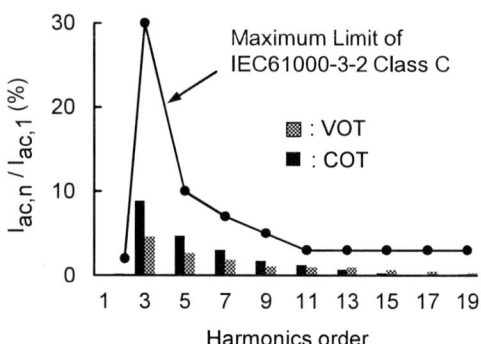

Fig. 9. Harmonic currents and IEC 61000-3-2 Class C standard.

found that compared with COT control, the harmonic currents is reduced by VOT control. The IEC standard is met with a huge margin in VOT control.

IV. CONCLUSION

A VOT control method is implemented to predictive digitally controlled boost-type BCM PFC converter in this paper. The derived variable on-time equation is very simple, and hence it takes short calculation time in DSP controller. Therefore, the VOT control is implemented to predictive control successfully. The THD of line current is significantly improved by 48.2 % compared with COT control. Hence, the PFC converter implementing the optimized VOT control can drive the LED lighting applications with high efficiency and low THD.

REFERENCES

[1] B. Wang, X. Ruan, K. Yao, and M. Xu, "A method of reducing the peak to average ratio of LED current for electrolytic capacitor-less ac–dc drivers," *IEEE Trans. Power Electron.*, vol. 25, no. 3, pp. 592-601, Mar. 2010.

[2] J. Wang, K. Yoshimura and F. Kurokawa, "Optimization design of novel zero-crossing point detection method for boundary current mode PFC converter," in *Proc. ICRERA,* pp. 1-5, Nov. 2015.

[3] J. Wang, H. Maruta, M. Matusnaga, and F. Kurokawa, "A novel predictive digital controlled sensorless PFC converter under the boundary conduction mode," *Journal of Power Electronics,* vol. 17, no. 1, pp. 1-10, Jan. 2017.

[4] L. Huber, B. Irving and M. Jovanovic, "Effect of valley switching and switching-frequency limitation on line-current distortions of DCM/CCM boundary boost PFC converters," *IEEE Trans. Power Electron.*, vol. 24, no.2, pp. 339-347, Feb. 2009.

[5] Z Liu, Z. Huang, F. C. Lee, Q. Li and Y. Yang, "Operation analysis of digital control based MHz totem-pole PFC with GaN device", *Workshop on Wide Bandgap Power Devices and Applications (WiPDA).*, pp. 281-286, Nov. 2015.

[6] C Gao, Y. Zhang, X. Zhang. H. Li, H. Chen and J. Wang, "Adaptive constant power control and power loss analysis of a MHz GaN-based ac-dc converter for low power applications", *Applied Power Electronics Conference and Exposition (APEC).*, pp. 1-8, Mar. 2017.

[7] J. Wang, K. Yoshimura and F. Kurokawa, "Zero-crossing point detection using differentiation circuit for boundary current mode PFC converter," in *Proc. IFEEC*, pp. 1-6, Nov. 2015.

[8] J. Wang, K. Yoshimura, H. Maruta, Y. Furukawa and F. Kurokawa, "Regulation characteristics of boundary conduction mode PFC converter with novel zero-crossing point detector," in *Proc. ICEMS*, pp. 1-5, Nov. 2016.

[9] J. W. Kim, H. S. Youn, and G. W. Moon, "A digitally controlled critical mode boost power factor corrector with optimized additional on time and reduced circulating losses," *IEEE. Trans. Power Electron.*, vol. 30, no. 6, pp. 3447-3456, Jun. 2015.

[10] J. Kim, J. Yi, and B. Cho, "Enhanced variable on-time control of critical conduction mode boost power factor correction converters," Journal of Power Electronics, vol. 14, no. 5, pp. 890-898, Sep. 2014.

[11] Z. Huang, Z. Liu, Q. Li and F. C. Lee, "Microcontroller-based MHz totem-pole PFC with critical mode control", *Energy Conversion Congress and Exposition (ECCE).*, pp. 1-8, Sep. 2016.

[12] Y. Lai, C. Yen, and K. Ho, "A family of predictive digital controlled PFC under boundary current mode control," *IEEE Trans. on Industrial Informatics*, vol. 8, pp. 448-458, Jul. 2012.

[13] Y. Su, C. Ni, and C. Chen, "Boundary Conduction Mode Controlled Power Factor Corrector with Line Voltage Recovery and Total Harmonic Distortion Improvement Techniques," I*EEE. Trans. On Power Electron.*, vol. 61, no. 7, Jul. 2014, pp. 3220-3231.

[14] J. Tsai, C. Chen, Y. Chen, C. Ni, and C. Chen, "Perturbation On-Time (POT) Technique in Power Factor Correction (PFC) Controller for Low Total Harmonic Distortion and High Power Factor," *IEEE. Trans. On Power Electron.*, vol. 28, no. 1, Jan. 2013, pp. 199-211.

[15] J. Moldaschl, J. Broulím, and L. Paločko, "Principle of power factor corrector with critical conduction mode," *IEEE Trans. Power Electron.*, vol. 25, no. 5, pp. 1103-1109, May. 2010.

Design of Multimode Battery Charger with Dynamic Voltage Tracking Control

Pang-Jung Liu*, Lin-Hao Chien, Song-Kai Lee and Ang-Tung Chen
Department of Electrical Engineering, National Taipei University of Technology, Taipei, Taiwan
*E-mail: pjliu@ntut.edu.tw

Abstract—This paper presents a multimode battery that is composed of a dc-dc buck converter and a charging circuit. The charging circuit can provide a ripple-free charging current and automatically switches among trickle current (TC), constant-current (CC), and constant-voltage (CV) stages depended on the battery voltage. The buck converter with a dynamic voltage tracking (DVT) control produces a dynamic supply voltage to closely follow the battery voltage, resulting in considerable efficiency improvement of the charging circuit. To reduce power loss of the buck converter, non-switching control and zero current detection are utilized to increase the efficiency of the battery charger at CV and TC stages. The battery charger was designed with a 0.35-μm CMOS technology. The simulation results are shown to verify the theoretical analysis of the proposed charger.

Keywords: Charger, dc-dc converter, dynamic voltage, constant voltage, constant current.

I. INTRODUCTION

Due to small size, low self-discharge and high energy density, Lithium-ion (Li-ion) batteries are widely used in battery-powered portable devices such as tablets and smart phones. Generally speaking, the charging process of Li-ion batteries includes constant-current (CC) stage, trickle current (TC) stage, and constant-voltage (CV) stage. To prevent battery damage, a small charging current is adopted at TC stage when the battery is discharged deeply [1]. CC stage is employed with a large constant charging current to decrease charging time. At CV stage, the charging current is reduced progressively to avoid overcharging and reach a rated voltage.

There are two categories of charger system: switching- and linear-type [2]-[5]. Although the switching-type charger has better power efficiency, it trades off accuracy for efficiency and it requires higher component counts [2]. On the other hand, a linear-type charger has ripple-freeness, high accuracy, and compact size. When the battery voltage is low, a linear-type charger suffers appreciable power dissipation. A variable supply voltage method had been presented to mitigate this issue of the linear-based charger

[2], [5]. The variable supply voltage is generated by a low dropout (LDO) regulator or operational amplifiers and resistors. They are effective methods to reduce the voltage value across the power transistor of the linear-type charger. But the power loss generated by the circuit in charge of variable supply voltage is also considerable when the battery voltage is much less than the input voltage.

A dynamic voltage tracking (DVT) control is presented to alleviate low charging efficiency under low battery voltage. The dynamic supply voltage is produced by a buck converter and is used as the voltage source for a charging circuit. Since the dynamic supply voltage closely follows the battery voltage, power consumption of the charging circuit is decreased dramatically. The buck converter also affects the overall efficiency of the proposed charger. Thus, non-switching control and zero current detection are employed in the buck converter to decrease power loss at CV and TC stages. In Section II and III, the proposed charger and its implementation are addressed. Section III shows the simulation results. Finally, conclusions are made in Section V.

II. PROPOSED MULTIMODE CHARGER WITH DYNAMIC VOLTAGE TRACKING (DVT) CONTROL

A. Concept of DVT control

Fig. 1(a) shows the schematic diagram of proposed charger composed of a dc-dc buck converter and a charging circuit. The charging circuit realizes TC, CC, and CV stages and switches among distinct operation stages depending on battery voltage V_{bat}. By sensing V_{bat}, the buck converter produces a dynamic supply voltage $V_{sup,dsv}$ for the charging circuit. $V_{sup,dsv}$ is marginally higher than V_{bat}, resulting in low power loss of the charging circuit during the whole charging process. Fig. 1(b) shows the theoretical waveforms of the charger with DVT control, where the voltage difference V_{diff} is between $V_{sup,dsv}$ and V_{bat}. To keep normal operation of the charging circuit, $V_{sup,dsv}$ must be larger than a voltage level, so V_{diff} is higher at the start of the charging process. At TC stage, V_{diff} decreases continuously. Moreover, to increase the

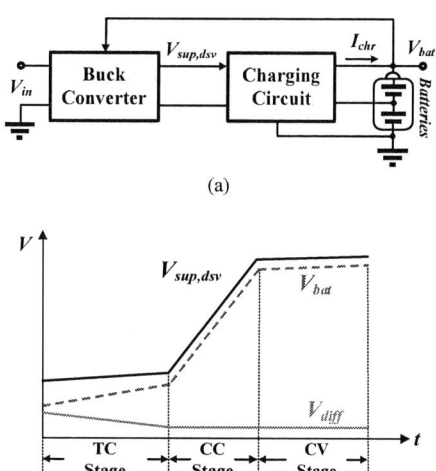

(a)

(b)

Fig. 1. (a) Proposed charger with DVT control and (b) theoretical waveforms.

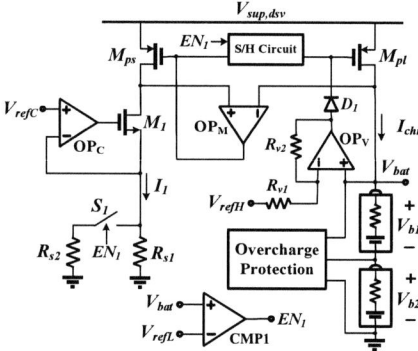

Fig. 2. Proposed charging circuit.

overall efficiency of the proposed charger, the power loss of the dc-dc converter should be minimized as low as possible. Thus, non-switching control and zero current detection are utilized to enhance the efficiency of the buck converter at CV and TC stages.

B. Multimode battery charging circuit

The schematic diagram of charging circuit is shown in Fig. 2. At the beginning of the charging process, when V_{bat} is smaller than a threshold voltage V_{refL} switch S_1 is turned off and diode D_1 is off because of low battery voltage. When the charging circuit operates at TC stage, current I_1 is equal to V_{refC}/R_{s1}. By using a current mirror (M_{ps} and M_{pl}), the charging current I_{chr} is enlarged to be 1000 times of I_1 and then charges the two batteries in series. To enhance the accuracy of the current mirror, transistor M_{ps} and operational amplifier OP_M are used to make the drain voltages of M_{pl} and M_{ps} to be the same.

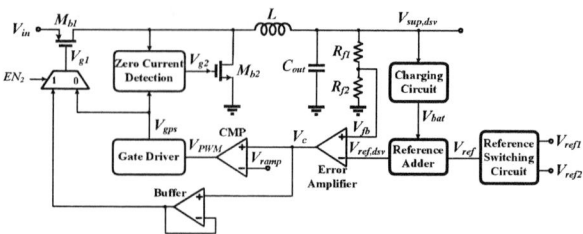

Fig. 3. Schematic diagram of buck converter with DVT and non-switching controls.

Once the charging process switches to CC stage, S_1 is turned on, leading to larger I_1 and I_{chr}. During the turn-on transition of S_1, a current spike in current I_1 could be generated. After flowing through the current mirror, I_{chr} could damage batteries. A sample-and-hold (S/H) circuit is employed to prevent this issue. When V_{bat} approaches a threshold voltage V_{refH}, the charging circuit enters CV stage. At CV stage, operational amplifier OP_V turns D_1 on to charge the parasitic capacitance at the gate terminal of M_{pl}, resulting in a progressively reducing charging current. Hence, TC, CC, and CV charging stages can be achieved.

C. Buck converter with DVT and non-switching controls

Fig. 3 shows the schematic diagram of the dc-dc buck converter with DVT and non-switching controls. To realize DVT control, a reference adder is employed to add V_{bat} and a reference signal V_{ref} with distinct weightings so as to generate an dynamic reference voltage $V_{ref,dsv}$. Since $V_{ref,dsv}$ is connected to the inverting input terminal of the error amplifier, the duty cycle value and the dynamic voltage $V_{sup,dsv}$ are adjusted. According to DVT control, $V_{sup,dsv}$ should be slightly larger than V_{bat}, but $V_{sup,dsv}$ must also be larger than a minimum voltage level for guaranteeing normal operation of the charging circuit. Moreover, due to applying different charging current values, the rising slopes of $V_{sup,dsv}$ in TC and CC modes are distinct. Hence, two reference voltages V_{ref1} and V_{ref2} are employed. V_{ref} is equal to V_{ref1} at TC stage while V_{ref} changes to V_{ref2} at CC stage. To preclude overshoot and ringing issues of $V_{sup,dsv}$ during reference interchanging transient, the reference switching circuit is utilized.

When the buck converter operates at TC stage, the inductor current will become negative, resulting in extra power consumption. Thus, to preclude output capacitor C_o from discharging energy, a zero-current detector is adopted to turn off power transistor M_{b2}. The voltage across M_{pl} of the charging circuit is decreased dramatically when V_{bat} reaches to a high voltage level. If the buck converter still uses a switching operation control, the overall efficiency of the proposed charger would be lower than that of the conventional linear charger. Moreover, the charging time at CV stage occupies a great portion of the total charging time [6]. Under the considerations, a non-switching operation control is utilized to keep M_{b2} off and M_{b1} on. Due to the

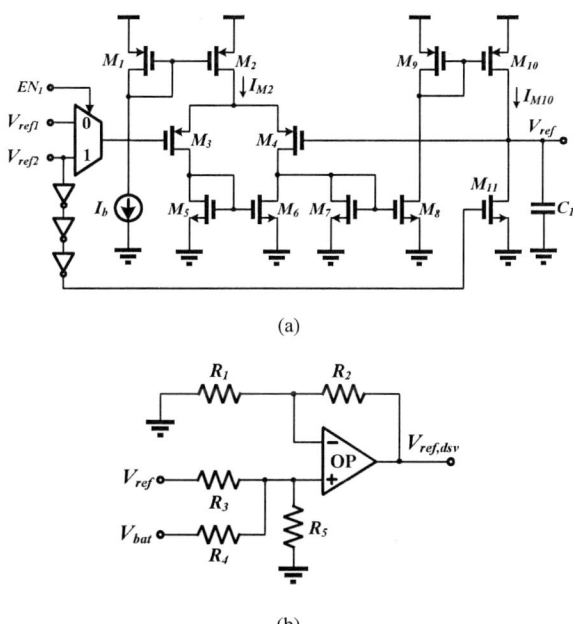

(a)

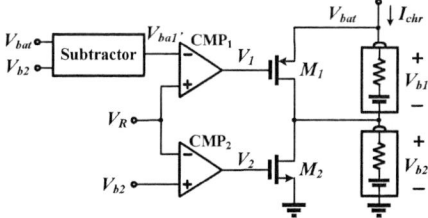

Fig. 5. Overcharge protection circuit.

(b)

Fig. 4. (a) Reference switching circuit and (b) reference adder.

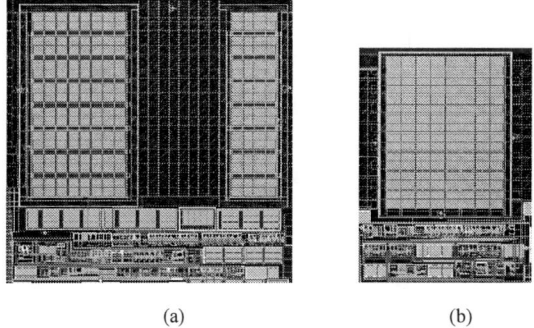

(a)   (b)

Fig. 6. Chip layouts of (a) buck converter and (b) charging circuit.

high state of signal EN_2, the value of gate voltage V_{g1} is determined by the error amplifier. On the other hand, since the V_c level is always higher than the amplitude of ramp signal V_{ramp}, gate voltage V_{g2} is kept at zero level. Thanks to the non-switching operation control and zero-current detector, the efficiency of the proposed charger can be maintained as high as possible.

III. CIRCUIT IMPLEMENTATION

A. Reference switching circuit and reference adder

To preclude ringing and overshoot in $V_{sup,dsv}$ during the start-up period and the interchanging transition between the two reference voltages, the reference switching circuit shown in Fig. 4(a) is adopted. Before the start-up process, the capacitor C_1 is discharged through transistor M_{11} due to a low level of V_{ref2}. This action ensures the reference voltage V_{ref} is reset from zero. After the buck converter is powered on, V_{ref2} is set to a predefined value. When V_{bat} is lower than V_{refL}, V_{ref1} is connected to the gate of transistor M_3. Most of current I_{M2} then passes through M_4 and M_6, leading to the rising drain voltage of M_7. Then, current I_{M10} charges capacitor C_1, so that V_{ref} rises slowly and linearly. Until V_{ref} approaches V_{ref1}, I_{M10} becomes zero and stops charging. The gate of M_3 is connected to V_{ref2} when the charging circuit switches to CC stage. Similarly, V_{ref} will be charged to V_{ref2} slowly. Fig. 4(b) shows the reference adder that is a noninverting amplifier with V_{ref} and V_{bat}.

B. Overcharge protection circuit

Although the two batteries are charged with the same current, the required charging times of the two batteries are distinct because of different battery parameters. Therefore, an overcharge protection circuit is needed as shown in Fig. 5. The V_{b1} information can be obtained with a subtractor. Two comparators CMP_1 and CMP_2 can determine whether the two battery voltages approach the rated voltage V_R or not. Transistors M_1 and M_2 are turned off before the two battery voltages reach V_R. Assuming V_{b2} reaches V_R first, V_3 is pulled up to a high level. Hence, M_2 is turned on and I_{chr} delivers energy to battery 1 and follows through M_2. On the other hand, the charging current delivers energy for battery 2 through M_1.

IV. SIMULATION RESULTS

To verify the circuit validity, a laboratory prototype was designed with an input voltage of 5.8-6.5 V, maximum charging current of 500 mA, dynamic supply voltage of 0.2-5.7V, and switching frequency of 1 MHz. The output inductance and capacitance were 4.7 μH and 10 μF, respectively. The battery charger was implemented 0.35-μm 2P4M CMOS technology. The chip layout of the buck converter and charging circuit are shown in Fig. 6. To shorten the simulation time, two capacitors are used as the batteries.

The 2018 International Power Electronics Conference

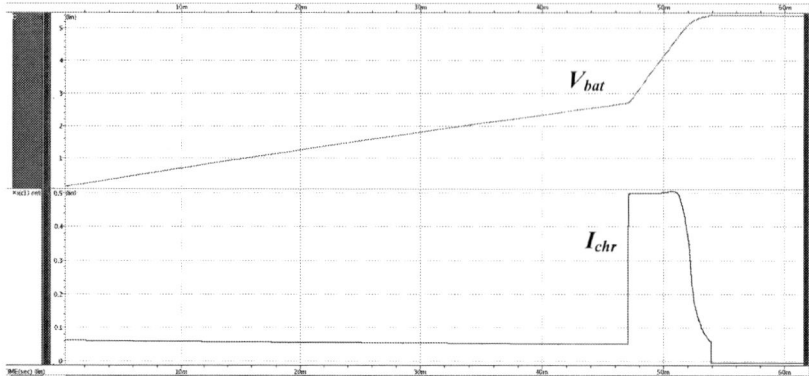

Fig. 7. Simulated waveforms of the charging circuit.

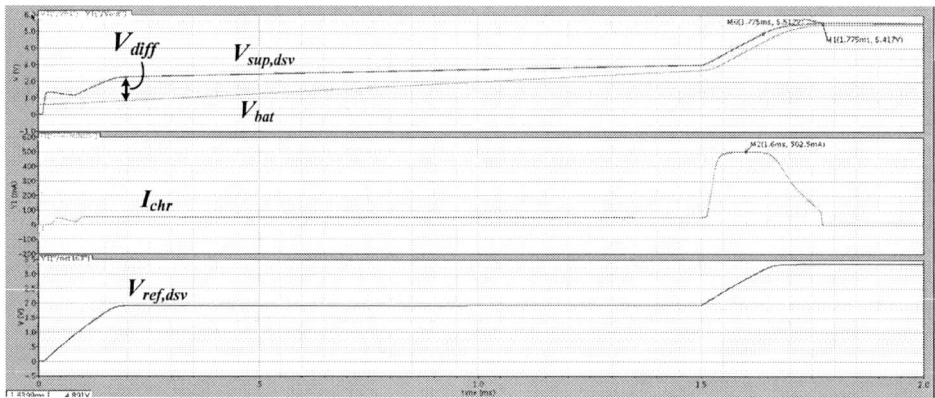

Fig. 8. Simulated waveforms of the proposed charger.

Fig. 7 shows the simulated waveforms of the charging circuit. The battery voltage is discharged to a low level before simulation, so the charging circuit operates at TC stage. Then, a small charging current I_{chr} starts to charge batteries and the battery voltage V_{bat} rises slowly. Once the charging circuit switches to CC stage, the charging current changes to 500 mA and V_{bat} then increases quickly. When the battery voltage reaches the rated voltage, the charging circuit enters CV stage and the charging current then gradually reduces to zero.

Fig. 8 shows the simulated waveforms of the proposed charger. Dynamic reference voltage $V_{ref,dsv}$ rises from zero linearly to preclude $V_{sup,dsv}$ from overshoot during the start-up period. At TC stage, $V_{ref,dsv}$ increases slowly, so does $V_{sup,dsv}$. Voltage difference V_{diff} between $V_{sup,dsv}$ and V_{bat} is larger at the beginning of the charging procedure and continues decreasing at TC stage. When the charging circuit switches to CC stage, the charging current changes to 500 mA, and V_{bat} and $V_{sup,dsv}$ rise quickly. Furthermore, V_{diff} is almost kept constant. At CV stage, $V_{sup,dsv}$ is maintained fixed while I_{chr} decreases toward zero.

V. CONCLUSION

This paper presents a battery charger consisted of a dc-dc buck converter with DVT control and a charging circuit. The charging circuit can automatically switch among TC, CC, and CV stages and provide a ripple-free charging current. To decrease extra power loss on the charging circuit, the buck converter with DVT control produces a dynamic supply voltage. Moreover, zero current detection and non-switching controls are used to reduce power consumption of the buck converter in TC and CV stages. The simulated results verify the theoretical analysis of the proposed charger.

ACKNOWLEDGMENT

This work was supported in part by Ministry of Science and Technology, Taiwan, under Grant MOST 105-2221-E-027-129 and 106-2221-E-027-082-MY3. The authors wish to thank National Chip Implementation Center, Taiwan, for chip fabrication.

REFERENCES

[1] K. Tae-Hoon, J. Jin-Beom, L. Baek-Haeng, S. Dong-Hyun, S. Hyun-Sik, K. Byoung-Hoon, and K. Hee-Jun, "Analytical study on low-frequency ripple effect of battery charging," in *Proc. Veh. Power Propulsion Conf.*, Oct. 2012, pp. 809–811.

[2] M. Chen and G. A. R. Rincon-Mora, "Accurate, compact, power efficient Li-ion battery charger circuit," *IEEE Trans. Circuits Syst. II, Exp. Briefs*, vol. 53, no. 11, pp. 1180-1184, Nov. 2006.

[3] C.-C. Tsai, C.-Y. Lin, Y.-S. Hwang, W.-T. Lee, and T.-Y. Lee, "A multimode LDO-based Li-ion battery charger in 0.35 μm CMOS technology," in *Proc. IEEE Asia-Pac. Conf. Circuits Syst.*, 2004, pp. 49-52.

[4] S.-H. Yang, J.-W. Liu, Y.-H. Wu, D.-S. Wang, and C.-C. Wang, "A high voltage battery charger with smooth charge mode transition in BCD process," in *Proc. IEEE ISCAS*, 2011, pp. 813-816.

[5] P. H. V. Quang, T. T. Ha, and J.-W. Lee, "A fully integrated multimode wireless power charger IC with adaptive supply control and built-in resistance compensation," *IEEE Trans. Ind. Electron.*, vol. 62, no. 2, pp. 1251-1260, Feb. 2015.

[6] P.-J. Liu, and C.-H. Yen, "A fast-charging switching-based charger with adaptive hybrid duty cycle control for multiple batteries," *IEEE Transactions on Power Electronics*, vol. 32, no. 3, pp. 1975-1983, Mar. 2017.

Dual-Slot Power-Pickup Structure for Contactless Strip Inductive Power Track System

Jia-You Lee, I-Lin Chen and Chien-Tzu Ko[*]

Department of Electrical Engineering, National Cheng Kung University, Tainan, Taiwan
*E-mail: N26054176@mail.ncku.edu.tw

Abstract— The paper aims to develop a dual-slot type power-pickup suitable for the inductive power track system for contactless power transmission. A new strip type inductive power track with high magnetic flux density and magnetic field uniformity is proposed. By means of equivalent magnetic circuit model analysis, a dual-slot type power-pickup with high magnetic flux linkage is designed to improve the power transmission capacity and efficiency. To verify the contactless transmission characteristics of the proposed strip type inductive power track system, an inductive power track with 1.2 m long is constructed. The SP_LS_C topology is chosen as the compound resonant circuit of a full bridge inverter. In accordance with the experimental results, the maximum output power of overall system is 775 W with transfer efficiency of 64.58%, and the maximum transmission efficiency is 72.52% at an output power of 289.52 W.

Keywords— *Contactless power transmission, dual-slot power-pickup structure, strip type inductive power track system.*

I. INTRODUCTION

With the maturity of industrial automation technology and the popularization of electric transport vehicle, contactless inductive power transmission techniques are not only well-developed in the industrial automation, but also using in particular environments such as dangerous, greatly precise, underwater and so on. However, because of the limited of coupling capability between the primary inductive power track and the secondary inductive power pickup, there is still much study to be need.

The common power tracks in automated production line system are shown in Fig. 1, and it can be divided into two types, linear and grounded [1-4]. These two types cannot set in specific circumstances where electrical contacts result in exploding or rusting. Therefore, the paper is concentrated on linear power track of contactless power transmission, which can transmit power without the brush or cable contact based on the theory of electromagnetic induction. A dual-slot power-pickup of the inductive power track System suitable for contactless power transmission has been develop.

Over the last few years, there has raised dramatically at research of contactless inductive power transmission techniques, and has been adopted by most panel, wafer manufacturing, semiconductor process industry, and so on. In addition, the international and domestic academia are researching on the contactless inductive power track system [5-7].

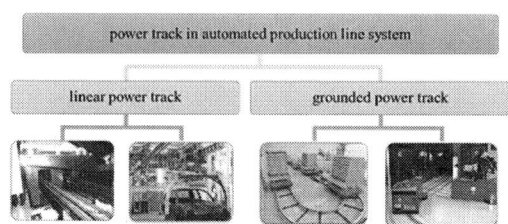

Fig. 1. Common power tracks [1-4].

II. INDUCTIVELY COUPLED STRUCTURE

A. Design of Strip Type Inductive Power Track

According to earlier research of inductive power track, it is noted that the circular inductive coil has better magnetic field intensity than the linear inductive coil of the tradition with the same line utilization and drive current, as shown in Fig. 2. In addition, based on

$$\phi = L I \tag{1}$$

where ϕ is magnetic field intensity, L is the inductance of coils, and I is the current in the coils. The current needs less in the circular inductive coil to produce same magnetic field intensity. Hence, this paper has chosen circular inductive coils as primary track coils.

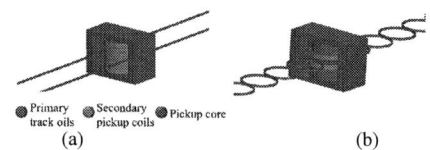

Fig. 2. Comparison of the inductive power track. (a) Linear power track. (b) Strip type power track.

For the same direction of magnetic field radiated from the inductive power track, the direction of the current in the coils is shown in Fig. 3(a). The winding method A is too ideal to make because when winding one coil to next coil, it will wind more half coil to the next. For that reason, a winding method B is shown in Fig. 3(b). The type is that finish one coil then to the next coil. As shown in Fig. 3(c), it is significant that the new proposed track wound by method B has more uniform the magnetic field than the ideal track wound by method A. Besides, the magnetic field on the top of the border between two coils would be offset because the magnetic field produced by the blue one is opposite to the field produced by the red one.

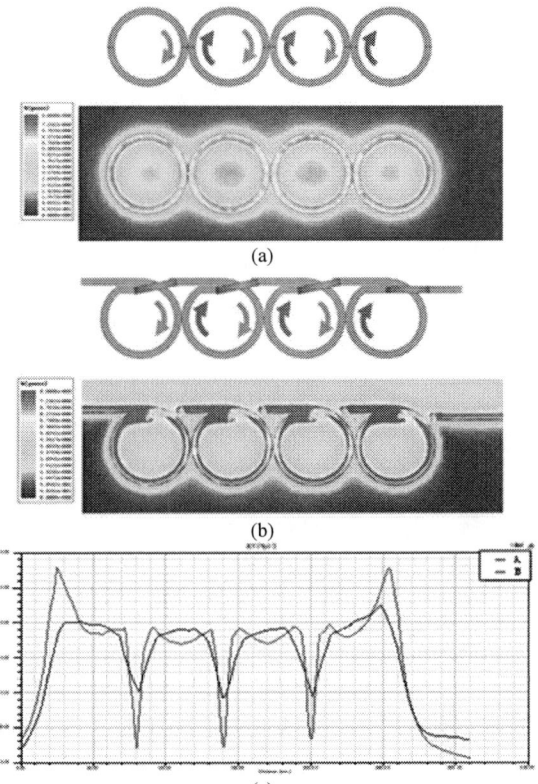

(a)

(b)

(c)

Fig. 3. Comparison of tracks with different winding method. (a) Winding method A. (b) Winding method B. (c) Comparison of magnetic field.

B. Design and Analysis of Power Pickup

Based on earlier study of power pickup, above all, the cross-sectional figure of a spherical core illustrates the direction of the magnetic field in the cylindrical core, as shown in Fig. 4(a). Hence, the magnetic flux flowing through the center column is equal to that through the two arms in the cross-sectional area. That is the core should be designed in consideration with the area A equaling to the sum of area B1 and area B2, as shown in Fig. 4(b).

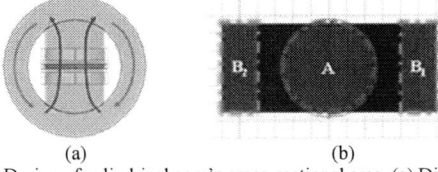

(a)   (b)

Fig. 4. Design of cylindrical core's cross-sectional area. (a) Direction of magnetic field. (b) Cross section of the core.

To avoid a long track bending down as a result of cable's weight in real applications, the left arm of power-pickup is removed and the cables need to be fastened on the braces to ensure smooth motion, as shown in Fig. 5. Then, since this study selects two single-layer power tracks as inductive power supply to power-pickup, the power-pickup needs the two slots to catch the energy easily. Owing to low magnetic permeability of two slots in the air, there is large amount of leakage flux in the gap. There are two methods to resolve these problems.

Fig. 5. An unclosed pickup with the track fixed on a brace.

At first, in view of the size of the power-pickup core and as previous mentioned, the coupling capability of the inductively coupled structure can be strong by thickening the width of the right arm. Because there is no gap in the right arm, thickening the right arm can decrease its equivalent magnetic reluctance, reinforcing the magnetic flux linkage. Second, between the two slots, it is replaced gaps with middle core. It can reduces the magnetic flux leakage between the orbital coils and pickup coils, because the magnetic permeability of middle core is 2000 to 8000 times better than air.

Besides, the shorter the semi-minor axis, the better core utilization is, and the higher coupling coefficient between the primary track and secondary pickup coils. As a result of having higher coupling coefficient, the cylindrical core is transformed into elliptically cylindrical core. Therefore, the proposed pickup structure is based on the structure with 4 cm semi-minor axis, 6 cm long semi-major axis, 3 cm wide right arm, and 5 cm width.

According to the above mentioned, the new power-pickup structures are proposed in Fig. 6 where left arm are removed because if the pickups have the left arm, the structures would be weak. It has been shown Type I, Type II and Type III theirs structures and magnetic simulations respectively, and Type I core structure has equivalent dimensions to Type II excluding Type II is smoother than Type I in the connection between the middle core and the right arm. Though it produces the magnetic reluctances of flux leakage, there is noticed that the magnetic flux density in the right arm of Type I is still higher than that of Type II. Therefore, there is new type, Type III, to improve the utilization of Type II core, but it is obscure to tell the difference of simulated results between Type I and Type III structure. Accordingly, the next section is comprehensively studied two types of structures via the magnetic equivalent circuit techniques, and the appropriate core structure of power pickup is designed.

C. Magnetic Equivalent Circuit Modeling of Pickup

On the basis of magnetic flux, the magnetomotive force $\mathscr{F}$ is defined as the product of the magnetic reluctance $\mathfrak{R}$ and the magnetic flux ϕ_c :

$$\mathscr{F} \equiv \mathfrak{R}\phi_C \tag{2}$$

Besides, the relational expression of the magnetic reluctance $\mathfrak{R}$ is given as

$$\mathfrak{R} = \frac{l_C}{\mu A} = \frac{l_C{}^2}{\mu_0 \mu_r V} \tag{3}$$

where the mean length l_C, the cross-sectional area A, the volume of the material V, and the permeability μ of the material, which is product of the permeability μ_0 of free space and the relative permeability μ_r of the material.

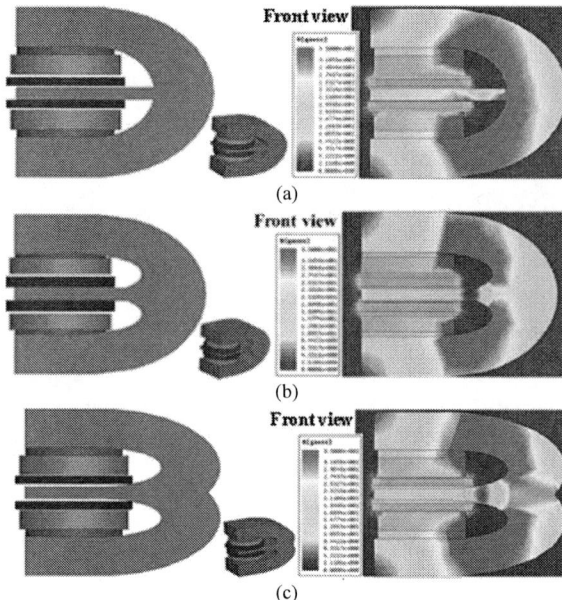

Fig. 6. Distinct structures and their magnetic simulations. (a) Type I structure. (b) Type II structure. (c) Type III structure.

By way of kirchhoff circuit laws to calculate the value of the magnetic flux, the coupling coefficient is computed, as given by

$$k = \frac{\phi_x}{\phi_{total}} = \frac{\phi_x}{\phi_x + \phi_{leak}} \tag{4}$$

The coupling coefficient k is known as the ratio of flux linkage ϕ_x to total flux produced by primary coils, and the flux leakage ϕ_{leak} is caused by the influence of core structures on parts of magnetic flux which is guided away from the flux linkage. Thus, when the value of k is close to 1, it represents the coupling capacity of the primary and the secondary coils is better. So the coupling coefficient has been the important basis for the pickup structure.

Apart from choosing the core structure between Type I and Type III structures, the proper dimensions of the core structure are designed based on the coupling coefficient. Fig. 7 illustrates the dimensions of Type I and Type III structure. In terms of the coupled structure, the magnetic equivalent circuit model is schematized in Fig. 8 where $\Re_{1A} \sim \Re_{6A}$ and $\Re_{1B} \sim \Re_{5B}$ are the magnetic reluctances of core, $\Re_{g1A} \sim \Re_{g2A}$ and $\Re_{g1B} \sim \Re_{g2B}$ are the magnetic reluctances of gaps, and $\Re_{leak1A} \sim \Re_{leak4A}$ and $\Re_{leak1B} \sim \Re_{leak2B}$ are the magnetic reluctances of flux leakage. With the simulations of the magnetic field, the main path of the flux leakage is around the gap between primary and secondary inductive coils, as shown in Fig. 9.

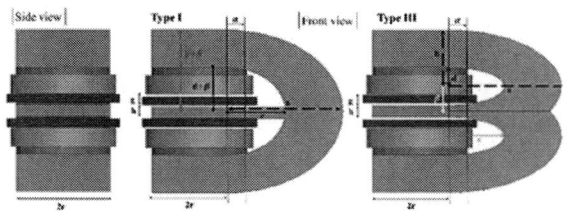

Fig. 7. The dimensions of Type I and Type III structure.

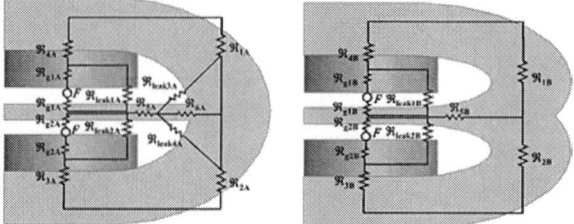

Fig. 8. Magnetic equivalent circuit model.

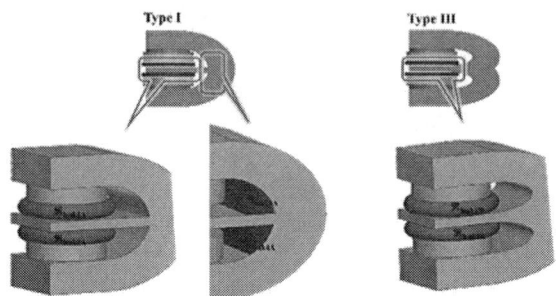

Fig. 9. Magnetic circuit model of the flux leakage.

In order to design the appropriate dimensions of the core structure, it is defined the size of a = 6 cm and c is a variable value to determine the right arm of the pickup core, because the width of the right arm of the pickup core have an high influence in the coupling coefficient of the power-pickup structure. Then, each magnetic reluctance can be given in Table I and Table II.

TABLE I
MAGNETIC RELUCTANCES OF TYPE I STRUCTURES

$\Re_{1A}(c)$	$\dfrac{l_{1A}^{2}}{\mu_0 \mu_r V_{1A}(c)}$
$\Re_{2A}(c)$	
$\Re_{3A}$	$\dfrac{b}{\mu_0 \mu_r \times \pi r^2}$
$\Re_{4A}$	
$\Re_{5A}$	$\dfrac{h}{\mu_r \times (2r)^2}$
$\Re_{6A}(c)$	$\dfrac{h}{\mu_r \times 2r \times (c - \alpha)}$
l_{1A}	$\displaystyle \int_{\alpha}^{a} \sqrt{1 + \left(\frac{\partial}{\partial x}(b + \beta)\sqrt{1 - \frac{x^2}{a^2}} \right)^2}\, dx + \alpha$
$V_{1A}(c)$	$\displaystyle 4r \int_{\alpha}^{a} (b + \beta)\sqrt{1 - \frac{x^2}{a^2}}\, dx - 4r \int_{\alpha}^{c}(d + \beta)\sqrt{1 - \frac{x^2}{c^2}}\, dx$
$\Re_{g1A}$	$\dfrac{g/2}{\mu_0 \times \pi r^2}$
$\Re_{g2A}$	
$\Re_{leak1A}$	$\dfrac{l_{mean}^{2}}{\mu_0 \times V_{leak1A}}$
$\Re_{leak2A}$	$= \dfrac{l_{mean}^{2}}{\mu_0 \times \int_{-1.5g}^{1.5g} \pi\left(r + \sqrt{(1.5g)^2 - y^2}\right)^2 dy - \pi r^2 \times 3g}$
$\Re_{leak3A}(c)$	$\dfrac{b}{\mu_0 \times [2r \times (c - \alpha)]}$
$\Re_{leak4A}(c)$	

TABLE II
MAGNETIC RELUCTANCES OF TYPE III STRUCTURES

$\mathfrak{R}_{1B}(c)$	$\dfrac{l_{1B}^2}{\mu_0 \mu_r V_{1B}(c)}$
$\mathfrak{R}_{2B}(c)$	
$\mathfrak{R}_{3B}$	$\dfrac{b}{\mu_0 \mu_r \times \pi r^2}$
$\mathfrak{R}_{4B}$	
$\mathfrak{R}_{5B}(c)$	$\dfrac{h}{\mu_r \times (2r)^2} + \dfrac{h}{\mu_r \times 2r \times (c-\alpha)}$
l_{1B}	$\dfrac{\int_\alpha^a \sqrt{1 + \left(\frac{\partial}{\partial x} b\sqrt{1 - \frac{x^2}{a^2}}\right)^2}\, dx + \alpha}{2}$ $+ \dfrac{\int_0^\beta \sqrt{1 + \left(\frac{\partial}{\partial y} a\sqrt{1 - \frac{y^2}{b^2}}\right)^2}\, dy + \alpha}{2}$
$V_{1B}(c)$	$4r\left(\int_\alpha^a b\sqrt{1 - \frac{x^2}{a^2}}\, dx - \int_\alpha^c d\sqrt{1 - \frac{x^2}{c^2}}\, dx\right)$ $-2r\int_\beta^b a\sqrt{1 - \frac{y^2}{b^2}}\, dy + 2r\int_0^\alpha b\sqrt{1 - \frac{x^2}{a^2}}\, dx - 2r\alpha\beta$
$\mathfrak{R}_{g1B}$	$\dfrac{g/2}{\mu_0 \times \pi r^2}$
$\mathfrak{R}_{g2B}$	
$\mathfrak{R}_{leak1B}$	$\dfrac{l_{mean}^2}{\mu_0 \times V_{leak1A}}$
$\mathfrak{R}_{leak2B}$	$= \dfrac{l_{mean}^2}{\mu_0 \times \int_{-1.5g}^{1.5g} \pi\left(r + \sqrt{(1.5g)^2 - y^2}\right)^2 dy - \pi r^2 \times 3g}$

Therefore, the magnetic equivalent circuit models of Type I and Type III structure are depicted in Fig. 10, and their coupling coefficients k_A and k_B representing Type I and Type III are given by

$$k_A = \frac{\phi_x}{\phi_x + \phi_{leak}} = \frac{\frac{1}{2}\mathfrak{R}_{leak1A}}{\frac{1}{2}\mathfrak{R}_{leak1A} + \mathfrak{R}_A} \tag{5}$$

and

$$k_B = \frac{\phi_x}{\phi_x + \phi_{leak}} = \frac{\frac{1}{2}\mathfrak{R}_{leak1B}}{\frac{1}{2}\mathfrak{R}_{leak1B} + \mathfrak{R}_B} \tag{6}$$

where the equivalent reluctance $\mathfrak{R}_A$ and $\mathfrak{R}_B$ are derived as

$$\mathfrak{R}_A = [(\mathfrak{R}_{2A}(c) + \mathfrak{R}_{3A}) \| (\mathfrak{R}_{1A}(c) + \mathfrak{R}_{4A})]$$

$$+ [\tfrac{1}{2}\mathfrak{R}_{leak3A}(c) \| \mathfrak{R}_{6A}(c)] + \mathfrak{R}_{5A} \tag{7}$$

and

$$\mathfrak{R}_B = [(\mathfrak{R}_{2B}(c) + \mathfrak{R}_{3B}) \| (\mathfrak{R}_{1B}(c) + \mathfrak{R}_{4B})]$$

$$+ \mathfrak{R}_{5B}(c) \tag{8}$$

Based on the above coupling capability of the calculation, it is obvious to design a proper inductively coupled structure. Then, we take those definition to simulations, as shown in Fig. 11 with the size of $r = 25$ mm, $a = 60$ mm, $b = 30$ mm, $d = 10$ mm, $h = 8$ mm, $g = 8$ mm, $\alpha = 10$ mm, $\beta = 14$ mm, and $c = 20\sim40$ mm.

Consequently, it is evident that the coupling coefficient of Type III is always higher than that of Type I. Simultaneously, it is thought over that to avoid the magnetic field of right arm is too high to pass through magnetic field due to the saturation of magnetic field and there is not much difference in coupling coefficient between two types of structure when c is wider than 30 mm. For these reasons, Type III structure with 30 mm width of right arm is chosen as the inductively coupled power pickup.

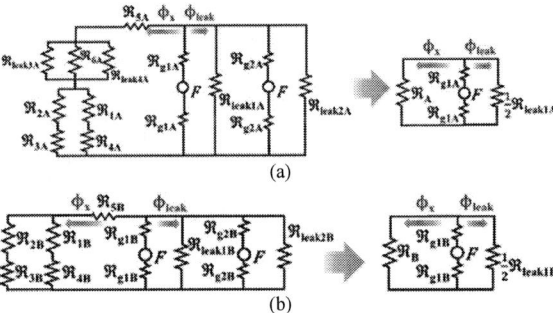

(a)

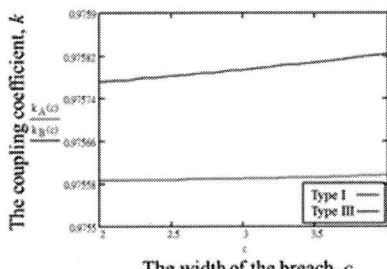

(b)

Fig. 10. The magnetic equivalent circuit models. (a) Type I structure. (b) Type III structure.

Fig. 11. Comparison of the coupling coefficients.

D. Analysis of Compound Resonant Circuits

The principle of contactless inductive power transmission is different from traditional transformer, and it is regarded as loosely coupled structure which the gap between primary and secondary coils leads to the value of coupling coefficient. Nevertheless, there is a wide gap causing low coupling coefficient in the inductive coupling structure for convenience of moving, and it results in low power transmission. In order to raise the efficiency of power transmission, the resonant tank are formed in the primary and secondary side by adding an additional capacitor. With resonant capacitor, the resonant tank at the primary side can increase the transmitted power, and the resonant tank at the secondary side can raise the acquired power. Accordingly, SP_LS_C topology is chosen to limit the current, as shown in Fig. 12, and their three resonant angular frequency are given by

$$\omega_1 = \frac{1}{\sqrt{L' C_{P2}}} \tag{9}$$

$$\omega_2 = \frac{1}{\sqrt{(L_{P1} + L_{P2})\, C_{P1}}} \tag{10}$$

2110

$$\omega_3 = \frac{1}{\sqrt{L_{P1}\,C_{P1}}} \tag{11}$$

where L' are derived as

$$L' = \frac{\omega^2\,L_{P1}\,L_{P2} - \frac{L_{P2}}{C_{P1}}}{\omega^2\,L_{P2} + \omega^2\,L_{P1} - \frac{1}{C_{P1}}} \tag{12}$$

To avoid the unlimited current and maintain the maximum power transfer capability, it is chosen to operate at the operating angular frequency ω equals to resonant angular frequency ω_3 because it is caused the circuit has a small impedance Z_1 that current is limit, as given by

$$Z_1 = \frac{1}{j\omega C_{P2}} \tag{13}$$

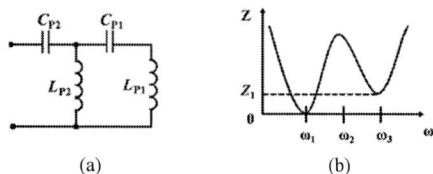

(a)   (b)

Fig. 12. SP$_L$S$_C$ topology. (a) Resonant circuit. (b) Impedance versus operating frequency.

III. DESIGN OF CONTACTLESS STRIP INDUCTIVE POWER TRACK SYSTEM

The framework of the proposed contactless inductive power transfer system is clarified in Fig. 13, and it is divided into three parts comprising primary circuits, inductively coupled structure, and secondary circuits.

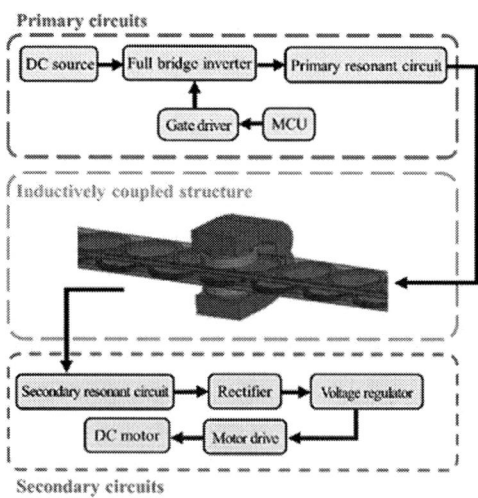

Fig. 13. Contactless inductive power transfer system.

In the primary circuits shown in Fig. 14(a), a full bridge inverter is controlled by driving signal from a micro-controller unit outputs a high-frequency square wave, and then the square wave is transformed into a sine wave with the resonant circuit of SP$_L$S$_C$ topology which can limit the current. It is noteworthy that the operating frequency of switches of the full bridge inverter have to adjust to the resonant frequency in primary resonant circuit, so the

power loss in the inductive power track is minimum. In the secondary circuits shown in Fig. 14(b), it is changed the AC to DC power by the rectifier, and then a voltage regulator adjusts to meet a proper voltage for the motor of the inductive power track in automated production line system.

As noted in the previous chapter, the inductively couple structure has been design that a 1.2 m long inductive power track is made of Litz wires, and the power-pickup made of several cores fastened with epoxy mucilage are the coils made of Litz wires winding on the core structure. The practical inductively coupled structure is manufactured, and its parameters are measured and demonstrated in Table III.

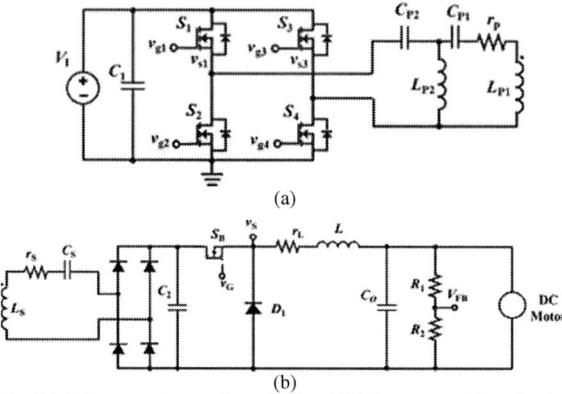

(a)

(b)

Fig. 14. Primary and secondary circuits. (a) Primary energizing circuits. (b) Secondary power pickup circuits.

TABLE III
PARAMETERS OF THE PRACTICAL PROPOSED INDUCTIVELY COUPLED STRUCTURE

The coils of the proposed strip-type power track	
Inductance L_{P1}	43.56 µH
Equivalent resistance r_P	0.64 Ω
Primary capacitance C_{P1}	1504 nF
The coils of the proposed power pickup	
Inductance L_S	331.02 µH
Equivalent resistance r_S	0.24 Ω
Secondary capacitance C_S	191 nF
Mutual inductance M	21.98 µH
Resonant frequency w_o	20 kHz

IV. EXPERIMENTAL RESULTS

The contactless strip inductive power track system with a dual-slot power-pickup structure is shown in Fig. 15. To consider the effect of load variation on the efficiency of power transmission in the system, Fig. 16 illustrates the connection between the range of load, output voltage and current with input voltage $V_1 = 40$ V. Then, we can multiply these results to get the output power and transmitting efficiency, as shown in Fig. 17. It indicates that the system has a highest efficiency and output power of transmission when R_L is between 11 Ω to 13 Ω.

2111

The 2018 International Power Electronics Conference

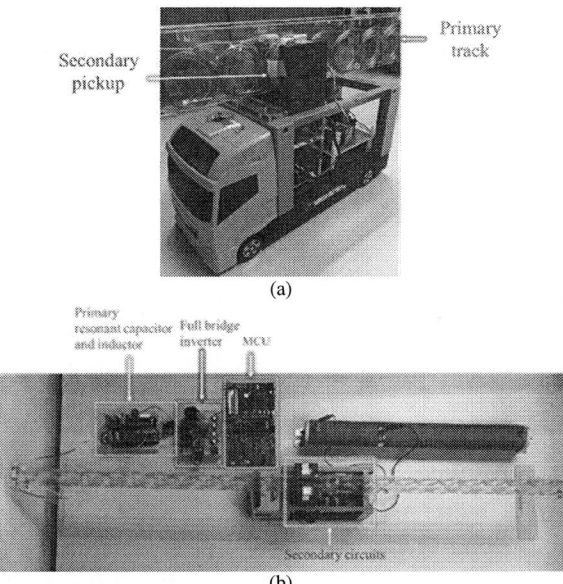

(a)

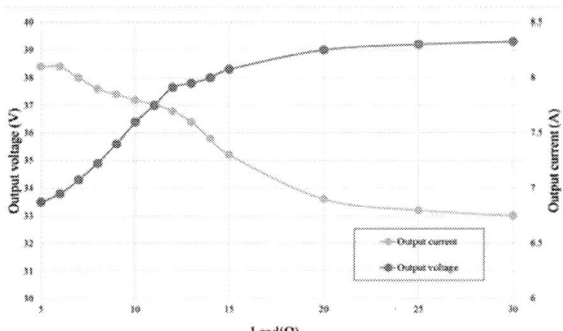

(b)

Fig. 15. Experiment Setup. (a) Actual inductively coupled structure. (b) Overall system.

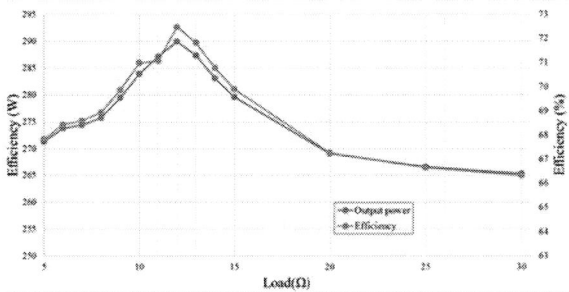

Fig. 16. Load versus output voltage and current with $V_1 = 40$ V.

Fig. 17. Load versus output power and transmitting efficiency with $V_1 = 40$ V.

Hence, we do more experiment in this area, as depicted in Fig. 18. It is obvious that a maximum efficiency happens at $R_L = 12.5$ Ω. The output voltage and current of load in an experiment with $V_1 = 40$ V, $I_1 = 9.98$ A, and 12.5 Ω load has a maximum efficiency 72.52% with output power 289.52 W. Fig. 19 demonstrates a maximum output power occurs at $R_L = 11.4$ Ω with input voltage $V_1 = 120$ V. Consequently, in an experiment with $V_1 = 120$ V, $I_1 = 10$ A, and 11.4 Ω load has a maximum output power 775 W with efficiency 64.58%.

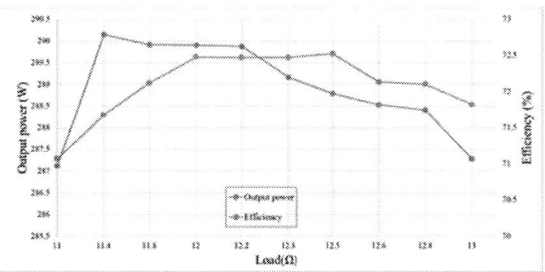

Fig. 18. Load between 11 Ω and 13 Ω versus output power and transmitting efficiency with $V_1 = 40$ V.

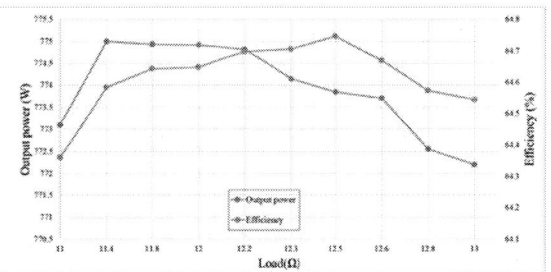

Fig. 19. Load between 11 Ω and 13 Ω versus output power and transmitting efficiency with $V_1 = 120$ V.

V. CONCLUSIONS

In the proposed system, the strip type inductive power track is made of circular coils tested the winding where the magnetic flux density is stronger and smoother at first. Next, the proposed structures of the power-pickup were analyzed for their magnetic simulations by Maxwell, and then their magnetic equivalent circuit models were established to observe so as to choose the structure with the best coupling coefficient. Furthermore, in the aspect of hardware, to develop a full bridge inverter with operating frequency controlled by microcontroller unit, the resonant circuits with SP_LS_C topology was used to provide a limit alternating current for the power track, while a rectifier and a buck converter were made for secondary output motor. Finally, the contactless strip inductive power track system, with a 1.2 m long strip type inductive power track and a 5 cm wide dual-slot type power-pickup, were fabricated. In accordance with the experimental results, the maximum power output of the systems is 775 W, and the maximum transmission efficiency is 72.52%.

REFERENCES

[1] Schoneberger, "Primove contactless and catenary-free operation," Bombardier Inc., 10832/SYS/09-2010/en, Canada, 2010.

[2] Kamen, "Contactless Power System," Vahle Corp., Germany, Nr. 9d/EN, Nov. 2008.

[3] "IPT charge for electric vehicles," Conductix-Wampfler delachaux group, Germany, KAT9200-0001-E, 2009.

[4] "Corporate profile," Daifuku Corp., Japan, CP13E, 2013.

[5] G. A. Covic and J. T. Boys, "Inductive power transfer," *Proc. IEEE*, vol. 101, no. 6, pp. 1276–1289, Jan. 2013.

[6] W. Zhang, S. C. Wong, C. K. Tse, and Q. Chen, "An optimized track length in roadway inductive power transfer systems," *IEEE J. Emerg. Sel. Topics Power Electron.*, vol. 2, no. 3, pp. 598–608, Sep. 2014.

[7] S. Y. Choi, B. W. Gu, S. Y. Jeong, and C. T. Rim, "Advances in wireless power transfer systems for roadway-powered electric vehicles," *IEEE J. Emerg. Sel. Topics Power Electron.*, vol. 3, no. 1, pp. 18–36, Mar. 2015.

2112

Discontinuous SVM Technique for Three-Leg VSI Fed Balanced/Unbalanced Two-Phase Loads

Supanut Charoensuksirikul[1] and Yuttana Kumsuwan[2*]
[1]Student of the Graduate School, Chiang Mai University, Chiang Mai, Thailand
[1,2]Department of Electrical Engineering, Faculty of Engineering, Chiang Mai University, Chiang Mai, Thailand
*E-mail: yt@eng.cmu.ac.th

Abstract—This paper presents a novel discontinuous space vector modulation (DSVM) technique to reduce the power switching loss by arranging the switching pattern of each an inverter leg for a three-leg voltage source inverter (VSI) fed balanced/unbalanced two-phase loads. This DSVM method provides a typical dwell time calculation as well as the four types of the discontinuous switching sequence selections, which results in optimization switching loss and significantly shares more balanced power, and thermal distributions of each an inverter leg. Especially, in the middle inverter leg, the proposed strategy can mitigate the switching loss up to 1/2 when compared with the conventional continuous SVM (CSVM) method. Calculation and simulation results are shown to verify the effectiveness of the proposed technique.

Keywords—Three-leg voltage source inverter (VSI), balanced/unbalanced two-phase loads, discontinuous space vector modulation (DSVM), switching loss.

I. INTRODUCTION

A typical three-leg VSI-fed balanced/unbalanced two-phase load is shown in Fig. 1. This inverter is widely used in domestic applications, such as symmetrical/asymmetrical two-phase induction motor, and ac servomotor drives [1]-[5]. However, one disadvantage of this inverter is that amplitude of the phase current in an inverter legs are unequal, resulting in unbalanced power and thermal distributions. Therefore, the reduction of switching loss operation is desirable for three-leg VSI-fed balanced/unbalanced impedance two-phase loads to avoid thermal problem and increase the efficiency of the inverter.

To overcome this problem, it can be done by a pulse width modulation (PWM) technique as known discontinuous modulation. By using a discontinuous PWM strategies [6], for the three-leg VSI with balanced two-phase loads, the number of switching step in three-leg can be reduced by 1/3 when compared with the continuous PWM with the same switching frequency. However, the switching loss of the inverter leg is unequal by 1/2, 1/4, and 1/4 for legs *A*, *B*, and *C*, respectively, due to unbalanced sector intervals and non-optimal switching sequence selections. In three-leg VSI-fed unbalanced impedance two-phase motor drive applications, it is possible to make the space vector PWM algorithm for

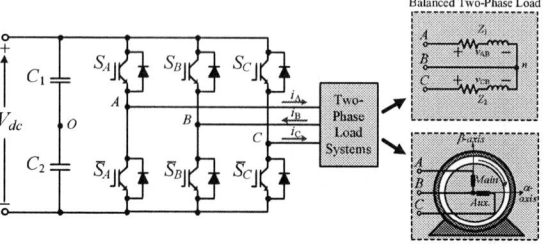

Fig. 1. Power circuit of the three-leg voltage source inverter (VSI) fed balanced/unbalanced two-phase loads.

continuous modulating waves [1], [3]-[5]. However, it still suffers from unbalanced current in the inverter legs and high current stress of the middle leg, causing unbalanced power losses.

In this paper, a novel discontinuous space vector modulation (DSVM) is developed for the three-leg VSI fed two-phase loads. The proposed DSVM can mitigate the disadvantages of the classical discontinuous PWMs to reduce the power losses at the middle leg as well as to improve the performance of the inverter. The simulation results are given in order to verify the performance of the proposed DSVM for balanced and unbalanced two-phase load applications.

II. OPERATION PRINCIPLE

Referring to Fig. 1, the topology fed two-phase loads is based on the three-leg VSI. Therefore, the switching states for three legs are totally 8 states, as the equivalent circuits are shown in Fig. 2. Among these switching states, they are same as the standard two-level three-leg VSI fed three-phase loads, unless there are only two-phase output voltages v_{AB} and v_{CB}, as summarized in Table I. In this table, considering the current flow through each inverter leg, it can be seen that the current flows through leg *B* results in the sum of the two output currents of leg *A* and leg *C*. For this reason, this leads to a higher switching loss in leg *B* compared with the two other legs, which is one of the drawbacks of this power conversion system. The switching loss in leg *B* is accordingly necessary to more take care and reduce from conventional systems [1]-[6], as its details are given in the following section.

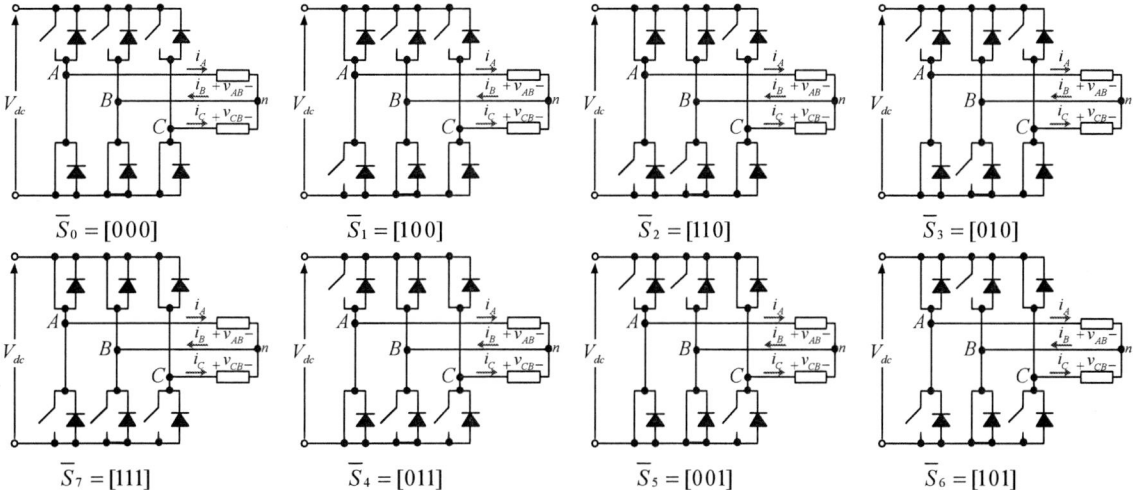

Fig. 2. Equivalent circuits of 8 switching states $\overline{S}_0$, $\overline{S}_1$, $\overline{S}_2$, $\overline{S}_3$, $\overline{S}_4$, $\overline{S}_5$, $\overline{S}_6$, and $\overline{S}_7$ for three-leg VSI fed two-phase loads.

III. PROPOSED DSVM TECHNIQUE

According to the proposed DSVM technique for three-leg VSI fed two-phase loads, its first contribution is the simplified SVM algorithm. Then, the generalized dwell time calculation for balanced/unbalanced two-phase loads is created. Finally, a novel switching sequence designs are significantly proposed in order to reduce the switching loss in the common leg (leg B).

A. Simplified SVM Algorithm

With the proposed simplified SVM algorithm for three-leg VSI fed balanced/unbalanced two-phase loads, the reference voltage in terms of the $\alpha\beta$ stationary frame can be expressed as

$$\vec{V}_{ref} = \frac{1}{\sqrt{2}}\left(S_A e^{j\frac{\delta}{2}}\cdot\overline{a}^0 + \sqrt{2}S_B e^{j\frac{\delta}{2}}\cdot\overline{a}^1 + S_C e^{j\frac{\delta}{2}}\cdot\overline{a}^2\right) \quad (1)$$

where $\overline{a} = e^{j3\pi/4}$. $\vec{V}_{ref}$ is the reference voltage in $\alpha\beta$ stationary frame of two-phase system, δ is the angle of freedom for setting balanced/unbalanced modulation.

Substituting 8 switching states (see Table I) into (1), the corresponding space vector diagram is represented by Fig. 3(a), which is consisted of 6 active voltage vectors ($\vec{V}_1,\vec{V}_2,\vec{V}_3,\vec{V}_4,\vec{V}_5,\vec{V}_6$) and 2 zero voltage vectors ($\vec{V}_0,\vec{V}_7$). For the balanced modulation scheme, where $\delta = 0$ rad, the space vector diagram is drawn by the voltage vectors with the black line. The trajectory of the reference voltage $\vec{V}_{ref}$ is constantly circle shape. By the way, when $\delta \neq 0$ ($\pi/6$ and $-\pi/6$ for example) for the unbalanced modulation, the space vector diagrams are drawn by the voltage vectors with the red and blue lines, respectively. Subsequent to the projection of the active voltage vectors excepting the $\vec{V}_3',\vec{V}_6'$ and $\vec{V}_3'',\vec{V}_6''$ into the

TABLE I
SWITCHING STATES CORRESPONDING TO RELATIVE TWO-PHASE
OUTPUT VOLTAGES AND CURRENT FLOW FOR THREE LEGS

States	Patterns of switching			Two-phase output voltages		Current flow		
	S_A	S_B	S_C	v_{AB}	v_{CB}	Leg A	Leg B	Leg C
$\overline{S}_0$	0	0	0	0	0	0	0	0
$\overline{S}_1$	1	0	0	V_{dc}	0	i_A	i_A+i_C	$-i_C$
$\overline{S}_2$	1	1	0	0	$-V_{dc}$	i_A	i_A+i_C	$-i_C$
$\overline{S}_3$	0	1	0	$-V_{dc}$	$-V_{dc}$	$-i_A$	i_A+i_C	$-i_C$
$\overline{S}_4$	0	1	1	$-V_{dc}$	0	$-i_A$	i_A+i_C	i_C
$\overline{S}_5$	0	0	1	0	V_{dc}	$-i_A$	i_A+i_C	i_C
$\overline{S}_6$	1	0	1	V_{dc}	V_{dc}	i_A	i_A+i_C	i_C
$\overline{S}_7$	1	1	1	0	0	0	0	0

$\alpha\beta$ axis, the consequent space vector diagram is shown in Fig. 3(b). This leads to the vertical and horizontal ellipse trajectories of the reference voltages for $\delta = \pi/6$ and $-\pi/6$ rad, respectively. Hereby, the summary of magnitudes and directions of the corresponding voltage vectors is also given in Table II.

B. Generalized Dwell Time Calculation

To determine the desired dwell times, the reference voltage vector is synthesized during a haft sampling time and its voltage vectors in each sector, as follow:

$$\vec{V}_{ref}\cdot\frac{T_s}{2} = \vec{V}_k\cdot t_k + \vec{V}_{k+1}\cdot t_{k+1} + \vec{V}_0\cdot T_0 \quad (2)$$

$$\frac{T_s}{2} = t_k + t_{k+1} + t_0 + t_7 \quad (3)$$

where T_s is the sampling time period. k is the number of sectors (I, II, III,…, VI).

From (2) and (3), the generalized dwell time

The 2018 International Power Electronics Conference

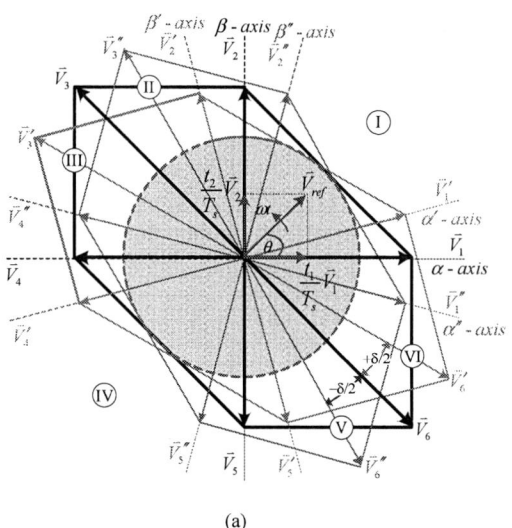
(a)

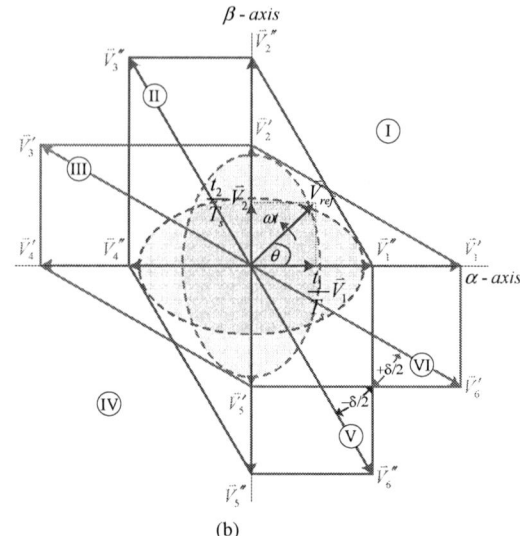
(b)

Fig. 3. Space vector diagrams of the proposed SVM algorithm for (a) initial balanced and unbalanced modulation schemes along with consequent balanced modulation scheme and (b) consequent unbalanced modulation scheme.

TABLE II
SUMMARY OF VOLTAGE SPACE VECTORS

Voltage space vectors	Switching states	Voltage space vector results of the proposed SVM algorithm		
		Balanced/ Unbalanced (Fig. 3(a))	Balanced (Fig. 3(a))	Unbalanced (Fig. 3(b))
$\vec{V}_0$	$\overline{S}_0$ [000]	0	0	0
$\vec{V}_7$	$\overline{S}_7$ [111]	0	0	0
$\vec{V}_1$	$\overline{S}_1$ [000]	$1/\sqrt{2}\,e^{j\delta/2}$	$1/\sqrt{2}\,e^{j0}$	$\cos(\pi/4+\delta/2)e^{j0}$
$\vec{V}_2$	$\overline{S}_2$ [111]	$1/\sqrt{2}\,e^{j(\pi/2+\delta/2)}$	$1/\sqrt{2}\,e^{j\pi/2}$	$\sin(\pi/4+\delta/2)e^{j\pi/2}$
$\vec{V}_3$	$\overline{S}_3$ [000]	$1e^{j(3\pi/4+\delta/2)}$	$1e^{j3\pi/4}$	$1e^{j(3\pi/4+\delta/2)}$
$\vec{V}_4$	$\overline{S}_4$ [111]	$1/\sqrt{2}\,e^{j(\pi+\delta/2)}$	$1/\sqrt{2}\,e^{j\pi}$	$\cos(\pi/4+\delta/2)e^{j\pi}$
$\vec{V}_5$	$\overline{S}_5$ [100]	$1/\sqrt{2}\,e^{j(3\pi/2+\delta/2)}$	$1/\sqrt{2}\,e^{j3\pi/2}$	$\sin(\pi/4+\delta/2)e^{j3\pi/2}$
$\vec{V}_6$	$\overline{S}_6$ [111]	$1e^{j(7\pi/4+\delta/2)}$	$1e^{j7\pi/4}$	$1e^{j(7\pi/4+\delta/2)}$

calculation for all sectors can be determined by

$$\begin{bmatrix} t_k \\ t_{k+1} \end{bmatrix} = 2M_a \cdot r(\theta) \cdot \frac{T_s}{2} \cdot \sqrt{\frac{2}{3}} \cdot \begin{bmatrix} \sin\dfrac{k\pi}{3} & \sin\dfrac{(k-1)\pi}{3} \\ -\sin\dfrac{(k-1)\pi}{3} & \sin\dfrac{(k+1)\pi}{3} \end{bmatrix} \begin{bmatrix} \cos\theta \\ \sin\theta \end{bmatrix} \quad (4)$$

where M_a is modulation index.

From (4), it can be explained that the unbalanced modulation scheme leads to the non-circle trajectory of the $\vec{V}_{ref}$ (see Fig. 3(b)). Based on the ellipse theory, the modulation index is identified by

$$M_a = \frac{V_{ref}}{r(\theta)}. \quad (5)$$

From (5), $0 \le M_a \le 1$, where $r(\theta)$ is the proportional ellipse expression, which is defined by

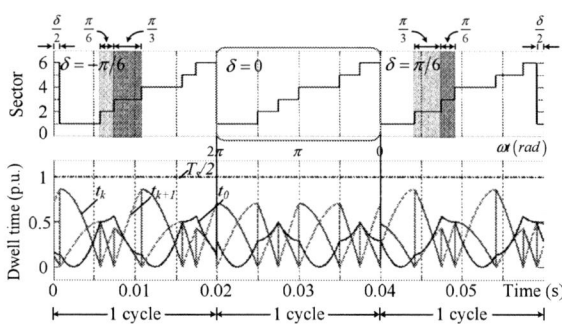

Fig. 4. Waveforms of sectors along with dwell times under balanced ($\delta = 0$) and unbalanced ($\delta = \pi/6, -\pi/6$) modulation schemes.

$$r(\theta) = \frac{\cos\left(\dfrac{\pi}{4}+\dfrac{\delta}{2}\right)\cdot\sin\left(\dfrac{\pi}{4}+\dfrac{\delta}{2}\right)}{\sqrt{2}\cdot\sqrt{\left(\cos\left(\dfrac{\pi}{4}+\dfrac{\delta}{2}\right)\cdot\sin\theta\right)^2 + \left(\sin\left(\dfrac{\pi}{4}+\dfrac{\delta}{2}\right)\cdot\cos\theta\right)^2}}. \quad (6)$$

With (4) to (6), the balanced modulation scheme ($\delta = 0$ rad) leads to $M_a = 2V_{ref}$ and $r(\theta) = 1/2$. Therefore, the dwell time calculation for the balanced modulation scheme can be given by

$$\begin{bmatrix} t_k \\ t_{k+1} \end{bmatrix} = 2V_{ref} \cdot \frac{T_s}{2} \cdot \sqrt{\frac{2}{3}} \cdot \begin{bmatrix} \sin\dfrac{k\pi}{3} & \sin\dfrac{(k-1)\pi}{3} \\ -\sin\dfrac{(k-1)\pi}{3} & \sin\dfrac{(k+1)\pi}{3} \end{bmatrix} \begin{bmatrix} \cos\theta \\ \sin\theta \end{bmatrix}. \quad (7)$$

Based on (4) and (7), it can be concluded that this paper can be proposed the generalized dwell time calculation for both balanced and unbalanced modulation schemes with the easier and straightforward algorithm using only the ellipse theory when compared to [1],[2],[4], and [6].

2115

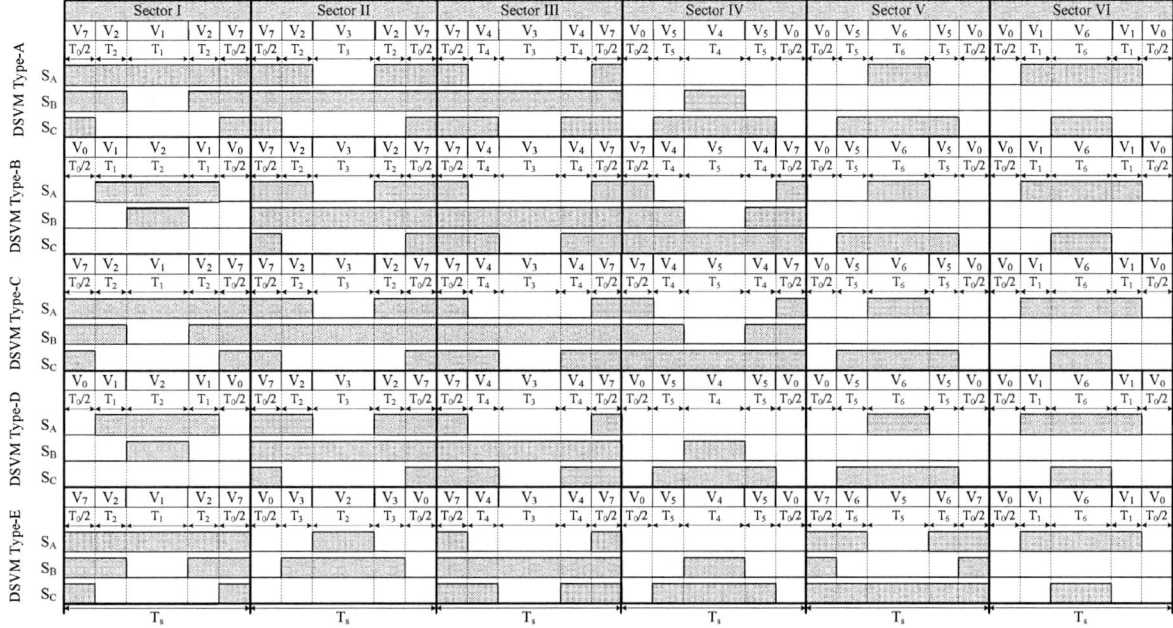

Fig. 5. Proposed DSVM switching sequence designs for Type-A, -B, -C, -D, and –E.

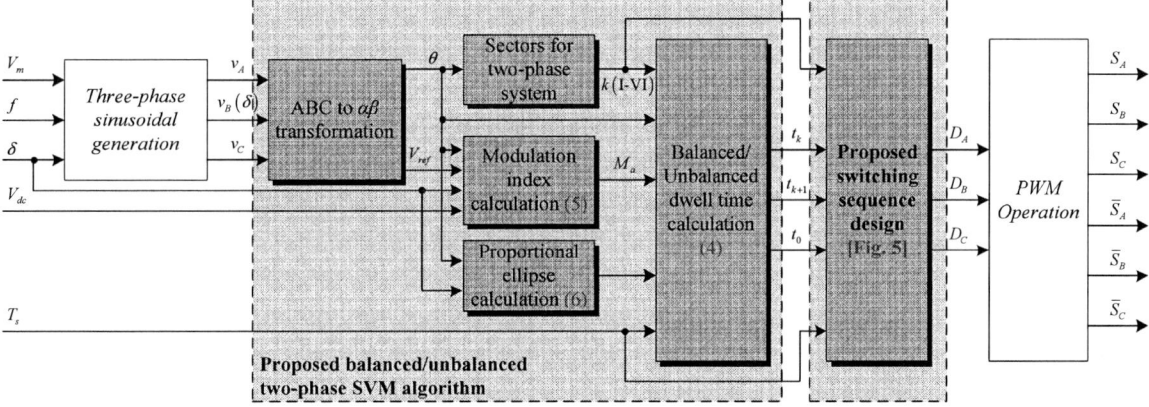

Fig. 6. Block diagram of the proposed DSVM technique for three-leg VSI fed balanced/unbalanced two-phase loads.

Fig. 4 shows the waveforms of the dwell time in order to exhibit the effect of the sector by δ, where the parameters are consistent with Table III to affirm the validity of the dwell times calculated by the proposed expression in balanced and unbalanced modulation schemes. Moreover, these dwell times can be converted to the duty cycles based on the proposed switching sequence design, as discussed in the next subsection.

C. Novel DSVM Switching Sequence Design

With a good candidate for the switching power loss reduction, which is depended on the switching pattern and the magnitude of the current through the switching device, this paper significantly designs a new switching sequences for the DSVM technique, as shown in Fig. 5, where its salient features are as follows:

1) Due to the high current stress on the middle leg, the maximum clamping range of π is designed to arrange for the duty cycle of phase B in order to optimize the switching power loss for this system.

2) The arrangement of the voltage vectors in sectors II, III, V, and VI are same for all types of the proposed DSVMs. That is $\vec{V}_7 \rightarrow \vec{V}_2 \rightarrow \vec{V}_3 \rightarrow \vec{V}_2 \rightarrow \vec{V}_7$ for sector II, $\vec{V}_7 \rightarrow \vec{V}_4 \rightarrow \vec{V}_3 \rightarrow \vec{V}_4 \rightarrow \vec{V}_7$ for sector III, $\vec{V}_0 \rightarrow \vec{V}_5 \rightarrow \vec{V}_6 \rightarrow \vec{V}_5 \rightarrow \vec{V}_0$ for sector V, and $\vec{V}_0 \rightarrow \vec{V}_1 \rightarrow \vec{V}_6 \rightarrow \vec{V}_1 \rightarrow \vec{V}_0$ for sector VI.

3) The proposed DSVMs type-C and type-D are designed to produce more sharing balance of the switching loss with the 25% switching loss reduction in leg A and leg C by swapping the clamping between phase A and phase C for sectors I and IV.

4) According to the proposed switching sequence design, the unbalanced algorithm dose not impact on the clamping duration for legs A, B, C.

The 2018 International Power Electronics Conference

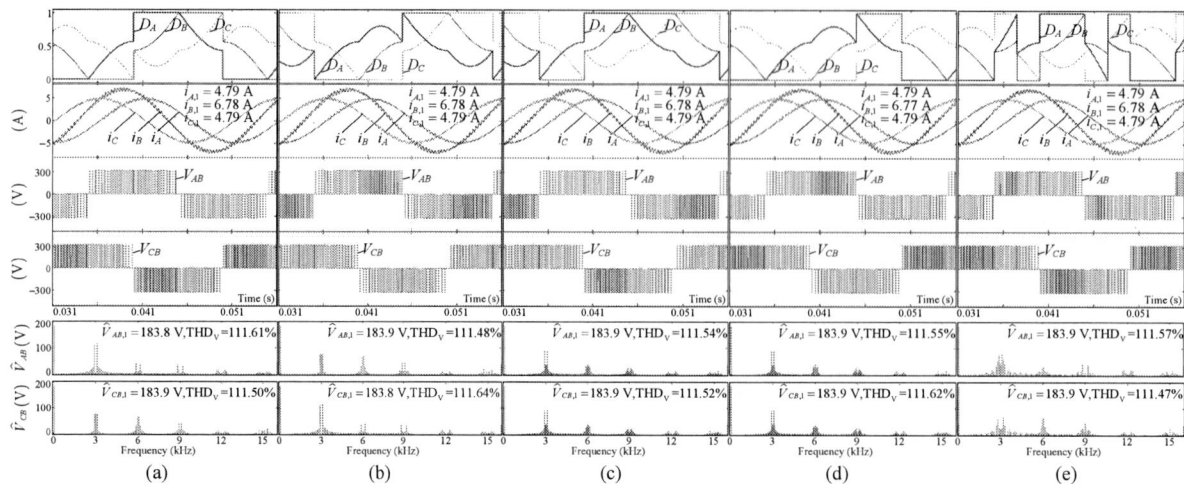

Fig. 7. Simulation results of the proposed DSVM schemes with $f = 50$ Hz, $f_{sw} = 3$ kHz, $M_a = 0.8$, and $\delta = 0$ rad. (Top to bottom) Three-phase duty cycles D_A, D_B, D_C, three-phase output currents i_A, i_B, i_C, two-phase output voltages v_{AB} and v_{CB}, and frequency spectrum of v_{AB} and v_{CB}, respectively. (a) Type-A. (b) Type-B. (c) Type-C. (d) Type-D. (e) Type-E.

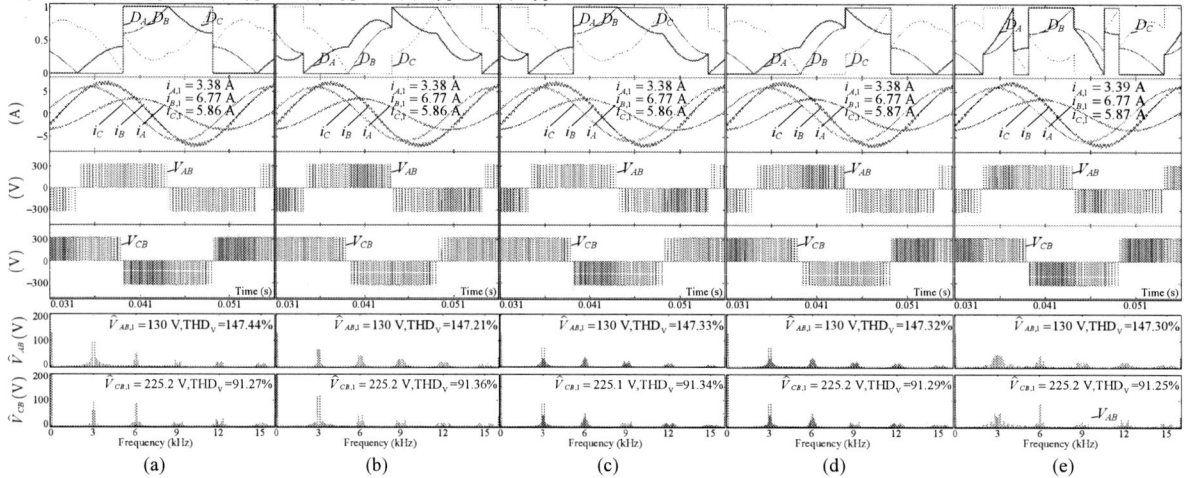

Fig. 8. Simulation results of the proposed DSVM schemes with $f = 50$ Hz, $f_{sw} = 3$ kHz, $M_a = 0.8$, and $\delta = \pi/6$ rad. (Top to bottom) Three-phase duty cycles D_A, D_B, D_C, three-phase output currents i_A, i_B, i_C, two-phase output voltages v_{AB} and v_{CB}, and frequency spectrum of v_{AB} and v_{CB}, respectively. (a) Type-A. (b) Type-B. (c) Type-C. (d) Type-D. (e) Type-E.

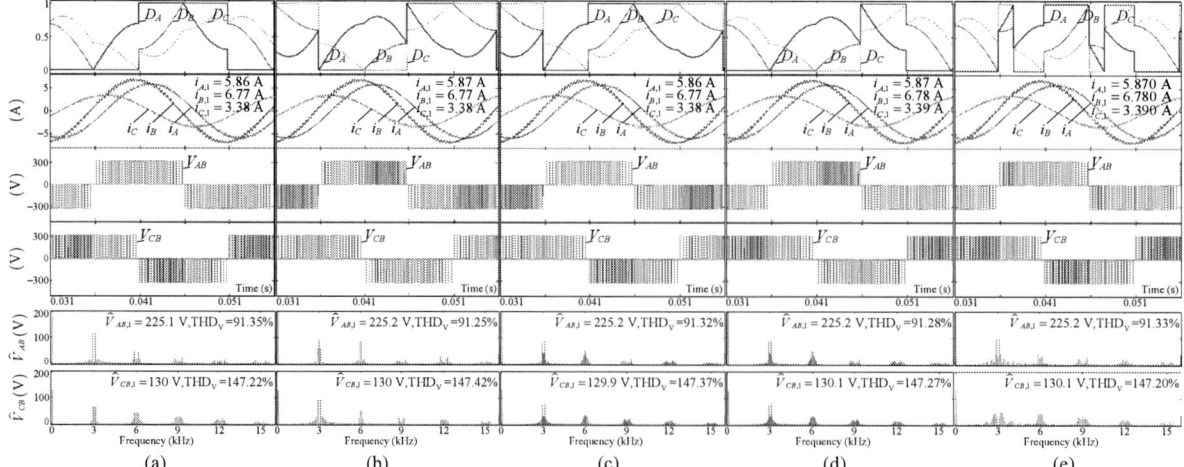

Fig. 9. Simulation results of the proposed DSVM schemes with $f = 50$ Hz, $f_{sw} = 3$ kHz, $M_a = 0.8$, and $\delta = -\pi/6$ rad. (Top to bottom) Three-phase duty cycles D_A, D_B, D_C, three-phase output currents i_A, i_B, i_C, two-phase output voltages v_{AB} and v_{CB}, and frequency spectrum of v_{AB} and v_{CB}, respectively. (a) Type-A. (b) Type-B. (c) Type-C. (d) Type-D. (e) Type-E.

2117

The 2018 International Power Electronics Conference

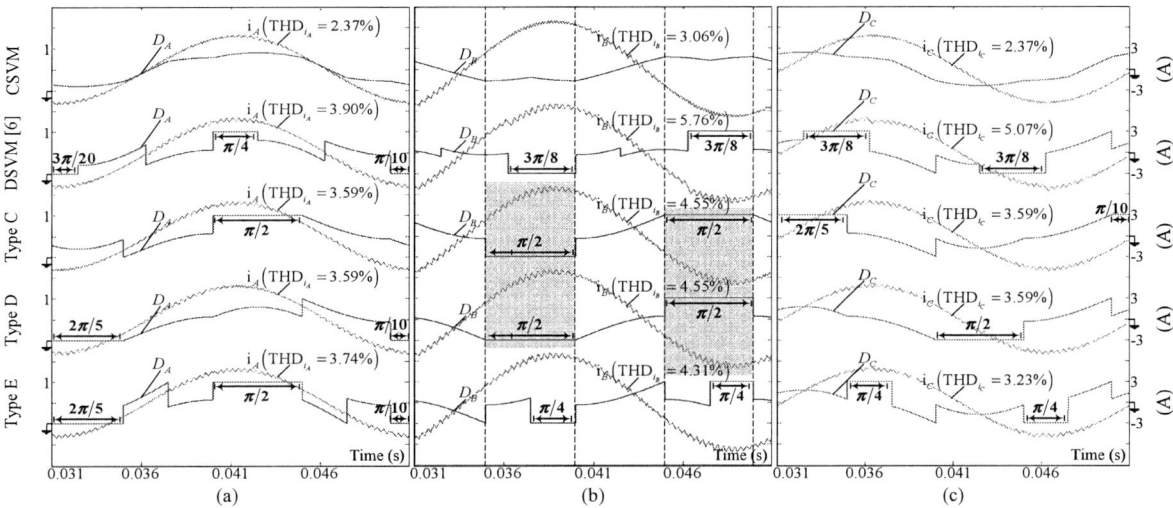

Fig. 10. Relative duty cycle and phase output current waveforms with $f = 50$ Hz, $f_{sw} = 3$ kHz, $M_a = 0.8$, and $\delta = 0$ rad in comparison between (top to bottom) the CSVM, DSVM [6], proposed DSVM Type-C, -D, and –E, respectively. (a) Leg A. (b) Leg B. (c) Leg C.

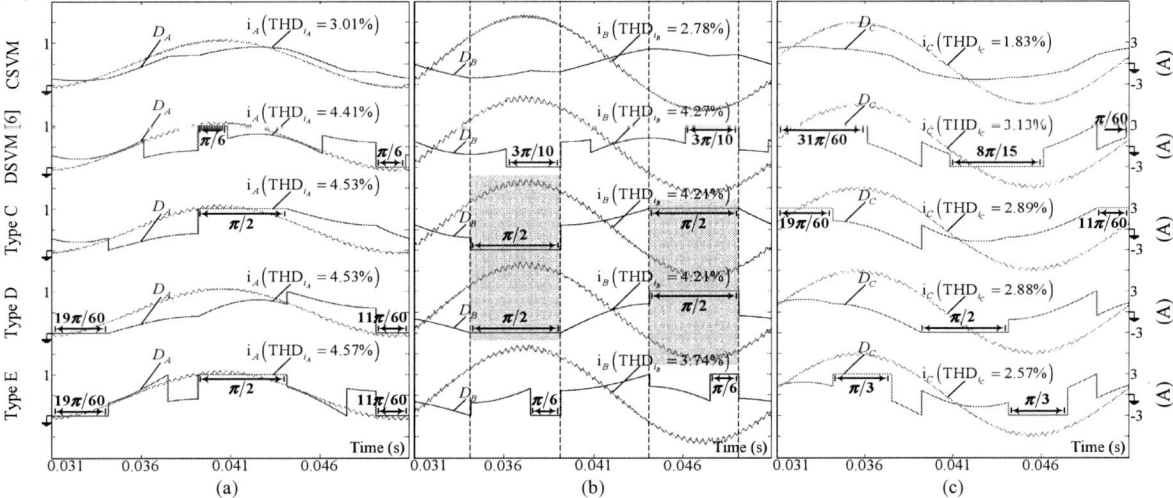

Fig. 11. Relative duty cycle and phase output current waveforms with $f = 50$ Hz, $f_{sw} = 3$ kHz, $M_a = 0.8$, and $\delta = \pi/6$ rad in comparison between (top to bottom) the CSVM, DSVM [6], proposed DSVM Type-C, -D, and –E, respectively. (a) Leg A. (b) Leg B. (c) Leg C.

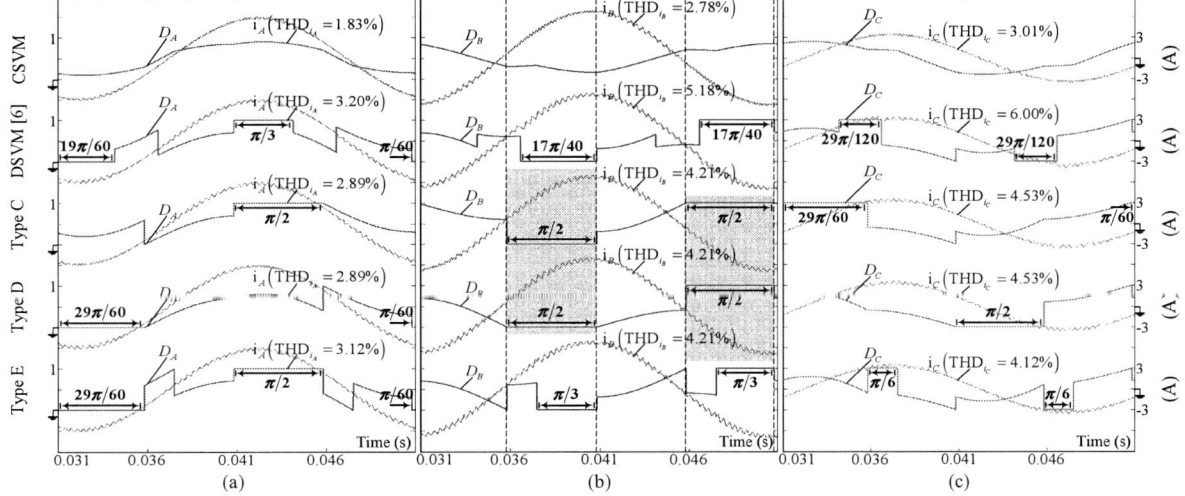

Fig. 12. Relative duty cycle and phase output current waveforms with $f = 50$ Hz, $f_{sw} = 3$ kHz, $M_a = 0.8$, and $\delta = -\pi/6$ rad in comparison between (top to bottom) the CSVM, DSVM [6], proposed DSVM Type-C, -D, and –E, respectively. (a) Leg A. (b) Leg B. (c) Leg C.

2118

TABLE III
SIMULATION PARAMETERS FOR PROPOSED DSVM TECHNIQUE FOR
THREE-LEG VSI FED BALANCED/UNBALANCED TWO-PHASE LOADS

Parameters		Values
Dc-link voltage	(V_{dc})	325 V
Fundamental frequency	(f)	50 Hz
Switching frequency	(f_{sw})	3 kHz
Modulation index	(M_a)	0.8
Angle of freedom	(δ)	$0, \pi/6, -\pi/6$ rad
Two-phase resistances	(R)	35 Ω
Two-phase inductances	(L)	50 mH

Therefore, the proposed DSVMs can also optimize the switching losses for the unbalanced algorithm.

5) In order to confirm the validity of the proposed dwell time calculation, the switching sequence design is also arranged as equivalent to the conventional three-phase DSVM, where it is called the proposed DSVM Type-E.

IV. SIMULATION RESULTS

To verify the validity and performance of the proposed DSVM technique for the three-leg VSI fed balanced and unbalanced two-phase loads, the simulation is implemented in the Matlab/Simulink software and the block diagram of the proposed DSVM algorithm is summarily depicted in Fig. 6. The set of simulation parameters are also given in Table III.

From Fig. 7, the inverter is loaded with a balanced two-phase inductive load with an angle of freedom $\delta = 0$ rad. Although duty cycles converted from (4) have different waveforms depend on switching sequence design in each type, degree interval clamped at leg B in 1 fundamental cycle is maximum by π rad in total and identical to all the proposed DSVM Type-A, -B, -C, and -D. This is due to the requirement of the switching power loss reduction; especially for leg B, which has a highest current stress leading to the highest switching loss compared with that of the other legs, as shown in the second traces. According to the principle, the two-phase output voltage waveforms V_{AB}, V_{CB} are displaced by $\pi/2$ rad and their spectrum of the voltage can be pointed out that their magnitudes result in around $0.707 M_a$ of the input dc voltage and their calculated THD$_V$ are nearby, which the dominant harmonics are centered around the switching frequency (3 kHz). As expected, the performance of the proposed DSVM Type-E is also stable and good as the others. This can confirm the performance and validity of the proposed DSVMs for the three-leg VSI fed the balanced two-phase load with the ability in the optimal switching power loss for this system.

In case of the unbalanced condition with $\delta = \pi/6$ rad, as shown in Fig. 8, it can be seen that the currents in leg B are still highest. However, the proposed DSVM Type-A, -B, -C, and -D are able to maximally reduce the switching power losses, as can be observed by the

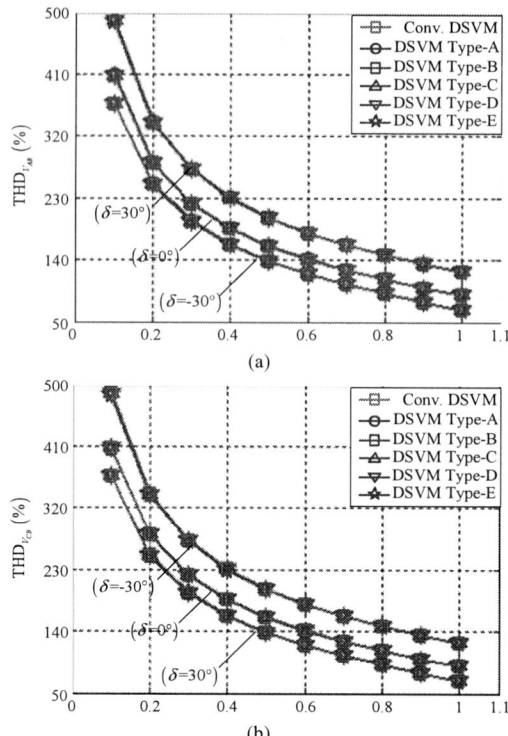

Fig. 13. THD$_V$ of the two-phase output voltages (a) v_{AB} and (b) v_{CB} in comparison between the conventional DSVM scheme [6] and the proposed DSVM schemes.

clamping ranges of the duty cycles for leg B maintaining at the π rad. They can also produce the stable output voltages (orthogonal displacement) and currents corresponding to the technical theory. Furthermore, the THD$_V$ profiles of the dual output voltages can well guarantee with good agreement of the order of the harmonics in accordance with the balanced condition (see Fig. 7).

In case of the unbalanced condition with $\delta = -\pi/6$ rad, as shown in Fig. 9, the results of the orthogonal two-phase output voltage and current profiles are switched each other from Fig. 8. However, the currents in leg B are highest, but the proposed DSVM Type-A, -B, -C, and -D can also handle and lead to the optimal switching loss reduction with the π rad clamping range in the duty cycle for leg B.

Figs. 10, 11, and 12 clearly show the three-phase duty cycles along with the corresponding three-phase output currents. It can be observed from the current ripples along with the THD$_i$ that the quality of the currents in leg B are worse compared with leg A and leg C under the balanced modulation scheme, as shown in Fig. 10. However, the quality of the currents in leg C is highest under the condition of $\delta = \pi/6$ rad and the quality of the currents in leg A is the best for $\delta = \pi/6$ rad, as shown in Fig. 11 and Fig. 12, respectively. Moreover, the tendency of the conventional DSVM [6] provides the lowermost

TABLE IV
COMPARISON OF SWITCHING LOSS REDUCTION

SVM schemes	Discontinuous interval (rad)			Proportional reduction of switching loss		
	Leg A	Leg B	Leg C	Leg A	Leg B	Leg C
CSVM	0	0	0	0	0	0
DSVM [6]	$\pi/2$	$3\pi/4$	$3\pi/4$	1/4	3/8	3/8
Proposed Type-A	π	π	0	1/2	1/2	0
Proposed Type-B	0	π	π	0	1/2	1/2
Proposed Type-C	$\pi/2$ (+)	π	$\pi/2$ (+)	1/4	1/2	1/4
Proposed Type-D	$\pi/2$ (-)	π	$\pi/2$ (-)	1/4	1/2	1/4
Proposed Type-E	π	$\pi/2$	$\pi/2$	1/2	1/4	1/4

current waveform qualities of legs A, B, and C compared to that of the CSVM and the proposed DSVM schemes.

For comparison the THD$_V$ of the two-phase output voltages v_{AB} shown in Fig. 13(a) and v_{CB} shown in Fig. 13(b) in case of the balanced $\delta = 0$ rad and unbalanced $\delta = \pi/6, -\pi/6$ rad conditions with the low to high values of the M_a, the THD$_V$ of the output voltage v_{AB} is the best in the case of the unbalanced $\delta = -\pi/6$ rad according to the spectrum harmonics (see Fig. 9), and the worst in the case of the unbalanced $\delta = \pi/6$ rad (see Fig. 8). For the THD$_V$ of the output voltage v_{CB}, it results in the contrary, as compared to that of the v_{AB}. However, the quality of the two-phase output voltages are identical in the condition of the $\delta = 0$ rad.

As shown in Table IV along with the corresponding results of Figs. 10 to 12, to evaluate the switching power loss reduction of the proposed DSVMs normalized by that of the CSVM and compared with the conventional DSVM [6], the desired switching loss reduction in leg B of the proposed DSVM Type-A, -B, -C, and -D can be satisfied by 50% from the CSVM, but that of the conventional DSVM [6] reduces only 12.5%. In addition, the proposed DSVMs type-C and type D can manage more sharing balance of the switching loss with the 25% switching loss reduction in leg A and leg C compared to the CSVM.

The adjustable speed drive application for the proposed DSVM type-C with an open-loop V/f control algorithm and the same parameters as [5] is shown in Fig. 14, where the turn ratio of the machine (auxiliary winding versus main winding) leads to the set of $\delta = -\pi/6$ rad. The two-phase output voltages and currents are performed in a good regulation with $\pi/2$ rad displacement of each phase, which can be confirmed by the XY plot trajectory of the two-phase output currents i_A and i_C. In the response of the machine, it is also smoothly started and accelerated the rotor speed from 0 rpm to 1,500 rpm with no-load torque, and then able to keep the stable behavior at the steady-state operation. It is evident that the thee-leg VSI can work well in the adjustable speed drive application for the unbalanced two-phase induction motor based on the proposed DSVM strategy.

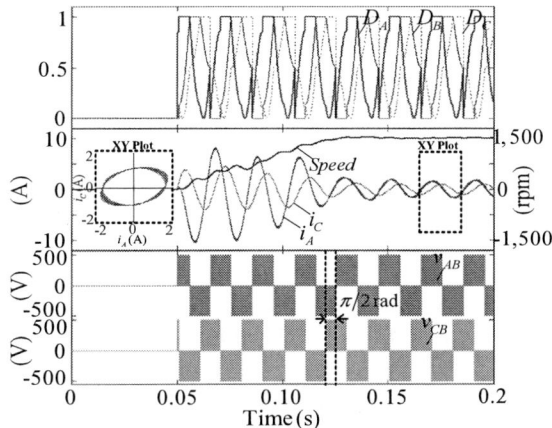

Fig. 14. Simulation results of two-phase unbalanced/induction motor load in an adjustable speed drive application.

V. CONCLUSION

This paper presents the DSVM technique for three-leg VSI fed balanced/unbalanced two-phase loads. The novelty of this paper is consisted of the simplified and straightforward SVM algorithm, which is based on only two-level SVM and ellipse principle, the generalized dwell time calculation for both balanced and unbalanced modulation schemes, and the discontinuous switching sequence designs for optimization of the switching loss reduction by maximum clamping of the switching in leg B with π rad or 50% switching loss reduction compared to the CSVM. Especially, the proposed DSVM Type-C and -D produce more balanced switching power loss distribution, when compared to that of the conventional DSVM [6].

ACKNOWLEDGMENT

The authors would like to thank financial supports from the graduate school and faculty of engineering, Chiang Mai University.

REFERENCES

[1] M.A. Jabbar, A.M. Khambadkone, and Z. Yanfeng, "Space-vector modulation in a two-phase induction motor drive for constant-power operation," *IEEE Trans. Ind. Electron.*, vol. 51, no. 5, pp. 1081-1088, Oct. 2004.

[2] D.C. Martins, L.C. Tomaselli, T.B. Lazzarin and I. Barbi, "Drive for a symmetrical two-phase induction machine using vector modulation," *IEEJ Trans. IA*, vol. 126, no.7, pp.835-840, 2006.

[3] C. Charumit and V. Kinnares, "Carrier-based unbalanced phase voltage space vector PWM strategy for asymmetrical parameter type two-phase induction motor drives," *Trans. Electr. Power Syst. Res.*, vol. 79, no. 7, pp. 1127–1135, Jul. 2009.

[4] V. Kinnares and C. Charumit, "Modulating functions of space vector PWM for three-leg VSI-fed unbalanced two-phase induction motors," *IEEE Trans. Power. Electron.*, vol. 24, no. 4, pp. 1135-1139, Apr. 2009.

[5] Y. Kumsuwan, S. Premrudeepreechacharn, and V. Kinnares, "A carrier-based unbalanced PWM method for four-leg voltage source inverter fed unsymmetrical two-phase induction motor," *IEEE Trans. Ind. Electron.*, vol. 60, no. 5, pp. 2031-2041, May 2013.

[6] C. Charumit and V. Kinnares, "Discontinuous SVPWM techniques of three-leg VSI-fed balanced two-phase loads for reduced switching losses and current ripple," *IEEE Trans. Power Electron.*, vol. 30, no. 4, pp. 2191-2204, Apr. 2015.

The 2018 International Power Electronics Conference

Reduction of Power Losses Based on Generalized Two-level PWM Algorithm for a Nine-switch VSI

Neerakorn Jarutus[1] and Yuttana Kumsuwan[2*]

[1]Student of the Graduate School, Chiang Mai University, Chiang Mai, Thailand
[1,2]Department of Electrical Engineering, Faculty of Engineering, Chiang Mai University, Chiang Mai, Thailand
*E-mail: yt@eng.cmu.ac.th

Abstract—The generalized two-level pulse-width-modulation (PWM) algorithm for the nine-switch voltage source inverter (VSI) is proposed for more simplicity of the implementation and significantly reducing the power losses compared with the conventional nine-switch PWM algorithm. In addition, a new power loss analysis method is also proposed to investigate the power loss reduction of the nine-switch VSI with the proposed PWM by synthesizing the voltage and current distribution models in line with the switching states. This method is simple and high accurate using a characteristic curve-based technique, and also remarkably able to be available for both conduction and switching loss analysis in only one point of view. The corresponding simulation results can well confirm the validity of the proposed PWM algorithm and the proposed power loss analysis method.

Keywords—Nine-switch voltage source inverter (VSI), two-level pulse width modulation (PWM), nine-switch PWM, conduction loss, switching loss.

I. INTRODUCTION

With an increasing demand for the use of the nine-switch voltage source inverter (VSI), as shown in Fig. 1, different pulse width modulation (PWM) algorithms [1]-[4] are continuously proposed in order to modulate the upper and lower VSIs of the nine-switch VSI for feeding the desired electrical energies into the dual three-phase loads. However, these conventional PWM algorithms fortuitously produce a double-switching operation in the coupling three middle switching devices. Then, the proposed generalized two-level PWM algorithm, which is very simpler to implementation, was firstly revealed in [5]. It leads to the cancellation of the switching in the middle switching devices and accordingly the reduction of the power losses in the nine-switch VSI.

In regard to the power loss analysis of the conversion systems, there are four groups categorized by the technique of the power loss determination as follows: the directly simulated/experimental technique [6], the linear [7] and non-linear [8] communication characteristic techniques for the switching power loss calculation, and the insulted-gate bipolar transistor (IGBT)/diode characteristic-curve-based technique [9]. Out of these techniques, the IGBT/diode characteristic-curve-based technique is agreeable with more straightforward, accurate, and uncomplicated way.

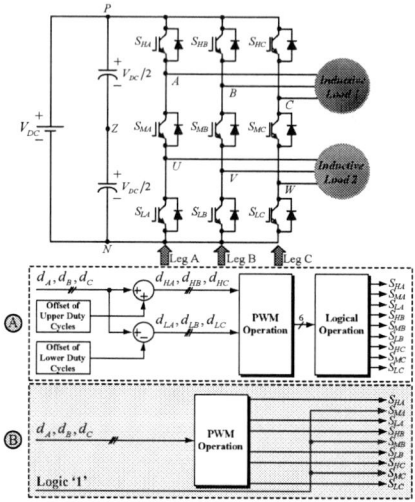

Fig. 1. Nine-switch voltage source inverter (VSI) modulated by either the conventional PWM algorithm [1] (block diagram Ⓐ) or the proposed generalized two-level PWM algorithm (block diagram Ⓑ).

For the power loss analysis of the nine-switch converter, it has received little attention in researches. In [1], the linear characteristic-curve-based interpolation for power loss analysis leads to the degraded of the mathematical switching loss model. Moreover, both analytical conduction and switching loss determinations are still unclear. In [2]-[4], the current distribution model is proposed, which works well with this unique nine-switch structure, to analyze the conduction power losses. However, this model is not available for the switching power loss calculation, which is also undefined and not in-depth analysis.

In this paper, the generalized two-level PWM algorithm for the nine-switch VSI is proposed to reduce the power losses in the conversion system from the conventional PWM algorithm. Subsequently, a novel power loss analysis method based on the characteristic-curve-based technique is also proposed to investigate the power loss reduction performance of the proposed PWM algorithm. Based on this method, the voltage and current distribution models are synthesized in the first time in order to be available for both conduction and switching loss analysis of the nine-switch VSI. It can be as well applied for any unique/common converter topologies.

2121

TABLE I
SWITCHING STATES PER LEG (LEG A) OF NINE-SWITCH INVERTER WITH
CONVENTIONAL PWM ALGORITHM [1]

States	Patterns of switching			Pole voltages	
	S_{HA}	S_{MA}	S_{LA}	v_{AN}	v_{UN}
[P]	1 (On)	1 (On)	0 (Off)	V_{DC}	V_{DC}
[Z]	1 (On)	0 (Off)	1 (On)	V_{DC}	0
[N]	0 (Off)	1 (On)	1 (On)	0	0

TABLE II
SWITCHING STATES PER LEG (LEG A) OF NINE-SWITCH INVERTER WITH
PROPOSED GENERALIZED TWO-LEVEL PWM ALGORITHM

States	Patterns of switching			Pole voltages	
	S_{HA}	S_{MA}	S_{LA}	v_{AN}	v_{UN}
[P]	1 (On)	1 (On)	0 (Off)	V_{DC}	V_{DC}
[N]	0 (Off)	1 (On)	1 (On)	0	0

II. DESCRIPTION OF CONVENTIONAL NINE-SWITCH PWM AND PROPOSED GENERALIZED TWO-LEVEL PWM ALGORITHMS FOR NINE-SWITCH VSI

Traditionally, the conventional PWM algorithm [1] (see block diagram Ⓐ of Fig. 1) is employed to modulated the nine-switch VSI by generating two sets of the three-phase duty cycle signals, d_{HA}, d_{HB}, d_{HC} and d_{LA}, d_{LB}, d_{LC}, consisted of the dc-offset components. Then, these duty cycles are entered into the PWM and logical operations to establish the gating signals for the switching devices. This significantly satisfies three switching states per leg, as listed in Table I.

For the proposed PWM algorithm (see block diagram Ⓑ of Fig. 1), it establishes a new switching function by specifying logical '1' to command the middle switches and two complement logical gating signals to command the upper and lower switches by generating only one set of the three-phase duty cycle signals, d_A, d_B, d_C. Therefore, it is easy and same as the use of the generalized two-level PWM algorithm. For this reason, there are remained only two switching states per leg, as shown in Table II.

III. PROPOSED POWER LOSS ANALYSIS MODEL

In this section, the novel power loss analysis method is also proposed in order to investigate the reduction of the power losses based on the proposed PWM algorithm in comparison to the conventional PWM algorithm [1].

Synthesizing the voltage and current distribution models, it is initialized by regarding the circuit operations of the nine-switch VSI. With the operations, it can be separated into two operating modes according to the direction of the sum of the output currents per leg $i_A + i_U$ and the relative switching states (see Table I and Table II), as shown in Table III, where ϕ is the power factor load angle. Mode ❶ denotes the positive duration of the sum of the output currents and Mode ❷ denotes the negative duration of the sum of the output currents.

TABLE III
EQUIVALENT CIRCUIT OPERATIONS AND CURRENT FLOW PATHS
PER LEG (LEG A)

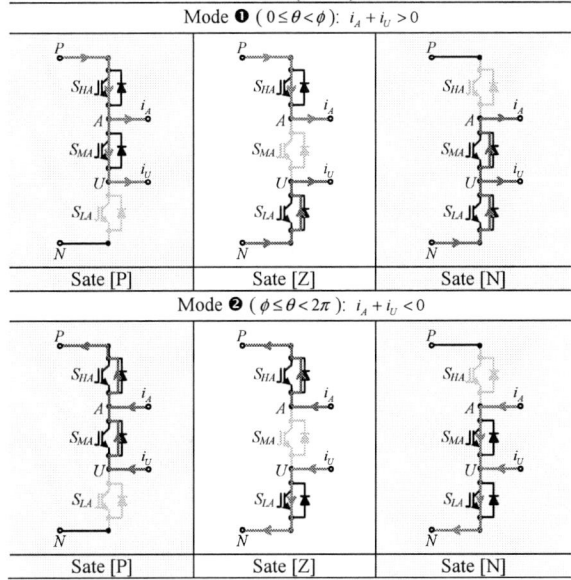

According to Table III, out of six equivalent circuits for the conventional PWM algorithm, there are only four equivalent circuits, which lack the equivalent circuits of state [Z], for the proposed PWM algorithm (see yellow shading area). Taking into account of all these circuit operations and their corresponding current flow paths, which are pointed out by the green line for the current through the IGBT and the red line for the current through the antiparallel diode, it can be clearly depicted the corresponding key principle waveforms, as shown in Fig. 2. Owing to the simplicity of the analysis, the output currents of the upper and lower VSIs are considered in terms of the fundamental components, such as the following output currents of leg A

$$i_{A,1} = I_{A,1} \sin(\omega t + \phi) \tag{1}$$

$$i_{U,1} = I_{U,1} \sin(\omega t + \phi) \tag{2}$$

where $i_{A,1}$ and $i_{U,1}$ are the fundamental current components of leg A for the upper and lower VSIs, respectively. $I_{A,1}$ and $I_{U,1}$ are the amplitudes of the fundamental current components $i_{A,1}$ and $i_{U,1}$, respectively.

Referring (1) and (2), it is realized by the common-frequency mode with the dual three-phase inductive loads of the nine-switch conversion system.

Using Table I to Table III, the proposed voltage and current distribution models for simultaneous conduction and switching loss analysis of the nine-switch VSI with the conventional PWM algorithm and the proposed PWM algorithm are further created, as shown in Figs. 3 and 4, respectively. Significantly, the detailed power loss analysis of the nine-switch VSI is then discussed as follows.

2122

The 2018 International Power Electronics Conference

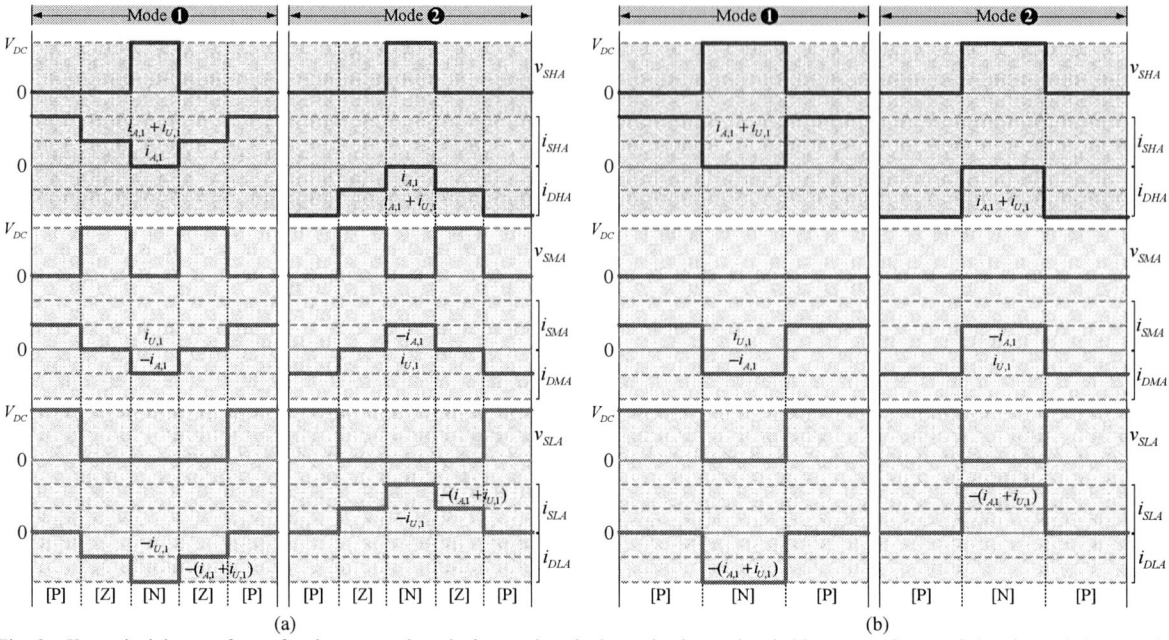

Fig. 2. Key principle waveforms for the proposed method to analyze both conduction and switching power losses of the nine-switch VSI with (a) conventional PWM algorithm [1] and (b) proposed PWM algorithm. (Top to bottom) Switching voltages v_{SHA}, v_{SMA}, v_{SLA}, IGBT currents i_{SHA}, i_{SMA}, i_{SLA}, and diode currents i_{DHA}, i_{DMA}, i_{DLA} of the upper, middle, and lower switching devices, respectively.

A. Conduction Power Losses

The factors of the proposed conduction power loss analysis in this paper are depended on as follows:

1) The characteristic curves of the collector-emitter voltage and current from the datasheet of the switching device, as based on the characteristic-curve-based technique.

2) The instantaneous current through the switching device, as based on the current distributions of the proposed models (see Figs. 3 and 4).

3) The PWM algorithm, as based on the pulse width and pattern of the duty cycle for the switching device.

With the characteristic-curve-based technique and using the switching devices of a 5SNA 1200G450300 Hipak 4.5-kV 1.2-kA module, as same as [9], the average conduction power losses of the IGBT can be obtained by $P_{C,IGBT} = (1.28) \cdot I_{C,Avg} + (0.00176) \cdot I_{C,Rms}^2$ and the average conduction power losses of the antiparallel diode can be obtained by $P_{C,Diode} = (1.76) \cdot I_{D,Avg} + (0.00126) \cdot I_{D,Rms}^2$.

Among these expressions, it is evident that the average conduction power losses of the IGBT and diode are depended on the average and RMS values of the current through the IGBT $I_{C,Avg}$, $I_{C,Rms}$ and diode $I_{D,Avg}$, $I_{D,Rms}$, respectively.

As discussed in aforementioned statement, the average and RMS currents for the conduction power loss analysis of the upper IGBT and diode with the conventional PWM algorithm for the nine-switch VSI can be determined by using the current distribution of the proposed model

given in Fig. 3 and their expressions are

$$
\begin{cases}
I_{SHA,Avg} = \dfrac{1}{2\pi} \displaystyle\int_{\phi}^{\pi+\phi} \left[i_{A,1} \dfrac{T_Z}{T_{SW}/2} + \left(i_{A,1} + i_{U,1} \right) \dfrac{T_P}{T_{SW}/2} \right] d\omega t \\[3mm]
I_{SHA,Rms}^2 = \dfrac{1}{2\pi} \displaystyle\int_{\phi}^{\pi+\phi} \left[i_{A,1}^2 \dfrac{T_Z}{T_{SW}/2} + \left(i_{A,1} + i_{U,1} \right)^2 \dfrac{T_P}{T_{SW}/2} \right] d\omega t
\end{cases}
\tag{3}
$$

and

$$
\begin{cases}
I_{DHA,Avg} = \dfrac{1}{2\pi} \displaystyle\int_{\pi+\phi}^{2\pi+\phi} \left[\left(-i_{A,1}\right) \dfrac{T_Z}{T_{SW}/2} + \left[-\left(i_{A,1} + i_{U,1} \right) \right] \dfrac{T_P}{T_{SW}/2} \right] d\omega t \\[3mm]
I_{DHA,Rms}^2 = \dfrac{1}{2\pi} \displaystyle\int_{\pi+\phi}^{2\pi+\phi} \left[\left(-i_{A,1}\right)^2 \dfrac{T_Z}{T_{SW}/2} + \left[-\left(i_{A,1} + i_{U,1} \right) \right]^2 \dfrac{T_P}{T_{SW}/2} \right] d\omega t
\end{cases}
\tag{4}
$$

respectively, where $I_{SHA,Avg}$, $I_{SHA,Rms}$ are the average and RMS currents of the upper IGBT, and $I_{DHA,Avg}$, $I_{DHA,Rms}$ are the average and RMS currents of the upper diode with the conventional PWM algorithm.

Accordingly, the average and RMS currents of the middle IGBT and diode can be formulated by

$$
\begin{cases}
I_{SMA,Avg} = \dfrac{1}{2\pi} \left[\displaystyle\int_{\phi}^{\pi+\phi} i_{U,1} \dfrac{T_P}{T_{SW}/2} d\omega t + \int_{\pi+\phi}^{2\pi+\phi} \left(-i_{A,1}\right) \dfrac{T_N}{T_{SW}/2} d\omega t \right] \\[3mm]
I_{SMA,Rms}^2 = \dfrac{1}{2\pi} \left[\displaystyle\int_{\phi}^{\pi+\phi} i_{U,1}^2 \dfrac{T_P}{T_{SW}/2} d\omega t + \int_{\pi+\phi}^{2\pi+\phi} \left(-i_{A,1}\right)^2 \dfrac{T_N}{T_{SW}/2} d\omega t \right]
\end{cases}
\tag{5}
$$

and

$$
\begin{cases}
I_{DMA,Avg} = \dfrac{1}{2\pi} \left[\displaystyle\int_{\phi}^{\pi+\phi} i_{A,1} \dfrac{T_N}{T_{SW}/2} d\omega t + \int_{\pi+\phi}^{2\pi+\phi} \left(-i_{U,1}\right) \dfrac{T_P}{T_{SW}/2} d\omega t \right] \\[3mm]
I_{DMA,Rms}^2 = \dfrac{1}{2\pi} \left[\displaystyle\int_{\phi}^{\pi+\phi} i_{A,1}^2 \dfrac{T_N}{T_{SW}/2} d\omega t + \int_{\pi+\phi}^{2\pi+\phi} \left(-i_{U,1}\right)^2 \dfrac{T_N}{T_{SW}/2} d\omega t \right]
\end{cases}
\tag{6}
$$

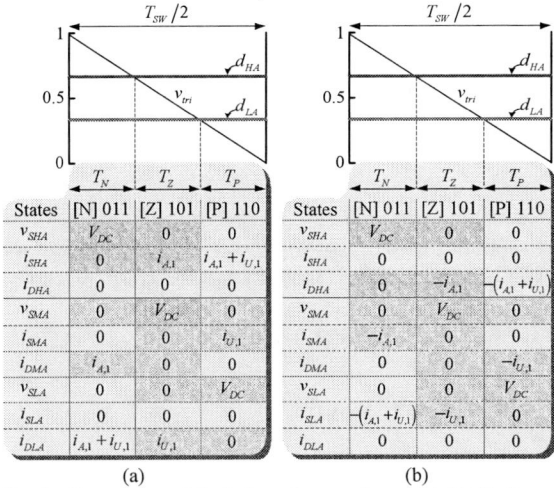

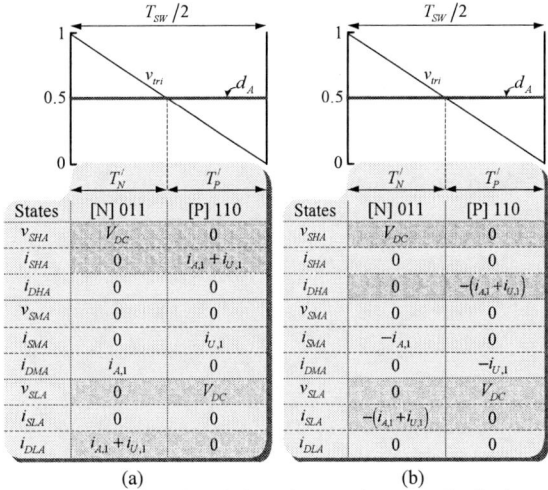

States	[N] 011	[Z] 101	[P] 110
v_{SHA}	V_{DC}	0	0
i_{SHA}	0	$i_{A,1}$	$i_{A,1}+i_{U,1}$
i_{DHA}	0	0	0
v_{SMA}	0	V_{DC}	0
i_{SMA}	0	0	$i_{U,1}$
i_{DMA}	$i_{A,1}$	0	0
v_{SLA}	0	0	V_{DC}
i_{SLA}	0	0	0
i_{DLA}	$i_{A,1}+i_{U,1}$	$i_{U,1}$	0

(a)

States	[N] 011	[Z] 101	[P] 110
v_{SHA}	V_{DC}	0	0
i_{SHA}	0	0	0
i_{DHA}	0	$-i_{A,1}$	$-(i_{A,1}+i_{U,1})$
v_{SMA}	0	V_{DC}	0
i_{SMA}	$-i_{A,1}$	0	0
i_{DMA}	0	0	$-i_{U,1}$
v_{SLA}	0	0	V_{DC}
i_{SLA}	$-(i_{A,1}+i_{U,1})$	$-i_{U,1}$	0
i_{DLA}	0	0	0

(b)

Fig. 3. Proposed model of the voltage and current distributions to analyze simultaneously the conduction and switching power losses of the nine-switch VSI with the conventional PWM algorithm [1]. (a) Mode ❶ ($0 \le \theta < \phi$). (b) Mode ❷ ($\phi \le \theta < 2\pi$).

States	[N] 011	[P] 110
v_{SHA}	V_{DC}	0
i_{SHA}	0	$i_{A,1}+i_{U,1}$
i_{DHA}	0	0
v_{SMA}	0	0
i_{SMA}	0	$i_{U,1}$
i_{DMA}	$i_{A,1}$	0
v_{SLA}	0	V_{DC}
i_{SLA}	0	0
i_{DLA}	$i_{A,1}+i_{U,1}$	0

(a)

States	[N] 011	[P] 110
v_{SHA}	V_{DC}	0
i_{SHA}	0	0
i_{DHA}	0	$-(i_{A,1}+i_{U,1})$
v_{SMA}	0	0
i_{SMA}	$-i_{A,1}$	0
i_{DMA}	0	$-i_{U,1}$
v_{SLA}	0	V_{DC}
i_{SLA}	$-(i_{A,1}+i_{U,1})$	0
i_{DLA}	0	0

(b)

Fig. 4. Proposed model of the voltage and current distributions to analyze simultaneously the conduction and switching power losses of the nine-switch VSI with the proposed PWM algorithm. (a) Mode ❶ ($0 \le \theta < \phi$). (b) Mode ❷ ($\phi \le \theta < 2\pi$).

respectively, where $I_{SMA,Avg}$, $I_{SMA,Rms}$ are the average and RMS currents of the middle IGBT, and $I_{DMA,Avg}$, $I_{DMA,Rms}$ are the average and RMS currents of the middle diode with the conventional PWM algorithm.

Subsequently, the average and RMS currents of the lower IGBT and diode can be calculated by

$$\begin{cases} I_{SLA,Avg} = \dfrac{1}{2\pi}\int_{\pi+\phi}^{2\pi+\phi}\left[\left[-\left(i_{A,1}+i_{U,1}\right)\right]\dfrac{T_N}{T_{SW}/2}+\left(-i_{U,1}\right)\dfrac{T_Z}{T_{SW}/2}\right]d\omega t \\ I_{SLA,Rms}^2 = \dfrac{1}{2\pi}\int_{\pi+\phi}^{2\pi+\phi}\left[\left[-\left(i_{A,1}+i_{U,1}\right)\right]^2\dfrac{T_N}{T_{SW}/2}+\left(-i_{U,1}\right)^2\dfrac{T_Z}{T_{SW}/2}\right]d\omega t \end{cases} \quad (7)$$

and

$$\begin{cases} I_{DLA,Avg} = \dfrac{1}{2\pi}\int_{\phi}^{\pi+\phi}\left[\left(i_{A,1}+i_{U,1}\right)\dfrac{T_N}{T_{SW}/2}+i_{U,1}\dfrac{T_Z}{T_{SW}/2}\right]d\omega t \\ I_{DLA,Rms}^2 = \dfrac{1}{2\pi}\int_{\phi}^{\pi+\phi}\left[\left(i_{A,1}+i_{U,1}\right)^2\dfrac{T_N}{T_{SW}/2}+i_{U,1}^2\dfrac{T_Z}{T_{SW}/2}\right]d\omega t \end{cases} \quad (8)$$

respectively, where $I_{SLA,Avg}$, $I_{SLA,Rms}$ are the average and RMS currents of the lower IGBT, and $I_{DLA,Avg}$, $I_{DLA,Rms}$ are the average and RMS currents of the lower diode with the conventional PWM algorithm.

In respect of (3) to (8), T_N, T_Z, T_P are the time intervals of the switching states [N], [Z], and [P], respectively, for the conventional PWM algorithm. T_{SW} is the switching time period, where $T_{SW}=1/f_{SW}=2(T_N+T_Z+T_P)$. By taking the concept of trigonometry, they can be found by deriving in terms of the duty cycles as $T_N=(1-d_{HA})T_{SW}/2$, $T_Z=(d_{HA}-d_{LA})T_{SW}/2$, and $T_P=d_{LA}T_{SW}/2$. Substituting (1) and (2) to (3) to (8), the average and RMS currents of the upper, middle, and lower IGBTs and diodes can be determined, as well as the average conduction power losses of all the IGBTs and

diodes of the nine-switch VSI with the conventional PWM algorithm.

Similarly, in part of the proposed PWM algorithm, the current distribution of the proposed model given in Fig. 4 is employed to find out the average and RMS currents of the upper IGBT and diode as expressed by

$$\begin{cases} I_{SHA,Avg}' = \dfrac{1}{2\pi}\int_{\phi}^{\pi+\phi}\left[\left(i_{A,1}+i_{U,1}\right)\dfrac{T_P'}{T_{SW}/2}\right]d\omega t \\ I_{SHA,Rms}'^2 = \dfrac{1}{2\pi}\int_{\phi}^{\pi+\phi}\left[\left(i_{A,1}+i_{U,1}\right)^2\dfrac{T_P'}{T_{SW}/2}\right]d\omega t \end{cases} \quad (9)$$

and

$$\begin{cases} I_{DHA,Avg}' = \dfrac{1}{2\pi}\int_{\pi+\phi}^{2\pi+\phi}\left[-\left(i_{A,1}+i_{U,1}\right)\right]\dfrac{T_P'}{T_{SW}/2}\,d\omega t \\ I_{DHA,Rms}'^2 = \dfrac{1}{2\pi}\int_{\pi+\phi}^{2\pi+\phi}\left[-\left(i_{A,1}+i_{U,1}\right)\right]^2\dfrac{T_P'}{T_{SW}/2}\,d\omega t \end{cases} \quad (10)$$

respectively, where $I_{SHA,Avg}'$, $I_{SHA,Rms}'$ are the average and RMS currents of the upper IGBT, and $I_{DHA,Avg}'$, $I_{DHA,Rms}'$ are the average and RMS currents of the upper diode with the proposed PWM algorithm.

Then, the average and RMS currents of the middle IGBT and diode can be calculated by

$$\begin{cases} I_{SMA,Avg}' = \dfrac{1}{2\pi}\left[\int_{\phi}^{\pi+\phi}i_{U,1}\dfrac{T_P'}{T_{SW}/2}d\omega t+\int_{\pi+\phi}^{2\pi+\phi}\left(-i_{A,1}\right)\dfrac{T_N'}{T_{SW}/2}d\omega t\right] \\ I_{SMA,Rms}'^2 = \dfrac{1}{2\pi}\left[\int_{\phi}^{\pi+\phi}i_{U,1}^2\dfrac{T_P'}{T_{SW}/2}d\omega t+\int_{\pi+\phi}^{2\pi+\phi}\left(-i_{A,1}\right)^2\dfrac{T_N'}{T_{SW}/2}d\omega t\right] \end{cases} \quad (11)$$

and

$$\begin{cases} I_{DMA,Avg}' = \dfrac{1}{2\pi}\left[\int_{\phi}^{\pi+\phi}i_{A,1}\dfrac{T_N'}{T_{SW}/2}d\omega t+\int_{\pi+\phi}^{2\pi+\phi}\left(-i_{U,1}\right)\dfrac{T_P'}{T_{SW}/2}d\omega t\right] \\ I_{DMA,Rms}'^2 = \dfrac{1}{2\pi}\left[\int_{\phi}^{\pi+\phi}i_{A,1}^2\dfrac{T_N'}{T_{SW}/2}d\omega t+\int_{\pi+\phi}^{2\pi+\phi}\left(-i_{U,1}\right)^2\dfrac{T_P'}{T_{SW}/2}d\omega t\right] \end{cases} \quad (12)$$

respectively, where $I'_{SMA,Avg}$, $I'_{SMA,Rms}$ are the average and RMS currents of the middle IGBT, and $I'_{DHA,Avg}$, $I'_{DHA,Rms}$ are the average and RMS currents of the middle diode with the proposed PWM algorithm.

Also, the average and RMS currents of the lower IGBT and diode can be calculated by

$$
\begin{cases}
I'_{SLA,Avg} = \dfrac{1}{2\pi} \displaystyle\int_{\pi+\phi}^{2\pi+\phi} \left[-(i_{A,1}+i_{U,1}) \right] \dfrac{T'_N}{T_{SW}/2} d\omega t \\[4mm]
I'^2_{SLA,Rms} = \dfrac{1}{2\pi} \displaystyle\int_{\pi+\phi}^{2\pi+\phi} \left[-(i_{A,1}+i_{U,1}) \right]^2 \dfrac{T'_N}{T_{SW}/2} d\omega t
\end{cases}
\tag{13}
$$

and

$$
\begin{cases}
I'_{DLA,Avg} = \dfrac{1}{2\pi} \displaystyle\int_{\phi}^{\pi+\phi} \left[(i_{A,1}+i_{U,1}) \dfrac{T'_N}{T_{SW}/2} \right] d\omega t \\[4mm]
I'^2_{DLA,Rms} = \dfrac{1}{2\pi} \displaystyle\int_{\phi}^{\pi+\phi} \left[(i_{A,1}+i_{U,1})^2 \dfrac{T'_N}{T_{SW}/2} \right] d\omega t
\end{cases}
\tag{14}
$$

respectively, where $I'_{SLA,Avg}$, $I'_{SLA,Rms}$ are the average and RMS currents of the lower IGBT, and $I'_{DLA,Avg}$, $I'_{DLA,Rms}$ are the average and RMS currents of the lower diode with the proposed PWM algorithm.

Based on (9) to (14), T'_N, T'_P are the time intervals of the switching states [N] and [P], respectively, for the proposed PWM algorithm. T_{SW} is the switching time period, where $T_{SW} = 1/f_{SW} = 2(T'_N + T'_P)$. By taking the concept of trigonometry, they can be found by deriving in terms of the duty cycles as $T'_N = (1-d_A)T_{SW}/2$ and $T'_P = d_A T_{SW}/2$. Replacing (1) and (2) to (9) to (14), the average and RMS currents of the upper, middle, and lower IGBTs and diodes can be determined, after that the average conduction power losses of all the IGBTs and diodes of the nine-switch VSI with the proposed PWM algorithm are also complete to be figured out.

Considering (3) to (14), the reduction of the conduction power loss with the proposed PWM algorithm is due to the lower average and RMS values of the currents for the upper and lower switches compared to the conventional PWM algorithm.

B. Switching Power Losses

The factors of the proposed switching power loss analysis in this paper are depended on as follows:

1) The characteristic curves of the switching energy loss from the datasheet of the switching device, as based on the characteristic-curve-based technique.
2) The instantaneous voltage across and current through the switching device during turns on and off or implication that changing the switching status, as based on the voltage and current distributions of the proposed models (see Figs. 3 and 4).

3) The PWM algorithm and the switching frequency, as based on the number of pulse and pattern of the duty cycle for the switching device.

For the characteristic-curve-based technique as [9] and the same switching device, the viewpoint of the instantaneous voltage across and current through the switching device during the changing of the switching states shown in the proposed voltage and current distribution models (see Fig. 3) can lead to the calculations of the switching energy losses of the upper IGBT and diode with the conventional PWM algorithm for the nine-switch VSI as follow:

$$
\begin{cases}
E_{SHA} = \displaystyle\sum_{t_1}^{t'_1} \left[1.1902 + 0.0061 i_{A,1} + 1.2791 \times 10^{-6} i^2_{A,1} \right] \\[4mm]
E_{DHA} = \displaystyle\sum_{t_2}^{t'_2} \left[0.375 - 0.00261 i_{A,1} - 5.25 \times 10^{-7} i^2_{A,1} \right]
\end{cases}
\tag{15}
$$

respectively, where E_{SHA}, E_{DHA} are the switching energy losses of the upper IGBT and diode with the conventional PWM algorithm.

Following the same procedure, the switching energy losses of the middle IGBT and diode can be formed by

$$
\begin{cases}
E_{SMA} = \displaystyle\sum_{t_1}^{t'_1} \left[1.1902 + 0.0061 i_{U,1} + 1.2791 \times 10^{-6} i^2_{U,1} \right] \\[3mm]
\qquad\quad + \displaystyle\sum_{t_2}^{t'_2} \left[1.1902 - 0.0061 i_{A,1} + 1.2791 \times 10^{-6} i^2_{A,1} \right] \\[3mm]
E_{DMA} = \displaystyle\sum_{t_1}^{t'_1} \left[0.375 + 0.00261 i_{A,1} - 5.25 \times 10^{-7} i^2_{A,1} \right] \\[3mm]
\qquad\quad + \displaystyle\sum_{t_2}^{t'_2} \left[0.375 - 0.00261 i_{U,1} - 5.25 \times 10^{-7} i^2_{U,1} \right]
\end{cases}
\tag{16}
$$

respectively, where E_{SMA}, E_{DMA} are the switching energy losses of the middle IGBT and diode with the conventional PWM algorithm.

Methodically, the switching energy losses of the lower IGBT and diode can be calculated by

$$
\begin{cases}
E_{SLA} = \displaystyle\sum_{t_2}^{t'_2} \left[1.1902 - 0.0061 i_{U,1} + 1.2791 \times 10^{-6} i^2_{U,1} \right] \\[4mm]
E_{DLA} = \displaystyle\sum_{t_1}^{t'_1} \left[0.375 + 0.00261 i_{U,1} - 5.25 \times 10^{-7} i^2_{U,1} \right]
\end{cases}
\tag{17}
$$

respectively, where E_{SLA}, E_{DLA} are the switching energy losses of the lower IGBT and diode with the conventional PWM algorithm.

According to (15) to (17), t_1, t'_1 is the time at the first and last pulses considered in mode ❶. t_2, t'_2 is the time at the first and last pulses considered in mode ❷. Dividing (15) to (17) by the fundamental time period and substituting (1) and (2) to them, the average switching power losses of all the IGBTs and diodes of the nine-switch VSI with the conventional PWM algorithm can be finished to be computed.

In the same way, to analysis of the switching power losses of the nine-switch VSI with the proposed PWM algorithm, the viewpoint of the instantaneous voltage across and current through the switching device during the changing of the switching states shown in the proposed voltage and current distribution models (see Fig. 4) results in the determinations of the switching energy losses of the upper IGBT and diode with the proposed PWM algorithm for the nine-switch VSI as follow:

$$\begin{cases} E'_{SHA} = \sum_{t_1}^{t'_1} \left[1.1902 + 0.0061\left(i_{A,1} + i_{U,1}\right) + 1.2791 \times 10^{-6}\left(i_{A,1} + i_{U,1}\right)^2 \right] \\ E'_{DHA} = \sum_{t_2}^{t'_2} \left[0.375 - 0.00261\left(i_{A,1} + i_{U,1}\right) - 5.25 \times 10^{-7}\left(i_{A,1} + i_{U,1}\right)^2 \right] \end{cases} \quad (18)$$

respectively, where E'_{SHA}, E'_{DHA} are the switching energy losses of the upper IGBT and diode with the proposed PWM algorithm.

In the next step, the switching energy losses of the middle IGBT and diode can be defined by

$$\begin{cases} E'_{SMA} = 0 \\ E'_{DMA} = 0 \end{cases} \quad (19)$$

respectively, where E'_{SMA}, E'_{DMA} are the switching energy losses of the middle IGBT and diode with the proposed PWM algorithm.

Finally, the switching energy losses of the lower IGBT and diode can be defined by

$$\begin{cases} E'_{SLA} = \sum_{t_1}^{t'_2} \left[1.1902 - 0.0061\left(i_{A,1} + i_{U,1}\right) + 1.2791 \times 10^{-6}\left(i_{A,1} + i_{U,1}\right)^2 \right] \\ E'_{DLA} = \sum_{t_1}^{t'_1} \left[0.375 + 0.00261\left(i_{A,1} + i_{U,1}\right) - 5.25 \times 10^{-7}\left(i_{A,1} + i_{U,1}\right)^2 \right] \end{cases} \quad (20)$$

respectively, where E'_{SLA}, E'_{DLA} are the switching energy losses of the lower IGBT and diode with the proposed PWM algorithm.

Also, dividing (18) to (20) by the fundamental time period and substituting (1) and (2) to them, the average switching power losses of all the IGBTs and diodes of the nine-switch VSI with the proposed PWM algorithm can be as well brought out in final.

Focusing on (15) to (20), it is pointed out that the reduction of the switching power loss in the nine-switch VSI with the proposed PWM algorithm is caused by the absence of the switching in the middle switch, as can be seen in (19), compared to the double switching of the switching frequency in the middle switch, as can be observed in (16), with the conventional PWM algorithm.

TABLE IV
SIMULATION AND EXPERIMENTAL PARAMETERS FOR VERIFICATION OF PROPOSED PWM ALGORITHM ALONG WITH PROPOSED POWER LOSS ANALYSIS METHOD

Parameters		Values	
Dc-link voltage	(V_{DC})	200	V
Fundamental frequency	(f)	50	Hz
Switching frequency	(f_{SW})	7.5	kHz
Modulation index	(m)	0.7	
Dual three-phase inductive loads (simulation)	(Z)	12.56∠30°	Ω
Dual three-phase inductive loads (experiment)	(Z)	40.49∠8.93°	Ω
Operating temperature		125	°C
IGBT module: a 5SNA 1200G450300 Hipak 4.5-kV 1.2-kA			

IV. SIMULATION AND EXPERIMENTAL RESULTS

To verify the reduction of the power losses using the proposed PWM algorithm for the nine-switch VSI in comparison to the conventional PWM algorithm [1] and investigate the validity of the proposed power loss analysis method, the simulation is carried out on the Matlab/Simulink software and the experimental prototype is also implemented with a 1-kW nine-switch VSI, where the used parameters are summarized in Table IV.

Figs. 5 and 6 show the results of the conventional PWM and the proposed PWM algorithms, respectively. The switching voltage across the middle switching device of the nine-switch VSI with the proposed PWM algorithm results in zero and no switching although there are the switching currents. This leads to the cancellation of the switching power loss in the middle switching device and the reduction of the switching power losses in the inverter from the conventional PWM algorithm, according to (15) to (20). Focusing on the switching currents, those of the conventional PWM algorithm have the $i_{A,1}$, $i_{A,1} + i_{U,1}$ components for the upper and lower switching devices and the $i_{A,1}$, $i_{U,1}$ components for the middle switching device. However, for the proposed PWM algorithm, the same components of the switching current for the middle switching device are contained, but those of the upper and lower switching devices contain only the $i_{A,1} + i_{U,1}$ component. Hereby, the conduction power loss of the middle switching device with the proposed PWM algorithm is nearby that of the conventional PWM algorithm but those of the upper and lower switching devices are reduced, corresponding to (3) to (14). From the above investigation, it is pointed out that these simulation results can well confirm the validity of the proposed technical power loss analysis and the power loss reduction performance of the proposed PWM algorithm for the nine-switch VSI.

For better verification, the zooming of two mode operations for the conventional PWM and the proposed PWM algorithms are shown in Figs. 7 and 8, respectively. These simulated waveforms are corresponded to the key principle waveforms (see Fig. 2). That means the patterns and profiles of the switching

The 2018 International Power Electronics Conference

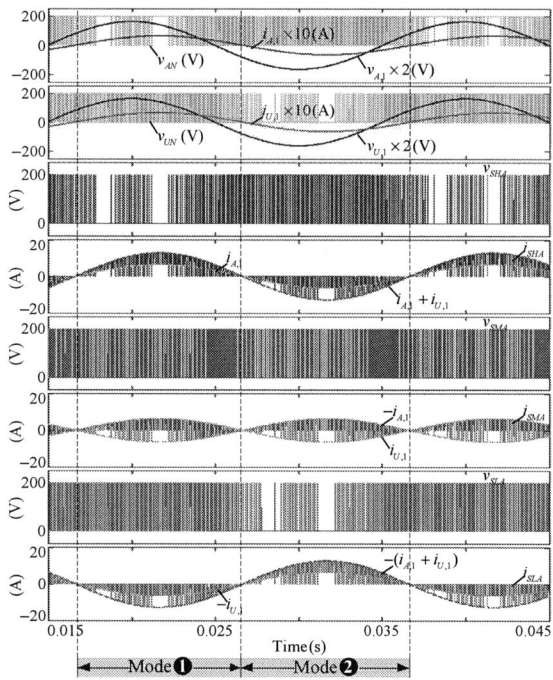

Fig. 5. Simulated waveforms of the conventional PWM algorithm [1] for the nine-switch VSI, consisting of pole voltages v_{AN}, v_{UN}, fundamental voltages $v_{A,1}$, $v_{U,1}$ and currents $i_{A,1}$, $i_{U,1}$ of upper and lower VSIs, respectively, switching voltages v_{SHA}, v_{SMA}, v_{SLA}, IGBT currents i_{SHA}, i_{SMA}, i_{SLA}, and diode currents i_{DHA}, i_{DMA}, i_{DLA} of upper, middle, and lower switching devices, respectively.

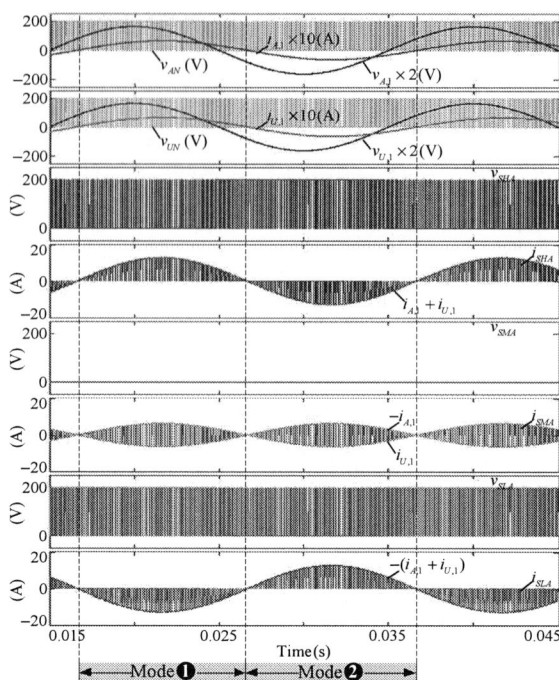

Fig. 6. Simulated waveforms of the proposed PWM algorithm for the nine-switch VSI, consisting of pole voltages v_{AN}, v_{UN}, fundamental voltages $v_{A,1}$, $v_{U,1}$ and currents $i_{A,1}$, $i_{U,1}$ of upper and lower VSIs, respectively, switching voltages v_{SHA}, v_{SMA}, v_{SLA}, IGBT currents i_{SHA}, i_{SMA}, i_{SLA}, and diode currents i_{DHA}, i_{DMA}, i_{DLA} of upper, middle, and lower switching devices, respectively.

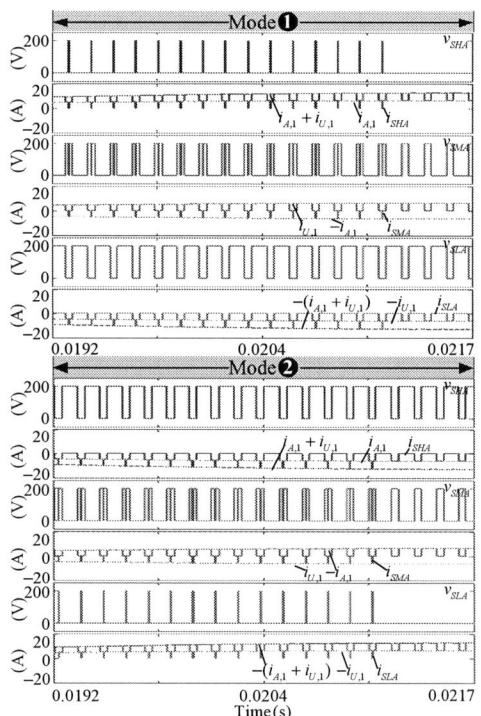

Fig. 7. Zooming in modes ❶ and ❷ of Fig. 5 to verify the proposed power loss analysis model for the conventional PWM algorithm [1].

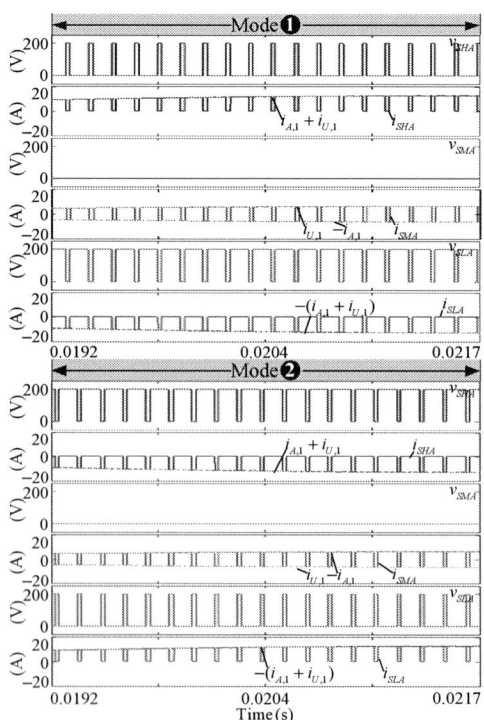

Fig. 8. Zooming in modes ❶ and ❷ of Fig. 6 to verify the proposed power loss analysis model for the proposed PWM algorithm.

2127

The 2018 International Power Electronics Conference

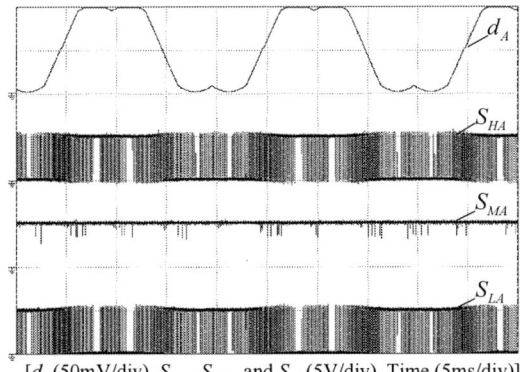

[d_A(50mV/div), S_{HA}, S_{MA}, and S_{LA}(5V/div), Time (5ms/div)]

Fig. 9. Experimental results of the duty cycle signal d_A and the corresponding gating pulse signals S_{HA}, S_{MA}, S_{LA} for the upper, middle, and lower switching devices, respectively, of the nine-switch VSI with the proposed PWM algorithm.

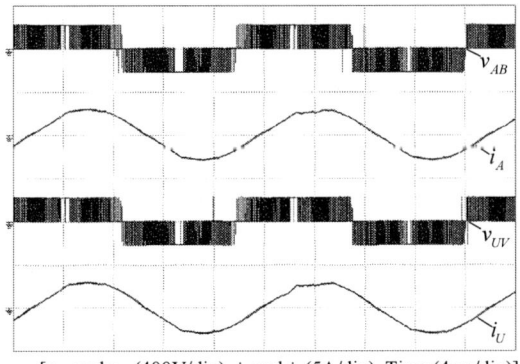

[v_{AB} and v_{UV} (400V/div), i_A and i_U (5A/div), Time (4ms/div)]

Fig. 10. Experimental results of the output line-to-line voltages v_{AB}, v_{UV} and the output phase currents i_A, i_U for the upper and lower VSIs, respectively, of the nine-switch VSI with the proposed PWM algorithm.

voltages and currents with the conventional PWM algorithm are directly in accordance with Fig. 2(a), which is consisted of three switching states ([P], [Z], and [N]). Additionally, those with the proposed PWM algorithm are also in good agreement with Fig. 2(b), which is consisted of two switching states ([P] and [N]). Therefore, this can upwards authenticate the accuracy of the proposed voltage and current distribution models corresponding to all mathematical models for analyzing the power loss reduction performance of the proposed PWM algorithm from the conventional PWM algorithm.

In real practical implementation of the proposed PWM algorithm, the lack of the switching in the middle switching device can be achieved, as shown in Fig. 9. The corresponding line-to-line output voltages and phase currents of the upper and lower VSIs, as shown in Fig. 10, perform a good stability and agreement with a symmetrical feature. Therefore, apart from the power loss reduction performance of the proposed PWM algorithm, it can also produce the maximum dc-link utilization, as equivalent to that of the standard two-level VSI.

V. CONCLUSION

This paper presents a generalized two-level PWM algorithm for modulating the nine-switch VSI. The development contributions of the proposed PWM algorithm from the conventional PWM algorithm are the simpler PWM algorithm and more reduction of the power losses, which results in the lack of the switching in the middle switching devices, in the nine-switch VSI. Therefore, in order to investigate such an attractive power loss reduction performance, a novel method of the power loss analysis is also presented, which is as well easy by basing on the characteristic-curve-based technique. With the straightforward method by synthesizing the voltage and current distribution models, the conduction and switching power losses can be simultaneously calculation in only one-stage point of view. The matching simulation results can well confirm the validity of the proposed power loss analysis method. This can further authenticate the power loss reduction performance of the proposed PWM algorithm as well. In addition, it is notable that the proposed power loss analysis method can be effectively applied for any unique/common converter topologies.

ACKNOWLEDGMENT

This work is supported in part by the Thailand Research Fund (TRF) under Grant PHD/0017/2559, and in part by the Graduate School and Faculty of Engineering, Chiang Mai University.

REFERENCES

[1] C. Liu, B. Wu, N. R. Zargari, D. Xu, and J. Wang, "A novel three-phase three-leg ac/ac converter using nine IGBTs," *IEEE Trans. Power Electron.*, vol. 24, no. 5, pp. 1151-1160, May 2009.

[2] Z. Qin, P. C. Loh, and F. Blaabjerg, "Application criteria for nine-switch power conversion systems with improved thermal performance," *IEEE Trans. Power Electron.*, vol. 30, no. 8, pp. 4608-4620, Aug. 2015.

[3] K. Ali, R. K. Surapaneni, P. Das, and S. K. Panda, "An SiC-MOSFET-based nine-switch single-stage three-phase AC-DC isolated converter," *IEEE Trans. Ind. Electron.*, vol. 64, no. 11, pp. 9083-9093, Nov. 2017.

[4] K. Ali, P. Das, and S. K. Panda, "A special application criterion of the nine-switch converter with reduced conduction loss," *IEEE Trans. Ind. Electron.*, vol. 65, no. 4, pp. 2853-2862, Apr. 2018.

[5] N. Jarutus and Y. kumsuwan, "Novel modulation strategy based on generalized two-level PWM theory for nine switch inverter with reduction of switching commutation," in *Proc. IEEE-IECON*, 2016, pp. 3529-3534.

[6] A. Stabile, C. Boccaletti, and A. J. M. Cardoso, "A power loss measurement method applied to static power converters," *IEEE Trans. Instrum. Meas.*, vol. 62, no. 2, pp. 344-352, Feb. 2013.

[7] A. M. Hava, R. J. Kerkman, and T. A. Lipo, "A high-performance generalized discontinuous PWM algorithm," *IEEE Trans. Ind. Applicat.*, vol. 34, no. 5, pp. 1059-1071, Sep./Oct. 1998.

[8] M. Bland, P. Wheeler, J. Clare, and L. Empringham, "Comparison of calculated and measured losses in direct ac-ac converters," in *Proc. IEEE-PESC*, 2001, pp. 1096-1101.

[9] A. K. Sadigh, V. Dargahi, and K. A. Corzine, "Analytical determination of conduction and switching power losses in flying-capacitor-based active neutral-point-clamped multilevel converter," *IEEE Trans. Power Electron.*, vol. 31, no. 8, pp. 5473-5494, Aug. 2016.

The 2018 International Power Electronics Conference

SiC-based three-phase Quasi-Z-Source Inverter versus the two-stage topology - a comparison

Kornel Wolski*, Mariusz Zdanowski and Jacek Rabkowski

Faculty of Electrical Engineering, Warsaw University of Technology, Warsaw, Poland

*E-mail: kornel.wolski@ee.pw.edu.pl

Abstract— The paper presents an in-depth experimental comparative study between a three-phase Quasi-Z-Source Inverter (qZSI) and a Boost Converter with Voltage-Source Inverter (BC+VSI). Both converters obtain identical output phase voltage value of 230 V RMS from DC input voltage in range of 400 V ÷ 550 V at a similar value of input inductor current ripple. The quality factors investigated are: power losses as well as both input inductor current and output voltage quality (considering both low- and high-frequency harmonics). Moreover, four variants of modulation methods are investigated. At switching frequency of 100 kHz results of experiments using SiC-based 6-kW laboratory models show that at identical conditions the BC+VSI configuration has lower losses and better quality of the output voltage, while qZSI has better quality of the input inductor current.

Keywords— SiC, qZSI, boost, comparison.

I. INTRODUCTION

Quasi-Z-Source Inverters (qZSIs), introduced in [1], are capable of boosting voltage in a single-stage configuration (Fig. 1(a)). As a result, there is one semiconductor switch less in comparison to the two-stage configuration (Fig. 1(b)) of a Boost Converter (BC) with a Voltage-Source Inverter (VSI). This suggests that total converter losses could be decreased due to lower semiconductor loss values; however, aside from number of switches, also values of currents and voltages of each switch and passive element must be considered. In [2] a comparative analysis was conducted for silicon IGBTs, which concluded that qZSI has efficiency comparable to BC+VSI and that it requires values of stored energy (size of passive elements) lower than the two-stage topology. In [3], on the other hand, the results for silicon IGBTs showed that for similar operating conditions qZSI has both better efficiency and better THD than BC+VSI. It is also pointed out that semiconductor voltage stress is a limiting factor for the single-stage converter and that it restricts the input voltage range. Aside from these two works, only comparisons between traditional Z-source inverters (ZSIs) and two-stage configurations were presented; however, due to discontinuous current in ZSI as opposed to in BC+VSI, these comparisons are not considered as comparisons between equivalent DC/AC converters.

This work is part of
National Science Centre project UMO-2015/19/B/ST7/00620.

In this work, an in-depth experimental comparison is conducted using two pairs of equivalent modulation methods and operating at several points throughout a specified input voltage range. A 6-kW SiC-based 100-kHz qZSI presented in [4] is tested at similar input/output operating points as an equivalent BC+VSI converter. Section II presents the main assumptions behind the comparison, while Section III shows laboratory setup. Section IV, V and VI are devoted to output voltage quality, input inductor current quality and losses, respectively. Section VII sums up and concludes the paper.

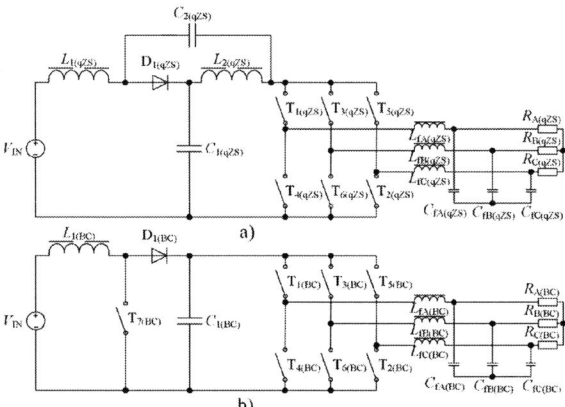

Fig. 1. Topologies of the compared converters: a) qZSI; b) BC+VSI.

II. INITIAL ASSUMPTIONS

Assumed boundary operating conditions result both from the aforementioned restriction of the input voltage range for qZSI due to semiconductor stress and from typical output parameters of photovoltaic modules. Considering the application of 1200-V SiC MOSFETs, input voltage V_{IN} of each of the DC-AC converters (qZSI or BC+VSI) is set to be in range of 400 V ÷ 550 V. Lower input voltage would cause higher voltage stress on the switches and higher input voltage would cause the qZSI to operate as a VSI (without shoot-through (ST) states) [1]. Impedance-network inductances L_1 and L_2, the input voltage range and input power as well as the number of ST states per period specify relative input inductor current ripple [4], which is kept the same in both

2129

of the compared configurations (<20%). Fig. 1 presents the considered single- and two-stage topologies along with the nomenclature used throughout the paper. Table I shows a summary of number of active and passive switches, which constitutes a first degree of indication of power density of the topologies.

Table II shows the considered operating parameters of the two DC-AC converters. The values from Table II necessitate the following parameter values of the elements from Fig. 1, which provides the second degree of indication of power density of the topologies – see Table III.

To obtain a fair comparison, identical values of gate resistance were applied and equivalent modulation methods for qZSI and VSI were used. The qZSI was designed in [4] to operate nominally with Minimum Switching Number (MSN) modulation method and the VSI-equivalent of it is a continuous variant of Space-Vector Modulation Pulse Width Modulation (SVMPWM). The adequacy of SVMPWM stems from the fact that ST states are seen from the perspective of the 3-phase bridge as zero states, and thus both methods have zero states outside and zero states inside of the active states, as shown in Fig. 2. Furthermore, two methods with a decreased amount of consecutive states were compared: the Minimum-Hard-Switching-Number (MHSN) method [5] as well as the Discontinuous Pulse-Width Modulation method in DPWMMAX variant [6]. Switching patterns for these can be observed in Fig. 3. for the first half of the first 60-degree sector of modulation.

All methods operate at maximum achievable value of modulation index M, which for the qZSI methods is limited by the value of boost factor B and for the VSI methods is simply equal to 1. Therefore, for qZSI only B was adjusted, while for the BC+VSI - the duty cycle coefficient D.

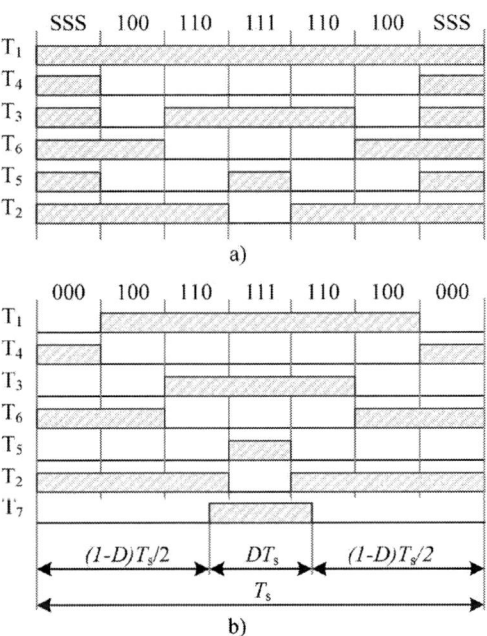

Fig. 2. Modulation schemes for a) MSN and b) SVMPWM methods in a first modulation sector.

TABLE I
NUMBERS OF ELEMENTS

Component	qZSI	BC+VSI
Transistor	6	7
Diode	1	1
Inductor	5	4
Capacitor	5	4

TABLE II
DESIGN PARAMETERS

Parameter	Value	Unit
V_{IN}	$400 \div 550$	V
$V_{A(RMS)}$	230	V
P_{OUT}	6	kW
f_S	100	kHz
$\Delta i_{L1\%\_max}$	20	%

TABLE III
SELECTED ACTIVE AND PASSIVE ELEMENTS

Element	qZSI	BC+VSI
$T_1 \div T_6$	C2M0080120D	C2M0080120D
T_7	-	C2M0080120D
D_1	C4D20120D	C4D20120D
L_1	250 μH	250 μH
L_2	250 μH	-
$L_{fA} \div L_{fC}$	100 μH	100 μH
C_1	30 μF	80 μF
C_2	30 μF	-
$C_{fA} \div C_{fC}$	0.47 μF	0.47 μF

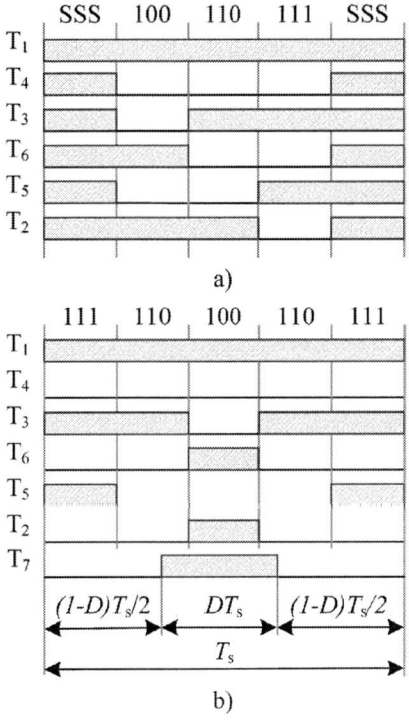

Fig. 3. Low-loss modulation schemes for a) MHSN and b) DPWMMAX methods in a first half of a first modulation sector.

III. EXPERIMENTAL SETUP

Both of the DC/AC topologies shown in Fig. 1 were realized with SiC diodes and MOSFETs to operate at nominal switching frequency of 100 kHz. High frequency design allowed to obtain relatively low dimensions for 6-kW converters [4]. Specifically for this paper a SiC-based 6-kW boost converter was built in a geometry compatible with an already available 6-kW SiC-based VSI. A 250-µH inductor based on the same core and litz wire as in case of qZSI [4] was utilized in the BC. Similar cooling solutions based on heatsink and a fan were applied for qZSI and for BC+VSI.

In case of both converters input voltage V_{IN} in the range of 400 V ÷ 550 V was obtained from three parallel-connected TDK-Lambda Gen600-5.5 power supplies and an auxiliary 24-V power supply was used for powering fans and gate drivers. TMS320F28335 digital signal processor was used for implementation of the four modulation methods and for generation of control signal for BC. A load consisting of three 26-Ω resistors was utilized.

As for measurements, Yokogawa WT1800 power analyser was connected to the outputs of GEN600-5.5 voltage sources and to the inputs of the star-connected three-phase load resistor configuration. Furthermore, oscilloscope DPO 5034B was utilized for measurements of input inductor current using TCP0030A current probe and output voltage across the A-phase load resistance using P5200 high-voltage differential probe.

The experiments were conducted by gradually increasing the input voltage throughout the established range of 400 V ÷ 550 V by 25-V steps and by tuning B and D parameters to obtain appropriate value of AC output voltage of 230 V RMS and thus output power in 26-Ω resistive load equal to 6 kW - see Table IV.

Duty cycle and modulation index for qZSI were calculated as follows:

$$D_{qZSI} = \frac{B-1}{2B} \tag{1}$$

$$M_{qZSI} = \frac{0.5(B+1)}{B} \tag{2}$$

TABLE IV
OPERATION PARAMETERS

Parameter	Value	Unit
V_{IN}	400 ÷ 550	V
B	1.99 ÷ 1.12	-
D_{qZSI}	0.249 ÷ 0.054	-
$V_{INV\text{-}qZSI}$	796 ÷ 616	V
M_{qZSI}	0.751 ÷ 0.946	-
D_{BC+VSI}	0.335 ÷ 0.059	-
$V_{INV\text{-}BC+VSI}$	580 ÷ 600	V
M_{BC+VSI}	1	-
$V_{A(RMS)}$	230	V
P_{OUT}	6	kW
$R_A ÷ R_C$	26	Ω
R_g	25	Ω
t_{dt}	330	ns
f_S	100	kHz
$\Delta i_{L\% \ max}$	20	%

Gate resistance values in both converters were set to 25 Ω, which necessitated deadtime value t_{dt} in both cases equal to 330 ns. Deadtime was also applied in qZSI, because omitting it due to the topology's immunity to phase short circuits would lead to increased switching losses [5]. The SiC MOSFETs were switching with frequency of 100 kHz, which allowed to keep the inductor ripple values below 20%.

As a result, two equivalent DC/AC converters were obtained with equally continuous input current, same inductor architecture and same values both at the input and in the output filter, same semiconductor switches and two very similar sets of modulation methods to control them.

IV. EXPERIMENTS – QUALITY OF OUTPUT VOLTAGE

A. qZSI (MSN) vs BC+VSI (SVMPWM)

Initially the comparison was conducted between the two equivalent modulation methods shown in Fig. 2 – MSN and SVMPWM, applied in qZSI and BC+VSI, respectively. To confirm proper operation of both DC/AC converters input and output waveforms were recorded and observed. Fig. 4 and Fig. 5 show the waveforms for the condition of $V_{IN} = 400$ V. The general conclusion from those figures is that the shapes of currents in inductors of boosting stages correlate with the shapes of the output sine waves.

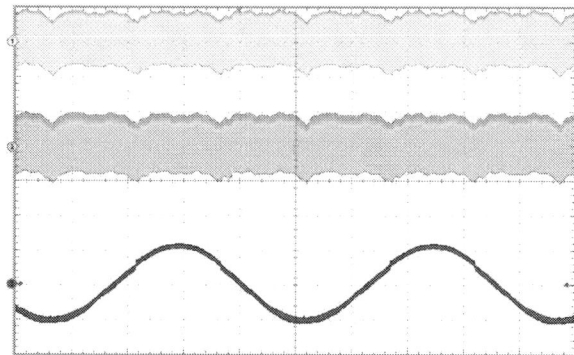

Fig. 4. Ripple of inductor current i_{L1} (yellow) and i_{L2} (light blue) waveforms and output voltage v_A (purple) waveform with qZSI operating with MSN at $V_{IN} = 400$ V. 4 A/div, 300 V/div, 5 ms/div.

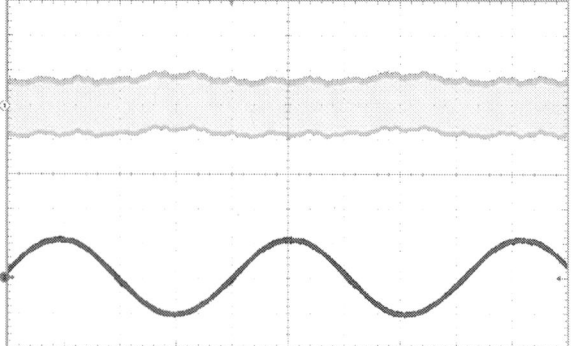

Fig. 5. Ripple of inductor current i_{L1} (yellow) waveform and output voltage v_A (purple) waveform with BC+VSI operating with SVMPWM at $V_{IN} = 400$ V. 4 A/div, 300 V/div, 5 ms/div.

In Fig. 4 it can be seen that the inductor current values of qZSI suddenly decrease when there is a discontinuity in the output sine wave, which occurs during entering and leaving a two-sector constant on-state of T_1 (Fig. 2(a)) and is additionally made worse by high value of deadtime t_{dt}. In case of BC+VSI in Fig. 5 the output sine wave does not feature distortions; however, the input inductor current still follows the general trend of the sine wave, showing a visible 50-Hz harmonic, as both systems operate in open-loop control. From this comparison results a superiority of BC+VSI with respect to qZSI. The difference stems from the fact that, even though from the perspective of output filter the state sequences shown in Fig. 2 are equivalent, from perspective of time they are not. This is the case firstly because the first "zero" state in qZSI has constant time and in BC+VSI it has not, and secondly, because the remaining state times also vary due to different values of M in qZSI and BC+VSI. Based on the waveforms seen in Fig. 4 and Fig. 5, FFT were realized and THD values were calculated using MATLAB based on the following formula:

$$THD_{v_A} - \frac{\sqrt{\sum_{n=2}^{nyquist} V^2_{(n)}}}{V_1} \qquad (3)$$

Formula (3) signifies that output voltage harmonics were calculated with respect to the output voltage fundamental. The upper limit of considered harmonic content was set at half of the sampling frequency in order to include all the vital switching-related harmonics. Based on (3), the following THD_{vA} values were obtained for the considered range of V_{IN}, as shown in Fig. 6.

Fig. 6 confirms what was stated based on Fig. 4 and Fig. 5 regarding shapes of the sine waves at $V_{IN} = 400$ V and shows that the situation looks similar throughout the entire input voltage range - BC+VSI with SVMPWM has better output waveform quality than qZSI with MSN.

Continuity in output voltage waveforms is directly related to repetitiveness of control pulse scheme. Constantly turned on transistor T_1 in sectors VI and I and constant-time shoot-through states (see Fig. 2) negatively influence that repetitiveness in case of qZSI.

B. qZSI (MHSN) vs BC+VSI (DPWMMAX)

Same tests were also conducted for the other set of considered modulation methods: MHSN and DPWMMAX (Fig. 3). Input/output waveforms for condition $V_{IN} = 400$ V for both of these methods are presented in Fig. 7 and Fig. 8. As in case of MSN and SVMPWM, the differences between MHSN and DPWMMAX regarding the quality of output voltage waveforms are clearly visible. However, in this case both waveforms feature certain distortions. The disruptions in MHSN waveform appear during changing of the switching sequence in a middle of a sector, as explained in [5], and those in DPWMMAX result from not switching T_1 for two sectors in a row. More high-frequency content can also be observed in case of MHSN. All these aspects were evaluated through calculations using (3) and the results are shown in Fig. 9. In that figure it can be seen that the total harmonic distortion is in fact much worse in case of MHSN at $V_{IN} -$ 400 V. The situation gradually improves with increase of V_{IN}, until the THD values become essentially equal.

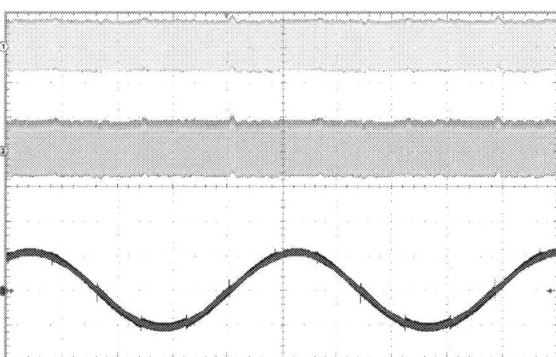

Fig. 7. Ripple of inductor current i_{L1} (yellow) and i_{L2} (light blue) waveforms and output voltage v_A (purple) waveform with qZSI operating with MHSN at $V_{IN} = 400$ V. 4 A/div, 300 V/div, 5 ms/div.

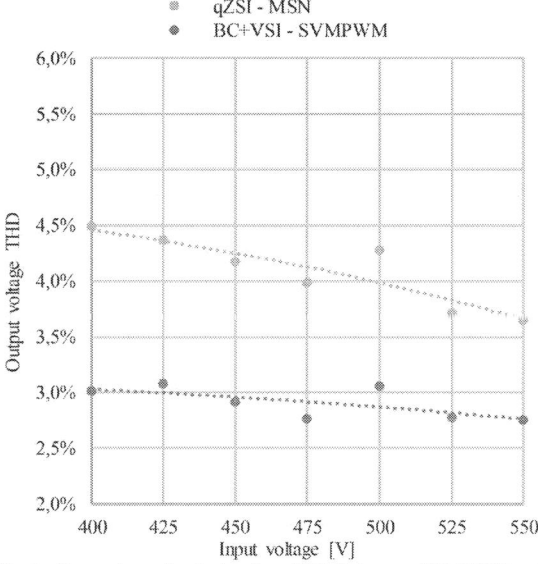

Fig. 6. Comparison of output voltage THD between qZSI (MSN) and BC+VSI (SVMPWM).

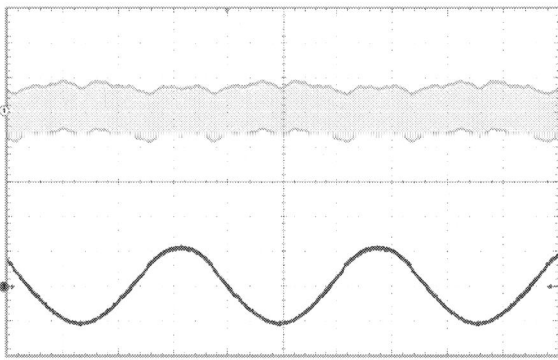

Fig. 8. Ripple of inductor current i_{L1} (yellow) waveform and output voltage v_A (purple) waveform with BC+VSI operating with DPWMMAX at $V_{IN} = 400$ V. 4 A/div, 300 V/div, 5 ms/div.

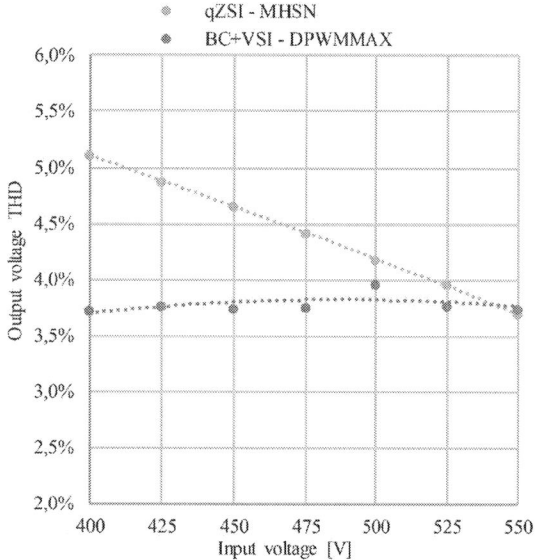

Fig. 9. Comparison of output voltage THD between qZSI (MHSN) and BC+VSI (DPWMMAX).

V. EXPERIMENTS – QUALITY OF INPUT INDUCTOR CURRENT

A. qZSI (MSN) vs BC+VSI (SVMPWM)

Currents from Fig. 4 and Fig. 5 were also magnified and showed in Fig. 10 and Fig. 11. From cursor measurements peak-to-peak values of the triangle waves were obtained and the following relative ripple calculations were conducted for the input inductors of qZSI and BC+VSI:

$$\Delta i_{L1\%.qZS} = \frac{i_{L1p-p.qZS}}{2I_{IN.qZS}} \qquad (4)$$

$$\Delta i_{L1\%.BC} = \frac{i_{L1p-p.BC}}{2I_{IN.BC}} \qquad (5)$$

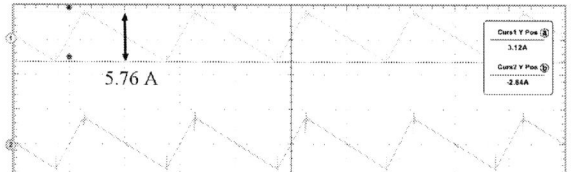

Fig. 10. High-frequency ripple of inductor current i_{L1} (yellow) and i_{L2} (light blue) waveforms with qZSI operating with MSN at V_{IN} = 400 V. 4 A/div, 5 μs/div.

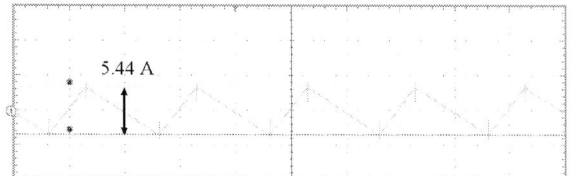

Fig. 11. High-frequency ripple of inductor current i_{L1} waveform with BC+VSI operating with SVMPWM at V_{IN} = 400 V. 4 A/div, 5 μs/div.

In accordance with the design data (Table II), maximum values of the parameters calculated in (4) and (5) for MSN and SVMPWM are 18% and the relative ripple value is similar in qZSI and BC+VSI.

High-frequency harmonics in an inductor current cause increased core losses due to the fact that hysteresis losses are proportional to frequency and eddy current losses in the core are proportional to frequency squared. High values of current harmonics at such high frequencies additionally increase the core losses by causing high values of induction swing ΔB at these frequencies. In case of a triangle wave the highest impact has the first harmonic, which can be calculated based on the magnified waveforms shown in Fig. 10 and Fig. 11.

$$i_{L1p-p.qZS(1)} = \frac{2i_{L1p-p.qZS}}{D_{qZS}^2(\frac{1}{D_{qZS}}-1)\pi^2} \qquad (6)$$

$$i_{L1p-p.BC(1)} = \frac{2i_{L1p-p.BC}}{D^2(\frac{1}{D}-1)\pi^2} \qquad (7)$$

These values change with input voltage; therefore, they are shown in a chart in Fig. 12. From Fig. 12 it can be seen that there is a difference in the peak-to-peak value of the first harmonic, even though the relative ripple in both cases is kept approximately equal. This difference is due to different duty cycle values in both converters and it can be as high as 1.3 A. It means that for the same core dimensions (which is the case here) core losses are higher when using qZSI than when using BC+VSI. The core losses are also higher in qZSI due to the fact that it features two inductors in the boosting stage, while BC+VSI - only one.

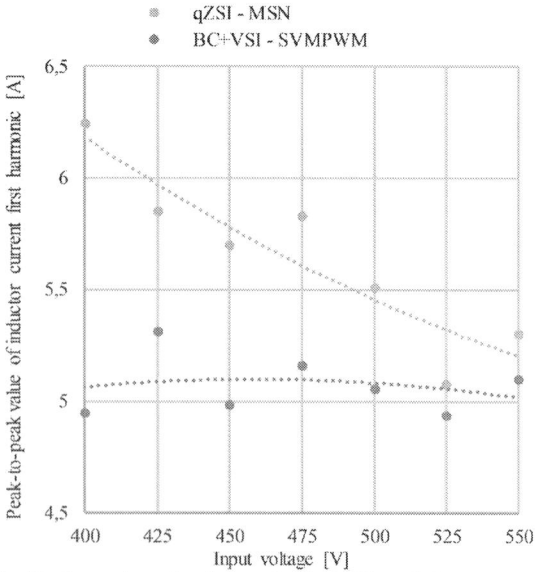

Fig. 12. Comparison of peak-to-peak values of input inductor current between qZSI (MSN) and BC+VSI (SVMPWM).

2133

As shown in Fig. 12, higher value of the current first harmonic characterize qZSI throughout the entire V_{IN} range, though the difference is diminishing as duty coefficients at V_{IN} = 550 V become almost equal (see Table IV).

In a similar manner to calculating a quality indicator for output voltage waveforms a quality indicator can be calculated for the input currents shown in Fig. 4 and Fig. 5. The following formula applies:

$$k_{T-L1} = \frac{\sqrt{\sum_{n=1}^{nyquist} I_{(n)RMS}^2}}{I_{DC}} \qquad (8)$$

This ripple factor is a measure of harmonic content in the input inductor with relation to the average value of input current. Same as with the voltage THD, harmonics up to the Nyquist frequency are considered. For the entire considered range of V_{IN} the appropriate chart is presented in Fig. 13. It is shown that even though input inductor current has worse quality in qZSI at V_{IN} = 400 V (as also visible in Fig. 4 and Fig. 5), for the most part of the input voltage the situation is reversed and SVMPWM features worse waveform quality. There seems to be a trade-off between the quality of waveforms at the input and at the output of the converters.

B. qZSI (MHSN) vs BC+VSI (DPWMMAX)

As it was the case with MSN and SVMPWM, also here the magnified high-frequency inductor current waveforms are shown along with the ripple value measurements – see Fig. 14 and Fig. 15. Based on (4) and (5) it was confirmed that the relative ripple values did not exceed the design constraint for qZSI and BC+VSI (17% and 16%, respectively).

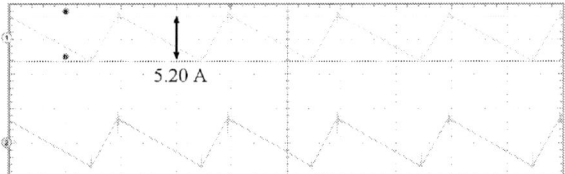

Fig. 14. High-frequency ripple of inductor current i_{L1} (yellow) and i_{L2} (light blue) waveforms with qZSI operating with MHSN at V_{IN} = 400 V. 4 A/div, 5 μs/div.

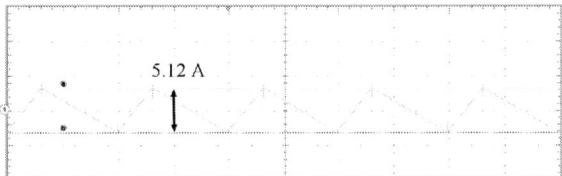

Fig. 15. High-frequency ripple of inductor current i_{L1} waveform with BC+VSI operating with DPWMMAX at V_{IN} = 400 V. 4 A/div, 5 μs/div.

Afterwards, using formulas (6) and (7), the peak-to-peak values of the first harmonic were calculated for entire range of V_{IN} and the results were presented in Fig. 16. A very similar dependency was observed as in the case of MSN and SVMPWM, which suggests that also in case of MHSN and DPWMMAX loss difference due to core losses may decrease with the value of input voltage V_{IN}.

As the final step in the analysis of current waveforms quality ripple factor for MHSN and DPWMMAX was calculated according to (8) and presented in Fig. 17. While looking at Fig. 7 and Fig. 8 alone it is hard to see why the ripple factor would be essentially the same in the two cases at V_{IN} = 400 V, it is also necessary to consider Fig. 16. At V_{IN} = 400 V in case of MHSN there is negligible low-frequency harmonic content, but significant high-frequency content. Meanwhile, in case of DPWMMAX there is some low-frequency and some high-frequency content.

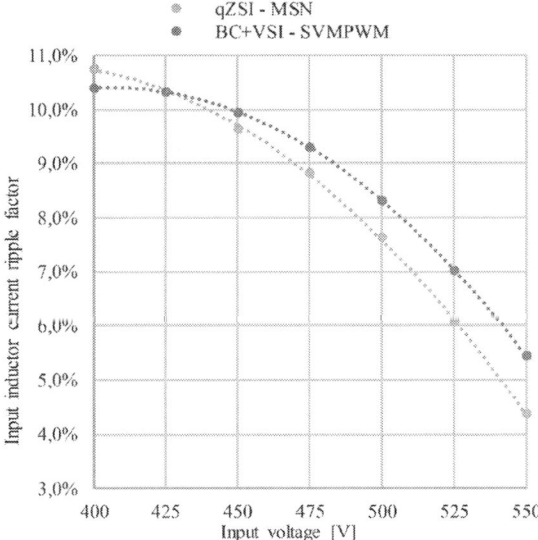

Fig. 13. Comparison of input inductor current ripple factor k_{T-L1} between qZSI (MSN) and BC+VSI (SVMPWM).

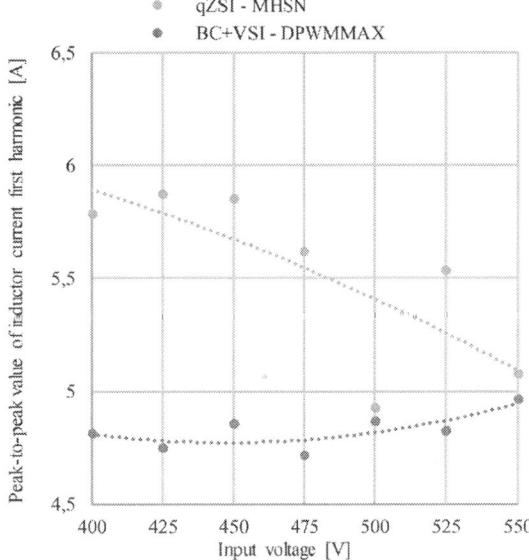

Fig. 16. Comparison of peak-to-peak values of input inductor current between qZSI (MHSN) and BC+VSI (DPWMMAX).

The 2018 International Power Electronics Conference

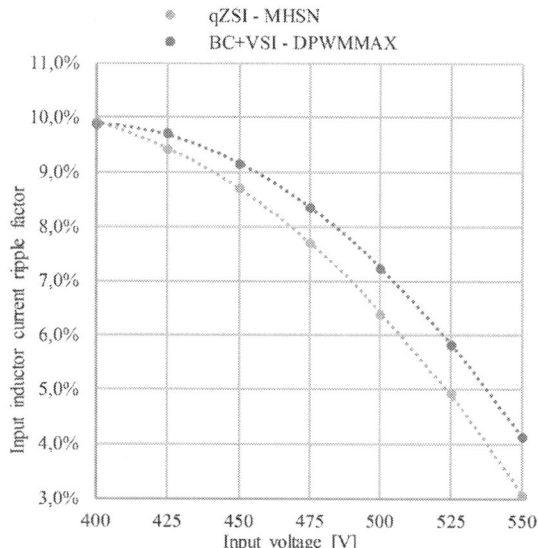

Fig. 17. Comparison of input inductor current ripple factor k_{T-L1} between qZSI (MHSN) and BC+VSI (DPWMMAX).

As V_{IN} increases, duty cycles of the two converters converge and the difference between high-frequency contents of both methods decreases, while DPWMMAX is still left with higher low-frequency harmonic content. This is why MHSN has better input inductor waveform quality, as indicated by the lower value of ripple factor.

VI. EXPERIMENTS – POWER LOSSES

A. qZSI (MSN) vs BC+VSI (SVMPWM)

The final stage of comparison of both converter topologies was to determine total losses of DC/AC conversion. For that Yokogawa WT1800 was used and the results for MSN and SVMPWM are provided in Fig. 18 and Fig. 19. It is clearly visible that at this operating point the VSI+BC topology has lower losses (dP) or, in other words, higher efficiency (Ef-cy). The difference is 90.4 W, which costs qZSI 1.42% of efficiency. Results for the entire range of the input voltage are presented in form of a chart in Fig. 20, where loss values are shown in percentage form, relative to the input power. Superiority of the BC+VSI topology is preserved throughout the whole range of input voltage. Aside from the aforementioned inductor losses, there are two reasons for the dominance of BC+VSI over qZSI. Firstly, the quasi-Z-source inverter operates at higher commutation voltage values of the bridge switches and, secondly, it features higher conduction losses of those switches during shoot-through states. The first one can be observed as V_{INV} parameter in Table IV, while for the second one it must be considered that during time of shoot-through state (Fig. 2(a), Fig. 3(a)) double the converter input inductor current is added to the bridge transistors currents, while in the equivalent methods (Fig. 2(b), Fig. 3(b)) only currents related to the zero state cause conduction losses.

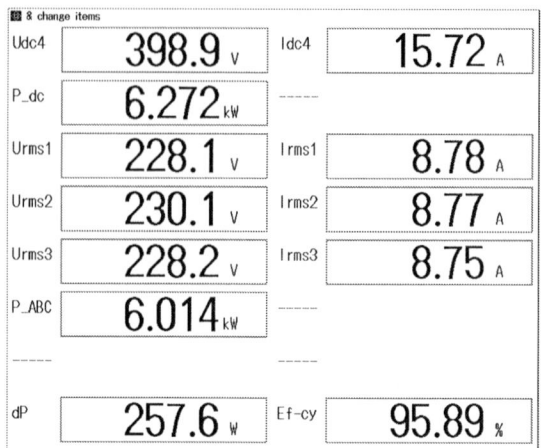

Fig. 18. Experimental results for qZSI with MSN modulation method at $V_{IN} \approx 400$ V.

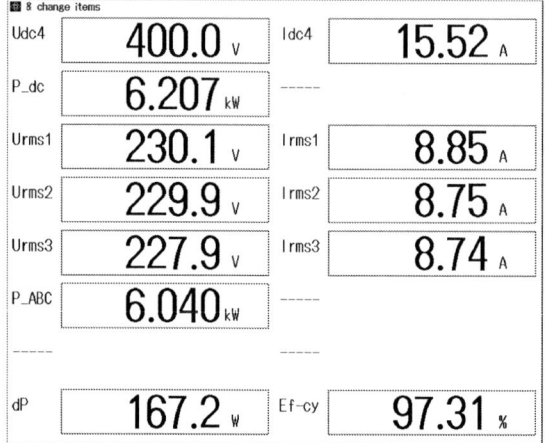

Fig. 19. Experimental results for BC+VSI with SVMPWM modulation method at $V_{IN} \approx 400$ V.

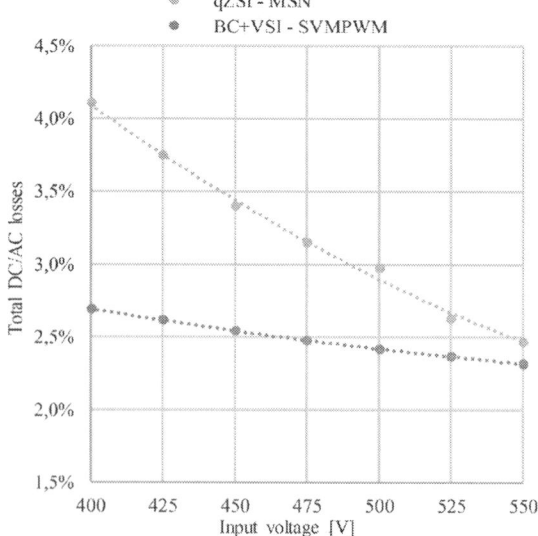

Fig. 20. Loss comparison between qZSI (MSN) and BC+VSI (SVMPWM) in the entire considered range of input voltage.

2135

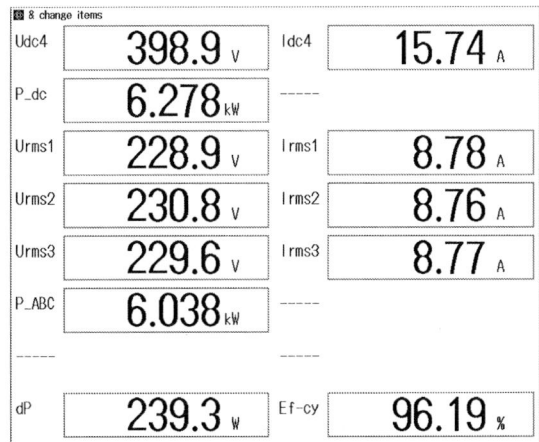

Fig. 21. Experimental results for qZSI with MHSN modulation method at $V_{IN} \approx 400$ V.

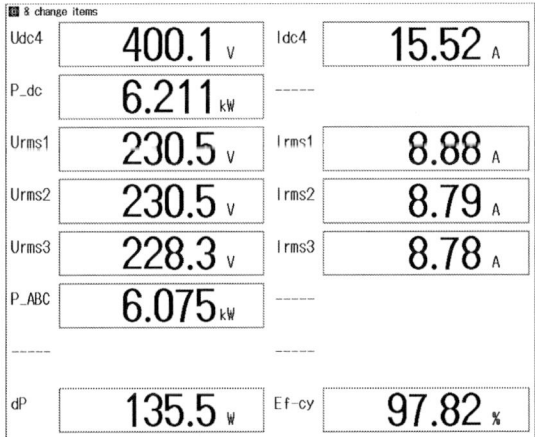

Fig. 22. Experimental results for BC+VSI with DPWMMAX modulation method at $V_{IN} \approx 400$ V.

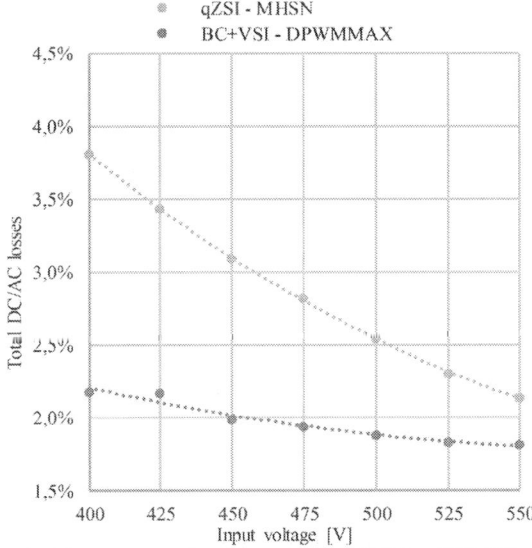

Fig. 23. Loss comparison between qZSI (MHSN) and BC+VSI (DPWMMAX) in the entire considered range of input voltage.

B. qZSI (MHSN) vs BC+VSI (DPWMMAX)

Loss values were also compared for the two methods with decreased amounts of switching losses – MHSN for qZSI and DPWMMAX for BC+VSI. The readings from Yokogawa WT1800 shown in Fig. 21 and Fig. 22 confirm that at $V_{IN} = 400$ V the tendency from the first set of modulation methods remains and the advantage of BC+VSI over qZSI is significant. The same can be said about the whole V_{IN} range, as shown in Fig. 23, where both plots from Fig. 20 seem to just have evenly lowered by deduction of certain amount of switching losses.

VII. CONCLUSIONS

Presented measurement results show that for a 6-kW three-phase DC/AC converter VSI+BC is a better solution than qZSI. With the same output LC filter THD of the output voltage is 2.8% ÷ 3.8% for VSI+BC, while 3.7% ÷ 5.1% for qZSI. With identical switching frequency, matching inductors and the same semiconductor switches total losses of VSI+BC are in range of 1.8% ÷ 2.7%, while for qZSI — 2.1% ÷ 4.1%. Even the minimization of amount of E_{on}-type transitions in qZSI by the use of MHSN method does not allow to obtain better efficiency than BC+VSI due to the fact that it is mainly the core losses in inductors, the commutation voltage value and the conduction losses in semiconductors that make qZSI non-competitive. In one of the three compared categories qZSI turns out to be a more favorable solution – it has better quality of input inductor current with less low-frequency harmonic content. However, the most important application-wise are the two categories in which qZSI performs worse than BC+VSI; therefore, using the single-stage three-phase DC/AC topology instead of the two-stage topology is not recommended, unless other factors (e.g. reliability) or different operating conditions are considered.

REFERENCES

[1] J. Anderson and F. Z. Peng, "Four quasi-Z-Source inverters," *2008 IEEE Power Electronics Specialists Conference*, Rhodes, 2008, pp. 2743-2749.

[2] A. Battiston et al., "Comparison Criteria for Electric Traction System Using Z-Source/Quasi Z-Source Inverter and Conventional Architectures," in *IEEE Journal of Emerging and Selected Topics in Power Electronics*, vol. 2, no. 3, pp. 467-476, Sept. 2014.

[3] A. Ayad, S. Hanafiah and R. Kennel, "A Comparison of Quasi-Z-Source Inverter and Traditional Two-Stage Inverter for Photovoltaic Application," *Proceedings of PCIM Europe 2015; International Exhibition and Conference for Power Electronics, Intelligent Motion, Renewable Energy and Energy Management*, Nuremberg, Germany, 2015, pp. 1-8.

[4] M. Zdanowski, D. Peftitsis, S. Piasecki and J. Rabkowski, "On the Design Process of a 6-kVA Quasi-Z-inverter Employing SiC Power Devices," in *IEEE Transactions on Power Electronics*, vol. 31, no. 11, pp. 7499-7508, Nov. 2016.

[5] K. Wolski, P. Majtczak and J. Rabkowski, "Minimum-Hard-Switching-Number (MHSN) Modulation Method for High-Frequency SiC-Based Impedance-Source Inverters," in *IEEE Transactions on Industrial Electronics*, vol. PP, no. 99, pp. 1-1.

[6] D. G. Holmes, T. A. Lipo, "Zero Space Vector Placement Modulation Strategies," in *Pulse Width Modulation for Power Converters: Principles and Practice*, Wiley-IEEE Press, 2003.

DC-side Circuit Implementation
of a Three-phase Inverter
for Balancing Phase-leg Capacitor Currents

Takashi Hirao[1]*, Keiji Wada[1]*, Toshihisa Shimizu[1]

1 Department of Electrical and Electronic Engineering, Tokyo Metropolitan University, Hachioji, Japan
*Email: thirao@mem.iee.or.jp, kj-wada@tmu.ac.jp

Abstract—**This paper proposes methods of balancing capacitor currents between phase legs in three-phase inverter circuits by changing the DC-side connection material or structure. The single-phase-leg unit consists of a half-bridge power module and DC-link capacitor that is connected directly to the module. Owing to this structure, it can be realized when there is a low stray inductance between the half-bridge module and DC-link capacitor. In the case of connecting a three-phase inverter, the stray inductance between the phase legs and the DC-link capacitor may lead to an imbalance in each capacitor current owing to DC-side resonance phenomena. The capacitor current imbalance may be due to the relationship between the switching and resonance frequencies. In order to balance the capacitor currents, this paper proposes two current balancing methods; applying (1) a graphite bus bar to increase the dumping resistance, and (2) a DC-side delta connection to geometrically balance the three phase legs. The experimental results obtained using a SiC MOSFET rated at 300 V, 4.6 A verify the suppression effectiveness of the DC-link capacitor current imbalance.**

Keywords—*Current imbalance, DC-link capacitor, resonance, stray inductance.*

I. INTRODUCTION

There have been studies aimed at improving of the passive components employed for power converter circuits, such as inductors and capacitors for next-generation circuits. Specifically, one of the key components of power-converter circuits is the DC-link capacitor, whose volume is an important factor in actual circuit implementation. Although the rules for determining capacitance are application specific, the required capacitance depends on the allowable voltage ripple. In applications where the low-voltage frequency ripple does not significantly affect the performance, the voltage ripple can be reduced by increasing the switching frequency. So a higher switching frequency can reduce the required DC capacitance. It has been reported that the capacitance of the DC-link capacitor can be reduced by increasing the switching frequency up to 100 kHz with a SiC device [1]. Normally, the DC voltage is maintained at a constant value, but the capacitor current waveform depends on the switching frequency, modulation scheme, and output power. In addition, the rms value of ripple current at high frequencies cannot be reduced by increasing the switching frequency [2], [3]. Recently, many studies focused on capacitor losses and reliabilities for power-converter circuits [4]–[11]. For

example, the influence of ripple current on reliability has been investigated [12]. A large ripple current that exceeds the rating of the capacitor may reduce its reliability and life time.

In the case involving the use of silicon IGBTs, a three-phase inverter structure with three single-phase-leg units has been used for motor-drive applications and uninterruptible power supplies (UPSs) [13]–[16]. The single-phase-leg unit consists of a half-bridge power module and a DC-link capacitor. Recently, even in the case of a SiC power device, three single-phase-leg units have been used, as shown in Fig. 1 [18]. Because the DC-link capacitor is connected close to the half-bridge module, the stray inductance between them can be minimized during circuit implementation. The DC-side stray inductance of the circuit may affect the surge voltage and switching loss [17]. The managed stray inductance contributes to the high-speed switching of SiC devices. However, there is a stray inductance on the DC-side circuit between the units, and it may lead to a resonance phenomenon with the DC-link capacitors for switching frequencies of over several tens of kHz [19], [20]. Furthermore, the resonance phenomenon may cause a current imbalance between the phase-leg units when the stray inductance between each phase-leg is unbalanced because of the difference of the resonance frequencies inside the DC-side circuit. Although the rms current of the DC-link capacitor does not depend on the switching frequency, it may increase because of both the resonance and imbalance phenomena. The large capacitor current may reduce the reliability of the capacitor [12].

The purpose of this paper is to present balancing methods of the DC-link capacitor current between phase-leg units of a three-phase pulse-width modulation (PWM) inverter. In an actual inverter circuit, an internal impedance of a DC power supply or a battery should be considered. However, in the case of high-switching operation, the switching frequency may be close to the resonant frequency of the inverter circuit. As a result, the internal impedance of the DC power supply is much larger than the internal impedance of the inverter circuit. In this paper, the DC power supply is assumed to be sufficiently larger than the internal impedance of the inverter circuit, and resonance phenomenon in the inverter circuit is analyzed. A graphite bus bar is applied to mitigate the resonance phenomena. In addition, a DC-

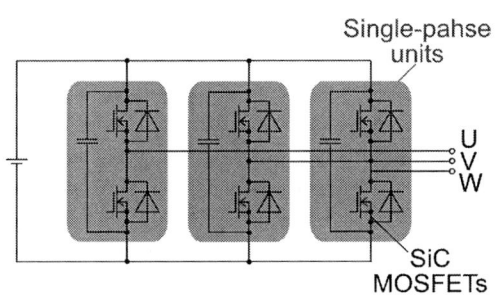

Fig. 1. Three-phase inverter structure with single-phase-leg units consisting of half-bridge power semiconductor modules and DC-link capacitors.

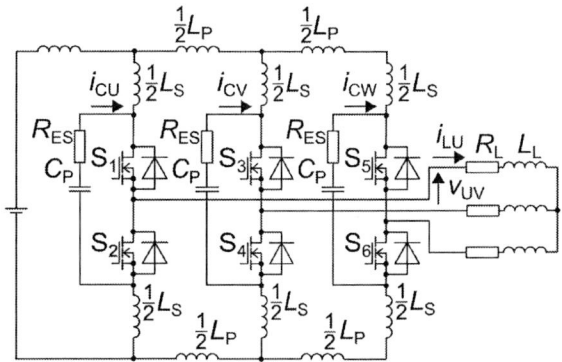

Fig. 3. Circuit configuration of conventional circuit implementation of three-phase inverter with three single-phase-leg units.

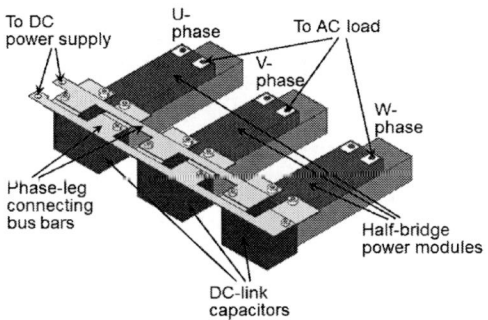

Fig. 2. Conventional circuit implementation of three-phase inverter with three single-phase-leg units.

side delta connection is also proposed to balance the three phase legs geometrically. It should be noted that the delta-type structure is not an AC-side but DC-side bus bar. The experimental results at 300 V and 4.6 A verify that the capacitor current imbalance between the phase-leg units can be suppressed by applying these methods.

II. CAPACITOR-CURRENT IMBALANCE PHENOMENON

Fig. 2 shows a conventional circuit of the three-phase inverter implementation of three single-phase-leg units with half-bridge power modules and DC-link capacitors. The units are connected using bus bars. In Fig. 2, the DC-link capacitor is connected to the nearest half-bridge power module, and they form a single-phase-leg unit. Phase-leg connecting bus bars connect the DC-side of the units. In order to reproduce it using a compact experimental set up, individual bus bars of the P- and N-side were used instead of the laminate bus bar.

Fig. 3 shows the circuit configuration of the three-phase inverter with three single-phase-leg units. In Fig. 3, S_1 - S_6 are MOSFETs, and C_P on each phase-leg is a DC-link capacitor. R_{ES} is their equivalent series resistance (ESR). There is a stray inductance L_S in series with each phase leg in the bus bar; there is also a stray inductance

L_P between each phase leg. These stray inductances are shown as $\frac{1}{2}L_S$ and $\frac{1}{2}L_P$, respectively, because the stray inductances are assumed to be equal for the P- and N-sides.

In order to analyze the DC-link capacitor current in each phase-leg, an equivalent circuit of the three-phase inverter is developed, as shown in Fig. 4(a). The stray inductances L_S and L_P are arranged in a ladder shape. Here, i_U, i_V and i_W show the pulse current sources with a switching frequency component of the drain current of MOSFETs S_1–S_6. The sinusoidal PWM (SPWM) is applied to control S_1–S_6. Fig. 5 shows the waveforms of i_U, i_V and i_W during the switching period. The cycle of i_U, i_V, and i_W is T_{SW}, which is the same as the switching period of S_1–S_6. $T_{SW} = 1/f_{SW}$ (f_{SW} is the switching frequency). The duty ratios of i_U, i_V and i_W are D_U, D_V, and D_W, respectively, which are the same as the switching duty ratios of S_1, S_3 and S_5, respectively. The peak-to-peak amplitudes of i_U, i_V, and i_W is equal to the load currents i_{LU}, i_{LV}, and i_{LW}, respectively. The k-th harmonic components of the switching frequency are expressed using the Fourier series of a pulse waveform

$$
\begin{bmatrix} i_U(k) \\ i_V(k) \\ i_W(k) \end{bmatrix} =
$$

$$
\begin{bmatrix} A_1(k, \omega_L t) & A_2(k, \omega_L t) \\ A_1(k, \omega_L t - \frac{2}{3}\pi) & A_2(k, \omega_L t - \frac{2}{3}\pi) \\ A_1(k, \omega_L t - \frac{4}{3}\pi) & A_2(k, \omega_L t - \frac{4}{3}\pi) \end{bmatrix}
\begin{bmatrix} \cos \omega_{SW} kt \\ \sin \omega_{SW} kt \end{bmatrix},
$$

(1)

where

$$
A_1(k, \theta) = \frac{\sin 2\pi k D(\theta)}{\pi k} i_L(\theta), \tag{2}
$$

$$
A_2(k, \theta) = \frac{\cos(2\pi k D(\theta)) - 1}{\pi k} i_L(\theta), \tag{3}
$$

$$
D(\theta) = \frac{1}{2}\{M \sin(2\pi f_L t - \theta) + 1\}, \tag{4}
$$

$$
i_L(\theta) = \sqrt{2} I_L \sin(2\pi f_L t - \theta - \phi) \tag{5}
$$

$\omega_{SW} = 2\pi f_{SW}$, $\omega_L = 2\pi f_L$ (f_L is load frequency), θ is the phase angle at the fundamental frequency of the load

The 2018 International Power Electronics Conference

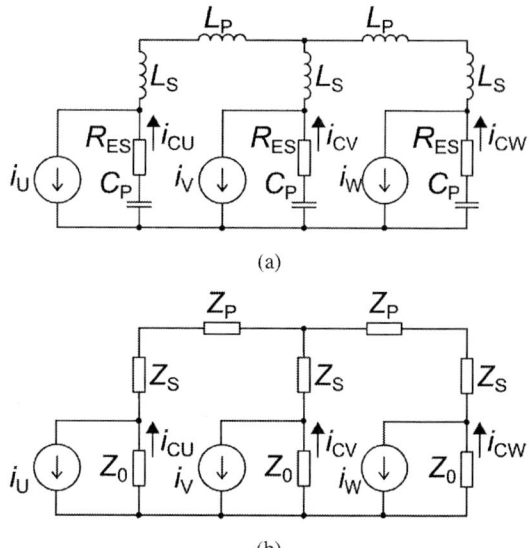

(a)

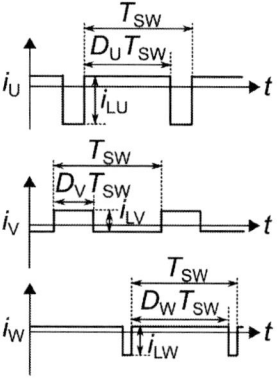

(b)

Fig. 4. Equivalent circuit focusing on high-frequency capacitor currents i_{CU}, i_{CV} and i_{CW} for conventional circuit implementation. (a) Ladder connection, and (b) simplified equivalent circuit.

Fig. 5. Waveforms of high-frequency current sources i_U, i_V and i_W.

(R_L, L_L), and I_L is the rms value of the load current, which is determined by R_L and L_L. Here, M is the modulation index, and ϕ is the phase difference at which the power factor is $\cos\phi$. $D(\theta)$ is the same as D_U, D_V, and D_W, and $i_L(\theta)$ is the same as i_{LU}, i_{LV}, and i_{LW}, as shown in Fig. 5.

Z_0, Z_S, and Z_P are defined as follows:

$$Z_0 = R_{ES} - j\frac{1}{k\omega_{SW}C_P}, \tag{6}$$

$$Z_S = jk\omega_{SW}L_S, \tag{7}$$

$$Z_P = jk\omega_{SW}L_P. \tag{8}$$

Fig. 4(b) shows the equivalent circuit using Z_0, Z_S, and Z_P. Because the equivalent circuit shown in Fig. 4(b) includes three current sources I_U, I_V, and I_W, the analysis was carried out using the superposition principle, i.e., the U-phase capacitor current I_{CU} is expressed by

the sum of the currents that constitute the single current source ($I_{CUU} + I_{CUV} + I_{CUW}$). The same applies to the V-phase capacitor current I_{CV} and the W-phase capacitor current I_{CW}. I_{CUU}, which assumes I_U as the single current source in the circuit, where I_V and I_W are considered as open circuits, is given by the following equation:

$$I_{CUU} = \frac{1}{Z_A}\left(Z_M + \frac{Z_0 Z_P}{Z_N}\right)I_U, \tag{9}$$

where

$$Z_A = 3(Z_0 + Z_S) + Z_P, \tag{10}$$

$$Z_M = Z_0 + 3Z_S + Z_P, \tag{11}$$

$$Z_N = Z_0 + Z_S + Z_P. \tag{12}$$

In the same way as I_{CUU}, I_{CUV}, which considers I_V as the single current source, and I_{CUW}, which considers I_W as the single current source, can be obtained. The sum of these values is the U-phase capacitor current I_{CU} (= $I_{CUU} + I_{CUV} + I_{CUW}$). The V-phase capacitor current I_{CV} and W-phase capacitor current I_{CW} can also be obtained by the following equation:

$$\begin{bmatrix} I_{CU} \\ I_{CV} \\ I_{CW} \end{bmatrix} = \begin{bmatrix} Z_{UU} & Z_{UV} & Z_{UW} \\ Z_{VU} & Z_{VV} & Z_{VU} \\ Z_{UU} & Z_{UV} & Z_{UW} \end{bmatrix} \begin{bmatrix} I_U \\ I_V \\ I_W \end{bmatrix}, \tag{13}$$

$$Z_{UU} = \frac{1}{Z_A}\left(Z_M + \frac{Z_0 Z_P}{Z_N}\right), \tag{14}$$

$$Z_{UV} = \frac{Z_0}{Z_A}, \tag{15}$$

$$Z_{UW} = \frac{Z_0(Z_0 + Z_S)}{Z_A Z_N}, \tag{16}$$

$$Z_{VU} = \frac{Z_0}{Z_A}, \tag{17}$$

$$Z_{VV} = \frac{Z_M}{Z_A}. \tag{18}$$

By using (1) and (13), the capacitor current considering the DC-side resonance can be calculated.

Fig. 6 shows the analytical results of the relationship between switching frequency and DC-link capacitor currents based on conventional circuit implementation. The analytical condition is that the input voltage is 300 V, load current 4.6 A, and load frequency 50 Hz. The capacitance C_P of the DC-link capacitor is set to 50 μF. The equivalent series resistance R_{ES} on each DC-link capacitor is set to 12 mΩ. The stray inductance of the bus bar is as follows: $L_S = 16.8$ nH and $L_P = 91.5$ nH. Because U- and W-phases are symmetrical positions, I_{CU} and I_{CW} are equal. The blue and red lines show the rms values I_{CU} and I_{CV} of the DC-link capacitor currents on the U- and V-phase legs, respectively. Because of the DC-side resonance phenomena between L_S, L_P, and C_P, the capacitor currents I_{CU} and I_{CV} increase. In addition, because the stray inductances L_{UW} and L_{UV} are different, there is a current imbalance between phase legs at switching frequencies from 10 kHz to 100 kHz. The resonant frequencies f_{rUW} and f_{rUV} are given in the

2139

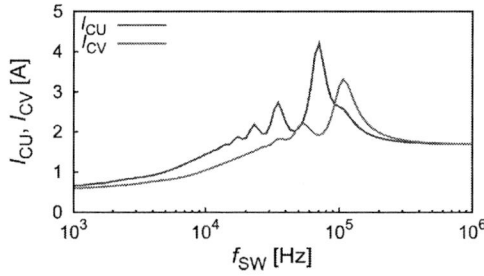

Fig. 6. Analytical switching frequency dependence of capacitor current in case of conventional circuit implementation.

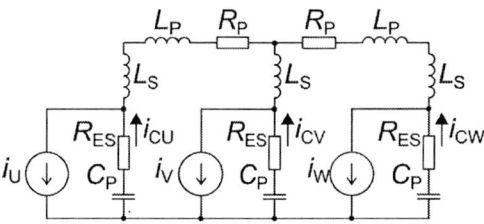

Fig. 7. Equivalent circuit focusing on high-frequency capacitor currents i_{CU}, i_{CV} and i_{CW} using graphite bus bar.

following equations:

$$f_{\mathrm{rUW}} = \frac{1}{2\pi\sqrt{L_{\mathrm{UW}}\left(\frac{1}{2}C_{\mathrm{P}}\right)}} = 68.4\,\mathrm{kHz}, \quad (19)$$

$$f_{\mathrm{rUV}} = \frac{1}{2\pi\sqrt{L_{\mathrm{UV}}\left(\frac{1}{2}C_{\mathrm{P}}\right)}} = 90.0\,\mathrm{kHz}. \quad (20)$$

where f_{rUW} is the joint inductance between the U- and W-phases, and f_{rUV} is the joint inductance between the U- and V-phases. For the imbalance to occur, f_{SW} is set to 70 kHz, where the current difference between I_{CU} and I_{CV} is a maximum. At the switching frequency, $I_{\mathrm{CU}} = 4.20$ A and $I_{\mathrm{CV}} = 1.92$ A, that is, a difference of 2.2 times was confirmed.

III. PRINCIPLE OF BALANCING METHODS

This paper proposes two current balancing methods; applying (1) a graphite bus bar, and (2) a delta-type structure. Principle of these methods are explained in this section.

A. Graphite Bus bar

In order to balance the currents on the DC-link capacitors, an increase in the damping resistance of the DC-side circuit was investigated. Because the electrical resistivity of graphite is 10^3 times that of copper, the use of graphite in a portion of the bus bar is proposed to increase the series resistance. Fig. 7 shows the equivalent circuit considering the bus bar resistance. In this equivalent circuit, the bus bar resistance R_{P} is added to the circuit shown in Fig. 4(a). In this case, (8) is replaced by the following equation:

$$Z_{\mathrm{P}} = R_{\mathrm{P}} + jk\omega_{\mathrm{SW}}L_{\mathrm{P}}. \quad (21)$$

Because R_{P} acts as a damping resistance, the capacitor current imbalance between phase legs may be suppressed.

B. Delta-type Structure

Another proposed approach to balancing the capacitor current between phase legs involves changing the bus bar structure. Fig. 8 shows the circuit implementation of a three-phase inverter with a delta-type structure of three single-phase-leg units. An additional bus bar directly

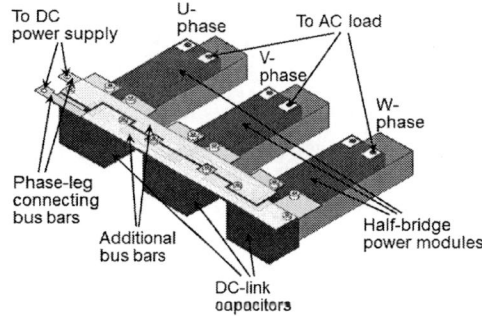

Fig. 8. Circuit implementation of three-phase inverter with delta-type structure of three single-phase-leg units.

connects the U- and W-phase legs, compared with those in Figs. 2 and 3.

Fig. 9 shows the circuit configuration of a three-phase inverter with a delta-type structure. The MOSFETs S_1–S_6 and the DC-link capacitors C_{P} are same as Fig. 3. L_{S}, L_{P} and L_{D} are the stray inductances, and L_{P} and L_{D} constitute the DC-side delta connection. As with L_{S} and L_{P}, the stray inductance is given as $\frac{1}{2}L_{\mathrm{D}}$ because the stray inductance is equal for the P- and N-sides. In this case, the stray inductances are balanced between the U-V, V-W, and U-W phases; then, the connection method is expected to balance the capacitor current.

As with conventional circuit implementation, the DC-link capacitor current on each phase leg is analyzed using a high-frequency equivalent circuit. Fig. 10(a) shows a high-frequency equivalent circuit using a delta-type structure. L_{D} and L_{P} constitute a delta connection. Here, i_{U}, i_{V} and i_{W} show the pulse current sources that equal the high-frequency components caused by S_1–S_6 as with the conventional circuit implementation. In order to analyze the capacitor current, the equivalent circuit is converted to a star connection, as shown in Fig. 10(b). Z_{D}, Z_{a} and Z_{b} are defined as follows:

$$Z_{\mathrm{D}} = jk\omega_{\mathrm{SW}}L_{\mathrm{D}}, \quad (22)$$

$$Z_{\mathrm{a}} = \frac{Z_{\mathrm{P}}Z_{\mathrm{D}}}{2Z_{\mathrm{P}} + Z_{\mathrm{D}}} + Z_{\mathrm{S}}, \quad (23)$$

$$Z_{\mathrm{b}} = \frac{Z_{\mathrm{P}}^2}{2Z_{\mathrm{P}} + Z_{\mathrm{D}}} + Z_{\mathrm{S}}. \quad (24)$$

The capacitor current I_{CU}, I_{CV} and I_{CW} on each phase

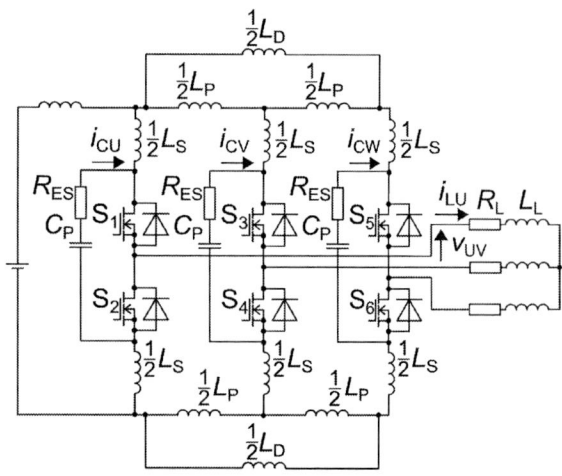

Fig. 9. Circuit configuration of three-phase inverter with delta-type structure of three single-phase-leg units.

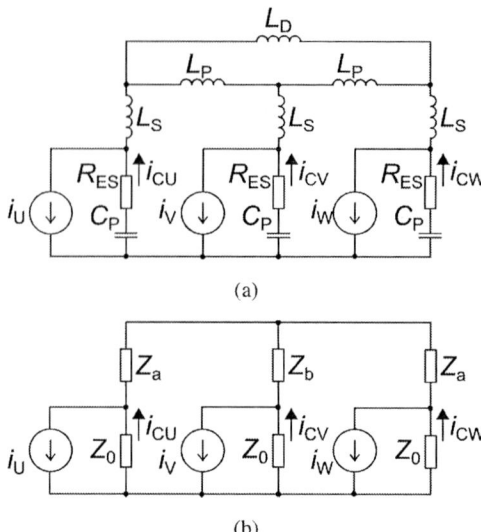

(a)

(b)

Fig. 10. Equivalent circuit focusing on high-frequency capacitor currents i_{CU}, i_{CV}, and i_{CW} using delta-type structure. (a) Delta connection, and (b) conversion to star connection.

leg can be obtained by the following equation:

$$\begin{bmatrix} I_{\mathrm{CU}} \\ I_{\mathrm{CV}} \\ I_{\mathrm{CW}} \end{bmatrix} = \begin{bmatrix} Z_{\mathrm{UU}}' & Z_{\mathrm{UV}}' & Z_{\mathrm{UW}}' \\ Z_{\mathrm{VU}}' & Z_{\mathrm{VV}}' & Z_{\mathrm{VU}}' \\ Z_{\mathrm{UU}}' & Z_{\mathrm{UV}}' & Z_{\mathrm{UW}}' \end{bmatrix} \begin{bmatrix} I_{\mathrm{U}} \\ I_{\mathrm{V}} \\ I_{\mathrm{W}} \end{bmatrix}, \quad (25)$$

$$Z_{\mathrm{UU}}' = \frac{Z_0{}^2 + Z_0(3Z_{\mathrm{a}} + Z_{\mathrm{b}}) + 2Z_{\mathrm{a}}Z_{\mathrm{b}} + Z_{\mathrm{a}}{}^2}{3Z_0{}^2 + 2Z_0(2Z_{\mathrm{a}} + Z_{\mathrm{b}}) + 2Z_{\mathrm{a}}Z_{\mathrm{b}} + Z_{\mathrm{a}}{}^2}, \quad (26)$$

$$Z_{\mathrm{UV}}' = \frac{Z_0}{3Z_0 + Z_{\mathrm{a}} + 2Z_{\mathrm{b}}}, \quad (27)$$

$$Z_{\mathrm{UW}}' = \frac{Z_0(Z_0 + Z_{\mathrm{b}})}{3Z_0{}^2 + 2Z_0(2Z_{\mathrm{a}} + Z_{\mathrm{b}}) + 2Z_{\mathrm{a}}Z_{\mathrm{b}} + Z_{\mathrm{a}}{}^2}, \quad (28)$$

$$Z_{\mathrm{VU}}' = \frac{Z_0(Z_0 + Z_{\mathrm{a}})}{3Z_0{}^2 + 2Z_0(2Z_{\mathrm{a}} + Z_{\mathrm{b}}) + 2Z_{\mathrm{a}}Z_{\mathrm{b}} + Z_{\mathrm{a}}{}^2}, \quad (29)$$

$$Z_{\mathrm{VV}}' = \frac{Z_0 + Z_{\mathrm{a}} + 2Z_{\mathrm{b}}}{3Z_0{}^2 + 2Z_0(2Z_{\mathrm{a}} + Z_{\mathrm{b}}) + 2Z_{\mathrm{a}}Z_{\mathrm{b}} + Z_{\mathrm{a}}{}^2}. \quad (30)$$

By using (1) and (25), the capacitor current can be calculated.

IV. EXPERIMENTAL VERIFICATION

A. Experimental Setup

In order to verify the effect on imbalance suppression, the circuits shown in Figs. 3 and 9 were implemented. Table I summarizes the circuit parameters employed in this experiment. The circuit parameters were the same as those in the analyses of the conventional circuit implementation shown in Fig. 6. The SiC MOSFET modules with SiC Schottky barrier diodes (SBDs) were used as switching devices S_1–S_6 to increase the switching frequency up to 100 kHz. The DC-link capacitors C_{P} were film capacitors that were installed close to the SiC MOSFET modules. The capacitance of each DC-link capacitor was set to $50\,\mu\mathrm{F}$; the value was not optimized because the purpose of this experiment is to observe the capacitor-current imbalance between the phase legs. The ESR R_{ES} on the capacitor is $12\,\mathrm{m}\Omega$. Because the influence of the large

TABLE I. EXPERIMENTAL CIRCUIT PARAMETERS

Switching devices $S_1 - S_6$	SiC MOSFET module with SiC-SBD, 1200 V, 120 A (Rohm, BSM120D12P2C005)
Phase-leg capacitors C_{P}	Film capacitor, $50\,\mu\mathrm{F}$, 600 V (Shizuki, MEC)
Equivalent series resistance R_{ES}	$12\,\mathrm{m}\Omega$
Input voltage	300 V
Input inductor	10 mH
Load	40 mH, 15 Ω
Switching frequency f_{SW}	$10 - 100$ kHz
Load frequency f_{L}	50 Hz
Modulation index M	0.9
Power factor $\cos\phi$	0.77

capacitor on the DC power supply is excluded, the 10-mH inductor was placed between the DC power supply and inverter circuit. The load frequency $f_{\mathrm{L}} = 50\,\mathrm{Hz}$, the modulation index $M = 0.9$, and the power factor $\cos\phi = 0.77$.

Table II summarizes the parameters of the bus bars used in the experiment. The parameters of the conventional circuit implementation are the same as the analyses shown in Fig. 6. Details of the parameters in this table are explained in the following sections.

B. Conventional Circuit Implementation

Fig. 11 shows experimental waveforms obtained using the conventional circuit implementation when the switch-

TABLE II. BUS BAR PARAMETERS

	Conventional	Proposed	
		Graphite	Delta-type
Stray inductance L_{S}	16.8 nH	16.8 nH	16.8 nH
Stray inductance L_{P}	91.5 nH	91.5 nH	91.5 nH
Stray inductance L_{D}	–	–	131.0 nH
Stray resistance R_{P}	0.18 mΩ	128 mΩ	0.18 mΩ

2141

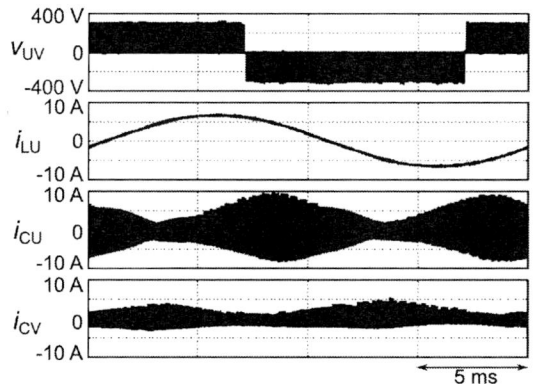

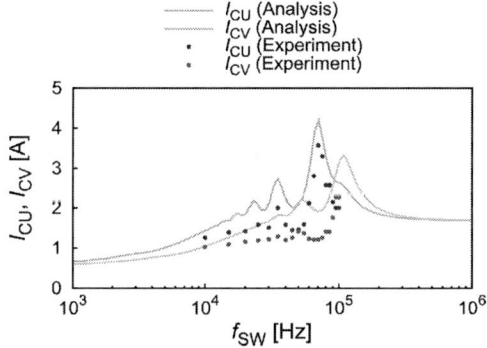

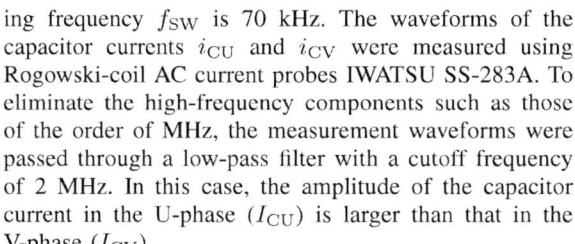

Fig. 11. Experimental waveforms using ladder-type bus bar when $f_{SW} = 70$ kHz.

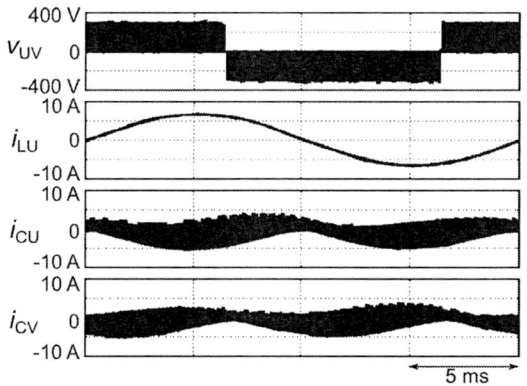

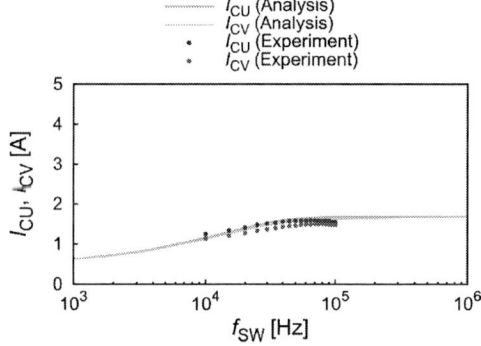

Fig. 13. Experimental waveforms using ladder-type bus bar when $f_{SW} = 70$ kHz.

Fig. 12. Analytical and experimental switching frequency dependence of capacitor current in case of conventional circuit implementation.

Fig. 14. Analytical and experimental switching frequency dependence of capacitor current in the case of graphite bus bar.

ing frequency f_{SW} is 70 kHz. The waveforms of the capacitor currents i_{CU} and i_{CV} were measured using Rogowski-coil AC current probes IWATSU SS-283A. To eliminate the high-frequency components such as those of the order of MHz, the measurement waveforms were passed through a low-pass filter with a cutoff frequency of 2 MHz. In this case, the amplitude of the capacitor current in the U-phase (I_{CU}) is larger than that in the V-phase (I_{CV}).

Fig. 12 shows the experimental results of the rms values for I_{CU} and I_{CV} in the case of conventional circuit implementation. The blue and red lines represent the analytical results shown in Fig. 6. The blue and red circles represent the experimental results for I_{CU} and I_{CV}, respectively. As with the analysis shown in Fig. 6, I_{CU} and I_{CV} increase depending on the switching frequency f_{SW}, whereas these should be constant regardless of f_{SW} in the situation without resonance. I_{CU} presents a peak for f_{SW} at around 68.4 kHz, which is calculated using (19). However, I_{CV} presents a different peak for f_{SW} at around 90.0 kHz, which is calculated using (20). In the experiment, when $f_{SW} = 70$ kHz, the imbalance between I_{CU} and I_{CV} was the maximum. In this situation, $I_{CU} = 3.57$ A and $I_{CV} = 1.21$ A, that is, I_{CU} is 3.0 times as large as I_{CV}.

C. Graphite Bus Bar

In this experiment, the graphite bus bar is only applied for the P-side. L_S and L_P are the same as in the conventional circuit implementation, as shown in Table II. The stray resistance R_P is 128 mΩ, which may act as a damping resistance for the DC-side resonance phenomena.

Fig. 13 shows experimental waveforms obtained using the graphite bus bar when the switching frequency f_{SW} is 70 kHz. The experimental results show the amplitudes of load voltage v_{UV} and load current i_{LU} that were the same as in conventional circuit implementation, as shown in Fig. 11. In this case, the amplitudes of I_{CU} and I_{CV} are the same, although I_{CU} is larger than I_{CV} with the conventional circuit implementation.

Fig. 14 shows the experimental results of the rms values for I_{CU} and I_{CV} in the case of the graphite bus bar. In this case, I_{CU} and I_{CV} increase depending on f_{SW}, but no resonance phenomena appeared at a certain switching frequency. For a f_{SW} of 70 kHz, at which frequency the capacitor-current imbalance was a maximum in the case of the conventional circuit implementation, I_{CU} and I_{CV} are 1.58 A and 1.50 A, respectively. The difference between I_{CU} and I_{CV} is suppressed to 1.05

The 2018 International Power Electronics Conference

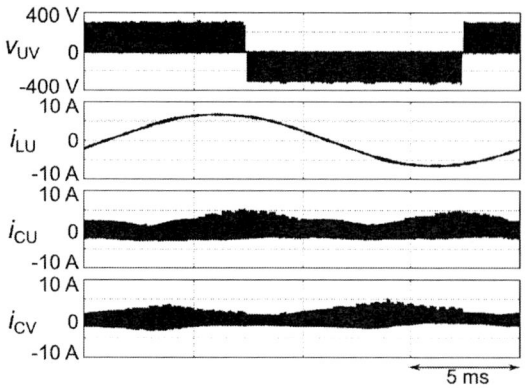

Fig. 15. Experimental waveforms obtained using delta-type structure when f_{SW} =70 kHz.

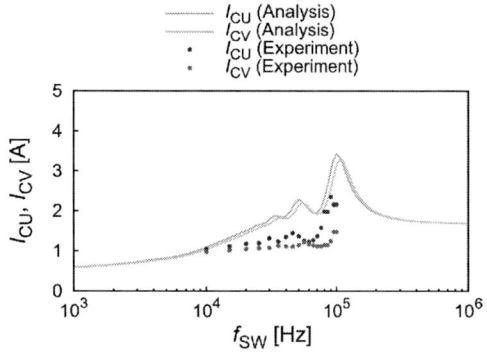

Fig. 16. Analytical switching frequency dependence of capacitor current in the case of delta-type structure.

times. Therefore, I_{CU} achieved a 56% reduction relative to the case of the conventional circuit implementation.

D. Delta-type Structure

In the case of the delta-type structure, L_S and L_P are the same as in the conventional circuit implementation, as shown in Table II. In addition, the stray inductance L_D of the additional bus bar between the U- and W-phases is set to 131.0 nH. L_D is larger than L_P because the distance between the U- and W-phase legs is longer than that between the U- and V-phases, and the V- and W-phases, as shown in Fig. 8. In the case involving the use of the delta-type structure, the resonant frequency $f_{rUW} = 96.0$ kHz and $f_{rUV} = 101.5$ kHz using the same calculation as (19) and (20). Because the two resonance frequencies f_{rUW} and f_{rUV} are close to each other, the capacitor currents between the phase legs are balanced.

Fig. 15 shows the experimental waveforms that were obtained using the delta-type structure when the switching frequency f_{SW} is 70 kHz. In this case, the amplitudes of I_{CU} and I_{CV} are the same, as with the graphite bus bar.

Fig. 16 shows the experimental result of the rms values for I_{CU} and I_{CV} in the case of the delta-type structure. The switching frequencies that have current

TABLE III.   CAPACITOR CURRENT COMPARISON WHEN $f_{SW} = 70$ kHz

	Conventional	Proposed	
		Graphite	Delta-type
U-phase capacitor current I_{CU}	3.57 A	1.58 A	1.37 A
V-phase capacitor current I_{CV}	1.21 A	1.50 A	1.12 A
Difference between I_{CU} and I_{CV}	3.0	1.05	1.05

peaks are shifted to higher region as follows:

$$f_{rUW} = \frac{1}{2\pi\sqrt{L_{UW}\left(\frac{1}{2}C_P\right)}} = 101.5\,\text{kHz}, \quad (31)$$

$$f_{rUV} = \frac{1}{2\pi\sqrt{L_{UV}\left(\frac{1}{2}C_P\right)}} = 96.0\,\text{kHz}. \quad (32)$$

As a result, at f_{SW} of 70 kHz, at which frequency the capacitor-current imbalance was a maximum in the case of the conventional circuit implementation, I_{CU} and I_{CV} are 1.37 A and 1.12 A, respectively. The difference between I_{CU} and I_{CV} is suppressed to 1.05 times. Therefore, I_{CU} achieved a 62% reduction relative to the conventional circuit implementation.

E. Discussion

Table III compares the capacitor current with each implementation when $f_{SW} = 70$ kHz. In the case of the conventional circuit implementation, I_{CU} is 3.0 times as large as I_{CV}, whereas it is 1.05 times in the case of the graphite bus bar and the delta-type structure. Compared with the conventional circuit implementation, the difference between the DC-link capacitor current for the U- and V- phases is small in the case of these two proposed methods. Therefore, these results show that the proposed methods can suppress the capacitor-current imbalance effectively. However, because there is an additional loss in the case of the graphite bus bar, the most suitable one depends on the application.

V. CONCLUSION

This paper proposed methods of balancing DC-link capacitor currents between phase-leg units in three-phase inverter circuits. Conventionally, there is a capacitor current imbalance between the phase-legs owing to the stray inductance in the bus bar. The proposed methods enable current balance by changing the DC-side connection material or structure without the use of an additional power semiconductor device or a special control scheme. The experimental results show that using the proposed structure the current difference between the phase legs was suppressed at f_{SW} of 70 kHz. As a result, the proposed methods can reduce the rms value of the capacitor current by 62%, which may lead to an improved reliability of the DC-link capacitor.

REFERENCES

[1] M. Eull, M. Preindl, and A. Emadi, "Analysis and design of a high efficiency, high power density three-phase silicon carbide inverter," in *Proc. IEEE ITEC*, June 2016.

2143

[2] P.A. Dahono, Y. Sato, and T. Kataoka, "Analysis and minimization of ripple components of input current and voltage of PWM inverters," *IEEE Trans. Ind. Appl.*, Vol. 32, No. 4, pp. 945–950, July/Aug. 1996.

[3] H. Wen, W. Xiao, X. Wen, and P. Armstrong, "Analysis and evaluation of DC-link capacitors for high-power-density electric vehicle drive systems," *IEEE Trans. Veh. Technol.*, Vol. 61, No. 7, pp. 2950–2964, Sept. 2012.

[4] G. I. Orfanoudakis, S. M. Sharkh, and M. A. Yuratich, "Capacitor size reduction for multiple inverter systems," in *Proc. IEEE RPG*, 2011.

[5] T. D. Nguyen, N. Patin, and G. Friedrich, "Extended double carrier PWM strategy dedicated to RMS current reduction in DC Link capacitors of three-phase inverters," *IEEE Trans. Power Electron.*, vol. 29, no. 1, pp. 396–406, Jan. 2014.

[6] T. Suzuki, H. Kabune, and N. Ito, "Capacitor current reducing control of the inverters for the dual winding motor," in *Proc. IEEE IECON*, 2015.

[7] T. Kaneko, K. Wada, and A. Tokumasu, "Switching-frequency ripple current reduction of DC-link capacitor for a single-phase charger," in *Proc. IEEE IFEEC*, 2015.

[8] K. Hasegawa, I. Omura, and S. Nishizawa, "Design and analysis of a new evaluation circuit for capacitors used in a high-power three-phase inverter," *IEEE Trans. Ind. Electron.*, vol. 63, no. 5, pp. 2679-2687, May 2016.

[9] H. Soliman, H. Wang, and F. Blaabjerg, "A review of the condition monitoring of capacitors in power electronic converters," *IEEE Trans. Ind. Appl.*, vol. 52, no. 6, pp. 4976–4989, Nov./Dec. 2016

[10] H. Nagasaki, P. Huang, and T. Shimizu, "Characterization of power capacitors under practical current condition using capacitor loss analyzer," in *Proc. IEEE ECCE*, 2016.

[11] X. Lyu, Y. Li, and D. Cao, "DC-Link RMS current reduction by increasing paralleled three-phase inverter module number for segmented traction drive," *IEEE J. Emerg. Sel. Topics Power Electron.*, vol. 5, no. 1, pp. 171–181, March 2017.

[12] H. Wang, and F. Blaabjerg, "Reliability of capacitors for DC-link applications in power electronic converters–an overview," *IEEE Trans. Ind. Appl.*, Vol. 50, No. 5, pp. 3569–3578, Sept/Oct. 2014.

[13] M. Mochizuki, M. Aoyama, and T. Yamada, "Large-capacity Inverters," *Fuji Electric Journal*, Vol. 78, No. 5, pp.377-380, Sept. 2005. (in Japanese)

[14] F. Hilpert, K. Brinkfeldt, and S. Arenz, "Modular integration of a 1200 V SiC inverter in a commercial vehicle wheelhub drivetrain," *in Proc. EDPC*, 2014.

[15] P. Brockerhoff, W. Schön, P. Blaha, P. Vclavek, and Y. Burkhardt, "Disc inverter in highly integrated 9-phase drivetrain for e-mobility," *in Proc. EPE*, 2015.

[16] H. Kamizuma, Y. Hattori, D. Matsumoto, A. Mima, T. Kawashima, Y. Mabuchi, and T. Ichikawa, "Low phase-to-phase-impedance busbar for modular power converter system," *in Proc. IEEE INTELEC*, pp. 772–776, 2015.

[17] M. Ando, and K. Wada, "Design of acceptable stray inductance based on scaling method for power electronics circuits," *IEEE J. Emerg. Sel. Topics Power Electron*, vol. 5, no. 1, pp. 568–575, Mar. 2017.

[18] K. Takao, T. Shinohe, T. Yamamoto, K. Hasegawa, and M. Ishida: "1200 V-360 A SiC power module with phase leg clustering concept for low parasitic inductance and high speed switching," in *Proc. IEEE CIPS* 2012.

[19] M. Nahrstaedt, and R. Zingel: "Novel high power IGBT inverter for a power conditioning system," *in Proc. EPE*, 2005.

[20] T. Hirao, K. Wada, and T. Shimizu, "Circulating resonant current between integrated half-bridge modules with capacitor for inverter circuit using SiC-MOSFET," *IEEE Trans. Ind. Appl.*, 2018.

The 2018 International Power Electronics Conference

A Three-Phase Hybrid Switched-Boost Inverter

Minh-Khai Nguyen[1*], Tan-Tai Tran[2], Hoan-Tien Luong[3], Kyoung-Won Lee[1], Youn-Ok Choi[1], and Geum-Bae Cho[1]

1 Dept. of Electrical Engineering, Chosun University, Gwangju, Korea
2 Dept. of Electrical Engineering, Chonnam National University, Gwangju, Korea
3 Dept. of Electrical and Electronics Engineering, HCMC University of Technology and Education,
Ho Chi Minh City, Vietnam
* E-mail: khaibk@ieee.org.

Abstract- **This study introduces a new topology of three-phase buck/boost inverter which can produce the ac voltage output higher than the input voltage source. The performance of proposed three-phase hybrid switched-boost inverter (hSBI) is similar to that of three-phase quasi Z-source inverter (qZSI). In comparison with the three-phase qZSI, the introduced three-phase hSBI uses one less inductor and one more diode. The comprehensive analysis in the operation modes, comprehensive comparisons with the three-phase qZSI, and simulation results are presented. Finally, the experimental test based on a laboratory prototype has been carried out to confirm practicality of the introduced three-phase hSBI.**

Keywords— Switched-boost inverter, three-phase topology, shoot through, impedance-source network, quasi-Z-source inverter.

I. INTRODUCTION

The voltage-source inverters (VSIs) [1], [2] are extensively used in industry for hybrid electric vehicles, uninterruptible power supplies, renewable energy systems, and electric machines. Nevertheless, the main drawback of VSI is that the maximum ac voltage obtained is less than the input voltage. To produce a desired ac voltage, an additional stage of the boost converter is demanded, which decreases the efficiency and rises the cost of two-stage power conversion. In addition, the conventional VSI cannot immune the shoot-through phenomenon.

To resolve the aforementioned demerits of VSI, the Z-source/quasi-Z-source inverters (ZS/qZSIs) has been presented in [3]-[7], where an impedance-source network was inserted between the H-bridge inverter and the input power source. Fig. 1 shows the three-phase qZSI with an LC output filter and a resistive load. The qZSI presents a buck/boost function in a one-stage power conversion processing. Furthermore, the reliability of the system is significantly improved because all switches in the same leg of H-bridge circuit can be turned on simultaneously, which is forbidden in VSI. However, the impedance-source in the qZSI with two capacitors and two inductors raises the cost and size of the power converter. To decrease the number of passive components and the cost of the power converter, the quasi-switched boost inverters were introduced in [8]. Although the qSBI reduces the passive elements, it uses one more active switch when compared with the qZSI.

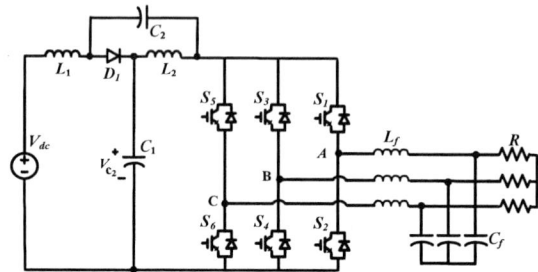

Fig. 1. Conventional three-phase qZSI.

Recently, the switched-boost network has been applied to the single-phase boost inverter [9] and the cascade H-bridge boost inverter [10]. The boost inverter based on switched-boost network has all inherit features of the qZSI such as shoot-though immunity and buck/boost voltage. In this study, a three-phase hybrid switched-boost inverter (hSBI) is suggested. The feature of the proposed hSBI is same as that of three-phase qZSI. Compared with the three-phase qZSI, the hSBI uses one less inductor and one more diode. The PWM control methods for three-phase qZSI can be applied to the proposed three-phase hSBI. The operating principles and comparison with the three-phase qZSI are presented. Simulation and experimental verifications has been provided to confirm the operating theory of the presented structure.

II. PROPOSED TOPOLOGY

The configuration of the introduced three-phase hSBI is indicated in Fig. 2. The introduced hSBI contains one DC source, six switches, two diodes, two capacitors, one boost inductor, and a three-phase filter (L_fC_f) connected to a three-phase resistor load. In the proposed inverter, the C_2 capacitor is connected in series to the phase C leg to cut out the DC component of the phase C voltage at the output side.

Similar to qZSI, the presented inverter has two operating modes as demonstrated in Fig. 3. Six switches are switched on at the same time; the D_1 and D_2 diodes are blocked in the shoot-through mode. The time in the shoot-through mode is $D.T$, where D represents the duty cycle and T represents the switching period. The line-to-line output voltage V_{AX} and V_{BX} is $-V_{C1}$. The inductor L_1 stores energy from the source. We have:

2145

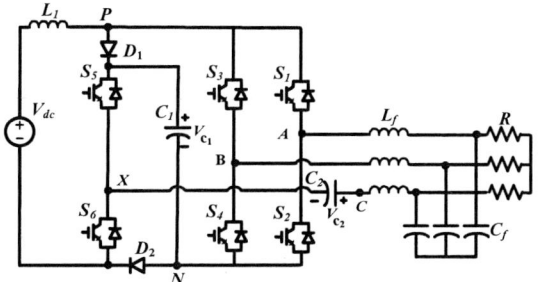

Fig. 2. Proposed three-phase hybrid switched-boost inverter.

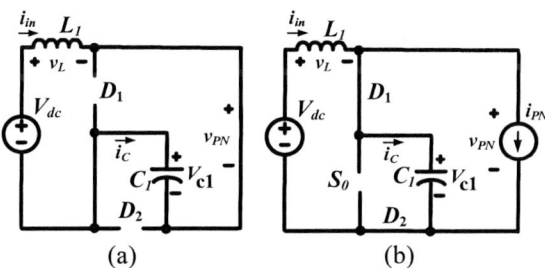

(a)   (b)

Fig. 3. Operating modes of introduced topology: (a) shoot-through, and (b) non-shoot-through.

$$\begin{cases} L_1 \dfrac{di_{L_1}}{dt} = V_{dc} \\ V_{AX} = V_{BX} = -V_{C1} \\ V_{AB} = 0 \\ V_{AC} = V_{BC} = -V_{C1} - V_{C2}. \end{cases} \quad (1)$$

In non-shoot-through mode, both D_1 and D_2 diodes are conducting as shown in Fig. 3(b). The inductor L_1 is discharged while the capacitor C_1 is charged. There is:

$$L_1 \frac{di_{L_1}}{dt} = V_{dc} - V_{C1} \quad (2)$$

Applying voltage balance law to L_1, (1) and (2) yield:

$$V_{C1} = \frac{1}{1-2D} V_{dc}. \quad (3)$$

During the shoot-through time interval $(D.T)$, the line-to-line output voltage V_{AX} and V_{BX} is not zero and equal to $-V_{C1}$. The voltage between nodes C and X is V_{C2} and calculated as

$$V_{C2} - DV_{C1} = \frac{D}{1-2D} V_{dc}. \quad (4)$$

The boost factor of the introduced hSBI is defined as

$$B = \frac{V_{C1}}{V_{dc}} = \frac{1}{1-2D}. \quad (5)$$

The output phase voltage peak is calculated as

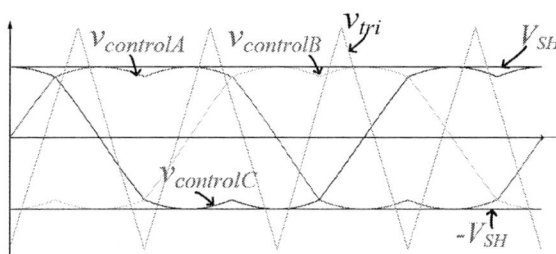

Fig. 4. The PWM method for the introduced inverter.

TABLE I
COMPARISON OF THE INTRODUCED TOPOLOGY AND
THE THREE PHASE-qZSI

	qZSI	Introduced Inverter
Capacitors	2	2
Inductors	2	1
Switches	6	6
Diodes	7	8
Shoot-through immunity	Yes	Yes
Input current	Continuous	Continuous

TABLE II
VOLTAGE STRESSES OF THE INTRODUCED TOPOLOGY AND
THE THREE-PHASE-qZSI

		Three Phase-qZSI	Introduced Topology
Capacitor voltage	V_{C1}	$\dfrac{1-D}{1-2D} V_{dc}$	$\dfrac{1}{1-2D} V_{dc}$
	V_{C2}	$\dfrac{D}{1-2D} V_{dc}$	$\dfrac{D}{1-2D} V_{dc}$
Diodes voltage stress		$\dfrac{1}{1-2D} V_{dc}$	
Switches voltage stress $(S_1$-$S_6)$		$\dfrac{1}{1-2D} V_{dc}$	

$$\hat{v}_{ph} = M \cdot B \cdot V_{dc} / 2, \quad (6)$$

where M is the index of modulation.

The PWM control methods for three-phase qZSI can be applied to the proposed three-phase hSBI. As an example, the PWM method [6] as indicated in Fig. 4 is applying for the introduced topology.

Table I presents a comparison of the components of the introduced hSBI and the three-phase qZSI. In comparison with the three-phase qZSI, the introduced structure only uses a inductor and one more diodes. Both inverters have the same following features as shoot-through immunity and continuous input current. The voltage stress comparison between the three-phase qZSI and the introduced topology is illustrated in Table II. From Table II, we can see that the capacitor voltages of the introduced topology are higher than capacitor voltages of the three-phase qZSI. The voltage stress across diodes and switches of the introduced topology are equal to that of the conventional three-phase qZSI.

TABLE III
EXPERIMENTAL PARAMETERS OF THE THREE-PHASE hSBI

Parameter	Value
Output line-to-line voltage	110 V (RMS)
Inductor (L_1) and Capacitor (C_1)	2 mH and 2200 μF
Capacitor (C_2)	4700 μF
Filter inductor (L_f) and Filter Capacitor (C_f)	3 mH and 10 μF
Resistive load (R)	40 Ω
MOSFETs (S_{1-6})	IRFP460
Diodes (D_a, D_b)	DSEP30-12AR
Switching frequency	10 KHz

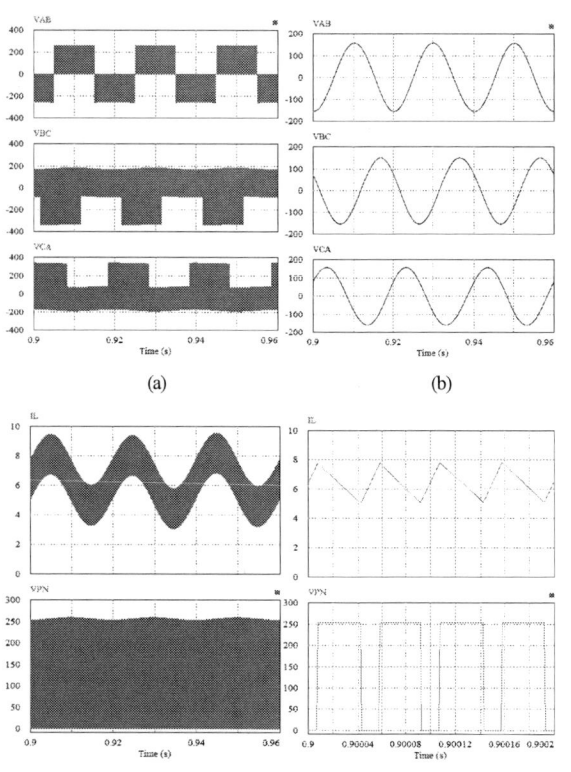

(a)

(b)

(c)

(d)

Fig. 5: Simulation results of introduced inverter. From top to bottom: (a)-(b) (a)-(b) output line to line voltages; (c)-(d) input current and DC-bus voltage.

III. SIMULATION RESULTS

The proposed three-phase hSBI is confirmed by PSIM simulation software. Table III gives the simulation parameters for the introduced topology. The input voltage is 96 V; the AC line-to-line voltage is 110 V_{rms}.

Fig. 5 presents the simulation results of the introduced topology when V_{dc} = 96 V. The simulated waveforms of the ac line-to-line voltages are illustrated in Fig. 5(a) and Fig. 5(b). The DC-link voltage V_{PN} is the square waveform and the peak value is 255 V as presented in Fig. 5(d). The average input current is 6.3 A. The high-frequency input ripple current from Fig. 5(d) is 2.6 A. The boost inductor operates with frequency of 20 kHz.

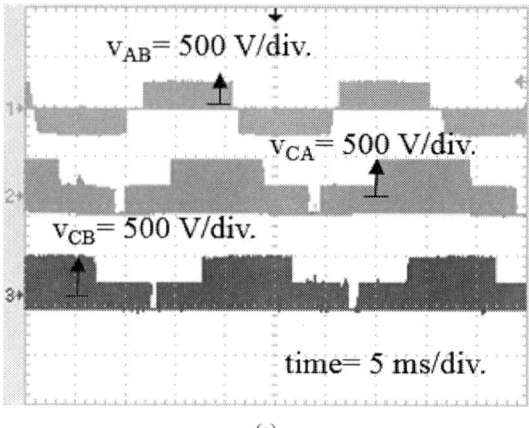

(a)

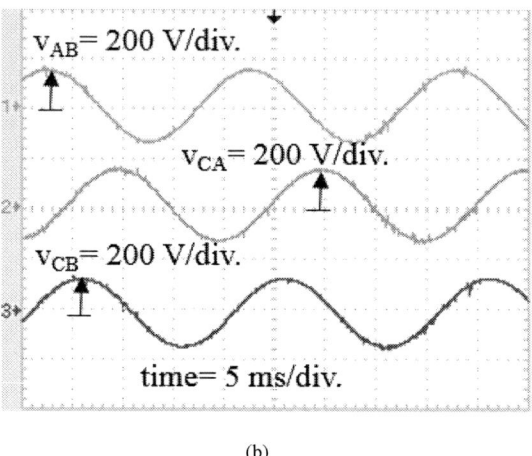

(b)

Fig. 6. Experimental output line-to-line voltage waveforms for the introduced topology at V_{dc} = 96 V. (a) Before filter and (b) after filter.

IV. EXPERIMENTAL RESULTS

A 400-W platform is set up to confirm the presented analyses of the introduced topology. The parameters of prototype are presented in Table III. The source voltage and the RMS value of ac voltage are 96 V and 110 V at 50 Hz, respectively.

Fig. 6 indicates the experimental waveforms of the output line-to-line voltages. From Fig. 7, we can see that the DC-link voltage is the pulse waveform and its peak value is 265 V. As shown in Fig. 6(b) and Fig. 7, the output voltage of 110 Vrms is produced from the dc voltage of 96 V. The high-frequency input ripple current from Fig. 7(d) is 2.5 A. The boost inductor current is continuous. The average inductor current is 3.3 A. The boost inductor operates with frequency of 20 kHz. As shown in Fig. 8, the capacitor C_1 and C_2 voltages are boosted to 265 V and 90 V, respectively. The RMS ac line-to-line voltage is 110 V. The THD of the ac voltage is 1.2 %. As indicated in Figs. 6 and 7, the experimental results match the presented analyses.

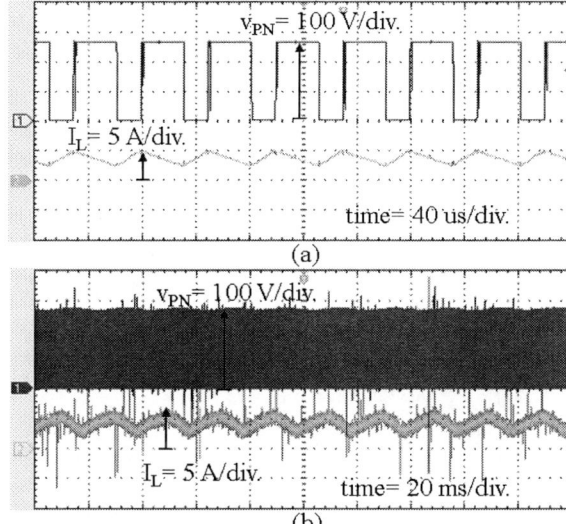

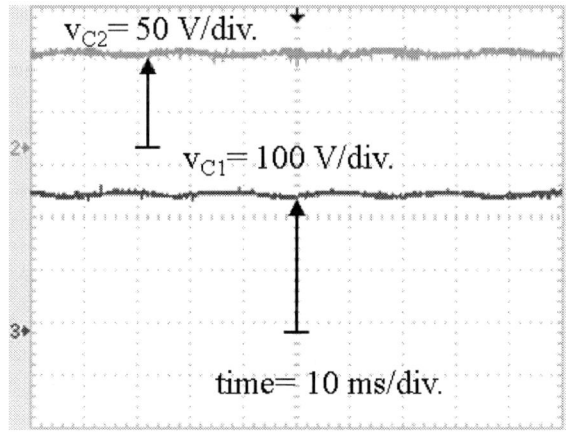

Fig. 7. Experimental waveforms of the introduced topology. (a)-(b) Top waveform: V_{PN} ; Bottom waveform: input current.

Fig. 8. Experimental results of the introduced topology. Top waveform: capacitor C_2 voltage; bottom waveform: capacitor C_1 voltage

V. CONCLUSION

A one-stage three-phase hSBI with boost voltage ability has been suggested in this study. The introduced structure has the following major features as shoot-through immunity and reduction in passive components. Compared with a three-phase qZSI, the proposed three-phase hSBI uses one less inductor and one more diode. Operating principle, steady-state analysis and simulation results for the introduced topology are presented. Experimental verifications are also carried out to verify the presented analyses of the introduced three-phase hSBI. Because of reducing passive component, the proposed three-phase hSBI can be replaced the three-phase qZSI in low-cost and low-power applications.

ACKNOWLEDGMENT

This work was supported by the Korea Institute of Energy Technology Evaluation and Planning (KETEP) and the Ministry of Trade, Industry& Energy (MOTIE) of the Republic of Korea (NO. 20164010201020).

REFERENCES

[1] M. Curkovic, K. Jezernik, R. Horvat, "FPGA-Based Predictive Sliding Mode Controller of a Three-Phase Inverter," *IEEE Trans. Ind. Electron.*, vol. 60, no. 2, pp. 637–644, Feb. 2013.

[2] T. Kerekes, R. Teodorescu, M. Liserre, C. Klumpner, and M. Sumner, "Evaluation of three-phase transformerless photovoltaic inverter topologies," *IEEE Trans. Power Electron.*, vol. 24, no. 9, pp. 2202–2211, 2009.

[3] F. Z. Peng, "Z-source inverter," *IEEE Trans. Ind. Appl.*, vol. 39, no. 2, pp. 504-510, Mar./Apr. 2003.

[4] J. Anderson, and F. Peng, "A class of quasi-Z-source inverters", *in Proc. IEEE IAS'08*, 2008, pp. 1 - 7.

[5] M. K. Nguyen, Y. C. Lim, and G. B. Cho, "Switched-inductor quasi-Z-source inverter," *IEEE Trans. Power Electron.*, vol. 26, no. 11, pp. 3183-3191, Nov. 2011.

[6] M. Shen, J. Wang, A. Joseph, F. Z. Peng, L. M. Tolbert, and D. J. Adams, "Constant boost control of the Z-source inverter to minimize current ripple and voltage stress," *IEEE Trans. Ind. Appl.*, vol. 42, no. 3, pp. 770-778, May/June 2006.

[7] Y. Liu, H. Abu-Rub, and B. Ge, "Z-source/quasi-Z-source inverters: derived networks, modulations, controls, and emerging applications to photovoltaic conversion," *IEEE Ind. Electron. Mag.*, vol. 8, no. 4, pp. 32–44, Dec. 2014.

[8] M. K. Nguyen, T. V. Le, S.J. Park, and Y. C. Lim, "A class of quasi-switched Boost inverters," *IEEE Trans. Ind. Election.*, vol. 62, no. 3, pp. 1526-1536, Mar. 2015.

[9] M. K. Nguyen, T. T. Tran, "A single-phase single-stage switched-boost inverter with four switches," *IEEE Trans. Power Electron*, vol. 33, no. 8, Aug. 2018.

[10] M. K. Nguyen and T. T. Tran, "Quasi cascaded H-bridge five-level boost inverter," *IEEE Trans. Ind. Election.*, vol. 64, no. 11, pp. 8525-8533, Nov. 2017.

The Effect of Built-in CR Snubber Capacitor into the Power Module

Ryotaro Hata[1] and Shigeki Nishiyama[1]
1 Murata Manufacturing Co., Ltd., Kyoto, Japan
*E-mail: ryotaro_hata@murata.com

Abstract— This paper presents the effect of built-in snubber capacitor into power module. SiC will be introduced to the power converter in various applications. Therefore, the switching frequency will get higher in order to increase output density. However, on high speed switching, the line inductance lead the surge voltage and ringing oscillation. These problems could be the problems in the circuit like EMI noise. One of the solution to reduce these issues is to apply CR-snubber capacitor into the circuit. Furthermore, if CR snubber capacitor is embedded to the power module, this effect get much higher than conventional method.

In this paper, we made built-in CR snubber capacitor into the power module and have tests to confirm this effect to reduce the above problems. Also we confirmed the effect difference from CR value, ambient temperature, switching frequency, line inductance and CR snubber capacitor position are evaluated quantitatively.

Keywords—Silicone Carbide, Ringing, Surge voltage, CR Snubber capacitor, Smoothing capacitor, Power Module.

I. Introduction

Silicon carbide (SiC) power devices have several advantages compared to the existing silicon (Si) one. The three main advantage of SiC are: high maximum operating temperature, low ON resistance and high switching frequency operation. Thanks to these advantages, SiC could be a potential replacement of the semiconductor components in several power electronics applications [1],[2]. It leads to reduce the output density of the power electronics applications. However, on high speed switching, the line inductance in the circuit lead higher surge voltage and ringing oscillation due to high parasitic inductance compared to existed Si power device. These problems could affect the misleading circuit operation and cause EMI noise, therefore, these problems must be resolved. One of the method to reduce these problems is to apply snubber capacitor into the circuit. Some papers have reported this effect to reduce them [3]. However, all papers should have reported the effect of snubber capacitor put outside power module. This paper report the effect of built-in snubber capacitor into the power module.

II. Experimental Ringing and Surge Supression

A. The power module

Fig.1 shows the power module we made and Table I shows the detail of power devices, SiC-SBD and SiC-MOSFET we used. This module was fabricated with the consideration for following items in order to get low ESL as much as possible,

1. To shorten current path
2. To diverge current flow
3. To make reciprocal current close.

The snubber capacitor is embedded in the power module.

TABLE I
POWER DEVICE DETAIL

Device	P/N	Supplier
SiC-MOSFET	CPM2-1200-0025B	CREE
SiC-SBD	PW5-1200-Z050B	CREE

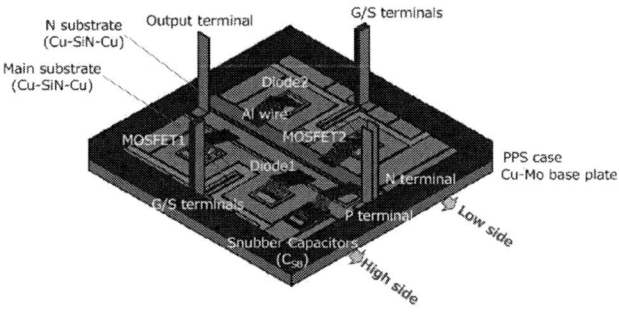

Fig.1. Outlook of power module with built-in snubber capacitor

B. Test circuit

Fig.2 shows the test circuit, double pulse circuit and Table II shows the parameter of each component.

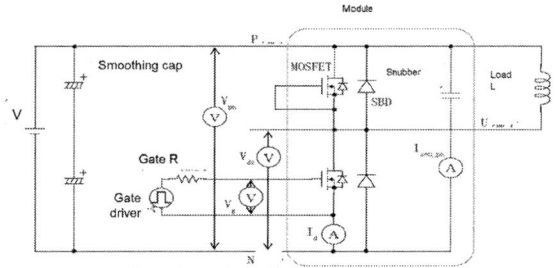

Fig.2. Test circuit (Double pulse circuit)

TABLE II

PARAMETER OF EACH COMPONENT

Device	Value
Voltage source	600V
Load current	90A
Smoothing cap.	1600uF/10nH
Snubber cap.	C:3.2-22.0nF/R:1.3-8.5 Ω
Gate resistance	0,2.2,4.7,10Ω
Gate voltage	+20V/-5V
Load Inductance	100uH

III. TEST RESULTS

We had several tests what kind of conditions would affect the reduction of surge voltage and ringing. We considered the following items in order to figure out them.

1. 1st peak voltage: Surge voltage

2. 2nd peak dumping ratio: 2nd peak voltage /1st-peak voltage (See Fig.3): Ringing

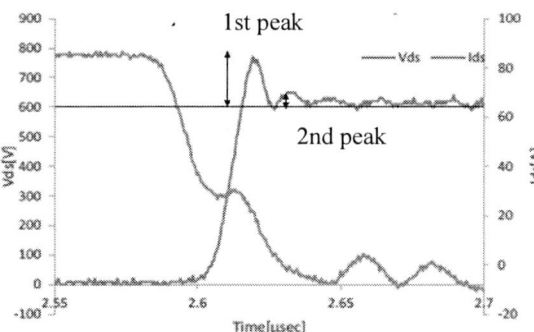

Fig.3. The definition of 1st peak and 2nd voltage

A. Variation of CR parameter of snubber capacitor

We tested with some parameters by various snubber capacitor's capacitance and resistance, then confirmed which parameter have a good effect to reduce the surge voltage and ringing. Fig.4 shows one of the result for the effect by applying the snubber capacitor. The capacitance and resistance was 3.2nF and 4.6 Ω respectively. Snubber capacitor could reduce the ringing although couldn't suppress the surge voltage.

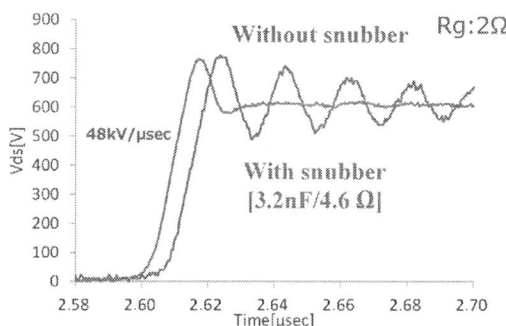

Fig.4. Turn-off result comparison between with and without snubber capacitor (3.2nF/4.6 Ω)

Fig.5 and Fig.6 show the results of the effect by snubber capacitor. The capacitance/resistance value were 12.0nF/4.3Ω in Fig.5 and 4.2nF/1.3Ω in Fig.6 respectively. From these results, we found the CR snubber capacitor would work only to reduce the ringing and not affect the suppression of surge voltage. Self-inductance of smoothing capacitor and line inductance must affect the behavior of the ringing. In Fig.6 the resistance is generally able to reduce the ringing. However, the low resistance could have bad effect for it. This is because the resonance was occurred in two path, between SiC-MOSFET/snubber capacitor and smoothing capacitor/ snubber capacitor.

We found 3-7nF and 4-9Ω have a good effect for the reduction of ringing.

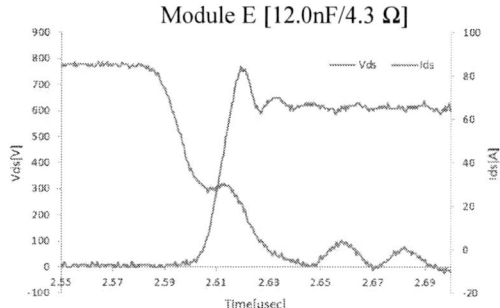

Fig.5. Vds and Ids behavior with 12.0nF/4.3Ω (turn-off)

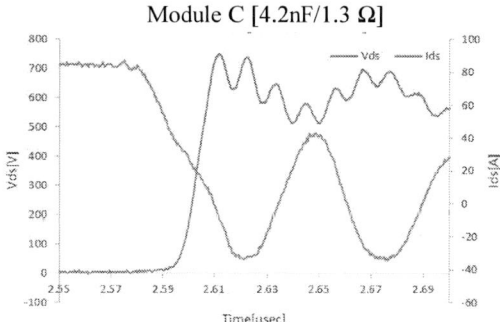

Fig.6. Vds and Ids behavior with 4.2nF/1.3Ω (turn-off)

2150

B. Influence by the temperature

We tested under different temperature, 100 ˚C, 150 ˚C and 200˚C. The temperature was changed by the hot plate where is set under the power device. Fig.7 shows surge voltage and damping ratio results under different temperature. 1st surge voltage under higher temperature could be suppressed as same as room temperature. The reason why the damping ration gets higher under high temperature is that this snubber capacitor have higher resistance under high temperature.

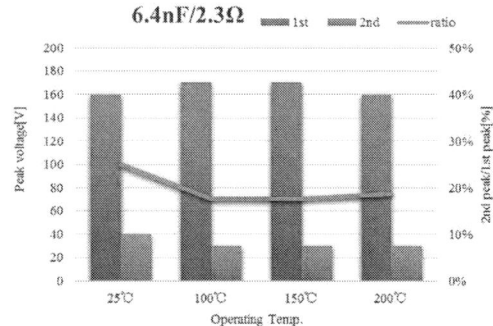

Fig.7. Surge voltage and damping ratio results under different temperature

C. Influence by the switching speed (Gate resistance)

We tested with different gate resistance, 0Ω, 2.0Ω, 4.7Ω and 10Ω. For the reference, the internal resistance of SiC-MOSFE is 1.1Ω. Fig.8 shows surge voltage and damping ratio results with different gate resistance (with 3.2nF/4.6Ω snubber capacitor). The snubber capacitor could work to reduce ringing with each gate resistance value (switching speed). Damping ratio were 10%-15%.

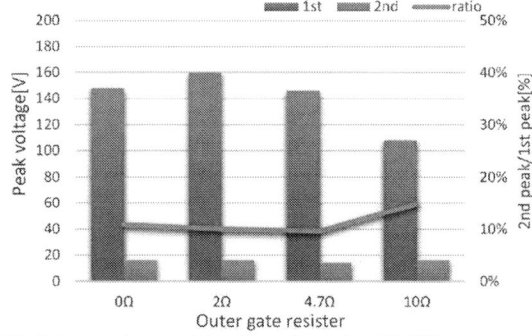

Fig.8. Surge voltage and damping ratio results with different gate resistance (snubber 3.2mF/4.6 Ω)

D. Influence of the line inductance between smoothing capacitor and power module

We tested with different line inductance between smoothing capacitor and power module, 7.5nH and 15nH. Fig.9 shows the surge voltage result with different line inductance. It was natural but higher line inductance cause higher surge voltage in both case with and without snubber capacitor. It becomes an evidence that snubber capacitor could not suppress the surge voltage, it is important for the suppression of surge voltage to reduce the line inductance between smoothing capacitor and power device. Fig.10 shows the damping ratio results with different line inductance. The ratio is almost same in all cases.

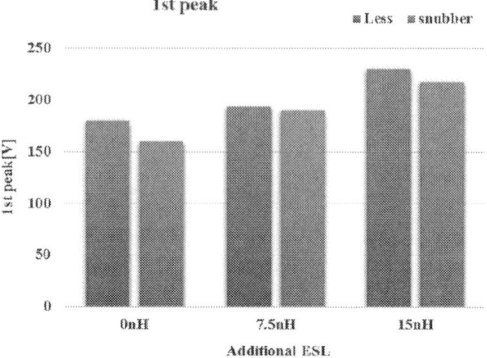

Fig.9. Surge voltage results with different line inductance

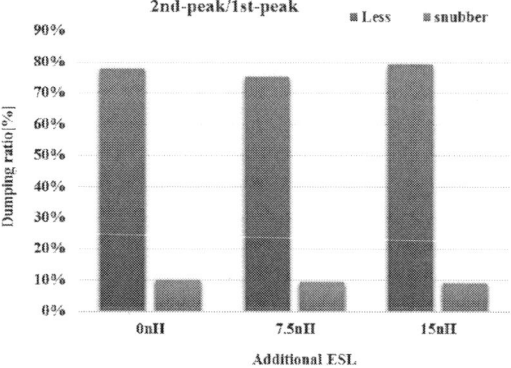

Fig.10. Damping ratio results with different line inductance

E. Influence by snubber capacitor position

We tested with different location to put the snubber capacitor in the circuit. Fig.11 shows each snubber capacitor position. We put it; Outside power module at 1. Smoothing capacitor side/ 2. Power module side and 3. Inside the power module (built-in snubber).

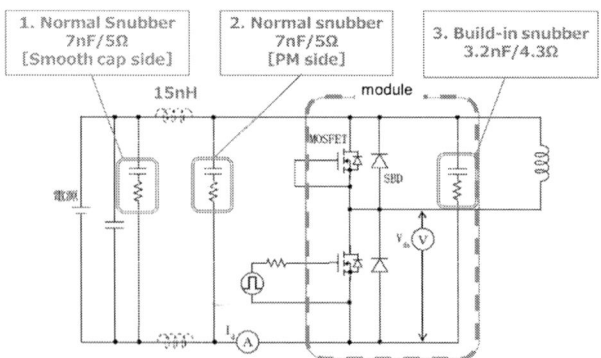

Fig.11. Snubber capacitor position in the circuit

2151

Fig.12 shows outlook of outside snubber capacitor module. The capacitance value is 7nF, the resistance value is 5Ω and the self-inductance value is 10nH. This module has 2 series capacitor for the solution of short failure and partial resistance to divide the loaded voltage.

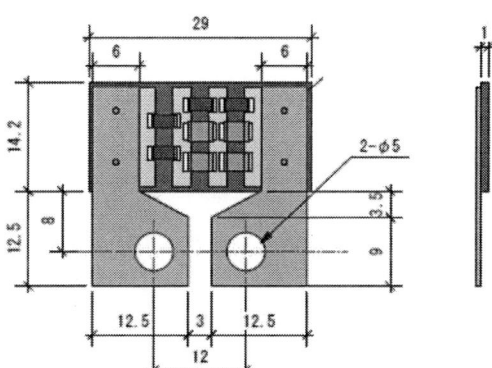

Fig.12. Outlook of outside snubber capacitor module

Fig.13 shows the result of the effect of snubber capacitor at different positions in the circuit. Snubber capacitor could not suppress the surge voltage regardless its position. In order to suppress it, much more capacitance would be needed and the capacitor size would be much bigger. The snubber capacitor position is so important to reduce the ringing. The effect difference is 25% between outside and inside power module. The capacitor positon inside the power module would not affect the effect of ringing reduction because it is affected only by capacitor resistance.

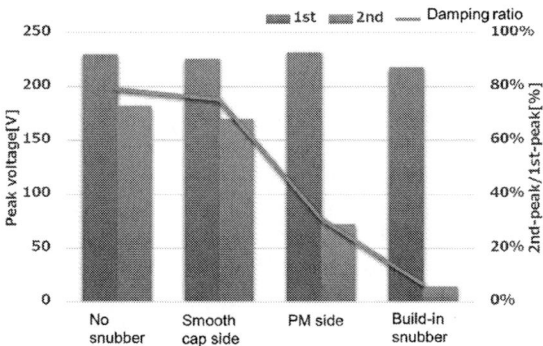

Fig.13. The result of the effect of snubber capacitor at different position in the circuit

IV. SUMMARY

In this article we described the effect of built-in snubber capacitor under different use. The capacitance and resistance value would be 3-7nF and 4-9 Ω respectively to reduce the ringing oscillation. The suppression of the surge voltage is affected by the line inductance and smoothing capacitor's ESL. Therefore, the parameter of snubber capacitor and smoothing capacitor are important to control the surge voltage and ringing oscillation.

In the next step, we will confirm this trend is applied in the different parameter of power module and circuit. Also we need to optimize the parameter of snubber and smoothing capacitor.

REFERENCES

[1] K.Mino, "Power Electronics Equipments Applying Novel SiC Power Semiconductor Modules", The 2014 International Power Electronics Conference.
[2] P.Friedrichs, "SiC Power devices complementing the silicon world –status and outlook". CIPS2016-9th International Conference on Integrated Power Electronics Systems.
[3] Y.Yamashita, "Ringing Suppression of Wiring Inductance by the RCD Snubber for MHz-Switching of SiC-MOSFET" *IEE Japan 2017(20-23・25-38・40・42-49), 147-152.*

AUTHOR INDEX

Aapro, Aapo 3156
Abdollahi, Hessamaldin..........................1719
Abe, Kazuyuki 1567
Abe, Kensho...767
Abe, Kodai 1741, 3890
Abe, Seiya 2360, 2370
Abe, Takashi 2176
Abrishamifar, Adib..................................2854
Abuogo, James 1125
Acharya, Anirudh Budnar 2630
Acharya, Sayan 3564
Adachi, Masakazu...................................2237
Afsharian, Jahangir 1537, 3797
Agarwal, Vivek.......................................3471
Agelidis, Vassilios G...............................3215
Agostinelli, Matteo 3140
Ahmad, Hamzeh J....................................3273
Aiso, Kohei...3186
Akagi, Hirofumi......................................2352
Akahane, Masashi 2774
Akama, Yousuke 1741
Akao, Naoki...1217
Akatsu, Kan....................................711, 3186
Alatise, O 1149
Alenius, Henrik 1704, 4205
Ali, Muhammad 528
Ali, Murad...2317
Allmeling, Jost............................422, 2199
Almér, Stefan...555
Alsofyani, Ibrahim Mohd.........................466
Alvarez, S..4009
Amano, Koki..94
Amei, Kenji..3182
Amin, Mohammad 759
Amrhein, Wolfgang.................................3640
An, Ronghui.................. 957, 1524, 3251, 3692, 3924
An, Zheng 4001
Andenna, M..3596
Andersen, A. E. Michael........................1351
Andersen, Michael A. E. 607, 4066
Ando, Akinobu 517
Ando, H..3665
Ando, Masato 1919
Ando, Takashi 3658
Ang, Simon S. 153
Antivachis, Michael...................................181
Antonini, Giulio......................................3588
Antonopoulos, Antonios 2335
Anurag, Anup...3564
Anyapo, Chan 3332
Aoyagi, Kazuki 2237
Aoyama, Masahiro 718, 753
Arai, Takuro...1997
Araumi, Ryunosuke 1877, 3658
Arimatsu, Kenji......................................1370

Arita, Hideaki 2796, 2820
Arrua, Silvia 1719
Asada, Kazunori 3658
Asama, Junichi 4016
Ashizaki, Yusuke 3450
Ashourloo, Mojtaba 2380
Aso, Shinji 3086
Aware, Mohan 1730
Ayano, Hideki 1080
Azad, A N M Wasekul............................. 2416
Azegami, Kazuya 3723
Azuma, S. 3665
Baba, Teppei 2283
Babasaki, Tadatoshi 207
Bach, Hoang Linh 2410
Baek, Jae-Il 108, 2365, 3100, 3533, 3538
Baek, Miran 1141
Bahat-Treidel, Eldad 3607
Bai, Baodong 2638
Baik, Jeong Min 3063
Bak, Yeongsu 1736, 4104
Bakran, Mark M. 2476
Bandyopadhyay, Soumya 1426
Barrena, Jon Andoni 759
Barrera-Cardenas, Rene........................... 3431
Bauer, Pavol 1426, 2630
Bauer, Walter 3640
Bayer, Christoph Friedrich 2410
Bellini, M. 4009
Berg, Matias 963, 4205
Bergveld, H.J. 267
Bertoldi, F. 488
Besselmann, Thomas............................... 555
Bezha, Minella 3170
Bhattacharya, Subhashish 3564, 3993
Bhowate, Apekshit 1730
Bhumkittipich, Krischonme 2430
Biela, J. 1896
Biela, Jürgen 1103, 1509, 2301, 3734
Bilal, Ahmad 2193
Bilsalam, A. 1622
Bin, Zhao 2692
Bixel, Paul 238
Blaabjerg, Frede 439, 746, 1183, 1246, 1711, 1788, 2512, 2604, 2743, 3123, 3164, 3357
Blanes, José M. 1435
Böcker, Jan 3607
Bojoi, R... 732
Bonyadi, R... 1149
Boroyevich, Dushan 790, 3705, 3749, 3985
Bortis, D. 4080
Bortis, Dominik 181
Boynov, K.O... 161
Braun, Michael 2848, 3074

AUTHOR INDEX

Büdel, Johannes 3034
Bui, M.X. 4174
Bunlaksananusorn, Chanin 2490
Burgos, Rolando 790, 3705, 3749, 3985
Cai, Kejun 3965
Cai, Panpan 3495
Cai, Xu 1004, 1491, 2245, 4162, 4220
Canales, F. 4009
Cao, Hu 1816, 3484
Cao, Pengpeng 2973
Cao, Qi 100
Cao, Wu 3002, 3010, 3015
Cardenas, Rene Alexander Barrera 1111
Carvalho, Kelly C. M. 3785
Castellazzi, Alberto 130, 2932
Ceballos, Salvador 3117
Celik, Mustafa 1680
Cha, Honnyong 927, 1046, 2619, 3134
Chae, Beomseok 1977
Chailloux, Thibaut 2153
Chang, Chen-Wei 1617
Chang, Chien-Hsuan 2860
Chang, Liuchen 815, 1472, 1793, 2505
Chang, Yung-Ruei 639, 883
Chanmontree, P. 1622
Chao, Yi-Hao 1145
Charalambous, Apollo 1634
Charoensuksirikul, Supanut 2113
Chattopadhyay, Ritwik 3564
Chazal, Hervé 2158
Chen, Ang-Tung 2102
Chen, Bo 1397
Chen, C. 142
Chen, Ching-Chen 1617
Chen, Ching-Jan 2086
Chen, Chuantong 1598
Chen, Dezhi 2638
Chen, Guan-Jung 1341
Chen, Guo 370
Chen, Hao 3112
Chen, I-Lin 2107
Chen, Jiangnan 1157, 1167
Chen, Jiann-Fuh 2653
Chen, Jie 1015, 1177
Chen, Kai-Hui 3081
Chen, Ke 1391
Chen, Kun-Feng 1341
Chen, Min 878
Chen, Minwu 2547
Chen, Nan 2335
Chen, Pingping 1118
Chen, Shen-Li 1145
Chen, Song 2153
Chen, Tang-Jung 1617
Chen, Tao 1872

Chen, Wan-Jung 3544
Chen, Wenjia 4213
Chen, Wenjie 1062, 2854, 3329
Chen, Wu 1504, 2496
Chen, Xiliang 3329
Chen, Xin 1015, 1177
Chen, Xingxing 1051, 3129, 3439
Chen, Yang 2785
Chen, Yangyang 560
Chen, Yaow-Ming 639, 883
Chen, Yenan 1118
Chen, Yen-Wen 2576
Chen, Yufeng 3383
Chen, Yu-Jen 275
Chen, Zhe 1758, 2708
Chen, Zhi 2997
Chen, Zhigang 3040
Cheng, Ching-Hsiang 2086
Cheng, Chun-An 2860
Cheng, Hung-Liang 2860
Cheng, Nie 2625
Cheng, Po-Tai 503, 1038, 2462, 3549
Cheng, Ran 3877
Cheng, Xiangpeng 2435, 3934
Chengbi, Zeng 2718
Chi, Yongning 1491
Chiba, Akira 3627
Chien, Lin-Hao 2102
Chiu, Huang-Jen 2092, 3151
Chiu, Hui-Lung 123
Chiu, Yi-Hao 1145
Cho, Geum-Bae 2145
Cho, In-Ho 3323
Cho, Shin-Young 1530
Cho, Young Joon 137
Cho, Younghoon 1403
Choe, Chanyang 1598
Choi, Byungcho 1465
Choi, Hyun-Jun 383
Choi, Jae Hyuk 1336
Choi, Jaeho 803
Choi, Joon-Ho 982, 1799
Choi, Seung-Hyun 4049
Choi, Sewan 256
Choi, Sung-Jin 1409
Choi, Youn-Ok 2145
Chou, Shih-Feng 1711
Chou, T.-C. 1912
Choudhury, Abhijit 3401
Chuai, Guoming 3025
Chung, Daewoong 1141
Chung, Henry S. H. 917
Chunkag, V. 1622
Collins, Caspar 1931
Cortes, Camilo 2193

AUTHOR INDEX

Corvasce, C. 3596
Cucala, Asuncion P. 2534
Cui, Shenghui 2250, 2484
Cui, Xiang 1125
Cvetkovic, Igor 790, 3985
Czyz, Piotr 396
D'arco, Salvatore 782, 2003
Da Silva, C. 267
Dahidah, Mohamed S A 3215
Dai, T. 1149
Dai, Wenjing.....................................1015
Daikoku, Akihiro 2796
Danqing, Liu 1376
Dao, Ngoc Dat...................................1212
Dauphin, Benjamin 3644
Davari, Pooya 746
Davletzhanova, Z 1149
De Doncker, Rik W. 375, 388, 598,
    1073, 2250, 2484, 2768, 3729, 3979
Decker, Simon 2848, 3074
Delaforge, Timothé 2158, 3820
Deng, Fujin 1758, 2708
Deng, Jinxin 2992
Deshpande, Prathamesh Pravin..................4186
Dieckerhoff, Sibylle............................3607
Dimarino, Christina 3985
Din, Zakiud 2262
Ding, Yong 815
Dinh, Nguyen Duy 363
Diouf, Fatou 2078
Dirksen, Daniel..................................2410
Divan, Deepak 4001
Doki, Shinji 1032, 1223, 1228, 1295, 1747, 2224
Dong, Hanjing 987
Dong, Mi 1771
Dong, Qinghua 459
Dong, Xiaofeng 4168
Dong, Zhen 459
Dong, Zheng 3768
Driesen, J. 488
Du, Chao 2204
Du, Xiaotong 1167, 2780
Du, Xizhou 1491
Du, Yan 1472, 2877
Du, Zhijiang.......................................84
Duarte, J. L. 946, 1067, 2697
Duarte, Jorge L. 1447, 3840
Dugal, F. 3596
Dujic, Drazen 422, 1484, 1498, 2170
Duong, Truong-Duy 982
Duque, C. A. 1067
Eberle, Wilson 927
Ekman, Jonas 3588
Elbaset, Adel A. 3945
Endegnanew, Atsede G. 2003

Endo, Hiroaki 4151
Endres, Tobias Maximilian 2410
Engelmann, Georges 3979
Enomoto, Bruno Yukio 3785
Eto, Haruhi 2097
Faiz, Muhammad Talib 528
Fajri, Poria 3223
Fan, Dongchen...............3002, 3010, 3015
Fan, Shengwen 977, 3040
Fan, Weiyan 1386, 1421
Fang, Jingyang....................... 337, 3910
Fang, Ran 4213
Fangfang, Luo 1282
Farkas, Gabor 137
Fayyaz, Asad 130
Felderer, Niklaus 2199
Feng, Chao 2058
Feng, Wei 3678
Ferdowsi, Mehdi 3223
Fernandez, Gabriel...............................3209
Fernandez-Cardador, Antonio......................2534
Fischer, F.......................................3596
Foo, Gilbert 1724
Formentini, A. 4034
Freijedo, Fracisco D. 1498
Friedrichs, Peter 3584
Fuchs, Simon 2301
Fujii, H. 1253
Fujii, Kansuke 3711
Fujii, Keisuke 1189
Fujii, Toshiyuki 2540, 3578
Fujimoto, Hiroshi 77, 663
Fujimoto, Kazuki.................................2047
Fujimoto, Yasutaka 571, 681
Fujimura, Akira 1080
Fujita, Atsushi 296
Fujita, Goro 363
Fujita, Hideaki...........626, 1854, 3813, 3940
Fujiwara, Hajime 1381
Fujiwara, Kazuya 3773
Fukuda, Hiroto 2938
Fukuda, Kenji 2558
Fukui, Tomoya 860
Fukuoka, T. 1240
Fukushima, Kentarou 2176
Fukushima, Takafumi 3478
Funabiki, Shigeyuki 2449
Funaki, Tsuyoshi................309, 2181, 3092
Funato, Hirohito 94, 2036
Funato, Hiroki 2073
Furukawa, Keita 3349
Furukawa, Kimihisa 3572
Furukawa, Yudai..................................4193
Furusho, Yasuaki 3711
Gan, Yiliang 1391

AUTHOR INDEX

Ganisetti, V. K. 2907
Gao, Feng 2016, 3383, 3965
Gao, Xiaonan 1661
Gao, Zhuo 3455
Garrigós, A. 1435
Gasim, Abdulaziz 2836
Gehlot, Deepak 3471
Geng, Hua 542
Geng, Yiwen 619
Gerada, C. 4034
Gheonjian, Anna 2078
Gietler, Harald 3140
Gohara, Hiromichi 2764
Gondo, Ryota 3490
Gong, Bing 3797
Gong, Chunying 1015, 1177
Gong, Z. 267
Gorodnichev, Anton 375
Goto, Akihisa 2449
Goto, Hiroki 3192
Goto, Kazuya 1315
Goto, Yasuyuki 809
Gou, Yating 1157, 1167
Grimm, Ferdinand 2895
Grossner, Ulrike 3588
Gruber, Wolfgang 3632, 3640, 4028
Gu, Lei 632
Gu, Qing 2963
Guajardo, Cristian Andres Garces 1854
Guan, Bo 1032
Guan, Yajuan 2668, 3678
Guan, Yueshi 614, 3780
Guangzhu, Wang 1376
Guerrero, Josep M. 1498, 2668, 3112
Guerrero, M. Josep 3678
Gui, Yonghao 2668
Guidi, Giuseppe 782, 2003
Guillod, Thomas 396
Gunji, Daisuke 663
Guo, Leilei 904
Guo, Yanjie 3338
Guozhao, Duan 2625
Gupta, K. 267
Gurpinar, Emre 130
Gutiérrez, R. 1435
Ha, Jung-Ik 565, 2500
Ha, Sang-Hyun 3466
Haga, Hitoshi 1370, 3890
Hagiwara, Makoto 3273
Hahashi, Yuji 4059
Haider, M. 4080
Halamicek, Michael 831
Halick, Mohamed 416
Hamabe, Yasumasa 1276
Hamada, Shizunori 227

Hamaguchi, Takumi 3507
Hamasaki, S. 1240
Hamasaki, Shin-Ichi 1217, 1276, 2938, 3237
Hameyer, Kay 740
Han, Byung-Moon 466
Han, Jung-Kyu 3107, 3533, 4049, 4054
Han, Pengcheng 1027, 2714
Han, Yang 3112
Hanajiri, Kensuke 663
Hanamoto, Tsuyoshi 1315, 1698
Hancioglu, Oguz Kaan 1680
Handa, Hiroyuki 3762
Handa, Yuuichi 4059
Hane, Yoshiki 2426
Hang, Lijun 1391, 2866
Hanju, Cha 1985
Hao, Liu 3484
Hao, Xiang 1478
Harnefors, Lennart 3684
Hartmann, S. 3596
Haruna, Junnosuke 94, 2036
Hasegawa, Kazunori 1938
Hasegawa, M. 3665
Hasegawa, Ryuta 2011
Hashempour, Mohammad M. 4198
Hashimoto, Kazuki 3757
Hasler, Jean-Philippe 3684
Hata, Katsuhiro 663
Hata, Ryotaro 2149
Hatakeyama, Tomoyuki 1991
Hataya, Morimasa 410
Hatipoglu, E. 3805
Hatsumi, Takuya 94
Hatta, Yoshiyuki 675
Hattori, Fumiya 2738
Hattori, Keisuke 3286
Haung-Jen, Chiu 645
Hayashi, Nobuo 866
Hayashi, Yuji 356
He, Wangpin 560
He, Xiaokun 1504, 2496
He, Xiaoqiong 1027, 2714
He, Yigang 2317
He, Yingjie 3439
Hendrix, M. A. M. 946, 2697
Heo, Jongwon 726
Hidaka, Yuki 2820
Higuchi, Keiichi 2764
Higuchi, Masato 3952
Higuchi, Shinichi 2216
Hikaru, Naruse 3418
Hikihara, T. 3665
Hikihara, Takashi 3654, 3757
Hiller, Marc 3074
Hillers, A. 1896

AUTHOR INDEX

Hillers, André 2301
Hilt, Oliver 3607
Hinz, Arne 598
Hirahara, Hideaki 1960
Hiraki, Eiji 410, 1602, 1610
Hirao, Takashi 2082, 2137
Hirase, Yuko 767
Hirayama, Katsutoshi 4193
Hirayama, Tadashi 3406
Hirokawa, Masahiko 1543, 4133
Hirokawa, Takayuki 296, 410
Hiromoto, Masayuki 3644
Hirose, Keiichi 593, 822
Hirose, Naoki 3791
Hiroshi, Tadano 3431
Hiroshige, Shinichi 3369
Hirota, Takashi 3952
Hoang, Tuan V. 1752
Hoda, Isao 2073
Hofmann, Viktor 2476
Hofmann, Wilfried 3243
Hojo, Masahide 3369
Holenstein, Thomas 3619
Holmes, D. G. 3670
Hong, Miao 2718
Hongpeng, Liu 1442, 2969
Honjo, Satoshi 2066
Hori, Motohito 3396
Hori, Yoichi 77, 663
Horie, Shunsuke 809
Horikoshi, Takahiro 1997
Hoshi, Nobukazu 971, 2660, 3855
Hou, Chung-Chuan 1617
Hou, Lijun 2901
Houran, Mohamad Abou 1062, 2854
Hsieh, Guan-Chyun 123
Hsieh, Hung-I 123
Hsieh, Yao-Ching 3151, 3544
Hsu, Chi-Hsuan 2653
Hu, Jiewen 3985
Hu, Jingxin 1073, 2250, 2484
Hu, Sheng 3052
Hu, Song 370
Hu, Xihong 614, 3780
Hu, Xing 2262
Huang, Bing-Siang 2092
Huang, Bo-Jia 3528
Huang, Chien-Chun 3151
Huang, Huazhen 1125
Huang, Jingjin 2980, 4157
Huang, Jingjing 1004, 2688, 2692
Huang, Jun-Xian 1626, 3081
Huang, Lang 1478
Huang, Pin Yu 2165
Huang, Ta-Wei 1626

Huang, Wen-Mei 2576
Huang, Xianjin 1131, 2051
Huang, Xiaoliang 84
Huang, Xuehao 3455
Huang, You-Chun 275
Huemer, Mario 3140
Hui, S. Y. Ron 889, 2552
Hung, Chun-Yao 2576
Hung, Shun-Kang 1575
Huo, Chongcan 987
Huo, Junya 1206, 1234
Hussein, Abdallah 130, 2932
Huynh, Dang Minh 3086
Hwang, Duck-Hwan 1403
Hwang, Seon-Ik 3323
Hwu, K.I. 851
Hyakutake, Y. 1253
Hyodo, Takashi 2589
Hyunsung, An 1985
Iannuzzi, Diego 2527
Ibuchi, Takaaki 309
Ichinose, H. 1240
Ide, Yuji 3896
Iijima, Ryuji 313, 1111
Iioka, Daisuke 2278
Ikari, Yuki 148
Ikeda, Hidehiro 1315
Ikeda, Yoshinari 3396
Ilves, Kalle 2335
Imai, Kazu 3363
Imai, Makoto 296, 410
Imamori, Satoshi 699
Imaoka, Jun 1087, 1095, 1554, 3773
Imoto, R. 2808
Imtiaz, Abu Saleh 2416
Imura, Takehiro 77, 663
Inaba, Tsuyoshi 4114
Inomata, Kentaro 3952
Inoue, Daisuke 2764
Inoue, Kaoru 1264, 2186, 4151
Inoue, Kent 348
Inoue, Masamichi 1228
Inoue, Takatoshi 1276
Inoue, Y. 704, 2808
Inoue, Yukinori 1189, 1289, 1329, 2802, 2814, 3197
Irino, Yusuke 244
Ise, Toshifumi 775, 2393, 3762, 3902
Ishibashi, Mikiya 1370
Ishibashi, Naoyuki 1543
Ishibashi, Taku 2292
Ishigaki, Shingo 227
Ishiguro, Takahiro 1997, 2011, 3304
Ishihara, Masataka 1610
Ishii, Y. 1834
Ishii, Yuki 1196

AUTHOR INDEX

Ishikawa, Hiroki...2176, 3412
Ishikawa, Kohsuke......................................2725
Ishikura, Yuki....................1087, 1095, 3717
Isobe, Eisuke..2042
Isobe, Takanori.................313, 1111, 3375, 3431
Isozaki, Keisaku...1364
Itaya, Yohei..3450
Ito, Kazuhiko..2540
Ito, Yasuaki.......................................1586, 2324
Ito, Yoichi...3086
Ito, Youichi...439
Itoh, Gimpei...1289
Itoh, Jun-Ichi...................69, 348, 534, 896, 1567,
    2229, 2237, 2519, 2596, 3349, 3797
Iwabuchi, Akio..439
Iwai, Akinobu...2066
Iwaji, Yoshitaka..1301
Iwasaki, Makoto...1666
Iwasaki, Tetsuya...3490
Iwata, Hiroki...3896
Iyasu, Seiji..4059
Iyoda, Isao..2914
Jacobs, Keijo..3292
Jaffar, Hanis Afiqah Binti..........................2956
Jain, Prashant..3471
Janah, Mounia...681
Jang, Duekjin...2619
Jang, Yu-Jin..1655, 3466
Jang, Yun...1736
Jangs, Yujin...1562
Jarutus, Neerakorn......................................2121
Jehle, Andreas...1509
Jennings, M...1149
Jeong, Seog Y...2564
Jeong, Si-Hoon..289
Jeong, Yeonho...................838, 2365, 2376
Jhang, Ying-Yi..3884
Jhou, Yu-Lin...1145
Ji, Guyuan...2921
Jia, Haiyang..998
Jia, Pengyu..977, 3040
Jia, Xu...3025
Jiacheng, Wang...2986
Jiajie, Zang...2986
Jiajie, Zhou..1442
Jian, Jun-Min..2653
Jiang, Jinhai..84
Jiang, Shuai..987
Jiang, Siyue..4168
Jiang, Yanfeng..2058
Jiang, Yongbin..3863
Jianhua, Wang..1282
Jianming, Xu...528
Jianqiao, Zhou..2986
Jianwen, Zhang...2986

Jiaxing, Liu...1376
Jikumaru, Takehiro.....................................177
Jimichi, T..1834
Jimichi, Takushi..3729
Jin, Nan...904
Jin, Zheming...2668
Jing, Lei..878
Jing, Lyu...2692
Jing, Yang..3383
Jingyu, Song...1282
Jing-Yuan, Lin...645
Jinjun, Liu...4181
Jinshui, Zhang...4181
Jisaki, Jun..3182
Joebges, Philipp..375
Jongudomkarn, Jonggrist............................3902
Jonishi, Akihiro..2774
Joryo, Satoshi...1202
Joseph, Anto...1358
Jumayev, S..161
Jung, Hanul...688
Jung, Hyun-Sam..911
Jung, Jae-Jung..3557
Jung, Jee-Hoon...................................289, 383
Jung, Jun-Hyung...3323
Jung, Si-Hoon...383
Jungmayr, Gerald..3640
Junior, Lourenço Matakas...........................3785
Jynu-Jhe, Jhang..645
Kada, Haruya..3890
Kadota, Mitsuhiro.......................................3572
Kai, Masahiko...1803
Kaicheng, Ding...4181
Kaipia, T...2948
Kaishakuji, Hikaru......................................2360
Kakigano, Hiroaki.............................583, 2956
Kamaeguchi, Koki.......................................410
Kamakura, Kousuke....................................2756
Kamejima, Takayoshi..................................3286
Kamiya, Naoki..1673
Kamiyama, Naosumi....................................1955
Kamoshida, Naoki.......................................1111
Kampeerawar, Warayut...............................3257
Kanai, Naoyuki...3396
Kanaya, Kazuhisa.......................................2011
Kanazawa, Yasuki.......................................2789
Kanchan, R. S...488
Kandula, Prasad..4001
Kaneko, Satoshi..3396
Kanetani, Kaisei..207
Kang, Dong-Hun..3030
Kang, Feel-Soon..2376
Kang, Kyoung-Suk......................................922
Kang, Tahyun..1977
Kang, Yong..2997

AUTHOR INDEX

Kanno, Junya 3299
Kano, Fumihisa 2036
Kanoda, Akihiko 3572
Kanzian, Marc...3140
Kapisch, E. B. 1067
Karami, Bagher...2854
Karppanen, J...2948
Kasai, Yuji..2036
Kashihara, Tatsuki 1741
Katayama, Tatsuji 1346
Kato, Hideaki 1580, 1586, 2324
Kato, Hirokazu..3478
Kato, Koji 439, 1370, 3086
Kato, Toshiji...........................1264, 2186, 4151
Katoh, Kaoru..233
Katoh, Shinji 2176
Katsuki, Akihiko 1543
Katsura, Seiichiro 669
Katsura, Shogo 767
Katsushi, Terazono 3431
Kawabata, Naoki......................................2887
Kawabata, Shuma 3406
Kawagoe, Natsuki 3490
Kawaguchi, Hironori 517
Kawaguchi, Jun'ichiro 1828
Kawaguchi, Yuki.......................................3572
Kawakami, Masaki....................................2756
Kawakami, Noriko....................................1346
Kawamura, Atsuo 318, 1649, 1687, 3916
Kawamura, Itsuo 3396
Kawamura, Kazuki 1567
Kawanishi, Kota..169
Kawashita, Jun...2042
Kayashima, Kazuya 1315
Kaymak, Murat..3729
Kazmi, Syed Muhammad Raza 4168
Ke, Junji 1125
Kennel, Ralph................... 1661, 2895, 3965
Kezuka, Nobutaka 227
Khan, Ashraf Ali 927
Khan, Faisal............................446, 2416
Khan, Muhammad Mansoor.......................528
Khan, Usman Ali 927
Khomfoi, Surin...1460
Khubchandani, Vasudha 845
Kiatsookkanatorn, Paiboon.......................2581
Kida, Masahiro.........................1586, 2324
Kido, Tatsuya 329
Kikuchi, Ryosuke 1877
Kikuchi, Takaaki 2292
Kikuchi, Takeshi 3578
Kikuma, Toshiaki 3299
Kim, Byeongwoo 256
Kim, Chong-Eun 108, 3538
Kim, Dong-Kwan 1655, 3466, 3538

Kim, Gun-Woo 838
Kim, Hansang 1465
Kim, Heung-Geun 927, 1046, 2619, 3134
Kim, Hideaki... 207
Kim, Hyeon-Sik 521
Kim, In-Dong 3229
Kim, Jae-Kuk... 3100
Kim, Jang-Mok 3323
Kim, Jin-Hak 1530
Kim, Jin-Young 3229
Kim, Jong-Woo 3107, 4049
Kim, Kangsan 256
Kim, Katherine A. 2092, 3063
Kim, Keon Young 4104
Kim, Keon-Woo..................108, 1562, 1655, 2365, 2376
Kim, Ki-Mok 2365
Kim, Myong Hwan 2500
Kim, Sanghun 2619
Kim, Sunju 3833
Kim, Yeonjung.. 1465
Kimura, Hideki 2036
Kimura, Mamoru 1991, 1997
Kimura, Noriyuki............1202, 1259, 2558, 2887, 2914
Kinoshita, Masahiro 3929
Kishimoto, Toshihiko 261
Kishita, Ken 1301
Kitagawa, Wataru 1847, 3507
Kitamura, Akio 2764
Kitamura, Toshinori 2660
Kiyoshi, Ohishi 1673
Kiyota, Kyohei 3182
Klammer, Bianca.................................... 3632
Ko, Chien-Tzu 2107
Kobayashi, Hiroyasu 2527, 3490
Kobayashi, Koji 1741
Kobayashi, Marika.................................. 2802
Kodaka, Wataru 2589
Kogai, Naoki 1364
Koizumi, Hirotaka................................... 4114
Kolar, J. W. 3805, 4080
Kolar, Johann W. 181, 396, 3619
Kolb, Johannes 2848
Komaru, Yuma 1329
Komatsu, Hiroyoshi 1346
Komatsu, Taiga 2820
Komatsu, Wilson 3785
Komeda, Shohei 3813
Kometani, Haruyuki 711
Kondo, Keiichiro 726, 2047, 2527, 3490
Kondo, Shota 1295
Kondo, Takeshi....................................... 4114
Kong, Wei 3460
Kongjeen, Yuttana 2430
Konishi, Akihiro 1602
Konno, Junya... 1692

AUTHOR INDEX

Konstantinou, Georgios 3117
Kopta, A. 3596
Kosaka, Takashi 3418
Koseki, K. 1162
Koseki, Takafumi 2042, 2309, 3257
Koshikizawa, Hiroyuki 1567
Kostov, Konstantin 2732
Kouketsu, Masaju 227
Kouno, Yusuke 2176
Kovacevic-Badstübner, Ivana 3588
Kowatari, Hiroki 2660
Koyama, Yushi 2011
Krismer, F. 3805
Krismer, Florian 396
Kubo, Hajime 483
Kubota, Hisao 1196
Kucka, Jakub 1904
Kumada, Keishirou 3396
Kumagai, Shuta 1264
Kumar, Ashish 3993
Kumar, Rajesh 2456
Kumar, S. Gautam 3471
Kumsuwan, Yuttana 2113, 2121
Kunomura, Ken 1803
Kuo, Chun-Ting 1145
Kuraishi, Daigo 3896
Kuraku, Nagendra Vara Prasad 2317
Kuring, Carsten 3607
Kurisaka, Masakatsu 4151
Kurita, Naoyuki 1991
Kurita, Nobuyuki 3640
Kurokawa, Fujio 826, 2097, 2283, 4193
Kurosawa, Nobuhito 1810
Kurumatani, Hiroki 669
Kusaka, Keisuke 69, 348, 2237, 3349
Kusumah, Ferdi Perdana 3870
Kuwata, Gen 177
Kwon, Min-Jun 114
Kyyrä, Jorma 2193, 3870
Lai, Jih-Sheng 3107, 4049
Lai, Jui-Hung 3081
Lan, Yuanliang 1167
Lana, A. 2948
Le, Hanh-Phuc 213
Le, Hoai Nam 2519
Lee, Byoung-Hee 838
Lee, Byung-Kwon 3030
Lee, Chan 688
Lee, Choongin 565
Lee, Dong-Choon 478, 1212
Lee, Hong-Hee 1752
Lee, Hyong Gun 1336
Lee, Il-Oun 1530
Lee, Jae-Bum 3100
Lee, Jia-You 657, 2107

Lee, Joon-Hee 3557
Lee, Junbae 1141
Lee, June-Hee 466
Lee, Jung-Yong 1403
Lee, Jun-Young 3030
Lee, Jusuk 1336
Lee, Kyo-Beum 466, 1736, 4104, 4109
Lee, Kyoung-Won 2145
Lee, Kyung-Hwan 2500
Lee, Min-Su 108
Lee, Minsub 1141
Lee, Nayoung 1562
Lee, Song-Kai 2102
Lee, T. L. 4198
Lee, Tzung-Lin 2576
Lee, Woo-Cheol 114
Lee, Woo-Seok 1530
Lee, Young-Dal 3466, 3538
Lehn, Peter W. 3203
Lei, Qin 2400, 3742
Leng, Darith 1764
Leubner, Martin 3243
Li, Bodong 878
Li, Chi 790, 3705
Li, Dongsheng 1301
Li, Fei 2611
Li, Fujian 2944
Li, Guanglei 1455
Li, Haijin 2270
Li, Haisi 3040
Li, Haoyu 2901
Li, Hong 2058
Li, Hongchang 337, 3910
Li, Jhih-Sian 3081
Li, Jia 2073
Li, Jianfeng 130
Li, Kaiyuan 1517, 1592
Li, Lei 1172
Li, Li 1771
Li, Ming 2973
Li, Mingshen 2668, 3678
Li, Pengcheng 3698
Li, Shufan 3338
Li, Sinan 889, 2552
Li, T.-Y. 1912
Li, Xiaodong 370
Li, Xiaolu Lucia 3768
Li, Xiaoqiang 3910
Li, Xingshuo 453
Li, Xinying 2646
Li, Yan 2245
Li, Yang 795, 1478
Li, Yangman 2901
Li, Yi-Chan 639, 883
Li, Yongdong 1010, 2386

AUTHOR INDEX

Li, Yong-Jyun 275
Li, Yunwei 3958
Li, Yunwei Ryan 1537
Li, Yuze 2997
Li, Zhenjie 84
Li, Zhenwei 998
Li, Zhiqing 100
Liang, Daniel 1943
Liang, Junrui 4122
Liang, Ning 1157
Liang, Wencai 1131
Liao, Chenglin 3338
Liao, Chih-Yi 657
Liao, Hsuan 2653
Liao, Jian-Tang 4233
Liao, Mengyan 1386, 1421
Liaw, C. M. 2907
Lim, Cheon-Yong 1655, 2376, 3533
Lim, Dae-Sik 1212
Lim, Kyungbae 803
Lim, Young-Cheol 982, 1799
Lin, Chang-Hua 1341, 1777
Lin, Cheng-Hung 2092
Lin, Fei 1131, 1816, 2051, 2058, 3484, 3495
Lin, Jin 3460
Lin, Jing-Yuan 3151
Lin, K.-E. 1912
Lin, Min 4133
Lin, Xiang 3460
Lin, Xiaolan 1027
Lin, Xuerui 1537
Lin, Yu-Hsiu 1575
Lin, Yu-Lin 1145
Lisha, Chen 3958
Liske, Andreas 2848
Liu, Baojin 1051, 2944, 3924
Liu, Bi 1872
Liu, Bo 542, 878
Liu, Chao 2245
Liu, Chunhui 3742
Liu, Cuicui 1157, 1167
Liu, Dong 1758, 2708
Liu, Fang 2611, 2992
Liu, Furong 3052
Liu, He 3215
Liu, Hwa-Dong 1341, 1777
Liu, Jia 775, 3902
Liu, Jiaxin 2016
Liu, Jinjun 957, 1051, 1524, 2435,
        2646, 2681, 3129, 3176, 3251, 3439, 3692,
        3924, 3934
Liu, Junwen 3863
Liu, Kangli 3010, 3015
Liu, Nianzhou 1010
Liu, Ning 2877

Liu, Pang-Jung 2102
Liu, Ruofei 2547
Liu, Shu 3052
Liu, Siqi 1491
Liu, Tao 1478
Liu, Teng 2681, 3176, 3934
Liu, Wei 3164
Liu, Wenzhao 3678
Liu, Xiaosheng 934
Liu, Xicai 1661, 3965
Liu, Xinbo 3455
Liu, Yifu 2400, 3742
Liu, Yu-Chen 2092
Liu, Yuping 1816
Liu, Zeng 957, 1524, 2435, 2681,
        3176, 3251, 3692, 3749, 3924
Liu, Zhiyuan 3495
Liu, Zipeng 2681, 3176
Lo, Jen-Hao 1145
Lomonova, E.A. 161
Lopez-Lopez, Alvaro J. 2534
Lotfi, Nima 3223
Lovison, Giorgio 77
Lu, David H. 2404
Lu, David Hongfei 3390
Lu, Kaiyuan 1183, 1246, 2842
Lu, M. Z. 2907
Lu, Shengli 3145
Lu, Shuai 3698
Lu, Y. 267
Luhtala, Roni 547, 2470, 3156
Lunglmayr, Michael 3140
Luo, Min 422, 2199
Luo, Rui 3129, 3439
Luo, Y. 267
Luong, Hoan-Tien 2145
Lyu, Jing 1004, 4162, 4220
Ma, Baohui 2882
Ma, Jie 1118
Ma, Ke 3877
Ma, Shaokang 542
Ma, Tianshu 2703
Ma, Yue 3717
Ma, Zhixun 917, 2688, 2692, 4157, 4162
Mabuchi, Yuichi 3572
Machavolu, Sawanth Krishna 753
Machida, Yuuki 2449
Maharjan, Laxman 1840
Makishima, Shingo 2047
Mannen, Tomoyuki 1414, 1866
Mantooth, H. Alan 153
Mao, Meiqin 815, 1472, 1793, 2505
Mariéthoz, Sébastien 2158, 3820
Marinescu, Radu-Florin 1822
Marroquí, D. 1435

AUTHOR INDEX

Martinez, Wilmar 2193
Maruta, Hidenori 826
Maruyama, Kouji 3396
März, Martin 2410
Masuda, Eisuke 309
Masuda, Mitsuru 88
Masuko, Toshitake 3723
Matsubayashi, Tatsushi 207
Matsuda, Akihiro 2329
Matsuda, Tomohiro 1972
Matsudate, Koki 2022
Matsui, Nobumasa 826, 2283
Matsui, Nobuyuki 3418
Matsui, Teruhisa 1803
Matsui, Yoshihiro 1080
Matsui, Yuto 1847, 3791
Matsuki, Yosuke 2224
Matsumori, Hiroaki 3357
Matsumoto, Satoshi 2360, 2370
Matsumoto, Takashi 2404
Matsumoto, Toshiaki 2011, 3304
Matsumoto, Yasuaki 517
Matsumoto, Yohei 233
Matsumura, Toshiro 809
Matsuo, Keisuke 169
Matsuse, Kouki 169
Mattsson, A. 2948
Mawby, P 1149
Mcgrath, B. P. 3670
Meng, Xin 957, 1549, 3251
Menzi, David 181
Mertens, Axel 1904
Messo, Tuomas 547, 963, 1704, 2470, 3156, 4205
Michihira, Masakazu 992, 3058
Michikoshi, Hisato 2558
Milovanovic, Stefan 1484
Min, Geon-Hong 2500
Minami, Masataka 992, 3058
Mino, Kazuaki 3717
Mira, Maria C. 1351
Mishima, Tomokazu 329, 872
Misra, Mitradatta 3884
Mitsantisuk, Chowarit 3332
Miura, Yushi 775, 2393, 3762
Miwa, Yoshihiro 404
Miyajima, Hiroki 1803
Miyama, Yoshihiro 711
Miyawaki, Satoshi 2738
Miyazaki, Toshimasa 1673
Mizumoto, Yuki 1810
Mizuno, Takayuki 169
Mizuno, Yuji 2283
Mizushima, Takuya 1543
Mocevic, Slavko 3985
Mochidate, Sae 1972

Mogorovic, Marko 2170
Moiannou, Tom 831
Mok, Hyung Soo 1336
Molinas, Marta 759
Moo, Chin-Sien 275, 3544
Moon, Gun-Woo 108, 838, 1562, 1655, 2365, 2376, 3100, 3466, 3533, 3538, 4049, 4054
Mori, Kazuhisa 233
Morimoto, Hiroaki 2540, 3265
Morimoto, S. 704, 2808
Morimoto, Shigeo ...1189, 1289, 1329, 2802, 2814, 3197
Morimoto, Shinya 2210
Morishima, Naoki 2540, 3450
Moriyama, Hiroyuki 1580, 1586, 2324
Morizane, Toshimitsu 1202, 1259, 2558, 2887, 2914
Mortimer, Benedict J. 598
Motegi, Shin-Ichi 992, 3058
Motohashi, Yuto 753
Motoyama, Hiromasa 356
Mouawad, Bassem 130
Mukaiyama, Naoki 2558
Müller-Hellmann, Adolf 598
Muni, Bishnu Prasad 3471
Murakami, Toshiyuki 575
Nabetani, Yoichi 2404
Nada, Kaho 3578
Nagai, Sakahisa 1687
Nagai, Satoshi 534
Nagao, S. 142
Nagaoka, Naoto 3170
Nagaoka, Shingo 118, 4139
Nagasaka, Kuniaki 1692
Nagashima, Takumi 3490
Nagira, Yoshiki 4016
Naina, Sagar 3046
Nakabayashi, Shigeaki 1692
Nakabayashi, Shigeyuki 517
Nakagawa, Hidehiko 767
Nakahara, Kengo 3237
Nakahara, Mizuki 3572
Nakai, Masanobu 3182
Nakajima, Mizuki 2750
Nakajima, Tatsuhito 1997, 3299
Nakamura, Fuminori 2329
Nakamura, Hideyo 1137
Nakamura, Kenji 2426
Nakamura, Kimikazu 4059
Nakamura, M 201
Nakamura, Masashi 471
Nakamura, Ritaka 495
Nakano, Hayato 2764
Nakano, Shigeki 2370
Nakao, Hiroshi 196
Nakao, Kazushige 148, 2914

AUTHOR INDEX

Nakao, Yuta 588
Nakashima, Yoshiyasu 196
Nakatsu, Kinya 2082
Nakazawa, Haruo 2404
Nakazawa, Y. 1253
Nakazawa, Yuji 244
Namba, Akihiro 2082
Nanamori, Kimihiro 2789
Naradhipa, Adhistira M. 3833
Narita, Takayoshi 1580, 1586, 2324
Narushima, Hiroki 693
Nashida, Norihiro 1137
Nasr, Miad 2380
Natori, Kenji 588, 1860
Nawaz, Muhammad 2335
Nazib, A. A. 3670
Nee, Hans-Peter 2732, 3292, 3684
Neubert, Markus 3979
Ngamroo, Issarachai 2287
Ngo, Tung 1724
Nguyen, Bang Le-Huy 1046, 3134
Nguyen, Hong-Quan 3426
Nguyen, Minh-Khai 982, 1799, 2145
Nguyen, Tien-The 1046, 3134
Nho, Eui-Cheol 922
Nicolae, Ileana-Diana 1822
Nicolae, Petre-Marian 1822
Nie, Jintong 2963
Niki, Toru 856
Nimura, Takumi 1295
Ninomiya, Tatsuya 2836
Nishikata, Shoji 4227
Nishimura, Yoshitaka 1137
Nishino, Taisei 1364
Nishiyama, Shigeki 2149
Nishizawa, Koroku 2229
Nishizawa, Shin-Ichi 1938
Niu, Haonan 3025
Niyomsatian, K. 4096
Noah, Mostafa 1087, 1095
Noda, Taku 2176
Noda, Yujiro 324
Noguchi, Toshihiko 718, 753
Noh, Seungjun 1598
Nomura, Naofumi 2216
Nomura, Shinichi 2022
Nonogaki, Midori 2292
Noro, Osamu 767
Norrga, Staffan 3292
Norum, Lars 2630
Noto, Yasuyuki 3711
Notohara, Yasuo 1301
Nuchnoi, S. 4096
Nugroho, Dannisworo S. 3855
Nussbaumer, Thomas 3619

Nuutinen, P. 2948
Obara, Hidemine 1649
Oda, Yoshiho 1586, 2324
Ogasawara, Satoshi 2589, 2725, 2796, 3315
Ogawa, Eri 2768
Ogawa, Kazuki 1580
Ogawa, Takuro 866
Ogawa, Tomoyuki 1828, 3265
Ogawa, Toru 2796
Ogino, Hiroshi 517
Oh, Sehoon 688
Ohashi, Hidetomo 2774
Ohdera, Fumiya 1322
Ohguchi, Hideki 699
Ohishi, Kiyoshi 1741, 3332, 3890, 3896
Ohji, Takahisa 3182
Ohnishi, Haruna 3273
Ohno, Takanobu 971
Ohno, Tatsuki 1649
Ohnuma, Naoto 233
Ohnuma, Takumi 1223
Ohnuma, Yoshiya 2738
Ohta, Kazuki 1223
Ohta, Takahiro 517
Ohtake, Asuka 3286
Ohyama, K. 1253
Ohyama, Kazuhiro 2921
Ohyama, Kazunobu 244
Oi, Kazunobu 1890
Oishi, Kazuki 3644
Oiwa, Takaaki 157, 4042
Oka, Toshiomi 2370
Okamoto, Kenkichiro 1095
Okazaki, Yuhei 2335
Okazawa, Toshio 2066
Oki, Yusuke 1828
Okitsu, Takashi 169
Okuda, Takafumi 3654, 3757
Okuno, Kengo 1586, 2324
Okuyama, Ryota 3450
Omori, Hideki 1202, 1259, 2558, 2887
Omori, Shuto 471
Omura, Ichiro 1938
Onishi, Hiroyuki 4139
Onishi, Masami 2082
Ono, Y. 4080
Onozawa, Yuichi 2768
Ooshima, Masahide 3613
Orikawa, Koji 2589, 2725, 3315
Ortiz-Gonzalez, J. 1149
Osawa, Akihiro 2764
Oshima, Takuya 4088
Osman, Ilham 3971
Ota, Ryosuke 3855
Ouaida, Rémy 2153

AUTHOR INDEX

Ouchi, Takayuki.................................250
Ouyang, Shaodi 1051, 3129
Ouyang, Ziwei 4066
Owaki, Daiki.....................................809
Paiboon, Supakorn.........................1642
Pairindra, Worapong.......................1460
Pan, Pengpeng 1504
Pan, Xuewei 1172
Panda, Sanjib Kumar 4186
Pang, Hui..2343
Papadopoulos, C.............................3596
Papini, L...4034
Paramalingam, Jan 2329
Parashar, Sanket 3993
Park, Hwa-Pyeong 289
Park, Jin-Hyuk 4104
Park, Jun H. 2564
Park, Kwon-Sik 922
Park, Moo-Hyun 1562, 3100, 3533
Park, Mu-Hyun 838
Park, Sang Uk 1336
Park, Sanghyeon 282
Partanen, J.....................................2948
Pasterczyk, Robert 2158
Patel, Prashant 3046
Patel, Utsav 3046
Pathmanathan, M. 488
Patwa, Premal 3046
Pauli, Florian 740
Pecharroman, Ramon R. 2534
Pei, Xuejun 2997
Peltoniemi, P..................................2948
Peng, Jinjie......................................939
Peng, Xu....................1027, 2714, 3020
Pengxiang, Zeng 4181
Pham, N. Ha 1414
Pidaparthy, Syam Kumar 1465
Pinomaa, P.....................................2948
Polmai, Sompob 1764, 2490
Pou, Josep 3117
Prabowo, Yos 3564
Prasanth, Sundararajan.....................416
Prodic, Aleksandar 831
Promyoo, Adisak 2871
Pueschel, Tilo...................................190
Pyrhonen, J.....................................161
Qi, Wenlong 889
Qian, Cheng 1472
Qian, Qinsong 3145
Qiao, Liang 3329
Qin, Zian 1925
Qiu, Maohang...................................878
Qiu, Zhifeng 939
Rabkowski, Jacek 2129
Radman, Karlo 3632

Radwan, Hamdy 3945
Rahimo, M. 3596, 4009
Rahman, Ahmad Arif Bin Abd 2956
Rahman, Faz 3971
Rahmati, Abdolreza 2854
Ramirez-Elizondo, Laura 1426
Ramos, Niño Christopher 3092
Ran, L. 1149
Ran, Li..1931
Rao, Eswar 3471
Rathore, Akshay Kumar 342, 2456
Reinikka, Tommi 1704
Remus, Nico 3243
Ren, Haijun 2714
Ren, Yu...3329
Rencz, Marta 137
Rengarajan, Satish 3564
Riar, Baljit 4074, 4145
Rietmann, Stefan 2301
Rim, Chun T. 2564
Risseh, Arash Edvin 2732
Rivas-Davila, Juan 282, 632, 3848
Robert, Mickaël 2158
Rodriguez-Diaz, Enrique 1498
Roes, M. G. L. 946, 2697
Roinila, Tomi 547, 1704, 1719, 2470, 3156, 4205
Romano, Daniele 3588
Roy, Sourov............................446, 2416
Ruan, Liheng 3010, 3015
Rubino, S. 732
Ruf, Andreas 740
Rygg, Atle 759
Sadakata, Hideki 410
Sagawa, Kouhei 2036
Saha, Tarak 4074, 4145
Saito, Tatsuhito 1828, 3265
Saito, Yota 1782
Saitoh, Hiroumi 2278
Sakabe, Tomoki 3058
Sakai, Kazuto 2826
Sakai, Ryosuke...............................2832
Sakai, Yoshikazu 4114
Sakawaki, Atsushi 244
Sakimoto, Kenichi 767
Sakiyama, Taiki 2186
Sakoda, Kenichi 860
Sakr, Nadim 2078
Sakuma, Kensuke 3522
Sakuraba, Tomokazu 2153
Sakurai, Seiya 3412
Samanta, Suvendu 342
Samermurn, S. 4096
Samizadeh, Mehdi 1062, 2854
Sanada, M. 704, 2808
Sanada, Masayuki ..1189, 1289, 1329, 2802, 2814, 3197

AUTHOR INDEX

Sangwongwanich, Ariya 2512
Sangwongwanich, S...4096
Sangwongwanich, Somboon 1642, 2581
Sannomiya, Kenta 1259
Sano, Kenichiro 3299
Sano, Toshiki 3896
Santi, Enrico 1719
Sasaki, Masahiro 2774
Sasaki, Masato 3344
Sasongko, Firman 416
Sathik, Mohamed 416
Sato, Fumihiro...250
Sato, Keisuke 3265
Sato, Kenji 3478
Sato, Mitsuru 118
Sato, Motoki..663
Sato, Takashi 3644
Sato, Yasuhiro 2042
Sato, Yukihiko 588, 1860, 1972, 3514, 3522
Satoh, Nobuo 2750
Sayed, Mahmoud A..3945
Schanen, Jean-Luc 2158
Schletz, Andreas 2410
Schülting, Philipp...388
Schweiker, Daniel...2848
Schweizer, Mario 555
Schwendemann, Rüdiger....................................3074
See, Kye Yak 2296
Sekiba, Yoichi 2176
Sekimoto, Morimitsu 866
Sekisue, Takayuki 2176
Sekiya, Hiroo 3650, 4127
Semwal, R. R. 1358
Senanayake, Thilak...313
Seng, Tan Chuan 416
Seo, Byuong-Jun 922
Seo, Gab-Su 213
Sera, Dezso..2512
Setiadi, Hadi..626
Settels, Sjef J...3840
Severson, Eric L..4020
Sewergin, Alexander 3979
Sha, Yilin 3329
Shabib, G. 3945
Shamseh, Mohammad Bani 3916
Shan, Zhenyu...977
Shang, Gao...1282
Shao, Chi...2866
Shao, Riming 1793
Sharma, Avinash 2456
Sharma, Sohit...1730
Shen, Yanfeng 1788, 1925
Shen, Yatao...815
Shen, Yecheng 2842
Shen, Zhan 1788, 1925

Sheng, Caiwang 1167
Shi, Gang 4220
Shi, Haixu 4168
Shi, Xiangyue 939
Shi, Yong 2877
Shibata, Naoya 3929
Shigeeda, Hidenori 2540
Shigematsu, Koichi 2176
Shigeuchi, Koji 3514, 3522
Shijo, Takuya 324
Shimada, Takae 250
Shimakage, Toyonari 2292
Shimamoto, Keita 2210
Shimao, Tohihiro 439
Shimaoka, Masahiro 1747
Shimizu, Toshihisa 302, 404, 2137, 2165, 3309, 3357
Shimizu, Toshimasa 1803
Shimomura, Shoji 2836
Shimono, Tomoyuki 675
Shimosato, Noboru 261, 3514, 3522
Shimoyama, A. 142
Shin, Sungyong 3418
Shinohara, Atsushi 1308, 1322
Shinohara, Hiroshi 1840
Shinshi, Tadahiko 4016
Shintani, Michihiro 3644
Shirai, Ryo 3309
Shirata, Kento 1137
Shiyuan, Yin 2625
Shoyama, Masahito 1095, 1554, 3773
Shujiang, Duan 2718
Shunsuke, Ohasi 3363
Shuto, Masao 699
Si, Yunpeng 2400, 3742
Sihvo, Jussi 2470
Sih-Yi, Lee 645
Silber, Siegfried 4028
Silventoinen, P. 2948
Simanjorang, Rejeki 416, 2296
Singh, Amit Kumar 4186
Singh, Vijay Kumar 1698
Son, Yung-Deug 3323
Song, Hongyu 3825
Song, Injong 803
Song, Kai 84
Song, Seung Min 3229
Song, Shuguang 1051, 3129, 3924
Song, Wensheng..1872
Song, Yang 3698
Song, Yipeng 746
Song, Yubo 3877
Soong, Boon-Hee 1517, 1592
Soong, Theodore 3203
Soontorntaweesub, Kittichot 1764
Spiliotis, K. 488

AUTHOR INDEX

Stieneker, Marco 598, 2484
Stock, Alexander 3034
Stojadinovic, Miloš....................................1103
Su, Huiling 795
Su, Jianhui 2877
Su, Yu-Chen 1038, 3549
Sudo, K. 1240
Suetake, A. 142
Suetsugu, Tadashi 4193
Sueuchi, Yuki 1955
Sugahara, Satoshi 2756
Sugahara, T. 142
Suganuma, K. 142
Suganuma, Katsuaki 1598
Sugihara, Yusuke 2789
Sugimoto, Hiroya 3627
Sugimoto, Kazushige 767
Sugiyama, Takashi 3578
Suh, Yongsug 1977
Sul, Seung-Ki 521, 911, 3557
Sumida, Hitoshi 2774
Sun, Bainan 607
Sun, Chuan 370
Sun, Haotian 2780
Sun, Jianning 2963
Sun, Kai 3460, 4168
Sun, Lejia 2882
Sun, Peng 1125
Sun, Shumin 1455
Sun, Weifeng 3145
Sun, Xiangdong 2204
Sun, Yongping 560
Sun, Yuchong 3650, 4127
Sung, Kyungmin 1364
Suntio, Teuvo 963
Supanyapong, S. 1622
Surakitbovorn, Kawin 632, 3848
Surinkaew, Tossaporn 2287
Suul, Jon Are 782, 2003
Suwa, Hiroshi 1997
Suwankawin, S. 4096
Suwankawin, Surapong 2871
Suzuki, Akio 1840
Suzuki, Dai 157
Suzuki, Hiromitsu 495
Suzuki, Kazuma 1847, 3501, 3507
Suzuki, Kenichiro 511
Suzuki, Toshiki 1586, 2324
Suzuki, Yuhei 3390
Suzumori, Hirofumi 2066
Tabata, Yoichiro 329
Tada, Makoto 1580
Tadano, Hiroshi 313, 1111, 3375
Tadano, Yugo 483, 1890
Taguchi, Masashi 826

Taguchi, Yoshiaki 3280
Taiyuan, Yin 2625
Tajima, Katsubumi 2832
Tajyuta, Toshihisa 1840
Takahashi, Akihiko 3896
Takahashi, Akiko 2449
Takahashi, Arata 1270
Takahashi, Isseki 575
Takahashi, Masaki 3186
Takahashi, R. 3665
Takahashi, Shotaro 3315
Takahashi, Tomohira 2796
Takahashi, Toshimichi 227
Takahashi, Yuki 3375
Takakura, Shotaro 1270
Takami, Hiroshi 471
Takamura, Kenya 1381
Takano, Sho 3390
Takasho, Kenta 1890
Takatori, Koji 4139
Takayanagi, Ryohei 3396
Takeda, Kodai 2309
Takemoto, Masatsugu 2589, 2725, 2796, 3315
Takenaka, Hiroshi 3304
Takeno, K. 201
Takenoiri, Shunji 2764
Takeshita, Takaharu 356, 1847, 3501,
    3507, 3791, 3945, 4088
Takeuchi, Norikazu 2292
Takeuchi, Yoko 1828, 3265
Takiguchi, Masashi 3723
Takimoto, Kazuyasu 3304
Takishima, Kenta 2826
Takubo, Hiromu 3390
Takuma, Shunsuke 2596
Takuno, Tsuguhiro 3578
Tamate, Michio 3315
Tan, Nguyen Anh 478
Tan, Siew-Chong 889
Tanaka, Akira 1960
Tanaka, Takaaki 2604
Tanaka, Takahide 2774
Tanaka, Toshihiko 324, 1381
Tanaka, Tsuguhiro 3929
Tanaka, Y. 1162
Tanemo, Masamichi 2022
Tang, Cheng-Yu 639
Tang, Houjun 528
Tang, Ye 3705
Tang, Yi 337, 428, 434, 3910
Taniguchi, Katsumi 3396
Taniguchi, Katsunori 1202
Taniguchi, Tomoisa 866
Tatsumi, Kazuto 1202
Tatsuta, Fujio 4227

AUTHOR INDEX

Tatte, Yogesh 1730
Tausif, Ali 3833
Tcai, Anatolii 4109
Techama, Pantarote 2490
Teerakawanich, Nithiphat................................3332
Teigelkötter, Johannes 3034
Tenconi, A. 732
Teraoka, Kenji 3086
Tey, Kuan-Chung..511
Thai, Van X. 2564
Thummala, Prasanth 4066
Tian, Mofan 998, 2785
Tian, Wei 1661
Tian, Xiaoyu 1771
Tian, Yanjun 1397
Tibola, Gabriel 1447
Tikka, V. 2948
Toba, Akio 1840
Toi, Takato 2229
Tokumaru, Syohei 2938
Tokusaki, Hiroyuki 2589
Tominaga, Isamu 1692
Tomita, Mutuwo..1295
Tong, Anping 1391, 2866
Tran, Hai N..3833
Tran, Tan-Tai...1799, 2145
Trescases, O. 267
Trescases, Olivier 2380
Tripathi, Ravi Nath 1698
Troppenz, Maria 3607
Trung, Tran Vu 1666
Tsai, Chang-Lin 3151
Tsai, Meng-Jiang...2462
Tsai, Men-Shen..1575
Tsai, Terng-Wei 639, 883
Tsai, Tsung-Lin 3151
Tsai, Yue-Ting...4198
Tse, Chi K. 3768
Tseng, King Jet.......................................1517, 1592
Tseng, Wei-Jing...1626
Tsuchiya, Taichiro 2329
Tsuji, Hitoshi 3717
Tsuji, M. 1240
Tsuji, Mineo...........................1217, 1276, 2938
Tsukakoshi, Masahiko 238
Tsumura, Akihiko 3490
Tsuno, Masahito 2558
Tsuruta, Ryoji 495
Tsuruta, Yukinori 318
Tsutsumi, Hirohiko 3723
Tu, Yiming.............. 2435, 2681, 3176, 3439, 3934
Tuji, Mineo 3237
Tumerdem, Ugur 1680
Tumurbaatar, Anudari......................................1972
Uchida, Junichi...1955

Uchida, Yuuki 2750
Uchino, Yuki 324
Uda, Ryosuke 3578
Udagawa, Ikuto 517, 1692
Ueda, Tetsuzo..3762
Uehara, H. 1253
Uematsu, Takeshi 118, 4139
Uemura, Takamasa 860
Ueno, Tsutomu 4151
Uesugi, Yuma 3412
Ueta, Hiroaki 1883
Umeda, Takashi 2814
Umetani, Kazuhiro.....................410, 1602, 1610
Unamuno, Eneko 759
Uno, Masatoshi 1782, 2030
Unterrieder, Christoph 3140
Ura, A. 704
Urabe, Shinichi 1782
Urata, Kazuki 302
Ute, Ryo 3773
Valente, G. 4034
Van De Ven, D.A.C. 267
Van Duivenbode, Jeroen 3840
Van Lam, Phi 571
Vasquez, C. Juan 3678
Vasquez, Juan C. 1498
Vass-Varnai, Andras 137
Veerachary, M. 845
Vemulapati, U. 3596, 4009
Vobecky, J. 3596
Vukadinovic, Nenad 831
Vyacheslav, Shkodyrev 1966
Wachi, Tsuneshisa 1997
Wada, Haruhisa...3286
Wada, Keiji 1414, 1866, 1919, 2137, 4059
Wakimoto, Hiroki..2404
Wang, Beibei 795
Wang, Bo 459
Wang, Can 1172
Wang, Chao 2386, 2901
Wang, Congling 3112
Wang, Dong 1183, 1246
Wang, Feng.......................................1157, 1167, 2882
Wang, Fusheng 2611, 2992
Wang, Gaolin 1206, 1234
Wang, Guoxin 1206
Wang, Hanyu 2997
Wang, Hao 2270
Wang, Haoyu 100, 3825
Wang, Hechao 1183, 1246
Wang, Hongjie 4074, 4145
Wang, Huai 1021, 1788, 1925, 2604, 2743, 3123
Wang, Huiying 1234
Wang, Jianing 2611
Wang, Jizhe 826, 2097

AUTHOR INDEX

Wang, Jun...3749, 3985
Wang, Kui..1010, 2386
Wang, Laili..2785, 3863
Wang, Liang..3958
Wang, Lifang...3338
Wang, Liwei...927
Wang, Meng...2992
Wang, Naizeng...998, 2785
Wang, Panrui..3383
Wang, Po-Wei...1617
Wang, Qiusheng..2421
Wang, Shike..1524, 3692
Wang, Shinn-Shyong...2086
Wang, Shitao..2866
Wang, Shunyu...3002
Wang, Wei..614, 3780
Wang, Wenjie...1391, 2866
Wang, Xiaolei...1455
Wang, Xiaoqing...878
Wang, Xiaoyang..453
Wang, Xiongfei................. 1711, 2673, 3164, 3357, 3684
Wang, Yanbo..1758, 2708
Wang, Yangyang..2505
Wang, Yi..1027, 1397, 3495
Wang, Yijie.....................614, 934, 3780, 3825
Wang, Youyun...2204
Wang, Yu-Chi...657
Wang, Yue...1455, 3863
Wang, Yuncheng..1177
Wang, Zhongxu..2743, 3123
Watanabe, Hiroki...896
Watanabe, Shoichiro...2042
Wei, Baoze..3678
Wei, Feng..1517, 1592
Wei, Jianzhao...2630
Wei, Juan...1131
Wei, Shilei...1397
Wei, Wang...1442, 2969
Wei, Xiaoguang...2343
Wei, Xiuqin..3650, 4127
Wei, Zhang..2969
Wellawatta, Thusitha Randima 1409
Wen, Huiqing...453
Wen, Po-Hsiang...3544
Wenbing, Li...1282
Wickramasinghe, Harith R. 3117
Wijaya, Febry Pandu...3490
Wikström, T..3596
Winter, Christian..388
Wolf, Mihaela...3607
Wolski, Kornel..2129
Wu, Bin..3797
Wu, Heng...2673
Wu, Hongfei..4168
Wu, Min..3863

Wu, Pei-Lin...1145
Wu, Ping-Heng 503, 3549
Wu, T.-F...1912
Wu, Tsai-Fu..3884
Wu, Tsung-Hsi..3544
Wu, Xiaojie..619
Wu, Xiaojun...3010, 3015
Wu, Ya'nan...2496
Wu, Zhiqian..1549
Würfl, Joachim...3607
Wyss, Jonas..3734
Xia, Meng..3484
Xia, Yongming..2842
Xiao, Chanjuan...1131
Xiao, Dan..3971
Xiao, Guochun...1549, 2944
Xiao, Jianfang...4157
Xiao, Xi...1966
Xiaoxi, Liu..2969
Xie, Jingwen...3069
Xie, Shaofeng..2547
Xie, Xiaogao..987
Xie, Zhen...2611, 2992
Xiong, Wei..939
Xu, Binci..2270
Xu, Cai..2986
Xu, Dehong..1118, 2270, 2569
Xu, Dewei David..1537, 3797
Xu, Dianguo 459, 560, 614, 934,
1206, 1234, 3780, 3825, 4213
Xu, Guangzhao...998
Xu, Huadian..2877
Xu, Jin..261, 3514, 3522
Xu, Peng...1478
Xu, Sheng..3002
Xu, Shuang...1793
Xu, Yin-Chi..3884
Xu, Yue..3985
Xuan, Yang...1478
Xuanjie, Gao...2718
Xue, Danhong...2435
Yabuuchi, Tatsushi...233
Yada, Tomoharu...1381
Yamada, Hiroaki...324, 1381
Yamada, Koji...169
Yamaguchi, Daiki...3940
Yamaguchi, Koji..1972
Yamaji, Masaharu...2774
Yamamoto, Aoto...2558
Yamamoto, Hidekazu..2750
Yamamoto, Kichiro......................................1308, 1322
Yamamoto, Masaya.......................................1782, 2030
Yamamoto, Masayoshi.....1087, 1095, 2738, 2789, 3344
Yamamoto, Ryo..4016
Yamamoto, Shu..1949, 1960

AUTHOR INDEX

Yamamoto, Yuuto 3197
Yamanaka, Daisuke 2329
Yamanaka, Kenji 3369
Yamashita, Hiroki 1196
Yamashita, Yoshinori 3490
Yamazaki, Katsumi...........................693, 699
Yamazaki, Masahiro 207
Yan, Qingzeng 619
Yan, Y.T. 851
Yan, Zhang 4181
Yanagisawa, Yuta 3762
Yang, Chang-Jun 3884
Yang, Cheng-Jhen 639, 883
Yang, Daoshu 1549
Yang, Dongsheng 3357
Yang, Geng 542
Yang, Hong-Tzer 4233
Yang, Hui-Chen 2296
Yang, Mei 3958
Yang, Ming 560
Yang, Peng 1966
Yang, Ping 3112
Yang, Renxin 4220
Yang, Sheng-Ming 651, 3426
Yang, Shunfeng 428
Yang, Shuying 2611
Yang, Xu 998, 1062, 1478, 2785, 2854, 3329
Yang, Ying 2973
Yang, Yongheng 439, 1021, 1788, 2512, 2743
Yang, Yugang 2703
Yang, Zebin 1157, 1167
Yang, Zhichang 2058
Yang, Zhihua 3797
Yang, Zhiqing 1073
Yang, Zhongping........... 1131, 1816, 2058, 3484, 3495
Yano, Junya 3723
Yao-Ching, Hsieh 645
Yaoqin, Jia 2441
Yasuda, Takumi 992
Yasuda, Yusuke 2082
Yaxin, Peng 416
Ye, Han 1504, 2496
Yeh, Shun-Hao 4233
Yelaverthi, Dorai Babu 4066
Yen, Chih-Ying 1145
Yenchamchalit, Kulsomsup.......................2430
Yi, Hao 2780, 2882
Yijie, Hou 2441
Yin, Shiyuan 1455
Yin, Taiyuan 1455
Yin, Zhijian 1021
Yin, Zhonggang 2204
Yingchun, Xu 2441
Yokokura, Yuki.................... 1673, 1741, 3890, 3896
Yokoyama, T. 3665

Yokoyama, Tomoki 1270, 1877, 1883, 2914, 3363, 3658
Yonezawa, Y. 3603
Yonezawa, Yu 196
Yoon, Bo-Kyung 3063
Yoshida, Souichi 2764
Yoshida, Yukihiro 2832
Yoshihara, Hidemasa 219
Yoshihara, Tohru 1997
Yoshikawa, Gaku 3280
Yoshimi, Daisuke 3952
Yoshimura, Eiji 767
Yoshino, Takuma 3363
Yoshino, Teruo 1692, 3916
Yoshioka, Yusuke 4151
Yoshizawa, Daisuke 238
You, Jiang 1386, 1421
You, Zih-Cing 651
Yu, Yong 459
Yuan, Huawei 889
Yuan, Liqiang 2963
Yuan, Xibo 619, 1634
Yuan, Yiqin 977, 3040
Yue, Wang 2625
Yui, Haiyan 699
Yukita, Kazuto 809
Zaijun, Wu 1282
Zaitsu, Toshiyuki 118, 4139
Zaman, Mohammad Shawkat 2380
Zanchetta, P. 4034
Zane, Regan 4066, 4074, 4145
Zdanowski, Mariusz 2129
Zeng, Pengxiang 2646
Zhang, Chen 4220
Zhang, Feili 1315
Zhang, Guoqiang 1206, 1234
Zhang, H. 142
Zhang, Hailong 3863
Zhang, Hao 1131, 1598
Zhang, Hongyang 3684
Zhang, Jianwen 1004
Zhang, Jianzhong 2262
Zhang, Le 3145
Zhang, Lei 3383
Zhang, Lifei 2703
Zhang, Meng 1966
Zhang, Qianfan 3025
Zhang, Runze 1816
Zhang, Shichong 2638
Zhang, Shu 614, 934, 3780
Zhang, Shuai 2944
Zhang, Tengfei 2980
Zhang, Wang 2625
Zhang, Xiaofang 2547

AUTHOR INDEX

Zhang, Xin 917, 953, 1004, 2688, 2692, 2980, 4157, 4162

Zhang, Xinan 1724

Zhang, Xing 2973, 2992

Zhang, Xueguang 4213

Zhang, Y. 946, 2697

Zhang, Yan 2646

Zhang, Yang 1177

Zhang, Yanping 2204

Zhang, Yaqian 2262

Zhang, Yi 2743, 3123

Zhang, Zhe 607, 1351, 3460

Zhang, Zhenbin 1661, 2895, 3965

Zhang, Zhigang 1157, 1167

Zhao, Chongyan 904

Zhao, Fangzhou 1549, 2944

Zhao, Fei 1172

Zhao, Jianfeng 3002, 3010, 3015

Zhao, Juan 2051

Zhao, Shengnan 795

Zhao, Tianshu 3020

Zhao, Tianyang 1172

Zhao, Yuanliang 3698

Zhao, Zhengming 2963

Zhao, Zhibin 1125

Zhao, Zhiqing 2714

Zheng, Deyou 2611

Zheng, Xuemei 2901

Zheng, Zedong 1010, 2386

Zhong, Wenxing 1118, 2569

Zhou, Dao 1758

Zhou, Dehong 428, 434

Zhou, Fulin 3257

Zhou, Jiuyang 2462

Zhou, Lei 2505

Zhou, Sheng-Zhi 370

Zhou, Victor 1943

Zhou, Yan 934

Zhou, Yimin 2547

Zhu, Cailing 3052

Zhu, Chunbo 84

Zhu, Helin 1336

Zhu, Junjie 3145

Zhu, Lianghong 1206, 1234

Zhu, Qingwei 3338

Zhu, Yanlin 2780

Zhu, Ye 2270

Zhujian, Ou 1376

Zhuo, Fang...............1157, 1167, 2780, 2882

Zhuyong, Li 2986

Zischler, Sigrid 2410

Zou, Yaohan 3455